Solutions Ma

to accompany

Thermodynamics
An Engineering Approach
Fourth Edition

Yunus A. Çengel
University of Nevada, Reno

Michael A. Boles
North Carolina State University

Boston Burr Ridge, IL Dubuque, IA Madison, WI New York San Francisco St. Louis
Bangkok Bogotá Caracas Kuala Lumpur Lisbon London Madrid Mexico City
Milan Montreal New Delhi Santiago Seoul Singapore Sydney Taipei Toronto

McGraw-Hill Higher Education

A Division of The McGraw-Hill Companies

Solutions Manual to accompany
THERMODYNAMICS: AN ENGINEERING APPROACH, FOURTH EDITION
YUNUS A. ÇENGEL AND MICHAEL A. BOLES

Published by McGraw-Hill Higher Education, an imprint of The McGraw-Hill Companies, Inc., 1221 Avenue of the Americas, New York, NY 10020.

This book is printed on acid-free paper.

3 4 5 6 7 8 9 0 BKM BKM 0 3 2

ISBN 0-07-245113-0

www.mhhe.com

Table of Contents

Preface

This manual is prepared as an aide to the instructors in correcting homework assignments, but it can also be used as a source of additional example problems for use in the classroom. With this in mind, all solutions are prepared in full detail in a systematic manner, using a word processor with an equation editor. The solutions are structured into the following sections to make it easy to locate information and to follow the solution procedure, as appropriate:

Solution - The problem is posed, and the quantities to be found are stated.
Assumptions - The significant assumptions in solving the problem are stated.
Properties - The material properties needed to solve the problem are listed.
Analysis - The problem is solved in a systematic manner, showing all steps.
Discussion - Comments are made on the results, as appropriate.

A sketch is included with most solutions to help the students visualize the physical problem, and also to enable the instructor to glance through several types of problems quickly, and to make selections easily.

Problems designated with the EES logo in the text are also solved with the EES software, and electronic solutions complete with parametric studies are available on the CD that accompanies the text. Comprehensive problems designated with the computer icon are solved using the EES software, and their solutions are placed at the *Instructor Manual* section of the *Online Learning Center* (OLC) at www.mhhe.com/cengel-boles. Access to solutions is limited to instructors only who adopted the text, and instructors may obtain their passwords for the OLC by contacting their McGraw-Hill Sales Representative at http://www.mhhe.com/catalogs/rep/.

Every effort is made to produce an error-free Solutions Manual. However, in a text of this magnitude, it is inevitable to have some, and we will appreciate hearing about them. We hope the text and this Manual serve their purpose in aiding with the instruction of Thermodynamics, and making the Thermodynamics experience of both the instructors and students a pleasant and fruitful one.

We acknowledge, with appreciation, the contributions of numerous users of the third edition of the book who took the time to report the typographical or calculational errors that they discovered. All of their suggestions have been incorporated. Special thanks are due to Dr. Mehmet Kanoglu who checked the accuracy of all of the solutions in this Manual.

Yunus A. Çengel
Michael A. Boles

June 2001

Chapter 1
BASIC CONCEPTS OF THERMODYNAMICS

Thermodynamics

1-1C Classical thermodynamics is based on experimental observations whereas statistical thermodynamics is based on the average behavior of large groups of particles.

1-2C On a downhill road the potential energy of the bicyclist is being converted to kinetic energy, and thus the bicyclist picks up speed. There is no creation of energy, and thus no violation of the conservation of energy principle.

1-3C There is no truth to his claim. It violates the second law of thermodynamics.

Mass, Force, and Acceleration

1-4C Pound-mass lbm is the mass unit in English system whereas pound-force lbf is the force unit. One pound-force is the force required to accelerate a mass of 32.174 lbm by 1 ft/s^2. In other words, the weight of a 1-lbm mass at sea level is 1 lbf.

1-5C Kg is the mass unit in the SI system whereas kg-force is a force unit. 1-kg-force is the force required to accelerate a 1-kg mass by 9.807 m/s^2. In other words, the weight of 1-kg mass at sea level is 1 kg-force.

1-6C There is no acceleration, thus the net force is zero in both cases.

1-7 A plastic tank is filled with water. The weight of the combined system is to be determined.

Assumptions The density of water is constant throughout.

Properties The density of water is given to be ρ = 1000 kg/m^3.

Analysis The mass of the water in the tank and the total mass are

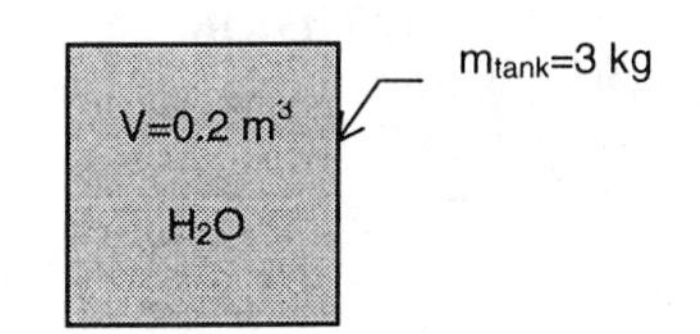

$$m_w = \rho V = (1000 \text{ kg/m}^3)(0.2 \text{ m}^3) = 200 \text{ kg}$$
$$m_{\text{total}} = m_w + m_{tank} = 200 + 3 = 203 \text{ kg}$$

Thus,

$$W = mg = (203 \text{ kg})(9.807 \text{ m/s}^2)\left(\frac{1\text{N}}{1\text{kg}\cdot\text{m/s}^2}\right) = \mathbf{1991\ N}$$

1-8 The interior dimensions of a room are given. The mass and weight of the air in the room are to be determined.

Assumptions The density of air is constant throughout the room.

Properties The density of air is given to be ρ = 1.16 kg/m^3.

Analysis The mass of the air in the room is

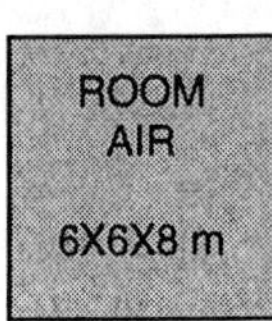

$$m = \rho V = (1.16 \text{ kg/m}^3)(6\times 6\times 8 \text{ m}^3) = \mathbf{334.1\ kg}$$

Thus,

$$W = mg = (334.1\text{kg})(9.807 \text{ m/s}^2)\left(\frac{1\text{ N}}{1\text{ kg}\cdot\text{m/s}^2}\right) = \mathbf{3277\ N}$$

1-9 The variation of gravitational acceleration above the sea level is given as a function of altitude. The height at which the weight of a body will decrease by 1% is to be determined.

Analysis The weight of a body at the elevation z can be expressed as

$$W = mg = m(9.807 - 3.32\times10^{-6}z)$$

In our case,

$$W = 0.99W_s = 0.99mg_s = 0.99(m)(9.807)$$

Substituting,

$$0.99(9.807) = (9.807 - 3.32\times10^{-6}z) \longrightarrow z = \mathbf{29{,}539m}$$

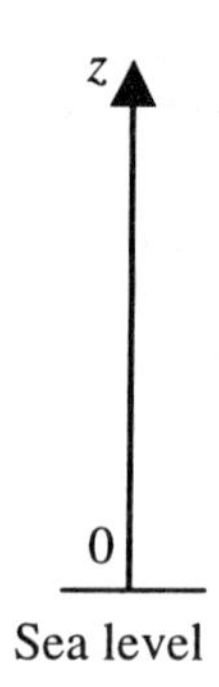

1-10E An astronaut took his scales with him to space. It is to be determined how much he will weigh on the spring and beam scales in space.

Analysis (*a*) A spring scale measures weight, which is the local gravitational force applied on a body:

$$W = mg = (150\text{lbm})(5.48\text{ft/s}^2)\left(\frac{1\text{lbf}}{32.174\text{lbm}\cdot\text{ft/s}^2}\right) = \mathbf{25.5lbf}$$

(*b*) A beam scale compares masses and thus is not affected by the variations in gravitational acceleration. The beam scale will read what it reads on earth,

$$W = \mathbf{150\ lbf}$$

1-11 The acceleration of an aircraft is given in *g*'s. The net upward force acting on a man in the aircraft is to be determined.

Analysis From the Newton's second law, the force applied is

$$F = ma = m(6\text{ g}) = (90\text{ kg})(6\times9.807\text{m/s}^2)\left(\frac{1\text{N}}{1\text{ kg}\cdot\text{m/s}^2}\right) = \mathbf{5296\ N}$$

1-12 [*Also solved by EES on enclosed CD*] A rock is thrown upward with a specified force. The acceleration of the rock is to be determined.

Analysis The weight of the rock is

$$W = mg = (5\text{kg})(9.79\text{m/s}^2)\left(\frac{1\text{N}}{1\text{kg}\cdot\text{m/s}^2}\right) = \mathbf{48.95N}$$

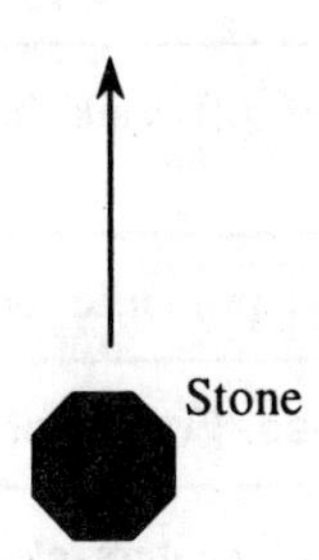

Then the net force that acts on the rock is

$$F_{net} = F_{up} - F_{down} = 150 - 48.95 = 101.05\ N$$

From the Newton's second law, the acceleration of the rock becomes

$$a = \frac{F}{m} = \frac{101.05\text{N}}{5\text{kg}}\left(\frac{1\text{kg}\cdot\text{m/s}^2}{1\text{N}}\right) = \mathbf{20.2m/s^2}$$

1-13 EES solution of this (and other comprehensive problems designated with the *computer icon*) is available to instructors at the *Instructor Manual* section of the *Online Learning Center* (OLC) at www.mhhe.com/cengel-boles. See the Preface for access information.

1-14 Gravitational acceleration g and thus the weight of bodies decreases with increasing elevation. The percent reduction in the weight of an airplane cruising at 13,000 m is to be determined.

Properties The gravitational acceleration g is given to be 9.807 m/s^2 at sea level and 9.767 m/s^2 at an altitude of 13,000 m.

Analysis Weigh is proportional to the gravitational acceleration g, and thus the percent reduction in weight is equivalent to the percent reduction in the gravitational acceleration, which is determined from

$$\%\text{Reduction in weight} = \%\text{Reduction in } g = \frac{\Delta g}{g}\times 100 = \frac{9.807 - 9.767}{9.807}\times 100 = \mathbf{0.41\%}$$

Therefore, the airplane and the people in it will weight 0.41% less at 13,000 m altitude.

Discussion Note that the weight loss at cruising altitudes is negligible.

Systems and Properties

1-15C The radiator should be analyzed as an open system since mass is crossing the boundaries of the system.

1-16C A can of soft drink should be analyzed as a closed system since no mass is crossing the boundaries of the system.

1-17C Intensive properties do not depend on the size (extent) of the system but extensive properties do.

State, Process, Forms of Energy

1-18C In electric heaters, electrical energy is converted to sensible internal energy.

1-19C The forms of energy involved are electrical energy and sensible internal energy. Electrical energy is converted to sensible internal energy, which is transferred to the water as heat.

1-20C The *macroscopic* forms of energy are those a system possesses as a whole with respect to some outside reference frame. The *microscopic* forms of energy, on the other hand, are those related to the molecular structure of a system and the degree of the molecular activity, and are independent of outside reference frames.

1-21C The sum of all forms of the energy a system possesses is called *total energy*. In the absence of magnetic, electrical and surface tension effects, the total energy of a system consists of the kinetic, potential, and internal energies.

1-22C The internal energy of a system is made up of sensible, latent, chemical and nuclear energies. The sensible internal energy is due to translational, rotational, and vibrational effects.

1-23C Thermal energy is the sensible and latent forms of internal energy, and it is referred to as heat in daily life.

1-24C For a system to be in thermodynamic equilibrium, the temperature has to be the same throughout but the pressure does not. However, there should be no unbalanced pressure forces present. The increasing pressure with depth in a fluid, for example, should be balanced by increasing weight.

1-25C A process during which a system remains almost in equilibrium at all times is called a quasi-equilibrium process. Many engineering processes can be approximated as being quasi-equilibrium. The work output of a device is maximum and the work input to a device is minimum when quasi-equilibrium processes are used instead of nonquasi-equilibrium processes.

1-26C A process during which the temperature remains constant is called isothermal; a process during which the pressure remains constant is called isobaric; and a process during which the volume remains constant is called isochoric.

1-27C The state of a simple compressible system is completely specified by two independent, intensive properties.

1-28C Yes, because temperature and pressure are two independent properties and the air in an isolated room is a simple compressible system.

1-29C A process is said to be steady-flow if it involves no changes with time anywhere within the system or at the system boundaries.

1-30 A 1000-MW power plant is powered by nuclear fuel. The amount of nuclear fuel consumed per year is to be determined

Assumptions **1** The power plant operates continuously. **2** The conversion efficiency of the power plant remains constant. **3** The nuclear fuel is uranium. **4** The uranium undergoes complete fission in the plant (this is not the case in practice).

Properties The complete fission of 1 kg of uranium-235 releases 6.73×10^{10} kJ/kg of heat (given in text).

Analysis Noting that the conversion efficiency is 30%, the amount of energy consumed by the power plant is

$$\text{Energy consumption rate} = \text{Power production/Efficiency}$$
$$= (1000\text{ MW})/0.3 = 3333\text{ MW} = 3.333 \times 10^6\text{ kJ/s}$$

$$\text{Annual energy consumption} = (\text{Energy consumtion rate})(1\text{ year})$$
$$= (3.333 \times 10^6\text{ kJ/s})(365 \times 24 \times 3600\text{ s/year})$$
$$= 1.051 \times 10^{14}\text{ kJ/year}$$

Noting that the complete fission of uranium-235 releases 6.73×10^{10} kJ/kg of heat, the amount of uranium that needs to be supplied to the power plant per year is

$$\text{Annual fuel consumption} = \frac{\text{Annual energy consumption}}{\text{Heating value of fuel}} = \frac{1.051 \times 10^{14}\text{ kJ/year}}{6.73 \times 10^{10}\text{ kJ/kg}}$$
$$= \mathbf{1562\ kg/year}$$

Therefore, this power plant will consume about one and a half tons of nuclear fuel per year.

1-31 A 1000-MW power plant is powered by burning coal. The amount of coal consumed per year is to be determined

Assumptions **1** The power plant operates continuously. **2** The conversion efficiency of the power plant remains constant.

Properties The heating value of the coal is given to be 28,000 kJ/kg.

Analysis Noting that the conversion efficiency is 30%, the amount of chemical energy consumed by the power plant is

$$\begin{aligned}\text{Energy consumption rate} &= \text{Power production/Efficiency}\\ &= (1000\ \text{MW})/0.3 = 3333\ \text{MW} = 3.333\times10^{6}\ \text{kJ/s}\end{aligned}$$

$$\begin{aligned}\text{Annual energy consumption} &= (\text{ Energy consumtion rate})(1\ \text{year})\\ &= (3.333\times10^{6}\ \text{kJ/s})(365\times24\times3600\ \text{s/year})\\ &= 1.051\times10^{14}\ \text{kJ/year}\end{aligned}$$

Noting that the heating value of the coal is 28,000 kJ/kg, the amount of coal that needs to be supplied to the power plant per year is

$$\begin{aligned}\text{Annual fuel consumption} &= \frac{\text{Annual energy consumption}}{\text{Heating value of fuel}} = \frac{1.051\times10^{14}\ \text{kJ/year}}{28{,}000\ \text{kJ/kg}}\\ &= 3.754\times10^{9}\ \text{kg/year}\\ &= \mathbf{3{,}754{,}000\ tons/year}\end{aligned}$$

Therefore, this power plant will consume almost 4 millions tons of coal per year.

Energy and Environment

1-32C Energy conversion pollutes the soil, the water, and the air, and the environmental pollution is a serious threat to vegetation, wild life, and human health. The emissions emitted during the combustion of fossil fuels are responsible for smog, acid rain, and global warming and climate change. The primary chemicals that pollute the air are hydrocarbons (HC, also referred to as volatile organic compounds, VOC), nitrogen oxides (NOx), and carbon monoxide (CO). The primary source of these pollutants is the motor vehicles.

1-33C Smog is the brown haze that builds up in a large stagnant air mass, and hangs over populated areas on calm hot summer days. Smog is made up mostly of ground-level ozone (O_3), but it also contains numerous other chemicals, including carbon monoxide (CO), particulate matter such as soot and dust, volatile organic compounds (VOC) such as benzene, butane, and other hydrocarbons. Ground-level ozone is formed when hydrocarbons and nitrogen oxides react in the presence of sunlight in hot calm days. Ozone irritates eyes and damage the air sacs in the lungs where oxygen and carbon dioxide are exchanged, causing eventual hardening of this soft and spongy tissue. It also causes shortness of breath, wheezing, fatigue, headaches, nausea, and aggravate respiratory problems such as asthma.

1-34C Fossil fuels include small amounts of sulfur. The sulfur in the fuel reacts with oxygen to form sulfur dioxide (SO_2), which is an air pollutant. The sulfur oxides and nitric oxides react with water vapor and other chemicals high in the atmosphere in the presence of sunlight to form sulfuric and nitric acids. The acids formed usually dissolve in the suspended water droplets in clouds or fog. These acid-laden droplets are washed from the air on to the soil by rain or snow. This is known as *acid rain.* It is called “rain” since it comes down with rain droplets.

As a result of acid rain, many lakes and rivers in industrial areas have become too acidic for fish to grow. Forests in those areas also experience a slow death due to absorbing the acids through their leaves, needles, and roots. Even marble structures deteriorate due to acid rain.

1-35C Carbon dioxide (CO_2), water vapor, and trace amounts of some other gases such as methane and nitrogen oxides act like a blanket and keep the earth warm at night by blocking the heat radiated from the earth. This is known as the *greenhouse effect.* The greenhouse effect makes life on earth possible by keeping the earth warm. But excessive amounts of these gases disturb the delicate balance by trapping too much energy, which causes the average temperature of the earth to rise and the climate at some localities to change. These undesirable consequences of the greenhouse effect are referred to as *global warming* or *global climate change.* The greenhouse effect can be reduced by reducing the net production of CO_2 by consuming less energy (for example, by buying energy efficient cars and appliances) and planting trees.

1-36C Carbon monoxide, which is a colorless, odorless, poisonous gas that deprives the body's organs from getting enough oxygen by binding with the red blood cells that would otherwise carry oxygen. At low levels, carbon monoxide decreases the amount of oxygen supplied to the brain and other organs and muscles, slows body reactions and reflexes, and impairs judgment. It poses a serious threat to people with heart disease because of the fragile condition of the circulatory system and to fetuses because of the oxygen needs of the developing brain. At high levels, it can be fatal, as evidenced by numerous deaths caused by cars that are warmed up in closed garages or by exhaust gases leaking into the cars.

1-37E A person trades in his Ford Taurus for a Ford Explorer. The extra amount of CO_2 emitted by the Explorer within 5 years is to be determined.

Assumptions The Explorer is assumed to use 940 gallons of gasoline a year compared to 715 gallons for Taurus.

Analysis The extra amount of gasoline the Explorer will use within 5 years is

$$\begin{aligned}\text{Extra Gasoline} &= \text{(Extra per year)(No. of years)}\\ &= (940 - 715\ \text{gal/yr})(5\ \text{yr})\\ &= 1125\ \text{gal}\end{aligned}$$

$$\begin{aligned}\text{Extra } CO_2 \text{ produced} &= \text{(Extra gallons of gasoline used)}(CO_2 \text{ emission per gallon})\\ &= (1125\ \text{gal})(19.7\ \text{lbm/gal})\\ &= \mathbf{22{,}163\ lbm\ CO_2}\end{aligned}$$

Discussion Note that the car we choose to drive has a significant effect on the amount of greenhouse gases produced.

1-38 A power plant that burns natural gas produces 0.59 kg of carbon dioxide (CO_2) per kWh. The amount of CO_2 production that is due to the refrigerators in a city is to be determined.

Assumptions The city uses electricity produced by a natural gas power plant.

Properties 0.59 kg of CO_2 is produced per kWh of electricity generated (given).

Analysis Noting that there are 200,000 households in the city and each household consumes 700 kWh of electricity for refrigeration, the total amount of CO_2 produced is

$$\begin{aligned}\text{Amount of CO}_2\text{ produced} &= (\text{Amount of electricity consumed})(\text{Amount of CO}_2\text{ per kWh})\\ &= (200{,}000\text{ household})(700\text{ kWh/household})(0.59\text{ kg/kWh})\\ &= 8.26\times10^7\ \text{CO}_2\text{ kg/year}\\ &= \mathbf{82{,}600\ CO_2\ ton/year}\end{aligned}$$

Therefore, the refrigerators in this city are responsible for the production of 82,600 tons of CO_2.

1-39 A power plant that burns coal, produces 1.1 kg of carbon dioxide (CO_2) per kWh. The amount of CO_2 production that is due to the refrigerators in a city is to be determined.

Assumptions The city uses electricity produced by a coal power plant.

Properties 1.1 kg of CO_2 is produced per kWh of electricity generated (given).

Analysis Noting that there are 200,000 households in the city and each household consumes 700 kWh of electricity for refrigeration, the total amount of CO_2 produced is

$$\begin{aligned}\text{Amount of CO}_2\text{ produced} &= (\text{Amount of electricity consumed})(\text{Amount of CO}_2\text{ per kWh})\\ &= (200{,}000\text{ household})(700\text{ kWh/household})(1.1\text{ kg/kWh})\\ &= 15.4\times10^7\ \text{CO}_2\text{ kg/year}\\ &= \mathbf{154{,}000\ CO_2\ ton/year}\end{aligned}$$

Therefore, the refrigerators in this city are responsible for the production of 154,000 tons of CO_2.

1-40E A household uses fuel oil for heating, and electricity for other energy needs. Now the household reduces its energy use by 20%. The reduction in the CO_2 production this household is responsible for is to be determined.

Properties The amount of CO_2 produced is 1.54 lbm per kWh and 26.4 lbm per gallon of fuel oil (given).

Analysis Noting that this household consumes 8000 kWh of electricity and 1500 gallons of fuel oil per year, the amount of CO_2 production this household is responsible for is

$$\begin{aligned}\text{Amount of CO}_2\text{ produced} &= (\text{Amount of electricity consumed})(\text{Amount of CO}_2\text{ per kWh})\\ &\quad+(\text{Amount of fuel oil consumed})(\text{Amount of CO}_2\text{ per gallon})\\ &= (8000\text{ kWh/yr})(1.54\text{ lbm/kWh})+(1500\text{ gal/yr})(26.4\text{ lbm/gal})\\ &= 51{,}920\text{ CO}_2\text{ lbm/year}\end{aligned}$$

Then reducing the electricity and fuel oil usage by 20% will reduce the annual amount of CO_2 production by this household by

$$\begin{aligned}\text{Reduction in CO}_2\text{ produced} &= (0.20)(\text{Current amount of CO}_2\text{ production})\\ &= (0.20)(51{,}920\text{ CO}_2\text{ kg/year})\\ &= \mathbf{10{,}384\ CO_2\ lbm/year}\end{aligned}$$

Therefore, any measure that saves energy also reduces the amount of pollution emitted to the environment.

1-41 A household has 2 cars, a natural gas furnace for heating, and uses electricity for other energy needs. The annual amount of NO_x emission to the atmosphere this household is responsible for is to be determined.

Properties The amount of NO_x produced is 7.1 g per kWh, 4.3 g per therm of natural gas, and 11 kg per car (given).

Analysis Noting that this household has 2 cars, consumes 1200 therms of natural gas, and 9,000 kWh of electricity per year, the amount of NO_x production this household is responsible for is

$$\begin{aligned}\text{Amount of NO}_x\text{ produced} &= (\text{No. of cars})(\text{Amount of NO}_x\text{ produced per car})\\ &\quad+(\text{Amount of electricity consumed})(\text{Amount of NO}_x\text{ per kWh})\\ &\quad+(\text{Amount of gas consumed})(\text{Amount of NO}_x\text{ per gallon})\\ &= (2\text{ cars})(11\text{ kg/car})+(9000\text{ kWh/yr})(0.0071\text{ kg/kWh})\\ &\qquad+(1200\text{ therms/yr})(0.0043\text{ kg/therm})\\ &= \mathbf{91.06\ NO_x\ kg/year}\end{aligned}$$

Discussion Any measure that saves energy will also reduce the amount of pollution emitted to the atmosphere.

Temperature

1-42C The zeroth law of thermodynamics states that two bodies are in thermal equilibrium if both have the same temperature reading, even if they are not in contact.

1-43C They are Celsius(°C) and Kelvin (K) in the SI, and Fahrenheit (°F) and Rankine (R) in the English system.

1-44C Probably, but not necessarily. The operation of these two thermometers is based on the thermal expansion of a fluid. If the thermal expansion coefficients of both fluids vary linearly with temperature, then both fluids will expand at the same rate with temperature, and both thermometers will always give identical readings. Otherwise, the two readings may deviate.

1-45 A temperature is given in °C. It is to be expressed in K.

Analysis The Kelvin scale is related to Celsius scale by

$$T(\text{K}) = T(°\text{C}) + 273$$

Thus,

$$T(\text{K}) = 37°\text{C} + 273 = \mathbf{310\ K}$$

1-46E A temperature is given in °C. It is to be expressed in °F, K, and R.

Analysis Using the conversion relations between the various temperature scales,

$$T(\text{K}) = T(°\text{C}) + 273 = 18°\text{C} + 273 = \mathbf{291\ K}$$
$$T(°\text{F}) = 1.8T(°\text{C}) + 32 = (1.8)(18) + 32 = \mathbf{64.4°F}$$
$$T(\text{R}) = T(°\text{F}) + 460 = 64.4 + 460 = \mathbf{524.4\ R}$$

1-47 A temperature change is given in °C. It is to be expressed in K.

Analysis This problem deals with temperature changes, which are identical in Kelvin and Celsius scales. Thus,

$$\Delta T(\text{K}) = \Delta T(°\text{C}) = \mathbf{15\ K}$$

1-48E A temperature change is given in °F. It is to be expressed in °C, K, and R.

Analysis This problem deals with temperature changes, which are identical in Rankine and Fahrenheit scales. Thus,

$$\Delta T(\text{R}) = \Delta T(°\text{F}) = \mathbf{27\ R}$$

The temperature changes in Celsius and Kelvin scales are also identical, and are related to the changes in Fahrenheit and Rankine scales by

$$\Delta T(\text{K}) = \Delta T(\text{R})/1.8 = 27/1.8 = \mathbf{15\ K}$$

and

$$\Delta T(°\text{C}) = \Delta T(\text{K}) = \mathbf{15°C}$$

1-49 Two systems having different temperatures and energy contents are brought in contact. The direction of heat transfer is to be determined.

Analysis Heat transfer occurs from warmer to cooler objects. Therefore, heat will be transferred from system B to system A until both systems reach the same temperature.

Pressure, Manometer, and Barometer

1-50C The pressure relative to the atmospheric pressure is called the *gage pressure*, and the pressure relative to an absolute vacuum is called *absolute pressure.*

1-51C The atmospheric air pressure which is the external pressure exerted on the skin decreases with increasing elevation. Therefore, the pressure is lower at higher elevations. As a result, the difference between the blood pressure in the veins and the air pressure outside increases. This pressure imbalance may cause some thin-walled veins such as the ones in the nose to burst, causing bleeding. The shortness of breath is caused by the lower air density at higher elevations, and thus lower amount of oxygen per unit volume.

1-52C No, the absolute pressure in a liquid of constant density does not double when the depth is doubled. It is the *gage pressure* that doubles when the depth is doubled.

1-53C If the lengths of the sides of the tiny cube suspended in water by a string are very small, the magnitudes of the pressures on all sides of the cube will be the same.

1-54C *Pascal's principle* states that *the pressure applied to a confined fluid increases the pressure throughout by the same amount.* This is a consequence of the pressure in a fluid remaining constant in the horizontal direction. An example of Pascal's principle is the operation of the hydraulic car jack.

1-55C The density of air at sea level is higher than the density of air on top of a high mountain. Therefore, the volume flow rates of the two fans running at identical speeds will be the same, but the mass flow rate of the fan at sea level will be higher.

1-56 The pressure in a vacuum chamber is measured by a vacuum gage. The absolute pressure in the chamber is to be determined.

Analysis The absolute pressure in the chamber is determined from

$$P_{\text{abs}} = P_{\text{atm}} - P_{\text{vac}} = 92 - 24 = \mathbf{68\ kPa}$$

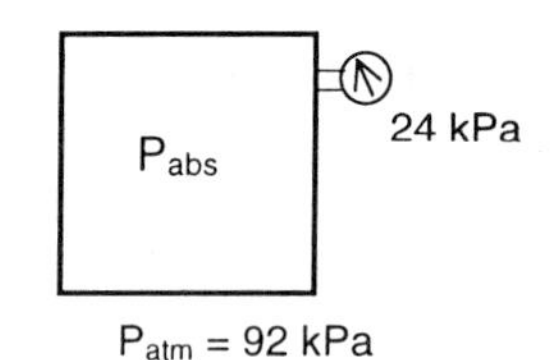

1-57E The pressure in a tank is measured with a manometer by measuring the differential height of the manometer fluid. The absolute pressure in the tank is to be determined for the cases of the manometer arm with the higher and lower fluid level being attached to the tank .

Assumptions The fluid in the manometer is incompressible.

Properties The specific gravity of the fluid is given to be ρ_s = 1.25. The density of water at 32°F is 62.4 lbm/ft^3 (Table A-3E).

Analysis The density of the fluid is obtained by multiplying its specific gravity by the density of water,

$$\rho = (\rho_s)(\rho_{H_2O}) = (1.25)(62.4\,\text{lbm/ft}^3) = 78.0\,\text{lbm/ft}^3$$

The pressure difference corresponding to a differential height of 28 in between the two arms of the manometer is

$$\Delta P = \rho g h = (78\,\text{lbm/ft}^3)(32.174\,\text{ft/s}^2)(28/12\,\text{ft})\left(\frac{1\,\text{lbf}}{32.174\,\text{lbm}\cdot\text{ft/s}^2}\right)\left(\frac{1\,\text{ft}^2}{144\,\text{in}^2}\right) = 1.26\,\text{psia}$$

Then the absolute pressures in the tank for the two cases become:

(*a*) The fluid level in the arm attached to the tank is higher (vacuum):

$$P_{abs} = P_{atm} - P_{vac} = 12.7 - 1.26 = \mathbf{11.44\ psia}$$

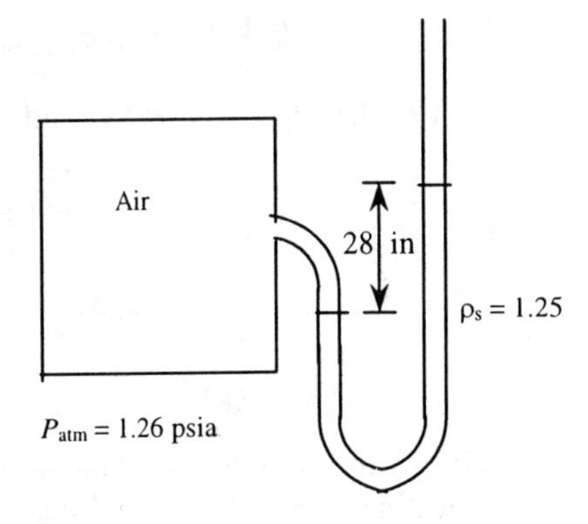

(*b*) The fluid level in the arm attached to the tank is lower:

$$P_{abs} = P_{gage} + P_{atm} = 12.7 + 1.26 = \mathbf{13.96\ psia}$$

Discussion Note that we can determine whether the pressure in a tank is above or below atmospheric pressure by simply observing the side of the manometer arm with the higher fluid level.

1-58 The pressure in a pressurized water tank is measured by a multi-fluid manometer. The gage pressure of air in the tank is to be determined.

Assumptions The air pressure in the tank is uniform (i.e., its variation with elevation is negligible due to its low density), and thus we can determine the pressure at the air-water interface.

Properties The densities of mercury, water, and oil are given to be 13,600, 1000, and 850 kg/m^3, respectively.

Analysis Starting with the pressure at point 1 at the air-water interface, and moving along the tube by adding (as we go down) or subtracting (as we go up) the ρgh terms until we reach point 2, and setting the result equal to P_{atm} since the tube is open to the atmosphere gives

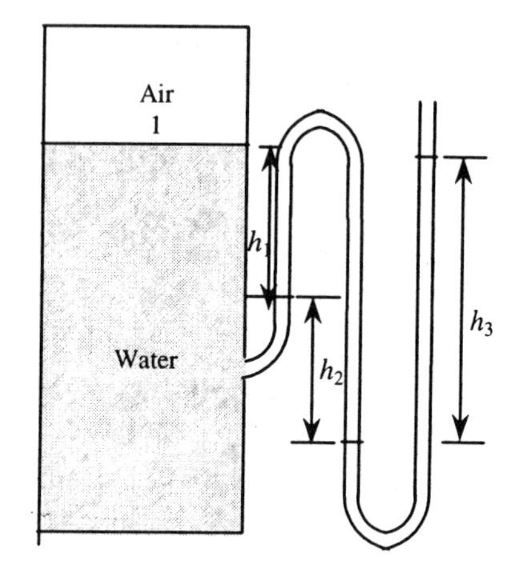

$$P_1 + \rho_{water} g h_1 + \rho_{oil} g h_2 - \rho_{mercury} g h_3 = P_{atm}$$

Solving for P_1,

$$P_1 = P_{atm} - \rho_{water} g h_1 - \rho_{oil} g h_2 + \rho_{mercury} g h_3$$

or,

$$P_1 - P_{atm} = g(\rho_{mercury} h_3 - \rho_{water} h_1 - \rho_{oil} h_2)$$

Noting that $P_{1,gage} = P_1 - P_{atm}$ and substituting,

$$P_{1,gage} = (9.81\text{ m/s}^2)[(13,600\text{ kg/m}^3)(0.46\text{ m}) - (1000\text{ kg/m}^3)(0.2\text{ m}) - (850\text{ kg/m}^3)(0.3\text{ m})+]\left(\frac{1\text{ N}}{1\text{ kg}\cdot\text{m/s}^2}\right)\left(\frac{1\text{ kPa}}{1000\text{ N/m}^2}\right)$$

$$= \mathbf{56.9\ kPa}$$

Discussion Note that jumping horizontally from one tube to the next and realizing that pressure remains the same in the same fluid simplifies the analysis greatly.

1-59 The barometric reading at a location is given in height of mercury column. The atmospheric pressure is to be determined.

Properties The density of mercury is given to be 13,600 kg/m^3.

Analysis The atmospheric pressure is determined directly from

$$P_{atm} = \rho gh$$

$$= (13,600\text{ kg/m}^3)(9.81\text{ m/s}^2)(0.750\text{ m})\left(\frac{1\text{ N}}{1\text{ kg}\cdot\text{m/s}^2}\right)\left(\frac{1\text{ kPa}}{1000\text{ N/m}^2}\right)$$

$$= 100.1\ \mathbf{kPa}$$

1-60 The gage pressure in a liquid at a certain depth is given. The gage pressure in the same liquid at a different depth is to be determined.

Assumptions The variation of the density of the liquid with depth is negligible.

Analysis The gage pressure at two different depths of a liquid can be expressed as

$$P_1 = \rho g h_1 \quad \text{and} \quad P_2 = \rho g h_2$$

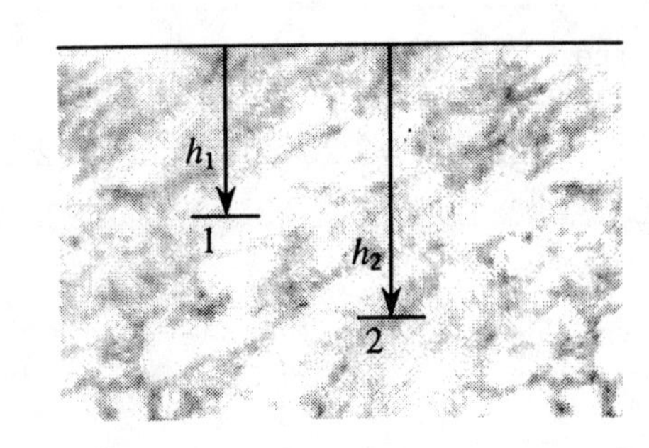

Taking their ratio,

$$\frac{P_2}{P_1} = \frac{\rho g h_2}{\rho g h_1} = \frac{h_2}{h_1}$$

Solving for P_2 and substituting gives

$$P_2 = \frac{h_2}{h_1} P_1 = \frac{12 \text{ m}}{3 \text{ m}} (28 \text{ kPa}) = \mathbf{112 \ kPa}$$

Discussion Note that the gage pressure in a given fluid is proportional to depth.

1-61 The absolute pressure in water at a specified depth is given. The local atmospheric pressure and the absolute pressure at the same depth in a different liquid are to be determined.

Assumptions The liquid and water are incompressible.

Properties The specific gravity of the fluid is given to be $\rho_s = 0.85$. We take the density of water to be 1000 kg/m^3. Then density of the liquid is obtained by multiplying its specific gravity by the density of water,

$$\rho = (\rho_s)(\rho_{H_2O}) = (0.85)(1000 \text{ kg/m}^3) = 850 \text{ kg/m}^3$$

Analysis (*a*) Knowing the absolute pressure, the atmospheric pressure can be determined from

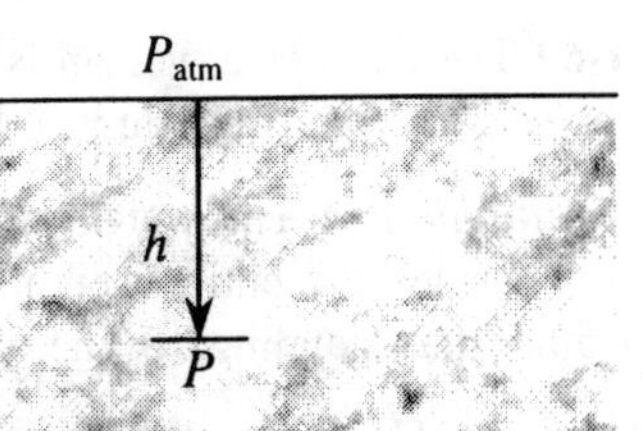

$$\begin{aligned} P_{atm} &= P - \rho g h \\ &= (145 \text{ kPa}) - (1000 \text{ kg/m}^3)(9.81 \text{ m/s}^2)(5 \text{ m})\left(\frac{1 \text{ kPa}}{1000 \text{ N/m}^2}\right) \\ &= \mathbf{96.0 \ kPa} \end{aligned}$$

(*b*) The absolute pressure at a depth of 5 m in the other liquid is

$$\begin{aligned} P &= P_{atm} + \rho g h \\ &= (96.0 \text{ kPa}) + (850 \text{ kg/m}^3)(9.81 \text{ m/s}^2)(5 \text{ m})\left(\frac{1\text{kPa}}{1000\text{N/m}^2}\right) \\ &= \mathbf{137.7 \ kPa} \end{aligned}$$

Discussion Note that at a given depth, the pressure in the lighter fluid is lower, as expected.

1-62E It is to be shown that 1 kgf/cm^2 = 14.223 psi .

Analysis Noting that 1 kgf = 9.80665 N, 1 N = 0.22481 lbf, and 1 in = 2.54 cm, we have

$$1\text{ kgf} = 9.80665\text{ N} = (9.80665\text{ N})\left(\frac{0.22481\text{ lbf}}{1\text{ N}}\right) = 2.20463\text{ lbf}$$

and

$$1\text{ kgf/cm}^2 = 2.20463\text{ lbf/cm}^2 = (2.20463\text{ lbf/cm}^2)\left(\frac{2.54\text{ cm}}{1\text{ in}}\right)^2 = 14.223\text{ lbf/in}^2 = \mathbf{14.223\text{ psi}}$$

1-63E The weight and the foot imprint area of a person are given. The pressures this man exerts on the ground when he stands on one and on both feet are to be determined.

Assumptions The weight of the person is distributed uniformly on foot imprint area.

Analysis The weight of the man is given to be 200 lbf. Noting that pressure is force per unit area, the pressure this man exerts on the ground is

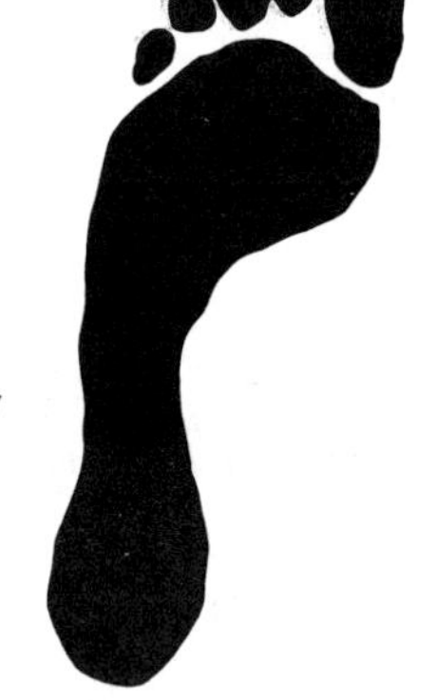

(*a*) On one foot: $$P = \frac{W}{A} = \frac{200\text{ lbf}}{36\text{ in}^2} = 5.56\text{ lbf/in}^2 = \mathbf{5.56\text{ psi}}$$

(*a*) On both feet: $$P = \frac{W}{2A} = \frac{200\text{ lbf}}{2\times 36\text{ in}^2} = 2.78\text{ lbf/in}^2 = \mathbf{2.78\text{ psi}}$$

Discussion Note that the pressure exerted on the ground (and on the feet) is reduced by half when the person stands on both feet.

1-64 The mass of a woman is given. The minimum imprint area per shoe needed to enable her to walk on the snow without sinking is to be determined.

Assumptions **1** The weight of the person is distributed uniformly on the imprint area of the shoes. **2** One foot carries the entire weight of a person during walking, and the shoe is sized for walking conditions (rather than standing). **3** The weight of the shoes is negligible.

Analysis The mass of the woman is given to be 70 kg. For a pressure of 0.5 kPa on the snow, the imprint area of one shoe must be

$$A = \frac{W}{P} = \frac{mg}{P} = \frac{(70\text{ kg})(9.81\text{ m/s}^2)}{0.5\text{ kPa}}\left(\frac{1\text{ N}}{1\text{ kg}\cdot\text{m/s}^2}\right) = \left(\frac{1\text{ kPa}}{1000\text{ N/m}^2}\right) = \mathbf{1.37\text{ m}^2}$$

Discussion This is a very large area for a shoe, and such shoes would be impractical to use. Therefore, some sinking of the snow should be allowed to have shoes of reasonable size.

1-65 The vacuum pressure reading of a tank is given. The absolute pressure in the tank is to be determined.

Properties The density of mercury is given to be $\rho = 13{,}590\ \text{kg/m}^3$.

Analysis The atmospheric (or barometric) pressure can be expressed as

$$
\begin{aligned}
P_{atm} &= \rho g h \\
&= (13{,}590\ \text{kg/m}^3)(9.807\ \text{m/s}^2)(0.755\ \text{m})\left(\frac{1\ \text{N}}{1\ \text{kg}\cdot\text{m/s}^2}\right)\left(\frac{1\ \text{kPa}}{1000\ \text{N/m}^2}\right) \\
&= 100.6\ \text{kPa}
\end{aligned}
$$

Then the absolute pressure in the tank becomes

$$P_{abs} = P_{atm} - P_{vac} = 100.6 - 30 = \mathbf{70.6\ kPa}$$

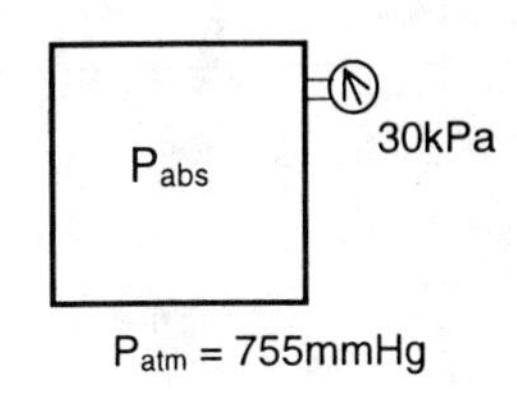

1-66E A pressure gage connected to a tank reads 50 psi. The absolute pressure in the tank is to be determined.

Pabs 50 psia

Properties The density of mercury is given to be $\rho = 848.4\ \text{lbm/ft}^3$.

Analysis The atmospheric (or barometric) pressure can be expressed as

$$
\begin{aligned}
P_{atm} &= \rho g h \\
&= (848.4\text{lbm/ft}^3)(32.174\text{ft/s}^2)(29.1/12\text{ft})\left(\frac{1\text{lbf}}{32.174\text{lbm}\cdot\text{ft/s}^2}\right)\left(\frac{1\text{ft}^2}{144\text{in}^2}\right) \\
&= 14.29\text{psia}
\end{aligned}
$$

Then the absolute pressure in the tank is

$$P_{abs} = P_{gage} + P_{atm} = 50 + 14.29 = \mathbf{64.29\ psia}$$

1-67 A pressure gage connected to a tank reads 500 kPa. The absolute pressure in the tank is to be determined.

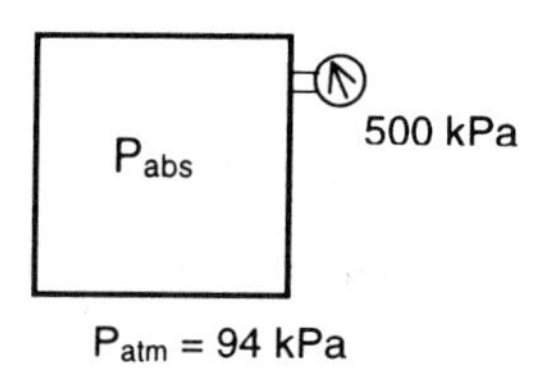

Analysis The absolute pressure in the tank is determined from

$$P_{\text{abs}} = P_{\text{gage}} + P_{\text{atm}} = 500 + 94 = \mathbf{594 kPa}$$

1-68 A mountain hiker records the barometric reading before and after a hiking trip. The vertical distance climbed is to be determined.

Assumptions The variation of air density and the gravitational acceleration with altitude is negligible.

Properties The density of air is given to be $\rho = 1.20$ kg/m^3.

Analysis Taking an air column between the top and the bottom of the mountain and writing a force balance per unit base area, we obtain

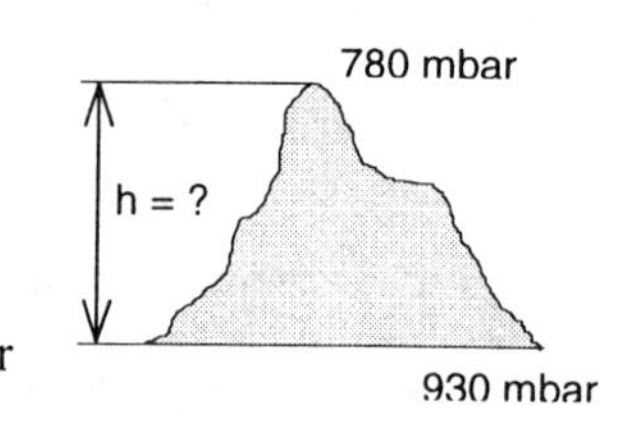

$$W_{\text{air}} / A = P_{\text{bottom}} - P_{\text{top}}$$

$$(\rho g h)_{\text{air}} = P_{\text{bottom}} - P_{\text{top}}$$

$$(1.20 \text{ kg/m}^3)(9.81 \text{ m/s}^2)(h)\left(\frac{1 \text{ N}}{1 \text{ kg}\cdot\text{m/s}^2}\right)\left(\frac{1 \text{ bar}}{100{,}000 \text{ N/m}^2}\right) = (0.930 - 0.780) \text{ bar}$$

It yields $\qquad h = \mathbf{1274\ m}$

which is also the distance climbed.

1-69 A barometer is used to measure the height of a building by recording reading at the bottom and at the top of the building. The height of the building is to be determined.

Assumptions The variation of air density with altitude is negligible.

Properties The density of air is given to be ρ = 1.18 kg/m³. The density of mercury is 13,600 kg/m³.

730 mmHg

h

755 mmHg

Analysis Atmospheric pressures at the top and at the bottom of the building are

$$\begin{aligned} P_{top} &= (\rho g h)_{top} \\ &= (13{,}600 \text{ kg/m}^3)(9.807 \text{ m/s}^2)(0.730 \text{ m})\left(\frac{1 \text{ N}}{1 \text{ kg}\cdot\text{m/s}^2}\right)\left(\frac{1 \text{ kPa}}{1000 \text{ N/m}^2}\right) \\ &= 97.36 \text{ kPa} \end{aligned}$$

$$\begin{aligned} P_{bottom} &= (\rho g h)_{bottom} \\ &= (13{,}600 \text{ kg/m}^3)(9.807 \text{ m/s}^2)(0.755 \text{ m})\left(\frac{1\text{N}}{1\text{kg}\cdot\text{m/s}^2}\right)\left(\frac{1 \text{ kPa}}{1000 \text{ N/m}^2}\right) \\ &= 100.70 \text{ kPa} \end{aligned}$$

Taking an air column between the top and the bottom of the building and writing a force balance per unit base area, we obtain

$$W_{air} / A = P_{bottom} - P_{top}$$

$$(\rho g h)_{air} = P_{bottom} - P_{top}$$

$$(1.18 \text{ kg/m}^3)(9.807 \text{ m/s}^2)(h)\left(\frac{1 \text{ N}}{1 \text{ kg}\cdot\text{m/s}^2}\right)\left(\frac{1 \text{ kPa}}{1000 \text{ N/m}^2}\right) = (100.70 - 97.36) \text{ kPa}$$

It yields $h = \mathbf{288.6\ m}$

which is also the height of the building.

1-70 EES solution of this (and other comprehensive problems designated with the *computer icon*) is available to instructors at the *Instructor Manual* section of the *Online Learning Center* (OLC) at www.mhhe.com/cengel-boles. See the Preface for access information.

1-71 A diver is moving at a specified depth from the water surface. The pressure exerted on the surface of the diver by water is to be determined.

Assumptions The variation of the density of water with depth is negligible.

Properties The specific gravity of sea water is given to be ρ_s = 1.03. We take the density of water to be 1000 kg/m^3.

Analysis The density of the sea water is obtained by multiplying its specific gravity by the density of water which is taken to be 1000 kg/m^3:

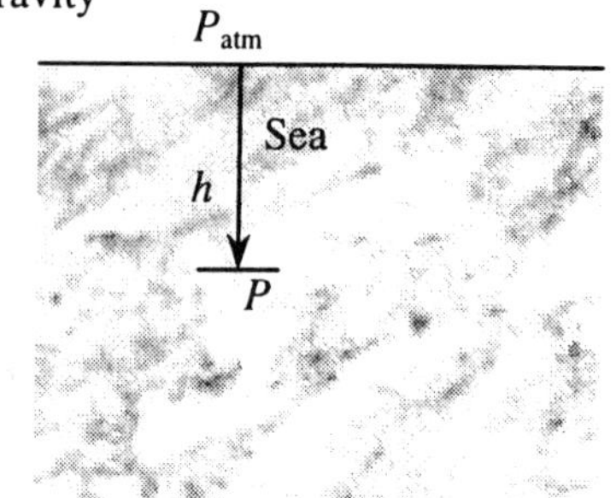

$$\rho = (\rho_s)(\rho_{H_2O}) = (1.03)(1000 \text{ kg/m}^3) = 1030 \ kg/m^3$$

The pressure exerted on a diver at 30 m below the free surface of the sea is the absolute pressure at that location:

$$\begin{aligned} P &= P_{atm} + \rho gh \\ &= (101\text{kPa}) + (1030\text{kg/m}^3)(9.807\text{m/s}^2)(30\text{m})\left(\frac{1\text{kPa}}{1000\text{N/m}^2}\right) \\ &= \mathbf{404.0\ kPa} \end{aligned}$$

1-72E A submarine is cruising at a specified depth from the water surface. The pressure exerted on the surface of the submarine by water is to be determined.

Assumptions The variation of the density of water with depth is negligible.

Properties The specific gravity of sea water is given to be ρ_s = 1.03. The density of water at 32°F is 62.4 lbm/ft^3 (Table A-3E).

Analysis The density of the sea water is obtained by multiplying its specific gravity by the density of water ,

$$\rho = (\rho_s)(\rho_{H_2O}) = (1.03)(62.4\ \text{lbm/ft}^3) = 64.27\ \text{lbm/ft}^3$$

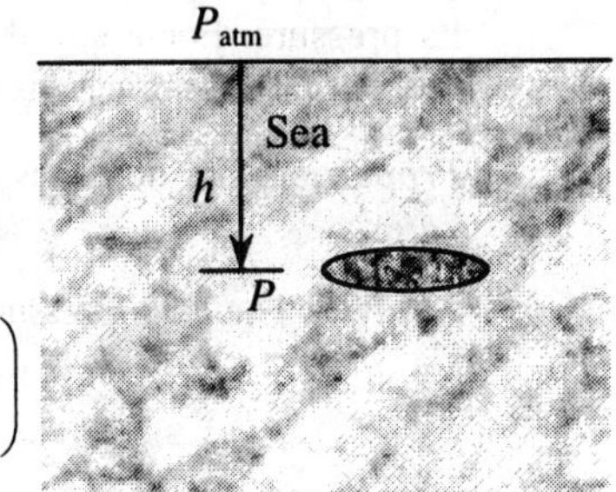

The pressure exerted on the surface of the submarine cruising 300 ft below the free surface of the sea is the absolute pressure at that location:

$$\begin{aligned}P &= P_{atm} + \rho gh \\ &= (14.7\,\text{psia}) + (64.27\,\text{lbm/ft}^3)(32.174\,\text{ft/s}^2)(300\,\text{ft})\left(\frac{1\,\text{lbf}}{32.174\,\text{lbm}\cdot\text{ft/s}^2}\right)\left(\frac{1\,\text{ft}^2}{144\,\text{in}^2}\right) \\ &= \mathbf{148.6\,psia}\end{aligned}$$

1-73 A gas contained in a vertical piston-cylinder device is pressurized by a spring and by the weight of the piston. The pressure of the gas is to be determined.

Analysis Drawing the free body diagram of the piston and balancing the vertical forces yield

$$PA = P_{atm}A + W + F_{spring}$$

Thus,

$$\begin{aligned}P &= P_{\text{atm}} + \frac{mg + F_{\text{spring}}}{A} \\ &= (95\ \text{kPa}) + \frac{(4\ \text{kg})(9.807\ \text{m/s}^2) + 60\ \text{N}}{35\times10^{-4}\,\text{m}^2}\left(\frac{1\ \text{kPa}}{1000\ \text{N/m}^2}\right) \\ &= \mathbf{123.4 kPa}\end{aligned}$$

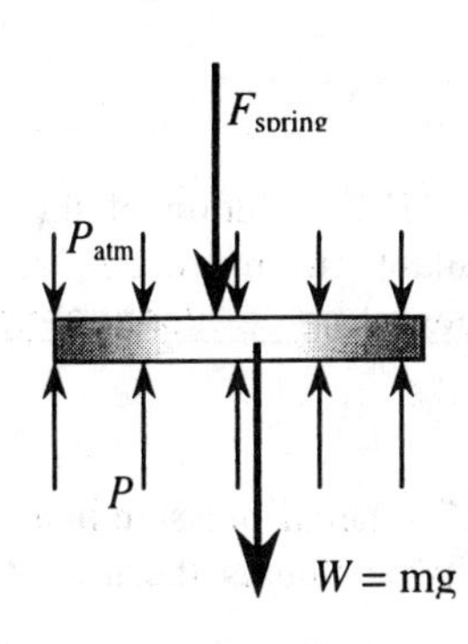

1-74 EES solution of this (and other comprehensive problems designated with the *computer icon*) is available to instructors at the *Instructor Manual* section of the *Online Learning Center* (OLC) at www.mhhe.com/cengel-boles. See the Preface for access information.

1-75 [*Also solved by EES on enclosed CD*] Both a gage and a manometer are attached to a gas to measure its pressure. For a specified reading of gage pressure, the difference between the fluid levels of the two arms of the manometer is to be determined for mercury and water.

Properties The densities of water and mercury are given to be ρ_{water} = 1000 kg/m^3 and be ρ_{Hg} = 13,600 kg/m^3.

Analysis The gage pressure is related to the vertical distance *h* between the two fluid levels by

$$P_{\text{gage}} = \rho g h \longrightarrow h = \frac{P_{\text{gage}}}{\rho g}$$

(*a*) For mercury,

$$h = \frac{P_{\text{gage}}}{\rho_{Hg} g} = \frac{80 \text{ kPa}}{(13600 \text{ kg/m}^3)(9.807 \text{ m/s}^2)}\left(\frac{1 \text{ kN/m}^2}{1 \text{ kPa}}\right)\left(\frac{1000 \text{ kg/m}\cdot\text{s}^2}{1 \text{ kN}}\right) = \mathbf{0.60\ m}$$

(*b*) For water,

$$h = \frac{P_{\text{gage}}}{\rho_{H_2O} g} = \frac{80 \text{ kPa}}{(1000 \text{ kg/m}^3)(9.807 \text{m/s}^2)}\left(\frac{1 \text{ kN/m}^2}{1 \text{ kPa}}\right)\left(\frac{1000 \text{ kg/m}\cdot\text{s}^2}{1 \text{ kN}}\right) = \mathbf{8.16\ m}$$

1-76 EES solution of this (and other comprehensive problems designated with the *computer icon*) is available to instructors at the *Instructor Manual* section of the *Online Learning Center* (OLC) at www.mhhe.com/cengel-boles. See the Preface for access information.

1-77 The air pressure in a tank is measured by an oil manometer. For a given oil-level difference between the two columns, the absolute pressure in the tank is to be determined.

Properties The density of oil is given to be ρ = 850 kg/m^3.

Analysis The absolute pressure in the tank is determined from

$$\begin{aligned} P &= P_{atm} + \rho g h \\ &= (98\text{kPa}) + (850\text{kg/m}^3)(9.807\text{m/s}^2)(0.45\text{m})\left(\frac{1\text{kPa}}{1000\text{N/m}^2}\right) \\ &= \mathbf{101.75kPa} \end{aligned}$$

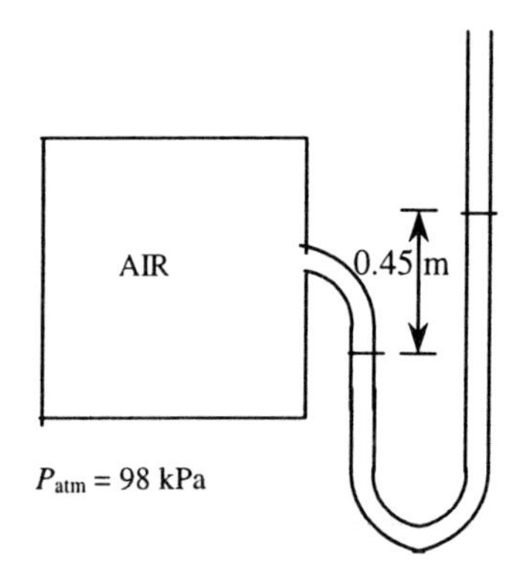

1-78 The air pressure in a duct is measured by a mercury manometer. For a given mercury-level difference between the two columns, the absolute pressure in the duct is to be determined.

Properties The density of mercury is given to be ρ = 13,600 kg/m³.

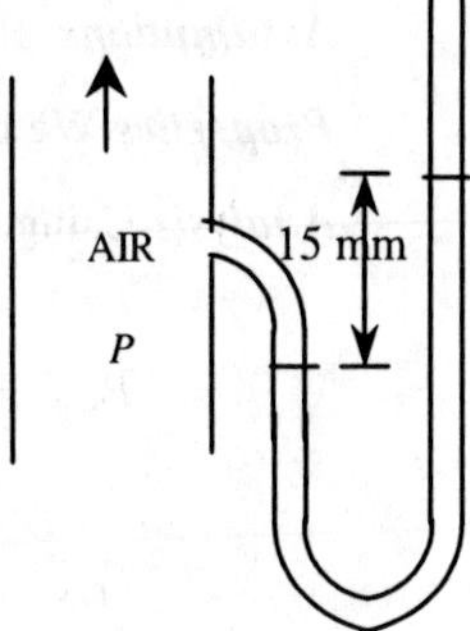

Analysis (*a*) The pressure in the duct is above atmospheric pressure since the fluid column on the duct side is at a lower level.

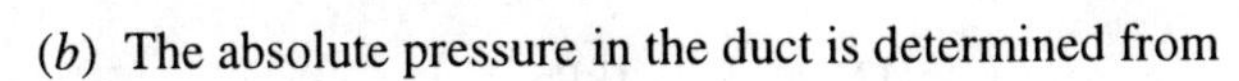

(*b*) The absolute pressure in the duct is determined from

$$P = P_{atm} + \rho gh$$
$$= (100 \text{ kPa}) + (13{,}600 \text{ kg/m}^3)(9.81 \text{ m/s}^2)(0.015 \text{ m})\left(\frac{1 \text{ N}}{1 \text{ kg}\cdot\text{m/s}^2}\right)\left(\frac{1 \text{ kPa}}{1000 \text{ N/m}^2}\right)$$
$$= \mathbf{102.0 \ kPa}$$

1-79 The air pressure in a duct is measured by a mercury manometer. For a given mercury-level difference between the two columns, the absolute pressure in the duct is to be determined.

Properties The density of mercury is given to be ρ = 13,600 kg/m³.

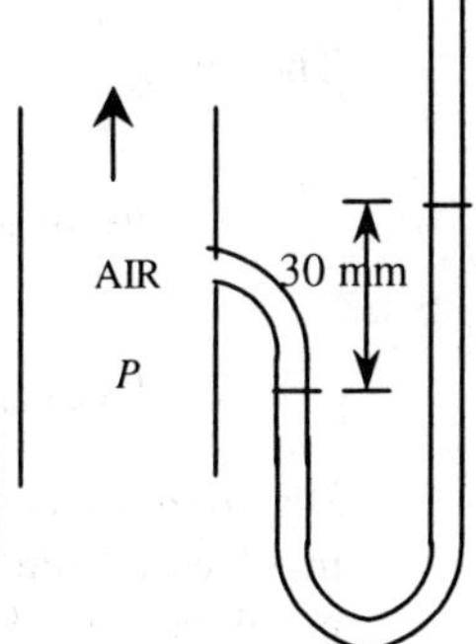

Analysis (*a*) The pressure in the duct is above atmospheric pressure since the fluid column on the duct side is at a lower level.

(*b*) The absolute pressure in the duct is determined from

$$P = P_{atm} + \rho gh$$
$$= (100 \text{ kPa}) + (13{,}600 \text{ kg/m}^3)(9.81 \text{ m/s}^2)(0.030 \text{ m})\left(\frac{1 \text{ N}}{1 \text{ kg}\cdot\text{m/s}^2}\right)\left(\frac{1 \text{ kPa}}{1000 \text{ N/m}^2}\right)$$
$$= \mathbf{104.0 \ kPa}$$

1-80E The systolic and diastolic pressures of a healthy person are given in mmHg. These pressures are to be expressed in kPa, psi, and meter water column.

Assumptions Both mercury and water are incompressible substances.

Properties We take the densities of water and mercury to be 1000 kg/m^3 and 13,600 kg/m^3, respectively.

Analysis Using the relation $P = \rho gh$ for gage pressure, the high and low pressures are expressed as

$$P_{\text{high}} = \rho g h_{\text{high}} = (13{,}600\ \text{kg/m}^3)(9.81\ \text{m/s}^2)(0.12\ \text{m})\left(\frac{1\ \text{N}}{1\ \text{kg}\cdot\text{m/s}^2}\right)\left(\frac{1\ \text{kPa}}{1000\ \text{N/m}^2}\right) = \mathbf{16.0\ kPa}$$

$$P_{\text{low}} = \rho g h_{\text{low}} = (13{,}600\ \text{kg/m}^3)(9.81\ \text{m/s}^2)(0.08\ \text{m})\left(\frac{1\ \text{N}}{1\ \text{kg}\cdot\text{m/s}^2}\right)\left(\frac{1\ \text{kPa}}{1000\ \text{N/m}^2}\right) = \mathbf{10.7\ kPa}$$

Noting that 1 psi = 6.895 kPa,

$$P_{\text{high}} = (16.0\ \text{Pa})\left(\frac{1\ \text{psi}}{6.895\ \text{kPa}}\right) = \mathbf{2.32\ psi} \quad \text{and} \quad P_{\text{low}} = (10.7\ \text{Pa})\left(\frac{1\ \text{psi}}{6.895\ \text{kPa}}\right) = \mathbf{1.55\ psi}$$

For a given pressure, the relation $P = \rho gh$ can be expressed for mercury and water as $P = \rho_{\text{water}} g h_{\text{water}}$ and $P = \rho_{\text{mercury}} g h_{\text{mercury}}$. Setting these two relations equal to each other and solving for water height gives

$$P = \rho_{\text{water}} g h_{\text{water}} = \rho_{\text{mercury}} g h_{\text{mercury}} \quad \rightarrow \quad h_{\text{water}} = \frac{\rho_{\text{mercury}}}{\rho_{\text{water}}} h_{\text{mercury}}$$

Therefore,

$$h_{\text{water, high}} = \frac{\rho_{\text{mercury}}}{\rho_{\text{water}}} h_{\text{mercury, high}} = \frac{13{,}600\ \text{kg/m}^3}{1000\ \text{kg/m}^3}(0.12\ \text{m}) = \mathbf{1.63\ m}$$

$$h_{\text{water, low}} = \frac{\rho_{\text{mercury}}}{\rho_{\text{water}}} h_{\text{mercury, low}} = \frac{13{,}600\ \text{kg/m}^3}{1000\ \text{kg/m}^3}(0.08\ \text{m}) = \mathbf{1.09\ m}$$

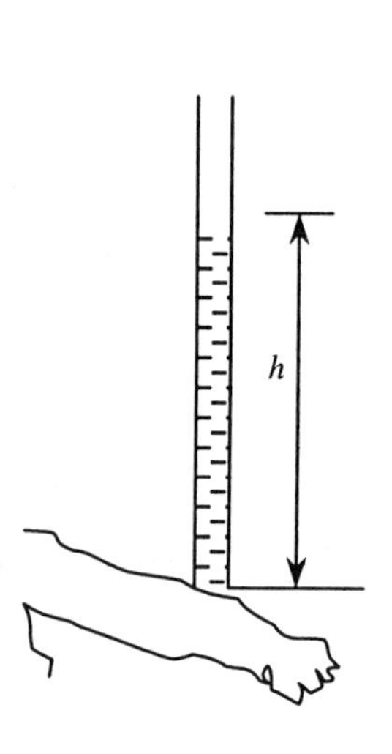

Discussion Note that measuring blood pressure with a "water" monometer would involve differential fluid heights higher than the person, and thus it is impractical. This problem shows why mercury is a suitable fluid for blood pressure measurement devices.

1-81 A vertical tube open to the atmosphere is connected to the vein in the arm of a person. The height that the blood will rise in the tube is to be determined.

Assumptions **1** The density of blood is constant. **2** The gage pressure of blood is 120 mmHg.

Properties The density of blood is given to be ρ = 1050 kg/m³.

Analysis For a given gage pressure, the relation $P = \rho g h$ can be expressed for mercury and blood as $P = \rho_{\text{blood}} g h_{\text{blood}}$ and $P = \rho_{\text{mercury}} g h_{\text{mercury}}$. Setting these two relations equal to each other we get

$$P = \rho_{\text{blood}} g h_{\text{blood}} = \rho_{\text{mercury}} g h_{\text{mercury}}$$

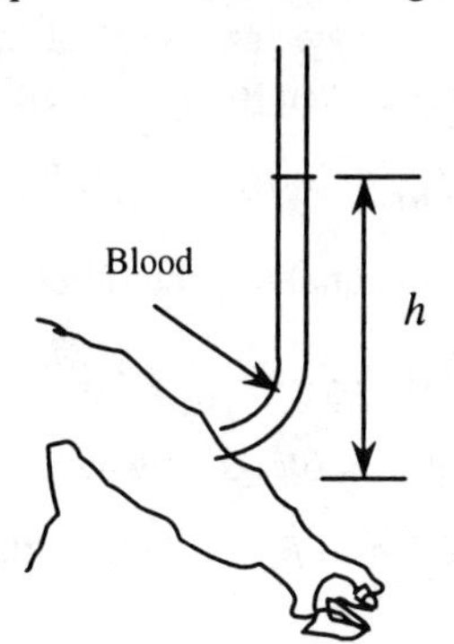

Solving for blood height and substituting gives

$$h_{\text{blood}} = \frac{\rho_{\text{mercury}}}{\rho_{\text{blood}}} h_{\text{mercury}} = \frac{13{,}600\ \text{kg/m}^3}{1050\ \text{kg/m}^3}(0.12\ \text{m}) = \mathbf{1.55\ m}$$

Discussion Note that the blood can rise about one and a half meters in a tube connected to the vein. This explains why IV tubes must be placed high to force a fluid into the vein of a patient.

1-82 A man is standing in water vertically while being completely submerged. The difference between the pressures acting on the head and on the toes is to be determined.

Assumptions Water is an incompressible substances, and thus the density does not change with depth.

Properties We take the density of water to be ρ =1000 kg/m³.

Analysis The pressures at the head and toes of the person can be expressed as

$$P_{\text{head}} = P_{\text{atm}} + \rho g h_{\text{head}} \quad \text{and} \quad P_{\text{toe}} = P_{\text{atm}} + \rho g h_{\text{toe}}$$

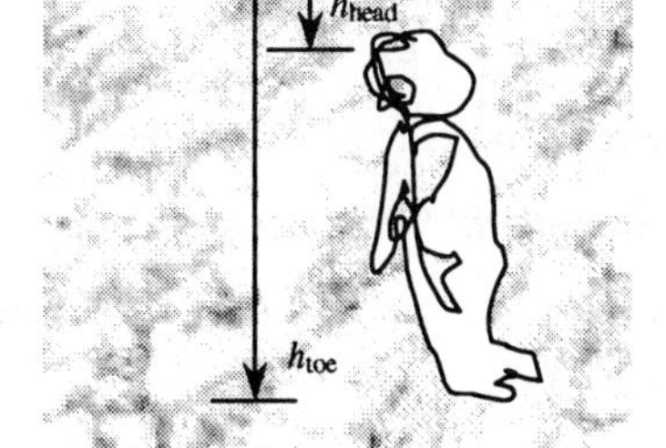

where h is the vertical distance of the location in water from the free surface. The pressure difference between the toes and the head is determined by subtracting the first relation above from the second,

$$P_{\text{toe}} - P_{\text{head}} = \rho g h_{\text{toe}} - \rho g h_{\text{head}} = \rho g (h_{\text{toe}} - h_{\text{head}})$$

Substituting,

$$P_{\text{toe}} - P_{\text{head}} = (1000\ \text{kg/m}^3)(9.81\ \text{m/s}^2)(1.80\ \text{m - 0})\left(\frac{1\text{N}}{1\text{kg}\cdot\text{m/s}^2}\right)\left(\frac{1\text{kPa}}{1000\,\text{N/m}^2}\right) = \mathbf{17.7\ kPa}$$

Discussion This problem can also be solved by noting that the atmospheric pressure (1 atm = 101.325 kPa) is equivalent to 10.3-m of water height, and finding the pressure that corresponds to a water height of 1.8 m.

1-83 Water is poured into the U-tube from one arm and ethyl alcohol from the other arm. The water column height in one arm and the ratio of the heights of the two fluids in the other arm are given. The height of each fluid in that arm is to be determined.

Assumptions Both water and ethyl alcohol are incompressible substances.

Properties The density of ethyl alcohol is given to be ρ = 790 kg/m^3. We take the density of water to be ρ =1000 kg/m^3.

Analysis The height of water column in the left arm of the monometer is given to be h_{w1} = 0.70 m. We let the height of water and alcohol in the right arm to be h_{w2} and h_a, respectively. Then, $h_a = 6h_{w2}$. Noting that both arms are open to the atmosphere, the pressure at the bottom of the U-tube can be expressed as

$$P_{\text{bottom}} = P_{\text{atm}} + \rho_w g h_{w1} \quad \text{and} \quad P_{\text{bottom}} = P_{\text{atm}} + \rho_w g h_{w2} + \rho_a g h_a$$

Setting them equal to each other and simplifying,

$$\rho_w g h_{w1} = \rho_w g h_{w2} + \rho_a g h_a \quad \rightarrow \quad \rho_w h_{w1} = \rho_w h_{w2} + \rho_a h_a \quad \rightarrow \quad h_{w1} = h_{w2} + (\rho_a / \rho_w) h_a$$

Noting that $h_a = 6h_{w2}$, the water and alcohol column heights in the second arm are determined to be

$$0.7 \text{ m} = h_{w2} + (790/1000)6h_{w2} \quad \rightarrow \quad h_{w2} = \mathbf{0.122 \text{ m}}$$

$$0.7 \text{ m} = 0.122 \text{ m} + (790/1000)h_a \quad \rightarrow \quad h_a = \mathbf{0.732 \text{ m}}$$

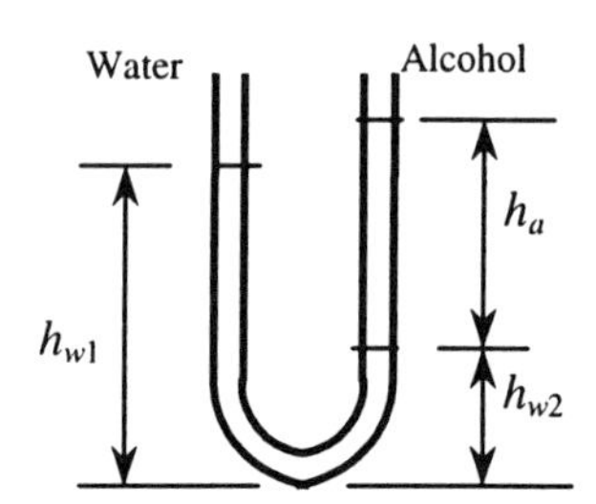

Discussion Note that the fluid height in the arm that contains alcohol is higher. This is expected since alcohol is lighter than water.

1-84 The hydraulic lift in a car repair shop is to lift cars. The fluid gage pressure that must be maintained in the reservoir is to be determined.

Assumptions The weight of the piston of the lift is negligible.

Analysis Pressure is force per unit area, and thus the gage pressure required is simply the ratio of the weight of the car to the area of the lift,

$$P_{\text{gage}} = \frac{W}{A} = \frac{mg}{\pi D^2/4} = \frac{(2000 \text{ kg})(9.81 \text{ m/s}^2)}{\pi (0.30 \text{ m})^2/4}\left(\frac{1 \text{ kN}}{1000 \text{ kg}\cdot\text{m/s}^2}\right) = 278 \text{ kN/m}^2 = \mathbf{278 \text{ kPa}}$$

Discussion Note that the pressure level in the reservoir can be reduced by using a piston with a larger area.

1-85 Fresh and seawater flowing in parallel horizontal pipelines are connected to each other by a double U-tube manometer. The pressure difference between the two pipelines is to be determined.

Assumptions **1** All the liquids are incompressible. **2** The effect of air column on pressure is negligible.

Properties The densities of seawater and mercury are given to be ρ_{sea} = 1035 kg/m^3 and ρ_{Hg} = 13,600 kg/m^3. We take the density of water to be ρ_w =1000 kg/m^3.

Analysis Starting with the pressure in the fresh water pipe (point 1) and moving along the tube by adding (as we go down) or subtracting (as we go up) the ρgh terms until we reach the sea water pipe (point 2), and setting the result equal to P_2 gives

$$P_1 + \rho_w g h_w - \rho_{Hg} g h_{Hg} - \rho_{air} g h_{air} + \rho_{sea} g h_{sea} = P_2$$

Rearranging and neglecting the effect of air column on pressure,

$$P_1 - P_2 = -\rho_w g h_w + \rho_{Hg} g h_{Hg} - \rho_{sea} g h_{sea} = g(\rho_{Hg} h_{Hg} - \rho_w h_w - \rho_{sea} h_{sea})$$

Substituting,

$$\begin{aligned} P_1 - P_2 &= (9.81\,\text{m/s}^2)[(13600\,\text{kg/m}^3)(0.1\,\text{m}) \\ &\quad - (1000\,\text{kg/m}^3)(0.6\,\text{m}) - (1035\,\text{kg/m}^3)(0.4\,\text{m})]\left(\frac{1\,\text{kN}}{1000\,\text{kg}\cdot\text{m/s}^2}\right) \\ &= 3.39\,\text{kN/m}^2 = \mathbf{3.39\,kPa} \end{aligned}$$

Therefore, the pressure in the fresh water pipe is 3.39 kPa higher than the pressure in the sea water pipe.

Discussion A 0.70-m high air column with a density of 1.2 kg/m^3 corresponds to a pressure difference of 0.008 kPa. Therefore, its effect on the pressure difference between the two pipes is negligible.

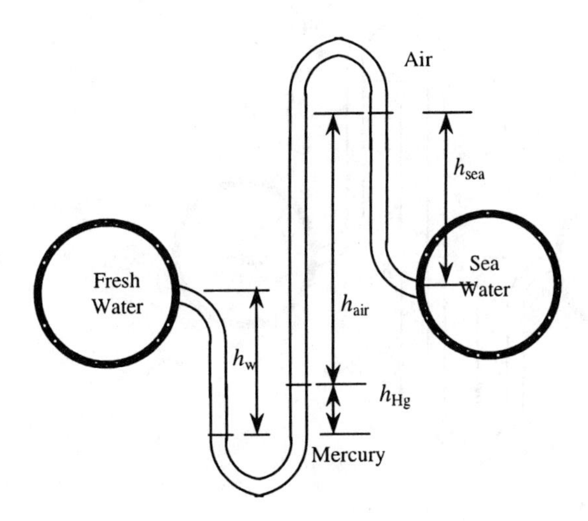

1-86 Fresh and seawater flowing in parallel horizontal pipelines are connected to each other by a double U-tube manometer. The pressure difference between the two pipelines is to be determined.

Assumptions All the liquids are incompressible.

Properties The densities of seawater and mercury are given to be ρ_{sea} = 1035 kg/m^3 and ρ_{Hg} = 13,600 kg/m^3. We take the density of water to be ρ_w =1000 kg/m^3. The specific gravity of oil is given to be 0.72, and thus its density is 720 kg/m^3.

Analysis Starting with the pressure in the fresh water pipe (point 1) and moving along the tube by adding (as we go down) or subtracting (as we go up) the ρgh terms until we reach the sea water pipe (point 2), and setting the result equal to P_2 gives

$$P_1 + \rho_w g h_w - \rho_{Hg} g h_{Hg} - \rho_{oil} g h_{oil} + \rho_{sea} g h_{sea} = P_2$$

Rearranging,

$$P_1 - P_2 = -\rho_w g h_w + \rho_{Hg} g h_{Hg} + \rho_{oil} g h_{oil} - \rho_{sea} g h_{sea}$$
$$= g(\rho_{Hg} h_{Hg} + \rho_{oil} h_{oil} - \rho_w h_w - \rho_{sea} h_{sea})$$

Substituting,

$$P_1 - P_2 = (9.81\,\text{m/s}^2)[(13600\,\text{kg/m}^3)(0.1\,\text{m}) + (720\,\text{kg/m}^3)(0.7\,\text{m}) - (1000\,\text{kg/m}^3)(0.6\,\text{m})$$
$$-(1035\,\text{kg/m}^3)(0.4\,\text{m})]\left(\frac{1\,\text{kN}}{1000\,\text{kg}\cdot\text{m/s}^2}\right)$$
$$= 8.34\,\text{kN/m}^2 = \mathbf{8.34\,kPa}$$

Therefore, the pressure in the fresh water pipe is 8.34 kPa higher than the pressure in the sea water pipe.

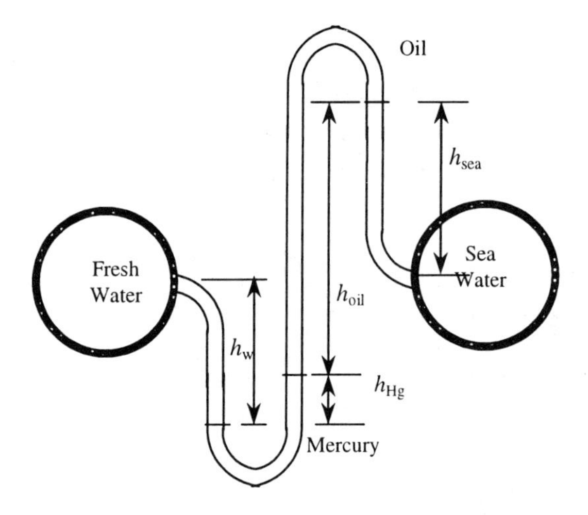

1-87E The pressure in a natural gas pipeline is measured by a double U-tube manometer with one of the arms open to the atmosphere. The absolute pressure in the pipeline is to be determined.

Assumptions **1** All the liquids are incompressible. **2** The effect of air column on pressure is negligible. **3** The pressure throughout the natural gas (including the tube) is uniform since its density is low.

Properties We take the density of water to be $\rho_w = 62.4$ lbm/ft³. The specific gravity of mercury is given to be 13.6, and thus its density is $\rho_{Hg} = 13.6 \times 62.4 = 848.6$ lbm/ft³.

Analysis Starting with the pressure at point 1 in the natural gas pipeline, and moving along the tube by adding (as we go down) or subtracting (as we go up) the ρgh terms until we reach the free surface of oil where the oil tube is exposed to the atmosphere, and setting the result equal to P_{atm} gives

$$P_1 - \rho_{Hg} g h_{Hg} - \rho_{water} g h_{water} = P_{atm}$$

Solving for P_1,

$$P_1 = P_{atm} + \rho_{Hg} g h_{Hg} + \rho_{water} g h_1$$

Substituting,

$$P = 14.2\,\text{psia} + (32.2\,\text{ft/s}^2)[(848.6\,\text{lbm/ft}^3)(6/12\,\text{ft}) + (62.4\,\text{lbm/ft}^3)(27/12\,\text{ft})]\left(\frac{1\,\text{lbf}}{32.2\,\text{lbm}\cdot\text{ft/s}^2}\right)\left(\frac{1\,\text{ft}^2}{144\,\text{in}^2}\right)$$
$$= \mathbf{18.1\,psia}$$

Discussion Note that jumping horizontally from one tube to the next and realizing that pressure remains the same in the same fluid simplifies the analysis greatly. Also, it can be shown that the 15-in high air column with a density of 0.075 lbm/ft³ corresponds to a pressure difference of 0.00065 psi. Therefore, its effect on the pressure difference between the two pipes is negligible.

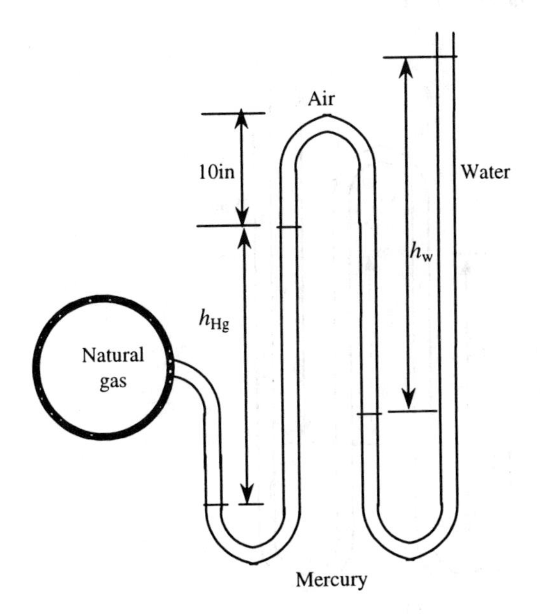

1-88E The pressure in a natural gas pipeline is measured by a double U-tube manometer with one of the arms open to the atmosphere. The absolute pressure in the pipeline is to be determined.

Assumptions **1** All the liquids are incompressible. **2** The pressure throughout the natural gas (including the tube) is uniform since its density is low.

Properties We take the density of water to be $\rho_w = 62.4$ lbm/ft^3. The specific gravity of mercury is given to be 13.6, and thus its density is $\rho_{Hg} = 13.6 \times 62.4 = 848.6$ lbm/ft^3. The specific gravity of oil is given to be 0.69, and thus its density is $\rho_{oil} = 0.69 \times 62.4 = 43.1$ lbm/ft^3.

Analysis Starting with the pressure at point 1 in the natural gas pipeline, and moving along the tube by adding (as we go down) or subtracting (as we go up) the ρgh terms until we reach the free surface of oil where the oil tube is exposed to the atmosphere, and setting the result equal to P_{atm} gives

$$P_1 - \rho_{Hg} g h_{Hg} + \rho_{oil} g h_{oil} - \rho_{water} g h_{water} = P_{atm}$$

Solving for P_1,

$$P_1 = P_{atm} + \rho_{Hg} g h_{Hg} + \rho_{water} g h_1 - \rho_{oil} g h_{oil}$$

Substituting,

$$P_1 = 14.2\text{psia} + (32.2\text{ ft/s}^2)[(848.6\text{lbm/ft}^3)(6/12\text{ ft}) + (62.4\text{lbm/ft}^3)(27/12\text{ ft})$$

$$-(43.1\text{lbm/ft}^3)(15/12\text{ ft})]\left(\frac{1\text{lbf}}{32.2\text{lbm}\cdot\text{ft/s}^2}\right)\left(\frac{1\text{ft}^2}{144\text{in}^2}\right)$$

$$= \mathbf{17.7\,psia}$$

Discussion Note that jumping horizontally from one tube to the next and realizing that pressure remains the same in the same fluid simplifies the analysis greatly.

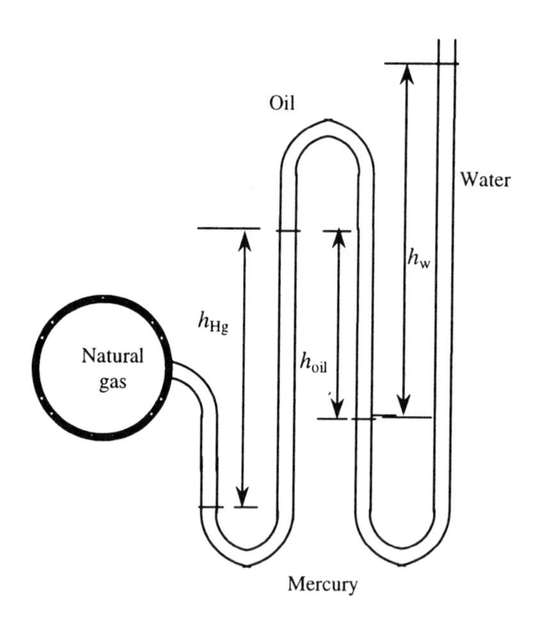

1-89 The gage pressure of air in a pressurized water tank is measured simultaneously by both a pressure gage and a manometer. The differential height h of the mercury column is to be determined.

Assumptions The air pressure in the tank is uniform (i.e., its variation with elevation is negligible due to its low density), and thus the pressure at the air-water interface is the same as the indicated gage pressure.

Properties We take the density of water to be $\rho_w = 1000\ \text{kg/m}^3$. The specific gravities of oil and mercury are given to be 0.72 and 13.6, respectively.

Analysis Starting with the pressure of air in the tank (point 1), and moving along the tube by adding (as we go down) or subtracting (as we go up) the $\rho g h$ terms until we reach the free surface of oil where the oil tube is exposed to the atmosphere, and setting the result equal to P_{atm} gives

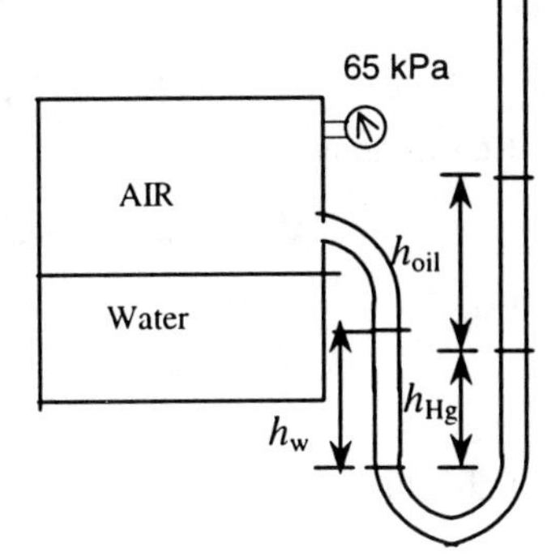

$$P_1 + \rho_w g h_w - \rho_{\text{Hg}} g h_{\text{Hg}} - \rho_{\text{oil}} g h_{\text{oil}} = P_{atm}$$

Rearranging,

$$P_1 - P_{\text{atm}} = \rho_{\text{oil}} g h_{\text{oil}} + \rho_{\text{Hg}} g h_{\text{Hg}} - \rho_w g h_w$$

or,

$$\frac{P_{1,\text{gage}}}{\rho_w g} = \rho_{\text{s,oil}} h_{\text{oil}} + \rho_{\text{s, Hg}} h_{\text{Hg}} - h_w$$

Substituting,

$$\left(\frac{65\ \text{kPa}}{(1000\ \text{kg/m}^3)(9.81\ \text{m/s}^2)} \right) \left(\frac{1000\ \text{kg} \cdot \text{m/s}^2}{1\ \text{kPa.} \cdot \text{m}^2} \right) = 0.72 \times (0.75\ \text{m}) + 13.6 \times h_{\text{Hg}} - 0.3\ \text{m}$$

Solving for h_{Hg} gives $h_{\text{Hg}} =$ **0.47 m**. Therefore, the differential height of the mercury column must be 47 cm.

Discussion Double instrumentation like this allows one to verify the measurement of one of the instruments by the measurement of another instrument.

1-90 The gage pressure of air in a pressurized water tank is measured simultaneously by both a pressure gage and a manometer. The differential height *h* of the mercury column is to be determined.

Assumptions The air pressure in the tank is uniform (i.e., its variation with elevation is negligible due to its low density), and thus the pressure at the air-water interface is the same as the indicated gage pressure.

Properties We take the density of water to be ρ_w =1000 kg/m^3. The specific gravities of oil and mercury are given to be 0.72 and 13.6, respectively.

Analysis Starting with the pressure of air in the tank (point 1), and moving along the tube by adding (as we go down) or subtracting (as we go up) the ρgh terms until we reach the free surface of oil where the oil tube is exposed to the atmosphere, and setting the result equal to P_{atm} gives

$$P_1 + \rho_w g h_w - \rho_{Hg} g h_{Hg} - \rho_{oil} g h_{oil} = P_{atm}$$

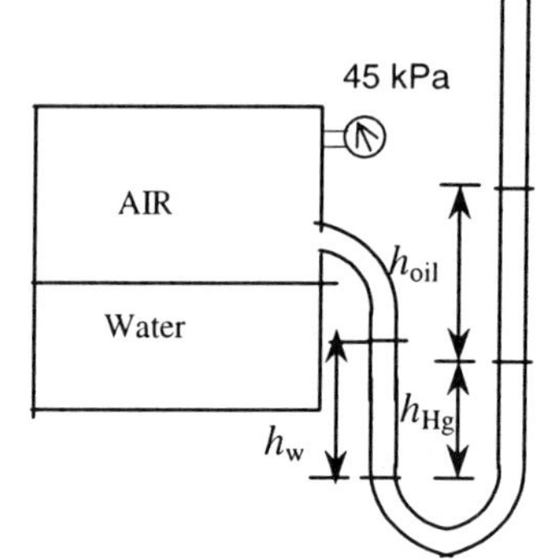

Rearranging,

$$P_1 - P_{atm} = \rho_{oil} g h_{oil} + \rho_{Hg} g h_{Hg} - \rho_w g h_w$$

or,

$$\frac{P_{1,gage}}{\rho_w g} = \rho_{s,oil} h_{oil} + \rho_{s,Hg} h_{Hg} - h_w$$

Substituting,

$$\frac{45 \text{ kPa}}{(1000 \text{ kg/m}^3)(9.81 \text{ m/s}^2)}\left(\frac{1000 \text{ kg}\cdot\text{m/s}^2}{1 \text{ kPa.}\cdot\text{m}^2}\right) = 0.72 \times (0.75 \text{ m}) + 13.6 \times h_{Hg} - 0.3 \text{ m}$$

Solving for h_{Hg} gives h_{Hg} = **0.32 m**. Therefore, the differential height of the mercury column must be 32 cm.

Discussion Double instrumentation like this allows one to verify the measurement of one of the instruments by the measurement of another instrument.

1-91 The top part of a water tank is divided into two compartments, and a fluid with an unknown density is poured into one side. The levels of the water and the liquid are measured. The density of the fluid is to be determined.

Assumptions **1** Both water and the added liquid are incompressible substances. **2** The added liquid does not mix with water.

Properties We take the density of water to be ρ =1000 kg/m^3.

Analysis Both fluids are open to the atmosphere. Noting that the pressure of both water and the added fluid is the same at the contact surface, the pressure at this surface can be expressed as

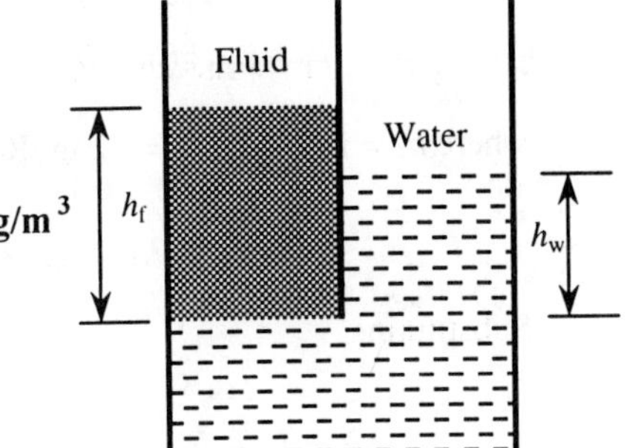

$$P_{\text{contact}} = P_{\text{atm}} + \rho_f g h_f = P_{\text{atm}} + \rho_w g h_w$$

Simplifying and solving for ρ_f gives

$$\rho_f g h_f = \rho_w g h_w \quad \rightarrow \quad \rho_f = \frac{h_w}{h_f}\rho_w = \frac{45 \text{ cm}}{80 \text{ cm}}(1000 \text{ kg/m}^3) = \mathbf{562.5 \text{ kg/m}^3}$$

Discussion Note that the added fluid is lighter than water as expected (a heavier fluid would sink in water).

1-92 A load on a hydraulic lift is to be raised by pouring oil from a thin tube. The height of oil in the tube required in order to raise that weight is to be determined.

Assumptions **1** The cylinders of the lift are vertical. **2** There are no leaks. **3** Atmospheric pressure act on both sides, and thus it can be disregarded.

Properties The density of oil is given to be ρ =780 kg/m^3.

Analysis Noting that pressure is force per unit area, the gage pressure in the fluid under the load is simply the ratio of the weight to the area of the lift,

$$P_{\text{gage}} = \frac{W}{A} = \frac{mg}{\pi D^2/4} = \frac{(500 \text{ kg})(9.81 \text{ m/s}^2)}{\pi(1.20 \text{ m})^2/4}\left(\frac{1 \text{ kN}}{1000 \text{ kg}\cdot\text{m/s}^2}\right) = 4.34 \text{ kN/m}^2 = 4.34 \text{ kPa}$$

The required oil height that will cause 4.34 kPa of pressure rise is

$$P_{\text{gage}} = \rho g h \quad \rightarrow \quad h = \frac{P_{\text{gage}}}{\rho g} = \frac{4.34 \text{ kN/m}^2}{(780 \text{ kg/m}^3)(9.81 \text{ m/s}^2)}\left(\frac{1000 \text{ kg}\cdot\text{m/s}^2}{1 \text{kN/m}^2}\right) = \mathbf{0.567 \text{ m}}$$

Therefore, a 500 kg load can be raised by this hydraulic lift by simply raising the oil level in the tube by 56.7 cm.

Discussion Note that large weights can be raised by little effort in hydraulic lift by making use of Pascal's principle.

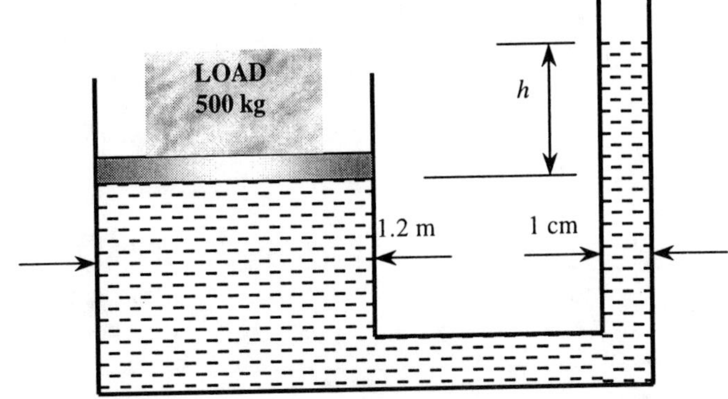

1-93E Two oil tanks are connected to each other through a mercury manometer. For a given differential height, the pressure difference between the two tanks is to be determined.

Assumptions **1** Both the oil and mercury are incompressible fluids. **2** The oils in both tanks have the same density.

Properties The densities of oil and mercury are given to be ρ_{oil} = 45 lbm/ft^3 and ρ_{Hg} = 848 lbm/ft^3.

Analysis Starting with the pressure at the bottom of tank 1 (where pressure is P_1) and moving along the tube by adding (as we go down) or subtracting (as we go up) the ρgh terms until we reach the bottom of tank 2 (where pressure is P_2) gives

$$P_1 + \rho_{oil} g(h_1 + h_2) - \rho_{Hg} gh_2 - \rho_{oil} gh_1 = P_2$$

where h_1 = 10 in and h_2 = 32 in. Rearranging and simplifying,

$$P_1 - P_2 = \rho_{Hg} gh_2 - \rho_{oil} gh_2 = (\rho_{Hg} - \rho_{oil})gh_2$$

Substituting,

$$\Delta P = P_1 - P_2 = (848 - 45 \text{ lbm/ft}^3)(32.2 \text{ ft/s}^2)(32/12 \text{ ft})\left(\frac{1 \text{lbf}}{32.2 \text{lbm} \cdot \text{ft/s}^2}\right)\left(\frac{1 \text{ft}^2}{144 \text{in}^2}\right) = \mathbf{14.9 \text{ psia}}$$

Therefore, the pressure in the left oil tank is 14.9 psia higher than the pressure in the right oil tank.

Discussion Note that large pressure differences can be measured conveniently by mercury manometers. If a water manometer were used in this case, the differential height would be over 30 ft.

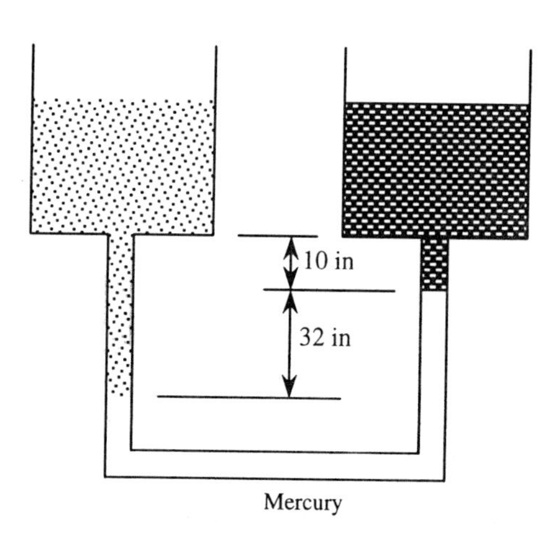

Problem Solving Techniques and EES

1-94C Despite the convenience and capability the engineering software packages offer, they are still just tools, and they will not replace the traditional engineering courses. They will simply cause a shift in emphasis in the course material from mathematics to physics. They are of great value in engineering practice, however, as engineers today rely on software packages for solving large and complex problems in a short time, and perform optimization studies efficiently.

1-95 Determine a positive real root of the following equation using EES:

$$2x^3 - 10x^{0.5} - 3x = -3$$

Solution by EES Software (Copy the following lines and paste on a blank EES screen to verify solution):

```
2*x^3-10*x^0.5-3*x = -3
```

Answer: x = 2.063 (using an initial guess of x=2)

1-96 Solve the following system of 2 equations with 2 unknowns using EES:

$$x^3 - y^2 = 7.75$$

$$3xy + y = 3.5$$

Solution by EES Software (Copy the following lines and paste on a blank EES screen to verify solution):

```
x^3-y^2=7.75
3*x*y+y=3.5
```

Answer x=2 y=0.5

1-97 Solve the following system of 3 equations with 3 unknowns using EES:

$$2x - y + z = 5$$

$$3x^2 + 2y = z + 2$$

$$xy + 2z = 8$$

Solution by EES Software (Copy the following lines and paste on a blank EES screen to verify solution):

```
2*x-y+z=5
3*x^2+2*y=z+2
x*y+2*z=8
```

Answer x=1.141, y=0.8159, z=3.535

1-98 Solve the following system of 3 equations with 3 unknowns using EES:

$$x^2y - z = 1$$

$$x - 3y^{0.5} + xz = -2$$

$$x + y - z = 2$$

Solution by EES Software (Copy the following lines and paste on a blank EES screen to verify solution):

```
x^2*y-z=1
x-3*y^0.5+x*z=-2
x+y-z=2
```

Answer x=1, y=1, z=0

Special Topic: Biological Systems

1-99C Metabolism refers to the chemical activity in the cells associated with the burning of foods. The basal metabolic rate is the metabolism rate of a resting person, which is 84 W for an average man.

1-100C The energy released during metabolism in humans is used to maintain the body temperature at 37°C.

1-101C The food we eat is not entirely metabolized in the human body. The fraction of metabolizable energy contents are 95.5% for carbohydrates, 77.5% for proteins, and 97.7% for fats. Therefore, the metabolizable energy content of a food is not the same as the energy released when it is burned in a bomb calorimeter.

1-102C Yes. Each body rejects the heat generated during metabolism, and thus serves as a heat source. For an average adult male it ranges from 84 W at rest to over 1000 W during heavy physical activity. Classrooms are designed for a large number of occupants, and thus the total heat dissipated by the occupants must be considered in the design of heating and cooling systems of classrooms.

1-103C 1 kg of natural fat contains almost 8 times the metabolizable energy of 1 kg of natural carbohydrates. Therefore, a person who fills his stomach with carbohydrates will satisfy his hunger without consuming too many calories.

1-104 Six people are fast dancing in a room, and there is a resistance heater in another identical room. The room that will heat up faster is to be determined.

Assumptions **1** The rooms are identical in every other aspect. **2** Half of the heat dissipated by people is in sensible form. **3** The people are of average size.

Properties An average fast dancing person dissipates 600 Cal/h of energy (sensible and latent) (Table 1-4).

Analysis Three couples will dissipate

$$E = (6\text{ persons})(600\text{ Cal/h.person})(4.1868\text{ kJ/Cal}) = 15{,}072\text{ kJ/h} = 4190\text{ W}$$

of energy. (About half of this is sensible heat). Therefore, the room with the **people dancing** will warm up much faster than the room with a 2-kW resistance heater.

1-105 Two men are identical except one jogs for 30 min while the other watches TV. The weight difference between these two people in one month is to be determined.

Assumptions The two people have identical metabolism rates, and are identical in every other aspect.

Properties An average 68-kg person consumes 540 Cal/h while jogging, and 72 Cal/h while watching TV (Table 1-4).

Analysis An 80-kg person who jogs 0.5 h a day will have jogged a total of 15 h a month, and will consume

$$\Delta E_{\text{consumed}} = [(540-72)\text{ Cal/h}](15\text{ h})\left(\frac{4.1868\text{kJ}}{1\text{Cal}}\right)\left(\frac{80\text{ kg}}{68\text{ kg}}\right) = 34{,}578\text{kJ}$$

more calories than the person watching TV. The metabolizable energy content of 1 kg of fat is 33,100 kJ. Therefore, the weight difference between these two people in 1-month will be

$$\Delta m_{fat} = \frac{\Delta E_{\text{consumed}}}{\text{Energy content of fat}} = \frac{34{,}578\text{kJ}}{33{,}100\text{kJ/kg}} = \mathbf{1.045kg}$$

1-106 A classroom has 30 students, each dissipating 100 W of sensible heat. It is to be determined if it is necessary to turn the heater on in the room to avoid cooling of the room.

Properties Each person is said to be losing sensible heat to the room air at a rate of 100 W.

Analysis We take the room is losing heat to the outdoors at a rate of

$$\dot{Q}_{loss} = (20{,}000\text{kJ/h})\left(\frac{1\text{h}}{3600\text{s}}\right) = 5.56\text{kW}$$

The rate of sensible heat gain from the students is

$$\dot{Q}_{gain} = (100\text{W/student})(30\text{students}) = 3000\text{W} = 3\text{kW}$$

which is less than the rate of heat loss from the room. Therefore, it is **necessary** to turn the heater on to prevent the room temperature from dropping.

1-107 A bicycling woman is to meet her entire energy needs by eating 30-g candy bars. The number of candy bars she needs to eat to bicycle for 1-h is to be determined.

Assumptions The woman meets her entire calorie needs from candy bars while bicycling.

Properties An average 68-kg person consumes 639 Cal/h while bicycling, and the energy content of a 20-g candy bar is 105 Cal (Tables 1-3 and 1-4).

Analysis Noting that a 20-g candy bar contains 105 Calories of metabolizable energy, a 30-g candy bar will contain

$$E_{candy} = (105\text{Cal})\left(\frac{30\text{g}}{20\text{g}}\right) = 157.5\text{Cal}$$

of energy. If this woman is to meet her entire energy needs by eating 30-g candy bars, she will need to eat

$$N_{candy} = \frac{639\text{ Cal/h}}{157.5\text{ Cal}} \cong \mathbf{4\ candy\ bars/h}$$

1-108 A 55-kg man eats 1-L of ice cream. The length of time this man needs to jog to burn off these calories is to be determined.

Assumptions The man meets his entire calorie needs from the ice cream while jogging.

Properties An average 68-kg person consumes 540 Cal/h while jogging, and the energy content of a 100-ml of ice cream is 110 Cal (Tables 1-3 and 1-4).

Analysis The rate of energy consumption of a 55-kg person while jogging is

$$\dot{E}_{consumed} = (540\text{Cal/h})\left(\frac{55\text{kg}}{68\text{kg}}\right) = 437\text{Cal/h}$$

Noting that a 100-ml serving of ice cream has 110 Cal of metabolizable energy, a 1-liter box of ice cream will have 1100 Calories. Therefore, it will take

$$\Delta t = \frac{1100\text{ Cal}}{437\text{ Cal / h}} = \mathbf{2.5\ h}$$

of jogging to burn off the calories from the ice cream.

1-109 A man with 20-kg of body fat goes on a hunger strike. The number of days this man can survive on the body fat alone is to be determined.

Assumptions **1** The person is an average male who remains in resting position at all times. **2** The man meets his entire calorie needs from the body fat alone.

Properties The metabolizable energy content of fat is 33,100 Cal/kg. An average resting person burns calories at a rate of 72 Cal/h (Table 1-4).

Analysis The metabolizable energy content of 20 kg of body fat is

$$E_{fat} = (33{,}100\text{kJ/kg})(20\text{kg}) = 662{,}000\text{kJ}$$

The person will consume

$$E_{consumed} = (72\text{Cal/h})(24\text{h})\left(\frac{4.1868\text{kJ}}{1\text{Cal}}\right) = 7235\text{kJ/day}$$

Therefore, this person can survive

$$\Delta t = \frac{662{,}000\text{ kJ}}{7235\text{ kJ / day}} = \mathbf{91.5\ days}$$

on his body fat alone. This result is not surprising since people are known to survive over 100 days without any food intake.

1-110 Two 50-kg women are identical except one eats her baked potato with 4 teaspoons of butter while the other eats hers plain every evening. The weight difference between these two woman in one year is to be determined.

Assumptions **1** These two people have identical metabolism rates, and are identical in every other aspect. **2** All the calories from the butter are converted to body fat.

Properties The metabolizable energy content of 1 kg of body fat is 33,100 kJ. The metabolizable energy content of 1 teaspoon of butter is 35 Calories (Table 1-3).

Analysis A person who eats 4 teaspoons of butter a day will consume

$$E_{\text{consumed}} = (35\text{Cal/teaspoon})(4\text{teaspoons/day})\left(\frac{365\text{days}}{1\text{year}}\right) = 51{,}100\text{Cal/year}$$

Therefore, the woman who eats her potato with butter will gain

$$m_{\text{fat}} = \frac{51{,}100\text{Cal}}{33{,}100\text{kJ/kg}}\left(\frac{4.1868\text{kJ}}{1\text{Cal}}\right) = \mathbf{6.5kg}$$

of additional body fat that year.

1-111 A woman switches from 1-L of regular cola a day to diet cola and 2 slices of apple pie. It is to be determined if she is now consuming more or less calories.

Properties The metabolizable energy contents are 300 Cal for a slice of apple pie, 87 Cal for a 200-ml regular cola, and 0 for the diet drink (Table 1-3).

Analysis The energy contents of 2 slices of apple pie and 1-L of cola are

$$E_{pie} = 2 \times (300\text{Cal}) = 600\text{Cal}$$
$$E_{cola} = 5 \times (87\text{Cal}) = 435\text{Cal}$$

Therefore, the woman is now consuming **more calories**.

1-112 A man switches from an apple a day to 200-ml of ice cream and 20-min walk every day. The amount of weight the person will gain or lose with the new diet is to be determined.

Assumptions All the extra calories are converted to body fat.

Properties The metabolizable energy contents are 70 Cal for a an apple and 220 Cal for a 200-ml serving of ice cream (Table 1-3). An average 68-kg man consumes 432 Cal/h while walking (Table 1-4). The metabolizable energy content of 1 kg of body fat is 33,100 kJ.

Analysis The person who switches from the apple to ice cream increases his calorie intake by

$$E_{extra} = 220 - 70 = 150\text{ Cal}$$

The amount of energy a 60-kg person uses during a 20-min walk is

$$E_{\text{consumed}} = (432\text{Cal/h})(20\text{ min})\left(\frac{1\text{ h}}{60\text{ min}}\right)\left(\frac{60\text{kg}}{68\text{kg}}\right) = 127\text{Cal}$$

Therefore, the man now has a net gain of 150 - 127 = 23 Cal per day, which corresponds to 23×30 = 690 Cal per month. Therefore, the man will gain

$$m_{\text{fat}} = \frac{690\text{Cal}}{33{,}100\text{kJ/kg}}\left(\frac{4.1868\text{kJ}}{1\text{Cal}}\right) = \mathbf{0.087kg}$$

of body fat per month with the new diet. (Without the exercise the man would gain 0.569 kg per month).

1-113 The average body temperature of the human body rises by 2°C during strenuous exercise. The increase in the thermal energy content of the body as a result is to be determined.

Properties The average specific heat of the human body is given to be 3.6 kJ/kg.°C.

Analysis The change in the sensible internal energy of the body is

$$\Delta U = mC\Delta T = (80\text{ kg})(3.6\text{ kJ/kg°C})(2\text{°C}) = \mathbf{576\ kJ}$$

as a result of body temperature rising 2°C during strenuous exercise.

1-114 An average American adult switches from drinking alcoholic beverages to drinking diet soda. The amount of weight the person will lose per year as a result of this switch is to be determined.

Assumptions **1** The diet and exercise habits of the person remain the same other than switching from alcoholic beverages to diet drinks. **2** All the excess calories from alcohol are converted to body fat.

Properties The metabolizable energy content of body fat is 33,100 Cal/kg (text).

Analysis when the person switches to diet drinks, he will consume 210 fewer Calories a day. Then the annual reduction in the calories consumed by the person becomes

$$\text{Reduction in energy intake:}\quad E_{\text{reduced}} = (210\ \text{Cal/day})(365\ \text{days/year}) = 76{,}650\ \text{Cal/year}$$

Therefore, assuming all the calories from the alcohol would be converted to body fat, the person who switches to diet drinks will lose

$$\text{Reduction in weight} = \frac{\text{Reduction in energy intake}}{\text{Enegy content of fat}} = \frac{E_{\text{reduced}}}{e_{\text{fat}}} = \frac{76{,}650\text{Cal/yr}}{33{,}100\text{kJ/kg}}\left(\frac{4.1868\text{kJ}}{1\text{Cal}}\right) = \mathbf{9.70\ kg/yr}$$

or about 21 pounds of body fat that year.

1-115 A person drinks a 12-oz beer, and then exercises on a treadmill. The time it will take to burn the calories from a 12-oz can of regular and light beer are to be determined.

Assumptions The drinks are completely metabolized by the body.

Properties The metabolizable energy contents of regular and light beer are 150 and 100 Cal, respectively. Exercising on a treadmill burns calories at an average rate of 700 Cal/h (given).

Analysis The exercising time it will take to burn off beer calories is determined directly from

(*a*) Regular beer: $$\Delta t_{\text{regular beer}} = \frac{150\text{Cal}}{700\text{Cal/h}} = 0.214\ \text{h} = \mathbf{12.9\ min}$$

(*b*) Light beer: $$\Delta t_{\text{light beer}} = \frac{100\ \text{Cal}}{700\text{Cal/h}} = 0.143\ \text{h} = \mathbf{8.6\ min}$$

1-116 A person has an alcoholic drink, and then exercises on a cross-country ski machine. The time it will take to burn the calories is to be determined for the cases of drinking a bloody mary and a martini.

Assumptions The drinks are completely metabolized by the body.

Properties The metabolizable energy contents of bloody mary and martini are 116 and 156 Cal, respectively. Exercising on a cross-country ski machine burns calories at an average rate of 600 Cal/h (given).

Analysis The exercising time it will take to burn off beer calories is determined directly from

(*a*) Bloody mary: $$\Delta t_{\text{Bloody Mary}} = \frac{116 \text{ Cal}}{600 \text{ Cal/h}} = 0.193 \text{ h} = \mathbf{11.6 \text{ min}}$$

(*b*) Martini: $$\Delta t_{\text{martini}} = \frac{156 \text{ Cal}}{600 \text{ Cal/h}} = 0.26 \text{ h} = \mathbf{15.6 \text{ min}}$$

1-117 A man and a woman have lunch at Burger King, and then shovel snow. The shoveling time it will take to burn off the lunch calories is to be determined for both.

Assumptions The food intake during lunch is completely metabolized by the body.

Properties The metabolizable energy contents of different foods are as given in the problem statement. Shoveling snow burns calories at a rate of 360 Cal/h for the woman and 480 Cal/h for the man (given).

Analysis The total calories consumed during lunch and the time it will take to burn them are determined for both the man and woman as follows:

Man: Lunch calories = 720+400+225 = 1345 Cal.

Shoveling time: $$\Delta t_{\text{shoveling, man}} = \frac{1345 \text{ Cal}}{480 \text{ Cal/h}} = \mathbf{2.80 \text{ h}}$$

Woman: Lunch calories = 330+400+0 = 730 Cal.

Shoveling time: $$\Delta t_{\text{shoveling, woman}} = \frac{730 \text{ Cal}}{360 \text{ Cal/h}} = \mathbf{2.03 \text{ h}}$$

1-118 Two friends have identical metabolic rates and lead identical lives, except they have different lunches. The weight difference between these two friends in a year is to be determined.

Assumptions **1** The diet and exercise habits of the people remain the same other than the lunch menus. **2** All the excess calories from the lunch are converted to body fat.

Properties The metabolizable energy content of body fat is 33,100 Cal/kg (text). The metabolizable energy contents of different foods are given in problem statement.

Analysis The person who has the double whopper sandwich consumes 1600 – 800 = 800 Cal more every day. The difference in calories consumed per year becomes

$$\text{Calorie consumption difference} = (800\ \text{Cal/day})(365\ \text{days/year}) = 292{,}000\ \text{Cal/year}$$

Therefore, assuming all the excess calories to be converted to body fat, the weight difference between the two persons after 1 year will be

$$\text{Weight difference} = \frac{\text{Calorie intake difference}}{\text{Enegy content of fat}} = \frac{\Delta E_{\text{intake}}}{e_{\text{fat}}} = \frac{292{,}000\text{Cal/yr}}{33{,}100\text{kJ/kg}}\left(\frac{4.1868\text{kJ}}{1\text{Cal}}\right) = \mathbf{36.9\ kg/yr}$$

or about 80 pounds of body fat per year.

1-119E A person eats dinner at a fast-food restaurant. The time it will take for this person to burn off the dinner calories by climbing stairs is to be determined.

Assumptions The food intake from dinner is completely metabolized by the body.

Properties The metabolizable energy contents are 270 Cal for regular roast beef, 410 Cal for big roast beef, and 150 Cal for the drink. Climbing stairs burns calories at a rate of 400 Cal/h (given).

Analysis The total calories consumed during dinner and the time it will take to burn them by climbing stairs are determined to be

$$\text{Dinner calories} = 270{+}410{+}150 = 830\ \text{Cal.}$$

$$\text{Stair climbing time:}\ \ \Delta t = \frac{830\ \text{Cal}}{400\ \text{Cal/h}} = \mathbf{2.08\ h}$$

1-120 Three people have different lunches. The person who consumed the most calories from lunch is to be determined.

Properties The metabolizable energy contents of different foods are 530 Cal for the Big Mac, 640 Cal for the whopper, 350 Cal for french fries, and 5 for each olive (given).

Analysis The total calories consumed by each person during lunch are:

Person 1: Lunch calories = 530 Cal
Person 2: Lunch calories = **640 Cal**
Person 3: Lunch calories = $350{+}5{\times}50 = 600$ Cal

Therefore, the person with the Whopper will consume the most calories.

1-121 A 100-kg man decides to lose 5 kg by exercising without reducing his calorie intake. The number of days it will take for this man to lose 5 kg is to be determined.

Assumptions **1** The diet and exercise habits of the person remain the same other than the new daily exercise program. **2** The entire calorie deficiency is met by burning body fat.

Properties The metabolizable energy content of body fat is 33,100 Cal/kg (text).

Analysis The energy consumed by an average 68-kg adult during fast-swimming, fast dancing, jogging, biking, and relaxing are 860, 600, 540, 639, and 72 Cal/h, respectively. The daily energy consumption of this 100-kg man is

$$[(860+600+540+639\text{Cal/h})(1\text{h})+(72\text{Cal/h})(20\text{h})]\left(\frac{100\text{kg}}{68\text{kg}}\right)=5999\text{Cal}$$

Therefore, this person burns 5999-3000 = 2999 more Calories than he takes in, which corresponds to

$$m_{\text{fat}}=\frac{2999\text{Cal}}{33{,}100\text{kJ/kg}}\left(\frac{4.1868\text{kJ}}{1\text{Cal}}\right)=0.38\text{kg}$$

of body fat per day. Thus it will take only

$$\Delta t=\frac{5\text{ kg}}{0.38\text{ kg/day}}=\mathbf{13.2\ days}$$

for this man to lose 5 kg.

1-122 The range of healthy weight for adults is usually expressed in terms of the *body mass index* (BMI) in SI units as $\text{BMI}=\dfrac{W(\text{kg})}{H^2(\text{m}^2)}$. This formula is to be converted to English units such that the weight is in pounds and the height in inches.

Analysis Noting that 1 kg = 2.2 lbm and 1 m =39.37 in, the weight in lbm must be divided by 2.2 to convert it to kg, and the height in inches must be divided by 39.37 to convert it to m before inserting them into the formula. Therefore,

$$\text{BMI}=\frac{W(\text{kg})}{H^2(\text{m}^2)}=\frac{W(\text{lbm})/2.2}{H^2(\text{in}^2)/(39.37)^2}=705\frac{W(\text{lbm})}{H^2(\text{in}^2)}$$

Every person can calculate their own BMI using either SI or English units, and determine if it is in the healthy range.

Review Problems

1-123 The gravitational acceleration changes with altitude. Accounting for this variation, the weights of a body at different locations are to be determined.

Analysis The weight of an 80-kg man at various locations is obtained by substituting the altitude z (values in m) into the relation

$$W = mg = (80\text{kg})(9.807 - 3.32 \times 10^{-6} z\,\text{m/s}^2)\left(\frac{1\text{N}}{1\text{kg}\cdot\text{m/s}^2}\right)$$

Sea level: (z = 0 m): W = 80×(9.807-3.32x10^{-6}×0) = 80×9.807 = **784.6 N**

Denver: (z = 1610 m): W = 80×(9.807-3.32x10^{-6}×1610) = 80×9.802 = **784.2 N**

Mt. Ev.: (z = 8848 m): W = 80×(9.807-3.32x10^{-6}×8848) = 80×9.778 = **782.2 N**

1-124 A man is considering buying a 12-oz steak for \$3.15, or a 320-g steak for \$2.80. The steak that is a better buy is to be determined.

Assumptions The steaks are of identical quality.

Analysis To make a comparison possible, we need to express the cost of each steak on a common basis. Let us choose 1 kg as the basis for comparison. Using proper conversion factors, the unit cost of each steak is determined to be

12 ounce steak: $\text{Unit Cost} = \left(\frac{\$3.15}{12\text{ oz}}\right)\left(\frac{16\text{ oz}}{1\text{ lbm}}\right)\left(\frac{1\text{ lbm}}{0.45359\text{ kg}}\right) = \mathbf{\$9.26/kg}$

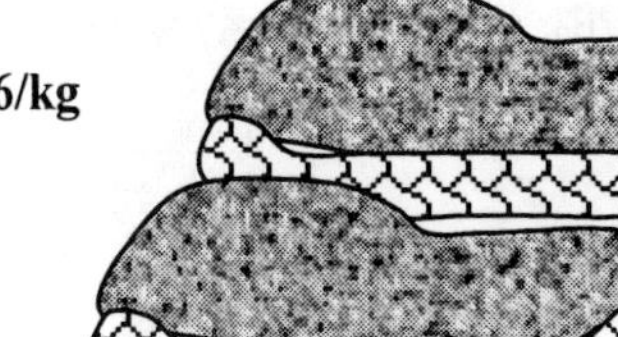

320 gram steak:

$$\text{Unit Cost} = \left(\frac{\$2.80}{320\text{ g}}\right)\left(\frac{1000\text{ g}}{1\text{ kg}}\right) = \mathbf{\$8.75/kg}$$

Therefore, the steak at the international market is a better buy.

1-125 The thrust developed by the jet engine of a Boeing 777 is given to be 85,000 pounds. This thrust is to be expressed in N and kgf.

Analysis Noting that 1 lbf = 4.448 N and 1 kgf = 9.807 N, the thrust developed can be expressed in two other units as

Thrust in N: $\text{Thrust} = (85{,}000\text{ lbf})\left(\frac{4.448\text{ N}}{1\text{ lbf}}\right) = \mathbf{378{,}080\ N}$

Thrust in kgf: $\text{Thrust} = (378{,}080\text{ N})\left(\frac{1\text{ kgf}}{9.807\text{ N}}\right) = \mathbf{38{,}552\ kgf}$

1-126E The efficiency of a refrigerator increases by 3% per °C rise in the minimum temperature. This increase is to be expressed per °F, K, and R rise in the minimum temperature.

Analysis The magnitudes of 1 K and 1°C are identical, so are the magnitudes of 1 R and 1°F. Also, a change of 1 K or 1°C in temperature corresponds to a change of 1.8 R or 1.8°F. Therefore, the increase in efficiency is

(a) **3%** for each K rise in temperature, and
(b), (c) 3/1.8 = **1.67%** for each R or °F rise in temperature.

1-127E The boiling temperature of water decreases by 3°C for each 1000 m rise in altitude. This decrease in temperature is to be expressed in °F, K, and R.

Analysis The magnitudes of 1 K and 1°C are identical, so are the magnitudes of 1 R and 1°F. Also, a change of 1 K or 1°C in temperature corresponds to a change of 1.8 R or 1.8°F. Therefore, the decrease in the boiling temperature is

(a) **3 K** for each 1000 m rise in altitude, and
(b), (c) 3×1.8 = **5.4°F = 5.4 R** for each 1000 m rise in altitude.

1-128E The average body temperature of a person rises by about 2°C during strenuous exercise. This increase in temperature is to be expressed in °F, K, and R.

Analysis The magnitudes of 1 K and 1°C are identical, so are the magnitudes of 1 R and 1°F. Also, a change of 1 K or 1°C in temperature corresponds to a change of 1.8 R or 1.8°F. Therefore, the rise in the body temperature during strenuous exercise is

(a) **2 K**
(b) 2×1.8 = **3.6°F**
(c) 2×1.8 = **3.6 R**

1-129E Hypothermia of 5°C is considered fatal. This fatal level temperature change of body temperature is to be expressed in °F, K, and R.

Analysis The magnitudes of 1 K and 1°C are identical, so are the magnitudes of 1 R and 1°F. Also, a change of 1 K or 1°C in temperature corresponds to a change of 1.8 R or 1.8°F. Therefore, the fatal level of hypothermia is

(a) **5 K**
(b) 5×1.8 = **9°F**
(c) 5×1.8 = **9 R**

1-130E A house is losing heat at a rate of 3000 kJ/h per °C temperature difference between the indoor and the outdoor temperatures. The rate of heat loss is to be expressed per °F, K, and R of temperature difference between the indoor and the outdoor temperatures.

Analysis The magnitudes of 1 K and 1°C are identical, so are the magnitudes of 1 R and 1°F. Also, a change of 1 K or 1°C in temperature corresponds to a change of 1.8 R or 1.8°F. Therefore, the rate of heat loss from the house is

(a) **3000 kJ/h** per K difference in temperature, and
(b), (c) 3000/1.8 = **1667 kJ/h** per R or °F rise in temperature.

1-131 The average temperature of the atmosphere is expressed as $T_{atm} = 288.15 - 6.5z$ where z is altitude in km. The temperature outside an airplane cruising at 12,000 m is to be determined.

Analysis Using the relation given, the average temperature of the atmosphere at an altitude of 12,000 m is determined to be

$$\begin{aligned} T_{atm} &= 288.15 - 6.5z \\ &= 288.15 - 6.5\times12 \\ &= \mathbf{210.15\ K = -63°C} \end{aligned}$$

Discussion This is the "average" temperature. The actual temperature at different times can be different.

1-132 A new "Smith" absolute temperature scale is proposed, and a value of 1000 S is assigned to the boiling point of water. The ice point on this scale, and its relation to the Kelvin scale are to be determined.

Analysis All linear absolute temperature scales read zero at absolute zero pressure, and are constant multiples of each other. For example, T(R) = 1.8 T(K). That is, multiplying a temperature value in K by 1.8 will give the same temperature in R.

The proposed temperature scale is an acceptable absolute temperature scale since it differs from the other absolute temperature scales by a constant only. The boiling temperature of water in the Kelvin and the Smith scales are 315.15 K and 1000 K, respectively. Therefore, these two temperature scales are related to each other by

$$T(S) = \frac{1000}{373.15} T(K) = \mathbf{2.6799\ T(K)}$$

K S
373.15 1000
0

The ice point of water on the Smith scale is

$$T(S)_{ice} = 2.6799\ T(K)_{ice} = 2.6799 \times 273.15 = \mathbf{732.0\ S}$$

1-133 An expression for the equivalent wind chill temperature is given in English units. It is to be converted to SI units.

Analysis The required conversion relations are 1 mph = 1.609 km/h and T(°F) = 1.8T(°C) + 32. The first thought that comes to mind is to replace T(°F) in the equation by its equivalent 1.8T(°C) + 32, and V in mph by 1.609 km/h, which is the "regular" way of converting units. However, the equation we have is not a regular dimensionally homogeneous equation, and thus the regular rules do not apply. The V in the equation is a constant whose value is equal to the numerical value of the velocity in mph. Therefore, if V is given in km/h, we should divide it by 1.609 to convert it to the desired unit of mph. That is,

$$T_{\text{equiv}}(°\text{F}) = 91.4 - [91.4 - T_{\text{ambient}}(°\text{F})][0.475 - 0.0203(V/1.609) + 0.304\sqrt{V/1.609}]$$

or

$$T_{\text{equiv}}(°\text{F}) = 91.4 - [91.4 - T_{\text{ambient}}(°\text{F})][0.475 - 0.0126V + 0.240\sqrt{V}]$$

where V is in km/h. Now the problem reduces to converting a temperature in °F to a temperature in °C, using the proper convection relation:

$$1.8T_{\text{equiv}}(°\text{C}) + 32 = 91.4 - [91.4 - (1.8T_{\text{ambient}}(°\text{C}) + 32)][0.475 - 0.0126V + 0.240\sqrt{V}]$$

which simplifies to

$$T_{\text{equiv}}(°\text{C}) = 33.0 - (33.0 - T_{\text{ambient}})(0.475 - 0.0126V + 0.240\sqrt{V})$$

where the ambient air temperature is in °C.

1-134 EES solution of this (and other comprehensive problems designated with the *computer icon*) is available to instructors at the *Instructor Manual* section of the *Online Learning Center* (OLC) at www.mhhe.com/cengel-boles. See the Preface for access information.

1-135 A water tank open to the atmosphere is initially filled with water. The plug near the bottom of a water tank is pulled out. The time it will take for the gage pressure at the bottom of the tank to drop to a specified value is to be determined.

Assumptions **1** The flow is uniform and incompressible. **2** The distance between the bottom of the tank and the center of the hole is negligible compared to the total water height.

Properties The density of water is taken to be ρ = 1000 kg/m³.

Analysis Noting that the gage pressure in a liquid is expressed as $P_{\text{gage}} = \rho g h$, the water height in the tank that will cause a gage pressure of 15.7 kPa is determined to be

$$h = \frac{P_{\text{gage}}}{\rho g} = \frac{15.7\text{ kPa}}{(1000\text{ kg/m}^3)(9.81\text{ m/s}^2)}\left(\frac{1000\text{ kg}\cdot\text{m/s}^2}{1\text{kN}}\right)\left(\frac{1\text{ kN/m}^2}{1\text{kPa}}\right) = 1.60\text{ m}$$

Now the question reduces to finding the time it takes for water level in the tank to drop to 1.6 m from 3.0 m.

We take the interior volume of the tank as the control volume. This is an unsteady flow problem since the properties (such as the amount of mass) within the control volume change with time. The conservation of mass relation for any system undergoing any process is given in the rate form as

$$\dot{m}_{\text{in}} - \dot{m}_{\text{out}} = \frac{dm_{\text{system}}}{dt} \quad (1)$$

During this process no mass enters the control volume ($\dot{m}_{\text{in}} = 0$), and the mass flow rate of discharged water can be expressed as

$$\dot{m}_{\text{out}} = (\rho \mathbf{V}_{ave} A)_{\text{out}} = \rho\sqrt{2gh}A_{\text{jet}} \quad (2)$$

where $A_{\text{jet}} = \pi D_{\text{jet}}^2 / 4$ is the cross-sectional area of the jet, which is constant. Noting that the density of water is constant, the mass of water in the tank at any time is

$$m_{\text{system}} = \rho V = \rho A_{\text{tank}} h \quad (3)$$

where A_{tank} is the base area of the tank. Substituting Eqs. (2) and (3) into (1) gives

$$-\rho\sqrt{2gh}A_{\text{jet}} = \frac{d(\rho A_{\text{tank}} h)}{dt} \rightarrow -\rho\sqrt{2gh}A_{\text{jet}} = \frac{\rho A_{\text{tank}}\, dh}{dt}$$

Canceling the densities and separating the variables give

$$dt = -\frac{A_{\text{tank}}}{A_{\text{jet}}}\frac{dh}{\sqrt{2gh}}$$

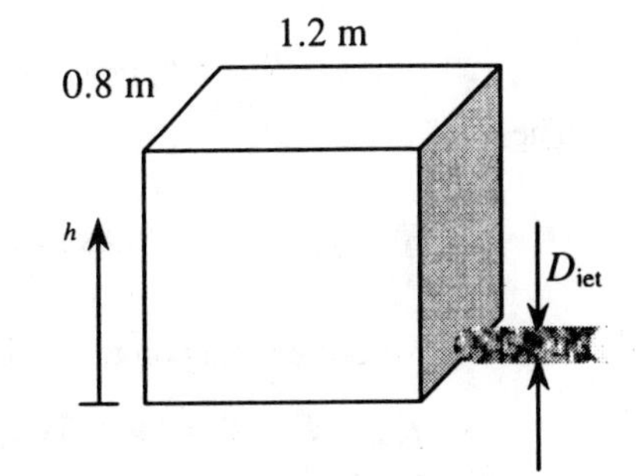

Integrating from $t = 0$ at which $h = h_0$ to $t = t$ at which $h = h_2$ gives

$$\int_0^t dt = -\frac{A_{\text{tank}}}{A_{\text{jet}}\sqrt{2g}}\int_{h_0}^{h_2}\frac{dh}{\sqrt{h}} \rightarrow t = \frac{A_{\text{tank}}}{A_{\text{jet}}}\frac{\sqrt{h_0} - \sqrt{h_2}}{\sqrt{g/2}}$$

Substituting, the time of discharge is determined to be

$$t = \frac{\sqrt{3\text{ m}} - \sqrt{1.6\text{ m}}}{\sqrt{(9.81\text{ m/s}^2)/2}}\left(\frac{0.8\text{ m}\times 1.2\text{ m}}{\pi(0.012\text{ m})^2/4}\right) = 1790\text{ s} = \mathbf{29.8\text{ min}}$$

Therefore, it will take about half an hour for the gage pressure at the bottom of the tank to drop to 15.7 kPa after the discharge hole is unplugged.

Discussion It can be shown that it will take 111 min for the tank to empty completely and the gage pressure at the bottom of the tank to drop to zero.

1-136 One section of the duct of an air-conditioning system is laid underwater. The upward force the water will exert on the duct is to be determined.

Assumptions **1** The diameter given is the outer diameter of the duct (or, the thickness of the duct material is negligible). **2** The weight of the duct and the air in is negligible.

Properties The density of air is given to be ρ = 1.30 kg/m³. We take the density of water to be 1000 kg/m³.

Analysis Noting that the weight of the duct and the air in it is negligible, the net upward force acting on the duct is the buoyancy force exerted by water. The volume of the underground section of the duct is

$$V = AL = (\pi D^2/4)L = [\pi(0.15\text{ m})^2/4](20\text{ m}) = 0.353\text{ m}^3$$

Then the buoyancy force becomes

$$F_B = \rho g V = (1000\text{ kg/m}^3)(9.81\text{ m/s}^2)(0.353\text{ m}^3)\left(\frac{1\text{ kN}}{1000\text{ kg}\cdot\text{m/s}^2}\right) = \mathbf{3.46\text{ kN}}$$

Discussion The upward force exerted by water on the duct is 3.46 kN, which is equivalent to the weight of a mass of 353 kg. Therefore, this force must be treated seriously.

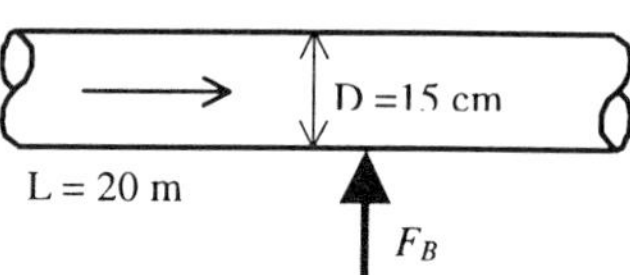

1-137 A helium balloon tied to the ground carries 2 people. The acceleration of the balloon when it is first released is to be determined.

Assumptions The weight of the cage and the ropes of the balloon is negligible.

Properties The density of air is given to be ρ = 1.16 kg/m³. The density of helium gas is 1/7th of this.

Analysis The buoyancy force acting on the balloon is

$$V_{balloon} = 4\pi r^3/3 = 4\pi(5m)^3/3 = 523.6m^3$$

$$F_B = \rho_{air} g V_{balloon}$$
$$= (1.16\text{kg/m}^3)(9.807\text{m/s}^2)(523.6\text{m}^3)\left(\frac{1\text{N}}{1\text{kg}\cdot\text{m/s}^2}\right) = 5956.5\text{N}$$

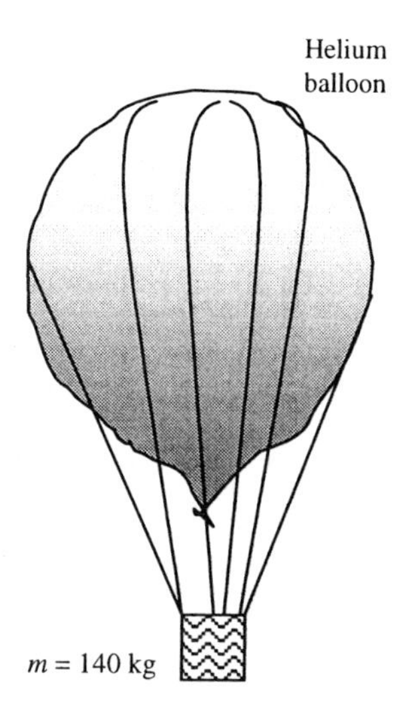

The total mass is

$$m_{He} = \rho_{He} V = \left(\frac{1.16}{7}\text{kg/m}^3\right)(523.6\text{m}^3) = 86.8\text{kg}$$

$$m_{total} = m_{He} + m_{people} = 86.8 + 2\times 70 = 226.8\text{kg}$$

The total weight is

$$W = m_{total}\, g = (226.8\text{kg})(9.807\text{m/s}^2)\left(\frac{1\text{N}}{1\text{kg}\cdot\text{m/s}^2}\right) = 2224.2\text{N}$$

Thus the net force acting on the balloon is

$$F_{net} = F_B - W = 5956.5 - 2224.2 = 3732.3\text{ N}$$

Then the acceleration becomes

$$a = \frac{F_{net}}{m_{total}} = \frac{3732.2\text{N}}{226.8\text{kg}}\left(\frac{1\text{kg}\cdot\text{m/s}^2}{1\text{N}}\right) = \mathbf{16.5m/s^2}$$

1-138 EES solution of this (and other comprehensive problems designated with the *computer icon*) is available to instructors at the *Instructor Manual* section of the *Online Learning Center* (OLC) at www.mhhe.com/cengel-boles. See the Preface for access information.

1-139 A balloon is filled with helium gas. The maximum amount of load the balloon can carry is to be determined.

Assumptions The weight of the cage and the ropes of the balloon is negligible.

Properties The density of air is given to be ρ = 1.16 kg/m^3. The density of helium gas is 1/7th of this.

Analysis In the limiting case, the net force acting on the balloon will be zero. That is, the buoyancy force and the weight will balance each other:

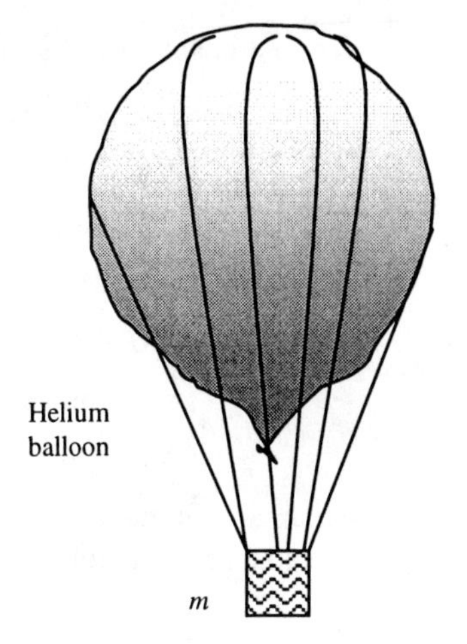

$$W = mg = F_B$$

$$m_{total} = \frac{F_B}{g} = \frac{5956.5 \text{ N}}{9.807 \text{ m/s}^2} = 607.4 \text{ kg}$$

Thus,

$$m_{\text{people}} = m_{\text{total}} - m_{\text{He}} = 607.4 - 86.8 = \mathbf{520.6\ kg}$$

1-140 The pressure in a steam boiler is given in kgf/cm^2. It is to be expressed in psi, kPa, atm, and bars.

Analysis We note that 1 atm = 1.03323 kgf/cm^2, 1 atm = 14.696 psi, 1 atm = 101.325 kPa, and 1 atm = 1.01325 bar (inner cover page of text). Then the desired conversions become:

In atm: $$P = (75 \text{ kgf/cm}^2)\left(\frac{1 \text{ atm}}{1.03323 \text{ kgf/cm}^2}\right) = 72.6 \text{ atm}$$

In psi: $$P = (75 \text{ kgf/cm}^2)\left(\frac{1 \text{ atm}}{1.03323 \text{ kgf/cm}^2}\right)\left(\frac{14.696 \text{ psi}}{1 \text{ atm}}\right) = \mathbf{1067\ psi}$$

In kPa: $$P = (75 \text{ kgf/cm}^2)\left(\frac{1 \text{ atm}}{1.03323 \text{ kgf/cm}^2}\right)\left(\frac{101.325 \text{ kPa}}{1 \text{ atm}}\right) = \mathbf{7355\ kPa}$$

In bars: $$P = (75 \text{ kgf/cm}^2)\left(\frac{1 \text{ atm}}{1.03323 \text{ kgf/cm}^2}\right)\left(\frac{1.01325 \text{ bar}}{1 \text{ atm}}\right) = \mathbf{73.55\ bar}$$

Discussion Note that the units atm, kgf/cm^2, and bar are almost identical to each other.

1-141 A barometer is used to measure the altitude of a plane relative to the ground. The barometric readings at the ground and in the plane are given. The altitude of the plane is to be determined.

Assumptions The variation of air density with altitude is negligible.

Properties The densities of air and mercury are given to be $\rho = 1.20\ \text{kg/m}^3$ and $\rho = 13{,}600\ \text{kg/m}^3$.

Analysis Atmospheric pressures at the location of the plane and the ground level are

$$\begin{aligned} P_{\text{plane}} &= (\rho g h)_{\text{plane}} \\ &= (13{,}600\ \text{kg/m}^3)(9.8\ \text{m/s}^2)(0.690\ \text{m})\left(\frac{1\ \text{N}}{1\ \text{kg}\cdot\text{m/s}^2}\right)\left(\frac{1\ \text{kPa}}{1000\ \text{N/m}^2}\right) \\ &= 91.96\ \text{kPa} \end{aligned}$$

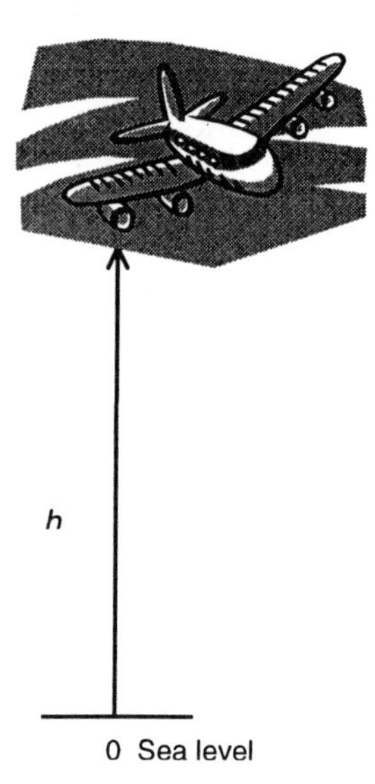

$$\begin{aligned} P_{\text{ground}} &= (\rho g h)_{\text{ground}} \\ &= (13{,}600\ \text{kg/m}^3)(9.8\ \text{m/s}^2)(0.753\ \text{m})\left(\frac{1\ \text{N}}{1\ \text{kg}\cdot\text{m/s}^2}\right)\left(\frac{1\ \text{kPa}}{1000\ \text{N/m}^2}\right) \\ &= 100.36\ \text{kPa} \end{aligned}$$

Taking an air column between the airplane and the ground and writing a force balance per unit base area, we obtain

$$W_{\text{air}} / A = P_{\text{ground}} - P_{\text{plane}}$$

$$(\rho g h)_{\text{air}} = P_{\text{ground}} - P_{\text{plane}}$$

$$(1.20\ \text{kg/m}^3)(9.8\ \text{m/s}^2)(h)\left(\frac{1\ \text{N}}{1\ \text{kg}\cdot\text{m/s}^2}\right)\left(\frac{1\ \text{kPa}}{1000\ \text{N/m}^2}\right) = (100.36 - 91.96)\ \text{kPa}$$

It yields $\quad h = \mathbf{714\ m}$

which is also the altitude of the airplane.

1-142 A 10-m high cylindrical container is filled with equal volumes of water and oil. The pressure difference between the top and the bottom of the container is to be determined.

Properties The density of water is given to be ρ = 1000 kg/m³. The specific gravity of oil is given to be 0.85.

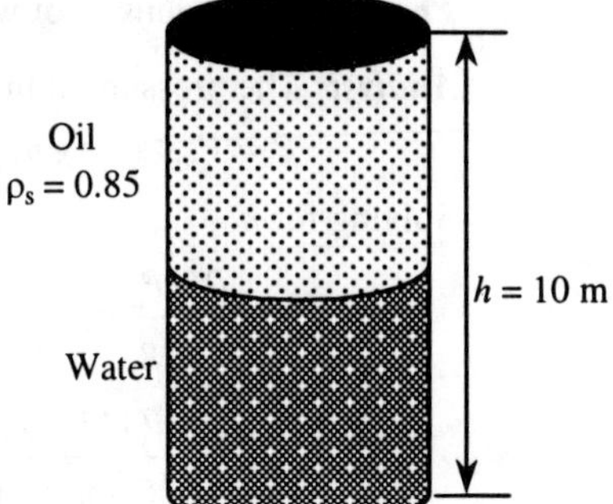

Analysis The density of the oil is obtained by multiplying its specific gravity by the density of water,

$$\rho = (\rho_s)(\rho_{H_2O}) = (0.85)(1000\text{ kg/m}^3) = 850\text{ kg/m}^3$$

The pressure difference between the top and the bottom of the cylinder is the sum of the pressure differences across the two fluids,

$$\begin{aligned}\Delta P_{total} &= \Delta P_{oil} + \Delta P_{water} = (\rho g h)_{oil} + (\rho g h)_{water} \\ &= \left[(850\text{kg/m}^3)(9.807\text{m/s}^2)(5\text{m}) + (1000\text{kg/m}^3)(9.807\text{m/s}^2)(5\text{m})\right]\left(\frac{1\text{kPa}}{1000\text{N/m}^2}\right) \\ &= \mathbf{90.7kPa}\end{aligned}$$

1-143 The pressure of a gas contained in a vertical piston-cylinder device is measured to be 500 kPa. The mass of the piston is to be determined.

Assumptions There is no friction between the piston and the cylinder.

Analysis Drawing the free body diagram of the piston and balancing the vertical forces yield

$$W = PA - P_{atm}A$$

$$mg = (P - P_{atm})A$$

$$(m)(9.807\text{m/s}^2) = (500 - 100\text{kPa})(30\times 10^{-4}\text{m}^2)\left(\frac{1000\text{kg/m}\cdot\text{s}^2}{1\text{kPa}}\right)$$

It yields $m = \mathbf{122.4\ kg}$

P_{atm}

P

$W = mg$

1-144 The gage pressure in a pressure cooker is maintained constant at 100 kPa by a petcock. The mass of the petcock is to be determined.

Assumptions There is no blockage of the pressure release valve.

Analysis Atmospheric pressure is acting on all surfaces of the petcock, which balances itself out. Therefore, it can be disregarded in calculations if we use the gage pressure as the cooker pressure. A force balance on the petcock ($\Sigma F_y = 0$) yields

$$W = P_{gage}A$$

$$m = \frac{P_{gage}A}{g} = \frac{(100kPa)(4\times 10^{-6}m^2)}{9.807m/s^2}\left(\frac{1000kg/m\cdot s^2}{1kPa}\right)$$

$$= \mathbf{0.0408kg}$$

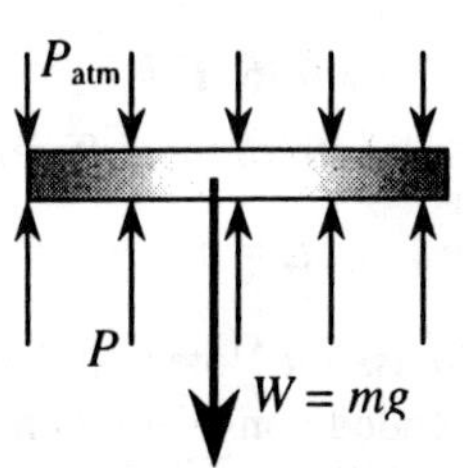

1-145 A glass tube open to the atmosphere is attached to a water pipe, and the pressure at the bottom of the tube is measured. It is to be determined how high the water will rise in the tube.

Properties The density of water is given to be $\rho = 1000\ \text{kg/m}^3$.

Analysis The pressure at the bottom of the tube can be expressed as

$$P = P_{atm} + (\rho g h)_{tube}$$

Solving for h,

$$h = \frac{P - P_{\text{atm}}}{\rho g}$$

$$= \frac{(115-92)\ \text{kPa}}{(1000\ \text{kg/m}^3)(9.8\ \text{m/s}^2)}\left(\frac{1\ \text{kg}\cdot\text{m/s}^2}{1\ \text{N}}\right)\left(\frac{1000\ \text{N/m}^2}{1\ \text{kPa}}\right)$$

$$= \mathbf{2.35\ m}$$

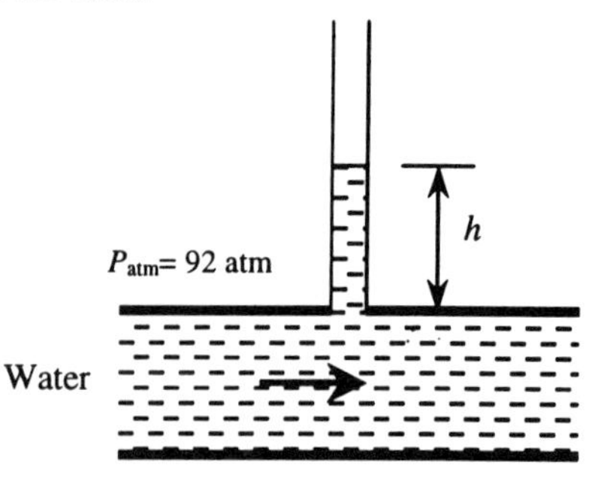

1-146 The average atmospheric pressure is given as $P_{atm} = 101.325(1 - 0.02256z)^{5.256}$ where z is the altitude in km. The atmospheric pressures at various locations are to be determined.

Analysis The atmospheric pressures at various locations are obtained by substituting the altitude z values in km into the relation

$$P_{atm} = 101.325(1 - 0.02256z)^{5.256}$$

Atlanta: (z = 0.306 km): $P_{\text{atm}} = 101.325(1 - 0.02256\times0.306)^{5.256} =$ **97.7 kPa**

Denver: (z = 1.610 km): $P_{\text{atm}} = 101.325(1 - 0.02256\times1.610)^{5.256} =$ **83.4 kPa**

M. City: (z = 2.309 km): $P_{\text{atm}} = 101.325(1 - 0.02256\times2.309)^{5.256} =$ **76.5 kPa**

Mt. Ev.: (z = 8.848 km): $P_{\text{atm}} = 101.325(1 - 0.02256\times8.848)^{5.256} =$ **31.4 kPa**

1-147 The air pressure in a duct is measured by an inclined manometer. For a given vertical level difference, the gage pressure in the duct and the length of the differential fluid column are to be determined.

Assumptions The manometer fluid is an incompressible substance.

Properties The density of the liquid is given to be $\rho = 0.81\ \text{kg/L} = 810\ \text{kg/m}^3$.

Analysis The gage pressure in the duct is determined from

$$P_{\text{gage}} = P_{\text{abs}} - P_{\text{atm}} = \rho g h$$

$$= (810\text{kg/m}^3)(9.81\text{m/s}^2)(0.08\text{m})\left(\frac{1\text{N}}{1\text{kg}\cdot\text{m/s}^2}\right)\left(\frac{1\text{Pa}}{1\text{N/m}^2}\right)$$

$$= \mathbf{636\ \ Pa}$$

The length of the differential fluid column is

$$L = h/\sin\theta = (8\text{cm})/\sin 35° = \mathbf{13.9\ cm}$$

Discussion Note that the length of the differential fluid column is extended considerably by inclining the manometer arm for better readability.

1-148E Equal volumes of water and ethyl alcohol are poured into a U-tube from different arms, and the alcohol side is pressurized until the contact surface of the two fluids moves to the bottom and the liquid levels in both arms become the same. The excess pressure applied on the alcohol side is to be determined.

Assumptions **1** Both water and alcohol are incompressible substances. **2** Alcohol does not mix with water. **3** The cross-sectional area of the U-tube is constant.

Properties The density of ethyl alcohol is given to be $\rho_a = 49.3$ lbm/ft³. We take the density of water to be $\rho_w = 62.4$ lbm/ft³.

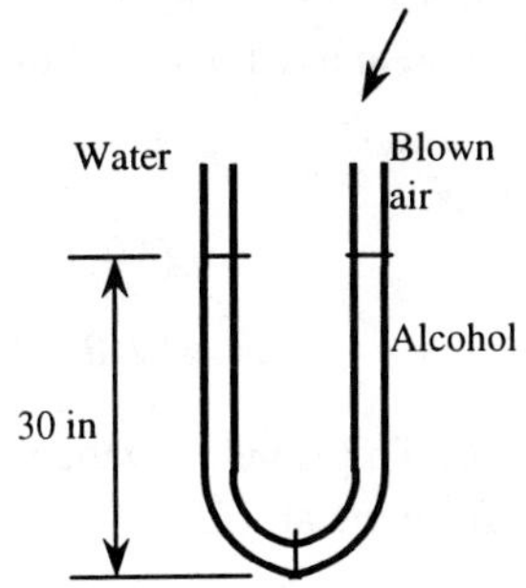

Water
Blown air
Alcohol
30 in

Analysis Noting that the pressure of both the water and the alcohol is the same at the contact surface, the pressure at this surface can be expressed as

$$P_{\text{contact}} = P_{\text{blow}} + \rho_a g h_a = P_{\text{atm}} + \rho_w g h_w$$

Noting that $h_a = h_w$ and rearranging,

$$\begin{aligned} P_{\text{gage,blow}} &= P_{\text{blow}} - P_{\text{atm}} = (\rho_w - \rho_a)gh \\ &= (62.4 - 49.3\ \text{lbm/ft}^3)(32.2\ \text{ft/s}^2)(30/12\ \text{ft})\left(\frac{1\,\text{lbf}}{32.2\,\text{lbm}\cdot\text{ft/s}^2}\right)\left(\frac{1\,\text{ft}^2}{144\,\text{in}^2}\right) \\ &= 0.227\,\textbf{psi} \end{aligned}$$

Discussion When the person stops blowing, the alcohol will rise and some water will flow into the right arm. It can be shown that when the curvature effects of the tube are disregarded, the differential height of water will be 23.7 in to balance 30-in of alcohol.

1-149 It is given that an IV fluid and the blood pressures balance each other when the bottle is at a certain height, and a certain gage pressure at the arm level is needed for sufficient flow rate. The gage pressure of the blood and elevation of the bottle required to maintain flow at the desired rate are to be determined.

Assumptions **1** The IV fluid is incompressible. **2** The IV bottle is open to the atmosphere.

Properties The density of the IV fluid is given to be $\rho = 1020 \text{ kg/m}^3$.

Analysis (*a*) Noting that the IV fluid and the blood pressures balance each other when the bottle is 1.2 m above the arm level, the gage pressure of the blood in the arm is simply equal to the gage pressure of the IV fluid at a depth of 1.2 m,

$$P_{\text{gage, arm}} = P_{\text{abs}} - P_{\text{atm}} = \rho g h_{\text{arm-bottle}}$$

$$= (1020 \text{ kg/m}^3)(9.81 \text{ m/s}^2)(1.20 \text{ m})\left(\frac{1 \text{kN}}{1000 \text{ kg} \cdot \text{m/s}^2}\right)\left(\frac{1 \text{ kPa}}{1 \text{ kN/m}^2}\right)$$

$$= \mathbf{12.0 \text{ kPa}}$$

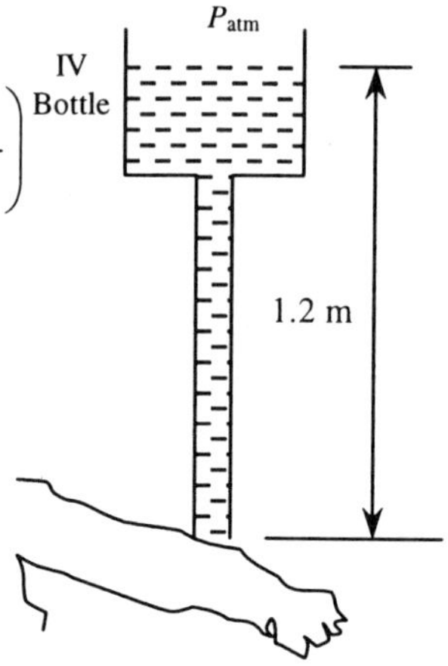

P_{atm}

IV Bottle

1.2 m

(*b*) To provide a gage pressure of 20 kPa at the arm level, the height of the bottle from the arm level is again determined from $P_{\text{gage, arm}} = \rho g h_{\text{arm-bottle}}$ to be

$$h_{\text{arm-bottle}} = \frac{P_{\text{gage, arm}}}{\rho g}$$

$$= \frac{20 \text{ kPa}}{(1020 \text{ kg/m}^3)(9.81 \text{ m/s}^2)}\left(\frac{1000 \text{ kg} \cdot \text{m/s}^2}{1 \text{ kN}}\right)\left(\frac{1 \text{ kN/m}^2}{1 \text{ kPa}}\right) = \mathbf{2.0 \text{ m}}$$

Discussion Note that the height of the reservoir can be used to control flow rates in gravity driven flows. When there is flow, the pressure drop in the tube due to friction should also be considered. This will result in raising the bottle a little higher to overcome pressure drop.

1-150 A gasoline line is connected to a pressure gage through a double-U manometer. For a given reading of the pressure gage, the gage pressure of the gasoline line is to be determined.

Assumptions **1** All the liquids are incompressible. **2** The effect of air column on pressure is negligible.

Properties The specific gravities of alcohol, mercury, and gasoline are given to be 0.79, 13.6, and 0.70, respectively. We take the density of water to be $\rho_w = 1000 \text{ kg/m}^3$.

Analysis Starting with the pressure indicated by the pressure gage and moving along the tube by adding (as we go down) or subtracting (as we go up) the ρgh terms until we reach the gasoline pipe, and setting the result equal to P_{gasoline} gives

$$P_{\text{gage}} - \rho_w g h_w + \rho_{\text{alcohol}} g h_{\text{alcohol}} - \rho_{\text{Hg}} g h_{\text{Hg}} - \rho_{\text{gasoline}} g h_{\text{gasoline}} = P_{\text{gasoline}}$$

Rearranging,

$$P_{\text{gasoline}} = P_{\text{gage}} - \rho_w g (h_w - \rho_{\text{s,alcohol}} h_{\text{alcohol}} + \rho_{\text{s,Hg}} h_{\text{Hg}} + \rho_{\text{s,gasoline}} h_{\text{gasoline}})$$

Substituting,

$$P_{\text{gasoline}} = 370 \text{ kPa} - (1000 \text{ kg/m}^3)(9.81 \text{ m/s}^2)[(0.45 \text{ m}) - 0.79(0.5 \text{ m}) + 13.6(0.1 \text{ m}) + 0.70(0.22 \text{ m})]$$
$$\times \left(\frac{1 \text{ kN}}{1000 \text{ kg}\cdot\text{m/s}^2}\right)\left(\frac{1 \text{ kPa}}{1 \text{ kN/m}^2}\right)$$
$$= \mathbf{354.6 \text{ kPa}}$$

Therefore, the pressure in the gasoline pipe is 15.4 kPa lower than the pressure reading of the pressure gage.

Discussion Note that sometimes the use of specific gravity offers great convenience in the solution of problems that involve several fluids.

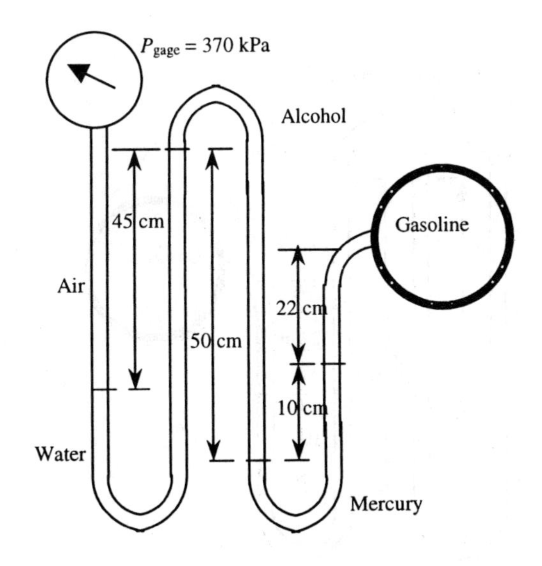

1-151 A gasoline line is connected to a pressure gage through a double-U manometer. For a given reading of the pressure gage, the gage pressure of the gasoline line is to be determined.

Assumptions **1** All the liquids are incompressible. **2** The effect of air column on pressure is negligible.

Properties The specific gravities of alcohol, mercury, and gasoline are given to be 0.79, 13.6, and 0.70, respectively. We take the density of water to be ρ_w = 1000 kg/m³.

Analysis Starting with the pressure indicated by the pressure gage and moving along the tube by adding (as we go down) or subtracting (as we go up) the ρgh terms until we reach the gasoline pipe, and setting the result equal to P_{gasoline} gives

$$P_{gage} - \rho_w gh_w + \rho_{\text{alcohol}} gh_{\text{alcohol}} - \rho_{\text{Hg}} gh_{\text{Hg}} - \rho_{\text{gasoline}} gh_{\text{gasoline}} = P_{\text{gasoline}}$$

Rearranging,

$$P_{\text{gasoline}} = P_{\text{gage}} - \rho_w g(h_w - \rho_{\text{s,alcohol}} h_{\text{s,alcohol}} + \rho_{\text{s,Hg}} h_{\text{Hg}} + \rho_{\text{s,gasoline}} h_{\text{s,gasoline}})$$

Substituting,

$$P_{\text{gasoline}} = 240\text{ kPa - }(1000\text{ kg/m}^3)(9.81\text{ m/s}^2)[(0.45\text{ m}) - 0.79(0.5\text{ m}) + 13.6(0.1\text{ m}) + 0.70(0.22\text{ m})]$$
$$\times\left(\frac{1\text{ kN}}{1000\text{ kg}\cdot\text{m/s}^2}\right)\left(\frac{1\text{ kPa}}{1\text{ kN/m}^2}\right)$$
$$= \mathbf{224.6\text{ kPa}}$$

Therefore, the pressure in the gasoline pipe is 15.4 kPa lower than the pressure reading of the pressure gage.

Discussion Note that sometimes the use of specific gravity offers great convenience in the solution of problems that involve several fluids.

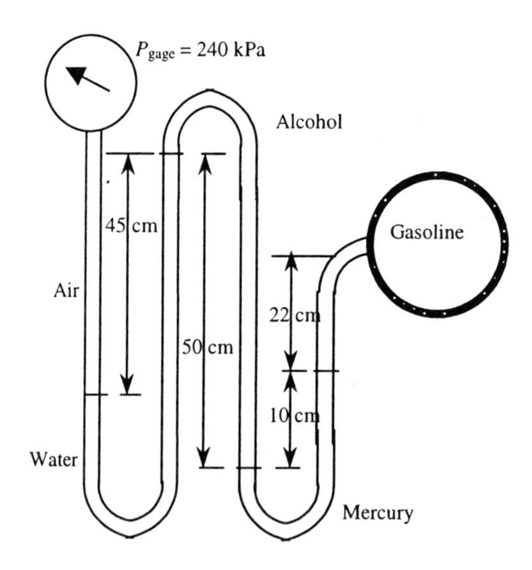

1-152E A water pipe is connected to a double-U manometer whose free arm is open to the atmosphere. The absolute pressure at the center of the pipe is to be determined.

Assumptions **1** All the liquids are incompressible. **2** The solubility of the liquids in each other is negligible.

Properties The specific gravities of alcohol, mercury, and oil are given to be 0.79, 13.6, and 0.80, respectively. We take the density of water to be $\rho_w = 62.4$ lbm/ft^3.

Analysis Starting with the pressure at the center of the water pipe, and moving along the tube by adding (as we go down) or subtracting (as we go up) the ρgh terms until we reach the free surface of oil where the oil tube is exposed to the atmosphere, and setting the result equal to P_{atm} gives

$$P_{\text{water pipe}} - \rho_{\text{water}} gh_{\text{water}} + \rho_{\text{alcohol}} gh_{\text{alcohol}} - \rho_{\text{Hg}} gh_{\text{Hg}} - \rho_{\text{oil}} gh_{\text{oil}} = P_{atm}$$

Solving for $P_{\text{water pipe}}$,

$$P_{\text{water pipe}} = P_{atm} + \rho_{\text{water}} g(h_{\text{water}} - \rho_{\text{s, alcohol}} h_{\text{alcohol}} + \rho_{\text{s, Hg}} h_{\text{Hg}} + \rho_{\text{s, oil}} h_{\text{oil}})$$

Substituting,

$$\begin{aligned} P_{\text{water pipe}} &= 14.2\,\text{psia} + (62.4\,\text{lbm/ft}^3)(32.2\,\text{ft/s}^2)[(35/12\,\text{ft}) - 0.79(60/12\,\text{ft}) + 13.6(15/12\,\text{ft}) \\ &\quad + 0.8(40/12\,\text{ft})] \times \left(\frac{1\,\text{lbf}}{32.2\,\text{lbm}\cdot\text{ft/s}^2}\right)\left(\frac{1\,\text{ft}^2}{144\,\text{in}^2}\right) \\ &= 22.3\,\textbf{psia} \end{aligned}$$

Therefore, the absolute pressure in the water pipe is 22.3 psia.

Discussion Note that jumping horizontally from one tube to the next and realizing that pressure remains the same in the same fluid simplifies the analysis greatly.

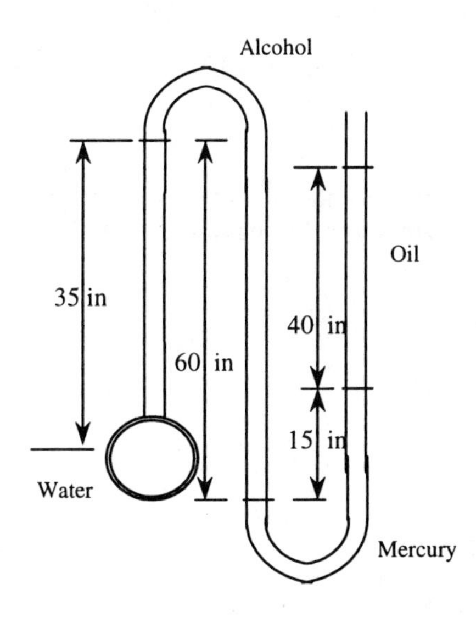

Fundamentals of Engineering (FE) Exam Problems

1-153 Consider a fish swimming 10 m below the free surface of water. The increase in the pressure exerted on the fish when it dives to a depth of 50 m below the free surface is

(a) 392 Pa (b) 9800 Pa (c) 40,000 Pa (d) 392,000 Pa (e) 490,000 Pa

Answer (*d*) 392,000 Pa

Solution Solved by EES Software. Solutions can be verified by copying-and-pasting the following lines on a blank EES screen. (Similar problems and their solutions can be obtained easily by modifying numerical values).

```
rho=1000 "kg/m3"
g=9.81 "m/s2"
z1=10 "m"
z2=50 "m"

DELTAP=rho*g*(z2-z1) "Pa"

"Some Wrong Solutions with Common Mistakes:"
W1_P=rho*g*(z2-z1)/1000        "dividing by 1000"
W2_P=rho*g*(z1+z2)   "adding depts instead of subtracting"
W3_P=rho*(z1+z2)   "not using g"
W4_P=rho*g*(0+z2)      "ignoring z1"
```

1-154 The atmospheric pressures at the top and the bottom of a building are read by a barometer to be 96.0 and 98.0 kPa. If the density of air is 1.2 kg/m^3, the height of the building is

(a) 14 m (b) 17 m (c) 142 m (d) 170 m (e) 210 m

Answer (d) 170 m

Solution Solved by EES Software. Solutions can be verified by copying-and-pasting the following lines on a blank EES screen. (Similar problems and their solutions can be obtained easily by modifying numerical values).

```
rho=1.2 "kg/m3"
g=9.81 "m/s2"
P1=96 "kPa"
P2=98 "kPa"
DELTAP=P2-P1 "kPa"

DELTAP=rho*g*h/1000 "kPa"

"Some Wrong Solutions with Common Mistakes:"
DELTAP=rho*W1_h/1000      "not using g"
DELTAP=g*W2_h/1000        "not using rho"
P2=rho*g*W3_h/1000          "ignoring P1"
P1=rho*g*W4_h/1000          "ignoring P2"
```

1-155 An apple loses 0.9 kJ of heat as it cools per °C drop in its temperature. The amount of heat loss from the apple per °F drop in its temperature is

(a) 0.25 kJ (b) 0.50 kJ (c) 1.00 kJ (d) 1.62 kJ (e) 0.81 kJ

Answer (b) 0.50 kJ

Solution Solved by EES Software. Solutions can be verified by copying-and-pasting the following lines on a blank EES screen. (Similar problems and their solutions can be obtained easily by modifying numerical values).

```
Q_perC=0.9 "kJ"
Q_perF=Q_perC/1.8 "kJ"

"Some Wrong Solutions with Common Mistakes:"
W1_Q=Q_perC*1.8 "multiplying instead of dividing"
W2_Q=Q_perC "setting them equal to each other"
```

1-156 Consider a 2-m deep swimming pool. The pressure difference between the top and bottom of the pool is

(a) 12.0 kPa (b) 19.6 kPa (c) 38.1 kPa (d) 50.8 kPa (e) 200 kPa

Answer (b) 19.6 kPa

Solution Solved by EES Software. Solutions can be verified by copying-and-pasting the following lines on a blank EES screen. (Similar problems and their solutions can be obtained easily by modifying numerical values).

```
rho=1000 "kg/m^3"
g=9.81 "m/s2"
z1=0 "m"
z2=2 "m"

DELTAP=rho*g*(z2-z1)/1000 "kPa"

"Some Wrong Solutions with Common Mistakes:"
W1_P=rho*(z1+z2)/1000    "not using g"
W2_P=rho*g*(z2-z1)/2000      "taking half of z"
W3_P=rho*g*(z2-z1) "not dividing by 1000"
```

1-157 At sea level, the weight of 1 kg mass in SI units is 9.81 N. The weight of 1 lbm mass in English units is

(a) 1 lbf (b) 9.81 lbf (c) 32.2 lbf (d) 0.1 lbf (e) 0.031 lbf

Answer (a) 1 lbf

Solution Solved by EES Software. Solutions can be verified by copying-and-pasting the following lines on a blank EES screen. (Similar problems and their solutions can be obtained easily by modifying numerical values).

```
m=1 "lbm"
g=32.2 "ft/s2"
W=m*g/32.2 "lbf"

"Some Wrong Solutions with Common Mistakes:"
gSI=9.81 "m/s2"
W1_W= m*gSI "Using wrong conversion"
W2_W= m*g "Using wrong conversion"
W3_W= m/gSI "Using wrong conversion"
W4_W= m/g "Using wrong conversion"
```

1-158 During a heating process, the temperature of an object rises by 10°C. This temperature rise is equivalent to a temperature rise of

(a) 10°F (b) 42°F (c) 18 K (d) 18 R (e) 283 K

Answer (d) 18 R

Solution Solved by EES Software. Solutions can be verified by copying-and-pasting the following lines on a blank EES screen. (Similar problems and their solutions can be obtained easily by modifying numerical values).

```
T_inC=10 "C"
T_inR=T_inC*1.8 "R"

"Some Wrong Solutions with Common Mistakes:"
W1_TinF=T_inC "F, setting C and F equal to each other"
W2_TinF=T_inC*1.8+32 "F, converting to F "
W3_TinK=1.8*T_inC "K, wrong conversion from C to K"
W4_TinK=T_inC+273 "K, converting to K"
```

1-159 ... 1-161 Design and Essay Problems

Chapter 2
PROPERTIES OF PURE SUBSTANCES

Pure Substances, Phase Change Processes, Property Diagrams

2-1C Yes. Because it has the same chemical composition throughout.

2-2C A liquid that is about to vaporize is saturated liquid; otherwise it is compressed liquid.

2-3C A vapor that is about to condense is saturated vapor; otherwise it is superheated vapor.

2-4C No.

2-5C No.

2-6C Yes. The saturation temperature of a pure substance depends on pressure. The higher the pressure, the higher the saturation or boiling temperature.

2-7C The temperature will also increase since the boiling or saturation temperature of a pure substance depends on pressure.

2-8C Because one cannot be varied while holding the other constant. In other words, when one changes, so does the other one.

2-9C At critical point the saturated liquid and the saturated vapor states are identical. At triple point the three phases of a pure substance coexist in equilibrium.

2-10C Yes.

2-11C Case (c) when the pan is covered with a heavy lid. Because the heavier the lid, the greater the pressure in the pan, and thus the greater the cooking temperature.

2-12C At supercritical pressures, there is no distinct phase change process. The liquid uniformly and gradually expands into a vapor. At subcritical pressures, there is always a distinct surface between the phases.

Property Tables

2-13C A given volume of water will boil at a higher temperature in a **tall and narrow pot** since the pressure at the bottom (and thus the corresponding saturation pressure) will be higher in that case.

2-14C A perfectly fitting pot and its lid often stick after cooking as a result of the vacuum created inside as the temperature and thus the corresponding saturation pressure inside the pan drops. An easy way of removing the lid is to reheat the food. When the temperature rises to boiling level, the pressure rises to atmospheric value and thus the lid will come right off.

2-15C The molar mass of gasoline (C_8H_{18}) is 114 kg/kmol, which is much larger than the molar mass of air that is 29 kg/kmol. Therefore, the gasoline vapor will settle down instead of rising even if it is at a much higher temperature than the surrounding air. As a result, the warm mixture of air and gasoline on top of an open gasoline can will most likely settle down instead of rising in a cooler environment

2-16C Ice can be made by evacuating the air in a water tank. During evacuation, vapor is also thrown out, and thus the vapor pressure in the tank drops, causing a difference between the vapor pressures at the water surface and in the tank. This pressure difference is the driving force of vaporization, and forces the liquid to evaporate. But the liquid must absorb the heat of vaporization before it can vaporize, and it absorbs it from the liquid and the air in the neighborhood, causing the temperature in the tank to drop. The process continues until water starts freezing. The process can be made more efficient by insulating the tank well so that the entire heat of vaporization comes essentially from the water.

2-17C Yes. Otherwise we can create energy by alternately vaporizing and condensing a substance.

2-18C No. Because in the thermodynamic analysis we deal with the changes in properties; and the changes are independent of the selected reference state.

2-19C The term h_{fg} represents the amount of energy needed to vaporize a unit mass of saturated liquid at a specified temperature or pressure. It can be determined from $h_{fg} = h_g - h_f$.

2-20C Yes; the higher the temperature the lower the h_{fg} value.

2-21C Quality is the fraction of vapor in a saturated liquid-vapor mixture. It has no meaning in the superheated vapor region.

2-22C Completely vaporizing 1 kg of saturated liquid at 1 atm pressure since the higher the pressure, the lower the h_{fg} .

2-23C Yes. It decreases with increasing pressure and becomes zero at the critical pressure.

2-24C No. Quality is a mass ratio, and it is not identical to the volume ratio.

2-25C The compressed liquid can be approximated as a saturated liquid at the given temperature. Thus $v_{T,P} \cong v_{f@T}$.

2-26 [*Also solved by EES on enclosed CD*] *Complete the following table for H_2O:*

T, °C	*P*, kPa	*v*, m³ / kg	Phase description
50	***12.349***	4.16	***Saturated mixture***
120.23	200	***0.8857***	Saturated vapor
250	400	***0.5951***	***Superheated vapor***
110	600	***0.001052***	***Compressed liquid***

2-27 EES solution of this (and other comprehensive problems designated with the *computer icon*) is available to instructors at the *Instructor Manual* section of the *Online Learning Center* (OLC) at www.mhhe.com/cengel-boles. See the Preface for access information.

2-28E *Complete the following table for H_2O:*

T, °F	P, psia	*u*, Btu / lbm	Phase description
300	***66.98***	782	***Saturated mixture***
267.26	40	***236.03***	Saturated liquid
500	120	***1174.2***	***Superheated vapor***
400	400	***374.27***	***Compressed liquid***

2-29 See the note above for Prob. 2-27.

2-30 *Complete the following table for* H_2O:

T, °C	P, kPa	*h*, kJ / kg	*x*	Phase description
120.23	200	***2046.0***	0.7	***Saturated mixture***
140	***361.3***	1800	***0.56***	***Saturated mixture***
177.69	950	***753.02***	0.0	***Saturated liquid***
80	500	***334.91***	- - -	***Compressed liquid***
350	800	3161.7	- - -	***Superheated vapor***

2-31 *Complete the following table for Refrigerant-134a*:

T, °C	P, kPa	*v*, m^3 / kg	Phase description
-8	500	***0.0007569***	***Compressed liquid***
30	***770.06***	0.022	***Saturated mixture***
2.48	320	***0.0632***	Saturated vapor
100	600	***0.04790***	***Superheated vapor***

2-32 *Complete the following table for Refrigerant-134a*:

T, °C	P, kPa	*u*, kJ / kg	Phase description
20	***571.6***	95	***Saturated mixture***
-12	***185.4***	***34.25***	Saturated liquid
86.88	400	300	***Superheated vapor***
8	600	***60.43***	***Compressed liquid***

2-33E *Complete the following table for Refrigerant-134a*:

T, °F	P, psia	*h*, Btu / lbm	*x*	Phase description
65.93	80	78	***0.581***	***Saturated mixture***
15	***29.756***	***68.83***	0.6	***Saturated mixture***
10	70	***14.66***	- - -	***Compressed liquid***
160	180	128.77	- - -	***Superheated vapor***
110	***161.04***	***115.96***	1.0	***Saturated vapor***

2-34 *Complete the following table for* H_2O:

T, °C	P, kPa	*v*, m^3 / kg	Phase description
140	***361.3***	0.48	***Saturated mixture***
170.43	800	***0.001115***	Saturated liquid
25	750	***0.001003***	***Compressed liquid***
500	***2709***	0.130	***Superheated vapor***

2-35 *Complete the following table for H_2O:*

T, °C	P, kPa	u, kJ / kg	Phase description
143.63	400	1825	***Saturated mixture***
220	***2318***	***2602.4***	Saturated vapor
190	2000	***806.19***	***Compressed liquid***
466.7	4000	3040	***Superheated vapor***

2-36E The temperature in a pressure cooker during cooking at sea level is measured to be 250°F. The absolute pressure inside the cooker and the effect of elevation on the answer are to be determined.

Assumptions Properties of pure water can be used to approximate the properties of juicy water in the cooker.

Properties The saturation pressure of water at 250°F is 29.82 psia (Table A-5E). The standard atmospheric pressure at sea level is 1 atm = 14.7 psia.

Analysis The absolute pressure in the cooker is simply the saturation pressure at the cooking temperature,

$$P_{\text{abs}} = P_{sat@250^\circ\text{F}} = \mathbf{29.82\ psia}$$

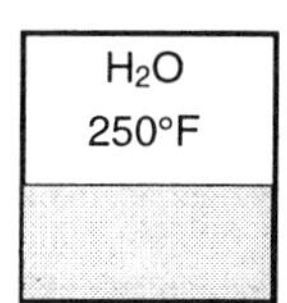

It is equivalent to

$$P_{\text{abs}} = 29.82\text{psia}\left(\frac{1\text{ atm}}{14.7\text{ psia}}\right) = \mathbf{2.03\,atm}$$

The elevation has **no effect** on the absolute pressure inside when the temperature is maintained constant at 250°F.

2-37E The local atmospheric pressure, and thus the boiling temperature, changes with the weather conditions. The change in the boiling temperature corresponding to a change of 0.3 in of mercury in atmospheric pressure is to be determined.

Properties The saturation pressures of water at 200 and 212°F are 11.529 and 14.698 psia, respectively (Table A-5E). One in. of mercury is equivalent to 1 inHg = 3.387 kPa = 0.491 psia (inner cover page).

Analysis A change of 0.3 in of mercury in atmospheric pressure corresponds to

$$\Delta P = (0.3\text{ inHg})\left(\frac{0.491\text{ psia}}{1\text{ inHg}}\right) = 0.147\text{psia}$$

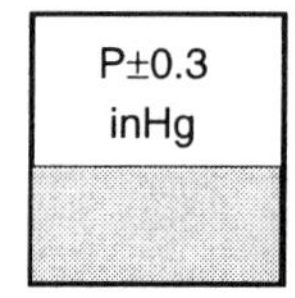

At about boiling temperature, the change in boiling temperature per 1 psia change in pressure is determined using data at 200 and 212°F to be

$$\frac{\Delta T}{\Delta P} = \frac{(212-200)^\circ\text{F}}{(14.698-11.529)\text{ psia}} = 3.787\ ^\circ\text{F/psia}$$

Then the change in saturation (boiling) temperature corresponding to a change of 0.147 psia becomes

$$\Delta T_{\text{boiling}} = (3.787\ ^\circ\text{F/psia})\Delta P = (3.787\ ^\circ\text{F/psia})(0.147\text{ psia}) = \mathbf{0.56^\circ F}$$

which is very small. Therefore, the effect of variation of atmospheric pressure on the boiling temperature is negligible.

2-38 A person cooks a meal in a pot that is covered with a well-fitting lid, and leaves the food to cool to the room temperature. It is to be determined if the lid will open or the pan will move up together with the lid when the person attempts to open the pan by lifting the lid up.

Assumptions **1** The local atmospheric pressure is 1 atm = 101.325 kPa. **2** The weight of the lid is small and thus its effect on the boiling pressure and temperature is negligible. **3** No air has leaked into the pan during cooling.

Properties The saturation pressure of water at 20°C is 2.339 kPa (Table A-4).

Analysis Noting that the weight of the lid is negligible, the reaction force F on the lid after cooling at the pan-lid interface can be determined from a force balance on the lid in the vertical direction to be

$$PA + F = P_{atm}A$$

or,

$$F = A(P_{atm} - P) = (\pi D^2/4)(P_{atm} - P)$$

$$= \frac{\pi(0.3\text{ m})^2}{4}(101{,}325 - 2339)\text{ Pa}$$

$$= 6997\text{ m}^2\text{Pa} = 6997\text{ N} \quad (\text{since } 1\text{ Pa} = 1\text{ N/m}^2)$$

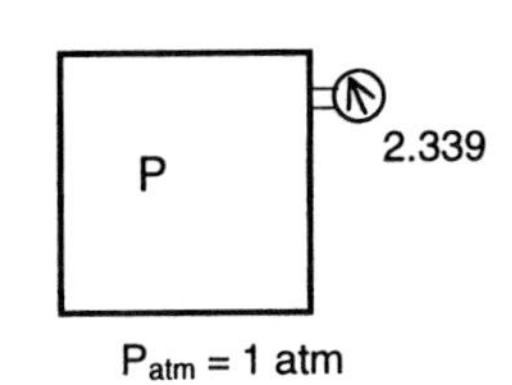

The weight of the pan and its contents is

$$W = mg = (8\text{ kg})(9.8\text{ m/s}^2) = 78\text{ N}$$

which is much less than the reaction force of 6997 N at the pan-lid interface. Therefore, the pan will **move up** together with the lid when the person attempts to open the pan by lifting the lid up. In fact, it looks like the lid will not open even if the mass of the pan and its contents is several hundred kg.

2-39 Water is boiled at sea level (1 atm pressure) in a pan placed on top of a 2-kW electric burner that transfers 60% of the heat generated to the water. The rate of evaporation of water is to be determined.

Properties The properties of water at 1 atm and thus at the saturation temperature of 100°C are h_{fg} = 2257 kJ/kg (Table A-4).

Analysis The net rate of heat transfer to the water is

$$\dot{Q} = 0.60 \times 3\text{ kW} = 1.8\text{ kW}$$

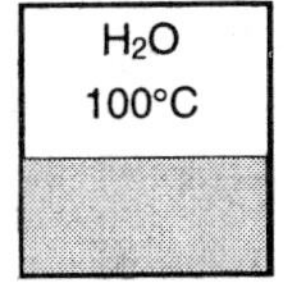

Noting that it takes 2257 kJ of energy to vaporize 1 kg of saturated liquid water, the rate of evaporation of water is determined to be

$$\dot{m}_{\text{evaporation}} = \frac{\dot{Q}}{h_{fg}} = \frac{1.8\text{ kJ/s}}{2257\text{ kJ/kg}} = \mathbf{0.80 \times 10^{-3}\ kg/s} = 2.87\text{ kg/h}$$

2-40 Water is boiled at 1500 m (84.5 kPa pressure) in a pan placed on top of a 2-kW electric burner that transfers 60% of the heat generated to the water. The rate of evaporation of water is to be determined.

Properties The properties of water at 84.5 kPa and thus at the saturation temperature of 95°C are h_{fg} = 2270.2 kJ/kg (Table A-4).

Analysis The net rate of heat transfer to the water is

H_2O
95°C

$$\dot{Q} = 0.60 \times 3 \text{ kW} = 1.8 \text{ kW}$$

Noting that it takes 2270.2 kJ of energy to vaporize 1 kg of saturated liquid water, the rate of evaporation of water is determined to be

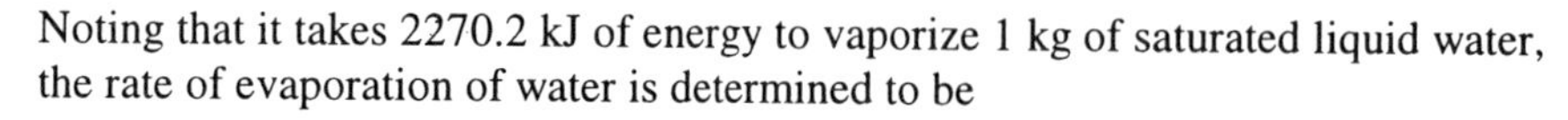

$$\dot{m}_{\text{evaporation}} = \frac{\dot{Q}}{h_{fg}} = \frac{1.8 \text{ kJ/s}}{2270.2 \text{ kJ/kg}} = \mathbf{0.79 \times 10^{-3} \text{ kg/s}} = 2.85 \text{ kg/h}$$

2-41 Water is boiled at 1 atm pressure in a pan placed on an electric burner. The water level drops by 10 cm in 30 min during boiling. The rate of heat transfer to the water is to be determined.

Properties The properties of water at 1 atm and thus at a saturation temperature of T_{sat} = 100°C are h_{fg} = 2257 kJ/kg and v_f = 0.001044 m^3/kg (Table A-4).

Analysis The rate of evaporation of water is

$$m_{\text{evap}} = \frac{V_{\text{evap}}}{v_f} = \frac{(\pi D^2/4)L}{v_f} = \frac{[\pi(0.2 \text{ m})^2/4](0.10 \text{ m})}{0.001044} = 3.009 \text{ kg}$$

H_2O
100°C

$$\dot{m}_{\text{evap}} = \frac{m_{\text{evap}}}{\Delta t} = \frac{3.009 \text{ kg}}{30 \times 60 \text{ s}} = 0.00167 \text{ kg/s}$$

Then the rate of heat transfer to water becomes

$$\dot{Q} = \dot{m}_{\text{evap}} h_{fg} = (0.00167 \text{ kg/s})(2257 \text{ kJ/kg}) = \mathbf{3.77 \text{ kW}}$$

2-42 Water is boiled at a location where the atmospheric pressure is 79.5 kPa in a pan placed on an electric burner. The water level drops by 10 cm in 30 min during boiling. The rate of heat transfer to the water is to be determined.

Properties The properties of water at 79.5 kPa are $T_{sat} = 93.2°C$, $h_{fg} = 2275$ kJ/kg and $v_f = 0.001038$ m^3/kg (Table A-5).

Analysis The rate of evaporation of water is

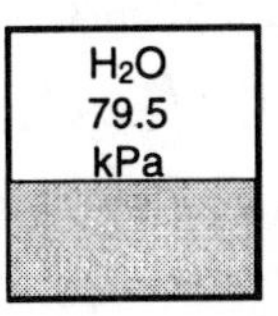

$$m_{evap} = \frac{V_{evap}}{v_f} = \frac{(\pi D^2/4)L}{v_f} = \frac{[\pi(0.2\text{ m})^2/4](0.10\text{ m})}{0.001038} = 3.027\text{ kg}$$

$$\dot{m}_{evap} = \frac{m_{evap}}{\Delta t} = \frac{3.027\text{ kg}}{30 \times 60\text{ s}} = 0.00168\text{ kg/s}$$

Then the rate of heat transfer to water becomes

$$\dot{Q} = \dot{m}_{evap} h_{fg} = (0.00168\text{ kg/s})(2275\text{ kJ/kg}) = \mathbf{3.82\text{ kW}}$$

2-43 Saturated steam at T_{sat} = 30°C condenses on the outer surface of a cooling tube at a rate of 45 kg/h. The rate of heat transfer from the steam to the cooling water is to be determined.

Assumptions **1** Steady operating conditions exist. **2** The condensate leaves the condenser as a saturated liquid at 30°C.

Properties The properties of water at the saturation temperature of 30°C are $h_{fg} = 2431$ kJ/kg.

Analysis Noting that 2431 kJ of heat is released as 1 kg of saturated vapor at 30°C condenses, the rate of heat transfer from the steam to the cooling water in the tube is determined directly from

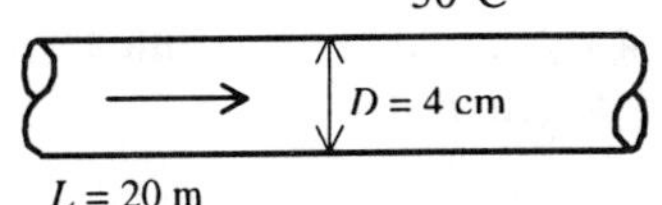

$$\dot{Q} = \dot{m}_{evap} h_{fg} = (45\text{ kg/h})(2431\text{ kJ/kg}) = 109{,}395\text{ kJ/h} = \mathbf{30.4\text{ kW}}$$

2-44 The average atmospheric pressure in Denver is 83.4 kPa. The boiling temperature of water in Denver is to be determined.

Analysis The boiling temperature of water in Denver is the saturation temperature corresponding to the atmospheric pressure in Denver, which is 83.4 kPa:

$$T = T_{sat@83.4\ kPa} = \mathbf{94.4°C}$$

2-45 The boiling temperature of water in a 5-cm deep pan is given. The boiling temperature in a 40-cm deep pan is to be determined.

Assumptions Both pans are full of water.

Properties The density of liquid water is approximately $\rho = 1000\ \text{kg/m}^3$.

Analysis The pressure at the bottom of the 5-cm pan is the saturation pressure corresponding to the boiling temperature of 98°C:

$$P = P_{\text{sat@98°C}} = 94.63\ \text{kPa}$$

The pressure difference between the bottoms of two pans is

$$\Delta P = \rho g h = (1000\text{kg/m}^3)(9.8\text{m/s}^2)(0.35\text{m})\left(\frac{1\text{kPa}}{1000\text{kg/m}\cdot\text{s}^2}\right) = 3.43\text{kPa}$$

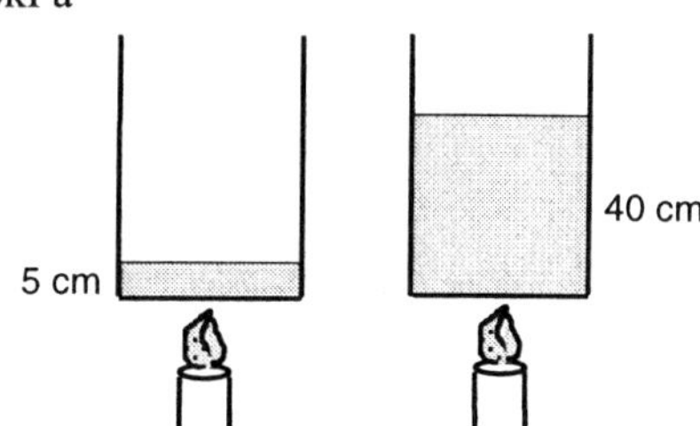

Then the pressure at the bottom of the 40-cm deep pan is

$$P = 94.63 + 3.43 = 98.06\ \text{kPa}$$

Then the boiling temperature becomes

$$T_{\text{boiling}} = T_{\text{sat@98.06kPa}} = \mathbf{99.0^\circ C}$$

2-46 A cooking pan is filled with water and covered with a 4-kg lid. The boiling temperature of water is to be determined.

Analysis The pressure in the pan is determined from a force balance on the lid,

$$PA = P_{atm}A + W$$

or,

$$\begin{aligned} P &= P_{atm} + \frac{mg}{A} \\ &= (101\text{kPa}) + \frac{(4\text{kg})(9.807\text{m/s}^2)}{\pi(0.1\text{m})^2}\left(\frac{1\text{kPa}}{1000\text{kg/m}\cdot\text{s}^2}\right) \\ &= 102.25\text{kPa} \end{aligned}$$

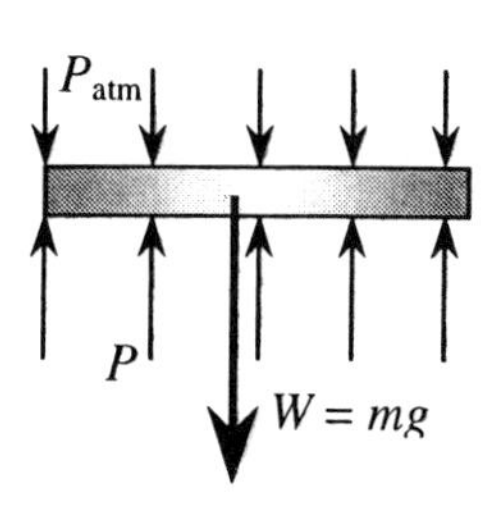

The boiling temperature is the saturation temperature corresponding to this pressure,

$$T = T_{sat\ @\ 102.25\ kPa} = \mathbf{100.2^\circ C}$$

2-47 EES solution of this (and other comprehensive problems designated with the *computer icon*) is available to instructors at the *Instructor Manual* section of the *Online Learning Center* (OLC) at www.mhhe.com/cengel-boles. See the Preface for access information.

2-48 A vertical piston-cylinder device is filled with water and covered with a 20-kg piston that serves as the lid. The boiling temperature of water is to be determined. √

Analysis The pressure in the cylinder is determined from a force balance on the piston,

$$PA = P_{atm}A + W$$

or,

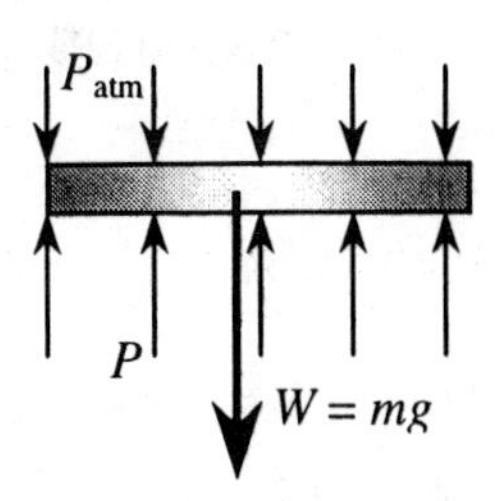

$$P = P_{atm} + \frac{mg}{A}$$
$$= (100\text{ kPa}) + \frac{(20\text{kg})(9.807\text{m/s}^2)}{0.01\text{m}^2}\left(\frac{1\text{ kPa}}{1000\text{kg/m}\cdot\text{s}^2}\right)$$
$$= 119.61\text{ kPa}$$

The boiling temperature is the saturation temperature corresponding to this pressure,

$$T = T_{sat\text{ @ }119.61\text{ }kPa} = \mathbf{104.6°C}$$

2-49 A rigid tank that is filled with saturated liquid-vapor mixture is heated. The temperature at which the liquid in the tank is completely vaporized is to be determined, and the *T-v* diagram is to be drawn.

Analysis This is a constant volume process (v = V/m = constant), and the specific volume is determined to be

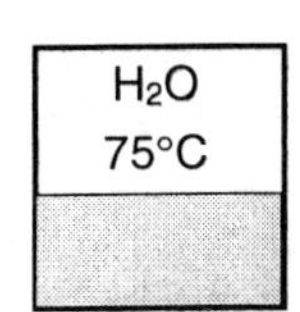

$$v = \frac{V}{m} = \frac{2.5\text{ m}^3}{5\text{ kg}} = 0.5\text{ m}^3\text{/kg}$$

When the liquid is completely vaporized the tank will contain saturated vapor only. Thus,

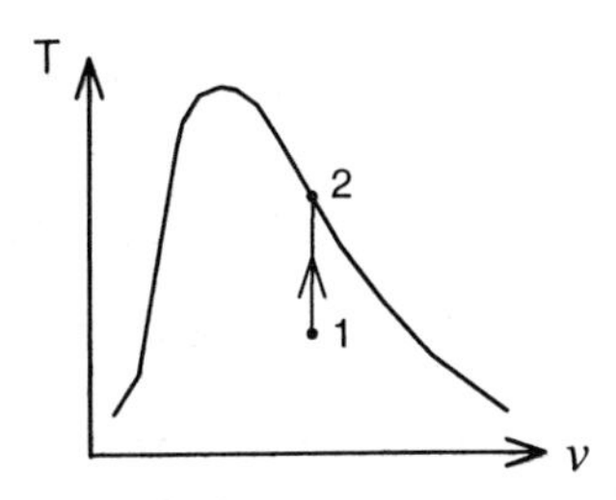

$$v_2 = v_g = 0.5\text{ m}^3\text{ /kg}$$

The temperature at this point is the temperature that corresponds to this v_g value,

$$T = T_{\text{sat@}v_g=0.5\text{ m}^3\text{/kg}} = \mathbf{140.7°C}$$

2-50 A rigid vessel is filled with refrigerant-134a. The total volume and the total internal energy are to be determined.

Properties The properties of R-134a at the given state are (Table A-13).

$$\left.\begin{array}{l} P = 900 \text{ kPa} \\ T = 80^\circ\text{C} \end{array}\right\} \begin{array}{l} u = 288.87 \text{ kJ/kg} \\ \text{v} = 0.02861 \text{ m}^3\text{/kg} \end{array}$$

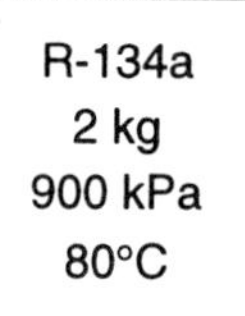

Analysis The total volume and internal energy are determined from

$$V = m\text{v} = (2 \text{ kg})(0.02861 \text{ m}^3\text{/kg}) = \mathbf{0.05722 \ m^3}$$

$$U = mu = (2 \text{ kg})(288.87 \text{ kJ/kg}) = \mathbf{577.74 \ kJ}$$

2-51E A rigid tank contains water at a specified pressure. The temperature, total enthalpy, and the mass of each phase are to be determined.

Analysis (a) The specific volume of the water is

$$\text{v} = \frac{V}{m} = \frac{5 \text{ ft}^3}{5 \text{ lbm}} = 1.0 \text{ ft}^3\text{/lbm}$$

At 20 psia, v_f = 0.01683 ft³/lbm and v_g = 20.09 ft³/lbm. Thus the tank contains saturated liquid-vapor mixture since $v_f < v < v_g$, and the temperature must be the saturation temperature at the specified pressure,

$$T = T_{sat\,@\,20\text{ psia}} = \mathbf{227.96^\circ F}$$

(b) The quality of the water and its total enthalpy are determined from

$$x = \frac{v - v_f}{v_{fg}} = \frac{1.0 - 0.01683}{20.09 - 0.01683} = 0.0490$$

$$h = h_f + xh_{fg} = 196.26 + 0.049 \times 960.1 = 243.3\text{Btu/lbm}$$

$$H = mh = (5\text{lbm})(243.3\text{Btu/lbm}) = \mathbf{1216.5Btu}$$

H_2O
5 lbm
20 psia

(c) The mass of each phase is determined from

$$m_g = xm_t = 0.049 \times 5 = \mathbf{0.245 \ lbm}$$

$$m_f = m_t + m_g = 5 - 0.245 = \mathbf{4.755 \ lbm}$$

2-52 A rigid vessel contains R-134a at specified temperature. The pressure, total internal energy, and the volume of the liquid phase are to be determined.

Analysis (*a*) The specific volume of the refrigerant is

$$v = \frac{V}{m} = \frac{0.5 \text{ m}^3}{10 \text{ kg}} = 0.05 \text{ m}^3/\text{kg}$$

At -20°C, $v_f = 0.0007361$ m³/kg and $v_g = 0.1464$ m³/kg. Thus the tank contains saturated liquid-vapor mixture since $v_f < v < v_g$, and the pressure must be the saturation pressure at the specified temperature,

$$P = P_{\text{sat @ }-20^\circ\text{C}} = \mathbf{132.99\ kPa}$$

(*b*) The quality of the refrigerant-134a and its total internal energy are determined from

R-134a
10 kg
-20°C

$$x = \frac{v - v_f}{v_{fg}} = \frac{0.05 - 0.0007361}{0.1464 - 0.0007361} = 0.338$$

$$u = u_f + x u_{fg} = 24.17 + 0.338 \times (215.84 - 24.17) = 88.95 \text{ kJ/kg}$$

$$U = mu = (10 \text{ kg})(88.95 \text{ kJ/kg}) = \mathbf{889.5\ \ kJ}$$

(*c*) The mass of the liquid phase and its volume are determined from

$$m_f = (1 - x)m_t = (1 - 0.338) \times 10 = 6.62 \text{ kg}$$

$$V_f = m_f v_f = (6.62 \text{ kg})(0.0007361 \text{ m}^3/\text{kg}) = \mathbf{0.00487\ m^3}$$

2-53 [*Also solved by EES on enclosed CD*] A piston-cylinder device contains a saturated liquid-vapor mixture of water at 800 kPa pressure. The mixture is heated at constant pressure until the temperature rises to 350°C. The initial temperature, the total mass of water, the final volume are to be determined, and the *P*-*v* diagram is to be drawn.

Analysis (a) Initially two phases coexist in equilibrium, thus we have a saturated liquid-vapor mixture. Then the temperature in the tank must be the saturation temperature at the specified pressure,

$$\text{T} = \text{T}_{\text{sat @ 800 kPa}} = \mathbf{170.43\ ^\circ C} \qquad \text{(Table A-5)}$$

(b) The total mass in this case can easily be determined by adding the mass of each phase,

$$m_f = \frac{V_f}{v_f} = \frac{0.1 \text{ m}^3}{0.001115 \text{ m}^3/\text{kg}} = 89.69 \text{ kg}$$

$$m_g = \frac{V_g}{v_g} = \frac{0.9 \text{ m}^3}{0.2404 \text{ m}^3/\text{kg}} = 3.74 \text{ kg}$$

$$m_t = m_f + m_g = 89.69 + 3.74 = \mathbf{93.43\ kg}$$

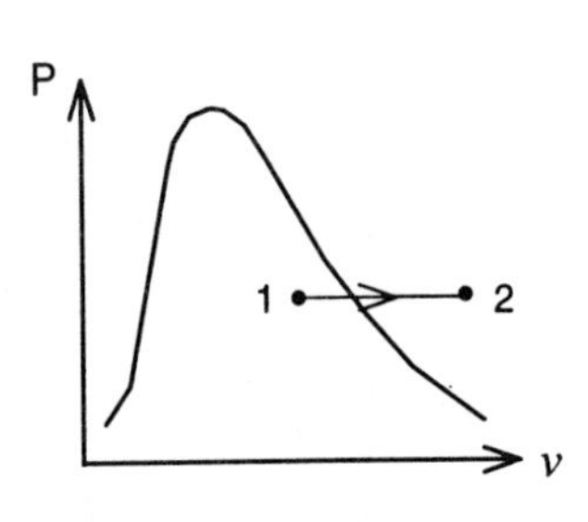

(c) At the final state water is superheated vapor, and its specific volume is

$$\left.\begin{array}{l} P_2 = 800 \text{ kPa} \\ T_2 = 350^\circ\text{C} \end{array}\right\} v_2 = 0.3544 \text{ m}^3/\text{kg} \qquad \text{(Table A-6)}$$

Then,

$$V_2 = m_t v_2 = (93.43 \text{ kg})(0.3544 \text{ m}^3/\text{kg}) = \mathbf{33.1\ m^3}$$

2-54 EES solution of this (and other comprehensive problems designated with the *computer icon*) is available to instructors at the *Instructor Manual* section of the *Online Learning Center* (OLC) at www.mhhe.com/cengel-boles. See the Preface for access information.

2-55E Superheated water vapor cools at constant volume until the temperature drops to 250°F. At the final state, the pressure, the quality, and the enthalpy are to be determined.

H_2O
180 psia
500 °F

Analysis This is a constant volume process (v = V/m = constant), and the initial specific volume is determined to be

$$\left.\begin{array}{l} P_1 = 180 \text{ psia} \\ T_1 = 500^\circ\text{F} \end{array}\right\} v_1 = 3.042 \text{ ft}^3/\text{lbm} \qquad \text{(Table A-6E)}$$

At 250°F, v_f = 0.017001 ft^3/lbm and v_g = 13.826 ft^3/lbm. Thus at the final state, the tank will contain saturated liquid-vapor mixture since $v_f < v < v_g$, and the final pressure must be the saturation pressure at the final temperature,

$$P = P_{sat\,@\,250^\circ\text{F}} = \mathbf{29.82\ psia}$$

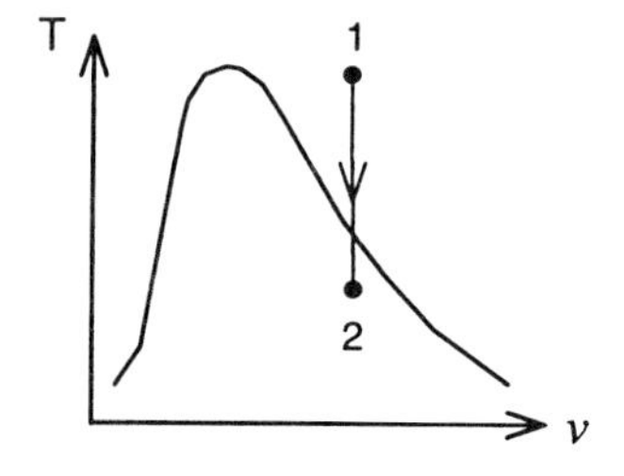

(*b*) The quality at the final state is determined from

$$x_2 = \frac{v_2 - v_f}{v_{fg}} = \frac{3.042 - 0.017001}{13.826 - 0.017001} = \mathbf{0.219}$$

(*c*) The enthalpy at the final state is determined from

$$h = h_f + xh_{fg} = 218.59 + 0.219 \times 945.6 = \mathbf{425.7\ Btu/lbm}$$

2-56 EES solution of this (and other comprehensive problems designated with the *computer icon*) is available to instructors at the *Instructor Manual* section of the *Online Learning Center* (OLC) at www.mhhe.com/cengel-boles. See the Preface for access information.

2-57 A piston-cylinder device that is initially filled with water is heated at constant pressure until all the liquid has vaporized. The mass of water, the final temperature, and the total enthalpy change are to be determined, and the *T*-*v* diagram is to be drawn.

Analysis Initially the cylinder contains compressed liquid (since $P > P_{atm\ @\ 25°C}$) that can be approximated as a saturated liquid at the specified temperature,

$$v_1 \cong v_{f\ @\ 25°C} = 0.001003\ \text{m}^3/\text{kg}$$

$$h_1 \cong h_{f\ @\ 25°C} = 104.89\ \text{kJ/kg}$$

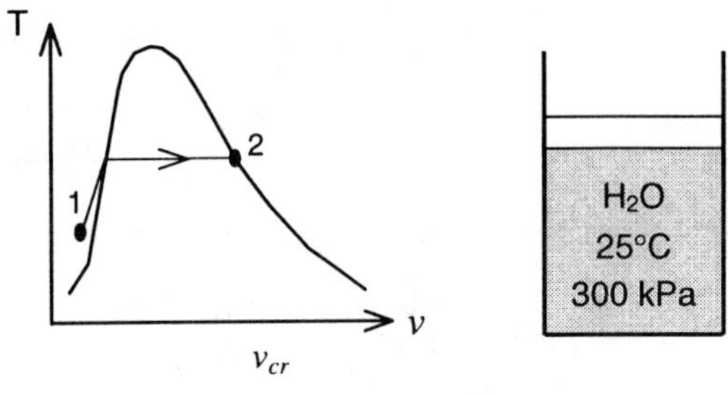

(*a*) The mass is determined from

$$m = \frac{V_1}{v_1} = \frac{0.050\ \text{m}^3}{0.001003\ \text{m}^3/\text{kg}} = \mathbf{49.85\ kg}$$

(*b*) At the final state, the cylinder contains saturated vapor and thus the final temperature must be the saturation temperature at the final pressure,

$$T = T_{sat\ @\ 300\ \text{kPa}} = \mathbf{133.55°\ C}$$

(*c*) The final enthalpy is $h_2 = h_{g\ @\ 300\ \text{kPa}} = 2725.3$ kJ/kg. Thus,

$$\Delta H = m(h_2 - h_1) = (49.85\ \text{kg})(2725.3 - 104.89)\ \text{kJ/kg} = \mathbf{130{,}627\ kJ}$$

2-58 A rigid vessel that contains a saturated liquid-vapor mixture is heated until it reaches the critical state. The mass of the liquid water and the volume occupied by the liquid at the initial state are to be determined.

Analysis This is a constant volume process ($v = V/m$ = constant) to the critical state, and thus the initial specific volume will be equal to the final specific volume, which is equal to the critical specific volume of water,

$$v_1 = v_2 = v_{cr} = 0.003155\ \text{m}^3/\text{kg} \qquad \text{(last row of Table A-4)}$$

The total mass is

$$m = \frac{V}{v} = \frac{0.5\ \text{m}^3}{0.003155\ \text{m}^3/\text{kg}} = 158.48\ \text{kg}$$

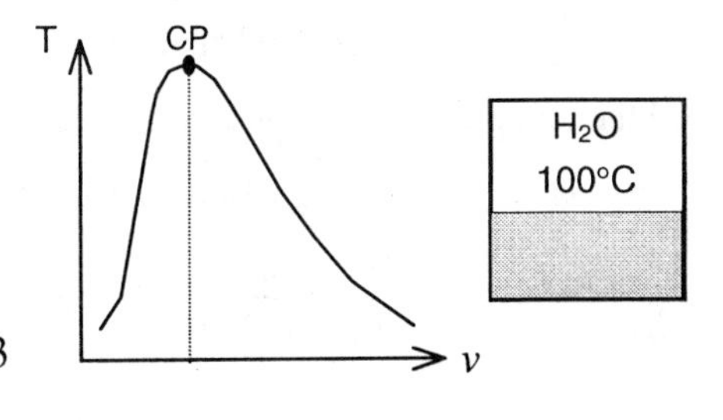

At 100°C, $v_f = 0.001044$ m³/kg and $v_g = 1.6729$ m³/kg. Then the quality of water at the initial state is

$$x_1 = \frac{v_1 - v_f}{v_{fg}} = \frac{0.003155 - 0.001044}{1.6729 - 0.001044} = 0.001263$$

Then the mass of the liquid phase and its volume at the initial state are determined from

$$m_f = (1 - x_1)m_t = (1 - 0.001263)(158.48) = \mathbf{158.28\ kg}$$

$$V_f = m_f v_f = (158.28\ \text{kg})(0.001044\ \text{m}^3/\text{kg}) = \mathbf{0.165\ m^3}$$

2-59 The properties of compressed liquid water at a specified state are to be determined using the compressed liquid tables, and also by using the saturated liquid approximation, and the results are to be compared.

Analysis Compressed liquid can be approximated as saturated liquid at the given temperature. Then from Table A-4,

$$\mathrm{T} = 100°\mathrm{C} \Rightarrow \quad \begin{aligned} v &\cong v_{f@100°C} = 0.001044\ \mathrm{m^3/kg} \quad &(0.76\%\ \text{error}) \\ u &\cong u_{f@100°C} = 418.94\ \mathrm{kJ/kg} \quad &(1.01\%\ \text{error}) \\ h &\cong h_{f@100°C} = 419.04\ \mathrm{kJ/kg} \quad &(2.61\%\ \text{error}) \end{aligned}$$

From compressed liquid table (Table A-7),

$$\left. \begin{array}{l} P = 15\mathrm{MPa} \\ T = 100°\mathrm{C} \end{array} \right\} \begin{array}{l} v = 0.0010361\ \mathrm{m^3/kg} \\ u = 414.74\ \mathrm{kJ/kg} \\ h = 430.28\ \mathrm{kJ/kg} \end{array}$$

The percent errors involved in the saturated liquid approximation are listed above in parentheses.

2-60 EES solution of this (and other comprehensive problems designated with the *computer icon*) is available to instructors at the *Instructor Manual* section of the *Online Learning Center* (OLC) at www.mhhe.com/cengel-boles. See the Preface for access information.

2-61E A rigid tank contains saturated liquid-vapor mixture of R-134a. The quality and total mass of the refrigerant are to be determined.

Analysis At 30 psia, $v_f = 0.01209\ \mathrm{ft^3/lbm}$ and $v_g = 1.5408\ \mathrm{ft^3/lbm}$. The volume occupied by the liquid and the vapor phases are

$$V_f = 1.5\ \mathrm{ft^3} \qquad \text{and} \qquad V_g = 13.5\ \mathrm{ft^3}$$

R-134a
15 ft^3
30 psia

Thus the mass of each phase is

$$m_f = \frac{V_f}{v_f} = \frac{1.5\ \mathrm{ft^3}}{0.01209\ \mathrm{ft^3/lbm}} = 124.1\ \mathrm{lbm}$$

$$m_g = \frac{V_g}{v_g} = \frac{13.5\ \mathrm{ft^3}}{1.5408\ \mathrm{ft^3/lbm}} = 8.76\ \mathrm{lbm}$$

Then the total mass and the quality of the refrigerant are

$$m_t = m_f + m_g = 124.1 + 8.76 = \mathbf{132.86\ lbm}$$

$$x = \frac{m_g}{m_t} = \frac{8.76}{132.86} = \mathbf{0.0659}$$

2-62 Superheated steam in a piston-cylinder device is cooled at constant pressure until half of the mass condenses. The final temperature and the volume change are to be determined, and the process should be shown on a *T-v* diagram.

Analysis (b) At the final state the cylinder contains saturated liquid-vapor mixture, and thus the final temperature must be the saturation temperature at the final pressure,

$$T = T_{\text{sat@1 MPa}} = \mathbf{179.91^\circ C}$$

(c) The quality at the final state is specified to be $x_2 = 0.5$.
The specific volumes at the initial and the final states are

$$\left.\begin{array}{l} P_1 = 1.0\text{ MPa} \\ T_1 = 300^\circ\text{C} \end{array}\right\} v_1 = 0.2579\text{ m}^3/\text{kg}$$

$$\left.\begin{array}{l} P_2 = 1.0\text{MPa} \\ x_2 = 0.5 \end{array}\right\} \begin{array}{l} v_2 = v_f + x_2 v_{fg} \\ = 0.001127 + 0.5 \times (0.19444 - 0.001127) \\ = 0.0978\text{ m}^3/\text{kg} \end{array}$$

Thus,

$$\Delta V = m(v_2 - v_1) = (0.8\text{ kg})(0.0978 - 0.2579)\text{ m}^3/\text{kg} = \mathbf{-0.128\text{ m}^3}$$

2-63 The water in a rigid tank is cooled until the vapor starts condensing. The initial pressure in the tank is to be determined.

Analysis This is a constant volume process ($v = V/m$ = constant), and the initial specific volume is equal to the final specific volume that is

$$v_1 = v_2 = v_{g@180^\circ C} = 0.19405\text{ m}^3/\text{kg}$$

since the vapor starts condensing at 180°C.
Then from Table A-6,

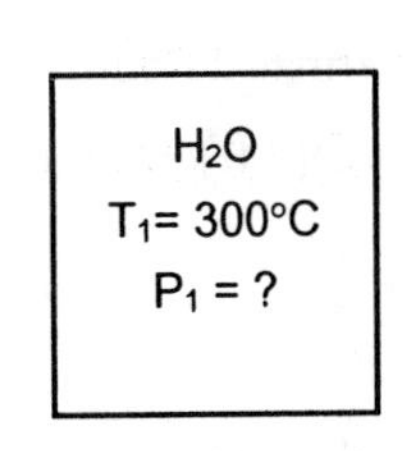

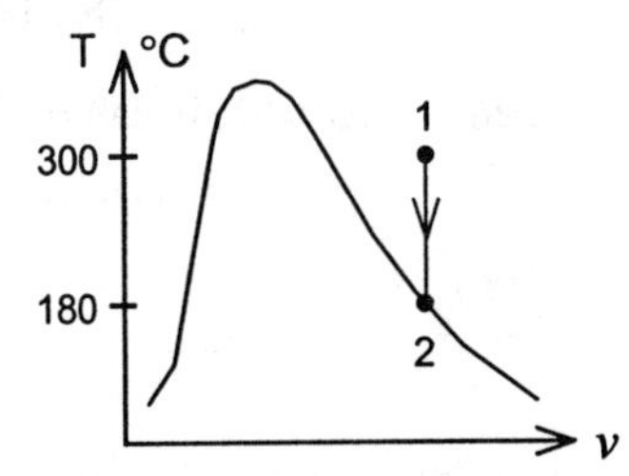

$$\left.\begin{array}{l} T_1 = 300^\circ\text{C} \\ v_1 = 0.194005\text{ m}^3/\text{kg} \end{array}\right\} P_1 = \mathbf{1.325\ MPa}$$

Ideal Gas

2-64C Propane (molar mass = 44.1 kg/kmol) poses a greater fire danger than methane (molar mass = 16 kg/kmol) since propane is heavier than air (molar mass = 29 kg/kmol), and it will settle near the floor. Methane, on the other hand, is lighter than air and thus it will rise and leak out.

2-65C A gas can be treated as an ideal gas when it is at a high temperature or low pressure relative to its critical temperature and pressure.

2-66C R_u is the universal gas constant that is the same for all gases whereas R is the specific gas constant that is different for different gases. These two are related to each other by $R = R_u/M$, where M is the molar mass of the gas.

2-67C Mass *m* is simply the amount of matter; molar mass *M* is the mass of one mole in grams or the mass of one kmol in kilograms. These two are related to each other by $m = NM$, where *N* is the number of moles.

2-68 A balloon is filled with helium gas. The mole number and the mass of helium in the balloon are to be determined.

Assumptions At specified conditions, helium behaves as an ideal gas.

Properties The universal gas constant is R_u = 8.314 kPa.m^3/kmol.K. The molar mass of helium is 4.0 kg/kmol (Table A-1).

Analysis The volume of the sphere is

$$V = \frac{4}{3}\pi r^3 = \frac{4}{3}\pi(3\text{ m})^3 = 113.1\text{ m}^3$$

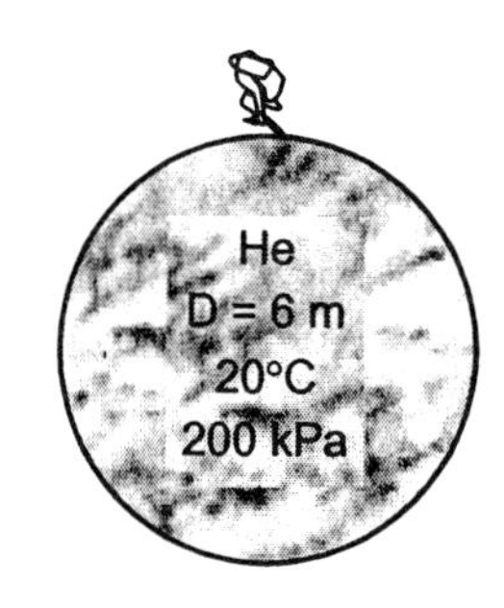

Assuming ideal gas behavior, the mole numbers of He is determined from

$$N = \frac{PV}{R_u T} = \frac{(200\text{ kPa})(113.1\text{ m}^3)}{(8.314\text{ kPa}\cdot\text{m}^3/\text{kmol}\cdot\text{K})(293\text{ K})} = \mathbf{9.28\ kmol}$$

Then the mass of He can be determined from

$$m = NM = (9.28\text{ kmol})(4.0\text{ kg/kmol}) = \mathbf{37.15\ kg}$$

2-69 EES solution of this (and other comprehensive problems designated with the *computer icon*) is available to instructors at the *Instructor Manual* section of the *Online Learning Center* (OLC) at www.mhhe.com/cengel-boles. See the Preface for access information.

2-70 An automobile tire is inflated with air. The pressure rise of air in the tire when the tire is heated and the amount of air that must be bled off to reduce the temperature to the original value are to be determined.

Assumptions **1** At specified conditions, air behaves as an ideal gas. **2** The volume of the tire remains constant.

Properties The gas constant of air is $R = 0.287$ kPa.m^3/kg.K (Table A-1).

Analysis Initially, the absolute pressure in the tire is

$$P_1 = P_g + P_{atm} = 210 + 100 = 310 \text{ kPa}$$

Treating air as an ideal gas and assuming the volume of the tire to remain constant, the final pressure in the tire can be determined from

$$\frac{P_1 V_1}{T_1} = \frac{P_2 V_2}{T_2} \longrightarrow P_2 = \frac{T_2}{T_1} P_1 = \frac{323 \text{ K}}{298 \text{ K}} (310 \text{ kPa}) = 336 \text{ kPa}$$

Tire

25°C

Thus the pressure rise is

$$\Delta P = P_2 - P_1 = 336 - 310 = \mathbf{26\ kPa}$$

The amount of air that needs to be bled off to restore pressure to its original value is

$$m_1 = \frac{P_1 V}{R T_1} = \frac{(310 \text{ kPa})(0.025 \text{ m}^3)}{(0.287 \text{ kPa} \cdot \text{m}^3 / \text{kg} \cdot \text{K})(298 \text{ K})} = 0.0906 \text{ kg}$$

$$m_2 = \frac{P_2 V}{R T_2} = \frac{(310 \text{ kPa})(0.025 \text{ m}^3)}{(0.287 \text{ kPa} \cdot \text{m}^3 / \text{kg} \cdot \text{K})(323 \text{ K})} = 0.0836 \text{ kg}$$

$$\Delta m = m_1 - m_2 = 0.0906 - 0.0836 = \mathbf{0.0070\ kg}$$

2-71E An automobile tire is under inflated with air. The amount of air that needs to be added to the tire to raise its pressure to the recommended value is to be determined.

Assumptions **1** At specified conditions, air behaves as an ideal gas. **2** The volume of the tire remains constant.

Properties The gas constant of air is $R = 0.3704$ psia.ft^3/lbm.R (Table A-1E).

Tire

0.53 ft^3

90°F

Analysis The initial and final absolute pressures in the tire are

$$P_1 = P_{g1} + P_{atm} = 20 + 14.6 = 34.6 \text{ psia}$$
$$P_2 = P_{g2} + P_{atm} = 30 + 14.6 = 44.6 \text{ psia}$$

Treating air as an ideal gas, the initial mass in the tire is

$$m_1 = \frac{P_1 V}{R T_1} = \frac{(34.6 \text{ psia})(0.53 \text{ ft}^3)}{(0.3704 \text{ psia} \cdot \text{ft}^3/\text{lbm} \cdot \text{R})(550 \text{ R})} = 0.0900 \text{ lbm}$$

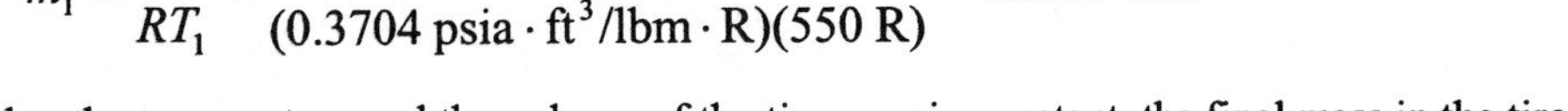

Noting that the temperature and the volume of the tire remain constant, the final mass in the tire becomes

$$m_2 = \frac{P_2 V}{R T_2} = \frac{(44.6 \text{ psia})(0.53 \text{ ft}^3)}{(0.3704 \text{ psia} \cdot \text{ft}^3/\text{lbm} \cdot \text{R})(550 \text{ R})} = 0.1160 \text{ lbm}$$

Thus the amount of air that needs to be added is

$$\Delta m = m_2 - m_1 = 0.1160 - 0.0900 = \mathbf{0.0260\ lbm}$$

2-72 The pressure and temperature of oxygen gas in a storage tank are given. The mass of oxygen in the tank is to be determined.

Assumptions At specified conditions, oxygen behaves as an ideal gas

Properties The gas constant of oxygen is $R = 0.2598$ kPa.m^3/kg.K (Table A-1).

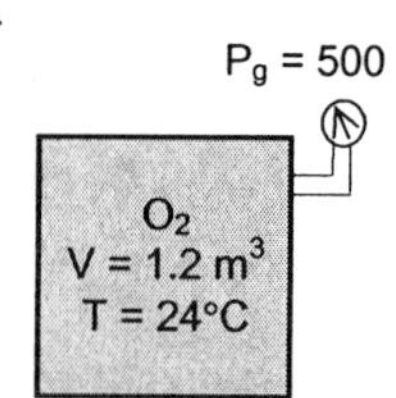

Analysis The absolute pressure of O_2 is

$$P = P_g + P_{atm} = 500 + 97 = 597 \text{ kPa}$$

Treating O_2 as an ideal gas, the mass of O_2 in tank is determined to be

$$m = \frac{PV}{RT} = \frac{(597 \text{ kPa})(1.2 \text{ m}^3)}{(0.2598 \text{ kPa} \cdot \text{m}^3/\text{kg} \cdot \text{K})(297 \text{ K})} = \mathbf{9.28 \text{ kg}}$$

2-73E A rigid tank contains slightly pressurized air. The amount of air that needs to be added to the tank to raise its pressure and temperature to the recommended values is to be determined. √

Assumptions **1** At specified conditions, air behaves as an ideal gas. **2** The volume of the tank remains constant.

Properties The gas constant of air is $R = 0.3704$ psia.ft^3/lbm.R (Table A-1E).

Analysis Treating air as an ideal gas, the initial volume and the final mass in the tank are determined to be

$$V = \frac{m_1 R T_1}{P_1} = \frac{(20 \text{ lbm})(0.3704 \text{ psia} \cdot \text{ft}^3/\text{lbm} \cdot \text{R})(530 \text{ R})}{20 \text{ psia}} = 196.3 \text{ ft}^3$$

$$m_2 = \frac{P_2 V}{R T_2} = \frac{(35 \text{ psia})(196.3 \text{ ft}^3)}{(0.3704 \text{ psia} \cdot \text{ft}^3/\text{lbm} \cdot \text{R})(550 \text{ R})} = 33.73 \text{ lbm}$$

Air, 20 lbm
20 psia
70°F

Thus the amount of air added is

$$\Delta m = m_2 - m_1 = 33.73 - 20.0 = \mathbf{13.73 \text{ lbm}}$$

2-74 A rigid tank contains air at a specified state. The gage pressure of the gas in the tank is to be determined..

Assumptions At specified conditions, air behaves as an ideal gas.

Properties The gas constant of air is $R = 0.287$ kPa.m^3/kg.K (Table A-1).

Analysis Treating air as an ideal gas, the absolute pressure in the tank is determined from

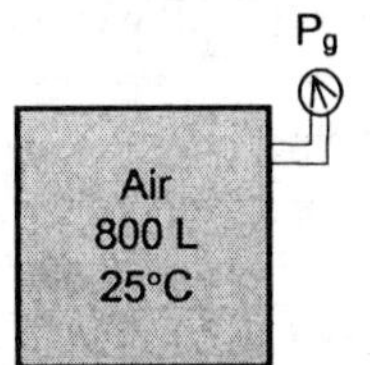

$$P_2 = \frac{mRT}{V} = \frac{(10\text{ kg})(0.287\text{ kPa}\cdot\text{m}^3/\text{kg}\cdot\text{K})(298\text{ K})}{0.8\text{ m}^3} = 1069.1\text{ kPa}$$

Thus the gage pressure is

$$P_g = P - P_{atm} = 1069.1 - 97 = \mathbf{972.1\ kPa}$$

2-75 Two rigid tanks connected by a valve to each other contain air at specified conditions. The volume of the second tank and the final equilibrium pressure when the valve is opened are to be determined.

Assumptions At specified conditions, air behaves as an ideal gas.

Properties The gas constant of air is $R = 0.287$ kPa.m^3/kg.K (Table A-1).

Analysis Let's call the first and the second tanks A and B. Treating air as an ideal gas, the volume of the second tank and the mass of air in the first tank are determined to be

$$V_B = \left(\frac{m_1 RT_1}{P_1}\right)_B = \frac{(5\text{kg})(0.287\text{kPa}\cdot\text{m}^3/\text{kg}\cdot\text{K})(308\text{K})}{200\text{kPa}} = \mathbf{2.21m^3}$$

$$m_A = \left(\frac{P_1 V}{RT_1}\right)_A = \frac{(500\text{kPa})(1.0\text{m}^3)}{(0.287\text{kPa}\cdot\text{m}^3/\text{kg}\cdot\text{K})(298\text{K})} = 5.846\text{kg}$$

Thus,

$$V = V_A + V_B = 1.0 + 2.21 = 3.21\text{ m}^3$$

$$m = m_A + m_B = 5.846 + 5.0 = 10.846\text{ kg}$$

Air
V=1 m^3
T=25°C
P=500 kPa

Air
m=5 kg
T=35°C
P=200 kPa

Then the final equilibrium pressure becomes

$$P_2 = \frac{mRT_2}{V} = \frac{(10.846\text{ kg})(0.287\text{ kPa}\cdot\text{m}^3/\text{kg}\cdot\text{K})(293\text{ K})}{3.21\text{ m}^3} = \mathbf{284.1\ kPa}$$

Compressibility Factor

2-76C It represent the deviation from ideal gas behavior. The further away it is from 1, the more the gas deviates from ideal gas behavior.

2-77C All gases have the same compressibility factor Z at the same reduced temperature and pressure.

2-78C Reduced pressure is the pressure normalized with respect to the critical pressure; and reduced temperature is the temperature normalized with respect to the critical temperature.

2-79 The specific volume of steam is to be determined using the ideal gas relation, the compressibility chart, and the steam tables. The errors involved in the first two approaches are also to be determined.

Properties The gas constant, the critical pressure, and the critical temperature of water are, from Table A-1,

$$R = 0.4615 \text{ kPa·m}^3\text{/kg·K}, \qquad T_{cr} = 647.3 \text{ K}, \qquad P_{cr} = 22.09 \text{ MPa}$$

Analysis (*a*) From the ideal gas equation of state,

$$v = \frac{RT}{P} = \frac{(0.4615 \text{ kPa}\cdot\text{m}^3\text{/kg}\cdot\text{K})(673 \text{ K})}{(10{,}000 \text{ kPa})} = \mathbf{0.03106 \ m^3/kg} \quad (17.6\%\text{error})$$

(*b*) From the compressibility chart (Fig. A-30),

H2O
10 MPa
400°C

$$\left.\begin{aligned} P_R &= \frac{P}{P_{cr}} = \frac{10 \text{ MPa}}{22.09 \text{ MPa}} = 0.453 \\ T_R &= \frac{T}{T_{cr}} = \frac{673 \text{ K}}{647.3 \text{ K}} = 1.04 \end{aligned}\right\} Z = 0.84$$

Thus,

$$v = (Z)(v_{\text{ideal}}) = (0.84)(0.03106 \text{ m}^3\text{/kg}) = \mathbf{0.02609 \ m^3/kg} \quad (1.2\%\text{error})$$

(*c*) From the superheated steam table (Table A-6),

$$\left.\begin{aligned} P &= 10 \text{ MPa} \\ T &= 400^\circ\text{C} \end{aligned}\right\} v = \mathbf{0.02641 \ m^3/kg}$$

2-80 EES solution of this (and other comprehensive problems designated with the *computer icon*) is available to instructors at the *Instructor Manual* section of the *Online Learning Center* (OLC) at www.mhhe.com/cengel-boles. See the Preface for access information.

2-81 The specific volume of R-134a is to be determined using the ideal gas relation, the compressibility chart, and the R-134a tables. The errors involved in the first two approaches are also to be determined. √

Properties The gas constant, the critical pressure, and the critical temperature of refrigerant-134a are, from Table A-1,

$$R = 0.08149 \text{ kPa·m}^3\text{/kg·K}, \qquad T_{cr} = 374.25 \text{ K}, \qquad P_{cr} = 4.067 \text{ MPa}$$

Analysis (*a*) From the ideal gas equation of state,

$$v = \frac{RT}{P} = \frac{(0.08149\text{kPa}\cdot\text{m}^3\text{/kg}\cdot\text{K})(413\text{K})}{1{,}400\text{kPa}} = \mathbf{0.02404m^3/kg} \quad (9.8\%\mathit{error})$$

(*b*) From the compressibility chart (Fig. A-30),

R-134a
1.4 MPa
140°C

$$\left.\begin{aligned} P_R &= \frac{P}{P_{cr}} = \frac{1.4 \text{ MPa}}{4.067 \text{ MPa}} = 0.344 \\ T_R &= \frac{T}{T_{cr}} = \frac{413 \text{ K}}{374.25 \text{ K}} = 1.104 \end{aligned}\right\} Z = 0.91$$

Thus,

$$v = (Z)(v_{\text{ideal}}) = (0.91)(0.02404 \text{ m}^3\text{/kg}) = \mathbf{0.02188 \ m^3/kg} \quad (0.05\%\text{error})$$

(*c*) From the superheated refrigerant table (Table A-13),

$$\left.\begin{aligned} P &= 1.4 \text{ MPa} \\ T &= 140^\circ\text{C} \end{aligned}\right\} v = \mathbf{0.02189m^3/kg}$$

2-82 The specific volume of nitrogen gas is to be determined using the ideal gas relation and the compressibility chart. The errors involved in these two approaches are also to be determined.

Properties The gas constant, the critical pressure, and the critical temperature of nitrogen are, from Table A-1,

$$R = 0.2968 \text{ kPa·m}^3/\text{kg·K}, \qquad T_{cr} = 126.2 \text{ K}, \qquad P_{cr} = 3.39 \text{ MPa}$$

Analysis (a) From the ideal gas equation of state,

$$v = \frac{RT}{P} = \frac{(0.2968 \text{ kPa}\cdot\text{m}^3/\text{kg}\cdot\text{K})(150 \text{ K})}{10{,}000 \text{ kPa}} = \mathbf{0.004452\ m^3/kg} \qquad (86.4\%\ \textit{error})$$

(*b*) From the compressibility chart (Fig. A-30),

$$\left.\begin{array}{l} P_R = \dfrac{P}{P_{cr}} = \dfrac{10 \text{ MPa}}{3.39 \text{ MPa}} = 2.95 \\ T_R = \dfrac{T}{T_{cr}} = \dfrac{150 \text{ K}}{126.2 \text{ K}} = 1.19 \end{array}\right\} Z = 0.54$$

N_2
10 MPa
150 K

Thus,

$$v = (Z)(v_{\text{ideal}}) = (0.54)(0.004452 \text{ m}^3/\text{kg}) = \mathbf{0.002404\ m^3/kg} \qquad (0.7\%\ \text{error})$$

2-83 The specific volume of steam is to be determined using the ideal gas relation, the compressibility chart, and the steam tables. The errors involved in the first two approaches are also to be determined.

Properties The gas constant, the critical pressure, and the critical temperature of water are, from Table A-1,

$$R = 0.4615 \text{ kPa·m}^3/\text{kg·K}, \qquad T_{cr} = 647.3 \text{ K}, \qquad P_{cr} = 22.09 \text{ MPa}$$

Analysis (*a*) From the ideal gas equation of state,

$$v = \frac{RT}{P} = \frac{(0.4615 \text{ kPa}\cdot\text{m}^3/\text{kg}\cdot\text{K})(498 \text{ K})}{1{,}600 \text{ kPa}} = \mathbf{0.14364\ m^3/kg} \qquad (8.1\%\ \textit{error})$$

(*b*) From the compressibility chart (Fig. A-30),

$$\left.\begin{array}{l} P_R = \dfrac{P}{P_{cr}} = \dfrac{1.6 \text{ MPa}}{22.09 \text{ MPa}} = 0.072 \\ T_R = \dfrac{T}{T_{cr}} = \dfrac{498 \text{ K}}{647.3 \text{ K}} = 0.769 \end{array}\right\} Z = 0.935$$

H_2O
1.6 MPa
225°C

Thus,

$$v = (Z)(v_{\text{ideal}}) = (0.935)(0.14364 \text{ m}^3/\text{kg}) = \mathbf{0.13430\ m^3/kg} \qquad (1.1\%\text{error})$$

(*c*) From the superheated steam table (Table A-6),

$$\left.\begin{array}{l} P = 1.6 \text{ MPa} \\ T = 225^\circ\text{C} \end{array}\right\} v = \mathbf{0.13287\ m^3/kg}$$

2-84E The temperature of R-134a is to be determined using the ideal gas relation, the compressibility chart, and the R-134a tables.

Properties The gas constant, the critical pressure, and the critical temperature of refrigerant-134a are, from Table A-1E,

$$R = 0.10517 \text{ psia}\cdot\text{ft}^3/\text{lbm}\cdot\text{R}, \qquad T_{cr} = 673.65 \text{ R}, \qquad P_{cr} = 590 \text{ psia}$$

Analysis (*a*) From the ideal gas equation of state,

$$T = \frac{Pv}{R} = \frac{(400 \text{ psia})(0.1386 \text{ ft}^3/\text{lbm})}{(0.10517 \text{ psia}\cdot\text{ft}^3/\text{lbm}\cdot\text{R})} = \mathbf{527.2\ R}$$

(*b*) From the compressibility chart (Fig. A-30a),

$$\left.\begin{aligned} P_R &= \frac{P}{P_{cr}} = \frac{400\text{psia}}{590\text{psia}} = 0.678 \\ v_R &= \frac{v_{\text{actual}}}{RT_{cr}/P_{cr}} = \frac{(0.1386 \text{ ft}^3/\text{lbm})(590 \text{ psia})}{(0.10517 \text{ psia}\cdot\text{ft}^3/\text{lbm}\cdot\text{R})(673.65 \text{ R})} = 1.15 \end{aligned}\right\} T_R = 1.03$$

Thus,

$$T = T_R T_{cr} = 1.03 \times 673.65 = \mathbf{693.9\ R}$$

(*c*) From the superheated refrigerant table (Table A-13E),

$$\left.\begin{aligned} P &= 400 \text{ psia} \\ v &= 0.1386 \text{ ft}^3/\text{lbm} \end{aligned}\right\} T = \mathbf{240^\circ F\ (700\ R)}$$

2-85 The pressure of R-134a is to be determined using the ideal gas relation, the compressibility chart, and the R-134a tables.

Properties The gas constant, the critical pressure, and the critical temperature of refrigerant-134a are, from Table A-1,

$$R = 0.08149 \text{ kPa·m}^3\text{/kg·K}, \qquad T_{cr} = 374.25 \text{ K}, \qquad P_{cr} = 4.067 \text{ MPa}$$

Analysis The specific volume of the refrigerant is

$$v = \frac{V}{m} = \frac{0.01677 \text{ m}^3}{1 \text{ kg}} = 0.01677 \text{ m}^3\text{/kg}$$

R-134a
0.01677 m^3/kg
110°C

(*a*) From the ideal gas equation of state,

$$P = \frac{RT}{v} = \frac{(0.08149 \text{ kPa} \cdot \text{m}^3/\text{kg} \cdot \text{K})(383 \text{ K})}{0.01677 \text{ m}^3/\text{kg}} = \mathbf{1861\ kPa}$$

(*b*) From the compressibility chart (Fig. A-30),

$$\left.\begin{aligned} T_R &= \frac{T}{T_{cr}} = \frac{383\text{K}}{374.25\text{K}} = 1.023 \\ v_R &= \frac{v_{\text{actual}}}{RT_{cr}/P_{cr}} = \frac{(0.01677 \text{ m}^3\text{/kg})(4067 \text{ kPa})}{(0.08149 \text{ kPa} \cdot \text{m}^3\text{/kg} \cdot \text{K})(374.25 \text{ K})} = 2.24 \end{aligned}\right\} P_R = 0.39$$

Thus,

$$P = P_R P_{cr} = 0.39 \times 4067 = \mathbf{1586\ kPa}$$

(*c*) From the superheated refrigerant table (Table A-13),

$$\left.\begin{aligned} T &= 110^\circ\text{C} \\ v &= 0.01677 \text{ m}^3\text{/kg} \end{aligned}\right\} \mathbf{P = 1600\ kPa}$$

2-86 Somebody claims that oxygen gas at a specified state can be treated as an ideal gas with an error less than 10%. The validity of this claim is to be determined.

Properties The critical pressure, and the critical temperature of oxygen are, from Table A-1,

$$T_{cr} = 154.8 \text{ K} \quad \text{and} \quad P_{cr} = 5.08 \text{ MPa}$$

Analysis From the compressibility chart (Fig. A-30),

$$\left.\begin{aligned} P_R &= \frac{P}{P_{cr}} = \frac{3 \text{ MPa}}{5.08 \text{ MPa}} = 0.591 \\ T_R &= \frac{T}{T_{cr}} = \frac{160 \text{ K}}{154.8 \text{ K}} = 1.034 \end{aligned}\right\} Z = 0.79$$

O_2
3 MPa
160 K

Then the error involved can be determined from

$$\text{Error} = \frac{v - v_{\text{ideal}}}{v} = 1 - \frac{1}{Z} = 1 - \frac{1}{0.79} = -26.6\%$$

Thus the claim is **false**.

2-87 The % error involved in treating CO_2 at a specified state as an ideal gas is to be determined.

Properties The critical pressure, and the critical temperature of CO_2 are, from Table A-1,

$$T_{cr} = 304.2 \text{ K} \quad \text{and} \quad P_{cr} = 7.39 \text{ MPa}$$

Analysis From the compressibility chart (Fig. A-30),

$$\left.\begin{aligned} P_R &= \frac{P}{P_{cr}} = \frac{3 \text{ MPa}}{7.39 \text{ MPa}} = 0.406 \\ T_R &= \frac{T}{T_{cr}} = \frac{283 \text{ K}}{304.2 \text{ K}} = 0.93 \end{aligned}\right\} Z = 0.80$$

CO_2
3 MPa
10°C

Then the error involved in treating CO_2 as an ideal gas is

$$\text{Error} = \frac{v - v_{\text{ideal}}}{v} = 1 - \frac{1}{Z} = 1 - \frac{1}{0.80} = -0.25 \text{ or } \mathbf{25.0\%}$$

2-88 The % error involved in treating CO_2 at a specified state as an ideal gas is to be determined.

Properties The critical pressure, and the critical temperature of CO_2 are, from Table A-1,

$$T_{cr} = 304.2 \text{ K} \quad \text{and} \quad P_{cr} = 7.39 \text{ MPa}$$

Analysis From the compressibility chart (Fig. A-30),

$$\left.\begin{aligned} P_R &= \frac{P}{P_{cr}} = \frac{5 \text{ MPa}}{7.39 \text{ MPa}} = 0.677 \\ T_R &= \frac{T}{T_{cr}} = \frac{350 \text{ K}}{304.2 \text{ K}} = 1.15 \end{aligned}\right\} Z = 0.84$$

CO_2
5 MPa
350 K

Then the error involved in treating CO_2 as an ideal gas is

$$\text{Error} = \frac{v - v_{\text{ideal}}}{v} = 1 - \frac{1}{Z} = 1 - \frac{1}{0.84} = -0.190 \text{ or } 19.0\%$$

Other Equations of State

2-89C The constant a represents the increase in pressure as a result of intermolecular forces; the constant b represents the volume occupied by the molecules. They are determined from the requirement that the critical isotherm has an inflection point at the critical point.

2-90 The pressure of nitrogen in a tank at a specified state is to be determined using the ideal gas, van der Waals, and Beattie-Bridgeman equations. The error involved in each case is to be determined.

Properties The gas constant, molar mass, critical pressure, and critical temperature of nitrogen are (Table A-1)

$$R = 0.2968 \text{ kPa·m}^3\text{/kg·K}, \quad M = 28.013 \text{ kg/kmol}, \quad T_{cr} = 126.2 \text{ K}, \quad P_{cr} = 3.39 \text{ MPa}$$

Analysis The specific volume of nitrogen is

$$v = \frac{V}{m} = \frac{3.27 \text{ m}^3}{100 \text{ kg}} = 0.0327 \text{ m}^3/\text{kg}$$

N_2
0.0327 m^3/kg
225 K

(*a*) From the ideal gas equation of state,

$$P = \frac{RT}{v} = \frac{(0.2968 \text{ kPa}\cdot\text{m}^3/\text{kg}\cdot\text{K})(225 \text{ K})}{0.0327 \text{ m}^3/\text{kg}} = \mathbf{2042 \text{ kPa}} \qquad (2.1\% \textit{ error})$$

(*b*) The van der Waals constants for nitrogen are determined from

$$a = \frac{27R^2T_{cr}^2}{64P_{cr}} = \frac{(27)(0.2968 \text{ kPa}\cdot\text{m}^3/\text{kg}\cdot\text{K})^2(126.2 \text{ K})^2}{(64)(3390 \text{ kPa})} = 0.175 \text{ m}^6\cdot\text{kPa/kg}^2$$

$$b = \frac{RT_{cr}}{8P_{cr}} = \frac{(0.2968 \text{ kPa}\cdot\text{m}^3/\text{kg}\cdot\text{K})(126.2 \text{ K})}{8 \times 3390 \text{ kPa}} = 0.00138 \text{ m}^3/\text{kg}$$

Then,

$$P = \frac{RT}{v-b} - \frac{a}{v^2} = \frac{0.2968 \times 225}{0.0327 - 0.00138} - \frac{0.175}{(0.0327)^2} = \mathbf{1969 \text{ kPa}} \ (1.6\% \text{ error})$$

(*c*) The constants in the Beattie-Bridgeman equation are

$$A = A_o\left(1 - \frac{a}{\bar{v}}\right) = 136.2315\left(1 - \frac{0.02617}{0.9160}\right) = 132.339$$

$$B = B_o\left(1 - \frac{b}{\bar{v}}\right) = 0.05046\left(1 - \frac{-0.00691}{0.9160}\right) = 0.05084$$

$$c = 4.2 \times 10^4 \text{ m}^3\cdot\text{K}^3\text{/kmol}$$

since $\bar{v} = Mv = (28.013 \text{ kg/kmol})(0.0327 \text{ m}^3/\text{kg}) = 0.9160 \text{ m}^3/\text{kmol}$. Substituting,

$$P = \frac{R_u T}{\bar{v}^2}\left(1 - \frac{c}{\bar{v}T^3}\right)(\bar{v} + B) - \frac{A}{\bar{v}^2}$$

$$= \frac{8.314 \times 225}{(0.9160)^2}\left(1 - \frac{4.2 \times 10^4}{0.9160 \times 225^3}\right)(0.9160 + 0.05084) - \frac{132.339}{(0.9160)^2} = \mathbf{1989 \text{ kPa}} \ (0.6\% \text{ error})$$

2-91 The temperature of steam in a tank at a specified state is to be determined using the ideal gas relation, van der Waals equation, and the steam tables.

Properties The gas constant, critical pressure, and critical temperature of steam are (Table A-1)

$$R = 0.4615 \text{ kPa·m}^3\text{/kg·K}, \qquad T_{cr} = 647.3 \text{ K}, \qquad P_{cr} = 22.09 \text{ MPa}$$

Analysis The specific volume of steam is

$$\text{v} = \frac{V}{m} = \frac{1 \text{ m}^3}{2.841 \text{ kg}} = 0.3520 \text{ m}^3/\text{kg}$$

H_2O
1 m³
2.841 kg
0.6 MPa

(*a*) From the ideal gas equation of state,

$$T = \frac{P\text{v}}{R} = \frac{(600 \text{ kPa})(0.352 \text{ m}^3/\text{kg})}{0.4615 \text{ kPa}\cdot\text{m}^3/\text{kg}\cdot\text{K}} = \mathbf{457.6\ K}$$

(*b*) The van der Waals constants for steam are determined from

$$a = \frac{27R^2T_{cr}^2}{64P_{cr}} = \frac{(27)(0.4615 \text{ kPa}\cdot\text{m}^3/\text{kg}\cdot\text{K})^2(647.3 \text{ K})^2}{(64)(22{,}090 \text{ kPa})} = 1.704 \text{ m}^6\cdot\text{kPa/kg}^2$$

$$b = \frac{RT_{cr}}{8P_{cr}} = \frac{(0.4615 \text{ kPa}\cdot\text{m}^3/\text{kg}\cdot\text{K})(647.3 \text{ K})}{8 \times 22{,}090 \text{ kPa}} = 0.00169 \text{ m}^3/\text{kg}$$

Then,

$$T = \frac{1}{R}\left(P + \frac{a}{\text{v}^2}\right)(\text{v} - b) = \frac{1}{0.4615}\left(600 + \frac{1.704}{(0.3520)^2}\right)(0.352 - 0.00169) = \mathbf{465.9\ K}$$

(*c*) From the superheated steam table (Tables A-6),

$$\left.\begin{array}{l} P = 0.6 \text{ MPa} \\ v = 0.352 \text{ m}^3/\text{kg} \end{array}\right\} T = \mathbf{200^\circ C} \quad (= 473 \text{ K})$$

2-92 EES solution of this (and other comprehensive problems designated with the *computer icon*) is available to instructors at the *Instructor Manual* section of the *Online Learning Center* (OLC) at www.mhhe.com/cengel-boles. See the Preface for access information.

2-93E The temperature of R-134a in a tank at a specified state is to be determined using the ideal gas relation, the van der Waals equation, and the refrigerant tables. √

Properties The gas constant, critical pressure, and critical temperature of R-134a are (Table A-1E)

$$R = 0.1052 \text{ psia·ft}^3\text{/lbm·R}, \qquad T_{cr} = 673.65 \text{ R}, \qquad P_{cr} = 590 \text{ psia}$$

Analysis (a) From the ideal gas equation of state,

$$T = \frac{Pv}{R} = \frac{(100 \text{ psia})(0.5388\,1 \text{ft}^3\text{/lbm})}{0.1052 \text{ psia}\cdot\text{ft}^3\text{/lbm}\cdot\text{R}} = \mathbf{512.2\ R}$$

(*b*) The van der Waals constants for the refrigerant are determined from

$$a = \frac{27R^2T_{cr}^2}{64P_{cr}} = \frac{(27)(0.1052 \text{ psia}\cdot\text{ft}^3\text{/lbm}\cdot\text{R})^2(673.65 \text{ R})^2}{(64)(590 \text{ psia})} = 3.591 \text{ ft}^6\cdot\text{psia/lbm}^2$$

$$b = \frac{RT_{cr}}{8P_{cr}} = \frac{(0.1052 \text{ psia}\cdot\text{ft}^3\text{/lbm}\cdot\text{R})(673.65 \text{ R})}{8\times 590 \text{ psia}} = 0.0150 \text{ ft}^3\text{/lbm}$$

Then,

$$T = \frac{1}{R}\left(P + \frac{a}{v^2}\right)(v-b) = \frac{1}{0.1052}\left(100 + \frac{3.591}{(0.5388)^2}\right)(0.5388 - 0.0150) = \mathbf{559.5\ R}$$

(c) From the superheated refrigerant table (Table A-13E),

$$\left.\begin{array}{l} P = 100\text{psia} \\ \text{v} = 0.5388\text{ft}^3\text{/lbm} \end{array}\right\} T = \mathbf{120^\circ F}\,(580\text{R})$$

2-94 [*Also solved by EES on enclosed CD*] The pressure of nitrogen in a tank at a specified state is to be determined using the ideal gas relation and the Beattie-Bridgeman equation. The error involved in each case is to be determined.

Properties The gas constant and molar mass of nitrogen are (Table A-1)

$$R = 0.2968 \text{ kPa·m}^3/\text{kg·K} \quad \text{and} \quad M = 28.013 \text{ kg/kmol}$$

Analysis (*a*) From the ideal gas equation of state,

$$P = \frac{RT}{\text{v}} = \frac{(0.2968 \text{ kPa}\cdot\text{m}^3/\text{kg}\cdot\text{K})(150 \text{ K})}{0.041884 \text{ m}^3/\text{kg}} = \mathbf{1063 \text{ kPa}} \qquad (6.3\% \textit{ error})$$

(*b*) The constants in the Beattie-Bridgeman equation are

$$A = A_o\left(1 - \frac{a}{\bar{\text{v}}}\right) = 136.2315\left(1 - \frac{0.02617}{1.1733}\right) = 133.193$$

$$B = B_o\left(1 - \frac{b}{\bar{\text{v}}}\right) = 0.05046\left(1 - \frac{-0.00691}{1.1733}\right) = 0.05076$$

$$c = 4.2 \times 10^4 \text{ m}^3 \cdot \text{K}^3/\text{kmol}$$

N_2
0.041884 m^3/kg
150 K

since $\bar{\text{v}} = M\text{v} = (28.013 \text{ kg/kmol})(0.041884 \text{ m}^3/\text{kg}) = 1.1733 \text{ m}^3/\text{kmol}$. Substituting,

$$P = \frac{R_u T}{\bar{\text{v}}^2}\left(1 - \frac{c}{\bar{\text{v}}T^3}\right)(\bar{\text{v}} + B) - \frac{A}{\bar{\text{v}}^2}$$

$$= \frac{8.314 \times 150}{(1.1733)^2}\left(1 - \frac{4.2 \times 10^4}{1.1733 \times 150^3}\right)(1.1733 + 0.05076) - \frac{133.193}{(1.1733)^2}$$

$$= \mathbf{1000.4 \text{ kPa}} \text{ (negligibleerror)}$$

2-95 EES solution of this (and other comprehensive problems designated with the *computer icon*) is available to instructors at the *Instructor Manual* section of the *Online Learning Center* (OLC) at www.mhhe.com/cengel-boles. See the Preface for access information.

Specific Heats, Δu and Δh of Ideal Gases

2-96C It can be used for any kind of process of an ideal gas.

2-97C It can be used for any kind of process of an ideal gas.

2-98C The desired result is obtained by multiplying the first relation by the molar mass M,

$$MC_p = MC_v + MR$$

or $$\overline{C}_p = \overline{C}_v + R_u$$

2-99C Very close, but no. Because the heat transfer during this process is $Q = mC_p\Delta T$, and C_p varies with temperature.

2-100C It can be either. The difference in temperature in both the K and °C scales is the same.

2-101C The energy required is $mC_p\Delta T$, which will be the same in both cases. This is because the C_p of an ideal gas does not vary with pressure.

2-102C The energy required is $mC_p\Delta T$, which will be the same in both cases. This is because the C_p of an ideal gas does not vary with volume.

2-103C For the constant pressure case. This is because the heat transfer to an ideal gas is $mC_p\Delta T$ at constant pressure, $mC_v\Delta T$ at constant volume, and C_p is always greater than C_v.

2-104 The enthalpy change of nitrogen gas during a heating process is to be determined using an empirical specific heat relation, constant specific heat at average temperature, and constant specific heat at room temperature.

Analysis (*a*) Using the empirical relation for $\overline{C}_p(T)$ from Table A-2c,

$$\overline{C}_p = a + bT + cT^2 + dT^3$$

where a = 28.90, b = -0.1571×10^{-2}, c = 0.8081×10^{-5}, and d = -2.873×10^{-9}. Then,

$$\begin{aligned}\Delta\bar{h} &= \int_1^2 \overline{C}_p(T)dT = \int_1^2 \left[a + bT + cT^2 + dT^3\right]dT \\ &= a(T_2 - T_1) + \tfrac{1}{2}b(T_2^2 + T_1^2) + \tfrac{1}{3}c(T_2^3 - T_1^3) + \tfrac{1}{4}d(T_2^4 - T_1^4) \\ &= 28.90(1000 - 600) - \tfrac{1}{2}(0.1571\times10^{-2})(1000^2 - 600^2) \\ &\quad + \tfrac{1}{3}(0.8081\times10^{-5})(1000^3 - 600^3) - \tfrac{1}{4}(2.873\times10^{-9})(1000^4 - 600^4) \\ &= 12{,}544 \text{ kJ/kmol}\end{aligned}$$

$$\Delta h = \frac{\Delta\bar{h}}{M} = \frac{12{,}544 \text{ kJ/kmol}}{28.013 \text{ kg/kmol}} = \mathbf{447.8 \text{ kJ/kg}} \quad (0.2\% \text{error})$$

(*b*) Using the constant C_p value from Table A-2b at the average temperature of 800 K,

$$C_{p,ave} = C_{p@800\text{K}} = 1.121 \text{ kJ/kg}\cdot\text{K}$$
$$\Delta h = C_{p,ave}(T_2 - T_1) = (1.121 \text{ kJ/kg}\cdot\text{K})(1000 - 600)\text{K} = \mathbf{448.4 \text{ kJ/kg}}$$

(*c*) Using the constant C_p value from Table A-2a at room temperature,

$$C_{p,ave} = C_{p@300\text{K}} = 1.039 \text{ kJ/kg}\cdot\text{K}$$
$$\Delta h = C_{p,ave}(T_2 - T_1) = (1.039 \text{ kJ/kg}\cdot\text{K})(1000 - 600)\text{K} = \mathbf{415.6 \text{ kJ/kg}}$$

2-105E The enthalpy change of oxygen gas during a heating process is to be determined using an empirical specific heat relation, constant specific heat at average temperature, and constant specific heat at room temperature.

Analysis (*a*) Using the empirical relation for $\overline{C}_p(T)$ from Table A-2Ec,

$$\overline{C}_p = a + bT + cT^2 + dT^3$$

where a = 6.085, b = 0.2017×10^{-2}, c = -0.05275×10^{-5}, and d = 0.05372×10^{-9}. Then,

$$\begin{aligned}
\Delta\bar{h} &= \int_1^2 \overline{C}_p(T)dT = \int_1^2 \left[a + bT + cT^2 + dT^3\right]dT \\
&= a(T_2 - T_1) + \tfrac{1}{2}b(T_2^2 + T_1^2) + \tfrac{1}{3}c(T_2^3 - T_1^3) + \tfrac{1}{4}d(T_2^4 - T_1^4) \\
&= 6.085(1500 - 800) + \tfrac{1}{2}(0.2017\times10^{-2})(1500^2 - 800^2) \\
&\quad - \tfrac{1}{3}(0.05275\times10^{-5})(1500^3 - 800^3) + \tfrac{1}{4}(0.05372\times10^{-9})(1500^4 - 800^4) \\
&= 5442.3\text{Btu/lbmol}
\end{aligned}$$

$$\Delta h = \frac{\Delta\bar{h}}{M} = \frac{5442.3\text{Btu/lbmol}}{31.999\text{lbm/lbmol}} = \mathbf{170.1Btu/lbm}$$

(*b*) Using the constant C_p value from Table A-2Eb at the average temperature of 1150 R,

$$C_{p,ave} = C_{p@1150\text{ R}} = 0.255\text{Btu/lbm}\cdot\text{R}$$

$$\Delta h = C_{p,ave}(T_2 - T_1) = (0.255\text{ Btu/lbm}\cdot\text{R})(1500 - 800)\text{ R} = \mathbf{178.5\ \ Btu/lbm}$$

(*c*) Using the constant C_p value from Table A-2Ea at room temperature,

$$C_{p,ave} = C_{p@537\text{ R}} = 0.219\text{ Btu / lbm}\cdot\text{R}$$

$$\Delta h = C_{p,ave}(T_2 - T_1) = (0.219\text{ Btu / lbm}\cdot\text{R})(1500 - 800)\text{R} = \mathbf{153.3\ Btu / lbm}$$

2-106 The internal energy change of hydrogen gas during a heating process is to be determined using an empirical specific heat relation, constant specific heat at average temperature, and constant specific heat at room temperature.

Analysis (*a*) Using the empirical relation for $\overline{C}_p(T)$ from Table A-2c and relating it to $\overline{C}_v(T)$,

$$\overline{C}_v(T) = \overline{C}_p - R_u = (a - R_u) + bT + cT^2 + dT^3$$

where a = 29.11, b = -0.1916×10^{-2}, c = 0.4003×10^{-5}, and d = -0.8704×10^{-9}. Then,

$$\begin{aligned}\Delta\overline{u} &= \int_1^2 \overline{C}_v(T)dT = \int_1^2 \left[(a - R_u) + bT + cT^2 + dT^3\right]dT \\ &= (a - R_u)(T_2 - T_1) + \tfrac{1}{2}b(T_2^2 + T_1^2) + \tfrac{1}{3}c(T_2^3 - T_1^3) + \tfrac{1}{4}d(T_2^4 - T_1^4) \\ &= (29.11 - 8.314)(1000 - 400) - \tfrac{1}{2}(0.1961\times10^{-2})(1000^2 - 400^2) \\ &\quad + \tfrac{1}{3}(0.4003\times10^{-5})(1000^3 - 400^3) - \tfrac{1}{4}(0.8704\times10^{-9})(1000^4 - 400^4) \\ &= 12{,}691 \text{ kJ/kmol}\end{aligned}$$

$$\Delta u = \frac{\Delta\overline{u}}{M} = \frac{12{,}691 \text{ kJ/kmol}}{2.016 \text{ kg/kmol}} = \mathbf{6295.3\ \ kJ/kg}$$

(*b*) Using a constant C_p value from Table A-2b at the average temperature of 700 K,

$$C_{v,ave} = C_{v@700\text{ K}} = 10.48 \text{ kJ/kg}\cdot\text{K}$$

$$\Delta u = C_{v,ave}(T_2 - T_1) = (10.48 \text{ kJ/kg}\cdot\text{K})(1000 - 400)\text{K} = \mathbf{6288\ \ kJ/kg}$$

(*c*) Using a constant C_p value from Table A-2a at room temperature,

$$C_{v,ave} = C_{v@300\text{ K}} = 10.183 \text{ kJ/kg}\cdot\text{K}$$

$$\Delta u = C_{v,ave}(T_2 - T_1) = (10.183 \text{ kJ/kg}\cdot\text{K})(1000 - 400)\text{K} = \mathbf{6110\ \ kJ/kg}$$

Special Topic: Vapor Pressure and Phase Equilibrium

2-107 A glass of water is left in a room. The vapor pressures at the free surface of the water and in the room far from the glass are to be determined.

Assumptions The water in the glass is at a uniform temperature.

Properties The saturation pressure of water is 2.339 kPa at 20°C, and 1.7051 kPa at 15°C (Table A-4).

Analysis The vapor pressure at the water surface is the saturation pressure of water at the water temperature,

$$P_{v,\text{ water surface}} = P_{\text{sat @ } T_\text{water}} = P_{\text{sat@15°C}} = \mathbf{1.7051\ kPa}$$

H₂O
15°C

Noting that the air in the room is not saturated, the vapor pressure in the room far from the glass is

$$P_{v,\text{ air}} = \phi P_{\text{sat @ } T_\text{air}} = \phi P_{\text{sat@20°C}} = (0.6)(2.339\text{ kPa}) = \mathbf{1.4034\ kPa}$$

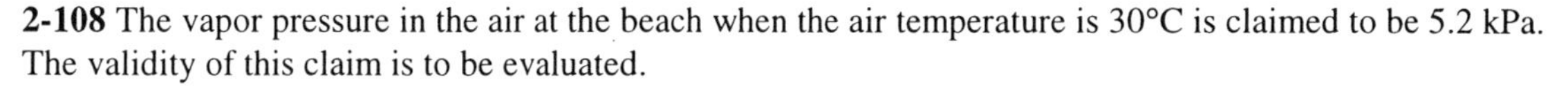

2-108 The vapor pressure in the air at the beach when the air temperature is 30°C is claimed to be 5.2 kPa. The validity of this claim is to be evaluated.

Properties The saturation pressure of water at 30°C is 4.246 kPa (Table A-4).

Analysis The maximum vapor pressure in the air is the saturation pressure of water at the given temperature, which is

$$P_{v,\text{ max}} = P_{\text{sat @ } T_\text{air}} = P_{\text{sat@30°C}} = \mathbf{4.246\ kPa}$$

which is less than the claimed value of 5.2 kPa. Therefore, the claim is **false**.

P_{atm} 30°C

WATER

2-109 The temperature and relative humidity of air over a swimming pool are given. The water temperature of the swimming pool when phase equilibrium conditions are established is to be determined.

Assumptions The temperature and relative humidity of air over the pool remain constant.

Properties The saturation pressure of water at 20°C is 2.339 kPa (Table A-4).

Analysis The vapor pressure of air over the swimming pool is

$$P_{v,\,\text{air}} = \phi P_{\text{sat}\,@\,T_{\text{air}}} = \phi P_{\text{sat@20°C}} = (0.4)(2.339\ \text{kPa}) = 0.9356\ \text{kPa}$$

Phase equilibrium will be established when the vapor pressure at the water surface equals the vapor pressure of air far from the surface. Therefore,

P_{atm} 20°C

POOL

$$P_{v,\,\text{water surface}} = P_{v,\,\text{air}} = 0.9356\ \text{kPa}$$

and

$$T_{\text{water}} = T_{\text{sat}\,@\,P_v} = T_{\text{sat}\,@\,0.9356\ \text{kPa}} = \mathbf{5.9°C}$$

Discussion Note that the water temperature drops to 5.9°C in an environment at 20°C when phase equilibrium is established.

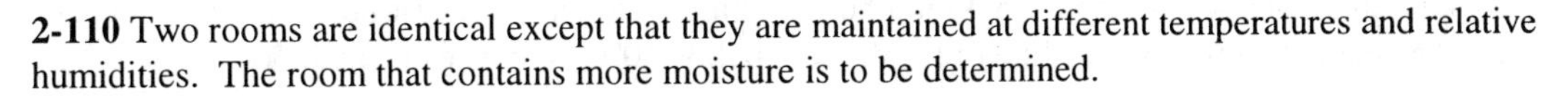

2-110 Two rooms are identical except that they are maintained at different temperatures and relative humidities. The room that contains more moisture is to be determined.

Properties The saturation pressure of water is 2.339 kPa at 20°C, and 4.246 kPa at 30°C (Table A-4).

Analysis The vapor pressures in the two rooms are

Room 1: $$P_{v1} = \phi_1 P_{\text{sat}\,@\,T_1} = \phi_1 P_{\text{sat@30°C}} = (0.4)(4.246\ \text{kPa}) = \mathbf{1.6984\ kPa}$$

Room 2: $$P_{v2} = \phi_2 P_{\text{sat}\,@\,T_2} = \phi_2 P_{\text{sat@20°C}} = (0.7)(2.339\ \text{kPa}) = \mathbf{1.6373\ kPa}$$

Therefore, room 1 at 30°C and 40% relative humidity contains more moisture.

2-111E A thermos bottle half-filled with water is left open to air in a room at a specified temperature and pressure. The temperature of water when phase equilibrium is established is to be determined.

Assumptions The temperature and relative humidity of air over the bottle remain constant.

Properties The saturation pressure of water at 70°F is 0.3632 psia (Table A-4E).

Analysis The vapor pressure of air in the room is

$$P_{v,\,\text{air}} = \phi P_{\text{sat @ } T_{\text{air}}} = \phi P_{\text{sat@70°F}} = (0.35)(0.3632 \text{ psia}) = 0.1271 \text{ psia}$$

Phase equilibrium will be established when the vapor pressure at the water surface equals the vapor pressure of air far from the surface. Therefore,

$$P_{v,\text{ water surface}} = P_{v,\text{ air}} = 0.1271 \text{ psia}$$

and

$$T_{\text{water}} = T_{\text{sat @ } P_v} = T_{\text{sat @ 0.1271 psia}} = \mathbf{41°F}$$

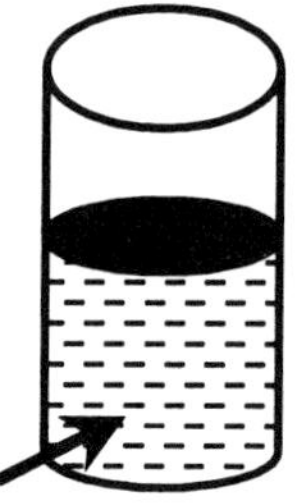

Discussion Note that the water temperature drops to 41°F in an environment at 70°F when phase equilibrium is established.

2-112 A person buys a supposedly cold drink in a hot and humid summer day, yet no condensation occurs on the drink. The claim that the temperature of the drink is below 10°C is to be evaluated.

Properties The saturation pressure of water at 35°C is 5.628 kPa (Table A-4).

Analysis The vapor pressure of air is

$$P_{v,\,\text{air}} = \phi P_{\text{sat @ } T_{\text{air}}} = \phi P_{\text{sat@35°C}} = (0.7)(5.628\text{ kPa}) = 3.9396\text{ kPa}$$

The saturation temperature corresponding to this pressure (called the dew-point temperature) is

$$T_{\text{sat}} = T_{\text{sat @ } P_v} = T_{\text{sat@3.9396 kPa}} = 28.6°\text{C}$$

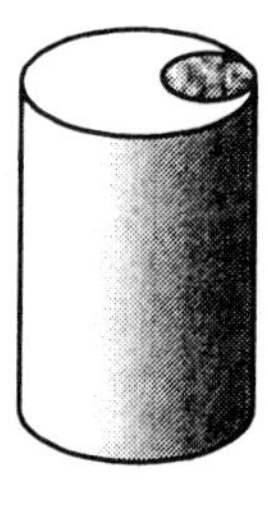

That is, the vapor in the air will condense at temperatures below 28.6°C. Noting that no condensation is observed on the can, the claim that the drink is at 10°C is **false**.

Review Problems

2-113 A smoking lounge that can accommodate 15 smokers is considered. The required minimum flow rate of air that needs to be supplied to the lounge and the diameter of the duct are to be determined.

Assumptions Infiltration of air into the smoking lounge is negligible.

Properties The minimum fresh air requirements for a smoking lounge is given to be 30 L/s per person.

Analysis The required minimum flow rate of air that needs to be supplied to the lounge is determined directly from

$$\dot{V}_{\text{air}} = \dot{V}_{\text{air per person}}(\text{No. of persons})$$
$$= (30\ \text{L/s}\cdot\text{person})(15\ \text{persons}) = 450\ \text{L/s} = \mathbf{0.45\ m^3/s}$$

Smoking Lounge

15 smokers
30 L/s person

The volume flow rate of fresh air can be expressed as

$$\dot{V} = \mathbf{V}A = \mathbf{V}(\pi D^2/4)$$

Solving for the diameter D and substituting,

$$D = \sqrt{\frac{4\dot{V}}{\pi \mathbf{V}}} = \sqrt{\frac{4(0.45\ \text{m}^3/\text{s})}{\pi(8\ \text{m/s})}} = \mathbf{0.268\ m}$$

Therefore, the diameter of the fresh air duct should be at least 26.8 cm if the velocity of air is not to exceed 8 m/s.

2-114 The minimum fresh air requirements of a residential building is specified to be 0.35 air changes per hour. The size of the fan that needs to be installed and the diameter of the duct are to be determined.

Analysis The volume of the building and the required minimum volume flow rate of fresh air are

$$V_{\text{room}} = (2.7\ \text{m})(200\ \text{m}^2) = 540\ \text{m}^3$$
$$\dot{V} = V_{\text{room}} \times \text{ACH} = (540\ \text{m}^3)(0.35/\text{h}) = 189\ \text{m}^3/\text{h} = 189{,}000\ \text{L/h} = \mathbf{3150\ \ L/min}$$

The volume flow rate of fresh air can be expressed as

$$\dot{V} = \mathbf{V}A = \mathbf{V}(\pi D^2/4)$$

Solving for the diameter D and substituting,

$$D = \sqrt{\frac{4\dot{V}}{\pi \mathbf{V}}} = \sqrt{\frac{4(189/3600\ \text{m}^3/\text{s})}{\pi(6\ \text{m/s})}} = \mathbf{0.106\ m}$$

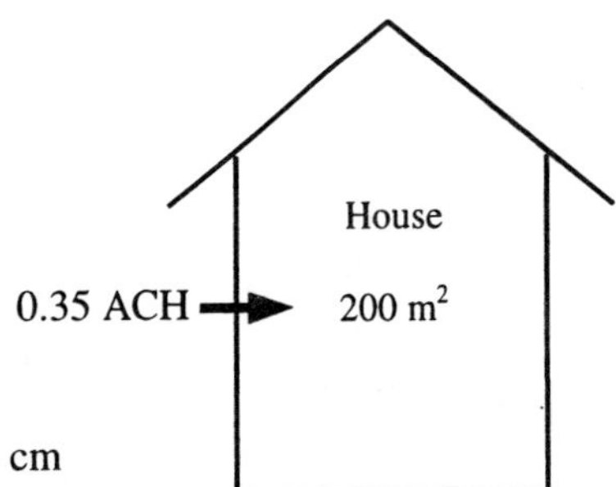

Therefore, the diameter of the fresh air duct should be at least 10.6 cm if the velocity of air is not to exceed 6 m/s.

2-115 The pressure in an automobile tire increases during a trip while its volume remains constant. The percent increase in the absolute temperature of the air in the tire is to be determined.

Assumptions **1** The volume of the tire remains constant. **2** Air is an ideal gas.

Properties The local atmospheric pressure is 90 kPa.

Analysis The absolute pressures in the tire before and after the trip are

$$P_1 = P_{\text{gage},1} + P_{\text{atm}} = 200 + 90 = 290 \text{ kPa}$$

$$P_2 = P_{\text{gage},2} + P_{\text{atm}} = 220 + 90 = 310 \text{ kPa}$$

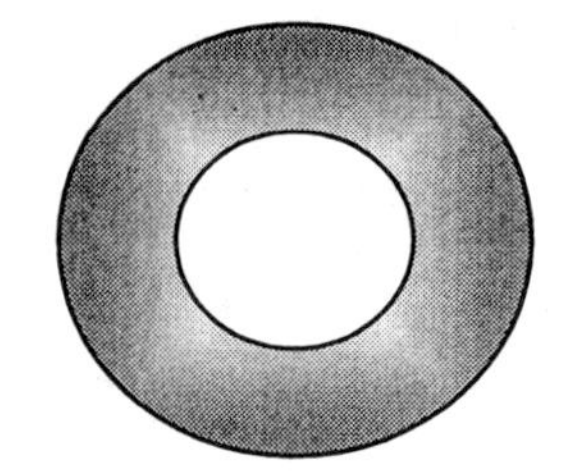

Noting that air is an ideal gas and the volume is constant, the ratio of absolute temperatures after and before the trip are

$$\frac{P_1 V_1}{T_1} = \frac{P_2 V_2}{T_2} \quad \rightarrow \quad \frac{T_2}{T_1} = \frac{P_2}{P_1} = \frac{310 \text{ kPa}}{290 \text{ kPa}} = 1.069$$

Therefore, the absolute temperature of air in the tire will increase by **6.9%** during this trip.

2-116 A hot air balloon with 3 people in its cage is hanging still in the air. The average temperature of the air in the balloon for two environment temperatures is to be determined.

Assumptions Air is an ideal gas.

Properties The gas constant of air is $R = 0.287$ kPa.m^3/kg.K.

Analysis The buoyancy force acting on the balloon is

$$V_{\text{balloon}} = 4\pi r^3/3 = 4\pi(10\text{m})^3/3 = 4189\text{m}^3$$

$$\rho_{\text{cool air}} = \frac{P}{RT} = \frac{90\text{kPa}}{(0.287\text{kPa}\cdot\text{m}^3/\text{kg}\cdot\text{K})(288\text{K})} = 1.089\text{kg/m}^3$$

$$\begin{aligned} F_B &= \rho_{\text{cool air}}\, g V_{\text{balloon}} \\ &= (1.089\text{ kg/m}^3)(9.8\text{ m/s}^2)(4189\text{ m}^3)\left(\frac{1\text{ N}}{1\text{kg}\cdot\text{m/s}^2}\right) = 44{,}700\text{ N} \end{aligned}$$

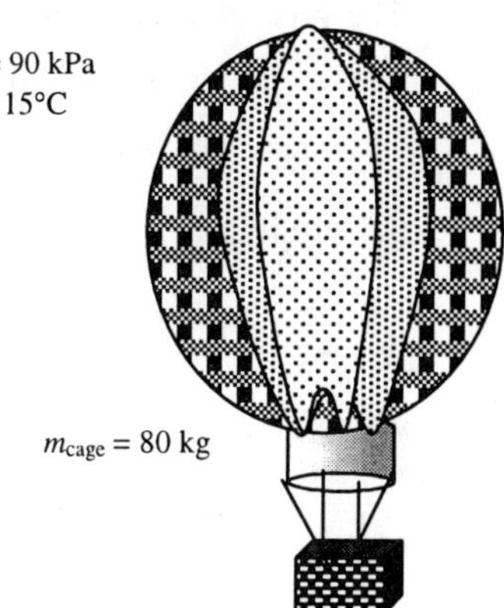

The vertical force balance on the balloon gives

$$\begin{aligned} F_B &= W_{\text{hot air}} + W_{\text{cage}} + W_{\text{people}} \\ &= (m_{\text{hot air}} + m_{\text{cage}} + m_{\text{people}})g \end{aligned}$$

Substituting,

$$44{,}700\text{ N} = (m_{\text{hot air}} + 80\text{ kg} + 195\text{ kg})(9.8\text{ m/s}^2)\left(\frac{1\text{ N}}{1\text{ kg}\cdot\text{m/s}^2}\right)$$

which gives

$$m_{\text{hot air}} = 4287\text{ kg}$$

Therefore, the average temperature of the air in the balloon is

$$T = \frac{PV}{mR} = \frac{(90\text{ kPa})(4189\text{ m}^3)}{(4287\text{ kg})(0.287\text{ kPa}\cdot\text{m}^3/\text{kg}\cdot\text{K})} = \mathbf{306.5\text{ K}}$$

Repeating the solution above for an atmospheric air temperature of 30°C gives **323.6 K** for the average air temperature in the balloon.

2-117 EES solution of this (and other comprehensive problems designated with the *computer icon*) is available to instructors at the *Instructor Manual* section of the *Online Learning Center* (OLC) at www.mhhe.com/cengel-boles. See the Preface for access information.

2-118 A hot air balloon with 2 people in its cage is about to take off. The average temperature of the air in the balloon for two environment temperatures is to be determined.

Assumptions Air is an ideal gas.

Properties The gas constant of air is R = 0.287 kPa.m^3/kg.K.

Analysis The buoyancy force acting on the balloon is

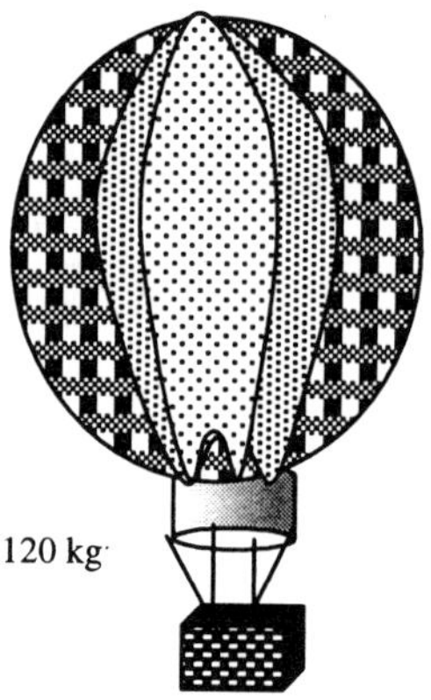

Hot air balloon
D = 18 m

P_{atm} = 93 kPa
T = 12°C

m_{cage} = 120 kg

$$V_{balloon} = 4\pi r^3/3 = 4\pi(9\text{ m})^3/3 = 3054\text{ m}^3$$

$$\rho_{coolair} = \frac{P}{RT} = \frac{93\text{ kPa}}{(0.287\text{ kPa}\cdot\text{m}^3/\text{kg}\cdot\text{K})(285\text{ K})} = 1.137\text{ kg/m}^3$$

$$F_B = \rho_{coolair}\, g V_{balloon}$$
$$= (1.137\text{ kg/m}^3)(9.8\text{ m/s}^2)(3054\text{ m}^3)\left(\frac{1\text{ N}}{1\text{ kg}\cdot\text{m/s}^2}\right) = 34{,}029\text{ N}$$

The vertical force balance on the balloon gives

$$F_B = W_{hot\,air} + W_{cage} + W_{people}$$
$$= (m_{hot\,air} + m_{cage} + m_{people})g$$

Substituting,

$$34{,}029\text{ N} = (m_{hot\,air} + 120\text{ kg} + 140\text{ kg})(9.8\text{ m/s}^2)\left(\frac{1\text{ N}}{1\text{ kg}\cdot\text{m/s}^2}\right)$$

which gives

$$m_{hot\,air} = 3212\text{ kg}$$

Therefore, the average temperature of the air in the balloon is

$$T = \frac{PV}{mR} = \frac{(93\text{ kPa})(3054\text{ m}^3)}{(3212\text{ kg})(0.287\text{ kPa}\cdot\text{m}^3/\text{kg}\cdot\text{K})} = \mathbf{308\text{ K}}$$

Repeating the solution above for an atmospheric air temperature of 25°C gives **323 K** for the average air temperature in the balloon.

2-119E Water in a pressure cooker boils at 260°F. The absolute pressure in the pressure cooker is to be determined.

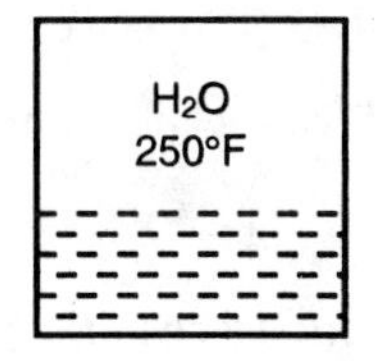

Analysis The absolute pressure in the pressure cooker is the saturation pressure that corresponds to the boiling temperature,

$$P = P_{\text{sat @ 260°F}} = \mathbf{35.42\ psia}$$

2-120 The refrigerant in a rigid tank is allowed to cool. The pressure at which the refrigerant starts condensing is to be determined, and the process is to be shown on a *P-v* diagram.

Analysis This is a constant volume process ($v = V/m$ = constant), and the specific volume is determined to be

$$v = \frac{V}{m} = \frac{0.07\ \text{m}^3}{1\ \text{kg}} = 0.07\ \text{m}^3/\text{kg}$$

When the refrigerant starts condensing, the tank will contain saturated vapor only. Thus,

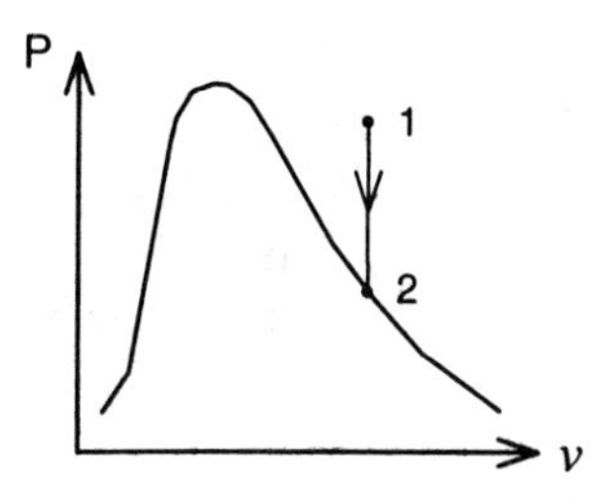

$$v_2 = v_g = 0.07\ \text{m}^3/\text{kg}$$

The pressure at this point is the pressure that corresponds to this v_g value,

$$P_2 = P_{\text{sat}@v_g=0.07\text{m}^3/\text{kg}} = \mathbf{0.29MPa}$$

2-121 The rigid tank contains saturated liquid-vapor mixture of water. The mixture is heated until it exists in a single phase. For a given tank volume, it is to be determined if the final phase is a liquid or a vapor.

Analysis This is a constant volume process ($v = V/m$ = constant), and thus the final specific volume will be equal to the initial specific volume,

H_2O
V = 4L
m = 2 kg
T = 50°C

$$v_2 = v_1$$

The critical specific volume of water is 0.003155 m³/kg. Thus if the final specific volume is smaller than this value, the water will exist as a liquid, otherwise as a vapor.

$$V = 4\ L \longrightarrow v = \frac{V}{m} = \frac{0.004\ \text{m}^3}{2\ \text{kg}} = 0.002\ \text{m}^3/\text{kg} < v_{cr}. \quad \text{Thus, } \textbf{liquid}.$$

$$V = 400\ L \longrightarrow v = \frac{V}{m} = \frac{0.4\ \text{m}^3}{2\ \text{kg}} = 0.2\ \text{m}^3/\text{kg} > v_{cr}. \quad \text{Thus, } \textbf{vapor}.$$

2-122 Superheated refrigerant-134a is cooled at constant pressure until it exists as a compressed liquid. The changes in total volume and internal energy are to be determined, and the process is to be shown on a *T-v* diagram.

Analysis The refrigerant is a superheated vapor at the initial state and a compressed liquid at the final state. From Tables A-13 and A-11,

$$\left.\begin{array}{l} P_1 = 0.8\text{MPa} \\ T_1 = 40^\circ\text{C} \end{array}\right\} \begin{array}{l} u_1 = 252.13\text{kJ/kg} \\ v_1 = 0.02691\text{m}^3/\text{kg} \end{array}$$

$$\left.\begin{array}{l} P_2 = 0.8\text{MPa} \\ T_2 = 20^\circ\text{C} \end{array}\right\} \begin{array}{l} u_2 \cong u_{f@20^\circ\text{C}} = 76.80\text{kJ/kg} \\ v_2 \cong v_{f@20^\circ\text{C}} = 0.0008157\text{m}^3/\text{kg} \end{array}$$

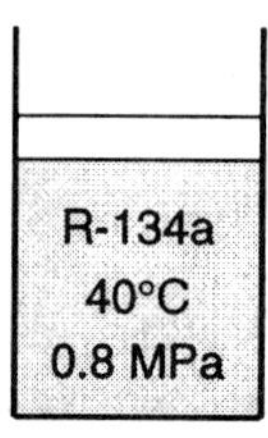

Thus,

$$\Delta V = m(v_2 - v_1) = (10\ \text{kg})(0.0008157 - 0.02691)\,\text{m}^3/\text{kg} = \mathbf{-0.261\ m^3}$$

and

$$\Delta U = m(u_2 - u_1) = (10\ \text{kg})(76.80 - 252.13)\,\text{kJ/kg} = \mathbf{-1753.3\ kJ}$$

2-123 Two rigid tanks that contain hydrogen at two different states are connected to each other. Now a valve is opened, and the two gases are allowed to mix while achieving thermal equilibrium with the surroundings. The final pressure in the tanks is to be determined.

Analysis Let's call the first and the second tanks A and B. Treating H_2 as an ideal gas, the total volume and the total mass of H_2 are

$$V = V_A + V_B = 0.5 + 0.5 = 1.0\text{m}^3$$

$$m_A = \left(\frac{P_1 V}{RT_1}\right)_A = \frac{(600\text{kPa})(0.5\text{m}^3)}{(4.124\text{kPa}\cdot\text{m}^3/\text{kg}\cdot\text{K})(293\text{K})} = 0.248 \text{ kg}$$

$$m_B = \left(\frac{P_1 V}{RT_1}\right)_B = \frac{(150\text{kPa})(0.5\text{m}^3)}{(4.124\text{kPa}\cdot\text{m}^3/\text{kg}\cdot\text{K})(303\text{K})} = 0.060 \text{ kg}$$

$$m = m_A + m_B = 0.248 + 0.060 = 0.308\text{kg}$$

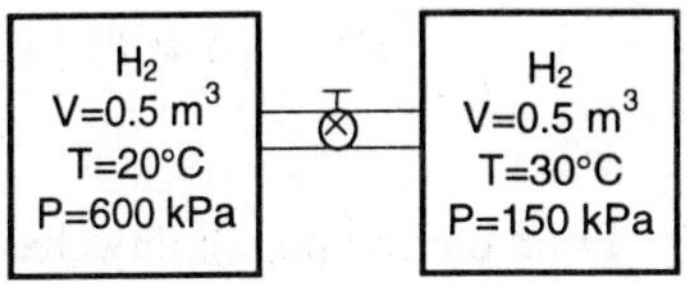

H2
V=0.5 m3
T=20°C
P=600 kPa

H2
V=0.5 m3
T=30°C
P=150 kPa

Then the final pressure can be determined from

$$P = \frac{mRT_2}{V} = \frac{(0.308 \text{ kg})(4.124 \text{ kPa}\cdot\text{m}^3/\text{kg}\cdot\text{K})(288 \text{ K})}{1.0 \text{ m}^3} = \mathbf{365.8 \text{ kPa}}$$

2-124 EES solution of this (and other comprehensive problems designated with the *computer icon*) is available to instructors at the *Instructor Manual* section of the *Online Learning Center* (OLC) at www.mhhe.com/cengel-boles. See the Preface for access information.

2-125 A large tank contains nitrogen at a specified temperature and pressure. Now some nitrogen is allowed to escape, and the temperature and pressure of nitrogen drop to new values. The amount of nitrogen that has escaped is to be determined.

Analysis Treating N_2 as an ideal gas, the initial and the final masses in the tank are determined to be

$$m_1 = \frac{P_1 V}{RT_1} = \frac{(800 \text{ kPa})(20 \text{ m}^3)}{(0.2968 \text{ kPa}\cdot\text{m}^3/\text{kg}\cdot\text{K})(298 \text{ K})} = 180.9 \text{ kg}$$

$$m_2 = \frac{P_2 V}{RT_2} = \frac{(600 \text{ kPa})(20 \text{ m}^3)}{(0.2968 \text{ kPa}\cdot\text{m}^3/\text{kg}\cdot\text{K})(293 \text{ K})} = 138.0 \text{ kg}$$

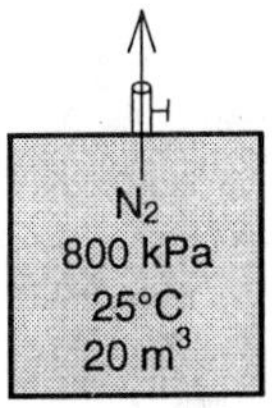

N2
800 kPa
25°C
20 m3

Thus the amount of N_2 that escaped is

$$\Delta m = m_1 - m_2 = 180.9 - 138.0 = \mathbf{42.9 \text{ kg}}$$

2-126 The temperature of steam in a tank at a specified state is to be determined using the ideal gas relation, the generalized chart, and the steam tables.

Properties The gas constant, the critical pressure, and the critical temperature of water are, from Table A-1,

$$R = 0.4615 \text{ kPa}\cdot\text{m}^3/\text{kg}\cdot\text{K}, \quad T_{cr} = 647.3 \text{ K}, \quad P_{cr} = 22.09 \text{ MPa}$$

Analysis (a) From the ideal gas equation of state,

$$P = \frac{RT}{\text{v}} = \frac{(0.4615 \text{ kPa}\cdot\text{m}^3/\text{kg}\cdot\text{K})(673\text{ K})}{0.02 \text{ m}^3/\text{kg}} = \mathbf{15{,}529 \ kPa}$$

(*b*) From the compressibility chart (Fig. A-30a),

$$\left.\begin{aligned} T_R &= \frac{T}{T_{cr}} = \frac{673\text{ K}}{647.3\text{ K}} = 1.040 \\ \text{v}_R &= \frac{v_{\text{actual}}}{RT_{cr}/P_{cr}} = \frac{(0.02\text{ m}^3/\text{kg})(22{,}090\text{ kPa})}{(0.4615\text{ kPa}\cdot\text{m}^3/\text{kg}\cdot\text{K})(647.3\text{ K})} = 1.48 \end{aligned}\right\} P_R = 0.57$$

H₂O
0.02 m³/kg
400°C

Thus,

$$P = P_R P_{cr} = 0.57 \times 22{,}090 = \mathbf{12{,}591 \ kPa}$$

(*c*) From the superheated steam table (Table A-6),

$$\left.\begin{aligned} T &= 400^\circ\text{C} \\ v &= 0.02 \text{ m}^3/\text{kg} \end{aligned}\right\} P = \mathbf{12{,}500 \ kPa}$$

2-127 One section of a tank is filled with saturated liquid R-134a while the other side is evacuated. The partition is removed, and the temperature and pressure in the tank are measured. The volume of the tank is to be determined.

Analysis The mass of the refrigerant contained in the tank is

$$m = \frac{V_1}{\text{v}_1} = \frac{0.01 \text{ m}^3}{0.0008454 \text{ m}^3/\text{kg}} = 11.83 \text{ kg}$$

since

$$v_1 = v_{f\,@\,0.8\text{ MPa}} = 0.0008454 \text{ m}^3/\text{kg}$$

R-134a
P=0.8 MPa
V=0.01 m³
Evacuated

At the final state (Table A-13),

$$\left.\begin{aligned} P_2 &= 200 \text{ kPa} \\ T_2 &= 25^\circ\text{C} \end{aligned}\right\} v_2 = 0.11625 \text{ m}^3/\text{kg}$$

Thus,

$$V_{\text{tank}} = V_2 = mv_2 = (11.83 \text{ kg})(0.11625 \text{ m}^3/\text{kg}) = \mathbf{1.375 \ m^3}$$

2-128 EES solution of this (and other comprehensive problems designated with the *computer icon*) is available to instructors at the *Instructor Manual* section of the *Online Learning Center* (OLC) at www.mhhe.com/cengel-boles. See the Preface for access information.

2-129 A propane tank contains 5 L of liquid propane at the ambient temperature. Now a leak develops at the top of the tank and propane starts to leak out. The temperature of propane when the pressure drops to 1 atm and the amount of heat transferred to the tank by the time the entire propane in the tank is vaporized are to be determined.

Properties The properties of propane at 1 atm are T_{sat} = -42.1°C, $\rho = 581 \text{ kg/m}^3$, and h_{fg} = 427.8 kJ/kg (Table A-3).

Analysis The temperature of propane when the pressure drops to 1 atm is simply the saturation pressure at that temperature,

$$T = T_{\text{sat @ 1 atm}} = \mathbf{-42.1°C}$$

The initial mass of liquid propane is

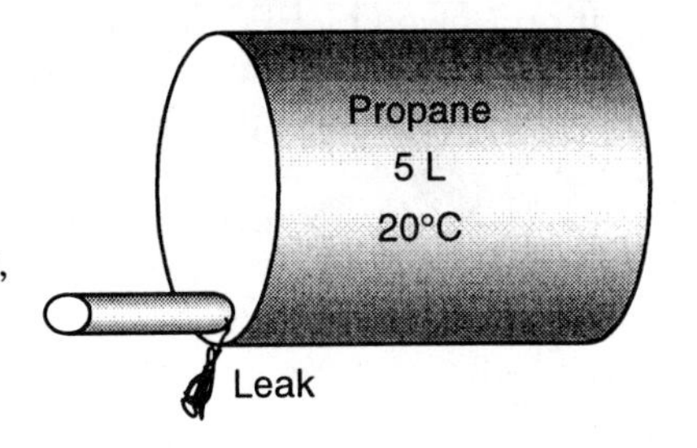

$$m = \rho V = (581 \text{ kg/m}^3)(0.005 \text{ m}^3) = 2.905 \text{ kg}$$

The amount of heat absorbed is simply the total heat of vaporization,

$$Q_{\text{absorbed}} = mh_{fg} = (2.905 \text{ kg})(427.8 \text{ kJ/kg}) = \mathbf{1243 \text{ kJ}}$$

2-130 An isobutane tank contains 5 L of liquid isobutane at the ambient temperature. Now a leak develops at the top of the tank and isobutane starts to leak out. The temperature of isobutane when the pressure drops to 1 atm and the amount of heat transferred to the tank by the time the entire isobutane in the tank is vaporized are to be determined.

Properties The properties of isobutane at 1 atm are T_{sat} = -11.7°C, $\rho = 593.8 \text{ kg/m}^3$, and h_{fg} = 367.1 kJ/kg (Table A-3).

Analysis The temperature of isobutane when the pressure drops to 1 atm is simply the saturation pressure at that temperature,

$$T = T_{\text{sat @ 1 atm}} = \mathbf{-11.7°C}$$

The initial mass of liquid isobutane is

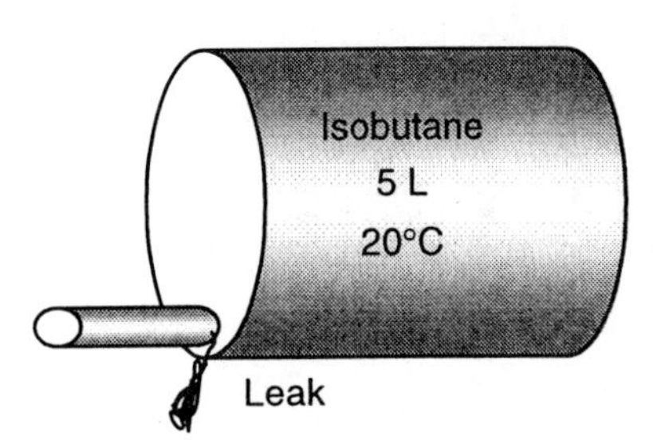

$$m = \rho V = (593.8 \text{ kg/m}^3)(0.005 \text{ m}^3) = 2.969 \text{ kg}$$

The amount of heat absorbed is simply the total heat of vaporization,

$$Q_{\text{absorbed}} = mh_{fg} = (2.969 \text{ kg})(367.1 \text{ kJ/kg}) = \mathbf{1090 \text{ kJ}}$$

Fundamentals of Engineering (FE) Exam Problems

2-131 A rigid tank contains 5 kg of an ideal gas at 4 atm and 40°C. Now a valve is opened, and half of mass of the gas is allowed to escape. If the final pressure in the tank is 1.5 atm, the final temperature in the tank is

(a) –38°C (b) 30°C (c) 40°C (d) 53°C (e) 144°C

Answer (a) –38°C

Solution Solved by EES Software. Solutions can be verified by copying-and-pasting the following lines on a blank EES screen. (Similar problems and their solutions can be obtained easily by modifying numerical values).

```
"When R=constant and V= constant, P1/P2=m1*T1/m2*T2"
m1=5 "kg"
P1=4 "atm"
P2=1.5 "atm"
T1=40+273 "K"

m2=0.5*m1 "kg"
P1/P2=m1*T1/(m2*T2)
T2_C=T2-273 "C"

"Some Wrong Solutions with Common Mistakes:"
P1/P2=m1*(T1-273)/(m2*W1_T2) "Using C instead of K"
P1/P2=m1*T1/(m1*(W2_T2+273)) "Disregarding the decrease in mass"
P1/P2=m1*T1/(m1*W3_T2) "Disregarding the decrease in mass, and not converting to deg. C"
W4_T2=(T1-273)/2 "Taking T2 to be half of T1 since half of the mass is discharged"
```

2-132 The pressure of an automobile tire is measured to be 200 kPa (gage) before a trip and 220 kPa (gage) after the trip at a location where the atmospheric pressure is 90 kPa. If the temperature of air in the tire before the trip is 25°C, the air temperature after the trip is

(a) 45.6°C (b) 54.8°C (c) 27.5°C (d) 26.7°C (e) 25.0°C

Answer (a) 45.6°C

Solution Solved by EES Software. Solutions can be verified by copying-and-pasting the following lines on a blank EES screen. (Similar problems and their solutions can be obtained easily by modifying numerical values).

```
"When R, V, and m are constant, P1/P2=T1/T2"
Patm=90
P1=200+Patm "kPa"
P2=220+Patm "kPa"
T1=25+273 "K"

P1/P2=T1/T2
T2_C=T2-273 "C"

"Some Wrong Solutions with Common Mistakes:"
P1/P2=(T1-273)/W1_T2 "Using C instead of K"
(P1-Patm)/(P2-Patm)=T1/(W2_T2+273) "Using gage pressure instead of absolute
pressure"
(P1-Patm)/(P2-Patm)=(T1-273)/W3_T2 "Making both of the mistakes above"
W4_T2=T1-273 "Assuming the temperature to remain constant"
```

2-133 A 500-m^3 rigid tank is filled with saturated liquid-vapor mixture of water at 200 kPa. If 20% of the mass is liquid and the 80% of the mass is vapor, the total mass in the tank is

(a) 705 kg (b) 500 kg (c) 258 kg (d) 635 kg (e) 2809 kg

Answer (a) 705 kg

Solution Solved by EES Software. Solutions can be verified by copying-and-pasting the following lines on a blank EES screen. (Similar problems and their solutions can be obtained easily by modifying numerical values).

```
V_tank=500 "m3"
P1=200 "kPa"
x=0.8

v_f=VOLUME(Steam_NBS, x=0,P=P1)
v_g=VOLUME(Steam_NBS, x=1,P=P1)
v=v_f+x*(v_g-v_f)
m=V_tank/v "kg"

"Some Wrong Solutions with Common Mistakes:"
R=0.4615 "kJ/kg.K"
T=TEMPERATURE(Steam_NBS,x=0,P=P1)
P1*V_tank=W1_m*R*(T+273) "Treating steam as ideal gas"
P1*V_tank=W2_m*R*T "Treating steam as ideal gas and using deg.C"
W3_m=V_tank "Taking the density to be 1 kg/m^3"
```

2-134 Water is boiled at 1 atm pressure in a coffee maker equipped with an immersion-type electric heating element. The coffee maker initially contains 1 kg of water. Once boiling started, it is observed that half of the water in the coffee maker evaporated in 25 minutes. If the heat loss from the coffee maker is negligible, the power rating of the heating element is

(a) 2.15 kW (b) 1.50 kW (c) 0.50 kW (d) 1.00 kW (e) 0.75 kW

Answer (e) 0.75 kW

Solution Solved by EES Software. Solutions can be verified by copying-and-pasting the following lines on a blank EES screen. (Similar problems and their solutions can be obtained easily by modifying numerical values).

```
m_1=1 "kg"
P=101.325 "kPa"
time=25*60 "s"
m_evap=0.5*m_1
Power*time=m_evap*h_fg "kJ"

h_f=ENTHALPY(Steam_NBS, x=0,P=P)
h_g=ENTHALPY(Steam_NBS, x=1,P=P)
h_fg=h_g-h_f

"Some Wrong Solutions with Common Mistakes:"
W1_Power*time=m_evap*h_g "Using h_g"
W2_Power*time/60=m_evap*h_g "Using minutes instead of seconds for time"
W3_Power=2*Power  "Assuming all the water evaporates"
```

2-135 A 1-m^3 rigid tank contains 10 kg of water (in any phase or phases) at 150°C. The pressure in the tank is

(a) 500 kPa (b) 475.8 kPa (c) 270.1 kPa (d) 1000 MPa (e) 1618 kPa

Answer (b) 475.8 kPa

Solution Solved by EES Software. Solutions can be verified by copying-and-pasting the following lines on a blank EES screen. (Similar problems and their solutions can be obtained easily by modifying numerical values).

```
V_tank=1 "m^3"
m=10 "kg"
v=V_tank/m
T=150 "C"

P=PRESSURE(Steam_NBS,v=v,T=T)

"Some Wrong Solutions with Common Mistakes:"
R=0.4615 "kJ/kg.K"
W1_P*V_tank=m*R*(T+273) "Treating steam as ideal gas"
W2_P*V_tank=m*R*T "Treating steam as ideal gas and using deg.C"
```

2-136 Water is boiling at 1 atm pressure in a stainless steel pan on an electric range. It is observed that 2 kg of liquid water evaporates in 30 minutes. The rate of heat transfer to the water is

(a) 2.51 kW (b) 2.32 kW (c) 2.97 kW (d) 0.47 kW (e) 3.12 kW

Answer (a) 2.51 kW

Solution Solved by EES Software. Solutions can be verified by copying-and-pasting the following lines on a blank EES screen. (Similar problems and their solutions can be obtained easily by modifying numerical values).

```
m_evap=2 "kg"
P=101.325 "kPa"
time=30*60 "s"
Q*time=m_evap*h_fg "kJ"

h_f=ENTHALPY(Steam_NBS, x=0,P=P)
h_g=ENTHALPY(Steam_NBS, x=1,P=P)
h_fg=h_g-h_f

"Some Wrong Solutions with Common Mistakes:"
W1_Q*time=m_evap*h_g "Using h_g"
W2_Q*time/60=m_evap*h_g "Using minutes instead of seconds for time"
W3_Q*time=m_evap*h_f "Using h_f"
```

2-137 Water is boiled in a pan on a stove at sea level. During 10 min of boiling, its is observed that 200 g of water has evaporated. Then the rate of heat transfer to the water is

(a) 0.84 kJ/min (b) 45.1 kJ/min (c) 41.8 kJ/min (d) 53.5 kJ/min (e) 225.7 kJ/min

Answer (b) 45.1 kJ/min

Solution Solved by EES Software. Solutions can be verified by copying-and-pasting the following lines on a blank EES screen. (Similar problems and their solutions can be obtained easily by modifying numerical values).

```
m_evap=0.2 "kg"
P=101.325 "kPa"
time=10 "min"
Q*time=m_evap*h_fg "kJ"

h_f=ENTHALPY(Steam_NBS, x=0,P=P)
h_g=ENTHALPY(Steam_NBS, x=1,P=P)
h_fg=h_g-h_f

"Some Wrong Solutions with Common Mistakes:"
W1_Q*time=m_evap*h_g "Using h_g"
W2_Q*time*60=m_evap*h_g "Using seconds instead of minutes for time"
W3_Q*time=m_evap*h_f "Using h_f"
```

2-138 A room is filled with saturated steam at 100°C. Now a 5-kg bowling ball at 25°C is brought to the room. Heat is transferred to the ball from the steam, and the temperature of the ball rises to 100°C while some steam condenses on the ball as it loses heat (but it still remains at 100°C). The specific heat of the ball can be taken to be 1.8 kJ/kg.°C. The mass of steam that condensed during this process is

(a) 80 g (b) 128 g (c) 299 g (d) 351 g (e) 405 g

Answer (c) 299 g

Solution Solved by EES Software. Solutions can be verified by copying-and-pasting the following lines on a blank EES screen. (Similar problems and their solutions can be obtained easily by modifying numerical values).

```
m_ball=5 "kg"
T=100 "C"
T1=25 "C"
T2=100 "C"
Cp=1.8 "kJ/kg.C"
Q=m_ball*Cp*(T2-T1)
Q=m_steam*h_fg "kJ"

h_f=ENTHALPY(Steam_NBS, x=0,T=T)
h_g=ENTHALPY(Steam_NBS, x=1,T=T)
h_fg=h_g-h_f

"Some Wrong Solutions with Common Mistakes:"
Q=W1m_steam*h_g "Using h_g"
Q=W2m_steam*4.18*(T2-T1) "Using m*C*DeltaT = Q for water"
Q=W3m_steam*h_f "Using h_f"
```

2-139 A rigid 5-m^3 rigid vessel contains steam at 20 MPa and 400°C. The mass of the steam is

(a) 5.0 kg (b) 0.322 kg (c) 322 kg (d) 503 kg (e) 680 kg

Answer (d) 503 kg

Solution Solved by EES Software. Solutions can be verified by copying-and-pasting the following lines on a blank EES screen. (Similar problems and their solutions can be obtained easily by modifying numerical values).

```
V=5 "m^3"
m=V/v1 "m^3/kg"
P1=20000 "kPa"
T1=400 "C"
v1=VOLUME(Steam_NBS,T=T1,P=P1)

"Some Wrong Solutions with Common Mistakes:"
R=0.4615 "kJ/kg.K"
P1*V=W1_m*R*(T1+273) "Treating steam as ideal gas"
P1*V=W2_m*R*T1 "Treating steam as ideal gas and using deg.C"
```

2-140 Consider a sealed can that is filled with refrigerant-134a. The contents of the can are at the room temperature of 25°C. Now a leak developes, and the pressure in the can drops to the local atmospheric pressure of 100 kPa. The temperature of the refrigerant in the can is expected to drop to (rounded to the nearest integer)

(a) 0°C (b) -26°C (c) -18°C (d) 3°C (e) 25°C

Answer (b) -26°C

Solution Solved by EES Software. Solutions can be verified by copying-and-pasting the following lines on a blank EES screen. (Similar problems and their solutions can be obtained easily by modifying numerical values).

```
T1=25 "C"
P2=100 "kPa"
T2=TEMPERATURE(R134a,x=0,P=P2)

"Some Wrong Solutions with Common Mistakes:"
W1_T2=T1 "Assuming temperature remains constant"
```

2-141 ... 2-147 Design and Essay Problems

2-141 A claim that fruits and vegetables are cooled by 6°C for each percentage point of weight loss as moisture during vacuum cooling is to be evaluated.

Analysis Assuming the fruits and vegetables are cooled from 30°C and 0°C, the average heat of vaporization can be taken to be 2466 kJ/kg, which is the value at 15°C, and the specific heat of products can be taken to be 4 kJ/kg.°C. Then the vaporization of 0.01 kg water will lower the temperature of 1 kg of produce by 24.66/4 = 6°C. Therefore, the vacuum cooled products will lose 1 percent moisture for each 6°C drop in temperature. Thus the claim is **reasonable**.

2-142 It is helium.

Chapter 3

ENERGY TRANSFER BY HEAT, WORK, AND MASS

Heat Transfer and Work

3-1C Energy can cross the boundaries of a closed system in two forms: heat and work.

3-2C The form of energy that crosses the boundary of a closed system because of a temperature difference is heat; all other forms are work.

3-3C An adiabatic process is a process during which there is no heat transfer. A system that does not exchange any heat with its surroundings is an adiabatic system.

3-4C It is a work interaction.

3-5C It is a work interaction since the electrons are crossing the system boundary, thus doing electrical work.

3-6C It is a heat interaction since it is due to the temperature difference between the sun and the room.

3-7C This is neither a heat nor a work interaction since no energy is crossing the system boundary. This is simply the conversion of one form of internal energy (chemical energy) to another form (sensible energy).

3-8C Point functions depend on the state only whereas the path functions depend on the path followed during a process. Properties of substances are point functions, heat and work are path functions.

3-9C The caloric theory is based on the assumption that heat is a fluid-like substance called the "caloric" which is a massless, colorless, odorless substance. It was abandoned in the middle of the nineteenth century after it was shown that there is no such thing as the caloric.

Boundary Work

3-10C It represents the boundary work for quasi-equilibrium processes.

3-11C Yes.

3-12C The area under the process curve, and thus the boundary work done, is greater in the constant pressure case.

3-13C $$1\ \text{kPa}\cdot\text{m}^3 = 1\ \text{k}(\text{N/m}^2)\cdot\text{m}^3 = 1\ \text{kN}\cdot\text{m} = 1\ \text{kJ}$$

3-14 Saturated water vapor in a cylinder is heated at constant pressure until its temperature rises to a specified value. The boundary work done during this process is to be determined.

Assumptions The process is quasi-equilibrium.

Properties Noting that the pressure remains constant during this process, the specific volumes at the initial and the final states are (Table A-4 through A-6)

$$\left.\begin{array}{l} P_1 = 200\text{ kPa} \\ \text{Sat.vapor} \end{array}\right\} v_1 = v_{g@200\text{kPa}} = 0.8857\text{ m}^3/\text{kg}$$

$$\left.\begin{array}{l} P_2 = 200\text{ kPa} \\ T_2 = 300^\circ\text{C} \end{array}\right\} v_2 = 1.3162\text{ m}^3/\text{kg}$$

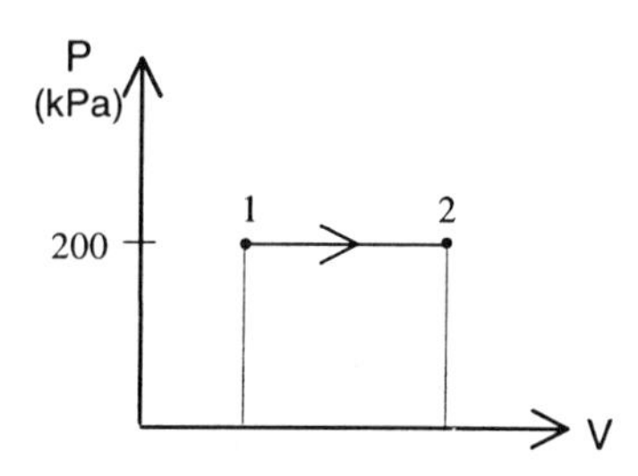

Analysis The boundary work is determined from its definition to be

$$W_{b,\text{out}} = \int_1^2 P dV = P(V_2 - V_1) = mP(v_2 - v_1)$$

$$= (5\text{kg})(200\text{kPa})(1.3162 - 0.8857)\text{ m}^3/\text{kg}\left(\frac{1\text{kJ}}{1\text{kPa}\cdot\text{m}^3}\right)$$

$$= \mathbf{430.5kJ}$$

Discussion The positive sign indicates that work is done by the system (work output).

3-15 Refrigerant-134a in a cylinder is heated at constant pressure until its temperature rises to a specified value. The boundary work done during this process is to be determined.

Assumptions The process is quasi-equilibrium.

Properties Noting that the pressure remains constant during this process, the specific volumes at the initial and the final states are (Table A-11 through A-13)

$$\left.\begin{array}{l} P_1 = 800\text{ kPa} \\ \text{Sat.liquid} \end{array}\right\} v_1 = v_{f@800\text{kPa}} = 0.0008454\text{m}^3/\text{kg}$$

$$\left.\begin{array}{l} P_2 = 800\text{kPa} \\ T_2 = 50^\circ\text{C} \end{array}\right\} v_2 = 0.02846\text{m}^3/\text{kg}$$

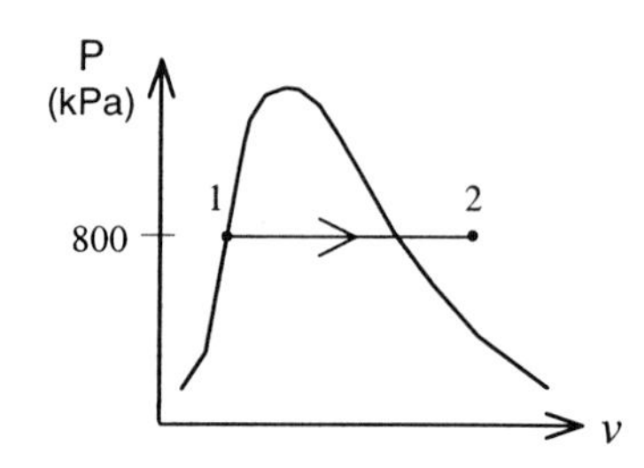

Analysis The boundary work is determined from its definition to be

$$m = \frac{V_1}{v_1} = \frac{0.2\text{ m}^3}{0.0008454\text{ m}^3/\text{kg}} = 236.6\text{ kg}$$

and

$$W_{b,\text{out}} = \int_1^2 P dV = P(V_2 - V_1) = mP(v_2 - v_1)$$

$$= (236.6\text{kg})(800\text{kPa})(0.02846 - 0.0008454)\text{m}^3/\text{kg}\left(\frac{1\text{kJ}}{1\text{kPa}\cdot\text{m}^3}\right)$$

$$= \mathbf{5227kJ}$$

Discussion The positive sign indicates that work is done by the system (work output).

3-16 EES solution of this (and other comprehensive problems designated with the *computer icon*) is available to instructors at the *Instructor Manual* section of the *Online Learning Center* (OLC) at www.mhhe.com/cengel-boles. See the Preface for access information.

3-17E Superheated water vapor in a cylinder is cooled at constant pressure until 70% of it condenses. The boundary work done during this process is to be determined.

Assumptions The process is quasi-equilibrium.

Properties Noting that the pressure remains constant during this process, the specific volumes at the initial and the final states are (Table A-4E through A-6E)

$$\left.\begin{array}{l} P_1 = 60 \text{ psia} \\ T_1 = 500^\circ\text{F} \end{array}\right\} v_1 = 9.399 \text{ft}^3/\text{lbm}$$

$$\left.\begin{array}{l} P_2 = 60 \text{ psia} \\ x_2 = 0.3 \end{array}\right\} v_2 = v_f + x_2 v_{fg}$$

$$= 0.017378 + 0.3(7.177 - 0.017378)$$

$$= 2.165 \text{ft}^3/\text{lbm}$$

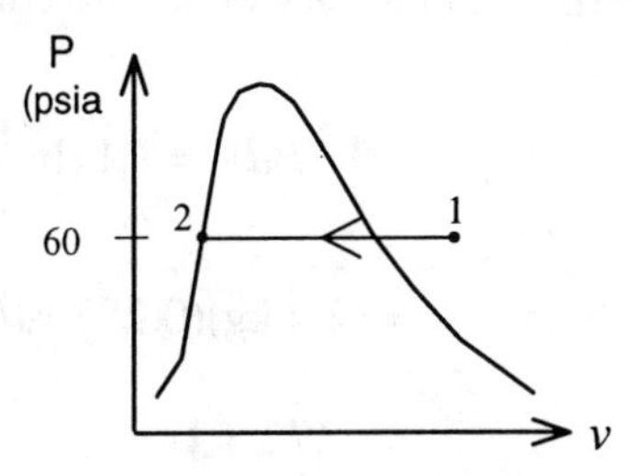

Analysis The boundary work is determined from its definition to be

$$W_{b,\text{out}} = \int_1^2 P dV = P(V_2 - V_1) = mP(v_2 - v_1)$$

$$= (12 \text{ lbm})(60 \text{ psia})(2.165 - 9.399) \text{ft}^3/\text{lbm}\left(\frac{1 \text{ Btu}}{5.4039 \text{ psia}\cdot\text{ft}^3}\right)$$

$$= \mathbf{-963.8 \text{ Btu}}$$

Discussion The negative sign indicates that work is done on the system (work input).

3-18 Air in a cylinder is compressed at constant temperature until its pressure rises to a specified value. The boundary work done during this process is to be determined.

Assumptions **1** The process is quasi-equilibrium. **2** Air is an ideal gas.

Properties The gas constant of air is R = 0.287 kJ/kg.K (Table A-1).

Analysis The boundary work is determined from its definition to be

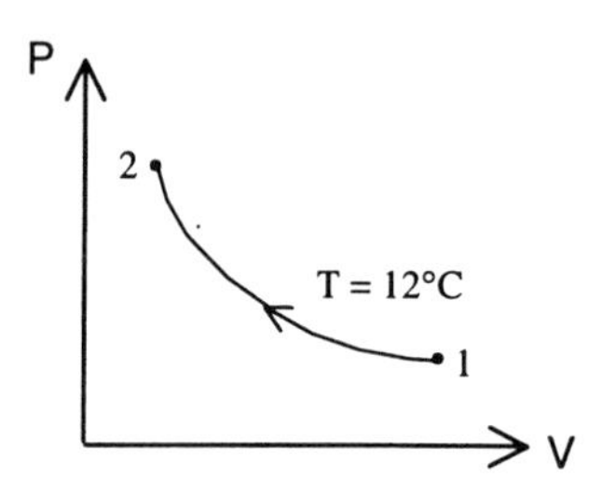

$$W_{b,out} = \int_1^2 PdV = P_1V_1 \ln\frac{V_2}{V_1} = mRT\ln\frac{P_1}{P_2}$$

$$= (2.4\ \text{kg})(0.287\ \text{kJ/kg}\cdot\text{K})(285\ \text{K})\ln\frac{150\ \text{kPa}}{600\ \text{kPa}}$$

$$= \mathbf{-272\ kJ}$$

Discussion The negative sign indicates that work is done on the system (work input).

3-19 Nitrogen gas in a cylinder is compressed at constant temperature until its pressure rises to a specified value. The boundary work done during this process is to be determined.

Assumptions **1** The process is quasi-equilibrium. **2** Nitrogen is an ideal gas.

Analysis The boundary work is determined from its definition to be

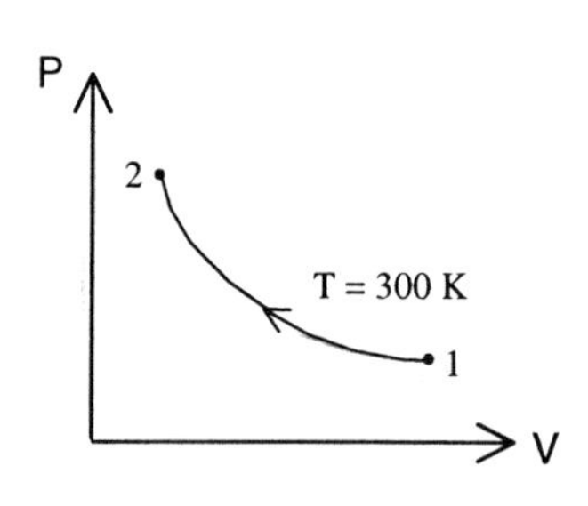

$$W_{b,out} = \int_1^2 PdV = P_1V_1 \ln\frac{V_2}{V_1} = P_1V_1 \ln\frac{P_1}{P_2}$$

$$= (150\text{kPa})(0.2\text{m}^3)\left(\ln\frac{150\text{kPa}}{800\text{kPa}}\right)\left(\frac{1\text{kJ}}{1\text{kPa}\cdot\text{m}^3}\right)$$

$$= \mathbf{-50.2kJ}$$

Discussion The negative sign indicates that work is done on the system (work input).

3-20 A gas in a cylinder is compressed to a specified volume in a process during which the pressure changes linearly with volume. The boundary work done during this process is to be determined by plotting the process on a P-V diagram and also by integration.

Assumptions The process is quasi-equilibrium.

Analysis (*a*) The pressure of the gas changes linearly with volume, and thus the process curve on a P-V diagram will be a straight line. The boundary work during this process is simply the area under the process curve, which is a trapezoidal. Thus,

$$P_1 = aV_1 + b = (-1200 \text{ kPa/m}^3)(0.42 \text{ m}^3) + (600 \text{ kPa}) = 96 \text{ kPa}$$

$$P_2 = aV_2 + b = (-1200 \text{ kPa/m}^3)(0.12 \text{ m}^3) + (600 \text{ kPa}) = 456 \text{ kPa}$$

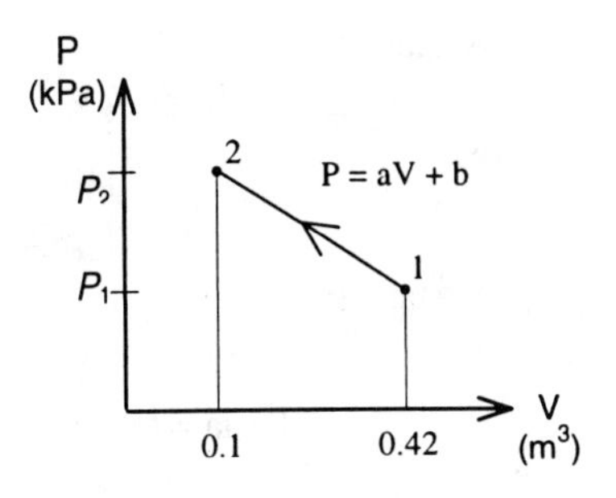

and

$$\begin{aligned} W_{b,\text{out}} &= \text{Area} = \frac{P_1 + P_2}{2}(V_2 - V_1) \\ &= \frac{(96+456)\text{kPa}}{2}(0.12-0.42)\text{m}^3\left(\frac{1\text{kJ}}{1\text{kPa}\cdot\text{m}^3}\right) \\ &= \mathbf{-82.8kJ} \end{aligned}$$

(*b*) The boundary work can also be determined by integration to be

$$\begin{aligned} W_{b,\text{out}} &= \int_1^2 P dV = \int_1^2 (aV + b)dV = a\frac{V_2^2 - V_1^2}{2} + b(V_2 - V_1) \\ &= (-1200 \text{ kPa/m}^3)\frac{(0.12^2 - 0.42^2)\text{m}^6}{2} + (600 \text{ kPa})(0.12 - 0.42)\text{m}^3 \\ &= \mathbf{-82.8 \text{ kJ}} \end{aligned}$$

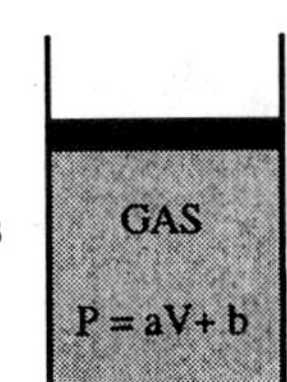

Discussion The negative sign indicates that work is done on the system (work input).

3-21E A gas in a cylinder is heated and is allowed to expand to a specified pressure in a process during which the pressure changes linearly with volume. The boundary work done during this process is to be determined.

Assumptions The process is quasi-equilibrium.

Analysis (*a*) The pressure of the gas changes linearly with volume, and thus the process curve on a P-V diagram will be a straight line. The boundary work during this process is simply the area under the process curve, which is a trapezoidal. Thus,

At state 1:

$$P_1 = aV_1 + b$$
$$15 \text{ psia} = (5 \text{ psia/ft}^3)(7 \text{ ft}^3) + b$$
$$b = -20 \text{ psia}$$

At state 2:

$$P_2 = aV_2 + b$$
$$100 \text{ psia} = (5 \text{ psia/ft}^3)V_2 + (-20 \text{ psia})$$
$$V_2 = 24 \text{ ft}^3$$

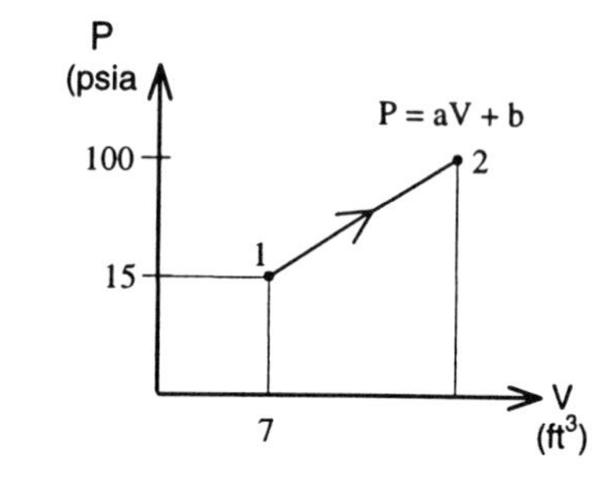

and,

$$W_{b,\text{out}} = \text{Area} = \frac{P_1 + P_2}{2}(V_2 - V_1) = \frac{(100+15)\text{psia}}{2}(24-7)\text{ft}^3\left(\frac{1 \text{Btu}}{5.4039 \text{psia}\cdot\text{ft}^3}\right)$$
$$= \mathbf{181\ Btu}$$

Discussion The positive sign indicates that work is done by the system (work output).

3-22 [*Also solved by EES on enclosed CD*] A gas in a cylinder expands polytropically to a specified volume. The boundary work done during this process is to be determined.

Assumptions The process is quasi-equilibrium.

Analysis The boundary work for this polytropic process can be determined directly from

$$P_2 = P_1\left(\frac{V_1}{V_2}\right)^n = (150 \text{ kPa})\left(\frac{0.03 \text{ m}^3}{0.2 \text{ m}^3}\right)^{1.3} = 12.74 \text{ kPa}$$

and,

$$W_{b,out} = \int_1^2 P dV = \frac{P_2V_2 - P_1V_1}{1-n}$$
$$= \frac{(12.74\times 0.2 - 150\times 0.03)\text{ kPa}\cdot\text{m}^3}{1-1.3}\left(\frac{1 \text{ kJ}}{1 \text{ kPa}\cdot\text{m}^3}\right)$$
$$= \mathbf{6.51\ kJ}$$

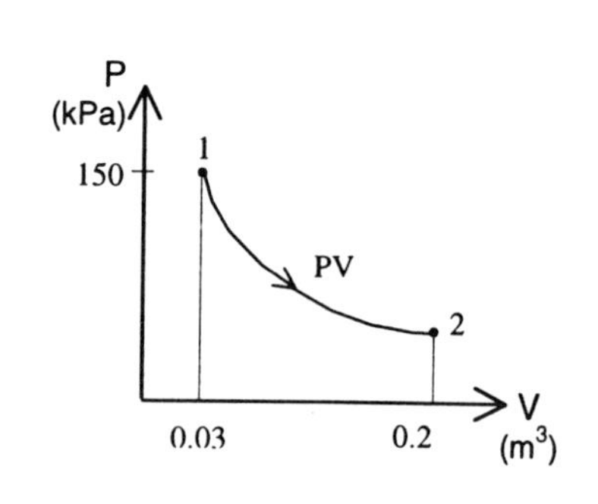

Discussion The positive sign indicates that work is done by the system (work output).

3-23 EES solution of this (and other comprehensive problems designated with the *computer icon*) is available to instructors at the *Instructor Manual* section of the *Online Learning Center* (OLC) at www.mhhe.com/cengel-boles. See the Preface for access information.

3-24 Nitrogen gas in a cylinder is compressed polytropically until the temperature rises to a specified value. The boundary work done during this process is to be determined.

Assumptions **1** The process is quasi-equilibrium. **2** Nitrogen is an ideal gas.

Analysis The boundary work for this polytropic process can be determined from

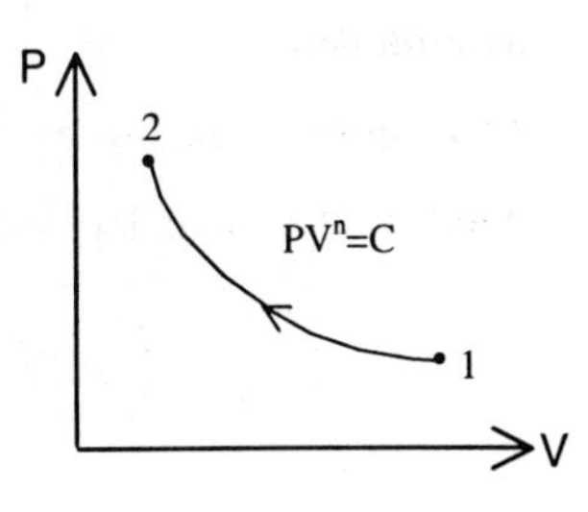

$$W_{b,\text{out}} = \int_1^2 P\,dV = \frac{P_2V_2 - P_1V_1}{1-n} = \frac{mR(T_2 - T_1)}{1-n}$$

$$= \frac{(2\text{kg})(0.2968\text{kJ/kg}\cdot\text{K})(360-300)\text{K}}{1-1.4}$$

$$= \mathbf{-89.0\ kJ}$$

Discussion The negative sign indicates that work is done on the system (work input).

3-25 [*Also solved by EES on enclosed CD*] A gas whose equation of state is $\bar{v}(P+10/\bar{v}^2) = R_uT$ expands in a cylinder isothermally to a specified volume. The unit of the quantity 10 and the boundary work done during this process are to be determined.

Assumptions The process is quasi-equilibrium.

Analysis (*a*) The term $10/\bar{v}^2$ must have pressure units since it is added to *P*. Thus the quantity 10 must have the unit kPa·m^6/kmol^2.

(*b*) The boundary work for this process can be determined from

P
T = 300 K
V
2
4

$$P = \frac{R_uT}{\bar{v}} - \frac{10}{\bar{v}^2} = \frac{R_uT}{V/N} - \frac{10}{(V/N)^2} = \frac{NR_uT}{V} - \frac{10N^2}{V^2}$$

and

$$W_{b,\text{out}} = \int_1^2 P\,dV = \int_1^2 \left(\frac{NR_uT}{V} - \frac{10N^2}{V^2}\right)dV = NR_uT\ln\frac{V_2}{V_1} + 10N^2\left(\frac{1}{V_2} - \frac{1}{V_1}\right)$$

$$= (0.5\text{ kmol})(8.314\text{ kJ/kmol}\cdot\text{K})(300\text{ K})\ln\frac{4\text{ m}^3}{2\text{ m}^3}$$

$$+ (10\text{ kPa}\cdot\text{m}^6/\text{kmol}^2)(0.5\text{kmol})^2\left(\frac{1}{4\text{ m}^3} - \frac{1}{2\text{ m}^3}\right)\left(\frac{1\text{ kJ}}{1\text{ kPa}\cdot\text{m}^3}\right)$$

$$= \mathbf{863\ kJ}$$

Discussion The positive sign indicates that work is done by the system (work output).

3-26 EES solution of this (and other comprehensive problems designated with the *computer icon*) is available to instructors at the *Instructor Manual* section of the *Online Learning Center* (OLC) at www.mhhe.com/cengel-boles. See the Preface for access information.

3-27 CO_2 gas in a cylinder is compressed until the volume drops to a specified value. The pressure changes during the process with volume as $P = aV^{-2}$. The boundary work done during this process is to be determined.

Assumptions The process is quasi-equilibrium.

Analysis The boundary work done during this process is determined from

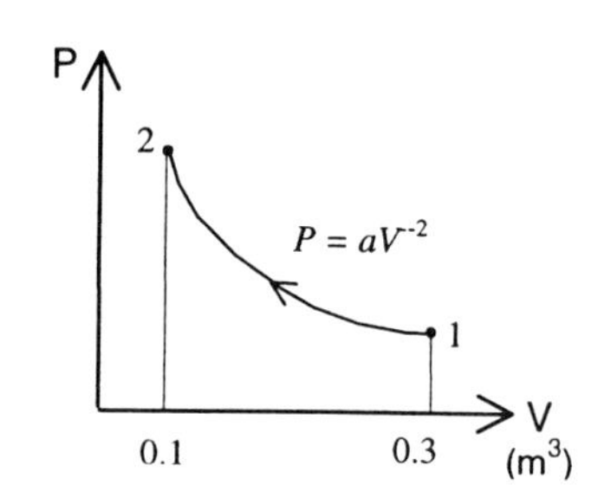

$$W_{b,\text{out}} = \int_1^2 P dV = \int_1^2 \left(\frac{a}{V^2}\right) dV = -a\left(\frac{1}{V_2} - \frac{1}{V_1}\right)$$

$$= -(8 \text{ kPa} \cdot \text{m}^6)\left(\frac{1}{0.1 \text{ m}^3} - \frac{1}{0.3 \text{ m}^3}\right)\left(\frac{1 \text{ kJ}}{1 \text{ kPa} \cdot \text{m}^3}\right)$$

$$= \mathbf{-53.3 \text{ kJ}}$$

Discussion The negative sign indicates that work is done on the system (work input).

3-28E Hydrogen gas in a cylinder equipped with a spring is heated. The gas expands and compresses the spring until its volume doubles. The final pressure, the boundary work done by the gas, and the work done against the spring are to be determined, and a *P-V* diagram is to be drawn.

Assumptions **1** The process is quasi-equilibrium. **2** Hydrogen is an ideal gas.

Analysis (*a*) When the volume doubles, the spring force and the final pressure of H_2 becomes

$$F_s = kx_2 = k\frac{\Delta V}{A} = (15{,}000\text{lbf/ft})\frac{15\text{ft}^3}{3\text{ft}^2} = 75{,}000\text{lbf}$$

$$P_2 = P_1 + \frac{F_s}{A} = (14.7\text{psia}) + \frac{75{,}000\text{lbf}}{3\text{ft}^2}\left(\frac{1\text{ft}^2}{144\text{in}^2}\right) = \mathbf{188.3 \text{ psia}}$$

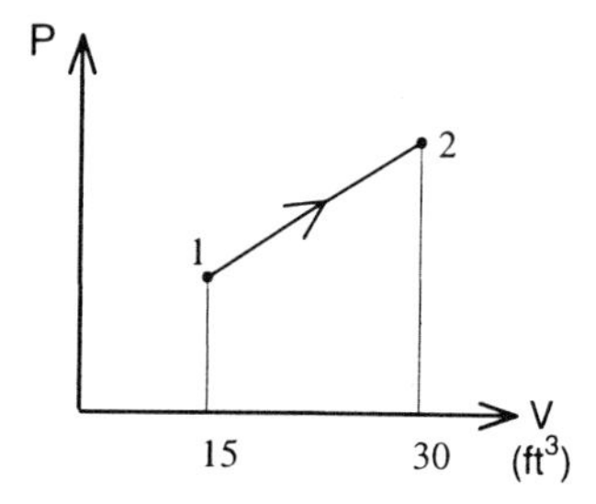

(*b*) The pressure of H_2 changes linearly with volume during this process, and thus the process curve on a *P-V* diagram will be a straight line. Then the boundary work during this process is simply the area under the process curve, which is a trapezoid. Thus,

$$W_{b,\text{out}} = \text{Area} = \frac{P_1 + P_2}{2}(V_2 - V_1)$$

$$= \frac{(188.3 + 14.7)\text{psia}}{2}(30 - 15)\text{ft}^3\left(\frac{1 \text{ Btu}}{5.40395 \text{ psia} \cdot \text{ft}^3}\right) = \mathbf{281.7 \text{ Btu}}$$

(*c*) If there were no spring, we would have a constant pressure process at P = 14.7 psia. The work done during this process would be

$$W_{b,\text{out,no spring}} = \int_1^2 P dV = P(V_2 - V_1)$$

$$= (14.7 \text{ psia})(30 - 15) \text{ ft}^3\left(\frac{1 \text{ Btu}}{5.40395 \text{ psia} \cdot \text{ft}^3}\right) = 40.8 \text{ Btu}$$

Thus,

$$W_{\text{spring}} = W_b - W_{b,\text{no spring}} = 281.7 - 40.8 = \mathbf{240.9 \text{ Btu}}$$

Discussion The positive sign for boundary work indicates that work is done by the system (work output).

3-29 Water in a cylinder equipped with a spring is heated and evaporated. The vapor expands until it compresses the spring 20 cm. The final pressure and temperature, and the boundary work done are to be determined, and the process is to be shown on a *P-V* diagram. √

Assumptions The process is quasi-equilibrium.

Analysis (*a*) The final pressure is determined from

$$P_3 = P_2 + \frac{F_s}{A} = P_2 + \frac{kx}{A} = (150\text{kPa}) + \frac{(100\text{kN/m})(0.2\text{m})}{0.1\text{m}^2}\left(\frac{1\text{kPa}}{1\text{kN/m}^2}\right) = \mathbf{350kPa}$$

The specific and total volumes at the three states are

$$\left.\begin{aligned} T_1 &= 25\,^\circ\text{C} \\ P_1 &= 150\text{kPa} \end{aligned}\right\} v_1 \cong v_{f@25^\circ\text{C}} = 0.001003\text{m}^3/\text{kg}$$

$$V_1 = mv_1 = (50\text{kg})(0.001003\text{m}^3/\text{kg}) = 0.05\text{m}^3$$

$$V_2 = 0.2\text{m}^3$$

$$V_3 = V_2 + x_{23}A_p = (0.2\text{m}^3) + (0.2\text{m})(0.1\text{m}^2) = 0.22\text{m}^3$$

$$v_3 = \frac{V_3}{m} = \frac{0.22\text{m}^3}{50\text{kg}} = 0.0044\ \text{m}^3/\text{kg}$$

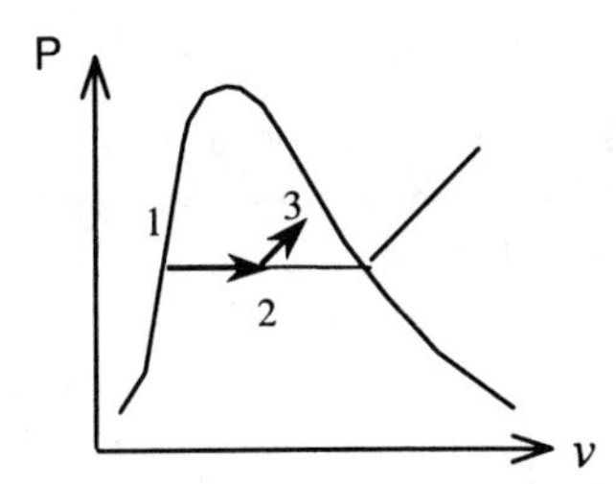

At 350 kPa, v_f = 0.0010 m^3/kg and v_g = 0.5243 m^3/kg. Noting that $v_f < v_3 < v_g$, the final state is a saturated mixture and thus the final temperature is

$$T_3 = T_{\text{sat@350 kPa}} = \mathbf{138.88\,^\circ C}$$

(*b*) The pressure remains constant during process 1-2 and changes linearly (a straight line) during process 2-3. Then the boundary work during this process is simply the total area under the process curve,

$$W_{b,\text{out}} = \text{Area} = P_1(V_2 - V_1) + \frac{P_2 + P_3}{2}(V_3 - V_2)$$

$$= \left((150\text{ kPa})(0.2 - 0.05)\text{m}^3 + \frac{(150 + 350)\text{kPa}}{2}(0.22 - 0.2)\text{m}^3\right)\left(\frac{1\text{ kJ}}{1\text{ kPa}\cdot\text{m}^3}\right)$$

$$= \mathbf{27.5\ \ kJ}$$

Discussion The positive sign indicates that work is done by the system (work output).

3-30 EES solution of this (and other comprehensive problems designated with the *computer icon*) is available to instructors at the *Instructor Manual* section of the *Online Learning Center* (OLC) at www.mhhe.com/cengel-boles. See the Preface for access information.

3-31 Refrigerant-134a in a cylinder equipped with a set of stops is heated and evaporated. The vapor expands until the piston hits the stops. The final temperature, and the boundary work done are to be determined, and the process is to be shown on a *P-V* diagram.

Assumptions The process is quasi-equilibrium.

Analysis (a) This a constant pressure process. Initially the system contains a saturated mixture, and thus the pressure is

$$P_2 = P_1 = P_{sat@-8^\circ C} = 217.04 \text{ kPa}$$

The specific volume of the refrigerant at the final state is

$$v_2 = \frac{V_2}{m} = \frac{0.4 \text{ m}^3}{10 \text{ kg}} = 0.04 \text{ m}^3/\text{kg}$$

At 217.04 kPa (or -8 °C), $v_f = 0.0007569 \text{ m}^3/\text{kg}$ and $v_g = 0.0919 \text{ m}^3/\text{kg}$. Noting that $v_f < v_2 < v_g$, the final state is a saturated mixture and thus the final temperature is

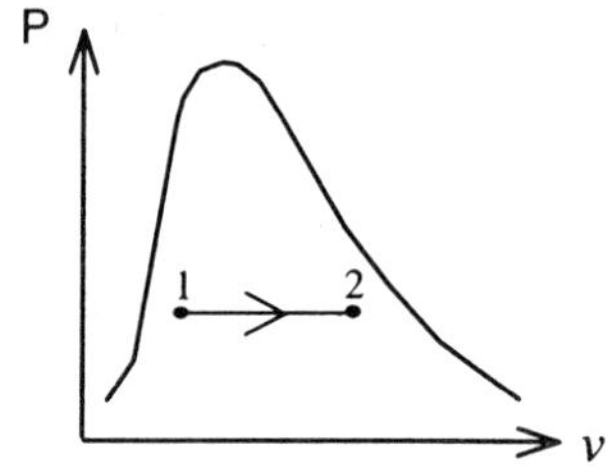

$$T_2 = T_{\text{sat@217.04kPa}} = \mathbf{-8^\circ C}$$

P

1 2

v

(*b*) The total initial volume is

$$V_1 = m_f v_f + m_g v_g = 8 \times 0.0007569 + 2 \times 0.0919 = 0.19 \text{ m}^3$$

Thus,

$$\begin{aligned} W_{b,\text{out}} &= \int_1^2 P dV = P(V_2 - V_1) \\ &= (217.04 \text{ kPa})(0.4 - 0.19)\text{m}^3 \left(\frac{1 \text{ kJ}}{1 \text{ kPa} \cdot \text{m}^3} \right) \\ &= \mathbf{45.6 \text{ kJ}} \end{aligned}$$

Discussion The positive sign indicates that work is done by the system (work output).

3-32 Saturated refrigerant-134a vapor in a cylinder is allowed to expand isothermally by gradually decreasing the pressure inside to 500 kPa. The boundary work done during this process is to be determined by using property data from the refrigerant tables, and by treating the refrigerant vapor as an ideal gas.

Assumptions The process is quasi-equilibrium.

Analysis From the refrigerant tables, the specific volume of the refrigerant at various pressures at 50°C are determined to be

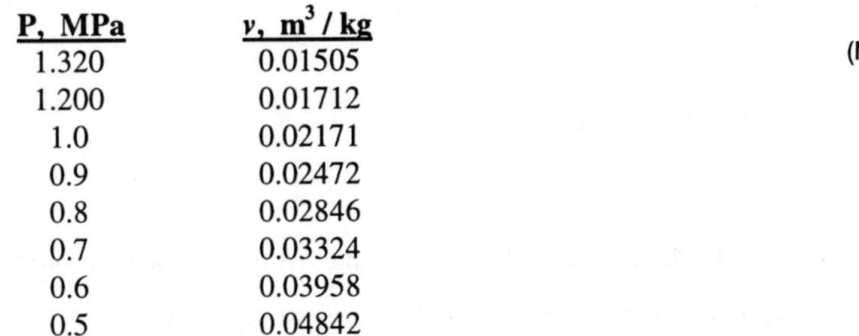

P, MPa	v, m^3 / kg
1.320	0.01505
1.200	0.01712
1.0	0.02171
0.9	0.02472
0.8	0.02846
0.7	0.03324
0.6	0.03958
0.5	0.04842

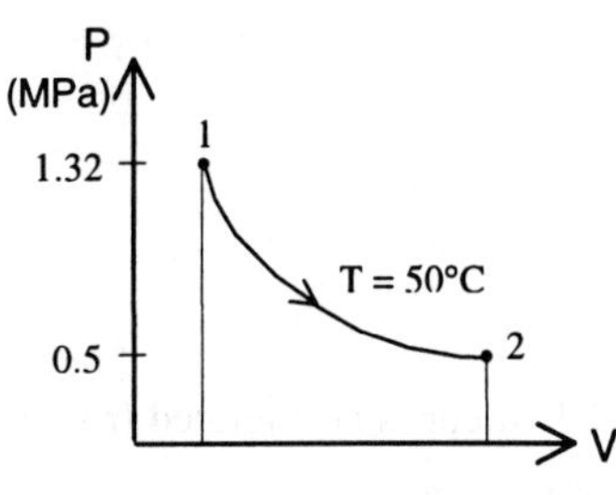

Plotting these on a P-V diagram and finding the area under the process curve, the boundary work during this isothermal process is determined to be

$$W_b = \mathbf{262.9\ kJ}$$

(b) Treating the refrigerant as an ideal gas, the boundary work for this isothermal process can be determined from

$$W_{b,\text{out}} = \int_1^2 PdV = P_1V_1 \ln\frac{V_2}{V_1} = mRT\ln\frac{P_1}{P_2}$$

$$= (10\ \text{kg})(0.08149\ \text{kJ/kg}\cdot\text{K})(323\ \text{K})\ln\frac{1320\ \text{kPa}}{500\ \text{kPa}}$$

$$= \mathbf{255.5\ \ kJ}$$

which is sufficiently close to the experimental value.

Discussion The positive sign indicates that work is done by the system (work output).

3-33 EES solution of this (and other comprehensive problems designated with the *computer icon*) is available to instructors at the *Instructor Manual* section of the *Online Learning Center* (OLC) at www.mhhe.com/cengel-boles. See the Preface for access information.

3-34 Several sets of pressure and volume data are taken as a gas expands. The boundary work done during this process is to be determined using the experimental data.

Assumptions The process is quasi-equilibrium.

Analysis Plotting the given data on a *P-V* diagram on a graph paper and evaluating the area under the process curve, the work done is determined to be **0.25 kJ**.

Other Forms of Work

3-35C The work done is the same, but the power is different.

3-36C The work done is the same, but the power is different.

3-37 A car is accelerated from rest to 100 km/h. The work needed to achieve this is to be determined.

Analysis The work needed to accelerate a body the change in kinetic energy of the body,

$$W_a = \frac{1}{2}m\left(\mathbf{V}_2^2 - \mathbf{V}_1^2\right) = \frac{1}{2}(800\text{kg})\left(\left(\frac{100{,}000\text{m}}{3600\text{s}}\right)^2 - 0\right)\left(\frac{1\text{kJ}}{1000\text{kg}\cdot\text{m}^2/\text{s}^2}\right) = \mathbf{308.6kJ}$$

3-38 A car is accelerated from 20 to 70 km/h on an uphill road. The work needed to achieve this is to be determined.

Analysis The total work required is the sum of the changes in potential and kinetic energies,

$$W_a = \frac{1}{2}m\left(\mathbf{V}_2^2 - \mathbf{V}_1^2\right) = \frac{1}{2}(2000\text{kg})\left(\left(\frac{70{,}000\text{m}}{3600\text{s}}\right)^2 - \left(\frac{20{,}000\text{m}}{3600\text{s}}\right)^2\right)\left(\frac{1\text{kJ}}{1000\text{kg}\cdot\text{m}^2/\text{s}^2}\right) = 347.2\text{kJ}$$

and,

$$W_g = mg(z_2 - z_1) = (2000\text{kg})(9.807\text{m/s}^2)(40\text{m})\left(\frac{1\text{kJ}}{1000\text{kg}\cdot\text{m}^2/\text{s}^2}\right) = 784.6\text{kJ}$$

Thus,

$$W_{\text{total}} = W_a + W_g = 347.2 + 784.6 = \mathbf{1131.8kJ}$$

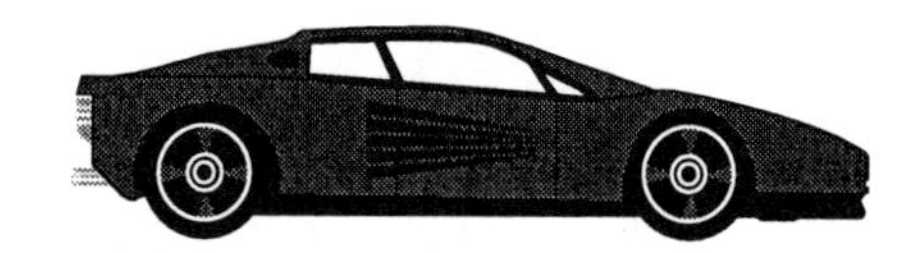

3-39E A engine of a car develops 450 hp at 3000 rpm. The torque transmitted through the shaft is to be determined.

Analysis The torque is determined from

$$T = \frac{\dot{W}_{sh}}{2\pi\dot{n}} = \frac{450 \text{ hp}}{2\pi(3000/60)/\text{s}}\left(\frac{550 \text{ lbf}\cdot\text{ft/s}}{1 \text{ hp}}\right) = \mathbf{787.8\ lbf\cdot ft}$$

3-40 A linear spring is elongated by 20 cm from its rest position. The work done is to be determined.

Analysis The spring work can be determined from

$$W_{spring} = \frac{1}{2}k(x_2^2 - x_1^2) = \frac{1}{2}(70 \text{ kN/m})(0.2^2 - 0)\text{m}^2 = 1.4 \text{ kN}\cdot\text{m} = \mathbf{1.4\ kJ}$$

3-41 The engine of a car develops 75 kW of power. The acceleration time of this car from rest to 85 km/h on a level road is to be determined.

Analysis The work needed to accelerate a body is the change in its kinetic energy,

$$W_a = \frac{1}{2}m(\mathrm{V}_2^2 - \mathrm{V}_1^2) = \frac{1}{2}(1500 \text{ kg})\left(\left(\frac{85{,}000 \text{ m}}{3600 \text{ s}}\right)^2 - 0\right)\left(\frac{1 \text{ kJ}}{1000 \text{ kg}\cdot\text{m}^2/\text{s}^2}\right) = 418.1 \text{ kJ}$$

Thus the time required is

$$\Delta t = \frac{W_a}{\dot{W}_a} = \frac{418.1 \text{ kJ}}{75 \text{ kJ/s}} = \mathbf{5.57\ s}$$

This answer is not realistic because part of the power will be used against the air drag, friction, and rolling resistance.

3-42 A ski lift is operating steadily at 10 km/h. The power required to operate and also to accelerate this ski lift from rest to the operating speed are to be determined.

Assumptions **1** Air drag and friction are negligible. **2** The average mass of each loaded chair is 250 kg. **3** The mass of chairs is small relative to the mass of people, and thus the contribution of returning empty chairs to the motion is disregarded (this provides a safety factor).

Analysis The lift is 1000 m long and the chairs are spaced 20 m apart. Thus at any given time there are 1000/20 = 50 chairs being lifted. Considering that the mass of each chair is 250 kg, the load of the lift at any given time is

$$\text{Load} = (50\text{ chairs})(250\text{ kg/chair}) = 12{,}500\text{ kg}$$

Neglecting the work done on the system by the returning empty chairs, the work needed to raise this mass by 200 m is

$$W_g = mg(z_2 - z_1) = (12{,}500\text{ kg})(9.81\text{ m/s}^2)(200\text{ m})\left(\frac{1\text{ kJ}}{1000\text{ kg}\cdot\text{m}^2/\text{s}^2}\right) = 24{,}525\text{ kJ}$$

At 10 km/h, it will take

$$\Delta t = \frac{\text{distance}}{\text{velocity}} = \frac{1\text{ km}}{10\text{ km/h}} = 0.1\text{ h} = 360\text{ s}$$

to do this work. Thus the power needed is

$$\dot{W}_g = \frac{W_g}{\Delta t} = \frac{24{,}525\text{ kJ}}{360\text{ s}} = \mathbf{68.1\text{ kW}}$$

The velocity of the lift during steady operation, and the acceleration during start up are

$$\mathbf{V} = (10\text{ km/h})\left(\frac{1\text{ m/s}}{3.6\text{ km/h}}\right) = 2.778\text{ m/s}$$

$$a = \frac{\Delta\mathbf{V}}{\Delta t} = \frac{2.778\text{ m/s - 0}}{5\text{ s}} = 0.556\text{ m/s}^2$$

During acceleration, the power needed is

$$\dot{W}_a = \frac{1}{2}m\left(\mathbf{V}_2^2 - \mathbf{V}_1^2\right)/\Delta t = \frac{1}{2}(12{,}500\text{ kg})\left((2.778\text{ m/s})^2 - 0\right)\left(\frac{1\text{ kJ/kg}}{1000\text{ m}^2/\text{s}^2}\right)/(5\text{ s}) = 9.6\text{ kW}$$

Assuming the power applied is constant, the acceleration will also be constant and the vertical distance traveled during acceleration will be

$$h = \frac{1}{2}at^2\sin\alpha = \frac{1}{2}at^2\frac{200\text{ m}}{1000\text{ m}} = \frac{1}{2}(0.556\text{ m/s}^2)(5\text{ s})^2(0.2) = 1.39\text{ m}$$

and

$$\dot{W}_g = mg(z_2 - z_1)/\Delta t = (12{,}500\text{ kg})(9.81\text{ m/s}^2)(1.39\text{ m})\left(\frac{1\text{ kJ/kg}}{1000\text{ kg}\cdot\text{m}^2/\text{s}^2}\right)/(5\text{ s}) = 34.1\text{ kW}$$

Thus,

$$\dot{W}_{\text{total}} = \dot{W}_a + \dot{W}_g = 9.6 + 34.1 = \mathbf{43.7\text{ kW}}$$

3-43 A car is to climb a hill in 10 s. The power needed is to be determined for three different cases.

Assumptions Air drag, friction, and rolling resistance are negligible.

Analysis The total power required for each case is the sum of the rates of changes in potential and kinetic energies. That is,

$$\dot{W}_{\text{total}} = \dot{W}_a + \dot{W}_g$$

(*a*) $\dot{W}_a = 0$ since the velocity is constant. Also, the vertical rise is h = (100 m)(sin 30°) = 50 m. Thus,

$$\dot{W}_g = mg(z_2 - z_1)/\Delta t = (2000\text{ kg})(9.81\text{ m/s}^2)(50\text{ m})\left(\frac{1\text{ kJ}}{1000\text{ kg}\cdot\text{m}^2/\text{s}^2}\right)/(10\text{ s}) = 98.1\text{ kW}$$

and

$$\dot{W}_{\text{total}} = \dot{W}_a + \dot{W}_g = 0 + 98.1 = \mathbf{98.1\ kW}$$

(*b*) The power needed to accelerate is

$$\dot{W}_a = \frac{1}{2}m\left(\mathrm{V}_2^2 - \mathrm{V}_1^2\right)/\Delta t = \frac{1}{2}(2000\text{kg})\left((30\text{m/s})^2 - 0\right)\left(\frac{1\text{kJ}}{1000\text{kg}\cdot\text{m}^2/\text{s}^2}\right)/(10\text{s}) = 90\text{kW}$$

and

$$\dot{W}_{\text{total}} = \dot{W}_a + \dot{W}_g = 90 + 98.1 = \mathbf{188.1\ kW}$$

(*c*) The power needed to decelerate is

$$\dot{W}_a = \frac{1}{2}m\left(\mathrm{V}_2^2 - \mathrm{V}_1^2\right)/\Delta t = \frac{1}{2}(2000\text{kg})\left((5\text{m/s})^2 - (35\text{m/s})^2\right)\left(\frac{1\text{kJ}}{1000\text{kg}\cdot\text{m}^2/\text{s}^2}\right)/(10\text{s}) = -120\text{kW}$$

and

$$\dot{W}_{\text{total}} = \dot{W}_a + \dot{W}_g = -120 + 98.1 = \mathbf{-21.9\ kW}\ \text{(breaking power)}$$

3-44 A damaged car is being towed by a truck. The extra power needed is to be determined for three different cases.

Assumptions Air drag, friction, and rolling resistance are negligible.

Analysis The total power required for each case is the sum of the rates of changes in potential and kinetic energies. That is,

$$\dot{W}_{\text{total}} = \dot{W}_a + \dot{W}_g$$

(*a*) Zero.

(*b*) $\dot{W}_a = 0$. Thus,

$$\dot{W}_{\text{total}} = \dot{W}_g = mg(z_2 - z_1)/\Delta t = mg\frac{\Delta z}{\Delta t} = mg\mathbf{V}_z = mg\mathbf{V}\sin 30^\circ$$

$$= (1200\text{kg})(9.8\text{m/s}^2)\left(\frac{50{,}000\text{m}}{3600\text{s}}\right)\left(\frac{1\text{kJ/kg}}{1000\text{m}^2/\text{s}^2}\right)(0.5) = \mathbf{81.7kW}$$

(*c*) $\dot{W}_g = 0$. Thus,

$$\dot{W}_{\text{total}} = \dot{W}_a = \frac{1}{2}m\left(\mathbf{V}_2^2 - \mathbf{V}_1^2\right)/\Delta t$$

$$= \frac{1}{2}(1200\text{kg})\left(\left(\frac{90{,}000\text{m}}{3600\text{s}}\right)^2 - 0\right)\left(\frac{1\text{kJ/kg}}{1000\text{m}^2/\text{s}^2}\right)/(12\text{s}) = \mathbf{31.25kW}$$

Conservation of Mass

3-45C Mass flow rate is the amount of mass flowing through a cross-section per unit time whereas the volume flow rate is the amount of volume flowing through a cross-section per unit time.

3-46C The amount of mass or energy entering a control volume does not have to be equal to the amount of mass or energy leaving during an unsteady-flow process.

3-47C Flow through a control volume is steady when it involves no changes with time at any specified position.

3-48C No, a flow with the same volume flow rate at the inlet and the exit is not necessarily steady (unless the density is constant). To be steady, the mass flow rate through the device must remain constant.

3-49E A garden hose is used to fill a water bucket. The volume and mass flow rates of water, the filling time, and the discharge velocity are to be determined.

Assumptions **1** Water is an incompressible substance. **2** Flow through the hose is steady. **3** There is no waste of water by splashing.

Properties We take the density of water to be 62.4 lbm/ft^3.

Analysis (*a*) The volume and mass flow rates of water are

$$\dot{V} = A\mathbf{V} = (\pi D^2/4)\mathbf{V} = [\pi(1/12\text{ ft})^2/4](8\text{ ft/s}) = \mathbf{0.04363\ ft^3/s}$$

$$\dot{m} = \rho\dot{V} = (62.4\text{ lbm/ft}^3)(0.04363\text{ ft}^3/\text{s}) = \mathbf{2.72\ lbm/s}$$

(*b*) The time it takes to fill a 20-gallon bucket is

$$\Delta t = \frac{V}{\dot{V}} = \frac{20\text{ gal}}{0.04363\text{ ft}^3/\text{s}}\left(\frac{1\text{ ft}^3}{7.4804\text{ gal}}\right) = \mathbf{61.3\ s}$$

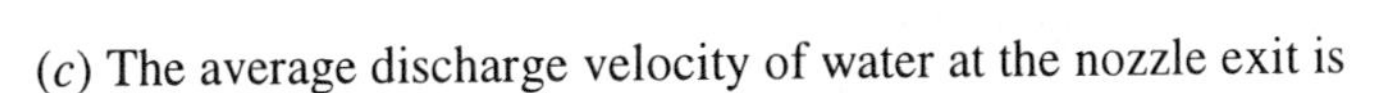

(*c*) The average discharge velocity of water at the nozzle exit is

$$\mathbf{V}_e = \frac{\dot{V}}{A_e} = \frac{\dot{V}}{\pi D_e^2/4} = \frac{0.04363\text{ ft}^3/\text{s}}{[\pi(0.5/12\text{ ft})^2/4]} = \mathbf{32\ ft/s}$$

Discussion Note that for a given flow rate, the average velocity is inversely proportional to the square of the velocity. Therefore, when the diameter is reduced by half, the velocity quadruples.

3-50 Air is accelerated in a nozzle. The mass flow rate and the exit area of the nozzle are to be determined.

Assumptions Flow through the nozzle is steady.

Properties The density of air is given to be 2.21 kg/m^3 at the inlet, and 0.762 kg/m^3 at the exit.

Analysis (a) The mass flow rate of air is determined from the inlet conditions to be

$$\dot{m} = \rho_1 A_1 V_1 = (2.21\ \text{kg/m}^3)(0.008\ \text{m}^2)(30\ \text{m/s}) = \mathbf{0.530\ kg/s}$$

$V_1 = 30$ m/s
$A_1 = 80$ cm^2
AIR
$V_2 = 180$ m/s

(*b*) There is only one inlet and one exit, and thus $\dot{m}_1 = \dot{m}_2 = \dot{m}$.
Then the exit area of the nozzle is determined to be

$$\dot{m} = \rho_2 A_2 \mathbf{V}_2 \longrightarrow A_2 = \frac{\dot{m}}{\rho_2 \mathbf{V}_2} = \frac{0.530\ \text{kg/s}}{(0.762\ \text{kg/m}^3)(180\ \text{m/s})} = 0.00387\ \text{m}^2 = \mathbf{38.7\ cm^2}$$

3-51 Air is expanded and is accelerated as it is heated by a hair dryer of constant diameter. The percent increase in the velocity of air as it flows through the drier is to be determined.

Assumptions Flow through the nozzle is steady.

Properties The density of air is given to be 1.20 kg/m^3 at the inlet, and 1.05 kg/m^3 at the exit.

Analysis There is only one inlet and one exit, and thus $\dot{m}_1 = \dot{m}_2 = \dot{m}$. Then,

$$\dot{m}_1 = \dot{m}_2$$

$$\rho_1 A\mathbf{V}_1 = \rho_2 A\mathbf{V}_2$$

$$\frac{\mathbf{V}_2}{\mathbf{V}_1} = \frac{\rho_1}{\rho_2} = \frac{1.20\ \text{kg/m}^3}{1.05\ \text{kg/m}^3} = 1.14 \quad \text{(or, and increase of } \mathbf{14\%}\text{)}$$

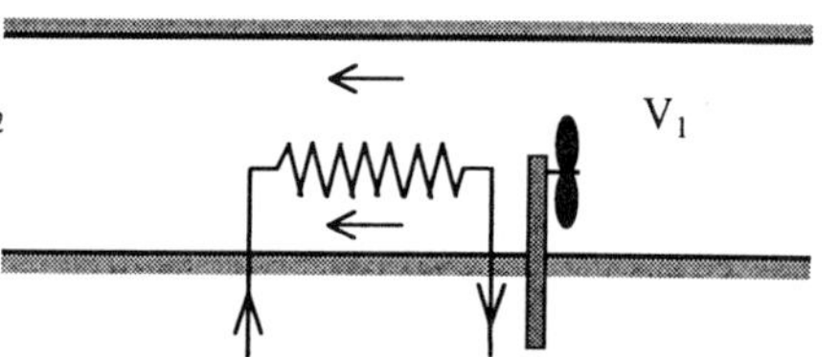

Therefore, the air velocity increases 14% as it flows through the hair drier.

3-52E The ducts of an air-conditioning system pass through an open area. The inlet velocity and the mass flow rate of air are to be determined.

Assumptions Flow through the air conditioning duct is steady.

Properties The density of air is given to be 0.078 lbm/ft^3 at the inlet.

Analysis The inlet velocity of air and the mass flow rate through the duct are

450 ft^3/min AIR D = 10 in

$$\mathrm{V}_1 = \frac{\dot{V}_1}{A_1} = \frac{\dot{V}_1}{\pi D^2/4} = \frac{450 \text{ ft}^3/\text{min}}{\pi(10/12 \text{ ft})^2/4} = \mathbf{825 \text{ ft/min}} = \mathbf{13.8 \text{ ft/s}}$$

$$\dot{m} = \rho_1 \dot{V}_1 = (0.078 \text{ lbm/ft}^3)(450 \text{ ft}^3/\text{min}) = 35.1 \text{ lbm/min} = \mathbf{0.585 lbm/s}$$

3-53 A rigid tank initially contains air at atmospheric conditions. The tank is connected to a supply line, and air is allowed to enter the tank until the density rises to a specified level. The mass of air that entered the tank is to be determined.

Properties The density of air is given to be 1.18 kg/m^3 at the beginning, and 7.20 kg/m^3 at the end.

Analysis We take the tank as the system, which is a control volume since mass crosses the boundary. The mass balance for this system can be expressed as

Mass balance: $m_{in} - m_{out} = \Delta m_{\text{system}} \rightarrow m_i = m_2 - m_1 = \rho_2 V - \rho_1 \mathrm{V}$

Substituting,

$$m_i = (\rho_2 - \rho_1)\mathrm{V} = [(7.20 - 1.18) \text{ kg/m}^3](1 \text{ m}^3) = 6.02 \text{ kg}$$

$\mathrm{V}_1 = 1$ m^3
$\rho_1 = 1.18$ kg/m^3

Therefore, 6.02 kg of mass entered the tank.

3-54 The ventilating fan of the bathroom of a building runs continuously. The mass of air "vented out" per day is to be determined.

Assumptions Flow through the fan is steady.

Properties The density of air in the building is given to be 1.20 kg/m^3.

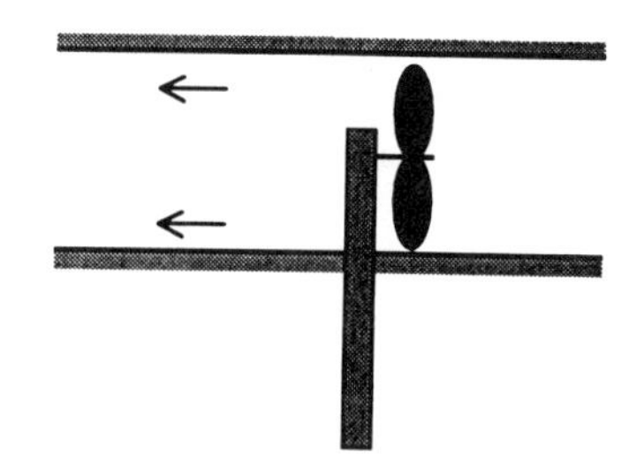

Analysis The mass flow rate of air vented out is

$$\dot{m}_{\text{air}} = \rho \dot{V}_{\text{air}} = (1.20\ \text{kg/m}^3)(0.030\ \text{m}^3/\text{s}) = 0.036\ \text{kg/s}$$

Then the mass of air vented out in 24 h becomes

$$m = \dot{m}_{\text{air}} \Delta t = (0.036\ \text{kg/s})(24 \times 3600\ \text{s}) = \mathbf{3110\ kg}$$

Discussion Note that more than 3 tons of air is vented out by a bathroom fan in one day.

3-55E Chickens are to be cooled by chilled water in an immersion chiller. The mass flow rate of chicken through the chiller is to be determined.

Assumptions Chickens are dropped into the chiller steadily.

Properties The average mass of a chicken is 4.5 lbm.

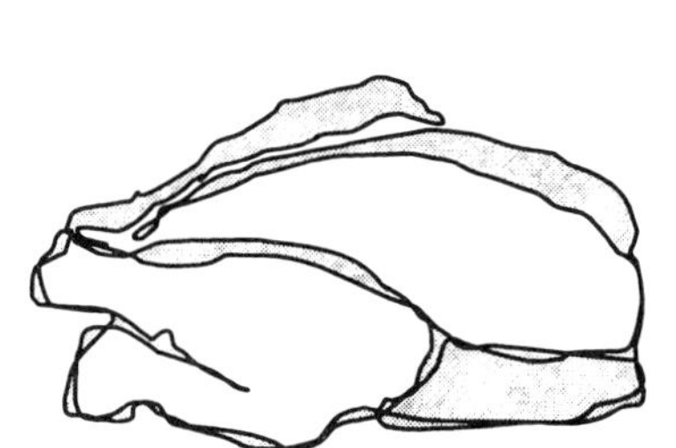

Analysis Chickens are dropped into the chiller at a rate of 500 per hour. Therefore, chickens can be considered to flow steadily through the chiller at a mass flow rate of

$$\dot{m}_{\text{chicken}} = (500\ \text{chicken/h})(4.5\ \text{lbm/chicken}) = 2250\ \text{kg/h} = \mathbf{0.625\ lbm/s}$$

Note that chicken can be treated conveniently as a "flowing fluid" in calculations.

3-56 A desktop computer is to be cooled by a fan at a high elevation where the air density is low. The mass flow rate of air through the fan and the diameter of the casing for a given velocity are to be determined.

Assumptions Flow through the fan is steady.

Properties The density of air at a high elevation is given to be 0.7 kg/m³.

Analysis The mass flow rate of air is

$$\dot{m}_{\text{air}} = \rho\dot{V}_{\text{air}} = (0.7\text{ kg/m}^3)(0.34\text{ m}^3/\text{min}) = 0.238\text{ kg/min} = \mathbf{0.0040\ kg/s}$$

If the mean velocity is 110 m/min, the diameter of the casing is

$$\dot{V} = A\mathbf{V} = \frac{\pi D^2}{4}\mathbf{V} \quad \rightarrow \quad D = \sqrt{\frac{4\dot{V}}{\pi\mathbf{V}}} = \sqrt{\frac{4(0.34\text{ m}^3/\text{min})}{\pi(110\text{ m/min})}} = \mathbf{0.063\ m}$$

Therefore, the diameter of the casing must be at least 6.3 cm to ensure that the mean velocity does not exceed 110 m/min.

Discussion This problem shows that engineering systems are sized to satisfy certain constraints imposed by certain considerations.

Flow Work and Energy Transfer by Mass

3-57C Energy can be transferred to or from a control volume as heat, various forms of work, and by mass.

3-58C Flow energy or flow work is the energy needed to push a fluid into or out of a control volume. Fluids at rest do not possess any flow energy.

3-59C Flowing fluids possess flow energy in addition to the forms of energy a fluid at rest possesses. The total energy of a fluid at rest consists of internal, kinetic, and potential energies. The total energy of a flowing fluid consists of internal, kinetic, potential, and flow energies.

3-60E Steam is leaving a pressure cooker at a specified pressure. The velocity, flow rate, the total and flow energies, and the rate of energy transfer by mass are to be determined.

Assumptions **1** The flow is steady, and the initial start-up period is disregarded. **2** The kinetic and potential energies are negligible, and thus they are not considered. **3** Saturation conditions exist within the cooker at all times so that steam leaves the cooker as a saturated vapor at 30 psia.

Properties The properties of saturated liquid water and water vapor at 30 psia are v_f = 0.017004 ft^3/lbm, v_g = 13.748 ft^3/lbm, u_g = 1088.0 Btu/lbm, and h_g = 1164.3 Btu/lbm (Table A-5E).

Analysis (*a*) Saturation conditions exist in a pressure cooker at all times after the steady operating conditions are established. Therefore, the liquid has the properties of saturated liquid and the exiting steam has the properties of saturated vapor at the operating pressure. The amount of liquid that has evaporated, the mass flow rate of the exiting steam, and the exit velocity are

$$m = \frac{\Delta V_{\text{liquid}}}{v_f} = \frac{0.4\text{ gal}}{0.017004\text{ ft}^3/\text{lbm}}\left(\frac{0.13368\text{ ft}^3}{1\text{ gal}}\right) = 3.145\text{ lbm}$$

$$\dot{m} = \frac{m}{\Delta t} = \frac{3.145\text{ lbm}}{45\text{ min}} = 0.0699\text{ lbm/min} = \mathbf{1.165\times 10^{-3}\ lbm/s}$$

$$\mathbf{V} = \frac{\dot{m}}{\rho_g A_c} = \frac{\dot{m}v_g}{A_c} = \frac{(1.165\times 10^{-3}\text{ lbm/s})(13.748\text{ ft}^3/\text{lbm})}{0.15\text{ in}^2}\left(\frac{144\text{ in}^2}{1\text{ ft}^2}\right) = \mathbf{15.4\ ft/s}$$

(*b*) Noting that $h = u + Pv$ and that the kinetic and potential energies are disregarded, the flow and total energies of the exiting steam are

$$e_{\text{flow}} = Pv = h - u = 1164.3 - 1088.0 = \mathbf{76.3\ Btu/lbm}$$

$$\theta = h + ke + pe \cong h = \mathbf{1164.3\ Btu/lbm}$$

Note that the kinetic energy in this case is ke = $\mathbf{V}^2/2$ = (15.4 ft/s)^2 = 237 ft^2/s^2 = 0.0095 Btu/lbm, which is very small compared to enthalpy.

(*c*) The rate at which energy is leaving the cooker by mass is simply the product of the mass flow rate and the total energy of the exiting steam per unit mass,

$$\dot{E}_{\text{mass}} = \dot{m}\theta = (1.165\times 10^{-3}\text{ lbm/s})(1164.3\text{ Btu/lbm}) = \mathbf{1.356\ Btu/s}$$

Discussion The numerical value of the energy leaving the cooker with steam alone does not mean much since this value depends on the reference point selected for enthalpy (it could even be negative). The significant quantity is the difference between the enthalpies of the exiting vapor and the liquid inside (which is h_{fg}) since it relates directly to the amount of energy supplied to the cooker.

3-61 Refrigerant-134a enters a compressor as a saturated vapor at a specified pressure, and leaves as superheated vapor at a specified rate. The rates of energy transfer by mass into and out of the compressor are to be determined.

Assumptions **1** The flow of the refrigerant through the compressor is steady. **2** The kinetic and potential energies are negligible, and thus they are not considered.

Properties The enthalpy of refrigerant-134a at the inlet and the exit are (Tables A-12 and A-13)

$$h_1 = h_{g\,@0.14\text{ MPa}} = 236.04\text{ kJ/kg}$$

$$\left.\begin{array}{l} P_2 = 0.8\text{ MPa} \\ T_2 = 50°\text{C} \end{array}\right\} h_2 = 284.39\text{ kJ/kg}$$

Analysis Noting that the total energy of a flowing fluid is equal to its enthalpy when the kinetic and potential energies are negligible, and that the rate of energy transfer by mass is equal to the product of the mass flow rate and the total energy of the fluid per unit mass, the rates of energy transfer by mass into and out of the compressor are

$$\dot{E}_{\text{mass, in}} = \dot{m}\theta_{in} = \dot{m}h_1 = (0.04\text{ kg/s})(236.04\text{ kJ/kg}) = 9.442\text{ kJ/s} = \mathbf{9.44\text{ kW}}$$

$$\dot{E}_{\text{mass, out}} = \dot{m}\theta_{out} = \dot{m}h_2 = (0.04\text{ kg/s})(284.39\text{ kJ/kg}) = 11.38\text{ kJ/s} = \mathbf{11.38\text{ kW}}$$

Discussion The numerical values of the energy entering or leaving a device by mass alone does not mean much since this value depends on the reference point selected for enthalpy (it could even be negative). The significant quantity here is the difference between the outgoing and incoming energy flow rates, which is

$$\Delta\dot{E}_{\text{mass}} = \dot{E}_{\text{mass, out}} - \dot{E}_{\text{mass, in}} = 11.38 - 9.44 = 1.94\text{ kW}$$

This quantity represents the rate of energy transfer to the refrigerant in the compressor.

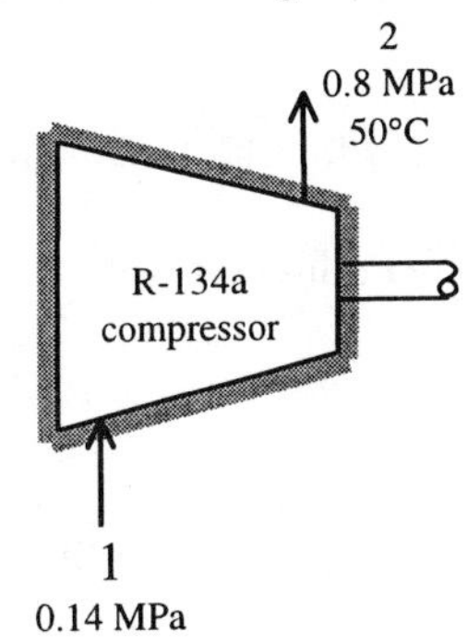

3-62 Warm air in a house is forced to leave by the infiltrating cold outside air at a specified rate. The net energy loss due to mass transfer is to be determined.

Assumptions **1** The flow of the air into and out of the house through the cracks is steady. **2** The kinetic and potential energies are negligible. **3** Air is an ideal gas with constant specific heats at room temperature.

Properties The gas constant of air is R = 0.287 kPa·m³/kg·K (Table A-1). The constant pressure specific heat of air at room temperature is C_p = 1.005 kJ/kg·°C (Table A-2).

Analysis The density of air at the indoor conditions and its mass flow rate are

$$\rho = \frac{P}{RT} = \frac{101.325\,\text{kPa}}{(0.287\,\text{kPa}\cdot\text{m}^3/\text{kg}\cdot\text{K})(24+273)\text{K}} = 1.189\,\text{kg/m}^3$$

$$\dot{m} = \rho\dot{V} = (1.189\,\text{kg/m}^3)(150\,\text{m}^3/\text{h}) = 178.35\,\text{kg/h} = 0.0495\,\text{kg/s}$$

Noting that the total energy of a flowing fluid is equal to its enthalpy when the kinetic and potential energies are negligible, and that the rate of energy transfer by mass is equal to the product of the mass flow rate and the total energy of the fluid per unit mass, the rates of energy transfer by mass into and out of the house by air re

$$\dot{E}_{\text{mass, in}} = \dot{m}\theta_{in} = \dot{m}h_1$$

$$\dot{E}_{\text{mass, out}} = \dot{m}\theta_{out} = \dot{m}h_2$$

The net energy loss by air infiltration is equal to the difference between the outgoing and incoming energy flow rates, which is

$$\begin{aligned}\Delta\dot{E}_{\text{mass}} &= \dot{E}_{\text{mass, out}} - \dot{E}_{\text{mass, in}} = \dot{m}(h_2 - h_1) = \dot{m}C_p(T_2 - T_1)\\ &= (0.0495\,\text{kg/s})(1.005\,\text{kJ/kg}\cdot°\text{C})(24 - 5)°\text{C} = 0.945\,\text{kJ/s}\\ &= \mathbf{0.945\,kW}\end{aligned}$$

This quantity represents the rate of energy transfer to the refrigerant in the compressor.

Discussion The rate of energy loss by infiltration will be less in reality since some air will leave the house before it is fully heated to 24°C.

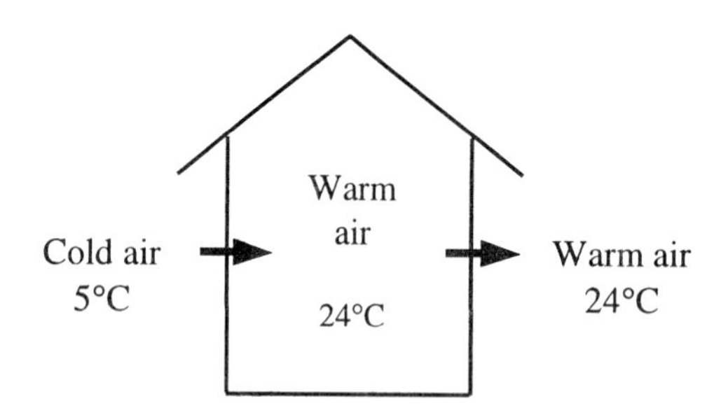

Special Topic: Mechanisms of Heat Transfer

3-63C The three mechanisms of heat transfer are conduction, convection, and radiation.

3-64C No. It is purely by radiation.

3-65C Diamond has a higher thermal conductivity than silver, and thus diamond is a better conductor of heat.

3-66C In forced convection, the fluid is forced to move by external means such as a fan, pump, or the wind. The fluid motion in natural convection is due to buoyancy effects only.

3-67C Emissivity is the ratio of the radiation emitted by a surface to the radiation emitted by a blackbody at the same temperature. Absorptivity is the fraction of radiation incident on a surface that is absorbed by the surface. The Kirchhoff's law of radiation states that the emissivity and the absorptivity of a surface are equal at the same temperature and wavelength.

3-68C A blackbody is an idealized body that emits the maximum amount of radiation at a given temperature, and that absorbs all the radiation incident on it. Real bodies emit and absorb less radiation than a blackbody at the same temperature.

3-69 The inner and outer surfaces of a brick wall are maintained at specified temperatures. The rate of heat transfer through the wall is to be determined.

Assumptions **1** Steady operating conditions exist since the surface temperatures of the wall remain constant at the specified values. **2** Thermal properties of the wall are constant.

Properties The thermal conductivity of the wall is given to be $k = 0.69$ W/m·°C.

Analysis Under steady conditions, the rate of heat transfer through the wall is

$$\dot{Q}_{\text{cond}} = kA\frac{\Delta T}{L} = (0.69\text{W/m}\cdot^\circ\text{C})(5\times 6\text{m}^2)\frac{(20-5)^\circ\text{C}}{0.3\text{m}} = \mathbf{1035W}$$

Brick Wall
Q = ?
30 cm
20°C
5°C

3-70 The inner and outer surfaces of a window glass are maintained at specified temperatures. The amount of heat transferred through the glass in 5 h is to be determined.

Assumptions **1** Steady operating conditions exist since the surface temperatures of the glass remain constant at the specified values. **2** Thermal properties of the glass are constant.

Properties The thermal conductivity of the glass is given to be $k = 0.78$ W/m·°C.

Analysis Under steady conditions, the rate of heat transfer through the glass by conduction is

Glass

$$\dot{Q}_{\text{cond}} = kA\frac{\Delta T}{L} = (0.78\text{W/m}\cdot^\circ\text{C})(2\times 2\text{m}^2)\frac{(10-3)^\circ\text{C}}{0.005\text{m}} = 4368\text{W}$$

Then the amount of heat transferred over a period of 5 h becomes

$$Q = \dot{Q}_{\text{cond}}\Delta t = (4.368\text{kJ/s})(5\times 3600\text{s}) = \mathbf{78{,}624kJ}$$

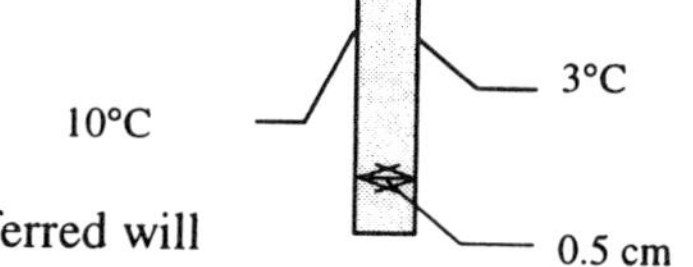

If the thickness of the glass is doubled to 1 cm, then the amount of heat transferred will go down by half to **39,312 kJ**.

3-71 EES solution of this (and other comprehensive problems designated with the *computer icon*) is available to instructors at the *Instructor Manual* section of the *Online Learning Center* (OLC) at www.mhhe.com/cengel-boles. See the Preface for access information.

3-72 Heat is transferred steadily to boiling water in the pan through its bottom. The inner surface temperature of the bottom of the pan is given. The temperature of the outer surface is to be determined.

Assumptions **1** Steady operating conditions exist since the surface temperatures of the pan remain constant at the specified values. **2** Thermal properties of the aluminum pan are constant.

Properties The thermal conductivity of the aluminum is given to be $k = 237$ W/m·°C.

Analysis The heat transfer surface area is

$$A = \pi r^2 = \pi(0.1\text{ m})^2 = 0.0314\text{ m}^2$$

Under steady conditions, the rate of heat transfer through the bottom of the pan by conduction is

$$\dot{Q} = kA\frac{\Delta T}{L} = kA\frac{T_2 - T_1}{L}$$

Substituting,

$$500\text{ W} = (237\text{ W/m}\cdot^\circ\text{C})(0.0314\text{ m}^2)\frac{T_2 - 105^\circ\text{C}}{0.004\text{ m}}$$

which gives $\quad T_2 = \mathbf{105.27\ ^\circ C}$

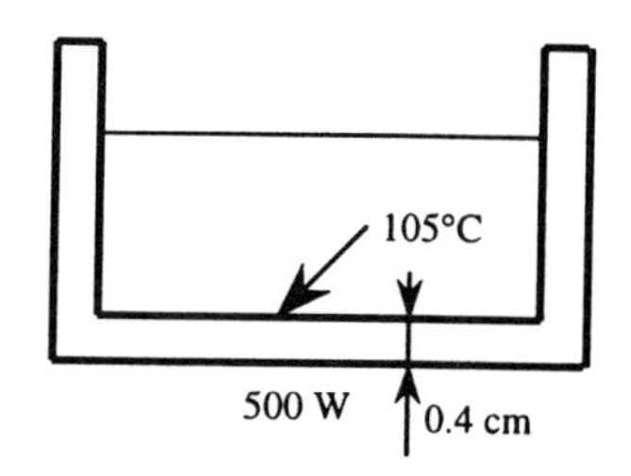

3-73 A person is standing in a room at a specified temperature. The rate of heat transfer between a person and the surrounding air by convection is to be determined.

Assumptions **1** Steady operating conditions exist. **2** Heat transfer by radiation is not considered. **3** The environment is at a uniform temperature.

Analysis The heat transfer surface area of the person is

$$A = (\pi D)h = \pi(0.3 \text{ m})(1.70 \text{ m}) = 1.60 \text{ m}^2$$

Under steady conditions, the rate of heat transfer by convection is

$$\dot{Q}_{\text{conv}} = hA\Delta T = (15 \text{W/m}^2\cdot{}^\circ\text{C})(1.60\text{m}^2)(34-20)^\circ\text{C} = \mathbf{336W}$$

3-74 A spherical ball whose surface is maintained at a temperature of 70°C is suspended in the middle of a room at 20°C. The total rate of heat transfer from the ball is to be determined.

Assumptions **1** Steady operating conditions exist since the ball surface and the surrounding air and surfaces remain at constant temperatures. **2** The thermal properties of the ball and the convection heat transfer coefficient are constant and uniform.

Properties The emissivity of the ball surface is given to be ε = 0.8.

Analysis The heat transfer surface area is

$$A = \pi D^2 = 3.14\text{x}(0.05 \text{ m})^2 = 0.007854 \text{ m}^2$$

Air
20°C
70°C
D = 5 cm
$\dot{Q}$

Under steady conditions, the rates of convection and radiation heat transfer are

$$\dot{Q}_{\text{conv}} = hA\Delta T = (15 \text{ W/m}^2\cdot{}^\circ\text{C})(0.007854 \text{ m}^2)(70-20)^\circ\text{C} = 5.89 \text{ W}$$

$$\dot{Q}_{\text{rad}} = \varepsilon\sigma A(T_s^4 - T_o^4)$$
$$= 0.8(0.007854 \text{ m}^2)(5.67\times10^{-8} \text{ W/m}^2\cdot\text{K}^4)[(343 \text{ K})^4 - (293 \text{ K})^4] = 2.31 \text{ W}$$

Therefore,

$$\dot{Q}_{\text{total}} = \dot{Q}_{\text{conv}} + \dot{Q}_{\text{rad}} = 5.89 + 2.31 = \mathbf{8.20W}$$

3-75 EES solution of this (and other comprehensive problems designated with the *computer icon*) is available to instructors at the *Instructor Manual* section of the *Online Learning Center* (OLC) at www.mhhe.com/cengel-boles. See the Preface for access information.

3-76 Hot air is blown over a flat surface at a specified temperature. The rate of heat transfer from the air to the plate is to be determined.

Assumptions **1** Steady operating conditions exist. **2** Heat transfer by radiation is not considered. **3** The convection heat transfer coefficient is constant and uniform over the surface.

Analysis Under steady conditions, the rate of heat transfer by convection is

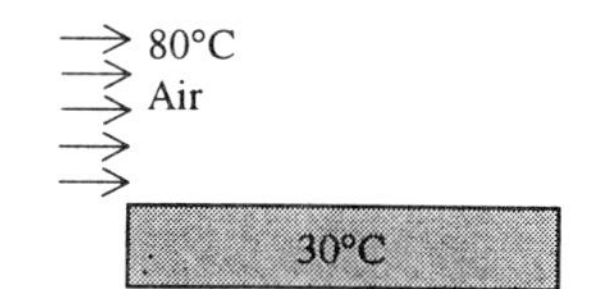

$$\dot{Q}_{\text{conv}} = hA\Delta T = (55\ \text{W/m}^2\cdot{}^\circ\text{C})(2\times 4\ \text{m}^2)(80-30)^\circ\text{C} = \mathbf{22{,}000\,W} = \mathbf{22\ kW}$$

3-77 A 1000-W iron is left on the iron board with its base exposed to the air at 20°C. The temperature of the base of the iron is to be determined in steady operation.

Assumptions **1** Steady operating conditions exist. **2** The thermal properties of the iron base and the convection heat transfer coefficient are constant and uniform. **3** The temperature of the surrounding surfaces is the same as the temperature of the surrounding air.

Properties The emissivity of the base surface is given to be $\varepsilon = 0.6$.

Analysis At steady conditions, the 1000 W of energy supplied to the iron will be dissipated to the surroundings by convection and radiation heat transfer. Therefore,

$$\dot{Q}_{\text{total}} = \dot{Q}_{\text{conv}} + \dot{Q}_{\text{rad}} = 1000\ \text{W}$$

where

$$\dot{Q}_{\text{conv}} = hA\Delta T = (35\ \text{W/m}^2\cdot\text{K})(0.02\ \text{m}^2)(T_s - 293\ \text{K}) = 0.7(T_s - 293\ \text{K})\ \text{W}$$

and

$$\dot{Q}_{\text{rad}} = \varepsilon\sigma A(T_s^4 - T_o^4) = 0.6(0.02\ \text{m}^2)(5.67\times 10^{-8}\ \text{W/m}^2\cdot\text{K}^4)[\text{T}_\text{s}^4 - (293\ \text{K})^4]$$
$$= 0.06804\times 10^{-8}[T_s^4 - (293\ \text{K})^4]\ \text{W}$$

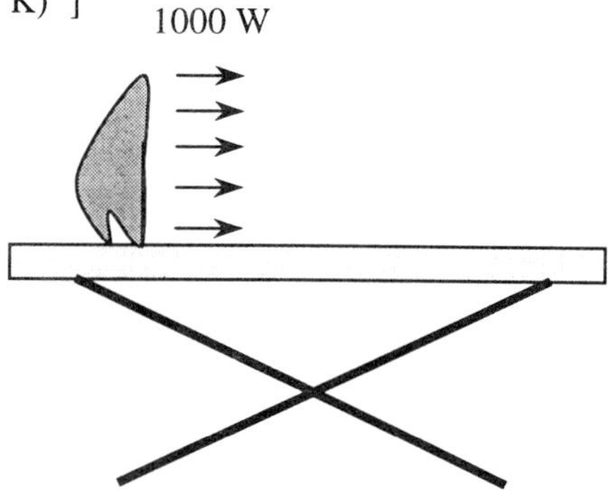

Substituting,

$$1000\ \text{W} = 0.7(T_s - 293\ \text{K}) + 0.06804\times 10^{-8}[T_s^4 - (293\ \text{K})^4]$$

Solving by trial and error gives

$$T_s = \mathbf{947\ K} = \mathbf{674^\circ C}$$

Discussion We note that the iron will dissipate all the energy it receives by convection and radiation when its surface temperature reaches 947 K.

3-78 The backside of the thin metal plate is insulated and the front side is exposed to solar radiation. The surface temperature of the plate is to be determined when it stabilizes.

Assumptions 1 Steady operating conditions exist. **2** Heat transfer through the insulated side of the plate is negligible. **3** The heat transfer coefficient is constant and uniform over the plate. **4** Heat loss by radiation is negligible.

Properties The solar absorptivity of the plate is given to be $\alpha = 0.6$.

Analysis When the heat loss from the plate by convection equals the solar radiation absorbed, the surface temperature of the plate can be determined from

$$\dot{Q}_{\text{solar absorbed}} = \dot{Q}_{\text{conv}}$$

$$\alpha\dot{Q}_{\text{solar}} = hA(T_s - T_o)$$

$$0.6 \times A \times 700 \text{ W/m}^2 = (50 \text{ W/m}^2 \cdot {}^\circ\text{C})A(T_s - 25)$$

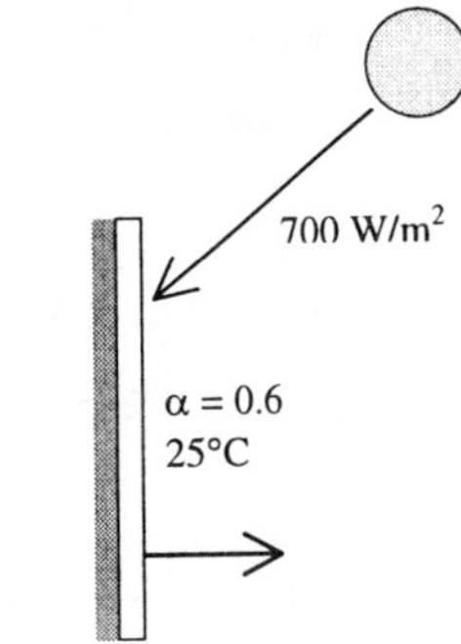

Canceling the surface area A and solving for T_s gives

$$T_s = \mathbf{33.4^\circ C}$$

3-79 EES solution of this (and other comprehensive problems designated with the *computer icon*) is available to instructors at the *Instructor Manual* section of the *Online Learning Center* (OLC) at www.mhhe.com/cengel-boles. See the Preface for access information.

3-80 A hot water pipe at 80°C is losing heat to the surrounding air at 5°C by natural convection with a heat transfer coefficient of 25 W/ m^2.°C. The rate of heat loss from the pipe by convection is to be determined.

Assumptions **1** Steady operating conditions exist. **2** Heat transfer by radiation is not considered. **3** The convection heat transfer coefficient is constant and uniform over the surface.

Analysis The heat transfer surface area is

$$A = (\pi D)L = 3.14\text{x}(0.05 \text{ m})(10 \text{ m}) = 1.571 \text{ m}^2$$

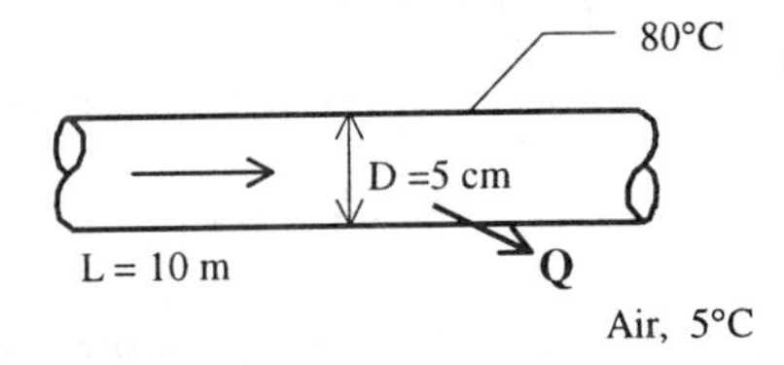

Under steady conditions, the rate of heat transfer by convection is

$$\dot{Q}_{\text{conv}} = hA\Delta T = (25 \text{ W/m}^2 \cdot {}^\circ\text{C})(1.571 \text{ m}^2)(80-5)^\circ\text{C} = \mathbf{2945 W} = 2.945 \text{ kW}$$

3-81 A spacecraft in space absorbs solar radiation while losing heat to deep space by thermal radiation. The surface temperature of the spacecraft is to be determined when steady conditions are reached..

Assumptions **1** Steady operating conditions exist since the surface temperatures of the wall remain constant at the specified values. **2** Thermal properties of the spacecraft are constant.

Properties The outer surface of a spacecraft has an emissivity of 0.8 and an absorptivity of 0.3.

Analysis When the heat loss from the outer surface of the spacecraft by radiation equals the solar radiation absorbed, the surface temperature can be determined from

$$\dot{Q}_{\text{solar absorbed}} = \dot{Q}_{\text{rad}}$$

$$\alpha\dot{Q}_{\text{solar}} = \varepsilon\sigma A(T_s^4 - T_{\text{space}}^4)$$

$$0.3\times A\times(1000\ \text{W/m}^2) = 0.8\times A\times(5.67\times10^{-8}\ \text{W/m}^2\cdot\text{K}^4)[\text{T}_s^4 - (0\ \text{K})^4]$$

Canceling the surface area A and solving for T_s gives

$$T_s = \mathbf{285\ K}$$

3-82 EES solution of this (and other comprehensive problems designated with the *computer icon*) is available to instructors at the *Instructor Manual* section of the *Online Learning Center* (OLC) at www.mhhe.com/cengel-boles. See the Preface for access information.

3-83 A hollow spherical iron container is filled with iced water at 0°C. The rate of heat loss from the sphere and the rate at which ice melts in the container are to be determined.

Assumptions **1** Steady operating conditions exist since the surface temperatures of the wall remain constant at the specified values. **2** Heat transfer through the shell is one-dimensional. **3** Thermal properties of the iron shell are constant. **4** The inner surface of the shell is at the same temperature as the iced water, 0°C.

Properties The thermal conductivity of iron is k = 80.2 W/m·°C (Table 3-1). The heat of fusion of water is at 1 atm is 333.7 kJ/kg.

Analysis This spherical shell can be approximated as a plate of thickness 0.4 cm and surface area

$$A = \pi D^2 = 3.14\text{x}(0.2\ \text{m})^2 = 0.126\ \text{m}^2$$

Then the rate of heat transfer through the shell by conduction is

$$\dot{Q}_{\text{cond}} = kA\frac{\Delta T}{L} = (80.2\text{W/m}\cdot^\circ\text{C})(0.126\text{m}^2)\frac{(5-0)^\circ\text{C}}{0.004\text{m}} = 12{,}632\text{W}$$

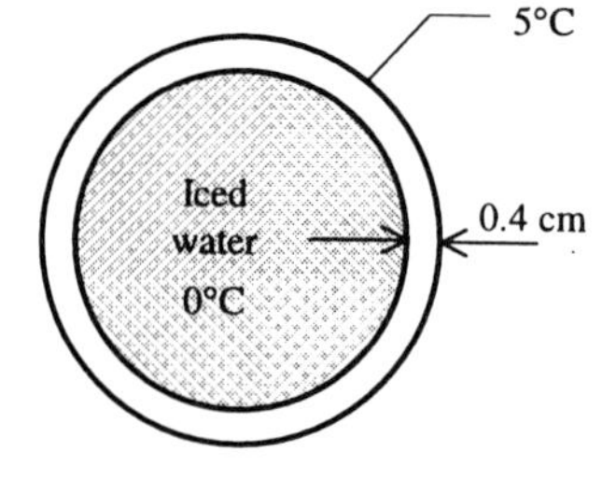

Considering that it takes 333.7 kJ of energy to melt 1 kg of ice at 0°C, the rate at which ice melts in the container can be determined from

$$\dot{m}_{\text{ice}} = \frac{\dot{Q}}{h_{if}} = \frac{12.632\text{kJ/s}}{333.7\text{kJ/kg}} = \mathbf{0.038kg/s}$$

Discussion We should point out that this result is slightly in error for approximating a curved wall as a plain wall. The error in this case is very small because of the large diameter to thickness ratio. For better accuracy, we could use the inner surface area (D = 19.2 cm) or the mean surface area (D = 19.6 cm) in the calculations.

3-84 The inner and outer glasses of a double pane window with a 1-cm air space are at specified temperatures. The rate of heat transfer through the window is to be determined.

Assumptions **1** Steady operating conditions exist since the surface temperatures of the glass remain constant at the specified values. **2** Heat transfer through the window is one-dimensional. **3** Thermal properties of the air are constant. **4** The air trapped between the two glasses is still, and thus heat transfer is by conduction only.

Properties The thermal conductivity of air at room temperature is $k = 0.026$ W/m.°C (Table 3-1).

Analysis Under steady conditions, the rate of heat transfer through the window by conduction is

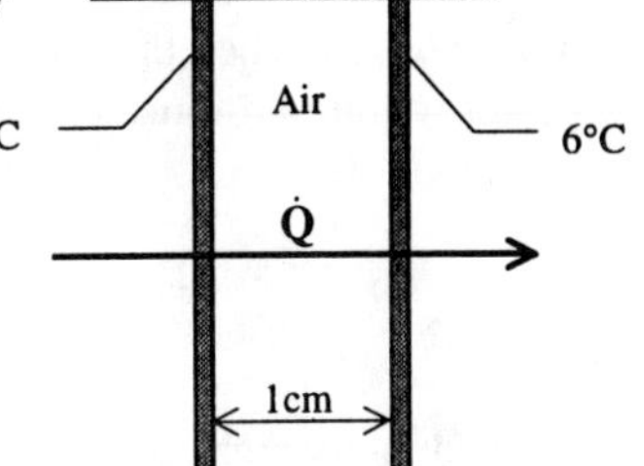

$$\dot{Q}_{\text{cond}} = kA\frac{\Delta T}{L} = (0.026\ \text{W/m}\cdot{}^\circ\text{C})(2\times 2\ \text{m}^2)\frac{(18-6)^\circ\text{C}}{0.01\ \text{m}} = \mathbf{125W}$$

$$= 0.125kW$$

3-85 Two surfaces of a flat plate are maintained at specified temperatures, and the rate of heat transfer through the plate is measured. The thermal conductivity of the plate material is to be determined.

Assumptions **1** Steady operating conditions exist since the surface temperatures of the plate remain constant at the specified values. **2** Heat transfer through the plate is one-dimensional. **3** Thermal properties of the plate are constant.

Analysis The thermal conductivity is determined directly from the steady one-dimensional heat conduction relation to be

Plate
2 cm
100°C
0°C
500 W/m²

$$\dot{Q} = kA\frac{T_1 - T_2}{L} \rightarrow k = \frac{(\dot{Q}/A)L}{T_1 - T_2} = \frac{(500\ \text{W/m}^2)(0.02\ \text{m})}{(100\text{-}0)^\circ\text{C}} = \mathbf{0.1\ W/m.^\circ C}$$

Review Problems

3-86 The weight of the cabin of an elevator is balanced by a counterweight. The power needed when the fully loaded cabin is rising , and when the empty cabin is descending at a constant speed are to be determined.

Assumptions **1** The weight of the cables is negligible. **2** The guide rails and pulleys are frictionless. **3** Air drag is negligible.

Analysis (*a*) When the cabin is fully loaded, half of the weight is balanced by the counterweight. The power required to raise the cabin at a constant speed of 2 m/s is

$$\dot{W} = \frac{mgz}{\Delta t} = mg\mathbf{V} = (400\text{kg})(9.8\text{m/s}^2)(2\text{m/s})\left(\frac{1\text{N}}{1\text{kg}\cdot\text{m/s}^2}\right)\left(\frac{1\text{kW}}{1000\text{N}\cdot\text{m/s}}\right) = \mathbf{7.84kW}$$

If no counterweight is used, the mass would double to 800 kg and the power would be 2×7.84 = **15.68 kW**.

(*b*) When the empty cabin is descending (and the counterweight is ascending) there is mass imbalance of 400-150 = 250 kg. The power required to raise this mass at a constant speed of 2 m/s is

$$\dot{W} = \frac{mgz}{\Delta t} = mg\mathbf{V} = (250\text{ kg})(9.81\text{ m/s}^2)(2\text{ m/s})\left(\frac{1\text{ N}}{1\text{ kg}\cdot\text{m/s}^2}\right)\left(\frac{1\text{ kW}}{1000\text{ N}\cdot\text{m/s}}\right) = \mathbf{4.9\text{ kW}}$$

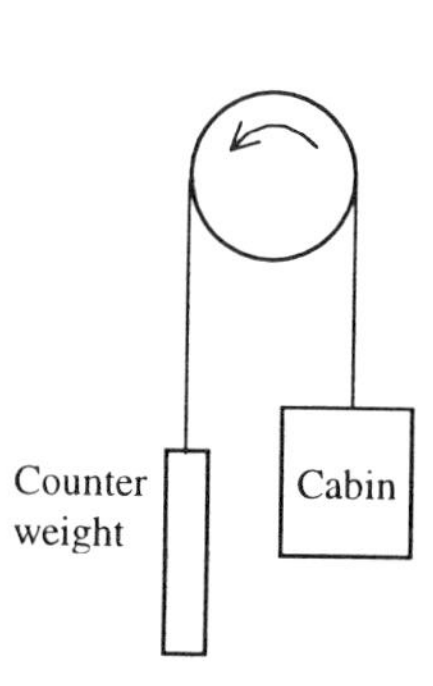

If a friction force of 1200 N develops between the cabin and the guide rails, we will need

$$\dot{W}_{\text{friction}} = \frac{F_{\text{friction}}\,z}{\Delta t} = F_{\text{friction}}\mathbf{V} = (1200\text{ N})(2\text{ m/s})\left(\frac{1\text{ kW}}{1000\text{ N}\cdot\text{m/s}}\right) = 2.4\text{ kW}$$

of additional power to combat friction which always acts in the opposite direction to motion. Therefore, the total power needed in this case is

$$\dot{W}_{\text{total}} = \dot{W} + \dot{W}_{\text{friction}} = 4.9 + 2.4 = \mathbf{7.3kW}$$

3-87 A cylinder equipped with an external spring is initially filled with air at a specified state. Heat is transferred to the air, and both the temperature and pressure rise. The total boundary work done by the air, and the amount of work done against the spring are to be determined, and the process is to be shown on a *P-v* diagram.

Assumptions **1** The process is quasi-equilibrium. **2** The spring is a linear spring.

Analysis (*a*) The pressure of the gas changes linearly with volume during this process, and thus the process curve on a *P-V* diagram will be a straight line. Then the boundary work during this process is simply the area under the process curve, which is a trapezoidal. Thus,

$$W_{b,out} = Area = \frac{P_1 + P_2}{2}(V_2 - V_1)$$

$$= \frac{(200 + 800)\text{kPa}}{2}(0.5 - 0.2)\text{m}^3\left(\frac{1 \text{ kJ}}{1 \text{ kPa} \cdot \text{m}^3}\right)$$

$$= \mathbf{150 \text{ kJ}}$$

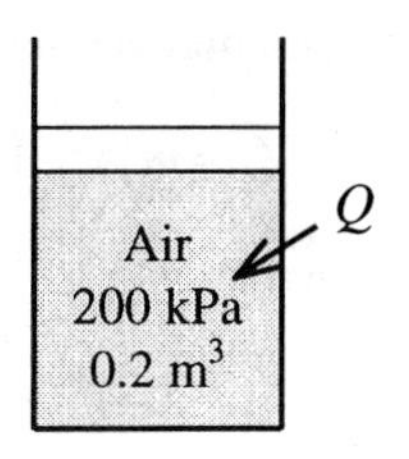

(*b*) If there were no spring, we would have a constant pressure process at P = 200 kPa. The work done during this process is

$$W_{\text{b,out,nospring}} = \int_1^2 PdV = P(V_2 - V_1)$$

$$= (200 \text{ kPa})(0.5 - 0.2)\text{m}^3\text{/kg}\left(\frac{1 \text{ kJ}}{1 \text{ kPa} \cdot \text{m}^3}\right) = 60 \text{ kJ}$$

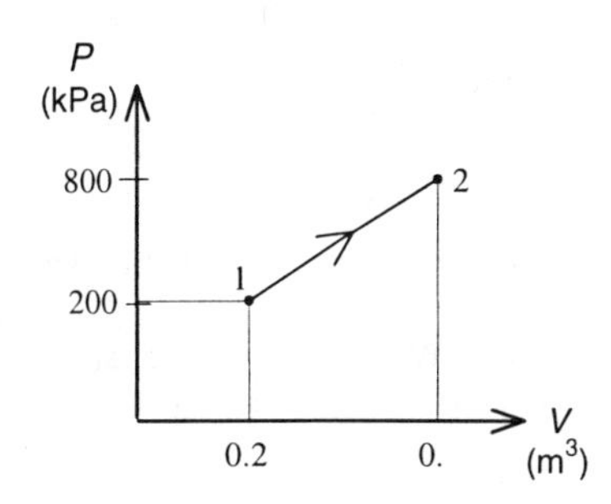

Thus,

$$W_{\text{spring}} = W_b - W_{\text{b,nospring}} = 150 - 60 = \mathbf{90 \text{ kJ}}$$

3-88 A cylinder equipped with a set of stops for the piston is initially filled with saturated liquid-vapor mixture of water a specified pressure. Heat is transferred to the water until the volume increases by 20%. The initial and final temperature, the mass of the liquid when the piston starts moving, and the work done during the process are to be determined, and the process is to be shown on a *P-v* diagram.

Assumptions The process is quasi-equilibrium.

Analysis (*a*) Initially the system is a saturated mixture at 100 kPa pressure, and thus the initial temperature is

$$T_1 = T_{sat\,@100\text{ kPa}} = \mathbf{99.63^\circ C}$$

The total initial volume is

$$V_1 = m_f v_f + m_g v_g = 2 \times 0.001043 + 3 \times 1.6940 = 5.084 \text{ m}^3$$

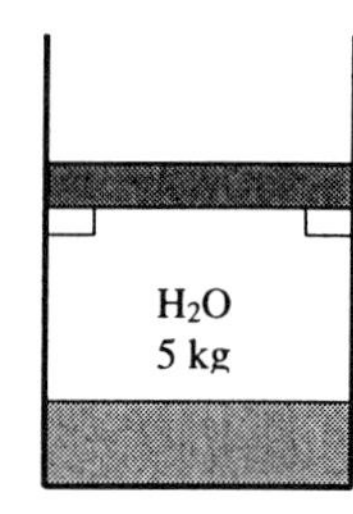

Then the total and specific volumes at the final state are

$$V_3 = 1.2V_1 = 1.2 \times 5.084 = 6.101 \text{ m}^3$$

$$v_3 = \frac{V_3}{m} = \frac{6.101 \text{ m}^3}{5 \text{ kg}} = 1.220 \text{ m}^3/\text{kg}$$

Thus,

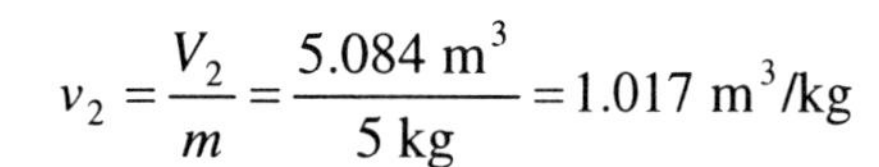

$$\left.\begin{array}{l} P_3 = 200 \text{ kPa} \\ v_3 = 1.220 \text{ m}^3/\text{kg} \end{array}\right\} T_3 = \mathbf{259.0^\circ C}$$

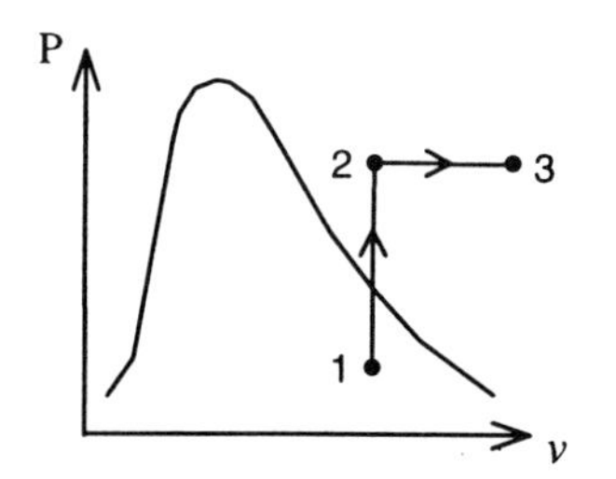

(*b*) When the piston first starts moving, $P_2 = 200$ kPa and $V_2 = V_1 = 5.084$ m^3. The specific volume at this state is

$$v_2 = \frac{V_2}{m} = \frac{5.084 \text{ m}^3}{5 \text{ kg}} = 1.017 \text{ m}^3/\text{kg}$$

which is greater than $v_g = 0.8857$ m^3/kg at 200 kPa. Thus **no liquid** is left in the cylinder when the piston starts moving.

(*c*) No work is done during process 1-2 since $V_1 = V_2$. The pressure remains constant during process 2-3 and the work done during this process is

$$W_b = \int_2^3 P dV = P_2(V_3 - V_2) = (200 \text{ kPa})(6.101 - 5.084)\text{m}^3 \left(\frac{1 \text{ kJ}}{1 \text{ kPa} \cdot \text{m}^3}\right) = \mathbf{203 \text{ kJ}}$$

3-89E A spherical balloon is initially filled with air at a specified state. The pressure inside is proportional to the square of the diameter. Heat is transferred to the air until the volume doubles. The work done is to be determined.

Assumptions **1** Air is an ideal gas. **2** The process is quasi-equilibrium.

Properties The gas constant of air is R = 0.06855 Btu/lbm.R (Table A-1E).

Analysis The dependence of pressure on volume can be expressed as

$$V = \frac{1}{6}\pi D^3 \longrightarrow D = \left(\frac{6V}{\pi}\right)^{1/3}$$

$$P \propto D^2 \longrightarrow P = kD^2 = k\left(\frac{6V}{\pi}\right)^{2/3}$$

AIR
10 lbm
30 psia
800 R

or,

$$k\left(\frac{6}{\pi}\right)^{2/3} = P_1V_1^{-2/3} = P_2V_2^{-2/3}$$

Also,

$$\frac{P_2}{P_1} = \left(\frac{V_2}{V_1}\right)^{2/3} = 2^{2/3} = 1.587$$

and

$$\frac{P_1V_1}{T_1} = \frac{P_2V_2}{T_2} \longrightarrow T_2 = \frac{P_2V_2}{P_1V_1}T_1 = 1.587 \times 2 \times (800\text{R}) = 2539\text{R}$$

Thus,

$$W_b = \int_1^2 PdV = \int_1^2 k\left(\frac{6V}{\pi}\right)^{2/3} dV = \frac{3k}{5}\left(\frac{6}{\pi}\right)^{2/3}\left(V_2^{5/3} - V_1^{5/3}\right) = \frac{3}{5}(P_2V_2 - P_1V_1)$$

$$= \frac{3}{5}mR(T_2 - T_1) = \frac{3}{5}(10 \text{ lbm})(0.06855 \text{ Btu/lbm}\cdot\text{R})(2539 - 800)\text{R} = \mathbf{715\ Btu}$$

3-90E EES solution of this (and other comprehensive problems designated with the *computer icon*) is available to instructors at the *Instructor Manual* section of the *Online Learning Center* (OLC) at www.mhhe.com/cengel-boles. See the Preface for access information.

3-91 A water tank open to the atmosphere is initially filled with water. The tank discharges to the atmosphere through a long pipe connected to a valve. The initial discharge velocity from the tank and the time required to empty the tank are to be determined. √

Assumptions **1** The flow is uniform and incompressible. **2** The draining pipe is horizontal. **3** The tank is considered to be empty when the water level drops to the center of the valve.

Analysis (*a*) Substituting the known quantities, the discharge velocity can be expressed as

$$\mathbf{V}=\sqrt{\frac{2gz}{1.5+fL/D}}=\sqrt{\frac{2gz}{1.5+0.015(100\text{ m})/(0.10\text{ m})}}=\sqrt{0.1212gz}$$

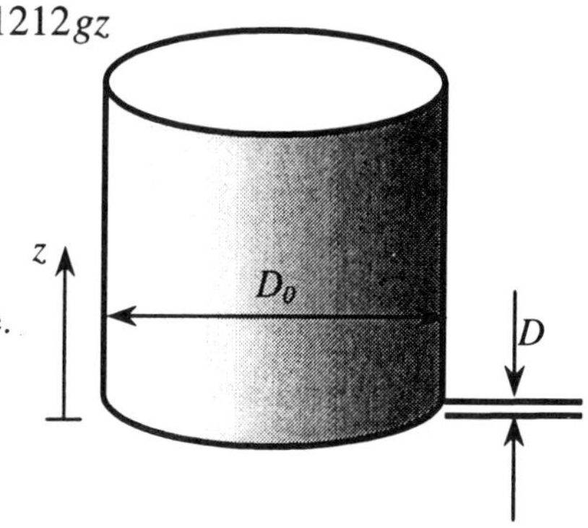

Then the initial discharge velocity becomes

$$\mathbf{V}_1=\sqrt{0.1212gz_1}=\sqrt{0.1212(9.81\text{ m/s}^2)(2\text{ m})}=\mathbf{1.54\ m/s}$$

where z is the water height relative to the center of the orifice at that time.

(*b*) The flow rate of water from the tank can be obtained by multiplying the discharge velocity by the pipe cross-sectional area,

$$\dot{V}=A_{\text{pipe}}\mathbf{V}_2=\frac{\pi D^2}{4}\sqrt{0.1212gz}$$

Then the amount of water that flows through the pipe during a differential time interval dt is

$$dV=\dot{V}dt=\frac{\pi D^2}{4}\sqrt{0.1212gz}\,dt \qquad (1)$$

which, from conservation of mass, must be equal to the decrease in the volume of water in the tank,

$$dV=A_{\tan k}(-dz)=-\frac{\pi D_0^2}{4}dz \qquad (2)$$

where dz is the change in the water level in the tank during dt. (Note that dz is a negative quantity since the positive direction of z is upwards. Therefore, we used $-dz$ to get a positive quantity for the amount of water discharged). Setting Eqs. (1) and (2) equal to each other and rearranging,

$$\frac{\pi D^2}{4}\sqrt{0.1212gz}\,dt=-\frac{\pi D_0^2}{4}dz \rightarrow dt=-\frac{D_0^2}{D^2}\frac{dz}{\sqrt{0.1212gz}}=-\frac{D_0^2}{D^2\sqrt{0.1212g}}z^{-\frac{1}{2}}dz$$

The last relation can be integrated easily since the variables are separated. Letting t_f be the discharge time and integrating it from $t=0$ when $z=z_1$ to $t=t_f$ when $z=0$ (completely drained tank) gives

$$\int_{t=0}^{t_f}dt=-\frac{D_0^2}{D^2\sqrt{0.1212g}}\int_{z=z_1}^{0}z^{-1/2}dz \rightarrow t_f=-\frac{D_0^2}{D^2\sqrt{0.1212g}}\left.\frac{z^{\frac{1}{2}}}{\frac{1}{2}}\right|_{z_1}^{0}=\frac{2D_0^2}{D^2\sqrt{0.1212g}}z_1^{\frac{1}{2}}$$

Simplifying and substituting the values given, the draining time is determined to be

$$t_f=\frac{2D_0^2}{D^2}\sqrt{\frac{z_1}{0.1212g}}=\frac{2(10\text{ m})^2}{(0.1\text{ m})^2}\sqrt{\frac{2\text{ m}}{0.1212(9.81\text{ m/s}^2)}}=25{,}940\text{ s}=\mathbf{7.21\,h}$$

Discussion The draining time can be shortened considerably by installing a pump in the pipe.

3-92 Milk is transported from Texas to California in a cylindrical tank. The amount of milk in the tank is to be determined.

Assumptions Milk is mostly water, and thus the properties of water can be used for milk.

Properties The density of milk is the same as that of water, $\rho = 1000$ kg/m^3.

Analysis Noting that the thickness of insulation is 0.05 m on all sides, the volume and mass of the milk in a full tank is determined to be

$$V_{\text{milk}} = (\pi D_i^2 / 4)L_i = [\pi(1.9\text{ m})^2 / 4](6.9\text{ m}) = 19.56\text{ m}^3$$

$$m_{\text{milk}} = \rho V_{\text{milk}} = (1000\text{ kg/m}^3)(19.56\text{ m}^3) = \mathbf{19{,}560\ \ kg}$$

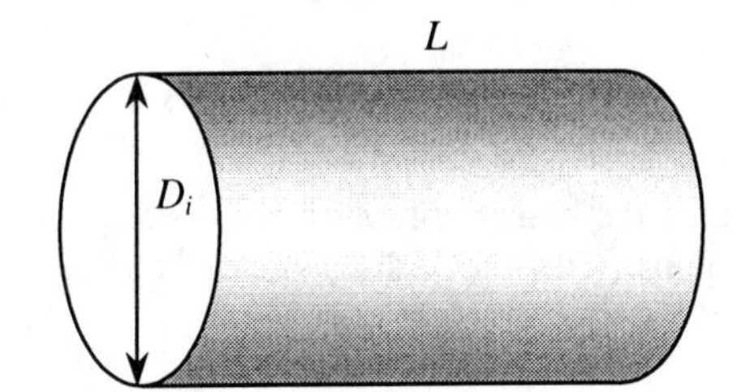

The volume of the milk in gallons is

$$V_{\text{milk}} = (19.56\text{ m}^3)\left(\frac{264.17\text{ gal}}{1\text{ m}^3}\right) = \mathbf{5167\ gal}$$

3-93 The rate of accumulation of water in a pool and the rate of discharge are given. The rate supply of water to the pool is to be determined.

Assumptions **1** Water is supplied and discharged steadily. **2** The rate of evaporation of water is negligible. **3** No water is supplied or removed through other means.

Analysis The conservation of mass principle applied to the pool requires that the rate of increase in the amount of water in the pool be equal to the difference between the rate of supply of water and the rate of discharge. That is,

$$\frac{dm_{\text{pool}}}{dt} = \dot{m}_i - \dot{m}_e \quad \rightarrow \quad \dot{m}_i = \frac{dm_{\text{pool}}}{dt} + \dot{m}_e \quad \rightarrow \quad \dot{V}_i = \frac{dV_{\text{pool}}}{dt} + \dot{V}_e$$

since the density of water is constant and thus the conservation of mass is equivalent to conservation of volume. The rate of discharge of water is

$$\dot{V}_e = A_e \mathbf{V}_e = (\pi D^2/4)\mathbf{V}_e = [\pi(0.05\text{ m})^2/4](5\text{ m/s}) = 0.00982\text{ m}^3/\text{s}$$

The rate of accumulation of water in the pool is equal to the cross-section of the pool times the rate at which the water level rises,

$$\frac{dV_{\text{pool}}}{dt} = A_{\text{cross-section}}\mathbf{V}_{\text{level}} = (3\text{ m}\times 4\text{ m})(0.015\text{ m/min}) = 0.18\text{ m}^3/\text{min} = 0.00300\text{ m}^3/\text{s}$$

Substituting, the rate at which water is supplied to the pool is determined to be

$$\dot{V}_i = \frac{dV_{\text{pool}}}{dt} + \dot{V}_e = 0.003 + 0.00982 = \mathbf{0.01282\text{ m}^3/\text{s}}$$

Therefore, water is supplied at a rate of 0.01282 m^3/s = 12.82 L/s.

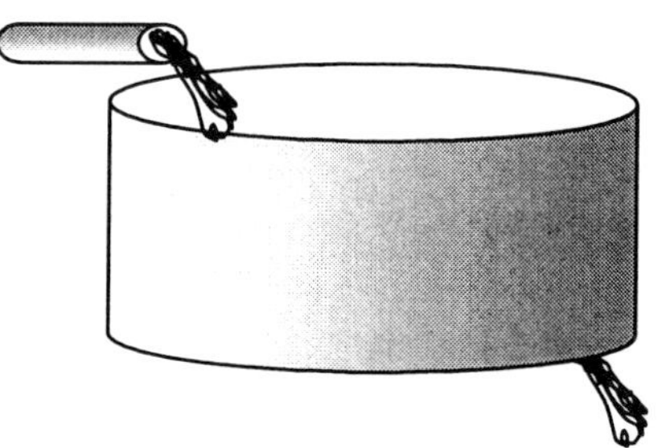

3-94 A fluid is flowing in a circular pipe. A relation is to be obtained for the average fluid velocity in therms of *V*(*r*), *R*, and *r*.

Analysis Choosing a circular ring of area dA = 2πrdr as our differential area, the mass flow rate through a cross-sectional area can be expressed as

$$\dot{m} = \int_A \rho \mathbf{V}(r)dA = \int_0^R \rho \mathbf{V}(r)2\pi r\,dr$$

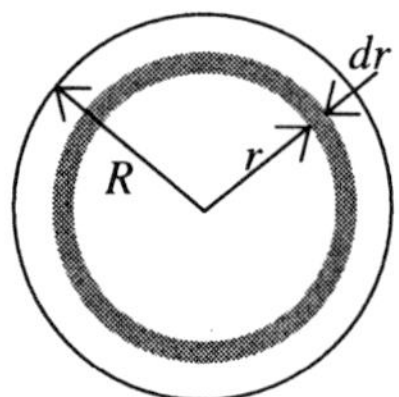

Setting this equal to and solving for V_{av},

$$\mathbf{V}_{av} = \frac{2}{R^2}\int_0^R \mathbf{V}(r)r\,dr$$

3-95 Air is accelerated in a nozzle. The density of air at the nozzle exit is to be determined.

Assumptions Flow through the nozzle is steady.

Properties The density of air is given to be 4.18 kg/m^3 at the inlet.

Analysis There is only one inlet and one exit, and thus $\dot{m}_1 = \dot{m}_2 = \dot{m}$. Then,

$$\dot{m}_1 = \dot{m}_2$$

$$\rho_1 A_1 \mathbf{V}_1 = \rho_2 A_2 \mathbf{V}_2$$

$$\rho_2 = \frac{A_1}{A_2}\frac{\mathbf{V}_1}{\mathbf{V}_2}\rho_1 = 2\frac{120 \text{ m/s}}{380 \text{ m/s}}(4.18 \text{ kg/m}^3) = \mathbf{2.64 \ kg/m^3}$$

1 AIR → 2

Discussion Note that the density of air decreases considerably despite a decrease in the cross-sectional area of the nozzle.

3-96 A long roll of large 1-Mn manganese steel plate is to be quenched in an oil bath at a specified rate. The mass flow rate of the plate is to be determined.

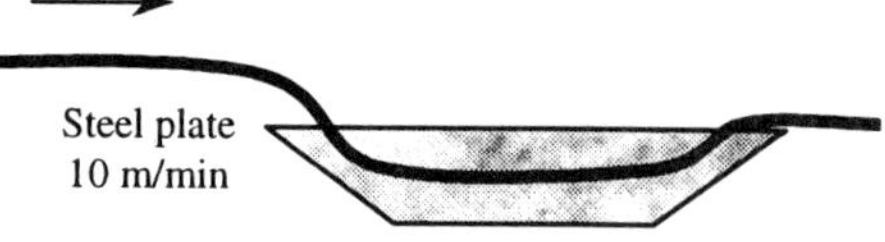

Assumptions The plate moves through the bath steadily.

Properties The density of steel plate is given to be $\rho = 7854\ \text{kg/m}^3$.

Analysis The mass flow rate of the sheet metal through the oil bath is

$$\dot{m} = \rho\dot{V} = \rho wtV = (7854\ \text{kg/m}^3)(1\ \text{m})(0.005\ \text{m})(10\ \text{m/min}) = 393\ \text{kg/min} = \mathbf{6.55\ kg/s}$$

Therefore, steel plate can be treated conveniently as a "flowing fluid" in calculations.

3-97 The air in a hospital room is to be replaced every 20 minutes. The minimum diameter of the duct is to be determined if the air velocity is not to exceed a certain value.

Assumptions **1** The volume occupied by the furniture etc in the room is negligible. **2** The incoming conditioned air does not mix with the air in the room.

Analysis The volume of the room is

$$V = (6\ \text{m})(5\ \text{m})(4\ \text{m}) = 120\ \text{m}^3$$

Hospital Room

6×5×4 m

To empty this air in 20 min, the volume flow rate must be

$$\dot{V} = \frac{V}{\Delta t} = \frac{120\ \text{m}^3}{20\times 60\ \text{s}} = 0.10\ \text{m}^3/\text{s}$$

If the mean velocity is 5 m/s, the diameter of the duct is

$$\dot{V} = A\mathbf{V} = \frac{\pi D^2}{4}\mathbf{V} \quad\rightarrow\quad D = \sqrt{\frac{4\dot{V}}{\pi\mathbf{V}}} = \sqrt{\frac{4(0.10\ \text{m}^3/\text{s})}{\pi(5\ \text{m/s})}} = \mathbf{0.16\ m}$$

Therefore, the diameter of the duct must be at least 0.16 m to ensure that the air in the room is exchanged completely within 20 min while the mean velocity does not exceed 5 m/s.

Discussion This problem shows that engineering systems are sized to satisfy certain constraints imposed by certain considerations.

3-98E A study quantifies the cost and benefits of enhancing IAQ by increasing the building ventilation. The net monetary benefit of installing an enhanced IAQ system to the employer per year is to be determined.

Assumptions The analysis in the report is applicable to this work place.

Analysis The report states that enhancing IAQ increases the productivity of a person by $90 per year, and decreases the cost of the respiratory illnesses by $39 a year while increasing the annual energy consumption by $6 and the equipment cost by about $4 a year. The net monetary benefit of installing an enhanced IAQ system to the employer per year is determined by adding the benefits and subtracting the costs to be

$$\text{Net benefit} = \text{Total benefits} - \text{total cost} = (90+39) - (6+4) = \$119/\text{year} \quad \text{(per person)}$$

The total benefit is determined by multiplying the benefit per person by the number of employees,

$$\text{Total net benefit} = \text{No. of employees} \times \text{Net benefit per person} = 120 \times \$119/\text{year} = \mathbf{\$14{,}280/year}$$

Discussion Note that the unseen savings in productivity and reduced illnesses can be very significant when they are properly quantified.

Fundamentals of Engineering (FE) Exam Problems

3-99 A frictionless piston-cylinder device and a rigid tank contain 1.2 kmol of an ideal gas at the same temperature, pressure and volume. Now heat is transferred, and the temperature of both systems is raised by 15°C. The amount of extra heat that must be supplied to the gas in the cylinder that is maintained at constant pressure is

(a) 0 kJ (b) 24 kJ (c) 78 kJ (d) 102 kJ (e) 150 kJ

Answer (e) 150 kJ

Solution Solved by EES Software. Solutions can be verified by copying-and-pasting the following lines on a blank EES screen. (Similar problems and their solutions can be obtained easily by modifying numerical values).

```
"Note that Cp-Cv=R, and thus Q_diff=m*R*dT=N*Ru*dT"

N=1.2 "kmol"
Ru=8.314 "kJ/kmol.K"
T_change=15
Q_diff=N*Ru*T_change

"Some Wrong Solutions with Common Mistakes:"
W1_Qdiff=0 "Assuming they are the same"
W2_Qdiff=Ru*T_change  "Not using mole numbers"
W3_Qdiff=Ru*T_change/N "Dividing by N instead of multiplying"
W4_Qdiff=N*Rair*T_change; Rair=0.287  "using Ru instead of R"
```

3-100 The specific heat of a material is given in a strange unit to be C = 3.60 kJ/kg.°F. The specific heat of this material in the SI units of kJ/kg.°C is

(a) 2.00 kJ/kg.°C (b) 3.20 kJ/kg.°C (c) 3.60 kJ/kg.°C (d) 4.80 kJ/kg.°C (e) 6.48 kJ/kg.°C

Answer (e) 6.48 kJ/kg.°C

Solution Solved by EES Software. Solutions can be verified by copying-and-pasting the following lines on a blank EES screen. (Similar problems and their solutions can be obtained easily by modifying numerical values).

```
C=3.60 "kJ/kg.F"
C_SI=C*1.8 "kJ/kg.C"

"Some Wrong Solutions with Common Mistakes:"
W1_C=C "Assuming they are the same"
W2_C=C/1.8 "Dividing by 1.8 instead of multiplying"
```

3-101 A 2-m^3 rigid tank contains nitrogen gas at 500 kPa and 300 K. Now heat is transferred to the nitrogen in the tank and the pressure of nitrogen rises to 800 kPa. The work done during this process is

(a) 600 kJ (b) 1000 kJ (c) 0 kJ (d) 500 kJ (e) 1600 kJ

Answer (b) 0 kJ

Solution Solved by EES Software. Solutions can be verified by copying-and-pasting the following lines on a blank EES screen. (Similar problems and their solutions can be obtained easily by modifying numerical values).

```
V=2 "m^3"
P1=500 "kPa"
T1=300 "K"
P2=800 "kPa"
W=0 "since constant volume"

"Some Wrong Solutions with Common Mistakes:"
R=0.297
W1_W=V*(P2-P1) "Using W=V*DELTAP"
W2_W=V*P1
W3_W=V*P2
W4_W=R*T1*ln(P1/P2)
```

3-102 A 2-m^3 cylinder contains nitrogen gas at 500 kPa and 300 K. Now the gas is compressed isothermally to a volume of 0.1 m^3. The work done on the gas during this compression process is

(a) 950 kJ (b) 0 kJ (c) 1610 kJ (d) 2996 kJ (e) 562 kJ

Answer (d) 2996 kJ

Solution Solved by EES Software. Solutions can be verified by copying-and-pasting the following lines on a blank EES screen. (Similar problems and their solutions can be obtained easily by modifying numerical values).

```
R=8.314/28
V1=2 "m^3"
V2=0.1 "m^3"
P1=500 "kPa"
T1=300 "K"
P1*V1=m*R*T1
W=m*R*T1* ln(V2/V1) "constant temperature"

"Some Wrong Solutions with Common Mistakes:"
W1_W=R*T1* ln(V2/V1)  "Forgetting m"
W2_W=P1*(V1-V2) "Using V*DeltaP"

P1*V1/T1=P2*V2/T1
W3_W=(V1-V2)*(P1+P2)/2 "Using P_ave*Delta V"
W4_W=P1*V1-P2*V2 "Using W=P1V1-P2V2"
```

3-103 A 2-kW electric resistance heater in a room is turned on and kept on for 30 min. The amount of energy transferred to the room by the heater is

(a) 1 kJ (b) 60 kJ (c) 1800 kJ (d) 3600 kJ (e) 7200 kJ

Answer (d) 3600 kJ

Solution Solved by EES Software. Solutions can be verified by copying-and-pasting the following lines on a blank EES screen. (Similar problems and their solutions can be obtained easily by modifying numerical values).

```
We= 2 "kJ/s"
time=30*60 "s"
We_total=We*time "kJ"

"Some Wrong Solutions with Common Mistakes:"
W1_Etotal=We*time/60 "using minutes instead of s"
W2_Etotal=We "ignoring time"
```

The following questions are based on the optional special topic of heat transfer:

3-104 A 12-cm high and 18-cm wide circuit board houses on its surface 100 closely spaced chips, each generating heat at a rate of 0.07 W and transferring it by convection to the surrounding air at 40°C. Heat transfer from the back surface of the board is negligible. If the convection heat transfer coefficient on the surface of the board is 10 $W/m^2.°C$ and radiation heat transfer is negligible, the average surface temperature of the chips is

(a) 72.4°C (b) 7.6°C (c) 40.7°C (d) 47.0°C (e) 68.2°C

Answer (a) 72.4°C

Solution Solved by EES Software. Solutions can be verified by copying-and-pasting the following lines on a blank EES screen. (Similar problems and their solutions can be obtained easily by modifying numerical values).

```
A=0.12*0.18 "m^2"
Q= 100*0.07 "W"
Tair=40 "C"
h=10 "W/m^2.C"
Q= h*A*(Ts-Tair) "W"

"Some Wrong Solutions with Common Mistakes:"
Q= h*(W1_Ts-Tair) "Not using area"
Q= h*2*A*(W2_Ts-Tair) "Using both sides of surfaces"
Q= h*A*(W3_Ts+Tair) "Adding temperatures instead of subtracting"
Q/100= h*A*(W4_Ts-Tair) "Considering 1 chip only"
```

3-105 A 50-cm-long, 0.2-cm-diameter electric resistance wire submerged in water is used to determine the boiling heat transfer coefficient in water at 1 atm experimentally. The surface temperature of the wire is measured to be 130°C when a wattmeter indicates the electric power consumption to be 4.1 kW. Then the heat transfer coefficient is

43500 W/m^2.°C (b) 137 W/m^2.°C (c) 68330 W/m^2.°C (d) 10038 W/m^2.°C (e) 37,540 W/m^2.°C

Answer (a) 43500 W/m^2.°C

Solution Solved by EES Software. Solutions can be verified by copying-and-pasting the following lines on a blank EES screen. (Similar problems and their solutions can be obtained easily by modifying numerical values).

```
L=0.5 "m"
D=0.002 "m"
A=pi*D*L "m^2"
We=4.1 "kW"
Ts=130 "C"
Tf=100 "C  (Boiling temperature of water at 1 atm)"
We= h*A*(Ts-Tf) "W"

"Some Wrong Solutions with Common Mistakes:"
We= W1_h*(Ts-Tf) "Not using area"
We= W2_h*(L*pi*D^2/4)*(Ts-Tf) "Using volume instead of area"
We= W3_h*A*Ts "Using Ts instead of temp difference"
```

3-106 A 3-m^2 hot black surface at 80°C is losing heat to the surrounding air at 25°C by convection with a convection heat transfer coefficient of 12 W/m^2.°C, and by radiation to the surrounding surfaces at 15°C. The total rate of heat loss from the surface is

(a) 1987 W (b) 2239 W (c) 2348 W (d) 3451 W (e) 3811 W

Answer (d) 3451 W

Solution Solved by EES Software. Solutions can be verified by copying-and-pasting the following lines on a blank EES screen. (Similar problems and their solutions can be obtained easily by modifying numerical values).

```
sigma=5.67E-8 "W/m^2.K^4"
eps=1
A=3 "m^2"
h_conv=12 "W/m^2.C"
Ts=80 "C"
Tf=25 "C"
Tsurr=15 "C"
Q_conv=h_conv*A*(Ts-Tf)  "W"
Q_rad=eps*sigma*A*((Ts+273)^4-(Tsurr+273)^4) "W"
Q_total=Q_conv+Q_rad "W"

"Some Wrong Solutions with Common Mistakes:"
W1_Ql=Q_conv "Ignoring radiation"
W2_Q=Q_rad "ignoring convection"
W3_Q=Q_conv+eps*sigma*A*(Ts^4-Tsurr^4) "Using C in radiation calculations"
W4_Q=Q_total/A "not using area"
```

3-107 Heat is transferred steadily through a 0.2-m thick 8 m by 4 m wall at a rate of 1.6 kW. The inner and outer surface temperatures of the wall are measured to be 15°C to 5°C. The average thermal conductivity of the wall is

(a) 0.001 W/m.°C (b) 0.5 W/m.°C (c) 1.0 W/m.°C (d) 2.0 W/m.°C (e) 5.0 W/m.°C

Answer (c) 1.0 W/m.°C

Solution Solved by EES Software. Solutions can be verified by copying-and-pasting the following lines on a blank EES screen. (Similar problems and their solutions can be obtained easily by modifying numerical values).

```
A=8*4 "m^2"
L=0.2 "m"
T1=15 "C"
T2=5 "C"
Q=1600 "W"
Q=k*A*(T1-T2)/L "W"

"Some Wrong Solutions with Common Mistakes:"
Q=W1_k*(T1-T2)/L "Not using area"
Q=W2_k*2*A*(T1-T2)/L "Using areas of both surfaces"
Q=W3_k*A*(T1+T2)/L "Adding temperatures instead of subtracting"
Q=W4_k*A*L*(T1-T2) "Multiplying by thickness instead of dividing by it"
```

3-108 The roof of an electrically heated house is 6 m long, 8 m wide, and 0.25 m thick. It is made of a flat layer of concrete whose thermal conductivity is 0.8 W/m.°C. During a certain winter night, the temperatures of the inner and outer surfaces of the roof are measured to be 15°C and 4°C, respectively. The average rate of heat loss through the roof that night was

(a) 35 W (b) 422 W (c) 3379 W (d) 2246 W (e) 1690 W

Answer (e) 1690 W

Solution Solved by EES Software. Solutions can be verified by copying-and-pasting the following lines on a blank EES screen. (Similar problems and their solutions can be obtained easily by modifying numerical values).

```
A=8*6 "m^2"
L=0.25 "m"
k=0.8 "W/m.C"
T1=15 "C"
T2=4 "C"
Q_cond=k*A*(T1-T2)/L "W"

"Some Wrong Solutions with Common Mistakes:"
W1_Q=k*(T1-T2)/L "Not using area"
W2_Q=k*2*A*(T1-T2)/L "Using areas of both surfaces"
W3_Q=k*A*(T1+T2)/L "Adding temperatures instead of subtracting"
W4_Q=k*A*L*(T1-T2) "Multiplying by thickness instead of dividing by it"
```

3-109 ... 3-111 Design and Essay Problems

Chapter 4
THE FIRST LAW OF THERMODYNAMICS

Closed System Energy Balance: General Systems

4-1C No. This is the case for adiabatic systems only.

4-2C Warmer. Because energy is added to the room air in the form of electrical work.

4-3C Warmer. If we take the room that contains the refrigerator as our system, we will see that electrical work is supplied to this room to run the refrigerator, which is eventually dissipated to the room as waste heat.

4-4C Energy can be transferred to or from a control volume as heat, various forms of work, and by mass transport.

4-5 Water is heated in a pan on top of a range while being stirred. The energy of the water at the end of the process is to be determined.

Assumptions The pan is stationary and thus the changes in kinetic and potential energies are negligible.

Analysis We take the water in the pan as our system. This is a closed system since no mass enters or leaves. Applying the energy balance on this system gives

$$\underbrace{E_{in} - E_{out}}_{\substack{\text{Net energy transfer} \\ \text{by heat, work, and mass}}} = \underbrace{\Delta E_{system}}_{\substack{\text{Change in internal, kinetic,} \\ \text{potential, etc. energies}}}$$

$$Q_{in} + W_{pw,in} - Q_{out} = \Delta U = U_2 - U_1$$

$$30\text{ kJ} + 0.5\text{ kJ} - 5\text{ kJ} = U_2 - 10\text{ kJ}$$

$$U_2 = \mathbf{35.5\ kJ}$$

Therefore, the final internal energy of the system is 35.5 kJ.

4-6E Water is heated in a cylinder on top of a range. The change in the energy of the water during this process is to be determined.

Assumptions The pan is stationary and thus the changes in kinetic and potential energies are negligible.

Analysis We take the water in the cylinder as the system. This is a closed system since no mass enters or leaves. Applying the energy balance on this system gives

$$\underbrace{E_{in} - E_{out}}_{\substack{\text{Net energy transfer} \\ \text{by heat, work, and mass}}} = \underbrace{\Delta E_{system}}_{\substack{\text{Change in internal, kinetic,} \\ \text{potential, etc. energies}}}$$

$$Q_{in} - W_{b,out} - Q_{out} = \Delta U = U_2 - U_1$$

$$65\text{ Btu} - 5\text{ Btu} - 8\text{ Btu} = \Delta U$$

$$\Delta U = U_2 - U_1 = \mathbf{52\ Btu}$$

Therefore, the energy content of the system increases by 52 Btu during this process.

4-7 A classroom is to be air-conditioned using window air-conditioning units. The cooling load is due to people, lights, and heat transfer through the walls and the windows. The number of 4-kW window air conditioning units required is to be determined.

Assumptions There are no heat dissipating equipment (such as computers, TVs, or ranges) in the room.

Analysis The total cooling load of the room is determined from

$$\dot{Q}_{cooling} = \dot{Q}_{lights} + \dot{Q}_{people} + \dot{Q}_{heat\ gain}$$

where

$$\dot{Q}_{lights} = 10 \times 100\text{ W} = 1\text{ kW}$$

$$\dot{Q}_{people} = 40 \times 360\text{ kJ/h} = 4\text{ kW}$$

$$\dot{Q}_{heat\ gain} = 15{,}000\text{ kJ/h} = 4.17\text{ kW}$$

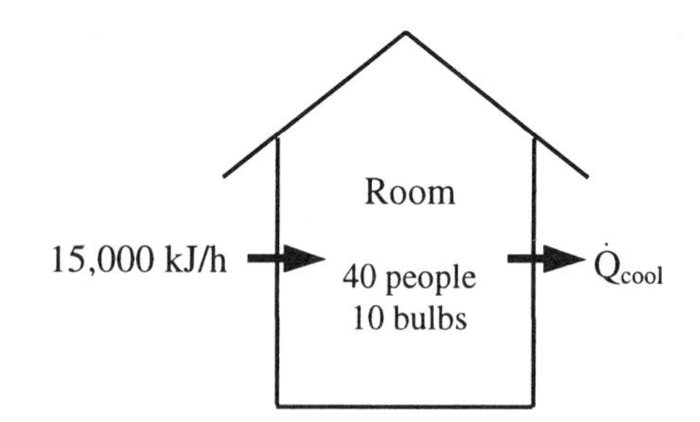

Substituting,

$$\dot{Q}_{cooling} = 1 + 4 + 4.17 = 9.17\text{ kW}$$

Thus the number of air-conditioning units required is

$$\frac{9.17\text{ kW}}{5\text{ kw/unit}} = 1.83 \longrightarrow \mathbf{2\ units}$$

4-8 An industrial facility is to replace its 40-W standard fluorescent lamps by their 34-W high efficiency counterparts. The amount of energy and money that will be saved a year as well as the simple payback period are to be determined.

Analysis The reduction in the total electric power consumed by the lighting as a result of switching to the high efficiency fluorescent is

Wattage reduction = (Wattage reduction per lamp)(Number of lamps)
= (40 - 34 W/lamp)(700 lamps)
= 4200 W

Then using the relations given earlier, the energy and cost savings associated with the replacement of the high efficiency fluorescent lamps are determined to be

Energy Savings = (Total wattage reduction)(Ballast factor)(Operating hours)
= (4.2 kW)(1.1)(2800 h/year)
= **12,936 kWh/year**

Cost Savings = (Energy savings)(Unit electricity cost)
= (12,936 kWh/year)($0.08/kWh)
= **$1035**

The implementation cost of this measure is simply the extra cost of the energy efficient fluorescent bulbs relative to standard ones, and is determined to be

Implementation Cost = (Cost difference of lamps)(Number of lamps)
= [($2.26-$1.77)/lamp](700 lamps)
= $343

This gives a simple payback period of

$$\text{Simple payback period} = \frac{\text{Implementation cost}}{\text{Annual cost savings}} = \frac{\$343}{\$1035\,/\,\text{year}} = \mathbf{0.33\ year}\ \ (4.0\ \text{months})$$

Discussion Note that if all the lamps were burned out today and are replaced by high-efficiency lamps instead of the conventional ones, the savings from electricity cost would pay for the cost differential in about 4 months. The electricity saved will also help the environment by reducing the amount of CO_2, CO, NO_x, etc. associated with the generation of electricity in a power plant.

4-9 The lighting energy consumption of a storage room is to be reduced by installing motion sensors. The amount of energy and money that will be saved as well as the simple payback period are to be determined.

Assumptions The electrical energy consumed by the ballasts is negligible.

Analysis The plant operates 12 hours a day, and thus currently the lights are on for the entire 12 hour period. The motion sensors installed will keep the lights on for 3 hours, and off for the remaining 9 hours every day. This corresponds to a total of 9×365 = 3285 off hours per year. Disregarding the ballast factor, the annual energy and cost savings become

Energy Savings = (Number of lamps)(Lamp wattage)(Reduction of annual operating hours)
= (24 lamps)(60 W/lamp)(3285 hours/year)
= **4730 kWh/year**

Cost Savings = (Energy Savings)(Unit cost of energy)
= (4730 kWh/year)($0.08/kWh)
= **$378/year**

The implementation cost of this measure is the sum of the purchase price of the sensor plus the labor,

Implementation Cost = Material + Labor = $32 + $40 = $72

This gives a simple payback period of

$$\text{Simple payback period} = \frac{\text{Implementation cost}}{\text{Annual cost savings}} = \frac{\$72}{\$378\,/\,\text{year}} = \mathbf{0.19\ year}\ \ (2.3\,\text{months})$$

Therefore, the motion sensor will pay for itself in about 2 months.

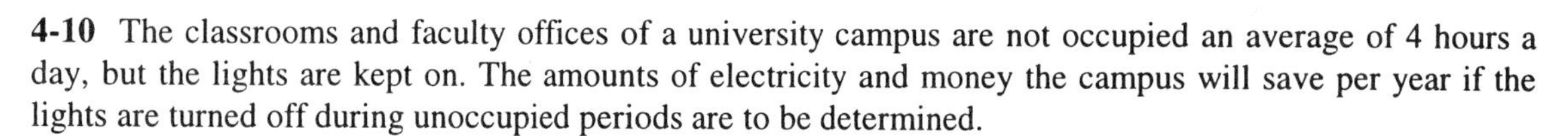

4-10 The classrooms and faculty offices of a university campus are not occupied an average of 4 hours a day, but the lights are kept on. The amounts of electricity and money the campus will save per year if the lights are turned off during unoccupied periods are to be determined.

Analysis The total electric power consumed by the lights in the classrooms and faculty offices is

$$\dot{E}_{\text{lighting, classroom}} = (\text{Power consumed per lamp}) \times (\text{No. of lamps}) = (200 \times 12 \times 110\ \text{W}) = 264{,}000 = 264\ \text{kW}$$

$$\dot{E}_{\text{lighting, offices}} = (\text{Power consumed per lamp}) \times (\text{No. of lamps}) = (400 \times 6 \times 110\ \text{W}) = 264{,}000 = 264\ \text{kW}$$

$$\dot{E}_{\text{lighting, total}} = \dot{E}_{\text{lighting, classroom}} + \dot{E}_{\text{lighting, offices}} = 264 + 264 = 528\ \text{kW}$$

Noting that the campus is open 240 days a year, the total number of unoccupied work hours per year is

Unoccupied hours = (4 hours/day)(240 days/year) = 960 h/yr

Then the amount of electrical energy consumed per year during unoccupied work period and its cost are

$$\text{Energy savings} = (\dot{E}_{\text{lighting, classroom}})(\text{Unoccupied hours}) = (528\ \text{kW})(960\ \text{h/yr}) = 506{,}880\ \text{kWh}$$

$$\text{Cost savings} = (\text{Energy savings})(\text{Unit cost of energy}) = (506{,}880\ \text{kWh})(\$0.082/\text{kWh}) = \mathbf{\$41{,}564/yr}$$

Discussion Note that simple conservation measures can result in significant energy and cost savings.

4-11 The radiator of a steam heating system is initially filled with superheated steam. The valves are closed, and steam is allowed to cool until the pressure drops to a specified value by transferring heat to the room. The amount of heat transfer is to be determined, and the process is to be shown on a *P-v* diagram.

Assumptions **1** The tank is stationary and thus the kinetic and potential energy changes are zero. **2** There are no work interactions.

Analysis We take the radiator as the system. This is a closed system since no mass enters or leaves. Noting that the volume of the system is constant and thus there is no boundary work, the energy balance for this stationary closed system can be expressed as

$$\underbrace{E_{in} - E_{out}}_{\substack{\text{Net energy transfer}\\ \text{by heat, work, and mass}}} = \underbrace{\Delta E_{system}}_{\substack{\text{Change in internal, kinetic,}\\ \text{potential, etc. energies}}}$$

$$-Q_{out} = \Delta U = m(u_2 - u_1) \quad (\text{since } W = \text{KE} = \text{PE} = 0)$$

$$Q_{out} = m(u_1 - u_2)$$

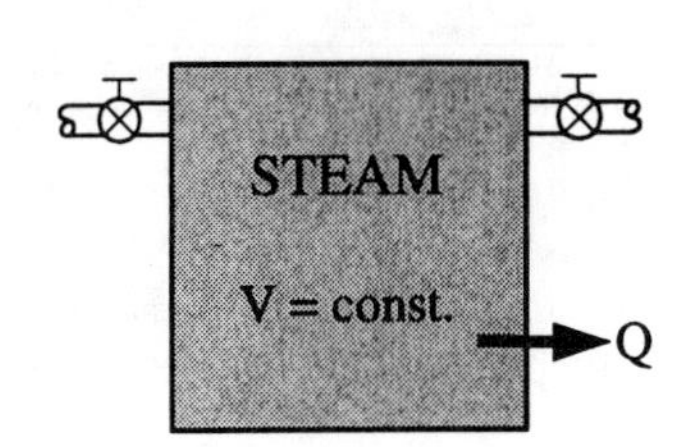

Using data from the steam tables (Tables A-4 through A-6), some properties are determined to be

$$\left.\begin{array}{l} P_1 = 300\text{kPa} \\ T_1 = 250^\circ\text{C} \end{array}\right\} \begin{array}{l} v_1 = 0.7964\text{m}^3/\text{kg} \\ u_1 = 2728.7\text{kJ/kg} \end{array}$$

$$P_2 = 100\text{kPa} \rightarrow \begin{array}{ll} v_f = 0.001043, & v_g = 1.6940\text{m}^3/\text{kg} \\ u_f = 417.36, & u_{fg} = 2088.7\text{kJ/kg} \end{array}$$

Noting that $v_1 = v_2$ and $v_f < v_2 < v_g$, the mass and the final internal energy becomes

$$m = \frac{V_1}{v_1} = \frac{0.020 \text{ m}^3}{0.7964 \text{ m}^3/\text{kg}} = 0.0251 \text{ kg}$$

$$x_2 = \frac{v_2 - v_f}{v_{fg}} = \frac{0.7964 - 0.001043}{1.6940 - 0.001043} = 0.470$$

$$u_2 = u_f + x_2 u_{fg} = 417.36 + (0.470 \times 2088.7) = 1399.0 \text{ kJ/kg}$$

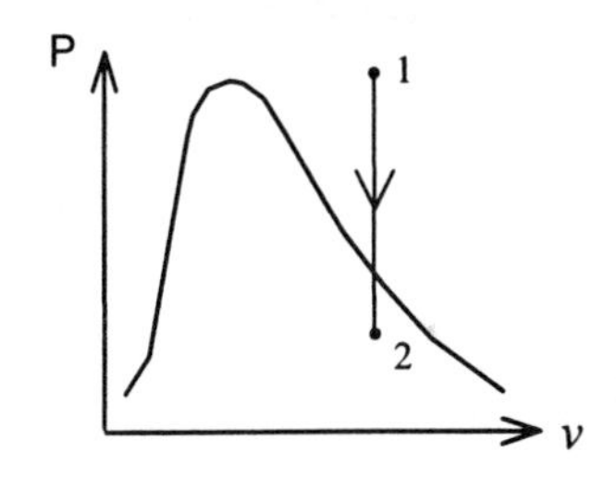

Substituting,

$$\begin{aligned} Q_{out} &= m(u_1 - u_2) \\ &= (0.0251 \text{ kg})(2728.7 - 1399.0) \text{ kJ/kg} \\ &= \mathbf{33.4 \text{ kJ}} \end{aligned}$$

4-12 A rigid tank is initially filled with superheated R-134a. Heat is transferred to the tank until the pressure inside rises to a specified value. The mass of the refrigerant and the amount of heat transfer are to be determined, and the process is to be shown on a *P-v* diagram. √

Assumptions **1** The tank is stationary and thus the kinetic and potential energy changes are zero. **2** There are no work interactions.

Analysis (*a*) We take the tank as the system. This is a closed system since no mass enters or leaves. Noting that the volume of the system is constant and thus there is no boundary work, the energy balance for this stationary closed system can be expressed as

$$\underbrace{E_{in} - E_{out}}_{\text{Net energy transfer by heat, work, and mass}} = \underbrace{\Delta E_{system}}_{\text{Change in internal, kinetic, potential, etc. energies}}$$

$$Q_{in} = \Delta U = m(u_2 - u_1) \quad (\text{since } W = \text{KE} = \text{PE} = 0)$$

R-134a
200 kPa

Using data from the refrigerant tables (Tables A-11 through A-13), the properties of R-134a are determined to be

$$\left.\begin{array}{l} P_1 = 200\text{kPa} \\ x_1 = 0.4 \end{array}\right\} \begin{array}{ll} v_f = 0.0007532, & v_g = 0.0993\text{m}^3/\text{kg} \\ u_f = 36.69, & u_g = 221.43\text{kJ/kg} \end{array}$$

$$v_1 = v_f + x_1 v_{fg} = 0.0007532 + [0.4 \times (0.0993 - 0.0007532)] = 0.04017\text{m}^3/\text{kg}$$

$$u_1 = u_f + x_1 u_{fg} = 36.69 + [0.4 \times (221.43 - 36.69)] = 110.59\text{kJ/kg}$$

$$\left.\begin{array}{l} P_2 = 800\text{kPa} \\ (v_2 = v_1) \end{array}\right\} u_2 = 349.82\text{kJ/kg (Superheated vapor)}$$

Then the mass of the refrigerant is determined to be

$$m = \frac{V_1}{v_1} = \frac{0.5\text{m}^3}{0.04017\text{m}^3/\text{kg}} = \mathbf{12.45kg}$$

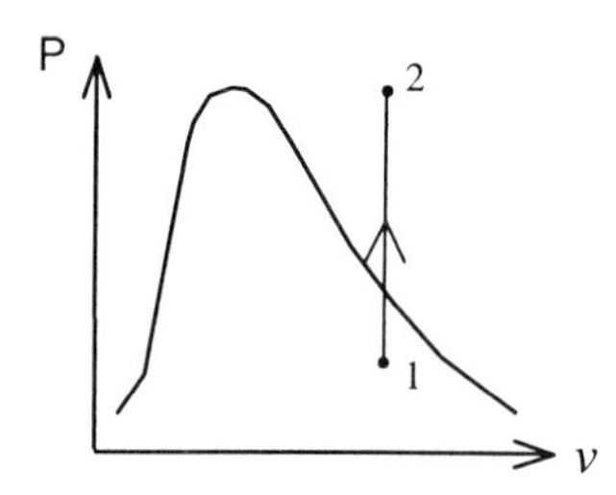

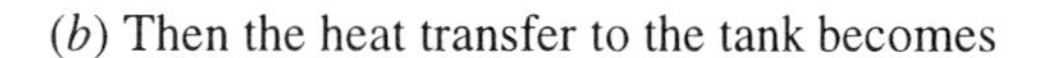

(*b*) Then the heat transfer to the tank becomes

$$\begin{aligned} Q_{in} &= m(u_2 - u_1) \\ &= (12.45\text{ kg})(349.82 - 110.59)\text{ kJ/kg} \\ &= \mathbf{2978\ kJ} \end{aligned}$$

4-13E A rigid tank is initially filled with saturated R-134a vapor. Heat is transferred from the refrigerant until the pressure inside drops to a specified value. The final temperature, the mass of the refrigerant that has condensed, and the amount of heat transfer are to be determined. Also, the process is to be shown on a *P-v* diagram.

Assumptions **1** The tank is stationary and thus the kinetic and potential energy changes are zero. **2** There are no work interactions.

Analysis (*a*) We take the tank as the system. This is a closed system since no mass enters or leaves. Noting that the volume of the system is constant and thus there is no boundary work, the energy balance for this stationary closed system can be expressed as

$$\underbrace{E_{in} - E_{out}}_{\substack{\text{Net energy transfer}\\ \text{by heat, work, and mass}}} = \underbrace{\Delta E_{system}}_{\substack{\text{Change in internal, kinetic,}\\ \text{potential, etc. energies}}}$$

$$-Q_{out} = \Delta U = m(u_2 - u_1) \quad (\text{since } W = \text{KE} = \text{PE} = 0)$$

$$Q_{out} = m(u_1 - u_2)$$

Using data from the refrigerant tables (Tables A-11 through A-13), the properties of R-134a are determined to be

$$\left.\begin{array}{l} P_1 = 120\text{psia} \\ \text{sat. vapor} \end{array}\right\} \begin{array}{l} v_1 = v_{g@120\text{psia}} = 0.3941\text{ft}^3/\text{lbm} \\ u_1 = u_{g@120\text{psia}} = 105.06\text{Btu/lbm} \end{array}$$

$$\left.\begin{array}{l} P_2 = 30\text{psia} \\ (v_2 = v_1) \end{array}\right\} \begin{array}{ll} v_f = 0.01209, & v_g = 1.5408\text{ft}^3/\text{lbm} \\ u_f = 16.24, & u_g = 95.40\text{Btu/lbm} \end{array}$$

R-134a
120 psia
Sat. vapor

The final state is saturated mixture. Thus,

$$T_2 = T_{sat\ @\ 30\ psia} = \mathbf{15.38\ °F}$$

(*b*) The total mass and the amount of refrigerant that has condensed are

$$m = \frac{V_1}{v_1} = \frac{20\text{ ft}^3}{0.3941\text{ ft}^3/\text{lbm}} = 50.8\text{ lbm}$$

$$x_2 = \frac{v_2 - v_f}{v_{fg}} = \frac{0.3941 - 0.01209}{1.5408 - 0.01209} = 0.250$$

$$m_f = (1 - x_2)m = (1 - 0.250)(50.8\text{ lbm}) = \mathbf{38.1\ lbm}$$

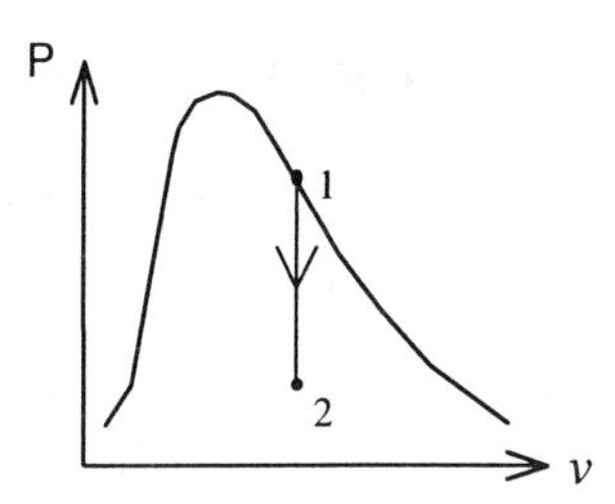

Also,

$$u_2 = u_f + x_2 u_{fg} = 16.24 + [0.250 \times (95.40 - 16.24)] = 36.03\text{ Btu/lbm}$$

(*c*) Substituting,

$$\begin{aligned} Q_{out} &= m(u_1 - u_2) \\ &= (50.8\text{ lbm})(105.06 - 36.03)\text{ Btu/lbm} \\ &= \mathbf{3507\ Btu} \end{aligned}$$

4-14 An insulated rigid tank is initially filled with a saturated liquid-vapor mixture of water. An electric heater in the tank is turned on, and the entire liquid in the tank is vaporized. The length of time the heater was kept on is to be determined, and the process is to be shown on a P-v diagram.

Assumptions **1** The tank is stationary and thus the kinetic and potential energy changes are zero. **2** The device is well-insulated and thus heat transfer is negligible. **3** The energy stored in the resistance wires, and the heat transferred to the tank itself is negligible.

Analysis We take the contents of the tank as the system. This is a closed system since no mass enters or leaves. Noting that the volume of the system is constant and thus there is no boundary work, the energy balance for this stationary closed system can be expressed as

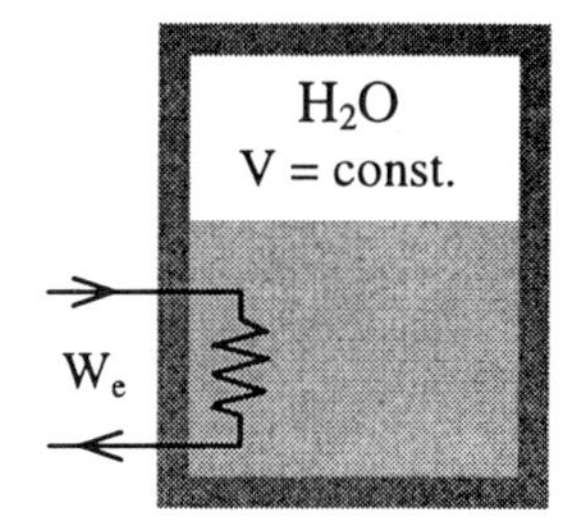

$$\underbrace{E_{in} - E_{out}}_{\text{Net energy transfer by heat, work, and mass}} = \underbrace{\Delta E_{system}}_{\text{Change in internal, kinetic, potential, etc. energies}}$$

$$W_{e,in} = \Delta U = m(u_2 - u_1) \quad (\text{since } Q = \text{KE} = \text{PE} = 0)$$

$$VI\Delta t = m(u_2 - u_1)$$

The properties of water are (Tables A-4 through A-6)

$$\left.\begin{array}{l} P_1 = 100\text{kPa} \\ x_1 = 0.25 \end{array}\right\} \begin{array}{l} v_f = 0.001043, \quad v_g = 1.6940\text{m}^3/\text{kg} \\ u_f = 417.36, \quad u_{fg} = 2088.7\text{kJ/kg} \end{array}$$

$$v_1 = v_f + x_1 v_{fg} = 0.001043 + [0.25 \times (1.6940 - 0.001043)] = 0.42428\text{m}^3/\text{kg}$$

$$u_1 = u_f + x_1 u_{fg} = 417.36 + (0.25 \times 2088.7) = 939.5\text{kJ/kg}$$

$$\left.\begin{array}{l} v_2 = v_1 = 0.42428\text{m}^3/\text{kg} \\ \text{sat.vapor} \end{array}\right\} u_2 = u_{g@0.42428\text{m}^3/\text{kg}} = 2556.7\text{kJ/kg}$$

Substituting,

$$(110\text{V})(8\text{A})\Delta t = (5\text{kg})(2556.7 - 939.5)\ \text{kJ/kg}\left(\frac{1000\text{VA}}{1\text{kJ/s}}\right)$$

$$\Delta t = 9189\text{s} \cong \mathbf{153.2min}$$

4-15 EES solution of this (and other comprehensive problems designated with the *computer icon*) is available to instructors at the *Instructor Manual* section of the *Online Learning Center* (OLC) at www.mhhe.com/cengel-boles. See the Preface for access information.

4-16 One part of an insulated tank contains compressed liquid while the other side is evacuated. The partition is then removed, and water is allowed to expand into the entire tank. The final temperature and the volume of the tank are to be determined.

Assumptions **1** The tank is stationary and thus the kinetic and potential energy changes are zero. **2** The tank is insulated and thus heat transfer is negligible. **3** There are no work interactions.

Analysis We take the entire contents of the tank as the system. This is a closed system since no mass enters or leaves. Noting that the volume of the system is constant and thus there is no boundary work, the energy balance for this stationary closed system can be expressed as

$$\underbrace{E_{in} - E_{out}}_{\substack{\text{Net energy transfer}\\ \text{by heat, work, and mass}}} = \underbrace{\Delta E_{system}}_{\substack{\text{Change in internal, kinetic,}\\ \text{potential, etc. energies}}}$$

$$0 = \Delta U = m(u_2 - u_1) \quad (\text{since } W = Q = \text{KE} = \text{PE} = 0)$$

$$u_1 = u_2$$

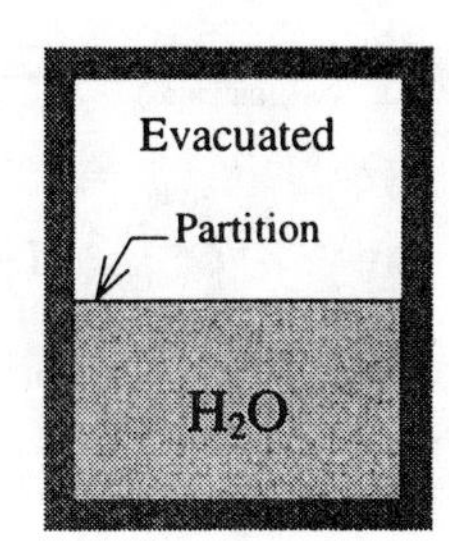

The properties of water are (Tables A-4 through A-6)

$$\left.\begin{array}{l} P_1 = 600\text{kPa} \\ T_1 = 60^\circ\text{C} \end{array}\right\} \begin{array}{l} v_1 \cong v_{f@60^\circ\text{C}} = 0.001017\text{m}^3/\text{kg} \\ u_1 \cong u_{f@60^\circ\text{C}} = 251.11\text{kJ/kg} \end{array}$$

We now assume the final state in the tank is saturated liquid-vapor mixture and determine quality. This assumption will be verified if we get a quality between 0 and 1.

$$\left.\begin{array}{l} P_2 = 10\text{kPa} \\ (u_2 = u_1) \end{array}\right\} \begin{array}{ll} v_f = 0.00101, & v_g = 14.67\ \text{m}^3/\text{kg} \\ u_f = 191.82, & u_{fg} = 2246.1\ \text{kJ/kg} \end{array}$$

$$x_2 = \frac{u_2 - u_f}{u_{fg}} = \frac{251.11 - 191.82}{2246.1} = 0.0264$$

Thus,

$$T_2 = = T_{sat\ @\ 10\ kPa} = \mathbf{45.81\ ^\circ C}$$

$$v_2 = v_f + x_2 v_{fg} = 0.0010 + [0.0264 \times (14.67 - 0.00101)] = 0.388\text{m}^3/\text{kg}$$

and,

$$V = m v_2 = (2.5\ \text{kg})(0.388\ \text{m}^3/\text{kg}) = \mathbf{0.97\ m^3}$$

4-17 EES solution of this (and other comprehensive problems designated with the *computer icon*) is available to instructors at the *Instructor Manual* section of the *Online Learning Center* (OLC) at www.mhhe.com/cengel-boles. See the Preface for access information.

4-18 A cylinder is initially filled with R-134a at a specified state. The refrigerant is cooled at constant pressure. The amount of heat loss is to be determined, and the process is to be shown on a *T-v* diagram.

Assumptions **1** The cylinder is stationary and thus the kinetic and potential energy changes are zero. **2** There are no work interactions involved other than the boundary work. **3** The thermal energy stored in the cylinder itself is negligible. **4** The compression or expansion process is quasi-equilibrium.

Analysis We take the contents of the cylinder as the system. This is a closed system since no mass enters or leaves. The energy balance for this stationary closed system can be expressed as

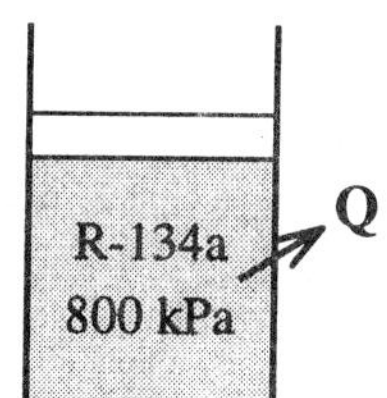

$$\underbrace{E_{in} - E_{out}}_{\substack{\text{Net energy transfer} \\ \text{by heat, work, and mass}}} = \underbrace{\Delta E_{system}}_{\substack{\text{Change in internal, kinetic,} \\ \text{potential, etc. energies}}}$$

$$-Q_{out} - W_{b,out} = \Delta U = m(u_2 - u_1) \quad \text{(since KE = PE = 0)}$$

$$-Q_{out} = m(h_2 - h_1)$$

since $\Delta U + W_b = \Delta H$ during a constant pressure quasi-equilibrium process. The properties of R-134a are (Tables A-11 through A-13)

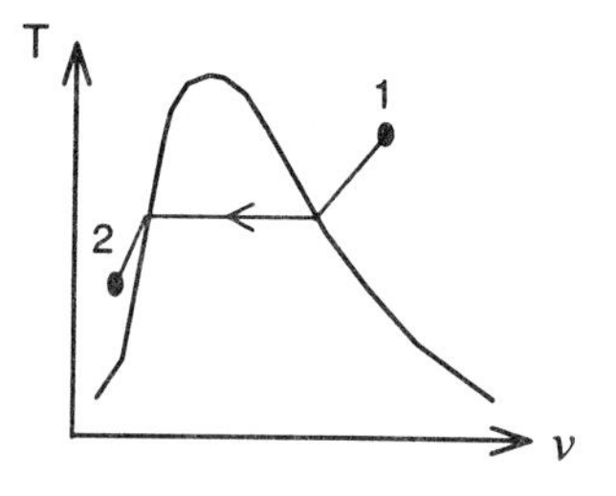

$$\left.\begin{array}{l} P_1 = 800\text{kPa} \\ T_1 = 60^\circ\text{C} \end{array}\right\} h_1 = 294.98\text{kJ/kg}$$

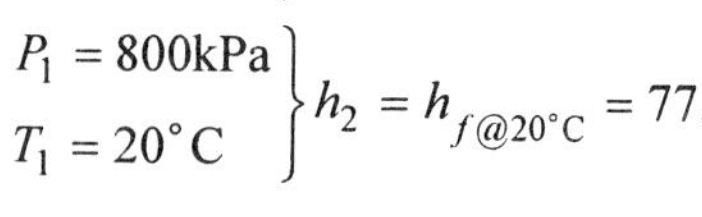

$$\left.\begin{array}{l} P_1 = 800\text{kPa} \\ T_1 = 20^\circ\text{C} \end{array}\right\} h_2 = h_{f@20^\circ\text{C}} = 77.26\text{kJ/kg}$$

Substituting,

$$Q_{out} = -\text{ (5 kg)(77.26 - 294.98) kJ/kg} = \mathbf{1089\ kJ}$$

4-19E A cylinder contains water initially at a specified state. The water is heated at constant pressure. The final temperature of the water is to be determined, and the process is to be shown on a *T-v* diagram.

Assumptions **1** The cylinder is stationary and thus the kinetic and potential energy changes are zero. **2** The thermal energy stored in the cylinder itself is negligible. **3** The compression or expansion process is quasi-equilibrium.

Analysis We take the contents of the cylinder as the system. This is a closed system since no mass enters or leaves. The energy balance for this stationary closed system can be expressed as

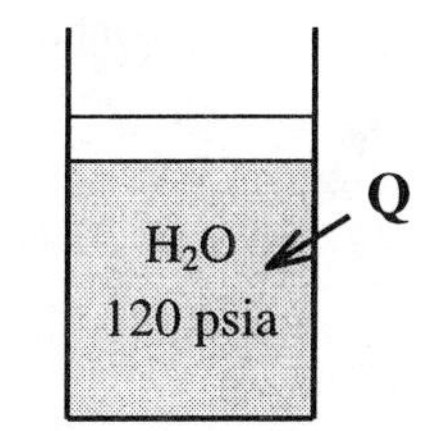

$$\underbrace{E_{\text{in}} - E_{\text{out}}}_{\substack{\text{Net energy transfer} \\ \text{by heat, work, and mass}}} = \underbrace{\Delta E_{\text{system}}}_{\substack{\text{Change in internal, kinetic,} \\ \text{potential, etc. energies}}}$$

$$Q_{\text{in}} - W_{b,\text{out}} = \Delta U = m(u_2 - u_1) \quad \text{(since KE = PE = 0)}$$

$$Q_{\text{in}} = m(h_2 - h_1)$$

since $\Delta U + W_b = \Delta H$ during a constant pressure quasi-equilibrium process. The properties of water are (Tables A-11E through A-13E)

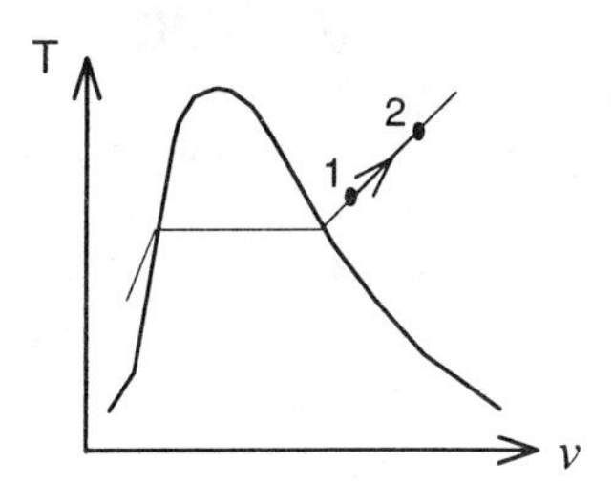

$$v_1 = \frac{V_1}{m} = \frac{2\text{ft}^3}{0.5\text{lbm}} = 4\text{ft}^3/\text{lbm}$$

$$\left.\begin{array}{l} P_1 = 120\text{psia} \\ v_1 = 4\text{ft}^3/\text{lbm} \end{array}\right\} h_1 = 1216.9\text{Btu/lbm}$$

Substituting,

$$200\text{ Btu} = (0.5\text{lbm})(h_2 - 1216.9)\text{Btu/lbm}$$

$$h_2 = 1616.9\text{ Btu/lbm}$$

Then,

$$\left.\begin{array}{l} P_2 = 120\text{ psia} \\ h_2 = 1616.9\text{ Btu/lbm} \end{array}\right\} T_2 = \mathbf{1161.4\,{}^\circ F}$$

4-20 A cylinder is initially filled with saturated liquid water at a specified pressure. The water is heated electrically as it is stirred by a paddle-wheel at constant pressure. The voltage of the current source is to be determined, and the process is to be shown on a *P-v* diagram.

Assumptions **1** The cylinder is stationary and thus the kinetic and potential energy changes are zero. **2** The cylinder is well-insulated and thus heat transfer is negligible. **3** The thermal energy stored in the cylinder itself is negligible. **4** The compression or expansion process is quasi-equilibrium.

Analysis We take the contents of the cylinder as the system. This is a closed system since no mass enters or leaves. The energy balance for this stationary closed system can be expressed as

$$\underbrace{E_{in} - E_{out}}_{\substack{\text{Net energy transfer}\\ \text{by heat, work, and mass}}} = \underbrace{\Delta E_{system}}_{\substack{\text{Change in internal, kinetic,}\\ \text{potential, etc. energies}}}$$

$$W_{e,in} + W_{pw,in} - W_{b,out} = \Delta U \quad (\text{since } Q = \text{KE} = \text{PE} = 0)$$

$$W_{e,in} + W_{pw,in} = m(h_2 - h_1)$$

$$(VI\Delta t) + W_{pw,in} = m(h_2 - h_1)$$

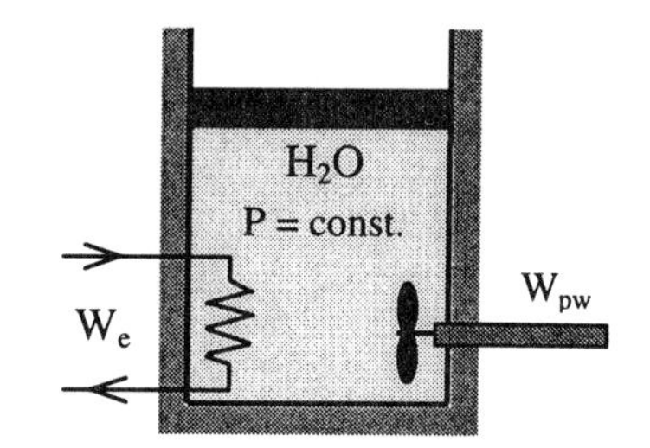

since $\Delta U + W_b = \Delta H$ during a constant pressure quasi-equilibrium process. The properties of water are (Tables A-4 through A-6)

$$\left.\begin{array}{l} P_1 = 150\text{kPa} \\ \text{sat.liquid} \end{array}\right\} \begin{array}{l} h_1 = h_{f@150\text{kPa}} = 467.11\text{kJ/kg} \\ v_1 = v_{f@150\text{kPa}} = 0.0010528\text{m}^3/\text{kg} \end{array}$$

$$\left.\begin{array}{l} P_2 = 150\text{kPa} \\ x_2 = 0.5 \end{array}\right\} h_2 = h_f + x_2 h_{fg} = 467.11 + (0.5 \times 2226.5) = 1580.36\text{kJ/kg}$$

$$m = \frac{V_1}{v_1} = \frac{0.005\text{m}^3}{0.0010528\text{m}^3/\text{kg}} = 4.75\text{kg}$$

Substituting,

$$VI\Delta t + (300\text{kJ}) = (4.75\text{kg})(1580.36 - 467.11)\text{kJ/kg}$$

$$VI\Delta t = 4988\text{kJ}$$

$$V = \frac{4988\text{kJ}}{(8\text{A})(45 \times 60\text{s})}\left(\frac{1000\text{VA}}{1\text{kJ/s}}\right) = \mathbf{230.9V}$$

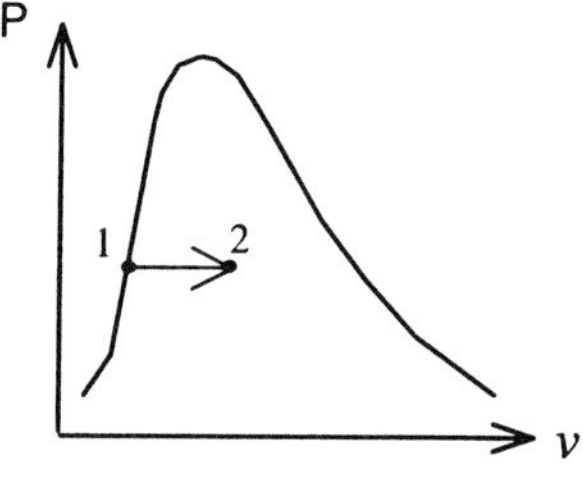

4-21 A cylinder is initially filled with steam at a specified state. The steam is cooled at constant pressure. The mass of the steam, the final temperature, and the amount of heat transfer are to be determined, and the process is to be shown on a *T-v* diagram.

Assumptions **1** The cylinder is stationary and thus the kinetic and potential energy changes are zero. **2** There are no work interactions involved other than the boundary work. **3** The thermal energy stored in the cylinder itself is negligible. **4** The compression or expansion process is quasi-equilibrium.

Analysis We take the contents of the cylinder as the system. This is a closed system since no mass enters or leaves. The energy balance for this stationary closed system can be expressed as

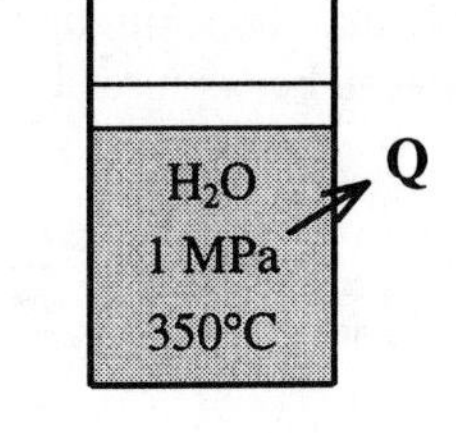

$$\underbrace{E_{in} - E_{out}}_{\substack{\text{Net energy transfer} \\ \text{by heat, work, and mass}}} = \underbrace{\Delta E_{system}}_{\substack{\text{Change in internal, kinetic,} \\ \text{potential, etc. energies}}}$$

$$-Q_{out} - W_{b,out} = \Delta U = m(u_2 - u_1) \quad \text{(since KE = PE = 0)}$$

$$-Q_{out} = m(h_2 - h_1)$$

since $\Delta U + W_b = \Delta H$ during a constant pressure quasi-equilibrium process. The properties of water are (Tables A-4 through A-6)

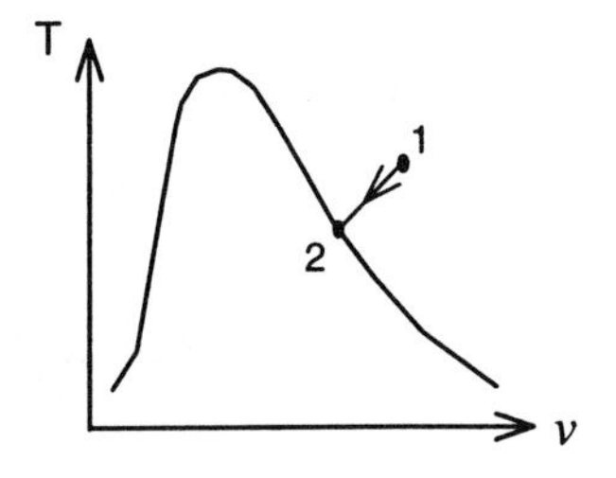

$$\left.\begin{array}{l} P_1 = 1\text{MPa} \\ T_2 = 350\,^\circ\text{C} \end{array}\right\} \begin{array}{l} v_1 = 0.2825\text{m}^3/\text{kg} \\ h_1 = 3157.7\text{kJ/kg} \end{array}$$

$$m = \frac{V_1}{v_1} = \frac{1.5\text{m}^3}{0.2825\text{m}^3/\text{kg}} = \mathbf{5.31kg}$$

(*b*) The final temperature is determined from

$$\left.\begin{array}{l} P_2 = 1\text{MPa} \\ sat.vapor \end{array}\right\} \begin{array}{l} T_2 = T_{sat@1\text{MPa}} = \mathbf{179.91\,^\circ C} \\ h_2 = h_{g@1\text{MPa}} = 2778.1\text{kJ/kg} \end{array}$$

(*c*) Substituting, the energy balance gives

$$Q_{out} = \text{-}\ (5.31\text{ kg})(2778.1 - 3157.7)\text{ kJ/kg} = \mathbf{2016\ kJ}$$

4-22 [*Also solved by EES on enclosed CD*] A cylinder equipped with an external spring is initially filled with steam at a specified state. Heat is transferred to the steam, and both the temperature and pressure rise. The final temperature, the boundary work done by the steam, and the amount of heat transfer are to be determined, and the process is to be shown on a *P-v* diagram.

Assumptions **1** The cylinder is stationary and thus the kinetic and potential energy changes are zero. **2** The thermal energy stored in the cylinder itself is negligible. **3** The compression or expansion process is quasi-equilibrium. **4** The spring is a linear spring.

Analysis We take the contents of the cylinder as the system. This is a closed system since no mass enters or leaves. Noting that the spring is not part of the system (it is external), the energy balance for this stationary closed system can be expressed as

H_2O
200 kPa
200°C
Q

$$\underbrace{E_{in} - E_{out}}_{\substack{\text{Net energy transfer}\\ \text{by heat, work, and mass}}} = \underbrace{\Delta E_{system}}_{\substack{\text{Change in internal, kinetic,}\\ \text{potential, etc. energies}}}$$

$$Q_{in} - W_{b,out} = \Delta U = m(u_2 - u_1) \quad \text{(since KE = PE = 0)}$$

$$Q_{in} = m(u_2 - u_1) + W_{b,out}$$

The properties of steam are (Tables A-4 through A-6)

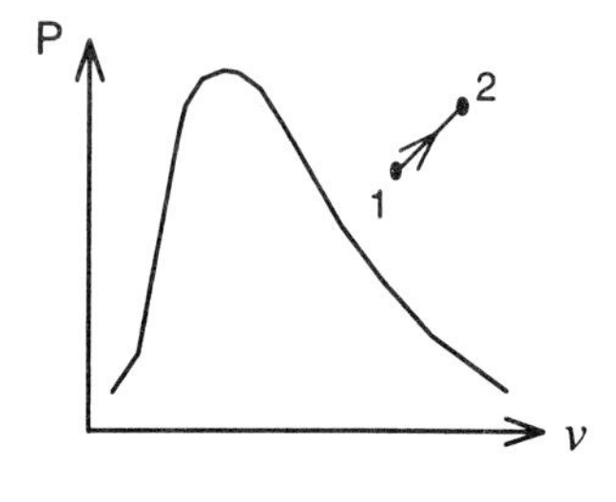

$$\left.\begin{array}{l} P_1 = 200\text{kPa} \\ T_1 = 200^\circ\text{C} \end{array}\right\} \begin{array}{l} v_1 = 1.0803\text{m}^3/\text{kg} \\ u_1 = 2654.4\text{kJ/kg} \end{array}$$

$$m = \frac{V_1}{v_1} = \frac{0.5\text{m}^3}{1.0803\text{m}^3/\text{kg}} = 0.463\text{kg}$$

$$v_2 = \frac{V_2}{m} = \frac{0.6\text{m}^3}{0.463\text{kg}} = 1.296\text{m}^3/\text{kg}$$

$$\left.\begin{array}{l} P_2 = 500\text{kPa} \\ v_2 = 1.296\text{m}^3/\text{kg} \end{array}\right\} \begin{array}{l} T_2 = \mathbf{1131\,^\circ C} \\ u_2 = 4321.9\text{kJ/kg} \end{array}$$

(*b*) The pressure of the gas changes linearly with volume, and thus the process curve on a P-V diagram will be a straight line. The boundary work during this process is simply the area under the process curve, which is a trapezoidal. Thus,

$$W_b = Area = \frac{P_1 + P_2}{2}(V_2 - V_1) = \frac{(200 + 500)\text{kPa}}{2}(0.6 - 0.5)\text{m}^3\left(\frac{1\text{kJ}}{1\text{kPa}\cdot\text{m}^3}\right) = \mathbf{35kJ}$$

(*c*) From the energy balance we have

$$Q_{in} = (0.463\text{ kg})(4321.9 - 2654.4)\text{kJ/kg} + 35\text{ kJ} = \mathbf{807\ kJ}$$

4-23 EES solution of this (and other comprehensive problems designated with the *computer icon*) is available to instructors at the *Instructor Manual* section of the *Online Learning Center* (OLC) at www.mhhe.com/cengel-boles. See the Preface for access information.

4-24 A cylinder equipped with a set of stops for the piston to rest on is initially filled with saturated water vapor at a specified pressure. Heat is transferred to water until the volume doubles. The final temperature, the boundary work done by the steam, and the amount of heat transfer are to be determined, and the process is to be shown on a *P-v* diagram.

Assumptions **1** The cylinder is stationary and thus the kinetic and potential energy changes are zero. **2** There are no work interactions involved other than the boundary work. **3** The thermal energy stored in the cylinder itself is negligible. **4** The compression or expansion process is quasi-equilibrium.

Analysis We take the contents of the cylinder as the system. This is a closed system since no mass enters or leaves. The energy balance for this stationary closed system can be expressed as

$$\underbrace{E_{in} - E_{out}}_{\substack{\text{Net energy transfer}\\ \text{by heat, work, and mass}}} = \underbrace{\Delta E_{system}}_{\substack{\text{Change in internal, kinetic,}\\ \text{potential, etc. energies}}}$$

$$Q_{in} - W_{b,out} = \Delta U = m(u_3 - u_1) \quad \text{(since KE = PE = 0)}$$

$$Q_{in} = m(u_3 - u_1) + W_{b,out}$$

300 kPa

H_2O
200 kPa
Sat. Vapor

The properties of steam are (Tables A-4 through A-6)

$$\left.\begin{array}{l} P_1 = 200\text{kPa} \\ sat.vapor \end{array}\right\} \begin{array}{l} v_1 = v_{g@200\text{kPa}} = 0.8857\text{m}^3/\text{kg} \\ u_1 = u_{g@200\text{kPa}} = 2529.5\text{kJ/kg} \end{array}$$

$$m = \frac{V_1}{v_1} = \frac{0.5\text{m}^3}{0.8857\text{m}^3/\text{kg}} = 0.5645\text{kg}$$

$$v_3 = \frac{V_3}{m} = \frac{1\text{m}^3}{0.5645\text{kg}} = 1.7715\text{m}^3/\text{kg}$$

$$\left.\begin{array}{l} P_3 = 300\text{kPa} \\ v_3 = 1.7715\text{m}^3/\text{kg} \end{array}\right\} \begin{array}{l} T_3 = \mathbf{878.9\,^\circ C} \\ u_3 = 3813.8\text{kJ/kg} \end{array}$$

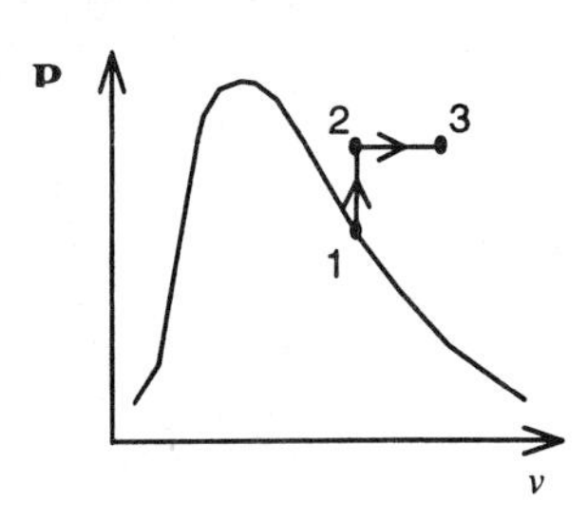

(*b*) The work done during process 1-2 is zero (since V = const) and the work done during the constant pressure process 2-3 is

$$W_{b,out} = \int_2^3 P dV = P(V_3 - V_2) = (300\text{ kPa})(1.0 - 0.5)\text{m}^3 \left(\frac{1\text{ kJ}}{1\text{ kPa}\cdot\text{m}^3}\right) = \mathbf{150\ kJ}$$

(*c*) Heat transfer is determined from the energy balance,

$$Q_{in} = m(u_3 - u_1) + W_{b,out}$$
$$= (0.5645\text{ kg})(3813.8 - 2529.5)\text{ kJ/kg} + 150\text{ kJ} = \mathbf{875.0\ kJ}$$

Closed System Energy Analysis: Ideal Gases

4-25C No, it isn't. This is because the first law relation $Q - W = \Delta U$ reduces to $W = 0$ in this case since the system is adiabatic ($Q = 0$) and $\Delta U = 0$ for the isothermal processes of ideal gases. Therefore, this adiabatic system cannot receive any net work at constant temperature.

4-26E The air in a rigid tank is heated until its pressure doubles. The volume of the tank and the amount of heat transfer are to be determined.

Assumptions **1** Air is an ideal gas since it is at a high temperature and low pressure relative to its critical point values of -141°C and 3.77 MPa. **2** The kinetic and potential energy changes are negligible, $\Delta pe \cong \Delta ke \cong 0$. **3** Constant specific heats at room temperature can be used for air. This assumption results in negligible error in heating and air-conditioning applications.

Properties The gas constant of air is R = 0.3704 psia.ft^3/lbm.R (Table A-1E).

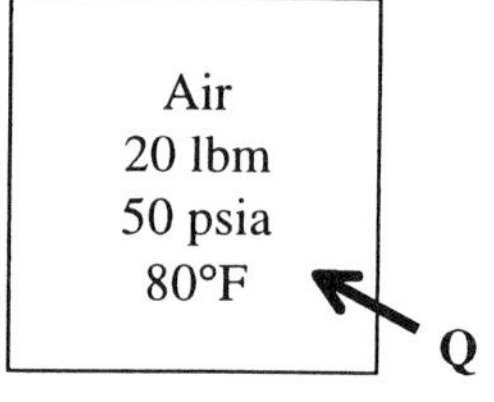

Analysis (*a*) The volume of the tank can be determined from the ideal gas relation,

$$V = \frac{mRT_1}{P_1} = \frac{(20\text{ lbm})(0.3704\text{ psia}\cdot\text{ft}^3/\text{lbm}\cdot\text{R})(540\text{ R})}{50\text{ psia}} = \mathbf{80.0\ ft^3}$$

(b) We take the air in the tank as our system. The energy balance for this stationary closed system can be expressed as

$$\underbrace{E_{in} - E_{out}}_{\substack{\text{Net energy transfer}\\ \text{by heat, work, and mass}}} = \underbrace{\Delta E_{system}}_{\substack{\text{Change in internal, kinetic,}\\ \text{potential, etc. energies}}}$$

$$Q_{in} = \Delta U$$

$$Q_{in} = m(u_2 - u_1) \cong mC_v(T_2 - T_1)$$

The final temperature of air is

$$\frac{P_1 V}{T_1} = \frac{P_2 V}{T_2} \longrightarrow T_2 = \frac{P_2}{P_1}T_1 = 2 \times (540\text{ R}) = 1080\text{ R}$$

The internal energies are (Table A-17E)

$$u_1 = u_{@\,540\text{ R}} = 92.04\text{ Btu / lbm}$$

$$u_2 = u_{@\,1080\text{ R}} = 186.93\text{ Btu / lbm}$$

Substituting,

$$Q_{in} = (20\text{ lbm})(186.93 - 92.04)\text{Btu/lbm} = \mathbf{1898\ Btu}$$

Alternative solutions The specific heat of air at the average temperature of T_{ave} = (540+1080)/2= 810 R = 350°F is, from Table A-2Eb, $C_{v,ave}$ = 0.175 Btu/lbm.R. Substituting,

$$Q_{in} = (20\text{ lbm})(0.175\text{ Btu/lbm.R})(1080 - 540)\text{ R} = \mathbf{1890\ Btu}$$

Discussion Both approaches resulted in almost the same solution in this case.

4-27 The hydrogen gas in a rigid tank is cooled until its temperature drops to 300 K. The final pressure in the tank and the amount of heat transfer are to be determined.

Assumptions **1** Hydrogen is an ideal gas since it is at a high temperature and low pressure relative to its critical point values of -240°C and 1.30 MPa. **2** The tank is stationary, and thus the kinetic and potential energy changes are negligible, $\Delta ke \cong \Delta pe \cong 0$.

Properties The gas constant of hydrogen is R = 4.124 kPa.m^3/kg.K (Table A-1). The constant volume specific heat of hydrogen at the average temperature of 400 K is , $C_{v,ave}$ = 10.352 kJ/kg.K (Table A-2).

Analysis (*a*) The final pressure of hydrogen can be determined from the ideal gas relation,

$$\frac{P_1 V}{T_1} = \frac{P_2 V}{T_2} \longrightarrow P_2 = \frac{T_2}{T_1} P_1 = \frac{300\text{ K}}{500\text{ K}}(250\text{ kPa}) = \mathbf{150\ kPa}$$

(*b*) We take the hydrogen in the tank as the system. This is a *closed system* since no mass enters or leaves. The energy balance for this stationary closed system can be expressed as

$$\underbrace{E_{in} - E_{out}}_{\substack{\text{Net energy transfer}\\ \text{by heat, work, and mass}}} = \underbrace{\Delta E_{system}}_{\substack{\text{Change in internal, kinetic,}\\ \text{potential, etc. energies}}}$$

$$-Q_{out} = \Delta U$$

$$Q_{out} = -\Delta U = -m(u_2 - u_1) \cong mC_v(T_1 - T_2)$$

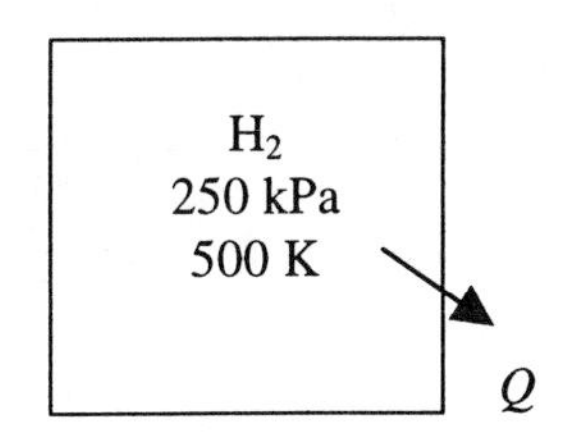

where

$$m = \frac{P_1 V}{RT_1} = \frac{(250\text{ kPa})(3.0\text{ m}^3)}{(4.124\text{ kPa}\cdot\text{m}^3/\text{kg}\cdot\text{K})(500\text{ K})} = 0.3637\text{ kg}$$

Substituting into the energy balance,

$$Q_{out} = (0.3637\text{ kg})(10.352\text{ kJ/kg}\cdot\text{ K})(500 - 300)\text{K} = \mathbf{753.0\ kJ}$$

4-28 A resistance heater is to raise the air temperature in the room from 7 to 23°C within 20 min. The required power rating of the resistance heater is to be determined. √

Assumptions **1** Air is an ideal gas since it is at a high temperature and low pressure relative to its critical point values of -141°C and 3.77 MPa. **2** The kinetic and potential energy changes are negligible, $\Delta ke \cong \Delta pe \cong 0$. **3** Constant specific heats at room temperature can be used for air. This assumption results in negligible error in heating and air-conditioning applications. **4** Heat losses from the room are negligible. **5** The room is air-tight so that no air leaks in and out during the process.

Properties The gas constant of air is R = 0.287 kPa.m^3/kg.K (Table A-1). Also, C_v = 0.718 kJ/kg.K for air at room temperature (Table A-2).

Analysis We take the air in the room to be the system. This is a closed system since no mass crosses the system boundary. The energy balance for this stationary constant-volume closed system can be expressed as

$$\underbrace{E_{\text{in}} - E_{\text{out}}}_{\substack{\text{Net energy transfer}\\ \text{by heat, work, and mass}}} = \underbrace{\Delta E_{\text{system}}}_{\substack{\text{Change in internal, kinetic,}\\ \text{potential, etc. energies}}}$$

$$W_{\text{e,in}} = \Delta U \cong mC_{v,ave}(T_2 - T_1) \text{ (since } Q = KE = PE = 0)$$

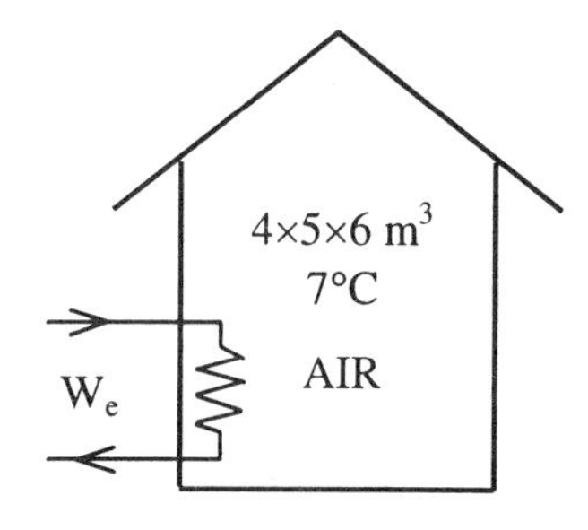

or,

$$\dot{W}_{e,in}\Delta t = mC_{v,\,ave}(T_2 - T_1)$$

The mass of air is

$$V = 4 \times 5 \times 6 = 120 \text{ m}^3$$

$$m = \frac{P_1V}{RT_1} = \frac{(100 \text{ kPa})(120 \text{ m}^3)}{(0.287 \text{ kPa}\cdot\text{m}^3/\text{kg}\cdot\text{K})(280 \text{ K})} = 149.3 \text{ kg}$$

Substituting, the power rating of the heater becomes

$$\dot{W}_{e,in} = \frac{(149.3 \text{ kg})(0.718 \text{ kJ/kg}\cdot^\circ\text{C})(23-7)^\circ\text{C}}{15 \times 60 \text{ s}} = \mathbf{1.91\ kW}$$

Discussion In practice, the pressure in the room will remain constant during this process rather than the volume, and some air will leak out as the air expands. As a result, the air in the room will undergo a constant pressure expansion process. Therefore, it is more proper to be conservative and to use ΔH instead of using ΔU in heating and air-conditioning applications.

4-29 A room is heated by a radiator, and the warm air is distributed by a fan. Heat is lost from the room. The time it takes for the air temperature to rise to 20°C is to be determined. √

Assumptions **1** Air is an ideal gas since it is at a high temperature and low pressure relative to its critical point values of -141°C and 3.77 MPa. **2** The kinetic and potential energy changes are negligible, $\Delta ke \cong \Delta pe \cong 0$. **3** Constant specific heats at room temperature can be used for air. This assumption results in negligible error in heating and air-conditioning applications. **4** The local atmospheric pressure is 100 kPa. **5** The room is air-tight so that no air leaks in and out during the process.

Properties The gas constant of air is R = 0.287 kPa.m^3/kg.K (Table A-1). Also, C_v = 0.718 kJ/kg.K for air at room temperature (Table A-2).

Analysis We take the air in the room to be the system. This is a closed system since no mass crosses the system boundary. The energy balance for this stationary constant-volume closed system can be expressed as

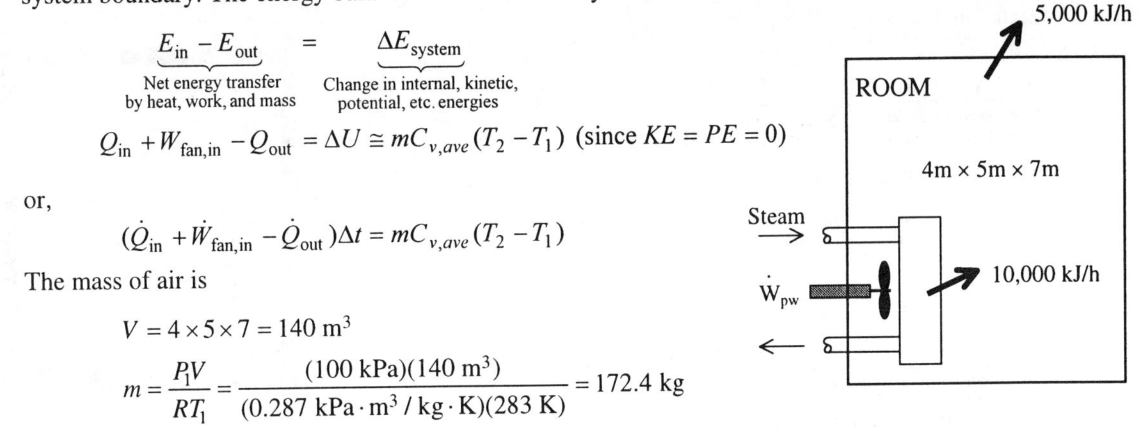

$$\underbrace{E_{in} - E_{out}}_{\substack{\text{Net energy transfer}\\ \text{by heat, work, and mass}}} = \underbrace{\Delta E_{system}}_{\substack{\text{Change in internal, kinetic,}\\ \text{potential, etc. energies}}}$$

$$Q_{in} + W_{fan,in} - Q_{out} = \Delta U \cong mC_{v,ave}(T_2 - T_1) \text{ (since } KE = PE = 0)$$

or,

$$(\dot{Q}_{in} + \dot{W}_{fan,in} - \dot{Q}_{out})\Delta t = mC_{v,ave}(T_2 - T_1)$$

The mass of air is

$$V = 4 \times 5 \times 7 = 140 \text{ m}^3$$

$$m = \frac{P_1 V}{RT_1} = \frac{(100 \text{ kPa})(140 \text{ m}^3)}{(0.287 \text{ kPa}\cdot\text{m}^3/\text{kg}\cdot\text{K})(283 \text{ K})} = 172.4 \text{ kg}$$

Using the C_v value at room temperature,

$$[(10{,}000 - 5{,}000)/3600 \text{ kJ/s} + 0.1 \text{ kJ/s}]\Delta t = (172.4\text{kg})(0.718 \text{ kJ/kg}\cdot^\circ\text{C})(20-10)^\circ\text{C}$$

It yields

$$\Delta t = \mathbf{831 \text{ s}}$$

Discussion In practice, the pressure in the room will remain constant during this process rather than the volume, and some air will leak out as the air expands. As a result, the air in the room will undergo a constant pressure expansion process. Therefore, it is more proper to be conservative and to using ΔH instead of use ΔU in heating and air-conditioning applications.

4-30 A student living in a room turns her 150-W fan on in the morning. The temperature in the room when she comes back 10 h later is to be determined. √

Assumptions **1** Air is an ideal gas since it is at a high temperature and low pressure relative to its critical point values of -141°C and 3.77 MPa. **2** The kinetic and potential energy changes are negligible, $\Delta ke \cong \Delta pe \cong 0$. **3** Constant specific heats at room temperature can be used for air. This assumption results in negligible error in heating and air-conditioning applications. **4** All the doors and windows are tightly closed, and heat transfer through the walls and the windows is disregarded.

Properties The gas constant of air is $R = 0.287$ kPa.m^3/kg.K (Table A-1). Also, $C_v = 0.718$ kJ/kg.K for air at room temperature (Table A-2).

Analysis We take the room as the system. This is a *closed system* since the doors and the windows are said to be tightly closed, and thus no mass crosses the system boundary during the process. The energy balance for this system can be expressed as

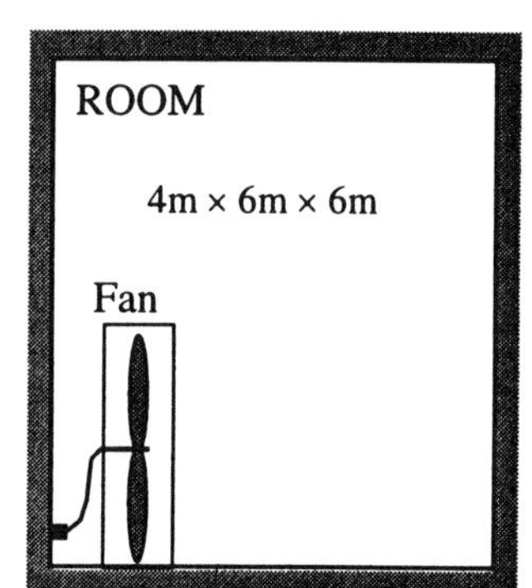

$$\underbrace{E_{in} - E_{out}}_{\substack{\text{Net energy transfer}\\ \text{by heat, work, and mass}}} = \underbrace{\Delta E_{\text{system}}}_{\substack{\text{Change in internal, kinetic,}\\ \text{potential, etc. energies}}}$$

$$W_{e,in} = \Delta U$$

$$W_{e,in} = m(u_2 - u_1) \cong mC_v(T_2 - T_1)$$

The mass of air is

$$V = 4 \times 6 \times 6 = 144 \text{ m}^3$$

$$m = \frac{P_1V}{RT_1} = \frac{(100 \text{ kPa})(144 \text{ m}^3)}{(0.287 \text{ kPa}\cdot\text{m}^3/\text{kg}\cdot\text{K})(288 \text{ K})} = 174.2 \text{ kg}$$

The electrical work done by the fan is

$$W_e = \dot{W}_e \Delta t = (0.15 \text{ kJ/s})(10 \times 3600 \text{ s}) = 5400 \text{ kJ}$$

Substituting and using the C_v value at room temperature,

$$5400 \text{ kJ} = (174.2 \text{ kg})(0.718 \text{ kJ/kg}\cdot°\text{C})(T_2 - 15)°\text{C}$$

$$T_2 = \mathbf{58.2°C}$$

4-31E A paddle wheel in an oxygen tank is rotated until the pressure inside rises to 20 psia while some heat is lost to the surroundings. The paddle wheel work done is to be determined.

Assumptions **1** Oxygen is an ideal gas since it is at a high temperature and low pressure relative to its critical point values of -181°F and 736 psia. **2** The kinetic and potential energy changes are negligible, $\Delta ke \cong \Delta pe \cong 0$. **3** The energy stored in the paddle wheel is negligible. **4** This is a rigid tank and thus its volume remains constant.

Properties The gas constant and molar mass of oxygen are R = 0.3353 psia.ft^3/lbm.R and M = 32 lbm/lbmol (Table A-1E). The specific heat of oxygen at the average temperature of T_{ave} = (735+540)/2= 638 R is $C_{v,ave}$ = 0.160 Btu/lbm.R (Table A-2E).

Analysis We take the oxygen in the tank as our system. This is a *closed system* since no mass enters or leaves. The energy balance for this system can be expressed as

$$\underbrace{E_{in} - E_{out}}_{\substack{\text{Net energy transfer}\\ \text{by heat, work, and mass}}} = \underbrace{\Delta E_{\text{system}}}_{\substack{\text{Change in internal, kinetic,}\\ \text{potential, etc. energies}}}$$

$$W_{pw,in} - Q_{out} = \Delta U$$

$$W_{pw,in} = Q_{out} + m(u_2 - u_1)$$

$$\cong Q_{out} + mC_v(T_2 - T_1)$$

O$_2$
14.7 psia
80°F
20 Btu

The final temperature and the mass of oxygen are

$$\frac{P_1 V}{T_1} = \frac{P_2 V}{T_2} \longrightarrow T_2 = \frac{P_2}{P_1} T_1 = \frac{20 \text{ psia}}{14.7 \text{ psia}}(540 \text{ R}) = 735 \text{ R}$$

$$m = \frac{P_1 V}{RT_1} = \frac{(14.7 \text{ psia})(10 \text{ ft}^3)}{(0.3353 \text{ psia} \cdot \text{ft}^3 / \text{lbmol} \cdot \text{R})(540 \text{ R})} = 0.812 \text{ lbm}$$

Substituting,

$W_{pw,in}$ = (20 Btu) + (0.812 lbm)(0.160 Btu/lbm.R)(735 - 540) R = **45.3 Btu**

Discussion Note that a fan actually causes the internal temperature of a confined space to rise. In fact, a 100-W fan supplies a room with as much energy as a 100-W resistance heater.

4-32 One part of an insulated rigid tank contains an ideal gas while the other side is evacuated. The final temperature and pressure in the tank are to be determined when the partition is removed.

Assumptions **1** The kinetic and potential energy changes are negligible, $\Delta ke \cong \Delta pe \cong 0$. **2** The tank is insulated and thus heat transfer is negligible.

Analysis We take the entire tank as the system. This is a *closed system* since no mass crosses the boundaries of the system. The energy balance for this system can be expressed as

$$\underbrace{E_{in} - E_{out}}_{\text{Net energy transfer by heat, work, and mass}} = \underbrace{\Delta E_{\text{system}}}_{\text{Change in internal, kinetic, potential, etc. energies}}$$

$$0 = \Delta U = m(u_2 - u_1)$$

$$u_2 = u_1$$

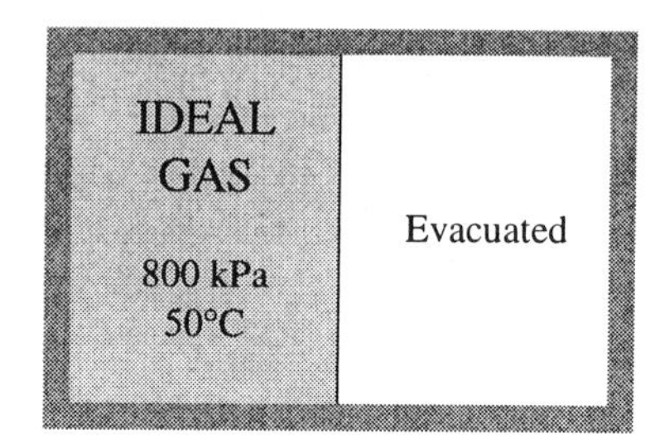

Therefore,

$$T_2 = T_1 = \mathbf{50°C}$$

Since $u = u(T)$ for an ideal gas. Then,

$$\frac{P_1 V_1}{T_1} = \frac{P_2 V_2}{T_2} \longrightarrow P_2 = \frac{V_1}{V_2} P_1 = \frac{1}{2}(800 \text{ kPa}) = \mathbf{400kPa}$$

4-33 A cylinder equipped with a set of stops for the piston to rest on is initially filled with helium gas at a specified state. The amount of heat that must be transferred to raise the piston is to be determined. √

Assumptions **1** Helium is an ideal gas with constant specific heats. **2** The kinetic and potential energy changes are negligible, $\Delta ke \cong \Delta pe \cong 0$. **3** There are no work interactions involved. **4** The thermal energy stored in the cylinder itself is negligible.

Properties The specific heat of helium at room temperature is C_v = 3.1156 kJ/kg.K (Table A-2).

Analysis We take the helium gas in the cylinder as the system. This is a closed system since no mass crosses the boundary of the system. The energy balance for this constant volume closed system can be expressed as

$$\underbrace{E_{in} - E_{out}}_{\substack{\text{Net energy transfer}\\\text{by heat, work, and mass}}} = \underbrace{\Delta E_{\text{system}}}_{\substack{\text{Change in internal, kinetic,}\\\text{potential, etc. energies}}}$$

$$Q_{in} = \Delta U = m(u_2 - u_1)$$

$$Q_{in} = m(u_2 - u_1) = mC_v(T_2 - T_1)$$

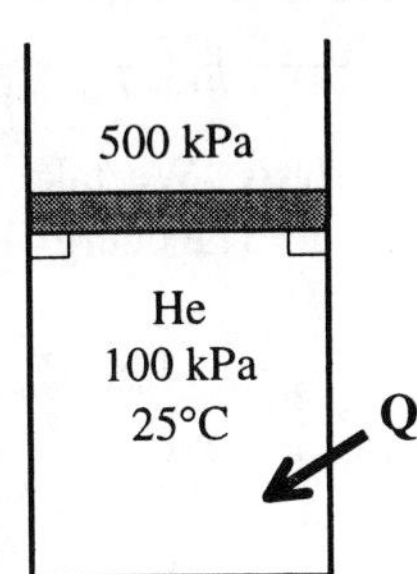

The final temperature of helium can be determined from the ideal gas relation to be

$$\frac{P_1 V}{T_1} = \frac{P_2 V}{T_2} \longrightarrow T_2 = \frac{P_2}{P_1} T_1 = \frac{500 \text{ kPa}}{100 \text{ kPa}} (298 \text{ K}) = 1490 \text{ K}$$

Substituting into the energy balance relation gives

Q_{in} = (0.5 kg)(3.1156 kJ/kg·K)(1490 - 298)K = **1857 kJ**

4-34 An insulated cylinder is initially filled with air at a specified state. A paddle-wheel in the cylinder stirs the air at constant pressure. The final temperature of air is to be determined.

Assumptions **1** Air is an ideal gas with variable specific heats. **2** The cylinder is stationary and thus the kinetic and potential energy changes are zero. **3** There are no work interactions involved other than the boundary work. **4** The cylinder is well-insulated and thus heat transfer is negligible. **5** The thermal energy stored in the cylinder itself and the paddle-wheel is negligible. **6** The compression or expansion process is quasi-equilibrium.

Properties The gas constant of air is $R = 0.287$ kPa.m^3/kg.K (Table A-1). Also, $C_p = 1.005$ kJ/kg.K for air at room temperature (Table A-2). The enthalpy of air at the initial temperature is

$$h_1 = h_{@298\text{ K}} = 298.18 \text{ kJ/kg}$$

Analysis We take the air in the cylinder as the system. This is a closed system since no mass enters or leaves. The energy balance for this stationary closed system can be expressed as

$$\underbrace{E_{in} - E_{out}}_{\substack{\text{Net energy transfer}\\ \text{by heat, work, and mass}}} = \underbrace{\Delta E_{\text{system}}}_{\substack{\text{Change in internal, kinetic,}\\ \text{potential, etc. energies}}}$$

$$W_{pw,in} - W_{b,out} = \Delta U$$

$$W_{pw,in} = m(h_2 - h_1)$$

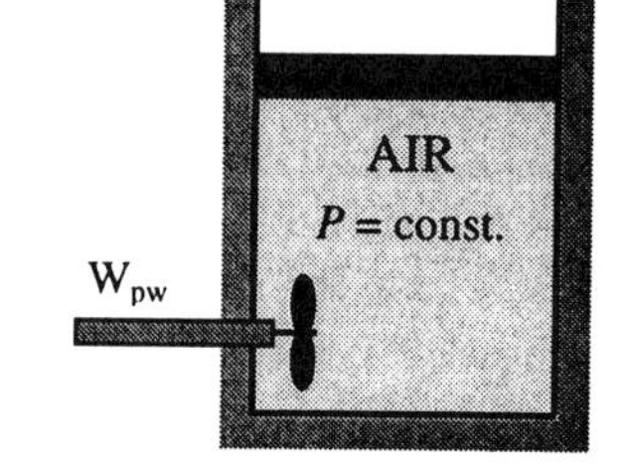

since $\Delta U + W_b = \Delta H$ during a constant pressure quasi-equilibrium process. The mass of air is

$$m = \frac{P_1 V}{RT_1} = \frac{(400 \text{ kPa})(0.1 \text{ m}^3)}{(0.287 \text{ kPa}\cdot\text{m}^3/\text{kg}\cdot\text{K})(298 \text{ K})} = 0.468 \text{ kg}$$

Substituting into the energy balance,

$$15 \text{ kJ} = (0.468 \text{ kg})(h_2 - 298.18 \text{ kJ/kg})$$
$$h_2 = 330.23 \text{ kJ/kg}$$

From Table A-17,

$$T_2 = \mathbf{329.9\ K}$$

Alternative solution Using specific heats at room temperature, $C_p = 1.005$ kJ/kg.°C, the final temperature is determined to be

$$W_{pw,in} = m(h_2 - h_1) \cong mC_p(T_2 - T_1)$$
$$15 \text{ kJ} = (0.468 \text{ kg})(1.005 \text{ kJ/kg.°C})(T_2 - 25)\text{°C}$$

which gives

$$T_2 = \mathbf{56.9°C}$$

4-35E A cylinder is initially filled with nitrogen gas at a specified state. The gas is cooled by transferring heat from it. The amount of heat transfer is to be determined.

Assumptions **1** The cylinder is stationary and thus the kinetic and potential energy changes are zero. **2** There are no work interactions involved other than the boundary work. **3** The thermal energy stored in the cylinder itself is negligible. **4** The compression or expansion process is quasi-equilibrium. **5** Nitrogen is an ideal gas with constant specific heats.

Properties The specific heat of nitrogen at the average temperature of T_{ave} = (700+ 140)/2 = 420°F is C_{pave} = 0.252 Btu/lbm.°F (Table A-2Eb).

Analysis We take the nitrogen gas in the cylinder as the system. This is a closed system since no mass enters or leaves. The energy balance for this closed system can be expressed as

$$\underbrace{E_{in} - E_{out}}_{\substack{\text{Net energy transfer} \\ \text{by heat, work, and mass}}} = \underbrace{\Delta E_{system}}_{\substack{\text{Change in internal, kinetic,} \\ \text{potential, etc. energies}}}$$

$$-Q_{out} - W_{b,out} = \Delta U = m(u_2 - u_1)$$

$$-Q_{out} = m(h_2 - h_1) = mC_p(T_2 - T_1)$$

N$_2$
50 psia
700°F
Q

since $\Delta U + W_b = \Delta H$ during a constant pressure quasi-equilibrium process.
The mass of nitrogen is

$$m = \frac{P_1 V}{RT_1} = \frac{(50 \text{ psia})(25 \text{ ft}^3)}{(0.3830 \text{ psia} \cdot \text{ft}^3 / \text{lbm} \cdot \text{R})(1160 \text{ R})} = 2.814 \text{ lbm}$$

Substituting,

Q_{out} = (2.814 lbm)(0.252 Btu/lbm.°F)(700 - 140)°F = **397 Btu**

4-36 A cylinder is initially filled with air at a specified state. Air is heated electrically at constant pressure, and some heat is lost in the process. The amount of electrical energy supplied is to be determined. √

Assumptions **1** The cylinder is stationary and thus the kinetic and potential energy changes are zero. **2** Air is an ideal gas with variable specific heats. **3** The thermal energy stored in the cylinder itself and the resistance wires is negligible. **4** The compression or expansion process is quasi-equilibrium.

Properties The initial and final enthalpies of air are (Table A-17)

$$h_1 = h_{@\,298\text{ K}} = 298.18\text{ kJ/kg}$$

$$h_2 = h_{@\,350\text{ K}} = 350.49\text{ kJ/kg}$$

Analysis We take the contents of the cylinder as the system. This is a closed system since no mass enters or leaves. The energy balance for this closed system can be expressed as

$$\underbrace{E_{in} - E_{out}}_{\substack{\text{Net energy transfer}\\\text{by heat, work, and mass}}} = \underbrace{\Delta E_{\text{system}}}_{\substack{\text{Change in internal, kinetic,}\\\text{potential, etc. energies}}}$$

$$W_{e,in} - Q_{out} - W_{b,out} = \Delta U$$

$$W_{e,in} = m(h_2 - h_1) + Q_{out}$$

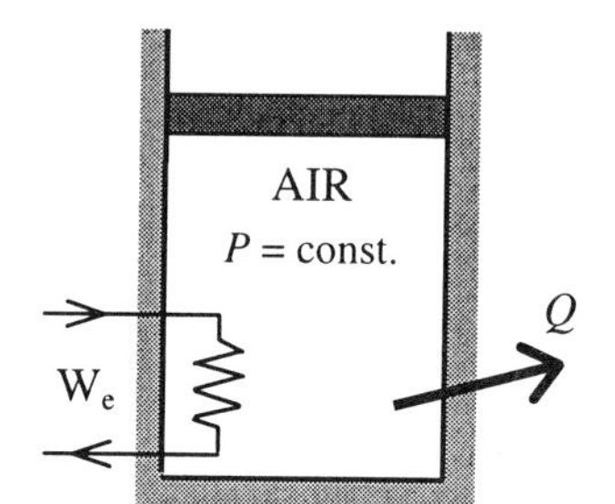

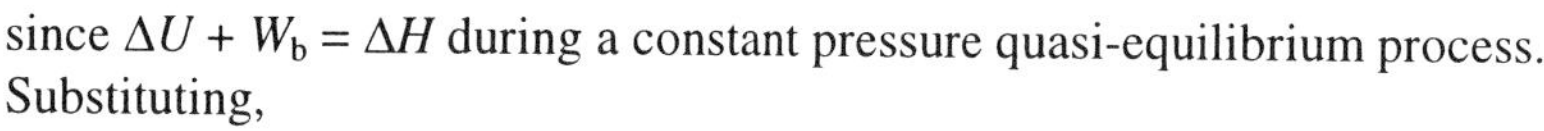
since $\Delta U + W_b = \Delta H$ during a constant pressure quasi-equilibrium process. Substituting,

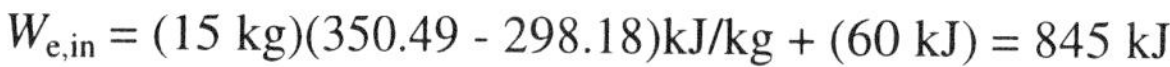
$$W_{e,in} = (15\text{ kg})(350.49 - 298.18)\text{kJ/kg} + (60\text{ kJ}) = 845\text{ kJ}$$

or,

$$W_{e,in} = (845\text{kJ})\left(\frac{1\text{ kWh}}{3600\text{ kJ}}\right) = \mathbf{0.235\ kWh}$$

Alternative solution The specific heat of air at the average temperature of T_{ave} = (25+ 77)/2 51°C = 324 K is, from Table A-2b, $C_{p,ave}$ = 1.0065 kJ/kg.°C. Substituting,

$$W_{e,in} = mC_p(T_2 - T_1) + Q_{out}$$
$$= (15\text{ kg})(1.0065\text{ kJ/kg.°C})(77 - 25)\text{°C} + 60\text{ kJ} = 845\text{ kJ}$$

or,

$$W_{e,in} = (845\text{ kJ})\left(\frac{1\text{ kWh}}{3600\text{ kJ}}\right) = \mathbf{0.235\ kWh}$$

Discussion Note that for small temperature differences, both approaches give the same result.

4-37 An insulated cylinder initially contains CO_2 at a specified state. The CO_2 is heated electrically for 10 min at constant pressure until the volume doubles. The electric current is to be determined.

Assumptions **1** The cylinder is stationary and thus the kinetic and potential energy changes are zero. **2** The CO_2 is an ideal gas with constant specific heats. **3** The thermal energy stored in the cylinder itself and the resistance wires is negligible. **4** The compression or expansion process is quasi-equilibrium.

Properties The gas constant and molar mass of CO_2 are R = 0.1889 kPa.m³/kg.K and M = 44 kg/kmol (Table A-1). The specific heat of CO_2 at the average temperature of T_{ave} = (300 + 600)/2 = 450 K is $C_{p,ave}$ = 0.978 kJ/kg.°C (Table A-2b).

Analysis We take the contents of the cylinder as the system. This is a closed system since no mass enters or leaves. The energy balance for this closed system can be expressed as

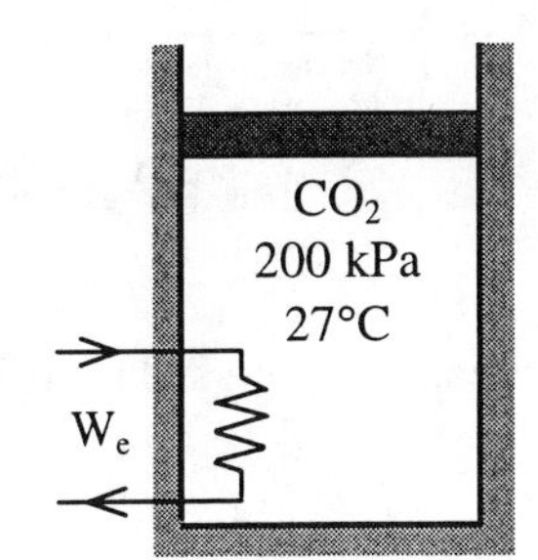

$$\underbrace{E_{in} - E_{out}}_{\substack{\text{Net energy transfer} \\ \text{by heat, work, and mass}}} = \underbrace{\Delta E_{system}}_{\substack{\text{Change in internal, kinetic,} \\ \text{potential, etc. energies}}}$$

$$W_{e,in} - W_{b,out} = \Delta U$$

$$W_{e,in} = m(h_2 - h_1) \cong mC_p(T_2 - T_1)$$

since $\Delta U + W_b = \Delta H$ during a constant pressure quasi-equilibrium process. The final temperature of CO_2 is

$$\frac{P_1 V_1}{T_1} = \frac{P_2 V_2}{T_2} \longrightarrow T_2 = \frac{P_2}{P_1}\frac{V_2}{V_1} T_1 = 1 \times 2 \times (300\text{K}) = 600\text{K}$$

The mass of CO_2 is

$$m = \frac{P_1 V_1}{RT_1} = \frac{(200 \text{ kPa})(0.3 \text{ m}^3)}{(0.1889 \text{ kPa} \cdot \text{m}^3/\text{kg} \cdot \text{K})(300 \text{ K})} = 1.059 \text{ kg}$$

Substituting,

$$W_{e,in} = (1.059 \text{ kg})(0.978 \text{ kJ/kg.K})(600 - 300)\text{K} = 311 \text{ kJ}$$

Then,

$$I = \frac{W_{e,in}}{V\Delta t} = \frac{311\text{kJ}}{(110\text{V})(10 \times 60\text{s})}\left(\frac{1000\text{VA}}{1\text{kJ/s}}\right) = \mathbf{4.71A}$$

4-38 A cylinder initially contains nitrogen gas at a specified state. The gas is compressed polytropically until the volume is reduced by one-half. The work done and the heat transfer are to be determined.

Assumptions **1** The cylinder is stationary and thus the kinetic and potential energy changes are zero. **2** The N_2 is an ideal gas with constant specific heats. **3** The thermal energy stored in the cylinder itself is negligible. **4** The compression or expansion process is quasi-equilibrium.

Properties The gas constant of N_2 are R = 0.2968 kPa.m^3/kg.K (Table A-1). The C_v value of N_2 at the average temperature (369+300)/2 = 335 K is 0.744 kJ/kg.K (Table A-2b).

Analysis We take the contents of the cylinder as the system. This is a closed system since no mass crosses the system boundary. The energy balance for this closed system can be expressed as

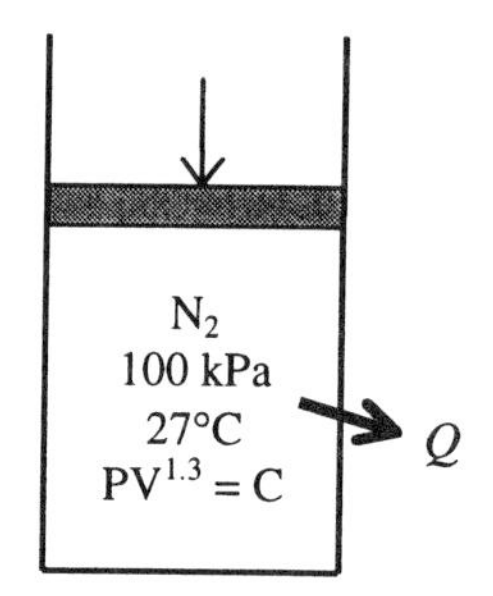

$$\underbrace{E_{in} - E_{out}}_{\substack{\text{Net energy transfer} \\ \text{by heat, work, and mass}}} = \underbrace{\Delta E_{\text{system}}}_{\substack{\text{Change in internal, kinetic,} \\ \text{potential, etc. energies}}}$$

$$W_{b,in} - Q_{out} = \Delta U = m(u_2 - u_1)$$

$$W_{b,in} - Q_{out} = mC_v(T_2 - T_1)$$

The final pressure and temperature of nitrogen are

$$P_2V_2^{1.3} = P_1V_1^{1.3} \longrightarrow P_2 = \left(\frac{V_1}{V_2}\right)^{1.3} P_1 = 2^{1.3}(100\text{kPa}) = 246.2\text{kPa}$$

$$\frac{P_1V_1}{T_1} = \frac{P_2V_2}{T_2} \longrightarrow T_2 = \frac{P_2}{P_1}\frac{V_2}{V_1}T_1 = \frac{246.2\text{kPa}}{100\text{kPa}} \times 0.5 \times (300\text{K}) = 369.3\text{K}$$

Then the boundary work for this polytropic process can be determined from

$$W_{b,in} = -\int_1^2 PdV = -\frac{P_2V_2 - P_1V_1}{1-n} = -\frac{mR(T_2 - T_1)}{1-n}$$

$$= -\frac{(0.8 \text{ kg})(0.2968 \text{ kJ/kg}\cdot\text{K})(369.3 - 300)\text{K}}{1-1.3} = \mathbf{54.8 \text{ kJ}}$$

Substituting into the energy balance gives

$$Q_{out} = W_{b,in} - mC_v(T_2 - T_1)$$
$$= 54.8 \text{ kJ} - (0.8 \text{ kg})(0.744 \text{ kJ/kg.K})(369.3 - 360)\text{K}$$
$$= \mathbf{13.6 \text{ kJ}}$$

4-39 EES solution of this (and other comprehensive problems designated with the *computer icon*) is available to instructors at the *Instructor Manual* section of the *Online Learning Center* (OLC) at www.mhhe.com/cengel-boles. See the Preface for access information.

4-40 It is observed that the air temperature in a room heated by electric baseboard heaters remains constant even though the heater operates continuously when the heat losses from the room amount to 8,000 kJ/h. The power rating of the heater is to be determined.

Assumptions **1** Air is an ideal gas since it is at a high temperature and low pressure relative to its critical point values of -141°C and 3.77 MPa. **2** The kinetic and potential energy changes are negligible, $\Delta ke \cong \Delta pe \cong 0$. **3** The temperature of the room is said to remain constant during this process.

Analysis We take the room as the system. This is a closed system since no mass crosses the boundary of the system. The energy balance for this system reduces to

$$\underbrace{E_{in} - E_{out}}_{\substack{\text{Net energy transfer} \\ \text{by heat, work, and mass}}} = \underbrace{\Delta E_{\text{system}}}_{\substack{\text{Change in internal, kinetic,} \\ \text{potential, etc. energies}}}$$

$$W_{e,in} - Q_{out} = \Delta U = 0$$

$$W_{e,in} = Q_{out}$$

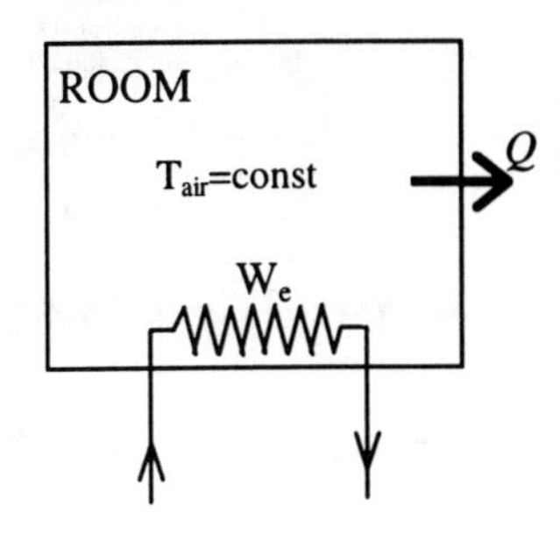

since $\Delta U = mC_v\Delta T = 0$ for isothermal processes of ideal gases. Thus,

$$\dot{W}_{e,in} = \dot{Q}_{out} = (6500 \text{ kJ/h})\left(\frac{1 \text{ kW}}{3600 \text{ kJ/h}}\right) = \mathbf{1.81\ kW}$$

4-41E A cylinder initially contains air at a specified state. Heat is transferred to the air, and air expands isothermally. The boundary work done is to be determined.

Assumptions **1** The cylinder is stationary and thus the kinetic and potential energy changes are zero. **2** The air is an ideal gas with constant specific heats. **3** The compression or expansion process is quasi-equilibrium.

Analysis We take the contents of the cylinder as the system. This is a closed system since no mass crosses the system boundary. The energy balance for this closed system can be expressed as

$$\underbrace{E_{in} - E_{out}}_{\substack{\text{Net energy transfer} \\ \text{by heat, work, and mass}}} = \underbrace{\Delta E_{\text{system}}}_{\substack{\text{Change in internal, kinetic,} \\ \text{potential, etc. energies}}}$$

$$Q_{in} - W_{b,out} = \Delta U = m(u_2 - u_1)$$

$$= mC_v(T_2 - T_1) = 0$$

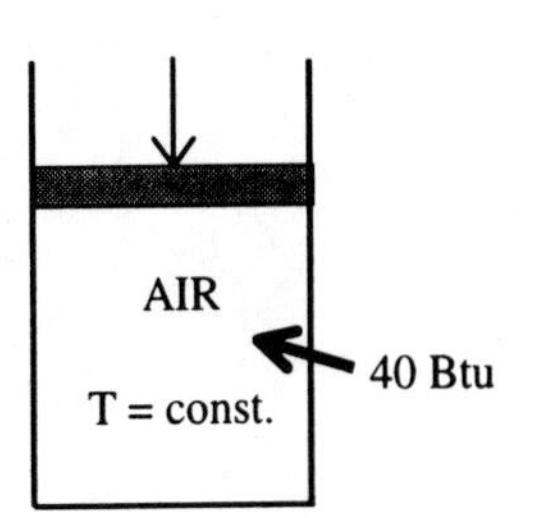

since $u = u(T)$ for ideal gases, and thus $u_2 = u_1$ when $T_1 = T_2$. Therefore,

$$W_{b,out} = Q_{in} = \mathbf{40\ Btu}$$

4-42 A cylinder initially contains argon gas at a specified state. The gas is stirred while being heated and expanding isothermally. The amount of heat transfer is to be determined.

Assumptions **1** The cylinder is stationary and thus the kinetic and potential energy changes are zero. **2** The air is an ideal gas with constant specific heats. **3** The compression or expansion process is quasi-equilibrium.

Analysis We take the contents of the cylinder as the system. This is a closed system since no mass crosses the system boundary. The energy balance for this closed system can be expressed as

$$\underbrace{E_{in} - E_{out}}_{\substack{\text{Net energy transfer} \\ \text{by heat, work, and mass}}} = \underbrace{\Delta E_{\text{system}}}_{\substack{\text{Change in internal, kinetic,} \\ \text{potential, etc. energies}}}$$

$$Q_{in} + W_{pw,in} - W_{b,out} = \Delta U = m(u_2 - u_1)$$

$$= mC_v(T_2 - T_1) = 0$$

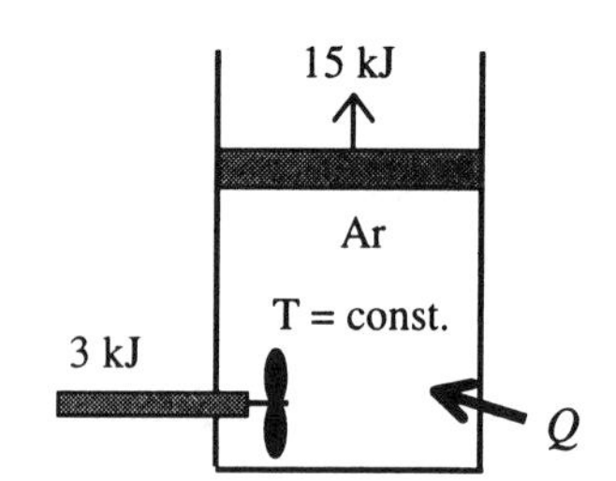

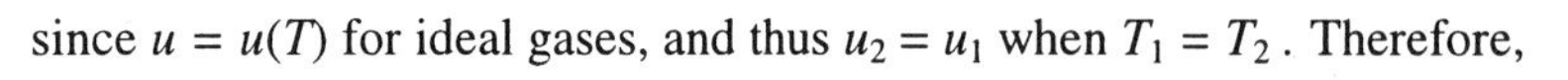

since $u = u(T)$ for ideal gases, and thus $u_2 = u_1$ when $T_1 = T_2$. Therefore,

$$Q_{in} = W_{b,out} - W_{pw,in} = 15 - 3 = \mathbf{12\ kJ}$$

4-43 A cylinder equipped with a set of stops for the piston is initially filled with air at a specified state. Heat is transferred to the air until the volume doubled. The work done by the air and the amount of heat transfer are to be determined, and the process is to be shown on a *P-v* diagram.

Assumptions **1** Air is an ideal gas with variable specific heats. **2** The kinetic and potential energy changes are negligible, $\Delta ke \cong \Delta pe \cong 0$. **3** The thermal energy stored in the cylinder itself is negligible. **4** The compression or expansion process is quasi-equilibrium.

Properties The gas constant of air is R = 0.287 kPa.m^3/kg.K (Table A-1).

Analysis We take the air in the cylinder as the system. This is a closed system since no mass crosses the boundary of the system. The energy balance for this closed system can be expressed as

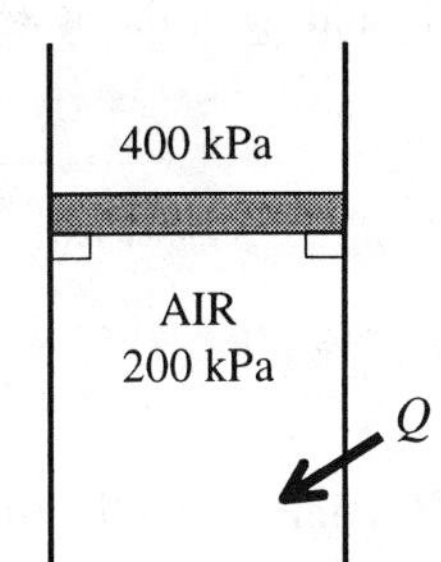

$$\underbrace{E_{in} - E_{out}}_{\substack{\text{Net energy transfer}\\ \text{by heat, work, and mass}}} = \underbrace{\Delta E_{\text{system}}}_{\substack{\text{Change in internal, kinetic,}\\ \text{potential, etc. energies}}}$$

$$Q_{in} - W_{b,out} = \Delta U = m(u_3 - u_1)$$

$$Q_{in} = m(u_3 - u_1) + W_{b,out}$$

The initial and the final volumes and the final temperature of air are

$$V_1 = \frac{mRT_1}{P_1} = \frac{(3\text{ kg})(0.287\text{ kPa}\cdot\text{m}^3/\text{kg}\cdot\text{K})(300\text{ K})}{200\text{ kPa}} = 1.29\text{ m}^3$$

$$V_3 = 2V_1 = 2\times 1.29 = 2.58\text{ m}^3$$

$$\frac{P_1V_1}{T_1} = \frac{P_3V_3}{T_3} \longrightarrow T_3 = \frac{P_3}{P_1}\frac{V_3}{V_1}T_1 = \frac{400\text{ kPa}}{200\text{ kPa}}\times 2\times(300\text{ K}) = 1200\text{ K}$$

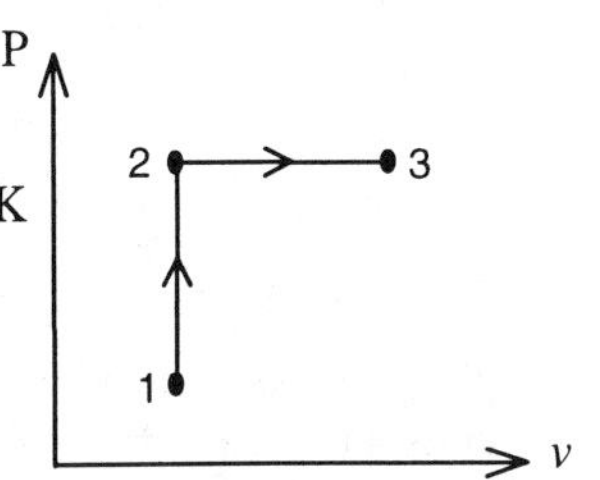

No work is done during process 1-2 since $V_1 = V_2$. The pressure remains constant during process 2-3 and the work done during this process is

$$W_{b,out} = \int_1^2 PdV = P_2(V_3 - V_2) = (400\text{ kPa})(2.58 - 1.29)\text{m}^3 = \mathbf{516\ kJ}$$

The initial and final internal energies of air are (Table A-17)

$$u_1 = u_{@\,300\text{ K}} = 214.07\text{ kJ/kg}$$

$$u_2 = u_{@\,1200\text{ K}} = 933.33\text{ kJ/kg}$$

Then from the energy balance,

$$Q_{\text{in}} = (3\text{ kg})(933.33 - 214.07)\text{kJ/kg} + 516\text{ kJ} = \mathbf{2674\ kJ}$$

Alternative solution The specific heat of air at the average temperature of T_{ave} = (300 + 1200)/2 = 750 K is, from Table A-2b, C_{vave} = 0.800 kJ/kg.K. Substituting,

$$Q_{in} = m(u_3 - u_1) + W_{b,out} \cong mC_v(T_3 - T_1) + W_{b,out}$$

$$Q_{\text{in}} = (3\text{ kg})(0.800\text{ kJ/kg.K})(1200 - 300)\text{ K} + 516\text{ kJ} = \mathbf{2676\ kJ}$$

4-44 [*Also solved by EES on enclosed CD*] A cylinder equipped with a set of stops on the top is initially filled with air at a specified state. Heat is transferred to the air until the piston hits the stops, and then the pressure doubles. The work done by the air and the amount of heat transfer are to be determined, and the process is to be shown on a *P-v* diagram.

Assumptions **1** Air is an ideal gas with variable specific heats. **2** The kinetic and potential energy changes are negligible, $\Delta ke \cong \Delta pe \cong 0$. **3** There are no work interactions involved. **3** The thermal energy stored in the cylinder itself is negligible.

Properties The gas constant of air is R = 0.287 kPa.m^3/kg.K (Table A-1).

Analysis We take the air in the cylinder to be the system. This is a closed system since no mass crosses the boundary of the system. The energy balance for this closed system can be expressed as

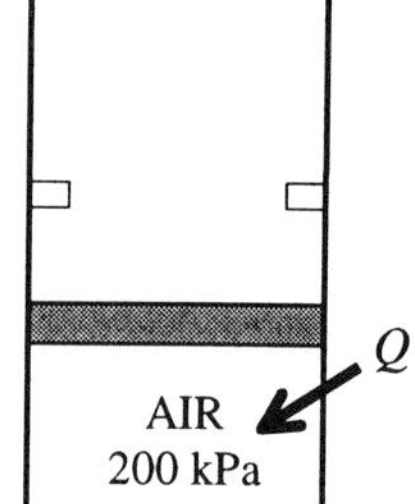

$$\underbrace{E_{in} - E_{out}}_{\substack{\text{Net energy transfer} \\ \text{by heat, work, and mass}}} = \underbrace{\Delta E_{\text{system}}}_{\substack{\text{Change in internal, kinetic,} \\ \text{potential, etc. energies}}}$$

$$Q_{in} - W_{b,out} = \Delta U = m(u_3 - u_1)$$

$$Q_{in} = m(u_3 - u_1) + W_{b,out}$$

The initial and the final volumes and the final temperature of air are determined from

$$V_1 = \frac{mRT_1}{P_1} = \frac{(3\text{ kg})(0.287\text{ kPa}\cdot\text{m}^3/\text{kg}\cdot\text{K})(300\text{ K})}{200\text{ kPa}} = 1.29\text{ m}^3$$

$$V_3 = 2V_1 = 2 \times 1.29 = 2.58\text{ m}^3$$

$$\frac{P_1 V_1}{T_1} = \frac{P_3 V_3}{T_3} \longrightarrow T_3 = \frac{P_3}{P_1}\frac{V_3}{V_1}T_1 = \frac{400\text{ kPa}}{200\text{ kPa}} \times 2 \times (300\text{ K}) = 1200\text{ K}$$

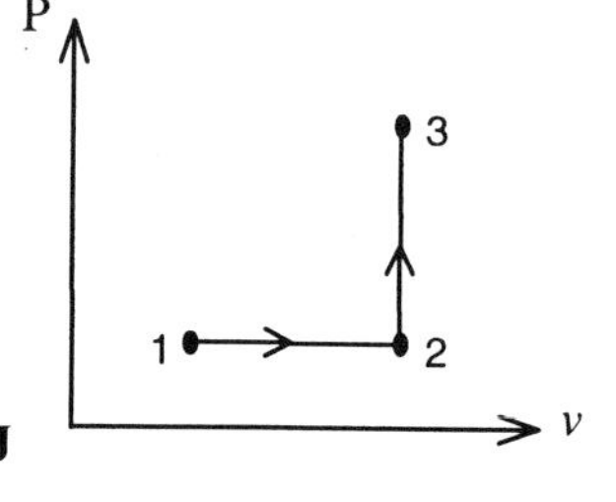

No work is done during process 2-3 since $V_2 = V_3$. The pressure remains constant during process 1-2 and the work done during this process is

$$W_b = \int_1^2 PdV = P_2(V_3 - V_2) = (200\text{kPa})(2.58 - 1.29)\text{m}^3\left(\frac{1\text{kJ}}{1\text{kPa}\cdot\text{m}^3}\right) = \mathbf{258kJ}$$

The initial and final internal energies of air are (Table A-17)

$$u_1 = u_{@\,300\text{ K}} = 214.07\text{ kJ/kg}$$

$$u_2 = u_{@\,1200\text{ K}} = 933.33\text{ kJ/kg}$$

Substituting,

$$Q_{\text{in}} = (3\text{ kg})(933.33 - 214.07)\text{kJ/kg} + 258\text{ kJ} = \mathbf{2416\text{ kJ}}$$

Alternative solution The specific heat of air at the average temperature of T_{ave} = (300 + 1200)/2 = 750 K is, from Table A-2b, C_{vave} = 0.800 kJ/kg.K. Substituting

$$Q_{in} = m(u_3 - u_1) + W_{b,out} \cong mC_v(T_3 - T_1) + W_{b,out}$$

$$= (3\text{ kg})(0.800\text{ kJ/kg.K})(1200 - 300)\text{ K} + 258\text{ kJ} = \mathbf{2418\text{ kJ}}$$

Closed System Energy Analysis: Solids and Liquids

4-45 A number of brass balls are to be quenched in a water bath at a specified rate. The rate at which heat needs to be removed from the water in order to keep its temperature constant is to be determined.

Assumptions **1** The thermal properties of the balls are constant. **2** The balls are at a uniform temperature before and after quenching. **3** The changes in kinetic and potential energies are negligible.

Properties The density and specific heat of the brass balls are given to be ρ = 8522 kg/m³ and C_p = 0.385 kJ/kg.°C.

Analysis We take a single ball as the system. The energy balance for this closed system can be expressed as

$$\underbrace{E_{in} - E_{out}}_{\substack{\text{Net energy transfer}\\ \text{by heat, work, and mass}}} = \underbrace{\Delta E_{\text{system}}}_{\substack{\text{Change in internal, kinetic,}\\ \text{potential, etc. energies}}}$$

$$-Q_{out} = \Delta U_{ball} = m(u_2 - u_1)$$

$$Q_{out} = mC(T_1 - T_2)$$

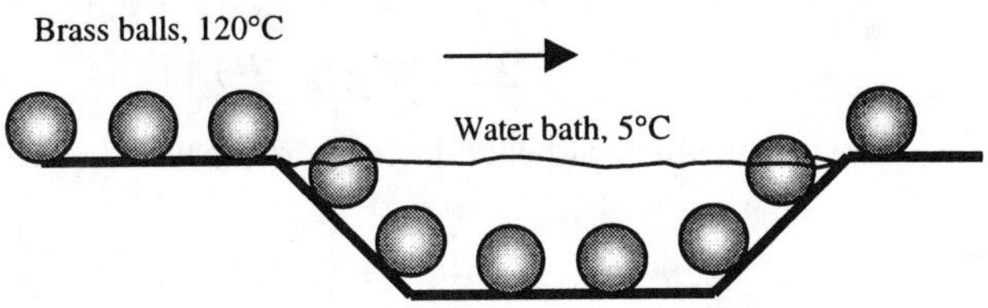

Brass balls, 120°C

Water bath, 5°C

The total amount of heat transfer from a ball is

$$m = \rho V = \rho\frac{\pi D^3}{6} = (8522\ \text{kg/m}^3)\frac{\pi(0.05\ \text{m})^3}{6} = 0.558\ \text{kg}$$

$$Q_{out} = mC(T_1 - T_2) = (0.558\ \text{kg})(0.385\ \text{kJ/kg.°C})(120 - 74)\text{°C} = 9.88\ \text{kJ/ball}$$

Then the rate of heat transfer from the balls to the water becomes

$$\dot{Q}_{total} = \dot{n}_{\text{ball}} Q_{\text{ball}} = (100\ \text{balls/min}) \times (9.88\ \text{kJ/ball}) = \mathbf{988\ kJ/min}$$

Therefore, heat must be removed from the water at a rate of 988 kJ/min in order to keep its temperature constant at 50°C since energy input must be equal to energy output for a system whose energy level remains constant. That is, $E_{in} = E_{out}$ when $\Delta E_{\text{system}} = 0$.

4-46 A number of aluminum balls are to be quenched in a water bath at a specified rate. The rate at which heat needs to be removed from the water in order to keep its temperature constant is to be determined.

Assumptions **1** The thermal properties of the balls are constant. **2** The balls are at a uniform temperature before and after quenching. **3** The changes in kinetic and potential energies are negligible.

Properties The density and specific heat of aluminum at the average temperature of (120+74)/2 = 97°C = 370 K are $\rho = 2700$ kg/m^3 and $C_p = 0.937$ kJ/kg.°C (Table A-3).

Analysis We take a single ball as the system. The energy balance for this closed system can be expressed as

$$\underbrace{E_{in} - E_{out}}_{\substack{\text{Net energy transfer}\\ \text{by heat, work, and mass}}} = \underbrace{\Delta E_{\text{system}}}_{\substack{\text{Change in internal, kinetic,}\\ \text{potential, etc. energies}}}$$

$$-Q_{out} = \Delta U_{ball} = m(u_2 - u_1)$$

$$Q_{out} = mC(T_1 - T_2)$$

Aluminum balls, 120°C

Water bath, 50°C

The total amount of heat transfer from a ball is

$$m = \rho V = \rho\frac{\pi D^3}{6} = (2700\text{ kg/m}^3)\frac{\pi(0.05\text{ m})^3}{6} = 0.1767\text{ kg}$$

$$Q_{out} = mC(T_1 - T_2) = (0.1767\text{ kg})(0.937\text{ kJ/kg.°C})(120 - 74)\text{°C} = 7.62\text{ kJ/ball}$$

Then the rate of heat transfer from the balls to the water becomes

$$\dot{Q}_{total} = \dot{n}_{\text{ball}}Q_{\text{ball}} = (100\text{ balls/min}) \times (7.62\text{ kJ/ball}) = \mathbf{762\ kJ/min}$$

Therefore, heat must be removed from the water at a rate of 762 kJ/min in order to keep its temperature constant at 50°C since energy input must be equal to energy output for a system whose energy level remains constant. That is, $E_{in} = E_{out}$ when $\Delta E_{\text{system}} = 0$.

4-47E A person shakes a canned of drink in a iced water to cool it. The mass of the ice that will melt by the time the canned drink is cooled to a specified temperature is to be determined.

Assumptions **1** The thermal properties of the drink are constant, and are taken to be the same as those of water. **2** The effect of agitation on the amount of ice melting is negligible. **3** The thermal energy capacity of the can itself is negligible, and thus it does not need to be considered in the analysis.

Properties The density and specific heat of water at the average temperature of (75+45)/2 = 60°F are ρ = 62.3 lbm/ft³, and C_p = 1.0 Btu/lbm.°F (Table A-3E). The heat of fusion of water is 143.5 Btu/lbm.

Analysis We take a canned drink as the system. The energy balance for this closed system can be expressed as

$$\underbrace{E_{in} - E_{out}}_{\substack{\text{Net energy transfer} \\ \text{by heat, work, and mass}}} = \underbrace{\Delta E_{\text{system}}}_{\substack{\text{Change in internal, kinetic,} \\ \text{potential, etc. energies}}}$$

$$-Q_{out} = \Delta U_{\text{canned drink}} = m(u_2 - u_1)$$

$$Q_{out} = mC(T_1 - T_2)$$

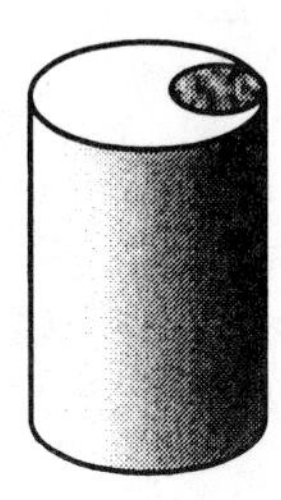

Noting that 1 gal = 128 oz and 1 ft³ = 7.48 gal = 957.5 oz, the total amount of heat transfer from a ball is

$$m = \rho V = (62.3\ \text{lbm/ft}^3)(12\ \text{oz/can})\left(\frac{1\ \text{ft}^3}{7.48\ \text{gal}}\right)\left(\frac{1\ \text{gal}}{128\ \text{fluid oz}}\right) = 0.781\ \text{lbm/can}$$

$$Q_{out} = mC(T_1 - T_2) = (0.781\ \text{lbm/can})(1.0\ \text{Btu/lbm.°F})(75-45)\text{°F} = 23.4\ \text{Btu/can}$$

Noting that the heat of fusion of water is 14.5 Btu/lbm, the amount of ice that will melt to cool the drink is

$$m_{ice} = \frac{Q_{out}}{h_{if}} = \frac{23.4\ \text{Btu/can}}{143.5\ \text{Btu/lbm}} = \mathbf{0.163\ lbm} \quad \text{(per can of drink)}$$

since heat transfer to the ice must be equal to heat transfer from the can.

Discussion The actual amount of ice melted will be greater since agitation will also cause some ice to melt.

4-48 An iron whose base plate is made of an aluminum alloy is turned on. The minimum time for the plate to reach a specified temperature is to be determined.

Assumptions **1** It is given that 85 percent of the heat generated in the resistance wires is transferred to the plate. **2** The thermal properties of the plate are constant. **3** Heat loss from the plate during heating is disregarded since the minimum heating time is to be determined. **4** There are no changes in kinetic and potential energies. **5** The plate is at a uniform temperature at the end of the process.

Properties The density and specific heat of the aluminum alloy plate are given to be ρ = 2770 kg/m^3 and C_p = 875 kJ/kg.°C.

Analysis The mass of the iron's base plate is

$$m = \rho V = \rho LA = (2770 \text{ kg/m}^3)(0.005 \text{ m})(0.03 \text{ m}^2) = 0.4155 \text{ kg}$$

Noting that only 85 percent of the heat generated is transferred to the plate, the rate of heat transfer to the iron's base plate is

$$\dot{Q}_{\text{in}} = 0.85 \times 1000 \text{ W} = 850 \text{ W}$$

We take plate to be the system. The energy balance for this closed system can be expressed as

$$\underbrace{E_{in} - E_{out}}_{\substack{\text{Net energy transfer} \\ \text{by heat, work, and mass}}} = \underbrace{\Delta E_{\text{system}}}_{\substack{\text{Change in internal, kinetic,} \\ \text{potential, etc. energies}}}$$

$$Q_{in} = \Delta U_{\text{plate}} = m(u_2 - u_1)$$

$$\dot{Q}_{in} \Delta t = mC(T_2 - T_1)$$

Air
22°C

IRON
1000 W

Solving for Δt and substituting,

$$\Delta t = \frac{mC\Delta T_{\text{plate}}}{\dot{Q}_{\text{in}}} = \frac{(0.4155 \text{ kg})(875 \text{ J/kg.°C})(140 - 22)\text{°C}}{850 \text{ J/s}} = \mathbf{50.5 \text{ s}}$$

which is the time required for the plate temperature to reach the specified temperature.

4-49 Stainless steel ball bearings leaving the oven at a specified uniform temperature at a specified rate are exposed to air and are cooled before they are dropped into the water for quenching. The rate of heat transfer from the ball bearing to the air is to be determined. √

Assumptions **1** The thermal properties of the bearing balls are constant. **2** The kinetic and potential energy changes of the balls are negligible. **3** The balls are at a uniform temperature at the end of the process

Properties The density and specific heat of the ball bearings are given to be ρ = 8085 kg/m³ and C_p = 0.480 kJ/kg.°C.

Analysis We take a single bearing ball as the system. The energy balance for this closed system can be expressed as

$$\underbrace{E_{in} - E_{out}}_{\substack{\text{Net energy transfer} \\ \text{by heat, work, and mass}}} = \underbrace{\Delta E_{\text{system}}}_{\substack{\text{Change in internal, kinetic,} \\ \text{potential, etc. energies}}}$$

$$-Q_{out} = \Delta U_{ball} = m(u_2 - u_1)$$

$$Q_{out} = mC(T_1 - T_2)$$

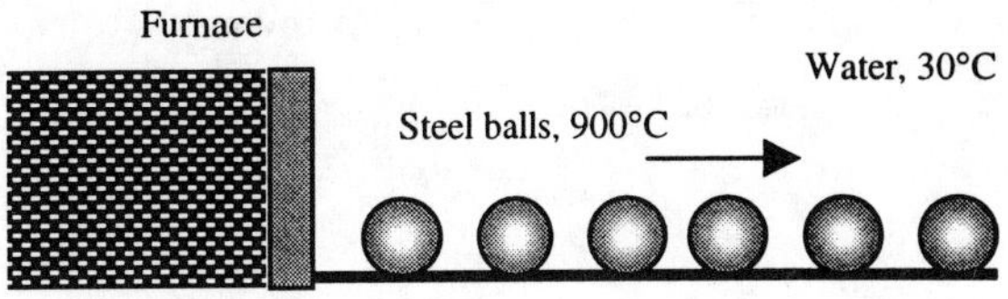

Furnace

Water, 30°C

Steel balls, 900°C

The total amount of heat transfer from a ball is

$$m = \rho V = \rho \frac{\pi D^3}{6} = (8085 \text{ kg/m}^3)\frac{\pi(0.012 \text{ m})^3}{6} = 0.007315 \text{ kg}$$

$$Q_{out} = mC(T_1 - T_2) = (0.007315 \text{ kg})(0.480 \text{ kJ/kg.°C})(900 - 850)\text{°C} = 0.1756 \text{ kJ/ball}$$

Then the rate of heat transfer from the balls to the air becomes

$$\dot{Q}_{total} = \dot{n}_{\text{ball}} Q_{\text{out (per ball)}} = (1400 \text{ balls/min}) \times (0.1756 \text{ kJ/ball}) = \mathbf{245.8 \text{ kJ/min} = 4.10 \text{ kW}}$$

Therefore, heat is lost to the air at a rate of 4.10 kW.

4-50 Carbon steel balls are to be annealed at a rate of 2500/h by heating them first and then allowing them to cool slowly in ambient air at a specified rate. The total rate of heat transfer from the balls to the ambient air is to be determined.

Assumptions **1** The thermal properties of the balls are constant. **2** There are no changes in kinetic and potential energies. **3** The balls are at a uniform temperature at the end of the process

Properties The density and specific heat of the balls are given to be ρ = 7833 kg/m^3 and C_p = 0.465 kJ/kg.°C.

Analysis We take a single ball as the system. The energy balance for this closed system can be expressed as

$$\underbrace{E_{in} - E_{out}}_{\substack{\text{Net energy transfer}\\ \text{by heat, work, and mass}}} = \underbrace{\Delta E_{\text{system}}}_{\substack{\text{Change in internal, kinetic,}\\ \text{potential, etc. energies}}}$$

$$-Q_{out} = \Delta U_{ball} = m(u_2 - u_1)$$

$$Q_{out} = mC(T_1 - T_2)$$

Furnace

Air, 35°C

Steel balls, 900°C

(*b*) The amount of heat transfer from a single ball is

$$m = \rho V = \rho \frac{\pi D^3}{6} = (7833\,\text{kg/m}^3)\frac{\pi(0.008\,\text{m})^3}{6} = 0.00210\,\text{kg}$$

$$Q_{out} = mC_p(T_1 - T_2) = (0.0021\,\text{kg})(0.465\,\text{kJ/kg.°C})(900 - 100)\text{°C} = 0.781\,\text{kJ (per ball)}$$

Then the total rate of heat transfer from the balls to the ambient air becomes

$$\dot{Q}_{out} = \dot{n}_{\text{ball}} Q_{out} = (2500\ \text{balls/h}) \times (0.781\ \text{kJ/ball}) = 1{,}953\ \text{kJ/h} = \mathbf{542\ W}$$

4-51 An electronic device is on for 5 minutes, and off for several hours. The temperature of the device at the end of the 4-min operating period is to be determined for the cases of operation with and without a heat sink.

Assumptions **1** The device and the heat sink are isothermal. **2** The thermal properties of the device and of the sink are constant. **3** Heat loss from the device during on time is disregarded since the highest possible temperature is to be determined.

Properties The specific heat of the device is given to be C_p = 850 J/kg.°C. The specific heat of aluminum at room temperature of 300 K is 902 J/kg.°C (Table A-3).

Analysis We take the device to be the system. Noting that electrical energy is supplied, the energy balance for this closed system can be expressed as

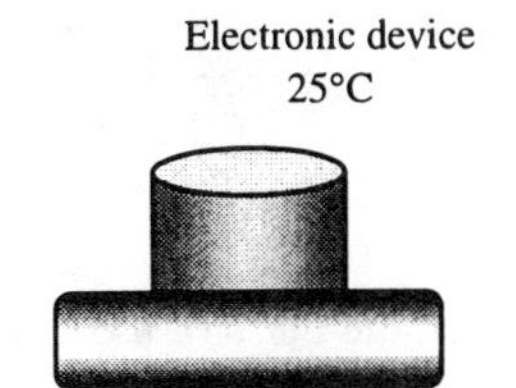

$$\underbrace{E_{in} - E_{out}}_{\substack{\text{Net energy transfer} \\ \text{by heat, work, and mass}}} = \underbrace{\Delta E_{\text{system}}}_{\substack{\text{Change in internal, kinetic,} \\ \text{potential, etc. energies}}}$$

$$W_{e,in} = \Delta U_{\text{device}} = m(u_2 - u_1)$$

$$\dot{W}_{e,in}\Delta t = mC(T_2 - T_1)$$

Substituting, the temperature of the device at the end of the process is determined to be

$$(30 \text{ J/s})(5 \times 60 \text{ s}) = (0.020 \text{ kg})(850 \text{ J/kg.°C})(T_2 - 25)\text{°C} \rightarrow T_2 = \mathbf{554°C} \text{ (without the heat sink)}$$

Case 2 When a heat sink is attached, the energy balance can be expressed as

$$W_{e,in} = \Delta U_{\text{device}} + \Delta U_{\text{heat sink}}$$

$$\dot{W}_{e,in}\Delta t = mC(T_2 - T_1)_{\text{device}} + mC(T_2 - T_1)_{\text{heat sink}}$$

Substituting, the temperature of the device-heat sink combination is determined to be

$$(30 \text{ J/s})(5 \times 60 \text{ s}) = (0.020 \text{ kg})(850 \text{ J/kg.°C})(T_2 - 25)\text{°C} + (0.200 \text{ kg})(902 \text{ J/kg.°C})(T_2 - 25)\text{°C}$$

$$T_2 = \mathbf{70.6°C} \text{ (with heat sink)}$$

Discussion These are the maximum temperatures. In reality, the temperatures will be lower because of the heat losses to the surroundings.

4-52 EES solution of this (and other comprehensive problems designated with the *computer icon*) is available to instructors at the *Instructor Manual* section of the *Online Learning Center* (OLC) at www.mhhe.com/cengel-boles. See the Preface for access information.

4-53 An egg is dropped into boiling water. The amount of heat transfer to the egg by the time it is cooked is to be determined.

Assumptions **1** The egg is spherical in shape with a radius of $r_0 = 2.75$ cm. **2** The thermal properties of the egg are constant. **3** Energy absorption or release associated with any chemical and/or phase changes within the egg is negligible. **4** There are no changes in kinetic and potential energies.

Properties The density and specific heat of the egg are given to be $\rho = 1020$ kg/m^3 and $C_p = 3.32$ kJ/kg.°C.

Analysis We take the egg as the system. This is a closes system since no mass enters or leaves the egg. The energy balance for this closed system can be expressed as

$$\underbrace{E_{in} - E_{out}}_{\substack{\text{Net energy transfer}\\ \text{by heat, work, and mass}}} = \underbrace{\Delta E_{\text{system}}}_{\substack{\text{Change in internal, kinetic,}\\ \text{potential, etc. energies}}}$$

$$Q_{in} = \Delta U_{egg} = m(u_2 - u_1) = mC(T_2 - T_1)$$

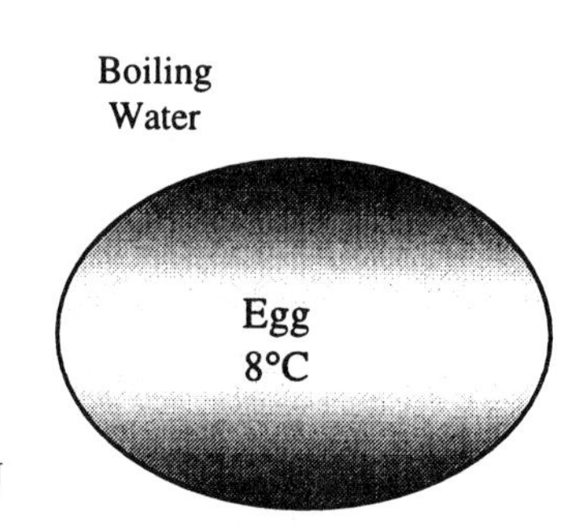

Then the mass of the egg and the amount of heat transfer become

$$m = \rho V = \rho \frac{\pi D^3}{6} = (1020\,\text{kg/m}^3)\frac{\pi(0.055\,\text{m})^3}{6} = 0.0889\,\text{kg}$$

$$Q_{in} = mC_p(T_2 - T_1) = (0.0889\,\text{kg})(3.32\,\text{kJ/kg.°C})(70 - 8)\text{°C} = \mathbf{18.3\,kJ}$$

4-54E Large brass plates are heated in an oven at a rate of 300/min. The rate of heat transfer to the plates in the oven is to be determined.

Assumptions **1** The thermal properties of the plates are constant. **2** The changes in kinetic and potential energies are negligible.

Properties The density and specific heat of the brass are given to be ρ = 532.5 lbm/ft³ and C_p = 0.091 Btu/lbm.°F.

Analysis We take the plate to be the system. The energy balance for this closed system can be expressed as

Plates
75°F

$$\underbrace{E_{in} - E_{out}}_{\substack{\text{Net energy transfer}\\ \text{by heat, work, and mass}}} = \underbrace{\Delta E_{\text{system}}}_{\substack{\text{Change in internal, kinetic,}\\ \text{potential, etc. energies}}}$$

$$Q_{in} = \Delta U_{\text{plate}} = m(u_2 - u_1) = mC(T_2 - T_1)$$

The mass of each plate and the amount of heat transfer to each plate is

$$m = \rho V = \rho LA = (532.5 \text{ lbm/ft}^3)[(1.2/12 \text{ ft})(2 \text{ ft})(2 \text{ ft})] = 213 \text{ lbm}$$

$$Q_{in} = mC(T_2 - T_1) = (213 \text{ lbm/plate})(0.091 \text{ Btu/lbm.°F})(1000 - 75)\text{°F} = 17{,}930 \text{ Btu/plate}$$

Then the total rate of heat transfer to the plates becomes

$$\dot{Q}_{total} = \dot{n}_{\text{plate}} Q_{\text{in, per plate}} = (300 \text{ plates/min}) \times (17{,}930 \text{ Btu/plate}) = \mathbf{5{,}379{,}000 \text{ Btu/min} = 89{,}650 \text{ Btu/s}}$$

4-55 Long cylindrical steel rods are heat-treated in an oven. The rate of heat transfer to the rods in the oven is to be determined.

Assumptions **1** The thermal properties of the rods are constant. **2** The changes in kinetic and potential energies are negligible.

Properties The density and specific heat of the steel rods are given to be ρ = 7833 kg/m³ and C_p = 0.465 kJ/kg.°C.

Analysis Noting that the rods enter the oven at a velocity of 3 m/min and exit at the same velocity, we can say that a 3-m long section of the rod is heated in the oven in 1 min. Then the mass of the rod heated in 1 minute is

$$m = \rho V = \rho LA = \rho L(\pi D^2/4) = (7833 \text{ kg/m}^3)(3 \text{ m})[\pi(0.1 \text{ m})^2/4] = 184.6 \text{ kg}$$

We take the 3-m section of the rod in the oven as the system. The energy balance for this closed system can be expressed as

$$\underbrace{E_{in} - E_{out}}_{\substack{\text{Net energy transfer}\\ \text{by heat, work, and mass}}} = \underbrace{\Delta E_{\text{system}}}_{\substack{\text{Change in internal, kinetic,}\\ \text{potential, etc. energies}}}$$

$$Q_{in} = \Delta U_{\text{rod}} = m(u_2 - u_1) = mC(T_2 - T_1)$$

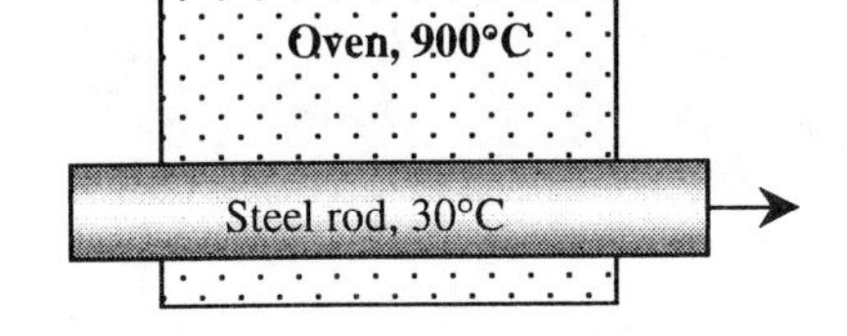

Substituting,

$$Q_{in} = mC(T_2 - T_1) = (184.6 \text{ kg})(0.465 \text{ kJ/kg.°C})(700 - 30)\text{°C} = 57{,}512 \text{ kJ}$$

Noting that this much heat is transferred in 1 min, the rate of heat transfer to the rod becomes

$$\dot{Q}_{in} = Q_{in}/\Delta t = (57{,}512 \text{ kJ})/(1 \text{ min}) = 57{,}512 \text{ kJ/min} = \mathbf{958.5 \text{ kW}}$$

Steady Flow Energy Balance: Nozzles and Diffusers

4-56C A steady-flow system involves no changes with time anywhere within the system or at the system boundaries

4-57C No.

4-58C It is mostly converted to internal energy as shown by a rise in the fluid temperature.

4-59C The kinetic energy of a fluid increases at the expense of the internal energy as evidenced by a decrease in the fluid temperature.

4-60C Heat transfer to the fluid as it flows through a nozzle is desirable since it will probably increase the kinetic energy of the fluid. Heat transfer from the fluid will decrease the exit velocity.

4-61 Air is accelerated in a nozzle from 30 m/s to 180 m/s. The mass flow rate, the exit temperature, and the exit area of the nozzle are to be determined.

Assumptions **1** This is a steady-flow process since there is no change with time. **2** Air is an ideal gas with constant specific heats. **3** Potential energy changes are negligible. **4** The device is adiabatic and thus heat transfer is negligible. **5** There are no work interactions.

Properties The gas constant of air is 0.287 kPa.m³/kg.K (Table A-1). The specific heat of air at the anticipated average temperature of 450 K is C_p = 1.02 kJ/kg.°C (Table A-2).

Analysis (*a*) There is only one inlet and one exit, and thus $\dot{m}_1 = \dot{m}_2 = \dot{m}$. Using the ideal gas relation, the specific volume and the mass flow rate of air are determined to be

$$v_1 = \frac{RT_1}{P_1} = \frac{(0.287\ \text{kPa}\cdot\text{m}^3/\text{kg}\cdot\text{K})(473\ \text{K})}{300\ \text{kPa}} = 0.4525\ \text{m}^3/\text{kg}$$

$$\dot{m} = \frac{1}{v_1}A_1\mathbf{V}_1 = \frac{1}{0.4525\text{m}^3/\text{kg}}(0.008\text{m}^2)(30\text{m/s}) = \mathbf{0.5304\ kg/s}$$

(*b*) We take nozzle as the system, which is a control volume since mass crosses the boundary. The energy balance for this steady-flow system can be expressed in the rate form as

$$\underbrace{\dot{E}_{in} - \dot{E}_{out}}_{\substack{\text{Rate of net energy transfer}\\ \text{by heat, work, and mass}}} = \underbrace{\Delta\dot{E}_{\text{system}}^{\nearrow 0\ \text{(steady)}}}_{\substack{\text{Rate of change in internal, kinetic,}\\ \text{potential, etc. energies}}} = 0$$

$$\dot{E}_{in} = \dot{E}_{out}$$

$$\dot{m}(h_1 + \mathbf{V}_1^2/2) = \dot{m}(h_2 + \mathbf{V}_2^2/2) \quad (\text{since } \dot{Q} \cong \dot{W} \cong \Delta\text{pe} \cong 0)$$

$$0 = h_2 - h_1 + \frac{\mathbf{V}_2^2 - \mathbf{V}_1^2}{2} \longrightarrow 0 = C_{p,ave}(T_2 - T_1) + \frac{\mathbf{V}_2^2 - \mathbf{V}_1^2}{2}$$

Substituting,

$$0 = (1.02\ \text{kJ/kg}\cdot\text{K})(T_2 - 200^\circ\text{C}) + \frac{(180\ \text{m/s})^2 - (30\ \text{m/s})^2}{2}\left(\frac{1\ \text{kJ/kg}}{1000\ \text{m}^2/\text{s}^2}\right)$$

It yields T_2 = **184.6°C**

(*c*) The specific volume of air at the nozzle exit is

$$v_2 = \frac{RT_2}{P_2} = \frac{(0.287\ \text{kPa}\cdot\text{m}^3/\text{kg}\cdot\text{K})(184.6 + 273\ \text{K})}{100\ \text{kPa}} = 1.313\ \text{m}^3/\text{kg}$$

$$\dot{m} = \frac{1}{v_2}A_2\mathbf{V}_2 \longrightarrow 0.5304\text{kg/s} = \frac{1}{1.313\text{m}^3/\text{kg}}A_2(180\text{m/s})$$

$$A_2 = 0.00387\ \text{m}^2 = \mathbf{38.7\ cm^2}$$

P_1 = 300 kPa
T_1 = 200°C
V_1 = 30 m/s
A_1 = 80 cm²
AIR
P_2 = 100 kPa
V_2 = 180 m/s

4-62 EES solution of this (and other comprehensive problems designated with the *computer icon*) is available to instructors at the *Instructor Manual* section of the *Online Learning Center* (OLC) at www.mhhe.com/cengel-boles. See the Preface for access information.

4-63 Steam is accelerated in a nozzle from a velocity of 80 m/s. The mass flow rate, the exit velocity, and the exit area of the nozzle are to be determined.

Assumptions **1** This is a steady-flow process since there is no change with time. **2** Potential energy changes are negligible. **3** There are no work interactions.

Properties From the steam tables (Table A-6)

$$\left.\begin{array}{l} P_1 = 5\ \text{MPa} \\ T_1 = 500^\circ\text{C} \end{array}\right\} \begin{array}{l} v_1 = 0.06857\ \text{m}^3/\text{kg} \\ h_1 = 3433.8\ \text{kJ/kg} \end{array}$$

and

$$\left.\begin{array}{l} P_2 = 2\text{MPa} \\ T_2 = 400^\circ\text{C} \end{array}\right\} \begin{array}{l} v_2 = 0.15120\text{m}^3/\text{kg} \\ h_2 = 3247.6\text{kJ/kg} \end{array}$$

90 kJ/s

1 Steam 2

Analysis (*a*) There is only one inlet and one exit, and thus $\dot{m}_1 = \dot{m}_2 = \dot{m}$.

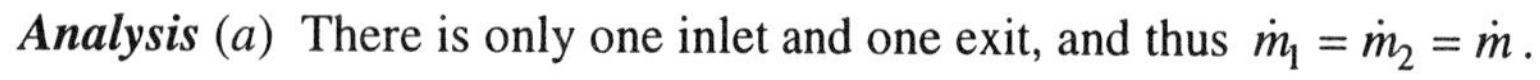

The mass flow rate of steam is

$$\dot{m} = \frac{1}{v_1}\mathbf{V}_1 A_1 = \frac{1}{0.06857\text{m}^3/\text{kg}}(80\ \text{m/s})(50 \times 10^{-4}\ \text{m}^2) = \mathbf{5.833\ kg/s}$$

(*b*) We take nozzle as the system, which is a control volume since mass crosses the boundary. The energy balance for this steady-flow system can be expressed in the rate form as

$$\underbrace{\dot{E}_{in} - \dot{E}_{out}}_{\substack{\text{Rate of net energy transfer} \\ \text{by heat, work, and mass}}} = \underbrace{\Delta \dot{E}_{\text{system}}^{\nearrow 0\ (\text{steady})}}_{\substack{\text{Rate of change in internal, kinetic,} \\ \text{potential, etc. energies}}} = 0$$

$$\dot{E}_{in} = \dot{E}_{out}$$

$$\dot{m}(h_1 + \mathbf{V}_1^2/2) = \dot{Q}_{\text{out}} + \dot{m}(h_2 + \mathbf{V}_2^2/2) \quad (\text{since } \dot{W} \cong \Delta\text{pe} \cong 0)$$

$$-\dot{Q}_{\text{out}} = \dot{m}\left(h_2 - h_1 + \frac{\mathbf{V}_2^2 - \mathbf{V}_1^2}{2}\right)$$

Substituting, the exit velocity of the steam is determined to be

$$-90\text{kJ/s} = (5.833\text{kg/s})\left(3247.6 - 3433.8 + \frac{\mathbf{V}_2^2 - (80\text{m/s})^2}{2}\left(\frac{1\text{kJ/kg}}{1000\text{m}^2/\text{s}^2}\right)\right)$$

It yields $\mathbf{V}_2 = \mathbf{589.9\ m/s}$

(*c*) The exit area of the nozzle is determined from

$$\dot{m} = \frac{1}{v_2}\mathbf{V}_2 A_2 \longrightarrow A_2 = \frac{\dot{m}v_2}{\mathbf{V}_2} = \frac{(5.833\ \text{kg/s})(0.1512\ \text{m}^3/\text{kg})}{589.9\ \text{m/s}} = \mathbf{15.0 \times 10^{-4}\ m^2}$$

4-64E Air is accelerated in a nozzle from 150 ft/s to 900 ft/s. The exit temperature of air and the exit area of the nozzle are to be determined.

Assumptions **1** This is a steady-flow process since there is no change with time. **2** Air is an ideal gas with variable specific heats. **3** Potential energy changes are negligible. **4** There are no work interactions.

Properties The enthalpy of air at the inlet is h_1 = 143.47 Btu/lbm (Table A-17E).

Analysis (a) There is only one inlet and one exit, and thus $\dot{m}_1 = \dot{m}_2 = \dot{m}$. We take nozzle as the system, which is a control volume since mass crosses the boundary. The energy balance for this steady-flow system can be expressed in the rate form as

$$\underbrace{\dot{E}_{in} - \dot{E}_{out}}_{\text{Rate of net energy transfer by heat, work, and mass}} = \underbrace{\Delta \dot{E}_{\text{system}}^{\nearrow 0 \text{ (steady)}}}_{\text{Rate of change in internal, kinetic, potential, etc. energies}} = 0$$

$$\dot{E}_{in} = \dot{E}_{out}$$

$$\dot{m}(h_1 + \mathbf{V}_1^2/2) = \dot{Q}_{\text{out}} + \dot{m}(h_2 + \mathbf{V}_2^2/2) \quad (\text{since } \dot{W} \cong \Delta\text{pe} \cong 0)$$

$$-\dot{Q}_{\text{out}} = \dot{m}\left(h_2 - h_1 + \frac{\mathbf{V}_2^2 - \mathbf{V}_1^2}{2}\right)$$

or,

$$h_2 = -q_{\text{out}} + h_1 - \frac{\mathbf{V}_2^2 - \mathbf{V}_1^2}{2}$$

$$= -6.5 \text{ Btu/lbm} + 143.47 \text{ Btu/lbm} - \frac{(900 \text{ ft/s})^2 - (150 \text{ ft/s})^2}{2}\left(\frac{1 \text{ Btu/lbm}}{25{,}037 \text{ ft}^2/\text{s}^2}\right)$$

$$= 121.2 \text{ Btu/lbm}$$

Thus, from Table A-17E, T_2 = **507 R**

(*b*) The exit area is determined from the conservation of mass relation,

$$\frac{1}{v_2}A_2\mathbf{V}_2 = \frac{1}{v_1}A_1\mathbf{V}_1 \longrightarrow A_2 = \frac{v_2}{v_1}\frac{\mathbf{V}_1}{\mathbf{V}_2}A_1 = \left(\frac{RT_2/P_2}{RT_1/P_1}\right)\frac{\mathbf{V}_1}{\mathbf{V}_2}A_1$$

$$A_2 = \frac{(508/14.7)(150\text{ft/s})}{(600/50)(900\text{ft/s})}(0.1\text{ft}^2) = \mathbf{0.048ft^2}$$

4-65 [*Also solved by EES on enclosed CD*] Steam is accelerated in a nozzle from a velocity of 40 m/s to 300 m/s. The exit temperature and the ratio of the inlet-to-exit area of the nozzle are to be determined.

Assumptions **1** This is a steady-flow process since there is no change with time. **2** Potential energy changes are negligible. **3** There are no work interactions. **4** The device is adiabatic and thus heat transfer is negligible.

Properties From the steam tables (Table A-6),

$$\left.\begin{array}{l} P_1 = 3\text{MPa} \\ T_1 = 400^\circ\text{C} \end{array}\right\} \begin{array}{l} v_1 = 0.09936\text{m}^3/\text{kg} \\ h_1 = 3230.9\text{kJ/kg} \end{array}$$

Analysis (*a*) There is only one inlet and one exit, and thus $\dot{m}_1 = \dot{m}_2 = \dot{m}$. We take nozzle as the system, which is a control volume since mass crosses the boundary. The energy balance for this steady-flow system can be expressed in the rate form as

$$\underbrace{\dot{E}_{in} - \dot{E}_{out}}_{\substack{\text{Rate of net energy transfer} \\ \text{by heat, work, and mass}}} = \underbrace{\Delta \dot{E}_{\text{system}}^{\nearrow 0 \text{ (steady)}}}_{\substack{\text{Rate of change in internal, kinetic,} \\ \text{potential, etc. energies}}} = 0$$

$$\dot{E}_{in} = \dot{E}_{out}$$

P₁ = 3 MPa, T₁ = 400°C, V₁ = 40 m/s → Steam → P₂ = 2.5 MPa, V₂ = 300 m/s

$$\dot{m}(h_1 + \mathbf{V}_1^2/2) = \dot{m}(h_2 + \mathbf{V}_2^2/2) \quad (\text{since } \dot{Q} \cong \dot{W} \cong \Delta\text{pe} \cong 0)$$

$$0 = h_2 - h_1 + \frac{\mathbf{V}_2^2 - \mathbf{V}_1^2}{2}$$

or,

$$h_2 = h_1 - \frac{\mathbf{V}_2^2 - \mathbf{V}_1^2}{2} = 3230.9 \text{ kJ/kg} - \frac{(300 \text{ m/s})^2 - (40 \text{ m/s})^2}{2}\left(\frac{1 \text{ kJ/kg}}{1000 \text{ m}^2/\text{s}^2}\right) = 3186.7 \text{ kJ/kg}$$

Thus,
$$\left.\begin{array}{l} P_2 = 2.5\text{MPa} \\ h_2 = 3186.7\text{kJ/kg} \end{array}\right\} \begin{array}{l} T_2 = \mathbf{376.7^\circ C} \\ v_2 = 0.1153\text{m}^3/\text{kg} \end{array}$$

(*b*) The ratio of the inlet to exit area is determined from the conservation of mass relation,

$$\frac{1}{v_2}A_2\mathbf{V}_2 = \frac{1}{v_1}A_1\mathbf{V}_1 \longrightarrow \frac{A_1}{A_2} = \frac{v_1}{v_2}\frac{\mathbf{V}_2}{\mathbf{V}_1} = \frac{(0.09936 \text{ m}^3/\text{kg})(300 \text{ m/s})}{(0.1153 \text{ m}^3/\text{kg})(40 \text{ m/s})} = \mathbf{6.46}$$

4-66 Air is accelerated in a nozzle from 120 m/s to 380 m/s. The exit temperature and pressure of air are to be determined.

Assumptions **1** This is a steady-flow process since there is no change with time. **2** Air is an ideal gas with variable specific heats. **3** Potential energy changes are negligible. **4** The device is adiabatic and thus heat transfer is negligible. **5** There are no work interactions.

Properties The enthalpy of air at the inlet temperature of 500 K is $h_1 = 503.02$ kJ/kg (Table A-17).

Analysis (a) There is only one inlet and one exit, and thus $\dot{m}_1 = \dot{m}_2 = \dot{m}$. We take nozzle as the system, which is a control volume since mass crosses the boundary. The energy balance for this steady-flow system can be expressed in the rate form as

$$\underbrace{\dot{E}_{in} - \dot{E}_{out}}_{\text{Rate of net energy transfer by heat, work, and mass}} = \underbrace{\Delta \dot{E}_{\text{system}}^{\nearrow 0 \text{ (steady)}}}_{\text{Rate of change in internal, kinetic, potential, etc. energies}} = 0$$

$$\dot{E}_{in} = \dot{E}_{out}$$

$$\dot{m}(h_1 + \mathbf{V}_1^2/2) = \dot{m}(h_2 + \mathbf{V}_2^2/2) \quad (\text{since } \dot{Q} \cong \dot{W} \cong \Delta\text{pe} \cong 0)$$

$$0 = h_2 - h_1 + \frac{\mathbf{V}_2^2 - \mathbf{V}_1^2}{2}$$

1 AIR → 2

or,

$$h_2 = h_1 - \frac{\mathbf{V}_2^2 - \mathbf{V}_1^2}{2} = 503.02\text{kJ/kg} - \frac{(380\text{m/s})^2 - (120\text{m/s})^2}{2}\left(\frac{1\text{kJ/kg}}{1000\text{m}^2/\text{s}^2}\right) = 438.02\text{kJ/kg}$$

Then from Table A-17 we read $T_2 =$ **436.5 K**

(b) The exit pressure is determined from the conservation of mass relation,

$$\frac{1}{v_2} A_2 \mathbf{V}_2 = \frac{1}{v_1} A_1 \mathbf{V}_1 \longrightarrow \frac{1}{RT_2/P_2} A_2 \mathbf{V}_2 = \frac{1}{RT_1/P_1} A_1 \mathbf{V}_1$$

Thus,

$$P_2 = \frac{A_1 T_2 \mathbf{V}_1}{A_2 T_1 \mathbf{V}_2} P_1 = \frac{2}{1}\frac{(436.5\text{ K})(120\text{ m/s})}{(500\text{ K})(380\text{ m/s})}(600\text{ kPa}) = \mathbf{330.8\ kPa}$$

4-67 Air is decelerated in a diffuser from 230 m/s to 30 m/s. The exit temperature of air and the exit area of the diffuser are to be determined.

Assumptions **1** This is a steady-flow process since there is no change with time. **2** Air is an ideal gas with variable specific heats. **3** Potential energy changes are negligible. **4** The device is adiabatic and thus heat transfer is negligible. **5** There are no work interactions.

Properties The gas constant of air is 0.287 kPa.m³/kg.K (Table A-1). The enthalpy of air at the inlet temperature of 400 K is $h_1 = 400.98$ kJ/kg (Table A-17).

Analysis (a) There is only one inlet and one exit, and thus $\dot{m}_1 = \dot{m}_2 = \dot{m}$. We take diffuser as the system, which is a control volume since mass crosses the boundary. The energy balance for this steady-flow system can be expressed in the rate form as

$$\underbrace{\dot{E}_{in} - \dot{E}_{out}}_{\text{Rate of net energy transfer by heat, work, and mass}} = \underbrace{\Delta \dot{E}_{system}^{\nearrow 0 \text{ (steady)}}}_{\text{Rate of change in internal, kinetic, potential, etc. energies}} = 0$$

$$\dot{E}_{in} = \dot{E}_{out}$$

$$\dot{m}(h_1 + \mathbf{V}_1^2/2) = \dot{m}(h_2 + \mathbf{V}_2^2/2) \quad (\text{since } \dot{Q} \cong \dot{W} \cong \Delta pe \cong 0)$$

$$0 = h_2 - h_1 + \frac{\mathbf{V}_2^2 - \mathbf{V}_1^2}{2}$$

1 AIR → 2

or,

$$h_2 = h_1 - \frac{\mathbf{V}_2^2 - \mathbf{V}_1^2}{2} = 400.98\text{kJ/kg} - \frac{(30\text{m/s})^2 - (230\text{m/s})^2}{2}\left(\frac{1\text{kJ/kg}}{1000\text{m}^2/\text{s}^2}\right) = 426.98\text{kJ/kg}$$

From Table A-17, $T_2 = \mathbf{425.6\ K}$

(b) The specific volume of air at the diffuser exit is

$$v_2 = \frac{RT_2}{P_2} = \frac{(0.287\text{kPa}\cdot\text{m}^3/\text{kg}\cdot\text{K})(425.6\text{K})}{(100\text{kPa})} = 1.221\text{m}^3/\text{kg}$$

From conservation of mass,

$$\dot{m} = \frac{1}{v_2}A_2\mathbf{V}_2 \longrightarrow A_2 = \frac{\dot{m}v_2}{\mathbf{V}_2} = \frac{(6000/3600\text{ kg/s})(1.221\text{ m}^3/\text{kg})}{30\text{ m/s}} = \mathbf{0.0678\ m^2}$$

4-68E Air is decelerated in a diffuser from 600 ft/s to a low velocity. The exit temperature and the exit velocity of air are to be determined.

Assumptions **1** This is a steady-flow process since there is no change with time. **2** Air is an ideal gas with variable specific heats. **3** Potential energy changes are negligible. **4** The device is adiabatic and thus heat transfer is negligible. **5** There are no work interactions.

Properties The enthalpy of air at the inlet temperature of 20°F is h_1 = 114.69 Btu/lbm (Table A-17E).

Analysis (*a*) There is only one inlet and one exit, and thus $\dot{m}_1 = \dot{m}_2 = \dot{m}$. We take diffuser as the system, which is a control volume since mass crosses the boundary. The energy balance for this steady-flow system can be expressed in the rate form as

$$\underbrace{\dot{E}_{in} - \dot{E}_{out}}_{\substack{\text{Rate of net energy transfer}\\ \text{by heat, work, and mass}}} = \underbrace{\Delta \dot{E}_{\text{system}}^{\nearrow 0 \text{ (steady)}}}_{\substack{\text{Rate of change in internal, kinetic,}\\ \text{potential, etc. energies}}} = 0$$

$$\dot{E}_{in} = \dot{E}_{out}$$

$$\dot{m}(h_1 + \mathbf{V}_1^2/2) = \dot{m}(h_2 + \mathbf{V}_2^2/2) \quad (\text{since } \dot{Q} \cong \dot{W} \cong \Delta\text{pe} \cong 0)$$

$$0 = h_2 - h_1 + \frac{\mathbf{V}_2^2 - \mathbf{V}_1^2}{2},$$

1 AIR → 2

or,

$$h_2 = h_1 - \frac{\mathbf{V}_2^2 - \mathbf{V}_1^2}{2} = 114.69\text{Btu/lbm} - \frac{0 - (600\text{ft/s})^2}{2}\left(\frac{1\text{Btu/lbm}}{25{,}037\text{ft}^2/\text{s}^2}\right) = 121.88\text{Btu/lbm}$$

From Table A-17E, T_2 = **510.0 R**

(*b*) The exit velocity of air is determined from the conservation of mass relation,

$$\frac{1}{v_2}A_2\mathbf{V}_2 = \frac{1}{v_1}A_1\mathbf{V}_1 \longrightarrow \frac{1}{RT_2/P_2}A_2\mathbf{V}_2 = \frac{1}{RT_1/P_1}A_1\mathbf{V}_1$$

Thus,

$$\mathbf{V}_2 = \frac{A_1T_2P_1}{A_2T_1P_2}\mathbf{V}_1 = \frac{1}{5}\frac{(510\text{ R})(13\text{ psia})}{(480\text{ R})(14.5\text{ psia})}(600\text{ ft/s}) = \mathbf{114.3\text{ ft/s}}$$

4-69E EES solution of this (and other comprehensive problems designated with the *computer icon*) is available to instructors at the *Instructor Manual* section of the *Online Learning Center* (OLC) at www.mhhe.com/cengel-boles. See the Preface for access information.

4-69 CO_2 gas is accelerated in a nozzle to 450 m/s. The inlet velocity and the exit temperature are to be determined.

Assumptions **1** This is a steady-flow process since there is no change with time. **2** CO_2 is an ideal gas with variable specific heats. **3** Potential energy changes are negligible. **4** The device is adiabatic and thus heat transfer is negligible. **5** There are no work interactions.

Properties The gas constant of CO_2 is 0.1889 kPa.m³/kg.K (Table A-1). The enthalpy of CO_2 at 500°C is $\bar{h}_1 = 30{,}797$ kJ/kmol (Table A-20).

Analysis (a) There is only one inlet and one exit, and thus $\dot{m}_1 = \dot{m}_2 = \dot{m}$. Using the ideal gas relation, the specific volume is determined to be

$$v_1 = \frac{RT_1}{P_1} = \frac{(0.1889 \text{ kPa}\cdot\text{m}^3/\text{kg}\cdot\text{K})(773 \text{ K})}{1000 \text{ kPa}} = 0.146 \text{ m}^3/\text{kg}$$

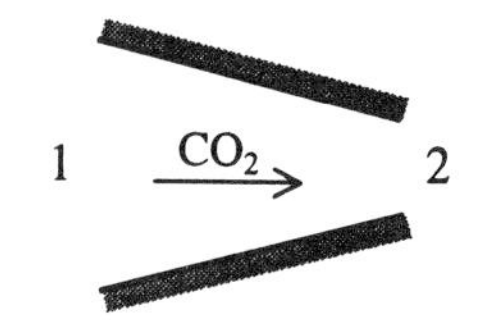

Thus,

$$\dot{m} = \frac{1}{v_1}A_1\mathbf{V}_1 \longrightarrow \mathbf{V}_1 = \frac{\dot{m}v_1}{A_1} = \frac{(6000/3600\text{kg/s})(0.146\text{m}^3/\text{kg})}{40\times 10^{-4}\text{m}^2} = \mathbf{60.8m/s}$$

(b) We take nozzle as the system, which is a control volume since mass crosses the boundary. The energy balance for this steady-flow system can be expressed in the rate form as

$$\underbrace{\dot{E}_{in} - \dot{E}_{out}}_{\substack{\text{Rate of net energy transfer}\\ \text{by heat, work, and mass}}} = \underbrace{\Delta \dot{E}_{\text{system}}^{\nearrow 0 \text{ (steady)}}}_{\substack{\text{Rate of change in internal, kinetic,}\\ \text{potential, etc. energies}}} = 0$$

$$\dot{E}_{in} = \dot{E}_{out}$$

$$\dot{m}(h_1 + \mathbf{V}_1^2/2) = \dot{m}(h_2 + \mathbf{V}_2^2/2) \quad (\text{since } \dot{Q} \cong \dot{W} \cong \Delta\text{pe} \cong 0)$$

$$0 = h_2 - h_1 + \frac{\mathbf{V}_2^2 - \mathbf{V}_1^2}{2}$$

Substituting,

$$\bar{h}_2 = \bar{h}_1 - \frac{\mathbf{V}_2^2 - \mathbf{V}_1^2}{2} \times M$$

$$= 30{,}797\text{kJ/kmol} - \frac{(450\text{m/s})^2 - (60.8\text{m/s})^2}{2}\left(\frac{1\text{kJ/kg}}{1000\text{m}^2/\text{s}^2}\right)(44\text{kg/kmol})$$

$$= 26{,}423\text{kJ/kmol}$$

Then the exit temperature of CO_2 from Table A-20 is obtained to be $T_2 = \mathbf{685.8\ K}$

4-70 R-134a is accelerated in a nozzle from a velocity of 20 m/s. The exit velocity of the refrigerant and the ratio of the inlet-to-exit area of the nozzle are to be determined.

Assumptions **1** This is a steady-flow process since there is no change with time. **2** Potential energy changes are negligible. **3** There are no work interactions. **4** The device is adiabatic and thus heat transfer is negligible.

Properties From the refrigerant tables (Table A-13)

$$\left.\begin{array}{l} P_1 = 700\text{kPa} \\ T_1 = 100^\circ\text{C} \end{array}\right\} \begin{array}{l} v_1 = 0.04064\text{m}^3/\text{kg} \\ h_1 = 338.19\text{kJ/kg} \end{array}$$

and

$$\left.\begin{array}{l} P_2 = 300\text{ kPa} \\ T_2 = 30^\circ\text{C} \end{array}\right\} \begin{array}{l} v_2 = 0.07767\text{ m}^3/\text{kg} \\ h_2 = 274.70\text{ kJ/kg} \end{array}$$

Analysis (*a*) There is only one inlet and one exit, and thus $\dot{m}_1 = \dot{m}_2 = \dot{m}$. We take nozzle as the system, which is a control volume since mass crosses the boundary. The energy balance for this steady-flow system can be expressed in the rate form as

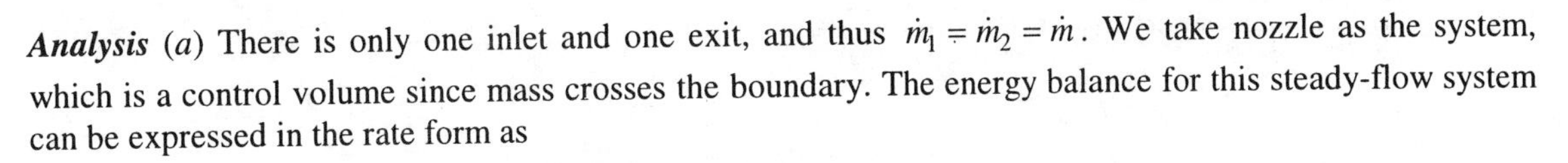

$$\underbrace{\dot{E}_{in} - \dot{E}_{out}}_{\substack{\text{Rate of net energy transfer} \\ \text{by heat, work, and mass}}} = \underbrace{\Delta \dot{E}_{\text{system}}^{\nearrow 0 \text{ (steady)}}}_{\substack{\text{Rate of change in internal, kinetic,} \\ \text{potential, etc. energies}}} = 0$$

$$\dot{E}_{in} = \dot{E}_{out}$$

$$\dot{m}(h_1 + \mathbf{V}_1^2/2) = \dot{m}(h_2 + \mathbf{V}_{21}^2/2) \quad (\text{since } \dot{Q} \cong \dot{W} \cong \Delta\text{pe} \cong 0)$$

$$0 = h_2 - h_1 + \frac{\mathbf{V}_2^2 - \mathbf{V}_1^2}{2}$$

Substituting,

$$0 = (274.70 - 338.19)\text{kJ/kg} + \frac{\mathbf{V}_2^2 - (20\text{m/s})^2}{2}\left(\frac{1\text{kJ/kg}}{1000\text{m}^2/\text{s}^2}\right)$$

It yields $\quad$ $\mathbf{V_2 = 356.9\ m/s}$

(*b*) The ratio of the inlet to exit area is determined from the conservation of mass relation,

$$\frac{1}{v_2}A_2\mathbf{V}_2 = \frac{1}{v_1}A_1\mathbf{V}_1 \longrightarrow \frac{A_1}{A_2} = \frac{v_2}{v_1}\frac{\mathbf{V}_1}{\mathbf{V}_2} = \frac{(0.04064\text{m}^3/\text{kg})(356.9\text{m/s})}{(0.07767\text{m}^3/\text{kg})(20\text{m/s})} = \mathbf{9.34}$$

4-71 Air is decelerated in a diffuser from 220 m/s. The exit velocity and the exit pressure of air are to be determined.

Assumptions **1** This is a steady-flow process since there is no change with time. **2** Air is an ideal gas with variable specific heats. **3** Potential energy changes are negligible. **4** There are no work interactions.

Properties The gas constant of air is 0.287 kPa.m^3/kg.K (Table A-1). The enthalpies are (Table A-17)

$$T_1 = 27°\text{C} = 300 \text{ K} \quad \rightarrow \quad h_1 = 300.19 \text{ kJ/kg}$$
$$T_2 = 42°\text{C} = 315 \text{ K} \quad \rightarrow \quad h_2 = 315.27 \text{ kJ/kg}$$

Analysis (*a*) There is only one inlet and one exit, and thus $\dot{m}_1 = \dot{m}_2 = \dot{m}$. We take diffuser as the system, which is a control volume since mass crosses the boundary. The energy balance for this steady-flow system can be expressed in the rate form as

$$\underbrace{\dot{E}_{in} - \dot{E}_{out}}_{\substack{\text{Rate of net energy transfer} \\ \text{by heat, work, and mass}}} = \underbrace{\Delta \dot{E}_{\text{system}}^{\nearrow 0 \text{ (steady)}}}_{\substack{\text{Rate of change in internal, kinetic,} \\ \text{potential, etc. energies}}} = 0$$

$$\dot{E}_{in} = \dot{E}_{out}$$

$$\dot{m}(h_1 + \mathbf{V}_1^2/2) = \dot{Q}_{out} + \dot{m}(h_2 + \mathbf{V}_2^2/2) \quad (\text{since } \dot{W} \cong \Delta\text{pe} \cong 0)$$

$$-\dot{Q}_{out} = \dot{m}\left(h_2 - h_1 + \frac{\mathbf{V}_2^2 - \mathbf{V}_1^2}{2}\right)$$

18 kJ/s

1 AIR 2

Substituting, the exit velocity of the air is determined to be

$$-18 \text{ kJ/s} = (2.5 \text{ kg/s})\left(315.27 - 300.19 + \frac{\mathbf{V}_2^2 - (220 \text{ m/s})^2}{2}\left(\frac{1 \text{ kJ/kg}}{1000 \text{ m}^2/\text{s}^2}\right)\right)$$

It yields $\mathbf{V_2 = 62.0 \text{ m/s}}$

(*b*) The exit pressure of air is determined from the conservation of mass and the ideal gas relations,

$$\dot{m} = \frac{1}{v_2} A_2 \mathbf{V}_2 \longrightarrow v_2 = \frac{A_2 \mathbf{V}_2}{\dot{m}} = \frac{(0.04\text{m}^2)(62\text{m/s})}{2.5\text{kg/s}} = 0.992\text{m}^3/\text{kg}$$

and

$$P_2 v_2 = RT_2 \longrightarrow P_2 = \frac{RT_2}{v_2} = \frac{(0.287\text{kPa}\cdot\text{m}^3/\text{kg}\cdot\text{K})(315\text{K})}{0.992\text{m}^3/\text{kg}} = \mathbf{91.1 \text{kPa}}$$

4-72 Nitrogen is decelerated in a diffuser from 200 m/s to a lower velocity. The exit velocity of nitrogen and the ratio of the inlet-to-exit area are to be determined.

Assumptions **1** This is a steady-flow process since there is no change with time. **2** Nitrogen is an ideal gas with variable specific heats. **3** Potential energy changes are negligible. **4** The device is adiabatic and thus heat transfer is negligible. **5** There are no work interactions.

Properties The molar mass of nitrogen is M = 28 kg/kmol (Table A-1). The enthalpies are (Table A-18)

$$T_1 = 7°\text{C} = 280\text{ K} \rightarrow \bar{h}_1 = 8141\text{ kJ/kmol}$$

$$T_2 = 22°\text{C} = 295\text{ K} \rightarrow \bar{h}_2 = 8580\text{ kJ/kmol}$$

Analysis (*a*) There is only one inlet and one exit, and thus $\dot{m}_1 = \dot{m}_2 = \dot{m}$. We take diffuser as the system, which is a control volume since mass crosses the boundary. The energy balance for this steady-flow system can be expressed in the rate form as

$$\underbrace{\dot{E}_{in} - \dot{E}_{out}}_{\substack{\text{Rate of net energy transfer}\\ \text{by heat, work, and mass}}} = \underbrace{\Delta \dot{E}_{\text{system}}^{\nearrow 0 \text{ (steady)}}}_{\substack{\text{Rate of change in internal, kinetic,}\\ \text{potential, etc. energies}}} = 0$$

$$\dot{E}_{in} = \dot{E}_{out}$$

$$\dot{m}(h_1 + \mathbf{V}_1^2/2) = \dot{m}(h_2 + \mathbf{V}_2^2/2) \quad (\text{since } \dot{Q} \cong \dot{W} \cong \Delta\text{pe} \cong 0)$$

$$0 = h_2 - h_1 + \frac{\mathbf{V}_2^2 - \mathbf{V}_1^2}{2} = \frac{\bar{h}_2 - \bar{h}_1}{M} + \frac{\mathbf{V}_2^2 - \mathbf{V}_1^2}{2},$$

1 N_2 2

Substituting,

$$0 = \frac{(8580 - 8141)\text{ kJ/kmol}}{28\text{ kJ/kmol}} + \frac{\mathbf{V}_2^2 - (200\text{m/s})^2}{2}\left(\frac{1\text{kJ/kg}}{1000\text{m}^2/\text{s}^2}\right)$$

It yields $\qquad$ $\mathbf{V_2 = 93.0}$ **m/s**

(*b*) The ratio of the inlet to exit area is determined from the conservation of mass relation,

$$\frac{1}{v_2}A_2\mathbf{V}_2 = \frac{1}{v_1}A_1\mathbf{V}_1 \longrightarrow \frac{A_1}{A_2} = \frac{v_1}{v_2}\frac{\mathbf{V}_2}{\mathbf{V}_1} = \left(\frac{RT_1/P_1}{RT_2/P_2}\right)\frac{\mathbf{V}_2}{\mathbf{V}_1}$$

or,

$$\frac{A_1}{A_2} = \left(\frac{T_1/P_1}{T_2/P_2}\right)\frac{\mathbf{V}_2}{\mathbf{V}_1} = \frac{(280\text{ K}/60\text{ kPa})(93.0\text{ m/s})}{(295\text{ K}/85\text{ kPa})(200\text{ m/s})} = \mathbf{0.625}$$

4-73 EES solution of this (and other comprehensive problems designated with the *computer icon*) is available to instructors at the *Instructor Manual* section of the *Online Learning Center* (OLC) at www.mhhe.com/cengel-boles. See the Preface for access information.

4-74 R-134a is decelerated in a diffuser from a velocity of 140 m/s. The exit velocity of R-134a and the mass flow rate of the R-134a are to be determined.

Assumptions **1** This is a steady-flow process since there is no change with time. **2** Potential energy changes are negligible. **3** There are no work interactions.

Properties From the R-134a tables (Tables A-11 through A-13)

$$\left.\begin{array}{l} P_1 = 700\text{kPa} \\ sat.vapor \end{array}\right\} \begin{array}{l} v_1 = 0.0292\text{m}^3/\text{kg} \\ h_1 = 261.85\text{kJ/kg} \end{array}$$

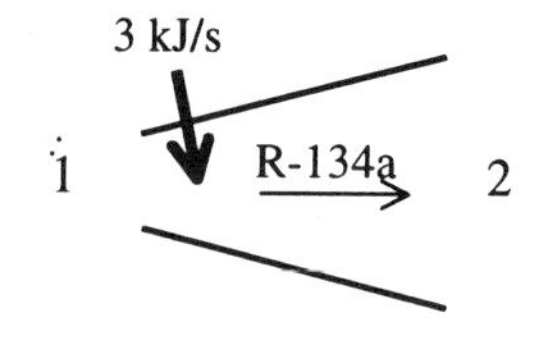

and

$$\left.\begin{array}{l} P_2 = 800\text{kPa} \\ T_2 = 40^\circ\text{C} \end{array}\right\} \begin{array}{l} v_2 = 0.02691\text{m}^3/\text{kg} \\ h_2 = 273.66\text{kJ/kg} \end{array}$$

Analysis (a) There is only one inlet and one exit, and thus $\dot{m}_1 = \dot{m}_2 = \dot{m}$. Then the exit velocity of R-134a is determined from the steady-flow mass balance to be

$$\frac{1}{v_2}A_2\mathbf{V}_2 = \frac{1}{v_1}A_1\mathbf{V}_1 \longrightarrow \mathbf{V}_2 = \frac{v_2}{v_1}\frac{A_1}{A_2}\mathbf{V}_1 = \frac{1}{1.8}\frac{0.02691\text{m}^3/\text{kg}}{0.02920\text{m}^3/\text{kg}}(140\text{m/s}) = \mathbf{71.7m/s}$$

(*b*) We take diffuser as the system, which is a control volume since mass crosses the boundary. The energy balance for this steady-flow system can be expressed in the rate form as

$$\underbrace{\dot{E}_{in} - \dot{E}_{out}}_{\text{Rate of net energy transfer by heat, work, and mass}} = \underbrace{\Delta\dot{E}_{\text{system}}^{\nearrow 0\ (\text{steady})}}_{\text{Rate of change in internal, kinetic, potential, etc. energies}} = 0$$

$$\dot{E}_{in} = \dot{E}_{out}$$

$$\dot{Q}_{\text{in}} + \dot{m}(h_1 + \mathbf{V}_1^2/2) = \dot{m}(h_2 + \mathbf{V}_2^2/2) \quad (\text{since } \dot{\text{W}} \cong \Delta\text{pe} \cong 0)$$

$$\dot{Q}_{\text{in}} = \dot{m}\left(h_2 - h_1 + \frac{\mathbf{V}_2^2 - \mathbf{V}_1^2}{2}\right)$$

Substituting, the mass flow rate of the refrigerant is determined to be

$$3\text{kJ/s} = \dot{m}\left(273.66 - 261.85 + \frac{(71.7\text{m/s})^2 - (140\text{m/s})^2}{2}\left(\frac{1\text{kJ/kg}}{1000\text{m}^2/\text{s}^2}\right)\right)$$

It yields $\dot{m} = \mathbf{0.655\ kg/s}$

Turbines and Compressors

4-75C Yes.

4-76C The volume flow rate at the compressor inlet will be greater than that at the compressor exit.

4-77C Yes. Because energy (in the form of shaft work) is being added to the air.

4-78C No.

4-79 Steam expands in a turbine. The change in kinetic energy, the power output, and the turbine inlet area are to be determined.

Assumptions **1** This is a steady-flow process since there is no change with time. **2** Potential energy changes are negligible. **3** The device is adiabatic and thus heat transfer is negligible.

Properties From the steam tables (Tables A-4 through 6)

$$\left.\begin{array}{l} P_1 = 10\text{MPa} \\ T_1 = 450^\circ\text{C} \end{array}\right\} \begin{array}{l} v_1 = 0.02975\text{m}^3/\text{kg} \\ h_1 = 3240.9\text{kJ/kg} \end{array}$$

and

$$\left.\begin{array}{l} P_2 = 10\text{ kPa} \\ x_2 = 0.92 \end{array}\right\} h_2 = h_f + x_2 h_{fg} = 191.83 + 0.92 \times 2392.8 = 2393.2\text{kJ/kg}$$

P1 = 10 MPa
T1 = 450°C
V1 = 80 m/s
STEAM
ṁ = 12 kg/s
Ẇ
P2 = 10 kPa
x2 = 0.92
V2 = 50 m/s

Analysis (a) The change in kinetic energy is determined from

$$\Delta ke = \frac{\mathbf{V}_2^2 - \mathbf{V}_1^2}{2} = \frac{(50\text{m/s})^2 - (80\text{m/s})^2}{2}\left(\frac{1\text{kJ/kg}}{1000\text{m}^2/\text{s}^2}\right) = \mathbf{-1.95kJ/kg}$$

(*b*) There is only one inlet and one exit, and thus $\dot{m}_1 = \dot{m}_2 = \dot{m}$. We take the turbine as the system, which is a control volume since mass crosses the boundary. The energy balance for this steady-flow system can be expressed in the rate form as

$$\underbrace{\dot{E}_{in} - \dot{E}_{out}}_{\substack{\text{Rate of net energy transfer} \\ \text{by heat, work, and mass}}} = \underbrace{\Delta \dot{E}_{\text{system}}^{\nearrow 0 \text{ (steady)}}}_{\substack{\text{Rate of change in internal, kinetic,} \\ \text{potential, etc. energies}}} = 0$$

$$\dot{E}_{in} = \dot{E}_{out}$$

$$\dot{m}(h_1 + \mathbf{V}_1^2/2) = \dot{W}_{\text{out}} + \dot{m}(h_2 + \mathbf{V}_2^2/2) \quad (\text{since } \dot{Q} \cong \Delta\text{pe} \cong 0)$$

$$\dot{W}_{\text{out}} = -\dot{m}\left(h_2 - h_1 + \frac{\mathbf{V}_2^2 - \mathbf{V}_1^2}{2}\right)$$

Then the power output of the turbine is determined by substitution to be

$$\dot{W}_{\text{out}} = -(12\text{ kg/s})(2393.2 - 3240.9 - 1.95)\text{kJ/kg} = \mathbf{10.2\ MW}$$

(*c*) The inlet area of the turbine is determined from the mass flow rate relation,

$$\dot{m} = \frac{1}{v_1}A_1\mathbf{V}_1 \longrightarrow A_1 = \frac{\dot{m}v_1}{\mathbf{V}_1} = \frac{(12\text{ kg/s})(0.02975\text{ m}^3/\text{kg})}{80\text{ m/s}} = \mathbf{0.00446\ m^2}$$

4-80 EES solution of this (and other comprehensive problems designated with the *computer icon*) is available to instructors at the *Instructor Manual* section of the *Online Learning Center* (OLC) at www.mhhe.com/cengel-boles. See the Preface for access information.

4-81 Steam expands in a turbine. The mass flow rate of steam for a power output of 5 MW is to be determined.

Assumptions **1** This is a steady-flow process since there is no change with time. **2** Kinetic and potential energy changes are negligible. **3** The device is adiabatic and thus heat transfer is negligible.

Properties From the steam tables (Tables A-4 through 6)

$$\left.\begin{array}{l} P_1 = 10\text{MPa} \\ T_1 = 400^\circ\text{C} \end{array}\right\} h_1 = 3096.5\text{kJ/kg}$$

$$\left.\begin{array}{l} P_2 = 20\text{kPa} \\ x_2 = 0.90 \end{array}\right\} h_2 = h_f + x_2 h_{fg} = 251.40 + 0.90 \times 2358.3 = 2373.9\text{kJ/kg}$$

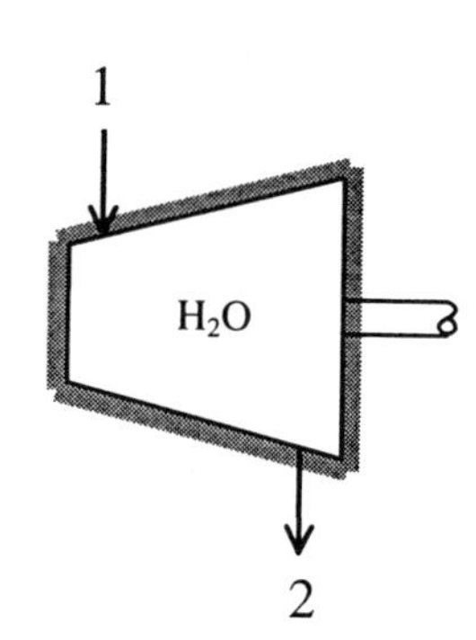

Analysis There is only one inlet and one exit, and thus $\dot{m}_1 = \dot{m}_2 = \dot{m}$. We take the turbine as the system, which is a control volume since mass crosses the boundary. The energy balance for this steady-flow system can be expressed in the rate form as

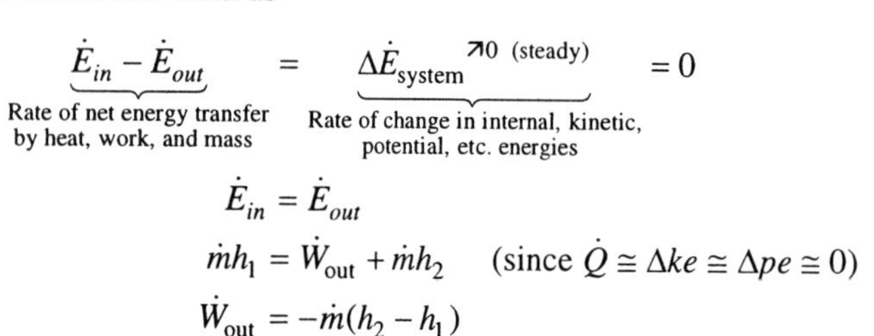

$$\underbrace{\dot{E}_{in} - \dot{E}_{out}}_{\text{Rate of net energy transfer by heat, work, and mass}} = \underbrace{\Delta \dot{E}_{\text{system}}^{\nearrow 0 \text{ (steady)}}}_{\text{Rate of change in internal, kinetic, potential, etc. energies}} = 0$$

$$\dot{E}_{in} = \dot{E}_{out}$$

$$\dot{m}h_1 = \dot{W}_{\text{out}} + \dot{m}h_2 \quad (\text{since } \dot{Q} \cong \Delta ke \cong \Delta pe \cong 0)$$

$$\dot{W}_{\text{out}} = -\dot{m}(h_2 - h_1)$$

Substituting, the required mass flow rate of the steam is determined to be

$$5000 \text{ kJ/s} = -\dot{m}(2373.9 - 3096.5) \text{ kJ/kg}$$

$$\dot{m} = \mathbf{6.919 \ kg/s}$$

4-82E Steam expands in a turbine. The rate of heat loss from the steam for a power output of 4 MW is to be determined.

Assumptions **1** This is a steady-flow process since there is no change with time. **2** Kinetic and potential energy changes are negligible.

Properties From the steam tables (Tables A-4 through 6)

$$\left.\begin{array}{l} P_1 = 1000\text{psia} \\ T_1 = 900^\circ\text{F} \end{array}\right\} h_1 = 1448.1\text{Btu/lbm}$$

$$\left.\begin{array}{l} P_2 = 5\text{psia} \\ sat.vapor \end{array}\right\} h_2 = 1131.0\text{Btu/lbm}$$

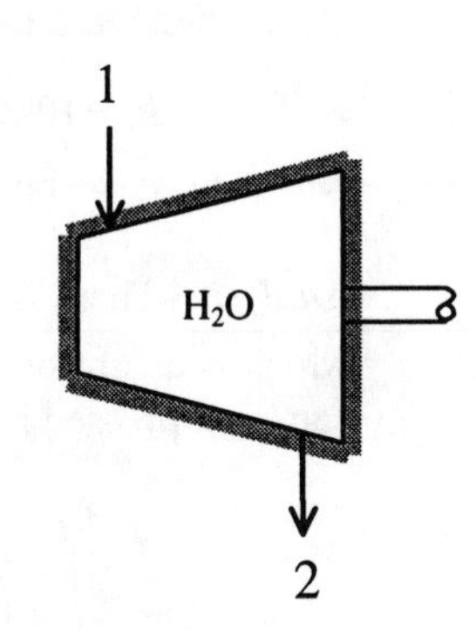

Analysis There is only one inlet and one exit, and thus $\dot{m}_1 = \dot{m}_2 = \dot{m}$. We take the turbine as the system, which is a control volume since mass crosses the boundary. The energy balance for this steady-flow system can be expressed in the rate form as

$$\underbrace{\dot{E}_{in} - \dot{E}_{out}}_{\substack{\text{Rate of net energy transfer} \\ \text{by heat, work, and mass}}} = \underbrace{\Delta \dot{E}_{\text{system}}^{\nearrow 0 \text{ (steady)}}}_{\substack{\text{Rate of change in internal, kinetic,} \\ \text{potential, etc. energies}}} = 0$$

$$\dot{E}_{in} = \dot{E}_{out}$$

$$\dot{m}h_1 = \dot{Q}_{\text{out}} + \dot{W}_{\text{out}} + \dot{m}h_2 \quad (\text{since } \Delta ke \cong \Delta pe \cong 0)$$

$$\dot{Q}_{\text{out}} = -\dot{m}(h_2 - h_1) - \dot{W}_{\text{out}}$$

Substituting,

$$\dot{Q}_{\text{out}} = -(45000/3600 \text{ lbm/s})(45000/3600 \text{ lbm/s})\text{Btu/lbm} - 4000 \text{ kJ/s}\left(\frac{1 \text{ Btu}}{1.055 \text{ kJ}}\right)$$

$$= \mathbf{172.3 \text{ Btu/s}}$$

4-83 Steam expands in a turbine. The exit temperature of the steam for a power output of 2 MW is to be determined.

Assumptions **1** This is a steady-flow process since there is no change with time. **2** Kinetic and potential energy changes are negligible. **3** The device is adiabatic and thus heat transfer is negligible.

Properties From the steam tables (Tables A-4 through 6)

$$\left.\begin{array}{l} P_1 = 10\text{MPa} \\ T_1 = 500^\circ\text{C} \end{array}\right\} h_1 = 3373.7\text{kJ/kg}$$

Analysis There is only one inlet and one exit, and thus $\dot{m}_1 = \dot{m}_2 = \dot{m}$. We take the turbine as the system, which is a control volume since mass crosses the boundary. The energy balance for this steady-flow system can be expressed in the rate form as

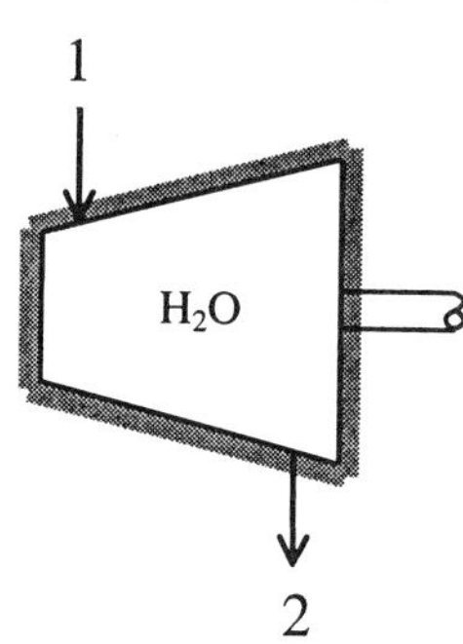

$$\underbrace{\dot{E}_{in} - \dot{E}_{out}}_{\substack{\text{Rate of net energy transfer} \\ \text{by heat, work, and mass}}} = \underbrace{\Delta \dot{E}_{\text{system}}^{\nearrow 0 \text{ (steady)}}}_{\substack{\text{Rate of change in internal, kinetic,} \\ \text{potential, etc. energies}}} = 0$$

$$\dot{E}_{in} = \dot{E}_{out}$$

$$\dot{m}h_1 = \dot{W}_{\text{out}} + \dot{m}h_2 \quad (\text{since } \dot{Q} \cong \Delta ke \cong \Delta pe \cong 0)$$

$$\dot{W}_{\text{out}} = \dot{m}(h_1 - h_2)$$

Substituting,

$$2000\text{kJ/s} = (3\text{kg/s})(3373.7 - h_2)\text{kJ/kg}$$

$$h_2 = 2707\text{kJ/kg}$$

Then the exit temperature becomes

$$\left.\begin{array}{l} P_2 = 20\text{kPa} \\ h_2 = 2707\text{kJ/kg} \end{array}\right\} T_2 = \mathbf{110.8^\circ C}$$

4-84 Argon gas expands in a turbine. The exit temperature of the argon for a power output of 250 kW is to be determined.

Assumptions **1** This is a steady-flow process since there is no change with time. **2** Potential energy changes are negligible. **3** The device is adiabatic and thus heat transfer is negligible. **4** Argon is an ideal gas with constant specific heats.

Properties The gas constant of Ar is R = 0.2081 kPa.m^3/kg.K. The constant pressure specific heat of Ar is C_p = 0.5203 kJ/kg· °C (Tables A-2a)

Analysis There is only one inlet and one exit, and thus $\dot{m}_1 = \dot{m}_2 = \dot{m}$. The inlet specific volume of argon and its mass flow rate are

$$v_1 = \frac{RT_1}{P_1} = \frac{(0.2081\text{kPa}\cdot\text{m}^3/\text{kg}\cdot\text{K})(723\text{K})}{900\text{kPa}} = 0.167\text{m}^3/\text{kg}$$

Thus,

$$\dot{m} = \frac{1}{v_1}A_1V_1 = \frac{1}{0.167\text{m}^3/\text{kg}}(0.006\text{m}^2)(80\text{m/s}) = 2.874\text{kg/s}$$

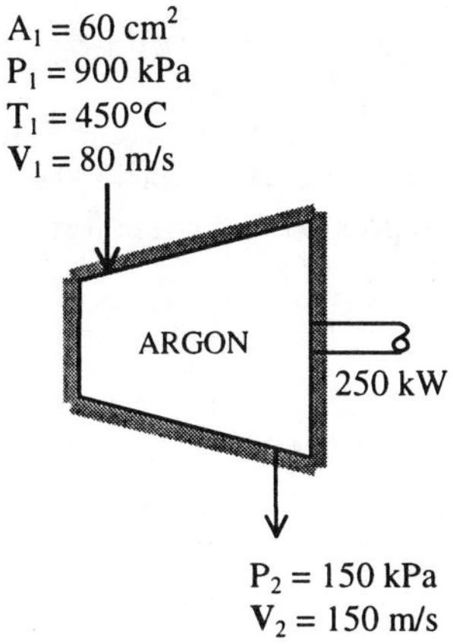

A$_1$ = 60 cm^2
P$_1$ = 900 kPa
T$_1$ = 450°C
V$_1$ = 80 m/s

ARGON

250 kW

P$_2$ = 150 kPa
V$_2$ = 150 m/s

We take the turbine as the system, which is a control volume since mass crosses the boundary. The energy balance for this steady-flow system can be expressed in the rate form as

$$\underbrace{\dot{E}_{in} - \dot{E}_{out}}_{\substack{\text{Rate of net energy transfer}\\ \text{by heat, work, and mass}}} = \underbrace{\Delta \dot{E}_{\text{system}}^{\nearrow 0\ (\text{steady})}}_{\substack{\text{Rate of change in internal, kinetic,}\\ \text{potential, etc. energies}}} = 0$$

$$\dot{E}_{in} = \dot{E}_{out}$$

$$\dot{m}(h_1 + \mathbf{V}_1^2/2) = \dot{W}_{\text{out}} + \dot{m}(h_2 + \mathbf{V}_2^2/2) \quad (\text{since } \dot{Q} \cong \Delta\text{pe} \cong 0)$$

$$\dot{W}_{\text{out}} = -\dot{m}\left(h_2 - h_1 + \frac{\mathbf{V}_2^2 - \mathbf{V}_1^2}{2}\right)$$

Substituting,

$$250\text{ kJ/s} = -(2.874\text{ kg/s})\left[(0.5203\text{ kJ/kg}\cdot{}^\circ\text{C})(T_2 - 450^\circ\text{C}) + \frac{(150\text{ m/s})^2 - (80\text{ m/s})^2}{2}\left(\frac{1\text{ kJ/kg}}{1000\text{ m}^2/\text{s}^2}\right)\right]$$

It yields $\qquad T_2 = \mathbf{267.3°C}$

4-85E Air expands in a turbine. The mass flow rate of air and the power output of the turbine are to be determined.

Assumptions **1** This is a steady-flow process since there is no change with time. **2** Potential energy changes are negligible. **3** The device is adiabatic and thus heat transfer is negligible. **4** Air is an ideal gas with constant specific heats.

Properties The gas constant of air is R = 0.3704 psia.ft³/lbm.R. The constant pressure specific heat of air at the average temperature of (900 + 300)/2 = 600°F is C_p = 0.25 Btu/lbm· °F (Tables A-2a)

Analysis (*a*) There is only one inlet and one exit, and thus $\dot{m}_1 = \dot{m}_2 = \dot{m}$. The inlet specific volume of air and its mass flow rate are

$$v_1 = \frac{RT_1}{P_1} = \frac{(0.3704\text{psia}\cdot\text{ft}^3/\text{lbm}\cdot\text{R})(1360\text{R})}{150\text{psia}} = 3.358\text{ft}^3/\text{lbm}$$

$$\dot{m} = \frac{1}{v_1}A_1V_1 = \frac{1}{3.358\text{ft}^3/\text{lbm}}(0.1\text{ft}^2)(350\text{ft/s}) = \mathbf{10.42lbm/s}$$

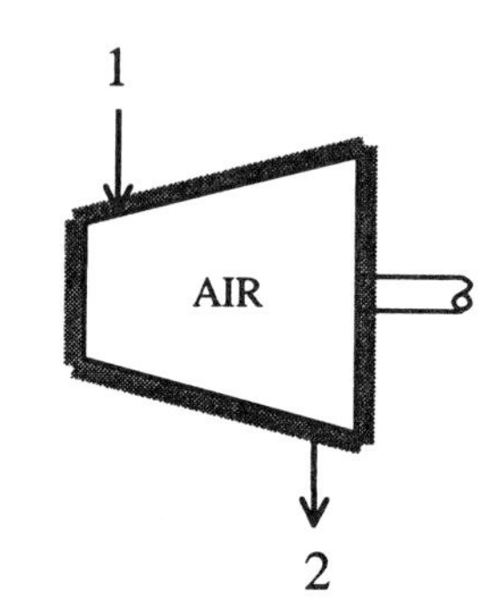

(*b*) We take the turbine as the system, which is a control volume since mass crosses the boundary. The energy balance for this steady-flow system can be expressed in the rate form as

$$\underbrace{\dot{E}_{in} - \dot{E}_{out}}_{\substack{\text{Rate of net energy transfer}\\ \text{by heat, work, and mass}}} = \underbrace{\Delta \dot{E}_{\text{system}}^{\nearrow 0 \text{ (steady)}}}_{\substack{\text{Rate of change in internal, kinetic,}\\ \text{potential, etc. energies}}} = 0$$

$$\dot{E}_{in} = \dot{E}_{out}$$

$$\dot{m}(h_1 + \mathbf{V}_1^2/2) = \dot{W}_{\text{out}} + \dot{m}(h_2 + \mathbf{V}_2^2/2) \quad (\text{since } \dot{Q} \cong \Delta\text{pe} \cong 0)$$

$$\dot{W}_{\text{out}} = -\dot{m}\left(h_2 - h_1 + \frac{\mathbf{V}_2^2 - \mathbf{V}_1^2}{2}\right) = -\dot{m}\left(C_p(T_2 - T_1) + \frac{\mathbf{V}_2^2 - \mathbf{V}_1^2}{2}\right)$$

Substituting,

$$\dot{W}_{\text{out}} = -(10.42\text{ lbm/s})\left[(0.250\text{ Btu/lbm}\cdot^\circ\text{F})(300 - 900)^\circ\text{F} + \frac{(700\text{ ft/s})^2 - (350\text{ ft/s})^2}{2}\left(\frac{1\text{ Btu/lbm}}{25{,}037\text{ ft}^2/\text{s}^2}\right)\right]$$

$$= 1486.5\text{ Btu/s} = \mathbf{1568\ kW}$$

4-86 Refrigerant-134a is compressed steadily by a compressor. The power input to the compressor and the volume flow rate of the refrigerant at the compressor inlet are to be determined.

Assumptions **1** This is a steady-flow process since there is no change with time. **2** Kinetic and potential energy changes are negligible. **3** The device is adiabatic and thus heat transfer is negligible.

Properties From the refrigerant tables (Tables A-11 through 13)

$$\left.\begin{array}{l} T_1 = -20^\circ\text{C} \\ sat.vapor \end{array}\right\} \begin{array}{l} v_1 = 0.1464\text{m}^3/\text{kg} \\ h_1 = 235.31\text{kJ/kg} \end{array}$$

$$\left.\begin{array}{l} P_2 = 0.7\text{MPa} \\ T_2 = 70^\circ\text{C} \end{array}\right\} h_2 = 307.01\text{kJ/kg}$$

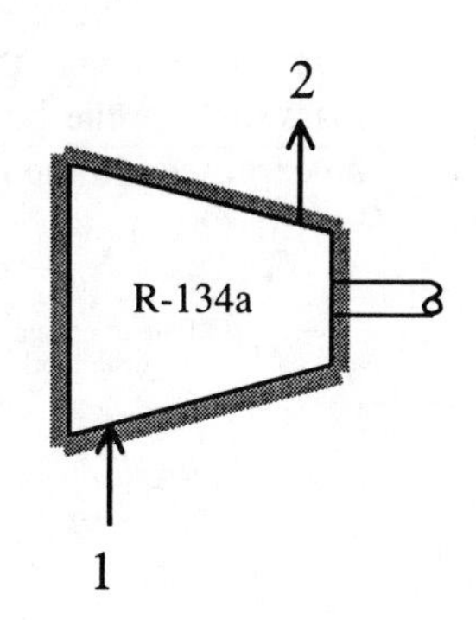

Analysis (*a*) There is only one inlet and one exit, and thus $\dot{m}_1 = \dot{m}_2 = \dot{m}$. We take the compressor as the system, which is a control volume since mass crosses the boundary. The energy balance for this steady-flow system can be expressed in the rate form as

$$\underbrace{\dot{E}_{in} - \dot{E}_{out}}_{\substack{\text{Rate of net energy transfer} \\ \text{by heat, work, and mass}}} = \underbrace{\Delta \dot{E}_{\text{system}}^{\nearrow 0 \text{ (steady)}}}_{\substack{\text{Rate of change in internal, kinetic,} \\ \text{potential, etc. energies}}} = 0$$

$$\dot{E}_{in} = \dot{E}_{out}$$

$$\dot{W}_{\text{in}} + \dot{m}h_1 = \dot{m}h_2 \quad (\text{since } \dot{Q} \cong \Delta ke \cong \Delta pe \cong 0)$$

$$\dot{W}_{\text{in}} = \dot{m}(h_2 - h_1)$$

Substituting,

$$\dot{W}_{in} = (1.2\text{kg/s})(307.01 - 235.31)\text{kJ/kg}$$
$$= \mathbf{86.04kJ/s}$$

(*b*) The volume flow rate of the refrigerant at the compressor inlet is

$$\dot{V}_1 = \dot{m}v_1 = (1.2\text{kg/s})(0.1464\text{m}^3/\text{kg}) = \mathbf{0.176m^3/s}$$

4-87 Air is compressed by a compressor. The mass flow rate of air through the compressor is to be determined.

Assumptions **1** This is a steady-flow process since there is no change with time. **2** Potential energy changes are negligible. **3** Air is an ideal gas with variable specific heats.

Properties The inlet and exit enthalpies of air are (Table A-17)

$$T_1 = 25°C = 298\ K \quad \rightarrow \quad h_1 = h_{@\ 298\ K} = 298.2\ kJ/kg$$
$$T_2 = 347°C = 620\ K \quad \rightarrow \quad h_2 = h_{@\ 620\ K} = 628.07\ kJ/kg$$

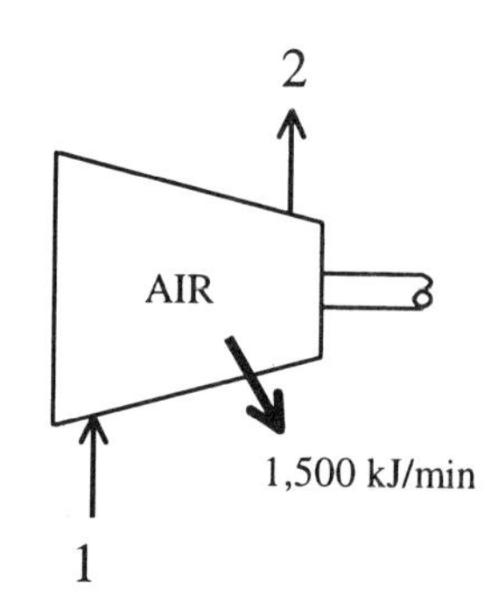

Analysis We take the compressor as the system, which is a control volume since mass crosses the boundary. The energy balance for this steady-flow system can be expressed in the rate form as

$$\underbrace{\dot{E}_{in} - \dot{E}_{out}}_{\text{Rate of net energy transfer by heat, work, and mass}} = \underbrace{\Delta \dot{E}_{system}}_{\text{Rate of change in internal, kinetic, potential, etc. energies}}^{\nearrow 0\ \text{(steady)}} = 0$$

$$\dot{E}_{in} = \dot{E}_{out}$$

$$\dot{W}_{in} + \dot{m}(h_1 + \mathbf{V}_1^2/2) = \dot{Q}_{out} + \dot{m}(h_2 + \mathbf{V}_2^2/2) \quad (\text{since } \Delta pe \cong 0)$$

$$\dot{W}_{in} - \dot{Q}_{out} = \dot{m}\left(h_2 - h_1 + \frac{\mathbf{V}_2^2 - \mathbf{V}_1^2}{2}\right)$$

Substituting, the mass flow rate is determined to be

$$250\ kJ/s - (1500/60\ kJ/s) = \dot{m}\left[628.07 - 298.2 + \frac{(90 m/s)^2 - 0}{2}\left(\frac{1 kJ/kg}{1000 m^2/s^2}\right)\right]$$

$$\dot{m} = \mathbf{0.674\ kg/s}$$

4-88E Air is compressed by a compressor. The mass flow rate of air through the compressor and the exit temperature of air are to be determined.

Assumptions **1** This is a steady-flow process since there is no change with time. **2** Kinetic and potential energy changes are negligible. **3** Air is an ideal gas with variable specific heats.

Properties The gas constant of air is R = 0.3704 psia.ft^3/lbm.R (Table A-1). The inlet enthalpy of air is (Table A-17E)

$$T_1 = 60°\text{F} = 520 \text{ R} \quad \rightarrow \quad h_1 = h_{@\,520\text{ R}} = 124.27 \text{ Btu/lbm}$$

Analysis (*a*) There is only one inlet and one exit, and thus $\dot{m}_1 = \dot{m}_2 = \dot{m}$. The inlet specific volume of air and its mass flow rate are

$$v_1 = \frac{RT_1}{P_1} = \frac{(0.3704\text{psia} \cdot \text{ft}^3/\text{lbm} \cdot \text{R})(520\text{R})}{14.7\text{psia}} = 13.1\text{ft}^3/\text{lbm}$$

$$\dot{m} = \frac{\dot{V}_1}{v_1} = \frac{5000 \text{ ft}^3/\text{min}}{13.1 \text{ ft}^3/\text{lbm}} = 381.7 \text{ lbm/min} = \mathbf{6.36 \text{ lbm/s}}$$

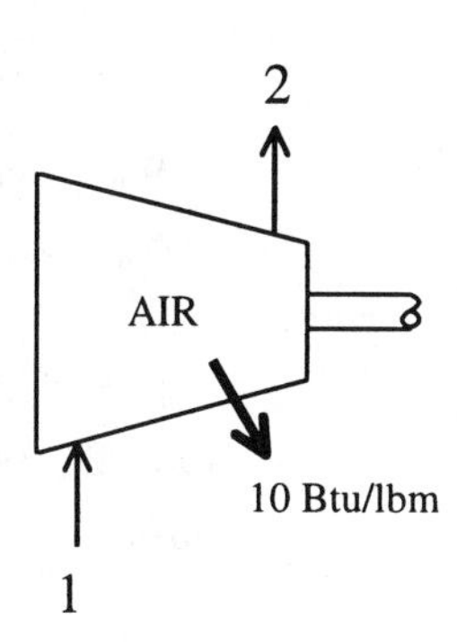

(*b*) We take the compressor as the system, which is a control volume since mass crosses the boundary. The energy balance for this steady-flow system can be expressed in the rate form as

$$\underbrace{\dot{E}_{in} - \dot{E}_{out}}_{\text{Rate of net energy transfer by heat, work, and mass}} = \underbrace{\Delta \dot{E}_{\text{system}}^{\nearrow 0 \text{ (steady)}}}_{\text{Rate of change in internal, kinetic, potential, etc. energies}} = 0$$

$$\dot{E}_{in} = \dot{E}_{out}$$

$$\dot{W}_{\text{in}} + \dot{m}h_1 = \dot{Q}_{\text{out}} + \dot{m}h_2 \quad (\text{since } \Delta\text{ke} \cong \Delta\text{pe} \cong 0)$$

$$\dot{W}_{\text{in}} - \dot{Q}_{\text{out}} = \dot{m}(h_2 - h_1)$$

Substituting,

$$(700\text{hp})\left(\frac{0.7068\text{Btu/s}}{1\text{hp}}\right) - (6.36 \text{ lbm/s}) \times (10\text{Btu/lbm}) = (6.36\text{lbm/s})(h_2 - 124.27\text{Btu/lbm})$$

$$h_2 = 192.06\text{Btu/lbm}$$

Then the exit temperature is determined from Table A-17E to be

$$T_2 = 801 \text{ R} = \mathbf{341°F}$$

4-89 EES solution of this (and other comprehensive problems designated with the *computer icon*) is available to instructors at the *Instructor Manual* section of the *Online Learning Center* (OLC) at www.mhhe.com/cengel-boles. See the Preface for access information.

4-90 Helium is compressed by a compressor. For a mass flow rate of 90 kg/min, the power input required is to be determined.

Assumptions **1** This is a steady-flow process since there is no change with time. **2** Kinetic and potential energy changes are negligible. **3** Helium is an ideal gas with constant specific heats.

Properties The constant pressure specific heat of helium is C_p = 5.1926 kJ/kg· K (Table A-2a).

Analysis There is only one inlet and one exit, and thus $\dot{m}_1 = \dot{m}_2 = \dot{m}$. We take the compressor as the system, which is a control volume since mass crosses the boundary. The energy balance for this steady-flow system can be expressed in the rate form as

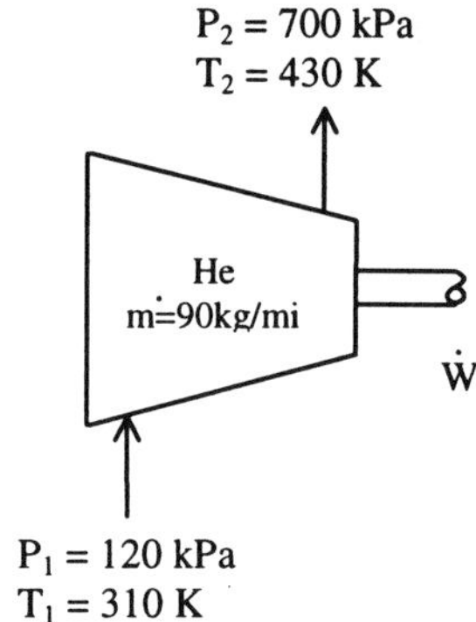

$$\underbrace{\dot{E}_{in} - \dot{E}_{out}}_{\text{Rate of net energy transfer by heat, work, and mass}} = \underbrace{\Delta \dot{E}_{system}}_{\text{Rate of change in internal, kinetic, potential, etc. energies}}^{\nearrow 0 \text{ (steady)}} = 0$$

$$\dot{E}_{in} = \dot{E}_{out}$$

$$\dot{W}_{in} + \dot{m}h_1 = \dot{Q}_{out} + \dot{m}h_2 \quad (\text{since } \Delta\text{ke} \cong \Delta\text{pe} \cong 0)$$

$$\dot{W}_{in} - \dot{Q}_{out} = \dot{m}(h_2 - h_1) = \dot{m}C_p(T_2 - T_1)$$

Thus,

$$\dot{W}_{in} = \dot{Q}_{out} + \dot{m}C_p(T_2 - T_1)$$
$$= (90/60\text{kg/s})(20 \text{ kJ/kg}) + (90/60\text{kg/s})(5.1926\text{kJ/kg}\cdot\text{K})(430 - 310)\text{K}$$
$$= \mathbf{965 kW}$$

4-91 CO_2 is compressed by a compressor. The volume flow rate of CO_2 at the compressor inlet and the power input to the compressor are to be determined.

Assumptions **1** This is a steady-flow process since there is no change with time. **2** Kinetic and potential energy changes are negligible. **3** Helium is an ideal gas with variable specific heats. **4** The device is adiabatic and thus heat transfer is negligible.

Properties The gas constant of CO_2 is R = 0.1889 kPa.m³/kg.K, and its molar mass is M = 44 kg/kmol (Table A-1). The inlet and exit enthalpies of CO_2 are (Table A-20)

$$T_1 = 300 \text{ K} \rightarrow \bar{h}_1 = 9{,}431 \text{ kJ/kmol}$$
$$T_2 = 450 \text{ K} \rightarrow \bar{h}_2 = 15{,}483 \text{ kJ/kmol}$$

Analysis (*a*) There is only one inlet and one exit, and thus $\dot{m}_1 = \dot{m}_2 = \dot{m}$. The inlet specific volume of air and its volume flow rate are

$$v_1 = \frac{RT_1}{P_1} = \frac{(0.1889\text{kPa}\cdot\text{m}^3/\text{kg}\cdot\text{K})(300\text{K})}{100\text{kPa}} = 0.5667\text{m}^3/\text{kg}$$

$$\dot{V} = \dot{m}v_1 = (0.5 \text{ kg/s})(0.5667 \text{ m}^3/\text{kg}) = \mathbf{0.283 \text{ m}^3/\text{s}}$$

(*b*) We take the compressor as the system, which is a control volume since mass crosses the boundary. The energy balance for this steady-flow system can be expressed in the rate form as

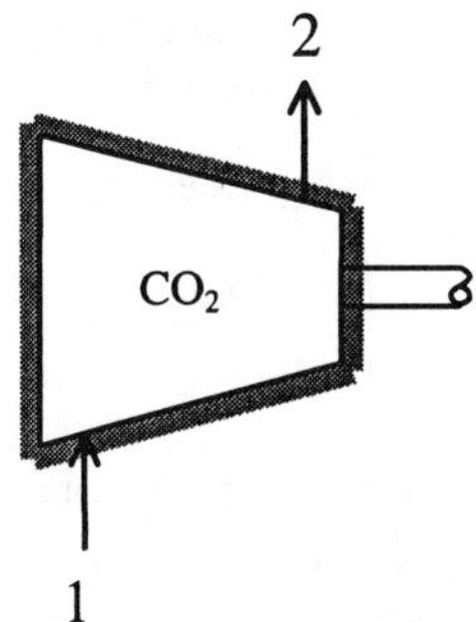

$$\underbrace{\dot{E}_{in} - \dot{E}_{out}}_{\text{Rate of net energy transfer by heat, work, and mass}} = \underbrace{\Delta \dot{E}_{system}^{\nearrow 0 \text{ (steady)}}}_{\text{Rate of change in internal, kinetic, potential, etc. energies}} = 0$$

$$\dot{E}_{in} = \dot{E}_{out}$$

$$\dot{W}_{in} + \dot{m}h_1 = \dot{m}h_2 \quad (\text{since } \dot{Q} \cong \Delta\text{ke} \cong \Delta\text{pe} \cong 0)$$
$$\dot{W}_{in} = \dot{m}(h_2 - h_1) = \dot{m}(\bar{h}_2 - \bar{h}_1)/M$$

$$\dot{W}_{in} = \frac{(0.5\text{kg/s})(15{,}483 - 9{,}431\text{kJ/kmol})}{44\text{kg/kmol}} = \mathbf{68.8kW}$$

Throttling Valves

4-92C Because usually there is a large temperature drop associated with the throttling process.

4-93C Yes.

4-94C No. Because air is an ideal gas and h = h(T) for ideal gases. Thus if h remains constant, so does the temperature.

4-95C If it remains in the liquid phase, no. But if some of the liquid vaporizes during throttling, then yes.

4-96 Refrigerant-134a is throttled by a valve. The temperature drop of the refrigerant and specific volume after expansion are to be determined. √

Assumptions **1** This is a steady-flow process since there is no change with time. **2** Kinetic and potential energy changes are negligible. **3** Heat transfer to or from the fluid is negligible. **4** There are no work interactions involved.

Properties The inlet enthalpy of R-134a is, from the refrigerant tables (Tables A-11 through 13),

$$\left.\begin{array}{l} P_1 = 0.8 MPa \\ sat.liquid \end{array}\right\} \begin{array}{l} T_1 = T_{sat} = 31.33^\circ C \\ h_1 = h_f = 93.42 kJ/kg \end{array}$$

Analysis There is only one inlet and one exit, and thus $\dot{m}_1 = \dot{m}_2 = \dot{m}$. We take the throttling valve as the system, which is a control volume since mass crosses the boundary. The energy balance for this steady-flow system can be expressed in the rate form as

$$\dot{E}_{in} - \dot{E}_{out} = \Delta \dot{E}_{system}^{\nearrow 0 \text{ (steady)}} = 0 \quad \rightarrow \quad \dot{E}_{in} = \dot{E}_{out} \quad \rightarrow \quad \dot{m}h_1 = \dot{m}h_2 \quad \rightarrow \quad h_1 = h_2$$

since $\dot{Q} \cong \dot{W} = \Delta ke \cong \Delta pe \cong 0$. Then,

$$\left.\begin{array}{l} P_2 = 0.14 \text{MPa} \\ (h_2 = h_1) \end{array}\right\} \begin{array}{l} h_f = 25.77 \text{kJ/kg}, \quad T_{sat} = -18.8^\circ \text{C} \\ h_g = 236.04 \text{kJ/kg} \end{array}$$

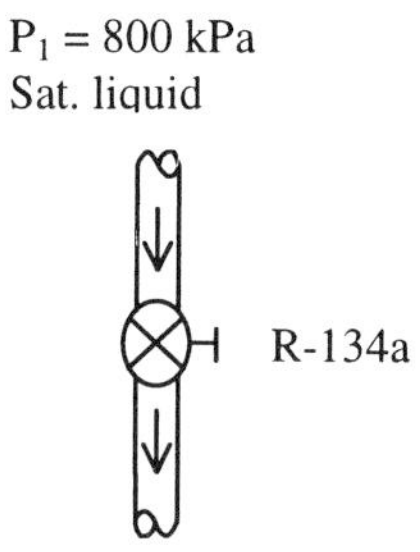

Obviously $h_f < h_2 < h_g$, thus the refrigerant exists as a saturated mixture at the exit state and thus $T_2 = T_{sat}$ = -18.8°C. Then the temperature drop becomes

$$\Delta T = T_2 - T_1 = -18.8 - 31.33 = \mathbf{-50.13^\circ C}$$

The quality at this state is determined from

$$x_2 = \frac{h_2 - h_f}{h_{fg}} = \frac{93.42 - 25.77}{210.27 - 25.77} = 0.322$$

Thus,

$$v_2 = v_f + x_2 v_{fg} = 0.0007381 + 0.322 \times 0.13876 = \mathbf{0.0454 m^3/kg}$$

4-97 [*Also solved by EES on enclosed CD*] Refrigerant-134a is throttled by a valve. The pressure and internal energy after expansion are to be determined.

Assumptions **1** This is a steady-flow process since there is no change with time. **2** Kinetic and potential energy changes are negligible. **3** Heat transfer to or from the fluid is negligible. **4** There are no work interactions involved.

Properties The inlet enthalpy of R-134a is, from the refrigerant tables (Tables A-11 through 13),

$$\left.\begin{array}{l} P_1 = 0.8\text{MPa} \\ T_1 = 25^\circ\text{C} \end{array}\right\} h_1 \cong h_{f@25^\circ\text{C}} = 84.33\text{kJ/kg}$$

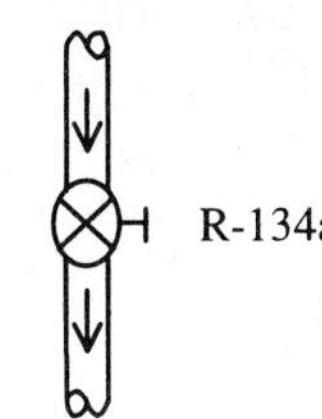

Analysis There is only one inlet and one exit, and thus $\dot{m}_1 = \dot{m}_2 = \dot{m}$. We take the throttling valve as the system, which is a control volume since mass crosses the boundary. The energy balance for this steady-flow system can be expressed in the rate form as

$$\dot{E}_{in} - \dot{E}_{out} = \Delta\dot{E}_{\text{system}}{}^{\nearrow 0\text{ (steady)}} = 0 \rightarrow \dot{E}_{in} = \dot{E}_{out} \rightarrow \dot{m}h_1 = \dot{m}h_2 \rightarrow h_1 = h_2$$

since $\dot{Q} \cong \dot{W} = \Delta ke \cong \Delta pe \cong 0$. Then,

$$\left.\begin{array}{l} T_2 = -20^\circ\text{C} \\ (h_2 = h_1) \end{array}\right\} \begin{array}{ll} h_f = 24.26\text{kJ/kg}, & u_f = 24.17\text{kJ/kg} \\ h_g = 235.31\text{kJ/kg} & u_g = 215.84\text{kJ/kg} \end{array}$$

Obviously $h_f < h_2 < h_g$, thus the refrigerant exists as a saturated mixture at the exit state, and thus

$$P_2 = P_{\text{sat @ -20°C}} = \mathbf{0.13299\ MPa}$$

Also, $$x_2 = \frac{h_2 - h_f}{h_{fg}} = \frac{84.33 - 24.26}{211.05} = 0.285$$

Thus, $$u_2 = u_f + x_2 u_{fg} = 24.17 + 0.285 \times (215.84 - 24.17) = \mathbf{78.8kJ/kg}$$

4-98 Steam is throttled by a well-insulated valve. The temperature drop of the steam after the expansion is to be determined.

Assumptions **1** This is a steady-flow process since there is no change with time. **2** Kinetic and potential energy changes are negligible. **3** Heat transfer to or from the fluid is negligible. **4** There are no work interactions involved.

Properties The inlet enthalpy of steam is (Tables A-6),

$$\left.\begin{array}{l} P_1 = 8\text{MPa} \\ T_1 = 500^\circ\text{C} \end{array}\right\} h_1 = 3398.3\text{kJ/kg}$$

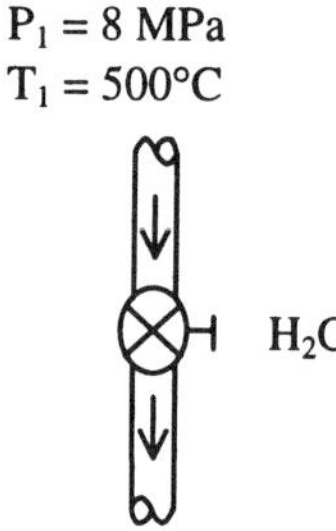

Analysis There is only one inlet and one exit, and thus $\dot{m}_1 = \dot{m}_2 = \dot{m}$. We take the throttling valve as the system, which is a control volume since mass crosses the boundary. The energy balance for this steady-flow system can be expressed in the rate form as

$$\dot{E}_{in} - \dot{E}_{out} = \Delta\dot{E}_{\text{system}}{}^{\nearrow 0\ (\text{steady})} = 0 \quad \rightarrow \quad \dot{E}_{in} = \dot{E}_{out} \quad \rightarrow \quad \dot{m}h_1 = \dot{m}h_2 \quad \rightarrow \quad h_1 = h_2$$

$P_2 = 6$ MPa

since $\dot{Q} \cong \dot{W} = \Delta ke \cong \Delta pe \cong 0$. Then the exit temperature of steam becomes

$$\left.\begin{array}{l} P_2 = 6\text{MPa} \\ (h_2 = h_1) \end{array}\right\} T_2 = \mathbf{490.1^\circ C}$$

4-99 EES solution of this (and other comprehensive problems designated with the *computer icon*) is available to instructors at the *Instructor Manual* section of the *Online Learning Center* (OLC) at www.mhhe.com/cengel-boles. See the Preface for access information.

4-100E High-pressure air is throttled to atmospheric pressure. The temperature of air after the expansion is to be determined.

Assumptions **1** This is a steady-flow process since there is no change with time. **2** Kinetic and potential energy changes are negligible. **3** Heat transfer to or from the fluid is negligible. **4** There are no work interactions involved. **5** Air is an ideal gas.

Analysis There is only one inlet and one exit, and thus $\dot{m}_1 = \dot{m}_2 = \dot{m}$. We take the throttling valve as the system, which is a control volume since mass crosses the boundary. The energy balance for this steady-flow system can be expressed in the rate form as

$P_1 = 200$ psia
$T_1 = 90$°F

$$\dot{E}_{in} - \dot{E}_{out} = \Delta\dot{E}_{\text{system}}{}^{\nearrow 0\ (\text{steady})} = 0 \quad \rightarrow \quad \dot{E}_{in} = \dot{E}_{out} \quad \rightarrow \quad \dot{m}h_1 = \dot{m}h_2 \quad \rightarrow \quad h_1 = h_2$$

Air

since $\dot{Q} \cong \dot{W} = \Delta ke \cong \Delta pe \cong 0$. For an ideal gas, $h = h(T)$.

Therefore,

$$T_2 = T_1 = \mathbf{90^\circ F}$$

$P_2 = 14.7$ psia

Mixing Chambers and Heat Exchangers

4-101C Yes, if the mixing chamber is losing heat to the surrounding medium.

4-102C Under the conditions of no heat and work interactions between the mixing chamber and the surrounding medium.

4-103C Under the conditions of no heat and work interactions between the heat exchanger and the surrounding medium.

4-104 A hot water stream is mixed with a cold water stream. For a specified mixture temperature, the mass flow rate of cold water is to be determined.

Assumptions **1** Steady operating conditions exist. **2** The mixing chamber is well-insulated so that heat loss to the surroundings is negligible. **3** Changes in the kinetic and potential energies of fluid streams are negligible. **4** Fluid properties are constant. **5** There are no work interactions.

Properties Noting that $T < T_{sat\ @\ 250\ kPa} = 127.44°C$, the water in all three streams exists as a compressed liquid, which can be approximated as a saturated liquid at the given temperature. Thus,

$$h_1 \cong h_{f\ @\ 80°C} = 334.91\ \text{kJ/kg}$$
$$h_2 \cong h_{f\ @\ 20°C} = 83.96\ \text{kJ/kg}$$
$$h_3 \cong h_{f\ @\ 42°C} = 175.92\ \text{kJ/kg}$$

Analysis We take the mixing chamber as the system, which is a control volume. The mass and energy balances for this steady-flow system can be expressed in the rate form as

Mass balance: $\dot{m}_{in} - \dot{m}_{out} = \Delta \dot{E}_{system}^{\nearrow 0\ \text{(steady)}} = 0 \longrightarrow \dot{m}_1 + \dot{m}_2 = \dot{m}_3$

Energy balance:

$$\underbrace{\dot{E}_{in} - \dot{E}_{out}}_{\substack{\text{Rate of net energy transfer}\\ \text{by heat, work, and mass}}} = \underbrace{\Delta \dot{E}_{system}^{\nearrow 0\ \text{(steady)}}}_{\substack{\text{Rate of change in internal, kinetic,}\\ \text{potential, etc. energies}}} = 0$$

$$\dot{E}_{in} = \dot{E}_{out}$$

$$\dot{m}_1 h_1 + \dot{m}_2 h_2 = \dot{m}_3 h_3 \quad (\text{since } \dot{Q} = \dot{W} = \Delta\text{ke} \cong \Delta\text{pe} \cong 0)$$

T₁ = 80°C, $\dot{m}_1$ = 0.5 kg/s; T₂ = 20°C, $\dot{m}_2$; H_2O (P = 250 kPa), T₃ = 42°C

Combining the two relations and solving for $\dot{m}_2$ gives

$$\dot{m}_1 h_1 + \dot{m}_2 h_2 = (\dot{m}_1 + \dot{m}_2) h_3$$

$$\dot{m}_2 = \frac{h_1 - h_3}{h_3 - h_2}\dot{m}_1$$

Substituting, the mass flow rate of cold water stream is determined to be

$$\dot{m}_2 = \frac{(334.91 - 175.92)\text{kJ/kg}}{(175.92 - 83.96)\text{kJ/kg}}(0.5\ \text{kg/s}) = \mathbf{0.864\ \ kg/s}$$

4-105 Liquid water is heated in a chamber by mixing it with superheated steam. For a specified mixing temperature, the mass flow rate of the steam is to be determined.

Assumptions **1** This is a steady-flow process since there is no change with time. **2** Kinetic and potential energy changes are negligible. **3** There are no work interactions. **4** The device is adiabatic and thus heat transfer is negligible.

Properties Noting that $T < T_{\text{sat @ 300 kPa}} = 133.55°\text{C}$, the cold water stream and the mixture exist as a compressed liquid, which can be approximated as a saturated liquid at the given temperature. Thus,

$$h_1 \cong h_{f\,@\,20°C} = 83.96 \text{ kJ/kg}$$
$$h_3 \cong h_{f\,@\,60°C} = 251.13 \text{ kJ/kg}$$

and

$$\left.\begin{array}{l} P_2 = 300\text{kPa} \\ T_2 = 300°\text{C} \end{array}\right\} h_2 = 3069.3\text{kJ/kg}$$

Analysis We take the mixing chamber as the system, which is a control volume since mass crosses the boundary. The mass and energy balances for this steady-flow system can be expressed in the rate form as

Mass balance: $\dot{m}_{in} - \dot{m}_{out} = \Delta\dot{m}_{\text{system}}{}^{\nearrow 0 \text{ (steady)}} = 0 \rightarrow \dot{m}_{in} = \dot{m}_{out} \rightarrow \dot{m}_1 + \dot{m}_2 = \dot{m}_3$

Energy balance:

$$\underbrace{\dot{E}_{in} - \dot{E}_{out}}_{\text{Rate of net energy transfer by heat, work, and mass}} = \underbrace{\Delta\dot{E}_{\text{system}}{}^{\nearrow 0 \text{ (steady)}}}_{\text{Rate of change in internal, kinetic, potential, etc. energies}} = 0$$

$$\dot{E}_{in} = \dot{E}_{out}$$

$$\dot{m}_1 h_1 + \dot{m}_2 h_2 = \dot{m}_3 h_3 \quad (\text{since } \dot{Q} \cong \dot{W} \cong \Delta\text{ke} \cong \Delta\text{pe} \cong 0)$$

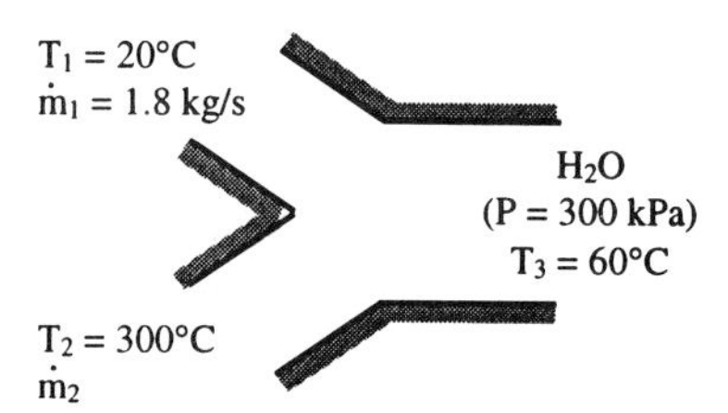

Combining the two, $\dot{m}_1 h_1 + \dot{m}_2 h_2 = (\dot{m}_1 + \dot{m}_2) h_3$

Solving for $\dot{m}_2$: $\dot{m}_2 = \dfrac{h_1 - h_3}{h_3 - h_2}\dot{m}_1$

Substituting,

$$\dot{m}_2 = \frac{83.96 - 251.13}{251.13 - 3069.3}(1.8 \text{ kg/s}) = \mathbf{0.107 \text{ kg/s}}$$

4-106 Feedwater is heated in a chamber by mixing it with superheated steam. If the mixture is saturated liquid, the ratio of the mass flow rates of the feedwater and the superheated vapor is to be determined.

Assumptions **1** This is a steady-flow process since there is no change with time. **2** Kinetic and potential energy changes are negligible. **3** There are no work interactions. **4** The device is adiabatic and thus heat transfer is negligible.

Properties Noting that $T < T_{sat\ @\ 800\ kPa} = 170.43°C$, the cold water stream and the mixture exist as a compressed liquid, which can be approximated as a saturated liquid at the given temperature. Thus,

$$h_1 \cong h_{f\ @\ 50°C} = 209.33\ \text{kJ/kg}$$
$$h_3 \cong h_{f\ @\ 800\ kPa} = 721.11\ \text{kJ/kg}$$

and

$$\left.\begin{array}{l} P_2 = 800\text{kPa} \\ T_2 = 200°\text{C} \end{array}\right\} h_2 = 2839.3\text{kJ/kg}$$

Analysis We take the mixing chamber as the system, which is a control volume since mass crosses the boundary. The mass and energy balances for this steady-flow system can be expressed in the rate form as

Mass balance: $$\dot{m}_{in} - \dot{m}_{out} = \Delta\dot{m}_{system}^{\nearrow 0\ (steady)} = 0 \rightarrow \dot{m}_{in} = \dot{m}_{out} \rightarrow \dot{m}_1 + \dot{m}_2 = \dot{m}_3$$

Energy balance:

$$\underbrace{\dot{E}_{in} - \dot{E}_{out}}_{\text{Rate of net energy transfer by heat, work, and mass}} = \underbrace{\Delta\dot{E}_{system}^{\nearrow 0\ (steady)}}_{\text{Rate of change in internal, kinetic, potential, etc. energies}} = 0$$

$$\dot{E}_{in} = \dot{E}_{out}$$

$$\dot{m}_1 h_1 + \dot{m}_2 h_2 = \dot{m}_3 h_3 \quad (\text{since } \dot{Q} \cong \dot{W} \cong \Delta\text{ke} \cong \Delta\text{pe} \cong 0)$$

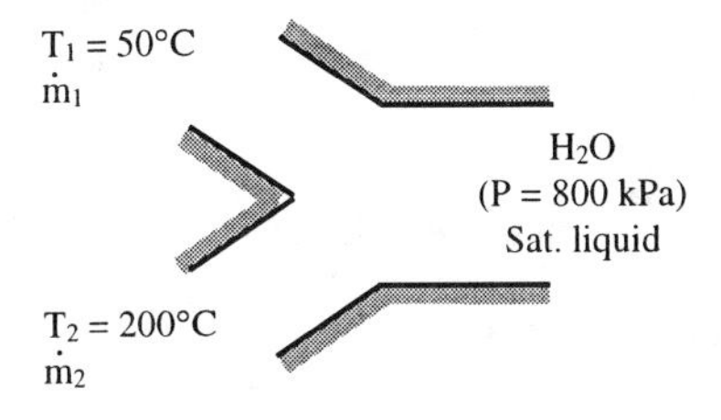

Combining the two, $$\dot{m}_1 h_1 + \dot{m}_2 h_2 = (\dot{m}_1 + \dot{m}_2) h_3$$

Dividing by $\dot{m}_2$ yields $$y h_1 + h_2 = (y+1) h_3$$

Solving for y: $$y = \frac{h_3 - h_2}{h_1 - h_3}$$

where $y = \dot{m}_1 / \dot{m}_2$ is the desired mass flow rate ratio. Substituting,

$$y = \frac{721.11 - 2839.3}{209.33 - 721.11} = \mathbf{4.14}$$

4-107E Liquid water is heated in a chamber by mixing it with saturated water vapor. If both streams enter at the same rate, the temperature and quality (if saturated) of the exit stream is to be determined.

Assumptions **1** This is a steady-flow process since there is no change with time. **2** Kinetic and potential energy changes are negligible. **3** There are no work interactions. **4** The device is adiabatic and thus heat transfer is negligible.

Properties From steam tables (Tables A-5 through A-6),

$$h_1 \cong h_{f\,@\,50°F} = 18.06 \text{ Btu/lbm}$$
$$h_2 = h_{g\,@\,50\text{ psia}} = 1174.4 \text{ Btu/lbm}$$

Analysis We take the mixing chamber as the system, which is a control volume since mass crosses the boundary. The mass and energy balances for this steady-flow system can be expressed in the rate form as

Mass balance: $\dot{m}_{in} - \dot{m}_{out} = \Delta \dot{m}_{system}^{\nearrow 0 \text{ (steady)}} = 0 \rightarrow \dot{m}_{in} = \dot{m}_{out} \rightarrow \dot{m}_1 + \dot{m}_2 = \dot{m}_3 = 2\dot{m} \rightarrow \dot{m}_1 = \dot{m}_2 = \dot{m}$

Energy balance:

$$\underbrace{\dot{E}_{in} - \dot{E}_{out}}_{\text{Rate of net energy transfer by heat, work, and mass}} = \underbrace{\Delta \dot{E}_{system}^{\nearrow 0 \text{ (steady)}}}_{\text{Rate of change in internal, kinetic, potential, etc. energies}} = 0$$

$$\dot{E}_{in} = \dot{E}_{out}$$

$$\dot{m}_1 h_1 + \dot{m}_2 h_2 = \dot{m}_3 h_3 \quad (\text{since } \dot{Q} \cong \dot{W} \cong \Delta\text{ke} \cong \Delta\text{pe} \cong 0)$$

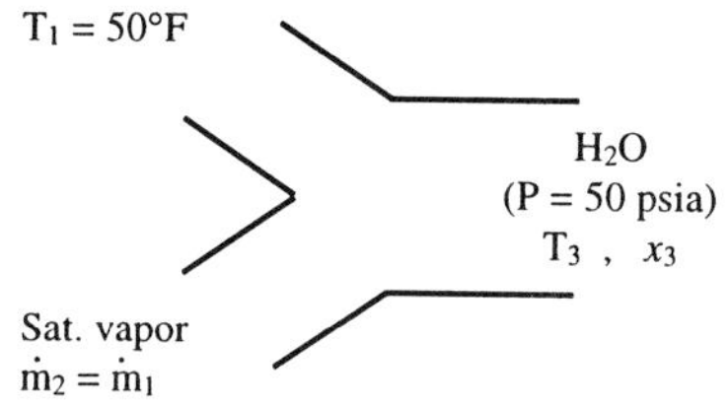

T₁ = 50°F

H₂O (P = 50 psia) T₃ , x₃

Sat. vapor $\dot{m}_2 = \dot{m}_1$

Combining the two gives $\dot{m}h_1 + \dot{m}h_2 = 2\dot{m}h_3$ or $h_3 = (h_1 + h_2)/2$

Substituting,

$$h_3 = (18.06 + 1174.4)/2 = 596.23 \text{ Btu/lbm}$$

At 50 psia, $h_f = 250.24$ Btu/lbm and $h_g = 1174.4$ Btu/lbm. Thus the exit stream is a saturated mixture since $h_f < h_3 < h_g$. Therefore,

$$T_3 = T_{sat\,@\,50\text{ psia}} = \mathbf{281.03°F}$$

and

$$x_3 = \frac{h_3 - h_f}{h_{fg}} = \frac{596.23 - 250.24}{924.2} = \mathbf{0.374}$$

4-108 Two streams of refrigerant-134a are mixed in a chamber. If the cold stream enters at twice the rate of the hot stream, the temperature and quality (if saturated) of the exit stream are to be determined. √

Assumptions **1** This is a steady-flow process since there is no change with time. **2** Kinetic and potential energy changes are negligible. **3** There are no work interactions. **4** The device is adiabatic and thus heat transfer is negligible.

Properties From R-134a tables (Tables A-11 through A-13),

$$h_1 \cong h_{f\,@\,12^\circ C} = 66.18 \text{ kJ/kg}$$
$$h_2 = h_{@\,1\text{ MPa},\,60^\circ C} = 291.36 \text{ kJ/kg}$$

Analysis We take the mixing chamber as the system, which is a control volume since mass crosses the boundary. The mass and energy balances for this steady-flow system can be expressed in the rate form as

Mass balance: $\dot{m}_{in} - \dot{m}_{out} = \Delta\dot{m}_{system}{}^{\nearrow 0\text{ (steady)}} = 0 \rightarrow \dot{m}_{in} = \dot{m}_{out} \rightarrow \dot{m}_1 + \dot{m}_2 = \dot{m}_3 = 3\dot{m}_2$ since $\dot{m}_1 = 2\dot{m}_2$

Energy balance:

$$\underbrace{\dot{E}_{in} - \dot{E}_{out}}_{\text{Rate of net energy transfer by heat, work, and mass}} = \underbrace{\Delta\dot{E}_{system}{}^{\nearrow 0\text{ (steady)}}}_{\text{Rate of change in internal, kinetic, potential, etc. energies}} = 0$$

$$\dot{E}_{in} = \dot{E}_{out}$$

$$\dot{m}_1 h_1 + \dot{m}_2 h_2 = \dot{m}_3 h_3 \quad (\text{since } \dot{Q} \cong \dot{W} \cong \Delta\text{ke} \cong \Delta\text{pe} \cong 0)$$

T_1 = 12°C, $\dot{m}_1 = 2\dot{m}_2$; T_2 = 60°C; R-134a (P = 1 MPa) T_3, x_3

Combining the two gives $\quad 2\dot{m}_2 h_1 + \dot{m}_2 h_2 = 3\dot{m}_2 h_3 \text{ or } h_3 = (2h_1 + h_2)/3$

Substituting,

$$h_3 = (2\times 66.18 + 291.36)/3 = 141.24 \text{ kJ/kg}$$

At 1 MPa, h_f = 105.29 kJ/kg and h_g = 267.97 kJ/kg. Thus the exit stream is a saturated mixture since $h_f < h_3 < h_g$. Therefore,

$$T_3 = T_{sat\,@\,1\text{ MPa}} = \mathbf{39.39^\circ C}$$

and

$$x_3 = \frac{h_3 - h_f}{h_{fg}} = \frac{141.24 - 105.29}{162.68} = \mathbf{0.221}$$

4-109 EES solution of this (and other comprehensive problems designated with the *computer icon*) is available to instructors at the *Instructor Manual* section of the *Online Learning Center* (OLC) at www.mhhe.com/cengel-boles. See the Preface for access information.

4-110 Refrigerant-134a is to be cooled by air in the condenser. For a specified volume flow rate of air, the mass flow rate of the refrigerant is to be determined.

Assumptions **1** This is a steady-flow process since there is no change with time. **2** Kinetic and potential energy changes are negligible. **3** There are no work interactions. **4** Heat loss from the device to the surroundings is negligible and thus heat transfer from the hot fluid is equal to the heat transfer to the cold fluid. **5** Air is an ideal gas with constant specific heats at room temperature.

Properties The gas constant of air is 0.287 kPa.m³/kg.K (Table A-1). The constant pressure specific heat of air is C_p = 1.005 kJ/kg· °C (Table A-2)..The enthalpies of the R-134a at the inlet and the exit states are (Tables A-11 through A-13)

$$\left.\begin{array}{l} P_3 = 1\text{MPa} \\ T_3 = 80^\circ\text{C} \end{array}\right\} h_3 = 313.20\text{kJ/kg}$$

$$\left.\begin{array}{l} P_4 = 1\text{MPa} \\ T_4 = 30^\circ\text{C} \end{array}\right\} h_4 \cong h_{f@30^\circ C} = 91.49\text{kJ/kg}$$

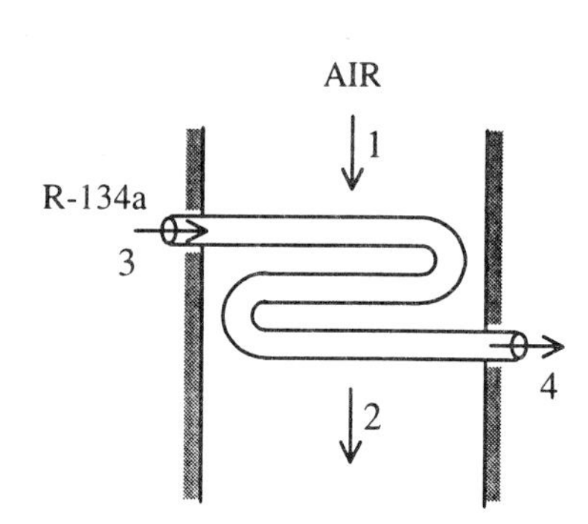

Analysis The inlet specific volume and the mass flow rate of air are

$$v_1 = \frac{RT_1}{P_1} = \frac{(0.287\text{kPa}\cdot\text{m}^3/\text{kg}\cdot\text{K})(300\text{K})}{100\text{kPa}} = 0.861\text{m}^3/\text{kg}$$

and

$$\dot{m} = \frac{\dot{V}_1}{v_1} = \frac{800\text{m}^3/\text{min}}{0.861\text{m}^3/\text{kg}} = 929.2\text{kg/min}$$

We take the entire heat exchanger as the system, which is a control volume. The mass and energy balances for this steady-flow system can be expressed in the rate form as

Mass balance (*for each fluid stream*):

$$\dot{m}_{in} - \dot{m}_{out} = \Delta\dot{m}_{\text{system}}^{\nearrow 0\ (\text{steady})} = 0 \rightarrow \dot{m}_{in} = \dot{m}_{out} \rightarrow \dot{m}_1 = \dot{m}_2 = \dot{m}_a \text{ and } \dot{m}_3 = \dot{m}_4 = \dot{m}_R$$

Energy balance (for the entire heat exchanger):

$$\underbrace{\dot{E}_{in} - \dot{E}_{out}}_{\substack{\text{Rate of net energy transfer} \\ \text{by heat, work, and mass}}} = \underbrace{\Delta\dot{E}_{\text{system}}^{\nearrow 0\ (\text{steady})}}_{\substack{\text{Rate of change in internal, kinetic,} \\ \text{potential, etc. energies}}} = 0$$

$$\dot{E}_{in} = \dot{E}_{out}$$

$$\dot{m}_1 h_1 + \dot{m}_3 h_3 = \dot{m}_2 h_2 + \dot{m}_4 h_4 \quad (\text{since } \dot{Q} = \dot{W} = \Delta\text{ke} \cong \Delta\text{pe} \cong 0)$$

Combining the two, $\dot{m}_a(h_2 - h_1) = \dot{m}_R(h_3 - h_4)$

Solving for $\dot{m}_R$: $\dot{m}_R = \frac{h_2 - h_1}{h_3 - h_4}\dot{m}_a \cong \frac{C_p(T_2 - T_1)}{h_3 - h_4}\dot{m}_a$

Substituting,

$$\dot{m}_R = \frac{(1.005\text{ kJ/kg}\cdot^\circ\text{C})(60 - 27)^\circ\text{C}}{(313.20 - 91.49)\text{ kJ/kg}}(929.2\text{ kg/min}) = \mathbf{139.0\ kg/min}$$

4-111E Refrigerant-134a is vaporized by air in the evaporator of an air-conditioner. For specified flow rates, the exit temperature of the air and the rate of heat transfer from the air are to be determined.

Assumptions **1** This is a steady-flow process since there is no change with time. **2** Kinetic and potential energy changes are negligible. **3** There are no work interactions. **4** Heat loss from the device to the surroundings is negligible and thus heat transfer from the hot fluid is equal to the heat transfer to the cold fluid. **5** Air is an ideal gas with constant specific heats at room temperature.

Properties The gas constant of air is 0.3704 psia.ft³/lbm.R (Table A-1E). The constant pressure specific heat of air is C_p = 0.240 Btu/lbm· °F (Table A-2E). The enthalpies of the R-134a at the inlet and the exit states are (Tables A-11E through A-13E)

$$\left.\begin{array}{l} P_3 = 20 \text{ psia} \\ x_3 = 0.3 \end{array}\right\} h_3 = h_f + x_3 h_{fg} = 10.89 + 0.3 \times 90.50 = 38.04 \text{ Btu/lbm}$$

$$\left.\begin{array}{l} P_4 = 20 \text{ psia} \\ \text{sat.vapor} \end{array}\right\} h_4 = h_{g\,@\,20\text{ psia}} = 101.39 \text{ Btu/lbm}$$

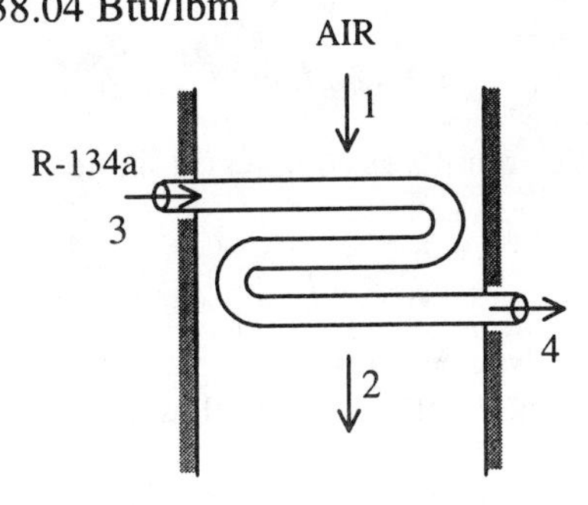

Analysis The inlet specific volume and the mass flow rate of air are

$$v_1 = \frac{RT_1}{P_1} = \frac{(0.3704\text{psia}\cdot\text{ft}^3/\text{lbm}\cdot\text{R})(550\text{R})}{14.7\text{psia}} = 13.86\text{ft}^3/\text{lbm}$$

and

$$\dot{m} = \frac{\dot{V}_1}{v_1} = \frac{200\text{ft}^3/\text{min}}{13.86\text{ft}^3/\text{lbm}} = 14.43\text{lbm/min}$$

We take the entire heat exchanger as the system, which is a control volume. The mass and energy balances for this steady-flow system can be expressed in the rate form as

Mass balance (for each fluid stream):

$$\dot{m}_{in} - \dot{m}_{out} = \Delta\dot{m}_{\text{system}}^{\nearrow 0 \text{ (steady)}} = 0 \rightarrow \dot{m}_{in} = \dot{m}_{out} \rightarrow \dot{m}_1 = \dot{m}_2 = \dot{m}_a \text{ and } \dot{m}_3 = \dot{m}_4 = \dot{m}_R$$

Energy balance (for the entire heat exchanger):

$$\underbrace{\dot{E}_{in} - \dot{E}_{out}}_{\substack{\text{Rate of net energy transfer} \\ \text{by heat, work, and mass}}} = \underbrace{\Delta\dot{E}_{\text{system}}^{\nearrow 0 \text{ (steady)}}}_{\substack{\text{Rate of change in internal, kinetic,} \\ \text{potential, etc. energies}}} = 0$$

$$\dot{E}_{in} = \dot{E}_{out}$$

$$\dot{m}_1 h_1 + \dot{m}_3 h_3 = \dot{m}_2 h_2 + \dot{m}_4 h_4 \quad (\text{since } \dot{Q} = \dot{W} = \Delta\text{ke} \cong \Delta\text{pe} \cong 0)$$

Combining the two, $\dot{m}_R(h_3 - h_4) = \dot{m}_a(h_2 - h_1) = \dot{m}_a C_p(T_2 - T_1)$

Solving for T_2: $T_2 = T_1 + \dfrac{\dot{m}_R(h_3 - h_4)}{\dot{m}_a C_p}$

Substituting, $T_2 = 90^\circ\text{F} + \dfrac{(4 \text{ lbm/min})(38.04 - 101.39)\text{Btu/lbm}}{(14.43 \text{ Btu/min})(0.24 \text{ Btu/lbm}\cdot^\circ\text{F})} = \mathbf{16.8^\circ F}$

(*b*) The rate of heat transfer from the air to the refrigerant is determined from the steady-flow energy balance applied to the air only. It yields

$$-\dot{Q}_{air,out} = \dot{m}_a(h_2 - h_1) = \dot{m}_a C_p (T_2 - T_1)$$

$$\dot{Q}_{air,out} = -(14.43\text{lbm/min})(0.24\text{Btu/lbm}\cdot^\circ\text{F})(16.8 - 90)^\circ\text{F}$$

$$= \mathbf{253.5 \text{ Btu/min}}$$

4-112 Refrigerant-134a is condensed in a water-cooled condenser. The mass flow rate of the cooling water required is to be determined.

Assumptions **1** This is a steady-flow process since there is no change with time. **2** Kinetic and potential energy changes are negligible. **3** There are no work interactions. **4** Heat loss from the device to the surroundings is negligible and thus heat transfer from the hot fluid is equal to the heat transfer to the cold fluid.

Properties The enthalpies of R-134a at the inlet and the exit states are (Tables A-5 and A-6)

$$\left.\begin{array}{l} P_3 = 800\text{kPa} \\ T_3 = 70^\circ\text{C} \end{array}\right\} h_3 = 305.50\text{kJ/kg}$$

$$\left.\begin{array}{l} P_4 = 800\text{kPa} \\ sat.liquid \end{array}\right\} h_4 = h_{f@800\text{kPa}} = 93.42\text{kJ/kg}$$

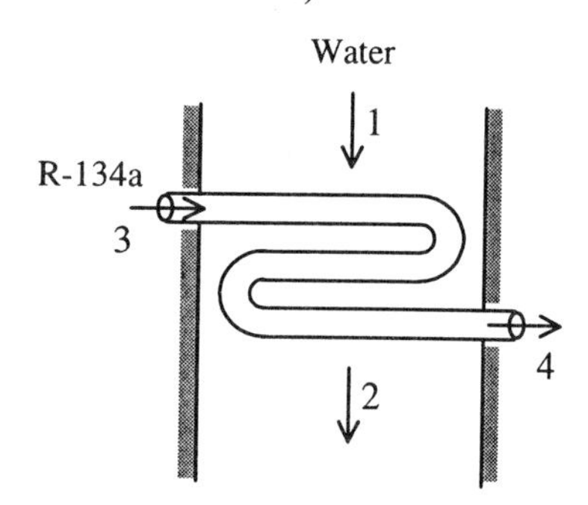

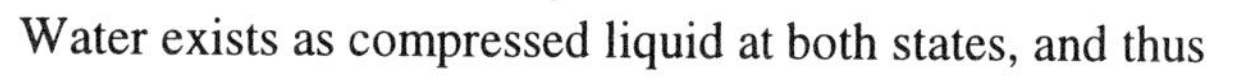

Water exists as compressed liquid at both states, and thus

$$h_1 \cong h_{f\,@\,15^\circ\text{C}} = 62.99 \text{ kJ/kg}$$
$$h_2 \cong h_{f\,@\,30^\circ\text{C}} = 125.79 \text{ kJ/kg}$$

Analysis We take the heat exchanger as the system, which is a control volume. The mass and energy balances for this steady-flow system can be expressed in the rate form as

Mass balance (for each fluid stream):

$$\dot{m}_{in} - \dot{m}_{out} = \Delta\dot{m}_{\text{system}}^{\nearrow 0\text{ (steady)}} = 0 \rightarrow \dot{m}_{in} = \dot{m}_{out} \rightarrow \dot{m}_1 = \dot{m}_2 = \dot{m}_w \quad \text{and} \quad \dot{m}_3 = \dot{m}_4 = \dot{m}_R$$

Energy balance (for the heat exchanger):

$$\underbrace{\dot{E}_{in} - \dot{E}_{out}}_{\substack{\text{Rate of net energy transfer}\\ \text{by heat, work, and mass}}} = \underbrace{\Delta\dot{E}_{\text{system}}^{\nearrow 0\text{ (steady)}}}_{\substack{\text{Rate of change in internal, kinetic,}\\ \text{potential, etc. energies}}} = 0$$

$$\dot{E}_{in} = \dot{E}_{out}$$

$$\dot{m}_1 h_1 + \dot{m}_3 h_3 = \dot{m}_2 h_2 + \dot{m}_4 h_4 \quad (\text{since } \dot{Q} = \dot{W} = \Delta\text{ke} \cong \Delta\text{pe} \cong 0)$$

Combining the two, $\qquad \dot{m}_w\left(h_2 - h_1\right) = \dot{m}_R\left(h_3 - h_4\right)$

Solving for $\dot{m}_w$: $\qquad \dot{m}_w = \dfrac{h_3 - h_4}{h_2 - h_1}\dot{m}_R$

Substituting,

$$\dot{m}_w = \frac{(305.50 - 93.42) \text{ kJ/kg}}{(125.79 - 62.99) \text{ kJ/kg}}(8 \text{ kg/min}) = \mathbf{27.0 \text{ kg/min}}$$

4-113E [*Also solved by EES on enclosed CD*] Air is heated in a steam heating system. For specified flow rates, the volume flow rate of air at the inlet is to be determined.

Assumptions **1** This is a steady-flow process since there is no change with time. **2** Kinetic and potential energy changes are negligible. **3** There are no work interactions. **4** Heat loss from the device to the surroundings is negligible and thus heat transfer from the hot fluid is equal to the heat transfer to the cold fluid. **5** Air is an ideal gas with constant specific heats at room temperature.

Properties The gas constant of air is 0.3704 psia.ft³/lbm.R (Table A-1E). The constant pressure specific heat of air is C_p = 0.240 Btu/lbm· °F (Table A-2E). The enthalpies of steam at the inlet and the exit states are (Tables A-4E through A-6E)

$$\left.\begin{array}{l} P_3 = 30\text{psia} \\ T_3 = 400^\circ\text{F} \end{array}\right\} h_3 = 1237.8\text{Btu/lbm}$$

$$\left.\begin{array}{l} P_4 = 25\text{psia} \\ T_4 = 212^\circ F \end{array}\right\} h_4 \cong h_{f@212^\circ\text{F}} = 180.16\text{Btu/lbm}$$

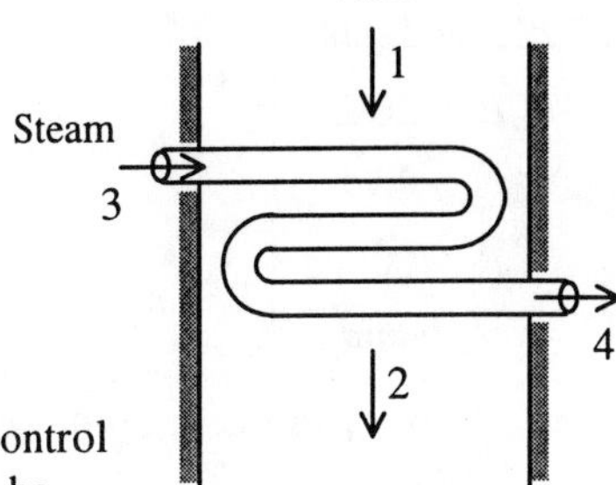

Analysis We take the entire heat exchanger as the system, which is a control volume. The mass and energy balances for this steady-flow system can be expressed in the rate form as

Mass balance (*for each fluid stream*):

$$\dot{m}_{in} - \dot{m}_{out} = \Delta\dot{m}_{\text{system}}^{\nearrow 0\ \text{(steady)}} = 0 \rightarrow \dot{m}_{in} = \dot{m}_{out} \rightarrow \dot{m}_1 = \dot{m}_2 = \dot{m}_a \text{ and } \dot{m}_3 = \dot{m}_4 = \dot{m}_s$$

Energy balance (for the entire heat exchanger):

$$\underbrace{\dot{E}_{in} - \dot{E}_{out}}_{\substack{\text{Rate of net energy transfer} \\ \text{by heat, work, and mass}}} = \underbrace{\Delta\dot{E}_{\text{system}}^{\nearrow 0\ \text{(steady)}}}_{\substack{\text{Rate of change in internal, kinetic,} \\ \text{potential, etc. energies}}} = 0$$

$$\dot{E}_{in} = \dot{E}_{out}$$

$$\dot{m}_1 h_1 + \dot{m}_3 h_3 = \dot{m}_2 h_2 + \dot{m}_4 h_4 \quad (\text{since } \dot{Q} = \dot{W} = \Delta\text{ke} \cong \Delta\text{pe} \cong 0)$$

Combining the two, $\dot{m}_a(h_2 - h_1) = \dot{m}_s(h_3 - h_4)$

Solving for $\dot{m}_a$: $\dot{m}_a = \dfrac{h_3 - h_4}{h_2 - h_1}\dot{m}_s \cong \dfrac{h_3 - h_4}{C_p(T_2 - T_1)}\dot{m}_s$

Substituting,

$$\dot{m}_a = \frac{(1237.8 - 180.16)\ \text{Btu/lbm}}{(0.240\ \text{Btu/lbm}\cdot{}^\circ\text{F})(130 - 80)^\circ\text{F}}(15\ \text{lbm/min}) = 1322\ \text{lbm/min} = 22.03\ \text{lbm/s}$$

Also, $\quad v_1 = \dfrac{RT_1}{P_1} = \dfrac{(0.3704\ \text{psia}\cdot\text{ft}^3/\text{lbm}\cdot\text{R})(540\ \text{R})}{14.7\ \text{psia}} = 13.61\ \text{ft}^3/\text{lbm}$

Then the volume flow rate of air at the inlet becomes

$$\dot{V}_1 = \dot{m}_a v_1 = (22.03\ \text{lbm/s})(13.61\ \text{ft}^3/\text{lbm}) = \mathbf{299.8\ ft^3/s}$$

4-114 Steam is condensed by cooling water in the condenser of a power plant. If the temperature rise of the cooling water is not to exceed 10°C, the minimum mass flow rate of the cooling water required is to be determined.

Assumptions **1** This is a steady-flow process since there is no change with time. **2** Kinetic and potential energy changes are negligible. **3** There are no work interactions. **4** Heat loss from the device to the surroundings is negligible and thus heat transfer from the hot fluid is equal to the heat transfer to the cold fluid. **5** Liquid water is an incompressible substance with constant specific heats at room temperature.

Properties The cooling water exists as compressed liquid at both states, and its specific heat at room temperature is C = 4.18 kJ/kg· °C (Table A-3). The enthalpies of the steam at the inlet and the exit states are (Tables A-5 and A-6)

$$\left.\begin{array}{l} P_3 = 20\text{kPa} \\ x_3 = 0.95 \end{array}\right\} h_3 = h_f + x_3 h_{fg} = 251.40 + 0.95 \times 2358.3 = 2491.8\text{kJ/kg}$$

$$\left.\begin{array}{l} P_4 = 20\text{kPa} \\ sat\,liquid \end{array}\right\} h_4 \cong h_{f@20kPa} = 251.40\text{kJ/kg}$$

Analysis We take the heat exchanger as the system, which is a control volume. The mass and energy balances for this steady-flow system can be expressed in the rate form as

Mass balance (for each fluid stream):

$$\dot{m}_{in} - \dot{m}_{out} = \Delta \dot{m}_{\text{system}}^{\nearrow 0\ (\text{steady})} = 0 \rightarrow \dot{m}_{in} = \dot{m}_{out} \rightarrow \dot{m}_1 = \dot{m}_2 = \dot{m}_w \quad \text{and} \quad \dot{m}_3 = \dot{m}_4 = \dot{m}_s$$

Energy balance (for the heat exchanger):

$$\underbrace{\dot{E}_{in} - \dot{E}_{out}}_{\substack{\text{Rate of net energy transfer} \\ \text{by heat, work, and mass}}} = \underbrace{\Delta \dot{E}_{\text{system}}^{\nearrow 0\ (\text{steady})}}_{\substack{\text{Rate of change in internal, kinetic,} \\ \text{potential, etc. energies}}} = 0$$

$$\dot{E}_{in} = \dot{E}_{out}$$

$$\dot{m}_1 h_1 + \dot{m}_3 h_3 = \dot{m}_2 h_2 + \dot{m}_4 h_4 \quad (\text{since } \dot{Q} = \dot{W} = \Delta\text{ke} \cong \Delta\text{pe} \cong 0)$$

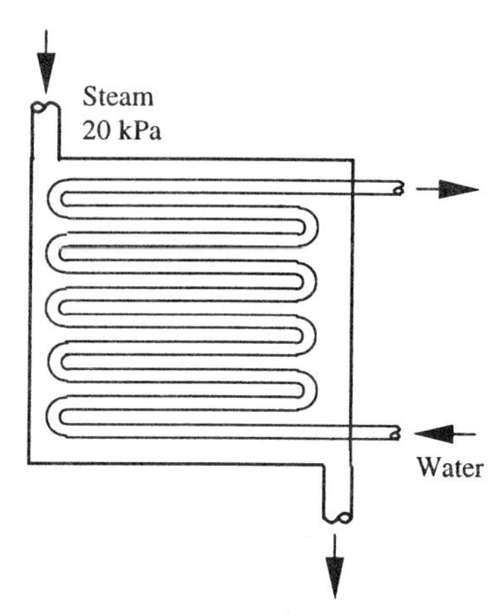

Combining the two, $\dot{m}_w(h_2 - h_1) = \dot{m}_s(h_3 - h_4)$

Solving for $\dot{m}_w$: $\dot{m}_w = \dfrac{h_3 - h_4}{h_2 - h_1}\dot{m}_s \cong \dfrac{h_3 - h_4}{C_p(T_2 - T_1)}\dot{m}_s$

Substituting,

$$\dot{m}_w = \frac{(2491.8 - 251.4)\text{kJ/kg}}{(4.18\text{kJ/kg}\cdot{}^\circ\text{C})(10^\circ\text{C})}(20{,}000/3600\text{kg/s}) = \mathbf{298 kg/s} = 17{,}866 \text{ kg/min}$$

4-115 Steam is condensed by cooling water in the condenser of a power plant. The rate of condensation of steam is to be determined.

Assumptions **1** Steady operating conditions exist. **2** The heat exchanger is well-insulated so that heat loss to the surroundings is negligible and thus heat transfer from the hot fluid is equal to the heat transfer to the cold fluid. **3** Changes in the kinetic and potential energies of fluid streams are negligible. **4** Fluid properties are constant.

Properties The heat of vaporization of water at 50°C is h_{fg} = 2382.7 kJ/kg and specific heat of cold water is C_p = 4.18 kJ/kg.°C (Tables A-3 and A-4).

Analysis We take the cold water tubes as the system, which is a control volume. The energy balance for this steady-flow system can be expressed in the rate form as

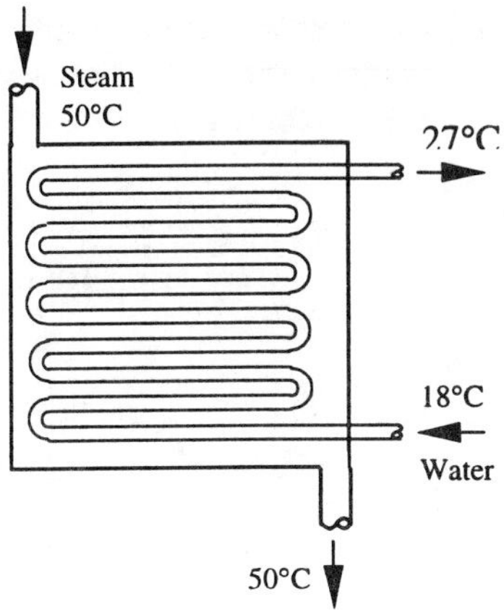

Steam 50°C

27°C

18°C

Water

50°C

$$\underbrace{\dot{E}_{in} - \dot{E}_{out}}_{\text{Rate of net energy transfer by heat, work, and mass}} = \underbrace{\Delta \dot{E}_{system}^{\nearrow 0 \text{ (steady)}}}_{\text{Rate of change in internal, kinetic, potential, etc. energies}} = 0$$

$$\dot{E}_{in} = \dot{E}_{out}$$

$$\dot{Q}_{in} + \dot{m}h_1 = \dot{m}h_2 \quad (\text{since } \Delta\text{ke} \cong \Delta\text{pe} \cong 0)$$

$$\dot{Q}_{in} = \dot{m}C_p(T_2 - T_1)$$

Then the heat transfer rate to the cooling water in the condenser becomes

$$\begin{aligned}\dot{Q} &= [\dot{m}C_p(T_{out} - T_{in})]_{\text{cooling water}} \\ &= (101\text{ kg/s})(4.18\text{ kJ/kg.°C})(27°\text{C} - 18°\text{C}) \\ &= 3800\text{ kJ/s}\end{aligned}$$

The rate of condensation of steam is determined to be

$$\dot{Q} = (\dot{m}h_{fg})_{\text{steam}} \longrightarrow \dot{m}_{\text{steam}} = \frac{\dot{Q}}{h_{fg}} = \frac{3800\text{ kJ/s}}{2382.7\text{kJ/kg}} = \mathbf{1.59\text{ kg/s}}$$

4-116 EES solution of this (and other comprehensive problems designated with the *computer icon*) is available to instructors at the *Instructor Manual* section of the *Online Learning Center* (OLC) at www.mhhe.com/cengel-boles. See the Preface for access information.

4-117 Water is heated in a heat exchanger by geothermal water. The rate of heat transfer to the water and the exit temperature of the geothermal water is to be determined.

Assumptions **1** Steady operating conditions exist. **2** The heat exchanger is well-insulated so that heat loss to the surroundings is negligible and thus heat transfer from the hot fluid is equal to the heat transfer to the cold fluid. **3** Changes in the kinetic and potential energies of fluid streams are negligible. **4** Fluid properties are constant.

Properties The specific heats of water and geothermal fluid are given to be 4.18 and 4.31 kJ/kg.°C, respectively.

Analysis We take the cold water tubes as the system, which is a control volume. The energy balance for this steady-flow system can be expressed in the rate form as

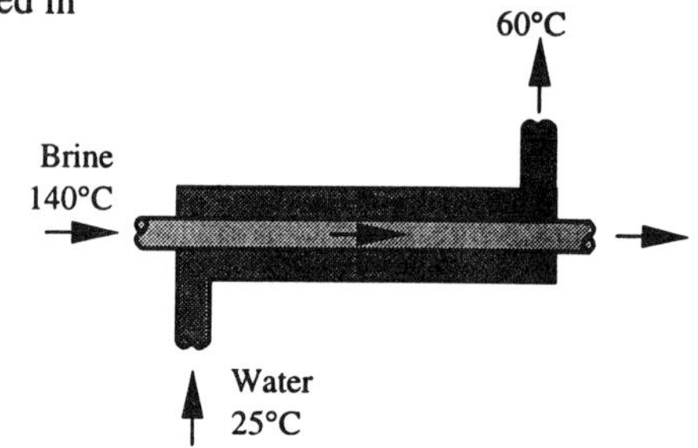

$$\underbrace{\dot{E}_{in} - \dot{E}_{out}}_{\substack{\text{Rate of net energy transfer} \\ \text{by heat, work, and mass}}} = \underbrace{\Delta \dot{E}_{system}^{\nearrow 0 \text{ (steady)}}}_{\substack{\text{Rate of change in internal, kinetic,} \\ \text{potential, etc. energies}}} = 0$$

$$\dot{E}_{in} = \dot{E}_{out}$$

$$\dot{Q}_{in} + \dot{m}h_1 = \dot{m}h_2 \quad (\text{since } \Delta\text{ke} \cong \Delta\text{pe} \cong 0)$$

$$\dot{Q}_{in} = \dot{m}C_p(T_2 - T_1)$$

Then the rate of heat transfer to the cold water in the heat exchanger becomes

$$\dot{Q} = [\dot{m}C_p(T_{out} - T_{in})]_{water} = (0.2\text{ kg/s})(4.18\text{ kJ/kg.°C})(60°\text{C} - 25°\text{C}) = \mathbf{29.26\text{ kW}}$$

Noting that heat transfer to the cold water is equal to the heat loss from the geothermal water, the outlet temperature of the geothermal water is determined from

$$\dot{Q} = [\dot{m}C_p(T_{in} - T_{out})]_{geot.water} \longrightarrow T_{out} = T_{in} - \frac{\dot{Q}}{\dot{m}C_p}$$

$$= 140°\text{C} - \frac{29.26\text{ kW}}{(0.3\text{ kg/s})(4.31\text{ kJ/kg.°C})} = \mathbf{117.4°C}$$

4-118 Ethylene glycol is cooled by water in a heat exchanger. The rate of heat transfer in the heat exchanger and the mass flow rate of water are to be determined.

Assumptions **1** Steady operating conditions exist. **2** The heat exchanger is well-insulated so that heat loss to the surroundings is negligible and thus heat transfer from the hot fluid is equal to the heat transfer to the cold fluid. **3** Changes in the kinetic and potential energies of fluid streams are negligible. **4** Fluid properties are constant.

Properties The specific heats of water and ethylene glycol are given to be 4.18 and 2.56 kJ/kg.°C, respectively.

Analysis (a) We take the ethylene glycol tubes as the system, which is a control volume. The energy balance for this steady-flow system can be expressed in the rate form as

Cold Water 20°C
Hot Glycol 80°C 2 kg/s
40°C

$$\underbrace{\dot{E}_{in} - \dot{E}_{out}}_{\substack{\text{Rate of net energy transfer} \\ \text{by heat, work, and mass}}} = \underbrace{\Delta \dot{E}_{system}^{\nearrow 0 \text{ (steady)}}}_{\substack{\text{Rate of change in internal, kinetic,} \\ \text{potential, etc. energies}}} = 0$$

$$\dot{E}_{in} = \dot{E}_{out}$$

$$\dot{m}h_1 = \dot{Q}_{out} + \dot{m}h_2 \quad (\text{since } \Delta\text{ke} \cong \Delta\text{pe} \cong 0)$$

$$\dot{Q}_{out} = \dot{m}C_p(T_1 - T_2)$$

Then the rate of heat transfer becomes

$$\dot{Q} = [\dot{m}C_p(T_{in} - T_{out})]_{glycol} = (2 \text{ kg/s})(2.56 \text{ kJ/kg.°C})(80°\text{C} - 40°\text{C}) = \mathbf{204.8 \text{ kW}}$$

(b) The rate of heat transfer from water must be equal to the rate of heat transfer to the glycol. Then,

$$\dot{Q} = [\dot{m}C_p(T_{out} - T_{in})]_{water} \longrightarrow \dot{m}_{water} = \frac{\dot{Q}}{C_p(T_{out} - T_{in})}$$

$$= \frac{204.8 \text{ kJ/s}}{(4.18 \text{ kJ/kg.°C})(55°\text{C} - 20°\text{C})} = \mathbf{1.4 \text{ kg/s}}$$

4-119 EES solution of this (and other comprehensive problems designated with the *computer icon*) is available to instructors at the *Instructor Manual* section of the *Online Learning Center* (OLC) at www.mhhe.com/cengel-boles. See the Preface for access information.

4-120 Oil is to be cooled by water in a thin-walled heat exchanger. The rate of heat transfer in the heat exchanger and the exit temperature of water is to be determined. √

Assumptions **1** Steady operating conditions exist. **2** The heat exchanger is well-insulated so that heat loss to the surroundings is negligible and thus heat transfer from the hot fluid is equal to the heat transfer to the cold fluid. **3** Changes in the kinetic and potential energies of fluid streams are negligible. **4** Fluid properties are constant.

Properties The specific heats of water and oil are given to be 4.18 and 2.20 kJ/kg.°C, respectively.

Analysis We take the oil tubes as the system, which is a control volume. The energy balance for this steady-flow system can be expressed in the rate form as

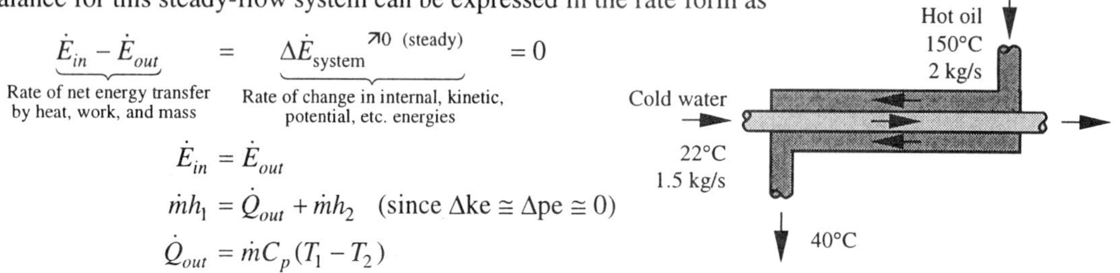

$$\underbrace{\dot{E}_{in} - \dot{E}_{out}}_{\text{Rate of net energy transfer by heat, work, and mass}} = \underbrace{\Delta \dot{E}_{\text{system}}^{\nearrow 0 \text{ (steady)}}}_{\text{Rate of change in internal, kinetic, potential, etc. energies}} = 0$$

$$\dot{E}_{in} = \dot{E}_{out}$$

$$\dot{m}h_1 = \dot{Q}_{out} + \dot{m}h_2 \quad (\text{since } \Delta\text{ke} \cong \Delta\text{pe} \cong 0)$$

$$\dot{Q}_{out} = \dot{m}C_p(T_1 - T_2)$$

Then the rate of heat transfer from the oil becomes

$$\dot{Q} = [\dot{m}C_p(T_{in} - T_{out})]_{\text{oil}} = (2\text{ kg/s})(2.2\text{ kJ/kg.°C})(150°\text{C} - 40°\text{C}) = \mathbf{484\ kW}$$

Noting that the heat lost by the oil is gained by the water, the outlet temperature of the water is determined from

$$\dot{Q} = [\dot{m}C_p(T_{out} - T_{in})]_{\text{water}} \longrightarrow T_{\text{out}} = T_{\text{in}} + \frac{\dot{Q}}{\dot{m}_{\text{water}}C_p} = 22°\text{C} + \frac{484\text{ kJ/s}}{(1.5\text{ kg/s})(4.18\text{ kJ/kg.°C})} = \mathbf{99.2°C}$$

4-121 Cold water is heated by hot water in a heat exchanger. The rate of heat transfer and the exit temperature of hot water are to be determined.

Assumptions **1** Steady operating conditions exist. **2** The heat exchanger is well-insulated so that heat loss to the surroundings is negligible and thus heat transfer from the hot fluid is equal to the heat transfer to the cold fluid. **3** Changes in the kinetic and potential energies of fluid streams are negligible. **4** Fluid properties are constant.

Properties The specific heats of cold and hot water are given to be 4.18 and 4.19 kJ/kg.°C, respectively.

Analysis We take the cold water tubes as the system, which is a control volume. The energy balance for this steady-flow system can be expressed in the rate form as

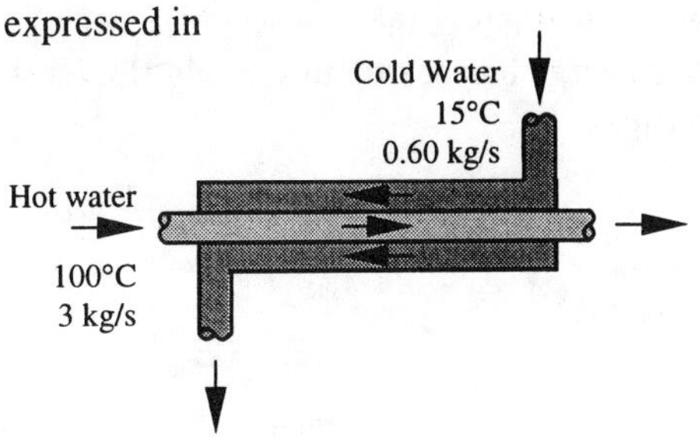

$$\underbrace{\dot{E}_{in} - \dot{E}_{out}}_{\text{Rate of net energy transfer by heat, work, and mass}} = \underbrace{\Delta \dot{E}_{system}^{\nearrow 0 \text{ (steady)}}}_{\text{Rate of change in internal, kinetic, potential, etc. energies}} = 0$$

$$\dot{E}_{in} = \dot{E}_{out}$$

$$\dot{Q}_{in} + \dot{m}h_1 = \dot{m}h_2 \quad (\text{since } \Delta\text{ke} \cong \Delta\text{pe} \cong 0)$$

$$\dot{Q}_{in} = \dot{m}C_p(T_2 - T_1)$$

Then the rate of heat transfer to the cold water in this heat exchanger becomes

$$\dot{Q} = [\dot{m}C_p(T_{out} - T_{in})]_{\text{cold water}} = (0.60\text{ kg/s})(4.18\text{ kJ/kg.°C})(45°\text{C} - 15°\text{C}) = \mathbf{75.24\ kW}$$

Noting that heat gain by the cold water is equal to the heat loss by the hot water, the outlet temperature of the hot water is determined to be

$$\dot{Q} = [\dot{m}C_p(T_{in} - T_{out})]_{\text{hot water}} \longrightarrow T_{out} = T_{in} - \frac{\dot{Q}}{\dot{m}C_p}$$

$$= 100°\text{C} - \frac{75.24\text{ kW}}{(3\text{ kg/s})(4.19\text{ kJ/kg.°C})} = \mathbf{94.0°C}$$

4-122 Air is preheated by hot exhaust gases in a cross-flow heat exchanger. The rate of heat transfer and the outlet temperature of the air are to be determined.

Assumptions **1** Steady operating conditions exist. **2** The heat exchanger is well-insulated so that heat loss to the surroundings is negligible and thus heat transfer from the hot fluid is equal to the heat transfer to the cold fluid. **3** Changes in the kinetic and potential energies of fluid streams are negligible. **4** Fluid properties are constant.

Properties The specific heats of air and combustion gases are given to be 1.005 and 1.10 kJ/kg.°C, respectively.

Analysis We take the exhaust pipes as the system, which is a control volume. The energy balance for this steady-flow system can be expressed in the rate form as

$$\underbrace{\dot{E}_{in} - \dot{E}_{out}}_{\substack{\text{Rate of net energy transfer}\\ \text{by heat, work, and mass}}} = \underbrace{\Delta \dot{E}_{system}^{\nearrow 0 \text{ (steady)}}}_{\substack{\text{Rate of change in internal, kinetic,}\\ \text{potential, etc. energies}}} = 0$$

$$\dot{E}_{in} = \dot{E}_{out}$$

$$\dot{m}h_1 = \dot{Q}_{out} + \dot{m}h_2 \quad (\text{since } \Delta\text{ke} \cong \Delta\text{pe} \cong 0)$$

$$\dot{Q}_{out} = \dot{m}C_p(T_1 - T_2)$$

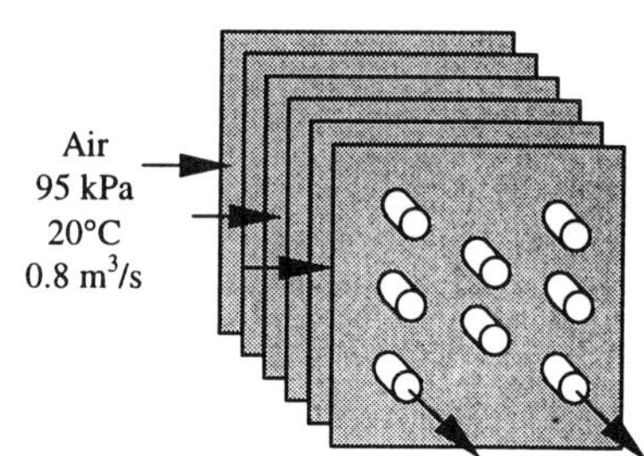

Then the rate of heat transfer from the exhaust gases becomes

$$\dot{Q} = [\dot{m}C_p(T_{in} - T_{out})]_{gas.} = (1.1 \text{ kg/s})(1.1 \text{ kJ/kg.°C})(180°\text{C} - 95°\text{C}) = \mathbf{102.85 \text{ kW}}$$

The mass flow rate of air is

$$\dot{m} = \frac{P\dot{V}}{RT} = \frac{(95 \text{ kPa})(0.8 \text{ m}^3/\text{s})}{(0.287 \text{ kPa.m}^3/\text{kg.K}) \times 293 \text{ K}} = 0.904 \text{ kg/s}$$

Noting that heat loss by the exhaust gases is equal to the heat gain by the air, the outlet temperature of the air becomes

$$\dot{Q} = \dot{m}C_p(T_{c,out} - T_{c,in}) \longrightarrow T_{c,out} = T_{c,in} + \frac{\dot{Q}}{\dot{m}C_p}$$

$$= 20°\text{C} + \frac{102.85 \text{ kW}}{(0.904 \text{ kg/s})(1.005 \text{ kJ/kg.°C})} = \mathbf{133.2°C}$$

4-123 Water is heated by hot oil in a heat exchanger. The rate of heat transfer in the heat exchanger and the outlet temperature of oil are to be determined.

Assumptions **1** Steady operating conditions exist. **2** The heat exchanger is well-insulated so that heat loss to the surroundings is negligible and thus heat transfer from the hot fluid is equal to the heat transfer to the cold fluid. **3** Changes in the kinetic and potential energies of fluid streams are negligible. **4** Fluid properties are constant.

Properties The specific heats of water and oil are given to be 4.18 and 2.3 kJ/kg.°C, respectively.

Analysis We take the cold water tubes as the system, which is a control volume. The energy balance for this steady-flow system can be expressed in the rate form as

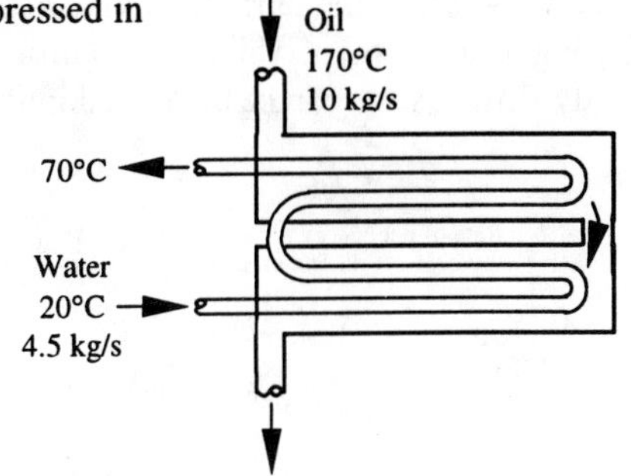

$$\underbrace{\dot{E}_{in} - \dot{E}_{out}}_{\text{Rate of net energy transfer by heat, work, and mass}} = \underbrace{\Delta \dot{E}_{system}^{\nearrow 0 \text{ (steady)}}}_{\text{Rate of change in internal, kinetic, potential, etc. energies}} = 0$$

$$\dot{E}_{in} = \dot{E}_{out}$$

$$\dot{Q}_{in} + \dot{m}h_1 = \dot{m}h_2 \quad (\text{since } \Delta\text{ke} \cong \Delta\text{pe} \cong 0)$$

$$\dot{Q}_{in} = \dot{m}C_p(T_2 - T_1)$$

Then the rate of heat transfer to the cold water in this heat exchanger becomes

$$\dot{Q} = [\dot{m}C_p(T_{out} - T_{in})]_{\text{water}} = (4.5 \text{ kg/s})(4.18 \text{ kJ/kg.°C})(70\text{°C} - 20\text{°C}) = \mathbf{940.5 \text{ kW}}$$

Noting that heat gain by the water is equal to the heat loss by the oil, the outlet temperature of the hot water is determined from

$$\dot{Q} = [\dot{m}C_p(T_{in} - T_{out})]_{\text{oil}} \longrightarrow T_{out} = T_{in} - \frac{\dot{Q}}{\dot{m}C_p} = 170\text{°C} - \frac{940.5 \text{ kW}}{(10 \text{ kg/s})(2.3 \text{ kJ/kg.°C})} = \mathbf{129.1\text{°C}}$$

4-124E Steam is condensed by cooling water in a condenser. The rate of heat transfer in the heat exchanger and the rate of condensation of steam are to be determined.

Assumptions **1** Steady operating conditions exist. **2** The heat exchanger is well-insulated so that heat loss to the surroundings is negligible and thus heat transfer from the hot fluid is equal to the heat transfer to the cold fluid. **3** Changes in the kinetic and potential energies of fluid streams are negligible. **4** Fluid properties are constant.

Properties The specific heat of water is 1.0 Btu/lbm.°F. The enthalpy of vaporization of water at 90°F is 1042.7 Btu/lbm (Table A-4E).

Analysis We take the tube-side of the heat exchanger where cold water is flowing as the system, which is a control volume. The energy balance for this steady-flow system can be expressed in the rate form as

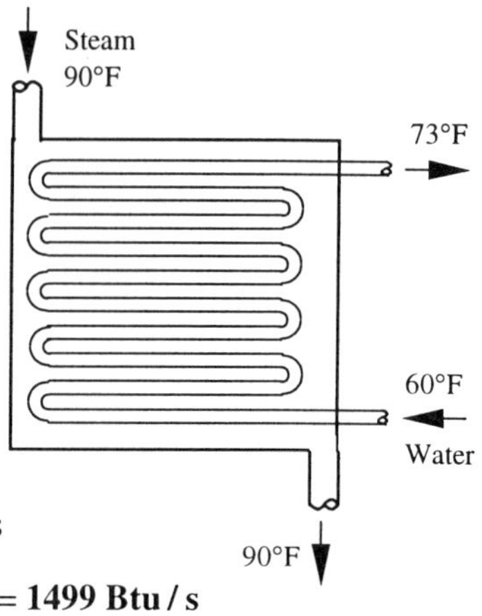

$$\underbrace{\dot{E}_{in} - \dot{E}_{out}}_{\text{Rate of net energy transfer by heat, work, and mass}} = \underbrace{\Delta \dot{E}_{\text{system}}^{\nearrow 0 \text{ (steady)}}}_{\text{Rate of change in internal, kinetic, potential, etc. energies}} = 0$$

$$\dot{E}_{in} = \dot{E}_{out}$$

$$\dot{Q}_{in} + \dot{m}h_1 = \dot{m}h_2 \quad (\text{since } \Delta\text{ke} \cong \Delta\text{pe} \cong 0)$$

$$\dot{Q}_{in} = \dot{m}C_p(T_2 - T_1)$$

Then the rate of heat transfer to the cold water in this heat exchanger becomes

$$\dot{Q} = [\dot{m}C_p(T_{out} - T_{in})]_{\text{water}} = (115.3 \text{ lbm/s})(1.0 \text{ Btu/lbm.°F})(73\text{°F} - 60\text{°F}) = \mathbf{1499 \text{ Btu/s}}$$

Noting that heat gain by the water is equal to the heat loss by the condensing steam, the rate of condensation of the steam in the heat exchanger is determined from

$$\dot{Q} = (\dot{m}h_{fg})_{steam} = \longrightarrow \dot{m}_{steam} = \frac{\dot{Q}}{h_{fg}} = \frac{1499 \text{ Btu/s}}{1042.7 \text{ Btu/lbm}} = \mathbf{1.44 \text{ lbm/s}}$$

Pipe and duct Flow

4-125 A desktop computer is to be cooled safely by a fan in hot environments and high elevations. The air flow rate of the fan and the diameter of the casing are to be determined.

Assumptions **1** Steady operation under worst conditions is considered. **2** Air is an ideal gas with constant specific heats. **3** Kinetic and potential energy changes are negligible.

Properties The specific heat of air at the average temperature of T_{ave} = (45+60)/2 = 52.5°C = 325.5 K is C_p = 1.0065 kJ/kg.°C (Table A-2b)

Analysis The fan selected must be able to meet the cooling requirements of the computer at worst conditions. Therefore, we assume air to enter the computer at 66.63 kPa and 45°C, and leave at 60°C.

We take the air space in the computer as the system, which is a control volume. The energy balance for this steady-flow system can be expressed in the rate form as

$$\underbrace{\dot{E}_{in} - \dot{E}_{out}}_{\substack{\text{Rate of net energy transfer} \\ \text{by heat, work, and mass}}} = \underbrace{\Delta \dot{E}_{\text{system}}^{\nearrow 0 \text{ (steady)}}}_{\substack{\text{Rate of change in internal, kinetic,} \\ \text{potential, etc. energies}}} = 0$$

$$\dot{E}_{in} = \dot{E}_{out}$$

$$\dot{Q}_{in} + \dot{m}h_1 = \dot{m}h_2 \quad (\text{since } \Delta\text{ke} \cong \Delta\text{pe} \cong 0)$$

$$\dot{Q}_{in} = \dot{m}C_p(T_2 - T_1)$$

Then the required mass flow rate of air to absorb heat at a rate of 60 W is determined to be

$$\dot{Q} = \dot{m}C_p(T_{out} - T_{in}) \rightarrow \dot{m} = \frac{\dot{Q}}{C_p(T_{out} - T_{in})} = \frac{60 \text{ W}}{(1006.5 \text{ J/kg.°C})(60 - 45)\text{°C}}$$

$$= 0.00397 \text{ kg/s} = 0.238 \text{ kg/min}$$

The density of air entering the fan at the exit and its volume flow rate are

$$\rho = \frac{P}{RT} = \frac{66.63 \text{ kPa}}{(0.287 \text{ kPa.m}^3\text{/kg.K})(60 + 273)\text{K}} = 0.6972 \text{ kg/m}^3$$

$$\dot{V} = \frac{\dot{m}}{\rho} = \frac{0.238 \text{ kg/min}}{0.6972 \text{ kg/m}^3} = \mathbf{0.341 \text{ m}^3/\text{min}}$$

For an average exit velocity of 110 m/min, the diameter of the casing of the fan is determined from

$$\dot{V} = A_c\mathbf{V} = \frac{\pi D^2}{4}\mathbf{V} \rightarrow D = \sqrt{\frac{4\dot{V}}{\pi\mathbf{V}}} = \sqrt{\frac{(4)(0.341 \text{ m}^3\text{/min})}{\pi(110 \text{ m/min})}} = 0.063 \text{ m} = \mathbf{6.3 \text{ cm}}$$

4-126 A desktop computer is to be cooled safely by a fan in hot environments and high elevations. The air flow rate of the fan and the diameter of the casing are to be determined.

Assumptions **1** Steady operation under worst conditions is considered. **2** Air is an ideal gas with constant specific heats. **3** Kinetic and potential energy changes are negligible.

Properties The specific heat of air at the average temperature of T_{ave} = (45+60)/2 = 52.5°C is C_p = 1.0065 kJ/kg.°C (Table A-3).

Analysis The fan selected must be able to meet the cooling requirements of the computer at worst conditions. Therefore, we assume air to enter the computer at 66.63 kPa and 45°C, and leave at 60°C.

We take the air space in the computer as the system, which is a control volume. The energy balance for this steady-flow system can be expressed in the rate form as

$$\underbrace{\dot{E}_{in} - \dot{E}_{out}}_{\text{Rate of net energy transfer by heat, work, and mass}} = \underbrace{\Delta \dot{E}_{system}^{\nearrow 0 \text{ (steady)}}}_{\text{Rate of change in internal, kinetic, potential, etc. energies}} = 0$$

$$\dot{E}_{in} = \dot{E}_{out}$$

$$\dot{Q}_{in} + \dot{m}h_1 = \dot{m}h_2 \quad (\text{since } \Delta\text{ke} \cong \Delta\text{pe} \cong 0)$$

$$\dot{Q}_{in} = \dot{m}C_p(T_2 - T_1)$$

Then the required mass flow rate of air to absorb heat at a rate of 100 W is determined to be

$$\dot{Q} = \dot{m}C_p(T_{out} - T_{in}) \rightarrow \dot{m} = \frac{\dot{Q}}{C_p(T_{out} - T_{in})} = \frac{100 \text{ W}}{(1006.5 \text{ J/kg.°C})(60 - 45)\text{°C}}$$
$$= 0.006624 \text{ kg/s} = 0.397 \text{ kg/min}$$

The density of air entering the fan at the exit and its volume flow rate are

$$\rho = \frac{P}{RT} = \frac{66.63 \text{ kPa}}{(0.287 \text{ kPa.m}^3\text{/kg.K})(60 + 273)\text{K}} = 0.6972 \text{ kg/m}^3$$

$$\dot{V} = \frac{\dot{m}}{\rho} = \frac{0.397 \text{ kg/min}}{0.6972 \text{ kg/m}^3} = \mathbf{0.57\ m^3/min}$$

For an average exit velocity of 110 m/min, the diameter of the casing of the fan is determined from

$$\dot{V} = A_c\mathbf{V} = \frac{\pi D^2}{4}\mathbf{V} \longrightarrow D = \sqrt{\frac{4\dot{V}}{\pi \mathbf{V}}} = \sqrt{\frac{(4)(0.57 \text{ m}^3\text{/min})}{\pi(110 \text{ m/min})}} = 0.081 \text{ m} = \mathbf{8.1\ cm}$$

4-127E Electronic devices mounted on a cold plate are cooled by water. The amount of heat generated by the electronic devices is to be determined.

Assumptions **1** Steady operating conditions exist. **2** About 15 percent of the heat generated is dissipated from the components to the surroundings by convection and radiation. **3** Kinetic and potential energy changes are negligible.

Properties The properties of water at room temperature are ρ = 62.1 lbm/ft^3 and C_p = 1.00 Btu/lbm.°F (Table A-3).

Analysis We take the tubes of the cold plate to be the system, which is a control volume. The energy balance for this steady-flow system can be expressed in the rate form as

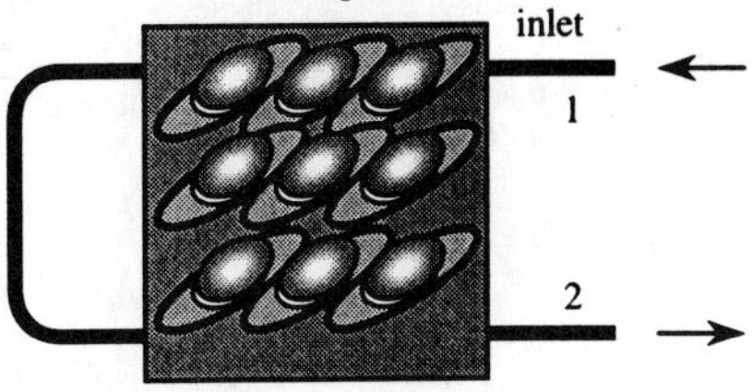

$$\underbrace{\dot{E}_{in} - \dot{E}_{out}}_{\text{Rate of net energy transfer by heat, work, and mass}} = \underbrace{\Delta \dot{E}_{system}^{\nearrow 0 \text{ (steady)}}}_{\text{Rate of change in internal, kinetic, potential, etc. energies}} = 0$$

$$\dot{E}_{in} = \dot{E}_{out}$$

$$\dot{Q}_{in} + \dot{m}h_1 = \dot{m}h_2 \quad (\text{since } \Delta\text{ke} \cong \Delta\text{pe} \cong 0)$$

$$\dot{Q}_{in} = \dot{m}C_p(T_2 - T_1)$$

Cold plate

Water inlet

1

2

Then mass flow rate of water and the rate of heat removal by the water are determined to be

$$\dot{m} = \rho A \mathbf{V} = \rho \frac{\pi D^2}{4} \mathbf{V} = (62.1\,\text{lbm/ft}^3)\frac{\pi(0.25/12\,\text{ft})^2}{4}(60\,\text{ft/min}) = 1.270\,\text{lbm/min} = 76.2\,\text{lbm/h}$$

$$\dot{Q} = \dot{m}C_p(T_{out} - T_{in}) = (76.2\,\text{lbm/h})(1.00\,\text{Btu/lbm.°F})(105 - 95)\text{°F} = 762\,\text{Btu/h}$$

which is 85 percent of the heat generated by the electronic devices. Then the total amount of heat generated by the electronic devices becomes

$$\dot{Q} = \frac{762\,\text{Btu/h}}{0.85} = \mathbf{896\,Btu/h = 263\,W}$$

4-128 A sealed electronic box is to be cooled by tap water flowing through channels on two of its sides. The mass flow rate of water and the amount of water used per year are to be determined.

Assumptions **1** Steady operating conditions exist. **2** Entire heat generated is dissipated by water. **3** Water is an incompressible substance with constant specific heats at room temperature. **4** Kinetic and potential energy changes are negligible.

Properties The specific heat of water at room temperature is $C_p = 4.18$ kJ/kg.°C (Table A-3).

Analysis We take the water channels on the sides to be the system, which is a control volume. The energy balance for this steady-flow system can be expressed in the rate form as

$$\underbrace{\dot{E}_{in} - \dot{E}_{out}}_{\text{Rate of net energy transfer by heat, work, and mass}} = \underbrace{\Delta \dot{E}_{system}}_{\text{Rate of change in internal, kinetic, potential, etc. energies}}^{\nearrow 0 \text{ (steady)}} = 0$$

$$\dot{E}_{in} = \dot{E}_{out}$$

$$\dot{Q}_{in} + \dot{m}h_1 = \dot{m}h_2 \quad (\text{since } \Delta \text{ke} \cong \Delta \text{pe} \cong 0)$$

$$\dot{Q}_{in} = \dot{m}C_p(T_2 - T_1)$$

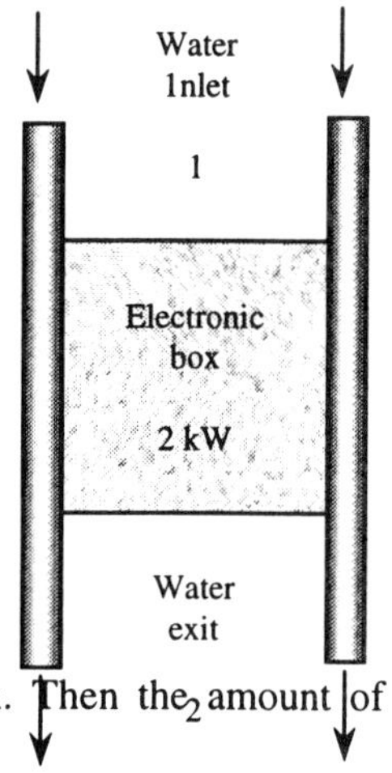

Water
Inlet
1
Electronic
box
2 kW
Water
exit
2

Then the mass flow rate of tap water flowing through the electronic box becomes

$$\dot{Q} = \dot{m}C_p\Delta T \longrightarrow \dot{m} = \frac{\dot{Q}}{C_p\Delta T} = \frac{2 \text{ kJ/s}}{(4.18 \text{ kJ/kg.°C})(4°\text{C})} = \mathbf{0.1196 \text{ kg/s}}$$

Therefore, 0.11962 kg of water is needed per second to cool this electronic box. Then the amount of cooling water used per year becomes

$$m = \dot{m}\Delta t = (0.1196 \text{ kg/s})(365 \text{ days/yr} \times 24 \text{ h/day} \times 3600 \text{ s/h})$$
$$= 3{,}772{,}000 \text{ kg/yr} = \mathbf{3{,}772 \text{ tons/yr}}$$

4-129 A sealed electronic box is to be cooled by tap water flowing through channels on two of its sides. The mass flow rate of water and the amount of water used per year are to be determined.

Assumptions **1** Steady operating conditions exist. **2** Entire heat generated is dissipated by water. **3** Water is an incompressible substance with constant specific heats at room temperature. **4** Kinetic and potential energy changes are negligible

Properties The specific heat of water at room temperature is $C_p = 4.18$ kJ/kg.°C (Table A-3).

Analysis We take the water channels on the sides to be the system, which is a control volume. The energy balance for this steady-flow system can be expressed in the rate form as

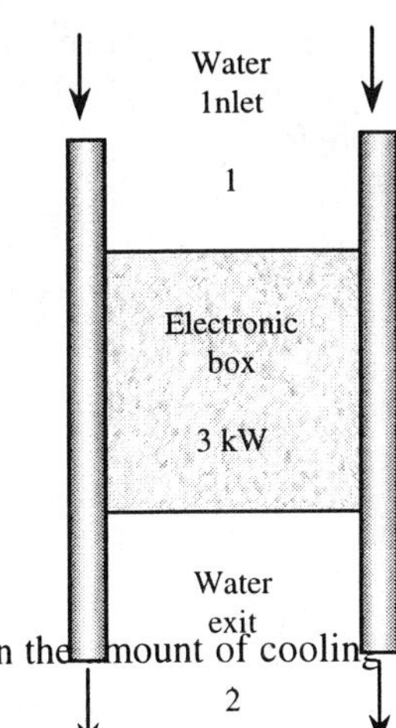

$$\underbrace{\dot{E}_{in} - \dot{E}_{out}}_{\text{Rate of net energy transfer by heat, work, and mass}} = \underbrace{\Delta \dot{E}_{system}}_{\text{Rate of change in internal, kinetic, potential, etc. energies}}^{\nearrow 0 \text{ (steady)}} = 0$$

$$\dot{E}_{in} = \dot{E}_{out}$$

$$\dot{Q}_{in} + \dot{m}h_1 = \dot{m}h_2 \quad (\text{since } \Delta\text{ke} \cong \Delta\text{pe} \cong 0)$$

$$\dot{Q}_{in} = \dot{m}C_p(T_2 - T_1)$$

Then the mass flow rate of tap water flowing through the electronic box becomes

$$\dot{Q} = \dot{m}C_p \Delta T \longrightarrow \dot{m} = \frac{\dot{Q}}{C_p \Delta T} = \frac{3 \text{ kJ/s}}{(4.18 \text{ kJ/kg.°C})(4\text{°C})} = \mathbf{0.1794\ kg/s}$$

Therefore, 0.1794 kg of water is needed per second to cool this electronic box. Then the amount of cooling water used per year becomes

$$m = \dot{m}\Delta t = (0.1794 \text{ kg/s})(365 \text{ days/yr} \times 24 \text{ h/day} \times 3600 \text{ s/h})$$

$$= 5{,}658{,}400 \text{ kg/yr} = \mathbf{5{,}658\ tons/yr}$$

4-130 A long roll of large 1-Mn manganese steel plate is to be quenched in an oil bath at a specified rate. The rate at which heat needs to be removed from the oil to keep its temperature constant is to be determined.

Assumptions **1** Steady operating conditions exist. **2** The thermal properties of the roll are constant. **3** Kinetic and potential energy changes are negligible

Properties The properties of the steel plate are given to be $\rho = 7854$ kg/m^3 and $C_p = 0.454$ kJ/kg.°C.

Analysis The mass flow rate of the sheet metal through the oil bath is

$$\dot{m} = \rho\dot{V} = \rho wtV = (7854\ \text{kg/m}^3)(2\ \text{m})(0.005\ \text{m})(10\ \text{m/min}) = 785.4\ \text{kg/min}$$

We take the volume occupied by the sheet metal in the oil bath to be the system, which is a control volume. The energy balance for this steady-flow system can be expressed I n the rate form as

$$\underbrace{\dot{E}_{in} - \dot{E}_{out}}_{\substack{\text{Rate of net energy transfer} \\ \text{by heat, work, and mass}}} = \underbrace{\Delta\dot{E}_{\text{system}}^{\nearrow 0\ \text{(steady)}}}_{\substack{\text{Rate of change in internal, kinetic,} \\ \text{potential, etc. energies}}} = 0$$

$$\dot{E}_{in} = \dot{E}_{out}$$

$$\dot{m}h_1 = \dot{Q}_{out} + \dot{m}h_2 \quad (\text{since } \Delta\text{ke} \cong \Delta\text{pe} \cong 0)$$

$$\dot{Q}_{out} = \dot{m}C_p(T_1 - T_2)$$

Oil bath
45°C
Steel plate
10 m/min

Then the rate of heat transfer from the sheet metal to the oil bath becomes

$$\begin{aligned}\dot{Q}_{out} &= \dot{m}C_p[T_{in} - T_{out}]_{\text{metal}} \\ &= (785.4\ \text{kg/min})(0.434\ \text{kJ/kg.°C})(820 - 51.1)\text{°C} \\ &= 262{,}090\ \text{kJ/min} = \mathbf{4368\ kW}\end{aligned}$$

This is the rate of heat transfer from the metal sheet to the oil, which is equal to the rate of heat removal from the oil since the oil temperature is maintained constant.

4-131 EES solution of this (and other comprehensive problems designated with the *computer icon*) is available to instructors at the *Instructor Manual* section of the *Online Learning Center* (OLC) at www.mhhe.com/cengel-boles. See the Preface for access information.

4-132 [*Also solved by EES on enclosed CD*] The components of an electronic device located in a horizontal duct of rectangular cross section are cooled by forced air. The heat transfer from the outer surfaces of the duct is to be determined.

Assumptions **1** Steady operating conditions exist. **2** Air is an ideal gas with constant specific heats at room temperature. **3** Kinetic and potential energy changes are negligible

Properties The gas constant of air is R = 0.287 kJ/kg.°C (Table A-1). The specific heat of air at room temperature is C_p = 1.005 kJ/kg.°C (Table A-2).

Analysis The density of air entering the duct and the mass flow rate are

$$\rho = \frac{P}{RT} = \frac{101.325 \text{ kPa}}{(0.287 \text{ kPa.m}^3/\text{kg.K})(30+273)\text{K}} = 1.165 \text{ kg/m}^3$$

$$\dot{m} = \rho\dot{V} = (1.165 \text{ kg/m}^3)(0.6 \text{ m}^3/\text{min}) = 0.700 \text{ kg/min}$$

We take the channel, excluding the electronic components, to be the system, which is a control volume. The energy balance for this steady-flow system can be expressed in the rate form as

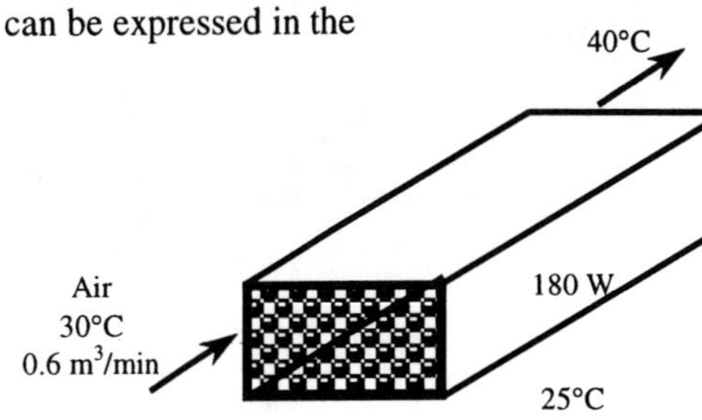

$$\underbrace{\dot{E}_{in} - \dot{E}_{out}}_{\text{Rate of net energy transfer by heat, work, and mass}} = \underbrace{\Delta\dot{E}_{system}^{\nearrow 0 \text{ (steady)}}}_{\text{Rate of change in internal, kinetic, potential, etc. energies}} = 0$$

$$\dot{E}_{in} = \dot{E}_{out}$$

$$\dot{Q}_{in} + \dot{m}h_1 = \dot{m}h_2 \quad (\text{since } \Delta\text{ke} \cong \Delta\text{pe} \cong 0)$$

$$\dot{Q}_{in} = \dot{m}C_p(T_2 - T_1)$$

Then the rate of heat transfer to the air passing through the duct becomes

$$\dot{Q}_{air} = [\dot{m}C_p(T_{out} - T_{in})]_{\text{air}} = (0.700/60 \text{ kg/s})(1.005 \text{ kJ/kg.°C})(40-30)\text{°C} = 0.117 \text{ kW} = 117 \text{ W}$$

The rest of the 180 W heat generated must be dissipated through the outer surfaces of the duct by natural convection and radiation,

$$\dot{Q}_{\text{external}} = \dot{Q}_{\text{total}} - \dot{Q}_{\text{internal}} = 180 - 117 = \mathbf{63\ W}$$

4-133 The components of an electronic device located in a horizontal duct of circular cross section is cooled by forced air. The heat transfer from the outer surfaces of the duct is to be determined.

Assumptions **1** Steady operating conditions exist. **2** Air is an ideal gas with constant specific heats at room temperature. **3** Kinetic and potential energy changes are negligible

Properties The gas constant of air is R = 0.287 kJ/kg.°C (Table A-1). The specific heat of air at room temperature is C_p = 1.005 kJ/kg.°C (Table A-2).

Analysis The density of air entering the duct and the mass flow rate are

$$\rho = \frac{P}{RT} = \frac{101.325 \text{ kPa}}{(0.287 \text{ kPa.m}^3/\text{kg.K})(30+273)\text{K}} = 1.165 \text{ kg/m}^3$$

$$\dot{m} = \rho\dot{V} = (1.165 \text{ kg/m}^3)(0.6 \text{ m}^3/\text{min}) = 0.700 \text{ kg/min}$$

We take the channel, excluding the electronic components, to be the system, which is a control volume. The energy balance for this steady-flow system can be expressed in the rate form as

$$\underbrace{\dot{E}_{in} - \dot{E}_{out}}_{\text{Rate of net energy transfer by heat, work, and mass}} = \underbrace{\Delta\dot{E}_{\text{system}}^{\nearrow 0 \text{ (steady)}}}_{\text{Rate of change in internal, kinetic, potential, etc. energies}} = 0$$

$$\dot{E}_{in} = \dot{E}_{out}$$

$$\dot{Q}_{in} + \dot{m}h_1 = \dot{m}h_2 \quad (\text{since } \Delta\text{ke} \cong \Delta\text{pe} \cong 0)$$

$$\dot{Q}_{in} = \dot{m}C_p(T_2 - T_1)$$

1
Air
30°C
0.6 m³/s
2

Then the rate of heat transfer to the air passing through the duct becomes

$$\dot{Q}_{air} = [\dot{m}C_p(T_{out} - T_{in})]_{\text{air}} = (0.700/60 \text{ kg/s})(1.005 \text{ kJ/kg.°C})(40-30)\text{°C} = 0.117 \text{ kW} = 117 \text{ W}$$

The rest of the 180 W heat generated must be dissipated through the outer surfaces of the duct by natural convection and radiation,

$$\dot{Q}_{\text{external}} = \dot{Q}_{\text{total}} - \dot{Q}_{\text{internal}} = 180 - 117 = \mathbf{63 \text{ W}}$$

4-134E Water is heated in a parabolic solar collector. The required length of parabolic collector is to be determined.

Assumptions **1** Steady operating conditions exist. **2** Heat loss from the tube is negligible so that the entire solar energy incident on the tube is transferred to the water. **3** Kinetic and potential energy changes are negligible

Properties The specific heat of water at room temperature is $C_p = 1.00$ Btu/lbm.°F (Table A-2E).

Analysis We take the thin aluminum tube to be the system, which is a control volume. The energy balance for this steady-flow system can be expressed in the rate form as

$$\underbrace{\dot{E}_{in} - \dot{E}_{out}}_{\substack{\text{Rate of net energy transfer} \\ \text{by heat, work, and mass}}} = \underbrace{\Delta \dot{E}_{\text{system}}^{\nearrow 0 \text{ (steady)}}}_{\substack{\text{Rate of change in internal, kinetic,} \\ \text{potential, etc. energies}}} = 0$$

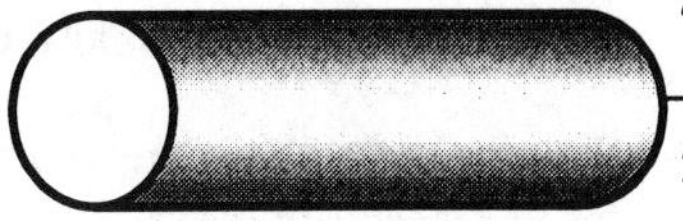

$$\dot{E}_{in} = \dot{E}_{out}$$

$$\dot{Q}_{in} + \dot{m}h_1 = \dot{m}h_2 \quad (\text{since } \Delta\text{ke} \cong \Delta\text{pe} \cong 0)$$

$$\dot{Q}_{in} = \dot{m}_{\text{water}} C_p (T_2 - T_1)$$

1
Water
55°F
4 lbm/s
2
200°F

Then the total rate of heat transfer to the water flowing through the tube becomes

$$\dot{Q}_{\text{total}} = \dot{m}C_p(T_e - T_i) = (4 \text{ lbm/s})(1.00 \text{ Btu/lbm.°F})(200 - 55)\text{°F} = 580 \text{ Btu/s} = 2{,}088{,}000 \text{ Btu/h}$$

The length of the tube required is

$$L = \frac{\dot{Q}_{\text{total}}}{\dot{Q}} = \frac{2{,}088{,}000 \text{ Btu/h}}{350 \text{ Btu/h.ft}} = \mathbf{5966 \text{ ft}}$$

4-135 Air enters a hollow-core printed circuit board. The exit temperature of the air is to be determined.

Assumptions **1** Steady operating conditions exist. **2** Air is an ideal gas with constant specific heats at room temperature. **3** The local atmospheric pressure is 1 atm. **4** Kinetic and potential energy changes are negligible.

Properties The gas constant of air is $R = 0.287$ kJ/kg.°C (Table A-1). The specific heat of air at room temperature is $C_p = 1.005$ kJ/kg.°C (Table A-2).

Analysis The density of air entering the duct and the mass flow rate are

$$\rho = \frac{P}{RT} = \frac{101.325 \text{ kPa}}{(0.287 \text{ kPa.m}^3/\text{kg.K})(32+273)\text{K}} = 1.16 \text{ kg/m}^3$$

$$\dot{m} = \rho\dot{V} = (1.16 \text{ kg/m}^3)(0.0008 \text{ m}^3/\text{s}) = 0.000928 \text{ kg/s}$$

We take the hollow core to be the system, which is a control volume. The energy balance for this steady-flow system can be expressed in the rate form as

$$\underbrace{\dot{E}_{in} - \dot{E}_{out}}_{\substack{\text{Rate of net energy transfer} \\ \text{by heat, work, and mass}}} = \underbrace{\Delta \dot{E}_{system}^{\nearrow 0 \text{ (steady)}}}_{\substack{\text{Rate of change in internal, kinetic,} \\ \text{potential, etc. energies}}} = 0$$

$$\dot{E}_{in} = \dot{E}_{out}$$

$$\dot{Q}_{in} + \dot{m}h_1 = \dot{m}h_2 \quad (\text{since } \Delta\text{ke} \cong \Delta\text{pe} \cong 0)$$

$$\dot{Q}_{in} = \dot{m}C_p(T_2 - T_1)$$

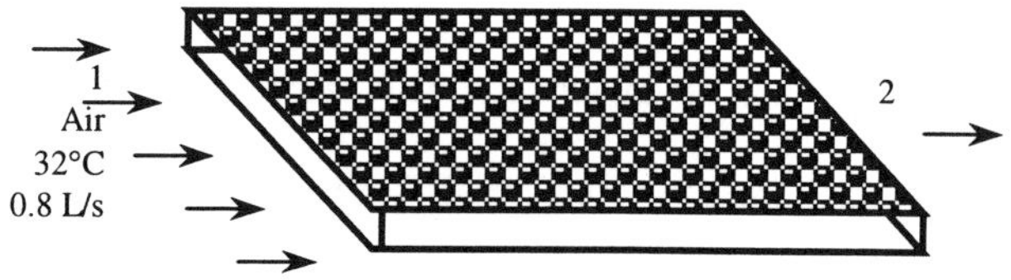

1
Air
32°C
0.8 L/s
2

Then the exit temperature of air leaving the hollow core becomes

$$\dot{Q}_{in} = \dot{m}C_p(T_2 - T_1) \rightarrow T_2 = T_1 + \frac{\dot{Q}_{in}}{\dot{m}C_p} = 32\text{ °C} + \frac{20 \text{ J/s}}{(0.000928 \text{ kg/s})(1005 \text{ J/kg.°C})} = \mathbf{53.4°C}$$

4-136 A computer is cooled by a fan blowing air through the case of the computer. The required flow rate of the air and the fraction of the temperature rise of air that is due to heat generated by the fan are to be determined.

Assumptions **1** Steady flow conditions exist. **2** Air is an ideal gas with constant specific heats. **3** The pressure of air is 1 atm. **4** Kinetic and potential energy changes are negligible

Properties The specific heat of air at room temperature is C_p = 1.005 kJ/kg.°C (Table A-2).

Analysis (*a*) We take the air space in the computer as the system, which is a control volume. The energy balance for this steady-flow system can be expressed in the rate form as

$$\underbrace{\dot{E}_{in} - \dot{E}_{out}}_{\substack{\text{Rate of net energy transfer}\\ \text{by heat, work, and mass}}} = \underbrace{\Delta \dot{E}_{system}^{\nearrow 0 \text{ (steady)}}}_{\substack{\text{Rate of change in internal, kinetic,}\\ \text{potential, etc. energies}}} = 0$$

$$\dot{E}_{in} = \dot{E}_{out}$$

$$\dot{Q}_{in} + \dot{W}_{in} + \dot{m}h_1 = \dot{m}h_2 \quad (\text{since } \Delta\text{ke} \cong \Delta\text{pe} \cong 0)$$

$$\dot{Q}_{in} + \dot{W}_{in} = \dot{m}C_p(T_2 - T_1)$$

Noting that the fan power is 25 W and the 8 PCBs transfer a total of 80 W of heat to air, the mass flow rate of air is determined to be

$$\dot{Q}_{in} + \dot{W}_{in} = \dot{m}C_p(T_e - T_i) \rightarrow \dot{m} = \frac{\dot{Q}_{in} + \dot{W}_{in}}{C_p(T_e - T_i)} = \frac{25 + (8 \times 10)\text{ W}}{(1005\text{ J/kg.°C})(10\text{°C})} = \mathbf{0.0104\ kg/s}$$

(*b*) The fraction of temperature rise of air that is due to the heat generated by the fan and its motor can be determined from

$$\dot{Q} = \dot{m}C_p\Delta T \rightarrow \Delta T = \frac{\dot{Q}}{\dot{m}C_p} = \frac{25\text{ W}}{(0.0104\text{ kg/s})(1005\text{ J/kg.°C})} = 2.4\text{°C}$$

$$f = \frac{2.4\text{°C}}{10\text{°C}} = 0.24 = \mathbf{24\%}$$

4-137 Hot water enters a pipe whose outer surface is exposed to cold air in a basement. The rate of heat loss from the water is to be determined.

Assumptions **1** Steady flow conditions exist. **2** Water is an incompressible substance with constant specific heats. **3** The changes in kinetic and potential energies are negligible.

Properties The properties of water at the average temperature of (90+88)/2 = 89°C are $\rho = 965\ \text{kg/m}^3$ and C_p = 4.21 kJ/kg.°C (Table A-3).

Analysis The mass flow rate of water is

$$\dot{m} = \rho A_c V = (965\ \text{kg/m}^3)\frac{\pi(0.04\ \text{m})^2}{4}(0.8\ \text{m/s}) = 0.970\ \text{kg/s}$$

We take the section of the pipe in the basement to be the system, which is a control volume. The energy balance for this steady-flow system can be expressed in the rate form as

$$\underbrace{\dot{E}_{in} - \dot{E}_{out}}_{\text{Rate of net energy transfer by heat, work, and mass}} = \underbrace{\Delta \dot{E}_{\text{system}}^{\nearrow 0\ \text{(steady)}}}_{\text{Rate of change in internal, kinetic, potential, etc. energies}} = 0$$

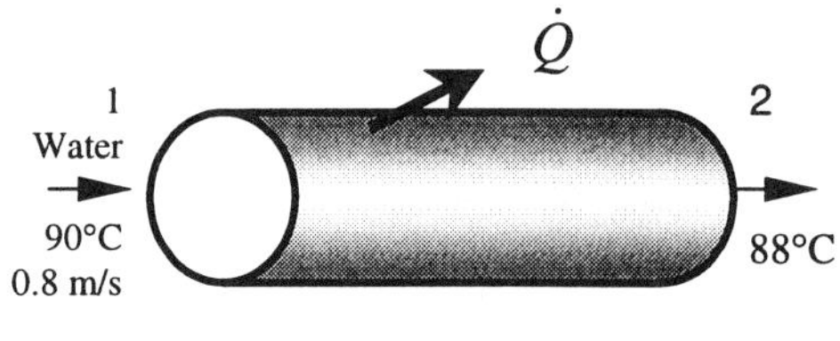

$$\dot{E}_{in} = \dot{E}_{out}$$

$$\dot{m}h_1 = \dot{Q}_{out} + \dot{m}h_2 \quad (\text{since } \Delta\text{ke} \cong \Delta\text{pe} \cong 0)$$

$$\dot{Q}_{out} = \dot{m}C_p(T_1 - T_2)$$

1
Water
90°C
0.8 m/s
$\dot{Q}$
2
88°C

Then the rate of heat transfer from the hot water to the surrounding air becomes

$$\begin{aligned}\dot{Q}_{out} &= \dot{m}C_p[T_{in} - T_{out}]_{\text{water}} \\ &= (0.970\ \text{kg/s})(4.21\ \text{kJ/kg.°C})(90 - 88)\text{°C} \\ &= \mathbf{8.17\ kW}\end{aligned}$$

3-138 EES solution of this (and other comprehensive problems designated with the *computer icon*) is available to instructors at the *Instructor Manual* section of the *Online Learning Center* (OLC) at www.mhhe.com/cengel-boles. See the Preface for access information.

4-139 A room is to be heated by an electric resistance heater placed in a duct in the room. The power rating of the electric heater and the temperature rise of air as it passes through the heater are to be determined.

Assumptions **1** Steady operating conditions exist. **2** Air is an ideal gas with constant specific heats at room temperature. **3** Kinetic and potential energy changes are negligible. **4** The heating duct is adiabatic, and thus heat transfer through it is negligible. **5** No air leaks in and out of the room.

Properties The gas constant of air is 0.287 kPa.m^3/kg.K (Table A-1). The specific heats of air at room temperature are $C_p = 1.005$ and $C_v = 0.718$ kJ/kg· K (Table A-2).

Analysis (*a*) The total mass of air in the room is

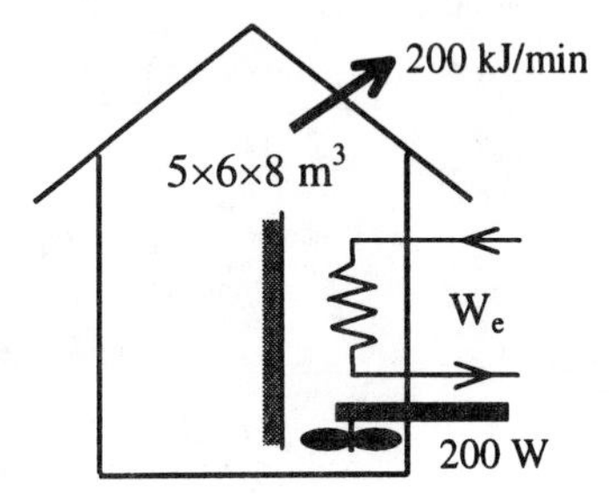

$$V = 5 \times 6 \times 8\text{m}^3 = 240\text{m}^3$$

$$m = \frac{P_1 V}{RT_1} = \frac{(98 \text{ kPa})(240 \text{ m}^3)}{(0.287 \text{ kPa}\cdot\text{m}^3/\text{kg}\cdot\text{K})(288 \text{ K})} = 284.6 \text{ kg}$$

We first take the *entire room* as our system, which is a closed system since no mass leaks in or out. The power rating of the electric heater is determined by applying the conservation of energy relation to this constant volume closed system:

$$\underbrace{E_{in} - E_{out}}_{\substack{\text{Net energy transfer}\\ \text{by heat, work, and mass}}} = \underbrace{\Delta E_{\text{system}}}_{\substack{\text{Change in internal, kinetic,}\\ \text{potential, etc. energies}}}$$

$$W_{e,in} + W_{fan,in} - Q_{out} = \Delta U \quad (\text{since } \Delta\text{KE} = \Delta\text{PE} = 0)$$

$$\Delta t\left(\dot{W}_{e,in} + \dot{W}_{fan,in} - \dot{Q}_{out}\right) = mC_{v,ave}\left(T_2 - T_1\right)$$

Solving for the electrical work input gives

$$\dot{W}_{e,in} = \dot{Q}_{out} - \dot{W}_{fan,in} + mC_v(T_2 - T_1)/\Delta t$$

$$= (200/60 \text{ kJ/s}) - (0.2 \text{ kJ/s}) + (284.6 \text{ kg})(0.718\text{kJ/kg}\cdot{}^\circ\text{C})(25-15)^\circ\text{C}/(15\times 60\text{s})$$

$$= \mathbf{5.40\ kW}$$

(*b*) We now take the *heating duct* as the system, which is a control volume since mass crosses the boundary. There is only one inlet and one exit, and thus $\dot{m}_1 = \dot{m}_2 = \dot{m}$. The energy balance for this adiabatic steady-flow system can be expressed in the rate form as

$$\underbrace{\dot{E}_{in} - \dot{E}_{out}}_{\substack{\text{Rate of net energy transfer}\\ \text{by heat, work, and mass}}} = \underbrace{\Delta \dot{E}_{\text{system}}^{\nearrow 0 \text{ (steady)}}}_{\substack{\text{Rate of change in internal, kinetic,}\\ \text{potential, etc. energies}}} = 0$$

$$\dot{E}_{in} = \dot{E}_{out}$$

$$\dot{W}_{e,in} + \dot{W}_{fan,in} + \dot{m}h_1 = \dot{m}h_2 \quad (\text{since } \dot{Q} = \Delta\text{ke} \cong \Delta\text{pe} \cong 0)$$

$$\dot{W}_{e,in} + \dot{W}_{fan,in} = \dot{m}(h_2 - h_1) = \dot{m}C_p(T_2 - T_1)$$

Thus,

$$\Delta T = T_2 - T_1 = \frac{\dot{W}_{e,in} + \dot{W}_{fan,in}}{\dot{m}C_p} = \frac{(5.40 + 0.2)\text{kJ/s}}{(50/60\text{kg/s})(1.005\text{kJ/kg}\cdot\text{K})} = \mathbf{6.7^\circ C}$$

4-140 A house is heated by an electric resistance heater placed in a duct. The power rating of the electric heater is to be determined.

Assumptions **1** This is a steady-flow process since there is no change with time. **2** Air is an ideal gas with constant specific heats at room temperature. **3** Kinetic and potential energy changes are negligible.

Properties The constant pressure specific heat of air at room temperature is C_p = 1.005 kJ/kg· K (Table A-2)

Analysis We take the *heating duct* as the system, which is a control volume since mass crosses the boundary. There is only one inlet and one exit, and thus $\dot{m}_1 = \dot{m}_2 = \dot{m}$. The energy balance for this steady-flow system can be expressed in the rate form as

$$\underbrace{\dot{E}_{in} - \dot{E}_{out}}_{\text{Rate of net energy transfer by heat, work, and mass}} = \underbrace{\Delta \dot{E}_{system}}_{\text{Rate of change in internal, kinetic, potential, etc. energies}}^{\nearrow 0 \text{ (steady)}} = 0$$

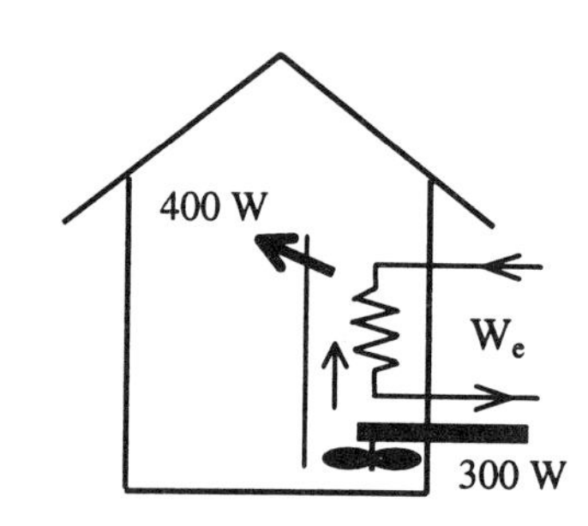

$$\dot{E}_{in} = \dot{E}_{out}$$

$$\dot{W}_{e,in} + \dot{W}_{fan,in} + \dot{m}h_1 = \dot{Q}_{out} + \dot{m}h_2 \quad (\text{since } \Delta ke \cong \Delta pe \cong 0)$$

$$\dot{W}_{e,in} + \dot{W}_{fan,in} = \dot{Q}_{out} + \dot{m}(h_2 - h_1)$$

$$= \dot{Q}_{out} + \dot{m}C_p(T_2 - T_1)$$

Substituting, the power rating of the heating element is determined to be

$$\dot{W}_{e,in} = \dot{Q}_{out} + \dot{m}C_p\Delta T - \dot{W}_{fan,in}$$

$$= (0.4 \text{ kJ/s}) + (0.6 \text{ kg/s})(1.005 \text{ kJ/kg}\cdot{}^\circ\text{C})(5^\circ\text{C}) - 0.3 \text{ kW}$$

$$= \mathbf{3.12\ kW}$$

4-141 A hair dryer consumes 1200 W of electric power when running. The inlet volume flow rate and the exit velocity of air are to be determined.

Assumptions **1** This is a steady-flow process since there is no change with time. **2** Air is an ideal gas with constant specific heats at room temperature. **3** Kinetic and potential energy changes are negligible. **4** The power consumed by the fan and the heat losses are negligible.

Properties The gas constant of air is 0.287 kPa.m^3/kg.K (Table A-1). The constant pressure specific heat of air at room temperature is C_p = 1.005 kJ/kg· K (Table A-2)

Analysis We take the *hair dryer* as the system, which is a control volume since mass crosses the boundary. There is only one inlet and one exit, and thus $\dot{m}_1 = \dot{m}_2 = \dot{m}$. The energy balance for this steady-flow system can be expressed in the rate form as

$$\underbrace{\dot{E}_{in} - \dot{E}_{out}}_{\text{Rate of net energy transfer by heat, work, and mass}} = \underbrace{\Delta \dot{E}_{system}^{\nearrow 0 \text{ (steady)}}}_{\text{Rate of change in internal, kinetic, potential, etc. energies}} = 0$$

$$\dot{E}_{in} = \dot{E}_{out}$$

$$\dot{W}_{e,in} + \dot{m}h_1 = \dot{m}h_2 \quad (\text{since } \dot{Q}_{out} \cong \Delta ke \cong \Delta pe \cong 0)$$

$$\dot{W}_{e,in} = \dot{m}(h_2 - h_1) = \dot{m}C_p(T_2 - T_1)$$

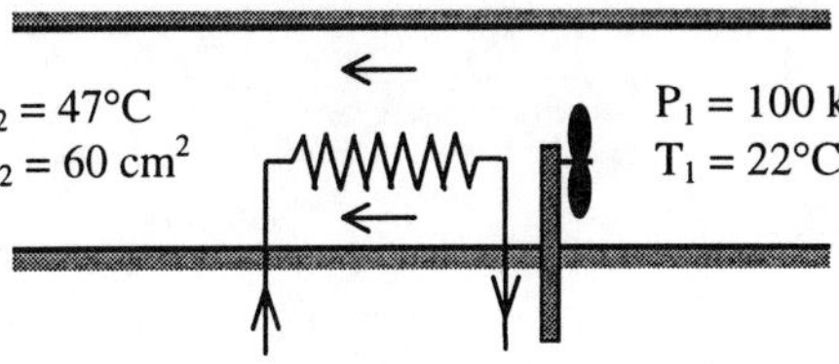

T$_2$ = 47°C
A$_2$ = 60 cm^2
P$_1$ = 100 kPa
T$_1$ = 22°C
W$_e$ = 1200 W

Substituting, the mass and volume flow rates of air are determined to be

$$\dot{m} = \frac{\dot{W}_{e,in}}{C_p(T_2 - T_1)} = \frac{1.2 \text{ kJ/s}}{(1.005 \text{kJ/kg}\cdot{}^\circ\text{C})(47-22)^\circ\text{C}} = 0.04776 \text{kg/s}$$

$$v_1 = \frac{RT_1}{P_1} = \frac{(0.287 \text{ kPa}\cdot\text{m}^3/\text{kg}\cdot\text{K})(295 \text{ K})}{(100 \text{ kPa})} = 0.8467 \text{ m}^3/\text{kg}$$

$$\dot{V}_1 = \dot{m}v_1 = (0.04776 \text{ kg/s})(0.8467 \text{ m}^3/\text{kg}) = \mathbf{0.0404 \text{ m}^3/\text{s}}$$

(*b*) The exit velocity of air is determined from the mass balance $\dot{m}_1 = \dot{m}_2 = \dot{m}$ to be

$$v_2 = \frac{RT_2}{P_2} = \frac{(0.287 \text{ kPa}\cdot\text{m}^3/\text{kg}\cdot\text{K})(320\text{K})}{(100 \text{ kPa})} = 0.9184 \text{ m}^3/\text{kg}$$

$$\dot{m} = \frac{1}{v_2}A_2\mathbf{V}_2 \longrightarrow \mathbf{V}_2 = \frac{\dot{m}v_2}{A_2} = \frac{(0.04776 \text{ kg/s})(0.9184 \text{ m}^3/\text{kg})}{60\times10^{-4}\text{m}^2} = \mathbf{7.31 \text{ m/s}}$$

4-142 EES solution of this (and other comprehensive problems designated with the *computer icon*) is available to instructors at the *Instructor Manual* section of the *Online Learning Center* (OLC) at www.mhhe.com/cengel-boles. See the Preface for access information.

4-143 The ducts of a heating system pass through an unheated area. The rate of heat loss from the air in the ducts is to be determined.

Assumptions **1** This is a steady-flow process since there is no change with time. **2** Air is an ideal gas with constant specific heats at room temperature. **3** Kinetic and potential energy changes are negligible. **4** There are no work interactions involved.

Properties The constant pressure specific heat of air at room temperature is C_p = 1.005 kJ/kg· K (Table A-2)

Analysis We take the *heating duct* as the system, which is a control volume since mass crosses the boundary. There is only one inlet and one exit, and thus $\dot{m}_1 = \dot{m}_2 = \dot{m}$. The energy balance for this steady-flow system can be expressed in the rate form as

$$\underbrace{\dot{E}_{in} - \dot{E}_{out}}_{\text{Rate of net energy transfer by heat, work, and mass}} = \underbrace{\Delta \dot{E}_{system}^{\nearrow 0 \text{ (steady)}}}_{\text{Rate of change in internal, kinetic, potential, etc. energies}} = 0$$

$$\dot{E}_{in} = \dot{E}_{out}$$

$$\dot{m}h_1 = \dot{Q}_{out} + \dot{m}h_2 \quad (\text{since } \dot{W} \cong \Delta\text{ke} \cong \Delta\text{pe} \cong 0)$$

$$\dot{Q}_{out} = \dot{m}(h_1 - h_2) = \dot{m}C_p(T_1 - T_2)$$

120 kg/min AIR

$\dot{Q}$

Substituting,

$$\dot{Q}_{out} = (120 \text{ kg/min})(1.005 \text{ kJ/kg·°C})(4°\text{C}) = \mathbf{482 \text{ kJ/min}}$$

4-144E The ducts of an air-conditioning system pass through an unconditioned area. The inlet velocity and the exit temperature of air are to be determined.

Assumptions **1** This is a steady-flow process since there is no change with time. **2** Air is an ideal gas with constant specific heats at room temperature. **3** Kinetic and potential energy changes are negligible. **4** There are no work interactions involved.

Properties The gas constant of air is 0.3704 psia.ft³/lbm.R (Table A-1E). The constant pressure specific heat of air at room temperature is C_p = 0.240 Btu/lbm.R (Table A-2E)

Analysis (*a*) The inlet velocity of air through the duct is

$$\mathbf{V}_1 = \frac{\dot{V}_1}{A_1} = \frac{\dot{V}_1}{\pi r^2} = \frac{450 \text{ ft}^3/\text{min}}{\pi (5/12 \text{ ft})^2} = \mathbf{825 \text{ ft/min}}$$

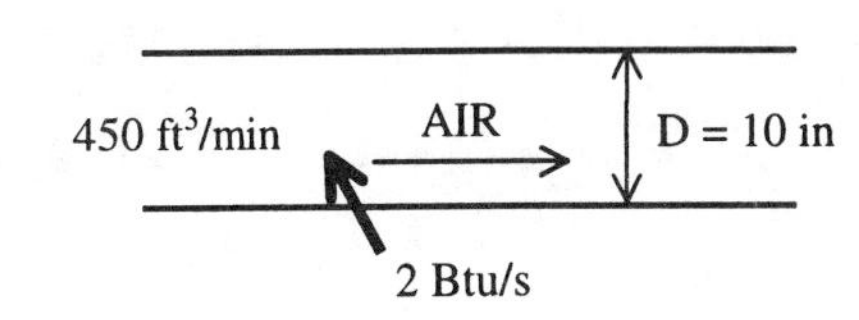

Then the mass flow rate of air becomes

$$v_1 = \frac{RT_1}{P_1} = \frac{(0.3704 \text{psia} \cdot \text{ft}^3/\text{lbm} \cdot \text{R})(510\text{R})}{(15\text{psia})} = 12.6\text{ft}^3/\text{lbm}$$

$$\dot{m} = \frac{\dot{V}_1}{v_1} = \frac{450\text{ft}^3/\text{min}}{12.6\text{ft}^3/\text{lbm}} = 35.7\text{lbm/min} = 0.595\text{lbm/s}$$

(*b*) We take the *air-conditioning duct* as the system, which is a control volume since mass crosses the boundary. There is only one inlet and one exit, and thus $\dot{m}_1 = \dot{m}_2 = \dot{m}$. The energy balance for this steady-flow system can be expressed in the rate form as

$$\underbrace{\dot{E}_{in} - \dot{E}_{out}}_{\substack{\text{Rate of net energy transfer} \\ \text{by heat, work, and mass}}} = \underbrace{\Delta \dot{E}_{\text{system}}^{\nearrow 0 \text{ (steady)}}}_{\substack{\text{Rate of change in internal, kinetic,} \\ \text{potential, etc. energies}}} = 0$$

$$\dot{E}_{in} = \dot{E}_{out}$$

$$\dot{Q}_{in} + \dot{m}h_1 = \dot{m}h_2 \quad (\text{since } \dot{W} \cong \Delta\text{ke} \cong \Delta\text{pe} \cong 0)$$

$$\dot{Q}_{in} = \dot{m}(h_2 - h_1) = \dot{m}C_p(T_2 - T_1)$$

Then the exit temperature of air becomes

$$T_2 = T_1 + \frac{\dot{Q}_{in}}{\dot{m}C_p} = 50^\circ\text{F} + \frac{2 \text{ Btu/s}}{(0.595 \text{ lbm/s})(0.24 \text{ Btu/lbm}\cdot^\circ\text{F})} = \mathbf{64.0^\circ F}$$

4-145 Water is heated by a 7-kW resistance heater as it flows through an insulated tube. The mass flow rate of water is to be determined.

Assumptions **1** This is a steady-flow process since there is no change with time. **2** Water is an incompressible substance with constant specific heats at room temperature. **3** Kinetic and potential energy changes are negligible. **4** The tube is adiabatic and thus heat losses are negligible.

Properties The specific heat of water at room temperature is $C = 4.18$ kJ/kg· °C (Table A-3).

Analysis We take the *water pipe* as the system, which is a control volume since mass crosses the boundary. There is only one inlet and one exit, and thus $\dot{m}_1 = \dot{m}_2 = \dot{m}$. The energy balance for this steady-flow system can be expressed in the rate form as

$$\underbrace{\dot{E}_{in} - \dot{E}_{out}}_{\text{Rate of net energy transfer by heat, work, and mass}} = \underbrace{\Delta \dot{E}_{system}^{\nearrow 0 \text{ (steady)}}}_{\text{Rate of change in internal, kinetic, potential, etc. energies}} = 0$$

$$\dot{E}_{in} = \dot{E}_{out}$$

$$\dot{W}_{e,in} + \dot{m}h_1 = \dot{m}h_2 \quad (\text{since } \dot{Q}_{out} \cong \Delta\text{ke} \cong \Delta\text{pe} \cong 0)$$

$$\dot{W}_{e,in} = \dot{m}(h_2 - h_1) = \dot{m}[C(T_2 - T_1) + v\Delta P^{\nearrow 0}] = \dot{m}C(T_2 - T_1)$$

Substituting, the mass flow rates of water is determined to be

$$\dot{m} = \frac{\dot{W}_{e,in}}{C(T_2 - T_1)} = \frac{7 \text{ kJ/s}}{(4.184 \text{ kJ/kg}\cdot{}^\circ\text{C})(75 - 20)^\circ\text{C}} = \mathbf{0.0304\ kg/s}$$

4-146 Steam pipes pass through an unheated area, and the temperature of steam drops as a result of heat losses. The mass flow rate of steam and the rate of heat loss from are to be determined. √

Assumptions **1** This is a steady-flow process since there is no change with time. **2** Kinetic and potential energy changes are negligible. **4** There are no work interactions involved.

Properties From the steam tables (Table A-6),

$$\left.\begin{array}{l} P_1 = 1\text{MPa} \\ T_1 = 250^\circ\text{C} \end{array}\right\} \begin{array}{l} v_1 = 0.2327\text{m}^3/\text{kg} \\ h_1 = 2942.6\text{kJ/kg} \end{array}$$

1 MPa
250°C
STEAM
800 kPa
200°C
$\dot{Q}$

and

$$\left.\begin{array}{l} P_2 = 800\text{kPa} \\ T_2 = 200^\circ\text{C} \end{array}\right\} h_2 = 2839.3\text{kJ/kg}$$

Analysis (*a*) The mass flow rate of steam is determined directly from

$$\dot{m} = \frac{1}{v_1} A_1 \mathbf{V}_1 = \frac{1}{0.2327\text{m}^3/\text{kg}} \left[\pi(0.06\text{m})^2\right](2\text{m/s}) = \mathbf{0.0972\ kg/s}$$

(*b*) We take the *steam pipe* as the system, which is a control volume since mass crosses the boundary. There is only one inlet and one exit, and thus $\dot{m}_1 = \dot{m}_2 = \dot{m}$. The energy balance for this steady-flow system can be expressed in the rate form as

$$\underbrace{\dot{E}_{in} - \dot{E}_{out}}_{\substack{\text{Rate of net energy transfer} \\ \text{by heat, work, and mass}}} = \underbrace{\Delta \dot{E}_{\text{system}}^{\nearrow 0\ \text{(steady)}}}_{\substack{\text{Rate of change in internal, kinetic,} \\ \text{potential, etc. energies}}} = 0$$

$$\dot{E}_{in} = \dot{E}_{out}$$

$$\dot{m}h_1 = \dot{Q}_{out} + \dot{m}h_2 \quad (\text{since } \dot{W} \cong \Delta\text{ke} \cong \Delta\text{pe} \cong 0)$$

$$\dot{Q}_{out} = \dot{m}(h_1 - h_2)$$

Substituting, the rate of heat loss is determined to be

$$\dot{Q}_{loss} = (0.0972\text{ kg/s})(2942.6 - 2839.3)\text{ kJ/kg} = \mathbf{10.04\ \ kJ/s}$$

Energy Balance for Charging and Discharging Processes

4-147 An evacuated bottle is surrounded by atmospheric air. A valve is opened, and air is allowed to fill the bottle. The amount of heat transfer through the wall of the bottle when thermal and mechanical equilibrium is established is to be determined.

Assumptions **1** This is an unsteady process since the conditions within the device are changing during the process, but it can be analyzed as a uniform-flow process since the state of fluid at the inlet remains constant. **2** Air is an ideal gas with variable specific heats. **3** Kinetic and potential energies are negligible. **4** There are no work interactions involved. **5** The direction of heat transfer is to the air in the bottle (will be verified).

Properties The gas constant of air is 0.287 kPa.m^3/kg.K (Table A-1).

Analysis We take the bottle as the system, which is a control volume since mass crosses the boundary. Noting that the microscopic energies of flowing and nonflowing fluids are represented by enthalpy h and internal energy u, respectively, the mass and energy balances for this uniform-flow system can be expressed as

Mass balance: $m_{in} - m_{out} = \Delta m_{\text{system}} \;\rightarrow\; m_i = m_2 \quad (\text{since } m_{out} = m_{initial} = 0)$

Energy balance:

$$\underbrace{E_{in} - E_{out}}_{\substack{\text{Net energy transfer} \\ \text{by heat, work, and mass}}} = \underbrace{\Delta E_{\text{system}}}_{\substack{\text{Change in internal, kinetic,} \\ \text{potential, etc. energies}}}$$

$$Q_{in} + m_i h_i = m_2 u_2 \quad (\text{since } W \cong E_{out} = E_{initial} = ke \cong pe \cong 0)$$

Combining the two balances:

$$Q_{in} = m_2 (u_2 - h_i)$$

where

$$m_2 = \frac{P_2 V}{RT_2} = \frac{(100 \text{ kPa})(0.008 \text{ m}^3)}{(0.287 \text{ kPa}\cdot\text{m}^3/\text{kg}\cdot\text{K})(290 \text{ K})} = 0.0096 \text{ kg}$$

$$T_i = T_2 = 290 \text{ K} \xrightarrow{\text{Table A-17}} \begin{array}{l} h_i = 290.16 \text{ kJ/kg} \\ u_2 = 206.91 \text{ kJ/kg} \end{array}$$

100 kPa
17°C

8 L
Evacuated

Substituting,

$$Q_{in} = (0.0096 \text{ kg})(206.91 - 290.16) \text{ kJ/kg} = -0.8 \text{ kJ} \quad \rightarrow \quad Q_{out} = \mathbf{0.8\ kJ}$$

Discussion The negative sign for heat transfer indicates that the assumed direction is wrong. Therefore, we reverse the direction.

4-148 An insulated rigid tank is evacuated. A valve is opened, and air is allowed to fill the tank until mechanical equilibrium is established. The final temperature in the tank is to be determined.

Assumptions **1** This is an unsteady process since the conditions within the device are changing during the process, but it can be analyzed as a uniform-flow process since the state of fluid at the inlet remains constant. **2** Air is an ideal gas with constant specific heats. **3** Kinetic and potential energies are negligible. **4** There are no work interactions involved. **5** The device is adiabatic and thus heat transfer is negligible.

Properties The specific heat ratio air at room temperature is $k = 1.4$ (Table A-2).

Analysis We take the tank as the system, which is a control volume since mass crosses the boundary. Noting that the microscopic energies of flowing and nonflowing fluids are represented by enthalpy h and internal energy u, respectively, the mass and energy balances for this uniform-flow system can be expressed as

Mass balance: $m_{in} - m_{out} = \Delta m_{\text{system}} \quad \rightarrow \quad m_i = m_2 \quad (\text{since } m_{out} = m_{initial} = 0)$

Energy balance:

$$\underbrace{E_{in} - E_{out}}_{\substack{\text{Net energy transfer}\\ \text{by heat, work, and mass}}} = \underbrace{\Delta E_{\text{system}}}_{\substack{\text{Change in internal, kinetic,}\\ \text{potential, etc. energies}}}$$

$$m_i h_i = m_2 u_2 \quad (\text{since } Q \cong W \cong E_{out} = E_{initial} = ke \cong pe \cong 0)$$

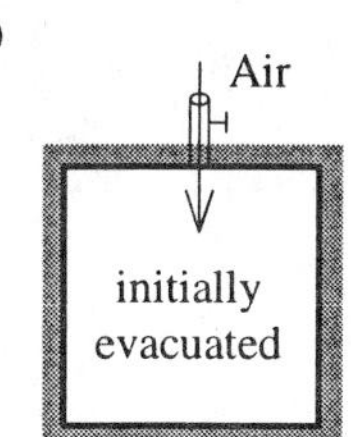

Combining the two balances:

$$u_2 = h_i \quad \rightarrow \quad C_v T_2 = C_p T_i \quad \rightarrow \quad T_2 = (C_p / C_v) T_i = k T_i$$

Substituting,

$$T_2 = 1.4 \times 290 \text{ K} = 406 \text{ K} = \mathbf{133^\circ C}$$

4-149 A rigid tank initially contains air at atmospheric conditions. The tank is connected to a supply line, and air is allowed to enter the tank until mechanical equilibrium is established. The mass of air that entered and the amount of heat transfer are to be determined.

Assumptions **1** This is an unsteady process since the conditions within the device are changing during the process, but it can be analyzed as a uniform-flow process since the state of fluid at the inlet remains constant. **2** Air is an ideal gas with variable specific heats. **3** Kinetic and potential energies are negligible. **4** There are no work interactions involved. **5** The direction of heat transfer is to the tank (will be verified).

Properties The gas constant of air is 0.287 kPa.m^3/kg.K (Table A-1). The properties of air are (Table A-17)

$$T_i = 295 \text{ K} \longrightarrow h_i = 295.17 \text{ kJ/kg}$$
$$T_1 = 295 \text{ K} \longrightarrow u_1 = 210.49 \text{ kJ/kg}$$
$$T_2 = 350 \text{ K} \longrightarrow u_2 = 250.02 \text{ kJ/kg}$$

Analysis (*a*) We take the tank as the system, which is a control volume since mass crosses the boundary. Noting that the microscopic energies of flowing and nonflowing fluids are represented by enthalpy *h* and internal energy *u*, respectively, the mass and energy balances for this uniform-flow system can be expressed as

Mass balance: $m_{in} - m_{out} = \Delta m_{\text{system}} \quad \rightarrow \quad m_i = m_2 - m_1$

Energy balance:

$$\underbrace{E_{in} - E_{out}}_{\substack{\text{Net energy transfer}\\ \text{by heat, work, and mass}}} = \underbrace{\Delta E_{\text{system}}}_{\substack{\text{Change in internal, kinetic,}\\ \text{potential, etc. energies}}}$$

$$Q_{in} + m_i h_i = m_2 u_2 - m_1 u_1 \quad (\text{since } W \cong ke \cong pe \cong 0)$$

The initial and the final masses in the tank are

$$m_1 = \frac{P_1 V}{RT_1} = \frac{(100 \text{ kPa})(2 \text{ m}^3)}{(0.287 \text{ kPa}\cdot\text{m}^3/\text{kg}\cdot\text{K})(295 \text{ K})} = 2.362 \text{ kg}$$

$$m_2 = \frac{P_2 V}{RT_2} = \frac{(600\text{kPa})(2\text{m}^3)}{(0.287\text{kPa}\cdot\text{m}^3/\text{kg}\cdot\text{K})(350\text{K})} = 11.946 \text{ kg}$$

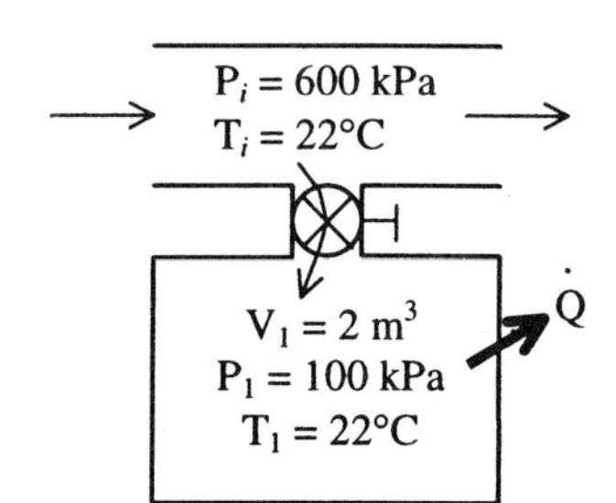

Then from the mass balance,

$$m_i = m_2 - m_1 = 11.946 - 2.362 = \mathbf{9.584 \text{ kg}}$$

(*b*) The heat transfer during this process is determined from

$$\begin{aligned} Q_{in} &= -m_i h_i + m_2 u_2 - m_1 u_1 \\ &= -(9.584 \text{ kg})(295.17 \text{ kJ/kg}) + (11.946 \text{ kg})(250.02 \text{ kJ/kg}) - (2.362 \text{ kg})(210.49 \text{ kJ/kg}) \\ &= -339 \text{ kJ} \rightarrow Q_{out} = \mathbf{339 \text{ kJ}} \end{aligned}$$

Discussion The negative sign for heat transfer indicates that the assumed direction is wrong. Therefore, we reversed the direction.

4-150 A rigid tank initially contains saturated R-134a vapor. The tank is connected to a supply line, and R-134a is allowed to enter the tank. The final temperature in the tank, the mass of R-134a that entered, and the heat transfer are to be determined.

Assumptions **1** This is an unsteady process since the conditions within the device are changing during the process, but it can be analyzed as a uniform-flow process since the state of fluid at the inlet remains constant. **2** Kinetic and potential energies are negligible. **3** There are no work interactions involved. **4** The direction of heat transfer is to the tank (will be verified).

Properties The properties of refrigerant are (Tables A-11 through A-13)

$$\left.\begin{array}{l} T_1 = 8^\circ\text{C} \\ x_1 = 0.6 \end{array}\right\} \begin{array}{l} v_1 = v_f + x_1 v_{fg} = 0.0007884 + 0.6 \times (0.0525 - 0.0007884) = 0.03182\text{m}^3/\text{kg} \\ u_1 = u_f + x_1 u_{fg} = 60.43 + 0.6 \times (231.46 - 60.43) = 163.05\text{kJ/kg} \end{array}$$

$$\left.\begin{array}{l} P_2 = 800\text{kPa} \\ sat.vapor \end{array}\right\} \begin{array}{l} v_2 = v_{g@800\text{kPa}} = 0.0255\text{m}^3/\text{kg} \\ u_2 = u_{g@800\text{kPa}} = 243.78\text{kJ/kg} \end{array}$$

$$\left.\begin{array}{l} P_i = 1.0\text{MPa} \\ T_i = 120^\circ C \end{array}\right\} h_i = 356.52\text{kJ/kg}$$

Analysis We take the tank as the system, which is a control volume since mass crosses the boundary. Noting that the microscopic energies of flowing and nonflowing fluids are represented by enthalpy *h* and internal energy *u*, respectively, the mass and energy balances for this uniform-flow system can be expressed as

Mass balance: $m_{in} - m_{out} = \Delta m_{\text{system}} \rightarrow m_i = m_2 - m_1$

Energy balance:

$$\underbrace{E_{in} - E_{out}}_{\substack{\text{Net energy transfer} \\ \text{by heat, work, and mass}}} = \underbrace{\Delta E_{\text{system}}}_{\substack{\text{Change in internal, kinetic,} \\ \text{potential, etc. energies}}}$$

$$Q_{in} + m_i h_i = m_2 u_2 - m_1 u_1 \quad (\text{since } W \cong ke \cong pe \cong 0)$$

(*a*) The tank contains saturated vapor at the final state at 800 kPa, and thus the final temperature is the saturation temperature at this pressure,

$$T_2 = T_{sat\ @\ 800\ kPa} = \mathbf{31.33^\circ C}$$

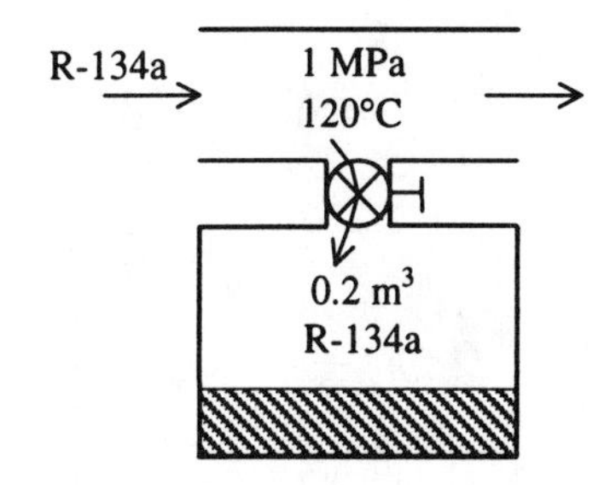

(*b*) The initial and the final masses in the tank are

$$m_1 = \frac{V}{v_1} = \frac{0.2\ m^3}{0.03182\ m^3/kg} = 6.29\ kg$$

$$m_2 = \frac{V}{v_2} = \frac{0.2\ m^3}{0.0255\ m^3/kg} = 7.84\ kg$$

Then from the mass balance

$$m_i = m_2 - m_1 = 7.84 - 6.29 = \mathbf{1.55\ kg}$$

(*c*) The heat transfer during this process is determined from the energy balance to be

$$\begin{aligned} Q_{in} &= -m_i h_i + m_2 u_2 - m_1 u_1 \\ &= -(1.55\text{ kg})(356.52\text{ kJ/kg}) + (7.84\text{ kg})(243.78\text{ kJ/kg}) - (6.29\text{ kg})(163.05\text{ kJ/kg}) \\ &= \mathbf{333\ kJ} \end{aligned}$$

4-151E A rigid tank initially contains saturated water vapor. The tank is connected to a supply line, and water vapor is allowed to enter the tank until one-half of the tank is filled with liquid water. The final pressure in the tank, the mass of steam that entered, and the heat transfer are to be determined.

Assumptions **1** This is an unsteady process since the conditions within the device are changing during the process, but it can be analyzed as a uniform-flow process since the state of fluid at the inlet remains constant. **2** Kinetic and potential energies are negligible. **3** There are no work interactions involved. **4** The direction of heat transfer is to the tank (will be verified).

Properties The properties of water are (Tables A-4E through A-6E)

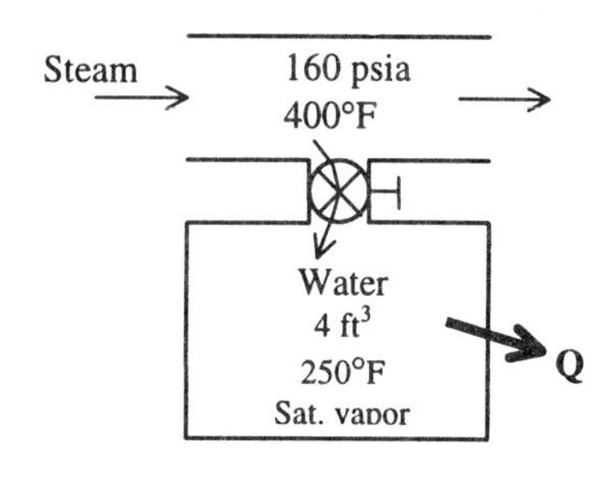

$$\left.\begin{array}{l} T_1 = 250^\circ\text{F} \\ sat.vapor \end{array}\right\} \begin{array}{l} v_1 = v_{g@250^\circ\text{F}} = 13.826\text{ft}^3/\text{lbm} \\ u_1 = u_{g@250^\circ\text{F}} = 1087.9\text{Btu/lbm} \end{array}$$

$$\left.\begin{array}{l} T_2 = 250^\circ F \\ sat.mixture \end{array}\right\} \begin{array}{ll} v_f = 0.017001, & v_g = 13.826\text{ft}^3/\text{lbm} \\ u_f = 218.49, & u_g = 1087.9\text{Btu/lbm} \end{array}$$

$$\left.\begin{array}{l} P_i = 160\text{psia} \\ T_i = 400^\circ\text{F} \end{array}\right\} h_i = 1217.8\text{Btu/lbm}$$

Analysis We take the tank as the system, which is a control volume since mass crosses the boundary. Noting that the microscopic energies of flowing and nonflowing fluids are represented by enthalpy h and internal energy u, respectively, the mass and energy balances for this uniform-flow system can be expressed as

Mass balance: $m_{in} - m_{out} = \Delta m_{\text{system}} \quad \rightarrow \quad m_i = m_2 - m_1$

Energy balance:

$$\underbrace{E_{in} - E_{out}}_{\substack{\text{Net energy transfer} \\ \text{by heat, work, and mass}}} = \underbrace{\Delta E_{\text{system}}}_{\substack{\text{Change in internal, kinetic,} \\ \text{potential, etc. energies}}}$$

$$Q_{in} + m_i h_i = m_2 u_2 - m_1 u_1 \quad (\text{since } W \cong ke \cong pe \cong 0)$$

(a) The tank contains saturated mixture at the final state at 250°F, and thus the exit pressure is the saturation pressure at this temperature,

$$P_2 = P_{sat@250^\circ F} = \mathbf{29.82\ psia}$$

(b) The initial and the final masses in the tank are

$$m_1 = \frac{V}{v_1} = \frac{4\text{ ft}^3}{13.826\text{ ft}^3/\text{lbm}} = 0.289\text{ lbm}$$

$$m_2 = m_f + m_g = \frac{V_f}{v_f} + \frac{V_g}{v_g} = \frac{2\text{ ft}^3}{0.017001\text{ ft}^3/\text{lbm}} + \frac{2\text{ ft}^3}{13.826\text{ ft}^3/\text{lbm}} = 117.64 + 0.14 = 117.78\text{ lbm}$$

Then from the mass balance

$$m_i = m_2 - m_1 = 117.78 - 0.289 = \mathbf{117.49\ lbm}$$

(c) The heat transfer during this process is determined from the energy balance to be

$$\begin{aligned} Q_{in} &= -m_i h_i + m_2 u_2 - m_1 u_1 \\ &= -(117.49\text{ lbm})(1217.8\text{ Btu/lbm}) + 25{,}855\text{ Btu} - (0.289\text{ lbm})(1087.9\text{ Btu/lbm}) \\ &= -117{,}539\text{ Btu} \rightarrow Q_{out} = \mathbf{117{,}539\ Btu} \end{aligned}$$

since $\quad U_2 = m_2 u_2 = m_f u_f + m_g u_g = 117.64 \times 218.49 + 0.14 \times 1087.9 = 25{,}855\text{ Btu}$

Discussion A negative result for heat transfer indicates that the assumed direction is wrong, and should be reversed.

4-152 A cylinder initially contains superheated steam. The cylinder is connected to a supply line, and is superheated steam is allowed to enter the cylinder until the volume doubles at constant pressure. The final temperature in the cylinder and the mass of the steam that entered are to be determined.

Assumptions **1** This is an unsteady process since the conditions within the device are changing during the process, but it can be analyzed as a uniform-flow process since the state of fluid at the inlet remains constant. **2** The expansion process is quasi-equilibrium. **3** Kinetic and potential energies are negligible. **3** There are no work interactions involved other than boundary work. **4** The device is insulated and thus heat transfer is negligible.

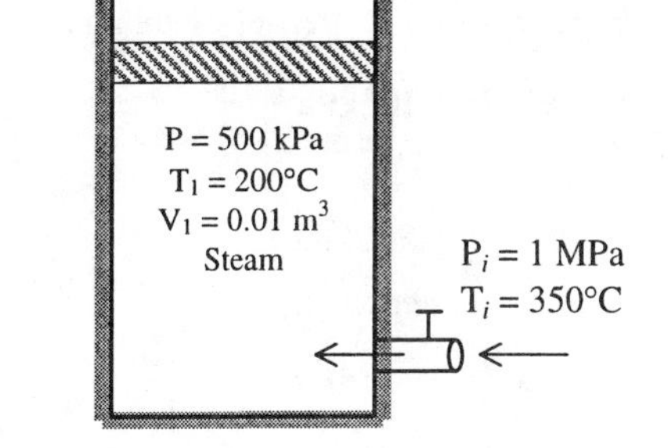

Properties The properties of steam are (Tables A-4 through A-6)

$$\left.\begin{array}{l} P_1 = 500\text{kPa} \\ T_1 = 200^\circ\text{C} \end{array}\right\} \begin{array}{l} v_1 = 0.4249\text{m}^3/\text{kg} \\ u_1 = 2642.9\text{kJ/kg} \end{array}$$

$$\left.\begin{array}{l} P_i = 1\text{MPa} \\ T_i = 350^\circ\text{C} \end{array}\right\} h_i = 3157.7\text{kJ/kg}$$

Analysis (*a*) We take the cylinder as the system, which is a control volume since mass crosses the boundary. Noting that the microscopic energies of flowing and nonflowing fluids are represented by enthalpy h and internal energy u, respectively, the mass and energy balances for this uniform-flow system can be expressed as

Mass balance: $\quad m_{in} - m_{out} = \Delta m_{\text{system}} \quad \rightarrow \quad m_i = m_2 - m_1$

Energy balance:

$$\underbrace{E_{in} - E_{out}}_{\substack{\text{Net energy transfer}\\ \text{by heat, work, and mass}}} = \underbrace{\Delta E_{\text{system}}}_{\substack{\text{Change in internal, kinetic,}\\ \text{potential, etc. energies}}}$$

$$m_i h_i = W_{b,out} + m_2 u_2 - m_1 u_1 \quad (\text{since } Q \cong ke \cong pe \cong 0)$$

Combining the two relations gives $\quad 0 = W_{b,out} - (m_2 - m_1)h_i + m_2 u_2 - m_1 u_1$

The boundary work done during this process is

$$W_{b,out} = \int_1^2 P d\text{V} = P(\text{V}_2 - \text{V}_1) = (500 \text{ kPa})(0.02 - 0.01)\text{m}^3 \left(\frac{1 \text{ kJ}}{1 \text{ kPa} \cdot \text{m}^3}\right) = 5 \text{ kJ}$$

The initial and the final masses in the cylinder are

$$m_1 = \frac{V_1}{v_1} = \frac{0.01 \text{ m}^3}{0.4249 \text{ m}^3/\text{kg}} = 0.0235 \text{ kg}$$

$$m_2 = \frac{V_2}{v_2} = \frac{0.02 \text{ m}^3}{v_2}$$

Substituting,

$$= 5 - \left(\frac{0.02}{v_2} - 0.0235\right)(3157.7) + \frac{0.02}{v_2} u_2 - (0.0235)(2642.9)$$

Then by trial and error, $\qquad T_2 = \mathbf{262.6°C}$ and $v_2 = 0.4865 \text{ m}^3/\text{kg}$

(*b*) The final mass in the cylinder is

$$m_2 = \frac{V_2}{v_2} = \frac{0.02 \text{ m}^3}{0.4865 \text{ m}^3/\text{kg}} = 0.0411 \text{ kg}$$

Then,

$$m_i = m_2 - m_1 = 0.0411 - 0.0235 = \mathbf{0.0176 \text{ kg}}$$

4-153 A cylinder initially contains saturated liquid-vapor mixture of water. The cylinder is connected to a supply line, and the steam is allowed to enter the cylinder until all the liquid is vaporized. The final temperature in the cylinder and the mass of the steam that entered are to be determined. √

Assumptions **1** This is an unsteady process since the conditions within the device are changing during the process, but it can be analyzed as a uniform-flow process since the state of fluid at the inlet remains constant. **2** The expansion process is quasi-equilibrium. **3** Kinetic and potential energies are negligible. **3** There are no work interactions involved other than boundary work. **4** The device is insulated and thus heat transfer is negligible.

Properties The properties of steam are (Tables A-4 through A-6)

$$\left.\begin{array}{l} P_1 = 300\text{kPa} \\ x_1 = 0.8 \end{array}\right\} h_1 = h_f + x_1 h_{fg} = 561.47 + 0.8 \times 2163.8 = 2292.51\text{kJ/kg}$$

$$\left.\begin{array}{l} P_2 = 300\text{kPa} \\ sat.vapor \end{array}\right\} h_2 = h_{g@300\text{kPa}} = 2725.3\text{kJ/kg}$$

$$\left.\begin{array}{l} P_i = 0.5\text{MPa} \\ T_i = 350^\circ\text{C} \end{array}\right\} h_i = 3167.7\text{kJ/kg}$$

(P = 300 kPa)
m_1 = 10 kg
H_2O

P_i = 0.5 MPa
T_i = 350°C

Analysis (*a*) The cylinder contains saturated vapor at the final state at a pressure of 300 kPa, thus the final temperature in the cylinder must be

$$T_2 = T_{\text{sat @ 300 kPa}} = \mathbf{133.6^\circ C}$$

(*b*) We take the cylinder as the system, which is a control volume since mass crosses the boundary. Noting that the microscopic energies of flowing and nonflowing fluids are represented by enthalpy h and internal energy u, respectively, the mass and energy balances for this uniform-flow system can be expressed as

Mass balance: $\quad m_{in} - m_{out} = \Delta m_{\text{system}} \quad \rightarrow \quad m_i = m_2 - m_1$

Energy balance:

$$\underbrace{E_{in} - E_{out}}_{\substack{\text{Net energy transfer} \\ \text{by heat, work, and mass}}} = \underbrace{\Delta E_{\text{system}}}_{\substack{\text{Change in internal, kinetic,} \\ \text{potential, etc. energies}}}$$

$$m_i h_i = W_{b,out} + m_2 u_2 - m_1 u_1 \quad (\text{since } Q \cong ke \cong pe \cong 0)$$

Combining the two relations gives $\quad 0 = W_{b,out} - (m_2 - m_1)h_i + m_2 u_2 - m_1 u_1$

or,

$$0 = -(m_2 - m_1)h_i + m_2 h_2 - m_1 h_1$$

since the boundary work and ΔU combine into ΔH for constant pressure expansion and compression processes. Solving for m_2 and substituting,

$$m_2 = \frac{h_i - h_1}{h_i - h_2} m_1 = \frac{(3167.7 - 2292.51)\text{kJ/kg}}{(3167.7 - 2725.3)\text{kJ/kg}}(10\text{kg}) = 19.78\text{kg}$$

Thus,

$$m_i = m_2 - m_1 = 19.78 - 10 = \mathbf{9.78\ kg}$$

4-154 A rigid tank initially contains saturated R-134a vapor. The tank is connected to a supply line, and R-134a is allowed to enter the tank. The mass of the R-134a that entered and the heat transfer are to be determined.

Assumptions **1** This is an unsteady process since the conditions within the device are changing during the process, but it can be analyzed as a uniform-flow process since the state of fluid at the inlet remains constant. **2** Kinetic and potential energies are negligible. **3** There are no work interactions involved. **4** The direction of heat transfer is to the tank (will be verified).

Properties The properties of refrigerant are (Tables A-11 through A-13)

$$\left.\begin{array}{l} P_1 = 1\text{ MPa} \\ sat.vapor \end{array}\right\} \begin{array}{l} v_1 = v_{g@1\text{ MPa}} = 0.0202\text{ m}^3/\text{kg} \\ u_1 = u_{g@1\text{ MPa}} = 247.77\text{ kJ/kg} \end{array}$$

$$\left.\begin{array}{l} P_2 = 1.2\text{ MPa} \\ sat.liquid \end{array}\right\} \begin{array}{l} v_2 = v_{f@1.2\text{ MPa}} = 0.0008928\text{ m}^3/\text{kg} \\ u_2 = u_{f@1.2\text{ MPa}} = 114.69\text{ kJ/kg} \end{array}$$

$$\left.\begin{array}{l} P_i = 1.2\text{ MPa} \\ T_i = 30^\circ\text{C} \end{array}\right\} h_i = h_{f@30^\circ\text{C}} = 91.49\text{ kJ/kg}$$

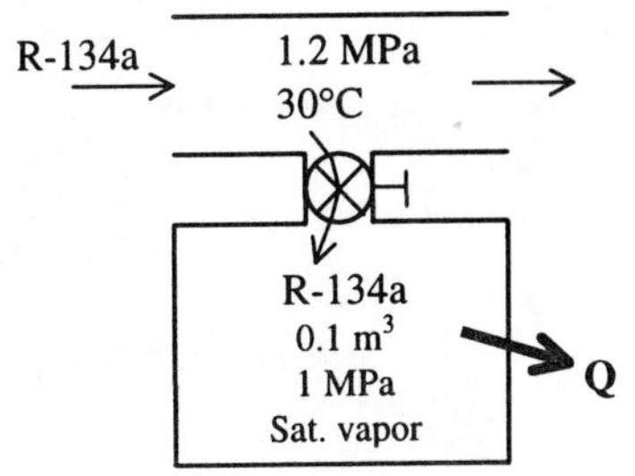

R-134a
1.2 MPa
30°C
R-134a
0.1 m³
1 MPa
Sat. vapor
Q

Analysis We take the tank as the system, which is a control volume since mass crosses the boundary. Noting that the microscopic energies of flowing and nonflowing fluids are represented by enthalpy h and internal energy u, respectively, the mass and energy balances for this uniform-flow system can be expressed as

Mass balance: $\quad m_{in} - m_{out} = \Delta m_{\text{system}} \quad \rightarrow \quad m_i = m_2 - m_1$

Energy balance:

$$\underbrace{E_{in} - E_{out}}_{\substack{\text{Net energy transfer}\\ \text{by heat, work, and mass}}} = \underbrace{\Delta E_{\text{system}}}_{\substack{\text{Change in internal, kinetic,}\\ \text{potential, etc. energies}}}$$

$$Q_{in} + m_i h_i = m_2 u_2 - m_1 u_1 \quad (\text{since } W \cong ke \cong pe \cong 0)$$

(*a*) The initial and the final masses in the tank are

$$m_1 = \frac{V_1}{v_1} = \frac{0.1\text{ m}^3}{0.0202\text{ m}^3/\text{kg}} = 4.95\text{ kg}$$

$$m_2 = \frac{V_2}{v_2} = \frac{0.1\text{ m}^3}{0.0008928\text{ m}^3/\text{kg}} = 112.01\text{ kg}$$

Then from the mass balance

$$m_i = m_2 - m_1 = 112.01 - 4.95 = \mathbf{107.06\text{ kg}}$$

(*c*) The heat transfer during this process is determined from the energy balance to be

$$\begin{aligned} Q_{in} &= -m_i h_i + m_2 u_2 - m_1 u_1 \\ &= -(107.06\text{ kg})(91.49\text{ kJ/kg}) + (112.01\text{ kg})(114.69\text{ kJ/kg}) - (4.95\text{ kg})(247.77\text{ kJ/kg}) \\ &= \mathbf{1825\text{ kJ}} \end{aligned}$$

4-155 A rigid tank initially contains saturated liquid water. A valve at the bottom of the tank is opened, and half of the mass in liquid form is withdrawn from the tank. The temperature in the tank is maintained constant. The amount of heat transfer is to be determined.

Assumptions **1** This is an unsteady process since the conditions within the device are changing during the process, but it can be analyzed as a uniform-flow process since the state of fluid leaving the device remains constant. **2** Kinetic and potential energies are negligible. **3** There are no work interactions involved. **4** The direction of heat transfer is to the tank (will be verified).

Properties The properties of water are (Tables A-4 through A-6)

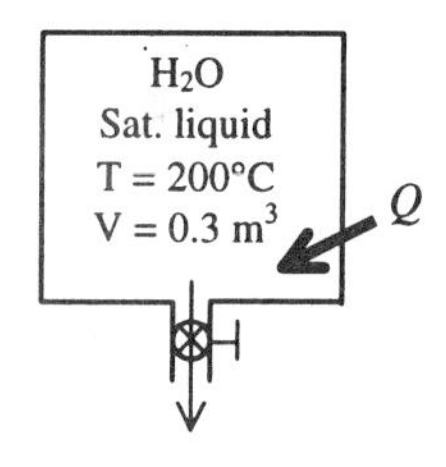

$$\left.\begin{array}{l} T_1 = 200^\circ\text{C} \\ sat.liquid \end{array}\right\} \begin{array}{l} v_1 = v_{f\,@\,200^\circ\text{C}} = 0.001157 \text{ m}^3\text{/kg} \\ u_1 = u_{f\,@\,200^\circ\text{C}} = 850.65 \text{ kJ/kg} \end{array}$$

$$\left.\begin{array}{l} T_e = 200^\circ\text{C} \\ sat.liquid \end{array}\right\} h_e = h_{f\,@\,200^\circ\text{C}} = 852.45 \text{ kJ/kg}$$

Analysis We take the tank as the system, which is a control volume since mass crosses the boundary. Noting that the microscopic energies of flowing and nonflowing fluids are represented by enthalpy h and internal energy u, respectively, the mass and energy balances for this uniform-flow system can be expressed as

Mass balance: $m_{in} - m_{out} = \Delta m_{\text{system}} \rightarrow m_e = m_1 - m_2$

Energy balance:

$$\underbrace{E_{in} - E_{out}}_{\substack{\text{Net energy transfer}\\ \text{by heat, work, and mass}}} = \underbrace{\Delta E_{\text{system}}}_{\substack{\text{Change in internal, kinetic,}\\ \text{potential, etc. energies}}}$$

$$Q_{in} = m_e h_e + m_2 u_2 - m_1 u_1 \quad (\text{since } W \cong ke \cong pe \cong 0)$$

The initial and the final masses in the tank are

$$m_1 = \frac{V_1}{v_1} = \frac{0.3\text{m}^3}{0.001157\text{m}^3\text{/kg}} = 259.3\text{kg}$$

$$m_2 = \tfrac{1}{2} m_1 = \tfrac{1}{2}(259.3\text{kg}) = 129.65\text{kg}$$

Then from the mass balance,

$$m_e = m_1 - m_2 = 259.3 - 129.65 = 129.65 \text{ kg}$$

Now we determine the final internal energy,

$$v_2 = \frac{V}{m_2} = \frac{0.3\text{m}^3}{129.65\text{kg}} = 0.002314\text{m}^3\text{/kg}$$

$$x_2 = \frac{v_2 - v_f}{v_{fg}} = \frac{0.002314 - 0.001157}{0.13736 - 0.001157} = 0.00849$$

$$\left.\begin{array}{l} T_2 = 200^\circ\text{C} \\ x_2 = 0.00849 \end{array}\right\} u_2 = u_f + x_2 u_{fg} = 850.65 + (0.00849)(1744.7) = 865.46 \text{ kJ/kg}$$

Then the heat transfer during this process is determined from the energy balance by substitution to be

$$Q = (129.65 \text{ kg})(852.45 \text{ kJ/kg}) + (129.65 \text{ kg})(865.46 \text{ kJ/kg}) - (259.3 \text{ kg})(850.65 \text{ kJ/kg})$$
$$= \mathbf{2153 \text{ kJ}}$$

4-156 A rigid tank initially contains saturated liquid-vapor mixture of refrigerant-134a. A valve at the bottom of the tank is opened, and liquid is withdrawn from the tank at constant pressure until no liquid remains inside. The amount of heat transfer is to be determined.

Assumptions **1** This is an unsteady process since the conditions within the device are changing during the process, but it can be analyzed as a uniform-flow process since the state of fluid leaving the device remains constant. **2** Kinetic and potential energies are negligible. **3** There are no work interactions involved. **4** The direction of heat transfer is to the tank (will be verified).

Properties The properties of R-134a are (Tables A-11 through A-13)

$$P_1 = 800\text{kPa} \rightarrow v_f = 0.0008454\text{m}^3/\text{kg},\ v_g = 0.0255\text{m}^3/\text{kg}$$
$$u_f = 92.75\text{kJ/kg},\ u_g = 243.78\text{kJ/kg}$$

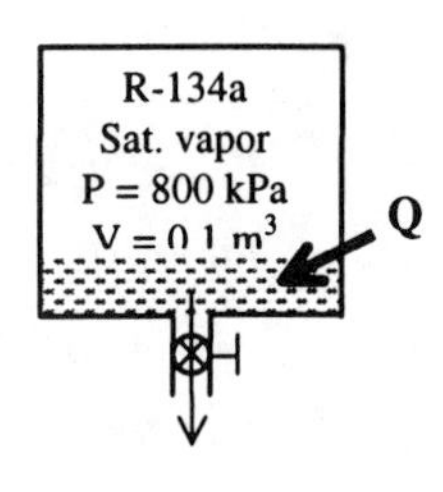

$$\left.\begin{array}{l} P_2 = 800\text{kPa} \\ sat.vapor \end{array}\right\} \begin{array}{l} v_2 = v_{g@800\text{kPa}} = 0.0255\text{m}^3/\text{kg} \\ u_2 = u_{g@800\text{kPa}} = 243.78\text{kJ/kg} \end{array}$$

$$\left.\begin{array}{l} P_e = 800\text{kPa} \\ sat.liquid \end{array}\right\} h_e = h_{f@800\text{kPa}} = 93.42\text{kJ/kg}$$

Analysis We take the tank as the system, which is a control volume since mass crosses the boundary. Noting that the microscopic energies of flowing and nonflowing fluids are represented by enthalpy h and internal energy u, respectively, the mass and energy balances for this uniform-flow system can be expressed as

Mass balance: $$m_{in} - m_{out} = \Delta m_{\text{system}} \rightarrow m_e = m_1 - m_2$$

Energy balance: $$\underbrace{E_{in} - E_{out}}_{\substack{\text{Net energy transfer} \\ \text{by heat, work, and mass}}} = \underbrace{\Delta E_{\text{system}}}_{\substack{\text{Change in internal, kinetic,} \\ \text{potential, etc. energies}}}$$

$$Q_{in} = m_e h_e + m_2 u_2 - m_1 u_1 \quad (\text{since } W \cong ke \cong pe \cong 0)$$

The initial mass, initial internal energy, and final mass in the tank are

$$m_1 = m_f + m_g = \frac{V_f}{v_f} + \frac{V_g}{v_g} = \frac{0.1 \times 0.4\text{m}^3}{0.0008454\text{m}^3/\text{kg}} + \frac{0.1 \times 0.6\text{m}^3}{0.0255\text{m}^3/\text{kg}} = 47.32 + 2.35 = 49.67\text{kg}$$

$$U_1 = m_1 u_1 = m_f u_f + m_g u_g = (47.32)(92.75) + (2.35)(243.78) = 4962\text{kJ}$$

$$m_2 = \frac{V}{v_2} = \frac{0.1\text{m}^3}{0.0255\text{m}^3/\text{kg}} = 3.92\text{kg}$$

Then from the mass and energy balances,

$$m_e = m_1 - m_2 = 49.67 - 3.92 = 45.75\text{ kg}$$

$$Q_{in} = (45.75\text{ kg})(93.42\text{ kJ/kg}) + (3.92\text{ kg})(243.78\text{ kJ/kg}) - 4962\text{ kJ} = \mathbf{267.6\ kJ}$$

4-157E A rigid tank initially contains saturated liquid-vapor mixture of R-134a. A valve at the top of the tank is opened, and vapor is allowed to escape at constant pressure until all the liquid in the tank disappears. The amount of heat transfer is to be determined.

Assumptions **1** This is an unsteady process since the conditions within the device are changing during the process, but it can be analyzed as a uniform-flow process since the state of fluid leaving the device remains constant. **2** Kinetic and potential energies are negligible. **3** There are no work interactions involved.

Properties The properties of R-134a are (Tables A-11E through A-13E)

$$P_1 = 100\text{psia}, \rightarrow v_f = 0.01332\text{ft}^3/\text{lbm},\ v_g = 0.4747\text{ft}^3/\text{lbm}$$
$$u_f = 36.75\text{Btu/lbm},\ u_g = 103.68\text{Btu/lbm}$$

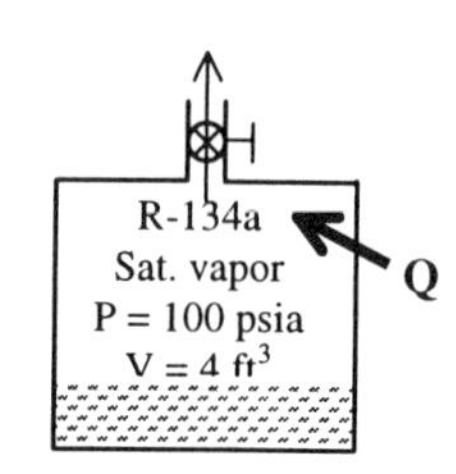

$$\left.\begin{array}{l} P_2 = 100\text{psia} \\ sat.vapor \end{array}\right\} \begin{array}{l} v_2 = v_{g@100\text{psia}} = 0.4747\text{ft}^3/\text{lbm} \\ u_2 = u_{g@100\text{psia}} = 103.68\text{Btu/lbm} \end{array}$$

$$\left.\begin{array}{l} P_e = 100\text{psia} \\ sat.vapor \end{array}\right\} h_e = h_{g@100\text{psia}} = 112.46\text{Btu/lbm}$$

Analysis We take the tank as the system, which is a control volume since mass crosses the boundary. Noting that the microscopic energies of flowing and nonflowing fluids are represented by enthalpy h and internal energy u, respectively, the mass and energy balances for this uniform-flow system can be expressed as

Mass balance: $m_{in} - m_{out} = \Delta m_{\text{system}} \rightarrow m_e = m_1 - m_2$

Energy balance:

$$\underbrace{E_{in} - E_{out}}_{\substack{\text{Net energy transfer} \\ \text{by heat, work, and mass}}} = \underbrace{\Delta E_{\text{system}}}_{\substack{\text{Change in internal, kinetic,} \\ \text{potential, etc. energies}}}$$

$$Q_{in} - m_e h_e = m_2 u_2 - m_1 u_1 \quad (\text{since } W \cong ke \cong pe \cong 0)$$

The initial mass, initial internal energy, and final mass in the tank are

$$m_1 = m_f + m_g = \frac{V_f}{v_f} + \frac{V_g}{v_g} = \frac{4 \times 0.2\text{ft}^3}{0.01332\text{ft}^3/\text{lbm}} + \frac{4 \times 0.8\text{ft}^3}{0.4747\text{ft}^3/\text{lbm}} = 60.06 + 6.74 = 66.8\text{lbm}$$

$$U_1 = m_1 u_1 = m_f u_f + m_g u_g = (60.06)(36.75) + (6.74)(103.68) = 2906\text{Btu}$$

$$m_2 = \frac{V}{v_2} = \frac{4\text{ft}^3}{0.4747\text{ft}^3/\text{lbm}} = 8.426\text{lbm}$$

Then from the mass and energy balances,

$$m_e = m_1 - m_2 = 66.8 - 8.426 = 58.374 \text{ lbm}$$

$$\begin{aligned} Q_{in} &= m_e h_e + m_2 u_2 - m_1 u_1 \\ &= (58.374 \text{ lbm})(112.46 \text{ Btu/lbm}) + (8.426 \text{ lbm})(103.68 \text{ Btu/lbm}) - 2906 \text{ Btu} \\ &= \mathbf{4532\ Btu} \end{aligned}$$

4-158 A rigid tank initially contains superheated steam. A valve at the top of the tank is opened, and vapor is allowed to escape at constant pressure until the temperature rises to 500°C. The amount of heat transfer is to be determined.

Assumptions **1** This is an unsteady process since the conditions within the device are changing during the process, but it can be analyzed as a uniform-flow process by using constant average properties for the steam leaving the tank. **2** Kinetic and potential energies are negligible. **3** There are no work interactions involved. **4** The direction of heat transfer is to the tank (will be verified).

Properties The properties of water are (Tables A-4 through A-6)

$$\left.\begin{array}{l} P_1 = 2\text{MPa} \\ T_1 = 300^\circ\text{C} \end{array}\right\} \begin{array}{l} v_1 = 0.12547\text{m}^3/\text{kg} \\ u_1 = 2772.6\text{kJ/kg}, \quad h_1 = 3023.5\text{kJ/kg} \end{array}$$

$$\left.\begin{array}{l} P_2 = 2\text{MPa} \\ T_2 = 500^\circ\text{C} \end{array}\right\} \begin{array}{l} v_2 = 0.17568\text{m}^3/\text{kg} \\ u_2 = 3116.2\text{kJ/kg}, \quad h_2 = 3467.6\text{kJ/kg} \end{array}$$

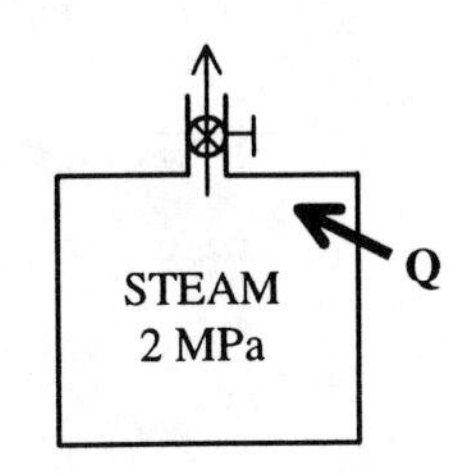

Analysis We take the tank as the system, which is a control volume since mass crosses the boundary. Noting that the microscopic energies of flowing and nonflowing fluids are represented by enthalpy h and internal energy u, respectively, the mass and energy balances for this uniform-flow system can be expressed as

Mass balance: $$m_{in} - m_{out} = \Delta m_{\text{system}} \quad \rightarrow \quad m_e = m_1 - m_2$$

Energy balance: $$\underbrace{E_{in} - E_{out}}_{\substack{\text{Net energy transfer} \\ \text{by heat, work, and mass}}} = \underbrace{\Delta E_{\text{system}}}_{\substack{\text{Change in internal, kinetic,} \\ \text{potential, etc. energies}}}$$

$$Q_{in} - m_e h_e = m_2 u_2 - m_1 u_1 \quad (\text{since } W \cong ke \cong pe \cong 0)$$

The state and thus the enthalpy of the steam leaving the tank is changing during this process. But for simplicity, we assume constant properties for the exiting steam at the average values. Thus,

$$h_e \cong \frac{h_1 + h_2}{2} = \frac{3023.5 + 3467.6 \text{ kJ/kg}}{2} = 3245.55 \text{ kJ/kg}$$

The initial and the final masses in the tank are

$$m_1 = \frac{V_1}{v_1} = \frac{0.2 \text{ m}^3}{0.12547 \text{ m}^3/\text{kg}} = 1.594 \text{ kg}$$

$$m_2 = \frac{V_2}{v_2} = \frac{0.2 \text{ m}^3}{0.17568 \text{ m}^3/\text{kg}} = 1.138 \text{ kg}$$

Then from the mass and energy balance relations,

$$m_e = m_1 - m_2 = 1.594 - 1.138 = 0.456 \text{ kg}$$

$$\begin{aligned} Q_{in} &= m_e h_e + m_2 u_2 - m_1 u_1 \\ &= (0.456 \text{ kg})(3245.55 \text{ kJ/kg}) + (1.138 \text{ kg})(3116.2 \text{ kJ/kg}) - (1.594 \text{ kg})(2772.6 \text{ kJ/kg}) \\ &= \mathbf{606.7 \text{ kJ}} \end{aligned}$$

4-159 A pressure cooker is initially half-filled with liquid water. If the pressure cooker is not to run out of liquid water for 1 h, the highest rate of heat transfer allowed is to be determined.

Assumptions **1** This is an unsteady process since the conditions within the device are changing during the process, but it can be analyzed as a uniform-flow process since the state of fluid leaving the device remains constant. **2** Kinetic and potential energies are negligible. **3** There are no work interactions involved.

Properties The properties of water are (Tables A-4 through A-6)

$$P_1 = 175 kPa \rightarrow v_f = 0.001057 \text{m}^3/\text{kg},\ v_g = 1.0036 \text{m}^3/\text{kg}$$
$$u_f = 486.8 \text{kJ/kg},\ u_g = 2524.9 \text{ kJ/kg}$$

$$\left.\begin{array}{l} P_2 = 175 \text{kPa} \\ sat.vapor \end{array}\right\} \begin{array}{l} v_2 = v_{g@175\text{kPa}} = 1.0036 \text{m}^3/\text{kg} \\ u_2 = u_{g@175\text{kPa}} = 2524.9 \text{kJ/kg} \end{array}$$

$$\left.\begin{array}{l} P_e = 175 \text{kPa} \\ sat.vapor \end{array}\right\} h_e = h_{g@175\text{kPa}} = 2700.6 \text{kJ/kg}$$

Pressure Cooker
4 L
175 kPa
$\dot{Q}$

Analysis We take the cooker as the system, which is a control volume since mass crosses the boundary. Noting that the microscopic energies of flowing and nonflowing fluids are represented by enthalpy h and internal energy u, respectively, the mass and energy balances for this uniform-flow system can be expressed as

Mass balance: $m_{in} - m_{out} = \Delta m_{\text{system}} \rightarrow m_e = m_1 - m_2$

Energy balance:
$$\underbrace{E_{in} - E_{out}}_{\substack{\text{Net energy transfer} \\ \text{by heat, work, and mass}}} = \underbrace{\Delta E_{\text{system}}}_{\substack{\text{Change in internal, kinetic,} \\ \text{potential, etc. energies}}}$$

$$Q_{in} - m_e h_e = m_2 u_2 - m_1 u_1 \quad (\text{since } W \cong ke \cong pe \cong 0)$$

The initial mass, initial internal energy, and final mass in the tank are

$$m_1 = m_f + m_g = \frac{V_f}{v_f} + \frac{V_g}{v_g} = \frac{0.002 \text{m}^3}{0.001057 \text{m}^3/\text{kg}} + \frac{0.002 \text{m}^3}{1.0036 \text{m}^3/\text{kg}} = 1.892 + 0.002 = 1.894 \text{kg}$$

$$U_1 = m_1 u_1 = m_f u_f + m_g u_g = (1.892)(486.8) + (0.002)(2524.9) = 926.1 \text{kJ}$$

$$m_2 = \frac{V}{v_2} = \frac{0.004 \text{m}^3}{1.0036 \text{m}^3/\text{kg}} = 0.004 \text{kg}$$

Then from the mass and energy balances,

$$m_e = m_1 - m_2 = 1.894 - 0.004 = 1.890 \text{ kg}$$

$$Q_{in} = m_e h_e + m_2 u_2 - m_1 u_1$$
$$= (1.890 \text{kg})(2700.6 \text{kJ/kg}) + (0.004 \text{kg})(2524.9 \text{kJ/kg}) - 926.1 \text{kJ} = 4188 \text{kJ}$$

Thus,

$$\dot{Q} = \frac{Q}{\Delta t} = \frac{4188 \text{ kJ}}{3600 \text{ s}} = \mathbf{1.163 \text{ kW}}$$

4-160 An insulated rigid tank initially contains helium gas at high pressure. A valve is opened, and half of the mass of helium is allowed to escape. The final temperature and pressure in the tank are to be determined.

Assumptions **1** This is an unsteady process since the conditions within the device are changing during the process, but it can be analyzed as a uniform-flow process by using constant average properties for the helium leaving the tank. **2** Kinetic and potential energies are negligible. **3** There are no work interactions involved. **4** The tank is insulated and thus heat transfer is negligible. **5** Helium is an ideal gas with constant specific heats.

Properties The specific heat ratio of helium is $k = 1.667$ (Table A-2).

Analysis We take the tank as the system, which is a control volume since mass crosses the boundary. Noting that the microscopic energies of flowing and nonflowing fluids are represented by enthalpy h and internal energy u, respectively, the mass and energy balances for this uniform-flow system can be expressed as

He
0.08 m^3
2 MPa
80°C

Mass balance: $$m_{in} - m_{out} = \Delta m_{\text{system}} \rightarrow m_e = m_1 - m_2$$

$$m_2 = \tfrac{1}{2} m_1 \text{ (given)} \longrightarrow m_e = m_2 = \tfrac{1}{2} m_1$$

Energy balance: $$\underbrace{E_{in} - E_{out}}_{\substack{\text{Net energy transfer}\\ \text{by heat, work, and mass}}} = \underbrace{\Delta E_{\text{system}}}_{\substack{\text{Change in internal, kinetic,}\\ \text{potential, etc. energies}}}$$

$$-m_e h_e = m_2 u_2 - m_1 u_1 \quad (\text{since } W \cong Q \cong ke \cong pe \cong 0)$$

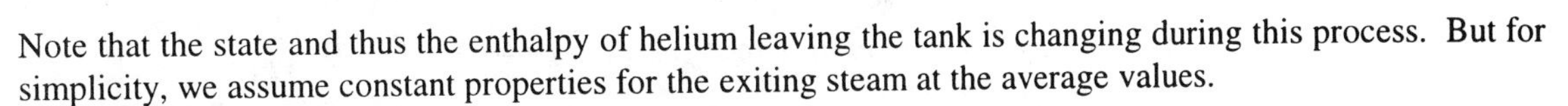

Note that the state and thus the enthalpy of helium leaving the tank is changing during this process. But for simplicity, we assume constant properties for the exiting steam at the average values.

Combining the mass and energy balances: $$0 = \tfrac{1}{2} m_1 h_e + \tfrac{1}{2} m_1 u_2 - m_1 u_1$$

Dividing by $2m_1$: $$0 = h_e + u_2 - 2u_1 \quad or \quad 0 = C_p \frac{T_1 + T_2}{2} + C_v T_2 - 2C_v T_1$$

Dividing by C_v: $$0 = k(T_1 + T_2) + 2T_2 - 4T_1 \quad \text{since } k = C_p / C_v$$

Solving for T_2: $$T_2 = \frac{(4-k)}{(2+k)} T_1 = \frac{(4-1.667)}{(2+1.667)}(353 \text{ K}) = \mathbf{225\ K}$$

The final pressure in the tank is

$$\frac{P_1 V}{P_2 V} = \frac{m_1 R T_1}{m_2 R T_2} \longrightarrow P_2 = \frac{m_2 T_2}{m_1 T_2} P_1 = \frac{1}{2}\frac{225}{353}(2000 \text{ kPa}) = \mathbf{637\ kPa}$$

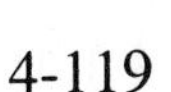

4-161E An insulated rigid tank equipped with an electric heater initially contains pressurized air. A valve is opened, and air is allowed to escape at constant temperature until the pressure inside drops to 30 psia. The amount of electrical work transferred is to be determined.

Assumptions **1** This is an unsteady process since the conditions within the device are changing during the process, but it can be analyzed as a uniform-flow process since the exit temperature (and enthalpy) of air remains constant. **2** Kinetic and potential energies are negligible. **3** The tank is insulated and thus heat transfer is negligible. **4** Air is an ideal gas with variable specific heats.

Properties The gas constant of air is R =0.3704 psia.ft^3/lbm.R (Table A-1E). The properties of air are (Table A-17E)

$$
\begin{aligned}
T_i &= 580 \text{ R} &\longrightarrow\quad h_i &= 138.66 \text{ Btu/lbm} \\
T_1 &= 580 \text{ R} &\longrightarrow\quad u_1 &= 98.90 \text{ Btu/lbm} \\
T_2 &= 580 \text{ R} &\longrightarrow\quad u_2 &= 98.90 \text{ Btu/lbm}
\end{aligned}
$$

Analysis We take the tank as the system, which is a control volume since mass crosses the boundary. Noting that the microscopic energies of flowing and nonflowing fluids are represented by enthalpy h and internal energy u, respectively, the mass and energy balances for this uniform-flow system can be expressed as

Mass balance: $m_{in} - m_{out} = \Delta m_{\text{system}} \quad \rightarrow \quad m_e = m_1 - m_2$

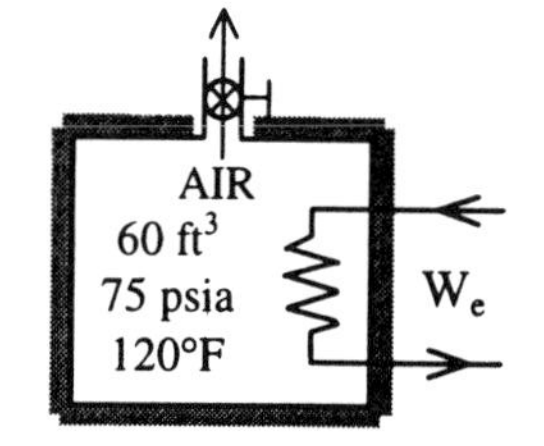

Energy balance:

$$\underbrace{E_{in} - E_{out}}_{\substack{\text{Net energy transfer}\\ \text{by heat, work, and mass}}} = \underbrace{\Delta E_{\text{system}}}_{\substack{\text{Change in internal, kinetic,}\\ \text{potential, etc. energies}}}$$

$$W_{e,in} - m_e h_e = m_2 u_2 - m_1 u_1 \quad (\text{since } Q \cong ke \cong pe \cong 0)$$

The initial and the final masses of air in the tank are

$$m_1 = \frac{P_1 V}{RT_1} = \frac{(75\text{psia})(60\text{ft}^3)}{(0.3704\text{psia}\cdot\text{ft}^3/\text{lbm}\cdot\text{R})(580\text{R})} = 20.95\text{lbm}$$

$$m_2 = \frac{P_2 V}{RT_2} = \frac{(30\text{psia})(60\text{ft}^3)}{(0.3704\text{psia}\cdot\text{ft}^3/\text{lbm}\cdot\text{R})(580\text{R})} = 8.38\text{lbm}$$

Then from the mass and energy balances,

$$m_e = m_1 - m_2 = 20.95 - 8.38 = 12.57 \text{ lbm}$$

$$
\begin{aligned}
W_{e,in} &= m_e h_e + m_2 u_2 - m_1 u_1 \\
&= (12.57 \text{ lbm})(138.66 \text{ Btu/lbm}) + (8.38 \text{ lbm})(98.90 \text{ Btu/lbm}) - (20.95 \text{ lbm})(98.90 \text{ Btu/lbm}) \\
&= \mathbf{500\ Btu}
\end{aligned}
$$

4-162 A vertical cylinder initially contains air at room temperature. Now a valve is opened, and air is allowed to escape at constant pressure and temperature until the volume of the cylinder goes down by half. The amount air that left the cylinder and the amount of heat transfer are to be determined.

Assumptions **1** This is an unsteady process since the conditions within the device are changing during the process, but it can be analyzed as a uniform-flow process since the exit temperature (and enthalpy) of air remains constant. **2** Kinetic and potential energies are negligible. **3** There are no work interactions. **4** Air is an ideal gas with constant specific heats. **5** The direction of heat transfer is to the cylinder (will be verified).

Properties The gas constant of air is R =0.287 kPa.m^3/kg.K.

Analysis (*a*) We take the cylinder as the system, which is a control volume since mass crosses the boundary. Noting that the microscopic energies of flowing and nonflowing fluids are represented by enthalpy h and internal energy u, respectively, the mass and energy balances for this uniform-flow system can be expressed as

Mass balance: $$m_{in} - m_{out} = \Delta m_{\text{system}} \quad \rightarrow \quad m_e = m_1 - m_2$$

Energy balance: $$\underbrace{E_{in} - E_{out}}_{\substack{\text{Net energy transfer} \\ \text{by heat, work, and mass}}} = \underbrace{\Delta E_{\text{system}}}_{\substack{\text{Change in internal, kinetic,} \\ \text{potential, etc. energies}}}$$

$$Q_{in} + W_{b,in} - m_e h_e = m_2 u_2 - m_1 u_1 \quad (\text{since } ke \cong pe \cong 0)$$

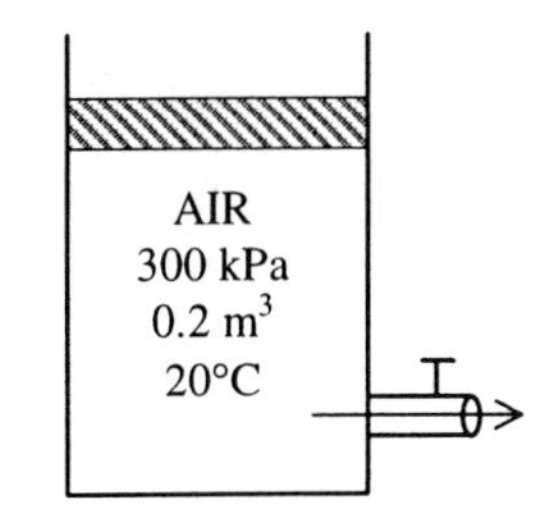

The initial and the final masses of air in the cylinder are

$$m_1 = \frac{P_1 V_1}{RT_1} = \frac{(300\text{kPa})(0.2\text{m}^3)}{(0.287\text{kPa}\cdot\text{m}^3/\text{kg}\cdot\text{K})(293\text{K})} = 0.714\text{kg}$$

$$m_2 = \frac{P_2 V_2}{RT_2} = \frac{(300\text{kPa})(0.1\text{m}^3)}{(0.287\text{kPa}\cdot\text{m}^3/\text{kg}\cdot\text{K})(293\text{K})} = 0.357\text{kg} = \tfrac{1}{2} m_1$$

Then from the mass balance,

$$m_e = m_1 - m_2 = 0.714 - 0.357 = \mathbf{0.357\ kg}$$

(*b*) This is a constant pressure process, and thus the W_b and the ΔU terms can be combined into Δh to yield

$$Q = m_e h_e + m_2 h_2 - m_1 h_1$$

Noting that the temperature of the air remains constant during this process, we have $h_i = h_1 = h_2 = h$. Also, $m_e = m_2 = \frac{1}{2} m_1$. Thus,

$$Q = \left(\tfrac{1}{2} m_1 + \tfrac{1}{2} m_1 - m_1\right) h = \mathbf{0}$$

4-163 A balloon is initially filled with helium gas at atmospheric conditions. The tank is connected to a supply line, and helium is allowed to enter the balloon until the pressure rises from 100 to 150 kPa. The final temperature in the balloon is to be determined.

Assumptions **1** This is an unsteady process since the conditions within the device are changing during the process, but it can be analyzed as a uniform-flow process since the state of fluid at the inlet remains constant. **2** Helium is an ideal gas with constant specific heats. **3** The expansion process is quasi-equilibrium. **4** Kinetic and potential energies are negligible. **5** There are no work interactions involved other than boundary work. **6** Heat transfer is negligible.

Properties The gas constant of helium is R = 2.0769 kJ/kg· K (Table A-1). The specific heats of helium are C_p = 5.1926 and C_v = 3.1156 kJ/kg· K (Table A-2a).

Analysis We take the cylinder as the system, which is a control volume since mass crosses the boundary. Noting that the microscopic energies of flowing and nonflowing fluids are represented by enthalpy h and internal energy u, respectively, the mass and energy balances for this uniform-flow system can be expressed as

Mass balance: $$m_{in} - m_{out} = \Delta m_{\text{system}} \quad \rightarrow \quad m_i = m_2 - m_1$$

Energy balance: $$\underbrace{E_{in} - E_{out}}_{\substack{\text{Net energy transfer} \\ \text{by heat, work, and mass}}} = \underbrace{\Delta E_{\text{system}}}_{\substack{\text{Change in internal, kinetic,} \\ \text{potential, etc. energies}}}$$

$$m_i h_i = W_{b,out} + m_2 u_2 - m_1 u_1 \quad (\text{since } Q \cong ke \cong pe \cong 0)$$

He
25°C
150 kPa

He
22°C
100 kPa

$$m_1 = \frac{P_1 V_1}{RT_1} = \frac{(100\text{kPa})(65\text{m}^3)}{(2.0769\text{kPa}\cdot\text{m}^3/\text{kg}\cdot\text{K})(295\text{K})} = 10.61\text{kg}$$

$$\frac{P_1}{P_2} = \frac{V_1}{V_2} \longrightarrow V_2 = \frac{P_2}{P_1}V_1 = \frac{150\text{kPa}}{100\text{kPa}}(65\text{m}^3) = 97.5\text{m}^3$$

$$m_2 = \frac{P_2 V_2}{RT_2} = \frac{(150\text{kPa})(97.5\text{m}^3)}{(2.0769\text{kPa}\cdot\text{m}^3/\text{kg}\cdot\text{K})(T_2\,\text{K})} = \frac{7041.74}{T_2}\text{kg}$$

Then from the mass balance,

$$m_i = m_2 - m_1 = \frac{7041.74}{T_2} - 10.61\text{ kg}$$

Noting that P varies linearly with V, the boundary work done during this process is

$$W_b = \frac{P_1 + P_2}{2}(V_2 - V_1) = \frac{(100+150)\text{kPa}}{2}(97.5-65)\text{m}^3 = 4062.5\text{kJ}$$

Using specific heats, the energy balance relation reduces to

$$W_{b,out} = m_i C_p T_i - m_2 C_v T_2 + m_1 C_v T_1$$

Substituting,

$$4062.5 = \left(\frac{7041.74}{T_2} - 10.61\right)(5.1926)(298) - \frac{7041.74}{T_2}(3.1156)T_2 + (10.61)(3.1156)(295)$$

It yields T_2 = **333.6 K**

4-164 A balloon is initially filled with pressurized helium gas. Now a valve is opened, and helium is allowed to escape until the pressure inside drops to atmospheric pressure. The final temperature of helium in the balloon and the amount helium that has escaped are to be determined.

Assumptions **1** This is an unsteady process since the conditions within the device are changing during the process, but it can be analyzed as a uniform-flow process by assuming exit properties to be constant at average conditions. **2** Kinetic and potential energies are negligible. **3** There are no work interactions other than boundary work. **4** Helium is an ideal gas with constant specific heats. **5** Heat transfer is negligible.

Properties The gas constant of helium is R =2.0769 kPa.m^3/kg.K (Table A-1). The specific heats of helium are C_p = 5.1926 and C_v = 3.1156 kJ/kg· K (Table A-2).

Analysis The properties of helium leaving the balloon are changing during this process. But we will treat them as a constant at the average temperature. Thus $T_e \cong (T_1 + T_2)/2$. Also $h = C_pT$ and $u = C_vT$.

We take the balloon as the system, which is a control volume since mass crosses the boundary. Noting that the microscopic energies of flowing and nonflowing fluids are represented by enthalpy h and internal energy u, the mass and energy balances for this uniform-flow system can be expressed as

Mass balance: $m_{in} - m_{out} = \Delta m_{\text{system}} \quad \rightarrow \quad m_e = m_1 - m_2$

Energy balance:

$$\underbrace{E_{in} - E_{out}}_{\substack{\text{Net energy transfer} \\ \text{by heat, work, and mass}}} = \underbrace{\Delta E_{\text{system}}}_{\substack{\text{Change in internal, kinetic,} \\ \text{potential, etc. energies}}}$$

$$W_{b,in} - m_e h_e = m_2 u_2 - m_1 u_1 \quad (\text{since } Q \cong ke \cong pe \cong 0)$$

or

$$W_{b,in} = m_e C_p \frac{T_1 + T_2}{2} + m_2 C_v T_2 - m_1 C_v T_1$$

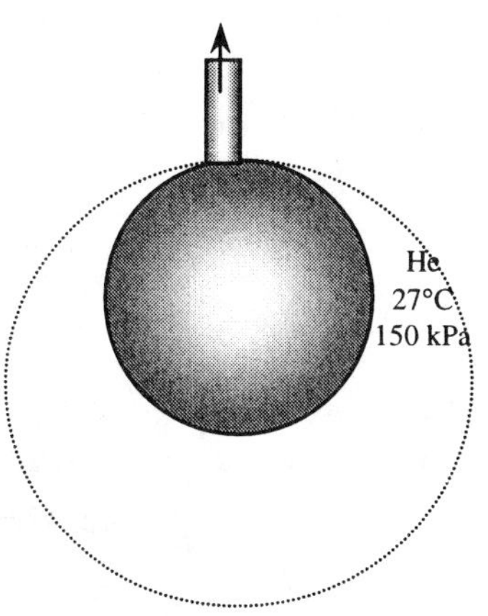

The initial and the final masses in the balloon are

$$m_1 = \frac{P_1 V_1}{RT_1} = \frac{(150\text{kPa})(10\text{m}^3)}{(2.0769\text{kPa}\cdot\text{m}^3/\text{kg}\cdot\text{K})(300\text{K})} = 2.4074\text{kg}$$

$$m_2 = \frac{P_2 V_2}{RT_2} = \frac{(100\text{kPa})(8.5\text{m}^3)}{(2.0769\text{kPa}\cdot\text{m}^3/\text{kg}\cdot\text{K})T_2} = \frac{409.264}{T_2}$$

Then from the mass balance,

$$m_e = m_1 - m_2 = 2.4074 - \frac{409.264}{T_2}$$

Noting that the pressure changes linearly with volume, the boundary work done during this process is

$$W_b = \frac{P_1 + P_2}{2}(V_2 - V_1) = \frac{(150+100)\text{kPa}}{2}(8.5-10)\text{m}^3 = -187.5\text{kJ}$$

Combining mass and energy balances and substituting,

$$-(-187.5) = \left(2.4074 - \frac{409.264}{T_2}\right)(5.1926)\frac{300+T_2}{2} + \frac{409.264}{T_2}(3.1156)T_2 - (2.4074)(3.1156)(300)$$

It yields $\quad T_2^2 - 56T_2 - 51{,}003.5 = 0 \quad \rightarrow \quad T_2 = \mathbf{256\ K}$

(b) The amount of helium that has escaped is

$$m_e = m_1 - m_2 = 2.4074 - \frac{409.264}{T_2} = 2.1074 - \frac{409.264}{256} = \mathbf{0.509\ \ kg}$$

4-165 EES solution of this (and other comprehensive problems designated with the *computer icon*) is available to instructors at the *Instructor Manual* section of the *Online Learning Center* (OLC) at www.mhhe.com/cengel-boles. See the Preface for access information.

4-166 A vertical piston-cylinder device equipped with an external spring initially contains superheated steam. Now a valve is opened, and steam is allowed to escape until the volume of the cylinder goes down by half. The initial and final masses of steam in the cylinder and the amount of heat transferred are to be determined.

Assumptions **1** This is an unsteady process since the conditions within the device are changing during the process, but it can be analyzed as a uniform-flow process by assuming the properties of steam that escape to be constant at average conditions. **2** Kinetic and potential energies are negligible. **3** The spring is a linear spring. **4** The direction of heat transfer is to the cylinder (will be verified).

Properties From the steam tables (Tables A-4 through A-6),

$$\left.\begin{array}{l} P_1 = 1\text{MPa} \\ T_1 = 250^\circ\text{C} \end{array}\right\} \begin{array}{l} v_1 = 0.2327\text{m}^3/\text{kg} \\ u_1 = 2709.9\text{kJ/kg}, \quad h_1 = 2942.6\text{kJ/kg} \end{array}$$

$$\left.\begin{array}{l} P_2 = 800\text{kPa} \\ sat.vapor \end{array}\right\} \begin{array}{l} v_2 = 0.2404\ \text{m}^3/\text{kg} \\ u_2 = 2576.8\ \text{kJ/kg}, \quad h_2 = 2769.1\text{kJ/kg} \end{array}$$

Analysis (*a*) We take the cylinder as the system, which is a control volume since mass crosses the boundary. Noting that the microscopic energies of flowing and nonflowing fluids are represented by enthalpy h and internal energy u, the mass and energy balances for this uniform-flow system can be expressed as

Mass balance: $$m_{in} - m_{out} = \Delta m_{\text{system}} \rightarrow m_e = m_1 - m_2$$

Energy balance: $$\underbrace{E_{in} - E_{out}}_{\substack{\text{Net energy transfer} \\ \text{by heat, work, and mass}}} = \underbrace{\Delta E_{\text{system}}}_{\substack{\text{Change in internal, kinetic,} \\ \text{potential, etc. energies}}}$$

$$Q_{in} + W_{b,in} - m_e h_e = m_2 u_2 - m_1 u_1 \quad (\text{since } ke \cong pe \cong 0)$$

The state and thus the enthalpy of the steam leaving the cylinder is changing during this process. But for simplicity, we assume constant properties for the exiting steam at the average values. Thus,

$$h_e \cong \frac{h_1 + h_2}{2} = \frac{2942.6 + 2769.1\ \text{kJ/kg}}{2} = 2855.9\ \text{kJ/kg}$$

The initial and the final masses in the tank are

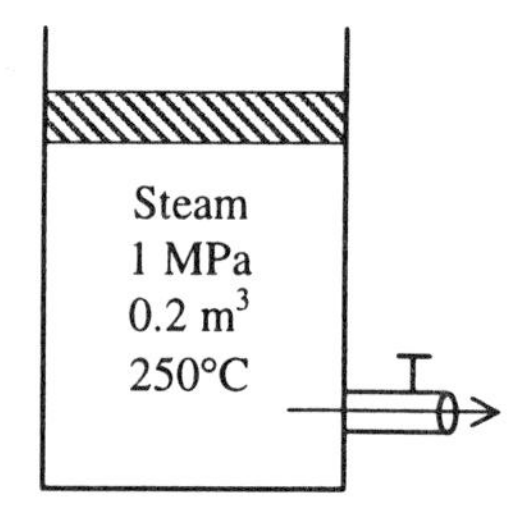

$$m_1 = \frac{V_1}{v_1} = \frac{0.2\ \text{m}^3}{0.2327\ \text{m}^3/\text{kg}} = \mathbf{0.859\ kg}$$

$$m_2 = \frac{V_2}{v_2} = \frac{0.1\ \text{m}^3}{0.2404\ \text{m}^3/\text{kg}} = \mathbf{0.416\ kg}$$

Then from the mass balance,

$$m_e = m_1 - m_2 = 0.859 - 0.416 = 0.443\ \text{kg}$$

(*b*) The boundary work done during this process is

$$W_{b,in} = \frac{P_1 + P_2}{2}(V_1 - V_2) = \frac{(1000 + 800)\text{kPa}}{2}(0.2 - 0.1)\text{m}^3 = 90\text{kJ}$$

Then the heat transfer during this process becomes

$$\begin{aligned} Q_{in} &= -W_{b,in} + m_e h_e + m_2 u_2 - m_1 u_1 \\ &= -90\text{kJ} + (0.443\text{kg})(2855.9\text{kJ/kg}) + (0.416\text{kg})(2576.8\text{kJ/kg}) - (0.859\text{kg})(2709.9\text{kJ/kg}) \\ &= -80.7\text{kJ} \rightarrow Q_{out} = \mathbf{80.7kJ} \end{aligned}$$

4-167 A vertical piston-cylinder device initially contains steam at a constant pressure of 300 kPa. Now a valve is opened, and steam is allowed to escape at constant temperature and pressure until the volume reduces to one-third. The mass of steam that escaped and the amount of heat transfer are to be determined.

Assumptions **1** This is an unsteady process since the conditions within the device are changing during the process, but it can be analyzed as a uniform-flow process since the properties of steam that escape remain constant. **2** Kinetic and potential energies are negligible. **3** There are no work interactions other than boundary work. **4** The direction of heat transfer is to the cylinder (will be verified).

Properties From the steam tables (Tables A-4 through A-6),

$$\left.\begin{array}{l} P_e = P_1 = P_2 = 300\text{kPa} \\ T_e = T_1 = T_2 = 250^\circ\text{C} \end{array}\right\} \begin{array}{l} h_e = 2967.6\text{kJ/kg} \\ v_1 = v_2 = 0.7964\text{m}^3\text{/kg} \\ u_1 = u_2 = 2728.7\text{kJ/kg} \end{array}$$

Analysis (*a*) We take the cylinder as the system, which is a control volume since mass crosses the boundary. Noting that the microscopic energies of flowing and nonflowing fluids are represented by enthalpy h and internal energy u, the mass and energy balances for this uniform-flow system can be expressed as

Mass balance: $$m_{in} - m_{out} = \Delta m_{\text{system}} \quad \rightarrow \quad m_e = m_1 - m_2$$

Energy balance: $$\underbrace{E_{in} - E_{out}}_{\substack{\text{Net energy transfer} \\ \text{by heat, work, and mass}}} = \underbrace{\Delta E_{\text{system}}}_{\substack{\text{Change in internal, kinetic,} \\ \text{potential, etc. energies}}}$$

$$Q_{in} + W_{b,in} - m_e h_e = m_2 u_2 - m_1 u_1 \quad (\text{since } ke \cong pe \cong 0)$$

or $$Q_{in} = m_e h_e + m_2 h_2 - m_1 h_1$$

since for a constant pressure process, the W_b and the ΔU terms can be combined into ΔH.

The initial and final masses of steam in the cylinder are

$$m_1 = \frac{V_1}{v_1} = \frac{0.3\text{ m}^3}{0.7964\text{ m}^3\text{/kg}} = 0.377\text{ kg}$$

$$m_2 = \frac{V_2}{v_2} = \frac{0.1\text{ m}^3}{0.7964\text{ m}^3\text{/kg}} = 0.126\text{ kg}$$

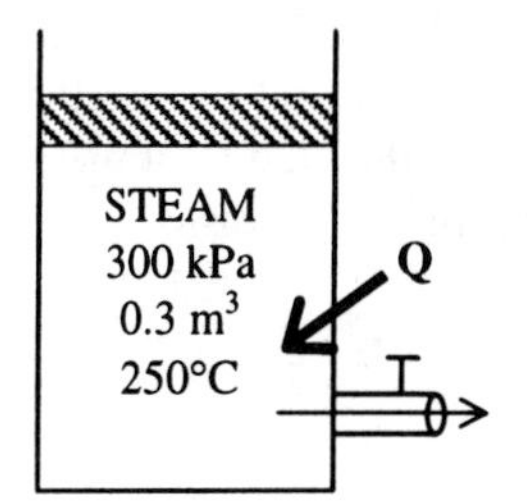

Then from the mass balance,

$$m_e = m_1 - m_2 = 0.377 - 0.126 = \mathbf{0.251\text{ kg}}$$

(*b*) Noting that $h_e = h_1 = h_2 = h$ and $m_e = m_1 - m_2$, the energy balance relation reduces to

$$\begin{aligned} Q_{in} &= m_e h_e + m_2 h_2 - m_1 h_1 \\ &= (m_e + m_2 - m_1)h \\ &= \mathbf{0} \end{aligned}$$

Therefore, there will be no heat transfer during this process.

Special Topic: Refrigeration and Freezing of Foods

4-168C The common kinds of microorganisms are bacteria, yeasts, molds, and viruses. The undesirable changes caused by microorganisms are off-flavors and colors, slime production, changes in the texture and appearances, and the spoilage of foods.

4-169C Microorganisms are the prime cause for the spoilage of foods. Refrigeration prevents or delays the spoilage of foods by reducing the rate of growth of microorganisms. Freezing extends the storage life of foods for months by preventing the growths of microorganisms.

4-170C Cooking kills the microorganisms in foods, and thus prevents spoilage of foods. It is important to raise the internal temperature of a roast in an oven above 70°C,.since most microorganisms, including some that cause diseases, may survive temperatures below 70°C.

4-171C The rate of freezing affects the size and number of ice crystals formed in foods during freezing. During slow freezing, ice crystals grow to a large size, damaging the walls of the cells and causing the loss of natural juices and sagging of foods. During fast freezing, a large number of ice crystals start forming at once, and thus the size of ice crystals is much smaller than that in slow freezing, causing minimal damage. Also, fast freezing forms a crust on the outer layer of the food product that prevents dehydration, and keeps all the flavoring agents sealed in the foods.

4-172C The latent heat of fusion of food products whose water content is known can be determined from $h_{\text{latent}} = 334a$ where a is the fraction of water in the food and 334 kJ/kg is the latent heat of fusion of water.

4-173C The specific heat of apples will be higher than that of apricots since the specific heats of food products are proportional to their water content, and the water content of apples (82%) is larger than that of apricots (70%).

4-174C About 70 percent of the beef carcass is water, and the carcass is cooled mostly by evaporative cooling as a result of moisture migration towards the surface where evaporation occurs. This may cause up to 2 percent of the total mass of the carcass to evaporate during an overnight chilling. This weight loss can be minimized by washing or spraying the carcass with water prior to cooling.

4-175C (a) The heat transfer coefficient during immersion cooling is much higher, and thus the cooling time during immersion chilling is much lower than that during forced air chilling. (b) The cool air chilling can cause a moisture loss of 1 to 2 percent while water immersion chilling can actually cause moisture absorption of 4 to 15 percent. (c) The chilled water circulated during immersion cooling encourages microbial growth, and thus immersion chilling is associated with more microbial growth. The problem can be minimized by adding chloride to the water.

4-176C Heat of respiration is the heat generated by fruits and vegetables due to their respiration during which glucose combines with O_2 to produce CO_2 and H_2O. This is an exothermic reaction that releases heat to surroundings.

4-177 A box of beef with a specified water content is to be frozen. The total amount of heat removed from the beef and the average rate of heat removal are to be determined.

Assumptions **1** The beef is at uniform temperatures at the beginning and at the end of the process. **2** The entire water content of the beef freezes during the process. **3** The thermal properties of beef are constant. **4** The freezing point of beef is -2°C.

Properties At a water content of 60 percent, the specific heats of fresh and frozen beef and the latent heat are determined from Siebel's formula to be

$$C_{p,\text{ fresh}} = 3.35a + 0.84 = 3.35 \times 0.60 + 0.84 = 2.85 \text{ kJ/kg.°C}$$

$$C_{p,\text{ frozen}} = 1.26a + 0.84 = 1.26 \times 0.60 + 0.84 = 1.60 \text{ kJ/kg.°C}$$

$$h_{\text{latent}} = 334a = 334 \times 0.60 = 200.4 \text{ kJ/kg}$$

Analysis (*a*) We take the beef as the system. The energy balance for this closed system can be expressed as

$$\underbrace{E_{in} - E_{out}}_{\substack{\text{Net energy transfer}\\ \text{by heat, work, and mass}}} = \underbrace{\Delta E_{\text{system}}}_{\substack{\text{Change in internal, kinetic,}\\ \text{potential, etc. energies}}}$$

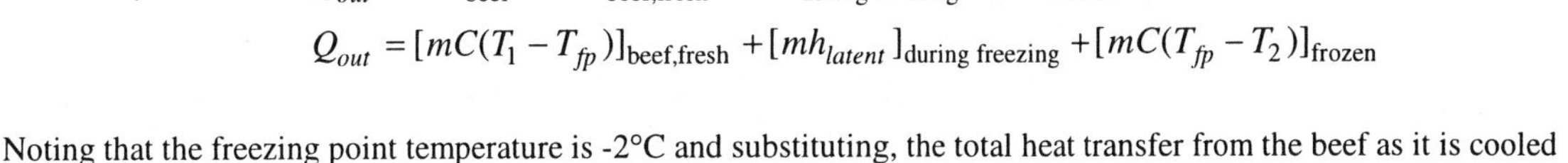

$$-Q_{out} = \Delta U_{\text{beef}} = \Delta U_{\text{beef,fresh}} + \Delta U_{\text{during freezing}} + \Delta U_{\text{frozen}}$$

$$Q_{out} = [mC(T_1 - T_{fp})]_{\text{beef,fresh}} + [mh_{latent}]_{\text{during freezing}} + [mC(T_{fp} - T_2)]_{\text{frozen}}$$

Noting that the freezing point temperature is -2°C and substituting, the total heat transfer from the beef as it is cooled from 6 to -20°C is determined to be

$$\begin{aligned} Q_{out} &= (35\text{ kg})(2.85\text{ kJ/kg.°C})[6-(-2)]\text{°C} + (35\text{ kg})(200.4\text{ kJ/kg}) + (35\text{ kg})(1.60\text{ kJ/kg.°C})[-2-(-20)]\text{°C} \\ &= 798 + 7014 + 1008 \\ &= \mathbf{8820\ kJ} \end{aligned}$$

(*b*) Noting that 8820 kJ of heat is removed from the beef in 3 h, the average rate of heat removal is determined to be

$$\dot{Q}_{ave} = \frac{Q_{out}}{\Delta t} = \frac{8820\text{ kJ}}{3\text{ h} \times (3600\text{ s/h})} = \mathbf{0.817\ kW}$$

4-178 A box of cherries with a specified water content is to be frozen. The total amount of heat removed from the cherries is to be determined.

Assumptions **1** The cherries are at uniform temperatures at the beginning and at the end of the process. **2** The entire water content of the cherries freezes during the process. **3** The thermal properties of cherries are constant

Properties . The freezing point of cherries is –1.8°C (Table A-3). At a water content of 77 percent, the specific heats of fresh and frozen cherries and the latent heat are determined from Siebel's formula to be

$$C_{p,\text{ fresh}} = 3.35a + 0.84 = 3.35 \times 0.77 + 0.84 = 3.42 \text{ kJ/kg.°C}$$

$$C_{p,\text{ frozen}} = 1.26a + 0.84 = 1.26 \times 0.77 + 0.84 = 1.81 \text{ kJ/kg.°C}$$

$$h_{\text{latent}} = 334a = 334 \times 0.77 = 257.2 \text{ kJ/kg}$$

Analysis We take the cherries as the system. The energy balance for this closed system can be expressed as

$$\underbrace{E_{in} - E_{out}}_{\substack{\text{Net energy transfer} \\ \text{by heat, work, and mass}}} = \underbrace{\Delta E_{\text{system}}}_{\substack{\text{Change in internal, kinetic,} \\ \text{potential, etc. energies}}}$$

$$-Q_{out} = \Delta U_{\text{beef}} = \Delta U_{\text{cherries,fresh}} + \Delta U_{\text{during freezing}} + \Delta U_{\text{frozen}}$$

$$Q_{out} = [mC(T_1 - T_{fp})]_{\text{cherries,fresh}} + [mh_{latent}]_{\text{during freezing}} + [mC(T_{fp} - T_2)]_{\text{frozen}}$$

Noting that the freezing point temperature is –1.8°C and substituting, the total heat transfer from the cherries as they are cooled from 8 to -20°C is determined to be

$$\begin{aligned} Q_{out} &= (50 \text{ kg})(3.42 \text{ kJ/kg.°C})[8-(-1.8)]\text{°C} + (50 \text{ kg})(257.2 \text{ kJ/kg}) + (50 \text{ kg})(1.81 \text{ kJ/kg.°C})[-1.8-(-20)]\text{°C} \\ &= 1676 + 12{,}860 + 1647 \\ &= \mathbf{16{,}183 \text{ kJ}} \end{aligned}$$

4-179 Fresh strawberries with a water content of 88% in nylon boxes are to be cooled to 4°C at a rate of 60 boxes/h. The rate of heat removal from the strawberries together with their boxes, and the percent error involved if the strawberry boxes were ignored in calculations are to be determined. √

Assumptions **1** Steady operating conditions exist. **2** The thermal properties of strawberries and boxes are constant.

Properties The specific heat of strawberries is given to be C_p = 3.89 kJ/kg.°C. The specific heat of nylon box is given to be C_p = 1.7 kJ/kg.°C. The heat of respiration of strawberries is given to be 0.21 W/kg.

Analysis Noting that the strawberries are cooled at a rate of 60 boxes per hour, the total amounts of strawberries and the box material cooled per hour are

$$m_{\text{strawberry}} = \text{(Mass per box)(Number of boxes per hour)} = (23\text{ kg/box})(60\text{ box}) = 1380\text{ kg}$$

$$m_{\text{box}} = \text{(Mass per box)(Number of boxes per hour)} = (0.8\text{ kg/box})(60\text{ box/h}) = 48\text{ kg}$$

Analysis We take the strawberries together with their boxes as the system. A practical way of considering the heat of respiration is to disregard the chemical changes that occur in the food, and to treat the heat of respiration as heat input. Then the energy balance for this closed system can be expressed as

$$\underbrace{E_{in} - E_{out}}_{\substack{\text{Net energy transfer}\\ \text{by heat, work, and mass}}} = \underbrace{\Delta E_{\text{system}}}_{\substack{\text{Change in internal, kinetic,}\\ \text{potential, etc. energies}}}$$

$$Q_{\text{respiration,in}} - Q_{out} = \Delta U_{\text{total}}$$

$$= \Delta U_{\text{strawberries}} + \Delta U_{\text{box}}$$

$$= [mC(T_2 - T_1)]_{\text{strawberries}} + [mC(T_2 - T_1)]_{\text{box}}$$

$$Q_{out} = Q_{\text{respiration,in}} - [mC(T_2 - T_1)]_{\text{strawberries}} - [mC(T_2 - T_1)]_{\text{box}}$$

$$= \dot{Q}_{\text{respiration,in}}\Delta t + [mC(T_1 - T_2)]_{\text{strawberries}} + [mC(T_1 - T_2)]_{\text{box}}$$

Q

Strawberries
30°C

Dividing by Δt, it can be expressed in the rate form as

$$\dot{Q}_{out} = (m_{\text{strawberries}}\dot{q}_{\text{respiration}}) + [mC(T_1 - T_2)/\Delta t]_{\text{strawberries}} + [mC(T_1 - T_2)/\Delta t]_{\text{box}}$$

Substituting, the rate of heat transfer is determined to be

$$\dot{Q}_{\text{out}} = (1380\text{ kg})(0.00021\text{ kW/kg}) + [(1380\text{ kg})(3.89\text{ kJ/kg.°C})(30-4)\text{°C}/1\text{ h}]$$
$$+ [(48\text{ kg})(1.7\text{ kJ/kg.°C})(30-4)\text{°C}]/(1\text{ h})$$
$$= 0.290\text{ kJ/s}\left(\frac{3600\text{ s}}{1\text{ h}}\right) + 139{,}573\text{ kJ/h} + 2122\text{ kJ/h}$$
$$= \mathbf{142{,}739\text{ kJ/h}} = 39.6\text{ kW}$$

The heat loss from the boxes constitute 2122/142,739 = 0.015 or just 1.5% of the total heat loss. Therefore, the error involved in ignoring the boxes in calculations is **negligible.**

Discussion Note that most of the cooling load is due to the removal of the latent heat during the phase change process. Also, the cooling load due to the box is negligible.

4-180 Lettuce is to be vacuum cooled in an insulated spherical vacuum chamber. The final pressure in the vacuum chamber and the amount of moisture removed from the lettuce are to be determined.

Assumptions **1** The thermal properties of lettuce are constant. **2** Heat transfer through the walls of the vacuum chamber is negligible. **3** All the air in the chamber is sucked out by the vacuum pump so that there is only vapor in the chamber at the end of the process.

Properties The specific heat of the lettuce is 4.02 kJ/kg.°C. The saturation pressure of water at 2°C is 0.716 kPa. At the average temperature of (24 + 2)/2 = 13°C, the latent heat of vaporization of water is h_{fg} = 2470.6 kJ/kg (Table A-4).

Analysis (*a*) The final vapor pressure in the vacuum chamber will be the saturation pressure at the final temperature of 2°C, which is determined from the saturated water tables to be

$$P_{\text{final}} = P_{\text{sat @ 2°C}} = \mathbf{0.716\ kPa}$$

Assuming all the air in the chamber is already sucked out by the vacuum pump, this vapor pressure will be equivalent to the final pressure in the chamber.

Vacuum Chamber

Lettuce 24°C

(*b*) We take the lettuce as the system. Disregarding the small amount of water that evaporates, the energy balance for this closed system can be expressed as

$$\underbrace{E_{in} - E_{out}}_{\substack{\text{Net energy transfer} \\ \text{by heat, work, and mass}}} = \underbrace{\Delta E_{\text{system}}}_{\substack{\text{Change in internal, kinetic,} \\ \text{potential, etc. energies}}}$$

$$-Q_{out} = \Delta U = [mC(T_2 - T_1)]_{\text{lettuce}}$$

Then the amount of heat transfer to cool the lettuce from 24 to 2°C becomes

$$Q_{\text{lettuce,out}} = [mC(T_1 - T_2)]_{\text{lettuce}} = (5000\text{ kg})(4.02\text{ kJ/kg.°C})(24 - 2)\text{°C} = 442{,}200\text{ kJ}$$

Each kg of water in lettuce absorbs 2470 kJ of heat as it evaporates. Disregarding any heat gain through the walls of the vacuum chamber, the amount of moisture removed is determined to be

$$Q_{\text{lettuce}} = m_{\text{evap}} h_{fg} \rightarrow m_{\text{evap}} = \frac{Q_{\text{lettuce}}}{h_{fg}} = \frac{442{,}200\text{ kJ}}{2470\text{ kJ/kg}} = \mathbf{179.0\ kg}$$

4-181 A box of shrimp is to be frozen in a freezer. The amount of heat that needs to be removed is to be determined.

Assumptions **1** The thermal properties of fresh and frozen shrimp are constant. **2** The entire water content of the shrimps freezes during the process.

Properties For shrimp, the freezing temperature is -2.2°C, the latent heat of fusion is 277 kJ/kg, the specific heat is 3.62 kJ/kg·°C above freezing and 1.89 kJ/kg·°C below freezing (given). The specific heat of polyethylene box is given to be 2.3 kJ/kg·°C.

Analysis We take the shrimp together with their boxes as the system. The energy balance for this system can be expressed as

$$\underbrace{E_{in} - E_{out}}_{\substack{\text{Net energy transfer}\\ \text{by heat, work, and mass}}} = \underbrace{\Delta E_{\text{system}}}_{\substack{\text{Change in internal, kinetic,}\\ \text{potential, etc. energies}}}$$

$$-Q_{out} = \Delta U_{\text{total}} = [\Delta U_{\text{fresh}} + \Delta U_{\text{during freezing}} + \Delta U_{\text{frozen}}]_{\text{shrimp}} + \Delta U_{\text{box}}$$

$$Q_{out} = -[\Delta U_{\text{fresh}} + \Delta U_{\text{during freezing}} + \Delta U_{\text{frozen}}]_{\text{shrimp}} - \Delta U_{\text{box}}$$

$$= [mC(T_1 - T_{fp})_{\text{fresh}} + mh_{latent} + mC(T_{fp} - T_2)_{\text{frozen}}]_{\text{shrimp}} + [mC(T_1 - T_2)]_{\text{box}}$$

Substituting, the total heat transfer is determined to be

$$\begin{aligned} Q_{\text{out}} &= (40\text{ kg})(3.62\text{ kJ/kg.°C})[8-(-2.2)]\text{°C}] + (40\text{ kg})(277\text{ kJ/kg}) \\ &\quad + (40\text{ kg})(1.89\text{ kJ/kg.°C})[-2.2-(-18)]\text{°C}] + (1.2\text{ kg})(2.3\text{ kJ/kg.°C})[8-(-18)\text{°C} \\ &= 1477 + 11{,}080 + 1195 + 72 \\ &= \mathbf{13{,}800\ kJ} \end{aligned}$$

Discussion Note that most of the cooling load (80 percent of it) is due to the removal of the latent heat during the phase change process. Also, the cooling load due to the box is negligible (less than 1 percent).

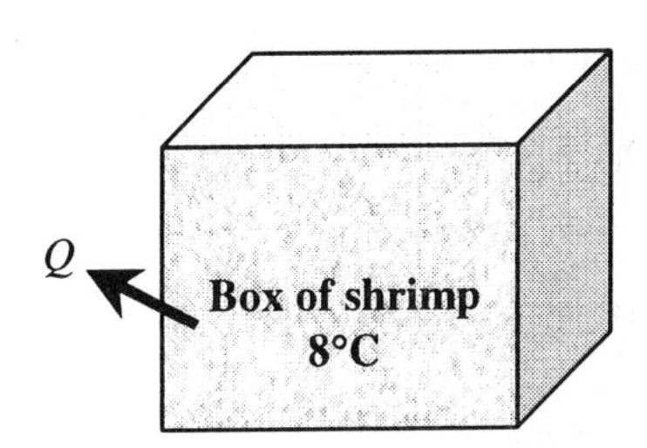

4-182 A box of shrimp is to be frozen in a freezer. The amount of heat that needs to be removed is to be determined.

Assumptions **1** The thermal properties of fresh and frozen shrimp are constant. **2** The entire water content of the shrimps freezes during the process. **3** The effect of the box is negligible.

Properties For shrimp, the freezing temperature is -2.2°C, the latent heat of fusion is 277 kJ/kg, the specific heat is 3.62 kJ/kg·°C above freezing and 1.89 kJ/kg·°C below freezing (given).

Analysis We take the shrimp within the boxes as the system. The energy balance for this system can be expressed as

$$\underbrace{E_{in} - E_{out}}_{\substack{\text{Net energy transfer} \\ \text{by heat, work, and mass}}} = \underbrace{\Delta E_{\text{system}}}_{\substack{\text{Change in internal, kinetic,} \\ \text{potential, etc. energies}}}$$

$$-Q_{out} = \Delta U_{\text{total}} = [\Delta U_{\text{fresh}} + \Delta U_{\text{during freezing}} + \Delta U_{\text{frozen}}]_{\text{shrimp}}$$

$$Q_{out} = -[\Delta U_{\text{fresh}} + \Delta U_{\text{during freezing}} + \Delta U_{\text{frozen}}]_{\text{shrimp}}$$

$$= [mC(T_1 - T_{fp})_{\text{fresh}} + mh_{latent} + mC(T_{fp} - T_2)_{\text{frozen}}]_{\text{shrimp}}$$

Substituting, the total heat transfer is determined to be

$$Q_{\text{out}} = (40\text{ kg})(3.62\text{ kJ/kg.°C})[8-(-2.2)]\text{°C}] + (40\text{ kg})(277\text{ kJ/kg})$$

$$+ (40\text{ kg})(1.89\text{ kJ/kg.°C})[-2.2-(-18)]\text{°C}]$$

$$= 1477 + 11{,}080 + 1195$$

$$= \mathbf{13{,}750\text{ kJ}}$$

Discussion Note that most of the cooling load (80 percent of it) is due to the removal of the latent heat during the phase change process. Also, it can be shown that the cooling load due to the box is negligible (less than 1 percent).

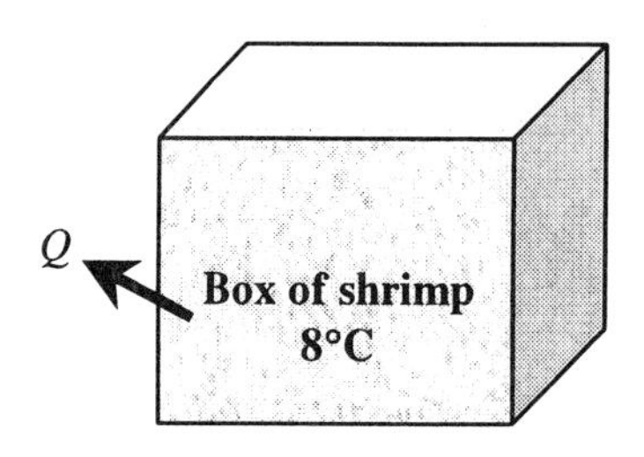

4-183E The infiltration rate of a cold storage room is given to be 0.4 air changes per hour (ACH). The total sensible infiltration load of the room is to be determined.

Assumptions **1** Air is an ideal gas with constant specific heats at room temperature. **2** Infiltrating air exfiltrates at the indoors temperature. **3** Infiltrating air is sufficiently dry so that the condensation of moisture in the air is disregarded. **4** The local atmospheric pressure is 14.7 psia.

Properties The gas constant of air is $R = 0.3704$ psia.ft³/lbm.R (Table A-1E). The specific heats of air at room temperature are $C_p = 0.240$ Btu/lbm.R and are $C_v = 0.171$ Btu/lbm.R (Table A-2E).

Analysis The density of air at the outdoor conditions is

$$\rho = \frac{P}{RT} = \frac{14.7\,\text{psia}^3}{(0.3704\ \text{psia.ft}^3/\text{lbm.R})(90+460)\text{R}} = 0.0722\ \text{lbm/ft}^3$$

Noting that the outdoor air will enter the room at a rate of 0.4 ACH, the mass flow rate of infiltrating air into the cold storage room is

$$\dot{m}_{\text{air}} = \rho V_{\text{room}} ACH = (0.0722\ \text{lbm/ft}^3)(12\times15\times30\ \text{ft}^3)(0.4\,\text{h}^{-1}) = 156\ \text{lbm/h}$$

We take the air that occupies the 0.4 room volume (156 lbm) as the system. Noting that this air is pushed out at constant pressure, boundary work is done. The energy balance for this fixed mass can be expressed as

$$\underbrace{E_{in} - E_{out}}_{\substack{\text{Net energy transfer}\\ \text{by heat, work, and mass}}} = \underbrace{\Delta E_{\text{system}}}_{\substack{\text{Change in internal, kinetic,}\\ \text{potential, etc. energies}}}$$

$$W_{b,in} - Q_{out} = \Delta U = m(u_2 - u_1)$$

$$-Q_{out} = m(h_2 - h_1)$$

$$Q_{out} = mC_p(T_1 - T_2)$$

Warm air 90°F → Cold storage room → Cold air 35°F

since $\Delta U + W_b = \Delta H$ during a constant pressure quasi-equilibrium process. It can also be expressed in the rate form as

$$\dot{Q}_{out} = \dot{m}C_p(T_1 - T_2)$$

Substituting, the rate of heat rejection from infiltrating air to the storage room, which is the sensible cooling load by infiltration, becomes

$$\dot{Q}_{\text{infiltration,sensible}} = (156\ \text{lbm/h})(0.240\ \text{Btu/lbm.°F})(90-35)\text{°F} = \mathbf{2059\ Btu/h}$$

Discussion Note that the refrigeration system of this cold storage room must be capable of removing heat at a rate of 2059 Btu/h to meet the sensible infiltration load. Of course, the total refrigeration capacity of the system will have to be larger to remove energy that entered by other mechanisms.

4-184 A banana cooling room is being analyzed. The minimum flow rate of air needed to cool bananas at a rate of 0.4°C/h while keeping the temperature rise of cool air under 2°C as it flows through the room is to be determined.

Assumptions **1** Steady operating conditions exist. **2** Thermal properties of air, bananas, and boxes are constant. **3** Air is an ideal gas with constant specific heats at room temperature. **4** The kinetic and potential energy changes are negligible.

Properties The specific heats of banana and the fiberboard are given to be 3.55 kJ/kg.°C and 1.7 kJ/kg.°C, respectively. The peak heat of respiration of bananas is given to be 0.3 W/kg. The density and specific heat of air are given to be 1.2 kg/m^3 and 1.0 kJ/kg.°C.

Analysis Noting that the banana room holds 432 boxes, the total mass of bananas and the boxes are determined to be

$$m_{\text{banana}} = (\text{Mass per box})(\text{Number of boxes}) = (19 \text{ kg / box})(864 \text{ box}) = 16{,}416 \text{ kg}$$

$$m_{\text{box}} = (\text{Mass per box})(\text{Number of boxes}) = (2.3 \text{ kg / box})(864 \text{ box}) = 1987 \text{ kg}$$

The total refrigeration load in this case is due to the heat of respiration, the cooling of the bananas and the boxes, and the heat gain through the walls, and is determined from

$$\dot{Q}_{\text{total}} = \dot{Q}_{\text{respiration}} + \dot{Q}_{\text{banana}} + \dot{Q}_{\text{box}} + \dot{Q}_{\text{wall}}$$

where

$$\dot{Q}_{\text{respiration}} = m_{\text{banana}}\dot{q}_{\text{respiration}} = (16{,}419 \text{ kg})(0.3 \text{ W / kg}) = 4925 \text{ W}$$

$$\dot{Q}_{\text{banana}} = (mC_p\Delta T / \Delta t)_{\text{banana}} = (16{,}416 \text{ kg})(3.55 \text{ kJ / kg.°C})(0.4 \text{ °C / h}) = 23{,}315 \text{ kJ / h} = 6476 \text{ W}$$

$$\dot{Q}_{\text{box}} = (mC_p\Delta T / \Delta t)_{\text{box}} = (1987 \text{ kg})(1.7 \text{ kJ / kg.°C})(0.4 \text{ °C / h}) = 1351 \text{ kJ / h} = 375 \text{ W}$$

$$\dot{Q}_{\text{wall}} = 1800 \text{ kJ / kg}) = 0.5 \text{ kW} = 500 \text{ W}$$

and the quantity $\Delta T / \Delta t$ is the rate of change of temperature of the products, and is given to be 0.4°C/W. Then the total rate of cooling becomes

$$\dot{Q}_{\text{total}} = \dot{Q}_{\text{respiration}} + \dot{Q}_{\text{banana}} + \dot{Q}_{\text{box}} + \dot{Q}_{\text{wall}} = 4925 + 6476 + 375 + 500 = 12{,}276 \text{ W}$$

All of these effects can be viewed as heat gain for the chilled air stream, which can be viewed as a steady stream of cool air that is heated as it flows in an imaginary duct passing through the room. The energy balance for this imaginary steady-flow system can be expressed in the rate form as

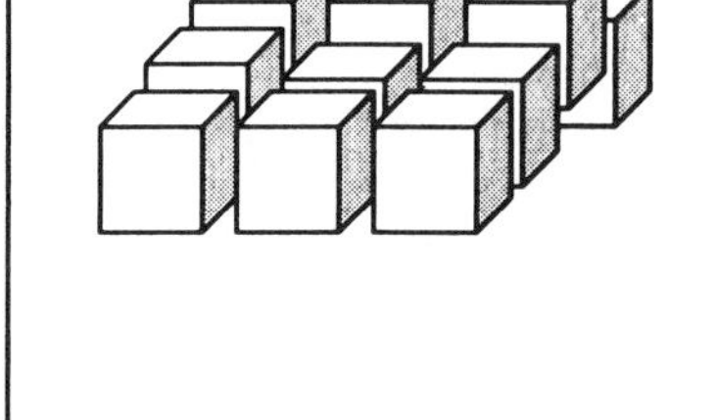

$$\underbrace{\dot{E}_{in} - \dot{E}_{out}}_{\substack{\text{Rate of net energy transfer}\\ \text{by heat, work, and mass}}} = \underbrace{\Delta \dot{E}_{\text{system}}^{\nearrow 0 \text{ (steady)}}}_{\substack{\text{Rate of change in internal, kinetic,}\\ \text{potential, etc. energies}}} = 0$$

$$\dot{E}_{in} = \dot{E}_{out}$$

$$\dot{Q}_{in} + \dot{m}h_1 = \dot{m}h_2 \quad (\text{since } \Delta\text{ke} \cong \Delta\text{pe} \cong 0)$$

$$\dot{Q}_{in} = \dot{Q}_{\text{total}} = \dot{m}C_p(T_2 - T_1)$$

The temperature rise of air is limited to 2.0 K as it flows through the load. Noting that air picks up heat at a rate of 12,276 W, the minimum mass and volume flow rates of air are determined to be

$$\dot{m}_{\text{air}} = \frac{\dot{Q}_{\text{air}}}{(C_p\Delta T)_{\text{air}}} = \frac{12{,}276 \text{ W}}{(1000 \text{ J / kg.°C})(2.0\text{°C})} = 6.14 \text{ kg / s}$$

$$\dot{V}_{\text{air}} = \frac{\dot{m}_{\text{air}}}{\rho_{\text{air}}} = \frac{6.14 \text{ kg / s}}{1.2 \text{ kg / m}^3} = \mathbf{5.12 \text{ m}^3 / \text{s}}$$

Therefore, the fan selected for the banana room must be large enough to circulate air at a rate of 5.12 m^3/s.

4-185 The chilling room of a meat plant with a capacity of 350 beef carcasses is considered. The cooling load and the air flow rate are to be determined.

Assumptions **1** Steady operating conditions exist. **2** Specific heats of beef carcass and air are constant. **3** Air is an ideal gas with constant specific. **4** The kinetic and potential energy changes are negligible.

Properties The density and specific heat of air are given to be 1.28 kg/m^3 and 1.0 kJ/kg·°C. The specific heat of beef carcass is given to be 3.14 kJ/kg·°C.

Analysis (a) The amount of beef mass that needs to be cooled per unit time is

$$\dot{m}_{beef} = (\text{Total beef mass cooled})/(\text{cooling time}) = (350\times 280\text{ kg/carcass})/(12\text{h}\times 3600\text{ s}) = 2.27\text{ kg/s}$$

The product refrigeration load can be viewed as the energy that needs to be removed from the beef carcass as it is cooled from 35 to 16°C at a rate of 2.27 kg/s. Then the energy balance for steadily flowing beef carcasses can be expressed in the rate form as

$$\underbrace{\dot{E}_{in} - \dot{E}_{out}}_{\substack{\text{Rate of net energy transfer}\\ \text{by heat, work, and mass}}} = \underbrace{\Delta\dot{E}_{\text{system}}^{\nearrow 0\text{ (steady)}}}_{\substack{\text{Rate of change in internal, kinetic,}\\ \text{potential, etc. energies}}} = 0$$

$$\dot{E}_{in} = \dot{E}_{out}$$

$$\dot{m}h_1 = \dot{Q}_{out} + \dot{m}h_2 \quad (\text{since } \Delta\text{ke} \cong \Delta\text{pe} \cong 0)$$

$$\dot{Q}_{out} = \dot{Q}_{\text{beef}} = \dot{m}_{\text{beef}} C_p (T_1 - T_2)$$

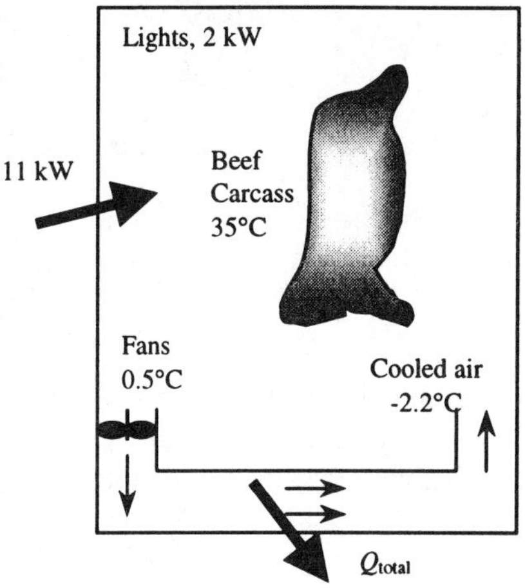

Substituting,

$$\dot{Q}_{beef} = (\dot{m}C_p\Delta T)_{beef} = (2.27\text{ kg/s})(3.14\text{ kJ/kg.°C})(35-16)\text{°C}$$
$$= 135\text{ kW}$$

Then the total refrigeration load of the chilling room becomes

$$\dot{Q}_{\text{total,chilling room}} = \dot{Q}_{beef} + \dot{Q}_{fan} + \dot{Q}_{lights} + \dot{Q}_{heat\ gain} = 135 + 22 + 2 + 11 = \mathbf{170\ kW}$$

(*b*) Heat is transferred to air at the rate determined above, and the temperature of air rises from -2.2°C to 0.5°C as a result. Therefore, the mass flow rate of air can be determined from the energy balance relation above to be

$$\dot{m}_{air} = \frac{\dot{Q}_{air}}{(C_p\Delta T)_{air}} = \frac{170\text{ kW}}{(1.0\text{ kJ/kg.°C})[0.5-(-2.2)\text{°C}]} = 63.0\text{ kg/s}$$

Then the volume flow rate of air becomes

$$\dot{V}_{air} = \frac{\dot{m}_{air}}{\rho_{air}} = \frac{63\text{ kg/s}}{1.28\text{ kg/m}^3} = \mathbf{49.2\ m^3/s}$$

4-186E Broccoli is to be vacuum cooled in an insulated spherical vacuum chamber. The final mass of broccoli after cooling is to be determined.

Assumptions **1** The thermal properties of broccoli are constant. **2** Heat transfer through the walls of the vacuum chamber is negligible. **3** All the air in the chamber is sucked out by the vacuum pump so that there is only vapor at the end.

Properties The specific heat of broccoli near freezing temperatures is 0.921 Btu/lbm.°F (Table A-3E). At the average temperature of (80 + 40)/2 = 60°F, the latent heat of vaporization of water is h_{fg} = 1059.6 Btu/lbm (Table A-4E).

Analysis We take the broccoli as the system. Disregarding the small amount of water that evaporates, the energy balance for this closed system can be expressed as

Vacuum Chamber

Broccoli 75°F

$$\underbrace{E_{in} - E_{out}}_{\substack{\text{Net energy transfer}\\ \text{by heat, work, and mass}}} = \underbrace{\Delta E_{\text{system}}}_{\substack{\text{Change in internal, kinetic,}\\ \text{potential, etc. energies}}}$$

$$-Q_{out} = \Delta U = [mC(T_2 - T_1)]_{\text{broccoli}}$$

The amount of heat transfer to cool 15,000 lbm of broccoli from 75 to 40°F is

$$\begin{aligned} Q_{\text{broccoli,out}} &= [mC(T_1 - T_2)]_{\text{broccoli}} \\ &= (15{,}000\ \text{lbm})(0.921\ \text{Btu/lbm.°F})(80 - 40)\text{°F} \\ &= 552{,}600\ \text{Btu} \end{aligned}$$

Noting that each kg of water in broccoli absorbs 1059.6 Btu of heat as it evaporates, and disregarding any heat gain through the walls of the vacuum chamber, the amount of moisture removed is determined to be

$$Q_{\text{broccoli}} = m_{\text{evap}} h_{fg} \rightarrow m_{\text{evap}} = \frac{Q_{\text{broccoli}}}{h_{fg}} = \frac{552{,}600\ \text{Btu}}{1059.6\ \text{Btu/lbm}} = 522\ \text{lbm}$$

Therefore, the final mass of the wet broccoli is

$$m_{\text{broccoli,final}} = m_{\text{broccoli,initial}} - m_{\text{evap}} = 15{,}000 - 522 = \mathbf{14{,}478\ lbm}$$

Discussion Initially, the wet broccoli contains 15,000×0.02 = 300 lbm of water. Therefore, broccoli loses 222 lbm of additional water from itself during vacuum cooling.

4-187 Cod fish in a polypropylene box is to be frozen in 4 h. The total amount of heat removed from the fish and the box and the average rate of heat removal are to be determined.

Assumptions **1** The specific heats of the box and the fish are constant. **2** The box and the fish are at uniform temperatures at the beginning and at the end of the process. **3** Evaporation from the fish is negligible.

Properties The specific heat of the cod fish is given to be 3.62 kJ/kg.°C above the freezing point of –2.2°C, and 1.89 kJ/kg.°C below the freezing point. The specific heat of the box is given to be 1.9 kJ/kg.°C. The latent heat of fish is given to be 277 kJ/kg.

Analysis (a) We take the fish and its box as the system. The energy balance for this fixed mass can be expressed as

$$\underbrace{E_{in} - E_{out}}_{\text{Net energy transfer by heat, work, and mass}} = \underbrace{\Delta E_{\text{system}}}_{\text{Change in internal, kinetic, potential, etc. energies}}$$

$$-Q_{out} = \Delta U = \Delta U_{\text{fish}} + \Delta U_{\text{box}}$$

$$Q_{out} = -[\Delta U_{\text{below freezing point}} + \Delta U_{\text{during freezing}} + \Delta U_{\text{below freezing point}}]_{\text{fish}} - \Delta U_{\text{box}}$$

or,

$$Q_{out} = m_{\text{fish}}[C_{\text{fresh}}(T_1 - T_{fp}) + h_{\text{latent}} + C_{\text{frozen}}(T_{fp} - T_2)]_{\text{fish}} + [mC(T_1 - T_2)]_{\text{box}}$$

Noting that T_{fp} = -2.2°C and substituting, the amount of heat removed as the cod fish and its box are cooled from 16 to -20°C becomes

$$\begin{aligned} Q_{\text{out}} &= (32\text{ kg})\{(3.62\text{ kJ/kg}\cdot{}^\circ\text{C})[16 - (-2.2)]^\circ\text{C} + 277\text{ kJ/kg} + (1.89\text{ kJ/kg}\cdot{}^\circ\text{C})(-2.2 - (-20)^\circ\text{C}]\\ &\quad + (1.6\text{ kg})(1.9\text{ kJ/kg}\cdot{}^\circ\text{C})[16 - (-20)]^\circ\text{C} = 0\\ &= 12{,}049 + 109\\ &= \mathbf{12{,}158\text{ kJ}} \end{aligned}$$

(b) Noting that 12,158 kJ of heat is removed in 4 h, the average rate of heat removal from the fish and its box is determined to be

$$\dot{Q}_{ave} = \frac{Q_{total}}{\Delta t} = \frac{12{,}158\text{ kJ}}{4\text{ h}} = \mathbf{3040\text{ kJ/h} = 0.844\text{ kW}}$$

Review Problems

4-188 A cylinder is initially filled with saturated R-134a vapor at a specified pressure. The refrigerant is heated both electrically and by heat transfer at constant pressure for 6 min. The electric current is to be determined, and the process is to be shown on a *T-v* diagram.

Assumptions **1** The cylinder is stationary and thus the kinetic and potential energy changes are negligible. **2** The thermal energy stored in the cylinder itself and the wires is negligible. **3** The compression or expansion process is quasi-equilibrium.

Analysis We take the contents of the cylinder as the system. This is a closed system since no mass enters or leaves. The energy balance for this stationary closed system can be expressed as

$$\underbrace{E_{in} - E_{out}}_{\substack{\text{Net energy transfer}\\ \text{by heat, work, and mass}}} = \underbrace{\Delta E_{\text{system}}}_{\substack{\text{Change in internal, kinetic,}\\ \text{potential, etc. energies}}}$$

$$Q_{in} + W_{e,in} - W_{b,out} = \Delta U \quad (\text{since } Q = \text{KE} = \text{PE} = 0)$$

$$Q_{in} + W_{e,in} = m(h_2 - h_1)$$

$$Q_{in} + (VI\Delta t) = m(h_2 - h_1)$$

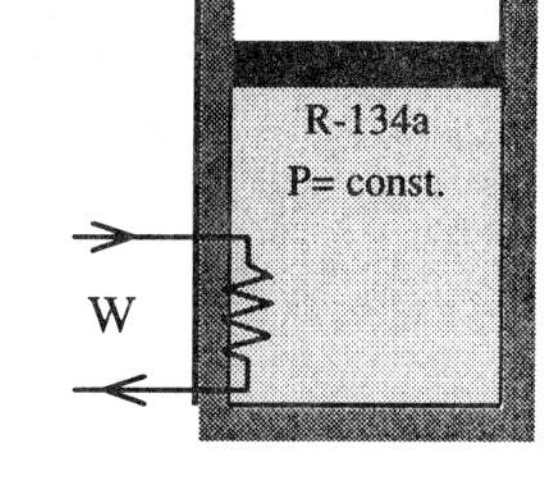

since $\Delta U + W_b = \Delta H$ during a constant pressure quasi-equilibrium process. The properties of R-134a are (Tables A-11 through A-13)

$$\left.\begin{array}{l} P_1 = 200\text{kPa} \\ \text{sat.vapor} \end{array}\right\} h_1 = h_{g@200\text{kPa}} = 241.30\text{kJ/kg}$$

$$\left.\begin{array}{l} P_2 = 200\text{kPa} \\ T_1 = 70^\circ\text{C} \end{array}\right\} h_2 = 314.02\text{kJ/kg}$$

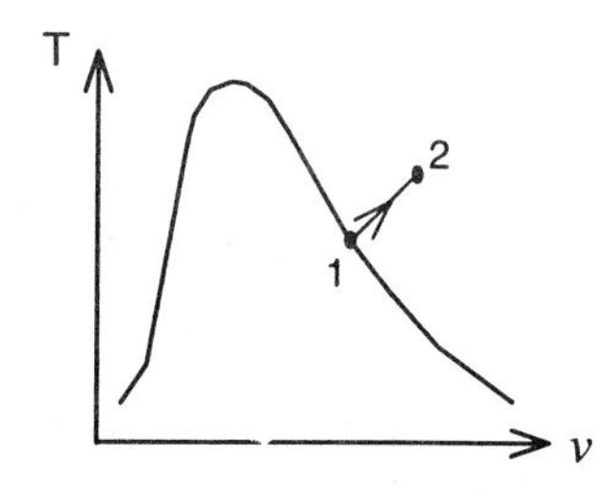

Substituting,

$$250{,}000\text{VA} + (110\text{V})(I)(6 \times 60\text{s}) = (12\text{kg})(314.02 - 241.30)\text{kJ/kg}\left(\frac{1000\text{VA}}{1\text{kJ/s}}\right)$$

$$I = \mathbf{15.72A}$$

4-189 A cylinder is initially filled with saturated liquid-vapor mixture of R-134a at a specified pressure. Heat is transferred to the cylinder until the refrigerant vaporizes completely at constant pressure. The initial volume, the work done, and the total heat transfer are to be determined, and the process is to be shown on a *P-v* diagram.

Assumptions **1** The cylinder is stationary and thus the kinetic and potential energy changes are negligible. **2** The thermal energy stored in the cylinder itself is negligible. **3** The compression or expansion process is quasi-equilibrium.

Analysis (*a*) Using property data from R-134a tables (Tables A-11 through A-13), the initial volume of the refrigerant is determined to be

$$\left.\begin{array}{l} P_1 = 200\text{kPa} \\ x_1 = 0.25 \end{array}\right\} \begin{array}{ll} v_f = 0.0007532, & v_g = 0.0993\text{m}^3/\text{kg} \\ u_f = 36.69, & u_g = 221.43\text{kJ/kg} \end{array}$$

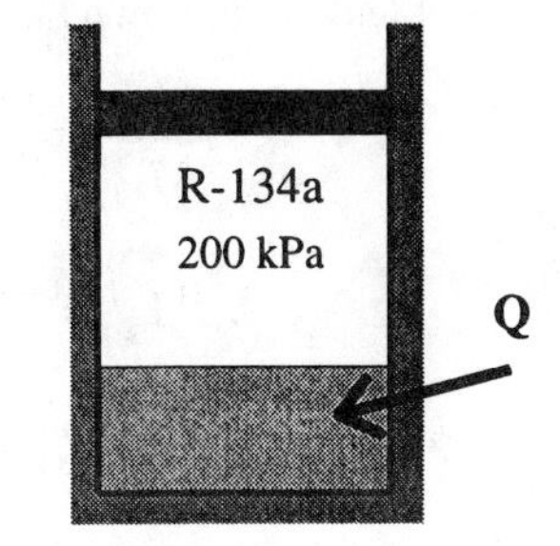

$$v_1 = v_f + x_1 v_{fg} = 0.0007532 + 0.25 \times (0.0993 - 0.0007532) = 0.02539\text{m}^3/\text{kg}$$

$$u_1 = u_f + x_1 u_{fg} = 36.69 + 0.25 \times (221.43 - 36.69) = 82.88\text{kJ/kg}$$

$$V_1 = mv_1 = (0.2\text{kg})(0.02539\text{m}^3/\text{kg}) = \mathbf{0.005078m^3}$$

(*b*) The work done during this constant pressure process is

$$\left.\begin{array}{l} P_2 = 200\text{kPa} \\ sat.vapor \end{array}\right\} \begin{array}{l} v_2 = v_{g@200\text{kPa}} = 0.0993\text{m}^3/\text{kg} \\ u_2 = u_{g@200\text{kPa}} = 221.43\text{kJ/kg} \end{array}$$

$$W_{b,out} = \int_1^2 PdV = P(V_2 - V_1) = mP(v_2 - v_1)$$

$$= (0.2\text{ kg})(200\text{ kPa})(0.0993 - 0.02539)\text{m}^3/\text{kg}\left(\frac{1\text{ kJ}}{1\text{ kPa}\cdot\text{m}^3}\right) = \mathbf{2.96\ kJ}$$

(*c*) We take the contents of the cylinder as the system. This is a closed system since no mass enters or leaves. The energy balance for this stationary closed system can be expressed as

$$\underbrace{E_{in} - E_{out}}_{\substack{\text{Net energy transfer} \\ \text{by heat, work, and mass}}} = \underbrace{\Delta E_{\text{system}}}_{\substack{\text{Change in internal, kinetic,} \\ \text{potential, etc. energies}}}$$

$$Q_{in} - W_{b,out} = \Delta U$$

$$Q_{in} = m(u_2 - u_1) + W_{b,out}$$

Substituting,

$$Q_{in} = (0.2\text{ kg})(221.43 - 82.88)\text{kJ/kg} + 2.96 = \mathbf{30.67\ kJ}$$

4-190 A cylinder is initially filled with helium gas at a specified state. Helium is compressed polytropically to a specified temperature and pressure. The heat transfer during the process is to be determined.

Assumptions **1** Helium is an ideal gas with constant specific heats. **2** The cylinder is stationary and thus the kinetic and potential energy changes are negligible. **3** The thermal energy stored in the cylinder itself is negligible. **4** The compression or expansion process is quasi-equilibrium.

Properties The gas constant of helium is R = 2.0769 kPa.m³/kg.K (Table A-1). Also, C_v = 3.1156 kJ/kg.K (Table A-2).

Analysis The mass of helium and the exponent n are determined to be

$$m = \frac{P_1 V_1}{RT_1} = \frac{(150\text{kPa})(0.5\text{m}^3)}{(2.0769\text{kPa}\cdot\text{m}^3/\text{kg}\cdot\text{K})(293\text{K})} = 0.123\text{kg}$$

$$\frac{P_1 V_1}{RT_1} = \frac{P_2 V_2}{RT_2} \longrightarrow V_2 = \frac{T_2 P_1}{T_1 P_2} V_1 = \frac{413\text{K}}{293\text{K}} \times \frac{150\text{kPa}}{400\text{kPa}} \times 0.5\text{m}^3 = 0.264\text{m}^3$$

$$P_2 V_2^n = P_1 V_1^n \longrightarrow \left(\frac{P_2}{P_1}\right) = \left(\frac{V_1}{V_2}\right)^n \longrightarrow \frac{400}{150} = \left(\frac{0.5}{0.264}\right)^n \longrightarrow n = 1.536$$

Then the boundary work for this polytropic process can be determined from

$$W_{b,in} = -\int_1^2 P dV = -\frac{P_2 V_2 - P_1 V_1}{1-n} = -\frac{mR(T_2 - T_1)}{1-n}$$

$$= -\frac{(0.123\text{kg})(2.0769\text{kJ/kg}\cdot\text{K})(413-293)\text{K}}{1-1.536} = \mathbf{57.2\ kJ}$$

We take the contents of the cylinder as the system. This is a closed system since no mass enters or leaves. Taking the direction of heat transfer to be to the cylinder, the energy balance for this stationary closed system can be expressed as

$$\underbrace{E_{in} - E_{out}}_{\substack{\text{Net energy transfer} \\ \text{by heat, work, and mass}}} = \underbrace{\Delta E_{system}}_{\substack{\text{Change in internal, kinetic,} \\ \text{potential, etc. energies}}}$$

$$Q_{in} + W_{b,in} = \Delta U = m(u_2 - u_1)$$

$$Q_{in} = m(u_2 - u_1) - W_{b,in}$$

$$= mC_v(T_2 - T_1) - W_{b,in}$$

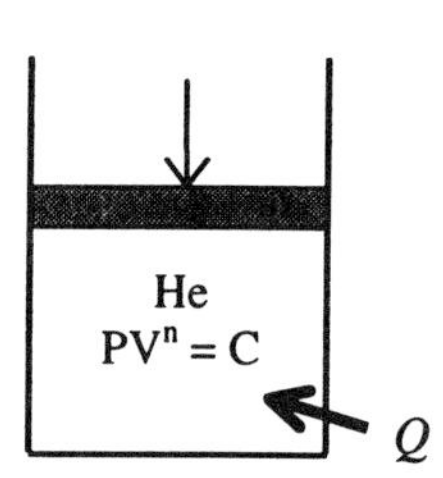

Substituting,

Q_{in} = (0.123 kg)(3.1156 kJ/kg· K)(413 - 293)K - (57.2 kJ) = **-11.2 kJ**

The negative sign indicates that heat is lost from the system.

4-191 A cylinder and a rigid tank initially contain the same amount of an ideal gas at the same state. The temperature of both systems is to be raised by the same amount. The amount of extra heat that must be transferred to the cylinder is to be determined.

Analysis In the absence of any work interactions, other than the boundary work, the ΔH and ΔU represent the heat transfer for ideal gases for constant pressure and constant volume processes, respectively. Thus the extra heat that must be supplied to the air maintained at constant pressure is

$$Q_{\text{in, extra}} = \Delta H - \Delta U = mC_p\Delta T - mC_v\Delta T = m(C_p - C_v)\Delta T = mR\Delta T$$

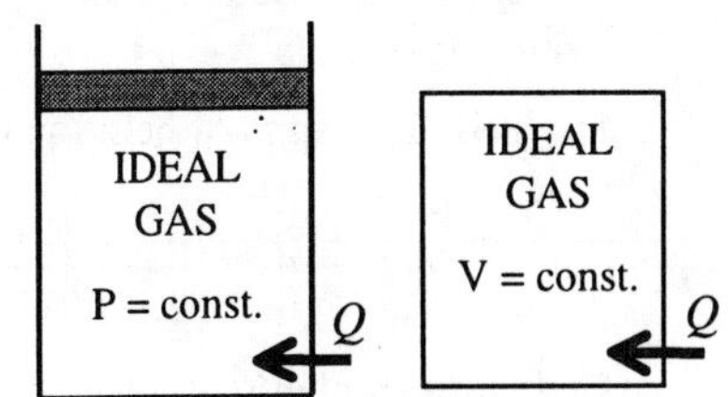

where

$$R = \frac{R_u}{M} = \frac{8.314\ \text{kJ/kmol}\cdot\text{K}}{25\ \text{kg/kmol}} = 0.3326\ \text{kJ/kg}\cdot\text{K}$$

Substituting,

$Q_{\text{in, extra}} = (12\ \text{kg})(0.3326\ \text{kJ/kg}\cdot\ \text{K})(15\ \text{K}) =$ **59.9 kJ**

4-192 The heating of a passive solar house at night is to be assisted by solar heated water. The length of time that the electric heating system would run that night with or without solar heating are to be determined.

Assumptions **1** Water is an incompressible substance with constant specific heats. **2** The energy stored in the glass containers themselves is negligible relative to the energy stored in water. **3** The house is maintained at 22°C at all times.

Properties The density and specific heat of water at room temperature are ρ = 1 kg/L and C = 4.18 kJ/kg· °C (Table A-3).

Analysis The total mass of water is

$$m_w = \rho V = (1\text{kg/L})(50 \times 20\text{L}) = 1000\text{kg}$$

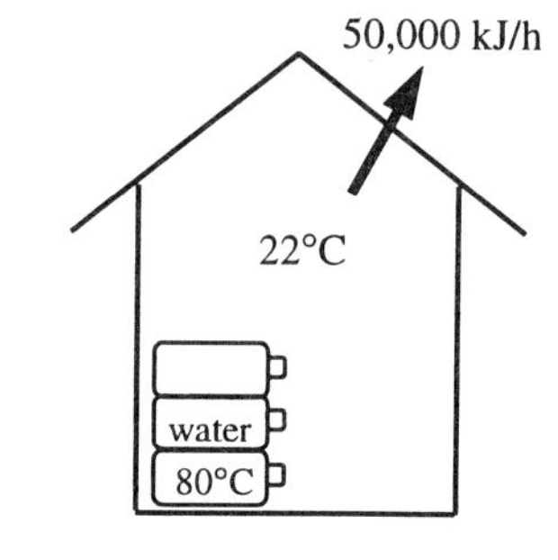

Taking the contents of the house, including the water as our system, the energy balance relation can be written as

$$\underbrace{E_{in} - E_{out}}_{\substack{\text{Net energy transfer}\\ \text{by heat, work, and mass}}} = \underbrace{\Delta E_{\text{system}}}_{\substack{\text{Change in internal, kinetic,}\\ \text{potential, etc. energies}}}$$

$$W_{e,in} - Q_{out} = \Delta U = (\Delta U)_{\text{water}} + (\Delta U)_{\text{air}}$$

$$= (\Delta U)_{\text{water}}$$

$$= mC(T_2 - T_1)_{\text{water}}$$

or,

$$\dot{W}_{e,in}\Delta t - Q_{out} = [mC(T_2 - T_1)]_{\text{water}}$$

Substituting,

$$(15 \text{ kJ/s})\Delta t \text{ - } (50{,}000 \text{ kJ/h})(10 \text{ h}) = (1000 \text{ kg})(4.18 \text{ kJ/kg· °C})(22 \text{ - } 80)\text{°C}$$

It gives

$$\Delta t = 17{,}170 \text{ s} = \mathbf{4.77\ h}$$

(b) If the house incorporated no solar heating, the energy balance relation above would simplify further to

$$\dot{W}_{e,in}\Delta t - Q_{out} = 0$$

Substituting,

$$(15 \text{ kJ/s})\Delta t \text{ - } (50{,}000 \text{ kJ/h})(10 \text{ h}) = 0$$

It gives

$$\Delta t = 33{,}333 \text{ s} = \mathbf{9.26\ h}$$

4-193 An electric resistance heater is immersed in water. The time it will take for the electric heater to raise the water temperature to a specified temperature is to be determined.

Assumptions **1** Water is an incompressible substance with constant specific heats. **2** The energy stored in the container itself and the heater is negligible. **3** Heat loss from the container is negligible.

Properties The density and specific heat of water at room temperature are ρ = 1 kg/L and C = 4.18 kJ/kg· °C (Table A-3).

Analysis Taking the water in the container as the system, the energy balance can be expressed as

$$\underbrace{E_{in} - E_{out}}_{\substack{\text{Net energy transfer} \\ \text{by heat, work, and mass}}} = \underbrace{\Delta E_{\text{system}}}_{\substack{\text{Change in internal, kinetic,} \\ \text{potential, etc. energies}}}$$

$$W_{e,in} = (\Delta U)_{\text{water}}$$

$$\dot{W}_{e,in} \Delta t = mC(T_2 - T_1)_{\text{water}}$$

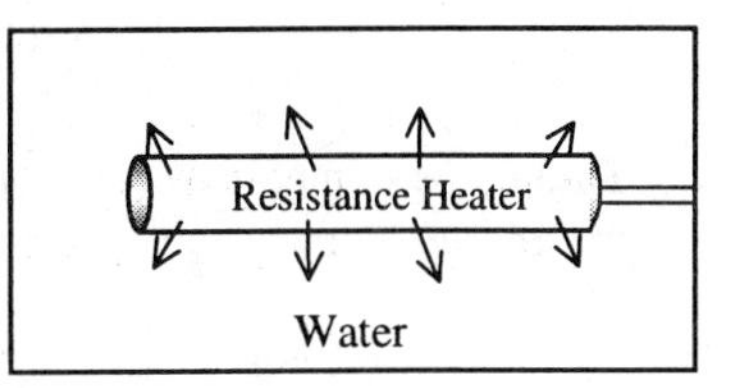

Substituting,

$$(800 \text{ J/s})\Delta t = (40 \text{ kg})(4180 \text{ J/kg} \cdot {}^\circ\text{C})(80 - 20){}^\circ\text{C}$$

Solving for Δt gives

$$\Delta t = \mathbf{12{,}540 \text{ s} = 209.0 \text{ min} = 3.483 \text{ h}}$$

4-194 One ton of liquid water at 80°C is brought into a room. The final equilibrium temperature in the room is to be determined.

Assumptions **1** The room is well insulated and well sealed. **2** The thermal properties of water and air are constant.

Properties The gas constant of air is R = 0.287 kPa.m^3/kg.K (Table A-1). The specific heat of water at room temperature is C = 4.18 kJ/kg·°C (Table A-3).

Analysis The volume and the mass of the air in the room are

$$V = 4\text{x}5\text{x}6 = 120 \text{ m}^3$$

$$m_{air} = \frac{P_1V_1}{RT_1} = \frac{(100\text{kPa})(120\text{m}^3)}{(0.2870\text{kPa}\cdot\text{m}^3/\text{kg}\cdot\text{K})(295\text{K})} = 141.7\text{kg}$$

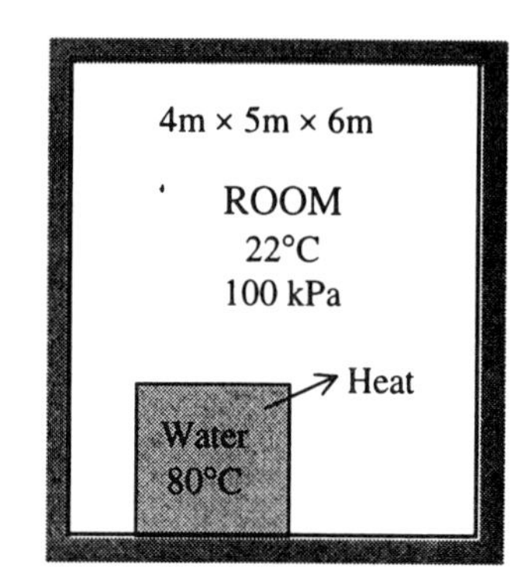

Taking the contents of the room, including the water, as our system, the energy balance can be written as

$$\underbrace{E_{in} - E_{out}}_{\text{Net energy transfer by heat, work, and mass}} = \underbrace{\Delta E_{\text{system}}}_{\text{Change in internal, kinetic, potential, etc. energies}} \rightarrow 0 = \Delta U = (\Delta U)_{\text{water}} + (\Delta U)_{\text{air}}$$

or $$[mC(T_2 - T_1)]_{\text{water}} + [mC_v(T_2 - T_1)]_{\text{air}} = 0$$

Substituting,

$$(1000\text{ kg})(4.180\text{ kJ/kg}\cdot^\circ\text{C})(T_f - 80)^\circ\text{C} + (147.7\text{ kg})(0.718\text{ kJ/kg}\cdot^\circ\text{C})(T_f - 22)^\circ\text{C} = 0$$

It gives T_f = **78.6°C**

where T_f is the final equilibrium temperature in the room.

4-195 A room is to be heated by 1 ton of hot water contained in a tank placed in the room. The minimum initial temperature of the water is to be determined if it to meet the heating requirements of this room for a 24-h period.

Assumptions **1** Water is an incompressible substance with constant specific heats. **2** Air is an ideal gas with constant specific heats. **3** The energy stored in the container itself is negligible relative to the energy stored in water. **4** The room is maintained at 20°C at all times. **5** The hot water is to meet the heating requirements of this room for a 24-h period.

Properties The specific heat of water at room temperature is $C = 4.18$ kJ/kg·°C (Table A-3).

Analysis Heat loss from the room during a 24-h period is

$$Q_{\text{loss}} = (10{,}000 \text{ kJ/h})(24 \text{ h}) = 240{,}000 \text{ kJ}$$

Taking the contents of the room, including the water, as our system, the energy balance can be written as

$$\underbrace{E_{in} - E_{out}}_{\substack{\text{Net energy transfer}\\ \text{by heat, work, and mass}}} = \underbrace{\Delta E_{\text{system}}}_{\substack{\text{Change in internal, kinetic,}\\ \text{potential, etc. energies}}} \rightarrow -Q_{\text{out}} = \Delta U = (\Delta U)_{\text{water}} + (\Delta U)_{\text{air}}^{\nearrow 0}$$

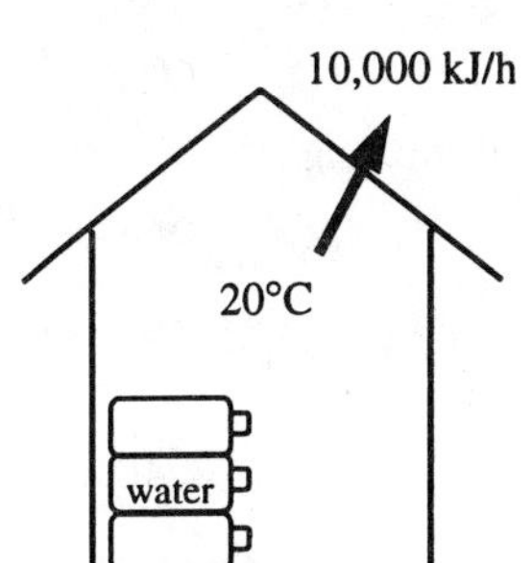

or

$$-Q_{\text{out}} = [mC(T_2 - T_1)]_{\text{water}}$$

Substituting,

$$-240{,}000 \text{ kJ} = (1000 \text{ kg})(4.18 \text{ kJ/kg·°C})(20 - T_1)$$

It gives

$$T_1 = \mathbf{77.4°C}$$

where T_1 is the temperature of the water when it is first brought into the room.

4-196 A sample of a food is burned in a bomb calorimeter, and the water temperature rises by 3.2°C when equilibrium is established. The energy content of the food is to be determined. √

Assumptions **1** Water is an incompressible substance with constant specific heats. **2** Air is an ideal gas with constant specific heats. **3** The energy stored in the reaction chamber is negligible relative to the energy stored in water. **4** The energy supplied by the mixer is negligible.

Properties The specific heat of water at room temperature is $C = 4.18$ kJ/kg· °C (Table A-3). The constant volume specific heat of air at room temperature is $C_v = 0.718$ kJ/kg· °C (Table A-2).

Analysis The chemical energy released during the combustion of the sample is transferred to the water as heat. Therefore, disregarding the change in the sensible energy of the reaction chamber, the energy content of the food is simply the heat transferred to the water. Taking the water as our system, the energy balance can be written as

$$\underbrace{E_{in} - E_{out}}_{\substack{\text{Net energy transfer} \\ \text{by heat, work, and mass}}} = \underbrace{\Delta E_{\text{system}}}_{\substack{\text{Change in internal, kinetic,} \\ \text{potential, etc. energies}}} \rightarrow Q_{in} = \Delta U$$

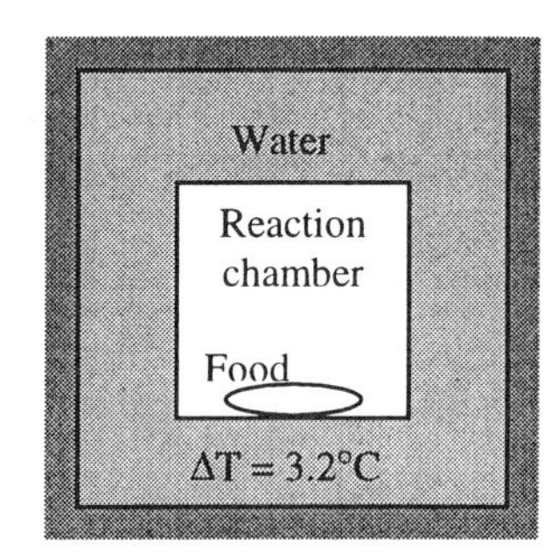

or

$$Q_{in} = (\Delta U)_{\text{water}} = [mC(T_2 - T_1)]_{\text{water}}$$

Substituting ,

$$Q_{\text{in}} = (3\text{ kg})(4.18\text{ kJ/kg}\cdot\text{°C})(3.2\text{°C}) = 40.13\text{ kJ}$$

for a 2-g sample. Then the energy content of the food per unit mass is

$$\frac{40.13\text{ kJ}}{2\text{ g}}\left(\frac{1000\text{ g}}{1\text{ kg}}\right) = \mathbf{20{,}060\ \ kJ/kg}$$

To make a rough estimate of the error involved in neglecting the thermal energy stored in the reaction chamber, we treat the entire mass within the chamber as air and determine the change in sensible internal energy:

$$(\Delta U)_{\text{chamber}} = [mC_v(T_2 - T_1)]_{\text{chamber}} = (0.102\text{ kg})(0.718\text{ kJ/kg}\cdot\text{°C})(3.2\text{°C}) = 0.23\text{ kJ}$$

which is less than 1% of the internal energy change of water. Therefore, it is reasonable to disregard the change in the sensible energy content of the reaction chamber in the analysis.

4-197 A man drinks one liter of cold water at 3°C in an effort to cool down. The drop in the average body temperature of the person under the influence of this cold water is to be determined.

Assumptions **1** Thermal properties of the body and water are constant. **2** The effect of metabolic heat generation and the heat loss from the body during that time period are negligible.

Properties The density of water is very nearly 1 kg/L, and the specific heat of water at room temperature is $C = 4.18$ kJ/kg· °C (Table A-3). The average specific heat of human body is given to be 3.6 kJ/kg.°C.

Analysis. The mass of the water is

$$m_w = \rho V = (1 \text{ kg/L})(1 \text{ L}) = 1 \text{ kg}$$

We take the man and the water as our system, and disregard any heat and mass transfer and chemical reactions. Of course these assumptions may be acceptable only for very short time periods, such as the time it takes to drink the water. Then the energy balance can be written as

$$\underbrace{E_{in} - E_{out}}_{\substack{\text{Net energy transfer} \\ \text{by heat, work, and mass}}} = \underbrace{\Delta E_{\text{system}}}_{\substack{\text{Change in internal, kinetic,} \\ \text{potential, etc. energies}}}$$

$$0 = \Delta U$$

$$0 = \Delta U_{\text{body}} + \Delta U_{\text{water}}$$

$$[mC_v(T_2 - T_1)]_{\text{body}} + [mC_v(T_2 - T_1)]_{\text{water}} = 0$$

Substituting,

$$(68 \text{ kg})(3.6 \text{ kJ/kg}\cdot^\circ\text{C})(T_f - 39)^\circ\text{C} + (1 \text{ kg})(4.18 \text{ kJ/kg}\cdot^\circ\text{C})(T_f - 3)^\circ\text{C} = 0$$

It gives

$$T_f = \mathbf{38.4°C}$$

Therefore, the average body temperature of this person should drop about half a degree Celsius.

4-198 A 0.2-L glass of water at 20°C is to be cooled with ice to 5°C. The amount of ice or cold water that needs to be added to the water is to be determined.

Assumptions **1** Thermal properties of the ice and water are constant. **2** Heat transfer to the glass is negligible. **3** There is no stirring by hand or a mechanical device (it will add energy).

Properties The density of water is 1 kg/L, and the specific heat of water at room temperature is C = 4.18 kJ/kg· °C (Table A-3). The specific heat of ice at about 0°C is C = 2.11 kJ/kg· °C (Table A-3). The melting temperature and the heat of fusion of ice at 1 atm are 0°C and 333.7 kJ/kg,.

Analysis (*a*) The mass of the water is

$$m_w = \rho V = (1\text{ kg/L})(0.2\text{ L}) = 0.2\text{kg}$$

We take the ice and the water as our system, and disregard any heat and mass transfer. This is a reasonable assumption since the time period of the process is very short. Then the energy balance can be written as

$$\underbrace{E_{in} - E_{out}}_{\substack{\text{Net energy transfer}\\ \text{by heat, work, and mass}}} = \underbrace{\Delta E_{\text{system}}}_{\substack{\text{Change in internal, kinetic,}\\ \text{potential, etc. energies}}}$$

$$0 = \Delta U$$

$$0 = \Delta U_{\text{ice}} + \Delta U_{\text{water}}$$

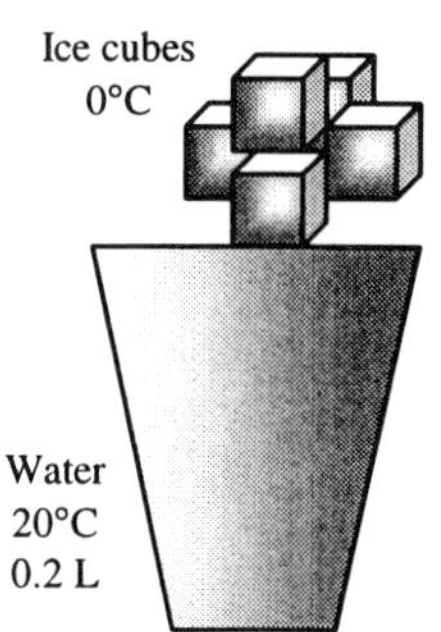

$$[mC(0^\circ\text{C} - T_1)_{\text{solid}} + mh_{if} + mC(T_2 - 0^\circ\text{C})_{\text{liquid}}]_{\text{ice}} + [mC(T_2 - T_1)]_{\text{water}} = 0$$

Noting that $T_{1,\text{ ice}}$ = 0°C and T_2 = 5°C and substituting gives

$$m[0 + 333.7\text{ kJ/kg} + (4.18\text{ kJ/kg}\cdot\ ^\circ\text{C})(4\text{-}0)^\circ\text{C}] + (0.2\text{ kg})(4.18\text{ kJ/kg}\cdot\ ^\circ\text{C})(4\text{-}20)^\circ\text{C} = 0$$

$$m = 0.0354\text{ kg} = \mathbf{35.4\ g}$$

(*b*) When $T_{1,\text{ ice}}$ = -8°C instead of 0°C, substituting gives

$$m[(2.11\text{ kJ/kg}\cdot\ ^\circ\text{C})[0\text{-}(\text{-}8)]^\circ\text{C} + 333.7\text{ kJ/kg} + (4.18\text{ kJ/kg}\cdot\ ^\circ\text{C})(4\text{-}0)^\circ\text{C}]$$
$$+ (0.2\text{ kg})(4.18\text{ kJ/kg}\cdot\ ^\circ\text{C})(4\text{-}20)^\circ\text{C} = 0$$

$$m = 0.0338\text{ kg} = \mathbf{33.8\ g}$$

Cooling with cold water can be handled the same way. All we need to do is replace the terms for ice by a term for cold water at 0°C:

$$(\Delta U)_{\text{coldwater}} + (\Delta U)_{\text{water}} = 0$$

$$[mC(T_2 - T_1)]_{\text{coldwater}} + [mC(T_2 - T_1)]_{\text{water}} = 0$$

Substituting,

$$[m_{\text{cold water}}(4.18\text{ kJ/kg}\cdot\ ^\circ\text{C})(5 - 0)^\circ\text{C}] + (0.2\text{ kg})(4.18\text{ kJ/kg}\cdot\ ^\circ\text{C})(4\text{-}20)^\circ\text{C} = 0$$

It gives $m = 0.6\text{ kg} = \mathbf{600\ g}$

Discussion Note that this is 17 times the amount of ice needed, and it explains why we use ice instead of water to cool drinks. Also, the temperature of ice does not seem to make a significant difference.

4-199 EES solution of this (and other comprehensive problems designated with the *computer icon*) is available to instructors at the *Instructor Manual* section of the *Online Learning Center* (OLC) at www.mhhe.com/cengel-boles. See the Preface for access information.

4-200 A 1- ton (1000 kg) of water is to be cooled in a tank by pouring ice into it. The final equilibrium temperature in the tank is to be determined.

Assumptions **1** Thermal properties of the ice and water are constant. **2** Heat transfer to the water tank is negligible. **3** There is no stirring by hand or a mechanical device (it will add energy).

Properties The specific heat of water at room temperature is $C = 4.18$ kJ/kg· °C, and the specific heat of ice at about 0°C is $C = 2.11$ kJ/kg· °C (Table A-3). The melting temperature and the heat of fusion of ice at 1 atm are 0°C and 333.7 kJ/kg..

Analysis We take the ice and the water as our system, and disregard any heat transfer between the system and the surroundings. Then the energy balance for this process can be written as

$$\underbrace{E_{in} - E_{out}}_{\text{Net energy transfer by heat, work, and mass}} = \underbrace{\Delta E_{\text{system}}}_{\text{Change in internal, kinetic, potential, etc. energies}}$$

$$0 = \Delta U$$

$$0 = \Delta U_{\text{ice}} + \Delta U_{\text{water}}$$

$$[mC(0^\circ\text{C} - T_1)_{\text{solid}} + mh_{if} + mC(T_2 - 0^\circ\text{C})_{\text{liquid}}]_{\text{ice}} + [mC(T_2 - T_1)]_{\text{water}} = 0$$

Substituting,

$$(80\text{ kg})\{(2.11\text{ kJ/kg}\cdot^\circ\text{C})[0-(-5)]^\circ\text{C} + 333.7\text{ kJ/kg} + (4.18\text{ kJ/kg}\cdot^\circ\text{C})(T_2 - 0)^\circ\text{C}\}$$
$$+(1000\text{ kg})(4.18\text{ kJ/kg}\cdot^\circ\text{C})(T_2 - 20)^\circ\text{C} = 0$$

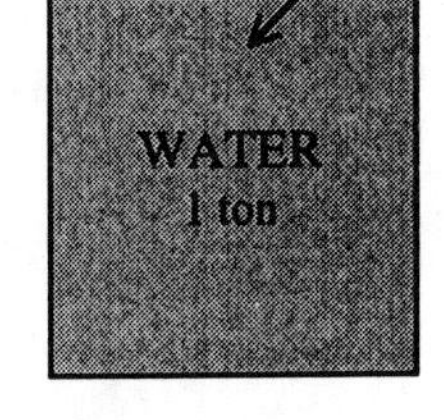

It gives $\qquad T_2 =$ **12.4°C**

which is the final equilibrium temperature in the tank.

4-201 An insulated cylinder initially contains a saturated liquid-vapor mixture of water at a specified temperature. The entire vapor in the cylinder is to be condensed isothermally by adding ice inside the cylinder. The amount of ice that needs to be added is to be determined.

Assumptions **1** Thermal properties of the ice are constant. **2** The cylinder is well-insulated and thus heat transfer is negligible. **3** There is no stirring by hand or a mechanical device (it will add energy).

Properties The specific heat of ice at about 0°C is C = 2.11 kJ/kg· °C (Table A-3). The melting temperature and the heat of fusion of ice at 1 atm are given to be 0°C and 333.7 kJ/kg.

Analysis (*a*) We take the contents of the cylinder (ice and saturated water) as our system, which is a closed system. Noting that the temperature and thus the pressure remains constant during this phase change process and thus $W_b + \Delta U = \Delta H$, the energy balance for this system can be written as

$$\underbrace{E_{in} - E_{out}}_{\text{Net energy transfer by heat, work, and mass}} = \underbrace{\Delta E_{\text{system}}}_{\text{Change in internal, kinetic, potential, etc. energies}}$$

$$W_{b,in} = \Delta U \rightarrow \Delta H = 0$$

$$\Delta H_{\text{ice}} + \Delta H_{\text{water}} = 0$$

$$[mC(0^\circ\text{C} - T_1)_{\text{solid}} + mh_{if} + mC(T_2 - 0^\circ\text{C})_{\text{liquid}}]_{\text{ice}} + [m(h_2 - h_1)]_{\text{water}} = 0$$

The properties of water at 100°C are (Table A-4)

$$v_f = 0.001044, \quad v_g = 1.6729 \text{ m}^3/\text{kg}$$
$$h_f = 419.04, \quad h_{fg} = 2257.0 \text{ kJ.kg}$$

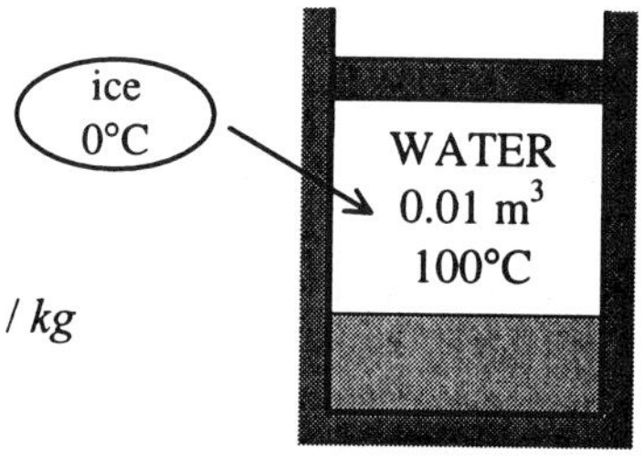

Then,

$$v_1 = v_f + x_1 v_{fg} = 0.001044 + 0.2 \times (1.6729 - 0.001044) = 0.3354 m^3 / kg$$
$$h_1 = h_f + x_1 h_{fg} = 419.04 + 0.2 \times 2257.0 = 870.4 kJ / kg$$
$$h_2 = h_{f@100°C} = 419.04 \text{ kJ/kg}$$

$$m_{steam} = \frac{V_1}{v_1} = \frac{0.01\, m^3}{0.3354\, m^3 / kg} = 0.0298\ kg$$

ice
0°C
WATER
0.01 m³
100°C

Noting that $T_{1,\text{ ice}}$ = 0°C and T_2 = 100°C and substituting gives

$$m[0 + 333.7 \text{ kJ/kg} + (4.18 \text{ kJ/kg·°C})(100\text{-}0)°\text{C}] + (0.0298 \text{ kg})(419.04 - 870.4) \text{ kJ/kg} = 0$$

m = 0.0179 kg = **17.9 g ice**

4-202 The cylinder of a steam engine initially contains saturated vapor of water at 100 kPa. The cylinder is cooled by pouring cold water outside of it, and some of the steam inside condenses. If the piston is stuck at its initial position, the friction force acting on the piston and the amount of heat transfer are to be determined.

Assumptions The device is air-tight so that no air leaks into the cylinder as the pressure drops.

Analysis We take the contents of the cylinder (the saturated liquid-vapor mixture) as the system, which is a closed system. Noting that the volume remains constant during this phase change process, the energy balance for this system can be expressed as

$$\underbrace{E_{in} - E_{out}}_{\substack{\text{Net energy transfer}\\ \text{by heat, work, and mass}}} = \underbrace{\Delta E_{\text{system}}}_{\substack{\text{Change in internal, kinetic,}\\ \text{potential, etc. energies}}}$$

$$-Q_{out} = \Delta U = m(u_2 - u_1)$$

The saturation properties of water at 100 kPa and at 30°C are (Tables A-4 and A-5)

$$P_1 = 100 \text{ kPa} \longrightarrow \begin{array}{l} v_f = 0.001043 \text{ m}^3/\text{kg}, \; v_g = 1.6940 \text{ m}^3/\text{kg} \\ u_f = 417.36 \text{ kJ/kg}, \quad u_g = 2506.1 \text{ kJ/kg} \end{array}$$

$$T_2 = 30°\text{C} \longrightarrow \begin{array}{l} v_f = 0.001004 \text{ m}^3/\text{kg}, \; v_g = 32.89 \text{ m}^3/\text{kg} \\ u_f = 125.78 \text{ kJ/kg}, \quad u_{fg} = 2290.8 \text{ kJ/kg} \\ P_{sat} = 4.246 \text{ kPa} \end{array}$$

Then,

$$P_2 = P_{sat\,@\,30°\text{C}} = 4.246 \text{ kPa}$$

$$v_1 = v_{g\,@\,100\text{ kPa}} = 1.694 \text{ m}^3/\text{kg}$$

$$u_1 = u_{g\,@\,100\text{ kPa}} = 2506.1 \text{ kJ/kg}$$

Cold water

Steam
0.05 m^3
100 kPa

and

$$m = \frac{V_1}{v_1} = \frac{0.05\text{m}^3}{1.6940\text{m}^3/\text{kg}} = 0.0295\text{kg}$$

$$v_2 = v_1 \longrightarrow x_2 = \frac{v_2 - v_f}{v_{fg}} = \frac{1.694 - 0.001}{32.89 - 0.001} = 0.05148$$

$$u_2 = u_f + x_2 u_{fg} = 125.78 + 0.05148 \times 2290.8 = 243.7\text{kJ/kg}$$

The friction force that develops at the piston-cylinder interface balances the force acting on the piston, and is equal to

$$F = A(P_1 - P_2) = (0.1 \text{ m}^3)(100 - 4.246)\text{kPa}\left(\frac{1000 \text{ N/m}^2}{1 \text{ kPa}}\right) = \mathbf{9575 \text{ N}}$$

The heat transfer is determined from the energy balance to be

$$\begin{aligned} Q_{out} &= m(u_1 - u_2) \\ &= (0.0295 \text{ kg})(2506.1 - 243.7)\text{kJ/kg} \\ &= \mathbf{66.7 \text{ kJ}} \end{aligned}$$

4-203 Water is boiled at sea level (1 atm pressure) in a coffee maker, and half of the water evaporates in 25 min. The power rating of the electric heating element and the time it takes to heat the cold water to the boiling temperature are to be determined. √

Assumptions **1** The electric power consumption by the heater is constant. **2** Heat losses from the coffee maker are negligible.

Properties The enthalpy of vaporization of water at the saturation temperature of 100°C is h_{fg} = 2257 kJ/kg (Table A-4). At an average temperature of (100+18)/2 = 59°C, the specific heat of water is C = 4.18 kJ/kg.°C, and the density is about 1 kg/L (Table A-3).

Analysis The density of water at room temperature is very nearly 1 kg/L, and thus the mass of 1 L water at 18°C is nearly 1 kg. Noting that the enthalpy of vaporization represents the amount of energy needed to vaporize a liquid at a specified temperature, the amount of electrical energy needed to vaporize 0.5 kg of water in 25 min is

$$W_e = \dot{W}_e \Delta t = mh_{fg} \rightarrow \dot{W}_e = \frac{mh_{fg}}{\Delta t} = \frac{(0.5\text{ kg})(2257\text{ kJ/kg})}{(25\times 60\text{ s})} = \mathbf{0.752\text{ kW}}$$

Therefore, the electric heater consumes (and transfers to water) 0.752 kW of electric power.

Noting that the specific heat of water at the average temperature of (18+100)/2 = 59°C is C = 4.18 kJ/kg·°C, the time it takes for the entire water to be heated from 18°C to 100°C is determined to be

$$W_e = \dot{W}_e \Delta t = mC\Delta T \rightarrow \Delta t = \frac{mC\Delta T}{\dot{W}_e} = \frac{(1\text{ kg})(4.18\text{ kJ/kg}\cdot{}^\circ\text{C})(100-18)^\circ\text{C}}{0.752\text{ kJ/s}} = 456\text{ s} = \mathbf{7.60\text{ min}}$$

Discussion We can also solve this problem using v_f data (instead of density), and h_f data instead of specific heat. At 100 °C, we have v_f = 0.001044 m^3/kg and h_f = 419.04 kJ/kg. At 18°C, we have h_f = 75.57 kJ/kg (Table A-4). The two results will be practically the same.

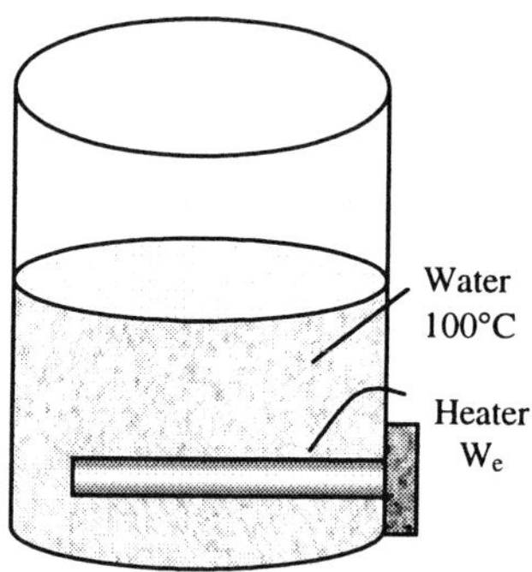

4-204 Water is boiled at a specified temperature by hot gases flowing through a stainless steel pipe submerged in water. The rate of evaporation of is to be determined.

Assumptions **1** Steady operating conditions exist. **2** Heat losses from the outer surfaces of the boiler are negligible.

Properties The enthalpy of vaporization of water at 150°C is h_{fg} = 2114.3 kJ/kg (Table A-4).

Analysis The rate of heat transfer to water is given to be 74 kJ/s. Noting that the enthalpy of vaporization represents the amount of energy needed to vaporize a unit mass of a liquid at a specified temperature, the rate of evaporation of water is determined to be

$$\dot{m}_{\text{evaporation}} = \frac{\dot{Q}_{\text{boiling}}}{h_{fg}} = \frac{74\text{ kJ/s}}{2114.3\text{kJ/kg}} = \mathbf{0.0350\text{ kg/s}}$$

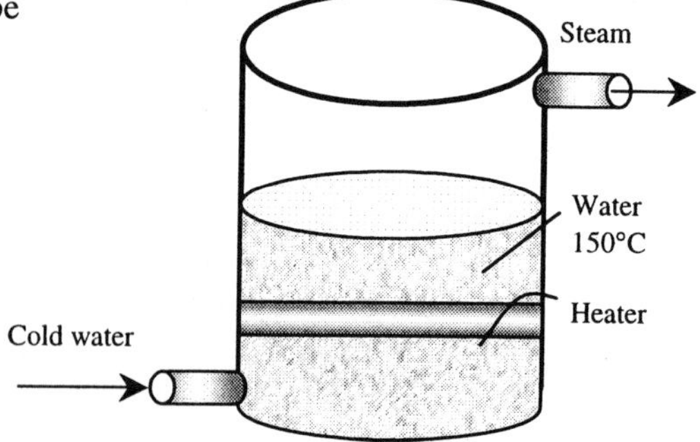

4-205 Cold water enters a steam generator at 20°C, and leaves as saturated vapor at T_{sat} = 100°C. The fraction of heat used to preheat the liquid water from 20°C to saturation temperature of 100°C is to be determined.

Assumptions **1** Steady operating conditions exist. **2** Heat losses from the steam generator are negligible. **3** The specific heat of water is constant at the average temperature.

Properties The heat of vaporization of water at 100°C is h_{fg} = 2257 kJ/kg (Table A-4), and the specific heat of liquid water at the average temperature of (20+100)/2 = 60°C is C = 4.18 kJ/kg.°C (Table A-3).

Analysis The heat of vaporization of water represents the amount of heat needed to vaporize a unit mass of liquid at a specified temperature. Using the average specific heat, the amount of heat transfer needed to preheat a unit mass of water from 20°C to 100°C is

$$q_{\text{preheating}} = C\Delta T = (4.18\ \text{kJ/kg·°C})(100-20)\text{°C} = 334.4\ \text{kJ/kg}$$

and

$$q_{\text{total}} = q_{\text{boiling}} + q_{\text{preheating}} = 2257 + 334.4 = 2591.4\ \text{kJ/kg}$$

Therefore, the fraction of heat used to preheat the water is

$$\text{Fraction to preheat} = \frac{q_{\text{preheating}}}{q_{\text{total}}} = \frac{334.4}{2591.4} = \mathbf{0.129}\ \text{(or } \mathbf{12.9\%}\text{)}$$

Steam

Water 100°C

Heater

Cold water 20°C

4-206 Cold water enters a steam generator at 20°C and is boiled, and leaves as saturated vapor at boiler pressure. The boiler pressure at which the amount of heat needed to preheat the water to saturation temperature that is equal to the heat of vaporization is to be determined.

Assumptions Heat losses from the steam generator are negligible.

Properties The enthalpy of liquid water at 20°C is 83.96 kJ/kg. Other properties needed to solve this problem are the heat of vaporization h_{fg} and the enthalpy of saturated liquid at the specified temperatures, and they can be obtained from Table A-4.

Analysis The heat of vaporization of water represents the amount of heat needed to vaporize a unit mass of liquid at a specified temperature, and Δh represents the amount of heat needed to preheat a unit mass of water from 20°C to the saturation temperature. Therefore,

$$q_{\text{preheating}} = q_{\text{boiling}}$$

$$(h_{f@T_{sat}} - h_{f@20°C}) = h_{fg@T_{sat}}$$

$$h_{f@T_{sat}} - 83.96\ \text{kJ/kg} = h_{fg@T_{sat}} \quad \rightarrow \quad h_{f@T_{sat}} - h_{fg@T_{sat}} = 83.96\ \text{kJ/kg}$$

The solution of this problem requires choosing a boiling temperature, reading h_f and h_{fg} at that temperature, and substituting the values into the relation above to see if it is satisfied. By trial and error, (Table A-4)

At 310°C: $h_{f@T_{sat}} - h_{fg@T_{sat}}$ = 1401.3 – 1326 = 75.3 kJ/kg

At 315°C: $h_{f@T_{sat}} - h_{fg@T_{sat}}$ = 1431.0 – 1283.5 = 147.5 kJ/kg

The temperature that satisfies this condition is determined from the two values above by interpolation to be 310.6°C. The saturation pressure corresponding to this temperature is **9.94 MPa.**

Steam

Water **P**

Heater

Cold water 20°C

4-207 Saturated steam at 1 atm pressure and thus at a saturation temperature of T_{sat} = 100°C condenses on a vertical plate maintained at 90°C by circulating cooling water through the other side. The rate of condensation of steam is to be determined.

Assumptions **1** Steady operating conditions exist. **2** The steam condenses and the condensate drips off at 100°C. (In reality, the condensate temperature will be between 90 and 100, and the cooling of the condensate a few °C should be considered if better accuracy is desired).

Properties The enthalpy of vaporization of water at 100°C is h_{fg} = 2257 kJ/kg (Table A-4).

Analysis The rate of heat transfer during this condensation process is given to be 180 kJ/s. Noting that the heat of vaporization of water represents the amount of heat released as a unit mass of vapor at a specified temperature condenses, the rate of condensation of steam is determined from

$$\dot{m}_{condensation} = \frac{\dot{Q}}{h_{fg}} = \frac{180 \text{ kJ/s}}{2257 \text{ kJ/kg}} = \mathbf{0.0798 \text{ kg/s}}$$

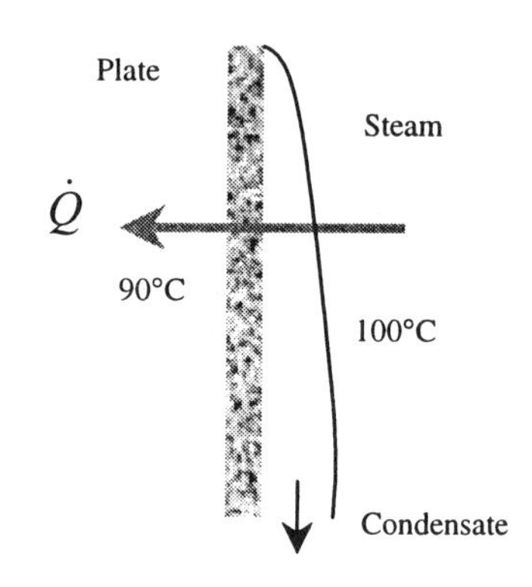

4-208 Water is boiled at T_{sat} = 100°C by an electric heater. The rate of evaporation of water is to be determined.

Assumptions **1** Steady operating conditions exist. **2** Heat losses from the outer surfaces of the water tank are negligible.

Properties The enthalpy of vaporization of water at 100°C is h_{fg} = 2257 kJ/kg (Table A-4).

Analysis Noting that the enthalpy of vaporization represents the amount of energy needed to vaporize a unit mass of a liquid at a specified temperature, the rate of evaporation of water is determined to be

$$\dot{m}_{evaporation} = \frac{\dot{W}_{e,boiling}}{h_{fg}} = \frac{5 \text{ kJ/s}}{2257 \text{ kJ/kg}} = \mathbf{0.00222 \text{ kg/s} = 7.98 \text{ kg/h}}$$

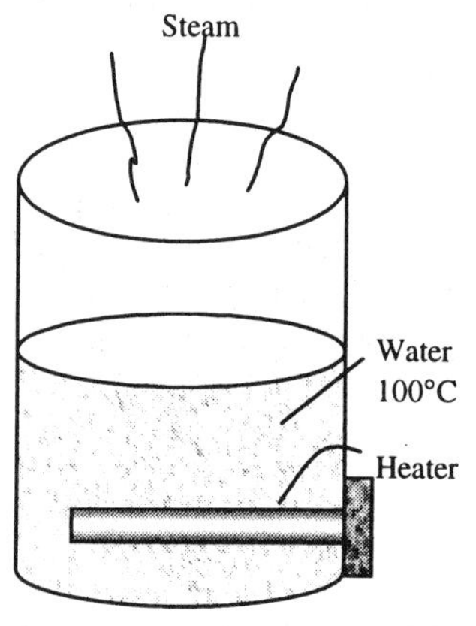

4-209 An insulated cylinder initially contains saturated liquid-vapor mixture of water at a specified temperature. A cold copper block is dropped into the cylinder. The final equilibrium temperature inside the cylinder and the final mass of water vapor are to be determined.

Assumptions **1** The copper block has a constant specific heat at the average temperature. **2** The cylinder is well-insulated and thus heat transfer is negligible. **3** Only part of the vapor in the cylinder condenses as a result of copper block being dropped into the cylinder, so that the temperature inside the cylinder remains constant (will be verified).

Properties The specific heat of copper at the average temperature of (30+120)/2 = 75°C is C = 0.391 kJ/kg· °C (Table A-3). The enthalpy of vaporization of water at 120°C is h_{fg} = 2202.6 kJ/kg (Table A-4).

Analysis We take the copper block as the system, which is closed system. Assuming the final temperature to be 120°C, the energy balance for copper can be expressed as

$$\underbrace{E_{in} - E_{out}}_{\substack{\text{Net energy transfer} \\ \text{by heat, work, and mass}}} = \underbrace{\Delta E_{\text{system}}}_{\substack{\text{Change in internal, kinetic,} \\ \text{potential, etc. energies}}}$$

$$Q_{in} = \Delta U = [mC(T_2 - T_1)]_{\text{copper}}$$

Cu 30°C

WATER 5 kg 120°C

Substituting,

$$Q_{\text{in}} = (5 \text{ kg})(0.391 \text{ kJ/kg.°C})(120 - 30)\text{°C} = 176.0 \text{ kJ}$$

This energy will come from the water vapor inside. Noting that the enthalpy of vaporization represents the amount of energy released when a unit mass of vapor condenses at a specified temperature, the amount of vapor that condenses to release 176 kJ of heat is

$$m_{\text{condensed}} = \frac{Q}{h_{fg}} = \frac{176 \text{ kJ}}{2202.6 \text{ kJ/kg}} = 0.080 \text{ kg}$$

Initially, the cylinder contained 1 kg of vapor. Then the amount of water vapor in the cylinder at the end of the process becomes

$$m_{\text{final}} = m_{\text{initial}} - m_{\text{condensed}} = 1.0 - 0.080 = \mathbf{0.920 \text{ kg}}$$

This verifies our assumption that the cylinder still contains some vapor at the end of the process. Then the final temperature in the cylinder must be the same as the initial temperature since saturation conditions still exist inside,

$$T_2 = T_1 = \mathbf{120°C}$$

4-210 The air pressure in a tire is measured before and after a trip. The temperature rise of air inside the tire is to be estimated.

Assumptions **1** Air is an ideal gas. **2** The volume of the tire remains constant. **3** No air leaks out of the tire during the trip.

Analysis Using the ideal gas relation between the two states, the final temperature in the tire is determined to be

Tire 25°C 200 kPa

$$\frac{P_1 V}{T_1} = \frac{P_2 V}{T_2} \rightarrow T_2 = \frac{P_2}{P_1} T_1 = \frac{(220 + 90) \text{ kPa}}{(200 + 90) \text{ kPa}} (25 + 273 \text{ K}) = 318.6 \text{ K} = 45.6°\text{C}$$

Therefore, the temperature rise of air in the tire during the trip is

$$\Delta T = T_2 - T_1 = 45.6 - 25 = \mathbf{20.6°\,C}$$

4-211 Two identical buildings in Los Angeles and Denver have the same infiltration rate. The ratio of the heat losses by infiltration at the two cities under identical conditions is to be determined.

Assumptions **1** Both buildings are identical and both are subjected to the same conditions except the atmospheric conditions. **2** Air is an ideal gas with constant specific heats at room temperature. **3** Steady flow conditions exist.

Analysis We can view infiltration as a steady stream of air that is heated as it flows in an imaginary duct passing through the building. The energy balance for this imaginary steady-flow system can be expressed in the rate form as

$$\underbrace{\dot{E}_{in} - \dot{E}_{out}}_{\text{Rate of net energy transfer by heat, work, and mass}} = \underbrace{\Delta \dot{E}_{\text{system}}^{\nearrow 0 \text{ (steady)}}}_{\text{Rate of change in internal, kinetic, potential, etc. energies}} = 0$$

$$\dot{E}_{in} = \dot{E}_{out}$$

$$\dot{Q}_{in} + \dot{m}h_1 = \dot{m}h_2 \quad (\text{since } \Delta\text{ke} \cong \Delta\text{pe} \cong 0)$$

$$\dot{Q}_{in} = \dot{m}C_p(T_2 - T_1) = \rho\dot{V}C_p(T_2 - T_1)$$

Los Angeles: 101 kPa
Denver: 83 kPa

Then the sensible infiltration heat loss (heat gain for the infiltrating air) can be expressed

$$\dot{Q}_{\text{infiltration}} = \dot{m}_{\text{air}} C_p (T_i - T_o) = \rho_{o,\text{air}} (ACH)(V_{\text{building}}) C_p (T_i - T_o)$$

where ACH is the infiltration volume rate in *air changes per hour*. Therefore, the infiltration heat loss is proportional to the density of air, and thus the ratio of infiltration heat losses at the two cities is simply the densities of outdoor air at those cities,

$$\text{Infiltration heat loss ratio} = \frac{\dot{Q}_{\text{infiltration, Los Angeles}}}{\dot{Q}_{\text{infiltration, Denver}}} = \frac{\rho_{o,\text{ air, Los Angeles}}}{\rho_{o,\text{ air, Denver}}}$$

$$= \frac{(P_0 / RT_0)_{\text{Los Angeles}}}{(P_0 / RT_0)_{\text{Denver}}} = \frac{P_{o,\text{ Los Angeles}}}{P_{0,\text{ Denver}}}$$

$$= \frac{101 \text{ kPa}}{83 \text{ kPa}} = \mathbf{1.22}$$

Therefore, the infiltration heat loss in Los Angeles will be 22% higher than that in Denver under identical conditions.

4-212 The ventilating fan of the bathroom of an electrically heated building in San Francisco runs continuously. The amount and cost of the heat "vented out" per month in winter are to be determined.

Assumptions **1** We take the atmospheric pressure to be 1 atm = 101.3 kPa since San Francisco is at sea level. **2** The building is maintained at 22°C at all times. **3** The infiltrating air is heated to 22°C before it exfiltrates. **4** Air is an ideal gas with constant specific heats at room temperature. **5** Steady flow conditions exist.

30 L/s
12.2°C
22°C

Properties The gas constant of air is R = 0.287 kPa.m^3/kg·K (Table A-1). The specific heat of air at room temperature is C_p = 1.005 kJ/kg·°C (Table A-2).

Analysis The density of air at the indoor conditions of 1 atm and 22°C is

$$\rho_o = \frac{P_o}{RT_o} = \frac{(101.3\,\text{kPa})}{(0.287\,\text{kPa.m}^3/\text{kg.K})(22+273\,\text{K})} = 1.20\,\text{kg/m}^3$$

Then the mass flow rate of air vented out becomes

$$\dot{m}_{\text{air}} = \rho\dot{V}_{\text{air}} = (1.20\,\text{kg/m}^3)(0.030\,\text{m}^3/\text{s}) = 0.036\,\text{kg/s}$$

We can view infiltration as a steady stream of air that is heated as it flows in an imaginary duct passing through the house. The energy balance for this imaginary steady-flow system can be expressed in the rate form as

$$\underbrace{\dot{E}_{in} - \dot{E}_{out}}_{\substack{\text{Rate of net energy transfer}\\ \text{by heat, work, and mass}}} = \underbrace{\Delta\dot{E}_{\text{system}}^{\nearrow 0\ \text{(steady)}}}_{\substack{\text{Rate of change in internal, kinetic,}\\ \text{potential, etc. energies}}} = 0$$

$$\dot{E}_{in} = \dot{E}_{out}$$

$$\dot{Q}_{in} + \dot{m}h_1 = \dot{m}h_2 \quad (\text{since } \Delta\text{ke} \cong \Delta\text{pe} \cong 0)$$

$$\dot{Q}_{in} = \dot{m}C_p(T_2 - T_1)$$

Noting that the indoor air vented out at 22°C is replaced by infiltrating outdoor air at 12.2°C, the sensible infiltration heat loss (heat gain for the infiltrating air) due to venting by fans can be expressed

$$\begin{aligned}\dot{Q}_{\text{loss by fan}} &= \dot{m}_{\text{air}}C_{\text{p}}(T_{\text{indoors}} - T_{\text{outdoors}})\\ &= (0.036\,\text{kg/s})(1.005\,\text{kJ/kg.°C})(22-12.2)\text{°C} = 0.355\,\text{kJ/s} = 0.355\,\text{kW}\end{aligned}$$

Then the amount and cost of the heat "vented out" per month (1 month = 30×24 = 720 h) becomes

Energy loss = $\dot{Q}_{\text{loss by fan}}\Delta t$ = (0.355 kW)(720 h/month) = **256 kWh/month**

Money loss = (Energy loss)(Unit cost of energy) = (256 kWh/month)($0.09/kWh) = **$23.0/month**

Discussion Note that the energy and money loss associated with ventilating fans can be very significant. Therefore, ventilating fans should be used with care.

4-213 Chilled air is to cool a room by removing the heat generated in a large insulated classroom by lights and students. The required flow rate of air that needs to be supplied to the room is to be determined.

Assumptions **1** The moisture produced by the bodies leave the room as vapor without any condensing, and thus the classroom has no latent heat load. **2** Heat gain through the walls and the roof is negligible. **4** Air is an ideal gas with constant specific heats at room temperature. **5** Steady operating conditions exist.

Properties The specific heat of air at room temperature is 1.005 kJ/kg·°C (Table A-2). The average rate of sensible heat generation by a person is given to be 60 W.

Analysis The rate of sensible heat generation by the people in the room and the total rate of sensible internal heat generation are

$$\dot{Q}_{\text{gen, sensible}} = \dot{q}_{\text{gen, sensible}}(\text{No. of people}) = (60\ \text{W/person})(150\ \text{persons}) = 9000\ \text{W}$$

$$\dot{Q}_{\text{total, sensible}} = \dot{Q}_{\text{gen, sensible}} + \dot{Q}_{\text{lighting}} = 9000 + 4000 = 13{,}000\ \text{W}$$

Both of these effects can be viewed as heat gain for the chilled air stream, which can be viewed as a steady stream of cool air that is heated as it flows in an imaginary duct passing through the room. The energy balance for this imaginary steady-flow system can be expressed in the rate form as

$$\underbrace{\dot{E}_{in} - \dot{E}_{out}}_{\substack{\text{Rate of net energy transfer} \\ \text{by heat, work, and mass}}} = \underbrace{\Delta \dot{E}_{\text{system}}}_{\substack{\text{Rate of change in internal, kinetic,} \\ \text{potential, etc. energies}}}^{\nearrow 0\ \text{(steady)}} = 0$$

$$\dot{E}_{in} = \dot{E}_{out}$$

$$\dot{Q}_{in} + \dot{m}h_1 = \dot{m}h_2 \quad (\text{since } \Delta\text{ke} \cong \Delta\text{pe} \cong 0)$$

$$\dot{Q}_{in} = \dot{Q}_{\text{total, sensible}} = \dot{m}C_p(T_2 - T_1)$$

Then the required mass flow rate of chilled air becomes

$$\dot{m}_{\text{air}} = \frac{\dot{Q}_{\text{total, sensible}}}{C_p \Delta T} = \frac{13\ \text{kJ/s}}{(1.005\ \text{kJ/kg·°C})(25-15)°\text{C}} = \mathbf{1.29\ kg/s}$$

Discussion The latent heat will be removed by the air-conditioning system as the moisture condenses outside the cooling coils.

4-214 Chickens are to be cooled by chilled water in an immersion chiller. The rate of heat removal from the chicken and the mass flow rate of water are to be determined.

Assumptions **1** Steady operating conditions exist. **2** The thermal properties of chickens and water are constant.

Properties The specific heat of chicken are given to be 3.54 kJ/kg.°C. The specific heat of water is 4.18 kJ/kg.°C (Table A-3).

Analysis (*a*) Chickens are dropped into the chiller at a rate of 500 per hour. Therefore, chickens can be considered to flow steadily through the chiller at a mass flow rate of

$$\dot{m}_{\text{chicken}} = (500\text{ chicken / h})(2.2\text{ kg / chicken}) = 1100\text{ kg / h} = 0.3056\text{ kg / s}$$

Taking the chicken flow stream in the chiller as the system, the energy balance for steadily flowing chickens can be expressed in the rate form as

$$\underbrace{\dot{E}_{in} - \dot{E}_{out}}_{\substack{\text{Rate of net energy transfer}\\ \text{by heat, work, and mass}}} = \underbrace{\Delta \dot{E}_{\text{system}}^{\nearrow 0\ \text{(steady)}}}_{\substack{\text{Rate of change in internal, kinetic,}\\ \text{potential, etc. energies}}} = 0$$

$$\dot{E}_{in} = \dot{E}_{out}$$

$$\dot{m}h_1 = \dot{Q}_{out} + \dot{m}h_2 \quad (\text{since } \Delta\text{ke} \cong \Delta\text{pe} \cong 0)$$

$$\dot{Q}_{out} = \dot{Q}_{\text{chicken}} = \dot{m}_{\text{chicken}} C_p (T_1 - T_2)$$

Then the rate of heat removal from the chickens as they are cooled from 15°C to 3°C becomes

$$\dot{Q}_{\text{chicken}} = (\dot{m}C_p\Delta T)_{\text{chicken}} = (0.3056\text{ kg/s})(3.54\text{ kJ/kg.° C})(15-3)\text{° C} = \mathbf{13.0\text{ kW}}$$

The chiller gains heat from the surroundings at a rate of 200 kJ/h = 0.0556 kJ/s. Then the total rate of heat gain by the water is

$$\dot{Q}_{\text{water}} = \dot{Q}_{\text{chicken}} + \dot{Q}_{\text{heat gain}} = 13.0 + 0.056 = 13.056\text{ kW}$$

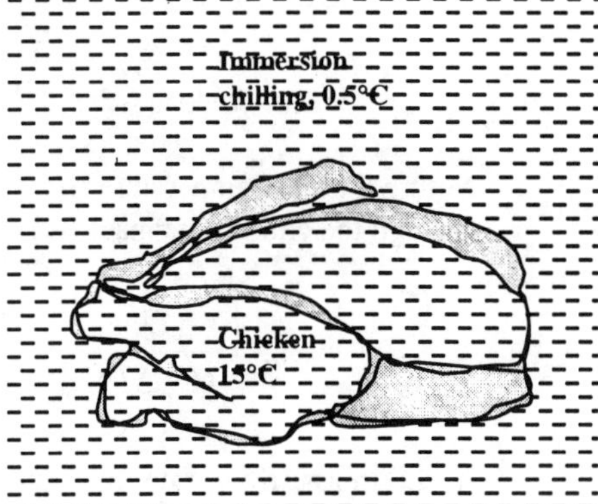

Noting that the temperature rise of water is not to exceed 2°C as it flows through the chiller, the mass flow rate of water must be at least

$$\dot{m}_{\text{water}} = \frac{\dot{Q}_{\text{water}}}{(C_p\Delta T)_{\text{water}}} = \frac{13.056\text{ kW}}{(4.18\text{ kJ/kg.° C})(2\text{° C})} = \mathbf{1.56\text{ kg/s}}$$

If the mass flow rate of water is less than this value, then the temperature rise of water will have to be more than 2°C.

4-215 Chickens are to be cooled by chilled water in an immersion chiller. The rate of heat removal from the chicken and the mass flow rate of water are to be determined.

Assumptions **1** Steady operating conditions exist. **2** The thermal properties of chickens and water are constant. **3** Heat gain of the chiller is negligible.

Properties The specific heat of chicken are given to be 3.54 kJ/kg.°C. The specific heat of water is 4.18 kJ/kg.°C (Table A-3).

Analysis (*a*) Chickens are dropped into the chiller at a rate of 500 per hour. Therefore, chickens can be considered to flow steadily through the chiller at a mass flow rate of

$$\dot{m}_{\text{chicken}} = (500\text{ chicken / h})(2.2\text{ kg / chicken}) = 1100\text{ kg / h} = 0.3056\text{ kg / s}$$

Taking the chicken flow stream in the chiller as the system, the energy balance for steadily flowing chickens can be expressed in the rate form as

$$\underbrace{\dot{E}_{in} - \dot{E}_{out}}_{\substack{\text{Rate of net energy transfer} \\ \text{by heat, work, and mass}}} = \underbrace{\Delta \dot{E}_{\text{system}}^{\nearrow 0\text{ (steady)}}}_{\substack{\text{Rate of change in internal, kinetic,} \\ \text{potential, etc. energies}}} = 0$$

$$\dot{E}_{in} = \dot{E}_{out}$$

$$\dot{m}h_1 = \dot{Q}_{out} + \dot{m}h_2 \quad (\text{since } \Delta\text{ke} \cong \Delta\text{pe} \cong 0)$$

$$\dot{Q}_{out} = \dot{Q}_{\text{chicken}} = \dot{m}_{\text{chicken}} C_p (T_1 - T_2)$$

Then the rate of heat removal from the chickens as they are cooled from 15°C to 3°C becomes

$$\dot{Q}_{\text{chicken}} = (\dot{m} C_p \Delta T)_{\text{chicken}} = (0.3056\text{ kg/s})(3.54\text{ kJ/kg.°C})(15-3)\text{°C} = \mathbf{13.0\text{ kW}}$$

Heat gain of the chiller from the surroundings is negligible. Then the total rate of heat gain by the water is

$$\dot{Q}_{\text{water}} = \dot{Q}_{\text{chicken}} = 13.0\text{ kW}$$

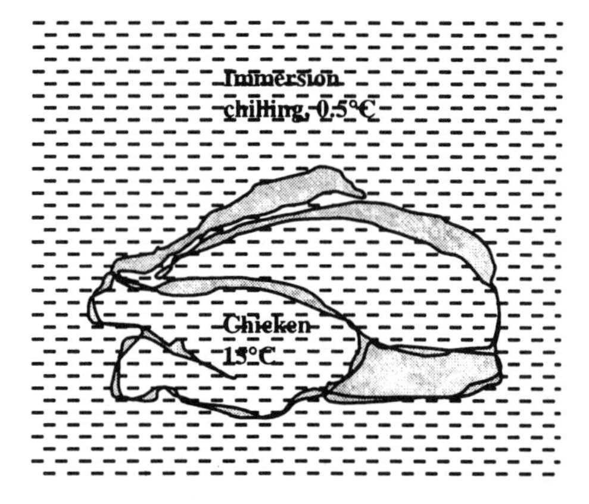

Noting that the temperature rise of water is not to exceed 2°C as it flows through the chiller, the mass flow rate of water must be at least

$$\dot{m}_{\text{water}} = \frac{\dot{Q}_{\text{water}}}{(C_p \Delta T)_{\text{water}}} = \frac{13.0\text{ kW}}{(4.18\text{ kJ/kg.°C})(2\text{°C})} = \mathbf{1.56\text{ kg/s}}$$

If the mass flow rate of water is less than this value, then the temperature rise of water will have to be more than 2°C.

4-216 A regenerator is considered to save heat during the cooling of milk in a dairy plant. The amounts of fuel and money such a generator will save per year are to be determined.

Assumptions 1 Steady operating conditions exist. **2** The properties of the milk are constant.

Properties The average density and specific heat of milk can be taken to be $\rho_{milk} \cong \rho_{water} = 1$ kg/L and $C_{p,\,milk}$= 3.79 kJ/kg.°C (Table A-3).

Analysis The mass flow rate of the milk is

$$\dot{m}_{milk} = \rho\dot{V}_{milk} = (1\text{ kg/L})(12\text{ L/s}) = 12\text{ kg/s} = 43{,}200\text{ kg/h}$$

Taking the pasteurizing section as the system, the energy balance for this steady-flow system can be expressed in the rate form as

$$\underbrace{\dot{E}_{in} - \dot{E}_{out}}_{\text{Rate of net energy transfer by heat, work, and mass}} = \underbrace{\Delta\dot{E}_{system}^{\nearrow 0\text{ (steady)}}}_{\text{Rate of change in internal, kinetic, potential, etc. energies}} = 0$$

$$\dot{E}_{in} = \dot{E}_{out}$$

$$\dot{Q}_{in} + \dot{m}h_1 = \dot{m}h_2 \quad (\text{since } \Delta\text{ke} \cong \Delta\text{pe} \cong 0)$$

$$\dot{Q}_{in} = \dot{m}_{milk}C_p(T_2 - T_1)$$

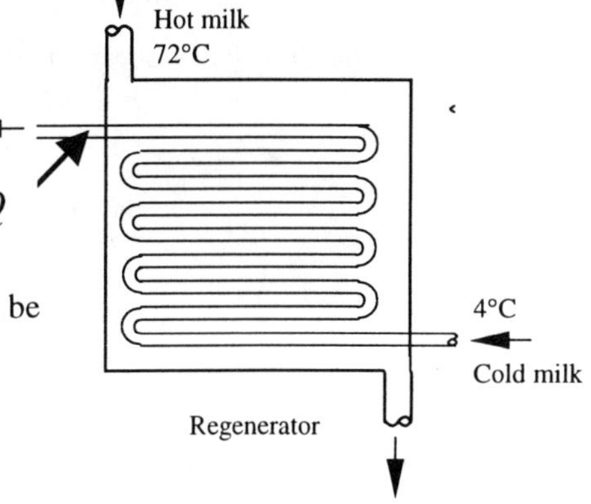

Therefore, to heat the milk from 4 to 72°C as being done currently, heat must be transferred to the milk at a rate of

$$\dot{Q}_{current} = [\dot{m}C_p(T_{pasturization} - T_{refrigeration})]_{milk}$$
$$= (12\text{ kg/s})(3.79\text{ kJ/kg.°C})(72-4)\text{°C} = 3093\text{ kJ/s}$$

The proposed regenerator has an effectiveness of ε = 0.82, and thus it will save 82 percent of this energy. Therefore,

$$\dot{Q}_{saved} = \varepsilon\dot{Q}_{current} = (0.82)(3093\text{ kJ/s}) = 2536\text{ kJ/s}$$

Noting that the boiler has an efficiency of $\eta_{boiler} = 0.82$, the energy savings above correspond to fuel savings of

$$\text{Fuel Saved} = \frac{\dot{Q}_{saved}}{\eta_{boiler}} = \frac{(2536\text{ kJ/s})}{(0.82)}\frac{(1\text{therm})}{(105{,}500\text{ kJ})} = 0.02931\text{ therm/s}$$

Noting that 1 year = 365×24=8760 h and unit cost of natural gas is \$0.52/therm, the annual fuel and money savings will be

Fuel Saved = (0.02931 therms/s)(8760×3600 s) = **924,450 therms/yr**

$$\text{Money saved} = (\text{Fuel saved})(\text{Unit cost of fuel})$$
$$= (924{,}450\text{ therm/yr})(\$0.52/\text{therm}) = \mathbf{\$480{,}700/yr}$$

4-217E A refrigeration system is to cool eggs by chilled air at a rate of 10,000 eggs per hour. The rate of heat removal from the eggs, the required volume flow rate of air, and the size of the compressor of the refrigeration system are to be determined. √

Assumptions **1** Steady operating conditions exist. **2** The eggs are at uniform temperatures before and after cooling. **3** The cooling section is well-insulated. **4** The properties of eggs are constant. **5** The local atmospheric pressure is 1 atm.

Properties The properties of the eggs are given to ρ = 67.4 lbm/ft^3 and C_p = 0.80 Btu/lbm.°F. The specific heat of air at room temperature C_p = 0.24 Btu/lbm. °F (Table A-2E). The gas constant of air is R = 0.3704 psia.ft^3/lbm.R (Table A-1E).

Analysis (*a*) Noting that eggs are cooled at a rate of 10,000 eggs per hour, eggs can be considered to flow steadily through the cooling section at a mass flow rate of

$$\dot{m}_{\text{egg}} = (10{,}000 \text{ eggs/h})(0.14 \text{ lbm/egg}) = 1400 \text{ lbm/h} = 0.3889 \text{ lbm/s}$$

Taking the egg flow stream in the cooler as the system, the energy balance for steadily flowing eggs can be expressed in the rate form as

$$\underbrace{\dot{E}_{in} - \dot{E}_{out}}_{\text{Rate of net energy transfer by heat, work, and mass}} = \underbrace{\Delta \dot{E}_{\text{system}}^{\nearrow 0 \text{ (steady)}}}_{\text{Rate of change in internal, kinetic, potential, etc. energies}} = 0$$

$$\dot{E}_{in} = \dot{E}_{out}$$

$$\dot{m}h_1 = \dot{Q}_{out} + \dot{m}h_2 \quad (\text{since } \Delta\text{ke} \cong \Delta\text{pe} \cong 0)$$

$$\dot{Q}_{out} = \dot{Q}_{\text{egg}} = \dot{m}_{\text{egg}} C_p (T_1 - T_2)$$

Egg
0.14 lbm

Air
34°F

Then the rate of heat removal from the eggs as they are cooled from 90°F to 50°F at this rate becomes

$$\dot{Q}_{\text{egg}} = (\dot{m}C_p\Delta T)_{\text{egg}} = (1400 \text{ lbm/h})(0.80 \text{ Btu/lbm.°F})(90-50)\text{°F} = \mathbf{44{,}800 \text{ Btu/h}}$$

(*b*) All the heat released by the eggs is absorbed by the refrigerated air since heat transfer through he walls of cooler is negligible, and the temperature rise of air is not to exceed 10°F. The minimum mass flow and volume flow rates of air are determined to be

$$\dot{m}_{air} = \frac{\dot{Q}_{air}}{(C_p\Delta T)_{air}} = \frac{44{,}800 \text{ Btu/h}}{(0.24 \text{ Btu/lbm.°F})(10\text{°F})} = 18{,}667 \text{ lbm/h}$$

$$\rho_{air} = \frac{P}{RT} = \frac{14.7 \text{ psia}}{(0.3704 \text{ psia.ft}^3/\text{lbm.R})(34+460)\text{R}} = 0.0803 \text{ lbm/ft}^3$$

$$\dot{V}_{air} = \frac{\dot{m}_{air}}{\rho_{air}} = \frac{18{,}667 \text{ lbm/h}}{0.0803 \text{ lbm/ft}^3} = \mathbf{232{,}500 \text{ ft}^3\text{/h}}$$

4-218 Dough is made with refrigerated water in order to absorb the heat of hydration and thus to control the temperature rise during kneading. The temperature to which the city water must be cooled before mixing with flour is to be determined to avoid temperature rise during kneading.

Assumptions **1** Steady operating conditions exist. **2** The dough is at uniform temperatures before and after cooling. **3** The kneading section is well-insulated. **4** The properties of water and dough are constant.

Properties The specific heats of the flour and the water are given to be 1.76 and 4.18 kJ/kg.°C, respectively. The heat of hydration of dough is given to be 15 kJ/kg.

Analysis It is stated that 2 kg of flour is mixed with 1 kg of water, and thus 3 kg of dough is obtained from each kg of water. Also, 15 kJ of heat is released for each kg of dough kneaded, and thus 3×15 = 45 kJ of heat is released from the dough made using 1 kg of water.

Taking the cooling section of water as the system, which is a steady-flow control volume, the energy balance for this steady-flow system can be expressed in the rate form as

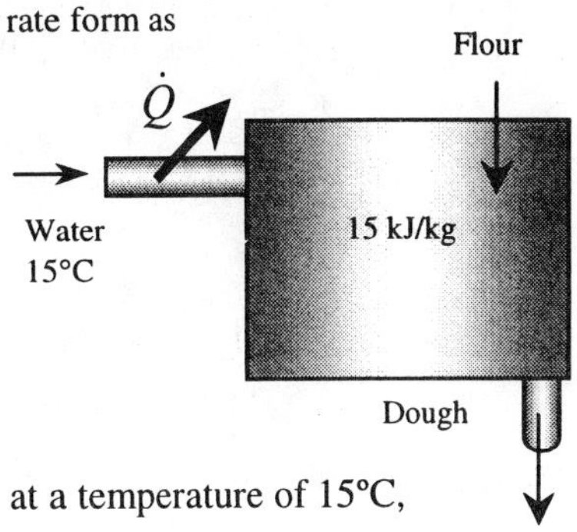

Flour

Q

15 kJ/kg

Water

15°C

Dough

$$\underbrace{\dot{E}_{in} - \dot{E}_{out}}_{\text{Rate of net energy transfer by heat, work, and mass}} = \underbrace{\Delta \dot{E}_{\text{system}}^{\nearrow 0 \text{ (steady)}}}_{\text{Rate of change in internal, kinetic, potential, etc. energies}} = 0$$

$$\dot{E}_{in} = \dot{E}_{out}$$

$$\dot{m}h_1 = \dot{Q}_{out} + \dot{m}h_2 \quad (\text{since } \Delta\text{ke} \cong \Delta\text{pe} \cong 0)$$

$$\dot{Q}_{out} = \dot{Q}_{\text{water}} = \dot{m}_{\text{water}} C_p (T_1 - T_2)$$

In order for water to absorb all the heat of hydration and end up at a temperature of 15°C, its temperature before entering the mixing section must be reduced to

$$Q_{in} = Q_{\text{dough}} = mC_p(T_2 - T_1) \rightarrow T_1 = T_2 - \frac{Q}{mC_p} = 15°\text{C} - \frac{45 \text{ kJ}}{(1 \text{ kg})(4.18 \text{ kJ/kg.°C})} = \mathbf{4.2°C}$$

That is, the water must be precooled to 4.2°C before mixing with the flour in order to absorb the entire heat of hydration.

4-219 Glass bottles are washed in hot water in an uncovered rectangular glass washing bath. The rates of heat and water mass that need to be supplied to the water are to be determined. √

Assumptions **1** Steady operating conditions exist. **2** The entire water body is maintained at a uniform temperature of 55°C. **3** Heat losses from the outer surfaces of the bath are negligible. **4** Water is an incompressible substance with constant properties.

Properties The specific heat of water at room temperature is C_p = 4.18 kJ/kg.°C. Also, the specific heat of glass is 0.80 kJ/kg.°C (Table A-3).

Analysis (*a*) The mass flow rate of glass bottles through the water bath in steady operation is

$$\dot{m}_{\text{bottle}} = m_{\text{bottle}} \times \text{Bottle flow rate} = (0.150 \text{ kg / bottle})(800 \text{ bottles / min}) = 120 \text{ kg / min} = 2 \text{ kg / s}$$

Taking the bottle flow section as the system, which is a steady-flow control volume, the energy balance for this steady-flow system can be expressed in the rate form as

$$\underbrace{\dot{E}_{in} - \dot{E}_{out}}_{\substack{\text{Rate of net energy transfer} \\ \text{by heat, work, and mass}}} = \underbrace{\Delta \dot{E}_{\text{system}}^{\nearrow 0 \text{ (steady)}}}_{\substack{\text{Rate of change in internal, kinetic,} \\ \text{potential, etc. energies}}} = 0$$

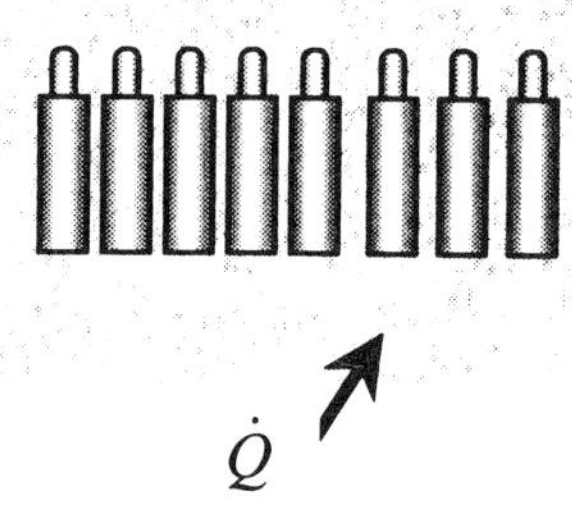

$$\dot{E}_{in} = \dot{E}_{out}$$

$$\dot{Q}_{in} + \dot{m}h_1 = \dot{m}h_2 \quad (\text{since } \Delta\text{ke} \cong \Delta\text{pe} \cong 0)$$

$$\dot{Q}_{in} = \dot{Q}_{\text{bottle}} = \dot{m}_{\text{water}} C_p (T_2 - T_1)$$

Then the rate of heat removal by the bottles as they are heated from 20 to 55°C is

$$\dot{Q}_{\text{bottle}} = \dot{m}_{\text{bottle}} C_p \Delta T = (2\text{kg/s})(0.8\text{kJ/kg.°C})(55 - 20)^\circ\text{C} = 56{,}000 \text{ W}$$

The amount of water removed by the bottles is

$$\dot{m}_{\text{water,out}} = (\text{Flow rate of bottles})(\text{Water removed per bottle})$$
$$= (800 \text{ bottles / min})(0.2 \text{ g/bottle}) = 160 \text{ g/min} = \mathbf{2.67 \times 10^{-3} \text{ kg/s}}$$

Noting that the water removed by the bottles is made up by fresh water entering at 15°C, the rate of heat removal by the water that sticks to the bottles is

$$\dot{Q}_{\text{water removed}} = \dot{m}_{\text{water removed}} C_p \Delta T = (2.67 \times 10^{-3} \text{ kg/s})(4180 \text{ J/kg·°C})(55 - 15)^\circ\text{C} = 446 \text{ W}$$

Therefore, the total amount of heat removed by the wet bottles is

$$\dot{Q}_{\text{total, removed}} = \dot{Q}_{\text{glass removed}} + \dot{Q}_{\text{water removed}} = 56{,}000 + 446 = \mathbf{56{,}446 \text{ W}}$$

Discussion In practice, the rates of heat and water removal will be much larger since the heat losses from the tank and the moisture loss from the open surface are not considered.

4-220 Glass bottles are washed in hot water in an uncovered rectangular glass washing bath. The rates of heat and water mass that need to be supplied to the water are to be determined. √

Assumptions **1** Steady operating conditions exist. **2** The entire water body is maintained at a uniform temperature of 50°C. **3** Heat losses from the outer surfaces of the bath are negligible. **4** Water is an incompressible substance with constant properties.

Properties The specific heat of water at room temperature is C_p = 4.18 kJ/kg.°C. Also, the specific heat of glass is 0.80 kJ/kg.°C (Table A-3).

Analysis (*a*) The mass flow rate of glass bottles through the water bath in steady operation is

$$\dot{m}_{\text{bottle}} = m_{\text{bottle}} \times \text{Bottle flow rate} = (0.150 \text{ kg / bottle})(800 \text{ bottles / min}) = 120 \text{ kg / min} = 2 \text{ kg / s}$$

Taking the bottle flow section as the system, which is a steady-flow control volume, the energy balance for this steady-flow system can be expressed in the rate form as

$$\underbrace{\dot{E}_{in} - \dot{E}_{out}}_{\substack{\text{Rate of net energy transfer} \\ \text{by heat, work, and mass}}} = \underbrace{\Delta \dot{E}_{\text{system}}^{\nearrow 0 \text{ (steady)}}}_{\substack{\text{Rate of change in internal, kinetic,} \\ \text{potential, etc. energies}}} = 0$$

$$\dot{E}_{in} = \dot{E}_{out}$$

$$\dot{Q}_{in} + \dot{m}h_1 = \dot{m}h_2 \quad (\text{since } \Delta\text{ke} \cong \Delta\text{pe} \cong 0)$$

$$\dot{Q}_{in} = \dot{Q}_{\text{bottle}} = \dot{m}_{\text{water}} C_p (T_2 - T_1)$$

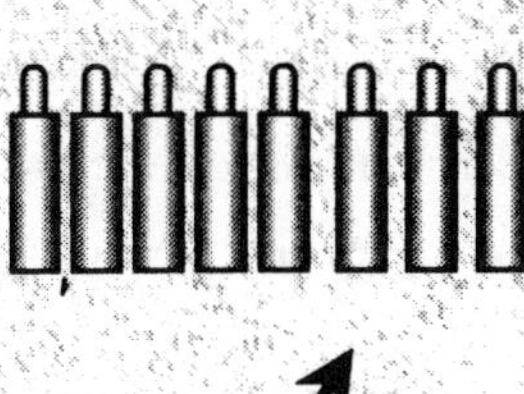

Then the rate of heat removal by the bottles as they are heated from 20 to 50°C is

$$\dot{Q}_{\text{bottle}} = \dot{m}_{\text{bottle}} C_p \Delta T = (2 \text{ kg/s})(0.8 \text{ kJ/kg.°C})(50 - 20)°\text{C} = 48{,}000 \text{ W}$$

The amount of water removed by the bottles is

$$\dot{m}_{\text{water,out}} = (\text{Flow rate of bottles})(\text{Water removed per bottle})$$
$$= (800 \text{ bottles / min})(0.2 \text{ g/bottle}) = 160 \text{ g/min} = \mathbf{2.67 \times 10^{-3} \text{ kg/s}}$$

Noting that the water removed by the bottles is made up by fresh water entering at 15°C, the rate of heat removal by the water that sticks to the bottles is

$$\dot{Q}_{\text{water removed}} = \dot{m}_{\text{water removed}} C_p \Delta T = (82.67 \times 10^{-3} \text{ kg/s})(4180 \text{ J/kg·°C})(50 - 15)°\text{C} = 391 \text{ W}$$

Therefore, the total amount of heat removed by the wet bottles is

$$\dot{Q}_{\text{total, removed}} = \dot{Q}_{\text{glass removed}} + \dot{Q}_{\text{water removed}} = 48{,}000 + 391 = \mathbf{48{,}391 \text{ W}}$$

Discussion In practice, the rates of heat and water removal will be much larger since the heat losses from the tank and the moisture loss from the open surface are not considered.

4-221 Long aluminum wires are extruded at a velocity of 10 m/min, and are exposed to atmospheric air. The rate of heat transfer from the wire is to be determined.

Assumptions **1** Steady operating conditions exist. **2** The thermal properties of the wire are constant.

Properties The properties of aluminum are given to be ρ = 2702 kg/m^3 and C_p = 0.896 kJ/kg.°C.

Analysis The mass flow rate of the extruded wire through the air is

$$\dot{m} = \rho\dot{V} = \rho(\pi r_0^2)V = (2702\ \text{kg/m}^3)\pi(0.0015\ \text{m})^2(10\ \text{m/min}) = 0.191\ \text{kg/min}$$

Taking the volume occupied by the extruded wire as the system, which is a steady-flow control volume, the energy balance for this steady-flow system can be expressed in the rate form as

$$\underbrace{\dot{E}_{in} - \dot{E}_{out}}_{\substack{\text{Rate of net energy transfer}\\ \text{by heat, work, and mass}}} = \underbrace{\Delta\dot{E}_{\text{system}}^{\nearrow 0\ \text{(steady)}}}_{\substack{\text{Rate of change in internal, kinetic,}\\ \text{potential, etc. energies}}} = 0$$

$$\dot{E}_{in} = \dot{E}_{out}$$

$$\dot{m}h_1 = \dot{Q}_{out} + \dot{m}h_2 \quad (\text{since } \Delta\text{ke} \cong \Delta\text{pe} \cong 0)$$

$$\dot{Q}_{out} = \dot{Q}_{\text{wire}} = \dot{m}_{\text{wire}}C_p(T_1 - T_2)$$

Then the rate of heat transfer from the wire to the air becomes

$$\dot{Q} = \dot{m}C_p[T(t) - T_\infty] = (0.191\ \text{kg/min})(0.896\ \text{kJ/kg.°C})(350-50)\text{°C} = 51.3\ \text{kJ/min} = \mathbf{0.856\ kW}$$

4-222 Long copper wires are extruded at a velocity of 10 m/min, and are exposed to atmospheric air. The rate of heat transfer from the wire is to be determined. √

Assumptions **1** Steady operating conditions exist. **2** The thermal properties of the wire are constant.

Properties The properties of copper are given to be ρ = 8950 kg/m^3 and C_p = 0.383 kJ/kg.°C.

Analysis The mass flow rate of the extruded wire through the air is

$$\dot{m} = \rho\dot{V} = \rho(\pi r_0^2)V = (8950\ \text{kg/m}^3)\pi(0.0015\ \text{m})^2(10\ \text{m/min}) = 0.633\ \text{kg/min}$$

Taking the volume occupied by the extruded wire as the system, which is a steady-flow control volume, the energy balance for this steady-flow system can be expressed in the rate form as

$$\underbrace{\dot{E}_{in} - \dot{E}_{out}}_{\substack{\text{Rate of net energy transfer}\\ \text{by heat, work, and mass}}} = \underbrace{\Delta\dot{E}_{\text{system}}^{\nearrow 0\ \text{(steady)}}}_{\substack{\text{Rate of change in internal, kinetic,}\\ \text{potential, etc. energies}}} = 0$$

$$\dot{E}_{in} = \dot{E}_{out}$$

$$\dot{m}h_1 = \dot{Q}_{out} + \dot{m}h_2 \quad (\text{since } \Delta\text{ke} \cong \Delta\text{pe} \cong 0)$$

$$\dot{Q}_{out} = \dot{Q}_{\text{wire}} = \dot{m}_{\text{wire}}C_p(T_1 - T_2)$$

Then the rate of heat transfer from the wire to the air becomes

$$\dot{Q} = \dot{m}C_p[T(t) - T_\infty] = (0.633\ \text{kg/min})(0.383\ \text{kJ/kg.°C})(350-50)\text{°C} = 72.7\ \text{kJ/min} = \mathbf{1.21\ kW}$$

4-223 Steam at a saturation temperature of T_{sat} = 40°C condenses on the outside of a thin horizontal tube. Heat is transferred to the cooling water that enters the tube at 25°C and exits at 35°C. The rate of condensation of steam is to be determined.

Assumptions **1** Steady operating conditions exist. **2** Water is an incompressible substance with constant properties at room temperature. **3** The changes in kinetic and potential energies are negligible.

Properties The properties of water at room temperature are ρ = 997 kg/m^3 and C_p = 4.18 kJ/kg.°C. The enthalpy of vaporization of water at a saturation temperature of 40°C is h_{fg} = 2406.7 kJ/kg (Table A-4).

Analysis The mass flow rate of water through the tube is

$$\dot{m}_{\text{water}} = \rho VA_c = (997\ \text{kg/m}^3)(2\ \text{m/s})[\pi(0.03\ \text{m})^2/4] = 1.409\ \text{kg/s}$$

Taking the volume occupied by the cold water in the tube as the system, which is a steady-flow control volume, the energy balance for this steady-flow system can be expressed in the rate form as

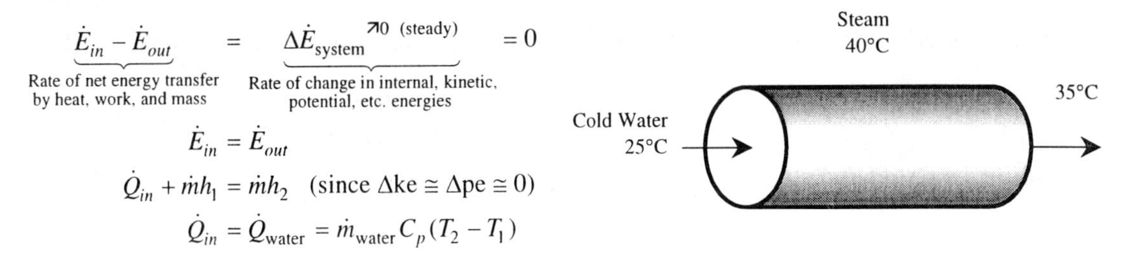

$$\underbrace{\dot{E}_{in} - \dot{E}_{out}}_{\text{Rate of net energy transfer by heat, work, and mass}} = \underbrace{\Delta \dot{E}_{\text{system}}^{\nearrow 0\ \text{(steady)}}}_{\text{Rate of change in internal, kinetic, potential, etc. energies}} = 0$$

$$\dot{E}_{in} = \dot{E}_{out}$$

$$\dot{Q}_{in} + \dot{m}h_1 = \dot{m}h_2 \quad (\text{since } \Delta\text{ke} \cong \Delta\text{pe} \cong 0)$$

$$\dot{Q}_{in} = \dot{Q}_{\text{water}} = \dot{m}_{\text{water}} C_p (T_2 - T_1)$$

Then the rate of heat transfer to the water and the rate of condensation become

$$\dot{Q} = \dot{m}C_p(T_{out} - T_{in}) = (1.409\ \text{kg/s})(4.18\ \text{kJ/kg}\cdot°\text{C})(35-25)°\text{C} = 58.9\ \text{kW}$$

$$\dot{Q} = \dot{m}_{\text{evap}} h_{fg} \rightarrow \dot{m}_{\text{evap}} = \frac{\dot{Q}}{h_{fg}} = \frac{58.9\ \text{kJ/s}}{2406.7\ \text{kJ/kg}} = \mathbf{0.0245\ kg/s}$$

4-224E Saturated steam at a saturation pressure of 0.95 psia and thus at a saturation temperature of T_{sat} = 100°F (Table A-4E) condenses on the outer surfaces of 144 horizontal tubes by circulating cooling water arranged in a 12 × 12 square array. The rate of heat transfer to the cooling water and the average velocity of the cooling water are to be determined.

Assumptions **1** Steady operating conditions exist. **2** The tubes are isothermal. **3** Water is an incompressible substance with constant properties at room temperature. **4** The changes in kinetic and potential energies are negligible.

Properties The properties of water at room temperature are ρ = 62.1 lbm/ft^3 and C_p = 1.00 Btu/lbm.°F. The enthalpy of vaporization of water at a saturation temperature of 100°F is h_{fg} = 1037 Btu/lbm (Table A-4E).

Analysis The rate of heat transfer from the steam to the cooling water is equal to the heat of vaporization released as the vapor condenses at the specified temperature,

$$\dot{Q} = \dot{m}h_{fg} = (6800 \text{ lbm/h})(1037 \text{ Btu/lbm}) = \mathbf{7{,}051{,}600\ Btu/h = 1959\ Btu/s}$$

All of this energy is transferred to the cold water. Therefore, the mass flow rate of cold water must be

$$\dot{Q} = \dot{m}_{\text{water}} C_p \Delta T \quad \rightarrow \quad \dot{m}_{\text{water}} = \frac{\dot{Q}}{C_p \Delta T} = \frac{1959 \text{ Btu/s}}{(1.00 \text{ Btu/lbm.°F})(8\text{°F})} = 244.8 \text{ lbm/s}$$

Then the average velocity of the cooling water through the 144 tubes becomes

$$\dot{m} = \rho A \mathbf{V} \quad \rightarrow \quad \mathbf{V} = \frac{\dot{m}}{\rho A} = \frac{\dot{m}}{\rho(n\pi D^2/4)} = \frac{244.8 \text{ lbm/s}}{(62.1 \text{ lbm/ft}^3)[144\pi(2/12 \text{ ft})^2/4]} = \mathbf{1.26\ ft/s}$$

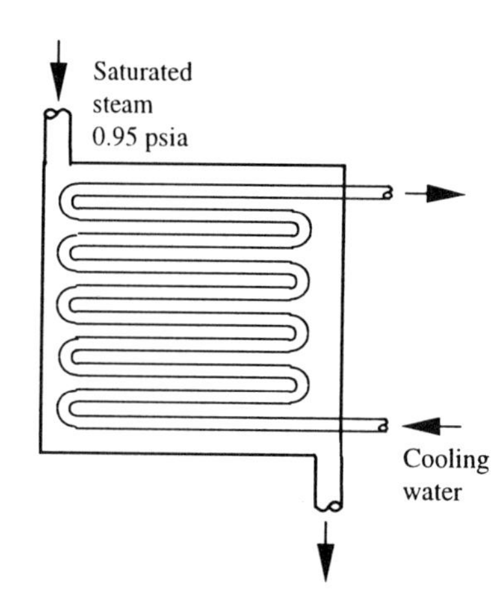

4-225 Saturated refrigerant-134a vapor at a saturation temperature of T_{sat} = 30°C condenses inside a tube. The rate of heat transfer from the refrigerant for the condensate exit temperatures of 16°C and 20°C are to be determined.

Assumptions **1** Steady flow conditions exist. **2** Kinetic and potential energy changes are negligible. **3** There are no work interactions involved.

Properties The properties of saturated refrigerant-134a at 30°C are h_f = 91.49 kJ/kg, are h_g = 263.50 kJ/kg, and are h_{fg} = 172.00 kJ/kg. The enthalpy of saturated liquid refrigerant at 16°C is h_f = 71.69 kJ/kg, (Table A-11).

Analysis We take the *tube and the refrigerant in it* as the system. This is a *control volume* since mass crosses the system boundary during the process. We note that there is only one inlet and one exit, and thus $\dot{m}_1 = \dot{m}_2 = \dot{m}$. Noting that heat is lost from the system, the energy balance for this steady-flow system can be expressed in the rate form as

$$\underbrace{\dot{E}_{in} - \dot{E}_{out}}_{\text{Rate of net energy transfer by heat, work, and mass}} = \underbrace{\Delta \dot{E}_{system}^{\nearrow 0 \text{ (steady)}}}_{\text{Rate of change in internal, kinetic, potential, etc. energies}} = 0$$

$$\dot{E}_{in} = \dot{E}_{out}$$

$$\dot{m}h_1 = \dot{Q}_{out} + \dot{m}h_2 \quad \text{(since } \Delta ke \cong \Delta pe \cong 0)$$

$$\dot{Q}_{out} = \dot{m}(h_1 - h_2)$$

Q_{out}

R-134a
30°C

where at the inlet state $h_1 = h_g$ = 263.50 kJ/kg. Then the rates of heat transfer during this condensation process for both cases become

Case 1: T_2 = 30°C: $h_2 = h_{f@30°C}$ = 91.49 kJ/kg.

$$\dot{Q}_{out} = (0.1 \text{ kg/min})(263.5 - 91.49) \text{ kJ/kg} = \mathbf{17.2 \text{ kW}}$$

Case 2: T_2 = 16°C: $h_2 \cong h_{f@16°C}$ =71.69 kJ/kg.

$$\dot{Q}_{out} = (0.1 \text{ kg/min})(263.5 - 71.69) \text{ kJ/kg} = \mathbf{19.2 \text{ kW}}$$

Discussion Note that the rate of heat removal is greater in the second case since the liquid is subcooled in that case.

4-226E A winterizing project is to reduce the infiltration rate of a house from 2.2 ACH to 1.1 ACH. The resulting cost savings are to be determined.

Assumptions **1** The house is maintained at 72°F at all times. **2** The latent heat load during the heating season is negligible. **3** The infiltrating air is heated to 72°F before it exfiltrates. **4** Air is an ideal gas with constant specific heats at room temperature. **5** The changes in kinetic and potential energies are negligible. **6** Steady flow conditions exist.

Properties The gas constant of air is 0.3704 psia.ft^3/lbm·R (Table A-1E). The specific heat of air at room temperature is 0.24 Btu/lbm·°F (Table A-2E).

Analysis The density of air at the outdoor conditions is

$$\rho_o = \frac{P_o}{RT_o} = \frac{13.5\text{ psia}}{(0.3704\text{ psia.ft}^3\text{/lbm.R})(496.5\text{ R})} = 0.0734\text{ lbm/ft}^3$$

The volume of the house is

$$V_{\text{building}} = (\text{Floor area})(\text{Height}) = (3000\text{ ft}^2)(9\text{ ft}) = 27{,}000\text{ ft}^3$$

We can view infiltration as a steady stream of air that is heated as it flows in an imaginary duct passing through the house. The energy balance for this imaginary steady-flow system can be expressed in the rate form as

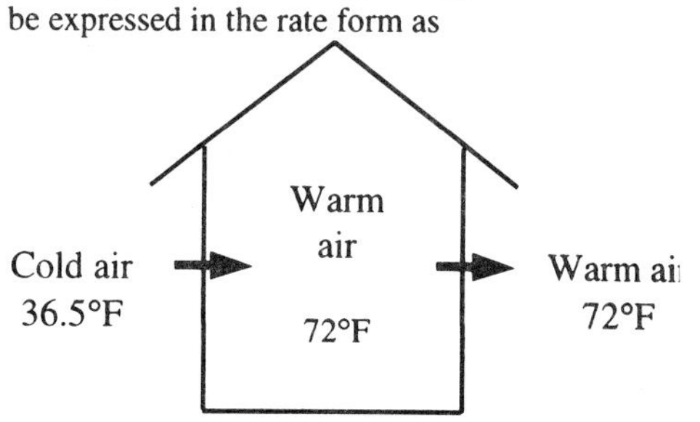

$$\underbrace{\dot{E}_{in} - \dot{E}_{out}}_{\substack{\text{Rate of net energy transfer}\\ \text{by heat, work, and mass}}} = \underbrace{\Delta \dot{E}_{\text{system}}^{\nearrow 0\text{ (steady)}}}_{\substack{\text{Rate of change in internal, kinetic,}\\ \text{potential, etc. energies}}} = 0$$

$$\dot{E}_{in} = \dot{E}_{out}$$

$$\dot{Q}_{in} + \dot{m}h_1 = \dot{m}h_2 \quad (\text{since } \Delta\text{ke} \cong \Delta\text{pe} \cong 0)$$

$$\dot{Q}_{in} = \dot{m}C_p(T_2 - T_1) = \rho\dot{V}C_p(T_2 - T_1)$$

The reduction in the infiltration rate is 2.2 – 1.1 = 1.1 ACH.
The reduction in the sensible infiltration heat load corresponding to it is

$$\dot{Q}_{\text{infiltration, saved}} = \rho_o C_p (ACH_{\text{saved}})(V_{\text{building}})(T_i - T_o)$$
$$= (0.0734\text{ lbm/ft}^3)(0.24\text{ Btu/lbm.°F})(1.1\text{/h})(27{,}000\text{ ft}^3)(72 - 36.5)\text{°F}$$
$$= 18{,}573\text{ Btu/h} = 0.18573\text{ therm/h}$$

since 1 therm = 100,000 Btu. The number of hours during a six month period is 6×30×24 = 4320 h. Noting that the furnace efficiency is 0.65 and the unit cost of natural gas is \$0.62/therm, the energy and money saved during the 6-month period are

$$\text{Energy savings} = (\dot{Q}_{\text{infiltration, saved}})(\text{No. of hours per year})/\text{Efficiency}$$
$$= (0.18573\text{ therm/h})(4320\text{ h/year})/0.65$$
$$= 1234\text{ therms/year}$$

$$\text{Cost savings} = (\text{Energy savings})(\text{Unit cost of energy})$$
$$= (1234\text{ therms/year})(\$0.62\text{/therm})$$
$$= \mathbf{\$765/year}$$

Therefore, reducing the infiltration rate by one-half will reduce the heating costs of this homeowner by \$765 per year.

4-227 Outdoors air at -10°C and 90 kPa enters the building at a rate of 35 L/s while the indoors is maintained at 20°C. The rate of sensible heat loss from the building due to infiltration is to be determined.

Assumptions **1** The house is maintained at 20°C at all times. **2** The latent heat load is negligible. **3** The infiltrating air is heated to 20°C before it exfiltrates. **4** Air is an ideal gas with constant specific heats at room temperature. **5** The changes in kinetic and potential energies are negligible. **6** Steady flow conditions exist.

Properties The gas constant of air is R = 0.287 kPa.m^3/kg·K (Table A-1). The specific heat of air at room temperature is C_p = 1.0 kJ/kg·°C (Table A-2).

Analysis The density of air at the outdoor conditions is

$$\rho_o = \frac{P_o}{RT_o} = \frac{90\text{ kPa}}{(0.287\text{ kPa.m}^3\text{/kg.K})(-10+273\text{ K})} = 1.19\text{ kg/m}^3$$

We can view infiltration as a steady stream of air that is heated as it flows in an imaginary duct passing through the building. The energy balance for this imaginary steady-flow system can be expressed in the rate form as

$$\underbrace{\dot{E}_{in} - \dot{E}_{out}}_{\text{Rate of net energy transfer by heat, work, and mass}} = \underbrace{\Delta \dot{E}_{system}^{\nearrow 0 \text{ (steady)}}}_{\text{Rate of change in internal, kinetic, potential, etc. energies}} = 0$$

$$\dot{E}_{in} = \dot{E}_{out}$$

$$\dot{Q}_{in} + \dot{m}h_1 = \dot{m}h_2 \quad (\text{since } \Delta\text{ke} \cong \Delta\text{pe} \cong 0)$$

$$\dot{Q}_{in} = \dot{m}C_p(T_2 - T_1)$$

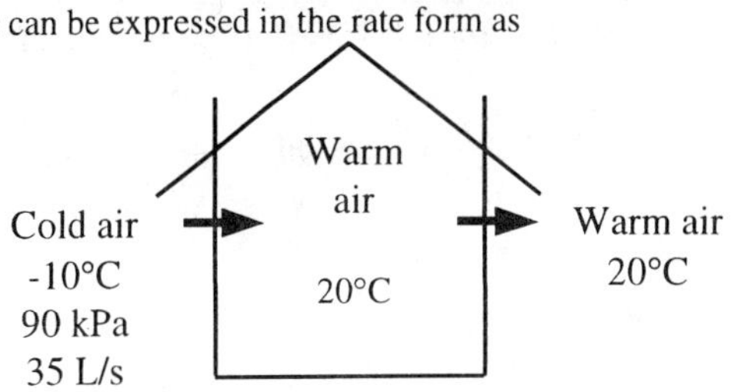

Then the sensible infiltration heat load corresponding to an infiltration rate of 35 L/s becomes

$$\dot{Q}_{\text{infiltration}} = \rho_o \dot{V}_{air} C_p (T_i - T_o)$$
$$= (1.19\text{ kg/m}^3)(0.035\text{ m}^3\text{/s})(1.005\text{ kJ/kg.°C})[20 - (-10)]°\text{C}$$
$$= \mathbf{1.256\ kW}$$

Therefore, sensible heat will be lost at a rate of 1.335 kJ/s due to infiltration.

4-228 The maximum flow rate of a standard shower head can be reduced from 13.3 to 10.5 L/min by switching to low-flow shower heads. The ratio of the hot-to-cold water flow rates and the amount of electricity saved by a family of four per year by replacing the standard shower heads by the low-flow ones are to be determined.

Assumptions **1** This is a steady-flow process since there is no change with time at any point and thus $\Delta m_{CV} = 0$ and $\Delta E_{CV} = 0$. **2** The kinetic and potential energies are negligible, $ke \cong pe \cong 0$. **3** Heat losses from the system are negligible and thus $\dot{Q} \cong 0$. **4** There are no work interactions involved. **5** .Showers operate at maximum flow conditions during the entire shower. **6** Each member of the household takes a 4-min shower every day. **7** Water is an incompressible substance with constant properties. **8** The efficiency of the electric water heater is 100%.

Properties The density and specific heat of water at room temperature are ρ = 1 kg/L and C = 4.18 kJ/kg.°C (Table A-3).

Analysis (*a*) We take the *mixing chamber* as the system. This is a *control volume* since mass crosses the system boundary during the process. We note that there are two inlets and one exit. The mass and energy balances for this steady-flow system can be expressed in the rate form as follows:

Mass balance:

$$\dot{m}_{in} - \dot{m}_{out} = \Delta \dot{m}_{system}^{\nearrow 0 \text{ (steady)}} = 0$$

$$\dot{m}_{in} = \dot{m}_{out} \rightarrow \dot{m}_1 + \dot{m}_2 = \dot{m}_3$$

Energy balance:

$$\underbrace{\dot{E}_{in} - \dot{E}_{out}}_{\text{Rate of net energy transfer by heat, work, and mass}} = \underbrace{\Delta \dot{E}_{system}^{\nearrow 0 \text{ (steady)}}}_{\text{Rate of change in internal, kinetic, potential, etc. energies}} = 0$$

$$\dot{E}_{in} = \dot{E}_{out}$$

$$\dot{m}_1 h_1 + \dot{m}_2 h_2 = \dot{m}_3 h_3 \quad (\text{since } \dot{Q} \cong 0,\ \dot{W} = 0,\ ke \cong pe \cong 0)$$

Mixture
3
Cold
water
1
Hot
water
2

Combining the mass and energy balances and rearranging,

$$\dot{m}_1 h_1 + \dot{m}_2 h_2 = (\dot{m}_1 + \dot{m}_2) h_3$$

$$\dot{m}_2 (h_2 - h_3) = \dot{m}_1 (h_3 - h_1)$$

Then the ratio of the mass flow rates of the hot water to cold water becomes

$$\frac{\dot{m}_2}{\dot{m}_1} = \frac{h_3 - h_1}{h_2 - h_3} = \frac{C(T_3 - T_1)}{C(T_2 - T_3)} = \frac{T_3 - T_1}{T_2 - T_3} = \frac{(42 - 15)°\text{C}}{(55 - 42)°\text{C}} = \mathbf{2.08}$$

(*b*) The low-flow heads will save water at a rate of

$$\dot{V}_{saved} = [(13.3 - 10.5)\ \text{L/min}](5\ \text{min/person.day})(4\ \text{persons})(365\ \text{days/yr}) = 20{,}440\ \text{L/year}$$

$$\dot{m}_{saved} = \rho \dot{V}_{saved} = (1\ \text{kg/L})(20{,}440\ \text{L/year}) = 20{,}440\ \text{kg/year}$$

Then the energy saved per year becomes

$$\begin{aligned} \text{Energy saved} &= \dot{m}_{saved} C \Delta T = (20{,}440\ \text{kg/year})(4.18\ \text{kJ/kg.°C})(42 - 15)°\text{C} \\ &= 2{,}307{,}000\ \text{kJ/year} \\ &= \mathbf{641\ kWh} \quad (\text{since 1 kWh} = 3600\ \text{kJ}) \end{aligned}$$

Therefore, switching to low-flow shower heads will save about 641 kWh of electricity per year.

4-229 EES solution of this (and other comprehensive problems designated with the *computer icon*) is available to instructors at the *Instructor Manual* section of the *Online Learning Center* (OLC) at www.mhhe.com/cengel-boles. See the Preface for access information.

4-230 A fan is powered by a 0.5 hp motor, and delivers air at a rate of 85 m³/min. The highest possible air velocity at the fan exit is to be determined.

Assumptions **1** This is a steady-flow process since there is no change with time at any point and thus $\Delta m_{CV} = 0$ and $\Delta E_{CV} = 0$. **2** The inlet velocity and the change in potential energy are negligible, $\mathbf{V}_1 \cong 0$ and $\Delta pe \cong 0$. **3** There are no heat and work interactions other than the electrical power consumed by the fan motor. **4** The efficiencies of the motor and the fan are 100% since best possible operation is assumed. **5** Air is an ideal gas with constant specific heats at room temperature.

Properties The density of air is given to be ρ = 1.18 kg/m³. The constant pressure specific heat of air at room temperature is C_p = 1.005 kJ/kg.°C (Table A-2).

Analysis We take the *fan-motor assembly* as the system. This is a *control volume* since mass crosses the system boundary during the process. We note that there is only one inlet and one exit, and thus $\dot{m}_1 = \dot{m}_2 = \dot{m}$.

The velocity of air leaving the fan will be highest when all of the entire electrical energy drawn by the motor is converted to kinetic energy, and the friction between the air layers is zero. In this best possible case, no energy will be converted to thermal energy, and thus the temperature change of air will be zero, $T_2 = T_1$. Then the energy balance for this steady-flow system can be expressed in the rate form as

$$\underbrace{\dot{E}_{in} - \dot{E}_{out}}_{\substack{\text{Rate of net energy transfer} \\ \text{by heat, work, and mass}}} = \underbrace{\Delta \dot{E}_{system}^{\nearrow 0 \text{ (steady)}}}_{\substack{\text{Rate of change in internal, kinetic,} \\ \text{potential, etc. energies}}} = 0$$

$$\dot{E}_{in} = \dot{E}_{out}$$

$$\dot{W}_{e,in} + \dot{m}h_1 = \dot{m}(h_2 + \mathbf{V}_2^2/2) \quad (\text{since } \mathbf{V}_1 \cong 0 \text{ and } \Delta\text{pe} \cong 0)$$

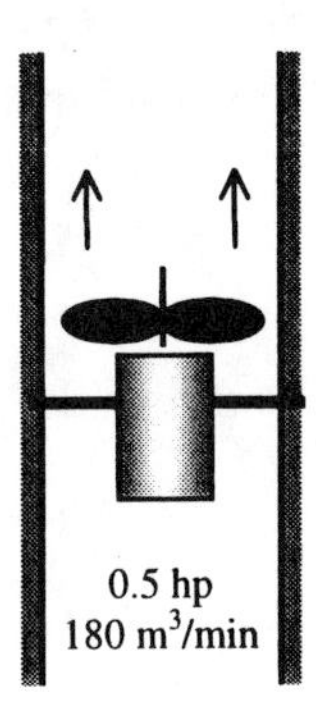

Noting that the temperature and thus enthalpy remains constant, the relation above simplifies further to

$$\dot{W}_{e,in} = \dot{m}\mathbf{V}_2^2/2$$

where

$$\dot{m} = \rho\dot{V} = (1.18 \text{ kg/m}^3)(85 \text{ m}^3/\text{min}) = 100.3 \text{ kg/min} = 1.67 \text{ kg/s}$$

Solving for $\mathbf{V}_2$ and substituting gives

$$\mathbf{V}_2 = \sqrt{\frac{2\dot{W}_{e,in}}{\dot{m}}} = \sqrt{\frac{2(0.5 \text{ hp})}{1.67 \text{ kg/s}}\left(\frac{745.7 \text{ W}}{1 \text{ hp}}\right)\left(\frac{1 \text{ m}^2/\text{s}^2}{1 \text{ W}}\right)} = \mathbf{21.1 \text{ m/s}}$$

Discussion In reality, the velocity will be less because of the inefficiencies of the motor and the fan.

4-231 The average air velocity in the circular duct of an air-conditioning system is not to exceed 10 m/s. If the fan converts 70 percent of the electrical energy into kinetic energy, the size of the fan motor needed and the diameter of the main duct are to be determined.

Assumptions **1** This is a steady-flow process since there is no change with time at any point and thus $\Delta m_{\rm CV} = 0$ and $\Delta E_{\rm CV} = 0$. **2** The inlet velocity is negligible, $\mathbf{V}_1 \cong 0$. **3** There are no heat and work interactions other than the electrical power consumed by the fan motor. **4** Air is an ideal gas with constant specific heats at room temperature.

Properties The density of air is given to be ρ = 1.20 kg/m³. The constant pressure specific heat of air at room temperature is C_p = 1.005 kJ/kg.°C (Table A-2).

Analysis We take the *fan-motor assembly* as the system. This is a *control volume* since mass crosses the system boundary during the process. We note that there is only one inlet and one exit, and thus $\dot{m}_1 = \dot{m}_2 = \dot{m}$. The change in the kinetic energy of air as it is accelerated from zero to .10 m/s at a rate of 180 m³/s is

$$\dot{m} = \rho\dot{V} = (1.20\,\text{kg/m}^3)(180\,\text{m}^3/\text{min}) = 216\,\text{kg/min} = 3.6\,\text{kg/s}$$

$$\Delta\dot{\text{KE}} = \dot{m}\frac{V_2^2 - V_1^2}{2} = (3.6\,\text{kg/s})\frac{(10\,\text{m/s})^2 - 0}{2}\left(\frac{1\,\text{kJ/kg}}{1000\,\text{m}^2/\text{s}^2}\right) = 0.18\,\text{kW}$$

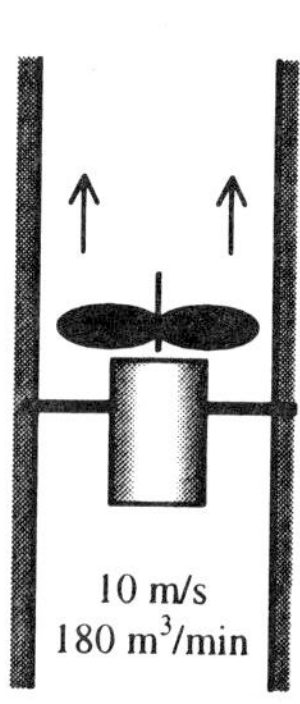

It is stated that this represents 70% of the electrical energy consumed by the motor. Then the total electrical power consumed by the motor is determined to be

$$0.7\dot{W}_{\text{motor}} = \Delta\dot{KE} \quad\rightarrow\quad \dot{W}_{\text{motor}} = \frac{\Delta\dot{KE}}{0.7} = \frac{0.18\text{ kW}}{0.7} = \mathbf{0.257\ kW}$$

The diameter of the main duct is

$$\dot{V} = \mathbf{V}A = \mathbf{V}(\pi D^2/4) \rightarrow D = \sqrt{\frac{4\dot{V}}{\pi\mathbf{V}}} = \sqrt{\frac{4(180\,\text{m}^3/\text{min})}{\pi(10\,\text{m/s})}\left(\frac{1\,\text{min}}{60\,\text{s}}\right)} = \mathbf{0.618\,m}$$

Therefore, the motor should have a rated power of at least 0.257 kW, and the diameter of the duct should be at least 61.8 cm

4-232 An evacuated bottle is surrounded by atmospheric air. A valve is opened, and air is allowed to fill the bottle. The amount of heat transfer through the wall of the bottle when thermal and mechanical equilibrium is established is to be determined.

Assumptions **1** This is an unsteady process since the conditions within the device are changing during the process, but it can be analyzed as a uniform-flow process since the state of fluid at the inlet remains constant. **2** Air is an ideal gas. **3** Kinetic and potential energies are negligible. **4** There are no work interactions involved. **5** The direction of heat transfer is to the air in the bottle (will be verified).

Analysis We take the bottle as the system, which is a control volume since mass crosses the boundary. Noting that the microscopic energies of flowing and nonflowing fluids are represented by enthalpy h and internal energy u, respectively, the mass and energy balances for this uniform-flow system can be expressed as

Mass balance: $m_{in} - m_{out} = \Delta m_{\text{system}} \rightarrow m_i = m_2 \quad (\text{since } m_{out} = m_{initial} = 0)$

Energy balance:

$$\underbrace{E_{in} - E_{out}}_{\substack{\text{Net energy transfer} \\ \text{by heat, work, and mass}}} = \underbrace{\Delta E_{\text{system}}}_{\substack{\text{Change in internal, kinetic,} \\ \text{potential, etc. energies}}}$$

$$Q_{in} + m_i h_i = m_2 u_2 \quad (\text{since } W \cong E_{out} = E_{initial} = ke \cong pe \cong 0)$$

Combining the two balances:

$$Q_{in} = m_2(u_2 - h_i) = m_2(C_v T_2 - C_p T_i)$$

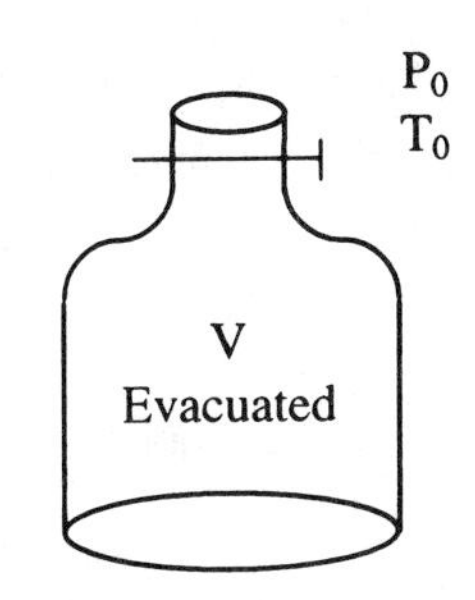

But $T_i = T_2 = T_0$ and $C_p - C_v = R$. Substituting,

$$Q_{in} = m_2(C_v - C_p)T_0 = -m_2 R T_0 = -\frac{P_0 V}{R T_0} R T_0 = -P_0 V$$

Therefore,

$$Q_{\text{out}} = \boldsymbol{P_0 V} \quad \text{(Heat is lost from the tank)}$$

4-233 An adiabatic air compressor is powered by a direct-coupled steam turbine, which is also driving a generator. The net power delivered to the generator is to be determined.

Assumptions **1** This is a steady-flow process since there is no change with time. **2** Kinetic and potential energy changes are negligible. **3** The devices are adiabatic and thus heat transfer is negligible. **4** Air is an ideal gas with variable specific heats.

Properties From the steam tables (Tables A-4 through 6)

$$\left.\begin{array}{l} P_3 = 12.5\text{MPa} \\ T_3 = 500^\circ\text{C} \end{array}\right\} h_3 = 3341.8\text{kJ/kg}$$

and

$$\left.\begin{array}{l} P_4 = 10\text{kPa} \\ x_4 = 0.92 \end{array}\right\} h_4 = h_f + x_4 h_{fg} = 191.83 + (0.92)(2392.8) = 2393.2\text{kJ/kg}$$

From the air table (Table A-17),

$$T_1 = 295\text{ K} \longrightarrow h_1 = 295.17\text{ kJ/kg}$$

$$T_2 = 620\text{ K} \longrightarrow h_2 = 628.07\text{ kJ/kg}$$

Analysis There is only one inlet and one exit for either device, and thus $\dot{m}_{in} = \dot{m}_{out} = \dot{m}$. We take either the turbine or the compressor as the system, which is a control volume since mass crosses the boundary. The energy balance for either steady-flow system can be expressed in the rate form as

$$\underbrace{\dot{E}_{in} - \dot{E}_{out}}_{\substack{\text{Rate of net energy transfer} \\ \text{by heat, work, and mass}}} = \underbrace{\Delta \dot{E}_{\text{system}}^{\nearrow 0 \text{ (steady)}}}_{\substack{\text{Rate of change in internal, kinetic,} \\ \text{potential, etc. energies}}} = 0$$

$$\dot{E}_{in} = \dot{E}_{out}$$

For the turbine and the compressor it becomes

Compressor: $\dot{W}_{\text{comp, in}} + \dot{m}_{\text{air}} h_1 = \dot{m}_{\text{air}} h_2 \rightarrow \dot{W}_{\text{comp, in}} = \dot{m}_{\text{air}}(h_2 - h_1)$

Turbine: $\dot{m}_{\text{steam}} h_3 = \dot{W}_{\text{turb, out}} + \dot{m}_{\text{steam}} h_4 \rightarrow \dot{W}_{\text{turb, out}} = \dot{m}_{\text{steam}}(h_3 - h_4)$

Substituting,

$$\dot{W}_{\text{comp,in}} = (10\text{kg/s})(628.07 - 295.17)\text{kJ/kg} = 3329\text{ kW}$$

$$\dot{W}_{\text{turb,out}} = (25\text{kg/s})(3341.8 - 2393.2)\text{kJ/kg} = 23{,}715\text{ kW}$$

Therefore,

$$\dot{W}_{\text{net,out}} = \dot{W}_{\text{turb,out}} - \dot{W}_{\text{comp,in}} = 23{,}715 - 3329 = \mathbf{20{,}386\text{ kW}}$$

4-234 Water is heated from 16°C to 43°C by an electric resistance heater placed in the water pipe as it flows through a showerhead steadily at a rate of 10 L/min. The electric power input to the heater, and the money that will be saved during a 10-min shower by installing a heat exchanger with an effectiveness of 0.50 are to be determined.

Assumptions **1** This is a steady-flow process since there is no change with time at any point within the system and thus $\Delta m_{CV} = 0$ and $\Delta E_{CV} = 0$,. **2** Water is an incompressible substance with constant specific heats. **3** The kinetic and potential energy changes are negligible, $\Delta ke \cong \Delta pe \cong 0$. **4** Heat losses from the pipe are negligible.

Properties The density and specific heat of water at room temperature are ρ = 1 kg/L and C = 4.18 kJ/kg· °C (Table A-3).

Analysis We take the pipe as the system. This is a *control volume* since mass crosses the system boundary during the process. We observe that there is only one inlet and one exit and thus $\dot{m}_1 = \dot{m}_2 = \dot{m}$. Then the energy balance for this steady-flow system can be expressed in the rate form as

$$\underbrace{\dot{E}_{in} - \dot{E}_{out}}_{\substack{\text{Rate of net energy transfer} \\ \text{by heat, work, and mass}}} = \underbrace{\Delta \dot{E}_{\text{system}}^{\nearrow 0 \text{ (steady)}}}_{\substack{\text{Rate of change in internal, kinetic,} \\ \text{potential, etc. energies}}} = 0 \rightarrow \dot{E}_{in} = \dot{E}_{out}$$

$$\dot{W}_{\text{e,in}} + \dot{m}h_1 = \dot{m}h_2 \quad (\text{since } \Delta\text{ke} \cong \Delta\text{pe} \cong 0)$$

$$\dot{W}_{\text{e,in}} = \dot{m}(h_2 - h_1) = \dot{m}[C(T_2 - T_1) + v(P_2 - P_1)^{\nearrow 0}] = \dot{m}C(T_2 - T_1)$$

where

$$\dot{m} = \rho\dot{V} = (1 \text{ kg/L})(10 \text{ L/min}) = 10 \text{ kg/min}$$

Substituting,

$$\dot{W}_{\text{e,in}} = (10/60 \text{kg/s})(4.18 \text{ kJ/kg}\cdot{}^\circ\text{C})(43-16)^\circ\text{C} = \mathbf{18.8 \text{ kW}}$$

WATER

16°C → 43°C

7 kW

The energy recovered by the heat exchanger is

$$\dot{Q}_{\text{saved}} = \varepsilon\dot{Q}_{\max} = \varepsilon\dot{m}C(T_{\max} - T_{\min})$$
$$= 0.5(10/60 \text{kg/s})(4.18 \text{ kJ/kg.}^\circ\text{C})(39-16)^\circ\text{C}$$
$$= 8.0 \text{kJ/s} = 8.0 \text{kW}$$

Therefore, 8.0 kW less energy is needed in this case, and the required electric power in this case reduces to

$$\dot{W}_{\text{in,new}} = \dot{W}_{\text{in,old}} - \dot{Q}_{\text{saved}} = 18.8 - 8.0 = \mathbf{10.8 \text{ kW}}$$

The money saved during a 10-min shower as a result of installing this heat exchanger is

$$(8.0 \text{ kW})(10/60 \text{ h})(8.5 \text{ cents/kWh}) = \mathbf{11.3 \text{ cents}}$$

4-235 EES solution of this (and other comprehensive problems designated with the *computer icon*) is available to instructors at the *Instructor Manual* section of the *Online Learning Center* (OLC) at www.mhhe.com/cengel-boles. See the Preface for access information.

4-236 [*Also solved by EES on enclosed CD*] Steam expands in a turbine steadily. The mass flow rate of the steam, the exit velocity, and the power output are to be determined.

Assumptions **1** This is a steady-flow process since there is no change with time. **2** Potential energy changes are negligible.

Properties From the steam tables (Tables A-4 through 6)

$$\left.\begin{array}{l} P_1 = 10 \text{ MPa} \\ T_1 = 550°\text{C} \end{array}\right\} \begin{array}{l} v_1 = 0.03564 \text{ m}^3/\text{kg} \\ h_1 = 3500.9 \text{ kJ/kg} \end{array}$$

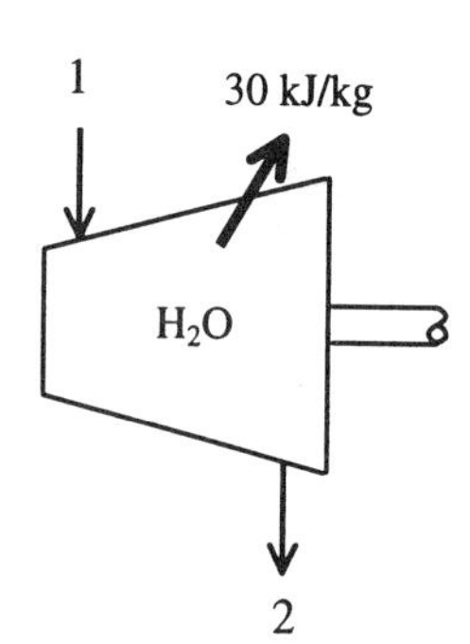

and

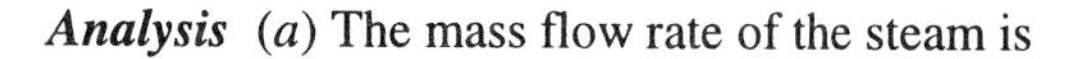

$$\left.\begin{array}{l} P_2 = 25 \text{ kPa} \\ x_2 = 0.95 \end{array}\right\} \begin{array}{l} v_2 = v_f + x_2 v_{fg} = 0.00102 + (0.95)(6.203) = 5.894 \text{ m}^3/\text{kg} \\ h_2 = h_f + x_2 h_{fg} = 271.93 + (0.95)(2346.3) = 2500.9 \text{ kJ/kg} \end{array}$$

Analysis (a) The mass flow rate of the steam is

$$\dot{m} = \frac{1}{v_1}\mathbf{V}_1 A_1 = \frac{1}{0.03564\text{m}^3/\text{kg}}(60\text{m/s})(0.015\text{m}^2) = \mathbf{25.3kg/s}$$

(b) There is only one inlet and one exit, and thus $\dot{m}_1 = \dot{m}_2 = \dot{m}$. Then the exit velocity is determined from

$$\dot{m} = \frac{1}{v_2}\mathbf{V}_2 A_2 \longrightarrow \mathbf{V}_2 = \frac{\dot{m} v_2}{A_2} = \frac{(25.3 \text{ kg/s})(5.894 \text{ m}^3/\text{kg})}{0.14 \text{ m}^2} = \mathbf{1065 \text{ m/s}}$$

(c) We take the turbine as the system, which is a control volume since mass crosses the boundary. The energy balance for this steady-flow system can be expressed in the rate form as

$$\underbrace{\dot{E}_{in} - \dot{E}_{out}}_{\substack{\text{Rate of net energy transfer}\\ \text{by heat, work, and mass}}} = \underbrace{\Delta \dot{E}_{\text{system}}^{\nearrow 0 \text{ (steady)}}}_{\substack{\text{Rate of change in internal, kinetic,}\\ \text{potential, etc. energies}}} = 0$$

$$\dot{E}_{in} = \dot{E}_{out}$$

$$\dot{m}(h_1 + \mathbf{V}_1^2/2) = \dot{W}_{\text{out}} + \dot{Q}_{\text{out}} + \dot{m}(h_2 + \mathbf{V}_2^2/2) \quad (\text{since } \Delta\text{pe} \cong 0)$$

$$\dot{W}_{\text{out}} = -\dot{Q}_{\text{out}} - \dot{m}\left(h_2 - h_1 + \frac{\mathbf{V}_2^2 - \mathbf{V}_1^2}{2}\right)$$

Then the power output of the turbine is determined by substituting to be

$$\dot{W}_{\text{out}} = -(25.3 \times 30)\text{kJ/s} - (25.3\text{kg/s})\left(2500.9 - 3500.9 + \frac{(1065\text{m/s})^2 - (60\text{m/s})^2}{2}\left(\frac{1\text{kJ/kg}}{1000\text{m}^2/\text{s}^2}\right)\right)$$

$$= \mathbf{10{,}240kW}$$

4-237 EES solution of this (and other comprehensive problems designated with the *computer icon*) is available to instructors at the *Instructor Manual* section of the *Online Learning Center* (OLC) at www.mhhe.com/cengel-boles. See the Preface for access information.

4-238E Refrigerant-134a is compressed steadily by a compressor. The mass flow rate of the refrigerant and the exit temperature are to be determined. √

Assumptions **1** This is a steady-flow process since there is no change with time. **2** Kinetic and potential energy changes are negligible. **3** The device is adiabatic and thus heat transfer is negligible.

Properties From the refrigerant tables (Tables A-11E through A-13E)

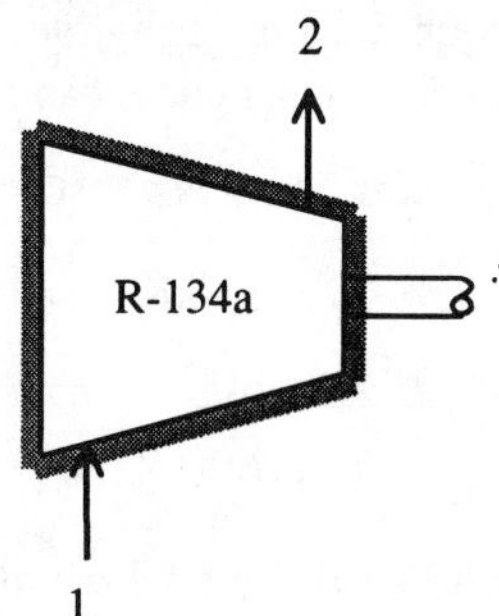

$$\left.\begin{array}{l} P_1 = 15 \text{ psia} \\ T_1 = 20^\circ\text{F} \end{array}\right\} \begin{array}{l} v_1 = 3.2468 \text{ ft}^3/\text{lbm} \\ h_1 = 106.34 \text{ Btu/lbm} \end{array}$$

Analysis (*a*) The mass flow rate of refrigerant is

$$\dot{m} = \frac{\dot{V}_1}{v_1} = \frac{10 \text{ ft}^3/\text{s}}{3.2468 \text{ ft}^3/\text{lbm}} = \mathbf{3.08 \text{ lbm/s}}$$

(*b*) There is only one inlet and one exit, and thus $\dot{m}_1 = \dot{m}_2 = \dot{m}$. We take the compressor as the system, which is a control volume since mass crosses the boundary. The energy balance for this steady-flow system can be expressed in the rate form as

$$\underbrace{\dot{E}_{in} - \dot{E}_{out}}_{\text{Rate of net energy transfer by heat, work, and mass}} = \underbrace{\Delta \dot{E}_{\text{system}}^{\nearrow 0 \text{ (steady)}}}_{\text{Rate of change in internal, kinetic, potential, etc. energies}} = 0$$

$$\dot{E}_{in} = \dot{E}_{out}$$

$$\dot{W}_{\text{in}} + \dot{m}h_1 = \dot{m}h_2 \quad (\text{since } \dot{Q} \cong \Delta ke \cong \Delta pe \cong 0)$$

$$\dot{W}_{\text{in}} = \dot{m}(h_2 - h_1)$$

Substituting,

$$(60 \text{ hp})\left(\frac{0.7068 \text{Btu/s}}{1 \text{hp}}\right) = (3.08 \text{lbm/s})(h_2 - 106.34)\text{Btu/lbm}$$

$$h_2 = 120.11 \text{Btu/lbm}$$

Then the exit temperature becomes

$$\left.\begin{array}{l} P_2 = 120 \text{psia} \\ h_2 = 120.11 \text{Btu/lbm} \end{array}\right\} T_2 = \mathbf{114.6^\circ F}$$

4-239 Air is preheated by the exhaust gases of a gas turbine in a regenerator. For a specified heat transfer rate, the exit temperature of air and the mass flow rate of exhaust gases are to be determined.

Assumptions **1** This is a steady-flow process since there is no change with time. **2** Kinetic and potential energy changes are negligible. **3** There are no work interactions. **4** Heat loss from the regenerator to the surroundings is negligible and thus heat transfer from the hot fluid is equal to the heat transfer to the cold fluid. **5** Exhaust gases can be treated as air. **6** Air is an ideal gas with variable specific heats.

Properties The gas constant of air is 0.287 kPa.m³/kg.K (Table A-1). The enthalpies of air are

$$T_1 = 550\text{ K} \rightarrow h_1 = 554.71\text{ kJ/kg}$$
$$T_3 = 800\text{ K} \rightarrow h_3 = 821.95\text{ kJ/kg}$$
$$T_4 = 600\text{ K} \rightarrow h_4 = 607.02\text{ kJ/kg}$$

Analysis (*a*) We take the *air side* of the heat exchanger as the system, which is a control volume since mass crosses the boundary. There is only one inlet and one exit, and thus $\dot{m}_1 = \dot{m}_2 = \dot{m}$. The energy balance for this steady-flow system can be expressed in the rate form as

$$\underbrace{\dot{E}_{in} - \dot{E}_{out}}_{\substack{\text{Rate of net energy transfer}\\ \text{by heat, work, and mass}}} = \underbrace{\Delta\dot{E}_{system}^{\nearrow 0\text{ (steady)}}}_{\substack{\text{Rate of change in internal, kinetic,}\\ \text{potential, etc. energies}}} = 0$$

$$\dot{E}_{in} = \dot{E}_{out}$$

$$\dot{Q}_{in} + \dot{m}_{air}h_1 = \dot{m}_{air}h_2 \quad (\text{since } \dot{W} = \Delta ke \cong \Delta pe \cong 0)$$

$$\dot{Q}_{in} = \dot{m}_{air}(h_2 - h_1)$$

Substituting,

$$3200\text{kJ/s} = (800/60\text{kg/s})(h_2 - 554.71\text{kJ/kg}) \rightarrow h_2 = 794.71 \text{ kJ/kg}$$

Then from Table A-17 we read $\quad T_2 = \mathbf{775.1\ K}$

(*b*) Treating the exhaust gases as an ideal gas, the mass flow rate of the exhaust gases is determined from the steady-flow energy relation applied only to the exhaust gases,

$$\dot{E}_{in} = \dot{E}_{out}$$

$$\dot{m}_{exhaust}h_3 = \dot{Q}_{out} + \dot{m}_{exhaust}h_4 \quad (\text{since } \dot{W} = \Delta ke \cong \Delta pe \cong 0)$$

$$\dot{Q}_{out} = \dot{m}_{exhaust}(h_3 - h_4)$$

$$3200\text{kJ/s} = \dot{m}_{exhaust}(821.95 - 607.02)\text{kJ/kg}$$

It yields

$$\dot{m}_{exhaust} = \mathbf{14.9\ kg/s}$$

4-240 Water is to be heated steadily from 20°C to 55°C by an electrical resistor inside an insulated pipe. The power rating of the resistance heater and the average velocity of the water are to be determined.

Assumptions **1** This is a steady-flow process since there is no change with time at any point within the system and thus $\Delta m_{CV} = 0$ and $\Delta E_{CV} = 0$. **2** Water is an incompressible substance with constant specific heats. **3** The kinetic and potential energy changes are negligible, $\Delta ke \cong \Delta pe \cong 0$. **4** The pipe is insulated and thus the heat losses are negligible.

Properties The density and specific heat of water at room temperature are ρ = 1000 kg/m³ and C = 4.18 kJ/kg· °C (Table A-3).

Analysis (*a*) We take the pipe as the system. This is a *control volume* since mass crosses the system boundary during the process. Also, there is only one inlet and one exit and thus $\dot{m}_1 = \dot{m}_2 = \dot{m}$. The energy balance for this steady-flow system can be expressed in the rate form as

$$\underbrace{\dot{E}_{in} - \dot{E}_{out}}_{\text{Rate of net energy transfer by heat, work, and mass}} = \underbrace{\Delta \dot{E}_{system}^{\nearrow 0 \text{ (steady)}}}_{\text{Rate of change in internal, kinetic, potential, etc. energies}} = 0$$

$$\dot{E}_{in} = \dot{E}_{out}$$

$$\dot{W}_{e,in} + \dot{m}h_1 = \dot{m}h_2 \quad (\text{since } \dot{Q}_{out} \cong \Delta\text{ke} \cong \Delta\text{pe} \cong 0)$$

$$\dot{W}_{e,in} = \dot{m}(h_2 - h_1) = \dot{m}[C(T_2 - T_1) + v\Delta P^{\nearrow 0}] = \dot{m}C(T_2 - T_1)$$

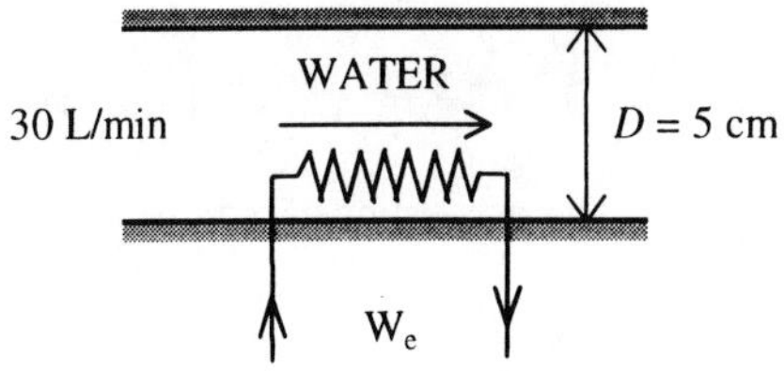

30 L/min | WATER | D = 5 cm | W_e

The mass flow rate of water through the pipe is

$$\dot{m} = \rho\dot{V}_1 = (1000 \text{ kg/m}^3)(0.030 \text{ m}^3/\text{min}) = 30 \text{ kg/min}$$

Therefore,

$$\dot{W}_{e,in} = \dot{m}C(T_2 - T_1) = (30/60 \text{ kg/s})(4.18 \text{ kJ/kg}\cdot^\circ\text{C})(55 - 20)^\circ\text{C} = \mathbf{73.2\ kW}$$

(*b*) The average velocity of water through the pipe is determined from

$$\mathbf{V}_1 = \frac{\dot{V}_1}{A_1} = \frac{\dot{V}}{\pi r^2} = \frac{0.030 \text{ m}^3/\text{min}}{\pi(0.025 \text{ m})^2} = \mathbf{15.3\ m/min}$$

4-241 Two rigid tanks that contain water at different states are connected by a valve. The valve is opened and the two tanks come to the same state at the temperature of the surroundings. The final pressure and the amount of heat transfer are to be determined.

Assumptions **1** The tanks are stationary and thus the kinetic and potential energy changes are zero. **2** The tank is insulated and thus heat transfer is negligible. **3** There are no work interactions.

Analysis We take the entire contents of the tank as the system. This is a closed system since no mass enters or leaves. Noting that the volume of the system is constant and thus there is no boundary work, the energy balance for this stationary closed system can be expressed as

$$\underbrace{E_{in} - E_{out}}_{\substack{\text{Net energy transfer}\\ \text{by heat, work, and mass}}} = \underbrace{\Delta E_{system}}_{\substack{\text{Change in internal, kinetic,}\\ \text{potential, etc. energies}}}$$

$$-Q_{out} = \Delta U = (\Delta U)_A + (\Delta U)_B \quad \text{(since } W = \text{KE} = \text{PE} = 0)$$

$$Q_{out} = -[U_{2,A+B} - U_{1,A} - U_{1,B}]$$

$$= -[m_{2,\text{total}} u_2 - (m_1 u_1)_A - (m_1 u_1)_B]$$

H2O
400 kPa
A
Q
H2O
200 kPa
B

The properties of water in each tank are (Tables A-4 through A-6)

Tank A:

$$\left.\begin{array}{l} P_1 = 400\text{kPa} \\ x_1 = 0.80 \end{array}\right\} \begin{array}{ll} v_f = 0.001084, & v_g = 0.4625\text{m}^3/\text{kg} \\ u_f = 604.31, & u_{fg} = 1949.3\text{kJ/kg} \end{array}$$

$$v_{1,A} = v_f + x_1 v_{fg} = 0.001084 + [0.8 \times (0.4625 - 0.001084)] = 0.3702\text{m}^3/\text{kg}$$

$$u_{1,A} = u_f + x_1 u_{fg} = 604.31 + (0.8 \times 1949.3) = 2163.75\text{kJ/kg}$$

Tank B:

$$\left.\begin{array}{l} P_1 = 200\text{ kPa} \\ T_1 = 250\,^\circ\text{C} \end{array}\right\} \begin{array}{l} v_{1,B} = 1.1988\text{ m}^3/\text{kg} \\ u_{1,B} = 2731.2\text{ kJ/kg} \end{array}$$

$$m_{1,A} = \frac{V_A}{v_{1,A}} = \frac{0.2\text{ m}^3}{0.3702\text{ m}^3/\text{kg}} = 0.540\text{ kg}$$

$$m_{1,B} = \frac{V_B}{v_{1,B}} = \frac{0.5\text{ m}^3}{1.1988\text{ m}^3/\text{kg}} = 0.417\text{ kg}$$

$$m_t = m_{1,A} + m_{1,B} = 0.540 + 0.417 = 0.957\text{ kg}$$

$$v_2 = \frac{V_t}{m_t} = \frac{0.7\text{ m}^3}{0.957\text{ kg}} = 0.731\text{ m}^3/\text{kg}$$

$$\left.\begin{array}{l} T_2 = 25\,^\circ\text{C} \\ v_2 = 0.731\text{ m}^3/\text{kg} \end{array}\right\} \begin{array}{ll} v_f = 0.001003, & v_g = 43.36\text{ m}^3/\text{kg} \\ u_f = 104.88, & u_{fg} = 2304.9\text{ kJ/kg} \end{array}$$

Thus at the final state the system will be a saturated liquid-vapor mixture since $v_f < v_2 < v_g$. Then the final pressure must be

$$P_2 = P_{sat\,@\,25\,^\circ C} = \mathbf{3.169\ kPa}$$

Also,

$$x_2 = \frac{v_2 - v_f}{v_{fg}} = \frac{0.731 - 0.001}{43.36 - 0.001} = 0.0168$$

$$u_2 = u_f + x_2 u_{fg} = 104.88 + (0.0168 \times 2304.9) = 143.60\text{kJ/kg}$$

Substituting,

$$Q_{out} = -[(0.957)(143.6) - (0.540)(2163.75) - (0.417)(2731.2)] = \mathbf{2170\ kJ}$$

4-242 EES solution of this (and other comprehensive problems designated with the *computer icon*) is available to instructors at the *Instructor Manual* section of the *Online Learning Center* (OLC) at www.mhhe.com/cengel-boles. See the Preface for access information.

4-243 A rigid tank filled with air is connected to a cylinder with zero clearance. The valve is opened, and air is allowed to flow into the cylinder. The temperature is maintained at 30°C at all times. The amount of heat transfer with the surroundings is to be determined.

Assumptions **1** Air is an ideal gas. **2** The kinetic and potential energy changes are negligible, $\Delta ke \cong \Delta pe \cong 0$. **3** There are no work interactions involved other than the boundary work.

Properties The gas constant of air is $R = 0.287$ kPa.m^3/kg.K (Table A-1).

Analysis We take the entire air in the tank and the cylinder to be the system. This is a closed system since no mass crosses the boundary of the system. The energy balance for this closed system can be expressed as

$$\underbrace{E_{in} - E_{out}}_{\substack{\text{Net energy transfer}\\ \text{by heat, work, and mass}}} = \underbrace{\Delta E_{\text{system}}}_{\substack{\text{Change in internal, kinetic,}\\ \text{potential, etc. energies}}}$$

$$Q_{in} - W_{b,out} = \Delta U = m(u_2 - u_1) = 0$$

$$Q_{in} = W_{b,out}$$

since $u = u(T)$ for ideal gases, and thus $u_2 = u_1$ when $T_1 = T_2$. The initial volume of air is

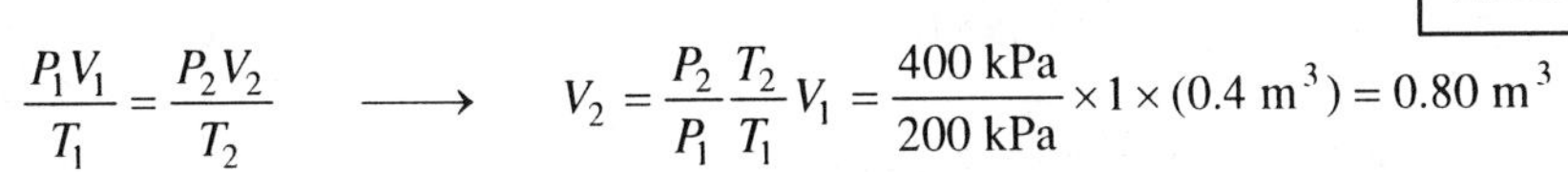

$$\frac{P_1 V_1}{T_1} = \frac{P_2 V_2}{T_2} \longrightarrow V_2 = \frac{P_2}{P_1}\frac{T_2}{T_1}V_1 = \frac{400 \text{ kPa}}{200 \text{ kPa}} \times 1 \times (0.4 \text{ m}^3) = 0.80 \text{ m}^3$$

The pressure at the piston face always remains constant at 200 kPa. Thus the boundary work done during this process is

$$W_{b,out} = \int_1^2 PdV = P_2(V_2 - V_1) = (200\text{kPa})(0.8 - 0.4)\text{m}^3\left(\frac{1\text{kJ}}{1\text{kPa}\cdot\text{m}^3}\right) = 80\text{kJ}$$

Therefore, the heat transfer is determined from the energy balance to be

$$W_{b,out} = Q_{in} = \mathbf{80\ kJ}$$

4-244 A well-insulated room is heated by a steam radiator, and the warm air is distributed by a fan. The average temperature in the room after 30 min is to be determined.

Assumptions **1** Air is an ideal gas with constant specific heats at room temperature. **2** The kinetic and potential energy changes are negligible. **3** The air pressure in the room remains constant and thus the air expands as it is heated, and some warm air escapes.

Properties The gas constant of air is $R = 0.287$ kPa.m^3/kg.K (Table A-1). Also, $C_p = 1.005$ kJ/kg.K for air at room temperature (Table A-2).

Analysis We first take the radiator as the system. This is a closed system since no mass enters or leaves. The energy balance for this closed system can be expressed as

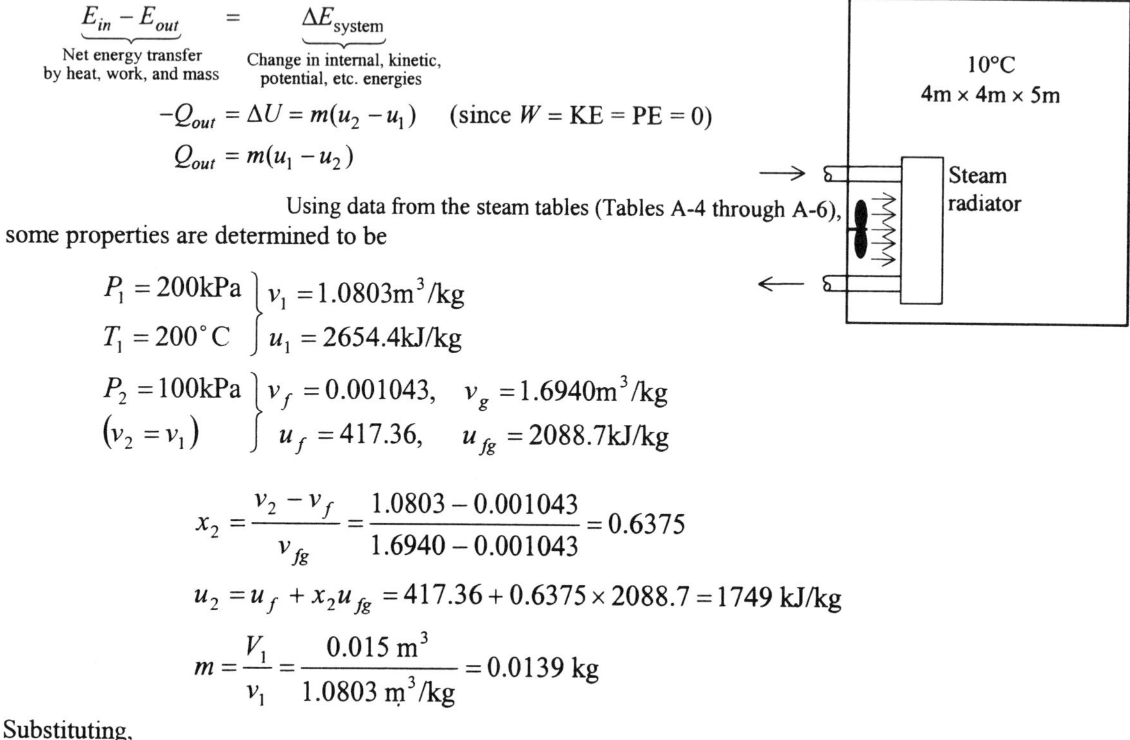

$$\underbrace{E_{in} - E_{out}}_{\substack{\text{Net energy transfer}\\ \text{by heat, work, and mass}}} = \underbrace{\Delta E_{\text{system}}}_{\substack{\text{Change in internal, kinetic,}\\ \text{potential, etc. energies}}}$$

$$-Q_{out} = \Delta U = m(u_2 - u_1) \quad (\text{since } W = \text{KE} = \text{PE} = 0)$$

$$Q_{out} = m(u_1 - u_2)$$

Using data from the steam tables (Tables A-4 through A-6), some properties are determined to be

$$\left.\begin{array}{l} P_1 = 200\text{kPa} \\ T_1 = 200^\circ\text{C} \end{array}\right\} \begin{array}{l} v_1 = 1.0803\text{m}^3/\text{kg} \\ u_1 = 2654.4\text{kJ/kg} \end{array}$$

$$\left.\begin{array}{l} P_2 = 100\text{kPa} \\ (v_2 = v_1) \end{array}\right\} \begin{array}{ll} v_f = 0.001043, & v_g = 1.6940\text{m}^3/\text{kg} \\ u_f = 417.36, & u_{fg} = 2088.7\text{kJ/kg} \end{array}$$

$$x_2 = \frac{v_2 - v_f}{v_{fg}} = \frac{1.0803 - 0.001043}{1.6940 - 0.001043} = 0.6375$$

$$u_2 = u_f + x_2 u_{fg} = 417.36 + 0.6375 \times 2088.7 = 1749 \text{ kJ/kg}$$

$$m = \frac{V_1}{v_1} = \frac{0.015 \text{ m}^3}{1.0803 \text{ m}^3/\text{kg}} = 0.0139 \text{ kg}$$

Substituting,

$$Q_{\text{out}} = (0.0139 \text{ kg})(2654.4 - 1749)\text{kJ/kg} = 12.6 \text{ kJ}$$

The volume and the mass of the air in the room are $V = 4\times4\times5 = 80$ m³ and

$$m_{air} = \frac{P_1 V_1}{RT_1} = \frac{(100\text{kPa})(80\text{m}^3)}{(0.2870\text{kPa}\cdot\text{m}^3/\text{kg}\cdot\text{K})(283\text{K})} = 98.5\text{kg}$$

The amount of fan work done in 30 min is

$$W_{fan,in} = \dot{W}_{fan,in}\Delta t = (0.120 \text{ kJ/s})(30 \times 60 \text{ s}) = 216 \text{ kJ}$$

We now take the air in the room as the system. The energy balance for this closed system is expressed as

$$E_{in} - E_{out} = \Delta E_{\text{system}}$$

$$Q_{in} + W_{fan,in} - W_{b,out} = \Delta U$$

$$Q_{in} + W_{fan,in} = \Delta H \cong mC_p(T_2 - T_1)$$

since the boundary work and ΔU combine into ΔH for a constant pressure expansion or compression process. It can also be expressed as

$$(\dot{Q}_{in} + \dot{W}_{fan,in})\Delta t = mC_{p,ave}(T_2 - T_1)$$

Substituting, (12.6 kJ) + (216 kJ) = (98.5 kg)(1.005 kJ/kg°C)(T_2 - 10)°C

which yields T_2 = **12.3°C**

Therefore, the air temperature in the room rises from 10°C to 12.3°C in 30 min.

4-245 A cylinder equipped with a set of stops for the piston is initially filled with saturated liquid-vapor mixture of water a specified pressure. Heat is transferred to the water until the volume increases by 20%. The initial and final temperature, the mass of the liquid when the piston starts moving, and the work done during the process are to be determined, and the process is to be shown on a *P*-*v* diagram.

Assumptions **1** The kinetic and potential energy changes are negligible, $\Delta ke \cong \Delta pe \cong 0$. **2** The thermal energy stored in the cylinder itself is negligible. **3** The compression or expansion process is quasi-equilibrium.

Analysis (*a*) Initially the system is a saturated mixture at 100 kPa pressure, and thus the initial temperature is

$$T_1 = T_{sat\,@\,100\text{ kPa}} = \mathbf{99.63^\circ C}$$

The total initial volume is

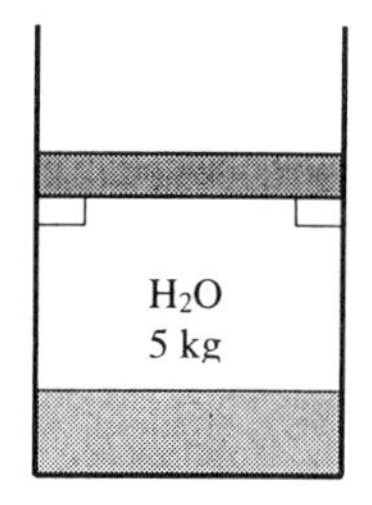

$$V_1 = m_f v_f + m_g v_g = 2 \times 0.001043 + 3 \times 1.6940 = 5.08\text{ m}^3$$

Then the total and specific volumes at the final state are

$$V_3 = 1.2V_1 = 1.2 \times 5.08 = 6.10\text{ m}^3$$

$$v_3 = \frac{V_3}{m} = \frac{6.10\text{ m}^3}{5\text{ kg}} = 1.22\text{ m}^3/\text{kg}$$

Thus,

$$\left.\begin{array}{l} P_3 = 200\text{ kPa} \\ v_3 = 1.22\text{ m}^3/\text{kg} \end{array}\right\} T_3 = \mathbf{259.0^\circ C}$$

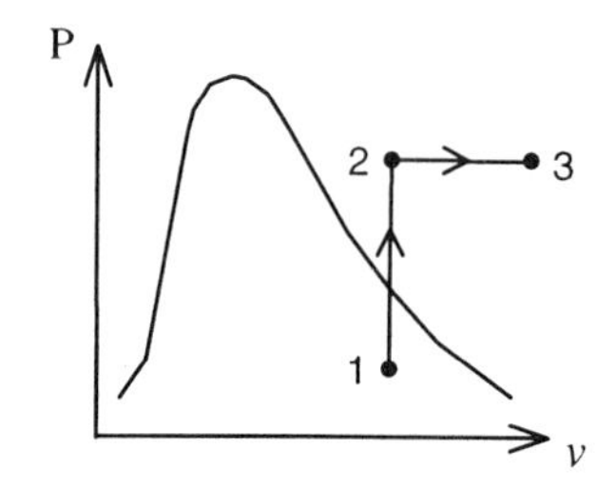

(*b*) When the piston first starts moving, $P_2 = 200$ kPa and $V_2 = V_1 = 5.08$ m^3. The specific volume at this state is

$$v_2 = \frac{V_2}{m} = \frac{5.08\text{ m}^3}{5\text{kg}} = 1.016\text{ m}^3/\text{kg}$$

which is greater than $v_g = 0.8857$ m^3/kg at 200 kPa. Thus **no liquid** is left in the cylinder when the piston starts moving.

(*c*) No work is done during process 1-2 since $V_1 = V_2$. The pressure remains constant during process 2-3 and the work done during this process is

$$W_b = \int_2^3 P dV = P_2(V_3 - V_2) = (200\text{ kPa})(6.10 - 5.08)\text{m}^3\left(\frac{1\text{ kJ}}{1\text{ kPa}\cdot\text{m}^3}\right) = \mathbf{204 kJ}$$

4-246 An insulated cylinder is divided into two parts. One side of the cylinder contains N_2 gas and the other side contains He gas at different states. The final equilibrium temperature in the cylinder when thermal equilibrium is established is to be determined for the cases of the piston being fixed and moving freely.

Assumptions **1** Both N_2 and He are ideal gases with constant specific heats. **2** The energy stored in the container itself is negligible. **3** The cylinder is well-insulated and thus heat transfer is negligible.

Properties The gas constants and the constant volume specific heats are R = 0.2968 kPa.m³/kg.K is C_v = 0.743 kJ/kg· °C for N_2, and R = 2.0769 kPa.m³/kg.K is C_v = 3.1156 kJ/kg· °C for He (Tables A-1 and A-2)

Analysis The mass of each gas in the cylinder is

$$m_{N_2} = \left(\frac{P_1V_1}{RT_1}\right)_{N_2} = \frac{(500\text{ kPa})(1\text{ m}^3)}{(0.2968\text{kPa}\cdot\text{m}^3/\text{kg}\cdot\text{K})(353\text{ K})} = 4.77\text{ kg}$$

$$m_{He} = \left(\frac{P_1V_1}{RT_1}\right)_{He} = \frac{(500\text{ kPa})(1\text{ m}^3)}{(2.0769\text{ kPa}\cdot\text{m}^3/\text{kg}\cdot\text{K})(298\text{ K})} = 0.808\text{ kg}$$

N_2	He
1 m³	1 m³
500 kPa	500 kPa
80°C	25°C

Taking the entire contents of the cylinder as our system, the 1st law relation can be written as

$$\underbrace{E_{in} - E_{out}}_{\substack{\text{Net energy transfer}\\ \text{by heat, work, and mass}}} = \underbrace{\Delta E_{\text{system}}}_{\substack{\text{Change in internal, kinetic,}\\ \text{potential, etc. energies}}}$$

$$0 = \Delta U = (\Delta U)_{N_2} + (\Delta U)_{\text{He}}$$

$$0 = [mC_v(T_2 - T_1)]_{N_2} + [mC_v(T_2 - T_1)]_{\text{He}}$$

Substituting,

$$(4.77\text{kg})(0.743\text{kJ/kg}\cdot^\circ\text{C})(T_f - 80)^\circ\text{C} + (0.808\text{kg})(3.1156\text{kJ/kg}\cdot^\circ\text{C})(T_f - 25)^\circ\text{C} = 0$$

It gives T_f = **57.2°C**

where T_f is the final equilibrium temperature in the cylinder.

The answer would be the **same** if the piston were not free to move since it would effect only pressure, and not the specific heats.

Discussion Using the relation PV = NR_uT, it can be shown that the total number of moles in the cylinder is 0.170 + 0.202 = 0.372 kmol, and the final pressure is 510.6 kPa.

4-247 An insulated cylinder is divided into two parts. One side of the cylinder contains N_2 gas and the other side contains He gas at different states. The final equilibrium temperature in the cylinder when thermal equilibrium is established is to be determined for the cases of the piston being fixed and moving freely.

Assumptions **1** Both N_2 and He are ideal gases with constant specific heats. **2** The energy stored in the container itself, except the piston, is negligible. **3** The cylinder is well-insulated and thus heat transfer is negligible. **4** Initially, the piston is at the average temperature of the two gases.

Properties The gas constants and the constant volume specific heats are $R = 0.2968$ kPa.m^3/kg.K is $C_v = 0.743$ kJ/kg·°C for N_2, and $R = 2.0769$ kPa.m^3/kg.K is $C_v = 3.1156$ kJ/kg·°C for He. (Tables A-1 and A-2). The specific heat of copper piston is $C = 0.386$ kJ/kg·°C (Table A-3).

Analysis The mass of each gas in the cylinder is

$$m_{N_2} = \left(\frac{P_1 V_1}{RT_1}\right)_{N_2} = \frac{(500\text{kPa})(1\text{m}^3)}{(0.2968\text{kPa}\cdot\text{m}^3/\text{kg}\cdot\text{K})(353\text{K})} = 4.77\text{kg}$$

$$m_{He} = \left(\frac{P_1 V_1}{RT_1}\right)_{He} = \frac{(500\text{kPa})(1\text{m}^3)}{(2.0769\text{kPa}\cdot\text{m}^3/\text{kg}\cdot\text{K})(353\text{K})} = 0.808\text{kg}$$

N_2
1 m^3
500 kPa
80°C

He
1 m^3
500 kPa
25°C

Coppe

Taking the entire contents of the cylinder as our system, the 1st law relation can be written as

$$\underbrace{E_{in} - E_{out}}_{\substack{\text{Net energy transfer}\\ \text{by heat, work, and mass}}} = \underbrace{\Delta E_{\text{system}}}_{\substack{\text{Change in internal, kinetic,}\\ \text{potential, etc. energies}}}$$

$$0 = \Delta U = (\Delta U)_{N_2} + (\Delta U)_{He} + (\Delta U)_{Cu}$$

$$0 = [mC_v(T_2 - T_1)]_{N_2} + [mC_v(T_2 - T_1)]_{He} + [mC(T_2 - T_1)]_{Cu}$$

where

$$T_{1,\text{Cu}} = (80 + 25)/2 = 52.5°\text{C}$$

Substituting,

$$(4.77\text{kg})(0.743\text{kJ/kg}\cdot°\text{C})(T_f - 80)°\text{C} + (0.808\text{kg})(3.1156\text{kJ/kg}\cdot°\text{C})(T_f - 25)°\text{C}$$
$$+ (5.0\text{kg})(0.386\text{kJ/kg}\cdot°\text{C})(T_f - 52.5)°\text{C} = 0$$

It gives

$$T_f = \mathbf{56.0°C}$$

where T_f is the final equilibrium temperature in the cylinder.

The answer would be the **same** if the piston were not free to move since it would effect only pressure, and not the specific heats.

4-248 EES solution of this (and other comprehensive problems designated with the *computer icon*) is available to instructors at the *Instructor Manual* section of the *Online Learning Center* (OLC) at www.mhhe.com/cengel-boles. See the Preface for access information.

4-249 A relation for the explosive energy of a fluid is given. A relation is to be obtained for the explosive energy of an ideal gas, and the value for air at a specified state is to be evaluated.

Analysis The explosive energy per unit volume is given as

$$e_{\text{explosion}} = \frac{u_1 - u_2}{v_1}$$

For an ideal gas, $u_1 - u_2 = C_v(T_1 - T_2)$

$$C_p - C_v = R$$

$$v_1 = \frac{RT_1}{P_1}$$

and thus

$$\frac{C_v}{R} = \frac{C_v}{C_p - C_v} = \frac{1}{C_p / C_v - 1} = \frac{1}{k-1}$$

Substituting,

$$e_{\text{explosion}} = \frac{C_v(T_1 - T_2)}{RT_1 / P_1} = \frac{P_1}{k-1}\left(1 - \frac{T_2}{T_1}\right)$$

which is the desired result.

Using the relation above, the total explosive energy of 20 m³ of air at 5 MPa and 100°C when the surroundings are at 20°C is determined to be

$$E_{\text{explosion}} = Ve_{\text{explosion}} = \frac{P_1 V_1}{k-1}\left(1 - \frac{T_2}{T_1}\right) = \frac{(5000kPa)(20m^3)}{1.4 - 1}\left(1 - \frac{293K}{373K}\right)\left(\frac{1kJ}{1kPa \cdot m^3}\right) = \mathbf{53{,}619kJ}$$

4-250 Using the relation for explosive energy given in the previous problem, the explosive energy of steam and its TNT equivalent at a specified state are to be determined.

Assumptions Steam condenses and becomes a liquid at room temperature after the explosion.

Properties The properties of steam at the initial and the final states are (Table A-4 through A-6)

$$\left.\begin{array}{l} P_1 = 10\text{MPa} \\ T_1 = 500^\circ\text{C} \end{array}\right\} \begin{array}{l} v_1 = 0.03279\text{m}^3/\text{kg} \\ u_1 = 3045.8\text{kJ/kg} \end{array}$$

$$\left.\begin{array}{l} T_2 = 25^\circ\text{C} \\ Comp.liquid \end{array}\right\} u_2 \cong u_{f@25^\circ\text{C}} = 104.88\text{kJ/kg}$$

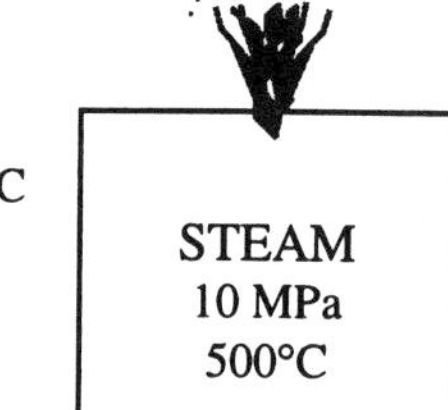

Analysis The mass of the steam is

$$m = \frac{V}{v_1} = \frac{20\ \text{m}^3}{0.03279\ \text{m}^3/\text{kg}} = 609.9\ \text{kg}$$

Then the total explosive energy of the steam is determined from

$$E_{\text{explosive}} = m(u_1 - u_2) = (609.9\text{kg})(3045.8 - 104.88)\text{kJ/kg} = \mathbf{1{,}793{,}667kJ}$$

which is equivalent to

$$\frac{1{,}793{,}667\ \text{kJ}}{3250\ \text{kJ/kg of TNT}} = \mathbf{552\ kg\ of\ TNT}$$

4-251 Solar energy is to be stored as sensible heat using phase-change materials, granite rocks, and water. The amount of heat that can be stored in a 5-m^3 = 5000 L space using these materials as the storage medium is to be determined.

Assumptions **1** The materials have constant properties at the specified values. **2** No allowance is made for voids, and thus the values calculated are the upper limits.

Analysis The amount of energy stored in a medium is simply equal to the increase in its internal energy, which, for incompressible substances, can be determined from $\Delta U = mC(T_2 - T_1)$.

(*a*) The latent heat of glaubers salts is given to be 329 kJ/L. Disregarding the sensible heat storage in this case, the amount of energy stored is becomes

$$\Delta U_{\text{salt}} = mh_{\text{if}} = (5000\text{ L})(329\text{ kJ/L}) = \mathbf{1{,}645{,}000\ kJ}$$

This value would be even larger if the sensible heat storage due to temperature rise is considered.

(*b*) The density of granite is 2700 kg/m^3 (Table A-3), and its specific heat is given to be C = 2.32 kJ/kg.°C. Then the amount of energy that can be stored in the rocks when the temperature rises by 20°C becomes

$$\Delta U_{\text{rock}} = \rho VC\Delta T = (2700\text{ kg/m}^3)(5\text{ m}^3)(2.32\text{ kJ/kg.°C})(20\text{°C}) = \mathbf{626{,}400\ kJ}$$

(*c*) The density of water is about 1000 kg/m^3 (Table A-3), and its specific heat is given to be C = 4.0 kJ/kg.°C. Then the amount of energy that can be stored in the water when the temperature rises by 20°C becomes

$$\Delta U_{\text{rock}} = \rho VC\Delta T = (1000\text{ kg/m}^3)(5\text{ m}^3)(4.0\text{ kJ/kg.°C})(20\text{°C}) = \mathbf{400{,}00\ kJ}$$

Discussion Note that the greatest amount of heat can be stored in phase-change materials essentially at constant temperature. Such materials are not without problems, however, and thus they are not widely used.

4-252 The feedwater of a steam power plant is preheated using steam extracted from the turbine. The ratio of the mass flow rates of the extracted seam the feedwater are to be determined.

Assumptions **1** This is a steady-flow process since there is no change with time. **2** Kinetic and potential energy changes are negligible. **3** There are no work interactions. **4** Heat loss from the device to the surroundings is negligible and thus heat transfer from the hot fluid is equal to the heat transfer to the cold fluid.

Properties The enthalpies of steam and feedwater at are (Tables A-4 through A-6)

$$\left.\begin{array}{l} P_1 = 1.2\text{MPa} \\ T_1 = 250^\circ\text{C} \end{array}\right\} h_1 = 2827.9\text{kJ/kg}$$

$$\left.\begin{array}{l} P_1 = 1\text{MPa} \\ sat\,liquid \end{array}\right\} \begin{array}{l} h_2 = h_{f@1\text{MPa}} = 762.81\text{kJ/kg} \\ T_2 = 179.91^\circ\text{C} \end{array}$$

and

$$\left.\begin{array}{l} P_3 = 2.5\text{MPa} \\ T_3 = 50^\circ\text{C} \end{array}\right\} h_3 \cong h_{f@50^\circ\text{C}} = 209.33\text{kJ/kg}$$

$$\left.\begin{array}{l} P_4 = 2.5\text{MPa} \\ T_4 = T_2 - 10 \cong 170^\circ\text{C} \end{array}\right\} h_4 \cong h_{f@170^\circ\text{C}} = 719.2\text{kJ/kg}$$

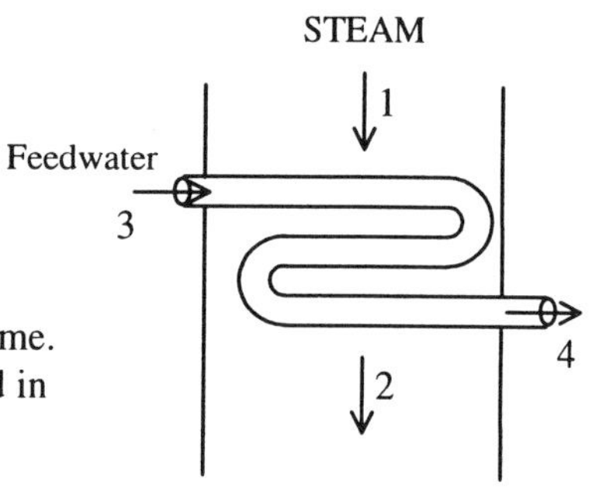

Analysis We take the heat exchanger as the system, which is a control volume. The mass and energy balances for this steady-flow system can be expressed in the rate form as

Mass balance (for each fluid stream):

$$\dot{m}_{in} - \dot{m}_{out} = \Delta\dot{m}_{\text{system}}^{\nearrow 0\text{ (steady)}} = 0 \rightarrow \dot{m}_{in} = \dot{m}_{out} \rightarrow \dot{m}_1 = \dot{m}_2 = \dot{m}_s \quad \text{and} \quad \dot{m}_3 = \dot{m}_4 = \dot{m}_{fw}$$

Energy balance (for the heat exchanger):

$$\underbrace{\dot{E}_{in} - \dot{E}_{out}}_{\substack{\text{Rate of net energy transfer} \\ \text{by heat, work, and mass}}} = \underbrace{\Delta\dot{E}_{\text{system}}^{\nearrow 0\text{ (steady)}}}_{\substack{\text{Rate of change in internal, kinetic,} \\ \text{potential, etc. energies}}} = 0$$

$$\dot{E}_{in} = \dot{E}_{out}$$

$$\dot{m}_1 h_1 + \dot{m}_3 h_3 = \dot{m}_2 h_2 + \dot{m}_4 h_4 \quad (\text{since } \dot{Q} = \dot{W} = \Delta\text{ke} \cong \Delta\text{pe} \cong 0)$$

Combining the two, $\dot{m}_s(h_2 - h_1) = \dot{m}_{fw}(h_3 - h_4)$

Dividing by $\dot{m}_{fw}$ and substituting,

$$\frac{\dot{m}_s}{\dot{m}_{fw}} = \frac{h_3 - h_4}{h_2 - h_1} = \frac{(719.2 - 209.33)\text{kJ/kg}}{(2827.9 - 762.81)\text{kJ/kg}} = \mathbf{0.247}$$

4-253 A building is to be heated by a 30-kW electric resistance heater placed in a duct inside. The time it takes to raise the interior temperature from 14°C to 24°C, and the average mass flow rate of air as it passes through the heater in the duct are to be determined.

Assumptions **1** Steady operating conditions exist. **2** Air is an ideal gas with constant specific heats at room temperature. **3** Kinetic and potential energy changes are negligible. **4** The heating duct is adiabatic, and thus heat transfer through it is negligible. **5** No air leaks in and out of the building.

Properties The gas constant of air is 0.287 kPa.m³/kg.K (Table A-1). The specific heats of air at room temperature are C_p = 1.005 and C_v = 0.718 kJ/kg· K (Table A-2).

Analysis (*a*) The total mass of air in the building is

$$m = \frac{P_1 V_1}{RT_1} = \frac{(95 \text{ kPa})(400 \text{ m}^3)}{(0.287 \text{ kPa}\cdot\text{m}^3/\text{kg}\cdot\text{K})(287 \text{ K})} = 461.3 \text{ kg}.$$

We first take the *entire building* as our system, which is a closed system since no mass leaks in or out. The time required to raise the air temperature to 24°C is determined by applying the energy balance to this constant volume closed system:

$$\underbrace{E_{in} - E_{out}}_{\substack{\text{Net energy transfer}\\ \text{by heat, work, and mass}}} = \underbrace{\Delta E_{\text{system}}}_{\substack{\text{Change in internal, kinetic,}\\ \text{potential, etc. energies}}}$$

$$W_{e,in} + W_{fan,in} - Q_{out} = \Delta U \quad (\text{since } \Delta\text{KE} = \Delta\text{PE} = 0)$$

$$\Delta t\left(\dot W_{e,in} + \dot W_{fan,in} - \dot Q_{out}\right) = mC_{v,ave}(T_2 - T_1)$$

Solving for Δt gives

$$\Delta t = \frac{mC_{v,ave}(T_2 - T_1)}{\dot W_{e,in} + \dot W_{fan,in} - \dot Q_{out}} = \frac{(461.3 \text{ kg})(0.718 \text{ kJ/kg}\cdot{}^\circ\text{C})(24-14)^\circ\text{C}}{(30 \text{ kJ/s}) + (0.25 \text{ kJ/s}) - (450/60 \text{ kJ/s})} = \mathbf{146 \ s}$$

(*b*) We now take the *heating duct* as the system, which is a control volume since mass crosses the boundary. There is only one inlet and one exit, and thus $\dot m_1 = \dot m_2 = \dot m$. The energy balance for this adiabatic steady-flow system can be expressed in the rate form as

$$\underbrace{\dot E_{in} - \dot E_{out}}_{\substack{\text{Rate of net energy transfer}\\ \text{by heat, work, and mass}}} = \underbrace{\Delta \dot E_{\text{system}}^{\nearrow 0 \text{ (steady)}}}_{\substack{\text{Rate of change in internal, kinetic,}\\ \text{potential, etc. energies}}} = 0$$

$$\dot E_{in} = \dot E_{out}$$

$$\dot W_{e,in} + \dot W_{fan,in} + \dot m h_1 = \dot m h_2 \quad (\text{since } \dot Q = \Delta\text{ke} \cong \Delta\text{pe} \cong 0)$$

$$\dot W_{e,in} + \dot W_{fan,in} = \dot m(h_2 - h_1) = \dot m C_p (T_2 - T_1)$$

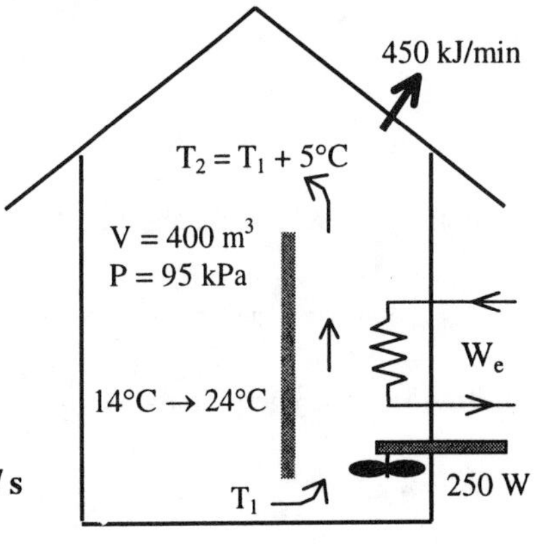

Thus,

$$\dot m = \frac{\dot W_{e,in} + \dot W_{fan,in}}{C_p \Delta T} = \frac{(30 + 0.25) \text{ kJ/s}}{(1.005 \text{ kJ/kg}\cdot{}^\circ\text{C})(5^\circ\text{C})} = \mathbf{6.02 \ kg/s}$$

450 kJ/min
$T_2 = T_1 + 5$°C
V = 400 m³
P = 95 kPa
14°C → 24°C
T_1
W_e
250 W

4-254 [*Also solved by EES on enclosed CD*] An insulated cylinder equipped with an external spring initially contains air. The tank is connected to a supply line, and air is allowed to enter the cylinder until its volume doubles. The mass of the air that entered and the final temperature in the cylinder are to be determined.

Assumptions **1** This is an unsteady process since the conditions within the device are changing during the process, but it can be analyzed as a uniform-flow process since the state of fluid at the inlet remains constant. **2** The expansion process is quasi-equilibrium. **3** Kinetic and potential energies are negligible. **4** The spring is a linear spring. **5** The device is insulated and thus heat transfer is negligible. **6** Air is an ideal gas with constant specific heats.

Properties The gas constant of air is R = 0.287 kJ/kg· K (Table A-1). The specific heats of air at room temperature are C_v = 0.718 and C_p = 1.005 kJ/kg· K (Table A-2a). Also, $u = C_vT$ and $h = C_pT$.

Analysis We take the cylinder as the system, which is a control volume since mass crosses the boundary. Noting that the microscopic energies of flowing and nonflowing fluids are represented by enthalpy h and internal energy u, respectively, the mass and energy balances for this uniform-flow system can be expressed as

Mass balance: $$m_{in} - m_{out} = \Delta m_{\text{system}} \rightarrow m_i = m_2 - m_1$$

Energy balance: $$\underbrace{E_{in} - E_{out}}_{\substack{\text{Net energy transfer}\\ \text{by heat, work, and mass}}} = \underbrace{\Delta E_{\text{system}}}_{\substack{\text{Change in internal, kinetic,}\\ \text{potential, etc. energies}}}$$

$$m_i h_i = W_{b,out} + m_2u_2 - m_1u_1 \quad (\text{since } Q \cong ke \cong pe \cong 0)$$

Combining the two relations, $$(m_2 - m_1)h_i = W_{b,out} + m_2u_2 - m_1u_1$$

or, $$(m_2 - m_1)C_pT_i = W_{b,out} + m_2C_vT_2 - m_1C_vT_1$$

F_{spring}

P = 200 kPa
T_1 = 22°C
V_1 = 0.2 m³
Air

P_i = 0.8 MPa
T_i = 22°C

The initial and the final masses in the tank are

$$m_1 = \frac{P_1V_1}{RT_1} = \frac{(200\text{kPa})(0.2\text{m}^3)}{(0.287\text{kPa}\cdot\text{m}^3/\text{kg}\cdot\text{K})(295\text{K})} = 0.472\text{kg}$$

$$m_2 = \frac{P_2V_2}{RT_2} = \frac{(600\text{kPa})(0.4\text{m}^3)}{(0.287\text{kPa}\cdot\text{m}^3/\text{kg}\cdot\text{K})T_2} = \frac{836.2}{T_2}$$

Then from the mass balance becomes $$m_i = m_2 - m_1 = \frac{836.2}{T_2} - 0.472$$

The spring is a linear spring, and thus the boundary work for this process can be determined from

$$W_b = Area = \frac{P_1 + P_2}{2}(V_2 - V_1) = \frac{(200+600)\text{kPa}}{2}(0.4-0.2)\text{m}^3 = 80\text{kJ}$$

Substituting into the energy balance, the final temperature of air T_2 is determined to be

$$-80 = -\left(\frac{836.2}{T_2} - 0.472\right)(1.005)(295) + \left(\frac{836.2}{T_2}\right)(0.718)(T_2) - (0.472)(0.718)(295)$$

It yields $$T_2 = \mathbf{344.1\ K}$$

Thus, $$m_2 = \frac{836.2}{T_2} = \frac{836.2}{344.1} = 2.430\text{ kg}$$

and $$m_i = m_2 - m_1 = 2.430 - 0.472 = \mathbf{1.958\ kg}$$

4-255 Pressurized air stored in a large cave is to be used to drive a turbine. The amount of work delivered by the turbine for specified turbine exit conditions is to be determined. √

Assumptions **1** This is an unsteady process since the conditions within the device are changing during the process, but it can be analyzed as a uniform-flow process since the exit temperature (and enthalpy) of air remains constant. **2** Kinetic and potential energies are negligible. **3** The system is insulated and thus heat transfer is negligible. **4** Air is an ideal gas with constant specific heats at room temperature.

Properties The gas constant of air is $R = 0.287$ kJ/kg· K (Table A-1). The specific heats of air at room temperature are $C_v = 0.718$ and $C_p = 1.005$ kJ/kg· K (Table A-2a). Also, $u = C_vT$ and $h = C_pT$.

Analysis We take the *cave* as the system, which is a control volume since mass crosses the boundary. Noting that the microscopic energies of flowing and nonflowing fluids are represented by enthalpy h and internal energy u, the mass and energy balances for this uniform-flow system can be expressed as

Mass balance: $m_{in} - m_{out} = \Delta m_{\text{system}} \rightarrow m_e = m_1 - m_2$

Energy balance:

$$\underbrace{E_{in} - E_{out}}_{\substack{\text{Net energy transfer}\\ \text{by heat, work, and mass}}} = \underbrace{\Delta E_{\text{system}}}_{\substack{\text{Change in internal, kinetic,}\\ \text{potential, etc. energies}}}$$

$$-m_e h_e = m_2u_2 - m_1u_1 \quad (\text{since } Q \cong W \cong ke \cong pe \cong 0)$$

Combining the two: $(m_1 - m_2)h_e + m_2u_2 - m_1u_1 = 0$

or,

$$\left(\frac{P_1V}{RT_1} - \frac{P_2V}{RT_2}\right)C_p\frac{T_1 + T_2}{2} + \frac{P_2V}{RT_2}C_vT_2 - \frac{P_1V}{RT_1}C_vT_1 = 0$$

Multiply by R/VC_v:

$$\left(\frac{P_1}{T_1} - \frac{P_2}{T_2}\right)k\frac{T_1 + T_2}{2} + P_2 - P_1 = 0$$

Substituting,

$$\left(\frac{500}{400} - \frac{300}{T_2}\right)(1.4)\frac{400 + T_2}{2} + 300 - 500 = 0$$

$$T_2^2 - 68.57T_2 - 96{,}000 = 0$$

It yields

$$T_2 = 346 \text{ K}$$

CAVE
10,000 m³
500 kPa
400 K
AIR
Air turbine
W
100 kPa
300 K

The initial and the final masses of air in the cave are determined to be

$$m_1 = \frac{P_1V}{RT_1} = \frac{(500\text{kPa})(10^4\text{ m}^3)}{(0.287\text{kPa}\cdot\text{m}^3/\text{kg}\cdot\text{K})(400\text{K})} = 43{,}554\text{kg}$$

$$m_2 = \frac{P_2V}{RT_2} = \frac{(300\text{kPa})(10^4\text{ m}^3)}{(0.287\text{kPa}\cdot\text{m}^3/\text{kg}\cdot\text{K})(346\text{K})} = 30{,}211\text{kg}$$

Then from the mass balance we get $m_e = m_1 - m_2 = 43{,}554 - 30{,}211 = 13{,}343$ kg

The average temperature at the turbine inlet is (400 + 346)/2 = 373 K. Taking the turbine as system and assuming the air properties at the turbine inlet to be constant at the average temperature, the turbine work output is determined from the steady-flow energy balance $\dot{E}_{in} = \dot{E}_{out}$ to be

$$m_ih_i - m_eh_e - W_{out} = 0$$

or

$$W_{out} = m_e(h_i - h_e)_{\text{turbine}} = (13{,}343\text{ kg})(373.7 - 300.19)\text{kJ/kg} = \mathbf{981 MJ}$$

4-256E Steam is decelerated in a diffuser from a velocity of 500 ft/s to 100 ft/s. The mass flow rate of steam, the rate of heat transfer, and the inlet area of the diffuser are to be determined.

Assumptions **1** This is a steady-flow process since there is no change with time. **2** Potential energy changes are negligible. **3** There are no work interactions.

Properties From the steam tables (Tables A-4E through A-6E)

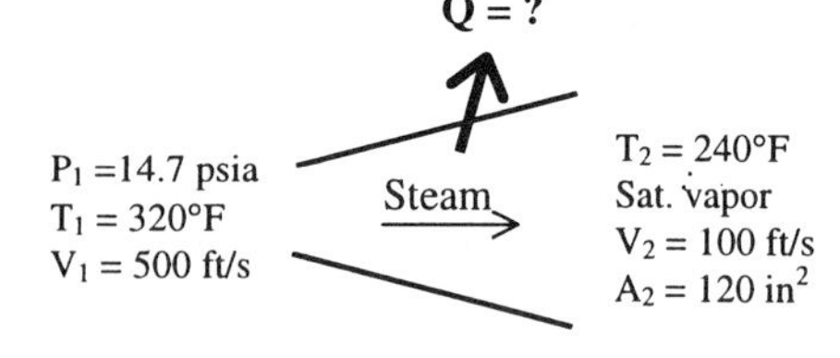

$$\left.\begin{array}{l} P_1 = 14.7\text{psia} \\ T_1 = 320^\circ\text{F} \end{array}\right\} \begin{array}{l} v_1 = 31.36\text{ft}^3/\text{lbm} \\ h_1 = 1202.1\text{Btu/lbm} \end{array}$$

and

$$\left.\begin{array}{l} T_2 = 240^\circ\text{F} \\ sat.vapor \end{array}\right\} \begin{array}{l} v_2 = 16.327\text{ft}^3/\text{lbm} \\ h_2 = 1160.7\text{Btu/lbm} \end{array}$$

Analysis (*a*) The mass flow rate of the steam can be determined from its definition to be

$$\dot{m} = \frac{1}{v_2}\mathbf{V}_2 A_2 = \frac{1}{16.327\text{ft}^3/\text{lbm}}(100\text{ft/s})(120/144\text{ft}^2) = \mathbf{5.104lbm/s}$$

(*b*) We take diffuser as the system, which is a control volume since mass crosses the boundary. The energy balance for this steady-flow system can be expressed in the rate form as

$$\underbrace{\dot{E}_{in} - \dot{E}_{out}}_{\substack{\text{Rate of net energy transfer} \\ \text{by heat, work, and mass}}} = \underbrace{\Delta \dot{E}_{\text{system}}^{\nearrow 0\ (\text{steady})}}_{\substack{\text{Rate of change in internal, kinetic,} \\ \text{potential, etc. energies}}} = 0$$

$$\dot{E}_{in} = \dot{E}_{out}$$

$$\dot{Q}_{\text{in}} + \dot{m}(h_1 + \mathbf{V}_1^2/2) = \dot{m}(h_2 + \mathbf{V}_1^2/2) \quad (\text{since } \dot{\text{W}} \cong \Delta\text{pe} \cong 0)$$

$$\dot{Q}_{\text{in}} = \dot{m}\left(h_2 - h_1 + \frac{\mathbf{V}_2^2 - \mathbf{V}_1^2}{2}\right)$$

Substituting,

$$\dot{Q}_{in} = (5.104\text{lbm/s})\left(1160.7 - 1202.1 + \frac{(100\text{ft/s})^2 - (500\text{ft/s})^2}{2}\left(\frac{1\text{Btu/lbm}}{25{,}037\text{ft}^2/\text{s}^2}\right)\right) = \mathbf{-235.8Btu/s}$$

(*c*) There is only one inlet and one exit, and thus $\dot{m}_1 = \dot{m}_2 = \dot{m}$. Then the inlet area of the diffuser becomes

$$\dot{m} = \frac{1}{v_1}\mathbf{V}_1 A_1 \longrightarrow A_1 = \frac{\dot{m}v_1}{\mathbf{V}_1} = \frac{(5.104\text{lbm/s})(31.36\text{ft}^3/\text{lbm})}{500\text{ft/s}} = \mathbf{0.320ft^2}$$

4-257 20% of the volume of a pressure cooker is initially filled with liquid water. Heat is transferred to the cooker at a rate of 400 W. The time it will take for the cooker to run out of liquid is to be determined.

Assumptions **1** This is an unsteady process since the conditions within the device are changing during the process, but it can be analyzed as a uniform-flow process since the state of fluid leaving the device remains constant. **2** Kinetic and potential energies are negligible. **3** There are no work interactions involved.

Properties The properties of water are (Tables A-4 through A-6)

$$P_1 = 200 \text{ kPa} \rightarrow \quad v_f = 0.001061 \text{ m}^3/\text{kg}, v_g = 0.8857 \text{ m}^3/\text{kg}$$
$$u_f = 504.49 \text{ kJ/kg}, u_g = 2529.5 \text{ kJ/kg}$$

$$\left.\begin{array}{l} P_2 = 200 \text{ kPa} \\ \text{sat.vapor} \end{array}\right\} \begin{array}{l} v_2 = v_{g\,@\,200\text{kPa}} = 0.8857 \text{ m}^3/\text{kg} \\ u_2 = u_{g\,@\,200\text{kPa}} = 2529.5 \text{ kJ/kg} \end{array}$$

$$\left.\begin{array}{l} P_e = 200 \text{ kPa} \\ \text{sat.vapor} \end{array}\right\} h_e = h_{g\,@\,200 \text{ kPa}} = 2706.7 \text{ kJ/kg}$$

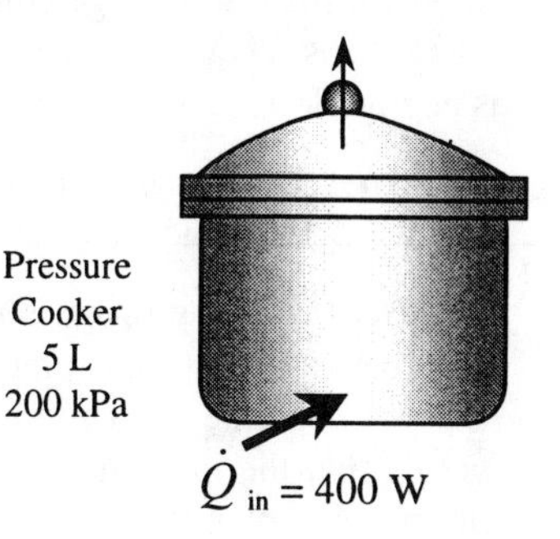

Analysis We take the cooker as the system, which is a control volume since mass crosses the boundary. Noting that the microscopic energies of flowing and nonflowing fluids are represented by enthalpy h and internal energy u, the mass and energy balances for this uniform-flow system can be expressed as

Mass balance: $\quad m_{in} - m_{out} = \Delta m_{\text{system}} \quad \rightarrow \quad m_e = m_1 - m_2$

Energy balance:

$$\underbrace{E_{in} - E_{out}}_{\substack{\text{Net energy transfer} \\ \text{by heat, work, and mass}}} = \underbrace{\Delta E_{\text{system}}}_{\substack{\text{Change in internal, kinetic,} \\ \text{potential, etc. energies}}}$$

$$Q_{in} - m_e h_e = m_2 u_2 - m_1 u_1 \quad (\text{since } W \cong ke \cong pe \cong 0)$$

The initial mass, initial internal energy, and final mass in the tank are

$$m_1 = m_f + m_g = \frac{V_f}{v_f} + \frac{V_g}{v_g} = \frac{0.001 \text{ m}^3}{0.001061 \text{ m}^3/\text{kg}} + \frac{0.004 \text{ m}^3}{0.8857 \text{ m}^3/\text{kg}} = 0.9425 + 0.0045 = 0.9470 \text{ kg}$$

$$U_1 = m_1 u_1 = m_f u_f + m_g u_g = (0.9425)(504.49) + (0.0045)(2529.5) = 486.86 \text{ kJ}$$

$$m_2 = \frac{V}{v_2} = \frac{0.005 \text{ m}^3}{0.8857 \text{ m}^3/\text{kg}} = 0.0056 \text{ kg}$$

From mass and energy balances,

$$m_e = m_1 - m_2 = 0.9470 - 0.0056 = 0.9414 \text{ kg}$$

and

$$\dot{Q}\Delta t = m_e h_e + m_2 u_2 - m_1 u_1$$
$$(0.4 \text{ kJ/s})\Delta t = (0.9414 \text{ kg})(2706.7 \text{ kJ/kg}) + (0.0056 \text{ kg})(2529.5 \text{ kJ/kg}) - 486.86 \text{ kJ}$$

It yields $\quad \Delta t = 5188 \text{ s} = \mathbf{1.44 \text{ h}}$

4-258 A balloon is initially filled with pressurized helium gas. Now a valve is opened, and helium is allowed to escape until the pressure inside drops to atmospheric pressure. The final temperature of helium in the balloon and the mass of helium that has escaped are to be determined.

Assumptions **1** This is an unsteady process since the conditions within the device are changing during the process, but it can be analyzed as a uniform-flow process by assuming the properties of helium that escape to be constant at average conditions. **2** Kinetic and potential energies are negligible. **3** There are no work interactions other than boundary work. **4** Helium is an ideal gas with constant specific heats. **5** Heat transfer is negligible.

Properties The gas constant of helium is R =2.0769 kPa.m^3/kg.K (Table A-1). The specific heats of helium are C_p = 5.1926 and C_v = 3.1156 kJ/kg· K (Table A-2).

Analysis (*a*) The properties of helium leaving the balloon are changing during this process. But we will treat them as a constant at the average temperature. Thus $T_e \cong (T_1 + T_2)/2$. Also $h = C_pT$ and $u = C_vT$.

We take the balloon as the system, which is a control volume since mass crosses the boundary. Noting that the microscopic energies of flowing and nonflowing fluids are represented by enthalpy h and internal energy u, the mass and energy balances for this uniform-flow system can be expressed as

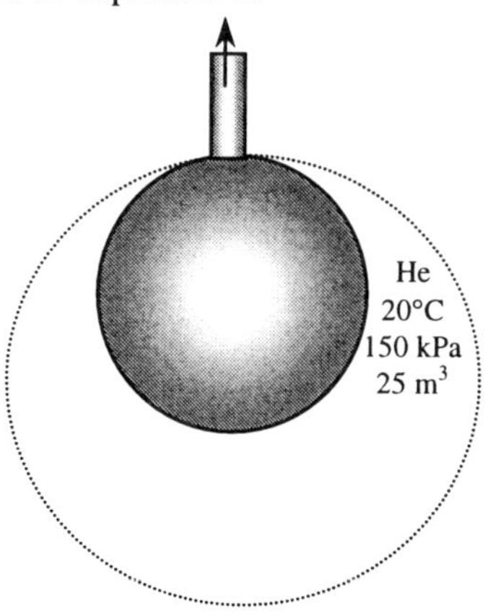

Mass balance: $$m_{in} - m_{out} = \Delta m_{\text{system}} \rightarrow m_e = m_1 - m_2$$

Energy balance: $$\underbrace{E_{in} - E_{out}}_{\substack{\text{Net energy transfer}\\ \text{by heat, work, and mass}}} = \underbrace{\Delta E_{\text{system}}}_{\substack{\text{Change in internal, kinetic,}\\ \text{potential, etc. energies}}}$$

$$W_{b,in} - m_e h_e = m_2u_2 - m_1u_1 \quad (\text{since } Q \cong ke \cong pe \cong 0)$$

or $$W_{b,in} = m_e C_p \frac{T_1 + T_2}{2} + m_2 C_v T_2 - m_1 C_v T_1$$

The final volume of helium is

$$P_1 = -100 + bV_1 \longrightarrow b = (P_1 + 100)/V_1 = (150 + 100)/25 = 10$$

$$P_2 = -100 + 10V_2 \longrightarrow V_2 = (P_2 + 100)/10 = (100 + 100)/10 = 20\text{m}^3$$

The initial and the final masses of helium in the balloon are

$$m_1 = \frac{P_1V_1}{RT_1} = \frac{(150\text{kPa})(25\text{m}^3)}{(2.0769\text{kPa}\cdot\text{m}^3/\text{kg}\cdot\text{K})(293\text{K})} = 6.162\text{kg}$$

$$m_2 = \frac{P_2V_2}{RT_2} = \frac{(100\text{kPa})(20\text{m}^3)}{(2.0769\text{kPa}\cdot\text{m}^3/\text{kg}\cdot\text{K})T_2} = \frac{962.974}{T_2}$$

Then from the mass balance we have $m_i = m_2 - m_1 = 6.162 - \dfrac{962.974}{T_2}$

The boundary work done during this process is

$$W_{b,in} = \frac{P_1 + P_2}{2}(V_1 - V_2) = \frac{(150 + 100)\text{kPa}}{2}(25 - 20)\text{m}^3 = 625\text{kJ}$$

Then from the energy balance,

$$W_{b,in} = m_e C_p \frac{T_1 + T_2}{2} + m_2 C_v T_2 - m_1 C_v T_1$$

$$625 = \left(6.162 - \frac{962.974}{T_2}\right)(5.1926)\frac{293 + T_2}{2} + \left(\frac{962.974}{T_2}\right)(3.1156)(T_2) - (6.162)(3.1156)(293)$$

It yields $$T_2^2 - 66.37T_2 - 45{,}789 = 0$$

Solving for T_2 yields $T_2 = \mathbf{249.7\ K}$

(*b*) The amount of helium that has escaped is

$$m_e = m_1 - m_2 = 6.162 - \frac{962.974}{T_2} = 6.162 - \frac{962.974}{249.7} = \mathbf{2.306\ kg}$$

4-259 EES solution of this (and other comprehensive problems designated with the *computer icon*) is available to instructors at the *Instructor Manual* section of the *Online Learning Center* (OLC) at www.mhhe.com/cengel-boles. See the Preface for access information.

Fundamentals of Engineering (FE) Exam Problems

4-260 A well-sealed room contains 80 kg of air at 200 kPa and 25°C. Now solar energy enters the room at an average rate of 1 kJ/s while a 100-W fan is turned on to circulate the air in the room. If heat transfer through the walls is negligible, the air temperature in the room in 30 min will be

(a) 28.3°C (b) 49.8°C (c) 52.5°C (d) 56.0°C (e) 59.5°C

Answer (e) 59.5°C

Solution Solved by EES Software. Solutions can be verified by copying-and-pasting the following lines on a blank EES screen. (Similar problems and their solutions can be obtained easily by modifying numerical values).

```
R=0.287 "kJ/kg.K"
Cv=0.718 "kJ/kg.K"
m=80 "kg"
P1=200 "kPa"
T1=25 "C"
Qsol=1 "kJ/s"
time=30*60 "s"
Wfan=0.1 "kJ/s"

"Applying energy balance E_in-E_out=dE_system gives"
time*(Wfan+Qsol)=m*Cv*(T2-T1)

"Some Wrong Solutions with Common Mistakes:"
Cp=1.005 "kJ/kg.K"
time*(Wfan+Qsol)=m*Cp*(W1_T2-T1)  "Using Cp instead of Cv "
time*(-Wfan+Qsol)=m*Cv*(W2_T2-T1) "Subtracting Wfan instead of adding"
time*Qsol=m*Cv*(W3_T2-T1) "Ignoring Wfan"
time*(Wfan+Qsol)/60=m*Cv*(W4_T2-T1) "Using min for time instead of s"
```

4-261 A 2-kW baseboard electric resistance heater in a vacant room is turned on and kept on for 15 min. The mass of the air in the room is 75 kg, and the room is tightly sealed so that no air can leak in or out. The temperature rise of air at the end of 15 min is

(a) 8.5°C (b) 12.4°C (c) 24.0°C (d) 33.4°C (e) 54.8°C

Answer (d) 33.4°C

Solution Solved by EES Software. Solutions can be verified by copying-and-pasting the following lines on a blank EES screen. (Similar problems and their solutions can be obtained easily by modifying numerical values).

```
R=0.287 "kJ/kg.K"
Cv=0.718 "kJ/kg.K"
m=75 "kg"
time=15*60 "s"
W_e=2 "kJ/s"

"Applying energy balance E_in-E_out=dE_system gives"
time*W_e=m*Cv*DELTAT "kJ"

"Some Wrong Solutions with Common Mistakes:"
Cp=1.005 "kJ/kg.K"
time*W_e=m*Cp*W1_DELTAT "Using Cp instead of Cv"
time*W_e/60=m*Cv*W2_DELTAT "Using min for time instead of s"
```

4-262 A room contains 25 kg of air at 100 kPa and 10°C. The room has a 250-W refrigerator (the refrigerator consumes 250 W of electricity when running), a 200-W TV, a 1-kW electric resistance heater, and a 100-W fan. During a cold winter day, it is observed that the refrigerator, the TV, the fan, and the electric resistance heater are running continuously but the air temperature in the room remains constant. The rate of heat loss from the room that day is

(a) 3600 kJ/h (b) 5220 kJ/h (c) 5580 kJ/h (d) 3780 kJ/h (e) 4680 kJ/h

Answer (c) 5580 kJ/h

Solution Solved by EES Software. Solutions can be verified by copying-and-pasting the following lines on a blank EES screen. (Similar problems and their solutions can be obtained easily by modifying numerical values).

```
R=0.287 "kJ/kg.K"
Cv=0.718 "kJ/kg.K"
m=25 "kg"
P_1=100 "kPa"
T_1=10 "C"

time=30*60 "s"
W_ref=0.250 "kJ/s"
W_TV=0.2 "kJ/s"
W_heater=1 "kJ/s"
W_fan=0.1 "kJ/s"

"Applying energy balance E_in-E_out=dE_system gives E_out=E_in since T=constant
and dE=0"
E_gain=W_ref+W_TV+W_heater+W_fan
Q_loss=E_gain*3600 "kJ/h"

"Some Wrong Solutions with Common Mistakes:"
E_gain1=-W_ref+W_TV+W_heater+W_fan "Subtracting Wrefrig instead of adding"
W1_Qloss=E_gain1*3600 "kJ/h"

E_gain2=W_ref+W_TV+W_heater-W_fan "Subtracting Wfan instead of adding"
W2_Qloss=E_gain2*3600 "kJ/h"

E_gain3=-W_ref+W_TV+W_heater-W_fan "Subtracting Wrefrig and Wfan instead of
adding"
W3_Qloss=E_gain3*3600 "kJ/h"

E_gain4=W_ref+W_heater+W_fan "Ignoring the TV"
W4_Qloss=E_gain4*3600 "kJ/h"
```

4-263 A piston-cylinder device contains 5 kg of air at 400 kPa and 30°C. During a quasi-equilibrium isothermal expansion process, 15 kJ of boundary work is done by the system, and 3 kJ of paddle-wheel work is done on the system. The heat transfer during this process is

(a) 12 kJ (b) 18 kJ (c) 2.4 kJ (d) 3.6 kJ (e) 60 kJ

Answer (a) 12 kJ

Solution Solved by EES Software. Solutions can be verified by copying-and-pasting the following lines on a blank EES screen. (Similar problems and their solutions can be obtained easily by modifying numerical values).

```
R=0.287 "kJ/kg.K"
Cv=0.718 "kJ/kg.K"
m=5 "kg"
P_1=400 "kPa"
T=30 "C"

Wout_b=15 "kJ"
Win_pw=3 "kJ"

"Noting that T=constant and thus dE_system=0, applying energy balance E_in-
E_out=dE_system gives"

Q_in+Win_pw-Wout_b=0

"Some Wrong Solutions with Common Mistakes:"

W1_Qin=Q_in/Cv "Dividing by Cv"
W2_Qin=Win_pw+Wout_b "Adding both quantities"
W3_Qin=Win_pw "Setting it equal to paddle-wheel work"
W4_Qin=Wout_b "Setting it equal to boundaru work"
```

4-264 A container equipped with a resistance heater and a mixer is initially filled with 3 kg of saturated water vapor at 100°C. Now the heater and the mixer are turned on; the steam is compressed, and there is heat loss to the surrounding air. At the end of the process, the temperature and pressure of steam in the container are measured to be 300°C and 0.5 MPa. The net energy transfer to the steam during this process is

(a) 196 kJ (b) 359 kJ (c) 561 kJ (d) 889 kJ (e) 1568 kJ

Answer (d) 889 kJ

Solution Solved by EES Software. Solutions can be verified by copying-and-pasting the following lines on a blank EES screen. (Similar problems and their solutions can be obtained easily by modifying numerical values).

```
m=3 "kg"
T1=100 "C"
x1=1 "saturated vapor"

P2=500 "kPa"
T2=300 "C"

u1=INTENERGY(Steam_NBS,T=T1,x=x1)
u2=INTENERGY(Steam_NBS,T=T2,P=P2)

"Noting that Eout=0 and dU_system=m*(u2-u1), applying energy balance E_in-
E_out=dE_system gives"
E_out=0
E_in=m*(u2-u1)

"Some Wrong Solutions with Common Mistakes:"
Cp_steam=1.8723 "kJ/kg.K"
Cv_steam=1.4108 "kJ/kg.K"
W1_Ein=m*Cp_Steam*(T2-T1) "Assuming ideal gas and using Cp"
W2_Ein=m*Cv_steam*(T2-T1) "Assuming ideal gas and using Cv"
W3_Ein=u2-u1 "Not using mass"

h1=ENTHALPY(Steam_NBS,T=T1,x=x1)
h2=ENTHALPY(Steam_NBS,T=T2,P=P2)
W4_Ein=m*(h2-h1) "Using enthalpy"
```

4-265 A 6-pack canned drink is to be cooled from 25°C to 3°C. The mass of each canned drink is 0.355 kg. The drinks can be treated as water, and the energy stored in the aluminum can itself is negligible. The amount of heat transfer from the 6 canned drinks is

(a) 33 kJ (b) 37 kJ (c) 47 kJ (d) 196 kJ (e) 223 kJ

Answer (d) 196 kJ

Solution Solved by EES Software. Solutions can be verified by copying-and-pasting the following lines on a blank EES screen. (Similar problems and their solutions can be obtained easily by modifying numerical values).

```
C=4.18 "kJ/kg.K"
m=6*0.355 "kg"
T1=25 "C"
T2=3 "C"
DELTAT=T2-T1 "C"

"Applying energy balance E_in-E_out=dE_system and noting that
dU_system=m*C*DELTAT gives"
-Q_out=m*C*DELTAT "kJ"

"Some Wrong Solutions with Common Mistakes:"
-W1_Qout=m*C*DELTAT/6 "Using one can only"
-W2_Qout=m*C*(T1+T2) "Adding temperatures instead of subtracting"
-W3_Qout=m*1.0*DELTAT  "Using specific heat of air or forgetting specific heat"
```

4-266 A glass of water with a mass of 0.3 kg at 20°C is to be cooled to 0°C by dropping ice cubes at 0°C into it. The latent heat of fusion of ice is 334 kJ/kg, and the specific heat of water is 4.18 kJ/kg.°C. The amount of ice that needs to be added is

(a) 24 g (b) 75 g (c) 124 g (d) 150 g (e) 300 g

Answer (b) 75 g

Solution Solved by EES Software. Solutions can be verified by copying-and-pasting the following lines on a blank EES screen. (Similar problems and their solutions can be obtained easily by modifying numerical values).

```
C=4.18 "kJ/kg.K"
h_melting=334 "kJ/kg.K"
m_w=0.3 "kg"
T1=20 "C"
T2=0 "C"
DELTAT=T2-T1 "C"
"Noting that there is no energy transfer with the surroundings and the latent heat of
melting of ice is transferred form the water, and applying energy balance E_in-
E_out=dE_system to ice+water gives"
dE_ice+dE_w=0
dE_ice=m_ice*h_melting
dE_w=m_w*C*DELTAT "kJ"

"Some Wrong Solutions with Common Mistakes:"
W1_mice*h_melting*(T1-T2)+m_w*C*DELTAT=0 "Multiplying h_latent by temperature
difference"
W2_mice=m_w "taking mass of water to be equal to the mass of ice"
```

4-267 A 2-kW electric resistance heater submerged in 5-kg water is turned on and kept on for 10 min. During the process, 300 kJ of heat is lost from the water. The temperature rise of water is

(a) 0.4°C (b) 43.1°C (c) 57.4°C (d) 71.8°C (e) 180.0°C

Answer (b) 43.1°C

Solution Solved by EES Software. Solutions can be verified by copying-and-pasting the following lines on a blank EES screen. (Similar problems and their solutions can be obtained easily by modifying numerical values).

```
C=4.18 "kJ/kg.K"
m=5 "kg"

Q_loss=300 "kJ"
time=10*60 "s"
W_e=2 "kJ/s"

"Applying energy balance E_in-E_out=dE_system gives"
time*W_e-Q_loss = dU_system
dU_system=m*C*DELTAT "kJ"

"Some Wrong Solutions with Common Mistakes:"
time*W_e = m*C*W1_T "Ignoring heat loss"
time*W_e+Q_loss = m*C*W2_T "Adding heat loss instead of subtracting"
time*W_e-Q_loss = m*1.0*W3_T "Using specific heat of air or not using specific heat"
```

4-268 1.2 kg of liquid water initially at 15°C is to be heated to 95°C in a teapot equipped with a 1200 W electric heating element inside. The specific heat of water can be taken to be 4.18 kJ/kg.°C, and the heat loss from the water during heating can be neglected. The time it takes to heat the water to the desired temperature is

(a) 4.6 min (b) 5.6 min (c) 6.7 min (d) 9.0 min (e) 11.4 min

Answer (b) 5.6 min

Solution Solved by EES Software. Solutions can be verified by copying-and-pasting the following lines on a blank EES screen. (Similar problems and their solutions can be obtained easily by modifying numerical values).

```
C=4.18 "kJ/kg.K"
m=1.2 "kg"
T1=15 "C"
T2=95 "C"
Q_loss=0 "kJ"
W_e=1.2 "kJ/s"

"Applying energy balance E_in-E_out=dE_system gives"
(time*60)*W_e-Q_loss = dU_system "time in minutes"
dU_system=m*C*(T2-T1) "kJ"

"Some Wrong Solutions with Common Mistakes:"
W1_time*60*W_e-Q_loss = m*C*(T2+T1) "Adding temps instead of subtracting"
W2_time*60*W_e-Q_loss = C*(T2-T1) "Not using mass"
```

4-269 An ordinary egg with a mass of 0.1 kg and a specific heat of 3.32 kJ/kg.°C is dropped into boiling water at 95°C. If the initial temperature of the egg is 5°C, the maximum amount of heat transfer to the egg is

(a) 12 kJ (b) 30 kJ (c) 24 kJ (d) 18 kJ (e) infinity

Answer (b) 30 kJ

Solution Solved by EES Software. Solutions can be verified by copying-and-pasting the following lines on a blank EES screen. (Similar problems and their solutions can be obtained easily by modifying numerical values).

```
C=3.32 "kJ/kg.K"
m=0.1 "kg"
T1=5 "C"
T2=95 "C"

"Applying energy balance E_in-E_out=dE_system gives"
E_in = dU_system
dU_system=m*C*(T2-T1) "kJ"

"Some Wrong Solutions with Common Mistakes:"
W1_Ein = m*C*T2  "Using T2 only"
W2_Ein=m*(ENTHALPY(Steam_NBS,T=T2,x=1)-ENTHALPY(Steam_NBS,T=T2,x=0))  "Using
h_fg"
```

4-270 An apple with an average mass of 0.15 kg and average specific heat of 3.65 kJ/kg.°C is cooled from 20°C to 5°C. The amount of heat transferred from the apple is

(a) 0.55 kJ (b) 54.8 kJ (c) 15 kJ (d) 8.21 kJ (e) 4.10 kJ

Answer (d) 8.21 kJ

Solution Solved by EES Software. Solutions can be verified by copying-and-pasting the following lines on a blank EES screen. (Similar problems and their solutions can be obtained easily by modifying numerical values).

```
C=3.65 "kJ/kg.K"
m=0.15 "kg"
T1=20 "C"
T2=5 "C"

"Applying energy balance E_in-E_out=dE_system gives"
-Q_out = dU_system
dU_system=m*C*(T2-T1) "kJ"

"Some Wrong Solutions with Common Mistakes:"
-W1_Qout =C*(T2-T1) "Not using mass"
-W2_Qout =m*C*(T2+T1) "adding temperatures"
```

4-271 Steam is accelerated by a nozzle steadily from a low velocity to a velocity of 250 m/s at a rate of 2 kg/s. If the temperature and pressure of the steam at the nozzle exit are 400°C and 2 MPa, the exit area of the nozzle is

(a) 4.0 cm^2 (b) 8.4 cm^2 (c) 10.2 cm^2 (d) 11.6 cm^2 (e) 12.1 cm^2

Answer (e) 12.1 cm^2

Solution Solved by EES Software. Solutions can be verified by copying-and-pasting the following lines on a blank EES screen. (Similar problems and their solutions can be obtained easily by modifying numerical values).

```
Vel_1=0 "m/s"
Vel_2=250 "m/s"
m=2 "kg/s"
T2=400 "C"
P2=2000 "kPa"

"The rate form of energy balance is E_dot_in - E_dot_out = DELTAE_dot_cv"
v2=VOLUME(Steam_NBS,T=T2,P=P2)
m=(1/v2)*A2*Vel_2  "A2 in m^2"

"Some Wrong Solutions with Common Mistakes:"
R=0.4615 "kJ/kg.K"
P2*v2ideal=R*(T2+273)
m=(1/v2ideal)*W1_A2*Vel_2 "assuming ideal gas"

P1*v2ideal=R*T2
m=(1/v2ideal)*W2_A2*Vel_2 "assuming ideal gas and using C"

m=W3_A2*Vel_2 "not using specific volume"
```

4-272 Steam enters a diffuser steadily at 0.5 MPa, 300°C, and 122 m/s at a rate of 3.5 kg/s. The inlet area of the diffuser is

(a) 15 cm^2 (b) 50 cm^2 (c) 105 cm^2 (d) 150 cm^2 (e) 190 cm^2

Answer (b) 50 cm^2

Solution Solved by EES Software. Solutions can be verified by copying-and-pasting the following lines on a blank EES screen. (Similar problems and their solutions can be obtained easily by modifying numerical values).

```
Vel_1=122 "m/s"
m=3.5 "kg/s"
T1=300 "C"
P1=500 "kPa"

"The rate form of energy balance is E_dot_in - E_dot_out = DELTAE_dot_cv"
v1=VOLUME(Steam_NBS,T=T1,P=P1)
m=(1/v1)*A*Vel_1  "A in m^2"

"Some Wrong Solutions with Common Mistakes:"
R=0.4615 "kJ/kg.K"
P1*v1ideal=R*(T1+273)
m=(1/v1ideal)*W1_A*Vel_1 "assuming ideal gas"

P1*v2ideal=R*T1
m=(1/v2ideal)*W2_A*Vel_1 "assuming ideal gas and using C"

m=W3_A*Vel_1 "not using specific volume"
```

4-273 An adiabatic heat exchanger is used to heat cold water at 15°C entering at a rate of 4 kg/s by hot air at 100°C entering also at rate of 4 kg/s. If the exit temperature of hot air is 20°C, the exit temperature of cold water is

(a) 29°C (b) 34°C (c) 52°C (d) 95°C (e) 100°C

Answer (b) 34°C

Solution Solved by EES Software. Solutions can be verified by copying-and-pasting the following lines on a blank EES screen. (Similar problems and their solutions can be obtained easily by modifying numerical values).

```
C_w=4.18 "kJ/kg-C"
Cp_air=1.005 "kJ/kg-C"
Tw1=15 "C"
m_dot_w=4 "kg/s"
Tair1=100 "C"
Tair2=20 "C"
m_dot_air=4 "kg/s"

"The rate form of energy balance for a steady-flow system is E_dot_in = E_dot_out"
m_dot_air*Cp_air*(Tair1-Tair2)=m_dot_w*C_w*(Tw2-Tw1)

"Some Wrong Solutions with Common Mistakes:"
(Tair1-Tair2)=(W1_Tw2-Tw1) "Equating temperature changes of fluids"
Cv_air=0.718 "kJ/kg.K"
m_dot_air*Cv_air*(Tair1-Tair2)=m_dot_w*C_w*(W2_Tw2-Tw1) "Using Cv for air"
W3_Tw2=Tair1 "Setting inlet temperature of hot fluid = exit temperature of cold fluid"
W4_Tw2=Tair2 "Setting exit temperature of hot fluid = exit temperature of cold fluid"
```

4-274 A heat exchanger is used to heat cold water at 15°C entering at a rate of 2 kg/s by hot air at 100°C entering at rate of 3 kg/s. The heat exchanger is not insulated, and is loosing heat at a rate of 40 kJ/s. If the exit temperature of hot air is 20°C, the exit temperature of cold water is

(a) 44°C (b) 49°C (c) 39°C (d) 72°C (e) 95°C

Answer (c) 39°C

Solution Solved by EES Software. Solutions can be verified by copying-and-pasting the following lines on a blank EES screen. (Similar problems and their solutions can be obtained easily by modifying numerical values).

```
C_w=4.18 "kJ/kg-C"
Cp_air=1.005 "kJ/kg-C"
Tw1=15 "C"
m_dot_w=2 "kg/s"
Tair1=100 "C"
Tair2=20 "C"
m_dot_air=3 "kg/s"
Q_loss=40 "kJ/s"

"The rate form of energy balance for a steady-flow system is E_dot_in = E_dot_out"
m_dot_air*Cp_air*(Tair1-Tair2)=m_dot_w*C_w*(Tw2-Tw1)+Q_loss

"Some Wrong Solutions with Common Mistakes:"
m_dot_air*Cp_air*(Tair1-Tair2)=m_dot_w*C_w*(W1_Tw2-Tw1) "Not considering Q_loss"
m_dot_air*Cp_air*(Tair1-Tair2)=m_dot_w*C_w*(W2_Tw2-Tw1)-Q_loss  "Taking heat loss
as heat gain"
(Tair1-Tair2)=(W3_Tw2-Tw1) "Equating temperature changes of fluids"
Cv_air=0.718 "kJ/kg.K"
m_dot_air*Cv_air*(Tair1-Tair2)=m_dot_w*C_w*(W4_Tw2-Tw1)+Q_loss "Using Cv for air"
```

4-275 An adiabatic heat exchanger is used to heat cold water at 20°C entering at a rate of 4 kg/s by hot water at 90°C entering at rate of 3 kg/s. If the exit temperature of hot water is 50°C, the exit temperature of cold water is

(a) 40°C (b) 50°C (c) 60°C (d) 75°C (e) 90°C

Answer (b) 50°C

Solution Solved by EES Software. Solutions can be verified by copying-and-pasting the following lines on a blank EES screen. (Similar problems and their solutions can be obtained easily by modifying numerical values).

```
C_w=4.18 "kJ/kg-C"
Tcold_1=20 "C"
m_dot_cold=4 "kg/s"
Thot_1=90 "C"
Thot_2=50 "C"
m_dot_hot=3 "kg/s"
Q_loss=0 "kJ/s"

"The rate form of energy balance for a steady-flow system is E_dot_in = E_dot_out"
m_dot_hot*C_w*(Thot_1-Thot_2)=m_dot_cold*C_w*(Tcold_2-Tcold_1)+Q_loss

"Some Wrong Solutions with Common Mistakes:"
Thot_1-Thot_2=W1_Tcold_2-Tcold_1 "Equating temperature changes of fluids"
W2_Tcold_2=90 "Taking exit temp of cold fluid=inlet temp of hot fluid"
```

4-276 In a shower, cold water at 10°C flowing at a rate of 5 kg/min is mixed with hot water at 60°C flowing at a rate of 2 kg/min. The exit temperature of the mixture will be

(a) 24.3°C (b) 35.0°C (c) 40.0°C (d) 44.3°C (e) 55.2°C

Answer (a) 24.3°C

Solution Solved by EES Software. Solutions can be verified by copying-and-pasting the following lines on a blank EES screen. (Similar problems and their solutions can be obtained easily by modifying numerical values).

```
C_w=4.18 "kJ/kg-C"
Tcold_1=10 "C"
m_dot_cold=5 "kg/min"
Thot_1=60 "C"
m_dot_hot=2 "kg/min"

"The rate form of energy balance for a steady-flow system is E_dot_in = E_dot_out"
m_dot_hot*C_w*Thot_1+m_dot_cold*C_w*Tcold_1=(m_dot_hot+m_dot_cold)*C_w*T
mix

"Some Wrong Solutions with Common Mistakes:"
W1_Tmix=(Tcold_1+Thot_1)/2 "Taking the average temperature of inlet fluids"
```

4-277 In a heating system, cold outdoor air at 15°C flowing at a rate of 5 kg/min is mixed adiabatically with heated air at 70°C flowing at a rate of 3 kg/min. The exit temperature of the mixture is

(a) 35.6°C (b) 41.7°C (c) 45.0°C (d) 52.9°C (e) 85°C

Answer (a) 35.6°C

Solution Solved by EES Software. Solutions can be verified by copying-and-pasting the following lines on a blank EES screen. (Similar problems and their solutions can be obtained easily by modifying numerical values).

```
C_air=1.005 "kJ/kg-C"
Tcold_1=15 "C"
m_dot_cold=5 "kg/min"
Thot_1=70 "C"
m_dot_hot=3 "kg/min"

"The rate form of energy balance for a steady-flow system is E_dot_in = E_dot_out"
m_dot_hot*C_air*Thot_1+m_dot_cold*C_air*Tcold_1=(m_dot_hot+m_dot_cold)*C_air
*Tmix

"Some Wrong Solutions with Common Mistakes:"
W1_Tmix=(Tcold_1+Thot_1)/2 "Taking the average temperature of inlet fluids"
```

4-278 Hot combustion gases (assumed to have the properties of air at room temperature) enter a gas turbine at 1 MPa and 1500 K at a rate of 0.1 kg/s, and exit at 0.2 MPa and 900 K. If heat is lost from the turbine to the surroundings at a rate of 15 kJ/s, the power output of the gas turbine is

(a) 15 kW (b) 30 kW (c) 45 kW (d) 60 kW (e) 75 kW

Answer (c) 45 kW

Solution Solved by EES Software. Solutions can be verified by copying-and-pasting the following lines on a blank EES screen. (Similar problems and their solutions can be obtained easily by modifying numerical values).

```
Cp_air=1.005 "kJ/kg-C"
T1=1500 "K"
T2=900 "K"
m_dot=0.1 "kg/s"
Q_dot_loss=15 "kJ/s"

"The rate form of energy balance for a steady-flow system is E_dot_in = E_dot_out"
W_dot_out+Q_dot_loss=m_dot*Cp_air*(T1-T2)

"Alternative: Variable specific heats - using EES data"
W_dot_outvariable+Q_dot_loss=m_dot*(ENTHALPY(Air,T=T1)-ENTHALPY(Air,T=T2))

"Some Wrong Solutions with Common Mistakes:"
W1_Wout=m_dot*Cp_air*(T1-T2) "Disregarding heat loss"
W2_Wout-Q_dot_loss=m_dot*Cp_air*(T1-T2) "Assuming heat gain instead of loss"
```

4-279 Steam expands in a turbine from 4 MPa and 500°C to 0.5 MPa and 250°C at a rate of 1740 kg/h. Heat is lost from the turbine at a rate of 12 kJ/s during the process. The power output of the turbine is

(a) 222 kW (b) 234 kW (c) 438 kW (d) 717 kW (e) 246 kW

Answer (a) 222 kW

Solution Solved by EES Software. Solutions can be verified by copying-and-pasting the following lines on a blank EES screen. (Similar problems and their solutions can be obtained easily by modifying numerical values).

```
T1=500 "C"
P1=4000 "kPa"

T2=250 "C"
P2=500 "kPa"

m_dot=1740/3600 "kg/s"
Q_dot_loss=12 "kJ/s"

h1=ENTHALPY(Steam_NBS,T=T1,P=P1)
h2=ENTHALPY(Steam_NBS,T=T2,P=P2)

"The rate form of energy balance for a steady-flow system is E_dot_in = E_dot_out"
W_dot_out+Q_dot_loss=m_dot*(h1-h2)

"Some Wrong Solutions with Common Mistakes:"
W1_Wout=m_dot*(h1-h2)  "Disregarding heat loss"
W2_Wout-Q_dot_loss=m_dot*(h1-h2)  "Assuming heat gain instead of loss"

u1=INTENERGY(Steam_NBS,T=T1,P=P1)
u2=INTENERGY(Steam_NBS,T=T2,P=P2)
W3_Wout+Q_dot_loss=m_dot*(u1-u2) "Using internal energy instead of enthalpy"
W4_Wout-Q_dot_loss=m_dot*(u1-u2) "Using internal energy and wrong direction for heat"
```

4-280 Steam is compressed by an adiabatic compressor from 0.2 MPa and 150°C to 2500 MPa and 250°C at a rate of 1.30 kg/s. The power input to the compressor is

(a) 144 kW (b) 302 kW (c) 393 kW (d) 717 kW (e) 901 kW

Answer (a) 144 kW

Solution Solved by EES Software. Solutions can be verified by copying-and-pasting the following lines on a blank EES screen. (Similar problems and their solutions can be obtained easily by modifying numerical values).

```
"Note: This compressor violates the 2nd law. Changing State 2 to 800 kPa and 350C
will correct this problem (it would give 511 kW)"

P1=200 "kPa"
T1=150 "C"

P2=2500 "kPa"
T2=250 "C"

m_dot=1.30 "kg/s"
Q_dot_loss=0 "kJ/s"

h1=ENTHALPY(Steam_NBS,T=T1,P=P1)
h2=ENTHALPY(Steam_NBS,T=T2,P=P2)

"The rate form of energy balance for a steady-flow system is E_dot_in = E_dot_out"
W_dot_in-Q_dot_loss=m_dot*(h2-h1)

"Some Wrong Solutions with Common Mistakes:"
W1_Win-Q_dot_loss=(h2-h1)/m_dot  "Dividing by mass flow rate instead of
multiplying"
W2_Win-Q_dot_loss=h2-h1  "Not considering mass flow rate"

u1=INTENERGY(Steam_NBS,T=T1,P=P1)
u2=INTENERGY(Steam_NBS,T=T2,P=P2)
W3_Win-Q_dot_loss=m_dot*(u2-u1) "Using internal energy instead of enthalpy"
W4_Win-Q_dot_loss=u2-u1 "Using internal energy and ignoring mass flow rate"
```

4-281 Refrigerant-134a is compressed by a compressor from the saturated vapor state at 0.12 MPa to 1.2 MPa and 70°C at a rate of 0.108 kg/s. The refrigerant is cooled at a rate of 1.30 kJ/s during compression. The power input to the compressor is

(a) 1.75 kW (b) 3.52 kW (c) 5.73 kW (d) 8.03 kW (e) 7.03 kW

Answer (d) 8.03 kW

Solution Solved by EES Software. Solutions can be verified by copying-and-pasting the following lines on a blank EES screen. (Similar problems and their solutions can be obtained easily by modifying numerical values).

```
P1=120 "kPa"
x1=1

P2=1200 "kPa"
T2=70 "C"

m_dot=0.108 "kg/s"
Q_dot_loss=1.30 "kJ/s"

h1=ENTHALPY(R134a,x=x1,P=P1)
h2=ENTHALPY(R134a,T=T2,P=P2)

"The rate form of energy balance for a steady-flow system is E_dot_in = E_dot_out"
W_dot_in-Q_dot_loss=m_dot*(h2-h1)

"Checking using properties from tables:"
h11=233.86
h22=298.96

Wtable-Q_dot_loss=m_dot*(h22-h11)

"Some Wrong Solutions with Common Mistakes:"
W1_Win+Q_dot_loss=m_dot*(h2-h1)  "Wrong direction for heat transfer"
W2_Win =m_dot*(h2-h1)  "Not considering heat loss"

u1=INTENERGY(R134a,x=x1,P=P1)
u2=INTENERGY(R134a,T=T2,P=P2)
W3_Win-Q_dot_loss=m_dot*(u2-u1) "Using internal energy instead of enthalpy"
W4_Win+Q_dot_loss=u2-u1 "Using internal energy and wrong direction for heat
transfer"
```

4-282 Refrigerant-134a expands in an adiabatic turbine from 1.2 MPa and 100°C to 0.18 MPa and 50°C at a rate of 1.25 kg/s. The power output of the turbine is

(a) 46.3 kW (b) 66.4 kW (c) 72.7 kW (d) 89.2 kW (e) 112.0 kW

Answer (a) 46.3 kW

Solution Solved by EES Software. Solutions can be verified by copying-and-pasting the following lines on a blank EES screen. (Similar problems and their solutions can be obtained easily by modifying numerical values).

```
P1=1200 "kPa"
T1=100 "C"

P2=180 "kPa"
T2=50 "C"

m_dot=1.25 "kg/s"
Q_dot_loss=0 "kJ/s"

h1=ENTHALPY(R134a,T=T1,P=P1)
h2=ENTHALPY(R134a,T=T2,P=P2)

"The rate form of energy balance for a steady-flow system is E_dot_in = E_dot_out"
-W_dot_out-Q_dot_loss=m_dot*(h2-h1)

"Checking using properties from tables:"
h11=332.47
h22=295.45

-Wtable-Q_dot_loss=m_dot*(h22-h11)

"Some Wrong Solutions with Common Mistakes:"
-W1_Wout-Q_dot_loss=(h2-h1)/m_dot  "Dividing by mass flow rate instead of
multiplying"
-W2_Wout-Q_dot_loss=h2-h1  "Not considering mass flow rate"

u1=INTENERGY(R134a,T=T1,P=P1)
u2=INTENERGY(R134a,T=T2,P=P2)
-W3_Wout-Q_dot_loss=m_dot*(u2-u1) "Using internal energy instead of enthalpy"
-W4_Wout-Q_dot_loss=u2-u1 "Using internal energy and ignoring mass flow rate"
```

4-283 Refrigerant-134a at 1.2 MPa and 90°C is throttled to a pressure of 0.8 MPa. The temperature of the refrigerant after throttling is

(a) 31.3°C (b) 56.1°C (c) 78.4°C (d) 85.1°C (e) 90.0°C

Answer (d) 85.1°C

Solution Solved by EES Software. Solutions can be verified by copying-and-pasting the following lines on a blank EES screen. (Similar problems and their solutions can be obtained easily by modifying numerical values).

```
P1=1200 "kPa"
T1=90 "C"

P2=800 "kPa"

h1=ENTHALPY(R134a,T=T1,P=P1)
T2=TEMPERATURE(R134a,h=h1,P=P2)

"Some Wrong Solutions with Common Mistakes:"
W1_T2=T1 "Assuming the temperature to remain constant"
W2_T2=TEMPERATURE(R134a,x=0,P=P2) "Taking the temperature to be the saturation temperature at P2"

u1=INTENERGY(R134a,T=T1,P=P1)
W3_T2=TEMPERATURE(R134a,u=u1,P=P2) "Assuming u=constant"

v1=VOLUME(R134a,T=T1,P=P1)
W4_T2=TEMPERATURE(R134a,v=v1,P=P2) "Assuming v=constant"
```

4-284 Air at 20°C and 5 atm is throttled by a valve to 2 atm. If the valve is adiabatic and the change in kinetic energy is negligible, the exit temperature of air will be

(a) 10°C (b) 14°C (c) 17°C (d) 20°C (e) 24°C

Answer (d) 20°C

Solution Solved by EES Software. Solutions can be verified by copying-and-pasting the following lines on a blank EES screen. (Similar problems and their solutions can be obtained easily by modifying numerical values).

```
"The temperature of an ideal gas remains constant during throttling, and thus T2=T1"
T1=20 "C"
P1=5 "atm"
P2=2 "atm"
T2=T1 "C"

"Some Wrong Solutions with Common Mistakes:"
W1_T2=T1*P1/P2 "Assuming v=constant and using C"
W2_T2=(T1+273)*P1/P2-273 "Assuming v=constant and using K"
W3_T2=T1*P2/P1 "Assuming v=constant and using C"
W4_T2=(T1+273)*P2/P1 "Assuming v=constant and using K"
```

4-285 Steam at 1 MPa and 300°C is throttled adiabatically to a pressure of 0.5 MPa. If the change in kinetic energy is negligible, the specific volume of the steam after throttling will be

(a) 0.2579 m^3/kg (b) 0.2327 m^3/kg (c) 0.3749 m^3/kg (d) 0.5165 m^3/kg (e) 0.5226 m^3/kg

Answer (d) 0.5165 m^3/kg

Solution Solved by EES Software. Solutions can be verified by copying-and-pasting the following lines on a blank EES screen. (Similar problems and their solutions can be obtained easily by modifying numerical values).

```
P1=1000 "kPa"
T1=300 "C"
P2=500 "kPa"

h1=ENTHALPY(Steam_NBS,T=T1,P=P1)
v2=VOLUME(Steam_NBS,h=h1,P=P2)

"Some Wrong Solutions with Common Mistakes:"
W1_v2=VOLUME(Steam_NBS,T=T1,P=P2) "Assuming the volume to remain constant"

u1=INTENERGY(Steam,T=T1,P=P1)
W2_v2=VOLUME(Steam_NBS,u=u1,P=P2) "Assuming u=constant"

W3_v2=VOLUME(Steam_NBS,T=T1,P=P2) "Assuming T=constant"
```

4-286 Air is to be heated steadily by an 8-kW electric resistance heater as it flows through an insulated duct. If the air enters at 50°C at a rate of 2 kg/s, the exit temperature of air will be

(a) 46.0°C (b) 50.0°C (c) 54.0°C (d) 55.4°C (e) 58.0°C

Answer (c) 54.0°C

Solution Solved by EES Software. Solutions can be verified by copying-and-pasting the following lines on a blank EES screen. (Similar problems and their solutions can be obtained easily by modifying numerical values).

```
Cp=1.005 "kJ/kg-C"
T1=50 "C"
m_dot=2 "kg/s"
W_dot_e=8 "kJ/s"

W_dot_e=m_dot*Cp*(T2-T1)

"Checking using data from EES table"
W_dot_e=m_dot*(ENTHALPY(Air,T=T_2table)-ENTHALPY(Air,T=T1))

"Some Wrong Solutions with Common Mistakes:"
Cv=0.718 "kJ/kg.K"
W_dot_e=Cp*(W1_T2-T1) "Not using mass flow rate"
W_dot_e=m_dot*Cv*(W2_T2-T1) "Using Cv"
W_dot_e=m_dot*Cp*W3_T2 "Ignoring T1"
```

4-287 Saturated water vapor at 40°C is to be condensed as it flows through a tube at a rate of 0.2 kg/s. The condensate leaves the tube as a saturated liquid at 40°C. The rate of heat transfer from the tube is

(a) 34 kJ/s (b) 268 kJ/s (c) 453 kJ/s (d) 481 kJ/s (e) 515 kJ/s

Answer (d) 481 kJ/s

Solution Solved by EES Software. Solutions can be verified by copying-and-pasting the following lines on a blank EES screen. (Similar problems and their solutions can be obtained easily by modifying numerical values).

```
T1=40 "C"
m_dot=0.2 "kg/s"

h_f=ENTHALPY(Steam_NBS,T=T1,x=0)
h_g=ENTHALPY(Steam_NBS,T=T1,x=1)
h_fg=h_g-h_f
Q_dot=m_dot*h_fg

"Some Wrong Solutions with Common Mistakes:"
W1_Q=m_dot*h_f "Using hf"
W2_Q=m_dot*h_g "Using hg"
W3_Q=h_fg "not using mass flow rate"
W4_Q=m_dot*(h_f+h_g) "Adding hf and hg"
```

4- 288 ... 2-294 Design and Essay Problems

Chapter 5
THE SECOND LAW OF THERMODYNAMICS

The Second Law of Thermodynamics and Thermal Energy Reservoirs

5-1C Water is not a fuel; thus the claim is false.

5-2C Transferring 5 kWh of heat to an electric resistance wire and producing 5 kWh of electricity.

5-3C An electric resistance heater which consumes 5 kWh of electricity and supplies 6 kWh of heat to a room.

5-4C Transferring 5 kWh of heat to an electric resistance wire and producing 6 kWh of electricity.

5-5C No. Heat cannot flow from a low-temperature medium to a higher temperature medium.

5-6C A thermal-energy reservoir is a body that can supply or absorb finite quantities of heat isothermally. Some examples are the oceans, the lakes, and the atmosphere.

5-7C Yes. Because the temperature of the oven remains constant no matter how much heat is transferred to the potatoes.

5-8C The surrounding air in the room that houses the TV set.

Heat Engines and Thermal Efficiency

5-9C No. Such an engine violates the Kelvin-Planck statement of the second law of thermodynamics.

5-10C Heat engines are cyclic devices that receive heat from a source, convert some of it to work, and reject the rest to a sink.

5-11C Method (b). With the heating element in the water, heat losses to the surrounding air are minimized, and thus the desired heating can be achieved with less electrical energy input.

5-12C No. Because 100% of the work can be converted to heat.

5-13C It is expressed as "No heat engine can exchange heat with a single reservoir, and produce an equivalent amount of work".

5-14C (a) No, (b) Yes. According to the second law, no heat engine can have and efficiency of 100%.

5-15C No. Such an engine violates the Kelvin-Planck statement of the second law of thermodynamics.

5-16C No. The Kelvin-Plank limitation applies only to heat engines; engines that receive heat and convert some of it to work.

5-17 The power output and thermal efficiency of a power plant are given. The rate of heat rejection is to be determined, and the result is to be compared to the actual case in practice.

Assumptions **1** The plant operates steadily. **2** Heat losses from the working fluid at the pipes and other components are negligible.

Analysis The rate of heat supply to the power plant is determined from the thermal efficiency relation,

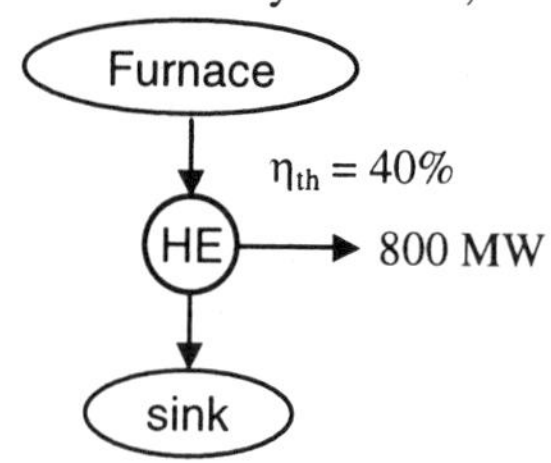

$$\dot{Q}_H = \frac{\dot{W}_{net,out}}{\eta_{th}} = \frac{600 \text{ MW}}{0.4} = 1500 \text{ MW}$$

The rate of heat transfer to the river water is determined from the first law relation for a heat engine,

$$\dot{Q}_L = \dot{Q}_H - \dot{W}_{net,out} = 1500 - 600 = \mathbf{900\ MW}$$

In reality the amount of heat rejected to the river will be **lower** since part of the heat will be lost to the surrounding air from the working fluid as it passes through the pipes and other components.

5-18 The rates of heat supply and heat rejection of a power plant are given. The power output and the thermal efficiency of this power plant are to be determined.

Assumptions **1** The plant operates steadily. **2** Heat losses from the working fluid at the pipes and other components are taken into consideration.

Analysis (*a*) The total heat rejected by this power plant is

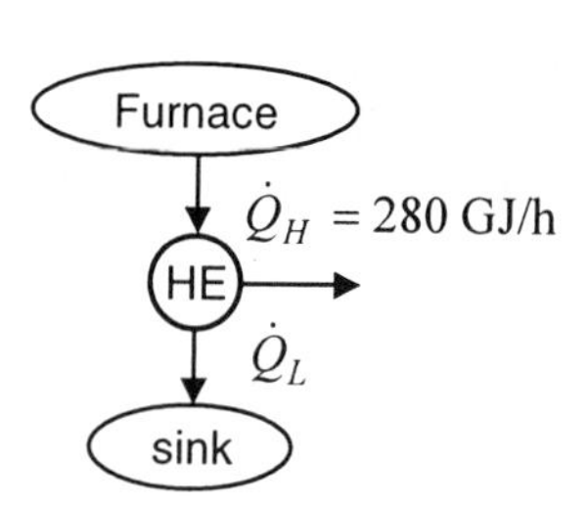

$$\dot{Q}_L = 145 + 8 = 153 \text{GJ/h}$$

Then the net power output of the plant becomes

$$\dot{W}_{net,out} = \dot{Q}_H - \dot{Q}_L = 280 - 153 = 127 \text{GJ/h} = \mathbf{35.3 MW}$$

(*b*) The thermal efficiency of the plant is determined from its definition,

$$\eta_{th} = \frac{\dot{W}_{net,out}}{\dot{Q}_H} = \frac{127 \text{GJ/h}}{280 \text{GJ/h}} = 0.454 = \mathbf{45.4\%}$$

5-19E The power output and thermal efficiency of a car engine are given. The rate of fuel consumption is to be determined.

Assumptions The car operates steadily.

Properties The heating value of the fuel is given to be 19,000 Btu/lbm.

Analysis This car engine is converting 28% of the chemical energy released during the combustion process into work. The amount of energy input required to produce a power output of 95 hp is determined from the definition of thermal efficiency to be

$$\dot{Q}_H = \frac{\dot{W}_{net,out}}{\eta_{th}} = \frac{95\text{hp}}{0.28}\left(\frac{2545\text{Btu/h}}{1\text{hp}}\right) = 863{,}482\text{Btu/h}$$

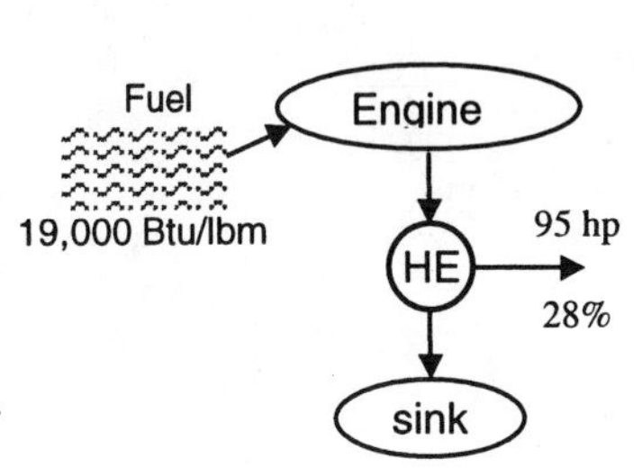

To supply energy at this rate, the engine must burn fuel at a rate of

$$\dot{m} = \frac{863{,}482\text{Btu/h}}{19{,}000\text{Btu/lbm}} = \mathbf{45.4lbm/h}$$

since 19,000 Btu of thermal energy is released for each lbm of fuel burned.

5-20 The power output and fuel consumption rate of a power plant are given. The thermal efficiency is to be determined.

Assumptions The plant operates steadily.

Properties The heating value of coal is given to be 30,000 kJ/kg.

Analysis The rate of heat supply to this power plant is

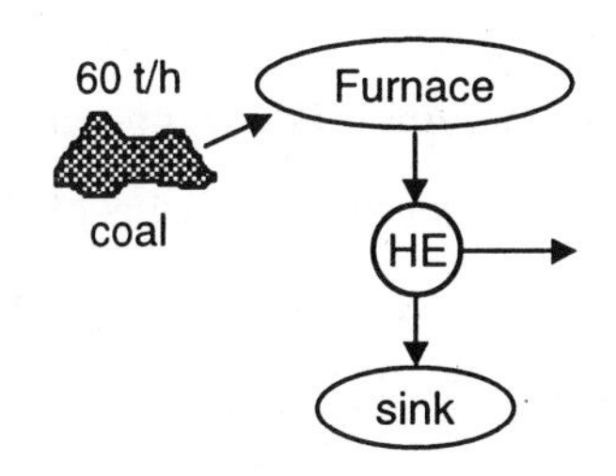

$$\dot{Q}_H = \dot{m}_{coal} u_{coal} = (60{,}000\text{kg/h})(30{,}000\text{kJ/kg}) = 1.8\times10^9\text{ kJ/h} = 500\text{MW}$$

Then the thermal efficiency of the plant becomes

$$\eta_{th} = \frac{\dot{W}_{net,out}}{\dot{Q}_H} = \frac{150\text{MW}}{500\text{MW}} = 0.300 = \mathbf{30.0\%}$$

5-21 The power output and fuel consumption rate of a car engine are given. The thermal efficiency of the engine is to be determined.

Assumptions The car operates steadily.

Properties The heating value of the fuel is given to be 44,000 kJ/kg.

Analysis The mass consumption rate of the fuel is

$$\dot{m}_{fuel} = (\rho\dot{V})_{fuel} = (0.8\text{ kg/L})(28\text{ L/h}) = 22.4\text{ kg/h}$$

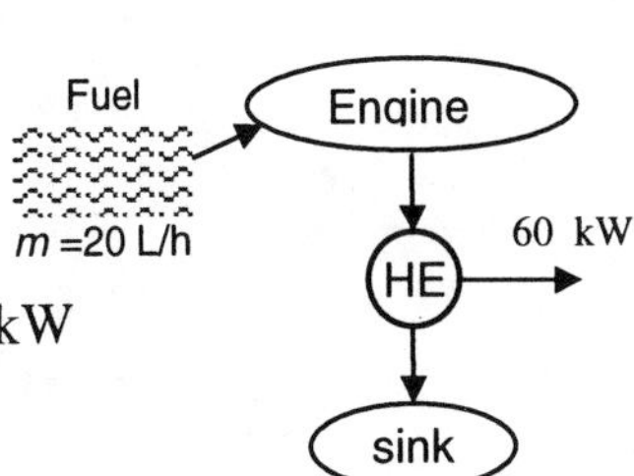

The rate of heat supply to the car is

$$\dot{Q}_H = \dot{m}_{coal} u_{coal} = (22.4\text{ kg/h})(44{,}000\text{kJ/kg}) = 985{,}600\text{ kJ/h} = 273.78\text{ kW}$$

Then the thermal efficiency of the car becomes

$$\eta_{th} = \frac{\dot{W}_{net,out}}{\dot{Q}_H} = \frac{60\text{ kW}}{273.78\text{ kW}} = 0.219 = \mathbf{21.9\%}$$

5-22E The power output and thermal efficiency of a solar pond power plant are given. The rate of solar energy collection is to be determined.

Assumptions The plant operates steadily.

Analysis The rate of solar energy collection or the rate of heat supply to the power plant is determined from the thermal efficiency relation to be

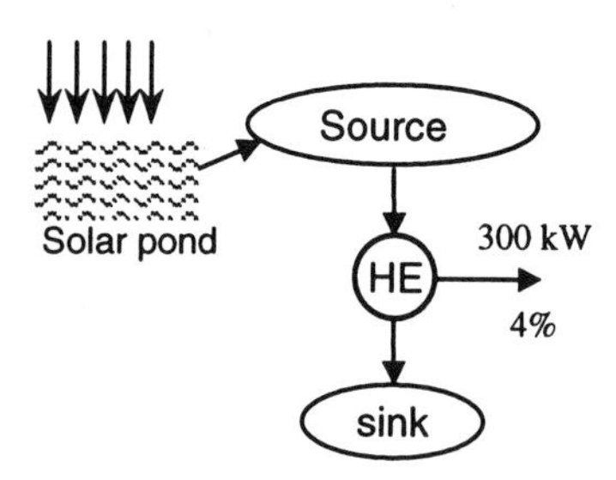

$$\dot{Q}_H = \frac{\dot{W}_{net,out}}{\eta_{th}} = \frac{200\text{ kW}}{0.04}\left(\frac{1\text{ Btu}}{1.055\text{ kJ}}\right)\left(\frac{3600\text{ s}}{1\text{ h}}\right) = 1.7\times\mathbf{10^7\ Btu/h}$$

5-23 The United States produces about 55 percent of its electricity from coal at a conversion efficiency of about 34 percent. The amount of heat rejected by the coal-fired power plants per year is to be determined.

Assumptions Power is generated continuously at an average rate of 400 MW.

Analysis The total amount of power generated by coal power plants is

$$\dot{W}_{\text{coal}} = 0.55\dot{W}_{\text{total}} = 0.55\times(400\text{ MW}) = 220\text{ MW}$$

Noting that the conversion efficiency is 34%, the rate of heat rejection is

$$\eta_{\text{th}} = \frac{\dot{W}_{\text{coal}}}{\dot{Q}_{\text{in}}} = \frac{\dot{W}_{\text{coal}}}{\dot{Q}_{\text{out}} + \dot{W}_{\text{coal}}} \rightarrow \dot{Q}_{\text{out}} = \frac{\dot{W}_{\text{coal}}}{\eta_{\text{th}}} - \dot{W}_{\text{coal}} = \frac{220}{0.34} - 220 = 427\text{ MW}$$

Coal
Furnace
HE
400,000 MW
34%
$\dot{Q}_{\text{out}}$
sink

Then the amount of heat rejected by the coal plants per year becomes

$$Q_{\text{out}} = \dot{Q}_{\text{out}}(1\text{ year}) = (427{,}000\text{ kJ/s})(365\times24\times3600\text{ s}) = \mathbf{1.35\times10^{13}\ kJ/year}$$

5-24 The projected power needs of the United States is to be met by building inexpensive but inefficient coal plants or by building expensive but efficient IGCC plants. The price of coal that will enable the IGCC plants to recover their cost difference from fuel savings in 5 years is to be determined.

Assumptions **1** Power is generated continuously by either plant at full capacity. **2** The time value of money (interest, inflation, etc.) is not considered.

Properties The heating value of the coal is given to be 28×10^6 kJ/ton.

Analysis For a power generation capacity of 150,000 MW, the construction costs of coal and IGCC plants and their difference are

$$\text{Construction cost}_{\text{coal}} = (150{,}000{,}000\text{ kW})(\$1300/\text{kW}) = \$195\times10^9$$

$$\text{Construction cost}_{\text{IGCC}} = (150{,}000{,}000\text{ kW})(\$1500/\text{kW}) = \$225\times10^9$$

$$\text{Construction cost difference} = \$225\times10^9 - \$195\times10^9 = \$30\times10^9$$

The amount of electricity produced by either plant in 5 years is

$$W_e = \dot{W}\Delta t = (150{,}000{,}000\text{ kW})(5\times365\times24\text{ h}) = 6.570\times10^{12}\text{ kWh}$$

The amount of fuel needed to generate a specified amount of power can be determined from

$$\eta = \frac{W_e}{Q_{\text{in}}} \rightarrow Q_{\text{in}} = \frac{W_e}{\eta} \quad \text{or} \quad m_{\text{fuel}} = \frac{Q_{\text{in}}}{\text{Heating value}} = \frac{W_e}{\eta(\text{Heating value})}$$

Then the amount of coal needed to generate this much electricity by each plant and their difference are

$$m_{\text{coal, coal plant}} = \frac{W_e}{\eta(\text{Heating value})} = \frac{6.570\times10^{12}\text{ kWh}}{(0.34)(28\times10^6\text{ kJ/ton})}\left(\frac{3600\text{ kJ}}{1\text{ kWh}}\right) = 2.484\times10^9\text{ tons}$$

$$m_{\text{coal, IGCC plant}} = \frac{W_e}{\eta(\text{Heating value})} = \frac{6.570\times10^{12}\text{ kWh}}{(0.45)(28\times10^6\text{ kJ/ton})}\left(\frac{3600\text{ kJ}}{1\text{ kWh}}\right) = 1.877\times10^9\text{ tons}$$

$$\Delta m_{\text{coal}} = m_{\text{coal, coal plant}} - m_{\text{coal, IGCC plant}} = 2.484\times10^9 - 1.877\times10^9 = 0.607\times10^9\text{ tons}$$

For Δm_{coal} to pay for the construction cost difference of \$30 billion, the price of coal should be

$$\text{Unit cost of coal} = \frac{\text{Construction cost difference}}{\Delta m_{\text{coal}}} = \frac{\$30\times10^9}{0.607\times10^9\text{ tons}} = \mathbf{\$49.4/ton}$$

Therefore, the IGCC plant becomes attractive when the price of coal is above \$49.4 per ton.

5-25 EES solution of this (and other comprehensive problems designated with the *computer icon*) is available to instructors at the *Instructor Manual* section of the *Online Learning Center* (OLC) at www.mhhe.com/cengel-boles. See the Preface for access information.

5-26 The projected power needs of the United States is to be met by building inexpensive but inefficient coal plants or by building expensive but efficient IGCC plants. The price of coal that will enable the IGCC plants to recover their cost difference from fuel savings in 3 years is to be determined.

Assumptions **1** Power is generated continuously by either plant at full capacity. **2** The time value of money (interest, inflation, etc.) is not considered.

Properties The heating value of the coal is given to be 28×10^6 kJ/ton.

Analysis For a power generation capacity of 150,000 MW, the construction costs of coal and IGCC plants and their difference are

$$\text{Construction cost}_{\text{coal}} = (150{,}000{,}000\text{ kW})(\$1300/\text{kW}) = \$195\times10^9$$

$$\text{Construction cost}_{\text{IGCC}} = (150{,}000{,}000\text{ kW})(\$1500/\text{kW}) = \$225\times10^9$$

$$\text{Construction cost difference} = \$225\times10^9 - \$195\times10^9 = \$30\times10^9$$

The amount of electricity produced by either plant in 3 years is

$$W_e = \dot{W}\Delta t = (150{,}000{,}000\text{ kW})(3\times365\times24\text{ h}) = 3.942\times10^{12}\text{ kWh}$$

The amount of fuel needed to generate a specified amount of power can be determined from

$$\eta = \frac{W_e}{Q_{\text{in}}} \rightarrow Q_{\text{in}} = \frac{W_e}{\eta} \quad \text{or} \quad m_{\text{fuel}} = \frac{Q_{\text{in}}}{\text{Heating value}} = \frac{W_e}{\eta(\text{Heating value})}$$

Then the amount of coal needed to generate this much electricity by each plant and their difference are

$$m_{\text{coal, coal plant}} = \frac{W_e}{\eta(\text{Heating value})} = \frac{3.942\times10^{12}\text{ kWh}}{(0.34)(28\times10^6\text{ kJ/ton})}\left(\frac{3600\text{ kJ}}{1\text{ kWh}}\right) = 1.491\times10^9\text{ tons}$$

$$m_{\text{coal, IGCC plant}} = \frac{W_e}{\eta(\text{Heating value})} = \frac{3.942\times10^{12}\text{ kWh}}{(0.45)(28\times10^6\text{ kJ/ton})}\left(\frac{3600\text{ kJ}}{1\text{ kWh}}\right) = 1.126\times10^9\text{ tons}$$

$$\Delta m_{\text{coal}} = m_{\text{coal, coal plant}} - m_{\text{coal, IGCC plant}} = 1.491\times10^9 - 1.126\times10^9 = 0.365\times10^9\text{ tons}$$

For Δm_{coal} to pay for the construction cost difference of \$30 billion, the price of coal should be

$$\text{Unit cost of coal} = \frac{\text{Construction cost difference}}{\Delta m_{\text{coal}}} = \frac{\$30\times10^9}{0.365\times10^9\text{ tons}} = \mathbf{\$82.2/ton}$$

Therefore, the IGCC plant becomes attractive when the price of coal is above \$82.2 per ton.

5-27 A wind turbine is rotating at 20 rpm under steady winds of 30 km/h. The power produced, the tip speed of the blade, and the revenue generated by the wind turbine per year are to be determined. √

Assumptions **1** Steady operating conditions exist. **2** The wind turbine operates continuously during the entire year at the specified conditions.

Properties The density of air is given to be $\rho = 1.20 \text{ kg/m}^3$.

Analysis (*a*) The blade span area and the mass flow rate of air through the turbine are

$$A = \pi D^2/4 = \pi(80 \text{ m})^2/4 = 5027 \text{ m}^2$$

$$\mathbf{V} = (30 \text{ km/h})\left(\frac{1000 \text{ m}}{1 \text{ km}}\right)\left(\frac{1 \text{ h}}{3600 \text{ s}}\right) = 8.333 \text{ m/s}$$

$$\dot{m} = \rho A\mathbf{V} = (1.2 \text{ kg/m}^3)(5027 \text{ m}^2)(8.333 \text{ m/s}) = 50{,}270 \text{ kg/s}$$

Noting that the kinetic energy of a unit mass is $\mathbf{V}^2/2$ and the wind turbine captures 35% of this energy, the power generated by this wind turbine becomes

$$\dot{W} = \eta\left(\frac{1}{2}\dot{m}\mathbf{V}^2\right) = (0.35)\frac{1}{2}(50{,}270 \text{ kg/s})(8.333 \text{ m/s})^2\left(\frac{1 \text{ kJ/kg}}{1000 \text{ m}^2/\text{s}^2}\right) = \mathbf{610.9 \text{ kW}}$$

(*b*) Noting that the tip of blade travels a distance of πD per revolution, the tip velocity of the turbine blade for an rpm of $\dot{n}$ becomes

$$\mathbf{V}_{\text{tip}} = \pi D\dot{n} = \pi(80 \text{ m})(20/\text{min}) = 5027 \text{ m/min} = 83.8 \text{ m/s} = \mathbf{302 \text{ km/h}}$$

(*c*) The amount of electricity produced and the revenue generated per year are

$$\text{Electricity produced} = \dot{W}\Delta t = (610.9 \text{ kW})(365 \times 24 \text{ h/year})$$
$$= 5.351 \times 10^6 \text{ kWh/year}$$

$$\text{Revenue generated} = (\text{Electricity produced})(\text{Unit price}) = (5.351 \times 10^6 \text{ kWh/year})(\$0.06/\text{kWh})$$
$$= \mathbf{\$321{,}100/year}$$

5-28 A wind turbine is rotating at 20 rpm under steady winds of 25 km/h. The power produced, the tip speed of the blade, and the revenue generated by the wind turbine per year are to be determined.

Assumptions **1** Steady operating conditions exist. **2** The wind turbine operates continuously during the entire year at the specified conditions.

Properties The density of air is given to be $\rho = 1.20\ \text{kg/m}^3$.

Analysis (*a*) The blade span area and the mass flow rate of air through the turbine are

$$A = \pi D^2/4 = \pi(80\ \text{m})^2/4 = 5027\ \text{m}^2$$

$$\mathbf{V} = (25\ \text{km/h})\left(\frac{1000\ \text{m}}{1\ \text{km}}\right)\left(\frac{1\ \text{h}}{3600\ \text{s}}\right) = 6.944\ \text{m/s}$$

$$\dot{m} = \rho A\mathbf{V} = (1.2\ \text{kg/m}^3)(5027\ \text{m}^2)(6.944\ \text{m/s}) = 41{,}890\ \text{kg/s}$$

Noting that the kinetic energy of a unit mass is $\mathbf{V}^2/2$ and the wind turbine captures 35% of this energy, the power generated by this wind turbine becomes

$$\dot{W} = \eta\left(\frac{1}{2}\dot{m}\mathbf{V}^2\right) = (0.35)\frac{1}{2}(41{,}890\ \text{kg/s})(6.944\ \text{m/s})^2\left(\frac{1\ \text{kJ/kg}}{1000\ \text{m}^2/\text{s}^2}\right) = \mathbf{353.5\ kW}$$

(*b*) Noting that the tip of blade travels a distance of πD per revolution, the tip velocity of the turbine blade for an rpm of $\dot{n}$ becomes

$$\mathbf{V}_{\text{tip}} = \pi D\dot{n} = \pi(80\ \text{m})(20/\text{min}) = 5027\ \text{m/min} = 83.8\ \text{m/s} = \mathbf{302\ km/h}$$

(*c*) The amount of electricity produced and the revenue generated per year are

$$\text{Electricity produced} = \dot{W}\Delta t = (353.5\ \text{kW})(365\times 24\ \text{h/year})$$
$$= 3.097\times 10^6\ \text{kWh/year}$$

$$\text{Revenue generated} = (\text{Electricity produced})(\text{Unit price}) = (3.097\times 10^6\ \text{kWh/year})(\$0.06/\text{kWh})$$
$$= \mathbf{\$185{,}780/year}$$

5-29E An OTEC power plant operates between the temperature limits of 86°F and 41°F. The cooling water experiences a temperature rise of 6°F in the condenser. The amount of power that can be generated by this OTEC plans is to be determined.

Assumptions **1** Steady operating conditions exist. **2** Water is an incompressible substance with constant properties.

Properties The density and specific heat of water are $\rho = 64.0$ lbm/ft³ and $C = 1.0$ Btu/lbm.°F, respectively.

Analysis The mass flow rate of the cooling water is

$$\dot{m}_{\text{water}} = \rho\dot{V}_{\text{water}} = (64.0\ \text{lbm/ft}^3)(13{,}300\ \text{gal/min})\left[\frac{1\ \text{ft}^3}{7.4804\ \text{gal}}\right] = 113{,}790\ \text{lbm/min} = 1897\ \text{lbm/s}$$

The rate of heat rejection to the cooling water is

$$\dot{Q}_{\text{out}} = \dot{m}_{\text{water}} C(T_{\text{out}} - T_{\text{in}}) = (1897\ \text{lbm/s})(1.0\ \text{Btu/lbm.°F})(6\text{°F}) = 11{,}380\ \text{Btu/s}$$

Noting that the thermal efficiency of this plant is 2.5%, the power generation is determined to be

$$\eta = \frac{\dot{W}}{\dot{Q}_{\text{in}}} = \frac{\dot{W}}{\dot{W} + \dot{Q}_{\text{out}}} \rightarrow 0.025 = \frac{\dot{W}}{\dot{W} + (11{,}380\ \text{Btu/s})} \rightarrow \dot{W} = 292\ \text{Btu/s} = \mathbf{308\ kW}$$

since 1 kW = 0.9478 Btu/s.

Energy Conversion Efficiencies

5-30 A hooded electric open burner and a gas burner are considered. The amount of the electrical energy used directly for cooking and the cost of energy per "utilized" kWh are to be determined. √

Analysis The efficiency of the electric heater is given to be 78 percent. Therefore, a burner that consumes 3-kW of electrical energy will supply

$\eta_{gas} = 38\%$

$\eta_{electric} = 73\%$

$$\dot{Q}_{\text{utilized}} = (\text{Energy input}) \times (\text{Efficiency}) = (3\text{ kW})(0.73) = \mathbf{2.19\ kW}$$

of useful energy. The unit cost of utilized energy is inversely proportional to the efficiency, and is determined from

$$\text{Cost of utilized energy} = \frac{\text{Cost of energy input}}{\text{Efficiency}} = \frac{\$0.07/\text{kWh}}{0.73} = \mathbf{\$0.096/kWh}$$

Noting that the efficiency of a gas burner is 38 percent, the energy input to a gas burner that supplies utilized energy at the same rate (2.19 kW) is

$$\dot{Q}_{\text{input, gas}} = \frac{\dot{Q}_{\text{utilized}}}{\text{Efficiency}} = \frac{2.19\text{ kW}}{0.38} = \mathbf{5.76\ kW}\ \ (= 19{,}660\text{ Btu/h})$$

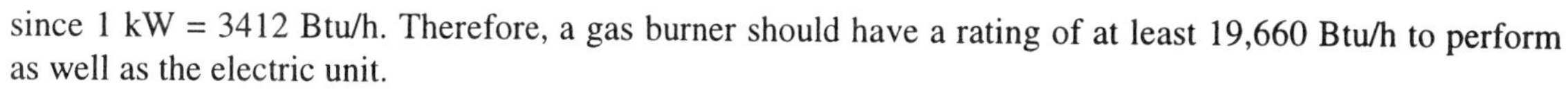

since 1 kW = 3412 Btu/h. Therefore, a gas burner should have a rating of at least 19,660 Btu/h to perform as well as the electric unit.

Noting that 1 therm = 29.3 kWh, the unit cost of utilized energy in the case of gas burner is determined the same way to be

$$\text{Cost of utilized energy} = \frac{\text{Cost of energy input}}{\text{Efficiency}} = \frac{\$0.60/(29.3\text{ kWh})}{0.38} = \mathbf{\$0.054/kWh}$$

which is about one-quarter of the unit cost of utilized electricity.

5-31 A worn out standard motor is replaced by a high efficiency one. The reduction in the internal heat gain due to the higher efficiency under full load conditions is to be determined.

Assumptions **1** The motor and the equipment driven by the motor are in the same room. **2** The motor operates at full load so that $f_{\text{load}} = 1$.

Analysis The heat generated by a motor is due to its inefficiency, and the difference between the heat generated by two motors that deliver the same shaft power is simply the difference between the electric power drawn by the motors,

$$\dot{W}_{\text{in, electric, standard}} = \dot{W}_{\text{shaft}} / \eta_{\text{motor}} = (75 \times 746 \text{ W})/0.91 = 61{,}484 \text{ W}$$

$$\dot{W}_{\text{in, electric, efficient}} = \dot{W}_{\text{shaft}} / \eta_{\text{motor}} = (75 \times 746 \text{ W})/0.954 = 58{,}648 \text{ W}$$

Then the reduction in heat generation becomes

$$\dot{Q}_{\text{reduction}} = \dot{W}_{\text{in, electric, standard}} - \dot{W}_{\text{in, electric, efficient}} = 61{,}484 - 58{,}648 = \mathbf{2836\ W}$$

5-32 An electric car is powered by an electric motor mounted in the engine compartment. The rate of heat supply by the motor to the engine compartment at full load conditions is to be determined.

Assumptions The motor operates at full load so that the load factor is 1.

Analysis The heat generated by a motor is due to its inefficiency, and is equal to the difference between the electrical energy it consumes and the shaft power it delivers,

$$\dot{W}_{\text{in, electric}} = \dot{W}_{\text{shaft}} / \eta_{\text{motor}} = (75 \text{ hp})/0.91 = 82.42 \text{ hp}$$

$$\dot{Q}_{\text{generation}} = \dot{W}_{\text{in, electric}} - \dot{W}_{\text{shaft out}} = 82.42 - 75 = 7.42 \text{ hp} = \mathbf{5.54\ kW}$$

since 1 hp = 0.746 kW.

Discussion Note that the electrical energy not converted to mechanical power is converted to heat.

5-33 A worn out standard motor is to be replaced by a high efficiency one. The amount of electrical energy and money savings as a result of installing the high efficiency motor instead of the standard one as well as the simple payback period are to be determined.

Assumptions The load factor of the motor remains constant at 0.75.

Analysis The electric power drawn by each motor and their difference can be expressed as

$$\dot{W}_{\text{electric in, standard}} = \dot{W}_{\text{shaft}} / \eta_{\text{standard}} = (\text{Power rating})(\text{Load factor}) / \eta_{\text{standard}}$$

$$\dot{W}_{\text{electric in, efficient}} = \dot{W}_{\text{shaft}} / \eta_{\text{efficient}} = (\text{Power rating})(\text{Load factor}) / \eta_{\text{efficient}}$$

$$\text{Power savings} = \dot{W}_{\text{electric in, standard}} - \dot{W}_{\text{electric in, efficient}}$$

$$= (\text{Power rating})(\text{Load factor})[1/\eta_{\text{standard}} - 1/\eta_{\text{efficient}}]$$

where η_{standard} is the efficiency of the standard motor, and $\eta_{\text{efficient}}$ is the efficiency of the comparable high efficiency motor. Then the annual energy and cost savings associated with the installation of the high efficiency motor are determined to be

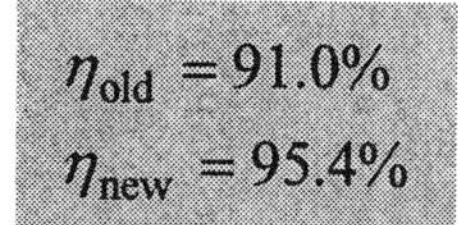

Energy Savings = (Power savings)(Operating Hours)
= (Power Rating)(Operating Hours)(Load Factor)($1/\eta_{\text{standard}} - 1/\eta_{\text{efficient}}$)
= (75 hp)(0.746 kW/hp)(4,368 hours/year)(0.75)(1/0.91 - 1/0.954)
= **9,290 kWh/year**

Cost Savings = (Energy savings)(Unit cost of energy)
= (9,290 kWh/year)($0.08/kWh)
= **$743/year**

The implementation cost of this measure consists of the excess cost the high efficiency motor over the standard one. That is,

Implementation Cost = Cost differential = $5,520 - $5,449 = $71

This gives a simple payback period of

$$\text{Simple payback period} = \frac{\text{Implementation cost}}{\text{Annual cost savings}} = \frac{\$71}{\$743/\text{year}} = \mathbf{0.096\ year}\ (\text{or } 1.1 \text{ months})$$

Therefore, the high-efficiency motor will pay for its cost differential in about one month.

5-34E The combustion efficiency of a furnace is raised from 0.7 to 0.8 by tuning it up. The annual energy and cost savings as a result of tuning up the boiler are to be determined. √

Assumptions The boiler operates at full load while operating.

Analysis The heat output of boiler is related to the fuel energy input to the boiler by

$$\text{Boiler output} = (\text{Boiler input})(\text{Combustion efficiency}) \quad \text{or} \quad \dot{Q}_{\text{out}} = \dot{Q}_{\text{in}} \eta_{\text{furnace}}$$

The current rate of heat input to the boiler is given to be $\dot{Q}_{\text{in, current}} = 3.6 \times 10^6$ Btu/h .
Then the rate of useful heat output of the boiler becomes

$$\dot{Q}_{\text{out}} = (\dot{Q}_{\text{in}} \eta_{\text{furnace}})_{\text{current}} = (3.6 \times 10^6 \text{ Btu/h})(0.7) = 2.52 \times 10^6 \text{ Btu/h}$$

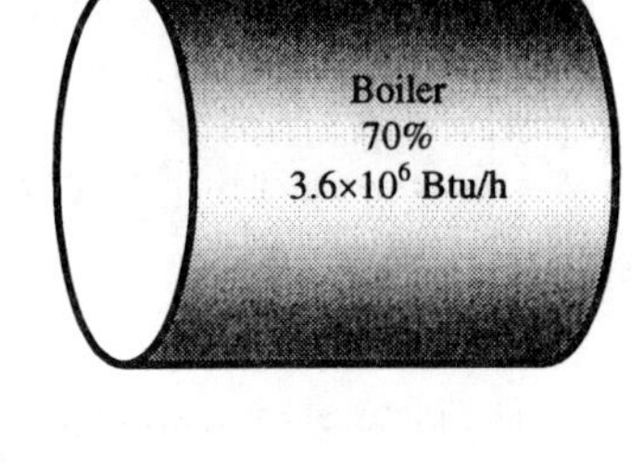

The boiler must supply useful heat at the same rate after the tune up. Therefore, the rate of heat input to the boiler after the tune up and the rate of energy savings become

$$\dot{Q}_{\text{in, new}} = \dot{Q}_{\text{out}} / \eta_{\text{furnace, new}} = (2.52 \times 10^6 \text{ Btu/h})/0.8 = 3.15 \times 10^6 \text{ Btu/h}$$

$$\dot{Q}_{\text{in, saved}} = \dot{Q}_{\text{in, current}} - \dot{Q}_{\text{in, new}} = 3.6 \times 10^6 - 3.15 \times 10^6 = 0.45 \times 10^6 \text{ Btu/h}$$

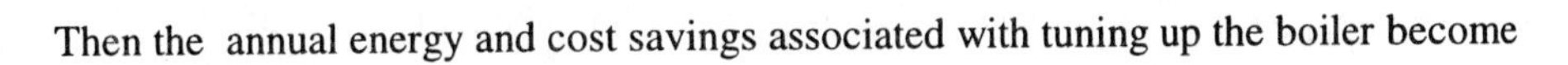
Then the annual energy and cost savings associated with tuning up the boiler become

$$\begin{aligned} \text{Energy Savings} &= \dot{Q}_{\text{in, saved}} (\text{Operation hours}) \\ &= (0.45 \times 10^6 \text{ Btu/h})(1500 \text{ h/year}) = \mathbf{675 \times 10^6 \text{ Btu/yr}} \end{aligned}$$

$$\begin{aligned} \text{Cost Savings} &= (\text{Energy Savings})(\text{Unit cost of energy}) \\ &= (675 \times 10^6 \text{ Btu/yr})(\$4.35 \text{ per } 10^6 \text{ Btu}) = \mathbf{\$2936/year} \end{aligned}$$

Discussion Notice that tuning up the boiler will save $2936 a year, which is a significant amount. The implementation cost of this measure is negligible if the adjustment can be made by in-house personnel. Otherwise it is worthwhile to have an authorized representative of the boiler manufacturer to service the boiler twice a year.

5-35E EES solution of this (and other comprehensive problems designated with the *computer icon*) is available to instructors at the *Instructor Manual* section of the *Online Learning Center* (OLC) at www.mhhe.com/cengel-boles. See the Preface for access information.

5-36 The gas space heating of a facility is to be supplemented by air heated in a liquid-to-air heat exchanger of a compressor. The amount of money that will be saved by diverting the compressor waste heat into the facility during the heating season is to be determined.

Assumptions The atmospheric pressure at that location is 1 atm.

Analysis The mass flow rate of air through the liquid-to-air heat exchanger is

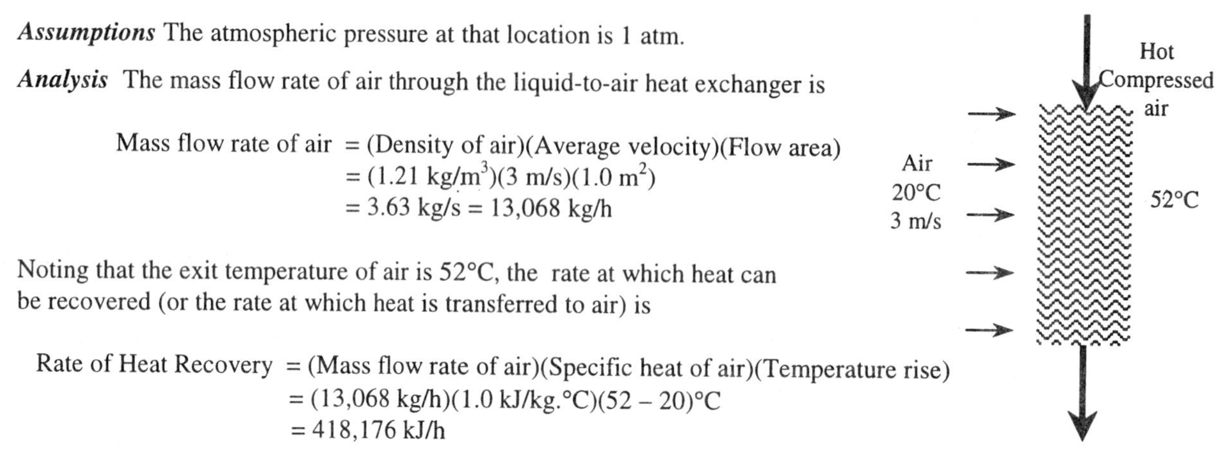

$$\begin{aligned}\text{Mass flow rate of air} &= \text{(Density of air)(Average velocity)(Flow area)}\\ &= (1.21\ \text{kg/m}^3)(3\ \text{m/s})(1.0\ \text{m}^2)\\ &= 3.63\ \text{kg/s} = 13{,}068\ \text{kg/h}\end{aligned}$$

Noting that the exit temperature of air is 52°C, the rate at which heat can be recovered (or the rate at which heat is transferred to air) is

$$\begin{aligned}\text{Rate of Heat Recovery} &= \text{(Mass flow rate of air)(Specific heat of air)(Temperature rise)}\\ &= (13{,}068\ \text{kg/h})(1.0\ \text{kJ/kg.°C})(52-20)\text{°C}\\ &= 418{,}176\ \text{kJ/h}\end{aligned}$$

The number of operating hours of this compressor during the heating season is

$$\text{Operating hours} = \text{(20 hours/day)(5 days/week)(26 weeks/year)} = 2600\ \text{hours/year}$$

Then the annual energy and cost savings become

$$\begin{aligned}\text{Energy Savings} &= \text{(Rate of Heat Recovery)(Annual Operating Hours)/Efficiency}\\ &= (418{,}176\ \text{kJ/h})(2600\ \text{h/year})/0.8\\ &= 1{,}359{,}100{,}000\ \text{kJ/year}\\ &= 12{,}882\ \text{therms/year}\end{aligned}$$

$$\begin{aligned}\text{Cost Savings} &= \text{(Energy savings)(Unit cost of energy saved)}\\ &= (12{,}882\ \text{therms/year})(\$0.50/\text{therm})\\ &= \$6441/\text{year}\end{aligned}$$

Discussion Notice that utilizing the waste heat from the compressor will save $6441 per year from the heating costs. The implementation of this measure requires the installation of an ordinary sheet metal duct from the outlet of the heat exchanger into the building. The installation cost associated with this measure is relatively low. Several manufacturing facilities already have this conservation system in place. A damper is used to direct the air into the building in winter and to the ambient in summer. Combined compressor/heat-recovery systems are available in the market for both air-cooled (greater than 50 hp) and water cooled (greater than 125 hp) systems.

5-37 Several people are working out in an exercise room. The rate of heat gain from people and the equipment is to be determined.

Assumptions The average rate of heat dissipated by people in an exercise room is 525 W.

Analysis The 8 weight lifting machines do not have any motors, and thus they do not contribute to the internal heat gain directly. The usage factors of the motors of the treadmills are taken to be unity since they are used constantly during peak periods. Noting that 1 hp = 746 W, the total heat generated by the motors is

$$\dot{Q}_{\text{motors}} = (\text{No. of motors}) \times \dot{W}_{\text{motor}} \times f_{\text{load}} \times f_{\text{usage}} / \eta_{\text{motor}}$$
$$= 4 \times (2.5 \times 746\text{ W}) \times 0.70 \times 1.0/0.77 = 6782\text{ W}$$

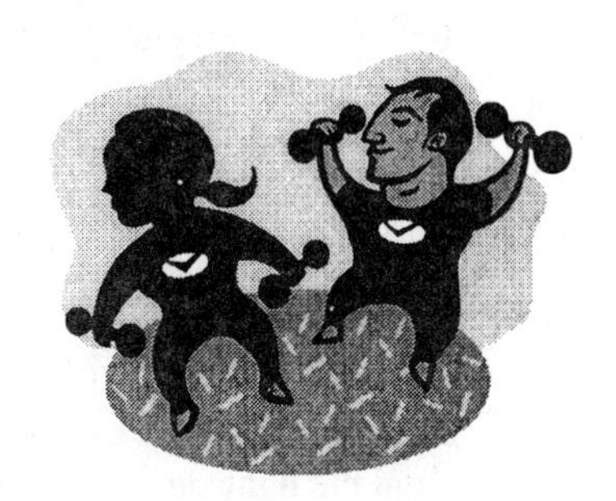

The heat gain from 14 people is

$$\dot{Q}_{\text{people}} = (\text{No. of people}) \times \dot{Q}_{\text{person}} = 14 \times (525\text{ W}) = 7350\text{ W}$$

Then the total rate of heat gain of the exercise room during peak period becomes

$$\dot{Q}_{\text{total}} = \dot{Q}_{\text{motors}} + \dot{Q}_{\text{people}} = 6782 + 7350 = \mathbf{14{,}132\text{ W}}$$

5-38 A classroom has a specified number of students, instructors, and fluorescent light bulbs. The rate of internal heat generation in this classroom is to be determined.

Assumptions **1** There is a mix of men, women, and children in the classroom. **2** The amount of light (and thus energy) leaving the room through the windows is negligible.

Properties The average rate of heat generation from people seated in a room/office is given to be 100 W.

Analysis The amount of heat dissipated by the lamps is equal to the amount of electrical energy consumed by the lamps, including the 10% additional electricity consumed by the ballasts. Therefore,

$$\dot{Q}_{\text{lighting}} = (\text{Energy consumed per lamp}) \times (\text{No. of lamps})$$
$$= (40\text{ W})(1.1)(18) = 792\text{ W}$$
$$\dot{Q}_{\text{people}} = (\text{No. of people}) \times \dot{Q}_{\text{person}} = 56 \times (100\text{ W}) = 5600\text{ W}$$

Then the total rate of heat gain (or the internal heat load) of the classroom from the lights and people become

$$\dot{Q}_{\text{total}} = \dot{Q}_{\text{lighting}} + \dot{Q}_{\text{people}} = 792 + 5600 = \mathbf{6392\ W}$$

5-39 A room is cooled by circulating chilled water through a heat exchanger, and the air is circulated through the heat exchanger by a fan. The contribution of the fan-motor assembly to the cooling load of the room is to be determined.

Assumptions The fan motor operates at full load so that $f_{\text{load}} = 1$.

Analysis The entire electrical energy consumed by the motor, including the shaft power delivered to the fan, is eventually dissipated as heat. Therefore, the contribution of the fan-motor assembly to the cooling load of the room is equal to the electrical energy it consumes,

$$\dot{Q}_{\text{internal generation}} = \dot{W}_{\text{in, electric}} = \dot{W}_{\text{shaft}} / \eta_{\text{motor}}$$
$$= (0.25\text{ hp})/0.54 = 0.463\text{ hp} = \mathbf{345\ W}$$

since 1 hp = 746 W.

Refrigerators and Heat Pumps

5-40C The difference between the two devices is one of purpose. The purpose of a refrigerator is to remove heat from a cold medium whereas the purpose of a heat pump is to supply heat to a warm medium.

5-41C The difference between the two devices is one of purpose. The purpose of a refrigerator is to remove heat from a refrigerated space whereas the purpose of an air-conditioner is remove heat from a living space.

5-42C No. Because the refrigerator consumes work to accomplish this task.

5-43C No. Because the heat pump consumes work to accomplish this task.

5-44C The coefficient of performance of a refrigerator represents the amount of heat removed from the refrigerated space for each unit of work supplied. It can be greater than unity.

5-45C The coefficient of performance of a heat pump represents the amount of heat supplied to the heated space for each unit of work supplied. It can be greater than unity.

5-46C No. The heat pump captures energy from a cold medium and carries it to a warm medium. It does not create it.

5-47C No. The refrigerator captures energy from a cold medium and carries it to a warm medium. It does not create it.

5-48C No device can transfer heat from a cold medium to a warm medium without requiring a heat or work input from the surroundings.

5-49C The violation of one statement leads to the violation of the other one, as shown in Sec. 5-4, and thus we conclude that the two statements are equivalent.

5-50 The COP and the refrigeration rate of a refrigerator are given. The power consumption and the rate of heat rejection are to be determined.

Assumptions The refrigerator operates steadily.

Analysis (*a*) Using the definition of the coefficient of performance, the power input to the refrigerator is determined to be

$$\dot{W}_{net,in} = \frac{\dot{Q}_L}{COP_R} = \frac{60 \text{ kJ/min}}{1.5} = 40 \text{ kJ/min} = \mathbf{0.67 \ kW}$$

Kitchen air

COP

R

$\dot{Q}_L$

cool space

(*b*) The heat transfer rate to the kitchen air is determined from the energy balance,

$$\dot{Q}_H = \dot{Q}_L + \dot{W}_{net,in} = 60 + 40 = \mathbf{100 \ kJ/min}$$

5-51 The power consumption and the cooling rate of an air conditioner are given. The COP and the rate of heat rejection are to be determined.

Assumptions The air conditioner operates steadily.

Analysis (*a*) The coefficient of performance of the air-conditioner (or refrigerator) is determined from its definition,

$$COP_R = \frac{\dot{Q}_L}{\dot{W}_{net,in}} = \frac{750 \text{ kJ/min}}{6 \text{ kW}}\left(\frac{1 \text{ kW}}{60 \text{ kJ/min}}\right) = \mathbf{2.08}$$

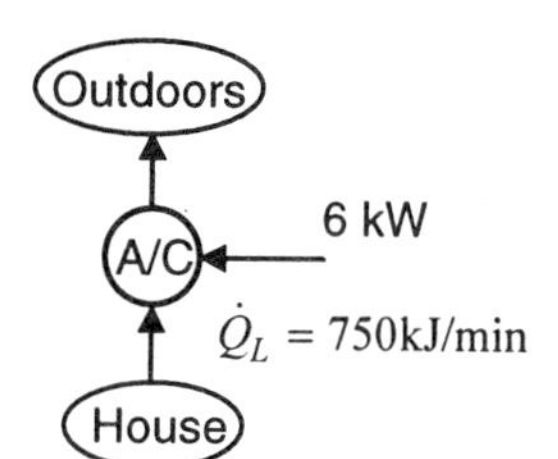

(*b*) The rate of heat discharge to the outside air is determined from the energy balance,

$$\dot{Q}_H = \dot{Q}_L + \dot{W}_{net,in} = (750 \text{ kJ/min}) + (6)(60 \text{ kJ/min}) = \mathbf{1110 \ kJ/min}$$

5-52 The COP and the refrigeration rate of a refrigerator are given. The power consumption of the refrigerator is to be determined.

Assumptions The refrigerator operates steadily.

Analysis Since the refrigerator runs one-fourth of the time and removes heat from the food compartment at an average rate of 800 kJ/h, the refrigerator removes heat at a rate of

$$\dot{Q}_L = 4(800 \text{ kJ/h}) = 3200 \text{ kJ/h}$$

when running. Thus the power the refrigerator draws when it is running is

$$\dot{W}_{net,in} = \frac{\dot{Q}_L}{COP_R} = \frac{3200 \text{kJ/h}}{2.2} = 1455 \text{ kJ/h} = \mathbf{0.40 \ kW}$$

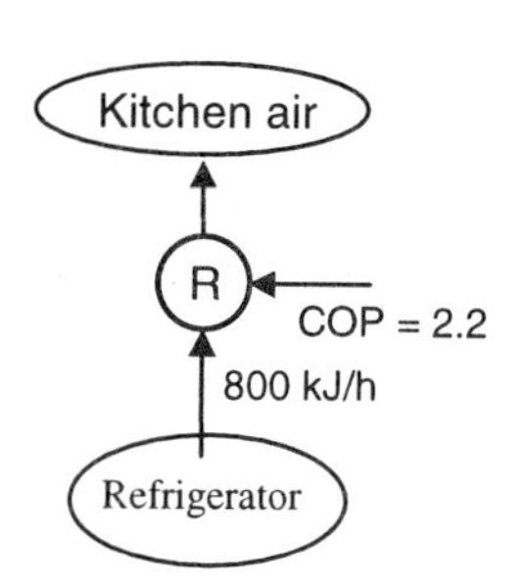

5-53E The COP and the refrigeration rate of an ice machine are given. The power consumption is to be determined.

Assumptions The ice machine operates steadily.

Analysis The cooling load of this ice machine is

$$\dot{Q}_L = \dot{m}q_L = (20 \text{lbm/h})(169 \text{Btu/lbm}) = 3380 \text{Btu/h}$$

Using the definition of the coefficient of performance, the power input to the ice machine system is determined to be

$$\dot{W}_{net,in} = \frac{\dot{Q}_L}{COP_R} = \frac{3380 \text{Btu/h}}{2.4}\left(\frac{1 \text{hp}}{2545 \text{Btu/h}}\right) = \mathbf{0.553 hp}$$

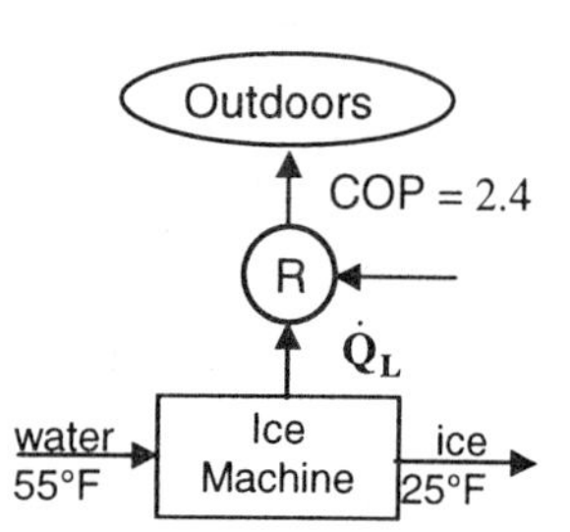

5-54 The COP and the power consumption of a refrigerator are given. The time it will take to cool 5 watermelons is to be determined.

Assumptions **1** The refrigerator operates steadily. **2** The heat gain of the refrigerator through its walls, door, etc. is negligible. **3** The watermelons are the only items in the refrigerator to be cooled.

Properties The specific heat of watermelons is given to be $C = 4.2$ kJ/kg.°C.

Analysis The total amount of heat that needs to be removed from the watermelons is

$$Q_L = (mC\Delta T)_{watermelons} = 5 \times (10\text{kg})(4.2\text{kJ/kg}\cdot^\circ\text{C})(20-8)^\circ\text{C} = 2520\text{kJ}$$

The rate at which this refrigerator removes heat is

$$\dot{Q}_L = (COP_R)(\dot{W}_{net,in}) = (2.5)(0.45\text{kW}) = 1.125\text{kW}$$

Kitchen air
R
450 W
COP = 2.5
cool space

That is, this refrigerator can remove 1.125 kJ of heat per second. Thus the time required to remove 2520 kJ of heat is

$$\Delta t = \frac{Q_L}{\dot{Q}_L} = \frac{2520\text{kJ}}{1.125\text{kJ/s}} = 2240\text{s} = \mathbf{37.3min}$$

This answer is optimistic since the refrigerated space will gain some heat during this process from the surrounding air, which will increase the work load. Thus, in reality, it will take longer to cool the watermelons.

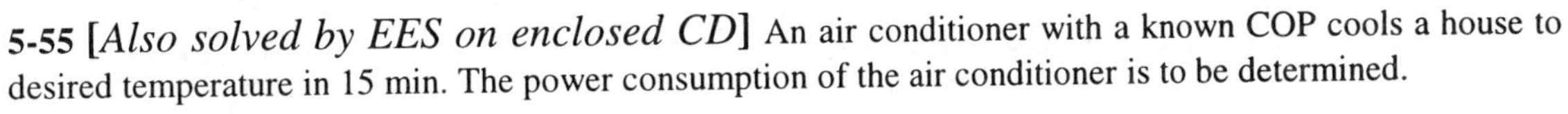

5-55 [*Also solved by EES on enclosed CD*] An air conditioner with a known COP cools a house to desired temperature in 15 min. The power consumption of the air conditioner is to be determined.

Assumptions **1** The air conditioner operates steadily. **2** The house is well-sealed so that no air leaks in or out during cooling. **3** Air is an ideal gas with constant specific heats at room temperature.

Properties The constant volume specific heat of air is given to be $C_v = 0.72$ kJ/kg.°C.

Analysis Since the house is well-sealed (constant volume), the total amount of heat that needs to be removed from the house is

$$Q_L = (mC_v\Delta T)_{House} = (800\text{kg})(0.72\text{kJ/kg}\cdot^\circ\text{C})(32-20)^\circ\text{C} = 6912\text{kJ}$$

This heat is removed in 15 minutes. Thus the average rate of heat removal from the house is

$$\dot{Q}_L = \frac{Q_L}{\Delta t} = \frac{6912\text{kJ}}{15 \times 60\text{s}} = 7.68\text{kW}$$

Outside
$\dot{Q}_H$
COP = 2.5
AC
32→20°C
House

Using the definition of the coefficient of performance, the power input to the air-conditioner is determined to be

$$\dot{W}_{net,in} = \frac{\dot{Q}_L}{COP_R} = \frac{7.68\text{kW}}{2.5} = \mathbf{3.07kW}$$

5-56 EES solution of this (and other comprehensive problems designated with the *computer icon*) is available to instructors at the *Instructor Manual* section of the *Online Learning Center* (OLC) at www.mhhe.com/cengel-boles. See the Preface for access information.

5-57 The heat removal rate of a refrigerator per kW of power it consumes is given. The COP and the rate of heat rejection are to be determined.

Assumptions The refrigerator operates steadily.

Analysis The coefficient of performance of the refrigerator is determined from its definition,

$$COP_R = \frac{\dot{Q}_L}{\dot{W}_{net,in}} = \frac{6500\text{ kJ/h}}{1\text{ kW}}\left(\frac{1\text{ kW}}{3600\text{ kJ/h}}\right) = \mathbf{1.81}$$

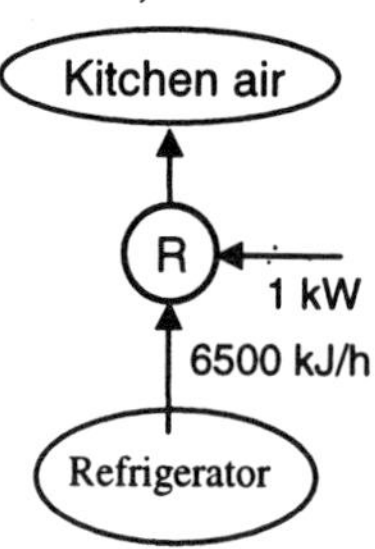

The rate of heat rejection to the surrounding air, per kW of power consumed, is determined from the energy balance,

$$\dot{Q}_H = \dot{Q}_L + \dot{W}_{net,in} = (6500\text{ kJ/h}) + (1)(3600\text{ kJ/h}) = \mathbf{10{,}100\ kJ/h}$$

5-58 The rate of heat supply of a heat pump per kW of power it consumes is given. The COP and the rate of heat absorption from the cold environment are to be determined.

Assumptions The heat pump operates steadily.

Analysis The coefficient of performance of the refrigerator is determined from its definition,

$$COP_{HP} = \frac{\dot{Q}_{HP}}{\dot{W}_{net,in}} = \frac{8000\text{ kJ/h}}{1\text{ kW}}\left(\frac{1\text{ kW}}{3600\text{ kJ/h}}\right) = \mathbf{2.22}$$

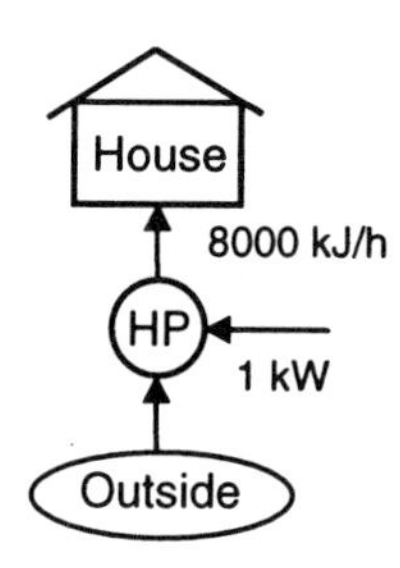

The rate of heat absorption from the surrounding air, per kW of power consumed, is determined from the energy balance,

$$\dot{Q}_L = \dot{Q}_H - \dot{W}_{net,in} = (8{,}000\text{ kJ/h}) - (1)(3600\text{ kJ/h}) = 4400\ \mathbf{kJ/h}$$

5-59 A house is heated by resistance heaters, and the amount of electricity consumed during a winter month is given. The amount of money that would be saved if this house were heated by a heat pump with a known COP is to be determined.

Assumptions The heat pump operates steadily.

Analysis The amount of heat the resistance heaters supply to the house is equal to he amount of electricity they consume. Therefore, to achieve the same heating effect, the house must be supplied with 1200 kWh of energy. A heat pump that supplied this much heat will consume electrical power in the amount of

$$\dot{W}_{net,in} = \frac{\dot{Q}_H}{COP_{HP}} = \frac{1200\text{ kWh}}{2.4} = 500\text{ kWh}$$

which represent a savings of 1200 – 500 = 700 kWh. Thus the homeowner would have saved

(700 kWh)(0.085 $/kWh) = **$59.50**

5-60E The rate of heat supply and the COP of a heat pump are given. The power consumption and the rate of heat absorption from the outside air are to be determined.

Assumptions The heat pump operates steadily.

Analysis (*a*) The power consumed by this heat pump can be determined from the definition of the coefficient of performance of a heat pump to be

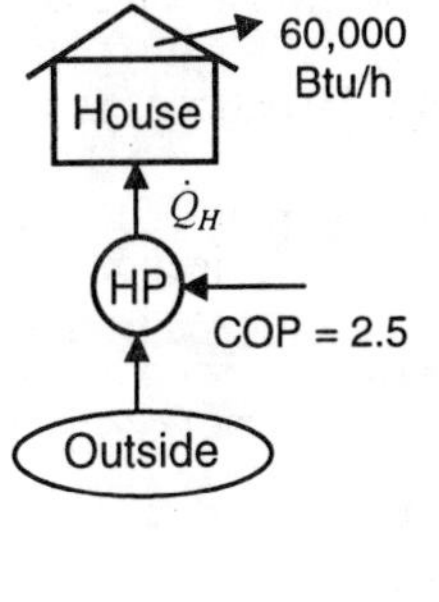

$$\dot{W}_{net,in} = \frac{\dot{Q}_H}{COP_{HP}} = \frac{60{,}000\text{Btu/h}}{2.5} = 24{,}000\text{Btu/h} = \mathbf{9.43hp}$$

(*b*) The rate of heat transfer from the outdoor air is determined from the conservation of energy principle,

$$\dot{Q}_L = \dot{Q}_H - \dot{W}_{net,in} = (60{,}000 - 24{,}000\text{Btu/h}) = \mathbf{36{,}000\ Btu/h}$$

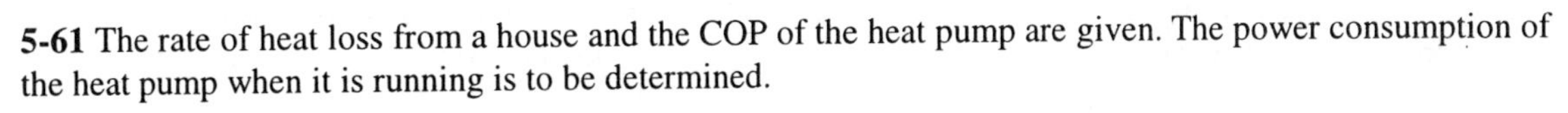

5-61 The rate of heat loss from a house and the COP of the heat pump are given. The power consumption of the heat pump when it is running is to be determined.

Assumptions The heat pump operates one-third of the time.

Analysis Since the heat pump runs one-third of the time and must supply heat to the house at an average rate of 15,000 kJ/h, the heat pump supplies heat at a rate of

$$\dot{Q}_H = 3(22{,}000\text{kJ/h}) = 66.000\text{ kJ/h}$$

when running. Thus the power the heat pump draws when it is running is

$$\dot{W}_{net,in} = \frac{\dot{Q}_H}{COP_{HP}} = \frac{66{,}000\text{ kJ/h}}{3.5}\left(\frac{1\text{ kW}}{3600\text{ kJ/h}}\right) = \mathbf{5.23\ kW}$$

5-62 The rate of heat loss, the rate of internal heat gain, and the COP of a heat pump are given. The power input to the heat pump is to be determined.

Assumptions The heat pump operates steadily.

Analysis The heating load of this heat pump system is the difference between the heat lost to the outdoors and the heat generated in the house from the people, lights, and appliances,

$$\dot{Q}_H = 60{,}000 - 4{,}000 = 56{,}000 \text{ kJ/h}$$

Using the definition of COP, the power input to the heat pump is determined to be

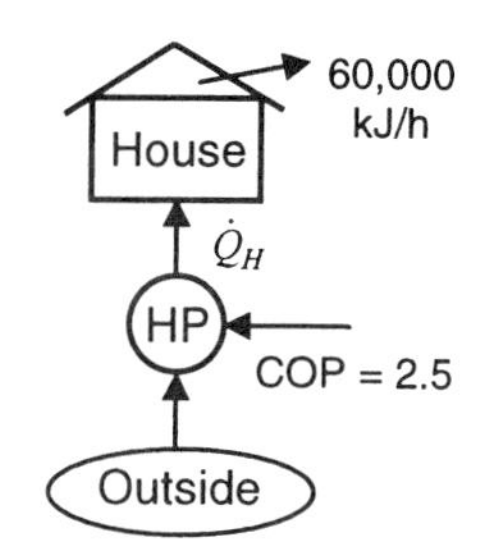

$$\dot{W}_{net,in} = \frac{\dot{Q}_H}{COP_{HP}} = \frac{56{,}000\text{kJ/h}}{2.5}\left(\frac{1\text{kW}}{3600\text{kJ/h}}\right) = \mathbf{6.22kW}$$

5-63 An office that is being cooled adequately by a 12,000 Btu/h window air-conditioner is converted to a computer room. The number of additional air-conditioners that need to be installed is to be determined.

Assumptions **1** The computer are operated by 4 adult men. **2** The computers consume 40 percent of their rated power at any given time.

Properties The average rate of heat generation from a person seated in a room/office is 100 W (given).

Analysis The amount of heat dissipated by the computers is equal to the amount of electrical energy they consume. Therefore,

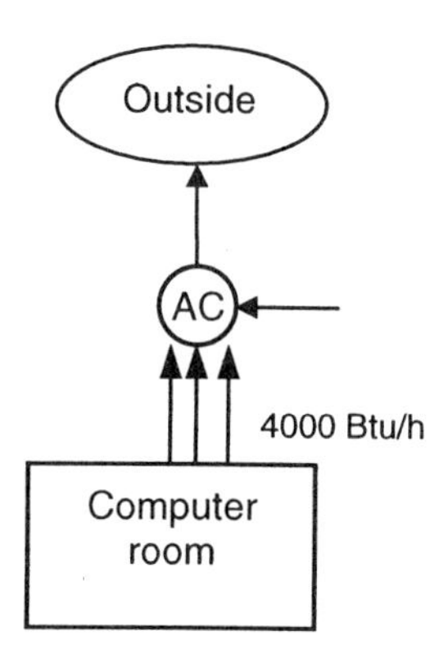

$$\dot{Q}_{\text{computers}} = (\text{Rated power}) \times (\text{Usage factor}) = (3.5 \text{ kW})(0.4) = 1.4 \text{ kW}$$

$$\dot{Q}_{\text{people}} = (\text{No. of people}) \times \dot{Q}_{\text{person}} = 4 \times (100 \text{ W}) = 400 \text{ W}$$

$$\dot{Q}_{\text{total}} = \dot{Q}_{\text{computers}} + \dot{Q}_{\text{people}} = 1400 + 400 = 1800 \text{ W} = 6142 \text{ Btu/h}$$

since 1 W = 3.412 Btu/h. Then noting that each available air conditioner provides 4,000 Btu/h cooling, the number of air-conditioners needed becomes

$$\text{No. of air conditioners} = \frac{\text{Cooling load}}{\text{Cooling capacity of A/C}} = \frac{6142 \text{ Btu/h}}{4000 \text{ Btu/h}}$$
$$= 1.5 \approx \mathbf{2\ Air\ conditioners}$$

5-64 A decision is to be made between a cheaper but inefficient air-conditioner and an expensive but efficient air-conditioner for a building. The better buy is to be determined.

Assumptions The two air conditioners are comparable in all aspects other than the initial cost and the efficiency.

Analysis The unit that will cost less during its lifetime is a better buy. The total cost of a system during its lifetime (the initial, operation, maintenance, etc.) can be determined by performing a life cycle cost analysis. A simpler alternative is to determine the simple payback period. The energy and cost savings of the more efficient air conditioner in this case is

$$
\begin{aligned}
\text{Energy savings} &= (\text{Annual energy usage of A}) - (\text{Annual energy usage of B}) \\
&= (\text{Annual cooling load})(1/\text{COP}_\text{A} - 1/\text{COP}_\text{B}) \\
&= (120{,}000 \text{ kWh/year})(1/3.2 - 1/5.0) \\
&= 13{,}500 \text{ kWh/year}
\end{aligned}
$$

Air Cond. A
COP = 3.2

$$
\begin{aligned}
\text{Cost savings} &= (\text{Energy savings})(\text{Unit cost of energy}) \\
&= (13{,}500 \text{ kWh/year})(\$0.10/\text{kWh}) = \mathbf{\$1350/year}
\end{aligned}
$$

The installation cost difference between the two air-conditioners is

$$
\text{Cost difference} = \text{Cost of B} - \text{cost of A} = 7000 - 5500 = \$1500
$$

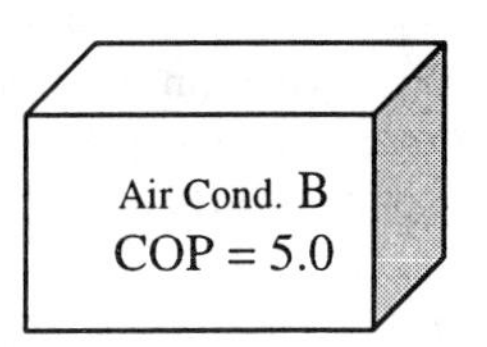

Therefore, the more efficient air-conditioner B will pay for the \$1500 cost differential in this case in about 1 year.

Discussion A cost conscious consumer will have no difficulty in deciding that the more expensive but more efficient air-conditioner B is clearly the better buy in this case since air conditioners last at least 15 years. But the decision would not be so easy if the unit cost of electricity at that location was much less than \$0.10/kWh, or if the annual air-conditioning load of the house was much less than 120,000 kWh.

Perpetual-Motion Machines

5-65C This device creates energy, and thus it is a PMM1.

5-66C This device creates energy, and thus it is a PMM1.

Reversible and Irreversible Processes

5-67C No. Because it involves heat transfer through a finite temperature difference.

5-68C Because reversible processes can be approached in reality, and they form the limiting cases. Work producing devices that operate on reversible processes deliver the most work, and work consuming devices that operate on reversible processes consume the least work.

5-69C When the compression process is non-quasiequilibrium, the molecules before the piston face cannot escape fast enough, forming a high pressure region in front of the piston. It takes more work to move the piston against this high pressure region.

5-70C When an expansion process is non-quasiequilibrium, the molecules before the piston face cannot follow the piston fast enough, forming a low pressure region behind the piston. The lower pressure that pushes the piston produces less work.

5-71C The irreversibilities that occur within the system boundaries are **internal** irreversibilities; those which occur outside the system boundaries are **external** irreversibilities.

5-72C A reversible expansion or compression process cannot involve unrestrained expansion or sudden compression, and thus it is quasi-equilibrium. A quasi-equilibrium expansion or compression process, on the other hand, may involve external irreversibilities (such as heat transfer through a finite temperature difference), and thus is not necessarily reversible.

The Carnot Cycle and Carnot's Principle

5-73C The four processes that make up the Carnot cycle are isothermal expansion, reversible adiabatic expansion, isothermal compression, and reversible adiabatic compression.

5-74C They are (1) the thermal efficiency of an irreversible heat engine is lower than the efficiency of a reversible heat engine operating between the same two reservoirs, and (2) the thermal efficiency of all the reversible heat engines operating between the same two reservoirs are equal.

5-75C False. The second Carnot principle states that no heat engine cycle can have a higher thermal efficiency than the Carnot cycle operating between the same temperature limits.

5-76C Yes. The second Carnot principle states that all reversible heat engine cycles operating between the same temperature limits have the thermal efficiency.

5-77C (a) No, (b) No. They would violate the Carnot principle.

Carnot Heat Engines

5-78C No.

5-79C The one that has a source temperature of 600°C. This is true because the higher the temperature at which heat is supplied to the working fluid of a heat engine, the higher the thermal efficiency.

5-80 The source and sink temperatures of a Carnot heat engine and the rate of heat supply are given. The thermal efficiency and the power output are to be determined.

Assumptions The Carnot heat engine operates steadily.

Analysis (a) The thermal efficiency of a Carnot heat engine depends on the source and the sink temperatures only, and is determined from

$$\eta_{th,C} = 1 - \frac{T_L}{T_H} = 1 - \frac{300 \text{ K}}{1000 \text{ K}} = 0.70 \quad \text{or} \quad \mathbf{70\%}$$

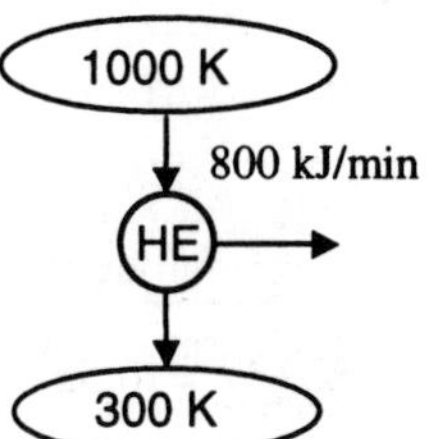

(b) The power output of this heat engine is determined from the definition of thermal efficiency,

$$\dot{W}_{net,out} = \eta_{th}\dot{Q}_H = (0.70)(800\text{kJ/min}) = 560\text{kJ/min} = \mathbf{9.33kW}$$

5-81 The sink temperature of a Carnot heat engine and the rates of heat supply and heat rejection are given. The source temperature and the thermal efficiency of the engine are to be determined.

Assumptions The Carnot heat engine operates steadily.

Analysis (a) For reversible cyclic devices we have $\left(\frac{Q_H}{Q_L}\right)_{rev} = \left(\frac{T_H}{T_L}\right)$

Thus the temperature of the source T_H must be

$$T_H = \left(\frac{Q_H}{Q_L}\right)_{rev} T_L = \left(\frac{650 \text{ kJ}}{200 \text{ kJ}}\right)(290 \text{ K}) = \mathbf{942.5 \text{ K}}$$

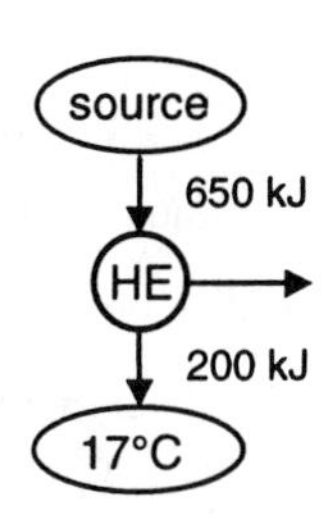

(*b*) The thermal efficiency of a Carnot heat engine depends on the source and the sink temperatures only, and is determined from

$$\eta_{th,C} = 1 - \frac{T_L}{T_H} = 1 - \frac{290 \text{ K}}{942.5 \text{ K}} = 0.69 \text{ or } \mathbf{69\%}$$

5-82 [*Also solved by EES on enclosed CD*] The source and sink temperatures of a heat engine and the rate of heat supply are given. The maximum possible power output of this engine is to be determined.

Assumptions The heat engine operates steadily.

Analysis The highest thermal efficiency a heat engine operating between two specified temperature limits can have is the Carnot efficiency, which is determined from

$$\eta_{th,\max} = \eta_{th,C} = 1 - \frac{T_L}{T_H} = 1 - \frac{298 \text{ K}}{823 \text{ K}} = 0.638 \quad \text{or} \quad \mathbf{63.8\%}$$

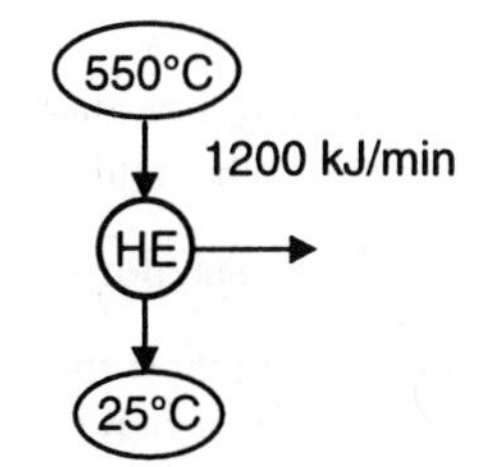

Then the maximum power output of this heat engine is determined from the definition of thermal efficiency to be

$$\dot{W}_{net,out} = \eta_{th}\dot{Q}_H = (0.638)(1200\text{kJ/min}) = 765.6kJ/\text{min} = \mathbf{12.8kW}$$

5-83 EES solution of this (and other comprehensive problems designated with the *computer icon*) is available to instructors at the *Instructor Manual* section of the *Online Learning Center* (OLC) at www.mhhe.com/cengel-boles. See the Preface for access information.

5-84E The sink temperature of a Carnot heat engine, the rate of heat rejection, and the thermal efficiency are given. The power output of the engine and the source temperature are to be determined.

Assumptions The Carnot heat engine operates steadily.

Analysis (a) The rate of heat input to this heat engine is determined from the definition of thermal efficiency,

$$\eta_{th} = 1 - \frac{\dot{Q}_L}{\dot{Q}_H} \longrightarrow 0.55 = 1 - \frac{800\text{Btu/min}}{\dot{Q}_H} \longrightarrow \dot{Q}_H = 1777.8\text{Btu/min}$$

Then the power output of this heat engine can be determined from

$$\dot{W}_{net,out} = \eta_{th}\dot{Q}_H = (0.55)(1777.8\text{Btu/min}) = 977.8\text{Btu/min} = \mathbf{23.1hp}$$

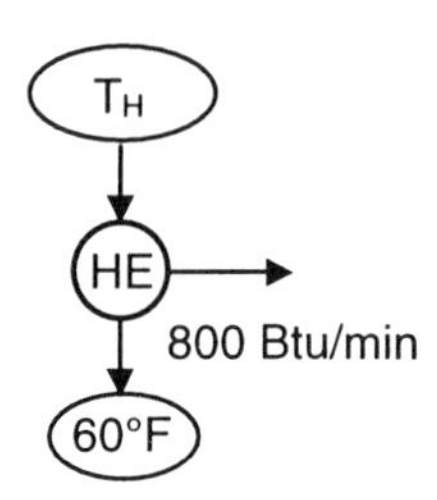

(*b*) For reversible cyclic devices we have $\left(\frac{\dot{Q}_H}{\dot{Q}_L}\right)_{rev} = \left(\frac{T_H}{T_L}\right)$

Thus the temperature of the source T_H must be

$$T_H = \left(\frac{\dot{Q}_H}{\dot{Q}_L}\right)_{rev} T_L = \left(\frac{1777.8\text{Btu/min}}{800\text{Btu/min}}\right)(520\text{R}) = \mathbf{1155.6R}$$

5-85 The source and sink temperatures of a OTEC (Ocean Thermal Energy Conversion) power plant are given. The maximum thermal efficiency is to be determined.

Assumptions The power plant operates steadily.

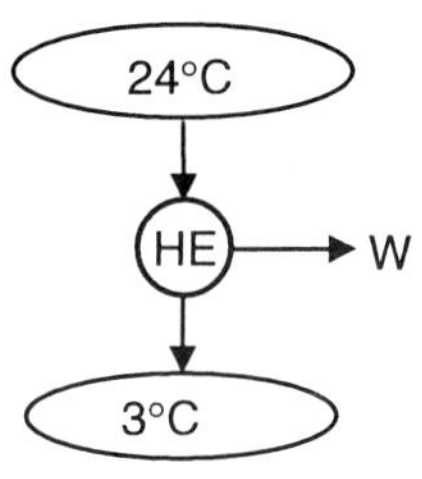

Analysis The highest thermal efficiency a heat engine operating between two specified temperature limits can have is the Carnot efficiency, which is determined from

$$\eta_{th,\max} = \eta_{th,C} = 1 - \frac{T_L}{T_H} = 1 - \frac{276\text{ K}}{297\text{ K}} = 0.071 \text{ or } \mathbf{7.1\%}$$

5-86 The source and sink temperatures of a geothermal power plant are given. The maximum thermal efficiency is to be determined.

Assumptions The power plant operates steadily.

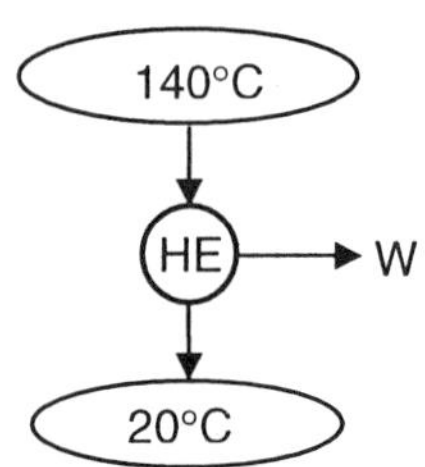

Analysis The highest thermal efficiency a heat engine operating between two specified temperature limits can have is the Carnot efficiency, which is determined from

$$\eta_{th,\max} = \eta_{th,C} = 1 - \frac{T_L}{T_H} = 1 - \frac{20 + 273\text{ K}}{140 + 273\text{ K}} = 0.291 \text{ or } \mathbf{29.1\%}$$

5-87 An inventor claims to have developed a heat engine. The inventor reports temperature, heat transfer, and work output measurements. The claim is to be evaluated.

Analysis The highest thermal efficiency a heat engine operating between two specified temperature limits can have is the Carnot efficiency, which is determined from

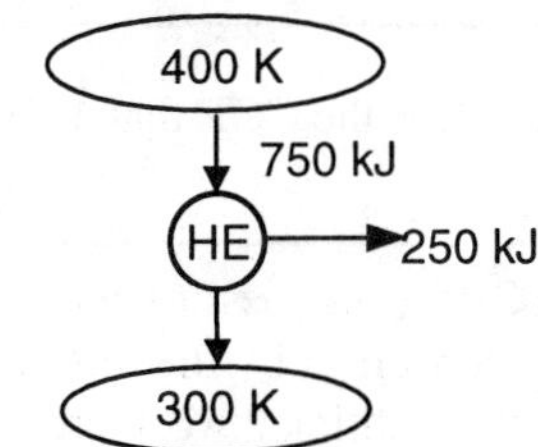

$$\eta_{th,\max} = \eta_{th,C} = 1 - \frac{T_L}{T_H} = 1 - \frac{300\text{ K}}{400\text{ K}} = 0.25 \quad or \quad \mathbf{25\%}$$

The actual thermal efficiency of the heat engine in question is

$$\eta_{th} = \frac{W_{net}}{Q_H} = \frac{250\text{ kJ}}{750\text{ kJ}} = 0.333 \text{ or } \mathbf{33.3\%}$$

which is greater than the maximum possible thermal efficiency. Therefore, this heat engine is a PMM2 and the claim is **false**.

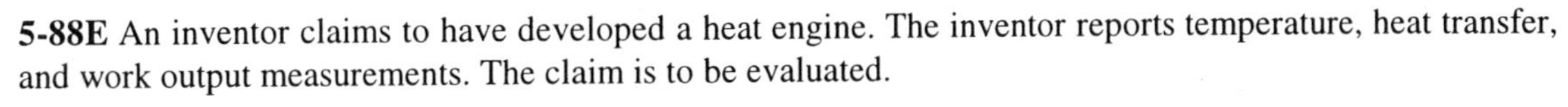

5-88E An inventor claims to have developed a heat engine. The inventor reports temperature, heat transfer, and work output measurements. The claim is to be evaluated.

Analysis The highest thermal efficiency a heat engine operating between two specified temperature limits can have is the Carnot efficiency, which is determined from

900 R
300Btu
HE
160 kJ
540 R

$$\eta_{th,\max} = \eta_{th,C} = 1 - \frac{T_L}{T_H} = 1 - \frac{540\text{ R}}{900\text{ R}} = 0.40 \quad or \quad \mathbf{40\%}$$

The actual thermal efficiency of the heat engine in question is

$$\eta_{th} = \frac{W_{net}}{Q_H} = \frac{160\text{ Btu}}{300\text{ Btu}} = 0.533 \quad \text{or} \quad \mathbf{53.3\%}$$

which is greater than the maximum possible thermal efficiency. Therefore, this heat engine is a PMM2 and the claim is **false**.

Carnot Refrigerators and Heat Pumps

5-89C By increasing T_L or by decreasing T_H.

5-90C It is the COP that a Carnot refrigerator would have, $COP_R = \dfrac{1}{T_H / T_L - 1}$.

5-91C No. At best (when everything is reversible), the increase in the work produced will be equal to the work consumed by the refrigerator. In reality, the work consumed by the refrigerator will always be greater than the additional work produced, resulting in a decrease in the thermal efficiency of the power plant.

5-92C No. At best (when everything is reversible), the increase in the work produced will be equal to the work consumed by the refrigerator. In reality, the work consumed by the refrigerator will always be greater than the additional work produced, resulting in a decrease in the thermal efficiency of the power plant.

5-93C Bad idea. At best (when everything is reversible), the increase in the work produced will be equal to the work consumed by the heat pump. In reality, the work consumed by the heat pump will always be greater than the additional work produced, resulting in a decrease in the thermal efficiency of the power plant.

5-94 The refrigerated space and the environment temperatures of a Carnot refrigerator and the power consumption are given. The rate of heat removal from the refrigerated space is to be determined.

Assumptions The Carnot refrigerator operates steadily.

Analysis The coefficient of performance of a Carnot refrigerator depends on the temperature limits in the cycle only, and is determined from

$$COP_{R,C} = \frac{1}{(T_H / T_L) - 1} = \frac{1}{(22 + 273\text{K})/(3 + 273\text{K}) - 1} = 14.5$$

The rate of heat removal from the refrigerated space is determined from the definition of the coefficient of performance of a refrigerator,

$$\dot{Q}_L = COP_R \times \dot{W}_{net,in} = (14.5)(2\text{ kW}) = 29.0\text{ kW} = \mathbf{1740\ kJ/min}$$

5-95 The refrigerated space and the environment temperatures for a refrigerator and the rate of heat removal from the refrigerated space are given. The minimum power input required is to be determined.

Assumptions The refrigerator operates steadily.

Analysis The power input to a refrigerator will be a minimum when the refrigerator operates in a reversible manner. The coefficient of performance of a reversible refrigerator depends on the temperature limits in the cycle only, and is determined from

$$COP_{R,rev} = \frac{1}{(T_H/T_L)-1} = \frac{1}{(25+273\text{K})/(-8+273\text{K})-1} = 8.03$$

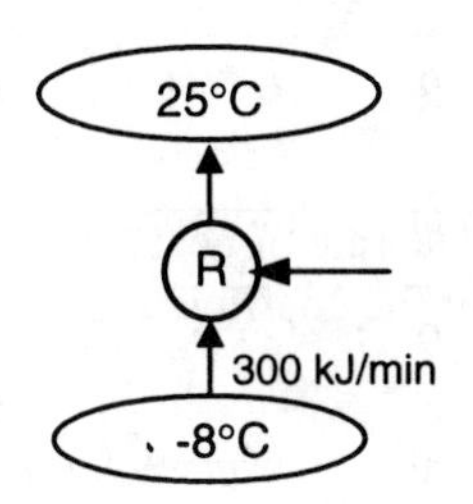

The power input to this refrigerator is determined from the definition of the coefficient of performance of a refrigerator,

$$\dot{W}_{net,in,\min} = \frac{\dot{Q}_L}{COP_{R,\max}} = \frac{300 \text{ kJ/min}}{8.03} = 37.36 \text{ kJ/min} = \mathbf{0.623\ kW}$$

5-96 The cooled space and the outdoors temperatures for a Carnot air-conditioner and the rate of heat removal from the air-conditioned room are given. The power input required is to be determined.

Assumptions The air-conditioner operates steadily.

Analysis The COP of a Carnot air conditioner (or Carnot refrigerator) depends on the temperature limits in the cycle only, and is determined from

$$COP_{R,C} = \frac{1}{(T_H/T_L)-1} = \frac{1}{(35+273\text{K})/(20+273\text{K})-1} = 19.5$$

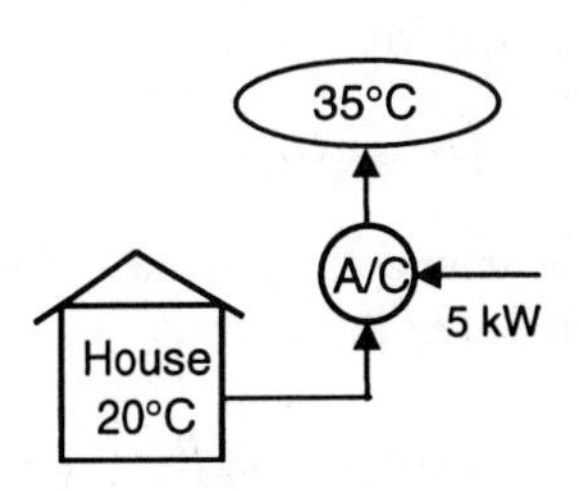

The power input to this refrigerator is determined from the definition of the coefficient of performance of a refrigerator,

$$\dot{W}_{net,in} = \frac{\dot{Q}_L}{COP_{R,\max}} = \frac{750 \text{ kJ/min}}{19.5} = 38.5 \text{ kJ/min} = \mathbf{0.64\ kW}$$

5-97E The cooled space and the outdoors temperatures for an air-conditioner and the power consumption are given. The maximum rate of heat removal from the air-conditioned space is to be determined.

Assumptions The air-conditioner operates steadily.

Analysis The rate of heat removal from a house will be a maximum when the air-conditioning system operates in a reversible manner. The coefficient of performance of a reversible air-conditioner (or refrigerator) depends on the temperature limits in the cycle only, and is determined from

$$COP_{R,rev} = \frac{1}{(T_H / T_L) - 1} = \frac{1}{(90 + 460\text{ R})/(72 + 460\text{ R}) - 1} = 29.6$$

The rate of heat removal from the house is determined from the definition of the coefficient of performance of a refrigerator,

$$\dot{Q}_L = COP_R \times \dot{W}_{net,in} = (29.6)(5\text{ hp})\left(\frac{42.41\text{ Btu/min}}{1\text{ hp}}\right) = \mathbf{6277\ Btu/min}$$

5-98 The refrigerated space temperature, the COP, and the power input of a Carnot refrigerator are given. The rate of heat removal from the refrigerated space and its temperature are to be determined.

Assumptions The refrigerator operates steadily.

Analysis (*a*) The rate of heat removal from the refrigerated space is determined from the definition of the COP of a refrigerator,

$$\dot{Q}_L = COP_R \times \dot{W}_{net,in} = (4.5)(0.5\text{kW}) = 2.25\text{kW} = \mathbf{135kJ/min}$$

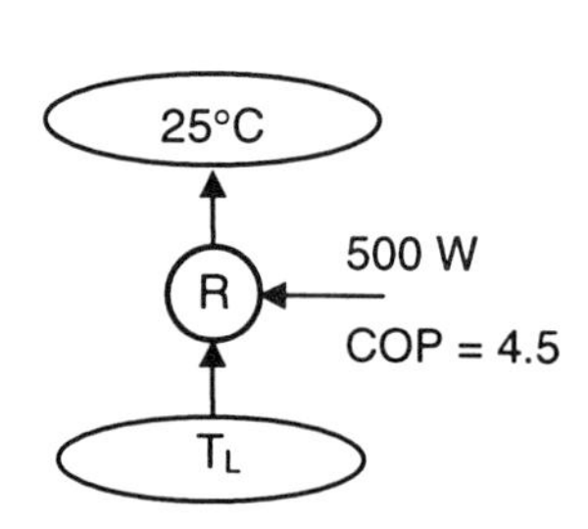

(*b*) The temperature of the refrigerated space T_L is determined from the coefficient of performance relation for a Carnot refrigerator,

$$COP_{R,rev} = \frac{1}{(T_H / T_L) - 1} \longrightarrow 4.5 = \frac{1}{(25 + 273\text{K})/T_L - 1}$$

It yields

$$T_L = 243.8\text{ K} = \mathbf{-29.2°C}$$

5-99 An inventor claims to have developed a refrigerator. The inventor reports temperature and COP measurements. The claim is to be evaluated.

Analysis The highest coefficient of performance a refrigerator can have when removing heat from a cool medium at -5°C to a warmer medium at 25°C is

$$COP_{R,\max} = COP_{R,rev} = \frac{1}{(T_H/T_L)-1} = \frac{1}{(25+273\text{K})/(-5+273\text{K})-1} = 8.9$$

25°C

COP=6.5

R

-5°C

The COP claimed by the inventor is below this maximum value, thus the claim is **reasonable**.

5-100 An experimentalist claims to have developed a refrigerator. The experimentalist reports temperature, heat transfer, and work input measurements. The claim is to be evaluated.

Analysis The highest coefficient of performance a refrigerator can have when removing heat from a cool medium at -30°C to a warmer medium at 25°C is

$$COP_{R,\max} = COP_{R,rev} = \frac{1}{(T_H/T_L)-1} = \frac{1}{(25+273\text{K})/(-30+273\text{K})-1} = 4.42$$

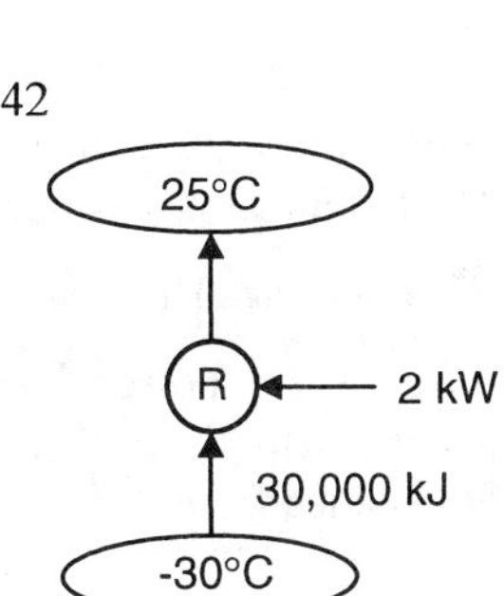

The work consumed by the actual refrigerator during this experiment is

$$W_{net,in} = \dot{W}_{net,in}\Delta t = (2\text{kJ/s})(20 \times 60\text{s}) = 2400\text{kJ}$$

Then the coefficient of performance of this refrigerator becomes

$$COP_R = \frac{Q_L}{W_{net,in}} = \frac{30{,}000 \text{ kJ}}{2400 \text{ kJ}} = 12.5$$

which is above the maximum value. Therefore, these measurements are **not reasonable**.

5-101E An air-conditioning system maintains a house at a specified temperature. The rate of heat gain of the house and the rate of internal heat generation are given. The maximum power input required is to be determined.

Assumptions The air-conditioner operates steadily.

Analysis The power input to an air-conditioning system will be a minimum when the air-conditioner operates in a reversible manner. The coefficient of performance of a reversible air-conditioner (or refrigerator) depends on the temperature limits in the cycle only, and is determined from

$$COP_{R,rev} = \frac{1}{(T_H/T_L)-1} = \frac{1}{(95+460\text{R})/(75+460\text{R})-1} = 26.75$$

95°F
A/C
House
75°F
750 kJ/min

The cooling load of this air-conditioning system is the sum of the heat gain from the outside and the heat generated within the house,

$$\dot{Q}_L = 750 + 150 = 900 \text{ Btu/min}$$

The power input to this refrigerator is determined from the definition of the coefficient of performance of a refrigerator,

$$\dot{W}_{net,in,\min} = \frac{\dot{Q}_L}{COP_{R,\max}} = \frac{900 \text{ Btu/min}}{26.75} = 33.6 \text{ Btu/min} = \mathbf{0.79 \text{ hp}}$$

5-102 A heat pump maintains a house at a specified temperature. The rate of heat loss of the house is given. The minimum power input required is to be determined.

Assumptions The heat pump operates steadily.

Analysis The power input to a heat pump will be a minimum when the heat pump operates in a reversible manner. The COP of a reversible heat pump depends on the temperature limits in the cycle only, and is determined from

$$COP_{HP,rev} = \frac{1}{1-(T_L/T_H)} = \frac{1}{1-(-5+273\text{K})/(24+273\text{K})} = 10.2$$

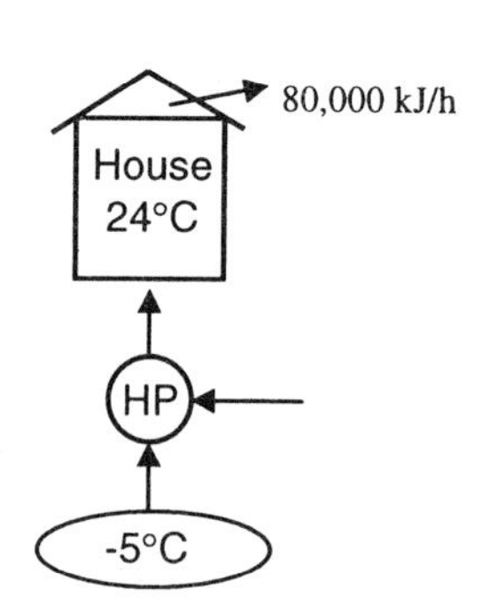

The required power input to this reversible heat pump is determined from the definition of the coefficient of performance to be

$$\dot{W}_{net,in,\min} = \frac{\dot{Q}_H}{COP_{HP}} = \frac{80{,}000\text{kJ/h}}{10.2}\left(\frac{1 \text{ h}}{3600 \text{ s}}\right) = \mathbf{2.18 \text{ kW}}$$

which is the *minimum* power input required.

5-103 A heat pump maintains a house at a specified temperature. The rate of heat loss of the house and the power consumption of the heat pump are given. It is to be determined if this heat pump can do the job.

Assumptions The heat pump operates steadily.

Analysis The power input to a heat pump will be a minimum when the heat pump operates in a reversible manner. The coefficient of performance of a reversible heat pump depends on the temperature limits in the cycle only, and is determined from

$$COP_{HP,rev} = \frac{1}{1-(T_L/T_H)} = \frac{1}{1-(2+273\text{K})/(22+273\text{K})} = 14.75$$

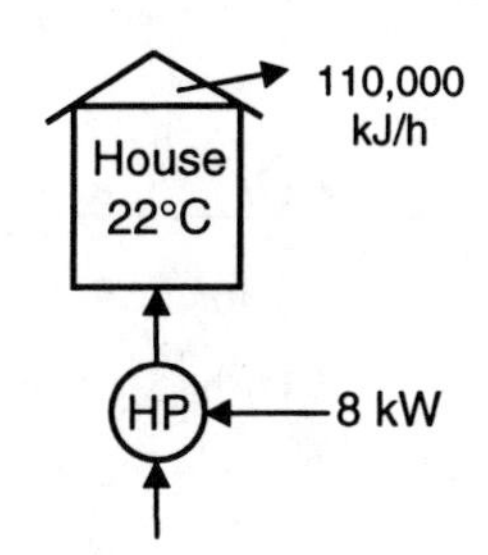

The required power input to this reversible heat pump is determined from the definition of the coefficient of performance to be

$$\dot{W}_{net,in,\min} = \frac{\dot{Q}_H}{COP_{HP}} = \frac{110{,}000\text{kJ/h}}{14.75}\left(\frac{1\text{h}}{3600\text{s}}\right) = \mathbf{2.07kW}$$

This heat pump is **powerful enough** since 8 kW > 2.07 kW.

5-104 A heat pump that consumes 5-kW of power when operating maintains a house at a specified temperature. The house is losing heat in proportion to the temperature difference between the indoors and the outdoors. The lowest outdoor temperature for which this heat pump can do the job is to be determined.

Assumptions The heat pump operates steadily.

Analysis Denoting the outdoor temperature by T_L, the heating load of this house can be expressed as

$$\dot{Q}_H = (5400\text{kJ/h}\cdot\text{K})(294 - T_L) = (1.5\text{kW/K})(294 - T_L)$$

The coefficient of performance of a Carnot heat pump depends on the temperature limits in the cycle only, and can be expressed as

$$COP_{HP} = \frac{1}{1-(T_L/T_H)} = \frac{1}{1-T_L/294\text{K}}$$

or, as

$$COP_{HP} = \frac{\dot{Q}_H}{\dot{W}_{net,in}} = \frac{(1.5\text{kW/K})(294 - T_L)}{6\text{kW}}$$

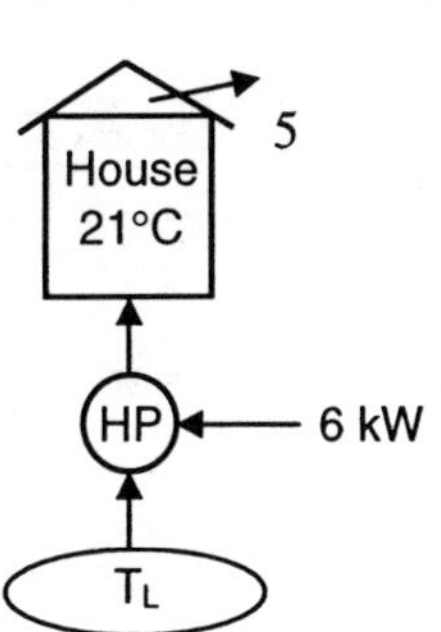

Equating the two relations above and solving for T_L, we obtain

T_L = 259.7 K = **-13.3°C**

5-105 A heat pump maintains a house at a specified temperature in winter. The maximum COPs of the heat pump for different outdoor temperatures are to be determined.

Analysis The coefficient of performance of a heat pump will be a maximum when the heat pump operates in a reversible manner. The coefficient of performance of a reversible heat pump depends on the temperature limits in the cycle only, and is determined for all three cases above to be

$$COP_{HP,rev} = \frac{1}{1-(T_L/T_H)} = \frac{1}{1-(10+273\text{K})/(20+273\text{K})} = \mathbf{29.3}$$

$$COP_{HP,rev} = \frac{1}{1-(T_L/T_H)} = \frac{1}{1-(-5+273\text{K})/(20+273\text{K})} = \mathbf{11.7}$$

$$COP_{HP,rev} = \frac{1}{1-(T_L/T_H)} = \frac{1}{1-(-30+273\text{K})/(20+273\text{K})} = \mathbf{5.86}$$

5-106E A heat pump maintains a house at a specified temperature. The rate of heat loss of the house is given. The minimum power inputs required for different source temperatures are to be determined.

Assumptions The heat pump operates steadily.

Analysis (*a*) The power input to a heat pump will be a minimum when the heat pump operates in a reversible manner. If the outdoor air at 25°F is used as the heat source, the COP of the heat pump and the required power input are determined to be

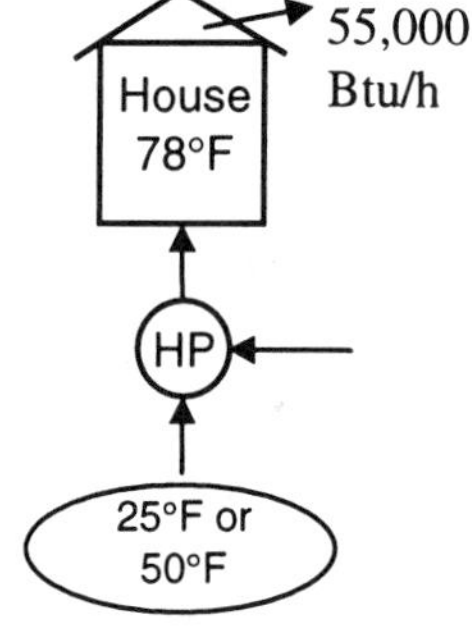

$$COP_{HP,\max} = COP_{HP,rev} = \frac{1}{1-(T_L/T_H)} = \frac{1}{1-(25+460\text{R})/(78+460\text{R})} = 10.15$$

and

$$\dot{W}_{net,in,\min} = \frac{\dot{Q}_H}{COP_{HP,\max}} = \frac{55{,}000 \text{ Btu/h}}{10.15}\left(\frac{1 \text{ hp}}{2545 \text{ Btu/h}}\right) = \mathbf{2.13 \text{ hp}}$$

(*b*) If the well-water at 50°F is used as the heat source, the COP of the heat pump and the required power input are determined to be

$$COP_{HP,\max} = COP_{HP,rev} = \frac{1}{1-(T_L/T_H)} = \frac{1}{1-(50+460\text{R})/(78+460\text{R})} = 19.2$$

and

$$\dot{W}_{net,in,\min} = \frac{\dot{Q}_H}{COP_{HP,\max}} = \frac{55{,}000 \text{ Btu/h}}{19.2}\left(\frac{1 \text{ hp}}{2545 \text{ Btu/h}}\right) = \mathbf{1.13 \text{ hp}}$$

5-107 A Carnot heat pump consumes 8-kW of power when operating, and maintains a house at a specified temperature. The average rate of heat loss of the house in a particular day is given. The actual running time of the heat pump that day, the heating cost, and the cost if resistance heating is used instead are to be determined.

Analysis (*a*) The coefficient of performance of this Carnot heat pump depends on the temperature limits in the cycle only, and is determined from

$$COP_{HP,rev} = \frac{1}{1-(T_L/T_H)} = \frac{1}{1-(2+273\text{K})/(20+273\text{K})} = 16.3$$

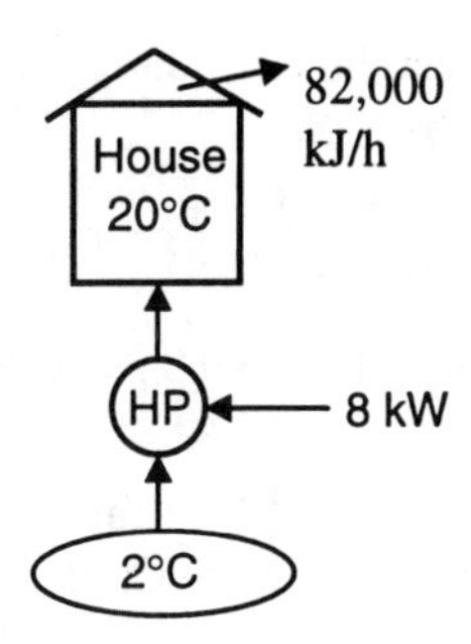

The amount of heat the house lost that day is

$$Q_H = \dot{Q}_H(1\text{day}) = (82{,}000\text{kJ/h})(24\text{h}) = 1{,}968{,}000\text{kJ}$$

Then the required work input to this Carnot heat pump is determined from the definition of the coefficient of performance to be

$$W_{net,in} = \frac{Q_H}{COP_{HP}} = \frac{1{,}968{,}000\text{ kJ}}{16.3} = 120{,}736\text{ kJ}$$

Thus the length of time the heat pump ran that day is

$$\Delta t = \frac{W_{net,in}}{\dot{W}_{net,in}} = \frac{120{,}736\text{ kJ}}{8\text{ kJ/s}} = 15{,}092\text{ s} = \mathbf{4.19\text{ h}}$$

(*b*) The total heating cost that day is

$$\text{Cost} = W \times \text{price} = (\dot{W}_{net,in} \times \Delta t)(\text{price}) = (8\text{kW})(4.19\text{h})(0.085\$/\text{kWh}) = \mathbf{\$2.85}$$

(*c*) If resistance heating were used, the entire heating load for that day would have to be met by electrical energy. Therefore, the heating system would consume 1,968,000 kJ of electricity that would cost

$$\text{New Cost} = Q_H \times \text{price} = (1{,}968{,}000\text{kJ})\left(\frac{1\text{kWh}}{3600\text{kJ}}\right)(0.085\$/\text{kWh}) = \mathbf{\$46.47}$$

5-108 A Carnot heat engine is used to drive a Carnot refrigerator. The maximum rate of heat removal from the refrigerated space and the total rate of heat rejection to the ambient air are to be determined.

Assumptions The heat engine and the refrigerator operate steadily.

Analysis (a) The highest thermal efficiency a heat engine operating between two specified temperature limits can have is the Carnot efficiency, which is determined from

$$\eta_{th,\max} = \eta_{th,C} = 1 - \frac{T_L}{T_H} = 1 - \frac{300\text{ K}}{1173\text{ K}} = 0.744$$

Then the maximum power output of this heat engine is determined from the definition of thermal efficiency to be

$$\dot{W}_{net,out} = \eta_{th}\dot{Q}_H = (0.744)(800\text{kJ/min}) = 595.2\text{kJ/min}$$

which is also the power input to the refrigerator, $\dot{W}_{net,in}$.

The rate of heat removal from the refrigerated space will be a maximum if a Carnot refrigerator is used. The COP of the Carnot refrigerator is

$$COP_{R,rev} = \frac{1}{(T_H/T_L)-1} = \frac{1}{(27+273\text{K})/(-5+273\text{K})-1} = 8.37$$

Then the rate of heat removal from the refrigerated space becomes

$$\dot{Q}_{L,R} = (COP_{R,rev})(\dot{W}_{net,in}) = (8.37)(595.2\text{kJ/min}) = \mathbf{4982kJ/min}$$

(b) The total rate of heat rejection to the ambient air is the sum of the heat rejected by the heat engine ($\dot{Q}_{L,HE}$) and the heat discarded by the refrigerator ($\dot{Q}_{H,R}$),

$$\dot{Q}_{L,HE} = \dot{Q}_{H,R} - \dot{W}_{net,out} = 800 - 595.2 = 204.8\text{ kJ/min}$$

$$\dot{Q}_{H,R} = \dot{Q}_{L,R} - \dot{W}_{net,in} = 4982 - 595.2 = 5577.2\text{ kJ/min}$$

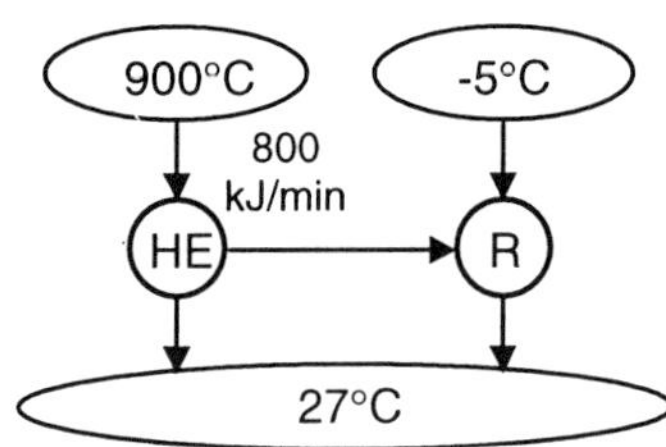

and

$$\dot{Q}_{Ambient} = \dot{Q}_{L,HE} + \dot{Q}_{H,R} = 204.8 + 5577.2 = \mathbf{5782\,kJ/min}$$

5-109E A Carnot heat engine is used to drive a Carnot refrigerator. The maximum rate of heat removal from the refrigerated space and the total rate of heat rejection to the ambient air are to be determined.

Assumptions The heat engine and the refrigerator operate steadily.

Analysis (*a*) The highest thermal efficiency a heat engine operating between two specified temperature limits can have is the Carnot efficiency, which is determined from

$$\eta_{th,\max} = \eta_{th,C} = 1 - \frac{T_L}{T_H} = 1 - \frac{540\text{R}}{2160\text{R}} = 0.75$$

Then the maximum power output of this heat engine is determined from the definition of thermal efficiency to be

$$\dot{W}_{net,out} = \eta_{th}\dot{Q}_H = (0.75)(700 \text{ Btu/min}) = 525 \text{ Btu/min}$$

which is also the power input to the refrigerator, $\dot{W}_{net,in}$.

The rate of heat removal from the refrigerated space will be a maximum if a Carnot refrigerator is used. The COP of the Carnot refrigerator is

$$COP_{R,rev} = \frac{1}{(T_H / T_L) - 1} = \frac{1}{(80 + 460\text{R})/(20 + 460\text{R}) - 1} = 8.0$$

Then the rate of heat removal from the refrigerated space becomes

$$\dot{Q}_{L,R} = (COP_{R,rev})(\dot{W}_{net,in}) = (8.0)(525 \text{ Btu/min}) = \mathbf{4200 \text{ Btu/min}}$$

(*b*) The total rate of heat rejection to the ambient air is the sum of the heat rejected by the heat engine ($\dot{Q}_{L,HE}$) and the heat discarded by the refrigerator ($\dot{Q}_{H,R}$),

$$\dot{Q}_{L,HE} = \dot{Q}_{H,R} - \dot{W}_{net,out} = 700 - 525 = 175 \text{ Btu/min}$$

$$\dot{Q}_{H,R} = \dot{Q}_{L,R} + \dot{W}_{net,in} = 4200 + 525 = 4725 \text{ Btu/min}$$

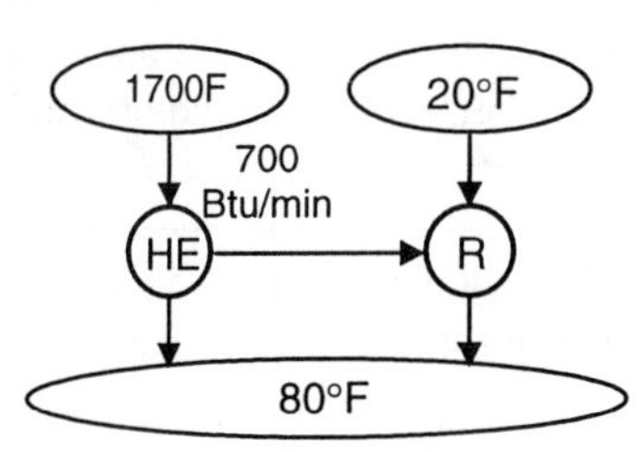

and

$$\dot{Q}_{Ambient} = \dot{Q}_{L,HE} + \dot{Q}_{H,R} = 175 + 4725 = \mathbf{4900 \text{ Btu/min}}$$

Special Topic: Household Refrigerators

5-110C It is a bad idea to overdesign the refrigeration system of a supermarket so that the entire air-conditioning needs of the store can be met by refrigerated air without installing any air-conditioning system. This is because the refrigerators cool the air to a much lower temperature than needed for air conditioning, and thus their efficiency is much lower, and their operating cost is much higher.

5-111C It is a bad idea to meet the entire refrigerator/freezer requirements of a store by using a large freezer that supplies sufficient cold air at -20°C instead of installing separate refrigerators and freezers . This is because the freezers cool the air to a much lower temperature than needed for refrigeration, and thus their efficiency is much lower, and their operating cost is much higher.

5-112C The energy consumption of a household refrigerator can be reduced by practicing good conservation measures such as (1) opening the refrigerator door the fewest times possible and for the shortest duration possible, (2) cooling the hot foods to room temperature first before putting them into the refrigerator, (3) cleaning the condenser coils behind the refrigerator, (4) checking the door gasket for air leaks, (5) avoiding unnecessarily low temperature settings, (6) avoiding excessive ice build-up on the interior surfaces of the evaporator, (7) using the power-saver switch that controls the heating coils that prevent condensation on the outside surfaces in humid environments, and (8) not blocking the air flow passages to and from the condenser coils of the refrigerator.

5-113C It is important to clean the condenser coils of a household refrigerator a few times a year since the dust that collects on them serves as insulation and slows down heat transfer. Also, it is important not to block air flow through the condenser coils since heat is rejected through them by natural convection, and blocking the air flow will interfere with this heat rejection process. A refrigerator cannot work unless it can reject the waste heat.

5-114C Today's refrigerators are much more efficient than those built in the past as a result of using smaller and higher efficiency motors and compressors, better insulation materials, larger coil surface areas, and better door seals.

5-115 A refrigerator consumes 300 W when running, and $83 worth of electricity per year under normal use. The fraction of the time the refrigerator will run in a year is to be determined.

Assumptions The electricity consumed by the light bulb is negligible.

Analysis The total amount of electricity the refrigerator uses a year is

$$\text{Total electric energy used} = W_{e,\text{total}} = \frac{\text{Total cost of energy}}{\text{Unit cost of energy}} = \frac{\$74/\text{year}}{\$0.07/\text{kWh}} = 1057 \text{ kWh/year}$$

The number of hours the refrigerator is on per year is

$$\text{Total operating hours} = \Delta t = \frac{W_{e,\text{total}}}{\dot{W}_e} = \frac{1057 \text{ kWh}}{0.3 \text{ kW}} = 3524 \text{ h/year}$$

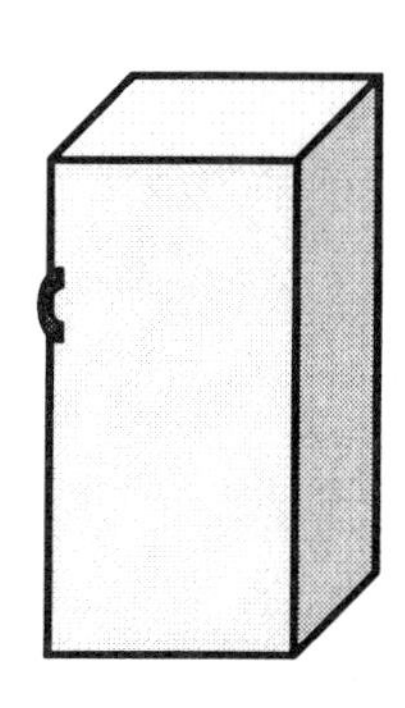

Noting that there are 365×24=8760 hours in a year, the fraction of the time the refrigerator is on during a year is determined to be

$$\text{Time fraction on} = \frac{\text{Total operating hours}}{\text{Total hours per year}} = \frac{3524/\text{year}}{8760 \text{ h/year}} = \mathbf{0.402}$$

Therefore, the refrigerator remained on 40.2% of the time.

5-116 The light bulb of a refrigerator is to be replaced by a \$25 energy efficient bulb that consumes less than half the electricity. It is to be determined if the energy savings of the efficient light bulb justify its cost.

Assumptions The new light bulb remains on the same number of hours a year.

Analysis The lighting energy saved a year by the energy efficient bulb is

$$\text{Lighting energy saved} = (\text{Lighting power saved})(\text{Operating hours})$$
$$= [(40-18)\text{W}](60\text{ h/year})$$
$$= 1320\text{ Wh} = 1.32\text{ kWh}$$

This means 1.32 kWh less heat is supplied to the refrigerated space by the light bulb, which must be removed from the refrigerated space. This corresponds to a refrigeration savings of

$$\text{Refrigeration energy saved} = \frac{\text{Lighting energy saved}}{\text{COP}} = \frac{1.32\text{ kWh}}{1.3} = 1.02\text{ kWh}$$

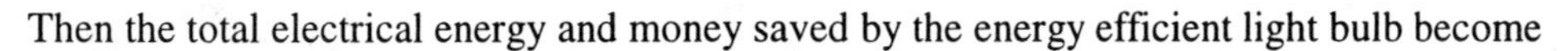

Then the total electrical energy and money saved by the energy efficient light bulb become

$$\text{Total energy saved} = (\text{Lighting} + \text{Refrigeration})\text{ energy saved} = 1.32 + 1.02 = 2.34\text{ kWh/year}$$
$$\text{Money saved} = (\text{Total energy saved})(\text{Unit cost of energy}) = (2.34\text{ kWh/year})(\$0.08/\text{kWh})$$
$$= \mathbf{\$0.19/year}$$

That is, the light bulb will save only 19 cents a year in energy costs, and it will take \$25/\$0.19 = 132 years for it to pay for itself from the energy it saves. Therefore, it is **not justified** in this case.

5-117 A person cooks twice a week and places the food into the refrigerator before cooling it first. The amount of money this person will save a year by cooling the hot foods to room temperature before refrigerating them is to be determined.

Assumptions **1** The heat stored in the pan itself is negligible. **2** The specific heat of the food is constant.

Properties The specific heat of food is $C = 3.90$ kJ/kg.°C (given).

Analysis The amount of hot food refrigerated per year is

$$m_{\text{food}} = (5\text{ kg/pan})(2\text{ pans/week})(52\text{ weeks/year}) = 520\text{ kg/year}$$

The amount of energy removed from food as it is unnecessarily cooled to room temperature in the refrigerator is

$$\text{Energy removed} = Q_{\text{out}} = m_{\text{food}} C\Delta T = (520\text{ kg/year})(3.90\text{ kJ/kg.°C})(95-20)\text{°C} = 152{,}100\text{ kJ/year}$$
$$\text{Energy saved} = E_{\text{saved}} = \frac{\text{Energy removed}}{\text{COP}} = \frac{152{,}100\text{ kJ/year}}{1.2}\left(\frac{1\text{ kWh}}{3600\text{ kJ}}\right) = 35.2\text{ kWh/year}$$
$$\text{Money saved} = (\text{Energy saved})(\text{Unit cost of energy}) = (35.2\text{ kWh/year})(\$0.10/\text{kWh}) = \mathbf{\$3.52/year}$$

Therefore, cooling the food to room temperature before putting it into the refrigerator will save about three and a half dollars a year.

5-118 The door of a refrigerator is opened 8 times a day, and half of the cool air inside is replaced by the warmer room air. The cost of the energy wasted per year as a result of opening the refrigerator door is to be determined for the cases of moist and dry air in the room. √

Assumptions **1** The room is maintained at 20°C and 95 kPa at all times. **2** Air is an ideal gas with constant specific heats at room temperature. **3** The moisture is condensed at an average temperature of 4°C. **4** Half of the air volume in the refrigerator is replaced by the warmer kitchen air each time the door is opened.

Properties The gas constant of air is R = 0.287 kPa.m^3/kg·K (Table A-1). The specific heat of air at room temperature is C_p = 1.005 kJ/kg·°C (Table A-2a). The heat of vaporization of water at 4°C is h_{fg} = 2492 kJ/kg.

Analysis The volume of the refrigerated air replaced each time the refrigerator is opened is 0.3 m^3 (half of the 0.6 m^3 air volume in the refrigerator). Then the total volume of refrigerated air replaced by room air per year is

$$\dot{V}_{\text{air, replaced}} = (0.3\text{ m}^3)(8/\text{day})(365\text{ days/year}) = 876\text{ m}^3/\text{year}$$

The density of air at the refrigerated space conditions of 95 kPa and 4°C and the mass of air replaced per year are

$$\rho_o = \frac{P_o}{RT_o} = \frac{95\text{ kPa}}{(0.287\text{ kPa.m}^3/\text{kg.K})(4+273\text{ K})} = 1.195\text{ kg/m}^3$$

$$m_{\text{air}} = \rho V_{\text{air}} = (1.195\text{ kg/m}^3)(876\text{ m}^3/\text{year}) = 1047\text{ kg/year}$$

The amount of moisture condensed and removed by the refrigerator is

$$m_{\text{moisture}} = m_{\text{air}}(\text{moisture removed per kg air}) = (1047\text{ kg air/year})(0.006\text{ kg/kg air}) = 6.28\text{ kg/year}$$

The sensible, latent, and total heat gains of the refrigerated space become

$$\begin{aligned} Q_{\text{gain,sensible}} &= m_{\text{air}} C_p (T_{\text{room}} - T_{\text{refrig}}) \\ &= (1047\text{ kg/year})(1.005\text{ kJ/kg.°C})(20-4)\text{°C} = 16{,}836\text{ kJ/year} \\ Q_{\text{gain,latent}} &= m_{\text{moisture}} h_{fg} \\ &= (6.28\text{ kg/year})(2492\text{ kJ/kg}) = 15{,}650\text{ kJ/year} \\ Q_{\text{gain,total}} &= Q_{\text{gain,sensible}} + Q_{\text{gain,latent}} = 16{,}836 + 15{,}650 = 32{,}486\text{ kJ/year} \end{aligned}$$

For a COP of 1.4, the amount of electrical energy the refrigerator will consume to remove this heat from the refrigerated space and its cost are

$$\text{Electrical energy used (total)} = \frac{Q_{\text{gain,total}}}{\text{COP}} = \frac{32{,}486\text{ kJ/year}}{1.4}\left(\frac{1\text{ kWh}}{3600\text{ kJ}}\right) = 6.45\text{ kWh/year}$$

$$\begin{aligned} \text{Cost of energy used (total)} &= (\text{Energy used})(\text{Unit cost of energy}) \\ &= (6.45\text{ kWh/year})(\$0.075/\text{kWh}) = \mathbf{\$0.48/year} \end{aligned}$$

If the room air is very dry and thus latent heat gain is negligible, then the amount of electrical energy the refrigerator will consume to remove the sensible heat from the refrigerated space and its cost become

$$\text{Electrical energy used (sensible)} = \frac{Q_{\text{gain,sensible}}}{\text{COP}} = \frac{16{,}836\text{ kJ/year}}{1.4}\left(\frac{1\text{ kWh}}{3600\text{ kJ}}\right) = 3.34\text{ kWh/year}$$

$$\begin{aligned} \text{Cost of energy used (sensible)} &= (\text{Energy used})(\text{Unit cost of energy}) \\ &= (3.34\text{ kWh/year})(\$0.075/\text{kWh}) = \mathbf{\$0.25/year} \end{aligned}$$

Review Problems

5-119 A Carnot heat engine cycle is executed in a steady-flow system with steam. The thermal efficiency and the mass flow rate of steam are given. The net power output of the engine is to be determined.

Assumptions All components operate steadily.

Properties The enthalpy of vaporization h_{fg} of water at 300°C is 1404.9 kJ/kg (Table A-4).

Analysis The enthalpy of vaporization h_{fg} at a given T or P represents the amount of heat transfer as 1 kg of a substance is converted from saturated liquid to saturated vapor at that T or P. Therefore, the rate of heat transfer to the steam during heat addition process is

$$\dot{Q}_H = \dot{m}h_{fg@300°C} = (5\text{kg/s})(1404.9\text{kJ/kg}) = 7025\text{kJ/s}$$

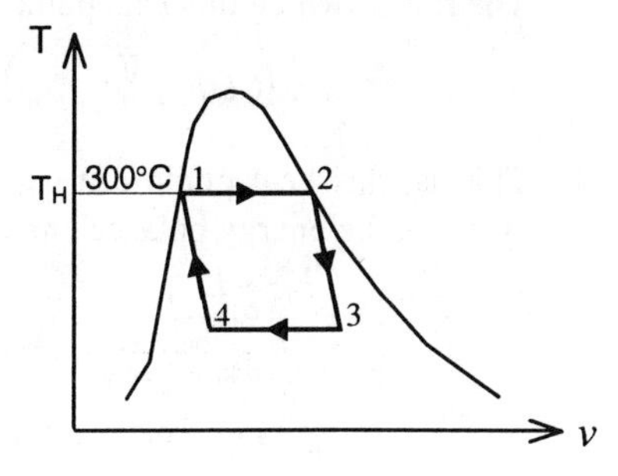

Then the power output of this heat engine becomes

$$\dot{W}_{net,out} = \eta_{th}\dot{Q}_H = (0.30)(7025\text{kW}) = \mathbf{2107.5kW}$$

5-120 A heat pump with a specified COP is to heat a house. The rate of heat loss of the house and the power consumption of the heat pump are given. The time it will take for the interior temperature to rise from 3°C to 22°C is to be determined.

Assumptions **1** Air is an ideal gas with constant specific heats at room temperature. **2** The house is well-sealed so that no air leaks in or out. **3** The COP of the heat pump remains constant during operation.

Properties The constant volume specific heat of air at room temperature is $C_v = 0.718$ kJ/kg.°C (Table A-2)

Analysis The house is losing heat at a rate of

$$\dot{Q}_{Loss} = 40{,}000 \text{ kJ/h} = 11.11 \text{ kJ/s}$$

The rate at which this heat pump supplies heat is

$$\dot{Q}_H = (COP_{HP})(\dot{W}_{net,in}) = (2.4)(8\text{kW}) = 19.2\text{kW}$$

That is, this heat pump can supply heat at a rate of 19.2 kJ/s. Taking the house as the system (a closed system), the energy balance can be written as

$$\underbrace{E_{in} - E_{out}}_{\substack{\text{Net energy transfer} \\ \text{by heat, work, and mass}}} = \underbrace{\Delta E_{\text{system}}}_{\substack{\text{Change in internal, kinetic,} \\ \text{potential, etc. energies}}}$$

$$Q_{in} - Q_{out} = \Delta U = m(u_2 - u_1)$$

$$Q_{in} - Q_{out} = mC_v(T_2 - T_1)$$

$$(\dot{Q}_{in} - \dot{Q}_{out})\Delta t = mC_v(T_2 - T_1)$$

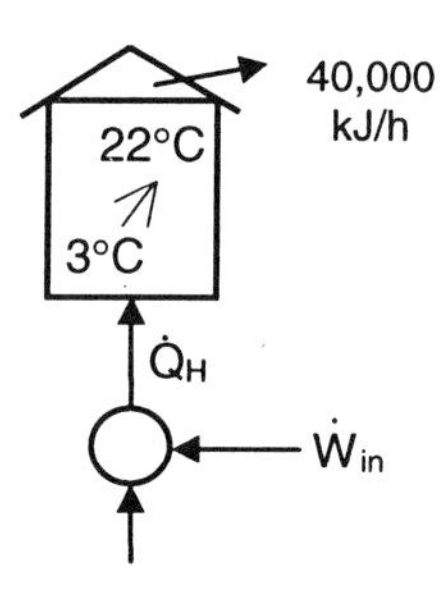

Substituting, $(19.2 - 11.11\text{kJ/s})\Delta t = (2000\text{kg})(0.718\text{kJ/kg}\cdot°\text{C})(22-3)°\text{C}$

Solving for Δt, it will take

$$\Delta t = 3373 \text{ s} = \mathbf{0.937 \ h}$$

for the temperature in the house to rise to 22°C.

5-121 The thermal efficiency and power output of a gas turbine are given. The rate of fuel consumption of the gas turbine is to be determined.

Assumptions Steady operating conditions exist.

Properties The density and heating value of the fuel are given to be 0.8 g/cm^3 and 46,000 kJ/kg, respectively.

Analysis This gas turbine is converting 21% of the chemical energy released during the combustion process into work. The amount of energy input required to produce a power output of 6,000 kW is determined from the definition of thermal efficiency to be

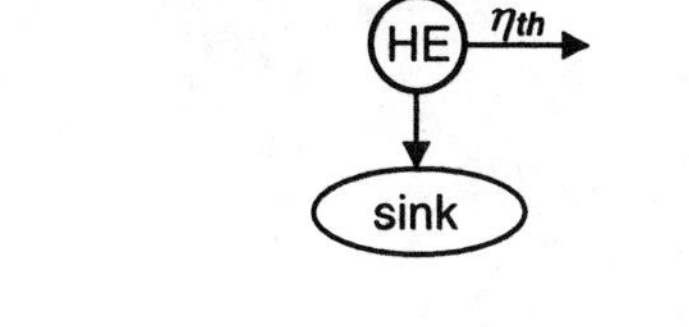

$$\dot{Q}_H = \frac{\dot{W}_{net,out}}{\eta_{th}} = \frac{6{,}000 \text{kJ, s}}{0.21} = 28{,}570 \text{ kJ/s}$$

To supply energy at this rate, the engine must burn fuel at a rate of

$$\dot{m} = \frac{28{,}570 \text{ kJ/s}}{46{,}000 \text{kJ/kg}} = 0.621 \text{ kg/s}$$

since 46,000 kJ of thermal energy is released for each kg of fuel burned. Then the volume flow rate of the fuel becomes

$$\dot{V} = \frac{\dot{m}}{\rho} = \frac{0.621 \text{ kg/s}}{0.8 \text{ kg/L}} = \mathbf{0.776\ \ L/s}$$

5-122 It is to be shown that $COP_{HP} = COP_R + 1$ for the same temperature and heat transfer terms.

Analysis Using the definitions of COPs, the desired relation is obtained to be

$$COP_{HP} = \frac{Q_H}{W_{net,in}} = \frac{Q_L + W_{net,in}}{W_{net,in}} = \frac{Q_L}{W_{net,in}} + 1 = \mathbf{COP_R + 1}$$

5-123 An air-conditioning system maintains a house at a specified temperature. The rate of heat gain of the house, the rate of internal heat generation, and the COP are given. The required power input is to be determined.

Assumptions Steady operating conditions exist.

Analysis The cooling load of this air-conditioning system is the sum of the heat gain from the outdoors and the heat generated in the house from the people, lights, and appliances:

$$\dot{Q}_L = 20{,}000 + 8{,}000 = 28{,}000 \text{ kJ/h}$$

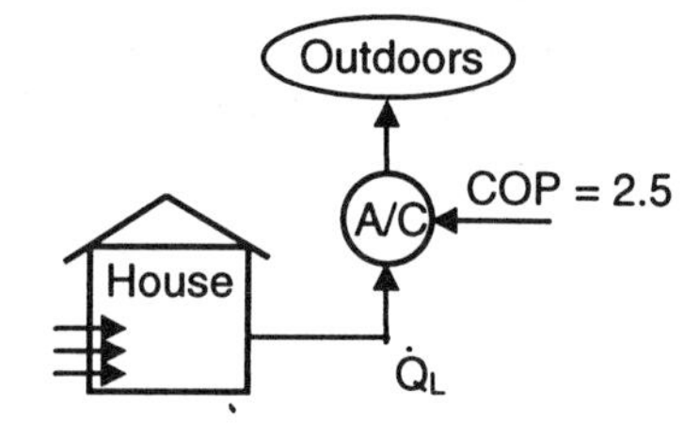

Using the definition of the coefficient of performance, the power input to the air-conditioning system is determined to be

$$\dot{W}_{net,in} = \frac{\dot{Q}_L}{COP_R} = \frac{28{,}000 \text{ kJ/h}}{2.5}\left(\frac{1 \text{ kW}}{3600 \text{ kJ/h}}\right) = \mathbf{3.11\ kW}$$

5-124 A Carnot heat engine cycle is executed in a closed system with a fixed mass of R-134a. The thermal efficiency of the cycle is given. The net work output of the engine is to be determined. √

Assumptions All components operate steadily.

Properties The enthalpy of vaporization of R-134a at 70°C is h_{fg} = 124.08 kJ/kg (Table A-11).

Analysis The enthalpy of vaporization h_{fg} at a given *T* or *P* represents the amount of heat transfer as 1 kg of a substance is converted from saturated liquid to saturated vapor at that *T* or *P*. Therefore, the amount of heat transfer to R-134a during the heat addition process of the cycle is

$$Q_H = mh_{fg@70^\circ C} = (0.01 \text{ kg})(124.08 \text{ kJ/kg}) = 1.24 \text{ kJ}$$

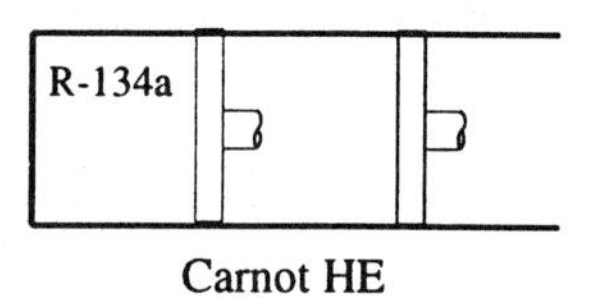

Carnot HE

Then the work output of this heat engine becomes

$$W_{\text{net,out}} = \eta_{th} Q_H = (0.15)(1.24 \text{ kJ}) = \mathbf{0.186\ kJ}$$

5-125 A heat pump with a specified COP and power consumption is used to heat a house. The time it takes for this heat pump to raise the temperature of a cold house to the desired level is to be determined.

Assumptions **1** Air is an ideal gas with constant specific heats at room temperature. **2** The heat loss of the house during the warp-up period is negligible. **3** The house is well-sealed so that no air leaks in or out.

Properties The constant volume specific heat of air at room temperature is C_v = 0.718 kJ/kg.°C.

Analysis Since the house is well-sealed (constant volume), the total amount of heat that needs to be supplied to the house is

$$Q_H = (mC_v\Delta T)_{house} = (1500\text{ kg})(0.718\text{ kJ/kg}\cdot^\circ\text{C})(22-7)^\circ\text{C} = 16{,}155\text{ kJ}$$

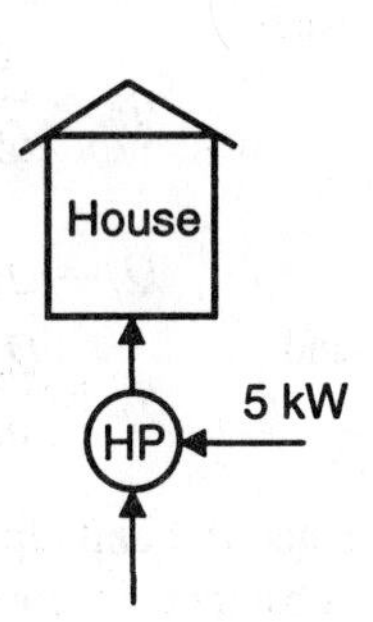

The rate at which this heat pump supplies heat is

$$\dot{Q}_H = COP_{HP}\dot{W}_{net,in} = (2.8)(5\text{ kW}) = 14\text{ kW}$$

That is, this heat pump can supply 14 kJ of heat per second. Thus the time required to supply 16,155 kJ of heat is

$$\Delta t = \frac{Q_H}{\dot{Q}_H} = \frac{16{,}155\text{ kJ}}{14\text{ kJ/s}} = 1154\text{ s} = \mathbf{19.2\ min}$$

5-126 A solar pond power plant operates by absorbing heat from the hot region near the bottom, and rejecting waste heat to the cold region near the top. The maximum thermal efficiency that the power plant can have is to be determined.

Analysis The highest thermal efficiency a heat engine operating between two specified temperature limits can have is the Carnot efficiency, which is determined from

$$\eta_{th,\max} = \eta_{th,C} = 1 - \frac{T_L}{T_H} = 1 - \frac{308\text{ K}}{353\text{ K}} = 0.127 \quad \text{or} \quad \mathbf{12.7\%}$$

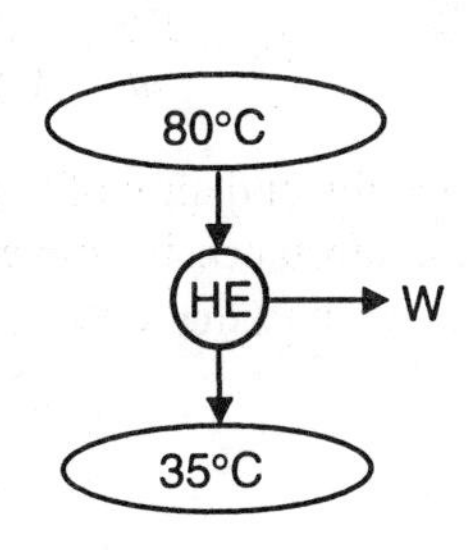

In reality, the temperature of the working fluid must be above 35°C in the condenser, and below 80°C in the boiler to allow for any effective heat transfer. Therefore, the maximum efficiency of the actual heat engine will be lower than the value calculated above.

5-127 A Carnot heat engine cycle is executed in a closed system with a fixed mass of steam. The net work output of the cycle and the ratio of sink and source temperatures are given. The low temperature in the cycle is to be determined.

Assumptions The engine is said to operate on the Carnot cycle, which is totally reversible.

Analysis The thermal efficiency of the cycle is

$$\eta_{th} = 1 - \frac{T_L}{T_H} = 1 - \frac{1}{2} = 0.5$$

Also,

$$\eta_{th} = \frac{W}{Q_H} \longrightarrow Q_H = \frac{W}{\eta_{th}} = \frac{25 \text{ kJ}}{0.5} = 50 \text{ kJ}$$

$$Q_L = Q_H - W = 50 - 25 = 25 \text{ kJ}$$

and

$$q_L = \frac{Q_L}{m} = \frac{25 \text{ kJ}}{0.0103 \text{ kg}} = 2427 \text{ kJ/kg} = h_{fg\,@\,T_L}$$

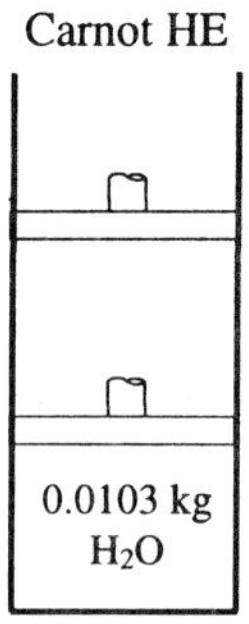

Carnot HE

0.0103 kg H_2O

since the enthalpy of vaporization h_{fg} at a given T or P represents the amount of heat transfer as 1 kg of a substance is converted from saturated liquid to saturated vapor at that T or P. Therefore, T_L is the temperature that corresponds to the h_{fg} value of 2427 kJ/kg, and is determined from the steam tables to be

$$T_L = \mathbf{31.5°C}$$

5-128 EES solution of this (and other comprehensive problems designated with the *computer icon*) is available to instructors at the *Instructor Manual* section of the *Online Learning Center* (OLC) at www.mhhe.com/cengel-boles. See the Preface for access information.

5-129 A Carnot refrigeration cycle is executed in a closed system with a fixed mass of R-134a. The net work input and the ratio of maximum-to-minimum temperatures are given. The minimum pressure in the cycle is to be determined.

Assumptions The refrigerator is said to operate on the reversed Carnot cycle, which is totally reversible.

Analysis The coefficient of performance of the cycle is

$$COP_R = \frac{1}{T_H / T_L - 1} = \frac{1}{1.2 - 1} = 5$$

Also,

$$COP_R = \frac{Q_L}{W_{in}} \longrightarrow Q_L = COP_R \times W_{in} = (5)(22\text{kJ}) = 110\text{kJ}$$

$$Q_H = Q_L + W = 110 + 22 = 132 \text{ kJ}$$

and

$$q_H = \frac{Q_H}{m} = \frac{132 \text{ kJ}}{0.96 \text{ kg}} = 137.5 \text{ kJ/kg} = h_{fg\,@\,T_H}$$

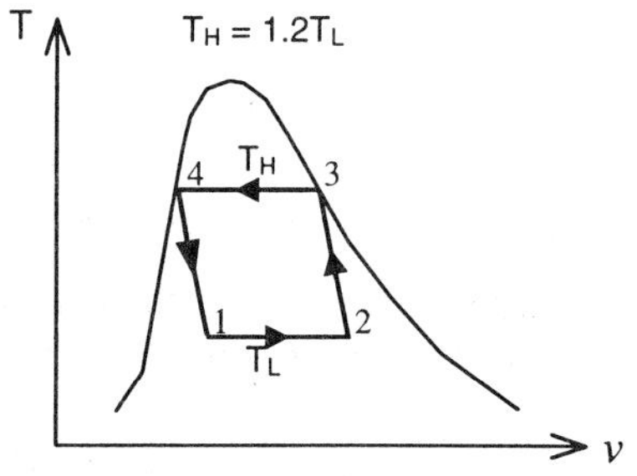

T

$T_H = 1.2T_L$

T_H

4

3

1

2

T_L

v

since the enthalpy of vaporization h_{fg} at a given T or P represents the amount of heat transfer per unit mass as a substance is converted from saturated liquid to saturated vapor at that T or P. Therefore, T_H is the temperature that corresponds to the h_{fg} value of 137.5 kJ/kg, and is determined from the R-134a tables to be

$$T_H \cong 61°\text{C} = 334\text{K}$$

Then,

$$T_L = \frac{T_H}{1.2} = \frac{334\text{K}}{1.2} = 278.3\text{K} \cong 5.3°\text{C}$$

Therefore,

$$P_{\min} = P_{sat@5.3°C} = \mathbf{0.354\ MPa}$$

5-130 EES solution of this (and other comprehensive problems designated with the *computer icon*) is available to instructors at the *Instructor Manual* section of the *Online Learning Center* (OLC) at www.mhhe.com/cengel-boles. See the Preface for access information.

5-131 Two Carnot heat engines operate in series between specified temperature limits. If the thermal efficiencies of both engines are the same, the temperature of the intermediate medium between the two engines is to be determined.

Assumptions The engines are said to operate on the Carnot cycle, which is totally reversible.

Analysis The thermal efficiency of the two Carnot heat engines can be expressed as

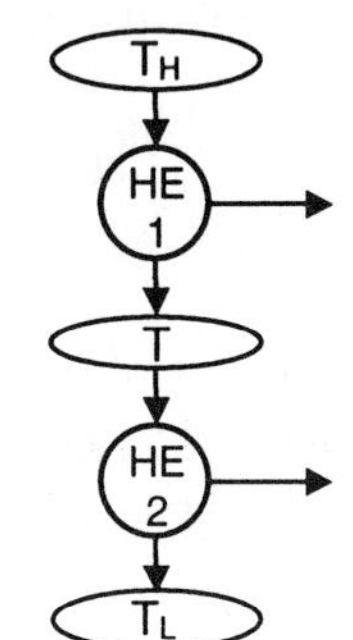

$$\eta_{th,I} = 1 - \frac{T}{T_H} \quad \text{and} \quad \eta_{th,II} = 1 - \frac{T_L}{T}$$

Equating,

$$1 - \frac{T}{T_H} = 1 - \frac{T_L}{T}$$

Solving for T,

$$T = \sqrt{T_H T_L} = \sqrt{(1200\text{ K})(300\text{ K})} = \mathbf{849\ K}$$

5-132 A performance of a refrigerator declines as the temperature of the refrigerated space decreases. The minimum amount of work needed to remove 1 kJ of heat from liquid helium at 3 K is to the determined.

Analysis The power input to a refrigerator will be a minimum when the refrigerator operates in a reversible manner. The coefficient of performance of a reversible refrigerator depends on the temperature limits in the cycle only, and is determined from

$$COP_{R,rev} = \frac{1}{(T_H / T_L) - 1} = \frac{1}{(300\text{K})/(3\text{K}) - 1} = 0.0101$$

The power input to this refrigerator is determined from the definition of the coefficient of performance of a refrigerator,

$$W_{net,in,\min} = \frac{Q_R}{COP_{R,\max}} = \frac{1\text{ kJ}}{0.0101} = \mathbf{99\ kJ}$$

5-133E A Carnot heat pump maintains a house at a specified temperature. The rate of heat loss from the house and the outdoor temperature are given. The COP and the power input are to be determined.

Analysis (*a*) The coefficient of performance of this Carnot heat pump depends on the temperature limits in the cycle only, and is determined from

$$COP_{HP,rev} = \frac{1}{1-(T_L/T_H)} = \frac{1}{1-(35-460\text{R})/(75+460\text{R})} = \mathbf{13.4}$$

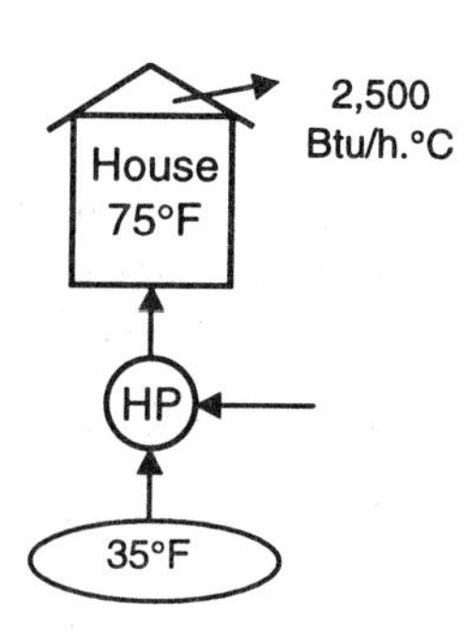

(*b*) The heating load of the house is

$$\dot{Q}_H = (2500\text{Btu/h}\cdot{}^\circ\text{F})(75-35)^\circ\text{F} = 100{,}000\text{Btu/h}$$

Then the required power input to this Carnot heat pump is determined from the definition of the coefficient of performance to be

$$\dot{W}_{net,in} = \frac{\dot{Q}_H}{COP_{HP}} = \frac{100{,}000\text{Btu/h}}{13.4}\left(\frac{1\text{hp}}{2545\text{Btu/h}}\right) = \mathbf{2.93hp}$$

5-134 A Carnot heat engine drives a Carnot refrigerator that removes heat from a cold medium at a specified rate. The rate of heat supply to the heat engine and the total rate of heat rejection to the environment are to be determined.

Analysis (*a*) The coefficient of performance of the Carnot refrigerator is

$$COP_{R,C} = \frac{1}{(T_H/T_L)-1} = \frac{1}{(300\text{K})/(258\text{K})-1} = 6.14$$

Then power input to the refrigerator becomes

$$\dot{W}_{net,in} = \frac{\dot{Q}_L}{COP_{R,C}} = \frac{400\text{ kJ/min}}{6.14} = 65.1\text{ kJ/min}$$

750 K
-15°C
$\dot{Q}_{H,HE}$
400 kJ/min
HE
R
$\dot{Q}_{L,HE}$
$\dot{Q}_{H,R}$
300 K

which is equal to the power output of the heat engine, $\dot{W}_{net,out}$.

The thermal efficiency of the Carnot heat engine is determined from

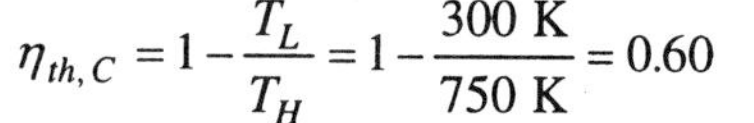

$$\eta_{th,C} = 1 - \frac{T_L}{T_H} = 1 - \frac{300\text{ K}}{750\text{ K}} = 0.60$$

Then the rate of heat input to this heat engine is determined from the definition of thermal efficiency to be

$$\dot{Q}_{H,HE} = \frac{\dot{W}_{net,out}}{\eta_{th,HE}} = \frac{65.1\text{ kJ/min}}{0.60} = \mathbf{108.5\ kJ/min}$$

(*b*) The total rate of heat rejection to the ambient air is the sum of the heat rejected by the heat engine ($\dot{Q}_{L,HE}$) and the heat discarded by the refrigerator ($\dot{Q}_{H,R}$),

$$\dot{Q}_{L,HE} = \dot{Q}_{H,HE} - \dot{W}_{net,out} = 108.5 - 65.1 = 43.4\text{kJ/min}$$

$$\dot{Q}_{H,R} = \dot{Q}_{L,R} + \dot{W}_{net,in} = 400 + 65.1 = 465.1\text{kJ/min}$$

and

$$\dot{Q}_{Ambient} = \dot{Q}_{L,HE} + \dot{Q}_{H,R} = 43.4 + 465.1 = \mathbf{508.5\ kJ/min}$$

5-135 EES solution of this (and other comprehensive problems designated with the *computer icon*) is available to instructors at the *Instructor Manual* section of the *Online Learning Center* (OLC) at www.mhhe.com/cengel-boles. See the Preface for access information.

5-136 Half of the work output of a Carnot heat engine is used to drive a Carnot heat pump that is heating a house. The minimum rate of heat supply to the heat engine is to be determined.

Assumptions Steady operating conditions exist.

Analysis The coefficient of performance of the Carnot heat pump is

$$COP_{HP,C} = \frac{1}{1-(T_L/T_H)} = \frac{1}{1-(2+273\text{K})/(22+273\text{K})} = 14.75$$

Then power input to the heat pump, which is supplying heat to the house at the same rate as the rate of heat loss, becomes

$$\dot{W}_{net,in} = \frac{\dot{Q}_H}{COP_{HP,C}} = \frac{62{,}000 \text{ kJ/h}}{14.75} = 4203 \text{ kJ/h}$$

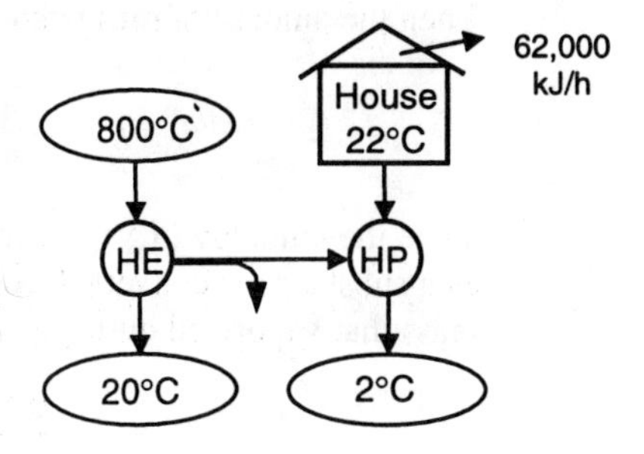

which is half the power produced by the heat engine. Thus the power output of the heat engine is

$$\dot{W}_{net,out} = 2\dot{W}_{net,in} = 2(4203 \text{ kJ/h}) = \mathbf{8406 \text{ kJ/h}}$$

To minimize the rate of heat supply, we must use a Carnot heat engine whose thermal efficiency is determined from

$$\eta_{th,C} = 1 - \frac{T_L}{T_H} = 1 - \frac{293 \text{ K}}{1073 \text{ K}} = 0.727$$

Then the rate of heat supply to this heat engine is determined from the definition of thermal efficiency to be

$$\dot{Q}_{H,HE} = \frac{\dot{W}_{net,out}}{\eta_{th,HE}} = \frac{8406 \text{ kJ/h}}{0.727} = \mathbf{11{,}560 \text{ kJ/h}}$$

5-137 A Carnot refrigeration cycle is executed in a closed system with a fixed mass of R-134a. The net work input and the maximum and minimum temperatures are given. The mass fraction of the refrigerant that vaporizes during the heat addition process, and the pressure at the end of the heat rejection process are to be determined.

Properties The enthalpy of vaporization of R-134a at -10°C is $h_{fg} = 204.46$ kJ/kg (Table A-12).

Analysis The coefficient of performance of the cycle is

$$COP_R = \frac{1}{T_H / T_L - 1} = \frac{1}{293/263 - 1} = 8.77$$

and

$$Q_L = COP_R \times W_{in} = (8.77)(12\text{kJ}) = 105.2\text{kJ}$$

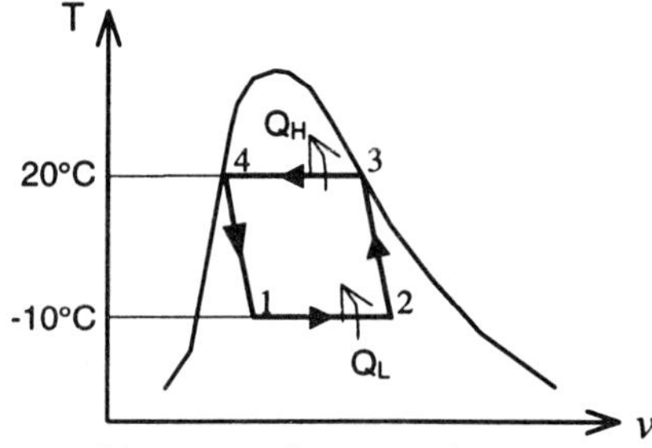

Then the amount of refrigerant that vaporizes during heat absorption is

$$Q_L = mh_{fg@T_L=-10^\circ C} \longrightarrow m = \frac{105.2\text{kJ}}{204.46\text{ kJ/kg}} = 0.514\text{ kg}$$

since the enthalpy of vaporization h_{fg} at a given T or P represents the amount of heat transfer per unit mass as a substance is converted from saturated liquid to saturated vapor at that T or P. Therefore, the fraction of mass that vaporized during heat addition process is

$$\frac{0.514\text{ kg}}{0.8\text{ kg}} = 0.642 \text{ or } \mathbf{64.2\%}$$

The pressure at the end of the heat rejection process is

$$P_4 = P_{sat@20^\circ C} = \mathbf{0.5716\ MPa}$$

5-138 A Carnot heat pump cycle is executed in a steady-flow system with R-134a flowing at a specified rate. The net power input and the ratio of the maximum-to-minimum temperatures are given. The ratio of the maximum to minimum pressures is to be determined.

Analysis The coefficient of performance of the cycle is

$$COP_{HP} = \frac{1}{1 - T_L/T_H} = \frac{1}{1 - 1/1.15} = 7.667$$

and

$$\dot{Q}_H = COP_{HP} \times \dot{W}_{in} = (7.667)(5 \text{ kW}) = 38.33 \text{ kJ/s}$$

$$q_H = \frac{\dot{Q}_H}{\dot{m}} = \frac{38.33 \text{kJ/s}}{0.264 \text{kg/s}} = 145.19 \text{ kJ/kg} = h_{fg\,@T_H}$$

since the enthalpy of vaporization h_{fg} at a given T or P represents the amount of heat transfer per unit mass as a substance is converted from saturated liquid to saturated vapor at that T or P. Therefore, T_H is the temperature that corresponds to the h_{fg} value of 145.19 kJ/kg, and is determined from the R-134a tables to be

$$T_H \cong 55^\circ\text{C} = 328 \text{ K}$$

and

$$P_{max} = P_{sat\,@55^\circ C} = 1.492 \text{ MPa}$$

Also,

$$T_L = \frac{T_H}{1.15} = \frac{328 \text{ K}}{1.15} = 285.2 \text{ K} \cong 12.2^\circ\text{C}$$

$$P_{min} = P_{sat\,@12.2^\circ C} = 0.446 \text{ MPa}$$

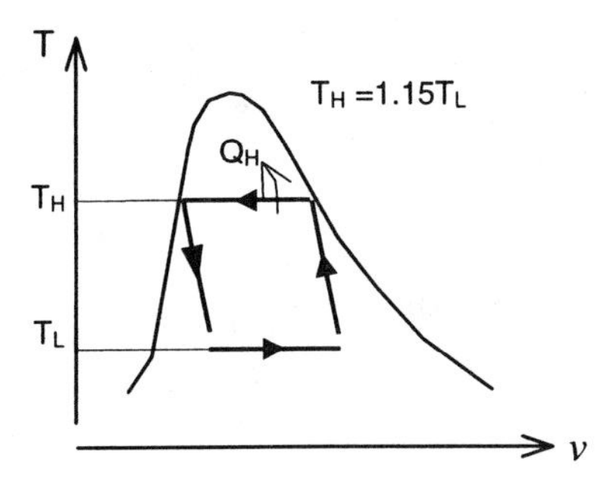

Then the ratio of the maximum to minimum pressures in the cycle is

$$\frac{P_{max}}{P_{min}} = \frac{1.492 \text{ MPa}}{0.446 \text{ MPa}} = \mathbf{3.34}$$

5-139 A Carnot heat engine is operating between specified temperature limits. The source temperature that will double the efficiency is to be determined.

Analysis Denoting the new source temperature by T_H^*, the thermal efficiency of the Carnot heat engine for both cases can be expressed as

$$\eta_{th,C} = 1 - \frac{T_L}{T_H} \quad \text{and} \quad \eta_{th,C}^* = 1 - \frac{T_L}{T_H^*} = 2\eta_{th,C}$$

Substituting,

$$1 - \frac{T_L}{T_H^*} = 2\left(1 - \frac{T_L}{T_H}\right)$$

Solving for T_H^*,

$$T_H^* = \frac{T_H T_L}{T_H - 2T_L}$$

which is the desired relation.

5-140 A Carnot cycle is analyzed for the case of temperature differences in the boiler and condenser. The ratio of overall temperatures for which the power output will be maximum, and an expression for the maximum net power output are to be determined.

Analysis It is given that $\dot{Q}_H = (hA)_H (T_H - T_H^*)$. Therefore,

$$\dot{W} = \eta_{th}\dot{Q}_H = \left(1 - \frac{T_L^*}{T_H^*}\right)(hA)_H (T_H - T_H^*) = \left(1 - \frac{T_L^*}{T_H^*}\right)(hA)_H \left(1 - \frac{T_H^*}{T_H}\right)$$

or,

$$\frac{\dot{W}}{(hA)_H T_H} = \left(1 - \frac{T_L^*}{T_H^*}\right)\left(1 - \frac{T_H^*}{T_H}\right) = (1 - r)x \qquad (1)$$

where we defined r and x as $r = T_L^*/T_H^*$ and $x = 1 - T_H^*/T_H$.

For a reversible cycle we also have

$$\frac{T_H^*}{T_L^*} = \frac{\dot{Q}_H}{\dot{Q}_L} \longrightarrow \frac{1}{r} = \frac{(hA)_H (T_H - T_H^*)}{(hA)_L (T_L^* - T_L)} = \frac{(hA)_H T_H (1 - T_H^*/T_H)}{(hA)_L T_H (T_L^*/T_H - T_L/T_H)}$$

But $\frac{T_L^*}{T_H} = \frac{T_L^*}{T_H^*}\frac{T_H^*}{T_H} = r(1-x)$. Substituting into above relation yields

$$\frac{1}{r} = \frac{(hA)_H x}{(hA)_L [r(1-x) - T_L/T_H]}$$

Solving for x,

$$x = \frac{r - T_L/T_H}{r[(hA)_H/(hA)_L + 1]} \qquad (2)$$

Substitute (2) into (1):

$$\dot{W} = (hA)_H T_H (1 - r)\frac{r - T_L/T_H}{r[(hA)_H/(hA)_L + 1]} \qquad (3)$$

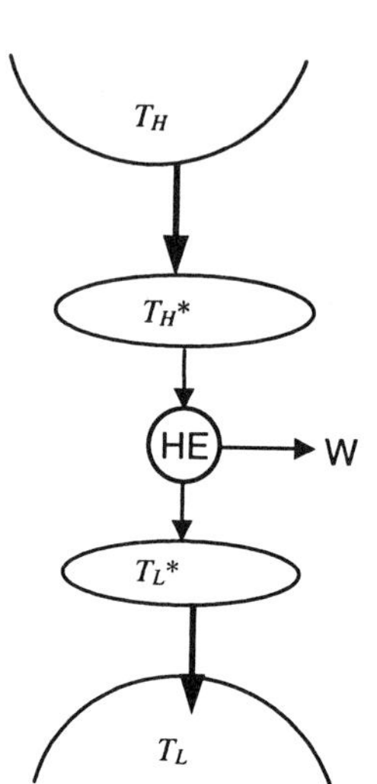

Taking the partial derivative $\frac{\partial \dot{W}}{\partial r}$ holding everything else constant and setting it equal to zero gives

$$r = \frac{T_L^*}{T_H^*} = \left(\frac{T_L}{T_H}\right)^{1/2} \qquad (4)$$

which is the desired relation. The maximum net power output in this case is determined by substituting (4) into (3). It simplifies to

$$\dot{W}_{\max} = \frac{(hA)_H T_H}{1 + (hA)_H/(hA)_L}\left\{1 - \left(\frac{T_L}{T_H}\right)^{1/2}\right\}^2$$

5-141 A decision is to be made between a cheaper but inefficient natural gas heater and an expensive but efficient natural gas heater for a house.

Assumptions The two heaters are comparable in all aspects other than the initial cost and efficiency.

Analysis Other things being equal, the logical choice is the heater that will cost less during its lifetime. The total cost of a system during its lifetime (the initial, operation, maintenance, etc.) can be determined by performing a life cycle cost analysis. A simpler alternative is to determine the simple payback period.

The annual heating cost is given to be $1200. Noting that the existing heater is 55% efficient, only 55% of that energy (and thus money) is delivered to the house, and the rest is wasted due to the inefficiency of the heater. Therefore, the monetary value of the heating load of the house is

$$\text{Cost of useful heat} = (55\%)(\text{Current annual heating cost}) = 0.55\times(\$1200/\text{yr})=\$660/\text{yr}$$

This is how much it would cost to heat this house with a heater that is 100% efficient. For heaters that are less efficient, the annual heating cost is determined by dividing $660 by the efficiency:

82% heater: Annual cost of heating = (Cost of useful heat)/Efficiency = ($660/yr)/0.82 = $805/yr

95% heater: Annual cost of heating = (Cost of useful heat)/Efficiency = ($660/yr)/0.95 = $695/yr

Annual cost savings with the efficient heater = 805 - 695 = $110

Excess initial cost of the efficient heater = 2700 - 1600 = $1100

Gas Heater

$\eta_1 = 82\%$
$\eta_2 = 95\%$

The simple payback period becomes

$$\text{Simple payback period} = \frac{\text{Excess initial cost}}{\text{Annaul cost savings}} = \frac{\$1100}{\$110/\text{yr}} = \mathbf{10\ years}$$

Therefore, the more efficient heater will pay for the $1100 cost differential in this case in 10 years, which is more than the 8-year limit. Therefore, the purchase of the cheater and less efficient heater is a better buy in this case.

5-142 Switching to energy efficient lighting reduces the electricity consumed for lighting as well as the cooling load in summer, but increases the heating load in winter. It is to be determined if switching to efficient lighting will increase or decrease the total energy cost of a building. √

Assumptions The light escaping through the windows is negligible so that the entire lighting energy becomes part of the internal heat generation.

Analysis (*a*) Efficient lighting reduces the amount of electrical energy used for lighting year-around as well as the amount of heat generation in the house since light is eventually converted to heat. As a result, the electrical energy needed to air condition the house is also reduced. Therefore, in summer, the total cost of energy use of the household definitely decreases.

(*b*) In winter, the heating system must make up for the reduction in the heat generation due to reduced energy used for lighting. The total cost of energy used in this case will still decrease if the cost of unit heat energy supplied by the heating system is less than the cost of unit energy provided by lighting.

The cost of 1 kWh heat supplied from lighting is \$0.08 since all the energy consumed by lamps is eventually converted to thermal energy. Noting that 1 therm = 29.3 kWh and the furnace is 80% efficient, the cost of 1 kWh heat supplied by the heater is

$$\begin{aligned}\text{Cost of 1 kWh heat supplied by furnace} &= (\text{Amount of useful energy}/\eta_{\text{furnace}})(\text{Price})\\ &= [(1\text{ kWh})/0.80](\$0.70/\text{therm})\left(\frac{1\text{ therm}}{29.3\text{ kWh}}\right)\\ &= \$0.030\ \text{(per kWh heat)}\end{aligned}$$

which is much less than \$0.08. Thus we conclude that switching to energy efficient lighting will **reduce** the total energy cost of this building both in summer and in winter.

Discussion To determine the amount of cost savings due to switching to energy efficient lighting, consider 10 h of operation of lighting in summer and in winter for 1 kW rated power for lighting.

Current lighting:
Lighting cost: (Energy used)(Unit cost)= (1 kW)(10 h)(\$0.08/kWh) = \$0.80
Increase in air conditioning cost: (Heat from lighting/COP)(unit cost) =(10 kWh/3.5)(\$0.08/kWh) = \$0.23
Decrease in the heating cost = [Heat from lighting/Eff](unit cost)=(10/0.8 kWh)(\$0.70/29.3/kWh) = 0.30

Total cost in summer = 0.80+0.23 = \$1.03; Total cost in winter = \$0.80-0.23 = 0.57.

Energy efficient lighting:
Lighting cost: (Energy used)(Unit cost)= (0.25 kW)(10 h)(\$0.08/kWh) = \$0.20
Increase in air conditioning cost: (Heat from lighting/COP)(unit cost) =(2.5 kWh/3.5)(\$0.08/kWh) = \$0.06
Decrease in the heating cost = [Heat from lighting/Eff](unit cost)=(2.5/0.8 kWh)(\$0.70/29.3/kWh) = \$0.07

Total cost in summer = 0.20+0.06 = \$0.26; Total cost in winter = \$0.20-0.07 = 0.13.

Note that during a day with 10 h of operation, the total energy cost decreases from \$1.03 to \$0.26 in summer, and from \$0.57 to \$0.13 in winter when efficient lighting is used.

5-143 The cargo space of a refrigerated truck is to be cooled from 25°C to an average temperature of 5°C. The time it will take for an 8-kW refrigeration system to precool the truck is to be determined.

Assumptions **1** The ambient conditions remain constant during precooling. **2** The doors of the truck are tightly closed so that the infiltration heat gain is negligible. **3** The air inside is sufficiently dry so that the latent heat load on the refrigeration system is negligible. **4** Air is an ideal gas with constant specific heats.

Properties The density of air is given to be 1.2 kg/m³, and its specific heat at the average temperature of 15°C is C_p = 1.0 kJ/kg·°C (Table A-2).

Analysis The mass of air in the truck is

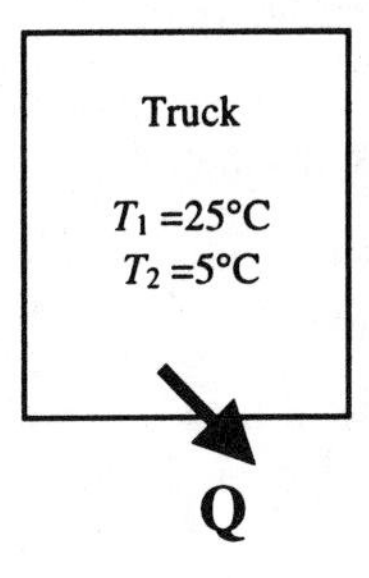

$$m_{\text{air}} = \rho_{\text{air}} V_{\text{truck}} = (1.2\,\text{kg/m}^3)(12\,\text{m} \times 2.3\,\text{m} \times 3.5\,\text{m}) = 116\,\text{kg}$$

The amount of heat removed as the air is cooled from 25 to 5°C

$$Q_{\text{cooling,air}} = (m\,C_p \Delta T)_{air} = (116\,\text{kg})(1.0\,\text{kJ/kg.}^\circ\text{C})(25-5)^\circ\text{C}$$
$$= 2{,}320\,\text{kJ}$$

Noting that *UA* is given to be 80 W/°C and the average air temperature in the truck during precooling is (25+5)/2 = 15°C, the average rate of heat gain by transmission is determined to be

$$\dot{Q}_{\text{transmission,ave}} = UA\Delta T = (80\,\text{W/}^\circ\text{C})(25-15)^\circ\text{C} = 800\text{ W} = 0.80\,\text{kJ/s}$$

Therefore, the time required to cool the truck from 25 to 5°C is determined to be

$$\dot{Q}_{\text{refrig.}}\Delta t = Q_{\text{cooling,air}} + \dot{Q}_{\text{transmission}}\Delta t \;\rightarrow \Delta t = \frac{Q_{\text{cooling,air}}}{\dot{Q}_{\text{refrig.}} - \dot{Q}_{\text{transmission}}} = \frac{2{,}320\,\text{kJ}}{(8-0.8)\text{ kJ/s}} = 322\,\text{s} \cong \mathbf{5.4\,min}$$

5-144 A refrigeration system is to cool bread loaves at a rate of 500 per hour by refrigerated air at -30°C. The rate of heat removal from the breads, the required volume flow rate of air, and the size of the compressor of the refrigeration system are to be determined.

Assumptions **1** Steady operating conditions exist. **2** The thermal properties of the bread loaves are constant. **3** The cooling section is well-insulated so that heat gain through its walls is negligible.

Properties The average specific and latent heats of bread are given to be 2.93 kJ/kg.°C and 109.3 kJ/kg, respectively. The gas constant of air is 0.287 kPa.m^3/kg.K (Table A-1), and the specific heat of air at the average temperature of (-30 + -22)/2 = -26°C ≈ 250 K is C_p =1.0 kJ/kg.°C (Table A-2).

Analysis (*a*) Noting that the breads are cooled at a rate of 500 loaves per hour, breads can be considered to flow steadily through the cooling section at a mass flow rate of

$$\dot{m}_{\text{bread}} = (500\text{ breads/h})(0.45\text{ kg/bread}) = 225\text{ kg/h} = 0.0625\text{ kg/s}$$

Then the rate of heat removal from the breads as they are cooled from 22°C to -10°C and frozen becomes

$$\dot{Q}_{\text{bread}} = (\dot{m}C_p\Delta T)_{\text{bread}} = (225\text{ kg/h})(2.93\text{ kJ/kg.°C})[(22-(-10)]\text{°C}$$
$$= 21{,}096\text{ kJ/h}$$

$$\dot{Q}_{\text{freezing}} = (\dot{m}h_{\text{latent}})_{\text{bread}} = (225\text{kg/h})(109.3\text{kJ/kg}) = 24{,}593\text{kJ/h}$$

and $$\dot{Q}_{\text{total}} = \dot{Q}_{\text{bread}} + \dot{Q}_{\text{freezing}} = 21{,}096 + 24{,}593 = \mathbf{45{,}689kJ/h}$$

Air
-30°C
Bread

(*b*) All the heat released by the breads is absorbed by the refrigerated air, and the temperature rise of air is not to exceed 8°C. The minimum mass flow and volume flow rates of air are determined to be

$$m_{air} = \frac{\dot{Q}_{air}}{(C_p\Delta T)_{air}} = \frac{45{,}689\text{ kJ/h}}{(1.0\text{ kJ/kg.°C})(8\text{°C})} = 5{,}711\text{ kg/h}$$

$$\rho = \frac{P}{RT} = \frac{101.3\text{ kPa}}{(0.287\text{ kPa.m}^3\text{/kg.K})(-30+273)\text{ K}} = 1.45\text{ kg/m}^3$$

$$\dot{V}_{air} = \frac{\dot{m}_{air}}{\rho_{air}} = \frac{5711\text{ kg/h}}{1.45\text{ kg/m}^3} = 3939\text{ m}^3\text{/h}$$

(*c*) For a COP of 1.2, the size of the compressor of the refrigeration system must be

$$\dot{W}_{\text{refrig}} = \frac{\dot{Q}_{\text{refrig}}}{COP} = \frac{45{,}689\text{kJ/h}}{1.2} = 38{,}074\text{kJ/h} = \mathbf{10.6\ kW}$$

5-145 The drinking water needs of a production facility with 20 employees is to be met by a bobbler type water fountain. The size of compressor of the refrigeration system of this water cooler is to be determined.

Assumptions **1** Steady operating conditions exist. **2** Water is an incompressible substance with constant properties at room temperature. **3** The cold water requirement is 0.4 L/h per person.

Properties The density and specific heat of water at room temperature are ρ = 1.0 kg/L and C = 4.18 kJ/kg.°C.C (Table A-3).

Analysis The refrigeration load in this case consists of the heat gain of the reservoir and the cooling of the incoming water. The water fountain must be able to provide water at a rate of

$$\dot{m}_{\text{water}} = \rho\dot{V}_{\text{water}} = (1\ \text{kg/L})(0.4\ \text{L/h}\cdot\text{person})(20\ \text{persons}) = 8.0\ \text{kg/h}$$

To cool this water from 22°C to 8°C, heat must removed from the water at a rate of

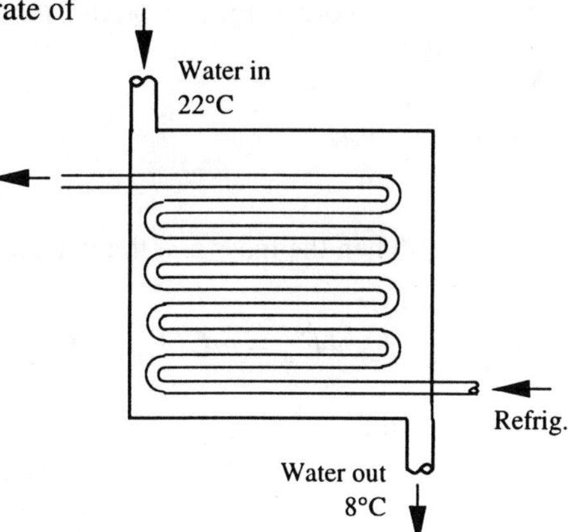

$$\dot{Q}_{\text{cooling}} = \dot{m}C_p(T_{\text{in}} - T_{\text{out}}) = (8.0\ \text{kg/h})(4.18\ \text{kJ/kg.°C})(22 - 8)\text{°C}$$
$$= 468\ \text{kJ/h} = 130\ \text{W} \quad (\text{since } 1\ \text{W} = 3.6\ \text{kJ/h})$$

Then total refrigeration load becomes

$$\dot{Q}_{\text{refrig, total}} = \dot{Q}_{\text{cooling}} + \dot{Q}_{\text{transfer}} = 130 + 45 = 175\ \text{W}$$

Noting that the coefficient of performance of the refrigeration system is 2.9, the required power input is

$$\dot{W}_{\text{refrig}} = \frac{\dot{Q}_{\text{refrig}}}{\text{COP}} = \frac{175\ \text{W}}{2.9} = \mathbf{60.3\ W}$$

Therefore, the power rating of the compressor of this refrigeration system must be at least 50.9 W to meet the cold water requirements of this office.

5-146 A washing machine uses $85/year worth of hot water heated by an electric water heater. The amount of hot water an average family uses per week is to be determined.

Assumptions **1** The electricity consumed by the motor of the washer is negligible. **2** Water is an incompressible substance with constant properties at room temperature.

Properties The density and specific heat of water at room temperature are ρ = 1.0 kg/L and C = 4.18 kJ/kg.°C (Table A-3).

Analysis The amount of electricity used to heat the water and the net amount transferred to water are

$$\text{Total energy used (electrical)} = \frac{\text{Total cost of energy}}{\text{Unit cost of energy}} = \frac{\$85/\text{year}}{\$0.082/\text{kWh}} = 1037\ \text{kWh/year}$$

$$\text{Total energy transfer to water} = \dot{E}_{\text{in}} = (\text{Efficiency})(\text{Total energy used}) = 0.91 \times 1037\ \text{kWh/year}$$

$$= 943.7\ \text{kWh/year} = (943.7\ \text{kWh/year})\left(\frac{3600\ \text{kJ}}{1\ \text{kWh}}\right)\left(\frac{1\ \text{year}}{52\ \text{weeks}}\right)$$

$$= 65{,}330\ \text{kJ/week}$$

Then the mass and the volume of hot water used per week become

$$\dot{E}_{\text{in}} = \dot{m}C(T_{\text{out}} - T_{\text{in}}) \rightarrow \dot{m} = \frac{\dot{E}_{\text{in}}}{C(T_{\text{out}} - T_{\text{in}})} = \frac{65{,}330\ \text{kJ/week}}{(4.18\ \text{kJ/kg.°C})(55 - 12)\text{°C}} = 411\ \text{kg/week}$$

and

$$\dot{V}_{\text{water}} = \frac{\dot{m}}{\rho} = \frac{441\ \text{kg/week}}{1\ \text{kg/L}} = \mathbf{441\ \ L/week}$$

Therefore, an average family uses 441 liters of hot water per week for washing clothes.

5-147E A washing machine uses \$33/year worth of hot water heated by a gas water heater. The amount of hot water an average family uses per week is to be determined.

Assumptions **1** The electricity consumed by the motor of the washer is negligible. **2** Water is an incompressible substance with constant properties at room temperature.

Properties The density and specific heat of water at room temperature are $\rho = 62.1$ lbm/ft^3 and $C = 1.00$ Btu/lbm.°F (Table A-3E).

Analysis The amount of electricity used to heat the water and the net amount transferred to water are

$$\text{Total energy used (gas)} = \frac{\text{Total cost of energy}}{\text{Unit cost of energy}} = \frac{\$33\text{/year}}{\$0.605\text{/therm}} = 54.55 \text{ therms/year}$$

$$\text{Total energy transfer to water} = \dot{E}_{\text{in}} = \text{(Efficiency)(Total energy used)} = 0.58 \times 54.55 \text{ therms/year}$$

$$= 31.6 \text{ therms/year} = (31.6 \text{ therms/year})\left(\frac{100{,}000 \text{ Btu}}{1 \text{ therm}}\right)\left(\frac{1 \text{ year}}{52 \text{ weeks}}\right)$$

$$= 60{,}770 \text{ Btu/week}$$

Then the mass and the volume of hot water used per week become

$$\dot{E}_{\text{in}} = \dot{m}C(T_{\text{out}} - T_{\text{in}}) \rightarrow \dot{m} = \frac{\dot{E}_{\text{in}}}{C(T_{\text{out}} - T_{\text{in}})} = \frac{60{,}770 \text{ Btu/week}}{(1.0 \text{ Btu/lbm.°F})(130 - 60)\text{°F}} = 868 \text{ lbm/week}$$

and

$$\dot{V}_{\text{water}} = \frac{\dot{m}}{\rho} = \frac{868 \text{ lbm/week}}{62.1 \text{ lbm/ft}^3} = (14.0 \text{ ft}^3)\left(\frac{7.4804 \text{ gal}}{1 \text{ ft}^3}\right) = \mathbf{105 \text{ gal/week}}$$

Therefore, an average family uses about 100 gallons of hot water per week for washing clothes.

5-148 [*Also solved by EES on enclosed CD*] A typical heat pump powered water heater costs about $800 more to install than a typical electric water heater. The number of years it will take for the heat pump water heater to pay for its cost differential from the energy it saves is to be determined.

Assumptions **1** The price of electricity remains constant. **2** Water is an incompressible substance with constant properties at room temperature. **3** Time value of money (interest, inflation) is not considered.

Properties The density and specific heat of water at room temperature are $\rho = 1.0$ kg/L and $C = 4.18$ kJ/kg.°C (Table A-3).

Analysis The amount of electricity used to heat the water and the net amount transferred to water are

$$\text{Total energy used (electrical)} = \frac{\text{Total cost of energy}}{\text{Unit cost of energy}} = \frac{\$390/\text{year}}{\$0.080/\text{kWh}} = 4875 \text{ kWh/year}$$

$$\text{Total energy transfer to water} = \dot{E}_{in} = \text{(Efficiency)(Total energy used)} = 0.9 \times 4875 \text{ kWh/year}$$
$$= 4388 \text{ kWh/year}$$

The amount of electricity consumed by the heat pump and its cost are

$$\text{Energy usage (of heat pump)} = \frac{\text{Energy transfer to water}}{\text{COP}_{HP}} = \frac{4388 \text{ kWh/year}}{2.2} = 1995 \text{ kWh/year}$$

$$\text{Energy cost (of heat pump)} = \text{(Energy usage)(Unit cost of energy)} = (1995 \text{ kWh/year})(\$0.08/\text{kWh})$$
$$= \$159.6/\text{year}$$

Then the money saved per year by the heat pump and the simple payback period become

$$\text{Money saved} = \text{(Energy cost of electric heater) - (Energy cost of heat pump)}$$
$$= \$390 - \$159.60 = \$230.40$$

$$\text{Simple payback period} = \frac{\text{Additional installation cost}}{\text{Money saved}} = \frac{\$800}{\$230.40/\text{year}} = \mathbf{3.5 \text{ years}}$$

Discussion The economics of heat pump water heater will be even better if the air in the house is used as the heat source for the heat pump in summer, and thus also serving as an air-conditioner.

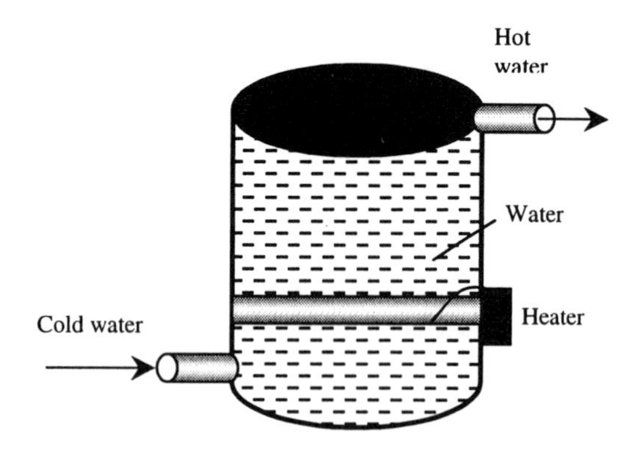

5-149 EES solution of this (and other comprehensive problems designated with the *computer icon*) is available to instructors at the *Instructor Manual* section of the *Online Learning Center* (OLC) at www.mhhe.com/cengel-boles. See the Preface for access information.

5-150E The energy contents, unit costs, and typical conversion efficiencies of various energy sources for use in water heaters are given. The lowest cost energy source is to be determined.

Assumptions The differences in installation costs of different water heaters are not considered.

Properties The energy contents, unit costs, and typical conversion efficiencies of different systems are given in the problem statement.

Analysis The unit cost of each Btu of useful energy supplied to the water heater by each system can be determined from

$$\text{Unit cost of useful energy} = \frac{\text{Unit cost of energy supplied}}{\text{Conversion efficiency}}$$

Substituting,

Natural gas heater:

$$\text{Unit cost of useful energy} = \frac{\$0.0060/\text{ft}^3}{0.55}\left(\frac{1\ \text{ft}^3}{1025\ \text{Btu}}\right) = \$10.6\times10^{-6}\ /\ \text{Btu}$$

Heating by oil heater:

$$\text{Unit cost of useful energy} = \frac{\$1.15/\text{gal}}{0.55}\left(\frac{1\ \text{gal}}{138{,}700\ \text{Btu}}\right) = \$15.1\times10^{-6}\ /\ \text{Btu}$$

Electric heater:

$$\text{Unit cost of useful energy} = \frac{\$0.084/\text{kWh})}{0.90}\left(\frac{1\ \text{kWh}}{3412\ \text{Btu}}\right) = \$27.4\times10^{-6}\ /\ \text{Btu}$$

Therefore, the lowest cost energy source for hot water heaters in this case is **natural gas**.

5-151E A home owner is considering three different heating systems for heating his house. The system with the lowest energy cost is to be determined.

Assumptions The differences in installation costs of different heating systems are not considered.

Properties The energy contents, unit costs, and typical conversion efficiencies of different systems are given in the problem statement.

Analysis The unit cost of each Btu of useful energy supplied to the house by each system can be determined from

$$\text{Unit cost of useful energy} = \frac{\text{Unit cost of energy supplied}}{\text{Conversion efficiency}}$$

Substituting,

Natural gas heater:

$$\text{Unit cost of useful energy} = \frac{\$0.62/\text{therm}}{0.87}\left(\frac{1\text{ therm}}{105{,}500\text{ kJ}}\right) = \$6.8\times10^{-6}\text{ / kJ}$$

Heating oil heater:

$$\text{Unit cost of useful energy} = \frac{\$1.25/\text{gal}}{0.87}\left(\frac{1\text{ gal}}{138{,}500\text{ kJ}}\right) = \$10.4\times10^{-6}\text{ / kJ}$$

Electric heater:

$$\text{Unit cost of useful energy} = \frac{\$0.09/\text{kWh})}{1.0}\left(\frac{1\text{ kWh}}{3600\text{ kJ}}\right) = \$25.0\times10^{-6}\text{ / kJ}$$

Therefore, the system with the lowest energy cost for heating the house is the **natural gas heater**.

5-152 A home owner is to choose between a high-efficiency natural gas furnace and a ground-source heat pump. The system with the lower energy cost is to be determined.

Assumptions The two heater are comparable in all aspects other than the cost of energy.

Analysis The unit cost of each kJ of useful energy supplied to the house by each system is

Natural gas furnace:

$$\text{Unit cost of useful energy} = \frac{(\$0.71/\text{therm})}{0.97}\left(\frac{1\ \text{therm}}{105{,}500\ \text{kJ}}\right) = \$6.94 \times 10^{-6}\ /\ \text{kJ}$$

Heat Pump System:

$$\text{Unit cost of useful energy} = \frac{(\$0.092/\text{kWh})}{3.5}\left(\frac{1\ \text{kWh}}{3600\ \text{kJ}}\right) = \$7.3 \times 10^{-6}\ /\ \text{kJ}$$

The two results are very close to each other, but the energy cost of **natural gas furnace** will be slightly lower.

5-153 The maximum flow rate of a standard shower head can be reduced from 13.3 to 10.5 L/min by switching to low-flow shower heads. The amount of oil and money a family of four will save per year by replacing the standard shower heads by the low-flow ones are to be determined.

Assumptions **1** Steady operating conditions exist. **2** Showers operate at maximum flow conditions during the entire shower. **3** Each member of the household takes a 5-min shower every day.

Properties The specific heat of water is C = 4.18 kJ/kg.°C and heating value of heating oil is 146,300 kJ/gal (given). The density of water is ρ = 1 kg/L.

Analysis The low-flow heads will save water at a rate of

$$\dot{V}_{\text{saved}} = [(13.3 - 10.5)\ \text{L/min}](6\ \text{min/person.day})(4\ \text{persons})(365\ \text{days/yr}) = 24{,}528\ \text{L/year}$$

$$\dot{m}_{\text{saved}} = \rho\dot{V}_{\text{saved}} = (1\ \text{kg/L})(24{,}528\ \text{L/year}) = 24{,}528\ \text{kg/year}$$

Then the energy, fuel, and money saved per year becomes

$$\text{Energy saved} = \dot{m}_{\text{saved}} C\Delta T = (24{,}528\ \text{kg/year})(4.18\ \text{kJ/kg.°C})(42 - 15)\text{°C} = 2{,}768{,}000\ \text{kJ/year}$$

$$\text{Fuel saved} = \frac{\text{Energy saved}}{(\text{Efficiency})(\text{Heating value of fuel})} = \frac{2{,}738{,}000\ \text{kJ/year}}{(0.65)(146{,}300\ \text{kJ/gal})} = \mathbf{29.1\ gal/year}$$

$$\text{Money saved} = (\text{Fuel saved})(\text{Unit cost of fuel}) = (29.1\ \text{gal/year})(\$1.20/\text{gal}) = \mathbf{\$34.9/year}$$

Therefore, switching to low-flow shower heads will save about $35 per year in energy costs..

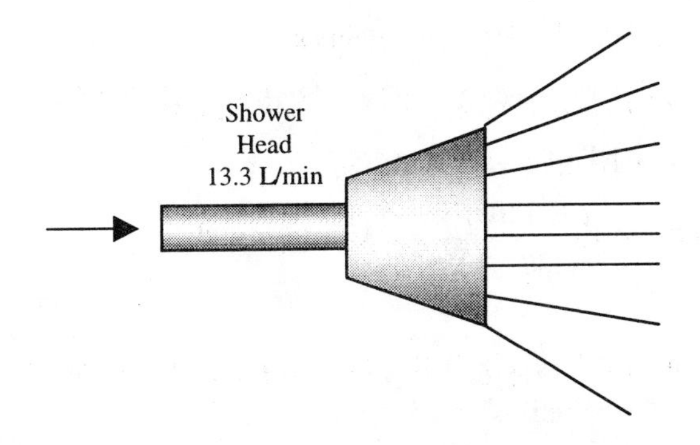

5-154 The heating and cooling costs of a poorly insulated house can be reduced by up to 30 percent by adding adequate insulation. The time it will take for the added insulation to pay for itself from the energy it saves is to be determined.

Assumptions It is given that the annual energy usage of a house is \$1200 a year, and 46% of it is used for heating and cooling. The cost of added insulation is given to be \$200.

Analysis The amount of money that would be saved per year is determined directly from

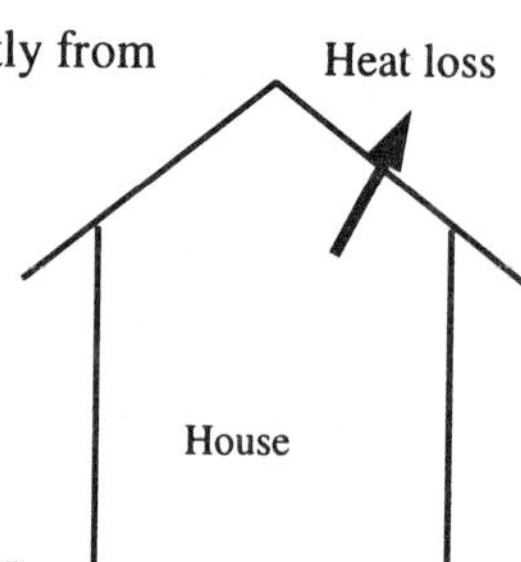

$$\text{Money saved } = (\$1200/\text{year})(0.46)(0.30) = \$166/\text{yr}$$

Then the simple payback period becomes

$$\text{Payback period} = \frac{\text{Cost}}{\text{Money saved}} = \frac{\$200}{\$166/\text{yr}} = \mathbf{1.2\ yr}$$

Therefore, the proposed measure will pay for itself in less than one and a half year.

5-155 The ventilating fans of a house discharge a houseful of warmed air in one hour (ACH = 1). For an average outdoor temperature of 5°C during the heating season, the cost of energy "vented out" by the fans in 1 h is to be determined.

Assumptions **1** Steady operating conditions exist. **2** The house is maintained at 22°C and 92 kPa at all times. **3** The infiltrating air is heated to 22°C before it is vented out. **4** Air is an ideal gas with constant specific heats at room temperature. **5** The volume occupied by the people, furniture, etc. is negligible.

Properties The gas constant of air is R = 0.287 kPa.m^3/kg·K (Table A-1). The specific heat of air at room temperature is C_p = 1.0 kJ/kg·°C (Table A-2a).

Analysis The density of air at the indoor conditions of 92 kPa and 22°C is

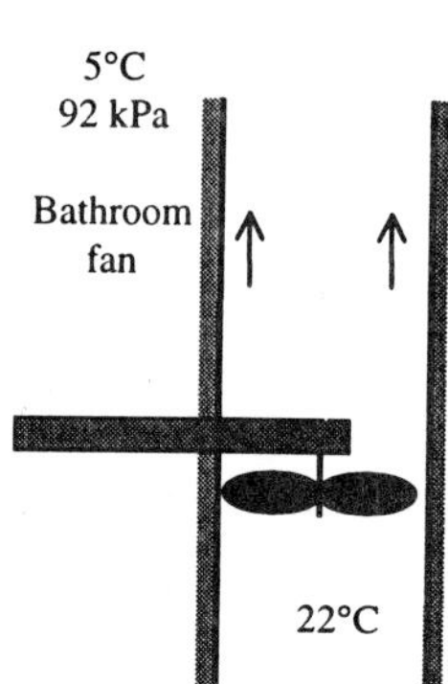

$$\rho_o = \frac{P_o}{RT_o} = \frac{92\ \text{kPa}}{(0.287\ \text{kPa.m}^3/\text{kg.K})(22+273\ \text{K})} = 1.087\ \text{kg/m}^3$$

Noting that the interior volume of the house is 200 × 2.8 = 560 m^3, the mass flow rate of air vented out becomes

$$\dot{m}_{\text{air}} = \rho\dot{V}_{\text{air}} = (1.087\ \text{kg/m}^3)(560\ \text{m}^3/\text{h}) = 608.7\ \text{kg/h} = 0.169\ \text{kg/s}$$

Noting that the indoor air vented out at 22°C is replaced by infiltrating outdoor air at 5°C, this corresponds to energy loss at a rate of

$$\dot{Q}_{\text{loss,fan}} = \dot{m}_{\text{air}} C_p (T_{\text{indoors}} - T_{\text{outdoors}})$$
$$= (0.169\ \text{kg/s})(1.0\ \text{kJ/kg.°C})(22-5)\text{°C} = 2.874\ \text{kJ/s} = 2.874\ \text{kW}$$

Then the amount and cost of the heat "vented out" per hour becomes

$$\text{Fuel energy loss} = \dot{Q}_{\text{loss,fan}}\Delta t/\eta_{furnace} = (2.874\ \text{kW})(720\ \text{h/month})/0.96 = \mathbf{2156\ kWh/month}$$

$$\text{Money loss} = (\text{Fuel energy loss})(\text{Unit cost of energy})$$

$$= (2156\ \text{kWh/month})(\$0.60/\text{therm})\frac{1\ \text{therm}}{29.3\ \text{kWh}} = \mathbf{\$44.1/month}$$

Discussion Note that the energy and money loss associated with ventilating fans can be very significant. Therefore, ventilating fans should be used sparingly.

5-156 The ventilating fans of a house discharge a houseful of air-conditioned air in one hour (ACH = 1). For an average outdoor temperature of 32°C during the cooling season, the cost of energy "vented out" by the fans in 1 h is to be determined.

Assumptions **1** Steady operating conditions exist. **2** The house is maintained at 22°C and 92 kPa at all times. **3** The infiltrating air is cooled to 22°C before it is vented out. **4** Air is an ideal gas with constant specific heats at room temperature. **5** The volume occupied by the people, furniture, etc. is negligible. **6** Latent heat load is negligible.

Properties The gas constant of air is R = 0.287 kPa.m³/kg·K (Table A-1). The specific heat of air at room temperature is C_p = 1.0 kJ/kg·°C (Table A-2a).

Analysis The density of air at the indoor conditions of 92 kPa and 22°C is

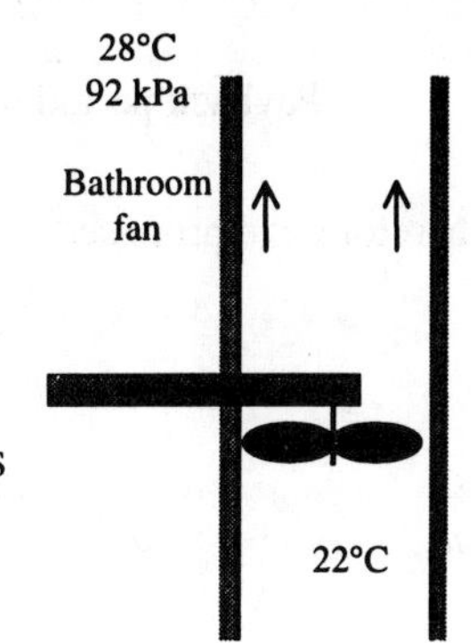

$$\rho_o = \frac{P_o}{RT_o} = \frac{92\ \text{kPa}}{(0.287\ \text{kPa.m}^3/\text{kg.K})(22+273\ \text{K})} = 1.087\ \text{kg/m}^3$$

Noting that the interior volume of the house is 200 × 2.8 = 560 m³, the mass flow rate of air vented out becomes

$$\dot{m}_{\text{air}} = \rho \dot{V}_{\text{air}} = (1.087\ \text{kg/m}^3)(560\ \text{m}^3/\text{h}) = 608.7\ \text{kg/h} = 0.169\ \text{kg/s}$$

Noting that the indoor air vented out at 22°C is replaced by infiltrating outdoor air at 28°C, this corresponds to energy loss at a rate of

$$\begin{aligned}\dot{Q}_{\text{loss,fan}} &= \dot{m}_{\text{air}} C_p (T_{\text{outdoors}} - T_{\text{indoors}}) \\ &= (0.169\ \text{kg/s})(1.0\ \text{kJ/kg.°C})(28-22)\text{°C} = 1.014\ \text{kJ/s} = 1.014\ \text{kW}\end{aligned}$$

Then the amount and cost of the electric energy "vented out" per hour becomes

$$\text{Electric energy loss} = \dot{Q}_{\text{loss,fan}} \Delta t / COP = (1.014\ \text{kW})(720\ \text{h/month})/3.2 = \mathbf{228\ kWh/month}$$

$$\text{Money loss} = (\text{Fuel energy loss})(\text{Unit cost of energy})$$
$$= (228\ \text{kWh/month})(\$0.10/\text{kWh}) = \mathbf{\$22.8/month}$$

Discussion Note that the energy and money loss associated with ventilating fans can be very significant. Therefore, ventilating fans should be used sparingly.

5-157 Caulking and weather-stripping doors and windows to reduce air leaks can reduce the energy use of a house by up to 10 percent. The time it will take for the caulking and weather-stripping to pay for itself from the energy it saves is to be determined.

Assumptions It is given that the annual energy usage of a house is $1100 a year, and the cost of caulking and weather-stripping a house is $50..

Analysis The amount of money that would be saved per year is determined directly from

$$\text{Money saved } = (\$1100/\text{year})(0.10) = \mathbf{\$110/yr}$$

Then the simple payback period becomes

$$\text{Payback period} = \frac{\text{Cost}}{\text{Money saved}} = \frac{\$50}{\$110/\text{yr}} = \mathbf{0.45\ yr}$$

Therefore, the proposed measure will pay for itself in less than half a year.

5-158 It is estimated that 570,000 barrels of oil would be saved per day if the thermostat setting in residences in winter were lowered by 6°F (3.3°C). The amount of money that would be saved per year is to be determined.

Assumptions The average heating season is given to be 180 days, and the cost of oil to be $20/barrel.

Analysis The amount of money that would be saved per year is determined directly from

$$(570{,}000\ \text{barrel/day})(180\ \text{days/year})(\$20/\text{barrel}) = \mathbf{\$2{,}052{,}000{,}000}$$

Therefore, the proposed measure will save more than 2-billion dollars a year in energy costs.

5-159 EES solution of this (and other comprehensive problems designated with the *computer icon*) is available to instructors at the *Instructor Manual* section of the *Online Learning Center* (OLC) at www.mhhe.com/cengel-boles. See the Preface for access information.

5-159 The maximum work will be obtained if a Carnot heat pump is used. The sink temperature of this heat engine will remain constant at 300 K but the source temperature will be decreasing from 350 K to 300 K. Then the thermal efficiency of the Carnot heat engine operating between pond and the ambient air can be expressed as

$$\eta_{th,C} = 1 - \frac{T_L}{T_H} = 1 - \frac{300}{T_H}$$

where T_H is a variable. The conservation of energy relation for the pond can be written in the differential form as

$$\partial Q_{pond} = mC\,dT_H$$

and

$$\partial Q_H = -\partial Q_{pond} = -mC\,dT_H = -(10^5\ \text{kg})(4.18\text{kJ/kg}\cdot\text{K})dT_H$$

Also,

$$\partial W_{net} = \eta_{th,C}\,\partial Q_H = -\left(1 - \frac{300}{T_H}\right)(10^5\ \text{kg})(4.18\text{kJ/kg}\cdot\text{K})\partial T_H$$

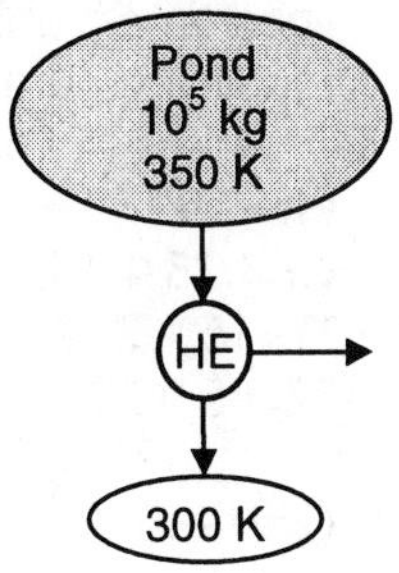

The total work output is obtained by integration,

$$W_{net} = \int_{300}^{350} \eta_{th,C}\,\partial Q_H = \int_{300}^{350}\left(1 - \frac{300}{T_H}\right)(10^5\ \text{kg})(4.18\text{kJ/kg}\cdot\text{K})dT_H$$

$$= 4.18\times10^5 \int_{300}^{350}\left(1 - \frac{300}{T_H}\right)dT_H = \mathbf{15.7\times10^5\ kJ}$$

which is the exact result. The values obtained by computer solution will approach this value as the temperature interval is decreased.

Fundamentals of Engineering (FE) Exam Problems

5-160 The label on a washing machine indicates that the washer will use $85 worth of hot water if the water is heated by a 90% efficiency electric heater at an electricity rate of $0.082/kWh. If the water is heated from 15°C to 55°C, the amount of hot water an average family uses per year is

(a) 14,800 kg (b) 17,200 kg (c) 20,100 kg (d) 22,300 kg (e) 24,800 kg

Answer (c) 20,100 kg

Solution Solved by EES Software. Solutions can be verified by copying-and-pasting the following lines on a blank EES screen. (Similar problems and their solutions can be obtained easily by modifying numerical values).

```
Eff=0.90
C=4.18 "kJ/kg-C"
T1=15 "C"
T2=55 "C"

Cost=85 "$"
Price=0.082 "$/kWh"
Ein=(85/0.082)*3600 "kJ"
Ein=m*C*(T2-T1)/Eff "kJ"

"Some Wrong Solutions with Common Mistakes:"
Ein=W1_m*C*(T2-T1)*Eff "Multiplying by Eff instead of dividing"
Ein=W2_m*C*(T2-T1) "Ignoring efficiency"
Ein=W3_m*(T2-T1)/Eff "Not using specific heat"
Ein=W4_m*C*(T2+T1)/Eff "Adding temperatures"
```

5-161 A 2.4-m high 200-m^2 house is maintained at 22°C by an air-conditioning system whose COP is 3.2. It is estimated that the kitchen, bath, and other ventilating fans of the house discharge a houseful of conditioned air once every hour. If the average outdoor temperature is 32°C, the density of air is 1.20 kg/m^3, and the unit cost of electricity is $0.10/kWh, the amount of money "vented out" by the fans in 10 hours is

(a) $0.50 (b) $1.60 (c) $5.00 (d) $11.00 (e) $16.00

Answer (a) $0.50

Solution Solved by EES Software. Solutions can be verified by copying-and-pasting the following lines on a blank EES screen. (Similar problems and their solutions can be obtained easily by modifying numerical values).

```
COP=3.2
T1=22 "C"
T2=32 "C"
Price=0.10 "$/kWh"
Cp=1.005 "kJ/kg-C"
rho=1.20 "kg/m^3"
V=2.4*200 "m^3"
m=rho*V
m_total=m*10
Ein=m_total*Cp*(T2-T1)/COP "kJ"
Cost=(Ein/3600)*Price

"Some Wrong Solutions with Common Mistakes:"
W1_Cost=(Price/3600)*m_total*Cp*(T2-T1)*COP "Multiplying by Eff instead of dividing"
W2_Cost=(Price/3600)*m_total*Cp*(T2-T1)  "Ignoring efficiency"
W3_Cost=(Price/3600)*m*Cp*(T2-T1)/COP "Using m instead of m_total"
W4_Cost=(Price/3600)*m_total*Cp*(T2+T1)/COP  "Adding temperatures"
```

5-162 The drinking water needs of an office are met by cooling tab water in a refrigerated water fountain from 22°C to 8°C at an average rate of 8 kg/h. If the COP of this refrigerator is 3.1, the required power input to this refrigerator is

(a) 28 W (b) 42 W (c) 88 W (d) 130 W (e) 403 W

Answer (b) 42 W

Solution Solved by EES Software. Solutions can be verified by copying-and-pasting the following lines on a blank EES screen. (Similar problems and their solutions can be obtained easily by modifying numerical values).

```
COP=3.1
Cp=4.18 "kJ/kg-C"

T1=22 "C"
T2=8 "C"
m_dot=8/3600 "kg/s"

Q_L=m_dot*Cp*(T1-T2)  "kW"
W_in=Q_L*1000/COP "W"

"Some Wrong Solutions with Common Mistakes:"
W1_Win=m_dot*Cp*(T1-T2) *1000*COP "Multiplying by COP instead of dividing"
W2_Win=m_dot*Cp*(T1-T2) *1000 "Not using COP"
W3_Win=m_dot*(T1-T2) *1000/COP "Not using specific heat"
W4_Win=m_dot*Cp*(T1+T2) *1000/COP "Adding temperatures"
```

5-163 In a hot summer day, the air in a well-sealed room is circulated by a 0.50-hp fan driven by a 65% efficient motor. (Note that the motor delivers 0.50 hp of net power to the fan). The rate of energy supply from the fan-motor assembly to the room is

(a) 0.769 kJ/s (b) 0.325 kJ/s (c) 0.574 kJ/s (d) 0.373 kJ/s (e) 0.242 kJ/s

Answer (c) 0.574 kJ/s

Solution Solved by EES Software. Solutions can be verified by copying-and-pasting the following lines on a blank EES screen. (Similar problems and their solutions can be obtained easily by modifying numerical values).

```
Eff=0.65
W_fan=0.50*0.7457 "kW"
E=W_fan/Eff  "kJ/s"

"Some Wrong Solutions with Common Mistakes:"
W1_E=W_fan*Eff "Multiplying by efficiency"
W2_E=W_fan "Ignoring efficiency"
W3_E=W_fan/Eff/0.7457 "Using hp instead of kW"
```

5-164 A heat pump is absorbing heat from the cold outdoors at 5°C and supplying heat to a house at 22°C at a rate of 18,000 kJ/h. If the power consumed by the heat pump is 2.5 kW, the coefficient of performance of the heat pump is

(a) 0.5 (b) 1.0 (c) 2.0 (d) 5.0 (e) 17.3

Answer (c) 2.0

Solution Solved by EES Software. Solutions can be verified by copying-and-pasting the following lines on a blank EES screen. (Similar problems and their solutions can be obtained easily by modifying numerical values).

```
TL=5 "C"
TH=22 "C"
QH=18000/3600 "kJ/s"
Win=2.5 "kW"
COP=QH/Win

"Some Wrong Solutions with Common Mistakes:"
W1_COP=Win/QH "Doing it backwards"
W2_COP=TH/(TH-TL) "Using temperatures in C"
W3_COP=(TH+273)/(TH-TL) "Using temperatures in K"
W4_COP=(TL+273)/(TH-TL) "Finding COP of refrigerator using temperatures in K"
```

5-165 A heat engine cycle is executed with steam in the saturation dome. The pressure of steam is 1 MPa during heat addition, and 0.5 MPa during heat rejection. The highest possible efficiency of this heat engine is

(a) 6.2% (b) 15.6% (c) 50.0% (d) 93.8% (e) 100%

Answer (a) 6.2%

Solution Solved by EES Software. Solutions can be verified by copying-and-pasting the following lines on a blank EES screen. (Similar problems and their solutions can be obtained easily by modifying numerical values).

```
PH=1000 "kPa"
PL=500 "kPa"

TH=TEMPERATURE(Steam_NBS,x=0,P=PH)
TL=TEMPERATURE(Steam_NBS,x=0,P=PL)

Eta_Carnot=1-(TL+273)/(TH+273)

"Some Wrong Solutions with Common Mistakes:"
W1_Eta_Carnot=1-PL/PH "Using pressures"
W2_Eta_Carnot=1-TL/TH "Using temperatures in C"
W3_Eta_Carnot=TL/TH "Using temperatures ratio"
```

5-166 A heat engine receives heat from a source at 1000°C and rejects the waste heat to a sink at 50°C. If heat is supplied to this engine at a rate of 100 kJ/s, the maximum power this heat engine can produce is

(a) 25.4 kW (b) 55.4 kW (c) 74.6 kW (d) 95.0 kW (e) 100.0 kW

Answer (c) 74.6 kW

Solution Solved by EES Software. Solutions can be verified by copying-and-pasting the following lines on a blank EES screen. (Similar problems and their solutions can be obtained easily by modifying numerical values).

```
TH=1000 "C"
TL=50 "C"
Q_in=100 "kW"

Eta=1-(TL+273)/(TH+273)
W_out=Eta*Q_in

"Some Wrong Solutions with Common Mistakes:"
W1_W_out=(1-TL/TH)*Q_in     "Using temperatures in C"
W2_W_out=Q_in               "Setting work equal to heat input"
W3_W_out=Q_in/Eta           "Dividing by efficiency instead of multiplying"
W4_W_out=(TL+273)/(TH+273)*Q_in  "Using temperature ratio"
```

5-167 A heat pump cycle is executed with R-134a under the saturation dome between the pressure limits of 2.0 MPa and 0.4 MPa. The maximum Coefficient of Performance of this heat pump is

(a) 1.2 (b) 3.7 (c) 5.0 (d) 5.8 (e) 7.1

Answer (d) 5.8

Solution Solved by EES Software. Solutions can be verified by copying-and-pasting the following lines on a blank EES screen. (Similar problems and their solutions can be obtained easily by modifying numerical values).

```
PH=2000 "kPa"
PL=400 "kPa"

TH=TEMPERATURE(R134a,x=0,P=PH) "C"
TL=TEMPERATURE(R134a,x=0,P=PL) "C"

COP_HP=(TH+273)/(TH-TL)

"Some Wrong Solutions with Common Mistakes:"
W1_COP=PH/(PH-PL)  "Using pressures"
W2_COP=TH/(TH-TL)  "Using temperatures in C"
W3_COP=TL/(TH-TL)  "Refrigeration COP using temperatures in C"
W4_COP=(TL+273)/(TH-TL)  "Refrigeration COP using temperatures in K"
```

5-168 A refrigeration cycle is executed with R-134a under the saturation dome between the pressure limits of 1.8 MPa and 0.2 MPa. If the power consumption of the refrigerator is 3 kW, the maximum rate of heat removal from the cooled space of this refrigerator is

(a) 0.42 kJ/s (b) 0.83 kJ/s (c) 3.0 kJ/s (d) 10.8 kJ/s (e) 13.8 kJ/s

Answer (d) 10.8 kJ/s

Solution Solved by EES Software. Solutions can be verified by copying-and-pasting the following lines on a blank EES screen. (Similar problems and their solutions can be obtained easily by modifying numerical values).

```
PH=1800 "kPa"
PL=200 "kPa"
W_in=3 "kW"

TH=TEMPERATURE(R134a,x=0,P=PH) "C"
TL=TEMPERATURE(R134a,x=0,P=PL) "C"

COP=(TL+273)/(TH-TL)
QL=W_in*COP "kW"

"Some Wrong Solutions with Common Mistakes:"
W1_QL=W_in*TL/(TH-TL)    "Using temperatures in C"
W2_QL=W_in               "Setting heat removal equal to power input"
W3_QL=W_in/COP           "Dividing by COP instead of multiplying"
W4_QL=W_in*(TH+273)/(TH-TL) "Using COP definition for Heat pump"
```

5-169 A heat pump with a COP of 2.8 is used to heat a perfectly sealed house (no air leaks). The entire mass within the house (air, furniture, etc.) is equivalent to 1200 kg of air. When running, the heat pump consumes electric power at a rate of 5 kW. The temperature of the house was 7°C when the heat pump was turned on. If heat transfer through the envelope of the house (walls, roof, etc.) is negligible, the length of time the heat pump must run to raise the temperature of the entire contents of the house to 22°C is

(a) 15.4 min (b) 44.5 min (c) 5.7 min (d) 21.4 min (e) 59.9 min

Answer (a) 15.4 min

Solution Solved by EES Software. Solutions can be verified by copying-and-pasting the following lines on a blank EES screen. (Similar problems and their solutions can be obtained easily by modifying numerical values).

```
COP=2.8
Cv=0.718 "kJ/kg.C"
m=1200 "kg"
T1=7 "C"
T2=22 "C"
QH=m*Cv*(T2-T1)
Win=5 "kW"
Win*time=QH/COP
time_min=time/60

"Some Wrong Solutions with Common Mistakes:"
Win*W1_time*60=m*Cv*(T2-T1) *COP "Multiplying by COP instead of dividing"
Win*W2_time*60=m*Cv*(T2-T1)        "Ignoring COP"
Win*W3_time=m*Cv*(T2-T1) /COP      "Finding time in seconds instead of minutes"
Win*W4_time*60=m*Cp*(T2-T1) /COP  "Using Cp instead of Cv"
Cp=1.005 "kJ/kg.K"
```

5-170 A heat engine cycle is executed with steam in the saturation dome between the pressure limits of 5 MPa and 2 MPa. If heat is supplied to the heat engine at a rate of 380 kJ/s, the maximum power output of this heat engine is

(a) 36.5 kW (b) 74.3 kW (c) 186.2 kW (d) 343.5 kW (e) 380.0 kW

Answer (a) 36.5 kW

Solution Solved by EES Software. Solutions can be verified by copying-and-pasting the following lines on a blank EES screen. (Similar problems and their solutions can be obtained easily by modifying numerical values).

```
PH=5000 "kPa"
PL=2000 "kPa"

Q_in=380 "kW"

TH=TEMPERATURE(Steam_NBS,x=0,P=PH)  "C"
TL=TEMPERATURE(Steam_NBS,x=0,P=PL) "C"

Eta=1-(TL+273)/(TH+273)
W_out=Eta*Q_in

"Some Wrong Solutions with Common Mistakes:"
W1_W_out=(1-TL/TH)*Q_in     "Using temperatures in C"
W2_W_out=(1-PL/PH)*Q_in     "Using pressures"
W3_W_out=Q_in/Eta     "Dividing by efficiency instead of multiplying"
W4_W_out=(TL+273)/(TH+273)*Q_in  "Using temperature ratio"
```

5-171 An air-conditioning system operating on the reversed Carnot cycle is required to remove heat from the house at a rate of 25 kJ/s to maintain its temperature constant at 20°C. If the temperature of the outdoors is 35°C, the power required to operate this air-conditioning system is

(a) 0.64 kW (b) 5.20 kW (c) 1.56 kW (d) 2.26 kW (e) 1.28 kW

Answer (e) 1.28 kW

Solution Solved by EES Software. Solutions can be verified by copying-and-pasting the following lines on a blank EES screen. (Similar problems and their solutions can be obtained easily by modifying numerical values).

```
TL=20 "C"
TH=35 "C"
QL=25 "kJ/s"
COP=(TL+273)/(TH-TL)
COP=QL/Win

"Some Wrong Solutions with Common Mistakes:"
QL=W1_Win*TL/(TH-TL)    "Using temperatures in C"
QL=W2_Win                  "Setting work equal to heat input"
QL=W3_Win/COP           "Dividing by COP instead of multiplying"
QL=W4_Win*(TH+273)/(TH-TL)   "Using COP of HP"
```

5-172 A refrigerator is removing heat from a cold medium at 3°C at a rate of 7200 kJ/h and rejecting the waste heat to a medium at 30°C. If the coefficient of performance of the refrigerator is 2, the power consumed by the refrigerator is

(a) 0.1 kW (b) 0.5 kW (c) 1.0 kW (d) 2.0 kW (e) 5.0 kW

Answer (c) 1.0 kW

Solution Solved by EES Software. Solutions can be verified by copying-and-pasting the following lines on a blank EES screen. (Similar problems and their solutions can be obtained easily by modifying numerical values).

```
TL=3 "C"
TH=30 "C"
QL=7200/3600 "kJ/s"
COP=2

QL=Win*COP

"Some Wrong Solutions with Common Mistakes:"
QL=W1_Win*(TL+273)/(TH-TL)  "Using Carnot COP"
QL=W2_Win                   "Setting work equal to heat input"
QL=W3_Win/COP               "Dividing by COP instead of multiplying"
QL=W4_Win*TL/(TH-TL)        "Using Carnot COP using C"
```

5-173 Two Carnot heat engines are operating in series such that the heat sink of the first engine serves as the heat source of the second one. If the source temperature of the first engine is 1800 K and the sink temperature of the second engine is 350 K and the thermal efficiencies of both engines are the same, the temperature of the intermediate reservoir is

(a) 1075 K (b) 794 K (c) 1000 K (d) 473 K (e) 1258 K

Answer (b) 794 K

Solution Solved by EES Software. Solutions can be verified by copying-and-pasting the following lines on a blank EES screen. (Similar problems and their solutions can be obtained easily by modifying numerical values).

```
TH=1800 "K"
TL=350 "K"

"Setting thermal efficiencies equal to each other:"
1-Tmid/TH=1-TL/Tmid

"Some Wrong Solutions with Common Mistakes:"
W1_Tmid=(TL+TH)/2 "Using average temperature"
```

5-174 Consider a Carnot refrigerator and a Carnot heat pump operating between the same two thermal energy reservoirs. If the COP of the refrigerator is 3.4, the COP of the heat pump is

(a) 1.7 (b) 2.4 (c) 3.4 (d) 4.4 (e) 5.0

Answer (d) 4.4

Solution Solved by EES Software. Solutions can be verified by copying-and-pasting the following lines on a blank EES screen. (Similar problems and their solutions can be obtained easily by modifying numerical values).

```
COP_R=3.4
COP_HP=COP_R+1

"Some Wrong Solutions with Common Mistakes:"
W1_COP=COP_R-1 "Subtracting 1 instead of adding 1"
W2_COP=COP_R "Setting COPs equal to each other"
```

5-175 A typical new household refrigerator consumes about 700 kWh of electricity per year, and has a coefficient of performance of 1.4. The amount of heat removed by this refrigerator from the refrigerated space per year is

(*a*) 1050 MJ/yr (*b*) 1800 MJ/yr (*c*) 2520 MJ/yr (*d*) 3528 MJ/yr (*e*) 6048 MJ/yr

Answer (*d*) 3528 MJ/yr

Solution Solved by EES Software. Solutions can be verified by copying-and-pasting the following lines on a blank EES screen. (Similar problems and their solutions can be obtained easily by modifying numerical values).

```
W_in=700*3.6 "MJ"
COP_R=1.4

QL=W_in*COP_R "MJ"

"Some Wrong Solutions with Common Mistakes:"
W1_QL=W_in*COP_R/3.6 "Not using the conversion factor"
W2_QL=W_in "Ignoring COP"
W3_QL=W_in/COP_R "Dividing by COP instead of multiplying"
```

5-176 A window air conditioner that consumes 2 kW of electricity when running and has a coefficient of performance of 4 is placed in the middle of a room, and is plugged in. The rate of cooling or heating this air conditioner will provide to the air in the room when running is

(*a*) 8 kJ/s, cooling (*b*) 2 kJ/s, cooling (*c*) 0.5kJ/s, heating (*d*) 2 kJ/s, heating (*e*) 8 kJ/s, heating

Answer (*d*) 2 kJ/s, heating

Solution Solved by EES Software. Solutions can be verified by copying-and-pasting the following lines on a blank EES screen. (Similar problems and their solutions can be obtained easily by modifying numerical values).

```
W_in=2 "kW"
COP=4

"From energy balance, heat supplied to the room is equal to electricity consumed,"

E_supplied=W_in "kJ/s, heating"

"Some Wrong Solutions with Common Mistakes:"
W1_E=-W_in "kJ/s, cooling"
W2_E=-COP*W_in "kJ/s, cooling"
W3_E=W_in/COP "kJ/s, heating"
W4_E=COP*W_in "kJ/s, heating"
```

5-177 · · · 5-186 Design and Essay Problems

Chapter 6
ENTROPY

Entropy and the Increase of Entropy Principle

6-1C Yes. Because we used the relation $(Q_H/T_H) = (Q_L/T_L)$ in the proof, which is the defining relation of absolute temperature.

6-2C No. The $\oint \delta Q$ represents the net heat transfer during a cycle, which could be positive.

6-3C Yes.

6-4C No. A system may reject more (or less) heat than it receives during a cycle. The steam in a steam power plant, for example, receives more heat than it rejects during a cycle.

6-5C No. A system may produce more (or less) work than it receives during a cycle. A steam power plant, for example, produces more work than it receives during a cycle, the difference being the net work output.

6-6C The entropy change will be the same for both cases since entropy is a property and it has a fixed value at a fixed state.

6-7C No. In general, that integral will have a different value for different processes. However, it will have the same value for all reversible processes.

6-8C Yes.

6-9C That integral should be performed along a reversible path to determine the entropy change.

6-10C No. An isothermal process can be irreversible. Example: A system that involves paddle-wheel work while losing an equivalent amount of heat.

6-11C The value of this integral is always larger for reversible processes.

6-12C No. Because the entropy of the surrounding air increases even more during that process, making the total entropy change positive.

6-13C It is possible to create entropy, but it is not possible to destroy it.

6-14C Sometimes.

6-15C Never.

6-16C Always.

6-17C Increase.

6-18C Increases.

6-19C Decreases.

6-20C Sometimes.

6-21C Yes. This will happen when the system is losing heat, and the decrease in entropy as a result of this heat loss is equal to the increase in entropy as a result of irreversibilities.

6-22C They are heat transfer, irreversibilities, and entropy transport with mass.

6-23C Greater than.

6-24 A rigid tank contains an ideal gas that is being stirred by a paddle wheel. The temperature of the gas remains constant as a result of heat transfer out. The entropy change of the gas is to be determined.

Assumptions The gas in the tank is given to be an ideal gas.

Analysis The temperature and the specific volume of the gas remain constant during this process. Therefore, the initial and the final states of the gas are the same. Then $s_2 = s_1$ since entropy is a property. Therefore,

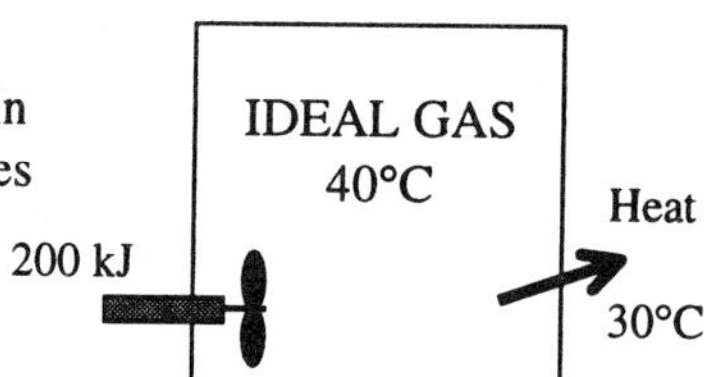

$$\Delta S_{sys} = \mathbf{0}$$

6-25 Air is compressed steadily by a compressor. The air temperature is maintained constant by heat rejection to the surroundings. The rate of entropy change of air is to be determined.

Assumptions **1** This is a steady-flow process since there is no change with time. **2** Kinetic and potential energy changes are negligible. **3** Air is an ideal gas. **4** The process involves no internal irreversibilities such as friction, and thus it is an isothermal, internally reversible process.

Properties Noting that $h = h(T)$ for ideal gases, we have $h_1 = h_2$ since $T_1 = T_2 = 25°C$.

Analysis We take the compressor as the system. Noting that the enthalpy of air remains constant, the energy balance for this steady-flow system can be expressed in the rate form as

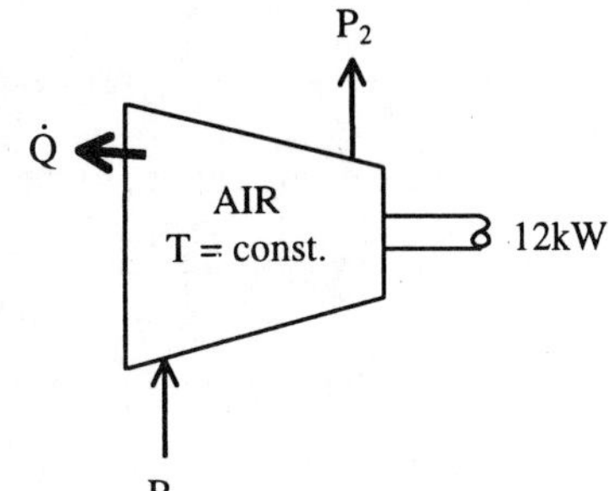

$$\underbrace{\dot{E}_{in} - \dot{E}_{out}}_{\text{Rate of net energy transfer by heat, work, and mass}} = \underbrace{\Delta \dot{E}_{system}}_{\text{Rate of change in internal, kinetic, potential, etc. energies}}^{\nearrow 0 \text{ (steady)}} = 0$$

$$\dot{E}_{in} = \dot{E}_{out}$$

$$\dot{W}_{in} = \dot{Q}_{out}$$

Therefore,

$$\dot{Q}_{out} = \dot{W}_{in} = 12 \text{ kW}$$

Noting that the process is assumed to be an isothermal and internally reversible process, the rate of entropy change of air is determined to be

$$\Delta \dot{S}_{air} = -\frac{\dot{Q}_{out,air}}{T_{sys}} = -\frac{12 \text{ kW}}{298 \text{ K}} = \mathbf{-0.0403 \ kW/K}$$

6-26 Heat is transferred isothermally from a source to the working fluid of a Carnot engine. The entropy change of the working fluid, the entropy change of the source, and the total entropy change during this process are to be determined.

Analysis (*a*) This is a reversible isothermal process, and the entropy change during such a process is given by

$$\Delta S = \frac{Q}{T}$$

Noting that heat transferred from the source is equal to the heat transferred to the working fluid, the entropy changes of the fluid and of the source become

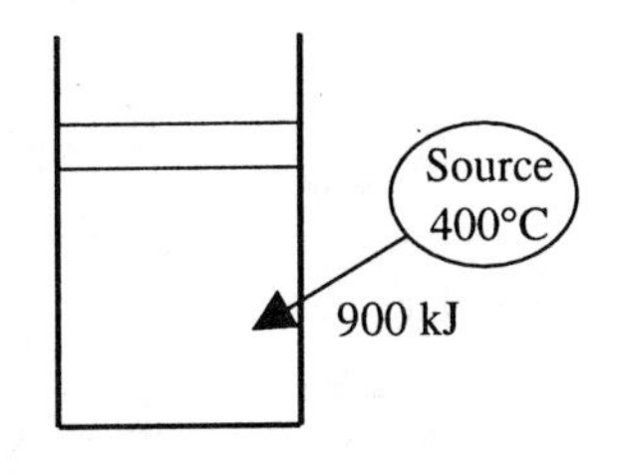

$$\Delta S_{fluid} = \frac{Q_{fluid}}{T_{fluid}} = \frac{Q_{in,fluid}}{T_{fluid}} = \frac{900 \text{ kJ}}{673 \text{ K}} = \mathbf{1.337 \ kJ/K}$$

(*b*)
$$\Delta S_{source} = \frac{Q_{source}}{T_{source}} = -\frac{Q_{\text{out, source}}}{T_{source}} = -\frac{900 \text{ kJ}}{673 \text{ K}} = \mathbf{-1.337 \ kJ/K}$$

(*c*) Thus the total entropy change of the process is

$$S_{gen} = \Delta S_{\text{total}} = \Delta S_{fluid} + \Delta S_{source} = 1.337 - 1.337 = \mathbf{0}$$

6-27 EES solution of this (and other comprehensive problems designated with the *computer icon*) is available to instructors at the *Instructor Manual* section of the *Online Learning Center* (OLC) at www.mhhe.com/cengel-boles. See the Preface for access information.

6-28E Heat is transferred isothermally from the working fluid of a Carnot engine to a heat sink. The entropy change of the working fluid is given. The amount of heat transfer, the entropy change of the sink, and the total entropy change during the process are to be determined.

Analysis (*a*) This is a reversible isothermal process, and the entropy change during such a process is given by

$$\Delta S = \frac{Q}{T}$$

Noting that heat transferred from the working fluid is equal to the heat transferred to the sink, the heat transfer become

$$Q_{fluid} = T_{fluid}\Delta S_{fluid} = (555\text{R})(-0.7\text{Btu/R}) = -388.5\text{Btu} \rightarrow Q_{fluid,out} = \mathbf{388.5\ Btu}$$

(*b*) The entropy change of the sink is determined from

$$\Delta S_{\text{sink}} = \frac{Q_{\text{sink,in}}}{T_{\text{sink}}} = \frac{388.5\text{ Btu}}{555\text{ R}} = \mathbf{0.7\ Btu/R}$$

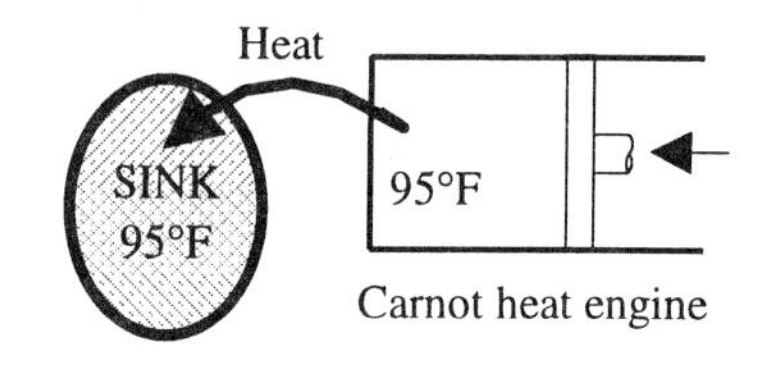

(*c*) Thus the total entropy change of the process is

$$S_{gen} = \Delta S_{\text{total}} = \Delta S_{fluid} + \Delta S_{\text{sink}} = -0.7 + 0.7 = \mathbf{0}$$

This is expected since all processes of the Carnot cycle are reversible processes, and no entropy is generated during a reversible process.

6-29 R-134a enters an evaporator as a saturated liquid-vapor at a specified pressure. Heat is transferred to the refrigerant from the cooled space, and the liquid is vaporized. The entropy change of the refrigerant, the entropy change of the cooled space, and the total entropy change for this process are to be determined.

Assumptions **1** Both the refrigerant and the cooled space involve no internal irreversibilities such as friction. **2** Any temperature change occurs within the wall of the tube, and thus both the refrigerant and the cooled space remain isothermal during this process. Thus it is an isothermal, internally reversible process.

Analysis Noting that both the refrigerant and the cooled space undergo reversible isothermal processes, the entropy change for them can be determined from

$$\Delta S = \frac{Q}{T}$$

(*a*) The pressure of the refrigerant is maintained constant. Therefore, the temperature of the refrigerant also remains constant at the saturation value,

$$T = T_{sat\ @\ 200\text{ kPa}} = -10.09^\circ\text{C} = 263\text{ K}$$

Then,

$$\Delta S_{\text{refrigerant}} = \frac{Q_{\text{refrigerant,in}}}{T_{\text{refrigerant}}} = \frac{120\text{ kJ}}{263\text{ K}} = \mathbf{0.456\ kJ/K}$$

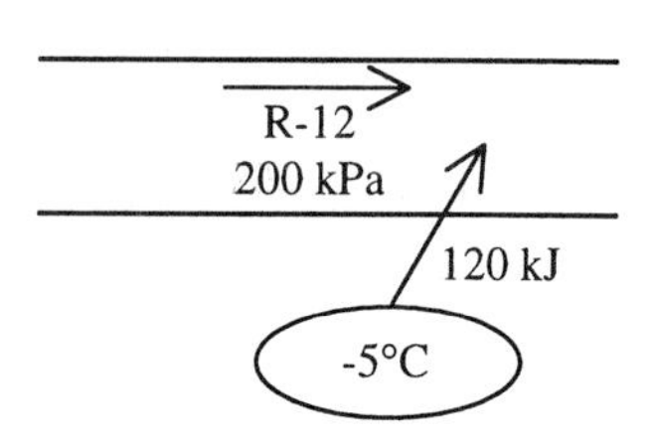

(*b*) Similarly,

$$\Delta S_{\text{space}} = -\frac{Q_{\text{space,out}}}{T_{\text{space}}} = -\frac{120\text{ kJ}}{268\text{ K}} = \mathbf{-0.448\ kJ/K}$$

(*c*) The total entropy change of the process is

$$S_{gen} = S_{\text{total}} = \Delta S_{\text{refrigerant}} + \Delta S_{\text{space}} = 0.456 - 0.448 = \mathbf{0.008\ kJ/K}$$

Entropy Changes of Pure Substances

6-30C Yes, because an internally reversible, adiabatic process involves no irreversibilities or heat transfer.

6-31 The radiator of a steam heating system is initially filled with superheated steam. The valves are closed, and steam is allowed to cool until the temperature drops to a specified value by transferring heat to the room. The entropy change of the steam during this process is to be determined. √

Analysis From the steam tables,

$$\left.\begin{array}{l} P_1 = 200\text{kPa} \\ T_1 = 200^\circ\text{C} \end{array}\right\} \begin{array}{l} v_1 = 1.0803\text{m}^3/\text{kg} \\ s_1 = 7.5066\text{kJ/kg}\cdot\text{K} \end{array}$$

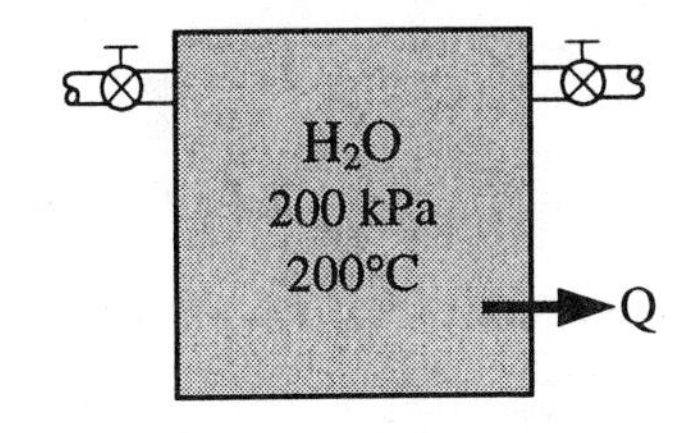

$$\left.\begin{array}{l} T_1 = 80^\circ\text{C} \\ v_2 = v_1 \end{array}\right\} x_2 = \frac{v_2 - v_f}{v_{fg}} = \frac{1.0803 - 0.001029}{3.407 - 0.001029} = 0.317$$

$$s_2 = s_f + x_2 s_{fg} = 1.0753 + (0.317)(6.5369)$$
$$= 3.1475\text{kJ/kg}\cdot\text{K}$$

The mass of the steam is

$$m = \frac{V}{v_1} = \frac{0.020\text{m}^3}{1.0803\text{m}^3/\text{kg}} = 0.0185\text{kg}$$

Then the entropy change of the steam during this process becomes

$$\Delta S = m(s_2 - s_1) = (0.0185\text{ kg})(3.1475 - 7.5066)\text{ kJ/kg}\cdot\text{K} = \mathbf{-0.0806\ kJ/K}$$

6-32 A rigid tank is initially filled with a saturated mixture of R-134a. Heat is transferred to the tank from a source until the pressure inside rises to a specified value. The entropy change of the refrigerant, entropy change of the source, and the total entropy change for this process are to be determined. √

Assumptions **1** The tank is stationary and thus the kinetic and potential energy changes are zero. **2** There are no work interactions.

Analysis (*a*) From the refrigerant tables (Tables A-11 through A-13),

$$\left.\begin{array}{l} P_1 = 200\text{kPa} \\ x_1 = 0.4 \end{array}\right\} \begin{array}{l} u_1 = u_f + x_1 u_{fg} = 36.69 + (0.4)(221.43 - 36.69) = 110.59\text{kJ/kg} \\ s_1 = s_f + x_1 s_{fg} = 0.1481 + (0.4)(0.9253 - 0.1481) = 0.4590\text{kJ/kg}\cdot\text{K} \\ v_1 = v_f + x_1 v_{fg} = 0.0007532 + (0.4)(0.0993 - 0.0007532) = 0.04017\text{m}^3\text{/kg} \end{array}$$

$$\left.\begin{array}{l} P_2 = 400\text{kPa} \\ v_2 = v_1 \end{array}\right\} \begin{array}{l} x_2 = \dfrac{v_2 - v_f}{v_{fg}} = \dfrac{0.04017 - 0.0007904}{0.0509 - 0.0007904} = 0.7859 \\ u_2 = u_f + x_2 u_{fg} = 61.69 + (0.7859)(231.97 - 61.69) = 195.51\text{kJ/kg} \\ s_2 = s_f + x_2 s_{fg} = 0.2399 + (0.7859)(0.9145 - 0.2399) = 0.7701\text{kJ/kg}\cdot\text{K} \end{array}$$

The mass of the refrigerant is

$$m = \frac{V}{v_1} = \frac{0.5\text{m}^3}{0.04017\text{m}^3\text{/kg}} = 12.45\text{ kg}$$

Then the entropy change of the refrigerant becomes

$$\Delta S_{system} = m(s_2 - s_1) = (12.45\text{kg})(0.7701 - 0.4590)\text{kJ/kg}\cdot\text{K} = \mathbf{3.873\ \ kJ/K}$$

(*b*) We take the tank as the system. This is a closed system since no mass enters or leaves. Noting that the volume of the system is constant and thus there is no boundary work, the energy balance for this stationary closed system can be expressed as

$$\underbrace{E_{in} - E_{out}}_{\substack{\text{Net energy transfer} \\ \text{by heat, work, and mass}}} = \underbrace{\Delta E_{\text{system}}}_{\substack{\text{Change in internal, kinetic,} \\ \text{potential, etc. energies}}}$$

$$Q_{in} = \Delta U = m(u_2 - u_1)$$

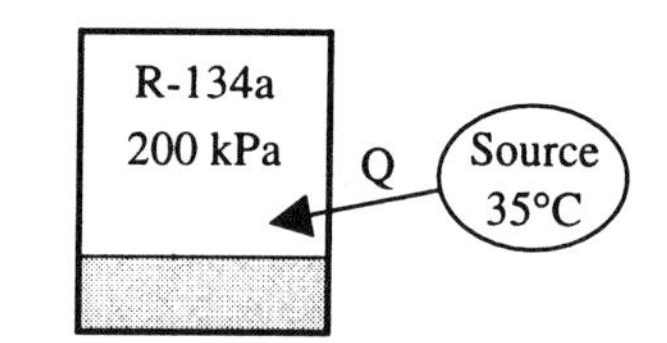

Substituting,

$$Q_{in} = m(u_2 - u_1) = (12.45\text{kg})(195.51 - 110.59) = 1057\text{ kJ}$$

The heat transfer for the source is equal in magnitude but opposite in direction. Therefore, $Q_{\text{source, out}} = -\,Q_{\text{tank, in}} = -\ 1057$ kJ, and

$$\Delta S_{\text{source}} = -\frac{Q_{\text{source,out}}}{T_{\text{source}}} = -\frac{1057\text{ kJ}}{308\text{ K}} = \mathbf{-3.432\ \ kJ/K}$$

(*c*) The total entropy change for this process is

$$\Delta S_{\text{total}} = \Delta S_{\text{system}} + \Delta S_{\text{source}} = 3.873 + (-3.432) = \mathbf{0.441\ \ kJ/K}$$

6-33 EES solution of this (and other comprehensive problems designated with the *computer icon*) is available to instructors at the *Instructor Manual* section of the *Online Learning Center* (OLC) at www.mhhe.com/cengel-boles. See the Preface for access information.

6-34 An insulated rigid tank contains a saturated liquid-vapor mixture of water at a specified pressure. An electric heater inside is turned on and kept on until all the liquid vaporized. The entropy change of the water during this process is to be determined.

Analysis From the steam tables (Tables A-4 through A-6)

$$\left.\begin{array}{l} P_1 = 100\text{kPa} \\ x_1 = 0.25 \end{array}\right\} \begin{array}{l} v_1 = v_f + x_1 v_{fg} = 0.001 + (0.25)(1.694 - 0.001) = 0.4243\text{m}^3/\text{kg} \\ s_1 = s_f + x_1 s_{fg} = 1.3026 + (0.25)(6.0568) = 2.8168\text{kJ/kg}\cdot\text{K} \end{array}$$

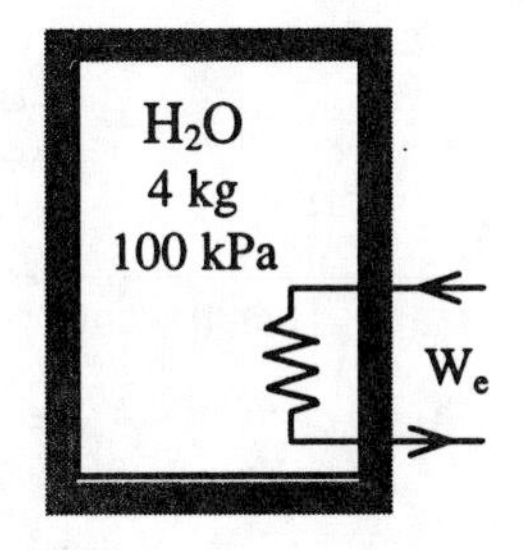

$$\left.\begin{array}{l} v_2 = v_1 \\ sat.vapor \end{array}\right\} s_2 = 6.8649\text{kJ/kg}\cdot\text{K}$$

Then the entropy change of the steam becomes

$$\Delta S = m(s_2 - s_1) = (4\text{ kg})(6.8649 - 2.8168)\text{ kJ/kg}\cdot\text{K} = \mathbf{16.19\ \ kJ/K}$$

6-35 [*Also solved by EES on enclosed CD*] A rigid tank is divided into two equal parts by a partition. One part is filled with compressed liquid water while the other side is evacuated. The partition is removed and water expands into the entire tank. The entropy change of the water during this process is to be determined.

Analysis The properties of the water are

$$\left.\begin{array}{l} P_1 = 300\text{kPa} \\ T_1 = 60^\circ\text{C} \end{array}\right\} \begin{array}{l} v_1 \cong v_{f@60^\circ\text{C}} = 0.001017\text{m}^3/\text{kg} \\ s_1 = s_{f@60^\circ\text{C}} = 0.8312\text{kJ/kg}\cdot\text{K} \end{array}$$

1.5 kg compressed liquid 300 kPa 60°C	Vacuum

Noting that $v_2 = 2v_1 = (2)(0.001017) = 0.002034\text{m}^3/\text{kg}$

$$\left.\begin{array}{l} P_2 = 15\text{kPa} \\ v_2 = 0.002034 \end{array}\right\} \begin{array}{l} x_2 = \dfrac{v_2 - v_f}{v_{fg}} = \dfrac{0.002034 - 0.001014}{10.02 - 0.001014} = 0.0001018 \\ s_2 = s_f + x_2 s_{fg} = 0.7549 + (0.0001018)(7.2536) = 0.7556\text{kJ/kg}\cdot\text{K} \end{array}$$

Then the entropy change of the water becomes

$$\Delta S = m(s_2 - s_1) = (1.5\text{kg})(0.7556 - 0.8312)\text{kJ/kg}\cdot\text{K} = \mathbf{-0.1134kJ/K}$$

6-36 EES solution of this (and other comprehensive problems designated with the *computer icon*) is available to instructors at the *Instructor Manual* section of the *Online Learning Center* (OLC) at www.mhhe.com/cengel-boles. See the Preface for access information.

6-37E A cylinder is initially filled with R-134a at a specified state. The refrigerant is cooled and condensed at constant pressure. The entropy change of refrigerant during this process is to be determined

Analysis From the refrigerant tables,

$$\left.\begin{array}{l} P_1 = 120 \text{ psia} \\ T_1 = 120^\circ\text{F} \end{array}\right\} s_1 = 0.2301 \text{ Btu/lbm}\cdot\text{R}$$

$$\left.\begin{array}{l} T_2 = 90^\circ\text{F} \\ P_2 = 120 \text{ psia} \end{array}\right\} s_2 \cong s_{f\,@\,90^\circ\text{F}} = 0.0836 \text{ Btu/lbm}\cdot\text{R}$$

Then the entropy change of the refrigerant becomes

$$\Delta S = m(s_2 - s_1) = (3 \text{ lbm})(0.0836 - 0.2301)\text{Btu/lbm}\cdot\text{R} = \mathbf{-0.4395\ \ Btu/R}$$

6-38 An insulated cylinder is initially filled with saturated liquid water at a specified pressure. The water is heated electrically at constant pressure. The entropy change of the water during this process is to be determined.

Assumptions **1** The kinetic and potential energy changes are negligible. **2** The cylinder is well-insulated and thus heat transfer is negligible. **3** The thermal energy stored in the cylinder itself is negligible. **4** The compression or expansion process is quasi-equilibrium.

Analysis From the steam tables,

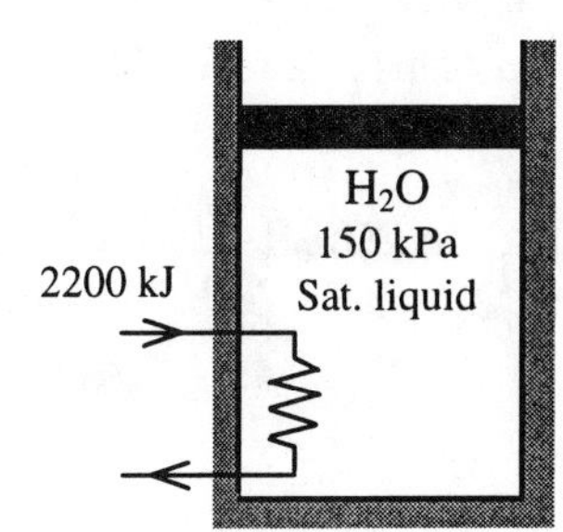

$$\left.\begin{array}{l} P_1 = 150\text{kPa} \\ \text{sat.liquid} \end{array}\right\} \begin{array}{l} v_1 = v_{f@150\text{kPa}} = 0.001053\text{m}^3/\text{kg} \\ h_1 = h_{f@150\text{kPa}} = 467.11\text{kJ/kg} \\ s_1 = s_{f@150\text{kPa}} = 1.4336\text{kJ/kg}\cdot\text{K} \end{array}$$

Also,

$$m = \frac{V}{v_1} = \frac{0.005\text{ m}^3}{0.001053\text{ m}^3/\text{kg}} = 4.748\text{ kg}$$

We take the contents of the cylinder as the system. This is a closed system since no mass enters or leaves. The energy balance for this stationary closed system can be expressed as

$$\underbrace{E_{in} - E_{out}}_{\substack{\text{Net energy transfer} \\ \text{by heat, work, and mass}}} = \underbrace{\Delta E_{\text{system}}}_{\substack{\text{Change in internal, kinetic,} \\ \text{potential, etc. energies}}}$$

$$W_{e,in} - W_{b,out} = \Delta U$$

$$W_{e,in} = m(h_2 - h_1)$$

since $\Delta U + W_b = \Delta H$ during a constant pressure quasi-equilibrium process. Solving for h_2,

$$h_2 = h_1 + \frac{W_{e,in}}{m} = 467.11 + \frac{2200\text{ kJ}}{4.748\text{ kg}} = 930.5\text{ kJ/kg}$$

Thus,

$$\left.\begin{array}{l} P_2 = 150\text{kPa} \\ h_2 = 930.5\text{kJ/kg} \end{array}\right\} \begin{array}{l} x_2 = \dfrac{h_2 - h_f}{h_{fg}} = \dfrac{930.5 - 467.11}{2226.5} = 0.208 \\ s_2 = s_f + x_2 s_{fg} = 1.4336 + (0.208)(5.7897) = 2.638\text{kJ/kg}\cdot\text{K} \end{array}$$

Then the entropy change of the water becomes

$$\Delta S = m(s_2 - s_1) = (4.748\text{kg})(2.638 - 1.4336)\text{kJ/kg}\cdot\text{K} = \mathbf{5.72\ \ kJ/K}$$

6-39 An insulated cylinder is initially filled with saturated R-134a vapor at a specified pressure. The refrigerant expands in a reversible manner until the pressure drops to a specified value. The final temperature in the cylinder and the work done by the refrigerant are to be determined.

Assumptions **1** The kinetic and potential energy changes are negligible. **2** The cylinder is well-insulated and thus heat transfer is negligible. **3** The thermal energy stored in the cylinder itself is negligible. **4** The process is stated to be reversible.

Analysis (*a*) This is a reversible adiabatic (i.e., isentropic) process, and thus $s_2 = s_1$. From the refrigerant tables,

$$\left.\begin{array}{l} P_1 = 0.8\text{MPa} \\ sat.vapor \end{array}\right\} \begin{array}{l} v_1 = v_{g@0.8\text{MPa}} = 0.0255\text{m}^3/\text{kg} \\ u_1 = u_{g@0.8\text{MPa}} = 243.78\text{kJ/kg} \\ s_1 = s_{g@0.8\text{MPa}} = 0.9066\text{kJ/kg}\cdot\text{K} \end{array}$$

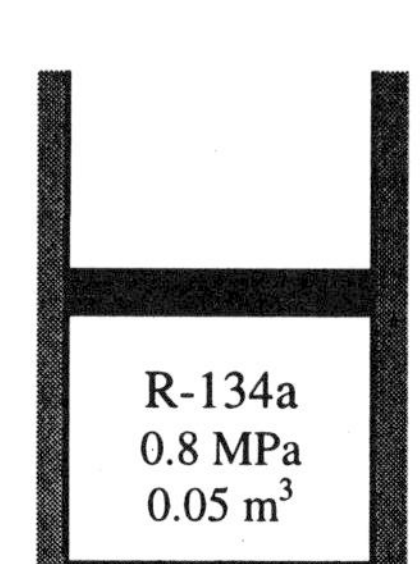

Also,

$$m = \frac{V}{v_1} = \frac{0.05\ \text{m}^3}{0.0255\ \text{m}^3/\text{kg}} = 1.961\ \text{kg}$$

and

$$\left.\begin{array}{l} P_2 = 0.4\text{MPa} \\ s_2 = s_1 \end{array}\right\} \begin{array}{l} x_2 = \dfrac{s_2 - s_f}{s_{fg}} = \dfrac{0.9066 - 0.2399}{0.9145 - 0.2399} = 0.988 \\ u_2 = u_f + x_2 u_{fg} = 61.69 + (0.988)(231.97 - 61.69) = 229.93\text{kJ/kg} \end{array}$$

$$T_2 = T_{sat@0.4\text{MPa}} = \mathbf{8.93^\circ C}$$

(*b*) We take the contents of the cylinder as the system. This is a closed system since no mass enters or leaves. The energy balance for this adiabatic closed system can be expressed as

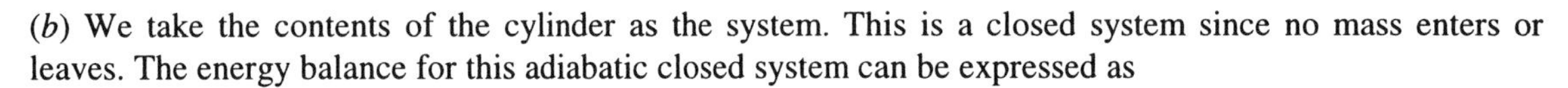

$$\underbrace{E_{in} - E_{out}}_{\substack{\text{Net energy transfer} \\ \text{by heat, work, and mass}}} = \underbrace{\Delta E_{\text{system}}}_{\substack{\text{Change in internal, kinetic,} \\ \text{potential, etc. energies}}}$$

$$-W_{b,out} = \Delta U$$

$$W_{b,out} = m(u_1 - u_2)$$

Substituting, the work done during this isentropic process is determined to be

$$W_{b,out} = m(u_1 - u_2) = (1.961\ \text{kg})(243.78 - 229.93)\ \text{kJ/kg} = \mathbf{27.16\ kJ}$$

6-40 EES solution of this (and other comprehensive problems designated with the *computer icon*) is available to instructors at the *Instructor Manual* section of the *Online Learning Center* (OLC) at www.mhhe.com/cengel-boles. See the Preface for access information.

6-41 Saturated Refrigerant-134a vapor at 140 kPa is compressed steadily by an adiabatic compressor. The minimum power input to the compressor is to be determined.

Assumptions **1** This is a steady-flow process since there is no change with time. **2** Kinetic and potential energy changes are negligible. **3** The device is adiabatic and thus heat transfer is negligible.

Analysis The power input to an adiabatic compressor will be a minimum when the compression process is reversible. For the reversible adiabatic process we have $s_2 = s_1$. From the refrigerant tables,

$$\left.\begin{array}{l} P_1 = 140 \text{ kPa} \\ sat.vapor \end{array}\right\} \begin{array}{l} v_1 = v_{g\,@140\text{kPa}} = 0.1395 \text{ m}^3\text{/kg} \\ h_1 = h_{g\,@140\text{kPa}} = 236.04 \text{ kJ/kg} \\ s_1 = s_{g\,@140kPa} = 0.9322 \text{ kJ/kg}\cdot\text{K} \end{array}$$

$$\left.\begin{array}{l} P_2 = 700 \text{ kPa} \\ s_2 = s_1 \end{array}\right\} h_2 = 269.23 \text{ kJ/kg}$$

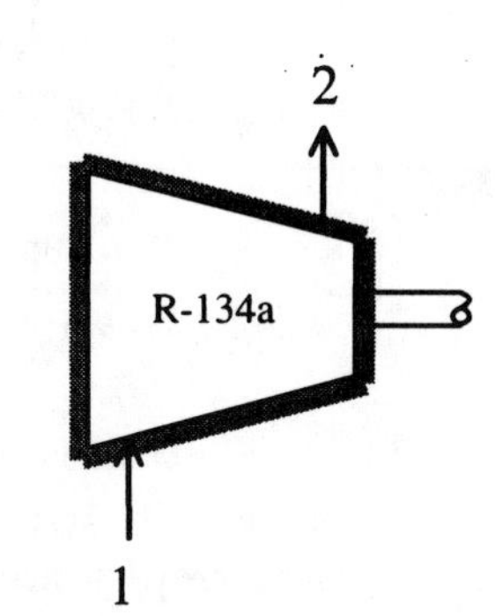

Also,

$$\dot{m} = \frac{\dot{V}_1}{v_1} = \frac{2 \text{ m}^3\text{/min}}{0.1395 \text{ m}^3\text{/kg}} = 14.34 \text{ kg/min} = 0.239 \text{ kg/s}$$

There is only one inlet and one exit, and thus $\dot{m}_1 = \dot{m}_2 = \dot{m}$. We take the compressor as the system, which is a control volume since mass crosses the boundary. The energy balance for this steady-flow system can be expressed in the rate form as

$$\underbrace{\dot{E}_{in} - \dot{E}_{out}}_{\text{Rate of net energy transfer by heat, work, and mass}} = \underbrace{\Delta \dot{E}_{\text{system}}^{\nearrow 0 \text{ (steady)}}}_{\text{Rate of change in internal, kinetic, potential, etc. energies}} = 0$$

$$\dot{E}_{in} = \dot{E}_{out}$$

$$\dot{W}_{\text{in}} + \dot{m}h_1 = \dot{m}h_2 \quad (\text{since } \dot{Q} \cong \Delta ke \cong \Delta pe \cong 0)$$

$$\dot{W}_{\text{in}} = \dot{m}(h_2 - h_1)$$

Substituting, the minimum power supplied to the compressor is determined to be

$$\dot{W}_{in} = (0.239 \text{kg/s})(269.23 - 236.04)\text{kJ/kg} = \mathbf{7.93 \ kW}$$

6-42E Steam expands in an adiabatic turbine. The maximum amount of work that can be done by the turbine is to be determined.

Assumptions **1** This is a steady-flow process since there is no change with time. **2** Kinetic and potential energy changes are negligible. **3** The device is adiabatic and thus heat transfer is negligible.

Analysis The work output of an adiabatic turbine is maximum when the expansion process is reversible. For the reversible adiabatic process we have $s_2 = s_1$. From the steam tables,

$$\left.\begin{array}{l} P_1 = 800\text{psia} \\ T_1 = 900^\circ\text{F} \end{array}\right\} \begin{array}{l} h_1 = 1455.6\text{Btu/lbm} \\ s_1 = 1.6408\text{Btu/lbm}\cdot\text{R} \end{array}$$

$$\left.\begin{array}{l} P_2 = 40\text{psia} \\ s_2 = s_1 \end{array}\right\} \begin{array}{l} x_2 = \dfrac{s_2 - s_f}{s_{fg}} = \dfrac{1.6408 - 0.39214}{1.2845} = 0.972 \\ h_2 = h_f + x_2 h_{fg} = 236.16 + (0.972)(933.8) = 1143.8\text{Btu/lbm} \end{array}$$

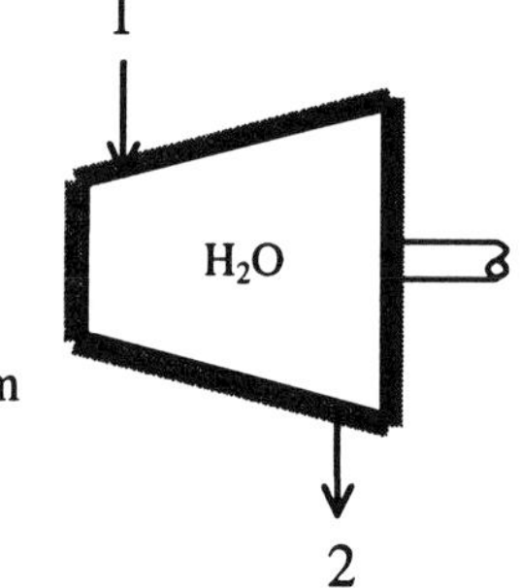

There is only one inlet and one exit, and thus $\dot{m}_1 = \dot{m}_2 = \dot{m}$. We take the turbine as the system, which is a control volume since mass crosses the boundary. The energy balance for this steady-flow system can be expressed in the rate form as

$$\underbrace{\dot{E}_{in} - \dot{E}_{out}}_{\text{Rate of net energy transfer by heat, work, and mass}} = \underbrace{\Delta \dot{E}_{\text{system}}^{\nearrow 0 \text{ (steady)}}}_{\text{Rate of change in internal, kinetic, potential, etc. energies}} = 0$$

$$\dot{E}_{in} = \dot{E}_{out}$$

$$\dot{m}h_1 = \dot{W}_{\text{out}} + \dot{m}h_2$$

$$\dot{W}_{\text{out}} = \dot{m}(h_1 - h_2)$$

Dividing by mass flow rate and substituting,

$$w_{out} = h_1 - h_2 = 1455.6 - 1143.8 = \mathbf{311.8\ Btu/lbm}$$

6-43 EES solution of this (and other comprehensive problems designated with the *computer icon*) is available to instructors at the *Instructor Manual* section of the *Online Learning Center* (OLC) at www.mhhe.com/cengel-boles. See the Preface for access information.

6-44 An insulated cylinder is initially filled with superheated steam at a specified state. The steam is compressed in a reversible manner until the pressure drops to a specified value. The work input during this process is to be determined. √

Assumptions **1** The kinetic and potential energy changes are negligible. **2** The cylinder is well-insulated and thus heat transfer is negligible. **3** The thermal energy stored in the cylinder itself is negligible. **4** The process is stated to be reversible.

Analysis This is a reversible adiabatic (i.e., isentropic) process, and thus $s_2 = s_1$. From the steam tables,

$$\left.\begin{array}{l} P_1 = 300 \text{ kPa} \\ T_1 = 150^\circ\text{C} \end{array}\right\} \begin{array}{l} v_1 = 0.6339 \text{ m}^3/\text{kg} \\ u_1 = 2570.8 \text{ kJ/kg} \\ s_1 = 7.0778 \text{ kJ/kg}\cdot\text{K} \end{array}$$

$$\left.\begin{array}{l} P_2 = 1 \text{ MPa} \\ s_2 = s_1 \end{array}\right\} u_2 = 2774.2 \text{ kJ/kg}$$

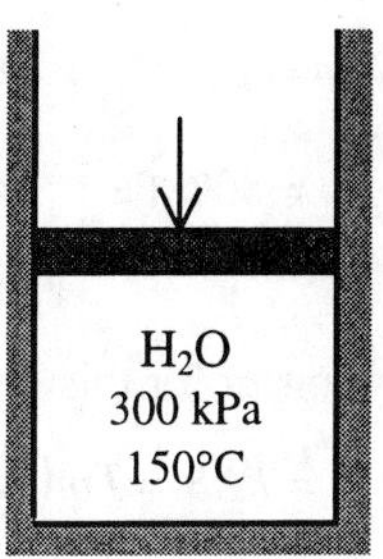

Also,

$$m = \frac{V}{v_1} = \frac{0.05 \text{ m}^3}{0.6339 \text{ m}^3/\text{kg}} = 0.0789 \text{ kg}$$

We take the contents of the cylinder as the system. This is a closed system since no mass enters or leaves. The energy balance for this adiabatic closed system can be expressed as

$$\underbrace{E_{in} - E_{out}}_{\substack{\text{Net energy transfer} \\ \text{by heat, work, and mass}}} = \underbrace{\Delta E_{\text{system}}}_{\substack{\text{Change in internal, kinetic,} \\ \text{potential, etc. energies}}}$$

$$W_{b,in} = \Delta U = m(u_2 - u_1)$$

Substituting, the work input during this adiabatic process is determined to be

$$W_{b,in} = m(u_2 - u_1) = (0.0789\text{kg})(2774.2 - 2570.8)\text{kJ/kg} = \mathbf{16.05 \ kJ}$$

6-45 EES solution of this (and other comprehensive problems designated with the *computer icon*) is available to instructors at the *Instructor Manual* section of the *Online Learning Center* (OLC) at www.mhhe.com/cengel-boles. See the Preface for access information.

6-46 A cylinder is initially filled with saturated water vapor at a specified temperature. Heat is transferred to the steam, and it expands in a reversible and isothermal manner until the pressure drops to a specified value. The heat transfer and the work output for this process are to be determined.

Assumptions **1** The kinetic and potential energy changes are negligible. **2** The cylinder is well-insulated and thus heat transfer is negligible. **3** The thermal energy stored in the cylinder itself is negligible. **4** The process is stated to be reversible and isothermal.

Analysis From the steam tables,

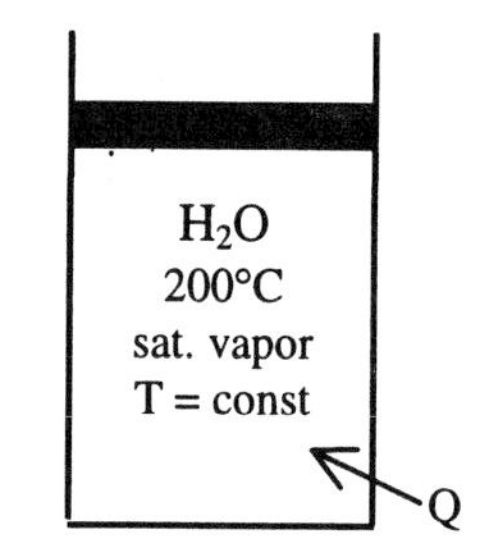

$$\left.\begin{array}{l} T_1 = 200^\circ\text{C} \\ sat.vapor \end{array}\right\} \begin{array}{l} u_1 = u_{g@200^\circ\text{C}} = 2595.3\text{kJ/kg} \\ s_1 = s_{g@200^\circ\text{C}} = 6.4323\text{kJ/kg}\cdot\text{K} \end{array}$$

$$\left.\begin{array}{l} P_2 = 800\text{kPa} \\ T_2 = T_1 \end{array}\right\} \begin{array}{l} u_2 = 2630.6\text{kJ/kg} \\ s_2 = 6.8158\text{kJ/kg}\cdot\text{K} \end{array}$$

The heat transfer for this reversible isothermal process can be determined from

$$Q = T\Delta S = Tm(s_2 - s_1) = (473\text{ K})(1.2\text{ kg})(6.8158 - 6.4323)\text{kJ/kg}\cdot\text{K} = \mathbf{217.7\ \ kJ}$$

We take the contents of the cylinder as the system. This is a closed system since no mass enters or leaves. The energy balance for this closed system can be expressed as

$$\underbrace{E_{in} - E_{out}}_{\substack{\text{Net energy transfer} \\ \text{by heat, work, and mass}}} = \underbrace{\Delta E_{\text{system}}}_{\substack{\text{Change in internal, kinetic,} \\ \text{potential, etc. energies}}}$$

$$Q_{in} - W_{b,out} = \Delta U = m(u_2 - u_1)$$

$$W_{b,out} = Q_{in} - m(u_2 - u_1)$$

Substituting, the work done during this process is determined to be

$$W_{b,out} = 217.7\text{ kJ} - (1.2\text{ kg})(2630.6 - 2595.3)\text{kJ/kg} = \mathbf{175.3\ \ kJ}$$

6-47 EES solution of this (and other comprehensive problems designated with the *computer icon*) is available to instructors at the *Instructor Manual* section of the *Online Learning Center* (OLC) at www.mhhe.com/cengel-boles. See the Preface for access information.

Entropy Change of Incompressible Substances

6-48C No, because entropy is not a conserved property.

6-49 A hot copper block is dropped into water in an insulated tank. The final equilibrium temperature of the tank and the total entropy change are to be determined.

Assumptions **1** Both the water and the copper block are incompressible substances with constant specific heats at room temperature. **2** The system is stationary and thus the kinetic and potential energies are negligible. **3** The tank is well-insulated and thus there is no heat transfer.

Properties The density and specific heat of water at 25°C are ρ = 997 kg/m³ and C_p = 4.18 kJ/kg.°C. The specific heat of copper at 27°C is C_p = 0.386 kJ/kg.°C (Table A-3).

Analysis We take the entire contents of the tank, water + copper block, as the *system*. This is a *closed system* since no mass crosses the system boundary during the process. The energy balance for this system can be expressed as

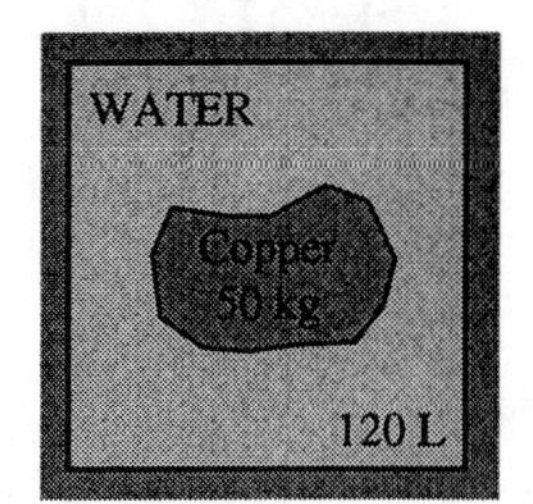

$$\underbrace{E_{in} - E_{out}}_{\substack{\text{Net energy transfer}\\ \text{by heat, work, and mass}}} = \underbrace{\Delta E_{\text{system}}}_{\substack{\text{Change in internal, kinetic,}\\ \text{potential, etc. energies}}}$$

$$0 = \Delta U$$

or,

$$\Delta U_{Cu} + \Delta U_{water} = 0$$

$$[mC(T_2 - T_1)]_{Cu} + [mC(T_2 - T_1)]_{water} = 0$$

where

$$m_{water} = \rho V = (997\text{kg/m}^3)(0.120\text{m}^3) = 120\text{kg}$$

Using specific heat values for copper and liquid water at room temperature and substituting,

$$(50\text{kg})(0.386\text{kJ/kg}\cdot^\circ\text{C})(T_2 - 80)^\circ\text{C} + (120\text{kg})(4.18\text{kJ/kg}\cdot^\circ\text{C})(T_2 - 25)^\circ\text{C} = 0$$

$$T_2 = \mathbf{27.0°C}$$

The entropy generated during this process is determined from

$$\Delta S_{\text{copper}} = mC_{ave} \ln\left(\frac{T_2}{T_1}\right) = (50\text{ kg})(0.386\text{ kJ/kg}\cdot\text{K})\ln\left(\frac{300.0\text{ K}}{353\text{ K}}\right) = -3.140\text{ kJ/K}$$

$$\Delta S_{\text{water}} = mC_{ave} \ln\left(\frac{T_2}{T_1}\right) = (120\text{ kg})(4.18\text{ kJ/kg}\cdot\text{K})\ln\left(\frac{300.0\text{ K}}{298\text{ K}}\right) = 3.355\text{ kJ/K}$$

Thus,

$$\Delta S_{total} = \Delta S_{copper} + \Delta S_{water} = -3.140 + 3.355 = \mathbf{0.215\ kJ/K}$$

6-50 A hot iron block is dropped into water in an insulated tank. The total entropy change during this process is to be determined.

Assumptions **1** Both the water and the iron block are incompressible substances with constant specific heats at room temperature. **2** The system is stationary and thus the kinetic and potential energies are negligible. **3** The tank is well-insulated and thus there is no heat transfer. **4** The water that evaporates, condenses back.

Properties The specific heat of water at 25°C is C_p = 4.18 kJ/kg.°C. The specific heat of iron at room temperature is C_p = 0.45 kJ/kg.°C (Table A-3).

Analysis We take the entire contents of the tank, water + iron block, as the *system*. This is a *closed system* since no mass crosses the system boundary during the process. The energy balance for this system can be expressed as

$$\underbrace{E_{in} - E_{out}}_{\substack{\text{Net energy transfer}\\ \text{by heat, work, and mass}}} = \underbrace{\Delta E_{\text{system}}}_{\substack{\text{Change in internal, kinetic,}\\ \text{potential, etc. energies}}}$$

$$0 = \Delta U$$

or,

$$\Delta U_{iron} + \Delta U_{water} = 0$$

$$[mC(T_2 - T_1)]_{iron} + [mC(T_2 - T_1)]_{water} = 0$$

Substituting,

$$(12 \text{ kg})(0.45 \text{ kJ/kg}\cdot\text{K})(T_2 - 350^\circ\text{C}) + (100 \text{ kg})(4.18 \text{ kJ/kg}\cdot\text{K})(T_2 - 22^\circ\text{C}) = 0$$

$$T_2 = \mathbf{26.2^\circ C}$$

The entropy generated during this process is determined from

$$\Delta S_{iron} = mC_{ave} \ln\left(\frac{T_2}{T_1}\right) = (12 \text{ kg})(0.45 \text{ kJ/kg}\cdot\text{K})\ln\left(\frac{299.2 \text{ K}}{623 \text{ K}}\right) = -3.96 \text{ kJ/K}$$

$$\Delta S_{water} = mC_{ave} \ln\left(\frac{T_2}{T_1}\right) = (100 \text{ kg})(4.18 \text{ kJ/kg}\cdot\text{K})\ln\left(\frac{299.2 \text{ K}}{295 \text{ K}}\right) = 5.91 \text{ kJ/K}$$

Thus,

$$S_{gen} = \Delta S_{total} = \Delta S_{iron} + \Delta S_{water} = -3.96 + 5.91 = \mathbf{1.95 \ kJ/K}$$

Discussion The results can be improved somewhat by using specific heats at average temperature.

6-51 An aluminum block is brought into contact with an iron block in an insulated enclosure. The final equilibrium temperature and the total entropy change for this process are to be determined.

Assumptions **1** Both the aluminum and the iron block are incompressible substances with constant specific heats. **2** The system is stationary and thus the kinetic and potential energies are negligible. **3** The system is well-insulated and thus there is no heat transfer.

Properties The specific heat of aluminum at the anticipated average temperature of 450 K is C_p = 0.973 kJ/kg.°C. The specific heat of iron at room temperature (the only value available in the tables) is C_p = 0.45 kJ/kg.°C (Table A-3).

Analysis We take the iron+aluminum blocks as the system, which is a closed system. The energy balance for this system can be expressed as

$$\underbrace{E_{in} - E_{out}}_{\substack{\text{Net energy transfer}\\\text{by heat, work, and mass}}} = \underbrace{\Delta E_{\text{system}}}_{\substack{\text{Change in internal, kinetic,}\\\text{potential, etc. energies}}}$$

$$0 = \Delta U$$

Iron	Aluminum
20 kg	20 kg
100°C	200°C

or,

$$\Delta U_{\text{alum}} + \Delta U_{iron} = 0$$

$$[mC(T_2 - T_1)]_{\text{alum}} + [mC(T_2 - T_1)]_{iron} = 0$$

Substituting,

$$(20\text{kg})(0.45\text{kJ/kg}\cdot\text{K})(T_2 - 100^\circ\text{C}) + (20\text{kg})(0.973\text{kJ/kg}\cdot\text{K})(T_2 - 200^\circ\text{C}) = 0$$

$$T_2 = \mathbf{168.4^\circ C} = 441.4\text{K}$$

The total entropy change for this process is determined from

$$\Delta S_{iron} = mC_{ave}\ln\left(\frac{T_2}{T_1}\right) = (20\text{kg})(0.45\text{kJ/kg}\cdot\text{K})\ln\left(\frac{441.4\text{K}}{373\text{K}}\right) = 1.515\text{kJ/K}$$

$$\Delta S_{alum} = mC_{ave}\ln\left(\frac{T_2}{T_1}\right) = (20\text{kg})(0.973\text{kJ/kg}\cdot\text{K})\ln\left(\frac{441.4\text{K}}{473\text{K}}\right) = -1.346\text{kJ/K}$$

Thus,

$$\Delta S_{total} = \Delta S_{iron} + \Delta S_{alum} = 1515 - 1.346 = \mathbf{0.169\ kJ/K}$$

6-52 EES solution of this (and other comprehensive problems designated with the *computer icon*) is available to instructors at the *Instructor Manual* section of the *Online Learning Center* (OLC) at www.mhhe.com/cengel-boles. See the Preface for access information.

6-53 An iron block and a copper block are dropped into a large lake. The total amount of entropy change when both blocks cool to the lake temperature is to be determined.

Assumptions **1** Both the water and the iron block are incompressible substances with constant specific heats at room temperature. **2** Kinetic and potential energies are negligible.

Properties The specific heats of iron and copper at room temperature are $C_{p,\text{ iron}}$ = 0.45 kJ/kg.°C and $C_{p,\text{copper}}$ = 0.386 kJ/kg.°C (Table A-3).

Analysis The thermal-energy capacity of the lake is very large, and thus the temperatures of both the iron and the copper blocks will drop to the lake temperature (15°C) when the thermal equilibrium is established. Then the entropy changes of the blocks become

$$\Delta S_{iron} = mC_{ave}\ln\left(\frac{T_2}{T_1}\right) = (50\text{kg})(0.45\text{kJ/kg}\cdot\text{K})\ln\left(\frac{288\text{K}}{353\text{K}}\right) = -4.579\text{kJ/K}$$

$$\Delta S_{copper} = mC_{ave}\ln\left(\frac{T_2}{T_1}\right) = (20\text{kg})(0.386\text{kJ/kg}\cdot\text{K})\ln\left(\frac{288\text{K}}{353\text{K}}\right) = -1.571\text{kJ/K}$$

We take both the iron and the copper blocks, as the *system*. This is a *closed system* since no mass crosses the system boundary during the process. The energy balance for this system can be expressed as

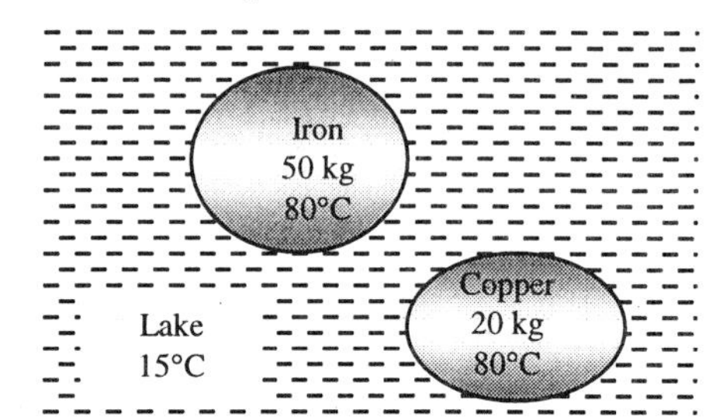

$$\underbrace{E_{in} - E_{out}}_{\substack{\text{Net energy transfer}\\\text{by heat, work, and mass}}} = \underbrace{\Delta E_{\text{system}}}_{\substack{\text{Change in internal, kinetic,}\\\text{potential, etc. energies}}}$$

$$-Q_{out} = \Delta U = \Delta U_{iron} + \Delta U_{copper}$$

or,

$$Q_{out} = [mC(T_1 - T_2)]_{iron} + [mC(T_1 - T_2)]_{copper} = 0$$

Substituting,

$$Q_{out} = (50\text{kg})(0.45\text{kJ/kg}\cdot\text{K})(353-288)\text{K} + (20\text{kg})(0.386\text{kJ/kg}\cdot\text{K})(353-288)\text{K}$$
$$= 1964\text{kJ}$$

Thus,

$$\Delta S_{lake} = \frac{Q_{lake,in}}{T_{lake}} = \frac{1964\text{kJ}}{288\text{K}} = 6.820\text{kJ/K}$$

Then the total entropy change for this process is

$$\Delta S_{total} = \Delta S_{iron} + \Delta S_{copper} + \Delta S_{lake} = -4.579 - 1.571 + 6.820 = \mathbf{0.670\ \ kJ/K}$$

Entropy Changes of Ideal Gases

6-54C For ideal gases, $C_p = C_v + R$ and

$$\frac{P_2 V_2}{T_2} = \frac{P_1 V_1}{T_1} \longrightarrow \frac{V_2}{V_1} = \frac{T_2 P_1}{T_1 P_2}$$

Thus,

$$\begin{aligned} s_2 - s_1 &= C_v \ln\left(\frac{T_2}{T_1}\right) + R\ln\left(\frac{V_2}{V_1}\right) \\ &= C_v \ln\left(\frac{T_2}{T_1}\right) + R\ln\left(\frac{T_2 P_1}{T_1 P_2}\right) \\ &= C_v \ln\left(\frac{T_2}{T_1}\right) + R\ln\left(\frac{T_2}{T_1}\right) - R\ln\left(\frac{P_2}{P_1}\right) \\ &= C_p \ln\left(\frac{T_2}{T_1}\right) - R\ln\left(\frac{P_2}{P_1}\right) \end{aligned}$$

6-55C For an ideal gas, $dh = C_p dT$ and $v = RT/P$. From the second Tds relation,

$$ds = \frac{dh}{T} - \frac{vdP}{T} = \frac{C_p dP}{T} - \frac{RT}{P}\frac{dP}{T} = C_p \frac{dT}{T} - R\frac{dP}{P}$$

Integrating,

$$s_2 - s_1 = C_p \ln\left(\frac{T_2}{T_1}\right) - R\ln\left(\frac{P_2}{P_1}\right)$$

since C_p is assumed to be constant.

6-56C No. The entropy of an ideal gas depends on the pressure as well as the temperature.

6-57C Setting $\Delta s = 0$ gives

$$C_p \ln\left(\frac{T_2}{T_1}\right) - R\ln\left(\frac{P_2}{P_1}\right) = 0 \longrightarrow \ln\left(\frac{T_2}{T_1}\right) = \frac{R}{C_p}\ln\left(\frac{P_2}{P_1}\right) \longrightarrow \frac{T_2}{T_1} = \left(\frac{P_2}{P_1}\right)^{R/C_p}$$

But

$$\frac{R}{C_p} = \frac{C_p - C_v}{C_p} = 1 - \frac{1}{k} = \frac{k-1}{k} \text{ since } k = C_p / C_v. \text{ Thus, } \frac{T_2}{T_1} = \left(\frac{P_2}{P_1}\right)^{(k-1)/k}$$

6-58C The P_r and v_r are called relative pressure and relative specific volume, respectively. They are derived for isentropic processes of ideal gases, and thus their use is limited to isentropic processes only.

6-59C The entropy of a gas *can* change during an isothermal process since entropy of an ideal gas depends on the pressure as well as the temperature.

6-60C The entropy change relations of an ideal gas simplify to

$\Delta s = C_p \ln(T_2/T_1)$ for a constant pressure process

and $\Delta s = C_v \ln(T_2/T_1)$ for a constant volume process.

Noting that $C_p > C_v$, the entropy change will be larger for a constant pressure process.

6-61 Oxygen gas is compressed from a specified initial state to a specified final state. The entropy change of oxygen during this process is to be determined for the case of constant specific heats.

Assumptions At specified conditions, oxygen can be treated as an ideal gas.

Properties The gas constant and molar mass of oxygen are R = 0.2598 kJ/kg.K and M = 32 kg/kmol (Table A-1).

Analysis The constant volume specific heat of oxygen at the average temperature is (Table A-2)

$$T_{ave} = \frac{298+560}{2} = 429\text{ K} \longrightarrow C_{v,ave} = 0.690\text{ kJ/kg}\cdot\text{K}$$

Thus,

$$s_2 - s_1 = C_{v,ave}\ln\frac{T_2}{T_1} + R\ln\frac{V_2}{V_1}$$

$$= (0.690\text{kJ/kg}\cdot\text{K})\ln\frac{560\text{K}}{298\text{K}} + (0.2598\text{kJ/kg}\cdot\text{K})\ln\frac{0.1\text{m}^3/\text{kg}}{0.8\text{m}^3/\text{kg}}$$

$$= \mathbf{-0.105 kJ/kg\cdot K}$$

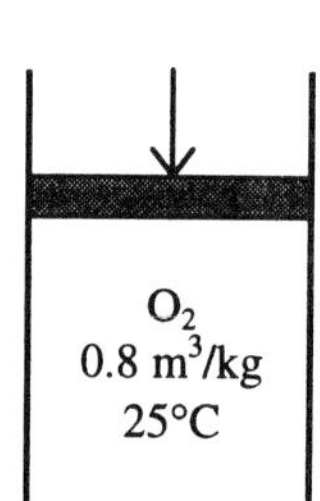

6-62 An insulated tank contains CO_2 gas at a specified pressure and volume. A paddle-wheel in the tank stirs the gas, and the pressure and temperature of CO_2 rises. The entropy change of CO_2 during this process is to be determined using constant specific heats.

Assumptions At specified conditions, CO_2 can be treated as an ideal gas with constant specific heats at room temperature.

Properties The specific heat of CO_2 is C_v = 0.657 kJ/kg.K (Table A-2).

Analysis Using the ideal gas relation, the entropy change is determined to be

$$\frac{P_2V}{T_2} = \frac{P_1V}{T_1} \longrightarrow \frac{T_2}{T_1} = \frac{P_2}{P_1} = \frac{120kPa}{100kPa} = 1.2$$

Thus,

$$\Delta S = m(s_2 - s_1) = m\left(C_{v,ave}\ln\frac{T_2}{T_1} + R\ln\frac{V_2}{V_1}^{\nearrow 0}\right) = mC_{v,ave}\ln\frac{T_2}{T_1}$$

$$= (2.7\text{ kg})(0.657\text{kJ/kg}\cdot\text{K})\ln(1.2)$$

$$= \mathbf{0.323\ kJ/K}$$

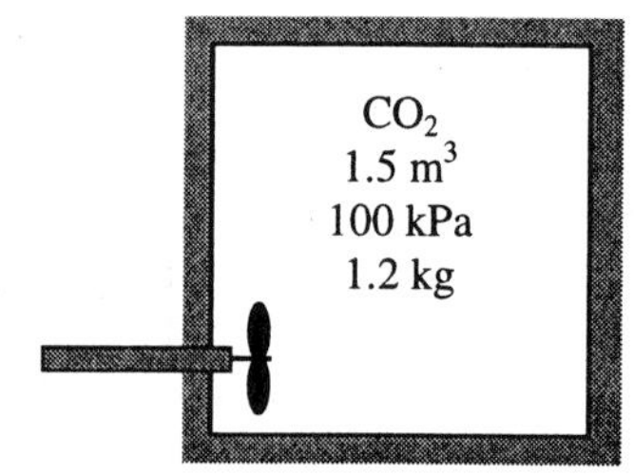

6-63 An insulated cylinder initially contains air at a specified state. A resistance heater inside the cylinder is turned on, and air is heated for 15 min at constant pressure. The entropy change of air during this process is to be determined for the cases of constant and variable specific heats. √

Assumptions At specified conditions, air can be treated as an ideal gas.

Properties The gas constant of air is $R = 0.287$ kJ/kg.K (Table A-1).

Analysis The mass of the air and the electrical work done during this process are

$$m = \frac{P_1 V_1}{RT_1} = \frac{(120\text{kPa})(0.3\text{m}^3)}{(0.287\text{kPa}\cdot\text{m}^3/\text{kg}\cdot\text{K})(290\text{K})} = 0.4325\text{kg}$$

$$W_{e,in} = \dot{W}_{e,in}\Delta t = (0.2\text{kJ/s})(15\times 60\text{s}) = 180\text{kJ}$$

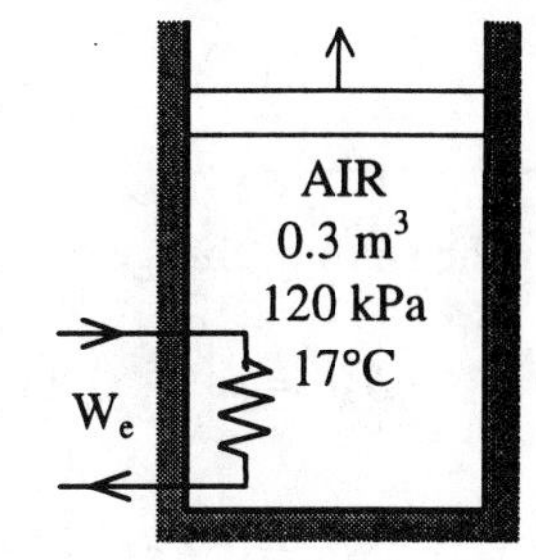

The energy balance for this stationary closed system can be expressed as

$$\underbrace{E_{in} - E_{out}}_{\substack{\text{Net energy transfer}\\ \text{by heat, work, and mass}}} = \underbrace{\Delta E_{\text{system}}}_{\substack{\text{Change in internal, kinetic,}\\ \text{potential, etc. energies}}}$$

$$W_{pw,in} - W_{b,out} = \Delta U$$

$$W_{pw,in} = m(h_2 - h_1) \cong C_p(T_2 - T_1)$$

since $\Delta U + W_b = \Delta H$ during a constant pressure quasi-equilibrium process.

(a) Using a constant C_p value at the anticipated average temperature of 450 K, the final temperature becomes

Thus, $$T_2 = T_1 + \frac{W_{e,in}}{mC_p} = 290\text{K} + \frac{180\text{kJ}}{(0.4325\text{kg})(1.02\text{kJ/kg}\cdot\text{K})} = 698\text{K}$$

Then the entropy change becomes

$$\Delta S_{sys} = m(s_2 - s_1) = m\left(C_{p,ave}\ln\frac{T_2}{T_1} - R\ln\frac{P_2}{P_1}^{\cancel{}0}\right) = mC_{p,ave}\ln\frac{T_2}{T_1}$$

$$= (0.4325\text{kg})(1.020\text{kJ/kg}\cdot\text{K})\ln\left(\frac{698\text{K}}{290\text{K}}\right)$$

$$= \mathbf{0.387kJ/K}$$

(b) Assuming variable specific heats,

$$W_{e,in} = m(h_2 - h_1) \longrightarrow h_2 = h_1 + \frac{W_{e,in}}{m} = 290.16\text{kJ/kg} + \frac{180\text{kJ}}{0.4325\text{kg}} = 706.34\text{kJ/kg}$$

From the air table (Table A-17, we read $s_2^\circ = 2.5628$ kJ/kg· K corresponding to this h_2 value. Then,

$$\Delta S_{sys} = m\left(s_2^\circ - s_1^\circ + R\ln\frac{P_2}{P_1}^{\cancel{}0}\right) = m(s_2^\circ - s_1^\circ)$$

$$= (0.4325\text{kg})(2.5628 - 1.66802)\text{kJ/kg}\cdot\text{K}$$

$$= \mathbf{0.387kJ/K}$$

6-64 A cylinder contains N_2 gas at a specified pressure and temperature. A gas is compressed polytropically until the volume is reduced by half. The entropy change of nitrogen during this process is to be determined.

Assumptions **1** At specified conditions, N_2 can be treated as an ideal gas. **2** Nitrogen has constant specific heats at room temperature.

Properties The gas constant of nitrogen is R = 0.297 kJ/kg.K (Table A-1). The constant volume specific heat of nitrogen at room temperature is C_v = 0.743 kJ/kg.K (Table A-2).

Analysis From the polytropic relation,

$$\frac{T_2}{T_1} = \left(\frac{v_1}{v_2}\right)^{n-1} \longrightarrow T_2 = T_1\left(\frac{v_1}{v_2}\right)^{n-1} = (300K)(2)^{1.3-1} = 369.3K$$

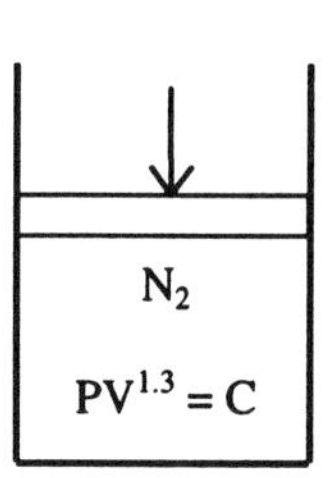

Then the entropy change of nitrogen becomes

$$\Delta S_{N_2} = m\left(C_{v,ave}\ln\frac{T_2}{T_1} + R\ln\frac{V_2}{V_1}\right)$$
$$= (1.2\text{ kg})\left((0.743\text{ kJ/kg}\cdot\text{K})\ln\frac{369.3\text{ K}}{300\text{ K}} + (0.297\text{ kJ/kg}\cdot\text{K})\ln(0.5)\right)$$
$$= \mathbf{-0.0617\ kJ/K}$$

6-65 EES solution of this (and other comprehensive problems designated with the *computer icon*) is available to instructors at the *Instructor Manual* section of the *Online Learning Center* (OLC) at www.mhhe.com/cengel-boles. See the Preface for access information.

6-66E A fixed mass of helium undergoes a process from one specified state to another specified state. The entropy change of helium is to be determined for the cases of reversible and irreversible processes.

Assumptions **1** At specified conditions, helium can be treated as an ideal gas. **2** Helium has constant specific heats at room temperature.

Properties The gas constant of helium is R = 0.4961 Btu/lbm.R (Table A-1E). The constant volume specific heat of helium is C_v = 0.753 Btu/lbm.R (Table A-2E).

Analysis From the ideal-gas entropy change relation,

$$\Delta S_{He} = m\left(C_{v,ave}\ln\frac{T_2}{T_1} + R\ln\frac{v_2}{v_1}\right)$$
$$= (15\text{ lbm})\left((0.753\text{ Btu/lbm}\cdot\text{R})\ln\frac{660\text{ R}}{540\text{ R}} + (0.4961\text{ Btu/lbm}\cdot\text{R})\ln\left(\frac{10\text{ ft}^3\text{/lbm}}{50\text{ ft}^3\text{/lbm}}\right)\right)$$
$$= \mathbf{-9.71\ Btu/R}$$

He
T_1 = 540 R
T_2 = 660 R

The entropy change will be the same for both cases.

6-67 Air is compressed in a piston-cylinder device in a reversible and isothermal manner. The entropy change of air and the work done are to be determined.

Assumptions **1** At specified conditions, air can be treated as an ideal gas. **2** The process is specified to be reversible.

Properties The gas constant of air is $R = 0.287$ kJ/kg.K (Table A-1).

Analysis (*a*) Noting that the temperature remains constant, the entropy change of air is determined from

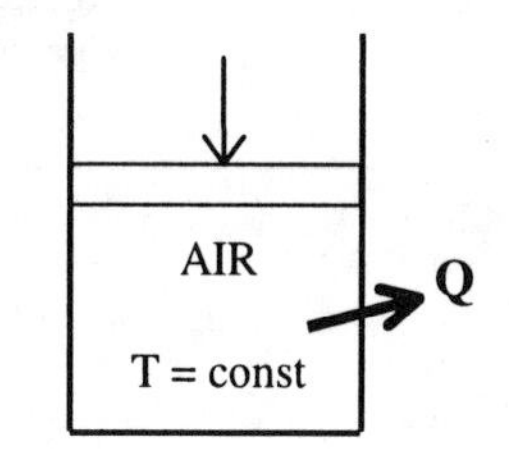

$$\Delta S_{air} = C_{p,ave} \ln\frac{T_2}{T_1}^{\nearrow 0} - R\ln\frac{P_2}{P_1} = -R\ln\frac{P_2}{P_1}$$

$$= -(0.287\text{kJ/kg}\cdot\text{K})\ln\left(\frac{400\text{kPa}}{90\text{kPa}}\right) = \mathbf{-0.428kJ/kg \cdot K}$$

Also, for a reversible isothermal process,

$$q = T\Delta s = (293\text{K})(-0.428\text{kJ/kg}\cdot\text{K}) = -125.4\text{kJ/kg} \rightarrow q_{out} = 125.4 \text{ kJ/kg}$$

(*b*) The work done during this process is determined from the closed system energy balance,

$$\underbrace{E_{in} - E_{out}}_{\substack{\text{Net energy transfer}\\ \text{by heat, work, and mass}}} = \underbrace{\Delta E_{\text{system}}}_{\substack{\text{Change in internal, kinetic,}\\ \text{potential, etc. energies}}}$$

$$W_{in} - Q_{out} = \Delta U = mC_v(T_2 - T_1) = 0$$

$$w_{in} = q_{out} = \mathbf{125.4\ kJ/kg}$$

6-68 Air is compressed steadily by a 5-kW compressor from one specified state to another specified state. The rate of entropy change of air is to be determined. √

Assumptions At specified conditions, air can be treated as an ideal gas. **2** Air has variable specific heats.

Properties The gas constant of air is $R = 0.287$ kJ/kg.K (Table A-1).

Analysis From the air table (Table A-17),

$$\left.\begin{array}{l} T_1 = 290 \text{ K} \\ P_1 = 100 \text{ kPa} \end{array}\right\} s_1^\circ = 1.66802 \text{ kJ/kg}\cdot\text{K}$$

$$\left.\begin{array}{l} T_2 = 440 \text{ K} \\ P_2 = 600 \text{ kPa} \end{array}\right\} s_2^\circ = 2.0887 \text{ kJ/kg}\cdot\text{K}$$

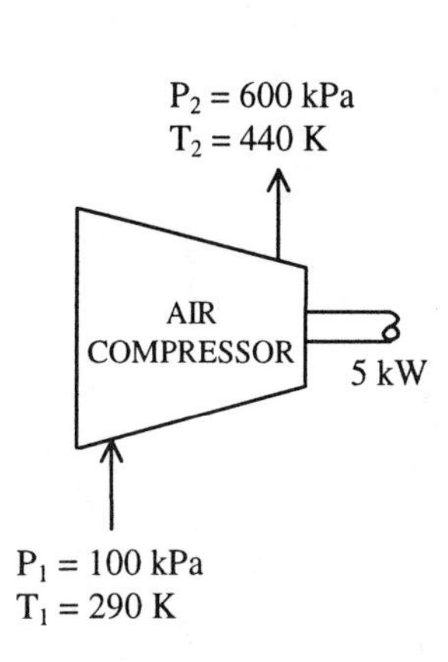

Then the rate of entropy change of air becomes

$$\Delta\dot{S}_{sys} = \dot{m}\left(s_2^\circ - s_1^\circ - R\ln\frac{P_2}{P_1}\right)$$

$$= (1.6/60\text{kg/s})\left(2.0887 - 1.66802 - (0.287\text{kJ/kg}\cdot\text{K})\ln\left(\frac{600\text{kPa}}{100\text{kPa}}\right)\right)$$

$$= \mathbf{-0.00250\ kW/K}$$

6-69 One side of a partitioned insulated rigid tank contains an ideal gas at a specified temperature and pressure while the other side is evacuated. The partition is removed, and the gas fills the entire tank. The total entropy change during this process is to be determined.

Assumptions The gas in the tank is given to be an ideal gas, and thus ideal gas relations apply.

Analysis Taking the entire rigid tank as the system, the energy balance can be expressed as

$$\underbrace{E_{in} - E_{out}}_{\substack{\text{Net energy transfer}\\ \text{by heat, work, and mass}}} = \underbrace{\Delta E_{\text{system}}}_{\substack{\text{Change in internal, kinetic,}\\ \text{potential, etc. energies}}}$$

$$0 = \Delta U = m(u_2 - u_1)$$

$$u_2 = u_1 \rightarrow T_2 = T_1$$

since $u = u(T)$ for an ideal gas. Then the entropy change of the gas becomes

$$\Delta S = N\left(\overline{C}_{v,ave} \ln\frac{T_2}{T_1}^{\nearrow 0} + R_u \ln\frac{V_2}{V_1}\right) = NR_u \ln\frac{V_2}{V_1}$$

$$= (5\text{kmol})(8.314\text{kJ/kmol}\cdot\text{K})\ln(2)$$

$$= \mathbf{28.81 kJ/K}$$

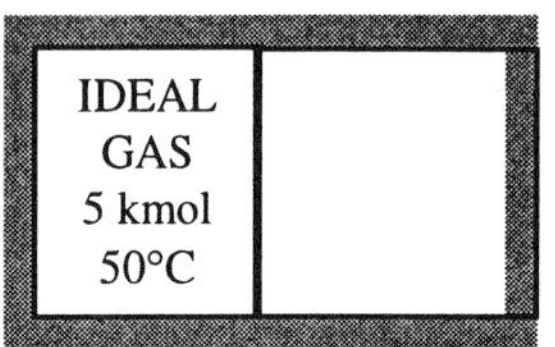

This also represents the **total entropy change** since the tank does not contain anything else, and there are no interactions with the surroundings.

6-70 Air is compressed in a piston-cylinder device in a reversible and adiabatic manner. The final temperature and the work are to be determined for the cases of constant and variable specific heats.

Assumptions **1** At specified conditions, air can be treated as an ideal gas. **2** The process is given to be reversible and adiabatic, and thus isentropic. Therefore, isentropic relations of ideal gases apply.

Properties The gas constant of air is R = 0.287 kJ/kg.K (Table A-1). The specific heat ratio of air at low to moderately high temperatures is k = 1.4 (Table A-2).

Analysis (a) Assuming constant specific heats, the ideal gas isentropic relations give

$$T_2 = T_1\left(\frac{P_2}{P_1}\right)^{(k-1)/k} = (290\text{K})\left(\frac{800\text{kPa}}{100\text{kPa}}\right)^{0.4/1.4} = \mathbf{525.3K}$$

Then,

$$T_{ave} = (290 + 525.3)/2 = 407.7\text{K} \longrightarrow C_{v,ave} = 0.727\text{kJ/kg}\cdot\text{K}$$

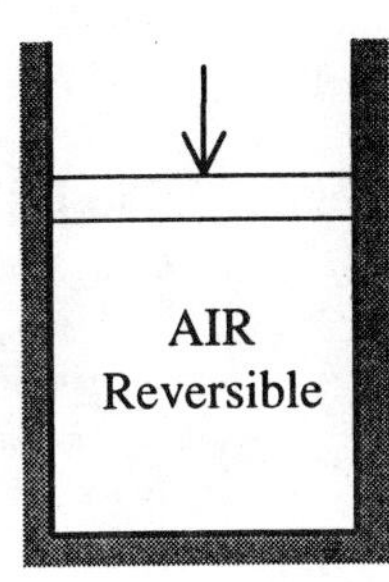

We take the air in the cylinder as the system. The energy balance for this stationary closed system can be expressed as

$$\underbrace{E_{in} - E_{out}}_{\substack{\text{Net energy transfer}\\ \text{by heat, work, and mass}}} = \underbrace{\Delta E_{\text{system}}}_{\substack{\text{Change in internal, kinetic,}\\ \text{potential, etc. energies}}}$$

$$W_{in} = \Delta U = m(u_2 - u_1) \cong mC_v(T_2 - T_1)$$

Thus,

$$w_{in} = C_{v,ave}(T_2 - T_1) = (0.727\text{kJ/kg}\cdot\text{K})(525.3 - 290)\text{K} = \mathbf{171.1kJ/kg}$$

(b) Assuming variable specific heats, the final temperature can be determined using the relative pressure data (Table A-17),

$$T_1 = 290\text{K} \longrightarrow \begin{matrix} P_{r_1} = 1.2311 \\ u_1 = 206.91\text{kJ/kg} \end{matrix}$$

and

$$P_{r_2} = \frac{P_2}{P_1}P_{r_1} = \frac{800\text{kPa}}{100\text{kPa}}(1.2311) = 9.849 \longrightarrow \begin{matrix} T_2 = \mathbf{522.4K} \\ u_2 = 376.16\text{kJ/kg} \end{matrix}$$

Then the work input becomes

$$w_{in} = u_2 - u_1 = (376.16 - 206.91)\text{kJ/kg} = \mathbf{169.25kJ/kg}$$

6-71 EES solution of this (and other comprehensive problems designated with the *computer icon*) is available to instructors at the *Instructor Manual* section of the *Online Learning Center* (OLC) at www.mhhe.com/cengel-boles. See the Preface for access information.

6-72 Helium gas is compressed in a piston-cylinder device in a reversible and adiabatic manner. The final temperature and the work are to be determined for the cases of the process taking place in a piston-cylinder device and a steady-flow compressor.

Assumptions **1** Helium is an ideal gas with constant specific heats. **2** The process is given to be reversible and adiabatic, and thus isentropic. Therefore, isentropic relations of ideal gases apply.

Properties The specific heats and the specific heat ratio of helium are C_v = 3.1156 kJ/kg.K, C_p = 5.1926 kJ/kg.K, and k = 1.667 (Table A-2).

Analysis (*a*) From the ideal gas isentropic relations,

$$T_2 = T_1\left(\frac{P_2}{P_1}\right)^{(k-1)/k} = (303\text{ K})\left(\frac{450\text{ kPa}}{90\text{ kPa}}\right)^{0.667/1.667} = \mathbf{576.9\ K}$$

He Rev.

He Rev.

2

1

(*a*) We take the air in the cylinder as the system. The energy balance for this stationary closed system can be expressed as

$$\underbrace{E_{in} - E_{out}}_{\substack{\text{Net energy transfer}\\ \text{by heat, work, and mass}}} = \underbrace{\Delta E_{\text{system}}}_{\substack{\text{Change in internal, kinetic,}\\ \text{potential, etc. energies}}}$$

$$W_{in} = \Delta U = m(u_2 - u_1) \cong mC_v(T_2 - T_1)$$

Thus,

$$w_{in} = C_v(T_2 - T_1) = (3.1156\text{kJ/kg}\cdot\text{K})(576.9 - 303)\text{K} = \mathbf{853.4 kJ/kg}$$

(*b*) If the process takes place in a steady-flow device, the final temperature will remain the same but the work done should be determined from an energy balance on this steady-flow device,

$$\underbrace{\dot{E}_{in} - \dot{E}_{out}}_{\substack{\text{Rate of net energy transfer}\\ \text{by heat, work, and mass}}} = \underbrace{\Delta \dot{E}_{\text{system}}^{\nearrow 0\text{ (steady)}}}_{\substack{\text{Rate of change in internal, kinetic,}\\ \text{potential, etc. energies}}} = 0$$

$$\dot{E}_{in} = \dot{E}_{out}$$

$$\dot{W}_{\text{in}} + \dot{m}h_1 = \dot{m}h_2$$

$$\dot{W}_{\text{in}} = \dot{m}(h_2 - h_1) \cong \dot{m}C_p(T_2 - T_1)$$

Thus,

$$w_{in} = C_p(T_2 - T_1) = (5.1926\text{ kJ/kg}\cdot\text{K})(576.9 - 303)\text{K} = \mathbf{1422.3\ kJ/kg}$$

6-73 An insulated rigid tank contains argon gas at a specified pressure and temperature. A valve is opened, and argon escapes until the pressure drops to a specified value. The final mass in the tank is to be determined.

Assumptions **1** At specified conditions, argon can be treated as an ideal gas. **2** The process is given to be reversible and adiabatic, and thus isentropic. Therefore, isentropic relations of ideal gases apply.

Properties The specific heat ratio of argon is $k = 1.667$ (Table A-2).

Analysis From the ideal gas isentropic relations,

$$T_2 = T_1\left(\frac{P_2}{P_1}\right)^{(k-1)/k} = (303\text{K})\left(\frac{150\text{kPa}}{450\text{kPa}}\right)^{0.667/1.667} = 195\text{K}$$

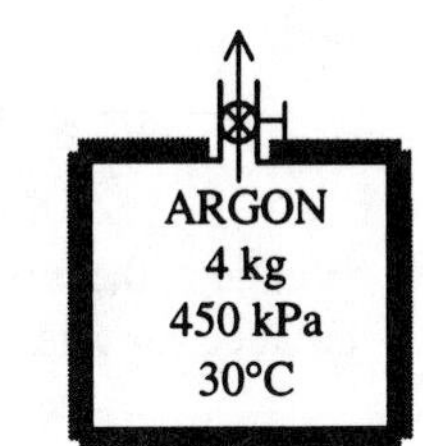

The final mass in the tank is determined from the ideal gas relation,

$$\frac{P_1 V}{P_2 V} = \frac{m_1 R T_1}{m_2 R T_2} \longrightarrow m_2 = \frac{P_2 T_1}{P_1 T_2} m_1 = \frac{(150\text{kPa})(303\text{K})}{(450\text{kPa})(195\text{K})}(4\text{kg}) = \mathbf{2.07kg}$$

6-74 EES solution of this (and other comprehensive problems designated with the *computer icon*) is available to instructors at the *Instructor Manual* section of the *Online Learning Center* (OLC) at www.mhhe.com/cengel-boles. See the Preface for access information.

6-75E Air is accelerated in an adiabatic nozzle. Disregarding irreversibilities, the exit velocity of air is to be determined.

Assumptions **1** Air is an ideal gas with variable specific heats. **2** The process is given to be reversible and adiabatic, and thus isentropic. Therefore, isentropic relations of ideal gases apply. **2** The nozzle operates steadily.

Analysis Assuming variable specific heats, the inlet and exit properties are determined to be

$$T_1 = 1000\text{R} \longrightarrow \begin{matrix} P_{r_1} = 12.30 \\ h_1 = 240.98\text{Btu/lbm} \end{matrix}$$

and

$$P_{r_2} = \frac{P_2}{P_1} P_{r_1} = \frac{12\text{psia}}{60\text{psia}}(12.30) = 2.46 \longrightarrow \begin{matrix} T_2 = 635.9\text{R} \\ h_2 = 152.11\text{Btu/lbm} \end{matrix}$$

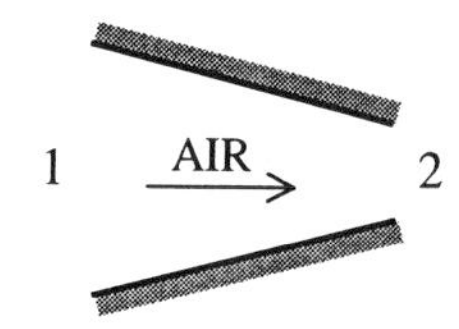

We take the nozzle as the system, which is a control volume. The energy balance for this steady-flow system can be expressed in the rate form as

$$\underbrace{\dot{E}_{in} - \dot{E}_{out}}_{\substack{\text{Rate of net energy transfer} \\ \text{by heat, work, and mass}}} = \underbrace{\Delta \dot{E}_{\text{system}}^{\nearrow 0 \text{ (steady)}}}_{\substack{\text{Rate of change in internal, kinetic,} \\ \text{potential, etc. energies}}} = 0$$

$$\dot{E}_{in} = \dot{E}_{out}$$

$$\dot{m}(h_1 + \mathbf{V}_1^2/2) = \dot{m}(h_2 + \mathbf{V}_2^2/2)$$

$$h_2 - h_1 + \frac{\mathbf{V}_2^2 - \mathbf{V}_1^2}{2} = 0$$

Therefore,

$$\mathbf{V}_2 = \sqrt{2(h_1 - h_2) + \mathbf{V}_1^2} = \sqrt{2(240.98 - 152.11)\text{Btu/lbm}\left(\frac{25{,}037\text{ft}^2/\text{s}^2}{1\text{Btu/lbm}}\right) + (200\text{ft/s})^2}$$

$$= \mathbf{2119ft/s}$$

6-76 Air is accelerated in an nozzle, and some heat is lost in the process. The exit temperature of air and the total entropy change during the process are to be determined. √

Assumptions **1** Air is an ideal gas with variable specific heats. **2** The nozzle operates steadily.

Analysis (*a*) Assuming variable specific heats, the inlet properties are determined to be,

$$T_1 = 350 \text{ K} \longrightarrow \begin{array}{l} h_1 = 350.49 \text{ kJ/kg} \\ s_1^\circ = 1.85708 / \text{kJ/kg} \cdot \text{K} \end{array} \quad \text{(Table A-17)}$$

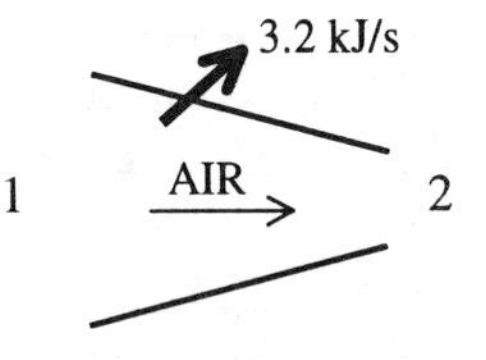

We take the nozzle as the system, which is a control volume. The energy balance for this steady-flow system can be expressed in the rate form as

$$\underbrace{\dot{E}_{in} - \dot{E}_{out}}_{\substack{\text{Rate of net energy transfer} \\ \text{by heat, work, and mass}}} = \underbrace{\Delta \dot{E}_{\text{system}}^{\nearrow 0 \text{ (steady)}}}_{\substack{\text{Rate of change in internal, kinetic,} \\ \text{potential, etc. energies}}} = 0$$

$$\dot{E}_{in} = \dot{E}_{out}$$

$$\dot{m}(h_1 + \mathbf{V}_1^2/2) = \dot{m}(h_2 + \mathbf{V}_2^2/2) + \dot{Q}_{out}$$

$$0 = q_{out} + h_2 - h_1 + \frac{\mathbf{V}_2^2 - \mathbf{V}_1^2}{2}$$

Therefore,

$$h_2 = h_1 - q_{out} - \frac{\mathbf{V}_2^2 - \mathbf{V}_1^2}{2} = 350.49 - 3.2 - \frac{(320\text{m/s})^2 - (50\text{m/s})^2}{2}\left(\frac{1\text{kJ/kg}}{1000\text{m}^2/\text{s}^2}\right)$$

$$= 297.34\text{kJ/kg}$$

At this h_2 value we read, from Table A-17, $T_2 = \mathbf{297.2\ K}$, $s_2^\circ = 1.6924 \text{ kJ/kg} \cdot \text{K}$

(*b*) The total entropy change is the sum of the entropy changes of the air and of the surroundings, and is determined from

$$\Delta s_{total} = \Delta s_{air} + \Delta s_{surr}$$

where

$$\Delta s_{air} = s_2^\circ - s_1^\circ - R \ln \frac{P_2}{P_1} = 1.6924 - 1.85708 - (0.287\text{kJ/kg} \cdot \text{K}) \ln \frac{85\text{kPa}}{280\text{kPa}} = 0.1775\text{kJ/kg} \cdot \text{K}$$

and

$$\Delta s_{surr} = \frac{q_{surr,in}}{T_{surr}} = \frac{3.2 \text{ kJ/kg}}{293 \text{ K}} = 0.0109 \text{ kJ/kg} \cdot \text{K}$$

Thus,

$$\Delta s_{total} = 0.1775 + 0.0109 = \mathbf{0.1884\ kJ/kg \cdot K}$$

6-77 EES solution of this (and other comprehensive problems designated with the *computer icon*) is available to instructors at the *Instructor Manual* section of the *Online Learning Center* (OLC) at www.mhhe.com/cengel-boles. See the Preface for access information.

Reversible Steady-Flow Work

6-78C The work associated with steady-flow devices is proportional to the specific volume of the gas. Cooling a gas during compression will reduce its specific volume, and thus the power consumed by the compressor.

6-79C Cooling the steam as it expands in a turbine will reduce its specific volume, and thus the work output of the turbine. Therefore, this is not a good proposal.

6-80C We would not support this proposal since the steady-flow work input to the pump is proportional to the specific volume of the liquid, and cooling will not affect the specific volume of a liquid significantly.

6-81 Liquid water is pumped reversibly to a specified pressure at a specified rate. The power input to the pump is to be determined.

Assumptions **1** Liquid water is an incompressible substance. **2** Kinetic and potential energy changes are negligible. **3** The process is reversible.

Properties The specific volume of saturated liquid water at 20 kPa is $v_1 = v_{f\,@\,20\,kPa} = 0.001017\ m^3/kg$ (Table A-5).

Analysis The power input to the pump can be determined directly from the steady-flow work relation for a liquid,

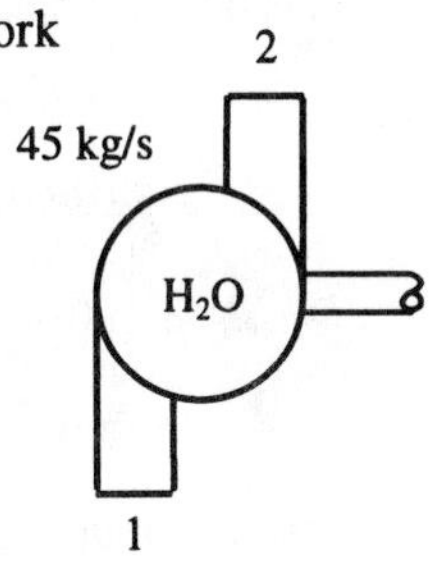

$$\dot{W}_{in} = \dot{m}\left(\int_1^2 v dP + \Delta ke^{\nearrow 0} + \Delta pe^{\nearrow 0}\right) = \dot{m}v_1(P_2 - P_1)$$

Substituting,

$$\dot{W}_{in} = (45\ kg/s)(0.001017\ m^3/kg)(6000 - 20)kPa\left(\frac{1\ kJ}{1\ kPa\cdot m^3}\right) = \mathbf{274\ kW}$$

6-82 Liquid water is to be pumped by a 10-kW pump at a specified rate. The highest pressure the water can be pumped to is to be determined.

Assumptions **1** Liquid water is an incompressible substance. **2** Kinetic and potential energy changes are negligible. **3** The process is assumed to be reversible since we will determine the limiting case.

Properties The specific volume of saturated liquid water at 20 kPa is $v_1 = v_{f\,@\,20\,kPa} = 0.001017\ m^3/kg$ (Table A-5).

Analysis The highest pressure the liquid can have at the pump exit can be determined from the reversible steady-flow work relation for a liquid,

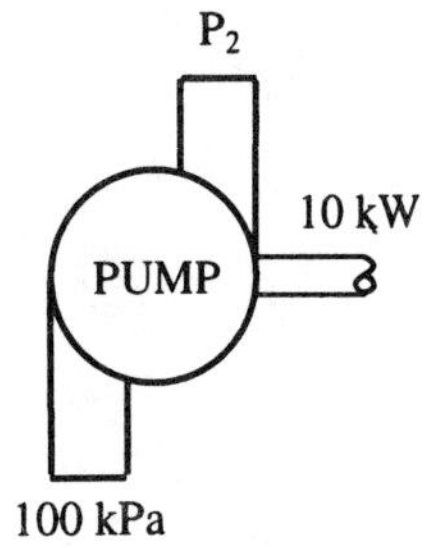

$$\dot{W}_{in} = \dot{m}\left(\int_1^2 v dP + \Delta ke^{\nearrow 0} + \Delta pe^{\nearrow 0}\right) = \dot{m}v_1(P_2 - P_1)$$

Thus,

$$10\ kJ/s = (5\ kg/s)(0.001017\ m^3/kg)(P_2 - 100)k\ Pa\left(\frac{1\ kJ}{1\ kPa\cdot m^3}\right)$$

It yields

$$P_2 = \mathbf{2100\ kPa}$$

6-83E Saturated refrigerant-134a vapor is to be compressed reversibly to a specified pressure. The power input to the compressor is to be determined, and it is also to be compared to the work input for the liquid case.

Assumptions **1** Liquid refrigerant is an incompressible substance. **2** Kinetic and potential energy changes are negligible. **3** The process is reversible. **4** The compressor is adiabatic.

Analysis The compression process is reversible and adiabatic, and thus isentropic, $s_1 = s_2$. Then the properties of the refrigerant are (Tables A-11E through A-13E)

$$\left.\begin{array}{l} P_1 = 20\text{psia} \\ sat.vapor \end{array}\right\} \begin{array}{l} h_1 = 101.39\text{Btu/lbm} \\ s_1 = 0.2227\text{Btu/lbm}\cdot\text{R} \end{array}$$

$$\left.\begin{array}{l} P_1 = 120\text{psia} \\ s_2 = s_1 \end{array}\right\} h_2 = 117.29\text{Btu/lbm}$$

The work input to this isentropic compressor is determined from the steady-flow energy balance to be

$$\underbrace{\dot{E}_{in} - \dot{E}_{out}}_{\text{Rate of net energy transfer by heat, work, and mass}} = \underbrace{\Delta \dot{E}_{\text{system}}^{\nearrow 0 \text{ (steady)}}}_{\text{Rate of change in internal, kinetic, potential, etc. energies}} = 0$$

$$\dot{E}_{in} = \dot{E}_{out}$$

$$\dot{W}_{\text{in}} + \dot{m}h_1 = \dot{m}h_2$$

$$\dot{W}_{\text{in}} = \dot{m}(h_2 - h_1)$$

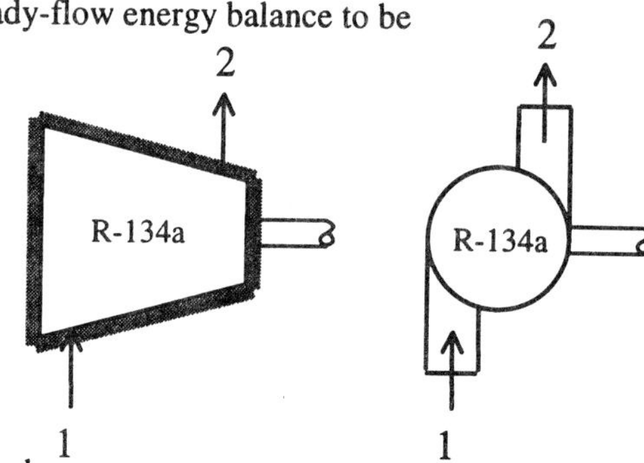

Thus, $w_{in} = h_2 - h_1 = 117.29 - 101.39 = \mathbf{15.9\ Btu/lbm}$

The pump work input is determined from the steady-flow work relation to be

$$w_{in} = \int_1^2 v\,dP + \Delta ke^{\nearrow 0} + \Delta pe^{\nearrow 0} = v_1(P_2 - P_1)$$

where $v_3 = v_{f\,@\,20\text{ psia}} = 0.01181$ ft^3/kg. Substituting,

$$w_{in} = (0.01181\text{ ft}^3\text{/lbm})(120 - 20)\text{ psia}\left(\frac{1\text{ Btu}}{5.4039\text{ psia}\cdot\text{ft}^3}\right) = \mathbf{0.2185\ Btu/lbm}$$

6-84 A steam power plant operates between the pressure limits of 10 MPa and 20 kPa. The ratio of the turbine work to the pump work is to be determined.

Assumptions **1** Liquid water is an incompressible substance. **2** Kinetic and potential energy changes are negligible. **3** The process is reversible. **4** The pump and the turbine are adiabatic.

Properties The specific volume of saturated liquid water at 20 kPa is $v_1 = v_{f\,@\,20\text{ kPa}} = 0.001017\text{ m}^3/\text{kg}$ (Table A-5).

Analysis Both the compression and expansion processes are reversible and adiabatic, and thus isentropic, $s_1 = s_2$ and $s_3 = s_4$. Then the properties of the steam are

$$\left.\begin{array}{l} P_4 = 20\text{ kPa} \\ sat.vapor \end{array}\right\} \begin{array}{l} h_4 = h_{g\,@\,20\text{kPa}} = 2609.7\text{ kJ/kg} \\ s_4 = s_{g\,@\,20\text{kPa}} = 7.9085\text{ kJ/kg}\cdot\text{K} \end{array}$$

$$\left.\begin{array}{l} P_3 = 10\text{ MPa} \\ s_3 = s_4 \end{array}\right\} h_3 = 4712.8\text{ kJ/kg}$$

Also, $v_1 = v_{f\,@\,20\text{ kPa}} = 0.001017\text{ m}^3/\text{kg}$.

The work output to this isentropic turbine is determined from the steady-flow energy balance to be

$$\underbrace{\dot{E}_{in} - \dot{E}_{out}}_{\substack{\text{Rate of net energy transfer} \\ \text{by heat, work, and mass}}} = \underbrace{\Delta\dot{E}_{\text{system}}^{\not{\,0}\text{ (steady)}}}_{\substack{\text{Rate of change in internal, kinetic,} \\ \text{potential, etc. energies}}} = 0$$

$$\dot{E}_{in} = \dot{E}_{out}$$

$$\dot{m}h_3 = \dot{m}h_4 + \dot{W}_{\text{out}}$$

$$\dot{W}_{\text{out}} = \dot{m}(h_3 - h_4)$$

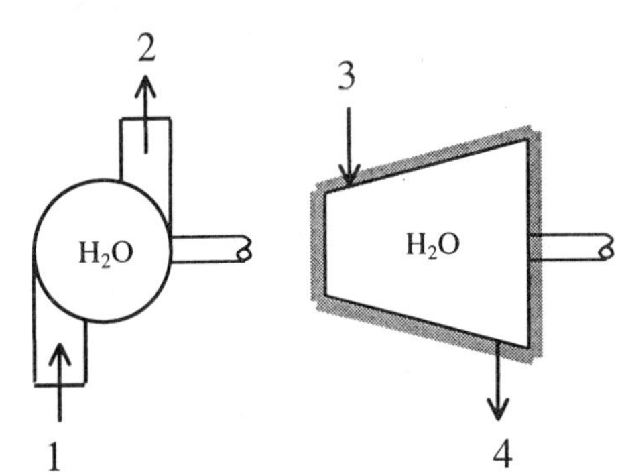

Substituting,

$$w_{turb,out} = h_3 - h_4 = 4712.8 - 2609.7 = 2103.1\text{ kJ/kg}$$

The pump work input is determined from the steady-flow work relation to be

$$w_{pump,in} = \int_1^2 v\,dP + \Delta ke^{\not{\,0}} + \Delta pe^{\not{\,0}} = v_1(P_2 - P_1)$$

$$= (0.001017\text{m}^3/\text{kg})(10{,}000 - 20)\text{kPa}\left(\frac{1\text{kJ}}{1\text{kPa}\cdot\text{m}^3}\right) = 10.15\text{kJ/kg}$$

Thus,

$$\frac{w_{turb,out}}{w_{pump,in}} = \frac{2103.1}{10.15} = \mathbf{207.2}$$

6-85 EES solution of this (and other comprehensive problems designated with the *computer icon*) is available to instructors at the *Instructor Manual* section of the *Online Learning Center* (OLC) at www.mhhe.com/cengel-boles. See the Preface for access information.

6-86 Liquid water is pumped by a 10-kW pump to a specified pressure at a specified level. The highest possible mass flow rate of water is to be determined.

Assumptions **1** Liquid water is an incompressible substance. **2** Kinetic energy changes are negligible, but potential energy changes may be significant. **3** The process is assumed to be reversible since we will determine the limiting case.

Properties The specific volume of liquid water is given to be $v_1 = 0.001 \text{ m}^3/\text{kg}$.

Analysis The highest mass flow rate will be realized when the entire process is reversible. Thus it is determined from the reversible steady-flow work relation for a liquid,

Thus,

$$\dot{W}_{in} = \dot{m}\left(\int_1^2 v dP + \Delta ke^{\not{0}} + \Delta pe\right) = \dot{m}\{v(P_2 - P_1) + g(z_2 - z_1)\}$$

$$7 \text{ kJ/s} = \dot{m}\left\{(0.001 \text{ m}^3/\text{kg})(3000 - 120)\text{kPa}\left(\frac{1 \text{ kJ}}{1 \text{ kPa}\cdot\text{m}^3}\right) + (9.8 \text{ m/s}^2)(10 \text{ m})\left(\frac{1 \text{ kJ/kg}}{1000 \text{ m}^2/\text{s}^2}\right)\right\}$$

It yields

$$\dot{m} = \mathbf{2.35 \text{ kg/s}}$$

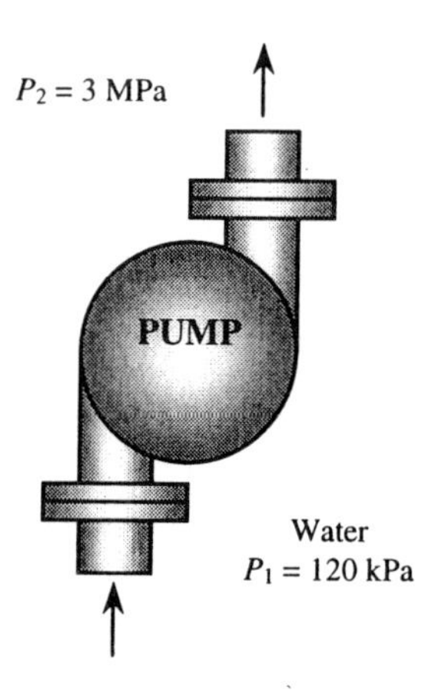

6-87E Helium gas is compressed from a specified state to a specified pressure at a specified rate. The power input to the compressor is to be determined for the cases of isentropic, polytropic, isothermal, and two-stage compression.

Assumptions **1** Helium is an ideal gas with constant specific heats. **2** The process is reversible. **3** Kinetic and potential energy changes are negligible.

Properties The gas constant of helium is R = 2.6805 psia.ft³/lbm.R = 0.4961 Btu/lbm.R (Table A-1). The specific heat ratio of helium is k = 1.667.

Analysis The mass flow rate of helium is

$$\dot{m} = \frac{P_1\dot{V}_1}{RT_1} = \frac{(14\text{psia})(5\text{ft}^3/\text{s})}{(2.6805\text{psia}\cdot\text{ft}^3/\text{lbm}\cdot\text{R})(530\text{R})} = 0.0493\text{lbm/s}$$

(*a*) Isentropic compression with k = 1.667:

$$\dot{W}_{comp,in} = \dot{m}\frac{kRT_1}{k-1}\left\{\left(\frac{P_2}{P_1}\right)^{(k-1)/k} - 1\right\}$$

$$= (0.0493\text{lbm/s})\frac{(1.667)(0.4961\text{Btu/lbm}\cdot\text{R})(530\text{R})}{1.667-1}\left\{\left(\frac{120\text{psia}}{14\text{psia}}\right)^{0.667/1.667} - 1\right\}$$

$$= 44.11\text{Btu/s}$$

$$= \mathbf{62.4hp} \quad \text{since 1 hp = 0.7068 Btu/s}$$

(*b*) Polytropic compression with n = 1.2:

$$\dot{W}_{comp,in} = \dot{m}\frac{nRT_1}{n-1}\left\{\left(\frac{P_2}{P_1}\right)^{(n-1)/n} - 1\right\}$$

$$= (0.0493\text{lbm/s})\frac{(1.2)(0.4961\text{Btu/lbm}\cdot\text{R})(530\text{R})}{1.2-1}\left\{\left(\frac{120\text{psia}}{14\text{psia}}\right)^{0.2/1.2} - 1\right\}$$

$$= 33.47\text{Btu/s}$$

$$= \mathbf{47.3hp} \quad \text{since 1 hp = 0.7068 Btu/s}$$

(*c*) Isothermal compression:

$$\dot{W}_{comp,in} = \dot{m}RT\ln\frac{P_2}{P_1} = (0.0493\text{lbm/s})(0.4961\text{Btu/lbm}\cdot\text{R})(530\text{R})\ln\frac{120\text{psia}}{14\text{psia}} = 27.83\text{Btu/s} = \mathbf{39.4hp}$$

(*d*) Ideal two-stage compression with intercooling (n = 1.2): In this case, the pressure ratio across each stage is the same, and its value is determined from

$$P_x = \sqrt{P_1P_2} = \sqrt{(14\text{psia})(120\text{psia})} = 41.0\text{psia}$$

The compressor work across each stage is also the same, thus total compressor work is twice the compression work for a single stage:

$$\dot{W}_{comp,in} = 2\dot{m}w_{comp,I} = 2\dot{m}\frac{nRT_1}{n-1}\left\{\left(\frac{P_x}{P_1}\right)^{(n-1)/n} - 1\right\}$$

$$= 2(0.0493\text{lbm/s})\frac{(1.2)(0.4961\text{Btu/lbm}\cdot\text{R})(530\text{R})}{1.2-1}\left\{\left(\frac{41\text{psia}}{14\text{psia}}\right)^{0.2/1.2} - 1\right\}$$

$$= 30.52\text{Btu/s}$$

$$= \mathbf{43.2hp} \quad \text{since 1 hp = 0.7068 Btu/s}$$

6-88E EES solution of this (and other comprehensive problems designated with the *computer icon*) is available to instructors at the *Instructor Manual* section of the *Online Learning Center* (OLC) at www.mhhe.com/cengel-boles. See the Preface for access information.

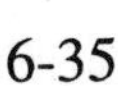

6-89 Nitrogen gas is compressed by a 10-kW compressor from a specified state to a specified pressure. The mass flow rate of nitrogen through the compressor is to be determined for the cases of isentropic, polytropic, isothermal, and two-stage compression.

Assumptions **1** Nitrogen is an ideal gas with constant specific heats. **2** The process is reversible. **3** Kinetic and potential energy changes are negligible.

Properties The gas constant of nitrogen is R = 0.297 kJ/kg.K (Table A-1). The specific heat ratio of Nitrogen is k = 1.4.

2

N_2
$\dot{m}$

10 kW

1

Analysis (*a*) Isentropic compression:

$$\dot{W}_{comp,in} = \dot{m}\frac{kRT_1}{k-1}\left\{(P_2/P_1)^{(k-1)/k} - 1\right\}$$

or,

$$10\text{kJ/s} = \dot{m}\frac{(1.4)(0.297\text{kJ/kg}\cdot\text{K})(300\text{K})}{1.4-1}\left\{(480\text{kPa}/80\text{kPa})^{0.4/1.4} - 1\right\}$$

It yields

$$\dot{m} = \mathbf{0.048\ kg/s}$$

(*b*) Polytropic compression with n = 1.3:

$$\dot{W}_{comp,in} = \dot{m}\frac{nRT_1}{n-1}\left\{(P_2/P_1)^{(n-1)/n} - 1\right\}$$

or,

$$10\text{kJ/s} = \dot{m}\frac{(1.3)(0.297\text{kJ/kg}\cdot\text{K})(300\text{K})}{1.3-1}\left\{(480\text{kPa}/80\text{kPa})^{0.3/1.3} - 1\right\}$$

It yields

$$\dot{m} = \mathbf{0.051\ kg/s}$$

(*c*) Isothermal compression:

$$\dot{W}_{comp,in} = \dot{m}RT\ln\frac{P_1}{P_2} \longrightarrow 10\text{kJ/s} = \dot{m}(0.297\text{kJ/kg}\cdot\text{K})(300\text{K})\ln\left(\frac{480\text{kPa}}{80\text{kPa}}\right)$$

It yields

$$\dot{m} = \mathbf{0.063\ kg/s}$$

(*d*) Ideal two-stage compression with intercooling (n = 1.3): In this case, the pressure ratio across each stage is the same, and its value is determined to be

$$P_x = \sqrt{P_1P_2} = \sqrt{(80\text{kPa})(480\text{kPa})} = 196\text{kPa}$$

The compressor work across each stage is also the same, thus total compressor work is twice the compression work for a single stage:

$$\dot{W}_{comp,in} = 2\dot{m}w_{comp,I} = 2\dot{m}\frac{nRT_1}{n-1}\left\{(P_x/P_1)^{(n-1)/n} - 1\right\}$$

or,

$$10\text{kJ/s} = 2\dot{m}\frac{(1.3)(0.297\text{kJ/kg}\cdot\text{K})(300\text{K})}{1.3-1}\left\{(196\text{kPa}/80\text{kPa})^{0.3/1.3} - 1\right\}$$

It yields

$$\dot{m} = \mathbf{0.056\ \ kg/s}$$

6-90 Water mist is to be sprayed into the air stream in the compressor to cool the air as the water evaporates and to reduce the compression power. The reduction in the exit temperature of the compressed air and the compressor power saved are to be determined.

Assumptions **1** Air is an ideal gas with variable specific heats. **2** The process is reversible. **3** Kinetic and potential energy changes are negligible. **3** Air is compressed isentropically. **4** Water vaporizes completely before leaving the compressor. **4** Air properties can be used for the air-vapor mixture.

Properties The gas constant of air is R = 0.287 kJ/kg.K (Table A-1). The specific heat ratio of air is k = 1.4. The inlet enthalpies of water and air are (Tables A-4 and A-17)

$$h_{w1} = h_{f@20°C} = 83.96 \text{ kJ/kg}, \; h_{fg@20°C} = 2454.1 \text{ kJ/kg and } h_{a1} = h_{@300\text{ K}} = 300.19 \text{ kJ/kg}$$

Analysis In the case of isentropic operation (thus no cooling or water spray), the exit temperature and the power input to the compressor are

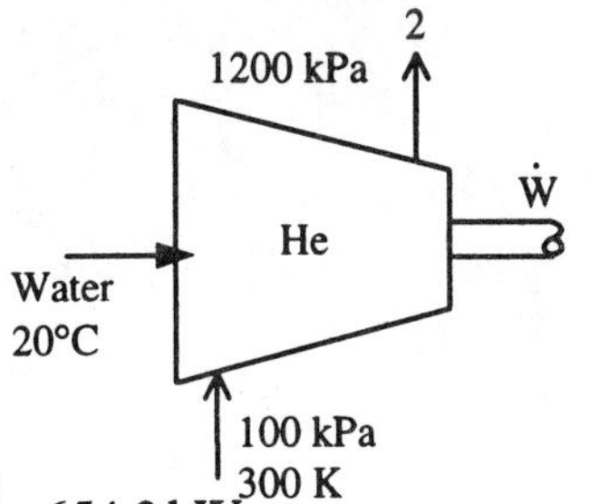

$$\frac{T_2}{T_1} = \left(\frac{P_2}{P_1}\right)^{(k-1)/k} \rightarrow T_2 = (300\text{ K})\left(\frac{1200\text{ kPa}}{100\text{ kPa}}\right)^{(1.4-1)/1.4} = 610.2\text{ K}$$

2
1200 kPa
W
He
Water
20°C
100 kPa
300 K

$$\dot{W}_{comp,in} = \dot{m}\frac{kRT_1}{k-1}\left\{(P_2/P_1)^{(k-1)/k} - 1\right\}$$

$$= (2.1\text{ kg/s})\frac{(1.4)(0.287\text{kJ/kg}\cdot\text{K})(300\text{K})}{1.4-1}\left\{(1200\text{ kPa}/100\text{ kPa})^{0.4/1.4} - 1\right\} = 654.3\text{ kW}$$

When water is sprayed, we first need to check the accuracy of the assumption that the water vaporizes completely in the compressor. In the limiting case, the compression will be isothermal at the compressor inlet temperature, and the water will be a saturated vapor. To avoid the complexity of dealing with two fluid streams and a gas mixture, we disregard water in the air stream (other than the mass flow rate), and assume air is cooled by an amount equal to the enthalpy change of water.

The rate of heat absorption of water as it evaporates at the inlet temperature completely is

$$\dot{Q}_{cooling,\max} = \dot{m}_w h_{fg@20°C} = (0.2\text{ kg/s})(2454.1\text{ kJ/kg}) = 490.8\text{ kW}$$

The minimum power input to the compressor is

$$\dot{W}_{comp,in,\min} = \dot{m}RT\ln\frac{P_2}{P_1} = (2.1\text{ kg/s})(0.287\text{ kJ/kg}\cdot\text{K})(300\text{ K})\ln\left(\frac{1200\text{ kPa}}{100\text{ kPa}}\right) = 449.3\text{ kW}$$

This corresponds to maximum cooling from the air since, at constant temperature, Δh = 0 and thus $\dot{Q}_{out} = \dot{W}_{in} = 449.1\text{ kW}$, which is close to 490.8 kW. Therefore, the assumption that all the water vaporizes is approximately valid. Then the reduction in required power input due to water spray becomes

$$\Delta\dot{W}_{comp,in} = \dot{W}_{\text{comp, isentropic}} - \dot{W}_{\text{comp, isothermal}} = 654.3 - 449.3 = \mathbf{205\ kW}$$

Discussion (can be ignored): At constant temperature, Δh = 0 and thus $\dot{Q}_{out} = \dot{W}_{in} = 449.1\,\text{kW}$ corresponds to maximum cooling from the air since, , which is less than 490.8 kW. Therefore, the assumption that all the water vaporizes is only roughly valid. As an alternative, we can assume the compression process to be polytropic and the water to be a saturated vapor at the compressor exit temperature, and disregard the remaining liquid. But in this case there is not a unique solution, and we will have to select either the amount of water or the exit temperature or the polytropic exponent to obtain a solution. Of course we can also tabulate the results for different cases, and then make a selection.

Sample Analysis: We take the compressor exit temperature to be T_2 = 200°C = 473 K. Then,

$$h_{w2} = h_{g@200°C} = 2793.2 \text{ kJ/kg and } h_{a2} = h_{@473\text{ K}} = 475.3 \text{ kJ/kg}$$

Then,

$$\frac{T_2}{T_1} = \left(\frac{P_2}{P_1}\right)^{(n-1)/n} \rightarrow \frac{473\text{ K}}{300\text{ K}} = \left(\frac{1200\text{ kPa}}{100\text{ kPa}}\right)^{(n-1)/n} \rightarrow n = 1.224$$

$$\dot{W}_{comp,in} = \dot{m}\frac{nRT_1}{n-1}\left\{(P_2/P_1)^{(n-1)/n} - 1\right\} = \dot{m}\frac{nR}{n-1}(T_2 - T_1)$$

$$= (2.1\text{ kg/s})\frac{(1.224)(0.287\text{ kJ/kg}\cdot\text{K})}{1.224-1}(473-300)\text{K} = 570\text{ kW}$$

Energy balance:

$$\dot{W}_{comp,in} - \dot{Q}_{out} = \dot{m}(h_2 - h_1) \rightarrow \dot{Q}_{out} = \dot{W}_{comp,in} - \dot{m}(h_2 - h_1)$$

$$= 569.7\text{ kW} - (2.1\text{ kg/s})(475.3 - 300.19) = 202.0\text{ kW}$$

Noting that this heat is absorbed by water, the rate at which water evaporates in the compressor becomes

$$\dot{Q}_{out,air} = \dot{Q}_{in,water} = \dot{m}_w(h_{w2} - h_{w1}) \rightarrow \dot{m}_w = \frac{\dot{Q}_{in,water}}{h_{w2} - h_{w1}} = \frac{202.0\text{ kJ/s}}{(2793.2 - 83.96)\text{ kJ/kg}} = 0.0746\text{ kg/s}$$

Then the reductions in the exit temperature and compressor power input become

$$\Delta T_2 = T_{2,\text{isentropic}} - T_{2,\text{water cooled}} = 610.2 - 473 = \mathbf{137.2°C}$$

$$\Delta\dot{W}_{comp,in} = \dot{W}_{comp,\text{isentropic}} - \dot{W}_{comp,\text{water cooled}} = 654.3 - 570 = \mathbf{84.3\text{ kW}}$$

Note that selecting a different compressor exit temperature T_2 will result in different values.

6-91 A water-injected compressor is used in a gas turbine power plant. It is claimed that the power output of a gas turbine will increase when water is injected into the compressor because of the increase in the mass flow rate of the gas (air + water vapor) through the turbine. This, however, is **not necessarily right** since the compressed air in this case enters the combustor at a low temperature, and thus it absorbs much more heat. In fact, the cooling effect will most likely dominate and cause the cyclic efficiency to drop.

Isentropic Efficiencies of Steady-Flow Devices

6-92C The ideal process for all three devices is the reversible adiabatic (i.e., isentropic) process. The adiabatic efficiencies of these devices are defined as

$$\eta_T = \frac{\text{actual work output}}{\text{insentropic work output}}, \eta_C = \frac{\text{insentropic work input}}{\text{actual work input}}, \text{and } \eta_N = \frac{\text{actual exit kinetic energy}}{\text{insentropic exit kinetic energy}}$$

6-93C No, because the isentropic process is not the model or ideal process for compressors that are cooled intentionally.

6-94C Yes. Because the entropy of the fluid must increase during an actual adiabatic process as a result of irreversibilities. Therefore, the actual exit state has to be on the right-hand side of the isentropic exit state

6-95 Steam enters an adiabatic turbine with an isentropic efficiency of 0.90 at a specified state with a specified mass flow rate, and leaves at a specified pressure. The turbine exit temperature and power output of the turbine are to be determined.

Assumptions **1** This is a steady-flow process since there is no change with time. **2** Kinetic and potential energy changes are negligible. **3** The device is adiabatic and thus heat transfer is negligible.

P_1 = 8 MPa
T_1 = 500°C

STEAM
TURBINE
η_T = 90%

P_2 = 30 kPa

Analysis (*a*) From the steam tables (Tables A-4 and A-6),

$$\left.\begin{array}{l} P_1 = 8\text{MPa} \\ T_1 = 500^\circ\text{C} \end{array}\right\} \begin{array}{l} h_1 = 3398.3\text{kJ/kg} \\ s_1 = 6.7240\text{kJ/kg}\cdot\text{K} \end{array}$$

$$\left.\begin{array}{l} P_{2s} = 30\text{kPa} \\ s_{2s} = s_1 \end{array}\right\} \begin{array}{l} x_{2s} = \dfrac{s_{2s} - s_f}{s_{fg}} = \dfrac{6.7240 - 0.9439}{6.8247} = 0.847 \\ h_{2s} = h_f + x_{2s}h_{fg} = 289.23 + (0.847)(2336.1) = 2267.9\text{kJ/kg} \end{array}$$

From the isentropic efficiency relation,

$$\eta_T = \frac{h_1 - h_{2a}}{h_1 - h_{2s}} \longrightarrow h_{2a} = h_1 - \eta_T(h_1 - h_{2s}) = 3398.3 - (0.9)(3398.3 - 2267.9) = 2380.9\text{kJ/kg}$$

Thus,

$$\left.\begin{array}{l} P_{2a} = 30\text{kPa} \\ h_{2a} = 2380.9\text{kJ/kg} \end{array}\right\} T_{2a} = T_{sat@30\text{kPa}} = \mathbf{69.10^\circ C}$$

(*b*) There is only one inlet and one exit, and thus $\dot{m}_1 = \dot{m}_2 = \dot{m}$. We take the actual turbine as the system, which is a control volume since mass crosses the boundary. The energy balance for this steady-flow system can be expressed in the rate form as

$$\underbrace{\dot{E}_{in} - \dot{E}_{out}}_{\substack{\text{Rate of net energy transfer}\\ \text{by heat, work, and mass}}} = \underbrace{\Delta \dot{E}_{\text{system}}^{\nearrow 0\ (\text{steady})}}_{\substack{\text{Rate of change in internal, kinetic,}\\ \text{potential, etc. energies}}} = 0$$

$$\dot{E}_{in} = \dot{E}_{out}$$

$$\dot{m}h_1 = \dot{W}_{a,\text{out}} + \dot{m}h_2 \quad (\text{since } \dot{Q} \cong \Delta ke \cong \Delta pe \cong 0)$$

$$\dot{W}_{a,\text{out}} = \dot{m}(h_1 - h_2)$$

Substituting,

$$\dot{W}_{a,out} = (3\text{kg/s})(3398.3 - 2380.9)\text{kJ/kg} = \mathbf{3052kW}$$

6-96 EES solution of this (and other comprehensive problems designated with the *computer icon*) is available to instructors at the *Instructor Manual* section of the *Online Learning Center* (OLC) at www.mhhe.com/cengel-boles. See the Preface for access information.

6-97 Steam enters an adiabatic turbine at a specified state, and leaves at a specified state. The mass flow rate of the steam and the isentropic efficiency are to be determined.

Assumptions **1** This is a steady-flow process since there is no change with time. **2** Potential energy changes are negligible. **3** The device is adiabatic and thus heat transfer is negligible.

Analysis (*a*) From the steam tables (Tables A-4 and A-6),

$$\left.\begin{array}{l} P_1 = 6\text{MPa} \\ T_1 = 600^\circ\text{C} \end{array}\right\} \begin{array}{l} h_1 = 3658.4\text{kJ/kg} \\ s_1 = 7.1677\text{kJ/kg}\cdot\text{K} \end{array}$$

$$\left.\begin{array}{l} P_2 = 50\text{kPa} \\ T_2 = 100^\circ\text{C} \end{array}\right\} h_{2a} = 2682.5\text{kJ/kg}$$

There is only one inlet and one exit, and thus $\dot{m}_1 = \dot{m}_2 = \dot{m}$. We take the actual turbine as the system, which is a control volume since mass crosses the boundary. The energy balance for this steady-flow system can be expressed in the rate form as

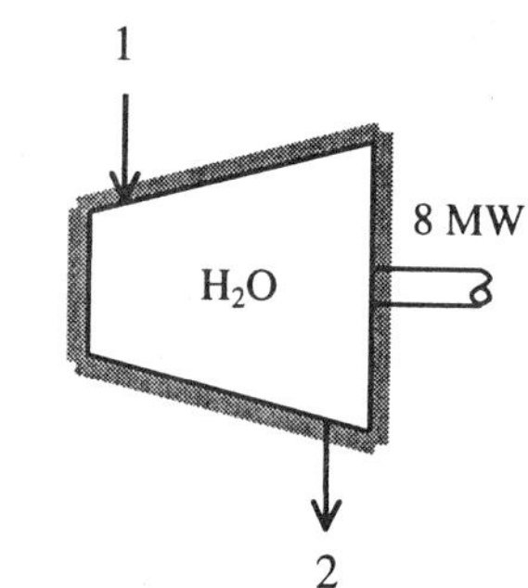

$$\underbrace{\dot{E}_{in} - \dot{E}_{out}}_{\substack{\text{Rate of net energy transfer} \\ \text{by heat, work, and mass}}} = \underbrace{\Delta \dot{E}_{\text{system}}^{\nearrow 0\text{ (steady)}}}_{\substack{\text{Rate of change in internal, kinetic,} \\ \text{potential, etc. energies}}} = 0$$

$$\dot{E}_{in} = \dot{E}_{out}$$

$$\dot{m}(h_1 + V_1^2/2) = \dot{W}_{a,\text{out}} + \dot{m}(h_2 + V_1^2/2) \quad (\text{since } \dot{Q} \cong \Delta\text{pe} \cong 0)$$

$$\dot{W}_{a,\text{out}} = -\dot{m}\left(h_2 - h_1 + \frac{V_2^2 - V_1^2}{2}\right)$$

Substituting, the mass flow rate of the steam is determined to be

$$8000\text{ kJ/s} = -\dot{m}\left(2682.5 - 3658.4 + \frac{(140\text{ m/s})^2 - (80\text{ m/s})^2}{2}\left(\frac{1\text{ kJ/kg}}{1000\text{ m}^2/\text{s}^2}\right)\right)$$

$$\dot{m} = \mathbf{8.25\ kg/s}$$

(*b*) The isentropic exit enthalpy of the steam and the power output of the isentropic turbine are

$$\left.\begin{array}{l} P_{2s} = 50\text{kPa} \\ s_{2s} = s_1 \end{array}\right\} \begin{array}{l} x_{2s} = \dfrac{s_{2s} - s_f}{s_{fg}} = \dfrac{7.1677 - 1.0910}{6.5029} = 0.934 \\ h_{2s} = h_f + x_{2s}h_{fg} = 340.49 + (0.934)(2305.4) = 2493.7\text{kJ/kg} \end{array}$$

and

$$\dot{W}_{s,out} = -\dot{m}\left(h_{2s} - h_1 + \left\{\left(V_2^2 - V_1^2\right)/2\right\}\right)$$

$$\dot{W}_{s,out} = -(8.25\text{ kg/s})\left(2493.7 - 3658.4 + \frac{(140\text{ m/s})^2 - (80\text{ m/s})^2}{2}\left(\frac{1\text{ kJ/kg}}{1000\text{ m}^2/\text{s}^2}\right)\right)$$

$$= 9554\text{ kW}$$

Then the isentropic efficiency of the turbine becomes

$$\eta_T = \frac{\dot{W}_a}{\dot{W}_s} = \frac{8000\text{ kW}}{9554\text{ kW}} = 0.837 = \mathbf{83.7\%}$$

6-98 Argon enters an adiabatic turbine at a specified state with a specified mass flow rate, and leaves at a specified pressure. The isentropic efficiency of the turbine is to be determined.

Assumptions **1** This is a steady-flow process since there is no change with time. **2** Kinetic and potential energy changes are negligible. **3** The device is adiabatic and thus heat transfer is negligible. **4** Argon is an ideal gas with constant specific heats.

Properties The specific heat ratio of argon is k = 1.667. The constant pressure specific heat of argon is C_p = 0.5203 kJ/kg.K (Table A-2).

Analysis There is only one inlet and one exit, and thus $\dot{m}_1 = \dot{m}_2 = \dot{m}$. We take the isentropic turbine as the system, which is a control volume since mass crosses the boundary. The energy balance for this steady-flow system can be expressed in the rate form as

$$\underbrace{\dot{E}_{in} - \dot{E}_{out}}_{\text{Rate of net energy transfer by heat, work, and mass}} = \underbrace{\Delta \dot{E}_{\text{system}}^{\nearrow 0 \text{ (steady)}}}_{\text{Rate of change in internal, kinetic, potential, etc. energies}} = 0$$

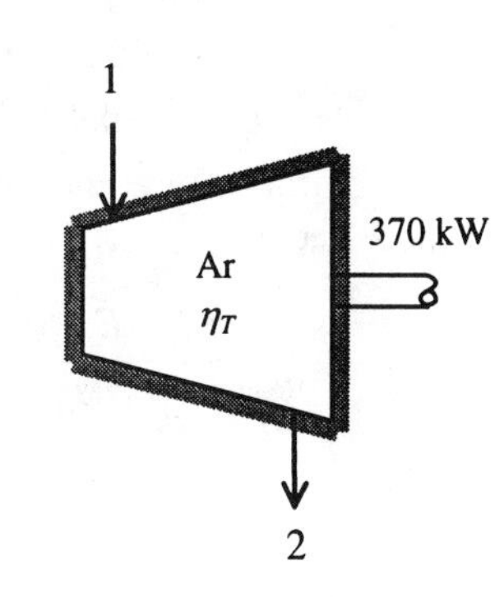

$$\dot{E}_{in} = \dot{E}_{out}$$

$$\dot{m}h_1 = \dot{W}_{s,\text{out}} + \dot{m}h_{2s} \quad (\text{since } \dot{Q} \cong \Delta ke \cong \Delta pe \cong 0)$$

$$\dot{W}_{s,\text{out}} = \dot{m}(h_1 - h_{2s})$$

From the isentropic relations,

$$T_{2s} = T_1 \left(\frac{P_{2s}}{P_1}\right)^{(k-1)/k} = (1073\text{K})\left(\frac{200\text{kPa}}{1500\text{kPa}}\right)^{0.667/1.667} = 479\text{K}$$

Then the power output of the isentropic turbine becomes

$$\dot{W}_{s,out} = \dot{m}C_p(T_1 - T_{2s}) = (80/60\text{kg/min})(0.5203\text{kJ/kg}\cdot\text{K})(1073 - 479) = 412.1\text{kW}$$

Then the isentropic efficiency of the turbine is determined from

$$\eta_T = \frac{\dot{W}_a}{\dot{W}_s} = \frac{370 \text{ kW}}{412.1 \text{ kW}} = 0.898 = \mathbf{89.8\%}$$

6-99E Combustion gases enter an adiabatic gas turbine with an isentropic efficiency of 86% at a specified state, and leave at a specified pressure. The work output of the turbine is to be determined.

Assumptions **1** This is a steady-flow process since there is no change with time. **2** Kinetic and potential energy changes are negligible. **3** The device is adiabatic and thus heat transfer is negligible. **4** Combustion gases can be treated as air that is an ideal gas with variable specific heats.

Analysis From the air table and isentropic relations,

$$T_1 = 2000\text{ R} \longrightarrow \begin{array}{l} h_1 = 504.71\text{ Btu/lbm} \\ P_{r_1} = 174.0 \end{array}$$

$$P_{r_2} = \left(\frac{P_2}{P_1}\right)P_{r_1} = \left(\frac{60\text{psia}}{120\text{psia}}\right)(174.0) = 87.0 \longrightarrow \text{h}_{2s} = 417.3\text{Btu/lbm}$$

There is only one inlet and one exit, and thus $\dot{m}_1 = \dot{m}_2 = \dot{m}$. We take the actual turbine as the system, which is a control volume since mass crosses the boundary. The energy balance for this steady-flow system can be expressed as

$$\underbrace{\dot{E}_{in} - \dot{E}_{out}}_{\text{Rate of net energy transfer by heat, work, and mass}} = \underbrace{\Delta \dot{E}_{\text{system}}^{\nearrow 0\text{ (steady)}}}_{\text{Rate of change in internal, kinetic, potential, etc. energies}} = 0$$

$$\dot{E}_{in} = \dot{E}_{out}$$

$$\dot{m}h_1 = \dot{W}_{a,\text{out}} + \dot{m}h_2 \quad (\text{since } \dot{Q} \cong \Delta ke \cong \Delta pe \cong 0)$$

$$\dot{W}_{a,\text{out}} = \dot{m}(h_1 - h_2)$$

Noting that $w_a = \eta_T w_s$, the work output of the turbine per unit mass is determined from

$$w_a = (0.86)(504.71 - 417.3)\text{Btu/lbm} = \mathbf{75.2Btu/lbm}$$

6-100 [*Also solved by EES on enclosed CD*] Refrigerant-134a enters an adiabatic compressor with an isentropic efficiency of 0.80 at a specified state with a specified volume flow rate, and leaves at a specified pressure. The compressor exit temperature and power input to the compressor are to be determined.

Assumptions **1** This is a steady-flow process since there is no change with time. **2** Kinetic and potential energy changes are negligible. **3** The device is adiabatic and thus heat transfer is negligible.

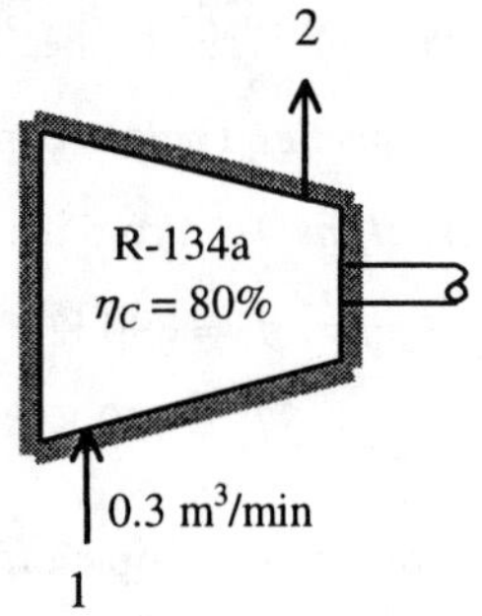

Analysis (a) From the refrigerant tables,

$$\left.\begin{array}{l} P_1 = 120\text{kPa} \\ sat.vapor \end{array}\right\} \begin{array}{l} h_1 = h_{g@120\text{kPa}} = 233.86\text{kJ/kg} \\ s_1 = s_{g@120\text{kPa}} = 0.9354\text{kJ/kg}\cdot\text{K} \\ v_1 = v_{g@120\text{kPa}} = 0.1614\text{m}^3\text{/kg} \end{array}$$

$$\left.\begin{array}{l} P_2 = 1\text{MPa} \\ s_{2s} = s_1 \end{array}\right\} h_{2s} = 277.84\text{kJ/kg}$$

2

R-134a

$\eta_C = 80\%$

0.3 m³/min

1

From the isentropic efficiency relation,

$$\eta_C = \frac{h_{2s} - h_1}{h_{2a} - h_1} \longrightarrow h_{2a} = h_1 + (h_{2s} - h_1)/\eta_C = 233.86 + (277.84 - 233.86)/0.80 = 288.84\text{kJ/kg}$$

Thus,

$$\left.\begin{array}{l} P_{2a} = 1\text{MPa} \\ h_{2a} = 288.84\text{kJ/kg} \end{array}\right\} T_{2a} = \mathbf{57.7^\circ C}$$

(b) The mass flow rate of the refrigerant is determined from

$$\dot{m} = \frac{\dot{V}_1}{v_1} = \frac{0.3/60\ \text{m}^3/\text{s}}{0.1614\ \text{m}^3/\text{kg}} = 0.031\ \text{kg/s}$$

There is only one inlet and one exit, and thus $\dot{m}_1 = \dot{m}_2 = \dot{m}$. We take the actual compressor as the system, which is a control volume since mass crosses the boundary. The energy balance for this steady-flow system can be expressed as

$$\underbrace{\dot{E}_{in} - \dot{E}_{out}}_{\text{Rate of net energy transfer by heat, work, and mass}} = \underbrace{\Delta \dot{E}_{\text{system}}^{\nearrow 0\ \text{(steady)}}}_{\text{Rate of change in internal, kinetic, potential, etc. energies}} = 0$$

$$\dot{E}_{in} = \dot{E}_{out}$$

$$\dot{W}_{a,\text{in}} + \dot{m}h_1 = \dot{m}h_2 \quad (\text{since } \dot{Q} \cong \Delta ke \cong \Delta pe \cong 0)$$

$$\dot{W}_{a,\text{in}} = \dot{m}(h_2 - h_1)$$

Substituting, the power input to the compressor becomes,

$$\dot{W}_{a,in} = (0.031\text{kg/s})(288.84 - 233.86)\text{kJ/kg} = \mathbf{1.70kW}$$

6-101 EES solution of this (and other comprehensive problems designated with the *computer icon*) is available to instructors at the *Instructor Manual* section of the *Online Learning Center* (OLC) at www.mhhe.com/cengel-boles. See the Preface for access information.

6-102 Air enters an adiabatic compressor with an isentropic efficiency of 84% at a specified state, and leaves at a specified temperature. The exit pressure of air and the power input to the compressor are to be determined.

Assumptions **1** This is a steady-flow process since there is no change with time. **2** Kinetic and potential energy changes are negligible. **3** The device is adiabatic and thus heat transfer is negligible. **4** Air is an ideal gas with variable specific heats.

Properties The gas constant of air is $R = 0.287$ kPa.m^3/kg.K (Table A-1)

Analysis (*a*) From the air table (Table A-17),

$$T_1 = 290\text{ K} \longrightarrow h_1 = 290.16\text{ kJ/kg}, \quad P_{r1} = 1.2311$$

$$T_2 = 530\text{ K} \longrightarrow h_{2a} = 533.98\text{ kJ/kg}$$

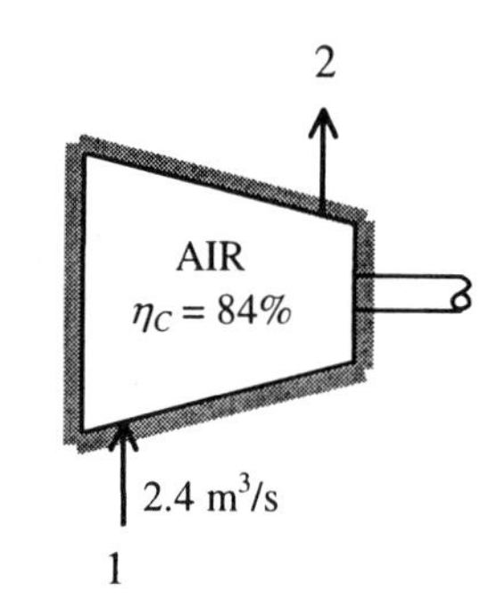

From the isentropic efficiency relation $\eta_C = \dfrac{h_{2s} - h_1}{h_{2a} - h_1}$,

$$h_{2s} = h_1 + \eta_C (h_{2a} - h_1)$$

$$= 290.16 + (0.84)(533.98 - 290.16) = 495.0\text{kJ/kg} \longrightarrow P_{r_2} = 7.951$$

Then from the isentropic relation ,

$$\frac{P_2}{P_1} = \frac{P_{r_2}}{P_{r_1}} \longrightarrow P_2 = \left(\frac{P_{r_2}}{P_{r_1}}\right) P_1 = \left(\frac{7.951}{1.2311}\right)(100\text{kPa}) = \mathbf{646kPa}$$

(*b*) There is only one inlet and one exit, and thus $\dot{m}_1 = \dot{m}_2 = \dot{m}$. We take the actual compressor as the system, which is a control volume since mass crosses the boundary. The energy balance for this steady-flow system can be expressed as

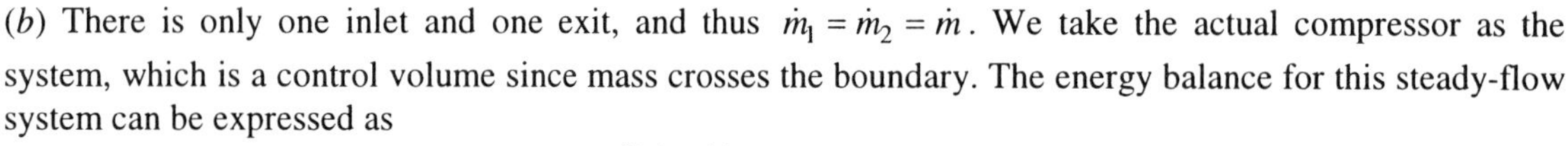

$$\underbrace{\dot{E}_{in} - \dot{E}_{out}}_{\text{Rate of net energy transfer by heat, work, and mass}} = \underbrace{\Delta \dot{E}_{\text{system}}^{\nearrow 0 \text{ (steady)}}}_{\text{Rate of change in internal, kinetic, potential, etc. energies}} = 0$$

$$\dot{E}_{in} = \dot{E}_{out}$$

$$\dot{W}_{a,\text{in}} + \dot{m}h_1 = \dot{m}h_2 \quad (\text{since } \dot{Q} \cong \Delta ke \cong \Delta pe \cong 0)$$

$$\dot{W}_{a,\text{in}} = \dot{m}(h_2 - h_1)$$

where $$\dot{m} = \frac{P_1 \dot{V}_1}{RT_1} = \frac{(100\text{ kPa})(2.4\text{ m}^3/\text{s})}{(0.287\text{ kPa}\cdot\text{m}^3/\text{kg}\cdot\text{K})(290\text{ K})} = 2.884\text{ kg/s}$$

Then the power input to the compressor is determined to be

$$\dot{W}_{a,out} = (2.884\text{ kg/s})(533.98 - 290.16)\text{kJ/kg} = \mathbf{703\ \ kW}$$

6-103 Air is compressed by an adiabatic compressor from a specified state to another specified state. The isentropic efficiency of the compressor and the exit temperature of air for the isentropic case are to be determined.

Assumptions **1** This is a steady-flow process since there is no change with time. **2** Kinetic and potential energy changes are negligible. **3** The device is adiabatic and thus heat transfer is negligible. **4** Air is an ideal gas with variable specific heats.

Analysis From the air table (Table A-17),

$$T_1 = 300 \text{ K} \longrightarrow h_1 = 300.19 \text{ kJ/kg}, \quad P_{r_1} = 1.386$$

$$T_2 = 550 \text{ K} \longrightarrow h_{2a} = 554.74 \text{ kJ/kg}$$

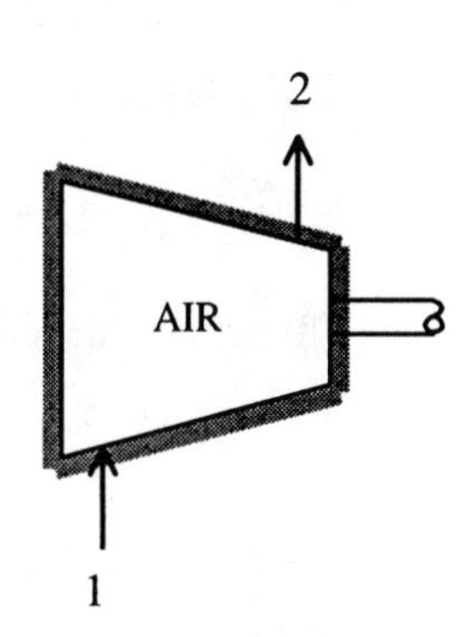

From the isentropic relation,

$$P_{r_2} = \left(\frac{P_2}{P_1}\right)P_{r_1} = \left(\frac{600\text{kPa}}{95\text{kPa}}\right)(1.386) = 8.754 \longrightarrow h_{2s} = 508.72\text{kJ/kg}$$

Then the isentropic efficiency becomes

$$\eta_C = \frac{h_{2s} - h_1}{h_{2a} - h_1} = \frac{508.72 - 300.19}{554.74 - 300.19} = 0.819 = \mathbf{81.9\%}$$

(b) If the process were isentropic, the exit temperature would be

$$h_{2s} = 508.72 \text{ kJ/kg} \longrightarrow T_{2s} = \mathbf{505.5 \text{ K}}$$

6-104E Argon enters an adiabatic compressor with an isentropic efficiency of 80% at a specified state, and leaves at a specified pressure. The exit temperature of argon and the work input to the compressor are to be determined.

Assumptions **1** This is a steady-flow process since there is no change with time. **2** Potential energy changes are negligible. **3** The device is adiabatic and thus heat transfer is negligible. **4** Argon is an ideal gas with constant specific heats.

Properties The specific heat ratio of argon is $k = 1.667$. The constant pressure specific heat of argon is $C_p = 0.1253$ Btu/lbm.R (Table A-2E).

Analysis (*a*) The isentropic exit temperature T_{2s} is determined from

$$T_{2s} = T_1\left(\frac{P_{2s}}{P_1}\right)^{(k-1)/k} = (550\text{R})\left(\frac{200\text{psia}}{20\text{psia}}\right)^{0.667/1.667} = 1381.9\text{R}$$

The actual kinetic energy change during this process is

$$\Delta ke_a = \frac{\mathbf{V}_2^2 - \mathbf{V}_1^2}{2} = \frac{(240\text{ft/s})^2 - (60\text{ft/s})^2}{2}\left(\frac{1\text{Btu/lbm}}{25{,}037\text{ft}^2/\text{s}^2}\right) = 1.08\text{Btu/lbm}$$

2

Ar

$\eta_C = 80\%$

1

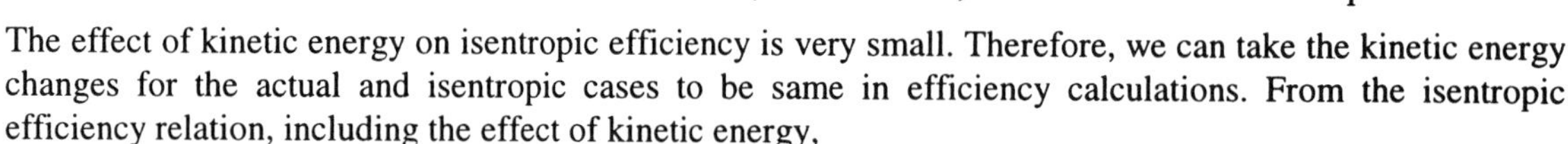

The effect of kinetic energy on isentropic efficiency is very small. Therefore, we can take the kinetic energy changes for the actual and isentropic cases to be same in efficiency calculations. From the isentropic efficiency relation, including the effect of kinetic energy,

$$\eta_C = \frac{w_s}{w_a} = \frac{(h_{2s} - h_1) + \Delta ke}{(h_{2a} - h_1) + \Delta ke} = \frac{C_p(T_{2s} - T_1) + \Delta ke_s}{C_p(T_{2a} - T_1) + \Delta ke_a} \longrightarrow 0.8 = \frac{0.1253(1381.9 - 550) + 1.08}{0.1253(T_{2a} - 550) + 1.08}$$

It yields $\quad T_{2a} = \mathbf{1592\ R}$

(*b*) There is only one inlet and one exit, and thus $\dot{m}_1 = \dot{m}_2 = \dot{m}$. We take the actual compressor as the system, which is a control volume since mass crosses the boundary. The energy balance for this steady-flow system can be expressed as

$$\underbrace{\dot{E}_{in} - \dot{E}_{out}}_{\substack{\text{Rate of net energy transfer}\\ \text{by heat, work, and mass}}} = \underbrace{\Delta \dot{E}_{\text{system}}^{\nearrow 0\ \text{(steady)}}}_{\substack{\text{Rate of change in internal, kinetic,}\\ \text{potential, etc. energies}}} = 0$$

$$\dot{E}_{in} = \dot{E}_{out}$$

$$\dot{W}_{a,\text{in}} + \dot{m}(h_1 + \mathbf{V}_1^2/2) = \dot{m}(h_2 + \mathbf{V}_2^2/2) \quad (\text{since } \dot{Q} \cong \Delta\text{pe} \cong 0)$$

$$\dot{W}_{a,\text{in}} = \dot{m}\left(h_2 - h_1 + \frac{\mathbf{V}_2^2 - \mathbf{V}_1^2}{2}\right) \rightarrow w_{a,\text{in}} = h_2 - h_1 + \Delta ke$$

Substituting, the work input to the compressor is determined to be

$$w_{a,in} = (0.1253\text{Btu/lbm}\cdot\text{R})(1592 - 550)\text{R} + 1.08\text{Btu/lbm} = \mathbf{131.6Btu/lbm}$$

6-105 CO_2 gas is compressed by an adiabatic compressor from a specified state to another specified state. The isentropic efficiency of the compressor is to be determined. √

Assumptions **1** This is a steady-flow process since there is no change with time. **2** Kinetic and potential energy changes are negligible. **3** The device is adiabatic and thus heat transfer is negligible. **4** CO_2 is an ideal gas with constant specific heats.

Properties At the average temperature of (300 + 450)/2 = 375 K, the constant pressure specific heat and the specific heat ratio of CO_2 are $k = 1.260$ and $C_p = 0.917$ kJ/kg.K (Table A-2).

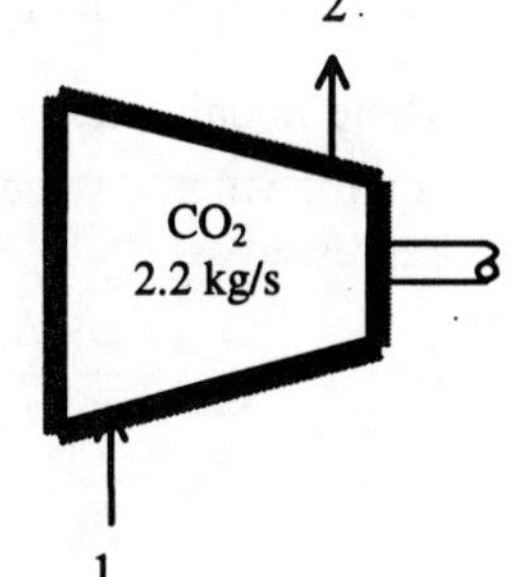

Analysis The isentropic exit temperature T_{2s} is

$$T_{2s} = T_1\left(\frac{P_{2s}}{P_1}\right)^{(k-1)/k} = (300\text{ K})\left(\frac{600\text{kPa}}{100\text{kPa}}\right)^{0.260/1.260} = 434.2\text{ K}$$

From the isentropic efficiency relation,

$$\eta_C = \frac{w_s}{w_a} = \frac{h_{2s}-h_1}{h_{2a}-h_1} = \frac{C_p(T_{2s}-T_1)}{C_p(T_{2a}-T_1)} = \frac{T_{2s}-T_1}{T_{2a}-T_1} = \frac{434.2-300}{450-300} = 0.895 = \mathbf{89.5\%}$$

6-106E Air is accelerated in a 90% efficient adiabatic nozzle from low velocity to a specified velocity. The exit temperature and pressure of the air are to be determined.

Assumptions **1** This is a steady-flow process since there is no change with time. **2** Potential energy changes are negligible. **3** The device is adiabatic and thus heat transfer is negligible. **4** Air is an ideal gas with variable specific heats.

Analysis From the air table (Table A-17),

$$T_1 = 1480 \text{ R} \longrightarrow h_1 = 363.89 \text{ Btu/lbm}, \qquad P_{r_1} = 53.04$$

There is only one inlet and one exit, and thus $\dot{m}_1 = \dot{m}_2 = \dot{m}$. We take the nozzle as the system, which is a control volume since mass crosses the boundary. The energy balance for this steady-flow system can be expressed as

$$\underbrace{\dot{E}_{in} - \dot{E}_{out}}_{\substack{\text{Rate of net energy transfer} \\ \text{by heat, work, and mass}}} = \underbrace{\Delta \dot{E}_{\text{system}}^{\nearrow 0 \text{ (steady)}}}_{\substack{\text{Rate of change in internal, kinetic,} \\ \text{potential, etc. energies}}} = 0$$

$$\dot{E}_{in} = \dot{E}_{out}$$

$$\dot{m}(h_1 + \mathbf{V}_1^2/2) = \dot{m}(h_2 + \mathbf{V}_2^2/2) \quad (\text{since } \dot{W} = \dot{Q} \cong \Delta\text{pe} \cong 0)$$

$$h_2 = h_1 - \frac{\mathbf{V}_2^2 - \mathbf{V}_1^{2 \nearrow 0}}{2}$$

Substituting, the exit temperature of air is determined to be

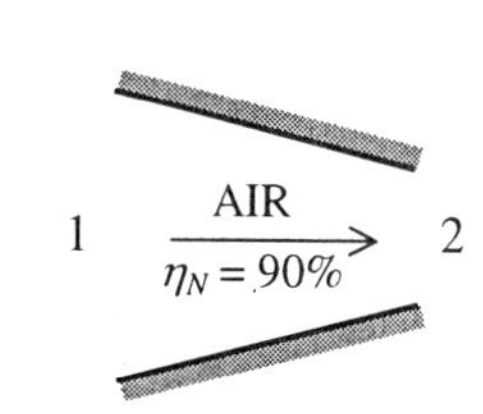

$$h_2 = 363.89 \text{kJ/kg} - \frac{(800 \text{ft/s})^2 - 0}{2}\left(\frac{1 \text{Btu/lbm}}{25{,}037 \text{ft}^2/\text{s}^2}\right) = 351.11 \text{Btu/lbm}$$

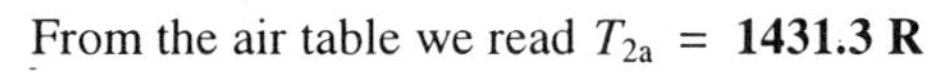

From the air table we read T_{2a} = **1431.3 R**

From the isentropic efficiency relation $\eta_N = \dfrac{(h_{2a} - h_1)}{(h_{2s} - h_1)}$,

$$h_{2s} = h_1 + (h_{2a} - h_1)/\eta_N = 363.89 + (351.11 - 363.89)/(0.90) = 349.69 \text{Btu/lbm} \longrightarrow P_{r_2} = 46.04$$

Then the exit pressure is determined from the isentropic relation to be

$$\frac{P_2}{P_1} = \frac{P_{r_2}}{P_{r_1}} \longrightarrow P_2 = \left(\frac{P_{r_2}}{P_{r_1}}\right)P_1 = \left(\frac{46.04}{53.04}\right)(60 \text{ psia}) = \mathbf{52.1 \text{ psia}}$$

6-107E EES solution of this (and other comprehensive problems designated with the *computer icon*) is available to instructors at the *Instructor Manual* section of the *Online Learning Center* (OLC) at www.mhhe.com/cengel-boles. See the Preface for access information.

6-108 Hot combustion gases are accelerated in a 92% efficient adiabatic nozzle from low velocity to a specified velocity. The exit velocity and the exit temperature are to be determined.

Assumptions **1** This is a steady-flow process since there is no change with time. **2** Potential energy changes are negligible. **3** The device is adiabatic and thus heat transfer is negligible. **4** Combustion gases can be treated as air that is an ideal gas with variable specific heats.

Analysis From the air table (Table A-17),

$$T_1 = 1020\text{ K} \longrightarrow h_1 = 1068.89\text{ kJ/kg}, \quad P_{r_1} = 123.4$$

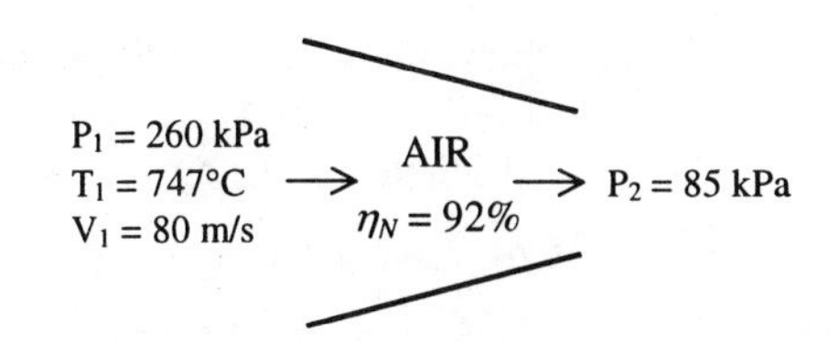

From the isentropic relation ,

$$P_{r_2} = \left(\frac{P_2}{P_1}\right)P_{r_1} = \left(\frac{85\text{kPa}}{260\text{kPa}}\right)(123.4) = 40.34 \longrightarrow h_{2s} = 783.92\text{kJ/kg}$$

There is only one inlet and one exit, and thus $\dot{m}_1 = \dot{m}_2 = \dot{m}$. We take the nozzle as the system, which is a control volume since mass crosses the boundary. The energy balance for this steady-flow system for the isentropic process can be expressed as

$$\underbrace{\dot{E}_{in} - \dot{E}_{out}}_{\substack{\text{Rate of net energy transfer} \\ \text{by heat, work, and mass}}} = \underbrace{\Delta \dot{E}_{\text{system}}^{\nearrow 0\text{ (steady)}}}_{\substack{\text{Rate of change in internal, kinetic,} \\ \text{potential, etc. energies}}} = 0$$

$$\dot{E}_{in} = \dot{E}_{out}$$

$$\dot{m}(h_1 + \mathbf{V}_1^2/2) = \dot{m}(h_{2s} + \mathbf{V}_{2s}^2/2) \quad (\text{since } \dot{W} = \dot{Q} \cong \Delta\text{pe} \cong 0)$$

$$h_{2s} = h_1 - \frac{\mathbf{V}_{2s}^2 - \mathbf{V}_1^2}{2}$$

Then the isentropic exit velocity becomes

$$\mathbf{V}_{2s} = \sqrt{\mathbf{V}_1^2 + 2(h_1 - h_{2s})} = \sqrt{(80\text{m/s})^2 + 2(1068.89 - 783.92)\text{kJ/kg}\left(\frac{1000\text{m}^2/\text{s}^2}{1\text{kJ/kg}}\right)} = 759.2\text{m/s}$$

Therefore,

$$\mathbf{V}_{2a} = \sqrt{\eta_N}\,\mathbf{V}_{2s} = \sqrt{0.92}(759.2\text{m/s}) = \mathbf{728.2m/s}$$

The exit temperature of air is determined from the steady-flow energy equation,

$$h_{2a} = 1068.89\text{kJ/kg} - \frac{(728.2\text{m/s})^2 - (80\text{m/s})^2}{2}\left(\frac{1\text{kJ/kg}}{1000\text{m}^2/\text{s}^2}\right) = 806.95\text{kJ/kg}$$

From the air table we read T_{2a} = **786.3 K**

Entropy Balance

6-109 Each member of a family of four take a 5-min shower every day. The amount of entropy generated by this family per year is to be determined.

Assumptions **1** Steady operating conditions exist. **2** The kinetic and potential energies are negligible. **3** Heat losses from the pipes and the mixing section are negligible and thus $\dot{Q} \cong 0$. **4** Showers operate at maximum flow conditions during the entire shower. **5** Each member of the household takes a 5-min shower every day. **6** Water is an incompressible substance with constant properties at room temperature. **7** The efficiency of the electric water heater is 100%.

Properties The density and specific heat of water at room temperature are ρ = 1 kg/L = 1000 kg/^{3} and C = 4.18 kJ/kg.°C (Table A-3).

Mixture
3

Cold
water
1

Hot
water
2

Analysis The mass flow rate of water at the shower head is

$$\dot{m} = \rho \dot{V} = (1 \text{ kg/L})(12 \text{ L/min}) = 12 \text{ kg/min}$$

The mass balance for the mixing chamber can be expressed in the rate form as

$$\dot{m}_{in} - \dot{m}_{out} = \Delta \dot{m}_{system}{}^{\nearrow 0 \text{ (steady)}} = 0 \rightarrow \dot{m}_{in} = \dot{m}_{out} \rightarrow \dot{m}_1 + \dot{m}_2 = \dot{m}_3$$

where the subscript 1 denotes the cold water stream, 2 the hot water stream, and 3 the mixture.

The rate of entropy generation during this process can be determined by applying the rate form of the entropy balance on a system that includes the electric water heater and the mixing chamber (the T-elbow). Noting that there is no entropy transfer associated with work transfer (electricity) and there is no heat transfer, the entropy balance for this steady-flow system can be expressed as

$$\underbrace{\dot{S}_{in} - \dot{S}_{out}}_{\substack{\text{Rate of net entropy transfer}\\ \text{by heat and mass}}} + \underbrace{\dot{S}_{gen}}_{\substack{\text{Rate of entropy}\\ \text{generation}}} = \underbrace{\Delta \dot{S}_{system}{}^{\nearrow 0 \text{ (steady)}}}_{\substack{\text{Rate of change}\\ \text{of entropy}}}$$

$$\dot{m}_1 s_1 + \dot{m}_2 s_2 - \dot{m}_3 s_3 + \dot{S}_{gen} = 0 \quad (\text{since } Q = 0 \text{ and work is entropy free})$$

$$\dot{S}_{gen} = \dot{m}_3 s_3 - \dot{m}_1 s_1 - \dot{m}_2 s_2$$

Noting from mass balance that $\dot{m}_1 + \dot{m}_2 = \dot{m}_3$ and $s_2 = s_1$ since hot water enters the system at the same temperature as the cold water, the rate of entropy generation is determined to be

$$\dot{S}_{gen} = \dot{m}_3 s_3 - (\dot{m}_1 + \dot{m}_2) s_1 = \dot{m}_3 (s_3 - s_1) = \dot{m}_3 C_p \ln \frac{T_3}{T_1}$$

$$= (12 \text{ kg/min})(4.18 \text{ kJ/kg.K}) \ln \frac{42 + 273}{15 + 273} = 4.495 \text{ kJ/min.K}$$

Noting that 4 people take a 5-min shower every day, the amount of entropy generated per year is

$$S_{gen} = (\dot{S}_{gen}) \Delta t (\text{No. of people})(\text{No. of days})$$

$$= (4.495 \text{ kJ/min.K})(5 \text{ min/person} \cdot \text{day})(4 \text{ persons})(365 \text{ days/year})$$

$$= \mathbf{32{,}814 \text{ kJ/K}} \quad (\text{per year})$$

Discussion *The value above represents the entropy generated within the water heater and the T-elbow in the absence of any heat losses. It does not include the entropy generated as the shower water at 42°C is discarded or cooled to the outdoor temperature. Also, an entropy balance on the mixing chamber alone (hot water entering at 55°C instead of 15°C) will exclude the entropy generated within the water heater.*

6-110 Steam is condensed by cooling water in the condenser of a power plant. The rate of condensation of steam and the rate of entropy generation are to be determined.

Assumptions **1** Steady operating conditions exist. **2** The heat exchanger is well-insulated so that heat loss to the surroundings is negligible and thus heat transfer from the hot fluid is equal to the heat transfer to the cold fluid. **3** Changes in the kinetic and potential energies of fluid streams are negligible. **4** Fluid properties are constant.

Properties The enthalpy and entropy of vaporization of water at 50°C are h_{fg} =2382.7 kJ/kg and s_{fg}= 7.3725 kJ/kg.K (Table A-4). The specific heat of water at room temperature is C_p = 4.18 kJ/kg.°C (Table A-3).

Analysis (*a*) We take the cold water tubes as the system, which is a control volume. The energy balance for this steady-flow system can be expressed in the rate form as

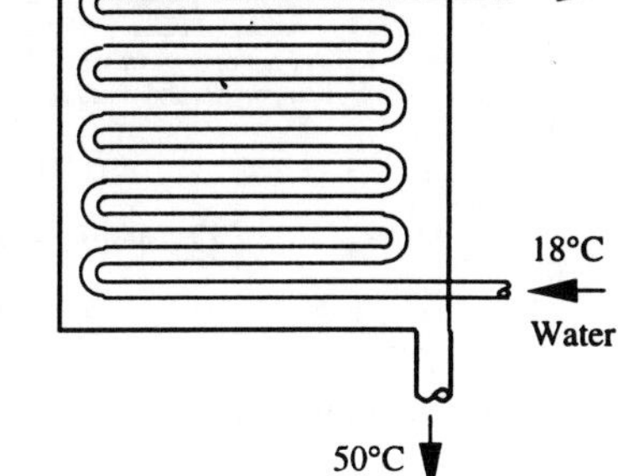

$$\underbrace{\dot{E}_{in} - \dot{E}_{out}}_{\text{Rate of net energy transfer by heat, work, and mass}} = \underbrace{\Delta \dot{E}_{system}^{\nearrow 0 \text{ (steady)}}}_{\text{Rate of change in internal, kinetic, potential, etc. energies}} = 0$$

$$\dot{E}_{in} = \dot{E}_{out}$$

$$\dot{Q}_{in} + \dot{m}h_1 = \dot{m}h_2 \quad (\text{since } \Delta\text{ke} \cong \Delta\text{pe} \cong 0)$$

$$\dot{Q}_{in} = \dot{m}C_p(T_2 - T_1)$$

Then the heat transfer rate to the cooling water in the condenser becomes

$$\dot{Q} = [\dot{m}C_p(T_{out} - T_{in})]_{\text{cooling water}}$$
$$= (101\,\text{kg/s})(4.18\,\text{kJ/kg.°C})(27°\text{C} - 18°\text{C}) = 3800\,\text{kJ/s}$$

The rate of condensation of steam is determined to be

$$\dot{Q} = (\dot{m}h_{fg})_{steam} \longrightarrow \dot{m}_{steam} = \frac{\dot{Q}}{h_{fg}} = \frac{3800\text{kJ/s}}{2382.7\text{ kJ/kg}} = \mathbf{1.595\ kg/s}$$

(*b*) The rate of entropy generation within the condenser during this process can be determined by applying the rate form of the entropy balance on the entire condenser. Noting that the condenser is well-insulated and thus heat transfer is negligible, the entropy balance for this steady-flow system can be expressed as

$$\underbrace{\dot{S}_{in} - \dot{S}_{out}}_{\text{Rate of net entropy transfer by heat and mass}} + \underbrace{\dot{S}_{gen}}_{\text{Rate of entropy generation}} = \underbrace{\Delta \dot{S}_{system}^{\nearrow 0 \text{ (steady)}}}_{\text{Rate of change of entropy}}$$

$$\dot{m}_1 s_1 + \dot{m}_3 s_3 - \dot{m}_2 s_2 - \dot{m}_4 s_4 + \dot{S}_{gen} = 0 \quad (\text{since } Q = 0)$$

$$\dot{m}_{\text{water}} s_1 + \dot{m}_{\text{steam}} s_3 - \dot{m}_{\text{water}} s_2 - \dot{m}_{\text{steam}} s_4 + \dot{S}_{gen} = 0$$

$$\dot{S}_{gen} = \dot{m}_{\text{water}}(s_2 - s_1) + \dot{m}_{\text{steam}}(s_4 - s_3)$$

Noting that water is an incompressible substance and steam changes from saturated vapor to saturated liquid, the rate of entropy generation is determined to be

$$\dot{S}_{gen} = \dot{m}_{\text{water}} C_p \ln\frac{T_2}{T_1} + \dot{m}_{\text{steam}}(s_f - s_g) = \dot{m}_{\text{water}} C_p \ln\frac{T_2}{T_1} - \dot{m}_{\text{steam}} s_{fg}$$

$$= (101\,\text{kg/s})(4.18\,\text{kJ/kg.K})\ln\frac{27+273}{18+273} - (1.595\,\text{kg/s})(7.3725\,\text{kJ/kg.K}) = 1.100\,\text{kW/K}$$

6-111 Water is heated in a heat exchanger by geothermal water. The rate of heat transfer to the water and the rate of entropy generation within the heat exchanger are to be determined.

Assumptions **1** Steady operating conditions exist. **2** The heat exchanger is well-insulated so that heat loss to the surroundings is negligible and thus heat transfer from the hot fluid is equal to the heat transfer to the cold fluid. **3** Changes in the kinetic and potential energies of fluid streams are negligible. **4** Fluid properties are constant.

Properties The specific heats of water and geothermal fluid are given to be 4.18 and 4.31 kJ/kg.°C, respectively.

Analysis (*a*) We take the cold water tubes as the system, which is a control volume. The energy balance for this steady-flow system can be expressed in the rate form as

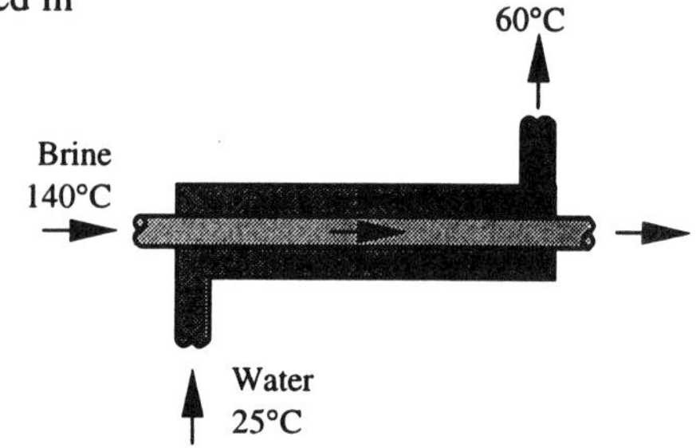

$$\underbrace{\dot{E}_{in} - \dot{E}_{out}}_{\text{Rate of net energy transfer by heat, work, and mass}} = \underbrace{\Delta \dot{E}_{\text{system}}^{\nearrow 0 \text{ (steady)}}}_{\text{Rate of change in internal, kinetic, potential, etc. energies}} = 0$$

$$\dot{E}_{in} = \dot{E}_{out}$$

$$\dot{Q}_{in} + \dot{m}h_1 = \dot{m}h_2 \quad (\text{since } \Delta\text{ke} \cong \Delta\text{pe} \cong 0)$$

$$\dot{Q}_{in} = \dot{m}C_p(T_2 - T_1)$$

Then the rate of heat transfer to the cold water in the heat exchanger becomes

$$\dot{Q}_{in,water} = [\dot{m}C_p(T_{out} - T_{in})]_{\text{water}} = (0.50\text{ kg/s})(4.18\text{ kJ/kg.°C})(60°\text{C} - 25°\text{C}) = \mathbf{73.15\text{ kW}}$$

Noting that heat transfer to the cold water is equal to the heat loss from the geothermal water, the outlet temperature of the geothermal water is determined from

$$\dot{Q}_{out} = [\dot{m}C_p(T_{in} - T_{out})]_{\text{geot.water}} \longrightarrow T_{out} = T_{in} - \frac{\dot{Q}_{out}}{\dot{m}C_p}$$

$$= 140°\text{C} - \frac{73.15\text{ kW}}{(0.75\text{ kg/s})(4.31\text{ kJ/kg.°C})} = \mathbf{117.4°C}$$

(*b*) The rate of entropy generation within the heat exchanger is determined by applying the rate form of the entropy balance on the entire heat exchanger:

$$\underbrace{\dot{S}_{in} - \dot{S}_{out}}_{\text{Rate of net entropy transfer by heat and mass}} + \underbrace{\dot{S}_{gen}}_{\text{Rate of entropy generation}} = \underbrace{\Delta \dot{S}_{\text{system}}^{\nearrow 0 \text{ (steady)}}}_{\text{Rate of change of entropy}}$$

$$\dot{m}_1 s_1 + \dot{m}_3 s_3 - \dot{m}_2 s_2 - \dot{m}_4 s_4 + \dot{S}_{gen} = 0 \quad (\text{since } Q = 0)$$

$$\dot{m}_{\text{water}} s_1 + \dot{m}_{\text{geo}} s_3 - \dot{m}_{\text{water}} s_2 - \dot{m}_{\text{geo}} s_4 + \dot{S}_{gen} = 0$$

$$\dot{S}_{gen} = \dot{m}_{\text{water}}(s_2 - s_1) + \dot{m}_{\text{geo}}(s_4 - s_3)$$

Noting that both fresh and geothermal water are incompressible substances, the rate of entropy generation is determined to be

$$\dot{S}_{gen} = \dot{m}_{\text{water}} C_p \ln\frac{T_2}{T_1} + \dot{m}_{\text{geo}} C_p \ln\frac{T_4}{T_3}$$

$$= (0.50\text{ kg/s})(4.18\text{ kJ/kg.K})\ln\frac{60+273}{25+273} + (0.75\text{ kg/s})(4.31\text{ kJ/kg.K})\ln\frac{117.4+273}{140+273} = \mathbf{0.050\text{ kW/K}}$$

6-112 Ethylene glycol is cooled by water in a heat exchanger. The rate of heat transfer and the rate of entropy generation within the heat exchanger are to be determined.

Assumptions **1** Steady operating conditions exist. **2** The heat exchanger is well-insulated so that heat loss to the surroundings is negligible and thus heat transfer from the hot fluid is equal to the heat transfer to the cold fluid. **3** Changes in the kinetic and potential energies of fluid streams are negligible. **4** Fluid properties are constant.

Properties The specific heats of water and ethylene glycol are given to be 4.18 and 2.56 kJ/kg.°C, respectively.

Analysis (a) We take the ethylene glycol tubes as the system, which is a control volume. The energy balance for this steady-flow system can be expressed in the rate form as

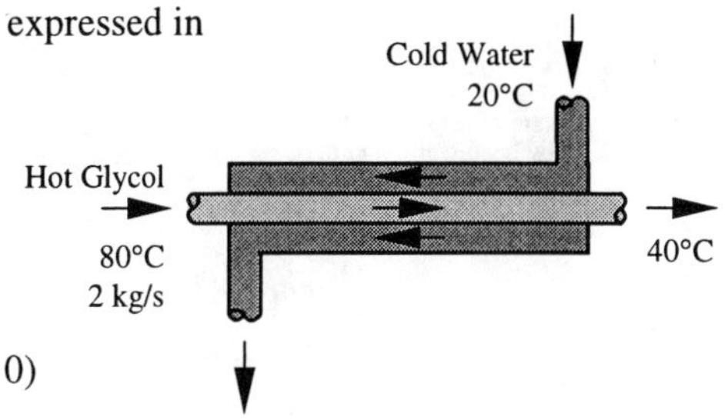

Cold Water
20°C
Hot Glycol
80°C
2 kg/s
40°C

$$\underbrace{\dot{E}_{in} - \dot{E}_{out}}_{\text{Rate of net energy transfer by heat, work, and mass}} = \underbrace{\Delta \dot{E}_{system}^{\nearrow 0 \text{ (steady)}}}_{\text{Rate of change in internal, kinetic, potential, etc. energies}} = 0$$

$$\dot{E}_{in} = \dot{E}_{out}$$

$$\dot{m}h_1 = \dot{Q}_{out} + \dot{m}h_2 \quad (\text{since } \Delta \text{ke} \cong \Delta \text{pe} \cong 0)$$

$$\dot{Q}_{out} = \dot{m}C_p(T_1 - T_2)$$

Then the rate of heat transfer becomes

$$\dot{Q}_{out} = [\dot{m}C_p(T_{in} - T_{out})]_{\text{glycol}} = (2 \text{ kg/s})(2.56 \text{ kJ/kg.°C})(80°\text{C} - 40°\text{C}) = \mathbf{204.8 \text{ kW}}$$

The rate of heat transfer from water must be equal to the rate of heat transfer to the glycol. Then,

$$\dot{Q}_{in} = [\dot{m}C_p(T_{out} - T_{in})]_{\text{water}} \longrightarrow \dot{m}_{\text{water}} = \frac{\dot{Q}_{in}}{C_p(T_{out} - T_{in})}$$

$$= \frac{204.8 \text{ kJ/s}}{(4.18 \text{ kJ/kg.°C})(55°\text{C} - 20°\text{C})} = \mathbf{1.4 \text{ kg/s}}$$

(*b*) The rate of entropy generation within the heat exchanger is determined by applying the rate form of the entropy balance on the entire heat exchanger:

$$\underbrace{\dot{S}_{in} - \dot{S}_{out}}_{\text{Rate of net entropy transfer by heat and mass}} + \underbrace{\dot{S}_{gen}}_{\text{Rate of entropy generation}} = \underbrace{\Delta \dot{S}_{\text{system}}^{\nearrow 0 \text{ (steady)}}}_{\text{Rate of change of entropy}}$$

$$\dot{m}_1 s_1 + \dot{m}_3 s_3 - \dot{m}_2 s_2 - \dot{m}_3 s_4 + \dot{S}_{gen} = 0 \quad (\text{since } Q = 0)$$

$$\dot{m}_{\text{glycol}} s_1 + \dot{m}_{\text{water}} s_3 - \dot{m}_{\text{glycol}} s_2 - \dot{m}_{\text{water}} s_4 + \dot{S}_{gen} = 0$$

$$\dot{S}_{gen} = \dot{m}_{\text{glycol}}(s_2 - s_1) + \dot{m}_{\text{water}}(s_4 - s_3)$$

Noting that both fluid streams are liquids (incompressible substances), the rate of entropy generation is determined to be

$$\dot{S}_{gen} = \dot{m}_{\text{glycol}} C_p \ln \frac{T_2}{T_1} + \dot{m}_{\text{water}} C_p \ln \frac{T_4}{T_3}$$

$$= (2 \text{ kg/s})(2.56 \text{ kJ/kg.K}) \ln \frac{40 + 273}{80 + 273} + (1.4 \text{ kg/s})(4.18 \text{ kJ/kg.K}) \ln \frac{55 + 273}{20 + 273} = \mathbf{0.0446 \text{ kW/K}}$$

6-113 Oil is to be cooled by water in a thin-walled heat exchanger. The rate of heat transfer and the rate of entropy generation within the heat exchanger are to be determined.

Assumptions **1** Steady operating conditions exist. **2** The heat exchanger is well-insulated so that heat loss to the surroundings is negligible and thus heat transfer from the hot fluid is equal to the heat transfer to the cold fluid. **3** Changes in the kinetic and potential energies of fluid streams are negligible. **4** Fluid properties are constant.

Properties The specific heats of water and oil are given to be 4.18 and 2.20 kJ/kg.°C, respectively.

Analysis We take the oil tubes as the system, which is a control volume. The energy balance for this steady-flow system can be expressed in the rate form as

$$\underbrace{\dot{E}_{in} - \dot{E}_{out}}_{\text{Rate of net energy transfer by heat, work, and mass}} = \underbrace{\Delta \dot{E}_{system}^{\nearrow 0\ \text{(steady)}}}_{\text{Rate of change in internal, kinetic, potential, etc. energies}} = 0$$

$$\dot{E}_{in} = \dot{E}_{out}$$

$$\dot{m}h_1 = \dot{Q}_{out} + \dot{m}h_2 \quad (\text{since } \Delta\text{ke} \cong \Delta\text{pe} \cong 0)$$

$$\dot{Q}_{out} = \dot{m}C_p(T_1 - T_2)$$

Hot oil
150°C
2 kg/s
Cold water
22°C
1.5 kg/s

Then the rate of heat transfer from the oil becomes

$$\dot{Q}_{out} = [\dot{m}C_p(T_{in} - T_{out})]_{\text{oil}} = (2\ \text{kg/s})(2.2\ \text{kJ/kg.°C})(150\text{°C} - 40\text{°C}) = \mathbf{484\ kW}$$

Noting that the heat lost by the oil is gained by the water, the outlet temperature of the water is determined from

$$Q = [mC_p(T_{out} - T_{in})]_{\text{water}} \longrightarrow T_{out} = T_{in} + \frac{Q}{mC_p}$$

$$= 22\text{°C} + \frac{484\ \text{kW}}{(1.5\ \text{kg/s})(4.18\ \text{kJ/kg.°C})} = 99.2\text{°C}$$

(*b*) The rate of entropy generation within the heat exchanger is determined by applying the rate form of the entropy balance on the entire heat exchanger:

$$\underbrace{\dot{S}_{in} - \dot{S}_{out}}_{\text{Rate of net entropy transfer by heat and mass}} + \underbrace{\dot{S}_{gen}}_{\text{Rate of entropy generation}} = \underbrace{\Delta \dot{S}_{system}^{\nearrow 0\ \text{(steady)}}}_{\text{Rate of change of entropy}}$$

$$\dot{m}_1 s_1 + \dot{m}_3 s_3 - \dot{m}_2 s_2 - \dot{m}_3 s_4 + \dot{S}_{gen} = 0 \quad (\text{since } Q = 0)$$

$$\dot{m}_{\text{oil}} s_1 + \dot{m}_{\text{water}} s_3 - \dot{m}_{\text{oil}} s_2 - \dot{m}_{\text{water}} s_4 + \dot{S}_{gen} = 0$$

$$\dot{S}_{gen} = \dot{m}_{\text{oil}}(s_2 - s_1) + \dot{m}_{\text{water}}(s_4 - s_3)$$

Noting that both fluid streams are liquids (incompressible substances), the rate of entropy generation is determined to be

$$\dot{S}_{gen} = \dot{m}_{\text{oil}} C_p \ln\frac{T_2}{T_1} + \dot{m}_{\text{water}} C_p \ln\frac{T_4}{T_3}$$

$$= (2\ \text{kg/s})(2.2\ \text{kJ/kg.K})\ln\frac{40+273}{150+273} + (1.5\ \text{kg/s})(4.18\ \text{kJ/kg.K})\ln\frac{99.2+273}{22+273} = \mathbf{0.132\ kW/K}$$

6-114 Cold water is heated by hot water in a heat exchanger. The rate of heat transfer and the rate of entropy generation within the heat exchanger are to be determined.

Assumptions **1** Steady operating conditions exist. **2** The heat exchanger is well-insulated so that heat loss to the surroundings is negligible and thus heat transfer from the hot fluid is equal to the heat transfer to the cold fluid. **3** Changes in the kinetic and potential energies of fluid streams are negligible. **4** Fluid properties are constant.

Properties The specific heats of cold and hot water are given to be 4.18 and 4.19 kJ/kg.°C, respectively.

Analysis We take the cold water tubes as the system, which is a control volume. The energy balance for this steady-flow system can be expressed in the rate form as

Cold Water 15°C 0.25 kg/s

Hot water 100°C 3 kg/s

$$\underbrace{\dot{E}_{in} - \dot{E}_{out}}_{\text{Rate of net energy transfer by heat, work, and mass}} = \underbrace{\Delta \dot{E}_{system}^{\nearrow 0 \text{ (steady)}}}_{\text{Rate of change in internal, kinetic, potential, etc. energies}} = 0$$

$$\dot{E}_{in} = \dot{E}_{out}$$

$$\dot{Q}_{in} + \dot{m}h_1 = \dot{m}h_2 \quad (\text{since } \Delta\text{ke} \cong \Delta\text{pe} \cong 0)$$

$$\dot{Q}_{in} = \dot{m}C_p(T_2 - T_1)$$

Then the rate of heat transfer to the cold water in this heat exchanger becomes

$$\dot{Q}_{in} = [\dot{m}C_p(T_{out} - T_{in})]_{\text{cold water}} = (0.25 \text{ kg/s})(4.18 \text{ kJ/kg.°C})(45°\text{C} - 15°\text{C}) = \mathbf{31.35 \text{ kW}}$$

Noting that heat gain by the cold water is equal to the heat loss by the hot water, the outlet temperature of the hot water is determined to be

$$\dot{Q} = [\dot{m}C_p(T_{in} - T_{out})]_{\text{hot water}} \longrightarrow T_{out} = T_{in} - \frac{\dot{Q}}{\dot{m}C_p}$$

$$= 100°\text{C} - \frac{31.35 \text{ kW}}{(3 \text{ kg/s})(4.19 \text{ kJ/kg.°C})} = 97.5°\text{C}$$

(*b*) The rate of entropy generation within the heat exchanger is determined by applying the rate form of the entropy balance on the entire heat exchanger:

$$\underbrace{\dot{S}_{in} - \dot{S}_{out}}_{\text{Rate of net entropy transfer by heat and mass}} + \underbrace{\dot{S}_{gen}}_{\text{Rate of entropy generation}} = \underbrace{\Delta \dot{S}_{system}^{\nearrow 0 \text{ (steady)}}}_{\text{Rate of change of entropy}}$$

$$\dot{m}_1 s_1 + \dot{m}_3 s_3 - \dot{m}_2 s_2 - \dot{m}_3 s_4 + \dot{S}_{gen} = 0 \quad (\text{since } Q = 0)$$

$$\dot{m}_{\text{cold}} s_1 + \dot{m}_{\text{hot}} s_3 - \dot{m}_{\text{cold}} s_2 - \dot{m}_{\text{hot}} s_4 + \dot{S}_{gen} = 0$$

$$\dot{S}_{gen} = \dot{m}_{\text{cold}}(s_2 - s_1) + \dot{m}_{\text{hot}}(s_4 - s_3)$$

Noting that both fluid streams are liquids (incompressible substances), the rate of entropy generation is determined to be

$$\dot{S}_{gen} = \dot{m}_{\text{cold}} C_p \ln\frac{T_2}{T_1} + \dot{m}_{\text{hot}} C_p \ln\frac{T_4}{T_3}$$

$$= (0.25 \text{ kg/s})(4.18 \text{ kJ/kg.K})\ln\frac{45+273}{15+273} + (3 \text{ kg/s})(4.19 \text{ kJ/kg.K})\ln\frac{97.5+273}{100+273} = \mathbf{0.0190 \text{ kW/K}}$$

6-115 Air is preheated by hot exhaust gases in a cross-flow heat exchanger. The rate of heat transfer, the outlet temperature of the air, and the rate of entropy generation are to be determined.

Assumptions **1** Steady operating conditions exist. **2** The heat exchanger is well-insulated so that heat loss to the surroundings is negligible and thus heat transfer from the hot fluid is equal to the heat transfer to the cold fluid. **3** Changes in the kinetic and potential energies of fluid streams are negligible. **4** Fluid properties are constant.

Properties The specific heats of air and combustion gases are given to be 1.005 and 1.10 kJ/kg.°C, respectively. The gas constant of air is R = 0.287 kJ/kg.K (Table A-1).

Analysis We take the exhaust pipes as the system, which is a control volume. The energy balance for this steady-flow system can be expressed in the rate form as

$$\underbrace{\dot{E}_{in} - \dot{E}_{out}}_{\substack{\text{Rate of net energy transfer}\\ \text{by heat, work, and mass}}} = \underbrace{\Delta \dot{E}_{\text{system}}^{\nearrow 0 \text{ (steady)}}}_{\substack{\text{Rate of change in internal, kinetic,}\\ \text{potential, etc. energies}}} = 0$$

$$\dot{E}_{in} = \dot{E}_{out}$$

$$\dot{m}h_1 = \dot{Q}_{out} + \dot{m}h_2 \quad (\text{since } \Delta\text{ke} \cong \Delta\text{pe} \cong 0)$$

$$\dot{Q}_{out} = \dot{m}C_p(T_1 - T_2)$$

Air
95 kPa
20°C
1.6 m³/s

Exhaust gases
2.2 kg/s, 95°C

Then the rate of heat transfer from the exhaust gases becomes

$$\dot{Q} = [\dot{m}C_p(T_{in} - T_{out})]_{\text{gas.}} = (2.2\text{ kg/s})(1.1\text{ kJ/kg.°C})(180\text{°C} - 95\text{°C}) = \mathbf{205.7\ kW}$$

The mass flow rate of air is

$$\dot{m} = \frac{P\dot{V}}{RT} = \frac{(95\text{ kPa})(1.6\text{ m}^3/\text{s})}{(0.287\text{ kPa.m}^3/\text{kg.K}) \times (293\text{ K})} = 1.808\text{ kg/s}$$

Noting that heat loss by the exhaust gases is equal to the heat gain by the air, the outlet temperature of the air becomes

$$\dot{Q} = \dot{m}C_p(T_{c,out} - T_{c,in}) \rightarrow T_{c,out} = T_{c,in} + \frac{\dot{Q}}{\dot{m}C_p} = 20\text{°C} + \frac{205.7\text{ kW}}{(1.808\text{ kg/s})(1.005\text{ kJ/kg.°C})} = \mathbf{133.2°C}$$

The rate of entropy generation within the heat exchanger is determined by applying the rate form of the entropy balance on the entire heat exchanger:

$$\underbrace{\dot{S}_{in} - \dot{S}_{out}}_{\substack{\text{Rate of net entropy transfer}\\ \text{by heat and mass}}} + \underbrace{\dot{S}_{gen}}_{\substack{\text{Rate of entropy}\\ \text{generation}}} = \underbrace{\Delta \dot{S}_{\text{system}}^{\nearrow 0 \text{ (steady)}}}_{\substack{\text{Rate of change}\\ \text{of entropy}}}$$

$$\dot{m}_1 s_1 + \dot{m}_3 s_3 - \dot{m}_2 s_2 - \dot{m}_3 s_4 + \dot{S}_{gen} = 0 \quad (\text{since } Q = 0)$$

$$\dot{m}_{\text{exhaust}} s_1 + \dot{m}_{\text{air}} s_3 - \dot{m}_{\text{exhaust}} s_2 - \dot{m}_{\text{air}} s_4 + \dot{S}_{gen} = 0$$

$$\dot{S}_{gen} = \dot{m}_{\text{exhaust}}(s_2 - s_1) + \dot{m}_{\text{air}}(s_4 - s_3)$$

Then the rate of entropy generation is determined to be

$$\dot{S}_{gen} = \dot{m}_{\text{exhaust}} C_p \ln\frac{T_2}{T_1} + \dot{m}_{\text{air}} C_p \ln\frac{T_4}{T_3}$$

$$= (2.2\text{ kg/s})(1.1\text{ kJ/kg.K})\ln\frac{95+273}{180+273} + (1.808\text{ kg/s})(1.005\text{ kJ/kg.K})\ln\frac{133.2+273}{20+273} = \mathbf{0.091\ kW/K}$$

6-116 Water is heated by hot oil in a heat exchanger. The outlet temperature of the oil and the rate of entropy generation within the heat exchanger are to be determined.

Assumptions **1** Steady operating conditions exist. **2** The heat exchanger is well-insulated so that heat loss to the surroundings is negligible and thus heat transfer from the hot fluid is equal to the heat transfer to the cold fluid. **3** Changes in the kinetic and potential energies of fluid streams are negligible. **4** Fluid properties are constant.

Properties The specific heats of water and oil are given to be 4.18 and 2.3 kJ/kg.°C, respectively.

Analysis (*a*) We take the cold water tubes as the system, which is a control volume. The energy balance for this steady-flow system can be expressed in the rate form as

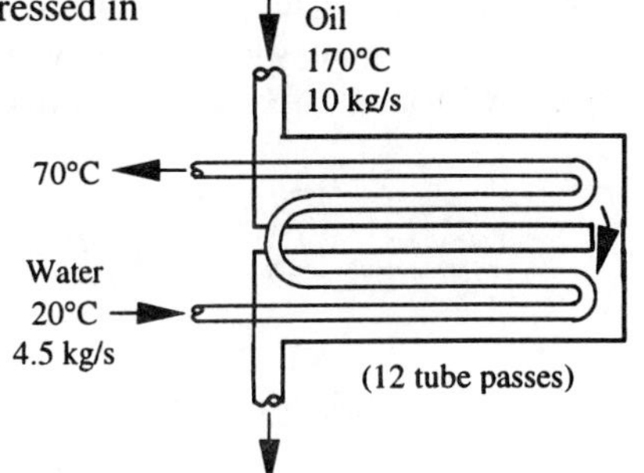

$$\underbrace{\dot{E}_{in} - \dot{E}_{out}}_{\text{Rate of net energy transfer by heat, work, and mass}} = \underbrace{\Delta \dot{E}_{\text{system}}^{\nearrow 0 \text{ (steady)}}}_{\text{Rate of change in internal, kinetic, potential, etc. energies}} = 0$$

$$\dot{E}_{in} = \dot{E}_{out}$$

$$\dot{Q}_{in} + \dot{m}h_1 = \dot{m}h_2 \quad (\text{since } \Delta\text{ke} \cong \Delta\text{pe} \cong 0)$$

$$\dot{Q}_{in} = \dot{m}C_p(T_2 - T_1)$$

Then the rate of heat transfer to the cold water in this heat exchanger becomes

$$\dot{Q} = [\dot{m}C_p(T_{out} - T_{in})]_{\text{water}} = (4.5 \text{ kg/s})(4.18 \text{ kJ/kg.°C})(70\text{°C} - 20\text{°C}) = 940.5 \text{ kW}$$

Noting that heat gain by the water is equal to the heat loss by the oil, the outlet temperature of the hot oil is determined from

$$\dot{Q} = [\dot{m}C_p(T_{in} - T_{out})]_{\text{oil}} \rightarrow T_{out} = T_{in} - \frac{\dot{Q}}{\dot{m}C_p} = 170\text{°C} - \frac{940.5 \text{ kW}}{(10 \text{ kg/s})(2.3 \text{ kJ/kg.°C})} = \mathbf{129.1°C}$$

(*b*) The rate of entropy generation within the heat exchanger is determined by applying the rate form of the entropy balance on the entire heat exchanger:

$$\underbrace{\dot{S}_{in} - \dot{S}_{out}}_{\text{Rate of net entropy transfer by heat and mass}} + \underbrace{\dot{S}_{gen}}_{\text{Rate of entropy generation}} = \underbrace{\Delta \dot{S}_{\text{system}}^{\nearrow 0 \text{ (steady)}}}_{\text{Rate of change of entropy}}$$

$$\dot{m}_1 s_1 + \dot{m}_3 s_3 - \dot{m}_2 s_2 - \dot{m}_3 s_4 + \dot{S}_{gen} = 0 \quad (\text{since } Q = 0)$$

$$\dot{m}_{\text{water}} s_1 + \dot{m}_{\text{oil}} s_3 - \dot{m}_{\text{water}} s_2 - \dot{m}_{\text{oil}} s_4 + \dot{S}_{gen} = 0$$

$$\dot{S}_{gen} = \dot{m}_{\text{water}}(s_2 - s_1) + \dot{m}_{\text{oil}}(s_4 - s_3)$$

Noting that both fluid streams are liquids (incompressible substances), the rate of entropy generation is determined to be

$$\dot{S}_{gen} = \dot{m}_{\text{water}} C_p \ln\frac{T_2}{T_1} + \dot{m}_{\text{oil}} C_p \ln\frac{T_4}{T_3}$$

$$= (4.5 \text{ kg/s})(4.18 \text{ kJ/kg.K})\ln\frac{70+273}{20+273} + (10 \text{ kg/s})(2.3 \text{ kJ/kg.K})\ln\frac{129.1+273}{170+273} = \mathbf{0.736 \text{ kW/K}}$$

6-117E Steam is condensed by cooling water in a condenser. The rate of heat transfer and the rate of entropy generation within the heat exchanger are to be determined.

Assumptions **1** Steady operating conditions exist. **2** The heat exchanger is well-insulated so that heat loss to the surroundings is negligible and thus heat transfer from the hot fluid is equal to the heat transfer to the cold fluid. **3** Changes in the kinetic and potential energies of fluid streams are negligible. **4** Fluid properties are constant.

Properties The specific heat of water is 1.0 Btu/lbm.°F (Table A-3E). The enthalpy and entropy of vaporization of water at 90°F are 1040.2 Btu/lbm and s_{fg} = 1.8966 Btu/lbm.R (Table A-4E).

Analysis We take the tube-side of the heat exchanger where cold water is flowing as the system, which is a control volume. The energy balance for this steady-flow system can be expressed in the rate form as

Steam 90°F; 73°F; 60°F Water; 90°F

$$\underbrace{\dot{E}_{in}-\dot{E}_{out}}_{\substack{\text{Rate of net energy transfer}\\ \text{by heat, work, and mass}}} = \underbrace{\Delta\dot{E}_{\text{system}}^{\nearrow 0\text{ (steady)}}}_{\substack{\text{Rate of change in internal, kinetic,}\\ \text{potential, etc. energies}}} = 0$$

$$\dot{E}_{in} = \dot{E}_{out}$$

$$\dot{Q}_{in} + \dot{m}h_1 = \dot{m}h_2 \quad (\text{since } \Delta\text{ke} \cong \Delta\text{pe} \cong 0)$$

$$\dot{Q}_{in} = \dot{m}C_p(T_2 - T_1)$$

Then the rate of heat transfer to the cold water in this heat exchanger becomes

$$\dot{Q} = [\dot{m}C_p(T_{out} - T_{in})]_{\text{water}} = (115.3\text{ lbm/s})(1.0\text{ Btu/lbm.°F})(73\text{°F} - 60\text{°F}) = \mathbf{1499\text{ Btu/s}}$$

Noting that heat gain by the water is equal to the heat loss by the condensing steam, the rate of condensation of the steam in the heat exchanger is determined from

$$\dot{Q} = (\dot{m}h_{fg})_{steam} = \longrightarrow \dot{m}_{steam} = \frac{\dot{Q}}{h_{fg}} = \frac{1499\text{ Btu/s}}{1040.2\text{ Btu/lbm}} = \mathbf{1.44\text{ lbm/s}}$$

(*b*) The rate of entropy generation within the heat exchanger is determined by applying the rate form of the entropy balance on the entire heat exchanger:

$$\underbrace{\dot{S}_{in}-\dot{S}_{out}}_{\substack{\text{Rate of net entropy transfer}\\ \text{by heat and mass}}} + \underbrace{\dot{S}_{gen}}_{\substack{\text{Rate of entropy}\\ \text{generation}}} = \underbrace{\Delta\dot{S}_{\text{system}}^{\nearrow 0\text{ (steady)}}}_{\substack{\text{Rate of change}\\ \text{of entropy}}}$$

$$\dot{m}_1 s_1 + \dot{m}_3 s_3 - \dot{m}_2 s_2 - \dot{m}_4 s_4 + \dot{S}_{gen} = 0 \quad (\text{since } Q = 0)$$

$$\dot{m}_{\text{water}} s_1 + \dot{m}_{\text{steam}} s_3 - \dot{m}_{\text{water}} s_2 - \dot{m}_{\text{steam}} s_4 + \dot{S}_{gen} = 0$$

$$\dot{S}_{gen} = \dot{m}_{\text{water}}(s_2 - s_1) + \dot{m}_{\text{steam}}(s_4 - s_3)$$

Noting that water is an incompressible substance and steam changes from saturated vapor to saturated liquid, the rate of entropy generation is determined to be

$$\dot{S}_{gen} = \dot{m}_{\text{water}} C_p \ln\frac{T_2}{T_1} + \dot{m}_{\text{steam}}(s_f - s_g) = \dot{m}_{\text{water}} C_p \ln\frac{T_2}{T_1} - \dot{m}_{\text{steam}} s_{fg}$$

$$= (115.3\text{ lbm/s})(1.0\text{ Btu/lbm.R})\ln\frac{73+460}{60+460} - (1.44\text{ lbm/s})(1.8966\text{ Btu/lbm.R}) = \mathbf{0.116\text{ Btu/s.R}}$$

6-118 Chickens are to be cooled by chilled water in an immersion chiller that is also gaining heat from the surroundings. The rate of heat removal from the chicken and the rate of entropy generation during this process are to be determined.

Assumptions **1** Steady operating conditions exist. **2** Thermal properties of chickens and water are constant. **3** The temperature of the surrounding medium is given to be 25°C.

Properties The specific heat of chicken is given to be 3.54 kJ/kg.°C. The specific heat of water at room temperature is 4.18 kJ/kg.°C (Table A-3).

Analysis (a) Chickens can be considered to flow steadily through the chiller at a mass flow rate of

$$\dot{m}_{\text{chicken}} = (250\text{ chicken/h})(2.2\text{ kg/chicken}) = 550\text{ kg/h} = 0.1528\text{ kg/s}$$

Taking the chicken flow stream in the chiller as the system, the energy balance for steadily flowing chickens can be expressed in the rate form as

$$\underbrace{\dot{E}_{in} - \dot{E}_{out}}_{\substack{\text{Rate of net energy transfer}\\ \text{by heat, work, and mass}}} = \underbrace{\Delta \dot{E}_{\text{system}}^{\nearrow 0\text{ (steady)}}}_{\substack{\text{Rate of change in internal, kinetic,}\\ \text{potential, etc. energies}}} = 0 \rightarrow \dot{E}_{in} = \dot{E}_{out}$$

$$\dot{m}h_1 = \dot{Q}_{out} + \dot{m}h_2 \quad (\text{since } \Delta\text{ke} \cong \Delta\text{pe} \cong 0)$$

$$\dot{Q}_{out} = \dot{Q}_{\text{chicken}} = \dot{m}_{\text{chicken}} C_p (T_1 - T_2)$$

Then the rate of heat removal from the chickens as they are cooled from 15°C to 3°C becomes

$$\dot{Q}_{\text{chicken}} = (\dot{m}C_p\Delta T)_{\text{chicken}} = (0.1528\text{ kg/s})(3.54\text{ kJ/kg.°C})(15-3)\text{°C} = \mathbf{6.49\text{ kW}}$$

The chiller gains heat from the surroundings as a rate of 150 kJ/h = 0.0417 kJ/s.
Then the total rate of heat gain by the water is

$$\dot{Q}_{\text{water}} = \dot{Q}_{\text{chicken}} + \dot{Q}_{\text{heat gain}} = 6.49 + 0.0417 = 6.532\text{ kW}$$

Noting that the temperature rise of water is not to exceed 2°C as it flows through the chiller, the mass flow rate of water must be at least

$$\dot{m}_{\text{water}} = \frac{\dot{Q}_{\text{water}}}{(C_p\Delta T)_{\text{water}}} = \frac{6.532\text{ kW}}{(4.18\text{ kJ/kg.°C})(2\text{°C})} = \mathbf{0.781\text{ kg/s}}$$

(*b*) The rate of entropy generation is determined by applying the entropy balance on an *extended system* that includes the chiller and the immediate surroundings so that boundary temperature is the surroundings temperature:

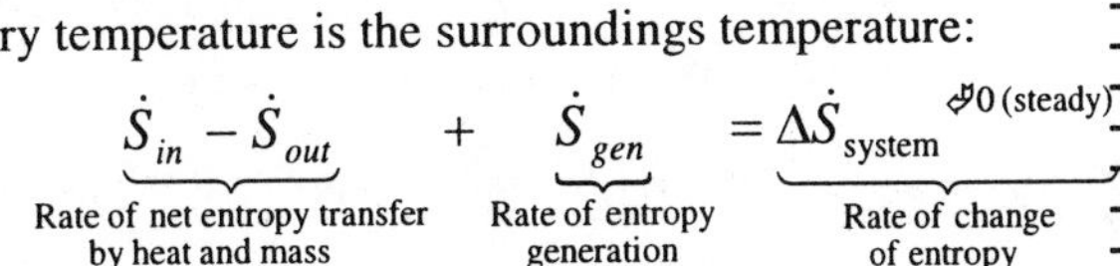

$$\underbrace{\dot{S}_{in} - \dot{S}_{out}}_{\substack{\text{Rate of net entropy transfer}\\ \text{by heat and mass}}} + \underbrace{\dot{S}_{gen}}_{\substack{\text{Rate of entropy}\\ \text{generation}}} = \underbrace{\Delta \dot{S}_{\text{system}}^{\nearrow 0\text{ (steady)}}}_{\substack{\text{Rate of change}\\ \text{of entropy}}}$$

$$\dot{m}_1 s_1 + \dot{m}_3 s_3 - \dot{m}_2 s_2 - \dot{m}_3 s_4 + \frac{\dot{Q}_{in}}{T_{surr}} + \dot{S}_{gen} = 0$$

$$\dot{m}_{\text{chicken}} s_1 + \dot{m}_{\text{water}} s_3 - \dot{m}_{\text{chicken}} s_2 - \dot{m}_{\text{water}} s_4 + \frac{\dot{Q}_{in}}{T_{surr}} + \dot{S}_{gen} = 0$$

$$\dot{S}_{gen} = \dot{m}_{\text{chicken}}(s_2 - s_1) + \dot{m}_{\text{water}}(s_4 - s_3) - \frac{\dot{Q}_{in}}{T_{surr}}$$

Noting that both streams are incompressible substances, the rate of entropy generation is determined to be

$$\dot{S}_{gen} = \dot{m}_{\text{chicken}} C_p \ln\frac{T_2}{T_1} + \dot{m}_{\text{water}} C_p \ln\frac{T_4}{T_3} - \frac{\dot{Q}_{in}}{T_{surr}}$$

$$= (0.1528\text{ kg/s})(3.54\text{ kJ/kg.K})\ln\frac{276}{288} + (0.781\text{ kg/s})(4.18\text{ kJ/kg.K})\ln\frac{275.5}{273.5} - \frac{0.0417\text{ kW}}{298\text{ K}} = \mathbf{0.000625\text{ kW/K}}$$

6-119 A regenerator is considered to save heat during the cooling of milk in a dairy plant. The amounts of fuel and money such a generator will save per year and the annual reduction in the rate of entropy generation are to be determined.

Assumptions 1 Steady operating conditions exist. **2** The properties of the milk are constant.

Properties The average density and specific heat of milk can be taken to be $\rho_{\text{milk}} \cong \rho_{\text{water}} = 1$ kg/L and $C_{p,\text{ milk}}$= 3.79 kJ/kg.°C (Table A-3).

Analysis The mass flow rate of the milk is

$$\dot{m}_{\text{milk}} = \rho\dot{V}_{\text{milk}} = (1\text{ kg/L})(12\text{ L/s}) = 12\text{ kg/s} = 43{,}200\text{ kg/h}$$

Taking the pasteurizing section as the system, the energy balance for this steady-flow system can be expressed in the rate form as

$$\underbrace{\dot{E}_{in} - \dot{E}_{out}}_{\substack{\text{Rate of net energy transfer}\\ \text{by heat, work, and mass}}} = \underbrace{\Delta\dot{E}_{\text{system}}^{\nearrow 0\text{ (steady)}}}_{\substack{\text{Rate of change in internal, kinetic,}\\ \text{potential, etc. energies}}} = 0 \quad \rightarrow \quad \dot{E}_{in} = \dot{E}_{out}$$

$$\dot{Q}_{in} + \dot{m}h_1 = \dot{m}h_2 \quad (\text{since } \Delta\text{ke} \cong \Delta\text{pe} \cong 0)$$

$$\dot{Q}_{in} = \dot{m}_{\text{milk}}C_p(T_2 - T_1)$$

Hot milk
72°C
$\dot{Q}$
4°C
Cold milk
Regenerator

Therefore, to heat the milk from 4 to 72°C as being done currently, heat must be transferred to the milk at a rate of

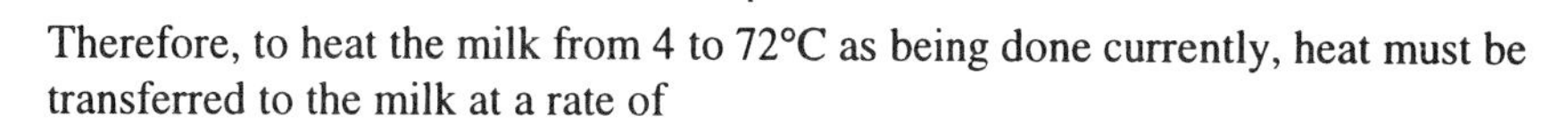

$$\dot{Q}_{\text{current}} = [\dot{m}C_{\text{p}}(T_{\text{pasturization}} - T_{\text{refrigeration}})]_{\text{milk}} = (12\text{kg/s})(3.79\text{kJ/kg.°C})(72-4)\text{°C} = 3093\text{kJ/s}$$

The proposed regenerator has an effectiveness of ε = 0.82, and thus it will save 82 percent of this energy. Therefore,

$$\dot{Q}_{\text{saved}} = \varepsilon\dot{Q}_{\text{current}} = (0.82)(3093\text{ kJ/s}) = 2536\text{ kJ/s}$$

Noting that the boiler has an efficiency of $\eta_{\text{boiler}} = 0.82$, the energy savings above correspond to fuel savings of

$$\text{Fuel Saved} = \frac{\dot{Q}_{\text{saved}}}{\eta_{\text{boiler}}} = \frac{(2536\text{ kJ/s})}{(0.82)}\frac{(1\text{therm})}{(105{,}500\text{ kJ})} = 0.02931\text{ therm/s}$$

Noting that 1 year = 365×24=8760 h and unit cost of natural gas is \$0.52/therm, the annual fuel and money savings will be

Fuel Saved = (0.02931 therms/s)(8760×3600 s) = **924,450 therms/yr**

$$\text{Money saved} = (\text{Fuel saved})(\text{Unit cost of fuel}) = (924{,}450\text{ therm/yr})(\$0.52/\text{therm}) = \mathbf{\$480{,}700/yr}$$

The rate of entropy generation during this process is determined by applying the rate form of the entropy balance on an *extended system* that includes the regenerator and the immediate surroundings so that the boundary temperature is the surroundings temperature, which we take to be the cold water temperature of 18°C.:

$$\underbrace{\dot{S}_{in} - \dot{S}_{out}}_{\substack{\text{Rate of net entropy transfer}\\ \text{by heat and mass}}} + \underbrace{\dot{S}_{gen}}_{\substack{\text{Rate of entropy}\\ \text{generation}}} = \underbrace{\Delta\dot{S}_{\text{system}}^{\nearrow 0\text{ (steady)}}}_{\substack{\text{Rate of change}\\ \text{of entropy}}} \rightarrow \dot{S}_{gen} = \dot{S}_{out} - \dot{S}_{in}$$

Disregarding entropy transfer associated with fuel flow, the only significant difference between the two cases is the reduction is the entropy transfer to water due to the reduction in heat transfer to water, and is determined to be

$$\dot{S}_{gen,\text{ reduction}} = \dot{S}_{out,\text{ reduction}} = \frac{\dot{Q}_{out,reduction}}{T_{surr}} = \frac{\dot{Q}_{saved}}{T_{surr}} = \frac{2536\text{ kJ/s}}{18+273} = 8.715\text{ kW/K}$$

$$S_{gen,\text{ reduction}} = \dot{S}_{gen,\text{ reduction}}\Delta t = (8.715\text{ kJ/s.K})(8760\times 3600\text{ s/year}) = \mathbf{2.75\times 10^8\ kJ/K}\text{ (per year)}$$

6-120 Stainless steel ball bearings leaving the oven at a uniform temperature of 900°C at a rate of 1400 /min are exposed to air and are cooled to 850°C before they are dropped into the water for quenching. The rate of heat transfer from the ball to the air and the rate of entropy generation due to this heat transfer are to be determined.

Assumptions **1** The thermal properties of the bearing balls are constant. **2** The kinetic and potential energy changes of the balls are negligible. **3** The balls are at a uniform temperature at the end of the process

Properties The density and specific heat of the ball bearings are given to be ρ = 8085 kg/m³ and C_p = 0.480 kJ/kg.°C.

Analysis (*a*) We take a single bearing ball as the system. The energy balance for this closed system can be expressed as

$$\underbrace{E_{in}-E_{out}}_{\substack{\text{Net energy transfer}\\ \text{by heat, work, and mass}}} = \underbrace{\Delta E_{\text{system}}}_{\substack{\text{Change in internal, kinetic,}\\ \text{potential, etc. energies}}}$$

$$-Q_{out} = \Delta U_{ball} = m(u_2-u_1)$$

$$Q_{out} = mC(T_1-T_2)$$

Furnace

Steel balls, 900°C

The total amount of heat transfer from a ball is

$$m=\rho V=\rho\frac{\pi D^3}{6}=(8085\ \text{kg/m}^3)\frac{\pi(0.012\ \text{m})^3}{6}=0.007315\ \text{kg}$$

$$Q_{out}=mC(T_1-T_2)=(0.007315\ \text{kg})(0.480\ \text{kJ/kg.°C})(900-850)\text{°C}=0.1756\ \text{kJ/ball}$$

Then the rate of heat transfer from the balls to the air becomes

$$\dot{Q}_{total}=\dot{n}_{\text{ball}}Q_{\text{out (per ball)}}=(1400\ \text{balls/min})\times(0.1756\ \text{kJ/ball})=\mathbf{245.8\ kJ/min=4.10\ kW}$$

Therefore, heat is lost to the air at a rate of 4.10 kW.

(*b*) We again take a single bearing ball as the system The entropy generated during this process can be determined by applying an entropy balance on an *extended system* that includes the ball and its immediate surroundings so that the boundary temperature of the extended system is at 30°C at all times:

$$\underbrace{S_{in}-S_{out}}_{\substack{\text{Net entropy transfer}\\ \text{by heat and mass}}}+\underbrace{S_{gen}}_{\substack{\text{Entropy}\\ \text{generation}}}=\underbrace{\Delta S_{\text{system}}}_{\substack{\text{Change}\\ \text{in entropy}}}$$

$$-\frac{Q_{out}}{T_b}+S_{gen}=\Delta S_{\text{system}}\ \rightarrow\ S_{gen}=\frac{Q_{out}}{T_b}+\Delta S_{\text{system}}$$

where

$$\Delta S_{\text{system}}=m(s_2-s_1)=mC_{av}\ln\frac{T_2}{T_1}=(0.007315\ \text{kg})(0.480\ \text{kJ/kg.K})\ln\frac{850+273}{900+273}=-0.0001530\ \text{kJ/K}$$

Substituting,

$$S_{gen}=\frac{Q_{out}}{T_b}+\Delta S_{\text{system}}=\frac{0.1756\ \text{kJ}}{303\ \text{K}}-0.0001530\ \text{kJ/K}=0.0004265\ \text{kJ/K}\ \ \text{(per ball)}$$

Then the rate of entropy generation becomes

$$\dot{S}_{gen}=S_{gen}\dot{n}_{ball}=(0.0004265\ \text{kJ/K}\cdot\text{ball})(1400\ \text{balls/min})=0.597\ \text{kJ/min.K}=\mathbf{0.00995\ kW/K}$$

6-121 Carbon steel balls are to be annealed at a rate of 2500/h by heating them first and then allowing them to cool slowly in ambient air at a specified rate. The total rate of heat transfer from the balls to the ambient air and the rate of entropy generation due to this heat transfer are to be determined.

Assumptions **1** The thermal properties of the balls are constant. **2** There are no changes in kinetic and potential energies. **3** The balls are at a uniform temperature at the end of the process

Properties The density and specific heat of the balls are given to be ρ = 7833 kg/m^3 and C_p = 0.465 kJ/kg.°C.

Analysis (*a*) We take a single ball as the system. The energy balance for this closed system can be expressed as

$$\underbrace{E_{in} - E_{out}}_{\substack{\text{Net energy transfer}\\ \text{by heat, work, and mass}}} = \underbrace{\Delta E_{\text{system}}}_{\substack{\text{Change in internal, kinetic,}\\ \text{potential, etc. energies}}}$$

$$-Q_{out} = \Delta U_{ball} = m(u_2 - u_1)$$

$$Q_{out} = mC(T_1 - T_2)$$

Furnace

Air, 35°C

Steel balls, 900°C

The amount of heat transfer from a single ball is

$$m = \rho V = \rho \frac{\pi D^3}{6} = (7833\ \text{kg/m}^3)\frac{\pi (0.008\ \text{m})^3}{6} = 0.00210\ \text{kg}$$

$$Q_{out} = mC_p(T_1 - T_2) = (0.0021\ \text{kg})(0.465\ \text{kJ/kg.°C})(900 - 100)°\text{C} = 781\ \text{J} = 0.781\ \text{kJ (per ball)}$$

Then the total rate of heat transfer from the balls to the ambient air becomes

$$\dot{Q}_{out} = \dot{n}_{\text{ball}} Q_{out} = (2500\ \text{balls/h}) \times (0.781\ \text{kJ/ball}) = 1{,}953\ \text{kJ/h} = \mathbf{542\ W}$$

(*b*) We again take a single ball as the system The entropy generated during this process can be determined by applying an entropy balance on an *extended system* that includes the ball and its immediate surroundings so that the boundary temperature of the extended system is at 35°C at all times:

$$\underbrace{S_{in} - S_{out}}_{\substack{\text{Net entropy transfer}\\ \text{by heat and mass}}} + \underbrace{S_{gen}}_{\substack{\text{Entropy}\\ \text{generation}}} = \underbrace{\Delta S_{\text{system}}}_{\substack{\text{Change}\\ \text{in entropy}}}$$

$$-\frac{Q_{out}}{T_b} + S_{gen} = \Delta S_{\text{system}} \rightarrow S_{gen} = \frac{Q_{out}}{T_b} + \Delta S_{\text{system}}$$

where

$$\Delta S_{\text{system}} = m(s_2 - s_1) = mC_{av} \ln\frac{T_2}{T_1} = (0.00210\ \text{kg})(0.465\ \text{kJ/kg.K})\ln\frac{100+273}{900+273} = -0.00112\ \text{kJ/K}$$

Substituting,

$$S_{gen} = \frac{Q_{out}}{T_b} + \Delta S_{\text{system}} = \frac{0.781\ \text{kJ}}{308\ \text{K}} - 0.00112\ \text{kJ/K} = 0.00142\ \text{kJ/K}\ \ \text{(per ball)}$$

Then the rate of entropy generation becomes

$$\dot{S}_{gen} = S_{gen}\dot{n}_{ball} = (0.00142\ \text{kJ/K}\cdot\text{ball})(2500\ \text{balls/h}) = 3.55\ \text{kJ/h.K} = \mathbf{0.986\ \ W/K}$$

6-122 An egg is dropped into boiling water. The amount of heat transfer to the egg by the time it is cooked and the amount of entropy generation associated with this heat transfer process are to be determined.

Assumptions **1** The egg is spherical in shape with a radius of $r_0 = 2.75$ cm. **2** The thermal properties of the egg are constant. **3** Energy absorption or release associated with any chemical and/or phase changes within the egg is negligible. **4** There are no changes in kinetic and potential energies.

Properties The density and specific heat of the egg are given to be $\rho = 1020$ kg/m^3 and $C_p = 3.32$ kJ/kg.°C.

Analysis We take the egg as the system. This is a closes system since no mass enters or leaves the egg. The energy balance for this closed system can be expressed as

$$\underbrace{E_{in} - E_{out}}_{\substack{\text{Net energy transfer}\\ \text{by heat, work, and mass}}} = \underbrace{\Delta E_{\text{system}}}_{\substack{\text{Change in internal, kinetic,}\\ \text{potential, etc. energies}}}$$

$$Q_{in} = \Delta U_{egg} = m(u_2 - u_1) = mC(T_2 - T_1)$$

Boiling Water

Egg 8°C

Then the mass of the egg and the amount of heat transfer become

$$m = \rho V = \rho \frac{\pi D^3}{6} = (1020\ \text{kg/m}^3)\frac{\pi (0.055\ \text{m})^3}{6} = 0.0889\ \text{kg}$$

$$Q_{in} = mC_p(T_2 - T_1) = (0.0889\ \text{kg})(3.32\ \text{kJ/kg.°C})(70 - 8)\text{°C} = \mathbf{18.3\ kJ}$$

We again take a single egg as the system The entropy generated during this process can be determined by applying an entropy balance on an *extended system* that includes the egg and its immediate surroundings so that the boundary temperature of the extended system is at 97°C at all times:

$$\underbrace{S_{in} - S_{out}}_{\substack{\text{Net entropy transfer}\\ \text{by heat and mass}}} + \underbrace{S_{gen}}_{\substack{\text{Entropy}\\ \text{generation}}} = \underbrace{\Delta S_{\text{system}}}_{\substack{\text{Change}\\ \text{in entropy}}}$$

$$\frac{Q_{in}}{T_b} + S_{gen} = \Delta S_{\text{system}} \rightarrow S_{gen} = -\frac{Q_{in}}{T_b} + \Delta S_{\text{system}}$$

where

$$\Delta S_{\text{system}} = m(s_2 - s_1) = mC_{av}\ln\frac{T_2}{T_1} = (0.0889\ \text{kg})(3.32\ \text{kJ/kg.K})\ln\frac{70+273}{8+273} = 0.0588\ \text{kJ/K}$$

Substituting,

$$S_{gen} = -\frac{Q_{in}}{T_b} + \Delta S_{\text{system}} = -\frac{18.3\ \text{kJ}}{370\ \text{K}} + 0.0588\ \text{kJ/K} = \mathbf{0.00961\ kJ/K}\ \ \text{(per egg)}$$

6-123E Large brass plates are heated in an oven at a specified rate. The rate of heat transfer to the plates in the oven and the rate of entropy generation associated with this heat transfer process are to be determined.

Assumptions **1** The thermal properties of the plates are constant. **2** The changes in kinetic and potential energies are negligible.

Properties The density and specific heat of the brass are given to be ρ = 532.5 lbm/ft³ and C_p = 0.091 Btu/lbm.°F.

Analysis We take the plate to be the system. The energy balance for this closed system can be expressed as

$$\underbrace{E_{in} - E_{out}}_{\substack{\text{Net energy transfer}\\ \text{by heat, work, and mass}}} = \underbrace{\Delta E_{\text{system}}}_{\substack{\text{Change in internal, kinetic,}\\ \text{potential, etc. energies}}}$$

Plates
75°F

$$Q_{in} = \Delta U_{\text{plate}} = m(u_2 - u_1) = mC(T_2 - T_1)$$

The mass of each plate and the amount of heat transfer to each plate is

$$m = \rho V = \rho LA = (532.5 \text{ lbm/ft}^3)[(1.2/12 \text{ ft})(2 \text{ ft})(2 \text{ ft})] = 213 \text{ lbm}$$

$$Q_{in} = mC(T_2 - T_1) = (213 \text{ lbm/plate})(0.091 \text{ Btu/lbm.°F})(1000 - 75)\text{°F} = 17{,}930 \text{ Btu/plate}$$

Then the total rate of heat transfer to the plates becomes

$$\dot{Q}_{total} = \dot{n}_{\text{plate}} Q_{\text{in, per plate}} = (450 \text{ plates/min}) \times (17{,}930 \text{ Btu/plate}) = \mathbf{8{,}069{,}000 \text{ Btu/min} = 134{,}500 \text{ Btu/s}}$$

We again take a single plate as the system The entropy generated during this process can be determined by applying an entropy balance on an *extended system* that includes the plate and its immediate surroundings so that the boundary temperature of the extended system is at 1300°F at all times:

$$\underbrace{S_{in} - S_{out}}_{\substack{\text{Net entropy transfer}\\ \text{by heat and mass}}} + \underbrace{S_{gen}}_{\substack{\text{Entropy}\\ \text{generation}}} = \underbrace{\Delta S_{\text{system}}}_{\substack{\text{Change}\\ \text{in entropy}}}$$

$$\frac{Q_{in}}{T_b} + S_{gen} = \Delta S_{\text{system}} \rightarrow S_{gen} = -\frac{Q_{in}}{T_b} + \Delta S_{\text{system}}$$

where

$$\Delta S_{\text{system}} = m(s_2 - s_1) = mC_{av} \ln\frac{T_2}{T_1} = (213 \text{ lbm})(0.091 \text{ Btu/lbm.R})\ln\frac{1000 + 460}{75 + 460} = 19.46 \text{ Btu/R}$$

Substituting,

$$S_{gen} = -\frac{Q_{in}}{T_b} + \Delta S_{\text{system}} = -\frac{17{,}930 \text{ Btu}}{1300 + 460 \text{ R}} + 19.46 \text{ Btu/R} = 9.272 \text{ Btu/R} \quad \text{(per plate)}$$

Then the rate of entropy generation becomes

$$\dot{S}_{gen} = S_{gen}\dot{n}_{ball} = (9.272 \text{ Btu/R} \cdot \text{plate})(450 \text{ plates/min}) = 4172 \text{ Btu/min.R} = \mathbf{69.5 \text{ Btu/s.R}}$$

6-124 Long cylindrical steel rods are heat-treated in an oven. The rate of heat transfer to the rods in the oven and the rate of entropy generation associated with this heat transfer process are to be determined.

Assumptions **1** The thermal properties of the rods are constant. **2** The changes in kinetic and potential energies are negligible.

Properties The density and specific heat of the steel rods are given to be ρ = 7833 kg/m³ and C_p = 0.465 kJ/kg.°C.

Analysis (*a*) Noting that the rods enter the oven at a velocity of 3 m/min and exit at the same velocity, we can say that a 3-m long section of the rod is heated in the oven in 1 min. Then the mass of the rod heated in 1 minute is

$$m = \rho V = \rho LA = \rho L(\pi D^2/4) = (7833\ \text{kg/m}^3)(3\ \text{m})[\pi(0.1\ \text{m})^2/4] = 184.6\ \text{kg}$$

We take the 3-m section of the rod in the oven as the system. The energy balance for this closed system can be expressed as

Oven, 900°C

Steel rod, 30°C

$$\underbrace{E_{in} - E_{out}}_{\substack{\text{Net energy transfer}\\ \text{by heat, work, and mass}}} = \underbrace{\Delta E_{\text{system}}}_{\substack{\text{Change in internal, kinetic,}\\ \text{potential, etc. energies}}}$$

$$Q_{in} = \Delta U_{\text{rod}} = m(u_2 - u_1) = mC(T_2 - T_1)$$

Substituting,

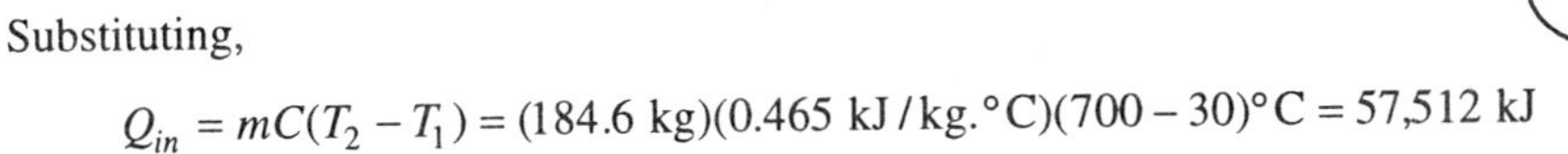

$$Q_{in} = mC(T_2 - T_1) = (184.6\ \text{kg})(0.465\ \text{kJ/kg.°C})(700 - 30)\text{°C} = 57{,}512\ \text{kJ}$$

Noting that this much heat is transferred in 1 min, the rate of heat transfer to the rod becomes

$$\dot{Q}_{in} = Q_{\text{in}}/\Delta t = (57{,}512\ \text{kJ})/(1\ \text{min}) = 57{,}512\ \text{kJ/min} = \mathbf{958.5\ kW}$$

(*b*) We again take the 3-m long section of the rod as the system The entropy generated during this process can be determined by applying an entropy balance on an *extended system* that includes the rod and its immediate surroundings so that the boundary temperature of the extended system is at 900°C at all times:

$$\underbrace{S_{in} - S_{out}}_{\substack{\text{Net entropy transfer}\\ \text{by heat and mass}}} + \underbrace{S_{gen}}_{\substack{\text{Entropy}\\ \text{generation}}} = \underbrace{\Delta S_{\text{system}}}_{\substack{\text{Change}\\ \text{in entropy}}}$$

$$\frac{Q_{in}}{T_b} + S_{gen} = \Delta S_{\text{system}} \rightarrow S_{gen} = -\frac{Q_{in}}{T_b} + \Delta S_{\text{system}}$$

where

$$\Delta S_{\text{system}} = m(s_2 - s_1) = mC_{av}\ln\frac{T_2}{T_1} = (184.6\ \text{kg})(0.465\ \text{kJ/kg.K})\ln\frac{700 + 273}{30 + 273} = 100.1\ \text{kJ/K}$$

Substituting,

$$S_{gen} = -\frac{Q_{in}}{T_b} + \Delta S_{\text{system}} = -\frac{57{,}512\ \text{kJ}}{900 + 273\ \text{K}} + 100.1\ \text{kJ/K} = 51.1\ \text{kJ/K}$$

Noting that this much entropy is generated in 1 min, the rate of entropy generation becomes

$$\dot{S}_{gen} = \frac{S_{gen}}{\Delta t} = \frac{51.1\ \text{kJ/K}}{1\ \text{min}} = 51.1\ \text{kJ/min.K} = \mathbf{0.85\ \ kW/K}$$

6-125 The inner and outer surfaces of a brick wall are maintained at specified temperatures. The rate of entropy generation within the wall is to be determined.

Assumptions Steady operating conditions exist since the surface temperatures of the wall remain constant at the specified values.

Analysis We take the wall to be the system, which is a closed system.

Under steady conditions, the rate form of the entropy balance for the wall simplifies to

Brick Wall

30 cm

20°C 5°C

$$\underbrace{\dot{S}_{in}-\dot{S}_{out}}_{\substack{\text{Rate of net entropy transfer}\\ \text{by heat and mass}}}+\underbrace{\dot{S}_{gen}}_{\substack{\text{Rate of entropy}\\ \text{generation}}}=\underbrace{\Delta\dot{S}_{\text{system}}^{\nearrow 0}}_{\substack{\text{Rate of change}\\ \text{of entropy}}}=0$$

$$\frac{\dot{Q}_{in}}{T_{b,in}}-\frac{\dot{Q}_{out}}{T_{b,out}}+\dot{S}_{\text{gen,wall}}=0$$

$$\frac{1035\text{ W}}{293\text{ K}}-\frac{1035\text{ W}}{278\text{ K}}+\dot{S}_{\text{gen,wall}}=0\ \rightarrow\ \dot{S}_{\text{gen,wall}}=\mathbf{0.191\ W/K}$$

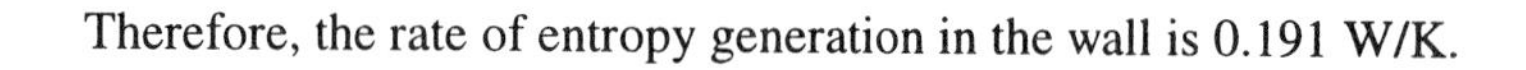

Therefore, the rate of entropy generation in the wall is 0.191 W/K.

6-126 A person is standing in a room at a specified temperature. The rate of entropy transfer from the body with heat is to be determined.

Assumptions Steady operating conditions exist.

Analysis Noting that Q/T represents entropy transfer with heat, the rate of entropy transfer from the body of the person accompanying heat transfer is

$$\dot{S}_{\text{transfer}}=\frac{\dot{Q}}{T}=\frac{336\text{ W}}{307\text{ K}}=\mathbf{1.094\ W/K}$$

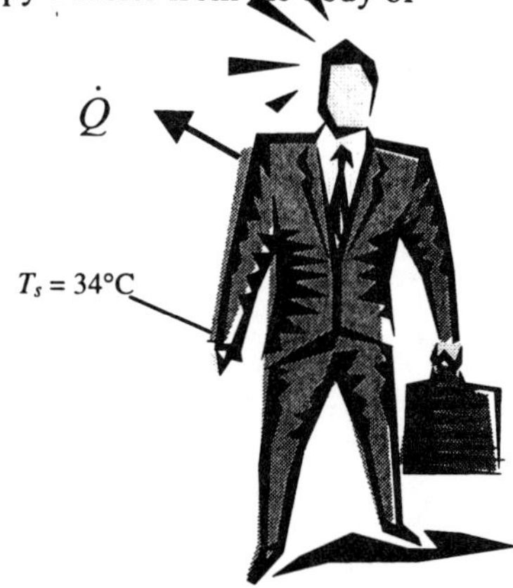

6-127 A 1000-W iron is left on the iron board with its base exposed to the air at 20°C. The rate of entropy generation is to be determined in steady operation.

Assumptions Steady operating conditions exist.

Analysis We take the iron to be the system, which is a closed system. Considering that the iron experiences no change in its properties in steady operation, including its entropy, the rate form of the entropy balance for the iron simplifies to

Iron
1000 W

$$\underbrace{\dot{S}_{in} - \dot{S}_{out}}_{\text{Rate of net entropy transfer by heat and mass}} + \underbrace{\dot{S}_{gen}}_{\text{Rate of entropy generation}} = \underbrace{\Delta \dot{S}_{\text{system}}^{\nearrow 0}}_{\text{Rate of change of entropy}} = 0$$

$$-\frac{\dot{Q}_{out}}{T_{b,out}} + \dot{S}_{\text{gen,iron}} = 0$$

Therefore,

$$\dot{S}_{\text{gen,iron}} = \frac{\dot{Q}_{out}}{T_{b,out}} = \frac{1000\ \text{W}}{673\ \text{K}} = \mathbf{1.486\ W/K}$$

The rate of total entropy generation during this process is determined by applying the entropy balance on an *extended system* that includes the iron and its immediate surroundings so that the boundary temperature of the extended system is at 20°C at all times. It gives

$$\dot{S}_{\text{gen,total}} = \frac{\dot{Q}_{out}}{T_{b,out}} = \frac{\dot{Q}}{T_{surr}} = \frac{1000\ \text{W}}{293\ \text{K}} = \mathbf{3.413\ W/K}$$

Discussion Note that only about one-third of the entropy generation occurs within the iron. The rest occurs in the air surrounding the iron as the temperature drops from 400°C to 20°C without serving any useful purpose.

6-128E A cylinder contains saturated liquid water at a specified pressure. Heat is transferred to liquid from a source and some liquid evaporates. The total entropy generation during this process is to be determined.

Assumptions **1** No heat loss occurs from the water to the surroundings during the process. **2** The pressure inside the cylinder and thus the water temperature remains constant during the process. **3** No irreversibilities occur within the cylinder during the process.

Analysis The pressure of the steam is maintained constant. Therefore, the temperature of the steam remains constant also at

$$T = T_{sat\,@\,20\text{ psia}} = 227.96°\text{F} = 688\text{ R}$$

Taking the contents of the cylinder as the system and noting that the temperature of water remains constant, the entropy change of the system during this isothermal, internally reversible process becomes

$$\Delta S_{\text{system}} = \frac{Q_{sys,in}}{T_{sys}} = \frac{600\text{ Btu}}{688\text{ R}} = 0.872\text{ Btu/R}$$

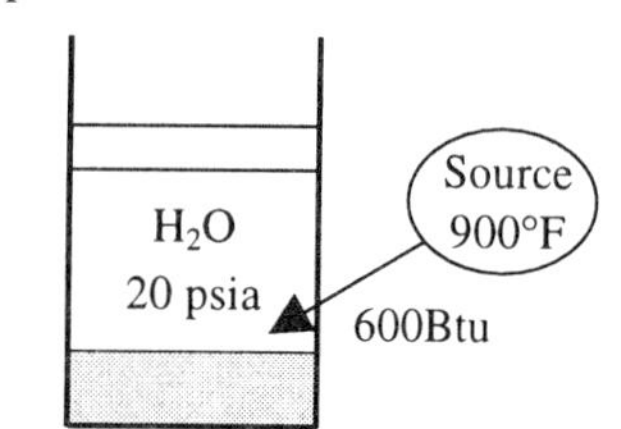

Similarly, the entropy change of the heat source is determined from

$$\Delta S_{\text{source}} = -\frac{Q_{\text{source,out}}}{T_{\text{source}}} = -\frac{600\text{ Btu}}{900+460\text{ R}} = -0.441\text{ Btu/R}$$

Now consider a combined system that includes the cylinder and the source. Noting that no heat or mass crosses the boundaries of this combined system, the entropy balance for it can be expressed as

$$\underbrace{S_{in} - S_{out}}_{\substack{\text{Net entropy transfer}\\ \text{by heat and mass}}} + \underbrace{S_{gen}}_{\substack{\text{Entropy}\\ \text{generation}}} = \underbrace{\Delta S_{\text{system}}}_{\substack{\text{Change}\\ \text{in entropy}}}$$

$$0 + S_{\text{gen,total}} = \Delta S_{\text{water}} + \Delta S_{\text{source}}$$

Therefore, the total entropy generated during this process is

$$S_{\text{gen,total}} = \Delta S_{\text{water}} + \Delta S_{\text{source}} = 0.872 - 0.441 = \mathbf{0.431\ Btu/R}$$

Discussion The entropy generation in this case is entirely due to the irreversible heat transfer through a finite temperature difference. We could also determine the total entropy generation by writing an energy balance on an extended system that includes the system and its immediate surroundings so that part of the boundary of the extended system, where heat transfer occurs, is at the source temperature.

6-129E Steam is decelerated in a diffuser from a velocity of 900 ft/s to 100 ft/s. The mass flow rate of steam and the rate of entropy generation are to be determined.

Assumptions **1** This is a steady-flow process since there is no change with time. **2** Potential energy changes are negligible. **3** There are no work interactions.

Properties The properties of steam at the inlet and the exit of the diffuser are (Tables A-4E through A-6E)

$$\left.\begin{array}{l} P_1 = 20 \text{ psia} \\ T_1 = 240^\circ\text{F} \end{array}\right\} \begin{array}{l} h_1 = 1162.3 \text{ Btu/lbm} \\ s_1 = 1.7405 \text{ Btu/lbm}\cdot\text{R} \end{array}$$

$$\left.\begin{array}{l} T_2 = 240^\circ\text{F} \\ sat.vapor \end{array}\right\} \begin{array}{l} h_2 = 1160.7 \text{ Btu/lbm} \\ s_2 = 1.7143 \text{ Btu/lbm}\cdot\text{R} \\ v_2 = 16.327 \text{ ft}^3\text{/lbm} \end{array}$$

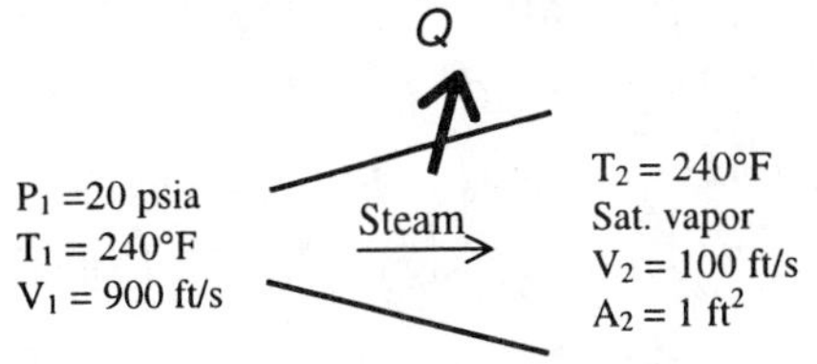

Analysis (a) The mass flow rate of the steam can be determined from its definition to be

$$\dot{m} = \frac{1}{v_2} A_2 \mathbf{V}_2 = \frac{1}{16.327\text{ft}^3\text{/lbm}} \left(1\text{ft}^2\right)\left(100\text{ft/s}\right) = \mathbf{6.125lbm/s}$$

(*b*) We take diffuser as the system, which is a control volume since mass crosses the boundary. The energy balance for this steady-flow system can be expressed in the rate form as

$$\underbrace{\dot{E}_{in} - \dot{E}_{out}}_{\substack{\text{Rate of net energy transfer} \\ \text{by heat, work, and mass}}} = \underbrace{\Delta \dot{E}_{\text{system}}^{\nearrow 0 \text{ (steady)}}}_{\substack{\text{Rate of change in internal, kinetic,} \\ \text{potential, etc. energies}}} = 0$$

$$\dot{E}_{in} = \dot{E}_{out}$$

$$\dot{m}(h_1 + \mathbf{V}_1^2/2) - \dot{Q}_{\text{out}} = \dot{m}(h_2 + \mathbf{V}_2^2/2) \quad (\text{since } \dot{W} \cong \Delta\text{pe} \cong 0)$$

$$\dot{Q}_{\text{out}} = -\dot{m}\left(h_2 - h_1 + \frac{\mathbf{V}_2^2 - \mathbf{V}_1^2}{2}\right)$$

Substituting, the rate of heat loss from the diffuser is determined to be

$$\dot{Q}_{out} = -(6.125\text{lbm/s})\left(1160.7 - 1162.3 + \frac{(100\text{ft/s})^2 - (900\text{ft/s})^2}{2}\left(\frac{1\text{Btu/lbm}}{25{,}037\text{ft}^2/\text{s}^2}\right)\right) = 107.66\text{Btu/s}$$

The rate of total entropy generation during this process is determined by applying the entropy balance on an *extended system* that includes the diffuser and its immediate surroundings so that the boundary temperature of the extended system is 77°F at all times. It gives

$$\underbrace{\dot{S}_{in} - \dot{S}_{out}}_{\substack{\text{Rate of net entropy transfer} \\ \text{by heat and mass}}} + \underbrace{\dot{S}_{gen}}_{\substack{\text{Rate of entropy} \\ \text{generation}}} = \underbrace{\Delta \dot{S}_{\text{system}}^{\nearrow 0}}_{\substack{\text{Rate of change} \\ \text{of entropy}}} = 0$$

$$\dot{m}s_1 - \dot{m}s_2 - \frac{\dot{Q}_{out}}{T_{b,surr}} + \dot{S}_{\text{gen}} = 0$$

Substituting, the total rate of entropy generation during this process becomes

$$\dot{S}_{gen} = \dot{m}(s_2 - s_1) + \frac{\dot{Q}_{out}}{T_{b,surr}} = (6.125 \text{ lbm/s})(1.7143 - 1.7405)\text{Btu/lbm}\cdot\text{R} + \frac{107.66 \text{ Btu/s}}{537 \text{ R}} = \mathbf{0.040 \ Btu/s\cdot R}$$

6-130 Steam expands in a turbine from a specified state to another specified state. The rate of entropy generation during this process is to be determined.

Assumptions **1** This is a steady-flow process since there is no change with time. **2** Kinetic and potential energy changes are negligible.

Properties From the steam tables (Tables A-4 through 6)

$$\left.\begin{array}{l} P_1 = 8\text{MPa} \\ T_1 = 450^\circ\text{C} \end{array}\right\} \begin{array}{l} h_1 = 3272.0\text{kJ/kg} \\ s_1 = 6.5551\text{kJ/kg}\cdot\text{K} \end{array}$$

$$\left.\begin{array}{l} P_2 = 50\text{kPa} \\ sat.vapor \end{array}\right\} \begin{array}{l} h_2 = 2645.9\text{kJ/kg} \\ s_2 = 7.5939\text{kJ/kg}\cdot\text{K} \end{array}$$

Analysis There is only one inlet and one exit, and thus $\dot{m}_1 = \dot{m}_2 = \dot{m}$. We take the turbine as the system, which is a control volume since mass crosses the boundary. The energy balance for this steady-flow system can be expressed in the rate form as

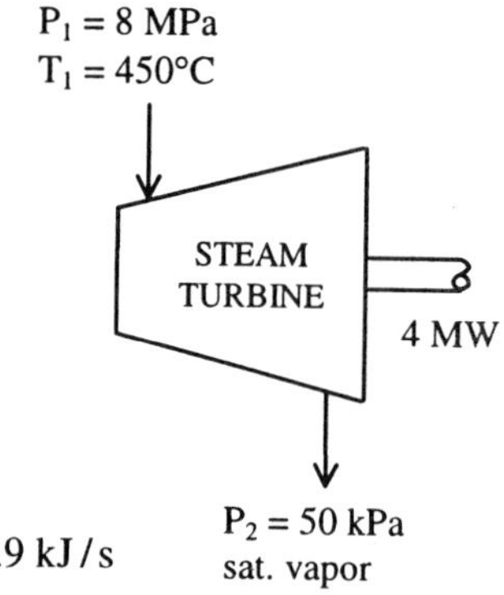

$$\underbrace{\dot{E}_{in} - \dot{E}_{out}}_{\substack{\text{Rate of net energy transfer} \\ \text{by heat, work, and mass}}} = \underbrace{\Delta\dot{E}_{\text{system}}^{\nearrow 0 \text{ (steady)}}}_{\substack{\text{Rate of change in internal, kinetic,} \\ \text{potential, etc. energies}}} = 0$$

$$\dot{E}_{in} = \dot{E}_{out}$$

$$\dot{m}h_1 = \dot{Q}_{\text{out}} + \dot{W}_{\text{out}} + \dot{m}h_2$$

$$\dot{Q}_{\text{out}} = \dot{m}(h_1 - h_2) - \dot{W}_{\text{out}}$$

P1 = 8 MPa
T1 = 450°C
STEAM TURBINE
4 MW
P2 = 50 kPa
sat. vapor

Substituting,

$$\dot{Q}_{out} = (25{,}000/3600 \text{ kg/s})(3272 - 2645.9) \text{ kJ/kg} - 4000 \text{ kJ/s} = 347.9 \text{ kJ/s}$$

The rate of total entropy generation during this process is determined by applying the entropy balance on an *extended system* that includes the turbine and its immediate surroundings so that the boundary temperature of the extended system is 25°C at all times. It gives

$$\underbrace{\dot{S}_{in} - \dot{S}_{out}}_{\substack{\text{Rate of net entropy transfer} \\ \text{by heat and mass}}} + \underbrace{\dot{S}_{gen}}_{\substack{\text{Rate of entropy} \\ \text{generation}}} = \underbrace{\Delta\dot{S}_{\text{system}}^{\nearrow 0}}_{\substack{\text{Rate of change} \\ \text{of entropy}}} = 0$$

$$\dot{m}s_1 - \dot{m}s_2 - \frac{\dot{Q}_{out}}{T_{b,surr}} + \dot{S}_{\text{gen}} = 0$$

Substituting, the rate of entropy generation during this process is determined to be

$$\dot{S}_{gen} = \dot{m}(s_2 - s_1) + \frac{\dot{Q}_{out}}{T_{b,surr}} = (25{,}000/3600\text{kg/s})(7.5939 - 6.5551)\text{kJ/kg}\cdot\text{K} + \frac{347.9\text{kW}}{298\text{K}} = \mathbf{8.38kW/K}$$

6-131 A hot water stream is mixed with a cold water stream. For a specified mixture temperature, the mass flow rate of cold water stream and the rate of entropy generation are to be determined.

Assumptions **1** Steady operating conditions exist. **2** The mixing chamber is well-insulated so that heat loss to the surroundings is negligible. **3** Changes in the kinetic and potential energies of fluid streams are negligible.

Properties Noting that $T < T_{sat\ @\ 200\ kPa} = 120.23°C$, the water in all three streams exists as a compressed liquid, which can be approximated as a saturated liquid at the given temperature. Thus from Table A-4,

$$\left.\begin{array}{l} P_1 = 200\text{kPa} \\ T_1 = 70^\circ\text{C} \end{array}\right\} \begin{array}{l} h_1 \cong h_{f@70^\circ C} = 292.88\text{kJ/kg} \\ s_1 \cong s_{f@70^\circ C} = 0.9549\text{kJ/kg}\cdot\text{K} \end{array}$$

$$\left.\begin{array}{l} P_2 = 200\text{kPa} \\ T_2 = 20^\circ\text{C} \end{array}\right\} \begin{array}{l} h_2 \cong h_{f@20^\circ C} = 83.96\text{kJ/kg} \\ s_2 \cong s_{f@20^\circ C} = 0.2966\text{kJ/kg}\cdot\text{K} \end{array}$$

$$\left.\begin{array}{l} P_3 = 200\text{kPa} \\ T_3 = 42^\circ\text{C} \end{array}\right\} \begin{array}{l} h_3 \cong h_{f@42^\circ C} = 175.92\text{kJ/kg} \\ s_3 \cong s_{f@42^\circ C} = 0.5990\text{kJ/kg}\cdot\text{K} \end{array}$$

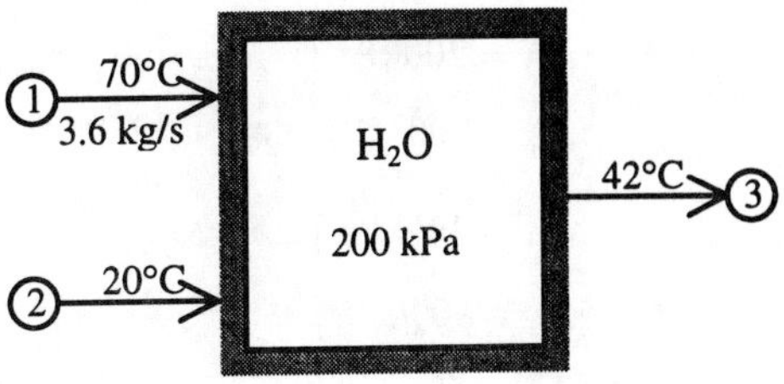

Analysis (*a*) We take the mixing chamber as the system, which is a control volume. The mass and energy balances for this steady-flow system can be expressed in the rate form as

Mass balance: $\dot{m}_{in} - \dot{m}_{out} = \Delta\dot{E}_{system}^{\nearrow 0\ (steady)} = 0 \longrightarrow \dot{m}_1 + \dot{m}_2 = \dot{m}_3$

Energy balance:

$$\underbrace{\dot{E}_{in} - \dot{E}_{out}}_{\substack{\text{Rate of net energy transfer} \\ \text{by heat, work, and mass}}} = \underbrace{\Delta\dot{E}_{system}^{\nearrow 0\ (steady)}}_{\substack{\text{Rate of change in internal, kinetic,} \\ \text{potential, etc. energies}}} = 0$$

$$\dot{E}_{in} = \dot{E}_{out}$$

$$\dot{m}h_1 + \dot{m}_2 h_2 = \dot{m}_3 h_3 \quad (\text{since } \dot{Q} = \dot{W} = \Delta\text{ke} \cong \Delta\text{pe} \cong 0)$$

Combining the two relations gives $\quad \dot{m}_1 h_1 + \dot{m}_2 h_2 = (\dot{m}_1 + \dot{m}_2) h_3$

Solving for $\dot{m}_2$ and substituting, the mass flow rate of cold water stream is determined to be

$$\dot{m}_2 = \frac{h_1 - h_3}{h_3 - h_2}\dot{m}_1 = \frac{(292.88 - 175.92)\text{kJ/kg}}{(175.92 - 83.96)\text{kJ/kg}}(3.6\ \text{kg/s}) = 4.58\,\mathbf{kg/s}$$

Also,

$$\dot{m}_3 = \dot{m}_1 + \dot{m}_2 = 3.6 + 4.58 = 8.18\ \text{kg/s}$$

(*b*) Noting that the mixing chamber is adiabatic and thus there is no heat transfer to the surroundings, the entropy balance of the steady-flow system (the mixing chamber) can be expressed as

$$\underbrace{\dot{S}_{in} - \dot{S}_{out}}_{\substack{\text{Rate of net entropy transfer} \\ \text{by heat and mass}}} + \underbrace{\dot{S}_{gen}}_{\substack{\text{Rate of entropy} \\ \text{generation}}} = \underbrace{\Delta\dot{S}_{system}^{\nearrow 0}}_{\substack{\text{Rate of change} \\ \text{of entropy}}} = 0$$

$$\dot{m}_1 s_1 + \dot{m}_2 s_2 - \dot{m}_3 s_3 + \dot{S}_{gen} = 0$$

Substituting, the total rate of entropy generation during this process becomes

$$\begin{aligned} \dot{S}_{gen} &= \dot{m}_3 s_3 - \dot{m}_2 s_2 - \dot{m}_1 s_1 \\ &= (4.09\text{kg/s})(0.599\text{kJ/kg}\cdot\text{K}) - (2.29\text{kg/s})(0.2966\text{kJ/kg}\cdot\text{K}) \\ &\quad - (1.8\text{kg/s})(0.9549\text{kJ/kg}\cdot\text{K}) \\ &= \mathbf{0.0519 kW/K} \end{aligned}$$

6-132 Liquid water is heated in a chamber by mixing it with superheated steam. For a specified mixing temperature, the mass flow rate of the steam and the rate of entropy generation are to be determined.

Assumptions **1** This is a steady-flow process since there is no change with time. **2** Kinetic and potential energy changes are negligible. **3** There are no work interactions.

Properties Noting that $T < T_{sat\ @\ 200\ kPa} = 120.23°C$, the cold water and the exit mixture streams exist as a compressed liquid, which can be approximated as a saturated liquid at the given temperature. From tables,

$$\left.\begin{array}{l} P_1 = 200\text{kPa} \\ T_1 \doteq 20°\text{C} \end{array}\right\} \begin{array}{l} h_1 \cong h_{f@20°\text{C}} = 83.96\text{kJ/kg} \\ s_1 \cong s_{f@20°\text{C}} = 0.2966\text{kJ/kg}\cdot\text{K} \end{array}$$

$$\left.\begin{array}{l} P_2 = 200\text{kPa} \\ T_2 = 300°\text{C} \end{array}\right\} \begin{array}{l} h_2 = 3071.8\text{kJ/kg} \\ s_2 = 7.8926\text{kJ/kg}\cdot\text{K} \end{array}$$

$$\left.\begin{array}{l} P_3 = 200\text{kPa} \\ T_3 = 60°\text{C} \end{array}\right\} \begin{array}{l} h_3 \cong h_{f@60°\text{C}} = 251.13\text{kJ/kg} \\ s_3 \cong s_{f@60°\text{C}} = 0.8312\text{kJ/kg}\cdot\text{K} \end{array}$$

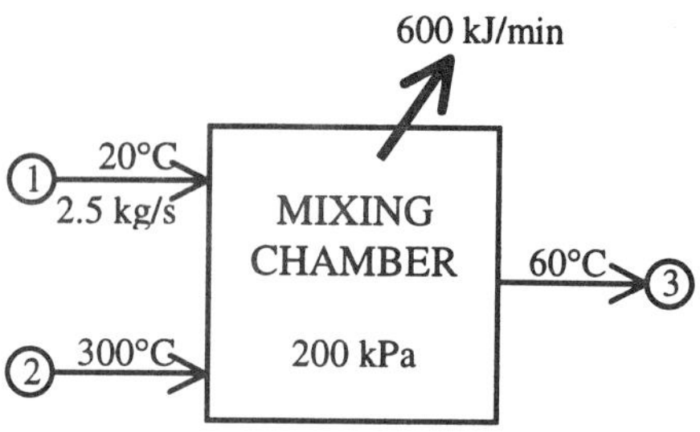

600 kJ/min
① 20°C 2.5 kg/s
② 300°C
MIXING CHAMBER
200 kPa
60°C ③

Analysis (a) We take the mixing chamber as the system, which is a control volume. The mass and energy balances for this steady-flow system can be expressed in the rate form as

Mass balance: $\dot{m}_{in} - \dot{m}_{out} = \Delta\dot{E}_{\text{system}}^{\nearrow 0\ \text{(steady)}} = 0 \longrightarrow \dot{m}_1 + \dot{m}_2 = \dot{m}_3$

Energy balance:

$$\underbrace{\dot{E}_{in} - \dot{E}_{out}}_{\text{Rate of net energy transfer by heat, work, and mass}} = \underbrace{\Delta\dot{E}_{\text{system}}^{\nearrow 0\ \text{(steady)}}}_{\text{Rate of change in internal, kinetic, potential, etc. energies}} = 0$$

$$\dot{E}_{in} = \dot{E}_{out}$$

$$\dot{m}_1 h_1 + \dot{m}_2 h_2 = \dot{Q}_{out} + \dot{m}_3 h_3$$

Combining the two relations gives $\dot{Q}_{out} = \dot{m}_1 h_1 + \dot{m}_2 h_2 - (\dot{m}_1 + \dot{m}_2)h_3 = \dot{m}_1(h_1 - h_3) + \dot{m}_2(h_2 - h_3)$

Solving for $\dot{m}_2$ and substituting, the mass flow rate of the superheated steam is determined to be

$$\dot{m}_2 = \frac{\dot{Q}_{out} - \dot{m}_1(h_1 - h_3)}{h_2 - h_3} = \frac{(600/60\text{kJ/s}) - (2.5\text{kg/s})(83.96 - 251.13)\text{kJ/kg}}{(3071.8 - 251.13)\text{kJ/kg}} = \mathbf{0.152kg/s}$$

Also, $\dot{m}_3 = \dot{m}_1 + \dot{m}_2 = 2.5 + 0.152 = 2.652\ \text{kg/s}$

(*b*) The rate of total entropy generation during this process is determined by applying the entropy balance on an *extended system* that includes the mixing chamber and its immediate surroundings so that the boundary temperature of the extended system is 25°C at all times. It gives

$$\underbrace{\dot{S}_{in} - \dot{S}_{out}}_{\text{Rate of net entropy transfer by heat and mass}} + \underbrace{\dot{S}_{gen}}_{\text{Rate of entropy generation}} = \underbrace{\Delta\dot{S}_{\text{system}}^{\nearrow 0}}_{\text{Rate of change of entropy}} = 0$$

$$\dot{m}_1 s_1 + \dot{m}_2 s_2 - \dot{m}_3 s_3 - \frac{\dot{Q}_{out}}{T_{b,surr}} + \dot{S}_{\text{gen}} = 0$$

Substituting, the rate of entropy generation during this process is determined to be

$$\begin{aligned}\dot{S}_{gen} &= \dot{m}_3 s_3 - \dot{m}_2 s_2 - \dot{m}_1 s_1 + \frac{\dot{Q}_{out}}{T_{b,surr}} \\ &= (2.652\text{kg/s})(0.8312\text{kJ/kg}\cdot\text{K}) - (0.152\text{kg/s})(7.8926\text{kJ/kg}\cdot\text{K}) \\ &\quad - (2.5\text{kg/s})(0.2966\text{kJ/kg}\cdot\text{K}) + \frac{(600/60\ \text{kJ/s})}{298\ \text{K}} \\ &= \mathbf{0.297kW/K}\end{aligned}$$

6-133 A rigid tank initially contains saturated liquid water. A valve at the bottom of the tank is opened, and half of mass in liquid form is withdrawn from the tank. The temperature in the tank is maintained constant. The amount of heat transfer and the entropy generation during this process are to be determined.

Assumptions **1** This is an unsteady process since the conditions within the device are changing during the process, but it can be analyzed as a uniform-flow process since the state of fluid leaving the device remains constant. **2** Kinetic and potential energies are negligible. **3** There are no work interactions involved. **4** The direction of heat transfer is to the tank (will be verified).

Properties The properties of water are (Tables A-4 through A-6)

$$\left.\begin{array}{l} T_1 = 200^\circ\text{C} \\ sat.liquid \end{array}\right\} \begin{array}{l} v_1 = v_{f@200^\circ\text{C}} = 0.001157\text{m}^3/\text{kg} \\ u_1 = u_{f@200^\circ\text{C}} = 850.65\text{kJ/kg} \\ s_1 = s_{f@200^\circ\text{C}} = 2.3309\text{kJ/kg}\cdot\text{K} \end{array}$$

$$\left.\begin{array}{l} T_e = 200^\circ\text{C} \\ sat.liquid \end{array}\right\} \begin{array}{l} h_e = h_{f@200^\circ\text{C}} = 852.45\text{kJ/kg} \\ s_e = s_{f@200^\circ\text{C}} = 2.3309\text{kJ/kg}\cdot\text{K} \end{array}$$

Analysis (*a*) We take the tank as the system, which is a control volume since mass crosses the boundary. Noting that the microscopic energies of flowing and nonflowing fluids are represented by enthalpy h and internal energy u, respectively, the mass and energy balances for this uniform-flow system can be expressed as

Mass balance: $m_{in} - m_{out} = \Delta m_{\text{system}} \quad \rightarrow \quad m_e = m_1 - m_2$

Energy balance:

$$\underbrace{E_{in} - E_{out}}_{\substack{\text{Net energy transfer}\\\text{by heat, work, and mass}}} = \underbrace{\Delta E_{\text{system}}}_{\substack{\text{Change in internal, kinetic,}\\\text{potential, etc. energies}}}$$

$$Q_{in} = m_e h_e + m_2 u_2 - m_1 u_1 \quad (\text{since } W \cong ke \cong pe \cong 0)$$

The initial and the final masses in the tank are

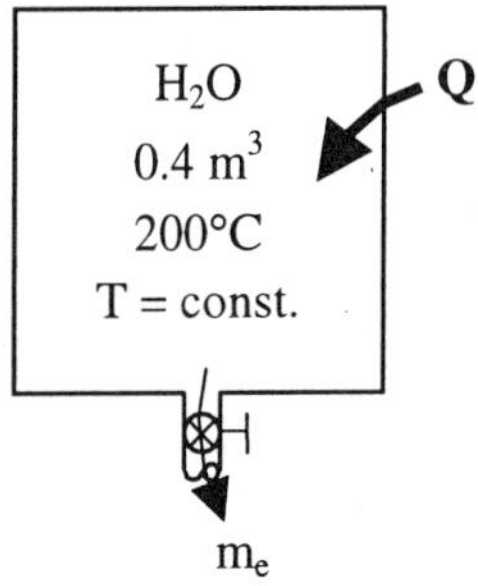

$$m_1 = \frac{V}{v_1} = \frac{0.4\text{m}^3}{0.001157\text{m}^3/\text{kg}} = 345.72\text{kg}$$

$$m_2 = \frac{1}{2}m_1 = \frac{1}{2}(345.72\text{kg}) = 172.86\text{kg} = m_e$$

Now we determine the final internal energy and entropy,

$$v_2 = \frac{V}{m_2} = \frac{0.4\text{m}^3}{172.86\text{kg}} = 0.002314\text{m}^3/\text{kg}$$

$$x_2 = \frac{v_2 - v_f}{v_{fg}} = \frac{0.002314 - 0.001157}{0.12736 - 0.001157} = 0.00917$$

$$\left.\begin{array}{l} T_2 = 200^\circ\text{C} \\ x_2 = 0.00917 \end{array}\right\} \begin{array}{l} u_2 = u_f + x_2 u_{fg} = 850.65 + (0.00917)(1744.7) = 866.65\text{kJ/kg} \\ s_2 = s_f + x_2 s_{fg} = 2.3309 + (0.00917)(4.1014) = 2.3685\text{kJ/kg}\cdot\text{K} \end{array}$$

The heat transfer during this process is determined by substituting these values into the energy balance equation,

$$\begin{aligned} Q_{in} &= m_e h_e + m_2 u_2 - m_1 u_1 \\ &= (172.86\text{kg})(852.45\text{kJ/kg}) + (172.86\text{kg})(866.65\text{kJ/kg}) \\ &\quad - (345.72\text{kg})(850.65\text{kJ/kg}) \\ &= \mathbf{3077kJ} \end{aligned}$$

(*b*) The total entropy generation is determined by considering a combined system that includes the tank and the heat source. Noting that no heat crosses the boundaries of this combined system and no mass enters, the entropy balance for it can be expressed as

$$\underbrace{S_{in} - S_{out}}_{\substack{\text{Net entropy transfer} \\ \text{by heat and mass}}} + \underbrace{S_{gen}}_{\substack{\text{Entropy} \\ \text{generation}}} = \underbrace{\Delta S_{\text{system}}}_{\substack{\text{Change} \\ \text{in entropy}}}$$

$$-m_e s_e + S_{\text{gen}} = \Delta S_{\text{tank}} + \Delta S_{\text{source}}$$

Therefore, the total entropy generated during this process is

$$\begin{aligned} S_{\text{gen}} &= m_e s_e + \Delta S_{\text{tank}} + \Delta S_{\text{source}} = m_e s_e + (m_2 s_2 - m_1 s_1) - \frac{Q_{\text{source,out}}}{T_{\text{source}}} \\ &= (172.86\text{kg})(2.3309\text{kJ/kg}\cdot\text{K}) + (172.86\text{kg})(2.3685\text{kJ/kg}\cdot\text{K}) \\ &\quad - (345.72\text{kg})(2.3309\text{kJ/kg}\cdot\text{K}) - \frac{3077\text{kJ}}{523\text{K}} \\ &= \mathbf{0.616kJ/K} \end{aligned}$$

6-134E An unknown mass of iron is dropped into water in an insulated tank while being stirred by a 200-W paddle wheel. Thermal equilibrium is established after 10 min. The mass of the iron block and the entropy generated during this process are to be determined.

Assumptions **1** Both the water and the iron block are incompressible substances with constant specific heats at room temperature. **2** The system is stationary and thus the kinetic and potential energy changes are zero. **3** The system is well-insulated and thus there is no heat transfer.

Properties The specific heats of water and the iron block at room temperature are $C_{p,\text{ water}}$ = 1.00 Btu/lbm.°F and $C_{p,\text{ iron}}$ = 0.107 Btu/lbm.°F (Table A-3E). The density of water at room temperature is 62.1 lbm/ft³.

Analysis We take the entire contents of the tank, water + iron block, as the system. This is a closed system since no mass crosses the system boundary during the process. The energy balance on the system can be expressed as

$$\underbrace{E_{in} - E_{out}}_{\substack{\text{Net energy transfer}\\ \text{by heat, work, and mass}}} = \underbrace{\Delta E_{\text{system}}}_{\substack{\text{Change in internal, kinetic,}\\ \text{potential, etc. energies}}}$$

$$W_{\text{pw,in}} = \Delta U$$

or,

$$W_{\text{pw,in}} = \Delta U_{\text{iron}} + \Delta U_{\text{water}}$$

$$W_{\text{pw,in}} = [mC(T_2 - T_1)]_{\text{iron}} + [mC(T_2 - T_1)]_{\text{water}}$$

where

$$m_{\text{water}} = \rho V = (62.1 \text{ lbm/ft}^3)(0.8 \text{ ft}^3) = 49.7 \text{ lbm}$$

$$W_{\text{pw}} = \dot{W}_{\text{pw}} \Delta t = (0.2 \text{ kJ/s})(10 \times 60 \text{ s})\left(\frac{1 \text{ Btu}}{1.055 \text{ kJ}}\right) = 113.7 \text{ Btu}$$

Using specific heat values for iron and liquid water and substituting,

$$113.7 \text{ Btu} = m_{iron}(0.107 \text{ Btu/lbm·°F})(75 - 185)\text{°F} + (49.7 \text{ lbm})(1.00 \text{ Btu/lbm·°F})(75 - 70)\text{°F}$$

$$m_{iron} = \mathbf{11.4\ lbm}$$

(*b*) Again we take the iron + water in the tank to be the system. Noting that no heat or mass crosses the boundaries of this combined system, the entropy balance for it can be expressed as

$$\underbrace{S_{in} - S_{out}}_{\substack{\text{Net entropy transfer}\\ \text{by heat and mass}}} + \underbrace{S_{gen}}_{\substack{\text{Entropy}\\ \text{generation}}} = \underbrace{\Delta S_{\text{system}}}_{\substack{\text{Change}\\ \text{in entropy}}}$$

$$0 + S_{\text{gen,total}} = \Delta S_{\text{iron}} + \Delta S_{\text{water}}$$

where

$$\Delta S_{iron} = mC_{ave} \ln\left(\frac{T_2}{T_1}\right) = (11.4\text{lbm})(0.107\text{Btu/lbm} \cdot \text{R})\ln\left(\frac{535\text{R}}{645\text{R}}\right) = -0.228\text{Btu/R}$$

$$\Delta S_{water} = mC_{ave} \ln\left(\frac{T_2}{T_1}\right) = (49.6\text{lbm})(1.0\text{Btu/lbm} \cdot \text{R})\ln\left(\frac{535\text{R}}{530\text{R}}\right) = 0.466\text{Btu/R}$$

Therefore, the total entropy generated during this process is

$$\Delta S_{total} = S_{gen,total} = \Delta S_{iron} + \Delta S_{water} = -0.228 + 0.466 = \mathbf{0.238\ Btu/R}$$

6-135E Air is compressed steadily by a compressor. The mass flow rate of air through the compressor and the rate of entropy generation are to be determined.

Assumptions **1** This is a steady-flow process since there is no change with time. **2** Potential energy changes are negligible. **3** Air is an ideal gas with variable specific heats.

Properties The inlet and exit enthalpies of air are (Table A-17)

$$\left.\begin{array}{l} T_1 = 520R \\ P_1 = 15\text{psia} \end{array}\right\} \begin{array}{l} h_1 = 124.27\text{Btu/lbm} \\ s_1^\circ = 0.59173\text{Btu/lbm}\cdot\text{R} \end{array}$$

$$\left.\begin{array}{l} T_2 = 1080\text{R} \\ P_2 = 150\text{psia} \end{array}\right\} \begin{array}{l} h_2 = 260.97\text{Btu/lbm} \\ s_2^\circ = 0.76964\text{Btu/lbm}\cdot\text{R} \end{array}$$

Analysis (*a*) We take the compressor as the system, which is a control volume since mass crosses the boundary. The energy balance for this steady-flow system can be expressed in the rate form as

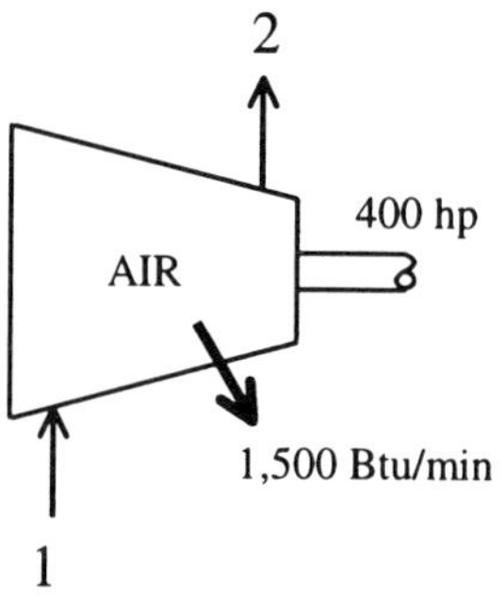

$$\underbrace{\dot{E}_{in} - \dot{E}_{out}}_{\substack{\text{Rate of net energy transfer} \\ \text{by heat, work, and mass}}} = \underbrace{\Delta \dot{E}_{\text{system}}^{\nearrow 0\ (\text{steady})}}_{\substack{\text{Rate of change in internal, kinetic,} \\ \text{potential, etc. energies}}} = 0$$

$$\dot{E}_{in} = \dot{E}_{out}$$

$$\dot{W}_{\text{in}} + \dot{m}(h_1 + \mathbf{V}_1^2/2) = \dot{Q}_{\text{out}} + \dot{m}(h_2 + \mathbf{V}_2^2/2) \quad (\text{since } \Delta\text{pe} \cong 0)$$

$$\dot{W}_{\text{in}} - \dot{Q}_{\text{out}} = \dot{m}\left(h_2 - h_1 + \frac{\mathbf{V}_2^2 - \mathbf{V}_1^2}{2}\right)$$

Substituting, the mass flow rate is determined to be

Thus, $$(400\text{hp})\left(\frac{0.7068\text{Btu/s}}{1\text{hp}}\right) - \frac{1500\text{Btu}}{60\text{s}} = \dot{m}\left(260.97 - 124.27 + \frac{(300\text{ft/s})^2}{2}\left(\frac{1\text{Btu/lbm}}{25{,}037\text{ft}^2/\text{s}^2}\right)\right)$$

It yields $$\dot{m} = \mathbf{1.861\ lbm/s}$$

(*b*) Again we take the compressor to be the system. Noting that no heat or mass crosses the boundaries of this combined system, the entropy balance for it can be expressed as

$$\underbrace{\dot{S}_{in} - \dot{S}_{out}}_{\substack{\text{Rate of net entropy transfer} \\ \text{by heat and mass}}} + \underbrace{\dot{S}_{gen}}_{\substack{\text{Rate of entropy} \\ \text{generation}}} = \underbrace{\Delta \dot{S}_{\text{system}}^{\nearrow 0}}_{\substack{\text{Rate of change} \\ \text{of entropy}}} = 0$$

$$\dot{m}s_1 - \dot{m}s_2 - \frac{\dot{Q}_{out}}{T_{b,surr}} + \dot{S}_{\text{gen}} = 0 \rightarrow \dot{S}_{gen} = \dot{m}(s_2 - s_1) + \frac{\dot{Q}_{out}}{T_{b,surr}}$$

where

$$\Delta \dot{S}_{air} = \dot{m}(s_2 - s_1) = \dot{m}\left(s_2^\circ - s_1^\circ - R\ln\frac{P_2}{P_1}\right)$$

$$= (1.861\ \text{lbm/s})\left(0.76964 - 0.59173 - (0.06855\ \text{Btu/lbm}\cdot\text{R})\ln\frac{150\ \text{psia}}{15\ \text{psia}}\right) = 0.0373\ \text{Btu/s}\cdot\text{R}$$

Substituting, the rate of entropy generation during this process is determined to be

$$\dot{S}_{gen} = \dot{m}(s_2 - s_1) + \frac{\dot{Q}_{out}}{T_{b,surr}} = 0.0373\,\text{Btu/s.R} + \frac{1500/60\ \text{Btu/s}}{520\ \text{R}} = \mathbf{0.0854\ kW/K}$$

6-136 Steam is accelerated in a nozzle from a velocity of 70 m/s to 320 m/s. The exit temperature and the rate of entropy generation are to be determined.

Assumptions **1** This is a steady-flow process since there is no change with time. **2** Potential energy changes are negligible. **3** There are no work interactions. **4** The device is adiabatic and thus heat transfer is negligible.

Properties From the steam tables (Table A-6),

$$\left.\begin{array}{l} P_1 = 3\text{MPa} \\ T_1 = 400^\circ\text{C} \end{array}\right\} \begin{array}{l} h_1 = 3230.9\text{kJ/kg} \\ s_1 = 6.9212\text{kJ/kg}\cdot\text{K} \\ v_1 = 0.09936\text{m}^3\text{/kg} \end{array}$$

Analysis (*a*) There is only one inlet and one exit, and thus $\dot{m}_1 = \dot{m}_2 = \dot{m}$. We take nozzle as the system, which is a control volume since mass crosses the boundary. The energy balance for this steady-flow system can be expressed in the rate form as

$$\underbrace{\dot{E}_{in} - \dot{E}_{out}}_{\substack{\text{Rate of net energy transfer} \\ \text{by heat, work, and mass}}} = \underbrace{\Delta \dot{E}_{\text{system}}^{\nearrow 0\ (\text{steady})}}_{\substack{\text{Rate of change in internal, kinetic,} \\ \text{potential, etc. energies}}} = 0$$

$$\dot{E}_{in} = \dot{E}_{out}$$

$$\dot{m}(h_1 + \mathbf{V}_1^2/2) = \dot{m}(h_2 + \mathbf{V}_2^2/2) \quad (\text{since } \dot{Q} \cong \dot{W} \cong \Delta\text{pe} \cong 0)$$

$$0 = h_2 - h_1 + \frac{\mathbf{V}_2^2 - \mathbf{V}_1^2}{2}$$

Substituting,,

or, $$h_2 = 3230.9\text{kJ/kg} - \frac{(320\text{m/s})^2 - (70\text{m/s})^2}{2}\left(\frac{1\text{kJ/kg}}{1000\text{m}^2/\text{s}^2}\right) = 3182.15\text{kJ/kg}$$

Thus,

$$\left.\begin{array}{l} P_2 = 2\text{MPa} \\ h_{2a} = 3182.15\text{kJ/kg} \end{array}\right\} \begin{array}{l} T_2 = \mathbf{370.4^\circ C} \\ s_2 = 7.0260\text{kJ/kg}\cdot\text{K} \end{array}$$

The mass flow rate of steam is

$$\dot{m} = \frac{1}{v_1} A_1 \mathbf{V}_1 = \frac{1}{0.09936\text{m}^3\text{/kg}}\left(7\times10^{-4}\text{m}^2\right)(70\text{m/s}) = 0.493\text{kg/s}$$

(*b*) Again we take the nozzle to be the system. Noting that no heat crosses the boundaries of this combined system, the entropy balance for it can be expressed as

$$\underbrace{\dot{S}_{in} - \dot{S}_{out}}_{\substack{\text{Rate of net entropy transfer} \\ \text{by heat and mass}}} + \underbrace{\dot{S}_{gen}}_{\substack{\text{Rate of entropy} \\ \text{generation}}} = \underbrace{\Delta \dot{S}_{\text{system}}^{\nearrow 0}}_{\substack{\text{Rate of change} \\ \text{of entropy}}} = 0$$

$$\dot{m}s_1 - \dot{m}s_2 + \dot{S}_{\text{gen}} = 0 \rightarrow \dot{S}_{gen} = \dot{m}(s_2 - s_1)$$

Substituting, the rate of entropy generation during this process is determined to be

$$\dot{S}_{gen} = \dot{m}(s_2 - s_1) = (0.493\text{kg/s})(7.0260 - 6.9212)\text{kJ/kg}\cdot\text{K} = \mathbf{0.0517kW/K}$$

Special Topic: Reducing the Cost of Compressed Air

6-137 The total installed power of compressed air systems in the US is estimated to be about 20 million horsepower. The amount of energy and money that will be saved per year if the energy consumed by compressors is reduced by 5 percent is to be determined.

Assumptions **1** The compressors operate at full load during one-third of the time on average, and are shut down the rest of the time. **2** The average motor efficiency is 85 percent.

Analysis The electrical energy consumed by compressors per year is

Energy consumed = (Power rating)(Load factor)(Annual Operating Hours)/Motor efficiency
= $(20\times10^6$ hp)(0.746 kW/hp)(1/3)(365×24 hours/year)/0.85
= 5.125×10^{10} kWh/year

Then the energy and cost savings corresponding to a 5% reduction in energy use for compressed air become

Energy Savings = (Energy consumed)(Fraction saved)
= $(5.125\times10^{10}$ kWh)(0.05)
= 2.563×10^9 kWh/year

Cost Savings = (Energy savings)(Unit cost of energy)
= $(2.563\times10^9$ kWh/year)($0.08/kWh)
= **$0.205×10⁹ /year**

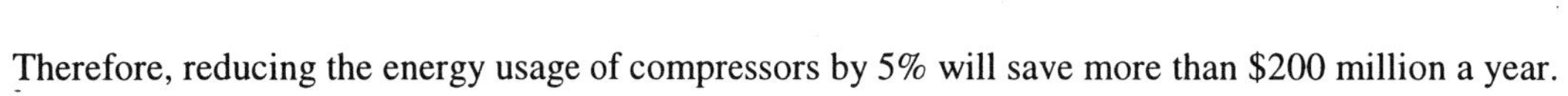

Therefore, reducing the energy usage of compressors by 5% will save more than $200 million a year.

6-138 The total energy used to compress air in the US is estimated to be 0.5×10^{15} kJ per year. About 20% of the compressed air is estimated to be lost by air leaks. The amount and cost of electricity wasted per year due to air leaks is to be determined.

Assumptions About 20% of the compressed air is lost by air leaks.

Analysis The electrical energy and money wasted by air leaks are

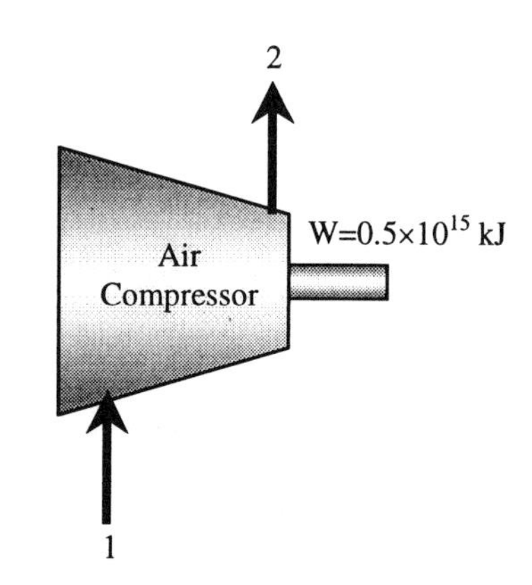

Energy wasted = (Energy consumed)(Fraction wasted)
= $(0.5\times10^{15}$ kJ)(1 kWh/3600 kJ)(0.20)
= **27.78×10^9 kWh/year**

Money wasted = (Energy wasted)(Unit cost of energy)
= $(27.78\times10^9$ kWh/year)($0.07/kWh)
= **$1.945×10⁹ /year**

Therefore, air leaks are costing almost $2 billion a year in electricity costs. The environment also suffers from this because of the pollution associated with the generation of this much electricity.

6-139 The compressed air requirements of a plant is being met by a 125 hp compressor that compresses air from 101.3 kPa to 900 kPa. The amount of energy and money saved by reducing the pressure setting of compressed air to 750 kPa is to be determined.

Assumptions **1** Air is an ideal gas with constant specific heats. **2** Kinetic and potential energy changes are negligible. **3** The load factor of the compressor is given to be 0.75. **4** The pressures given are absolute pressure rather than gage pressure.

Properties The specific heat ratio of air is k = 1.4 (Table A-2).

Analysis The electrical energy consumed by this compressor per year is

Energy consumed = (Power rating)(Load factor)(Annual Operating Hours)/Motor efficiency
= (125 hp)(0.746 kW/hp)(0.75)(3500 hours/year)/0.88
= 278,160 kWh/year

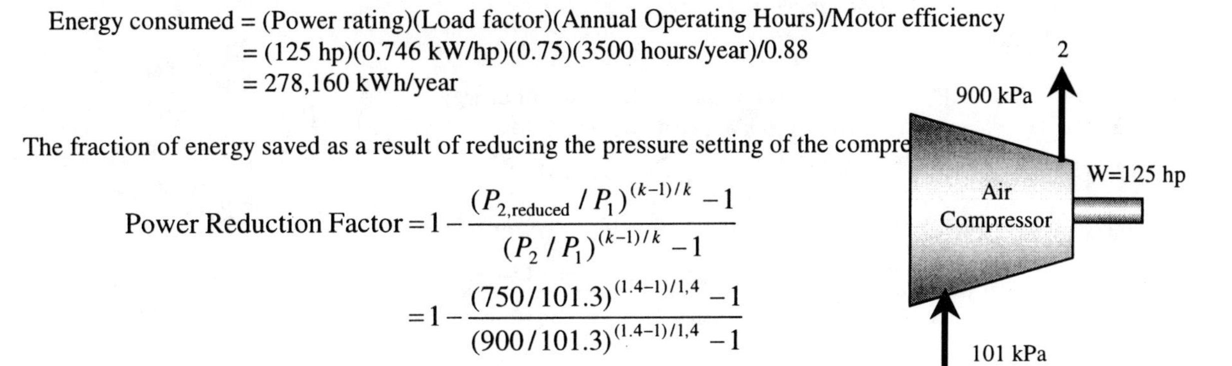

The fraction of energy saved as a result of reducing the pressure setting of the compressor is

$$\text{Power Reduction Factor} = 1 - \frac{(P_{2,\text{reduced}} / P_1)^{(k-1)/k} - 1}{(P_2 / P_1)^{(k-1)/k} - 1}$$

$$= 1 - \frac{(750/101.3)^{(1.4-1)/1,4} - 1}{(900/101.3)^{(1.4-1)/1,4} - 1}$$

$$= 0.1093$$

That is, reducing the pressure setting will result in about 11 percent savings from the energy consumed by the compressor and the associated cost. Therefore, the energy and cost savings in this case become

Energy Savings = (Energy consumed)(Power reduction factor)
= (278,160 kWh/year)(0.1093)
= 30,410 kWh/year

Cost Savings = (Energy savings)(Unit cost of energy)
= (30,410 kWh/year)($0.085/kWh)
= $2585/year

Therefore, reducing the pressure setting by 150 kPa will result in annual savings of 30.410 kWh that is worth $2585 in this case.

Discussion Some applications require very low pressure compressed air. In such cases the need can be met by a blower instead of a compressor. Considerable energy can be saved in this manner, since a blower requires a small fraction of the power needed by a compressor for a specified mass flow rate.

6-140 A 150 hp compressor in an industrial facility is housed inside the production area where the average temperature during operating hours is 25°C. The amounts of energy and money saved as a result of drawing cooler outside air to the compressor instead of using the inside air are to be determined.

Assumptions **1** Air is an ideal gas with constant specific heats. **2** Kinetic and potential energy changes are negligible.

Analysis The electrical energy consumed by this compressor per year is

Energy consumed = (Power rating)(Load factor)(Annual Operating Hours)/Motor efficiency
= (150 hp)(0.746 kW/hp)(0.85)(6000 hours/year)/0.9
= 634,100 kWh/year

Also,

Cost of Energy = (Energy consumed)(Unit cost of energy)
= (634,100 kWh/year)($0.07/kWh)
= $44,390/year

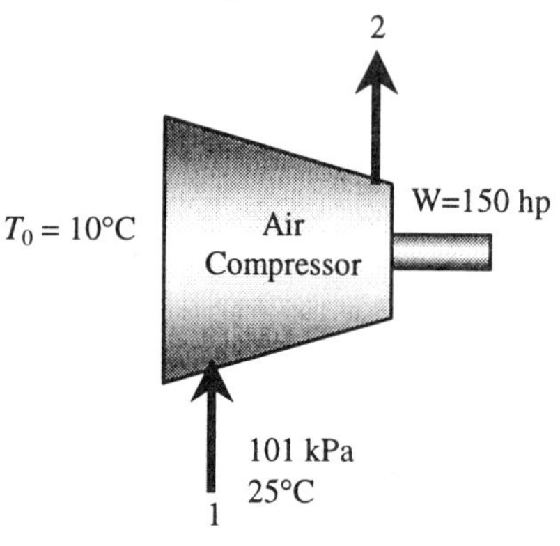

The fraction of energy saved as a result of drawing in cooler outside air is

$$\text{Power Reduction Factor} = 1 - \frac{T_{\text{outside}}}{T_{\text{inside}}} = 1 - \frac{10+273}{25+273} = 0.0503$$

That is, drawing in air which is 15°C cooler will result in 5.03 percent savings from the energy consumed by the compressor and the associated cost. Therefore, the energy and cost savings in this case become

Energy Savings = (Energy consumed)(Power reduction factor)
= (634,100 kWh/year)(0.0503)
= 31,895 kWh/year

Cost Savings = (Energy savings)(Unit cost of energy)
= (31,895 kWh/year)($0.07/kWh)
= $2233/year

Therefore, drawing air in from the outside will result in annual savings of 31,895 kWh, which is worth $2233 in this case.

Discussion The price of a typical 150 hp compressor is much lower than $50,000. Therefore, it is interesting to note that the cost of energy a compressor uses a year may be more than the cost of the compressor itself.

The implementation of this measure requires the installation of an ordinary sheet metal or PVC duct from the compressor intake to the outside. The installation cost associated with this measure is relatively low, and the pressure drop in the duct in most cases is negligible. About half of the manufacturing facilities we have visited, especially the newer ones, have the duct from the compressor intake to the outside in place, and they are already taking advantage of the savings associated with this measure.

6-141 The compressed air requirements of the facility during 60 percent of the time can be met by a 25 hp reciprocating compressor instead of the existing 100 hp compressor. The amounts of energy and money saved as a result of switching to the 25 hp compressor during 60 percent of the time are to be determined. √

Analysis Noting that 1 hp = 0.746 kW, the electrical energy consumed by each compressor per year is determined from

$$\text{(Energy consumed)}_{\text{Large}} = \text{(Power)(Hours)}[(\text{LFxTF}/\eta_{\text{motor}})_{\text{Unloaded}} + (\text{LFxTF}/\eta_{\text{motor}})_{\text{Loaded}}]$$
$$= (100\text{ hp})(0.746\text{ kW/hp})(3800\text{ hours/year})[0.35\times0.6/0.82+0.90\times0.4/0.9]$$
$$= 185{,}990\text{ kWh/year}$$

$$\text{(Energy consumed)}_{\text{Small}} = \text{(Power)(Hours)}[(\text{LFxTF}/\eta_{\text{motor}})_{\text{Unloaded}} + (\text{LFxTF}/\eta_{\text{motor}})_{\text{Loaded}}]$$
$$= (25\text{ hp})(0.746\text{ kW/hp})(3800\text{ hours/year})[0.0\times0.15+0.95\times0.85]/0.88$$
$$= 65{,}031\text{ kWh/year}$$

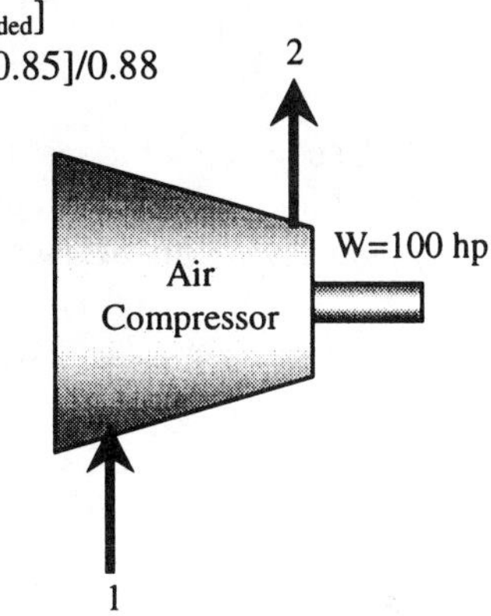

Therefore, the energy and cost savings in this case become

$$\text{Energy Savings} = \text{(Energy consumed)}_{\text{Large}} - \text{(Energy consumed)}_{\text{Small}}$$
$$= 185{,}990 - 65{,}031\text{ kWh/year}$$
$$= \mathbf{120{,}959\text{ kWh/year}}$$

$$\text{Cost Savings} = \text{(Energy savings)(Unit cost of energy)}$$
$$= (120{,}959\text{ kWh/year})(\$0.075/\text{kWh})$$
$$= \mathbf{\$9{,}072/year}$$

Discussion Note that utilizing a small compressor during the times of reduced compressed air requirements and shutting down the large compressor will result in annual savings of 120,959 kWh, which is worth $9,072 in this case.

6-142 A facility stops production for one hour every day, including weekends, for lunch break, but the 125 hp compressor is kept operating. If the compressor consumes 35 percent of the rated power when idling, the amounts of energy and money saved per year as a result of turning the compressor off during lunch break are to be determined.

Analysis It seems like the compressor in this facility is kept on unnecessarily for one hour a day and thus 365 hours a year, and the idle factor is 0.35. Then the energy and cost savings associated with turning the compressor off during lunch break are determined to be

$$\text{Energy Savings} = \text{(Power Rating)(Turned Off Hours)(Idle Factor)}/\eta_{\text{motor}}$$
$$= (125\text{ hp})(0.746\text{ kW/hp})(365\text{ hours/year})(0.35)/0.84$$
$$= \mathbf{14{,}182\text{ kWh/year}}$$

$$\text{Cost Savings} = \text{(Energy savings)(Unit cost of energy)}$$
$$= (14{,}182\text{ kWh/year})(\$0.09/\text{kWh})$$
$$= \mathbf{\$1{,}276/year}$$

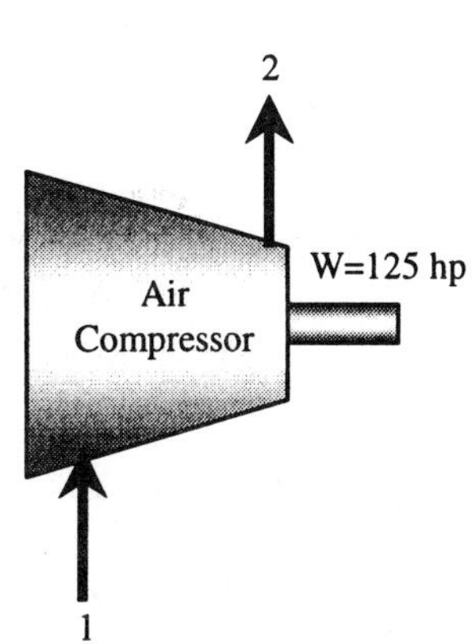

Discussion Note that the simple practice of turning the compressor off during lunch break will save this facility $1,276 a year in energy costs. There are also side benefits such as extending the life of the motor and the compressor, and reducing the maintenance costs.

6-143 It is determined that 40 percent of the energy input to the compressor is removed from the compressed air as heat in the aftercooler with a refrigeration unit whose COP is 3.5. The amounts of the energy and money saved per year as a result of cooling the compressed air before it enters the refrigerated dryer are to be determined.

Assumptions The compressor operates at full load when operating.

Analysis Noting that 40 percent of the energy input to the compressor is removed by the aftercooler, the rate of heat removal from the compressed air in the aftercooler under full load conditions is

$$\begin{aligned}\dot{Q}_{\text{aftercooling}} &= \text{(Rated Power of Compressor)(Load Factor)(Aftercooling Fraction)} \\ &= (150\text{ hp})(0.746\text{ kW/hp})(1.0)(0.4) \\ &= 44.76\text{ kW}\end{aligned}$$

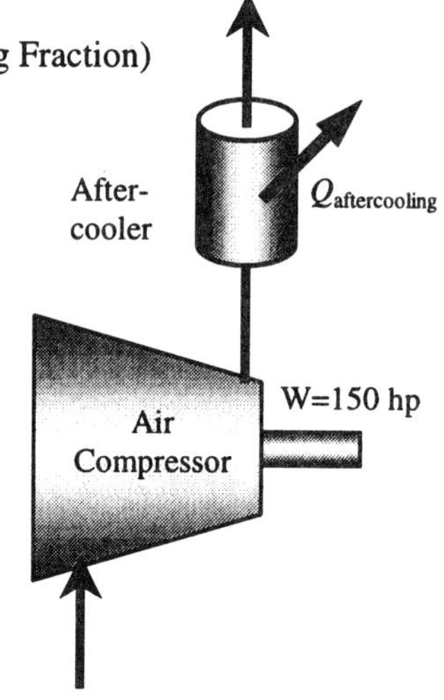

The compressor is said to operate at full load for 1600 hours a year, and the COP of the refrigeration unit is 3.5. Then the energy and cost savings associated with this measure become

$$\begin{aligned}\text{Energy Savings} &= (\dot{Q}_{\text{aftercooling}})\text{(Annual Operating Hours)/COP} \\ &= (44.76\text{ kW})(1600\text{ hours/year})/3.5 \\ &= \mathbf{20{,}462\text{ kWh/year}}\end{aligned}$$

$$\begin{aligned}\text{Cost Savings} &= \text{(Energy savings)(Unit cost of energy saved)} \\ &= (20{,}462\text{ kWh/year})(\$0.08/\text{kWh}) \\ &= \mathbf{\$1{,}637/year}\end{aligned}$$

Discussion Note that the aftercooler will save this facility 20,462 kWh of electrical energy worth $1,637 per year. The actual savings will be less than indicated above since we have not considered the power consumed by the fans and/or pumps of the aftercooler. However, if the heat removed by the aftercooler is utilized for some useful purpose such as space heating or process heating, then the actual savings will be much more.

6-144 The motor of a 150 hp compressor is burned out and is to be replaced by either a 93% efficient standard motor or a 96.2% efficient high efficiency motor. It is to be determined if the savings from the high efficiency motor justify the price differential.

Assumptions **1** The compressor operates at full load when operating. **2** The life of the motors is 10 years. **3** There are no rebates involved. **4** The price of electricity remains constant.

Analysis The energy and cost savings associated with the installation of the high efficiency motor in this case are determined to be

Energy Savings = (Power Rating)(Operating Hours)(Load Factor)$(1/\eta_{standard} - 1/\eta_{efficient})$
= (150 hp)(0.746 kW/hp)(4,368 hours/year)(1.0)(1/0.930 - 1/0.962)
= **17,483 kWh/year**

Cost Savings = (Energy savings)(Unit cost of energy)
= (17,483 kWh/year)($0.075/kWh)
= **$1311/year**

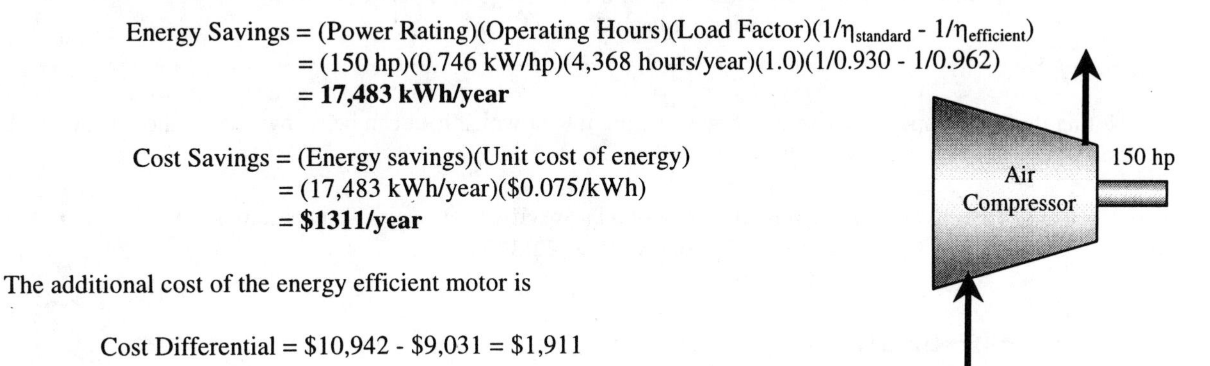

The additional cost of the energy efficient motor is

Cost Differential = $10,942 - $9,031 = $1,911

Discussion The money saved by the high efficiency motor will pay for this cost difference in $1,911/$1311 = 1.5 years, and will continue saving the facility money for the rest of the 10 years of its lifetime. Therefore, the use of the high efficiency motor is recommended in this case even in the absence of any incentives from the local utility company.

6-145 The compressor of a facility is being cooled by air in a heat-exchanger. This air is to be used to heat the facility in winter. The amount of money that will be saved by diverting the compressor waste heat into the facility during the heating season is to be determined.

Assumptions The compressor operates at full load when operating.

Analysis Assuming operation at sea level and taking the density of air to be 1.2 kg/m^3, the mass flow rate of air through the liquid-to-air heat exchanger is determined to be

Mass flow rate of air = (Density of air)(Average velocity)(Flow area)
= (1.2 kg/m^3)(3 m/s)(1.0 m^2)
= 3.6 kg/s = 12,960 kg/h

Noting that the temperature rise of air is 32°C, the rate at which heat can be recovered (or the rate at which heat is transferred to air) is

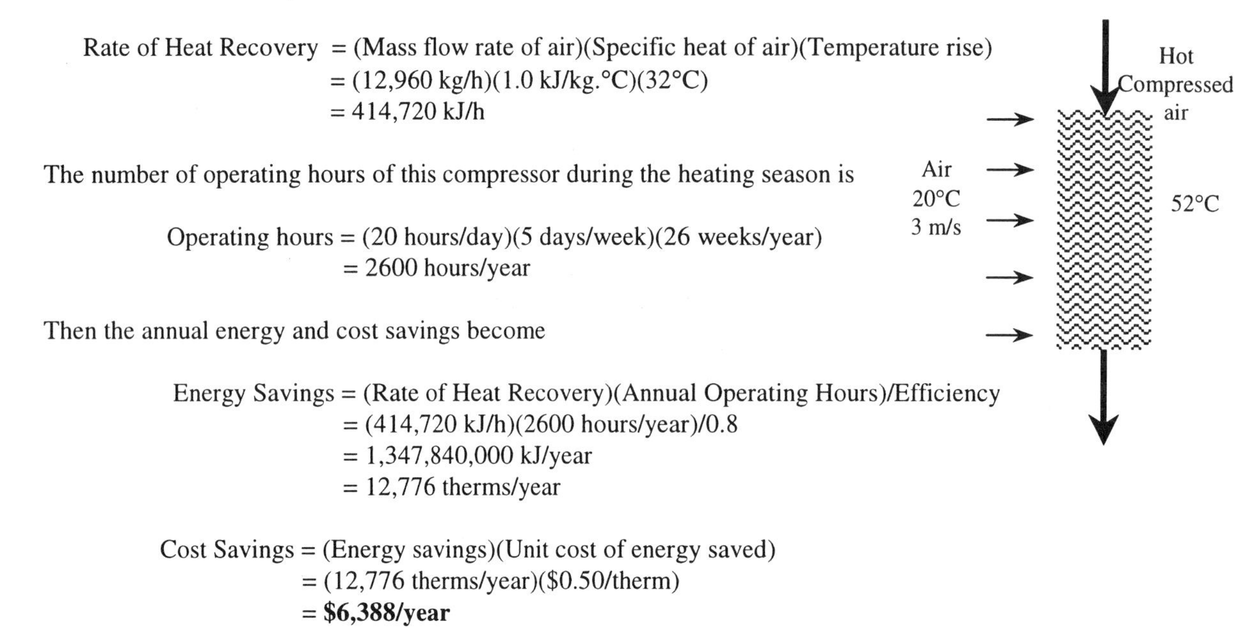

Rate of Heat Recovery = (Mass flow rate of air)(Specific heat of air)(Temperature rise)
= (12,960 kg/h)(1.0 kJ/kg.°C)(32°C)
= 414,720 kJ/h

The number of operating hours of this compressor during the heating season is

Operating hours = (20 hours/day)(5 days/week)(26 weeks/year)
= 2600 hours/year

Then the annual energy and cost savings become

Energy Savings = (Rate of Heat Recovery)(Annual Operating Hours)/Efficiency
= (414,720 kJ/h)(2600 hours/year)/0.8
= 1,347,840,000 kJ/year
= 12,776 therms/year

Cost Savings = (Energy savings)(Unit cost of energy saved)
= (12,776 therms/year)($0.50/therm)
= **$6,388/year**

Therefore, utilizing the waste heat from the compressor will save $6,388 per year from the heating costs.

Discussion The implementation of this measure requires the installation of an ordinary sheet metal duct from the outlet of the heat exchanger into the building. The installation cost associated with this measure is relatively low. A few of the manufacturing facilities we have visited already have this conservation system in place. A damper is used to direct the air into the building in winter and to the ambient in summer.

Combined compressor/heat-recovery systems are available in the market for both air-cooled (greater than 50 hp) and water cooled (greater than 125 hp) systems.

6-146 The compressed air lines in a facility are maintained at a gage pressure of 850 kPa at a location where the atmospheric pressure is 85.6 kPa. There is a 5-mm diameter hole on the compressed air line. The energy and money saved per year by sealing the hole on the compressed air line.

Assumptions **1** Air is an ideal gas with constant specific heats. **2** Kinetic and potential energy changes are negligible.

Properties The gas constant of air is R = 0.287 kJ/kg.K (Table A-1). The specific heat ratio of air is k = 1.4.

Analysis Disregarding any pressure losses and noting that the absolute pressure is the sum of the gage pressure and the atmospheric pressure, the work needed to compress a unit mass of air at 15°C from the atmospheric pressure of 85.6 kPa to 850+85.6 = 935.6 kPa is determined to be

$$w_{\text{comp, in}} = \frac{kRT_1}{\eta_{\text{comp}}(k-1)}\left[\left(\frac{P_2}{P_1}\right)^{(k-1)/k} - 1\right]$$

$$= \frac{(1.4)(0.287\text{ kJ/kg.K})(288\text{ K})}{(0.8)(1.4-1)}\left[\left(\frac{935.6\text{ kPa}}{85.6\text{ kPa}}\right)^{(1.4-1)/1.4} - 1\right]$$

$$= 354.5\text{ kJ/kg}$$

P_{atm} = 85.6 kPa, 15°C

Air leak

Compressed air line
850 kPa, 25°C

The cross-sectional area of the 5-mm diameter hole is

$$A = \pi D^2/4 = \pi(5\times10^{-3}\text{ m})^2/4 = 19.63\times10^{-6}\text{ m}^2$$

Noting that the line conditions are T_0 = 298 K and P_0 = 935.6 kPa, the mass flow rate of the air leaking through the hole is determined to be

$$\dot{m}_{\text{air}} = C_{\text{loss}}\left(\frac{2}{k+1}\right)^{1/(k-1)}\frac{P_0}{RT_0}A\sqrt{kR\left(\frac{2}{k+1}\right)T_0}$$

$$= (0.65)\left(\frac{2}{1.4+1}\right)^{1/(1.4-1)}\frac{935.6\text{ kPa}}{(0.287\text{ kPa.m}^3/\text{kg.K})(298\text{ K})}(19.63\times10^{-6}\text{ m}^2)$$

$$\times\sqrt{(1.4)(0.287\text{ kJ/kg.K})\left(\frac{1000\text{ m}^2/\text{s}^2}{1\text{ kJ/kg}}\right)\left(\frac{2}{1.4+1}\right)(298\text{ K})}$$

$$= 0.02795\text{ kg/s}$$

Then the power wasted by the leaking compressed air becomes

$$\text{Power wasted} = \dot{m}_{\text{air}}w_{\text{comp,in}} = (0.02795\text{ kg/s})(354.5\text{ kJ/kg}) = 9.91\text{ kW}$$

Noting that the compressor operates 5200 hours a year and the motor efficiency is 0.93, the annual energy and cost savings resulting from repairing this leak are determined to be

Energy Savings = (Power wasted)(Annual operating hours)/Motor efficiency
= (9.91 kW)(5200 hours/year)/0.93
= 55,400 kWh/year

Cost Savings = (Energy savings)(Unit cost of energy)
= (55,400 kWh/year)($0.072/kWh)
= $3989/year

Therefore, the facility will save 55,400 kWh of electricity that is worth $3989 a year when this air leak is sealed.

Review Problems

6-147 It is to be shown that the difference between the steady-flow and boundary works is the flow energy.

Analysis The total differential of flow energy Pv can be expressed as

$$d(Pv) = Pdv + vdP = \delta w_b - \delta w_{flow} = \delta(w_b - w_{flow})$$

Therefore, the difference between the reversible steady-flow work and the reversible boundary work is the flow energy.

6-148E An insulated rigid can initially contains R-134a at a specified state. A crack develops, and refrigerant escapes slowly. The final mass in the can is to be determined when the pressure inside drops to a specified value.

Assumptions **1** The can is well-insulated and thus heat transfer is negligible. **2** The refrigerant that remains in the can underwent a reversible adiabatic process.

Analysis Noting that for a reversible adiabatic (i.e., isentropic) process, $s_1 = s_2$, the properties of the refrigerant in the can are

$$\left.\begin{array}{l} P_1 = 120 \text{ psia} \\ T_1 = 80^\circ\text{F} \end{array}\right\} s_1 \cong s_{f @ 80^\circ\text{F}} = 0.0774 \text{ Btu/lbm}\cdot\text{R}$$

$$\left.\begin{array}{l} P_2 = 30 \text{ psia} \\ s_2 = s_1 \end{array}\right\} \begin{array}{l} x_2 = \dfrac{s_2 - s_f}{s_{fg}} = \dfrac{0.0774 - 0.0364}{0.2209 - 0.0364} = 0.2222 \\ v_2 = v_f + x_2 v_{fg} = 0.01209 + (0.2222)(1.5408 - 0.01209) = 0.3518 \text{ ft}^3\text{/lbm} \end{array}$$

Thus the final mass of the refrigerant in the can is

$$m = \frac{V}{v_1} = \frac{1.2 \text{ ft}^3}{0.3518 \text{ ft}^3\text{/lbm}} = \mathbf{3.411 \text{ lbm}}$$

6-149 An insulated rigid tank is connected to a piston-cylinder device with zero clearance that is maintained at constant pressure. A valve is opened, and some steam in the tank is allowed to flow into the cylinder. The final temperatures in the tank and the cylinder are to be determined.

Assumptions **1** Both the tank and cylinder are well-insulated and thus heat transfer is negligible. **2** The water that remains in the tank underwent a reversible adiabatic process. **3** The thermal energy stored in the tank and cylinder themselves is negligible. **4** The system is stationary and thus kinetic and potential energy changes are negligible.

Analysis (a) The steam in tank A undergoes a reversible, adiabatic process, and thus $s_2 = s_1$. From the steam tables,

$$\left.\begin{array}{l} P_1 = 500\text{kPa} \\ sat.vapor \end{array}\right\} \begin{array}{l} v_1 = v_{g@500\text{kPa}} = 0.3749\text{m}^3/\text{kg} \\ u_1 = u_{g@500\text{kPa}} = 2561.2\text{kJ/kg} \\ s_1 = s_{g@500\text{kPa}} = 6.8213\text{kJ/kg}\cdot\text{K} \end{array}$$

$$\left.\begin{array}{l} P_2 = 150\text{kPa} \\ s_2 = s_1 \\ (sat.mixture) \end{array}\right\} \begin{array}{l} T_{2,A} = T_{sat@150\text{kPa}} = \mathbf{111.37^\circ C} \\ x_{2,A} = \dfrac{s_{2,A} - s_f}{s_{fg}} = \dfrac{6.8213 - 1.4336}{5.7897} = 0.9306 \\ v_{2,A} = v_f + x_{2,A} v_{fg} = 0.001053 + (0.9306)(1.1593 - 0.001053) = 1.079\text{ m}^3/\text{kg} \\ u_{2,A} = u_f + x_{2,A} u_{fg} = 466.94 + (0.9306)(2052.7\text{ kJ/kg}) = 2377.2\text{ kJ/kg} \end{array}$$

The initial and the final masses in tank A are

$$m_{1,A} = \frac{V_A}{v_{1,A}} = \frac{0.4\text{m}^3}{0.3749\text{m}^3/\text{kg}} = 1.067\text{kg} \; and \; m_{2,A} = \frac{V_A}{v_{2,A}} = \frac{0.4\text{m}^3}{1.079\text{m}^3/\text{kg}} = 0.371\text{kg}$$

Thus,

$$m_{2,B} = m_{1,A} - m_{2,B} = 1.067 - 0.371 = 0.696\text{ kg}$$

(*b*) The boundary work done during this process is

$$W_{b,out} = \int_1^2 PdV = P_B\left(V_{2,B} - 0\right) = P_B m_{2,B} v_{2,B}$$

Taking the contents of both the tank and the cylinder to be the system, the energy balance for this closed system can be expressed as

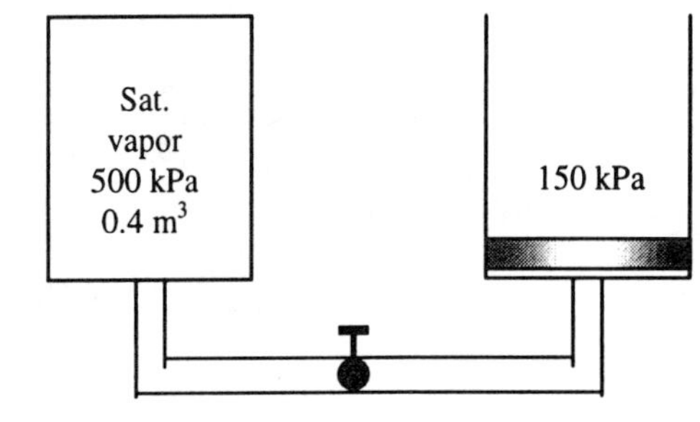

$$\underbrace{E_{in} - E_{out}}_{\substack{\text{Net energy transfer} \\ \text{by heat, work, and mass}}} = \underbrace{\Delta E_{\text{system}}}_{\substack{\text{Change in internal, kinetic,} \\ \text{potential, etc. energies}}}$$

$$-W_{\text{b,out}} = \Delta U = (\Delta U)_A + (\Delta U)_B$$

$$W_{b,out} + (\Delta U)_A + (\Delta U)_B = 0$$

or,

$$P_B m_{2,B} v_{2,B} + (m_2u_2 - m_1u_1)_A + (m_2u_2)_B = 0$$

$$m_{2,B} h_{2,B} + (m_2u_2 - m_1u_1)_A = 0$$

Thus,

$$h_{2,B} = \frac{(m_1u_1 - m_2u_2)_A}{m_{2,B}} = \frac{(1.067)(2561.2) - (0.371)(2377.2)}{0.696} = 2659.3 kJ/kg$$

At 150 kPa, $h_f = 467.11$ and $h_g = 2693.6$ kJ/kg. Thus at the final state, the cylinder will contain a saturated liquid-vapor mixture since $h_f < h_2 < h_g$. Therefore,

$$T_{2,B} = T_{sat\,@\,150\,kPa} = \mathbf{111.37^\circ C}$$

6-150 One ton of liquid water at 80°C is brought into a room. The final equilibrium temperature in the room and the entropy change during this process are to be determined.

Assumptions **1** The room is well insulated and well sealed. **2** The thermal properties of water and air are constant at room temperature. **3** The system is stationary and thus the kinetic and potential energy changes are zero. **4** There are no work interactions involved.

Properties The gas constant of air is R = 0.287 kPa.m^3/kg.K (Table A-1). The specific heat of water at room temperature is C = 4.18 kJ/kg·°C (Table A-3). For air is C_v = 0.718 kJ/kg·°C at room temperature.

Analysis The volume and the mass of the air in the room are

V = 4x5x6 = 120 m³

$$m_{air} = \frac{P_1 V_1}{RT_1} = \frac{(100\text{kPa})(120\text{m}^3)}{(0.2870\text{kPa}\cdot\text{m}^3/\text{kg}\cdot\text{K})(295\text{K})} = 141.7\text{kg}$$

4m × 5m × 6m

ROOM
22°C
100 kPa

Water
80°C

Heat

Taking the contents of the room, including the water, as our system, the energy balance can be written as

$$\underbrace{E_{in} - E_{out}}_{\substack{\text{Net energy transfer}\\\text{by heat, work, and mass}}} = \underbrace{\Delta E_{\text{system}}}_{\substack{\text{Change in internal, kinetic,}\\\text{potential, etc. energies}}} \rightarrow 0 = \Delta U = (\Delta U)_{\text{water}} + (\Delta U)_{\text{air}}$$

or
$$[mC(T_2 - T_1)]_{\text{water}} + [mC_v(T_2 - T_1)]_{\text{air}} = 0$$

Substituting,

$$(1000\text{ kg})(4.18\text{ kJ/kg}\cdot{}^\circ\text{C})(T_f - 80)^\circ\text{C} + (141.7\text{ kg})(0.718\text{ kJ/kg}\cdot{}^\circ\text{C})(T_f - 22)^\circ\text{C} = 0$$

It gives the final equilibrium temperature in the room to be

$$T_f = \mathbf{78.6°C}$$

(*b*) Considering that the system is well-insulated and no mass is entering and leaving, the total entropy change during this process is the sum of the entropy changes of water and the room air,

$$\Delta S_{\text{total}} = S_{\text{gen}} = \Delta S_{\text{air}} + \Delta S_{\text{water}}$$

where

$$\Delta S_{air} = mC_v \ln\frac{T_2}{T_1} + mR\ln\frac{V_2}{V_1}^{\nearrow 0} = (141.7\text{ kg})(0.718\text{kJ/kg}\cdot\text{K})\ln\frac{351.6\text{K}}{295\text{K}} = 17.86\text{kJ/K}$$

$$\Delta S_{water} = mC\ln\frac{T_2}{T_1} = (1000\text{kg})(4.18\text{ kJ/kg}\cdot\text{K})\ln\frac{351.6\text{ K}}{353\text{ K}} = -16.61\text{ kJ/K}$$

Substituting, the total entropy change is determined to be

$$\Delta S_{\text{total}} = 17.86 - 16.61 = \mathbf{1.25\ kJ/K}$$

6-151E A cylinder initially filled with helium gas at a specified state is compressed polytropically to a specified temperature and pressure. The entropy changes of the helium and the surroundings are to be determined, and it is to be assessed if the process is reversible, irreversible, or impossible.

Assumptions **1** Helium is an ideal gas with constant specific heats. **2** The cylinder is stationary and thus the kinetic and potential energy changes are negligible. **3** The thermal energy stored in the cylinder itself is negligible. **4** The compression or expansion process is quasi-equilibrium.

Properties The gas constant of helium is R = 2.6805 psia.ft^3/lbm.R = 0.4961 Btu/lbm.R (Table A-1E). The specific heats of helium are C_v = 0.753 and C_v = 1.25 Btu/lbm.R (Table A-2E).

Analysis (*a*) The mass of helium is

$$m = \frac{P_1 V_1}{RT_1} = \frac{(25\text{psia})(15\text{ft}^3)}{(2.6805\text{psia}\cdot\text{ft}^3/\text{lbm}\cdot\text{R})(530\text{R})} = 0.264\text{lbm}$$

HELIUM
15 ft^3
PV^n = const
Q

Then the entropy change of helium becomes

$$\Delta S_{sys} = \Delta S_{helium} = m\left(C_{p,ave}\ln\frac{T_2}{T_1} - R\ln\frac{P_2}{P_1}\right)$$

$$= (0.264\text{lbm})\left[(1.25\text{Btu/lbm}\cdot\text{R})\ln\frac{760\text{R}}{530\text{R}} - (0.4961\text{Btu/lbm}\cdot\text{R})\ln\frac{70\text{psia}}{25\text{psia}}\right] = \mathbf{-0.016Btu/R}$$

(*b*) The exponent n and the boundary work for this polytropic process are determined to be

$$\frac{P_1 V_1}{T_1} = \frac{P_2 V_2}{T_2} \longrightarrow V_2 = \frac{T_2}{T_1}\frac{P_1}{P_2}V_1 = \frac{(760\text{R})(25\text{psia})}{(530\text{R})(70\text{psia})}(15\text{ft}^3) = 7.682\text{ft}^3$$

$$P_2 V_2^n = P_1 V_1^n \longrightarrow \left(\frac{P_2}{P_1}\right) = \left(\frac{V_1}{V_2}\right)^n \longrightarrow \left(\frac{70}{25}\right) = \left(\frac{15}{7.682}\right)^n \longrightarrow n = 1.539$$

Then the boundary work for this polytropic process can be determined from

$$W_{b,in} = -\int_1^2 P\,dV = -\frac{P_2 V_2 - P_1 V_1}{1-n} = -\frac{mR(T_2 - T_1)}{1-n}$$

$$= -\frac{(0.264\text{lbm})(0.4961\text{Btu/lbm}\cdot\text{R})(760-530)\text{R}}{1-1.539} = 55.9\text{Btu}$$

We take the helium in the cylinder as the system, which is a closed system. Taking the direction of heat transfer to be from the cylinder, the energy balance for this stationary closed system can be expressed as

$$\underbrace{E_{in} - E_{out}}_{\substack{\text{Net energy transfer}\\ \text{by heat, work, and mass}}} = \underbrace{\Delta E_{\text{system}}}_{\substack{\text{Change in internal, kinetic,}\\ \text{potential, etc. energies}}}$$

$$-Q_{out} + W_{b,in} = \Delta U = m(u_2 - u_1)$$

$$-Q_{out} = m(u_2 - u_1) - W_{b,in}$$

$$Q_{out} = W_{b,in} - mC_v(T_2 - T_1)$$

Substituting,

$$Q_{out} = 55.9\text{ Btu} - (0.264\text{ lbm})(0.753\text{ Btu/lbm}\cdot\text{R})(760-530)\text{R} = 10.2\text{ Btu}$$

Noting that the surroundings undergo a reversible isothermal process, its entropy change becomes

$$\Delta S_{surr} = \frac{Q_{surr,in}}{T_{surr}} = \frac{10.2\text{ Btu}}{530\text{ R}} = \mathbf{0.019\ Btu/R}$$

(*c*) Noting that the system+surroundings combination can be treated as an isolated system,

$$\Delta S_{total} = \Delta S_{sys} + \Delta S_{surr} = -0.016 + 0.019 = 0.003\text{Btu/R} > 0$$

Therefore, the process is **irreversible**.

6-152 Air is compressed steadily by a compressor from a specified state to a specified pressure. The minimum power input required is to be determined for the cases of adiabatic and isothermal operation.

Assumptions **1** This is a steady-flow process since there is no change with time. **2** Kinetic and potential energy changes are negligible. **3** Air is an ideal gas with variable specific heats. **4** The process is reversible since the work input to the compressor will be minimum when the compression process is reversible.

Properties The gas constant of air is R = 0.287 kJ/kg.K (Table A-1).

Analysis (*a*) For the adiabatic case, the process will be reversible and adiabatic (i.e., isentropic), thus the isentropic relations are applicable.

$$T_1 = 290\text{K} \longrightarrow P_{r_1} = 1.2311 \text{ and } h_1 = 290.16 \text{ kJ/kg}$$

and

$$P_{r_2} = \frac{P_2}{P_1} P_{r_1} = \frac{700 \text{ kPa}}{100 \text{ kPa}}(1.2311) = 8.6177 \rightarrow \begin{array}{l} T_2 = 503.3 \text{ K} \\ h_2 = 506.45 \text{ kJ/kg} \end{array}$$

2
AIR
Rev.
1

2
T = const
1

The energy balance for the compressor, which is a steady-flow system, can be expressed in the rate form as

$$\underbrace{\dot{E}_{in} - \dot{E}_{out}}_{\text{Rate of net energy transfer by heat, work, and mass}} = \underbrace{\Delta \dot{E}_{\text{system}}^{\nearrow 0 \text{ (steady)}}}_{\text{Rate of change in internal, kinetic, potential, etc. energies}} = 0$$

$$\dot{E}_{in} = \dot{E}_{out}$$

$$\dot{W}_{\text{in}} + \dot{m}h_1 = \dot{m}h_2 \rightarrow \dot{W}_{\text{in}} = \dot{m}(h_2 - h_1)$$

Substituting, the power input to the compressor is determined to be

$$\dot{W}_{in} = (5/60 \text{ kg/s})(506.45 - 290.16)\text{kJ/kg} = 18.0 \textbf{ kW}$$

(*b*) In the case of the reversible isothermal process, the steady-flow energy balance becomes

$$\dot{E}_{in} = \dot{E}_{out} \rightarrow \dot{W}_{in} + \dot{m}h_1 - \dot{Q}_{out} = \dot{m}h_2 \rightarrow \dot{W}_{\text{in}} = \dot{Q}_{out} + \dot{m}(h_2 - h_1)^{\nearrow 0} = \dot{Q}_{out}$$

since $h = h(T)$ for ideal gases, and thus the enthalpy change in this case is zero. Also, for a reversible isothermal process,

$$\dot{Q}_{out} = \dot{m}T(s_1 - s_2) = -\dot{m}T(s_2 - s_1)$$

where

$$s_2 - s_1 = \left(s_2^\circ - s_1^\circ\right)^{\nearrow 0} - R\ln\frac{P_2}{P_1} = -R\ln\frac{P_2}{P_1} = -(0.287\text{kJ/kg}\cdot\text{K})\ln\frac{700\text{kPa}}{100\text{kPa}} = -0.5585\text{kJ/kg}\cdot\text{K}$$

Substituting, the power input for the reversible isothermal case becomes

$$\dot{W}_{in} = -(5/60 \text{ kg/s})(290 \text{ K})(-0.5585 \text{ kJ/kg}\cdot\text{K}) = 13.5 \textbf{ kW}$$

6-153 Air is compressed in a two-stage ideal compressor with intercooling. For a specified mass flow rate of air, the power input to the compressor is to be determined, and it is to be compared to the power input to a single-stage compressor. √

Assumptions **1** The compressor operates steadily. **2** Kinetic and potential energies are negligible. **3** The compression process is reversible adiabatic, and thus isentropic. **4** Air is an ideal gas with constant specific heats at room temperature.

Properties The gas constant of air is R =0.287 kPa.m^3/kg.K (Table A-1). The specific heat ratio of air is k = 1.4 (Table A-2)

Analysis The intermediate pressure between the two stages is

$$P_x = \sqrt{P_1 P_2} = \sqrt{(100\text{kPa})(900\text{kPa})} = 300\text{kPa}$$

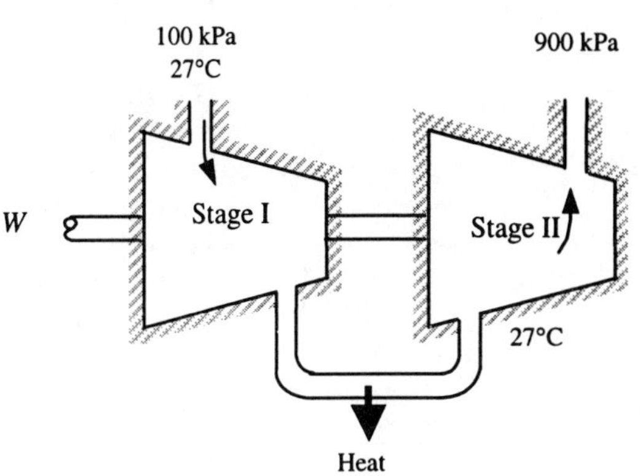

The compressor work across each stage is the same, thus total compressor work is twice the compression work for a single stage:

$$w_{comp,in} = (2)(w_{comp,in,I}) = 2\frac{kRT_1}{k-1}\left((P_x/P_1)^{(k-1)/k} - 1\right)$$

$$= 2\frac{(1.4)(0.287\text{kJ/kg}\cdot\text{K})(300\text{K})}{1.4-1}\left(\left(\frac{300\text{kPa}}{100\text{kPa}}\right)^{0.4/1.4} - 1\right)$$

$$= 222.2\text{kJ/kg}$$

and

$$\dot{W}_{in} = \dot{m}w_{comp,in} = (0.02\text{kg/s})(222.2\text{kJ/kg}) = \mathbf{4.44kW}$$

The work input to a single-stage compressor operating between the same pressure limits would be

$$w_{\text{comp,in}} = \frac{kRT_1}{k-1}\left((P_2/P_1)^{(k-1)/k} - 1\right) = \frac{(1.4)(0.287\text{kJ/kg}\cdot\text{K})(300\text{K})}{1.4-1}\left(\left(\frac{900\text{kPa}}{100\text{kPa}}\right)^{0.4/1.4} - 1\right) = 263.2\text{kJ/kg}$$

and

$$\dot{W}_{in} = \dot{m}w_{comp,in} = (0.02\text{kg/s})(263.2\text{kJ/kg}) = \mathbf{5.26\ kW}$$

Discussion Note that the power consumption of the compressor decreases significantly by using 2-stage compression with intercooling.

6-154 A three-stage compressor with two stages of intercooling is considered. The two intermediate pressures that will minimize the work input are to be determined in terms of the inlet and exit pressures.

Analysis The work input to this three-stage compressor with intermediate pressures P_x and P_y and two intercoolers can be expressed as

$$
\begin{aligned}
w_{comp} &= w_{comp,\text{I}} + w_{comp,\text{II}} + w_{comp,\text{III}} \\
&= \frac{nRT_1}{n-1}\left(1-(P_x/P_1)^{(n-1)/n}\right) + \frac{nRT_1}{n-1}\left(1-(P_y/P_x)^{(n-1)/n}\right) + \frac{nRT_1}{n-1}\left(1-(P_x/P_1)^{(n-1)/n}\right) \\
&= \frac{nRT_1}{n-1}\left(1-(P_x/P_1)^{(n-1)/n} + 1-(P_y/P_x)^{(n-1)/n} + 1-(P_x/P_1)^{(n-1)/n}\right) \\
&= \frac{nRT_1}{n-1}\left(3-(P_x/P_1)^{(n-1)/n} - (P_y/P_x)^{(n-1)/n} - (P_x/P_1)^{(n-1)/n}\right)
\end{aligned}
$$

The P_x and P_y values that will minimize the work input are obtained by taking the partial differential of w with respect to P_x and P_y, and setting them equal to zero:

$$\frac{\partial w}{\partial P_x} = 0 \longrightarrow -\frac{n-1}{n}\left(\frac{1}{P_1}\right)\left(\frac{P_x}{P_1}\right)^{\frac{n-1}{n}-1} + \frac{n-1}{n}\left(\frac{1}{P_y}\right)\left(\frac{P_x}{P_y}\right)^{-\frac{n-1}{n}-1} = 0$$

$$\frac{\partial w}{\partial P_y} = 0 \longrightarrow -\frac{n-1}{n}\left(\frac{1}{P_x}\right)\left(\frac{P_y}{P_x}\right)^{\frac{n-1}{n}-1} + \frac{n-1}{n}\left(\frac{1}{P_2}\right)\left(\frac{P_y}{P_2}\right)^{-\frac{n-1}{n}-1} = 0$$

Simplifying,

$$\frac{1}{P_1}\left(\frac{P_x}{P_1}\right)^{-\frac{1}{n}} = \frac{1}{P_y}\left(\frac{P_x}{P_y}\right)^{-\frac{2n-1}{n}} \longrightarrow \frac{1}{P_1^n}\left(\frac{P_1}{P_x}\right) = \frac{1}{P_y^n}\left(\frac{P_x}{P_y}\right)^{1-2n} \longrightarrow P_x^{2(1-n)} = \left(P_1 P_y\right)^{1-n}$$

$$\frac{1}{P_x}\left(\frac{P_y}{P_x}\right)^{-\frac{1}{n}} = \frac{1}{P_2}\left(\frac{P_y}{P_2}\right)^{-\frac{2n-1}{n}} \longrightarrow \frac{1}{P_x^n}\left(\frac{P_x}{P_y}\right) = \frac{1}{P_2^n}\left(\frac{P_y}{P_2}\right)^{1-2n} \longrightarrow P_y^{2(1-n)} = \left(P_x P_2\right)^{1-n}$$

which yield

$$\mathbf{P_x^2 = P_1\sqrt{P_x P_2} \longrightarrow P_x = \left(P_1^2 P_2\right)^{1/3}}$$

$$\mathbf{P_y^2 = P_2\sqrt{P_1 P_y} \longrightarrow P_y = \left(P_1 P_2^2\right)^{1/3}}$$

6-155 Steam expands in a two-stage adiabatic turbine from a specified state to specified pressure. Some steam is extracted at the end of the first stage. The power output of the turbine is to be determined for the cases of 100% and 88% isentropic efficiencies.

Assumptions **1** This is a steady-flow process since there is no change with time. **2** Kinetic and potential energy changes are negligible. **3** The turbine is adiabatic and thus heat transfer is negligible.

Properties From the steam tables (Tables A-4 through 6)

$$\left.\begin{array}{l} P_1 = 7\text{MPa} \\ T_1 = 500^\circ\text{C} \end{array}\right\} \begin{array}{l} h_1 = 3410.3\text{kJ/kg} \\ s_1 = 6.7975\text{kJ/kg}\cdot\text{K} \end{array}$$

$$\left.\begin{array}{l} P_2 = 1\text{MPa} \\ s_2 = s_1 \end{array}\right\} h_2 = 2879.4 \text{ kJ / kg}$$

$$\left.\begin{array}{l} P_3 = 50\text{kPa} \\ s_3 = s_1 \end{array}\right\} \begin{array}{l} x_{3s} = \dfrac{s_{3s} - s_f}{s_{fg}} = \dfrac{6.7975 - 1.0910}{6.5029} = 0.8775 \\ h_{3s} = h_f + x_{3s} h_{fg} = 340.49 + (0.8775)(2305.4) = 2363.5\text{kJ/kg} \end{array}$$

Analysis (a) The mass flow rate through the second stage is

$$\dot{m}_3 = 0.9\dot{m}_1 = (0.9)(15\text{kg/s}) = 13.5\text{kg/s}$$

We take the entire turbine, including the connection part between the two stages, as the system, which is a control volume since mass crosses the boundary. Noting that one fluid stream enters the turbine and two fluid streams leave, the energy balance for this steady-flow system can be expressed in the rate form as

$$\underbrace{\dot{E}_{in} - \dot{E}_{out}}_{\substack{\text{Rate of net energy transfer} \\ \text{by heat, work, and mass}}} = \underbrace{\Delta\dot{E}_{\text{system}}^{\nearrow 0 \text{ (steady)}}}_{\substack{\text{Rate of change in internal, kinetic,} \\ \text{potential, etc. energies}}} = 0$$

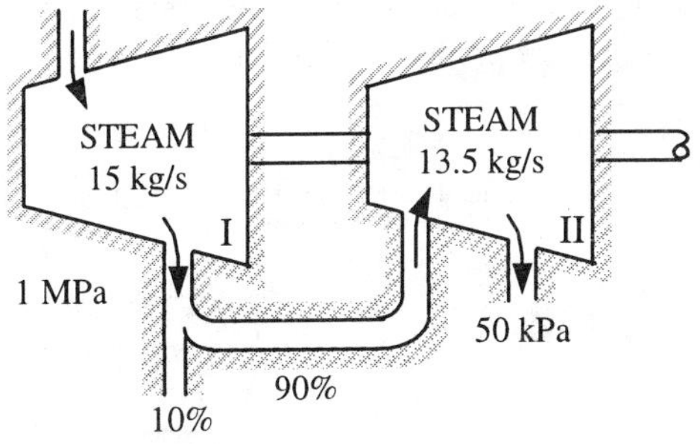

$$\dot{E}_{in} = \dot{E}_{out}$$

$$\dot{m}_1 h_1 = (\dot{m}_1 - \dot{m}_3)h_2 + \dot{W}_{\text{out}} + \dot{m}_3 h_3$$

$$\dot{W}_{\text{out}} = \dot{m}_1 h_1 - (\dot{m}_1 - \dot{m}_3)h_2 - \dot{m}_3 h_3$$

$$= \dot{m}_1(h_1 - h_2) + \dot{m}_3(h_2 - h_3)$$

7 MPa
500°C
STEAM
15 kg/s
I
STEAM
13.5 kg/s
II
1 MPa
50 kPa
90%
10%

Substituting, the power output of the turbine is

$$\dot{W}_{out} = (15 \text{ kg/s})(3410.3 - 2879.4)\text{kJ/kg} + (13.5 \text{ kg})(2879.4 - 2363.5)\text{kJ/kg} = \mathbf{14{,}930 \text{ kW}}$$

(*b*) If the turbine has an adiabatic efficiency of 88%, then the power output becomes

$$\dot{W}_a = \eta_T \dot{W}_s = (0.88)(14{,}928 \text{ kW}) = \mathbf{13{,}140 \text{ kW}}$$

6-156 Steam expands in an 84% efficient two-stage adiabatic turbine from a specified state to a specified pressure. Steam is reheated between the stages. For a given power output, the mass flow rate of steam through the turbine is to be determined.

Assumptions **1** This is a steady-flow process since there is no change with time. **2** Kinetic and potential energy changes are negligible. **3** The turbine is adiabatic and thus heat transfer is negligible.

Properties From the steam tables (Tables A-4 through 6)

$$\left.\begin{array}{l} P_1 = 8 \text{ MPa} \\ T_1 = 500^\circ\text{C} \end{array}\right\} \begin{array}{l} h_1 = 3398.3 \text{ kJ/kg} \\ s_1 = 6.7240 \text{ kJ/kg}\cdot\text{K} \end{array}$$

$$\left.\begin{array}{l} P_{2s} = 2 \text{ MPa} \\ s_{2s} = s_1 \end{array}\right\} h_{2s} = 3000.3 \text{ kJ/kg}$$

$$\left.\begin{array}{l} P_3 = 2 \text{ MPa} \\ T_3 = 500^\circ\text{C} \end{array}\right\} \begin{array}{l} h_3 = 3467.6 \text{ kJ/kg} \\ s_3 = 7.4317 \text{ kJ/kg}\cdot\text{K} \end{array}$$

$$\left.\begin{array}{l} P_{4s} = 100 \text{ kPa} \\ s_{4s} = s_3 \end{array}\right\} h_{4s} = 2704.2 \text{ kJ/kg}$$

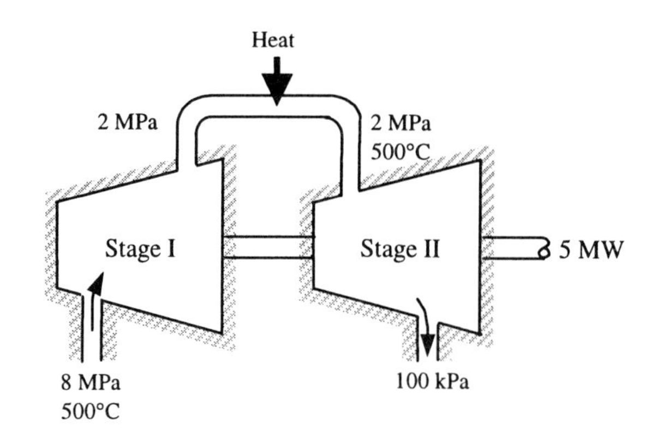

Analysis The power output of the actual turbine is given to be 80 MW. Then the power output for the isentropic operation becomes

$$\dot{W}_{s,out} = \dot{W}_{a,out} / \eta_T = (80{,}000 \text{ kW})/0.84 = 95{,}240 \text{ kW}$$

We take the entire turbine, excluding the reheat section, as the system, which is a control volume since mass crosses the boundary. The energy balance for this steady-flow system in isentropic operation can be expressed in the rate form as

$$\underbrace{\dot{E}_{in} - \dot{E}_{out}}_{\substack{\text{Rate of net energy transfer} \\ \text{by heat, work, and mass}}} = \underbrace{\Delta \dot{E}_{\text{system}}^{\nearrow 0 \text{ (steady)}}}_{\substack{\text{Rate of change in internal, kinetic,} \\ \text{potential, etc. energies}}} = 0$$

$$\dot{E}_{in} = \dot{E}_{out}$$

$$\dot{m}h_1 + \dot{m}h_3 = \dot{m}h_{2s} + \dot{m}h_{4s} + \dot{W}_{\text{s,out}}$$

$$\dot{W}_{\text{s,out}} = \dot{m}[(h_1 - h_{2s}) + (h_3 - h_{4s})]$$

Substituting,

$$95{,}240 \text{ kJ/s} = \dot{m}\left[(3398.3 - 3000.3) + (3467.6 - 2704.2)\text{kJ/kg}\right]$$

which gives

$$\dot{m} = 82.\mathbf{0} \ \mathbf{kg/s}$$

6-157 Refrigerant-134a is compressed by a 0.5-kW adiabatic compressor from a specified state to another specified state. The isentropic efficiency, the volume flow rate at the inlet, and the maximum flow rate at the compressor inlet are to be determined.

Assumptions **1** This is a steady-flow process since there is no change with time. **2** Kinetic and potential energy changes are negligible. **3** The device is adiabatic and thus heat transfer is negligible.

Properties From the R-134a tables (Tables A-11 through A-13)

$$\left.\begin{array}{l} P_1 = 140\text{kPa} \\ T_1 = -10^\circ\text{C} \end{array}\right\} \begin{array}{l} v_1 = 0.14549\text{m}^3/\text{kg} \\ h_1 = 243.40\text{kJ/kg} \\ s_1 = 0.9606\text{kJ/kg}\cdot\text{K} \end{array}$$

$$\left.\begin{array}{l} P_2 = 700\text{kPa} \\ T_2 = 60^\circ\text{C} \end{array}\right\} \text{h}_2 = 296.69\text{kJ/kg}$$

$$\left.\begin{array}{l} P_2 = 700\text{kPa} \\ s_{2s} = s_1 \end{array}\right\} h_{2s} = 278.06\text{kJ/kg}$$

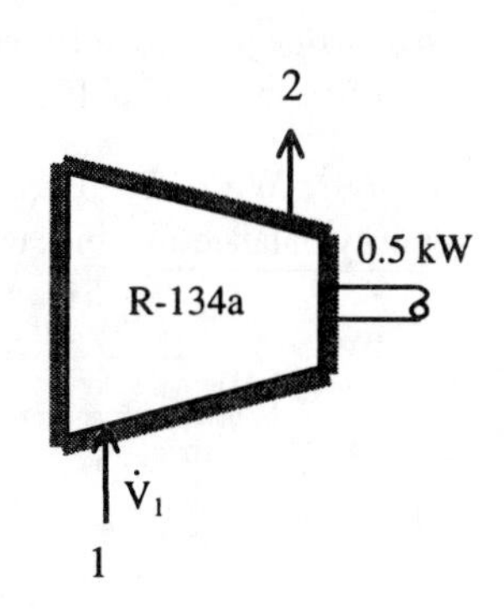

Analysis (a) The isentropic efficiency is determined from its definition,

$$\eta_C = \frac{h_{2s} - h_1}{h_{2a} - h_1} = \frac{278.06 - 243.40}{296.69 - 243.40} = 0.650 = \mathbf{65.0\%}$$

(*b*) There is only one inlet and one exit, and thus $\dot{m}_1 = \dot{m}_2 = \dot{m}$. We take the actual compressor as the system, which is a control volume. The energy balance for this steady-flow system can be expressed as

$$\underbrace{\dot{E}_{in} - \dot{E}_{out}}_{\substack{\text{Rate of net energy transfer} \\ \text{by heat, work, and mass}}} = \underbrace{\Delta \dot{E}_{\text{system}}^{\nearrow 0 \text{ (steady)}}}_{\substack{\text{Rate of change in internal, kinetic,} \\ \text{potential, etc. energies}}} = 0$$

$$\dot{E}_{in} = \dot{E}_{out}$$

$$\dot{W}_{a,\text{in}} + \dot{m}h_1 = \dot{m}h_2 \quad (\text{since } \dot{Q} \cong \Delta ke \cong \Delta pe \cong 0)$$

$$\dot{W}_{a,\text{in}} = \dot{m}(h_2 - h_1)$$

Then the mass and volume flow rates of the refrigerant are determined to be

$$\dot{m} = \frac{\dot{W}_{a,in}}{h_{2a} - h_1} = \frac{0.5\text{kJ/s}}{(296.69 - 243.40)\text{kJ/kg}} = 0.0094\text{kg/s}$$

$$\dot{V}_1 = \dot{m}v_1 = (0.0094\text{kg/s})(0.14549\text{m}^3/\text{kg}) = 0.00137\text{m}^3/\text{s} = \mathbf{82L/min}$$

(*c*) The volume flow rate will be a maximum when the process is isentropic, and it is determined similarly from the steady-flow energy equation applied to the isentropic process. It gives

$$\dot{m}_{\max} = \frac{\dot{W}_{s,in}}{h_{2s} - h_1} = \frac{0.5\text{kJ/s}}{(278.06 - 243.40)\text{kJ/kg}} = 0.0144 \text{ kg/s}$$

$$\dot{V}_{1,\max} = \dot{m}_{\max} v_1 = (0.0144 \text{ kg/s})(0.14549 \text{ m}^3/\text{kg}) = 0.00210 \text{ m}^3/\text{s} = \mathbf{126 \ L/min}$$

Discussion Note that the raising the isentropic efficiency of the compressor to 100% would increase the volumetric flow rate by more than 50%.

6-158E Helium is accelerated by a 94% efficient nozzle from a low velocity to 1000 ft/s. The pressure and temperature at the nozzle inlet are to be determined.

Assumptions **1** This is a steady-flow process since there is no change with time. **2** Helium is an ideal gas with constant specific heats. **3** Potential energy changes are negligible. **4** The device is adiabatic and thus heat transfer is negligible.

Properties The specific heat ratio of helium is k = 1.667. The constant pressure specific heat of helium is 1.25 Btu/lbm.R (Table A-2E).

Analysis We take nozzle as the system, which is a control volume since mass crosses the boundary. The energy balance for this steady-flow system can be expressed in the rate form as

$$\underbrace{\dot{E}_{in} - \dot{E}_{out}}_{\text{Rate of net energy transfer by heat, work, and mass}} = \underbrace{\Delta \dot{E}_{\text{system}}^{\nearrow 0 \text{ (steady)}}}_{\text{Rate of change in internal, kinetic, potential, etc. energies}} = 0$$

$$\dot{E}_{in} = \dot{E}_{out}$$

$$\dot{m}(h_1 + \mathbf{V}_1^2/2) = \dot{m}(h_2 + \mathbf{V}_2^2/2) \quad (\text{since } \dot{Q} \cong \dot{W} \cong \Delta\text{pe} \cong 0)$$

$$0 = h_2 - h_1 + \frac{\mathbf{V}_2^2 - \mathbf{V}_1^2}{2} \longrightarrow 0 = C_{p,ave}(T_2 - T_1) + \frac{\mathbf{V}_2^2 - \mathbf{V}_1^2}{2}$$

1 HELIUM $\eta_N = 94\%$ 2

Solving for T_1 and substituting,

$$T_1 = T_{2a} + \frac{\mathbf{V}_{2s}^2 - \mathbf{V}_1^{2\nearrow 0}}{2C_p} = 180^\circ\text{F} + \frac{(1000\text{ft/s})^2}{2(1.25\text{Btu/lbm}\cdot\text{R})}\left(\frac{1\text{Btu/lbm}}{25{,}037\text{ft}^2/\text{s}^2}\right) = \mathbf{196.0^\circ F = 656R}$$

From the isentropic efficiency relation,

$$\eta_N = \frac{h_{2a} - h_1}{h_{2s} - h_1} = \frac{C_P(T_{2a} - T_1)}{C_P(T_{2s} - T_1)}$$

or,

$$T_{2s} = T_1 + (T_{2a} - T_1)/\eta_N = 656 + (640 - 656)/(0.94) = 639\text{R}$$

From the isentropic relation, $\dfrac{T_{2s}}{T_1} = \left(\dfrac{P_2}{P_1}\right)^{(k-1)/k}$

$$P_1 = P_2\left(\frac{T_1}{T_{2s}}\right)^{k/(k-1)} = (14\text{ psia})\left(\frac{656\text{ R}}{639\text{ R}}\right)^{1.667/0.667} = \mathbf{14.9\ psia}$$

6-159 [*Also solved by EES on enclosed CD*] An adiabatic compressor is powered by a direct-coupled steam turbine, which also drives a generator. The net power delivered to the generator and the rate of entropy generation are to be determined.

Assumptions **1** This is a steady-flow process since there is no change with time. **2** Kinetic and potential energy changes are negligible. **3** The devices are adiabatic and thus heat transfer is negligible. **4** Air is an ideal gas with variable specific heats.

Properties From the steam tables (Tables A-4 through 6) and air table (Table A-17),

$$T_1 = 295 \text{ K} \longrightarrow h_1 = 295.17 \text{kJ/kg}, \; s_1^\circ = 1.68515 \text{ kJ/kg}\cdot\text{K}$$

$$T_2 = 620 \text{ K} \longrightarrow h_1 = 628.07 \text{kJ/kg}, \; s_2^\circ = 2.44356 \text{ kJ/kg}\cdot\text{K}$$

$$\left.\begin{array}{l} P_3 = 12.5\text{MPa} \\ T_3 = 500^\circ\text{C} \end{array}\right\} \begin{array}{l} h_3 = 3341.8 \text{kJ/kg} \\ s_3 = 6.4618 \text{kJ/kg}\cdot\text{K} \end{array}$$

$$\left.\begin{array}{l} P_4 = 10\text{kPa} \\ x_4 = 0.92 \end{array}\right\} \begin{array}{l} h_4 = h_f + x_4 h_{fg} = 191.83 + (0.92)(2392.8) = 2393.2 \text{kJ/kg} \\ s_4 = s_f + x_4 s_{fg} = 0.6493 + (0.92)(7.5009) = 7.5501 \text{kJ/kg}\cdot\text{K} \end{array}$$

Analysis There is only one inlet and one exit for either device, and thus $\dot{m}_{in} = \dot{m}_{out} = \dot{m}$. We take either the turbine or the compressor as the system, which is a control volume since mass crosses the boundary. The energy balance for either steady-flow system can be expressed in the rate form as

$$\underbrace{\dot{E}_{in} - \dot{E}_{out}}_{\substack{\text{Rate of net energy transfer} \\ \text{by heat, work, and mass}}} = \underbrace{\Delta \dot{E}_{\text{system}}^{\nearrow 0 \text{ (steady)}}}_{\substack{\text{Rate of change in internal, kinetic,} \\ \text{potential, etc. energies}}} = 0$$

$$\dot{E}_{in} = \dot{E}_{out}$$

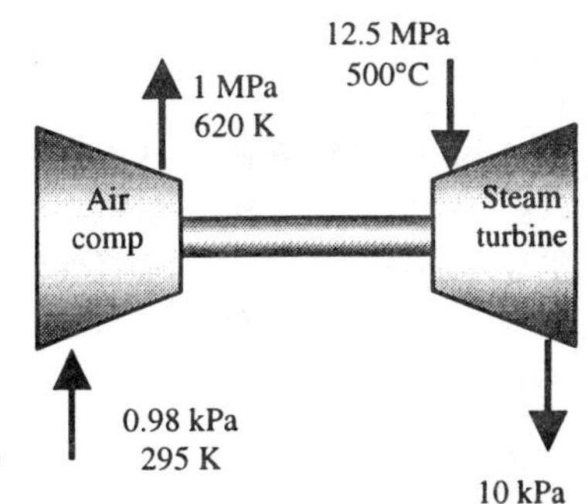

For the turbine and the compressor it becomes

Compressor: $\dot{W}_{\text{comp, in}} + \dot{m}_{\text{air}} h_1 = \dot{m}_{\text{air}} h_2 \rightarrow \dot{W}_{\text{comp, in}} = \dot{m}_{\text{air}}(h_2 - h_1)$

Turbine: $\dot{m}_{\text{steam}} h_3 = \dot{W}_{\text{turb, out}} + \dot{m}_{\text{steam}} h_4 \rightarrow \dot{W}_{\text{turb, out}} = \dot{m}_{\text{steam}}(h_3 - h_4)$

Substituting,

$$\dot{W}_{\text{comp,in}} = (10 \text{kg/s})(628.07 - 295.17) \text{kJ/kg} = 3329 \text{ kW}$$

$$\dot{W}_{\text{turb,out}} = (25 \text{kg/s})(3341.8 - 2393.2) \text{kJ/kg} = 23{,}715 \text{ kW}$$

Therefore,

$$\dot{W}_{\text{net,out}} = \dot{W}_{\text{turb,out}} - \dot{W}_{\text{comp,in}} = 23{,}715 - 3329 = \mathbf{20{,}386 \text{ kW}}$$

Noting that the system is adiabatic, the total rate of entropy change (or generation) during this process is the sum of the entropy changes of both fluids,

$$\dot{S}_{\text{gen}} = \dot{m}_{air}(s_2 - s_1) + \dot{m}_{steam}(s_4 - s_3)$$

where

$$\dot{m}_{air}(s_2 - s_1) = \dot{m}\left(s_2^\circ - s_1^\circ - R\ln\frac{P_2}{P_1}\right) = (10 \text{ kg/s})\left(2.44356 - 1.68515 - 0.287\ln\frac{1000\text{kPa}}{98\text{kPa}}\right)\text{kJ/kg}\cdot\text{K} = 0.92 \text{ kW/K}$$

$$\dot{m}_{steam}(s_4 - s_3) = (25 \text{kg/s})(7.5501 - 6.4618) \text{kJ/kg}\cdot\text{K} = 27.2 \text{ kW/K}$$

Substituting, the total rate of entropy generation is determined to be

$$\dot{S}_{gen,total} = \dot{S}_{gen,comp} + \dot{S}_{gen,turb} = 0.92 + 27.2 = \mathbf{28.12 \text{ kW/K}}$$

6-160 EES solution of this (and other comprehensive problems designated with the *computer icon*) is available to instructors at the *Instructor Manual* section of the *Online Learning Center* (OLC) at www.mhhe.com/cengel-boles. See the Preface for access information.

6-161 Two identical bodies at different temperatures are connected to each other through a heat engine. It is to be shown that the final common temperature of the two bodies will be $T_f = \sqrt{T_1 T_2}$ when the work output of the heat engine is maximum.

Analysis For maximum power production, the entropy generation must be zero. Taking the source, the sink, and the heat engine as our system, which is adiabatic, and noting that the entropy change for cyclic devices is zero, the entropy generation for this system can be expressed as

$$S_{gen} = (\Delta S)_{\text{source}} + (\Delta S)_{\text{engine}}^{\nearrow 0} + (\Delta S)_{\text{sink}} = 0$$

$$mC\ln\frac{T_f}{T_1} + 0 + mC\ln\frac{T_f}{T_2} = 0$$

$$\ln\frac{T_f}{T_1} + \ln\frac{T_f}{T_2} = 0 \longrightarrow \ln\frac{T_f}{T_1}\frac{T_f}{T_2} = 0 \longrightarrow T_f^2 = T_1 T_2$$

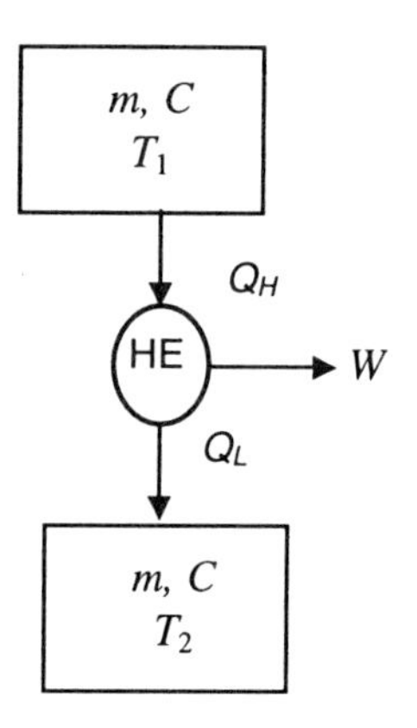

and thus

$$T_f = \sqrt{T_1 T_2}$$

for maximum power production.

6-162 The pressure in a hot water tank rises to 2 MPa, and the tank explodes. The explosion energy of the water is to be determined, and expressed in terms of its TNT equivalence.

Assumptions **1** The expansion process during explosion is isentropic. **2** Kinetic and potential energy changes are negligible. **3** Heat transfer with the surroundings during explosion is negligible.

Properties The explosion energy of TNT is 3250 kJ/kg. From the steam tables (Tables A-4 through 6)

$$\left.\begin{array}{l} P_1 = 2\text{ MPa} \\ \textit{Sat.liquid} \end{array}\right\} \begin{array}{l} v_1 = v_{f@2\text{MPa}} = 0.001177\text{ m}^3/\text{kg} \\ u_1 = u_{f@2\text{MPa}} = 906.44\text{ kJ/kg} \\ s_1 = s_{f@2\text{MPa}} = 2.4474\text{ kJ/kg}\cdot\text{K} \end{array}$$

$$\left.\begin{array}{l} P_2 = 100\text{ kPa} \\ s_2 = s_1 \end{array}\right\} \begin{array}{l} u_f = 417.36, \quad u_{fg} = 2088.7\text{ kJ/kg} \\ s_f = 1.3026, \quad s_{fg} = 6.0568\text{ kJ/kg}\cdot\text{K} \end{array}$$

Water
Tank
2 MPa

$$x_2 = \frac{s_2 - s_f}{s_{fg}} = \frac{2.4474 - 1.3026}{6.0568} = 0.189$$

$$u_2 = u_f + x_2 u_{fg} = 417.36 + (0.189)(2088.7) = 812.1\text{ kJ/kg}$$

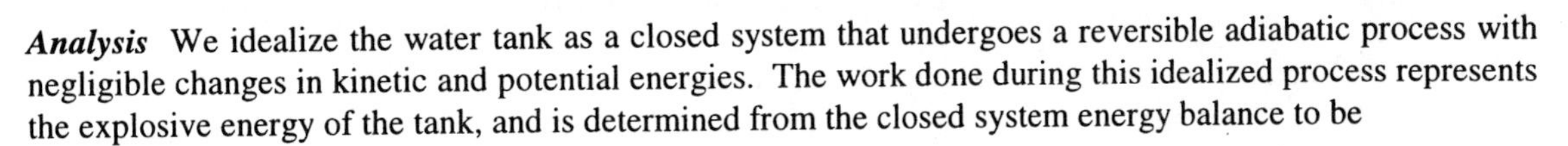

Analysis We idealize the water tank as a closed system that undergoes a reversible adiabatic process with negligible changes in kinetic and potential energies. The work done during this idealized process represents the explosive energy of the tank, and is determined from the closed system energy balance to be

$$\underbrace{E_{in} - E_{out}}_{\substack{\text{Net energy transfer} \\ \text{by heat, work, and mass}}} = \underbrace{\Delta E_{\text{system}}}_{\substack{\text{Change in internal, kinetic,} \\ \text{potential, etc. energies}}}$$

$$-W_{b,out} = \Delta U = m(u_2 - u_1)$$

$$E_{\exp} = W_{b,out} = m(u_1 - u_2)$$

where

$$m = \frac{V}{v_1} = \frac{0.1\text{m}^3}{0.001177\text{m}^3/\text{kg}} = 85.0\text{kg}$$

Substituting,

$$E_{\exp} = (85.0\text{ kg})(906.44 - 812.1)\text{kJ/kg} = 8019\text{ kJ}$$

which is equivalent to

$$m_{TNT} = \frac{8019\text{ kJ}}{3250\text{ kJ / kg}} = \mathbf{2.47\ kg\ TNT}$$

6-163 A 0.2-L canned drink explodes at a pressure of 1 MPa. The explosive energy of the drink is to be determined, and expressed in terms of its TNT equivalence. √

Assumptions **1** The expansion process during explosion is isentropic. **2** Kinetic and potential energy changes are negligible. **3** Heat transfer with the surroundings during explosion is negligible. **4** The drink can be treated as pure water.

Properties The explosion energy of TNT is 3250 kJ/kg. From the steam tables (Tables A-4 through 6)

$$\left.\begin{array}{l} P_1 = 1\text{ MPa} \\ Comp.liquid \end{array}\right\} \begin{array}{l} v_1 = v_{f@1\text{MPa}} = 0.001127\text{ m}^3\text{/kg} \\ u_1 = u_{f@1\text{MPa}} = 761.68\text{ kJ/kg} \\ s_1 = s_{f@1\text{MPa}} = 2.1387\text{ kJ/kg}\cdot\text{K} \end{array}$$

COLA
1 MPa

$$\left.\begin{array}{l} P_2 = 100\text{ kPa} \\ s_2 = s_1 \end{array}\right\} \begin{array}{l} u_f = 417.36, \quad u_{fg} = 2088.7\text{ kJ/kg} \\ s_f = 1.3026, \quad s_{fg} = 6.0568\text{ kJ/kg}\cdot\text{K} \end{array}$$

$$x_2 = \frac{s_2 - s_f}{s_{fg}} = \frac{2.1387 - 1.3026}{6.0568} = 0.138$$

$$u_2 = u_f + x_2 u_{fg} = 417.36 + (0.138)(2088.7) = 705.6\text{ kJ/kg}$$

Analysis We idealize the canned drink as a closed system that undergoes a reversible adiabatic process with negligible changes in kinetic and potential energies. The work done during this idealized process represents the explosive energy of the can, and is determined from the closed system energy balance to be

$$\underbrace{E_{in} - E_{out}}_{\substack{\text{Net energy transfer}\\ \text{by heat, work, and mass}}} = \underbrace{\Delta E_{\text{system}}}_{\substack{\text{Change in internal, kinetic,}\\ \text{potential, etc. energies}}}$$

$$-W_{b,out} = \Delta U = m(u_2 - u_1)$$

$$E_{\text{exp}} = W_{b,out} = m(u_1 - u_2)$$

where

$$m = \frac{V}{v_1} = \frac{0.0002\text{m}^3}{0.001127\text{m}^3\text{/kg}} = 0.177\text{kg}$$

Substituting,

$$E_{\text{exp}} = (0.177\text{kg})(761.68 - 705.6)\text{kJ/kg} = \mathbf{9.9\ kJ}$$

which is equivalent to

$$m_{TNT} = \frac{9.9\text{ kJ}}{3250\text{ kJ/kg}} = \mathbf{0.00305\ kgTNT}$$

Review Problems

6-164 The validity of the Clausius inequality is to be demonstrated using a reversible and an irreversible heat engine operating between the same temperature limits.

Analysis Consider two heat engines, one reversible and one irreversible, both operating between a high-temperature reservoir at T_H and a low-temperature reservoir at T_L. Both heat engines receive the same amount of heat, Q_H. The reversible heat engine rejects heat in the amount of Q_L, and the irreversible one in the amount of $Q_{L,\ irrev} = Q_L + Q_{diff}$, where Q_{diff} is a positive quantity since the irreversible heat engine produces less work. Noting that Q_H and Q_L are transferred at constant temperatures of T_H and T_L, respectively, the cyclic integral of $\delta Q/T$ for the reversible and irreversible heat engine cycles become

$$\oint\left(\frac{\delta Q}{T}\right)_{rev} = \int\frac{\delta Q_H}{T_H} - \int\frac{\delta Q_L}{T_L} = \frac{1}{T_H}\int\delta Q_H - \frac{1}{T_L}\int\delta Q_L = \frac{Q_H}{T_H} - \frac{Q_L}{T_L} = 0$$

since $(Q_H/T_H) = (Q_L/T_L)$ for reversible cycles. Also,

$$\oint\left(\frac{\delta Q}{T}\right)_{irr} = \frac{Q_H}{T_H} - \frac{Q_{L,irrev}}{T_L} = \frac{Q_H}{T_H} - \frac{Q_L}{T_L} - \frac{Q_{diff}}{T_L} = -\frac{Q_{diff}}{T_L} < 0$$

since Q_{diff} is a positive quantity. Thus, $\oint\left(\frac{\delta Q}{T}\right) \le 0$.

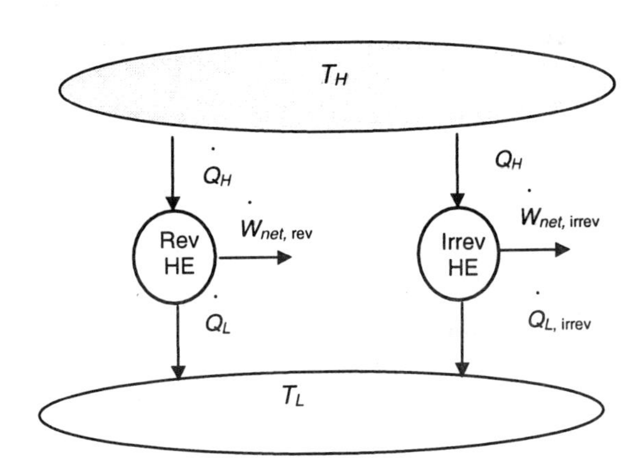

6-165 The inner and outer surfaces of a window glass are maintained at specified temperatures. The amount of heat transfer through the glass and the amount of entropy generation within the glass in 5 h are to be determined

Assumptions **1** Steady operating conditions exist since the surface temperatures of the glass remain constant at the specified values. **2** Thermal properties of the glass are constant.

Analysis The amount of heat transfer over a period of 5 h is

$$Q = \dot{Q}_{cond}\Delta t = (3.2\text{ kJ/s})(5\times 3600\text{ s}) = \mathbf{57,600\ kJ}$$

Glass

10°C

3°C

We take the glass to be the system, which is a closed system. Under steady conditions, the rate form of the entropy balance for the glass simplifies to

$$\underbrace{\dot{S}_{in}-\dot{S}_{out}}_{\substack{\text{Rate of net entropy transfer}\\ \text{by heat and mass}}} + \underbrace{\dot{S}_{gen}}_{\substack{\text{Rate of entropy}\\ \text{generation}}} = \underbrace{\Delta\dot{S}_{\text{system}}^{\nearrow 0}}_{\substack{\text{Rate of change}\\ \text{of entropy}}} = 0$$

$$\frac{\dot{Q}_{in}}{T_{b,in}} - \frac{\dot{Q}_{out}}{T_{b,out}} + \dot{S}_{\text{gen,glass}} = 0$$

$$\frac{3200\text{ W}}{283\text{ K}} - \frac{3200\text{ W}}{276\text{ K}} + \dot{S}_{\text{gen, wall}} = 0 \rightarrow \dot{S}_{\text{gen,glass}} = \mathbf{0.287\ W/K}$$

Then the amount of entropy generation over a period of 5 h becomes

$$S_{\text{gen,glass}} = \dot{S}_{\text{gen,glass}}\Delta t = (0.287\text{ W/K})(5\times 3600\text{ s}) = \mathbf{5160\ J/K}$$

6-166 Two rigid tanks that contain water at different states are connected by a valve. The valve is opened and steam flows from tank A to tank B until the pressure in tank A drops to a specified value. Tank B loses heat to the surroundings. The final temperature in each tank and the entropy generated during this process are to be determined.

Assumptions **1** Tank A is insulated, and thus heat transfer is negligible. **2** The water that remains in tank A undergoes a reversible adiabatic process. **3** The thermal energy stored in the tanks themselves is negligible. **4** The system is stationary and thus kinetic and potential energy changes are negligible. **5** There are no work interactions.

Analysis (*a*) The steam in tank A undergoes a reversible, adiabatic process, and thus $s_2 = s_1$. From the steam tables (Tables A-4 through A-6),

Tank A:

$$\left.\begin{array}{l} P_1 = 400\text{kPa} \\ x_1 = 0.8 \end{array}\right\} \begin{array}{l} v_{1,A} = v_f + x_1 v_{fg} = 0.001084 + (0.8)(0.4625 - 0.001084) = 0.3702\text{m}^3/\text{kg} \\ u_{1,A} = u_f + x_1 u_{fg} = 604.31 + (0.8)(1949.3) = 2163.75\text{kJ/kg} \\ s_{1,A} = s_f + x_1 s_{fg} = 1.7766 + (0.8)(5.1193) = 5.8720\text{kJ/kg}\cdot\text{K} \end{array}$$

$$T_{2,A} = T_{sat@300kPa} = \mathbf{133.55^\circ C}$$

$$\left.\begin{array}{l} P_1 = 300\text{kPa} \\ s_2 = s_1 \\ (sat.mixture) \end{array}\right\} \begin{array}{l} x_{2,A} = \dfrac{s_{2,A} - s_f}{s_{fg}} = \dfrac{5.8720 - 1.6718}{5.3201} = 0.790 \\ v_{2,A} = v_f + x_{2,A} v_{fg} = 0.001073 + (0.790)(0.6058 - 0.001073) = 0.479\text{m}^3/\text{kg} \\ u_{2,A} = u_f + x_{2,A} u_{fg} = 561.15 + (0.790)(1982.4\text{kJ/kg}) = 2127.2\text{kJ/kg} \end{array}$$

Tank B:

$$\left.\begin{array}{l} P_1 = 200\text{kPa} \\ T_1 = 250^\circ\text{C} \end{array}\right\} \begin{array}{l} v_{1,B} = 1.1988\text{m}^3/\text{kg} \\ u_{1,B} = 2731.2\text{kJ/kg} \\ s_{1,B} = 7.7086\text{kJ/kg}\cdot\text{K} \end{array}$$

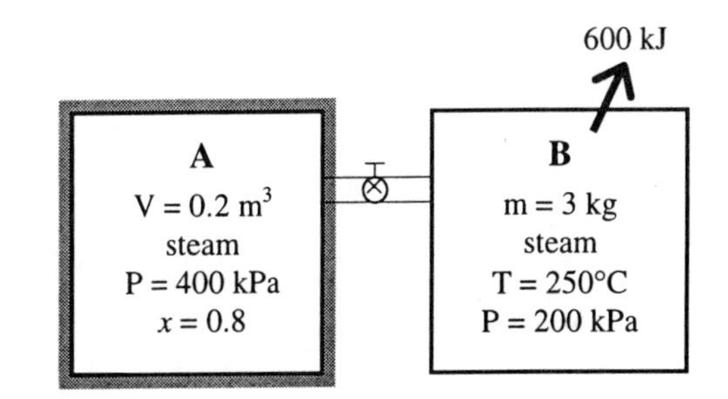

The initial and the final masses in tank A are

$$m_{1,A} = \frac{V_A}{v_{1,A}} = \frac{0.2\text{ m}^3}{0.3702\text{ m}^3/\text{kg}} = 0.540\text{ kg}$$

and

$$m_{2,A} = \frac{V_A}{v_{2,A}} = \frac{0.2\text{ m}^3}{0.479\text{ m}^3/\text{kg}} = 0.418\text{ kg}$$

Thus, 0.540 - 0.418 = 0.122 kg of mass flows into tank B. Then,

$$m_{2,B} = m_{1,B} - 0.122 = 3 + 0.122 = 3.122\text{ kg}$$

The final specific volume of steam in tank B is determined from

$$v_{2,B} = \frac{V_B}{m_{2,B}} = \frac{(m_1 v_1)_B}{m_{2,B}} = \frac{(3\text{kg})(1.1988\text{m}^3/\text{kg})}{3.122\text{m}^3} = 1.152\text{m}^3/\text{kg}$$

We take the entire contents of both tanks as the system, which is a closed system. The energy balance for this stationary closed system can be expressed as

$$\underbrace{E_{in} - E_{out}}_{\substack{\text{Net energy transfer}\\ \text{by heat, work, and mass}}} = \underbrace{\Delta E_{\text{system}}}_{\substack{\text{Change in internal, kinetic,}\\ \text{potential, etc. energies}}}$$

$$-Q_{out} = \Delta U = (\Delta U)_A + (\Delta U)_B \quad (\text{since } W = \text{KE} = \text{PE} = 0)$$

$$-Q_{out} = (m_2 u_2 - m_1 u_1)_A + (m_2 u_2 - m_1 u_1)_B$$

Substituting,

$$-600 = \{(0.418)(2127..2) - (0.540)(2163.8)\} + \{(3.122)u_{2,B} - (3)(2731.2)\}$$

$$u_{2,B} = 2521.7 \text{kJ/kg}$$

Thus,

$$\left.\begin{array}{l} v_{2,B} = 1.152 \text{m}^3/\text{kg} \\ u_{2,B} = 2521.7 \text{kJ/kg} \end{array}\right\} \begin{array}{l} T_{2,B} = \mathbf{113^\circ C} \\ s_{2,B} = 7.225 \text{kJ/kg} \cdot \text{K} \end{array}$$

(*b*) The total entropy generation during this process is determined by applying the entropy balance on an *extended system* that includes both tanks and their immediate surroundings so that the boundary temperature of the extended system is the temperature of the surroundings at all times. It gives

$$\underbrace{S_{in} - S_{out}}_{\substack{\text{Net entropy transfer} \\ \text{by heat and mass}}} + \underbrace{S_{gen}}_{\substack{\text{Entropy} \\ \text{generation}}} = \underbrace{\Delta S_{\text{system}}}_{\substack{\text{Change} \\ \text{in entropy}}}$$

$$-\frac{Q_{out}}{T_{b,surr}} + S_{\text{gen}} = \Delta S_{\text{A}} + \Delta S_{\text{B}}$$

Rearranging and substituting, the total entropy generated during this process is determined to be

$$S_{gen} = \Delta S_A + \Delta S_B + \frac{Q_{out}}{T_{b,surr}} = (m_2 s_2 - m_1 s_1)_A + (m_2 s_2 - m_1 s_1)_B + \frac{Q_{out}}{T_{b,surr}}$$

$$= \{(0.418)(5.872) - (0.540)(5.872)\} + \{(3.122)(7.225) - (3)(7.7086)\} + \frac{600 \text{kJ}}{273 \text{K}}$$

$$= \mathbf{0.912 kJ/K}$$

6-167 Heat is transferred steadily to boiling water in a pan through its bottom. The rate of entropy generation within the bottom plate is to be determined.

Assumptions Steady operating conditions exist since the surface temperatures of the pan remain constant at the specified values.

Analysis We take the bottom of the pan to be the system, which is a closed system. Under steady conditions, the rate form of the entropy balance for this system can be expressed as

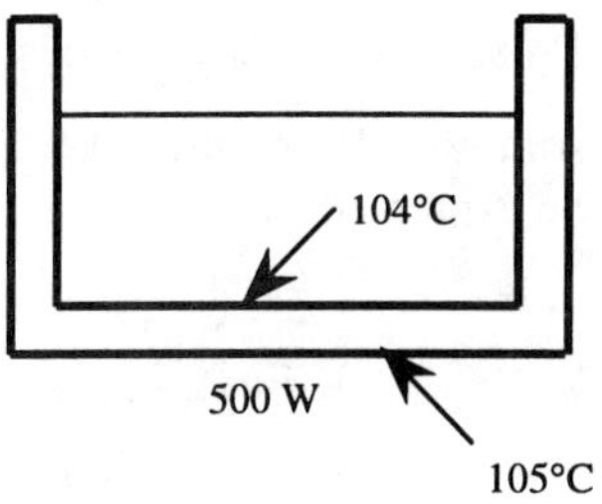

$$\underbrace{\dot{S}_{in} - \dot{S}_{out}}_{\text{Rate of net entropy transfer by heat and mass}} + \underbrace{\dot{S}_{gen}}_{\text{Rate of entropy generation}} = \underbrace{\Delta \dot{S}_{\text{system}}^{\nearrow 0}}_{\text{Rate of change of entropy}} = 0$$

$$\frac{\dot{Q}_{in}}{T_{b,in}} - \frac{\dot{Q}_{out}}{T_{b,out}} + \dot{S}_{\text{gen,system}} = 0$$

$$\frac{800 \text{ W}}{378 \text{ K}} - \frac{800 \text{ W}}{377 \text{ K}} + \dot{S}_{\text{gen,system}} = 0 \rightarrow \dot{S}_{\text{gen,system}} = \mathbf{0.0056\ W/K}$$

104°C

500 W

105°C

Discussion Note that there is a small temperature drop across the bottom of the pan, and thus a small amount of entropy generation.

6-168 An electric resistance heater is immersed in water. The time it will take for the electric heater to raise the water temperature to a specified temperature and the entropy generated during this process are to be determined.

Assumptions **1** Water is an incompressible substance with constant specific heats. **2** The energy stored in the container itself and the heater is negligible. **3** Heat loss from the container is negligible.

Properties The specific heat of water at room temperature is $C = 4.18$ kJ/kg· °C (Table A-3).

Analysis Taking the water in the container as the system, which is a closed system, the energy balance can be expressed as

$$\underbrace{E_{in} - E_{out}}_{\substack{\text{Net energy transfer} \\ \text{by heat, work, and mass}}} = \underbrace{\Delta E_{\text{system}}}_{\substack{\text{Change in internal, kinetic,} \\ \text{potential, etc. energies}}}$$

$$W_{e,in} = (\Delta U)_{\text{water}}$$

$$\dot{W}_{e,in} \Delta t = mC(T_2 - T_1)_{\text{water}}$$

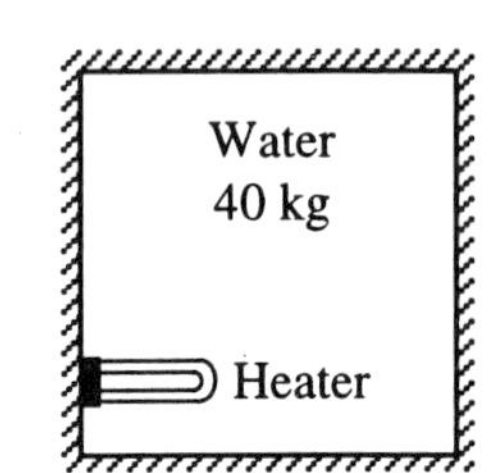

Substituting,

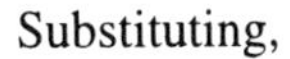

$$(800 \text{ J/s})\Delta t = (40 \text{ kg})(4180 \text{ J/kg· °C})(80 - 20)\text{°C}$$

Solving for Δt gives

$$\Delta t = \mathbf{12{,}540\ s = 209\ min = 3.48\ h}$$

Again we take the water in the tank to be the system. Noting that no heat or mass crosses the boundaries of this system and the energy and entropy contents of the heater are negligible, the entropy balance for it can be expressed as

$$\underbrace{S_{in} - S_{out}}_{\substack{\text{Net entropy transfer} \\ \text{by heat and mass}}} + \underbrace{S_{gen}}_{\substack{\text{Entropy} \\ \text{generation}}} = \underbrace{\Delta S_{\text{system}}}_{\substack{\text{Change} \\ \text{in entropy}}}$$

$$0 + S_{\text{gen}} = \Delta S_{\text{water}}$$

Therefore, the entropy generated during this process is

$$S_{gen} = \Delta S_{\text{water}} = mC \ln \frac{T_2}{T_1} = (40 \text{ kg})(4.18 \text{ kJ/kg} \cdot \text{K}) \ln \frac{353 \text{ K}}{293 \text{ K}} = \mathbf{31.15\ kJ/K}$$

6-169 A hot water pipe at a specified temperature is losing heat to the surrounding air at a specified rate. The rate of entropy generation in the surrounding air due to this heat transfer are to be determined.

Assumptions Steady operating conditions exist.

Analysis We take the air in the vicinity of the pipe (excluding the pipe) as our system, which is a closed system.. The system extends from the outer surface of the pipe to a distance at which the temperature drops to the surroundings temperature. In steady operation, the rate form of the entropy balance for this system can be expressed as

$$\underbrace{\dot{S}_{in} - \dot{S}_{out}}_{\text{Rate of net entropy transfer by heat and mass}} + \underbrace{\dot{S}_{gen}}_{\text{Rate of entropy generation}} = \underbrace{\Delta \dot{S}_{\text{system}}^{\;\nearrow 0}}_{\text{Rate of change of entropy}} = 0$$

$$\frac{\dot{Q}_{in}}{T_{b,in}} - \frac{\dot{Q}_{out}}{T_{b,out}} + \dot{S}_{\text{gen,system}} = 0$$

$$\frac{2200\text{ W}}{353\text{ K}} - \frac{2200\text{ W}}{278\text{ K}} + \dot{S}_{\text{gen,system}} = 0 \rightarrow \dot{S}_{\text{gen,system}} = \mathbf{1.68\ W/K}$$

80°C

Q

Air, 5°C

6-170 The feedwater of a steam power plant is preheated using steam extracted from the turbine. The ratio of the mass flow rates of the extracted steam to the feedwater and entropy generation per unit mass of feedwater are to be determined.

Assumptions **1** This is a steady-flow process since there is no change with time. **2** Kinetic and potential energy changes are negligible. **3** Heat loss from the device to the surroundings is negligible.

Properties The properties of steam and feedwater are (Tables A-4 through A-6)

$$\left.\begin{array}{l} P_1 = 1\text{ MPa} \\ T_1 = 200^\circ\text{C} \end{array}\right\} \begin{array}{l} h_1 = 2827.9\text{ kJ/kg} \\ s_1 = 6.6940\text{ kJ/kg}\cdot\text{K} \end{array}$$

$$\left.\begin{array}{l} P_2 = 1\text{MPa} \\ sat.liquid \end{array}\right\} \begin{array}{l} h_2 = h_{f@1\text{ MPa}} = 762.81\text{ kJ/kg} \\ s_2 = s_{f@1\text{ MPa}} = 2.1387\text{ kJ/kg}\cdot\text{K} \\ T_2 = 179.91^\circ\text{C} \end{array}$$

$$\left.\begin{array}{l} P_3 = 2.5\text{ MPa} \\ T_3 = 50^\circ\text{C} \end{array}\right\} \begin{array}{l} h_3 \cong h_{f@50^\circ\text{C}} = 209.33\text{ kJ/kg} \\ s_3 \cong s_{f@50^\circ\text{C}} = 0.7038\text{ kJ/kg}\cdot\text{K} \end{array}$$

$$\left.\begin{array}{l} P_4 = 2.5\text{ MPa} \\ T_4 = T_2 - 10^\circ\text{C} \cong 170^\circ\text{C} \end{array}\right\} \begin{array}{l} h_4 \cong h_{f@170^\circ\text{C}} = 719.21\text{ kJ/kg} \\ s_4 \cong s_{f@170^\circ\text{C}} = 2.0419\text{ kJ/kg}\cdot\text{K} \end{array}$$

Steam from turbine ① 1 MPa 200°C; Feedwater ③ 2.5 MPa; ④; ② sat. liquid

Analysis (*a*) We take the heat exchanger as the system, which is a control volume. The mass and energy balances for this steady-flow system can be expressed in the rate form as follows:

Mass balance (for each fluid stream):

$$\dot{m}_{in} - \dot{m}_{out} = \Delta\dot{m}_{\text{system}}^{\nearrow 0\text{ (steady)}} = 0 \rightarrow \dot{m}_{in} = \dot{m}_{out} \rightarrow \dot{m}_1 = \dot{m}_2 = \dot{m}_s \quad\text{and}\quad \dot{m}_3 = \dot{m}_4 = \dot{m}_{fw}$$

Energy balance (for the heat exchanger):

$$\underbrace{\dot{E}_{in} - \dot{E}_{out}}_{\substack{\text{Rate of net energy transfer}\\\text{by heat, work, and mass}}} = \underbrace{\Delta\dot{E}_{\text{system}}^{\nearrow 0\text{ (steady)}}}_{\substack{\text{Rate of change in internal, kinetic,}\\\text{potential, etc. energies}}} = 0$$

$$\dot{E}_{in} = \dot{E}_{out}$$

$$\dot{m}_1 h_1 + \dot{m}_3 h_3 = \dot{m}_2 h_2 + \dot{m}_4 h_4 \quad (\text{since } \dot{Q} = \dot{W} = \Delta\text{ke} \cong \Delta\text{pe} \cong 0)$$

Combining the two, $\dot{m}_s(h_2 - h_1) = \dot{m}_{fw}(h_3 - h_4)$

Dividing by $\dot{m}_{fw}$ and substituting,

$$\frac{\dot{m}_s}{\dot{m}_{fw}} = \frac{h_4 - h_3}{h_1 - h_2} = \frac{(719.2 - 209.33)\text{kJ/kg}}{(2827.9 - 762.81)\text{kJ/kg}} = \mathbf{0.247}$$

(*b*) The total entropy change (or entropy generation) during this process per unit mass of feedwater can be determined from an entropy balance expressed in the rate form as

$$\underbrace{\dot{S}_{in} - \dot{S}_{out}}_{\substack{\text{Rate of net entropy transfer}\\\text{by heat and mass}}} + \underbrace{\dot{S}_{gen}}_{\substack{\text{Rate of entropy}\\\text{generation}}} = \underbrace{\Delta\dot{S}_{\text{system}}^{\nearrow 0}}_{\substack{\text{Rate of change}\\\text{of entropy}}} = 0$$

$$\dot{m}_1 s_1 - \dot{m}_2 s_2 + \dot{m}_3 s_3 - \dot{m}_4 s_4 + \dot{S}_{\text{gen}} = 0$$

$$\dot{m}_s(s_1 - s_2) + \dot{m}_{fw}(s_3 - s_4) + \dot{S}_{\text{gen}} = 0$$

$$\frac{\dot{S}_{gen}}{\dot{m}_{fw}} = \frac{\dot{m}_s}{\dot{m}_{fw}}(s_2 - s_1) + (s_4 - s_3) = (0.247)(2.1387 - 6.694) + (2.0419 - 0.7038) = \mathbf{0.213kW/(K \cdot kgfw)}$$

6-171 EES solution of this (and other comprehensive problems designated with the *computer icon*) is available to instructors at the *Instructor Manual* section of the *Online Learning Center* (OLC) at www.mhhe.com/cengel-boles. See the Preface for access information.

6-172E A rigid tank initially contains saturated R-134a vapor. The tank is connected to a supply line, and is charged until the tank contains saturated liquid at a specified pressure. The mass of R-134a that entered the tank, the heat transfer with the surroundings at 120°F, and the entropy generated during this process are to be determined.

Assumptions **1** This is an unsteady process since the conditions within the device are changing during the process, but it can be analyzed as a uniform-flow process since the state of fluid at the inlet remains constant. **2** Kinetic and potential energies are negligible. **3** There are no work interactions involved. **4** The direction of heat transfer is to the tank (will be verified).

Properties The properties of R-134a are (Tables A-11 through A-13)

$$\left.\begin{array}{l} P_1 = 120\text{psia} \\ sat.vapor \end{array}\right\} \begin{array}{l} v_1 = v_{g@120\text{psia}} = 0.3941\text{ft}^3/\text{lbm} \\ u_1 = u_{g@120\text{psia}} = 105.06\text{Btu/lbm} \\ s_1 = s_{g@120\text{psia}} = 0.2165\text{Btu/lbm}\cdot\text{R} \end{array}$$

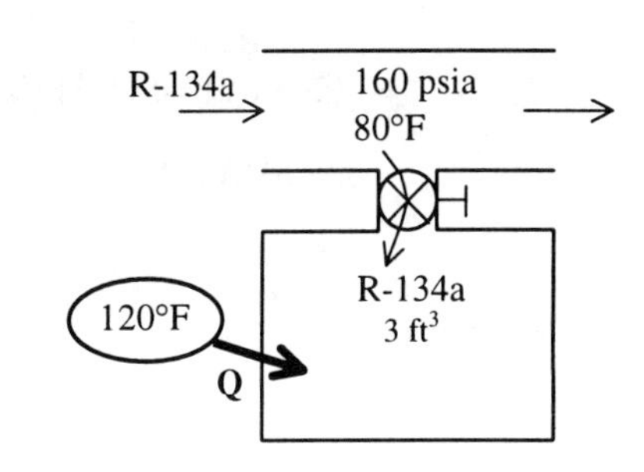

$$\left.\begin{array}{l} P_2 = 140\text{psia} \\ sat.liquid \end{array}\right\} \begin{array}{l} v_2 = v_{f@140\text{psia}} = 0.01386\text{ft}^3/\text{lbm} \\ u_2 = u_{f@140\text{psia}} = 44.07\text{Btu/lbm} \\ s_2 = s_{f@140\text{psia}} = 0.0902\text{Btu/lbm}\cdot\text{R} \end{array}$$

$$\left.\begin{array}{l} P_i = 160\text{psia} \\ T_i = 80^\circ\text{F} \end{array}\right\} \begin{array}{l} h_i \cong h_{f@80^\circ\text{F}} = 37.27\text{ Btu/lbm} \\ s_i \cong s_{f@80^\circ\text{F}} = 0.0774\text{ Btu/lbm}\cdot\text{R} \end{array}$$

Analysis (*a*) We take the tank as the system, which is a control volume since mass crosses the boundary. Noting that the microscopic energies of flowing and nonflowing fluids are represented by enthalpy h and internal energy u, respectively, the mass and energy balances for this uniform-flow system can be expressed as

Mass balance: $m_{in} - m_{out} = \Delta m_{\text{system}} \rightarrow m_i = m_2 - m_1$

Energy balance:

$$\underbrace{E_{in} - E_{out}}_{\substack{\text{Net energy transfer} \\ \text{by heat, work, and mass}}} = \underbrace{\Delta E_{\text{system}}}_{\substack{\text{Change in internal, kinetic,} \\ \text{potential, etc. energies}}}$$

$$Q_{in} + m_i h_i = m_2 u_2 - m_1 u_1 \quad (\text{since } W \cong ke \cong pe \cong 0)$$

The initial and the final masses in the tank are

$$m_1 = \frac{V}{v_1} = \frac{3\text{ ft}^3}{0.3941\text{ ft}^3/\text{lbm}} = 7.61\text{ lbm}$$

$$m_2 = \frac{V}{v_2} = \frac{3\text{ ft}^3}{0.01386\text{ ft}^3/\text{lbm}} = 216.45\text{ lbm}$$

Then from the mass balance,

$$m_i = m_2 - m_1 = 216.45 - 7.61 = \mathbf{208.84\text{ lbm}}$$

(*b*) The heat transfer during this process is determined from the energy balance to be

$$\begin{aligned} Q_{in} &= -m_i h_i + m_2 u_2 - m_1 u_1 \\ &= -(208.84\text{lbm})(37.27\text{Btu/lbm}) + (216.45\text{lbm})(44.07\text{Btu/lbm}) - (7.61\text{lbm})(105.06\text{Btu/lbm}) \\ &= \mathbf{956\text{Btu}} \end{aligned}$$

(*c*) The entropy generated during this process is determined by applying the entropy balance on an *extended system* that includes the tank and its immediate surroundings so that the boundary temperature of the extended system is the temperature of the surroundings at all times. The entropy balance for it can be expressed as

$$\underbrace{S_{in} - S_{out}}_{\substack{\text{Net entropy transfer} \\ \text{by heat and mass}}} + \underbrace{S_{gen}}_{\substack{\text{Entropy} \\ \text{generation}}} = \underbrace{\Delta S_{\text{system}}}_{\substack{\text{Change} \\ \text{in entropy}}}$$

$$\frac{Q_{in}}{T_{b,in}} + m_i s_i + S_{\text{gen}} = \Delta S_{\text{tank}} = m_2 s_2 - m_1 s_1$$

Therefore, the total entropy generated during this process is

$$S_{gen} = -m_i s_i + (m_2 s_2 - m_1 s_1) - \frac{Q_{\text{in}}}{T_{\text{b,in}}}$$

$$= -(208.84)(0.0774) + (216.45)(0.0902) - (7.61)(0.2165) - \frac{956\text{Btu}}{580\text{R}}$$

$$= \mathbf{0.0637 Btu/R}$$

6-173 It is to be shown that for thermal energy reservoirs, the entropy change relation $\Delta S = mC\ln(T_2/T_1)$ reduces to $\Delta S = Q/T$ as $T_2 \rightarrow T_1$.

Analysis Consider a thermal energy reservoir of mass m, specific heat C, and initial temperature T_1. Now heat, in the amount of Q, is transferred to this reservoir. The first law and the entropy change relations for this reservoir can be written as

$$Q = mC(T_2 - T_1) \longrightarrow mC = \frac{Q}{T_2 - T_1}$$

and

$$\Delta S = mC\ln\frac{T_2}{T_1} = Q\frac{\ln(T_2/T_1)}{T_2 - T_1}$$

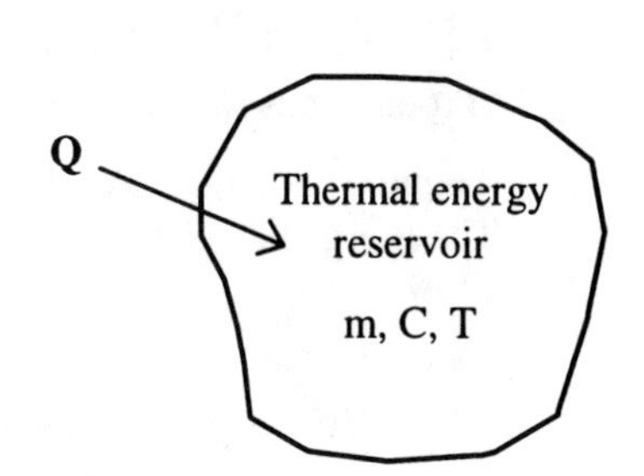

Taking the limit as $T_2 \rightarrow T_1$ by applying the L'Hospital's rule,

$$\Delta S = Q\frac{1/T_1}{1} = \frac{Q}{T_1}$$

which is the desired result.

6-174 The inner and outer glasses of a double pane window are at specified temperatures. The rates of entropy transfer through both sides of the window and the rate of entropy generation within the window are to be determined.

Assumptions Steady operating conditions exist since the surface temperatures of the glass remain constant at the specified values.

Analysis The entropy flows associated with heat transfer through the left and right glasses are

$$\dot{S}_{left} = \frac{\dot{Q}_{left}}{T_{left}} = \frac{110\text{ W}}{291\text{ K}} = \mathbf{0.378\ W/K}$$

$$\dot{S}_{right} = \frac{\dot{Q}_{right}}{T_{right}} = \frac{110\text{ W}}{279\text{ K}} = \mathbf{0.394\ W/K}$$

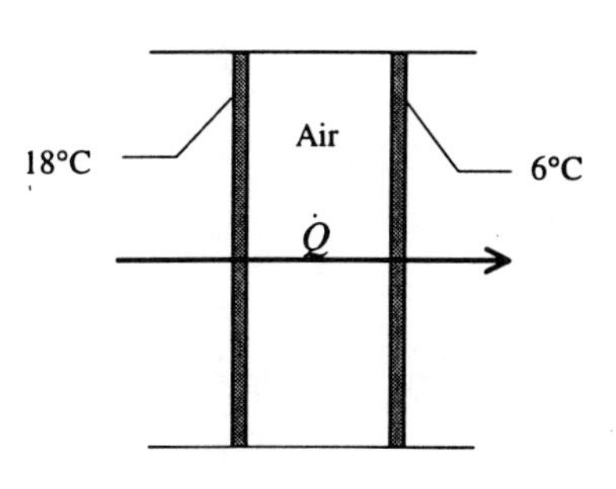

We take the double pane window as the system, which is a closed system. In steady operation, the rate form of the entropy balance for this system can be expressed as

$$\underbrace{\dot{S}_{in} - \dot{S}_{out}}_{\text{Rate of net entropy transfer by heat and mass}} + \underbrace{\dot{S}_{gen}}_{\text{Rate of entropy generation}} = \underbrace{\Delta\dot{S}_{\text{system}}^{\nearrow 0}}_{\text{Rate of change of entropy}} = 0$$

$$\frac{\dot{Q}_{in}}{T_{b,in}} - \frac{\dot{Q}_{out}}{T_{b,out}} + \dot{S}_{\text{gen,system}} = 0$$

$$\frac{110\text{ W}}{291\text{ K}} - \frac{110\text{ W}}{279\text{ K}} + \dot{S}_{\text{gen,system}} = 0 \rightarrow \dot{S}_{\text{gen,system}} = \mathbf{0.016\ W/K}$$

6-175 A well-insulated room is heated by a steam radiator, and the warm air is distributed by a fan. The average temperature in the room after 30 min, the entropy changes of steam and air, and the entropy generated during this process are to be determined.

Assumptions **1** Air is an ideal gas with constant specific heats at room temperature. **2** The kinetic and potential energy changes are negligible. **3** The air pressure in the room remains constant and thus the air expands as it is heated, and some warm air escapes.

Properties The gas constant of air is R = 0.287 kPa.m³/kg.K (Table A-1). Also, C_p = 1.005 kJ/kg.K for air at room temperature (Table A-2).

Analysis We first take the radiator as the system. This is a closed system since no mass enters or leaves. The energy balance for this closed system can be expressed as

$$\underbrace{E_{in} - E_{out}}_{\substack{\text{Net energy transfer}\\ \text{by heat, work, and mass}}} = \underbrace{\Delta E_{\text{system}}}_{\substack{\text{Change in internal, kinetic,}\\ \text{potential, etc. energies}}}$$

$$-Q_{out} = \Delta U = m(u_2 - u_1) \quad (\text{since } W = \text{KE} = \text{PE} = 0)$$

$$Q_{out} = m(u_1 - u_2)$$

10°C
4m x 4m x 5m

Steam radiator

Using data from the steam tables (Tables A-4 through A-6), some properties are determined to be

$$\left.\begin{array}{l} P_1 = 200\text{kPa} \\ T_1 = 200^\circ\text{C} \end{array}\right\} \begin{array}{l} v_1 = 1.0803\text{m}^3/\text{kg} \\ u_1 = 2654.4\text{kJ/kg} \\ s_1 = 7.5066\text{kJ/kg.K} \end{array}$$

$$\left.\begin{array}{l} P_2 = 100\text{kPa} \\ (v_2 = v_1) \end{array}\right\} \begin{array}{ll} v_f = 0.001043, & v_g = 1.6940\text{m}^3/\text{kg} \\ u_f = 417.46, & u_{fg} = 2088.7\text{kJ/kg} \\ s_f = 1.3026, & s_{fg} = 6.0568\text{ kJ/kg.K} \end{array}$$

$$x_2 = \frac{v_2 - v_f}{v_{fg}} = \frac{1.0803 - 0.001043}{1.6940 - 0.001043} = 0.6375$$

$$u_2 = u_f + x_2 u_{fg} = 417.36 + 0.6375 \times 2088.7 = 1748.9 \text{ kJ/kg}$$

$$m = \frac{V_1}{v_1} = \frac{0.015\text{m}^3}{1.0803\text{m}^3/\text{kg}} = 0.0139\text{kg}$$

Substituting,

$$Q_{\text{out}} = (0.0139 \text{ kg})(2654.4 - 1749.0)\text{kJ/kg} = 12.6 \text{ kJ}$$

The volume and the mass of the air in the room are V = 4×4×5 = 80 m³ and

$$m_{air} = \frac{P_1 V_1}{RT_1} = \frac{(100\text{kPa})(80\text{ m}^3)}{(0.2870\text{kPa}\cdot\text{m}^3/\text{kg}\cdot\text{K})(283\text{K})} = 98.5\text{kg}$$

The amount of fan work done in 30 min is

$$W_{fan,in} = \dot{W}_{fan,in}\Delta t = (0.120 \text{ kJ/s})(30 \times 60 \text{ s}) = 216 \text{ kJ}$$

We now take the air in the room as the system. The energy balance for this closed system is expressed as

$$E_{in} - E_{out} = \Delta E_{\text{system}}$$
$$Q_{in} + W_{fan,in} - W_{b,out} = \Delta U$$
$$Q_{in} + W_{fan,in} = \Delta H \cong mC_p(T_2 - T_1)$$

since the boundary work and ΔU combine into ΔH for a constant pressure expansion or compression process.

Substituting, $\quad$ (12.6 kJ) + (216 kJ) = (98.5 kg)(1.005 kJ/kg°C)(T_2 - 10)°C

which yields $\quad$ T_2 = **12.3°C**

Therefore, the air temperature in the room rises from 10°C to 12.3°C in 30 mi.

(*b*) The entropy change of the steam is

$$\Delta S_{steam} = m(s_2 - s_1) = (0.0139\text{kg})(5.1638 - 7.5066)\text{kJ/kg}\cdot\text{K} = \mathbf{-0.0326kJ/K}$$

(*c*) Noting that air expands at constant pressure, the entropy change of the air in the room is

$$\Delta S_{air} = mC_p \ln\frac{T_2}{T_1} - mR\ln\frac{P_2}{P_1}^{\nearrow 0} = (98.5\text{kg})(1.005\text{kJ/kg}\cdot\text{K})\ln\frac{285.3\text{ K}}{283\text{ K}} = \mathbf{0.8013\ \ kJ/K}$$

We take the air in the room (including the steam radiator) as our system, which is a closed system. Noting that no heat or mass crosses the boundaries of this system, the entropy balance for it can be expressed as

$$\underbrace{S_{in} - S_{out}}_{\substack{\text{Net entropy transfer}\\ \text{by heat and mass}}} + \underbrace{S_{gen}}_{\substack{\text{Entropy}\\ \text{generation}}} = \underbrace{\Delta S_{\text{system}}}_{\substack{\text{Change}\\ \text{in entropy}}}$$
$$0 + S_{\text{gen}} = \Delta S_{\text{steam}} + \Delta S_{\text{air}}$$

Substituting, the entropy generated during this process is determined to be

$$S_{gen} = \Delta S_{steam} + \Delta S_{air} = -0.0326 + 0.8013 = \mathbf{0.7687\ \ kJ/K}$$

6-176 The heating of a passive solar house at night is to be assisted by solar heated water. The length of time that the electric heating system would run that night and the amount of entropy generated that night are to be determined.

Assumptions **1** Water is an incompressible substance with constant specific heats. **2** The energy stored in the glass containers themselves is negligible relative to the energy stored in water. **3** The house is maintained at 22°C at all times.

Properties The density and specific heat of water at room temperature are ρ = 1 kg/L and C = 4.18 kJ/kg· °C (Table A-3).

Analysis The total mass of water is

$$m_w = \rho V = (1\text{kg/L})(50 \times 20\text{L}) = 1000\text{kg}$$

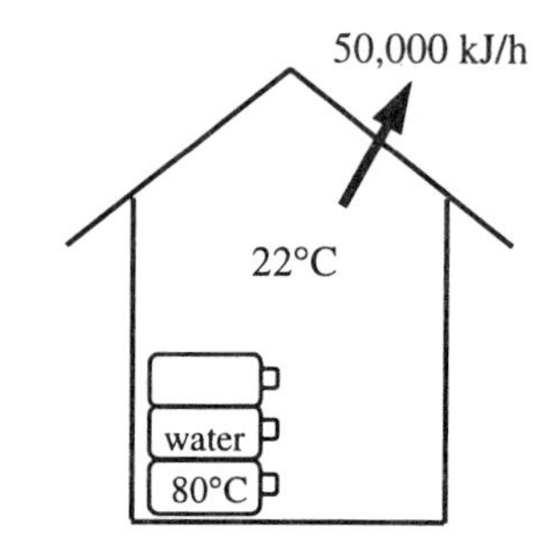

Taking the contents of the house, including the water as our system, the energy balance relation can be written as

$$\underbrace{E_{in} - E_{out}}_{\substack{\text{Net energy transfer} \\ \text{by heat, work, and mass}}} = \underbrace{\Delta E_{\text{system}}}_{\substack{\text{Change in internal, kinetic,} \\ \text{potential, etc. energies}}}$$

$$W_{e,in} - Q_{out} = \Delta U = (\Delta U)_{\text{water}} + (\Delta U)_{\text{air}}$$
$$= (\Delta U)_{\text{water}}$$
$$= mC(T_2 - T_1)_{\text{water}}$$

or,

$$\dot{W}_{e,in}\Delta t - Q_{out} = [mC(T_2 - T_1)]_{\text{water}}$$

Substituting,

$$(15 \text{ kJ/s})\Delta t \text{ - } (50{,}000 \text{ kJ/h})(10 \text{ h}) = (1000 \text{ kg})(4.18 \text{ kJ/kg}\cdot\,^\circ\text{C})(22 - 80)^\circ\text{C}$$

It gives

$$\Delta t = 17{,}170 \text{ s} = \mathbf{4.77 \text{ h}}$$

We take the house as the system, which is a closed system. The entropy generated during this process is determined by applying the entropy balance on an *extended system* that includes the house and its immediate surroundings so that the boundary temperature of the extended system is the temperature of the surroundings at all times. The entropy balance for the extended system can be expressed as

$$\underbrace{S_{in} - S_{out}}_{\substack{\text{Net entropy transfer} \\ \text{by heat and mass}}} + \underbrace{S_{gen}}_{\substack{\text{Entropy} \\ \text{generation}}} = \underbrace{\Delta S_{\text{system}}}_{\substack{\text{Change} \\ \text{in entropy}}}$$

$$-\frac{Q_{out}}{T_{b,out}} + S_{\text{gen}} = \Delta S_{\text{water}} + \Delta S_{\text{air}}^{\nearrow 0} = \Delta S_{\text{water}}$$

since the state of air in the house remains unchanged. Then the entropy generated during the 10-h period that night is

$$S_{gen} = \Delta S_{\text{water}} + \frac{Q_{out}}{T_{b,out}} = \left(mC\ln\frac{T_2}{T_1}\right)_{water} + \frac{Q_{out}}{T_{surr}}$$
$$= (1000 \text{ kg})(4.18 \text{ kJ/kg}\cdot\text{K})\ln\frac{295 \text{ K}}{353 \text{ K}} + \frac{500{,}000 \text{ kJ}}{276 \text{ K}}$$
$$= -750 + 1811 = \mathbf{1061 \text{ kJ/K}}$$

6-177E A steel container that is filled with hot water is allowed to cool to the ambient temperature. The total entropy generated during this process is to be determined.

Assumptions **1** Both the water and the steel tank are incompressible substances with constant specific heats at room temperature. **2** The system is stationary and thus the kinetic and potential energy changes are zero. **3** Specific heat of iron can be used for steel. **4** There are no work interactions involved.

Properties The specific heats of water and the iron at room temperature are $C_{p,\text{ water}}$ = 1.00 Btu/lbm.°F and $C_{p,\text{ iron}}$ = 0.107 Btu/lbm.°C (Table A-3E). The density of water at room temperature is 62.1 lbm/ft³.

Analysis The mass of the water is

$$m_{water} = \rho V = (62.1 \text{ lbm/ft}^3)(15 \text{ ft}^3) = 931.5 \text{ lbm}$$

We take the steel container and the water in it as the system, which is a closed system. The energy balance on the system can be expressed as

$$\underbrace{E_{in} - E_{out}}_{\substack{\text{Net energy transfer} \\ \text{by heat, work, and mass}}} = \underbrace{\Delta E_{\text{system}}}_{\substack{\text{Change in internal, kinetic,} \\ \text{potential, etc. energies}}}$$

$$-Q_{\text{out}} = \Delta U = \Delta U_{\text{container}} + \Delta U_{\text{water}}$$

$$= [mC(T_2 - T_1)]_{\text{container}} + [mC(T_2 - T_1)]_{\text{water}}$$

Steel

WATER
120°F

Q
70°F

Substituting, the heat loss to the surrounding air is determined to be

$$Q_{out} = [mC(T_1 - T_2)]_{\text{container}} + [mC(T_1 - T_2)]_{\text{water}}$$

$$= (75 \text{ lbm})(0.107 \text{ Btu/lbm}\cdot^\circ\text{F})(120 - 70)^\circ\text{F} + (931.5 \text{ lbm})(1.00 \text{ Btu/lbm}\cdot^\circ\text{F})(120 - 70)^\circ\text{F}$$

$$= 46{,}976 \text{ Btu}$$

We again take the container and the water In it as the system. The entropy generated during this process is determined by applying the entropy balance on an *extended system* that includes the container and its immediate surroundings so that the boundary temperature of the extended system is the temperature of the surrounding air at all times. The entropy balance for the extended system can be expressed as

$$\underbrace{S_{in} - S_{out}}_{\substack{\text{Net entropy transfer} \\ \text{by heat and mass}}} + \underbrace{S_{gen}}_{\substack{\text{Entropy} \\ \text{generation}}} = \underbrace{\Delta S_{\text{system}}}_{\substack{\text{Change} \\ \text{in entropy}}}$$

$$-\frac{Q_{out}}{T_{b,out}} + S_{\text{gen}} = \Delta S_{\text{container}} + \Delta S_{\text{water}}$$

where

$$\Delta S_{\text{container}} = mC_{ave} \ln\frac{T_2}{T_1} = (75 \text{ lbm})(0.107 \text{ Btu/lbm}\cdot\text{R})\ln\frac{530 \text{ R}}{580 \text{ R}} = -0.72 \text{ Btu/R}$$

$$\Delta S_{\text{water}} = mC_{ave} \ln\frac{T_2}{T_1} = (931.5 \text{ lbm})(1.00 \text{ Btu/lbm}\cdot\text{R})\ln\frac{530 \text{ R}}{580 \text{ R}} = -83.98 \text{ Btu/R}$$

Therefore, the total entropy generated during this process is

$$S_{gen} = \Delta S_{\text{container}} + \Delta S_{\text{water}} + \frac{Q_{out}}{T_{b,out}} = -0.72 - 83.98 + \frac{46{,}976 \text{ Btu}}{70 + 460 \text{ R}} = \mathbf{3.93\ \ Btu/R}$$

6-178 Refrigerant-134a is vaporized by air in the evaporator of an air-conditioner. For specified flow rates, the exit temperature of air and the rate of entropy generation are to be determined for the cases of an insulated and uninsulated evaporator. √

Assumptions **1** This is a steady-flow process since there is no change with time. **2** Kinetic and potential energy changes are negligible. **3** There are no work interactions. **4** Air is an ideal gas with constant specific heats at room temperature.

Properties The gas constant of air is 0.287 kPa.m^3/kg.K (Table A-1). The constant pressure specific heat of air at room temperature is C_p = 1.005 kJ/kg.K (Table A-2). The properties of R-134a at the inlet and the exit states are (Tables A-11 through A-13)

$$\left.\begin{array}{l} P_1 = 120\text{ kPa} \\ x_1 = 0.3 \end{array}\right\} \begin{array}{l} h_1 = h_f + x_1 h_{fg} = 21.32 + 0.3 \times 212.54 = 85.08\text{ kJ/kg} \\ s_1 = s_f + x_1 s_{fg} = 0.0879 + 0.3(0.9354 - 0.0879) = 0.3422\text{ kJ/kg}\cdot\text{K} \end{array}$$

$$\left.\begin{array}{l} T_2 = 120\text{ kPa} \\ \text{sat.vapor} \end{array}\right\} \begin{array}{l} h_2 = h_{g@120\text{ kPa}} = 233.86\text{ kJ/kg} \\ s_2 = h_{g@120\text{ kPa}} = 0.9354\text{ kJ/kg}\cdot\text{K} \end{array}$$

Analysis (*a*) The mass flow rate of air is

$$\dot{m}_{air} = \frac{P_3\dot{V}_3}{RT_3} = \frac{(100\text{kPa})(6\text{m}^3/\text{min})}{(0.287\text{kPa}\cdot\text{m}^3/\text{kg}\cdot\text{K})(300\text{K})} = 6.97\text{kg/min}$$

We take the entire heat exchanger as the system, which is a control volume. The mass and energy balances for this steady-flow system can be expressed in the rate form as

Mass balance (*for each fluid stream*):

$$\dot{m}_{in} - \dot{m}_{out} = \Delta\dot{m}_{\text{system}}^{\nearrow 0\text{ (steady)}} = 0 \rightarrow \dot{m}_{in} = \dot{m}_{out} \rightarrow \dot{m}_1 = \dot{m}_2 = \dot{m}_{air} \text{ and } \dot{m}_3 = \dot{m}_4 = \dot{m}_R$$

Energy balance (for the entire heat exchanger):

$$\underbrace{\dot{E}_{in} - \dot{E}_{out}}_{\substack{\text{Rate of net energy transfer}\\ \text{by heat, work, and mass}}} = \underbrace{\Delta\dot{E}_{\text{system}}^{\nearrow 0\text{ (steady)}}}_{\substack{\text{Rate of change in internal, kinetic,}\\ \text{potential, etc. energies}}} = 0$$

$$\dot{E}_{in} = \dot{E}_{out}$$

$$\dot{m}_1 h_1 + \dot{m}_3 h_3 = \dot{m}_2 h_2 + \dot{m}_4 h_4 \quad (\text{since } \dot{Q} = \dot{W} = \Delta\text{ke} \cong \Delta\text{pe} \cong 0)$$

6 m^3/min AIR ③
R-134a ①
2 kg/min
② sat. vapor ④

Combining the two, $\dot{m}_R(h_2 - h_1) = \dot{m}_{air}(h_3 - h_4) = \dot{m}_{air} C_p (T_3 - T_4)$

Solving for T_4, $T_4 = T_3 - \dfrac{\dot{m}_R(h_2 - h_1)}{\dot{m}_{air} C_p}$

Substituting, $T_4 = 27°\text{C} - \dfrac{(2\text{ kg/min})(233.86 - 85.08)\text{ kJ/kg}}{(6.97\text{ kg/min})(1.005\text{ kJ/kg}\cdot\text{K})} = \mathbf{-15.5°C} = 257.5\text{ K}$

Noting that the condenser is well-insulated and thus heat transfer is negligible, the entropy balance for this steady-flow system can be expressed as

$$\underbrace{\dot{S}_{in} - \dot{S}_{out}}_{\substack{\text{Rate of net entropy transfer}\\ \text{by heat and mass}}} + \underbrace{\dot{S}_{gen}}_{\substack{\text{Rate of entropy}\\ \text{generation}}} = \underbrace{\Delta\dot{S}_{\text{system}}^{\nearrow 0\text{ (steady)}}}_{\substack{\text{Rate of change}\\ \text{of entropy}}}$$

$$\dot{m}_1 s_1 + \dot{m}_3 s_3 - \dot{m}_2 s_2 - \dot{m}_4 s_4 + \dot{S}_{gen} = 0 \quad (\text{since } Q = 0)$$

$$\dot{m}_R s_1 + \dot{m}_{\text{air}} s_3 - \dot{m}_R s_2 - \dot{m}_{\text{air}} s_4 + \dot{S}_{gen} = 0$$

or,

$$\dot{S}_{gen} = \dot{m}_R(s_2 - s_1) + \dot{m}_{air}(s_4 - s_3)$$

where

$$s_4 - s_3 = C_p \ln\frac{T_4}{T_3} - R\ln\frac{P_4^{\nearrow 0}}{P_3} = C_p \ln\frac{T_4}{T_3} = (1.005\ \text{kJ/kg}\cdot\text{K})\ln\frac{257.5\ \text{K}}{300\ \text{K}} = -0.154\ \text{kJ/kg}\cdot\text{K}$$

Substituting,

$$\begin{aligned}\dot{S}_{gen} &= (2\ \text{kg/min})(0.9354 - 0.3422\ \text{kJ/kg}\cdot\text{K}) + (6.97\ \text{kg/min})(0.154\ \text{kJ/kg}\cdot\text{K}) \\ &= 0.113\ \text{kJ/min}\cdot\text{K} \\ &= \mathbf{0.00188\ kW/K}\end{aligned}$$

(*b*) When there is a heat gain from the surroundings at a rate of 30 kJ/min, the steady-flow energy equation reduces to

$$\dot{Q}_{in} = \dot{m}_R(h_2 - h_1) + \dot{m}_{air}C_p(T_4 - T_3)$$

Solving for T_4, $$T_4 = T_3 + \frac{\dot{Q}_{in} - \dot{m}_R(h_2 - h_1)}{\dot{m}_{air}C_p}$$

Substituting, $$T_4 = 27°\text{C} + \frac{(30\ \text{kJ/min}) - (2\ \text{kg/min})(233.86 - 85.08)\ \text{kJ/kg}}{(6.97\ \text{kg/min})(1.005\ \text{kJ/kg}\cdot\text{K})} = \mathbf{-11.2°C}$$

The entropy generation in this case is determined by applying the entropy balance on an *extended system* that includes the evaporator and its immediate surroundings so that the boundary temperature of the extended system is the temperature of the surrounding air at all times. The entropy balance for the extended system can be expressed as

$$\underbrace{\dot{S}_{in} - \dot{S}_{out}}_{\substack{\text{Rate of net entropy transfer}\\ \text{by heat and mass}}} + \underbrace{\dot{S}_{gen}}_{\substack{\text{Rate of entropy}\\ \text{generation}}} = \underbrace{\Delta\dot{S}_{system}^{\nearrow 0\ (\text{steady})}}_{\substack{\text{Rate of change}\\ \text{of entropy}}}$$

$$\frac{Q_{in}}{T_{b,out}} + \dot{m}_1 s_1 + \dot{m}_3 s_3 - \dot{m}_2 s_2 - \dot{m}_4 s_4 + \dot{S}_{gen} = 0$$

$$\frac{Q_{in}}{T_{surr}} + \dot{m}_R s_1 + \dot{m}_{air} s_3 - \dot{m}_R s_2 - \dot{m}_{air} s_4 + \dot{S}_{gen} = 0$$

or $$\dot{S}_{gen} = \dot{m}_R(s_2 - s_1) + \dot{m}_{air}(s_4 - s_3) - \frac{\dot{Q}_{in}}{T_0}$$

where $$s_4 - s_3 = C_p \ln\frac{T_4}{T_3} - R\ln\frac{P_4^{\nearrow 0}}{P_3} = (1.005\ \text{kJ/kg}\cdot\text{K})\ln\frac{261.8\ \text{K}}{300\ \text{K}} = -0.137\ \text{kJ/kg}\cdot\text{K}$$

Substituting,

$$\begin{aligned}\dot{S}_{gen} &= (2\text{kg/min})(0.9354 - 0.3422)\text{kJ/kg}\cdot\text{K} + (6.97\text{kg/min})(-0.137\text{kJ/kg}\cdot\text{K}) - \frac{30\text{kJ/min}}{305\text{K}} \\ &= 0.13315\ \text{kJ/min}\cdot\text{K} \\ &= \mathbf{0.00222\ kW/K}\end{aligned}$$

Discussion Note that the rate of entropy generation in the second case is greater because of the irreversibility associated with heat transfer between the evaporator and the surrounding air.

6-179 A room is to be heated by hot water contained in a tank placed in the room. The minimum initial temperature of the water needed to meet the heating requirements of this room for a 24-h period and the entropy generated are to be determined.

Assumptions **1** Water is an incompressible substance with constant specific heats. **2** Air is an ideal gas with constant specific heats. **3** The energy stored in the container itself is negligible relative to the energy stored in water. **4** The room is maintained at 20°C at all times. **5** The hot water is to meet the heating requirements of this room for a 24-h period.

Properties The specific heat of water at room temperature is $C = 4.18$ kJ/kg· °C (Table A-3).

Analysis Heat loss from the room during a 24-h period is

$$Q_{\text{loss}} = (10{,}000 \text{ kJ/h})(24 \text{ h}) = 240{,}000 \text{ kJ}$$

Taking the contents of the room, including the water, as our system, the energy balance can be written as

$$\underbrace{E_{in} - E_{out}}_{\substack{\text{Net energy transfer}\\ \text{by heat, work, and mass}}} = \underbrace{\Delta E_{\text{system}}}_{\substack{\text{Change in internal, kinetic,}\\ \text{potential, etc. energies}}} \rightarrow -Q_{\text{out}} = \Delta U = (\Delta U)_{\text{water}} + (\Delta U)_{\text{air}}^{\nearrow 0}$$

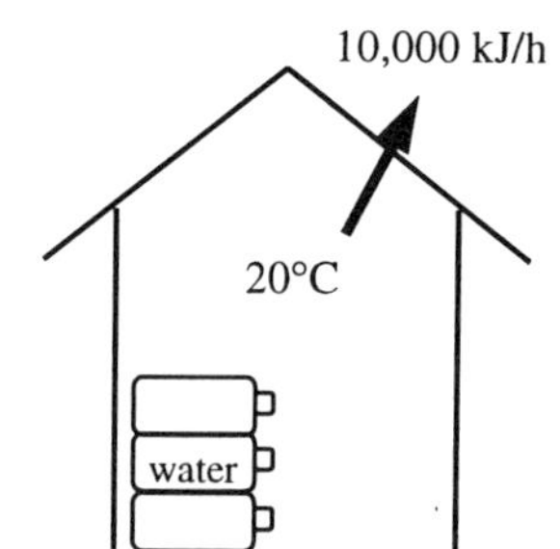

or

$$-Q_{\text{out}} = [mC(T_2 - T_1)]_{\text{water}}$$

Substituting,

$$-240{,}000 \text{ kJ} = (1500 \text{ kg})(4.18 \text{ kJ/kg· °C})(20 - T_1)$$

It gives

$$T_1 = \mathbf{58.3°C}$$

where T_1 is the temperature of the water when it is first brought into the room.

(*b*) We take the house as the system, which is a closed system. The entropy generated during this process is determined by applying the entropy balance on an *extended system* that includes the house and its immediate surroundings so that the boundary temperature of the extended system is the temperature of the surroundings at all times. The entropy balance for the extended system can be expressed as

$$\underbrace{S_{in} - S_{out}}_{\substack{\text{Net entropy transfer}\\ \text{by heat and mass}}} + \underbrace{S_{gen}}_{\substack{\text{Entropy}\\ \text{generation}}} = \underbrace{\Delta S_{\text{system}}}_{\substack{\text{Change}\\ \text{in entropy}}}$$

$$-\frac{Q_{out}}{T_{b,out}} + S_{\text{gen}} = \Delta S_{\text{water}} + \Delta S_{\text{air}}^{\nearrow 0} = \Delta S_{\text{water}}$$

since the state of air in the house (and thus its entropy) remains unchanged. Then the entropy generated during the 24 h period becomes

$$S_{gen} = \Delta S_{\text{water}} + \frac{Q_{out}}{T_{b,out}} = \left(mC \ln \frac{T_2}{T_1} \right)_{water} + \frac{Q_{out}}{T_{surr}}$$

$$= (1500 \text{ kg})(4.18 \text{ kJ/kg} \cdot \text{K}) \ln \frac{293 \text{ K}}{331.3 \text{ K}} + \frac{240{,}000 \text{ kJ}}{278 \text{ K}}$$

$$= -770.3 + 863.3 = \mathbf{93.0 \text{ kJ/K}}$$

6-180 An insulated cylinder is divided into two parts. One side of the cylinder contains N_2 gas and the other side contains He gas at different states. The final equilibrium temperature in the cylinder and the entropy generated are to be determined for the cases of the piston being fixed and moving freely.

Assumptions **1** Both N_2 and He are ideal gases with constant specific heats. **2** The energy stored in the container itself is negligible. **3** The cylinder is well-insulated and thus heat transfer is negligible.

Properties The gas constants and the constant volume specific heats are R = 0.2968 kPa.m³/kg.K, C_v = 0.743 kJ/kg·°C and C_p =1.039 kJ/kg·°C for N_2, and R = 2.0769 kPa.m³/kg.K, C_v = 3.1156 kJ/kg·°C, and C_p = 5.1926 kJ/kg·°C for He (Tables A-1 and A-2)

Analysis The mass of each gas in the cylinder is

$$m_{N_2} = \left(\frac{P_1 V_1}{RT_1}\right)_{N_2} = \frac{(500\text{kPa})(1\text{m}^3)}{(0.2968\text{kPa}\cdot\text{m}^3/\text{kg}\cdot\text{K})(353\text{K})} = 4.77\text{kg}$$

$$m_{He} = \left(\frac{P_1 V_1}{RT_1}\right)_{He} = \frac{(500\text{kPa})(1\text{m}^3)}{(2.0769\text{kPa}\cdot\text{m}^3/\text{kg}\cdot\text{K})(298\text{K})} = 0.808\text{kg}$$

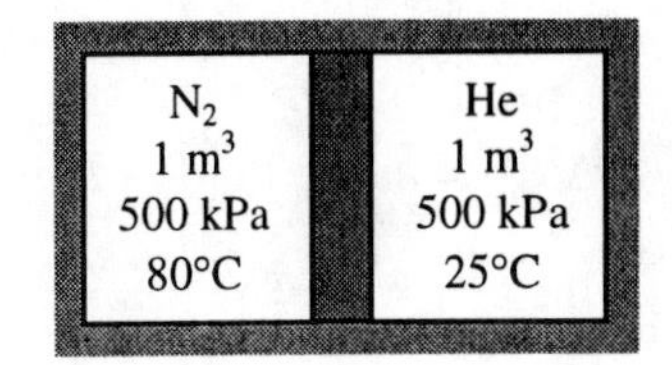

Taking the entire contents of the cylinder as our system, the 1st law relation can be written as

$$\underbrace{E_{in} - E_{out}}_{\substack{\text{Net energy transfer}\\ \text{by heat, work, and mass}}} = \underbrace{\Delta E_{\text{system}}}_{\substack{\text{Change in internal, kinetic,}\\ \text{potential, etc. energies}}}$$

$$0 = \Delta U = (\Delta U)_{N_2} + (\Delta U)_{He}$$

$$0 = [mC_v(T_2 - T_1)]_{N_2} + [mC_v(T_2 - T_1)]_{He}$$

Substituting,

$$(4.77\text{kg})(0.743\text{kJ/kg}\cdot^\circ\text{C})(T_f - 80)^\circ\text{C} + (0.808\text{kg})(3.1156\text{kJ/kg}\cdot^\circ\text{C})(T_f - 25)^\circ\text{C} = 0$$

It gives T_f = **57.2°C**

where T_f is the final equilibrium temperature in the cylinder.

The answer would be the **same** if the piston were not free to move since it would effect only pressure, and not the specific heats.

(*b*) We take the entire cylinder as our system, which is a closed system. Noting that the cylinder is well-insulated and thus there is no heat transfer, the entropy balance for this closed system can be expressed as

$$\underbrace{S_{in} - S_{out}}_{\substack{\text{Net entropy transfer}\\ \text{by heat and mass}}} + \underbrace{S_{gen}}_{\substack{\text{Entropy}\\ \text{generation}}} = \underbrace{\Delta S_{\text{system}}}_{\substack{\text{Change}\\ \text{in entropy}}}$$

$$0 + S_{gen} = \Delta S_{N_2} + \Delta S_{He}$$

But first we determine the final pressure in the cylinder:

$$N_{total} = N_{N_2} + N_{He} = \left(\frac{m}{M}\right)_{N_2} + \left(\frac{m}{M}\right)_{He} = \frac{4.77\text{kg}}{28\text{kg/kmol}} + \frac{0.808\text{kg}}{4\text{kg/kmol}} = 0.372\text{kmol}$$

$$P_2 = \frac{N_{total} R_u T}{V_{total}} = \frac{(0.372\text{kmol})(8.314\text{kPa}\cdot\text{m}^3/\text{kmol}\cdot\text{K})(330.2\text{K})}{2\text{m}^3} = 510.6\text{kPa}$$

Then,

$$\Delta S_{N_2} = m\left(C_p \ln\frac{T_2}{T_1} - R\ln\frac{P_2}{P_1}\right)_{N_2}$$

$$= (4.77\text{kg})\left[(1.039\text{kJ/kg}\cdot\text{K})\ln\frac{330.2\text{K}}{353\text{K}} - (0.2968\text{kJ/kg}\cdot\text{K})\ln\frac{510.6\text{kPa}}{500\text{kPa}}\right] = -0.361\text{kJ/K}$$

$$\Delta S_{He} = m\left(C_p \ln\frac{T_2}{T_1} - R\ln\frac{P_2}{P_1}\right)_{He}$$

$$= (0.808\text{kg})\left[(5.1926\text{kJ/kg}\cdot\text{K})\ln\frac{330.2\text{K}}{298\text{K}} - (2.0769\text{kJ/kg}\cdot\text{K})\ln\frac{510.6\text{kPa}}{500\text{kPa}}\right] = 0.395\text{kJ/K}$$

$$S_{gen} = \Delta S_{N_2} + \Delta S_{He} = -0.361 + 0.395 = \mathbf{0.034kJ/K}$$

If the piston were not free to move, we would still have T_2 = 330.2 K but the volume of each gas would remain constant in this case:

$$\Delta S_{N_2} = m\left(C_v \ln\frac{T_2}{T_1} - R\ln\frac{V_2}{V_1}^{\nearrow 0}\right)_{N_2} = (4.77\text{ kg})(0.743\text{ kJ/kg}\cdot\text{K})\ln\frac{330.2\text{ K}}{353\text{ K}} = -0.237\text{ kJ/K}$$

$$\Delta S_{He} = m\left(C_v \ln\frac{T_2}{T_1} - R\ln\frac{V_2}{V_1}^{\nearrow 0}\right)_{He} = (0.808\text{ kg})(3.1156\text{ kJ/kg}\cdot\text{K})\ln\frac{330.2\text{ K}}{298\text{ K}} = 0.258\text{ kJ/K}$$

$$S_{gen} = \Delta S_{N_2} + \Delta S_{He} = -0.237 + 0.258 = \mathbf{0.021\ kJ/K}$$

6-181 EES solution of this (and other comprehensive problems designated with the *computer icon*) is available to instructors at the *Instructor Manual* section of the *Online Learning Center* (OLC) at www.mhhe.com/cengel-boles. See the Preface for access information.

6-182 An insulated cylinder is divided into two parts. One side of the cylinder contains N_2 gas and the other side contains He gas at different states. The final equilibrium temperature in the cylinder and the entropy generated are to be determined for the cases of the piston being fixed and moving freely.

Assumptions **1** Both N_2 and He are ideal gases with constant specific heats. **2** The energy stored in the container itself, except the piston, is negligible. **3** The cylinder is well-insulated and thus heat transfer is negligible. **4** Initially, the piston is at the average temperature of the two gases.

Properties The gas constants and the constant volume specific heats are R = 0.2968 kPa.m³/kg.K, C_v = 0.743 kJ/kg·°C and C_p =1.039 kJ/kg·°C for N_2, and R = 2.0769 kPa.m³/kg.K, C_v = 3.1156 kJ/kg·°C, and C_p = 5.1926 kJ/kg·°C for He (Tables A-1 and A-2). The specific heat of the copper at room temperature is C = 0.386 kJ/kg·°C (Table A-3).

Analysis The mass of each gas in the cylinder is

$$m_{N_2} = \left(\frac{P_1V_1}{RT_1}\right)_{N_2} = \frac{(500\text{kPa})(1\text{m}^3)}{(0.2968\text{kPa}\cdot\text{m}^3/\text{kg}\cdot\text{K})(353\text{K})} = 4.77\text{kg}$$

$$m_{He} = \left(\frac{P_1V_1}{RT_1}\right)_{He} = \frac{(500\text{kPa})(1\text{m}^3)}{(2.0769\text{kPa}\cdot\text{m}^3/\text{kg}\cdot\text{K})(353\text{K})} = 0.808\text{kg}$$

N_2 1 m³ 500 kPa 80°C | He 1 m³ 500 kPa 25°C | Coppe

Taking the entire contents of the cylinder as our system, the 1st law relation can be written as

$$\underbrace{E_{in} - E_{out}}_{\substack{\text{Net energy transfer}\\ \text{by heat, work, and mass}}} = \underbrace{\Delta E_{\text{system}}}_{\substack{\text{Change in internal, kinetic,}\\ \text{potential, etc. energies}}}$$

$$0 = \Delta U = (\Delta U)_{N_2} + (\Delta U)_{He} + (\Delta U)_{Cu}$$

$$0 = [mC_v(T_2 - T_1)]_{N_2} + [mC_v(T_2 - T_1)]_{He} + [mC(T_2 - T_1)]_{Cu}$$

where

$$T_{1,\,Cu} = (80 + 25)/2 = 52.5°\text{C}$$

Substituting,

$$(4.77\text{kg})(0.743\text{kJ/kg}\cdot°\text{C})(T_f - 80)°\text{C} + (0.808\text{kg})(3.1156\text{kJ/kg}\cdot°\text{C})(T_f - 25)°\text{C} + (5.0\text{kg})(0.386\text{kJ/kg}\cdot°\text{C})(T_f - 52.5)°\text{C} = 0$$

It gives

$$T_f = \mathbf{56.0°C}$$

where T_f is the final equilibrium temperature in the cylinder.

The answer would be the **same** if the piston were not free to move since it would effect only pressure, and not the specific heats.

(*b*) We take the entire cylinder as our system, which is a closed system. Noting that the cylinder is well-insulated and thus there is no heat transfer, the entropy balance for this closed system can be expressed as

$$\underbrace{S_{in} - S_{out}}_{\substack{\text{Net entropy transfer}\\ \text{by heat and mass}}} + \underbrace{S_{gen}}_{\substack{\text{Entropy}\\ \text{generation}}} = \underbrace{\Delta S_{\text{system}}}_{\substack{\text{Change}\\ \text{in entropy}}}$$

$$0 + S_{gen} = \Delta S_{N_2} + \Delta S_{He} + \Delta S_{piston}$$

But first we determine the final pressure in the cylinder:

$$N_{total} = N_{N_2} + N_{He} = \left(\frac{m}{M}\right)_{N_2} + \left(\frac{m}{M}\right)_{He} = \frac{4.77\text{kg}}{28\text{kg/kmol}} + \frac{0.808\text{kg}}{4\text{kg/kmol}} = 0.372\text{kmol}$$

$$P_2 = \frac{N_{total}R_uT}{V_{total}} = \frac{(0.372\text{kmol})(8.314\text{kPa}\cdot\text{m}^3/\text{kmol}\cdot\text{K})(329\text{K})}{2\text{m}^3} = 508.8\text{kPa}$$

Then,

$$\Delta S_{N_2} = m\left(C_p \ln\frac{T_2}{T_1} - R\ln\frac{P_2}{P_1}\right)_{N_2}$$

$$= (4.77\text{kg})\left[(1.039\text{kJ/kg}\cdot\text{K})\ln\frac{329\text{K}}{353\text{K}} - (0.2968\text{kJ/kg}\cdot\text{K})\ln\frac{508.8\text{kPa}}{500\text{kPa}}\right] = -0.374\text{kJ/K}$$

$$\Delta S_{He} = m\left(C_p \ln\frac{T_2}{T_1} - R\ln\frac{P_2}{P_1}\right)_{He}$$

$$= (0.808\text{kg})\left[(5.1926\text{kJ/kg}\cdot\text{K})\ln\frac{329\text{ K}}{298\text{ K}} - (2.0769\text{kJ/kg}\cdot\text{K})\ln\frac{508.8\text{kPa}}{500\text{kPa}}\right] = 0.386\text{kJ/K}$$

$$\Delta S_{\text{piston}} = \left(mC\ln\frac{T_2}{T_1}\right)_{\text{piston}} = (5\text{kg})(0.386\text{kJ/kg}\cdot\text{K})\ln\frac{329\text{K}}{325.5\text{K}} = 0.021\text{kJ/K}$$

$$S_{gen} = \Delta S_{N_2} + \Delta S_{He} + \Delta S_{\text{piston}} = -0.374 + 0.386 + 0.021 = \mathbf{0.033\ kJ/K}$$

If the piston were not free to move, we would still have T_2 = 329 K but the volume of each gas would remain constant in this case:

$$\Delta S_{N_2} = m\left(C_v \ln\frac{T_2}{T_1} - R\ln\frac{V_2}{V_1}^{\nearrow 0}\right)_{N_2} = (4.77\text{kg})(0.743\text{kJ/kg}\cdot\text{K})\ln\frac{329\text{K}}{353\text{K}} = -0.250\text{kJ/K}$$

$$\Delta S_{He} = m\left(C_v \ln\frac{T_2}{T_1} - R\ln\frac{V_2}{V_1}^{\nearrow 0}\right)_{He} = (0.808\text{kg})(3.1156\text{kJ/kg}\cdot\text{K})\ln\frac{329\text{ K}}{298\text{ K}} = 0.249\text{ kJ/K}$$

$$S_{gen} = \Delta S_{N_2} + \Delta S_{He} + \Delta S_{piston} = -0.250 + 0.249 + 0.021 = \mathbf{0.020kJ/K}$$

6-183 An insulated rigid tank equipped with an electric heater initially contains pressurized air. A valve is opened, and air is allowed to escape at constant temperature until the pressure inside drops to a specified value. The amount of electrical work done during this process and the total entropy change are to be determined.

Assumptions **1** This is an unsteady process since the conditions within the device are changing during the process, but it can be analyzed as a uniform-flow process since the exit temperature (and enthalpy) of air remains constant. **2** Kinetic and potential energies are negligible. **3** The tank is insulated and thus heat transfer is negligible. **4** Air is an ideal gas with variable specific heats.

Properties The gas constant of air is R =0.287 kPa.m³/kg.K (Table A-1). The properties of air are (Table A-17)

$$T_e = 330 \text{ K} \longrightarrow h_e = 330.34 \text{ kJ/kg}$$
$$T_1 = 330 \text{ K} \longrightarrow u_1 = 235.61 \text{ kJ/kg}$$
$$T_2 = 330 \text{ K} \longrightarrow u_2 = 235.61 \text{ kJ/kg}$$

Analysis We take the tank as the system, which is a control volume since mass crosses the boundary. Noting that the microscopic energies of flowing and nonflowing fluids are represented by enthalpy h and internal energy u, respectively, the mass and energy balances for this uniform-flow system can be expressed as

Mass balance: $$m_{in} - m_{out} = \Delta m_{\text{system}} \quad \rightarrow \quad m_e = m_1 - m_2$$

Energy balance: $$\underbrace{E_{in} - E_{out}}_{\substack{\text{Net energy transfer} \\ \text{by heat, work, and mass}}} = \underbrace{\Delta E_{\text{system}}}_{\substack{\text{Change in internal, kinetic,} \\ \text{potential, etc. energies}}}$$

$$W_{e,in} - m_e h_e = m_2 u_2 - m_1 u_1 \quad (\text{since } Q \cong ke \cong pe \cong 0)$$

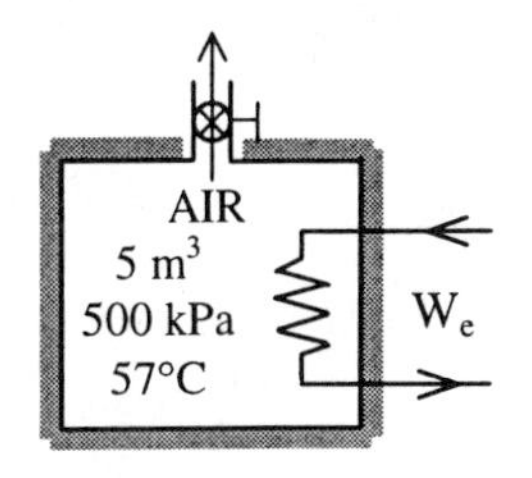

The initial and the final masses of air in the tank are

$$m_1 = \frac{P_1 V}{R T_1} = \frac{(500 \text{ kPa})(5 \text{ m}^3)}{(0.287 \text{ kPa} \cdot \text{m}^3/\text{kg} \cdot \text{K})(330 \text{ K})} = 26.40 \text{kg}$$

$$m_2 = \frac{P_2 V}{R T_2} = \frac{(200 \text{ kPa})(5 \text{ m}^3)}{(0.287 \text{ kPa} \cdot \text{m}^3/\text{kg} \cdot \text{K})(330 \text{ K})} = 10.56 \text{ kg}$$

Then from the mass and energy balances,

$$m_e = m_1 - m_2 = 26.40 - 10.56 = 15.84 \text{ kg}$$

$$\begin{aligned} W_{e,in} &= m_e h_e + m_2 u_2 - m_1 u_1 \\ &= (15.84 \text{ kg})(330.34 \text{ kJ/kg}) + (10.56 \text{ kg})(235.61 \text{ kJ/kg}) - (26.40 \text{ kg})(235.61 \text{ kJ/kg}) \\ &= \mathbf{1501 \text{ kJ}} \end{aligned}$$

(*b*) The total entropy change, or the total entropy generation within the tank boundaries is determined from an entropy balance on the tank expressed as

$$\underbrace{S_{in} - S_{out}}_{\substack{\text{Net entropy transfer} \\ \text{by heat and mass}}} + \underbrace{S_{gen}}_{\substack{\text{Entropy} \\ \text{generation}}} = \underbrace{\Delta S_{\text{system}}}_{\substack{\text{Change} \\ \text{in entropy}}}$$

$$-m_e s_e + S_{\text{gen}} = \Delta S_{\text{tank}}$$

or,

$$\begin{aligned} S_{\text{gen}} &= m_e s_e + \Delta S_{\text{tank}} = m_e s_e + (m_2 s_2 - m_1 s_1) \\ &= (m_1 - m_2) s_e + (m_2 s_2 - m_1 s_1) = m_2 (s_2 - s_e) - m_1 (s_1 - s_e) \end{aligned}$$

Assuming a constant average pressure of (500 + 200)/2 = 350 kPa for the exit stream, the entropy changes are determined to be

$$s_2 - s_e = C_p \ln \frac{T_2}{T_e}^{\nearrow 0} - R \ln \frac{P_2}{P_e} = -R \ln \frac{P_2}{P_e} = -(0.287 \text{kJ/kg} \cdot \text{K}) \ln \frac{200 \text{kPa}}{350 \text{kPa}} = 0.1606 \text{kJ/kg} \cdot \text{K}$$

$$s_1 - s_e = C_p \ln \frac{T_1}{T_e}^{\nearrow 0} - R \ln \frac{P_2}{P_e} = -R \ln \frac{P_1}{P_e} = -(0.287 \text{kJ/kg} \cdot \text{K}) \ln \frac{500 \text{kPa}}{350 \text{kPa}} = -0.1024 \text{kJ/kg} \cdot \text{K}$$

Therefore, the total entropy generated within the tank during this process is

$$S_{gen} = (10.56 \text{ kg})(0.1606 \text{ kJ/kg} \cdot \text{K}) - (26.40 \text{ kg})(-0.1024 \text{kJ/kg} \cdot \text{K}) = \mathbf{4.40 \ kJ/K}$$

6-184 A 1- ton (1000 kg) of water is to be cooled in a tank by pouring ice into it. The final equilibrium temperature in the tank and the entropy generation are to be determined.

Assumptions **1** Thermal properties of the ice and water are constant. **2** Heat transfer to the water tank is negligible. **3** There is no stirring by hand or a mechanical device (it will add energy).

Properties The specific heat of water at room temperature is $C = 4.18$ kJ/kg· °C, and the specific heat of ice at about 0°C is $C = 2.11$ kJ/kg· °C (Table A-3). The melting temperature and the heat of fusion of ice at 1 atm are 0°C and 333.7 kJ/kg..

Analysis (*a*) We take the ice and the water as the system, and disregard any heat transfer between the system and the surroundings. Then the energy balance for this process can be written as

$$\underbrace{E_{in} - E_{out}}_{\substack{\text{Net energy transfer}\\ \text{by heat, work, and mass}}} = \underbrace{\Delta E_{\text{system}}}_{\substack{\text{Change in internal, kinetic,}\\ \text{potential, etc. energies}}}$$

$$0 = \Delta U$$

$$0 = \Delta U_{\text{ice}} + \Delta U_{\text{water}}$$

$$[mC(0^\circ\text{C} - T_1)_{\text{solid}} + mh_{if} + mC(T_2 - 0^\circ\text{C})_{\text{liquid}}]_{\text{ice}} + [mC(T_2 - T_1)]_{\text{water}} = 0$$

Substituting,

$$(80 \text{ kg})\{(2.11 \text{ kJ/kg}\cdot^\circ\text{C})[0-(-5)]^\circ\text{C} + 333.7 \text{ kJ/kg} + (4.18 \text{ kJ/kg}\cdot^\circ\text{C})(T_2 - 0)^\circ\text{C}\}$$
$$+(1000 \text{ kg})(4.18 \text{ kJ/kg}\cdot^\circ\text{C})(T_2 - 20)^\circ\text{C} = 0$$

It gives $T_2 = \mathbf{12.4°C}$

which is the final equilibrium temperature in the tank.

ice
-5°C
80 kg

WATER
1 ton

(*b*) We take the ice and the water as our system, which is a closed system. Considering that the tank is well-insulated and thus there is no heat transfer, the entropy balance for this closed system can be expressed as

$$\underbrace{S_{in} - S_{out}}_{\substack{\text{Net entropy transfer}\\ \text{by heat and mass}}} + \underbrace{S_{gen}}_{\substack{\text{Entropy}\\ \text{generation}}} = \underbrace{\Delta S_{\text{system}}}_{\substack{\text{Change}\\ \text{in entropy}}}$$

$$0 + S_{\text{gen}} = \Delta S_{ice} + \Delta S_{water}$$

where

$$\Delta S_{water} = \left(mC\ln\frac{T_2}{T_1}\right)_{water} = (1000\text{kg})(4.18\text{kJ/kg}\cdot\text{K})\ln\frac{285.4\text{K}}{293\text{K}} = -109.9 \text{ kJ/K}$$

$$\Delta S_{ice} = (\Delta S_{solid} + \Delta S_{melting} + \Delta S_{liquid})_{ice}$$

$$= \left(\left(mC\ln\frac{T_2}{T_1}\right)_{solid} + \frac{mh_{if}}{T_{melting}} + \left(mC\ln\frac{T_2}{T_1}\right)_{liquid}\right)_{ice}$$

$$= (80\text{kg})\left((2.11\text{kJ/kg}\cdot\text{K})\ln\frac{273\text{K}}{268\text{K}} + \frac{333.7\text{kJ/kg}}{273\text{K}} + (4.18\text{kJ/kg}\cdot\text{K})\ln\frac{285.4\text{K}}{273\text{K}}\right)$$

$$= 115.8\text{kJ/K}$$

Then,

$$S_{gen} = \Delta S_{water} + \Delta S_{ice} = -109.9 + 115.8 = \mathbf{5.9\ kJ/K}$$

6-185 An insulated cylinder initially contains a saturated liquid-vapor mixture of water at a specified temperature. The entire vapor in the cylinder is to be condensed isothermally by adding ice inside the cylinder. The amount of ice added and the entropy generation are to be determined.

Assumptions **1** Thermal properties of the ice are constant. **2** The cylinder is well-insulated and thus heat transfer is negligible. **3** There is no stirring by hand or a mechanical device (it will add energy).

Properties The specific heat of ice at about 0°C is C = 2.11 kJ/kg· °C (Table A-3). The melting temperature and the heat of fusion of ice at 1 atm are 0°C and 333.7 kJ/kg.

Analysis (*a*) We take the contents of the cylinder (ice and saturated water) as our system, which is a closed system. Noting that the temperature and thus the pressure remains constant during this phase change process and thus $W_b + \Delta U = \Delta H$, the energy balance for this system can be written as

$$\underbrace{E_{in} - E_{out}}_{\substack{\text{Net energy transfer}\\ \text{by heat, work, and mass}}} = \underbrace{\Delta E_{\text{system}}}_{\substack{\text{Change in internal, kinetic,}\\ \text{potential, etc. energies}}}$$

$$W_{b,in} = \Delta U \quad \rightarrow \quad \Delta H = 0$$

$$\Delta H_{\text{ice}} + \Delta H_{\text{water}} = 0$$

$$[mC(0^\circ\text{C} - T_1)_{\text{solid}} + mh_{if} + mC(T_2 - 0^\circ\text{C})_{\text{liquid}}]_{\text{ice}} + [m(h_2 - h_1)]_{\text{water}} = 0$$

The properties of water at 100°C are (Table A-4)

$$v_f = 0.001044, \quad v_g = 1.6729 \text{ m}^3/\text{kg}$$
$$h_f = 419.04, \quad h_{fg} = 2257.0 \text{ kJ.kg}$$

Then,

$$v_1 = v_f + x_1 v_{fg} = 0.001044 + (0.2)(1.6729 - 0.001044) = 0.3354 \text{ m}^3/\text{kg}$$
$$h_1 = h_f + x_1 h_{fg} = 419.04 + (0.2)(2257.0) = 870.4 \text{ kJ/kg}$$
$$s_1 = s_f + x_1 s_{fg} = 1.3069 + (0.2)(6.048) = 2.5165 \text{ kJ/kg}\cdot\text{K}$$

$$h_2 = h_{f@100^\circ\text{C}} = 419.04 \text{ kJ/kg}$$
$$s_2 = s_{f@100^\circ\text{C}} = 1.3069 \text{ kJ/kg}\cdot\text{K}$$

$$m_{steam} = \frac{V_1}{v_1} = \frac{0.02 \text{ m}^3}{0.3354 \text{ m}^3/\text{kg}} = 0.0596 \text{ kg}$$

Noting that $T_{1,\text{ ice}}$ = -5°C and T_2 = 100°C and substituting gives

m\{(2.11 kJ/kg.K)[0-(-5)] + 333.7 kJ/kg + (4.18 kJ/kg· °C)(100-0)°C\} + (0.0596 kg)(419.04 – 870.4) kJ/kg= 0

m = 0.0353 kg = **35.3 g ice**

(*b*) We take the ice and the steam as our system, which is a closed system. Considering that the tank is well-insulated and thus there is no heat transfer, the entropy balance for this closed system can be expressed as

$$\underbrace{S_{in} - S_{out}}_{\substack{\text{Net entropy transfer} \\ \text{by heat and mass}}} + \underbrace{S_{gen}}_{\substack{\text{Entropy} \\ \text{generation}}} = \underbrace{\Delta S_{\text{system}}}_{\substack{\text{Change} \\ \text{in entropy}}}$$

$$0 + S_{\text{gen}} = \Delta S_{ice} + \Delta S_{steam}$$

where

$$\Delta S_{steam} = m(s_2 - s_1) = (0.0596 \text{ kg})(1.3069 - 2.5165)\text{kJ/kg} \cdot \text{K} = -0.0721 \text{ kJ/K}$$

$$\Delta S_{ice} = \left(\Delta S_{solid} + \Delta S_{melting} + \Delta S_{liquid}\right)_{ice} = \left(\left(mC \ln \frac{T_2}{T_1} \right)_{solid} + \frac{mh_{ig}}{T_{melting}} + \left(mC \ln \frac{T_2}{T_1} \right)_{liquid} \right)_{ice}$$

$$= (0.0353 \text{ kg}) \left((2.11 \text{ kJ/kg} \cdot \text{K}) \ln \frac{273 \text{ K}}{268 \text{ K}} + \frac{333.7 \text{kJ/kg}}{273 \text{K}} + (4.18 \text{ kJ/kg} \cdot \text{K}) \ln \frac{373 \text{ K}}{273 \text{ K}} \right)$$

$$= 0.0906 \text{ kJ/K}$$

Then,

$$S_{gen} = \Delta S_{steam} + \Delta S_{ice} = -0.0721 + 0.0906 = \mathbf{0.0185 \ kJ/K}$$

6-186 An evacuated bottle is surrounded by atmospheric air. A valve is opened, and air is allowed to fill the bottle. The amount of heat transfer through the wall of the bottle when thermal and mechanical equilibrium is established and the amount of entropy generated are to be determined.

Assumptions **1** This is an unsteady process since the conditions within the device are changing during the process, but it can be analyzed as a uniform-flow process since the state of fluid at the inlet remains constant. **2** Air is an ideal gas. **3** Kinetic and potential energies are negligible. **4** There are no work interactions involved. **5** The direction of heat transfer is to the air in the bottle (will be verified).

Properties The gas constant of air is 0.287 kPa.m^3/kg.K (Table A-1).

Analysis We take the bottle as the system, which is a control volume since mass crosses the boundary. Noting that the microscopic energies of flowing and nonflowing fluids are represented by enthalpy h and internal energy u, respectively, the mass and energy balances for this uniform-flow system can be expressed as

Mass balance: $m_{in} - m_{out} = \Delta m_{\text{system}} \quad \to \quad m_i = m_2 \quad (\text{since } m_{out} = m_{initial} = 0)$

Energy balance:

$$\underbrace{E_{in} - E_{out}}_{\substack{\text{Net energy transfer}\\ \text{by heat, work, and mass}}} = \underbrace{\Delta E_{\text{system}}}_{\substack{\text{Change in internal, kinetic,}\\ \text{potential, etc. energies}}}$$

$$Q_{in} + m_i h_i = m_2 u_2 \quad (\text{since } W \cong E_{out} = E_{initial} = ke \cong pe \cong 0)$$

Combining the two balances:

$$Q_{in} = m_2(u_2 - h_i)$$

where

$$m_2 = \frac{P_2 V}{RT_2} = \frac{(100\text{kPa})(0.005\text{m}^3)}{(0.287\text{kPa}\cdot\text{m}^3/\text{kg}\cdot\text{K})(290\text{K})} = 0.0060\text{kg}$$

$$T_i = T_2 = 290\text{K} \xrightarrow{\text{Table A-17}} \begin{array}{l} h_i = 290.16\text{kJ/kg} \\ u_2 = 206.91\text{kJ/kg} \end{array}$$

10 kPa
17°C
5 L
Evacuated

Substituting,

$$Q_{in} = (0.0060 \text{ kg})(206.91 - 290.16) \text{ kJ/kg} = -0.5 \text{ kJ} \quad \to \quad Q_{out} = \mathbf{0.5 \text{ kJ}}$$

Note that the negative sign for heat transfer indicates that the assumed direction is wrong. Therefore, we reverse the direction.

The entropy generated during this process is determined by applying the entropy balance on an *extended system* that includes the bottle and its immediate surroundings so that the boundary temperature of the extended system is the temperature of the surroundings at all times. The entropy balance for it can be expressed as

$$\underbrace{S_{in} - S_{out}}_{\substack{\text{Net entropy transfer}\\ \text{by heat and mass}}} + \underbrace{S_{gen}}_{\substack{\text{Entropy}\\ \text{generation}}} = \underbrace{\Delta S_{\text{system}}}_{\substack{\text{Change}\\ \text{in entropy}}}$$

$$m_i s_i - \frac{Q_{out}}{T_{b,in}} + S_{\text{gen}} = \Delta S_{\text{tank}} = m_2 s_2 - m_1 s_1^{\nearrow 0} = m_2 s_2$$

Therefore, the total entropy generated during this process is

$$S_{gen} = -m_i s_i + m_2 s_2 + \frac{Q_{\text{out}}}{T_{\text{b,out}}} = m_2(s_2 - s_i)^{\nearrow 0} + \frac{Q_{\text{out}}}{T_{\text{b,out}}} = \frac{Q_{out}}{T_{surr}} = \frac{0.5\text{kJ}}{290\text{K}} = \mathbf{0.0017kJ/K}$$

6-187 (a) Water is heated from 16°C to 43°C by an electric resistance heater placed in the water pipe as it flows through a showerhead steadily at a rate of 10 L/min. The electric power input to the heater and the rate of entropy generation are to be determined. The reduction in power input and entropy generation as a result of installing a 50% efficient regenerator are also to be determined.

Assumptions **1** This is a steady-flow process since there is no change with time at any point within the system and thus $\Delta m_{CV} = 0$ and $\Delta E_{CV} = 0$. **2** Water is an incompressible substance with constant specific heats. **3** The kinetic and potential energy changes are negligible, $\Delta ke \cong \Delta pe \cong 0$. **4** Heat losses from the pipe are negligible.

Properties The density of water is given to be ρ = 1 kg/L. The specific heat of water at room temperature is $C = 4.18$ kJ/kg· °C (Table A-3).

Analysis We take the pipe as the system. This is a *control volume* since mass crosses the system boundary during the process. We observe that there is only one inlet and one exit and thus $\dot{m}_1 = \dot{m}_2 = \dot{m}$. Then the energy balance for this steady-flow system can be expressed in the rate form as

$$\underbrace{\dot{E}_{in} - \dot{E}_{out}}_{\text{Rate of net energy transfer by heat, work, and mass}} = \underbrace{\Delta \dot{E}_{\text{system}}^{\nearrow 0 \text{ (steady)}}}_{\text{Rate of change in internal, kinetic, potential, etc. energies}} = 0 \rightarrow \dot{E}_{in} = \dot{E}_{out}$$

$$\dot{W}_{\text{e,in}} + \dot{m}h_1 = \dot{m}h_2 \quad (\text{since } \Delta\text{ke} \cong \Delta\text{pe} \cong 0)$$

$$\dot{W}_{\text{e,in}} = \dot{m}(h_2 - h_1) = \dot{m}C(T_2 - T_1)$$

WATER

16°C → 43°C

where

$$\dot{m} = \rho\dot{V} = (1 \text{ kg/L})(10 \text{ L/min}) = 10 \text{ kg/min}$$

Substituting,

$$\dot{W}_{\text{e,in}} = (10/60\text{kg/s})(4.18 \text{ kJ/kg}\cdot^\circ\text{C})(43-16)^\circ\text{C} = \mathbf{18.8\ kW}$$

The rate of entropy generation in the heating section during this process is determined by applying the entropy balance on the heating section. Noting that this is a steady-flow process and heat transfer from the heating section is negligible,

$$\underbrace{\dot{S}_{in} - \dot{S}_{out}}_{\text{Rate of net entropy transfer by heat and mass}} + \underbrace{\dot{S}_{gen}}_{\text{Rate of entropy generation}} = \underbrace{\Delta \dot{S}_{\text{system}}^{\nearrow 0}}_{\text{Rate of change of entropy}} = 0$$

$$\dot{m}s_1 - \dot{m}s_2 + \dot{S}_{\text{gen}} = 0$$

$$\dot{S}_{\text{gen}} = \dot{m}(s_2 - s_1)$$

Noting that water is an incompressible substance and substituting,

$$\dot{S}_{gen} = \dot{m}C\ln\frac{T_2}{T_1} = (10/60 \text{ kg/s})(4.18 \text{ kJ/kg}\cdot\text{K})\ln\frac{316 \text{ K}}{289 \text{ K}} = \mathbf{0.0622\ kJ/K}$$

(*b*) The energy recovered by the heat exchanger is

$$\begin{aligned}\dot{Q}_{\text{saved}} &= \varepsilon\dot{Q}_{\max} = \varepsilon\dot{m}C(T_{\max} - T_{\min}) \\ &= 0.5(10/60\text{kg/s})(4.18 \text{ kJ/kg}^\circ\text{C})(39-16)^\circ\text{C} \\ &= 8.0\text{kJ/s} = 8.0\text{kW}\end{aligned}$$

Therefore, 8.0 kW less energy is needed in this case, and the required electric power in this case reduces to

$$\dot{W}_{\text{in,new}} = \dot{W}_{\text{in,old}} - \dot{Q}_{\text{saved}} = 18.8 - 8.0 = \mathbf{10.8\ kW}$$

Taking the cold water stream in the heat exchanger as our control volume (a steady-flow system), the temperature at which the cold water leaves the heat exchanger and enters the electric resistance heating section is determined to be

$$\dot{Q} = \dot{m}C(T_{c,out} - T_{c,in})$$

Substituting,

$$8 \text{ kJ/s} = (10/60 \text{ kg/s})(4.184 \text{ kJ/kg}\cdot^\circ\text{C})(T_{c,out} - 16^\circ\text{C})$$

It yields

$$T_{c,out} = 27.5^\circ\text{C} = 300.5 \text{ K}$$

The rate of entropy generation in the heating section in this case is determined similarly to be

$$\dot{S}_{gen} = \dot{m}C\ln\frac{T_2}{T_1} = (10/60 \text{ kg/s})(4.18 \text{ kJ/kg}\cdot\text{K})\ln\frac{316 \text{ K}}{300.5 \text{ K}} = \mathbf{0.0351 kJ/K}$$

Thus the reduction in the rate of entropy generation within the heating section is

$$\dot{S}_{reduction} = 0.0622 - 0.0350 = \mathbf{0.0272 \ \ kW/K}$$

6-188 EES solution of this (and other comprehensive problems designated with the *computer icon*) is available to instructors at the *Instructor Manual* section of the *Online Learning Center* (OLC) at www.mhhe.com/cengel-boles. See the Preface for access information.

Fundamentals of Engineering (FE) Exam Problems

6-189 Steam is condensed at a constant temperature of 30°C as it flows through the condenser of a power plant by rejecting heat at a rate of 55 MW. The rate of entropy change of steam as it flows through the condenser is

(a) –1.83 MW/K (b) –0.18 MW/K (c) 0 MW/K (d) 0.56 MW/K (e) 1.22 MW/K

Answer (b) –0.18 MW/K

Solution Solved by EES Software. Solutions can be verified by copying-and-pasting the following lines on a blank EES screen. (Similar problems and their solutions can be obtained easily by modifying numerical values).

```
T1=30 "C"
Q_out=55 "MW"

S_change=-Q_out/(T1+273) "MW/K"

"Some Wrong Solutions with Common Mistakes:"
W1_S_change=0 "Assuming no change"
W2_S_change=Q_out/T1 "Using temperature in C"
W3_S_change=Q_out/(T1+273) "Wrong sign"
W4_S_change=-s_fg "Taking entropy of vaporization"
s_fg=(ENTROPY(Steam_NBS,T=T1,x=1)-ENTROPY(Steam_NBS,T=T1,x=0))
```

6-190 Steam is compressed from 8 MPa and 300°C to 10 MPa isentropically. The final temperature of the steam is

(a) 290°C (b) 300°C (c) 320°C (d) 330°C (e) 340°C

Answer (d) 330°C

Solution Solved by EES Software. Solutions can be verified by copying-and-pasting the following lines on a blank EES screen. (Similar problems and their solutions can be obtained easily by modifying numerical values).

```
P1=8000 "kPa"
T1=300 "C"

P2=10000 "kPa"
s2=s1

s1=ENTROPY(Steam_NBS,T=T1,P=P1)
T2=TEMPERATURE(Steam_NBS,s=s2,P=P2)

"Some Wrong Solutions with Common Mistakes:"
W1_T2=T1 "Assuming temperature remains constant"
W2_T2=TEMPERATURE(Steam_NBS,x=0,P=P2) "Saturation temperature at P2"
W3_T2=TEMPERATURE(Steam_NBS,x=0,P=P2) "Saturation temperature at P1"
```

6-191 An apple with an average mass of 0.15 kg and average specific heat of 3.65 kJ/kg.°C is cooled from 20°C to 5°C. The entropy change of the apple is

(a) –0.0288 kJ/K (b) –0.192 kJ/K (c) -0.526 kJ/K (d) 0 kJ/K (e) 0.657 kJ/K

Answer (a) –0.0288 kJ/K

Solution Solved by EES Software. Solutions can be verified by copying-and-pasting the following lines on a blank EES screen. (Similar problems and their solutions can be obtained easily by modifying numerical values).

```
C=3.65 "kJ/kg.K"
m=0.15 "kg"
T1=20 "C"
T2=5 "C"

S_change=m*C*ln((T2+273)/(T1+273))

"Some Wrong Solutions with Common Mistakes:"
W1_S_change=C*ln((T2+273)/(T1+273)) "Not using mass"
W2_S_change=m*C*ln(T2/T1) "Using C"
W3_S_change=m*C*(T2-T1) "Using Wrong relation"
```

6-192 A piston-cylinder device contains 5 kg of saturated water vapor at 3.5 MPa. Now heat is rejected from the cylinder at constant pressure until the water vapor completely condenses so that the cylinder contains saturated liquid at 3.5 MPa at the end of the process. The entropy change of the system during this process is

(a) 0 kJ/K (b) -3.4 kJ/K (c) -10.5 kJ/K (d) -17.0 kJ/K (e) -17.5 kJ/K

Answer (d) -17.0 kJ/K

Solution Solved by EES Software. Solutions can be verified by copying-and-pasting the following lines on a blank EES screen. (Similar problems and their solutions can be obtained easily by modifying numerical values).

```
P1=3500 "kPa"
m=5 "kg"

s_fg=(ENTROPY(Steam_NBS,P=P1,x=1)-ENTROPY(Steam_NBS,P=P1,x=0))

S_change=-m*s_fg "kJ/K"
```

6-193 Helium gas is compressed from 1 atm and 25°C to a pressure of 10 atm adiabatically. The lowest temperature of helium after compression is

(a) 25°C (b) 63°C (c) 250°C (d) 384°C (e) 476°C

Answer (e) 476°C

Solution Solved by EES Software. Solutions can be verified by copying-and-pasting the following lines on a blank EES screen. (Similar problems and their solutions can be obtained easily by modifying numerical values).

```
k=1.667
P1=101.325 "kPa"
T1=25 "C"

P2=10*101.325 "kPa"
"s2=s1"

"The exit temperature will be lowest for isentropic compression,"
T2=(T1+273)*(P2/P1)^((k-1)/k) "K"
T2_C= T2-273 "C"

"Some Wrong Solutions with Common Mistakes:"
W1_T2=T1 "Assuming temperature remains constant"
W2_T2=T1*(P2/P1)^((k-1)/k) "Using C instead of K"
W3_T2=(T1+273)*(P2/P1)-273 "Assuming T is proportional to P"
W4_T2=T1*(P2/P1) "Assuming T is proportional to P, using C"
```

6-194 Steam expands in an adiabatic turbine from 10 MPa and 500°C to 0.1 MPa at a rate of 8 kg/s. If steam leaves the turbine as saturated vapor, the power output of the turbine is

(a) 5586 kW (b) 698 kW (c) 2136 kW (d) 12,452 kW (e) 26,990 kW

Answer (a) 5586 kW

Solution Solved by EES Software. Solutions can be verified by copying-and-pasting the following lines on a blank EES screen. (Similar problems and their solutions can be obtained easily by modifying numerical values).

```
P1=10000 "kPa"
T1=500 "C"

P2=100 "kPa"
x2=1

m=8 "kg/s"

h1=ENTHALPY(Steam_NBS,T=T1,P=P1)
h2=ENTHALPY(Steam_NBS,x=x2,P=P2)

W_out=m*(h1-h2)

"Some Wrong Solutions with Common Mistakes:"
s1=ENTROPY(Steam_NBS,T=T1,P=P1)
h2s=ENTHALPY(Steam_NBS, s=s1,P=P2)
W1_Wout=m*(h1-h2s) "Assuming isentropic expansion"
```

6-195 Argon gas expands in an adiabatic turbine from 3 MPa and 750°C to 0.2 MPa at a rate of 5 kg/s. The maximum power output of the turbine is

(a) 1.06 MW (b) 1.29 MW (c) 1.43 MW (d) 1.76 MW (e) 2.08 MW

Answer (d) 1.76 MW

Solution Solved by EES Software. Solutions can be verified by copying-and-pasting the following lines on a blank EES screen. (Similar problems and their solutions can be obtained easily by modifying numerical values).

```
Cp=0.5203
k=1.667
P1=3000 "kPa"
T1=750 "C"
m=5 "kg/s"
P2=200 "kPa"
"s2=s1"

T2=(T1+750)*(P2/P1)^((k-1)/k)

W_max=m*Cp*(T1-T2)

"Some Wrong Solutions with Common Mistakes:"
Cv=0.2081"kJ/kg.K"
W1_Wmax=m*Cv*(T1-T2) "Using Cv"

T22=T1*(P2/P1)^((k-1)/k) "Using C instead of K"
W2_Wmax=m*Cp*(T1-T22)

W3_Wmax=Cp*(T1-T2) "Not using mass flow rate"

T24=T1*(P2/P1) "Assuming T is proportional to P, using C"
W4_Wmax=m*Cp*(T1-T24)
```

6-196 A unit mass of a substance undergoes an irreversible process from state 1 to state 2 while gaining heat from the surroundings at temperature T in the amount of q. If the entropy of the substance is s_1 at state 1, and s_2 at state 2, the entropy change of the substance Δs during this process is

(a) $\Delta s < s_2 - s_1$ (b) $\Delta s > s_2 - s_1$ (c) $\Delta s = s_2 - s_1$ (d) $\Delta s = s_2 - s_1 + q/T$ (e) $\Delta s > s_2 - s_1 + q/T$

Answer (c) $\Delta s = s_2 - s_1$

6-197 A unit mass of an ideal gas at temperature T undergoes a reversible isothermal process from pressure P_1 to pressure P_2 while loosing heat to the surroundings at temperature T in the amount of q. If the gas constant of the gas is R, the entropy change of the gas Δs during this process is

(a) $\Delta s = R \ln(P_2/P_1)$ (b) $\Delta s = R \ln(P_2/P_1)$- q/T (c) $\Delta s = R \ln(P_1/P_2)$ (d) $\Delta s = R \ln(P_1/P_2)$-q/T
(e) $\Delta s = 0$

Answer (c) $\Delta s = R \ln(P_1/P_2)$

6-198 Air is compressed from room conditions to a specified pressure in a reversible manner by two compressors: one isothermal and the other adiabatic. If the entropy change of air is Δs_{isot} during the reversible isothermal compression, and Δs_{adia} during the reversible adiabatic compression, the correct statement regarding entropy change of air per unit mass is

(a) $\Delta s_{isot} = \Delta s_{adia} = 0$ (b) $\Delta s_{isot} = \Delta s_{adia} > 0$ (c) $\Delta s_{adia} > 0$ (d) $\Delta s_{isot} < 0$ (e) $\Delta s_{isot} = 0$

Answer (d) $\Delta s_{isot} < 0$

6-199 Helium gas is compressed from 20°C and 6.20 m^3/kg to 0.775 m^3/kg in a reversible adiabatic manner. The temperature of helium after compression is (corrected and checked)

(a) 160°C (b) 80°C (c) 400°C (d) 46°C (e) 900°C

Answer (e) 900°C

Solution Solved by EES Software. Solutions can be verified by copying-and-pasting the following lines on a blank EES screen. (Similar problems and their solutions can be obtained easily by modifying numerical values).

```
k=1.667
v1=6.20 "m^3/kg"
T1=20 "C"

v2=0.775 "m^3/kg"
"s2=s1"

"The exit temperature is determined from isentropic compression relation,"
T2=(T1+273)*(v1/v2)^(k-1) "K"
T2_C= T2-273 "C"

"Some Wrong Solutions with Common Mistakes:"
W1_T2=T1 "Assuming temperature remains constant"
W2_T2=T1*(v1/v2)^(k-1) "Using C instead of K"
W3_T2=(T1+273)*(v1/v2)-273 "Assuming T is proportional to v"
W4_T2=T1*(v1/v2) "Assuming T is proportional to v, using C"
```

6-200 Heat is lost through a plane wall steadily at a rate of 600 W. If the inner and outer surface temperatures of the wall are 20°C and 5°C, respectively, the rate of entropy generation within the wall is

(a) 0.11 W/K (b) 4.21 W/K (c) 2.10 W/K (d) 42.1 W/K (e) 90.0 W/K

Answer (a) 0.11 W/K

Solution Solved by EES Software. Solutions can be verified by copying-and-pasting the following lines on a blank EES screen. (Similar problems and their solutions can be obtained easily by modifying numerical values).

```
Q=600 "W"
T1=20+273 "K"
T2=5+273 "K"

"Entropy balance S_in - S_out + S_gen= DS_system for the wall for steady operation
gives"
Q/T1-Q/T2+S_gen=0  "W/K"

"Some Wrong Solutions with Common Mistakes:"
Q/(T1+273)-Q/(T2+273)+W1_Sgen=0 "Using C instead of K"
W2_Sgen=Q/((T1+T2)/2)  "Using avegage temperature in K"
W3_Sgen=Q/((T1+T2)/2-273)  "Using avegage temperature in C"
W4_Sgen=Q/(T1-T2+273)  "Using temperature difference in K"
```

6-201 Air is compressed steadily and adiabatically from 20°C and 100 kPa to 200°C and 500 kPa. Assuming constant specific heats for air at room temperature, the isentropic efficiency of the compressor is (corrected)

(a) 0.67 (b) 0.78 (c) 0.89 (d) 0.95 (e) 1.00

Answer (d) 0.95

Solution Solved by EES Software. Solutions can be verified by copying-and-pasting the following lines on a blank EES screen. (Similar problems and their solutions can be obtained easily by modifying numerical values).

```
Cp=1.005 "kJ/kg.K"
k=1.4
P1=100 "kPa"
T1=20 "C"

P2=500 "kPa"
T2=200 "C"
T2s=(T1+273)*(P2/P1)^((k-1)/k)-273
Eta_comp=(Cp*(T2s-T1))/(Cp*(T2-T1))

"Some Wrong Solutions with Common Mistakes:"
T2sW1=T1*(P2/P1)^((k-1)/k) "Using C instead of K in finding T2s"
W1_Eta_comp=(Cp*(T2sW1-T1))/(Cp*(T2-T1))

W2_Eta_comp=T2s/T2 "Using wrong definition for isentropic efficiency, and using C"
W3_Eta_comp=(T2s+273)/(T2+273) "Wrong definition for isentr. efficiency, with K"
```

6-202 Argon gas expands in an adiabatic turbine steadily from 500°C and 800 kPa to 80 kPa at a rate of 2.5 kg/s. For an isentropic efficiency of 80%, the power produced by the turbine is

(a) 194 kW (b) 291 kW (c) 484 kW (d) 363 kW (e) 605 kW

Answer (c) 484 kW

Solution Solved by EES Software. Solutions can be verified by copying-and-pasting the following lines on a blank EES screen. (Similar problems and their solutions can be obtained easily by modifying numerical values).

```
Cp=0.5203 "kJ/kg-K"
k=1.667
m=2.5 "kg/s"
T1=500 "C"
P1=800 "kPa"

P2=80 "kPa"

T2s=(T1+273)*(P2/P1)^((k-1)/k)-273

Eta_turb=0.8
Eta_turb=(Cp*(T2-T1))/(Cp*(T2s-T1))

W_out=m*Cp*(T1-T2)

"Some Wrong Solutions with Common Mistakes:"
T2sW1=T1*(P2/P1)^((k-1)/k) "Using C instead of K to find T2s"
Eta_turb=(Cp*(T2W1-T1))/(Cp*(T2sW1-T1))
W1_Wout=m*Cp*(T1-T2W1)

Eta_turb=(Cp*(T2s-T1))/(Cp*(T2W2-T1)) "Using wrong definition for isentropic
efficiency, and using C"
W2_Wout=m*Cp*(T1-T2W2)

W3_Wout=Cp*(T1-T2) "Not using mass flow rate"

Cv=0.3122 "kJ/kg.K"
W4_Wout=m*Cv*(T1-T2) "Using Cv instead of Cp"
```

6-203 Water enters a pump steadily at 100 kPa at a rate of 50 L/s and leaves at 900 kPa. The flow velocities at the inlet and the exit are the same, but the pump exit where the discharge pressure is measured is 6.1 m above the inlet section. The minimum power input to the pump is

(a) 37 kW (b) 40 kW (c) 43 kW (d) 52 kW (e) 74 kW

Answer (c) 43 kW

Solution Solved by EES Software. Solutions can be verified by copying-and-pasting the following lines on a blank EES screen. (Similar problems and their solutions can be obtained easily by modifying numerical values).

```
V=0.050 "m^3/s"
g=9.81 "m/s^2"
h=6.1 "m"
P1=100 "kPa"
T1=20 "C"

P2=900 "kPa"

"Pump power input is minimum when compression is reversible and thus w=v(P2-
P1)+Dpe"

v1=VOLUME(Steam_NBS,T=T1,P=P1)
m=V/v1
W_min=m*v1*(P2-P1)+m*g*h/1000 "kPa.m^3/s=kW"

"(The effect of 6.1 m elevation difference turns out to be small)"

"Some Wrong Solutions with Common Mistakes:"
W1_Win=m*v1*(P2-P1) "Disregarding potential energy"
W2_Win=m*v1*(P2-P1)-m*g*h/1000 "Subtracting potential energy instead of adding"
W3_Win=m*v1*(P2-P1)+m*g*h "Not using the conversion factor 1000 in PE term"
W4_Win=m*v1*(P2+P1)+m*g*h/1000 "Adding pressures instead of subtracting"
```

6-204 Air at 15°C is compressed steadily and isothermally from 100 kPa to 700 kPa at a rate of 0.12 kg/s. The minimum power input to the compressor is

(a) 1.0 kW (b) 11.2 kW (c) 25.8 kW (d) 19.3 kW (e) 161 kW

Answer (d) 19.3 kW

Solution Solved by EES Software. Solutions can be verified by copying-and-pasting the following lines on a blank EES screen. (Similar problems and their solutions can be obtained easily by modifying numerical values).

```
Cp=1.005 "kJ/kg.K"
R=0.287 "kJ/kg.K"
Cv=0.718 "kJ/kg.K"
k=1.4

P1=100 "kPa"
T=15 "C"
m=0.12 "kg/s"

P2=700 "kPa"

Win=m*R*(T+273)*ln(P2/P1)

"Some Wrong Solutions with Common Mistakes:"
W1_Win=m*R*T*ln(P2/P1) "Using C instead of K"
W2_Win=m*T*(P2-P1) "Using wrong relation"
W3_Win=R*(T+273)*ln(P2/P1) "Not using mass flow rate"
```

6-205 Air is to be compressed steadily and isentropically from 1 atm to 16 atm by a two-stage compressor. To minimize the total compression work, the intermediate pressure between the two stages must be

(a) 2 atm (b) 4 atm (c) 8 atm (d) 10 atm (e) 12 atm

Answer (b) 4 atm

Solution Solved by EES Software. Solutions can be verified by copying-and-pasting the following lines on a blank EES screen. (Similar problems and their solutions can be obtained easily by modifying numerical values).

```
P1=1 "atm"
P2=16 "atm"

P_mid=SQRT(P1*P2)

"Some Wrong Solutions with Common Mistakes:"
W1_P=(P1+P2)/2 "Using average pressure"
W2_P=P1*P2/2 "Half of product"
```

6-206 Helium gas enters an adiabatic nozzle steadily at 500°C and 600 kPa with a low velocity, and exits at a pressure of 90 kPa. The highest possible velocity of helium gas at the nozzle exit is

(a) 1475 m/s (b) 1660 m/s (c) 1830 m/s (d) 2066 m/s (e) 3040 m/s

Answer (d) 2066 m/s

Solution Solved by EES Software. Solutions can be verified by copying-and-pasting the following lines on a blank EES screen. (Similar problems and their solutions can be obtained easily by modifying numerical values).

```
k=1.667
Cp=5.1926 "kJ/kg.K"
Cv=3.1156 "kJ/kg.K"
T1=500 "C"
P1=600 "kPa"
Vel1=0

P2=90 "kPa"
"s2=s1 for maximum exit velocity"

"The exit velocity will be highest for isentropic expansion,"
T2=(T1+273)*(P2/P1)^((k-1)/k)-273 "C"

"Energy balance for this case is h+ke=constant for the fluid stream (Q=W=pe=0)"
(0.5*Vel1^2)/1000+Cp*T1=(0.5*Vel2^2)/1000+Cp*T2

"Some Wrong Solutions with Common Mistakes:"
T2a=T1*(P2/P1)^((k-1)/k) "Using C for temperature"
(0.5*Vel1^2)/1000+Cp*T1=(0.5*W1_Vel2^2)/1000+Cp*T2a

T2b=T1*(P2/P1)^((k-1)/k) "Using Cv"
(0.5*Vel1^2)/1000+Cv*T1=(0.5*W2_Vel2^2)/1000+Cv*T2b

T2c=T1*(P2/P1)^k "Using wrong relation"
(0.5*Vel1^2)/1000+Cp*T1=(0.5*W3_Vel2^2)/1000+Cp*T2c
```

6-207 Combustion gases with a specific heat ratio of 1.3 enter an adiabatic nozzle steadily at 600°C and 800 kPa with a low velocity, and exit at a pressure of 85 kPa. The lowest possible temperature of combustion gases at the nozzle exit is

(a) 64°C (b) 187°C (c) 247°C (d) 316°C (e) 358°C

Answer (c) 247°C

Solution Solved by EES Software. Solutions can be verified by copying-and-pasting the following lines on a blank EES screen. (Similar problems and their solutions can be obtained easily by modifying numerical values).

```
k=1.3
T1=600 "C"
P1=800 "kPa"
P2=85 "kPa"

"Nozzle exit temperature will be lowest for isentropic operation"
T2=(T1+273)*(P2/P1)^((k-1)/k)-273

"Some Wrong Solutions with Common Mistakes:"
W1_T2=T1*(P2/P1)^((k-1)/k) "Using C for temperature"
W2_T2=(T1+273)*(P2/P1)^((k-1)/k) "Not converting the answer to C"
W3_T2=T1*(P2/P1)^k "Using wrong relation"
```

6-208 Steam enters an adiabatic turbine steadily at 400°C and 3 MPa, and leaves at 50 kPa. The highest possible percentage of mass of steam that condenses at the turbine exit and leaves the turbine as a liquid is

(a) 6% (b) 10% (c) 15% (d) 21% (e) 0%

Answer (b) 10%

Solution Solved by EES Software. Solutions can be verified by copying-and-pasting the following lines on a blank EES screen. (Similar problems and their solutions can be obtained easily by modifying numerical values).

```
P1=3000 "kPa"
T1=400 "C"

P2=50 "kPa"
s2=s1

s1=ENTROPY(Steam_NBS,T=T1,P=P1)
x2=QUALITY(Steam_NBS,s=s2,P=P2)
misture=1-x2

"Checking x2 using data from table"
x2_table=(6.9212-1.091)/6.5029
```

6-209 Liquid water enters an adiabatic piping system at 15°C at a rate of 5 kg/s. If the water temperature rises by 0.5°C during flow due to friction, the rate of entropy generation in the pipe is

(a) 36 W/K (b) 29 W/K (c) 685 W/K (d) 920 W/K (e) 8370 W/K

Answer (a) 36 W/K

Solution Solved by EES Software. Solutions can be verified by copying-and-pasting the following lines on a blank EES screen. (Similar problems and their solutions can be obtained easily by modifying numerical values).

```
Cp=4.18 "kJ/kg.K"
m=5 "kg/s"
T1=15 "C"
T2=15.5 "C"
S_gen=m*Cp*ln((T2+273)/(T1+273)) "kW/K"

"Some Wrong Solutions with Common Mistakes:"
W1_Sgen=m*Cp*ln(T2/T1) "Using deg. C"
W2_Sgen=Cp*ln(T2/T1) "Not using mass flow rate with deg. C"
W3_Sgen=Cp*ln((T2+273)/(T1+273)) "Not using mass flow rate with deg. C"
```

6-210 Liquid water is to be compressed by a pump whose isentropic efficiency is 75 percent from 0.2 MPa to 5 MPa at a rate of 0.15 m^3/min. The required power input to this pump is

(*a*) 4.8 kW (*b*) 6.4 kW (*c*) 9.0 kW (*d*) 16.0 kW (*e*) 12.0 kW

Answer (*d*) 16.0 kW

Solution Solved by EES Software. Solutions can be verified by copying-and-pasting the following lines on a blank EES screen. (Similar problems and their solutions can be obtained easily by modifying numerical values).

```
V=0.15/60 "m^3/s"
rho=1000 "kg/m^3"
v1=1/rho
m=rho*V "kg/s"
P1=200 "kPa"
Eta_pump=0.75
P2=5000 "kPa"

"Reversible pump power input is w =mv(P2-P1) = V(P2-P1)"

W_rev=m*v1*(P2-P1) "kPa.m^3/s=kW"
W_pump=W_rev/Eta_pump

"Some Wrong Solutions with Common Mistakes:"
W1_Wpump=W_rev*Eta_pump "Multiplying by efficiency"
W2_Wpump=W_rev "Disregarding efficiency"
W3_Wpump=m*v1*(P2+P1)/Eta_pump "Adding pressures instead of subtracting"
```

6-211 Steam enters an adiabatic turbine at 4 MPa and 500°C at a rate of 15 kg/s, and exits at 0.2 MPa and 300°C. The rate of entropy generation in the turbine is

(*a*) 0.8 kW/K (*b*) 1.2 kW/K (*c*) 12.0 kW/K (*d*) 15.1 kW/K (*e*) 17.4 kW/K

Answer (*c*) 12.0 kW/K

Solution Solved by EES Software. Solutions can be verified by copying-and-pasting the following lines on a blank EES screen. (Similar problems and their solutions can be obtained easily by modifying numerical values).

```
P1=4000 "kPa"
T1=500 "C"
m=15 "kg/s"
P2=200 "kPa"
T2=300 "C"
s1=ENTROPY(Steam_NBS,T=T1,P=P1)
s2=ENTROPY(Steam_NBS,T=T2,P=P2)
S_gen=m*(s2-s1) "kW/K"

"Some Wrong Solutions with Common Mistakes:"
W1_Sgen=0 "Assuming isentropic expansion"
```

6-212 Helium gas is compressed steadily from 90 kPa and 25°C to 600 kPa at a rate of 2 kg/min by an adiabatic compressor. If the compressor consumes 70 kW of power while operating, the isentropic efficiency of this compressor is

(*a*) 56.7% (*b*) 83.7% (*c*) 75.4% (*d*) 92.1% (*e*) 100.0%

Answer (*b*) 83.7%

Solution Solved by EES Software. Solutions can be verified by copying-and-pasting the following lines on a blank EES screen. (Similar problems and their solutions can be obtained easily by modifying numerical values).

```
Cp=5.1926 "kJ/kg-K"
Cv=3.1156 "kJ/kg.K"
k=1.667
m=2/60 "kg/s"
T1=25 "C"
P1=90 "kPa"
P2=600 "kPa"
W_comp=70 "kW"
T2s=(T1+273)*(P2/P1)^((k-1)/k)-273
W_s=m*Cp*(T2s-T1)
Eta_comp=W_s/W_comp

"Some Wrong Solutions with Common Mistakes:"
T2sA=T1*(P2/P1)^((k-1)/k) "Using C instead of K"
W1_Eta_comp=m*Cp*(T2sA-T1)/W_comp
W2_Eta_comp=m*Cv*(T2s-T1)/W_comp "Using Cv instead of Cp"
```

6-213 ... 6-216 Design and Essay Problems

Chapter 7
EXERGY – A MEASURE OF WORK POTENTIAL

Exergy, Irreversibility, Reversible Work, and Second-Law Efficiency

7-1C Reversible work differs from the useful work by irreversibilities. For reversible processes both are identical. $W_u = W_{rev} - I$.

7-2C Reversible work and irreversibility are identical for processes that involve no actual useful work.

7-3C The dead state.

7-4C Yes; exergy is a function of the state of the surroundings as well as the state of the system.

7-5C Useful work differs from the actual work by the surroundings work. They are identical for systems that involve no surroundings work such as steady-flow systems.

7-6C Yes.

7-7C No, not necessarily. The well with the higher temperature will have a higher exergy.

7-8C The system that is at the temperature of the surroundings has zero exergy. But the system that is at a lower temperature than the surroundings has some exergy since we can run a heat engine between these two temperature levels.

7-9C They would be identical.

7-10C The second-law efficiency is a measure of the performance of a device relative to its performance under reversible conditions. It differs from the first law efficiency in that it is not a conversion efficiency.

7-11C No. The power plant that has a lower thermal efficiency may have a higher second-law efficiency.

7-12C No. The refrigerator that has a lower COP may have a higher second-law efficiency.

7-13C A processes with $W_{rev} = 0$ is reversible if it involves no actual useful work. Otherwise it is irreversible.

7-14C Yes.

7-15 Windmills are to be installed at a location with steady winds to generate power. The minimum number of windmills that need to be installed is to be determined.

Assumptions Air is at standard conditions of 1 atm and 25°C

Properties The gas constant of air is 0.287 kPa.^{3}/kg.K.

Analysis The exergy or work potential of the blowing air is the kinetic energy it possesses,

$$\text{Exergy} = ke_i = \frac{\mathbf{V}_1^2}{2} = \frac{(12\ \text{m/s})^2}{2}\left(\frac{1\ \text{kJ/kg}}{1000\ \text{m}^2/\text{s}^2}\right) = 0.072\ \text{kJ/kg}$$

At standard atmospheric conditions (25°C, 101 kPa), the density and the mass flow rate of air are

$$\rho = \frac{P}{RT} = \frac{101\ \text{kPa}}{(0.287\ \text{kPa}\cdot\text{m}^3/\text{kg}\cdot\text{K})(298\ \text{K})} = 1.18\ \text{m}^3/\text{kg}$$

and

$$\dot{m} = \rho A V_1 = \rho\frac{\pi D^2}{4}\mathbf{V}_1 = (1.18\ \text{kg/m}^3)(\pi/4)(10\ \text{m})^2(12\ \text{m/s}) = 1{,}112\ \text{kg/s}$$

Thus,

$$\textit{Available Power} = \dot{m}\,ke_1 = (1{,}112\ \text{kg/s})(0.072\ \text{kJ/kg}) = 80.1\ \text{kW}$$

The minimum number of windmills that needs to be installed is

$$N = \frac{\dot{W}_{total}}{\dot{W}} = \frac{600\ \text{kW}}{80.1\ \text{kW}} = 7.49 \cong \mathbf{8\ windmills}$$

7-16 Water is to be pumped to a high elevation lake at times of low electric demand for use in a hydroelectric turbine at times of high demand. For a specified energy storage capacity, the minimum amount of water that needs to be stored in the lake is to be determined.

Assumptions The evaporation of water from the lake is negligible.

Analysis The exergy or work potential of the water is the potential energy it possesses,

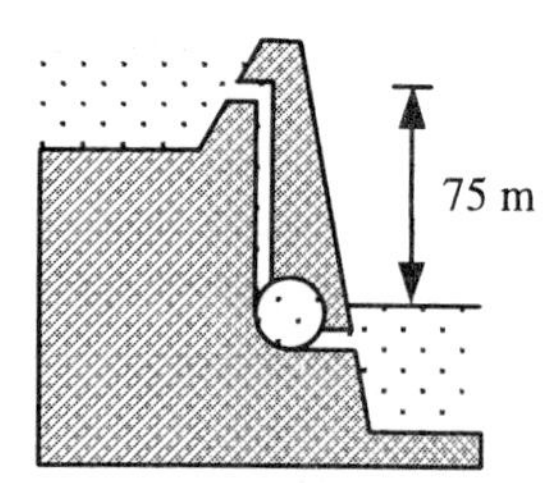

$$\text{Exergy} = \text{PE} = \text{mgh}$$

Thus,

$$m = \frac{PE}{gh} = \frac{5\times10^6\ \text{kWh}}{(9.8\ \text{m/s}^2)(75\ \text{m})}\left(\frac{3600\ \text{s}}{1\ \text{h}}\right)\left(\frac{1000\ \text{m}^2/\text{s}^2}{1\ \text{kW}\cdot\text{s/kg}}\right) = \mathbf{2.45\times10^{10}\ kg}$$

7-17 A heat reservoir at a specified temperature can supply heat at a specified rate. The exergy of this heat supplied is to be determined.

Analysis The exergy of the supplied heat, in the rate form, is the amount of power that would be produced by a reversible heat engine,

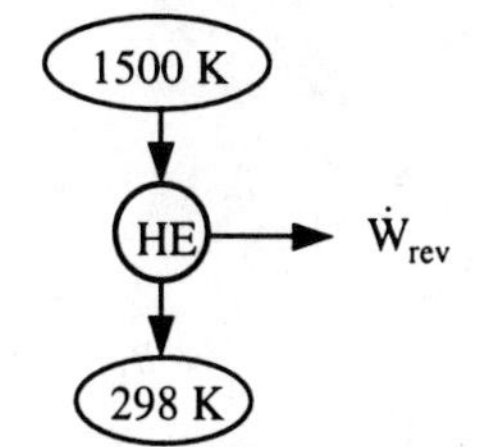

$$\eta_{th,\max} = \eta_{th,rev} = 1 - \frac{T_0}{T_H} = 1 - \frac{298 \text{ K}}{1500 \text{ K}} = 0.8013$$

$$\begin{aligned}\text{Exergy} &= \dot{W}_{\max,out} = \dot{W}_{rev,out} = \eta_{th,rev}\dot{Q}_{in} \\ &= (0.8013)(150{,}000/3600 \text{ kJ/s}) = \mathbf{33.4\ kW}\end{aligned}$$

7-18 [*Also solved by EES on enclosed CD*] A heat engine receives heat from a source at a specified temperature at a specified rate, and rejects the waste heat to a sink. For a given power output, the reversible power, the rate of irreversibility, and the 2nd law efficiency are to be determined.

Analysis (a) The reversible power is the power produced by a reversible heat engine operating between the specified temperature limits,

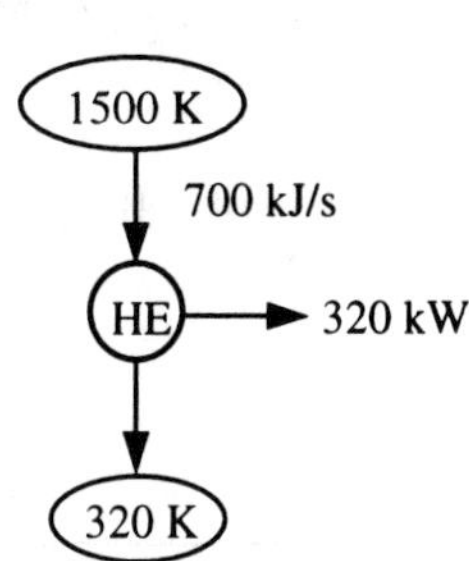

$$\eta_{th,\max} = \eta_{th,rev} = 1 - \frac{T_L}{T_H} = 1 - \frac{320 \text{ K}}{1500 \text{ K}} = 0.7867$$

$$\dot{W}_{rev,out} = \eta_{th,rev}\dot{Q}_{in} = (0.7867)(700 \text{ kJ/s}) = \mathbf{550.7\ kW}$$

(b) The irreversibility rate is the difference between the reversible power and the actual power output:

$$\dot{I} = \dot{W}_{rev,out} - \dot{W}_{u,out} = 550.7 - 320 = \mathbf{230.7\ kW}$$

(c) The second law efficiency is determined from its definition,

$$\eta_{II} = \frac{W_{u,out}}{W_{rev,out}} = \frac{320 \text{ kW}}{550.7 \text{ kW}} = \mathbf{58.1\%}$$

7-19 EES solution of this (and other comprehensive problems designated with the *computer icon*) is available to instructors at the *Instructor Manual* section of the *Online Learning Center* (OLC) at www.mhhe.com/cengel-boles. See the Preface for access information.

7-20E The thermal efficiency and the second-law efficiency of a heat engine are given. The source temperature is to be determined.

Analysis From the definition of the second law efficiency,

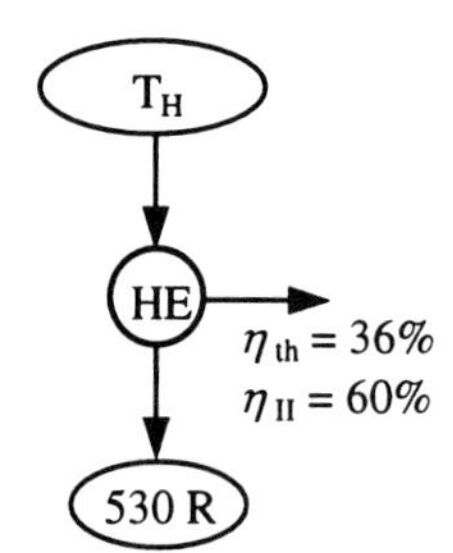

$$\eta_{II} = \frac{\eta_{th}}{\eta_{th,rev}} \longrightarrow \eta_{th,rev} = \frac{\eta_{th}}{\eta_{II}} = \frac{0.36}{0.60} = 0.60$$

Thus,

$$\eta_{th,rev} = 1 - \frac{T_L}{T_H} \longrightarrow T_H = T_L / (1 - \eta_{th,rev}) = (530\text{ R})/0.40 = \mathbf{1325\ R}$$

7-21 A body contains a specified amount of thermal energy at a specified temperature. The amount that can be converted to work is to be determined.

Analysis The amount of heat that can be converted to work is simply the amount that a reversible heat engine can convert to work,

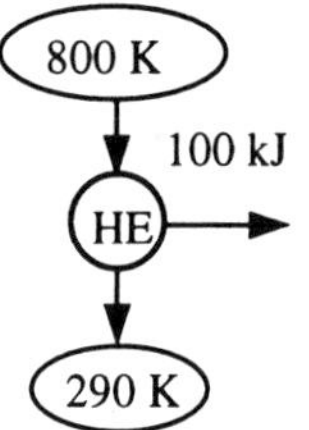

$$\eta_{th,rev} = 1 - \frac{T_0}{T_H} = 1 - \frac{298\text{ K}}{800\text{ K}} = 0.734$$

$$W_{\max,out} = W_{rev,out} = \eta_{th,rev} Q_{in}$$
$$= (0.734)(100\text{ kJ}) = \mathbf{73.4\ kJ}$$

7-22 The thermal efficiency of a heat engine operating between specified temperature limits is given. The second-law efficiency of a engine is to be determined.

Analysis The thermal efficiency of a reversible heat engine operating between the same temperature reservoirs is

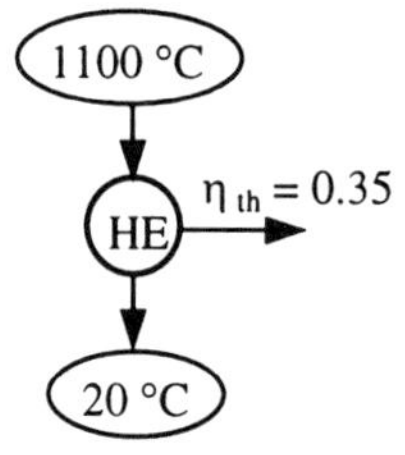

$$\eta_{th,rev} = 1 - \frac{T_0}{T_H} = 1 - \frac{293\text{ K}}{1100\text{ K}} = 0.734$$

Thus,

$$\eta_{II} = \frac{\eta_{th}}{\eta_{th,rev}} = \frac{0.35}{0.734} = \mathbf{47.6\%}$$

7-23 A house is maintained at a specified temperature by electric resistance heaters. The reversible work for this heating process and irreversibility are to be determined.

Analysis The reversible work is the minimum work required to accomplish this process, and the irreversibility is the difference between the reversible work and the actual electrical work consumed. The actual power input is

$$\dot{W}_{in} = \dot{Q}_{out} = \dot{Q}_H = 80{,}000 \text{ kJ/h} = 22.22 \text{ kW}$$

The COP of a reversible heat pump operating between the specified temperature limits is

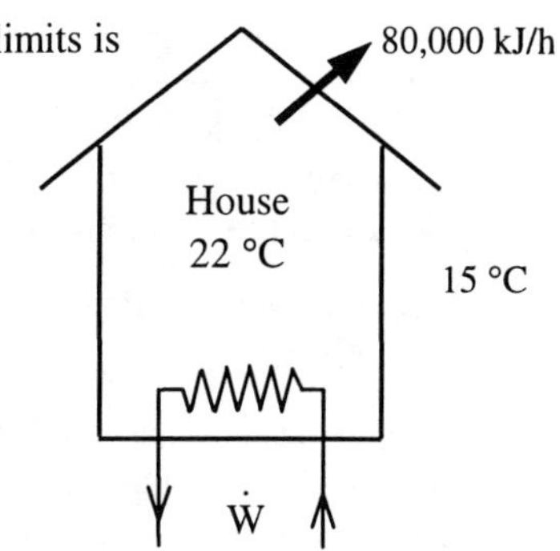

$$COP_{HP,rev} = \frac{1}{1-T_L/T_H} = \frac{1}{1-288/295} = 47.14$$

Thus,

$$\dot{W}_{rev,in} = \frac{\dot{Q}_H}{COP_{HP,rev}} = \frac{22.22 \text{ kW}}{42.14} = \mathbf{0.53 \text{ kW}}$$

and

$$I = \dot{W}_{u,in} - \dot{W}_{rev,in} = 22.22 - 0.53 = \mathbf{21.69 \text{ kW}}$$

7-24E A freezer is maintained at a specified temperature by removing heat from it at a specified rate. The power consumption of the freezer is given. The reversible power, irreversibility, and the second-law efficiency are to be determined.

Analysis (a) The reversible work is the minimum work required to accomplish this task, which is the work that a reversible refrigerator operating between the specified temperature limits would consume,

$$COP_{R,rev} = \frac{1}{T_H/T_L - 1} = \frac{1}{535/480-1} = 8.73$$

$$\dot{W}_{rev,in} = \frac{\dot{Q}_L}{COP_{R,rev}} = \frac{75 \text{ Btu/min}}{8.73}\left(\frac{1 \text{ hp}}{42.41 \text{ Btu/min}}\right) = \mathbf{0.20 \text{ hp}}$$

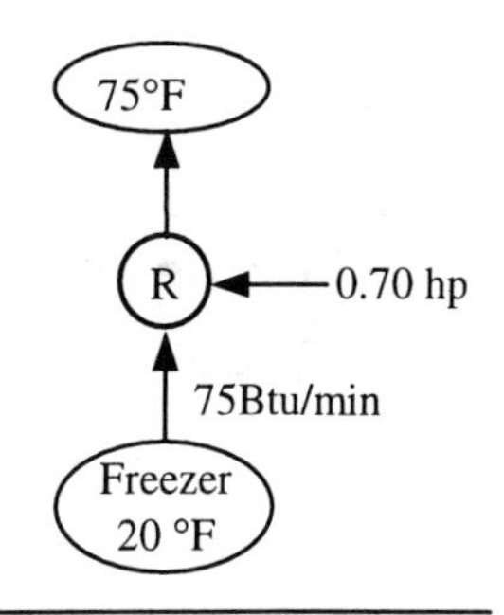

(b) The irreversibility is the difference between the reversible work and the actual electrical work consumed,

$$\dot{I} = \dot{W}_{u,in} - \dot{W}_{rev,in} = 0.70 - 0.20 = \mathbf{0.50 \text{ hp}}$$

(c) The second law efficiency is determined from its definition,

$$\eta_{II} = \frac{\dot{W}_{rev}}{\dot{W}_u} = \frac{-0.20 \text{ hp}}{-0.7 \text{ hp}} = \mathbf{28.9\%}$$

7-25 It is to be shown that the power produced by a wind turbine is proportional to the cube of the wind velocity and the square of the blade span diameter.

Analysis The power produced by a wind turbine is proportional to the kinetic energy of the wind, which is equal to the product of the kinetic energy of air per unit mass and the mass flow rate of air through the blade span area. Therefore,

$$\begin{aligned}\text{Wind power} &= \text{(Efficiency)(Kinetic energy)(Mass flow rate of air)} \\ &= \eta_{\text{wind}} \frac{\mathbf{V}^2}{2}(\rho A \mathbf{V}) = \eta_{\text{wind}} \frac{\mathbf{V}^2}{2} \rho \frac{\pi D^2}{4} \mathbf{V} \\ &= \eta_{\text{wind}} \rho \frac{\pi \mathbf{V}^3 D^2}{8} = (\text{Constant})\mathbf{V}^3 D^2\end{aligned}$$

which completes the proof that wind power is proportional to the cube of the wind velocity and to the square of the blade span diameter.

Second-Law Analysis of Closed Systems

7-26C Yes.

7-27C Yes, it can. For example, the 1st law efficiency of a reversible heat engine operating between the temperature limits of 300 K and 1000 K is 70%. However, the second law efficiency of this engine, like all reversible devices, is 100%.

7-28 A cylinder initially contains air at atmospheric conditions. Air is compressed to a specified state and the useful work input is measured. The exergy of the air at the initial and final states, and the minimum work input to accomplish this compression process, and the second-law efficiency are to be determined

Assumptions **1** Air is an ideal gas with constant specific heats. **2** The kinetic and potential energies are negligible.

Properties The gas constant of air is R = 0.287 kPa.m^3/kg.K (Table A-1). The specific heats of air at the average temperature of (298+423)/2=360. K are C_p = 1.009 kJ/kg· K and C_v = 0.722 kJ/kg· K (Table A-2).

Analysis (a) We realize that $X_1 = \Phi_1 = \mathbf{0}$ since air initially is at the dead state. The mass of air is

$$m = \frac{P_1V_1}{RT_1} = \frac{(100\text{ kPa})(0.002\text{ m}^3)}{(0.287\text{ kPa}\cdot\text{m}^3/\text{kg}\cdot\text{K})(298\text{ K})} = 0.00234\text{ kg}$$

Also, $$\frac{P_2V_2}{T_2} = \frac{P_1V_1}{T_1} \longrightarrow V_2 = \frac{P_1T_2}{P_2T_1}V_1 = \frac{(100\text{ kPa})(423\text{ K})}{(600\text{ kPa})(298\text{ K})}(2\text{ L}) = 0.473\text{ L}$$

and

$$\begin{aligned} s_2 - s_0 &= C_{p,ave}\ln\frac{T_2}{T_0} - R\ln\frac{P_2}{P_0} \\ &= (1.009\text{ kJ/kg}\cdot\text{K})\ln\frac{423\text{ K}}{198\text{ K}} - (0.287\text{ kJ/kg}\cdot\text{K})\ln\frac{600\text{ kPa}}{100\text{ kPa}} \\ &= -0.1608\text{ kJ/kg}\cdot\text{K} \end{aligned}$$

AIR
V_1 = 2 L
P_1 = 100 kPa
T_1 = 25°C

Thus, the exergy of air at the final state is

$$\begin{aligned} X_2 = \Phi_2 &= m\left[C_{v,ave}(T_2 - T_0) - T_0(s_2 - s_0)\right] + P_0(V_2 - V_0) \\ &= (0.00234\text{ kg})\left[(0.722\text{ kJ/kg}\cdot\text{K})(423 - 298)\text{K} - (298\text{ K})(-0.1608\text{ kJ/kg}\cdot\text{K})\right] \\ &\quad +(100\text{ kPa})(0.000473 - 0.002)\text{m}^3[\text{kJ/m}^3\cdot\text{kPa}] \\ &= \mathbf{0.171\text{ kJ}} \end{aligned}$$

(*b*) The minimum work input is the reversible work input, which can be determined from the exergy balance by setting the exergy destruction equal to zero,

$$\underbrace{X_{in} - X_{out}}_{\substack{\text{Net exergy transfer}\\ \text{by heat, work, and mass}}} \underbrace{- X_{\text{destroyed}}^{\nearrow 0\text{ (reversible)}}}_{\substack{\text{Exergy}\\ \text{destruction}}} = \underbrace{\Delta X_{\text{system}}}_{\substack{\text{Change}\\ \text{in exergy}}}$$

$$\begin{aligned} W_{rev,in} &= X_2 - X_1 \\ &= 0.171 - 0 = \mathbf{0.171\text{ kJ}} \end{aligned}$$

(c) The second-law efficiency of this process is

$$\eta_{II} = \frac{W_{u,in}}{W_{rev,in}} = \frac{0.171\text{ kJ}}{1.2\text{ kJ}} = \mathbf{14.3\%}$$

7-29 A cylinder is initially filled with R-134a at a specified state. The refrigerant is cooled and condensed at constant pressure. The exergy of the refrigerant at the initial and final states, and the exergy destroyed during this process are to be determined. √

Assumptions The kinetic and potential energies are negligible.

Properties From the refrigerant tables (Tables A-11 through A-13),

$$\left.\begin{array}{l} P_1 = 0.8\text{ MPa} \\ T_1 = 50°C \end{array}\right\} \begin{array}{l} v_1 = 0.02846\text{ m}^3/\text{kg} \\ u_1 = 261.62\text{ kJ/kg} \\ s_1 = 0.9711\text{ kJ/kg}\cdot\text{K} \end{array}$$

$$\left.\begin{array}{l} P_2 = 0.8\text{ MPa} \\ T_2 = 30°C \end{array}\right\} \begin{array}{l} v_2 \cong v_{f@30°C} = 0.0008417\text{ m}^3/\text{kg} \\ u_2 \cong u_{f@30°C} = 90.84\text{ kJ/kg} \\ s_2 \cong s_{f@30°C} = 0.3396\text{ kJ/kg}\cdot\text{K} \end{array}$$

$$\left.\begin{array}{l} P_0 = 0.1\text{ MPa} \\ T_0 = 30°C \end{array}\right\} \begin{array}{l} v_0 = 0.24216\text{ m}^3/\text{kg} \\ u_0 = 254.54\text{ kJ/kg} \\ s_0 = 1.1122\text{ kJ/kg}\cdot\text{K} \end{array}$$

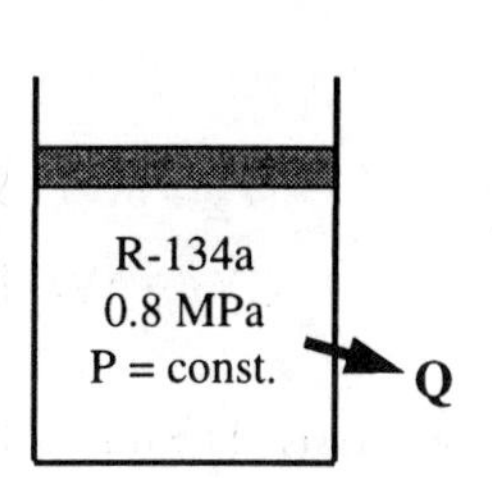

Analysis (*a*) From the closed system exergy relation,

$$\begin{aligned} X_1 = \Phi_1 &= m\{(u_1 - u_0) - T_0(s_1 - s_0) + P_0(v_1 - v_0)\} \\ &= (5\text{ kg})\{(261.62 - 254.54)\text{ kJ/kg} - (303\text{ K})(0.9711 - 1.1122)\text{ kJ/kg}\cdot\text{K} \\ &\quad + (100\text{ kPa})(0.02846 - 0.24216)\text{m}^3\text{/kg}[\text{kJ/m}^3\cdot\text{kPa}]\} \\ &= \mathbf{142.3\ kJ} \end{aligned}$$

and,

$$\begin{aligned} X_2 = \Phi_2 &= m\{(u_2 - u_0) - T_0(s_2 - s_0) + P_0(v_2 - v_0)\} \\ &= (5\text{ kg})\{(90.84 - 254.54)\text{ kJ/kg} - (303\text{ K})(0.3396 - 1.1122)\text{ kJ/kg}\cdot\text{K} \\ &\quad + (100\text{ kPa})(0.0008417 - 0.24216)\text{m}^3\text{/kg}[\text{kJ/m}^3\cdot\text{kPa}]\} \\ &= \mathbf{231.1\ kJ} \end{aligned}$$

(*b*) The reversible work input, which represents the minimum work input $W_{rev,in}$ in this case can be determined from the exergy balance by setting the exergy destruction equal to zero,

$$\underbrace{X_{in} - X_{out}}_{\substack{\text{Net exergy transfer} \\ \text{by heat, work, and mass}}} \underbrace{- X_{destroyed}^{\nearrow 0\text{ (reversible)}}}_{\substack{\text{Exergy} \\ \text{destruction}}} = \underbrace{\Delta X_{system}}_{\substack{\text{Change} \\ \text{in exergy}}}$$

$$W_{rev,in} = X_2 - X_1 = 231.1 - 142.3 = 88.8\text{ kJ}$$

Noting that the process involves only boundary work, the useful work input during this process is simply the boundary work in excess of the work done by the surrounding air,

$$\begin{aligned} W_{u,in} &= W_{in} - W_{surr,in} = W_{in} - P_0(V_1 - V_2) = P(V_1 - V_2) - P_0 m(v_1 - v_2) \\ &= m(P - P_0)(v_1 - v_2) \\ &= (5\text{ kg})(800 - 100\text{ kPa})(0.02846 - 0.0008417\text{ m}^3/\text{kg})\left(\frac{1\text{ kJ}}{1\text{ kPa}\cdot\text{m}^3}\right) = 96.7\text{ kJ} \end{aligned}$$

Knowing both the actual useful and reversible work inputs, the exergy destruction or irreversibility that is the difference between the two is determined from its definition to be

$$X_{destroyed} = I = W_{u,in} - W_{rev,in} = 96.7 - 88.8 = \mathbf{7.9\ kJ}$$

7-30 The radiator of a steam heating system is initially filled with superheated steam. The valves are closed, and steam is allowed to cool until the pressure drops to a specified value by transferring heat to the room. The amount of heat transfer to the room and the maximum amount of heat that can be supplied to the room are to be determined.

Assumptions Kinetic and potential energies are negligible.

Properties (a) From the steam tables,

STEAM
20 L
$P_1 = 200$ kPa
$T_1 = 200°C$
Q_{out}

$$\left.\begin{array}{l} P_1 = 200\text{ kPa} \\ T_1 = 200°C \end{array}\right\} \begin{array}{l} v_1 = 1.0803\text{ m}^3/\text{kg} \\ u_1 = 2654.4\text{ kJ/kg} \\ s_1 = 7.5066\text{ kJ/kg}\cdot\text{K} \end{array}$$

$$\left.\begin{array}{l} T_2 = 80°C \\ (v_2 = v_1) \end{array}\right\} \begin{array}{l} x_2 = \dfrac{v_2 - v_f}{v_{fg}} = \dfrac{1.0803 - 0.001029}{3.407 - 0.001029} = 0.3169 \\ u_2 = u_f + x_2 u_{fg} = 334.86 + 0.3169 \times (2482.2 - 334.86) = 1015.35\text{ kJ/kg} \\ s_2 = s_f + x_2 s_{fg} = 1.0753 + 0.3169 \times 6.5369 = 3.1468\text{ kJ/kg}\cdot\text{K} \end{array}$$

Analysis The mass of the steam is

$$m = \frac{V}{v_1} = \frac{0.020\text{ m}^3}{1.0803\text{ m}^3/\text{kg}} = 0.01851\text{ kg}$$

The amount of heat transfer to the room is determined from an energy balance on the radiator expressed as

$$\underbrace{E_{in} - E_{out}}_{\substack{\text{Net energy transfer}\\ \text{by heat, work, and mass}}} = \underbrace{\Delta E_{\text{system}}}_{\substack{\text{Change in internal, kinetic,}\\ \text{potential, etc. energies}}}$$

$$-Q_{out} = \Delta U = m(u_2 - u_1) \quad (\text{since } W = \text{KE} = \text{PE} = 0)$$

$$Q_{out} = m(u_1 - u_2)$$

or,

$$Q_{out} = (0.01851\text{ kg})(2654.4 - 1015.35)\text{ kJ/kg} = \mathbf{30.3\text{ kJ}}$$

(*b*) The reversible work output, which represents the maximum work output $W_{\text{rev,out}}$ in this case can be determined from the exergy balance by setting the exergy destruction equal to zero,

$$\underbrace{X_{in} - X_{out}}_{\substack{\text{Net exergy transfer}\\ \text{by heat, work, and mass}}} \underbrace{- X_{\text{destroyed}}^{\nearrow 0\text{ (reversible)}}}_{\substack{\text{Exergy}\\ \text{destruction}}} = \underbrace{\Delta X_{\text{system}}}_{\substack{\text{Change}\\ \text{in exergy}}}$$

$$-W_{rev,out} = X_2 - X_1 \rightarrow W_{rev,out} = X_1 - X_2 = \Phi_1 - \Phi_2$$

Substituting the closed system exergy relation, the reversible work during this process is determined to be

$$\begin{aligned} W_{rev,out} &= m\left[(u_1 - u_2) - T_0(s_1 - s_2) + P_0(v_1 \overset{\nearrow 0}{} - v_2)\right] \\ &= m\left[(u_1 - u_2) - T_0(s_1 - s_2)\right] \\ &= (0.01851\text{ kg})\left[(2654.4 - 1015.35)\text{kJ/kg} - (273\text{ K})(7.5066 - 3.1468)\text{kJ/kg}\cdot\text{K}\right] = 8.308\text{ kJ} \end{aligned}$$

When this work is supplied to a reversible heat pump, it will supply the room heat in the amount of

$$Q_H = COP_{HP,rev} W_{rev} = \frac{W_{rev}}{1 - T_L/T_H} = \frac{8.308\text{ kJ}}{1 - 273/294} = \mathbf{116.3\text{ kJ}}$$

Discussion Note that the amount of heat supplied to the room can be increased by about 3 times by eliminating the irreversibility associated with the irreversible heat transfer process.

7-31 EES solution of this (and other comprehensive problems designated with the *computer icon*) is available to instructors at the *Instructor Manual* section of the *Online Learning Center* (OLC) at www.mhhe.com/cengel-boles. See the Preface for access information.

7-32E An insulated rigid tank contains saturated liquid-vapor mixture of water at a specified pressure. An electric heater inside is turned on and kept on until all the liquid is vaporized. The exergy destruction and the second-law efficiency are to be determined.

Assumptions Kinetic and potential energies are negligible.

Properties From the steam tables (Tables A-4 through A-6)

$$\left.\begin{array}{l} P_1 = 15\text{ psia} \\ x_1 = 0.25 \end{array}\right\} \begin{array}{l} v_1 = v_f + x_1 v_{fg} = 0.016723 + 0.25 \times (26.29 - 0.016723) = 6.585\text{ ft}^3/\text{lbm} \\ u_1 = s_f + x_1 u_{fg} = 181.147 + 0.25 \times 896.8 = 405.34\text{ Btu/lbm} \\ s_1 = s_f + x_1 s_{fg} = 0.31367 + 0.25 \times 1.4414 = 0.674\text{ Btu/lbm}\cdot\text{R} \end{array}$$

$$\left.\begin{array}{l} v_2 = v_1 \\ sat.vapor \end{array}\right\} \begin{array}{l} u_2 = u_{g\,@\,v_g} = 6.586\text{ ft}^3/\text{lbm} = 1099.68\text{ Btu/lbm} \\ s_2 = s_{g\,@\,v_g} = 6.586\text{ ft}^3/\text{lbm} = 1.637\text{ Btu/lbm}\cdot\text{R} \end{array}$$

Analysis (*a*) The irreversibility can be determined from its definition $X_{\text{destroyed}} = T_0 S_{\text{gen}}$ where the entropy generation is determined from an entropy balance on the tank, which is an insulated closed system,

$$\underbrace{S_{in} - S_{out}}_{\substack{\text{Net entropy transfer} \\ \text{by heat and mass}}} + \underbrace{S_{gen}}_{\substack{\text{Entropy} \\ \text{generation}}} = \underbrace{\Delta S_{\text{system}}}_{\substack{\text{Change} \\ \text{in entropy}}}$$

$$S_{\text{gen}} = \Delta S_{\text{system}} = m(s_2 - s_1)$$

Substituting,

$$X_{\text{destroyed}} = T_0 S_{gen} = mT_0(s_2 - s_1)$$
$$= (4\text{ lbm})(535\text{ R})(1.637 - 0.674)\text{Btu/lbm}\cdot\text{R} = \mathbf{2060.8\ Btu}$$

(*b*) Noting that V = constant during this process, the W and W_u are identical and are determined from the energy balance on the closed system energy equation,

$$\underbrace{E_{in} - E_{out}}_{\substack{\text{Net energy transfer} \\ \text{by heat, work, and mass}}} = \underbrace{\Delta E_{\text{system}}}_{\substack{\text{Change in internal, kinetic,} \\ \text{potential, etc. energies}}}$$

$$W_{e,in} = \Delta U = m(u_2 - u_1)$$

or,

$$W_{e,in} = (4\text{ lbm})(1099.68 - 405.34)\text{Btu/lbm} = 2777.4\text{ Btu}$$

Then the reversible work during this process and the second-law efficiency become

$$W_{rev,in} = W_{u,in} - X_{\text{destroyed}} = 2777.4 - 2060.8 = 716.6\text{ Btu}$$

Thus,

$$\eta_{II} = \frac{W_{rev}}{W_u} = \frac{716.6\text{ Btu}}{2777.4\text{ Btu}} = \mathbf{25.8\%}$$

7-33 A rigid tank is divided into two equal parts by a partition. One part is filled with compressed liquid while the other side is evacuated. The partition is removed and water expands into the entire tank. The exergy destroyed during this process is to be determined.

Assumptions Kinetic and potential energies are negligible.

Analysis The properties of the water are

$$\left.\begin{array}{l} P_1 = 300\text{ kPa} \\ T_1 = 60°C \end{array}\right\} \begin{array}{l} v_1 \cong v_{f\,@\,60°C} = 0.001017\text{ m}^3/\text{kg} \\ u_1 \cong u_{f\,@\,60°C} = 251.11\text{ kJ/kg} \\ s_1 \cong s_{f\,@\,60°C} = 0.8312\text{ kJ/kg}\cdot\text{K} \end{array}$$

1.5 kg 300 kPa 60°C WATER	Vacuum

Noting that $v_2 = 2v_1 = 2\times 0.001017 = 0.002034\text{ m}^3/\text{kg}$,

$$\left.\begin{array}{l} P_2 = 15\text{ kPa} \\ v_2 = 0.002034 \end{array}\right\} \begin{array}{l} x_2 = \dfrac{v_2 - v_f}{v_{fg}} = \dfrac{0.002034 - 0.001014}{10.02 - 0.001014} = 0.0001017 \\ u_2 = u_f + x_2 u_{fg} = 225.92 + 0.0001017\times 2222.8 = 226.15\text{ kJ/kg} \\ s_2 = s_f + x_2 s_{fg} = 0.7549 + 0.0001017\times 7.2536 = 0.7556\text{ kJ/kg}\cdot\text{K} \end{array}$$

Taking the direction of heat transfer to be *to* the tank, the energy balance on this closed system becomes

$$\underbrace{E_{in} - E_{out}}_{\substack{\text{Net energy transfer}\\ \text{by heat, work, and mass}}} = \underbrace{\Delta E_{\text{system}}}_{\substack{\text{Change in internal, kinetic,}\\ \text{potential, etc. energies}}}$$

$$Q_{in} = \Delta U = m(u_2 - u_1)$$

or,

$$Q_{in} = (1.5\text{ kg})(226.15 - 251.11)\text{kJ/kg} = -37.4\text{ kJ} \rightarrow Q_{out} = 37.4\text{ kJ}$$

The irreversibility can be determined from its definition $X_{\text{destroyed}} = T_0 S_{\text{gen}}$ where the entropy generation is determined from an entropy balance on an *extended system* that includes the tank and its immediate surroundings so that the boundary temperature of the extended system is the temperature of the surroundings at all times,

$$\underbrace{S_{in} - S_{out}}_{\substack{\text{Net entropy transfer}\\ \text{by heat and mass}}} + \underbrace{S_{gen}}_{\substack{\text{Entropy}\\ \text{generation}}} = \underbrace{\Delta S_{\text{system}}}_{\substack{\text{Change}\\ \text{in entropy}}}$$

$$-\frac{Q_{out}}{T_{b,out}} + S_{\text{gen}} = \Delta S_{\text{system}} = m(s_2 - s_1)$$

$$S_{\text{gen}} = m(s_2 - s_1) + \frac{Q_{out}}{T_{surr}}$$

Substituting,

$$X_{\text{destroyed}} = T_0 S_{gen} = T_0\left(m(s_2 - s_1) + \frac{Q_{out}}{T_{surr}}\right)$$

$$= (298\text{ K})\left[(1.5\text{ kg})(0.7556 - 0.8312)\text{kJ/kg}\cdot\text{K} + \frac{37.4\text{ kJ}}{298\text{ K}}\right]$$

$$= \mathbf{3.61\ kJ}$$

7-34 EES solution of this (and other comprehensive problems designated with the *computer icon*) is available to instructors at the *Instructor Manual* section of the *Online Learning Center* (OLC) at www.mhhe.com/cengel-boles. See the Preface for access information.

7-35 An insulated cylinder is initially filled with saturated liquid water at a specified pressure. The water is heated electrically at constant pressure. The minimum work by which this process can be accomplished and the exergy destroyed are to be determined.

Assumptions **1** The kinetic and potential energy changes are negligible. **2** The cylinder is well-insulated and thus heat transfer is negligible. **3** The thermal energy stored in the cylinder itself is negligible. **4** The compression or expansion process is quasi-equilibrium.

Analysis (a) From the steam tables,

$$\left.\begin{array}{l} P_1 = 150\text{ kPa} \\ \text{sat.liquid} \end{array}\right\} \begin{array}{l} u_1 = u_{f\,@\,150\text{ kPa}} = 466.94\text{ kJ/kg} \\ v_1 = v_{f\,@\,150\text{ kPa}} = 0.001053\text{ m}^3\text{/kg} \\ h_1 = h_{f\,@\,150\text{ kPa}} = 467.11\text{ kJ/kg} \\ s_1 = s_{f\,@\,150\text{ kPa}} = 1.4336\text{ kJ/kg}\cdot\text{K} \end{array}$$

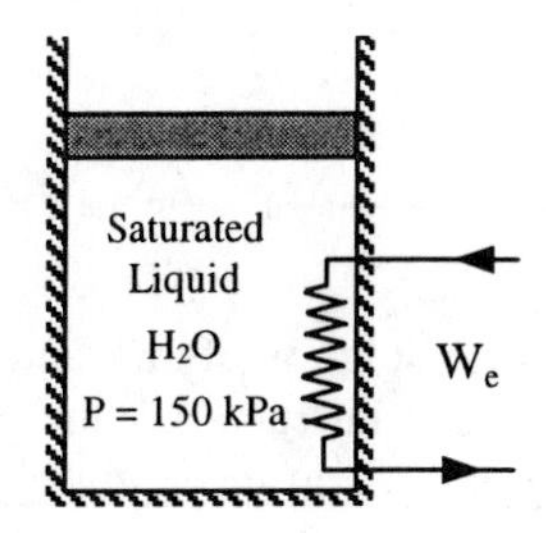

The mass of the steam is

$$m = \frac{V}{v_1} = \frac{0.002\text{ m}^3}{0.001053\text{ m}^3\text{/kg}} = 1.899\text{ kg}$$

We take the contents of the cylinder as the system. This is a closed system since no mass enters or leaves. The energy balance for this stationary closed system can be expressed as

$$\underbrace{E_{in} - E_{out}}_{\substack{\text{Net energy transfer} \\ \text{by heat, work, and mass}}} = \underbrace{\Delta E_{\text{system}}}_{\substack{\text{Change in internal, kinetic,} \\ \text{potential, etc. energies}}}$$

$$W_{e,in} - W_{b,out} = \Delta U$$

$$W_{e,in} = m(h_2 - h_1)$$

since $\Delta U + W_b = \Delta H$ during a constant pressure quasi-equilibrium process. Solving for h_2,

$$h_2 = h_1 + \frac{W_{e,in}}{m} = 467.11 + \frac{2200\text{ kJ}}{1.899\text{ kg}} = 1625.6\text{ kJ/kg}$$

Thus,

$$\left.\begin{array}{l} P_2 = 150\text{ kPa} \\ h_2 = 1625.6\text{ kJ/kg} \end{array}\right\} \begin{array}{l} x_2 = \dfrac{h_2 - h_f}{h_{fg}} = \dfrac{1625.6 - 467.11}{2226.5} = 0.520 \\ s_2 = s_f + x_2 s_{fg} = 1.4336 + 0.520 \times 5.7897 = 4.4442\text{ kJ/kg}\cdot\text{K} \\ u_2 = u_f + x_2 u_{fg} = 466.94 + 0.520 \times (2519.7 - 466.94) = 1534.38\text{ kJ/kg} \\ v_2 = v_f + x_2 v_{fg} = 0.001053 + 0.520 \times (1.1593 - 0.001053) = 0.6033\text{ m}^3\text{/kg} \end{array}$$

The reversible work input, which represents the minimum work input $W_{\text{rev,in}}$ in this case can be determined from the exergy balance by setting the exergy destruction equal to zero,

$$\underbrace{X_{in} - X_{out}}_{\substack{\text{Net exergy transfer} \\ \text{by heat, work, and mass}}} \underbrace{- X_{\text{destroyed}}^{\nearrow 0\text{ (reversible)}}}_{\substack{\text{Exergy} \\ \text{destruction}}} = \underbrace{\Delta X_{\text{system}}}_{\substack{\text{Change} \\ \text{in exergy}}} \rightarrow W_{rev,in} = X_2 - X_1$$

Substituting the closed system exergy relation, the reversible work input during this process is determined to be

$$\begin{aligned} W_{rev,in} &= -m[(u_1 - u_2) - T_0(s_1 - s_2) + P_0(v_1 - v_2)] \\ &= -(1.899\text{ kg})\{(466.94 - 1534.38)\text{ kJ/kg} - (298\text{ K})(1.4336 - 4.4442)\text{ kJ/kg}\cdot\text{K} \\ &\quad + (100\text{ kPa})(0.001053 - 0.6033)\text{m}^3\text{/kg}[\text{kJ/kPa}\cdot\text{m}^3]\} \\ &= \mathbf{437.7\ kJ} \end{aligned}$$

(b) The exergy destruction (or irreversibility) associated with this process can be determined from its definition $X_{\text{destroyed}} = T_0 S_{\text{gen}}$ where the entropy generation is determined from an entropy balance on the cylinder, which is an insulated closed system,

$$\underbrace{S_{in} - S_{out}}_{\substack{\text{Net entropy transfer}\\ \text{by heat and mass}}} + \underbrace{S_{gen}}_{\substack{\text{Entropy}\\ \text{generation}}} = \underbrace{\Delta S_{\text{system}}}_{\substack{\text{Change}\\ \text{in entropy}}}$$

$$S_{\text{gen}} = \Delta S_{\text{system}} = m(s_2 - s_1)$$

Substituting,

$$X_{\text{destroyed}} = T_0 S_{gen} = mT_0(s_2 - s_1) = (298 \text{ K})(1.899 \text{ kg})(4.4442 - 1.4336)\text{kJ/kg}\cdot\text{K} = \mathbf{1705\ kJ}$$

7-36 EES solution of this (and other comprehensive problems designated with the *computer icon*) is available to instructors at the *Instructor Manual* section of the *Online Learning Center* (OLC) at www.mhhe.com/cengel-boles. See the Preface for access information.

7-37 An insulated cylinder is initially filled with saturated R-134a vapor at a specified pressure. The refrigerant expands in a reversible manner until the pressure drops to a specified value. The change in the exergy of the refrigerant during this process and the reversible work are to be determined.

Assumptions **1** The kinetic and potential energy changes are negligible. **2** The cylinder is well-insulated and thus heat transfer is negligible. **3** The thermal energy stored in the cylinder itself is negligible. **4** The process is stated to be reversible.

Analysis This is a reversible adiabatic (i.e., isentropic) process, and thus $s_2 = s_1$. From the refrigerant tables,

$$\left.\begin{array}{l} P_1 = 0.8\text{ MPa} \\ \text{sat.vapor} \end{array}\right\} \begin{array}{l} v_1 = v_{g\,@\,0.8\text{ MPa}} = 0.0255\text{ m}^3/\text{kg} \\ u_1 = u_{g\,@\,0.8\text{ MPa}} = 243.78\text{ kJ/kg} \\ s_1 = s_{g\,@\,0.8\text{ MPa}} = 0.9068\text{ kJ/kg}\cdot\text{K} \end{array}$$

R-134a
0.8 MPa
Reversible

The mass of the refrigerant is

$$m = \frac{V}{v_1} = \frac{0.08\text{ m}^3}{0.0255\text{ m}^3/\text{kg}} = 3.137\text{ kg}$$

$$\left.\begin{array}{l} P_2 = 0.4\text{ MPa} \\ s_2 = s_1 \end{array}\right\} \begin{array}{l} x_2 = \dfrac{s_2 - s_f}{s_{fg}} = \dfrac{0.9066 - 0.2399}{0.9145 - 0.2399} = 0.988 \\ v_2 = v_f + x_2 v_{fg} = 0.0007904 + 0.988 \times (0.0509 - 0.0007904) = 0.05030\text{ m}^3/\text{kg} \\ u_2 = u_f + x_2 u_{fg} = 61.69 + 0.988 \times (231.97 - 61.69) = 229.93\text{ kJ/kg} \end{array}$$

The reversible work output, which represents the maximum work output $W_{\text{rev,out}}$ can be determined from the exergy balance by setting the exergy destruction equal to zero,

$$\underbrace{X_{in} - X_{out}}_{\substack{\text{Net exergy transfer} \\ \text{by heat, work, and mass}}} - \underbrace{X_{\text{destroyed}}^{\nearrow 0\ (\text{reversible})}}_{\substack{\text{Exergy} \\ \text{destruction}}} = \underbrace{\Delta X_{\text{system}}}_{\substack{\text{Change} \\ \text{in exergy}}}$$

$$-W_{rev,out} = X_2 - X_1 \rightarrow W_{rev,out} = X_1 - X_2 = \Phi_1 - \Phi_2$$

Therefore, the change in exergy and the reversible work are identical in this case. Using the definition of the closed system exergy and substituting, the reversible work is determined to be

$$W_{rev,out} = \Phi_1 - \Phi_2 = m\left[(u_1 - u_2) - T_0(s_1 - s_2)^{\nearrow 0} + P_0(v_1 - v_2)\right] = m[(u_1 - u_2) + P_0(v_1 - v_2)]$$

$$= (3.137\text{ kg})[(243.78 - 229.93)\text{ kJ/kg} + (100\text{ kPa})(0.0255 - 0.05030)\text{m}^3/\text{kg}[\text{kJ/kPa}\cdot\text{m}^3]$$

$$= \mathbf{35.7\ kJ}$$

7-38E Oxygen gas is compressed from a specified initial state to a final specified state. The reversible work and the increase in the exergy of the oxygen during this process are to be determined.

Assumptions At specified conditions, oxygen can be treated as an ideal gas with constant specific heats.

Properties The gas constant of oxygen is R = 0.06206 Btu/lbm.R (Table A-1). The constant-volume specific heat of oxygen at the average temperature is

$$T_{av} = (T_1 + T_2)/2 = (75 + 525)/2 = 300°\text{F} \longrightarrow C_{v,ave} = 0.164 \text{ Btu/lbm}\cdot\text{R}$$

O_2
12 ft^3/lbm
75°F

Analysis The entropy change of oxygen is

$$\begin{aligned} s_2 - s_1 &= C_{av} \ln\left(\frac{T_2}{T_1}\right) + R\ln\left(\frac{v_2}{v_1}\right) \\ &= (0.164 \text{ Btu/lbm}\cdot\text{R})\ln\left(\frac{985\text{ R}}{535\text{ R}}\right) + (0.06206 \text{ Btu/lbm}\cdot\text{R})\ln\left(\frac{1.5\text{ ft}^3\text{/lbm}}{12\text{ ft}^3\text{/lbm}}\right) \\ &= -0.02894 \text{ Btu/lbm}\cdot\text{R} \end{aligned}$$

The reversible work input, which represents the minimum work input $W_{\text{rev,in}}$ in this case can be determined from the exergy balance by setting the exergy destruction equal to zero,

$$\underbrace{X_{in} - X_{out}}_{\substack{\text{Net exergy transfer} \\ \text{by heat, work, and mass}}} \underbrace{- X_{\text{destroyed}}^{\nearrow 0 \text{ (reversible)}}}_{\substack{\text{Exergy} \\ \text{destruction}}} = \underbrace{\Delta X_{\text{system}}}_{\substack{\text{Change} \\ \text{in exergy}}} \rightarrow W_{rev,in} = X_2 - X_1$$

Therefore, the change in exergy and the reversible work are identical in this case. Substituting the closed system exergy relation, the reversible work input during this process is determined to be

$$\begin{aligned} w_{rev,in} &= \phi_2 - \phi_1 = -[(u_1 - u_2) - T_0(s_1 - s_2) + P_0(v_1 - v_2)] \\ &= -\{(0.164 \text{ Btu/lbm}\cdot\text{R})(535\text{ - }985)\text{R} - (535\text{ R})(0.02894 \text{ Btu/lbm}\cdot\text{R}) \\ &\quad + (14.7\text{ psia})(12 - 1.5)\text{ft}^3\text{/lbm}[\text{Btu/5.4039 psia}\cdot\text{ft}^3]\} \\ &= \mathbf{60.7\ Btu/lbm} \end{aligned}$$

Also, the increase in the exergy of oxygen is

$$\phi_2 - \phi_1 = w_{rev,in} = \mathbf{60.7\ \ Btu/lbm}$$

7-39 An insulated tank contains CO_2 gas at a specified pressure and volume. A paddle-wheel in the tank stirs the gas, and the pressure and temperature of CO_2 rises. The actual paddle-wheel work and the minimum paddle-wheel work by which this process can be accomplished are to be determined.

Assumptions **1** At specified conditions, CO_2 can be treated as an ideal gas with constant specific heats at the average temperature. **2** The surroundings temperature is 298 K.

Analysis (a) The initial and final temperature of CO_2 are

$$T_1 = \frac{P_1 V_1}{mR} = \frac{(100\,\text{kPa})(1.2\,\text{m}^3)}{(2.13\,\text{kg})(0.1889\,\text{kPa}\cdot\text{m}^3/\text{kg}\cdot\text{K})} = 298.2\,\text{K}$$

$$T_2 = \frac{P_2 V_2}{mR} = \frac{(120\,\text{kPa})(1.2\,\text{m}^3)}{(2.13\,\text{kg})(0.1889\,\text{kPa}\cdot\text{m}^3/\text{kg}\cdot\text{K})} = 357.9\,\text{K}$$

$$T_{av} = (T_1 + T_2)/2 = (298.2 + 357.9)/2 = 328\,\text{K} \longrightarrow C_{v,ave} = 0.684\,\text{kJ/kg}\cdot\text{K}$$

1.2 m³
2.13 kg
CO_2
140 kPa
W_{pw}

The actual paddle-wheel work done is determined from the energy balance on the CO gas in the tank,

We take the contents of the cylinder as the system. This is a closed system since no mass enters or leaves. The energy balance for this stationary closed system can be expressed as

$$\underbrace{E_{in} - E_{out}}_{\substack{\text{Net energy transfer}\\ \text{by heat, work, and mass}}} = \underbrace{\Delta E_{\text{system}}}_{\substack{\text{Change in internal, kinetic,}\\ \text{potential, etc. energies}}}$$

$$W_{pw,in} = \Delta U = mC_v(T_2 - T_1)$$

or,

$$W_{pw,in} = (2.13\text{ kg})(0.684\text{ kJ/kg}\cdot\text{K})(357.9 - 298.2)\text{K} = \mathbf{87.0\ kJ}$$

(b) The minimum paddle-wheel work with which this process can be accomplished is the reversible work, which can be determined from the exergy balance by setting the exergy destruction equal to zero,

$$\underbrace{X_{in} - X_{out}}_{\substack{\text{Net exergy transfer}\\ \text{by heat, work, and mass}}} \underbrace{- X_{\text{destroyed}}^{\nearrow 0\ \text{(reversible)}}}_{\substack{\text{Exergy}\\ \text{destruction}}} = \underbrace{\Delta X_{\text{system}}}_{\substack{\text{Change}\\ \text{in exergy}}} \rightarrow W_{rev,in} = X_2 - X_1$$

Substituting the closed system exergy relation, the reversible work input for this process is determined to be

$$\begin{aligned} W_{rev,in} &= m\left[(u_2 - u_1) - T_0(s_2 - s_1) + P_0(\text{v}_2^{\nearrow 0} - \text{v}_1)\right] \\ &= m\left[C_{v,av}(T_2 - T_1) - T_0(s_2 - s_1)\right] \\ &= (2.13\,\text{kg})\left[(0.684\,\text{kJ/kg}\cdot\text{K})(357.9 - 298)\text{K} - (298)(0.1253\,\text{kJ/kg}\cdot\text{K})\right] \\ &= \mathbf{7.74\ kJ} \end{aligned}$$

since

$$s_2 - s_1 = C_{v,av}\ln\frac{T_2}{T_1} + R\ln\frac{\text{v}_2^{\nearrow 0}}{\text{v}_1} = (0.684\,\text{kJ/kg}\cdot\text{K})\ln\left(\frac{357.9\,\text{K}}{298\,\text{K}}\right) = 0.1253\,\text{kJ/kg}\cdot\text{K}$$

7-40 An insulated cylinder initially contains air at a specified state. A resistance heater inside the cylinder is turned on, and air is heated for 15 min at constant pressure. The exergy destruction during this process is to be determined.

Assumptions Air is an ideal gas with variable specific heats.

Properties The gas constant of air is $R = 0.287$ kJ/kg.K (Table A-1).

Analysis The mass of the air and the electrical work done during this process are

$$m = \frac{P_1 V_1}{RT_1} = \frac{(120 \text{ kPa})(0.03 \text{ m}^3)}{(0.287 \text{ kPa}\cdot\text{m}^3/\text{kg}\cdot\text{K})(300 \text{ K})} = 0.0418 \text{ kg}$$

$$W_e = \dot{W}_e \Delta t = (-0.05 \text{ kJ/s})(5 \times 60 \text{ s}) = -15 \text{ kJ}$$

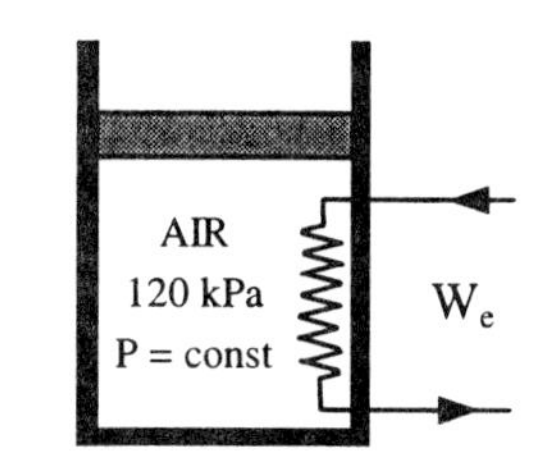

Also,

$$T_1 = 300 \text{ K} \longrightarrow h_1 = 300.19 \text{ kJ/kg} \quad \text{and} \quad s_1^\circ = 1.70202 \text{ kJ/kg}\cdot\text{K}$$

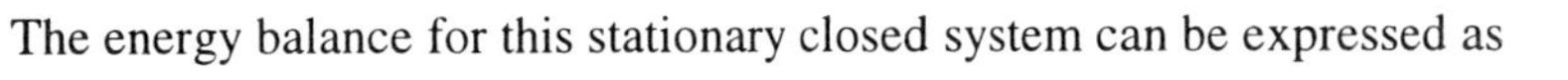

The energy balance for this stationary closed system can be expressed as

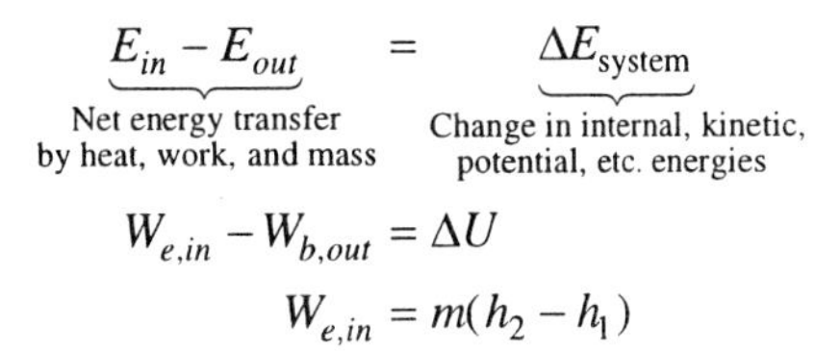

$$\underbrace{E_{in} - E_{out}}_{\substack{\text{Net energy transfer}\\ \text{by heat, work, and mass}}} = \underbrace{\Delta E_{\text{system}}}_{\substack{\text{Change in internal, kinetic,}\\ \text{potential, etc. energies}}}$$

$$W_{e,in} - W_{b,out} = \Delta U$$

$$W_{e,in} = m(h_2 - h_1)$$

since $\Delta U + W_b = \Delta H$ during a constant pressure quasi-equilibrium process. Thus,

$$h_2 = h_1 + \frac{W_{e,in}}{m} = 300.19 + \frac{15 \text{ kJ}}{0.0418 \text{ kg}} = 659.04 \text{ kJ/kg} \longrightarrow T_2 = 650 \text{ K}$$

$$s_2^\circ = 2.49364 \text{ kJ/kg}\cdot\text{K}$$

Also,

$$s_2 - s_1 = s_2^\circ - s_1^\circ - R\ln\left(\frac{P_2}{P_1}\right)^{\nearrow 0} = s_2^\circ - s_1^\circ = 2.49364 - 1.70202 = 0.79162 \text{ kJ/kg}\cdot\text{K}$$

The exergy destruction (or irreversibility) associated with this process can be determined from its definition $X_{\text{destroyed}} = T_0 S_{\text{gen}}$ where the entropy generation is determined from an entropy balance on the cylinder, which is an insulated closed system,

$$\underbrace{S_{in} - S_{out}}_{\substack{\text{Net entropy transfer}\\ \text{by heat and mass}}} + \underbrace{S_{gen}}_{\substack{\text{Entropy}\\ \text{generation}}} = \underbrace{\Delta S_{\text{system}}}_{\substack{\text{Change}\\ \text{in entropy}}}$$

$$S_{\text{gen}} = \Delta S_{\text{system}} = m(s_2 - s_1)$$

Substituting,

$$X_{\text{destroyed}} = T_0 S_{gen} = mT_0(s_2 - s_1) == (300 \text{ K})(0.0418 \text{ kg})(0.79162 \text{ kJ/kg}\cdot\text{K}) = \mathbf{9.9 \text{ kJ}}$$

7-41 A fixed mass of helium undergoes a process from a specified state to another specified state. The increase in the useful energy potential of helium is to be determined.

He
5 kg
288 K

Assumptions **1** At specified conditions, helium can be treated as an ideal gas.
2 Helium has constant specific heats at room temperature.

Properties The gas constant of helium is R = 2.0769 kJ/kg.K (Table A-1). The constant volume specific heat of helium is C_v = 3.1156 kJ/kg.K (Table A-2).

Analysis From the ideal-gas entropy change relation,

$$s_2 - s_1 = C_{v,av} \ln\frac{T_2}{T_1} + R\ln\frac{v_2}{v_1}$$

$$= (3.1156 \text{ kJ/kg}\cdot\text{K})\ln\frac{353 \text{ K}}{288 \text{ K}} + (2.0769 \text{ kJ/kg}\cdot\text{K})\ln\frac{0.5 \text{ m}^3/\text{kg}}{3 \text{ m}^3/\text{kg}} = -3.087 \text{ kJ/kg}\cdot\text{K}$$

The increase in the useful potential of helium during this process is simply the increase in exergy,

$$\Phi_2 - \Phi_1 = -m[(u_1 - u_2) - T_0(s_1 - s_2) + P_0(v_1 - v_2)]$$

$$= -(8 \text{ kg})\{(3.1156 \text{ kJ/kg}\cdot\text{K})(288 - 353)\text{ K} - (298 \text{ K})(3.087 \text{ kJ/kg}\cdot\text{K})$$

$$+ (100 \text{ kPa})(0.5 - 3)\text{m}^3/\text{kg}[\text{kJ/kPa}\cdot\text{m}^3]\}$$

$$= \mathbf{10{,}980 \text{ kJ}}$$

7-42 One side of a partitioned insulated rigid tank contains argon gas at a specified temperature and pressure while the other side is evacuated. The partition is removed, and the gas fills the entire tank. The exergy destroyed during this process is to be determined.

Assumptions Argon is an ideal gas with constant specific heats, and thus ideal gas relations apply.

Analysis Taking the entire rigid tank as the system, the energy balance can be expressed as

$$\underbrace{E_{in} - E_{out}}_{\substack{\text{Net energy transfer}\\ \text{by heat, work, and mass}}} = \underbrace{\Delta E_{\text{system}}}_{\substack{\text{Change in internal, kinetic,}\\ \text{potential, etc. energies}}}$$

$$0 = \Delta U = m(u_2 - u_1)$$

$$u_2 = u_1 \quad\rightarrow\quad T_2 = T_1$$

Argon
300 kPa
70°C

Vacuum

since $u = u(T)$ for an ideal gas.

The exergy destruction (or irreversibility) associated with this process can be determined from its definition $X_{\text{destroyed}} = T_0 S_{\text{gen}}$ where the entropy generation is determined from an entropy balance on the entire tank, which is an insulated closed system,

$$\underbrace{S_{in} - S_{out}}_{\substack{\text{Net entropy transfer}\\ \text{by heat and mass}}} + \underbrace{S_{gen}}_{\substack{\text{Entropy}\\ \text{generation}}} = \underbrace{\Delta S_{\text{system}}}_{\substack{\text{Change}\\ \text{in entropy}}}$$

$$S_{\text{gen}} = \Delta S_{\text{system}} = m(s_2 - s_1)$$

where

$$\Delta S_{\text{system}} = m(s_2 - s_1) = m\left(C_{v,ave} \ln\frac{T_2}{T_1}^{\nearrow 0} + R\ln\frac{V_2}{V_1}\right) = mR\ln\frac{V_2}{V_1}$$

$$= (3 \text{ kg})(0.208 \text{ kJ/kg}\cdot\text{K})\ln(2) = 0.433 \text{ kJ/K}$$

Substituting,

$$X_{\text{destroyed}} = T_0 S_{gen} = mT_0(s_2 - s_1) = (298 \text{ K})(0.433 \text{ kJ/K}) = \mathbf{129 \text{ kJ}}$$

7-43E A hot copper block is dropped into water in an insulated tank. The final equilibrium temperature of the tank and the work potential wasted during this process are to be determined.

Assumptions **1** Both the water and the copper block are incompressible substances with constant specific heats at room temperature. **2** The system is stationary and thus the kinetic and potential energies are negligible. **3** The tank is well-insulated and thus there is no heat transfer.

Properties The density and specific heat of water at the anticipated average temperature of 90°F are ρ = 62.1 lbm/ft^3 and C_p = 1.00 Btu/lbm.°F. The specific heat of copper at the anticipated average temperature of 100°F is C_p = 0.0925 Btu/lbm.°F (Table A-3).

Analysis We take the entire contents of the tank, water + copper block, as the *system*, which is a closed system. The energy balance for this system can be expressed as

$$\underbrace{E_{in} - E_{out}}_{\substack{\text{Net energy transfer}\\ \text{by heat, work, and mass}}} = \underbrace{\Delta E_{\text{system}}}_{\substack{\text{Change in internal, kinetic,}\\ \text{potential, etc. energies}}}$$

$$0 = \Delta U$$

Water 75°F

Copper 200°F

or,

$$\Delta U_{Cu} + \Delta U_{water} = 0$$

$$[mC(T_2 - T_1)]_{Cu} + [mC(T_2 - T_1)]_{water} = 0$$

where

$$m_w = \rho V = (62.1 \text{ lbm/ft}^3)(1 \text{ ft}^3) = 62.1 \text{ lbm}$$

Substituting,

$$(70 \text{ lbm})(0.0925 \text{ Btu/lbm}\cdot°\text{F})(T_2 - 200°\text{F}) + (62.1 \text{ lbm})(1.0 \text{ Btu/lbm}\cdot°\text{F})(T_2 - 75°\text{F}) = 0$$

$$T_2 = \mathbf{86.8°F} = 546.8 \text{ R}$$

The wasted work potential is equivalent to the exergy destruction (or irreversibility), and it can be determined from its definition $X_{\text{destroyed}} = T_0 S_{\text{gen}}$ where the entropy generation is determined from an entropy balance on the system, which is an insulated closed system,

$$\underbrace{S_{in} - S_{out}}_{\substack{\text{Net entropy transfer}\\ \text{by heat and mass}}} + \underbrace{S_{gen}}_{\substack{\text{Entropy}\\ \text{generation}}} = \underbrace{\Delta S_{\text{system}}}_{\substack{\text{Change}\\ \text{in entropy}}}$$

$$S_{\text{gen}} = \Delta S_{\text{system}} = \Delta S_{\text{water}} + \Delta S_{\text{copper}}$$

where

$$\Delta S_{copper} = mC_{ave} \ln\left(\frac{T_2}{T_1}\right) = (70 \text{ lbm})(0.092 \text{ Btu/lbm}\cdot\text{R}) \ln\left(\frac{546.8 \text{ R}}{640 \text{ R}}\right) = -1.212 \text{ Btu/R}$$

$$\Delta S_{water} = mC_{ave} \ln\left(\frac{T_2}{T_1}\right) = (62 \text{ lbm})(1.0 \text{ Btu/lbm}\cdot\text{R}) \ln\left(\frac{546.8 \text{ R}}{535 \text{ R}}\right) = 1.353 \text{ Btu/R}$$

Substituting,

$$X_{\text{destroyed}} = (535 \text{ R})(-1.212 + 1.353) \text{Btu/R} = \mathbf{75.4 \text{ Btu}}$$

7-44 A hot iron block is dropped into water in an insulated tank that is stirred by a paddle-wheel. The mass of the iron block and the exergy destroyed during this process are to be determined. √

Assumptions **1** Both the water and the iron block are incompressible substances with constant specific heats at room temperature. **2** The system is stationary and thus the kinetic and potential energies are negligible. **3** The tank is well-insulated and thus there is no heat transfer.

Properties The density and specific heat of water at 25°C are ρ = 997 kg/m^3 and C_p = 4.18 kJ/kg.°F. The specific heat of iron at room temperature (the only value available in the tables) is C_p = 0.45 kJ/kg.°C (Table A-3).

Analysis We take the entire contents of the tank, water + iron block, as the system, which is a closed system. The energy balance for this system can be expressed as

100 L
20°C
Iron
85°C
Water
W$_{pw}$

$$\underbrace{E_{in} - E_{out}}_{\substack{\text{Net energy transfer}\\ \text{by heat, work, and mass}}} = \underbrace{\Delta E_{\text{system}}}_{\substack{\text{Change in internal, kinetic,}\\ \text{potential, etc. energies}}}$$

$$W_{pw,in} = \Delta U = \Delta U_{iron} + \Delta U_{water}$$

$$W_{pw,in} = [mC(T_2 - T_1)]_{iron} + [mC(T_2 - T_1)]_{water}$$

where

$$m_{water} = \rho V = (997\ \text{kg/m}^3)(0.1\ \text{m}^3) = 99.7\ \text{kg}$$

$$W_{pw} = \dot{W}_{pw,in}\Delta t = (0.2\ \text{kJ/s})(20 \times 60\ \text{s}) = 240\ \text{kJ}$$

Substituting,

$$240\ \text{kJ} = m_{iron}(0.45\ \text{kJ/kg}\cdot{}^\circ\text{C})(24-85)^\circ\text{C} + (99.7\ \text{kg})(4.18\ \text{kJ/kg}\cdot{}^\circ\text{C})(24-20)^\circ\text{C}$$

$$m_{iron} = \mathbf{52.0\ kg}$$

(*b*) The exergy destruction (or irreversibility) can be determined from its definition $X_{\text{destroyed}} = T_0 S_{\text{gen}}$ where the entropy generation is determined from an entropy balance on the system, which is an insulated closed system,

$$\underbrace{S_{in} - S_{out}}_{\substack{\text{Net entropy transfer}\\ \text{by heat and mass}}} + \underbrace{S_{gen}}_{\substack{\text{Entropy}\\ \text{generation}}} = \underbrace{\Delta S_{\text{system}}}_{\substack{\text{Change}\\ \text{in entropy}}}$$

$$S_{\text{gen}} = \Delta S_{\text{system}} = \Delta S_{\text{iron}} + \Delta S_{\text{water}}$$

where

$$\Delta S_{iron} = mC_{ave}\ln\left(\frac{T_2}{T_1}\right) = (52.0\ \text{kg})(0.45\ \text{kJ/kg}\cdot\text{K})\ln\left(\frac{297\ \text{K}}{358\ \text{K}}\right) = -4.371\ \text{kJ/K}$$

$$\Delta S_{water} = mC_{ave}\ln\left(\frac{T_2}{T_1}\right) = (99.7\ \text{kg})(4.18\ \text{kJ/kg}\cdot\text{K})\ln\left(\frac{297\ \text{K}}{293\ \text{K}}\right) = 5.651\ \text{kJ/K}$$

Substituting,

$$X_{\text{destroyed}} = T_0 S_{gen} = (293\ \text{K})(-4.371 + 5.651)\ \text{kJ/K} = \mathbf{375.0\ kJ}$$

7-45 An iron block and a copper block are dropped into a large lake where they cool to lake temperature. The amount of work that could have been produced is to be determined.

Assumptions **1** The iron and copper blocks and water are incompressible substances with constant specific heats at room temperature. **2** Kinetic and potential energies are negligible.

Properties The specific heats of iron and copper at room temperature are $C_{p,\text{ iron}}$ = 0.45 kJ/kg.°C and $C_{p,\text{copper}}$ = 0.386 kJ/kg.°C (Table A-3).

Analysis The thermal-energy capacity of the lake is very large, and thus the temperatures of both the iron and the copper blocks will drop to the lake temperature (15°C) when the thermal equilibrium is established.

We take both the iron and the copper blocks as the system, which is a closed system. The energy balance for this system can be expressed as

$$\underbrace{E_{in}-E_{out}}_{\substack{\text{Net energy transfer}\\ \text{by heat, work, and mass}}} = \underbrace{\Delta E_{\text{system}}}_{\substack{\text{Change in internal, kinetic,}\\ \text{potential, etc. energies}}}$$

$$-Q_{out} = \Delta U = \Delta U_{iron} + \Delta U_{copper}$$

or,

$$Q_{out} = [mC(T_1 - T_2)]_{iron} + [mC(T_1 - T_2)]_{copper}$$

Substituting,

$$Q_{out} = (50\text{ kg})(0.45\text{ kJ/kg}\cdot\text{K})(353-288)\text{K} + (20\text{ kg})(0.386\text{ kJ/kg}\cdot\text{K})(353-288)\text{K}$$
$$= 1964\text{ kJ}$$

The work that could have been produced is equal to the wasted work potential. It is equivalent to the exergy destruction (or irreversibility), and it can be determined from its definition $X_{\text{destroyed}} = T_0 S_{\text{gen}}$. The entropy generation is determined from an entropy balance on an *extended system* that includes the blocks and the water in their immediate surroundings so that the boundary temperature of the extended system is the temperature of the lake water at all times,

$$\underbrace{S_{in}-S_{out}}_{\substack{\text{Net entropy transfer}\\ \text{by heat and mass}}} + \underbrace{S_{gen}}_{\substack{\text{Entropy}\\ \text{generation}}} = \underbrace{\Delta S_{\text{system}}}_{\substack{\text{Change}\\ \text{in entropy}}}$$

$$-\frac{Q_{out}}{T_{b,out}} + S_{\text{gen}} = \Delta S_{\text{system}} = \Delta S_{\text{iron}} + \Delta S_{\text{copper}}$$

$$S_{\text{gen}} = \Delta S_{\text{iron}} + \Delta S_{\text{copper}} + \frac{Q_{out}}{T_{lake}}$$

where

$$\Delta S_{iron} = mC_{ave}\ln\left(\frac{T_2}{T_1}\right) = (50\text{ kg})(0.45\text{ kJ/kg}\cdot\text{K})\ln\left(\frac{288\text{ K}}{353\text{ K}}\right) = -4.579\text{ kJ/K}$$

$$\Delta S_{copper} = mC_{ave}\ln\left(\frac{T_2}{T_1}\right) = (20\text{ kg})(0.386\text{ kJ/kg}\cdot\text{K})\ln\left(\frac{288\text{ K}}{353\text{ K}}\right) = -1.571\text{ kJ/K}$$

Substituting,

$$X_{\text{destroyed}} = T_0 S_{gen} = (293\text{ K})\left(-4.579 - 1.571 + \frac{1964\text{ kJ}}{288\text{ K}}\right)\text{kJ/K} = \mathbf{196\ kJ}$$

7-46E A rigid tank is initially filled with saturated mixture of R-134a. Heat is transferred to the tank from a source until the pressure inside rises to a specified value. The amount of heat transfer to the tank from the source and the exergy destroyed are to be determined.

Assumptions **1** The tank is stationary and thus the kinetic and potential energy changes are zero. **2** There is no heat transfer with the environment.

Properties From the refrigerant tables (Tables A-11 through A-13),

$$\left.\begin{array}{l} P_1 = 30\text{ psia} \\ x_1 = 0.4 \end{array}\right\} \begin{array}{l} u_1 = u_f + x_1 u_{fg} = 16.24 + 0.4\times(95.40 - 16.24) = 47.90\text{ Btu / lbm} \\ s_1 = s_f + x_1 s_{fg} = 0.0364 + 0.4\times(0.2209 - 0.0364) = 0.1102\text{ Btu / lbm}\cdot\text{R} \\ \text{v}_1 = \text{v}_f + x_1 \text{v}_{fg} = 0.01209 + 0.4\times(1.5408 - 0.01209) = 0.6236\text{ ft}^3\text{ / lbm} \end{array}$$

$$\left.\begin{array}{l} P_2 = 60\text{ psia} \\ (\text{v}_2 = \text{v}_1) \end{array}\right\} \begin{array}{l} x_2 = \dfrac{\text{v}_2 - \text{v}_f}{\text{v}_{fg}} = \dfrac{0.6236 - 0.01270}{0.7887 - 0.01270} = 0.7872 \\ s_2 = s_f + x_2 s_{fg} = 0.0584 + 0.7872\times(0.2183 - 0.0584) = 0.1843\text{ Btu/lbm}\cdot\text{R} \\ u_2 = u_f + x_2 u_{fg} = 27.10 + 0.7872\times(99.96 - 27.10) = 84.46\text{ Btu/lbm} \end{array}$$

Analysis (*a*) The mass of the refrigerant is

$$m = \frac{V}{\text{v}_1} = \frac{12\text{ ft}^3}{0.6236\text{ ft}^3\text{ / lbm}} = 19.24\text{ lbm}$$

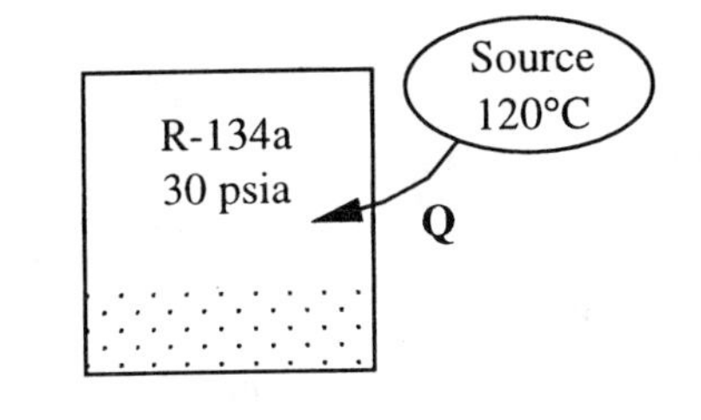

We take the tank as the system, which is a closed system. The energy balance for this stationary closed system can be expressed as

$$\underbrace{E_{in} - E_{out}}_{\substack{\text{Net energy transfer}\\ \text{by heat, work, and mass}}} = \underbrace{\Delta E_{\text{system}}}_{\substack{\text{Change in internal, kinetic,}\\ \text{potential, etc. energies}}}$$

$$Q_{in} = \Delta U = m(u_2 - u_1)$$

Substituting,

$$Q_{in} = m(u_2 - u_1) = (19.24\text{ lbm})(84.46 - 47.90)\text{ Btu/lbm} = \mathbf{703.4\ Btu}$$

(*b*) The exergy destruction (or irreversibility) can be determined from its definition $X_{\text{destroyed}} = T_0 S_{\text{gen}}$. The entropy generation is determined from an entropy balance on an *extended system* that includes the tank and the region in its immediate surroundings so that the boundary temperature of the extended system where heat transfer occurs is the source temperature,

$$\underbrace{S_{in} - S_{out}}_{\substack{\text{Net entropy transfer}\\ \text{by heat and mass}}} + \underbrace{S_{gen}}_{\substack{\text{Entropy}\\ \text{generation}}} = \underbrace{\Delta S_{\text{system}}}_{\substack{\text{Change}\\ \text{in entropy}}}$$

$$\frac{Q_{in}}{T_{b,in}} + S_{\text{gen}} = \Delta S_{\text{system}} = m(s_2 - s_1),$$

$$S_{\text{gen}} = m(s_2 - s_1) - \frac{Q_{in}}{T_{source}}$$

Substituting,

$$X_{\text{destroyed}} = T_0 S_{gen} = (535\text{ R})\left[(19.24\text{ lbm})(0.1843 - 0.1102)\text{Btu/lbm}\cdot\text{R} - \frac{703.4\text{ Btu}}{580\text{ R}}\right] = \mathbf{114\ Btu}$$

7-47 Chickens are to be cooled by chilled water in an immersion chiller that is also gaining heat from the surroundings. The rate of heat removal from the chicken and the rate of exergy destruction during this process are to be determined.

Assumptions **1** Steady operating conditions exist. **2** Thermal properties of chickens and water are constant. **3** The temperature of the surrounding medium is 25°C.

Properties The specific heat of chicken is given to be 3.54 kJ/kg.°C. The specific heat of water at room temperature is 4.18 kJ/kg.°C (Table A-3).

Analysis (a) Chickens are dropped into the chiller at a rate of 500 per hour. Therefore, chickens can be considered to flow steadily through the chiller at a mass flow rate of

$$\dot{m}_{\text{chicken}} = (500 \text{ chicken / h})(2.2 \text{ kg / chicken}) = 1100 \text{ kg / h} = 0.3056 \text{ kg / s}$$

Taking the chicken flow stream in the chiller as the system, the energy balance for steadily flowing chickens can be expressed in the rate form as

$$\underbrace{\dot{E}_{in} - \dot{E}_{out}}_{\substack{\text{Rate of net energy transfer} \\ \text{by heat, work, and mass}}} = \underbrace{\Delta \dot{E}_{\text{system}}^{\nearrow 0 \text{ (steady)}}}_{\substack{\text{Rate of change in internal, kinetic,} \\ \text{potential, etc. energies}}} = 0 \rightarrow \dot{E}_{in} = \dot{E}_{out}$$

$$\dot{m}h_1 = \dot{Q}_{out} + \dot{m}h_2 \quad (\text{since } \Delta\text{ke} \cong \Delta\text{pe} \cong 0)$$

$$\dot{Q}_{out} = \dot{Q}_{\text{chicken}} = \dot{m}_{\text{chicken}} C_p (T_1 - T_2)$$

Then the rate of heat removal from the chickens as they are cooled from 15°C to 3°C becomes

$$\dot{Q}_{\text{chicken}} = (\dot{m}C_p\Delta T)_{\text{chicken}} = (0.3056 \text{ kg/s})(3.54 \text{ kJ/kg.°C})(15-3)\text{°C} = \mathbf{13.0 \text{ kW}}$$

The chiller gains heat from the surroundings as a rate of 200 kJ/h = 0.0556 kJ/s. Then the total rate of heat gain by the water is

$$\dot{Q}_{\text{water}} = \dot{Q}_{\text{chicken}} + \dot{Q}_{\text{heat gain}} = 13.0 + 0.056 = 13.056 \text{ kW}$$

Noting that the temperature rise of water is not to exceed 2°C as it flows through the chiller, the mass flow rate of water must be at least

$$\dot{m}_{\text{water}} = \frac{\dot{Q}_{\text{water}}}{(C_p\Delta T)_{\text{water}}} = \frac{13.056 \text{ kW}}{(4.18 \text{ kJ/kg.°C})(2\text{°C})} = \mathbf{1.56 \text{ kg/s}}$$

(*b*) The exergy destruction can be determined from its definition $X_{\text{destroyed}} = T_0 S_{\text{gen}}$ The rate of entropy generation during this chilling process is determined by applying the rate form of the entropy balance on an *extended system* that includes the chiller and the immediate surroundings so that the boundary temperature is the surroundings temperature:

$$\underbrace{\dot{S}_{in} - \dot{S}_{out}}_{\substack{\text{Rate of net entropy transfer} \\ \text{by heat and mass}}} + \underbrace{\dot{S}_{gen}}_{\substack{\text{Rate of entropy} \\ \text{generation}}} = \underbrace{\Delta \dot{S}_{\text{system}}^{\nearrow 0 \text{ (steady)}}}_{\substack{\text{Rate of change} \\ \text{of entropy}}}$$

$$\dot{m}_1 s_1 + \dot{m}_3 s_3 - \dot{m}_2 s_2 - \dot{m}_3 s_4 + \frac{\dot{Q}_{in}}{T_{surr}} + \dot{S}_{gen} = 0$$

$$\dot{m}_{\text{chicken}} s_1 + \dot{m}_{\text{water}} s_3 - \dot{m}_{\text{chicken}} s_2 - \dot{m}_{\text{water}} s_4 + \frac{\dot{Q}_{in}}{T_{surr}} + \dot{S}_{gen} = 0$$

$$\dot{S}_{gen} = \dot{m}_{\text{chicken}} (s_2 - s_1) + \dot{m}_{\text{water}} (s_4 - s_3) - \frac{\dot{Q}_{in}}{T_{surr}}$$

Noting that both streams are incompressible substances, the rate of entropy generation is determined to be

$$\begin{aligned}\dot{S}_{gen} &= \dot{m}_{\text{chicken}} C_p \ln\frac{T_2}{T_1} + \dot{m}_{\text{water}} C_p \ln\frac{T_4}{T_3} - \frac{\dot{Q}_{in}}{T_{surr}} \\ &= (0.3056 \text{ kg/s})(3.54 \text{ kJ/kg.K})\ln\frac{276}{288} + (1.56 \text{ kg/s})(4.18 \text{ kJ/kg.K})\ln\frac{275.5}{273.5} - \frac{0.0556 \text{ kW}}{298 \text{ K}} \\ &= \mathbf{0.00128 \text{ kW/K}}\end{aligned}$$

Finally,

$$\dot{X}_{\text{destroyed}} = T_0 \dot{S}_{gen} = (298 \text{ K})(0.00128 \text{ kW/K}) = \mathbf{0.381 \text{ kW/K}}$$

7-48 An egg is dropped into boiling water. The amount of heat transfer to the egg by the time it is cooked and the amount of exergy destruction associated with this heat transfer process are to be determined.

Assumptions **1** The egg is spherical in shape with a radius of $r_0 = 2.75$ cm. **2** The thermal properties of the egg are constant. **3** Energy absorption or release associated with any chemical and/or phase changes within the egg is negligible. **4** There are no changes in kinetic and potential energies. **5** The temperature of the surrounding medium is 25°C.

Properties The density and specific heat of the egg are given to be $\rho = 1020$ kg/m^3 and $C_p = 3.32$ kJ/kg.°C.

Analysis We take the egg as the system. This is a closed system since no mass enters or leaves the egg. The energy balance for this closed system can be expressed as

$$\underbrace{E_{in} - E_{out}}_{\text{Net energy transfer by heat, work, and mass}} = \underbrace{\Delta E_{\text{system}}}_{\text{Change in internal, kinetic, potential, etc. energies}}$$

$$Q_{in} = \Delta U_{egg} = m(u_2 - u_1) = mC(T_2 - T_1)$$

Then the mass of the egg and the amount of heat transfer become

$$m = \rho V = \rho \frac{\pi D^3}{6} = (1020\,\text{kg/m}^3)\frac{\pi(0.055\,\text{m})^3}{6} = 0.0889\,\text{kg}$$

$$Q_{in} = mC_p(T_2 - T_1) = (0.0889\,\text{kg})(3.32\,\text{kJ/kg.°C})(70 - 8)\text{°C} = \mathbf{18.3\ kJ}$$

The exergy destruction can be determined from its definition $X_{\text{destroyed}} = T_0 S_{\text{gen}}$ The entropy generated during this process can be determined by applying an entropy balance on an *extended system* that includes the egg and its immediate surroundings so that the boundary temperature of the extended system is at 97°C at all times:

$$\underbrace{S_{in} - S_{out}}_{\text{Net entropy transfer by heat and mass}} + \underbrace{S_{gen}}_{\text{Entropy generation}} = \underbrace{\Delta S_{\text{system}}}_{\text{Change in entropy}}$$

$$\frac{Q_{in}}{T_b} + S_{gen} = \Delta S_{\text{system}} \quad \rightarrow \quad S_{gen} = -\frac{Q_{in}}{T_b} + \Delta S_{\text{system}}$$

where

$$\Delta S_{\text{system}} = m(s_2 - s_1) = mC_{av} \ln\frac{T_2}{T_1} = (0.0889\ \text{kg})(3.32\ \text{kJ/kg.K})\ln\frac{70+273}{8+273} = 0.0588\ \text{kJ/K}$$

Substituting,

$$S_{gen} = -\frac{Q_{in}}{T_b} + \Delta S_{\text{system}} = -\frac{18.3\,\text{kJ}}{370\,\text{K}} + 0.0588\,\text{kJ/K} = 0.00934\ \text{kJ/K}\quad \text{(per egg)}$$

Finally,

$$\dot{X}_{\text{destroyed}} = T_0 \dot{S}_{gen} = (298\ \text{K})(0.00934\,\text{kJ/K}) = \mathbf{2.78\ kJ}$$

7-49 Stainless steel ball bearings leaving the oven at a uniform temperature of 900°C at a rate of 1400 /min are exposed to air and are cooled to 850°C before they are dropped into the water for quenching. The rate of heat transfer from the ball to the air and the rate of exergy destruction due to this heat transfer are to be determined.

Assumptions **1** The thermal properties of the bearing balls are constant. **2** The kinetic and potential energy changes of the balls are negligible. **3** The balls are at a uniform temperature at the end of the process

Properties The density and specific heat of the ball bearings are given to be ρ = 8085 kg/m³ and C_p = 0.480 kJ/kg.°C.

Analysis We take a single bearing ball as the system. The energy balance for this closed system can be expressed as

$$\underbrace{E_{in} - E_{out}}_{\substack{\text{Net energy transfer}\\ \text{by heat, work, and mass}}} = \underbrace{\Delta E_{\text{system}}}_{\substack{\text{Change in internal, kinetic,}\\ \text{potential, etc. energies}}}$$

$$-Q_{out} = \Delta U_{ball} = m(u_2 - u_1)$$

$$Q_{out} = mC(T_1 - T_2)$$

The total amount of heat transfer from a ball is

$$m = \rho V = \rho\frac{\pi D^3}{6} = (8085\ \text{kg/m}^3)\frac{\pi(0.012\ \text{m})^3}{6} = 0.007315\ \text{kg}$$

$$Q_{out} = mC(T_1 - T_2) = (0.007315\ \text{kg})(0.480\ \text{kJ/kg.°C})(900 - 850)\text{°C} = 0.1756\ \text{kJ/ball}$$

Then the rate of heat transfer from the balls to the air becomes

$$\dot{Q}_{total} = \dot{n}_{\text{ball}}Q_{\text{out (per ball)}} = (1400\ \text{balls/min})\times(0.1756\ \text{kJ/ball}) = \mathbf{245.8\ kJ/min} = \mathbf{4.10\ kW}$$

Therefore, heat is lost to the air at a rate of 4.10 kW.

(*b*) The exergy destruction can be determined from its definition $X_{\text{destroyed}} = T_0 S_{\text{gen}}$ The entropy generated during this process can be determined by applying an entropy balance on an *extended system* that includes the ball and its immediate surroundings so that the boundary temperature of the extended system is at 30°C at all times:

$$\underbrace{S_{in} - S_{out}}_{\substack{\text{Net entropy transfer}\\ \text{by heat and mass}}} + \underbrace{S_{gen}}_{\substack{\text{Entropy}\\ \text{generation}}} = \underbrace{\Delta S_{\text{system}}}_{\substack{\text{Change}\\ \text{in entropy}}}$$

$$-\frac{Q_{out}}{T_b} + S_{gen} = \Delta S_{\text{system}} \rightarrow S_{gen} = \frac{Q_{out}}{T_b} + \Delta S_{\text{system}}$$

where

$$\Delta S_{\text{system}} = m(s_2 - s_1) = mC_{av}\ln\frac{T_2}{T_1} = (0.007315\ \text{kg})(0.480\ \text{kJ/kg.K})\ln\frac{850+273}{900+273} = -0.0001530\ \text{kJ/K}$$

Substituting,

$$S_{gen} = \frac{Q_{out}}{T_b} + \Delta S_{\text{system}} = \frac{0.1756\ \text{kJ}}{303\ \text{K}} - 0.0001530\ \text{kJ/K} = 0.0004265\ \text{kJ/K}\quad\text{(per ball)}$$

Then the rate of entropy generation becomes

$$\dot{S}_{gen} = S_{gen}\dot{n}_{ball} = (0.0004265\ \text{kJ/K}\cdot\text{ball})(1400\ \text{balls/min}) = 0.597\ \text{kJ/min.K} = \mathbf{0.00995\ kW/K}$$

Finally,

$$\dot{X}_{\text{destroyed}} = T_0\dot{S}_{gen} = (303\ \text{K})(0.00995\ \text{kW/K}) = \mathbf{3.01\ kW/K}$$

7-50 Carbon steel balls are to be annealed at a rate of 2500/h by heating them first and then allowing them to cool slowly in ambient air at a specified rate. The total rate of heat transfer from the balls to the ambient air and the rate of exergy destruction due to this heat transfer are to be determined.

Assumptions **1** The thermal properties of the balls are constant. **2** There are no changes in kinetic and potential energies. **3** The balls are at a uniform temperature at the end of the process.

Properties The density and specific heat of the balls are given to be ρ = 7833 kg/m^3 and C_p = 0.465 kJ/kg.°C.

Analysis (*a*) We take a single ball as the system. The energy balance for this closed system can be expressed as

$$\underbrace{E_{in} - E_{out}}_{\substack{\text{Net energy transfer}\\\text{by heat, work, and mass}}} = \underbrace{\Delta E_{\text{system}}}_{\substack{\text{Change in internal, kinetic,}\\\text{potential, etc. energies}}}$$

$$-Q_{out} = \Delta U_{ball} = m(u_2 - u_1)$$

$$Q_{out} = mC(T_1 - T_2)$$

The amount of heat transfer from a single ball is

$$m = \rho V = \rho\frac{\pi D^3}{6} = (7833\ \text{kg/m}^3)\frac{\pi(0.008\ \text{m})^3}{6} = 0.00210\ \text{kg}$$

$$Q_{out} = mC_p(T_1 - T_2) = (0.0021\ \text{kg})(0.465\ \text{kJ/kg.°C})(900-100)\text{°C} = 781\ \text{J} = 0.781\ \text{kJ (per ball)}$$

Then the total rate of heat transfer from the balls to the ambient air becomes

$$\dot{Q}_{out} = \dot{n}_{\text{ball}} Q_{out} = (1200\ \text{balls/h}) \times (0.781\ \text{kJ/ball}) = 936\ \text{kJ/h} = \mathbf{260\ \ W}$$

(*b*) The exergy destruction (or irreversibility) can be determined from its definition $X_{\text{destroyed}} = T_0 S_{\text{gen}}$ The entropy generated during this process can be determined by applying an entropy balance on an *extended system* that includes the ball and its immediate surroundings so that the boundary temperature of the extended system is at 35°C at all times:

$$\underbrace{S_{in} - S_{out}}_{\substack{\text{Net entropy transfer}\\\text{by heat and mass}}} + \underbrace{S_{gen}}_{\substack{\text{Entropy}\\\text{generation}}} = \underbrace{\Delta S_{\text{system}}}_{\substack{\text{Change}\\\text{in entropy}}}$$

$$-\frac{Q_{out}}{T_b} + S_{gen} = \Delta S_{\text{system}} \rightarrow S_{gen} = \frac{Q_{out}}{T_b} + \Delta S_{\text{system}}$$

where

$$\Delta S_{\text{system}} = m(s_2 - s_1) = mC_{av}\ln\frac{T_2}{T_1} = (0.00210\ \text{kg})(0.465\ \text{kJ/kg.K})\ln\frac{100+273}{900+273} = -0.00112\ \text{kJ/K}$$

Substituting,

$$S_{gen} = \frac{Q_{out}}{T_b} + \Delta S_{\text{system}} = \frac{0.781\ \text{kJ}}{308\ \text{K}} - 0.00112\ \text{kJ/K} = 0.00142\ \text{kJ/K}\quad\text{(per ball)}$$

Then the rate of entropy generation becomes

$$\dot{S}_{gen} = S_{gen}\dot{n}_{ball} = (0.00142\ \text{kJ/K}\cdot\text{ball})(1200\ \text{balls/h}) = 1.704\ \text{kJ/h.K} = 0.000473\ \ \text{kW/K}$$

Finally,

$$\dot{X}_{\text{destroyed}} = T_0\dot{S}_{gen} = (308\ \text{K})(0.000473\ \text{kW/K}) = 0.146\ \text{kW} = \mathbf{146\ \ W}$$

Second-Law Analysis of Control Volumes

7-51 Steam is throttled from a specified state to a specified pressure. The wasted work potential during this throttling process is to be determined.

Assumptions **1** This is a steady-flow process since there is no change with time. **2** Kinetic and potential energy changes are negligible. **3** The temperature of the surroundings is given to be 25°C. **4** Heat transfer is negligible.

Properties The properties of steam before and after the throttling process are (Tables A-4 through A-6)

$$\left.\begin{array}{l} P_1 = 10\text{ MPa} \\ T_1 = 450°\text{C} \end{array}\right\} \begin{array}{l} h_1 = 3240.9\text{ kJ/kg} \\ s_1 = 6.4190\text{ kJ/kg}\cdot\text{K} \end{array}$$

$$\left.\begin{array}{l} P_2 = 8\text{ MPa} \\ h_2 = h_1 \end{array}\right\} s_2 = 6.5105\text{ kJ/kg}\cdot\text{K}$$

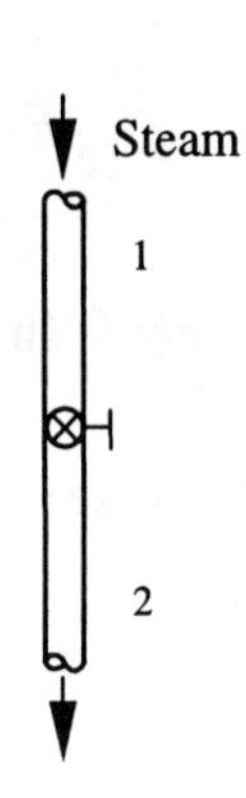

Analysis The wasted work potential is equivalent to the exergy destruction (or irreversibility). It can be determined from an exergy balance or directly from its definition $X_{\text{destroyed}} = T_0 S_{\text{gen}}$ where the entropy generation is determined from an entropy balance on the device, which is an adiabatic steady-flow system,

$$\underbrace{\dot{S}_{in} - \dot{S}_{out}}_{\substack{\text{Rate of net entropy transfer} \\ \text{by heat and mass}}} + \underbrace{\dot{S}_{gen}}_{\substack{\text{Rate of entropy} \\ \text{generation}}} = \underbrace{\Delta\dot{S}_{\text{system}}^{\nearrow 0}}_{\substack{\text{Rate of change} \\ \text{of entropy}}} = 0$$

$$\dot{m}s_1 - \dot{m}s_2 + \dot{S}_{\text{gen}} = 0 \rightarrow \dot{S}_{\text{gen}} = \dot{m}(s_2 - s_1) \text{ or } s_{\text{gen}} = s_2 - s_1$$

Substituting,

$$x_{\text{destroyed}} = T_0 s_{gen} = T_0(s_2 - s_1) = (298\text{ K})(6.5105 - 6.4190)\text{ kJ/kg}\cdot\text{K} = \mathbf{27.3\text{ kJ/kg}}$$

Discussion Note that 27.3 kJ/kg of work potential is wasted during this throttling process.

7-52 [*Also solved by EES on enclosed CD*] Air is compressed steadily by an 8-kW compressor from a specified state to another specified state. The increase in the exergy of air and the rate of exergy destruction are to be determined.

Assumptions **1** Air is an ideal gas with variable specific heats. **2** Kinetic and potential energy changes are negligible.

Properties The gas constant of air is R = 0.287 kJ/kg.K (Table A-1). From the air table (Table A-17)

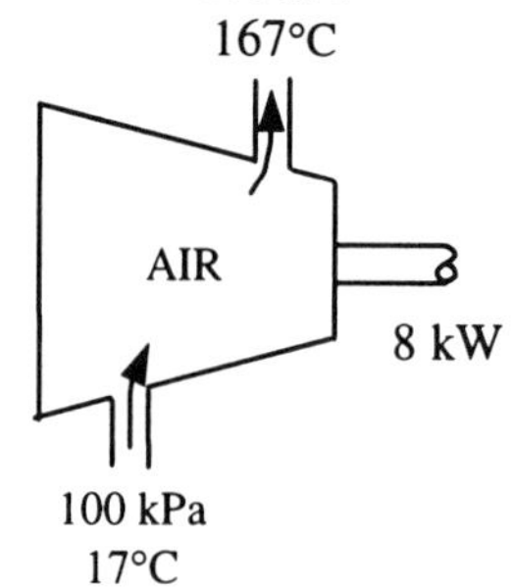

$$T_1 = 290\text{ K} \longrightarrow \begin{array}{l} h_1 = 290.16\text{ kJ/kg} \\ s_1^o = 1.66802\text{ kJ/kg}\cdot\text{K} \end{array}$$

$$T_2 = 440\text{ K} \longrightarrow \begin{array}{l} h_2 = 441.61\text{ kJ/kg} \\ s_2^o = 2.0887\text{ kJ/kg}\cdot\text{K} \end{array}$$

Analysis The increase in exergy is the difference between the exit and inlet flow exergies,

$$\text{Increase in exergy} = \psi_2 - \psi_1 = [(h_2 - h_1) + \Delta ke^{\not 0} + \Delta pe^{\not 0} - T_0(s_2 - s_1)] = (h_2 - h_1) - T_0(s_2 - s_1)$$

where

$$\begin{aligned} s_2 - s_1 &= (s_2^o - s_1^o) - R\ln\frac{P_2}{P_1} \\ &= (2.0887 - 1.66802)\text{kJ/kg}\cdot\text{K} - (0.287\text{ kJ/kg}\cdot\text{K})\ln\frac{600\text{ kPa}}{100\text{ kPa}} \\ &= -0.09356\text{ kJ/kg}\cdot\text{K} \end{aligned}$$

Substituting,

$$\text{Increase in exergy} = \psi_2 - \psi_1 = \left[(441.61 - 290.16)\text{kJ/kg} - (290\text{ K})(-0.09356\text{ kJ/kg}\cdot\text{K})\right] = \mathbf{178.6\ kJ/kg}$$

Then the reversible power input becomes

$$\dot{W}_{rev,in} = \dot{m}(\psi_2 - \psi_1) = (2.1/60\text{ kg/s})(178.6\text{ kJ/kg}) = \mathbf{6.25\ kW}$$

(*b*) The rate of exergy destruction (or irreversibility) is determined from its definition,

$$\dot{X}_{\text{destroyed}} = \dot{W}_{in} - \dot{W}_{rev,in} = 8 - 6.25 = \mathbf{1.75\ kW}$$

Discussion Note that 1.75 kW of power input is wasted during this compression process.

7-53 EES solution of this (and other comprehensive problems designated with the *computer icon*) is available to instructors at the *Instructor Manual* section of the *Online Learning Center* (OLC) at www.mhhe.com/cengel-boles. See the Preface for access information.

7-54 Refrigerant-124a is throttled from a specified state to a specified pressure. The reversible work and the exergy destroyed during this process are to be determined.

Assumptions **1** This is a steady-flow process since there is no change with time. **2** Kinetic and potential energy changes are negligible. **3** Heat transfer is negligible.

Properties The properties of R-134a before and after the throttling process are (Tables A-11 through A-13)

$$\left.\begin{array}{l} P_1 = 1\,\text{MPa} \\ T_1 = 100°\text{C} \end{array}\right\} \begin{array}{l} h_1 = 334.82\,\text{kJ/kg} \\ s_1 = 1.1000\,\text{kJ/kg}\cdot\text{K} \end{array}$$

$$\left.\begin{array}{l} P_2 = 0.8\,\text{MPa} \\ h_2 = h_1 \end{array}\right\} s_2 = 1.1166\,\text{kJ/kg}\cdot\text{K}$$

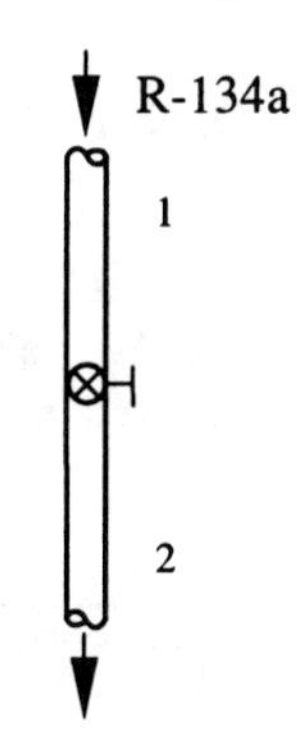

Analysis The exergy destruction (or irreversibility) can be determined from an exergy balance or directly from its definition $X_{\text{destroyed}} = T_0 S_{\text{gen}}$ where the entropy generation is determined from an entropy balance on the system, which is an adiabatic steady-flow device,

$$\underbrace{\dot{S}_{in} - \dot{S}_{out}}_{\substack{\text{Rate of net entropy transfer} \\ \text{by heat and mass}}} + \underbrace{\dot{S}_{gen}}_{\substack{\text{Rate of entropy} \\ \text{generation}}} = \underbrace{\Delta \dot{S}_{\text{system}}^{\nearrow 0}}_{\substack{\text{Rate of change} \\ \text{of entropy}}} = 0$$

$$\dot{m}s_1 - \dot{m}s_2 + \dot{S}_{\text{gen}} = 0 \rightarrow \dot{S}_{\text{gen}} = \dot{m}(s_2 - s_1) \quad \text{or} \quad s_{\text{gen}} = s_2 - s_1$$

Substituting,

$$x_{\text{destroyed}} = T_0 s_{gen} = T_0(s_2 - s_1) = (303\,\text{K})(1.1166 - 1.1000)\text{kJ/kg}\cdot\text{K} = \mathbf{5.03\ kJ/kg}$$

This process involves no actual work, and thus the reversible work and irreversibility are identical,

$$x_{destroyed} = w_{rev,out} - w_{act,out}^{\nearrow 0} \rightarrow w_{rev,out} = x_{destroyed} = \mathbf{5.03\ kJ/kg}$$

Discussion Note that 4.95 kJ/kg of work potential is wasted during this throttling process.

7-55 EES solution of this (and other comprehensive problems designated with the *computer icon*) is available to instructors at the *Instructor Manual* section of the *Online Learning Center* (OLC) at www.mhhe.com/cengel-boles. See the Preface for access information.

7-56 Air is accelerated in a nozzle while losing some heat to the surroundings. The exit temperature of air and the exergy destroyed during the process are to be determined.

Assumptions **1** Air is an ideal gas with variable specific heats. **2** The nozzle operates steadily.

Properties The gas constant of air is R = 0.287 kJ/kg.K (Table A-1). The properties of air at the nozzle inlet are (Table A-17)

$$T_1 = 360\text{ K} \longrightarrow \begin{array}{l} h_1 = 360.58\text{ kJ/kg} \\ s_1^o = 1.88543\text{ kJ/kg}\cdot\text{K} \end{array}$$

Analysis (a) We take the nozzle as the system, which is a control volume. The energy balance for this steady-flow system can be expressed in the rate form as

$$\underbrace{\dot{E}_{in} - \dot{E}_{out}}_{\substack{\text{Rate of net energy transfer} \\ \text{by heat, work, and mass}}} = \underbrace{\Delta \dot{E}_{\text{system}}^{\nearrow 0\text{ (steady)}}}_{\substack{\text{Rate of change in internal, kinetic,} \\ \text{potential, etc. energies}}} = 0$$

$$\dot{E}_{in} = \dot{E}_{out}$$

$$\dot{m}(h_1 + \mathbf{V}_1^2/2) = \dot{m}(h_2 + \mathbf{V}_2^2/2) + \dot{Q}_{out}$$

$$0 = q_{out} + h_2 - h_1 + \frac{\mathbf{V}_2^2 - \mathbf{V}_1^2}{2}$$

4 kJ/kg

50 m/s → AIR → 300 m/s

Therefore,

$$h_2 = h_1 - q_{out} - \frac{\mathbf{V}_2^2 - \mathbf{V}_1^2}{2} = 360.58 - 4 - \frac{(300\text{ m/s})^2 - (50\text{ m/s})^2}{2}\left(\frac{1\text{ kJ/kg}}{1000\text{ m}^2/\text{s}^2}\right) = 312.83\text{ kJ/kg}$$

At this h_2 value we read, from Table A-17, $T_2 = \mathbf{312.5\ K = 39.5°C}$ and $s_2^0 = 1.74302\text{ kJ/kg}\cdot\text{K}$

(b) The exergy destroyed during a process can be determined from an exergy balance or directly from its definition $X_{\text{destroyed}} = T_0 S_{gen}$ where the entropy generation S_{gen} is determined from an entropy balance on an *extended system* that includes the device and its immediate surroundings so that the boundary temperature of the extended system is T_{surr} at all times. It gives

$$\underbrace{\dot{S}_{in} - \dot{S}_{out}}_{\substack{\text{Rate of net entropy transfer} \\ \text{by heat and mass}}} + \underbrace{\dot{S}_{gen}}_{\substack{\text{Rate of entropy} \\ \text{generation}}} = \underbrace{\Delta \dot{S}_{\text{system}}^{\nearrow 0}}_{\substack{\text{Rate of change} \\ \text{of entropy}}} = 0$$

$$\dot{m}s_1 - \dot{m}s_2 - \frac{\dot{Q}_{out}}{T_{b,surr}} + \dot{S}_{\text{gen}} = 0 \rightarrow \dot{S}_{gen} = \dot{m}(s_2 - s_1) + \frac{\dot{Q}_{out}}{T_{surr}}$$

where

$$\Delta s_{air} = s_2^o - s_1^o - R\ln\frac{P_2}{P_1} = (1.74302 - 1.88543)\text{ kJ/kg}\cdot\text{K} - (0.287\text{ kJ/kg}\cdot\text{K})\ln\frac{95\text{ kPa}}{300\text{ kPa}} = 0.1876\text{ kJ/kg}\cdot\text{K}$$

Substituting, the entropy generation and exergy destruction per unit mass of air are determined to be

$$x_{\text{destroyed}} = T_0 s_{gen} = T_{surr} s_{gen} = T_0\left(s_2 - s_1 + \frac{q_{surr}}{T_{surr}}\right) = (290\text{ K})\left(0.1876\text{ kJ/kg}\cdot\text{K} + \frac{4\text{ kJ/kg}}{290\text{ K}}\right) = \mathbf{58.4\ kJ/kg}$$

Alternative solution The exergy destroyed during a process can be determined from an exergy balance applied on the *extended system* that includes the device and its immediate surroundings so that the boundary temperature of the extended system is environment temperature T_0 (or T_{surr}) at all times. Noting that exergy transfer with heat is zero when the temperature at the point of transfer is the environment temperature, the exergy balance for this steady-flow system can be expressed as

$$\underbrace{\dot{X}_{in} - \dot{X}_{out}}_{\substack{\text{Rate of net exergy transfer} \\ \text{by heat, work, and mass}}} \; - \underbrace{\dot{X}_{destroyed}}_{\substack{\text{Rate of exergy} \\ \text{destruction}}} = \underbrace{\Delta \dot{X}_{system}^{\nearrow 0\ (steady)}}_{\substack{\text{Rate of change} \\ \text{of exergy}}} = 0$$

$$\dot{X}_{destroyed} = \dot{X}_{in} - \dot{X}_{out} = \dot{m}\psi_1 - \dot{m}\psi_2 = \dot{m}(\psi_1 - \psi_2)$$
$$= \dot{m}[(h_1 - h_2) - T_0(s_1 - s_2) - \Delta ke - \Delta pe^{\nearrow 0}]$$
$$= \dot{m}[T_0(s_2 - s_1) - (h_2 - h_1 + \Delta ke)]$$
$$= \dot{m}[T_0(s_2 - s_1) + q_{out}] \quad \text{since, from energy balance,} -q_{out} = h_2 - h_1 + \Delta ke$$
$$= T_0\left(\dot{m}(s_2 - s_1) + \frac{\dot{Q}_{out}}{T_0}\right) = T_0 \dot{S}_{gen}$$

Therefore, the two approaches for the determination of exergy destruction are identical.

7-57 EES solution of this (and other comprehensive problems designated with the *computer icon*) is available to instructors at the *Instructor Manual* section of the *Online Learning Center* (OLC) at www.mhhe.com/cengel-boles. See the Preface for access information.

7-58 Steam is decelerated in a diffuser. The mass flow rate of steam and the wasted work potential during the process are to be determined.

Assumptions **1** The diffuser operates steadily. **2** The changes in potential energies are negligible.

Properties The properties of steam at the inlet and the exit of the diffuser are (Tables A-4 through A-6)

$$\left.\begin{array}{l} P_1 = 10\ \text{kPa} \\ T_1 = 50°\text{C} \end{array}\right\} \begin{array}{l} h_1 = 2592.6\ \text{kJ/kg} \\ s_1 = 8.1749\ \text{kJ/kg}\cdot\text{K} \end{array}$$

$$\left.\begin{array}{l} T_2 = 50°\text{C} \\ \text{sat.vapor} \end{array}\right\} \begin{array}{l} h_2 = 2592.1\ \text{kJ/kg} \\ s_2 = 8.0763\ \text{kJ/kg}\cdot\text{K} \\ v_2 = 12.03\ \text{m}^3/\text{kg} \end{array}$$

300 m/s → H_2O → 50 m/s

Analysis (a) The mass flow rate of the steam is

$$\dot{m} = \frac{1}{v_2} A_2 \mathbf{V}_2 = \frac{1}{12.03\ \text{m}^3/\text{kg}}(2\ \text{m}^2)(50\ \text{m/s}) = \mathbf{8.31\ kg/s}$$

(*b*) We take the diffuser to be the system, which is a control volume. Assuming the direction of heat transfer to be from the stem, the energy balance for this steady-flow system can be expressed in the rate form as

$$\underbrace{\dot{E}_{in} - \dot{E}_{out}}_{\substack{\text{Rate of net energy transfer} \\ \text{by heat, work, and mass}}} = \underbrace{\Delta \dot{E}_{\text{system}}^{\nearrow 0\ \text{(steady)}}}_{\substack{\text{Rate of change in internal, kinetic,} \\ \text{potential, etc. energies}}} = 0$$

$$\dot{E}_{in} = \dot{E}_{out}$$

$$\dot{m}(h_1 + \mathbf{V}_1^2/2) = \dot{m}(h_2 + \mathbf{V}_2^2/2) + \dot{Q}_{out}$$

$$\dot{Q}_{out} = -\dot{m}\left(h_2 - h_1 + \frac{\mathbf{V}_2^2 - \mathbf{V}_1^2}{2}\right)$$

Substituting,

$$\dot{Q}_{out} = -(8.31\ \text{kg/s})\left[2592.1 - 2592.6 + \frac{(50\ \text{m/s})^2 - (300\ \text{m/s})^2}{2}\left(\frac{1\ \text{kJ/kg}}{1000\ \text{m}^2/\text{s}^2}\right)\right] = 367.7\ \text{kJ/s}$$

The wasted work potential is equivalent to exergy destruction. The exergy destroyed during a process can be determined from an exergy balance or directly from its definition $X_{\text{destroyed}} = T_0 S_{gen}$ where the entropy generation S_{gen} is determined from an entropy balance on an *extended system* that includes the device and its immediate surroundings so that the boundary temperature of the extended system is T_{surr} at all times. It gives

$$\underbrace{\dot{S}_{in} - \dot{S}_{out}}_{\substack{\text{Rate of net entropy transfer} \\ \text{by heat and mass}}} + \underbrace{\dot{S}_{gen}}_{\substack{\text{Rate of entropy} \\ \text{generation}}} = \underbrace{\Delta \dot{S}_{\text{system}}^{\nearrow 0}}_{\substack{\text{Rate of change} \\ \text{of entropy}}} = 0$$

$$\dot{m}s_1 - \dot{m}s_2 - \frac{\dot{Q}_{out}}{T_{b,surr}} + \dot{S}_{\text{gen}} = 0 \rightarrow \dot{S}_{gen} = \dot{m}(s_2 - s_1) + \frac{\dot{Q}_{out}}{T_{surr}}$$

Substituting, the exergy destruction is determined to be

$$\dot{X}_{\text{destroyed}} = T_0 \dot{S}_{gen} = T_0\left(\dot{m}(s_2 - s_1) + \frac{\dot{Q}_{out}}{T_0}\right)$$

$$= (298\ \text{K})\left((8.31\ \text{kg/s})(8.0763 - 8.1749)\text{kJ/kg}\cdot\text{K} + \frac{367.7\ \text{kW}}{298\ \text{K}}\right) = \mathbf{123.5\ kW}$$

7-59E Air is compressed steadily by a compressor from a specified state to another specified state. The minimum power input required for the compressor is to be determined.

Assumptions **1** Air is an ideal gas with variable specific heats. **2** Kinetic and potential energy changes are negligible.

100 psia
480°F

AIR
15 lbm/min

14.7 psia
60°F

Properties The gas constant of air is R = 0.06855 Btu/lbm.R (Table A-1E). From the air table (Table A-17E)

$$T_1 = 520\text{ R} \longrightarrow h_1 = 124.16\text{ Btu/lbm}$$
$$s_1^{\circ} = 0.59173\text{ Btu/lbm}\cdot\text{R}$$
$$T_2 = 940\text{ R} \longrightarrow h_2 = 226.11\text{ Btu/lbm}$$
$$s_2^{\circ} = 0.73509\text{ Btu/lbm}\cdot\text{R}$$

Analysis The reversible (or minimum) power input is determined from the rate form of the exergy balance applied on the compressor and setting the exergy destruction term equal to zero,

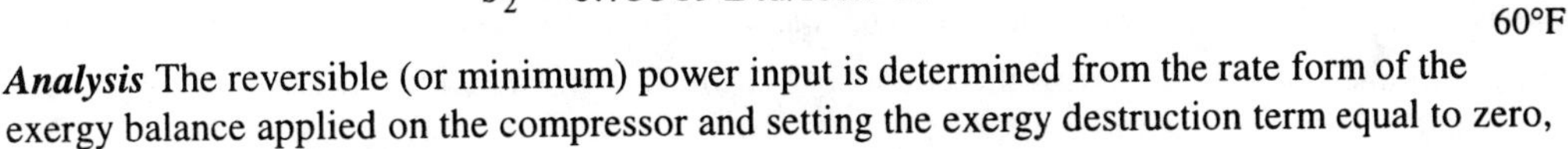

$$\underbrace{\dot{X}_{in} - \dot{X}_{out}}_{\text{Rate of net exergy transfer by heat, work, and mass}} - \underbrace{\dot{X}_{\text{destroyed}}^{\nearrow 0\text{ (reversible)}}}_{\text{Rate of exergy destruction}} = \underbrace{\Delta \dot{X}_{\text{system}}^{\nearrow 0\text{ (steady)}}}_{\text{Rate of change of exergy}} = 0$$

$$\dot{X}_{in} = \dot{X}_{out}$$
$$\dot{m}\psi_1 + \dot{W}_{rev,in} = \dot{m}\psi_2$$
$$\dot{W}_{rev,in} = \dot{m}(\psi_2 - \psi_1) = \dot{m}[(h_2 - h_1) - T_0(s_2 - s_1) + \Delta ke^{\nearrow 0} + \Delta pe^{\nearrow 0}]$$

where

$$\Delta s_{air} = s_2^{\circ} - s_1^{\circ} - R\ln\frac{P_2}{P_1}$$
$$= (0.73509 - 0.59173)\text{ Btu/lbm}\cdot\text{R} - (0.06855\text{ Btu/lbm}\cdot\text{R})\ln\frac{100\text{ psia}}{14.7\text{ psia}}$$
$$= 0.01193\text{ Btu/lbm}\cdot\text{R}$$

Substituting,

$$\dot{W}_{rev,in} = (22/60\text{ lbm/s})[(226.11 - 124.27)\text{Btu/lbm} - (520\text{ R})(0.01193\text{ Btu/lbm}\cdot\text{R})]$$
$$= 35.1\text{ Btu/s} = \mathbf{49.6\ hp}$$

Discussion Note that this is the minimum power input needed for this compressor.

7-60 Steam expands in a turbine from a specified state to another specified state. The actual power output of the turbine is given. The reversible power output and the second-law efficiency are to be determined.

Assumptions **1** This is a steady-flow process since there is no change with time. **2** Kinetic and potential energy changes are negligible. **3** The temperature of the surroundings is given to be 25°C.

Properties From the steam tables (Tables A-4 through A-6)

$$\left.\begin{array}{l} P_1 = 6\text{ MPa} \\ T_1 = 600°\text{C} \end{array}\right\} \begin{array}{l} h_1 = 3658.4\text{ kJ/kg} \\ s_1 = 7.1677\text{ kJ/kg}\cdot\text{K} \end{array}$$

$$\left.\begin{array}{l} P_2 = 50\text{ kPa} \\ T_2 = 100°\text{C} \end{array}\right\} \begin{array}{l} h_2 = 2682.5\text{ kJ/kg} \\ s_2 = 7.6947\text{ kJ/kg}\cdot\text{K} \end{array}$$

Analysis There is only one inlet and one exit, and thus $\dot{m}_1 = \dot{m}_2 = \dot{m}$. We take the turbine as the system, which is a control volume since mass crosses the boundary. The energy balance for this steady-flow system can be expressed in the rate form as

$$\underbrace{\dot{E}_{in} - \dot{E}_{out}}_{\text{Rate of net energy transfer by heat, work, and mass}} = \underbrace{\Delta \dot{E}_{\text{system}}^{\nearrow 0\text{ (steady)}}}_{\text{Rate of change in internal, kinetic, potential, etc. energies}} = 0$$

$$\dot{E}_{in} = \dot{E}_{out}$$

$$\dot{m}(h_1 + \mathbf{V}_1^2/2) = \dot{W}_{\text{out}} + \dot{m}(h_2 + \mathbf{V}_2^2/2)$$

$$\dot{W}_{\text{out}} = \dot{m}\left[h_1 - h_2 + \frac{\mathbf{V}_1^2 - \mathbf{V}_2^2}{2}\right]$$

80 m/s
6 MPa
600°C
STEAM
5 MW
50 kPa
100°C
140 m/s

Substituting,

$$5000\text{ kJ/s} = \dot{m}\left(3658.4 - 2682.5 + \frac{(80\text{ m/s})^2 - (140\text{ m/s})^2}{2}\left(\frac{1\text{ kJ/kg}}{1000\text{ m}^2/\text{s}^2}\right)\right)$$

$$\dot{m} = 5.158\text{ kg/s}$$

The reversible (or maximum) power output is determined from the rate form of the exergy balance applied on the turbine and setting the exergy destruction term equal to zero,

$$\underbrace{\dot{X}_{in} - \dot{X}_{out}}_{\text{Rate of net exergy transfer by heat, work, and mass}} \underbrace{- \dot{X}_{\text{destroyed}}^{\nearrow 0\text{ (reversible)}}}_{\text{Rate of exergy destruction}} = \underbrace{\Delta \dot{X}_{\text{system}}^{\nearrow 0\text{ (steady)}}}_{\text{Rate of change of exergy}} = 0$$

$$\dot{X}_{in} = \dot{X}_{out}$$

$$\dot{m}\psi_1 = \dot{W}_{rev,out} + \dot{m}\psi_2$$

$$\dot{W}_{rev,out} = \dot{m}(\psi_1 - \psi_2) = \dot{m}[(h_1 - h_2) - T_0(s_1 - s_2) - \Delta ke^{\nearrow 0} - \Delta pe^{\nearrow 0}]$$

Substituting,

$$\dot{W}_{rev,out} = \dot{m}[(h_1 - h_2) - T_0(s_1 - s_2)]$$

$$= (5.158\text{ kg/s})[3658.4 - 2682.5 - (298\text{ K})(7.1677 - 7.6947)\text{ kJ/kg}\cdot\text{K}] = \mathbf{5844\text{ kW}}$$

(*b*) The second-law efficiency of a turbine is the ratio of the actual work output to the reversible work,

$$\eta_{II} = \frac{\dot{W}_{out}}{\dot{W}_{rev,out}} = \frac{5\text{ MW}}{5.844\text{ MW}} = \mathbf{85.6\%}$$

Discussion Note that 14.4% percent of the work potential of the steam is wasted as it flows through the turbine during this process.

7-61 Steam is throttled from a specified state to a specified pressure. The decrease in the exergy of the steam during this throttling process is to be determined.

Assumptions **1** This is a steady-flow process since there is no change with time. **2** Kinetic and potential energy changes are negligible. **3** The temperature of the surroundings is given to be 25°C. **4** Heat transfer is negligible.

Properties The properties of steam before and after throttling are (Tables A-4 through A-6)

$$\left.\begin{array}{l} P_1 = 9\text{ MPa} \\ T_1 = 500°\text{C} \end{array}\right\} \begin{array}{l} h_1 = 3386.1\text{ kJ/kg} \\ s_1 = 6.6576\text{ kJ/kg}\cdot\text{K} \end{array}$$

$$\left.\begin{array}{l} P_2 = 7\text{ MPa} \\ h_2 = h_1 \end{array}\right\} s_2 = 6.7651\text{ kJ/kg}\cdot\text{K}$$

Steam

1

2

Analysis The decrease in exergy is of the steam is the difference between the inlet and exit flow exergies,

$$\begin{aligned} \text{Decrease in exergy} &= \psi_1 - \psi_2 = -[\Delta h^{\nearrow 0} - \Delta ke^{\nearrow 0} - \Delta pe^{\nearrow 0} - T_0(s_1 - s_2)] = T_0(s_2 - s_1) \\ &= (298\text{ K})(6.7651 - 6.6576)\text{kJ/kg}\cdot\text{K} \\ &= \mathbf{32.0\ kJ/kg} \end{aligned}$$

Discussion Note that 32.0 kJ/kg of work potential is wasted during this throttling process.

7-62 Combustion gases expand in a turbine from a specified state to another specified state. The exergy of the gases at the inlet and the reversible work output of the turbine are to be determined. √

Assumptions **1** This is a steady-flow process since there is no change with time. **2** Potential energy changes are negligible. **3** The temperature of the surroundings is given to be 25°C. **4** The combustion gases are ideal gases with constant specific heats.

Properties The constant pressure specific heat and the specific heat ratio are given to be C_p = 1.15 kJ/kg.K and k = 1.3. The gas constant R is determined from

$$R = C_p - C_v = C_p - C_p / k = C_p(1 - 1/k) = (1.15 \text{ kJ/kg}\cdot\text{K})(1 - 1/1.3) = 0.265 \text{ kJ/kg}\cdot\text{K}$$

Analysis (*a*) The exergy of the gases at the turbine inlet is simply the flow exergy,

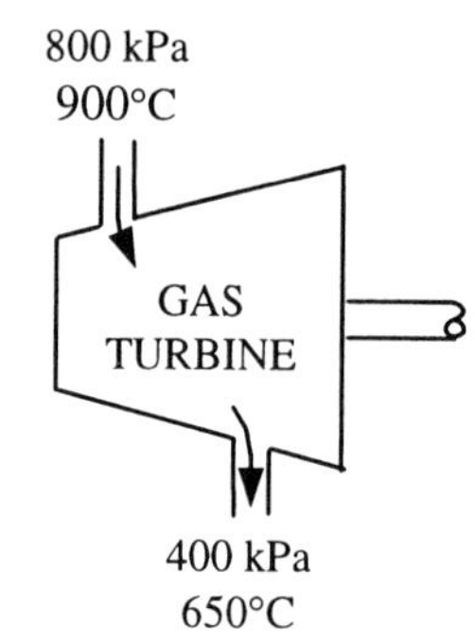

$$\psi_1 = h_1 - h_0 - T_0(s_1 - s_0) + \frac{\mathbf{V}_1^2}{2} + gz_1^{\nearrow 0}$$

where

$$\begin{aligned} s_1 - s_0 &= C_p \ln\frac{T_1}{T_0} - R\ln\frac{P_1}{P_0} \\ &= (1.15 \text{ kJ/kg}\cdot\text{K})\ln\frac{1173\text{ K}}{298\text{ K}} - (0.265 \text{ kJ/kg}\cdot\text{K})\ln\frac{800\text{ kPa}}{100\text{ kPa}} \\ &= 1.025 \text{ kJ/kg}\cdot\text{K} \end{aligned}$$

Thus,

$$\begin{aligned} \psi_1 &= (1.15 \text{ kJ/kg.K})(900 - 25)^\circ\text{C} - (298\text{ K})(1.025 \text{ kJ/kg}\cdot\text{K}) + \frac{(100\text{ m/s})^2}{2}\left(\frac{1\text{ kJ/kg}}{1000\text{ m}^2/\text{s}^2}\right) \\ &= \mathbf{705.8 \text{ kJ/kg}} \end{aligned}$$

(*b*) The reversible (or maximum) work output is determined from an exergy balance applied on the turbine and setting the exergy destruction term equal to zero,

$$\underbrace{\dot{X}_{in} - \dot{X}_{out}}_{\substack{\text{Rate of net exergy transfer} \\ \text{by heat, work, and mass}}} - \underbrace{\dot{X}_{\text{destroyed}}^{\nearrow 0 \text{ (reversible)}}}_{\substack{\text{Rate of exergy} \\ \text{destruction}}} = \underbrace{\Delta\dot{X}_{\text{system}}^{\nearrow 0 \text{ (steady)}}}_{\substack{\text{Rate of change} \\ \text{of exergy}}} = 0$$

$$\dot{X}_{in} = \dot{X}_{out}$$

$$\dot{m}\psi_1 = \dot{W}_{rev,in} + \dot{m}\psi_2$$

$$\dot{W}_{rev,out} = \dot{m}(\psi_1 - \psi_2) = \dot{m}[(h_1 - h_2) - T_0(s_1 - s_2) - \Delta ke - \Delta pe^{\nearrow 0}]$$

where

$$\Delta ke = \frac{\mathbf{V}_2^2 - \mathbf{V}_1^2}{2} = \frac{(220\text{ m/s})^2 - (100\text{ m/s})^2}{2}\left(\frac{1\text{ kJ/kg}}{1000\text{ m}^2/\text{s}^2}\right) = 19.2 \text{ kJ/kg}$$

and

$$\begin{aligned} s_2 - s_1 &= C_p \ln\frac{T_2}{T_1} - R\ln\frac{P_2}{P_1} \\ &= (1.15 \text{ kJ/kg}\cdot\text{K})\ln\frac{923\text{ K}}{1173\text{ K}} - (0.265 \text{ kJ/kg}\cdot\text{K})\ln\frac{400\text{ kPa}}{800\text{ kPa}} \\ &= -0.09196 \text{ kJ/kg}\cdot\text{K} \end{aligned}$$

Then the reversible work output on a unit mass basis becomes

$$\begin{aligned} w_{rev,out} &= h_1 - h_2 + T_0(s_2 - s_1) - \Delta ke = C_p(T_1 - T_2) + T_0(s_2 - s_1) - \Delta ke \\ &= (1.15 \text{ kJ/kg}\cdot\text{K})(900 - 650)^\circ\text{C} + (298\text{ K})(-0.09196 \text{ kJ/kg}\cdot\text{K}) - 19.2 \text{ kJ/kg} = \mathbf{240.9 \text{ kJ/kg}} \end{aligned}$$

7-63E Refrigerant-134a enters an adiabatic compressor with an isentropic efficiency of 0.80 at a specified state with a specified volume flow rate, and leaves at a specified pressure. The actual power input and the second-law efficiency to the compressor are to be determined.

Assumptions **1** This is a steady-flow process since there is no change with time. **2** Kinetic and potential energy changes are negligible. **3** The device is adiabatic and thus heat transfer is negligible.

Properties From the refrigerant tables (Tables A-11E through A-13E)

$$\left.\begin{array}{l} P_1 = 20\text{ psia} \\ \text{sat.vapor} \end{array}\right\} \begin{array}{l} h_1 = h_{g\,@\,20\text{ psia}} = 101.39\text{ Btu/lbm} \\ s_1 = s_{g\,@\,20\text{ psia}} = 0.2227\text{ Btu/lbm}\cdot\text{R} \\ \text{v}_1 = \text{v}_{g\,@\,20\text{ psia}} = 2.2661\text{ ft}^3\text{/lbm} \end{array}$$

$$\left.\begin{array}{l} P_2 = 100\text{ psia} \\ s_{2s} = s_1 \end{array}\right\} h_{2s} = 115.64\text{ Btu/lbm}$$

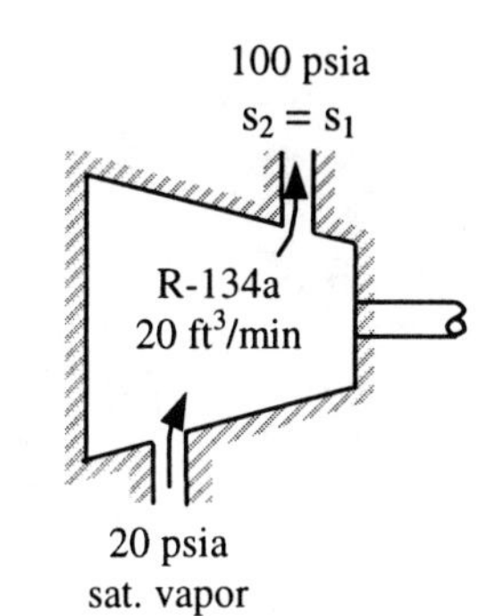

Analysis From the isentropic efficiency relation,

$$h_c = \frac{h_{2s} - h_1}{h_{2a} - h_1} \longrightarrow h_{2a} = h_1 + (h_{2s} - h_1)/h_c$$
$$= 101.39 + (115.64 - 101.39)/0.80$$
$$= 119.20\ Btu/lbm$$

Also,

$$\dot{m} = \frac{\dot{V}_1}{v_1} = \frac{20/60\text{ ft}^3/\text{s}}{2.2661\text{ ft}^3/\text{lbm}} = 0.1471\text{ lbm/s}$$

There is only one inlet and one exit, and thus $\dot{m}_1 = \dot{m}_2 = \dot{m}$. We take the actual compressor as the system, which is a control volume. The energy balance for this steady-flow system can be expressed as

$$\underbrace{\dot{E}_{in} - \dot{E}_{out}}_{\substack{\text{Rate of net energy transfer} \\ \text{by heat, work, and mass}}} = \underbrace{\Delta \dot{E}_{\text{system}}^{\nearrow 0\text{ (steady)}}}_{\substack{\text{Rate of change in internal, kinetic,} \\ \text{potential, etc. energies}}} = 0$$

$$\dot{E}_{in} = \dot{E}_{out}$$
$$\dot{W}_{a,\text{in}} + \dot{m}h_1 = \dot{m}h_2 \quad (\text{since } \dot{Q} \cong \Delta ke \cong \Delta pe \cong 0)$$
$$\dot{W}_{a,\text{in}} = \dot{m}(h_2 - h_1)$$

Substituting, the actual power input to the compressor becomes

$$\dot{W}_{a,in} = (0.1471\text{ lbm/s})(119.20 - 101.39)\text{ Btu/lbm}\left(\frac{1\text{ hp}}{0.7068\text{ Btu/s}}\right) = \mathbf{3.71\ hp}$$

(*b*) The entropy of the refrigerant at the compressor exit state (P_2 = 100 psia, h_2 = 119.20 Btu/lbm) is s_2 = 0.2291 Btu/lbm· R. The reversible (or minimum) power input is determined from the exergy balance applied on the compressor and setting the exergy destruction term equal to zero,

$$\underbrace{\dot{X}_{in} - \dot{X}_{out}}_{\substack{\text{Rate of net exergy transfer} \\ \text{by heat, work, and mass}}} \underbrace{- \dot{X}_{\text{destroyed}}^{\nearrow 0\text{ (reversible)}}}_{\substack{\text{Rate of exergy} \\ \text{destruction}}} = \underbrace{\Delta \dot{X}_{\text{system}}^{\nearrow 0\text{ (steady)}}}_{\substack{\text{Rate of change} \\ \text{of exergy}}} = 0$$

$$\dot{X}_{in} = \dot{X}_{out}$$
$$\dot{W}_{rev,in} + \dot{m}\psi_1 = \dot{m}\psi_2$$
$$\dot{W}_{rev,in} = \dot{m}(\psi_2 - \psi_1) = \dot{m}[(h_2 - h_1) - T_0(s_2 - s_1) + \Delta ke^{\nearrow 0} + \Delta pe^{\nearrow 0}]$$

Substituting,

$$\dot{W}_{rev,in} = (0.1471 \text{ lbm/s})\left[(119.20 - 101.39) \text{ Btu/lbm} - (535 \text{ R})(0.2291 - 0.2227) \text{ Btu/lbm} \cdot \text{R}\right]$$
$$= 2.116 \text{ Btu/s} = \mathbf{2.99\ hp} \qquad (\text{since } 1 \text{ hp} = 0.7068 \text{ Btu/s})$$

Thus,

$$\eta_{II} = \frac{\dot{W}_{rev,in}}{\dot{W}_{act,in}} = \frac{2.99 \text{ hp}}{3.71 \text{ hp}} = \mathbf{80.6\%}$$

7-64 Refrigerant-134a is compressed by an adiabatic compressor from a specified state to another specified state. The isentropic efficiency and the second-law efficiency of the compressor are to be determined.

Assumptions **1** This is a steady-flow process since there is no change with time. **2** Kinetic and potential energy changes are negligible. **3** The device is adiabatic and thus heat transfer is negligible.

Properties From the refrigerant tables (Tables A-11E through A-13E)

$$\left.\begin{array}{l} P_1 = 140\text{ kPa} \\ T_1 = -10°\text{C} \end{array}\right\} \begin{array}{l} h_1 = 243.40\text{ kJ/kg} \\ s_1 = 0.9606\text{ kJ/kg}\cdot\text{K} \\ v_1 = 0.14549\text{ m}^3\text{/kg} \end{array}$$

$$\left.\begin{array}{l} P_2 = 700\text{ kPa} \\ T_2 = 60°\text{C} \end{array}\right\} \begin{array}{l} h_2 = 296.69\text{ kJ/kg} \\ s_2 = 1.0182\text{ kJ/kg}\cdot\text{K} \end{array}$$

$$\left.\begin{array}{l} P_{2s} = 700\text{ kPa} \\ s_{2s} = s_1 \end{array}\right\} h_{2s} = 278.06\text{ kJ/kg}$$

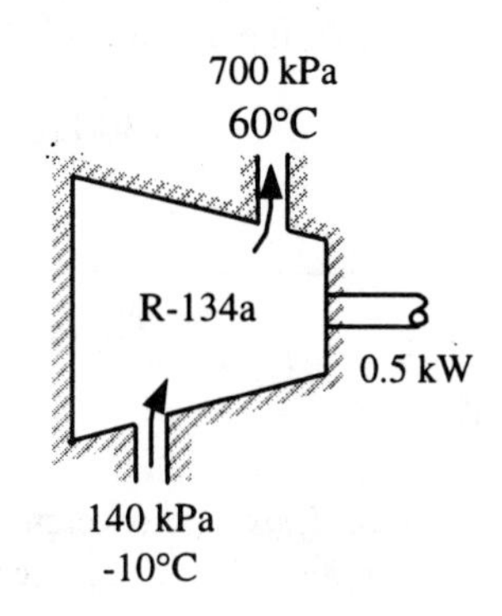

Analysis (*a*) The isentropic efficiency is

$$\eta_c = \frac{h_{2s} - h_1}{h_{2a} - h_1} = \frac{278.06 - 243.40}{296.69 - 243.40} = 0.650 = \mathbf{65.0\%}$$

(*b*) There is only one inlet and one exit, and thus $\dot{m}_1 = \dot{m}_2 = \dot{m}$. We take the actual compressor as the system, which is a control volume. The energy balance for this steady-flow system can be expressed as

$$\underbrace{\dot{E}_{in} - \dot{E}_{out}}_{\substack{\text{Rate of net energy transfer} \\ \text{by heat, work, and mass}}} = \underbrace{\Delta \dot{E}_{\text{system}}^{\nearrow 0\text{ (steady)}}}_{\substack{\text{Rate of change in internal, kinetic,} \\ \text{potential, etc. energies}}} = 0$$

$$\dot{E}_{in} = \dot{E}_{out}$$

$$\dot{W}_{a,\text{in}} + \dot{m}h_1 = \dot{m}h_2 \quad (\text{since } \dot{Q} \cong \Delta ke \cong \Delta pe \cong 0)$$

$$\dot{W}_{a,\text{in}} = \dot{m}(h_2 - h_1)$$

Then the mass flow rate of the refrigerant becomes

$$\dot{m} = \frac{\dot{W}_{a,in}}{h_1 - h_{2a}} = \frac{0.5\text{ kJ/s}}{(296.69 - 243.40)\text{kJ/kg}} = 0.00938\text{ kg/s}$$

The reversible (or minimum) power input is determined from the exergy balance applied on the compressor and setting the exergy destruction term equal to zero,

$$\underbrace{\dot{X}_{in} - \dot{X}_{out}}_{\substack{\text{Rate of net exergy transfer} \\ \text{by heat, work, and mass}}} \underbrace{- \dot{X}_{\text{destroyed}}^{\nearrow 0\text{ (reversible)}}}_{\substack{\text{Rate of exergy} \\ \text{destruction}}} = \underbrace{\Delta \dot{X}_{\text{system}}^{\nearrow 0\text{ (steady)}}}_{\substack{\text{Rate of change} \\ \text{of exergy}}} = 0$$

$$\dot{X}_{in} = \dot{X}_{out}$$

$$\dot{W}_{rev,in} + \dot{m}\psi_1 = \dot{m}\psi_2$$

$$\dot{W}_{rev,in} = \dot{m}(\psi_2 - \psi_1) = \dot{m}[(h_2 - h_1) - T_0(s_2 - s_1) + \Delta ke^{\nearrow 0} + \Delta pe^{\nearrow 0}]$$

Substituting,

$$\dot{W}_{rev,in} = (0.00938\text{ kg/s})\left[(296.69 - 243.40)\text{kJ/kg} - (300\text{ K})(1.018 - 0.9606)\text{kJ/kg}\cdot\text{K}\right] = 0.338\text{ kW}$$

and

$$\eta_{II} = \frac{\dot{W}_{rev,in}}{\dot{W}_{a,in}} = \frac{0.338\text{ kW}}{0.5\text{ kW}} = \mathbf{67.7\%}$$

7-65 Air is compressed steadily by a compressor from a specified state to another specified state. The increase in the exergy of air and the rate of exergy destruction are to be determined.

Assumptions **1** Air is an ideal gas with variable specific heats. **2** Kinetic and potential energy changes are negligible.

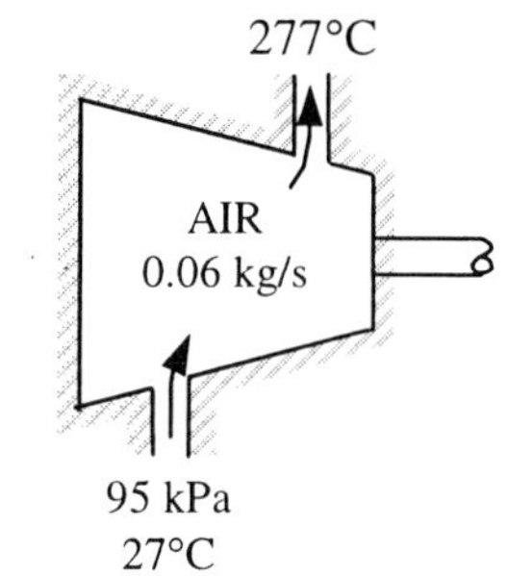

Properties The gas constant of air is R = 0.287 kJ/kg.K (Table A-1).
From the air table (Table A-17)

$$T_1 = 300 \text{ K} \longrightarrow h_1 = 300.19 \text{ kJ/kg}$$
$$s_1^o = 1.702 \text{ kJ/kg} \cdot \text{K}$$

$$T_2 = 550 \text{ K} \longrightarrow h_2 = 555.74 \text{ kJ/kg}$$
$$s_2^o = 2.318 \text{ kJ/kg} \cdot \text{K}$$

Analysis The reversible (or minimum) power input is determined from the rate form of the exergy balance applied on the compressor and setting the exergy destruction term equal to zero,

$$\underbrace{\dot{X}_{in} - \dot{X}_{out}}_{\substack{\text{Rate of net exergy transfer} \\ \text{by heat, work, and mass}}} \underbrace{- \dot{X}_{\text{destroyed}}^{\nearrow 0 \text{ (reversible)}}}_{\substack{\text{Rate of exergy} \\ \text{destruction}}} = \underbrace{\Delta \dot{X}_{\text{system}}^{\nearrow 0 \text{ (steady)}}}_{\substack{\text{Rate of change} \\ \text{of exergy}}} = 0$$

$$\dot{X}_{in} = \dot{X}_{out}$$
$$\dot{m}\psi_1 + \dot{W}_{rev,in} = \dot{m}\psi_2$$
$$\dot{W}_{rev,in} = \dot{m}(\psi_2 - \psi_1) = \dot{m}[(h_2 - h_1) - T_0(s_2 - s_1) + \Delta ke^{\nearrow 0} + \Delta pe^{\nearrow 0}]$$

where

$$\begin{aligned} s_2 - s_1 &= s_2^o - s_1^o - R \ln \frac{P_2}{P_1} \\ &= (2.318 - 1.702) \text{ kJ/kg} \cdot \text{K} - (0.287 \text{ kJ/kg} \cdot \text{K}) \ln \frac{600 \text{ kPa}}{95 \text{ kPa}} \\ &= 0.0870 \text{ kJ/kg} \cdot \text{K} \end{aligned}$$

Substituting,

$$\dot{W}_{rev,in} = (0.06 \text{ kg/s})[(555.74 - 300.19)\text{kJ/kg - }(298 \text{ K})(0.0870 \text{ kJ/kg} \cdot \text{K})] = \mathbf{13.7 \text{ kW}}$$

Discussion Note that a minimum of 13.7 kW of power input is needed for this compression process.

7-66 EES solution of this (and other comprehensive problems designated with the *computer icon*) is available to instructors at the *Instructor Manual* section of the *Online Learning Center* (OLC) at www.mhhe.com/cengel-boles. See the Preface for access information.

7-67 Argon enters an adiabatic compressor at a specified state, and leaves at another specified state. The reversible power input and irreversibility are to be determined.

Assumptions **1** This is a steady-flow process since there is no change with time. **2** Potential energy changes are negligible. **3** The device is adiabatic and thus heat transfer is negligible. **4** Argon is an ideal gas with constant specific heats.

Properties For argon, the gas constant is R = 0.2081 kJ/kg.K; the specific heat ratio is k = 1.667; the constant pressure specific heat is C_p = 0.5203 kJ/kg.K (Table A-2).

Analysis The mass flow rate, the entropy change, and the kinetic energy change of argon during this process are

$$v_1 = \frac{RT_1}{P_1} = \frac{(0.2081\ \text{kPa}\cdot\text{m}^3/\text{kg}\cdot\text{K})(303\ \text{K})}{(120\ \text{kPa})} = 0.5255\ \text{m}^3/kg$$

$$\dot{m} = \frac{1}{v_1}\mathbf{A}_1\mathbf{V}_1 = \frac{1}{0.5255\ \text{m}^3/\text{kg}}(0.0130\ \text{m}^2)(20\ \text{m/s}) = 0.495\ \text{kg/s}$$

$$\begin{aligned} s_2 - s_1 &= C_p \ln\frac{T_2}{T_1} - R\ln\frac{P_2}{P_1} \\ &= (0.5203\ \text{kJ/kg}\cdot\text{K})\ln\frac{803\ \text{K}}{303\ \text{K}} - (0.2081\ \text{kJ/kg}\cdot\text{K})\ln\frac{1200\ \text{kPa}}{120\ \text{kPa}} \\ &= 0.02793\ \text{kJ/kg}\cdot\text{K} \end{aligned}$$

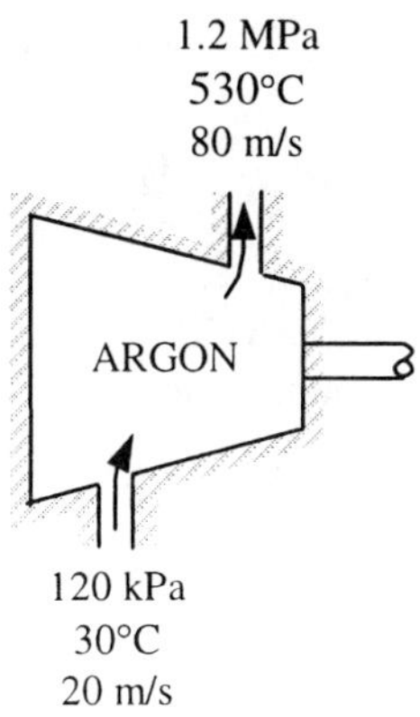

1.2 MPa
530°C
80 m/s

ARGON

120 kPa
30°C
20 m/s

and

$$\Delta ke = \frac{\mathbf{V}_2^2 - \mathbf{V}_1^2}{2} = \frac{(80\ \text{m/s})^2 - (20\ \text{m/s})^2}{2}\left(\frac{1\ \text{kJ/kg}}{1000\ \text{m}^2/\text{s}^2}\right) = 3.0\ \text{kJ/kg}$$

The reversible (or minimum) power input is determined from the rate form of the exergy balance applied on the compressor, and setting the exergy destruction term equal to zero,

$$\underbrace{\dot{X}_{in} - \dot{X}_{out}}_{\substack{\text{Rate of net exergy transfer}\\ \text{by heat, work, and mass}}} - \underbrace{\dot{X}_{\text{destroyed}}^{\nearrow 0\ \text{(reversible)}}}_{\substack{\text{Rate of exergy}\\ \text{destruction}}} = \underbrace{\Delta\dot{X}_{\text{system}}^{\nearrow 0\ \text{(steady)}}}_{\substack{\text{Rate of change}\\ \text{of exergy}}} = 0$$

$$\dot{X}_{in} = \dot{X}_{out}$$

$$\dot{m}\psi_1 + \dot{W}_{rev,in} = \dot{m}\psi_2$$

$$\dot{W}_{rev,in} = \dot{m}(\psi_2 - \psi_1) = \dot{m}[(h_2 - h_1) - T_0(s_2 - s_1) + \Delta ke + \Delta pe^{\nearrow 0}]$$

Substituting,

$$\begin{aligned} \dot{W}_{rev,in} &= \dot{m}\left[C_p(T_2 - T_1) - T_0(s_2 - s_1) + \Delta ke\right] \\ &= (0.495\ \text{kg/s})\left[(0.5203\ \text{kJ/kg}\cdot\text{K})(530 - 30)\text{K} - (298\ \text{K})(0.02793\ \text{kJ/kg}\cdot\text{K}) + 3.0\right] \\ &= \mathbf{126\ kW} \end{aligned}$$

The exergy destruction (or irreversibility) can be determined from an exergy balance or directly from its definition $X_{\text{destroyed}} = T_0 S_{\text{gen}}$ where the entropy generation is determined from an entropy balance on the system, which is an adiabatic steady-flow device,

$$\underbrace{\dot{S}_{in} - \dot{S}_{out}}_{\substack{\text{Rate of net entropy transfer}\\ \text{by heat and mass}}} + \underbrace{\dot{S}_{gen}}_{\substack{\text{Rate of entropy}\\ \text{generation}}} = \underbrace{\Delta\dot{S}_{\text{system}}^{\nearrow 0}}_{\substack{\text{Rate of change}\\ \text{of entropy}}} = 0$$

$$\dot{m}s_1 - \dot{m}s_2 + \dot{S}_{\text{gen}} = 0 \rightarrow \dot{S}_{\text{gen}} = \dot{m}(s_2 - s_1)$$

Substituting,

$$\dot{X}_{\text{destroyed}} = T_0\dot{m}(s_2 - s_1) = (298\ \text{K})(0.495\ \text{kg/s})(0.02793\ \text{kJ/kg}\cdot\text{K}) = \mathbf{4.12\ kW}$$

7-68 Steam expands in a turbine steadily at a specified rate from a specified state to another specified state. The power potential of the steam at the inlet conditions and the reversible power output are to be determined.

Assumptions **1** This is a steady-flow process since there is no change with time. **2** Kinetic and potential energy changes are negligible. **3** The temperature of the surroundings is given to be 25°C.

8 MPa
450°C
STEAM
15,000 kg/h
50 kPa
sat. vapor

Properties From the steam tables (Tables A-4 through 6)

$$\left.\begin{array}{l} P_1 = 8\text{ MPa} \\ T_1 = 450°\text{C} \end{array}\right\} \begin{array}{l} h_1 = 3272.0\text{ kJ/kg} \\ s_1 = 6.5551\text{ kJ/kg}\cdot\text{K} \end{array}$$

$$\left.\begin{array}{l} P_2 = 50\text{ kPa} \\ \text{sat.vapor} \end{array}\right\} \begin{array}{l} h_2 = 2645.9\text{ kJ/kg} \\ s_2 = 7.5939\text{ kJ/kg}\cdot\text{K} \end{array}$$

$$\left.\begin{array}{l} P_0 = 100\text{ kPa} \\ T_0 = 25°\text{C} \end{array}\right\} \begin{array}{l} h_0 \cong h_{f\,@\,25°\text{C}} = 104.89\text{ kJ/kg} \\ s_0 \cong s_{f\,@\,25°\text{C}} = 0.3674\text{ kJ/kg}\cdot\text{K} \end{array}$$

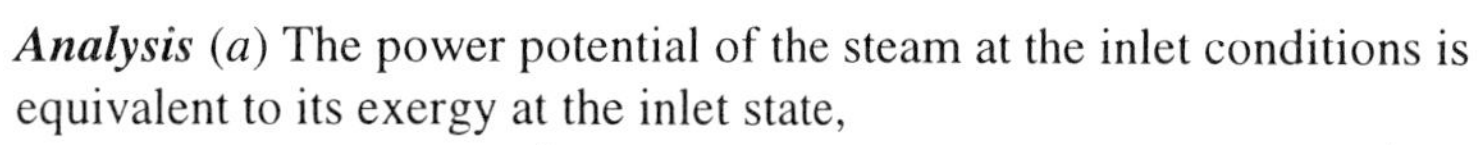

Analysis (*a*) The power potential of the steam at the inlet conditions is equivalent to its exergy at the inlet state,

$$\dot{\Psi} = \dot{m}\psi_1 = \dot{m}\left(h_1 - h_0 - T_0(s_1 - s_0) + \frac{\mathbf{V}_1^{2\,\nearrow 0}}{2} + gz_1^{\nearrow 0}\right) = \dot{m}\left(h_1 - h_0 - T_0(s_1 - s_0)\right)$$

$$= (15{,}000/3600\text{ kg/s})\left[(3272 - 104.89)\text{kJ/kg} - (298\text{ K})(6.5551 - 0.3674)\text{kJ/kg}\cdot\text{K}\right]$$

$$= \mathbf{5513\ kW}$$

(*b*) The power output of the turbine if there were no irreversibilities is the reversible power, is determined from the rate form of the exergy balance applied on the turbine and setting the exergy destruction term equal to zero,

$$\underbrace{\dot{X}_{in} - \dot{X}_{out}}_{\substack{\text{Rate of net exergy transfer} \\ \text{by heat, work, and mass}}} \underbrace{- \dot{X}_{\text{destroyed}}^{\nearrow 0\ (\text{reversible})}}_{\substack{\text{Rate of exergy} \\ \text{destruction}}} = \underbrace{\Delta \dot{X}_{\text{system}}^{\nearrow 0\ (\text{steady})}}_{\substack{\text{Rate of change} \\ \text{of exergy}}} = 0$$

$$\dot{X}_{in} = \dot{X}_{out}$$

$$\dot{m}\psi_1 = \dot{W}_{rev,out} + \dot{m}\psi_2$$

$$\dot{W}_{rev,out} = \dot{m}(\psi_1 - \psi_2) = \dot{m}[(h_1 - h_2) - T_0(s_1 - s_2) - \Delta ke^{\nearrow 0} - \Delta pe^{\nearrow 0}]$$

Substituting,

$$\dot{W}_{rev,out} = \dot{m}[(h_1 - h_2) - T_0(s_1 - s_2)]$$

$$= (15{,}000/3600\text{ kg/s})\left[(3272 - 2645.9)\text{ kJ/kg} - (298\text{ K})(6.5551 - 7.5939)\text{ kJ/kg}\cdot\text{K}\right]$$

$$= \mathbf{3899\ kW}$$

7-69E Air is compressed steadily by a 400-hp compressor from a specified state to another specified state while being cooled by the ambient air. The mass flow rate of air and the part of input power that is used to just overcome the irreversibilities are to be determined.

Assumptions **1** Air is an ideal gas with variable specific heats. **2** Potential energy changes are negligible. **3** The temperature of the surroundings is given to be 60°F.

Properties The gas constant of air is R = 0.06855 Btu/lbm.R (Table A-1E). From the air table (Table A-17E)

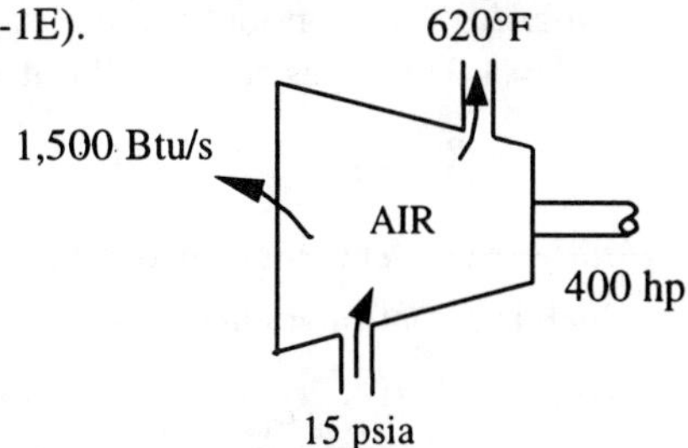

$$\left.\begin{array}{l} T_1 = 520\,\text{R} \\ P_1 = 15\,\text{psia} \end{array}\right\} \begin{array}{l} h_1 = 124.27\ \text{Btu/lbm} \\ s_1^{\text{o}} = 0.59173\ \text{Btu/lbm}\cdot\text{R} \end{array}$$

$$\left.\begin{array}{l} T_2 = 1080\,\text{R} \\ P_2 = 150\,\text{psia} \end{array}\right\} \begin{array}{l} h_2 = 260.97\ \text{Btu/lbm} \\ s_1^{\text{o}} = 0.76964\ \text{Btu/lbm}\cdot\text{R} \end{array}$$

Analysis (*a*) There is only one inlet and one exit, and thus $\dot{m}_1 = \dot{m}_2 = \dot{m}$. We take the actual compressor as the system, which is a control volume. The energy balance for this steady-flow system can be expressed as

$$\underbrace{\dot{E}_{in} - \dot{E}_{out}}_{\text{Rate of net energy transfer by heat, work, and mass}} = \underbrace{\Delta \dot{E}_{\text{system}}^{\nearrow 0\ (\text{steady})}}_{\text{Rate of change in internal, kinetic, potential, etc. energies}} = 0$$

$$\dot{E}_{in} = \dot{E}_{out}$$

$$\dot{W}_{a,\text{in}} + \dot{m}(h_1 + \mathbf{V}_1^2/2) = \dot{m}(h_2 + \mathbf{V}_2^2/2) + \dot{Q}_{out} \rightarrow \dot{W}_{a,\text{in}} - \dot{Q}_{out} = \dot{m}\left(h_2 - h_1 + \frac{\mathbf{V}_2^2 - \mathbf{V}_1^2}{2}\right)$$

Substituting, the mass flow rate of the refrigerant becomes

$$(400\ \text{hp})\left(\frac{0.7068\ \text{Btu/s}}{1\ \text{hp}}\right) - (1500/60\ \text{Btu/s}) = \dot{m}\left(260.97 - 124.27 + \frac{(350\ \text{ft/s})^2 - 0}{2}\frac{1\ \text{Btu/lbm}}{25{,}037\ \text{ft}^2/\text{s}^2}\right)$$

It yields $\dot{m} = \mathbf{1.852\ lbm/s}$

(*b*) The portion of the power output that is used just to overcome the irreversibilities is equivalent to exergy destruction, which can be determined from an exergy balance or directly from its definition $X_{\text{destroyed}} = T_0 S_{gen}$ where the entropy generation S_{gen} is determined from an entropy balance on an *extended system* that includes the device and its immediate surroundings. It gives

$$\underbrace{\dot{S}_{in} - \dot{S}_{out}}_{\text{Rate of net entropy transfer by heat and mass}} + \underbrace{\dot{S}_{gen}}_{\text{Rate of entropy generation}} = \underbrace{\Delta \dot{S}_{\text{system}}^{\nearrow 0}}_{\text{Rate of change of entropy}} = 0$$

$$\dot{m}s_1 - \dot{m}s_2 - \frac{\dot{Q}_{out}}{T_{b,surr}} + \dot{S}_{\text{gen}} = 0 \rightarrow \dot{S}_{gen} = \dot{m}(s_2 - s_1) + \frac{\dot{Q}_{out}}{T_0}$$

where

$$s_2 - s_1 = s_2^0 - s_1^0 - R\ln\frac{P_2}{P_1} = (0.76964 - 0.59173)\ \text{Btu/lbm} - (0.06855\ \text{Btu/lbm.R})\ln\frac{150\ \text{psia}}{15\ \text{psia}}$$

$$= 0.02007\ \ \text{Btu/lbm.R}$$

Substituting, the exergy destruction is determined to be

$$\dot{X}_{\text{destroyed}} = T_0 \dot{S}_{gen} = T_0\left(\dot{m}(s_2 - s_1) + \frac{\dot{Q}_{out}}{T_0}\right)$$

$$= (520\ \text{R})\left((1.852\ \text{lbm/s})(0.02007\ \text{Btu/lbm}\cdot\text{R}) + \frac{+1500/60\ \text{Btu/s}}{520\ \text{R}}\right)\left(\frac{1\ \text{hp}}{5.4039\ \text{Btu/s}}\right) = \mathbf{8.21\ hp}$$

7-70 Hot combustion gases are accelerated in an adiabatic nozzle. The exit velocity and the decrease in the exergy of the gases are to be determined.

Assumptions **1** This is a steady-flow process since there is no change with time. **2** Potential energy changes are negligible. **3** The device is adiabatic and thus heat transfer is negligible. **4** The combustion gases are ideal gases with constant specific heats.

Properties The constant pressure specific heat and the specific heat ratio are given to be C_p = 1.15 kJ/kg.K and k = 1.3. The gas constant R is determined from

$$R = C_p - C_v = C_p - C_p / k = C_p(1 - 1/k) = (1.15 \text{ kJ/kg}\cdot\text{K})(1 - 1/1.3) = 0.265 \text{ kJ/kg}\cdot\text{K}$$

Analysis (*a*) There is only one inlet and one exit, and thus $\dot{m}_1 = \dot{m}_2 = \dot{m}$. We take the nozzle as the system, which is a control volume. The energy balance for this steady-flow system can be expressed as

$$\underbrace{\dot{E}_{in} - \dot{E}_{out}}_{\substack{\text{Rate of net energy transfer} \\ \text{by heat, work, and mass}}} = \underbrace{\Delta \dot{E}_{\text{system}}^{\nearrow 0 \text{ (steady)}}}_{\substack{\text{Rate of change in internal, kinetic,} \\ \text{potential, etc. energies}}} = 0$$

$$\dot{E}_{in} = \dot{E}_{out}$$

$$\dot{m}(h_1 + \mathbf{V}_1^2/2) = \dot{m}(h_2 + \mathbf{V}_2^2/2) \quad (\text{since } \dot{W} = \dot{Q} \cong \Delta\text{pe} \cong 0)$$

$$h_2 = h_1 - \frac{\mathbf{V}_2^2 - \mathbf{V}_1^2}{2}$$

260 kPa, 747°C, 80 m/s → Combustion gases → 70 kPa, 500°C

Then the exit velocity becomes

$$V_2 = \sqrt{2C_p(T_1 - T_2) + \mathbf{V}_1^2}$$

$$= \sqrt{2(1.15 \text{ kJ/kg}\cdot\text{K})(747 - 500)K\left(\frac{1000 \text{ m}^2/\text{s}^2}{1 \text{ kJ/kg}}\right) + (80 \text{ m/s})^2} = \mathbf{758 \text{ m/s}}$$

(*b*) The decrease in exergy of combustion gases is simply the difference between the initial and final values of flow exergy, and is determined to be

$$\psi_1 - \psi_2 = w_{rev} = h_1 - h_2 - \Delta ke - \Delta pe^{\nearrow 0} + T_0(s_2 - s_1) = C_p(T_1 - T_2) + T_0(s_2 - s_1) - \Delta ke$$

where

$$\Delta ke = \frac{\mathbf{V}_2^2 - \mathbf{V}_1^2}{2} = \frac{(758 \text{ m/s})^2 - (80 \text{ m/s})^2}{2}\left(\frac{1 \text{ kJ/kg}}{1000 \text{ m}^2/\text{s}^2}\right) = 284.1 \text{ kJ/kg}$$

and

$$s_2 - s_1 = C_p \ln\frac{T_2}{T_1} - R\ln\frac{P_2}{P_1}$$

$$= (1.15 \text{ kJ/kg}\cdot\text{K})\ln\frac{773 \text{ K}}{1020 \text{ K}} - (0.265 \text{ kJ/kg}\cdot\text{K})\ln\frac{70 \text{ kPa}}{260 \text{ kPa}}$$

$$= 0.02886 \text{ kJ/kg}\cdot\text{K}$$

Substituting,

$$\text{Decrease in exergy} = \psi_1 - \psi_2$$

$$= (1.15 \text{ kJ/kg}\cdot\text{K})(747 - 500)°\text{C} + (293 \text{ K})(0.02886 \text{ kJ/kg}\cdot\text{K}) - 284.1 \text{ kJ/kg}$$

$$= \mathbf{8.40 \text{ kJ/kg}}$$

7-71 Steam is accelerated in an adiabatic nozzle. The exit velocity of the steam, the isentropic efficiency, and the exergy destroyed within the nozzle are to be determined.

Assumptions **1** The nozzle operates steadily. **2** The changes in potential energies are negligible.

Properties The properties of steam at the inlet and the exit of the nozzle are (Tables A-4 through A-6)

$$\left.\begin{array}{l} P_1 = 7\text{ MPa} \\ T_1 = 500°\text{C} \end{array}\right\} \begin{array}{l} h_1 = 3410.3\text{ kJ/kg} \\ s_1 = 6.7975\text{ kJ/kg}\cdot\text{K} \end{array}$$

$$\left.\begin{array}{l} P_2 = 5\text{ MPa} \\ T_2 = 450°\text{C} \end{array}\right\} \begin{array}{l} h_2 = 3316.2\text{ kJ/kg} \\ s_2 = 6.8186\text{ kJ/kg}\cdot\text{K} \end{array}$$

$$\left.\begin{array}{l} P_{2s} = 5\text{ MPa} \\ s_{2s} = s_1 \end{array}\right\} h_{2s} = 3301.5\text{ kJ/kg}$$

7 MPa
500°C
70 m/s
STEAM
5 MPa
450°C

Analysis We take the nozzle to be the system, which is a control volume. The energy balance for this steady-flow system can be expressed in the rate form as

$$\underbrace{\dot{E}_{in} - \dot{E}_{out}}_{\substack{\text{Rate of net energy transfer} \\ \text{by heat, work, and mass}}} = \underbrace{\Delta\dot{E}_{\text{system}}^{\nearrow 0\ (\text{steady})}}_{\substack{\text{Rate of change in internal, kinetic,} \\ \text{potential, etc. energies}}} = 0$$

$$\dot{E}_{in} = \dot{E}_{out}$$

$$\dot{m}(h_1 + \mathbf{V}_1^2/2) = \dot{m}(h_2 + \mathbf{V}_2^2/2) \quad (\text{since } \dot{W} = \dot{Q} \cong \Delta\text{pe} \cong 0)$$

$$0 = h_2 - h_1 + \frac{\mathbf{V}_2^2 - \mathbf{V}_1^2}{2}$$

Then the exit velocity becomes

$$\mathbf{V}_2 = \sqrt{2(h_1 - h_2) + \mathbf{V}_1^2} = \sqrt{2(3410.3 - 3316.2)\text{ kJ/kg}\left(\frac{1000\text{ m}^2/\text{s}^2}{1\text{ kJ/kg}}\right) + (70\text{ m/s})^2} = \mathbf{439.4\ m/s}$$

(*b*) The exit velocity for the isentropic case is determined from

$$\mathbf{V}_{2s} = \sqrt{2(h_1 - h_{2s}) + \mathbf{V}_1^2} = \sqrt{2(3410.3 - 3301.5)\text{ kJ/kg}\left(\frac{1000\text{ m}^2/\text{s}^2}{1\text{ kJ/kg}}\right) + (70\text{ m/s})^2} = \mathbf{471.7\ m/s}$$

Thus,

$$\eta_N = \frac{\mathbf{V}_2^2/2}{\mathbf{V}_{2s}^2/2} = \frac{(439.4\text{ m/s})^2/2}{(471.7\text{ m/s})^2/2} = \mathbf{86.8\%}$$

(*c*) The exergy destroyed during a process can be determined from an exergy balance or directly from its definition $X_{\text{destroyed}} = T_0 S_{gen}$ where the entropy generation S_{gen} is determined from an entropy balance on the actual nozzle. It gives

$$\underbrace{\dot{S}_{in} - \dot{S}_{out}}_{\substack{\text{Rate of net entropy transfer} \\ \text{by heat and mass}}} + \underbrace{\dot{S}_{gen}}_{\substack{\text{Rate of entropy} \\ \text{generation}}} = \underbrace{\Delta\dot{S}_{\text{system}}^{\nearrow 0}}_{\substack{\text{Rate of change} \\ \text{of entropy}}} = 0$$

$$\dot{m}s_1 - \dot{m}s_2 + \dot{S}_{\text{gen}} = 0 \rightarrow \dot{S}_{gen} = \dot{m}(s_2 - s_1) \text{ or } s_{gen} = s_2 - s_1$$

Substituting, the exergy destruction in the nozzle on a unit mass basis is determined to be

$$x_{\text{destroyed}} = T_0 s_{gen} = T_0(s_2 - s_1) = (298\text{ K})(6.8186 - 6.7975)\text{kJ/kg}\cdot\text{K} = \mathbf{6.26\ kJ/kg}$$

7-72 CO_2 gas is compressed steadily by a compressor from a specified state to another specified state. The power input to the compressor if the process involved no irreversibilities is to be determined.

Assumptions **1** This is a steady-flow process since there is no change with time. **2** Kinetic and potential energy changes are negligible. **3** The device is adiabatic and thus heat transfer is negligible. **4** CO_2 is an ideal gas with constant specific heats.

Properties At the average temperature of (300 + 450)/2 = 375 K, the constant pressure specific heat and the specific heat ratio of CO_2 are $k = 1.261$ and $C_p = 0.917$ kJ/kg.K (Table A-2).

Analysis The reversible (or minimum) power input is determined from the exergy balance applied on the compressor, and setting the exergy destruction term equal to zero,

$$\underbrace{\dot{X}_{in} - \dot{X}_{out}}_{\substack{\text{Rate of net exergy transfer}\\ \text{by heat, work, and mass}}} - \underbrace{\dot{X}_{destroyed}^{\nearrow 0 \text{ (reversible)}}}_{\substack{\text{Rate of exergy}\\ \text{destruction}}} = \underbrace{\Delta \dot{X}_{system}^{\nearrow 0 \text{ (steady)}}}_{\substack{\text{Rate of change}\\ \text{of exergy}}} = 0$$

$$\dot{X}_{in} = \dot{X}_{out}$$

$$\dot{m}\psi_1 + \dot{W}_{rev,in} = \dot{m}\psi_2$$

$$\dot{W}_{rev,in} = \dot{m}(\psi_2 - \psi_1)$$

$$= \dot{m}[(h_2 - h_1) - T_0(s_2 - s_1) + \Delta ke^{\nearrow 0} + \Delta pe^{\nearrow 0}]$$

600 kPa
450 K

CO_2
0.2 kg/s

100 kPa
300 K

where

$$s_2 - s_1 = C_p \ln\frac{T_2}{T_1} - R\ln\frac{P_2}{P_1}$$

$$= (0.9175 \text{ kJ/kg}\cdot\text{K})\ln\frac{450 \text{ K}}{300 \text{ K}} - (0.1889 \text{ kJ/kg}\cdot\text{K})\ln\frac{600 \text{ kPa}}{100 \text{ kPa}}$$

$$= 0.03335 \text{ kJ/kg}\cdot\text{K}$$

Substituting,

$$\dot{W}_{rev,in} = (0.2 \text{ kg/s})[(0.917 \text{ kJ/kg}\cdot\text{K})(450 - 300)\text{ K} - (298 \text{ K})(0.03335 \text{ kJ/kg}\cdot\text{K})] = \mathbf{25.5 \text{ kW}}$$

Discussion Note that a minimum of 25.5 kW of power input is needed for this compressor.

7-73E A hot water stream is mixed with a cold water stream. For a specified mixture temperature, the mass flow rate of cold water stream and the rate of exergy destruction are to be determined.

Assumptions **1** Steady operating conditions exist. **2** The mixing chamber is well-insulated so that heat loss to the surroundings is negligible. **3** Changes in the kinetic and potential energies of fluid streams are negligible.

Properties Noting that that $T < T_{sat@50\ psia} = 281.03°F$, the water in all three streams exists as a compressed liquid, which can be approximated as a saturated liquid at the given temperature. Thus from Table A-4E,

$$\left.\begin{array}{l} P_1 = 50\text{ psia} \\ T_1 = 160°\text{F} \end{array}\right\} \begin{array}{l} h_1 \cong h_{f@160°F} = 127.96\text{ Btu/lbm} \\ s_1 \cong s_{f@160°F} = 0.2313\text{ Btu/lbm}\cdot\text{R} \end{array}$$

$$\left.\begin{array}{l} P_2 = 50\text{ psia} \\ T_2 = 70°\text{F} \end{array}\right\} \begin{array}{l} h_2 \cong h_{f@70°F} = 38.09\text{ Btu/lbm} \\ s_2 \cong s_{f@70°F} = 0.07463\text{ Btu/lbm}\cdot\text{R} \end{array}$$

$$\left.\begin{array}{l} P_3 = 50\text{ psia} \\ T_3 = 110°\text{F} \end{array}\right\} \begin{array}{l} h_3 \cong h_{f@110°F} = 78.02\text{ Btu/lbm} \\ s_3 \cong s_{f@110°F} = 0.1473\text{ Btu/lbm}\cdot\text{R} \end{array}$$

Analysis (*a*) We take the mixing chamber as the system, which is a control volume. The mass and energy balances for this steady-flow system can be expressed in the rate form as

Mass balance: $$\dot{m}_{in} - \dot{m}_{out} = \Delta\dot{m}_{system}^{\nearrow 0\ (steady)} = 0 \longrightarrow \dot{m}_1 + \dot{m}_2 = \dot{m}_3$$

Energy balance:

$$\underbrace{\dot{E}_{in} - \dot{E}_{out}}_{\substack{\text{Rate of net energy transfer}\\ \text{by heat, work, and mass}}} = \underbrace{\Delta\dot{E}_{system}^{\nearrow 0\ (steady)}}_{\substack{\text{Rate of change in internal, kinetic,}\\ \text{potential, etc. energies}}} = 0$$

$$\dot{E}_{in} = \dot{E}_{out}$$

$$\dot{m}h_1 + \dot{m}_2 h_2 = \dot{m}_3 h_3 \quad (\text{since } \dot{Q} = \dot{W} = \Delta\text{ke} \cong \Delta\text{pe} \cong 0)$$

Combining the two relations gives $\quad \dot{m}_1 h_1 + \dot{m}_2 h_2 = (\dot{m}_1 + \dot{m}_2)h_3$

Solving for $\dot{m}_2$ and substituting, the mass flow rate of cold water stream is determined to be

$$\dot{m}_2 = \frac{h_1 - h_3}{h_3 - h_2}\dot{m}_1 = \frac{(127.96 - 78.02)\text{Btu/lbm}}{(78.02 - 38.09)\text{Btu/lbm}}(4.0\text{ lbm/s}) = \mathbf{5.0\ lbm/s}$$

Also,

$$\dot{m}_3 = \dot{m}_1 + \dot{m}_2 = 4 + 5 = 9\text{ lbm/s}$$

(*b*) The exergy destroyed during a process can be determined from an exergy balance or directly from its definition $X_{destroyed} = T_0 S_{gen}$ where the entropy generation S_{gen} is determined from an entropy balance on the mixing chamber. It gives

$$\underbrace{\dot{S}_{in} - \dot{S}_{out}}_{\substack{\text{Rate of net entropy transfer}\\ \text{by heat and mass}}} + \underbrace{\dot{S}_{gen}}_{\substack{\text{Rate of entropy}\\ \text{generation}}} = \underbrace{\Delta\dot{S}_{system}^{\nearrow 0}}_{\substack{\text{Rate of change}\\ \text{of entropy}}} = 0$$

$$\dot{m}_1 s_1 + \dot{m}_2 s_2 - \dot{m}_3 s_3 + \dot{S}_{gen} = 0 \rightarrow \dot{S}_{gen} = \dot{m}_3 s_3 - \dot{m}_1 s_1 - \dot{m}_2 s_2$$

Substituting, the exergy destruction is determined to be

$$\begin{aligned} \dot{X}_{destroyed} &= T_0 \dot{S}_{gen} = T_0(\dot{m}_3 s_3 - \dot{m}_2 s_2 - \dot{m}_1 s_1) \\ &= (535\text{ R})(9.0 \times 0.1473 - 4.0 \times 0.2313 - 5.0 \times 0.07463)\text{Btu/s}\cdot\text{R} \\ &= \mathbf{14.6\ Btu/s} \end{aligned}$$

7-74 Liquid water is heated in a chamber by mixing it with superheated steam. For a specified mixing temperature, the mass flow rate of the steam and the rate of exergy destruction are to be determined.

Assumptions **1** This is a steady-flow process since there is no change with time. **2** Kinetic and potential energy changes are negligible. **3** There are no work interactions.

Properties Noting that $T < T_{sat\,@\,200\,kPa} = 120.23°C$, the cold water and the exit mixture streams exist as a compressed liquid, which can be approximated as a saturated liquid at the given temperature. From tables,

$$\left.\begin{array}{l} P_1 = 200\text{kPa} \\ T_1 = 20°\text{C} \end{array}\right\} \begin{array}{l} h_1 \cong h_{f\,@\,20°\text{C}} = 83.96\text{kJ/kg} \\ s_1 \cong s_{f\,@\,20°\text{C}} = 0.2966\text{kJ/kg}\cdot\text{K} \end{array}$$

$$\left.\begin{array}{l} P_2 = 200\text{kPa} \\ T_2 = 300°\text{C} \end{array}\right\} \begin{array}{l} h_2 = 3071.8\text{kJ/kg} \\ s_2 = 7.8926\text{kJ/kg}\cdot\text{K} \end{array}$$

$$\left.\begin{array}{l} P_3 = 200\text{kPa} \\ T_3 = 60°\text{C} \end{array}\right\} \begin{array}{l} h_3 \cong h_{f\,@\,60°\text{C}} = 251.13\text{kJ/kg} \\ s_3 \cong s_{f\,@\,60°\text{C}} = 0.8312\text{kJ/kg}\cdot\text{K} \end{array}$$

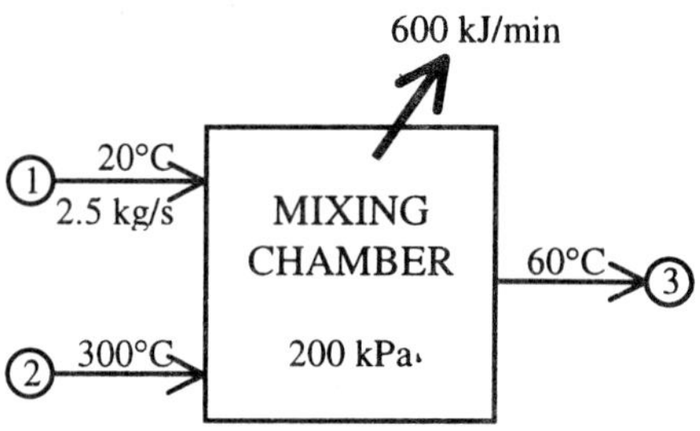

Analysis (a) We take the mixing chamber as the system, which is a control volume. The mass and energy balances for this steady-flow system can be expressed in the rate form as

Mass balance: $\dot{m}_{in} - \dot{m}_{out} = \Delta\dot{m}_{\text{system}}^{\nearrow 0\text{ (steady)}} = 0 \longrightarrow \dot{m}_1 + \dot{m}_2 = \dot{m}_3$

Energy balance:

$$\underbrace{\dot{E}_{in} - \dot{E}_{out}}_{\substack{\text{Rate of net energy transfer} \\ \text{by heat, work, and mass}}} = \underbrace{\Delta\dot{E}_{\text{system}}^{\nearrow 0\text{ (steady)}}}_{\substack{\text{Rate of change in internal, kinetic,} \\ \text{potential, etc. energies}}} = 0$$

$$\dot{E}_{in} = \dot{E}_{out}$$

$$\dot{m}h_1 + \dot{m}_2 h_2 = \dot{Q}_{out} + \dot{m}_3 h_3$$

Combining the two relations gives $\dot{Q}_{out} = \dot{m}_1 h_1 + \dot{m}_2 h_2 - (\dot{m}_1 + \dot{m}_2)h_3 = \dot{m}_1(h_1 - h_3) + \dot{m}_2(h_2 - h_3)$

Solving for $\dot{m}_2$ and substituting, the mass flow rate of the superheated steam is determined to be

$$\dot{m}_2 = \frac{\dot{Q}_{out} - \dot{m}_1(h_1 - h_3)}{h_2 - h_3} = \frac{(600/60\text{kJ/s}) - (2.5\text{kg/s})(83.96 - 251.13)\text{kJ/kg}}{(3071.8 - 251.13)\text{kJ/kg}} = \mathbf{0.152kg/s}$$

Also, $\dot{m}_3 = \dot{m}_1 + \dot{m}_2 = 2.5 + 0.152 = 2.652\text{ kg/s}$

(*b*) The exergy destroyed during a process can be determined from an exergy balance or directly from its definition $X_{\text{destroyed}} = T_0 S_{gen}$ where the entropy generation S_{gen} is determined from an entropy balance on an *extended system* that includes the mixing chamber and its immediate surroundings. It gives

$$\underbrace{\dot{S}_{in} - \dot{S}_{out}}_{\substack{\text{Rate of net entropy transfer} \\ \text{by heat and mass}}} + \underbrace{\dot{S}_{gen}}_{\substack{\text{Rate of entropy} \\ \text{generation}}} = \underbrace{\Delta\dot{S}_{\text{system}}^{\nearrow 0}}_{\substack{\text{Rate of change} \\ \text{of entropy}}} = 0$$

$$\dot{m}_1 s_1 + \dot{m}_2 s_2 - \dot{m}_3 s_3 - \frac{\dot{Q}_{out}}{T_{b,surr}} + \dot{S}_{\text{gen}} = 0 \rightarrow \dot{S}_{\text{gen}} = \dot{m}_3 s_3 - \dot{m}_1 s_1 - \dot{m}_2 s_2 + \frac{\dot{Q}_{out}}{T_0}$$

Substituting, the exergy destruction is determined to be

$$\begin{aligned} \dot{X}_{\text{destroyed}} &= T_0 \dot{S}_{gen} = T_0\left(\dot{m}_3 s_3 - \dot{m}_2 s_2 - \dot{m}_1 s_1 + \frac{\dot{Q}_{out}}{T_{b,surr}}\right) \\ &= (298\text{ K})(2.652\times0.8312 - 0.152\times7.8926 - 2.5\times0.2966 + 10/298)\text{kW/K} \\ &= \mathbf{88.5\,kW} \end{aligned}$$

7-75 Refrigerant-134a is vaporized by air in the evaporator of an air-conditioner. For specified flow rates, the exit temperature of air and the rate of exergy destruction are to be determined for the cases of insulated and uninsulated evaporator.

Assumptions **1** This is a steady-flow process since there is no change with time. **2** Kinetic and potential energy changes are negligible. **3** There are no work interactions. **4** Air is an ideal gas with constant specific heats at room temperature.

Properties The gas constant of air is 0.287 kPa.m^3/kg.K (Table A-1). The constant pressure specific heat of air at room temperature is C_p = 1.005 kJ/kg.K (Table A-2). The properties of R-134a at the inlet and the exit states are (Tables A-11 through A-13)

$$\left.\begin{array}{l} P_1 = 120\text{ kPa} \\ x_1 = 0.3 \end{array}\right\} \begin{array}{l} h_1 = h_f + x_1 h_{fg} = 21.32 + 0.3 \times 212.54 = 85.08\text{ kJ/kg} \\ s_1 = s_f + x_1 s_{fg} = 0.0879 + 0.3(0.9354 - 0.0879) = 0.3422\text{ kJ/kg}\cdot\text{K} \end{array}$$

$$\left.\begin{array}{l} T_2 = 120\text{ kPa} \\ \text{sat.vapor} \end{array}\right\} \begin{array}{l} h_2 = h_{g\,@\,120\text{ kPa}} = 233.86\text{ kJ/kg} \\ s_2 = s_{g\,@\,120\text{ kPa}} = 0.9354\text{ kJ/kg}\cdot\text{K} \end{array}$$

Analysis Air at specified conditions can be treated as an ideal gas with specific heats at room temperature. The properties of the refrigerant are

$$\dot{m}_{air} = \frac{P_3\dot{V}_3}{RT_3} = \frac{(100\text{kPa})(6\text{m}^3/\text{min})}{(0.287\text{kPa}\cdot\text{m}^3/\text{kg}\cdot\text{K})(300\text{K})} = 6.97\text{kg/min}$$

(a) We take the entire heat exchanger as the system, which is a control volume. The mass and energy balances for this steady-flow system can be expressed in the rate form as:

Mass balance (*for each fluid stream*):

$$\dot{m}_{in} - \dot{m}_{out} = \Delta\dot{m}_{\text{system}}^{\nearrow 0\text{ (steady)}} = 0 \rightarrow \dot{m}_{in} = \dot{m}_{out} \rightarrow \dot{m}_1 = \dot{m}_2 = \dot{m}_{air} \text{ and } \dot{m}_3 = \dot{m}_4 = \dot{m}_R$$

Energy balance (for the entire heat exchanger):

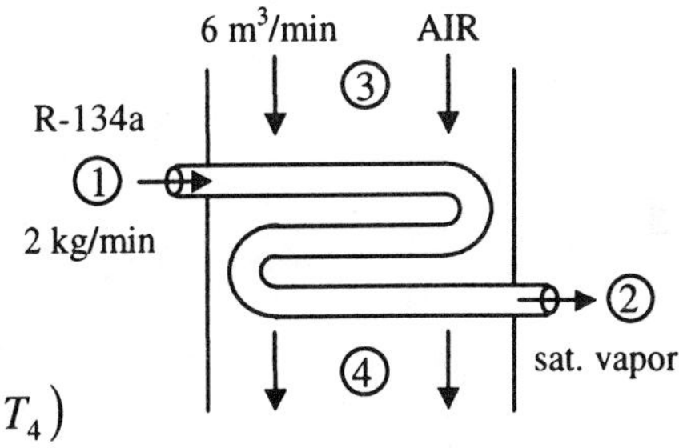

$$\underbrace{\dot{E}_{in} - \dot{E}_{out}}_{\substack{\text{Rate of net energy transfer} \\ \text{by heat, work, and mass}}} = \underbrace{\Delta\dot{E}_{\text{system}}^{\nearrow 0\text{ (steady)}}}_{\substack{\text{Rate of change in internal, kinetic,} \\ \text{potential, etc. energies}}} = 0$$

$$\dot{E}_{in} = \dot{E}_{out}$$

$$\dot{m}_1 h_1 + \dot{m}_3 h_3 = \dot{m}_2 h_2 + \dot{m}_4 h_4 \quad (\text{since } \dot{Q} = \dot{W} = \Delta\text{ke} \cong \Delta\text{pe} \cong 0)$$

6 m^3/min AIR
③
R-134a
①
2 kg/min
②
④
sat. vapor

Combining the two, $\dot{m}_R(h_2 - h_1) = \dot{m}_{air}(h_3 - h_4) = \dot{m}_{air}C_p(T_3 - T_4)$

Solving for T_4, $T_4 = T_3 - \dfrac{\dot{m}_R(h_2 - h_1)}{\dot{m}_{air}C_p}$

Substituting, $T_4 = 27°\text{C} - \dfrac{(2\text{ kg/min})(233.86 - 85.08)\text{ kJ/kg}}{(6.97\text{ kg/min})(1.005\text{ kJ/kg}\cdot\text{K})} = \mathbf{-15.5°C} = 257.5\text{ K}$

The exergy destroyed during a process can be determined from an exergy balance or directly from its definition $X_{\text{destroyed}} = T_0 S_{gen}$ where the entropy generation S_{gen} is determined from an entropy balance on the evaporator. Noting that the condenser is well-insulated and thus heat transfer is negligible, the entropy balance for this steady-flow system can be expressed as

$$\underbrace{\dot{S}_{in} - \dot{S}_{out}}_{\substack{\text{Rate of net entropy transfer} \\ \text{by heat and mass}}} + \underbrace{\dot{S}_{gen}}_{\substack{\text{Rate of entropy} \\ \text{generation}}} = \underbrace{\Delta \dot{S}_{\text{system}}^{\nearrow 0 \text{ (steady)}}}_{\substack{\text{Rate of change} \\ \text{of entropy}}}$$

$$\dot{m}_1 s_1 + \dot{m}_3 s_3 - \dot{m}_2 s_2 - \dot{m}_4 s_4 + \dot{S}_{gen} = 0 \quad (\text{since } Q = 0)$$

$$\dot{m}_R s_1 + \dot{m}_{\text{air}} s_3 - \dot{m}_R s_2 - \dot{m}_{\text{air}} s_4 + \dot{S}_{gen} = 0$$

or,

$$\dot{S}_{gen} = \dot{m}_R (s_2 - s_1) + \dot{m}_{air} (s_4 - s_3)$$

where $s_4 - s_3 = C_p \ln \frac{T_4}{T_3} - R \ln \frac{P_4^{\nearrow 0}}{P_3} = C_p \ln \frac{T_4}{T_3} = (1.005 \text{ kJ/kg} \cdot \text{K}) \ln \frac{257.5 \text{ K}}{300 \text{ K}} = 0.154 \text{ kJ/kg} \cdot \text{K}$

Substituting, the exergy destruction is determined to be

$$\begin{aligned} \dot{X}_{\text{destroyed}} &= T_0 \dot{S}_{gen} = T_0 [\dot{m}_R (s_2 - s_1) + \dot{m}_{air} (s_4 - s_3)] \\ &= (305 \text{ K})[(2 \text{ kg/min})(0.9354 - 0.3422) \text{kJ/kg} \cdot \text{K} + (6.97 \text{ kg/min})(-0.154 \text{ kJ/kg} \cdot \text{K})] \\ &= 34.5 \text{ kJ/min} = \mathbf{0.57 \text{ kW}} \end{aligned}$$

(*b*) When there is a heat gain from the surroundings, the steady-flow energy equation reduces to

$$\dot{Q}_{in} = \dot{m}_R (h_2 - h_1) + \dot{m}_{air} C_p (T_4 - T_3)$$

Solving for T_4, $\quad T_4 = T_3 + \frac{\dot{Q}_{in} - \dot{m}_R (h_2 - h_1)}{\dot{m}_{air} C_p}$

Substituting, $\quad T_4 = 27°\text{C} + \frac{(30 \text{ kJ/min}) - (2 \text{ kg/min})(233.86 - 85.08) \text{ kJ/kg}}{(6.97 \text{ kg/min})(1.005 \text{ kJ/kg} \cdot \text{K})} = \mathbf{-11.2°C}$

The exergy destroyed during a process can be determined from an exergy balance or directly from its definition $X_{\text{destroyed}} = T_0 S_{gen}$ where the entropy generation S_{gen} is determined from an entropy balance on an *extended system* that includes the evaporator and its immediate surroundings. It gives

$$\underbrace{\dot{S}_{in} - \dot{S}_{out}}_{\substack{\text{Rate of net entropy transfer} \\ \text{by heat and mass}}} + \underbrace{\dot{S}_{gen}}_{\substack{\text{Rate of entropy} \\ \text{generation}}} = \underbrace{\Delta \dot{S}_{\text{system}}^{\nearrow 0 \text{ (steady)}}}_{\substack{\text{Rate of change} \\ \text{of entropy}}}$$

$$\frac{Q_{in}}{T_{b,in}} + \dot{m}_1 s_1 + \dot{m}_3 s_3 - \dot{m}_2 s_2 - \dot{m}_4 s_4 + \dot{S}_{gen} = 0$$

$$\frac{Q_{in}}{T_0} + \dot{m}_R s_1 + \dot{m}_{\text{air}} s_3 - \dot{m}_R s_2 - \dot{m}_{\text{air}} s_4 + \dot{S}_{gen} = 0$$

or $\quad \dot{S}_{gen} = \dot{m}_R (s_2 - s_1) + \dot{m}_{air} (s_4 - s_3) - \frac{\dot{Q}_{in}}{T_0}$

where $s_4 - s_3 = C_p \ln \frac{T_4}{T_3} - R \ln \frac{P_4^{\nearrow 0}}{P_3} = (1.005 \text{ kJ/kg} \cdot \text{K}) \ln \frac{261.8 \text{ K}}{300 \text{ K}} = -0.137 \text{ kJ/kg} \cdot \text{K}$

Substituting, the exergy destruction is determined to be

$$\begin{aligned} \dot{X}_{\text{destroyed}} &= T_0 \dot{S}_{gen} = T_0 \left[\dot{m}_R (s_2 - s_1) + \dot{m}_{air} (s_4 - s_3) - \frac{\dot{Q}_{in}}{T_0} \right] \\ &= (305 \text{ K}) \left[(2 \text{kg/min})(0.9354 - 0.3422) \text{kJ/kg} \cdot \text{K} + (6.97 \text{kg/min})(-0.137 \text{kJ/kg} \cdot \text{K}) - \frac{30 \text{kJ/min}}{305 \text{K}} \right] \\ &= 40.6 \text{ kJ/min} = \mathbf{0.67 \text{ kW}} \end{aligned}$$

7-76 A rigid tank initially contains saturated R-134a vapor. The tank is connected to a supply line, and R-134a is allowed to enter the tank. The mass of the R-134a that entered the tank and the exergy destroyed during this process are to be determined.

Assumptions **1** This is an unsteady process since the conditions within the device are changing during the process, but it can be analyzed as a uniform-flow process since the state of fluid at the inlet remains constant. **2** Kinetic and potential energies are negligible. **3** There are no work interactions involved. **4** The direction of heat transfer is to the tank (will be verified).

Properties The properties of refrigerant are (Tables A-11 through A-13)

$$\left.\begin{array}{l} P_1 = 1\text{ MPa} \\ \text{sat.vapor} \end{array}\right\} \begin{array}{l} v_1 = v_{g\,@\,1\,\text{MPa}} = 0.0202\text{ m}^3/\text{kg} \\ u_1 = u_{g\,@\,1\,\text{MPa}} = 247.77\text{ kJ/kg} \\ s_1 = s_{g\,@\,1\,\text{MPa}} = 0.9043\text{ kJ/kg}\cdot\text{K} \end{array}$$

$$\left.\begin{array}{l} T_2 = 1.2\text{ MPa} \\ \text{sat.liquid} \end{array}\right\} \begin{array}{l} v_2 = v_{f\,@\,1.2\,\text{MPa}} = 0.0008928\text{ m}^3/\text{kg} \\ u_2 = u_{f\,@\,1.2\,\text{MPa}} = 114.69\text{ kJ/kg} \\ s_2 = s_{f\,@\,1.2\,\text{MPa}} = 0.4164\text{ kJ/kg}\cdot\text{K} \end{array}$$

$$\left.\begin{array}{l} P_i = 1.4\text{ MPa} \\ T_i = 30°\text{C} \end{array}\right\} \begin{array}{l} h_i = 91.49\text{ kJ/kg} \\ s_i = 0.3396\text{ kJ/kg}\cdot\text{K} \end{array}$$

R-134a
1.4 MPa
30°C
R-134a
0.1 m³
1 MPa
Sat. vapor
Q

Analysis We take the tank as the system, which is a control volume. Noting that the microscopic energies of flowing and nonflowing fluids are represented by enthalpy h and internal energy u, respectively, the mass and energy balances for this uniform-flow system can be expressed as

Mass balance: $m_{in} - m_{out} = \Delta m_{\text{system}} \rightarrow m_i = m_2 - m_1$

Energy balance:

$$\underbrace{E_{in} - E_{out}}_{\substack{\text{Net energy transfer} \\ \text{by heat, work, and mass}}} = \underbrace{\Delta E_{\text{system}}}_{\substack{\text{Change in internal, kinetic,} \\ \text{potential, etc. energies}}}$$

$$Q_{in} + m_i h_i = m_2 u_2 - m_1 u_1 \quad (\text{since } W \cong ke \cong pe \cong 0)$$

(*a*) The initial and the final masses in the tank are

$$m_1 = \frac{V_1}{v_1} = \frac{0.1\text{ m}^3}{0.0202\text{ m}^3/\text{kg}} = 4.95\text{ kg}$$

$$m_2 = \frac{V_2}{v_2} = \frac{0.1\text{ m}^3}{0.0008928\text{ m}^3/\text{kg}} = 112.01\text{ kg}$$

Then from the mass balance

$$m_i = m_2 - m_1 = 112.01 - 4.95 = \mathbf{107.06\ kg}$$

The heat transfer during this process is determined from the energy balance to be

$$\begin{aligned} Q_{in} &= -m_i h_i + m_2 u_2 - m_1 u_1 \\ &= -(107.06\text{kg})(91.49\text{kJ/kg}) + (112.01\text{kg})(114.69\text{kJ/kg}) - (4.95\text{kg})(247.77\text{kJ/kg}) \\ &= 1825\text{kJ} \end{aligned}$$

(*b*) The exergy destroyed during a process can be determined from an exergy balance or directly from its definition $X_{\text{destroyed}} = T_0 S_{gen}$. The entropy generation S_{gen} in this case is determined from an entropy balance on an *extended system* that includes the tank and its immediate surroundings so that the boundary temperature of the extended system is the surroundings temperature T_{surr} at all times. It gives

$$\underbrace{S_{in} - S_{out}}_{\substack{\text{Net entropy transfer} \\ \text{by heat and mass}}} + \underbrace{S_{gen}}_{\substack{\text{Entropy} \\ \text{generation}}} = \underbrace{\Delta S_{\text{system}}}_{\substack{\text{Change} \\ \text{in entropy}}}$$

$$\frac{Q_{in}}{T_{b,in}} + m_i s_i + S_{gen} = \Delta S_{\text{tank}} = (m_2 s_2 - m_1 s_1)_{\text{tank}}$$

$$S_{gen} = m_2 s_2 - m_1 s_1 - m_i s_i - \frac{Q_{in}}{T_0}$$

Substituting, the exergy destruction is determined to be

$$X_{\text{destroyed}} = T_0 S_{gen} = T_0 \left[m_2 s_2 - m_1 s_1 - m_i s_i - \frac{Q_{in}}{T_0} \right]$$

$$= (318 \text{ K})[112.00 \times 0.4164 - 4.95 \times 0.9043 - 107.05 \times 0.3396 - (1{,}824.8 \text{ kJ})/(318 \text{ K})]$$

$$= \mathbf{21.6 \ kJ}$$

7-77 A rigid tank initially contains saturated liquid water. A valve at the bottom of the tank is opened, and half of mass in liquid form is withdrawn from the tank. The temperature in the tank is maintained constant. The amount of heat transfer, the reversible work, and the exergy destruction during this process are to be determined.

Assumptions **1** This is an unsteady process since the conditions within the device are changing during the process, but it can be analyzed as a uniform-flow process since the state of fluid leaving the device remains constant. **2** Kinetic and potential energies are negligible. **3** There are no work interactions involved. **4** The direction of heat transfer is to the tank (will be verified).

Properties The properties of water are (Tables A-4 through A-6)

$$\left.\begin{array}{l} T_1 = 200^\circ\text{C} \\ sat.liquid \end{array}\right\} \begin{array}{l} v_1 = v_{f@200^\circ\text{C}} = 0.001157\text{m}^3/\text{kg} \\ h_1 = h_{f@200^\circ\text{C}} = 850.65\text{kJ/kg} \\ s_1 = s_{f@200^\circ\text{C}} = 2.3309\text{kJ/kg}\cdot\text{K} \end{array}$$

$$\left.\begin{array}{l} T_e = 200^\circ\text{C} \\ sat.liquid \end{array}\right\} \begin{array}{l} h_e = h_{f@200^\circ\text{C}} = 852.45\text{kJ/kg} \\ s_e = s_{f@200^\circ\text{C}} = 2.3309\text{kJ/kg}\cdot\text{K} \end{array}$$

Analysis We take the tank as the system, which is a control volume since mass crosses the boundary. Noting that the microscopic energies of flowing and nonflowing fluids are represented by enthalpy h and internal energy u, respectively, the mass and energy balances for this uniform-flow system can be expressed as

Mass balance: $$m_{in} - m_{out} = \Delta m_{\text{system}} \quad \rightarrow \quad m_e = m_1 - m_2$$

Energy balance: $$\underbrace{E_{in} - E_{out}}_{\substack{\text{Net energy transfer} \\ \text{by heat, work, and mass}}} = \underbrace{\Delta E_{\text{system}}}_{\substack{\text{Change in internal, kinetic,} \\ \text{potential, etc. energies}}}$$

$$Q_{in} = m_e h_e + m_2 u_2 - m_1 u_1 \quad (\text{since } W \cong ke \cong pe \cong 0)$$

The initial and the final masses in the tank are

H₂O 0.4 m³ 200°C T = const. — Q — mₑ

$$m_1 = \frac{V}{v_1} = \frac{0.4\text{m}^3}{0.001157\text{m}^3/\text{kg}} = 345.72\text{kg}$$

$$m_2 = \frac{1}{2} m_1 = \frac{1}{2}(345.72\text{kg}) = 172.86\text{kg} = m_e$$

Now we determine the final internal energy and entropy,

$$v_2 = \frac{V}{m_2} = \frac{0.4\text{m}^3}{172.86\text{kg}} = 0.002314\text{m}^3/\text{kg}$$

$$x_2 = \frac{v_2 - v_f}{v_{fg}} = \frac{0.002314 - 0.001157}{0.13736 - 0.001157} = 0.00917$$

$$\left.\begin{array}{l} T_2 = 200^\circ\text{C} \\ x_2 = 0.00917 \end{array}\right\} \begin{array}{l} u_2 = u_f + x_2 u_{fg} = 850.65 + (0.00917)(1744.7) = 866.65\text{kJ/kg} \\ s_2 = s_f + x_2 s_{fg} = 2.3309 + (0.00917)(4.1014) = 2.3685\text{kJ/kg}\cdot\text{K} \end{array}$$

The heat transfer during this process is determined by substituting these values into the energy balance equation,

$$\begin{aligned}Q_{in} &= m_e h_e + m_2 u_2 - m_1 u_1 \\ &= (172.86\text{kg})(852.45\text{kJ/kg}) + (172.86\text{kg})(866.65\text{kJ/kg}) \\ &\quad - (345.72\text{kg})(850.65\text{kJ/kg}) \\ &= \mathbf{3077 kJ}\end{aligned}$$

(*b*) The exergy destroyed during a process can be determined from an exergy balance or directly from its definition $X_{\text{destroyed}} = T_0 S_{gen}$. The entropy generation S_{gen} in this case is determined from an entropy balance on an *extended system* that includes the tank and the region between the tank and the source so that the boundary temperature of the extended system at the location of heat transfer is the source temperature T_{source} at all times. It gives

$$\underbrace{S_{in} - S_{out}}_{\substack{\text{Net entropy transfer} \\ \text{by heat and mass}}} + \underbrace{S_{gen}}_{\substack{\text{Entropy} \\ \text{generation}}} = \underbrace{\Delta S_{\text{system}}}_{\substack{\text{Change} \\ \text{in entropy}}}$$

$$\frac{Q_{in}}{T_{b,in}} - m_e s_e + S_{gen} = \Delta S_{\text{tank}} = (m_2 s_2 - m_1 s_1)_{\text{tank}}$$

$$S_{gen} = m_2 s_2 - m_1 s_1 + m_e s_e - \frac{Q_{in}}{T_{\text{source}}}$$

Substituting, the exergy destruction is determined to be

$$\begin{aligned}X_{\text{destroyed}} &= T_0 S_{gen} = T_0 \left[m_2 s_2 - m_1 s_1 + m_e s_e - \frac{Q_{in}}{T_{\text{source}}} \right] \\ &= (298\text{ K})[172.86 \times 2.3309 + 172.86 \times 2.3685 - 345.72 \times 2.3309 + (-3077\text{ kJ})/(523\text{ K})] \\ &= \mathbf{183.6\ kJ}\end{aligned}$$

For processes that involve no actual work, the reversible work output and exergy destruction are identical. Therefore,

$$X_{\text{destroyed}} = W_{rev,out} - W_{act,out} \quad \rightarrow \quad W_{rev,out} = X_{\text{destroyed}} = \mathbf{183.6\ kJ}$$

7-78E An insulated rigid tank equipped with an electric heater initially contains pressurized air. A valve is opened, and air is allowed to escape at constant temperature until the pressure inside drops to 30 psia. The amount of electrical work done and the exergy destroys are to be determined.

Assumptions **1** This is an unsteady process since the conditions within the device are changing during the process, but it can be analyzed as a uniform-flow process since the exit temperature (and enthalpy) of air remains constant. **2** Kinetic and potential energies are negligible. **3** The tank is insulated and thus heat transfer is negligible. **4** Air is an ideal gas with variable specific heats. **5** The environment temperature is given to be 70°F.

Properties The gas constant of air is R =0.3704 psia.ft³/lbm.R (Table A-1E). The properties of air are (Table A-17E)

$$T_e = 600 \text{ R} \longrightarrow h_e = 143.47 \text{ Btu/lbm}$$
$$T_1 = 600 \text{ R} \longrightarrow u_1 = 102.34 \text{ Btu/lbm}$$
$$T_2 = 600 \text{ R} \longrightarrow u_2 = 102.34 \text{ Btu/lbm}$$

Analysis We take the tank as the system, which is a control volume. Noting that the microscopic energies of flowing and nonflowing fluids are represented by enthalpy h and internal energy u, respectively, the mass and energy balances for this uniform-flow system can be expressed as

Mass balance: $m_{in} - m_{out} = \Delta m_{\text{system}} \rightarrow m_e = m_1 - m_2$

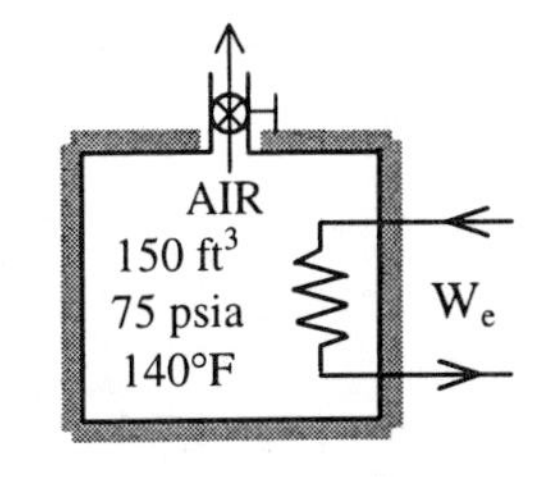

Energy balance:

$$\underbrace{E_{in} - E_{out}}_{\substack{\text{Net energy transfer}\\ \text{by heat, work, and mass}}} = \underbrace{\Delta E_{\text{system}}}_{\substack{\text{Change in internal, kinetic,}\\ \text{potential, etc. energies}}}$$

$$W_{e,in} - m_e h_e = m_2 u_2 - m_1 u_1 \quad (\text{since } Q \cong ke \cong pe \cong 0)$$

The initial and the final masses of air in the tank are

$$m_1 = \frac{P_1 V}{RT_1} = \frac{(75 \text{ psia})(150 \text{ ft}^3)}{(0.3704 \text{ psia}\cdot\text{ft}^3/\text{lbm}\cdot\text{R})(600 \text{ R})} = 50.62 \text{ lbm}$$

$$m_2 = \frac{P_2 V}{RT_2} = \frac{(30 \text{ psia})(150 \text{ ft}^3)}{(0.3704 \text{ psia}\cdot\text{ft}^3/\text{lbm}\cdot\text{R})(600 \text{ R})} = 20.25 \text{ lbm}$$

Then from the mass and energy balances,

$$m_e = m_1 - m_2 = 50.62 - 20.25 = 30.37 \text{ lbm}$$

$$\begin{aligned} W_{e,in} &= m_e h_e + m_2 u_2 - m_1 u_1 \\ &= (30.37 \text{ lbm})(143.47 \text{ Btu/lbm}) + (20.25 \text{ lbm})(102.34 \text{ Btu/lbm}) - (50.62 \text{ lbm})(102.34 \text{ Btu/lbm}) \\ &= \mathbf{1249 \text{ Btu}} \end{aligned}$$

(*b*) The exergy destroyed during a process can be determined from an exergy balance or directly from its definition $X_{\text{destroyed}} = T_0 S_{gen}$ where the entropy generation S_{gen} is determined from an entropy balance on the insulated tank. It gives

$$\underbrace{S_{in} - S_{out}}_{\substack{\text{Net entropy transfer}\\ \text{by heat and mass}}} + \underbrace{S_{gen}}_{\substack{\text{Entropy}\\ \text{generation}}} = \underbrace{\Delta S_{\text{system}}}_{\substack{\text{Change}\\ \text{in entropy}}}$$

$$-m_e s_e + S_{gen} = \Delta S_{\text{tank}} = (m_2 s_2 - m_1 s_1)_{\text{tank}}$$

$$\begin{aligned} S_{gen} &= m_2 s_2 - m_1 s_1 + m_e s_e \\ &= m_2 s_2 - m_1 s_1 + (m_1 - m_2) s_e \\ &= m_2 (s_2 - s_e) - m_1 (s_1 - s_e) \end{aligned}$$

Assuming a constant average pressure of (75 + 30) / 2 = 52.5 psia for the exit stream, the entropy changes are determined to be

$$s_2 - s_e = C_p \ln \frac{T_2}{T_e}^{\nearrow 0} - R \ln \frac{P_2}{P_e} = -(0.06855\ \text{Btu/lbm} \cdot \text{R}) \ln \frac{30\ \text{psia}}{52.5\ \text{psia}} = 0.03836\ \text{Btu/lbm} \cdot \text{R}$$

$$s_1 - s_e = C_p \ln \frac{T_1}{T_e}^{\nearrow 0} - R \ln \frac{P_1}{P_e} = -(0.06855\ \text{Btu/lbm} \cdot \text{R}) \ln \frac{75\ \text{psia}}{52.5\ \text{psia}} = -0.02445\ \text{Btu/lbm} \cdot \text{R}$$

Substituting, the exergy destruction is determined to be

$$\begin{aligned} X_{\text{destroyed}} &= T_0 S_{gen} = T_0 [m_2 (s_2 - s_e) - m_1 (s_1 - s_e)] \\ &= (530\ \text{R}) [(20.25\ \text{lbm})(0.03836\ \text{Btu/lbm} \cdot \text{R}) - (50.62\ \text{lbm})(-0.02445\ \text{Btu/lbm} \cdot \text{R})] \\ &= \mathbf{1{,}068\ Btu} \end{aligned}$$

7-79 A rigid tank initially contains saturated liquid-vapor mixture of refrigerant-134a. A valve at the bottom of the tank is opened, and liquid is withdrawn from the tank at constant pressure until no liquid remains inside. The final mass in the tank and the reversible work associated with this process are to be determined.

Assumptions **1** This is an unsteady process since the conditions within the device are changing during the process. It can be analyzed as a uniform-flow process since the state of fluid leaving the device remains constant. **2** Kinetic and potential energies are negligible. **3** There are no work interactions involved. **4** The direction of heat transfer is to the tank (will be verified).

Properties The properties of R-134a are (Tables A-11 through A-13)

R-134a
800 kPa
P = const.
Source
50°C
Q
m_e

$$P_1 = 800\text{kPa} \rightarrow v_f = 0.0008454\text{m}^3/\text{kg}, v_g = 0.0255\text{m}^3/\text{kg}$$
$$u_f = 92.75\text{kJ/kg}, \quad u_g = 243.78\text{kJ/kg}$$
$$s_f = 0.3459\text{kJ/kg.K}, \quad s_g = 0.09066\text{kJ/kg.K}$$

$$\left.\begin{array}{l} P_2 = 800 \text{ kPa} \\ \text{sat.vapor} \end{array}\right\} \begin{array}{l} v_2 = v_{g\,@\,800\text{ kPa}} = 0.0255 \text{ m}^3/\text{kg} \\ u_2 = u_{g\,@\,800\text{ kPa}} = 243.78 \text{ kJ/kg} \\ s_2 = s_{g\,@\,800\text{ kPa}} = 0.9066 \text{ kJ/kg}\cdot\text{K} \end{array}$$

$$\left.\begin{array}{l} P_e = 800\text{kPa} \\ sat.liquid \end{array}\right\} \begin{array}{l} h_e = h_{f\,@800\text{kPa}} = 93.42\text{kJ/kg} \\ s_e = s_{f\,@800\text{kPa}} = 0.3459\text{kJ/kg.K} \end{array}$$

Analysis We take the tank as the system, which is a control volume. Noting that the microscopic energies of flowing and nonflowing fluids are represented by enthalpy *h* and internal energy *u*, respectively, the mass and energy balances for this uniform-flow system can be expressed as

Mass balance: $$m_{in} - m_{out} = \Delta m_{\text{system}} \quad \rightarrow \quad m_e = m_1 - m_2$$

Energy balance: $$\underbrace{E_{in} - E_{out}}_{\substack{\text{Net energy transfer} \\ \text{by heat, work, and mass}}} = \underbrace{\Delta E_{\text{system}}}_{\substack{\text{Change in internal, kinetic,} \\ \text{potential, etc. energies}}}$$

$$Q_{in} = m_e h_e + m_2 u_2 - m_1 u_1 \quad (\text{since } W \cong ke \cong pe \cong 0)$$

The initial mass, initial internal energy, and final mass in the tank are

$$m_1 = m_f + m_g = \frac{V_f}{v_f} + \frac{V_g}{v_g} = \frac{0.1 \times 0.2 \text{ m}^3}{0.0008454 \text{ m}^3/\text{kg}} + \frac{0.1 \times 0.8 \text{ m}^3}{0.0255 \text{ m}^3/\text{kg}} = 23.657 + 3.137 = 26.794 \text{ kg}$$

$$U_1 = m_1 u_1 = m_f u_f + m_g u_g = 23.657 \times 92.75 + 3.137 \times 243.78 = 2{,}958.9 \text{ kJ}$$
$$S_1 = m_1 s_1 = m_f s_f + m_g s_g = 23.657 \times 0.3459 + 3.137 \times 0.9066 = 11.027 \text{ kJ/K}$$

$$m_2 = \frac{V}{v_2} = \frac{0.1 \text{ m}^3}{0.0255 \text{ m}^3/\text{kg}} = \mathbf{3.922 \text{ kg}}$$

Then from the mass and energy balances,

$$m_e = m_1 - m_2 = 26.794 - 3.922 = 22.872 \text{ kg}$$

$$Q_{in} = (22.872 \text{ kg})(93.42 \text{ kJ/kg}) + (3.922 \text{ kg})(243.78 \text{ kJ/kg}) - 2{,}958.9 \text{ kJ} = 133.9 \text{ kJ}$$

(*b*) This process involves no actual work, thus the reversible work and exergy generation are identical since $X_{\text{destroyed}} = W_{rev,out} - W_{act,out} \rightarrow W_{rev,out} = X_{\text{destroyed}}$.

The exergy destroyed during a process can be determined from an exergy balance or directly from its definition $X_{\text{destroyed}} = T_0 S_{gen}$. The entropy generation S_{gen} in this case is determined from an entropy balance on an *extended system* that includes the tank and the region between the tank and the heat source so that the boundary temperature of the extended system at the location of heat transfer is the source temperature T_{source} at all times. It gives

$$\underbrace{S_{in} - S_{out}}_{\substack{\text{Net entropy transfer} \\ \text{by heat and mass}}} + \underbrace{S_{gen}}_{\substack{\text{Entropy} \\ \text{generation}}} = \underbrace{\Delta S_{\text{system}}}_{\substack{\text{Change} \\ \text{in entropy}}}$$

$$\frac{Q_{in}}{T_{b,in}} - m_e s_e + S_{gen} = \Delta S_{\text{tank}} = (m_2 s_2 - m_1 s_1)_{\text{tank}}$$

$$S_{gen} = m_2 s_2 - m_1 s_1 + m_e s_e - \frac{Q_{in}}{T_{\text{source}}}$$

Substituting,

$$W_{rev,out} = X_{\text{destroyed}} = T_0 S_{gen} = T_0 \left[m_2 s_2 - m_1 s_1 + m_e s_e - \frac{Q_{in}}{T_{\text{source}}} \right]$$

$$= (298 \text{ K})[3.922 \times 0.9066 - 11.027 + 22.872 \times 0.3459 - 133.9/323)]$$

$$= \mathbf{7.62 \ kJ}$$

That is, 7.62 kJ of work could have been produced during this process.

7-80 A cylinder initially contains helium gas at a specified pressure and temperature. A valve is opened, and helium is allowed to escape until its volume decreases by half. The work potential of the helium at the initial state and the exergy destroyed during the process are to be determined.

Assumptions **1** This is an unsteady process since the conditions within the device are changing during the process, but it can be analyzed as a uniform-flow process by using constant average properties for the helium leaving the tank. **2** Kinetic and potential energies are negligible. **3** There are no work interactions involved other than boundary work. **4** The tank is insulated and thus heat transfer is negligible. **5** Helium is an ideal gas with constant specific heats.

Properties The gas constant of helium is R = 2.0769 kPa.m³/kg.K = 2.0769 kJ/kg.K. The constant pressure specific heat of helium is C_p = 5.1926 kJ/kg.K (Table A-2).

Analysis (a) From the ideal gas relation, the initial and the final masses in the cylinder are determined to be

$$m_1 = \frac{P_1 V}{RT_1} = \frac{(300 \text{ kPa})(0.1 \text{ m}^3)}{(2.0769 \text{ kPa}\cdot\text{m}^3/\text{kg}\cdot\text{K})(293 \text{ K})} = 0.0493 \text{ kg}$$

$$m_e = m_2 = m_1/2 = 0.0493/2 = 0.02465 \text{ kg}$$

The work potential of helium at the initial state is simply the initial exergy of helium, and is determined from the closed-system exergy relation,

$$\Phi_1 = m_1\phi = m_1\left[(u_1 - u_0) - T_0(s_1 - s_0) + P_0(v_1 - v_0)\right]$$

where

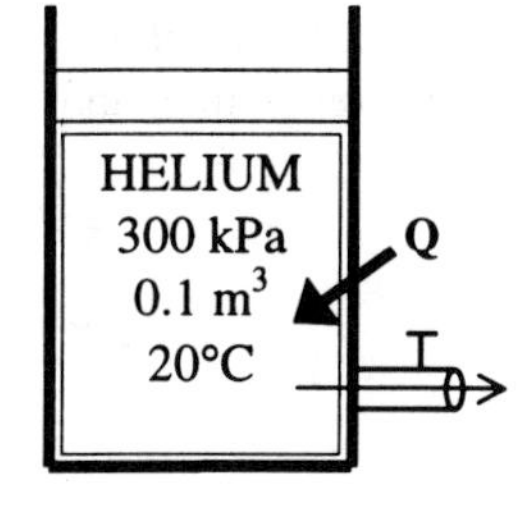

$$v_1 = \frac{RT_1}{P_1} = \frac{(2.0769 \text{ kPa}\cdot\text{m}^3/\text{kg}\cdot\text{K})(293 \text{ K})}{300 \text{ kPa}} = 2.0284 \text{ m}^3/\text{kg}$$

$$v_0 = \frac{RT_0}{P_0} = \frac{(2.0769 \text{ kPa}\cdot\text{m}^3/\text{kg}\cdot\text{K})(293 \text{ K})}{95 \text{ kPa}} = 6.405 \text{ m}^3/\text{kg}$$

and

$$\begin{aligned} s_1 - s_0 &= C_p \ln\frac{T_1}{T_0} - R\ln\frac{P_1}{P_0} \\ &= (5.1926 \text{ kJ/kg}\cdot\text{K})\ln\frac{293 \text{ K}}{293 \text{ K}} - (2.0769 \text{ kJ/kg}\cdot\text{K})\ln\frac{300 \text{ kPa}}{100 \text{ kPa}} \\ &= -2.38 \text{ kJ/kg}\cdot\text{K} \end{aligned}$$

Thus,

$$\begin{aligned} \Phi_1 &= (0.0493 \text{ kg})\{(3.1156 \text{ kJ/kg}\cdot\text{K})(20-20)^\circ\text{C} - (293 \text{ K})(-2.338 \text{ kJ/kg}\cdot\text{K}) \\ &\quad + (95 \text{ kPa})(2.0284 - 6.405)\text{m}^3/\text{kg}[\text{kJ/kPa}\cdot\text{m}^3]\} \\ &= \mathbf{13.3 \text{ kJ}} \end{aligned}$$

(*b*) We take the cylinder as the system, which is a control volume. Noting that the microscopic energies of flowing and nonflowing fluids are represented by enthalpy h and internal energy u, respectively, the mass and energy balances for this uniform-flow system can be expressed as

Mass balance: $$m_{in} - m_{out} = \Delta m_{\text{system}} \rightarrow m_e = m_1 - m_2$$

Energy balance: $$\underbrace{E_{in} - E_{out}}_{\substack{\text{Net energy transfer} \\ \text{by heat, work, and mass}}} = \underbrace{\Delta E_{\text{system}}}_{\substack{\text{Change in internal, kinetic,} \\ \text{potential, etc. energies}}}$$

$$Q_{in} - m_e h_e + W_{b,in} = m_2 u_2 - m_1 u_1$$

Combining the two relations gives

$$\begin{aligned} Q_{in} &= (m_1 - m_2)h_e + m_2u_2 - m_1u_1 - W_{b,in} \\ &= (m_1 - m_2)h_e + m_2h_2 - m_1h_1 \\ &= (m_1 - m_2 + m_2 - m_1)h_1 \\ &= 0 \end{aligned}$$

since the boundary work and ΔU combine into ΔH for constant pressure expansion and compression processes.

The exergy destroyed during a process can be determined from an exergy balance or directly from its definition $X_{\text{destroyed}} = T_0S_{gen}$ where the entropy generation S_{gen} can be determined from an entropy balance on the cylinder. Noting that the pressure and temperature of helium in the cylinder are maintained constant during this process and heat transfer is zero, it gives

$$\underbrace{S_{in} - S_{out}}_{\substack{\text{Net entropy transfer}\\ \text{by heat and mass}}} + \underbrace{S_{gen}}_{\substack{\text{Entropy}\\ \text{generation}}} = \underbrace{\Delta S_{\text{system}}}_{\substack{\text{Change}\\ \text{in entropy}}}$$

$$\begin{aligned} -m_es_e + S_{gen} &= \Delta S_{\text{cylinder}} = (m_2s_2 - m_1s_1)_{\text{cylinder}} \\ S_{gen} &= m_2s_2 - m_1s_1 + m_es_e \\ &= m_2s_2 - m_1s_1 + (m_1 - m_2)s_e \\ &= (m_2 - m_1 + m_1 - m_2)s_1 \\ &= 0 \end{aligned}$$

since the initial, final, and the exit states are identical and thus $s_e = s_2 = s_1$. Therefore, this discharge process is reversible, and

$$X_{\text{destroyed}} = T_0S_{gen} = \mathbf{0}$$

7-81 A rigid tank initially contains saturated R-134a vapor at a specified pressure. The tank is connected to a supply line, and R-134a is allowed to enter the tank. The amount of heat transfer with the surroundings and the exergy destruction are to be determined.

Assumptions **1** This is an unsteady process since the conditions within the device are changing during the process, but it can be analyzed as a uniform-flow process since the state of fluid at the inlet remains constant. **2** Kinetic and potential energies are negligible. **3** There are no work interactions involved. **4** The direction of heat transfer is from the tank (will be verified).

Properties The properties of refrigerant are (Tables A-11 through A-13)

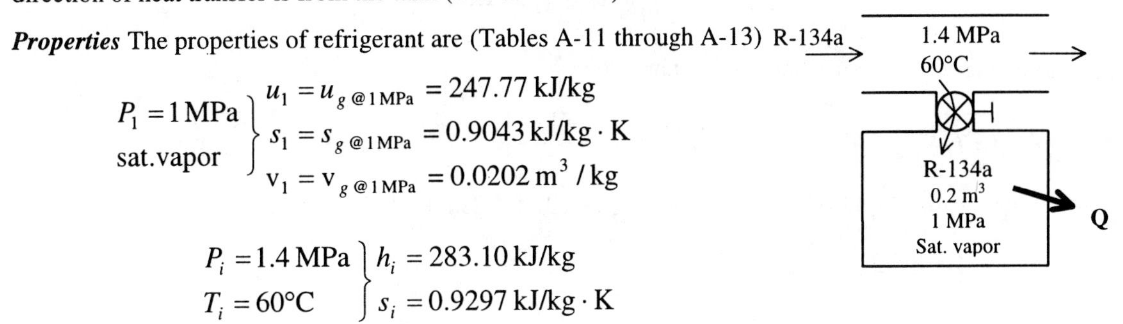

$$\left.\begin{array}{l} P_1 = 1\,\text{MPa} \\ \text{sat.vapor} \end{array}\right\} \begin{array}{l} u_1 = u_{g\,@\,1\,\text{MPa}} = 247.77\ \text{kJ/kg} \\ s_1 = s_{g\,@\,1\,\text{MPa}} = 0.9043\ \text{kJ/kg}\cdot\text{K} \\ \text{v}_1 = \text{v}_{g\,@\,1\,\text{MPa}} = 0.0202\ \text{m}^3/\text{kg} \end{array}$$

$$\left.\begin{array}{l} P_i = 1.4\,\text{MPa} \\ T_i = 60°\text{C} \end{array}\right\} \begin{array}{l} h_i = 283.10\ \text{kJ/kg} \\ s_i = 0.9297\ \text{kJ/kg}\cdot\text{K} \end{array}$$

Analysis (*a*) We take the tank as the system, which is a control volume since mass crosses the boundary. Noting that the microscopic energies of flowing and nonflowing fluids are represented by enthalpy *h* and internal energy *u*, respectively, the mass and energy balances for this uniform-flow system can be expressed as

Mass balance: $m_{in} - m_{out} = \Delta m_{\text{system}} \quad \rightarrow \quad m_i = m_2 - m_1$

Energy balance:

$$\underbrace{E_{in} - E_{out}}_{\substack{\text{Net energy transfer} \\ \text{by heat, work, and mass}}} = \underbrace{\Delta E_{\text{system}}}_{\substack{\text{Change in internal, kinetic,} \\ \text{potential, etc. energies}}}$$

$$m_i h_i - Q_{out} = m_2 u_2 - m_1 u_1 \quad (\text{since } W \cong ke \cong pe \cong 0)$$

The initial and the final masses in the tank are

$$m_1 = \frac{V}{v_1} = \frac{0.2\ \text{m}^3}{0.0202\ \text{m}^3/\text{kg}} = 9.9\ \text{kg}$$

$$m_2 = m_f + m_g = \frac{V_f}{v_f} + \frac{V_g}{v_g} = \frac{0.1\ \text{m}^3}{0.0008928\ \text{m}^3/\text{kg}} + \frac{0.1\ \text{m}^3}{0.0166\ \text{m}^3/\text{kg}} = 112.00 + 6.02 = 118.02\ \text{kg}$$

$$U_2 = m_2 u_2 = m_f u_f + m_g u_g = 112.00 \times 114.69 + 6.02 \times 251.03 = 14{,}357\ \text{kJ}$$

$$S_2 = m_2 s_2 = m_f s_f + m_g s_g = 112.00 \times 0.4164 + 6.02 \times 0.9023 = 52.07\ \text{kJ/K}$$

Then from the mass and energy balances,

$$m_i = m_2 - m_1 = 118.02 - 9.9 = \mathbf{108.12\ kg}$$

The heat transfer during this process is determined from the energy balance to be

$$Q_{out} = m_i h_i - m_2 u_2 + m_1 u_1 = 108.12 \times 283.10 - 14{,}357 + 9.9 \times 247.77 = \mathbf{18{,}705\ kJ}$$

(*b*) The exergy destroyed during a process can be determined from an exergy balance or directly from its definition $X_{\text{destroyed}} = T_0 S_{gen}$. The entropy generation S_{gen} in this case is determined from an entropy balance on an *extended system* that includes the cylinder and its immediate surroundings so that the boundary temperature of the extended system is the surroundings temperature T_{surr} at all times. It gives

$$\underbrace{S_{in} - S_{out}}_{\substack{\text{Net entropy transfer}\\ \text{by heat and mass}}} + \underbrace{S_{gen}}_{\substack{\text{Entropy}\\ \text{generation}}} = \underbrace{\Delta S_{\text{system}}}_{\substack{\text{Change}\\ \text{in entropy}}}$$

$$-\frac{Q_{out}}{T_{b,out}} + m_i s_i + S_{gen} = \Delta S_{\text{tank}} = (m_2 s_2 - m_1 s_1)_{\text{tank}}$$

$$S_{gen} = m_2 s_2 - m_1 s_1 - m_i s_i + \frac{Q_{out}}{T_0}$$

Substituting, the exergy destruction is determined to be

$$X_{\text{destroyed}} = T_0 S_{gen} = T_0 \left[m_2 s_2 - m_1 s_1 - m_i s_i + \frac{Q_{out}}{T_0} \right]$$

$$= (298\text{ K})[52.07 - 9.9 \times 0.9043 - 108.12 \times 0.9297 + 18{,}705/298]$$

$$= \mathbf{1{,}599\ kJ}$$

7-82 An insulated cylinder initially contains saturated liquid-vapor mixture of water. The cylinder is connected to a supply line, and the steam is allowed to enter the cylinder until all the liquid is vaporized. The amount of steam that entered the cylinder and the exergy destroyed are to be determined.

Assumptions **1** This is an unsteady process since the conditions within the device are changing during the process, but it can be analyzed as a uniform-flow process since the state of fluid at the inlet remains constant. **2** The expansion process is quasi-equilibrium. **3** Kinetic and potential energies are negligible. **4** The device is insulated and thus heat transfer is negligible.

Properties The properties of steam are (Tables A-4 through A-6)

$$\left.\begin{array}{l} P_1 = 300\text{ kPa} \\ x_1 = 4/5 = 0.8 \end{array}\right\} \begin{array}{l} h_1 = h_f + x_1 h_{fg} = 561.47 + 0.8\times 2163.8 = 2292.51\text{ kJ/kg} \\ s_1 = s_f + x_1 s_{fg} = 1.6718 + 0.8\times 5.3201 = 5.9279\text{ kJ/kg}\cdot\text{K} \end{array}$$

$$\left.\begin{array}{l} P_2 = 300\text{ kPa} \\ \text{sat.vapor} \end{array}\right\} \begin{array}{l} h_2 = h_{g\,@\,300\text{ kPa}} = 2725.3\text{ kJ/kg} \\ s_2 = s_{g\,@\,300\text{ kPa}} = 6.9919\text{ kJ/kg}\cdot\text{K} \end{array}$$

$$\left.\begin{array}{l} P_i = 1\text{ MPa} \\ T_i = 400°\text{C} \end{array}\right\} \begin{array}{l} h_i = 3263.9\text{ kJ/kg} \\ s_i = 7.4651\text{ kJ/kg}\cdot\text{K} \end{array}$$

Analysis (*a*) We take the cylinder as the system, which is a control volume. Noting that the microscopic energies of flowing and nonflowing fluids are represented by enthalpy *h* and internal energy *u*, respectively, the mass and energy balances for this unsteady-flow system can be expressed as

Mass balance: $$m_{in} - m_{out} = \Delta m_{\text{system}} \rightarrow m_i = m_2 - m_1$$

Energy balance: $$\underbrace{E_{in} - E_{out}}_{\substack{\text{Net energy transfer} \\ \text{by heat, work, and mass}}} = \underbrace{\Delta E_{\text{system}}}_{\substack{\text{Change in internal, kinetic,} \\ \text{potential, etc. energies}}}$$

$$m_i h_i = W_{b,out} + m_2 u_2 - m_1 u_1 \quad (\text{since } Q \cong ke \cong pe \cong 0)$$

Combining the two relations gives $0 = W_{b,out} - (m_2 - m_1)h_i + m_2 u_2 - m_1 u_1$

or, $$0 = -(m_2 - m_1)h_i + m_2 h_2 - m_1 h_1$$

since the boundary work and ΔU combine into ΔH for constant pressure expansion and compression processes. Solving for m_2 and substituting,

$$m_2 = \frac{h_i - h_1}{h_i - h_2} m_1 = \frac{(3263.9 - 2292.51)\text{ kJ/kg}}{(3263.9 - 2725.3)\text{ kJ/kg}}(15\text{ kg}) = 27.05\text{ kg}$$

H$_2$O
300 kPa
P = const.
1 MPa
400°C

Thus,

$$m_i = m_2 - m_1 = 27.05 - 15 = \mathbf{12.05\ kg}$$

(*b*) The exergy destroyed during a process can be determined from an exergy balance or directly from its definition $X_{\text{destroyed}} = T_0 S_{gen}$ where the entropy generation S_{gen} is determined from an entropy balance on the insulated cylinder,

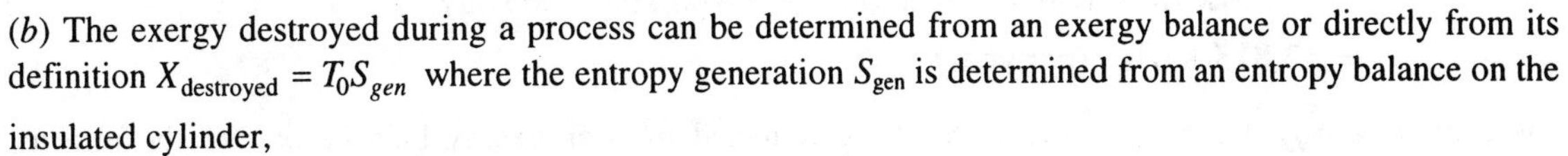

$$m_i s_i + S_{gen} = \Delta S_{\text{system}} = m_2 s_2 - m_1 s_1 \rightarrow S_{gen} = m_2 s_2 - m_1 s_1 - m_i s_i$$

Substituting, the exergy destruction is determined to be

$$X_{\text{destroyed}} = T_0 S_{gen} = T_0[m_2 s_2 - m_1 s_1 - m_i s_i]$$
$$= (298\text{ K})(27.05\times 6.9919 - 15\times 5.9279 - 12.05\times 7.4651) = \mathbf{3{,}057\ kJ}$$

7-83 Each member of a family of four take a shower every day. The amount of exergy destroyed by this family per year is to be determined.

Assumptions **1** Steady operating conditions exist. **2** The kinetic and potential energies are negligible. **3** Heat losses from the pipes, mixing section are negligible and thus $\dot{Q} \cong 0$. **4** Showers operate at maximum flow conditions during the entire shower. **5** Each member of the household takes a shower every day. **6** Water is an incompressible substance with constant properties at room temperature. **7** The efficiency of the electric water heater is 100%.

Properties The density and specific heat of water are at room temperature are ρ = 1 kg/L = 1000 kg/^{3} and C = 4.18 kJ/kg.°C (Table A-3).

Analysis The mass flow rate of water at the shower head is

$$\dot{m} = \rho\dot{V} = (1\,\text{kg/L})(10\,\text{L/min}) = 10\,\text{kg/min}$$

The mass balance for the mixing chamber can be expressed in the rate form as

$$\dot{m}_{in} - \dot{m}_{out} = \Delta\dot{m}_{\text{system}}^{\nearrow 0\ (\text{steady})} = 0 \quad \rightarrow \quad \dot{m}_{in} = \dot{m}_{out} \quad \rightarrow \dot{m}_1 + \dot{m}_2 = \dot{m}_3$$

where the subscript 1 denotes the cold water stream, 2 the hot water stream, and 3 the mixture.

The rate of entropy generation during this process can be determined by applying the rate form of the entropy balance on a system that includes the electric water heater and the mixing chamber (the T-elbow). Noting that there is no entropy transfer associated with work transfer (electricity) and there is no heat transfer, the entropy balance for this steady-flow system can be expressed as

$$\underbrace{\dot{S}_{in} - \dot{S}_{out}}_{\substack{\text{Rate of net entropy transfer}\\ \text{by heat and mass}}} + \underbrace{\dot{S}_{gen}}_{\substack{\text{Rate of entropy}\\ \text{generation}}} = \underbrace{\Delta\dot{S}_{\text{system}}^{\nearrow 0\,(\text{steady})}}_{\substack{\text{Rate of change}\\ \text{of entropy}}}$$

$$\dot{m}_1 s_1 + \dot{m}_2 s_2 - \dot{m}_3 s_3 + \dot{S}_{gen} = 0 \quad (\text{since } Q = 0 \text{ and work is entropy free})$$

$$\dot{S}_{gen} = \dot{m}_3 s_3 - \dot{m}_1 s_1 - \dot{m}_2 s_2$$

Noting from mass balance that $\dot{m}_1 + \dot{m}_2 = \dot{m}_3$ and $s_2 = s_1$ since hot water enters the system at the same temperature as the cold water, the rate of entropy generation is determined to be

$$\dot{S}_{gen} = \dot{m}_3 s_3 - (\dot{m}_1 + \dot{m}_2)s_1 = \dot{m}_3(s_3 - s_1) = \dot{m}_3 C_p \ln\frac{T_3}{T_1}$$

$$= (10\,\text{kg/min})(4.18\,\text{kJ/kg.K})\ln\frac{42+273}{15+273} = 3.746\,\text{kJ/min.K}$$

Noting that 4 people take a 6-min shower every day, the amount of entropy generated per year is

$$S_{gen} = (\dot{S}_{gen})\Delta t(\text{No. of people})(\text{No. of days})$$

$$= (3.746\,\text{kJ/min.K})(6\,\text{min/person}\cdot\text{day})(4\,\text{persons})(365\,\text{days/year})$$

$$= \mathbf{32{,}815\ kJ/K}\ \ (\text{per year})$$

The exergy destroyed during a process can be determined from an exergy balance or directly from its definition $X_{\text{destroyed}} = T_0 S_{gen}$,

$$X_{\text{destroyed}} = T_0 S_{gen} = (298\,\text{K})(32{,}815\,\text{kJ/K}) = \mathbf{9{,}779{,}000\,kJ}$$

Discussion The value above represents the exergy destroyed within the water heater and the T-elbow in the absence of any heat losses. It does not include the exergy destroyed as the shower water at 42°C is discarded or cooled to the outdoor temperature. Also, an entropy balance on the mixing chamber alone (hot water entering at 55°C instead of 15°C) will exclude the exergy destroyed within the water heater.

7-84 Air is compressed in a steady-flow device isentropically. The work done, the exit exergy of compressed air, and the exergy of compressed air after it is cooled to ambient temperature are to be determined.

Assumptions **1** Air is an ideal gas with constant specific heats at room temperature. **2** The process is given to be reversible and adiabatic, and thus isentropic. Therefore, isentropic relations of ideal gases apply. **3** The environment temperature and pressure are given to be 300 K and 100 kPa. **4** The kinetic and potential energies are negligible.

Properties The gas constant of air is R = 0.287 kJ/kg.K (Table A-1). The constant pressure specific heat and specific heat ratio of air at room temperature are C_p = 1.005 kJ/kg.K and k = 1.4 (Table A-2).

Analysis (a) From the constant specific heats ideal gas isentropic relations,

$$T_2 = T_1\left(\frac{P_2}{P_1}\right)^{(k-1)/k} = (300\text{K})\left(\frac{1000\text{kPa}}{100\text{kPa}}\right)^{0.4/1.4} = 579.2\text{K}$$

For a steady-flow isentropic compression process, the work in put is determined from

$$w_{comp,in} = \frac{kRT_1}{k-1}\left\{(P_2/P_1)^{(k-1)/k} - 1\right\}$$
$$= \frac{(1.4)(0.287\text{kJ/kg}\cdot\text{K})(300\text{K})}{1.4-1}\left\{(1000/100)^{0.4/1.4} - 1\right\}$$
$$= \mathbf{280.5\ kJ/kg}$$

(*b*) The exergy of air at the compressor exit is simply the flow exergy at the exit state,

$$\psi_2 = h_2 - h_0 - T_0(s_2 - s_0)^{\nearrow 0} + \frac{\mathbf{V}_2^2}{2}^{\nearrow 0} + gz_2^{\nearrow 0} \quad \text{(since the proccess 0 - 2 is isentropic)}$$
$$= C_p(T_2 - T_0)$$
$$= (1.005\ \text{kJ/kg.K})(579.2 - 300)\text{K} = \mathbf{280.6\ kJ/kg}$$

which is the same as the compressor work input. This is not surprising since the compression process is reversible.

(*c*) The exergy of compressed air at 1 MPa after it is cooled to 300 K is again the flow exergy at that state,

$$\psi_3 = h_3 - h_0 - T_0(s_3 - s_0) + \frac{\mathbf{V}_3^2}{2}^{\nearrow 0} + gz_3^{\nearrow 0}$$
$$= C_p(T_3 - T_0)^{\nearrow 0} - T_0(s_3 - s_0) \quad (\text{since } T_3 = T_0 = 300\text{ K})$$
$$= -T_0(s_3 - s_0)$$

where

$$s_3 - s_0 = C_p \ln\frac{T_3}{T_0}^{\nearrow 0} - R\ln\frac{P_3}{P_0} = -R\ln\frac{P_3}{P_0} = -(0.287\ \text{kJ/kg}\cdot\text{K})\ln\frac{1000\ \text{kPa}}{100\ \text{kPa}} = -0.661\ \text{kJ/kg.K}$$

Substituting,

$$\psi_3 = -(300\ \text{K})(-0.661\ \text{kJ/kg.K}) = \mathbf{198\ kJ/kg}$$

Note that the exergy of compressed air decreases from 280.6 to 198 as it is cooled to ambient temperature.

7-85 Cold water is heated by hot water in a heat exchanger. The rate of heat transfer and the rate of exergy destruction within the heat exchanger are to be determined.

Assumptions **1** Steady operating conditions exist. **2** The heat exchanger is well-insulated so that heat loss to the surroundings is negligible and thus heat transfer from the hot fluid is equal to the heat transfer to the cold fluid. **3** Changes in the kinetic and potential energies of fluid streams are negligible. **4** Fluid properties are constant. **5** The temperature of the environment is 25°C.

Properties The specific heats of cold and hot water are given to be 4.18 and 4.19 kJ/kg.°C, respectively.

Analysis We take the cold water tubes as the system, which is a control volume. The energy balance for this steady-flow system can be expressed in the rate form as

$$\underbrace{\dot{E}_{in} - \dot{E}_{out}}_{\text{Rate of net energy transfer by heat, work, and mass}} = \underbrace{\Delta \dot{E}_{\text{system}}^{\nearrow 0\ \text{(steady)}}}_{\text{Rate of change in internal, kinetic, potential, etc. energies}} = 0$$

$$\dot{E}_{in} = \dot{E}_{out}$$

$$\dot{Q}_{in} + \dot{m}h_1 = \dot{m}h_2 \quad (\text{since } \Delta\text{ke} \cong \Delta\text{pe} \cong 0)$$

$$\dot{Q}_{in} = \dot{m}C_p(T_2 - T_1)$$

Cold Water
15°C
0.25 kg/s
Hot water
100°C
3 kg/s

Then the rate of heat transfer to the cold water in this heat exchanger becomes

$$\dot{Q}_{in} = [\dot{m}C_p(T_{out} - T_{in})]_{\text{cold water}} = (0.25\ \text{kg/s})(4.18\ \text{kJ/kg.°C})(45°\text{C} - 15°\text{C}) = \mathbf{31.35\ kW}$$

Noting that heat gain by the cold water is equal to the heat loss by the hot water, the outlet temperature of the hot water is determined to be

$$\dot{Q} = [\dot{m}C_p(T_{in} - T_{out})]_{\text{hot water}} \longrightarrow T_{out} = T_{in} - \frac{\dot{Q}}{\dot{m}C_p}$$

$$= 100°\text{C} - \frac{31.35\ \text{kW}}{(3\ \text{kg/s})(4.19\ \text{kJ/kg.°C})} = \mathbf{97.5°C}$$

(*b*) The rate of entropy generation within the heat exchanger is determined by applying the rate form of the entropy balance on the entire heat exchanger:

$$\underbrace{\dot{S}_{in} - \dot{S}_{out}}_{\text{Rate of net entropy transfer by heat and mass}} + \underbrace{\dot{S}_{gen}}_{\text{Rate of entropy generation}} = \underbrace{\Delta \dot{S}_{\text{system}}^{\nearrow 0\ \text{(steady)}}}_{\text{Rate of change of entropy}}$$

$$\dot{m}_1 s_1 + \dot{m}_3 s_3 - \dot{m}_2 s_2 - \dot{m}_3 s_4 + \dot{S}_{gen} = 0 \quad (\text{since } Q = 0)$$

$$\dot{m}_{\text{cold}} s_1 + \dot{m}_{\text{hot}} s_3 - \dot{m}_{\text{cold}} s_2 - \dot{m}_{\text{hot}} s_4 + \dot{S}_{gen} = 0$$

$$\dot{S}_{gen} = \dot{m}_{\text{cold}}(s_2 - s_1) + \dot{m}_{\text{hot}}(s_4 - s_3)$$

Noting that both fluid streams are liquids (incompressible substances), the rate of entropy generation is determined to be

$$\dot{S}_{gen} = \dot{m}_{\text{cold}} C_p \ln\frac{T_2}{T_1} + \dot{m}_{\text{hot}} C_p \ln\frac{T_4}{T_3}$$

$$= (0.25\ \text{kg/s})(4.18\ \text{kJ/kg.K})\ln\frac{45+273}{15+273} + (3\ \text{kg/s})(4.19\ \text{kJ/kg.K})\ln\frac{97.5+273}{100+273} = \mathbf{0.0190\ kW/K}$$

The exergy destroyed during a process can be determined from an exergy balance or directly from its definition $X_{\text{destroyed}} = T_0 S_{gen}$,

$$X_{\text{destroyed}} = T_0 S_{gen} = (298\ \text{K})(0.019\ \text{kW/K}) = \mathbf{5.66\ kW}$$

7-86 Air is preheated by hot exhaust gases in a cross-flow heat exchanger. The rate of heat transfer and the outlet temperature of the air are to be determined.

Assumptions **1** Steady operating conditions exist. **2** The heat exchanger is well-insulated so that heat loss to the surroundings is negligible and thus heat transfer from the hot fluid is equal to the heat transfer to the cold fluid. **3** Changes in the kinetic and potential energies of fluid streams are negligible. **4** Fluid properties are constant.

Properties The specific heats of air and combustion gases are given to be 1.005 and 1.10 kJ/kg.°C, respectively. The gas constant of air is $R = 0.287$ kJ/kg.K (Table A-1).

Analysis We take the exhaust pipes as the system, which is a control volume. The energy balance for this steady-flow system can be expressed in the rate form as

$$\underbrace{\dot{E}_{in} - \dot{E}_{out}}_{\substack{\text{Rate of net energy transfer} \\ \text{by heat, work, and mass}}} = \underbrace{\Delta \dot{E}_{\text{system}}^{\nearrow 0 \text{ (steady)}}}_{\substack{\text{Rate of change in internal, kinetic,} \\ \text{potential, etc. energies}}} = 0$$

$$\dot{E}_{in} = \dot{E}_{out}$$

$$\dot{m}h_1 = \dot{Q}_{out} + \dot{m}h_2 \quad (\text{since } \Delta\text{ke} \cong \Delta\text{pe} \cong 0)$$

$$\dot{Q}_{out} = \dot{m}C_p(T_1 - T_2)$$

Air
95 kPa
20°C
0.8 m³/s

Exhaust gases
1.1 kg/s, 95°C

Then the rate of heat transfer from the exhaust gases becomes

$$\dot{Q} = [\dot{m}C_p(T_{in} - T_{out})]_{\text{gas.}} = (1.1 \text{ kg/s})(1.1 \text{ kJ/kg.°C})(180\text{°C} - 95\text{°C}) = \mathbf{102.85 \text{ kW}}$$

The mass flow rate of air is

$$\dot{m} = \frac{P\dot{V}}{RT} = \frac{(95 \text{ kPa})(0.8 \text{ m}^3/\text{s})}{(0.287 \text{ kPa.m}^3/\text{kg.K}) \times 293 \text{ K}} = 0.904 \text{ kg/s}$$

Noting that heat loss by exhaust gases is equal to the heat gain by the air, the air exit temperature becomes

$$\dot{Q} = \dot{m}C_p(T_{c,out} - T_{c,in}) \rightarrow T_{c,out} = T_{c,in} + \frac{\dot{Q}}{\dot{m}C_p} = 20\text{°C} + \frac{102.85 \text{ kW}}{(0.904 \text{ kg/s})(1.005 \text{ kJ/kg.°C})} = \mathbf{133.2\text{°C}}$$

Extra: The rate of entropy generation within the heat exchanger is determined by applying the rate form of the entropy balance on the entire heat exchanger:

$$\underbrace{\dot{S}_{in} - \dot{S}_{out}}_{\substack{\text{Rate of net entropy transfer} \\ \text{by heat and mass}}} + \underbrace{\dot{S}_{gen}}_{\substack{\text{Rate of entropy} \\ \text{generation}}} = \underbrace{\Delta \dot{S}_{\text{system}}^{\nearrow 0 \text{ (steady)}}}_{\substack{\text{Rate of change} \\ \text{of entropy}}}$$

$$\dot{m}_1 s_1 + \dot{m}_3 s_3 - \dot{m}_2 s_2 - \dot{m}_3 s_4 + \dot{S}_{gen} = 0 \quad (\text{since } Q = 0)$$

$$\dot{m}_{\text{exhaust}} s_1 + \dot{m}_{\text{air}} s_3 - \dot{m}_{\text{exhaust}} s_2 - \dot{m}_{\text{air}} s_4 + \dot{S}_{gen} = 0$$

$$\dot{S}_{gen} = \dot{m}_{\text{exhaust}}(s_2 - s_1) + \dot{m}_{\text{air}}(s_4 - s_3)$$

Noting that both fluid streams are liquids (incompressible substances), the rate of entropy generation is

$$\dot{S}_{gen} = \dot{m}_{\text{exhaust}} C_p \ln\frac{T_2}{T_1} + \dot{m}_{\text{air}} C_p \ln\frac{T_4}{T_3}$$

$$= (1.1 \text{kg/s})(1.1 \text{kJ/kg.K})\ln\frac{95 + 273}{180 + 273} + (0.904 \text{kg/s})(1.005 \text{kJ/kg.K})\ln\frac{133.2 + 273}{20 + 273} = \mathbf{0.0453 \text{ kW/K}}$$

The exergy destroyed during a process can be determined from an exergy balance or directly from its definition $X_{\text{destroyed}} = T_0 S_{gen}$,

$$X_{\text{destroyed}} = T_0 S_{gen} = (293 \text{ K})(0.0453 \text{kW/K}) = \mathbf{13.3 \text{ kW}}$$

7-87 Water is heated by hot oil in a heat exchanger. The outlet temperature of the oil and the rate of exergy destruction within the heat exchanger are to be determined.

Assumptions **1** Steady operating conditions exist. **2** The heat exchanger is well-insulated so that heat loss to the surroundings is negligible and thus heat transfer from the hot fluid is equal to the heat transfer to the cold fluid. **3** Changes in the kinetic and potential energies of fluid streams are negligible. **4** Fluid properties are constant.

Properties The specific heats of water and oil are given to be 4.18 and 2.3 kJ/kg.°C, respectively.

Analysis We take the cold water tubes as the system, which is a control volume. The energy balance for this steady-flow system can be expressed in the rate form as

$$\underbrace{\dot{E}_{in} - \dot{E}_{out}}_{\text{Rate of net energy transfer by heat, work, and mass}} = \underbrace{\Delta \dot{E}_{system}^{\nearrow 0 \text{ (steady)}}}_{\text{Rate of change in internal, kinetic, potential, etc. energies}} = 0$$

$$\dot{E}_{in} = \dot{E}_{out}$$

$$\dot{Q}_{in} + \dot{m}h_1 = \dot{m}h_2 \quad (\text{since } \Delta\text{ke} \cong \Delta\text{pe} \cong 0)$$

$$\dot{Q}_{in} = \dot{m}C_p(T_2 - T_1)$$

Oil
170°C
10 kg/s
70°C
Water
20°C
4.5 kg/s
(12 tube passes)

Then the rate of heat transfer to the cold water in this heat exchanger becomes

$$\dot{Q} = [\dot{m}C_p(T_{out} - T_{in})]_{\text{water}} = (4.5 \text{ kg/s})(4.18 \text{ kJ/kg.°C})(70°\text{C} - 20°\text{C}) = 940.5 \text{ kW}$$

Noting that heat gain by the water is equal to the heat loss by the oil, the outlet temperature of the hot water is determined from

$$\dot{Q} = [\dot{m}C_p(T_{in} - T_{out})]_{\text{oil}} \rightarrow T_{out} = T_{in} - \frac{\dot{Q}}{\dot{m}C_p} = 170°\text{C} - \frac{940.5 \text{ kW}}{(10 \text{ kg/s})(2.3 \text{ kJ/kg.°C})} = \mathbf{129.1°C}$$

(*b*) The rate of entropy generation within the heat exchanger is determined by applying the rate form of the entropy balance on the entire heat exchanger:

$$\underbrace{\dot{S}_{in} - \dot{S}_{out}}_{\text{Rate of net entropy transfer by heat and mass}} + \underbrace{\dot{S}_{gen}}_{\text{Rate of entropy generation}} = \underbrace{\Delta \dot{S}_{system}^{\nearrow 0 \text{ (steady)}}}_{\text{Rate of change of entropy}}$$

$$\dot{m}_1 s_1 + \dot{m}_3 s_3 - \dot{m}_2 s_2 - \dot{m}_3 s_4 + \dot{S}_{gen} = 0 \quad (\text{since } Q = 0)$$

$$\dot{m}_{\text{water}} s_1 + \dot{m}_{\text{oil}} s_3 - \dot{m}_{\text{water}} s_2 - \dot{m}_{\text{oil}} s_4 + \dot{S}_{gen} = 0$$

$$\dot{S}_{gen} = \dot{m}_{\text{water}}(s_2 - s_1) + \dot{m}_{\text{oil}}(s_4 - s_3)$$

Noting that both fluid streams are liquids (incompressible substances), the rate of entropy generation is determined to be

$$\dot{S}_{gen} = \dot{m}_{\text{water}} C_p \ln\frac{T_2}{T_1} + \dot{m}_{\text{oil}} C_p \ln\frac{T_4}{T_3}$$

$$= (4.5 \text{ kg/s})(4.18 \text{ kJ/kg.K})\ln\frac{70+273}{20+273} + (10 \text{ kg/s})(2.3 \text{ kJ/kg.K})\ln\frac{129.1+273}{170+273} = \mathbf{0.736 \text{ kW/K}}$$

The exergy destroyed during a process can be determined from an exergy balance or directly from its definition $X_{\text{destroyed}} = T_0 S_{gen}$,

$$X_{\text{destroyed}} = T_0 S_{gen} = (298 \text{ K})(0.736 \text{ kW/K}) = \mathbf{219 \text{ kW}}$$

7-88E Steam is condensed by cooling water in a condenser. The rate of heat transfer and the rate of exergy destruction within the heat exchanger are to be determined. (Note: the given condensation rate of steam is to be ignored).

Assumptions **1** Steady operating conditions exist. **2** The heat exchanger is well-insulated so that heat loss to the surroundings is negligible and thus heat transfer from the hot fluid is equal to the heat transfer to the cold fluid. **3** Changes in the kinetic and potential energies of fluid streams are negligible. **4** Fluid properties are constant. **5** The temperature of the environment is 77°F.

Properties The specific heat of water is 1.0 Btu/lbm.°F (Table A-3E). The enthalpy and entropy of vaporization of water at 90°F are 1040.2 Btu/lbm and s_{fg} = 1.8966 Btu/lbm.R (Table A-4E).

Analysis We take the tube-side of the heat exchanger where cold water is flowing as the system, which is a control volume. The energy balance for this steady-flow system can be expressed in the rate form as

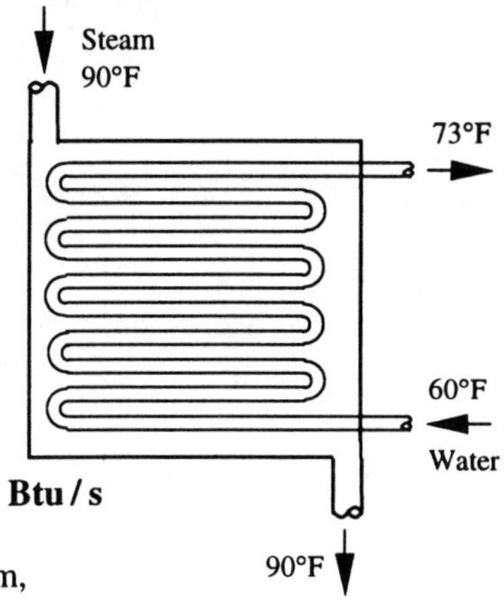

$$\underbrace{\dot{E}_{in} - \dot{E}_{out}}_{\substack{\text{Rate of net energy transfer}\\ \text{by heat, work, and mass}}} = \underbrace{\Delta \dot{E}_{system}^{\nearrow 0 \text{ (steady)}}}_{\substack{\text{Rate of change in internal, kinetic,}\\ \text{potential, etc. energies}}} = 0$$

$$\dot{E}_{in} = \dot{E}_{out}$$

$$\dot{Q}_{in} + \dot{m}h_1 = \dot{m}h_2 \quad (\text{since } \Delta\text{ke} \cong \Delta\text{pe} \cong 0)$$

$$\dot{Q}_{in} = \dot{m}C_p(T_2 - T_1)$$

Steam
90°F
73°F
60°F
Water
90°F

Then the rate of heat transfer to the cold water in this heat exchanger becomes

$$\dot{Q} = [\dot{m}C_p(T_{out} - T_{in})]_{\text{water}} = (115.3 \text{ lbm/s})(1.0 \text{ Btu/lbm.°F})(73°\text{F} - 60°\text{F}) = \mathbf{1499 \text{ Btu/s}}$$

Noting that heat gain by the water is equal to the heat loss by the condensing steam, the rate of condensation of the steam in the heat exchanger is determined from

$$\dot{Q} = (\dot{m}h_{fg})_{steam} \longrightarrow \dot{m}_{steam} = \frac{\dot{Q}}{h_{fg}} = \frac{1499 \text{ Btu/s}}{1040.2 \text{ Btu/lbm}} = \mathbf{1.44 \text{ lbm/s}}$$

(*b*) The rate of entropy generation within the heat exchanger is determined by applying the rate form of the entropy balance on the entire heat exchanger:

$$\underbrace{\dot{S}_{in} - \dot{S}_{out}}_{\substack{\text{Rate of net entropy transfer}\\ \text{by heat and mass}}} + \underbrace{\dot{S}_{gen}}_{\substack{\text{Rate of entropy}\\ \text{generation}}} = \underbrace{\Delta \dot{S}_{system}^{\nearrow 0 \text{ (steady)}}}_{\substack{\text{Rate of change}\\ \text{of entropy}}}$$

$$\dot{m}_1 s_1 + \dot{m}_3 s_3 - \dot{m}_2 s_2 - \dot{m}_4 s_4 + \dot{S}_{gen} = 0 \quad (\text{since } Q = 0)$$

$$\dot{m}_{\text{water}} s_1 + \dot{m}_{\text{steam}} s_3 - \dot{m}_{\text{water}} s_2 - \dot{m}_{\text{steam}} s_4 + \dot{S}_{gen} = 0$$

$$\dot{S}_{gen} = \dot{m}_{\text{water}}(s_2 - s_1) + \dot{m}_{\text{steam}}(s_4 - s_3)$$

Noting that water is an incompressible substance and steam changes from saturated vapor to saturated liquid, the rate of entropy generation is determined to be

$$\dot{S}_{gen} = \dot{m}_{\text{water}} C_p \ln\frac{T_2}{T_1} + \dot{m}_{\text{steam}}(s_f - s_g) = \dot{m}_{\text{water}} C_p \ln\frac{T_2}{T_1} - \dot{m}_{\text{steam}} s_{fg}$$

$$= (115.3 \text{ lbm/s})(1.0 \text{ Btu/lbm.R})\ln\frac{73 + 460}{60 + 460} - (1.44 \text{ lbm/s})(1.8966 \text{ Btu/lbm.R}) = \mathbf{0.116 \text{ Btu/s.R}}$$

The exergy destroyed during a process can be determined from an exergy balance or directly from its definition $X_{\text{destroyed}} = T_0 S_{gen}$,

$$X_{\text{destroyed}} = T_0 S_{gen} = (537 \text{ R})(0.116 \text{ Btu/s.R}) = \mathbf{62.3 \text{ Btu/s}}$$

Review Problems

7-89 The inner and outer surfaces of a brick wall are maintained at specified temperatures. The rate of exergy destruction is to be determined.

Assumptions **1** Steady operating conditions exist since the surface temperatures of the wall remain constant at the specified values. **2** The environment temperature is given to be $T_0 = 0°C$.

Analysis We take the wall to be the system, which is a closed system. Under steady conditions, the rate form of the entropy balance for the wall simplifies to

Brick Wall

$\dot{Q}$

30 cm

20°C

5°C

$$\underbrace{\dot{S}_{in} - \dot{S}_{out}}_{\text{Rate of net entropy transfer by heat and mass}} + \underbrace{\dot{S}_{gen}}_{\text{Rate of entropy generation}} = \underbrace{\Delta\dot{S}_{\text{system}}^{\;\nearrow 0}}_{\text{Rate of change of entropy}} = 0$$

$$\frac{\dot{Q}_{in}}{T_{b,in}} - \frac{\dot{Q}_{out}}{T_{b,out}} + \dot{S}_{\text{gen,wall}} = 0$$

$$\frac{900\text{ W}}{293\text{ K}} - \frac{900\text{ W}}{278\text{ K}} + \dot{S}_{\text{gen,wall}} = 0 \rightarrow \dot{S}_{\text{gen,wall}} = \mathbf{0.166\ W/K}$$

The exergy destroyed during a process can be determined from an exergy balance or directly from its definition $X_{\text{destroyed}} = T_0 S_{gen}$,

$$\dot{X}_{\text{destroyed}} = T_0 \dot{S}_{gen} = (273\text{ K})(0.166\ \text{W/K}) = \mathbf{45.3\ W}$$

7-90 A 1000-W iron is left on the iron board with its base exposed to air. The rate of exergy destruction in steady operation is to be determined.

Assumptions Steady operating conditions exist.

Analysis The rate of total entropy generation during this process is determined by applying the entropy balance on an *extended system* that includes the iron and its immediate surroundings so that the boundary temperature of the extended system is 20°C at all times. It gives

$$\underbrace{\dot{S}_{in} - \dot{S}_{out}}_{\text{Rate of net entropy transfer by heat and mass}} + \underbrace{\dot{S}_{gen}}_{\text{Rate of entropy generation}} = \underbrace{\Delta \dot{S}_{\text{system}}^{\nearrow 0}}_{\text{Rate of change of entropy}} = 0$$

$$-\frac{\dot{Q}_{out}}{T_{b,out}} + \dot{S}_{\text{gen}} = 0$$

Therefore,

$$\dot{S}_{\text{gen}} = \frac{\dot{Q}_{out}}{T_{b,out}} = \frac{\dot{Q}}{T_0} = \frac{1000 \text{ W}}{293 \text{ K}} = \mathbf{3.413 \text{ W/K}} \qquad (*)$$

The exergy destroyed during a process can be determined from an exergy balance or directly from its definition $X_{\text{destroyed}} = T_0 S_{gen}$,

$$\dot{X}_{\text{destroyed}} = T_0 \dot{S}_{gen} = (293 \text{ K})(3.413 \text{ W/K}) = \mathbf{1000 \text{ W}}$$

Discussion The rate of entropy generation within the iron can be determined by performing an entropy balance on the iron alone (it gives 2.21 W/K). Therefore, about one-third of the entropy generation and thus exergy destruction occurs within the iron. The rest occurs in the air surrounding the iron as the temperature drops from 180°C to 20°C without serving any useful purpose.

7-91 The heating of a passive solar house at night is to be assisted by solar heated water. The amount of heating this water will provide to the house at night and the exergy destruction during this heat transfer process are to be determined.

Assumptions **1** Water is an incompressible substance with constant specific heats. **2** The energy stored in the glass containers themselves is negligible relative to the energy stored in water. **3** The house is maintained at 22°C at all times. **4** The outdoor temperature is given to be 5°C.

Properties The density and specific heat of water at room temperature are ρ = 1 kg/L and C = 4.18 kJ/kg· °C (Table A-3).

Analysis The total mass of water is

$$m_w = \rho V = (1\text{kg/L})(500\text{L}) = 500\text{kg}$$

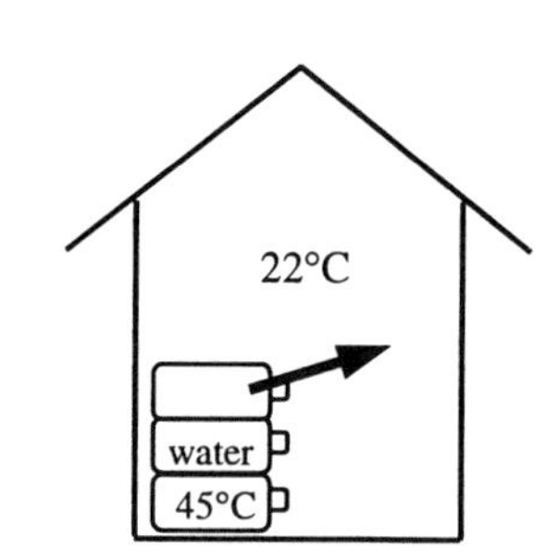

The amount of heat this water storage system can provide is determined from an energy balance on the 500-L water storage system

$$\underbrace{E_{in} - E_{out}}_{\substack{\text{Net energy transfer} \\ \text{by heat, work, and mass}}} = \underbrace{\Delta E_{\text{system}}}_{\substack{\text{Change in internal, kinetic,} \\ \text{potential, etc. energies}}}$$

$$-Q_{out} = \Delta U_{\text{system}} = mC(T_2 - T_1)_{\text{water}}$$

Substituting,

$$Q_{out} = (500\text{ kg})(4.18\text{ kJ/kg}\cdot°\text{C})(45 - 22)°\text{C} = 48{,}070\text{ kJ}$$

The entropy generated during this process is determined by applying the entropy balance on an *extended system* that includes the water and its immediate surroundings so that the boundary temperature of the extended system is the environment temperature at all times. It gives

$$\underbrace{S_{in} - S_{out}}_{\substack{\text{Net entropy transfer} \\ \text{by heat and mass}}} + \underbrace{S_{gen}}_{\substack{\text{Entropy} \\ \text{generation}}} = \underbrace{\Delta S_{\text{system}}}_{\substack{\text{Change} \\ \text{in entropy}}}$$

$$-\frac{Q_{out}}{T_{b,out}} + S_{\text{gen}} = \Delta S_{\text{water}}$$

Substituting,

$$S_{gen} = \Delta S_{\text{water}} + \frac{Q_{out}}{T_{b,out}} = \left(mC\ln\frac{T_2}{T_1}\right)_{water} + \frac{Q_{out}}{T_{room}}$$

$$= (500\text{kg})(4.18\text{kJ/kg}\cdot\text{K})\ln\frac{295\text{K}}{318\text{K}} + \frac{48{,}070\text{kJ}}{295\text{ K}}$$

$$= 6.040\text{ kJ/K}$$

The exergy destroyed during a process can be determined from an exergy balance or directly from its definition $X_{\text{destroyed}} = T_0 S_{gen}$,

$$X_{\text{destroyed}} = T_0 S_{gen} = (278\text{ K})(6.04\text{ kJ/K}) = \mathbf{1679\text{ kJ}}$$

7-92 The inner and outer surfaces of a window glass are maintained at specified temperatures. The amount of exergy destruction in 5 h is to be determined

Assumptions Steady operating conditions exist since the surface temperatures of the glass remain constant at the specified values.

Analysis We take the glass to be the system, which is a closed system. Under steady conditions, the rate form of the entropy balance for the glass simplifies to

$$\underbrace{\dot{S}_{in} - \dot{S}_{out}}_{\text{Rate of net entropy transfer by heat and mass}} + \underbrace{\dot{S}_{gen}}_{\text{Rate of entropy generation}} = \underbrace{\Delta \dot{S}_{\text{system}}^{\nearrow 0}}_{\text{Rate of change of entropy}} = 0$$

$$\frac{\dot{Q}_{in}}{T_{b,in}} - \frac{\dot{Q}_{out}}{T_{b,out}} + \dot{S}_{\text{gen,glass}} = 0$$

$$\frac{3200 \text{ W}}{283 \text{ K}} - \frac{3200 \text{ W}}{276 \text{ K}} + \dot{S}_{\text{gen,wall}} = 0 \quad \rightarrow \quad \dot{S}_{\text{gen,glass}} = \mathbf{0.2868 \ W/K}$$

Glass

10°C

3°C

Then the amount of entropy generation over a period of 5 h becomes

$$S_{\text{gen,glass}} = \dot{S}_{\text{gen,glass}} \Delta t = (0.2868 \ \text{W/K})(5 \times 3600 \text{ s}) = \mathbf{5162 \ J/K}$$

The exergy destroyed during a process can be determined from an exergy balance or directly from its definition $X_{\text{destroyed}} = T_0 S_{gen}$,

$$X_{\text{destroyed}} = T_0 S_{gen} = (278 \text{ K})(5.162 \text{ kJ/K}) = \mathbf{1435 \ kJ}$$

Discussion The total entropy generated during this process can be determined by applying the entropy balance on an *extended system* that includes the glass and its immediate surroundings on both sides so that the boundary temperature of the extended system is the room temperature on one side and the environment temperature on the other side at all times. Using this value of entropy generation will give the total exergy destroyed during the process, including the temperature gradient zones on both sides of the window.

7-93 Heat is transferred steadily to boiling water in the pan through its bottom. The inner and outer surface temperatures of the bottom of the pan are given. The rate of exergy destruction within the bottom plate is to be determined.

Assumptions Steady operating conditions exist since the surface temperatures of the pan remain constant at the specified values.

Analysis We take the bottom of the pan to be the system, which is a closed system. Under steady conditions, the rate form of the entropy balance for this system can be expressed as

$$\underbrace{\dot{S}_{in} - \dot{S}_{out}}_{\substack{\text{Rate of net entropy transfer} \\ \text{by heat and mass}}} + \underbrace{\dot{S}_{gen}}_{\substack{\text{Rate of entropy} \\ \text{generation}}} = \underbrace{\Delta \dot{S}_{\text{system}}^{\;\nearrow 0}}_{\substack{\text{Rate of change} \\ \text{of entropy}}} = 0$$

$$\frac{\dot{Q}_{in}}{T_{b,in}} - \frac{\dot{Q}_{out}}{T_{b,out}} + \dot{S}_{\text{gen,system}} = 0$$

$$\frac{800\text{ W}}{378\text{ K}} - \frac{800\text{ W}}{377\text{ K}} + \dot{S}_{\text{gen,system}} = 0 \;\rightarrow\; \dot{S}_{\text{gen,system}} = \mathbf{0.00561\ W/K}$$

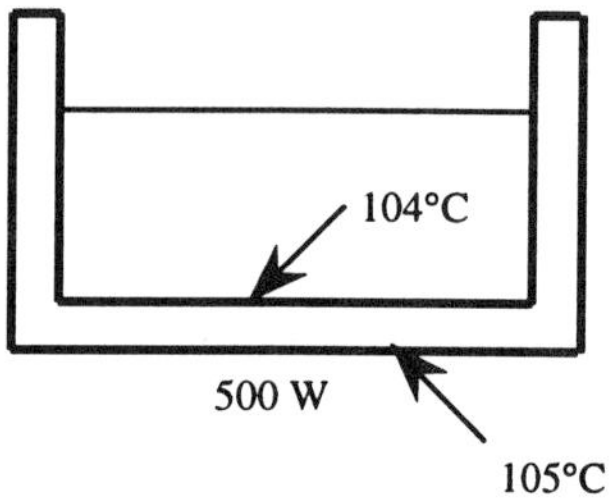

The exergy destroyed during a process can be determined from an exergy balance or directly from its definition $X_{\text{destroyed}} = T_0 S_{gen}$,

$$\dot{X}_{\text{destroyed}} = T_0 \dot{S}_{gen} = (298\text{ K})(0.00561\text{ W/K}) = \mathbf{1.67\ W}$$

7-94 A elevation, base area, and the depth of a crater lake are given. The maximum amount of electricity that can be generated by a hydroelectric power plant is to be determined.

Assumptions The evaporation of water from the lake is negligible.

Analysis The exergy or work potential of the water is the potential energy it possesses relative to the ground level,

$$\text{Exergy} = \text{PE} = \text{mgh}$$

Therefore,

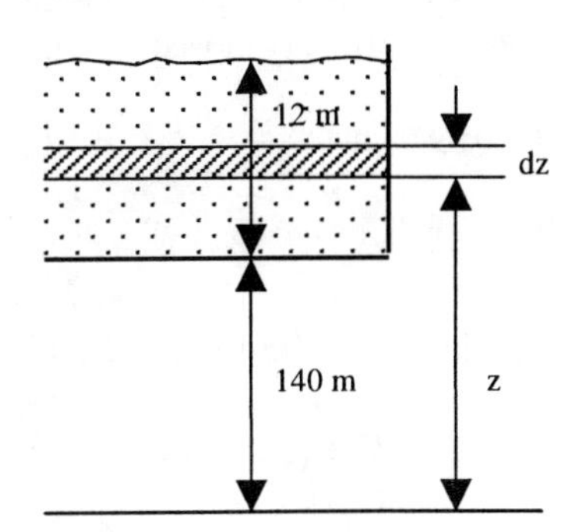

$$\text{Exergy} = PE = \int dPE = \int gz dm = \int gz(\rho A dz)$$

$$= \rho A g \int_{z_1}^{z_2} z dz = \rho A g (z_2^2 - z_1^2)/2$$

$$= 0.5(1000\ \text{kg/m}^3)(2 \times 10^4\ \text{m}^2)(9.81\ \text{m/s}^2)$$

$$\times \left((152\ \text{m})^2 - (140\ \text{m}^2)\right)\left(\frac{1\ \text{h}}{3600\ \text{s}}\right)\left(\frac{1\ \text{kJ/kg}}{1000\ \text{m}^2/\text{s}^2}\right)$$

$$= \mathbf{9.55 \times 10^4\ kWh}$$

7-95E The 2nd-law efficiency of a refrigerator and the refrigeration rate are given. The power input to the refrigerator is to be determined.

Analysis From the definition of the second law efficiency, the COP of the refrigerator is determined to be

$$COP_{R,rev} = \frac{1}{T_H / T_L - 1} = \frac{1}{535/495 - 1} = 12.375$$

$$\eta_{II} = \frac{COP_{R,rev}}{COP_R} \longrightarrow COP_R = \eta_{II} COP_{R,rev} = 0.45 \times 12.375 = 5.57$$

75°F

$\eta_{II} = 0.45$

R

200 Btu/min

35°F

Thus the power input is

$$\dot{W}_{in} = \frac{\dot{Q}_L}{COP_R} = \frac{200\ \text{Btu/min}}{5.57}\left(\frac{1\ \text{hp}}{42.41\ \text{Btu/min}}\right) = \mathbf{0.85\ hp}$$

7-96 Writing energy and entropy balances, a relation for the reversible work is to be obtained for a closed system that exchanges heat with surroundings at T_0 in the amount of Q_0 as well as a heat reservoir at temperature T_R in the amount Q_R.

Assumptions Kinetic and potential changes are negligible.

Analysis We take the direction of heat transfers to be to the system (heat input) and the direction of work transfer to be from the system (work output). The result obtained is still general since quantities wit opposite directions can be handled the same way by using negative signs. The energy and entropy balances for this stationary closed system can be expressed as

Energy balance: $E_{in} - E_{out} = \Delta E_{\text{system}} \rightarrow Q_0 + Q_R - W = U_2 - U_1 \longrightarrow W = U_1 - U_2 + Q_0 + Q_R \quad (1)$

Entropy balance: $S_{in} - S_{out} + S_{gen} = \Delta S_{\text{system}} \rightarrow S_{gen} = (S_2 - S_1) + \dfrac{-Q_R}{T_R} + \dfrac{-Q_0}{T_0} \quad (2)$

Solving for Q_0 from (2) and substituting in (1) yields

$$W = (U_1 - U_2) - T_0(S_1 - S_2) - Q_R\left(1 - \frac{T_0}{T_R}\right) - T_0 S_{gen}$$

Source
T_R
System
Q_R

The useful work relation for a closed system is obtained from

$$W_u = W - W_{surr} = (U_1 - U_2) - T_0(S_1 - S_2) - Q_R\left(1 - \frac{T_0}{T_R}\right) - T_0 S_{gen} - P_0(V_2 - V_1)$$

Then the reversible work relation is obtained by substituting $S_{gen} = 0$,

$$W_{rev} = (U_1 - U_2) - T_0(S_1 - S_2) + P_0(V_1 - V_2) - Q_R\left(1 - \frac{T_0}{T_R}\right)$$

A positive result for W_{rev} indicates work output, and a negative result work input. Also, the Q_R is a positive quantity for heat transfer to the system, and a negative quantity for heat transfer from the system.

7-97 Writing energy and entropy balances, a relation for the reversible work is to be obtained for a steady-flow system that exchanges heat with surroundings at T_0 at a rate of $\dot{Q}_0$ as well as a heat reservoir at temperature T_R in the amount $\dot{Q}_R$.

Analysis We take the direction of heat transfers to be to the system (heat input) and the direction of work transfer to be from the system (work output). The result obtained is still general since quantities wit opposite directions can be handled the same way by using negative signs. The energy and entropy balances for this stationary closed system can be expressed as

Energy balance: $\dot{E}_{in} - \dot{E}_{out} = \Delta\dot{E}_{system} \rightarrow \dot{E}_{in} = \dot{E}_{out}$

$$\dot{Q}_0 + \dot{Q}_R - \dot{W} = \Sigma \dot{m}_e (h_e + \frac{\mathbf{V}_e^2}{2} + gz_e) - \Sigma \dot{m}_i (h_i + \frac{\mathbf{V}_i^2}{2} + gz_i)$$

or

$$\dot{W} = \Sigma \dot{m}_i (h_i + \frac{\mathbf{V}_i^2}{2} + gz_i) - \Sigma \dot{m}_e (h_e + \frac{\mathbf{V}_e^2}{2} + gz_e) + \dot{Q}_0 + \dot{Q}_R \quad (1)$$

System

Entropy balance: $\dot{S}_{in} - \dot{S}_{out} + \dot{S}_{gen} = \Delta\dot{S}_{system} \rightarrow \dot{S}_{gen} = \dot{S}_{out} - \dot{S}_{in}$

$$\dot{S}_{gen} = \Sigma \dot{m}_e s_e - \Sigma \dot{m}_i s_i + \frac{-\dot{Q}_R}{T_R} + \frac{-\dot{Q}_0}{T_0} \quad (2)$$

Solving for $\dot{Q}_0$ from (2) and substituting in (1) yields

$$\dot{W} = \Sigma \dot{m}_i (h_i + \frac{\mathbf{V}_i^2}{2} + gz_i - T_0 s_i) - \Sigma \dot{m}_e (h_e + \frac{\mathbf{V}_e^2}{2} + gz_e - T_0 s_e) - T_0 \dot{S}_{gen} - \dot{Q}_R \left(1 - \frac{T_0}{T_R}\right)$$

Then the reversible work relation is obtained by substituting $S_{gen} = 0$,

$$\dot{W}_{rev} = \Sigma \dot{m}_i (h_i + \frac{\mathbf{V}_i^2}{2} + gz_i - T_0 s_i) - \Sigma \dot{m}_e (h_e + \frac{\mathbf{V}_e^2}{2} + gz_e - T_0 s_e) - \dot{Q}_R \left(1 - \frac{T_0}{T_R}\right)$$

A positive result for W_{rev} indicates work output, and a negative result work input. Also, the Q_R is a positive quantity for heat transfer to the system, and a negative quantity for heat transfer from the system.

7-98 Writing energy and entropy balances, a relation for the reversible work is to be obtained for a uniform-flow system that exchanges heat with surroundings at T_0 in the amount of Q_0 as well as a heat reservoir at temperature T_R in the amount Q_R.

Assumptions Kinetic and potential changes are negligible.

Analysis We take the direction of heat transfers to be to the system (heat input) and the direction of work transfer to be from the system (work output). The result obtained is still general since quantities wit opposite directions can be handled the same way by using negative signs. The energy and entropy balances for this stationary closed system can be expressed as

Energy balance: $E_{in} - E_{out} = \Delta E_{\text{system}}$

$$Q_0 + Q_R - W = \Sigma m_e(h_e + \frac{\mathbf{V}_e^2}{2} + gz_e) - \Sigma m_i(h_i + \frac{\mathbf{V}_i^2}{2} + gz_i) + (U_2 - U_1)_{cv}$$

or, $$W = \Sigma m_i(h_i + \frac{\mathbf{V}_i^2}{2} + gz_i) - \Sigma m_e(h_e + \frac{\mathbf{V}_e^2}{2} + gz_e) - (U_2 - U_1)_{cv} + Q_0 + Q_R \quad (1)$$

Entropy balance: $S_{in} - S_{out} + S_{gen} = \Delta S_{\text{system}}$

$$S_{gen} = (S_2 - S_1)_{cv} + \Sigma m_e s_e - \Sigma m_i s_i + \frac{-Q_R}{T_R} + \frac{-Q_0}{T_0} \quad (2)$$

Solving for Q_0 from (2) and substituting in (1) yields

$$W = \Sigma m_i(h_i + \frac{\mathbf{V}_i^2}{2} + gz_i - T_0 s_i) - \Sigma m_e(h_e + \frac{\mathbf{V}_e^2}{2} + gz_e - T_0 s_e)$$
$$+ [(U_1 - U_2) - T_0(S_1 - S_2)]_{cv} - T_0 S_{gen} - Q_R\left(1 - \frac{T_0}{T_R}\right)$$

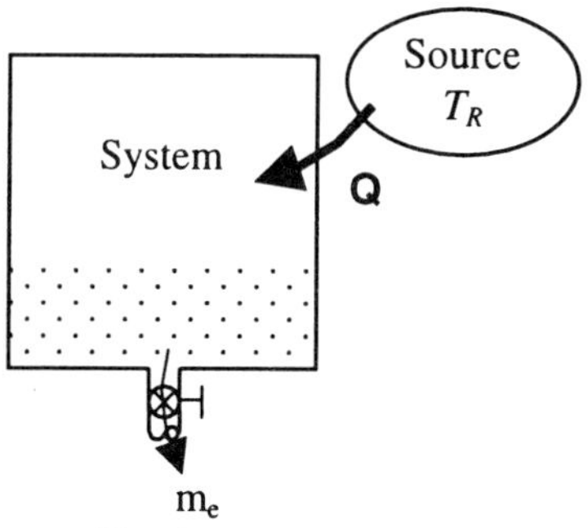

The useful work relation for a closed system is obtained from

$$W_u = W - W_{surr} = \Sigma m_i(h_i + \frac{\mathbf{V}_i^2}{2} + gz_i - T_0 s_i) - \Sigma m_e(h_e + \frac{\mathbf{V}_e^2}{2} + gz_e - T_0 s_e)$$
$$+ [(U_1 - U_2) - T_0(S_1 - S_2)]_{cv} - T_0 S_{gen} - Q_R\left(1 - \frac{T_0}{T_R}\right) - P_0(V_2 - V_1)$$

Then the reversible work relation is obtained by substituting $S_{gen} = 0$,

$$W_{rev} = \Sigma m_i(h_i + \frac{\mathbf{V}_i^2}{2} + gz_i - T_0 s_i) - \Sigma m_e(h_e + \frac{\mathbf{V}_e^2}{2} + gz_e - T_0 s_e)$$
$$+ [(U_1 - U_2) - T_0(S_1 - S_2) + P_0(V_1 - V_2)]_{cv} - Q_R\left(1 - \frac{T_0}{T_R}\right)$$

A positive result for W_{rev} indicates work output, and a negative result work input. Also, the Q_R is a positive quantity for heat transfer to the system, and a negative quantity for heat transfer from the system.

7-99 An electric resistance heater is immersed in water. The time it will take for the electric heater to raise the water temperature to a specified temperature, the minimum work input, and the exergy destroyed during this process are to be determined.

Assumptions **1** Water is an incompressible substance with constant specific heats. **2** The energy stored in the container itself and the heater is negligible. **3** Heat loss from the container is negligible. **4** The environment temperature is given to be $T_0 = 20°C$.

Properties The specific heat of water at room temperature is $C = 4.18$ kJ/kg·°C (Table A-3).

Analysis Taking the water in the container as the system, which is a closed system, the energy balance can be expressed as

$$\underbrace{E_{in} - E_{out}}_{\substack{\text{Net energy transfer}\\ \text{by heat, work, and mass}}} = \underbrace{\Delta E_{\text{system}}}_{\substack{\text{Change in internal, kinetic,}\\ \text{potential, etc. energies}}}$$

$$W_{e,in} = (\Delta U)_{\text{water}}$$

$$\dot{W}_{e,in}\Delta t = mC(T_2 - T_1)_{\text{water}}$$

Water
40 kg
Heater

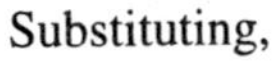

Substituting,

$$(800 \text{ J/s})\Delta t = (40 \text{ kg})(4184 \text{ J/kg·°C})(80 - 20)°C$$

Solving for Δt gives

$$\Delta t = \mathbf{12{,}552 \text{ s} = 209.2 \text{ min} = 3.487 \text{ h}}$$

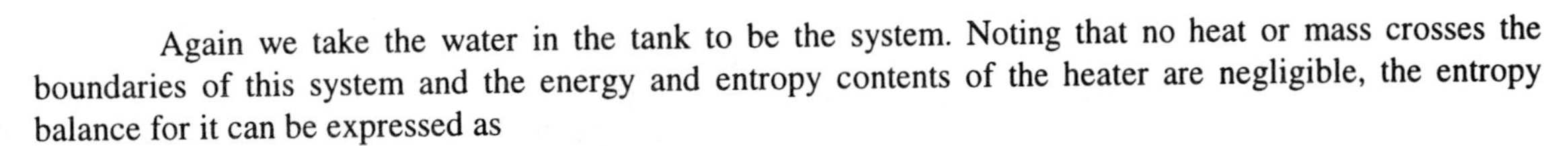

Again we take the water in the tank to be the system. Noting that no heat or mass crosses the boundaries of this system and the energy and entropy contents of the heater are negligible, the entropy balance for it can be expressed as

$$\underbrace{S_{in} - S_{out}}_{\substack{\text{Net entropy transfer}\\ \text{by heat and mass}}} + \underbrace{S_{gen}}_{\substack{\text{Entropy}\\ \text{generation}}} = \underbrace{\Delta S_{\text{system}}}_{\substack{\text{Change}\\ \text{in entropy}}}$$

$$0 + S_{\text{gen}} = \Delta S_{\text{water}}$$

Therefore, the entropy generated during this process is

$$S_{gen} = \Delta S_{\text{water}} = mC\ln\frac{T_2}{T_1} = (40\text{kg})(4.184\text{kJ/kg}\cdot\text{K})\ln\frac{353\text{K}}{293\text{K}} = \mathbf{31.18kJ/K}$$

The exergy destroyed during a process can be determined from an exergy balance or directly from its definition $X_{\text{destroyed}} = T_0 S_{gen}$,

$$X_{\text{destroyed}} = T_0 S_{gen} = (293 \text{ K})(31.18 \text{ kJ/K}) = \mathbf{9136 \text{ kJ}}$$

The actual work input for this process is

$$W_{act,in} = \dot{W}_{act,in}\Delta t = (0.8 \text{ kJ/s})(12{,}552 \text{ s}) = 10{,}042 \text{ kJ}$$

Then the reversible (or minimum required)work input becomes

$$W_{rev,in} = W_{act,in} - X_{\text{destroyed}} = 10{,}042 - 9136 = \mathbf{906 \text{ kJ}}$$

7-100 A hot water pipe at a specified temperature is losing heat to the surrounding air at a specified rate. The rate at which the work potential is wasted during this process is to be determined.

Assumptions Steady operating conditions exist.

Analysis We take the air in the vicinity of the pipe (excluding the pipe) as our system, which is a closed system.. The system extends from the outer surface of the pipe to a distance at which the temperature drops to the surroundings temperature. In steady operation, the rate form of the entropy balance for this system can be expressed as

$$\underbrace{\dot{S}_{in} - \dot{S}_{out}}_{\text{Rate of net entropy transfer by heat and mass}} + \underbrace{\dot{S}_{gen}}_{\text{Rate of entropy generation}} = \underbrace{\Delta \dot{S}_{\text{system}}^{\nearrow 0}}_{\text{Rate of change of entropy}} = 0$$

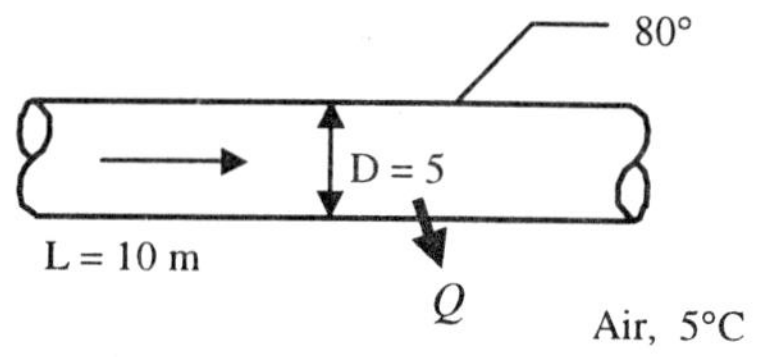

$$\frac{\dot{Q}_{in}}{T_{b,in}} - \frac{\dot{Q}_{out}}{T_{b,out}} + \dot{S}_{\text{gen,system}} = 0$$

$$\frac{45\text{ W}}{353\text{ K}} - \frac{45\text{ W}}{278\text{ K}} + \dot{S}_{\text{gen,system}} = 0 \quad \rightarrow \quad \dot{S}_{\text{gen,system}} = 0.0344\text{ W/K}$$

80°

D = 5

L = 10 m

Q

Air, 5°C

The exergy destroyed during a process can be determined from an exergy balance or directly from its definition $X_{\text{destroyed}} = T_0 S_{gen}$,

$$\dot{X}_{\text{destroyed}} = \dot{T}_0 S_{gen} = (278\text{ K})(0.0344\text{ W/K}) = \mathbf{9.56\ W}$$

7-101 Two rigid tanks that contain water at different states are connected by a valve. The valve is opened and steam flows from tank A to tank B until the pressure in tank A drops to a specified value. Tank B loses heat to the surroundings. The final temperature in each tank and the work potential wasted during this process are to be determined.

Assumptions **1** Tank A is insulated and thus heat transfer is negligible. **2** The water that remains in tank A undergoes a reversible adiabatic process. **3** The thermal energy stored in the tanks themselves is negligible. **4** The system is stationary and thus kinetic and potential energy changes are negligible. **5** There are no work interactions.

Analysis (a) The steam in tank A undergoes a reversible, adiabatic process, and thus $s_2 = s_1$. From the steam tables (Tables A-4 through A-6),

Tank A :

$$\left.\begin{array}{l} P_1 = 400\text{kPa} \\ x_1 = 0.8 \end{array}\right\} \begin{array}{l} v_{1,A} = v_f + x_1 v_{fg} = 0.001084 + (0.8)(0.4625 - 0.001084) = 0.3702\text{m}^3/\text{kg} \\ u_{1,A} = u_f + x_1 u_{fg} = 604.31 + (0.8)(1949.3) = 2163.75\text{kJ/kg} \\ s_{1,A} = s_f + x_1 s_{fg} = 1.7766 + (0.8)(5.1193) = 5.8720\text{kJ/kg}\cdot\text{K} \end{array}$$

$$\left.\begin{array}{l} P_2 = 300\text{kPa} \\ s_2 = s_1 \\ (sat.mixture) \end{array}\right\} \begin{array}{l} T_{2,A} = T_{sat@300kPa} = \mathbf{133.55^\circ C} \\ x_{2,A} = \dfrac{s_{2,A} - s_f}{s_{fg}} = \dfrac{5.8720 - 1.6718}{5.3201} = 0.790 \\ v_{2,A} = v_f + x_{2,A} v_{fg} = 0.001073 + (0.790)(0.6058 - 0.001073) = 0.479\text{m}^3/\text{kg} \\ u_{2,A} = u_f + x_{2,A} u_{fg} = 561.15 + (0.790)(1982.4\text{kJ/kg}) = 2127.2\text{kJ/kg} \end{array}$$

TankB :

$$\left.\begin{array}{l} P_1 = 200\text{kPa} \\ T_1 = 250^\circ\text{C} \end{array}\right\} \begin{array}{l} v_{1,B} = 1.1988\text{m}^3/\text{kg} \\ u_{1,B} = 2731.2\text{kJ/kg} \\ s_{1,B} = 7.7086\text{kJ/kg}\cdot\text{K} \end{array}$$

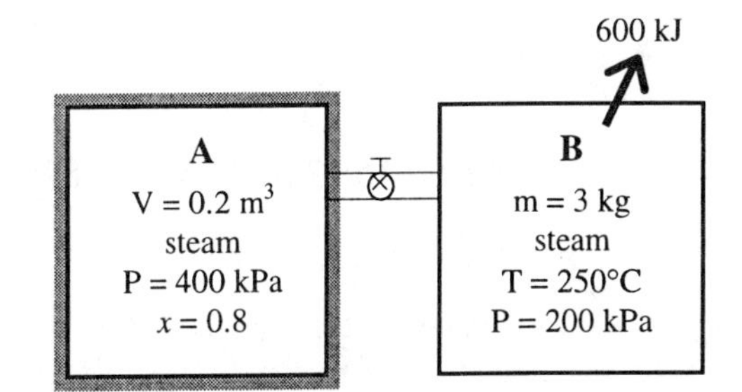

The initial and the final masses in tank A are

$$m_{1,A} = \frac{V_A}{v_{1,A}} = \frac{0.2\text{ m}^3}{0.3702\text{ m}^3/\text{kg}} = 0.540\text{ kg}$$

and

$$m_{2,A} = \frac{V_A}{v_{2,A}} = \frac{0.2\text{ m}^3}{0.479\text{ m}^3/\text{kg}} = 0.418\text{ kg}$$

Thus, 0.540 - 0.418 = 0.122 kg of mass flows into tank B. Then,

$$m_{2,B} = m_{1,B} - 0.122 = 3 + 0.122 = 3.122\text{ kg}$$

The final specific volume of steam in tank B is determined from

$$v_{2,B} = \frac{V_B}{m_{2,B}} = \frac{(m_1 v_1)_B}{m_{2,B}} = \frac{(3\text{kg})(1.1988\text{m}^3/\text{kg})}{3.122\text{m}^3} = 1.152\text{m}^3/\text{kg}$$

We take the entire contents of both tanks as the system, which is a closed system. The energy balance for this stationary closed system can be expressed as

$$\underbrace{E_{in} - E_{out}}_{\substack{\text{Net energy transfer}\\ \text{by heat, work, and mass}}} = \underbrace{\Delta E_{\text{system}}}_{\substack{\text{Change in internal, kinetic,}\\ \text{potential, etc. energies}}}$$

$$-Q_{out} = \Delta U = (\Delta U)_A + (\Delta U)_B \quad (\text{since } W = \text{KE} = \text{PE} = 0)$$

$$-Q_{out} = (m_2u_2 - m_1u_1)_A + (m_2u_2 - m_1u_1)_B$$

Substituting,

$$-600 = \{(0.418)(2127..2) - (0.540)(2163.8)\} + \{(3.122)u_{2,B} - (3)(2731.2)\}$$

$$u_{2,B} = 2521.7 \text{kJ/kg}$$

Thus,

$$\left.\begin{array}{l} v_{2,B} = 1.152 \text{m}^3\text{/kg} \\ u_{2,B} = 2521.7 \text{kJ/kg} \end{array}\right\} \begin{array}{l} T_{2,B} = \mathbf{113^\circ C} \\ s_{2,B} = 7.225 \text{ kJ/kg} \cdot \text{K} \end{array}$$

(*b*) The total entropy generation during this process is determined by applying the entropy balance on an *extended system* that includes both tanks and their immediate surroundings so that the boundary temperature of the extended system is the temperature of the surroundings at all times. It gives

$$\underbrace{S_{in} - S_{out}}_{\substack{\text{Net entropy transfer}\\ \text{by heat and mass}}} + \underbrace{S_{gen}}_{\substack{\text{Entropy}\\ \text{generation}}} = \underbrace{\Delta S_{\text{system}}}_{\substack{\text{Change}\\ \text{in entropy}}}$$

$$-\frac{Q_{out}}{T_{b,surr}} + S_{\text{gen}} = \Delta S_{\text{A}} + \Delta S_{\text{B}}$$

Rearranging and substituting, the total entropy generated during this process is determined to be

$$S_{gen} = \Delta S_A + \Delta S_B + \frac{Q_{out}}{T_{b,surr}} = (m_2s_2 - m_1s_1)_A + (m_2s_2 - m_1s_1)_B + \frac{Q_{out}}{T_{b,surr}}$$

$$= \{(0.418)(5.872) - (0.540)(5.872)\} + \{(3.122)(7.225) - (3)(7.7086)\} + \frac{600\text{kJ}}{273\text{K}}$$

$$= \mathbf{0.912 \ kJ/K}$$

The work potential wasted is equivalent to the exergy destroyed during a process, which can be determined from an exergy balance or directly from its definition $X_{\text{destroyed}} = T_0 S_{gen}$,

$$X_{\text{destroyed}} = T_0 S_{gen} = (273 \text{ K})(0.912 \text{ kJ/K}) = \mathbf{249 \ kJ}$$

7-102E A cylinder initially filled with helium gas at a specified state is compressed polytropically to a specified temperature and pressure. The actual work consumed and the minimum useful work input needed are to be determined.

Assumptions **1** Helium is an ideal gas with constant specific heats. **2** The cylinder is stationary and thus the kinetic and potential energy changes are negligible. **3** The thermal energy stored in the cylinder itself is negligible. **4** The compression or expansion process is quasi-equilibrium. **5** The environment temperature is 70°F.

Properties The gas constant of helium is R = 2.6805 psia.ft³/lbm.R = 0.4961 Btu/lbm.R (Table A-1E). The specific heats of helium are C_v = 0.753 and C_v = 1.25 Btu/lbm.R (Table A-2E).

Analysis (a) Helium at specified conditions can be treated as an ideal gas. The mass of helium is

$$m = \frac{P_1 V_1}{RT_1} = \frac{(25\text{ psia})(15\text{ ft}^3)}{(2.6805\text{ psia}\cdot\text{ft}^3/\text{lbm}\cdot\text{R})(530\text{ R})} = 0.264\text{ lbm}$$

HELIUM
15 ft³
PVⁿ = const
Q

The exponent n and the boundary work for this polytropic process are determined to be

$$\frac{P_1 V_1}{T_1} = \frac{P_2 V_2}{T_2} \longrightarrow V_2 = \frac{T_2}{T_1}\frac{P_1}{P_2}V_1 = \frac{(760\text{ R})(25\text{ psia})}{(530\text{ R})(70\text{ psia})}(15\text{ ft}^3) = 7.682\text{ ft}^3$$

$$P_2 V_2^n = P_1 V_1^n \longrightarrow \left(\frac{P_2}{P_1}\right) = \left(\frac{V_1}{V_2}\right)^n \longrightarrow \left(\frac{70}{25}\right) = \left(\frac{15}{7.682}\right)^n \longrightarrow n = 1.539$$

Then the boundary work for this polytropic process can be determined from

$$W_{b,in} = -\int_1^2 P\,dV = -\frac{P_2 V_2 - P_1 V_1}{1-n} = -\frac{mR(T_2 - T_1)}{1-n}$$

$$= -\frac{(0.264\text{lbm})(0.4961\text{Btu/lbm}\cdot\text{R})(760-530)\text{R}}{1-1.539} = 55.9\text{Btu}$$

Also,

$$W_{surr,in} = -P_0(V_2 - V_1) = -(14.7\text{ psia})(7.682-15)\text{ft}^3\left(\frac{1\text{ Btu}}{5.4039\text{ psia}\cdot\text{ft}^3}\right) = 19.9\text{ Btu}$$

Thus,

$$W_{u,in} = W_{b,in} - W_{surr,in} = 55.9 - 19.9 = \mathbf{36.0\ Btu}$$

(*b*) We take the helium in the cylinder as the system, which is a closed system. Taking the direction of heat transfer to be from the cylinder, the energy balance for this stationary closed system can be expressed as

$$\underbrace{E_{in} - E_{out}}_{\substack{\text{Net energy transfer}\\ \text{by heat, work, and mass}}} = \underbrace{\Delta E_{\text{system}}}_{\substack{\text{Change in internal, kinetic,}\\ \text{potential, etc. energies}}}$$

$$-Q_{out} + W_{b,in} = \Delta U = m(u_2 - u_1)$$

$$-Q_{out} = m(u_2 - u_1) - W_{b,in}$$

$$Q_{out} = W_{b,in} - mC_v(T_2 - T_1)$$

Substituting,

$$Q_{out} = 55.9\text{ Btu} - (0.264\text{ lbm})(0.753\text{ Btu/lbm}\cdot\text{R})(760-530)\text{R} = 10.2\text{ Btu}$$

The total entropy generation during this process is determined by applying the entropy balance on an *extended system* that includes the cylinder and its immediate surroundings so that the boundary temperature of the extended system is the temperature of the surroundings at all times. It gives

$$\underbrace{S_{in} - S_{out}}_{\substack{\text{Net entropy transfer} \\ \text{by heat and mass}}} + \underbrace{S_{gen}}_{\substack{\text{Entropy} \\ \text{generation}}} = \underbrace{\Delta S_{\text{system}}}_{\substack{\text{Change} \\ \text{in entropy}}}$$

$$-\frac{Q_{out}}{T_{b,surr}} + S_{\text{gen}} = \Delta S_{\text{sys}}$$

where the entropy change of helium is

$$\Delta S_{sys} = \Delta S_{helium} = m\left(C_{p,ave} \ln\frac{T_2}{T_1} - R\ln\frac{P_2}{P_1}\right)$$

$$= (0.264 \text{ lbm})\left[(1.25 \text{ Btu/lbm}\cdot\text{R})\ln\frac{760 \text{ R}}{530 \text{ R}} - (0.4961 \text{ Btu/lbm}\cdot\text{R})\ln\frac{70 \text{ psia}}{25 \text{ psia}}\right]$$

$$= -0.0159 \text{ Btu/R}$$

Rearranging and substituting, the total entropy generated during this process is determined to be

$$S_{gen} = \Delta S_{helium} + \frac{Q_{out}}{T_0} = (-0.0159 \text{ Btu/R}) + \frac{10.2 \text{ Btu}}{530 \text{ R}} = 0.003345 \text{ Btu/R}$$

The work potential wasted is equivalent to the exergy destroyed during a process, which can be determined from an exergy balance or directly from its definition $X_{\text{destroyed}} = T_0 S_{gen}$,

$$X_{\text{destroyed}} = T_0 S_{gen} = (530 \text{ R})(0.003345 \text{ Btu/R}) = \mathbf{1.77\ Btu}$$

The minimum work with which this process could be accomplished is the reversible work input, $W_{\text{rev, in}}$. which can be determined directly from

$$W_{rev,in} = W_{act,in} - X_{\text{destroyed}} = 36.0 - 1.77 = \mathbf{34.23\ Btu}$$

Discussion The reversible work input, which represents the minimum work input $W_{\text{rev,in}}$ in this case can be determined from the exergy balance by setting the exergy destruction term equal to zero,

$$\underbrace{X_{in} - X_{out}}_{\substack{\text{Net exergy transfer} \\ \text{by heat, work, and mass}}} \underbrace{- X_{\text{destroyed}}^{\nearrow 0 \text{ (reversible)}}}_{\substack{\text{Exergy} \\ \text{destruction}}} = \underbrace{\Delta X_{\text{system}}}_{\substack{\text{Change} \\ \text{in exergy}}} \rightarrow W_{rev,in} = X_2 - X_1$$

Substituting the closed system exergy relation, the reversible work input during this process is determined to be

$$W_{rev} = (U_2 - U_1) - T_0(S_2 - S_1) + P_0(V_2 - V_1)$$

$$= (0.264 \text{ lbm})(0.753 \text{ Btu/lbm}\cdot\text{R})(300-70)°\text{F} - (530 \text{ R})(-0.0159 \text{ Btu/R})$$

$$+ (14.7 \text{ psia})(7.682 - 15) \text{ ft}^3 \text{ [Btu/5.4039 psia}\cdot\text{ft}^3]$$

$$= \mathbf{34.24\ Btu}$$

7-103 A well-insulated room is heated by a steam radiator, and the warm air is distributed by a fan. The average temperature in the room after 30 min, the entropy changes of steam and air, and the exergy destruction during this process are to be determined.

Assumptions **1** Air is an ideal gas with constant specific heats at room temperature. **2** The kinetic and potential energy changes are negligible. **3** The air pressure in the room remains constant and thus the air expands as it is heated, and some warm air escapes. **4** The environment temperature is given to be T_0 = 10°C.

Properties The gas constant of air is R = 0.287 kPa.m³/kg.K (Table A-1). Also, C_p = 1.005 kJ/kg.K for air at room temperature (Table A-2).

Analysis We first take the radiator as the system. This is a closed system since no mass enters or leaves. The energy balance for this closed system can be expressed as

$$\underbrace{E_{in} - E_{out}}_{\text{Net energy transfer by heat, work, and mass}} = \underbrace{\Delta E_{\text{system}}}_{\text{Change in internal, kinetic, potential, etc. energies}}$$

$$-Q_{out} = \Delta U = m(u_2 - u_1) \quad \text{(since } W = \text{KE} = \text{PE} = 0\text{)}$$

$$Q_{out} = m(u_1 - u_2)$$

10°C
4m × 4m × 5m
Steam
radiator

Using data from the steam tables (Tables A-4 through A-6), some properties are determined to be

$$\left.\begin{array}{l} P_1 = 200 \text{ kPa} \\ T_1 = 200^\circ\text{C} \end{array}\right\} \begin{array}{l} v_1 = 1.0803 \text{ m}^3/\text{kg} \\ u_1 = 2654.4 \text{ kJ/kg} \end{array}$$

$$\left.\begin{array}{l} P_2 = 100 \text{ kPa} \\ (v_2 = v_1) \end{array}\right\} \begin{array}{l} v_f = 0.001043, \quad v_g = 1.6940 \text{ m}^3/\text{kg} \\ u_f = 417.36, \quad u_{fg} = 2088.7 \text{ kJ/kg} \end{array}$$

$$x_2 = \frac{v_2 - v_f}{v_{fg}} = \frac{1.0803 - 0.001043}{1.6940 - 0.001043} = 0.6375$$

$$u_2 = u_f + x_2 u_{fg} = 417.36 + 0.6375 \times 2088.7 = 1749.0\text{kJ/kg}$$

$$m = \frac{V_1}{v_1} = \frac{0.015\text{m}^3}{1.0803\text{m}^3/\text{kg}} = 0.0139\text{kg}$$

Substituting,

$$Q_{out} = (0.0139 \text{ kg})(2654.4 - 1749.0)\text{kJ/kg} = 12.6 \text{ kJ}$$

The volume and the mass of the air in the room are V = 4×4×5 = 80 m³ and

$$m_{air} = \frac{P_1 V_1}{RT_1} = \frac{(100 \text{ kPa})(80 \text{ m}^3)}{(0.2870 \text{ kPa}\cdot\text{m}^3/\text{kg}\cdot\text{K})(283 \text{ K})} = 98.5 \text{ kg}$$

The amount of fan work done in 30 min is

$$W_{fan,in} = \dot{W}_{fan,in}\Delta t = (0.120 \text{ kJ/s})(30 \times 60 \text{ s}) = 216 \text{ kJ}$$

We now take the air in the room as the system. The energy balance for this closed system is expressed as

$$E_{in} - E_{out} = \Delta E_{\text{system}}$$

$$Q_{in} + W_{fan,in} - W_{b,out} = \Delta U$$

$$Q_{in} + W_{fan,in} = \Delta H \cong mC_p(T_2 - T_1)$$

since the boundary work and ΔU combine into ΔH for a constant pressure expansion or compression process. It can also be expressed as

$$(\dot{Q}_{in} + \dot{W}_{fan,in})\Delta t = mC_{p,ave}(T_2 - T_1)$$

Substituting, (12.6 kJ) + (216 kJ) = (98.5 kg)(1.005 kJ/kg°C)(T_2 - 10)°C

which yields T_2 = **12.3°C**

Therefore, the air temperature in the room rises from 10°C to 12.3°C in 30 mi.

(*b*) The entropy change of the steam is

$$\Delta S_{steam} = m(s_2 - s_1) = (0.0139 \text{ kg})(5.1638 - 7.5066)\text{kJ/kg}\cdot\text{K} = \mathbf{-0.0326\ kJ/K}$$

(*c*) Noting that air expands at constant pressure, the entropy change of the air in the room is

$$\Delta S_{air} = mC_p \ln\frac{T_2}{T_1} - mR\ln\frac{P_2}{P_1}^{\nearrow 0} = (98.5 \text{ kg})(1.005 \text{ kJ/kg}\cdot\text{K})\ln\frac{285.3 \text{ K}}{283 \text{ K}} = \mathbf{0.801\ kJ/K}$$

(*d*) We take the contents of the room (including the steam radiator) as our system, which is a closed system. Noting that no heat or mass crosses the boundaries of this system, the entropy balance for it can be expressed as

$$\underbrace{S_{in} - S_{out}}_{\substack{\text{Net entropy transfer}\\ \text{by heat and mass}}} + \underbrace{S_{gen}}_{\substack{\text{Entropy}\\ \text{generation}}} = \underbrace{\Delta S_{\text{system}}}_{\substack{\text{Change}\\ \text{in entropy}}}$$

$$0 + S_{\text{gen}} = \Delta S_{\text{steam}} + \Delta S_{\text{air}}$$

Substituting, the entropy generated during this process is determined to be

$$S_{gen} = \Delta S_{steam} + \Delta S_{air} = -0.0326 + 0.801 = 0.768 \text{ kJ/K}$$

The exergy destroyed during a process can be determined from an exergy balance or directly from its definition $X_{\text{destroyed}} = T_0 S_{gen}$,

$$X_{\text{destroyed}} = T_0 S_{gen} = (283 \text{ K})(0.768 \text{ kJ/K}) = \mathbf{217\ kJ}$$

Alternative Solution In the solution above, we assumed the air pressure in the room to remain constant. This is an extreme case, and it is commonly used in practice since it gives higher results for heat loads, and thus allows the designer to be conservative results. The other extreme is to assume the house to be airtight, and thus the volume of the air in the house to remain constant as the air is heated. There is no expansion in this case and thus boundary work, and C_v is used in energy change relation instead of C_p. It gives the following results:

$$T_2 = 12.3°\text{C}$$

$$\Delta S_{steam} = m(s_2 - s_1) = (0.0139 \text{ kg})(5.1638 - 7.5066)\text{kJ/kg}\cdot\text{K} = -0.0326 \text{ kJ/K}$$

$$\Delta S_{air} = mC_v \ln\frac{T_2}{T_1} + mR\ln\frac{V_2}{V_1}^{\nearrow 0} = (98.5 \text{ kg})(0.718 \text{ kJ/kg}\cdot\text{K})\ln\frac{286.2 \text{ K}}{283 \text{ K}} = 0.795 \text{ kJ/K}$$

$$S_{gen} = \Delta S_{steam} + \Delta S_{air} = -0.0326 + 0.795 = 0.7625 \text{ kJ/K}$$

and

$$X_{\text{destroyed}} = T_0 S_{gen} = (283 \text{ K})(0.7625 \text{ kJ/K}) = 216 \text{ kJ}$$

The actual value in practice will be between these two limits.

7-104 The heating of a passive solar house at night is to be assisted by solar heated water. The length of time that the electric heating system would run that night, the exergy destruction, and the minimum work input required that night are to be determined.

Assumptions **1** Water is an incompressible substance with constant specific heats. **2** The energy stored in the glass containers themselves is negligible relative to the energy stored in water. **3** The house is maintained at 22°C at all times. **4** The environment temperature is given to be $T_0 = 5°C$.

Properties The density and specific heat of water at room temperature are ρ = 1 kg/L and C = 4.18 kJ/kg· °C (Table A-3).

Analysis The total mass of water is

$$m_w = \rho V = (1\text{kg/L})(50 \times 20\text{L}) = 1000\text{kg}$$

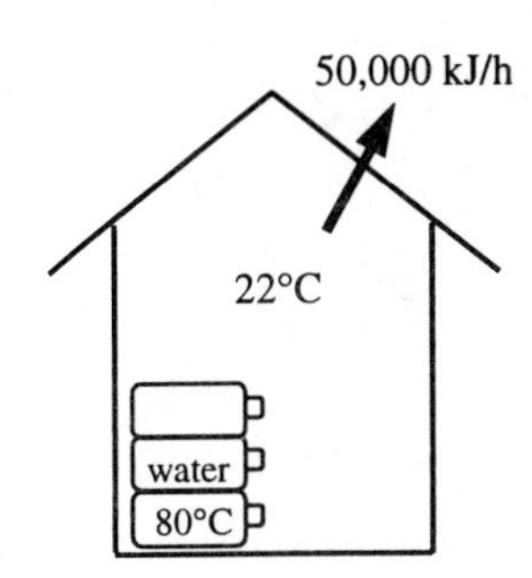

Taking the contents of the house, including the water as our system, the energy balance relation can be written as

$$\underbrace{E_{in} - E_{out}}_{\substack{\text{Net energy transfer}\\ \text{by heat, work, and mass}}} = \underbrace{\Delta E_{\text{system}}}_{\substack{\text{Change in internal, kinetic,}\\ \text{potential, etc. energies}}}$$

$$W_{e,in} - Q_{out} = \Delta U = (\Delta U)_{\text{water}} + (\Delta U)_{\text{air}}^{\nearrow 0}$$
$$= (\Delta U)_{\text{water}} = mC(T_2 - T_1)_{\text{water}}$$

or, $$\dot{W}_{e,in}\Delta t - Q_{out} = [mC(T_2 - T_1)]_{\text{water}}$$

Substituting,

$$(15 \text{ kJ/s})\Delta t - (50{,}000 \text{ kJ/h})(10 \text{ h}) = (1000 \text{ kg})(4.18 \text{ kJ/kg·°C})(22 - 80)°\text{C}$$

It gives

$$\Delta t = 17{,}170 \text{ s} = \mathbf{4.77 \text{ h}}$$

We take the house as the system, which is a closed system. The entropy generated during this process is determined by applying the entropy balance on an *extended system* that includes the house and its immediate surroundings so that the boundary temperature of the extended system is the temperature of the surroundings at all times. The entropy balance for the extended system can be expressed as

$$\underbrace{S_{in} - S_{out}}_{\substack{\text{Net entropy transfer}\\ \text{by heat and mass}}} + \underbrace{S_{gen}}_{\substack{\text{Entropy}\\ \text{generation}}} = \underbrace{\Delta S_{\text{system}}}_{\substack{\text{Change}\\ \text{in entropy}}}$$

$$-\frac{Q_{out}}{T_{b,out}} + S_{\text{gen}} = \Delta S_{\text{water}} + \Delta S_{\text{air}}^{\nearrow 0} = \Delta S_{\text{water}}$$

since the state of air in the house remains unchanged. Then the entropy generated during the 10-h period that night is

$$S_{gen} = \Delta S_{\text{water}} + \frac{Q_{out}}{T_{b,out}} = \left(mC\ln\frac{T_2}{T_1}\right)_{water} + \frac{Q_{out}}{T_0}$$
$$= (1000 \text{ kg})(4.18 \text{ kJ/kg}\cdot\text{K})\ln\frac{295 \text{ K}}{353 \text{ K}} + \frac{500{,}000 \text{ kJ}}{278 \text{ K}} = 1048 \text{ kJ/K}$$

The exergy destroyed during a process can be determined from an exergy balance or directly from its definition $X_{\text{destroyed}} = T_0 S_{gen}$,

$$X_{\text{destroyed}} = T_0 S_{gen} = (278 \text{ K})(1048 \text{ kJ/K}) = \mathbf{291{,}400 \text{ kJ}}$$

The actual work input during this process is

$$W_{act,in} = \dot{W}_{act,in}\Delta t = (15 \text{ kJ/s})(17{,}170 \text{ s}) = 257{,}550 \text{ kJ}$$

The minimum work with which this process could be accomplished is the reversible work input, $W_{rev,\ in}$. which can be determined directly from

$$W_{rev,in} = W_{act,in} - X_{\text{destroyed}} = 257{,}550 - 291{,}400 = \text{-}33{,}850 \text{ kJ}$$

$$\rightarrow\ W_{rev,out} = 33{,}850 \text{ kJ} = \mathbf{9.40\ kWh}$$

That is, 9.40 kWh of electricity could be *generated* while heating the house by the solar heated water (instead of consuming electricity) if the process was done reversibly.

7-105 Steam expands in a two-stage adiabatic turbine from a specified state to specified pressure. Some steam is extracted at the end of the first stage. The wasted power potential is to be determined.

Assumptions **1** This is a steady-flow process since there is no change with time. **2** Kinetic and potential energy changes are negligible. **3** The turbine is adiabatic and thus heat transfer is negligible. **4** The environment temperature is given to be $T_0 = 25°C$.

Analysis The wasted power potential is equivalent to the rate of exergy destruction during a process, which can be determined from an exergy balance or directly from its definition $X_{destroyed} = T_0 S_{gen}$.

The total rate of entropy generation during this process is determined by taking the entire turbine, which is a control volume, as the system and applying the entropy balance. Noting that this is a steady-flow process and there is no heat transfer,

$$\underbrace{\dot{S}_{in} - \dot{S}_{out}}_{\text{Rate of net entropy transfer by heat and mass}} + \underbrace{\dot{S}_{gen}}_{\text{Rate of entropy generation}} = \underbrace{\Delta \dot{S}_{system}^{\nearrow 0}}_{\text{Rate of change of entropy}} = 0$$

$$\dot{m}_1 s_1 - \dot{m}_2 s_2 - \dot{m}_3 s_3 + \dot{S}_{gen} = 0$$

$$\dot{m}_1 s_1 - 0.1\dot{m}_1 s_2 - 0.9\dot{m}_1 s_3 + \dot{S}_{gen} = 0 \rightarrow \dot{S}_{gen} = \dot{m}_1[0.9s_3 + 0.1s_2 - s_1]$$

and

$$X_{destroyed} = T_0 S_{gen} = T_0 \dot{m}_1[0.9s_3 + 0.1s_2 - s_1]$$

From the steam tables (Tables A-4 through 6)

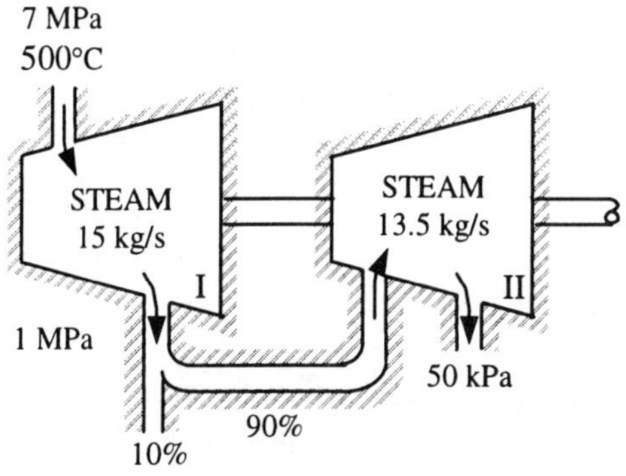

$$\left.\begin{array}{l} P_1 = 7\text{ MPa} \\ T_1 = 500°\text{C} \end{array}\right\} \begin{array}{l} h_1 = 3410.3\text{ kJ/kg} \\ s_1 = 6.7975\text{ kJ/kg}\cdot\text{K} \end{array}$$

$$\left.\begin{array}{l} P_2 = 1\text{ MPa} \\ s_{2s} = s_1 \end{array}\right\} h_{2s} = 2879.4\text{ kJ/kg}$$

7 MPa
500°C
STEAM
15 kg/s
I
STEAM
13.5 kg/s
II
1 MPa
50 kPa
10%
90%

and,

$$\eta_T = \frac{h_1 - h_2}{h_1 - h_{2s}} \longrightarrow \begin{array}{l} h_2 = h_1 - \eta_T(h_1 - h_{2s}) \\ = 3410.3 - 0.88(3410.3 - 2879.4) \\ = 2943.1\text{ kJ/kg} \end{array}$$

$$\left.\begin{array}{l} P_2 = 1\text{ MPa} \\ h_2 = 2943.1\text{ kJ/kg} \end{array}\right\} s_2 = 6.9256\text{ kJ/kg}\cdot\text{K}$$

$$\left.\begin{array}{l} P_3 = 50\text{ kPa} \\ s_{3s} = s_1 \end{array}\right\} \begin{array}{l} x_{3s} = \dfrac{s_{3s} - s_f}{s_{fg}} = \dfrac{6.7975 - 1.0910}{6.5029} = 0.8775 \\ h_{3s} = h_f + x_{3s}h_{fg} = 340.49 + 0.8775 \times 2305.4 = 2363.5\text{ kJ/kg} \end{array}$$

and,

$$\eta_T = \frac{h_1 - h_3}{h_1 - h_{3s}} \longrightarrow \begin{array}{l} h_3 = h_1 - \eta_T(h_1 - h_{3s}) \\ = 3410.3 - 0.88(3410.3 - 2363.5) \\ = 2489.1\text{ kJ/kg} \end{array}$$

$$\left.\begin{array}{l} P_3 = 50\text{ kPa} \\ h_3 = 2489.1\text{ kJ/kg} \end{array}\right\} \begin{array}{l} x_3 = \dfrac{h_3 - h_f}{h_{fg}} = \dfrac{2489.1 - 340.49}{2305.4} = 0.932 \\ s_3 = s_f + x_3 s_{fg} = 1.091 + 0.932 \times 6.5029 = 7.152\text{ kJ/kg}\cdot\text{K} \end{array}$$

Substituting, the wasted work potential is determined to be

$$X_{destroyed} = T_0 S_{gen} = (298\text{ K})(15\text{ kg/s})(0.9 \times 7.152 + 0.1 \times 6.9256 - 6.7975)\text{kJ/kg} = \mathbf{1483\ kW}$$

7-106 Steam expands in a two-stage adiabatic turbine from a specified state to another specified state. Steam is reheated between the stages. For a given power output, the reversible power output and the rate of exergy destruction are to be determined.

Assumptions **1** This is a steady-flow process since there is no change with time. **2** Kinetic and potential energy changes are negligible. **3** The turbine is adiabatic and thus heat transfer is negligible. **4** The environment temperature is given to be $T_0 = 25°C$.

Heat
2 MPa
350°C
2 MPa
500°C
Stage I
Stage II
5 MW
8 MPa
500°C
30 kPa
x = 97%

Properties From the steam tables (Tables A-4 through 6)

$$\left.\begin{array}{l} P_1 = 8\text{ MPa} \\ T_1 = 500°\text{C} \end{array}\right\} \begin{array}{l} h_1 = 3398.3\text{ kJ/kg} \\ s_1 = 6.7240\text{ kJ/kg}\cdot\text{K} \end{array}$$

$$\left.\begin{array}{l} P_2 = 2\text{ MPa} \\ T_2 = 350°\text{C} \end{array}\right\} \begin{array}{l} h_2 = 3137.0\text{ kJ/kg} \\ s_2 = 6.9563\text{ kJ/kg}\cdot\text{K} \end{array}$$

$$\left.\begin{array}{l} P_3 = 2\text{ MPa} \\ T_3 = 500°\text{C} \end{array}\right\} \begin{array}{l} h_3 = 3467.6\text{ kJ/kg} \\ s_3 = 7.4317\text{ kJ/kg}\cdot\text{K} \end{array}$$

$$\left.\begin{array}{l} P_4 = 30\text{ kPa} \\ x_4 = 0.97 \end{array}\right\} \begin{array}{l} h_4 = h_f + x_4 h_{fg} = 289.23 + 0.97\times 2336.1 = 2555.25\text{ kJ/kg} \\ s_4 = s_f + x_4 s_{fg} = 0.9439 + 0.97\times 6.8247 = 7.5639\text{ kJ/kg}\cdot\text{K} \end{array}$$

Analysis We take the entire turbine, excluding the reheat section, as the system, which is a control volume. The energy balance for this steady-flow system can be expressed in the rate form as

$$\underbrace{\dot{E}_{in} - \dot{E}_{out}}_{\substack{\text{Rate of net energy transfer}\\ \text{by heat, work, and mass}}} = \underbrace{\Delta \dot{E}_{\text{system}}^{\nearrow 0\text{ (steady)}}}_{\substack{\text{Rate of change in internal, kinetic,}\\ \text{potential, etc. energies}}} = 0$$

$$\dot{E}_{in} = \dot{E}_{out}$$

$$\dot{m}h_1 + \dot{m}h_3 = \dot{m}h_{2s} + \dot{m}h_{4s} + \dot{W}_{\text{s,out}}$$

$$\dot{W}_{\text{s,out}} = \dot{m}[(h_1 - h_2) + (h_3 - h_4)]$$

Substituting, the mass flow rate of the steam is determined from the steady-flow energy equation applied to the actual process,

$$\dot{m} = \frac{W_{out}}{h_1 - h_2 + h_3 - h_4} = \frac{5000\text{ kJ/s}}{(3398.3 - 3137.0 + 3467.6 - 2555.25)\text{kJ/kg}} = \mathbf{4.26\ kg/s}$$

The reversible (or maximum) power output is determined from the rate form of the exergy balance applied on the turbine and setting the exergy destruction term equal to zero,

$$\underbrace{\dot{X}_{in} - \dot{X}_{out}}_{\substack{\text{Rate of net exergy transfer}\\ \text{by heat, work, and mass}}} \underbrace{- \dot{X}_{\text{destroyed}}^{\nearrow 0\text{ (reversible)}}}_{\substack{\text{Rate of exergy}\\ \text{destruction}}} = \underbrace{\Delta \dot{X}_{\text{system}}^{\nearrow 0\text{ (steady)}}}_{\substack{\text{Rate of change}\\ \text{of exergy}}} = 0$$

$$\dot{X}_{in} = \dot{X}_{out}$$

$$\dot{m}\psi_1 + \dot{m}\psi_3 = \dot{m}\psi_2 + \dot{m}\psi_4 + \dot{W}_{rev,out}$$

$$\begin{aligned} \dot{W}_{rev,out} &= \dot{m}(\psi_1 - \psi_2) + \dot{m}(\psi_3 - \psi_4) \\ &= \dot{m}[(h_1 - h_2) + T_0(s_2 - s_1) - \Delta ke^{\nearrow 0} - \Delta pe^{\nearrow 0}] \\ &\quad + \dot{m}[(h_3 - h_4) + T_0(s_4 - s_3) - \Delta ke^{\nearrow 0} - \Delta pe^{\nearrow 0}] \end{aligned}$$

Then the reversible power becomes

$$
\begin{aligned}
\dot{W}_{rev,out} &= \dot{m}\left[h_1 - h_2 + h_3 - h_4 + T_0(s_2 - s_1 + s_4 - s_3)\right] \\
&= (4.26\text{ kg/s})[(3398.3 - 3137.0 + 3467.6 - 2555.25)\text{kJ/kg} \\
&\quad +(298\text{ K})(6.9563 - 6.7240 + 7.5639 - 7.4317)\text{kJ/kg}\cdot\text{K}] \\
&= \mathbf{5463\ kW}
\end{aligned}
$$

Then the rate of exergy destruction is determined from its definition,

$$
\dot{X}_{\text{destroyed}} = \dot{W}_{rev,out} - \dot{W}_{out} = 5463 - 5000 = \mathbf{463\ kW}
$$

7-107 One ton of liquid water at 80°C is brought into a room. The final equilibrium temperature in the room and the entropy generated are to be determined.

Assumptions **1** The room is well insulated and well sealed. **2** The thermal properties of water and air are constant at room temperature. **3** The system is stationary and thus the kinetic and potential energy changes are zero. **4** There are no work interactions involved.

Properties The gas constant of air is R = 0.287 kPa.m^3/kg.K (Table A-1). The specific heat of water at room temperature is C = 4.18 kJ/kg·°C (Table A-3).

Analysis The volume and the mass of the air in the room are

$$V = 4\text{x}5\text{x}6 = 120\ \text{m}^3$$

$$m_{air} = \frac{P_1 V_1}{RT_1} = \frac{(100\ \text{kPa})(120\ \text{m}^3)}{(0.2870\ \text{kPa}\cdot\text{m}^3/\text{kg}\cdot\text{K})(295\ \text{K})} = 141.7\ \text{kg}$$

4m x 5m x 6m

ROOM

22°C

100 kPa

Water

80°C

Heat

Taking the contents of the room, including the water, as our system, the energy balance can be written as

$$\underbrace{E_{in} - E_{out}}_{\substack{\text{Net energy transfer}\\ \text{by heat, work, and mass}}} = \underbrace{\Delta E_{\text{system}}}_{\substack{\text{Change in internal, kinetic,}\\ \text{potential, etc. energies}}} \rightarrow 0 = \Delta U = (\Delta U)_{\text{water}} + (\Delta U)_{\text{air}}$$

or

$$[mC(T_2 - T_1)]_{\text{water}} + [mC_v(T_2 - T_1)]_{\text{air}} = 0$$

Substituting,

$$(1000\ \text{kg})(4.18\ \text{kJ/kg}\cdot{}^\circ\text{C})(T_f - 80)^\circ\text{C} + (141.7\ \text{kg})(0.718\ \text{kJ/kg}\cdot{}^\circ\text{C})(T_f - 22)^\circ\text{C} = 0$$

It gives the final equilibrium temperature in the room to be

$$T_f = \mathbf{78.6°C}$$

(*b*) We again take the room and the water in it as the system, which is a closed system. Considering that the system is well-insulated and no mass is entering and leaving, the entropy balance for this system can be expressed as

$$\underbrace{S_{in} - S_{out}}_{\substack{\text{Net entropy transfer}\\ \text{by heat and mass}}} + \underbrace{S_{gen}}_{\substack{\text{Entropy}\\ \text{generation}}} = \underbrace{\Delta S_{\text{system}}}_{\substack{\text{Change}\\ \text{in entropy}}}$$

$$0 + S_{\text{gen}} = \Delta S_{\text{air}} + \Delta S_{\text{water}}$$

where

$$\Delta S_{air} = mC_v \ln\frac{T_2}{T_1} + mR\ln\frac{V_2}{V_1}^{\nearrow 0} = (141.7\ \text{kg})(0.718\ \text{kJ/kg}\cdot\text{K})\ln\frac{351.6\ \text{K}}{295\ \text{K}} = 17.86\ \text{kJ/K}$$

$$\Delta S_{water} = mC\ln\frac{T_2}{T_1} = (1000\ \text{kg})(4.18\ \text{kJ/kg}\cdot\text{K})\ln\frac{351.6\ \text{K}}{353\ \text{K}} = -16.63\ \text{kJ/K}$$

Substituting, the entropy generation is determined to be

$$S_{gen} = 17.86 - 16.63 = 1.23\ \text{kJ/K}$$

The exergy destroyed during a process can be determined from an exergy balance or directly from its definition $X_{\text{destroyed}} = T_0 S_{gen}$,

$$X_{\text{destroyed}} = T_0 S_{gen} = (293\ \text{K})(1.23\ \text{kJ/K}) = \mathbf{360\ kJ}$$

(c) The work potential (the maximum amount of work that can be produced) during a process is simply the reversible work output. Noting that the actual work for this process is zero, it becomes

$$X_{\text{destroyed}} = W_{rev,out} - W_{act,out} \rightarrow W_{rev,out} = X_{\text{destroyed}} = \mathbf{360\ kJ}$$

7-108 An insulated cylinder is divided into two parts. One side of the cylinder contains N_2 gas and the other side contains He gas at different states. The final equilibrium temperature in the cylinder and the wasted work potential are to be determined for the cases of piston being fixed and moving freely.

Assumptions **1** Both N_2 and He are ideal gases with constant specific heats. **2** The energy stored in the container itself is negligible. **3** The cylinder is well-insulated and thus heat transfer is negligible.

Properties The gas constants and the constant volume specific heats are R = 0.2968 kPa.m^3/kg.K is C_v = 0.743 kJ/kg· °C for N_2, and R = 2.0769 kPa.m^3/kg.K is C_v = 3.1156 kJ/kg· °C for He (Tables A-1 and A-2)

Analysis The mass of each gas in the cylinder is

$$m_{N_2} = \left(\frac{P_1V_1}{RT_1}\right)_{N_2} = \frac{(500\text{kPa})(1\text{m}^3)}{(0.2968\text{kPa}\cdot\text{m}^3/\text{kg}\cdot\text{K})(353\text{ K})} = 4.77\text{ kg}$$

$$m_{He} = \left(\frac{P_1V_1}{RT_1}\right)_{He} = \frac{(500\text{kPa})(1\text{m}^3)}{(2.0769\text{kPa}\cdot\text{m}^3/\text{kg}\cdot\text{K})(298\text{ K})} = 0.808\text{ kg}$$

N_2 1 m^3 500 kPa 80°C	He 1 m^3 500 kPa 25°C

Taking the entire contents of the cylinder as our system, the 1st law relation can be written as

$$\underbrace{E_{in} - E_{out}}_{\substack{\text{Net energy transfer}\\ \text{by heat, work, and mass}}} = \underbrace{\Delta E_{\text{system}}}_{\substack{\text{Change in internal, kinetic,}\\ \text{potential, etc. energies}}}$$

$$0 = \Delta U = (\Delta U)_{N_2} + (\Delta U)_{He}$$

$$0 = [mC_v(T_2 - T_1)]_{N_2} + [mC_v(T_2 - T_1)]_{He}$$

Substituting,

$$(4.77\text{kg})(0.743\text{kJ/kg}\cdot{}^\circ\text{C})(T_f - 80)^\circ\text{C} + (0.808\text{kg})(3.1156\text{kJ/kg}\cdot{}^\circ\text{C})(T_f - 25)^\circ\text{C} = 0$$

It gives $\quad$ $\mathbf{T_f = 57.2°C}$

where T_f is the final equilibrium temperature in the cylinder.

The answer would be the **same** if the piston were not free to move since it would effect only pressure, and not the specific heats.

(*b*) We take the entire cylinder as our system, which is a closed system. Noting that the cylinder is well-insulated and thus there is no heat transfer, the entropy balance for this closed system can be expressed as

$$\underbrace{S_{in} - S_{out}}_{\substack{\text{Net entropy transfer}\\ \text{by heat and mass}}} + \underbrace{S_{gen}}_{\substack{\text{Entropy}\\ \text{generation}}} = \underbrace{\Delta S_{\text{system}}}_{\substack{\text{Change}\\ \text{in entropy}}}$$

$$0 + S_{\text{gen}} = \Delta S_{N_2} + \Delta S_{He}$$

But first we determine the final pressure in the cylinder:

$$N_{total} = N_{N_2} + N_{He} = \left(\frac{m}{M}\right)_{N_2} + \left(\frac{m}{M}\right)_{He} = \frac{4.77\text{kg}}{28\text{kg/kmol}} + \frac{0.808\text{kg}}{4\text{kg/kmol}} = 0.372\text{kmol}$$

$$P_2 = \frac{N_{total}R_uT}{V_{total}} = \frac{(0.372\text{kmol})(8.314\text{kPa}\cdot\text{m}^3/\text{kmol}\cdot\text{K})(330.2\text{K})}{2\text{m}^3} = 510.6\text{kPa}$$

Then,

$$\Delta S_{N_2} = m\left(C_p \ln\frac{T_2}{T_1} - R\ln\frac{P_2}{P_1}\right)_{N_2}$$

$$= (4.77\text{kg})\left[(1.039\text{kJ/kg}\cdot\text{K})\ln\frac{330.2\text{K}}{353\text{K}} - (0.2968\text{kJ/kg}\cdot\text{K})\ln\frac{510.6\text{kPa}}{500\text{kPa}}\right] = -0.361\text{kJ/K}$$

$$\Delta S_{He} = m\left(C_p \ln\frac{T_2}{T_1} - R\ln\frac{P_2}{P_1}\right)_{He}$$

$$= (0.808\text{kg})\left[(5.1926\text{kJ/kg}\cdot\text{K})\ln\frac{330.2\text{K}}{298\text{ K}} - (2.0769\text{kJ/kg}\cdot\text{K})\ln\frac{510.6\text{kPa}}{500\text{kPa}}\right] = 0.395\text{kJ/K}$$

$$S_{gen} = \Delta S_{N_2} + \Delta S_{He} = -0.361 + 0.395 = 0.034\text{kJ/K}$$

The wasted work potential is equivalent to the exergy destroyed during a process, and it can be determined from an exergy balance or directly from its definition $X_{\text{destroyed}} = T_0 S_{gen}$,

$$X_{\text{destroyed}} = T_0 S_{gen} = (298\text{ K})(0.034\text{ kJ/K}) = \mathbf{10.1\text{ kJ}}$$

If the piston were not free to move, we would still have T_2 = 330.2 K but the volume of each gas would remain constant in this case:

$$\Delta S_{N_2} = m\left(C_v \ln\frac{T_2}{T_1} - R\ln\frac{V_2}{V_1}^{\nearrow 0}\right)_{N_2} = (4.77\text{kg})(0.743\text{kJ/kg}\cdot\text{K})\ln\frac{330.2\text{ K}}{353\text{ K}} = -0.237\text{ kJ/K}$$

$$\Delta S_{He} = m\left(C_v \ln\frac{T_2}{T_1} - R\ln\frac{V_2}{V_1}^{\nearrow 0}\right)_{He} = (0.808\text{kg})(3.1156\text{kJ/kg}\cdot\text{K})\ln\frac{330.2\text{K}}{298\text{ K}} = 0.258\text{ kJ/K}$$

$$S_{gen} = \Delta S_{N_2} + \Delta S_{He} = -0.237 + 0.258 = \mathbf{0.021\text{ kJ/K}}$$

and

$$X_{\text{destroyed}} = T_0 S_{gen} = (298\text{ K})(0.021\text{kJ/K}) = \mathbf{6.26\text{ kJ}}$$

7-109 An insulated cylinder is divided into two parts. One side of the cylinder contains N_2 gas and the other side contains He gas at different states. The final equilibrium temperature in the cylinder and the wasted work potential are to be determined for the cases of piston being fixed and moving freely. √

Assumptions **1** Both N_2 and He are ideal gases with constant specific heats. **2** The energy stored in the container itself, except the piston, is negligible. **3** The cylinder is well-insulated and thus heat transfer is negligible. **4** Initially, the piston is at the average temperature of the two gases.

Properties The gas constants and the constant volume specific heats are R = 0.2968 kPa.m³/kg.K is C_v = 0.743 kJ/kg· °C for N_2, and R = 2.0769 kPa.m³/kg.K is C_v = 3.1156 kJ/kg· °C for He (Tables A-1 and A-2). The specific heat of copper piston is C = 0.386 kJ/kg· °C (Table A-3).

Analysis The mass of each gas in the cylinder is

$$m_{N_2} = \left(\frac{P_1V_1}{RT_1}\right)_{N_2} = \frac{(500\text{kPa})(1\text{m}^3)}{(0.2968\text{kPa}\cdot\text{m}^3/\text{kg}\cdot\text{K})(353\text{K})} = 4.77\text{kg}$$

$$m_{He} = \left(\frac{P_1V_1}{RT_1}\right)_{He} = \frac{(500\text{kPa})(1\text{m}^3)}{(2.0769\text{kPa}\cdot\text{m}^3/\text{kg}\cdot\text{K})(353\text{K})} = 0.808\text{kg}$$

Taking the entire contents of the cylinder as our system, the 1st law relation can be written as

$$\underbrace{E_{in} - E_{out}}_{\substack{\text{Net energy transfer}\\ \text{by heat, work, and mass}}} = \underbrace{\Delta E_{\text{system}}}_{\substack{\text{Change in internal, kinetic,}\\ \text{potential, etc. energies}}}$$

$$0 = \Delta U = (\Delta U)_{\text{N}_2} + (\Delta U)_{\text{He}} + (\Delta U)_{\text{Cu}}$$

$$0 = [mC_v(T_2 - T_1)]_{\text{N}_2} + [mC_v(T_2 - T_1)]_{\text{He}} + [mC(T_2 - T_1)]_{\text{Cu}}$$

where

$$T_{1,\text{Cu}} = (80 + 25)/2 = 52.5°\text{C}$$

Substituting,

$$(4.77\text{kg})(0.743\text{kJ/kg}\cdot°\text{C})(T_f - 80)°\text{C} + (0.808\text{kg})(3.1156\text{kJ/kg}\cdot°\text{C})(T_f - 25)°\text{C}$$
$$+ (5.0\text{kg})(0.386\text{kJ/kg}\cdot°\text{C})(T_f - 52.5)°\text{C} = 0$$

It gives

$$T_f = \mathbf{56.0°C}$$

where T_f is the final equilibrium temperature in the cylinder.

The answer would be the **same** if the piston were not free to move since it would effect only pressure, and not the specific heats.

(*b*) We take the entire cylinder as our system, which is a closed system. Noting that the cylinder is well-insulated and thus there is no heat transfer, the entropy balance for this closed system can be expressed as

$$\underbrace{S_{in} - S_{out}}_{\substack{\text{Net entropy transfer}\\ \text{by heat and mass}}} + \underbrace{S_{gen}}_{\substack{\text{Entropy}\\ \text{generation}}} = \underbrace{\Delta S_{\text{system}}}_{\substack{\text{Change}\\ \text{in entropy}}}$$

$$0 + S_{\text{gen}} = \Delta S_{N_2} + \Delta S_{He} + \Delta S_{\text{piston}}$$

But first we determine the final pressure in the cylinder:

$$N_{total} = N_{N_2} + N_{He} = \left(\frac{m}{M}\right)_{N_2} + \left(\frac{m}{M}\right)_{He} = \frac{4.77\text{kg}}{28\text{kg/kmol}} + \frac{0.808\text{kg}}{4\text{kg/kmol}} = 0.372\text{kmol}$$

$$P_2 = \frac{N_{total}R_uT}{V_{total}} = \frac{(0.372\text{kmol})(8.314\text{kPa}\cdot\text{m}^3/\text{kmol}\cdot\text{K})(329\text{K})}{2\text{m}^3} = 508.8\text{kPa}$$

Then,

$$\Delta S_{N_2} = m\left(C_p \ln\frac{T_2}{T_1} - R\ln\frac{P_2}{P_1}\right)_{N_2}$$

$$= (4.77\text{kg})\left[(1.039\text{kJ/kg}\cdot\text{K})\ln\frac{329\text{K}}{353\text{K}} - (0.2968\text{kJ/kg}\cdot\text{K})\ln\frac{508.8\text{kPa}}{500\text{kPa}}\right] = -0.374\text{kJ/K}$$

$$\Delta S_{He} = m\left(C_p \ln\frac{T_2}{T_1} - R\ln\frac{P_2}{P_1}\right)_{He}$$

$$= (0.808\text{kg})\left[(5.1926\text{kJ/kg}\cdot\text{K})\ln\frac{329\text{K}}{353\text{K}} - (2.0769\text{kJ/kg}\cdot\text{K})\ln\frac{508.8\text{kPa}}{500\text{kPa}}\right] = 0.386\text{kJ/K}$$

$$\Delta S_{\text{piston}} = \left(mC\ln\frac{T_2}{T_1}\right)_{\text{piston}} = (5\text{kg})(0.386\text{kJ/kg}\cdot\text{K})\ln\frac{329\text{K}}{325.5\text{K}} = 0.021\text{kJ/K}$$

$$S_{gen} = \Delta S_{N_2} + \Delta S_{He} + \Delta S_{\text{piston}} = -0.374 + 0.386 + 0.021 = \mathbf{0.0334\ \ kJ/K}$$

The wasted work potential is equivalent to the exergy destroyed during a process, and it can be determined from an exergy balance or directly from its definition $X_{\text{destroyed}} = T_0 S_{gen}$,

$$X_{\text{destroyed}} = T_0 S_{gen} = (298\text{ K})(0.033\text{ kJ/K}) = \mathbf{9.83\ kJ}$$

If the piston were not free to move, we would still have T_2 = 330.2 K but the volume of each gas would remain constant in this case:

$$\Delta S_{N_2} = m\left(C_v \ln\frac{T_2}{T_1} - R\ln\frac{V_2}{V_1}^{\nearrow 0}\right)_{N_2} = (4.77\text{kg})(0.743\text{kJ/kg}\cdot\text{K})\ln\frac{329\text{K}}{353\text{K}} = -0.250\text{kJ/K}$$

$$\Delta S_{He} = m\left(C_v \ln\frac{T_2}{T_1} - R\ln\frac{V_2}{V_1}^{\nearrow 0}\right)_{He} = (0.808\text{kg})(3.1156\text{kJ/kg}\cdot\text{K})\ln\frac{329\text{K}}{353\text{K}} = 0.249\text{kJ/K}$$

$$S_{gen} = \Delta S_{N_2} + \Delta S_{He} + \Delta S_{piston} = -0.250 + 0.249 + 0.021 = \mathbf{0.020kJ/K}$$

and

$$X_{\text{destroyed}} = T_0 S_{gen} = (298\text{ K})(0.020\text{ kJ/K}) = \mathbf{6.0\ kJ}$$

7-110E Argon enters an adiabatic turbine at a specified state with a specified mass flow rate, and leaves at a specified pressure. The isentropic efficiency of turbine is to be determined.

Assumptions **1** This is a steady-flow process since there is no change with time. **2** Kinetic and potential energy changes are negligible. **3** The device is adiabatic and thus heat transfer is negligible. **4** Argon is an ideal gas with constant specific heats.

Properties The specific heat ratio of argon is k = 1.667. The constant pressure specific heat of argon is C_p = 0.1253 Btu/lbm.R (Table A-2E).

Analysis There is only one inlet and one exit, and thus $\dot{m}_1 = \dot{m}_2 = \dot{m}$. We take the isentropic turbine as the system, which is a control volume since mass crosses the boundary. The energy balance for this steady-flow system can be expressed in the rate form as

$$\underbrace{\dot{E}_{in} - \dot{E}_{out}}_{\text{Rate of net energy transfer by heat, work, and mass}} = \underbrace{\Delta \dot{E}_{\text{system}}^{\nearrow 0\ (\text{steady})}}_{\text{Rate of change in internal, kinetic, potential, etc. energies}} = 0$$

$$\dot{E}_{in} = \dot{E}_{out}$$

$$\dot{m}h_1 = \dot{W}_{s,\text{out}} + \dot{m}h_{2s} \quad (\text{since } \dot{Q} \cong \Delta ke \cong \Delta pe \cong 0)$$

$$\dot{W}_{s,\text{out}} = \dot{m}(h_1 - h_{2s})$$

1

370 kW

Ar

η_T

2

From the isentropic relations,

$$T_{2s} = T_1\left(\frac{P_{2s}}{P_1}\right)^{(k-1)/k} = (1960\ \text{R})\left(\frac{30\ \text{psia}}{200\ \text{psia}}\right)^{0.667/1.667} = 917.5\ \text{R}$$

Then the power output of the isentropic turbine becomes

$$\dot{W}_{s,out} = \dot{m}C_p(T_1 - T_{2s})$$

$$= (40\ \text{lbm/min})(0.1253\ \text{Btu/lbm}\cdot\text{R})(1960 - 917.5)\text{R}\left(\frac{1\ \text{hp}}{42.41\ \text{Btu/min}}\right)$$

$$= 123.2\ \text{hp}$$

Then the isentropic efficiency of the turbine is determined from

$$\eta_T = \frac{\dot{W}_{a,out}}{\dot{W}_{s,out}} = \frac{95\ \text{hp}}{123.2\ \text{hp}} = 0.771 = \mathbf{77.1\%}$$

(*b*) Using the steady-flow energy balance relation $\dot{W}_{a,out} = \dot{m}C_p(T_1 - T_2)$ above, the actual turbine exit temperature is determined to be

$$T_2 = T_1 - \frac{\dot{W}_{a,out}}{\dot{m}C_p} = 1500 - \frac{95\ \text{hp}}{(40\ \text{lbm/min})(0.1253\ \text{Btu/lbm}\cdot\text{R})}\left(\frac{42.41\ \text{Btu/min}}{1\ \text{hp}}\right) = 696.1°\text{F} = 1156.1\ \text{R}$$

The entropy generation during this process can be determined from an entropy balance on the turbine,

$$\underbrace{\dot{S}_{in} - \dot{S}_{out}}_{\text{Rate of net entropy transfer by heat and mass}} + \underbrace{\dot{S}_{gen}}_{\text{Rate of entropy generation}} = \underbrace{\Delta \dot{S}_{\text{system}}^{\nearrow 0}}_{\text{Rate of change of entropy}} = 0$$

$$\dot{m}s_1 - \dot{m}s_2 + \dot{S}_{\text{gen}} = 0$$

$$\dot{S}_{\text{gen}} = \dot{m}(s_2 - s_1)$$

where

$$s_2 - s_1 = C_p \ln\frac{T_2}{T_1} - R\ln\frac{P_2}{P_1} = (0.1253\ \text{Btu/lbm}\cdot\text{R})\ln\frac{1156.1\ \text{R}}{1960\ \text{R}} - (0.04971\ \text{Btu/lbm}\cdot\text{R})\ln\frac{30\ \text{psia}}{200\ \text{psia}}$$

$$= 0.02816\ \text{Btu/lbm.R}$$

The exergy destroyed during a process can be determined from an exergy balance or directly from its definition $X_{\text{destroyed}} = T_0 S_{gen}$,

$$\dot{X}_{\text{destroyed}} = T_0 \dot{S}_{gen} = \dot{m} T_0 (s_2 - s_1)$$

$$= (40\ \text{lbm/min})(537\ \text{R})(0.02816\ \text{Btu/lbm}\cdot\text{R})\left(\frac{1\ \text{hp}}{42.41\ \text{Btu/min}}\right) = 14.3\ \text{hp}$$

Then the reversible power and second-law efficiency become

$$\dot{W}_{rev,out} = \dot{W}_{a,out} + \dot{X}_{\text{destroyed}} = 95 + 14.3 = 109.3\ \text{hp}$$

and

$$\eta_{II} = \frac{\dot{W}}{\dot{W}_{rev}} = \frac{95\ \text{hp}}{109.3\ \text{hp}} = \mathbf{86.9\%}$$

7-111 [*Also solved by EES on enclosed CD*] The feedwater of a steam power plant is preheated using steam extracted from the turbine. The ratio of the mass flow rates of the extracted steam and the feedwater are to be determined.

Assumptions **1** This is a steady-flow process since there is no change with time. **2** Kinetic and potential energy changes are negligible. **3** Heat loss from the device to the surroundings is negligible and thus heat transfer from the hot fluid is equal to the heat transfer to the cold fluid.

Properties The properties of steam and feedwater are (Tables A-4 through A-6)

$$\left.\begin{array}{l} P_1 = 1\text{MPa} \\ T_1 = 200^\circ\text{C} \end{array}\right\} \begin{array}{l} h_1 = 2827.9\text{kJ/kg} \\ s_1 = 6.6940\text{kJ/kg}\cdot\text{K} \end{array}$$

$$\left.\begin{array}{l} P_2 = 1\text{MPa} \\ sat.liquid \end{array}\right\} \begin{array}{l} h_2 = h_{f@1\text{MPa}} = 762.81\text{kJ/kg} \\ s_2 = s_{f@1\text{MPa}} = 2.1387\text{kJ/kg}\cdot\text{K} \\ T_2 = 179.91^\circ\text{C} \end{array}$$

$$\left.\begin{array}{l} P_3 = 2.5\text{MPa} \\ T_3 = 50^\circ\text{C} \end{array}\right\} \begin{array}{l} h_3 \cong h_{f@50^\circ\text{C}} = 209.33\text{kJ/kg} \\ s_3 \cong s_{f@50^\circ\text{C}} = 0.7038\text{kJ/kg}\cdot\text{K} \end{array}$$

$$\left.\begin{array}{l} P_4 = 2.5\text{MPa} \\ T_4 = T_2 - 10^\circ\text{C} \cong 170^\circ\text{C} \end{array}\right\} \begin{array}{l} h_4 \cong h_{f@170^\circ\text{C}} = 719.2\text{kJ/kg} \\ s_4 \cong s_{f@170^\circ\text{C}} = 2.0419\text{kJ/kg}\cdot\text{K} \end{array}$$

Steam from turbine ① 1 MPa 200°C; Feedwater ③ 2.5 MPa; ④; ② sat. liquid

Analysis (*a*) We take the heat exchanger as the system, which is a control volume. The mass and energy balances for this steady-flow system can be expressed in the rate form as follows:

Mass balance (for each fluid stream):

$$\dot{m}_{in} - \dot{m}_{out} = \Delta\dot{m}_{system}^{\nearrow 0\ (\text{steady})} = 0 \rightarrow \dot{m}_{in} = \dot{m}_{out} \rightarrow \dot{m}_1 = \dot{m}_2 = \dot{m}_s \quad \text{and} \quad \dot{m}_3 = \dot{m}_4 = \dot{m}_{fw}$$

Energy balance (for the heat exchanger):

$$\underbrace{\dot{E}_{in} - \dot{E}_{out}}_{\substack{\text{Rate of net energy transfer}\\ \text{by heat, work, and mass}}} = \underbrace{\Delta\dot{E}_{system}^{\nearrow 0\ (\text{steady})}}_{\substack{\text{Rate of change in internal, kinetic,}\\ \text{potential, etc. energies}}} = 0$$

$$\dot{E}_{in} = \dot{E}_{out}$$

$$\dot{m}_1 h_1 + \dot{m}_3 h_3 = \dot{m}_2 h_2 + \dot{m}_4 h_4 \quad (\text{since } \dot{Q} = \dot{W} = \Delta\text{ke} \cong \Delta\text{pe} \cong 0)$$

Combining the two, $\dot{m}_s (h_2 - h_1) = \dot{m}_{fw} (h_3 - h_4)$

Dividing by $\dot{m}_{fw}$ and substituting,

$$\frac{\dot{m}_s}{\dot{m}_{fw}} = \frac{h_3 - h_4}{h_2 - h_1} = \frac{(209.33 - 719.2)\text{kJ/kg}}{(762.81 - 2827.9)\text{kJ/kg}} = \mathbf{0.247}$$

(*b*) The entropy generation during this process per unit mass of feedwater can be determined from an entropy balance on the feedwater heater expressed in the rate form as

$$\underbrace{\dot{S}_{in} - \dot{S}_{out}}_{\substack{\text{Rate of net entropy transfer}\\ \text{by heat and mass}}} + \underbrace{\dot{S}_{gen}}_{\substack{\text{Rate of entropy}\\ \text{generation}}} = \underbrace{\Delta\dot{S}_{system}^{\nearrow 0}}_{\substack{\text{Rate of change}\\ \text{of entropy}}} = 0$$

$$\dot{m}_1 s_1 - \dot{m}_2 s_2 + \dot{m}_3 s_3 - \dot{m}_4 s_4 + \dot{S}_{gen} = 0$$

$$\dot{m}_s (s_1 - s_2) + \dot{m}_{fw} (s_3 - s_4) + \dot{S}_{gen} = 0$$

$$\frac{\dot{S}_{gen}}{\dot{m}_{fw}} = \frac{\dot{m}_s}{\dot{m}_{fw}}(s_2 - s_1) + (s_4 - s_3) = (0.247)(2.1387 - 6.694) + (2.0419 - 0.7038) = \mathbf{0.213\ kJ/K \cdot kgfw}$$

Noting that this process involves no actual work, the reversible work and exergy destruction become equivalent since $X_{\text{destroyed}} = W_{rev,out} - W_{act,out} \rightarrow W_{rev,out} = X_{\text{destroyed}}$. The exergy destroyed during a process can be determined from an exergy balance or directly from its definition $X_{\text{destroyed}} = T_0 S_{gen}$,

$$X_{\text{destroyed}} = T_0 S_{gen} = (298\text{ K})(0.213\text{ kJ/K} \cdot \text{kgfw}) = \mathbf{63.5\ kJ/kgfeedwater}$$

7-112 EES solution of this (and other comprehensive problems designated with the *computer icon*) is available to instructors at the *Instructor Manual* section of the *Online Learning Center* (OLC) at www.mhhe.com/cengel-boles. See the Preface for access information.

7-113 A 1-ton (1000 kg) of water is to be cooled in a tank by pouring ice into it. The final equilibrium temperature in the tank and the exergy destruction are to be determined.

Assumptions **1** Thermal properties of the ice and water are constant. **2** Heat transfer to the water tank is negligible. **3** There is no stirring by hand or a mechanical device (it will add energy).

Properties The specific heat of water at room temperature is $C = 4.18$ kJ/kg· °C, and the specific heat of ice at about 0°C is $C = 2.11$ kJ/kg· °C (Table A-3). The melting temperature and the heat of fusion of ice at 1 atm are 0°C and 333.7 kJ/kg..

Analysis (*a*) We take the ice and the water as the system, and disregard any heat transfer between the system and the surroundings. Then the energy balance for this process can be written as

$$\underbrace{E_{in} - E_{out}}_{\substack{\text{Net energy transfer}\\ \text{by heat, work, and mass}}} = \underbrace{\Delta E_{\text{system}}}_{\substack{\text{Change in internal, kinetic,}\\ \text{potential, etc. energies}}}$$

$$0 = \Delta U$$

$$0 = \Delta U_{\text{ice}} + \Delta U_{\text{water}}$$

$$[mC(0^\circ\text{C} - T_1)_{\text{solid}} + mh_{if} + mC(T_2 - 0^\circ\text{C})_{\text{liquid}}]_{\text{ice}} + [mC(T_2 - T_1)]_{\text{water}} = 0$$

Substituting,

$$(80\text{ kg})\{(2.11\text{ kJ/kg·}^\circ\text{C})[0-(-5)]^\circ\text{C} + 333.7\text{ kJ/kg} + (4.18\text{ kJ/kg·}^\circ\text{C})(T_2 - 0)^\circ\text{C}\}$$
$$+(1000\text{ kg})(4.18\text{ kJ/kg·}^\circ\text{C})(T_2 - 20)^\circ\text{C} = 0$$

ice
-5°C
80 kg

WATER
1 ton

It gives $T_2 =$ **12.4°C**

which is the final equilibrium temperature in the tank.

(*b*) We take the ice and the water as our system, which is a closed system .Considering that the tank is well-insulated and thus there is no heat transfer, the entropy balance for this closed system can be expressed as

$$\underbrace{S_{in} - S_{out}}_{\substack{\text{Net entropy transfer}\\ \text{by heat and mass}}} + \underbrace{S_{gen}}_{\substack{\text{Entropy}\\ \text{generation}}} = \underbrace{\Delta S_{\text{system}}}_{\substack{\text{Change}\\ \text{in entropy}}}$$

$$0 + S_{\text{gen}} = \Delta S_{ice} + \Delta S_{water}$$

where

$$\Delta S_{water} = \left(mC\ln\frac{T_2}{T_1}\right)_{water} = (1000\text{kg})(4.18\text{kJ/kg}\cdot\text{K})\ln\frac{285.4\text{K}}{293\text{K}} = -110.0\text{kJ/K}$$

$$\Delta S_{ice} = (\Delta S_{solid} + \Delta S_{melting} + \Delta S_{liquid})_{ice}$$

$$= \left(\left(mC\ln\frac{T_2}{T_1}\right)_{solid} + \frac{mh_{ig}}{T_{melting}} + \left(mC\ln\frac{T_2}{T_1}\right)_{liquid}\right)_{ice}$$

$$= (80\text{kg})\left((2.11\text{kJ/kg}\cdot\text{K})\ln\frac{273\text{K}}{268\text{K}} + \frac{333.7\text{kJ/kg}}{273\text{K}} + (4.18\text{kJ/kg}\cdot\text{K})\ln\frac{285.4\text{K}}{273\text{K}}\right)$$

$$= 115.8\text{kJ/K}$$

Then,

$$S_{gen} = \Delta S_{water} + \Delta S_{ice} = -110.0 + 115.8 = 5.8\text{ kJ/K}$$

The exergy destroyed during a process can be determined from an exergy balance or directly from its definition $X_{\text{destroyed}} = T_0 S_{gen}$,

$$X_{\text{destroyed}} = T_0 S_{gen} = (293\text{ K})(5.80\text{ kJ/K}) = \mathbf{1699\ kJ}$$

7-114 An evacuated bottle is surrounded by atmospheric air. A valve is opened, and air is allowed to fill the bottle. The amount of heat transfer through the wall of the bottle when thermal and mechanical equilibrium is established and the amount of exergy destroyed are to be determined.

Assumptions **1** This is an unsteady process since the conditions within the device are changing during the process, but it can be analyzed as a uniform-flow process since the state of fluid at the inlet remains constant. **2** Air is an ideal gas. **3** Kinetic and potential energies are negligible. **4** There are no work interactions involved. **5** The direction of heat transfer is to the air in the bottle (will be verified).

Properties The gas constant of air is 0.287 kPa.m^3/kg.K (Table A-1).

Analysis We take the bottle as the system, which is a control volume since mass crosses the boundary. Noting that the microscopic energies of flowing and nonflowing fluids are represented by enthalpy h and internal energy u, respectively, the mass and energy balances can be expressed as

Mass balance: $m_{in} - m_{out} = \Delta m_{\text{system}} \rightarrow m_i = m_2 \quad (\text{since } m_{out} = m_{initial} = 0)$

Energy balance:

$$\underbrace{E_{in} - E_{out}}_{\substack{\text{Net energy transfer}\\ \text{by heat, work, and mass}}} = \underbrace{\Delta E_{\text{system}}}_{\substack{\text{Change in internal, kinetic,}\\ \text{potential, etc. energies}}}$$

$$Q_{in} + m_i h_i = m_2 u_2 \quad (\text{since } W \cong E_{out} = E_{initial} = ke \cong pe \cong 0)$$

Combining the two balances:

$$Q_{in} = m_2 (u_2 - h_i)$$

where

$$m_2 = \frac{P_2 V}{R T_2} = \frac{(100 \text{ kPa})(0.012 \text{ m}^3)}{(0.287 \text{ kPa}\cdot\text{m}^3/\text{kg}\cdot\text{K})(290 \text{ K})} = 0.0144 \text{ kg}$$

$$T_i = T_2 = 290 \text{ K} \xrightarrow{\text{Table A-17}} \begin{array}{l} h_i = 290.16 \text{ kJ/kg} \\ u_2 = 206.91 \text{ kJ/kg} \end{array}$$

100kPa
17°C

12 L
Evacuated

Substituting,

Q_{in} = (0.0144 kg)(206.91 - 290.16) kJ/kg = - 1.2 kJ $\rightarrow$ Q_{out} = **1.2 kJ**

Note that the negative sign for heat transfer indicates that the assumed direction is wrong. Therefore, we reversed the direction.

The entropy generated during this process is determined by applying the entropy balance on an *extended system* that includes the bottle and its immediate surroundings so that the boundary temperature of the extended system is the temperature of the surroundings at all times. The entropy balance for it can be expressed as

$$\underbrace{S_{in} - S_{out}}_{\substack{\text{Net entropy transfer}\\ \text{by heat and mass}}} + \underbrace{S_{gen}}_{\substack{\text{Entropy}\\ \text{generation}}} = \underbrace{\Delta S_{\text{system}}}_{\substack{\text{Change}\\ \text{in entropy}}}$$

$$m_i s_i - \frac{Q_{out}}{T_{b,in}} + S_{\text{gen}} = \Delta S_{\text{tank}} = m_2 s_2 - m_1 s_1^{\nearrow 0} = m_2 s_2$$

Therefore, the total entropy generated during this process is

$$S_{gen} = -m_i s_i + m_2 s_2 + \frac{Q_{\text{out}}}{T_{\text{b,out}}} = m_2 (s_2 - s_i)^{\nearrow 0} + \frac{Q_{\text{out}}}{T_{\text{b,out}}} = \frac{Q_{out}}{T_{surr}} = \frac{1.2 \text{ kJ}}{290 \text{ K}} = 0.00415 \text{ kJ/K}$$

The exergy destroyed during a process can be determined from an exergy balance or directly from its definition $X_{\text{destroyed}} = T_0 S_{gen}$,

$$X_{\text{destroyed}} = T_0 S_{gen} = (290 \text{ K})(0.00415 \text{ kJ/K}) = \mathbf{1.2 \text{ kJ}}$$

7-115 A heat engine operates between two tanks filled with air at different temperatures. The maximum work that can be produced and the final temperatures of the tanks are to be determined.

Assumptions Air is an ideal gas with constant specific heats at room temperature.

Properties The gas constant of air is 0.287 kPa.m^3/kg.K (Table A-1). The constant volume specific heat of air at room temperature is C_v = 0.718 kJ/kg.K (Table A-2).

Analysis For maximum power production, the entropy generation must be zero. We take the two tanks (the heat source and heat sink) and the heat engine as the system. Noting that the system involves no heat and mass transfer and that the entropy change for cyclic devices is zero, the entropy balance can be expressed as

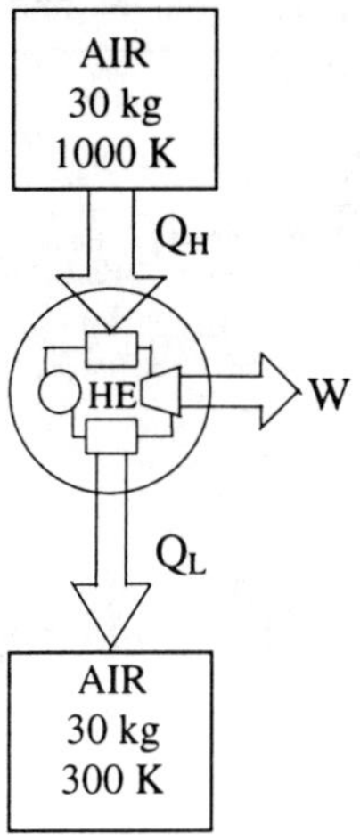

$$\underbrace{S_{in} - S_{out}}_{\substack{\text{Net entropy transfer}\\ \text{by heat and mass}}} + \underbrace{S_{gen}^{\nearrow 0}}_{\substack{\text{Entropy}\\ \text{generation}}} = \underbrace{\Delta S_{\text{system}}}_{\substack{\text{Change}\\ \text{in entropy}}}$$

$$0 + S_{gen}^{\nearrow 0} = \Delta S_{\text{tank,source}} + \Delta S_{\text{tank,sink}} + \Delta S_{\text{heat engine}}^{\nearrow 0}$$

$$\Delta S_{\text{tank,source}} + \Delta S_{\text{tank,sink}} = 0$$

$$\left(mC_v \ln\frac{T_2}{T_1} + mR\ln\frac{V_2}{V_1}^{\nearrow 0} \right)_{\text{source}} + \left(mC_v \ln\frac{T_2}{T_1} + mR\ln\frac{V_2}{V_1}^{\nearrow 0} \right)_{\text{sink}} = 0$$

$$\ln\frac{T_2}{T_{1A}}\frac{T_2}{T_{1B}} = 0 \longrightarrow T_2^2 = T_{1A}T_{1B}$$

AIR
30 kg
1000 K

Q_H

HE

W

Q_L

AIR
30 kg
300 K

where T_{1A} and T_{1B} are the initial temperatures of the source and the sink, respectively, and T_2 is the common final temperature. Therefore, the final temperature of the tanks for maximum power production is

$$T_2 = \sqrt{T_{1A}T_{1B}} = \sqrt{(1000\text{ K})(300\text{ K})} = \mathbf{547.7\ K}$$

The energy balance $E_{in} - E_{out} = \Delta E_{\text{system}}$ for the source and sink can be expressed as follows:

Source: $\quad -Q_{\text{source,out}} = \Delta U = mC_v(T_2 - T_{1A}) \rightarrow Q_{\text{source,out}} = mC_v(T_{1A} - T_2)$

$$Q_{\text{source,out}} = mC_v(T_{1A} - T_2) = (30\text{ kg})(0.718\text{ kJ/kg}\cdot\text{K})(1000 - 547.7)\text{K} = 9743\text{ kJ}$$

Sink: $\quad Q_{\text{sink,in}} = mC_v(T_2 - T_{1A}) = (30\text{ kg})(0.718\text{ kJ/kg}\cdot\text{K})(547.7 - 300)\text{K} = 5336\text{ kJ}$

Then the work produced in this case becomes

$$W_{\text{max,out}} = Q_H - Q_L = Q_{\text{source,out}} - Q_{\text{sink,in}} = 9743 - 5336 = \mathbf{4407\ kJ}$$

Therefore, a maximum of 4407 kJ of work can be produced during this process.

7-116 A heat engine operates between two constant-pressure cylinders filled with air at different temperatures. The maximum work that can be produced and the final temperatures of the cylinders are to be determined.

Assumptions Air is an ideal gas with constant specific heats at room temperature.

Properties The gas constant of air is 0.287 kPa.m^3/kg.K (Table A-1). The constant pressure specific heat of air at room temperature is $C_p = 1.005$ kJ/kg.K (Table A-2).

Analysis For maximum power production, the entropy generation must be zero. We take the two cylinders (the heat source and heat sink) and the heat engine as the system. Noting that the system involves no heat and mass transfer and that the entropy change for cyclic devices is zero, the entropy balance can be expressed as

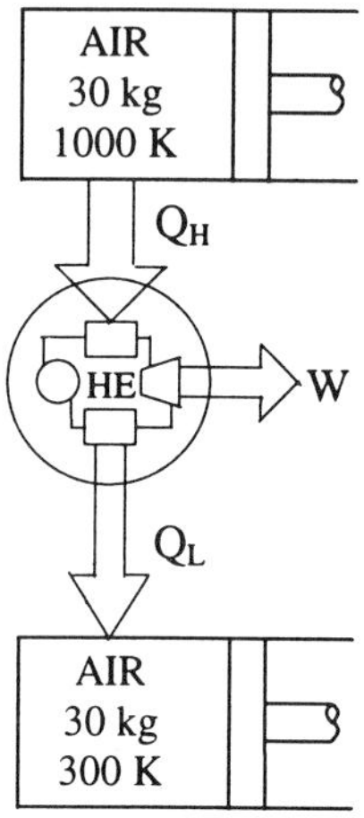

$$\underbrace{S_{in} - S_{out}}_{\substack{\text{Net entropy transfer}\\ \text{by heat and mass}}} + \underbrace{S_{gen}^{\nearrow 0}}_{\substack{\text{Entropy}\\ \text{generation}}} = \underbrace{\Delta S_{\text{system}}}_{\substack{\text{Change}\\ \text{in entropy}}}$$

$$0 + S_{gen}^{\nearrow 0} = \Delta S_{\text{cylinder,source}} + \Delta S_{\text{cylinder,sink}} + \Delta S_{\text{heat engine}}^{\nearrow 0}$$

$$\Delta S_{\text{cylinder,source}} + \Delta S_{\text{cylinder,sink}} = 0$$

$$\left(mC_v \ln\frac{T_2}{T_1} - mR\ln\frac{P_2}{P_1}^{\nearrow 0} \right)_{source} + 0 + \left(mC_v \ln\frac{T_2}{T_1} - mR\ln\frac{P_2}{P_1}^{\nearrow 0} \right)_{\text{sink}} = 0$$

$$\ln\frac{T_2}{T_{1A}}\frac{T_2}{T_{1B}} = 0 \longrightarrow T_2^2 = T_{1A}T_{1B}$$

AIR
30 kg
1000 K
Q_H
HE
W
Q_L
AIR
30 kg
300 K

where T_{1A} and T_{1B} are the initial temperatures of the source and the sink, respectively, and T_2 is the common final temperature. Therefore, the final temperature of the tanks for maximum power production is

$$T_2 = \sqrt{T_{1A}T_{1B}} = \sqrt{(1000 \text{ K})(300 \text{ K})} = \mathbf{547.7\ K}$$

The energy balance $E_{in} - E_{out} = \Delta E_{\text{system}}$ for the source and sink can be expressed as follows:

Source: $\quad -Q_{\text{source,out}} + W_{b,in} = \Delta U \rightarrow Q_{\text{source,out}} = \Delta H = mC_p(T_{1A} - T_2)$

$$Q_{\text{source,out}} = mC_p(T_{1A} - T_2) = (30\text{ kg})(1.005\text{ kJ/kg}\cdot\text{K})(1000 - 547.7)\text{K} = 13{,}640\text{ kJ}$$

Sink: $\quad Q_{\text{sink,in}} - W_{b,out} = \Delta U \rightarrow Q_{\text{sink,in}} = \Delta H = mC_p(T_2 - T_{1A})$

$$Q_{\text{sink,in}} = mC_v(T_2 - T_{1B}) = (30\text{ kg})(1.005\text{ kJ/kg}\cdot\text{K})(547.7 - 300)\text{K} = 7470\text{ kJ}$$

Then the work produced becomes

$$W_{\text{max,out}} = Q_H - Q_L = Q_{\text{source,out}} - Q_{\text{sink,in}} = 13{,}640 - 7470 = \mathbf{6170\ kJ}$$

Therefore, a maximum of 6170 kJ of work can be produced during this process

7-117 A pressure cooker is initially half-filled with liquid water. It is kept on the heater for 30 min. The amount water that remained in the cooker and the exergy destroyed are to be determined.

Assumptions **1** This is an unsteady process since the conditions within the device are changing during the process, but it can be analyzed as a uniform-flow process since the state of water vapor leaving the cooker remains constant. **2** Kinetic and potential energies are negligible. **3** Heat loss from the cooker is negligible.

Properties The properties of water are (Tables A-4 through A-6)

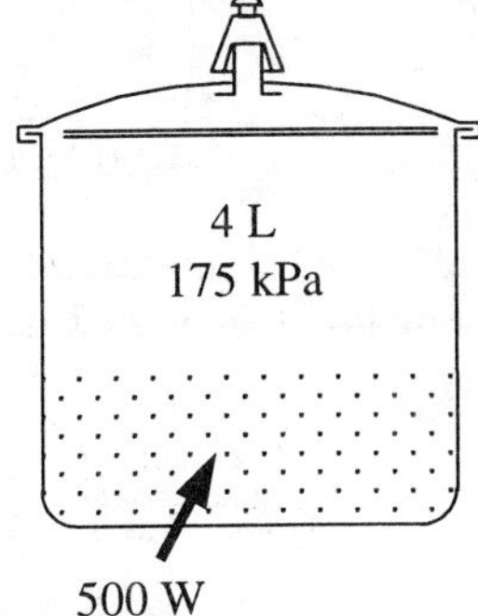

$$P_1 = 175\ kPa \quad \rightarrow \quad \begin{array}{l} v_f = 0.001057\ \text{m}^3/\text{kg},\ v_g = 1.0036\ \text{m}^3/\text{kg} \\ u_f = 486.8\ \text{kJ/kg},\ u_g = 2524.9\ \text{kJ/kg} \\ s_f = 1.4849\ \text{kJ/kg.K},\ s_g = 7.1717\ \text{kJ/kg.K} \end{array}$$

$$\left.\begin{array}{l} P_e = 175\ \text{kPa} \\ \text{sat.vapor} \end{array}\right\} \begin{array}{l} h_e = h_{g\ @\ 175\ \text{kPa}} = 2700.6\ \text{kJ/kg} \\ s_e = s_{g\ @\ 175\ \text{kPa}} = 7.1717\ \text{kJ/kg}\cdot\text{K} \end{array}$$

Analysis We take the cooker as the system, which is a control volume since mass crosses the boundary. Noting that the microscopic energies of flowing and nonflowing fluids are represented by enthalpy h and internal energy u, respectively, the mass and energy balances for this uniform-flow system can be expressed as

Mass balance: $$m_{in} - m_{out} = \Delta m_{\text{system}} \quad \rightarrow \quad m_e = m_1 - m_2$$

Energy balance:

$$\underbrace{E_{in} - E_{out}}_{\substack{\text{Net energy transfer} \\ \text{by heat, work, and mass}}} = \underbrace{\Delta E_{\text{system}}}_{\substack{\text{Change in internal, kinetic,} \\ \text{potential, etc. energies}}}$$

$$W_{e,in} - m_e h_e = m_2 u_2 - m_1 u_1 \quad (\text{since } Q \cong ke \cong pe \cong 0)$$

The initial mass, initial internal energy, initial entropy, and final mass in the tank are

$$V_f = V_g = 2\ \text{L} = 0.002\ \text{m}^3$$

$$m_1 = m_f + m_g = \frac{V_f}{v_f} + \frac{V_g}{v_g} = \frac{0.002\ \text{m}^3}{0.001057\ \text{m}^3/\text{kg}} + \frac{0.002\ \text{m}^3}{1.0036\ \text{m}^3/\text{kg}} = 1.892 + 0.002 = 1.8945\ \text{kg}$$

$$U_1 = m_1 u_1 = m_f u_f + m_g u_g = 1.892 \times 486.8 + 0.002 \times 2524.9 = 926.1\ \text{kJ}$$

$$S_1 = m_1 s_1 = m_f s_f + m_g s_g = 1.892 \times 1.4849 + 0.002 \times 7.1717 = 2.8238\ \text{kJ/K}$$

$$m_2 = \frac{V}{v_2} = \frac{0.004\ \text{m}^3}{v_2}$$

The amount of electrical energy supplied during this process is

$$W_{e,in} = \dot{W}_{e,in} \Delta t = (0.5\ \text{kJ/s})(30 \times 60\ \text{s}) = 900\ \text{kJ}$$

Then from the mass and energy balances,

$$m_e = m_1 - m_2 = 1.894 - \frac{0.004}{v_2}$$

$$900\ \text{kJ} = (1.894 - \frac{0.004}{v_2})(2700.6\ \text{kJ/kg}) + (\frac{0.004}{v_2})(u_2) - 926.1\ \text{kJ}$$

Substituting $u_2 = u_f + x_2 u_{fg}$ and $v_2 = v_f + x_2 v_{fg}$, and solving for x_2 yields

$$x_2 = 0.00163$$

Thus,

$$v_2 = v_f + x_2 v_{fg} = 0.001057 + 0.00163 \times (1.0036 - 0.001057) = 0.00269 \text{ m}^3/\text{kg}$$

$$s_2 = s_f + x_2 s_{fg} = 1.4849 + 0.00163 \times 5.6868 = 1.494 \text{ kJ/kg} \cdot \text{K}$$

and

$$m_2 = \frac{V}{v_2} = \frac{0.004 \text{ m}^3}{0.00269 \text{ m}^3/\text{kg}} = \mathbf{1.487 \text{ kg}}$$

(*b*) The entropy generated during this process is determined by applying the entropy balance on the cooker. Noting that there is no heat transfer and some mass leaves, the entropy balance can be expressed as

$$\underbrace{S_{in} - S_{out}}_{\substack{\text{Net entropy transfer} \\ \text{by heat and mass}}} + \underbrace{S_{gen}}_{\substack{\text{Entropy} \\ \text{generation}}} = \underbrace{\Delta S_{\text{system}}}_{\substack{\text{Change} \\ \text{in entropy}}}$$

$$-m_e s_e + S_{\text{gen}} = \Delta S_{\text{sys}} = m_2 s_2 - m_1 s_1$$

$$S_{\text{gen}} = m_e s_e + m_2 s_2 - m_1 s_1$$

The exergy destroyed during a process can be determined from an exergy balance or directly from its definition $X_{\text{destroyed}} = T_0 S_{gen}$. Using the S_{gen} relation obtained above and substituting,

$$\begin{aligned} X_{\text{destroyed}} &= T_0 S_{gen} = T_0 (m_e s_e + m_2 s_2 - m_1 s_1) \\ &= (298 \text{ K})[(1.894 - 1.487) \times 7.1717 + 1.487 \times 1.494 - 2.8238] \\ &= \mathbf{690 \text{ kJ}} \end{aligned}$$

7-118 A pressure cooker is initially half-filled with liquid water. Heat is transferred to the cooker for 30 min. The amount water that remained in the cooker and the exergy destroyed are to be determined.

Assumptions **1** This is an unsteady process since the conditions within the device are changing during the process, but it can be analyzed as a uniform-flow process since the state of water vapor leaving the cooker remains constant. **2** Kinetic and potential energies are negligible. **3** Heat loss from the cooker is negligible.

Properties The properties of water are (Tables A-4 through A-6)

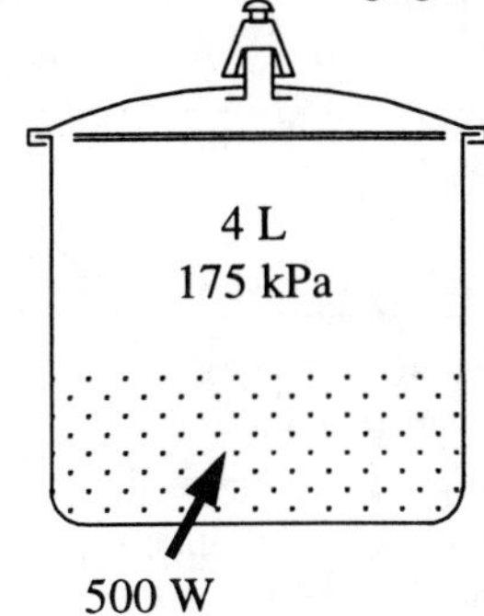

$$P_1 = 175\ kPa \ \rightarrow\ v_f = 0.001057\ \text{m}^3/\text{kg},\ v_g = 1.0036\ \text{m}^3/\text{kg}$$
$$u_f = 486.8\ \text{kJ/kg},\ u_g = 2524.9\ \text{kJ/kg}$$
$$s_f = 1.4849\ \text{kJ/kg.K},\ s_g = 7.1717\ \text{kJ/kg.K}$$

$$\left.\begin{array}{l} P_e = 175\ \text{kPa} \\ \text{sat.vapor} \end{array}\right\} \begin{array}{l} h_e = h_{g\ @\ 175\ \text{kPa}} = 2700.6\ \text{kJ/kg} \\ s_e = s_{g\ @\ 175\ \text{kPa}} = 7.1717\ \text{kJ/kg}\cdot\text{K} \end{array}$$

Analysis We take the cooker as the system, which is a control volume since mass crosses the boundary. Noting that the microscopic energies of flowing and nonflowing fluids are represented by enthalpy h and internal energy u, respectively, the mass and energy balances for this uniform-flow system can be expressed as

Mass balance: $\quad m_{in} - m_{out} = \Delta m_{\text{system}} \ \rightarrow\ m_e = m_1 - m_2$

Energy balance:

$$\underbrace{E_{in} - E_{out}}_{\substack{\text{Net energy transfer} \\ \text{by heat, work, and mass}}} = \underbrace{\Delta E_{\text{system}}}_{\substack{\text{Change in internal, kinetic,} \\ \text{potential, etc. energies}}}$$

$$Q_{in} - m_e h_e = m_2 u_2 - m_1 u_1 \quad (\text{since } W \cong ke \cong pe \cong 0)$$

The initial mass, initial internal energy, initial entropy, and final mass in the tank are

$$V_f = V_g = 2\ \text{L} = 0.002\ \text{m}^3$$

$$m_1 = m_f + m_g = \frac{V_f}{v_f} + \frac{V_g}{v_g} = \frac{0.002\ \text{m}^3}{0.001057\ \text{m}^3/\text{kg}} + \frac{0.002\ \text{m}^3}{1.0036\ \text{m}^3/\text{kg}} = 1.892 + 0.002 = 1.8945\ \text{kg}$$

$$U_1 = m_1 u_1 = m_f u_f + m_g u_g = 1.892 \times 486.8 + 0.002 \times 2524.9 = 926.1\ \text{kJ}$$
$$S_1 = m_1 s_1 = m_f s_f + m_g s_g = 1.892 \times 1.4849 + 0.002 \times 7.1717 = 2.8238\ \text{kJ/K}$$

$$m_2 = \frac{V}{v_2} = \frac{0.004\ \text{m}^3}{v_2}$$

The amount of heat transfer during this process is

$$Q = \dot{Q}\Delta t = (0.5\ \text{kJ/s})(30 \times 60\ \text{s}) = 900\ \text{kJ}$$

Then from the mass and energy balances,

$$m_e = m_1 - m_2 = 1.894 - \frac{0.004}{v_2}$$

$$900\ \text{kJ} = (1.894 - \frac{0.004}{v_2})(2700.6\ \text{kJ/kg}) + (\frac{0.004}{v_2})(u_2) - 926.1\ \text{kJ}$$

Substituting $u_2 = u_f + x_2 u_{fg}$ and $v_2 = v_f + x_2 v_{fg}$, and solving for x_2 yields

$$x_2 = 0.00163$$

Thus,

$$v_2 = v_f + x_2 v_{fg} = 0.001057 + 0.00163 \times (1.0036 - 0.001057) = 0.00269\ \text{m}^3/\text{kg}$$

$$s_2 = s_f + x_2 s_{fg} = 1.4849 + 0.00163 \times 5.6868 = 1.494\ \text{kJ/kg}\cdot\text{K}$$

and

$$m_2 = \frac{V}{v_2} = \frac{0.004\ \text{m}^3}{0.00269\ \text{m}^3/\text{kg}} = \mathbf{1.487\ kg}$$

(*b*) The entropy generated during this process is determined by applying the entropy balance on an *extended system* that includes the cooker and its immediate surroundings so that the boundary temperature of the extended system at the location of heat transfer is the heat source temperature, T_{source} = 180°C at all times. The entropy balance for it can be expressed as

$$\underbrace{S_{in} - S_{out}}_{\substack{\text{Net entropy transfer}\\ \text{by heat and mass}}} + \underbrace{S_{gen}}_{\substack{\text{Entropy}\\ \text{generation}}} = \underbrace{\Delta S_{\text{system}}}_{\substack{\text{Change}\\ \text{in entropy}}}$$

$$\frac{Q_{in}}{T_{b,in}} - m_e s_e + S_{\text{gen}} = \Delta S_{\text{sys}} = m_2 s_2 - m_1 s_1$$

$$S_{\text{gen}} = m_e s_e + m_2 s_2 - m_1 s_1 - \frac{Q_{in}}{T_{\text{source}}}$$

The exergy destroyed during a process can be determined from an exergy balance or directly from its definition $X_{\text{destroyed}} = T_0 S_{gen}$. Using the S_{gen} relation obtained above and substituting,

$$\begin{aligned} X_{\text{destroyed}} &= T_0 S_{gen} = T_0 (m_e s_e + m_2 s_2 - m_1 s_1) \\ &= (298\ \text{K})\left[(1.894 - 1.487) \times 7.1717 + 1.487 \times 1.494 - 2.8238 - 900/453\right] \\ &= \mathbf{97.8\ kJ} \end{aligned}$$

Note that the exergy destroyed is much less when heat is supplied from a heat source rather than an electric resistance heater.

7-119 A heat engine operates between a nitrogen tank and an argon cylinder at different temperatures. The maximum work that can be produced and the final temperatures are to be determined.

Assumptions Nitrogen and argon are ideal gases with constant specific heats at room temperature.

Properties The constant volume specific heat of nitrogen at room temperature is C_v = 0.743 kJ/kg.K. The constant pressure specific heat of argon at room temperature is C_p = 0.5203 kJ/kg.K (Table A-2).

Analysis For maximum power production, the entropy generation must be zero. We take the tank, the cylinder (the heat source and the heat sink) and the heat engine as the system. Noting that the system involves no heat and mass transfer and that the entropy change for cyclic devices is zero, the entropy balance can be expressed as

$$\underbrace{S_{in} - S_{out}}_{\substack{\text{Net entropy transfer} \\ \text{by heat and mass}}} + \underbrace{S_{gen}^{\nearrow 0}}_{\substack{\text{Entropy} \\ \text{generation}}} = \underbrace{\Delta S_{\text{system}}}_{\substack{\text{Change} \\ \text{in entropy}}}$$

$$0 + S_{gen}^{\nearrow 0} = \Delta S_{\text{tank,source}} + \Delta S_{\text{cylinder,sink}} + \Delta S_{\text{heat engine}}^{\nearrow 0}$$

$$(\Delta S)_{source} + (\Delta S)_{sink} = 0$$

$$\left(mC_v \ln\frac{T_2}{T_1} - mR\ln\frac{V_2}{V_1}^{\not 0} \right)_{source} + 0 + \left(mC_p \ln\frac{T_2}{T_1} - mR\ln\frac{P_2}{P_1}^{\not 0} \right)_{\text{sink}} = 0$$

N2, 20 kg, 1000 K; QH; HE; W; QL; Ar, 10 kg, 300 K

Substituting,

$$(20 \text{ kg})(0.743 \text{ kJ/kg}\cdot\text{K})\ln\frac{T_2}{1000 \text{ K}} + (10 \text{ kg})(0.5203 \text{ kJ/kg}\cdot\text{K})\ln\frac{T_2}{300 \text{ K}} = 0$$

Solving for T_2 yields

$$T_2 = \mathbf{731 \ K}$$

where T_2 is the common final temperature of the tanks for maximum power production.

The energy balance $E_{in} - E_{out} = \Delta E_{\text{system}}$ for the source and sink can be expressed as follows:

Source: $\quad -Q_{\text{source,out}} = \Delta U = mC_v(T_2 - T_{1A}) \rightarrow Q_{\text{source,out}} = mC_v(T_{1A} - T_2)$

$$Q_{\text{source,out}} = mC_v(T_{1A} - T_2) = (20 \text{ kg})(0.743 \text{ kJ/kg}\cdot\text{K})(1000 - 731)\text{K} = 3997 \text{ kJ}$$

Sink: $\quad Q_{\text{sink,in}} - W_{b,out} = \Delta U \rightarrow Q_{\text{sink,in}} = \Delta H = mC_p(T_2 - T_{1A})$

$$Q_{\text{sink,in}} = mC_v(T_2 - T_{1A}) = (10 \text{ kg})(0.5203 \text{ kJ/kg}\cdot\text{K})(731 - 300)\text{K} = 2242 \text{ kJ}$$

Then the work produced becomes

$$W_{\text{max,out}} = Q_H - Q_L = Q_{\text{source,out}} - Q_{\text{sink,in}} = 3997 - 2242 = \mathbf{1755 \ kJ}$$

Therefore, a maximum of 1755 kJ of work can be produced during this process

7-120 A heat engine operates between a tank and a cylinder filled with air at different temperatures. The maximum work that can be produced and the final temperatures are to be determined.

Assumptions Air is an ideal gas with constant specific heats at room temperature.

Properties The specific heats of air are $C_v = 0.718$ kJ/kg.K and $C_p = 1.005$ kJ/kg.K (Table A-2).

Analysis For maximum power production, the entropy generation must be zero. We take the tank, the cylinder (the heat source and the heat sink) and the heat engine as the system. Noting that the system involves no heat and mass transfer and that the entropy change for cyclic devices is zero, the entropy balance can be expressed as

$$\underbrace{S_{in} - S_{out}}_{\substack{\text{Net entropy transfer}\\ \text{by heat and mass}}} + \underbrace{S_{gen}^{\nearrow 0}}_{\substack{\text{Entropy}\\ \text{generation}}} = \underbrace{\Delta S_{\text{system}}}_{\substack{\text{Change}\\ \text{in entropy}}}$$

$$0 + S_{gen}^{\nearrow 0} = \Delta S_{\text{tank,source}} + \Delta S_{\text{cylinder,sink}} + \Delta S_{\text{heat engine}}^{\nearrow 0}$$

$$(\Delta S)_{source} + (\Delta S)_{\text{sink}} = 0$$

$$\left(mC_v \ln\frac{T_2}{T_1} - mR\ln\frac{V_2}{V_1}^{\nearrow 0} \right)_{source} + 0 + \left(mC_p \ln\frac{T_2}{T_1} - mR\ln\frac{P_2}{P_1}^{\nearrow 0} \right)_{\text{sink}} = 0$$

$$\ln\frac{T_2}{T_{1A}} + \frac{C_p}{C_v}\ln\frac{T_2}{T_{1B}} = 0 \longrightarrow \frac{T_2}{T_{1A}}\left(\frac{T_2}{T_{1B}}\right)^k = 1 \longrightarrow T_2 = \left(T_{1A}T_{1B}^k\right)^{1/(k+1)}$$

where T_{1A} and T_{1B} are the initial temperatures of the source and the sink, respectively, and T_2 is the common final temperature. Therefore, the final temperature of the tanks for maximum power production is

$$T_2 = \left((800\text{ K})(290\text{ K})^{1.4}\right)^{\frac{1}{2.4}} = \mathbf{443\ K}$$

Source: $\quad -Q_{\text{source,out}} = \Delta U = mC_v(T_2 - T_{1A}) \rightarrow Q_{\text{source,out}} = mC_v(T_{1A} - T_2)$

$$Q_{\text{source,out}} = mC_v(T_{1A} - T_2) = (20\text{ kg})(0.718\text{ kJ/kg}\cdot\text{K})(800 - 443)\text{K} = 5127\text{ kJ}$$

Sink: $\quad Q_{\text{sink,in}} - W_{b,out} = \Delta U \rightarrow Q_{\text{sink,in}} = \Delta H = mC_p(T_2 - T_{1A})$

$$Q_{\text{sink,in}} = mC_v(T_2 - T_{1A}) = (20\text{ kg})(1.005\text{ kJ/kg}\cdot\text{K})(443 - 290)\text{K} = 3075\text{ kJ}$$

Then the work produced becomes

$$W_{\text{max,out}} = Q_H - Q_L = Q_{\text{source,out}} - Q_{\text{sink,in}} = 5127 - 3075 = \mathbf{2052\ kJ}$$

Therefore, a maximum of 2052 kJ of work can be produced during this process

7-121 Using an incompressible substance as an example, it is to be demonstrated if closed system and flow exergies can be negative.

Analysis The availability of a closed system cannot be negative. However, the flow availability can be negative at low pressures. A closed system has zero availability at dead state, and positive availability at any other state since we can always produce work when there is a pressure or temperature differential.

To see that the flow availability can be negative, consider an incompressible substance. The flow availability can be written as

$$\begin{aligned}\psi &= h - h_0 + T_0(s - s_0) \\ &= (u - u_0) + v(P - P_0) + T_0(s - s_0) \\ &= \xi + v(P - P_0)\end{aligned}$$

The closed system availability ξ is always positive or zero, and the flow availability can be negative when P << P_0.

7-122 A relation for the second-law efficiency of a heat engine operating between a heat source and a heat sink at specified temperatures is to be obtained.

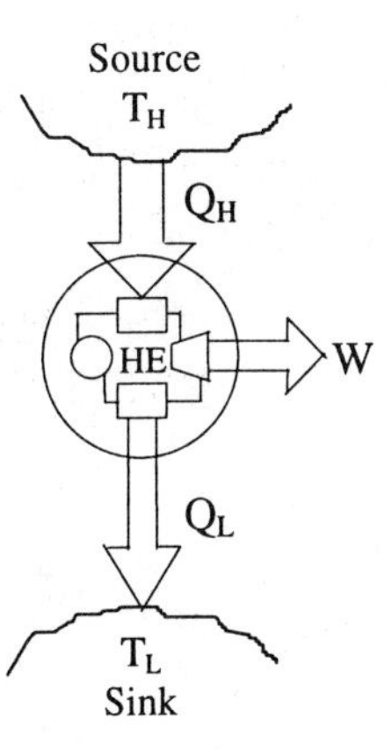

Analysis The second-law efficiency is defined as the ratio of the availability recovered to availability supplied during a process. The work W produced is the availability recovered. The decrease in the availability of the heat supplied Q_H is the availability supplied or invested.

Therefore,

$$\eta_{II} = \frac{W}{\left(1 - \frac{T_0}{T_H}\right)Q_H - \left(1 - \frac{T_0}{T_L}\right)(Q_H - W)}$$

Note that the first term in the denominator is the availability of heat supplied to the heat engine whereas the second term is the availability of the heat rejected by the heat engine. The difference between the two is the availability consumed during the process.

7-123E Large brass plates are heated in an oven at a rate of 300/min. The rate of heat transfer to the plates in the oven and the rate of exergy destruction associated with this heat transfer process are to be determined.

Assumptions **1** The thermal properties of the plates are constant. **2** The changes in kinetic and potential energies are negligible. **3** The environment temperature is 75°F.

Properties The density and specific heat of the brass are given to be ρ = 532.5 lbm/ft^3 and C_p = 0.091 Btu/lbm.°F.

Analysis We take the plate to be the system. The energy balance for this closed system can be expressed as

$$\underbrace{E_{in} - E_{out}}_{\substack{\text{Net energy transfer}\\ \text{by heat, work, and mass}}} = \underbrace{\Delta E_{\text{system}}}_{\substack{\text{Change in internal, kinetic,}\\ \text{potential, etc. energies}}}$$

$$Q_{in} = \Delta U_{\text{plate}} = m(u_2 - u_1) = mC(T_2 - T_1)$$

The mass of each plate and the amount of heat transfer to each plate is

$$m = \rho V = \rho LA = (532.5 \text{ lbm/ft}^3)[(1.2/12 \text{ ft})(2 \text{ ft})(2 \text{ ft})] = 213 \text{ lbm}$$

$$Q_{in} = mC(T_2 - T_1) = (213 \text{ lbm/plate})(0.091 \text{ Btu/lbm.°F})(1000 - 75)\text{°F} = 17{,}930 \text{ Btu/plate}$$

Then the total rate of heat transfer to the plates becomes

$$\dot{Q}_{total} = \dot{n}_{\text{plate}} Q_{\text{in, per plate}} = (300 \text{ plates/min}) \times (17{,}930 \text{ Btu/plate}) = \mathbf{5{,}379{,}000 \text{ Btu/min} = 89{,}650 \text{ Btu/s}}$$

We again take a single plate as the system The entropy generated during this process can be determined by applying an entropy balance on an *extended system* that includes the plate and its immediate surroundings so that the boundary temperature of the extended system is at 1300°F at all times:

$$\underbrace{S_{in} - S_{out}}_{\substack{\text{Net entropy transfer}\\ \text{by heat and mass}}} + \underbrace{S_{gen}}_{\substack{\text{Entropy}\\ \text{generation}}} = \underbrace{\Delta S_{\text{system}}}_{\substack{\text{Change}\\ \text{in entropy}}}$$

$$\frac{Q_{in}}{T_b} + S_{gen} = \Delta S_{\text{system}} \rightarrow S_{gen} = -\frac{Q_{in}}{T_b} + \Delta S_{\text{system}}$$

where

$$\Delta S_{\text{system}} = m(s_2 - s_1) = mC_{av} \ln\frac{T_2}{T_1} = (213 \text{ lbm})(0.091 \text{ Btu/lbm.R})\ln\frac{1000+460}{75+460} = 19.46 \text{ Btu/R}$$

Substituting,

$$S_{gen} = -\frac{Q_{in}}{T_b} + \Delta S_{\text{system}} = -\frac{17{,}930 \text{ Btu}}{1300+460 \text{ R}} + 19.46 \text{ Btu/R} = 9.272 \text{ Btu/R} \quad \text{(per plate)}$$

Then the rate of entropy generation becomes

$$\dot{S}_{gen} = S_{gen}\dot{n}_{ball} = (9.272 \text{ Btu/R}\cdot\text{plate})(300 \text{ plates/min}) = 2781 \text{ Btu/min.R} = \mathbf{46.35 \text{ Btu/s.R}}$$

The exergy destroyed during a process can be determined from an exergy balance or directly from its definition $X_{\text{destroyed}} = T_0 S_{gen}$,

$$\dot{X}_{\text{destroyed}} = T_0 \dot{S}_{gen} = (535 \text{ R})(46.35 \text{ Btu/s.R}) = \mathbf{24{,}797 \text{ Btu/s}}$$

7-124 Long cylindrical steel rods are heat-treated in an oven. The rate of heat transfer to the rods in the oven and the rate of exergy destruction associated with this heat transfer process are to be determined.

Assumptions **1** The thermal properties of the rods are constant. **2** The changes in kinetic and potential energies are negligible. **3** The environment temperature is 30°C.

Properties The density and specific heat of the steel rods are given to be ρ = 7833 kg/m^3 and C_p = 0.465 kJ/kg.°C.

Analysis Noting that the rods enter the oven at a velocity of 3 m/min and exit at the same velocity, we can say that a 3-m long section of the rod is heated in the oven in 1 min. Then the mass of the rod heated in 1 minute is

$$m = \rho V = \rho LA = \rho L(\pi D^2/4) = (7833\ \text{kg/m}^3)(3\ \text{m})[\pi(0.1\ \text{m})^2/4] = 184.6\ \text{kg}$$

We take the 3-m section of the rod in the oven as the system. The energy balance for this closed system can be expressed as

$$\underbrace{E_{in} - E_{out}}_{\substack{\text{Net energy transfer}\\ \text{by heat, work, and mass}}} = \underbrace{\Delta E_{\text{system}}}_{\substack{\text{Change in internal, kinetic,}\\ \text{potential, etc. energies}}}$$

$$Q_{in} = \Delta U_{\text{rod}} = m(u_2 - u_1) = mC(T_2 - T_1)$$

Substituting,

$$Q_{in} = mC(T_2 - T_1) = (184.6\ \text{kg})(0.465\ \text{kJ/kg.°C})(700 - 30)\text{°C} = 57{,}512\ \text{kJ}$$

Noting that this much heat is transferred in 1 min, the rate of heat transfer to the rod becomes

$$\dot{Q}_{in} = Q_{\text{in}}/\Delta t = (57{,}512\ \text{kJ})/(1\ \text{min}) = 57{,}512\ \text{kJ/min} = \mathbf{958.5\ kW}$$

We again take the 3-m long section of the rod as the system The entropy generated during this process can be determined by applying an entropy balance on an *extended system* that includes the rod and its immediate surroundings so that the boundary temperature of the extended system is at 900°C at all times:

$$\underbrace{S_{in} - S_{out}}_{\substack{\text{Net entropy transfer}\\ \text{by heat and mass}}} + \underbrace{S_{gen}}_{\substack{\text{Entropy}\\ \text{generation}}} = \underbrace{\Delta S_{\text{system}}}_{\substack{\text{Change}\\ \text{in entropy}}}$$

$$\frac{Q_{in}}{T_b} + S_{gen} = \Delta S_{\text{system}} \rightarrow S_{gen} = -\frac{Q_{in}}{T_b} + \Delta S_{\text{system}}$$

where

$$\Delta S_{\text{system}} = m(s_2 - s_1) = mC_{av}\ln\frac{T_2}{T_1} = (184.6\ \text{kg})(0.465\ \text{kJ/kg.K})\ln\frac{700 + 273}{30 + 273} = 100.1\ \text{kJ/K}$$

Substituting,

$$S_{gen} = -\frac{Q_{in}}{T_b} + \Delta S_{\text{system}} = -\frac{57{,}512\ \text{kJ}}{900 + 273\ \text{R}} + 100.1\ \text{kJ/K} = 51.1\ \text{kJ/K}$$

Noting that this much entropy is generated in 1 min, the rate of entropy generation becomes

$$\dot{S}_{gen} = \frac{S_{gen}}{\Delta t} = \frac{51.1\ \text{kJ/K}}{1\ \text{min}} = 51.1\ \text{kJ/min.K} = \mathbf{0.852\ \ kW/K}$$

The exergy destroyed during a process can be determined from an exergy balance or directly from its definition $X_{\text{destroyed}} = T_0 S_{gen}$,

$$\dot{X}_{\text{destroyed}} = T_0 \dot{S}_{gen} = (298\ \text{K})(0.852\ \text{kW/K}) = \mathbf{254\ \ kW}$$

7-125 Steam is condensed by cooling water in the condenser of a power plant. The rate of condensation of steam and the rate of exergy destruction are to be determined.

Assumptions **1** Steady operating conditions exist. **2** The heat exchanger is well-insulated so that heat loss to the surroundings is negligible and thus heat transfer from the hot fluid is equal to the heat transfer to the cold fluid. **3** Changes in the kinetic and potential energies of fluid streams are negligible. **4** Fluid properties are constant.

Properties The enthalpy and entropy of vaporization of water at 50°C are h_{fg} =2382.7 kJ/kg and s_{fg}= 7.3725 kJ/kg.K (Table A-4). The specific heat of water at room temperature is $C_p = 4.18$ kJ/kg.°C (Table A-3).

Analysis (*a*) We take the cold water tubes as the system, which is a control volume. The energy balance for this steady-flow system can be expressed in the rate form as

$$\underbrace{\dot{E}_{in} - \dot{E}_{out}}_{\text{Rate of net energy transfer by heat, work, and mass}} = \underbrace{\Delta \dot{E}_{\text{system}}^{\nearrow 0 \text{ (steady)}}}_{\text{Rate of change in internal, kinetic, potential, etc. energies}} = 0$$

$$\dot{E}_{in} = \dot{E}_{out}$$

$$\dot{Q}_{in} + \dot{m}h_1 = \dot{m}h_2 \quad (\text{since } \Delta\text{ke} \cong \Delta\text{pe} \cong 0)$$

$$\dot{Q}_{in} = \dot{m}C_p(T_2 - T_1)$$

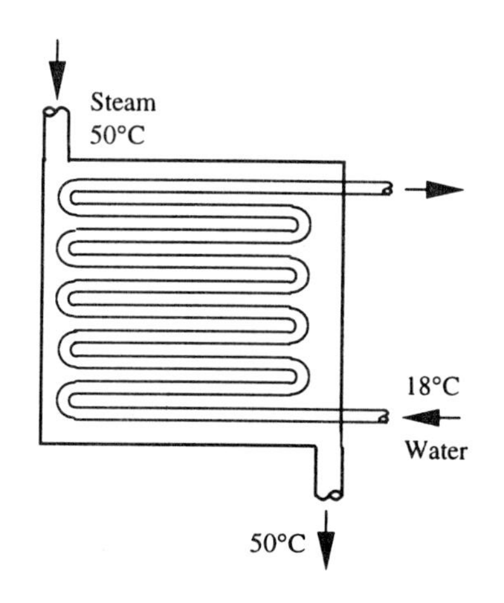

Then the heat transfer rate to the cooling water in the condenser becomes

$$\dot{Q} = [\dot{m}C_p(T_{out} - T_{in})]_{\text{cooling water}}$$
$$= (140 \text{ kg/s})(4.18 \text{ kJ/kg.°C})(27°\text{C} - 18°\text{C}) = 5267 \text{ kJ/s}$$

The rate of condensation of steam is determined to be

$$\dot{Q} = (\dot{m}h_{fg})_{steam} \longrightarrow \dot{m}_{steam} = \frac{\dot{Q}}{h_{fg}} = \frac{5267 \text{kJ/s}}{2382.7 \text{ kJ/kg}} = \mathbf{2.210 \ kg/s}$$

(*b*) The rate of entropy generation within the condenser during this process can be determined by applying the rate form of the entropy balance on the entire condenser. Noting that the condenser is well-insulated and thus heat transfer is negligible, the entropy balance for this steady-flow system can be expressed as

$$\underbrace{\dot{S}_{in} - \dot{S}_{out}}_{\text{Rate of net entropy transfer by heat and mass}} + \underbrace{\dot{S}_{gen}}_{\text{Rate of entropy generation}} = \underbrace{\Delta \dot{S}_{\text{system}}^{\nearrow 0 \text{ (steady)}}}_{\text{Rate of change of entropy}}$$

$$\dot{m}_1 s_1 + \dot{m}_3 s_3 - \dot{m}_2 s_2 - \dot{m}_4 s_4 + \dot{S}_{gen} = 0 \quad (\text{since } Q = 0)$$

$$\dot{m}_{\text{water}} s_1 + \dot{m}_{\text{steam}} s_3 - \dot{m}_{\text{water}} s_2 - \dot{m}_{\text{steam}} s_4 + \dot{S}_{gen} = 0$$

$$\dot{S}_{gen} = \dot{m}_{\text{water}}(s_2 - s_1) + \dot{m}_{\text{steam}}(s_4 - s_3)$$

Noting that water is an incompressible substance and steam changes from saturated vapor to saturated liquid, the rate of entropy generation is determined to be

$$\dot{S}_{gen} = \dot{m}_{\text{water}} C_p \ln\frac{T_2}{T_1} + \dot{m}_{\text{steam}}(s_f - s_g) = \dot{m}_{\text{water}} C_p \ln\frac{T_2}{T_1} - \dot{m}_{\text{steam}} s_{fg}$$

$$= (140 \text{ kg/s})(4.18 \text{ kJ/kg.K})\ln\frac{27+273}{18+273} - (2.21 \text{ kg/s})(7.3725 \text{ kJ/kg.K}) = 1.532 \text{ kW/K}$$

Then the exergy destroyed can be determined directly from its definition $X_{\text{destroyed}} = T_0 S_{gen}$ to be

$$\dot{X}_{\text{destroyed}} = T_0 \dot{S}_{gen} = (291 \text{ K})(1.532 \text{ kW/K}) = \mathbf{446 \ kW}$$

7-126 Water is heated in a heat exchanger by geothermal water. The rate of heat transfer to the water and the rate of exergy destruction within the heat exchanger are to be determined.

Assumptions **1** Steady operating conditions exist. **2** The heat exchanger is well-insulated so that heat loss to the surroundings is negligible and thus heat transfer from the hot fluid is equal to the heat transfer to the cold fluid. **3** Changes in the kinetic and potential energies of fluid streams are negligible. **4** Fluid properties are constant. **5** The environment temperature is 25°C.

Properties The specific heats of water and geothermal fluid are given to be 4.18 and 4.31 kJ/kg.°C, respectively.

Analysis (*a*) We take the cold water tubes as the system, which is a control volume. The energy balance for this steady-flow system can be expressed in the rate form as

$$\underbrace{\dot{E}_{in} - \dot{E}_{out}}_{\text{Rate of net energy transfer by heat, work, and mass}} = \underbrace{\Delta \dot{E}_{\text{system}}^{\nearrow 0 \text{ (steady)}}}_{\text{Rate of change in internal, kinetic, potential, etc. energies}} = 0$$

$$\dot{E}_{in} = \dot{E}_{out}$$

$$\dot{Q}_{in} + \dot{m}h_1 = \dot{m}h_2 \quad (\text{since } \Delta\text{ke} \cong \Delta\text{pe} \cong 0)$$

$$\dot{Q}_{in} = \dot{m}C_p(T_2 - T_1)$$

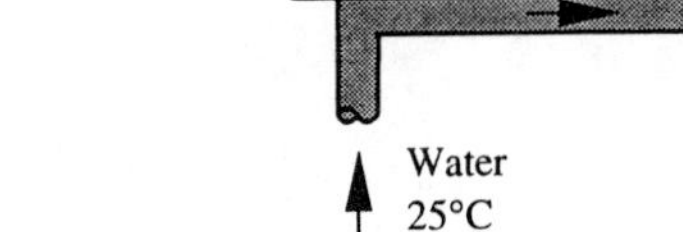

Then the rate of heat transfer to the cold water in the heat exchanger becomes

$$\dot{Q}_{in,water} = [\dot{m}C_p(T_{out} - T_{in})]_{\text{water}} = (0.2 \text{ kg/s})(4.18 \text{ kJ/kg.°C})(60°\text{C} - 25°\text{C}) = \mathbf{29.26 \text{ kW}}$$

Noting that heat transfer to the cold water is equal to the heat loss from the geothermal water, the outlet temperature of the geothermal water is determined from

$$\dot{Q}_{out} = [\dot{m}C_p(T_{in} - T_{out})]_{\text{geo}} \longrightarrow T_{out} = T_{in} - \frac{\dot{Q}_{out}}{\dot{m}C_p} = 140°\text{C} - \frac{29.26 \text{ kW}}{(0.3 \text{ kg/s})(4.31 \text{ kJ/kg.°C})} = \mathbf{117.4°C}$$

(*b*) The rate of entropy generation within the heat exchanger is determined by applying the rate form of the entropy balance on the entire heat exchanger:

$$\underbrace{\dot{S}_{in} - \dot{S}_{out}}_{\text{Rate of net entropy transfer by heat and mass}} + \underbrace{\dot{S}_{gen}}_{\text{Rate of entropy generation}} = \underbrace{\Delta \dot{S}_{\text{system}}^{\nearrow 0 \text{ (steady)}}}_{\text{Rate of change of entropy}}$$

$$\dot{m}_1 s_1 + \dot{m}_3 s_3 - \dot{m}_2 s_2 - \dot{m}_4 s_4 + \dot{S}_{gen} = 0 \quad (\text{since } Q = 0)$$

$$\dot{m}_{\text{water}} s_1 + \dot{m}_{\text{geo}} s_3 - \dot{m}_{\text{water}} s_2 - \dot{m}_{\text{geo}} s_4 + \dot{S}_{gen} = 0$$

$$\dot{S}_{gen} = \dot{m}_{\text{water}}(s_2 - s_1) + \dot{m}_{\text{geo}}(s_4 - s_3)$$

Noting that both fresh and geothermal water are incompressible substances, the rate of entropy generation is determined to be

$$\dot{S}_{gen} = \dot{m}_{\text{water}} C_p \ln\frac{T_2}{T_1} + \dot{m}_{\text{geo}} C_p \ln\frac{T_4}{T_3}$$

$$= (0.2 \text{ kg/s})(4.18 \text{ kJ/kg.K})\ln\frac{60+273}{25+273} + (0.3 \text{ kg/s})(4.31 \text{ kJ/kg.K})\ln\frac{117.3+273}{140+273} = \mathbf{0.0197 \text{ kW/K}}$$

The exergy destroyed during a process can be determined from an exergy balance or directly from its definition $X_{\text{destroyed}} = T_0 S_{gen}$,

$$\dot{X}_{\text{destroyed}} = T_0 \dot{S}_{gen} = (298 \text{ K})(0.0197 \text{ kW/K}) = \mathbf{5.87 \text{ kW}}$$

7-127 Ethylene glycol is cooled by water in a heat exchanger. The rate of heat transfer and the rate of exergy destruction within the heat exchanger are to be determined.

Assumptions **1** Steady operating conditions exist. **2** The heat exchanger is well-insulated so that heat loss to the surroundings is negligible and thus heat transfer from the hot fluid is equal to the heat transfer to the cold fluid. **3** Changes in the kinetic and potential energies of fluid streams are negligible. **4** Fluid properties are constant. **5** The environment temperature is 20°C.

Properties The specific heats of water and ethylene glycol are given to be 4.18 and 2.56 kJ/kg.°C, respectively.

Analysis (*a*) We take the ethylene glycol tubes as the system, which is a control volume. The energy balance for this steady-flow system can be expressed in the rate form as

$$\underbrace{\dot{E}_{in} - \dot{E}_{out}}_{\substack{\text{Rate of net energy transfer}\\ \text{by heat, work, and mass}}} = \underbrace{\Delta \dot{E}_{\text{system}}^{\nearrow 0\ \text{(steady)}}}_{\substack{\text{Rate of change in internal, kinetic,}\\ \text{potential, etc. energies}}} = 0$$

$$\dot{E}_{in} = \dot{E}_{out}$$

$$\dot{m}h_1 = \dot{Q}_{out} + \dot{m}h_2 \quad (\text{since } \Delta\text{ke} \cong \Delta\text{pe} \cong 0)$$

$$\dot{Q}_{out} = \dot{m}C_p(T_1 - T_2)$$

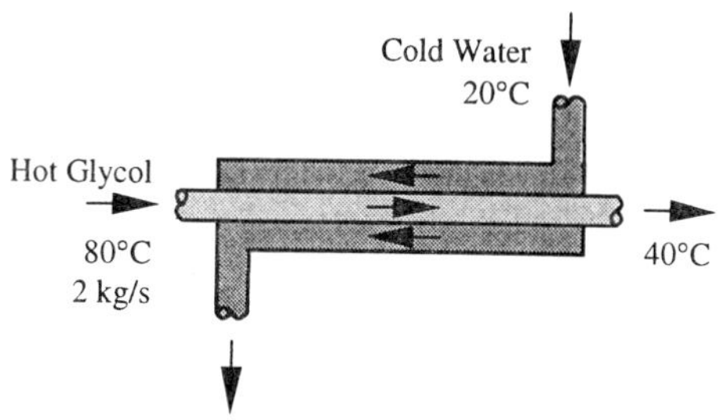

Cold Water 20°C

Hot Glycol 80°C 2 kg/s

40°C

Then the rate of heat transfer becomes

$$\dot{Q}_{out} = [\dot{m}C_p(T_{in} - T_{out})]_{\text{glycol}} = (2\text{ kg/s})(2.56\text{ kJ/kg.°C})(80°\text{C} - 40°\text{C}) = \mathbf{204.8\ kW}$$

The rate of heat transfer from water must be equal to the rate of heat transfer to the glycol. Then,

$$\dot{Q}_{in} = [\dot{m}C_p(T_{out} - T_{in})]_{\text{water}} \longrightarrow \dot{m}_{\text{water}} = \frac{\dot{Q}_{in}}{C_p(T_{out} - T_{in})} = \frac{204.8\text{ kJ/s}}{(4.18\text{ kJ/kg.°C})(55-20)\text{°C}} = \mathbf{1.4\ kg/s}$$

(*b*) The rate of entropy generation within the heat exchanger is determined by applying the rate form of the entropy balance on the entire heat exchanger:

$$\underbrace{\dot{S}_{in} - \dot{S}_{out}}_{\substack{\text{Rate of net entropy transfer}\\ \text{by heat and mass}}} + \underbrace{\dot{S}_{gen}}_{\substack{\text{Rate of entropy}\\ \text{generation}}} = \underbrace{\Delta \dot{S}_{\text{system}}^{\nearrow 0\ \text{(steady)}}}_{\substack{\text{Rate of change}\\ \text{of entropy}}}$$

$$\dot{m}_1 s_1 + \dot{m}_3 s_3 - \dot{m}_2 s_2 - \dot{m}_3 s_4 + \dot{S}_{gen} = 0 \quad (\text{since } Q = 0)$$

$$\dot{m}_{\text{glycol}} s_1 + \dot{m}_{\text{water}} s_3 - \dot{m}_{\text{glycol}} s_2 - \dot{m}_{\text{water}} s_4 + \dot{S}_{gen} = 0$$

$$\dot{S}_{gen} = \dot{m}_{\text{glycol}}(s_2 - s_1) + \dot{m}_{\text{water}}(s_4 - s_3)$$

Noting that both fluid streams are liquids (incompressible substances), the rate of entropy generation is determined to be

$$\dot{S}_{gen} = \dot{m}_{\text{glycol}} C_p \ln\frac{T_2}{T_1} + \dot{m}_{\text{water}} C_p \ln\frac{T_4}{T_3}$$

$$= (2\text{ kg/s})(2.56\text{ kJ/kg.K})\ln\frac{40+273}{80+273} + (1.4\text{ kg/s})(4.18\text{ kJ/kg.K})\ln\frac{55+273}{20+273} = \mathbf{0.0446\ kW/K}$$

The exergy destroyed during a process can be determined from an exergy balance or directly from its definition $X_{\text{destroyed}} = T_0 S_{gen}$,

$$\dot{X}_{\text{destroyed}} = T_0 \dot{S}_{gen} = (293\text{ K})(0.0446\text{ kW/K}) = \mathbf{13.1\ kW}$$

7-128 Oil is to be cooled by water in a thin-walled heat exchanger. The rate of heat transfer and the rate of exergy destruction within the heat exchanger are to be determined.

Assumptions **1** Steady operating conditions exist. **2** The heat exchanger is well-insulated so that heat loss to the surroundings is negligible and thus heat transfer from the hot fluid is equal to the heat transfer to the cold fluid. **3** Changes in the kinetic and potential energies of fluid streams are negligible. **4** Fluid properties are constant.

Properties The specific heats of water and oil are given to be 4.18 and 2.20 kJ/kg.°C, respectively.

Analysis We take the oil tubes as the system, which is a control volume. The energy balance for this steady-flow system can be expressed in the rate form as

Hot oil
150°C
2 kg/s
Cold water
22°C
1.5 kg/s

$$\underbrace{\dot{E}_{in} - \dot{E}_{out}}_{\text{Rate of net energy transfer by heat, work, and mass}} = \underbrace{\Delta \dot{E}_{\text{system}}^{\nearrow 0 \text{ (steady)}}}_{\text{Rate of change in internal, kinetic, potential, etc. energies}} = 0$$

$$\dot{E}_{in} = \dot{E}_{out}$$

$$\dot{m}h_1 = \dot{Q}_{out} + \dot{m}h_2 \quad (\text{since } \Delta\text{ke} \cong \Delta\text{pe} \cong 0)$$

$$\dot{Q}_{out} = \dot{m}C_p(T_1 - T_2)$$

Then the rate of heat transfer from the oil becomes

$$\dot{Q}_{out} = [\dot{m}C_p(T_{in} - T_{out})]_{\text{oil}} = (2 \text{ kg/s})(2.2 \text{ kJ/kg.°C})(150\text{°C} - 40\text{°C}) = \mathbf{484 \text{ kW}}$$

Noting that heat lost by the oil is gained by the water, the outlet temperature of water is determined from

$$Q = [mC_p(T_{out} - T_{in})]_{\text{water}} \longrightarrow T_{out} = T_{in} + \frac{Q}{mC_p}$$

$$= 22\text{°C} + \frac{484 \text{ kW}}{(1.5 \text{ kg/s})(4.18 \text{ kJ/kg.°C})} = 99.2\text{°C}$$

(*b*) The rate of entropy generation within the heat exchanger is determined by applying the rate form of the entropy balance on the entire heat exchanger:

$$\underbrace{\dot{S}_{in} - \dot{S}_{out}}_{\text{Rate of net entropy transfer by heat and mass}} + \underbrace{\dot{S}_{gen}}_{\text{Rate of entropy generation}} = \underbrace{\Delta \dot{S}_{\text{system}}^{\nearrow 0 \text{ (steady)}}}_{\text{Rate of change of entropy}}$$

$$\dot{m}_1 s_1 + \dot{m}_3 s_3 - \dot{m}_2 s_2 - \dot{m}_3 s_4 + \dot{S}_{gen} = 0 \quad (\text{since } Q = 0)$$

$$\dot{m}_{\text{oil}} s_1 + \dot{m}_{\text{water}} s_3 - \dot{m}_{\text{oil}} s_2 - \dot{m}_{\text{water}} s_4 + \dot{S}_{gen} = 0$$

$$\dot{S}_{gen} = \dot{m}_{\text{oil}}(s_2 - s_1) + \dot{m}_{\text{water}}(s_4 - s_3)$$

Noting that both fluid streams are liquids (incompressible substances), the rate of entropy generation is determined to be

$$\dot{S}_{gen} = \dot{m}_{\text{oil}} C_p \ln\frac{T_2}{T_1} + \dot{m}_{\text{water}} C_p \ln\frac{T_4}{T_3}$$

$$= (2 \text{ kg/s})(2.2 \text{ kJ/kg.K})\ln\frac{40+273}{150+273} + (1.5 \text{ kg/s})(4.18 \text{ kJ/kg.K})\ln\frac{99.2+273}{22+273} = \mathbf{0.132 \text{ kW/K}}$$

The exergy destroyed during a process can be determined from an exergy balance or directly from its definition $X_{\text{destroyed}} = T_0 S_{gen}$,

$$\dot{X}_{\text{destroyed}} = T_0 \dot{S}_{gen} = (295 \text{ K})(0.132 \text{ kW/K}) = \mathbf{38.9 \text{ kW}}$$

7-129 A regenerator is considered to save heat during the cooling of milk in a dairy plant. The amounts of fuel and money such a generator will save per year and the rate of exergy destruction within the regenerator are to be determined.

Assumptions 1 Steady operating conditions exist. **2** The properties of the milk are constant. **5** The environment temperature is 18°C.

Properties The average density and specific heat of milk can be taken to be $\rho_{milk} \cong \rho_{water} = 1$ kg/L and $C_{p,milk} = 3.79$ kJ/kg.°C (Table A-3).

Analysis The mass flow rate of the milk is

$$\dot{m}_{milk} = \rho \dot{V}_{milk} = (1\ \text{kg/L})(12\ \text{L/s}) = 12\ \text{kg/s} = 43{,}200\ \text{kg/h}$$

Taking the pasteurizing section as the system, the energy balance for this steady-flow system can be expressed in the rate form as

$$\underbrace{\dot{E}_{in} - \dot{E}_{out}}_{\substack{\text{Rate of net energy transfer}\\ \text{by heat, work, and mass}}} = \underbrace{\Delta \dot{E}_{system}^{\nearrow 0\ \text{(steady)}}}_{\substack{\text{Rate of change in internal, kinetic,}\\ \text{potential, etc. energies}}} = 0 \rightarrow \dot{E}_{in} = \dot{E}_{out}$$

$$\dot{Q}_{in} + \dot{m}h_1 = \dot{m}h_2 \quad (\text{since } \Delta\text{ke} \cong \Delta\text{pe} \cong 0)$$

$$\dot{Q}_{in} = \dot{m}_{milk} C_p (T_2 - T_1)$$

Therefore, to heat the milk from 4 to 72°C as being done currently, heat must be transferred to the milk at a rate of

$$\dot{Q}_{current} = [\dot{m}C_p(T_{pasturization} - T_{refrigeration})]_{milk} = (12\,\text{kg/s})(3.79\,\text{kJ/kg.°C})(72-4)\text{°C} = 3093\,\text{kJ/s}$$

The proposed regenerator has an effectiveness of $\varepsilon = 0.82$, and thus it will save 82 percent of this energy. Therefore,

$$\dot{Q}_{saved} = \varepsilon \dot{Q}_{current} = (0.82)(3093\ \text{kJ/s}) = 2536\ \text{kJ/s}$$

Noting that the boiler has an efficiency of $\eta_{boiler} = 0.82$, the energy savings above correspond to fuel savings of

$$\text{Fuel Saved} = \frac{\dot{Q}_{saved}}{\eta_{boiler}} = \frac{(2536\ \text{kJ/s})}{(0.82)} \frac{(1\text{therm})}{(105{,}500\ \text{kJ})} = 0.02931\,\text{therm/s}$$

Noting that 1 year = 365×24=8760 h and unit cost of natural gas is \$0.52/therm, the annual fuel and money savings will be

Fuel Saved = (0.02931 therms/s)(8760×3600 s) = **924,450 therms/yr**

$$\text{Money saved} = (\text{Fuel saved})(\text{Unit cost of fuel}) = (924{,}450\ \text{therm/yr})(\$0.52/\text{therm}) = \mathbf{\$480{,}700/yr}$$

The rate of entropy generation during this process is determined by applying the rate form of the entropy balance on an *extended system* that includes the regenerator and the immediate surroundings so that the boundary temperature is the surroundings temperature, which we take to be the cold water temperature of 18°C.:

$$\underbrace{\dot{S}_{in} - \dot{S}_{out}}_{\substack{\text{Rate of net entropy transfer}\\ \text{by heat and mass}}} + \underbrace{\dot{S}_{gen}}_{\substack{\text{Rate of entropy}\\ \text{generation}}} = \underbrace{\Delta \dot{S}_{system}^{\nearrow 0\ \text{(steady)}}}_{\substack{\text{Rate of change}\\ \text{of entropy}}} \rightarrow \dot{S}_{gen} = \dot{S}_{out} - \dot{S}_{in}$$

Disregarding entropy transfer associated with fuel flow, the only significant difference between the two cases is the reduction is the entropy transfer to water due to the reduction in heat transfer to water, and is determined to be

$$\dot{S}_{gen,\text{ reduction}} = \dot{S}_{out,\text{ reduction}} = \frac{\dot{Q}_{out,reduction}}{T_{surr}} = \frac{\dot{Q}_{saved}}{T_{surr}} = \frac{2536 \text{ kJ/s}}{18+273} = 8.715 \text{ kW/K}$$

$$S_{gen,\text{ reduction}} = \dot{S}_{gen,\text{ reduction}} \Delta t = (8.715 \text{ kJ/s.K})(8760 \times 3600 \text{ s/year}) = \mathbf{2.75 \times 10^8 \text{ kJ/K}} \text{ (per year)}$$

The exergy destroyed during a process can be determined from an exergy balance or directly from its definition $X_{\text{destroyed}} = T_0 S_{gen}$,

$$X_{\text{destroyed, reduction}} = T_0 S_{\text{gen, reduction}} = (291 \text{ K})(2.75 \times 10^8 \text{ kJ/K}) = \mathbf{8.00 \times 10^{10} \text{ kJ}} \text{ (per year)}$$

Fundamentals of Engineering (FE) Exam Problems

7-130 Heat is lost through a plane wall steadily at a rate of 600 W. If the inner and outer surface temperatures of the wall are 20°C and 5°C, respectively, and the environment temperature is 0°C, the rate of exergy destruction within the wall is (done with EES)

(a) 30 W (b) 1150 W (c) 573 W (d) 24,570 W (e) 0 W

Answer (a) 30 W

Solution Solved by EES Software. Solutions can be verified by copying-and-pasting the following lines on a blank EES screen. (Similar problems and their solutions can be obtained easily by modifying numerical values).

```
Q=600 "W"
T1=20 "C"
T2=5 "C"
To=0 "C"
"Entropy balance S_in - S_out + S_gen= DS_system for the wall for steady operation
gives"
Q/(T1+273)-Q/(T2+273)+S_gen=0  "W/K"
X_dest=(To+273)*S_gen "W"

"Some Wrong Solutions with Common Mistakes:"
Q/T1-Q/T2+Sgen1=0; W1_Xdest=(To+273)*Sgen1  "Using C instead of K in Sgen"
Sgen2=Q/((T1+T2)/2); W2_Xdest=(To+273)*Sgen2  "Using avegage temperature in C
for Sgen"
Sgen3=Q/((T1+T2)/2+273); W3_Xdest=(To+273)*Sgen3  "Using avegage temperature
in K"
W4_Xdest=To*S_gen  "Using C for To"
```

7-131 Liquid water enters an adiabatic piping system at 15°C at a rate of 5 kg/s. It is observed that the water temperature rises by 0.5°C in the pipe due to friction. If the environment temperature is also 15°C, the rate of exergy destruction in the pipe is

(a) 8.36 kW (b) 10.4 kW (c) 197 kW (d) 265 kW (e) 2410 kW

Answer (b) 10.4 kW

Solution Solved by EES Software. Solutions can be verified by copying-and-pasting the following lines on a blank EES screen. (Similar problems and their solutions can be obtained easily by modifying numerical values).

```
Cp=4.18 "kJ/kg.K"
m=5 "kg/s"
T1=15 "C"
T2=15.5 "C"
To=15 "C"

S_gen=m*Cp*ln((T2+273)/(T1+273)) "kW/K"
X_dest=(To+273)*S_gen "kW"

"Some Wrong Solutions with Common Mistakes:"
W1_Xdest=(To+273)*m*Cp*ln(T2/T1) "Using deg. C in Sgen"
W2_Xdest=To*m*Cp*ln(T2/T1) "Using deg. C in Sgen and To"
W3_Xdest=(To+273)*Cp*ln(T2/T1) "Not using mass flow rate with deg. C"
W4_Xdest=(To+273)*Cp*ln((T2+273)/(T1+273)) "Not using mass flow rate with K"
```

7-132 A heat engine receives heat from a source at 1200 K at a rate of 500 kJ/s and rejects the waste heat to a sink at 300 K. If the power output of the engine is 200 kW, the second-law efficiency of this heat engine is

(a) 35% (b) 40% (c) 53% (d) 75% (e) 100%

Answer (c) 53%

Solution Solved by EES Software. Solutions can be verified by copying-and-pasting the following lines on a blank EES screen. (Similar problems and their solutions can be obtained easily by modifying numerical values).

```
Qin=500 "kJ/s"
W=200 "kW"
TL=300 "K"
TH=1200 "K"

Eta_rev=1-TL/TH
Eta_th=W/Qin
Eta_II=Eta_th/Eta_rev

"Some Wrong Solutions with Common Mistakes:"
W1_Eta_II=Eta_th1/Eta_rev; Eta_th1=1-W/Qin "Wrong relation for thermal efficiency"
W2_Eta_II=Eta_th "Taking second-law efficiency to be thermal efficiency"
W3_Eta_II=Eta_rev "Taking second-law efficiency to be reversible efficiency"
W4_Eta_II=Eta_th*Eta_rev "Multiplying thermal and rev. efficiencies instead of
dividing"
```

7-133 A water reservoir contains 100 tons of water at an average elevation of 60 m. The maximum amount of electric power that can be generated from this water is

(a) 8 kWh (b) 16 kWh (c) 1630 kWh (d) 16,300 kWh (e) 58,800 kWh

Answer (b) 16 kWh

Solution Solved by EES Software. Solutions can be verified by copying-and-pasting the following lines on a blank EES screen. (Similar problems and their solutions can be obtained easily by modifying numerical values).

```
m=100000 "kg"
h=60 "m"
g=9.81 "m/s^2"

"Maximum power is simply the potential energy change,"
W_max=m*g*h/1000 "kJ"
W_max_kWh=W_max/3600 "kWh"

"Some Wrong Solutions with Common Mistakes:"
W1_Wmax =m*g*h/3600 "Not using the conversion factor 1000"
W2_Wmax =m*g*h/1000 "Obtaining the result in kJ instead of kWh"
W3_Wmax =m*g*h*3.6/1000 "Using worng conversion factor"
W4_Wmax =m*h/3600"Not using g and the factor 1000 in calculations"
```

7-134 A house is maintained at 22°C in winter by electric resistance heaters. If the outdoor temperature is 5°C, the second-law efficiency of the resistance heaters is

(a) 0% (b) 5.8% (c) 34% (d) 77% (e) 100%

Answer (b) 5.8%

Solution Solved by EES Software. Solutions can be verified by copying-and-pasting the following lines on a blank EES screen. (Similar problems and their solutions can be obtained easily by modifying numerical values).

```
TL=5+273 "K"
TH=22+273 "K"
To=TL

COP_rev=TH/(TH-TL)
COP=1
Eta_II=COP/COP_rev

"Some Wrong Solutions with Common Mistakes:"
W1_Eta_II=COP/COP_rev1; COP_rev1=TL/(TH-TL) "Using wrong relation for COP_rev"
W2_Eta_II=1-(TL-273)/(TH-273) "Taking second-law efficiency to be reversible thermal
efficiency with C for temp"
W3_Eta_II=COP_rev "Taking second-law efficiency to be reversible COP"
W4_Eta_II=COP_rev2/COP; COP_rev2=(TL-273)/(TH-TL) "Using C in COP_rev relation
instead of K, and reversing"
```

7-135 A 10-kg solid whose specific heat is 2.8 kJ/kg.°C is at a uniform temperature of -10°C. For an environment temperature of 25°C, the exergy content of this solid is

(a) Less than zero (b) 0 kJ (c) 22.3 kJ (d) 62.5 kJ (e) 980 kJ

Answer (d) 62.5 kJ

Solution Solved by EES Software. Solutions can be verified by copying-and-pasting the following lines on a blank EES screen. (Similar problems and their solutions can be obtained easily by modifying numerical values).

```
m=10 "kg"
Cp=2.8 "kJ/kg.K"
T1=-10+273 "K"
To=25+273 "K"

"Exergy content of a fixed mass is x1=u1-uo-To*(s1-so)+Po*(v1-vo)"
 ex=m*(Cp*(T1-To)-To*Cp*ln(T1/To))

"Some Wrong Solutions with Common Mistakes:"
W1_ex=m*Cp*(To-T1) "Taking the energy content as the exergy content"
W2_ex=m*(Cp*(T1-To)+To*Cp*ln(T1/To)) "Using + for the second term instead of -"
W3_ex=Cp*(T1-To)-To*Cp*ln(T1/To) "Using exergy content per unit mass"
W4_ex=0 "Taking the exergy content to be zero"
```

7- 136 Keeping the limitations imposed by the second-law of thermodynamics in mind, choose the wrong statement below:

(a) A heat engine cannot have a thermal efficiency of 100%.
(b) For all reversible processes, the second-law efficiency is 100%.
(c) The second-law efficiency of a heat engine cannot be greater than its thermal efficiency.
(d) The second-law efficiency of a process is 100% if no entropy is generated during that process.
(e) The coefficient of performance of a refrigerator can be greater than 1.

Answer (c) The second-law efficiency of a heat engine cannot be greater than its thermal efficiency.

7-137 A furnace can supply heat steadily at a 1200 K at a rate of 800 kJ/s. The maximum amount of power that can be produced by using the heat supplied by this furnace in an environment at 300 K is

(a) 100 kW (b) 200 kW (c) 400 kW (d) 600 kW (e) 800 kW

Answer (d) 600 kW

Solution Solved by EES Software. Solutions can be verified by copying-and-pasting the following lines on a blank EES screen. (Similar problems and their solutions can be obtained easily by modifying numerical values).

```
Q_in=800 "kJ/s"
TL=300 "K"
TH=1200 "K"

W_max=Q_in*(1-TL/TH) "kW"

"Some Wrong Solutions with Common Mistakes:"
W1_Wmax=W_max/2 "Taking half of Wmax"
W2_Wmax=Q_in/(1-TL/TH) "Dividing by efficiency instead of multiplying by it"
W3_Wmax =Q_in*TL/TH "Using wrong relation"
W4_Wmax=Q_in "Assuming entire heat input is converted to work"
```

7-138 Air is throttled from 50°C and 800 kPa to a pressure of 200 kPa at a rate of 0.5 kg/s in an environment at 25°C. The change in kinetic energy is negligible, and no heat transfer occurs during the process. The power potential wasted during this process is

(a) 0 (b) 0.20 kW (c) 47 kW (d) 59 kW (e) 119 kW

Answer (d) 59 kW

Solution Solved by EES Software. Solutions can be verified by copying-and-pasting the following lines on a blank EES screen. (Similar problems and their solutions can be obtained easily by modifying numerical values).

```
R=0.287 "kJ/kg.K"
Cp=1.005 "kJ/kg.K"
m=0.5 "kg/s"
T1=50+273 "K"
P1=800 "kPa"
To=25 "C"
P2=200 "kPa"
"Temp. of ideal gases remain constant during throttling since h=const and h=h(T)"
T2=T1

ds=Cp*ln(T2/T1)-R*ln(P2/P1)
X_dest=(To+273)*m*ds "kW"

"Some Wrong Solutions with Common Mistakes:"
W1_dest=0 "Assuming no loss"
W2_dest=(To+273)*ds "Not using mass flow rate"
W3_dest=To*m*ds "Using C for To instead of K"
W4_dest=m*(P1-P2) "Using wrong relations"
```

7-139 Steam enters a turbine steadily at 3 MPa and 450°C and exits at 0.2 MPa and 150°C in an environment at 25°C. The decrease in the exergy of the steam as it flows through the turbine is

(a) 58 kJ/kg (b) 517 kJ/kg (c) 575 kJ/kg (d) 580 kJ/kg (e) 634 kJ/kg

Answer (e) 634 kJ/kg

Solution Solved by EES Software. Solutions can be verified by copying-and-pasting the following lines on a blank EES screen. (Similar problems and their solutions can be obtained easily by modifying numerical values).

```
P1=3000 "kPa"
T1=450 "C"

P2=200 "kPa"
T2=150 "C"
To=25 "C"

h1=ENTHALPY(Steam_NBS,T=T1,P=P1)
s1=ENTROPY(Steam_NBS,T=T1,P=P1)

h2=ENTHALPY(Steam_NBS,T=T2,P=P2)
s2=ENTROPY(Steam_NBS,T=T2,P=P2)

"Exergy change of s fluid stream is Dx=h2-h1-To(s2-s1)"

-Dx=h2-h1-(To+273)*(s2-s1)

"Some Wrong Solutions with Common Mistakes:"
-W1_Dx=0 "Assuming no exergy destruction"
-W2_Dx=h2-h1 "Using enthalpy change"
-W3_Dx=h2-h1-To*(s2-s1) "Using C for To instead of K"
-W4_Dx=(h2+(T2+273)*s2)-(h1+(T1+273)*s1) "Using wrong relations for exergy"
```

7- 140 ... 7- 143 Design and Essay Problems

Chapter 8
GAS POWER CYCLES

Actual and Ideal Cycles, Carnot cycle, Air-Standard Assumptions

8-1C The Carnot cycle is not suitable as an ideal cycle for all power producing devices because it cannot be approximated using the hardware of actual power producing devices.

8-2C It is less than the thermal efficiency of a Carnot cycle.

8-3C It represents the net work on both diagrams.

8-4C The cold air standard assumptions involves the additional assumption that air can be treated as an ideal gas with constant specific heats at room temperature.

8-5C Under the air standard assumptions, the combustion process is modeled as a heat addition process, and the exhaust process as a heat rejection process.

8-6C The air standard assumptions are: (1) the working fluid is air which behaves as an ideal gas, (2) all the processes are internally reversible, (3) the combustion process is replaced by the heat addition process, and (4) the exhaust process is replaced by the heat rejection process which returns the working fluid to its original state.

8-7C The clearance volume is the minimum volume formed in the cylinder whereas the displacement volume is the volume displaced by the piston as the piston moves between the top dead center and the bottom dead center.

8-8C It is the ratio of the maximum to minimum volumes in the cylinder.

8-9C The MEP is the fictitious pressure which, if acted on the piston during the entire power stroke, would produce the same amount of net work as that produced during the actual cycle.

8-10C Yes.

8-11C Assuming no accumulation of carbon deposits on the piston face, the compression ratio will remain the same (otherwise it will increase). The mean effective pressure, on the other hand, will decrease as a car gets older as a result of wear and tear.

8-12C The SI and CI engines differ from each other in the way combustion is initiated; by a spark in SI engines, and by compressing the air above the self-ignition temperature of the fuel in CI engines.

8-13C Stroke is the distance between the TDC and the BDC, bore is the diameter of the cylinder, TDC is the position of the piston when it forms the smallest volume in the cylinder, and clearance volume is the minimum volume formed in the cylinder.

8-14 The four processes of an air-standard cycle are described. The cycle is to be shown on *P-v* and *T-s* diagrams, and the net work output and the thermal efficiency are to be determined.

Assumptions **1** The air-standard assumptions are applicable. **2** Kinetic and potential energy changes are negligible. **3** Air is an ideal gas with variable specific heats.

Properties The properties of air are given in Table A-17.

Analysis (*b*) The properties of air at various states are

$$T_1 = 300\text{K} \longrightarrow \begin{array}{l} h_1 = 300.19\text{kJ/kg} \\ P_{r_1} = 1.386 \end{array}$$

$$P_{r_2} = \frac{P_2}{P_1} P_{r_1} = \frac{800\text{kPa}}{100\text{kPa}}(1.386) = 11.088 \longrightarrow \begin{array}{l} u_2 = 389.22\text{kJ/kg} \\ T_2 = 539.8\text{K} \end{array}$$

$$T_3 = 1800\text{K} \longrightarrow \begin{array}{l} u_3 = 1487.2\text{kJ/kg} \\ P_{r_3} = 1310 \end{array}$$

$$\frac{P_3 v_3}{T_3} = \frac{P_2 v_2}{T_2} \longrightarrow P_3 = \frac{T_3}{T_2} P_2 = \frac{1800\text{K}}{539.8\text{K}}(800\text{kPa}) = 2668\text{kPa}$$

$$P_{r_4} = \frac{P_4}{P_3} P_{r_3} = \frac{100\text{kPa}}{2668\text{kPa}}(1310) = 49.10 \longrightarrow h_4 = 828.1\text{kJ/kg}$$

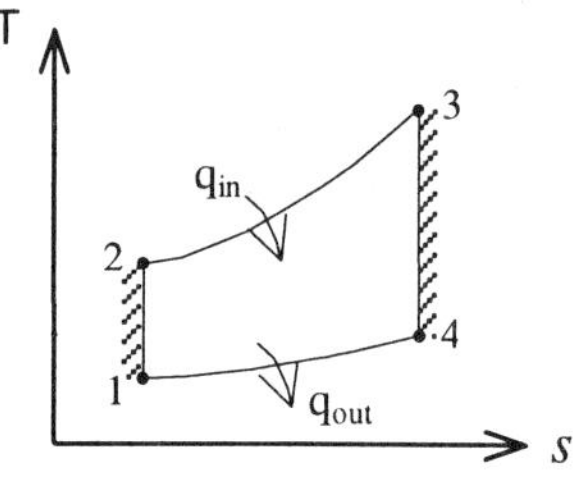

From energy balances,

$$q_{in} = u_3 - u_2 = 1487.2 - 389.2 = 1098.0 \text{ kJ/kg}$$

$$q_{out} = h_4 - h_1 = 828.1 - 300.19 = 527.9 \text{ kJ/kg}$$

$$w_{net,out} = q_{in} - q_{out} = 1098.0 - 527.9 = \mathbf{570.1 \text{ kJ/kg}}$$

(*c*) Then the thermal efficiency becomes

$$\eta_{th} = \frac{w_{net,out}}{q_{in}} = \frac{570.1 \text{ kJ/kg}}{1098.0 \text{ kJ/kg}} = \mathbf{51.9\%}$$

8-15 EES solution of this (and other comprehensive problems designated with the *computer icon*) is available to instructors at the *Instructor Manual* section of the *Online Learning Center* (OLC) at www.mhhe.com/cengel-boles. See the Preface for access information.

8-16 The four processes of an air-standard cycle are described. The cycle is to be shown on *P-v* and *T-s* diagrams, and the maximum temperature in the cycle and the thermal efficiency are to be determined.

Assumptions **1** The air-standard assumptions are applicable. **2** Kinetic and potential energy changes are negligible. **3** Air is an ideal gas with constant specific heats.

Properties The properties of air at room temperature are C_p = 1.005 kJ/kg.K, C_v = 0.718 kJ/kg· K, and k = 1.4 (Table A-2).

Analysis (*b*) From the ideal gas isentropic relations and energy balance,

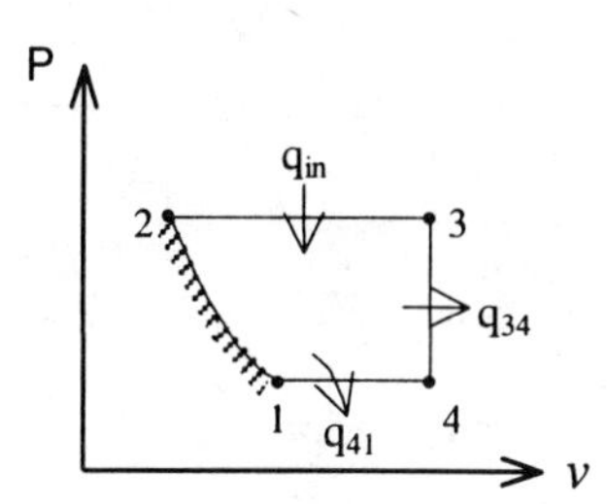

$$T_2 = T_1 \left(\frac{P_2}{P_1} \right)^{(k-1)/k} = (300\text{K}) \left(\frac{1000\text{kPa}}{100\text{kPa}} \right)^{0.4/1.4} = 579.2\text{K}$$

$$q_{in} = h_3 - h_2 = C_p (T_3 - T_2)$$

or,

$$2200 \text{ kJ/kg} = (1.005 \text{ kJ/kg} \cdot \text{K})(T_3 - 579.2)$$

$$T_{\text{max}} = T_3 = \mathbf{2768 \ K}$$

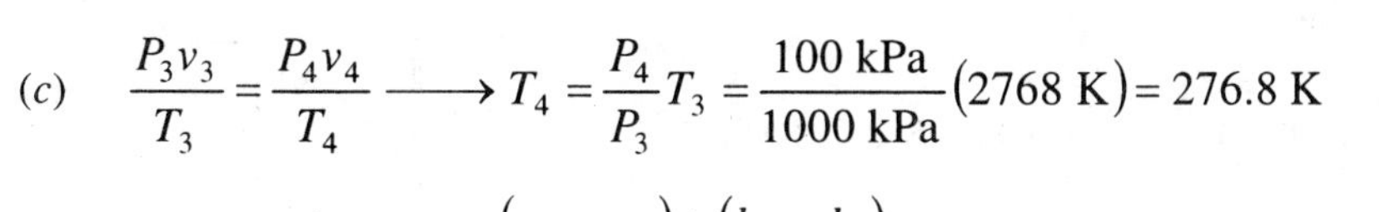

(*c*) $$\frac{P_3 v_3}{T_3} = \frac{P_4 v_4}{T_4} \longrightarrow T_4 = \frac{P_4}{P_3} T_3 = \frac{100 \text{ kPa}}{1000 \text{ kPa}} (2768 \text{ K}) = 276.8 \text{ K}$$

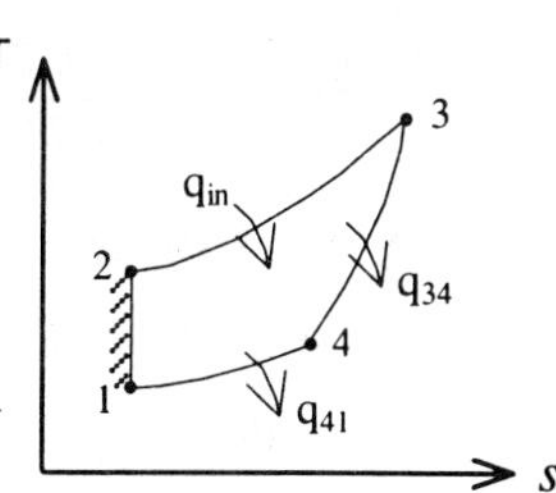

$$\begin{aligned} q_{out} &= q_{34,out} + q_{41,out} = (u_3 - u_4) + (h_4 - h_1) \\ &= C_v (T_3 - T_4) + C_p (T_4 - T_1) \\ &= (0.718 \text{ kJ/kg} \cdot \text{K})(2768 - 276.8)\text{K} + (1.005 \text{ kJ/kg} \cdot \text{K})(276.8 - 300)\text{K} \\ &= 1766 \text{ kJ/kg} \end{aligned}$$

$$\eta_{th} = 1 - \frac{q_{out}}{q_{in}} = 1 - \frac{1766 \text{ kJ/kg}}{2200 \text{ kJ/kg}} = \mathbf{19.7\%}$$

8-17E The four processes of an air-standard cycle are described. The cycle is to be shown on *P-v* and *T-s* diagrams, and the total heat input and the thermal efficiency are to be determined.

Assumptions **1** The air-standard assumptions are applicable. **2** Kinetic and potential energy changes are negligible. **3** Air is an ideal gas with variable specific heats.

Properties The properties of air are given in Table A-17E.

Analysis (*b*) The properties of air at various states are

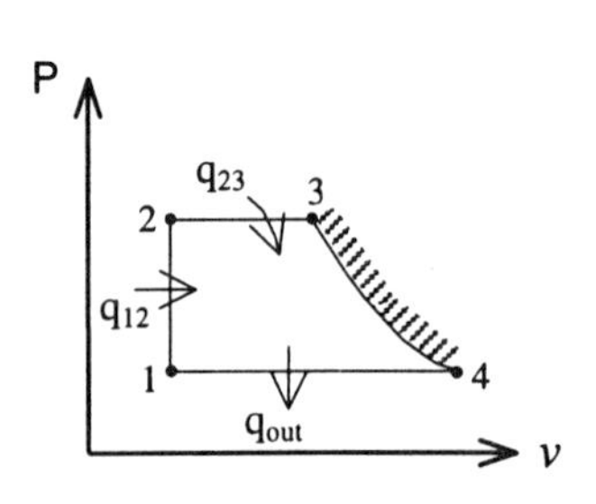

$$T_1 = 540 \text{ R} \longrightarrow \begin{aligned} u_1 &= 92.04 \text{ Btu/lbm} \\ h_1 &= 129.06 \text{ Btu/lbm} \end{aligned}$$

$$q_{\text{in},12} = u_2 - u_1 \longrightarrow \begin{aligned} u_2 &= u_1 + q_{\text{in},12} = 92.04 + 300 = 392.04 \text{Btu/lbm} \\ T_2 &= 2116\text{R},\ h_2 = 537.1 \text{Btu/lbm} \end{aligned}$$

$$\frac{P_2 v_2}{T_2} = \frac{P_1 v_1}{T_1} \longrightarrow P_2 = \frac{T_2}{T_1} P_1 = \frac{2116\text{R}}{540\text{R}}(14.7\text{psia}) = 57.6\text{psia}$$

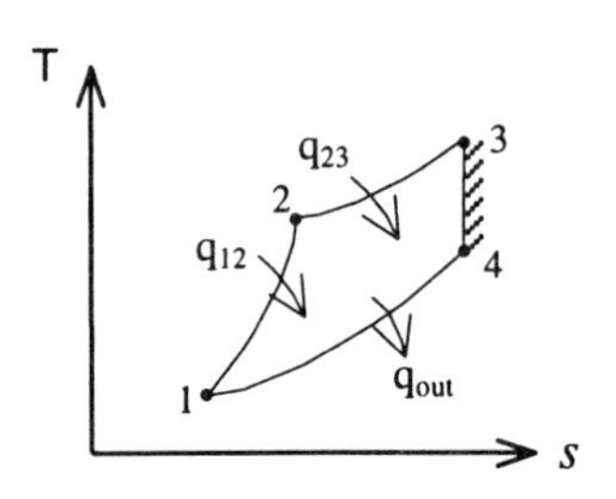

$$T_3 = 3200\text{R} \longrightarrow \begin{aligned} h_3 &= 849.48 \text{Btu/lbm} \\ P_{r_3} &= 1242 \end{aligned}$$

$$P_{r_4} = \frac{P_4}{P_3} P_{r_3} = \frac{14.7\text{psia}}{57.6\text{psia}}(1242) = 317.0 \longrightarrow h_4 = 593.22 \text{Btu/lbm}$$

From energy balance,

$$q_{23,in} = h_3 - h_2 = 849.48 - 537.1 = 312.38 \text{ Btu/lbm}$$

$$q_{in} = q_{12,in} + q_{23,in} = 300 + 312.38 = \mathbf{612.38 \text{ Btu/lbm}}$$

$$q_{out} = h_4 - h_1 = 593.22 - 129.06 = 464.16 \text{ Btu/lbm}$$

(*c*) Then the thermal efficiency becomes

$$\eta_{th} = 1 - \frac{q_{out}}{q_{in}} = 1 - \frac{464.16 \text{ Btu/lbm}}{612.38 \text{ Btu/lbm}} = \mathbf{24.2\%}$$

8-18E The four processes of an air-standard cycle are described. The cycle is to be shown on *P-v* and *T-s* diagrams, and the total heat input and the thermal efficiency are to be determined.

Assumptions **1** The air-standard assumptions are applicable. **2** Kinetic and potential energy changes are negligible. **3** Air is an ideal gas with constant specific heats.

Properties The properties of air at room temperature are C_p = 0.240 Btu/lbm.R, C_v = 0.171 Btu/lbm.R, and k = 1.4 (Table A-2E).

Analysis (*b*)

$$q_{in,12} = u_2 - u_1 = C_v(T_2 - T_1)$$

$$300\text{Btu/lbm} = (0.171\text{Btu/lbm.R})(T_2 - 540)\text{R}$$

$$T_2 = 2294\text{R}$$

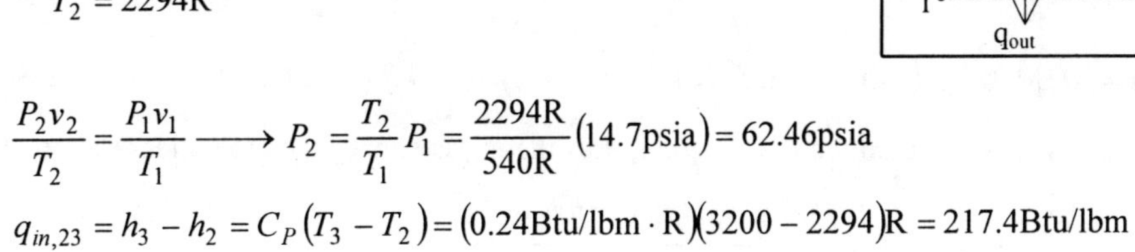

$$\frac{P_2 v_2}{T_2} = \frac{P_1 v_1}{T_1} \longrightarrow P_2 = \frac{T_2}{T_1} P_1 = \frac{2294\text{R}}{540\text{R}}(14.7\text{psia}) = 62.46\text{psia}$$

$$q_{in,23} = h_3 - h_2 = C_P(T_3 - T_2) = (0.24\text{Btu/lbm}\cdot\text{R})(3200 - 2294)\text{R} = 217.4\text{Btu/lbm}$$

Process 3-4 is isentropic:

$$T_4 = T_3\left(\frac{P_4}{P_3}\right)^{(k-1)/k} = (3200\text{R})\left(\frac{14.7\text{psia}}{62.46\text{psia}}\right)^{0.4/1.4} = 2117\text{R}$$

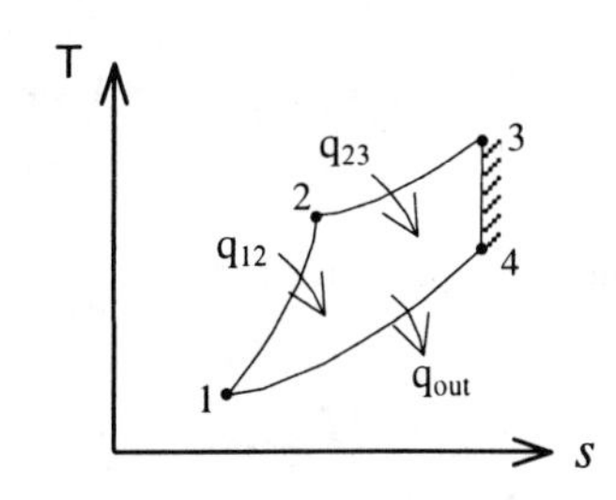

$$q_{in} = q_{in,12} + q_{in,23} = 300 + 217.4 = \mathbf{517.4 Btu/lbm}$$

$$q_{out} = h_4 - h_1 = C_p(T_4 - T_1) = (0.240\text{Btu/lbm.R})(2117 - 540)$$

$$= 378.5\text{Btu/lbm}$$

(c) $$\eta_{th} = 1 - \frac{q_{out}}{q_{in}} = 1 - \frac{378.5 \text{ Btu/lbm}}{517.4 \text{ Btu/lbm}} = \mathbf{26.8\%}$$

8-19 The three processes of an air-standard cycle are described. The cycle is to be shown on *P-v* and *T-s* diagrams, and the heat rejected and the thermal efficiency are to be determined. √

Assumptions **1** The air-standard assumptions are applicable. **2** Kinetic and potential energy changes are negligible. **3** Air is an ideal gas with constant specific heats.

Properties The properties of air at room temperature are C_p = 1.005 kJ/kg.K, C_v = 0.718 kJ/kg· K, and k = 1.4 (Table A-2).

Analysis (*b*)

$$T_2 = T_1\left(\frac{P_2}{P_1}\right)^{(k-1)/k} = (300\text{K})\left(\frac{1000\text{kPa}}{100\text{kPa}}\right)^{0.4/1.4} = 579.2\text{K}$$

$$Q_{in} = m(h_3 - h_2) = mC_p(T_3 - T_2)$$

or,

$$2.76 \text{ kJ} = (0.0015 \text{ kg})(1.005 \text{ kJ/kg}\cdot\text{K})(T_3 - 579.2) \longrightarrow T_3 = 2410 \text{ K}$$

Process 3-1 is a straight line on the P-*v* diagram, thus the w_{31} is simply the area under the process curve,

$$w_{31} = area = \frac{P_3 + P_1}{2}(v_1 - v_3) = \frac{P_3 + P_1}{2}\left(\frac{RT_1}{P_1} - \frac{RT_3}{P_3}\right)$$

$$= \left(\frac{1000 + 100\text{kPa}}{2}\right)\left(\frac{300\text{K}}{100\text{kPa}} - \frac{2410\text{K}}{1000\text{kPa}}\right)(0.287\text{kJ/kg}\cdot\text{K}) = 93.1\text{kJ/kg}$$

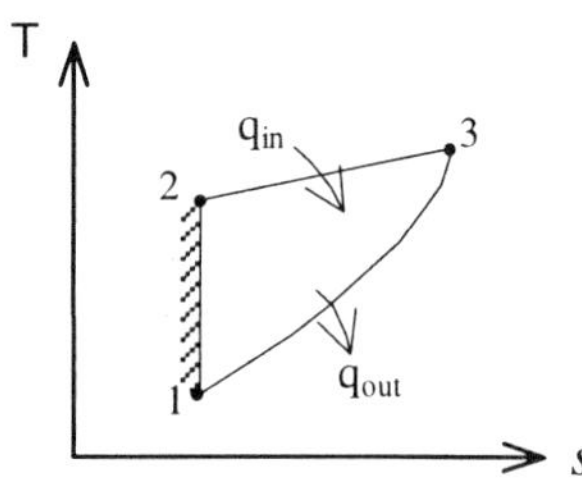

Energy balance for process 3-1 gives

$$E_{in} - E_{out} = \Delta E_{system}$$

$$-Q_{31,out} - W_{31,out} = m(u_1 - u_3)$$

$$Q_{31,out} = -mw_{31,out} - mC_v(T_1 - T_3) = -m[w_{31,out} + C_v(T_1 - T_3)]$$

$$= -(0.0015 \text{ kg})[93.1 + (0.718\text{kJ/kg}\cdot\text{K})(300 - 2410)\text{K}] = \mathbf{2.133 \text{ kJ}}$$

(*c*)

$$\eta_{th} = 1 - \frac{Q_{out}}{Q_{in}} = 1 - \frac{2.133 \text{ kJ}}{2.76 \text{ kJ}} = \mathbf{22.7\%}$$

8-20 The three processes of an air-standard cycle are described. The cycle is to be shown on *P-v* and *T-s* diagrams, and the net work per cycle and the thermal efficiency are to be determined. √

Assumptions **1** The air-standard assumptions are applicable. **2** Kinetic and potential energy changes are negligible. **3** Air is an ideal gas with variable specific heats.

Properties The properties of air are given in Table A-17.

Analysis (*b*) The properties of air at various states are

$$T_1 = 290K \longrightarrow \begin{array}{l} u_1 = 206.91\text{kJ/kg} \\ h_1 = 290.16\text{kJ/kg} \end{array}$$

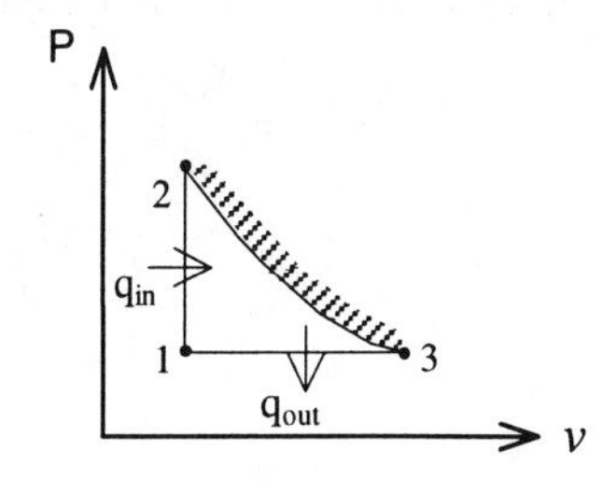

$$\frac{P_2 v_2}{T_2} = \frac{P_1 v_1}{T_1} \longrightarrow T_2 = \frac{P_2}{P_1} T_1 = \frac{380\text{kPa}}{95\text{kPa}}(290\text{K}) = 1160\text{K}$$

$$\longrightarrow u_2 = 897.91\text{kJ/kg},\ P_{r_2} = 207.2$$

$$P_{r_3} = \frac{P_3}{P_2} P_{r_2} = \frac{95\text{kPa}}{380\text{kPa}}(207.2) = 51.8 \longrightarrow h_3 = 840.38\text{kJ/kg}$$

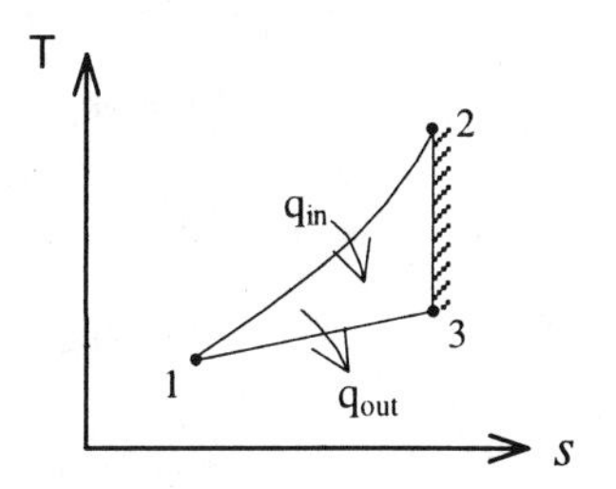

$$Q_{in} = m(u_2 - u_1) = (0.003\text{kg})(897.91 - 206.91)\text{kJ/kg} = 2.073\text{kJ}$$

$$Q_{out} = m(h_3 - h_1) = (0.003\text{kg})(840.38 - 290.16)\text{kJ/kg} = 1.651\text{kJ}$$

$$W_{net,out} = Q_{in} - Q_{out} = 2.073 - 1.651 = \mathbf{0.422kJ}$$

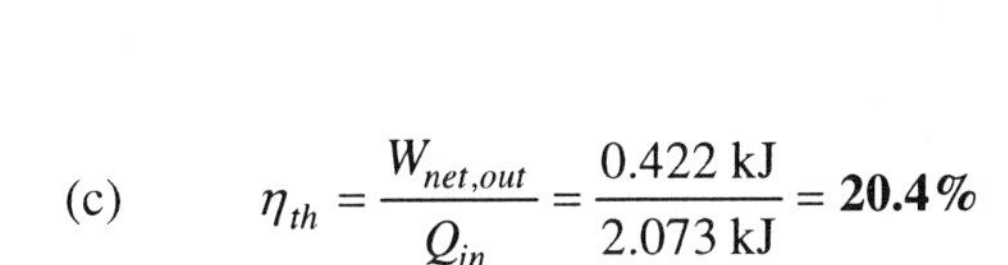

(c) $$\eta_{th} = \frac{W_{net,out}}{Q_{in}} = \frac{0.422\text{ kJ}}{2.073\text{ kJ}} = \mathbf{20.4\%}$$

8-21 The three processes of an air-standard cycle are described. The cycle is to be shown on *P-v* and *T-s* diagrams, and the net work per cycle and the thermal efficiency are to be determined.

Assumptions **1** The air-standard assumptions are applicable. **2** Kinetic and potential energy changes are negligible. **3** Air is an ideal gas with constant specific heats.

Properties The properties of air at room temperature are C_p = 1.005 kJ/kg.K, C_v = 0.718 kJ/kg· K, and k = 1.4 (Table A-2).

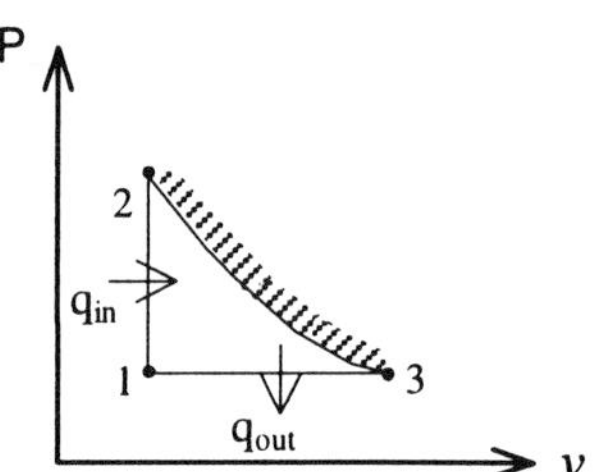

Analysis (*b*) From the isentropic relations and energy balance,

$$\frac{P_2 v_2}{T_2} = \frac{P_1 v_1}{T_1} \longrightarrow T_2 = \frac{P_2}{P_1} T_1 = \frac{380\text{kPa}}{95\text{kPa}}(290\text{K}) = 1160\text{K}$$

$$T_3 = T_2 \left(\frac{P_3}{P_2}\right)^{(k-1)/k} = (1160\text{K})\left(\frac{95\text{kPa}}{380\text{kPa}}\right)^{0.4/1.4} = 780.6\text{K}$$

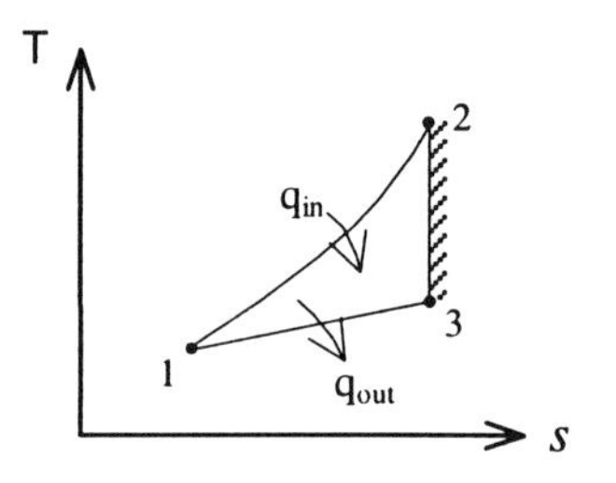

$$Q_{in} = m(u_2 - u_1) = mC_v(T_2 - T_1)$$
$$= (0.003\text{kg})(0.718\text{kJ/kg}\cdot\text{K})(1160 - 290)\text{K} = 1.87\text{kJ}$$

$$Q_{out} = m(h_3 - h_1) = mC_p(T_3 - T_1)$$
$$= (0.003\text{kg})(1.005\text{kJ/kg}\cdot\text{K})(780.6 - 290)\text{K} = 1.48\text{kJ}$$

$$W_{net,out} = Q_{in} - Q_{out} = 1.87 - 1.48 = \mathbf{0.39kJ}$$

(*c*)

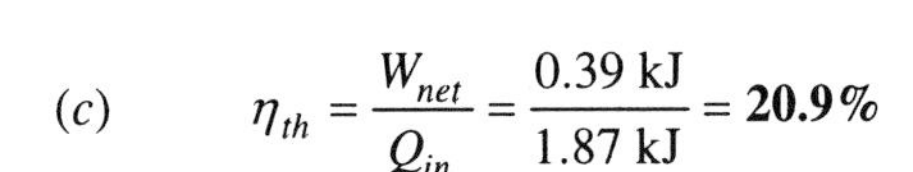

$$\eta_{th} = \frac{W_{net}}{Q_{in}} = \frac{0.39\text{ kJ}}{1.87\text{ kJ}} = \mathbf{20.9\%}$$

8-22 A Carnot cycle with the specified temperature limits is considered. The net work output per cycle is to be determined.

Assumptions Air is an ideal gas with constant specific heats.

Properties The properties of air at room temperature are C_p = 1.005 kJ/kg.K, C_v = 0.718 kJ/kg· K, and k = 1.4 (Table A-2).

Analysis The minimum pressure in the cycle is P_3 and the maximum pressure is P_1. Then,

$$\frac{T_2}{T_3} = \left(\frac{P_2}{P_3}\right)^{(k-1)/k}$$

or,

$$P_2 = P_3\left(\frac{T_2}{T_3}\right)^{k/(k-1)} = (20\text{kPa})\left(\frac{1000\text{K}}{300\text{K}}\right)^{1.4/0.4} = 1352\text{kPa}$$

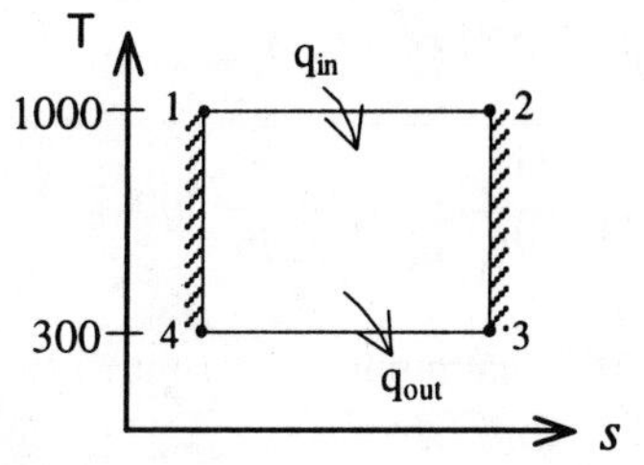

T
1000
300
1
2
3
4
q_{in}
q_{out}
s

The heat input is determined from

$$s_2 - s_1 = C_p \ln\frac{T_2}{T_1}^{\nearrow 0} - R\ln\frac{P_2}{P_1} = -(0.287\text{kJ/kg}\cdot\text{K})\ln\frac{1352\text{ kPa}}{1800\text{ kPa}} = 0.08205\text{ kJ/kg}\cdot\text{K}$$

$$Q_{in} = mT_H(s_2 - s_1) = (0.003\text{ kg})(1000\text{ K})(0.08205\text{ kJ/kg}\cdot\text{K}) = 0.246\text{ kJ}$$

Then,

$$\eta_{th} = 1 - \frac{T_L}{T_H} = 1 - \frac{300\text{ K}}{1000\text{ K}} = 70.0\%$$

$$W_{net,out} = \eta_{th}Q_{in} = (0.70)(0.246\text{ kJ}) = \mathbf{0.172\ \ kJ}$$

8-23 A Carnot cycle with specified temperature limits is considered. The maximum pressure in the cycle, the heat transfer to the working fluid, and the mass of the working fluid are to be determined. √

Assumptions Air is an ideal gas with variable specific heats.

Analysis (*a*) In a Carnot cycle, the maximum pressure occurs at the beginning of the expansion process, which is state 1.

$$T_1 = 1200\text{ K} \longrightarrow P_{r_1} = 238$$
$$T_4 = 350\text{ K} \longrightarrow P_{r_4} = 2.379 \qquad \text{(Table A-17)}$$

$$P_1 = \frac{P_{r_1}}{P_{r_4}} P_4 = \frac{238}{2.379}(300\text{kPa}) = \mathbf{30{,}013kPa} = P_{\max}$$

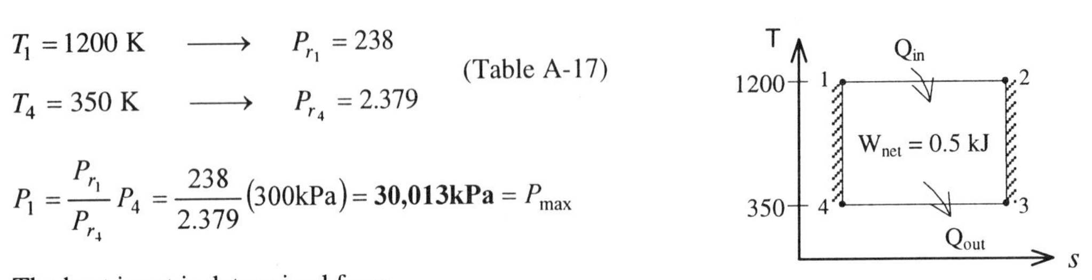

(*b*) The heat input is determined from

$$\eta_{th} = 1 - \frac{T_L}{T_H} = 1 - \frac{350\text{K}}{1200\text{K}} = 70.83\%$$

$$Q_{in} = W_{net,out} / \eta_{th} = (0.5\text{kJ})/(0.7083) = \mathbf{0.706kJ}$$

(*c*) The mass of air is

$$s_4 - s_3 = \left(s_4^\circ - s_3^\circ\right)^{\not 0} - R\ln\frac{P_4}{P_3} = -(0.287\text{kJ/kg}\cdot\text{K})\ln\frac{300\text{kPa}}{150\text{kPa}}$$
$$= -0.199\text{kJ/kg}\cdot K = s_1 - s_2$$

$$w_{net,out} = (s_2 - s_1)(T_H - T_L) = (0.199\text{kJ/kg}\cdot\text{K})(1200 - 350)K = 169.15\text{kJ/kg}$$

$$m = \frac{W_{net,out}}{w_{net,out}} = \frac{0.5\text{kJ}}{169.15\text{kJ/kg}} = \mathbf{0.00296kg}$$

8-24 A Carnot cycle with specified temperature limits is considered. The maximum pressure in the cycle, the heat transfer to the working fluid, and the mass of the working fluid are to be determined.

Assumptions Helium is an ideal gas with constant specific heats.

Properties The properties of helium at room temperature are $R = 2.0769$ kJ/kg.K and $k = 1.667$ (Table A-2).

Analysis (*a*) In a Carnot cycle, the maximum pressure occurs at the beginning of the expansion process, which is state 1.

$$\frac{T_1}{T_4} = \left(\frac{P_1}{P_4}\right)^{(k-1)/k}$$

or,

$$P_1 = P_4\left(\frac{T_1}{T_4}\right)^{k/(k-1)} = (300\text{kPa})\left(\frac{1200\text{K}}{350\text{K}}\right)^{1.667/0.667} = 6524\text{kPa}$$

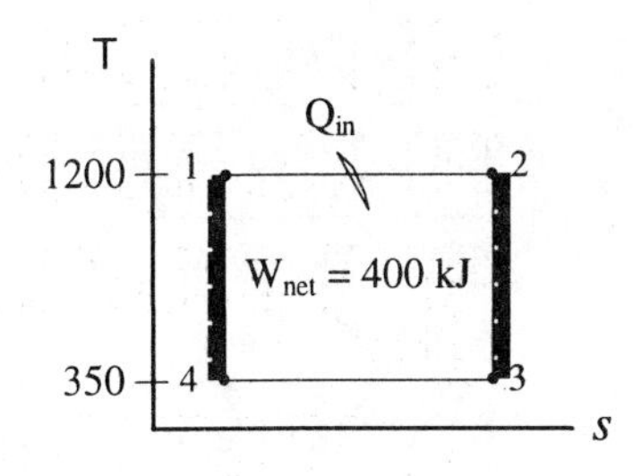

(*b*) The heat input is determined from

$$\eta_{th} = 1 - \frac{T_L}{T_H} = 1 - \frac{350\text{K}}{1200\text{K}} = 70.83\%$$

$$Q_{in} = W_{net,out} / \eta_{th} = (0.5\text{kJ})/(0.7083) = \mathbf{0.706kJ}$$

(*c*) The mass of helium is determined from

$$s_4 - s_3 = C_p \ln\frac{T_4}{T_3}^{\nearrow 0} - R\ln\frac{P_4}{P_3} = -(2.0769\ \text{kJ/kg}\cdot\text{K})\ln\frac{300\ \text{kPa}}{150\ \text{kPa}}$$

$$= -1.4396\ \text{kJ/kg}\cdot\text{K} = s_1 - s_2$$

$$w_{net,out} = (s_2 - s_1)(T_H - T_L) = (1.4396\ \text{kJ/kg}\cdot\text{K})(1200 - 350)\text{K} = 1223.7\ \text{kJ/kg}$$

$$m = \frac{W_{net,out}}{w_{net,out}} = \frac{0.5\ \text{kJ}}{1223.7\ \text{kJ/kg}} = \mathbf{0.000409\ kg}$$

Otto Cycle

8-25C The four processes that make up the Otto cycle are (1) isentropic compression, (2) v = constant heat addition, (3) isentropic expansion, and (4) v = constant heat rejection.

8-26C The ideal Otto cycle involves external irreversibilities, and thus it has a lower thermal efficiency.

8-27C For actual four-stroke engines, the rpm is twice the number of thermodynamic cycles; for two-stroke engines, it is equal to the number of thermodynamic cycles.

8-28C They are analyzed as closed system processes because no mass crosses the system boundaries during any of the processes.

8-29C It increases with both of them.

8-30C Because high compression ratios cause engine knock.

8-31C The thermal efficiency will be the highest for argon because it has the highest specific heat ratio, k = 1.667.

8-32C The fuel is injected into the cylinder in both engines, but it is ignited with a spark plug in gasoline engines.

8-33 An ideal Otto cycle with air as the working fluid has a compression ratio of 8. The pressure and temperature at the end of the heat addition process, the net work output, the thermal efficiency, and the mean effective pressure for the cycle are to be determined.

Assumptions **1** The air-standard assumptions are applicable. **2** Kinetic and potential energy changes are negligible. **3** Air is an ideal gas with variable specific heats.

Properties The properties of air are given in Table A-17.

Analysis (*a*) Process 1-2: isentropic compression.

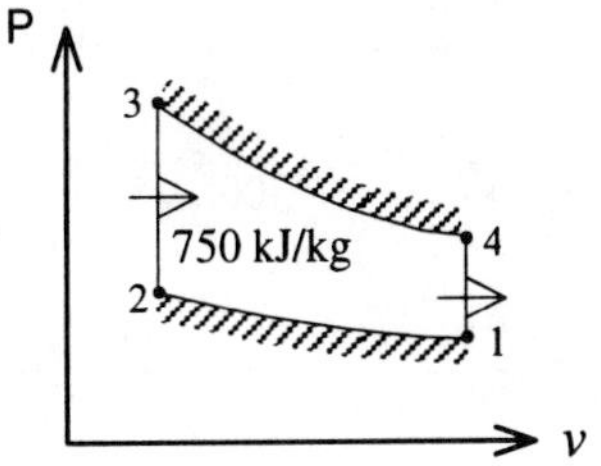

$$T_1 = 300 \text{ K} \longrightarrow \begin{array}{l} u_1 = 214.07 \text{ kJ/kg} \\ v_{r_1} = 621.2 \end{array}$$

$$v_{r_2} = \frac{v_2}{v_1} v_{r_1} = \frac{1}{r} v_{r_1} = \frac{1}{8}(621.2) = 77.65$$

$$\longrightarrow \begin{array}{l} T_2 = 673.1\text{K} \\ u_2 = 491.2\text{kJ/kg} \end{array}$$

$$\frac{P_2 v_2}{T_2} = \frac{P_1 v_1}{T_1} \longrightarrow P_2 = \frac{v_1}{v_2}\frac{T_2}{T_1} P_1 = (8)\left(\frac{673.1\text{K}}{300\text{K}}\right)(95\text{kPa}) = 1705\text{kPa}$$

Process 2-3: v = constant heat addition.

$$q_{23,\text{in}} = u_3 - u_2 \longrightarrow u_3 = u_2 + q_{23,in} = 491.2 + 750 = 1241.2 \text{ kJ/kg} \longrightarrow \begin{array}{l} T_3 = \mathbf{1539 \text{ K}} \\ v_{r_3} = 6.588 \end{array}$$

$$\frac{P_3 v_3}{T_3} = \frac{P_2 v_2}{T_2} \longrightarrow P_3 = \frac{T_3}{T_2} P_2 = \left(\frac{1539\text{K}}{673.1\text{K}}\right)(1705\text{kPa}) = \mathbf{3898kPa}$$

(*b*) Process 3-4: isentropic expansion.

$$v_{r_4} = \frac{v_1}{v_2} v_{r_3} = r v_{r_3} = (8)(6.588) = 52.70 \longrightarrow \begin{array}{l} T_4 = 774.5\text{K} \\ u_4 = 571.69\text{kJ/kg} \end{array}$$

Process 4-1: v = constant heat rejection.

$$q_{\text{out}} = u_4 - u_1 = 571.69 - 214.07 = 357.62 \text{ kJ/kg}$$

$$w_{net,out} = q_{in} - q_{out} = 750 - 357.62 = \mathbf{392.38 \text{ kJ/kg}}$$

(*c*)
$$\eta_{\text{th}} = \frac{w_{\text{net,out}}}{q_{\text{in}}} = \frac{392.38\text{kJ/kg}}{750\text{kJ/kg}} = \mathbf{52.3\%}$$

(*d*)
$$v_1 = \frac{RT_1}{P_1} = \frac{(0.287\text{kPa}\cdot\text{m}^3/\text{kg}\cdot\text{K})(300\text{K})}{95\text{kPa}} = 0.906\text{m}^3/\text{kg} = v_{\max}$$

$$v_{\min} = v_2 = \frac{v_{\max}}{r}$$

$$MEP = \frac{w_{\text{net,out}}}{v_1 - v_2} = \frac{w_{\text{net,out}}}{v_1(1-1/r)} = \frac{392.38\text{kJ/kg}}{(0.906\text{m}^3/\text{kg})(1-1/8)}\left(\frac{\text{kPa}\cdot\text{m}^3}{\text{kJ}}\right) = \mathbf{495.0kPa}$$

8-34 EES solution of this (and other comprehensive problems designated with the *computer icon*) is available to instructors at the *Instructor Manual* section of the *Online Learning Center* (OLC) at www.mhhe.com/cengel-boles. See the Preface for access information.

8-35 An ideal Otto cycle with air as the working fluid has a compression ratio of 8. The pressure and temperature at the end of the heat addition process, the net work output, the thermal efficiency, and the mean effective pressure for the cycle are to be determined.

Assumptions **1** The air-standard assumptions are applicable. **2** Kinetic and potential energy changes are negligible. **3** Air is an ideal gas with constant specific heats.

Properties The properties of air at room temperature are C_p = 1.005 kJ/kg· K, C_v = 0.718 kJ/kg· K, and k = 1.4 (Table A-2).

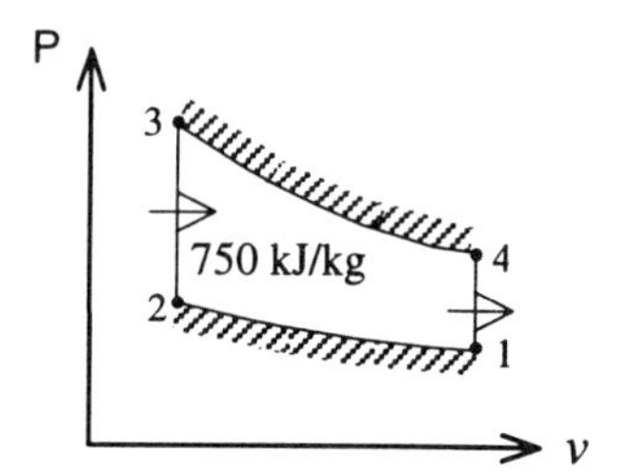

Analysis (*a*) Process 1-2: isentropic compression.

$$T_2 = T_1\left(\frac{v_1}{v_2}\right)^{k-1} = (300\text{K})(8)^{0.4} = 689\text{K}$$

$$\frac{P_2 v_2}{T_2} = \frac{P_1 v_1}{T_1} \longrightarrow P_2 = \frac{v_1}{v_2}\frac{T_2}{T_1}P_1 = (8)\left(\frac{689\text{K}}{300\text{K}}\right)(95\text{kPa}) = 1745\text{kPa}$$

Process 2-3: v = constant heat addition.

$$q_{23,\text{in}} = u_3 - u_2 = C_v(T_3 - T_2)$$
$$750\text{kJ/kg} = (0.718\text{kJ/kg}\cdot\text{K})(T_3 - 689)\text{K}$$
$$T_3 = \mathbf{1734K}$$

$$\frac{P_3 v_3}{T_3} = \frac{P_2 v_2}{T_2} \longrightarrow P_3 = \frac{T_3}{T_2}P_2 = \left(\frac{1734\text{K}}{689\text{K}}\right)(1745\text{kPa}) = \mathbf{4392kPa}$$

(*b*) Process 3-4: isentropic expansion.

$$T_4 = T_3\left(\frac{v_3}{v_4}\right)^{k-1} = (1734\text{K})\left(\frac{1}{8}\right)^{0.4} = 755\text{K}$$

Process 4-1: v = constant heat rejection.

$$q_{\text{out}} = u_4 - u_1 = C_v(T_4 - T_1) = (0.718\text{kJ/kg}\cdot\text{K})(755 - 300)\text{K} = 327\text{kJ/kg}$$

$$w_{\text{net,out}} = q_{\text{in}} - q_{\text{out}} = 750 - 327 = \mathbf{423kJ/kg}$$

(*c*)
$$\eta_{\text{th}} = \frac{w_{\text{net,out}}}{q_{\text{in}}} = \frac{423\text{ kJ/kg}}{750\text{ kJ/kg}} = \mathbf{56.4\%}$$

(*d*)
$$v_1 = \frac{RT_1}{P_1} = \frac{(0.287\text{kPa}\cdot\text{m}^3/\text{kg}\cdot\text{K})(300\text{K})}{95\text{kPa}} = 0.906\text{m}^3/\text{kg} = v_{\max}$$

$$v_{\min} = v_2 = \frac{v_{\max}}{r}$$

$$MEP = \frac{w_{\text{net,out}}}{v_1 - v_2} = \frac{w_{\text{net,out}}}{v_1(1 - 1/r)} = \frac{423\text{kJ/kg}}{(0.906\text{m}^3/\text{kg})(1 - 1/8)}\left(\frac{\text{kPa}\cdot\text{m}^3}{\text{kJ}}\right) = \mathbf{534kPa}$$

8-36 An ideal Otto cycle with air as the working fluid has a compression ratio of 9.5. The highest pressure and temperature in the cycle, the amount of heat transferred, the thermal efficiency, and the mean effective pressure are to be determined.

Assumptions **1** The air-standard assumptions are applicable. **2** Kinetic and potential energy changes are negligible. **3** Air is an ideal gas with constant specific heats.

Properties The properties of air at room temperature are C_p = 1.005 kJ/kg· K, C_v = 0.718 kJ/kg· K, and k = 1.4 (Table A-2).

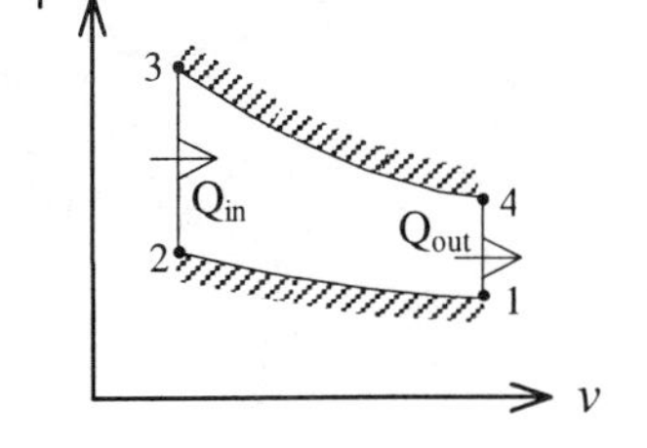

Analysis (*a*) Process 1-2: isentropic compression.

$$T_2 = T_1\left(\frac{v_1}{v_2}\right)^{k-1} = (290\text{K})(9.5)^{0.4} = 713.7\text{K}$$

$$\frac{P_2 v_2}{T_2} = \frac{P_1 v_1}{T_1} \longrightarrow P_2 = \frac{v_1}{v_2}\frac{T_2}{T_1}P_1 = (9.5)\left(\frac{713.7\text{K}}{290\text{K}}\right)(100\text{kPa}) = 2338\text{kPa}$$

Process 3-4: isentropic expansion.

$$T_3 = T_4\left(\frac{v_4}{v_3}\right)^{k-1} = (800\text{K})(9.5)^{0.4} = \mathbf{1969K}$$

Process 2-3: v = constant heat addition.

$$\frac{P_3 v_3}{T_3} = \frac{P_2 v_2}{T_2} \longrightarrow P_3 = \frac{T_3}{T_2}P_2 = \left(\frac{1969\text{K}}{713.7\text{K}}\right)(2338\text{kPa}) = \mathbf{6449kPa}$$

(*b*)

$$m = \frac{P_1 V_1}{RT_1} = \frac{(100\text{kPa})(0.0006\text{m}^3)}{(0.287\text{kPa}\cdot\text{m}^3/\text{kg}\cdot\text{K})(290\text{K})} = 7.21\times10^{-4}\text{kg}$$

$$Q_{\text{in}} = m(u_3 - u_2) = mC_v(T_3 - T_2) = (7.21\times10^{-4}\text{kg})(0.718\text{kJ/kg}\cdot\text{K})(1969 - 713.7)\text{K} = \mathbf{0.650kJ}$$

(c) Process 4-1: v = constant heat rejection.

$$Q_{\text{out}} = m(u_4 - u_1) = mC_v(T_4 - T_1) = -(7.21\times10^{-4}\text{kg})(0.718\text{kJ/kg}\cdot\text{K})(800 - 290)\text{K} = \mathbf{0.264kJ}$$

$$W_{\text{net}} = Q_{\text{in}} - Q_{\text{out}} = 0.650 - 0.264 = 0.386\text{ kJ}$$

$$\eta_{\text{th}} = \frac{W_{\text{net,out}}}{Q_{\text{in}}} = \frac{0.386\text{ kJ}}{0.650\text{ kJ}} = \mathbf{59.4\%}$$

(*d*)

$$V_{\min} = V_2 = \frac{V_{\max}}{r}$$

$$MEP = \frac{W_{net,out}}{V_1 - V_2} = \frac{W_{net,out}}{V_1(1-1/r)} = \frac{0.386\text{kJ}}{(0.0006\text{m}^3)(1-1/9.5)}\left(\frac{\text{kPa}\cdot\text{m}^3}{\text{kJ}}\right) = \mathbf{719kPa}$$

8-37 An Otto cycle with air as the working fluid has a compression ratio of 9.5. The highest pressure and temperature in the cycle, the amount of heat transferred, the thermal efficiency, and the mean effective pressure are to be determined.

Assumptions **1** The air-standard assumptions are applicable. **2** Kinetic and potential energy changes are negligible. **3** Air is an ideal gas with constant specific heats.

Properties The properties of air at room temperature are $C_p = 1.005$ kJ/kg· K, $C_v = 0.718$ kJ/kg· K, and $k = 1.4$ (Table A-2).

Analysis (*a*) Process 1-2: isentropic compression.

$$T_2 = T_1\left(\frac{v_1}{v_2}\right)^{k-1} = (290\text{K})(9.5)^{0.4} = 713.7\text{K}$$

$$\frac{P_2 v_2}{T_2} = \frac{P_1 v_1}{T_1} \longrightarrow P_2 = \frac{v_1}{v_2}\frac{T_2}{T_1}P_1 = (9.5)\left(\frac{713.7\text{K}}{290\text{K}}\right)(100\text{kPa}) = 2338\text{kPa}$$

Process 3-4: polytropic expansion.

$$m = \frac{P_1 V_1}{RT_1} = \frac{(100\text{ kPa})(0.0006\text{ m}^3)}{(0.287\text{ kPa}\cdot\text{m}^3/\text{kg}\cdot\text{K})(290\text{ K})} = 7.209\times10^{-4}\text{ kg}$$

$$T_3 = T_4\left(\frac{v_4}{v_3}\right)^{n-1} = (800\text{K})(9.5)^{0.35} = \mathbf{1759\ K}$$

$$W_{34} = \frac{mR(T_4 - T_3)}{1-n} = \frac{(7.209\times10^{-4})(0.287\text{ kJ/kg}\cdot\text{K})(800-1759)\text{K}}{1-1.35} = 0.567\text{ kJ}$$

Then energy balance for process 3-4 gives

$$E_{in} - E_{out} = \Delta E_{system}$$

$$Q_{34,in} - W_{34,out} = m(u_4 - u_3)$$

$$Q_{34,in} = m(u_4 - u_3) + W_{34,out} = mC_v(T_4 - T_3) + W_{34,out}$$

$$Q_{34,in} = (7.209\times10^{-4}\text{kg})(0.718\text{kJ/kg}\cdot\text{K})(800-1759)\text{K} + 0.567\text{kJ} = 0.071\text{kJ}$$

That is, 0.071 kJ of heat is added to the air during the expansion process (This is not realistic, and probably is due to assuming constant specific heats at room temperature).

(*b*) Process 2-3: v = constant heat addition.

$$\frac{P_3 v_3}{T_3} = \frac{P_2 v_2}{T_2} \longrightarrow P_3 = \frac{T_3}{T_2}P_2 = \left(\frac{1759\text{K}}{713.7\text{K}}\right)(2338\text{kPa}) = \mathbf{5762kPa}$$

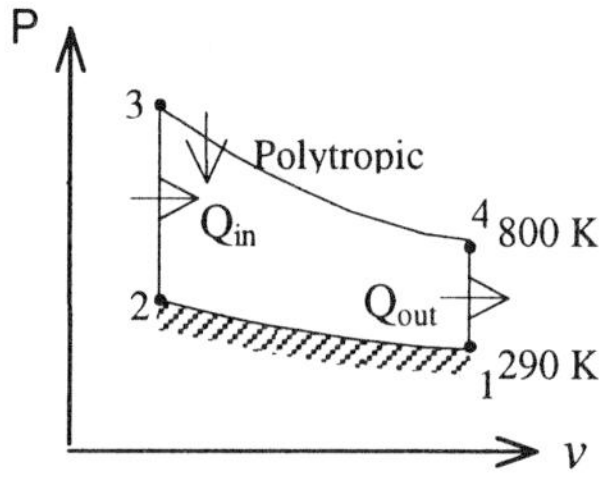

$$Q_{23,in} = m(u_3 - u_2) = mC_v(T_3 - T_2)$$

$$Q_{23,in} = (7.209\times10^{-4}\text{kg})(0.718\text{kJ/kg}\cdot\text{K})(1759 - 713.7)K = \mathbf{0.541kJ}$$

Therefore,

$$Q_{in} = Q_{23,in} + Q_{34,in} = 0.541 + 0.071 = 0.612\text{ kJ}$$

(*c*) Process 4-1: v = constant heat rejection.

$$Q_{out} = m(u_4 - u_1) = mC_v(T_4 - T_1) = (7.209\times10^{-4}\text{kg})(0.718\text{kJ/kg}\cdot\text{K})(800-290)K = \mathbf{0.264kJ}$$

$$W_{net,out} = Q_{in} - Q_{out} = 0.612 - 0.264 = 0.348\text{ kJ}$$

$$\eta_{th} = \frac{W_{net,out}}{Q_{in}} = \frac{0.348 \text{ kJ}}{0.612 \text{ kJ}} = \mathbf{56.9\%}$$

(*d*) $$V_{\min} = V_2 = \frac{V_{\max}}{r}$$

$$MEP = \frac{W_{net,out}}{V_1 - V_2} = \frac{W_{net,out}}{V_1(1-1/r)} = \frac{0.348\text{kJ}}{(0.0006\text{m}^3)(1-1/9.5)}\left(\frac{\text{kPa}\cdot\text{m}^3}{\text{kJ}}\right) = \mathbf{648kPa}$$

8-38E An ideal Otto cycle with air as the working fluid has a compression ratio of 8. The amount of heat transferred to the air during the heat addition process, the thermal efficiency, and the thermal efficiency of a Carnot cycle operating between the same temperature limits are to be determined.

Assumptions **1** The air-standard assumptions are applicable. **2** Kinetic and potential energy changes are negligible. **3** Air is an ideal gas with variable specific heats.

Properties The properties of air are given in Table A-17E.

Analysis (a) Process 1-2: isentropic compression.

$$T_1 = 540 \text{ R} \longrightarrow \begin{array}{l} u_1 = 92.04 \text{ Btu/lbm} \\ v_{r_1} = 144.32 \end{array}$$

$$v_{r_2} = \frac{v_2}{v_1} v_{r_2} = \frac{1}{r} v_{r_2} = \frac{1}{8}(144.32) = 18.04 \longrightarrow u_2 = 211.28 \text{Btu/lbm}$$

Process 2-3: v = constant heat addition.

$$T_3 = 2400\text{R} \longrightarrow \begin{array}{l} u_3 = 452.70 \text{ Btu/lbm} \\ v_{r_3} = 2.419 \end{array}$$

$$q_{in} = u_3 - u_2 = 452.70 - 211.28 = 241.42 \textbf{ Btu/lbm}$$

(b) Process 3-4: isentropic expansion.

$$v_{r_4} = \frac{v_4}{v_3} v_{r_3} = r v_{r_3} = (8)(2.419) = 19.35 \longrightarrow u_4 = 205.54 \text{ Btu/lbm}$$

Process 4-1: v = constant heat rejection.

$$q_{out} = u_4 - u_1 = 205.54 - 92.04 = 113.50 \text{ Btu/lbm}$$

$$\eta_{th} = 1 - \frac{q_{out}}{q_{in}} = 1 - \frac{113.50 \text{ Btu/lbm}}{241.42 \text{ Btu/lbm}} = 47.0\%$$

(c)

$$\eta_{th,C} = 1 - \frac{T_H}{T_L} = 1 - \frac{540 \text{ R}}{2400 \text{ R}} = \mathbf{77.5\%}$$

8-39E An ideal Otto cycle with argon as the working fluid has a compression ratio of 8. The amount of heat transferred to the argon during the heat addition process, the thermal efficiency, and the thermal efficiency of a Carnot cycle operating between the same temperature limits are to be determined. √

Assumptions **1** The air-standard assumptions are applicable with argon as the working fluid. **2** Kinetic and potential energy changes are negligible. **3** Argon is an ideal gas with constant specific heats.

Properties The properties of argon are C_p = 0.1253 Btu/lbm.R, C_v = 0.0756 Btu/lbm.R, and k = 1.667 (Table A-2E).

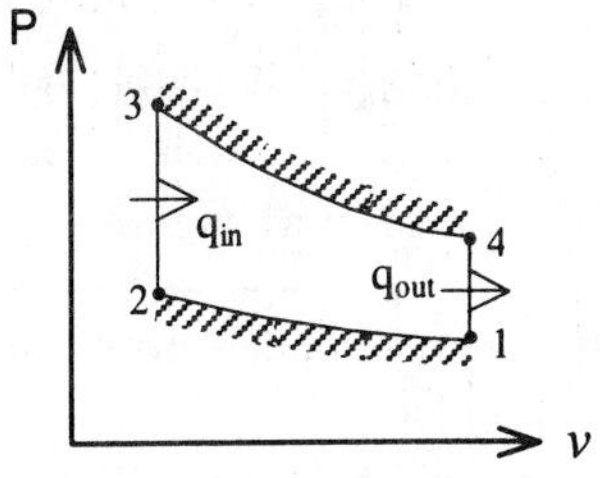

Analysis (*a*) Process 1-2: isentropic compression.

$$T_2 = T_1\left(\frac{v_1}{v_2}\right)^{k-1} = (540\text{ R})(8)^{0.667} = 2161\text{ R}$$

P, 3, q_{in}, 4, q_{out}, 2, 1, v

Process 2-3: v = constant heat addition.

$$q_{in} = u_3 - u_2 = C_v(T_3 - T_2) = (0.0756\text{ Btu/lbm.R})(2400 - 2161)\text{ R} = \mathbf{18.07\ \ Btu/lbm.R}$$

(*b*) Process 3-4: isentropic expansion.

$$T_4 = T_3\left(\frac{v_3}{v_4}\right)^{k-1} = (2400\text{R})\left(\frac{1}{8}\right)^{0.667} = 600\text{R}$$

Process 4-1: v = constant heat rejection.

$$q_{out} = u_4 - u_1 = C_v(T_4 - T_1) = (0.0756\text{ Btu/lbm.R})(600 - 540)\text{R} = 4.536\text{ Btu/lbm}$$

$$\eta_{th} = 1 - \frac{q_{out}}{q_{in}} = 1 - \frac{4.536\text{ Btu/lbm}}{18.07\text{ Btu/lbm}} = \mathbf{74.9\%}$$

(*c*)
$$\eta_{th,C} = 1 - \frac{T_H}{T_L} = 1 - \frac{540\text{ R}}{2400\text{ R}} = \mathbf{77.5\%}$$

Diesel Cycle

8-40C A diesel engine differs from the gasoline engine in the way combustion is initiated. In diesel engines combustion is initiated by compressing the air above the self-ignition temperature of the fuel whereas it is initiated by a spark plug in a gasoline engine.

8-41C The Diesel cycle differs from the Otto cycle in the heat addition process only; it takes place at constant volume in the Otto cycle, but at constant pressure in the Diesel cycle.

8-42C The gasoline engine.

8-43C Diesel engines operate at high compression ratios because the diesel engines do not have the engine knock problem.

8-44C Cutoff ratio is the ratio of the cylinder volumes after and before the combustion process. As the cutoff ratio decreases, the efficiency of the diesel cycle increases.

8-45 An air-standard Diesel cycle with a compression ratio of 16 and a cutoff ratio of 2 is considered. The temperature after the heat addition process, the thermal efficiency, and the mean effective pressure are to be determined.

Assumptions **1** The air-standard assumptions are applicable. **2** Kinetic and potential energy changes are negligible. **3** Air is an ideal gas with variable specific heats.

Properties The properties of air are given in Table A-17.

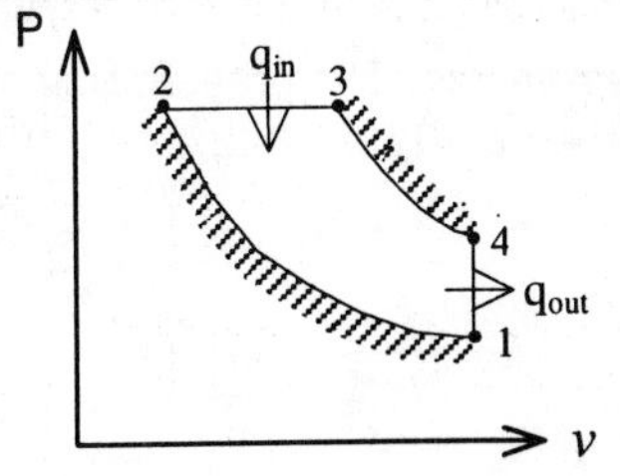

Analysis (*a*) Process 1-2: isentropic compression.

$$T_1 = 300\text{ K} \longrightarrow \begin{array}{l} u_1 = 214.07\text{ kJ/kg} \\ v_{r_1} = 621.2 \end{array}$$

$$v_{r_2} = \frac{v_2}{v_1} v_{r_1} = \frac{1}{r} v_{r_1} = \frac{1}{16}(621.2) = 38.825 \longrightarrow \begin{array}{l} T_2 = 862.4\text{ K} \\ h_2 = 890.9\text{ kJ/kg} \end{array}$$

Process 2-3: P = constant heat addition.

$$\frac{P_3 v_3}{T_3} = \frac{P_2 v_2}{T_2} \longrightarrow T_3 = \frac{v_3}{v_2} T_2 = 2T_2 = (2)(862.4\text{ K}) = \mathbf{1724.8\text{ K}}$$

$$\longrightarrow \begin{array}{l} h_3 = 1910.6\text{ kJ/kg} \\ v_{r_3} = 4.546 \end{array}$$

(*b*) $$q_{in} = h_3 - h_2 = 1910.6 - 890.9 = 1019.7\text{ kJ/kg}$$

Process 3-4: isentropic expansion.

$$v_{r_4} = \frac{v_4}{v_3} v_{r_3} = \frac{v_4}{2v_2} v_{r_3} = \frac{r}{2} v_{r_3} = \frac{16}{2}(4.546) = 36.37 \longrightarrow u_4 = 659.7\text{kJ/kg}$$

Process 4-1: v = constant heat rejection.

$$q_{out} = u_4 - u_1 = 659.7 - 214.07 = 445.63\text{ kJ/kg}$$

$$\eta_{th} = 1 - \frac{q_{out}}{q_{in}} = 1 - \frac{445.63\text{ kJ/kg}}{1019.7\text{ kJ/kg}} = \mathbf{56.3\%}$$

(*c*) $$w_{net,out} = q_{in} - q_{out} = 1019.7 - 445.63 = 574.07\text{kJ/kg}$$

$$v_1 = \frac{RT_1}{P_1} = \frac{(0.287\text{kPa}\cdot\text{m}^3/\text{kg}\cdot\text{K})(300\text{K})}{95\text{kPa}} = 0.906\text{m}^3/\text{kg} = v_{max}$$

$$v_{min} = v_2 = \frac{v_{max}}{r}$$

$$MEP = \frac{w_{net,out}}{v_1 - v_2} = \frac{w_{net,out}}{v_1(1-1/r)} = \frac{574.07\text{kJ/kg}}{(0.906\text{m}^3/\text{kg})(1-1/16)}\left(\frac{\text{kPa}\cdot\text{m}^3}{\text{kJ}}\right) = \mathbf{675.9\text{kPa}}$$

8-46 An air-standard Diesel cycle with a compression ratio of 16 and a cutoff ratio of 2 is considered. The temperature after the heat addition process, the thermal efficiency, and the mean effective pressure are to be determined.

Assumptions **1** The air-standard assumptions are applicable. **2** Kinetic and potential energy changes are negligible. **3** Air is an ideal gas with constant specific heats.

Properties The properties of air at room temperature are C_p = 1.005 kJ/kg· K, C_v = 0.718 kJ/kg· K, and k = 1.4 (Table A-2).

Analysis (*a*) Process 1-2: isentropic compression.

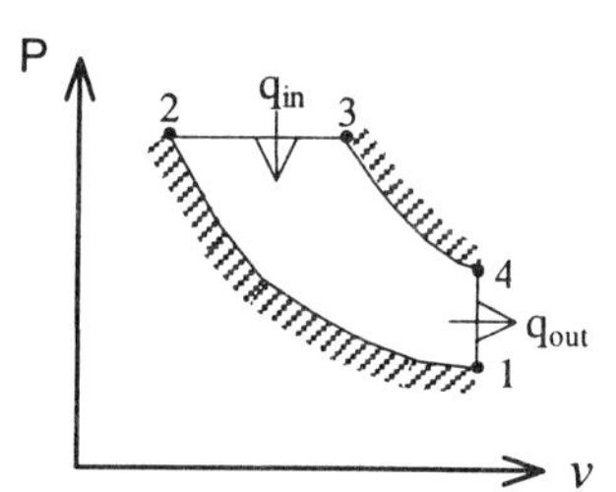

$$T_2 = T_1\left(\frac{v_1}{v_2}\right)^{k-1} = (300\text{K})(16)^{0.4} = 909.4\text{K}$$

Process 2-3: P = constant heat addition.

$$\frac{P_3 v_3}{T_3} = \frac{P_2 v_2}{T_2} \longrightarrow T_3 = \frac{v_3}{v_2}T_2 = 2T_2 = (2)(909.4\text{K}) = \mathbf{1818.8K}$$

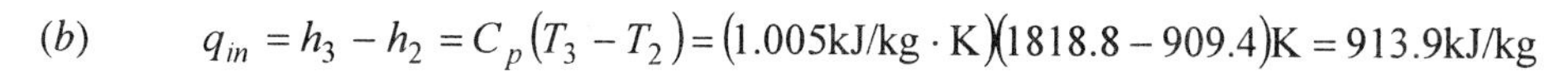

(*b*) $$q_{in} = h_3 - h_2 = C_p(T_3 - T_2) = (1.005\text{kJ/kg}\cdot\text{K})(1818.8 - 909.4)\text{K} = 913.9\text{kJ/kg}$$

Process 3-4: isentropic expansion.

$$T_4 = T_3\left(\frac{v_3}{v_4}\right)^{k-1} = T_3\left(\frac{2v_2}{v_4}\right)^{k-1} = (1818.8\text{K})\left(\frac{2}{16}\right)^{0.4} = 791.7\text{K}$$

Process 4-1: v = constant heat rejection.

$$q_{out} = u_4 - u_1 = C_v(T_4 - T_1) = (0.718\text{kJ/kg}\cdot\text{K})(791.7 - 300)\text{K} = 353\text{kJ/kg}$$

$$\eta_{th} = 1 - \frac{q_{out}}{q_{in}} = 1 - \frac{353\text{kJ/kg}}{913.9\text{kJ/kg}} = \mathbf{61.4\%}$$

(*c*) $$w_{net.out} = q_{in} - q_{out} = 913.9 - 353 = 560.9\text{kJ/kg}$$

$$v_1 = \frac{RT_1}{P_1} = \frac{(0.287\text{kPa}\cdot\text{m}^3/\text{kg}\cdot\text{K})(300\text{K})}{95\text{kPa}} = 0.906\text{m}^3/\text{kg} = v_{max}$$

$$v_{min} = v_2 = \frac{v_{max}}{r}$$

$$MEP = \frac{w_{net,out}}{v_1 - v_2} = \frac{w_{net,out}}{v_1(1 - 1/r)} = \frac{560.9\text{kJ/kg}}{(0.906\text{m}^3/\text{kg})(1 - 1/16)}\left(\frac{\text{kPa}\cdot\text{m}^3}{\text{kJ}}\right) = \mathbf{660.4kPa}$$

8-47E An air-standard Diesel cycle with a compression ratio of 18.2 is considered. The cutoff ratio, the heat rejection per unit mass, and the thermal efficiency are to be determined. √

Assumptions **1** The air-standard assumptions are applicable. **2** Kinetic and potential energy changes are negligible. **3** Air is an ideal gas with variable specific heats.

Properties The properties of air are given in Table A-17E.

Analysis (*a*) Process 1-2: isentropic compression.

$$T_1 = 540 \text{ R} \longrightarrow \begin{array}{l} u_1 = 92.04 \text{ Btu/lbm} \\ v_{r_1} = 144.32 \end{array}$$

$$v_{r_2} = \frac{v_2}{v_1} v_{r_1} = \frac{1}{r} v_{r_1} = \frac{1}{18.2}(144.32) = 7.93 \longrightarrow \begin{array}{l} T_2 = 1623.6\text{R} \\ h_2 = 402.05\text{Btu/lbm} \end{array}$$

Process 2-3: P = constant heat addition.

$$\frac{P_3 v_3}{T_3} = \frac{P_2 v_2}{T_2} \longrightarrow \frac{v_3}{v_2} = \frac{T_3}{T_2} = \frac{3000 \text{ R}}{1623.6 \text{ R}} = 1.848$$

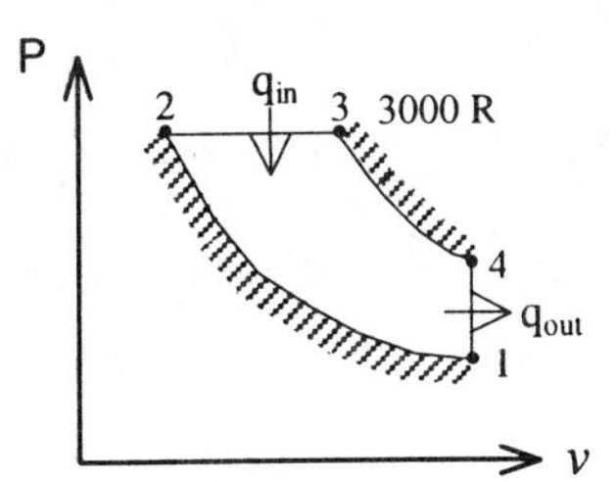

(*b*)
$$T_3 = 3000 \text{ R} \longrightarrow \begin{array}{l} h_3 = 790.68 \text{ Btu/lbm} \\ v_{r_3} = 1.180 \end{array}$$

$$q_{in} = h_3 - h_2 = 790.68 - 402.05 = 388.63 \text{ Btu/lbm}$$

Process 3-4: isentropic expansion.

$$v_{r_4} = \frac{v_4}{v_3} v_{r_3} = \frac{v_4}{1.848 v_2} v_{r_3} = \frac{r}{1.848} v_{r_3} = \frac{18.2}{1.848}(1.180) = 11.621 \longrightarrow u_4 = 250.91\text{Btu/lbm}$$

Process 4-1: v = constant heat rejection.

$$q_{out} = u_4 - u_1 = 250.91 - 92.04 = \mathbf{158.87 \ Btu/lbm}$$

(*c*)
$$\eta_{th} = 1 - \frac{q_{out}}{q_{in}} = 1 - \frac{158.87 \text{ Btu/lbm}}{388.63 \text{ Btu/lbm}} = \mathbf{59.1\%}$$

8-48E An air-standard Diesel cycle with a compression ratio of 18.2 is considered. The cutoff ratio, the heat rejection per unit mass, and the thermal efficiency are to be determined.

Assumptions **1** The air-standard assumptions are applicable. **2** Kinetic and potential energy changes are negligible. **3** Air is an ideal gas with constant specific heats.

Properties The properties of air at room temperature are C_p = 0.240 Btu/lbm.R, C_v = 0.171 Btu/lbm.R, and $k = 1.4$ (Table A-2E).

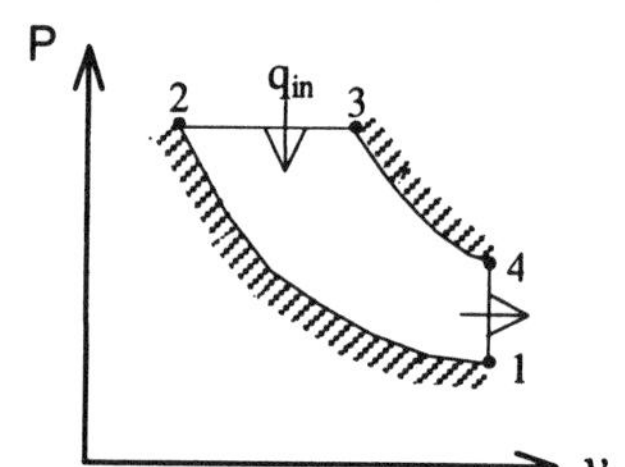

Analysis (*a*) Process 1-2: isentropic compression.

$$T_2 = T_1\left(\frac{v_1}{v_2}\right)^{k-1} = (540\text{R})(18.2)^{0.4} = 1724\text{R}$$

Process 2-3: P = constant heat addition.

$$\frac{P_3 v_3}{T_3} = \frac{P_2 v_2}{T_2} \longrightarrow \frac{v_3}{v_2} = \frac{T_3}{T_2} = \frac{3000\text{ R}}{1724\text{ R}} = \mathbf{1.741}$$

(*b*) $$q_{in} = h_3 - h_2 = C_p(T_3 - T_2) = (0.240\text{ Btu/lbm.R})(3000 - 1724)\text{R} = 306\text{ Btu/lbm}$$

Process 3-4: isentropic expansion.

$$T_4 = T_3\left(\frac{v_3}{v_4}\right)^{k-1} = T_3\left(\frac{1.741 v_2}{v_4}\right)^{k-1} = (3000\text{ R})\left(\frac{1.741}{18.2}\right)^{0.4} = 1173\text{ R}$$

Process 4-1: v = constant heat rejection.

$$q_{out} = u_4 - u_1 = C_v(T_4 - T_1)$$
$$= (0.171\text{ Btu/lbm.R})(1173 - 540)\text{R} = \mathbf{108\ Btu/lbm}$$

(*c*) $$\eta_{th} = 1 - \frac{q_{out}}{q_{in}} = 1 - \frac{108\text{ Btu/lbm}}{306\text{ Btu/lbm}} = \mathbf{64.6\%}$$

8-49 An ideal diesel engine with air as the working fluid has a compression ratio of 20. The thermal efficiency and the mean effective pressure are to be determined.

Assumptions **1** The air-standard assumptions are applicable. **2** Kinetic and potential energy changes are negligible. **3** Air is an ideal gas with constant specific heats.

Properties The properties of air at room temperature are C_p = 1.005 kJ/kg· K, C_v = 0.718 kJ/kg· K, and $k = 1.4$ (Table A-2).

Analysis (*a*) Process 1-2: isentropic compression.

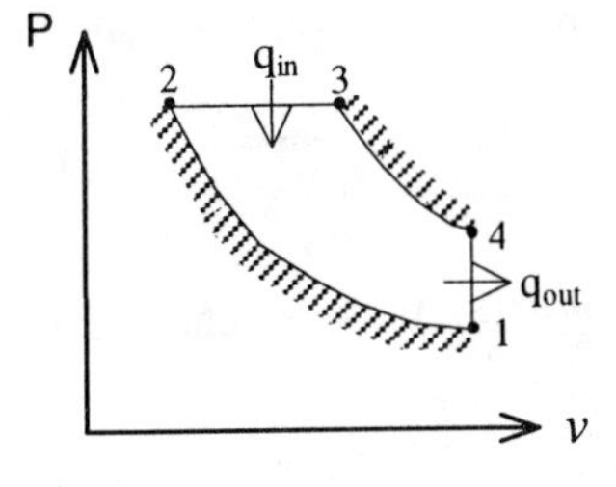

$$T_2 = T_1\left(\frac{V_1}{V_2}\right)^{k-1} = (293\text{K})(20)^{0.4} = 971.1\text{K}$$

Process 2-3: P = constant heat addition.

$$\frac{P_3V_3}{T_3} = \frac{P_2V_2}{T_2} \longrightarrow \frac{V_3}{V_2} = \frac{T_3}{T_2} = \frac{2200\text{ K}}{971.1\text{ K}} = 2.265$$

Process 3-4: isentropic expansion.

$$T_4 = T_3\left(\frac{V_3}{V_4}\right)^{k-1} = T_3\left(\frac{2.265V_2}{V_4}\right)^{k-1} = T_3\left(\frac{2.265}{r}\right)^{k-1} = (2200\text{ K})\left(\frac{2.265}{20}\right)^{0.4} = 920.6\text{ K}$$

$$q_{in} = h_3 - h_2 = C_p(T_3 - T_2) = (1.005\text{ kJ/kg}\cdot\text{K})(2200 - 971.1)\text{K} = 1235\text{ kJ/kg}$$

$$q_{out} = u_4 - u_1 = C_v(T_4 - T_1) = (0.718\text{ kJ/kg}\cdot\text{K})(920.6 - 293)\text{K} = 450.6\text{ kJ/kg}$$

$$w_{net,out} = q_{in} - q_{out} = 1235 - 450.6 = 784.4\text{ kJ/kg}$$

$$\eta_{th} = \frac{w_{net,out}}{q_{in}} = \frac{784.4\text{ kJ/kg}}{1235\text{ kJ/kg}} = \mathbf{63.5\%}$$

(*b*) $$v_1 = \frac{RT_1}{P_1} = \frac{(0.287\text{kPa}\cdot\text{m}^3/\text{kg}\cdot\text{K})(293\text{K})}{95\text{kPa}} = 0.885\text{m}^3/\text{kg} = v_{\max}$$

$$v_{\min} = v_2 = \frac{v_{\max}}{r}$$

$$MEP = \frac{w_{net,out}}{v_1 - v_2} = \frac{w_{net,out}}{v_1(1-1/r)} = \frac{784.4\text{kJ/kg}}{(0.885\text{m}^3/\text{kg})(1-1/20)}\left(\frac{\text{kPa}\cdot\text{m}^3}{\text{kJ}}\right) = \mathbf{933kPa}$$

8-50 A diesel engine with air as the working fluid has a compression ratio of 20. The thermal efficiency and the mean effective pressure are to be determined. √

Assumptions **1** The air-standard assumptions are applicable. **2** Kinetic and potential energy changes are negligible. **3** Air is an ideal gas with constant specific heats.

Properties The properties of air at room temperature are C_p = 1.005 kJ/kg· K, C_v = 0.718 kJ/kg· K, and k = 1.4 (Table A-2).

Analysis (a) Process 1-2: isentropic compression.

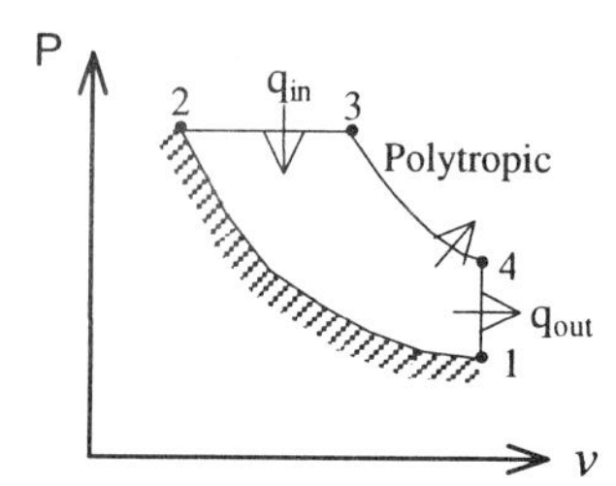

$$T_2 = T_1\left(\frac{V_1}{V_2}\right)^{k-1} = (293\text{K})(20)^{0.4} = 971.1\text{ K}$$

Process 2-3: P = constant heat addition.

$$\frac{P_3V_3}{T_3} = \frac{P_2V_2}{T_2} \longrightarrow \frac{V_3}{V_2} = \frac{T_3}{T_2} = \frac{2200\text{ K}}{971.1\text{ K}} = 2.265$$

Process 3-4: polytropic expansion.

$$T_4 = T_3\left(\frac{V_3}{V_4}\right)^{n-1} = \text{T}_3\left(\frac{2.265\text{V}_2}{\text{V}_4}\right)^{\text{n}-1} = \text{T}_3\left(\frac{2.265}{\text{r}}\right)^{\text{n}-1} = (2200\text{ K})\left(\frac{2.265}{20}\right)^{0.35} = 1026\text{ K}$$

$$q_{in} = h_3 - h_2 = C_p(T_3 - T_2) = (1.005\text{ kJ/kg}\cdot\text{K})(2200 - 971.1)\text{ K} = 1235\text{ kJ/kg}$$

$$q_{out} = u_4 - u_1 = C_v(T_4 - T_1) = (0.718\text{ kJ/kg}\cdot\text{K})(1026 - 293)\text{ K} = 526.3\text{ kJ/kg}$$

Note that q_{out} in this case does not represent the entire heat rejected since some heat is also rejected during the polytropic process, which is determined from an energy balance on process 3-4:

$$w_{34,out} = \frac{R(T_4 - T_3)}{1-n} = \frac{(0.287\text{kJ/kg}\cdot\text{K})(1026 - 2200)\text{ K}}{1 - 1.35} = 963\text{ kJ/kg}$$

$$E_{\text{in}} - E_{\text{out}} = \Delta E_{\text{system}}$$

$$q_{34,in} - w_{34,out} = u_4 - u_3 \longrightarrow q_{34,in} = w_{34,out} + C_v(T_4 - T_3)$$
$$= 963\text{ kJ/kg} + (0.718\text{ kJ/kg}\cdot\text{K})(1026 - 2200)\text{ K}$$
$$= 120.1\text{ kJ/kg}$$

which means that 120.1 kJ/kg of heat is transferred to the combustion gases during the expansion process. This is unrealistic since the gas is at a much higher temperature than the surroundings, and a hot gas loses heat during polytropic expansion. The cause of this unrealistic result is the constant specific heat assumption. If we were to use u data from the air table, we would obtain

$$q_{34,in} = w_{34,out} + (u_4 - u_3) = 963 + (781.3 - 1872.4) = -128.1\text{ kJ/kg}$$

which is a heat loss as expected. Then q_{out} becomes

$$q_{out} = q_{34,out} + q_{41,out} = 128.1 + 526.3 = 654.4\text{ kJ/kg}$$

and

$$w_{net,out} = q_{in} - q_{out} = 1235 - 654.4 = 580.6\text{ kJ/kg}$$

$$\eta_{th} = \frac{w_{net,out}}{q_{in}} = \frac{580.6\text{ kJ/kg}}{1235\text{ kJ/kg}} = 47.0\%$$

(c) $$v_1 = \frac{RT_1}{P_1} = \frac{(0.287\text{ kPa}\cdot\text{m}^3/\text{kg}\cdot\text{K})(293\text{ K})}{95\text{ kPa}} = 0.885\text{ m}^3/\text{kg} = v_{\max}$$

$$v_{\min} = v_2 = \frac{v_{\max}}{r}$$

$$MEP = \frac{w_{net,out}}{v_1 - v_2} = \frac{w_{net,out}}{v_1(1-1/r)} = \frac{580.6\text{ kJ/kg}}{(0.885\text{ m}^3/\text{kg})(1-1/20)}\left(\frac{1\text{ kPa}\cdot\text{m}^3}{\text{kJ}}\right) = \mathbf{691\text{ kPa}}$$

8-51 EES solution of this (and other comprehensive problems designated with the *computer icon*) is available to instructors at the *Instructor Manual* section of the *Online Learning Center* (OLC) at www.mhhe.com/cengel-boles. See the Preface for access information.

8-52 A four-cylinder ideal diesel engine with air as the working fluid has a compression ratio of 17 and a cutoff ratio of 2.2. The power the engine will deliver at 1500 rpm is to be determined.

Assumptions **1** The air-standard assumptions are applicable. **2** Kinetic and potential energy changes are negligible. **3** Air is an ideal gas with constant specific heats.

Properties The properties of air at room temperature are C_p = 1.005 kJ/kg· K, C_v = 0.718 kJ/kg· K, and k = 1.4 (Table A-2).

Analysis Process 1-2: isentropic compression.

$$T_2 = T_1\left(\frac{V_1}{V_2}\right)^{k-1} = (300\text{K})(17)^{0.4} = 931.8\text{K}$$

Process 2-3: P = constant heat addition.

$$\frac{P_3 v_3}{T_3} = \frac{P_2 v_2}{T_2} \longrightarrow T_3 = \frac{v_3}{v_2}T_2 = 2.2T_2 = (2.2)(931.8\text{K}) = 2050\text{K}$$

Process 3-4: isentropic expansion.

$$T_4 = T_3\left(\frac{V_3}{V_4}\right)^{n-1} = T_3\left(\frac{2.2V_2}{V_4}\right)^{n-1} = T_3\left(\frac{2.2}{r}\right)^{n-1} = (2050\text{K})\left(\frac{2.2}{17}\right)^{0.4} = 904.8\text{K}$$

$$m = \frac{P_1V_1}{RT_1} = \frac{(97\text{ kPa})(0.003\text{ m}^3)}{(0.287\text{ kPa}\cdot\text{m}^3/\text{kg}\cdot\text{K})(300\text{ K})} = 3.380\times10^{-3}\text{kg}$$

$$Q_{in} = m(h_3 - h_2) = mC_p(T_3 - T_2)$$
$$= (3.380\times10^{-3}\text{kg})(1.005\text{ kJ/kg}\cdot\text{K})(2050 - 931.8)\text{K} = 3.798\text{ kJ}$$

$$Q_{out} = m(u_4 - u_1) = mC_v(T_4 - T_1)$$
$$= (3.380\times10^{-3}\text{kg})(0.718\text{ kJ/kg}\cdot\text{K})(904.8 - 300)\text{K} = 1.468\text{ kJ}$$

$$W_{net,out} = Q_{in} - Q_{out} = 3.798 - 1.468 = 2.330\text{ kJ/rev}$$

$$\dot{W}_{net,out} = \dot{n}W_{net,out} = (1500/60\text{ rev/s})(2.330\text{ kJ/rev}) = \mathbf{58.2\ kW}$$

Discussion Note that for 2-stroke engines, 1 thermodynamic cycle is equivalent to 1 mechanical cycle (and thus revolutions).

8-53 A four-cylinder ideal diesel engine with nitrogen as the working fluid has a compression ratio of 17 and a cutoff ratio of 2.2. The power the engine will deliver at 1500 rpm is to be determined.

Assumptions **1** The air-standard assumptions are applicable with nitrogen as the working fluid. **2** Kinetic and potential energy changes are negligible. **3** Nitrogen is an ideal gas with constant specific heats.

Properties The properties of nitrogen at room temperature are C_p = 1.039 kJ/kg· K, C_v = 0.743 kJ/kg· K, and k = 1.4 (Table A-2).

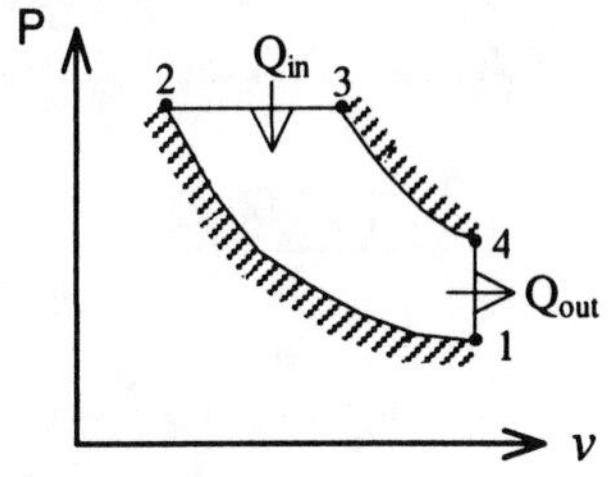

Analysis Process 1-2: isentropic compression.

$$T_2 = T_1\left(\frac{V_1}{V_2}\right)^{k-1} = (300\text{K})(17)^{0.4} = 931.8\text{K}$$

Process 2-3: P = constant heat addition.

$$\frac{P_3 v_3}{T_3} = \frac{P_2 v_2}{T_2} \longrightarrow T_3 = \frac{v_3}{v_2}T_2 = 2.2T_2 = (2.2)(931.8\text{K}) = 2050\text{K}$$

Process 3-4: isentropic expansion.

$$T_4 = T_3\left(\frac{V_3}{V_4}\right)^{n-1} = T_3\left(\frac{2.2V_2}{V_4}\right)^{n-1} = T_3\left(\frac{2.2}{r}\right)^{n-1} = (2050\text{K})\left(\frac{2.2}{17}\right)^{0.4} = 904.8\text{K}$$

$$m = \frac{P_1V_1}{RT_1} = \frac{(97\text{ kPa})(0.003\text{ m}^3)}{(0.2968\text{ kPa}\cdot\text{m}^3/\text{kg}\cdot\text{K})(300\text{ K})} = 3.268\times10^{-3}\text{kg}$$

$$Q_{in} = m(h_3 - h_2) = mC_p(T_3 - T_2)$$
$$= (3.268\times10^{-3}\text{kg})(1.039\text{ kJ/kg}\cdot\text{K})(2050 - 931.8)\text{K} = 3.797\text{ kJ}$$

$$Q_{out} = m(u_4 - u_1) = mC_v(T_4 - T_1)$$
$$= (3.268\times10^{-3}\text{kg})(0.743\text{ kJ/kg}\cdot\text{K})(904.8 - 300)\text{K} = 1.469\text{ kJ}$$

$$W_{net,out} = Q_{in} - Q_{out} = 3.797 - 1.469 = 2.328\text{ kJ/rev}$$

$$\dot{W}_{net,out} = \dot{n}W_{net,out} = (1500/60\text{rev/s})(2.328\text{ kJ/rev}) = \mathbf{58.2\ kW}$$

Discussion Note that for 2-stroke engines, 1 thermodynamic cycle is equivalent to 1 mechanical cycle (and thus revolutions).

8-54 [*Also solved by EES on enclosed CD*] An ideal dual cycle with air as the working fluid has a compression ratio of 14. The fraction of heat transferred at constant volume and the thermal efficiency of the cycle are to be determined.

Assumptions **1** The air-standard assumptions are applicable. **2** Kinetic and potential energy changes are negligible. **3** Air is an ideal gas with variable specific heats.

Properties The properties of air are given in Table A-17.

Analysis (a) Process 1-2: isentropic compression.

$$T_1 = 300 \text{ K} \longrightarrow \begin{array}{l} u_1 = 214.07 \text{ kJ/kg} \\ v_{r_1} = 621.2 \end{array}$$

$$v_{r_2} = \frac{v_2}{v_1} v_{r_1} = \frac{1}{r} v_{r_1} = \frac{1}{14}(621.2) = 44.37 \longrightarrow \begin{array}{l} T_2 = 823.1\text{K} \\ u_2 = 611.2\text{kJ/kg} \end{array}$$

Process 2-x, x-3: heat addition,

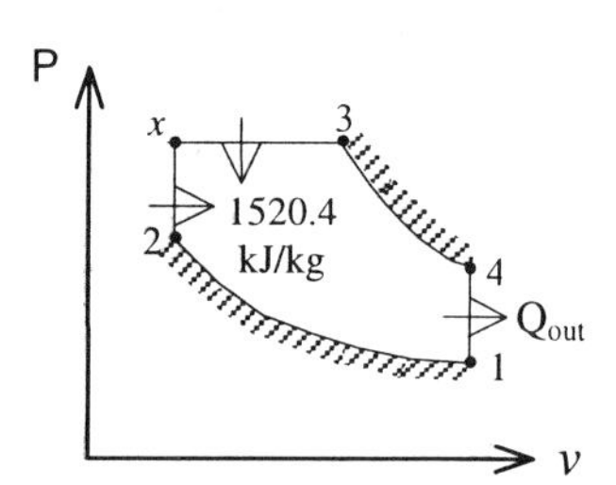

$$T_3 = 2200\text{K} \longrightarrow \begin{array}{l} h_3 = 2503.2\text{kJ/kg} \\ v_{r_3} = 2.012 \end{array}$$

$$q_{in} = q_{x-2,in} + q_{3-x,in} = (u_x - u_2) + (h_3 - h_x)$$
$$1520.4 = (u_x - 611.2) + (2503.2 - h_x)$$

By trial and error, we get T_x = 1300 K and h_x = 1395.97, u_x = 1022.82 kJ /kg.

Thus,

$$q_{2-x,in} = u_x - u_2 = 1022.82 - 611.2 = 411.62 \text{ kJ/kg}$$

and

$$ratio = \frac{q_{2-x,in}}{q_{in}} = \frac{411.62 \text{ kJ/kg}}{1520.4 \text{ kJ/kg}} = \mathbf{27.1\%}$$

(b) $$\frac{P_3 v_3}{T_3} = \frac{P_x v_x}{T_x} \longrightarrow \frac{v_3}{v_x} = \frac{T_3}{T_x} = \frac{2200 \text{ K}}{1300 \text{ K}} = 1.692 = r_c$$

$$v_{r_4} = \frac{v_4}{v_3} v_{r_3} = \frac{v_4}{1.692 v_2} v_{r_3} = \frac{r}{1.692} v_{r_3} = \frac{14}{1.692}(2.012) = 16.648 \longrightarrow u_4 = 886.3\text{kJ/kg}$$

Process 4-1: v = constant heat rejection.

$$q_{out} = u_4 - u_1 = 886.3 - 214.07 = 672.23 \text{ kJ/kg}$$

$$\eta_{th} = 1 - \frac{q_{out}}{q_{in}} = 1 - \frac{672.23 \text{ kJ/kg}}{1520.4 \text{ kJ/kg}} = \mathbf{55.8\%}$$

8-55 EES solution of this (and other comprehensive problems designated with the *computer icon*) is available to instructors at the *Instructor Manual* section of the *Online Learning Center* (OLC) at www.mhhe.com/cengel-boles. See the Preface for access information.

8-56 An ideal dual cycle with air as the working fluid has a compression ratio of 14. The fraction of heat transferred at constant volume and the thermal efficiency of the cycle are to be determined.

Assumptions **1** The air-standard assumptions are applicable. **2** Kinetic and potential energy changes are negligible. **3** Air is an ideal gas with constant specific heats.

Properties The properties of air at room temperature are C_p = 1.005 kJ/kg· K, C_v = 0.718 kJ/kg· K, and k = 1.4 (Table A-2).

Analysis (a) Process 1-2: isentropic compression.

$$T_2 = T_1\left(\frac{v_1}{v_2}\right)^{k-1} = (300\text{K})(14)^{0.4} = 862\text{K}$$

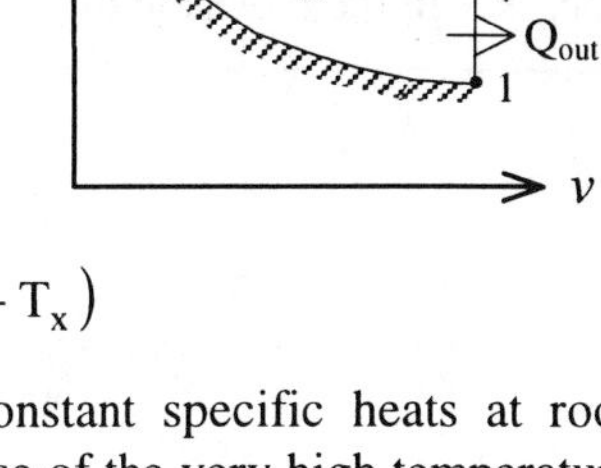

Process 2-x, x-3: heat addition,

$$q_{in} = q_{2-x,in} + q_{3-x,in} = (u_x - u_2) + (h_3 - h_x)$$
$$= C_v(T_x - T_2) + C_p(T_3 - T_x)$$

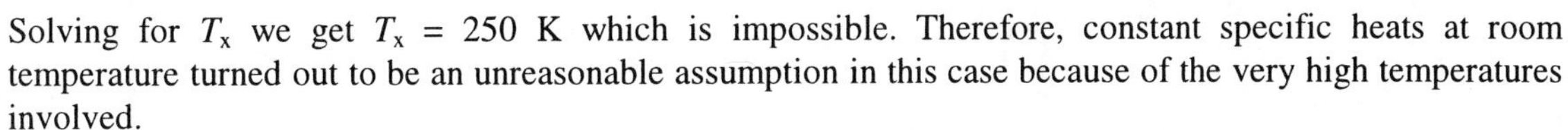

$$1520.4\text{kJ/kg} = (0.718\text{kJ/kg}\cdot\text{K})(\text{T}_\text{x} - 862) + (1.005\text{kJ/kg}\cdot\text{K})(2200 - \text{T}_\text{x})$$

Solving for T_x we get T_x = 250 K which is impossible. Therefore, constant specific heats at room temperature turned out to be an unreasonable assumption in this case because of the very high temperatures involved.

Stirling and Ericsson Cycles

8-57C The efficiencies of the Carnot and the Stirling cycles would be the same, the efficiency of the Otto cycle would be less.

8-58C The efficiencies of the Carnot and the Ericsson cycles would be the same, the efficiency of the Diesel cycle would be less.

8-59C The Stirling cycle.

8-60C The two isentropic processes of the Carnot cycle are replaced by two constant pressure regeneration processes in the Ericsson cycle.

8-61E An ideal Ericsson engine with helium as the working fluid operates between the specified temperature and pressure limits. The thermal efficiency of the cycle, the heat transfer rate in the regenerator, and the power delivered are to be determined.

Assumptions Helium is an ideal gas with constant specific heats.

Properties The gas constant and the specific heat of helium at room temperature are $R = 0.4961$ Btu/lbm.R and $C_p = 1.25$ Btu/lbm· R (Table A-2).

Analysis (*a*) The thermal efficiency of this totally reversible cycle is determined from

$$\eta_{th} = 1 - \frac{T_L}{T_H} = 1 - \frac{550\text{ R}}{3000\text{ R}} = \mathbf{81.67\%}$$

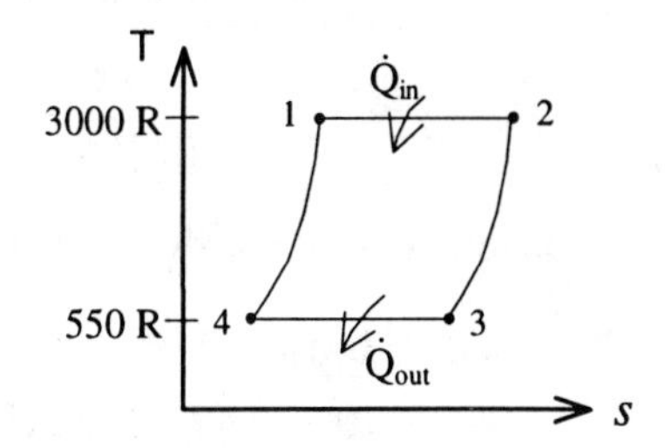

(*b*) The amount of heat transferred in the regenerator is

$$\begin{aligned}\dot{Q}_{regen} &= \dot{Q}_{41,in} = \dot{m}(h_1 - h_4) = \dot{m}C_p(T_1 - T_4)\\ &= (8\text{lbm/s})(1.25\text{Btu/lbm}\cdot\text{R})(3000 - 550)R\\ &= \mathbf{24{,}500 Btu/s}\end{aligned}$$

(*c*) The net power output is determined from

$$s_2 - s_1 = C_p \ln\frac{T_2}{T_1}^{\nearrow 0} - R\ln\frac{P_2}{P_1} = -(0.4961\text{Btu/lbm}\cdot\text{R})\ln\frac{25\text{psia}}{200\text{psia}} = 1.0316\text{Btu/lbm}\cdot\text{R}$$

$$\dot{Q}_{in} = \dot{m}T_H(s_2 - s_1) = (8\text{lbm/s})(3000\text{R})(1.0316\text{Btu/lbm}\cdot\text{R}) = 24{,}758\text{Btu/s}$$

$$\dot{W}_{net,out} = \eta_{th}\dot{Q}_{in} = (0.8167)(24{,}758) = \mathbf{20{,}220\ \ Btu/s}$$

8-62 An ideal steady-flow Ericsson engine with air as the working fluid is considered. The maximum pressure in the cycle, the net work output, and the thermal efficiency of the cycle are to be determined.

Assumptions Air is an ideal gas.

Properties The gas constant of air is $R = 0.287$ kJ/kg.K (Table A-1).

Analysis (*a*) The entropy change during process 3-4 is

$$s_4 - s_3 = -\frac{q_{34,out}}{T_0} = -\frac{150\text{ kJ/kg}}{300\text{ K}} = -0.5\text{ kJ/kg}\cdot\text{K}$$

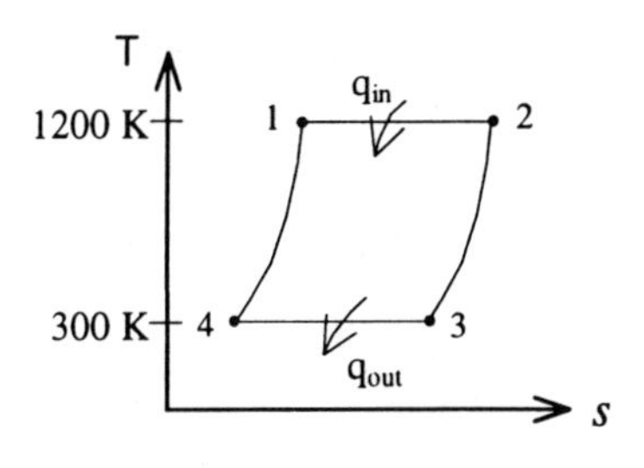

and

$$\begin{aligned}s_4 - s_3 &= C_p \ln\frac{T_4}{T_3}^{\nearrow 0} - R\ln\frac{P_4}{P_3}\\ &= -(0.287\text{kJ/kg}\cdot\text{K})\ln\frac{P_4}{120\text{kPa}} = -0.5\text{kJ/kg}\cdot\text{K}\end{aligned}$$

It yields $\qquad P_4 = \mathbf{685.2\ kPa}$

(*b*) For reversible cycles, $\dfrac{q_{out}}{q_{in}} = \dfrac{T_L}{T_H} \longrightarrow q_{in} = \dfrac{T_H}{T_L}q_{out} = \dfrac{1200\text{ K}}{300\text{ K}}(150\text{ kJ/kg}) = \mathbf{600\ \ kJ/kg}$

Thus, $\quad w_{net,out} = q_{in} - q_{out} = 600 - 150 = \mathbf{450\ \ kJ/kg}$

(*c*) The thermal efficiency of this totally reversible cycle is determined from

$$\eta_{th} = 1 - \frac{T_L}{T_H} = 1 - \frac{300\text{ K}}{1200\text{ K}} = \mathbf{75.0\%}$$

8-63 An ideal Stirling engine with helium as the working fluid operates between the specified temperature and pressure limits. The thermal efficiency of the cycle, the amount of heat transfer in the regenerator, and the work output per cycle are to be determined.

Assumptions Helium is an ideal gas with constant specific heats.

Properties The gas constant and the specific heat of helium at room temperature are R = 2.0769 kJ/kg.K, C_v = 3.1156 kJ/kg.K and C_p = 5.1926 kJ/kg.K (Table A-2).

Analysis (*a*) The thermal efficiency of this totally reversible cycle is determined from

$$\eta_{th} = 1 - \frac{T_L}{T_H} = 1 - \frac{300 \text{ K}}{2000 \text{ K}} = \mathbf{85.0\%}$$

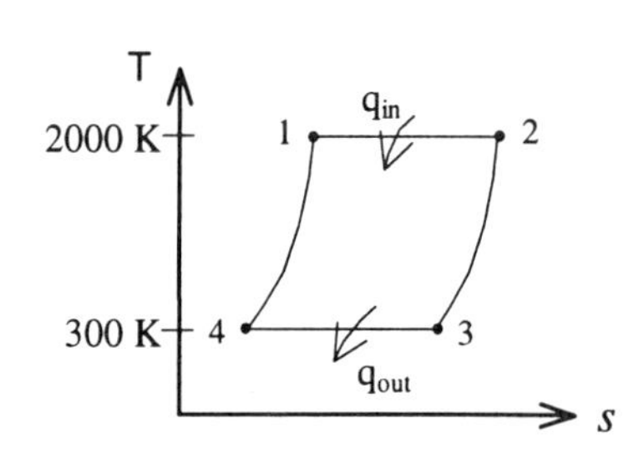

(*b*) The amount of heat transferred in the regenerator is

$$\begin{aligned}\dot{Q}_{regen} &= \dot{Q}_{41,in} = \dot{m}(u_1 - u_4) = \dot{m}C_v(T_1 - T_4) \\ &= (0.12 \text{ kg})(3.1156 \text{kJ/kg}\cdot\text{K})(2000 - 300)\text{K} \\ &= \mathbf{635.6 \ kJ}\end{aligned}$$

(*c*) The net work output is determined from

$$\frac{P_3 v_3}{T_3} = \frac{P_1 v_1}{T_1} \longrightarrow \frac{v_3}{v_1} = \frac{T_3 P_1}{T_1 P_3} = \frac{(300 \text{ K})(3000 \text{ kPa})}{(2000 \text{ K})(150 \text{ kPa})} = 3 = \frac{v_2}{v_1}$$

$$s_2 - s_1 = C_v \ln\frac{T_2}{T_1}^{\nearrow 0} + R\ln\frac{v_2}{v_1} = (2.0769 \text{kJ/kg}\cdot\text{K})\ln(3) = 2.282 \text{kJ/kg}\cdot\text{K}$$

$$Q_{in} = mT_H(s_2 - s_1) = (0.12 \text{ kg})(2000 \text{ K})(2.282 \text{ kJ/kg}\cdot\text{K}) = 547.6 \text{ kJ}$$

$$W_{net,out} = \eta_{th} Q_{in} = (0.85)(547.6 \text{ kJ}) = \mathbf{465.5 \ kJ}$$

Ideal and Actual Gas-Turbine (Brayton) Cycles

8-64C In gas turbine engines a gas is compressed, and thus the compression work requirements are very large since the steady-flow work is proportional to the specific volume.

8-65C They are (1) isentropic compression (in a compressor), (2) P = constant heat addition, (3) isentropic expansion (in a turbine), and (4) P = constant heat rejection.

8-66C For fixed maximum and minimum temperatures, (a) the thermal efficiency increases with pressure ratio, (b) the net work first increases with pressure ratio, reaches a maximum, and then decreases.

8-67C Back work ratio is the ratio of the compressor (or pump) work input to the turbine work output. It is usually between 0.40 and 0.6 for gas turbine engines.

8-68C As a result of turbine and compressor inefficiencies, (a) the back work ratio increases, and (b) the thermal efficiency decreases.

8-69E A simple ideal Brayton cycle with air as the working fluid has a pressure ratio of 10. The air temperature at the compressor exit, the back work ratio, and the thermal efficiency are to be determined.

Assumptions **1** Steady operating conditions exist. **2** The air-standard assumptions are applicable. **3** Kinetic and potential energy changes are negligible. **4** Air is an ideal gas with variable specific heats.

Properties The properties of air are given in Table A-17E.

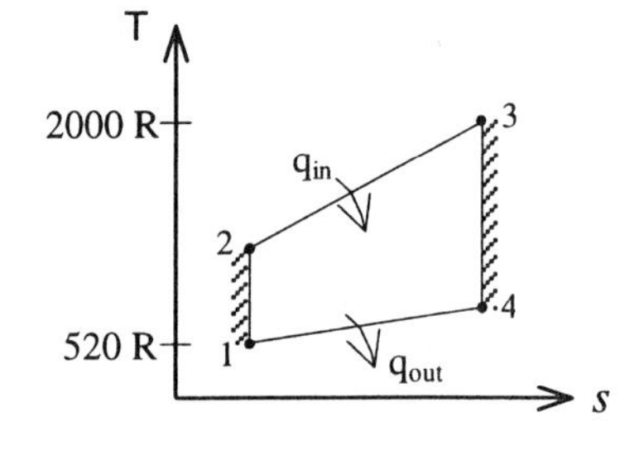

Analysis (*a*) Noting that process 1-2 is isentropic,

$$T_1 = 520 \text{ R} \longrightarrow \begin{array}{l} h_1 = 124.27 \text{ Btu/lbm} \\ P_{r_1} = 1.2147 \end{array}$$

$$P_{r_2} = \frac{P_2}{P_1} P_{r_1} = (10)(1.2147) = 12.147 \longrightarrow \begin{array}{l} T_2 = \mathbf{996.5R} \\ h_2 = 240.11\text{Btu/lbm} \end{array}$$

(*b*) Process 3-4 is isentropic, and thus

$$T_3 = 2000\text{R} \longrightarrow \begin{array}{l} h_3 = 504.71\text{Btu/lbm} \\ P_{r_3} = 174.0 \end{array}$$

$$P_{r_4} = \frac{P_4}{P_3} P_{r_3} = \left(\frac{1}{10}\right)(174.0) = 17.4 \longrightarrow h_4 = 265.83\text{Btu/lbm}$$

$$w_{C,in} = h_2 - h_1 = 240.11 - 124.27 = 115.84\text{Btu/lbm}$$

$$w_{T,out} = h_3 - h_4 = 504.71 - 265.83 = 238.88\text{Btu/lbm}$$

Then the back-work ratio becomes

$$r_{bw} = \frac{w_{C,in}}{w_{T,out}} = \frac{115.84 \text{ Btu/lbm}}{238.88 \text{ Btu/lbm}} = \mathbf{48.5\%}$$

(*c*)

$$q_{in} = h_3 - h_2 = 504.71 - 240.11 = 264.60 \text{ Btu/lbm}$$

$$w_{net,out} = w_{T,out} - w_{C,in} = 238.88 - 115.84 = 123.04 \text{ Btu/lbm}$$

$$\eta_{th} = \frac{w_{net,out}}{q_{in}} = \frac{123.04 \text{ Btu/lbm}}{264.60 \text{ Btu/lbm}} = \mathbf{46.5\%}$$

8-70 [*Also solved by EES on enclosed CD*] A simple Brayton cycle with air as the working fluid has a pressure ratio of 8. The air temperature at the turbine exit, the net work output, and the thermal efficiency are to be determined.

Assumptions **1** Steady operating conditions exist. **2** The air-standard assumptions are applicable. **3** Kinetic and potential energy changes are negligible. **4** Air is an ideal gas with variable specific heats.

Properties The properties of air are given in Table A-17.

Analysis (*a*) Noting that process 1-2s is isentropic,

$$T_1 = 310 \text{ K} \longrightarrow \begin{array}{l} h_1 = 310.24 \text{ kJ/kg} \\ P_{r_1} = 1.5546 \end{array}$$

$$P_{r_2} = \frac{P_2}{P_1} P_{r_1} = (8)(1.5546) = 12.44 \longrightarrow h_{2s} = 562.58 \text{ kJ/kg and } T_{2s} = 557.25 \text{ K}$$

$$\eta_C = \frac{h_{2s} - h_1}{h_2 - h_1} \longrightarrow h_2 = h_1 + \frac{h_{2s} - h_1}{\eta_C}$$

$$= 310.24 + \frac{562.58 - 310.24}{0.75} = 646.7 \text{ kJ/kg}$$

$$T_3 = 1160\text{K} \longrightarrow \begin{array}{l} h_3 = 1230.92\text{kJ/kg} \\ P_{r_3} = 207.2 \end{array}$$

$$P_{r_4} = \frac{P_4}{P_3} P_{r_3} = \left(\frac{1}{8}\right)(207.2) = 25.90 \longrightarrow h_{4s} = 692.19 \text{ kJ/kg and } T_{4s} = 680.3 \text{ K}$$

$$\eta_T = \frac{h_3 - h_4}{h_3 - h_{4s}} \longrightarrow h_4 = h_3 - \eta_T (h_3 - h_{4s})$$

$$= 1230.92 - (0.82)(1230.92 - 692.19)$$

$$= 789.16\text{kJ/kg}$$

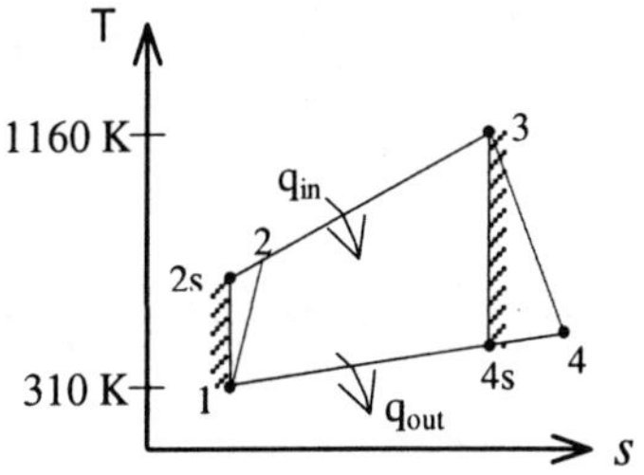

Thus, $T_4 = \mathbf{770.1 \text{ K}}$

(*b*) $q_{in} = h_3 - h_2 = 1230.92 - 646.7 = 584.2\text{kJ/kg}$

$q_{out} = h_4 - h_1 = 789.16 - 310.24 = 478.92\text{kJ/kg}$

$w_{net,out} = w_{in} - w_{out} = 584.2 - 478.92 = \mathbf{105.3kJ/kg}$

(*c*) $$\eta_{th} = \frac{w_{net,out}}{q_{in}} = \frac{105.3\text{kJ/kg}}{584.2\text{kJ/kg}} = \mathbf{18.0\%}$$

8-71 EES solution of this (and other comprehensive problems designated with the *computer icon*) is available to instructors at the *Instructor Manual* section of the *Online Learning Center* (OLC) at www.mhhe.com/cengel-boles. See the Preface for access information.

8-72 A simple Brayton cycle with air as the working fluid has a pressure ratio of 8. The air temperature at the turbine exit, the net work output, and the thermal efficiency are to be determined.

Assumptions **1** Steady operating conditions exist. **2** The air-standard assumptions are applicable. **3** Kinetic and potential energy changes are negligible. **4** Air is an ideal gas with constant specific heats.

Properties The properties of air at room temperature are $C_p = 1.005$ kJ/kg· K and $k = 1.4$ (Table A-2).

Analysis (*a*) Using the compressor and turbine efficiency relations,

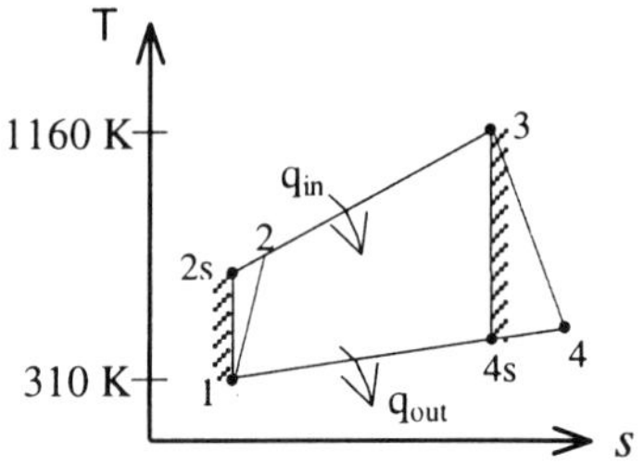

$$T_{2s} = T_1\left(\frac{P_2}{P_1}\right)^{(k-1)/k} = (310\text{K})(8)^{0.4/1.4} = 561.5\text{K}$$

$$T_{4s} = T_3\left(\frac{P_4}{P_3}\right)^{(k-1)/k} = (1160\text{K})\left(\frac{1}{8}\right)^{0.4/1.4} = 640.4\text{K}$$

$$\eta_C = \frac{h_{2s}-h_1}{h_2-h_1} = \frac{C_p(T_{2s}-T_1)}{C_p(T_2-T_1)} \longrightarrow T_2 = T_1 + \frac{T_{2s}-T_1}{\eta_C} = 310 + \frac{561.5-310}{0.75} = 645.3\text{K}$$

$$\eta_T = \frac{h_3-h_{4s}}{h_3-h_4} = \frac{C_p(T_3-T_{4s})}{C_p(T_3-T_4)} \longrightarrow T_4 = T_3 - \eta_T(T_3-T_{4s}) = 1160-(0.82)(1160-640.4) = \mathbf{733.9K}$$

(*b*)

$$q_{in} = h_3 - h_2 = C_p(T_3 - T_2) = (1.005\text{kJ/kg}\cdot\text{K})(1160-645.3)\text{K} = 517.3\text{kJ/kg}$$

$$q_{out} = h_4 - h_1 = C_p(T_4 - T_1) = (1.005\text{kJ/kg}\cdot\text{K})(733.9-310)\text{K} = 426.0\text{kJ/kg}$$

$$w_{net,out} = w_{in} - w_{out} = 517.3 - 426.0 = \mathbf{91.3kJ/kg}$$

(*c*)

$$\eta_{th} = \frac{w_{net,out}}{q_{in}} = \frac{91.3\text{ kJ/kg}}{517.3\text{ kJ/kg}} = \mathbf{17.6\%}$$

8-73 A gas turbine power plant that operates on the simple Brayton cycle with air as the working fluid has a specified pressure ratio. The required mass flow rate of air is to be determined for two cases.

Assumptions **1** Steady operating conditions exist. **2** The air-standard assumptions are applicable. **3** Kinetic and potential energy changes are negligible. **4** Air is an ideal gas with constant specific heats.

Properties The properties of air at room temperature are C_p = 1.005 kJ/kg· K and k = 1.4 (Table A-2).

Analysis (*a*) Using the isentropic relations,

$$T_{2s} = T_1\left(\frac{P_2}{P_1}\right)^{(k-1)/k} = (300\text{ K})(12)^{0.4/1.4} = 610.2\text{ K}$$

$$T_{4s} = T_3\left(\frac{P_4}{P_3}\right)^{(k-1)/k} = (1000\text{ K})\left(\frac{1}{12}\right)^{0.4/1.4} = 491.7\text{ K}$$

$$w_{s,C,in} = h_{2s} - h_1 = C_p(T_{2s} - T_1) = (1.005\text{ kJ/kg}\cdot\text{K})(610.2 - 300)\text{K} = 311.75\text{ kJ/kg}$$

$$w_{s,T,out} = h_3 - h_{4s} = C_p(T_3 - T_{4s}) = (1.005\text{ kJ/kg}\cdot\text{K})(1000 - 491.7)\text{K} = 510.84\text{ kJ/kg}$$

$$w_{s,net,out} = w_{s,T,out} - w_{s,C,in} = 510.84 - 311.75 = 199.09\text{ kJ/kg}$$

$$\dot{m}_s = \frac{\dot{W}_{net,out}}{w_{s,net,out}} = \frac{90{,}000\text{ kJ/s}}{199.09\text{ kJ/kg}} = \mathbf{452.1\ kg/s}$$

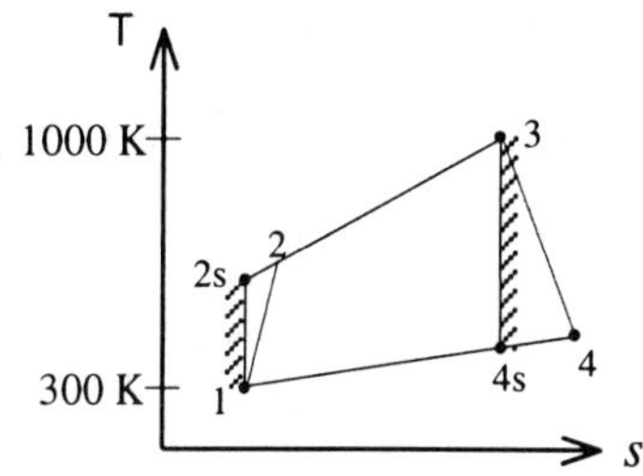

(*b*) The net work output is determined to be

$$w_{a,net,out} = w_{a,T,out} - w_{a,C,in} = \eta_T w_{s,T,out} - w_{s,C,in}/\eta_C$$
$$= (0.80)(510.84) - 311.75/0.80 = 18.98\text{kJ/kg}$$

$$\dot{m}_a = \frac{\dot{W}_{net,out}}{w_{a,net,out}} = \frac{90{,}000\text{ kJ/s}}{18.98\text{ kJ/kg}} = \mathbf{4742\ kg/s}$$

8-74 A stationary gas-turbine power plant operates on a simple ideal Brayton cycle with air as the working fluid. The power delivered by this plant is to be determined assuming constant and variable specific heats.

Assumptions **1** Steady operating conditions exist. **2** The air-standard assumptions are applicable. **3** Kinetic and potential energy changes are negligible. **4** Air is an ideal gas.

Analysis (*a*) Assuming constant specific heats,

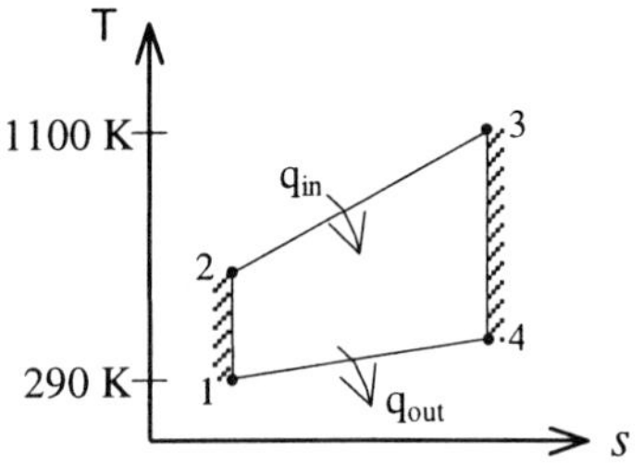

$$T_{2s} = T_1\left(\frac{P_2}{P_1}\right)^{(k-1)/k} = (290\text{K})(8)^{0.4/1.4} = 525.3\text{K}$$

$$T_{4s} = T_3\left(\frac{P_4}{P_3}\right)^{(k-1)/k} = (1100\text{K})\left(\frac{1}{8}\right)^{0.4/1.4} = 607.2\text{ K}$$

T, 1100 K, 290 K, 1, 2, 3, 4, q_{in}, q_{out}, s

$$\eta_{th} = 1 - \frac{q_{out}}{q_{in}} = 1 - \frac{C_p(T_4 - T_1)}{C_p(T_3 - T_2)} = 1 - \frac{T_4 - T_1}{T_3 - T_2} = 1 - \frac{607.2 - 290}{1100 - 525.3} = 0.448$$

$$\dot{W}_{net,out} = \eta_T \dot{Q}_{in} = (0.448)(35{,}000\text{ kW}) = \mathbf{15{,}680\ kW}$$

(*b*) Assuming variable specific heats (Table A-17),

$$T_1 = 290\text{ K} \longrightarrow \begin{matrix} h_1 = 290.16\text{ kJ/kg} \\ P_{r_1} = 1.2311 \end{matrix}$$

$$P_{r_2} = \frac{P_2}{P_1}P_{r_1} = (8)(1.2311) = 9.8488 \longrightarrow h_2 = 526.12\text{ kJ/kg}$$

$$T_3 = 1100\text{ K} \longrightarrow \begin{matrix} h_3 = 1161.07\text{ kJ/kg} \\ P_{r_3} = 167.1 \end{matrix}$$

$$P_{r_4} = \frac{P_4}{P_3}P_{r_3} = \left(\frac{1}{8}\right)(167.1) = 20.89 \longrightarrow h_4 = 651.37\text{ kJ/kg}$$

$$\eta_{th} = 1 - \frac{q_{out}}{q_{in}} == 1 - \frac{h_4 - h_1}{h_3 - h_2} = 1 - \frac{651.37 - 290.16}{1161.07 - 526.11} = 0.431$$

$$\dot{W}_{net,out} = \eta_T \dot{Q}_{in} = (0.431)(35{,}000\text{ kW}) = \mathbf{15{,}085\ kW}$$

8-75 An actual gas-turbine power plant operates at specified conditions. The fraction of the turbine work output used to drive the compressor and the thermal efficiency are to be determined. √

Assumptions **1** Steady operating conditions exist. **2** The air-standard assumptions are applicable. **3** Kinetic and potential energy changes are negligible. **4** Air is an ideal gas with variable specific heats.

Properties The properties of air are given in Table A-17.

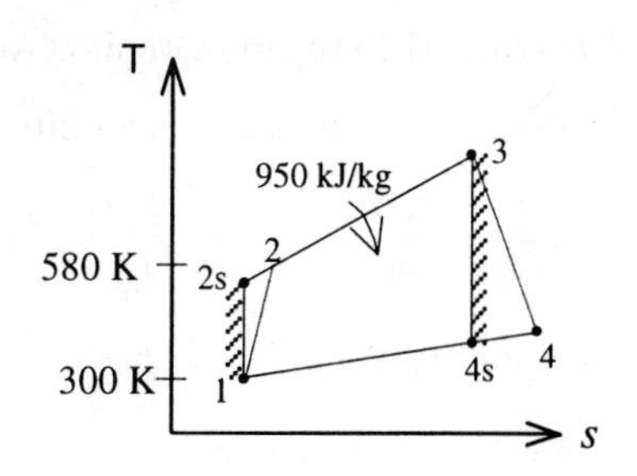

Analysis (a) Using the isentropic relations,

$$T_1 = 300 \text{ K} \longrightarrow h_1 = 300.19 \text{ kJ/kg}$$

$$T_2 = 580 \text{ K} \longrightarrow h_2 = 586.04 \text{ kJ/kg}$$

$$r_p = \frac{P_2}{P_1} = \frac{700}{100} = 7$$

$$q_{in} = h_3 - h_2 \longrightarrow h_3 = 950 + 586.04 = 1536.04 \text{kJ/kg}$$

$$\longrightarrow P_{r_3} = 474.11$$

$$P_{r_4} = \frac{P_4}{P_3} P_{r_3} = \left(\frac{1}{7}\right)(474.11) = 67.73 \longrightarrow h_{4s} = 905.83 \text{kJ/kg}$$

$$w_{C,in} = h_2 - h_1 = 586.04 - 300.19 = 285.85 \text{kJ/kg}$$

$$w_{T,out} = \eta_T (h_3 - h_{4s}) = (0.86)(1536.04 - 905.83) = 542.0 \text{kJ/kg}$$

Thus, $$r_{bw} = \frac{w_{C,in}}{w_{T,out}} = \frac{285.85 \text{ kJ/kg}}{542.0 \text{ kJ/kg}} = \mathbf{52.7\%}$$

(b) $$w_{net.out} = w_{T,out} - w_{C,in} = 542.0 - 285.85 = 256.15 \text{ kJ/kg}$$

$$\eta_{th} = \frac{w_{net,out}}{q_{in}} = \frac{256.15 \text{ kJ/kg}}{950 \text{ kJ/kg}} = \mathbf{27.0\%}$$

8-76 A gas-turbine power plant operates at specified conditions. The fraction of the turbine work output used to drive the compressor and the thermal efficiency are to be determined.

Assumptions **1** Steady operating conditions exist. **2** The air-standard assumptions are applicable. **3** Kinetic and potential energy changes are negligible. **4** Air is an ideal gas with constant specific heats.

Properties The properties of air at room temperature are C_p = 1.005 kJ/kg· K and k = 1.4 (Table A-2).

Analysis (a) Using constant specific heats,

$$r_p = \frac{P_2}{P_1} = \frac{700}{100} = 7$$

$$q_{in} = h_3 - h_2 = C_p(T_3 - T_2)$$

$$\longrightarrow T_3 = T_2 + q_{in}/C_p$$

$$= 580\text{K} + (950\text{kJ/kg})/(1.005\text{kJ/kg}\cdot\text{K})$$

$$= 1525.3\text{K}$$

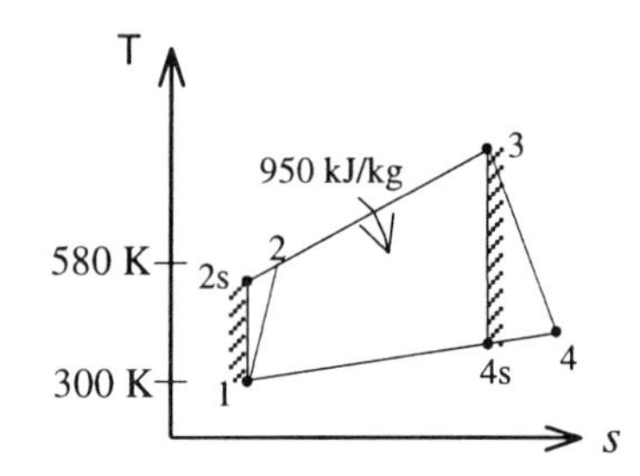

$$T_{4s} = T_3\left(\frac{P_4}{P_3}\right)^{(k-1)/k} = (1525.3\text{K})\left(\frac{1}{7}\right)^{0.4/1.4} = 874.8\text{K}$$

$$w_{C,in} = h_2 - h_1 = C_p(T_2 - T_1) = (1.005\text{kJ/kg}\cdot\text{K})(580 - 300)\text{K} = 281.4\text{kJ/kg}$$

$$w_{T,out} = \eta_T(h_3 - h_{4s}) = \eta_T C_p(T_3 - T_{4s}) = (0.86)(1.005\text{kJ/kg}\cdot\text{K})(1525.3 - 874.8)\text{K} = 562.2\text{kJ/kg}$$

Thus, $$r_{bw} = \frac{w_{C,in}}{w_{T,out}} = \frac{281.4 \text{ kJ/kg}}{562.2 \text{ kJ/kg}} = \mathbf{50.1\%}$$

(b) $$w_{net,out} = w_{T,out} - w_{C,in} = 562.2 - 281.4 = 280.8 \text{ kJ/kg}$$

$$\eta_{th} = \frac{w_{net,out}}{q_{in}} = \frac{280.8 \text{ kJ/kg}}{950 \text{ kJ/kg}} = \mathbf{29.6\%}$$

8-77E A gas-turbine power plant operates on a simple Brayton cycle with air as the working fluid. The net power output of the plant is to be determined.

Assumptions **1** Steady operating conditions exist. **2** The air-standard assumptions are applicable. **3** Kinetic and potential energy changes are negligible. **4** Air is an ideal gas with variable specific heats.

Properties The properties of air are given in Table A-17E.

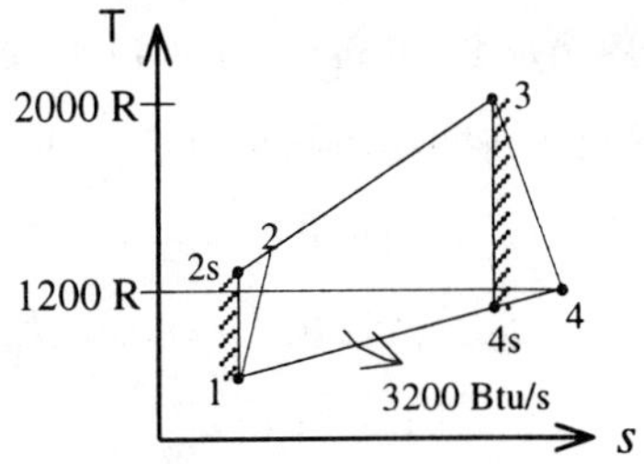

Analysis Using variable specific heats for air,

$$T_3 = 2000 \text{ R} \longrightarrow h_3 = 504.71 \text{ Btu/lbm}$$

$$T_4 = 1200 \text{ R} \longrightarrow h_4 = 291.30 \text{ Btu/lbm}$$

$$r_p = \frac{P_2}{P_1} = \frac{120}{15} = 8$$

$$\dot{Q}_{out} = \dot{m}(h_4 - h_1) \longrightarrow h_1 = 291.30 - 6400/40 = 131.30 \text{ Btu/lbm}$$

$$\longrightarrow P_{r_1} = 1.474$$

$$P_{r_2} = \frac{P_2}{P_1} P_{r_1} = (8)(1.474) = 11.79 \longrightarrow h_{2s} = 238.07 \text{Btu/lbm}$$

$$\dot{W}_{C,in} = \dot{m}(h_{2s} - h_1)/\eta_C = (40 \text{lbm/s})(238.07 - 131.30)/(0.80) = 5339 \text{Btu/s}$$

$$\dot{W}_{T,out} = \dot{m}(h_3 - h_4) = (40 \text{lbm/s})(504.71 - 291.30) = 8536 \text{Btu/s}$$

$$\dot{W}_{net,out} = \dot{W}_{T,out} - \dot{W}_{C,in} = 8536 - 5339 = 3197 \text{Btu/s} = \mathbf{3373 kW}$$

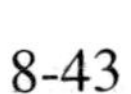

8-78E A gas-turbine power plant operates on a simple Brayton cycle with air as the working fluid. The compressor efficiency for which the power plant produces zero net work is to be determined.

Assumptions **1** Steady operating conditions exist. **2** The air-standard assumptions are applicable. **3** Kinetic and potential energy changes are negligible. **4** Air is an ideal gas with variable specific heats.

Properties The properties of air are given in Table A-17E.

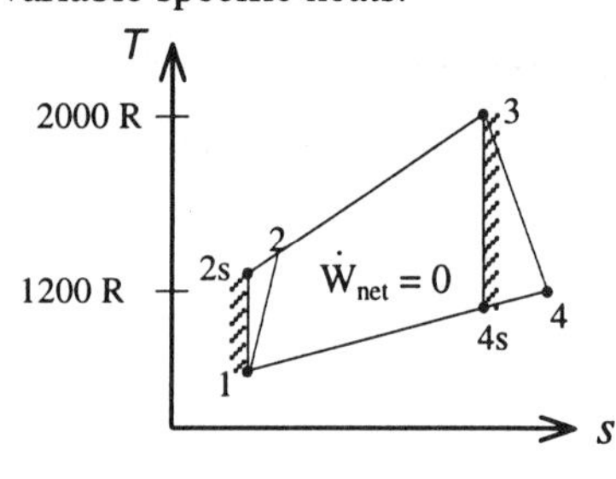

Analysis Using variable specific heats,

$$T_3 = 2000 \text{ R} \longrightarrow h_3 = 504.71 \text{ Btu/lbm}$$

$$T_4 = 1200 \text{ R} \longrightarrow h_4 = 291.30 \text{ Btu/lbm}$$

$$r_p = \frac{P_2}{P_1} = \frac{120}{15} = 8$$

$$\dot{Q}_{out} = \dot{m}(h_4 - h_1) \longrightarrow h_1 = 291.30 - 6400/40 = 131.30\text{Btu/lbm} \longrightarrow P_{r_1} = 1.474$$

$$P_{r_2} = \frac{P_2}{P_1} P_{r_1} = (8)(1.474) = 11.79 \longrightarrow h_{2s} = 238.07\text{Btu/lbm}$$

Then, $\dot{W}_{C,in} = \dot{W}_{T,out} \longrightarrow \dot{m}(h_{2s} - h_1)/\eta_C = \dot{m}(h_3 - h_4)$

$$\eta_C = \frac{h_{2s} - h_1}{h_3 - h_4} = \frac{238.07 - 131.30}{504.71 - 291.30} = \mathbf{50.0\%}$$

8-79 A 15-MW gas-turbine power plant operates on a simple Brayton cycle with air as the working fluid. The mass flow rate of air through the cycle is to be determined.

Assumptions **1** Steady operating conditions exist. **2** The air-standard assumptions are applicable. **3** Kinetic and potential energy changes are negligible. **4** Air is an ideal gas with variable specific heats.

Properties The properties of air are given in Table A-17.

Analysis Using variable specific heats,

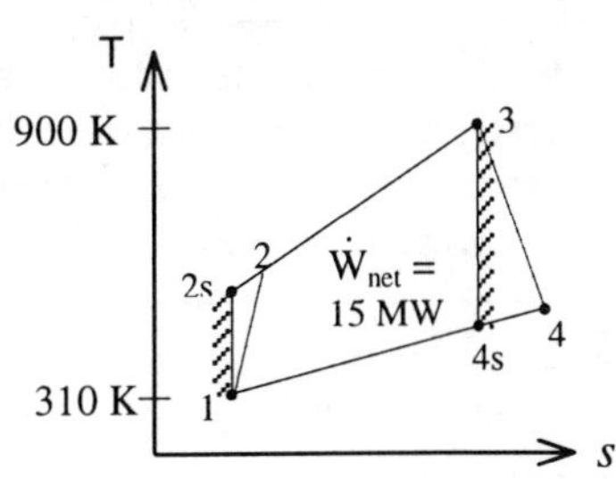

$$T_1 = 310\text{ K} \longrightarrow \begin{array}{l} h_1 = 310.24\text{ kJ/kg} \\ P_{r_1} = 1.5546 \end{array}$$

$$P_{r_2} = \frac{P_2}{P_1} P_{r_1} = (8)(1.5546) = 12.44 \longrightarrow h_{2s} = 562.26\text{kJ/kg}$$

$$T_3 = 900\text{K} \longrightarrow \begin{array}{l} h_3 = 932.93\text{kJ/kg} \\ P_{r_3} = 75.29 \end{array}$$

$$P_{r_4} = \frac{P_4}{P_3} P_{r_3} = \left(\frac{1}{8}\right)(75.29) = 9.411 \longrightarrow h_{4s} = 519.32\text{kJ/kg}$$

$$\begin{aligned} w_{net,out} &= w_{T,out} - w_{C,in} = \eta_T (h_3 - h_{4s}) - (h_{2s} - h_1)/\eta_C \\ &= (0.86)(932.93 - 519.32) - (562.26 - 310.24)/(0.80) = 40.68\text{kJ/kg} \end{aligned}$$

and

$$\dot{m} = \frac{\dot{W}_{net,out}}{w_{net,out}} = \frac{15{,}000\text{ kJ/s}}{40.68\text{ kJ/kg}} = \mathbf{368.7\ kg/s}$$

8-80 A 15-MW gas-turbine power plant operates on a simple Brayton cycle with air as the working fluid. The mass flow rate of air through the cycle is to be determined.

Assumptions **1** Steady operating conditions exist. **2** The air-standard assumptions are applicable. **3** Kinetic and potential energy changes are negligible. **4** Air is an ideal gas with constant specific heats.

Properties The properties of air at room temperature are C_p = 1.005 kJ/kg· K and k = 1.4 (Table A-2).

Analysis Using constant specific heats,

$$T_{2s} = T_1 \left(\frac{P_2}{P_1}\right)^{(k-1)/k} = (310\text{ K})(8)^{0.4/1.4} = 561.5\text{K}$$

$$T_{4s} = T_3 \left(\frac{P_4}{P_3}\right)^{(k-1)/k} = (900\text{K})\left(\frac{1}{8}\right)^{0.4/1.4} = 496.8\text{K}$$

$$\begin{aligned} w_{net,out} &= w_{T,out} - w_{C,in} = \eta_T C_p (T_3 - T_{4s}) - C_p (T_{2s} - T_1)/\eta_C \\ &= (1.005\text{kJ/kg}\cdot\text{K})[(0.86)(900 - 496.8) - (561.5 - 310)/(0.80)]\text{K} \\ &= 32.5\text{kJ/kg} \end{aligned}$$

and

$$\dot{m} = \frac{\dot{W}_{net,out}}{w_{net,out}} = \frac{15{,}000\text{ kJ/s}}{32.5\text{ kJ/kg}} = \mathbf{461.5\ kg/s}$$

Brayton Cycle with Regeneration

8-81C Regeneration increases the thermal efficiency of a Brayton cycle by capturing some of the waste heat from the exhaust gases and preheating the air before it enters the combustion chamber.

8-82C Yes. At very high compression ratios, the gas temperature at the turbine exit may be lower than the temperature at the compressor exit. Therefore, if these two streams are brought into thermal contact in a regenerator, heat will flow to the exhaust gases instead of from the exhaust gases. As a result, the thermal efficiency will decrease.

8-83C The extent to which a regenerator approaches an ideal regenerator is called the effectiveness ε, and is defined as $\varepsilon = q_{regen,\ act} / q_{regen,\ max}$.

8-84C (b) turbine exit.

8-85C The steam injected increases the mass flow rate through the turbine and thus the power output. This, in turn, increases the thermal efficiency since $\eta = W / Q_{in}$ and W increases while Q_{in} remains constant. Steam can be obtained by utilizing the hot exhaust gases.

8-86E A car is powered by a gas turbine with a pressure ratio of 4. The thermal efficiency of the car and the mass flow rate of air for a net power output of 135 hp are to be determined.

Assumptions **1** Steady operating conditions exist. **2** Air is an ideal gas with variable specific heats. **3** The ambient air is 540 R and 14.5 psia. **4** The effectiveness of the regenerator is 0.9, and the isentropic efficiencies for both the compressor and the turbine are 80%. **5** The combustion gases can be treated as air.

Properties The properties of air at the compressor and turbine inlet temperatures can be obtained from Table A-17E.

Analysis The gas turbine cycle with regeneration can be analyzed as follows:

$$T_1 = 540\text{R} \longrightarrow \begin{array}{l} h_1 = 129.06\text{Btu/lbm} \\ P_{r_1} = 1.386 \end{array}$$

$$P_{r_2} = \frac{P_2}{P_1} P_{r_1} = (4)(1.386) = 5.544 \longrightarrow h_{2s} = 192.0\text{Btu/lbm}$$

$$T_3 = 2160\text{R} \longrightarrow \begin{array}{l} h_3 = 549.35\text{Btu/lbm} \\ P_{r_3} = 230.12 \end{array}$$

$$P_{r_4} = \frac{P_4}{P_3} P_{r_3} = \left(\frac{1}{4}\right)(230.12) = 57.53 \longrightarrow h_{4s} = 372.2\text{Btu/lbm}$$

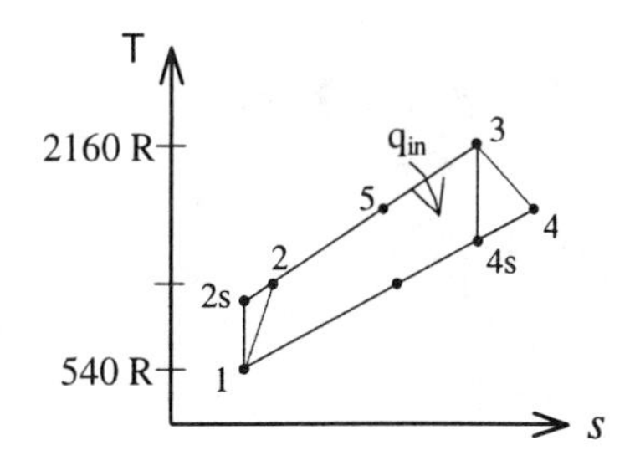

and

$$\eta_{\text{comp}} = \frac{h_{2s} - h_1}{h_2 - h_1} \rightarrow 0.80 = \frac{192.0 - 129.06}{h_2 - 129.06} \rightarrow h_2 = 207.74 \text{ kJ/kg}$$

$$\eta_{\text{turb}} = \frac{h_3 - h_4}{h_3 - h_{4s}} \rightarrow 0.80 = \frac{549.35 - h_4}{549.35 - 372.2} \rightarrow h_4 = 407.63 \text{ kJ/kg}$$

Then the thermal efficiency of the gas turbine cycle becomes

$$\dot{q}_{\text{regen}} = \varepsilon(h_4 - h_2) = 0.9(407.63 - 207.74) = 179.9 \text{ Btu/lbm}$$

$$q_{in} = (h_3 - h_2) - q_{regen} = (549.35 - 207.74) - 179.9 \text{ kJ/} = 161.7 \text{ Btu/s}$$

$$w_{net,out} = w_{T,out} - w_{C,in} = (h_3 - h_4) - (h_2 - h_1) = (549.35 - 407.63) - (207.74 - 129.06) = \mathbf{63.0\ Btu/lbm}$$

$$\eta_{th} = \frac{w_{net,out}}{q_{in}} = \frac{63.0 \text{ Btu/lbm}}{161.7 \text{ Btu/lbm}} = \mathbf{0.39\%}$$

Finally, the mass flow rate of air through the turbine becomes

$$\dot{m}_{steam} = \frac{\dot{W}_{net}}{w_{net}} = \frac{135 \text{ hp}}{63.0 \text{ Btu/lbm}}\left(\frac{0.7068 \text{ Btu/s}}{1 \text{ hp}}\right) = \mathbf{1.51\ lbm/s}$$

8-87 [*Also solved by EES on enclosed CD*] The thermal efficiency and power output of an actual gas turbine are given. The isentropic efficiency of the turbine and of the compressor, and the thermal efficiency of the gas turbine modified with a regenerator are to be determined.

Assumptions **1** Air is an ideal gas with variable specific heats. **2** Kinetic and potential energy changes are negligible. **3** The mass flow rates of air and of the combustion gases are the same, and the properties of combustion gases are the same as those of air.

Properties The properties of air are given in Table A-17.

Analysis The properties at various states are

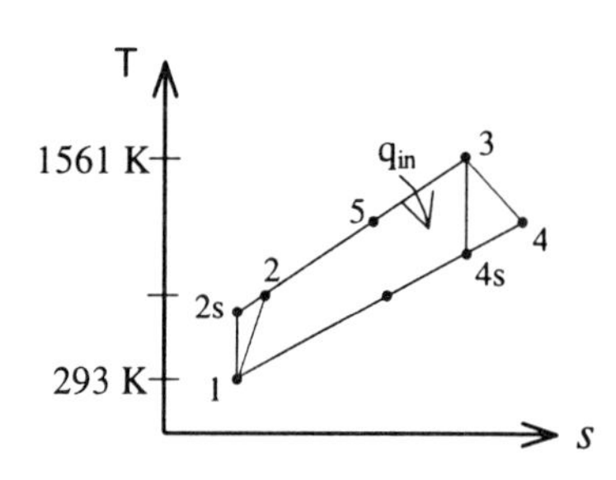

$$T_1 = 20°\text{C} = 293\text{K} \longrightarrow \begin{array}{l} h_1 = 293.2\text{kJ/kg} \\ P_{r_1} = 1.2765 \end{array}$$

$$P_{r_2} = \frac{P_2}{P_1} P_{r_1} = (14.7)(1.2765) = 18.765 \longrightarrow h_{2s} = 643.3\text{kJ/kg}$$

$$T_3 = 1288°\text{C} = 1561\,\text{K} \longrightarrow \begin{array}{l} h_3 = 1710.0\text{kJ/kg} \\ P_{r_3} = 712.5 \end{array}$$

$$P_{r_4} = \frac{P_4}{P_3} P_{r_3} = \left(\frac{1}{14.7}\right)(712.5) = 48.47 \longrightarrow h_{4s} = 825.23\text{kJ/kg}$$

The net work output and the heat input per unit mass are

$$w_{net} = \frac{\dot{W}_{net}}{\dot{m}} = \frac{159{,}000\,\text{kW}}{1{,}536{,}000\,\text{kg/h}}\left(\frac{3600\,\text{s}}{1\,\text{h}}\right) = 372.66\,\text{kJ/kg}$$

$$q_{in} = \frac{w_{net}}{\eta_{th}} = \frac{372.66\,\text{kJ/kg}}{0.359} = 1038.0\,\text{kJ/kg}$$

$$q_{in} = h_3 - h_2 \;\rightarrow\; h_2 = h_3 - q_{in} = 1710 - 1038 = 672.0\,\text{kJ/kg}$$

$$q_{out} = q_{in} - w_{net} = 1038.0 - 372.66 = 665.34\,\text{kJ/kg}$$

$$q_{out} = h_4 - h_1 \;\rightarrow\; h_4 = q_{out} + h_1 = 665.34 + 293.2 = 958.54\,\text{kJ/kg} \;\rightarrow\; T_4 = 650°\text{C}$$

Then the compressor and turbine efficiencies become

$$\eta_T = \frac{h_3 - h_4}{h_3 - h_{4s}} = \frac{1710 - 958.54}{1710 - 825.23} = \mathbf{0.849}$$

$$\eta_C = \frac{h_{2s} - h_1}{h_2 - h_1} = \frac{643.3 - 293.2}{672 - 293.2} = \mathbf{0.924}$$

When a regenerator is added, the new heat input and the thermal efficiency become

$$q_{regen} = \varepsilon(h_4 - h_2) = (0.80)(958.54 - 672.0) = 286.54\,\text{kJ/kg}$$

$$q_{in,new} = q_{in} - q_{regen} = 1038 - 286.54 = 751.46\,\text{kJ/kg}$$

$$\eta_{th,new} = \frac{w_{net}}{q_{in,new}} = \frac{372.66\,\text{kJ/kg}}{751.46\,\text{kJ/kg}} = \mathbf{0.496}$$

Discussion Note an 80% efficient regenerator would increase the thermal efficiency of this gas turbine from 35.9% to 49.6%.

8-88 EES solution of this (and other comprehensive problems designated with the *computer icon*) is available to instructors at the *Instructor Manual* section of the *Online Learning Center* (OLC) at www.mhhe.com/cengel-boles. See the Preface for access information.

8-89 An ideal Brayton cycle with regeneration is considered. The effectiveness of the regenerator is 100%. The net work output and the thermal efficiency of the cycle are to be determined.

Assumptions **1** The air standard assumptions are applicable. **2** Air is an ideal gas with variable specific heats. **3** Kinetic and potential energy changes are negligible.

Properties The properties of air are given in Table A-17.

Analysis Noting that this is an ideal cycle and thus the compression and expansion processes are isentropic, we have

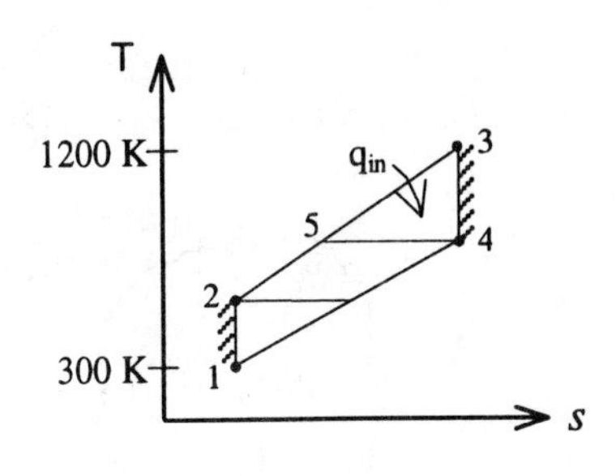

$$T_1 = 300\text{K} \longrightarrow \begin{matrix} h_1 = 300.19\text{kJ/kg} \\ P_{r_1} = 1.386 \end{matrix}$$

$$P_{r_2} = \frac{P_2}{P_1} P_{r_1} = (10)(1.386) = 13.86 \longrightarrow h_2 = 579.87\text{kJ/kg}$$

$$T_3 = 1200\text{K} \longrightarrow \begin{matrix} h_3 = 1277.79\text{kJ/kg} \\ P_{r_3} = 238 \end{matrix}$$

$$P_{r_4} = \frac{P_4}{P_3} P_{r_3} = \left(\frac{1}{10}\right)(238) = 23.8 \longrightarrow h_4 = 675.85\text{kJ/kg}$$

$$w_{C,in} = h_2 - h_1 = 579.87 - 300.19 = 279.68\text{kJ/kg}$$

$$w_{T,out} = h_3 - h_4 = 1277.79 - 675.85 = 601.94\text{kJ/kg}$$

Thus,

$$w_{net} = w_{T,out} - w_{C,in} = 601.94 - 279.68 = \mathbf{322.26kJ/kg}$$

Also, $\varepsilon = 100\% \longrightarrow h_5 = h_4 = 675.85 \text{ kJ/kg}$

$$q_{in} = h_3 - h_5 = 1277.79 - 675.85 = 601.94 \text{ kJ/kg}$$

and

$$\eta_{th} = \frac{w_{net}}{q_{in}} = \frac{322.26 \text{ kJ/kg}}{601.94 \text{ kJ/kg}} = \mathbf{53.5\%}$$

8-90 EES solution of this (and other comprehensive problems designated with the *computer icon*) is available to instructors at the *Instructor Manual* section of the *Online Learning Center* (OLC) at www.mhhe.com/cengel-boles. See the Preface for access information.

8-91 An ideal Brayton cycle with regeneration is considered. The effectiveness of the regenerator is 100%. The net work output and the thermal efficiency of the cycle are to be determined.

Assumptions **1** The air standard assumptions are applicable. **2** Air is an ideal gas with constant specific heats at room temperature. **3** Kinetic and potential energy changes are negligible.

Properties The properties of air at room temperature are C_p = 1.005 kJ/kg.K and k = 1.4 (Table A-2a).

Analysis Noting that this is an ideal cycle and thus the compression and expansion processes are isentropic, we have

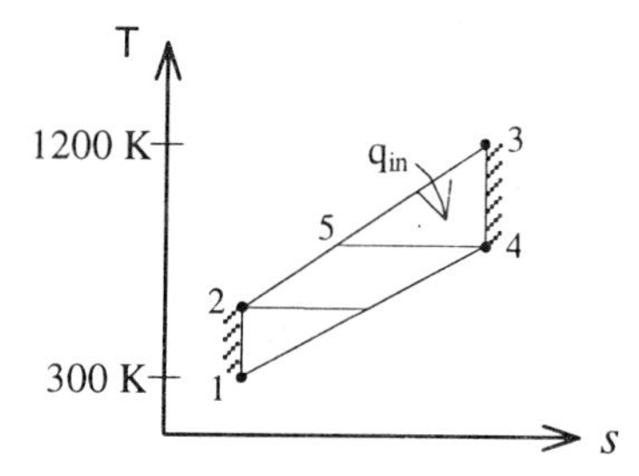

$$T_2 = T_1\left(\frac{P_2}{P_1}\right)^{(k-1)/k} = (300\text{K})(10)^{0.4/1.4} = 579.2\text{K}$$

$$T_4 = T_3\left(\frac{P_4}{P_3}\right)^{(k-1)/k} = (1200\text{K})\left(\frac{1}{10}\right)^{0.4/1.4} = 621.5\text{K}$$

$$\varepsilon = 100\% \longrightarrow T_5 = T_4 = 621.5\text{K and } T_6 = T_2 = 579.2\text{K}$$

$$\eta_{th} = 1 - \frac{q_{out}}{q_{in}} = 1 - \frac{C_p(T_6 - T_1)}{C_p(T_3 - T_5)} = 1 - \frac{T_6 - T_1}{T_3 - T_5} = 1 - \frac{579.2 - 300}{1200 - 621.5} = \mathbf{0.517}$$

$$(\text{or, } \eta_{th} = 1 - \left(\frac{T_1}{T_3}\right) r_p^{(k-1)/k} = 1 - \left(\frac{300}{1200}\right)(10)^{(1.4-1)/1.4} = 0.517)$$

Then,

$$\begin{aligned} w_{\text{net}} &= w_{\text{turb, out}} - w_{\text{comp, in}} = (h_3 - h_4) - (h_2 - h_1) \\ &= C_p[(T_3 - T_4) - (T_2 - T_1)] \\ &= (1.005 \text{ kJ/kg.K})[(1200 - 621.5) - (579.2 - 300)]\text{K} \\ &= \mathbf{300.8 \text{ kJ/kg}} \end{aligned}$$

or,

$$\begin{aligned} w_{net} &= \eta_{th} q_{in} = \eta_{th}(h_3 - h_5) = \eta_{th} C_p (T_3 - T_5) \\ &= (0.517)(1.005 \text{ kJ/kg.K})(1200 - 621.5) \\ &= 300.6 \text{ kJ/kg} \end{aligned}$$

8-92 A Brayton cycle with regeneration using air as the working fluid is considered. The air temperature at the turbine exit, the net work output, and the thermal efficiency are to be determined.

Assumptions **1** The air standard assumptions are applicable. **2** Air is an ideal gas with variable specific heats. **3** Kinetic and potential energy changes are negligible.

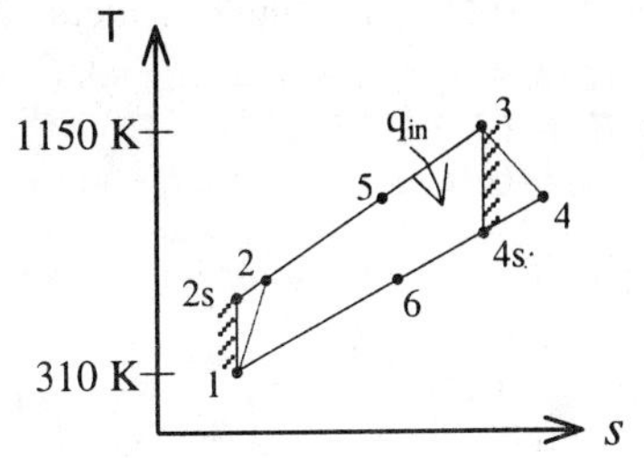

Properties The properties of air are given in Table A-17.

Analysis (a) The properties of air at various states are

T
1150 K
310 K
s
q_{in}
1, 2s, 2, 5, 3, 4s, 4, 6

$$T_1 = 310\text{K} \longrightarrow \begin{array}{l} h_1 = 310.24\text{kJ/kg} \\ P_{r_1} = 1.5546 \end{array}$$

$$P_{r_2} = \frac{P_2}{P_1} P_{r_1} = (7)(1.5546) = 10.88 \longrightarrow h_{2s} = 541.26 \text{ kJ/kg}$$

$$\eta_C = \frac{h_{2s} - h_1}{h_2 - h_1} \longrightarrow h_2 = h_1 + (h_{2s} - h_1)/\eta_C = 310.24 + (541.26 - 310.24)/(0.75) = 618.26 \text{ kJ/kg}$$

$$T_3 = 1150\text{K} \longrightarrow \begin{array}{l} h_3 = 1219.25\text{kJ/kg} \\ P_{r_3} = 200.15 \end{array}$$

$$P_{r_4} = \frac{P_4}{P_3} P_{r_3} = \left(\frac{1}{7}\right)(200.15) = 28.59 \longrightarrow h_{4s} = 711.80 \text{ kJ/kg}$$

$$\eta_T = \frac{h_3 - h_4}{h_3 - h_{4s}} \longrightarrow h_4 = h_3 - \eta_T (h_3 - h_{4s}) = 1219.25 - (0.82)(1219.25 - 711.80) = 803.14 \text{ kJ/kg}$$

Thus, $T_4 = \mathbf{782.8\ K}$

(b)

$$w_{net} = w_{T,out} - w_{C,in} = (h_3 - h_4) - (h_2 - h_1)$$
$$= (1219.25 - 803.14) - (618.26 - 310.24) = \mathbf{108.09\ kJ/kg}$$

(c)

$$\varepsilon = \frac{h_5 - h_2}{h_4 - h_2} \longrightarrow h_5 = h_2 + \varepsilon(h_4 - h_2)$$
$$= 618.26 + (0.65)(803.14 - 618.26)$$
$$= 738.43 \text{ kJ/kg}$$

Then,

$$q_{in} = h_3 - h_5 = 1219.25 - 738.43 = 480.82 \text{ kJ/kg}$$

$$\eta_{th} = \frac{w_{net}}{q_{in}} = \frac{108.09 \text{ kJ/kg}}{480.82 \text{ kJ/kg}} = \mathbf{22.5\%}$$

8-93 A stationary gas-turbine power plant operating on an ideal regenerative Brayton cycle with air as the working fluid is considered. The power delivered by this plant is to be determined for two cases.

Assumptions **1** The air standard assumptions are applicable. **2** Air is an ideal gas. **3** Kinetic and potential energy changes are negligible.

Properties When assuming constant specific heats, the properties of air at room temperature are C_p = 1.005 kJ/kg.K and k = 1.4 (Table A-2a). When assuming variable specific heats, the properties of air are obtained from Table A-17.

Analysis (*a*) Assuming constant specific heats,

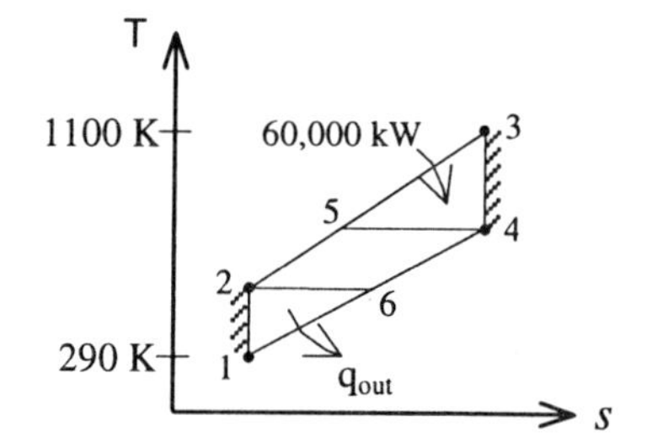

$$T_2 = T_1\left(\frac{P_2}{P_1}\right)^{(k-1)/k} = (290\text{K})(8)^{0.4/1.4} = 525.3 \text{ K}$$

$$T_4 = T_3\left(\frac{P_4}{P_3}\right)^{(k-1)/k} = (1100\text{K})\left(\frac{1}{8}\right)^{0.4/1.4} = 607.2 \text{ K}$$

$$\varepsilon = 100\% \longrightarrow T_5 = T_4 = 607.2 \text{ K and } T_6 = T_2 = 525.3 \text{ K}$$

$$\eta_{th} = 1 - \frac{q_{out}}{q_{in}} = 1 - \frac{C_p(T_6 - T_1)}{C_p(T_3 - T_5)} = 1 - \frac{T_6 - T_1}{T_3 - T_5} = 1 - \frac{525.3 - 290}{1100 - 607.2} = 0.5225$$

$$\dot{W}_{net} = \eta_T \dot{Q}_{in} = (0.5225)(90{,}000 \text{ kW}) = \mathbf{47{,}027 \text{ kW}}$$

(*b*) Assuming variable specific heats,

$$T_1 = 290\text{K} \longrightarrow \begin{array}{l} h_1 = 290.16 \text{ kJ/kg} \\ P_{r_1} = 1.2311 \end{array}$$

$$P_{r_2} = \frac{P_2}{P_1} P_{r_1} = (8)(1.2311) = 9.8488 \longrightarrow h_2 = 526.12 \text{ kJ/kg}$$

$$T_3 = 1100\text{K} \longrightarrow \begin{array}{l} h_3 = 1161.07 \text{ kJ/kg} \\ P_{r_3} = 167.1 \end{array}$$

$$P_{r_4} = \frac{P_4}{P_3} P_{r_3} = \left(\frac{1}{8}\right)(167.1) = 20.89 \longrightarrow h_4 = 651.37 \text{ kJ/kg}$$

$$\varepsilon = 100\% \longrightarrow h_5 = h_4 = 651.37 \text{ kJ/kg and } h_6 = h_2 = 526.12 \text{ kJ/kg}$$

$$\eta_{th} = 1 - \frac{q_{out}}{q_{in}} = 1 - \frac{h_6 - h_1}{h_3 - h_5} = 1 - \frac{526.12 - 290.16}{1161.07 - 651.37} = 0.5371$$

$$\dot{W}_{net} = \eta_T \dot{Q}_{in} = (0.5371)(90{,}000\text{kW}) = \mathbf{48{,}335 \text{ kW}}$$

8-94 A regenerative gas-turbine engine using air as the working fluid is considered. The amount of heat transfer in the regenerator and the thermal efficiency are to be determined.

Assumptions **1** The air standard assumptions are applicable. **2** Air is an ideal gas with variable specific heats. **3** Kinetic and potential energy changes are negligible.

Properties The properties of air are given in Table A-17.

Analysis (*a*) The properties at various states are

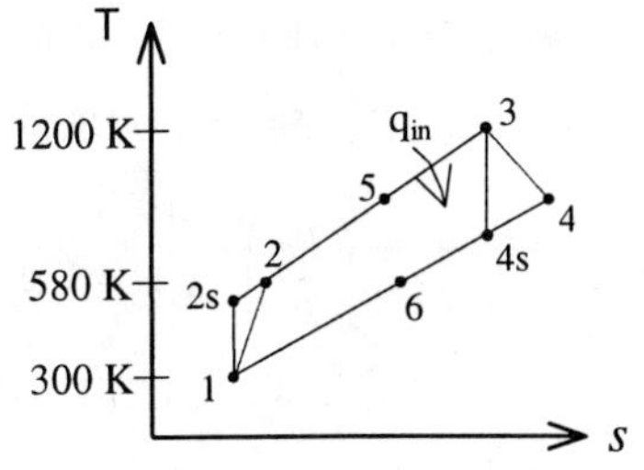

$$r_p = P_2 / P_1 = 800/100 = 8$$

$$T_1 = 300\text{K} \longrightarrow h_1 = 300.19\text{kJ/kg}$$
$$T_2 = 580\text{K} \longrightarrow h_2 = 586.04\text{kJ/kg}$$
$$T_3 = 1200\text{K} \longrightarrow \text{h}_3 = 1277.79\text{kJ/kg}$$
$$\text{P}_{\text{r}_3} = 238.0$$

$$P_{r_4} = \frac{P_4}{P_3} P_{r_3} = \left(\frac{1}{8}\right)(238.0) = 29.75 \longrightarrow h_{4s} = 719.75\text{kJ/kg}$$

$$\eta_T = \frac{h_3 - h_4}{h_3 - h_{4s}} \longrightarrow h_4 = h_3 - \eta_T (h_3 - h_{4s})$$
$$= 1277.79 - (0.86)(1277.79 - 719.75)$$
$$= 797.88 \text{ kJ/kg}$$

$$q_{regen} = \varepsilon(h_4 - h_2) = (0.72)(797.88 - 586.04) = \mathbf{152.5kJ/kg}$$

(*b*)
$$w_{net} = w_{T,out} - w_{C,in} = (h_3 - h_4) - (h_2 - h_1)$$
$$= (1277.79 - 797.88) - (586.04 - 300.19) = 194.06 \text{ kJ/kg}$$

$$q_{in} = (h_3 - h_2) - q_{regen} = (1277.79 - 586.04) - 152.52 = 539.23 \text{ kJ/kg}$$

$$\eta_{th} = \frac{w_{net}}{q_{in}} = \frac{194.06 \text{ kJ/kg}}{539.23 \text{ kJ/kg}} = \mathbf{36.0\%}$$

8-95 A regenerative gas-turbine engine using air as the working fluid is considered. The amount of heat transfer in the regenerator and the thermal efficiency are to be determined. √

Assumptions **1** The air standard assumptions are applicable. **2** Air is an ideal gas with constant specific heats. **3** Kinetic and potential energy changes are negligible.

Properties The properties of air at room temperature are C_p = 1.005 kJ/kg.K and k = 1.4 (Table A-2a).

Analysis (a) Using the isentropic relations and turbine efficiency,

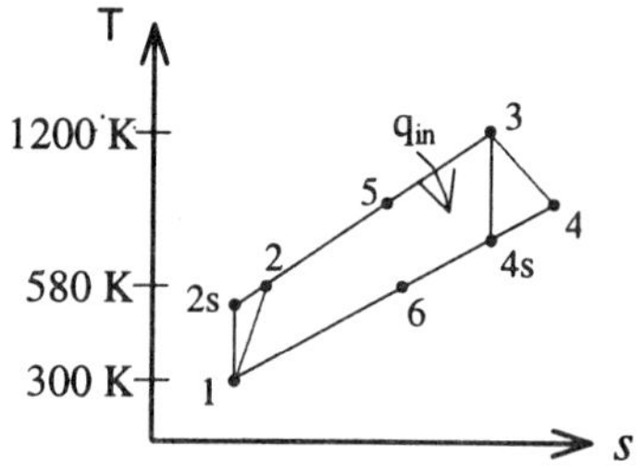

$$r_p = P_2 / P_1 = 800/100 = 8$$

$$T_{4s} = T_3\left(\frac{P_4}{P_3}\right)^{(k-1)/k} = (1200\text{K})\left(\frac{1}{8}\right)^{0.4/1.4} = 662.5\text{K}$$

$$\eta_T = \frac{h_3 - h_4}{h_3 - h_{4s}} = \frac{C_p(T_3 - T_4)}{C_p(T_3 - T_{4s})} \longrightarrow T_4 = T_3 - \eta_T(T_3 - T_{4s})$$

$$= 1200 - (0.86)(1200 - 662.5)$$

$$= 737.8\text{K}$$

$$q_{regen} = \varepsilon(h_4 - h_2) = \varepsilon C_p(T_4 - T_2)$$

$$= (0.72)(1.005\text{kJ/kg}\cdot\text{K})(737.8 - 580)\text{K} = \mathbf{114.2kJ/kg}$$

(b)

$$w_{net} = w_{T,out} - w_{C,in} = C_p(T_3 - T_s) - C_p(T_2 - T_1)$$

$$= (1.005\text{kJ/kg}\cdot\text{K})[(1200 - 737.8) - (580 - 300)]\text{K} = 183.1\text{kJ/kg}$$

$$q_{in} = (h_3 - h_2) - q_{regen} = C_p(T_3 - T_2) - q_{regen}$$

$$= (1.005\text{kJ/kg}\cdot\text{K})(1200 - 580)\text{K} - 114.2 = 508.9\text{kJ/kg}$$

$$\eta_{th} = \frac{w_{net}}{q_{in}} = \frac{183.1\text{kJ/kg}}{508.9\text{kJ/kg}} = \mathbf{36.0\%}$$

8-96 A regenerative gas-turbine engine using air as the working fluid is considered. The amount of heat transfer in the regenerator and the thermal efficiency are to be determined.

Assumptions **1** The air standard assumptions are applicable. **2** Air is an ideal gas with variable specific heats. **3** Kinetic and potential energy changes are negligible.

Properties The properties of air are given in Table A-17.

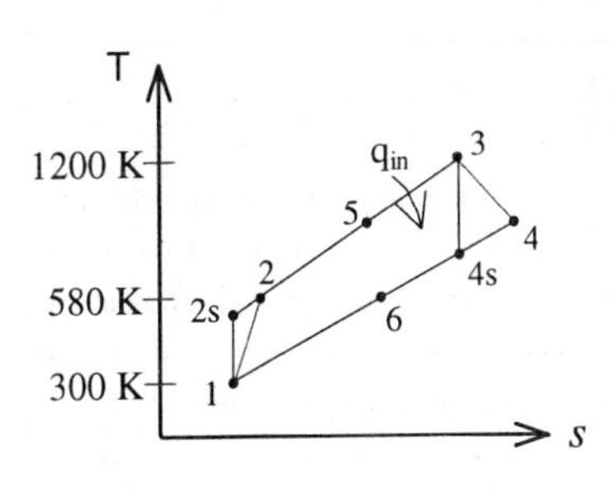

Analysis (*a*) The properties of air at various states are

$$r_p = P_2 / P_1 = 800/100 = 8$$

$$T_1 = 300\text{K} \longrightarrow h_1 = 300.19\text{kJ/kg}$$
$$T_2 = 580\text{K} \longrightarrow h_2 = 586.04\text{kJ/kg}$$
$$T_3 = 1200\text{K} \longrightarrow \text{h}_3 = 1277.79\text{kJ/kg}$$
$$\text{P}_{\text{r}_3} = 238.0$$

$$P_{r_4} = \frac{P_4}{P_3} P_{r_3} = \left(\frac{1}{8}\right)(238.0) = 29.75 \longrightarrow h_{4s} = 719.75\text{kJ/kg}$$

$$\eta_T = \frac{h_3 - h_4}{h_3 - h_{4s}} \longrightarrow h_4 = h_3 - \eta_T (h_3 - h_{4s})$$
$$= 1277.79 - (0.86)(1277.79 - 719.75)$$
$$= 797.88 \text{ kJ/kg}$$

$$q_{regen} = \varepsilon(h_3 - h_2) = (0.70)(797.88 - 586.04) = \mathbf{148.3kJ/kg}$$

(*b*)
$$w_{net} = w_{T,out} - w_{C,in} = (h_3 - h_4) - (h_2 - h_1)$$
$$= (1277.79 - 797.88) - (586.04 - 300.19) = 194.06 \text{ kJ/kg}$$

$$q_{in} = (h_3 - h_2) - q_{regen} = (1277.79 - 586.04) - 148.3 = 543.5 \text{ kJ/kg}$$

$$\eta_{th} = \frac{w_{net}}{q_{in}} = \frac{194.06 \text{ kJ/kg}}{543.5 \text{ kJ/kg}} = \mathbf{35.7\%}$$

Brayton Cycle with Intercooling, Reheating, and Regeneration

8-97C As the number of compression and expansion stages are increased and regeneration is employed, the ideal Brayton cycle will approach the Ericsson cycle.

8-98C (a) decrease, (b) decrease, and (c) decrease.

8-99C (a) increase, (b) decrease, and (c) decrease.

8-100C (a) increase, (b) decrease, (c) decrease, and (d) increase.

8-101C (a) increase, (b) decrease, (c) increase, and (d) decrease.

8-102C Because the steady-flow work is proportional to the specific volume of the gas. Intercooling decreases the average specific volume of the gas during compression, and thus the compressor work. Reheating increases the average specific volume of the gas, and thus the turbine work output.

8-103C (c) The Carnot (or Ericsson) cycle efficiency.

8-104 An ideal gas-turbine cycle with two stages of compression and two stages of expansion is considered. The back work ratio and the thermal efficiency of the cycle are to be determined for the cases of with and without a regenerator.

Assumptions **1** The air standard assumptions are applicable. **2** Air is an ideal gas with variable specific heats. **3** Kinetic and potential energy changes are negligible.

Properties The properties of air are given in Table A-17.

Analysis (*a*) The work inputs to each stage of compressor are identical, so are the work outputs of each stage of the turbine since this is an ideal cycle. Then,

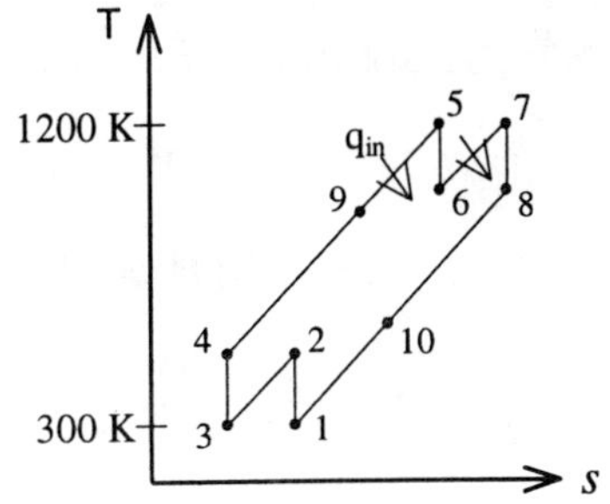

$$T_1 = 300\text{K} \longrightarrow \begin{array}{l} h_1 = 300.19\text{kJ/kg} \\ P_{r_1} = 1.386 \end{array}$$

$$P_{r_2} = \frac{P_2}{P_1} P_{r_1} = (3)(1.386) = 4.158 \longrightarrow h_2 = h_4 = 411.26\text{kJ/kg}$$

$$T_5 = 1200\text{K} \longrightarrow \begin{array}{l} h_5 = h_7 = 1277.79\text{kJ/kg} \\ P_{r_5} = 238 \end{array}$$

$$P_{r_6} = \frac{P_6}{P_5} P_{r_5} = \left(\frac{1}{3}\right)(238) = 79.33 \longrightarrow h_6 = h_8 = 946.36\text{kJ/kg}$$

$$w_{C,in} = 2(h_2 - h_1) = 2(411.26 - 300.19) = 222.14\text{kJ/kg}$$

$$w_{T,out} = 2(h_5 - h_6) = 2(1277.79 - 946.36) = 662.86\text{kJ/kg}$$

Thus,

$$r_{bw} = \frac{w_{C,in}}{w_{T,out}} = \frac{222.14\text{kJ/kg}}{662.86\text{kJ/kg}} = \mathbf{33.5\%}$$

$$q_{in} = (h_5 - h_4) + (h_7 - h_6) = (1277.79 - 411.26) + (1277.79 - 946.36) = 1197.96\text{kJ/kg}$$

$$w_{net} = w_{T,out} - w_{C,in} = 662.86 - 222.14 = 440.72\text{kJ/kg}$$

$$\eta_{th} = \frac{w_{net}}{q_{in}} = \frac{440.72\text{kJ/kg}}{1197.96\text{kJ/kg}} = \mathbf{36.8\%}$$

(*b*) When a regenerator is used, r_{bw} remains the same. The thermal efficiency in this case becomes

$$q_{regen} = \varepsilon(h_8 - h_4) = (0.75)(946.36 - 411.26) = 401.33\text{kJ/kg}$$

$$q_{in} = q_{in,old} - q_{regen} = 1197.96 - 401.33 = 796.63\text{kJ/kg}$$

$$\eta_{th} = \frac{w_{net}}{q_{in}} = \frac{440.72\text{kJ/kg}}{796.63\text{kJ/kg}} = \mathbf{55.3\%}$$

8-105 A gas-turbine cycle with two stages of compression and two stages of expansion is considered. The back work ratio and the thermal efficiency of the cycle are to be determined for the cases of with and without a regenerator.

Assumptions **1** The air standard assumptions are applicable. **2** Air is an ideal gas with variable specific heats. **3** Kinetic and potential energy changes are negligible.

Properties The properties of air are given in Table A-17.

Analysis (*a*) The work inputs to each stage of compressor are identical, so are the work outputs of each stage of the turbine. Then,

$$T_1 = 300\text{K} \longrightarrow h_1 = 300.19\text{kJ/kg}$$
$$P_{r_1} = 1.386$$

$$P_{r_2} = \frac{P_2}{P_1} P_{r_1} = (3)(1.386) = 4.158 \longrightarrow h_{2s} = h_{4s} = 411.26\text{kJ/kg}$$

$$\eta_C = \frac{h_{2s} - h_1}{h_2 - h_1} \longrightarrow h_2 = h_4 = h_1 + (h_{2s} - h_1)/\eta_C$$
$$= 300.19 + (411.26 - 300.19)/(0.80)$$
$$= 439.03\text{kJ/kg}$$

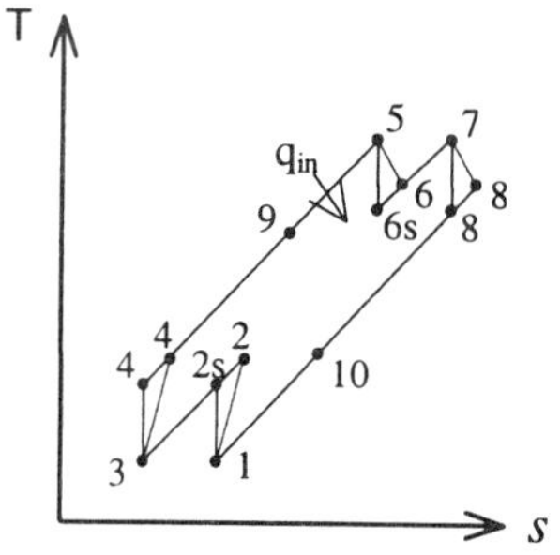

$$T_5 = 1200\text{K} \longrightarrow h_5 = h_7 = 1277.79\text{kJ/kg}$$
$$P_{r_5} = 238$$

$$P_{r_6} = \frac{P_6}{P_5} P_{r_5} = \left(\frac{1}{3}\right)(238) = 79.33 \longrightarrow h_6 = h_8 = 946.36\text{kJ/kg}$$

$$\eta_T = \frac{h_5 - h_6}{h_5 - h_{6s}} \longrightarrow h_6 = h_8 = h_5 - \eta_T (h_5 - h_{6s})$$
$$= 1277.79 - (0.85)(1277.79 - 946.36)$$
$$= 996.07\text{kJ/kg}$$

$$w_{C,in} = 2(h_2 - h_1) = 2(439.03 - 300.19) = 277.68 \text{ kJ/kg}$$
$$w_{T,out} = 2(h_5 - h_6) = 2(1277.79 - 996.07) = 563.44 \text{ kJ/kg}$$

Thus, $$r_{bw} = \frac{w_{C,in}}{w_{T,out}} = \frac{277.68 \text{ kJ/kg}}{563.44 \text{ kJ/kg}} = \mathbf{49.3\%}$$

$$q_{in} = (h_5 - h_4) + (h_7 - h_6) = (1277.79 - 439.03) + (1277.79 - 996.07) = 1120.48\text{kJ/kg}$$
$$w_{net} = w_{T,out} - w_{C,in} = 563.44 - 277.68 = 285.76 \text{ kJ/kg}$$

$$\eta_{th} = \frac{w_{net}}{q_{in}} = \frac{285.76 \text{ kJ/kg}}{1120.48 \text{ kJ/kg}} = \mathbf{25.5\%}$$

(*b*) When a regenerator is used, r_{bw} remains the same. The thermal efficiency in this case becomes

$$q_{regen} = \varepsilon(h_8 - h_4) = (0.75)(996.07 - 439.03) = 417.78\text{kJ/kg}$$
$$q_{in} = q_{in,old} - q_{regen} = 1120.48 - 417.78 = 702.70\text{kJ/kg}$$

$$\eta_{th} = \frac{w_{net}}{q_{in}} = \frac{285.76 \text{ kJ/kg}}{702.70 \text{ kJ/kg}} = \mathbf{40.7\%}$$

8-106 A regenerative gas-turbine cycle with two stages of compression and two stages of expansion is considered. The minimum mass flow rate of air needed to develop a specified net power output is to be determined.

Assumptions **1** The air standard assumptions are applicable. **2** Air is an ideal gas with variable specific heats. **3** Kinetic and potential energy changes are negligible.

Properties The properties of air are given in Table A-17.

Analysis The mass flow rate will be a minimum when the cycle is ideal. That is, the turbine and the compressors are isentropic, the regenerator has an effectiveness of 100%, and the compression ratios across each compression or expansion stage are identical. In our case it is $r_p = \sqrt{9} = 3$. Then the work inputs to each stage of compressor are identical, so are the work outputs of each stage of the turbine.

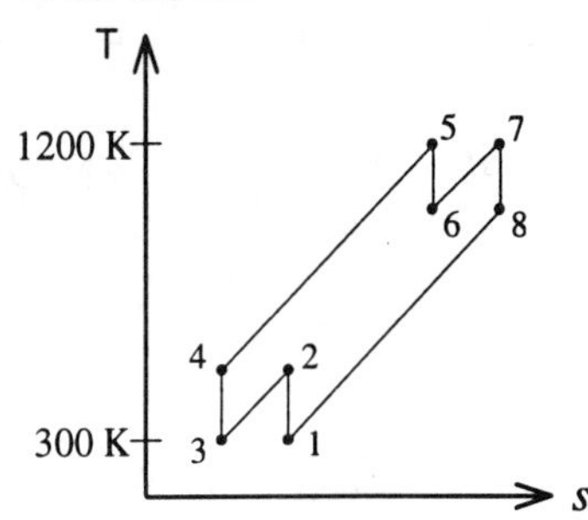

$$T_1 = 300\text{K} \longrightarrow h_1 = 300.19\text{kJ/kg}$$
$$P_{r_1} = 1.386$$

$$P_{r_2} = \frac{P_2}{P_1} P_{r_1} = (3)(1.386) = 4.158 \longrightarrow h_2 = h_4 = 411.26\text{kJ/kg}$$

$$T_5 = 1200K \longrightarrow h_5 = h_7 = 1277.79\text{kJ/kg}$$
$$P_{r_5} = 238$$

$$P_{r_6} = \frac{P_6}{P_5} P_{r_5} = \left(\frac{1}{3}\right)(238) = 79.33 \longrightarrow h_6 = h_8 = 946.36\text{kJ/kg}$$

$$w_{C,in} = 2(h_2 - h_1) = 2(411.26 - 300.19) = 222.14\text{kJ/kg}$$

$$w_{T,out} = 2(h_5 - h_6) = 2(1277.79 - 946.36) = 662.86\text{kJ/kg}$$

$$w_{net} = w_{T,out} - w_{C,in} = 662.86 - 222.14 = 440.72\text{kJ/kg}$$

$$\dot{m} = \frac{\dot{W}_{net}}{w_{net}} = \frac{90{,}000\text{ kJ/s}}{440.72\text{ kJ/kg}} = \mathbf{204.2\ kg/s}$$

8-107 A regenerative gas-turbine cycle with two stages of compression and two stages of expansion is considered. The minimum mass flow rate of air needed to develop a specified net power output is to be determined.

Assumptions **1** Argon is an ideal gas with constant specific heats. **2** Kinetic and potential energy changes are negligible.

Properties The properties of argon at room temperature are C_p = 0.5203 kJ/kg.K and k = 1.667 (Table A-2a).

Analysis The mass flow rate will be a minimum when the cycle is ideal. That is, the turbine and the compressors are isentropic, the regenerator has an effectiveness of 100%, and the compression ratios across each compression or expansion stage are identical. In our case it is $r_p = \sqrt{9} = 3$. Then the work inputs to each stage of compressor are identical, so are the work outputs of each stage of the turbine.

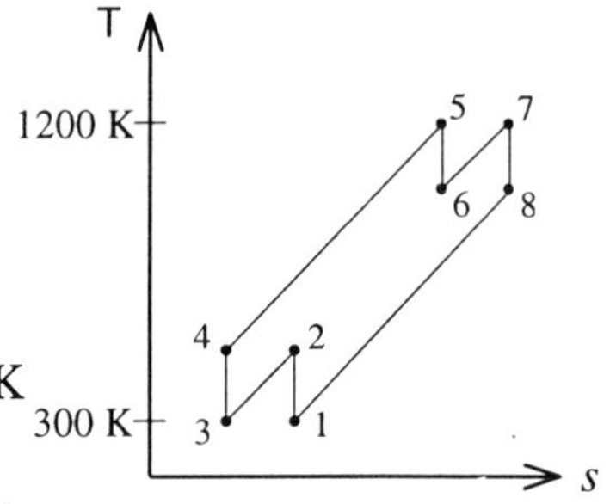

$$T_2 = T_1\left(\frac{P_2}{P_1}\right)^{(k-1)/k} = (300\text{K})(3)^{0.667/1.667} = 465.6\text{K}$$

$$T_6 = T_5\left(\frac{P_6}{P_5}\right)^{(k-1)/k} = (1200\text{K})\left(\frac{1}{3}\right)^{0.667/1.667} = 773.2\text{K}$$

$$w_{C,in} = 2(h_2 - h_1) = 2C_p(T_2 - T_1) = 2(0.5203\text{kJ/kg}\cdot\text{K})(465.6 - 300)\text{K} = 172.3\text{kJ/kg}$$

$$w_{T,out} = 2(h_5 - h_6) = 2C_p(T_5 - T_6) = 2(0.5203\text{kJ/kg}\cdot\text{K})(1200 - 773.2)\text{K} = 444.1\text{kJ/kg}$$

$$w_{net} = w_{T,out} - w_{C,in} = 444.1 - 172.3 = 271.8\text{kJ/kg}$$

$$\dot{m} = \frac{\dot{W}_{net}}{w_{net}} = \frac{90{,}000\text{ kJ/s}}{271.8\text{ kJ/kg}} = \mathbf{331.1\ kg/s}$$

T
1200 K
300 K
s
1
2
3
4
5
6
7
8

Jet-Propulsion Cycles

8-108C The power developed from the thrust of the engine is called the propulsive power. It is equal to thrust times the aircraft velocity.

8-109C The ratio of the propulsive power developed and the rate of heat input is called the propulsive efficiency. It is determined by calculating these two quantities separately, and taking their ratio.

8-110C It reduces the exit velocity, and thus the thrust.

8-111E A turbojet engine operating on an ideal cycle is flying at an altitude of 20,000 ft. The pressure at the turbine exit, the velocity of the exhaust gases, and the propulsive efficiency are to be determined.

Assumptions **1** Steady operating conditions exist. **2** The air standard assumptions are applicable. **3** Air is an ideal gas with constant specific heats at room temperature. **4** Kinetic and potential energies are negligible, except at the diffuser inlet and the nozzle exit. **5** The turbine work output is equal to the compressor work input.

Properties The properties of air at room temperature are C_p = 0.24 Btu/lbm.R and k = 1.4 (Table A-2Ea).

Analysis (*a*) For convenience, we assume the aircraft is stationary and the air is moving towards the aircraft at a velocity of $\mathbf{V}_1$ = 900 ft/s. Ideally, the air will leave the diffuser with a negligible velocity ($\mathbf{V}_2 \cong 0$).

Diffuser:

$$\dot{E}_{in} - \dot{E}_{out} = \Delta\dot{E}_{system}^{\nearrow 0 \text{ (steady)}}$$

$$\dot{E}_{in} = \dot{E}_{out}$$

$$h_1 + V_1^2/2 = h_2 + V_2^2/2$$

$$0 = h_2 - h_1 + \frac{\mathbf{V}_2^{2\nearrow 0} - \mathbf{V}_1^2}{2}$$

$$0 = C_p(T_2 - T_1) - \mathbf{V}_1^2/2$$

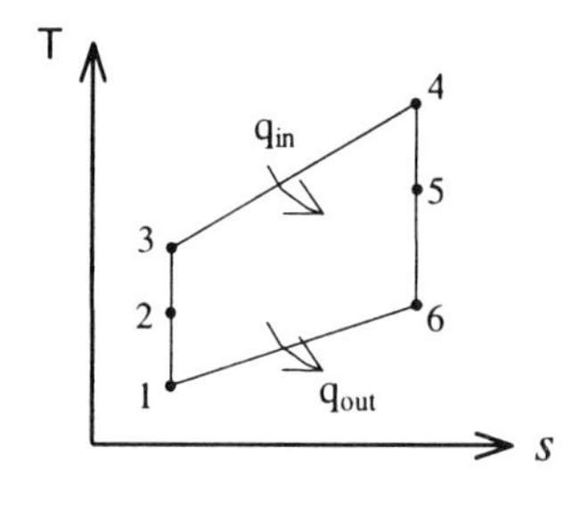

$$T_2 = T_1 + \frac{\mathbf{V}_1^2}{2C_p} = 470 + \frac{(900\text{ ft/s})^2}{(2)(0.24\text{ Btu/lbm}\cdot\text{R})}\left(\frac{1\text{ Btu/lbm}}{25,037\text{ ft}^2/\text{s}^2}\right) = 537.4\text{ R}$$

$$P_2 = P_1\left(\frac{T_2}{T_1}\right)^{k/(k-1)} = (7\text{ psia})\left(\frac{537.3\text{ R}}{470\text{ R}}\right)^{1.4/0.4} = 11.19\text{ psia}$$

Compressor:

$$P_3 = P_4 = (r_p)(P_2) = (13)(11.19\text{ psia}) = 145.5\text{ psia}$$

$$T_3 = T_2\left(\frac{P_3}{P_2}\right)^{(k-1)/k} = (537.4\text{ R})(13)^{0.4/1.4} = 1118.3\text{ R}$$

Turbine:

$$w_{comp,in} = w_{turb,out} \longrightarrow h_3 - h_2 = h_4 - h_5 \longrightarrow C_p(T_3 - T_2) = C_p(T_4 - T_5)$$

or,

$$T_5 = T_4 - T_3 + T_2 = 2400 - 1118.3 + 537.4 = 1819.1\text{ R}$$

$$P_5 = P_4\left(\frac{T_5}{T_4}\right)^{k/(k-1)} = (145.5\text{ psia})\left(\frac{1819.1\text{ R}}{2400\text{ R}}\right)^{1.4/0.4} = \mathbf{55.2\ psia}$$

(*b*) Nozzle:

$$T_6 = T_5\left(\frac{P_6}{P_5}\right)^{(k-1)/k} = (1819.1\ \text{R})\left(\frac{7\ \text{psia}}{55.2\ \text{psia}}\right)^{0.4/1.4} = 1008.6\ \text{R}$$

$$\dot{E}_{\text{in}} - \dot{E}_{\text{out}} = \Delta\dot{E}_{\text{system}}^{\ \nearrow 0\ \text{(steady)}}$$

$$\dot{E}_{in} = \dot{E}_{out}$$

$$h_5 + V_5^2/2 = h_6 + V_6^2/2$$

$$0 = h_6 - h_5 + \frac{\mathbf{V}_6^2 - \mathbf{V}_5^{2\ \nearrow 0}}{2} \longrightarrow 0 = C_p(T_6 - T_5) + \mathbf{V}_6^2/2$$

or,

$$\mathbf{V}_6 = \sqrt{(2)(0.240\text{Btu/lbm}\cdot\text{R})(1819.1 - 1008.6)\text{R}\left(\frac{25{,}037\ \text{ft}^2/\text{s}^2}{1\ \text{Btu/lbm}}\right)} = \mathbf{3121\ ft/s}$$

(*c*) The propulsive efficiency is the ratio of the propulsive work to the heat input,

$$w_p = (\mathbf{V}_{exit} - \mathbf{V}_{inlet})\mathbf{V}_{aircraft}$$

$$= (3121 - 900)\text{ft/s}(900\ \text{m/s})\left(\frac{1\ \text{Btu/lbm}}{25{,}037\ \text{ft}^2/\text{s}^2}\right) = 79.8\ \text{Btu/lbm}$$

$$q_{in} = h_4 - h_3 = C_p(T_4 - T_3) = (0.24\ \text{Btu/lbm}\cdot\text{R})(2400 - 1118.3)\text{R} = 307.6\ \text{Btu/lbm}$$

$$\eta_p = \frac{w_p}{q_{in}} = \frac{79.8\ \text{Btu/lbm}}{307.6\ \text{Btu/lbm}} = \mathbf{25.9\%}$$

8-112E A turbojet engine operating on an ideal cycle is flying at an altitude of 20,000 ft. The pressure at the turbine exit, the velocity of the exhaust gases, and the propulsive efficiency are to be determined.

Assumptions **1** Steady operating conditions exist. **2** The air standard assumptions are applicable. **3** Air is an ideal gas with variable specific heats. **4** Kinetic and potential energies are negligible, except at the diffuser inlet and the nozzle exit. **5** The turbine work output is equal to the compressor work input.

Properties The properties of air are given in Table A-17E.

Analysis (a) For convenience, we assume the aircraft is stationary and the air is moving towards the aircraft at a velocity of $\mathbf{V}_1$= 900 ft/s. Ideally, the air will leave the diffuser with a negligible velocity ($\mathbf{V}_2 \cong 0$).

Diffuser: $T_1 = 470\text{R} \longrightarrow h_1 = 112.20 \text{ Btu/lbm}$

$$P_{r_1} = 0.8548$$

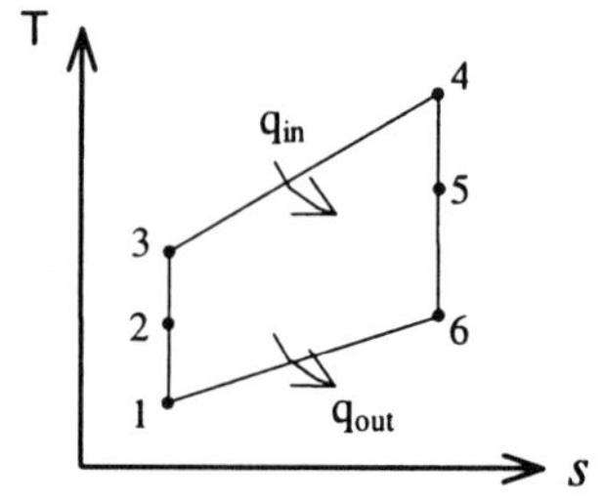

$$\dot{E}_{in} - \dot{E}_{out} = \Delta\dot{E}_{system}^{\nearrow 0 \text{ (steady)}}$$

$$\dot{E}_{in} = \dot{E}_{out}$$

$$h_1 + V_1^2/2 = h_2 + V_2^2/2$$

$$0 = h_2 - h_1 + \frac{\mathbf{V}_2^{2 \nearrow 0} - \mathbf{V}_1^2}{2}$$

$$h_2 = h_1 + \frac{\mathbf{V}_1^2}{2} = 112.20 + \frac{(900 \text{ ft/s})^2}{2}\left(\frac{1 \text{ Btu/lbm}}{25{,}037 \text{ ft}^2/\text{s}^2}\right) = 128.48 \text{ Btu/lbm}$$

$$\longrightarrow P_{r_2} = 1.3698$$

$$P_2 = P_1\left(\frac{P_{r_2}}{P_{r_1}}\right) = (7 \text{ psia})\left(\frac{1.3698}{0.8548}\right) = 11.22 \text{ psia}$$

Compressor:

$$P_3 = P_4 = (r_p)(P_2) = (13)(11.22 \text{ psia}) = 145.8 \text{ psia}$$

$$P_{r_3} = \left(\frac{P_3}{P_2}\right)P_{r_2} = \left(\frac{145.8}{11.22}\right)(1.368) = 17.80 \longrightarrow h_3 = 267.56 \text{ Btu/lbm}$$

Turbine: $T_4 = 2400 \text{ R} \longrightarrow$ $h_4 = 617.22 \text{ Btu/lbm}$, $P_{r_4} = 367.6$

$$w_{comp,in} = w_{turb,out} \longrightarrow h_3 - h_2 = h_4 - h_5$$

or,

$$h_5 = h_4 - h_3 + h_2 = 617.22 - 267.56 + 128.48 = 478.14 \text{ Btu/lbm} \longrightarrow P_{r_5} = 142.7$$

$$P_5 = P_4\left(\frac{P_{r_5}}{P_{r_4}}\right) = (145.8 \text{ psia})\left(\frac{142.7}{367.6}\right) = \mathbf{56.6 \text{ psia}}$$

(*b*) Nozzle:

$$P_{r_6} = P_{r_5}\left(\frac{P_6}{P_5}\right) = (142.7)\left(\frac{7\text{ psia}}{56.6\text{ psia}}\right) = 17.66 \longrightarrow h_6 = 266.93\text{ Btu/lbm}$$

$$\dot{E}_{\text{in}} - \dot{E}_{\text{out}} = \Delta\dot{E}_{\text{system}}^{\;\nearrow 0\text{ (steady)}}$$

$$\dot{E}_{in} = \dot{E}_{out}$$

$$h_5 + V_5^2/2 = h_6 + V_6^2/2$$

$$0 = h_6 - h_5 + \frac{\mathbf{V}_6^2 - \mathbf{V}_5^{2\;\nearrow 0}}{2}$$

or,

$$\mathbf{V}_6 = \sqrt{2(h_5 - h_6)} = \sqrt{(2)(478.14 - 266.93)\text{Btu/lbm}\left(\frac{25{,}037\text{ ft}^2/\text{s}^2}{1\text{ Btu/lbm}}\right)} = \mathbf{3252\ ft/s}$$

(*c*) The propulsive efficiency is the ratio of the propulsive work to the heat input,

$$w_p = (\mathbf{V}_{exit} - \mathbf{V}_{inlet})\mathbf{V}_{aircraft}$$

$$= (3252 - 900\text{ ft/s})(900\text{ ft/s})\left(\frac{1\text{ Btu/lbm}}{25{,}037\text{ ft}^2/\text{s}^2}\right) = 84.55\text{ Btu/lbm}$$

$$q_{in} = h_4 - h_3 = 617.22 - 267.56 = 349.66\text{ Btu/lbm}$$

$$\eta_p = \frac{w_p}{q_{in}} = \frac{84.55\text{ Btu/lbm}}{349.66\text{ Btu/lbm}} = \mathbf{24.2\%}$$

8-113 A turbojet aircraft flying at an altitude of 9150 m is operating on the ideal jet propulsion cycle. The velocity of exhaust gases, the propulsive power developed, and the rate of fuel consumption are to be determined.

Assumptions **1** Steady operating conditions exist. **2** The air standard assumptions are applicable. **3** Air is an ideal gas with constant specific heats at room temperature. **4** Kinetic and potential energies are negligible, except at the diffuser inlet and the nozzle exit. **5** The turbine work output is equal to the compressor work input.

Properties The properties of air at room temperature are C_p = 1.005 kJ/kg.K and k = 1.4 (Table A-2a).

Analysis (*a*) We assume the aircraft is stationary and the air is moving towards the aircraft at a velocity of $\mathbf{V}_1$ = 320 m/s. Ideally, the air will leave the diffuser with a negligible velocity ($\mathbf{V}_2 \cong 0$).

Diffuser:

$$\dot{E}_{in} - \dot{E}_{out} = \Delta\dot{E}_{system}^{\nearrow 0\ (steady)}$$

$$\dot{E}_{in} = \dot{E}_{out}$$

$$h_1 + V_1^2/2 = h_2 + V_2^2/2$$

$$0 = h_2 - h_1 + \frac{\mathbf{V}_2^{2\nearrow 0} - \mathbf{V}_1^2}{2}$$

$$0 = C_p(T_2 - T_1) - \mathbf{V}_1^2/2$$

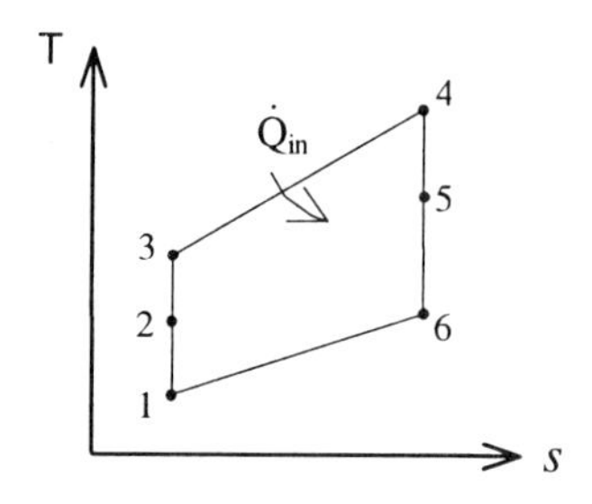

$$T_2 = T_1 + \frac{\mathbf{V}_1^2}{2C_p} = 241 + \frac{(320\text{m/s})^2}{(2)(1.005\text{kJ/kg}\cdot\text{K})}\left(\frac{1\text{kJ/kg}}{1000\text{m}^2/\text{s}^2}\right) = 291.9\text{K}$$

$$P_2 = P_1\left(\frac{T_2}{T_1}\right)^{k/(k-1)} = (32\text{kPa})\left(\frac{291.9\text{K}}{241\text{K}}\right)^{1.4/0.4} = 62.6\text{kPa}$$

Compressor:

$$P_3 = P_4 = (r_p)(P_2) = (12)(62.6\text{kPa}) = 751.2\text{kPa}$$

$$T_3 = T_2\left(\frac{P_3}{P_2}\right)^{(k-1)/k} = (291.9\text{K})(12)^{0.4/1.4} = 593.7\text{K}$$

Turbine:

$$w_{comp,in} = w_{turb,out} \longrightarrow h_3 - h_2 = h_4 - h_5 \longrightarrow C_p(T_3 - T_2) = C_p(T_4 - T_5)$$

or,

$$T_5 = T_4 - T_3 + T_2 = 1400 - 593.7 + 291.9 = 1098.2\text{K}$$

Nozzle:

$$T_6 = T_4\left(\frac{P_6}{P_4}\right)^{(k-1)/k} = (1400\text{K})\left(\frac{32\text{kPa}}{751.2\text{kPa}}\right)^{0.4/1.4} = 568.2\text{K}$$

$$\dot{E}_{\text{in}} - \dot{E}_{\text{out}} = \Delta\dot{E}_{\text{system}}^{\nearrow 0\ \text{(steady)}}$$

$$\dot{E}_{in} = \dot{E}_{out}$$

$$h_5 + V_5^2/2 = h_6 + V_6^2/2$$

$$0 = h_6 - h_5 + \frac{\mathbf{V}_6^2 - \mathbf{V}_5^{2\nearrow 0}}{2} \longrightarrow 0 = C_p(T_6 - T_5) + \mathbf{V}_6^2/2$$

or,

$$\mathbf{V}_6 = \sqrt{(2)(1.005\text{kJ/kg}\cdot\text{K})(1098.2 - 568.2)\text{K}\left(\frac{1000\text{m}^2/\text{s}^2}{1\text{kJ/kg}}\right)} = \mathbf{1032m/s}$$

(*b*) $$\dot{W}_p = \dot{m}(\mathbf{V}_{exit} - \mathbf{V}_{inlet})\mathbf{V}_{aircraft}$$

$$= (60\text{ kg/s})(1032 - 320)\text{m/s}(320\text{ m/s})\left(\frac{1\text{ kJ/kg}}{1000\text{ m}^2/\text{s}^2}\right) = \mathbf{13{,}670\ kW}$$

(*c*) $$\dot{Q}_{in} = \dot{m}(h_4 - h_3) = \dot{m}C_p(T_4 - T_3) = (60\text{ kg/s})(1.005\text{ kJ/kg}\cdot\text{K})(1400 - 593.7)\text{K}$$
$$= 48{,}620\text{ kJ/s}$$

$$\dot{m}_{fuel} = \frac{\dot{Q}_{in}}{HV} = \frac{48{,}620\text{ kJ/s}}{42{,}700\text{ kJ/kg}} = \mathbf{1.14\ \ kg/s}$$

8-114 A turbojet aircraft is flying at an altitude of 9150 m. The velocity of exhaust gases, the propulsive power developed, and the rate of fuel consumption are to be determined.

Assumptions **1** Steady operating conditions exist. **2** The air standard assumptions are applicable. **3** Air is an ideal gas with constant specific heats at room temperature. **4** Kinetic and potential energies are negligible, except at the diffuser inlet and the nozzle exit.

Properties The properties of air at room temperature are C_p = 1.005 kJ/kg.K and k = 1.4 (Table A-2a).

Analysis (*a*) For convenience, we assume the aircraft is stationary and the air is moving towards the aircraft at a velocity of $\mathbf{V}_1$ = 320 m/s. Ideally, the air will leave the diffuser with a negligible velocity ($\mathbf{V}_2 \cong 0$).

Diffuser:

$$\dot{E}_{in} - \dot{E}_{out} = \Delta\dot{E}_{system}^{\nearrow 0 \text{ (steady)}}$$

$$\dot{E}_{in} = \dot{E}_{out}$$

$$h_1 + V_1^2/2 = h_2 + V_2^2/2$$

$$0 = h_2 - h_1 + \frac{\mathbf{V}_2^{2\nearrow 0} - \mathbf{V}_1^2}{2}$$

$$0 = C_p(T_2 - T_1) - \mathbf{V}_1^2/2$$

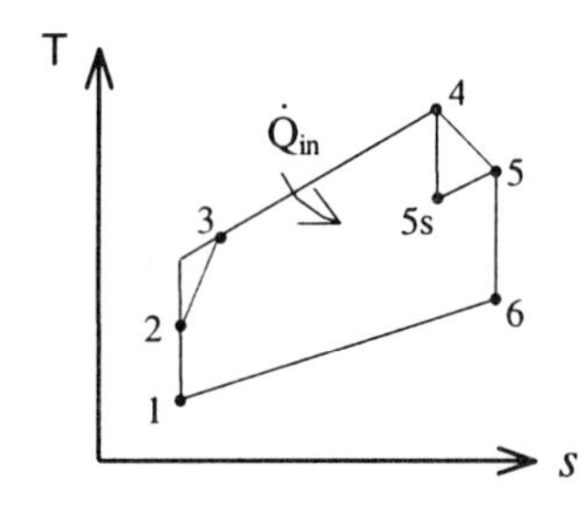

$$T_2 = T_1 + \frac{\mathbf{V}_1^2}{2C_p} = 241 + \frac{(320\text{m/s})^2}{(2)(1.005\text{kJ/kg}\cdot\text{K})}\left(\frac{1\text{kJ/kg}}{1000\text{m}^2/\text{s}^2}\right) = 291.9\text{K}$$

$$P_2 = P_1\left(\frac{T_2}{T_1}\right)^{k/(k-1)} = (32\text{kPa})\left(\frac{291.9\text{K}}{241\text{K}}\right)^{1.4/0.4} = 62.6\text{kPa}$$

Compressor:

$$P_3 = P_4 = (r_p)(P_2) = (12)(62.6\text{kPa}) = 751.2\text{kPa}$$

$$T_3 = T_2\left(\frac{P_3}{P_2}\right)^{(k-1)/k} = (291.9\text{K})(12)^{0.4/1.4} = 593.7\text{K}$$

$$\eta_C = \frac{h_{3s} - h_2}{h_3 - h_2} = \frac{C_p(T_{3s} - T_2)}{C_p(T_3 - T_2)} \longrightarrow T_3 = T_2 + (T_{3s} - T_2)/\eta_C = 291.9 + (593.7 - 291.9)/(0.80) = 669.2\text{K}$$

Turbine:

$$w_{comp,in} = w_{turb,out} \longrightarrow h_3 - h_2 = h_4 - h_5 \longrightarrow C_p(T_3 - T_2) = C_p(T_4 - T_5)$$

or,

$$T_5 = T_4 - T_3 + T_2 = 1400 - 669.2 + 291.9 = 1022.7\text{K}$$

$$\eta_T = \frac{h_4 - h_5}{h_4 - h_{5s}} = \frac{C_p(T_4 - T_5)}{C_p(T_4 - T_{5s})} \longrightarrow \text{T}_{5s} = \text{T}_4 - (\text{T}_4 - \text{T}_5)/\eta_T = 1400 - (1400 - 1022.7)/0.85 = 956.1\text{ K}$$

$$P_5 = P_4\left(\frac{T_{5s}}{T_4}\right)^{k/(k-1)} = (751.2\text{kPa})\left(\frac{956.1\text{K}}{1400\text{K}}\right)^{1.4/0.4} = 197.7\text{kPa}$$

Nozzle:

$$T_6 = T_5 \left(\frac{P_6}{P_5} \right)^{(k-1)/k} = (1022.7 \text{ K}) \left(\frac{32 \text{ kPa}}{197.7 \text{ kPa}} \right)^{0.4/1.4} = 607.8 \text{ K}$$

$$\dot{E}_{\text{in}} - \dot{E}_{\text{out}} = \Delta \dot{E}_{\text{system}}^{\;\not{0} \text{ (steady)}}$$

$$\dot{E}_{in} = \dot{E}_{out}$$

$$h_5 + V_5^2/2 = h_6 + V_6^2/2$$

$$0 = h_6 - h_5 + \frac{\mathbf{V}_6^2 - \mathbf{V}_5^{2\;\not{0}}}{2} \longrightarrow 0 = C_p(T_6 - T_5) + \mathbf{V}_6^2/2$$

or,

$$\mathbf{V}_6 = \sqrt{(2)(1.005 \text{ kJ/kg} \cdot \text{K})(1022.7 - 607.8)\text{K} \left(\frac{1000 \text{ m}^2/\text{s}^2}{1 \text{ kJ/kg}} \right)} = \mathbf{913.2 \text{ m/s}}$$

(*b*)

$$\dot{W}_p = \dot{m}(\mathbf{V}_{exit} - \mathbf{V}_{inlet})\mathbf{V}_{aircraft}$$

$$= (60 \text{ kg/s})(913.2 - 320)\text{m/s}(320 \text{ m/s}) \left(\frac{1 \text{ kJ/kg}}{1000 \text{ m}^2/\text{s}^2} \right) = \mathbf{11{,}390 \text{ kW}}$$

(*c*) $\dot{Q}_{in} = \dot{m}(h_4 - h_3) = \dot{m}C_p(T_4 - T_3) = (60 \text{ kg/s})(1.005 \text{ kJ/kg} \cdot \text{K})(1400 - 669.2)\text{K} = 44{,}067 \text{ kJ/s}$

$$\dot{m}_{fuel} = \frac{\dot{Q}_{in}}{HV} = \frac{44{,}067 \text{ kJ/s}}{42{,}700 \text{ kJ/kg}} = \mathbf{1.03 \text{ kg/s}}$$

8-115 A turbojet aircraft that has a pressure rate of 12 is stationary on the ground. The force that must be applied on the brakes to hold the plane stationary is to be determined.

Assumptions **1** Steady operating conditions exist. **2** The air standard assumptions are applicable. **3** Air is an ideal gas with variable specific heats. **4** Kinetic and potential energies are negligible, except at the nozzle exit.

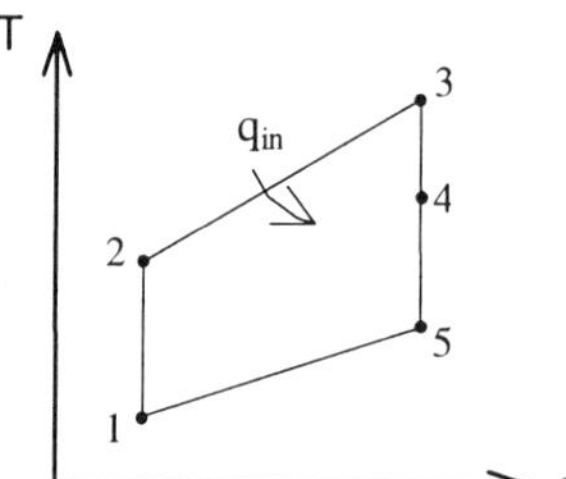

Properties The properties of air are given in Table A17.

Analysis (*a*) Using variable specific heats for air,

Compressor: $T_1 = 300 \text{ K} \longrightarrow h_1 = 300.19 \text{ kJ/kg}$, $P_{r_1} = 1.386$

$$P_{r_2} = \frac{P_2}{P_1} P_{r_1} = (12)(1.386) = 16.63 \longrightarrow h_2 = 610.65 \text{ kJ/kg}$$

$$\dot{Q}_{in} = \dot{m}_{fuel} \times HV = (0.2 \text{ kg/s})(42{,}700 \text{ kJ/kg}) = 8540 \text{ kJ/s}$$

$$q_{in} = \frac{\dot{Q}_{in}}{\dot{m}} = \frac{8540 \text{ kJ/s}}{10 \text{ kg/s}} = 854 \text{ kJ/kg}$$

$$q_{in} = h_3 - h_2 \longrightarrow h_3 = h_2 + q_{in} = 610.65 + 854 = 1464.65 \text{ kJ/kg}$$

$$\longrightarrow P_{r_3} = 396.27$$

Turbine:

$$w_{comp,in} = w_{turb,out} \longrightarrow h_3 - h_2 = h_4 - h_5$$

or,

$$h_4 = h_3 - h_2 + h_1 = 1464.65 - 610.65 + 300.19 = 741.17 \text{ kJ/kg}$$

Nozzle:

$$P_{r_5} = P_{r_3}\left(\frac{P_5}{P_3}\right) = (396.27)\left(\frac{1}{12}\right) = 33.02 \longrightarrow h_5 = 741.79 \text{kJ/kg}$$

$$\dot{E}_{in} - \dot{E}_{out} = \Delta \dot{E}_{system}^{\nearrow 0 \text{ (steady)}}$$

$$\dot{E}_{in} = \dot{E}_{out}$$

$$h_4 + V_4^2/2 = h_5 + V_5^2/2$$

$$0 = h_5 - h_4 + \frac{\mathbf{V}_5^2 - \mathbf{V}_4^{2\nearrow 0}}{2}$$

or,

$$\mathbf{V}_5 = \sqrt{2(h_4 - h_5)} = \sqrt{(2)(1154.19 - 741.17)\text{kJ/kg}\left(\frac{1000 \text{ m}^2/\text{s}^2}{1 \text{ kJ/kg}}\right)} = \mathbf{908.9 \text{ m/s}}$$

$$\text{Brake force} = \text{Thrust} = \dot{m}(\mathbf{V}_{exit} - \mathbf{V}_{inlet}) = (10 \text{ kg/s})(908.9 - 0)\text{m/s}\left(\frac{1\text{N}}{1\text{kg} \cdot \text{m/s}^2}\right) = \mathbf{9089 \text{ N}}$$

8-116 EES solution of this (and other comprehensive problems designated with the *computer icon*) is available to instructors at the *Instructor Manual* section of the *Online Learning Center* (OLC) at www.mhhe.com/cengel-boles. See the Preface for access information.

8-117 Air enters a turbojet engine. The thrust produced by this turbojet engine is to be determined.

Assumptions **1** Steady operating conditions exist. **2** The air standard assumptions are applicable. **3** Air is an ideal gas with variable specific heats. **4** Kinetic and potential energies are negligible, except at the diffuser inlet and the nozzle exit.

Properties The properties of air are given in Table A-17.

Analysis We assume the aircraft is stationary and the air is moving towards the aircraft at a velocity of $\mathbf{V}_1$ = 300 m/s. Taking the entire engine as our control volume and writing the steady-flow energy balance yield

$$T_1 = 280 \text{ K} \longrightarrow h_1 = 280.13 \text{ kJ/kg}$$

$$T_2 = 700 \text{ K} \longrightarrow h_2 = 713.27 \text{ kJ/kg}$$

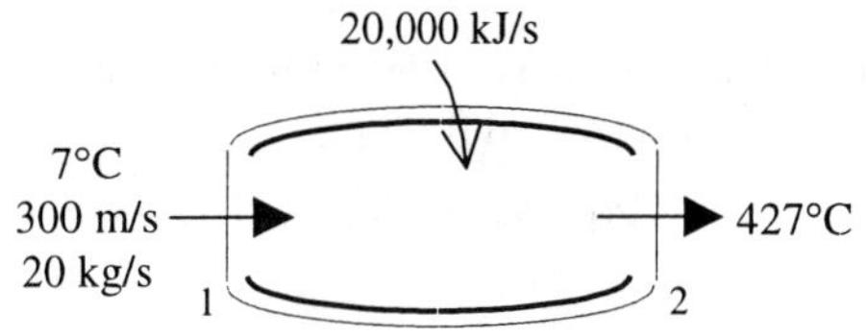

20,000 kJ/s
7°C
300 m/s
20 kg/s
1
2
427°C

$$\dot{E}_{\text{in}} - \dot{E}_{\text{out}} = \Delta \dot{E}_{\text{system}}^{\nearrow 0 \text{ (steady)}}$$

$$\dot{E}_{in} = \dot{E}_{out}$$

$$\dot{Q}_{in} + \dot{m}(h_1 + V_1^2/2) = \dot{m}(h_2 + V_2^2/2)$$

$$\dot{Q}_{in} = \dot{m}\left(h_2 - h_1 + \frac{\mathbf{V}_2^2 - \mathbf{V}_1^2}{2}\right)$$

$$20{,}000\text{kJ/s} = (20\text{kg/s})\left[713.27 - 280.13 + \frac{\mathbf{V}_2^2 - (300\text{m/s})^2}{2}\left(\frac{1\text{kJ/kg}}{1000\text{m}^2/\text{s}^2}\right)\right]$$

It gives $\mathbf{V}_2 = 1106$ m/s

Thus,

$$F_p = \dot{m}(\mathbf{V}_2 - \mathbf{V}_1) = (20\text{kg/s})(1106 - 300)\text{m/s} = \mathbf{16{,}120N}$$

Second-Law Analysis of Gas Power Cycles

8-118 The total exergy destruction associated with the Otto cycle described in Prob. 8-33 and the exergy at the end of the power stroke are to be determined.

Analysis From Prob. 8-33, $q_{in} = 750$, $q_{out} = 357.62$ kJ/kg, $T_1 = 300$ K, and $T_4 = 774.5$ K.

The total exergy destruction associated with this Otto cycle is determined from

$$x_{\text{destroyed}} = T_0\left(\frac{q_{out}}{T_L} - \frac{q_{in}}{T_H}\right) = (300\text{ K})\left(\frac{357.62\text{ kJ/kg}}{300\text{ K}} - \frac{750\text{ kJ/kg}}{2000\text{ K}}\right) = \mathbf{245.12\ kJ/kg}$$

Noting that state 4 is identical to the state of the surroundings, the exergy at the end of the power stroke (state 4) is determined from

$$\phi_4 = (u_4 - u_0) - T_0(s_4 - s_0) + P_0(v_4 - v_0)$$

where

$$u_4 - u_0 = u_4 - u_1 = q_{out} = 357.62\text{kJ/kg}$$

$$v_4 - v_0 = v_4 - v_1 = 0$$

$$s_4 - s_0 = s_4 - s_1 = s_4^\circ - s_1^\circ - R\ln\frac{P_4}{P_1} = s_4^\circ - s_1^\circ - R\ln\frac{T_4 v_1}{T_1 v_4} = s_4^\circ - s_1^\circ - R\ln\frac{T_4}{T_1}$$

$$= 2.6823 - 1.70203 - (0.287\text{kJ/kg}\cdot\text{K})\ln\frac{774.5\text{ K}}{300\text{ K}} = 0.7081\text{ kJ/kg}\cdot\text{K}$$

Thus,

$$\phi_4 = (357.62\text{ kJ/kg}) - (300\text{ K})(0.7081\text{ kJ/kg}\cdot\text{K}) + 0 = \mathbf{145.2\ kJ/kg}$$

8-119 The total exergy destruction associated with the Diesel cycle described in Prob. 8-45 and the exergy at the end of the compression stroke are to be determined.

Analysis From Prob. 8-45, $q_{in} = 1019.7$, $q_{out} = 445.63$ kJ/kg, $T_1 = 300$ K, $v_1 = 0.906$ m^3/kg, and $v_2 = v_1 / r = 0.906 / 12 = 0.0566$ m^3/kg.

The total exergy destruction associated with this Otto cycle is determined from

$$x_{\text{destroyed}} = T_0\left(\frac{q_{out}}{T_L} - \frac{q_{in}}{T_H}\right) = (300\text{ K})\left(\frac{445.63\text{ kJ/kg}}{300\text{ K}} - \frac{1019.7\text{ kJ/kg}}{2000\text{ K}}\right) = \mathbf{292.7\ kJ/kg}$$

Noting that state 1 is identical to the state of the surroundings, the exergy at the end of the compression stroke (state 2) is determined from

$$\phi_2 = (u_2 - u_0) - T_0(s_2 - s_0) + P_0(v_2 - v_0)$$

$$= (u_2 - u_1) - T_0(s_2 - s_1) + P_0(v_2 - v_1)$$

$$= (643.3 - 214.07) - 0 + (95\text{ kPa})(0.0566 - 0.906)\text{m}^3\text{/kg}\left(\frac{1\text{ kJ}}{1\text{ kPa}\cdot\text{m}^3}\right)$$

$$= \mathbf{348.6\ kJ/kg}$$

8-120E The exergy destruction associated with the heat rejection process of the Diesel cycle described in Prob. 8-47E and the exergy at the end of the expansion stroke are to be determined.

Analysis From Prob. 8-47E, q_{out} = 158.9 Btu/lbm, T_1 = 540 R, T_4 = 1420.6 R, and $v_4 = v_1$.

At $T_{ave} = (T_4 + T_1)/2 = (1420.6 + 540)/2 = 980.3$ R, we have $C_{v,\,ave}$ = 0.180 Btu/lbm· R. The entropy change during process 4-1 is

$$s_1 - s_4 = C_v \ln\frac{T_1}{T_4} + R\ln\frac{v_1}{v_4}^{\nearrow 0} = (0.180\ \text{Btu/lbm}\cdot\text{R})\ln\frac{540\ \text{R}}{1420.6\ \text{R}} = -0.1741\ \text{Btu/lbm}\cdot\text{R}$$

Thus,

$$x_{\text{destroyed},41} = T_0\left(s_1 - s_4 + \frac{q_{R,41}}{T_R}\right) = (540\text{R})\left(-0.1741\ \text{Btu/lbm}\cdot\text{R} + \frac{158.9\ \text{Btu/lbm}}{540\ \text{R}}\right) = \mathbf{64.9\ Btu/lbm}$$

Noting that state 4 is identical to the state of the surroundings, the exergy at the end of the power stroke (state 4) is determined from

$$\phi_4 = (u_4 - u_0) - T_0(s_4 - s_0) + P_0(v_4 - v_0)$$

where

$$u_4 - u_0 = u_4 - u_1 = q_{out} = 158.9\ \text{Btu/lbm}\cdot\text{R}$$

$$v_4 - v_0 = v_4 - v_1 = 0$$

$$s_4 - s_0 = s_4 - s_1 = 0.1741\ \text{Btu/lbm}\cdot\text{R}$$

Thus,

$$\phi_4 = (158.9\ \text{Btu/lbm}) - (540\text{R})(0.1741\ \text{Btu/lbm}\cdot\text{R}) + 0 = \mathbf{64.9\ Btu/lbm}$$

Analysis Note that the exergy at state 4 is identical to the exergy destruction for the process 4-1 since state 1 is identical to the dead state, and the entire exergy at state 4 is wasted during process 4-1.

8-121 The exergy destruction associated with each of the processes of the Brayton cycle described in Prob. 8-70 is to be determined.

Analysis From Prob. 8-70, q_{in} = 584.62 kJ/kg, q_{out} = 478.92 kJ/kg, and

$$T_1 = 310\text{K} \longrightarrow s_1^\circ = 1.73498\text{kJ/kg}\cdot\text{K}$$

$$h_2 = 646.3\text{kJ/kg} \longrightarrow s_2^\circ = 2.47256\text{kJ/kg}\cdot\text{K}$$

$$\text{T}_3 = 1160\text{K} \longrightarrow \text{s}_3^\circ = 3.13916\text{kJ/kg}\cdot\text{K}$$

$$h_4 = 789.16\text{kJ/kg} \longrightarrow s_4^\circ = 2.67602\text{kJ/kg}\cdot\text{K}$$

Thus,

$$x_{\text{destroyed},12} = T_0 s_{gen,12} = T_0(s_2 - s_1) = T_0\left(s_2^\circ - s_1^\circ - R\ln\frac{P_2}{P_1}\right) =$$

$$= (290\text{ K})(2.47256 - 1.73498 - (0.287\text{kJ/kg}\cdot\text{K})\ln(8)) = \mathbf{40.83\ kJ/kg}$$

$$x_{\text{destroyed},23} = T_0 s_{gen,23} = T_0\left(s_3 - s_2 + \frac{q_{R,23}}{T_R}\right) = T_0\left(s_3^\circ - s_2^\circ - R\ln\frac{P_3}{P_2}^{\nearrow 0} + \frac{-q_{in}}{T_H}\right)$$

$$= (290\text{ K})\left(3.13916 - 2.47256 - \frac{584.62\text{ kJ/kg}}{1600\text{ K}}\right) = \mathbf{87.35\ kJ/kg}$$

$$x_{\text{destroyed},34} = T_0 s_{gen,34} = T_0(s_4 - s_3) = T_0\left(s_4^\circ - s_3^\circ - R\ln\frac{P_4}{P_3}\right) =$$

$$= (290\text{ K})(2.67602 - 3.13916 - (0.287\text{ kJ/kg}\cdot\text{K})\ln(1/8)) = \mathbf{38.76\ kJ/kg}$$

$$x_{\text{destroyed},41} = T_0 s_{gen,41} = T_0\left(s_1 - s_4 + \frac{q_{R,41}}{T_R}\right) = T_0\left(s_1^\circ - s_4^\circ - R\ln\frac{P_1}{P_4}^{\nearrow 0} + \frac{q_{out}}{T_L}\right)$$

$$= (290\text{ K})\left(1.73498 - 2.67602 + \frac{478.92\text{ kJ/kg}}{310\text{ K}}\right) = \mathbf{206.0\ kJ/kg}$$

8-122 The total exergy destruction associated with the Brayton cycle described in Prob. 8-89 and the exergy at the exhaust gases at the turbine exit are to be determined.

Analysis From Prob. 8-89, q_{in} = 601.94, q_{out} = 279.68 kJ/kg, and h_6 = 579.87 kJ/kg.

The total exergy destruction associated with this Otto cycle is determined from

$$x_{\text{destroyed}} = T_0\left(\frac{q_{out}}{T_L} - \frac{q_{in}}{T_H}\right) = (300\text{ K})\left(\frac{279.68\text{ kJ/kg}}{300\text{ K}} - \frac{601.94\text{ kJ/kg}}{1600\text{ K}}\right) = \mathbf{166.8\ kJ/kg}$$

Noting that $h_0 = h_{@\,300\text{ K}}$ = 300.19 kJ/kg, the stream exergy at the exit of the regenerator (state 6) is determined from

$$\phi_6 = (h_6 - h_0) - T_0(s_6 - s_0) + \frac{\mathbf{V}_6^2}{2}^{\nearrow 0} + gz_6^{\nearrow 0}$$

where

$$s_6 - s_0 = s_6 - s_1 = s_6^\circ - s_1^\circ - R\ln\frac{P_6}{P_1}^{\nearrow 0} = 2.36275 - 1.70203 = 0.66072\text{ kJ/kg}\cdot\text{K}$$

Thus,

$$\phi_6 = 579.87 - 300.19 - (300\text{ K})(0.66072\text{ kJ/kg}\cdot\text{K}) = \mathbf{81.5\ kJ/kg}$$

8-123 EES solution of this (and other comprehensive problems designated with the *computer icon*) is available to instructors at the *Instructor Manual* section of the *Online Learning Center* (OLC) at www.mhhe.com/cengel-boles. See the Preface for access information.

8-124 The exergy destruction associated with each of the processes of the Brayton cycle described in Prob. 8-94 and the exergy at the end of the exhaust gases at the exit of the regenerator are to be determined.

Analysis From Prob. 8-94, q_{in} = 480.82 kJ/kg, q_{out} = 372.73 kJ/kg, and

$$T_1 = 310\text{K} \longrightarrow s_1^\circ = 1.73498\text{kJ/kg}\cdot\text{K}$$

$$h_2 = 618.26 \text{ kJ/kg} \longrightarrow s_2^\circ = 2.42763 \text{ kJ/kg}\cdot\text{K}$$

$$T_3 = 1150 \text{ K} \longrightarrow s_3^\circ = 3.1290\text{kJ/kg}\cdot\text{K}$$

$$h_4 = 803.14\text{kJ/kg} \longrightarrow s_4^\circ = 2.69407 \text{ kJ/kg}\cdot\text{K}$$

$$h_5 = 738.43\text{kJ/kg} \longrightarrow s_5^\circ = 2.60815 \text{ kJ/kg}\cdot\text{K}$$

and, from an energy balance on the heat exchanger,

$$h_5 - h_2 = h_4 - h_6 \longrightarrow h_6 = 803.14 - 738.43 + 618.26 = 682.97 \text{ kJ/kg}$$

$$\longrightarrow s_6^\circ = 2.52861 \text{ kJ/kg}\cdot\text{K}$$

Thus,

$$x_{\text{destroyed},12} = T_0 s_{gen,12} = T_0(s_2 - s_1) = T_0\left(s_2^\circ - s_1^\circ - R\ln\frac{P_2}{P_1}\right) =$$

$$= (310\text{ K})(2.42763 - 1.73498 - (0.287\text{ kJ/kg}\cdot\text{K})\ln(7)) = \mathbf{41.59\ kJ/kg}$$

$$x_{\text{destroyed},34} = T_0 s_{gen,34} = T_0(s_4 - s_3) = T_0\left(s_4^\circ - s_3^\circ - R\ln\frac{P_4}{P_3}\right) =$$

$$= (310\text{ K})(2.69407 - 3.1290 - (0.287\text{kJ/kg}\cdot\text{K})\ln(1/7)) = \mathbf{38.30\ kJ/kg}$$

$$x_{\text{destroyed,regen}} = T_0 s_{gen,regen} = T_0[(s_5 - s_2) + (s_6 - s_4)] = T_0[(s_5^\circ - s_2^\circ) + (s_6^\circ - s_4^\circ)]$$

$$= (310\text{ K})(2.6082 - 2.4276 + 2.5286 - 2.69407) = \mathbf{4.68\ kJ/kg}$$

$$x_{\text{destroyed},53} = T_0 s_{gen,53} = T_0\left(s_3 - s_5 + \frac{q_{R,53}}{T_R}\right) = T_0\left(s_3^\circ - s_5^\circ - R\ln\frac{P_3}{P_5}^{\nearrow 0} + \frac{q_{in}}{T_H}\right)$$

$$= (310\text{ K})\left(3.1290 - 2.6082 + \frac{480.82\text{ kJ/kg}}{1260\text{ K}}\right) = \mathbf{43.2\ kJ/kg}$$

$$x_{\text{destroyed},61} = T_0 s_{gen,61} = T_0\left(s_1 - s_6 + \frac{q_{R,61}}{T_R}\right) = T_0\left(s_1^\circ - s_6^\circ - R\ln\frac{P_1}{P_6}^{\nearrow 0} + \frac{q_{out}}{T_L}\right)$$

$$= (310\text{ K})\left(1.73498 - 2.5286 + \frac{372.73\text{ kJ/kg}}{310\text{ K}}\right) = \mathbf{126.7\ kJ/kg}$$

Noting that $h_0 = h_{@\ 310\text{ K}}$ = 310.24 kJ/kg, the stream exergy at the exit of the regenerator (state 6) is determined from

$$\phi_6 = (h_6 - h_0) - T_0(s_6 - s_0) + \frac{\mathbf{V}_6^2}{2}^{\nearrow 0} + gz_6^{\nearrow 0}$$

where

$$s_6 - s_0 = s_6 - s_1 = s_6^\circ - s_1^\circ - R\ln\frac{P_6}{P_1}^{\nearrow 0} = 2.5286 - 1.73498 = 0.7936 \text{ kJ/kg}\cdot\text{K}$$

Thus,

$$\phi_6 = 682.97 - 310.24 - (310\text{ K})(0.7986\text{ kJ/kg}\cdot\text{K}) = \mathbf{126.7\ kJ/kg}$$

Review Problems

8-125 A turbocharged four-stroke V-16 diesel engine produces 4000 hp at 1050 rpm. The amount of power produced per cylinder per mechanical and per thermodynamic cycle is to be determined.

Analysis Noting that there are 16 cylinders and each thermodynamic cycle corresponds to 2 mechanical cycles, we have

(*a*)

$$\begin{aligned} w_{\text{mechanical}} &= \frac{\text{Total power produced}}{(\text{No. of cylinders})(\text{No. of mechanical cycles})} \\ &= \frac{4000\text{hp}}{(16\,\text{cylinders})(1050\,\text{rev/min})}\left(\frac{42.41\,\text{Btu/min}}{1\,\text{hp}}\right) \\ &= \mathbf{10.1\ Btu/cyl \cdot mech\ cycle} \quad (= 10.7\,\text{kJ/cyl}\cdot\text{mech cycle}) \end{aligned}$$

(*b*)

$$\begin{aligned} w_{\text{thermodynamic}} &= \frac{\text{Total power produced}}{(\text{No. of cylinders})(\text{No. of thermodynamic cycles})} \\ &= \frac{4000\text{hp}}{(16\,\text{cylinders})(1050/2\,\text{rev/min})}\left(\frac{42.41\,\text{Btu/min}}{1\,\text{hp}}\right) \\ &= \mathbf{20.2\ Btu/cyl \cdot therm\ cycle} \quad (= 21.3\,\text{kJ/cyl}\cdot\text{therm cycle}) \end{aligned}$$

8-126 A simple ideal Brayton cycle operating between the specified temperature limits is considered. The pressure ratio for which the compressor and the turbine exit temperature of air are equal is to be determined.

Assumptions **1** Steady operating conditions exist. **2** The air-standard assumptions are applicable. **3** Kinetic and potential energy changes are negligible. **4** Air is an ideal gas with constant specific heats.

Properties The specific heat ratio of air is k =1.4 (Table A-2).

Analysis We treat air as an ideal gas with constant specific heats. Using the isentropic relations, the temperatures at the compressor and turbine exit can be expressed as

$$T_2 = T_1\left(\frac{P_2}{P_1}\right)^{(k-1)/k} = T_1 (r_p)^{(k-1)/k}$$

$$T_4 = T_3\left(\frac{P_4}{P_3}\right)^{(k-1)/k} = T_3\left(\frac{1}{r_p}\right)^{(k-1)/k}$$

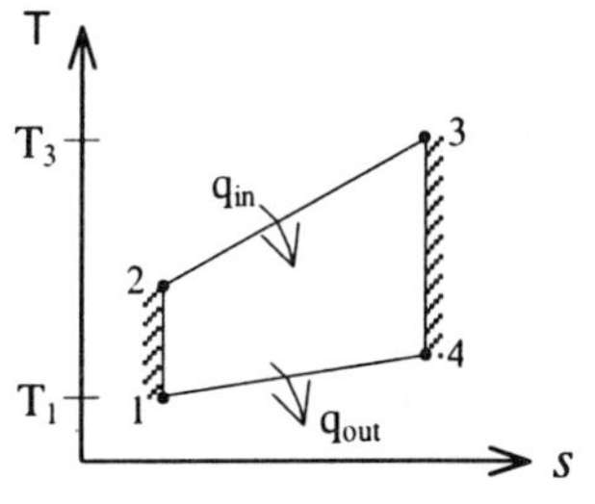

T, T₃, T₁, 1, 2, 3, 4, q_{in}, q_{out}, s

Setting $T_2 = T_4$ and solving for r_p gives

$$r_p = \left(\frac{T_3}{T_1}\right)^{k/2(k-1)} = \left(\frac{1800\text{ K}}{300\text{ K}}\right)^{1.4/0.8} = \mathbf{23.0}$$

Therefore, the compressor and turbine exit temperatures will be equal when the compression ratio is 23.

8-127 The four processes of an air-standard cycle are described. The cycle is to be shown on *P-v* and *T-s* diagrams, and the net work output and the thermal efficiency are to be determined.

Assumptions **1** The air-standard assumptions are applicable. **2** Kinetic and potential energy changes are negligible. **3** Air is an ideal gas with variable specific heats.

Properties The properties of air are given in Table A-17.

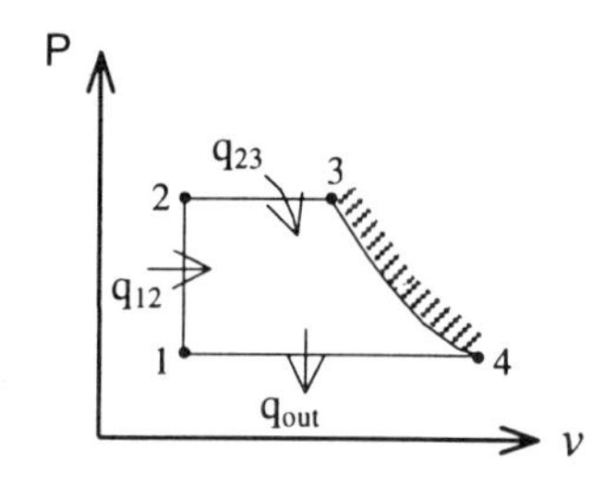

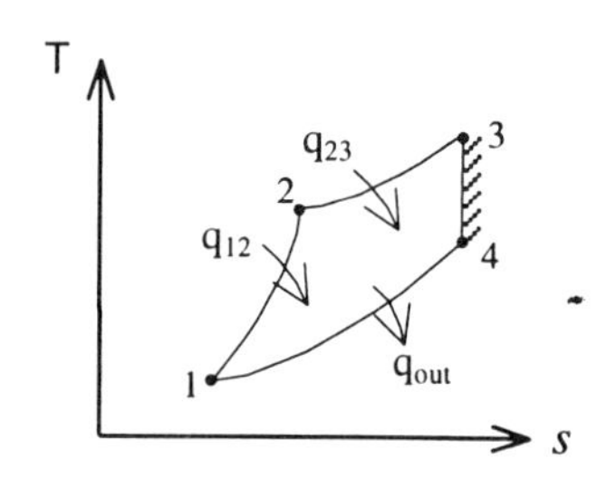

Analysis (b) We treat air as an ideal gas with variable specific heats,

$$T_1 = 300\text{K} \longrightarrow u_1 = 214.07\text{kJ/kg}$$
$$h_1 = 300.19\text{kJ/kg}$$

$$\frac{P_2 v_2}{T_2} = \frac{P_1 v_1}{T_1} \longrightarrow T_2 = \frac{P_2}{P_1} T_1 = \left(\frac{300\text{kPa}}{100\text{kPa}}\right)(300\text{K})$$
$$= 900\text{K} \longrightarrow u_2 = 674.58\text{kJ/kg}$$
$$h_2 = 932.93\text{kJ/kg}$$

$$T_3 = 1300K \longrightarrow u_3 = 1022.82\text{kJ/kg}$$
$$h_3 = 1395.97\text{kJ/kg},\ P_{r3} = 330.9$$

$$P_{r_4} = \frac{P_4}{P_3} P_{r_3} = \left(\frac{100\text{kPa}}{300\text{kPa}}\right)(330.9) = 110.3$$
$$\longrightarrow h_4 = 1036.46\text{kJ/kg}$$

$$q_{in} = q_{12,in} + q_{23,in} = (u_2 - u_1) + (h_3 - h_2)$$
$$= (674.58 - 214.07) + (1395.97 - 932.93)$$
$$= 923.55\text{kJ/kg}$$

$$q_{out} = h_4 - h_1 = 1036.46 - 300.19 = 736.27\text{kJ/kg}$$

$$w_{net} = q_{in} - q_{out} = 923.55 - 736.27 = \mathbf{187.28kJ/kg}$$

(c) $$\eta_{th} = \frac{w_{net}}{q_{in}} = \frac{187.28\ \text{kJ/kg}}{923.55\ \text{kJ/kg}} = \mathbf{20.3\%}$$

8-128 All four processes of an air-standard cycle are described. The cycle is to be shown on *P-v* and *T-s* diagrams, and the net work output and the thermal efficiency are to be determined.

Assumptions **1** The air-standard assumptions are applicable. **2** Kinetic and potential energy changes are negligible. **3** Air is an ideal gas with constant specific heats.

Properties The properties of air at room temperature are C_p = 1.005 kJ/kg.K, C_v = 0.718 kJ/kg· K, and k = 1.4 (Table A-2).

Analysis (*b*) Process 3-4 is isentropic:

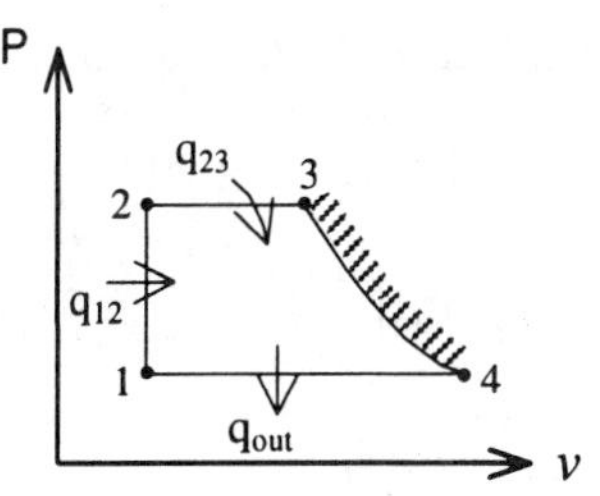

$$T_4 = T_3\left(\frac{P_4}{P_3}\right)^{(k-1)/k} = (1300\text{K})\left(\frac{1}{3}\right)^{0.4/1.4} = 949.8\text{K}$$

$$\frac{P_2 v_2}{T_2} = \frac{P_1 v_1}{T_1} \longrightarrow T_2 = \frac{P_2}{P_1}T_1 = \left(\frac{300\text{kPa}}{100\text{kPa}}\right)(300\text{K}) = 900\text{K}$$

$$\begin{aligned} q_{in} &= q_{12,in} + q_{23,in} = (u_2 - u_1) + (h_3 - h_2) = C_v(T_2 - T_1) + C_p(T_3 - T_2) \\ &= (0.718\text{kJ/kg}\cdot\text{K})(900-300)\text{K} + (1.005\text{kJ/kg}\cdot\text{K})(1300-900)\text{K} \\ &= 832.8\text{kJ/kg} \end{aligned}$$

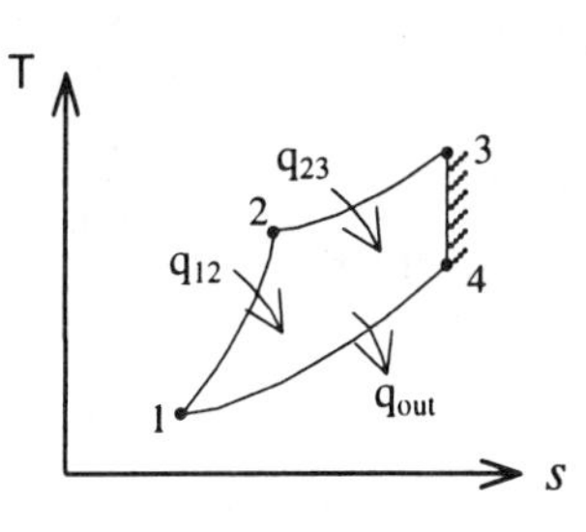

$$\begin{aligned} q_{out} &= h_4 - h_1 = C_p(T_4 - T_1) = (1.005\text{kJ/kg}\cdot\text{K})(949.8-300)\text{K} \\ &= 653\text{kJ/kg} \end{aligned}$$

$$w_{net} = q_{in} - q_{out} = 832.8 - 653 = \mathbf{179.8kJ/kg}$$

(*c*) $$\eta_{th} = \frac{w_{net}}{q_{in}} = \frac{179.8\text{ kJ/kg}}{832.8\text{ kJ/kg}} = \mathbf{21.6\%}$$

8-129 The three processes of an air-standard cycle are described. The cycle is to be shown on *P-v* and *T-s* diagrams, and the maximum temperature in the cycle and the thermal efficiency are to be determined.

Assumptions **1** The air-standard assumptions are applicable. **2** Kinetic and potential energy changes are negligible. **3** Air is an ideal gas with variable specific heats.

Properties The properties of air are given in Table A-17.

Analysis (*b*) We treat air as an ideal gas with variable specific heats,

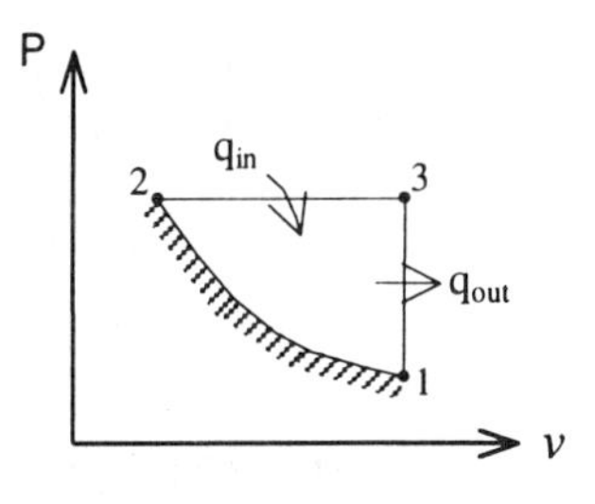

$$T_1 = 300\text{K} \longrightarrow \begin{aligned} u_1 &= 214.07\text{kJ/kg} \\ P_{r_1} &= 1.386 \end{aligned}$$

$$P_{r_2} = \frac{P_2}{P_1}P_{r_1} = \left(\frac{700\text{kPa}}{100\text{kPa}}\right)(1.386) = 9.702 \longrightarrow h_2 = 523.90\text{kJ/kg}$$

$$\frac{P_3 v_3}{T_3} = \frac{P_1 v_1}{T_1} \longrightarrow T_{\max} = T_3 = \frac{P_3}{P_1}T_1 = \left(\frac{700\text{kPa}}{100\text{kPa}}\right)(300\text{K}) = \mathbf{2100K}$$

$$T_3 = 2100\text{K} \longrightarrow \begin{aligned} u_3 &= 1775.3\text{kJ/kg} \\ h_3 &= 2377.7\text{ kJ/kg} \end{aligned}$$

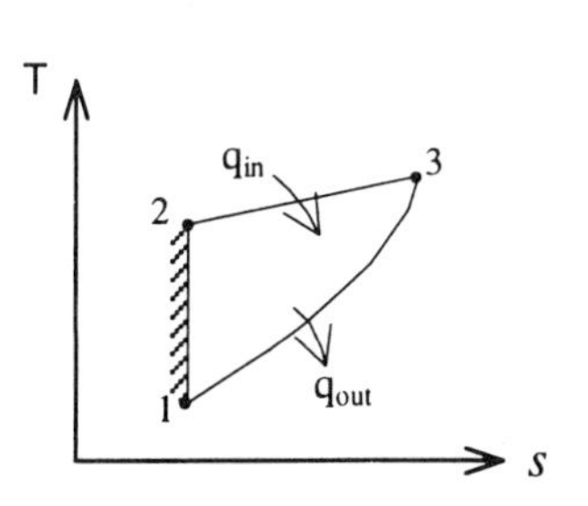

(*c*) $$q_{in} = h_3 - h_2 = 2377.7 - 523.9 = 1853.8\text{ kJ/kg}$$

$$q_{out} = u_3 - u_1 = 1775.3 - 214.07 = 1561.23\text{kJ/kg}$$

$$\eta_{th} = 1 - \frac{q_{out}}{q_{in}} = 1 - \frac{1561.23\text{kJ/kg}}{1853.8\text{kJ/kg}} = \mathbf{15.8\%}$$

8-130 All three processes of an air-standard cycle are described. The cycle is to be shown on *P-v* and *T-s* diagrams, and the maximum temperature in the cycle and the thermal efficiency are to be determined.

Assumptions **1** The air-standard assumptions are applicable. **2** Kinetic and potential energy changes are negligible. **3** Air is an ideal gas with constant specific heats.

Properties The properties of air at room temperature are C_p = 1.005 kJ/kg.K, C_v = 0.718 kJ/kg· K, and k = 1.4 (Table A-2).

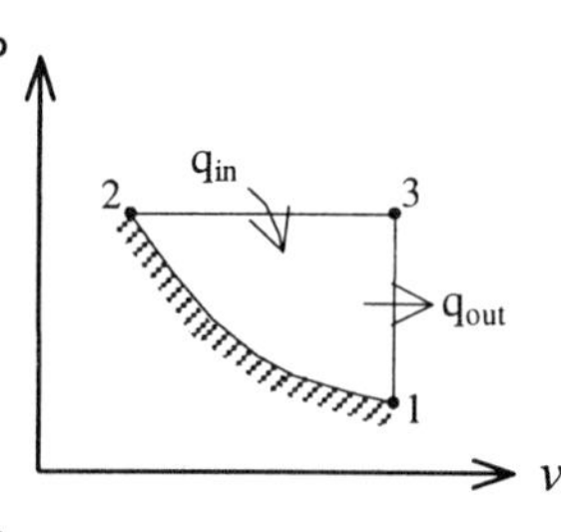

Analysis (*b*) We treat air as an ideal gas with constant specific heats.

Process 1-2 is isentropic:

$$T_2 = T_1\left(\frac{P_2}{P_1}\right)^{(k-1)/k} = (300\text{K})\left(\frac{700\text{kPa}}{100\text{kPa}}\right)^{0.4/1.4} = 523.1\text{K}$$

$$\frac{P_3 v_3}{T_3} = \frac{P_1 v_1}{T_1} \longrightarrow T_{\max} = T_3 = \frac{P_3}{P_1}T_1 = \left(\frac{700\text{kPa}}{100\text{kPa}}\right)(300\text{K}) = \mathbf{2100K}$$

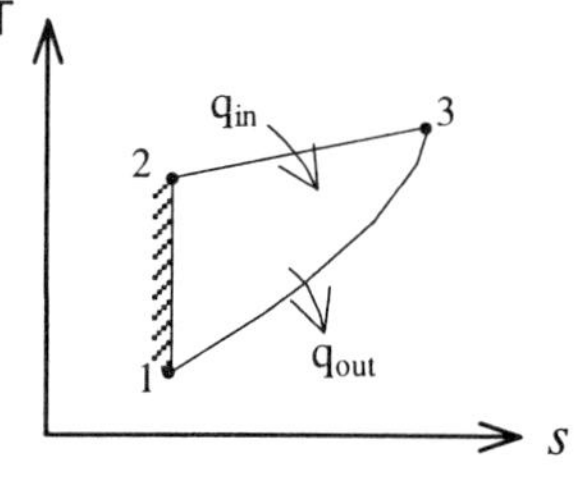

(*c*) $q_{in} = h_3 - h_2 = C_p(T_3 - T_2) = (1.005\text{kJ/kg}\cdot\text{K})(2100 - 523.1)\text{K} = 1584.8\text{kJ/kg}$

$q_{out} = u_3 - u_1 = C_v(T_3 - T_1) = (0.718\text{kJ/kg}\cdot\text{K})(2100 - 300)\text{K} = 1292.4\text{kJ/kg}$

$$\eta_{th} = 1 - \frac{q_{out}}{q_{in}} = 1 - \frac{1292.4\text{kJ/kg}}{1584.8\text{kJ/kg}} = \mathbf{18.5\%}$$

8-131 A Carnot cycle executed in a closed system uses air as the working fluid. The net work output per cycle is to be determined.

Assumptions **1** Air is an ideal gas with variable specific heats.

Analysis (a) The maximum temperature is determined from

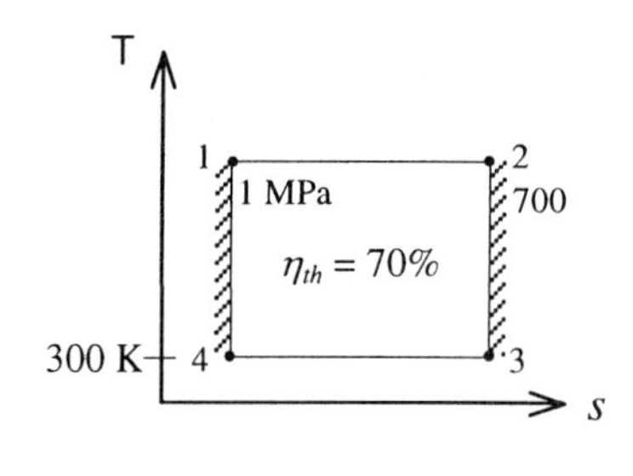

$$\eta_{th} = 1 - \frac{T_L}{T_H} \longrightarrow 0.70 = 1 - \frac{300\text{ K}}{T_H} \longrightarrow T_H = 1000\text{K}$$

$$s_2 - s_1 = s_2^\circ - s_1^\circ - R\ln\frac{P_2}{P_1} = -(0.287\text{kJ/kg}\cdot\text{K})\ln\frac{700\text{kPa}}{1000\text{kPa}}$$
$$= 0.1204\text{kJ/kg}\cdot\text{K}$$

$$W_{net} = m(s_2 - s_1)(T_H - T_L)$$
$$= (0.0015\text{ kg})(0.1024\text{kJ/kg}\cdot\text{K})(1000 - 300)\text{K}$$
$$= \mathbf{0.107\ kJ}$$

8-132 [*Also solved by EES on enclosed CD*] A four-cylinder spark-ignition engine with a compression ratio of 8 is considered. The amount of heat supplied per cylinder, the thermal efficiency, and the rpm for a net power output of 60 kW are to be determined.

Assumptions **1** The air-standard assumptions are applicable. **2** Kinetic and potential energy changes are negligible. **3** Air is an ideal gas with variable specific heats.

Properties The properties of air are given in Table A-17.

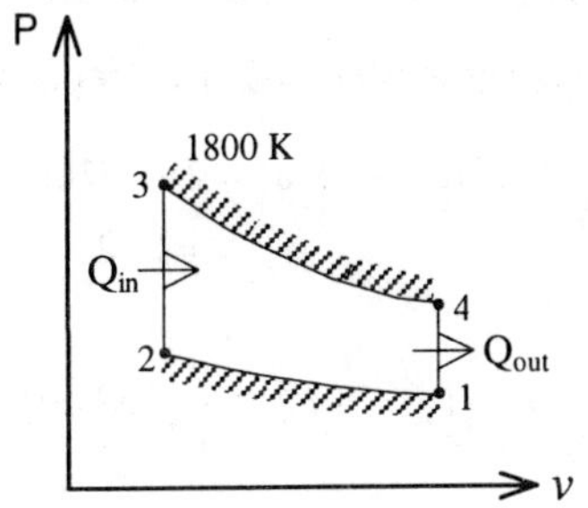

Analysis (*a*) Process 1-2: isentropic compression.

$$T_1 = 290K \longrightarrow u_1 = 206.91\text{kJ/kg}$$
$$v_{r_1} = 676.1$$

$$v_{r_2} = \frac{v_2}{v_1}v_{r_1} = \frac{1}{r}v_{r_1} = \frac{1}{8}(676.1) = 84.51$$
$$\longrightarrow u_2 = 475.11\text{kJ/kg}$$

P
1800 K
3
Q_{in}
4
2
Q_{out}
1
v

Process 2-3: v = constant heat addition.

$$T_3 = 1800\text{K} \longrightarrow u_3 = 1487.2\text{kJ/kg}$$
$$v_{r_3} = 3.994$$

$$m = \frac{P_1V_1}{RT_1} = \frac{(98\text{kPa})(0.0006\text{m}^3)}{(0.287\text{kPa}\cdot\text{m}^3/\text{kg}\cdot\text{K})(290\text{K})} = 7.065\times10^{-4}\text{kg}$$

$$Q_{in} = m(u_3 - u_2) = (7.065\times10^{-4}\text{kg})(1487.2 - 475.11)\text{kJ/kg} = \mathbf{0.715kJ}$$

(*b*) Process 3-4: isentropic expansion.

$$v_{r_4} = \frac{v_4}{v_3}v_{r_3} = rv_{r_3} = (8)(3.994) = 31.95 \longrightarrow u_4 = 693.23\text{ kJ/kg}$$

Process 4-1: v = constant heat rejection.

$$Q_{out} = m(u_4 - u_1) = (7.065\times10^{-4}\text{kg})(693.23 - 206.91)\text{kJ/kg} = \mathbf{0.344\ kJ}$$

$$W_{net} = Q_{in} - Q_{out} = 0.715 - 0.344 = 0.371\text{ kJ}$$

$$\eta_{th} = \frac{W_{net}}{Q_{in}} = \frac{0.371\text{kJ}}{0.715\text{kJ}} = \mathbf{51.9\%}$$

(*c*)
$$\dot{n} = \frac{\dot{W}_{net}}{n_{cyl}W_{net,cyl}} = \frac{60\text{kJ/s}}{4\times0.371\text{ kJ}}\left(\frac{60\text{s}}{1\text{min}}\right) = \mathbf{2426\ rpm}$$

Discussion Note for the ideal Otto cycle, a thermodynamic cycle is equivalent to a mechanical cycle (a revolution) (In actual 4-storke engines, 2 revolutions correspond to 1 thermodynamic cycle).

8-133 EES solution of this (and other comprehensive problems designated with the *computer icon*) is available to instructors at the *Instrúctor Manual* section of the *Online Learning Center* (OLC) at www.mhhe.com/cengel-boles. See the Preface for access information.

8-134 An ideal Otto cycle with air as the working fluid with a compression ratio of 9.2 is considered. The amount of heat transferred to the air, the net work output, the thermal efficiency, and the mean effective pressure are to be determined.

Assumptions **1** The air-standard assumptions are applicable. **2** Kinetic and potential energy changes are negligible. **3** Air is an ideal gas with variable specific heats.

Properties The properties of air are given in Table A-17.

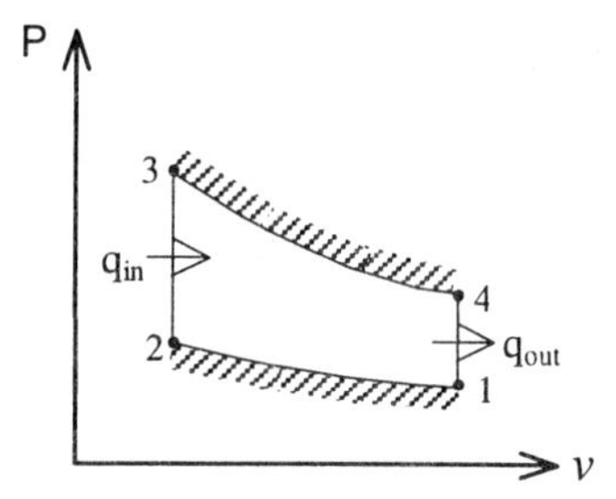

Analysis (*a*) Process 1-2: isentropic compression.

$$T_1 = 300\text{K} \longrightarrow u_1 = 214.07\text{kJ/kg}$$
$$v_{r_1} = 621.2$$

$$v_{r_2} = \frac{v_2}{v_1} v_{r_1} = \frac{1}{r} v_{r_1} = \frac{1}{9.2}(621.2) = 67.52 \longrightarrow T_2 = 708.3K$$
$$u_2 = 518.9kJ/kg$$

$$\frac{P_2 v_2}{T_2} = \frac{P_1 v_1}{T_1} \longrightarrow P_2 = \frac{v_1}{v_2}\frac{T_2}{T_1} P_1 = (9.2)\left(\frac{708.3\text{K}}{300\text{K}}\right)(98\text{kPa}) = 2129\text{kPa}$$

Process 2-3: v = constant heat addition.

$$\frac{P_3 v_3}{T_3} = \frac{P_2 v_2}{T_2} \longrightarrow T_3 = \frac{P_3}{P_2} T_2 = 2T_2 = (2)(708.3) = 1416.6\text{K} \longrightarrow u_3 = 1128.7\text{kJ/kg}$$
$$v_{r_3} = 8.593$$

$$q_{in} = u_3 - u_2 = 1128.7 - 518.9 = \mathbf{609.8kJ/kg}$$

(*b*) Process 3-4: isentropic expansion.

$$v_{r_4} = \frac{v_4}{v_3} v_{r_3} = r v_{r_3} = (9.2)(8.593) = 79.06 \longrightarrow u_4 = 487.75\text{kJ/kg}$$

Process 4-1: v = constant heat rejection.

$$q_{out} = u_4 - u_1 = 487.75 - 214.07 = 273.7 \text{ kJ/kg}$$
$$w_{net} = q_{in} - q_{out} = 609.8 - 273.7 = \mathbf{336.1 \text{ kJ/kg}}$$

(*c*)
$$\eta_{th} = \frac{w_{net}}{q_{in}} = \frac{336.1 \text{ kJ/kg}}{609.8 \text{ kJ/kg}} = \mathbf{55.1\%}$$

(*d*)
$$v_{\max} = v_1 = \frac{RT_1}{P_1} = \frac{(0.287\text{kPa}\cdot\text{m}^3/\text{kg}\cdot\text{K})(300\text{K})}{98\text{kPa}} = 0.879\text{m}^3/\text{kg}$$

$$v_{\min} = v_2 = \frac{v_{\max}}{r}$$

$$MEP = \frac{w_{net}}{v_1 - v_2} = \frac{w_{net}}{v_1(1 - 1/r)} = \frac{336.1\text{kJ/kg}}{(0.879\text{m}^3/\text{kg})(1 - 1/9.2)}\left(\frac{1\text{kPa}\cdot\text{m}^3}{1\text{kJ}}\right) = \mathbf{429kPa}$$

8-135 An ideal Otto cycle with air as the working fluid with a compression ratio of 9.2 is considered. The amount of heat transferred to the air, the net work output, the thermal efficiency, and the mean effective pressure are to be determined.

Assumptions **1** The air-standard assumptions are applicable. **2** Kinetic and potential energy changes are negligible. **3** Air is an ideal gas with constant specific heats.

Properties The properties of air at room temperature are C_p = 1.005 kJ/kg.K, C_v = 0.718 kJ/kg· K, and k = 1.4 (Table A-2).

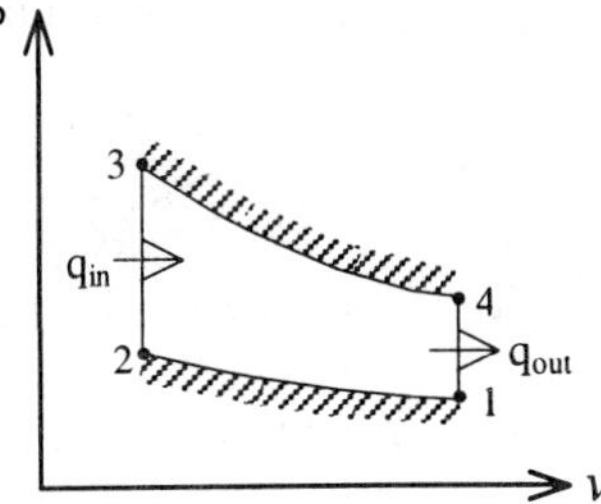

Analysis (a) Process 1-2 is isentropic compression:

$$T_2 = T_1\left(\frac{v_1}{v_2}\right)^{k-1} = (300\text{K})(9.2)^{0.4} = 728.8\text{K}$$

$$\frac{P_2 v_2}{T_2} = \frac{P_1 v_1}{T_1} \longrightarrow P_2 = \frac{v_1}{v_2}\frac{T_2}{T_1}P_1 = (9.2)\left(\frac{728.8\text{K}}{300\text{K}}\right)(98\text{kPa}) = 2190\text{kPa}$$

Process 2-3: v = constant heat addition.

$$\frac{P_3 v_3}{T_3} = \frac{P_2 v_2}{T_2} \longrightarrow T_3 = \frac{P_3}{P_2}T_2 = 2T_2 = (2)(728.8) = 1457.6\text{K}$$

$$q_{in} = u_3 - u_2 = C_v(T_3 - T_2) = (0.718\text{kJ/kg}\cdot\text{K})(1457.6 - 728.8)\text{K} = \mathbf{523.3kJ/kg}$$

(*b*) Process 3-4: isentropic expansion.

$$T_4 = T_3\left(\frac{v_3}{v_4}\right)^{k-1} = (1457.6\text{K})\left(\frac{1}{9.2}\right)^{0.4} = 600.0\text{K}$$

Process 4-1: v = constant heat rejection.

$$q_{out} = u_4 - u_1 = C_v(T_4 - T_1) = (0.718\text{kJ/kg}\cdot\text{K})(600 - 300)\text{K} = 215.4\text{kJ/kg}$$

$$w_{net} = q_{in} - q_{out} = 523.3 - 215.4 = \mathbf{307.9kJ/kg}$$

(*c*) $$\eta_{th} = \frac{w_{net}}{q_{in}} = \frac{307.9\text{ kJ/kg}}{523.3\text{ kJ/kg}} = \mathbf{58.8\%}$$

(*d*) $$v_{max} = v_1 = \frac{RT_1}{P_1} = \frac{(0.287\text{kPa}\cdot\text{m}^3/\text{kg}\cdot\text{K})(300\text{K})}{98\text{kPa}} = 0.879\text{m}^3/\text{kg}$$

$$v_{min} = v_2 = \frac{v_{max}}{r}$$

$$MEP = \frac{w_{net}}{v_1 - v_2} = \frac{w_{net}}{v_1(1-1/r)} = \frac{307.9\text{kJ/kg}}{(0.879\text{m}^3/\text{kg})(1-1/9.2)}\left(\frac{1\text{kPa}\cdot\text{m}^3}{1\text{kJ}}\right) = \mathbf{393kPa}$$

8-136 An engine operating on the ideal diesel cycle with air as the working fluid is considered. The pressure at the beginning of the heat-rejection process, the net work per cycle, and the mean effective pressure are to be determined. √

Assumptions **1** The air-standard assumptions are applicable. **2** Kinetic and potential energy changes are negligible. **3** Air is an ideal gas with variable specific heats.

Properties The properties of air are given in Table A-17.

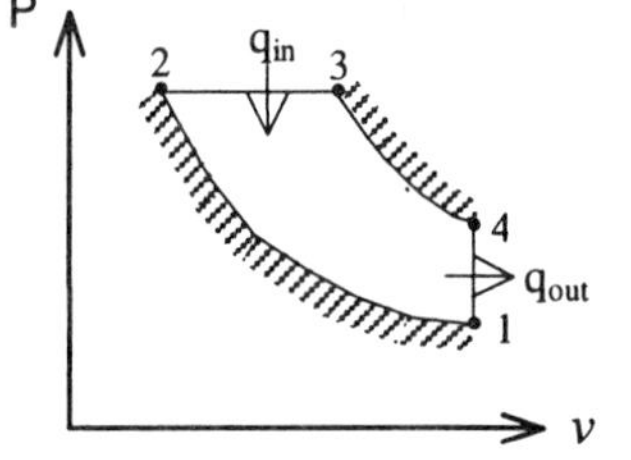

Analysis (a) The compression and the cutoff ratios are

$$r = \frac{V_1}{V_2} = \frac{1200 \text{ cm}^3}{75 \text{ cm}^3} = 16 \qquad r_c = \frac{V_3}{V_2} = \frac{150 \text{ cm}^3}{75 \text{ cm}^3} = 2$$

Process 1-2: isentropic compression.

$$T_1 = 290K \longrightarrow u_1 = 206.91\text{kJ/kg}$$
$$v_{r_1} = 676.1$$

$$v_{r_2} = \frac{v_2}{v_1} v_{r_1} = \frac{1}{r} v_{r_1} = \frac{1}{16}(676.1) = 42.256 \longrightarrow T_2 = 837.3\text{K}$$
$$h_2 = 863.03\text{kJ/kg}$$

Process 2-3: P = constant heat addition.

$$\frac{P_3 v_3}{T_3} = \frac{P_2 v_2}{T_2} \longrightarrow T_3 = \frac{v_3}{v_2} T_2 = 2T_2 = (2)(837.3) = 1674.6 \text{ K}$$
$$\longrightarrow h_3 = 1848.9\text{kJ/kg}$$
$$v_{r_3} = 5.002$$

Process 3-4: isentropic expansion.

$$v_{r_4} = \frac{v_4}{v_3} v_{r_3} = \frac{v_4}{2v_2} v_{r_3} = \frac{r}{2} v_{r_3} = \left(\frac{16}{2}\right)(5.002) = 40.016 \longrightarrow T_4 = 853.4\text{K}$$
$$u_4 = 636.00\text{kJ/kg}$$

Process 4-1: v = constant heat rejection.

$$\frac{P_4 v_4}{T_4} = \frac{P_1 v_1}{T_1} \longrightarrow P_4 = \frac{T_4}{T_1} P_1 = \left(\frac{853.4 \text{ K}}{290 \text{ K}}\right)(100 \text{ kPa}) = \mathbf{294.3 \text{ kPa}}$$

(b) $$m = \frac{P_1 V_1}{RT_1} = \frac{(100\text{kPa})(0.0012\text{m}^3)}{(0.287\text{kPa}\cdot\text{m}^3/\text{kg}\cdot\text{K})(290\text{K})} = 1.442 \times 10^{-3}\text{kg}$$

$$Q_{in} = m(h_3 - h_2) = (1.442 \times 10^{-3}\text{kg})(1848.9 - 863.08) = 1.422\text{kJ}$$

$$Q_{out} = m(u_4 - u_1) = (1.442 \times 10^{-3}\text{kg})(636.00 - 206.91)\text{kJ/kg} = 0.619\text{kJ}$$

$$W_{net} = Q_{in} - Q_{out} = 1.422 - 0.619 = \mathbf{0.803 \text{ kJ}}$$

(c) $$MEP = \frac{W_{net}}{V_1 - V_2} = \frac{W_{net}}{V_1(1 - 1/r)} = \frac{0.803 \text{ kJ}}{(0.0012\text{m}^3)(1 - 1/16)}\left(\frac{1 \text{ kPa}\cdot\text{m}^3}{1 \text{ kJ}}\right) = \mathbf{714 \text{ kPa}}$$

8-137 An engine operating on the ideal diesel cycle with argon as the working fluid is considered. The pressure at the beginning of the heat-rejection process, the net work per cycle, and the mean effective pressure are to be determined.

Assumptions **1** The air-standard assumptions are applicable. **2** Kinetic and potential energy changes are negligible. **3** Argon is an ideal gas with constant specific heats.

Properties The properties of argon at room temperature are C_p = 0.5203 kJ/kg.K, C_v = 0.3122 kJ/kg· K, and k = 1.667 (Table A-2).

Analysis (*a*) The compression and the cutoff ratios are

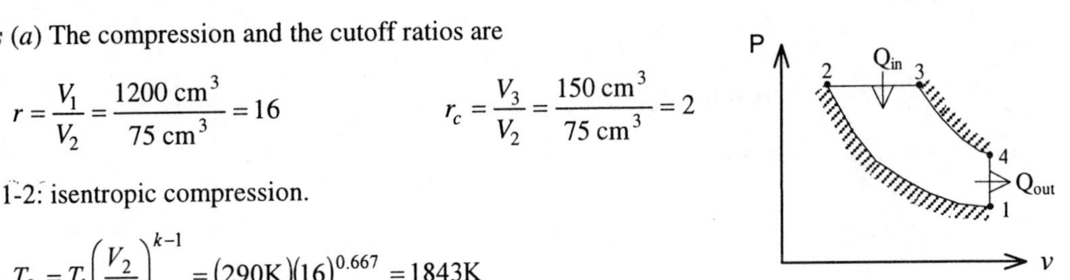

$$r = \frac{V_1}{V_2} = \frac{1200 \text{ cm}^3}{75 \text{ cm}^3} = 16 \qquad r_c = \frac{V_3}{V_2} = \frac{150 \text{ cm}^3}{75 \text{ cm}^3} = 2$$

Process 1-2: isentropic compression.

$$T_2 = T_1\left(\frac{V_2}{V_1}\right)^{k-1} = (290\text{K})(16)^{0.667} = 1843\text{K}$$

Process 2-3: P = constant heat addition.

$$\frac{P_3 v_3}{T_3} = \frac{P_2 v_2}{T_2} \longrightarrow T_3 = \frac{v_3}{v_2} T_2 = 2T_2 = (2)(1843) = 3686\text{K}$$

Process 3-4: isentropic expansion.

$$T_4 = T_3\left(\frac{V_3}{V_4}\right)^{k-1} = T_3\left(\frac{2V_2}{V_4}\right)^{k-1} = T_3\left(\frac{2}{r}\right)^{k-1} = (3686\text{K})\left(\frac{2}{16}\right)^{0.667} = 920.9\text{K}$$

Process 4-1: v = constant heat rejection.

$$\frac{P_4 v_4}{T_4} = \frac{P_1 v_1}{T_1} \longrightarrow P_4 = \frac{T_4}{T_1} P_1 = \left(\frac{920.9 \text{ K}}{290 \text{ K}}\right)(100 \text{ kPa}) = \mathbf{317.6 \text{ kPa}}$$

(*b*)
$$m = \frac{P_1 V_1}{R T_1} = \frac{(100\text{kPa})(0.0012\text{m}^3)}{(0.2081\text{kPa}\cdot\text{m}^3/\text{kg}\cdot\text{K})(290\text{K})} = 1.988 \times 10^{-3}\text{kg}$$

$$Q_{in} = m(h_3 - h_2) = mC_p(T_3 - T_2) = (1.988 \times 10^{-3}\text{kg})(0.5203 \text{ kJ/kg}\cdot\text{K})(3686 - 1843)\text{K} = 1.906 \text{ kJ}$$

$$Q_{out} = m(u_4 - u_1) = mC_v(T_4 - T_1) = (1.988 \times 10^{-3}\text{kg})(0.3122 \text{ kJ/kg}\cdot\text{K})(920.9 - 290)\text{K} = 0.392 \text{ kJ}$$

$$W_{net} = Q_{in} - Q_{out} = 1.906 - 0.392 = \mathbf{1.514 \text{ kJ}}$$

(*c*)
$$MEP = \frac{W_{net}}{V_1 - V_2} = \frac{W_{net}}{V_1(1 - 1/r)} = \frac{1.514 \text{ kJ}}{(0.0012 \text{ m}^3)(1 - 1/16)}\left(\frac{1 \text{ kPa}\cdot\text{m}^3}{1 \text{ kJ}}\right) = \mathbf{1346 \text{ kPa}}$$

8-138E An ideal dual cycle with air as the working fluid with a compression ratio of 12 is considered. The thermal efficiency of the cycle is to be determined. √

Assumptions **1** The air-standard assumptions are applicable. **2** Kinetic and potential energy changes are negligible. **3** Air is an ideal gas with constant specific heats.

Properties The properties of air at room temperature are C_p = 0.240 Btu/lbm.R, C_v = 0.171 Btu/lbm.R, and k = 1.4 (Table A-2E).

Analysis (*a*) The mass of air is

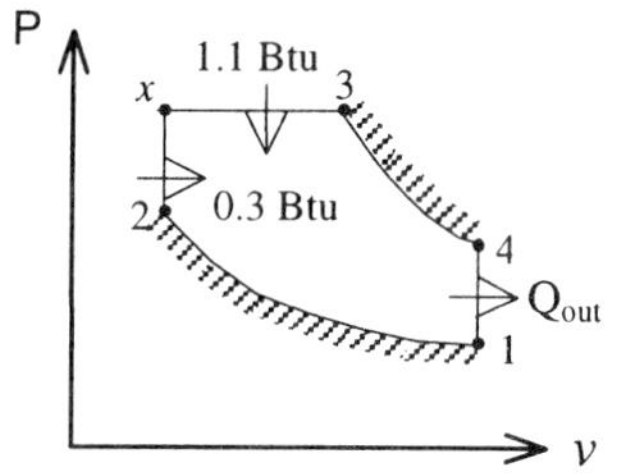

$$m = \frac{P_1 V_1}{R T_1} = \frac{(14.7\text{psia})(75/1728\text{ft}^3)}{(0.3704\text{psia}\cdot\text{ft}^3/\text{lbm}\cdot\text{R})(550\text{R})} = 3.132\times10^{-3}\text{ lbm}$$

P
1.1 Btu
x
3
0.3 Btu
2
4
Q_{out}
1
v

Process 1-2: isentropic compression.

$$T_2 = T_1\left(\frac{V_1}{V_2}\right)^{k-1} = (550\text{R})(12)^{0.4} = 1486\text{R}$$

Process 2-x: v = *constant* heat addition,

$$Q_{2-x,in} = m(u_x - u_2) = mC_v(T_x - T_2)$$
$$0.3\text{Btu} = (3.132\times10^{-3}\text{lbm})(0.171\text{Btu/lbm}\cdot\text{R})(T_x - 1486)\text{R} \longrightarrow T_x = 2046\text{R}$$

Process x-3: P = *constant* heat addition.

$$Q_{x-3,in} = m(h_3 - h_x) = mC_p(T_3 - T_x)$$
$$1.1\text{Btu} = (3.132\times10^{-3}\text{lbm})(0.240\text{Btu/lbm}\cdot\text{R})(T_3 - 2046)\text{R} \longrightarrow T_3 = 3509\text{ R}$$

$$\frac{P_3 V_3}{T_3} = \frac{P_x V_x}{T_x} \longrightarrow r_c = \frac{V_3}{V_x} = \frac{T_3}{T_x} = \frac{3509\text{ R}}{2046\text{ R}} = 1.715$$

Process 3-4: isentropic expansion.

$$T_4 = T_3\left(\frac{V_3}{V_4}\right)^{k-1} = T_3\left(\frac{1.715V_1}{V_4}\right)^{k-1} = T_3\left(\frac{1.715}{r}\right)^{k-1} = (3509\text{ R})\left(\frac{1.715}{12}\right)^{0.4} = 1611\text{ R}$$

Process 4-1: v = constant heat rejection.

$$Q_{out} = m(u_4 - u_1) = mC_v(T_4 - T_1)$$
$$= (3.132\times10^{-3}\text{lbm})(0.171\text{ Btu/lbm}\cdot\text{R})(1611 - 550)\text{R} = 0.568\text{ Btu}$$

$$\eta_{th} = 1 - \frac{Q_{out}}{Q_{in}} = 1 - \frac{0.568\text{ Btu}}{1.4\text{ Btu}} = \mathbf{59.4\%}$$

8-139 An ideal Stirling cycle with air as the working fluid is considered. The maximum pressure in the cycle and the net work output are to be determined.

Assumptions **1** The air-standard assumptions are applicable. **2** Kinetic and potential energy changes are negligible. **3** Air is an ideal gas with constant specific heats.

Properties The properties of air at room temperature are $R = 0.287$ kJ/kg.K, $C_p = 1.005$ kJ/kg.K, $C_v = 0.718$ kJ/kg· K, and $k = 1.4$ (Table A-2).

Analysis (*a*) The entropy change during process 1-2 is

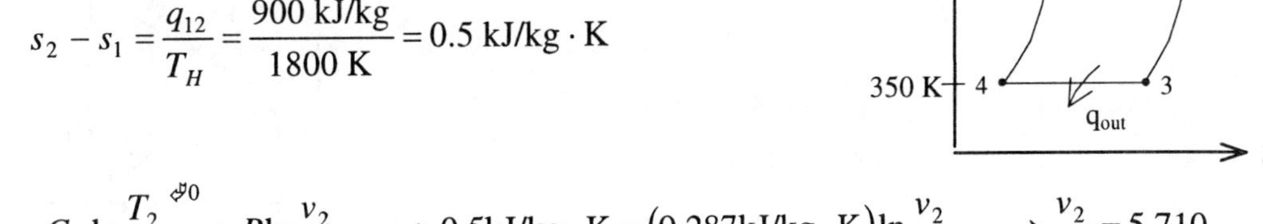

$$s_2 - s_1 = \frac{q_{12}}{T_H} = \frac{900 \text{ kJ/kg}}{1800 \text{ K}} = 0.5 \text{ kJ/kg} \cdot \text{K}$$

and

$$s_2 - s_1 = C_v \ln \frac{T_2}{T_1}^{\nearrow 0} + R \ln \frac{v_2}{v_1} \longrightarrow 0.5 \text{kJ/kg} \cdot \text{K} = (0.287 \text{kJ/kg} \cdot \text{K}) \ln \frac{v_2}{v_1} \longrightarrow \frac{v_2}{v_1} = 5.710$$

$$\frac{P_3 v_3}{T_3} = \frac{P_1 v_1}{T_1} \longrightarrow P_1 = P_3 \frac{v_3}{v_1} \frac{T_1}{T_3} = P_3 \frac{v_2}{v_1} \frac{T_1}{T_3} = (200 \text{ kPa})(5.710)\left(\frac{1800 \text{ K}}{350 \text{ K}}\right) = \mathbf{5873 \text{ kPa}}$$

(*b*)

$$w_{net} = \eta_{th} q_{in} = \left(1 - \frac{T_L}{T_H}\right) q_{in} = \left(1 - \frac{350 \text{ K}}{1800 \text{ K}}\right)(900 \text{ kJ/kg}) = \mathbf{725 \text{ kJ/kg}}$$

8-140 A simple ideal Brayton cycle with air as the working fluid is considered. The changes in the net work output per unit mass and the thermal efficiency are to be determined. √

Assumptions **1** The air-standard assumptions are applicable. **2** Kinetic and potential energy changes are negligible. **3** Air is an ideal gas with variable specific heats.

Properties The properties of air are given in Table A-17.

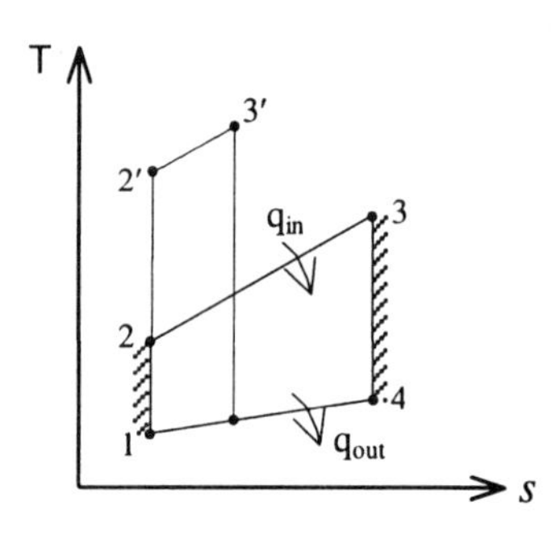

Analysis (*a*) The properties at various states are

$$T_1 = 300 \text{ K} \longrightarrow \begin{array}{l} h_1 = 300.19 \text{ kJ/kg} \\ P_{r_1} = 1.386 \end{array}$$

$$T_3 = 1300 \text{ K} \longrightarrow \begin{array}{l} h_3 = 1395.97 \text{ kJ/kg} \\ P_{r_3} = 330.9 \end{array}$$

For $r_p = 6$,

$$P_{r_2} = \frac{P_2}{P_1} P_{r_1} = (6)(1.386) = 8.316 \longrightarrow h_2 = 501.40\text{kJ/kg}$$

$$P_{r_4} = \frac{P_4}{P_3} P_{r_3} = \left(\frac{1}{6}\right)(330.9) = 55.15 \longrightarrow h_4 = 855.3\text{kJ/kg}$$

$$\begin{aligned} q_{in} &= h_3 - h_2 = 1395.97 - 501.40 = 894.57\text{kJ/kg} \\ q_{out} &= h_4 - h_1 = 855.3 - 300.19 = 555.11\text{kJ/kg} \\ w_{net} &= q_{in} - q_{out} = 894.57 - 555.11 = 339.46\text{kJ/kg} \end{aligned}$$

$$\eta_{th} = \frac{w_{net}}{q_{in}} = \frac{339.46\text{kJ/kg}}{894.57\text{kJ/kg}} = 37.9\%$$

For $r_p = 12$,

$$P_{r_2} = \frac{P_2}{P_1} P_{r_1} = (12)(1.386) = 16.63 \longrightarrow h_2 = 610.6\text{kJ/kg}$$

$$P_{r_4} = \frac{P_4}{P_3} P_{r_3} = \left(\frac{1}{12}\right)(330.9) = 27.58 \longrightarrow h_4 = 704.6\text{kJ/kg}$$

$$\begin{aligned} q_{in} &= h_3 - h_2 = 1395.97 - 610.60 = 785.37\text{kJ/kg} \\ q_{out} &= h_4 - h_1 = 704.6 - 300.19 = 404.41\text{kJ/kg} \\ w_{net} &= q_{in} - q_{out} = 785.37 - 404.41 = 380.96\text{kJ/kg} \end{aligned}$$

$$\eta_{th} = \frac{w_{net}}{q_{in}} = \frac{380.96\text{kJ/kg}}{785.37\text{kJ/kg}} = 48.5\%$$

Thus,

(*a*) $\Delta w_{net} = 380.96 - 339.46 = \mathbf{41.5\ kJ/kg}$ (increase)

(*b*) $\Delta \eta_{th} = 48.5\% - 37.9\% = \mathbf{10.6\%}$ (increase)

8-141 A simple ideal Brayton cycle with air as the working fluid is considered. The changes in the net work output per unit mass and the thermal efficiency are to be determined.

Assumptions **1** The air-standard assumptions are applicable. **2** Kinetic and potential energy changes are negligible. **3** Air is an ideal gas with constant specific heats.

Properties The properties of air at room temperature are R = 0.287 kJ/kg.K, C_p = 1.005 kJ/kg.K, C_v = 0.718 kJ/kg· K, and k = 1.4 (Table A-2).

Analysis Processes 1-2 and 3-4 are isentropic. Therefore, For $r_p = 6$,

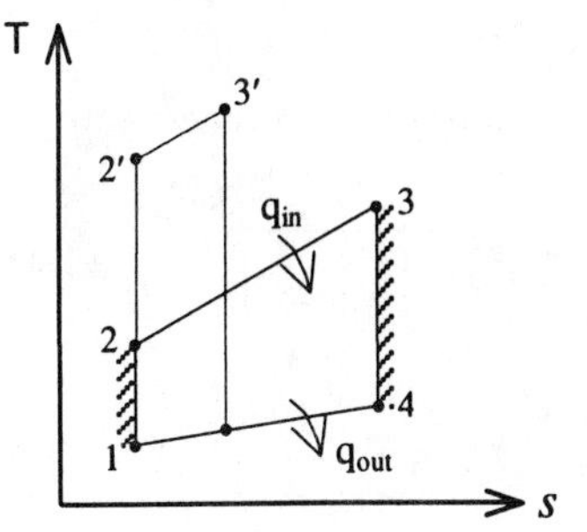

$$T_2 = T_1\left(\frac{P_2}{P_1}\right)^{(k-1)/k} = (300\text{K})(6)^{0.4/1.4} = 500.6\text{K}$$

$$T_4 = T_3\left(\frac{P_4}{P_3}\right)^{(k-1)/k} = (1300\text{K})\left(\frac{1}{6}\right)^{0.4/1.4} = 779.1\text{K}$$

$$\begin{aligned} q_{in} &= h_3 - h_2 = C_p(T_3 - T_2) \\ &= (1.005\text{kJ/kg}\cdot\text{K})(1300 - 500.6)\text{K} = 803.4\text{kJ/kg} \end{aligned}$$

$$\begin{aligned} q_{out} &= h_4 - h_1 = C_p(T_4 - T_1) \\ &= (1.005\text{kJ/kg}\cdot\text{K})(779.1 - 300)\text{K} = 481.5\text{kJ/kg} \end{aligned}$$

$$w_{net} = q_{in} - q_{out} = 803.4 - 481.5 = 321.9\text{kJ/kg}$$

$$\eta_{th} = \frac{w_{net}}{q_{in}} = \frac{321.9\text{kJ/kg}}{803.4\text{kJ/kg}} = 40.1\%$$

For $r_p = 12$,

$$T_2 = T_1\left(\frac{P_2}{P_1}\right)^{(k-1)/k} = (300\text{K})(12)^{0.4/1.4} = 610.2\text{K}$$

$$T_4 = T_3\left(\frac{P_4}{P_3}\right)^{(k-1)/k} = (1300\text{K})\left(\frac{1}{12}\right)^{0.4/1.4} = 639.2\text{K}$$

$$\begin{aligned} q_{in} &= h_3 - h_2 = C_p(T_3 - T_2) \\ &= (1.005\text{kJ/kg}\cdot\text{K})(1300 - 610.2)\text{K} = 693.2\text{kJ/kg} \end{aligned}$$

$$\begin{aligned} q_{out} &= h_4 - h_1 = C_p(T_4 - T_1) \\ &= (1.005\text{kJ/kg}\cdot\text{K})(639.2 - 300)\text{K} = 340.9\text{kJ/kg} \end{aligned}$$

$$w_{net} = q_{in} - q_{out} = 693.2 - 340.9 = 352.3\text{kJ/kg}$$

$$\eta_{th} = \frac{w_{net}}{q_{in}} = \frac{352.3\text{kJ/kg}}{693.2\text{kJ/kg}} = 50.8\%$$

Thus,

(*a*) $\Delta w_{net} = 352.3 - 321.9 = \mathbf{30.4\ kJ/kg}$ (increase)

(*b*) $\Delta\eta_{th} = 50.8\% - 40.1\% = \mathbf{10.7\%}$ (increase)

8-142 A regenerative Brayton cycle with helium as the working fluid is considered. The thermal efficiency and the required mass flow rate of helium are to be determined for 100 percent and 80 percent isentropic efficiencies for both the compressor and the turbine.

Assumptions **1** The air-standard assumptions are applicable. **2** Kinetic and potential energy changes are negligible. **3** Helium is an ideal gas with constant specific heats.

Properties The properties of helium at room temperature are C_p = 5.1926 kJ/kg.K and k = 1.667 (Table A-2).

Analysis (a) Assuming $\eta_T = \eta_C$ = 100%,

$$T_{2s} = T_1\left(\frac{P_2}{P_1}\right)^{(k-1)/k} = (300\text{ K})(8)^{0.667/1.667} = 689.4\text{ K}$$

$$T_{4s} = T_3\left(\frac{P_4}{P_3}\right)^{(k-1)/k} = (1800\text{ K})\left(\frac{1}{8}\right)^{0.667/1.667} = 783.3\text{ K}$$

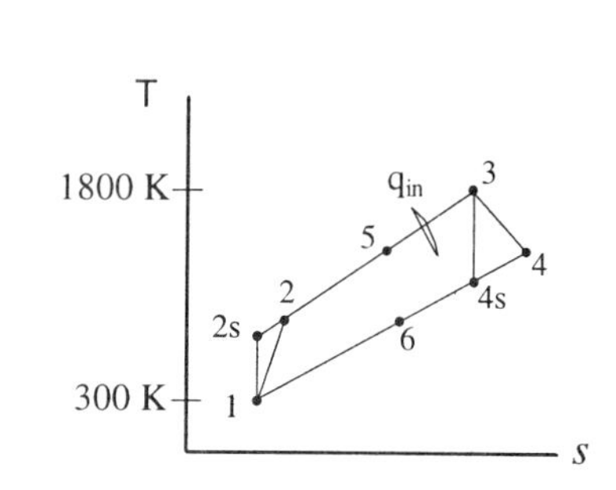

$$\varepsilon = \frac{h_5 - h_2}{h_4 - h_2} = \frac{C_p(T_5 - T_2)}{C_p(T_4 - T_2)} \longrightarrow T_5 = T_2 + \varepsilon(T_4 - T_2)$$
$$= 689.4 + (0.75)(783.3 - 689.4)$$
$$= 759.8\text{ K}$$

$$w_{net} = w_{T,out} - w_{C,in} = (h_3 - h_4) - (h_2 - h_1) = C_p[(T_3 - T_4) - (T_2 - T_1)]$$
$$= (5.1926\text{ kJ/kg}\cdot\text{K})[(1800 - 783.3) - (689.4 - 300)]\text{K} = 3257.3\text{kJ/kg}$$

$$\dot{m} = \frac{\dot{W}_{net}}{w_{net}} = \frac{450{,}000\text{ kJ/s}}{3257.3\text{ kJ/kg}} = \mathbf{13.82\ kg/s}$$

$$q_{in} = h_3 - h_5 = C_p(T_3 - T_5) = (5.1926\text{ kJ/kg}\cdot\text{K})(1800 - 759.8)\text{K} = 5401.3\text{ kJ/kg}$$

$$\eta_{th} = \frac{w_{net}}{q_{in}} = \frac{3257.3\text{ kJ/kg}}{5401.3\text{ kJ/kg}} = \mathbf{60.3\%}$$

(*b*) Assuming $\eta_T = \eta_C = 80\%$,

$$T_{2s} = T_1\left(\frac{P_2}{P_1}\right)^{(k-1)/k} = (300\text{K})(8)^{0.667/1.667} = 689.4\text{K}$$

$$\eta_C = \frac{h_{2s} - h_1}{h_2 - h_1} = \frac{C_p(T_{2s} - T_1)}{C_p(T_2 - T_1)} \longrightarrow T_2 = T_1 + (T_{2s} - T_1)/\eta_C$$
$$= 300 + (689.4 - 300)/(0.80)$$
$$= 786.8\text{K}$$

$$T_{4s} = T_3\left(\frac{P_4}{P_3}\right)^{(k-1)/k} = (1800\text{K})\left(\frac{1}{8}\right)^{0.667/1.667} = 783.3\text{K}$$

$$\eta_T = \frac{h_3 - h_4}{h_3 - h_{4s}} = \frac{C_p(T_3 - T_4)}{C_p(T_3 - T_{4s})} \longrightarrow T_4 = T_3 - \eta_T(T_3 - T_{4s})$$
$$= 1800 - (0.80)(1800 - 783.3)$$
$$= 986.6\text{K}$$

$$\varepsilon = \frac{h_5 - h_2}{h_4 - h_2} = \frac{C_p(T_5 - T_2)}{C_p(T_4 - T_2)} \longrightarrow T_5 = T_2 + \varepsilon(T_4 - T_2)$$
$$= 786.8 + (0.75)(986.6 - 786.8)$$
$$= 936.7\text{K}$$

$$w_{net} = w_{T,out} - w_{C,in} = (h_3 - h_4) - (h_2 - h_1) = C_p[(T_3 - T_4) - (T_2 - T_1)]$$
$$= (5.1926 \text{ kJ/kg}\cdot\text{K})[(1800 - 986.6) - (786.8 - 300)]\text{K} = 1695.9 \text{ kJ/kg}$$

$$\dot{m} = \frac{\dot{W}_{net}}{w_{net}} = \frac{45{,}000 \text{ kJ/s}}{1695.9 \text{ kJ/kg}} = \mathbf{26.5\ \ kg/s}$$

$$q_{in} = h_3 - h_5 = C_p(T_3 - T_5) = (5.1926 \text{ kJ/kg}\cdot\text{K})(1800 - 936.7)\text{K} = 4482.8\text{kJ/kg}$$

$$\eta_{th} = \frac{w_{net}}{q_{in}} = \frac{1695.9 \text{ kJ/kg}}{4482.8 \text{ kJ/kg}} = \mathbf{37.8\%}$$

8-143 A regenerative gas-turbine engine operating with two stages of compression and two stages of expansion is considered. The back work ratio and the thermal efficiency are to be determined.

Assumptions **1** The air-standard assumptions are applicable. **2** Kinetic and potential energy changes are negligible. **3** Air is an ideal gas with constant specific heats.

Properties The properties of air at room temperature are C_p = 1.005 kJ/kg.K, C_v = 0.718 kJ/kg· K, and k = 1.4 (Table A-2).

Analysis The work inputs to each stage of compressor are identical, so are the work outputs of each stage of the turbine.

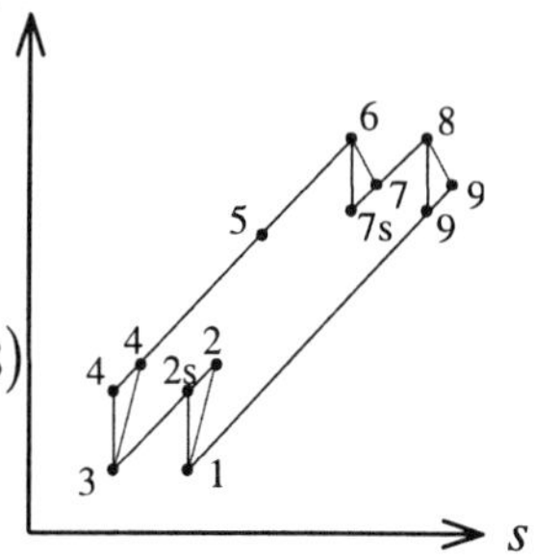

$$T_{4s} = T_{2s} = T_1\left(\frac{P_2}{P_1}\right)^{(k-1)/k} = (300\text{K})(3.5)^{0.4/1.4} = 429.1\text{K}$$

$$\eta_C = \frac{h_{2s}-h_1}{h_2-h_1} = \frac{C_p(T_{2s}-T_1)}{C_p(T_2-T_1)} \longrightarrow T_4 = T_2 = T_1 + (T_{2s}-T_1)/\eta_C$$
$$= 300 + (429.1-300)/(0.78)$$
$$= 465.5 \text{ K}$$

$$T_{9s} = T_{7s} = T_6\left(\frac{P_7}{P_6}\right)^{(k-1)/k} = (1200\text{K})\left(\frac{1}{3.5}\right)^{0.4/1.4} = 838.9\text{K}$$

$$\eta_T = \frac{h_6-h_7}{h_6-h_{7s}} = \frac{C_p(T_6-T_7)}{C_p(T_6-T_{7s})} \longrightarrow T_9 = T_7 = T_6 - \eta_T(T_6-T_{7s})$$
$$= 1200 - (0.86)(1200-838.9)$$
$$= 889.5\text{K}$$

$$\varepsilon = \frac{h_5-h_4}{h_9-h_4} = \frac{C_p(T_5-T_4)}{C_p(T_9-T_4)} \longrightarrow T_5 = T_4 + \varepsilon(T_9-T_4)$$
$$= 465.5 + (0.72)(889.5-465.5)$$
$$= 770.8\text{K}$$

$$w_{C,in} = 2(h_2-h_1) = 2C_p(T_2-T_1) = 2(1.005\text{kJ/kg}\cdot\text{K})(465.5-300)\text{K} = 332.7\text{kJ/kg}$$

$$w_{T,out} = 2(h_6-h_7) = 2C_p(T_6-T_7) = 2(1.005\text{kJ/kg}\cdot\text{K})(1200-889.5)\text{K} = 624.1\text{kJ/kg}$$

Thus,

$$r_{bw} = \frac{w_{C,in}}{w_{T,out}} = \frac{332.7\text{ kJ/kg}}{624.1\text{ kJ/kg}} = \mathbf{53.3\%}$$

$$q_{in} = (h_6-h_5)+(h_8-h_7) = C_p[(T_6-T_5)+(T_8-T_7)]$$
$$= (1.005\text{kJ/kg}\cdot\text{K})[(1200-770.8)+(1200-889.5)]\text{K} = 743.4\text{kJ/kg}$$

$$w_{net} = w_{T,out} - w_{C,in} = 624.1 - 332.7 = 291.4\text{kJ/kg}$$

$$\eta_{th} = \frac{w_{net}}{q_{in}} = \frac{291.4\text{kJ/kg}}{743.4\text{kJ/kg}} = \mathbf{39.2\%}$$

8-144 EES solution of this (and other comprehensive problems designated with the *computer icon*) is available to instructors at the *Instructor Manual* section of the *Online Learning Center* (OLC) at www.mhhe.com/cengel-boles. See the Preface for access information.

8-145 A regenerative gas-turbine engine operating with two stages of compression and two stages of expansion is considered. The back work ratio and the thermal efficiency are to be determined.

Assumptions **1** The air-standard assumptions are applicable. **2** Kinetic and potential energy changes are negligible. **3** Helium is an ideal gas with constant specific heats.

Properties The properties of helium at room temperature are C_p = 5.1926 kJ/kg.K and k = 1.667 (Table A-2).

Analysis The work inputs to each stage of compressor are identical, so are the work outputs of each stage of the turbine.

$$T_{4s} = T_{2s} = T_1\left(\frac{P_2}{P_1}\right)^{(k-1)/k} = (300\text{K})(3.5)^{0.667/1.667} = 495.2\text{K}$$

$$\eta_C = \frac{h_{2s} - h_1}{h_2 - h_1} = \frac{C_p(T_{2s} - T_1)}{C_p(T_2 - T_1)} \longrightarrow T_4 = T_2 = T_1 + (T_{2s} - T_1)/\eta_C$$
$$= 300 + (495.2 - 300)/(0.78)$$
$$= 550.3\text{K}$$

$$T_{9s} = T_{7s} = T_6\left(\frac{P_7}{P_6}\right)^{(k-1)/k} = (1200\text{K})\left(\frac{1}{3.5}\right)^{0.667/1.667} = 726.9\text{K}$$

$$\eta_T = \frac{h_6 - h_7}{h_6 - h_{7s}} = \frac{C_p(T_6 - T_7)}{C_p(T_6 - T_{7s})} \longrightarrow T_9 = T_7 = T_6 - \eta_T(T_6 - T_{7s})$$
$$= 1200 - (0.86)(1200 - 726.9)$$
$$= 793.1\text{K}$$

$$\varepsilon = \frac{h_5 - h_4}{h_9 - h_4} = \frac{C_p(T_5 - T_4)}{C_p(T_9 - T_4)} \longrightarrow T_5 = T_4 + \varepsilon(T_9 - T_4)$$
$$= 550.3 + (0.72)(793.1 - 550.3)$$
$$= 725.1\text{K}$$

$$w_{C,in} = 2(h_2 - h_1) = 2C_p(T_2 - T_1) = 2(5.1926\text{kJ/kg}\cdot\text{K})(550.3 - 300)\text{K} = 2599.4\text{kJ/kg}$$

$$w_{T,out} = 2(h_6 - h_7) = 2C_p(T_6 - T_7) = 2(5.1926\text{kJ/kg}\cdot\text{K})(1200 - 793.1)\text{K} = 4225.7\text{kJ/kg}$$

Thus,
$$r_{bw} = \frac{w_{C,in}}{w_{T,out}} = \frac{2599.4\text{ kJ/kg}}{4225.7\text{ kJ/kg}} = \mathbf{61.5\%}$$

$$q_{in} = (h_6 - h_5) + (h_8 - h_7) = C_p[(T_6 - T_5) + (T_8 - T_7)]$$
$$= (5.1926\text{kJ/kg}\cdot\text{K})[(1200 - 725.1) + (1200 - 793.1)]\text{K} = 4578.8\text{kJ/kg}$$

$$w_{net} = w_{T,out} - w_{C,in} = 4225.7 - 2599.4 = 1626.3\text{kJ/kg}$$

$$\eta_{th} = \frac{w_{net}}{q_{in}} = \frac{1626.3\text{kJ/kg}}{4578.8\text{kJ/kg}} = \mathbf{35.5\%}$$

8-146 An ideal regenerative Brayton cycle is considered. The pressure ratio that maximizes the thermal efficiency of the cycle is to be determined, and to be compared with the pressure ratio that maximizes the cycle net work.

Analysis Using the isentropic relations, the temperatures at the compressor and turbine exit can be expressed as

$$T_2 = T_1\left(\frac{P_2}{P_1}\right)^{(k-1)/k} = T_1\left(r_p\right)^{(k-1)/k}$$

$$T_4 = T_3\left(\frac{P_4}{P_3}\right)^{(k-1)/k} = T_3\left(\frac{1}{r_p}\right)^{(k-1)/k} = T_3 r_p^{(1-k)/k}$$

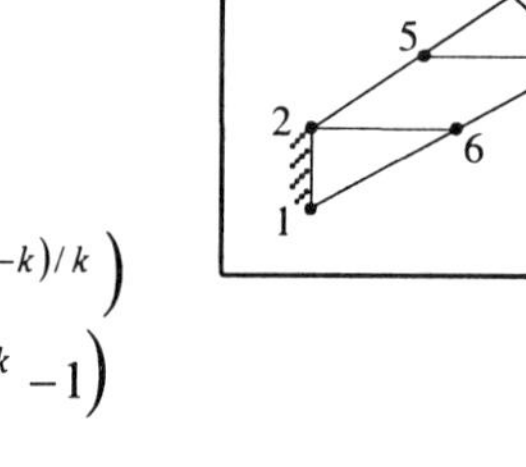

Then,

$$q_{in} = h_3 - h_5 = C_p\left(T_3 - T_5\right) = C_p\left(T_3 - T_4\right) = C_p T_3\left(1 - r_p^{(1-k)/k}\right)$$

$$q_{out} = h_6 - h_1 = C_p\left(T_6 - T_1\right) = C_p\left(T_2 - T_1\right) = C_p T_1\left(r_p^{(k-1)/k} - 1\right)$$

$$w_{net} = q_{in} - q_{out} = C_p\left(T_3 - T_3 r_p^{(1-k)/k} - T_1 r_p^{(k-1)/k} + T_1\right)$$

To maximize the net work, we must have

$$\frac{\partial w_{net}}{\partial r_p} = C_p\left(-\frac{1-k}{k}T_3 r_p^{(1-k)/k} - 1 - \frac{k-1}{k}T_1 r_p^{(k-1)/k} - 1\right) = 0$$

Solving for r_p gives

$$r_p = \left(\frac{T_1}{T_3}\right)^{k/2(1-k)}$$

Similarly,

$$\eta_{th} = 1 - \frac{q_{out}}{q_{in}} = 1 - \frac{C_p T_1\left(r_p^{(k-1)/k} - 1\right)}{C_p T_3\left(1 - r_p^{(1-k)/k}\right)}$$

which simplifies to

$$\eta_{th} = 1 - \frac{T_1}{T_3} r_p^{(k-1)/k}$$

When $r_p = 1$, the thermal efficiency becomes $\eta_{th} = 1 - T_1/T_3$, which is the Carnot efficiency. Therefore, the efficiency is a maximum when $r_p = 1$, and must decrease as r_p increases for the fixed values of T_1 and T_3. Note that the compression ratio cannot be less than 1, and the factor

$$r_p^{(k-1)/k}$$

is always greater than 1 for $r_p > 1$. Also note that the net work $w_{net} = 0$ for $r_p = 1$. This being the case, the pressure ratio for maximum thermal efficiency, which is $r_p = 1$, is always less than the pressure ratio for maximum work.

8-147 An ideal gas-turbine cycle with one stage of compression and two stages of expansion and regeneration is considered. The thermal efficiency of the cycle as a function of the compressor pressure ratio and the high-pressure turbine to compressor inlet temperature ratio is to be determined, and to be compared with the efficiency of the standard regenerative cycle.

Analysis The T-s diagram of the cycle is as shown in the figure. If the overall pressure ratio of the cycle is r_p, which is the pressure ratio across the compressor, then the pressure ratio across each turbine stage in the ideal case becomes $\sqrt{r_p}$. Using the isentropic relations, the temperatures at the compressor and turbine exit can be expressed as

$$T_5 = T_2 = T_1\left(\frac{P_2}{P_1}\right)^{(k-1)/k} = T_1(r_p)^{(k-1)/k}$$

$$T_7 = T_4 = T_3\left(\frac{P_4}{P_3}\right)^{(k-1)/k} = T_3\left(\frac{1}{\sqrt{r_p}}\right)^{(k-1)/k} = T_3 r_p^{(1-k)/2k}$$

$$T_6 = T_5\left(\frac{P_6}{P_5}\right)^{(k-1)/k} = T_5\left(\frac{1}{\sqrt{r_p}}\right)^{(k-1)/k} = T_2 r_p^{(1-k)/2k} = T_1 r_p^{(k-1)/k} r_p^{(1-k)/2k} = T_1 r_p^{(k-1)/2k}$$

Then,

$$q_{in} = h_3 - h_7 = C_p(T_3 - T_7) = C_p T_3\left(1 - r_p^{(1-k)/2k}\right)$$

$$q_{out} = h_6 - h_1 = C_p(T_6 - T_1) = C_p T_1\left(r_p^{(k-1)/2k} - 1\right)$$

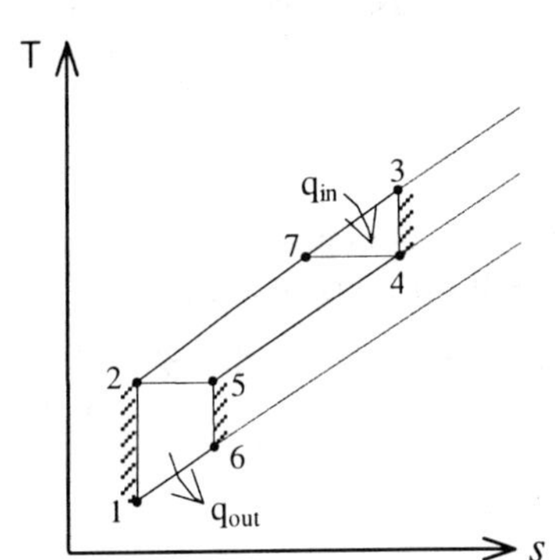

and thus

$$\eta_{th} = 1 - \frac{q_{out}}{q_{in}} = 1 - \frac{C_p T_1\left(r_p^{(k-1)/2k} - 1\right)}{C_p T_3\left(1 - r_p^{(1-k)/2k}\right)}$$

which simplifies to

$$\eta_{th} = 1 - \frac{T_1}{T_3} r_p^{(k-1)/2k}$$

The thermal efficiency of the single stage ideal regenerative cycle is given as

$$\eta_{th} = 1 - \frac{T_1}{T_3} r_p^{(k-1)/k}$$

Therefore, the regenerative cycle with two stages of expansion has a higher thermal efficiency than the standard regenerative cycle with a single stage of expansion for any given value of the pressure ratio r_p.

8-148 through 8-159 EES solution of these problems are available to instructors at the *Instructor Manual* section of the *Online Learning Center* (OLC) at www.mhhe.com/cengel-boles. See the Preface for access information.

Fundamentals of Engineering (FE) Exam Problems.

8-160 An Otto cycle with air as the working fluid has a compression ratio of 9.5. Under cold air standard conditions, the thermal efficiency of this cycle is

(a) 24% (b) 41% (c) 52% (d) 59% (e) 63%

Answer (d) 59%

Solution Solved by EES Software. Solutions can be verified by copying-and-pasting the following lines on a blank EES screen. (Similar problems and their solutions can be obtained easily by modifying numerical values).

```
r=9.5
k=1.4

Eta_Otto=1-1/r^(k-1)

"Some Wrong Solutions with Common Mistakes:"
W1_Eta = 1/r "Taking efficiency to be 1/r"
W2_Eta = 1/r^(k-1)  "Using incorrect relation"
W3_Eta =  1-1/r^(k1-1); k1=1.667 "Using wrong k value"
```

8-161 For specified limits for the maximum and minimum temperatures, the ideal cycle with the lowest thermal efficiency is

(a) Carnot (b) Stirling (c) Ericsson (d) Otto (e) All are the same

Answer (d) Otto

8-162 A Carnot cycle operates between the temperatures limits of 300 K and 1500 K, and produces 600 kW of net power. The rate of entropy change of the working fluid during the heat addition process is

(a) 0 (b) 0.4 kW/K (c) 0.5 kW/K (d) 2.0 kW/K (e) 5.0 kW/K

Answer (c) 0.5 kW/K

Solution Solved by EES Software. Solutions can be verified by copying-and-pasting the following lines on a blank EES screen. (Similar problems and their solutions can be obtained easily by modifying numerical values).

```
TL=300 "K"
TH=1500 "K"
Wnet=600 "kJ/s"
Wnet= (TH-TL)*DS

"Some Wrong Solutions with Common Mistakes:"
W1_DS = Wnet/TH "Using TH instead of TH-TL"
W2_DS = Wnet/TL "Using TL instead of TH-TL"
W3_DS =  Wnet/(TH+TL) "Using TH+TL  instead of TH-TL"
```

8-163 Air in an ideal Diesel cycle is compressed from 3 L to 0.15 L, and then it expands during the constant pressure heat addition process to 0.30 L. Under cold air standard conditions, the thermal efficiency of this cycle is

(a) 35% (b) 44% (c) 65% (d) 70% (e) 82%

Answer (c) 65%

Solution Solved by EES Software. Solutions can be verified by copying-and-pasting the following lines on a blank EES screen. (Similar problems and their solutions can be obtained easily by modifying numerical values).

```
V1=3 "L"
V2= 0.15 "L"
V3= 0.30 "L"
r=V1/V2
rc=V3/V2
k=1.4
Eta_Diesel=1-(1/r^(k-1))*(rc^k-1)/k/(rc-1)

"Some Wrong Solutions with Common Mistakes:"
W1_Eta = 1-(1/r1^(k-1))*(rc^k-1)/k/(rc-1); r1=V1/V3 "Wrong r value"
W2_Eta = 1-Eta_Diesel  "Using incorrect relation"
W3_Eta =  1-(1/r^(k1-1))*(rc^k1-1)/k1/(rc-1); k1=1.667 "Using wrong k value"
W4_Eta = 1-1/r^(k-1) "Using Otto cycle efficiency"
```

8-164 Helium gas in an ideal Otto cycle is compressed from 20°C and 2 L to 0.25 L, and its temperature increases by an additional 800°C during the heat addition process. The temperature of helium before the expansion process is

(a) 1700°C (b) 1440°C (c) 1240°C (d) 880°C (e) 820°C

Answer (a) 1700°C

Solution Solved by EES Software. Solutions can be verified by copying-and-pasting the following lines on a blank EES screen. (Similar problems and their solutions can be obtained easily by modifying numerical values).

```
k=1.667
V1=2
V2=0.25
r=V1/V2

T1=20+273 "K"
T2=T1*r^(k-1)
T3=T2+800-273 "C"

"Some Wrong Solutions with Common Mistakes:"
W1_T3 =T22+800-273; T22=T1*r^(k1-1); k1=1.4 "Using wrong k value"
W2_T3 = T3+273  "Using K instead of C"
W3_T3 = T1+800-273 "Disregarding temp rise during compression"
W4_T3 = T222+800-273; T222=(T1-273)*r^(k-1) "Using C for T1 instead of K"
```

8-165 In an ideal Otto cycle, air is compressed from 1.20 kg/m^3 and 2.2 L to 0.26 L, and the net work output of the cycle is 440 kJ/kg. The mean effective pressure (MEP) for this cycle is

(a) 612 kPa (b) 599 kPa (c) 528 kPa (d) 416 kPa (e) 367 kPa

Answer (b) 599 kPa

Solution Solved by EES Software. Solutions can be verified by copying-and-pasting the following lines on a blank EES screen. (Similar problems and their solutions can be obtained easily by modifying numerical values).

```
rho1=1.20 "kg/m^3"
k=1.4
V1=2.2
V2=0.26
m=rho1*V1/1000 "kg"
w_net=440 "kJ/kg"

Wtotal=m*w_net
MEP=Wtotal/((V1-V2)/1000)

"Some Wrong Solutions with Common Mistakes:"
W1_MEP = w_net/((V1-V2)/1000) "Disregarding mass"
W2_MEP = Wtotal/(V1/1000)  "Using V1 instead of V1-V2"
W3_MEP = (rho1*V2/1000)*w_net/((V1-V2)/1000);  "Finding mass using V2 instead of V1"
W4_MEP = Wtotal/((V1+V2)/1000) "Adding V1 and V2 instead of subtracting"
```

8-166 In an ideal Brayton cycle, air is compressed from 100 kPa and 25°C to 700 kPa. Under cold air standard conditions, the thermal efficiency of this cycle is

(a) 43% (b) 52% (c) 57% (d) 70% (e) 84%

Answer (a) 43%

Solution Solved by EES Software. Solutions can be verified by copying-and-pasting the following lines on a blank EES screen. (Similar problems and their solutions can be obtained easily by modifying numerical values).

```
P1=100 "kPa"
P2=700 "kPa"
T1=25+273 "K"

rp=P2/P1
k=1.4

Eta_Brayton=1-1/rp^((k-1)/k)

"Some Wrong Solutions with Common Mistakes:"
W1_Eta = 1/rp "Taking efficiency to be 1/rp"
W2_Eta = 1/rp^((k-1)/k)  "Using incorrect relation"
W3_Eta = 1-1/rp^((k1-1)/k1); k1=1.667 "Using wrong k value"
```

8-167 Consider an ideal Brayton cycle executed between the pressure limits of 1200 kPa and 100 kPa and temperature limits of 20°C and 1000°C with argon as the working fluid. The net work output of the cycle is

(a) 68 kJ/kg (b) 93 kJ/kg (c) 158 kJ/kg (d) 186 kJ/kg (e) 310 kJ/kg

Answer (c) 158 kJ/kg

Solution Solved by EES Software. Solutions can be verified by copying-and-pasting the following lines on a blank EES screen. (Similar problems and their solutions can be obtained easily by modifying numerical values).

```
P1=100 "kPa"
P2=1200 "kPa"
T1=20+273 "K"
T3=1000+273 "K"

rp=P2/P1
k=1.667
Cp=0.5203 "kJ/kg.K"
Cv=0.3122 "kJ/kg.K"

T2=T1*rp^((k-1)/k)
q_in=Cp*(T3-T2)
Eta_Brayton=1-1/rp^((k-1)/k)

w_net=Eta_Brayton*q_in

"Some Wrong Solutions with Common Mistakes:"
W1_wnet = (1-1/rp^((k-1)/k))*qin1;  qin1=Cv*(T3-T2) "Using Cv instead of Cp"
W2_wnet = (1-1/rp^((k-1)/k))*qin2;  qin2=1.005*(T3-T2)  "Using Cp of air instead of argon"
W3_wnet = (1-1/rp^((k1-1)/k1))*Cp*(T3-T22); T22=T1*rp^((k1-1)/k1); k1=1.4  "Using k of air instead of argon"
W4_wnet = (1-1/rp^((k-1)/k))*Cp*(T3-T222); T222=(T1-273)*rp^((k-1)/k) "Using C for T1 instead of K"
```

8-168 An ideal Brayton cycle has a net work output of 150 kJ/kg and a backwork ratio of 0.4. If both the turbine and the compressor had an isentropic efficiency of 80%, the net work output of the cycle would be

(a) 50 kJ/kg (b) 75 kJ/kg (c) 98 kJ/kg (d) 120 kJ/kg (e) 188 kJ/kg

Answer (b) 75 kJ/kg

Solution Solved by EES Software. Solutions can be verified by copying-and-pasting the following lines on a blank EES screen. (Similar problems and their solutions can be obtained easily by modifying numerical values).

```
wcomp/wturb=0.4
wturb-wcomp=150 "kJ/kg"
Eff=0.8
w_net=Eff*wturb-wcomp/Eff

"Some Wrong Solutions with Common Mistakes:"
W1_wnet = Eff*wturb-wcomp*Eff "Making a mistake in Wnet relation"
W2_wnet = (wturb-wcomp)/Eff "Using a wrong relation"
W3_wnet = wturb/eff-wcomp*Eff "Using a wrong relation"
```

8-169 In an ideal Brayton cycle, air is compressed from 100 kPa and 25°C to 1 MPa, and then heated to 1200°C before entering the turbine. Under cold air standard conditions, the air temperature at the turbine exit is

(a) 490°C (b) 515°C (c) 622°C (d) 763°C (e) 895°C

Answer (a) 490°C

Solution Solved by EES Software. Solutions can be verified by copying-and-pasting the following lines on a blank EES screen. (Similar problems and their solutions can be obtained easily by modifying numerical values).

```
P1=100 "kPa"
P2=1000 "kPa"
T1=25+273 "K"
T3=1200+273 "K"

rp=P2/P1
k=1.4

T4=T3*(1/rp)^((k-1)/k)-273

"Some Wrong Solutions with Common Mistakes:"
W1_T4 = T3/rp "Using wrong relation"
W2_T4 = (T3-273)/rp "Using wrong relation"
W3_T4 = T4+273  "Using K instead of C"
W4_T4 = T1+800-273 "Disregarding temp rise during compression"
```

8-170 In an ideal Brayton cycle with regeneration, argon gas is compressed from 100 kPa and 25°C to 500 kPa, and then heated to 1200°C before entering the turbine. The highest temperature that argon can be heated in the regenerator is

(a) 1200°C (b) 774°C (c) 630°C (d) 567°C (e) 501°C

Answer (e) 501°C

Solution Solved by EES Software. Solutions can be verified by copying-and-pasting the following lines on a blank EES screen. (Similar problems and their solutions can be obtained easily by modifying numerical values).

```
k=1.667
Cp=0.5203 "kJ/kg.K"

P1=100 "kPa"
P2=500 "kPa"
T1=25+273 "K"
T3=1200+273 "K"

"The highest temperature that argon can be heated in the regenerator is the turbine
exit temperature,"
rp=P2/P1
T2=T1*rp^((k-1)/k)
T4=T3/rp^((k-1)/k)-273

"Some Wrong Solutions with Common Mistakes:"
W1_T4 = T3/rp "Using wrong relation"
W2_T4 = (T3-273)/rp^((k-1)/k) "Using C instead of K for T3"
W3_T4 = T4+273  "Using K instead of C"
W4_T4 = T2-273 "Taking compressor exit temp as the answer"
```

8-171 In an ideal Brayton cycle with regeneration, air is compressed from 80 kPa and 10°C to 400 kPa and 175°C, is heated to 450°C in the regenerator, and then further heated to 1000°C before entering the turbine. Under cold air standard conditions, the effectiveness of the regenerator is

(a) 33% (b) 44% (c) 62% (d) 77% (e) 89%

Answer (d) 77%

Solution Solved by EES Software. Solutions can be verified by copying-and-pasting the following lines on a blank EES screen. (Similar problems and their solutions can be obtained easily by modifying numerical values).

```
k=1.4
Cp=1.005 "kJ/kg.K"
P1=80 "kPa"
P2=400 "kPa"
T1=10+273 "K"
T2=175+273 "K"
T3=1000+273 "K"
T5=450+273 "K"

"The highest temperature that the gas can be heated in the regenerator is the
turbine exit temperature,"
rp=P2/P1
T2check=T1*rp^((k-1)/k) "Checking the given value of T2. It checks."
T4=T3/rp^((k-1)/k)
Effective=(T5-T2)/(T4-T2)

"Some Wrong Solutions with Common Mistakes:"
W1_eff = (T5-T2)/(T3-T2) "Using wrong relation"
W2_eff = (T5-T2)/(T44-T2); T44=(T3-273)/rp^((k-1)/k) "Using C instead of K for T3"
W3_eff = (T5-T2)/(T444-T2);  T444=T3/rp "Using wrong relation for T4"
```

8-172 Consider a gas turbine that has a pressure ratio of 7 and operates on the Brayton cycle with regeneration between the temperature limits of 20°C and 900°C. If the specific heat ratio of the working fluid is 1.3, the highest thermal efficiency this gas turbine can have is

(a) 40% (b) 46% (c) 61% (d) 63% (e) 97%

Answer (c) 61%

Solution Solved by EES Software. Solutions can be verified by copying-and-pasting the following lines on a blank EES screen. (Similar problems and their solutions can be obtained easily by modifying numerical values).

```
k=1.3
rp=7
T1=20+273 "K"
T3=900+273 "K"
Eta_regen=1-(T1/T3)*rp^((k-1)/k)

"Some Wrong Solutions with Common Mistakes:"
W1_Eta = 1-((T1-273)/(T3-273))*rp^((k-1)/k) "Using C for temperatures instead of K"
W2_Eta = (T1/T3)*rp^((k-1)/k)  "Using incorrect relation"
W3_Eta = 1-(T1/T3)*rp^((k1-1)/k1); k1=1.4 "Using wrong k value (the one for air)"
```

8-173 An ideal gas turbine cycle with many stages of compression and expansion and a regenerator of 100 percent effectiveness has an overall pressure ratio of 10. Air enters every stage of compressor at 290 K, and every stage of turbine at 1200 K. The thermal efficiency of this gas-turbine cycle is

(a) 36% (b) 40% (c) 52% (d) 64% (e) 76%

Answer (e) 76%

Solution Solved by EES Software. Solutions can be verified by copying-and-pasting the following lines on a blank EES screen. (Similar problems and their solutions can be obtained easily by modifying numerical values).

```
k=1.4
rp=10
T1=290 "K"
T3=1200 "K"

Eff=1-T1/T3

"Some Wrong Solutions with Common Mistakes:"
W1_Eta = 100
W2_Eta = 1-1/rp^((k-1)/k) "Using incorrect relation"
W3_Eta = 1-(T1/T3)*rp^((k-1)/k) "Using wrong relation"
W4_Eta = T1/T3 "Using wrong relation"
```

8-174 Air enters a turbojet engine at 200 m/s at a rate of 20 kg/s, and exits at 800 m/s relative to the aircraft. The thrust developed by the engine is

(a) 6 kN (b) 12 kN (c) 16 kN (d) 20 kN (e) 34 kN

Answer (b) 12 kN

Solution Solved by EES Software. Solutions can be verified by copying-and-pasting the following lines on a blank EES screen. (Similar problems and their solutions can be obtained easily by modifying numerical values).

```
Vel1=200 "m/s"
Vel2=800 "m/s"

Thrust=m*(Vel2-Vel1)/1000 "kN"
m= 20 "kg/s"

"Some Wrong Solutions with Common Mistakes:"
W1_thrust = (Vel2-Vel1)/1000 "Disregarding mass flow rate"
W2_thrust = m*Vel2/1000 "Using incorrect relation"
```

8-175 · · · 8-180 Design and Essay Problems.

Chapter 9
VAPOR AND COMBINED POWER CYCLES

Carnot Vapor Cycle

9-1C Because excessive moisture in steam causes erosion on the turbine blades. The highest moisture content allowed is about 10%.

9-2C The Carnot cycle is not a realistic model for steam power plants because (1) limiting the heat transfer processes to two-phase systems to maintain isothermal conditions severely limits the maximum temperature that can be used in the cycle, (2) the turbine will have to handle steam with a high moisture content which causes erosion, and (3) it is not practical to design a compressor that will handle two phases.

9-3E A steady-flow Carnot engine with water as the working fluid operates at specified conditions. The thermal efficiency, the quality at the end of the heat rejection process, and the net work output are to be determined.

Assumptions **1** Steady operating conditions exist. **2** Kinetic and potential energy changes are negligible.

Analysis (*a*) We note that

$$T_H = T_{\text{sat @ 120 psia}} = 341.3°\text{F} = 801.3 \text{ R}$$
$$T_L = T_{\text{sat @ 14.7 psia}} = 212°\text{F} = 672 \text{ R}$$

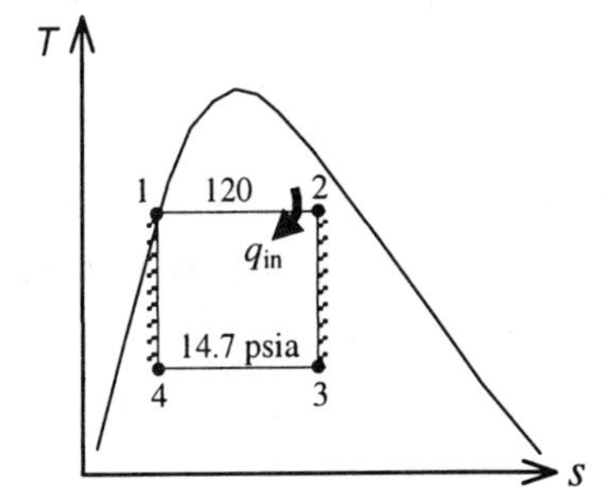

and

$$\eta_{\text{th,C}} = 1 - \frac{T_L}{T_H} = 1 - \frac{672 \text{ R}}{801.3 \text{ R}} = \mathbf{16.1\%}$$

(*b*) Noting that $s_4 = s_1 = s_{f\text{ @ 120 psia}} = 0.49201$ Btu/lbm R,

$$x_4 = \frac{s_4 - s_f}{s_{fg}} = \frac{0.49201 - 0.31212}{1.4446} = \mathbf{0.1245}$$

(*c*) The enthalpies before and after the heat addition process are

$$h_1 = h_{f\text{ @ 120 psia}} = 312.67 \text{ Btu/lbm}$$
$$h_2 = h_f + x_2 h_{fg} = 312.67 + (0.95)(878.5) = 1147.25 \text{ Btu/lbm}$$

Thus,

$$q_{\text{in}} = h_2 - h_1 = 1147.25 - 312.67 = 834.58 \text{ Btu/lbm}$$

and,

$$w_{\text{net}} = \eta_{\text{th}} q_{\text{in}} = (0.161)(834.58 \text{ Btu/lbm}) = \mathbf{134.4 \text{ Btu/lbm}}$$

9-4 A steady-flow Carnot engine with water as the working fluid operates at specified conditions. The thermal efficiency, the amount of heat rejected, and the net work output are to be determined.

Assumptions **1** Steady operating conditions exist. **2** Kinetic and potential energy changes are negligible.

Analysis (*a*) Noting that T_H = 250°C = 523 K and $T_L = T_{\text{sat @ 20 kPa}}$ = 60.06°C = 333.1 K, the thermal efficiency becomes

$$\eta_{\text{th,C}} = 1 - \frac{T_L}{T_H} = 1 - \frac{333.1\text{ K}}{523\text{ K}} = \mathbf{36.3\%}$$

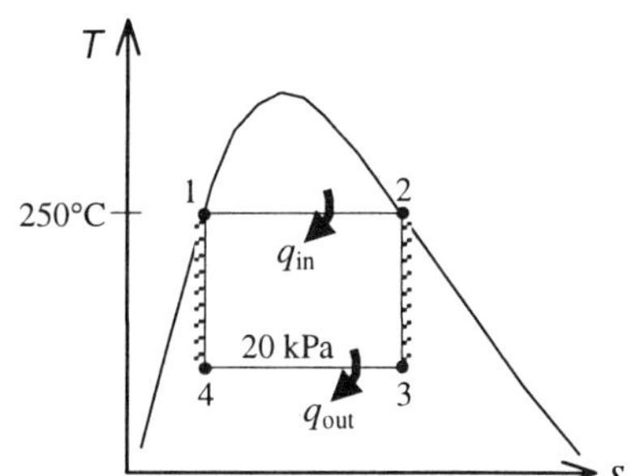

(*b*) The heat supplied during this cycle is simply the enthalpy of vaporization ,

$$q_{\text{in}} = h_{fg@\ 250°C} = 1716.2\text{ kJ/kg}$$

Thus,

$$q_{\text{out}} = q_L = \frac{T_L}{T_H} q_{\text{in}} = \left(\frac{333.1\text{ K}}{523\text{ K}}\right)(1716.2\text{ kJ/kg}) = \mathbf{1093.1\ kJ/kg}$$

(*c*) The net work output of this cycle is

$$w_{\text{net}} = \eta_{\text{th}} q_{\text{in}} = (0.363)(1716.2\text{ kJ/kg}) = \mathbf{623.0\ kJ/kg}$$

9-5 A steady-flow Carnot engine with water as the working fluid operates at specified conditions. The thermal efficiency, the amount of heat rejected, and the net work output are to be determined.

Assumptions **1** Steady operating conditions exist. **2** Kinetic and potential energy changes are negligible.

Analysis (*a*) Noting that T_H = 250°C = 523 K and $T_L = T_{\text{sat @ 10 kPa}}$ = 45.81°C = 318.8 K, the thermal efficiency becomes

$$\eta_{\text{th, C}} = 1 - \frac{T_L}{T_H} = 1 - \frac{318.8\text{ K}}{523\text{ K}} = \mathbf{39.0\%}$$

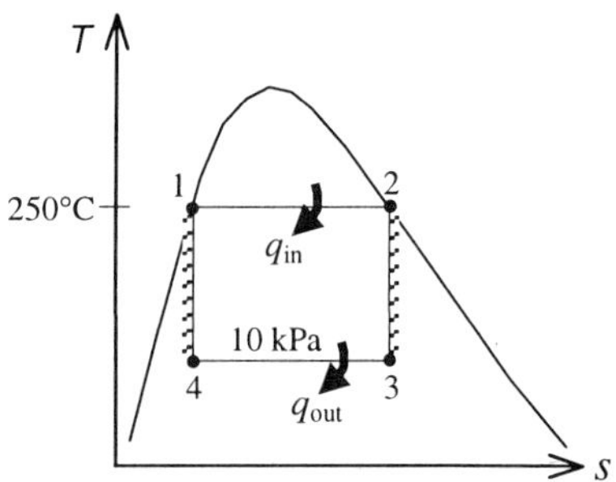

(*b*) The heat supplied during this cycle is simply the enthalpy of vaporization ,

$$q_{\text{in}} = h_{fg@\ 250°C} = 1716.2\text{ kJ/kg}$$

Thus,

$$q_{\text{out}} = q_L = \frac{T_L}{T_H} q_{\text{in}} = \left(\frac{318.8\text{K}}{523\text{ K}}\right)(1716.2\text{ kJ/kg}) = \mathbf{1046.1\ kJ/kg}$$

(*c*) The net work output of this cycle is

$$w_{\text{net}} = \eta_{\text{th}} q_{\text{in}} = (0.390)(1716.2\text{ kJ/kg}) = \mathbf{669.3\ kJ/kg}$$

9-6 A steady-flow Carnot engine with water as the working fluid operates at specified conditions. The thermal efficiency, the pressure at the turbine inlet, and the net work output are to be determined.

Assumptions **1** Steady operating conditions exist. **2** Kinetic and potential energy changes are negligible.

Analysis (*a*) The thermal efficiency is determined from

$$\eta_{\text{th,C}} = 1 - \frac{T_L}{T_H} = 1 - \frac{60 + 273\text{ K}}{350 + 273\text{ K}} = \mathbf{46.5\%}$$

(*b*) Note that $s_2 = s_3 = s_f + x_3 s_{fg}$

$$= 0.8312 + 0.891 \times 7.0784 = 7.138 \text{ kJ/kg K}$$

Thus ,

$$\left.\begin{array}{l} T_2 = 350\,^\circ\text{C} \\ s_2 = 7.138 \text{ kJ/kg} \cdot \text{K} \end{array}\right\} P_2 = 1.40 \text{ MPa}$$

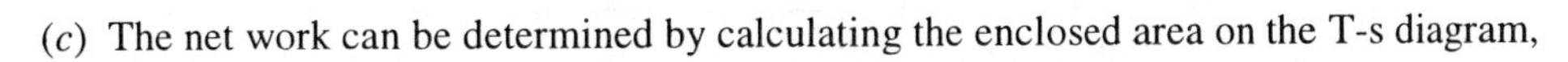

(*c*) The net work can be determined by calculating the enclosed area on the T-s diagram,

$$s_4 = s_f + x_4 s_{fg} = 0.8312 + (0.1)(7.0784) = 1.539 \text{ kJ/kg} \cdot \text{K}$$

Thus,

$$w_{\text{net}} = \text{Area} = (T_H - T_L)(s_3 - s_4) = (350 - 60)(7.138 - 1.539) = \mathbf{1624 \text{ kJ/kg}}$$

The Simple Rankine Cycle

9-7C The four processes that make up the simple ideal cycle are (1) Isentropic compression in a pump, (2) P = constant heat addition in a boiler, (3) Isentropic expansion in a turbine, and (4) P = constant heat rejection in a condenser.

9-8C Heat rejected decreases; everything else increases.

9-9C Heat rejected decreases; everything else increases.

9-10C The pump work remains the same, the moisture content decreases, everything else increases.

9-11C The actual vapor power cycles differ from the idealized ones in that the actual cycles involve friction and pressure drops in various components and the piping, and heat loss to the surrounding medium from these components and piping.

9-12C The boiler exit pressure will be (a) lower than the boiler inlet pressure in actual cycles, and (b) the same as the boiler inlet pressure in ideal cycles.

9-13C We would reject this proposal because $w_{turb} = h_1 - h_2 - q_{out}$, and any heat loss from the steam will adversely affect the turbine work output.

9-14C Yes, because the saturation temperature of steam at 10 kPa is 45.81°C, which is much higher than the temperature of the cooling water.

9-15 A steam power plant operates on a simple ideal Rankine cycle between the specified pressure limits. The thermal efficiency of the cycle and the net power output of the plant are to be determined.

Assumptions **1** Steady operating conditions exist. **2** Kinetic and potential energy changes are negligible.

Analysis (*a*) From the steam tables (Tables A-4, A-5, and A-6),

$$h_1 = h_{f\,@\,50\text{ kPa}} = 340.49\text{ kJ/kg}$$

$$v_1 = v_{f\,@\,50\text{ kPa}} = 0.001030\text{ m}^3\text{/kg}$$

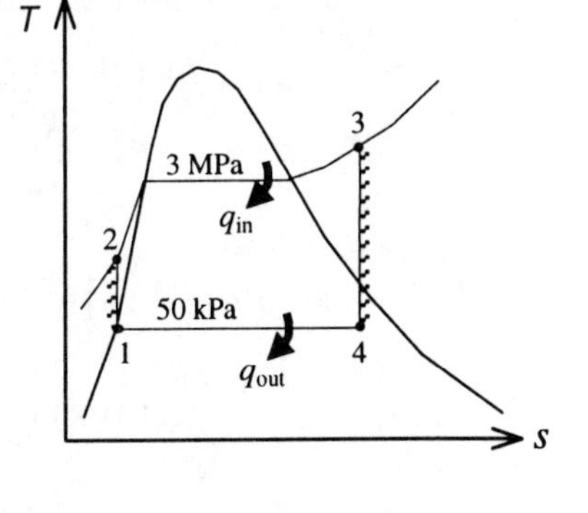

$$\begin{aligned} w_{p,\text{in}} &= v_1(P_2 - P_1) \\ &= (0.001030\text{ m}^3\text{/kg})(3000 - 50)\text{kPa}\left(\frac{1\text{ kJ}}{1\text{ kPa}\cdot\text{m}^3}\right) \\ &= 3.04\text{ kJ/kg} \end{aligned}$$

$$h_2 = h_1 + w_{p,\text{in}} = 340.49 + 3.04 = 343.53\text{ kJ/kg}$$

$$\left.\begin{aligned} P_3 &= 3\text{ MPa} \\ T_3 &= 400\,°\text{C} \end{aligned}\right\} \begin{aligned} h_3 &= 3230.9\text{ kJ/kg} \\ s_3 &= 6.9212\text{ kJ/kg}\cdot\text{K} \end{aligned}$$

$$\left.\begin{aligned} P_4 &= 50\text{ kPa} \\ s_4 &= s_3 \end{aligned}\right\} x_4 = \frac{s_4 - s_f}{s_{fg}} = \frac{6.9212 - 1.0910}{6.5029} = 0.8966$$

$$\begin{aligned} h_4 &= h_f + x_4 h_{fg} = 340.49 + (0.8966)(2305.4) \\ &= 2407.5\text{ kJ/kg} \end{aligned}$$

Thus,

$$\begin{aligned} q_{\text{in}} &= h_3 - h_2 = 3230.9 - 343.53 = 2887.37\text{ kJ/kg} \\ q_{\text{out}} &= h_4 - h_1 = 2407.5 - 340.49 = 2067.01\text{ kJ/kg} \\ w_{\text{net}} &= q_{\text{in}} - q_{\text{out}} = 2887.37 - 2067.01 = 820.36\text{ kJ/kg} \end{aligned}$$

and

$$\eta_{\text{th}} = 1 - \frac{q_{\text{out}}}{q_{\text{in}}} = 1 - \frac{2067.01}{2887.73} = \mathbf{28.4\%}$$

(*b*) $\dot{W}_{\text{net}} = \dot{m}w_{\text{net}} = (60\text{ kg/s})(820.36\text{ kJ/kg}) = \mathbf{49.2\text{ MW}}$

9-16 A steam power plant that operates on a simple ideal Rankine cycle is considered. The quality of the steam at the turbine exit, the thermal efficiency of the cycle, and the mass flow rate of the steam are to be determined.

Assumptions **1** Steady operating conditions exist. **2** Kinetic and potential energy changes are negligible.

Analysis (*a*) From the steam tables (Tables A-4, A-5, and A-6),

$$h_1 = h_{f\,@\,10\text{ kPa}} = 191.83\text{ kJ/kg}$$
$$v_1 = v_{f\,@\,10\text{ kPa}} = 0.00101\text{ m}^3\text{/kg}$$

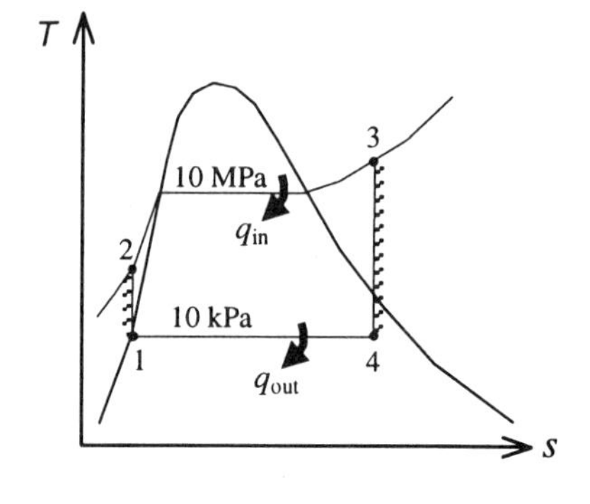

$$\begin{aligned} w_{p,\text{in}} &= v_1(P_2 - P_1) \\ &= (0.00101\text{ m}^3\text{/kg})(10{,}000 - 10\text{ kPa})\left(\frac{1\text{ kJ}}{1\text{ kPa}\cdot\text{m}^3}\right) \\ &= 10.09\text{kJ/kg} \end{aligned}$$
$$h_2 = h_1 + w_{p,\text{in}} = 191.83 + 10.09 = 201.92\text{ kJ/kg}$$

$$\left.\begin{array}{l} P_3 = 10\text{ MPa} \\ T_3 = 500\,°\text{C} \end{array}\right\} \begin{array}{l} h_3 = 3373.7\text{ kJ/kg} \\ s_3 = 6.5966\text{ kJ/kg}\cdot\text{K} \end{array}$$

$$\left.\begin{array}{l} P_4 = 10\text{ kPa} \\ s_4 = s_3 \end{array}\right\} x_4 = \frac{s_4 - s_f}{s_{fg}} = \frac{6.5966 - 0.6493}{7.5009} = 0.793$$
$$h_4 = h_f + x_4 h_{fg} = 191.83 + (0.793)(2392.8) = 2089.3\text{ kJ/kg}$$

(*b*)

$$q_{\text{in}} = h_3 - h_2 = 3373.7 - 201.92 = 3171.78\text{ kJ/kg}$$
$$q_{\text{out}} = h_4 - h_1 = 2089.3 - 191.83 = 1897.47\text{ kJ/kg}$$
$$w_{\text{net}} = q_{\text{in}} - q_{\text{out}} = 3171.78 - 1897.47 = 1274.31\text{ kJ/kg}$$

and

$$\eta_{\text{th}} = \frac{w_{\text{net}}}{q_{\text{in}}} = \frac{1274.31\text{ kJ/kg}}{3171.78\text{ kJ/kg}} = \mathbf{40.2\%}$$

(*c*)

$$\dot{m} = \frac{\dot{W}_{\text{net}}}{w_{\text{net}}} = \frac{210{,}000\text{kJ/s}}{1274.31\text{kJ/kg}} = \mathbf{165\text{ kg/s}}$$

9-17 A steam power plant that operates on a simple nonideal Rankine cycle is considered. The quality of the steam at the turbine exit, the thermal efficiency of the cycle, and the mass flow rate of the steam are to be determined.

Assumptions **1** Steady operating conditions exist. **2** Kinetic and potential energy changes are negligible.

Analysis (*a*) From the steam tables (Tables A-4, A-5, and A-6),

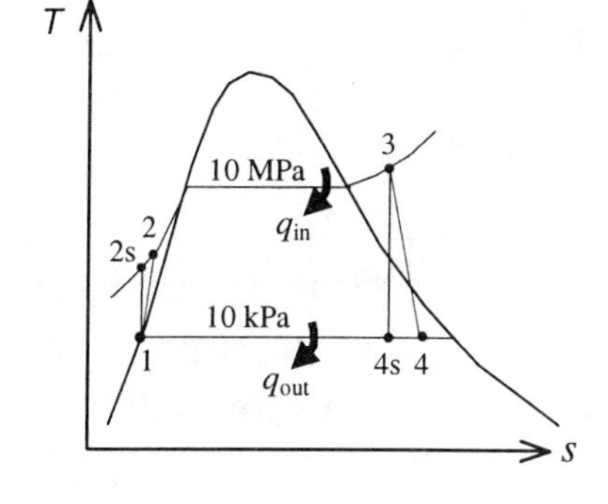

$$h_1 = h_{f\,@\,10\text{ kPa}} = 191.83\text{ kJ/kg}$$
$$v_1 = v_{f\,@\,10\text{ kPa}} = 0.00101\text{ m}^3\text{/kg}$$

$$\begin{aligned} w_{p,\text{in}} &= v_1(P_2 - P_1)/\eta_p \\ &= \left(0.00101\text{ m}^3\text{/kg}\right)\left(10{,}000 - 10\text{ kPa}\right)\left(\frac{1\text{ kJ}}{1\text{ kPa}\cdot\text{m}^3}\right)/(0.85) \\ &= 11.87\text{ kJ/kg} \end{aligned}$$
$$h_2 = h_1 + w_{p,\text{in}} = 191.83 + 11.87 = 203.70\text{ kJ/kg}$$

$$\left.\begin{array}{l} P_3 = 10\text{ MPa} \\ T_3 = 500\,^\circ\text{C} \end{array}\right\} \begin{array}{l} h_3 = 3373.7\text{ kJ/kg} \\ s_3 = 6.5966\text{ kJ/kg}\cdot\text{K} \end{array}$$

$$\left.\begin{array}{l} P_{4s} = 10\text{ kPa} \\ s_{4s} = s_3 \end{array}\right\} x_{4s} = \frac{s_{4s} - s_f}{s_{fg}} = \frac{6.5966 - 0.6493}{7.5009} = 0.793$$
$$h_{4s} = h_f + x_4 h_{fg} = 191.83 + (0.793)(2392.8) = 2089.3\text{ kJ/kg}$$

$$\eta_T = \frac{h_3 - h_4}{h_3 - h_{4s}} \longrightarrow h_4 = h_3 - \eta_T(h_3 - h_{4s})$$
$$= 3373.7 - (0.85)(3373.7 - 2089.3) = 2281.96\text{ kJ/kg}$$

$$\left.\begin{array}{l} P_4 = 10\text{ kPa} \\ h_4 = 2281.96\text{ kJ/kg} \end{array}\right\} x_4 = \mathbf{0.874}$$

(*b*)

$$q_{\text{in}} = h_3 - h_2 = 3373.7 - 203.70 = 3170.0\text{ kJ/kg}$$
$$q_{\text{out}} = h_4 - h_1 = 2281.96 - 191.83 = 2090.13\text{ kJ/kg}$$
$$w_{\text{net}} = q_{\text{in}} - q_{\text{out}} = 3170.0 - 2090.13 = 1079.87\text{ kJ/kg}$$

and

$$\eta_{\text{th}} = \frac{w_{\text{net}}}{q_{\text{in}}} = \frac{1079.87\text{ kJ/kg}}{3170.0\text{ kJ/kg}} = \mathbf{34.1\%}$$

(*c*)

$$\dot{m} = \frac{\dot{W}_{\text{net}}}{w_{\text{net}}} = \frac{210{,}000\text{ kJ/s}}{1079.87\text{ kJ/kg}} = \mathbf{194.5\text{ kg/s}}$$

9-18E A steam power plant that operates on a simple ideal Rankine cycle between the specified pressure limits is considered. The minimum turbine inlet temperature, the rate of heat input in the boiler, and the thermal efficiency of the cycle are to be determined.

Assumptions **1** Steady operating conditions exist. **2** Kinetic and potential energy changes are negligible.

Analysis (*a*) From the steam tables (Tables A-4E, A-5E, and A-6E),

$$h_1 = h_{f\,@\,2\text{ psia}} = 94.02\text{ Btu/lbm}$$
$$v_1 = v_{f\,@\,2\text{ psia}} = 0.01623\text{ ft}^3/\text{lbm}$$

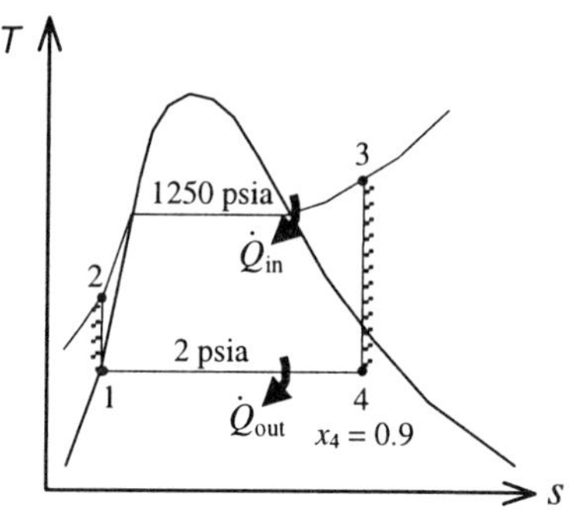

$$\begin{aligned} w_{p,\text{in}} &= v_1(P_2 - P_1) \\ &= (0.01623\text{ ft}^3/\text{lbm})(1250 - 2\text{ psia})\left(\frac{1\text{ Btu}}{5.4039\text{ psia}\cdot\text{ft}^3}\right) \\ &= 3.75\text{ Btu/lbm} \end{aligned}$$

$$h_2 = h_1 + w_{p,\text{in}} = 94.02 + 3.75 = 97.77\text{ Btu/lbm}$$

$$h_4 = h_f + x_4 h_{fg} = 94.02 + (0.9)(1022.1) = 1013.91\text{ Btu/lbm}$$
$$s_4 = s_f + x_4 s_{fg} = 0.17499 + (0.9)(1.7448) = 1.7453\text{ Btu/lbm}\cdot\text{R}$$

$$\left.\begin{array}{l} P_3 = 1250\text{ psia} \\ s_3 = s_4 \end{array}\right\} \begin{array}{l} h_3 = 1695.74\text{ Btu/lbm} \\ T_3 = \mathbf{1340.7°F} \end{array}$$

(*b*) $$\dot{Q}_{\text{in}} = \dot{m}(h_3 - h_2) = (75\text{ lbm/s})(1695.74 - 97.77) = \mathbf{119{,}848\text{ Btu/s}}$$

(*c*) $$\dot{Q}_{\text{out}} = \dot{m}(h_4 - h_1) = (75\text{ lbm/s})(1013.91 - 94.02) = 68{,}992\text{ Btu/s}$$

$$\eta_{th} = 1 - \frac{\dot{Q}_{\text{out}}}{\dot{Q}_{\text{in}}} = 1 - \frac{68{,}992\text{ Btu/s}}{119{,}848\text{ Btu/s}} = \mathbf{42.4\%}$$

9-19E A steam power plant operates on a simple nonideal Rankine cycle between the specified pressure limits. The minimum turbine inlet temperature, the rate of heat input in the boiler, and the thermal efficiency of the cycle are to be determined.

Assumptions **1** Steady operating conditions exist. **2** Kinetic and potential energy changes are negligible.

Analysis (*a*) From the steam tables (Tables A-4E, A-5E, and A-6E),

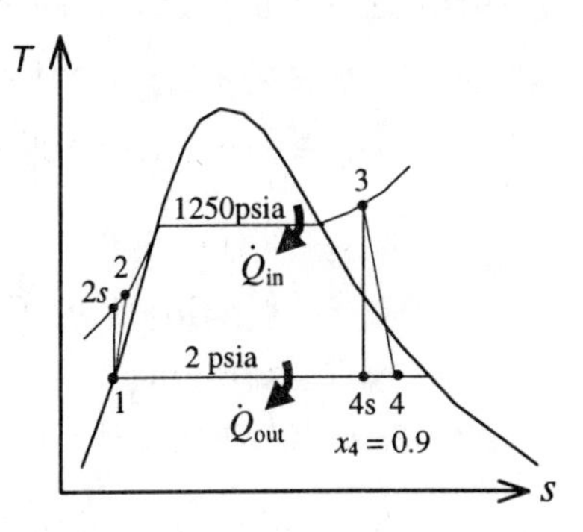

$$h_1 = h_{f\,@\,2\text{ psia}} = 94.02 \text{ Btu/lbm}$$

$$v_1 = v_{f\,@\,2\text{ psia}} = 0.01623 \text{ ft}^3\text{/lbm}$$

$$\begin{aligned} w_{p,\text{in}} &= v_1(P_2 - P_1)/\eta_P \\ &= (0.01623 \text{ ft}^3\text{/lbm})(1250 - 2 \text{ psia})\left(\frac{1 \text{ Btu}}{5.4039 \text{ psia}\cdot\text{ft}^3}\right)/0.85 \\ &= 4.41 \text{ Btu/lbm} \end{aligned}$$

$$h_2 = h_1 + w_{p,\text{in}} = 94.02 + 4.41 = 98.43 \text{ Btu/lbm}$$

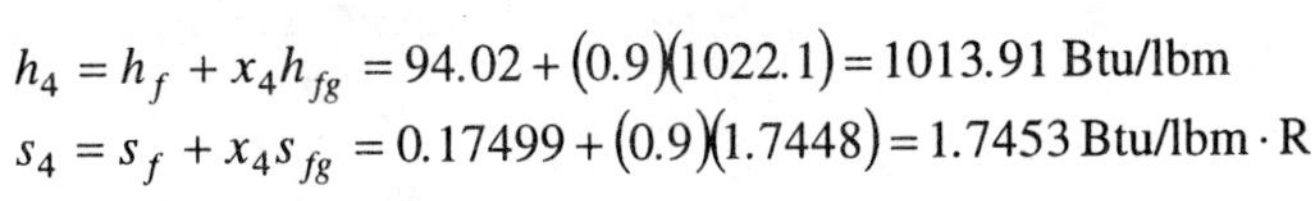

$$h_4 = h_f + x_4 h_{fg} = 94.02 + (0.9)(1022.1) = 1013.91 \text{ Btu/lbm}$$

$$s_4 = s_f + x_4 s_{fg} = 0.17499 + (0.9)(1.7448) = 1.7453 \text{ Btu/lbm}\cdot\text{R}$$

The turbine inlet temperature is determined by trial and error ,

Try 1:
$$\left.\begin{array}{l} P_3 = 1250 \text{ psia} \\ T_3 = 900°\text{F} \end{array}\right\} \begin{array}{l} h_3 = 1438.4 \text{ Btu/lbm} \\ s_3 = 1.582 \text{ Btu/lbm.R} \end{array}$$

$$x_{4s} = \frac{s_{4s} - s_f}{s_{fg}} = \frac{s_3 - s_f}{s_{fg}} = \frac{1.582 - 0.17499}{1.7448} = 0.8064$$

$$h_{4s} = h_f + x_{4s} h_{fg} = 94.02 + (0.8064)(1022.1) = 918.2 \text{ Btu/lbm}$$

$$\eta_T = \frac{h_3 - h_4}{h_3 - h_{4s}} = \frac{1438.4 - 1013.91}{1438.4 - 918.2} = 0.8160$$

Try 2:
$$\left.\begin{array}{l} P_3 = 1250 \text{ psia} \\ T_3 = 1000°\text{F} \end{array}\right\} \begin{array}{l} h_3 = 1498.2 \text{ Btu/lbm} \\ s_3 = 1.6244 \text{ Btu/lbm.R} \end{array}$$

$$x_{4s} = \frac{s_{4s} - s_f}{s_{fg}} = \frac{s_3 - s_f}{s_{fg}} = \frac{1.6244 - 0.17499}{1.7448} = 0.8307$$

$$h_{4s} = h_f + x_{4s} h_{fg} = 94.02 + (0.8307)(1022.1) = 943.1 \text{ Btu/lbm}$$

$$\eta_T = \frac{h_3 - h_4}{h_3 - h_{4s}} = \frac{1498.2 - 1013.91}{1498.2 - 943.1} = 0.8724$$

By linear interpolation, at $\eta_T = 0.85$ we obtain $T_3 =$ **960.3°F**. Also, $h_3 = 1474.4$ Btu/lbm.

(*b*) $\dot{Q}_{\text{in}} = \dot{m}(h_3 - h_2) = (75 \text{ lbm/s})(1474.4 - 98.43) = \mathbf{103{,}200\ Btu/s}$

(*c*) $\dot{Q}_{\text{out}} = \dot{m}(h_4 - h_1) = (75 \text{ lbm/s})(1013.91 - 94.02) = 69{,}000 \text{ Btu/s}$

$$\eta_{th} = 1 - \frac{\dot{Q}_{\text{out}}}{\dot{Q}_{\text{in}}} = 1 - \frac{69{,}000 \text{ Btu/s}}{103{,}200 \text{ Btu/s}} = \mathbf{33.1\%}$$

9-20 A 300-MW coal-fired steam power plant operates on a simple ideal Rankine cycle between the specified pressure limits. The overall plant efficiency and the required rate of the coal supply are to be determined.

Assumptions **1** Steady operating conditions exist. **2** Kinetic and potential energy changes are negligible.

Analysis (*a*) From the steam tables (Tables A-4, A-5, and A-6),

$$h_1 = h_{f\,@\,25\text{ kPa}} = 271.93\text{ kJ/kg}$$
$$v_1 = v_{f\,@\,25\text{ kPa}} = 0.00102\text{ m}^3\text{/kg}$$

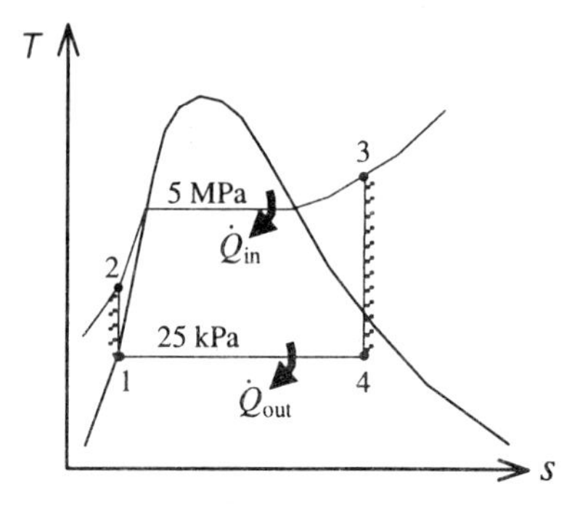

$$\begin{aligned} w_{p,\text{in}} &= v_1(P_2 - P_1) \\ &= \left(0.00102\text{ m}^3\text{/kg}\right)\left(5{,}000 - 25\text{ kPa}\right)\left(\frac{1\text{ kJ}}{1\text{ kPa}\cdot\text{m}^3}\right) \\ &= 5.07\text{ kJ/kg} \end{aligned}$$
$$h_2 = h_1 + w_{p,\text{in}} = 271.93 + 5.07 = 277.0\text{ kJ/kg}$$

$$\left.\begin{array}{l} P_3 = 5\text{ MPa} \\ T_3 = 450°\text{C} \end{array}\right\} \begin{array}{l} h_3 = 3316.2\text{ kJ/kg} \\ s_3 = 6.8186\text{ kJ/kg}\cdot\text{K} \end{array}$$

$$\left.\begin{array}{l} P_4 = 25\text{ kPa} \\ s_4 = s_3 \end{array}\right\} x_4 = \frac{s_4 - s_f}{s_{fg}} = \frac{6.8186 - 0.8931}{6.9383} = 0.8540$$
$$h_4 = h_f + x_4 h_{fg} = 271.93 + (0.8540)(2346.3) = 2275.7\text{ kJ/kg}$$

The thermal efficiency is determined from

$$q_{\text{in}} = h_3 - h_2 = 3316.2 - 277.0 = 3039.2\text{ kJ/kg}$$
$$q_{\text{out}} = h_4 - h_1 = 2275.7 - 271.93 = 2003.8\text{ kJ/kg}$$

and

$$\eta_{th} = 1 - \frac{q_{\text{out}}}{q_{\text{in}}} = 1 - \frac{2003.8}{3039.2} = 34.1\%$$

Thus,

$$\eta_{\text{overall}} = \eta_{th} \times \eta_{\text{comb}} \times \eta_{\text{gen}} = (0.341)(0.75)(0.96) = \mathbf{24.6\%}$$

(*b*) Then the required rate of coal supply becomes

$$\dot{Q}_{\text{in}} = \frac{\dot{W}_{\text{net}}}{\eta_{\text{overall}}} = \frac{300{,}000\text{ kJ/s}}{0.246} = 1{,}219{,}500\text{ kJ/s}$$

and

$$\dot{m}_{\text{coal}} = \frac{\dot{Q}_{\text{in}}}{C_{\text{coal}}} = \frac{1{,}219{,}500\text{ kJ/s}}{29{,}300\text{ kJ/kg}}\left(\frac{1\text{ ton}}{1000\text{ kg}}\right) = 0.04162\text{ tons/s} = \mathbf{149.8\text{ tons/h}}$$

9-21 A solar-pond power plant that operates on a simple ideal Rankine cycle with refrigerant-134a as the working fluid is considered. The thermal efficiency of the cycle and the power output of the plant are to be determined.

Assumptions **1** Steady operating conditions exist. **2** Kinetic and potential energy changes are negligible.

Analysis (*a*) From the refrigerant tables (Tables A-11, A-12, and A-13),

$$h_1 = h_{f\,@\,0.7\text{ MPa}} = 86.78 \text{ kJ/kg}$$

$$v_1 = v_{f\,@\,0.7\text{ MPa}} = 0.0008328 \text{ m}^3\text{/kg}$$

$$w_{p,\text{in}} = v_1(P_2 - P_1)$$

$$= (0.0008328 \text{ m}^3\text{/kg})(1600 - 700 \text{ kPa})\left(\frac{1 \text{ kJ}}{1 \text{ kPa}\cdot\text{m}^3}\right)$$

$$= 0.75 \text{ kJ/kg}$$

$$h_2 = h_1 + w_{p,\text{in}} = 86.78 + 0.75 = 87.53 \text{ kJ/kg}$$

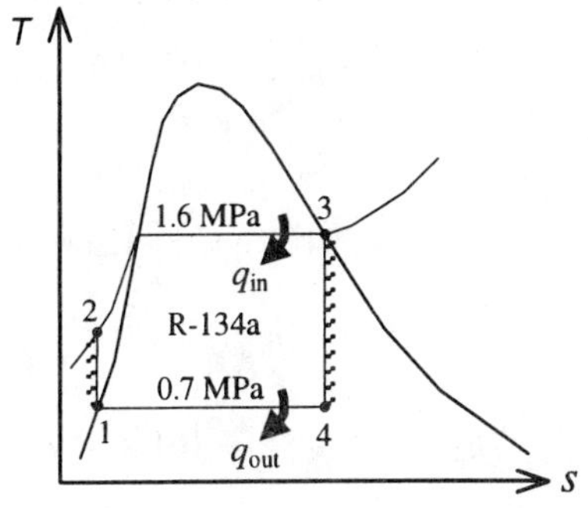

$$\left.\begin{array}{l} P_3 = 1.6 \text{ MPa} \\ \text{sat.vapor} \end{array}\right\} \begin{array}{l} h_3 = h_{g\,@\,1.6\text{ MPa}} = 275.33 \text{ kJ/kg} \\ s_3 = s_{g\,@\,1.6\text{ MPa}} = 0.8982 \text{ kJ/kg}\cdot\text{K} \end{array}$$

$$\left.\begin{array}{l} P_4 = 0.7 \text{ MPa} \\ s_4 = s_3 \end{array}\right\} x_4 = \frac{s_4 - s_f}{s_{fg}} = \frac{0.8982 - 0.3242}{0.9080 - 0.3242} = 0.983$$

$$h_4 = h_f + x_4 h_{fg} = 86.78 + (0.983)(175.07) = 258.87 \text{ kJ/kg}$$

Thus ,

$$q_{\text{in}} = h_3 - h_2 = 275.33 - 87.53 = 187.80 \text{ kJ/kg}$$

$$q_{\text{out}} = h_4 - h_1 = 258.87 - 86.78 = 172.09 \text{ kJ/kg}$$

$$w_{\text{net}} = q_{\text{in}} - q_{\text{out}} = 187.80 - 172.09 = 15.71 \text{ kJ/kg}$$

and

$$\eta_{th} = \frac{w_{\text{net}}}{q_{\text{in}}} = \frac{15.71 \text{ kJ/kg}}{187.80 \text{ kJ/kg}} = \mathbf{8.4\%}$$

(*b*) $\dot{W}_{\text{net}} = \dot{m} w_{\text{net}} = (6 \text{ kg/s})(15.71 \text{ kJ/kg}) = \mathbf{94.26 \text{ kW}}$

9-22 A steam power plant operates on a simple ideal Rankine cycle between the specified pressure limits. The thermal efficiency of the cycle, the mass flow rate of the steam, and the temperature rise of the cooling water are to be determined.

Assumptions **1** Steady operating conditions exist. **2** Kinetic and potential energy changes are negligible.

Analysis (*a*) From the steam tables (Tables A-4, A-5, and A-6),

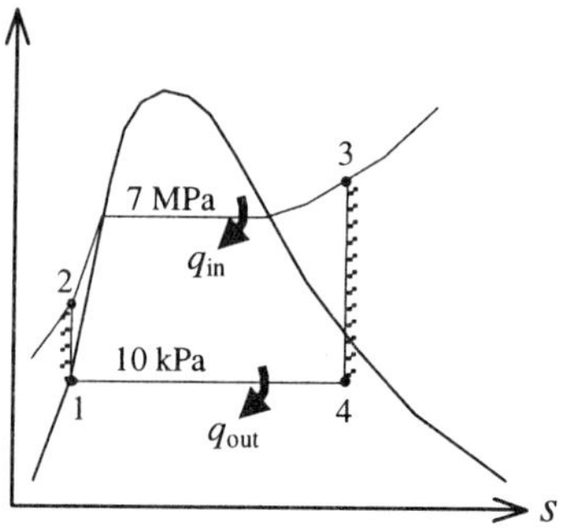

$$h_1 = h_{f @ 10\text{ kPa}} = 191.83 \text{ kJ/kg}$$

$$v_1 = v_{f @ 10\text{ kPa}} = 0.00101 \text{ m}^3\text{/kg}$$

$$w_{p,\text{in}} = v_1(P_2 - P_1)$$

$$= (0.00101 \text{ m}^3\text{/kg})(7{,}000 - 10 \text{ kPa})\left(\frac{1 \text{ kJ}}{1 \text{ kPa}\cdot\text{m}^3}\right)$$

$$= 7.06 \text{ kJ/kg}$$

$$h_2 = h_1 + w_{p,\text{in}} = 191.83 + 7.06 = 198.89 \text{ kJ/kg}$$

$$\left.\begin{array}{l} P_3 = 7 \text{ MPa} \\ T_3 = 500°\text{C} \end{array}\right\} \begin{array}{l} h_3 = 3410.5 \text{ kJ/kg} \\ s_3 = 6.7975 \text{ kJ/kg}\cdot\text{K} \end{array}$$

$$\left.\begin{array}{l} P_4 = 10 \text{ kPa} \\ s_4 = s_3 \end{array}\right\} x_4 = \frac{s_4 - s_f}{s_{fg}} = \frac{6.7975 - 0.6493}{7.5009} = 0.820$$

$$h_4 = h_f + x_4 h_{fg} = 191.83 + (0.820)(2392.8) = 2153.93 \text{ kJ/kg}$$

Thus,

$$q_{\text{in}} = h_3 - h_2 = 3410.5 - 198.89 = 3211.61 \text{ kJ/kg}$$

$$q_{\text{out}} = h_4 - h_1 = 2153.93 - 191.83 = 1962.10 \text{ kJ/kg}$$

$$w_{\text{net}} = q_{\text{in}} - q_{\text{out}} = 3211.61 - 1962.10 = 1249.51 \text{ kJ/kg}$$

and

$$\eta_{th} = \frac{w_{\text{net}}}{q_{\text{in}}} = \frac{1249.51 \text{ kJ/kg}}{3211.61 \text{ kJ/kg}} = \mathbf{38.9\%}$$

(*b*)

$$\dot{m} = \frac{\dot{W}_{\text{net}}}{w_{\text{net}}} = \frac{45{,}000 \text{ kJ/s}}{1249.51 \text{ kJ/kg}} = \mathbf{36.0 \text{ kg/s}}$$

(*c*) The rate of heat rejection to the cooling water and its temperature rise are

$$\dot{Q}_{\text{out}} = \dot{m}q_{\text{out}} = (36.0 \text{ kg/s})(1962.1 \text{ kJ/kg}) = 70{,}636 \text{ kJ/s}$$

$$\Delta T_{\text{cooling water}} = \frac{\dot{Q}_{\text{out}}}{(\dot{m}C)_{\text{cooling water}}} = \frac{70{,}636 \text{ kJ/s}}{(2000 \text{ kg/s})(4.18 \text{ kJ/kg}\cdot°\text{C})} = \mathbf{8.45°C}$$

9-23 A steam power plant operates on a simple nonideal Rankine cycle between the specified pressure limits. The thermal efficiency of the cycle, the mass flow rate of the steam, and the temperature rise of the cooling water are to be determined.

Assumptions **1** Steady operating conditions exist. **2** Kinetic and potential energy changes are negligible.

Analysis (*a*) From the steam tables (Tables A-4, A-5, and A-6),

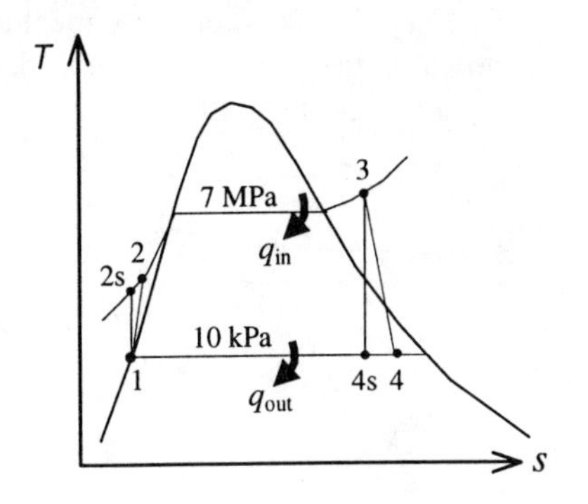

$$h_1 = h_{f@\,10\text{ kPa}} = 191.83\text{ kJ/kg}$$

$$v_1 = v_{f@\,10\text{ kPa}} = 0.00101\text{ m}^3\text{/kg}$$

$$\begin{aligned} w_{p,\text{in}} &= v_1(P_2 - P_1)/\eta_p \\ &= (0.00101\text{ m}^3\text{/kg})(7{,}000 - 10\text{ kPa})\left(\frac{1\text{ kJ}}{1\text{ kPa}\cdot\text{m}^3}\right)/(0.87) \\ &= 8.11\text{ kJ/kg} \end{aligned}$$

$$h_2 = h_1 + w_{p,\text{in}} = 191.83 + 8.11 = 199.94\text{ kJ/kg}$$

$$\left.\begin{array}{l} P_3 = 7\text{ MPa} \\ T_3 = 500°\text{C} \end{array}\right\} \begin{array}{l} h_3 = 3410.5\text{ kJ/kg} \\ s_3 = 6.7975\text{ kJ/kg}\cdot\text{K} \end{array}$$

$$\left.\begin{array}{l} P_4 = 10\text{ kPa} \\ s_4 = s_3 \end{array}\right\} x_4 = \frac{s_4 - s_f}{s_{fg}} = \frac{6.7975 - 0.6493}{7.5009} = 0.820$$

$$h_{4s} = h_f + x_4 h_{fg} = 191.83 + (0.820)(2392.8) = 2153.93\text{ kJ/kg}$$

$$\eta_T = \frac{h_3 - h_4}{h_3 - h_{4s}} \longrightarrow h_4 = h_3 - \eta_T(h_3 - h_{4s}) = 3410.5 - (0.87)(3410.5 - 2153.93) = 2317.28\text{ kJ/kg}$$

Thus,

$$q_{in} = h_3 - h_2 = 3410.5 - 199.94 = 3210.56\text{ kJ/kg}$$

$$q_{out} = h_4 - h_1 = 2317.28 - 191.83 = 2125.45\text{ kJ/kg}$$

$$w_{net} = q_{in} - q_{out} = 3210.56 - 2125.45 = 1085.11\text{ kJ/kg}$$

and

$$\eta_{th} = \frac{w_{net}}{q_{in}} = \frac{1085.11\text{ kJ/kg}}{3210.56\text{ kJ/kg}} = \mathbf{33.8\%}$$

(*b*)

$$\dot{m} = \frac{\dot{W}_{net}}{w_{net}} = \frac{45{,}000\text{ kJ/s}}{1085.11\text{ kJ/kg}} = \mathbf{41.5\text{ kg/s}}$$

(*c*) The rate of heat rejection to the cooling water and its temperature rise are

$$\dot{Q}_{out} = \dot{m}q_{out} = (41.5\text{ kg/s})(2125.45\text{ kJ/kg}) = 88{,}143\text{ kJ/s}$$

$$\Delta T_{\text{cooling water}} = \frac{\dot{Q}_{out}}{(\dot{m}C)_{\text{cooling water}}} = \frac{88{,}143\text{ kJ/s}}{(2000\text{ kg/s})(4.18\text{ kJ/kg}\cdot°\text{C})} = \mathbf{10.54°C}$$

The Reheat Rankine Cycle

9-24C The pump work remains the same, the moisture content decreases, everything else increases.

9-25C The T-s diagram of the ideal Rankine cycle with 3 stages of reheat is shown on the side. The cycle efficiency will increase as the number of reheating stages increases.

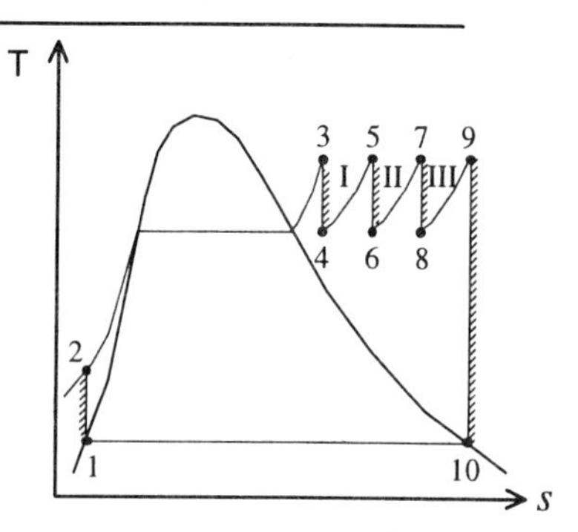

9-26C The thermal efficiency of the simple ideal Rankine cycle will probably be higher since the average temperature at which heat is added will be higher in this case.

9-27 [*Also solved by EES on enclosed CD*] A steam power plant that operates on the ideal reheat Rankine cycle is considered. The turbine work output and the thermal efficiency of the cycle are to be determined.

Assumptions **1** Steady operating conditions exist. **2** Kinetic and potential energy changes are negligible.

Analysis From the steam tables (Tables A-4, A-5, and A-6),

$$h_1 = h_{f\,@\,20\text{ kPa}} = 251.40\text{ kJ/kg}$$

$$v_1 = v_{f\,@\,20\text{ kPa}} = 0.001017\text{ m}^3\text{/kg}$$

$$\begin{aligned} w_{p,\text{in}} &= v_1(P_2 - P_1) \\ &= (0.001017\text{ m}^3\text{/kg})(8{,}000 - 20\text{ kPa})\left(\frac{1\text{ kJ}}{1\text{ kPa}\cdot\text{m}^3}\right) \\ &= 8.12\text{ kJ/kg} \end{aligned}$$

$$h_2 = h_1 + w_{p,\text{in}} = 251.40 + 8.12 = 259.52\text{ kJ/kg}$$

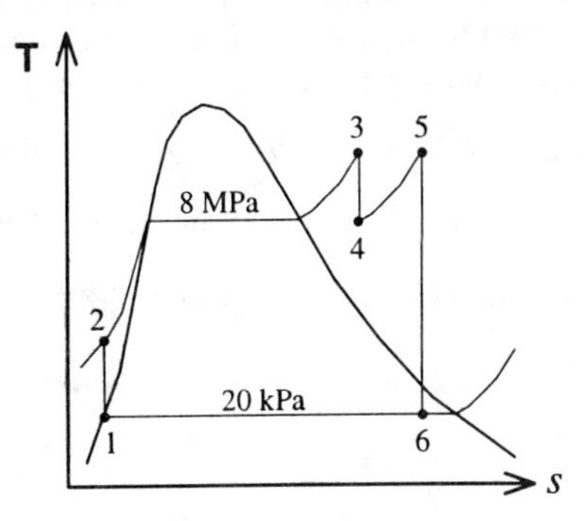

$$\left.\begin{array}{l} P_3 = 8\text{ MPa} \\ T_3 = 500°\text{C} \end{array}\right\} \begin{array}{l} h_3 = 3398.3\text{ kJ/kg} \\ s_3 = 6.7240\text{ kJ/kg}\cdot\text{K} \end{array}$$

$$\left.\begin{array}{l} P_4 = 3\text{ MPa} \\ s_4 = s_3 \end{array}\right\} h_4 = 3104.1\text{ kJ/kg}$$

$$\left.\begin{array}{l} P_5 = 3\text{ MPa} \\ T_5 = 500°\text{C} \end{array}\right\} \begin{array}{l} h_5 = 3456.5\text{ kJ/kg} \\ s_5 = 7.2338\text{ kJ/kg}\cdot\text{K} \end{array}$$

$$\left.\begin{array}{l} P_6 = 20\text{ kPa} \\ s_6 = s_5 \end{array}\right\} \begin{array}{l} x_6 = \dfrac{s_6 - s_f}{s_{fg}} = \dfrac{7.2338 - 0.8320}{7.0766} = 0.9046 \\ h_6 = h_f + x_6 h_{fg} = 251.40 + (0.9046)(2358.3) = 2384.7\text{ kJ/kg} \end{array}$$

The turbine work output and the thermal efficiency are determined from

$$w_{\text{T,out}} = (h_3 - h_4) + (h_5 - h_6) = 3398.3 - 3104.1 + 3456.5 - 2384.7 = \mathbf{1366\ kJ/kg}$$

and

$$q_{\text{in}} = (h_3 - h_2) + (h_5 - h_4) = 3398.3 - 259.52 + 3456.5 - 3104.1 = 3491.2\text{ kJ/kg}$$

$$w_{\text{net}} = w_{T,out} - w_{p,\text{in}} = 1366 - 8.12 = 1357.88\text{ kJ/kg}$$

Thus,

$$\eta_{th} = \frac{w_{\text{net}}}{q_{\text{in}}} = \frac{1357.88\text{ kJ/kg}}{3491.2\text{ kJ/kg}} = \mathbf{38.9\%}$$

9-28 EES solution of this (and other comprehensive problems designated with the *computer icon*) is available to instructors at the *Instructor Manual* section of the *Online Learning Center* (OLC) at www.mhhe.com/cengel-boles. See the Preface for access information.

9-29 A steam power plant that operates on a reheat Rankine cycle is considered. The quality (or temperature, if superheated) of the steam at the turbine exit, the thermal efficiency of the cycle, and the mass flow rate of the steam are to be determined.

Assumptions **1** Steady operating conditions exist. **2** Kinetic and potential energy changes are negligible.

Analysis (*a*) From the steam tables (Tables A-4, A-5, and A-6),

$$h_1 = h_{f\,@\,10\text{ kPa}} = 191.83 \text{ kJ/kg}$$

$$v_1 = v_{f\,@\,10\text{ kPa}} = 0.00101 \text{ m}^3\text{/kg}$$

$$\begin{aligned} w_{p,\text{in}} &= v_1(P_2 - P_1)/\eta_p \\ &= \left(0.00101 \text{ m}^3\text{/kg}\right)\left(10{,}000 - 10 \text{ kPa}\right)\left(\frac{1 \text{ kJ}}{1 \text{ kPa}\cdot\text{m}^3}\right)/(0.95) \\ &= 10.62 \text{ kJ/kg} \end{aligned}$$

$$h_2 = h_1 + w_{p,\text{in}} = 191.83 + 10.62 = 202.45 \text{ kJ/kg}$$

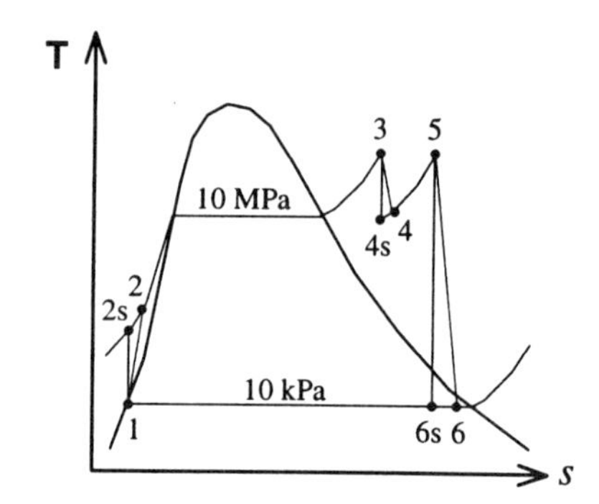

$$\left.\begin{array}{l} P_3 = 10 \text{ MPa} \\ T_3 = 500°\text{C} \end{array}\right\} \begin{array}{l} h_3 = 3373.7 \text{ kJ/kg} \\ s_3 = 6.5966 \text{ kJ/kg}\cdot\text{K} \end{array}$$

$$\left.\begin{array}{l} P_{4s} = 1 \text{ MPa} \\ s_{4s} = s_3 \end{array}\right\} h_{4s} = 2782.5 \text{ kJ/kg}$$

$$\begin{aligned} \eta_T = \frac{h_3 - h_4}{h_3 - h_{4s}} \longrightarrow h_4 &= h_3 - \eta_T\left(h_3 - h_{4s}\right) \\ &= 3373.7 - (0.80)(3373.7 - 2782.5) = 2900.74 \text{ kJ/kg} \end{aligned}$$

$$\left.\begin{array}{l} P_5 = 1 \text{ MPa} \\ T_5 = 500°\text{C} \end{array}\right\} \begin{array}{l} h_5 = 3478.5 \text{ kJ/kg} \\ s_5 = 7.7622 \text{ kJ/kg}\cdot\text{K} \end{array}$$

$$\left.\begin{array}{l} P_{6s} = 10 \text{ kPa} \\ s_{6s} = s_5 \end{array}\right\} \begin{array}{l} x_{6s} = \dfrac{s_{6s} - s_f}{s_{fg}} = \dfrac{7.7622 - 0.6493}{7.5009} = 0.948 \ \text{(at turbine exit)} \\ h_{6s} = h_f + x_{6s}h_{fg} = 191.83 + (0.948)(2392.8) = 2460.2 \text{ kJ/kg} \end{array}$$

$$\begin{aligned} \eta_T = \frac{h_5 - h_6}{h_5 - h_{6s}} \longrightarrow h_6 &= h_5 - \eta_T\left(h_5 - h_{6s}\right) \\ &= 3478.5 - (0.80)(3478.5 - 2460.2) \\ &= 2663.86 \text{ kJ/kg} > h_g \ \text{(superheated vapor)} \end{aligned}$$

From steam tables at 10 kPa we read T_6 = **87.5°C**.

(*b*) $$w_{\text{T,out}} = (h_3 - h_4) + (h_5 - h_6) = 3373.7 - 2900.74 + 3478.5 - 2663.86 = 1287.6 \text{ kJ/kg}$$

$$q_{\text{in}} = (h_3 - h_2) + (h_5 - h_4) = 3373.7 - 202.45 + 3478.5 - 2900.74 = 3749.0 \text{ kJ/kg}$$

$$w_{\text{net}} = w_{\text{T,out}} - w_{\text{p,in}} = 1287.6 - 10.62 = 1277.0 \text{ kJ/kg}$$

Thus the thermal efficiency is $$\eta_{th} = \frac{w_{\text{net}}}{q_{\text{in}}} = \frac{1277.0 \text{ kJ/kg}}{3749.0 \text{ kJ/kg}} = \mathbf{34.1\%}$$

(*c*) The mass flow rate of the steam is $$\dot{m} = \frac{\dot{W}_{\text{net}}}{w_{\text{net}}} = \frac{80{,}000 \text{ kJ/s}}{1277.0 \text{ kJ/kg}} = \mathbf{62.6 \text{ kg/s}}$$

9-30 A steam power plant that operates on the ideal reheat Rankine cycle is considered. The quality (or temperature, if superheated) of the steam at the turbine exit, the thermal efficiency of the cycle, and the mass flow rate of the steam are to be determined.

Assumptions **1** Steady operating conditions exist. **2** Kinetic and potential energy changes are negligible.

Analysis (*a*) From the steam tables (Tables A-4, A-5, and A-6),

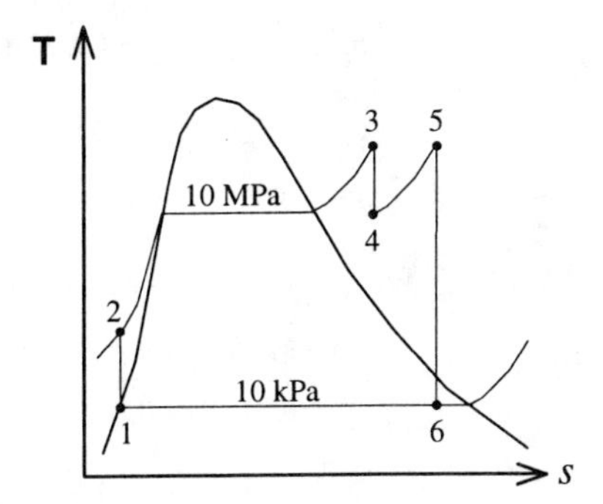

$$h_1 = h_{f\,@\,10\text{ kPa}} = 191.83 \text{ kJ/kg}$$

$$v_1 = v_{f\,@\,10\text{ kPa}} = 0.00101 \text{ m}^3\text{/kg}$$

$$\begin{aligned} w_{p,\text{in}} &= v_1(P_2 - P_1) \\ &= (0.00101 \text{ m}^3\text{/kg})(10{,}000 - 10 \text{ kPa})\left(\frac{1 \text{ kJ}}{1 \text{ kPa}\cdot\text{m}^3}\right) \\ &= 10.09 \text{ kJ/kg} \end{aligned}$$

$$h_2 = h_1 + w_{p,\text{in}} = 191.83 + 10.09 = 201.92 \text{ kJ/kg}$$

$$\left.\begin{aligned} P_3 &= 10 \text{ MPa} \\ T_3 &= 500°\text{C} \end{aligned}\right\} \begin{aligned} h_3 &= 3373.7 \text{ kJ/kg} \\ s_3 &= 6.5966 \text{ kJ/kg}\cdot\text{K} \end{aligned}$$

$$\left.\begin{aligned} P_4 &= 1 \text{ MPa} \\ s_4 &= s_3 \end{aligned}\right\} h_4 = 2782.5 \text{ kJ/kg}$$

$$\left.\begin{aligned} P_5 &= 1 \text{ MPa} \\ T_5 &= 500°\text{C} \end{aligned}\right\} \begin{aligned} h_5 &= 3478.5 \text{ kJ/kg} \\ s_5 &= 7.7622 \text{ kJ/kg}\cdot\text{K} \end{aligned}$$

$$\left.\begin{aligned} P_6 &= 10 \text{ kPa} \\ s_6 &= s_5 \end{aligned}\right\} \begin{aligned} x_6 &= \frac{s_6 - s_f}{s_{fg}} = \frac{7.7622 - 0.6493}{7.5009} = \mathbf{0.948} \text{ (at turbine exit)} \\ h_6 &= h_f + x_6 h_{fg} = 191.83 + (0.948)(2392.8) = 2460.2 \text{ kJ/kg} \end{aligned}$$

(*b*)

$$w_{\text{T,out}} = (h_3 - h_4) + (h_5 - h_6) = 3373.7 - 2782.5 + 3478.5 - 2460.2 = 1609.5 \text{ kJ/kg}$$

$$q_{\text{in}} = (h_3 - h_2) + (h_5 - h_4) = 3373.7 - 201.92 + 3478.5 - 2782.5 = 3867.78 \text{ kJ/kg}$$

$$w_{\text{net}} = w_{\text{T,out}} - w_{p,\text{in}} = 1609.5 - 10.09 = 1599.41 \text{ kJ/kg}$$

Thus the thermal efficiency is

$$\eta_{th} = \frac{w_{\text{net}}}{q_{\text{in}}} = \frac{1599.41 \text{ kJ/kg}}{3867.78 \text{ kJ/kg}} = \mathbf{41.4\%}$$

(*c*) The mass flow rate of the steam is

$$\dot{m} = \frac{\dot{W}_{\text{net}}}{w_{\text{net}}} = \frac{80{,}000 \text{ kJ/s}}{1599.41 \text{ kJ/kg}} = \mathbf{50.0 \text{ kg/s}}$$

9-31E A steam power plant that operates on the ideal reheat Rankine cycle is considered. The pressure at which reheating takes place, the net power output, the thermal efficiency, and the minimum mass flow rate of the cooling water required are to be determined.

Assumptions **1** Steady operating conditions exist. **2** Kinetic and potential energy changes are negligible.

Analysis (*a*) From the steam tables (Tables A-4E, A-5E, and A-6E),

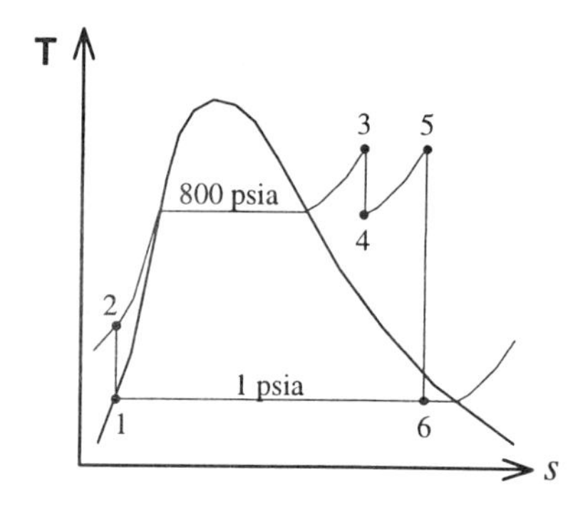

$$h_1 = h_{\text{sat@ 1 psia}} = 69.74 \text{ Btu/lbm}$$

$$v_1 = v_{\text{sat@ 1 psia}} = 0.016136 \text{ ft}^3/\text{lbm}$$

$$T_1 = T_{\text{sat@ 1 psia}} = 101.70°\text{F}$$

$$\begin{aligned} w_{p,\text{in}} &= v_1(P_2 - P_1) \\ &= (0.016136 \text{ ft}^3/\text{lbm})(800 - 1 \text{ psia})\left(\frac{1 \text{ Btu}}{5.4039 \text{ psia}\cdot\text{ft}^3}\right) \\ &= 2.39 \text{ Btu/lbm} \end{aligned}$$

$$h_2 = h_1 + w_{p,\text{in}} = 69.74 + 2.39 = 72.13 \text{ Btu/lbm}$$

$$\left.\begin{aligned} P_3 &= 800 \text{ psia} \\ T_3 &= 900°\text{F} \end{aligned}\right\} \begin{aligned} h_3 &= 1455.6 \text{ Btu/lbm} \\ s_3 &= 1.6408 \text{ Btu/lbm}\cdot\text{R} \end{aligned}$$

$$\left.\begin{aligned} &s_4 = s_3 \\ &(\text{sat.vapor}) \end{aligned}\right\} \begin{aligned} h_4 &= h_{g@s_g=s_4} = 1178.9 \text{ Btu/lbm} \\ P_4 &= P_{\text{sat}@s_g=s_4} = \mathbf{62.81\ psia} \quad \text{(the reheat pressure)} \end{aligned}$$

$$\left.\begin{aligned} P_5 &= 62.81 \text{ psia} \\ T_5 &= 800°\text{F} \end{aligned}\right\} \begin{aligned} h_5 &= 1431.1 \text{ Btu/lbm} \\ s_5 &= 1.8977 \text{ Btu/lbm}\cdot\text{R} \end{aligned}$$

$$\left.\begin{aligned} P_6 &= 1 \text{ psia} \\ s_6 &= s_5 \end{aligned}\right\} \begin{aligned} x_6 &= \frac{s_6 - s_f}{s_{fg}} = \frac{1.8977 - 0.13266}{1.8453} = 0.9565 \\ h_6 &= h_f + x_6 h_{fg} = 69.74 + (0.9565)(1036) = 1060.7 \text{ Btu/lbm} \end{aligned}$$

(*b*) $$q_{\text{in}} = (h_3 - h_2) + (h_5 - h_4) = 1455.6 - 72.13 + 1431.1 - 1178.9 = 1635.7 \text{ Btu/lbm}$$

$$q_{\text{out}} = h_6 - h_1 = 1060.7 - 69.74 = 991.0 \text{ Btu/lbm}$$

Thus,

$$\eta_{th} = 1 - \frac{q_{\text{out}}}{q_{\text{in}}} = 1 - \frac{991.0 \text{ Btu/lbm}}{1635.7 \text{ Btu/lbm}} = \mathbf{39.4\%}$$

(*c*) The mass flow rate of the cooling water will be minimum when it is heated to the temperature of the steam in the condenser, which is 101.7°F,

$$\dot{Q}_{\text{out}} = \dot{Q}_{\text{in}} - \dot{W}_{\text{net}} = (1 - \eta_{th})\dot{Q}_{\text{in}} = (1 - 0.394)(6 \times 10^4 \text{ Btu/s}) = 3.636 \times 10^4 \text{ Btu/s}$$

$$\dot{m}_{\text{cool}} = \frac{\dot{Q}_{\text{out}}}{C\Delta T} = \frac{3.636 \times 10^4 \text{ Btu/s}}{(1.0 \text{ Btu/lbm}\cdot°\text{F})(101.7 - 45)°\text{F}} = \mathbf{641.3\ lbm/s}$$

9-32 A steam power plant that operates on an ideal reheat Rankine cycle between the specified pressure limits is considered. The pressure at which reheating takes place, the total rate of heat input in the boiler, and the thermal efficiency of the cycle are to be determined.

Assumptions **1** Steady operating conditions exist. **2** Kinetic and potential energy changes are negligible.

Analysis (*a*) From the steam tables (Tables A-4, A-5, and A-6),

$$h_1 = h_{sat@\ 10\ kPa} = 191.83\ \text{kJ/kg}$$

$$v_1 = v_{sat@\ 10\ kPa} = 0.00101\ \text{m}^3/\text{kg}$$

$$w_{p,in} = v_1(P_2 - P_1)$$
$$= (0.00101\ \text{m}^3/\text{kg})(9{,}000 - 10\ \text{kPa})\left(\frac{1\ \text{kJ}}{1\ \text{kPa}\cdot\text{m}^3}\right)$$
$$= 9.08\ \text{kJ/kg}$$

$$h_2 = h_1 + w_{p,in} = 191.83 + 9.08 = 200.91\ \text{kJ/kg}$$

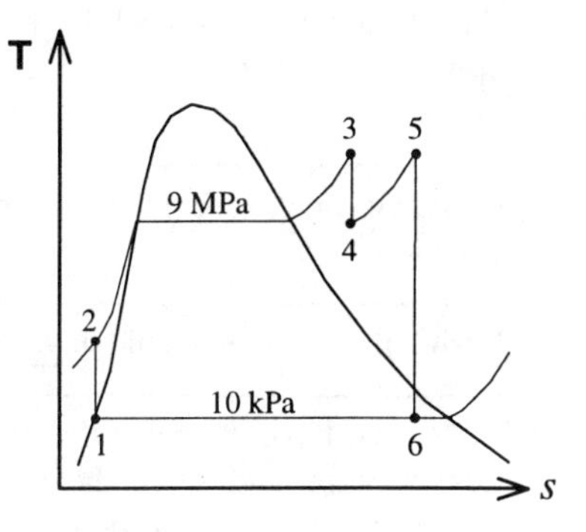

$$\left.\begin{array}{l} P_3 = 9\ \text{MPa} \\ T_3 = 500°\text{C} \end{array}\right\} \begin{array}{l} h_3 = 3386.1\ \text{kJ/kg} \\ s_3 = 6.6576\ \text{kJ/kg}\cdot\text{K} \end{array}$$

$$\left.\begin{array}{l} P_6 = 10\ \text{kPa} \\ s_6 = s_5 \end{array}\right\} \begin{array}{l} h_6 = h_f + x_6 h_{fg} = 191.83 + (0.90)(2392.8) = 2345.4\ \text{kJ/kg} \\ s_6 = s_f + x_6 s_{fg} = 0.6493 + (0.90)(7.5009) = 7.4001\ \text{kJ/kg}\cdot\text{K} \end{array}$$

$$\left.\begin{array}{l} T_5 = 500°\text{C} \\ s_5 = s_6 \end{array}\right\} \begin{array}{l} P_5 = \mathbf{2.146\ MPa}\ (\text{the reheat pressure}) \\ h_5 = 3466.0\ \text{kJ/kg} \end{array}$$

$$\left.\begin{array}{l} P_4 = 2.146\ \text{MPa} \\ s_4 = s_3 \end{array}\right\} h_4 = 2979.5\ \text{kJ/kg}$$

(*b*) The rate of heat supply is

$$\dot{Q}_{in} = \dot{m}[(h_3 - h_2) + (h_5 - h_4)]$$
$$= (25\ \text{kJ/s})(3386.1 - 200.91 + 3466 - 2979.5)\text{kJ/kg} = \mathbf{91{,}792\ \ kJ/s}$$

(*c*) The thermal efficiency is determined from

$$\dot{Q}_{out} = \dot{m}(h_6 - h_1) = (25\ \text{kJ/s})(2345.4 - 191.83)\text{kJ/kg} = 53{,}839\ \text{kJ/s}$$

Thus,

$$\eta_{th} = 1 - \frac{\dot{Q}_{out}}{\dot{Q}_{in}} = 1 - \frac{53{,}839\ \text{kJ/s}}{91{,}792\ \text{kJ/s}} = \mathbf{41.3\%}$$

Regenerative Rankine Cycle

9-33C Moisture content remains the same, everything else decreases.

9-34C This is a smart idea because we waste little work potential but we save a lot from the heat input. The extracted steam has little work potential left, and most of its energy would be part of the heat rejected anyway. Therefore, by regeneration, we utilize a considerable amount of heat by sacrificing little work output.

9-35C In open feedwater heaters, the two fluids actually mix, but in closed feedwater heaters there is no mixing.

9-36C Both cycles would have the same efficiency.

9-37C To have the same thermal efficiency as the Carnot cycle, the cycle must receive and reject heat isothermally. Thus the liquid should be brought to the saturated liquid state at the boiler pressure isothermally, and the steam must be a saturated vapor at the turbine inlet. This will require an infinite number of heat exchangers (feedwater heaters), as shown on the T-s diagram.

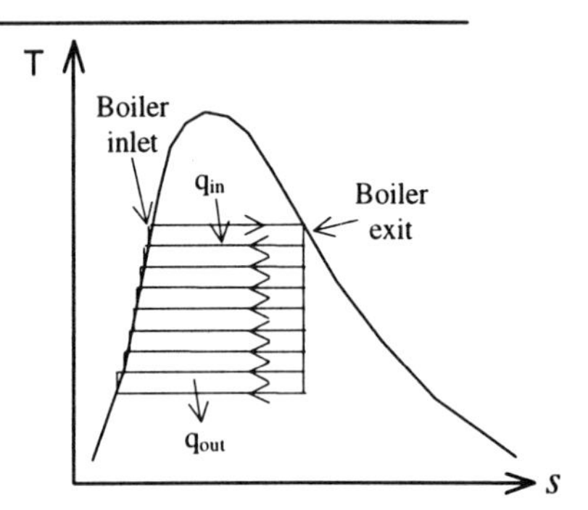

9-38 A steam power plant that operates on an ideal regenerative Rankine cycle with an open feedwater heater is considered. The net work output per kg of steam and the thermal efficiency of the cycle are to be determined.

Assumptions **1** Steady operating conditions exist. **2** Kinetic and potential energy changes are negligible.

Analysis (*a*) From the steam tables (Tables A-4, A-5, and A-6),

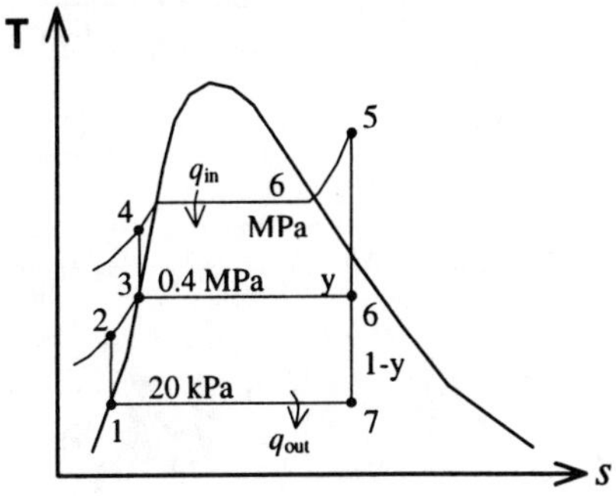

$$h_1 = h_{f\,@\,20\text{ kPa}} = 251.40 \text{ kJ/kg}$$

$$v_1 = v_{f\,@\,20\text{ kPa}} = 0.001017 \text{ m}^3\text{/kg}$$

$$w_{pI,\text{in}} = v_1(P_2 - P_1) = (0.001017 \text{ m}^3\text{/kg})(400 - 20 \text{ kPa})\left(\frac{1 \text{ kJ}}{1 \text{ kPa}\cdot\text{m}^3}\right) = 0.39 \text{ kJ/kg}$$

$$h_2 = h_1 + w_{pI,\text{in}} = 251.40 + 0.39 = 251.79 \text{ kJ/kg}$$

$$\left.\begin{array}{l} P_3 = 0.4 \text{ MPa} \\ \text{sat.liquid} \end{array}\right\} \begin{array}{l} h_3 = h_{f\,@\,0.4\text{ MPa}} = 604.74 \text{ kJ/kg} \\ v_3 = v_{f\,@\,0.4\text{ MPa}} = 0.001084 \text{ m}^3\text{/kg} \end{array}$$

$$w_{pII,\text{in}} = v_3(P_4 - P_3) = (0.001084 \text{ m}^3\text{/kg})(6000 - 400 \text{ kPa})\left(\frac{1 \text{ kJ}}{1 \text{ kPa}\cdot\text{m}^3}\right) = 6.07 \text{ kJ/kg}$$

$$h_4 = h_3 + w_{pII,\text{in}} = 604.74 + 6.07 = 610.81 \text{ kJ/kg}$$

$$\left.\begin{array}{l} P_5 = 6 \text{ MPa} \\ T_5 = 450°\text{C} \end{array}\right\} \begin{array}{l} h_5 = 3301.8 \text{ kJ/kg} \\ s_5 = 6.7193 \text{ kJ/kg}\cdot\text{K} \end{array}$$

$$\left.\begin{array}{l} P_6 = 0.4 \text{ MPa} \\ s_6 = s_5 \end{array}\right\} \begin{array}{l} x_6 = \dfrac{s_6 - s_f}{s_{fg}} = \dfrac{6.7193 - 1.7766}{5.1193} = 0.9655 \\ h_6 = h_f + x_6 h_{fg} = 604.74 + (0.9655)(2133.8) = 2664.93 \text{ kJ/kg} \end{array}$$

$$\left.\begin{array}{l} P_7 = 20 \text{ kPa} \\ s_7 = s_5 \end{array}\right\} \begin{array}{l} x_7 = \dfrac{s_7 - s_f}{s_{fg}} = \dfrac{6.7193 - 0.8320}{7.0766} = 0.8319 \\ h_7 = h_f + x_7 h_{fg} = 251.40 + (0.8319)(2358.3) = 2213.3 \text{ kJ/kg} \end{array}$$

The fraction of steam extracted is determined from the steady-flow energy balance equation applied to the feedwater heater. Noting that $\dot{Q} \cong \dot{W} \cong \Delta ke \cong \Delta pe \cong 0$,

$$\dot{E}_{in} - \dot{E}_{out} = \Delta\dot{E}_{system}^{\;\nearrow 0\,(\text{steady})} = 0$$

$$\dot{E}_{in} = \dot{E}_{out}$$

$$\sum \dot{m}_i h_i = \sum \dot{m}_e h_e \longrightarrow \dot{m}_6 h_6 + \dot{m}_2 h_2 = \dot{m}_3 h_3 \longrightarrow y h_6 + (1 - y)h_2 = 1(h_3)$$

where y is the fraction of steam extracted from the turbine ($= \dot{m}_6 / \dot{m}_3$). Solving for y,

$$y = \frac{h_3 - h_2}{h_6 - h_2} = \frac{604.74 - 251.79}{2664.93 - 251.79} = 0.1463$$

Then,

$$q_{\text{in}} = h_5 - h_4 = 3301.8 - 610.81 = 2691.0 \text{ kJ/kg}$$

$$q_{\text{out}} = (1 - y)(h_7 - h_1) = (1 - 0.1463)(2213.3 - 251.4) = 1674.9 \text{ kJ/kg}$$

and

$$w_{\text{net}} = q_{\text{in}} - q_{\text{out}} = 2691.0 - 1674.9 = \mathbf{1016.1\ kJ/kg}$$

(b) The thermal efficiency is determined from

$$\eta_{th} = 1 - \frac{q_{\text{out}}}{q_{\text{in}}} = 1 - \frac{1674.9\ \text{kJ/kg}}{2691.0\ \text{kJ/kg}} = \mathbf{37.8\%}$$

9-39 A steam power plant that operates on an ideal regenerative Rankine cycle with a closed feedwater heater is considered. The net work output per kg of steam and the thermal efficiency of the cycle are to be determined.

Assumptions **1** Steady operating conditions exist. **2** Kinetic and potential energy changes are negligible.

Analysis (*a*) From the steam tables (Tables A-4, A-5, and A-6),

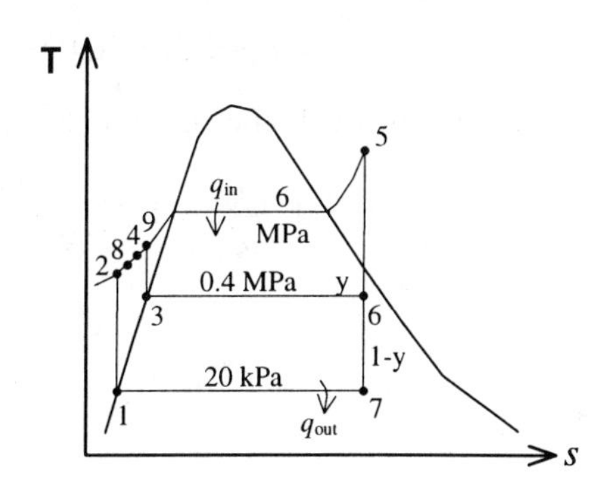

$$h_1 = h_{f\,@\,20\text{ kPa}} = 251.40 \text{ kJ/kg}$$

$$v_1 = v_{f\,@\,20\text{ kPa}} = 0.001017 \text{ m}^3\text{/kg}$$

$$\begin{aligned} w_{pI,in} &= v_1(P_2 - P_1) \\ &= (0.001017 \text{ m}^3\text{/kg})(6000 - 20 \text{ kPa})\left(\frac{1 \text{ kJ}}{1 \text{ kPa}\cdot\text{m}^3}\right) \\ &= 6.08 \text{ kJ/kg} \end{aligned}$$

$$h_2 = h_1 + w_{pI,in} = 251.40 + 6.08 = 257.48 \text{ kJ/kg}$$

$$\left.\begin{array}{l} P_3 = 0.4 \text{ MPa} \\ \text{sat.liquid} \end{array}\right\} \begin{array}{l} h_3 = h_{f\,@\,0.4\text{ MPa}} = 604.74 \text{ kJ/kg} \\ v_3 = v_{f\,@\,0.4\text{ MPa}} = 0.001084 \text{ m}^3\text{/kg} \end{array}$$

$$\begin{aligned} w_{pII,in} &= v_3(P_9 - P_3) \\ &= (0.001084 \text{ m}^3\text{/kg})(6000 - 400 \text{ kPa})\left(\frac{1 \text{ kJ}}{1 \text{ kPa}\cdot\text{m}^3}\right) \\ &= 6.07 \text{ kJ/kg} \end{aligned}$$

$$h_9 = h_3 + w_{pII,in} = 604.74 + 6.07 = 610.81 \text{ kJ/kg}$$

$$h_8 = h_3 + v_3(P_8 - P_3) = h_9 = 610.81 \text{ kJ/kg}$$

Also, $h_4 = h_9 = h_8 = 610.81$ kJ/kg since the two fluid streams which are being mixed have the same enthalpy.

$$\left.\begin{array}{l} P_5 = 6 \text{ MPa} \\ T_5 = 450°\text{C} \end{array}\right\} \begin{array}{l} h_5 = 3301.8 \text{ kJ/kg} \\ s_5 = 6.7193 \text{ kJ/kg}\cdot\text{K} \end{array}$$

$$\left.\begin{array}{l} P_6 = 0.4 \text{ MPa} \\ s_6 = s_5 \end{array}\right\} \begin{array}{l} x_6 = \dfrac{s_6 - s_f}{s_{fg}} = \dfrac{6.7193 - 1.7766}{5.1193} = 0.9655 \\ h_6 = h_f + x_6 h_{fg} = 604.74 + (0.9655)(2133.8) = 2664.93 \text{ kJ/kg} \end{array}$$

$$\left.\begin{array}{l} P_7 = 20 \text{ kPa} \\ s_7 = s_5 \end{array}\right\} \begin{array}{l} x_7 = \dfrac{s_7 - s_f}{s_{fg}} = \dfrac{6.7193 - 0.8320}{7.0766} = 0.8319 \\ h_7 = h_f + x_7 h_{fg} = 251.40 + (0.8319)(2358.3) = 2213.3 \text{ kJ/kg} \end{array}$$

The fraction of steam extracted is determined from the steady-flow energy balance equation applied to the feedwater heater. Noting that $\dot{Q} \cong \dot{W} \cong \Delta ke \cong \Delta pe \cong 0$,

$$\dot{E}_{in} - \dot{E}_{out} = \Delta \dot{E}_{system}^{\;\nearrow 0\text{ (steady)}} = 0$$

$$\dot{E}_{in} = \dot{E}_{out}$$

$$\sum \dot{m}_i h_i = \sum \dot{m}_e h_e \longrightarrow \dot{m}_2(h_8 - h_2) = \dot{m}_6(h_6 - h_3) \longrightarrow (1-y)(h_8 - h_2) = y(h_6 - h_3)$$

where y is the fraction of steam extracted from the turbine ($= \dot{m}_6 / \dot{m}_5$). Solving for y,

$$y = \frac{h_8 - h_2}{(h_6 - h_3) + (h_8 - h_2)} = \frac{610.81 - 257.48}{2664.93 - 604.64 + 610.81 - 257.48} = 0.1464$$

Then,

$$q_{\text{in}} = h_5 - h_4 = 3301.8 - 610.81 = 2691.0 \text{ kJ/kg}$$
$$q_{\text{out}} = (1-y)(h_7 - h_1) = (1-0.1464)(2213.3 - 251.4) = 1674.7 \text{ kJ/kg}$$

and

$$w_{\text{net}} = q_{\text{in}} - q_{\text{out}} = 2691.0 - 1674.7 = \mathbf{1016.3\ kJ/kg}$$

(*b*) The thermal efficiency is determined from

$$\eta_{th} = 1 - \frac{q_{\text{out}}}{q_{\text{in}}} = 1 - \frac{1674.7 \text{ kJ/kg}}{2691.0 \text{ kJ/kg}} = \mathbf{37.8\%}$$

9-40 A steam power plant operates on an ideal regenerative Rankine cycle with two open feedwater heaters. The net power output of the power plant and the thermal efficiency of the cycle are to be determined.

Assumptions **1** Steady operating conditions exist. **2** Kinetic and potential energy changes are negligible.

Analysis

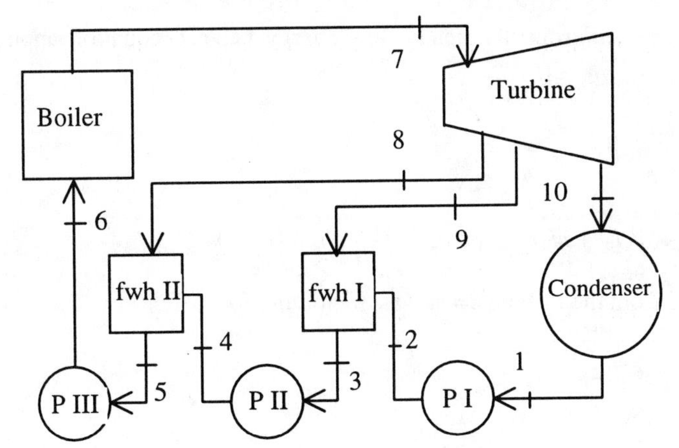

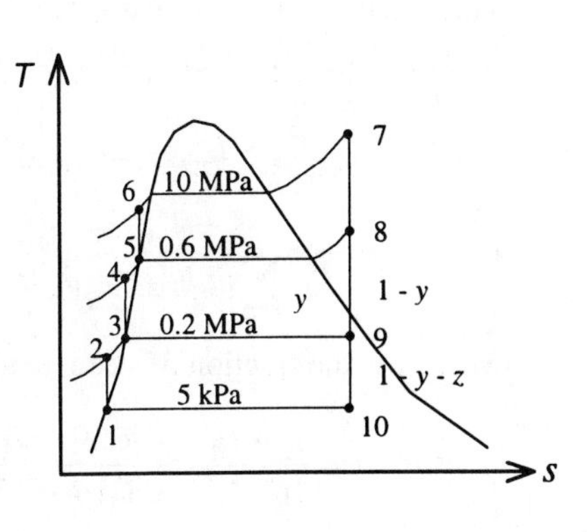

(*a*) From the steam tables (Tables A-4, A-5, and A-6),

$$h_1 = h_{f @ 5\text{ kPa}} = 137.82 \text{ kJ/kg}$$

$$v_1 = v_{f @ 5\text{ kPa}} = 0.001005 \text{ m}^3\text{/kg}$$

$$w_{pI,\text{in}} = v_1(P_2 - P_1) = (0.001005 \text{ m}^3\text{/kg})(200 - 5 \text{ kPa})\left(\frac{1 \text{ kJ}}{1 \text{ kPa}\cdot\text{m}^3}\right) = 0.20 \text{ kJ/kg}$$

$$h_2 = h_1 + w_{pI,\text{in}} = 137.82 + 0.20 = 138.02 \text{ kJ/kg}$$

$$\left.\begin{array}{l} P_3 = 0.2 \text{ MPa} \\ \text{sat.liquid} \end{array}\right\} \begin{array}{l} h_3 = h_{f @ 0.2\text{ MPa}} = 504.70 \text{ kJ/kg} \\ v_3 = v_{f @ 0.2\text{ MPa}} = 0.001061 \text{ m}^3\text{/kg} \end{array}$$

$$w_{pII,\text{in}} = v_3(P_4 - P_3) = (0.001061 \text{ m}^3\text{/kg})(600 - 200 \text{ kPa})\left(\frac{1 \text{ kJ}}{1 \text{ kPa}\cdot\text{m}^3}\right)$$
$$= 0.42 \text{ kJ/kg}$$

$$h_4 = h_3 + w_{pII,\text{in}} = 504.70 + 0.42 = 505.12 \text{ kJ/kg}$$

$$\left.\begin{array}{l} P_5 = 0.6 \text{ MPa} \\ \text{sat.liquid} \end{array}\right\} \begin{array}{l} h_5 = h_{f @ 0.6\text{ MPa}} = 670.56 \text{ kJ/kg} \\ v_5 = v_{f @ 0.6\text{ MPa}} = 0.001101 \text{ m}^3\text{/kg} \end{array}$$

$$w_{pIII,\text{in}} = v_5(P_6 - P_5) = (0.001101 \text{ m}^3\text{/kg})(10{,}000 - 600 \text{ kPa})\left(\frac{1 \text{ kJ}}{1 \text{ kPa}\cdot\text{m}^3}\right)$$
$$= 10.35 \text{ kJ/kg}$$

$$h_6 = h_5 + w_{pIII,\text{in}} = 670.56 + 10.35 = 680.91 \text{ kJ/kg}$$

$$\left.\begin{array}{l} P_7 = 10 \text{ MPa} \\ T_7 = 600°\text{C} \end{array}\right\} \begin{array}{l} h_7 = 3625.3 \text{ kJ/kg} \\ s_7 = 6.9029 \text{ kJ/kg}\cdot\text{K} \end{array}$$

$$\left.\begin{array}{l} P_8 = 0.6 \text{ MPa} \\ s_8 = s_7 \end{array}\right\} h_8 = 2821.4 \text{ kJ/kg}$$

$$\left.\begin{array}{l} P_9 = 0.2 \text{ MPa} \\ s_9 = s_7 \end{array}\right\} \begin{array}{l} x_9 = \dfrac{s_9 - s_f}{s_{fg}} = \dfrac{6.9029 - 1.5301}{5.5970} = 0.9599 \\ h_9 = h_f + x_9 h_{fg} = 504.70 + (0.9599)(2201.9) \\ = 2618.3 \text{ kJ/kg} \end{array}$$

$$\left.\begin{array}{l} P_{10} = 5 \text{ kPa} \\ s_{10} = s_7 \end{array}\right\} \begin{array}{l} x_{10} = \dfrac{s_{10} - s_f}{s_{fg}} = \dfrac{6.9029 - 0.4764}{7.9187} = 0.8116 \\ h_{10} = h_f + x_{10} h_{fg} = 137.82 + (0.8116)(2423.7) = 2104.8 \text{ kJ/kg} \end{array}$$

The fraction of steam extracted is determined from the steady-flow energy balance equation applied to the feedwater heaters. Noting that $\dot{Q} \cong \dot{W} \cong \Delta ke \cong \Delta pe \cong 0$,

$$\dot{E}_{in} - \dot{E}_{out} = \Delta \dot{E}_{system}^{\,\nearrow 0 \text{ (steady)}} = 0$$

FWH-2:

$$\dot{E}_{in} = \dot{E}_{out}$$

$$\sum \dot{m}_i h_i = \sum \dot{m}_e h_e \longrightarrow \dot{m}_8 h_8 + \dot{m}_4 h_4 = \dot{m}_5 h_5 \longrightarrow y h_8 + (1-y) h_4 = 1(h_5)$$

where y is the fraction of steam extracted from the turbine ($= \dot{m}_8 / \dot{m}_5$). Solving for y,

$$y = \frac{h_5 - h_4}{h_8 - h_4} = \frac{670.56 - 505.12}{2821.4 - 505.12} = 0.0714$$

FWH-1: $\sum \dot{m}_i h_i = \sum \dot{m}_e h_e \longrightarrow \dot{m}_9 h_9 + \dot{m}_2 h_2 = \dot{m}_3 h_3 \longrightarrow z h_9 + (1 - y - z) h_2 = (1 - y) h_3$

where z is the fraction of steam extracted from the turbine ($= \dot{m}_9 / \dot{m}_5$) at the second stage. Solving for z,

$$z = \frac{h_3 - h_2}{h_9 - h_2}(1 - y) = \frac{504.7 - 138.02}{2618.3 - 138.02}(1 - 0.0714) = 0.137$$

Then,

$$q_{\text{in}} = h_7 - h_6 = 3625.3 - 680.91 = 2944.4 \text{ kJ/kg}$$

$$q_{\text{out}} = (1 - y - z)(h_{10} - h_1) = (1 - 0.0714 - 0.137)(2104.8 - 137.82) = 1557.1 \text{ kJ/kg}$$

$$w_{\text{net}} = q_{\text{in}} - q_{\text{out}} = 2944.4 - 1557.1 = 1387.3 \text{ kJ/kg}$$

and

$$\dot{W}_{\text{net}} = \dot{m} w_{\text{net}} = (27 \text{ kg/s})(1387.3 \text{ kJ/kg}) = 37{,}457 \text{ kW} \cong \mathbf{37.5 \text{ MW}}$$

(*b*) $$\eta_{th} = 1 - \frac{q_{\text{out}}}{q_{\text{in}}} = 1 - \frac{1557.1 \text{ kJ/kg}}{2944.4 \text{ kJ/kg}} = \mathbf{47.1\%}$$

9-41 [*Also solved by EES on enclosed CD*] A steam power plant operates on an ideal regenerative Rankine cycle with two feedwater heaters, one closed and one open. The mass flow rate of steam through the boiler for a net power output of 250 MW and the thermal efficiency of the cycle are to be determined.

Assumptions **1** Steady operating conditions exist. **2** Kinetic and potential energy changes are negligible.

Analysis (*a*) From the steam tables (Tables A-4, A-5, and A-6),

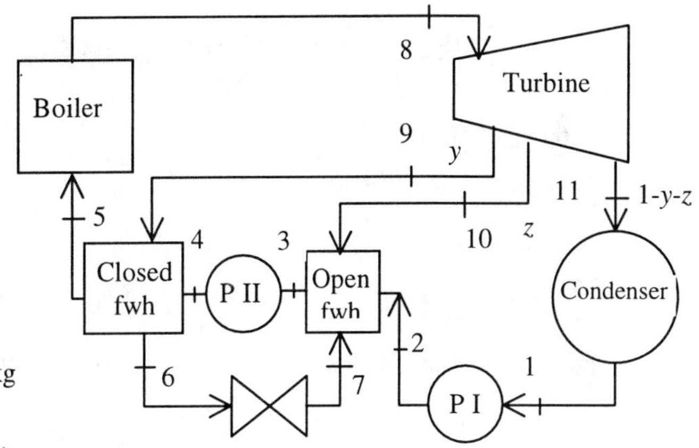

$$h_1 = h_{f\,@\,10\text{ kPa}} = 191.83 \text{ kJ/kg}$$
$$v_1 = v_{f\,@\,10\text{ kPa}} = 0.00101 \text{ m}^3\text{/kg}$$

$$w_{pI,\text{in}} = v_1(P_2 - P_1) = \left(0.00101 \text{ m}^3\text{/kg}\right)(300 - 10 \text{ kPa})\left(\frac{1 \text{ kJ}}{1 \text{ kPa}\cdot\text{m}^3}\right) = 0.29 \text{ kJ/kg}$$
$$h_2 = h_1 + w_{pI,\text{in}} = 191.83 + 0.29 = 192.12 \text{ kJ/kg}$$

$$\left.\begin{array}{l} P_3 = 0.3 \text{ MPa} \\ \text{sat.liquid} \end{array}\right\} \begin{array}{l} h_3 = h_{f\,@\,0.3\text{ MPa}} = 561.47 \text{ kJ/kg} \\ v_3 = v_{f\,@\,0.3\text{ MPa}} = 0.001073 \text{ m}^3\text{/kg} \end{array}$$

$$w_{pII,\text{in}} = v_3(P_4 - P_3) = \left(0.001073 \text{ m}^3\text{/kg}\right)(12{,}500 - 300 \text{ kPa})\left(\frac{1 \text{ kJ}}{1 \text{ kPa}\cdot\text{m}^3}\right) = 13.09 \text{ kJ/kg}$$
$$h_4 = h_3 + w_{pII,\text{in}} = 561.47 + 13.09 = 574.56 \text{ kJ/kg}$$

$$\left.\begin{array}{l} P_6 = 0.8 \text{ MPa} \\ \text{sat.liquid} \end{array}\right\} \begin{array}{l} h_6 = h_7 = h_{f\,@\,0.8\text{ MPa}} = 721.11 \text{ kJ/kg} \\ v_6 = v_{f\,@\,0.8\text{ MPa}} = 0.001115 \text{ m}^3\text{/kg} \end{array}$$

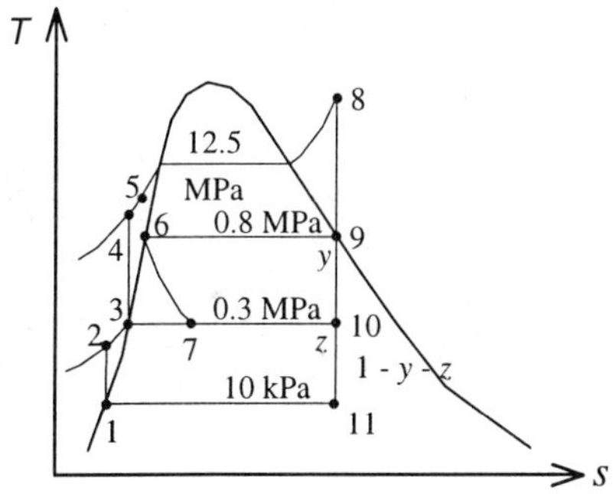

$$T_6 = T_5 \rightarrow h_5 = h_6 + v_6(P_5 - P_6) = 721.11 + \left(0.001115 \text{ m}^3\text{/kg}\right)(12{,}500 - 800 \text{ kPa})\left(\frac{1 \text{ kJ}}{1 \text{ kPa}\cdot\text{m}^3}\right) = 734.16 \text{ kJ/kg}$$

$$\left.\begin{array}{l} P_8 = 12.5 \text{ MPa} \\ T_8 = 550°\text{C} \end{array}\right\} \begin{array}{l} h_8 = 3475.2 \text{ kJ/kg} \\ s_8 = 6.6290 \text{ kJ/kg}\cdot\text{K} \end{array}$$

$$\left.\begin{array}{l} P_9 = 0.8 \text{ MPa} \\ s_9 = s_8 \end{array}\right\} \begin{array}{l} x_9 = \dfrac{s_9 - s_f}{s_{fg}} = \dfrac{6.6290 - 2.0462}{4.6166} = 0.9927 \\ h_9 = h_f + x_9 h_{fg} = 721.11 + (0.9927)(2048.0) = 2754.1 \text{ kJ/kg} \end{array}$$

$$\left.\begin{array}{l} P_{10} = 0.3 \text{ MPa} \\ s_{10} = s_8 \end{array}\right\} \begin{array}{l} x_{10} = \dfrac{s_{10} - s_f}{s_{fg}} = \dfrac{6.6290 - 1.6718}{5.3201} = 0.9318 \\ h_{10} = h_f + x_{10} h_{fg} = 561.47 + (0.9318)(2163.8) = 2577.7 \text{ kJ/kg} \end{array}$$

$$\left.\begin{array}{l} P_{11} = 10 \text{ kPa} \\ s_{11} = s_8 \end{array}\right\} \begin{array}{l} x_{11} = \dfrac{s_{11} - s_f}{s_{fg}} = \dfrac{6.6290 - 0.6493}{7.5009} = 0.7972 \\ h_{11} = h_f + x_{11} h_{fg} = 191.83 + (0.7972)(2392.8) = 2099.4 \text{ kJ/kg} \end{array}$$

The fraction of steam extracted is determined from the steady-flow energy balance equation applied to the feedwater heaters. Noting that $\dot{Q} \cong \dot{W} \cong \Delta ke \cong \Delta pe \cong 0$,

$$\dot{E}_{in} - \dot{E}_{out} = \Delta\dot{E}_{system}{}^{\nearrow 0\,(steady)} = 0$$
$$\dot{E}_{in} = \dot{E}_{out}$$
$$\sum \dot{m}_i h_i = \sum \dot{m}_e h_e \longrightarrow \dot{m}_9(h_9 - h_6) = \dot{m}_5(h_5 - h_4) \longrightarrow y(h_9 - h_6) = (h_5 - h_4)$$

where y is the fraction of steam extracted from the turbine ($= \dot{m}_{10} / \dot{m}_5$). Solving for y,

$$y = \frac{h_5 - h_4}{h_9 - h_6} = \frac{734.16 - 574.56}{2754.1 - 721.11} = 0.0785$$

For the open FWH,

$$\dot{E}_{in} - \dot{E}_{out} = \Delta\dot{E}_{system}{}^{\nearrow 0\,(steady)} = 0$$
$$\dot{E}_{in} = \dot{E}_{out}$$
$$\sum \dot{m}_i h_i = \sum \dot{m}_e h_e \longrightarrow \dot{m}_7 h_7 + \dot{m}_2 h_2 + \dot{m}_{10} h_{10} = \dot{m}_3 h_3 \longrightarrow y h_7 + (1 - y - z)h_2 + z h_{10} = (1)h_3$$

where z is the fraction of steam extracted from the turbine ($= \dot{m}_9 / \dot{m}_5$) at the second stage. Solving for z,

$$z = \frac{(h_3 - h_2) - y(h_7 - h_2)}{h_{10} - h_2} = \frac{561.47 - 192.12 - (0.0785)(721.11 - 192.12)}{2577.7 - 192.12} = 0.1374$$

Then,

$$q_{in} = h_8 - h_5 = 3475.2 - 734.16 = 2741.0 \text{ kJ/kg}$$
$$q_{out} = (1 - y - z)(h_{11} - h_1) = (1 - 0.0785 - 0.1374)(2099.4 - 191.83) = 1495.7 \text{ kJ/kg}$$
$$w_{net} = q_{in} - q_{out} = 2741.0 - 1495.7 = 1245.3 \text{ kJ/kg}$$

and

$$\dot{m} = \frac{\dot{W}_{net}}{w_{net}} = \frac{250{,}000 \text{ kJ/s}}{1245.3 \text{ kJ/kg}} = \mathbf{200.8\ kg/s}$$

(*b*)
$$\eta_{th} = 1 - \frac{q_{out}}{q_{in}} = 1 - \frac{1495.7 \text{ kJ/kg}}{2741.0 \text{ kJ/kg}} = \mathbf{45.4\%}$$

9-42 EES solution of this (and other comprehensive problems designated with the *computer icon*) is available to instructors at the *Instructor Manual* section of the *Online Learning Center* (OLC) at www.mhhe.com/cengel-boles. See the Preface for access information.

9-43 A steam power plant operates on an ideal reheat-regenerative Rankine cycle with an open feedwater heater. The mass flow rate of steam through the boiler and the thermal efficiency of the cycle are to be determined.

Assumptions **1** Steady operating conditions exist. **2** Kinetic and potential energy changes are negligible.

Analysis (*a*) From the steam tables (Tables A-4, A-5, and A-6),

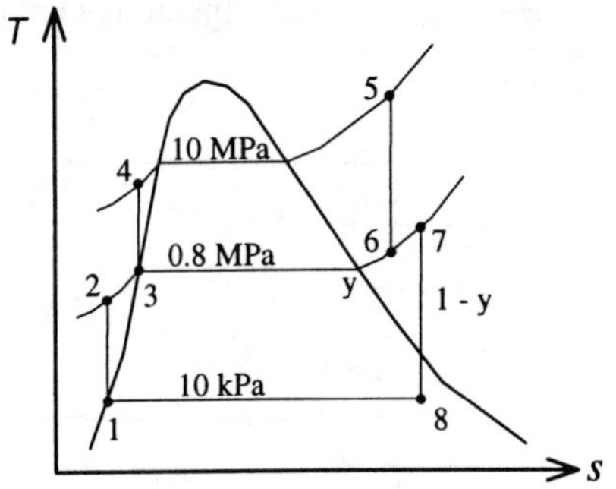

$$h_1 = h_{f\,@\,10\text{ kPa}} = 191.83\text{ kJ/kg}$$
$$v_1 = v_{f\,@\,10\text{ kPa}} = 0.00101\text{ m}^3\text{/kg}$$
$$w_{pI,\text{in}} = v_1(P_2 - P_1) = (0.00101\text{ m}^3\text{/kg})(800 - 10\text{ kPa})\left(\frac{1\text{ kJ}}{1\text{ kPa}\cdot\text{m}^3}\right) = 0.80\text{ kJ/kg}$$
$$h_2 = h_1 + w_{pI,\text{in}} = 191.83 + 0.80 = 192.63\text{ kJ/kg}$$

$$\left.\begin{array}{l} P_3 = 0.8\text{ MPa} \\ \text{sat.liquid} \end{array}\right\} \begin{array}{l} h_3 = h_{f\,@\,0.8\text{ MPa}} = 721.11\text{ kJ/kg} \\ v_3 = v_{f\,@\,0.8\text{ MPa}} = 0.001115\text{ m}^3\text{/kg} \end{array}$$

$$w_{pII,\text{in}} = v_3(P_4 - P_3) = (0.001115\text{ m}^3\text{/kg})(10{,}000 - 800\text{ kPa})\left(\frac{1\text{ kJ}}{1\text{ kPa}\cdot\text{m}^3}\right) = 10.26\text{ kJ/kg}$$
$$h_4 = h_3 + w_{pII,\text{in}} = 721.11 + 10.26 = 731.37\text{ kJ/kg}$$

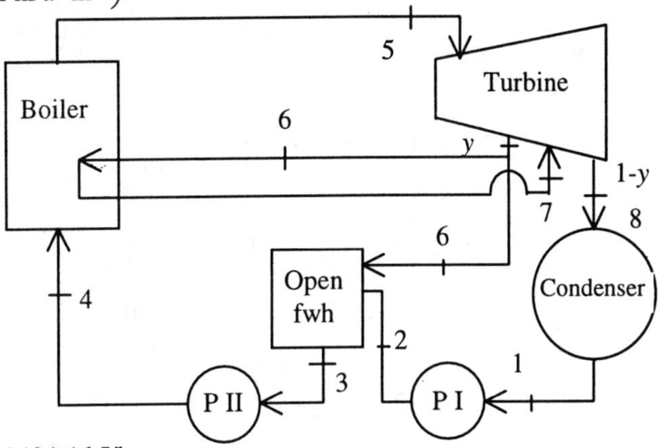

$$\left.\begin{array}{l} P_5 = 10\text{ MPa} \\ T_5 = 550°\text{C} \end{array}\right\} \begin{array}{l} h_5 = 3500.9\text{ kJ/kg} \\ s_5 = 6.7561\text{ kJ/kg}\cdot\text{K} \end{array}$$

$$\left.\begin{array}{l} P_6 = 0.8\text{ MPa} \\ s_6 = s_5 \end{array}\right\} h_6 = 2811.9\text{ kJ/kg}$$

$$\left.\begin{array}{l} P_7 = 0.8\text{ MPa} \\ T_7 = 500°\text{C} \end{array}\right\} \begin{array}{l} h_7 = 3480.6\text{ kJ/kg} \\ s_7 = 7.8673\text{ kJ/kg}\cdot\text{K} \end{array}$$

$$\left.\begin{array}{l} P_8 = 10\text{ kPa} \\ s_8 = s_7 \end{array}\right\} \begin{array}{l} x_8 = \dfrac{s_8 - s_f}{s_{fg}} = \dfrac{7.8673 - 0.6493}{7.5009} = 0.9623 \\ h_8 = h_f + x_8 h_{fg} = 191.83 + (0.9623)(2392.8) = 2494.4\text{ kJ/kg} \end{array}$$

The fraction of steam extracted is determined from the steady-flow energy balance equation applied to the feedwater heaters. Noting that $\dot{Q} \cong \dot{W} \cong \Delta ke \cong \Delta pe \cong 0$,

$$\dot{E}_{\text{in}} - \dot{E}_{\text{out}} = \Delta\dot{E}_{\text{system}}^{\;\nearrow 0\text{ (steady)}} = 0 \rightarrow \dot{E}_{\text{in}} = \dot{E}_{\text{out}}$$
$$\sum \dot{m}_i h_i = \sum \dot{m}_e h_e \longrightarrow \dot{m}_6 h_6 + \dot{m}_2 h_2 = \dot{m}_3 h_3 \longrightarrow y h_6 + (1-y)h_2 = 1(h_3)$$

where y is the fraction of steam extracted from the turbine ($= \dot{m}_6 / \dot{m}_3$). Solving for y,

$$y = \frac{h_3 - h_2}{h_6 - h_2} = \frac{721.11 - 192.63}{2711.9 - 192.63} = 0.2018$$

Then, $q_{\text{in}} = (h_5 - h_4) + (1-y)(h_7 - h_6) = (3500.9 - 731.37) + (1 - 0.2098)(3480.6 - 2811.9) = 3297.9\text{ kJ/kg}$

$$q_{\text{out}} = (1-y)(h_8 - h_1) = (1 - 0.2098)(2494.4 - 191.83) = 1819.5\text{ kJ/kg}$$
$$w_{\text{net}} = q_{\text{in}} - q_{\text{out}} = 3297.9 - 1819.5 = 1478.5\text{ kJ/kg}$$

and $$\dot{m} = \frac{\dot{W}_{\text{net}}}{w_{\text{net}}} = \frac{80{,}000\text{ kJ/s}}{1478.4\text{ kJ/kg}} = \mathbf{54.1\text{ kg/s}}$$

(*b*) $$\eta_{th} = \frac{w_{\text{net}}}{q_{\text{in}}} = \frac{1478.4\text{ kJ/kg}}{3297.9\text{ kJ/kg}} = \mathbf{44.8\%}$$

9-44 A steam power plant operates on an ideal reheat-regenerative Rankine cycle with a closed feedwater heater. The mass flow rate of steam through the boiler and the thermal efficiency of the cycle are to be determined.

Assumptions **1** Steady operating conditions exist. **2** Kinetic and potential energy changes are negligible.

Analysis

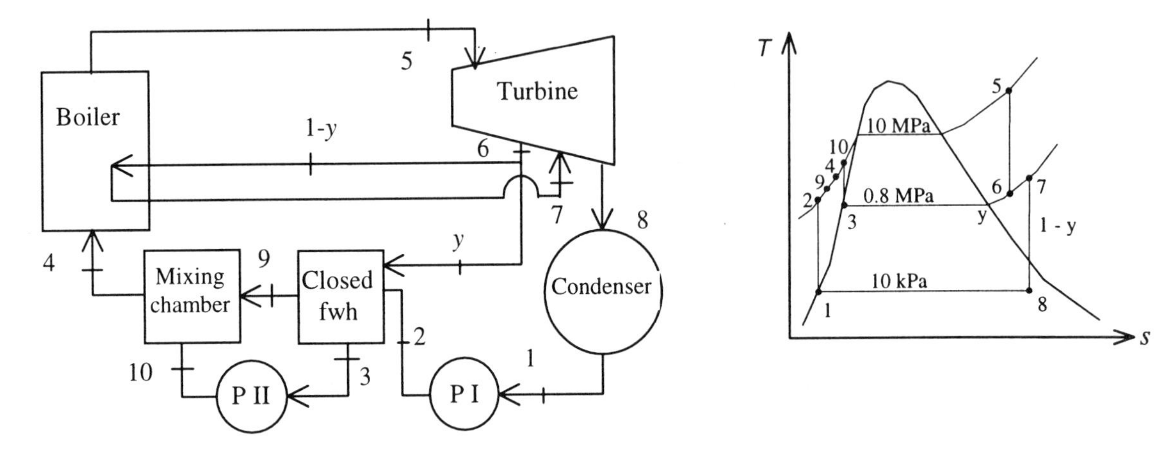

(*a*) From the steam tables (Tables A-4, A-5, and A-6),

$$h_1 = h_{f\,@\,10\text{ kPa}} = 191.83 \text{ kJ/kg}$$

$$v_1 = v_{f\,@\,10\text{ kPa}} = 0.00101 \text{ m}^3\text{/kg}$$

$$w_{pI,\text{in}} = v_1(P_2 - P_1) = (0.00101 \text{ m}^3\text{/kg})(10{,}000 - 10 \text{ kPa})\left(\frac{1 \text{ kJ}}{1 \text{ kPa}\cdot\text{m}^3}\right)$$
$$= 10.09 \text{ kJ/kg}$$

$$h_2 = h_1 + w_{pI,\text{in}} = 191.83 + 10.09 = 201.92 \text{ kJ/kg}$$

$$\left.\begin{array}{l} P_3 = 0.8 \text{ MPa} \\ \text{sat.liquid} \end{array}\right\} \begin{array}{l} h_3 = h_{f\,@\,0.8\text{ MPa}} = 721.11 \text{ kJ/kg} \\ v_3 = v_{f\,@\,0.8\text{ MPa}} = 0.001115 \text{ m}^3\text{/kg} \end{array}$$

$$w_{pII,\text{in}} = v_3(P_4 - P_3) = (0.001115 \text{ m}^3\text{/kg})(10{,}000 - 800 \text{ kPa})\left(\frac{1 \text{ kJ}}{1 \text{ kPa}\cdot\text{m}^3}\right)$$
$$= 10.26 \text{ kJ/kg}$$

$$h_4 = h_3 + w_{pII,\text{in}} = 721.11 + 10.26 = 731.37 \text{ kJ/kg}$$

Also, $h_4 = h_9 = h_{10} = 731.37$ kJ/kg since the two fluid streams that are being mixed have the same enthalpy.

$$\left.\begin{array}{l} P_5 = 10 \text{ MPa} \\ T_5 = 550°\text{C} \end{array}\right\} \begin{array}{l} h_5 = 3500.9 \text{ kJ/kg} \\ s_5 = 6.7561 \text{ kJ/kg} \cdot \text{K} \end{array}$$

$$\left.\begin{array}{l} P_6 = 0.8 \text{ MPa} \\ s_6 = s_5 \end{array}\right\} h_6 = 2811.9 \text{ kJ/kg}$$

$$\left.\begin{array}{l} P_7 = 0.8 \text{ MPa} \\ T_7 = 500°\text{C} \end{array}\right\} \begin{array}{l} h_7 = 3480.6 \text{ kJ/kg} \\ s_7 = 7.8673 \text{ kJ/kg} \cdot \text{K} \end{array}$$

$$\left.\begin{array}{l} P_8 = 10 \text{ kPa} \\ s_8 = s_7 \end{array}\right\} \begin{array}{l} x_8 = \dfrac{s_8 - s_f}{s_{fg}} = \dfrac{7.8673 - 0.6493}{7.5009} = 0.9623 \\ h_8 = h_f + x_8 h_{fg} = 191.83 + (0.9622)(2392.8) = 2494.4 \text{ kJ/kg} \end{array}$$

The fraction of steam extracted is determined from the steady-flow energy balance equation applied to the feedwater heaters. Noting that $\dot{Q} \cong \dot{W} \cong \Delta ke \cong \Delta pe \cong 0$,

$$\dot{E}_{\text{in}} - \dot{E}_{\text{out}} = \Delta \dot{E}_{\text{system}}^{\nearrow 0 \text{ (steady)}} = 0$$

$$\dot{E}_{\text{in}} = \dot{E}_{\text{out}}$$

$$\sum \dot{m}_i h_i = \sum \dot{m}_e h_e \longrightarrow \dot{m}_2 (h_9 - h_2) = \dot{m}_3 (h_6 - h_3) \longrightarrow (1 - y)(h_9 - h_2) = y(h_6 - h_3)$$

where y is the fraction of steam extracted from the turbine ($= \dot{m}_3 / \dot{m}_4$). Solving for y,

$$y = \frac{h_9 - h_2}{(h_6 - h_3) + (h_9 - h_2)} = \frac{731.37 - 201.92}{2811.9 - 721.11 + 731.37 - 201.92} = 0.2021$$

Then,

$$q_{\text{in}} = (h_5 - h_4) + (1 - y)(h_7 - h_6) = (3500.9 - 731.37) + (1 - 0.2021)(3480.6 - 2811.9) = 3303.1 \text{ kJ/kg}$$

$$q_{\text{out}} = (1 - y)(h_8 - h_1) = (1 - 0.2021)(2494.4 - 191.83) = 1837.2 \text{ kJ/kg}$$

$$w_{\text{net}} = q_{\text{in}} - q_{\text{out}} = 3303.1 - 1837.2 = 1465.9 \text{ kJ/kg}$$

and

$$\dot{m} = \frac{\dot{W}_{\text{net}}}{w_{\text{net}}} = \frac{120{,}000 \text{ kJ/s}}{1465.9 \text{ kJ/kg}} = \mathbf{54.6 \ kg/s}$$

(*b*) $$\eta_{\text{th}} = 1 - \frac{q_{\text{out}}}{q_{\text{in}}} = 1 - \frac{1837.2 \text{ kJ/kg}}{3303.1 \text{ kJ/kg}} = \mathbf{44.4\%}$$

9-45E A steam power plant operates on an ideal reheat-regenerative Rankine cycle with one reheater and two open feedwater heaters. The mass flow rate of steam through the boiler, the net power output of the plant, and the thermal efficiency of the cycle are to be determined.

Assumptions **1** Steady operating conditions exist. **2** Kinetic and potential energy changes are negligible.

Analysis

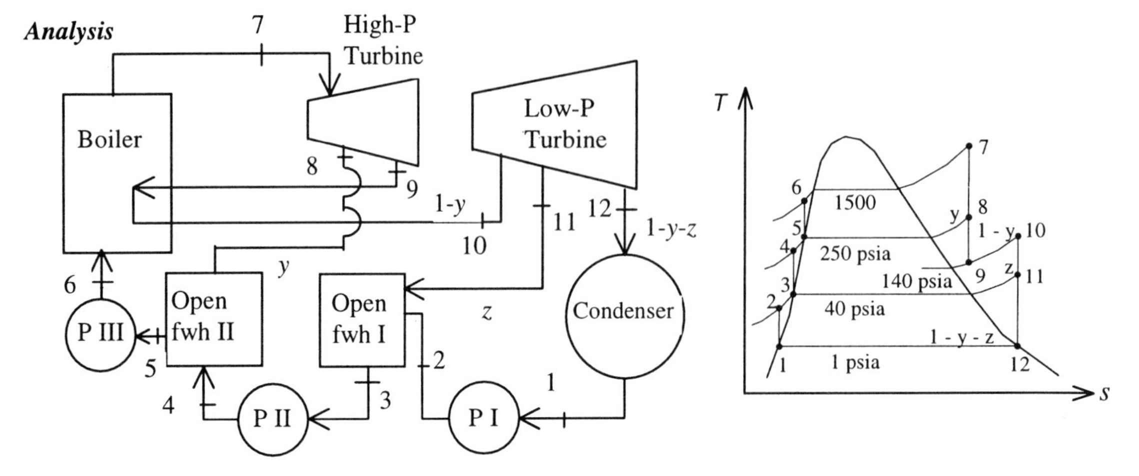

(*a*) From the steam tables (Tables A-4E, A-5E, and A-6E),

$$h_1 = h_{f\,@\,1\text{ psia}} = 69.74 \text{ Btu/lbm}$$

$$v_1 = v_{f\,@\,1\text{ psia}} = 0.016136 \text{ ft}^3\text{/lbm}$$

$$w_{pI,\text{in}} = v_1(P_2 - P_1) = \left(0.016136 \text{ ft}^3\text{/lbm}\right)(40 - 1 \text{ psia})\left(\frac{1 \text{ Btu}}{5.4039 \text{ psia}\cdot\text{ft}^3}\right) = 0.12 \text{ Btu/lbm}$$

$$h_2 = h_1 + w_{pI,\text{in}} = 69.74 + 0.12 = 69.86 \text{ Btu/lbm}$$

$$\left.\begin{array}{l} P_3 = 40 \text{ psia} \\ \text{sat.liquid} \end{array}\right\} \begin{array}{l} h_3 = h_{f\,@\,40\text{ psia}} = 236.16 \text{ Btu/lbm} \\ v_3 = v_{f\,@\,40\text{ psia}} = 0.017146 \text{ ft}^3\text{/lbm} \end{array}$$

$$w_{pII,\text{in}} = v_3(P_4 - P_3) = \left(0.017146 \text{ ft}^3\text{/lbm}\right)(250 - 40 \text{ psia})\left(\frac{1 \text{ Btu}}{5.4039 \text{ psia}\cdot\text{ft}^3}\right) = 0.67 \text{ Btu/lbm}$$

$$h_4 = h_3 + w_{pII,\text{in}} = 236.16 + 0.67 = 236.83 \text{ Btu/lbm}$$

$$\left.\begin{array}{l} P_5 = 250 \text{ psia} \\ \text{sat.liquid} \end{array}\right\} \begin{array}{l} h_5 = h_{f\,@\,250\text{ psia}} = 376.20 \text{ Btu/lbm} \\ v_5 = v_{f\,@\,250\text{ psia}} = 0.018653 \text{ ft}^3\text{/lbm} \end{array}$$

$$w_{pIII,\text{in}} = v_5(P_6 - P_5) = \left(0.018653 \text{ ft}^3\text{/lbm}\right)(1500 - 250 \text{ psia})\left(\frac{1 \text{ Btu}}{5.4039 \text{ psia}\cdot\text{ft}^3}\right) = 4.31 \text{ Btu/lbm}$$

$$h_6 = h_5 + w_{pIII,\text{in}} = 376.20 + 4.31 = 380.51 \text{ Btu/lbm}$$

$$\left.\begin{array}{l} P_7 = 1500 \text{ psia} \\ T_7 = 1100°\text{F} \end{array}\right\} \begin{array}{l} h_7 = 1550.3 \text{ Btu/lbm} \\ s_7 = 1.6399 \text{ Btu/lbm}\cdot\text{R} \end{array}$$

$$\left.\begin{array}{l} P_8 = 250 \text{ psia} \\ s_8 = s_7 \end{array}\right\} h_8 = 1308.5 \text{ Btu/lbm}$$

$$\left.\begin{array}{l} P_9 = 140 \text{ psia} \\ s_9 = s_7 \end{array}\right\} h_9 = 1234.3 \text{ Btu/lbm}$$

$$\left.\begin{array}{l} P_{10} = 140 \text{ psia} \\ T_{10} = 1000°\text{F} \end{array}\right\} \begin{array}{l} h_{10} = 1531.0 \text{ Btu/lbm} \\ s_{10} = 1.8827 \text{ Btu/lbm} \cdot \text{R} \end{array}$$

$$\left.\begin{array}{l} P_{11} = 40 \text{ psia} \\ s_{11} = s_{10} \end{array}\right\} h_{11} = 1356.2 \text{ Btu/lbm}$$

$$\left.\begin{array}{l} P_{12} = 1 \text{ psia} \\ s_{12} = s_{10} \end{array}\right\} \begin{array}{l} x_{12} = \dfrac{s_{12} - s_f}{s_{fg}} = \dfrac{1.8827 - 0.13266}{1.8453} = 0.9484 \\ h_{12} = h_f + x_{12} h_{fg} = 69.74 + (0.9484)(1036.0) \\ \qquad = 1052.3 \text{ Btu/lbm} \end{array}$$

The fraction of steam extracted is determined from the steady-flow energy balance equation applied to the feedwater heaters. Noting that $\dot{Q} \cong \dot{W} \cong \Delta ke \cong \Delta pe \cong 0$,

$$\dot{E}_{in} - \dot{E}_{out} = \Delta \dot{E}_{system}^{\nearrow 0 \text{ (steady)}} = 0$$

FWH-2:

$$\dot{E}_{in} = \dot{E}_{out}$$

$$\sum \dot{m}_i h_i = \sum \dot{m}_e h_e \longrightarrow \dot{m}_8 h_8 + \dot{m}_4 h_4 = \dot{m}_5 h_5 \longrightarrow y h_8 + (1 - y) h_4 = 1(h_5)$$

where y is the fraction of steam extracted from the turbine ($= \dot{m}_8 / \dot{m}_5$). Solving for y,

$$y = \frac{h_5 - h_4}{h_8 - h_4} = \frac{376.20 - 236.83}{1308.5 - 236.83} = 0.1300$$

$$\dot{E}_{in} - \dot{E}_{out} = \Delta \dot{E}_{system}^{\nearrow 0 \text{ (steady)}} = 0$$

FWH-1

$$\dot{E}_{in} = \dot{E}_{out}$$

$$\sum \dot{m}_i h_i = \sum \dot{m}_e h_e \longrightarrow \dot{m}_{11} h_{11} + \dot{m}_2 h_2 = \dot{m}_3 h_3 \longrightarrow z h_{11} + (1 - y - z) h_2 = (1 - y) h_3$$

where z is the fraction of steam extracted from the turbine ($= \dot{m}_9 / \dot{m}_5$) at the second stage. Solving for z,

$$z = \frac{h_3 - h_2}{h_{11} - h_2}(1 - y) = \frac{236.16 - 69.86}{1356.2 - 69.86}(1 - 0.1300) = 0.1125$$

Then,

$$q_{in} = h_7 - h_6 + (1 - y)(h_{10} - h_9) = 1550.3 - 380.51 + (1 - 0.1300)(1531.0 - 1234.3) = 1427.9 \text{ Btu/lbm}$$

$$q_{out} = (1 - y - z)(h_{12} - h_1) = (1 - 0.1300 - 0.1125)(1052.3 - 69.74) = 744.3 \text{ Btu/lbm}$$

$$w_{net} = q_{in} - q_{out} = 1427.9 - 744.3 = 683.6 \text{ Btu/lbm}$$

and

$$\dot{m} = \frac{\dot{Q}_{in}}{q_{in}} = \frac{6 \times 10^5 \text{ Btu/s}}{1427.9 \text{ Btu/lbm}} = \mathbf{420.2\ lbm/s}$$

(*b*)

$$\dot{W}_{net} = \dot{m} w_{net} = (420.2 \text{ lbm/s})(683.6 \text{ Btu/lbm})\left(\frac{1.055 \text{ kJ}}{1 \text{ Btu}}\right) = \mathbf{303.0\ MW}$$

(*c*)

$$\eta_{th} = 1 - \frac{q_{out}}{q_{in}} = 1 - \frac{744.3 \text{ Btu/lbm}}{1427.9 \text{ Btu/lbm}} = \mathbf{47.9\%}$$

Second-Law Analysis of Vapor Power Cycles

9-46 The exergy destructions associated with each of the processes of the Rankine cycle described in Prob. 9-15 are to be determined for the specified source and sink temperatures.

Assumptions **1** Steady operating conditions exist. **2** Kinetic and potential energy changes are negligible.

Analysis From Problem 9-15,

$$s_1 = s_2 = s_{f\,@\,50\text{ kPa}} = 1.0910 \text{ kJ/kg}\cdot\text{K}$$
$$s_3 = s_4 = 6.9212 \text{ kJ/kg}\cdot\text{K}$$
$$q_{in} = 2887.37 \text{ kJ/kg}$$
$$q_{\text{out}} = 2067.01 \text{ kJ/kg}$$

Processes 1-2 and 3-4 are isentropic. Thus, $i_{12} = \mathbf{0}$ and $i_{34} = \mathbf{0}$. Also,

$$x_{\text{destroyed},23} = T_0\left(s_3 - s_2 + \frac{q_{R,23}}{T_R}\right) = (290\text{ K})\left(6.9212 - 1.0910 + \frac{-2887.37\text{ kJ/kg}}{1500\text{ K}}\right) = \mathbf{1132.5\ kJ/kg}$$

$$x_{\text{destroyed},41} = T_0\left(s_1 - s_4 + \frac{q_{R,41}}{T_R}\right) = (290\text{ K})\left(1.0910 - 6.9212 + \frac{2067.01\text{ kJ/kg}}{290\text{ K}}\right) = \mathbf{376.3\ kJ/kg}$$

9-47 The exergy destructions associated with each of the processes of the Rankine cycle described in Prob. 9-16 are to be determined for the specified source and sink temperatures.

Assumptions **1** Steady operating conditions exist. **2** Kinetic and potential energy changes are negligible.

Analysis From Problem 9-16,

$$s_1 = s_2 = s_{f\,@\,10\text{ kPa}} = 0.6493 \text{ kJ / kg}\cdot\text{K}$$
$$s_3 = s_4 = 6.5966 \text{ kJ / kg}\cdot\text{K}$$
$$q_{\text{in}} = 3171.78 \text{ kJ / kg}$$
$$q_{\text{out}} = 1897.47 \text{ kJ / kg}$$

Processes 1-2 and 3-4 are isentropic. Thus, $i_{12} = \mathbf{0}$ and $i_{34} = \mathbf{0}$. Also,

$$x_{\text{destroyed},23} = T_0\left(s_3 - s_2 + \frac{q_{R,23}}{T_R}\right) = (290\text{ K})\left(6.5966 - 0.6493 + \frac{-3171.78\text{ kJ/kg}}{1500\text{ K}}\right) = \mathbf{1111.5\ kJ/kg}$$

$$x_{\text{destroyed},41} = T_0\left(s_1 - s_4 + \frac{q_{R,41}}{T_R}\right) = (290\text{ K})\left(0.6493 - 6.5966 + \frac{1897.47\text{ kJ/kg}}{290\text{ K}}\right) = \mathbf{172.8\ kJ/kg}$$

9-48 The exergy destruction associated with the heat rejection process in Prob. 9-22 is to be determined for the specified source and sink temperatures. The exergy of the steam at the boiler exit is also to be determined.

Assumptions **1** Steady operating conditions exist. **2** Kinetic and potential energy changes are negligible.

Analysis From Problem 9-22,

$$s_1 = s_2 = s_{f\,@\,10\,\text{kPa}} = 0.6493\ \text{kJ/kg}\cdot\text{K}$$
$$s_3 = s_4 = 6.7975\ \text{kJ/kg}\cdot\text{K}$$
$$h_3 = 3410.5\ \text{kJ/kg}$$
$$q_{out} = 1962.1\ \text{kJ/kg}$$

The exergy destruction associated with the heat rejection process is

$$x_{\text{destroyed},41} = T_0\left(s_1 - s_4 + \frac{q_{R,41}}{T_R}\right) = (290\ \text{K})\left(0.6493 - 6.7975 + \frac{1962.1\ \text{kJ/kg}}{290\ \text{K}}\right) = \mathbf{179.1\ kJ/kg}$$

The exergy of the steam at the boiler exit is simply the flow exergy,

$$\psi_3 = (h_3 - h_0) - T_0(s_3 - s_0) + \frac{\mathbf{V}_3^2}{2}^{\nearrow 0} + qz_3^{\nearrow 0}$$
$$= (h_3 - h_0) - T_0(s_3 - s_0)$$

where
$$h_0 = h_{@(290\ \text{K},\ 100\ \text{kPa})} \cong h_{f\,@\,290\ \text{K}} = 71.38\ \text{kJ/kg}$$
$$s_0 = s_{@(290\ \text{K},100\ \text{kPa})} \cong s_{f\,@\,290\ \text{K}} = 0.2533\ \text{kJ/kg}\cdot\text{K}$$

Thus, $\psi_3 = (3410.5 - 71.38)\ \text{kJ/kg} - (290\ \text{K})(6.7975 - 0.2533)\ \text{kJ/kg}\cdot\text{K} = \mathbf{1441.3\ \ kJ/kg}$

9-49 The exergy destructions associated with each of the processes of the reheat Rankine cycle described in Prob. 9-27 are to be determined for the specified source and sink temperatures.

Assumptions **1** Steady operating conditions exist. **2** Kinetic and potential energy changes are negligible.

Analysis From Problem 9-27,

$$s_1 = s_2 = s_{f\,@\,20\,\text{kPa}} = 0.8320\ \text{kJ/kg}\cdot\text{K}$$
$$s_3 = s_4 = 6.7240\ \text{kJ/kg}\cdot\text{K}$$
$$s_5 = s_6 = 7.2338\ \text{kJ/kg}\cdot\text{K}$$
$$q_{23,\text{in}} = 3398.3 - 259.52 = 3138.8\ \text{kJ/kg}$$
$$q_{45,\text{in}} = 3465.5 - 3104.1 = 352.4\ \text{kJ/kg}$$
$$q_{out} = h_6 - h_1 = 2384.7 - 251.4 = 2133.3\ \text{kJ/kg}$$

Processes 1-2, 3-4, and 5-6 are isentropic. Thus, $i_{12} = i_{34} = i_{56} = \mathbf{0}$. Also,

$$x_{\text{destroyed},23} = T_0\left(s_3 - s_2 + \frac{q_{R,23}}{T_R}\right) = (300\ \text{K})\left(6.7240 - 0.8320 + \frac{-3138.8\ \text{kJ/kg}}{1800\ \text{K}}\right) = \mathbf{1244\ kJ/kg}$$

$$x_{\text{destroyed},45} = T_0\left(s_5 - s_4 + \frac{q_{R,45}}{T_R}\right) = (300\ \text{K})\left(7.2338 - 6.7240 + \frac{-352.4\ \text{kJ/kg}}{1800\ \text{K}}\right) = \mathbf{94.2\ kJ/kg}$$

$$x_{\text{destroyed},61} = T_0\left(s_1 - s_6 + \frac{q_{R,61}}{T_R}\right) = (300\ \text{K})\left(0.8320 - 7.2338 + \frac{2133.3\ \text{kJ/kg}}{300\ \text{K}}\right) = \mathbf{212.8\ kJ/kg}$$

9-50 EES solution of this (and other comprehensive problems designated with the *computer icon*) is available to instructors at the *Instructor Manual* section of the *Online Learning Center* (OLC) at www.mhhe.com/cengel-boles. See the Preface for access information.

9-51 The exergy destruction associated with the heat addition process and the expansion process in Prob. 9-29 are to be determined for the specified source and sink temperatures. The exergy of the steam at the boiler exit is also to be determined.

Assumptions **1** Steady operating conditions exist. **2** Kinetic and potential energy changes are negligible.

Analysis From Problem 9-29,

$$s_1 = s_2 = s_{f\ @\ 10\ \text{kPa}} = 0.6493\ \text{kJ/kg}\cdot\text{K}$$
$$s_3 = 6.5966\ \text{kJ/kg}\cdot\text{K}$$
$$s_4 = 6.8405\ \text{kJ/kg}\cdot\text{K}\quad (P_4 = 1\ \text{MPa},\ h_4 = 2900.74\ \text{kJ/kg})$$
$$s_5 = 7.7622\ \text{kJ/kg}\cdot\text{K}$$
$$s_6 = 8.3799\ \text{kJ/kg}\cdot\text{K}\quad (P_6 = 10\ \text{kPa},\ h_6 = 2663.9\ \text{kJ/kg})$$
$$h_3 = 3373.7\ \text{kJ/kg}$$
$$q_{in} = 3749.0\ \text{kJ/kg}$$

The exergy destruction associated with the combined pumping and the heat addition processes is

$$x_{\text{destroyed}} = T_0\left(s_3 - s_1 + s_5 - s_4 + \frac{q_{R,15}}{T_R}\right) = (290\ \text{K})\left(6.5966 - 0.6493 + 7.7622 - 6.8405 + \frac{-3749.0\ \text{kJ/kg}}{1500\ \text{K}}\right)$$
$$= 1267.2\ \text{kJ/kg}$$

The exergy destruction associated with the pumping process is

$$x_{\text{destroyed},12} \cong w_{p,a} - w_{p,s} = w_{p,a} - v\Delta P = 10.62 - 10.09 = 0.53\text{kJ/kg}$$

Thus,

$$x_{\text{destroyed, heating}} = x_{\text{destroyed}} - x_{\text{destroyed},12} = 1267.2 - 0.5 = \mathbf{1266.7kJ/kg}$$

The exergy destruction associated with the expansion process is

$$x_{\text{destroyed},34} = T_0\left((s_4 - s_3) + (s_6 - s_5) + \frac{q_{R,36}^{\nearrow 0}}{T_R}\right) = (290\ \text{K})(6.8405 - 6.5966 + 8.3799 - 7.7622)\text{kJ/kg}\cdot\text{K}$$
$$= \mathbf{249.9\ \ kJ/kg}$$

The exergy of the steam at the boiler exit is determined from

$$\psi_3 = (h_3 - h_0) - T_0(s_3 - s_0) + \frac{\mathbf{V}_3^{2\ \nearrow 0}}{2} + gz_3^{\nearrow 0}$$
$$= (h_3 - h_0) - T_0(s_3 - s_0)$$

where

$$h_0 = h_{@\ (290\ \text{K},\ 100\ \text{kPa})} \cong h_{f\ @\ 290\ \text{K}} = 71.38\ \text{kJ/kg}$$
$$s_0 = s_{@\ (290\ \text{K},\ 100\ \text{kPa})} \cong s_{f\ @\ 290\ \text{K}} = 0.2533\ \text{kJ/kg}\cdot\text{K}$$

Thus,

$$\psi_3 = (3373.7 - 71.38)\ \text{kJ/kg} - (290\ \text{K})(6.5966 - 0.2533)\ \text{kJ/kg}\cdot\text{K} = \mathbf{1462.8\ \ kJ/kg}$$

9-52 The exergy destruction associated with the regenerative cycle described in Prob. 9-38 is to be determined for the specified source and sink temperatures.

Assumptions **1** Steady operating conditions exist. **2** Kinetic and potential energy changes are negligible.

Analysis From Problem 9-38, q_{in} = 2691 kJ/kg and q_{out} = 1674.9 kJ/kg. Then the exergy destruction associated with this regenerative cycle is

$$x_{\text{destroyed},cycle} = T_0\left(\frac{q_{\text{out}}}{T_L} - \frac{q_{\text{in}}}{T_H}\right) = (290\text{ K})\left(\frac{1674.9\text{ kJ/kg}}{290\text{ K}} - \frac{2691\text{ kJ/kg}}{1500\text{ K}}\right) = \mathbf{1155\ kJ/kg}$$

9-53 The exergy destruction associated with the reheating and regeneration processes described in Prob. 9-43 are to be determined for the specified source and sink temperatures.

Assumptions **1** Steady operating conditions exist. **2** Kinetic and potential energy changes are negligible.

Analysis From Problem 9-43 and the steam tables,

$$\begin{aligned} y &= 0.2098 \\ s_3 &= s_{f\,@\,0.8\text{ MPa}} = 2.0462\text{ kJ/kg}\cdot\text{K} \\ s_5 &= s_6 = 6.7561\text{ kJ/kg}\cdot\text{K} \\ s_7 &= 7.8673\text{ kJ/kg}\cdot\text{K} \\ s_1 &= s_2 = s_{f\,@\,10\text{ kPa}} = 0.6493\text{ kJ/kg}\cdot\text{K} \\ q_{\text{reheat}} &= h_7 - h_6 = 3480.6 - 2811.9 = 668.7\text{ kJ/kg} \end{aligned}$$

Then the exergy destruction associated with reheat and regeneration processes are

$$\begin{aligned} x_{\text{destroyed},reheat} &= T_0\left(s_7 - s_6 + \frac{q_{R,67}}{T_R}\right) \\ &= (290\text{ K})\left(7.8673 - 6.7561 + \frac{-668.7\text{ kJ/kg}}{1800\text{ K}}\right) = \mathbf{214.5\ kJ/kg} \end{aligned}$$

$$\begin{aligned} x_{\text{destroyed},regen} &= T_0 s_{\text{gen}} = T_0\left(\sum m_e s_e - \sum m_i s_i + \frac{q_{\text{surr}}}{T_0}^{\nearrow 0}\right) = T_0\left(s_3 - y s_6 - (1-y)s_2\right) \\ &= (290\text{ K})[2.0462 - (0.2098)(6.7561) - (1 - 0.2098)(0.6493)] = \mathbf{33.6\ kJ/kg} \end{aligned}$$

Cogeneration

9-54C The utilization factor of a cogeneration plant is the ratio of the energy utilized for a useful purpose to the total energy supplied. It could be unity for a plant that does not produce any power.

9-55C No. A cogeneration plant may involve throttling, friction, and heat transfer through a finite temperature difference, and still have a utilization factor of unity.

9-56C Yes, if the cycle involves no irreversibilities such as throttling, friction, and heat transfer through a finite temperature difference.

9-57C Cogeneration is the production of more than one useful form of energy from the same energy source. Regeneration is the transfer of heat from the working fluid at some stage to the working fluid at some other stage.

9-58 A cogeneration plant is to generate power and process heat. Part of the steam extracted from the turbine at a relatively high pressure is used for process heating. The net power produced and the utilization factor of the plant are to be determined.

Assumptions **1** Steady operating conditions exist. **2** Kinetic and potential energy changes are negligible.

Analysis

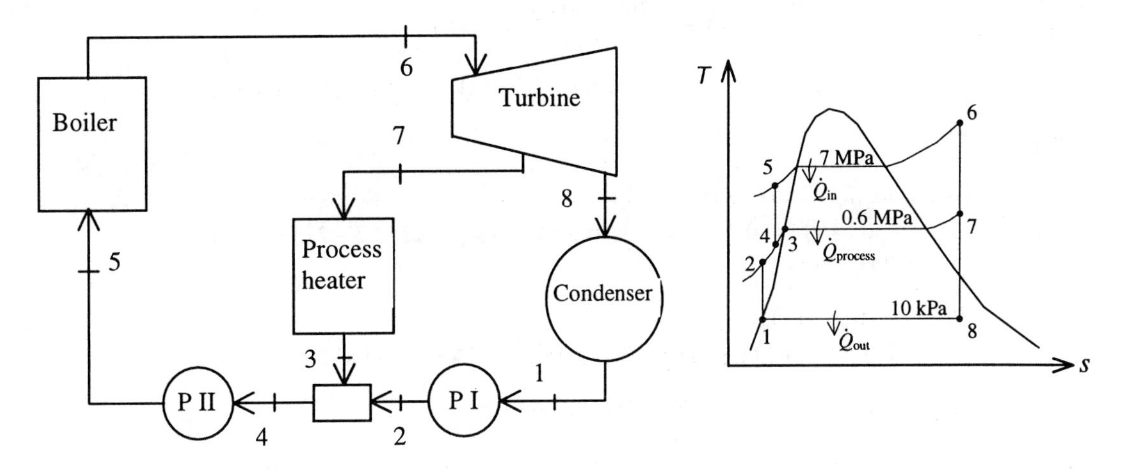

From the steam tables (Tables A-4, A-5, and A-6),

$$h_1 = h_{f\ @\ 10\ \text{kPa}} = 191.83\ \text{kJ/kg}$$
$$v_1 = v_{f\ @\ 10\ \text{kPa}} = 0.00101\ \text{m}^3/\text{kg}$$

$$\begin{aligned} w_{pI,in} &= v_1(P_2 - P_1) \\ &= (0.00101\ \text{m}^3/\text{kg})(600 - 10\ \text{kPa})\left(\frac{1\ \text{kJ}}{1\ \text{kPa}\cdot\text{m}^3}\right) \\ &= 0.60\ \text{kJ/kg} \end{aligned}$$
$$h_2 = h_1 + w_{pI,in} = 191.83 + 0.60 = 192.43\ \text{kJ/kg}$$

$$h_3 = h_{f\ @\ 0.6\ \text{MPa}} = 670.56\ \text{kJ/kg}$$

Mixing chamber:

$$\dot{E}_{in} - \dot{E}_{out} = \Delta\dot{E}_{system}{}^{\nearrow 0\ (\text{steady})} = 0$$
$$\dot{E}_{in} = \dot{E}_{out}$$
$$\sum \dot{m}_i h_i = \sum \dot{m}_e h_e \longrightarrow \dot{m}_4 h_4 = \dot{m}_2 h_2 + \dot{m}_3 h_3$$

or,
$$h_4 = \frac{\dot{m}_2 h_2 + \dot{m}_3 h_3}{\dot{m}_4} = \frac{(22.50)(192.43) + (7.50)(670.56)}{30} = 311.96\ \text{kJ/kg}$$
$$v_4 \cong v_{f\ @\ h_f = 311.96\ \text{kJ/kg}} = 0.001026\ \text{m}^3/\text{kg}$$

$$\begin{aligned} w_{pII,in} &= v_4(P_5 - P_4) \\ &= (0.001026\ \text{m}^3/\text{kg})(7000 - 600\ \text{kPa})\left(\frac{1\ \text{kJ}}{1\ \text{kPa}\cdot\text{m}^3}\right) \\ &= 6.57\ \text{kJ/kg} \end{aligned}$$
$$h_5 = h_4 + w_{pII,in} = 311.96 + 6.57 = 318.53\ \text{kJ/kg}$$

$$\left.\begin{array}{l} P_6 = 7 \text{ MPa} \\ T_6 = 500°\text{C} \end{array}\right\} \begin{array}{l} h_6 = 3410.3 \text{ kJ/kg} \\ s_6 = 6.7975 \text{ kJ/kg}\cdot\text{K} \end{array}$$

$$\left.\begin{array}{l} P_7 = 0.6 \text{ MPa} \\ s_7 = s_6 \end{array}\right\} h_7 = 2773.7 \text{ kJ/kg}$$

$$\left.\begin{array}{l} P_8 = 10 \text{ kPa} \\ s_8 = s_6 \end{array}\right\} \begin{array}{l} x_8 = \dfrac{s_8 - s_f}{s_{fg}} = \dfrac{6.7975 - 0.6493}{7.5009} = 0.8197 \\ h_8 = h_f + x_8 h_{fg} = 191.83 + (0.8197)(2392.8) = 2153.2 \text{ kJ/kg} \end{array}$$

Then,

$$\begin{aligned} \dot{W}_{\text{T,out}} &= \dot{m}_6(h_6 - h_7) + \dot{m}_8(h_7 - h_8) \\ &= (30 \text{ kg/s})(3410.3 - 2773.7)\text{kJ/kg} + (22.5 \text{ kg/s})(2773.7 - 2153.2)\text{kJ/kg} \\ &= 33{,}059 \text{ kW} \end{aligned}$$

$$\begin{aligned} \dot{W}_{\text{p,in}} &= \dot{m}_1 w_{\text{pI,in}} + \dot{m}_4 w_{\text{pII,in}} \\ &= (22.5 \text{ kg/s})(0.60 \text{ kJ/kg}) + (30 \text{ kg/s})(6.57 \text{ kJ/kg}) = 211 \text{ kW} \end{aligned}$$

$$\dot{W}_{\text{net}} = \dot{W}_{\text{T,out}} - \dot{W}_{\text{p,in}} = 33{,}059 - 211 = \mathbf{32{,}848 \text{ kW}}$$

Also,

$$\dot{Q}_{\text{process}} = \dot{m}_7(h_7 - h_3) = (7.5 \text{ kg/s})(2773.7 - 670.56) \text{ kJ/kg} = 15{,}774 \text{ kW}$$

$$\dot{Q}_{\text{in}} = \dot{m}_5(h_6 - h_5) = (30 \text{ kg/s})(3410.3 - 318.53) = 92{,}753 \text{ kW}$$

and

$$\varepsilon_u = \frac{\dot{W}_{\text{net}} + \dot{Q}_{\text{process}}}{\dot{Q}_{\text{in}}} = \frac{32{,}848 + 15{,}774}{92{,}753} = \mathbf{52.4\%}$$

9-59E A large food-processing plant requires steam at a relatively high pressure, which is extracted from the turbine of a cogeneration plant. The rate of heat transfer to the boiler and the power output of the cogeneration plant are to be determined.

Assumptions **1** Steady operating conditions exist. **2** Kinetic and potential energy changes are negligible.

Analysis

(*a*) From the steam tables (Tables A-4E, A-5E, and A-6E),

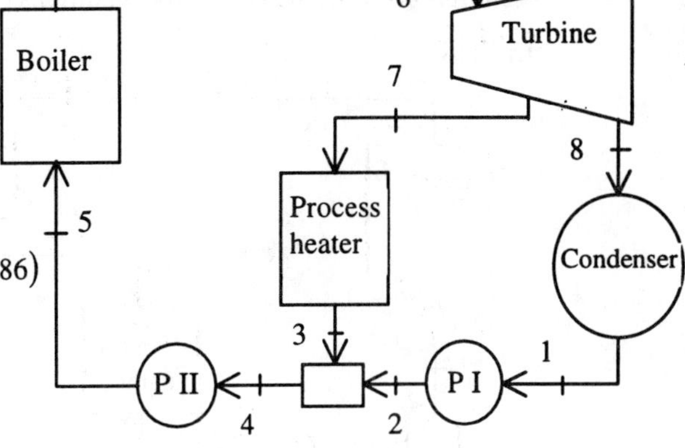

$$h_1 = h_{f\ @\ 2\ \text{psia}} = 94.02\ \text{Btu/lbm}$$
$$v_1 = v_{f\ @\ 2\ \text{psia}} = 0.01623\ \text{ft}^3/\text{lbm}$$

$$\begin{aligned} w_{\text{pI,in}} &= v_1(P_2 - P_1)/\eta_p \\ &= \left(0.01623\ \text{ft}^3/\text{lbm}\right)(80 - 2\ \text{psia})\left(\frac{1\ \text{Btu}}{5.4039\ \text{psia}\cdot\text{ft}^3}\right)/(0.86) \\ &= 0.27\ \text{Btu/lbm} \end{aligned}$$

$$h_2 = h_1 + w_{\text{pI,in}} = 94.02 + 0.27 = 94.29\ \text{Btu/lbm}$$

$$h_3 = h_{f\ @\ 80\ \text{psia}} = 282.21\ \text{Btu/lbm}$$

Mixing chamber:

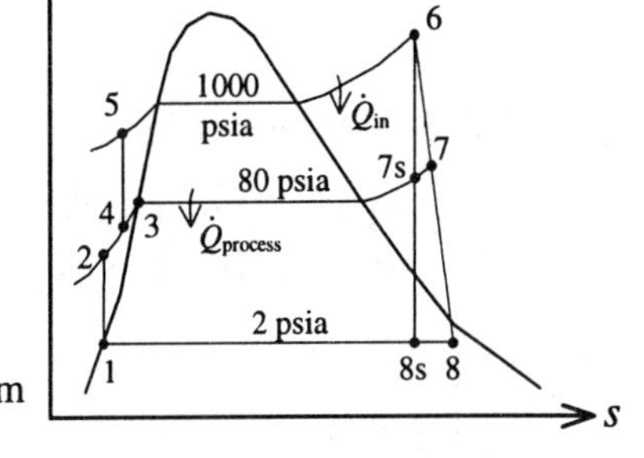

$$\dot{E}_{in} - \dot{E}_{out} = \Delta\dot{E}_{system}^{\nearrow 0\ (\text{steady})} = 0$$
$$\dot{E}_{in} = \dot{E}_{out}$$
$$\sum \dot{m}_i h_i = \sum \dot{m}_e h_e \longrightarrow \dot{m}_4 h_4 = \dot{m}_2 h_2 + \dot{m}_3 h_3$$

or,

$$h_4 = \frac{\dot{m}_2 h_2 + \dot{m}_3 h_3}{\dot{m}_4} = \frac{(3)(94.29) + (2)(282.21)}{5} = 169.46\ \text{Btu/lbm}$$

$$v_4 \cong v_{f\ @\ h_f = 169.46\ \text{Btu/lbm}} = 0.01664\ \text{ft}^3/\text{lbm}$$

$$\begin{aligned} w_{\text{pII,in}} &= v_4(P_5 - P_4)/\eta_p \\ &= \left(0.01664\ \text{ft}^3/\text{lbm}\right)(1000 - 80\ \text{psia})\left(\frac{1\ \text{Btu}}{5.4039\ \text{psia}\cdot\text{ft}^3}\right)/(0.86) \\ &= 3.29\ \text{Btu/lbm} \end{aligned}$$

$$h_5 = h_4 + w_{pII,\text{in}} = 169.46 + 3.29 = 172.75\ \text{Btu/lbm}$$

$$\left.\begin{array}{l} P_6 = 1000\ \text{psia} \\ T_6 = 1000°\text{F} \end{array}\right\} \begin{array}{l} h_6 = 1505.9\ \text{Btu/lbm} \\ s_6 = 1.653\ \text{Btu/lbm}\cdot\text{R} \end{array}$$

$$\left.\begin{array}{l} P_{7s} = 80\ \text{psia} \\ s_{7s} = s_6 \end{array}\right\} h_{7s} = 1208.8\ \text{Btu/lbm}$$

$$\left.\begin{array}{l} P_{8s} = 2\ \text{psia} \\ s_{8s} = s_6 \end{array}\right\} \begin{array}{l} x_{8s} = \dfrac{s_{8s} - s_f}{s_{fg}} = \dfrac{1.6530 - 0.17499}{1.7448} = 0.8471 \\ h_{8s} = h_f + x_{8s} h_{fg} = 94.02 + (0.8471)(1022.1) = 959.84\ \text{Btu/lbm} \end{array}$$

Then, $\dot{Q}_{\text{in}} = \dot{m}_5(h_6 - h_5) = (5\ \text{lbm/s})(1505.9 - 172.75)\text{Btu/lbm} = \mathbf{6666 Btu/s}$

(*b*)
$$\begin{aligned} \dot{W}_{\text{T,out}} &= \eta_T \dot{W}_{T,s} = \eta_T\left[\dot{m}_6(h_6 - h_{7s}) + \dot{m}_8(h_{7s} - h_{8s})\right] \\ &= (0.86)\left[(5\ \text{lbm/s})(1505.9 - 1208.8)\ \text{Btu/lbm} + (3\ \text{lbm/s})(1208.8 - 959.84)\ \text{Btu/lbm}\right] \\ &= 1920\ \text{Btu/s} = \mathbf{2026\ kW} \end{aligned}$$

9-60 A cogeneration plant has two modes of operation. In the first mode, all the steam leaving the turbine at a relatively high pressure is routed to the process heater. In the second mode, 60 percent of the steam is routed to the process heater and remaining is expanded to the condenser pressure. The power produced and the rate at which process heat is supplied in the first mode, and the power produced and the rate of process heat supplied in the second mode are to be determined.

Assumptions **1** Steady operating conditions exist. **2** Kinetic and potential energy changes are negligible.

Analysis

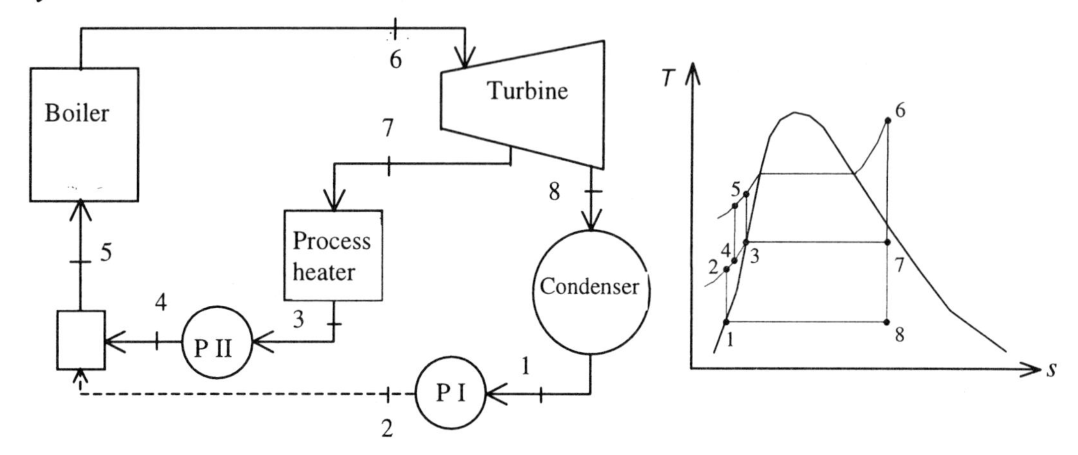

(*a*) From the steam tables (Tables A-4, A-5, and A-6),

$$h_1 = h_{f\ @\ 20\ \text{kPa}} = 251.40\ \text{kJ/kg}$$
$$v_1 = v_{f\ @\ 10\ \text{kPa}} = 0.001017\ \text{m}^3/\text{kg}$$

$$\begin{aligned} w_{\text{pI,in}} &= v_1(P_2 - P_1) \\ &= (0.001017\ \text{m}^3/\text{kg})(10{,}000 - 20\ \text{kPa})\left(\frac{1\ \text{kJ}}{1\ \text{kPa}\cdot\text{m}^3}\right) \\ &= 10.15\ \text{kJ/kg} \end{aligned}$$
$$h_2 = h_1 + w_{\text{pI,in}} = 251.40 + 10.15 = 261.55\ \text{kJ/kg}$$

$$h_3 = h_{f\ @\ 0.5\ \text{MPa}} = 640.23\ \text{kJ/kg}$$
$$v_3 = v_{f\ @\ 0.5\ \text{MPa}} = 0.001093\ \text{m}^3/\text{kg}$$

$$\begin{aligned} w_{\text{pII,in}} &= v_3(P_4 - P_3) \\ &= (0.001093\ \text{m}^3/\text{kg})(10{,}000 - 500\ \text{kPa})\left(\frac{1\ \text{kJ}}{1\ \text{kPa}\cdot\text{m}^3}\right) \\ &= 10.38\ \text{kJ/kg} \end{aligned}$$
$$h_4 = h_3 + w_{\text{pII,in}} = 640.23 + 10.38 = 650.61\ \text{kJ/kg}$$

Mixing chamber:

$$\dot{E}_{in} - \dot{E}_{out} = \Delta\dot{E}_{system}^{\nearrow 0\ \text{(steady)}} = 0 \rightarrow \dot{E}_{in} = \dot{E}_{out}$$
$$\sum \dot{m}_i h_i = \sum \dot{m}_e h_e \longrightarrow \dot{m}_5 h_5 = \dot{m}_2 h_2 + \dot{m}_4 h_4$$

or, $$h_5 = \frac{\dot{m}_2 h_2 + \dot{m}_4 h_4}{\dot{m}_5} = \frac{(2)(261.55) + (3)(650.61)}{5} = 495.0\ \text{kJ/kg}$$

$$\left.\begin{array}{l} P_6 = 10 \text{ MPa} \\ T_6 = 450°\text{C} \end{array}\right\} \begin{array}{l} h_6 = 3240.9 \text{ kJ/kg} \\ s_6 = 6.4190 \text{ kJ/kg} \cdot \text{K} \end{array}$$

$$\left.\begin{array}{l} P_7 = 0.5 \text{ MPa} \\ s_7 = s_6 \end{array}\right\} \begin{array}{l} x_7 = \dfrac{s_7 - s_f}{s_{fg}} = \dfrac{6.4190 - 1.8607}{4.9606} = 0.9189 \\ h_7 = h_f + x_7 h_{fg} = 640.23 + (0.9189)(2108.5) = 2577.7 \text{ kJ/kg} \end{array}$$

$$\left.\begin{array}{l} P_8 = 10 \text{ kPa} \\ s_8 = s_6 \end{array}\right\} \begin{array}{l} x_8 = \dfrac{s_8 - s_f}{s_{fg}} = \dfrac{6.4190 - 0.6493}{7.5009} = 0.7692 \\ h_8 = h_f + x_8 h_{fg} = 191.83 + (0.7692)(2392.8) = 2032.4 \text{ kJ/kg} \end{array}$$

When the entire steam is routed to the process heater,

$$\dot{W}_{\text{T,out}} = \dot{m}_6 (h_6 - h_7) = (5 \text{ kg/s})(3240.9 - 2577.7) \text{kJ/kg} = \mathbf{3316 \ kW}$$

$$\dot{Q}_{\text{process}} = \dot{m}_7 (h_7 - h_3) = (5 \text{ kg/s})(2577.7 - 640.23) \text{kJ/kg} = \mathbf{9687 \ kW}$$

(*b*) When only 60% of the steam is routed to the process heater,

$$\begin{aligned} \dot{W}_{\text{T,out}} &= \dot{m}_6 (h_6 - h_7) + \dot{m}_8 (h_7 - h_8) \\ &= (5 \text{ kg/s})(3240.9 - 2577.7) \text{ kJ/kg} + (2 \text{ kg/s})(2577.7 - 2032.4) \text{ kJ/kg} \\ &= \mathbf{4407 \ kW} \end{aligned}$$

$$\dot{Q}_{\text{process}} = \dot{m}_7 (h_7 - h_3) = (3 \text{ kg/s})(2577.7 - 640.23) \text{ kJ/kg} = \mathbf{5812 \ kW}$$

9-61 A cogeneration plant modified with regeneration is to generate power and process heat. The mass flow rate of steam through the boiler for a net power output of 15 MW is to be determined.

Assumptions **1** Steady operating conditions exist. **2** Kinetic and potential energy changes are negligible.

Analysis

From the steam tables (Tables A-4, A-5, and A-6),

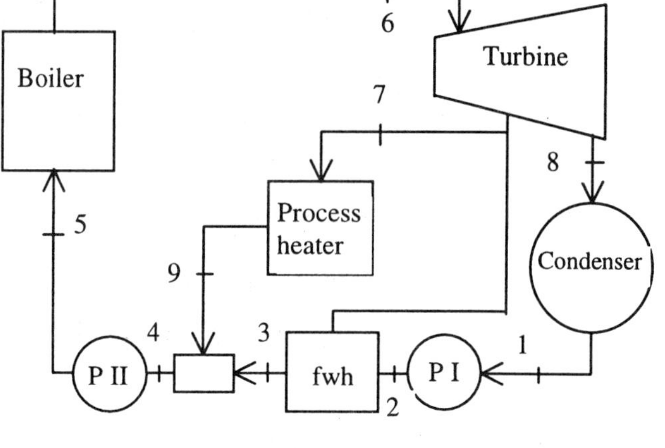

$$h_1 = h_{f\ @\ 10\ \text{kPa}} = 191.83\ \text{kJ/kg}$$
$$v_1 = v_{f\ @\ 10\ \text{kPa}} = 0.00101\ \text{m}^3/\text{kg}$$

$$\begin{aligned} w_{\text{pI,in}} &= v_1(P_2 - P_1) \\ &= (0.00101\ \text{m}^3/\text{kg})(400 - 10\ \text{kPa})\left(\frac{1\ \text{kJ}}{1\ \text{kPa}\cdot\text{m}^3}\right) \\ &= 0.39\ \text{kJ/kg} \end{aligned}$$

$$h_2 = h_1 + w_{\text{pI,in}} = 191.83 + 0.39 = 192.22\ \text{kJ/kg}$$
$$h_3 = h_4 = h_9 = h_{f\ @\ 0.4\ \text{MPa}} = 604.74\ \text{kJ/kg}$$
$$v_4 = v_{f\ @\ 0.4\ \text{MPa}} = 0.001084\ \text{m}^3/\text{kg}$$

$$\begin{aligned} w_{\text{pII,in}} &= v_4(P_5 - P_4) \\ &= (0.001084\ \text{m}^3/\text{kg})(6000 - 400\ \text{kPa})\left(\frac{1\ \text{kJ}}{1\ \text{kPa}\cdot\text{m}^3}\right) \\ &= 6.07\ \text{kJ/kg} \end{aligned}$$

$$h_4 = h_3 + w_{\text{pII,in}} = 604.74 + 6.07 = 610.81\ \text{kJ/kg}$$

$$\left.\begin{array}{l} P_6 = 6\ \text{MPa} \\ T_6 = 450°\text{C} \end{array}\right\} \begin{array}{l} h_6 = 3301.8\ \text{kJ/kg} \\ s_6 = 6.7193\ \text{kJ/kg}\cdot\text{K} \end{array}$$

$$\left.\begin{array}{l} P_7 = 0.4\ \text{MPa} \\ s_7 = s_6 \end{array}\right\} \begin{array}{l} x_7 = \dfrac{s_7 - s_f}{s_{fg}} = \dfrac{6.7193 - 1.7766}{5.1193} = 0.9655 \\ h_7 = h_f + x_7 h_{fg} = 604.74 + (0.9655)(2133.8) = 2664.9\ \text{kJ/kg} \end{array}$$

$$\left.\begin{array}{l} P_8 = 10\ \text{kPa} \\ s_8 = s_6 \end{array}\right\} \begin{array}{l} x_8 = \dfrac{s_8 - s_f}{s_{fg}} = \dfrac{6.7193 - 0.6493}{7.5009} = 0.8092 \\ h_8 = h_f + x_8 h_{fg} = 191.83 + (0.8092)(2392.8) = 2128.1\ \text{kJ/kg} \end{array}$$

Then, per kg of steam flowing through the boiler, we have

$$\begin{aligned} w_{\text{T,out}} &= (h_6 - h_7) + 0.4(h_7 - h_8) \\ &= (3301.8 - 2664.9)\ \text{kJ/kg} + (0.4)(2664.9 - 2128.1)\ \text{kJ/kg} \\ &= 851.62\ \text{kJ/kg} \end{aligned}$$

$$\begin{aligned} w_{\text{p,in}} &= 0.4 w_{\text{pI,in}} + w_{\text{pII,in}} \\ &= (0.4)(0.39\ \text{kJ/kg}) + (6.07\ \text{kJ/kg}) \\ &= 6.23\ \text{kJ/kg} \end{aligned}$$

$$w_{\text{net}} = w_{\text{T,out}} - w_{\text{p,in}} = 851.62 - 6.23 = 845.39\ \text{kJ/kg}$$

Thus,

$$\dot{m} = \frac{\dot{W}_{\text{net}}}{w_{\text{net}}} = \frac{15{,}000\ \text{kJ/s}}{845.39\ \text{kJ/kg}} = \mathbf{17.7\ kg/s}$$

9-62 EES solution of this (and other comprehensive problems designated with the *computer icon*) is available to instructors at the *Instructor Manual* section of the *Online Learning Center* (OLC) at www.mhhe.com/cengel-boles. See the Preface for access information.

9-63E A cogeneration plant is to generate power while meeting the process steam requirements for a certain industrial application. The net power produced, the rate of process heat supply, and the utilization factor of this plant are to be determined.

Assumptions **1** Steady operating conditions exist. **2** Kinetic and potential energy changes are negligible.

Analysis

(*a*) From the steam tables (Tables A-4E, A-5E, and A-6E),

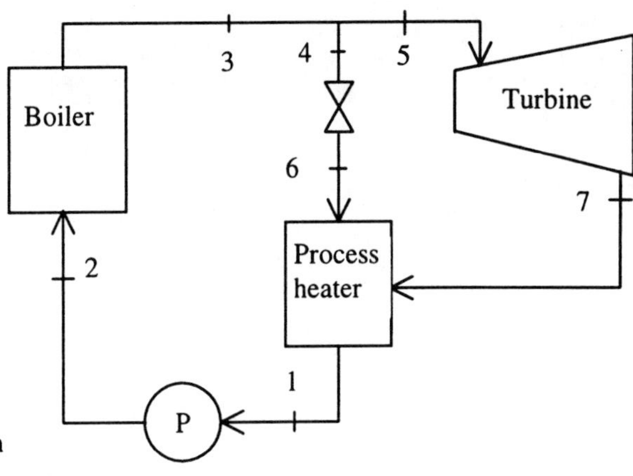

$$h_1 \cong h_{f\ @\ 240°F} = 208.44\ \text{Btu/lbm}$$
$$h_2 \cong h_1$$

$$\left.\begin{array}{l} P_3 = 800\ \text{psia} \\ T_3 = 900°F \end{array}\right\} \begin{array}{l} h_3 = 1455.6\ \text{Btu/lbm} \\ s_3 = s_5 = s_7 = 1.6408\ \text{Btu/lbm}\cdot\text{R} \end{array}$$

$$h_3 = h_4 = h_5 = h_6$$

$$\left.\begin{array}{l} P_7 = 120\ \text{psia} \\ s_7 = s_3 \end{array}\right\} h_7 = 1235.0\ \text{Btu/lbm}$$

$$\dot{W}_{\text{net}} = \dot{m}_5(h_5 - h_7) = (10\text{lbm/s})(1455.6 - 1235.0)\ \text{Btu/lbm}$$
$$= 2206\ \text{Btu/s} = \mathbf{2327.3\ kW}$$

(*b*) $$\dot{Q}_{\text{process}} = \sum \dot{m}_e h_e - \sum \dot{m}_i h_i = \dot{m}_1 h_1 - \dot{m}_6 h_6 - \dot{m}_7 h_7$$
$$= (15)(208.44) - (5)(1455.6) - (10)(1235.0)$$
$$= \mathbf{-16{,}501\ Btu/s}$$

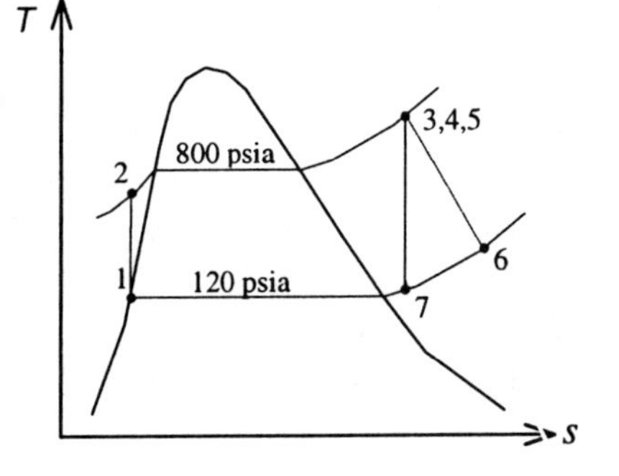

(*c*) $\varepsilon_u = \mathbf{1}$ since all the energy is utilized.

Combined Gas-Vapor Power Cycles

9-64C The energy source of the steam is the waste energy of the exhausted combustion gases.

9-65C Because the combined gas-steam cycle takes advantage of the desirable characteristics of the gas cycle at high temperature, and those of steam cycle at low temperature, and combines them. The result is a cycle that is more efficient than either cycle executed operated alone.

9-66 A combined gas-steam power cycle is considered. The topping cycle is a gas-turbine cycle and the bottoming cycle is a simple ideal Rankine cycle. The mass flow rate of the steam, the net power output, and the thermal efficiency of the combined cycle are to be determined.

Assumptions **1** Steady operating conditions exist. **2** Kinetic and potential energy changes are negligible. **3** Air is an ideal gas with constant specific heats.

Properties The properties of air at room temperature are C_p = 1.005 kJ/kg K and k = 1.4 (Table A-2).

Analysis (*a*) The analysis of gas cycle yields

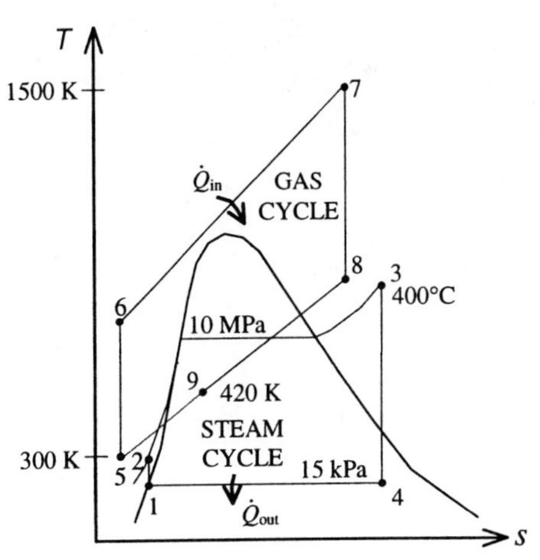

$$T_6 = T_5\left(\frac{P_6}{P_5}\right)^{(k-1)/k} = (300\text{ K})(16)^{0.4/1.4} = 662.5\text{ K}$$

$$\dot{Q}_{in} = \dot{m}_{air}(h_7 - h_6) = \dot{m}_{air}C_p(T_7 - T_6)$$
$$= (14\text{ kg/s})(1.005\text{ kJ/kg}\cdot\text{K})(1500 - 662.5)\text{ K} = 11{,}784\text{ kW}$$

$$\dot{W}_{C,gas} = \dot{m}_{air}(h_6 - h_5) = \dot{m}_{air}C_p(T_6 - T_5)$$
$$= (14\text{ kg/s})(1.005\text{ kJ/kg}\cdot\text{K})(662.5 - 300)\text{ K} = 5{,}100\text{ kW}$$

$$T_8 = T_7\left(\frac{P_8}{P_7}\right)^{(k-1)/k} = (1500\text{K})\left(\frac{1}{16}\right)^{0.4/1.4} = 679.3\text{ K}$$

$$\dot{W}_{T,gas} = \dot{m}_{air}(h_7 - h_8) = \dot{m}_{air}C_p(T_7 - T_8)$$
$$= (14\text{ kg/s})(1.005\text{ kJ/kg}\cdot\text{K})(1500 - 679.3)\text{ K} = 11{,}547\text{ kW}$$

$$\dot{W}_{net,gas} = \dot{W}_{T,gas} - \dot{W}_{C,gas} = 11{,}547 - 5{,}100 = 6447\text{ kW}$$

From the steam tables (Tables A-4, A-5, and A-6),

$$h_1 = h_{f\,@\,15\text{ kPa}} = 225.94\text{ kJ/kg}$$
$$v_1 = v_{f\,@\,15\text{ kPa}} = 0.001014\text{ m}^3\text{/kg}$$

$$w_{pI,in} = v_1(P_2 - P_1)$$
$$= (0.001014\text{ m}^3\text{/kg})(10{,}000 - 15\text{ kPa})\left(\frac{1\text{ kJ}}{1\text{ kPa}\cdot\text{m}^3}\right)$$
$$= 10.12\text{ kJ/kg}$$
$$h_2 = h_1 + w_{pI,in} = 225.94 + 10.12 = 236.06\text{ kJ/kg}$$

$$\left.\begin{array}{l}P_3 = 10\text{ MPa}\\ T_3 = 400°\text{C}\end{array}\right\}\begin{array}{l}h_3 = 3096.5\text{ kJ/kg}\\ s_3 = 6.2120\text{ kJ/kg}\cdot\text{K}\end{array}$$

$$\left.\begin{array}{l}P_4 = 15\text{ kPa}\\ s_4 = s_3\end{array}\right\}\begin{array}{l}x_4 = \dfrac{s_4 - s_f}{s_{fg}} = \dfrac{6.2120 - 0.7549}{7.2536} = 0.7523\\ h_4 = h_f + x_4 h_{fg} = 225.94 + (0.7523)(2373.1) = 2011.3\text{ kJ/kg}\end{array}$$

Noting that $\dot{Q} \cong \dot{W} \cong \Delta ke \cong \Delta pe \cong 0$ for the heat exchanger, the steady-flow energy balance equation yields

$$\dot{E}_{in} - \dot{E}_{out} = \Delta\dot{E}_{system}^{\nearrow 0\text{ (steady)}} = 0$$
$$\dot{E}_{in} = \dot{E}_{out}$$
$$\sum \dot{m}_i h_i = \sum \dot{m}_e h_e \longrightarrow \dot{m}_s(h_3 - h_2) = \dot{m}_{air}(h_8 - h_9)$$
$$\dot{m}_s = \frac{h_8 - h_9}{h_3 - h_2}\dot{m}_{air} = \frac{C_p(T_8 - T_9)}{h_3 - h_2}\dot{m}_{air} = \frac{(1.005\text{ kJ/kg}\cdot\text{K})(679.3 - 420)\text{ K}}{(3096.5 - 236.06)\text{ kJ/kg}}(14\text{ kg/s}) = \mathbf{1.275\ kg/s}$$

(*b*)

$$\dot{W}_{T,\text{steam}} = \dot{m}_s(h_3 - h_4) = (1.275\ \text{kg/s})(3096.5 - 2011.3)\ \text{kJ/kg} = 1384\ \text{kW}$$

$$\dot{W}_{p,\text{steam}} = \dot{m}_s w_p = (1.275\ \text{kg/s})(10.12\ \text{kJ/kg}) = 13\ \text{kW}$$

$$\dot{W}_{\text{net,steam}} = \dot{W}_{T,\text{steam}} - \dot{W}_{p,\text{steam}} = 1384 - 13 = 1371\ \text{kW}$$

and

$$\dot{W}_{\text{net}} = \dot{W}_{\text{net, steam}} + \dot{W}_{\text{net, gas}} = 1371 + 6447 = \mathbf{7818\ kW}$$

(*c*)

$$\eta_{th} = \frac{\dot{W}_{\text{net}}}{\dot{Q}_{\text{in}}} = \frac{7818\ \text{kW}}{11{,}784\ \text{kW}} = \mathbf{66.3\%}$$

9-67 [*Also solved by EES on enclosed CD*] A 450-MW combined gas-steam power plant is considered. The topping cycle is a gas-turbine cycle and the bottoming cycle is an ideal Rankine cycle with an open feedwater heater. The mass flow rate of air to steam, the required rate of heat input in the combustion chamber, and the thermal efficiency of the combined cycle are to be determined.

Assumptions **1** Steady operating conditions exist. **2** Kinetic and potential energy changes are negligible. **3** Air is an ideal gas with variable specific heats.

Analysis (*a*) The analysis of gas cycle yields (Table A-17)

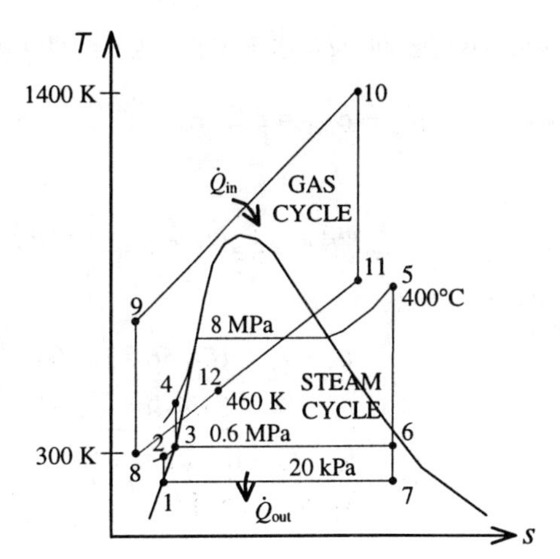

$$T_8 = 300 \text{ K} \longrightarrow h_8 = 300.19 \text{ kJ/kg}$$
$$P_{r_8} = 1.386$$

$$P_{r_9} = \frac{P_9}{P_8} P_{r_8} = (14)(1.386) = 19.40 \longrightarrow h_9 = 635.5 \text{ kJ/kg}$$

$$T_{10} = 1400 \text{ K} \longrightarrow h_{10} = 1515.42 \text{ kJ/kg}$$
$$P_{r_{10}} = 450.5$$

$$P_{r_{11}} = \frac{P_{11}}{P_{10}} P_{r_{10}} = \left(\frac{1}{14}\right)(450.5) = 32.18 \longrightarrow h_{11} = 735.8 \text{ kJ/kg}$$

$$T_{12} = 460 \text{ K} \longrightarrow h_{12} = 462.02 \text{ kJ/kg}$$

From the steam tables (Tables A-4, A-5, A-6),

$$h_1 = h_{f\ @\ 20\ \text{kPa}} = 251.40 \text{ kJ/kg}$$
$$v_1 = v_{f\ @\ 20\ \text{kPa}} = 0.001017 \text{ m}^3\text{/kg}$$

$$\begin{aligned} w_{\text{pI,in}} &= v_1(P_2 - P_1) \\ &= (0.001017 \text{ m}^3\text{/kg})(600 - 20 \text{ kPa})\left(\frac{1 \text{ kJ}}{1 \text{ kPa} \cdot \text{m}^3}\right) \\ &= 0.59 \text{ kJ/kg} \end{aligned}$$

$$h_2 = h_1 + w_{\text{pI,in}} = 251.40 + 0.59 = 251.99 \text{ kJ/kg}$$

$$h_3 = h_{f\ @\ 0.6\ \text{MPa}} = 670.56 \text{ kJ/kg}$$
$$v_3 = v_{f\ @\ 0.6\ \text{MPa}} = 0.001101 \text{ m}^3\text{/kg}$$

$$\begin{aligned} w_{\text{pII,in}} &= v_3(P_4 - P_3) \\ &= (0.001104 \text{ m}^3\text{/kg})(8{,}000 - 600 \text{ kPa})\left(\frac{1 \text{ kJ}}{1 \text{ kPa} \cdot \text{m}^3}\right) \\ &= 8.15 \text{ kJ/kg} \end{aligned}$$

$$h_4 = h_3 + w_{\text{pI,in}} = 670.56 + 8.15 = 678.71 \text{ kJ/kg}$$

$$\left.\begin{array}{l} P_5 = 8 \text{ MPa} \\ T_5 = 400°\text{C} \end{array}\right\} \begin{array}{l} h_5 = 3138.3 \text{ kJ/kg} \\ s_5 = 6.3634 \text{ kJ/kg} \cdot \text{K} \end{array}$$

$$\left.\begin{array}{l} P_6 = 0.6 \text{ MPa} \\ s_6 = s_5 \end{array}\right\} \begin{array}{l} x_6 = \dfrac{s_6 - s_f}{s_{fg}} = \dfrac{6.3634 - 1.9312}{4.8288} = 0.9179 \\ h_6 = h_f + x_6 h_{fg} = 670.56 + (0.9179)(2086.3) = 2585.6 \text{ kJ/kg} \end{array}$$

$$\left.\begin{array}{l} P_7 = 20 \text{ kPa} \\ s_7 = s_5 \end{array}\right\} \begin{array}{l} x_7 = \dfrac{s_7 - s_f}{s_{fg}} = \dfrac{6.3634 - 0.8320}{7.0766} = 0.7816 \\ h_7 = h_f + x_7 h_{fg} = 251.40 + (0.7816)(2358.3) = 2094.8 \text{ kJ/kg} \end{array}$$

Noting that $\dot{Q} \cong \dot{W} \cong \Delta ke \cong \Delta pe \cong 0$ for the heat exchanger, the steady-flow energy balance equation yields

$$\dot{E}_{in} - \dot{E}_{out} = \Delta\dot{E}_{system}^{\nearrow 0\,(steady)} = 0$$

$$\dot{E}_{in} = \dot{E}_{out}$$

$$\sum \dot{m}_i h_i = \sum \dot{m}_e h_e \longrightarrow \dot{m}_s(h_5 - h_4) = \dot{m}_{air}(h_{11} - h_{12})$$

$$\frac{\dot{m}_{air}}{\dot{m}_s} = \frac{h_5 - h_4}{h_{11} - h_{12}} = \frac{3138.3 - 678.71}{735.80 - 462.02} = \mathbf{9.0\ \ kg\ air / kg\ steam}$$

(*b*) Noting that $\dot{Q} \cong \dot{W} \cong \Delta ke \cong \Delta pe \cong 0$ for the open FWH, the steady-flow energy balance equation yields

$$\dot{E}_{in} - \dot{E}_{out} = \Delta\dot{E}_{system}^{\nearrow 0\,(steady)} = 0$$

$$\dot{E}_{in} = \dot{E}_{out}$$

$$\sum \dot{m}_i h_i = \sum \dot{m}_e h_e \longrightarrow \dot{m}_2 h_2 + \dot{m}_6 h_6 = \dot{m}_3 h_3 \longrightarrow y h_6 + (1-y)h_2 = (1)h_3$$

Thus,

$$y = \frac{h_3 - h_2}{h_6 - h_2} = \frac{670.56 - 251.99}{2585.6 - 251.99} = 0.1794 \quad (\text{the fraction of steam extracted})$$

$$w_T = h_5 - h_6 + (1-y)(h_6 - h_7)$$
$$= 3138.3 - 2585.6 + (1 - 0.1794)(2585.6 - 2094.8) = 955.5 \text{ kJ/kg}$$

$$w_{net,steam} = w_T - w_{p,in} = w_T - (1-y)w_{p,I} - w_{p,II}$$
$$= 955.5 - (1 - 0.1794)(0.59) - 8.15 = 946.9 \text{ kJ/kg}$$

$$w_{net,gas} = w_T - w_{C,in} = (h_{10} - h_{11}) - (h_9 - h_8)$$
$$= 1515.42 - 735.8 - (635.5 - 300.19) = 444.31 \text{ kJ/kg}$$

The net work output per unit mass of gas is

$$w_{net} = w_{net,gas} + \tfrac{1}{9} w_{net,steam} = 444.3 + \tfrac{1}{9}(946.9) = 549.51 \text{ kJ/kg}$$

$$\dot{m}_{air} = \frac{\dot{W}_{net}}{w_{net}} = \frac{450{,}000 \text{ kJ/s}}{549.51 \text{ kJ/kg}} = 819 \text{ kg/s}$$

and

$$\dot{Q}_{in} = \dot{m}_{air}(h_{10} - h_9) - (819 \text{ kg/s})(1515.42 - 635.5) \text{ kJ/kg} = \mathbf{720{,}655\ \ kW}$$

(*c*)

$$\eta_{th} = \frac{\dot{W}_{net}}{\dot{Q}_{in}} = \frac{450{,}000 \text{ kW}}{720{,}655 \text{ kW}} = \mathbf{62.4\%}$$

9-68 EES solution of this (and other comprehensive problems designated with the *computer icon*) is available to instructors at the *Instructor Manual* section of the *Online Learning Center* (OLC) at www.mhhe.com/cengel-boles. See the Preface for access information.

9-69 A 450-MW combined gas-steam power plant is considered. The topping cycle is a gas-turbine cycle and the bottoming cycle is a nonideal Rankine cycle with an open feedwater heater. The mass flow rate of air to steam, the required rate of heat input in the combustion chamber, and the thermal efficiency of the combined cycle are to be determined.

Assumptions **1** Steady operating conditions exist. **2** Kinetic and potential energy changes are negligible. **3** Air is an ideal gas with variable specific heats.

Analysis (*a*) Using the properties of air from Table A-17, the analysis of gas cycle yields

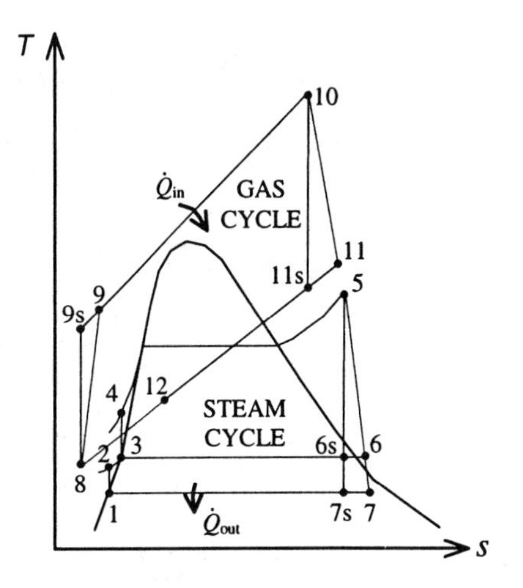

$$T_8 = 300 \text{ K} \longrightarrow h_8 = 300.19 \text{ kJ/kg}$$
$$P_{r_8} = 1.386$$

$$P_{r_9} = \frac{P_9}{P_8} P_{r_8} = (14)(1.386) = 19.40 \longrightarrow h_{9s} = 635.5 \text{ kJ/kg}$$

$$\eta_C = \frac{h_{9s} - h_8}{h_9 - h_8} \longrightarrow h_9 = h_8 + (h_{9s} - h_8)/\eta_C$$
$$= 300.19 + (635.5 - 300.19)/(0.82)$$
$$= 709.1 \text{ kJ/kg}$$

$$T_{10} = 1400 \text{ K} \longrightarrow h_{10} = 1515.42 \text{ kJ/kg}$$
$$P_{r_{10}} = 450.5$$

$$P_{r_{11}} = \frac{P_{11}}{P_{10}} P_{r_{10}} = \left(\frac{1}{14}\right)(450.5) = 32.18 \longrightarrow h_{11s} = 735.8 \text{ kJ/kg}$$

$$\eta_T = \frac{h_{10} - h_{11}}{h_{10} - h_{11s}} \longrightarrow h_{11} = h_{10} - \eta_T (h_{10} - h_{11s})$$
$$= 1515.42 - (0.86)(1515.42 - 735.8)$$
$$= 844.95 \text{ kJ/kg}$$

$$T_{12} = 460 \text{ K} \longrightarrow h_{12} = 462.02 \text{ kJ/kg}$$

From the steam tables (Tables A-4, A-5, and A-6),

$$h_1 = h_{f\text{ @ 20 kPa}} = 251.40 \text{ kJ/kg}$$
$$v_1 = v_{f\text{ @ 20 kPa}} = 0.001017 \text{ m}^3\text{/kg}$$

$$w_{\text{pI,in}} = v_1(P_2 - P_1)$$
$$= (0.001017 \text{ m}^3\text{/kg})(600 - 20 \text{ kPa})\left(\frac{1 \text{ kJ}}{1 \text{ kPa} \cdot \text{m}^3}\right)$$
$$= 0.59 \text{ kJ/kg}$$

$$h_2 = h_1 + w_{\text{pI,in}} = 251.40 + 0.59 = 251.99 \text{ kJ/kg}$$

$$h_3 = h_{f\text{ @ 0.6 MPa}} = 670.56 \text{ kJ/kg}$$
$$v_3 = v_{f\text{ @ 0.6 MPa}} = 0.001101 \text{ m}^3\text{/kg}$$

$$w_{\text{pII,in}} = v_3(P_4 - P_3)$$
$$= (0.001101 \text{ m}^3\text{/kg})(8{,}000 - 600 \text{ kPa})\left(\frac{1 \text{ kJ}}{1 \text{ kPa} \cdot \text{m}^3}\right)$$
$$= 8.15 \text{ kJ/kg}$$

$$h_4 = h_3 + w_{\text{pI,in}} = 670.56 + 8.15 = 678.71 \text{ kJ/kg}$$

$$\left.\begin{array}{l} P_5 = 8 \text{ MPa} \\ T_5 = 400°\text{C} \end{array}\right\} \begin{array}{l} h_5 = 3138.3 \text{ kJ/kg} \\ s_5 = 6.3634 \text{ kJ/kg}\cdot\text{K} \end{array}$$

$$\left.\begin{array}{l} P_6 = 0.6 \text{ MPa} \\ s_{6s} = s_5 \end{array}\right\} \begin{array}{l} x_{6s} = \dfrac{s_{6s} - s_f}{s_{fg}} = \dfrac{6.3634 - 1.9312}{4.8288} = 0.9179 \\ h_{6s} = h_f + x_{6s} h_{fg} = 670.56 + (0.9179)(2086.3) = 2585.6 \text{ kJ/kg} \end{array}$$

$$\eta_T = \frac{h_5 - h_6}{h_5 - h_{6s}} \longrightarrow h_6 = h_5 - \eta_T (h_5 - h_{6s}) = 3138.3 - (0.86)(3138.3 - 2585.6) = 2663.0 \text{ kJ/kg}$$

$$\left.\begin{array}{l} P_7 = 20 \text{ kPa} \\ s_7 = s_5 \end{array}\right\} \begin{array}{l} x_7 = \dfrac{s_7 - s_f}{s_{fg}} = \dfrac{6.3634 - 0.8320}{7.0766} = 0.7816 \\ h_7 = h_f + x_7 h_{fg} = 251.40 + (0.7816)(2358.3) = 2094.8 \text{ kJ/kg} \end{array}$$

Noting that $\dot{Q} \cong \dot{W} \cong \Delta ke \cong \Delta pe \cong 0$ for the heat exchanger, the steady-flow energy balance equation yields

$$\dot{E}_{in} - \dot{E}_{out} = \Delta \dot{E}_{system}{}^{\nearrow 0 \text{ (steady)}} = 0$$

$$\dot{E}_{in} = \dot{E}_{out}$$

$$\sum \dot{m}_i h_i = \sum \dot{m}_e h_e \longrightarrow \dot{m}_s (h_5 - h_4) = \dot{m}_{air} (h_{11} - h_{12})$$

$$\frac{\dot{m}_{air}}{\dot{m}_s} = \frac{h_5 - h_4}{h_{11} - h_{12}} = \frac{3138.3 - 678.71}{844.95 - 462.02} = \mathbf{6.423 \text{ kg air / kg steam}}$$

(*b*) Noting that $\dot{Q} \cong \dot{W} \cong \Delta ke \cong \Delta pe \cong 0$ for the open FWH, the steady-flow energy balance equation yields

$$\dot{E}_{in} - \dot{E}_{out} = \Delta \dot{E}_{system}{}^{\nearrow 0 \text{ (steady)}} = 0 \rightarrow \dot{E}_{in} = \dot{E}_{out}$$

$$\sum \dot{m}_i h_i = \sum \dot{m}_e h_e \longrightarrow \dot{m}_2 h_2 + \dot{m}_6 h_6 = \dot{m}_3 h_3 \longrightarrow y h_6 + (1 - y) h_2 = (1) h_3$$

Thus,

$$y = \frac{h_3 - h_2}{h_6 - h_2} = \frac{670.56 - 251.99}{2663.0 - 251.99} = 0.1736 \quad \text{(the fraction of steam extracted)}$$

$$\begin{aligned} w_T &= \eta_T [h_5 - h_{6s} + (1 - y)(h_{6s} - h_7)] \\ &= (0.86)[3138.3 - 2585.6 + (1 - 0.1736)(2585.6 - 2094.8)] = 824.1 \text{ kJ/kg} \end{aligned}$$

$$\begin{aligned} w_{net,steam} &= w_T - w_{p,in} = w_T - (1 - y) w_{p,I} - w_{p,II} \\ &= 824.1 - (1 - 0.1736)(0.59) - 8.15 = 815.5 \text{kJ/kg} \end{aligned}$$

$$\begin{aligned} w_{net,gas} &= w_T - w_{C,in} = (h_{10} - h_{11}) - (h_9 - h_8) \\ &= 1515.42 - 844.95 - (709.1 - 300.19) = 261.56 \text{ kJ/kg} \end{aligned}$$

The net work output per unit mass of gas is

$$w_{net} = w_{net,gas} + \tfrac{1}{6.423} w_{net,steam} = 261.56 + \tfrac{1}{6.423}(815.5) = 388.53 \text{ kJ/kg}$$

$$\dot{m}_{air} = \frac{\dot{W}_{net}}{w_{net}} = \frac{450{,}000 \text{ kJ/s}}{388.53 \text{ kJ/kg}} = \mathbf{1158 \text{ kg/s}}$$

and $\quad \dot{Q}_{in} = \dot{m}_{air}(h_{10} - h_9) = (1158 \text{kg/s})(1515.42 - 709.1) \text{ kJ/kg} = \mathbf{933{,}71 \text{ kW}}$

(*c*) $\quad \eta_{th} = \dfrac{\dot{W}_{net}}{\dot{Q}_{in}} = \dfrac{450{,}000 \text{ kW}}{933{,}719 \text{ kW}} = \mathbf{48.2\%}$

9-70 EES solution of this problem is available to instructors at the *Instructor Manual* section of the *Online Learning Center* (OLC) at www.mhhe.com/cengel-boles. See the Preface for access information.

Special Topic: Binary Vapor Cycles

9-71C Binary power cycle is a cycle which is actually a combination of two cycles; one in the high temperature region, and the other in the low temperature region. Its purpose is to increase thermal efficiency.

9-72C Consider the heat exchanger of a binary power cycle. The working fluid of the topping cycle (cycle A) enters the heat exchanger at state 1 and leaves at state 2. The working fluid of the bottoming cycle (cycle B) enters at state 3 and leaves at state 4. Neglecting any changes in kinetic and potential energies, and assuming the heat exchanger is well-insulated, the steady-flow energy balance relation yields

$$\dot{E}_{in} - \dot{E}_{out} = \Delta\dot{E}_{system}^{\;\nearrow 0\,(steady)} = 0$$

$$\dot{E}_{in} = \dot{E}_{out}$$

$$\sum \dot{m}_e h_e = \sum \dot{m}_i h_i$$

$$\dot{m}_A h_2 + \dot{m}_B h_4 = \dot{m}_A h_1 + \dot{m}_B h_3 \;\; or \;\; \dot{m}_A (h_2 - h_1) = \dot{m}_B (h_3 - h_4)$$

Thus,

$$\frac{\dot{m}_A}{\dot{m}_B} = \frac{h_3 - h_4}{h_2 - h_1}$$

9-73C Steam is not an ideal fluid for vapor power cycles because its critical temperature is low, its saturation dome resembles an inverted V, and its condenser pressure is too low.

9-74C Because mercury has a high critical temperature, relatively low critical pressure, but a very low condenser pressure. It is also toxic, expensive, and has a low enthalpy of vaporization.

9-75C In binary vapor power cycles, both cycles are vapor cycles. In the combined gas-steam power cycle, one of the cycles is a gas cycle.

Review Problems

9-76 It is to be demonstrated that the thermal efficiency of a combined gas-steam power plant η_{cc} can be expressed as $\eta_{cc} = \eta_g + \eta_s - \eta_g\eta_s$ where $\eta_g = W_g / Q_{in}$ and $\eta_s = W_s / Q_{g,out}$ are the thermal efficiencies of the gas and steam cycles, respectively, and the efficiency of a combined cycle is to be obtained.

Analysis The thermal efficiencies of gas, steam, and combined cycles can be expressed as

$$\eta_{cc} = \frac{W_{total}}{Q_{in}} = 1 - \frac{Q_{out}}{Q_{in}}$$

$$\eta_g = \frac{W_g}{Q_{in}} = 1 - \frac{Q_{g,out}}{Q_{in}}$$

$$\eta_s = \frac{W_s}{Q_{g,out}} = 1 - \frac{Q_{out}}{Q_{g,out}}$$

where Q_{in} is the heat supplied to the gas cycle, where Q_{out} is the heat rejected by the steam cycle, and where $Q_{g,out}$ is the heat rejected from the gas cycle and supplied to the steam cycle.

Using the relations above, the expression $\eta_g + \eta_s - \eta_g\eta_s$ can be expressed as

$$\begin{aligned}
\eta_g + \eta_s - \eta_g\eta_s &= \left(1 - \frac{Q_{g,out}}{Q_{in}}\right) + \left(1 - \frac{Q_{out}}{Q_{g,out}}\right) - \left(1 - \frac{Q_{g,out}}{Q_{in}}\right)\left(1 - \frac{Q_{out}}{Q_{g,out}}\right) \\
&= 1 - \frac{Q_{g,out}}{Q_{in}} + 1 - \frac{Q_{out}}{Q_{g,out}} - 1 + \frac{Q_{g,out}}{Q_{in}} + \frac{Q_{out}}{Q_{g,out}} - \frac{Q_{out}}{Q_{in}} \\
&= 1 - \frac{Q_{out}}{Q_{in}} \\
&= \eta_{cc}
\end{aligned}$$

Therefore, the proof is complete. Using the relation above, the thermal efficiency of the given combined cycle is determined to be

$$\eta_{cc} = \eta_g + \eta_s - \eta_g\eta_s = 0.4 + 0.30 - 0.40 \times 0.30 = \mathbf{0.58}$$

9-77 The thermal efficiency of a combined gas-steam power plant η_{cc} can be expressed in terms of the thermal efficiencies of the gas and the steam turbine cycles as $\eta_{cc} = \eta_g + \eta_s - \eta_g\eta_s$. It is to be shown that the value of η_{cc} is greater than either of η_g or η_s.

Analysis By factoring out terms, the relation $\eta_{cc} = \eta_g + \eta_s - \eta_g\eta_s$ can be expressed as

$$\eta_{cc} = \eta_g + \eta_s - \eta_g\eta_s = \eta_g + \underbrace{\eta_s(1-\eta_g)}_{\substack{\text{Positive since} \\ \eta_g < 1}} > \eta_g$$

or

$$\eta_{cc} = \eta_g + \eta_s - \eta_g\eta_s = \eta_s + \underbrace{\eta_g(1-\eta_s)}_{\substack{\text{Positive since} \\ \eta_s < 1}} > \eta_s$$

Thus we conclude that the combined cycle is more efficient than either of the gas turbine or steam turbine cycles alone.

9-78 A steam power plant operating on the ideal Rankine cycle with reheating is considered. The reheat pressures of the cycle are to be determined for the cases of single and double reheat.

Assumptions **1** Steady operating conditions exist. **2** Kinetic and potential energy changes are negligible.

Analysis (*a*) Single Reheat: From the steam tables (Tables A-4, A-5, and A-6),

$$\left.\begin{array}{l} P_6 = 10 \text{ kPa} \\ x_6 = 0.88 \end{array}\right\} \begin{array}{l} h_6 = h_f + x_6 h_{fg} = 191.83 + (0.88)(2392.8) = 2297.5 \text{ kJ/kg} \\ s_6 = s_f + x_6 s_{fg} = 0.6493 + (0.88)(7.5009) = 7.250 \text{ kJ/kg}\cdot\text{K} \end{array}$$

$$\left.\begin{array}{l} T_5 = 600°\text{C} \\ s_5 = s_6 \end{array}\right\} P_5 = \mathbf{5.098\ MPa}$$

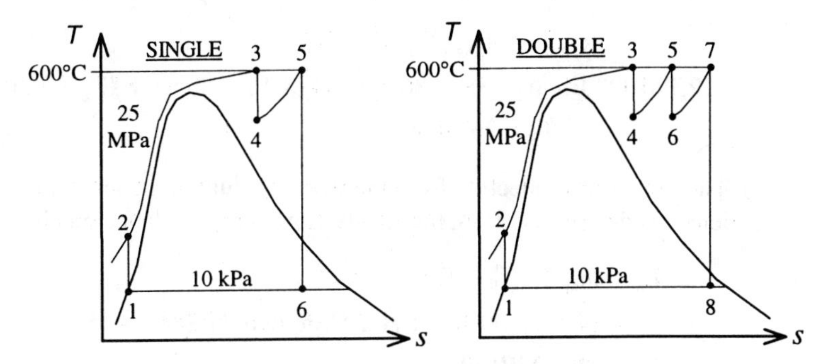

(*b*) Double Reheat :

$$\left.\begin{array}{l} P_3 = 25 \text{ MPa} \\ T_3 = 600°\text{C} \end{array}\right\} s_3 = 6.3602 \text{ kJ/kg}\cdot\text{K}$$

$$\begin{array}{l} P_4 = P_x \\ s_4 = s_3 \end{array} \quad \text{and} \quad \begin{array}{l} P_5 = P_x \\ T_5 = 600°\text{C} \end{array}$$

Any pressure P_x selected between the limits of 25 MPa and 5.098 MPa will satisfy the requirements, and can be used for the double reheat pressure.

9-79E A geothermal power plant operating on the simple Rankine cycle using an organic fluid as the working fluid is considered. The exit temperature of the geothermal water from the vaporizer, the rate of heat rejection from the working fluid in the condenser, the mass flow rate of geothermal water at the preheater, and the thermal efficiency of the Level I cycle of this plant are to be determined.

Assumptions **1** Steady operating conditions exist. **2** Kinetic and potential energy changes are negligible.

Analysis (*a*) The exit temperature of geothermal water from the vaporizer is determined from the steady-flow energy balance on the geothermal water (brine),

$$\dot{Q}_{\text{brine}} = \dot{m}_{\text{brine}} C_p (T_2 - T_1)$$

$$-22{,}790{,}000 \text{ Btu/h} = (384{,}286 \text{ lbm/h})(1.03 \text{ Btu/lbm}\cdot{}^\circ\text{F})(T_2 - 325^\circ\text{F})$$

$$T_2 = \mathbf{267.4^\circ F}$$

(*b*) The rate of heat rejection from the working fluid to the air in the condenser is determined from the steady-flow energy balance on air,

$$\dot{Q}_{\text{air}} = \dot{m}_{\text{air}} C_p (T_9 - T_8)$$

$$= (4{,}195{,}100 \text{ lbm/h})(0.24 \text{ Btu/lbm}\cdot{}^\circ\text{F})(84.5 - 55^\circ\text{F})$$

$$= \mathbf{29.7 \text{ MBtu/h}}$$

(*c*) The mass flow rate of geothermal water at the preheater is determined from the steady-flow energy balance on the geothermal water,

$$\dot{Q}_{\text{geo}} = \dot{m}_{\text{geo}} C_p (T_{\text{out}} - T_{\text{in}})$$

$$-11{,}140{,}000 \text{ Btu/h} = \dot{m}_{\text{geo}} (1.03 \text{ Btu/lbm}\cdot{}^\circ\text{F})(154.0 - 221.8^\circ\text{F})$$

$$\dot{m}_{\text{geo}} = \mathbf{187{,}120 \text{ lbm/h}}$$

(*d*) The rate of heat input is

$$\dot{Q}_{\text{in}} = \dot{Q}_{\text{vaporizer}} + \dot{Q}_{\text{reheater}} = 22{,}790{,}000 + 11{,}140{,}000$$

$$= 33{,}930{,}000 \text{ Btu/h}$$

and

$$\dot{W}_{\text{net}} = 1271 - 200 = 1071 \text{ kW}$$

Then,

$$\eta_{th} = \frac{\dot{W}_{\text{net}}}{\dot{Q}_{\text{in}}} = \frac{1071 \text{ kW}}{33{,}930{,}000 \text{ Btu/h}}\left(\frac{3412.14 \text{ Btu}}{1 \text{ kWh}}\right) = \mathbf{10.8\%}$$

9-80 A steam power plant operates on the simple ideal Rankine cycle. The turbine inlet temperature, the net power output, the thermal efficiency, and the minimum mass flow rate of the cooling water required are to be determined.

Assumptions **1** Steady operating conditions exist. **2** Kinetic and potential energy changes are negligible.

Analysis (*a*) From the steam tables (Tables A-4, A-5, and A-6),

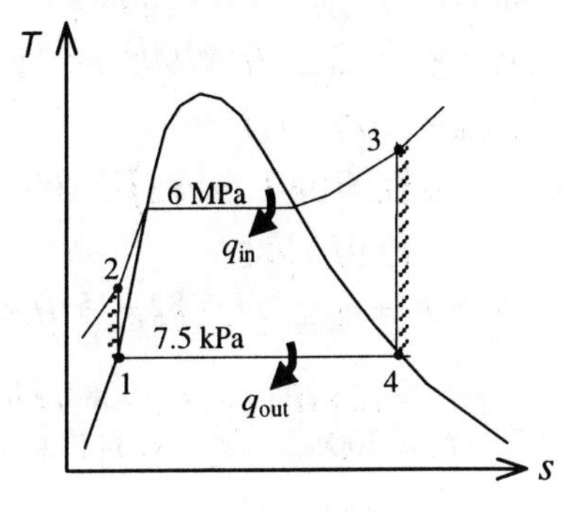

$$h_1 = h_{f\,@\,7.5\text{ kPa}} = 168.79\text{ kJ/kg}$$
$$v_1 = v_{f\,@\,7.5\text{ kPa}} = 0.001008\text{ m}^3\text{/kg}$$
$$T_1 = T_{\text{sat}\,@\,7.5\text{ kPa}} = 40.29°\text{C}$$

$$\begin{aligned} w_{p,in} &= v_1(P_2 - P_1) \\ &= (0.001008\text{ m}^3\text{/kg})(6{,}000 - 7.5\text{ kPa})\left(\frac{1\text{ kJ}}{1\text{ kPa}\cdot\text{m}^3}\right) \\ &= 6.04\text{ kJ/kg} \end{aligned}$$

$$h_2 = h_1 + w_{p,in} = 168.79 + 6.04 = 174.8\text{ kJ/kg}$$

$$h_4 = h_{g\,@\,7.5\text{ kPa}} = 2574.8\text{ kJ/kg}$$
$$s_4 = s_{g\,@\,7.5\text{ kPa}} = 8.2515\text{ kJ/kg}$$

$$\left.\begin{array}{l} P_3 = 6\text{ MPa} \\ s_3 = s_4 \end{array}\right\} \begin{array}{l} h_3 = 4856.1\text{ kJ/kg} \\ T_3 = \mathbf{1092.4°C} \end{array}$$

(*b*)
$$q_{in} = h_3 - h_2 = 4856.1 - 174.8 = 4681.3\text{ kJ / kg}$$
$$q_{out} = h_4 - h_1 = 2574.8 - 168.79 = 2406\text{ kJ / kg}$$
$$w_{net} = q_{in} - q_{out} = 4681.3 - 2406.0 = 2275.3\text{ kJ / kg}$$

and

$$\eta_{th} = \frac{w_{net}}{q_{in}} = \frac{2275.3\text{ kJ / kg}}{4681.3\text{ kJ / kg}} = \mathbf{48.6\%}$$

Thus,

$$\dot{W}_{net} = \eta_{th}\dot{Q}_{in} = (0.486)(60{,}000\text{kJ/s}) = \mathbf{29{,}160\ kJ/s}$$

(*c*) The mass flow rate of the cooling water will be minimum when it is heated to the temperature of the steam in the condenser, which is 40.29°C,

$$\dot{Q}_{out} = \dot{Q}_{in} - \dot{W}_{net} = 60{,}000 - 29{,}160 = 30{,}840\text{ kJ/s}$$

$$\dot{m}_{cool} = \frac{\dot{Q}_{out}}{C\Delta T} = \frac{30{,}840\text{ kJ/s}}{(4.18\text{ kJ/kg}\cdot°\text{C})(40.29 - 18\,°\text{C})} = \mathbf{331.0\ \ kg/s}$$

9-81 A steam power plant operating on an ideal Rankine cycle with two stages of reheat is considered. The thermal efficiency of the cycle and the mass flow rate of the steam are to be determined.

Assumptions **1** Steady operating conditions exist. **2** Kinetic and potential energy changes are negligible.

Analysis (*a*) From the steam tables (Tables A-4, A-5, and A-6),

$$h_1 = h_{f\ @\ 5\text{ kPa}} = 137.82 \text{ kJ/kg}$$
$$v_1 = v_{f\ @\ 5\text{ kPa}} = 0.001005 \text{ m}^3\text{/kg}$$
$$w_{p,in} = v_1(P_2 - P_1) = (0.001005 \text{ m}^3\text{/kg})(15{,}000 - 5 \text{ kPa})\left(\frac{1 \text{ kJ}}{1 \text{ kPa}\cdot\text{m}^3}\right) = 15.07 \text{ kJ/kg}$$
$$h_2 = h_1 + w_{p,in} = 137.82 + 15.07 = 152.89 \text{ kJ/kg}$$

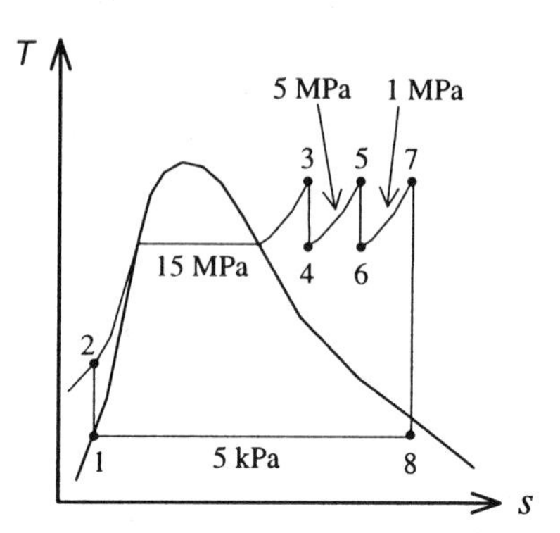

$$\left.\begin{array}{l} P_3 = 15 \text{ MPa} \\ T_3 = 500°\text{C} \end{array}\right\} \begin{array}{l} h_3 = 3308.6 \text{ kJ/kg} \\ s_3 = 6.3443 \text{ kJ/kg}\cdot\text{K} \end{array}$$

$$\left.\begin{array}{l} P_4 = 5 \text{ MPa} \\ s_4 = s_3 \end{array}\right\} h_4 = 3005.7 \text{ kJ/kg}$$

$$\left.\begin{array}{l} P_5 = 5 \text{ MPa} \\ T_5 = 500°\text{C} \end{array}\right\} \begin{array}{l} h_5 = 3433.8 \text{ kJ/kg} \\ s_5 = 6.9759 \text{ kJ/kg}\cdot\text{K} \end{array}$$

$$\left.\begin{array}{l} P_6 = 1 \text{ MPa} \\ s_6 = s_5 \end{array}\right\} h_6 = 2970.7 \text{ kJ/kg}$$

$$\left.\begin{array}{l} P_7 = 1 \text{ MPa} \\ T_7 = 500°\text{C} \end{array}\right\} \begin{array}{l} h_7 = 3478.5 \text{ kJ/kg} \\ s_7 = 7.7622 \text{ kJ/kg}\cdot\text{K} \end{array}$$

$$\left.\begin{array}{l} P_8 = 5 \text{ kPa} \\ s_8 = s_7 \end{array}\right\} \begin{array}{l} x_8 = \dfrac{s_8 - s_f}{s_{fg}} = \dfrac{7.7622 - 0.4764}{7.9187} = 0.920 \\ h_8 = h_f + x_8 h_{fg} = 137.82 + (0.920)(2423.7) = 2367.8 \text{ kJ/kg} \end{array}$$

Then,

$$q_{in} = (h_3 - h_2) + (h_5 - h_4) + (h_7 - h_6)$$
$$= 3308.6 - 152.89 + 3433.8 - 3005.7 + 3478.5 - 2970.7 = 4091.6 \text{ kJ/kg}$$
$$q_{out} = h_8 - h_1 = 2367.8 - 137.82 = 2230.0 \text{ kJ/kg}$$
$$w_{net} = q_{in} - q_{out} = 4091.6 - 2230.0 = 1861.6 \text{ kJ/kg}$$

Thus,

$$\eta_{th} = \frac{w_{net}}{q_{in}} = \frac{1861.6 \text{ kJ/kg}}{4091.6 \text{ kJ/kg}} = \mathbf{45.5\%}$$

(*b*)

$$\dot{m} = \frac{\dot{W}_{net}}{w_{net}} = \frac{120{,}000 \text{ kJ/s}}{1861.6 \text{ kJ/kg}} = \mathbf{64.5 \text{ kg/s}}$$

9-82 An 150-MW steam power plant operating on a regenerative Rankine cycle with an open feedwater heater is considered. The mass flow rate of steam through the boiler, the thermal efficiency of the cycle, and the irreversibility associated with the regeneration process are to be determined.

Assumptions **1** Steady operating conditions exist. **2** Kinetic and potential energy changes are negligible.

Analysis

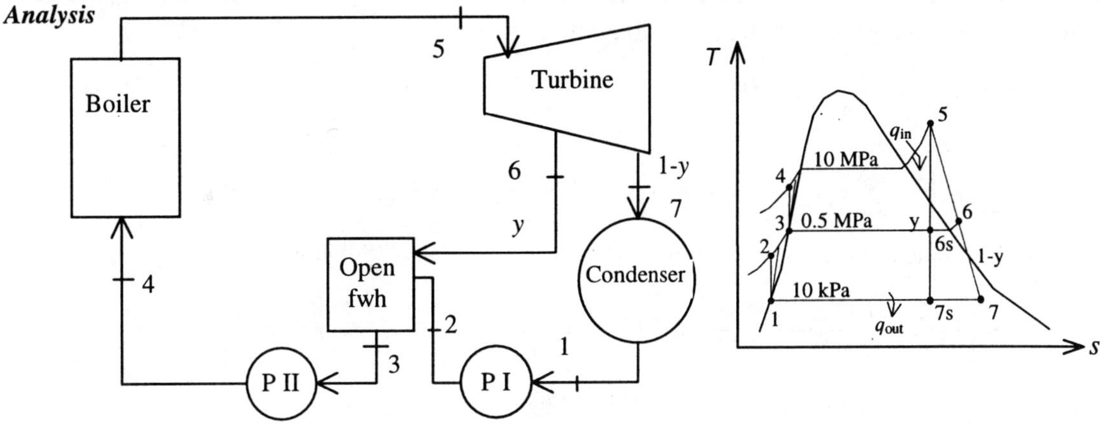

(*a*) From the steam tables (Tables A-4, A-5, and A-6),

$$h_1 = h_{f\ @\ 10\ \text{kPa}} = 191.83\ \text{kJ/kg}$$
$$v_1 = v_{f\ @\ 10\ \text{kPa}} = 0.00101\ \text{m}^3\text{/kg}$$

$$\begin{aligned} w_{\text{pI,in}} &= v_1(P_2 - P_1)/\eta_p \\ &= (0.00101\ \text{m}^3\text{/kg})(500 - 10\ \text{kPa})\left(\frac{1\ \text{kJ}}{1\ \text{kPa}\cdot\text{m}^3}\right)/(0.95) \\ &= 0.52\ \text{kJ/kg} \end{aligned}$$
$$h_2 = h_1 + w_{\text{pI,in}} = 191.83 + 0.52 = 192.35\ \text{kJ/kg}$$

$$\left.\begin{array}{l} P_3 = 0.5\ \text{MPa} \\ \text{sat.liquid} \end{array}\right\} \begin{array}{l} h_3 = h_{f\ @\ 0.5\ \text{MPa}} = 640.23\ \text{kJ/kg} \\ v_3 = v_{f\ @\ 0.5\ \text{MPa}} = 0.001093\ \text{m}^3\text{/kg} \end{array}$$

$$\begin{aligned} w_{\text{pII,in}} &= v_3(P_4 - P_3)/\eta_p \\ &= (0.001039\ \text{m}^3\text{/kg})(10{,}000 - 500\ \text{kPa})\left(\frac{1\ \text{kJ}}{1\ \text{kPa}\cdot\text{m}^3}\right)/(0.95) \\ &= 10.39\ \text{kJ/kg} \end{aligned}$$
$$h_4 = h_3 + w_{\text{pII,in}} = 640.23 + 10.39 = 650.62\ \text{kJ/kg}$$

$$\left.\begin{array}{l} P_5 = 10\ \text{MPa} \\ T_5 = 500°\text{C} \end{array}\right\} \begin{array}{l} h_5 = 3373.7\ \text{kJ/kg} \\ s_5 = 6.5966\ \text{kJ/kg}\cdot\text{K} \end{array}$$

$$\left.\begin{array}{l} P_{6s} = 0.5\ \text{MPa} \\ s_{6s} = s_5 \end{array}\right\} \begin{array}{l} x_{6s} = \dfrac{s_{6s} - s_f}{s_{fg}} = \dfrac{6.5966 - 1.8607}{4.9606} = 0.9547 \\ h_{6s} = h_f + x_{6s}h_{fg} = 640.23 + (0.9547)(2108.5) \\ \quad = 2653.2\ \text{kJ/kg} \end{array}$$

$$\eta_T = \frac{h_5 - h_6}{h_5 - h_{6s}} \longrightarrow \begin{array}{l} h_6 = h_5 - \eta_T(h_5 - h_{6s}) \\ \quad = 3373.7 - (0.80)(3373.7 - 2653.2) \\ \quad = 2797.3\ \text{kJ/kg} \end{array}$$

$$\left.\begin{array}{l} P_{7s} = 10\ \text{kPa} \\ s_{7s} = s_5 \end{array}\right\} \begin{array}{l} x_{7s} = \dfrac{s_{7s} - s_f}{s_{fg}} = \dfrac{6.5966 - 0.6493}{7.5009} = 0.7929 \\ h_{7s} = h_f + x_{7s}h_{fg} = 191.83 + (0.7929)(2392.8) \\ \quad = 2089.1\ \text{kJ/kg} \end{array}$$

$$\eta_T = \frac{h_5 - h_7}{h_5 - h_{7s}} \longrightarrow h_7 = h_5 - \eta_T(h_5 - h_{7s}) = 3373.7 - (0.80)(3373.7 - 2089.1) = 2346.0 \text{ kJ/kg}$$

The fraction of steam extracted is determined from the steady-flow energy balance equation applied to the feedwater heaters. Noting that $\dot{Q} \cong \dot{W} \cong \Delta ke \cong \Delta pe \cong 0$,

$$\dot{E}_{in} - \dot{E}_{out} = \Delta\dot{E}_{system}^{\nearrow 0 \text{ (steady)}} = 0$$

$$\dot{E}_{in} = \dot{E}_{out}$$

$$\sum \dot{m}_i h_i = \sum \dot{m}_e h_e \longrightarrow \dot{m}_6 h_6 + \dot{m}_2 h_2 = \dot{m}_3 h_3 \longrightarrow y h_6 + (1-y)h_2 = 1(h_3)$$

where y is the fraction of steam extracted from the turbine ($= \dot{m}_6 / \dot{m}_3$). Solving for y,

$$y = \frac{h_3 - h_2}{h_6 - h_2} = \frac{640.23 - 192.35}{2797.3 - 192.35} = 0.1719$$

Then,

$$q_{in} = h_5 - h_4 = 3373.7 - 650.62 = 2723.1 \text{ kJ/kg}$$

$$q_{out} = (1-y)(h_7 - h_1) = (1 - 0.1719)(2346.0 - 191.83) = 1783.9 \text{ kJ/kg}$$

$$w_{net} = q_{in} - q_{out} = 2723.1 - 1783.9 = 939.2 \text{ kJ/kg}$$

and

$$\dot{m} = \frac{\dot{W}_{net}}{w_{net}} = \frac{150{,}000 \text{ kJ/s}}{939.2 \text{ kJ/kg}} = \mathbf{159.7 \text{ kg/s}}$$

(*b*) The thermal efficiency is determined from

$$\eta_{th} = 1 - \frac{q_{out}}{q_{in}} = 1 - \frac{1783.9 \text{ kJ/kg}}{2723.1 \text{ kJ/kg}} = \mathbf{34.5\%}$$

Also,

$$\left.\begin{array}{l} P_6 = 0.5 \text{ MPa} \\ h_6 = 2797.8 \text{ kJ/kg} \end{array}\right\} s_6 = 6.9308 \text{ kJ/kg} \cdot \text{K}$$

$$s_3 = s_{f \text{ @ } 0.5 \text{ MPa}} = 1.8607 \text{ kJ/kg} \cdot \text{K}$$

$$s_2 = s_1 = s_{f \text{ @ } 10 \text{ kPa}} = 0.6493 \text{ kJ/kg} \cdot \text{K}$$

Then the irreversibility (or exergy destruction) associated with this regeneration process is

$$i_{regen} = T_0 s_{gen} = T_0\left(\sum m_e s_e - \sum m_i s_i + \frac{q_{surr}}{T_L}^{\nearrow 0}\right) = T_0[s_3 - y s_6 - (1-y)s_2]$$

$$= (303 \text{ K})[1.8607 - (0.1719)(6.9308) - (1 - 0.1719)(0.6493)] = \mathbf{39.9 \text{ kJ/kg}}$$

9-83 An 150-MW steam power plant operating on an ideal regenerative Rankine cycle with an open feedwater heater is considered. The mass flow rate of steam through the boiler, the thermal efficiency of the cycle, and the irreversibility associated with the regeneration process are to be determined.

Assumptions **1** Steady operating conditions exist. **2** Kinetic and potential energy changes are negligible.

Analysis

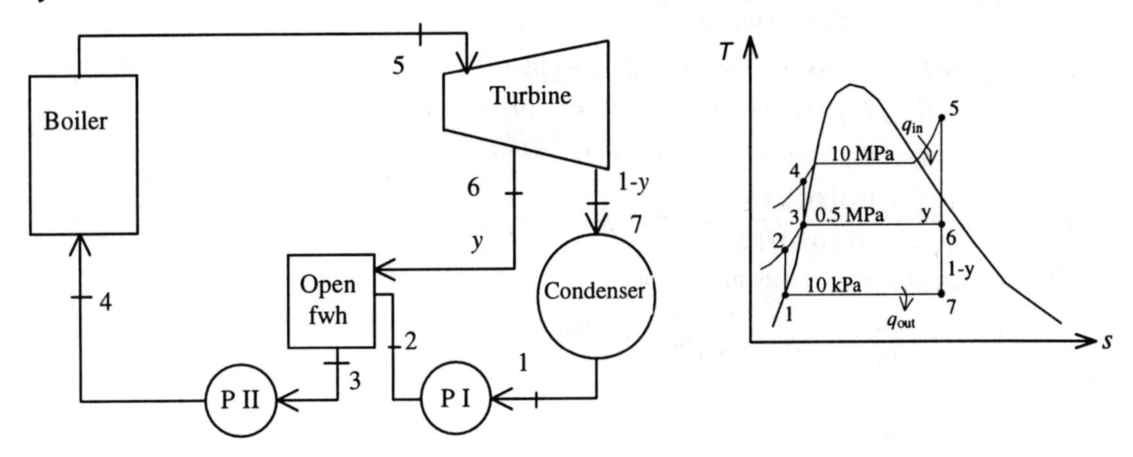

(*a*) From the steam tables (Tables A-4, A-5, and A-6),

$$h_1 = h_{f\,@\,10\text{ kPa}} = 191.83\text{ kJ/kg}$$
$$v_1 = v_{f\,@\,10\text{ kPa}} = 0.00101\text{ m}^3\text{/kg}$$

$$w_{pI,in} = v_1(P_2 - P_1) = (0.00101\text{ m}^3\text{/kg})(500 - 10\text{ kPa})\left(\frac{1\text{ kJ}}{1\text{ kPa}\cdot\text{m}^3}\right) = 0.50\text{ kJ/kg}$$

$$h_2 = h_1 + w_{pI,in} = 191.83 + 0.50 = 192.33\text{ kJ/kg}$$

$$\left.\begin{array}{l} P_3 = 0.5\text{ MPa} \\ \text{sat.liquid} \end{array}\right\} \begin{array}{l} h_3 = h_{f\,@\,0.5\text{ MPa}} = 640.23\text{ kJ/kg} \\ v_3 = v_{f\,@\,0.5\text{ MPa}} = 0.001093\text{ m}^3\text{/kg} \end{array}$$

$$w_{pII,in} = v_3(P_4 - P_3) = (0.001039\text{ m}^3\text{/kg})(10{,}000 - 500\text{ kPa})\left(\frac{1\text{ kJ}}{1\text{ kPa}\cdot\text{m}^3}\right) = 9.87\text{ kJ/kg}$$

$$h_4 = h_3 + w_{pII,in} = 640.23 + 9.87 = 650.10\text{ kJ/kg}$$

$$\left.\begin{array}{l} P_5 = 10\text{ MPa} \\ T_5 = 500°\text{C} \end{array}\right\} \begin{array}{l} h_5 = 3373.7\text{ kJ/kg} \\ s_5 = 6.5966\text{ kJ/kg}\cdot\text{K} \end{array}$$

$$\left.\begin{array}{l} P_6 = 0.5\text{ MPa} \\ s_6 = s_5 \end{array}\right\} \begin{array}{l} x_6 = \dfrac{s_6 - s_f}{s_{fg}} = \dfrac{6.5966 - 1.8607}{4.9606} = 0.9547 \\ h_6 = h_f + x_6 h_{fg} = 640.23 + (0.9547)(2108.5) = 2653.2\text{ kJ/kg} \end{array}$$

$$\left.\begin{array}{l} P_7 = 10\text{ kPa} \\ s_7 = s_5 \end{array}\right\} \begin{array}{l} x_7 = \dfrac{s_7 - s_f}{s_{fg}} = \dfrac{6.5966 - 0.6493}{7.5009} = 0.7929 \\ h_7 = h_f + x_7 h_{fg} = 191.83 + (0.7929)(2392.8) = 2089.1\text{ kJ/kg} \end{array}$$

The fraction of steam extracted is determined from the steady-flow energy equation applied to the feedwater heaters. Noting that $\dot{Q} \cong \dot{W} \cong \Delta ke \cong \Delta pe \cong 0$,

$$\dot{E}_{in} - \dot{E}_{out} = \Delta \dot{E}_{system}^{\nearrow 0\,(steady)} = 0 \rightarrow \dot{E}_{in} = \dot{E}_{out}$$

$$\sum \dot{m}_i h_i = \sum \dot{m}_e h_e \longrightarrow \dot{m}_6 h_6 + \dot{m}_2 h_2 = \dot{m}_3 h_3 \longrightarrow y h_6 + (1-y) h_2 = 1(h_3)$$

where y is the fraction of steam extracted from the turbine ($= \dot{m}_6 / \dot{m}_3$). Solving for y,

$$y = \frac{h_3 - h_2}{h_6 - h_2} = \frac{640.23 - 192.33}{2653.2 - 192.33} = 0.1820$$

Then,.

$$q_{in} = h_5 - h_4 = 3373.7 - 650.10 = 2723.6 \text{ kJ/kg}$$

$$q_{out} = (1-y)(h_7 - h_1) = (1 - 0.1820)(2089.1 - 191.83) = 1552.0 \text{ kJ/kg}$$

$$w_{net} = q_{in} - q_{out} = 2723.6 - 1552.0 = 1171.6 \text{ kJ/kg}$$

and

$$\dot{m} = \frac{\dot{W}_{net}}{w_{net}} = \frac{150{,}000 \text{ kJ/s}}{1171.6 \text{ kJ/kg}} = \mathbf{128.0\ kg/s}$$

(*b*) The thermal efficiency is determined from

$$\eta_{th} = 1 - \frac{q_{out}}{q_{in}} = 1 - \frac{1552.0 \text{ kJ/kg}}{2723.6 \text{ kJ/kg}} = \mathbf{43.0\%}$$

Also,

$$s_6 = s_5 = 6.5966 \text{ kJ/kg} \cdot \text{K}$$

$$s_3 = s_{f\ @\ 0.5\ MPa} = 1.8607 \text{ kJ/kg} \cdot \text{K}$$

$$s_2 = s_1 = s_{f\ @\ 10\ kPa} = 0.6493 \text{ kJ/kg} \cdot \text{K}$$

Then the irreversibility (or exergy destruction) associated with this regeneration process is

$$i_{regen} = T_0 s_{gen} = T_0 \left(\sum m_e s_e - \sum m_i s_i + \frac{q_{surr}}{T_L}^{\nearrow 0} \right) = T_0 \left[s_3 - y s_6 - (1-y) s_2 \right]$$

$$= (303 \text{ K})[1.8607 - (0.1820)(6.5966) - (1 - 0.1820)(0.6493)] = \mathbf{39.1\ kJ/kg}$$

9-84 An ideal reheat-regenerative Rankine cycle with one open feedwater heater is considered. The fraction of steam extracted for regeneration and the thermal efficiency of the cycle are to be determined.

Assumptions **1** Steady operating conditions exist. **2** Kinetic and potential energy changes are negligible.

Analysis

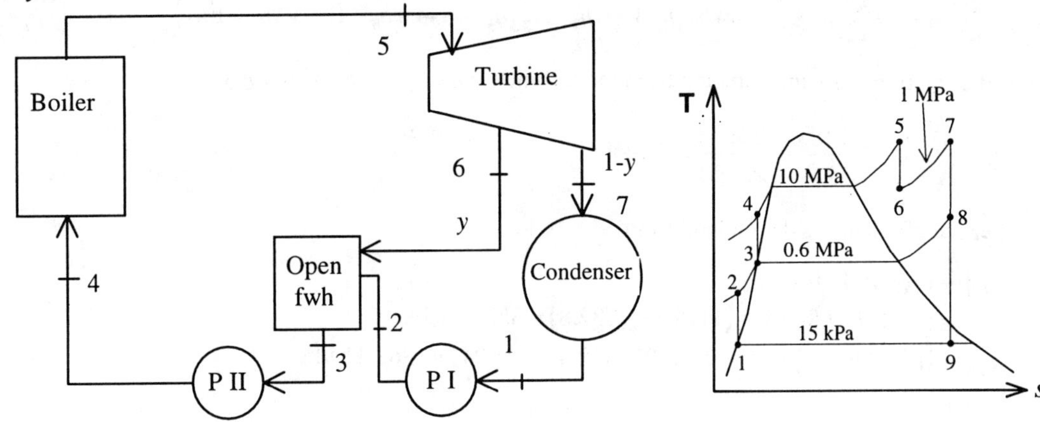

(*a*) From the steam tables (Tables A-4, A-5, and A-6),

$$h_1 = h_{f\ @\ 15\ \text{kPa}} = 225.94\ \text{kJ/kg}$$
$$v_1 = v_{f\ @\ 15\ \text{kPa}} = 0.001014\ \text{m}^3/\text{kg}$$

$$\begin{aligned} w_{\text{pI,in}} &= v_1(P_2 - P_1) \\ &= (0.001014\ \text{m}^3/\text{kg})(600 - 15\ \text{kPa})\left(\frac{1\ \text{kJ}}{1\ \text{kPa}\cdot\text{m}^3}\right) \\ &= 0.59\ \text{kJ/kg} \end{aligned}$$
$$h_2 = h_1 + w_{\text{pI,in}} = 225.94 + 0.59 = 226.53\ \text{kJ/kg}$$

$$\left.\begin{array}{l} P_3 = 0.6\ \text{MPa} \\ \text{sat.liquid} \end{array}\right\} \begin{array}{l} h_3 = h_{f\ @\ 0.6\ \text{MPa}} = 670.56\ \text{kJ/kg} \\ v_3 = v_{f\ @\ 0.6\ \text{MPa}} = 0.001101\ \text{m}^3/\text{kg} \end{array}$$

$$\begin{aligned} w_{\text{pII,in}} &= v_3(P_4 - P_3) \\ &= (0.001101\ \text{m}^3/\text{kg})(10{,}000 - 600\ \text{kPa})\left(\frac{1\ \text{kJ}}{1\ \text{kPa}\cdot\text{m}^3}\right) \\ &= 10.35\ \text{kJ/kg} \end{aligned}$$
$$h_4 = h_3 + w_{\text{pII,in}} = 670.56 + 10.35 = 680.91\ \text{kJ/kg}$$

$$\left.\begin{array}{l} P_5 = 10\ \text{MPa} \\ T_5 = 500°\text{C} \end{array}\right\} \begin{array}{l} h_5 = 3373.7\ \text{kJ/kg} \\ s_5 = 6.5966\ \text{kJ/kg}\cdot\text{K} \end{array}$$

$$\left.\begin{array}{l} P_6 = 1.0\ \text{MPa} \\ s_6 = s_5 \end{array}\right\} h_6 = 2782.8\ \text{kJ/kg}$$

$$\left.\begin{array}{l} P_7 = 1.0\ \text{MPa} \\ T_7 = 500°\text{C} \end{array}\right\} \begin{array}{l} h_7 = 3478.5\ \text{kJ/kg} \\ s_7 = 7.7622\ \text{kJ/kg}\cdot\text{K} \end{array}$$

$$\left.\begin{array}{l} P_8 = 0.6\ \text{MPa} \\ s_8 = s_7 \end{array}\right\} h_8 = 3309.5\ \text{kJ/kg}$$

$$\left.\begin{array}{l} P_9 = 15\ \text{kPa} \\ s_9 = s_7 \end{array}\right\} \begin{array}{l} x_9 = \dfrac{s_9 - s_f}{s_{fg}} = \dfrac{7.7622 - 0.7549}{7.2536} = 0.9660 \\ h_9 = h_f + x_9 h_{fg} = 225.94 + (0.9660)(2373.1) = 2518.4\ \text{kJ/kg} \end{array}$$

The fraction of steam extracted is determined from the steady-flow energy balance equation applied to the feedwater heaters. Noting that $\dot{Q} \cong \dot{W} \cong \Delta ke \cong \Delta pe \cong 0$,

$$\dot{E}_{in} - \dot{E}_{out} = \Delta \dot{E}_{system}{}^{\nearrow 0 \text{ (steady)}} = 0 \rightarrow \dot{E}_{in} = \dot{E}_{out}$$

$$\sum \dot{m}_i h_i = \sum \dot{m}_e h_e \longrightarrow \dot{m}_8 h_8 + \dot{m}_2 h_2 = \dot{m}_3 h_3 \longrightarrow y h_8 + (1-y) h_2 = 1(h_3)$$

where y is the fraction of steam extracted from the turbine ($= \dot{m}_8 / \dot{m}_3$). Solving for y,

$$y = \frac{h_3 - h_2}{h_8 - h_2} = \frac{670.56 - 226.53}{3309.5 - 226.53} = \mathbf{0.1440}$$

(*b*) The thermal efficiency is determined from

$$\begin{aligned} q_{in} &= (h_5 - h_4) + (h_7 - h_6) \\ &= (3373.7 - 680.91) + (3478.5 - 2782.8) = 3388.5 \text{ kJ/kg} \\ q_{out} &= (1 - y)(h_9 - h_1) = (1 - 0.1440)(2518.4 - 225.94) = 1962.3 \text{ kJ/kg} \end{aligned}$$

and

$$\eta_{th} = 1 - \frac{q_{out}}{q_{in}} = 1 - \frac{1962.3 \text{ kJ/kg}}{3388.5 \text{ kJ/kg}} = \mathbf{42.1\%}$$

9-85 A nonideal reheat-regenerative Rankine cycle with one open feedwater heater is considered. The fraction of steam extracted for regeneration and the thermal efficiency of the cycle are to be determined.

Assumptions **1** Steady operating conditions exist. **2** Kinetic and potential energy changes are negligible.

Analysis

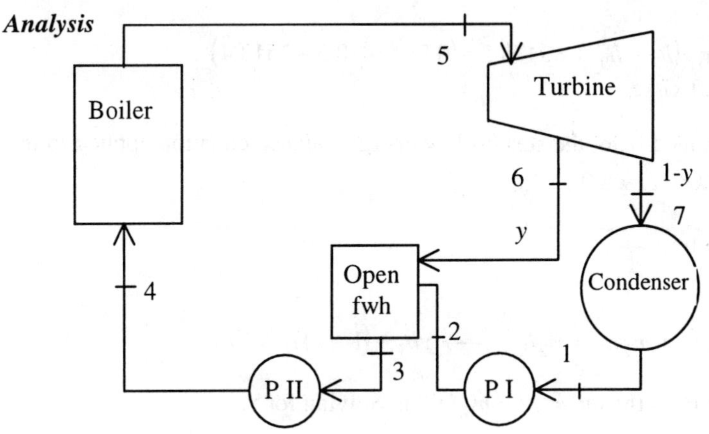

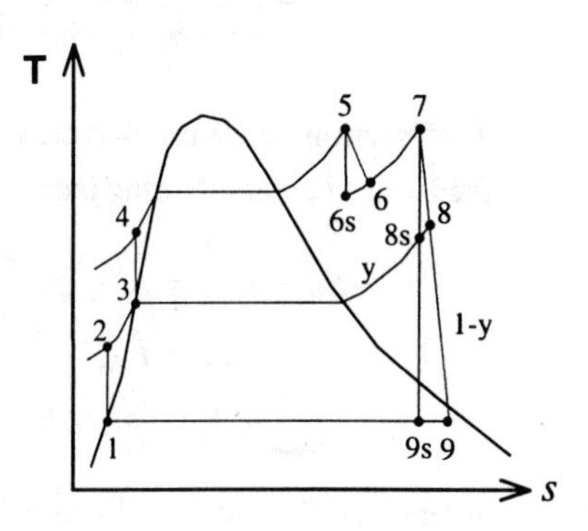

(*a*) From the steam tables (Tables A-4, A-5, and A-6),

$$h_1 = h_{f@\,15\text{ kPa}} = 225.94 \text{ kJ/kg}$$
$$v_1 = v_{f@\,15\text{ kPa}} = 0.001014 \text{ m}^3\text{/kg}$$

$$\begin{aligned} w_{pI,\text{in}} &= v_1(P_2 - P_1) \\ &= (0.001014 \text{ m}^3\text{/kg})(600 - 15 \text{ kPa})\left(\frac{1 \text{ kJ}}{1 \text{ kPa}\cdot\text{m}^3}\right) \\ &= 0.59 \text{ kJ/kg} \end{aligned}$$
$$h_2 = h_1 + w_{pI,\text{in}} = 225.94 + 0.59 = 226.53 \text{ kJ/kg}$$

$$\left.\begin{array}{l} P_3 = 0.6 \text{ MPa} \\ \text{sat. liquid} \end{array}\right\} \begin{array}{l} h_3 = h_{f@\,0.6\text{ MPa}} = 670.56 \text{ kJ/kg} \\ v_3 = v_{f@\,0.6\text{ MPa}} = 0.001101 \text{ m}^3\text{/kg} \end{array}$$

$$\begin{aligned} w_{\text{pII,in}} &= v_3(P_4 - P_3) \\ &= (0.001101 \text{ m}^3\text{/kg})(10{,}000 - 600 \text{ kPa})\left(\frac{1 \text{ kJ}}{1 \text{ kPa}\cdot\text{m}^3}\right) \\ &= 10.35 \text{ kJ/kg} \end{aligned}$$
$$h_4 = h_3 + w_{\text{pII,in}} = 670.56 + 10.35 = 680.91 \text{ kJ/kg}$$

$$\left.\begin{array}{l} P_5 = 10 \text{ MPa} \\ T_5 = 500°\text{C} \end{array}\right\} \begin{array}{l} h_5 = 3373.7 \text{ kJ/kg} \\ s_5 = 6.5966 \text{ kJ/kg}\cdot\text{K} \end{array}$$

$$\left.\begin{array}{l} P_{6s} = 1.0 \text{ MPa} \\ s_{6s} = s_5 \end{array}\right\} h_{6s} = 2782.8 \text{ kJ/kg}$$

$$\begin{aligned} \eta_T = \frac{h_5 - h_6}{h_5 - h_{6s}} \longrightarrow h_6 &= h_5 - \eta_T(h_5 - h_{6s}) \\ &= 3373.7 - (0.84)(3373.7 - 2782.8) \\ &= 2877.3 \text{ kJ/kg} \end{aligned}$$

$$\left.\begin{array}{l} P_7 = 1.0 \text{ MPa} \\ T_7 = 500°\text{C} \end{array}\right\} \begin{array}{l} h_7 = 3478.5 \text{ kJ/kg} \\ s_7 = 7.7622 \text{ kJ/kg}\cdot\text{K} \end{array}$$

$$\left.\begin{array}{l} P_{8s} = 0.6 \text{ MPa} \\ s_{8s} = s_7 \end{array}\right\} h_{8s} = 3309.5 \text{ kJ/kg}$$

$$\begin{aligned} \eta_T = \frac{h_7 - h_8}{h_7 - h_{8s}} \longrightarrow h_8 &= h_7 - \eta_T(h_7 - h_{8s}) = 3478.5 - (0.84)(3478.5 - 3309.5) \\ &= 3336.5 \text{ kJ/kg} \end{aligned}$$

$$\left.\begin{array}{l} P_{9s} = 15 \text{ kPa} \\ s_{9s} = s_7 \end{array}\right\} \begin{array}{l} x_{9s} = \dfrac{s_{9s} - s_f}{s_{fg}} = \dfrac{7.7622 - 0.7549}{7.2536} = 0.9660 \\ h_{9s} = h_f + x_{9s} h_{fg} = 225.94 + (0.9660)(2373.1) = 2518.4 \text{ kJ/kg} \end{array}$$

$$\eta_T = \frac{h_7 - h_9}{h_7 - h_{9s}} \longrightarrow h_9 = h_7 - \eta_T (h_7 - h_{9s}) = 3478.5 - (0.84)(3478.5 - 2518.4)$$
$$= 2672.0 \text{ kJ/kg}$$

The fraction of steam extracted is determined from the steady-flow energy balance equation applied to the feedwater heaters. Noting that $\dot{Q} \cong \dot{W} \cong \Delta ke \cong \Delta pe \cong 0$,

$$\dot{E}_{in} - \dot{E}_{out} = \Delta \dot{E}_{system}^{\nearrow 0 \text{ (steady)}} = 0$$
$$\dot{E}_{in} = \dot{E}_{out}$$
$$\sum \dot{m}_i h_i = \sum \dot{m}_e h_e \longrightarrow \dot{m}_8 h_8 + \dot{m}_2 h_2 = \dot{m}_3 h_3 \longrightarrow y h_8 + (1 - y) h_2 = 1(h_3)$$

where y is the fraction of steam extracted from the turbine ($= \dot{m}_8 / \dot{m}_3$). Solving for y,

$$y = \frac{h_3 - h_2}{h_8 - h_2} = \frac{670.56 - 226.53}{3336.5 - 226.53} = \mathbf{0.1428}$$

(b) The thermal efficiency is determined from

$$q_{in} = (h_5 - h_4) + (h_7 - h_6)$$
$$= (3373.7 - 680.91) + (3478.5 - 2877.3) = 3294.0 \text{ kJ/kg}$$
$$q_{out} = (1 - y)(h_9 - h_1) = (1 - 0.1428)(2672.0 - 225.94) = 2096.8 \text{ kJ/kg}$$

and

$$\eta_{th} = 1 - \frac{q_{out}}{q_{in}} = 1 - \frac{2096.8 \text{ kJ/kg}}{3294.0 \text{ kJ/kg}} = \mathbf{36.3\%}$$

9-86 A steam power plant operates on an ideal reheat-regenerative Rankine cycle with one reheater and two feedwater heaters, one open and one closed. The fraction of steam extracted from the turbine for the open feedwater heater, the thermal efficiency of the cycle, and the net power output are to be determined.

Assumptions **1** Steady operating conditions exist. **2** Kinetic and potential energy changes are negligible.

Analysis

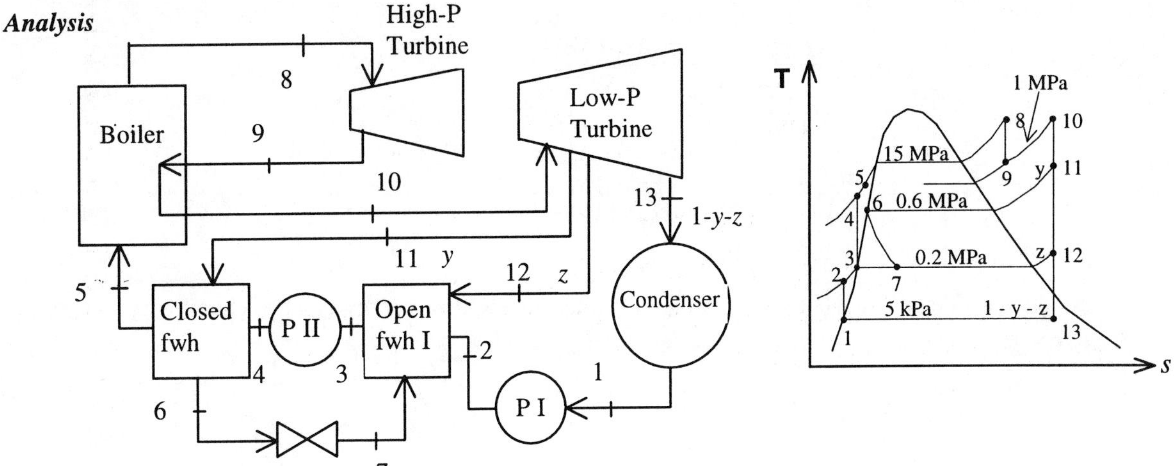

(*a*) From the steam tables (Tables A-4, A-5, and A-6),

$$h_1 = h_{f\ @\ 5\ \text{kPa}} = 137.82\ \text{kJ/kg}$$
$$v_1 = v_{f\ @\ 5\ \text{kPa}} = 0.001005\ \text{m}^3\text{/kg}$$

$$\begin{aligned} w_{\text{pI,in}} &= v_1(P_2 - P_1) \\ &= (0.001005\ \text{m}^3\text{/kg})(200 - 5\ \text{kPa})\left(\frac{1\ \text{kJ}}{1\ \text{kPa}\cdot\text{m}^3}\right) \\ &= 0.20\ \text{kJ/kg} \end{aligned}$$

$$h_2 = h_1 + w_{\text{pI,in}} = 137.82 + 0.20 = 138.02\ \text{kJ/kg}$$

$$\left.\begin{array}{l} P_3 = 0.2\ \text{MPa} \\ \text{sat.liquid} \end{array}\right\} \begin{array}{l} h_3 = h_{f\ @\ 0.2\ \text{MPa}} = 504.70\ \text{kJ/kg} \\ v_3 = v_{f\ @\ 0.2\ \text{MPa}} = 0.001061\ \text{m}^3\text{/kg} \end{array}$$

$$\begin{aligned} w_{\text{pII,in}} &= v_3(P_4 - P_3) \\ &= (0.001061\ \text{m}^3\text{/kg})(15{,}000 - 200\ \text{kPa})\left(\frac{1\ \text{kJ}}{1\ \text{kPa}\cdot\text{m}^3}\right) \\ &= 15.70\ \text{kJ/kg} \end{aligned}$$

$$h_4 = h_3 + w_{\text{pII,in}} = 504.70 + 15.70 = 520.40\ \text{kJ/kg}$$

$$\left.\begin{array}{l} P_6 = 0.6\ \text{MPa} \\ \text{sat.liquid} \end{array}\right\} \begin{array}{l} h_6 = h_7 = h_{f\ @\ 0.6\ \text{MPa}} = 670.56\ \text{kJ/kg} \\ v_6 = v_{f\ @\ 0.6\ \text{MPa}} = 0.001101\ \text{m}^3\text{/kg} \end{array}$$

$$\begin{aligned} T_6 = T_5 \longrightarrow h_5 &= h_6 + v_6(P_5 - P_6) \\ &= 670.56 + (0.001101\ \text{m}^3\text{/kg})(15{,}000 - 600\ \text{kPa})\left(\frac{1\ \text{kJ}}{1\ \text{kPa}\cdot\text{m}^3}\right) \\ &= 686.41\ \text{kJ/kg} \end{aligned}$$

$$\left.\begin{array}{l} P_8 = 15\ \text{MPa} \\ T_8 = 600°\text{C} \end{array}\right\} \begin{array}{l} h_8 = 3582.3\ \text{kJ/kg} \\ s_8 = 6.6776\ \text{kJ/kg}\cdot\text{K} \end{array}$$

$$\left.\begin{array}{l} P_9 = 1.0\ \text{MPa} \\ s_9 = s_8 \end{array}\right\} h_9 = 2820.3\ \text{kJ/kg}$$

$$\left.\begin{array}{l} P_{10} = 1.0\ \text{MPa} \\ T_{10} = 500°\text{C} \end{array}\right\} \begin{array}{l} h_{10} = 3478.5\ \text{kJ/kg} \\ s_{10} = 7.7622\ \text{kJ/kg}\cdot\text{K} \end{array}$$

$$\left.\begin{array}{l} P_{11} = 0.6 \text{ MPa} \\ s_{11} = s_{10} \end{array}\right\} h_{11} = 3309.5 \text{ kJ/kg}$$

$$\left.\begin{array}{l} P_{12} = 0.2 \text{ MPa} \\ s_{12} = s_{10} \end{array}\right\} h_{12} = 3000.4 \text{ kJ/kg}$$

$$\left.\begin{array}{l} P_{13} = 5 \text{ kPa} \\ s_{13} = s_{10} \end{array}\right\} \begin{array}{l} x_{13} = \dfrac{s_{13} - s_f}{s_{fg}} = \dfrac{7.7622 - 0.4764}{7.9187} = 0.9201 \\ h_{13} = h_f + x_{13} h_{fg} = 137.82 + (0.9201)(2423.7) = 2367.9 \text{ kJ/kg} \end{array}$$

The fraction of steam extracted is determined from the steady-flow energy balance equation applied to the feedwater heaters. Noting that $\dot{Q} \cong \dot{W} \cong \Delta ke \cong \Delta pe \cong 0$,

$$\dot{E}_{in} - \dot{E}_{out} = \Delta \dot{E}_{system}^{\nearrow 0 \text{ (steady)}} = 0$$

$$\dot{E}_{in} = \dot{E}_{out}$$

$$\sum \dot{m}_i h_i = \sum \dot{m}_e h_e \longrightarrow \dot{m}_{11}(h_{11} - h_6) = \dot{m}_5 (h_5 - h_4) \longrightarrow y(h_{11} - h_6) = (h_5 - h_4)$$

where y is the fraction of steam extracted from the turbine ($= \dot{m}_{11} / \dot{m}_5$). Solving for y,

$$y = \frac{h_5 - h_4}{h_{11} - h_6} = \frac{686.41 - 520.40}{3309.5 - 670.56} = 0.0629$$

For the open FWH,

$$\dot{E}_{in} - \dot{E}_{out} = \Delta \dot{E}_{system}^{\nearrow 0 \text{ (steady)}} = 0$$

$$\dot{E}_{in} = \dot{E}_{out}$$

$$\sum \dot{m}_i h_i = \sum \dot{m}_e h_e$$

$$\dot{m}_7 h_7 + \dot{m}_2 h_2 + \dot{m}_{12} h_{12} = \dot{m}_3 h_3$$

$$y h_7 + (1 - y - z) h_2 + z h_{12} = (1) h_3$$

where z is the fraction of steam extracted from the turbine ($= \dot{m}_{12} / \dot{m}_5$) at the second stage. Solving for z,

$$z = \frac{(h_3 - h_2) - y(h_7 - h_2)}{h_{12} - h_2} = \frac{504.70 - 138.02 - (0.0629)(670.56 - 138.02)}{3000.4 - 138.02} = \mathbf{0.1164}$$

(*b*)

$$q_{in} = (h_8 - h_5) + (h_{10} - h_9) = (3582.3 - 686.41) + (3478.5 - 2820.3) = 3554.1 \text{ kJ/kg}$$

$$q_{out} = (1 - y - z)(h_{13} - h_1) = (1 - 0.0629 - 0.1164)(2367.9 - 137.82) = 1830.2 \text{ kJ/kg}$$

$$w_{net} = q_{in} - q_{out} = 3554.1 - 1830.2 = 1723.9 \text{ kJ/kg}$$

and

$$\eta_{th} = 1 - \frac{q_{out}}{q_{in}} = 1 - \frac{1830.2 \text{ kJ/kg}}{3554.1 \text{ kJ/kg}} = \mathbf{48.5\%}$$

(*c*)

$$\dot{W}_{net} = \dot{m} w_{net} = (42 \text{ kg/s})(1723.9 \text{ kJ/kg}) = \mathbf{72{,}404 \text{ kW}}$$

9-87 A cogeneration power plant is modified with reheat and that produces 3 MW of power and supplies 7 MW of process heat. The rate of heat input in the boiler and the fraction of steam extracted for process heating are to be determined.

Assumptions **1** Steady operating conditions exist. **2** Kinetic and potential energy changes are negligible.

Analysis (*a*) From the steam tables (Tables A-4, A-5, and A-6),

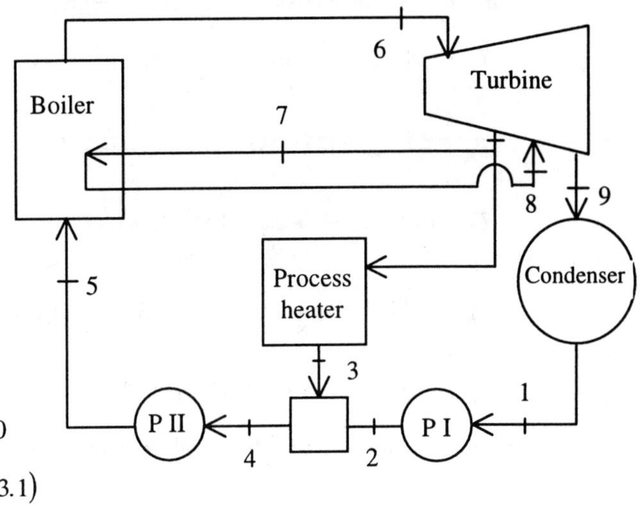

$$h_1 = h_{f\,@\,15\text{ kPa}} = 225.94\text{ kJ/kg}$$
$$h_2 \cong h_1$$
$$h_3 = h_{f\,@\,120°\text{C}} = 503.71\text{ kJ/kg}$$

$$\left.\begin{array}{l} P_6 = 8\text{ MPa} \\ T_6 = 500°\text{C} \end{array}\right\} \begin{array}{l} h_6 = 3398.3\text{ kJ/kg} \\ s_6 = 6.7240\text{ kJ/kg}\cdot\text{K} \end{array}$$

$$\left.\begin{array}{l} P_7 = 1\text{ MPa} \\ s_7 = s_6 \end{array}\right\} h_7 = 2842.8\text{ kJ/kg}$$

$$\left.\begin{array}{l} P_8 = 1\text{ MPa} \\ T_8 = 500°\text{C} \end{array}\right\} \begin{array}{l} h_8 = 3478.5\text{ kJ/kg} \\ s_8 = 7.7622\text{ kJ/kg}\cdot\text{K} \end{array}$$

$$\left.\begin{array}{l} P_9 = 15\text{ kPa} \\ s_9 = s_8 \end{array}\right\} \begin{array}{l} x_9 = \dfrac{s_9 - s_f}{s_{fg}} = \dfrac{7.7622 - 0.7549}{7.2536} = 0.9660 \\ h_9 = h_f + x_9 h_{fg} = 225.94 + (0.9660)(2373.1) \\ \quad = 2518.4\text{ kJ/kg} \end{array}$$

The mass flow rate through the process heater is

$$\dot{m}_3 = \frac{\dot{Q}_{\text{process}}}{h_7 - h_3} = \frac{7{,}000\text{ kJ/s}}{(2842.8 - 503.71)\text{ kJ/kg}} = 2.993\text{ kg/s}$$

Also,

$$\dot{W}_T = \dot{m}_6(h_6 - h_7) + \dot{m}_9(h_8 - h_9) = \dot{m}_6(h_6 - h_7) + (\dot{m}_6 - 2.993)(h_8 - h_9)$$

or,

$$3{,}000\text{ kJ/s} = \dot{m}_6(3398.3 - 2842.8) + (\dot{m}_6 - 2.993)(3478.5 - 2518.4)$$

It yields $\dot{m}_6 = 3.875\text{ kg/s}$

and $\dot{m}_9 = \dot{m}_6 - \dot{m}_3 = 3.875 - 2.993 = 0.882\text{ kg/s}$

Mixing chamber:

$$\dot{E}_{\text{in}} - \dot{E}_{\text{out}} = \Delta\dot{E}_{\text{system}}^{\nearrow 0\text{ (steady)}} = 0$$
$$\dot{E}_{\text{in}} = \dot{E}_{\text{out}}$$
$$\sum \dot{m}_i h_i = \sum \dot{m}_e h_e \longrightarrow \dot{m}_4 h_4 = \dot{m}_2 h_2 + \dot{m}_3 h_3$$

or,

$$h_4 \cong h_5 = \frac{\dot{m}_2 h_2 + \dot{m}_3 h_3}{\dot{m}_4} = \frac{(0.882)(225.94) + (2.993)(503.71)}{3.875} = 440.5\text{ kJ/kg}$$

Then,

$$\begin{aligned} \dot{Q}_{\text{in}} &= \dot{m}_6(h_6 - h_5) + \dot{m}_8(h_8 - h_7) \\ &= (3.875\text{ kg/s})(3398.3 - 440.5\text{ kJ/kg}) + (0.882\text{ kg/s})(3478.5 - 2842.8\text{ kJ/kg}) \\ &= \mathbf{12{,}022\ kW} \end{aligned}$$

(*b*) The fraction of steam extracted for process heating is $y = \dfrac{\dot{m}_3}{\dot{m}_{\text{total}}} = \dfrac{2.993\text{ kg/s}}{3.875\text{ kg/s}} = \mathbf{77.2\%}$

9-88 A combined gas-steam power plant is considered. The topping cycle is an ideal gas-turbine cycle and the bottoming cycle is an ideal reheat Rankine cycle. The mass flow rate of air in the gas-turbine cycle, the rate of total heat input, and the thermal efficiency of the combined cycle are to be determined.

Assumptions **1** Steady operating conditions exist. **2** Kinetic and potential energy changes are negligible. **3** Air is an ideal gas with variable specific heats.

Analysis (*a*) The analysis of gas cycle yields

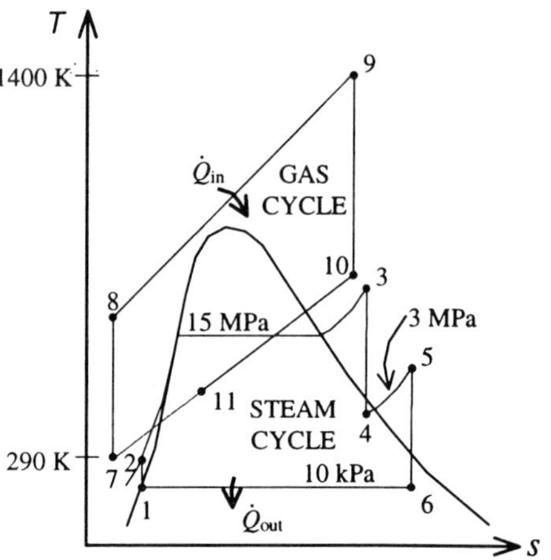

$$T_7 = 290 \text{ K} \longrightarrow h_7 = 290.16 \text{ kJ/kg}$$
$$P_{r_7} = 1.2311$$

$$P_{r_8} = \frac{P_8}{P_7} P_{r_7} = (8)(1.2311) = 9.849 \longrightarrow h_8 = 526.12 \text{ kJ/kg}$$

$$T_9 = 1400 \text{ K} \longrightarrow h_9 = 1515.42 \text{ kJ/kg}$$
$$P_{r_9} = 450.5$$

$$P_{r_{10}} = \frac{P_{10}}{P_9} P_{r_9} = \left(\frac{1}{8}\right)(450.5) = 56.3 \longrightarrow h_{10} = 860.35 \text{ kJ/kg}$$

$$T_{11} = 520 \text{ K} \longrightarrow h_{11} = 523.63 \text{ kJ/kg}$$

From the steam tables (Tables A-4, A-5, and A-6),

$$h_1 = h_{f\,@\,10\text{ kPa}} = 191.83 \text{ kJ/kg}$$
$$v_1 = v_{f\,@\,10\text{ kPa}} = 0.00101 \text{ m}^3\text{/kg}$$

$$\begin{aligned} w_{\text{pI,in}} &= v_1(P_2 - P_1) \\ &= (0.00101 \text{ m}^3\text{/kg})(15{,}000 - 10 \text{ kPa})\left(\frac{1 \text{ kJ}}{1 \text{ kPa}\cdot\text{m}^3}\right) \\ &= 15.14 \text{ kJ/kg} \end{aligned}$$

$$h_2 = h_1 + w_{\text{pI,in}} = 191.83 + 15.14 = 206.97 \text{ kJ/kg}$$

$$\left.\begin{aligned} P_3 &= 15 \text{ MPa} \\ T_3 &= 450°\text{C} \end{aligned}\right\} \begin{aligned} h_3 &= 3156.2 \text{ kJ/kg} \\ s_3 &= 6.1404 \text{ kJ/kg}\cdot\text{K} \end{aligned}$$

$$\left.\begin{aligned} P_4 &= 3 \text{ MPa} \\ s_4 &= s_3 \end{aligned}\right\} \begin{aligned} x_4 &= \frac{s_4 - s_f}{s_{fg}} = \frac{6.1404 - 2.6457}{3.5412} = 0.9869 \\ h_4 &= h_f + x_4 h_{fg} = 1008.42 + (0.9869)(1795.7) = 2780.6 \text{ kJ/kg} \end{aligned}$$

$$\left.\begin{aligned} P_5 &= 3 \text{ MPa} \\ T_5 &= 500°\text{C} \end{aligned}\right\} \begin{aligned} h_5 &= 3456.3 \text{ kJ/kg} \\ s_5 &= 7.2338 \text{ kJ/kg}\cdot\text{K} \end{aligned}$$

$$\left.\begin{aligned} P_6 &= 10 \text{ kPa} \\ s_6 &= s_5 \end{aligned}\right\} \begin{aligned} x_6 &= \frac{s_6 - s_f}{s_{fg}} = \frac{7.2338 - 0.6493}{7.5009} = 0.8778 \\ h_6 &= h_f + x_6 h_{fg} = 191.83 + (0.8778)(2392.8) = 2292.2 \text{ kJ/kg} \end{aligned}$$

Noting that $\dot{Q} \cong \dot{W} \cong \Delta ke \cong \Delta pe \cong 0$ for the heat exchanger, the steady-flow energy balance equation yields

$$\dot{E}_{\text{in}} - \dot{E}_{\text{out}} = \Delta \dot{E}_{\text{system}}^{\nearrow 0 \text{ (steady)}} = 0$$

$$\dot{E}_{\text{in}} = \dot{E}_{\text{out}}$$

$$\sum \dot{m}_i h_i = \sum \dot{m}_e h_e \longrightarrow \dot{m}_s (h_3 - h_2) = \dot{m}_{\text{air}} (h_{10} - h_{11})$$

$$\dot{m}_{\text{air}} = \frac{h_3 - h_2}{h_{10} - h_{11}} \dot{m}_s = \frac{3156.2 - 206.97}{860.35 - 523.63}(30 \text{ kg/s}) = \mathbf{262.8 \text{ kg/s}}$$

(*b*) $\dot{Q}_{in} = \dot{Q}_{air} + \dot{Q}_{reheat} = \dot{m}_{air}(h_9 - h_8) + \dot{m}_{reheat}(h_5 - h_4)$
$= (262.8\text{ kg/s})(1515.42 - 526.12)\text{ kJ/kg} + (30\text{ kg/s})(3456.3 - 2780.6)\text{kJ/kg} = 280{,}259\text{ kW}$
$\cong \mathbf{2.80 \times 10^5\ kW}$

(*c*) $\dot{Q}_{out} = \dot{Q}_{out,air} + \dot{Q}_{out,steam} = \dot{m}_{air}(h_{11} - h_7) + \dot{m}_s(h_6 - h_1)$
$= (262.8\text{ kg/s})(523.63 - 290.16)\text{ kJ/kg} + (30\text{ kg/s})(2292.2 - 191.83)\text{ kJ/kg} = 124{,}367\text{ kW}$

$$\eta_{th} = 1 - \frac{\dot{Q}_{out}}{\dot{Q}_{in}} = 1 - \frac{124{,}367\ \text{kW}}{280{,}259\ \text{kW}} = \mathbf{55.6\%}$$

9-89 A combined gas-steam power plant is considered. The topping cycle is a gas-turbine cycle and the bottoming cycle is a nonideal reheat Rankine cycle. The mass flow rate of air in the gas-turbine cycle, the rate of total heat input, and the thermal efficiency of the combined cycle are to be determined.

Assumptions **1** Steady operating conditions exist. **2** Kinetic and potential energy changes are negligible. **3** Air is an ideal gas with variable specific heats.

Analysis (*a*) The analysis of gas cycle yields (Table A-17)

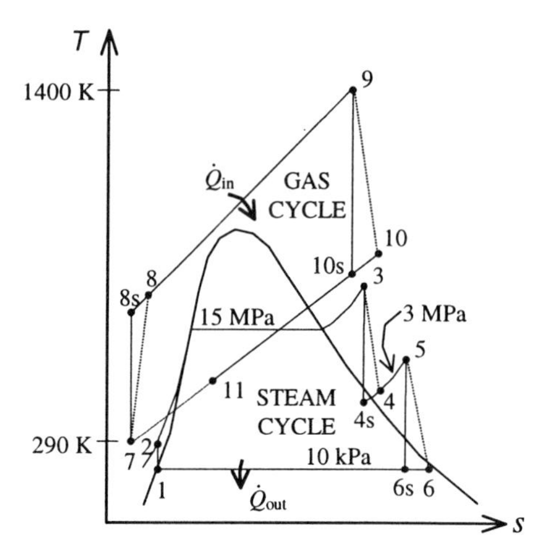

$$T_7 = 290 \text{ K} \longrightarrow h_7 = 290.16 \text{ kJ/kg}, \quad P_{r_7} = 1.2311$$

$$P_{r_{8s}} = \frac{P_{8s}}{P_7} P_{r_7} = (8)(1.2311) = 9.849 \longrightarrow h_{8s} = 526.12 \text{ kJ/kg}$$

$$\eta_C = \frac{h_{8s} - h_7}{h_8 - h_7} \longrightarrow h_8 = h_7 + (h_{8s} - h_7)/\eta_C = 290.16 + (526.12 - 290.16)/(0.80) = 585.1 \text{ kJ/kg}$$

$$T_9 = 1400 \text{ K} \longrightarrow h_9 = 1515.42 \text{ kJ/kg}, \quad P_{r_9} = 450.5$$

$$P_{r_{10s}} = \frac{P_{10s}}{P_9} P_{r_9} = \left(\frac{1}{8}\right)(450.5) = 56.3 \longrightarrow h_{10s} = 860.35 \text{ kJ/kg}$$

$$\eta_T = \frac{h_9 - h_{10}}{h_9 - h_{10s}} \longrightarrow h_{10} = h_9 - \eta_T (h_9 - h_{10s}) = 1515.42 - (0.85)(1515.42 - 860.35) = 958.6 \text{ kJ/kg}$$

$$T_{11} = 520 \text{ K} \longrightarrow h_{11} = 523.63 \text{ kJ/kg}$$

From the steam tables (Tables A-4, A-5, and A-6),

$$h_1 = h_{f\,@\,10\text{ kPa}} = 191.83 \text{ kJ/kg}$$
$$v_1 = v_{f\,@\,10\text{ kPa}} = 0.00101 \text{ m}^3/\text{kg}$$

$$w_{\text{pI,in}} = v_1 (P_2 - P_1) = (0.00101 \text{ m}^3/\text{kg})(15{,}000 - 10 \text{ kPa})\left(\frac{1 \text{ kJ}}{1 \text{ kPa}\cdot\text{m}^3}\right) = 15.14 \text{ kJ/kg}$$

$$h_2 = h_1 + w_{\text{pI,in}} = 191.83 + 15.14 = 206.97 \text{ kJ/kg}$$

$$\left.\begin{array}{l} P_3 = 15 \text{ MPa} \\ T_3 = 450°\text{C} \end{array}\right\} \begin{array}{l} h_3 = 3156.2 \text{ kJ/kg} \\ s_3 = 6.1404 \text{ kJ/kg}\cdot\text{K} \end{array}$$

$$\left.\begin{array}{l} P_4 = 3 \text{ MPa} \\ s_{4s} = s_3 \end{array}\right\} \begin{array}{l} x_{4s} = \dfrac{s_{4s} - s_f}{s_{fg}} = \dfrac{6.1404 - 2.6457}{3.5412} = 0.9869 \\ h_{4s} = h_f + x_{4s} h_{fg} = 1008.42 + (0.9869)(1795.7) = 2780.6 \text{ kJ/kg} \end{array}$$

$$\eta_T = \frac{h_3 - h_4}{h_3 - h_{4s}} \longrightarrow h_4 = h_3 - \eta_T (h_3 - h_{4s}) = 3156.2 - (0.85)(3156.2 - 2780.6) = 2836.9 \text{ kJ/kg}$$

$$\left.\begin{array}{l} P_5 = 3\text{ MPa} \\ T_5 = 500°\text{C} \end{array}\right\} \begin{array}{l} h_5 = 3456.3\text{ kJ/kg} \\ s_5 = 7.2338\text{ kJ/kg}\cdot\text{K} \end{array}$$

$$\left.\begin{array}{l} P_6 = 10\text{ kPa} \\ s_{6s} = s_5 \end{array}\right\} \begin{array}{l} x_{6s} = \dfrac{s_{6s} - s_f}{s_{fg}} = \dfrac{7.2338 - 0.6493}{7.5009} = 0.8778 \\ h_{6s} = h_f + x_{6s} h_{fg} = 191.83 + (0.8778)(2392.8) = 2292.2\text{ kJ/kg} \end{array}$$

$$\eta_T = \frac{h_5 - h_6}{h_5 - h_{6s}} \longrightarrow h_6 = h_5 - \eta_T (h_5 - h_{6s}) = 3456.3 - (0.85)(3456.3 - 2292.2) = 2466.8\text{ kJ/kg}$$

Noting that $\dot{Q} \cong \dot{W} \cong \Delta ke \cong \Delta pe \cong 0$ for the heat exchanger, the steady-flow energy balance equation yields

$$\dot{E}_{in} - \dot{E}_{out} = \Delta \dot{E}_{system}^{\nearrow 0\text{ (steady)}} = 0$$

$$\dot{E}_{in} = \dot{E}_{out}$$

$$\sum \dot{m}_i h_i = \sum \dot{m}_e h_e \longrightarrow \dot{m}_s (h_3 - h_2) = \dot{m}_{air} (h_{10} - h_{11})$$

$$\dot{m}_{air} = \frac{h_3 - h_2}{h_{10} - h_{11}} \dot{m}_s = \frac{3156.2 - 206.97}{958.60 - 523.63}(30\text{ kg/s}) = \mathbf{203.4\ kg/s}$$

(*b*) $\dot{Q}_{in} = \dot{Q}_{air} + \dot{Q}_{reheat} = \dot{m}_{air}(h_9 - h_8) + \dot{m}_{reheat}(h_5 - h_4)$
$= (203.4\text{ kg/s})(1515.42 - 585.1)\text{ kJ/kg} + (30\text{ kg/s})(3456.3 - 2836.9)\text{ kJ/kg} = \mathbf{207{,}809\ kW}$

(*c*) $\dot{Q}_{out} = \dot{Q}_{out,air} + \dot{Q}_{out,steam} = \dot{m}_{air}(h_{10} - h_{11}) + \dot{m}_s (h_6 - h_1)$
$= (203.4\text{ kg/s})(958.6 - 523.63)\text{ kJ/kg} + (30\text{ kg/s})(2466.8 - 191.83)\text{ kJ/kg} = \mathbf{156{,}722\ kW}$

$$\eta_{th} = 1 - \frac{\dot{Q}_{out}}{\dot{Q}_{in}} = 1 - \frac{156{,}722\text{ kW}}{207{,}809\text{ kW}} = \mathbf{24.6\%}$$

9-90 It is to be shown that the exergy destruction associated with a simple ideal Rankine cycle can be expressed as $x_{\text{destroyed}} = q_{in}(\eta_{\text{th,Carnot}} - \eta_{th})$, where η_{th} is efficiency of the Rankine cycle and $\eta_{\text{th, Carnot}}$ is the efficiency of the Carnot cycle operating between the same temperature limits.

Analysis The exergy destruction associated with a cycle is given on a unit mass basis as

$$x_{\text{destroyed}} = T_0 \sum \frac{q_R}{T_R}$$

where the direction of q_{in} is determined with respect to the reservoir (positive if to the reservoir and negative if from the reservoir). For a cycle that involves heat transfer only with a source at T_{H} and a sink at T_0, the irreversibility becomes

$$x_{\text{destroyed}} = T_0\left(\frac{q_{\text{out}}}{T_0} - \frac{q_{\text{in}}}{T_H}\right) = q_{\text{out}} - \frac{T_0}{T_H} q_{\text{in}} = q_{\text{in}}\left(\frac{q_{\text{out}}}{q_{\text{in}}} - \frac{T_0}{T_H}\right) = q_{\text{in}}\left[(1 - \eta_{th}) - (1 - \eta_{th,C})\right] = q_{\text{in}}(\eta_{th,C} - \eta_{th})$$

9-91 A cogeneration plant is to produce power and process heat. There are two turbines in the cycle: a high-pressure turbine and a low-pressure turbine. The temperature, pressure, and mass flow rate of steam at the inlet of high-pressure turbine are to be determined.

Assumptions **1** Steady operating conditions exist. **2** Kinetic and potential energy changes are negligible.

Analysis From the steam tables (Tables A-4, A-5, and A-6),

$$\left.\begin{array}{l} P_4 = 1.4 \text{ MPa} \\ \text{sat. vapor} \end{array}\right\} \begin{array}{l} h_4 = h_{g\,@\,1.4\text{ MPa}} = 2790.0 \text{ kJ/kg} \\ s_4 = s_{g\,@\,1.4\text{ MPa}} = 6.4693 \text{ kJ/kg}\cdot\text{K} \end{array}$$

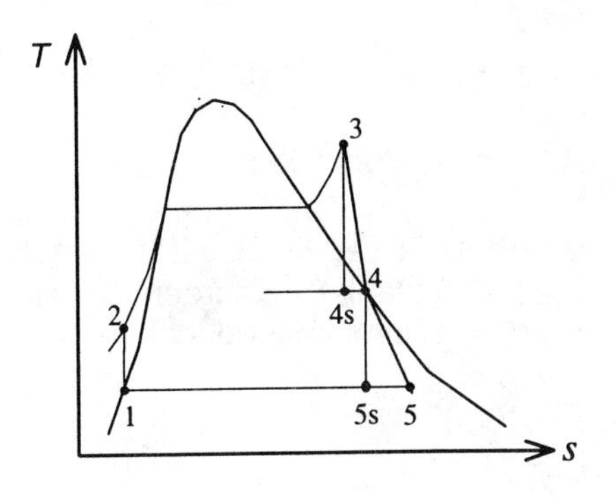

$$\left.\begin{array}{l} P_5 = 10 \text{ kPa} \\ s_{5s} = s_4 \end{array}\right\} \begin{array}{l} x_{5s} = \dfrac{s_{4s} - s_f}{s_{fg}} = \dfrac{6.4693 - 0.6493}{7.5009} = 0.776 \\ h_{5s} = h_f + x_{5s} h_{fg} \\ = 191.83 + (0.776)(2392.8) = 2048.4 \text{ kJ/kg} \end{array}$$

$$\eta_T = \frac{h_4 - h_5}{h_4 - h_{5s}} \longrightarrow h_5 = h_4 - \eta_T (h_4 - h_{5s}) = 2790.0 - (0.60)(2790.0 - 2048.4) = 2345 \text{ kJ/kg}$$

and

$$w_{\text{turb, low}} = h_4 - h_5 = 2790 - 2345 = 445 \text{ kJ / kg}$$

$$\dot{m}_{\text{low turb}} = \frac{\dot{W}_{\text{turb, II}}}{w_{\text{turb, low}}} = \frac{800 \text{ kJ / s}}{445 \text{ kJ / kg}} = 1.8 \text{ kg / s} = 108 \text{ kg / min}$$

Therefore ,

$$\dot{m}_{\text{total}} = 1000 + 108 = 1108 \text{ kg / min} = \mathbf{18.47 \text{ kg / s}}$$

$$w_{\text{turb, high}} = \frac{\dot{W}_{\text{turb, }I}}{\dot{m}_{\text{high, turb}}} = \frac{1000 \text{ kJ / s}}{18.47 \text{ kg / s}} = 54.1 \text{ kJ / kg} = h_3 - h_4$$

$$h_3 = w_{\text{turb, high}} + h_4 = 54.1 + 2790 = 2844.1 \text{ kJ / kg}$$

$$\eta_T = \frac{h_3 - h_4}{h_3 - h_{4s}} \longrightarrow h_{4s} = h_3 - (h_3 - h_4)/\eta_T = 2844.1 - (2844.1 - 2790.0)/(0.75) = 2772.0 \text{ kJ/kg}$$

$$\left.\begin{array}{l} P_{4s} = 1.4 \text{ MPa} \\ s_{4s} = s_3 \end{array}\right\} \begin{array}{l} x_{4s} = \dfrac{h_{4s} - h_f}{h_{fg}} = \dfrac{2772.0 - 830.3}{1957.7} = 0.992 \\ s_{4s} = s_f + x_{4s} s_{fg} = 2.2842 + (0.992)(4.185) = 6.436 \text{ kJ/kg}\cdot\text{K} \end{array}$$

Then from the tables or the software, the turbine inlet temperature and pressure becomes

$$\left.\begin{array}{l} h_3 = 2844.1 \text{ kJ/kg} \\ s_3 = 6.436 \text{ kJ/kg}\cdot\text{K} \end{array}\right\} \begin{array}{l} P_3 = \mathbf{2 \text{ MPa}} \\ T_3 = \mathbf{228°C} \end{array}$$

9-92 · · · 9-98 EES solution of these problems are available to instructors at the *Instructor Manual* section of the *Online Learning Center* (OLC) at www.mhhe.com/cengel-boles. See the Preface for access information.

Fundamentals of Engineering (FE) Exam Problems

9-99 Consider a steady-flow Carnot cycle with water as the working fluid executed under the saturation dome between the pressure limits of 10 MPa and 10 kPa. Water changes from saturated liquid to saturated vapor during the heat addition process. The net work output of for this cycle is

(a) 194 kJ/kg (b) 342 kJ/kg (c) 598 kJ/kg (d) 719 kJ/kg (e) 1123 kJ/kg

Answer (c) 598 kJ/kg

Solution Solved by EES Software. Solutions can be verified by copying-and-pasting the following lines on a blank EES screen. (Similar problems and their solutions can be obtained easily by modifying numerical values).

```
P1=10000 "kPa"
P2=10 "kPa"

h_fg=ENTHALPY(Steam_NBS,x=1,P=P1)-ENTHALPY(Steam_NBS,x=0,P=P1)
T1=TEMPERATURE(Steam_NBS,x=0,P=P1)+273
T2=TEMPERATURE(Steam_NBS,x=0,P=P2)+273

q_in=h_fg
Eta_Carnot=1-T2/T1
w_net=Eta_Carnot*q_in

"Some Wrong Solutions with Common Mistakes:"
W1_work = Eta1*q_in; Eta1=T2/T1 "Taking Carnot efficiency to be T2/T1"
W2_work = Eta2*q_in; Eta2=1-(T2-273)/(T1-273) "Using C instead of K"
W3_work = Eta_Carnot*ENTHALPY(Steam_NBS,x=1,P=P1) "Using h_g instead of h_fg"
W4_work = Eta_Carnot*q2; q2=ENTHALPY(Steam_NBS,x=1,P=P2)-
ENTHALPY(Steam_NBS,x=0,P=P2) "Using h_fg at P2"
```

9-100 A simple ideal Rankine cycle operates between the pressure limits of 10 kPa and 3 MPa, with a turbine inlet temperature of 600°C. Disregarding the pump work, the cycle efficiency is

(a) 24% (b) 37% (c) 52% (d) 63% (e) 71%

Answer (b) 37%

Solution Solved by EES Software. Solutions can be verified by copying-and-pasting the following lines on a blank EES screen. (Similar problems and their solutions can be obtained easily by modifying numerical values).

```
P1=10 "kPa"
P2=3000 "kPa"
P3=P2
P4=P1

T3=600 "C"
s4=s3

h1=ENTHALPY(Steam_NBS,x=0,P=P1)
v1=VOLUME(Steam_NBS,x=0,P=P1)
w_pump=v1*(P2-P1) "kJ/kg"
h2=h1+w_pump

h3=ENTHALPY(Steam_NBS,T=T3,P=P3)
s3=ENTROPY(Steam_NBS,T=T3,P=P3)
h4=ENTHALPY(Steam_NBS,s=s4,P=P4)

q_in=h3-h2
q_out=h4-h1
Eta_th=1-q_out/q_in

"Some Wrong Solutions with Common Mistakes:"
W1_Eff = q_out/q_in "Using wrong relation"
W2_Eff = 1-(h44-h1)/(h3-h2); h44 = ENTHALPY(Steam_NBS,x=1,P=P4) "Using h_g for h4"
W3_Eff = 1-(T1+273)/(T3+273); T1=TEMPERATURE(Steam_NBS,x=0,P=P1) "Using Carnot
efficiency"
W4_Eff = (h3-h4)/q_in  "Disregarding pump work"
```

9-101 A simple ideal Rankine cycle operates between the pressure limits of 10 kPa and 3 MPa, with a turbine inlet temperature of 600°C. The mass fraction of steam that condenses at the turbine exit is

(a) 4% (b) 6% (c) 9% (d) 11% (e) 17%

Answer (c) 9%

Solution Solved by EES Software. Solutions can be verified by copying-and-pasting the following lines on a blank EES screen. (Similar problems and their solutions can be obtained easily by modifying numerical values).

```
P1=10 "kPa"
P2=3000 "kPa"
P3=P2
P4=P1

T3=600 "C"
s4=s3

h3=ENTHALPY(Steam_NBS,T=T3,P=P3)
s3=ENTROPY(Steam_NBS,T=T3,P=P3)
h4=ENTHALPY(Steam_NBS,s=s4,P=P4)
x4=QUALITY(Steam_NBS,s=s4,P=P4)
moisture=1-x4

"Some Wrong Solutions with Common Mistakes:"
W1_moisture = x4 "Taking quality as moisture"
W2_moisture = 0 "Assuming superheated vapor"
```

9-102 A steam power plant operates on the simple ideal Rankine cycle between the pressure limits of 10 kPa and 15 MPa, with a turbine inlet temperature of 600°C. The rate of heat transfer in the boiler is 800 kJ/s. Disregarding the pump work, the power output of this plant is

(a) 215 kW (b) 254 kW (c) 315 kW (d) 346 kW (e) 800 kW

Answer (d) 346 kW

Solution Solved by EES Software. Solutions can be verified by copying-and-pasting the following lines on a blank EES screen. (Similar problems and their solutions can be obtained easily by modifying numerical values).

```
P1=10 "kPa"
P2=15000 "kPa"
P3=P2
P4=P1

T3=600 "C"
s4=s3

Q_rate=800 "kJ/s"
m=Q_rate/q_in

h1=ENTHALPY(Steam_NBS,x=0,P=P1)
h2=h1  "pump work is neglected"
"v1=VOLUME(Steam_NBS,x=0,P=P1)
w_pump=v1*(P2-P1)
h2=h1+w_pump"

h3=ENTHALPY(Steam_NBS,T=T3,P=P3)
s3=ENTROPY(Steam_NBS,T=T3,P=P3)
h4=ENTHALPY(Steam_NBS,s=s4,P=P4)

q_in=h3-h2
W_turb=m*(h3-h4)

"Some Wrong Solutions with Common Mistakes:"
W1_power = Q_rate "Assuming all heat is converted to power"
W2_power = h3-h4 "Not using mass flow rate"
W3_power = Q_rate*Carnot; Carnot = 1-(T1+273)/(T3+273);
T1=TEMPERATURE(Steam_NBS,x=0,P=P1) "Using Carnot efficiency"
W4_power = m*(h3-h44); h44 = ENTHALPY(Steam_NBS,x=1,P=P4) "Taking h4=h_g"
```

9-103 Consider a combined gas-steam power plant. Water for the steam cycle is heated in a well-insulated heat exchanger by the exhaust gases that enter at 800 K at a rate of 60 kg/s and leave at 400 K. Water enters the heat exchanger at 200°C and 8 MPa and leaves at 350°C and 8 MPa. If the exhaust gases are treated as air with constant specific heats at room temperature, the mass flow rate of water through the heat exchanger becomes

(a) 11 kg/s (b) 24 kg/s (c) 46 kg/s (d) 53 kg/s (e) 60 kg/s

Answer (a) 11 kg/s

Solution Solved by EES Software. Solutions can be verified by copying-and-pasting the following lines on a blank EES screen. (Similar problems and their solutions can be obtained easily by modifying numerical values).

```
m_gas=60 "kg/s"
Cp=1.005 "kJ/kg.K"
T3=800 "K"
T4=400 "K"
Q_gas=m_gas*Cp*(T3-T4)

P1=8000 "kPa"
T1=200 "C"

P2=8000 "kPa"
T2=350 "C"

h1=ENTHALPY(Steam_NBS,T=T1,P=P1)
h2=ENTHALPY(Steam_NBS,T=T2,P=P2)
Q_steam=m_steam*(h2-h1)

Q_gas=Q_steam

"Some Wrong Solutions with Common Mistakes:"
m_gas*Cp*(T3 -T4)=W1_msteam*4.18*(T2-T1) "Assuming no evaporation of liquid water"
m_gas*Cv*(T3 -T4)=W2_msteam*(h2-h1); Cv=0.718 "Using Cv for air instead of Cp"
W3_msteam = m_gas "Taking the mass flow rates of two fluids to be equal"
m_gas*Cp*(T3 -T4)=W4_msteam*(h2-h11); h11=ENTHALPY(Steam_NBS,x=0,P=P1) "Taking h1=hf@P1"
```

9-104 An ideal reheat Rankine cycle operates between the pressure limits of 20 kPa and 8 MPa, with reheat occurring at 3 MPa. The temperature of steam at the inlets of both turbines is 500°C, and the enthalpy of steam is 3104 kJ/kg at the exit of the high-pressure turbine, and 2385 kJ/kg at the exit of the low-pressure turbine. Disregarding the pump work, the cycle efficiency is

(a) 26% (b) 30% (c) 35% (d) 39% (e) 43%

Answer (d) 39%

Solution Solved by EES Software. Solutions can be verified by copying-and-pasting the following lines on a blank EES screen. (Similar problems and their solutions can be obtained easily by modifying numerical values).

```
P1=20 "kPa"
P2=8000 "kPa"
P3=P2
P4=3000 "kPa"
P5=P4
P6=P1

T3=500 "C"
T5=500 "C"
s4=s3
s6=s5

h1=ENTHALPY(Steam_NBS,x=0,P=P1)
h2=h1

h44=3104 "kJ/kg - for checking given data"
h66=2385 "kJ/kg - for checking given data"

h3=ENTHALPY(Steam_NBS,T=T3,P=P3)
s3=ENTROPY(Steam_NBS,T=T3,P=P3)
h4=ENTHALPY(Steam_NBS,s=s4,P=P4)

h5=ENTHALPY(Steam_NBS,T=T5,P=P5)
s5=ENTROPY(Steam_NBS,T=T5,P=P5)
h6=ENTHALPY(Steam_NBS,s=s6,P=P6)
q_in=(h3-h2)+(h5-h4)
q_out=h6-h1
Eta_th=1-q_out/q_in

"Some Wrong Solutions with Common Mistakes:"
W1_Eff = q_out/q_in "Using wrong relation"
W2_Eff = 1-q_out/(h3-h2) "Disregarding heat input during reheat"
W3_Eff = 1-(T1+273)/(T3+273); T1=TEMPERATURE(Steam_NBS,x=0,P=P1) "Using Carnot
efficiency"
W4_Eff = 1-q_out/(h5-h2)  "Using wrong relation for q_in"
```

9-105 Pressurized feedwater in a steam power plant is to be heated in an ideal open feedwater heater that operates at a pressure of 0.4 MPa with steam extracted from the turbine. If the enthalpy of feedwater is 252 kJ/kg and the enthalpy of extracted steam is 2665 kJ/kg, the mass fraction of steam extracted from the turbine is

(a) 5% (b) 9% (c) 15% (d) 20% (e) 25%

Answer (c) 15%

Solution Solved by EES Software. Solutions can be verified by copying-and-pasting the following lines on a blank EES screen. (Similar problems and their solutions can be obtained easily by modifying numerical values).

```
h_feed=252 "kJ/kg"
h_extracted=2665 "kJ/kg"
P3=400 "kPa"
h3=ENTHALPY(Steam_NBS,x=0,P=P3)

"Energy balance on the FWH"
h3=x_ext*h_extracted+(1-x_ext)*h_feed

"Some Wrong Solutions with Common Mistakes:"
W1_ext = h_feed/h_extracted "Using wrong relation"
W2_ext = h3/(h_extracted-h_feed) "Using wrong relation"
W3_ext = h_feed/(h_extracted-h_feed) "Using wrong relation"
```

9-106 Consider a steam power plant that operates on the regenerative Rankine cycle with one open feedwater heater. The enthalpy of the steam is 3374 kJ/kg at the turbine inlet, 2797 kJ/kg at the location of bleeding, and 2346 kJ/kg at the turbine exit. The net power output of the plant is 80 MW, and the fraction of steam bled off the turbine for regeneration is 0.172. If the pump work is negligible, the mass flow rate of steam at the turbine inlet is

(a) 78 kg/s (b) 84 kg/s (c) 122 kg/s (d) 288 kg/s (e) 452 kg/s

Answer (b) 84 kg/s

Solution Solved by EES Software. Solutions can be verified by copying-and-pasting the following lines on a blank EES screen. (Similar problems and their solutions can be obtained easily by modifying numerical values).

```
h_in=3374 "kJ/kg"
h_out=2346 "kJ/kg"
h_feed=252 "kJ/kg"
h_extracted=2797 "kJ/kg"
Wnet_out=80000 "kW"
x_bleed=0.172

w_turb=(h_in-h_extracted)+(1-x_bleed)*(h_extracted-h_out)
m=Wnet_out/w_turb

"Some Wrong Solutions with Common Mistakes:"
W1_mass = Wnet_out/(h_in-h_out) "Disregarding extraction of steam"
W2_mass = Wnet_out/(x_bleed*(h_in-h_out)) "Assuming steam is extracted at turbine
inlet"
W3_mass = Wnet_out/(h_in-h_out-x_bleed*h_extracted) "Using wrong relation"
```

9-107 Consider a simple ideal Rankine cycle. If the condenser pressure is lowered while keeping turbine inlet state the same, (select the correct statement)

(a) the turbine work output will decrease.
(b) the amount of heat rejected will decrease.
(c) the cycle efficiency will decrease.
(d) the moisture content at turbine exit will decrease.
(e) the pump work input will decrease.

Answer (b) the amount of heat rejected will decrease.

9-108 Consider a simple ideal Rankine cycle with fixed boiler and condenser pressures. If the steam is superheated to a higher temperature, (select the correct statement)

(a) the turbine work output will decrease.
(b) the amount of heat rejected will decrease.
(c) the cycle efficiency will decrease.
(d) the moisture content at turbine exit will decrease.
(e) the amount of heat input will decrease.

Answer (d) the moisture content at turbine exit will decrease.

9-109 Consider a simple ideal Rankine cycle with fixed boiler and condenser pressures . If the cycle is modified with reheating, (select the correct statement)

(a) the turbine work output will decrease.
(b) the amount of heat rejected will decrease.
(c) the pump work input will decrease.
(d) the moisture content at turbine exit will decrease.
(e) the amount of heat input will decrease.

Answer (d) the moisture content at turbine exit will decrease.

9-110 Consider a simple ideal Rankine cycle with fixed boiler and condenser pressures . If the cycle is modified with regeneration that involves one open feed water heater, (select the correct statement per unit mass of steam flowing through the boiler)

(a) the turbine work output will decrease.
(b) the amount of heat rejected will increase.
(c) the cycle thermal efficiency will decrease.
(d) the quality of steam at turbine exit will decrease.
(e) the amount of heat input will increase.

Answer (a) the turbine work output will decrease.

9-111 Consider a cogeneration power plant modified with regeneration. Steam enters the turbine at 6 MPa and 450°C at a rate of 20 kg/s and expands to a pressure of 0.4 MPa. At this pressure, 60% of the steam is extracted from the turbine, and the remainder expands to a pressure of 10 kPa. Part of the extracted steam is used to heat feedwater in an open feedwater heater. The rest of the extracted steam is used for process heating and leaves the process heater as a saturated liquid at 0.4 MPa. It is subsequently mixed with the feedwater leaving the feedwater heater, and the mixture is pumped to the boiler pressure. The steam in the condenser is cooled and condensed by the cooling water from a nearby river, which enters the adiabatic condenser at a rate of 463 kg/s.

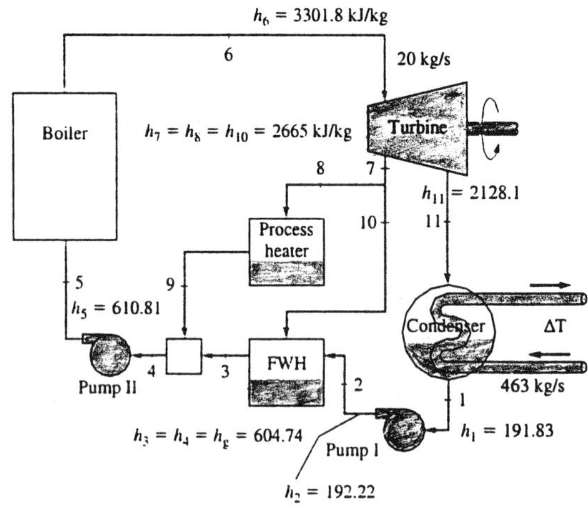

FIGURE P9–111

(I) The total power output of the turbine is

(*a*) 17.0 MW (*b*) 8.4 MW (*c*) 12.2 MW (*d*) 20.0 MW (*e*) 3.4 MW

Answer (*a*) 17.0 MW

(II) The temperature rise of the cooling water from the river in the condenser is

(*a*) 8.0°C (*b*) 5.2°C (*c*) 9.6°C (*d*) 12.9°C (*e*) 16.2°C

Answer (*a*) 8.0°C

(III) The mass flow rate of steam through the process heater is

(*a*) 1.6 kg/s (*b*) 3.8 kg/s (*c*) 5.2 kg/s (*d*) 7.6 kg/s (*e*) 10.4 kg/s

Answer (*e*) 10.4 kg/s

(IV) The rate of heat supply from the process heater per unit mass of steam passing through it is

(*a*) 246 kJ/kg (*b*) 893 kJ/kg (*c*) 1344 kJ/kg (*d*) 1891 kJ/kg (*e*) 2060 kJ/kg

Answer (*e*) 2060 kJ/kg

(V) The rate of heat transfer to the steam in the boiler is

(*a*) 26.0 MJ/s (*b*) 53.8 MJ/s (*c*) 39.5 MJ/s (*d*) 62.8 MJ/s (*e*) 125.4 MJ/s

Answer (*b*) 53.8 MJ/s

Solution Solved by EES Software. Solutions can be verified by copying-and-pasting the following lines on a blank EES screen. (Similar problems and their solutions can be obtained easily by modifying numerical values).

Note: The solution given below also evaluates all enthalpies given on the figure.

```
P1=10 "kPa"
P11=P1
P2=400 "kPa"
P3=P2; P4=P2; P7=P2; P8=P2; P9=P2; P10=P2
P5=6000 "kPa"
P6=P5

T6=450 "C"
```

```
m_total=20 "kg/s"
m7=0.6*m_total
m_cond=0.4*m_total
C=4.18 "kJ/kg.K"
m_cooling=463 "kg/s"
s7=s6
s11=s6

h1=ENTHALPY(Steam_NBS,x=0,P=P1)
v1=VOLUME(Steam_NBS,x=0,P=P1)
w_pump=v1*(P2-P1)
h2=h1+w_pump

h3=ENTHALPY(Steam_NBS,x=0,P=P3)
h4=h3; h9=h3

v4=VOLUME(Steam_NBS,x=0,P=P4)
w_pump2=v4*(P5-P4)
h5=h4+w_pump2

h6=ENTHALPY(Steam_NBS,T=T6,P=P6)
s6=ENTROPY(Steam_NBS,T=T6,P=P6)

h7=ENTHALPY(Steam_NBS,s=s7,P=P7)
h8=h7; h10=h7

h11=ENTHALPY(Steam_NBS,s=s11,P=P11)

W_turb=m_total*(h6-h7)+m_cond*(h7-h11)
m_cooling*C*T_rise=m_cond*(h11-h1)
m_cond*h2+m_feed*h10=(m_cond+m_feed)*h3
m_process=m7-m_feed
q_process=h8-h9
Q_in=m_total*(h6-h5)
```

9-112 · · · 9-121 Design and Essay Problems

Chapter 10
REFRIGERATION CYCLES

The Reversed Carnot Cycle

10-1C Because the compression process involves the compression of a liquid-vapor mixture which requires a compressor that will handle two phases, and the expansion process involves the expansion of high-moisture content refrigerant.

10-2 A steady-flow Carnot refrigeration cycle with refrigerant-134a as the working fluid is considered. The coefficient of performance, the amount of heat absorbed from the refrigerated space, and the net work input are to be determined.

Assumptions **1** Steady operating conditions exist. **2** Kinetic and potential energy changes are negligible.

Analysis (*a*) Noting that T_H = 30°C = 303 K and $T_L = T_{\text{sat @ 120 kPa}}$ = -22.36°C = 250.6 K, the COP of this Carnot refrigerator is determined from

$$\text{COP}_{\text{R,C}} = \frac{1}{T_H / T_L - 1} = \frac{1}{(303\text{ K})/(250.6\text{ K}) - 1} = \mathbf{4.78}$$

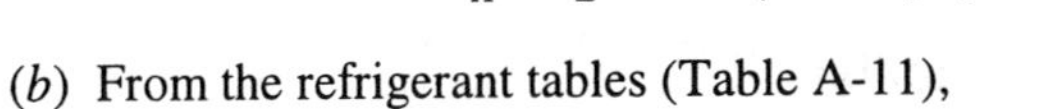

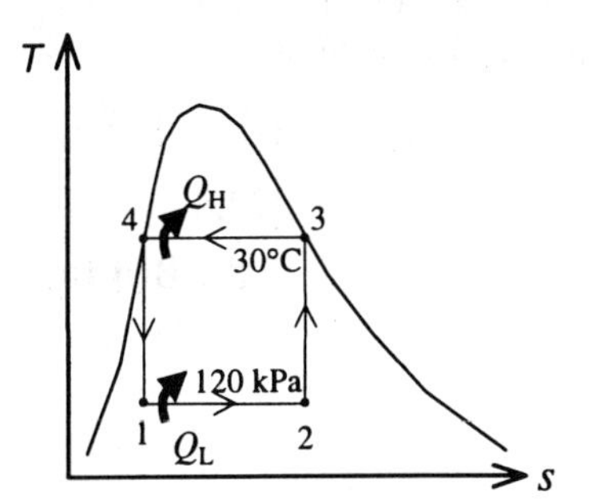

(*b*) From the refrigerant tables (Table A-11),

$$h_3 = h_{g\,@\,30°\text{C}} = 263.50\text{ kJ/kg}$$
$$h_4 = h_{f\,@\,30°\text{C}} = 91.49\text{ kJ/kg}$$

Thus,

$$q_H = h_3 - h_4 = 263.50 - 91.49 = 172.01\text{ kJ/kg}$$

and

$$\frac{q_H}{q_L} = \frac{T_H}{T_L} \longrightarrow q_L = \frac{T_L}{T_H} q_H = \left(\frac{250.6\text{ K}}{303\text{ K}}\right)(172.01\text{ kJ/kg}) = \mathbf{142.3\ kJ/kg}$$

(*c*) The net work input is determined from

$$w_{\text{net}} = q_H - q_L = 172.01 - 142.3 = \mathbf{29.71\ kJ/kg}$$

10-3E A steady-flow Carnot refrigeration cycle with refrigerant-134a as the working fluid is considered. The coefficient of performance, the quality at the beginning of the heat-absorption process, and the net work input are to be determined.

Assumptions **1** Steady operating conditions exist. **2** Kinetic and potential energy changes are negligible.

Analysis (*a*) Noting that $T_H = T_{sat\ @\ 90\ psia} = 72.83°F = 532.8$ R and $T_L = T_{sat\ @\ 30\ psia} = 15.38°F = 475.4$ R.

$$\mathrm{COP}_{R,C} = \frac{1}{T_H/T_L - 1} = \frac{1}{(532.8\ \mathrm{R})/(475.4\ \mathrm{R}) - 1} = \mathbf{8.28}$$

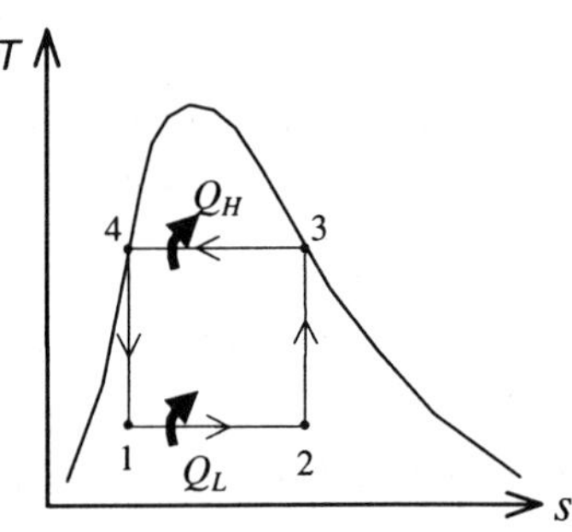

(*b*) Process 4-1 is isentropic, and thus

$$s_1 = s_4 = \left(s_f + x_4 s_{fg}\right)_{@\ 90\ \mathrm{psia}} = 0.0729 + (0.05)(0.2172 - 0.0729)$$
$$= 0.0801\ \mathrm{Btu/lbm \cdot R}$$

$$x_1 = \left(\frac{s_1 - s_f}{s_{fg}}\right)_{@\ 30\ \mathrm{psia}} = \frac{0.0801 - 0.0364}{0.2209 - 0.0364} = \mathbf{0.237}$$

(*c*) Remembering that on a *T-s* diagram the area enclosed represents the net work, and $s_3 = s_{g\ @\ 90\ psia} =$ 0.2172 Btu/lbm· R,

$$w_{\mathrm{net,in}} = (T_H - T_L)(s_3 - s_4)$$
$$= [72.83 - (-15.38)\ \mathrm{R}](0.2172 - 0.0801)\ \mathrm{Btu/lbm \cdot R}$$
$$= \mathbf{7.88\ Btu/lbm}$$

Ideal and Actual Vapor-Compression Cycles

10-4C Yes; the throttling process is an internally irreversible process.

10-5C To make the ideal vapor-compression refrigeration cycle more closely approximate the actual cycle.

10-6C No. Assuming the water is maintained at 10°C in the evaporator, the evaporator pressure will be the saturation pressure corresponding to this pressure, which is 1.2 kPa. It is not practical to design refrigeration or air-conditioning devices that involve such extremely low pressures.

10-7C Allowing a temperature difference of 10°C for effective heat transfer, the condensation temperature of the refrigerant should be 25°C. The saturation pressure corresponding to 25°C is 0.67 MPa. Therefore, the recommended pressure would be 0.7 MPa.

10-8C The area enclosed by the cyclic curve on a T-s diagram represents the net work input for the reversed Carnot cycle, but not so for the ideal vapor-compression refrigeration cycle. This is because the latter cycle involves an irreversible process for which the process path is not known.

10-9C The cycle that involves saturated liquid at 30°C will have a higher COP because, judging from the T-s diagram, it will require a smaller work input for the same refrigeration capacity.

10-10C The minimum temperature that the refrigerant can be cooled to before throttling is the temperature of the sink (the cooling medium) since heat is transferred from the refrigerant to the cooling medium.

10-11 An ideal vapor-compression refrigeration cycle with refrigerant-134a as the working fluid is considered. The rate of heat removal from the refrigerated space, the power input to the compressor, the rate of heat rejection to the environment, and the COP are to be determined.

Assumptions **1** Steady operating conditions exist. **2** Kinetic and potential energy changes are negligible.

Analysis (*a*) In an ideal vapor-compression refrigeration cycle, the compression process is isentropic, the refrigerant enters the compressor as a saturated vapor at the evaporator pressure, and leaves the condenser as saturated liquid at the condenser pressure. From the refrigerant tables (Tables A-12 and A-13),

$$\left.\begin{array}{l} P_1 = 120 \text{ kPa} \\ \text{sat.vapor} \end{array}\right\} \begin{array}{l} h_1 = h_{g\,@\,120\text{ kPa}} = 233.86 \text{ kJ/kg} \\ s_1 = s_{g\,@\,120\text{ kPa}} = 0.9354 \text{ kJ/kg}\cdot\text{K} \end{array}$$

$$\left.\begin{array}{l} P_2 = 0.7 \text{ MPa} \\ s_2 = s_1 \end{array}\right\} h_2 = 270.22 \text{ kJ/kg } (T_2 = 34.6°\text{C})$$

$$\left.\begin{array}{l} P_3 = 0.7 \text{ MPa} \\ \text{sat.liquid} \end{array}\right\} h_3 = h_{f\,@\,0.7\text{ MPa}} = 86.78 \text{ kJ/kg}$$

$$h_4 \cong h_3 = 86.78 \text{ kJ/kg (throttling)}$$

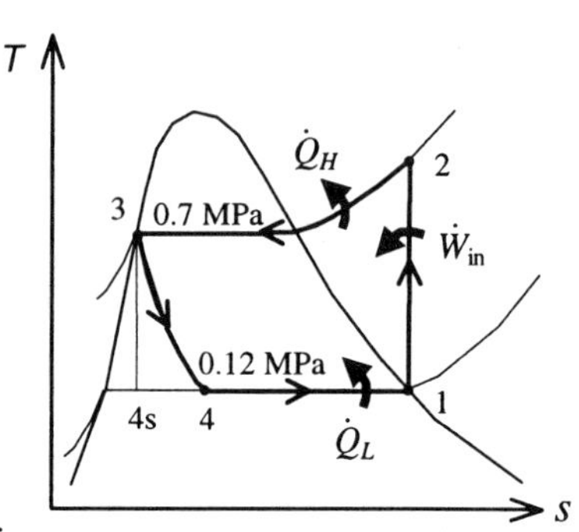

Then the rate of heat removal from the refrigerated space and the power input to the compressor are determined from

$$\dot{Q}_L = \dot{m}(h_1 - h_4) = (0.05 \text{ kg/s})(233.86 - 86.78) \text{ kJ/kg} = \mathbf{7.35 \ kW}$$

and

$$\dot{W}_{\text{in}} = \dot{m}(h_2 - h_1) = (0.05 \text{ kg/s})(270.22 - 233.86) \text{ kJ/kg} = \mathbf{1.82 \ kW}$$

(*b*) The rate of heat rejection to the environment is determined from

$$\dot{Q}_H = \dot{Q}_L + \dot{W}_{\text{in}} = 7.35 + 1.82 = \mathbf{9.17 \ kW}$$

(*c*) The COP of the refrigerator is determined from its definition,

$$\text{COP}_\text{R} = \frac{\dot{Q}_L}{\dot{W}_{\text{in}}} = \frac{7.35 \text{ kW}}{1.82 \text{ kW}} = \mathbf{4.04}$$

10-12 An ideal vapor-compression refrigeration cycle with refrigerant-134a as the working fluid is considered. The rate of heat removal from the refrigerated space, the power input to the compressor, the rate of heat rejection to the environment, and the COP are to be determined.

Assumptions **1** Steady operating conditions exist. **2** Kinetic and potential energy changes are negligible.

Analysis (a) In an ideal vapor-compression refrigeration cycle, the compression process is isentropic, the refrigerant enters the compressor as a saturated vapor at the evaporator pressure, and leaves the condenser as saturated liquid at the condenser pressure. From the refrigerant tables (Tables A-12 and A-13),

$$\left.\begin{array}{l} P_1 = 120 \text{ kPa} \\ \text{sat.vapor} \end{array}\right\} \begin{array}{l} h_1 = h_{g\text{ @ 120 kPa}} = 233.86 \text{ kJ/kg} \\ s_1 = s_{g\text{ @ 120 kPa}} = 0.9354 \text{ kJ/kg}\cdot\text{K} \end{array}$$

$$\left.\begin{array}{l} P_2 = 0.8 \text{ MPa} \\ s_2 = s_1 \end{array}\right\} h_2 = 273.04 \text{ kJ/kg } (T_2 = 39.4°\text{C})$$

$$\left.\begin{array}{l} P_3 = 0.8 \text{ MPa} \\ \text{sat.liquid} \end{array}\right\} h_3 = h_{f\text{ @ 0.7 MPa}} = 93.42 \text{ kJ/kg}$$

$$h_4 \cong h_3 = 93.42 \text{ kJ/kg (throttling)}$$

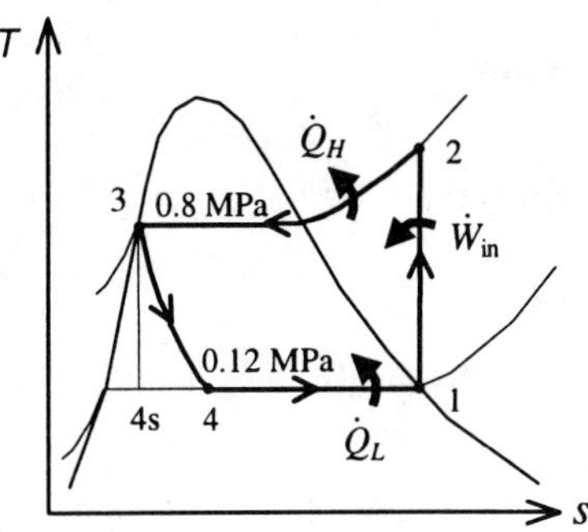

Then the rate of heat removal from the refrigerated space and the power input to the compressor are determined from

$$\dot{Q}_L = \dot{m}(h_1 - h_4) = (0.05 \text{ kg/s})(233.86 - 93.42) \text{ kJ/kg} = \mathbf{7.02 \text{ kW}}$$

and

$$\dot{W}_{in} = \dot{m}(h_2 - h_1) = (0.05 \text{ kg/s})(273.04 - 233.86) \text{ kJ/kg} = \mathbf{1.96 \text{ kW}}$$

(b) The rate of heat rejection to the environment is determined from

$$\dot{Q}_H = \dot{Q}_L + \dot{W}_{in} = 7.02 + 1.96 = \mathbf{8.98 \text{ kW}}$$

(c) The COP of the refrigerator is determined from its definition,

$$\text{COP}_R = \frac{\dot{Q}_L}{\dot{W}_{in}} = \frac{7.02 \text{ kW}}{1.96 \text{ kW}} = \mathbf{3.58}$$

10-13 An ideal vapor-compression refrigeration cycle with refrigerant-134a as the working fluid is considered. The throttling valve in the cycle is replaced by an isentropic turbine. The percentage increase in the COP and in the rate of heat removal from the refrigerated space due to this replacement are to be determined.

Assumptions **1** Steady operating conditions exist. **2** Kinetic and potential energy changes are negligible.

Analysis If the throttling valve in the previous problem is replaced by an isentropic turbine, we would have $s_{4s} = s_3 = s_{f\ @\ 0.7\ \mathrm{MPa}} = 0.3242$ kJ/kg· K, and the enthalpy at the turbine exit would be

$$x_{4s} = \left(\frac{s_3 - s_f}{s_{fg}}\right)_{@\ 120\ \mathrm{kPa}} = \frac{0.3242 - 0.0879}{0.9354 - 0.0879} = 0.279$$

$$h_{4s} = \left(h_f + x_{4s} h_{fg}\right)_{@\ 120\ \mathrm{kPa}} = 21.32 + (0.279)(212.54) = 80.62\ \mathrm{kJ/kg}$$

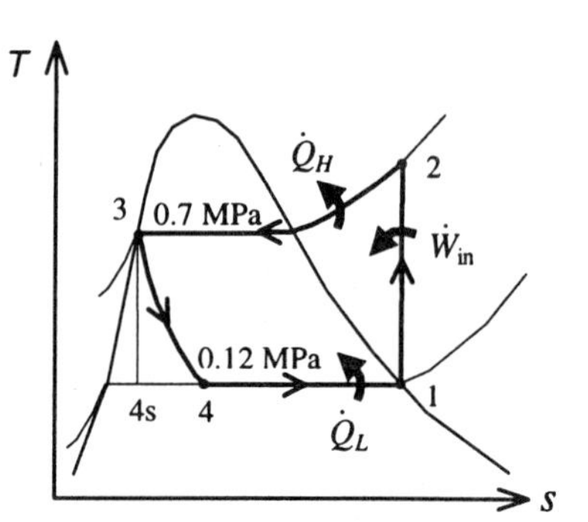

Then,

$$\dot{Q}_L = \dot{m}(h_1 - h_{4s}) = (0.05\ \mathrm{kg/s})(233.86 - 80.62)\ \mathrm{kJ/kg} = 7.66\ \mathrm{kW}$$

and

$$\mathrm{COP_R} = \frac{\dot{Q}_L}{\dot{W}_{\mathrm{in}}} = \frac{7.66\ \mathrm{kW}}{1.82\ \mathrm{kW}} = 4.21$$

Then the percentage increase in $\dot{Q}$ and COP becomes

$$\text{Increase in } \dot{Q}_L = \frac{\Delta \dot{Q}_L}{\dot{Q}_L} = \frac{7.66 - 7.35}{7.35} = \mathbf{4.2\%}$$

$$\text{Increase in } \mathrm{COP_R} = \frac{\Delta \mathrm{COP_R}}{\mathrm{COP_R}} = \frac{4.21 - 4.04}{4.04} = \mathbf{4.2\%}$$

10-14 [*Also solved by EES on enclosed CD*] An ideal vapor-compression refrigeration cycle with refrigerant-134a as the working fluid is considered. The quality of the refrigerant at the end of the throttling process, the COP, and the power input to the compressor are to be determined.

Assumptions **1** Steady operating conditions exist. **2** Kinetic and potential energy changes are negligible.

Analysis (*a*) In an ideal vapor-compression refrigeration cycle, the compression process is isentropic, the refrigerant enters the compressor as a saturated vapor at the evaporator pressure, and leaves the condenser as saturated liquid at the condenser pressure. From the refrigerant tables (Tables A-12 and A-13),

$$\left.\begin{array}{l} P_1 = 140 \text{ kPa} \\ \text{sat. vapor} \end{array}\right\} \begin{array}{l} h_1 = h_{g\text{ @ 140 kPa}} = 236.04 \text{ kJ/kg} \\ s_1 = s_{g\text{ @ 140 kPa}} = 0.9322 \text{ kJ/kg}\cdot\text{K} \end{array}$$

$$\left.\begin{array}{l} P_2 = 0.8 \text{ MPa} \\ s_2 = s_1 \end{array}\right\} h_2 = 272.05 \text{ kJ/kg}$$

$$\left.\begin{array}{l} P_3 = 0.8 \text{ MPa} \\ \text{sat. liquid} \end{array}\right\} h_3 = h_{f\text{ @ 0.8 MPa}} = 93.42 \text{ kJ/kg}$$

$$h_4 \cong h_3 = 93.42 \text{ kJ/kg (throttling)}$$

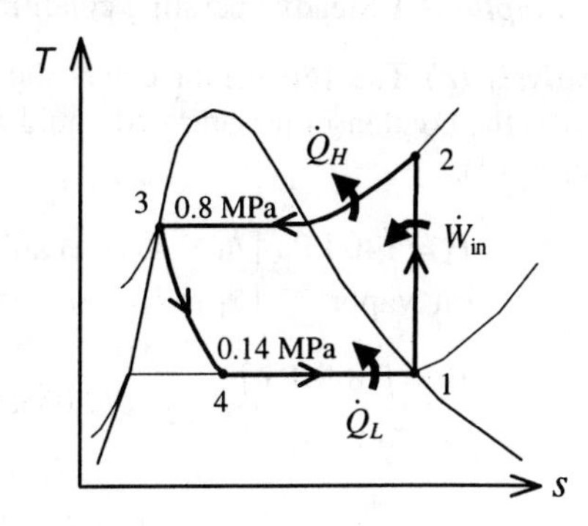

The quality of the refrigerant at the end of the throttling process is

$$x_4 = \left(\frac{h_4 - h_f}{h_{fg}}\right)_{\text{@ 140 kPa}} = \frac{93.42 - 25.77}{210.27} = \mathbf{0.322}$$

(*b*) The COP of the refrigerator is determined from its definition,

$$\text{COP}_R = \frac{q_L}{w_{in}} = \frac{h_1 - h_4}{h_2 - h_1} = \frac{236.04 - 93.42}{272.05 - 236.04} = \mathbf{3.96}$$

(*c*) The power input to the compressor is determined from

$$\dot{W}_{in} = \frac{\dot{Q}_L}{\text{COP}_R} = \frac{5 \text{ kW}}{3.96} = \mathbf{1.26\ kW}$$

10-15 EES solution of this (and other comprehensive problems designated with the *computer icon*) is available to instructors at the *Instructor Manual* section of the *Online Learning Center* (OLC) at www.mhhe.com/cengel-boles. See the Preface for access information.

10-16 A nonideal vapor-compression refrigeration cycle with refrigerant-134a as the working fluid is considered. The quality of the refrigerant at the end of the throttling process, the COP, the power input to the compressor, and the irreversibility rate associated with the compression process are to be determined.

Assumptions **1** Steady operating conditions exist. **2** Kinetic and potential energy changes are negligible.

Analysis (*a*) The refrigerant enters the compressor as a saturated vapor at the evaporator pressure, and leaves the condenser as saturated liquid at the condenser pressure. From the refrigerant tables (Tables A-12 and A-13),

$$\left.\begin{array}{l} P_1 = 140 \text{ kPa} \\ \text{sat. vapor} \end{array}\right\} \begin{array}{l} h_1 = h_{g\,@\,140\text{ kPa}} = 236.04 \text{ kJ/kg} \\ s_1 = s_{g\,@\,140\text{ kPa}} = 0.9322 \text{ kJ/kg}\cdot\text{K} \end{array}$$

$$\left.\begin{array}{l} P_2 = 0.8 \text{ MPa} \\ s_{2s} = s_1 \end{array}\right\} h_{2s} = 272.05 \text{ kJ/kg}$$

$$\eta_C = \frac{h_{2s} - h_1}{h_2 - h_1} \longrightarrow h_2 = h_1 + (h_{2s} - h_1)/\eta_C = 236.04 + (272.05 - 236.04)/(0.85) = 278.40 \text{ kJ/kg}$$

$$\left.\begin{array}{l} P_3 = 0.8 \text{ MPa} \\ \text{sat. liquid} \end{array}\right\} h_3 = h_{f\,@\,0.8\text{ MPa}} = 93.42 \text{ kJ/kg}$$

$$h_4 \cong h_3 = 93.42 \text{ kJ/kg (throttling)}$$

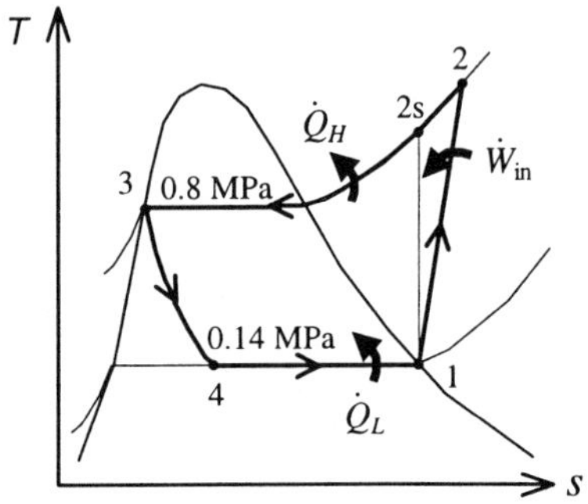

The quality of the refrigerant at the end of the throttling process is

$$x_4 = \left(\frac{h_4 - h_f}{h_{fg}}\right)_{@\,140\text{ kPa}} = \frac{93.42 - 25.77}{210.27} = \mathbf{0.322}$$

(*b*) The COP of the refrigerator is determined from its definition,

$$\text{COP}_\text{R} = \frac{q_L}{w_{in}} = \frac{h_1 - h_4}{h_2 - h_1} = \frac{236.04 - 93.42}{278.40 - 236.04} = \mathbf{3.37}$$

(*c*) The power input to the compressor is determined from

$$\dot{W}_{in} = \frac{\dot{Q}_L}{\text{COP}_R} = \frac{5 \text{ kW}}{3.37} = \mathbf{1.48\ kW}$$

The exergy destruction associated with the compression process is determined from

$$\dot{X}_{destroyed} = T_0 \dot{S}_{gen} = T_0 \dot{m}\left(s_2 - s_1 + \frac{q_{surr} \nearrow^{0}}{T_0}\right) = T_0 \dot{m}(s_2 - s_1)$$

where

$$\dot{m} = \frac{\dot{Q}_L}{q_L} = \frac{\dot{Q}_L}{h_1 - h_4} = \frac{5 \text{ kJ/s}}{(236.04 - 93.42) \text{ kJ/kg}} = 0.03506 \text{ kg/s}$$

$$\left.\begin{array}{l} P_2 = 0.8 \text{ MPa} \\ h_2 = 278.40 \text{ kJ/kg} \end{array}\right\} s_2 = 0.9522 \text{ kJ/kg}\cdot\text{K}$$

Thus, $\dot{X}_{destroyed} = (298 \text{ K})(0.03506 \text{ kg/s})(0.9522 - 0.9322) \text{ kJ/kg}\cdot\text{K} = \mathbf{0.209\ kW}$

10-17 A refrigerator with refrigerant-134a as the working fluid is considered. The rate of heat removal from the refrigerated space, the power input to the compressor, the isentropic efficiency of the compressor, and the COP of the refrigerator are to be determined.

Assumptions **1** Steady operating conditions exist. **2** Kinetic and potential energy changes are negligible.

Analysis (*a*) From the refrigerant tables (Tables A-12 and A-13),

$$\left.\begin{array}{l} P_1 = 0.14 \text{ MPa} \\ T_1 = -10°\text{C} \end{array}\right\} \begin{array}{l} h_1 = 243.40 \text{ kJ/kg} \\ s_1 = 0.9606 \text{ kJ/kg}\cdot\text{K} \end{array}$$

$$\left.\begin{array}{l} P_2 = 0.7 \text{ MPa} \\ T_2 = 50°\text{C} \end{array}\right\} h_2 = 286.35 \text{ kJ/kg}$$

$$\left.\begin{array}{l} P_{2s} = 0.7 \text{ MPa} \\ s_{2s} = s_1 \end{array}\right\} h_{2s} = 278.06 \text{ kJ/kg}$$

$$\left.\begin{array}{l} P_3 = 0.65 \text{ MPa} \\ T_3 = 24°\text{C} \end{array}\right\} h_3 = h_{f\text{ @ }24°\text{C}} = 82.90 \text{ kJ/kg}$$

$$h_4 \cong h_3 = 82.90 \text{ kJ/kg (throttling)}$$

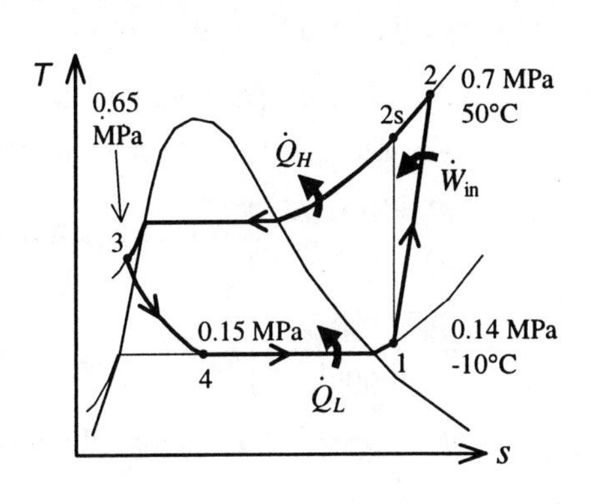

Then the rate of heat removal from the refrigerated space and the power input to the compressor are determined from

$$\dot{Q}_L = \dot{m}(h_1 - h_4) = (0.12 \text{ kg/s})(243.40 - 82.90) \text{ kJ/kg} = \mathbf{19.3 \text{ kW}}$$

and

$$\dot{W}_{\text{in}} = \dot{m}(h_2 - h_1) = (0.12 \text{ kg/s})(286.35 - 243.40) \text{ kJ/kg} = \mathbf{5.15 \text{ kW}}$$

(*b*) The adiabatic efficiency of the compressor is determined from

$$\eta_C = \frac{h_{2s} - h_1}{h_2 - h_1} = \frac{278.06 - 243.40}{286.35 - 243.40} = \mathbf{80.7\%}$$

(*c*) The COP of the refrigerator is determined from its definition,

$$\text{COP}_\text{R} = \frac{\dot{Q}_L}{\dot{W}_{\text{in}}} = \frac{19.3 \text{kW}}{5.15 \text{kW}} = \mathbf{3.75}$$

10-18E An ice-making machine operates on the ideal vapor-compression refrigeration cycle, using refrigerant-134a as the working fluid. The power input to the ice machine is to be determined.

Assumptions **1** Steady operating conditions exist. **2** Kinetic and potential energy changes are negligible.

Analysis In an ideal vapor-compression refrigeration cycle, the compression process is isentropic, the refrigerant enters the compressor as a saturated vapor at the evaporator pressure, and leaves the condenser as saturated liquid at the condenser pressure. From the refrigerant tables (Tables A-12 and A-13),

$$\left.\begin{array}{l} P_1 = 20 \text{ psia} \\ \text{sat. vapor} \end{array}\right\} \begin{array}{l} h_1 = h_{g\ @\ 20\text{ psia}} = 101.39 \text{ Btu/lbm} \\ s_1 = s_{g\ @\ 20\text{ psia}} = 0.2227 \text{ Btu/lbm}\cdot\text{R} \end{array}$$

$$\left.\begin{array}{l} P_2 = 100 \text{ psia} \\ s_2 = s_1 \end{array}\right\} h_2 = 115.64 \text{ Btu/lbm}$$

$$\left.\begin{array}{l} P_3 = 100 \text{ psia} \\ \text{sat. liquid} \end{array}\right\} h_3 = h_{f\ @\ 100\text{ psia}} = 36.99 \text{ Btu/lbm}$$

$$h_4 \cong h_3 = 36.99 \text{ Btu/lbm (throttling)}$$

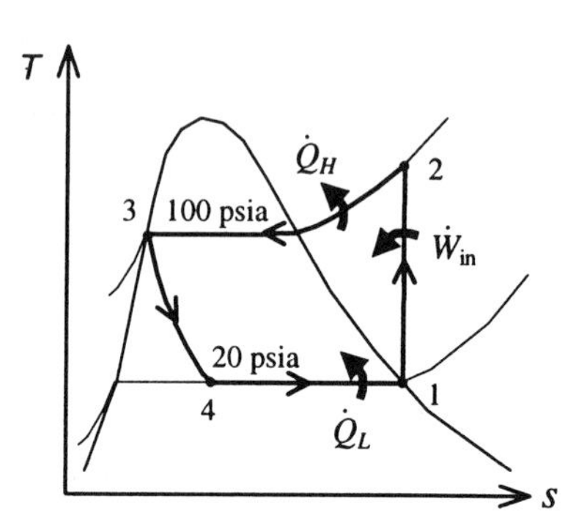

The cooling load of this refrigerator is

$$\dot{Q}_L = \dot{m}_{ice}(\Delta h)_{ice} = (20/3600 \text{ lbm/s})(169 \text{ Btu/lbm}) = 0.9389 \text{ Btu/s}$$

Then the mass flow rate of the refrigerant and the power input become

$$\dot{m}_R = \frac{\dot{Q}_L}{h_1 - h_4} = \frac{0.9389 \text{ Btu/s}}{(101.39 - 36.99) \text{ Btu/lbm}} = 0.0146\text{l bm/s}$$

and

$$\dot{W}_{in} = \dot{m}_R(h_2 - h_1) = (0.0146 \text{ lbm/s})(115.64 - 101.39) \text{ Btu/lbm}\left(\frac{1 \text{ hp}}{0.7068 \text{ Btu/s}}\right) = \mathbf{0.294\ hp}$$

10-19 A refrigerator with refrigerant-134a as the working fluid is considered. The power input to the compressor, the rate of heat removal from the refrigerated space, and the pressure drop and the rate of heat gain in the line between the evaporator and the compressor are to be determined.

Assumptions **1** Steady operating conditions exist. **2** Kinetic and potential energy changes are negligible.

Analysis (*a*) From the refrigerant tables (Tables A-12 and A-13),

$$\left.\begin{array}{l} P_1 = 140 \text{ kPa} \\ T_1 = -10°\text{C} \end{array}\right\} \begin{array}{l} h_1 = 243.40 \text{ kJ/kg} \\ s_1 = 0.9606 \text{ kJ/kg}\cdot\text{K} \\ v_1 = 0.14549 \text{ m}^3/\text{kg} \end{array}$$

$$\left.\begin{array}{l} P_2 = 1.0 \text{ MPa} \\ s_{2s} = s_1 \end{array}\right\} h_{2s} = 286.04 \text{ kJ/kg}$$

$$\left.\begin{array}{l} P_3 = 0.95 \text{ MPa} \\ T_3 = 30°\text{C} \end{array}\right\} h_3 \cong h_{f\ @\ 30°\text{C}} = 91.49 \text{ kJ/kg}$$

$$h_4 \cong h_3 = 91.49 \text{ kJ/kg (throttling)}$$

$$\left.\begin{array}{l} T_5 = -18.5°\text{C} \\ \text{sat. vapor} \end{array}\right\} \begin{array}{l} P_5 = 0.14187 \text{ MPa} \\ h_5 = 236.23 \text{ kJ/kg} \end{array}$$

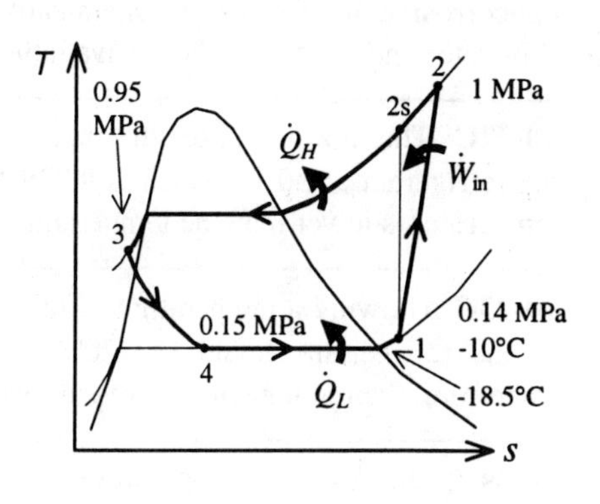

Then the mass flow rate of the refrigerant and the power input becomes

$$\dot{m} = \frac{\dot{V}_1}{v_1} = \frac{0.3/60 \text{ m}^3/\text{s}}{0.14549 \text{ m}^3/\text{kg}} = 0.0344 \text{ kg/s}$$

$$\dot{W}_{\text{in}} = \dot{m}(h_{2s} - h_1)/\eta_C = (0.0344 \text{ kg/s})[(286.04 - 243.40) \text{ kJ/kg}]/(0.78) = \mathbf{1.88 \text{ kW}}$$

(*b*) The rate of heat removal from the refrigerated space is

$$\dot{Q}_L = \dot{m}(h_5 - h_4) = (0.0344 \text{ kg/s})(236.23 - 91.49) \text{ kJ/kg} = \mathbf{4.98 \text{ kW}}$$

(*c*) The pressure drop and the heat gain in the line between the evaporator and the compressor are

$$\Delta P = P_5 - P_1 = 141.87 - 140 = \mathbf{1.87}$$

and

$$\dot{Q}_{\text{gain}} = \dot{m}(h_1 - h_5) = (0.0344 \text{ kg/s})(243.40 - 236.23) \text{ kJ/kg} = \mathbf{0.247 \text{ kW}}$$

10-20 EES solution of this (and other comprehensive problems designated with the *computer icon*) is available to instructors at the *Instructor Manual* section of the *Online Learning Center* (OLC) at www.mhhe.com/cengel-boles. See the Preface for access information.

Selecting the Right Refrigerant

10-21C The desirable characteristics of a refrigerant are to have an evaporator pressure which is above the atmospheric pressure, and a condenser pressure which corresponds to a saturation temperature above the temperature of the cooling medium. Other desirable characteristics of a refrigerant include being nontoxic, noncorrosive, nonflammable, chemically stable, having a high enthalpy of vaporization (minimizes the mass flow rate) and, of course, being available at low cost.

10-22C The minimum pressure that the refrigerant needs to be compressed to is the saturation pressure of the refrigerant at 30°C, which is **0.770 MPa**. At lower pressures, the refrigerant will have to condense at temperatures lower than the temperature of the surroundings, which cannot happen.

10-23C Allowing a temperature difference of 10°C for effective heat transfer, the evaporation temperature of the refrigerant should be -20°C. The saturation pressure corresponding to -20°C is 0.133 MPa. Therefore, the recommended pressure would be 0.12 MPa.

10-24 A refrigerator that operates on the ideal vapor-compression cycle with refrigerant-134a is considered. Reasonable pressures for the evaporator and the condenser are to be selected.

Assumptions **1** Steady operating conditions exist. **2** Kinetic and potential energy changes are negligible.

Analysis Allowing a temperature difference of 10°C for effective heat transfer, the evaporation and condensation temperatures of the refrigerant should be -20°C and 35°C, respectively. The saturation pressures corresponding to these temperatures are 0.133 MPa and 0.887 MPa. Therefore, the recommended evaporator and condenser pressures are **0.133 MPa** and **0.887 MPa**, respectively.

10-25 A heat pump that operates on the ideal vapor-compression cycle with refrigerant-134a is considered. Reasonable pressures for the evaporator and the condenser are to be selected.

Assumptions **1** Steady operating conditions exist. **2** Kinetic and potential energy changes are negligible.

Analysis Allowing a temperature difference of 10°C for effective heat transfer, the evaporation and condensation temperatures of the refrigerant should be 0°C and 32°C, respectively. The saturation pressures corresponding to these temperatures are 0.29 MPa and 0.77 MPa. Therefore, the recommended evaporator and condenser pressures are **0.29 MPa** and **0.815 MPa**, respectively.

Heat Pump Systems

10-26C A heat pump system is more cost effective in Miami because of the low heating loads and high cooling loads at that location.

10-27C A water-source heat pump extracts heat from water instead of air. Water-source heat pumps have higher COPs than the air-source systems because the temperature of water is higher than the temperature of air in winter.

10-28E A heat pump that operates on the ideal vapor-compression cycle with refrigerant-134a is considered. The power input to the heat pump and the electric power saved by using a heat pump instead of a resistance heater are to be determined.

Assumptions **1** Steady operating conditions exist. **2** Kinetic and potential energy changes are negligible.

Analysis In an ideal vapor-compression refrigeration cycle, the compression process is isentropic, the refrigerant enters the compressor as a saturated vapor at the evaporator pressure, and leaves the condenser as saturated liquid at the condenser pressure. From the refrigerant tables (Tables A-12E and A-13E),

$$\left.\begin{array}{l} P_1 = 50 \text{ psia} \\ \text{sat. vapor} \end{array}\right\} \begin{array}{l} h_1 = h_{g\ @\ 50\text{ psia}} = 107.43 \text{ Btu/lbm} \\ s_1 = s_{g\ @\ 50\text{ psia}} = 0.2189 \text{ Btu/lbm}\cdot\text{R} \end{array}$$

$$\left.\begin{array}{l} P_2 = 120 \text{ psia} \\ s_2 = s_1 \end{array}\right\} h_2 = 115.15 \text{ Btu/lbm}$$

$$\left.\begin{array}{l} P_3 = 120 \text{ psia} \\ \text{sat. liquid} \end{array}\right\} h_3 = h_{f\ @\ 120\text{ psia}} = 40.91 \text{ Btu/lbm}$$

$$h_4 \cong h_3 = 40.91 \text{ Btu/lbm (throttling)}$$

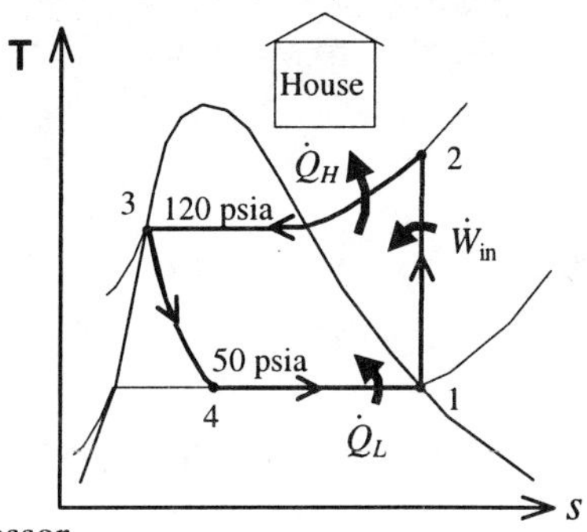

The mass flow rate of the refrigerant and the power input to the compressor are determined from

$$\dot{m} = \frac{\dot{Q}_H}{q_H} = \frac{\dot{Q}_H}{h_2 - h_3} = \frac{60{,}000/3600 \text{ Btu/s}}{(115.15 - 40.91) \text{ Btu/lbm}} = 0.225 \text{ lbm/s}$$

and

$$\dot{W}_{\text{in}} = \dot{m}(h_2 - h_1) = (0.225 \text{ kg/s})(115.15 - 107.43) \text{ Btu/lbm} = 1.73 \text{ Btu/s}$$
$$= \mathbf{2.45\ hp} \text{ since } 1 \text{ hp} = 0.7068 \text{ Btu/s}$$

The electrical power required without the heat pump is

$$\dot{W}_e = \dot{Q}_H = (60{,}000/3600 \text{ Btu/s})\left(\frac{1 \text{ hp}}{0.7068 \text{ Btu/s}}\right) = 23.58 \text{ hp}$$

Thus,

$$\dot{W}_{\text{saved}} = \dot{W}_e - \dot{W}_{\text{in}} = 23.58 - 2.45 = \mathbf{21.13\ hp}$$
$$= 15.76 \text{ kW since } 1 \text{ hp} = 0.7457 \text{ kW}$$

10-29 A heat pump that operates on the ideal vapor-compression cycle with refrigerant-134a is considered. The power input to the heat pump is to be determined.

Assumptions **1** Steady operating conditions exist. **2** Kinetic and potential energy changes are negligible.

Analysis In an ideal vapor-compression refrigeration cycle, the compression process is isentropic, the refrigerant enters the compressor as a saturated vapor at the evaporator pressure, and leaves the condenser as saturated liquid at the condenser pressure. From the refrigerant tables (Tables A-12 and A-13),

$$\left.\begin{array}{l} P_1 = 320\text{ kPa} \\ \text{sat. vapor} \end{array}\right\} \begin{array}{l} h_1 = h_{g\,@320\text{ kPa}} = 248.66\text{ kJ/kg} \\ s_1 = s_{g\,@\,320\text{ kPa}} = 0.9177\text{ kJ/kg}\cdot\text{K} \end{array}$$

$$\left.\begin{array}{l} P_2 = 1.4\text{ MPa} \\ s_2 = s_1 \end{array}\right\} h_2 = 279.14\text{ kJ/kg}$$

$$\left.\begin{array}{l} P_3 = 1.4\text{ MPa} \\ \text{sat. liquid} \end{array}\right\} h_3 = h_{f\,@\,1.4\text{ MPa}} = 125.26\text{ kJ/kg}$$

$$h_4 \cong h_3 = 125.26\text{ kJ/kg} \quad (\text{throttling})$$

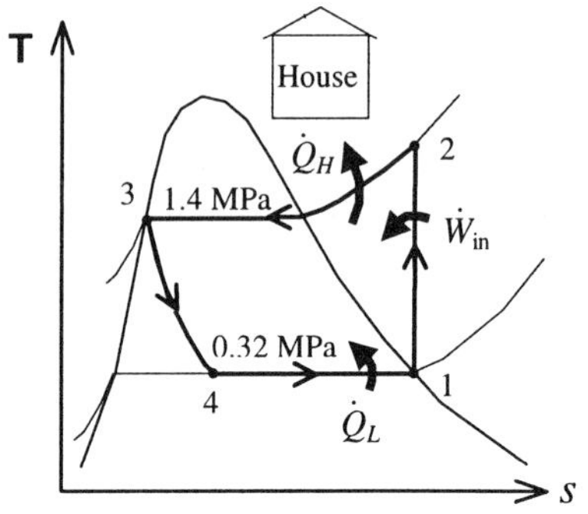

T
House
$\dot{Q}_H$
2
3
1.4 MPa
$\dot{W}_{in}$
0.32 MPa
4
1
$\dot{Q}_L$
s

The heating load of this heat pump is determined from

$$\dot{Q}_H = [\dot{m}C(T_2 - T_1)]_{\text{water}} = (0.24\text{ kg/s})(4.18\text{ kJ/kg}\cdot{}^\circ\text{C})(54 - 15)^\circ\text{C} = 39.12\text{ kW}$$

and

$$\dot{m}_R = \frac{\dot{Q}_H}{q_H} = \frac{\dot{Q}_H}{h_2 - h_3} = \frac{39.12\text{ kJ/s}}{(279.14 - 125.26)\text{ kJ/kg}} = 0.254\text{ kg/s}$$

Then,

$$\dot{W}_{\text{in}} = \dot{m}_R(h_2 - h_1) = (0.254\text{ kg/s})(279.14 - 248.66)\text{ kJ/kg} = \mathbf{7.75\ kW}$$

10-30 A heat pump with refrigerant-134a as the working fluid heats a house by using underground water as the heat source. The power input to the heat pump, the rate of heat absorption from the water, and the increase in electric power input if an electric resistance heater is used instead of a heat pump are to be determined.

Assumptions **1** Steady operating conditions exist. **2** Kinetic and potential energy changes are negligible.

Analysis (*a*) From the refrigerant tables (Tables A-12 and A-13),

$$\left.\begin{array}{l} P_1 = 280\text{ kPa} \\ T_1 = 0°\text{C} \end{array}\right\} h_1 = 247.64\text{ kJ/kg}$$

$$\left.\begin{array}{l} P_2 = 1.0\text{ MPa} \\ T_2 = 60°\text{C} \end{array}\right\} h_2 = 291.36\text{ kJ/kg}$$

$$\left.\begin{array}{l} P_3 = 1.0\text{ MPa} \\ T_3 = 30°\text{C} \end{array}\right\} h_3 \cong h_{f\ @\ 30°\text{C}} = 91.49\text{ kJ/kg}$$

$$h_4 \cong h_3 = 91.49\text{ kJ/kg}\ (\text{throttling})$$

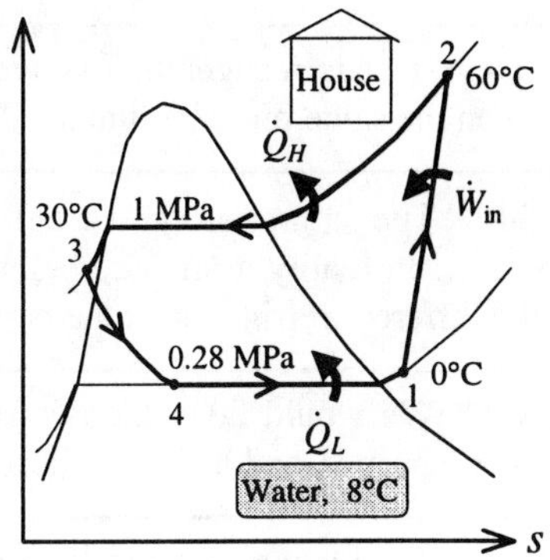

The mass flow rate of the refrigerant is

$$\dot{m}_R = \frac{\dot{Q}_H}{q_H} = \frac{\dot{Q}_H}{h_2 - h_3} = \frac{60{,}000/3{,}600\text{ kJ/s}}{(291.36 - 91.49)\text{ kJ/kg}} = 0.0834\text{ kg/s}$$

Then the power input to the compressor becomes

$$\dot{W}_{\text{in}} = \dot{m}(h_2 - h_1) = (0.0834\text{ kg/s})(291.36 - 247.64)\text{ kJ/kg} = \mathbf{3.65\ kW}$$

(*b*) The rate of hat absorption from the water is

$$\dot{Q}_L = \dot{m}(h_1 - h_4) = (0.0834\text{ kg/s})(247.64 - 91.49)\text{ kJ/kg} = \mathbf{13.02\ kW}$$

(*c*) The electrical power required without the heat pump is

$$\dot{W}_e = \dot{Q}_H = 60{,}000/3600\text{ kJ/s} = 16.67\text{ kW}$$

Thus,

$$\dot{W}_{\text{increase}} = \dot{W}_e - \dot{W}_{\text{in}} = 16.67 - 3.65 = \mathbf{13.02\ kW}$$

10-31 EES solution of this (and other comprehensive problems designated with the *computer icon*) is available to instructors at the *Instructor Manual* section of the *Online Learning Center* (OLC) at www.mhhe.com/cengel-boles. See the Preface for access information.

Innovative Refrigeration Systems

10-32C Performing the refrigeration in stages is called cascade refrigeration. In cascade refrigeration, two or more refrigeration cycles operate in series. Cascade refrigerators are more complex and expensive, but they have higher COP's, they can incorporate two or more different refrigerants, and they can achieve much lower temperatures.

10-33C Cascade refrigeration systems have higher COPs than the ordinary refrigeration systems operating between the same pressure limits.

10-34C The saturation pressure of refrigerant-134a at -32°C is 77 kPa, which is below the atmospheric pressure. In reality a pressure below this value should be used. Therefore, a cascade refrigeration system with a different refrigerant at the bottoming cycle is recommended in this case.

10-35C We would favor the two-stage compression refrigeration system with a flash chamber since it is simpler, cheaper, and has better heat transfer characteristics.

10-36C Yes, by expanding the refrigerant in stages in several throttling devices.

10-37C To take advantage of the cooling effect by throttling from high pressures to low pressures.

10-38 A two-stage cascade refrigeration system is considered. Each stage operates on the ideal vapor-compression cycle with refrigerant-134a as the working fluid. The mass flow rate of refrigerant through the lower cycle, the rate of heat removal from the refrigerated space, the power input to the compressor, and the COP of this cascade refrigerator are to be determined.

Assumptions **1** Steady operating conditions exist. **2** Kinetic and potential energy changes are negligible. **3** The heat exchanger is adiabatic.

Analysis (*a*) Each stage of the cascade refrigeration cycle is said to operate on the ideal vapor compression refrigeration cycle. Thus the compression process is isentropic, and the refrigerant enters the compressor as a saturated vapor at the evaporator pressure. Also, the refrigerant leaves the condenser as a saturated liquid at the condenser pressure. The enthalpies of the refrigerant at all 8 states are determined from the refrigerant tables (Tables A-11, A-12, and A-13) to be

$$h_1 = 236.04 \text{ kJ/kg}, \quad h_2 = 257.39 \text{ kJ/kg}$$
$$h_3 = 62.00 \text{ kJ/kg}, \quad h_4 = 62.00 \text{ kJ/kg}$$
$$h_5 = 252.32 \text{ kJ/kg}, \quad h_6 = 266.59 \text{ kJ/kg}$$
$$h_7 = 93.42 \text{ kJ/kg}, \quad h_8 = 93.42 \text{ kJ/kg}$$

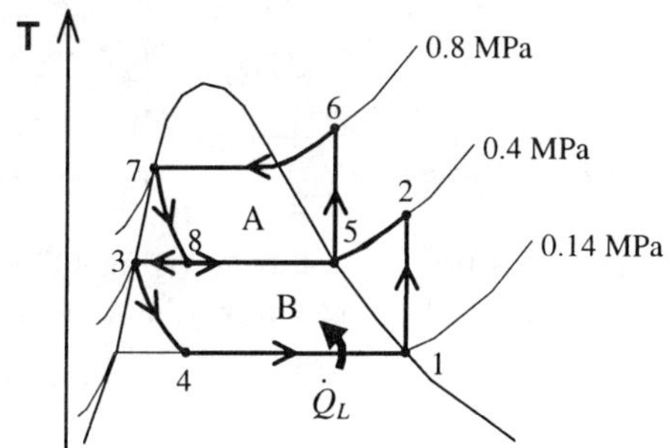

The mass flow rate of the refrigerant through the lower cycle is determined from an energy balance on the heat exchanger:

$$\dot{E}_{\text{in}} - \dot{E}_{\text{out}} = \Delta\dot{E}_{\text{system}}^{\nearrow 0 \text{ (steady)}} = 0$$
$$\dot{E}_{\text{in}} = \dot{E}_{\text{out}}$$
$$\sum \dot{m}_e h_e = \sum \dot{m}_i h_i$$
$$\dot{m}_A (h_5 - h_8) = \dot{m}_B (h_2 - h_3)$$
$$\dot{m}_B = \frac{h_5 - h_8}{h_2 - h_3}\dot{m}_A = \frac{252.32 - 93.42}{257.39 - 62.00}(0.24 \text{ kg/s}) = \mathbf{0.1952 \text{ kg/s}}$$

(*b*) The rate of heat removed by a cascade cycle is the rate of heat absorption in the evaporator of the lowest stage. The power input to a cascade cycle is the sum of the power inputs to all of the compressors:

$$\dot{Q}_L = \dot{m}_B (h_1 - h_4) = (0.1952 \text{ kg/s})(236.04 - 62.00) \text{ kJ/kg} = \mathbf{34.0 \text{ kW}}$$

$$\begin{aligned}\dot{W}_{\text{in}} &= \dot{W}_{\text{compI,in}} + \dot{W}_{\text{compII,in}} = \dot{m}_A (h_6 - h_5) + \dot{m}_B (h_2 - h_1) \\ &= (0.24 \text{ kg/s})(266.59 - 252.32) \text{ kJ/kg} + (0.1952 \text{ kg/s})(257.39 - 236.04) \text{ kJ/kg} \\ &= \mathbf{7.59 \text{ kW}}\end{aligned}$$

(*c*) The COP of this refrigeration system is determined from its definition,

$$\text{COP}_\text{R} = \frac{\dot{Q}_L}{\dot{W}_{\text{net,in}}} = \frac{34.0\text{kW}}{7.59\text{kW}} = \mathbf{4.48}$$

10-39 A two-stage cascade refrigeration system is considered. Each stage operates on the ideal vapor-compression cycle with refrigerant-134a as the working fluid. The mass flow rate of refrigerant through the lower cycle, the rate of heat removal from the refrigerated space, the power input to the compressor, and the COP of this cascade refrigerator are to be determined.

Assumptions **1** Steady operating conditions exist. **2** Kinetic and potential energy changes are negligible. **3** The heat exchanger is adiabatic.

Analysis (*a*) Each stage of the cascade refrigeration cycle is said to operate on the ideal vapor compression refrigeration cycle. Thus the compression process is isentropic, and the refrigerant enters the compressor as a saturated vapor at the evaporator pressure. Also, the refrigerant leaves the condenser as a saturated liquid at the condenser pressure. The enthalpies of the refrigerant at all 8 states are determined from the refrigerant tables (Tables A-11, A-12, and A-13) to be

$$h_1 = 236.04\ \text{kJ/kg},\quad h_2 = 262.07\ \text{kJ/kg}$$
$$h_3 = 71.33\ \text{kJ/kg},\quad h_4 = 71.33\ \text{kJ/kg}$$
$$h_5 = 256.07\ \text{kJ/kg},\quad h_6 = 265.72\ \text{kJ/kg}$$
$$h_7 = 93.42\ \text{kJ/kg},\quad h_8 = 93.42\ \text{kJ/kg}$$

The mass flow rate of the refrigerant through the lower cycle is determined from an energy balance on the heat exchanger:

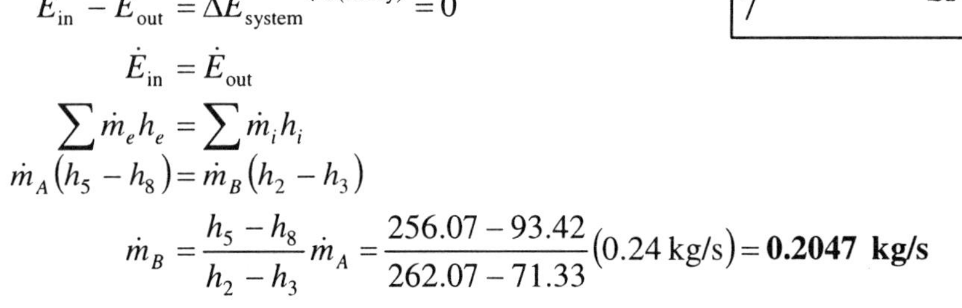

$$\dot{E}_{in} - \dot{E}_{out} = \Delta\dot{E}_{system}^{\nearrow 0\,(steady)} = 0$$
$$\dot{E}_{in} = \dot{E}_{out}$$
$$\sum \dot{m}_e h_e = \sum \dot{m}_i h_i$$
$$\dot{m}_A(h_5 - h_8) = \dot{m}_B(h_2 - h_3)$$
$$\dot{m}_B = \frac{h_5 - h_8}{h_2 - h_3}\dot{m}_A = \frac{256.07 - 93.42}{262.07 - 71.33}(0.24\ \text{kg/s}) = \mathbf{0.2047\ kg/s}$$

(*b*) The rate of heat removed by a cascade cycle is the rate of heat absorption in the evaporator of the lowest stage. The power input to a cascade cycle is the sum of the power inputs to all of the compressors:

$$\dot{Q}_L = \dot{m}_B(h_1 - h_4) = (0.2047\ \text{kg/s})(236.04 - 71.33)\ \text{kJ/kg} = \mathbf{33.71\ kW}$$
$$\dot{W}_{in} = \dot{W}_{compI,in} + \dot{W}_{compII,in} = \dot{m}_A(h_6 - h_5) + \dot{m}_B(h_2 - h_1)$$
$$= (0.24\ \text{kg/s})(265.72 - 256.07)\ \text{kJ/kg} + (0.2047\ \text{kg/s})(262.07 - 236.04)\ \text{kJ/kg}$$
$$= \mathbf{7.64\ kW}$$

(*c*) The COP of this refrigeration system is determined from its definition,

$$\text{COP}_R = \frac{\dot{Q}_L}{\dot{W}_{net,in}} = \frac{33.71\text{kW}}{7.64\text{kW}} = \mathbf{4.41}$$

10-40 [*Also solved by EES on enclosed CD*] A two-stage compression refrigeration system with refrigerant-134a as the working fluid is considered. The fraction of the refrigerant that evaporates as it is throttled to the flash chamber, the rate of heat removed from the refrigerated space, and the COP are to be determined.

Assumptions **1** Steady operating conditions exist. **2** Kinetic and potential energy changes are negligible. **3** The flash chamber is adiabatic.

Analysis (*a*) The enthalpies of the refrigerant at several states are determined from the refrigerant tables (Tables A-11, A-12, and A-13) to be

$$h_1 = 236.04 \text{ kJ/kg}, \quad h_2 = 262.07 \text{ kJ/kg}$$
$$h_3 = 256.07 \text{ kJ/kg},$$
$$h_5 = 105.29 \text{ kJ/kg}, \quad h_6 = 105.29 \text{ kJ/kg}$$
$$h_7 = 71.33 \text{ kJ/kg}, \quad h_8 = 71.33 \text{ kJ/kg}$$

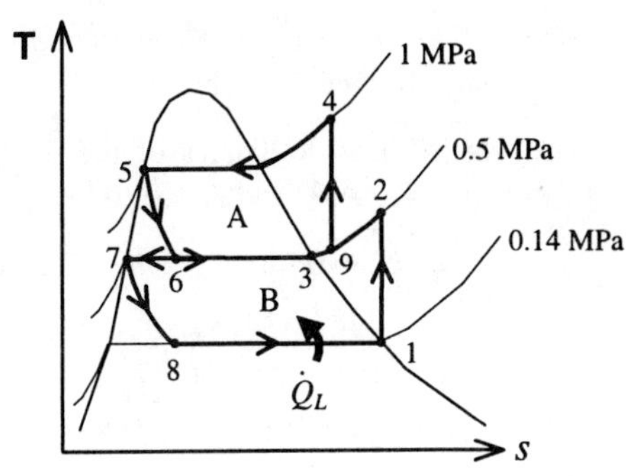

The fraction of the refrigerant that evaporates as it is throttled to the flash chamber is simply the quality at state 6,

$$x_6 = \frac{h_6 - h_f}{h_{fg}} = \frac{105.29 - 71.33}{184.74} = \mathbf{0.1838}$$

(*b*) The enthalpy at state 9 is determined from an energy balance on the mixing chamber:

$$\dot{E}_{\text{in}} - \dot{E}_{\text{out}} = \Delta \dot{E}_{\text{system}}^{\nearrow 0 \text{ (steady)}} = 0$$
$$\dot{E}_{\text{in}} = \dot{E}_{\text{out}}$$
$$\sum \dot{m}_e h_e = \sum \dot{m}_i h_i$$
$$(1)h_9 = x_6 h_3 + (1 - x_6)h_2$$
$$h_9 = (0.1838)(256.07) + (1 - 0.1838)(262.07) = 260.97 \text{ kJ/kg}$$

also,

$$\left.\begin{array}{l} P_4 = 1 \text{ MPa} \\ s_4 = s_9 = 0.9285 \text{ kJ/kg} \cdot \text{K} \end{array}\right\} h_4 = 275.64 \text{ kJ/kg}$$

Then the rate of heat removed from the refrigerated space and the compressor work input per unit mass of refrigerant flowing through the condenser are

$$\dot{m}_B = (1 - x_6)\dot{m}_A = (1 - 0.1838)(0.25 \text{ kg/s}) = 0.20405 \text{ kg/s}$$

$$\dot{Q}_L = \dot{m}_B (h_1 - h_8) = (0.20405 \text{ kg/s})(236.04 - 71.33) \text{ kJ/kg} = \mathbf{33.61 \text{ kW}}$$

$$\dot{W}_{\text{in}} = \dot{W}_{\text{compI,in}} + \dot{W}_{\text{compII,in}} = \dot{m}_A (h_4 - h_9) + \dot{m}_B (h_2 - h_1)$$
$$= (0.25 \text{ kg/s})(275.64 - 260.97) \text{ kJ/kg} + (0.20405 \text{ kg/s})(262.07 - 236.04) \text{ kJ/kg}$$
$$= 8.98 \text{ kW}$$

(*c*) The coefficient of performance is determined from

$$\text{COP}_\text{R} = \frac{\dot{Q}_L}{\dot{W}_{\text{net, in}}} = \frac{33.61 \text{ kW}}{8.98 \text{ kW}} = \mathbf{3.74}$$

10-41 EES solution of this (and other comprehensive problems designated with the *computer icon*) is available to instructors at the *Instructor Manual* section of the *Online Learning Center* (OLC) at www.mhhe.com/cengel-boles. See the Preface for access information.

10-42 [*Also solved by EES on enclosed CD*] A two-stage compression refrigeration system with refrigerant-134a as the working fluid is considered. The fraction of the refrigerant that evaporates as it is throttled to the flash chamber, the rate of heat removed from the refrigerated space, and the COP are to be determined.

Assumptions **1** Steady operating conditions exist. **2** Kinetic and potential energy changes are negligible. **3** The flash chamber is adiabatic.

Analysis (*a*) The enthalpies of the refrigerant at several states are determined from the refrigerant tables (Tables A-11, A-12, and A-13) to be

$$h_1 = 236.04 \text{ kJ/kg}, \quad h_2 = 252.71 \text{ kJ/kg}$$
$$h_3 = 248.66 \text{ kJ/kg},$$
$$h_5 = 105.29 \text{ kJ/kg}, \quad h_6 = 105.29 \text{ kJ/kg}$$
$$h_7 = 53.31 \text{ kJ/kg}, \quad h_8 = 53.31 \text{ kJ/kg}$$

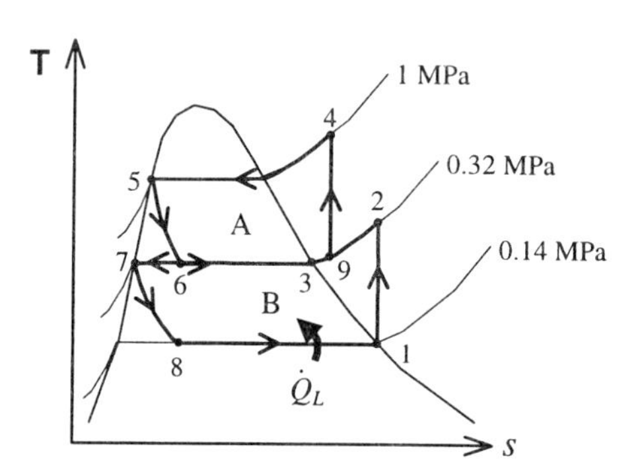

The fraction of the refrigerant that evaporates as it is throttled to the flash chamber is simply the quality at state 6,

$$x_6 = \frac{h_6 - h_f}{h_{fg}} = \frac{105.29 - 53.31}{195.35} = \mathbf{0.266}$$

(*b*) The enthalpy at state 9 is determined from an energy balance on the mixing chamber:

$$\dot{E}_{\text{in}} - \dot{E}_{\text{out}} = \Delta \dot{E}_{\text{system}}^{\nearrow 0 \text{ (steady)}} = 0$$
$$\dot{E}_{\text{in}} = \dot{E}_{\text{out}}$$
$$\sum \dot{m}_e h_e = \sum \dot{m}_i h_i$$
$$(1)h_9 = x_6 h_3 + (1 - x_6)h_2$$
$$h_9 = (0.266)(248.66) + (1 - 0.266)(252.71) = 251.63 \text{ kJ/kg}$$

also,

$$\left. \begin{array}{l} P_4 = 1 \text{ MPa} \\ s_4 = s_9 = 0.9283 \text{ kJ/kg} \cdot \text{K} \end{array} \right\} h_4 = 275.58 \text{ kJ/kg}$$

Then the rate of heat removed from the refrigerated space and the compressor work input per unit mass of refrigerant flowing through the condenser are

$$\dot{m}_B = (1 - x_6)\dot{m}_A = (1 - 0.266)(0.25 \text{ kg/s}) = 0.1835 \text{ kg/s}$$

$$\dot{Q}_L = \dot{m}_B (h_1 - h_8) = (0.1835 \text{ kg/s})(236.04 - 53.31) \text{ kJ/kg} = \mathbf{33.53 \text{ kW}}$$

$$\begin{aligned} \dot{W}_{\text{in}} &= \dot{W}_{\text{compI,in}} + \dot{W}_{\text{compII,in}} = \dot{m}_A (h_4 - h_9) + \dot{m}_B (h_2 - h_1) \\ &= (0.25 \text{ kg/s})(275.58 - 251.63) \text{ kJ/kg} + (0.1835 \text{ kg/s})(252.71 - 236.04) \text{ kJ/kg} \\ &= 9.05 \text{ kW} \end{aligned}$$

(*c*) The coefficient of performance is determined from

$$\text{COP}_\text{R} = \frac{\dot{Q}_L}{\dot{W}_{\text{net, in}}} = \frac{33.53 \text{ kW}}{9.05 \text{ kW}} = \mathbf{3.71}$$

Gas Refrigeration Cycles

10-43C The ideal gas refrigeration cycle is identical to the Brayton cycle, except it operates in the reversed direction.

10-44C The reversed Stirling cycle is identical to the Stirling cycle, except it operates in the reversed direction. Remembering that the Stirling cycle is a totally reversible cycle, the reversed Stirling cycle is also totally reversible, and thus its COP is

$$\text{COP}_{\text{R,Stirling}} = \frac{1}{T_H / T_L - 1}$$

10-45C In the ideal gas refrigeration cycle, the heat absorption and the heat rejection processes occur at constant pressure instead of at constant temperature.

10-46C In aircraft cooling, the atmospheric air is compressed by a compressor, cooled by the surrounding air, and expanded in a turbine. The cool air leaving the turbine is then directly routed to the cabin.

10-47C No; because $h = h(T)$ for ideal gases, and the temperature of air will not drop during a throttling ($h_1 = h_2$) process.

10-48C By regeneration.

10-49 An ideal-gas refrigeration cycle with air as the working fluid is considered. The maximum and minimum temperatures in the cycle, the COP, and the rate of refrigeration are to be determined.

Assumptions **1** Steady operating conditions exist. **2** Air is an ideal gas with variable specific heats. **3** Kinetic and potential energy changes are negligible.

Analysis (*a*) We assume both the turbine and the compressor to be isentropic, the turbine inlet temperature to be the temperature of the surroundings, and the compressor inlet temperature to be the temperature of the refrigerated space. From the air table (Table A-17),

$$T_1 = 250 \text{ K} \longrightarrow \begin{array}{l} h_1 = 250.05 \text{ kJ/kg} \\ P_{r_1} = 0.7329 \end{array}$$

$$T_1 = 300 \text{ K} \longrightarrow \begin{array}{l} h_3 = 300.19 \text{ kJ/kg} \\ P_{r_3} = 1.386 \end{array}$$

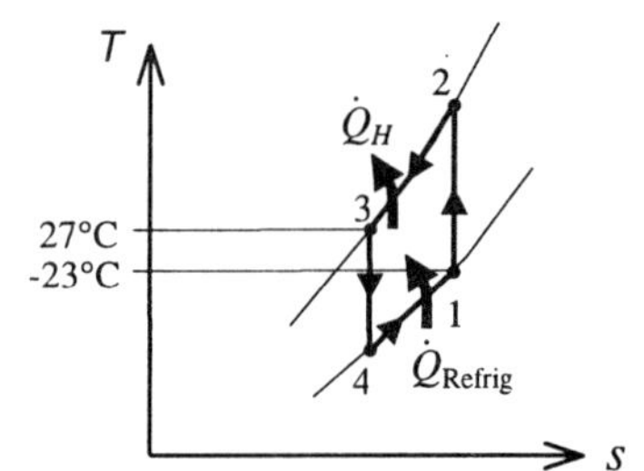

Thus,

$$P_{r_2} = \frac{P_2}{P_1} P_{r_1} = (3)(0.7329) = 2.1987 \longrightarrow T_2 = T_{\max} = \mathbf{342.2 \text{ K}}$$
$$h_2 = 342.60 \text{ kJ/kg}$$

$$P_{r_4} = \frac{P_4}{P_3} P_{r_3} = \left(\frac{1}{3}\right)(1.386) = 0.462 \longrightarrow T_4 = T_{\min} = \mathbf{219.0 \text{ K}}$$
$$h_4 = 218.97 \text{ kJ/kg}$$

(*b*) The COP of this ideal gas refrigeration cycle is determined from

$$\text{COP}_\text{R} = \frac{q_L}{w_{\text{net,in}}} = \frac{q_L}{w_{\text{comp,in}} - w_{\text{turb,out}}}$$

where

$$q_L = h_1 - h_4 = 250.05 - 218.97 = 31.08 \text{ kJ/kg}$$
$$w_{\text{comp,in}} = h_2 - h_1 = 342.60 - 250.05 = 92.55 \text{ kJ/kg}$$
$$w_{\text{turb,out}} = h_3 - h_4 = 300.19 - 218.97 = 81.22 \text{ kJ/kg}$$

Thus, $$\text{COP}_\text{R} = \frac{31.08}{92.55 - 81.22} = \mathbf{2.74}$$

(*c*) The rate of refrigeration is determined to be

$$\dot{Q}_{\text{refrig}} = \dot{m}(q_L) = (0.08 \text{ kg/s})(31.08 \text{ kJ/kg}) = \mathbf{2.49 \text{ kJ/s}}$$

10-50 [*Also solved by EES on enclosed CD*] An ideal-gas refrigeration cycle with air as the working fluid is considered. The rate of refrigeration, the net power input, and the COP are to be determined.

Assumptions **1** Steady operating conditions exist. **2** Air is an ideal gas with variable specific heats. **3** Kinetic and potential energy changes are negligible.

Analysis (*a*) We assume both the turbine and the compressor to be isentropic, the turbine inlet temperature to be the temperature of the surroundings, and the compressor inlet temperature to be the temperature of the refrigerated space. From the air table (Table A-17),

$$T_1 = 285\text{ K} \longrightarrow \begin{array}{l} h_1 = 285.14\text{ kJ/kg} \\ P_{r_1} = 1.1584 \end{array}$$

$$T_1 = 320\text{ K} \longrightarrow \begin{array}{l} h_3 = 320.29\text{ kJ/kg} \\ P_{r_3} = 1.7375 \end{array}$$

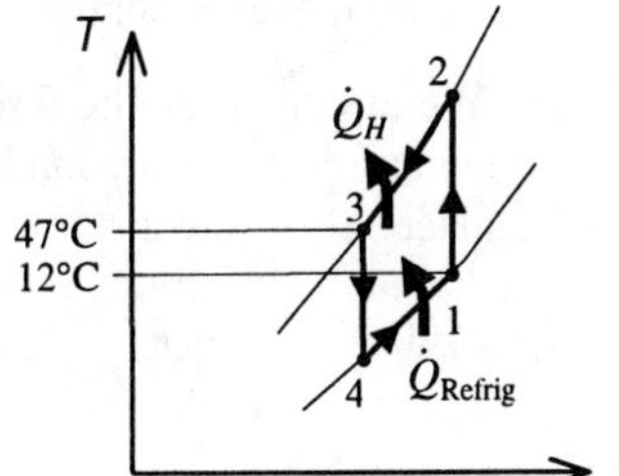

Thus,

$$P_{r_2} = \frac{P_2}{P_1} P_{r_1} = \left(\frac{250}{50}\right)(1.1584) = 5.792 \longrightarrow \begin{array}{l} T_2 = 450.4\text{ K} \\ h_2 = 452.17\text{ kJ/kg} \end{array}$$

$$P_{r_4} = \frac{P_4}{P_3} P_{r_3} = \left(\frac{50}{250}\right)(1.7375) = 0.3475 \longrightarrow \begin{array}{l} T_4 = 201.8\text{ K} \\ h_4 = 201.76\text{ kJ/kg} \end{array}$$

Then the rate of refrigeration is

$$\dot{Q}_{\text{refrig}} = \dot{m}(q_L) = \dot{m}(h_1 - h_4) = (0.08\text{ kg/s})(285.14 - 201.76)\text{kJ/kg} = \mathbf{6.67\ kW}$$

(*b*) The net power input is determined from

$$\dot{W}_{\text{net, in}} = \dot{W}_{\text{comp, in}} - \dot{W}_{\text{turb, out}}$$

where

$$\dot{W}_{\text{comp,in}} = \dot{m}(h_2 - h_1) = (0.08\text{ kg/s})(452.17 - 285.14)\text{kJ/kg} = 13.36\text{ kW}$$
$$\dot{W}_{\text{turb,out}} = \dot{m}(h_3 - h_4) = (0.08\text{ kg/s})(320.29 - 201.76)\text{kJ/kg} = 9.48\text{ kW}$$

Thus, $\dot{W}_{\text{net, in}} = 13.36 - 9.48 = \mathbf{3.88\ kW}$

(*c*) The COP of this ideal gas refrigeration cycle is determined from

$$\text{COP}_\text{R} = \frac{\dot{Q}_L}{\dot{W}_{\text{net, in}}} = \frac{6.67\text{ kW}}{3.88\text{ kW}} = \mathbf{1.72}$$

10-51 EES solution of this (and other comprehensive problems designated with the *computer icon*) is available to instructors at the *Instructor Manual* section of the *Online Learning Center* (OLC) at www.mhhe.com/cengel-boles. See the Preface for access information.

10-52E An ideal-gas refrigeration cycle with air as the working fluid is considered. The rate of refrigeration, the net power input, and the COP are to be determined.

Assumptions **1** Steady operating conditions exist. **2** Air is an ideal gas with variable specific heats. **3** Kinetic and potential energy changes are negligible.

Analysis (*a*) We assume both the turbine and the compressor to be isentropic, the turbine inlet temperature to be the temperature of the surroundings, and the compressor inlet temperature to be the temperature of the refrigerated space. From the air table (Table A-17E),

$$T_1 = 500 \text{ R} \longrightarrow \begin{array}{l} h_1 = 119.48 \text{ Btu/lbm} \\ P_{r_1} = 1.0590 \end{array}$$

$$T_1 = 580 \text{ R} \longrightarrow \begin{array}{l} h_3 = 138.66 \text{ Btu/lbm} \\ P_{r_3} = 1.7800 \end{array}$$

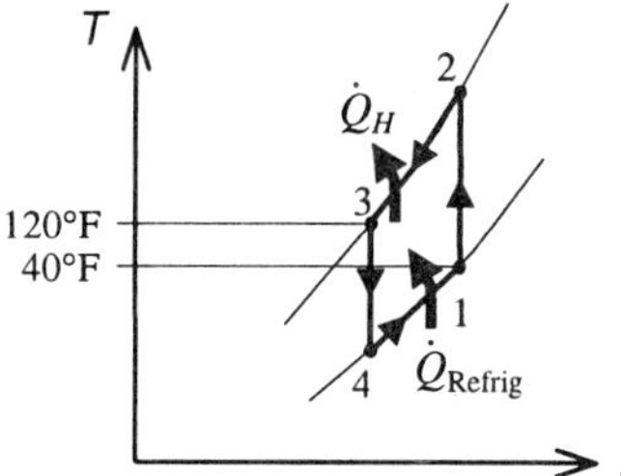

Thus,

$$P_{r_2} = \frac{P_2}{P_1} P_{r_1} = \left(\frac{30}{10}\right)(1.0590) = 3.177 \longrightarrow \begin{array}{l} T_2 = 683.9 \text{ R} \\ h_2 = 163.68 \text{ Btu/lbm} \end{array}$$

$$P_{r_4} = \frac{P_4}{P_3} P_{r_3} = \left(\frac{10}{30}\right)(1.7800) = 0.5933 \longrightarrow \begin{array}{l} T_4 = 423.4 \text{ R} \\ h_4 = 101.14 \text{ Btu/lbm} \end{array}$$

Then the rate of refrigeration is

$$\dot{Q}_{\text{refrig}} = \dot{m}(q_L) = \dot{m}(h_1 - h_4) = (0.5 \text{ lbm/s})(119.48 - 101.14) \text{ Btu/lbm} = \mathbf{9.17 \ Btu/s}$$

(*b*) The net power input is determined from

$$\dot{W}_{\text{net, in}} = \dot{W}_{\text{comp, in}} - \dot{W}_{\text{turb, out}}$$

where

$$\dot{W}_{\text{comp,in}} = \dot{m}(h_2 - h_1) = (0.5 \text{ lbm/s})(163.68 - 119.48) \text{ Btu/lbm} = 22.10 \text{ Btu/s}$$
$$\dot{W}_{\text{turb,out}} = \dot{m}(h_3 - h_4) = (0.5 \text{ lbm/s})(138.66 - 101.14) \text{ Btu/lbm} = 18.79 \text{ Btu/s}$$

Thus, $\dot{W}_{\text{net, in}} = 22.10 - 18.76 = 3.34 \text{ Btu/s} = \mathbf{4.73 \ hp}$

(*c*) The COP of this ideal gas refrigeration cycle is determined from

$$\text{COP}_\text{R} = \frac{\dot{Q}_L}{\dot{W}_{\text{net, in}}} = \frac{9.17 \text{ Btu/s}}{3.34 \text{ Btu/s}} = \mathbf{2.75}$$

10-53 [*Also solved by EES on enclosed CD*] An ideal-gas refrigeration cycle with air as the working fluid is considered. The rate of refrigeration, the net power input, and the COP are to be determined.

Assumptions **1** Steady operating conditions exist. **2** Air is an ideal gas with variable specific heats. **3** Kinetic and potential energy changes are negligible.

Analysis (*a*) We assume the turbine inlet temperature to be the temperature of the surroundings, and the compressor inlet temperature to be the temperature of the refrigerated space. From the air table (Table A-17),

$$T_1 = 285 \text{ K} \longrightarrow \begin{array}{l} h_1 = 285.14 \text{ kJ/kg} \\ P_{r_1} = 1.1584 \end{array}$$

$$T_1 = 320 \text{ K} \longrightarrow \begin{array}{l} h_3 = 320.29 \text{ kJ/kg} \\ P_{r_3} = 1.7375 \end{array}$$

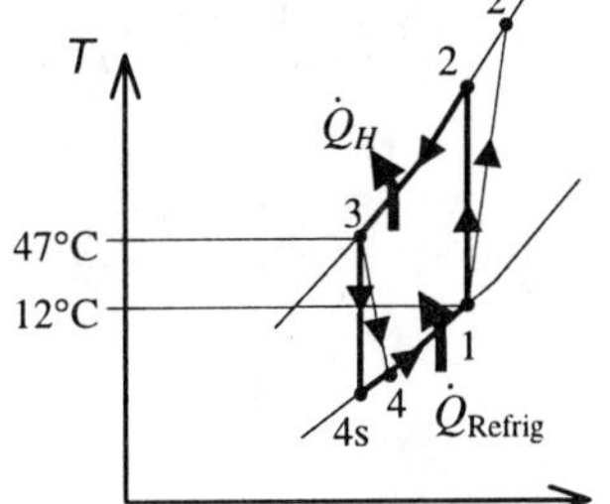

Thus,

$$P_{r_2} = \frac{P_2}{P_1} P_{r_1} = \left(\frac{250}{50}\right)(1.1584) = 5.792 \longrightarrow \begin{array}{l} T_{2s} = 450.4 \text{ K} \\ h_{2s} = 452.17 \text{ kJ/kg} \end{array}$$

$$P_{r_4} = \frac{P_4}{P_3} P_{r_3} = \left(\frac{50}{250}\right)(1.7375) = 0.3475 \longrightarrow \begin{array}{l} T_{4s} = 201.8 \text{ K} \\ h_{4s} = 201.76 \text{ kJ/kg} \end{array}$$

Also,

$$\eta_T = \frac{h_3 - h_4}{h_3 - h_{4s}} \longrightarrow \begin{aligned} h_4 &= h_3 - \eta_T (h_3 - h_{4s}) \\ &= 320.29 - (0.85)(320.29 - 201.76) \\ &= 219.54 \text{ kJ/kg} \end{aligned}$$

Then the rate of refrigeration is

$$\dot{Q}_{\text{refrig}} = \dot{m}(q_L) = \dot{m}(h_1 - h_4) = (0.08 \text{ kg/s})(285.14 - 219.54) \text{ kJ/kg} = \mathbf{5.25 \text{ kW}}$$

(*b*) The net power input is determined from

$$\dot{W}_{\text{net, in}} = \dot{W}_{\text{comp, in}} - \dot{W}_{\text{turb, out}}$$

where

$$\begin{aligned} \dot{W}_{\text{comp,in}} &= \dot{m}(h_2 - h_1) = \dot{m}(h_{2s} - h_1)/\eta_C \\ &= (0.08 \text{ kg/s})[(452.17 - 285.14) \text{ kJ/kg}]/(0.80) = 16.70 \text{ kW} \end{aligned}$$

$$\dot{W}_{\text{turb,out}} = \dot{m}(h_3 - h_4) = (0.08 \text{ kg/s})(320.29 - 219.54) \text{ kJ/kg} = 8.06 \text{ kW}$$

Thus, $\dot{W}_{\text{net, in}} = 16.70 - 8.06 = \mathbf{8.64 \text{ kW}}$

(*c*) The COP of this ideal gas refrigeration cycle is determined from

$$\text{COP}_R = \frac{\dot{Q}_L}{\dot{W}_{\text{net, in}}} = \frac{5.25 \text{ kW}}{8.64 \text{ kW}} = \mathbf{0.61}$$

10-54 A gas refrigeration cycle with helium as the working fluid is considered. The minimum temperature in the cycle, the COP, and the mass flow rate of the helium are to be determined.

Assumptions **1** Steady operating conditions exist. **2** Helium is an ideal gas with constant specific heats. **3** Kinetic and potential energy changes are negligible.

Properties The properties of helium are C_p = 5.1926 kJ/kg· K and k = 1.667 (Table A-2).

Analysis (*a*) From the isentropic relations,

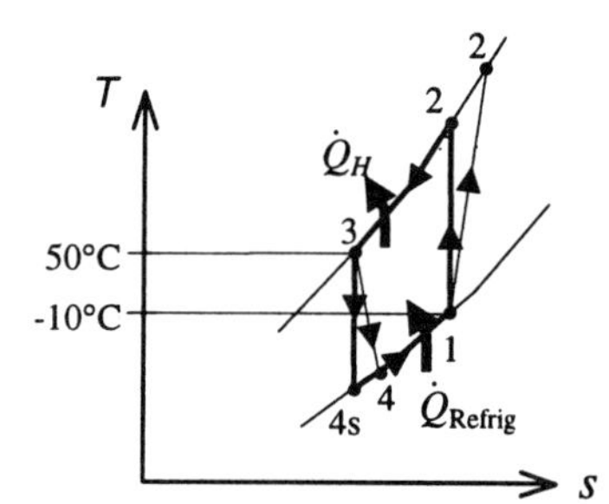

$$T_{2s} = T_1\left(\frac{P_2}{P_1}\right)^{(k-1)/k} = (263\text{K})(3)^{0.667/1.667} = 408.2\text{K}$$

$$T_{4s} = T_3\left(\frac{P_4}{P_3}\right)^{(k-1)/k} = (323\text{K})\left(\frac{1}{3}\right)^{0.667/1.667} = 208.1\text{K}$$

and

$$\eta_T = \frac{h_3 - h_4}{h_3 - h_{4s}} = \frac{T_3 - T_4}{T_3 - T_{4s}} \longrightarrow T_4 = T_3 - \eta_T(T_3 - T_{4s}) = 323 - (0.82)(323 - 208.1) = \mathbf{228.8\ K} = T_{\min}$$

$$\eta_C = \frac{h_{2s} - h_1}{h_2 - h_1} = \frac{T_{2s} - T_1}{T_2 - T_1} \longrightarrow T_2 = T_1 + (T_{2s} - T_1)/\eta_C = 263 + (408.2 - 263)/(0.82) = 440.1\text{ K}$$

(*b*) The COP of this ideal gas refrigeration cycle is determined from

$$\begin{aligned}\text{COP}_\text{R} &= \frac{q_L}{w_{\text{net,in}}} = \frac{q_L}{w_{\text{comp,in}} - w_{\text{turb,out}}} \\ &= \frac{h_1 - h_4}{(h_2 - h_1) - (h_3 - h_4)} \\ &= \frac{T_1 - T_4}{(T_2 - T_1) - (T_3 - T_4)} \\ &= \frac{263 - 228.8}{(440.1 - 263) - (323 - 228.8)} = \mathbf{0.413}\end{aligned}$$

(*c*) The mass flow rate of helium is determined from

$$\dot{m} = \frac{\dot{Q}_{\text{refrig}}}{q_L} = \frac{\dot{Q}_{\text{refrig}}}{h_1 - h_4} = \frac{\dot{Q}_{\text{refrig}}}{C_p(T_1 - T_4)} = \frac{12\text{ kJ/s}}{(5.1926\text{ kJ/kg}\cdot\text{K})(263 - 228.8)\text{K}} = \mathbf{0.0676\ kg/s}$$

10-55 An ideal-gas refrigeration cycle with air as the working fluid is considered. The lowest temperature that can be obtained by this cycle, the COP, and the mass flow rate of air are to be determined.

Assumptions **1** Steady operating conditions exist. **2** Air is an ideal gas with constant specific heats. **3** Kinetic and potential energy changes are negligible.

Properties The properties of air at room temperature are $C_p = 1.005$ kJ/kg· K and $k = 1.4$ (Table A-2).

Analysis (*a*) The lowest temperature in the cycle occurs at the turbine exit. From the isentropic relations,

$$T_2 = T_1\left(\frac{P_2}{P_1}\right)^{(k-1)/k} = (266\text{ K})(4)^{0.4/1.4} = 395.3\text{ K} = 122.3°\text{C}$$

$$T_5 = T_4\left(\frac{P_5}{P_4}\right)^{(k-1)/k} = (258\text{ K})\left(\frac{1}{4}\right)^{0.4/1.4} = 173.6\text{ K} = \mathbf{-99.4°C} = T_{\min}$$

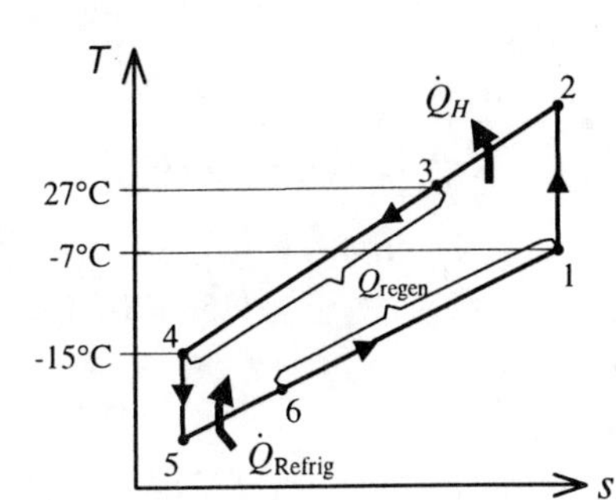

(*b*) From an energy balance on the regenerator,

$$\dot{E}_{in} - \dot{E}_{out} = \Delta\dot{E}_{system}^{\;\nearrow 0\text{ (steady)}} = 0$$

$$\dot{E}_{in} = \dot{E}_{out}$$

$$\sum \dot{m}_e h_e = \sum \dot{m}_i h_i \longrightarrow \dot{m}(h_3 - h_4) = \dot{m}(h_1 - h_6)$$

or,

$$\dot{m}C_p(T_3 - T_4) = \dot{m}C_p(T_1 - T_6) \longrightarrow T_3 - T_4 = T_1 - T_6$$

or,

$$T_6 = T_1 - T_3 + T_4 = (-7°\text{C}) - 27°\text{C} + (-15°\text{C}) = -49°\text{C}$$

Then the COP of this ideal gas refrigeration cycle is determined from

$$\begin{aligned}\text{COP}_R &= \frac{q_L}{w_{net,in}} = \frac{q_L}{w_{comp,in} - w_{turb,out}}\\ &= \frac{h_6 - h_5}{(h_2 - h_1) - (h_4 - h_5)}\\ &= \frac{T_6 - T_5}{(T_2 - T_1) - (T_4 - T_5)}\\ &= \frac{-49°\text{C} - (-99.4°\text{C})}{[122.3 - (-7)]°\text{C} - [-15 - (-99.4)]°\text{C}} = \mathbf{1.12}\end{aligned}$$

(*c*) The mass flow rate is determined from

$$\dot{m} = \frac{\dot{Q}_{refrig}}{q_L} = \frac{\dot{Q}_{refrig}}{h_6 - h_5} = \frac{\dot{Q}_{refrig}}{C_p(T_6 - T_5)} = \frac{12\text{ kJ/s}}{(1.005\text{ kJ/kg}\cdot°\text{C})[-49 - (-99.4)]°\text{C}} = \mathbf{0.237\ kg/s}$$

10-56 An ideal-gas refrigeration cycle with air as the working fluid is considered. The lowest temperature that can be obtained by this cycle, the COP, and the mass flow rate of air are to be determined.

Assumptions **1** Steady operating conditions exist. **2** Air is an ideal gas with constant specific heats. **3** Kinetic and potential energy changes are negligible.

Properties The properties of air at room temperature are C_p = 1.005 kJ/kg· K and k = 1.4 (Table A-2).

Analysis (*a*) The lowest temperature in the cycle occurs at the turbine exit. From the isentropic relations,

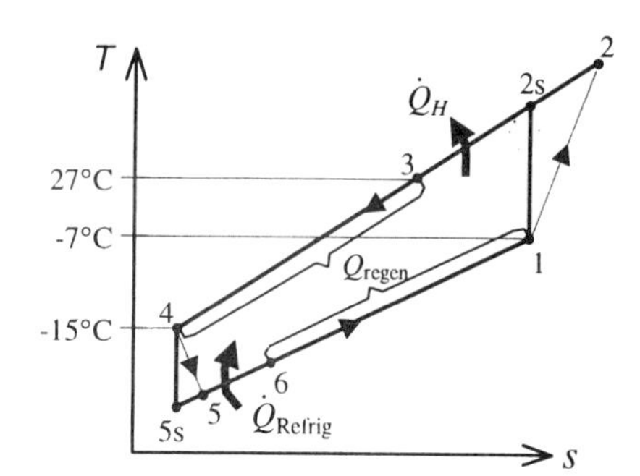

$$T_{2s} = T_1\left(\frac{P_2}{P_1}\right)^{(k-1)/k} = (266\text{ K})(4)^{0.4/1.4} = 395.3\text{ K} = 122.3°\text{C}$$

$$T_{5s} = T_4\left(\frac{P_5}{P_4}\right)^{(k-1)/k} = (258\text{ K})\left(\frac{1}{4}\right)^{0.4/1.4} = 173.6\text{ K} = -99.4°\text{C}$$

and

$$\eta_T = \frac{h_4 - h_5}{h_4 - h_{5s}} = \frac{T_4 - T_5}{T_4 - T_{5s}} \longrightarrow T_5 = T_4 - \eta_T(T_4 - T_{5s}) = -15 - (0.80)(-15-(-99.4)) = \mathbf{-82.5°C} = T_{\min}$$

$$\eta_C = \frac{h_{2s} - h_1}{h_2 - h_1} = \frac{T_{2s} - T_1}{T_2 - T_1} \longrightarrow T_2 = T_1 + (T_{2s} - T_1)/\eta_C = -7 + (122.3 - (-7))/(0.75) = 165.4°\text{C}$$

(*b*) From an energy balance on the regenerator,

$$\dot{E}_{\text{in}} - \dot{E}_{\text{out}} = \Delta\dot{E}_{\text{system}}^{\;\not{0}\text{ (steady)}} = 0 \rightarrow \dot{E}_{\text{in}} = \dot{E}_{\text{out}}$$

$$\sum \dot{m}_e h_e = \sum \dot{m}_i h_i \longrightarrow \dot{m}(h_3 - h_4) = \dot{m}(h_1 - h_6)$$

or,

$$\dot{m}C_p(T_3 - T_4) = \dot{m}C_p(T_1 - T_6) \longrightarrow T_3 - T_4 = T_1 - T_6$$

or,

$$T_6 = T_1 - T_3 + T_4 = (-7°\text{C}) - 27°\text{C} + (-15°\text{C}) = -49°\text{C}$$

Then the COP of this ideal gas refrigeration cycle is determined from

$$\text{COP}_\text{R} = \frac{q_L}{w_{\text{net,in}}} = \frac{q_L}{w_{\text{comp,in}} - w_{\text{turb,out}}} = \frac{h_6 - h_5}{(h_2 - h_1) - (h_4 - h_5)} = \frac{T_6 - T_5}{(T_2 - T_1) - (T_4 - T_5)} = \frac{-49°\text{C} - (-82.5°\text{C})}{[165.4 - (-7)]°\text{C} - [-15 - (-82.5)]°\text{C}} = \mathbf{0.32}$$

(*c*) The mass flow rate is determined from

$$\dot{m} = \frac{\dot{Q}_{\text{refrig}}}{q_L} = \frac{\dot{Q}_{\text{refrig}}}{h_6 - h_5} = \frac{\dot{Q}_{\text{refrig}}}{C_p(T_6 - T_5)} = \frac{12\text{ kJ/s}}{(1.005\text{ kJ/kg}\cdot°\text{C})[-49 - (-82.5)]°\text{C}} = \mathbf{0.356\ kg/s}$$

Absorption Refrigeration Systems

10-57C Absorption refrigeration is the kind of refrigeration that involves the absorption of the refrigerant during part of the cycle. In absorption refrigeration cycles, the refrigerant is compressed in the liquid phase instead of in the vapor form.

10-58C The main advantage of absorption refrigeration is its being economical in the presence of an inexpensive heat source. Its disadvantages include being expensive, complex, and requiring an external heat source.

10-59C In absorption refrigeration, water can be used as the refrigerant in air conditioning applications since the temperature of water never needs to fall below the freezing point.

10-60C The fluid in the absorber is cooled to maximize the refrigerant content of the liquid; the fluid in the generator is heated to maximize the refrigerant content of the vapor.

10-61C The coefficient of performance of absorption refrigeration systems is defined as

$$\mathrm{COP_R} = \frac{\text{desiredoutput}}{\text{requiredinput}} = \frac{Q_L}{Q_{\text{gen}} + W_{\text{pump,in}}} \cong \frac{Q_L}{Q_{\text{gen}}}$$

10-62C The rectifier separates the water from NH_3 and returns it to the generator. The regenerator transfers some heat from the water-rich solution leaving the generator to the NH_3-rich solution leaving the pump.

10-63 The COP of an absorption refrigeration system that operates at specified conditions is given. It is to be determined whether the given COP value is possible.

Analysis The maximum COP that this refrigeration system can have is

$$\mathrm{COP_{R,max}} = \left(1 - \frac{T_0}{T_s}\right)\left(\frac{T_L}{T_0 - T_L}\right) = \left(1 - \frac{300\text{ K}}{403\text{ K}}\right)\left(\frac{268}{300 - 268}\right) = 2.14$$

which is slightly greater than 2. Thus the claim is **possible**, but not probable.

10-64 The conditions at which an absorption refrigeration system operates are specified. The maximum COP this absorption refrigeration system can have is to be determined.

Analysis The maximum COP that this refrigeration system can have is

$$\mathrm{COP_{R,max}} = \left(1 - \frac{T_0}{T_s}\right)\left(\frac{T_L}{T_0 - T_L}\right) = \left(1 - \frac{298\text{ K}}{383\text{ K}}\right)\left(\frac{253}{298 - 253}\right) = \mathbf{1.25}$$

10-65 The conditions at which an absorption refrigeration system operates are specified. The maximum rate at which this system can remove heat from the refrigerated space is to be determined.

Analysis The maximum COP that this refrigeration system can have is

$$\text{COP}_{\text{R,max}} = \left(1 - \frac{T_0}{T_s}\right)\left(\frac{T_L}{T_0 - T_L}\right) = \left(1 - \frac{298\text{ K}}{403\text{ K}}\right)\left(\frac{243}{298 - 243}\right) = 1.15$$

Thus,

$$\dot{Q}_{\text{L,max}} = COP_{\text{R,max}}\dot{Q}_{\text{gen}} = (1.15)(5\times10^5\text{ kJ/h}) = \mathbf{5.75\times10^5\ kJ/h}$$

10-66E The conditions at which an absorption refrigeration system operates are specified. The COP is also given. The maximum rate at which this system can remove heat from the refrigerated space is to be determined.

Analysis For a COP = 0.7, the rate at which this system can remove heat from the refrigerated space is

$$\dot{Q}_L = \text{COP}_\text{R}\dot{Q}_{\text{gen}} = (0.55)(10^5\text{ Btu/h}) = \mathbf{0.55\times10^5\ Btu/h}$$

Special Topic: Thermoelectric Power Generation and Refrigeration Systems

---

10-67C The circuit that incorporates both thermal and electrical effects is called a thermoelectric circuit.

10-68C When two wires made from different metals joined at both ends (junctions) forming a closed circuit and one of the joints is heated, a current flows continuously in the circuit. This is called the Seebeck effect. When a small current is passed through the junction of two dissimilar wires, the junction is cooled. This is called the Peltier effect.

10-69C No.

10-70C No.

10-71C Yes.

10-72C When a thermoelectric circuit is broken, the current will cease to flow, and we can measure the voltage generated in the circuit by a voltmeter. The voltage generated is a function of the temperature difference, and the temperature can be measured by simply measuring voltages.

10-73C The performance of thermoelectric refrigerators improves considerably when semiconductors are used instead of metals.

10-74C The efficiency of a thermoelectric generator is limited by the Carnot efficiency because a thermoelectric generator fits into the definition of a heat engine with electrons serving as the working fluid.

10-75E A thermoelectric generator that operates at specified conditions is considered. The maximum thermal efficiency this thermoelectric generator can have is to be determined.

Analysis The maximum thermal efficiency of this thermoelectric generator is the Carnot efficiency,

$$\eta_{th,\max} = \eta_{th,\text{Carnot}} = 1 - \frac{T_L}{T_H} = 1 - \frac{550\ \text{R}}{800\ \text{R}} = \mathbf{31.3\%}$$

10-76 A thermoelectric generator that operates at specified conditions is considered. The maximum COP this thermoelectric generator can have and the minimum required power input are to be determined.

Analysis The maximum COP of this thermoelectric refrigerator is the COP of a Carnot refrigerator operating between the same temperature limits,

$$\text{COP}_{\max} = \text{COP}_{\text{R,Carnot}} = \frac{1}{(T_H/T_L)-1} = \frac{1}{(293\text{ K})/(268\text{ K})-1} = 10.72$$

Thus,

$$\dot{W}_{\text{in,min}} = \frac{\dot{Q}_L}{\text{COP}_{\max}} = \frac{130\text{ W}}{10.72} = \mathbf{12.1\text{ W}}$$

10-77 A thermoelectric cooler that operates at specified conditions with a given COP is considered. The required power input to the thermoelectric cooler is to be determined.

Analysis The required power input is determined from the definition of COP_R,

$$\text{COP}_\text{R} = \frac{\dot{Q}_L}{\dot{W}_{\text{in}}} \longrightarrow \dot{W}_{\text{in}} = \frac{\dot{Q}_L}{\text{COP}_\text{R}} = \frac{180\text{ W}}{0.15} = \mathbf{1200\text{ W}}$$

10-78E A thermoelectric cooler that operates at specified conditions with a given COP is considered. The required power input to the thermoelectric cooler is to be determined.

Analysis The required power input is determined from the definition of COP_R,

$$\text{COP}_\text{R} = \frac{\dot{Q}_L}{\dot{W}_{\text{in}}} \longrightarrow \dot{W}_{\text{in}} = \frac{\dot{Q}_L}{\text{COP}_\text{R}} = \frac{35\text{ Btu/min}}{0.15} = 233.3\text{ Btu/min} = \mathbf{5.5\text{ hp}}$$

10-79 A thermoelectric refrigerator powered by a car battery cools 9 canned drinks in 12 h. The average COP of this refrigerator is to be determined.

Assumptions Heat transfer through the walls of the refrigerator is negligible.

Properties The properties of canned drinks are the same as those of water at room temperature, ρ = 1 kg/L and C_p = 4.18 kJ/kg. °C (Table A-3).

Analysis The cooling rate of the refrigerator is simply the rate of decrease of the energy of the canned drinks,

$$m = \rho V = 9 \times (1\text{ kg/L})(0.350\text{ L}) = 3.15\text{ kg}$$

$$Q_{\text{cooling}} = mC\Delta T = (3.15\text{ kg})(4.18\text{ kJ/kg}\cdot{}^\circ\text{C})(25 - 3)^\circ\text{C} = 290\text{ kJ}$$

$$\dot{Q}_{\text{cooling}} = \frac{Q_{\text{cooling}}}{\Delta t} = \frac{290\text{ kJ}}{12 \times 3600\text{ s}} = 0.00671\text{ kW} = 6.71\text{ W}$$

The electric power consumed by the refrigerator is

$$\dot{W}_{\text{in}} = VI = (12\text{ V})(3\text{ A}) = 36\text{ W}$$

Then the COP of the refrigerator becomes

$$\text{COP} = \frac{\dot{Q}_{\text{cooling}}}{\dot{W}_{\text{in}}} = \frac{6.71\text{ W}}{36\text{ W}} = \mathbf{0.186} \approx 0.20$$

10-80E A thermoelectric cooler is said to cool a 12-oz drink or to heat a cup of coffee in about 15 min. The average rate of heat removal from the drink, the average rate of heat supply to the coffee, and the electric power drawn from the battery of the car are to be determined.

Assumptions Heat transfer through the walls of the refrigerator is negligible.

Properties The properties of canned drinks are the same as those of water at room temperature, C_p = 1.0 Btu/lbm.°F (Table A-3E).

Analysis (*a*) The average cooling rate of the refrigerator is simply the rate of decrease of the energy content of the canned drinks,

$$Q_{\text{cooling}} = mC_p\Delta T = (0.771\,\text{lbm})(1.0\ \text{Btu/lbm}\cdot{}^\circ\text{F})(78-38)^\circ\text{F} = 30.84\ \text{Btu}$$

$$\dot{Q}_{\text{cooling}} = \frac{Q_{\text{cooling}}}{\Delta t} = \frac{30.84\ \text{Btu}}{15\times 60\,\text{s}}\left(\frac{1055\,\text{J}}{1\,\text{Btu}}\right) = \mathbf{36.2\ W}$$

(*b*) The average heating rate of the refrigerator is simply the rate of increase of the energy content of the canned drinks,

$$Q_{\text{heating}} = mC_p\Delta T = (0.771\,\text{lbm})(1.0\ \text{Btu/lbm}\cdot{}^\circ\text{F})(130-75)^\circ\text{F} = 42.4\ \text{Btu}$$

$$\dot{Q}_{\text{heating}} = \frac{Q_{\text{heating}}}{\Delta t} = \frac{42.4\ \text{Btu}}{15\times 60\,\text{s}}\left(\frac{1055\,\text{J}}{1\,\text{Btu}}\right) = \mathbf{49.7\ W}$$

(*c*) The electric power drawn from the car battery during cooling and heating is

$$\dot{W}_{\text{in,cooling}} = \frac{\dot{Q}_{\text{cooling}}}{\text{COP}_{\text{cooling}}} = \frac{36.2\ \text{W}}{0.2} = \mathbf{181\ W}$$

$$\text{COP}_{\text{heating}} = \text{COP}_{\text{cooling}} + 1 = 0.2 + 1 = 1.2$$

$$\dot{W}_{\text{in,heating}} = \frac{\dot{Q}_{\text{heating}}}{\text{COP}_{\text{heating}}} = \frac{49.7\ \text{W}}{1.2} = \mathbf{41.4\ W}$$

10-81 The maximum power a thermoelectric generator can produce is to be determined.

Analysis The maximum thermal efficiency this thermoelectric generator can have is

$$\eta_{\text{th,max}} = 1 - \frac{T_L}{T_H} = 1 - \frac{303\ \text{K}}{353\ \text{K}} = 0.142$$

Thus,

$$\dot{W}_{\text{out,max}} = \eta_{\text{th,max}}\dot{Q}_{\text{in}} = (0.142)(10^6\ \text{kJ/h}) = 141{,}643\ \text{kJ/h} = \mathbf{39.3\,kW}$$

Review Problems

10-82 A steady-flow Carnot refrigeration cycle with refrigerant-134a as the working fluid is considered. The COP, the condenser and evaporator pressures, and the net work input are to be determined.

Assumptions **1** Steady operating conditions exist. **2** Kinetic and potential energy changes are negligible.

Analysis (*a*) The COP of this refrigeration cycle is determined from

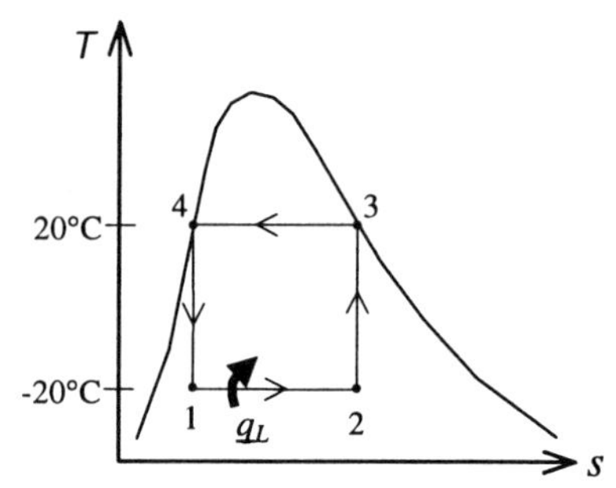

$$\mathrm{COP_{R,C}} = \frac{1}{(T_H/T_L)-1} = \frac{1}{(293\ \mathrm{K})/(253\ \mathrm{K})-1} = \mathbf{6.33}$$

(*b*) The condenser and evaporative pressures are (Table A-11)

$$P_{\text{evap}} = P_{\text{sat @ }-20°\text{C}} = \mathbf{0.13299\ kPa}$$
$$P_{\text{cond}} = P_{\text{sat @ }20°\text{C}} = \mathbf{0.57160\ kPa}$$

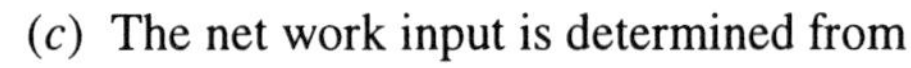

(*c*) The net work input is determined from

$$h_1 = (h_f + x_1 h_{fg})_{@-20°\text{C}} = 24.26 + (0.2)(211.05) = 66.5\ \text{kJ/kg}$$
$$h_2 = (h_f + x_2 h_{fg})_{@-20°\text{C}} = 24.26 + (0.85)(211.05) = 203.7\ \text{kJ/kg}$$

$$q_L = h_2 - h_1 = 203.7 - 66.5 = 137.2\ \text{kJ/kg}$$

and

$$w_{\text{net, in}} = \frac{q_L}{\mathrm{COP_R}} = \frac{137.2\ \text{kJ/kg}}{6.33} = \mathbf{21.7\ kJ/kg}$$

10-83 A large refrigeration plant that operates on the ideal vapor-compression cycle with refrigerant-134a as the working fluid is considered. The mass flow rate of the refrigerant, the power input to the compressor, and the mass flow rate of the cooling water are to be determined.

Assumptions **1** Steady operating conditions exist. **2** Kinetic and potential energy changes are negligible.

Analysis In an ideal vapor-compression refrigeration cycle, the compression process is isentropic, the refrigerant enters the compressor as a saturated vapor at the evaporator pressure, and leaves the condenser as saturated liquid at the condenser pressure. From the refrigerant tables (Tables A-12 and A-13),

$$\left.\begin{array}{l} P_1 = 120\text{ kPa} \\ \text{sat. vapor} \end{array}\right\} \begin{array}{l} h_1 = h_{g\,@\,120\text{ kPa}} = 233.86\text{ kJ/kg} \\ s_1 = s_{g\,@\,120\text{ kPa}} = 0.9354\text{ kJ/kg}\cdot\text{K} \end{array}$$

$$\left.\begin{array}{l} P_2 = 0.7\text{ MPa} \\ s_2 = s_1 \end{array}\right\} h_2 = 270.22\text{ kJ/kg} \quad (T_2 = 34.6°\text{C})$$

$$\left.\begin{array}{l} P_3 = 0.7\text{ MPa} \\ \text{sat. liquid} \end{array}\right\} h_3 = h_{f\,@\,0.7\text{ MPa}} = 86.78\text{ kJ/kg}$$

$$h_4 \cong h_3 = 86.78\text{ kJ/kg} \quad (\text{throttling})$$

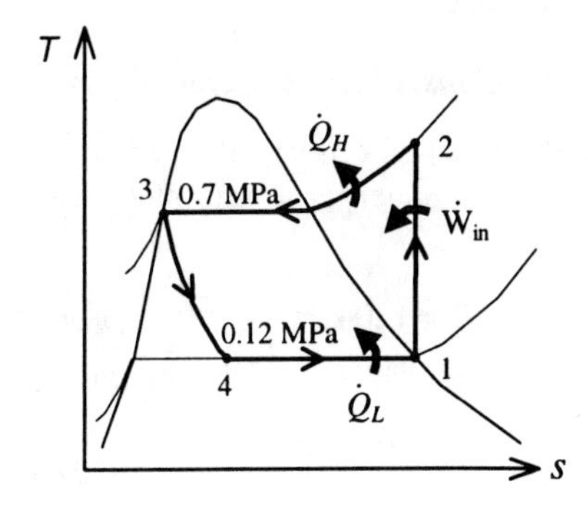

The mass flow rate of the refrigerant is determined from

$$\dot{m} = \frac{\dot{Q}_L}{h_1 - h_4} = \frac{100\text{ kJ/s}}{(233.86 - 86.78)\text{ kJ/kg}} = \mathbf{0.68\ kg/s}$$

(*b*) The power input to the compressor is

$$\dot{W}_{\text{in}} = \dot{m}(h_2 - h_1) = (0.68\text{ kg/s})(270.22 - 233.86)\text{ kJ/kg} = \mathbf{24.7\ kW}$$

(*c*) The mass flow rate of the cooling water is determined from

$$\dot{Q}_H = \dot{m}(h_2 - h_3) = (0.68\text{ kg/s})(270.22 - 86.78)\text{ kJ/kg} = 124.7\text{ kW}$$

and

$$\dot{m}_{\text{cooling}} = \frac{\dot{Q}_H}{(C_p \Delta T)_{\text{water}}} = \frac{124.7\text{ kJ/s}}{(4.18\text{ kJ/kg}\cdot°\text{C})(8°\text{C})} = \mathbf{3.73\ kg/s}$$

10-84 EES solution of this (and other comprehensive problems designated with the *computer icon*) is available to instructors at the *Instructor Manual* section of the *Online Learning Center* (OLC) at www.mhhe.com/cengel-boles. See the Preface for access information.

10-85 A large refrigeration plant operates on the vapor-compression cycle with refrigerant-134a as the working fluid. The mass flow rate of the refrigerant, the power input to the compressor, the mass flow rate of the cooling water, and the rate of exergy destruction associated with the compression process are to be determined.

Assumptions **1** Steady operating conditions exist. **2** Kinetic and potential energy changes are negligible.

Analysis (*a*) The refrigerant enters the compressor as a saturated vapor at the evaporator pressure, and leaves the condenser as saturated liquid at the condenser pressure. From the refrigerant tables (Tables A-12 and A-13),

$$\left.\begin{array}{l} P_1 = 120\text{ kPa} \\ \text{sat. vapor} \end{array}\right\} \begin{array}{l} h_1 = h_{g\ @\ 120\text{ kPa}} = 233.86\text{ kJ/kg} \\ s_1 = s_{g\ @\ 120\text{ kPa}} = 0.9354\text{ kJ/kg}\cdot\text{K} \end{array}$$

$$\left.\begin{array}{l} P_2 = 0.7\text{ MPa} \\ s_{2s} = s_1 \end{array}\right\} h_{2s} = 270.22\text{ kJ/kg}\ \ (T_{2s} = 34.6°\text{C})$$

$$\left.\begin{array}{l} P_3 = 0.7\text{ MPa} \\ \text{sat. liquid} \end{array}\right\} h_3 = h_{f\ @\ 0.7\text{ MPa}} = 86.78\text{ kJ/kg}$$

$$h_4 \cong h_3 = 86.78\text{ kJ/kg}\ \ (\text{throttling})$$

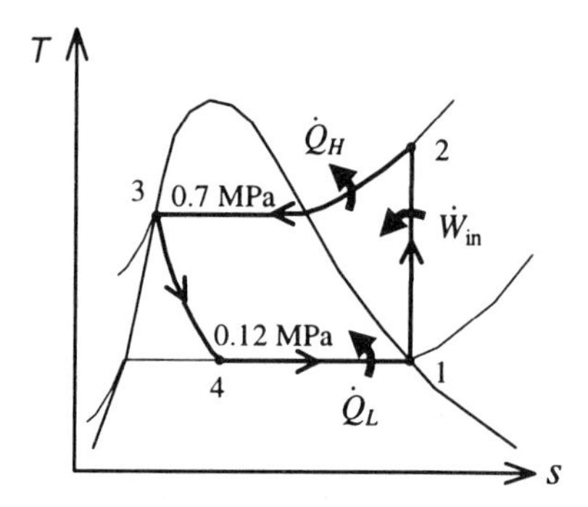

The mass flow rate of the refrigerant is determined from

$$\dot{m} = \frac{\dot{Q}_L}{h_1 - h_4} = \frac{100\text{ kJ/s}}{(233.86 - 86.78)\text{ kJ/kg}} = \mathbf{0.68\ kg/s}$$

(*b*) The actual enthalpy at the compressor exit is

$$\eta_C = \frac{h_{2s} - h_1}{h_2 - h_1} \longrightarrow h_2 = h_1 + (h_{2s} - h_1)/\eta_C = 233.86 + (270.22 - 233.86)/(0.75) = 282.34\text{ kJ/kg}$$

Thus,

$$\dot{W}_{in} = \dot{m}(h_2 - h_1) = (0.68\text{ kg/s})(282.34 - 233.86)\text{ kJ/kg} = \mathbf{33.0\ kW}$$

(*c*) The mass flow rate of the cooling water is determined from

$$\dot{Q}_H = \dot{m}(h_2 - h_3) = (0.68\text{ kg/s})(282.34 - 86.78)\text{ kJ/kg} = 133.0\text{ kW}$$

and

$$\dot{m}_{cooling} = \frac{\dot{Q}_H}{(C_p \Delta T)_{water}} = \frac{133.0\text{ kJ/s}}{(4.18\text{ kJ/kg}\cdot°\text{C})(8°\text{C})} = \mathbf{3.98\ kg/s}$$

The exergy destruction associated with this adiabatic compression process is determined from

$$\dot{X}_{destroyed} = T_0 \dot{S}_{gen} = T_0 \dot{m}(s_2 - s_1)$$

where

$$\left.\begin{array}{l} P_2 = 0.7\text{ MPa} \\ h_2 = 282.34\text{ kJ/kg} \end{array}\right\} s_2 = 0.9741\text{ kJ/kg}\cdot\text{K}$$

Thus, $\dot{X}_{destroyed} = (298\text{ K})(0.68\text{ kg/s})(0.9741 - 0.9354)\text{ kJ/kg}\cdot\text{K} = \mathbf{7.84\ kW}$

10-86 A heat pump that operates on the ideal vapor-compression cycle with refrigerant-134a as the working fluid is used to heat a house. The rate of heat supply to the house, the volume flow rate of the refrigerant at the compressor inlet, and the COP of this heat pump are to be determined.

Assumptions **1** Steady operating conditions exist. **2** Kinetic and potential energy changes are negligible.

Analysis (*a*) In an ideal vapor-compression refrigeration cycle, the compression process is isentropic, the refrigerant enters the compressor as a saturated vapor at the evaporator pressure, and leaves the condenser as saturated liquid at the condenser pressure. From the refrigerant tables (Tables A-12 and A-13),

$$\left.\begin{array}{l} P_1 = 240\text{ kPa} \\ \text{sat. vapor} \end{array}\right\} \begin{array}{l} h_1 = h_{g\ @\ 240\text{ kPa}} = 244.09\text{ kJ/kg} \\ s_1 = s_{g\ @\ 240\text{ kPa}} = 0.9222\text{ kJ/kg}\cdot\text{K} \\ v_1 = v_{g\ @\ 240\text{ kPa}} = 0.0834\text{ m}^3\text{/kg} \end{array}$$

$$\left.\begin{array}{l} P_2 = 0.9\text{ MPa} \\ s_2 = s_1 \end{array}\right\} h_2 = 271.41\text{ kJ/kg}$$

$$\left.\begin{array}{l} P_3 = 0.9\text{ MPa} \\ \text{sat. liquid} \end{array}\right\} h_3 = h_{f\ @\ 0.9\text{ MPa}} = 99.56\text{ kJ/kg}$$

$$h_4 \cong h_3 = 99.56\text{ kJ/kg}\ \text{(throttling)}$$

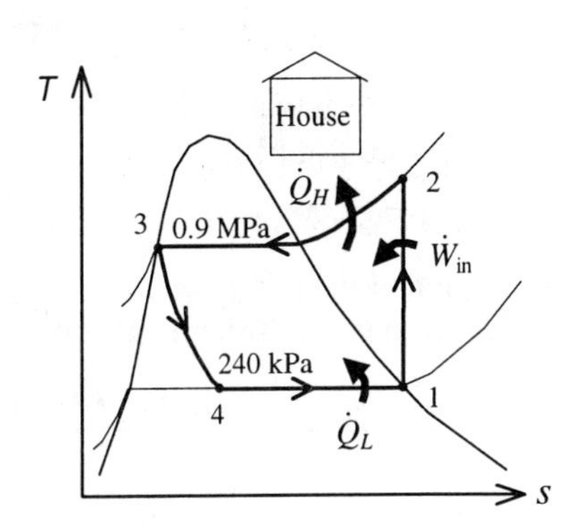

The rate of heat supply to the house is determined from

$$\dot{Q}_H = \dot{m}(h_2 - h_3) = (0.24\text{ kg/s})(271.41 - 99.56)\text{ kJ/kg} = \mathbf{41.24\ kW}$$

(*b*) The volume flow rate of the refrigerant at the compressor inlet is

$$\dot{V}_1 = \dot{m}v_1 = (0.24\text{ kg/s})(0.0834\text{ m}^3\text{/kg}) = \mathbf{0.02002\ m^3/s}$$

(*c*) The COP of t his heat pump is determined from

$$\text{COP}_\text{R} = \frac{q_L}{w_\text{in}} = \frac{h_2 - h_3}{h_2 - h_1} = \frac{271.41 - 99.56}{271.41 - 244.09} = \mathbf{6.29}$$

10-87 A relation for the COP of the two-stage refrigeration system with a flash chamber shown in Fig. 10-12 is to be derived.

Analysis The coefficient of performance is determined from

$$\text{COP}_\text{R} = \frac{q_L}{w_\text{in}}$$

where

$$q_L = (1 - x_6)(h_1 - h_8)\ \text{ with }\ x_6 = \frac{h_6 - h_f}{h_{fg}}$$

$$w_\text{in} = w_\text{compI,in} + w_\text{compII,in} = (1 - x_6)(h_2 - h_1) + (1)(h_4 - h_9)$$

10-88 A two-stage compression refrigeration system using refrigerant-134a as the working fluid is considered. The fraction of the refrigerant that evaporates as it is throttled to the flash chamber, the amount of heat removed from the refrigerated space, the compressor work, and the COP are to be determined.

Assumptions **1** Steady operating conditions exist. **2** Kinetic and potential energy changes are negligible. **3** The flashing chamber is adiabatic.

Analysis (*a*) The enthalpies of the refrigerant at several states are determined from the refrigerant tables to be (Tables A-11, A-12, and A-13)

$$h_1 = 236.04 \text{ kJ/kg}, \quad h_2 = 257.4 \text{ kJ/kg}$$
$$h_3 = 252.30 \text{ kJ/kg},$$
$$h_5 = 93.42 \text{ kJ/kg}, \quad h_6 = 93.42 \text{ kJ/kg}$$
$$h_7 = 62.00 \text{ kJ/kg}, \quad h_8 = 62.00 \text{ kJ/kg}$$

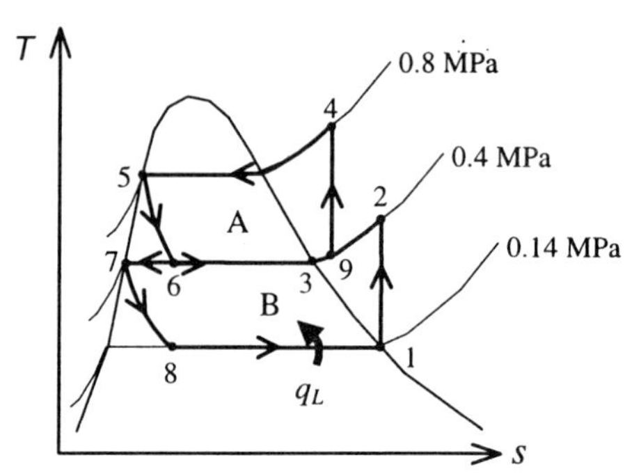

The fraction of the refrigerant that evaporates as it is throttled to the flash chamber is simply the quality at state 6,

$$x_6 = \frac{h_6 - h_f}{h_{fg}} = \frac{93.42 - 62.00}{190.32} = \mathbf{0.165}$$

(*b*) The enthalpy at state 9 is determined from an energy balance on the mixing chamber:

$$\dot{E}_{in} - \dot{E}_{out} = \Delta \dot{E}_{system}^{\;\nearrow 0 \text{ (steady)}} = 0 \rightarrow \dot{E}_{in} = \dot{E}_{out}$$
$$\sum \dot{m}_e h_e = \sum \dot{m}_i h_i$$
$$(1)h_9 = x_6 h_3 + (1 - x_6)h_2$$
$$h_9 = (0.165)(252.30) + (1 - 0.165)(257.4) = 256.6 \text{ kJ/kg}$$

Also,

$$\left.\begin{array}{l} P_4 = 0.8 \text{ MPa} \\ s_4 = s_9 = 0.9295 \text{ kJ/kg} \cdot \text{K} \end{array}\right\} h_4 = 271.22 \text{ kJ/kg}$$

Then the amount of heat removed from the refrigerated space and the compressor work input per unit mass of refrigerant flowing through the condenser are

$$q_L = (1 - x_6)(h_1 - h_8) = (1 - 0.165)(236.04 - 62.00) \text{ kJ/kg} = 145.3 \text{ kJ/kg}$$

$$\begin{aligned} w_{in} &= w_{compI,in} + w_{compII,in} = (1 - x_6)(h_2 - h_1) + (1)(h_4 - h_9) \\ &= (1 - 0.165)(257.4 - 236.04) \text{ kJ/kg} + (271.22 - 256.6) \text{ kJ/kg} \\ &= \mathbf{32.5 \text{ kJ/kg}} \end{aligned}$$

(*c*) The coefficient of performance is determined from

$$\text{COP}_R = \frac{q_L}{w_{in}} = \frac{145.3 \text{ kJ/kg}}{32.5 \text{ kJ/kg}} = \mathbf{4.47}$$

10-89 An aircraft on the ground is to be cooled by a gas refrigeration cycle operating with air on an open cycle. The temperature of the air leaving the turbine is to be determined.

Assumptions **1** Steady operating conditions exist. **2** Air is an ideal gas with constant specific heats at room temperature. **3** Kinetic and potential energy changes are negligible.

Properties The specific heat ratio of air at room temperature is $k = 1.4$ (Table A-2).

Analysis Assuming the turbine to be isentropic, the air temperature at the turbine exit is determined from

$$T_4 = T_3\left(\frac{P_4}{P_3}\right)^{(k-1)/k} = (343\text{ K})\left(\frac{100\text{ kPa}}{250\text{ kPa}}\right)^{0.4/1.4} = 264\text{ K} = \mathbf{-9.0°C}$$

10-90 A regenerative gas refrigeration cycle with helium as the working fluid is considered. The temperature of the helium at the turbine inlet, the COP of the cycle, and the net power input required are to be determined.

Assumptions **1** Steady operating conditions exist. **2** Helium is an ideal gas with constant specific heats at room temperature. **3** Kinetic and potential energy changes are negligible.

Properties The properties of helium are $C_p = 5.1926$ kJ/kg· K and $k = 1.667$ (Table A-2).

Analysis (*a*) The temperature of the helium at the turbine inlet is determined from an energy balance on the regenerator,

$$\dot{E}_{in} - \dot{E}_{out} = \Delta\dot{E}_{system}^{\;\nearrow 0\text{ (steady)}} = 0$$

$$\dot{E}_{in} = \dot{E}_{out}$$

$$\sum \dot{m}_e h_e = \sum \dot{m}_i h_i \longrightarrow \dot{m}(h_3 - h_4) = \dot{m}(h_1 - h_6)$$

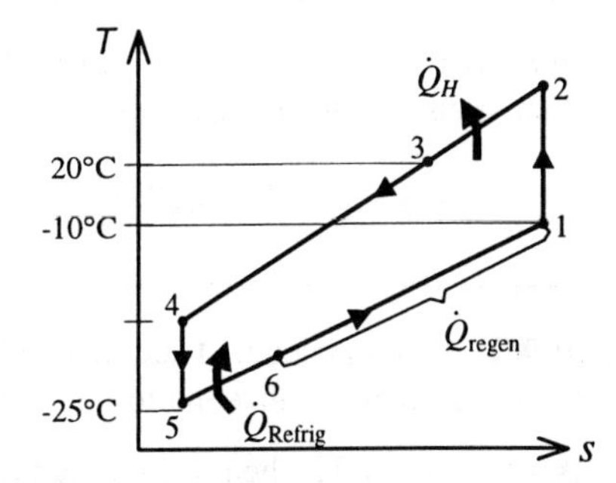

or,

$$\dot{m}C_p(T_3 - T_4) = \dot{m}C_p(T_1 - T_6) \longrightarrow T_3 - T_4 = T_1 - T_6$$

Thus,

$$T_4 = T_3 - T_1 + T_6 = 20°\text{C} - (-10°\text{C}) + (-25°\text{C}) = \mathbf{5°C} = 278\text{ K}$$

(*b*) From the isentropic relations,

$$T_2 = T_1\left(\frac{P_2}{P_1}\right)^{(k-1)/k} = (263\text{ K})(3)^{0.667/1.667} = 408.2\text{ K} = 135.2°\text{C}$$

$$T_5 = T_4\left(\frac{P_5}{P_4}\right)^{(k-1)/k} = (278\text{ K})\left(\frac{1}{3}\right)^{0.667/1.667} = 179.1\text{ K} = -93.9°\text{C}$$

Then the COP of this ideal gas refrigeration cycle is determined from

$$\text{COP}_R = \frac{q_L}{w_{net,in}} = \frac{q_L}{w_{comp,in} - w_{turb,out}} = \frac{h_6 - h_5}{(h_2 - h_1) - (h_4 - h_5)}$$

$$= \frac{T_6 - T_5}{(T_2 - T_1) - (T_4 - T_5)} = \frac{-25°\text{C} - (-93.9°\text{C})}{[135.2 - (-10)]°\text{C} - [5 - (-93.9)]°\text{C}} = \mathbf{1.49}$$

(*c*) The net power input is determined from

$$\dot{W}_{net,in} = \dot{W}_{comp,in} - \dot{W}_{turb,out} = \dot{m}[(h_2 - h_1) - (h_4 - h_5)] = \dot{m}C_p[(T_2 - T_1) - (T_4 - T_5)]$$

$$= (0.45\text{ kg/s})(5.1926\text{ kJ/kg}\cdot°\text{C})([135.2 - (-10)] - [5 - (-93.9)]) = \mathbf{108.2\text{ kW}}$$

10-91 An absorption refrigeration system operating at specified conditions is considered. The minimum rate of heat supply required is to be determined.

Analysis The maximum COP that this refrigeration system can have is

$$\text{COP}_{\text{R,max}} = \left(1 - \frac{T_0}{T_s}\right)\left(\frac{T_L}{T_0 - T_L}\right) = \left(1 - \frac{298\text{K}}{358\text{K}}\right)\left(\frac{263}{298 - 263}\right) = 1.26$$

Thus,

$$\dot{Q}_{\text{gen, min}} = \frac{\dot{Q}_L}{\text{COP}_{\text{R, max}}} = \frac{12\text{ kW}}{1.26} = \mathbf{9.53\ kW}$$

10-92 EES solution of this (and other comprehensive problems designated with the *computer icon*) is available to instructors at the *Instructor Manual* section of the *Online Learning Center* (OLC) at www.mhhe.com/cengel-boles. See the Preface for access information.

10-93 A house is cooled adequately by a 3.5 ton air-conditioning unit. The rate of heat gain of the house when the air-conditioner is running continuously is to be determined.

Assumptions **1** The heat gain includes heat transfer through the walls and the roof, infiltration heat gain, solar heat gain, internal heat gain, etc. **2** Steady operating conditions exist.

Analysis Noting that 1 ton of refrigeration is equivalent to a cooling rate of 211 kJ/min, the rate of heat gain of the house in steady operation is simply equal to the cooling rate of the air-conditioning system,

$$\dot{Q}_{\text{heat gain}} = \dot{Q}_{\text{cooling}} = (3.5\text{ ton})(211\text{ kJ/min}) = 738.5\text{ kJ/min} = \mathbf{44{,}310\ kJ/h}$$

10-94E A room is cooled adequately by a 5000 Btu/h window air-conditioning unit. The rate of heat gain of the room when the air-conditioner is running continuously is to be determined.

Assumptions **1** The heat gain includes heat transfer through the walls and the roof, infiltration heat gain, solar heat gain, internal heat gain, etc. **2** Steady operating conditions exist.

Analysis The rate of heat gain of the room in steady operation is simply equal to the cooling rate of the air-conditioning system,

$$\dot{Q}_{\text{heat gain}} = \dot{Q}_{\text{cooling}} = \mathbf{5{,}000\ Btu/h}$$

10-95 A heat pump water heater has a COP of 2.2 and consumes 2 kW when running. It is to be determined if this heat pump can be used to meet the cooling needs of a room by absorbing heat from it.

Assumptions The COP of the heat pump remains constant whether heat is absorbed from the outdoor air or room air.

Analysis The COP of the heat pump is given to be 2.2. Then the COP of the air-conditioning system becomes

$$\text{COP}_{\text{air-cond}} = \text{COP}_{\text{heat pump}} - 1 = 2.2 - 1 = 1.2$$

Then the rate of cooling (heat absorption from the air) becomes

$$\dot{Q}_{\text{cooling}} = \text{COP}_{\text{air-cond}} \dot{W}_{in} = (1.2)(2\text{ kW}) = 2.4\text{ kW} = 8640\text{ kJ/h}$$

since 1 kW = 3600 kJ/h. We conclude that this heat pump **can meet** the cooling needs of the room since its cooling rate is greater than the rate of heat gain of the room.

10-96 A vortex tube receives compressed air at 500 kPa and 300 K, and supplies 25 percent of it as cold air and the rest as hot air. The COP of the vortex tube is to be compared to that of a reversed Brayton cycle for the same pressure ratio; the exit temperature of the hot fluid stream and the COP are to be determined; and it is to be shown if this process violates the second law.

Assumptions **1** The vortex tube is adiabatic. **2** Air is an ideal gas with constant specific heats at room temperature. **3** Steady operating conditions exist.

Properties The gas constant of air is 0.287 kJ/kg.K (Table A-1). The specific heat of air at room temperature is C_p = 1.005 kJ/kg.K (Table A-2). The enthalpy of air at absolute temperature T can be expressed in terms of specific heats as $h = C_p T$.

Analysis (*a*) The COP of the vortex tube is much lower than the COP of a reversed Brayton cycle of the same pressure ratio since the vortex tube involves *vortices*, which are highly irreversible. Owing to this irreversibility, the minimum temperature that can be obtained by the vortex tube is not as low as the one that can be obtained by the revered Brayton cycle.

(*b*) We take the vortex tube as the system. This is a steady flow system with one inlet and two exits, and it involves no heat or work interactions. Then the steady-flow energy balance equation for this system $\dot{E}_{in} = \dot{E}_{out}$ for a unit mass flow rate at the inlet $(\dot{m}_1 = 1\text{ kg/s})$ can be expressed as

$$\dot{m}_1 h_1 = \dot{m}_2 h_2 + \dot{m}_3 h_3$$
$$\dot{m}_1 C_p T_1 = \dot{m}_2 C_p T_2 + \dot{m}_3 C_p T_3$$
$$1 C_p T_1 = 0.25 C_p T_2 + 0.75 C_p T_3$$

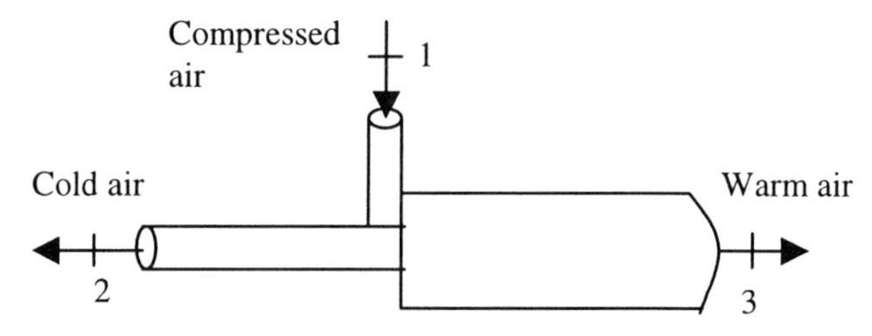

Canceling C_p and solving for T_3 gives

$$T_3 = \frac{T_1 - 0.25T_2}{0.75} = \frac{300 - 0.25 \times 278}{0.75} = \mathbf{307.3\ K}$$

Therefore, the hot air stream will leave the vortex tube at an average temperature of 307.3 K.

(*c*) The entropy balance for this steady flow system $\dot{S}_{in} - \dot{S}_{out} + \dot{S}_{gen} = 0$ can be expressed as with one inlet and two exits, and it involves no heat or work interactions. Then the steady-flow energy balance equation for this system for a unit mass flow rate at the inlet $(\dot{m}_1 = 1\text{ kg/s})$ can be expressed

$$\begin{aligned}
\dot{S}_{gen} &= \dot{S}_{out} - \dot{S}_{in} \\
&= \dot{m}_2 s_2 + \dot{m}_3 s_3 - \dot{m}_1 s_1 = \dot{m}_2 s_2 + \dot{m}_3 s_3 - (\dot{m}_2 + \dot{m}_3) s_1 \\
&= \dot{m}_2 (s_2 - s_1) + \dot{m}_3 (s_3 - s_1) \\
&= 0.25(s_2 - s_1) + 0.75(s_3 - s_1) \\
&= 0.25\left(C_p \ln\frac{T_2}{T_1} - R\ln\frac{P_2}{P_1}\right) + 0.75\left(C_p \ln\frac{T_3}{T_1} - R\ln\frac{P_3}{P_1}\right)
\end{aligned}$$

Substituting the known quantities, the rate of entropy generation is determined to be

$$\begin{aligned}
\dot{S}_{gen} &= 0.25\left((1.005\text{ kJ/kg.K})\ln\frac{278\text{ K}}{300\text{ K}} - (0.287\text{ kJ/kg.K})\ln\frac{100\text{ kPa}}{500\text{ kPa}}\right) \\
&\quad + 0.75\left((1.005\text{ kJ/kg.K})\ln\frac{307.3\text{ K}}{300\text{ K}} - (0.287\text{ kJ/kg.K})\ln\frac{100\text{ kPa}}{500\text{ kPa}}\right) \\
&= 0.461\text{ kW/K} > 0
\end{aligned}$$

which is a positive quantity. Therefore, this process **satisfies** the 2nd law of thermodynamics.

(*d*) For a unit mass flow rate at the inlet ($\dot{m}_1 = 1 \text{ kg/s}$), the cooling rate and the power input to the compressor are determined to

$$\dot{Q}_{cooling} = \dot{m}_c(h_1 - h_c) = \dot{m}_c C_p(T_1 - T_c)$$
$$= (0.25 \text{ kg/s})(1.005 \text{ kJ/kg.K})(300 - 278)\text{K} = 5.53 \text{ kW}$$

$$\dot{W}_{comp,in} = \frac{\dot{m}_0 R T_0}{(k-1)\eta_{comp}}\left[\left(\frac{P_1}{P_0}\right)^{(k-1)/k} - 1\right]$$

$$= \frac{(1 \text{ kg/s})(0.287 \text{ kJ/kg.K})(300 \text{ K})}{(1.4-1)0.80}\left[\left(\frac{535 \text{ kPa}}{100 \text{ kPa}}\right)^{(1.4-1)/1.4} - 1\right] = 165.4 \text{ kW}$$

Then the COP of the vortex refrigerator becomes

$$\text{COP} = \frac{\dot{Q}_{cooling}}{\dot{W}_{comp, in}} = \frac{5.53 \text{ kW}}{165.4 \text{ kW}} = \mathbf{0.033}$$

The COP of a Carnot refrigerator operating between the same temperature limits of 300 K and 278 K is

$$\text{COP}_{Carnot} = \frac{T_L}{T_H - T_L} = \frac{278 \text{ K}}{(300 - 278) \text{ K}} = \mathbf{12.6}$$

Discussion Note that the COP of the vortex refrigerator is a small fraction of the COP of a Carnot refrigerator operating between the same temperature limits.

10-97 A vortex tube receives compressed air at 600 kPa and 300 K, and supplies 25 percent of it as cold air and the rest as hot air. The COP of the vortex tube is to be compared to that of a reversed Brayton cycle for the same pressure ratio; the exit temperature of the hot fluid stream and the COP are to be determined; and it is to be shown if this process violates the second law.

Assumptions **1** The vortex tube is adiabatic. **2** Air is an ideal gas with constant specific heats at room temperature. **3** Steady operating conditions exist.

Properties The gas constant of air is 0.287 kJ/kg.K (Table A-1). The specific heat of air at room temperature is C_p = 1.005 kJ/kg.K (Table A-2). The enthalpy of air at absolute temperature T can be expressed in terms of specific heats as $h = C_p T$.

Analysis (*a*) The COP of the vortex tube is much lower than the COP of a reversed Brayton cycle of the same pressure ratio since the vortex tube involves *vortices*, which are highly irreversible. Owing to this irreversibility, the minimum temperature that can be obtained by the vortex tube is not as low as the one that can be obtained by the revered Brayton cycle.

(*b*) We take the vortex tube as the system. This is a steady flow system with one inlet and two exits, and it involves no heat or work interactions. Then the steady-flow energy balance equation for this system $\dot{E}_{\text{in}} = \dot{E}_{\text{out}}$ for a unit mass flow rate at the inlet $(\dot{m}_1 = 1\text{ kg/s})$ can be expressed as

$$\dot{m}_1 h_1 = \dot{m}_2 h_2 + \dot{m}_3 h_3$$
$$\dot{m}_1 C_p T_1 = \dot{m}_2 C_p T_2 + \dot{m}_3 C_p T_3$$
$$1 C_p T_1 = 0.25 C_p T_2 + 0.75 C_p T_3$$

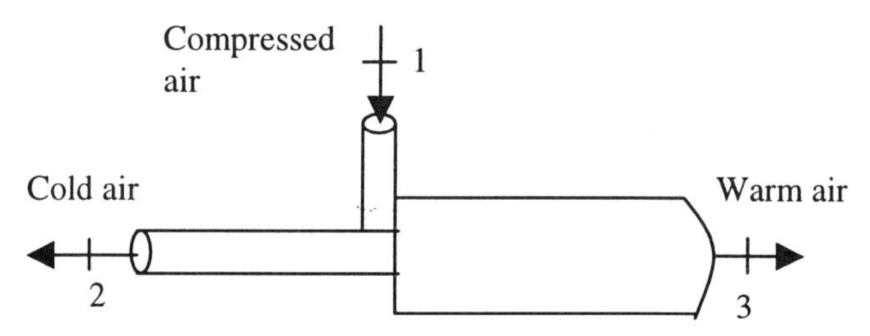

Canceling C_p and solving for T_3 gives

$$T_3 = \frac{T_1 - 0.25T_2}{0.75} = \frac{300 - 0.25 \times 278}{0.75} = \mathbf{307.3\ K}$$

Therefore, the hot air stream will leave the vortex tube at an average temperature of 307.3 K.

(*c*) The entropy balance for this steady flow system $\dot{S}_{\text{in}} - \dot{S}_{\text{out}} + \dot{S}_{\text{gen}} = 0$ can be expressed as with one inlet and two exits, and it involves no heat or work interactions. Then the steady-flow energy balance equation for this system for a unit mass flow rate at the inlet $(\dot{m}_1 = 1\text{ kg/s})$ can be expressed

$$\begin{aligned}
\dot{S}_{\text{gen}} &= \dot{S}_{\text{out}} - \dot{S}_{\text{in}} \\
&= \dot{m}_2 s_2 + \dot{m}_3 s_3 - \dot{m}_1 s_1 = \dot{m}_2 s_2 + \dot{m}_3 s_3 - (\dot{m}_2 + \dot{m}_3) s_1 \\
&= \dot{m}_2 (s_2 - s_1) + \dot{m}_3 (s_3 - s_1) \\
&= 0.25(s_2 - s_1) + 0.75(s_3 - s_1) \\
&= 0.25\left(C_p \ln\frac{T_2}{T_1} - R\ln\frac{P_2}{P_1}\right) + 0.75\left(C_p \ln\frac{T_3}{T_1} - R\ln\frac{P_3}{P_1}\right)
\end{aligned}$$

Substituting the known quantities, the rate of entropy generation is determined to be

$$\begin{aligned}
\dot{S}_{\text{gen}} &= 0.25\left((1.005\text{ kJ/kg.K})\ln\frac{278\text{ K}}{300\text{ K}} - (0.287\text{ kJ/kg.K})\ln\frac{100\text{ kPa}}{600\text{ kPa}}\right) \\
&\quad + 0.75\left((1.005\text{ kJ/kg.K})\ln\frac{307.3\text{ K}}{300\text{ K}} - (0.287\text{ kJ/kg.K})\ln\frac{100\text{ kPa}}{600\text{ kPa}}\right) \\
&= 0.513\ \text{kW/K} > 0
\end{aligned}$$

which is a positive quantity. Therefore, this process **satisfies** the 2nd law of thermodynamics.

(*d*) For a unit mass flow rate at the inlet ($\dot{m}_1 = 1\ \text{kg/s}$), the cooling rate and the power input to the compressor are determined to

$$\dot{Q}_{\text{cooling}} = \dot{m}_c(h_1 - h_c) = \dot{m}_c C_p(T_1 - T_c)$$
$$= (0.25\ \text{kg/s})(1.005\ \text{kJ/kg.K})(300 - 278)\text{K} = 5.53\ \text{kW}$$

$$\dot{W}_{\text{comp,in}} = \frac{\dot{m}_0 R T_0}{(k-1)\eta_{\text{comp}}}\left[\left(\frac{P_1}{P_0}\right)^{(k-1)/k} - 1\right]$$

$$= \frac{(1\ \text{kg/s})(0.287\ \text{kJ/kg.K})(300\ \text{K})}{(1.4-1)0.80}\left[\left(\frac{635\ \text{kPa}}{100\ \text{kPa}}\right)^{(1.4-1)/1.4} - 1\right] = 187.2\ \text{kW}$$

Then the COP of the vortex refrigerator becomes

$$\text{COP} = \frac{\dot{Q}_{\text{cooling}}}{\dot{W}_{\text{comp, in}}} = \frac{5.53\ \text{kW}}{187.2\ \text{kW}} = \mathbf{0.030}$$

The COP of a Carnot refrigerator operating between the same temperature limits of 300 K and 278 K is

$$\text{COP}_{\text{Carnot}} = \frac{T_L}{T_H - T_L} = \frac{278\ \text{K}}{(300 - 278)\ \text{K}} = \mathbf{12.6}$$

Discussion Note that the COP of the vortex refrigerator is a small fraction of the COP of a Carnot refrigerator operating between the same temperature limits.

10-98, 10-99 EES solution of this (and other comprehensive problems designated with the *computer icon*) is available to instructors at the *Instructor Manual* section of the *Online Learning Center* (OLC) at www.mhhe.com/cengel-boles. See the Preface for access information.

Fundamentals of Engineering (FE) Exam Problems

10-100 Consider a heat pump that operates on the reversed Carnot cycle with R-134a as the working fluid executed under the saturation dome between the pressure limits of 120 kPa and 900 kPa. R-134a changes from saturated vapor to saturated liquid during the heat rejection process. The net work input for this cycle is

(a) 31 kJ/kg (b) 38 kJ/kg (c) 49 kJ/kg (d) 888 kJ/kg (e) 1133 kJ/kg

Answer (a) 31 kJ/kg

Solution Solved by EES Software. Solutions can be verified by copying-and-pasting the following lines on a blank EES screen. (Similar problems and their solutions can be obtained easily by modifying numerical values).

```
P1=900 "kPa"
P2=120 "kPa"

h_fg=ENTHALPY(R134a,x=1,P=P1)-ENTHALPY(R134a,x=0,P=P1)
TH=TEMPERATURE(R134a,x=0,P=P1)+273
TL=TEMPERATURE(R134a,x=0,P=P2)+273
q_H=h_fg
COP=TH/(TH-TL)
w_net=q_H/COP

"Some Wrong Solutions with Common Mistakes:"
W1_work = q_H/COP1; COP1=TL/(TH-TL) "Using COP of regrigerator"
W2_work = q_H/COP2; COP2=(TH-273)/(TH-TL) "Using C instead of K"
W3_work = h_fg3/COP; h_fg3= ENTHALPY(R134a,x=1,P=P2)-ENTHALPY(R134a,x=0,P=P2)
"Using h_fg at P2"
W4_work = q_H*TL/TH "Using the wrong relation"
```

10-101 A refrigerator removes heat from a refrigerated space at –5°C at a rate of 0.35 kJ/s and rejects it to an environment at 20°C. The minimum required power input is

(a) 30 W (b) 33 W (c) 56 W (d) 124 W (e) 350 W

Answer (b) 33 W

Solution Solved by EES Software. Solutions can be verified by copying-and-pasting the following lines on a blank EES screen. (Similar problems and their solutions can be obtained easily by modifying numerical values).

```
TH=20+273
TL=-5+273
Q_L=0.35 "kJ/s"
COP_max=TL/(TH-TL)
w_min=Q_L/COP_max

"Some Wrong Solutions with Common Mistakes:"
W1_work = Q_L/COP1; COP1=TH/(TH-TL) "Using COP of heat pump"
W2_work = Q_L/COP2; COP2=(TH-273)/(TH-TL) "Using C instead of K"
W3_work = Q_L*TL/TH "Using the wrong relation"
W4_work = Q_L "Taking the rate of refrigeration as power input"
```

10-102 A refrigerator operates on the ideal vapor compression refrigeration cycle with R-134a as the working fluid between the pressure limits of 140 kPa and 800 kPa. If the rate of heat removal from the refrigerated space is 25 kJ/s, the mass flow rate of the refrigerant is

(a) 0.10 kg/s (b) 0.15 kg/s (c) 0.18 kg/s (d) 0.25 kg/s (e) 0.38 kg/s

Answer (c) 0.18 kg/s

Solution Solved by EES Software. Solutions can be verified by copying-and-pasting the following lines on a blank EES screen. (Similar problems and their solutions can be obtained easily by modifying numerical values).

```
P1=140 "kPa"
P2=800 "kPa"
P3=P2
P4=P1
s2=s1

Q_refrig=25 "kJ/s"
m=Q_refrig/(h1-h4)

h1=ENTHALPY(R134a,x=1,P=P1)
s1=ENTROPY(R134a,x=1,P=P1)

h2=ENTHALPY(R134a,s=s2,P=P2)
h3=ENTHALPY(R134a,x=0,P=P3)
h4=h3

"Some Wrong Solutions with Common Mistakes:"
W1_mass = Q_refrig/(h2-h1) "Using wrong enthalpies, for W_in"
W2_mass = Q_refrig/(h2-h3) "Using wrong enthalpies, for Q_H"
W3_mass = Q_refrig/(h1-h44); h44=ENTHALPY(R134a,x=0,P=P4)  "Using wrong enthalpy
h4 (at P4)"
W4_mass = Q_refrig/h_fg; h_fg=ENTHALPY(R134a,x=1,P=P2) -
ENTHALPY(R134a,x=0,P=P2)  "Using h_fg at P2"
```

10-103 A heat pump operates on the ideal vapor compression refrigeration cycle with R-134a as the working fluid between the pressure limits of 0.32 MPa and 1.2 MPa. If the mass flow rate of the refrigerant is 0.193 kg/s, the rate of heat supply by the heat pump to the heated space is

(a) 3.3 kW (b) 23 kW (c) 26 kW (d) 31 kW (e) 45 kW

Answer (d) 31 kW

Solution Solved by EES Software. Solutions can be verified by copying-and-pasting the following lines on a blank EES screen. (Similar problems and their solutions can be obtained easily by modifying numerical values).

```
P1=320 "kPa"
P2=1200 "kPa"
P3=P2
P4=P1
s2=s1

m=0.193 "kg/s"
Q_supply=m*(h2-h3) "kJ/s"

h1=ENTHALPY(R134a,x=1,P=P1)
s1=ENTROPY(R134a,x=1,P=P1)

h2=ENTHALPY(R134a,s=s2,P=P2)
h3=ENTHALPY(R134a,x=0,P=P3)
h4=h3

"Some Wrong Solutions with Common Mistakes:"
W1_Qh = m*(h2-h1) "Using wrong enthalpies, for W_in"
W2_Qh = m*(h1-h4) "Using wrong enthalpies, for Q_L"
W3_Qh = m*(h22-h4); h22=ENTHALPY(R134a,x=1,P=P2)  "Using wrong enthalpy h2 (hg
at P2)"
W4_Qh = m*h_fg; h_fg=ENTHALPY(R134a,x=1,P=P1) - ENTHALPY(R134a,x=0,P=P1)
"Using h_fg at P1"
```

10-104 An ideal vapor compression refrigeration cycle with R-134a as the working fluid operates between the pressure limits of 140 kPa and 900 kPa. The mass fraction of the refrigerant that is in the liquid phase at the inlet of the evaporator is

(a) 0.75 (b) 0.65 (c) 0.50 (d) 0.35 (e) 0.25

Answer (b) 0.65

Solution Solved by EES Software. Solutions can be verified by copying-and-pasting the following lines on a blank EES screen. (Similar problems and their solutions can be obtained easily by modifying numerical values).

```
P1=140 "kPa"
P2=900 "kPa"
P3=P2
P4=P1

h1=ENTHALPY(R134a,x=1,P=P1)
h3=ENTHALPY(R134a,x=0,P=P3)
h4=h3
x4=QUALITY(R134a,h=h4,P=P4)
liquid=1-x4

"Some Wrong Solutions with Common Mistakes:"
W1_liquid = x4 "Taking quality as liquid content"
W2_liquid = 0 "Assuming superheated vapor"
W3_liquid = 1-x4s; x4s=QUALITY(R134a,s=s3,P=P4) "Assuming isentropic expansion"
s3=ENTROPY(R134a,x=0,P=P3)
```

10-105 Consider a heat pump that operates on the ideal vapor compression refrigeration cycle with R-134a as the working fluid between the pressure limits of 0.32 MPa and 1.2 MPa. The coefficient of performance of this heat pump is

(a) 0.17 (b) 1.2 (c) 3.1 (d) 4.9 (e) 5.9

Answer (e) 5.9

Solution Solved by EES Software. Solutions can be verified by copying-and-pasting the following lines on a blank EES screen. (Similar problems and their solutions can be obtained easily by modifying numerical values).

```
P1=320 "kPa"
P2=1200 "kPa"
P3=P2
P4=P1
s2=s1
h1=ENTHALPY(R134a,x=1,P=P1)
s1=ENTROPY(R134a,x=1,P=P1)
h2=ENTHALPY(R134a,s=s2,P=P2)
h3=ENTHALPY(R134a,x=0,P=P3)
h4=h3
COP_HP=qH/Win
Win=h2-h1
qH=h2-h3

"Some Wrong Solutions with Common Mistakes:"
W1_COP = (h1-h4)/(h2-h1) "COP of refrigerator"
W2_COP = (h1-h4)/(h2-h3) "Using wrong enthalpies, QL/QH"
W3_COP = (h22-h3)/(h22-h1); h22=ENTHALPY(R134a,x=1,P=P2) "Using wrong enthalpy
h2 (hg at P2)"
```

10-106 An ideal gas refrigeration cycle using air as the working fluid operates between the pressure limits of 100 kPa and 300 kPa. Air is cooled to 35°C before entering the turbine. The lowest temperature of this cycle is

(a) –48°C (b) -22°C (c) 5°C (d) 10°C (e) 22°C

Answer (a) –48°C

Solution Solved by EES Software. Solutions can be verified by copying-and-pasting the following lines on a blank EES screen. (Similar problems and their solutions can be obtained easily by modifying numerical values).

```
k=1.4
P1= 100 "kPa"
P2=300 "kPa"
T3=35+273 "K"
T4=T3*(P1/P2)^((k-1)/k) – 273 "T_min is the turbine exit temperature"

"Some Wrong Solutions with Common Mistakes:"
W1_Tmin = (T3-273)*(P1/P2)^((k-1)/k) "Using C instead of K"
W2_Tmin = T3*(P1/P2)^((k-1)) - 273 "Using wrong exponent"
W3_Tmin = T3*(P1/P2)^k - 273 "Using wrong exponent"
```

10-107 Consider an ideal gas refrigeration cycle using helium as the working fluid. Helium enters the compressor at 100 kPa and –10°C and is compressed to 250 kPa. Helium is then cooled to 20°C before it enters the turbine. For a mass flow rate of 0.2 kg/s, the net power input required is

(a) 9.3 kW (b) 27.6 kW (c) 48.8 kW (d) 93.5 kW (e) 119 kW

Answer (b) 27.6 kW

Solution Solved by EES Software. Solutions can be verified by copying-and-pasting the following lines on a blank EES screen. (Similar problems and their solutions can be obtained easily by modifying numerical values).

```
k=1.667
Cp=5.1926 "kJ/kg.K"
P1= 100 "kPa"
T1=-10+273 "K"
P2=250 "kPa"
T3=20+273 "K"
m=0.2 "kg/s"

"Mimimum temperature is the turbine exit temperature"
T2=T1*(P2/P1)^((k-1)/k)
T4=T3*(P1/P2)^((k-1)/k)
W_netin=m*Cp*((T2-T1)-(T3-T4))

"Some Wrong Solutions with Common Mistakes:"
W1_Win = m*Cp*((T22-T1)-(T3-T44)); T22=T1*P2/P1; T44=T3*P1/P2 "Using wrong
relations for temps"
W2_Win = m*Cp*(T2-T1) "Ignoring turbine work"
W3_Win=m*1.005*((T2B-T1)-(T3-T4B)); T2B=T1*(P2/P1)^((kB-1)/kB); T4B=T3*(P1/P2)^((kB-
1)/kB); kB=1.4 "Using air properties"
W4_Win=m*Cp*((T2A-(T1-273))-(T3-273-T4A)); T2A=(T1-273)*(P2/P1)^((k-1)/k); T4A=(T3-
273)*(P1/P2)^((k-1)/k) "Using C instead of K"
```

10-108 An absorption air-conditioning system is to remove heat from the conditioned space at 20°C at a rate of 120 kJ/s while operating in an environment at 35°C. Heat is to be supplied from a geothermal source at 160°C. The minimum rate of heat supply required is

(a) 7.9 kJ/s (b) 15 kJ/s (c) 21 kJ/s (d) 58 kJ/s (e) 115 kJ/s

Answer (c) 21 kJ/s

Solution Solved by EES Software. Solutions can be verified by copying-and-pasting the following lines on a blank EES screen. (Similar problems and their solutions can be obtained easily by modifying numerical values).

```
TL=20+273 "K"
Q_refrig=120 "kJ/s"
To=35+273 "K"
Ts=160+273 "K"

COP_max=(1-To/Ts)*(TL/(To-TL))
Q_in=Q_refrig/COP_max

"Some Wrong Solutions with Common Mistakes:"
W1_Qin = Q_refrig "Taking COP = 1"
W2_Qin = Q_refrig/COP2; COP2=TL/(Ts-TL) "Wrong COP expression"
W3_Qin = Q_refrig/COP3; COP3=(1-To/Ts)*(Ts/(To-TL)) "Wrong COP expression"
W4_Qin = Q_refrig*COP_max "Multiplying by COP instead of dividing"
```

10-109 Consider a refrigerator that operates on the vapor compression refrigeration cycle with R-134a as the working fluid. The refrigerant enters the compressor as saturated vapor at 140 kPa, and exits at 800 kPa and 60°C, and leaves the condenser as saturated liquid at 800 kPa. The coefficient of performance of this refrigerator is

(a) 0.41 (b) 1.0 (c) 1.8 (d) 2.5 (e) 3.4

Answer (d) 2.5

Solution Solved by EES Software. Solutions can be verified by copying-and-pasting the following lines on a blank EES screen. (Similar problems and their solutions can be obtained easily by modifying numerical values).

```
P1=140 "kPa"
P2=800 "kPa"
T2=60 "C"
P3=P2
P4=P1

h1=ENTHALPY(R134a,x=1,P=P1)
s1=ENTROPY(R134a,x=1,P=P1)

h2=ENTHALPY(R134a,T=T2,P=P2)
h3=ENTHALPY(R134a,x=0,P=P3)
h4=h3

COP_R=qL/Win
Win=h2-h1
qL=h1-h4

"Some Wrong Solutions with Common Mistakes:"
W1_COP = (h2-h3)/(h2-h1) "COP of heat pump"
W2_COP = (h1-h4)/(h2-h3)  "Using wrong enthalpies, QL/QH"
W3_COP = (h1-h4)/(h2s-h1); h2s=ENTHALPY(R134a,s=s1,P=P2)  "Assuming isentropic compression"
```

10-110 · · · 10-119 Design and Essay Problems

Chapter 11
THERMODYNAMIC PROPERTY RELATIONS

Partial Derivatives and Associated Relations

11-1C

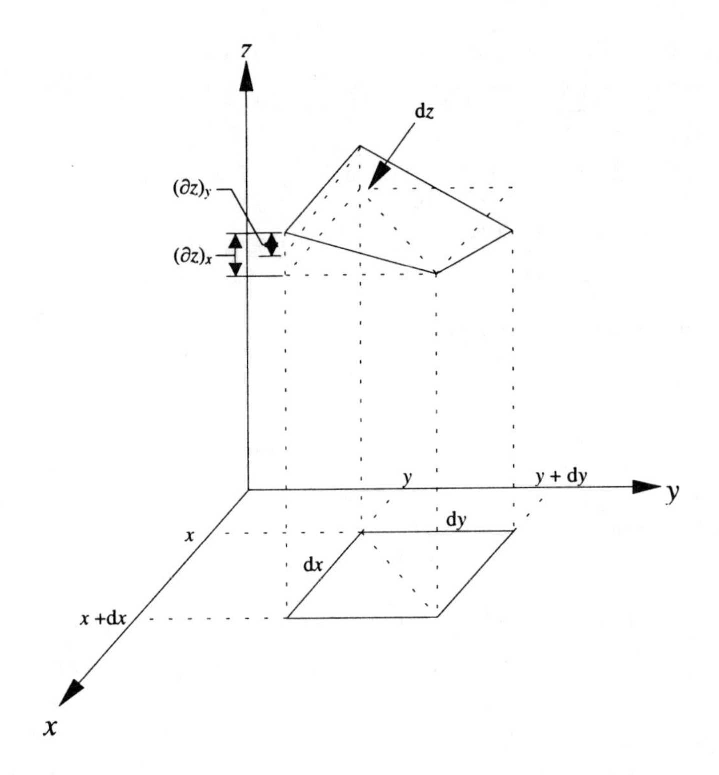

$$\partial x \equiv dx$$
$$\partial y \equiv dy$$
$$dz = (\partial z)_x + (\partial z)_y$$

11-2C For functions that depend on one variable, they are identical. For functions that depend on two or more variable, the partial differential represents the change in the function with one of the variables as the other variables are held constant. The ordinary differential for such functions represents the total change as a result of differential changes in all variables.

11-3C (*a*) $(\partial x)_y = dx$; (*b*) $(\partial z)_y \leq dz$; and (*c*) $dz = (\partial z)_x + (\partial z)_y$

11-4C Only when $(\partial z/\partial x)_y = 0$. That is, when z does not depend on y and thus $z = z(x)$.

11-5C It indicates that z does not depend on y. That is, $z = z(x)$.

11-6C Yes.

11-7C Yes.

11-8 Air at a specified temperature and specific volume is considered. The changes in pressure corresponding to a certain increase of different properties are to be determined.

Assumptions Air is an ideal gas

Properties The gas constant of air is R = 0.287 kPa· m^3/kg· K (Table A-1).

Analysis An ideal gas equation can be expressed as $P = RT/v$. Noting that R is a constant and $P = P(T, v)$,

$$dP = \left(\frac{\partial P}{\partial T}\right)_v dT + \left(\frac{\partial P}{\partial v}\right)_T dv = \frac{RdT}{v} - \frac{RTdv}{v^2}$$

(a) The change in T can be expressed as $dT \cong \Delta T = 400 \times 0.01 = 4.0$ K. At v = constant,

$$(dP)_v = \frac{RdT}{v} = \frac{(0.287\text{kPa}\cdot\text{m}^3/\text{kg}\cdot\text{K})(4.0\text{K})}{0.90\text{m}^3/\text{kg}} = \mathbf{1.276\ kPa}$$

(b) The change in v can be expressed as $dv \cong \Delta v = 0.90 \times 0.01 = 0.009$ m^3/kg. At T = constant,

$$(dP)_T = -\frac{RTdv}{v^2} = \frac{(0.287\text{kPa}\cdot\text{m}^3/\text{kg}\cdot\text{K})(400\text{K})(0.009\text{m}^3)}{(0.90\text{m}^3/\text{kg})^2} = \mathbf{-1.276\ kPa}$$

(c) When both v and T increases by 1%, the change in P becomes

$$dP = (dP)_v + (dP)_T = 1.276 + (-1.276) = \mathbf{0}$$

Thus the changes in T and v balance each other.

11-9 Helium at a specified temperature and specific volume is considered. The changes in pressure corresponding to a certain increase of different properties are to be determined.

Assumptions Helium is an ideal gas

Properties The gas constant of helium is R = 2.0769 kPa· m^3/kg· K (Table A-1).

Analysis An ideal gas equation can be expressed as $P = RT/v$. Noting that R is a constant and $P = P(T, v)$,

$$dP = \left(\frac{\partial P}{\partial T}\right)_v dT + \left(\frac{\partial P}{\partial v}\right)_T dv = \frac{RdT}{v} - \frac{RTdv}{v^2}$$

(a) The change in T can be expressed as $dT \cong \Delta T = 400 \times 0.01 = 4.0$ K. At v = constant,

$$(dP)_v = \frac{RdT}{v} = \frac{(2.0769\text{kPa}\cdot\text{m}^3/\text{kg}\cdot\text{K})(4.0\text{K})}{0.90\text{m}^3/\text{kg}} = \mathbf{9.231 kPa}$$

(b) The change in v can be expressed as $dv \cong \Delta v = 0.90 \times 0.01 = 0.009$ m^3/kg. At T = constant,

$$(dP)_T = -\frac{RTdv}{v^2} = \frac{(2.0769\text{kPa}\cdot\text{m}^3/\text{kg}\cdot\text{K})(400\text{K})(0.009\text{m}^3)}{(0.90\text{m}^3/\text{kg})^2} = \mathbf{-9.231 kPa}$$

(c) When both v and T increases by 1%, the change in P becomes

$$dP = (dP)_v + (dP)_T = 1.276 + (-1.276) = \mathbf{0}$$

Thus the changes in T and v balance each other.

11-10 It is to be proven for an ideal gas that the P = constant lines on a T-v diagram are straight lines and that the high pressure lines are steeper than the low-pressure lines.

Analysis (*a*) For an ideal gas $Pv = RT$ or $T = Pv/R$. Taking the partial derivative of T with respect to v holding P constant yields

$$\left(\frac{\partial T}{\partial v}\right)_P = \frac{P}{R}$$

which remains constant at P = constant. Thus the derivative $(\partial T/\partial v)_P$, which represents the slope of the P = const. lines on a T-v diagram, remains constant. That is, the P = const. lines are straight lines on a T-v diagram.

(*b*) The slope of the P = const. lines on a T-v diagram is equal to P/R, which is proportional to P. Therefore, the high pressure lines are steeper than low pressure lines on the T-v diagram.

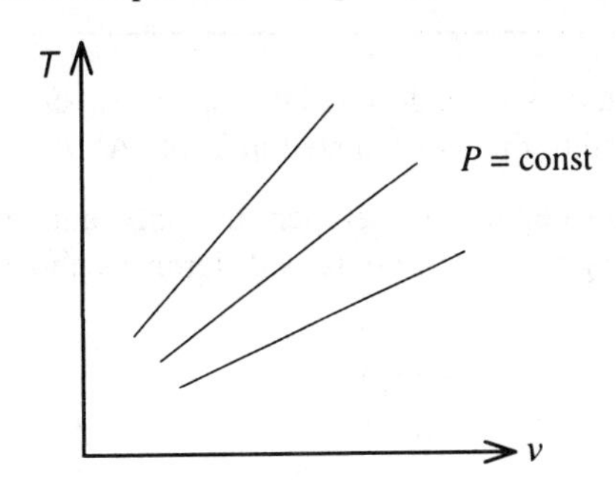

11-11 A relation is to be derived for the slope of the v = constant lines on a T-P diagram for a gas that obeys the van der Waals equation of state.

Analysis The van der Waals equation of state can be expressed as

$$T = \frac{1}{R}\left(P + \frac{a}{v^2}\right)(v - b)$$

Taking the derivative of T with respect to P holding v constant,

$$\left(\frac{\partial T}{\partial P}\right)_v = \frac{1}{R}(1 + 0)(v - b) = \frac{v - \mathbf{b}}{\mathbf{R}}$$

which is the slope of the v = constant lines on a T-P diagram.

11-12 Nitrogen gas at a specified state is considered. The C_p and C_v of the nitrogen are to be determined using Table A-18, and to be compared to the values listed in Table A-2*b*.

Analysis The C_p and C_v of ideal gases depends on temperature only, and are expressed as $C_p(T) = dh(T)/dT$ and $C_v(T) = du(T)/dT$. Approximating the differentials as differences about 400 K, the C_p and C_v values are determined to be

$$C_p(400\text{K}) = \left(\frac{dh(T)}{dT}\right)_{T=400\text{K}} \cong \left(\frac{\Delta h(T)}{\Delta T}\right)_{T\cong 400\text{K}}$$

$$= \frac{h(410\text{K}) - h(390\text{K})}{(410 - 390)\text{K}}$$

$$= \frac{(11{,}932 - 11{,}347)/28.0\text{kJ/kg}}{(410 - 390)\text{K}}$$

$$= \mathbf{1.045 kJ/kg \cdot K}$$

(Compare: Table A-2b at 400 K $\rightarrow$ C_p = 1.044 kJ/kg· K)

$$C_v(400\text{K}) = \left(\frac{du(T)}{dT}\right)_{T=400\text{K}} \cong \left(\frac{\Delta u(T)}{\Delta T}\right)_{T\cong 400\text{K}}$$

$$= \frac{u(410\text{K}) - u(390\text{K})}{(410 - 390)\text{K}}$$

$$= \frac{(8{,}523 - 8{,}104)/28.0\text{kJ/kg}}{(410 - 390)\text{K}}$$

$$= \mathbf{0.748 kJ/kg \cdot K}$$

(Compare: Table A-2b at 400 K $\rightarrow$ C_v = 0.747 kJ/kg· K)

11-13E Nitrogen gas at a specified state is considered. The C_p and C_v of the nitrogen are to be determined using Table A-18E, and to be compared to the values listed in Table A-2E*b*.

Analysis The C_p and C_v of ideal gases depends on temperature only, and are expressed as $C_p(T) = dh(T)/dT$ and $C_v(T) = du(T)/dT$. Approximating the differentials as differences about 600 R, the C_p and C_v values are determined to be

$$C_p(600\text{R}) = \left(\frac{dh(T)}{dT}\right)_{T=600\text{R}} \cong \left(\frac{\Delta h(T)}{\Delta T}\right)_{T\cong 600\text{R}}$$

$$= \frac{h(620\text{R}) - h(580\text{R})}{(620-580)\text{R}}$$

$$= \frac{(4{,}307.1 - 4{,}028.7)/28.0\text{Btu/lbm}}{(620-580)\text{R}}$$

$$= \mathbf{0.249 Btu/lbm \cdot R}$$

(Compare: Table A-2E*b* at 600 R → C_p = 0.248 Btu/lbm· R)

$$C_v(600\text{R}) = \left(\frac{du(T)}{dT}\right)_{T=600\text{R}} \cong \left(\frac{\Delta u(T)}{\Delta T}\right)_{T\cong 600\text{R}}$$

$$= \frac{u(620\text{R}) - u(580\text{R})}{(620-580)\text{R}}$$

$$= \frac{(3{,}075.9 - 2{,}876.9)/28.0\text{Btu/lbm}}{(620-580)\text{R}}$$

$$= \mathbf{0.178 Btu/lbm \cdot R}$$

Compare: Table A-2Eb at 600 R → C_v = 0.178 Btu/lbm· R)

11-14 The state of an ideal gas is altered slightly. The change in the specific volume of the gas is to be determined using differential relations and the ideal-gas relation at each state.

Assumptions The gas is air and air is an ideal gas.

Properties The gas constant of air is $R = 0.287$ kPa· m³/kg· K (Table A-1).

Analysis (*a*) The changes in T and P can be expressed as

$$dT \cong \Delta T = (404 - 400)\text{ K} = 4\text{ K}$$
$$dP \cong \Delta P = (98 - 100)\text{ kPa} = -2\text{ kPa}$$

The ideal gas relation $Pv = RT$ can be expressed as $v = RT/P$. Note that R is a constant and $v = v(T, P)$. Applying the total differential relation and using average values for T and P,

$$dv = \left(\frac{\partial v}{\partial T}\right)_P dT + \left(\frac{\partial v}{\partial P}\right)_T dP = \frac{RdT}{P} - \frac{RT\,dP}{P^2}$$
$$= (0.287\text{kPa}\cdot\text{m}^3/\text{kg}\cdot\text{K})\left(\frac{4\text{K}}{99\text{kPa}} - \frac{(402\text{K})(-2\text{kPa})}{(99\text{kPa})^2}\right)$$
$$= (0.0116\text{m}^3/\text{kg}) + (0.0235\text{m}^3/\text{kg}) = \mathbf{0.0351m^3/kg}$$

(*b*) Using the ideal gas relation at each state,

$$v_1 = \frac{RT_1}{P_1} = \frac{(0.287\text{ kPa}\cdot\text{m}^3/\text{kg}\cdot\text{K})(400\text{ K})}{100\text{ kPa}} = 1.1480\text{ m}^3/\text{kg}$$

$$v_2 = \frac{RT_2}{P_2} = \frac{(0.287\text{ kPa}\cdot\text{m}^3/\text{kg}\cdot\text{K})(404\text{ K})}{98\text{ kPa}} = 1.1831\text{ m}^3/\text{kg}$$

Thus,

$$\Delta v = v_2 - v_1 = 1.1831 - 1.1480 = \mathbf{0.0351\ m^3/kg}$$

The two results are identical.

11-15 Using the equation of state $P(v\text{-}a) = RT$, the cyclic relation, and the reciprocity relation at constant v are to be verified.

Analysis (*a*) This equation of state involves three variables P, v, and T. Any two of these can be taken as the independent variables, with the remaining one being the dependent variable. Replacing x, y, and z by P, v, and T, the cyclic relation can be expressed as

$$\left(\frac{\partial P}{\partial v}\right)_T \left(\frac{\partial v}{\partial T}\right)_P \left(\frac{\partial T}{\partial P}\right)_v = -1$$

where

$$P = \frac{RT}{v-a} \longrightarrow \left(\frac{\partial P}{\partial v}\right)_T = \frac{-RT}{(v-a)^2} = -\frac{P}{v-a}$$

$$v = \frac{RT}{P} + a \longrightarrow \left(\frac{\partial v}{\partial T}\right)_P = \frac{R}{P}$$

$$T = \frac{P(v-a)}{R} \longrightarrow \left(\frac{\partial T}{\partial P}\right)_v = \frac{v-a}{R}$$

Substituting,

$$\left(\frac{\partial P}{\partial v}\right)_T \left(\frac{\partial v}{\partial T}\right)_P \left(\frac{\partial T}{\partial P}\right)_v = \left(-\frac{P}{v-a}\right)\left(\frac{R}{P}\right)\left(\frac{v-a}{R}\right) = -1$$

which is the desired result.

(*b*) The reciprocity rule for this gas at v = constant can be expressed as

$$\left(\frac{\partial P}{\partial T}\right)_v = \frac{1}{(\partial T / \partial P)_v}$$

$$T = \frac{P(v-a)}{R} \longrightarrow \left(\frac{\partial T}{\partial P}\right)_v = \frac{v-a}{R}$$

$$P = \frac{RT}{v-a} \longrightarrow \left(\frac{\partial P}{\partial T}\right)_v = \frac{R}{v-a}$$

We observe that the first differential is the inverse of the second one. Thus the proof is complete.

The Maxwell Relations

11-16 The validity of the last Maxwell relation for refrigerant-134a at a specified state is to be verified.

Analysis We do not have exact analytical property relations for refrigerant-134a, and thus we need to replace the differential quantities in the last Maxwell relation with the corresponding finite quantities. Using property values from the tables about the specified state,

$$\left(\frac{\partial s}{\partial P}\right)_T \stackrel{?}{=} -\left(\frac{\partial v}{\partial T}\right)_P$$

$$\left(\frac{\Delta s}{\Delta P}\right)_{T=80°C} \stackrel{?}{\cong} -\left(\frac{\Delta v}{\Delta T}\right)_{P=1200kPa}$$

$$\left(\frac{s_{1400kPa} - s_{1000kPa}}{(1400-1000)kPa}\right)_{T=80°C} \stackrel{?}{\cong} -\left(\frac{v_{100°C} - v_{60°C}}{(100-60)°C}\right)_{P=1200kPa}$$

$$\frac{(0.9997-1.0405)\text{kJ/kg}\cdot\text{K}}{(1400-1000)\text{kPa}} \stackrel{?}{\cong} -\frac{(0.02244-0.01835)\text{m}^3\text{/kg}}{(100-60)°\text{C}}$$

$$-1.02\times10^{-4}\text{m}^3\text{/kg}\cdot\text{K} \cong -1.0225\times10^{-4}\text{m}^3\text{/kg}\cdot\text{K}$$

since kJ ≡ kPa· m³, and K ≡ °C for temperature differences. Thus the last Maxwell relation is satisfied.

11-17 EES solution of this (and other comprehensive problems designated with the *computer icon*) is available to instructors at the *Instructor Manual* section of the *Online Learning Center* (OLC) at www.mhhe.com/cengel-boles. See the Preface for access information.

11-18E The validity of the last Maxwell relation for refrigerant-134a at a specified state is to be verified.

Analysis We do not have exact analytical property relations for steam, and thus we need to replace the differential quantities in the last Maxwell relation with the corresponding finite quantities. Using property values from the tables about the specified state,

$$\left(\frac{\partial s}{\partial P}\right)_T \stackrel{?}{=} -\left(\frac{\partial v}{\partial T}\right)_P$$

$$\left(\frac{\Delta s}{\Delta P}\right)_{T=1000°F} \stackrel{?}{\cong} -\left(\frac{\Delta v}{\Delta T}\right)_{P=450psia}$$

$$\left(\frac{s_{500psia} - s_{400psia}}{(500-400)psia}\right)_{T=1000°F} \stackrel{?}{\cong} -\left(\frac{v_{1200°F} - v_{800°F}}{(1200-800)°F}\right)_{P=450psia}$$

$$\frac{(1.7371-1.7632)\text{Btu/lbm}\cdot\text{R}}{(500-400)\text{psia}} \stackrel{?}{\cong} -\frac{(2.172-1.6077)\text{ft}^3\text{/lbm}}{(1200-800)°\text{F}}$$

$$-1.410\times10^{-3}\text{ft}^3\text{/lbm}\cdot\text{R} \cong -1.411\times10^{-3}\text{ft}^3\text{/lbm}\cdot\text{R}$$

since 1 Btu ≡ 5.4039 psia· ft³, and R ≡ °F for temperature differences. Thus the fourth Maxwell relation is satisfied.

11-19 Using the Maxwell relations, a relation for $(\partial s/\partial P)_T$ for a gas whose equation of state is $P(v-b) = RT$ is to be obtained.

Analysis This equation of state can be expressed as $v = \frac{RT}{P} + b$. Then,

$$\left(\frac{\partial v}{\partial T}\right)_P = \frac{R}{P}$$

From the fourth Maxwell relation,

$$\left(\frac{\partial s}{\partial P}\right)_T = -\left(\frac{\partial v}{\partial T}\right)_P = -\frac{\mathbf{R}}{\mathbf{P}}$$

11-20 Using the Maxwell relations, a relation for $(\partial s/\partial v)_T$ for a gas whose equation of state is $(P-a/v^2)(v-b) = RT$ is to be obtained.

Analysis This equation of state can be expressed as $P = \frac{RT}{v-b} + \frac{a}{v^2}$. Then,

$$\left(\frac{\partial P}{\partial T}\right)_v = \frac{R}{v-b}$$

From the third Maxwell relation,

$$\left(\frac{\partial s}{\partial v}\right)_T = \left(\frac{\partial P}{\partial T}\right)_v = \frac{\mathbf{R}}{\mathbf{v-b}}$$

11-21 Using the Maxwell relations and the ideal-gas equation of state, a relation for $(\partial s/\partial v)_T$ for an ideal gas is to be obtained.

Analysis The ideal gas equation of state can be expressed as $P = \frac{RT}{v}$. Then,

$$\left(\frac{\partial P}{\partial T}\right)_v = \frac{R}{v}$$

From the third Maxwell relation,

$$\left(\frac{\partial s}{\partial v}\right)_T = \left(\frac{\partial P}{\partial T}\right)_v = \frac{\mathbf{R}}{\mathbf{v}}$$

The Clapeyron Equation

11-22C It enables us to determine the enthalpy of vaporization from h_{fg} at a given temperature from the *P*, *v*, *T* data alone.

11-23C It is exact.

11-24C It is assumed that $v_{fg} \cong v_g \cong RT/P$, and $h_{fg} \cong$ constant for small temperature intervals.

11-25 Using the Clapeyron equation, the enthalpy of vaporization of refrigerant-134a at a specified temperature is to be estimated and to be compared to the tabulated data.

Analysis From the Clapeyron equation,

$$\begin{aligned} h_{fg} &= Tv_{fg}\left(\frac{dP}{dT}\right)_{\text{sat}} \\ &\cong T(v_g - v_f)_{@40^\circ\text{C}}\left(\frac{\Delta P}{\Delta T}\right)_{\text{sat},40^\circ\text{C}} \\ &= T(v_g - v_f)_{@40^\circ\text{C}}\left(\frac{P_{\text{sat}@42^\circ\text{C}} - P_{\text{sat}@38^\circ\text{C}}}{42^\circ\text{C} - 38^\circ\text{C}}\right) \\ &= (40 + 273.15\text{K})(0.0199 - 0.0008714\text{m}^3/\text{kg})\left(\frac{(1072.0 - 962.98)\text{kPa}}{4\text{K}}\right) \\ &= \mathbf{162.41 kJ/kg} \end{aligned}$$

The tabulated value of h_{fg} at 40°C is **162.05 kJ/kg**.

11-26 EES solution of this (and other comprehensive problems designated with the *computer icon*) is available to instructors at the *Instructor Manual* section of the *Online Learning Center* (OLC) at www.mhhe.com/cengel-boles. See the Preface for access information.

11-27 Using the Clapeyron equation, the enthalpy of vaporization of steam at a specified pressure is to be estimated and to be compared to the tabulated data.

Analysis From the Clapeyron equation,

$$\begin{aligned} h_{fg} &= Tv_{fg}\left(\frac{dP}{dT}\right)_{sat} \\ &\cong T(v_g - v_f)_{@200\text{kPa}}\left(\frac{\Delta P}{\Delta T}\right)_{\text{sat},200\text{kPa}} \\ &= T_{\text{sat}@200\text{kPa}}(v_g - v_f)_{@200\text{kPa}}\left(\frac{(225 - 175)\text{kPa}}{T_{\text{sat}@225\text{kPa}} - T_{\text{sat}@175\text{kPa}}}\right) \\ &= (120.23 + 273.15\text{K})(0.8857 - 0.001061\text{m}^3/\text{kg})\left(\frac{50\text{kPa}}{(124.00 - 116.06)^\circ\text{C}}\right) \\ &= \mathbf{2191.4 kJ/kg} \end{aligned}$$

The tabulated value of h_{fg} at 200 kPa is **2201.9 kJ/kg**.

11-28 The h_{fg} and s_{fg} of steam at a specified temperature are to be calculated using the Clapeyron equation and to be compared to the tabulated data.

Analysis From the Clapeyron equation,

$$h_{fg} = Tv_{fg}\left(\frac{dP}{dT}\right)_{sat}$$

$$\cong T(v_g - v_f)_{@120°C}\left(\frac{\Delta P}{\Delta T}\right)_{sat,120°C}$$

$$= T(v_g - v_f)_{@120°C}\left(\frac{P_{sat@125°C} - P_{sat@115°C}}{125°C - 115°C}\right)$$

$$= (120 + 273.15\text{K})(0.8919 - 0.00106\text{m}^3/\text{kg})\left(\frac{(232.1 - 169.06)\text{kPa}}{10\text{K}}\right)$$

$$= \mathbf{2207.9 kJ/kg}$$

Also,

$$s_{fg} = \frac{h_{fg}}{T} = \frac{2207.9\text{ kJ/kg}}{(120 + 273.15)\text{ K}} = \mathbf{5.6159\ kJ/kg \cdot K}$$

The tabulated values at 120°C are h_{fg} = **2202.6 kJ/kg** and s_{fg} = **5.6020 kJ/kg· K**.

11-29E [*Also solved by EES on enclosed CD*] The h_{fg} of refrigerant-134a at a specified temperature is to be calculated using the Clapeyron equation and Clapeyron-Clausius equation and to be compared to the tabulated data.

Analysis (*a*) From the Clapeyron equation,

$$h_{fg} = Tv_{fg}\left(\frac{dP}{dT}\right)_{sat}$$

$$\cong T(v_g - v_f)_{@\ 50°F}\left(\frac{\Delta P}{\Delta T}\right)_{sat,\ 50°F}$$

$$= T(v_g - v_f)_{@\ 50°F}\left(\frac{P_{sat@\ 60°F} - P_{sat@\ 40°F}}{60°F - 40°F}\right)$$

$$= (50 + 459.67\text{ R})(0.7871 - 0.01270\ \text{ft}^3/\text{lbm})\left(\frac{(72.092 - 49.738)\text{ psia}}{20\text{ R}}\right)$$

$$= 441.1\ \text{psia}\cdot\text{ft}^3/\text{lbm} = \mathbf{81.64\ Btu/lbm}\ (0.2\%\ \text{error})$$

since 1 Btu = 5.4039 psia· ft^3.

(*b*) From the Clapeyron-Clausius equation,

$$ln\left(\frac{P_2}{P_1}\right)_{sat} \cong \frac{h_{fg}}{R}\left(\frac{1}{T_1} - \frac{1}{T_2}\right)_{sat}$$

$$\ln\left(\frac{72.092\ \text{psia}}{49.738\ \text{psia}}\right) \cong \frac{h_{fg}}{0.01946\ \text{Btu/lbm}\cdot\text{R}}\left(\frac{1}{40 + 459.67\ \text{R}} - \frac{1}{60 + 459.67\ \text{R}}\right)$$

$$h_{fg} = \mathbf{93.8\ Btu/lbm}\ (15.1\%\ \text{error})$$

The tabulated value of h_{fg} at 50°F is **81.46 Btu/lbm**.

General Relations for du, dh, ds, C_v, and C_p

11-30C Yes, through the relation

$$\left(\frac{\partial C_p}{\partial P}\right)_T = -T\left(\frac{\partial^2 v}{\partial T^2}\right)_P$$

11-31 It is to be shown that the enthalpy of an ideal gas is a function of temperature only and that for an incompressible substance it also depends on pressure.

Analysis The change in enthalpy is expressed as

$$dh = C_p dT + \left(v - T\left(\frac{\partial v}{\partial T}\right)_P\right)dP$$

For an ideal gas $v = RT/P$. Then,

$$v - T\left(\frac{\partial v}{\partial T}\right)_P = v - T\left(\frac{R}{P}\right) = v - v = 0$$

Thus,

$$dh = C_p dT$$

To complete the proof we need to show that C_p is not a function of P either. This is done with the help of the relation

$$\left(\frac{\partial C_p}{\partial P}\right)_T = -T\left(\frac{\partial^2 v}{\partial T^2}\right)_P$$

For an ideal gas,

$$\left(\frac{\partial v}{\partial T}\right)_P = \frac{R}{P} \text{ and } \left(\frac{\partial^2 v}{\partial T^2}\right)_P \doteq \left(\frac{\partial (R/P)}{\partial T}\right)_P = 0$$

Thus,

$$\left(\frac{\partial C_p}{\partial P}\right)_T = 0$$

Therefore we conclude that the enthalpy of an ideal gas is a function of temperature only.

For an incompressible substance v = constant and thus $\partial v/\partial T = 0$. Then,

$$dh = C_p dT + v dP$$

Therefore we conclude that the enthalpy of an incompressible substance is a function of temperature and pressure.

11-32 General expressions for Δu, Δh, and Δs for a gas that obeys the van der Waals equation of state for an isothermal process are to be derived.

Analysis (a) For an isothermal process $dT = 0$ and the general relation for Δu reduces to

$$\Delta u = u_2 - u_1 = \int_{T_1}^{T_2} C_v dT + \int_{v_1}^{v_2}\left(T\left(\frac{\partial P}{\partial T}\right)_v - P\right)dv = \int_{v_1}^{v_2}\left(T\left(\frac{\partial P}{\partial T}\right)_v - P\right)dv$$

The van der Waals equation of state can be expressed as

$$P = \frac{RT}{v-b} - \frac{a}{v^2} \longrightarrow \left(\frac{\partial P}{\partial T}\right)_v = \frac{R}{v-b}$$

Thus,

$$T\left(\frac{\partial P}{\partial T}\right)_v - P = \frac{RT}{v-b} - \frac{RT}{v-b} + \frac{a}{v^2} = \frac{a}{v^2}$$

Substituting,

$$\Delta u = \int_{v_1}^{v_2} \frac{a}{v^2} dv = a\left(\frac{1}{v_1} - \frac{1}{v_2}\right)$$

(b) The enthalpy change Δh is related to Δu through the relation

$$\Delta h = \Delta u + P_2 v_2 - P_1 v_1$$

where

$$Pv = \frac{RTv}{v-b} - \frac{a}{v}$$

Thus,

$$P_2 v_2 - P_1 v_1 = RT\left(\frac{v_2}{v_2 - b} - \frac{v_1}{v_1 - b}\right) + a\left(\frac{1}{v_1} - \frac{1}{v_2}\right)$$

Substituting,

$$\Delta h = 2a\left(\frac{1}{v_1} - \frac{1}{v_2}\right) + RT\left(\frac{v_2}{v_2 - b} - \frac{v_1}{v_1 - b}\right)$$

(c) For an isothermal process $dT = 0$ and the general relation for Δs reduces to

$$\Delta s = s_2 - s_1 = \int_{T_1}^{T_2} \frac{C_v}{T} dT + \int_{v_1}^{v_2}\left(\frac{\partial P}{\partial T}\right)_v dv = \int_{v_1}^{v_2}\left(\frac{\partial P}{\partial T}\right)_v dv$$

Substituting $(\partial P/\partial T)_v = R/(v - b)$,

$$\Delta s = \int_{v_1}^{v_2} \frac{R}{v-b} dv = R\ln\frac{v_2 - b}{v_1 - b}$$

11-33 General expressions for Δu, Δh, and Δs for a gas whose equation of state is $P(v\text{-}a) = RT$ for an isothermal process are to be derived.

Analysis *(a)* A relation for Δu is obtained from the general relation

$$\Delta u = u_2 - u_1 = \int_{T_1}^{T_2} C_v dT + \int_{v_1}^{v_2} \left(T\left(\frac{\partial P}{\partial T}\right)_v - P \right) dv$$

The equation of state for the specified gas can be expressed as

$$P = \frac{RT}{v-a} \longrightarrow \left(\frac{\partial P}{\partial T}\right)_v = \frac{R}{v-a}$$

Thus,

$$T\left(\frac{\partial P}{\partial T}\right)_v - P = \frac{RT}{v-a} - P = P - P = 0$$

Substituting, $\qquad \Delta u = \int_{T_1}^{T_2} C_v dT$

(b) A relation for Δh is obtained from the general relation

$$\Delta h = h_2 - h_1 = \int_{T_1}^{T_2} C_P dT + \int_{P_1}^{P_2} \left(v - T\left(\frac{\partial v}{\partial T}\right)_P \right) dP$$

The equation of state for the specified gas can be expressed as

$$v = \frac{RT}{P} + a \longrightarrow \left(\frac{\partial v}{\partial T}\right)_P = \frac{R}{P}$$

Thus,

$$v - T\left(\frac{\partial v}{\partial T}\right)_P = v - T\frac{R}{P} = v - (v-a) = a$$

Substituting,

$$\Delta h = \int_{T_1}^{T_2} C_P dT + \int_{P_1}^{P_2} a dP = \int_{T_1}^{T_2} C_P dT + a(P_2 - P_1)$$

(c) A relation for Δs is obtained from the general relation

$$\Delta s = s_2 - s_1 = \int_{T_1}^{T_2} \frac{C_P}{T} dT - \int_{P_1}^{P_2} \left(\frac{\partial v}{\partial T}\right)_P dP$$

Substituting $(\partial v/\partial T)_P = R/T$,

$$\Delta s = \int_{T_1}^{T_2} \frac{C_P}{T} dT - \int_{P_1}^{P_2} \left(\frac{R}{P}\right)_P dP = \int_{T_1}^{T_2} \frac{C_P}{T} dT - R\ln\frac{P_2}{P_1}$$

For an isothermal process $dT = 0$ and these relations reduce to

$$\Delta u = 0,\ \Delta h = a(P_2 - P_1),\ \text{and}\ \Delta s = -R\ln\frac{P_2}{P_1}$$

11-34 General expressions for $(\partial u/\partial P)_T$ and $(\partial h/\partial v)_T$ in terms of P, v, and T only are to be derived.

Analysis The general relation for du is

$$du = C_v dT + \left(T\left(\frac{\partial P}{\partial T}\right)_v - P\right)dv$$

Differentiating each term in this equation with respect to P at T = constant yields

$$\left(\frac{\partial u}{\partial P}\right)_T = 0 + \left(T\left(\frac{\partial P}{\partial T}\right)_v - P\right)\left(\frac{\partial v}{\partial P}\right)_T = T\left(\frac{\partial P}{\partial T}\right)_v\left(\frac{\partial v}{\partial P}\right)_T - P\left(\frac{\partial v}{\partial P}\right)_T$$

Using the properties P, T, v, the cyclic relation can be expressed as

$$\left(\frac{\partial P}{\partial T}\right)_v\left(\frac{\partial T}{\partial v}\right)_P\left(\frac{\partial v}{\partial P}\right)_T = -1 \longrightarrow \left(\frac{\partial P}{\partial T}\right)_v\left(\frac{\partial v}{\partial P}\right)_T = -\left(\frac{\partial v}{\partial T}\right)_P$$

Substituting, we get

$$\left(\frac{\partial u}{\partial P}\right)_T = -T\left(\frac{\partial v}{\partial T}\right)_P - P\left(\frac{\partial v}{\partial P}\right)_T$$

The general relation for dh is

$$dh = C_P dT + \left(v - T\left(\frac{\partial v}{\partial T}\right)_P\right)dP$$

Differentiating each term in this equation with respect to v at T = constant yields

$$\left(\frac{\partial h}{\partial v}\right)_T = 0 + \left(v - T\left(\frac{\partial v}{\partial T}\right)_P\right)\left(\frac{\partial P}{\partial v}\right)_T = v\left(\frac{\partial P}{\partial v}\right)_T - T\left(\frac{\partial v}{\partial T}\right)_P\left(\frac{\partial P}{\partial v}\right)_T$$

Using the properties v, T, P, the cyclic relation can be expressed as

$$\left(\frac{\partial v}{\partial T}\right)_P\left(\frac{\partial T}{\partial P}\right)_v\left(\frac{\partial P}{\partial v}\right)_T = -1 \longrightarrow \left(\frac{\partial v}{\partial T}\right)_P\left(\frac{\partial P}{\partial v}\right)_T = -\left(\frac{\partial T}{\partial P}\right)_v$$

Substituting, we get

$$\left(\frac{\partial h}{\partial v}\right)_T = v\left(\frac{\partial P}{\partial v}\right)_T + T\left(\frac{\partial T}{\partial P}\right)_v$$

11-35 Expressions for the specific heat difference C_p-C_v for three substances are to be derived.

Analysis The general relation for the specific heat difference C_p - C_v is

$$C_P - C_v = -T\left(\frac{\partial v}{\partial T}\right)_P^2 \left(\frac{\partial P}{\partial v}\right)_T$$

(*a*) For an ideal gas $Pv = RT$. Then.

$$v = \frac{RT}{P} \longrightarrow \left(\frac{\partial v}{\partial T}\right)_P = \frac{R}{P}$$

$$P = \frac{RT}{v} \longrightarrow \left(\frac{\partial P}{\partial v}\right)_T = -\frac{RT}{v^2} = -\frac{P}{v}$$

Substituting.

$$C_P - C_v = -T\left(-\frac{P}{v}\right)^2\left(\frac{R}{P}\right) = \frac{TR}{Pv}R = \mathbf{R}$$

(*b*) For a van der Waals gas $\left(P + \frac{a}{v^2}\right)(v - b) = RT$. Then.

$$T = \frac{1}{R}\left(P + \frac{a}{v^2}\right)(v - b) \longrightarrow \left(\frac{\partial T}{\partial v}\right)_P = \frac{1}{R}\left(-\frac{2a}{v^3}\right)(v - b) + \frac{1}{R}\left(P + \frac{a}{v^2}\right)$$

$$= \frac{2a(b - v)}{Rv^3} + \frac{T}{v - b}$$

Inverting.

$$\left(\frac{\partial v}{\partial T}\right)_P = \frac{1}{\dfrac{2a(b - v)}{Rv^3} + \dfrac{T}{v - b}}$$

Also.

$$P = \frac{RT}{v - b} - \frac{a}{v^2} \longrightarrow \left(\frac{\partial P}{\partial v}\right)_T = -\frac{RT}{(v - b)^2} + \frac{2a}{v^3}$$

Substituting.

$$C_P - C_v = T\left(\frac{1}{\dfrac{2a(b - v)}{Rv^3} + \dfrac{T}{v - b}}\right)^2\left(\frac{RT}{(v - b)^2} - \frac{2a}{v^3}\right)$$

(*c*) For an incompressible substance v = constant and thus $(\partial v/\partial T)_P = 0$. Therefore.

$$C_P - C_v = \mathbf{0}$$

11-36 The specific heat difference C_p-C_v for liquid water at 20 MPa and 60°C is to be estimated.

Analysis The specific heat difference C_P - C_v is given as

$$C_p - C_v = -T\left(\frac{\partial v}{\partial T}\right)_P^2\left(\frac{\partial P}{\partial v}\right)_T$$

Approximating differentials by differences about the specified state,

$$C_p - C_v \cong -T\left(\frac{\Delta v}{\Delta T}\right)_{P=20\text{MPa}}^2\left(\frac{\Delta P}{\Delta v}\right)_{T=60°\text{C}}$$

$$= -(60+273.15\text{ K})\left(\frac{v_{80°\text{C}} - v_{40°\text{C}}}{(80-40)°\text{C}}\right)_{P=20\text{MPa}}^2\left(\frac{(30-10)\text{MPa}}{v_{30\text{MPa}} - v_{10\text{MPa}}}\right)_{T=60°\text{C}}$$

$$= -(333.15\text{K})\left(\frac{(0.0010199-0.0009992)\text{m}^3/\text{kg}}{40\text{K}}\right)^2\left(\frac{20{,}000\text{kPa}}{(0.0010042-0.0010127)\text{m}^3/\text{kg}}\right)$$

$$= 0.2099\text{kPa}\cdot\text{m}^3/\text{kg}\cdot\text{K} = \mathbf{0.2099\ \ kJ/kg\cdot K}$$

11-37E The specific heat difference C_p-C_v for liquid water at 1000 psia and 150°F is to be estimated.

Analysis The specific heat difference C_P - C_v is given as

$$C_p - C_v = -T\left(\frac{\partial v}{\partial T}\right)_P^2\left(\frac{\partial P}{\partial v}\right)_T$$

Approximating differentials by differences about the specified state,

$$C_p - C_v \cong -T\left(\frac{\Delta v}{\Delta T}\right)_{P=1000\text{psia}}^2\left(\frac{\Delta P}{\Delta v}\right)_{T=150°\text{F}}$$

$$= -(150+460)\left(\frac{v_{200°\text{F}} - v_{100°\text{F}}}{(200-100)°\text{F}}\right)_{P=1000\text{psia}}^2\left(\frac{(1500-500)\text{psia}}{v_{1500\text{psia}} - v_{500\text{psia}}}\right)_{T=150°\text{F}}$$

$$= -(610\text{R})\left(\frac{(0.016580-0.016082)\text{ft}^3/\text{lbm}}{100\text{R}}\right)^2\left(\frac{1{,}000\text{psia}}{(0.016268-0.016318)\text{ft}^3/\text{lbm}}\right)$$

$$= 0.30256\text{psia}\cdot\text{ft}^3/\text{lbm}\cdot\text{R} = \mathbf{0.0560Btu/lbm\cdot R}\quad(1\text{Btu} = 5.4039\text{psia}\cdot\text{ft}^3)$$

11-38 Relations for the volume expansivity β and the isothermal compressibility α for an ideal gas and for a gas whose equation of state is $P(v-a) = RT$ are to be obtained.

Analysis The volume expansivity and isothermal compressibility are expressed as

$$\beta = \frac{1}{v}\left(\frac{\partial v}{\partial T}\right)_P \quad \text{and} \quad \alpha = -\frac{1}{v}\left(\frac{\partial v}{\partial P}\right)_T$$

(*a*) For an ideal gas $v = RT/P$. Thus,

$$\left(\frac{\partial v}{\partial T}\right)_P = \frac{R}{P} \longrightarrow \beta = \frac{1}{v}\frac{R}{P} = \mathbf{\frac{1}{T}}$$

$$\left(\frac{\partial v}{\partial P}\right)_T = -\frac{RT}{P^2} = -\frac{v}{P} \longrightarrow \alpha = -\frac{1}{v}\left(-\frac{v}{P}\right) = \mathbf{\frac{1}{P}}$$

(*b*) For a gas whose equation of state is $v = RT/P + a$,

$$\left(\frac{\partial v}{\partial T}\right)_P = \frac{R}{P} \longrightarrow \beta = \frac{1}{v}\frac{R}{P} = \mathbf{\frac{R}{RT + aP}}$$

$$\left(\frac{\partial v}{\partial P}\right)_T = -\frac{RT}{P^2} = -\frac{v-a}{P} \longrightarrow \alpha = -\frac{1}{v}\left(-\frac{v-a}{P}\right) = \mathbf{\frac{v-a}{Pv}}$$

11-39 The volume expansivity β and the isothermal compressibility α of refrigerant-134a at 200 kPa and 30°C are to be estimated.

Analysis The volume expansivity and isothermal compressibility are expressed as

$$\beta = \frac{1}{v}\left(\frac{\partial v}{\partial T}\right)_P \quad \text{and} \quad \alpha = -\frac{1}{v}\left(\frac{\partial v}{\partial P}\right)_T$$

Approximating differentials by differences about the specified state,

$$\beta \cong \frac{1}{v}\left(\frac{\Delta v}{\Delta T}\right)_{P=200\text{kPa}} = \frac{1}{v}\left(\frac{v_{40°\text{C}} - v_{20°\text{C}}}{(40-20)°\text{C}}\right)_{P=200\text{kPa}}$$

$$= \frac{1}{0.11856\text{m}^3/\text{kg}}\left(\frac{(0.12311 - 0.11394)\text{m}^3/\text{kg}}{20\text{K}}\right) = \mathbf{0.00387K^{-1}}$$

and

$$\alpha \cong -\frac{1}{v}\left(\frac{\Delta v}{\Delta P}\right)_{T=30°\text{C}} = -\frac{1}{v}\left(\frac{v_{240\text{kPa}} - v_{180\text{kPa}}}{(240-180)\text{kPa}}\right)_{T=30°\text{C}}$$

$$= -\frac{1}{0.11856\text{m}^3/\text{kg}}\left(\frac{(0.09794 - 0.13230)\text{m}^3/\text{kg}}{60\text{kPa}}\right) = \mathbf{0.00483kPa^{-1}}$$

The Joule-Thomson Coefficient

11-40C It represents the variation of temperature with pressure during a throttling process.

11-41C The line that passes through the peak points of the constant enthalpy lines on a *T-P* diagram is called the inversion line. The maximum inversion temperature is the highest temperature a fluid can be cooled by throttling.

11-42C No. The temperature may even increase as a result of throttling.

11-43C Yes.

11-44C No. Helium is an ideal gas and $h = h(T)$ for ideal gases. Therefore, the temperature of an ideal gas remains constant during a throttling (h = constant) process.

11-45 The equation of state of a gas is given to be $P(v-a) = RT$. It is to be determined if it is possible to cool this gas by throttling.

Analysis The equation of state of this gas can be expressed as

$$v = \frac{RT}{P} + a \longrightarrow \left(\frac{\partial v}{\partial T}\right)_P = \frac{R}{P}$$

Substituting into the Joule-Thomson coefficient relation,

$$\mu = -\frac{1}{C_p}\left(v - T\left(\frac{\partial v}{\partial T}\right)_P\right) = -\frac{1}{C_p}\left(v - T\frac{R}{P}\right) = -\frac{1}{C_p}(v - v + a) = -\frac{a}{C_p} < 0$$

Therefore, this gas **cannot** be cooled by throttling since μ is always a negative quantity.

11-46 Relations for the Joule-Thompson coefficient and the inversion temperature for a gas whose equation of state is $(P+a/v^2)\,v = RT$ are to be obtained.

Analysis The equation of state of this gas can be expressed as

$$T = \frac{v}{R}\left(P + \frac{a}{v^2}\right) \longrightarrow \left(\frac{\partial T}{\partial v}\right)_P = \frac{v}{R}\left(-\frac{2a}{v^3}\right) + \frac{1}{R}\left(P + \frac{a}{v^2}\right) = -\frac{2av}{Rv^2} - \frac{T}{v} = \frac{RTv - 2a}{Rv^2}$$

Substituting into the Joule-Thomson coefficient relation,

$$\mu = -\frac{1}{C_p}\left(v - T\left(\frac{\partial v}{\partial T}\right)_P\right) = -\frac{1}{C_p}\left(v - \frac{Rv^2}{RTv - 2a}\right) = -\frac{2av}{C_p(2a - RTv)}$$

The temperature at $\mu = 0$ is the inversion temperature,

$$\mu = -\frac{2av}{C_p(2a - RTv)} = 0 \quad \longrightarrow \quad v = 0$$

Thus the line of $v = 0$ is the inversion line. Since it is not physically possible to have $v = 0$, this gas does not have an inversion line.

11-47 The Joule-Thompson coefficient of steam at two states is to be estimated.

Analysis (*a*) The enthalpy of steam at 3 MPa and 300°C is h = 2993.5 kJ/kg. Approximating differentials by differences about the specified state, the Joule-Thomson coefficient is expressed as

$$\mu = \left(\frac{\partial T}{\partial P}\right)_h \cong \left(\frac{\Delta T}{\Delta P}\right)_{h=2993.5\text{kJ/kg}}$$

Considering a throttling process from 3.5 MPa to 2.5 MPa at h = 2993.5 kJ/kg, the Joule-Thomson coefficient is determined to be

$$\mu = \left(\frac{T_{3.5\text{ MPa}} - T_{2.5\text{ MPa}}}{(3.5-2.5)\text{ MPa}}\right)_{h=2993.5\text{ kJ/kg}} = \frac{(306.3-294.1)°\text{C}}{(3.5-2.5)\text{ MPa}} = \mathbf{12.2\ °C/MPa}$$

(*b*) The enthalpy of steam at 6 MPa and 500°C is h = 3422.2 kJ/kg. Approximating differentials by differences about the specified state, the Joule-Thomson coefficient is expressed as

$$\mu = \left(\frac{\partial T}{\partial P}\right)_h \cong \left(\frac{\Delta T}{\Delta P}\right)_{h=3422.2\text{kJ/kg}}$$

Considering a throttling process from 7.0 MPa to 5.0 MPa at h = 3422.2 kJ/kg, the Joule-Thomson coefficient is determined to be

$$\mu = \left(\frac{T_{7.0\text{ MPa}} - T_{5.0\text{ MPa}}}{(7.0-5.0)\text{ MPa}}\right)_{h=3422.2\text{ kJ/kg}} = \frac{(504.9-495.1)°\text{C}}{(7.0-5.0)\text{ MPa}} = \mathbf{4.9\ °C/MPa}$$

11-48E [*Also solved by EES on enclosed CD*] The Joule-Thompson coefficient of nitrogen at two states is to be estimated.

Analysis (*a*) The enthalpy of nitrogen at 200 psia and 500 R is, from EES data, h = 185.8 Btu/lbm. Approximating differentials by differences about the specified state, the Joule-Thomson coefficient is expressed as

$$\mu = \left(\frac{\partial T}{\partial P}\right)_h \cong \left(\frac{\Delta T}{\Delta P}\right)_{h=185.8\ \text{Btu/lbm}}$$

Considering a throttling process from 201 psia to 199 psia at h = 185.8 Btu/lbm, the Joule-Thomson coefficient is determined to be

$$\mu = \left(\frac{T_{199\ \text{psia}} - T_{201\ \text{psia}}}{(199-201)\ \text{psia}}\right)_{h=185.8\ \text{Btu/lbm}} = \frac{(499.969 - 500.031)\ \text{R}}{(199-201)\ \text{psia}} = \mathbf{0.03145\ R/psia}$$

(*b*) The enthalpy of nitrogen at 2000 psia and 400 R is h = 140.9 Btu/lbm. Approximating differentials by differences about the specified state, the Joule-Thomson coefficient is expressed as

$$\mu = \left(\frac{\partial T}{\partial P}\right)_h \cong \left(\frac{\Delta T}{\Delta P}\right)_{h=140.9\text{Btu/lbm}}$$

Considering a throttling process from 2001 psia to 1999 psia at h = 140.9 Btu/lbm, the Joule-Thomson coefficient is determined to be

$$\mu = \left(\frac{T_{1999\ \text{psia}} - T_{2001\ \text{psia}}}{(1999-2001)\ \text{psia}}\right)_{h=177.78\ \text{Btu/lbm}} = \frac{(399.978 - 400.022)\ \text{R}}{(1999-2001)\ \text{psia}} = \mathbf{0.022243\ R/psia}$$

11-49E EES solution of this (and other comprehensive problems designated with the *computer icon*) is available to instructors at the *Instructor Manual* section of the *Online Learning Center* (OLC) at www.mhhe.com/cengel-boles. See the Preface for access information.

11-50 The Joule-Thompson coefficient of refrigerant-134a at a specified state is to be estimated.

Analysis The enthalpy of refrigerant-134a at 0.5 MPa and T = 80°C is h = 319.96 kJ/kg. Approximating differentials by differences about the specified state, the Joule-Thomson coefficient is expressed as

$$\mu = \left(\frac{\partial T}{\partial P}\right)_h \cong \left(\frac{\Delta T}{\Delta P}\right)_{h=319.96\text{kJ/kg}}$$

Considering a throttling process from 0.6 MPa to 0.4 MPa at h = 319.96 kJ/kg, the Joule-Thomson coefficient is determined to be

$$\mu = \left(\frac{T_{0.6\text{MPa}} - T_{0.4\text{MPa}}}{(0.6-0.4)\text{MPa}}\right)_{h=319.96\text{kJ/kg}} = \frac{(81.26-78.72)^\circ\text{C}}{(0.6-0.4)\text{MPa}} = \mathbf{12.7^\circ C/MPa}$$

11-51 Steam is throttled slightly from 1 MPa and 300°C. It is to be determined if the temperature of the steam will increase, decrease, or remain the same during this process.

Analysis The enthalpy of steam at 1 MPa and T = 300°C is h = 3051.2 kJ/kg. Now consider a throttling process from this state to 0.8 MPa, which is the next lowest pressure listed in the tables. The temperature of the steam at the end of this throttling process will be

$$\left.\begin{array}{l} P = 0.8\text{MPa} \\ h = 3051.2\text{kJ/kg} \end{array}\right\} T_2 = 297.51^\circ\text{C}$$

Therefore, the temperature will **decrease**.

The Δh, Δu, and Δs of Real Gases

11-52C It is the variation of enthalpy with pressure at a fixed temperature.

11-53C As P_R approaches zero, the gas approaches ideal gas behavior. As a result, the deviation from ideal gas behavior diminishes.

11-54C So that a single chart can be used for all gases instead of a single particular gas.

11-55 The enthalpy of nitrogen at 175 K and 8 MPa is to be determined using data from the ideal-gas nitrogen table and the generalized enthalpy departure chart.

Analysis (*a*) From the ideal gas table of nitrogen (Table A-18) we read

$$h = 5083.75 \text{ kJ/kmol} = \mathbf{181.48 \text{ kJ/kg}} \qquad (M_{N_2} = 28.013 \text{ kg/kmol})$$

N_2
175 K
8 MPa

at the specified temperature. This value involves 44.6% error.

(*b*) The enthalpy departure of nitrogen at the specified state is determined from the generalized chart to be

$$\left.\begin{aligned} T_R &= \frac{T}{T_{cr}} = \frac{175}{126.2} = 1.387 \\ P_R &= \frac{P}{P_{cr}} = \frac{8}{3.39} = 2.360 \end{aligned}\right\} \longrightarrow Z_h = \frac{(\bar{h}_{\text{ideal}} - \bar{h})_{T,P}}{R_u T_{cr}} = 1.6$$

and

Thus,

$$\bar{h} = \bar{h}_{\text{ideal}} - Z_h R_u T_{cr} = 5083.75 - [(1.6)(8.314)(126.2)] = 3405.0 \text{kJ/kmol}$$

or,

$$h = \frac{\bar{h}}{M} = \frac{3405.0 \text{kJ/kmol}}{28.013 \text{kg/kmol}} = \mathbf{121.6 \text{ kJ/kg}} \text{ (3.1\%error)}$$

11-56E The enthalpy of nitrogen at 400 R and 2000 psia is to be determined using data from the ideal-gas nitrogen table and the generalized enthalpy departure chart.

Analysis (*a*) From the ideal gas table of nitrogen (Table A-18E) we read

$$h = 2777.0 \text{ Btu/lbmol} = \mathbf{99.18 \text{ Btu/lbm}} \qquad (M_{N_2} = 28 \text{ lbm/lbmol})$$

at the specified temperature. This value involves 44.2% error.

N$_2$
400 R
2000 psia

(*b*) The enthalpy departure of nitrogen at the specified state is determined from the generalized chart to be

$$\left.\begin{array}{l} T_R = \dfrac{T}{T_{cr}} = \dfrac{400}{227.1} = 1.761 \\ P_R = \dfrac{P}{P_{cr}} = \dfrac{2000}{492} = 4.065 \end{array}\right\} \longrightarrow Z_h = \frac{(\bar{h}_{\text{ideal}} - \bar{h})_{T,P}}{R_u T_{cr}} = 1.18$$

and

Thus,

$$\bar{h} = \bar{h}_{\text{ideal}} - Z_h R_u T_{cr} = 2777.0 - [(1.18)(1.986)(227.1)] = 2244.8 \text{ Btu/lbmol}$$

or,

$$h = \frac{\bar{h}}{M} = \frac{2244.8 \text{ Btu/lbmol}}{28 \text{ lbm/lbmol}} = \mathbf{80.17 \text{ Btu/lbm}} \qquad (54.9\% \text{ error})$$

11-57 The errors involved in the enthalpy and internal energy of CO_2 at 350 K and 10 MPa if it is assumed to be an ideal gas are to be determined.

Analysis (*a*) The enthalpy departure of CO_2 at the specified state is determined from the generalized chart to be

$$\left.\begin{array}{l} T_R = \dfrac{T}{T_{cr}} = \dfrac{350}{304.2} = 1.151 \\ P_R = \dfrac{P}{P_{cr}} = \dfrac{10}{7.39} = 1.353 \end{array}\right\} \longrightarrow Z_h = \frac{(\bar{h}_{\text{ideal}} - \bar{h})_{T,P}}{R_u T_{cr}} = 1.5$$

and

CO$_2$
350 K
10 MPa

Thus,

$$\bar{h} = \bar{h}_{\text{ideal}} - Z_h R_u T_{cr} = 11{,}351 - [(1.5)(8.314)(304.2)] = 7{,}557 \text{ kJ/kmol}$$

and,

$$\text{Error} = \frac{(\bar{h}_{\text{ideal}} - \bar{h})_{T,P}}{\bar{h}} = \frac{11{,}351 - 7{,}557}{7{,}557} = \mathbf{50.2\%}$$

(*b*) At the calculated T_R and P_R the compressibility factor is determined from the compressibility chart to be Z = 0.65. Then using the definition of enthalpy, the internal energy is determined to be

$$\bar{u} = \bar{h} - P\bar{v} = \bar{h} - ZR_u T = 7557 - [(0.65)(8.314)(350)] = 5{,}666 \text{ kJ/kmol}$$

and,

$$\text{Error} = \frac{\bar{u}_{\text{ideal}} - \bar{u}}{\bar{u}} = \frac{8{,}439 - 5{,}666}{5{,}666} = \mathbf{48.9\%}$$

11-58 The enthalpy and entropy changes of nitrogen during a process are to be determined assuming ideal gas behavior and using generalized charts.

Analysis (*a*) Using data from the ideal gas property table of nitrogen (Table A-18),

$$(\bar{h}_2 - \bar{h}_1)_{\text{ideal}} = \bar{h}_{2,\text{ideal}} - \bar{h}_{1,\text{ideal}} = 9306 - 6{,}537 = \mathbf{2769\ kJ/kmol}$$

and

$$(\bar{s}_2 - \bar{s}_1)_{\text{ideal}} = \bar{s}_2^\circ - \bar{s}_1^\circ - R_u \ln\frac{P_2}{P_1} = 193.562 - 183.289 - 8.314 \times \ln\frac{12}{6} = \mathbf{4.510\ kJ/kmol \cdot K}$$

(*b*) The enthalpy and entropy departures of nitrogen at the specified states are determined from the generalized charts to be

$$\left.\begin{aligned} T_{R1} &= \frac{T_1}{T_{cr}} = \frac{225}{126.2} = 1.783 \\ P_{R1} &= \frac{P_1}{P_{cr}} = \frac{6}{3.39} = 1.770 \end{aligned}\right\} \longrightarrow Z_{h1} = 0.6 \text{ and } Z_{s1} = 0.25$$

and

$$\left.\begin{aligned} T_{R2} &= \frac{T_2}{T_{cr}} = \frac{320}{126.2} = 2.536 \\ P_{R2} &= \frac{P_2}{P_{cr}} = \frac{12}{3.39} = 2.540 \end{aligned}\right\} \longrightarrow Z_{h2} = 0.4 \text{ and } Z_{s2} = 0.15$$

Substituting,

$$\begin{aligned} \bar{h}_2 - \bar{h}_1 &= R_u T_{cr}(Z_{h1} - Z_{h2}) + (\bar{h}_2 - \bar{h}_1)_{\text{ideal}} \\ &= (8.314)(126.2)(0.6 - 0.4) + 2769 = \mathbf{2979\ kJ/kmol} \end{aligned}$$

$$\begin{aligned} \bar{s}_2 - \bar{s}_1 &= R_u(Z_{s1} - Z_{s2}) + (\bar{s}_2 - \bar{s}_1)_{\text{ideal}} \\ &= (8.314)(0.25 - 0.15) + 4.510 = \mathbf{5.341\ kJ/kmol \cdot K} \end{aligned}$$

11-59 The enthalpy and entropy changes of CO_2 during a process are to be determined assuming ideal gas behavior and using generalized charts.

Analysis (*a*) Using data from the ideal gas property table of CO_2 (Table A-20),

$$(\bar{h}_2 - \bar{h}_1)_{\text{ideal}} = \bar{h}_{2,\text{ideal}} - \bar{h}_{1,\text{ideal}} = 8,697 - 7,627 = 1,070\ \text{kJ/kmol}$$

$$(\bar{s}_2 - \bar{s}_1)_{\text{ideal}} = \bar{s}_2^\circ - \bar{s}_1^\circ - R_u \ln\frac{P_2}{P_1} = 211.376 - 207.337 - 8.314 \times \ln\frac{12}{7} = -0.442\ \text{kJ/kmol}\cdot\text{K}$$

Thus,

$$(h_2 - h_1)_{\text{ideal}} = \frac{(\bar{h}_2 - \bar{h}_1)_{\text{ideal}}}{M} = \frac{1,070\ \text{kJ/kmol}}{44\ \text{kg/kmol}} = \mathbf{24.32\ kJ/kg}$$

$$(s_2 - s_1)_{\text{ideal}} = \frac{(\bar{s}_2 - \bar{s}_1)_{\text{ideal}}}{M} = \frac{-0.442\ \text{kJ/kmol}}{44\ \text{kg/kmol}} = \mathbf{-0.0100\ kJ/kg\cdot K}$$

(*b*) The enthalpy and entropy departures of CO_2 at the specified states are determined from the generalized charts to be

250 K, 7 MPa → CO_2 → 280 K, 12 MPa

$$\left.\begin{aligned} T_{R1} &= \frac{T_1}{T_{cr}} = \frac{250}{304.2} = 0.822 \\ P_{R1} &= \frac{P_1}{P_{cr}} = \frac{7}{7.39} = 0.947 \end{aligned}\right\} \longrightarrow Z_{h1} = 5.5 \text{ and } Z_{s1} = 5.3$$

and

$$\left.\begin{aligned} T_{R2} &= \frac{T_2}{T_{cr}} = \frac{280}{304.2} = 0.920 \\ P_{R2} &= \frac{P_2}{P_{cr}} = \frac{12}{7.39} = 1.624 \end{aligned}\right\} \longrightarrow Z_{h2} = 5.0 \text{ and } Z_{s2} = 4.2$$

Thus,

$$h_2 - h_1 = RT_{cr}(Z_{h1} - Z_{h2}) + (h_2 - h_1)_{\text{ideal}} = (0.1889)(304.2)(5.5 - 5.0) + 24.32 = \mathbf{53.05 kJ/kg}$$

$$s_2 - s_1 = R(Z_{s1} - Z_{s2}) + (s_2 - s_1)_{\text{ideal}} = (0.1889)(5.3 - 4.2) - 0.010 = \mathbf{0.198 kJ/kg\cdot K}$$

11-60 Methane is compressed adiabatically by a steady-flow compressor. The required power input to the compressor is to be determined using the generalized charts.

Assumptions **1** Steady operating conditions exist. **2** Kinetic and potential energy changes are negligible.

Analysis The steady-flow energy balance equation for this compressor can be expressed as

$$\dot{E}_{in} - \dot{E}_{out} = \Delta\dot{E}_{system}^{\nearrow 0\text{ (steady)}} = 0$$

$$\dot{E}_{in} = \dot{E}_{out}$$

$$\dot{W}_{C,in} + \dot{m}h_1 = \dot{m}h_2$$

$$\dot{W}_{C,in} = \dot{m}(h_2 - h_1)$$

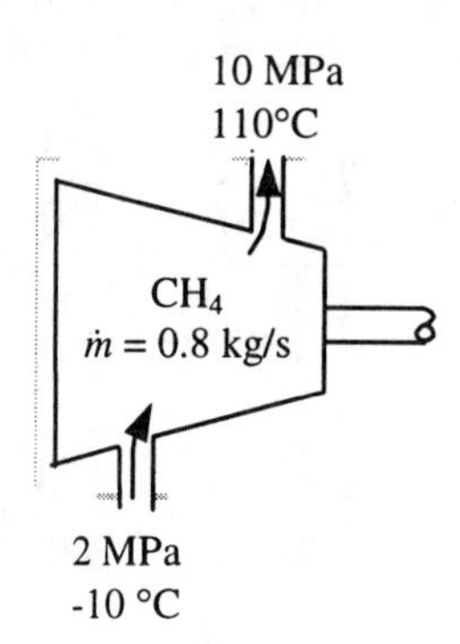

The enthalpy departures of CH_4 at the specified states are determined from the generalized charts to be

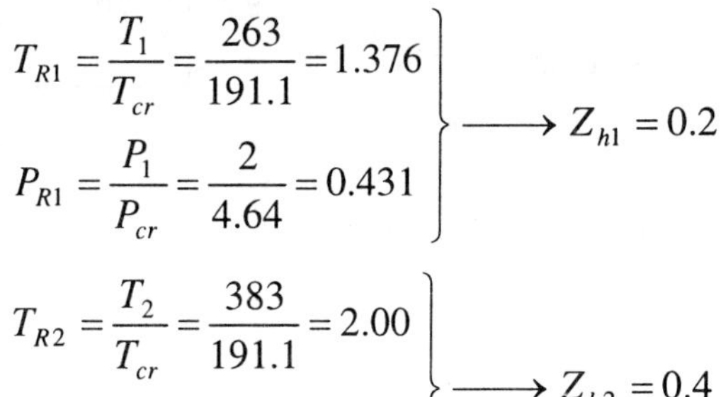

$$\left.\begin{aligned} T_{R1} &= \frac{T_1}{T_{cr}} = \frac{263}{191.1} = 1.376 \\ P_{R1} &= \frac{P_1}{P_{cr}} = \frac{2}{4.64} = 0.431 \end{aligned}\right\} \longrightarrow Z_{h1} = 0.2$$

and

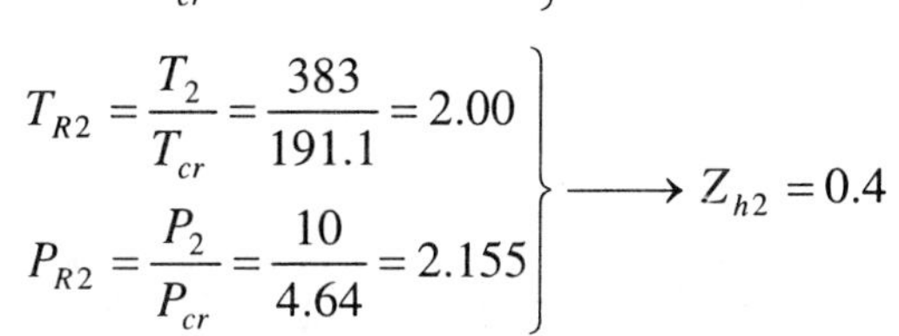

$$\left.\begin{aligned} T_{R2} &= \frac{T_2}{T_{cr}} = \frac{383}{191.1} = 2.00 \\ P_{R2} &= \frac{P_2}{P_{cr}} = \frac{10}{4.64} = 2.155 \end{aligned}\right\} \longrightarrow Z_{h2} = 0.4$$

Thus,

$$\begin{aligned} h_2 - h_1 &= RT_{cr}(Z_{h1} - Z_{h2}) + (h_2 - h_1)_{\text{ideal}} \\ &= (0.5182)(191.1)(0.2 - 0.4) + 2.2537(110 - (-10)) = 250.6\text{kJ/kg} \end{aligned}$$

Substituting,

$$\dot{W}_{C,in} = (0.8\text{kg/s})(250.6\text{kJ/kg}) = \mathbf{201\ kW}$$

11-61 [*Also solved by EES on enclosed CD*] Propane is compressed isothermally by a piston-cylinder device. The work done and the heat transfer are to be determined using the generalized charts.

Assumptions **1**The compression process is quasi-equilibrium. **2** Kinetic and potential energy changes are negligible.

Analysis (*a*) The enthalpy departure and the compressibility factors of propane at the initial and the final states are determined from the generalized charts to be

$$\left.\begin{aligned} T_{R1} &= \frac{T_1}{T_{cr}} = \frac{373}{370} = 1.008 \\ P_{R1} &= \frac{P_1}{P_{cr}} = \frac{1}{4.26} = 0.235 \end{aligned}\right\} \longrightarrow Z_{h1} = 0.28 \text{ and } Z_1 = 0.92$$

Propane
1 MPa
100 °C
Q

and

$$\left.\begin{aligned} T_{R2} &= \frac{T_2}{T_{cr}} = \frac{373}{370} = 1.008 \\ P_{R2} &= \frac{P_2}{P_{cr}} = \frac{4}{4.26} = 0.939 \end{aligned}\right\} \longrightarrow Z_{h2} = 1.8 \text{ and } Z_2 = 0.50$$

Treating propane as a real gas with $Z_{ave} = (Z_1+Z_2)/2 = (0.92 + 0.50)/2 = 0.71$,

$$Pv = ZRT \cong Z_{\text{ave}} RT = C = \text{constant}$$

Then the boundary work becomes

$$\begin{aligned} w_{b,\text{in}} &= -\int_1^2 P dv = -\int_1^2 \frac{C}{v} dv = -C \ln\frac{v_2}{v_1} = Z_{\text{ave}} RT \ln\frac{Z_2 RT / P_2}{Z_1 RT / P_1} = -Z_{\text{ave}} RT \ln\frac{Z_2 P_1}{Z_1 P_2} \\ &= -(0.71)(0.1885 \text{kJ/kg} \cdot \text{K})(373 \text{K}) \ln\frac{(0.50)(1)}{(0.92)(4)} = \mathbf{99.6 kJ/kg} \end{aligned}$$

Also,

$$h_2 - h_1 = RT_{cr}(Z_{h1} - Z_{h2}) + (h_2 - h_1)_{\text{ideal}} = (0.1885)(370)(0.28 - 1.8) + 0 = -106 \text{kJ/kg}$$

$$u_2 - u_1 = (h_2 - h_1) - R(Z_2 T_2 - Z_1 T_1) = -106 - (0.1885)[(0.5)(373) - (0.92)(373)] = -76.5 \text{kJ/kg}$$

Then the heat transfer for this process is determined from the closed system energy balance to be

$$E_{\text{in}} - E_{\text{out}} = \Delta E_{\text{system}}$$

$$q_{\text{in}} + w_{b,\text{in}} = \Delta u = u_2 - u_1$$

$$q_{\text{in}} = (u_2 - u_1) - w_{b,\text{in}} = -76.5 - 99.6 = -176.1 \text{kJ/kg} \rightarrow q_{\text{out}} = \mathbf{176.1 kJ/kg}$$

11-62 EES solution of this (and other comprehensive problems designated with the *computer icon*) is available to instructors at the *Instructor Manual* section of the *Online Learning Center* (OLC) at www.mhhe.com/cengel-boles. See the Preface for access information.

11-63E Propane is compressed isothermally by a piston-cylinder device. The work done and the heat transfer are to be determined using the generalized charts.

Assumptions **1**The compression process is quasi-equilibrium. **2** Kinetic and potential energy changes are negligible. **3** The device is well-insulated and thus heat transfer is negligible

Analysis (*a*) The enthalpy departure and the compressibility factors of propane at the initial and the final states are determined from the generalized charts to be

$$\left.\begin{aligned} T_{R1} &= \frac{T_1}{T_{cr}} = \frac{660}{665.9} = 0.991 \\ P_{R1} &= \frac{P_1}{P_{cr}} = \frac{200}{617} = 0.324 \end{aligned}\right\} \longrightarrow Z_{h1} = 0.37 \text{ and } Z_1 = 0.88$$

and

$$\left.\begin{aligned} T_{R2} &= \frac{T_2}{T_{cr}} = \frac{660}{665.9} = 0.991 \\ P_{R2} &= \frac{P_2}{P_{cr}} = \frac{800}{617} = 1.297 \end{aligned}\right\} \longrightarrow Z_{h2} = 4.2 \text{ and } Z_2 = 0.22$$

Propane
200 psia
200 °F
Q

Treating propane as a real gas with $Z_{ave} = (Z_1+Z_2)/2 = (0.88 + 0.22)/2 = 0.55$,

$$Pv = ZRT \cong Z_{ave}RT = C = \text{constant}$$

Then the boundary work becomes

$$w_{b,\text{in}} = -\int_1^2 P dv = -\int_1^2 \frac{C}{v} dv = -C\ln\frac{v_2}{v_1} = -Z_{ave}RT\ln\frac{Z_2RT/P_2}{Z_1RT/P_1} = -Z_{ave}RT\ln\frac{Z_2P_1}{Z_1P_2}$$

$$= -(0.55)(0.04504\text{Btu/lbm}\cdot\text{R})(660\text{R})\ln\frac{(0.22)(200)}{(0.88)(800)} = \mathbf{45.3Btu/lbm}$$

Also,

$$h_2 - h_1 = RT_{cr}(Z_{h1} - Z_{h2}) + (h_2 - h_1)_{\text{ideal}}^{\nearrow 0}$$
$$= (0.04504)(665.9)(0.37 - 4.2) + 0 = -114.9\text{Btu/lbm}$$

$$u_2 - u_1 = (h_2 - h_1) - R(Z_2T_2 - Z_1T_1)$$
$$= -114.9 - (0.04504)[(0.22)(660) - (0.88)(660)] = -95.3\text{Btu/lbm}$$

Then the heat transfer for this process is determined from the closed system energy balance equation to be

$$E_{\text{in}} - E_{\text{out}} = \Delta E_{\text{system}}$$

$$q_{\text{in}} + w_{b,\text{in}} = \Delta u = u_2 - u_1$$

$$q_{\text{in}} = (u_2 - u_1) - w_{b,\text{in}} = -95.3 - 45.3 = -140.6\text{Btu/lbm} \rightarrow q_{\text{out}} = \mathbf{140.6Btu/lbm}$$

11-64 Propane is compressed isothermally by a piston-cylinder device. The exergy destruction associated with this process is to be determined.

Assumptions **1**The compression process is quasi-equilibrium. **2** Kinetic and potential energy changes are negligible.

Properties The gas constant of propane is $R = 0.1885$ kJ/kg.K (Table A-1).

Analysis The exergy destruction is determined from its definition $x_{\text{destroyed}} = T_0 s_{\text{gen}}$ where the entropy generation is determined from an entropy balance on the contents of the cylinder. It gives

$$S_{in} - S_{out} + S_{gen} = \Delta S_{system}$$

$$-\frac{Q_{out}}{T_{b,surr}} + S_{gen} = m(s_2 - s_1) \quad \rightarrow \quad s_{gen} = (s_2 - s_1) + \frac{q_{out}}{T_{surr}}$$

where

$$\Delta s_{\text{sys}} = s_2 - s_1 = R(Z_{s1} - Z_{s2}) + (s_2 - s_1)_{\text{ideal}}$$

$$(s_2 - s_1)_{\text{ideal}} = C_p \ln\frac{T_2}{T_1}^{\nearrow 0} - R\ln\frac{P_2}{P_1} = 0 - 0.1885\ln\frac{4}{1} = -0.261 \text{ kJ/kg}\cdot\text{K}$$

$$\left.\begin{aligned} T_{R1} &= \frac{T_1}{T_{cr}} = \frac{373}{370} = 1.008 \\ P_{R1} &= \frac{P_1}{P_{cr}} = \frac{1}{4.26} = 0.235 \end{aligned}\right\} \longrightarrow Z_{s1} = 0.21$$

and

$$\left.\begin{aligned} T_{R2} &= \frac{T_2}{T_{cr}} = \frac{373}{370} = 1.008 \\ P_{R2} &= \frac{P_2}{P_{cr}} = \frac{4}{4.26} = 0.939 \end{aligned}\right\} \longrightarrow Z_{s2} = 1.5$$

Thus,

$$\Delta s_{\text{sys}} = s_2 - s_1 = (0.1885)(0.21 - 1.5) - 0.261 = -0.504 \text{kJ/kg}\cdot\text{K}$$

and

$$x_{\text{destroyed}} = T_0 s_{\text{gen}} = T_0\left((s_2 - s_1) + \frac{q_{\text{out}}}{T_{\text{surr}}}\right) = (303\text{K})\left(-0.504 + \frac{176.1\text{kJ/kg}}{303\text{K}}\right)\text{kJ/kg}\cdot\text{K} = \mathbf{23.4 kJ/kg}$$

11-65 Carbon dioxide passes through an adiabatic nozzle. The exit velocity is to be determined using the generalized enthalpy departure chart.

Assumptions **1**Steady operating conditions exist. **2** Kinetic and potential energy changes are negligible. **3** The nozzle is adiabatic and thus heat transfer is negligible

Properties The gas constant of CO_2 is 0.1889 kJ/kg.K (Table A-1).

Analysis The steady-flow energy balance equation for this nozzle can be expressed as

$$\dot{E}_{in} - \dot{E}_{out} = \Delta\dot{E}_{system}^{\;\nearrow 0\,(steady)} = 0$$

$$\dot{E}_{in} = \dot{E}_{out}$$

$$h_1 + (\mathbf{V}_1^2/2)^{\nearrow 0} = h_2 + (\mathbf{V}_2^2/2)$$

$$\mathbf{V}_2 = \sqrt{2(h_1 - h_2)}$$

P_1 = 8 MPa
T_1 = 450 K
CO_2
P_2 = 2 MPa
T_2 = 350 K

The enthalpy departures of CO_2 at the specified states are determined from the generalized enthalpy departure chart to be

$$\left.\begin{aligned} T_{R1} &= \frac{T_1}{T_{cr}} = \frac{450}{304.2} = 1.48 \\ P_{R1} &= \frac{P_1}{P_{cr}} = \frac{8}{7.39} = 1.08 \end{aligned}\right\} \longrightarrow Z_{h1} = 0.55$$

and

$$\left.\begin{aligned} T_{R2} &= \frac{T_2}{T_{cr}} = \frac{350}{304.2} = 1.15 \\ P_{R2} &= \frac{P_2}{P_{cr}} = \frac{2}{7.39} = 0.27 \end{aligned}\right\} \longrightarrow Z_{h2} = 0.20$$

Thus,

$$\begin{aligned} h_2 - h_1 &= RT_{cr}(Z_{h1} - Z_{h2}) + (h_2 - h_1)_{ideal} \\ &= (0.1889)(304.2)(0.55 - 0.2) + (11{,}351 - 15{,}483)/44 = -73.8 \text{ kJ/kg} \end{aligned}$$

Substituting,

$$\mathbf{V}_2 = \sqrt{2(73.8 \text{ kJ/kg})\left(\frac{1000 \text{ m}^2/\text{s}^2}{1 \text{ kJ/kg}}\right)} = \mathbf{384\ m/s}$$

11-66 EES solution of this (and other comprehensive problems designated with the *computer icon*) is available to instructors at the *Instructor Manual* section of the *Online Learning Center* (OLC) at www.mhhe.com/cengel-boles. See the Preface for access information.

11-67 A paddle-wheel placed in a well-insulated rigid tank containing oxygen is turned on. The final pressure in the tank and the paddle-wheel work done during this process are to be determined.

Assumptions **1**The tank is well-insulated and thus heat transfer is negligible. **2** Kinetic and potential energy changes are negligible.

Properties The gas constant of O_2 is R = 0.2598 kJ/kg.K (Table A-1).

Analysis (*a*) The compressibility factor of oxygen at the initial state is determined from the generalized chart to be

$$\left.\begin{aligned} T_{R1} &= \frac{T_1}{T_{cr}} = \frac{220}{154.8} = 1.42 \\ P_{R1} &= \frac{P_1}{P_{cr}} = \frac{10}{5.08} = 1.97 \end{aligned}\right\} \longrightarrow Z_1 = 0.80 \text{ and } Z_{h1} = 1.15$$

O_2

220 K
10 MPa

Then,

$$Pv = ZRT \longrightarrow v_1 = \frac{ZRT_1}{P_1} = \frac{(0.80)(0.2598 \text{ kPa}\cdot\text{m}^3/\text{kg}\cdot\text{K})(220\text{ K})}{10{,}000 \text{ kPa}} = 0.00457 \text{ m}^3/\text{kg}$$

$$m = \frac{V}{v_1} = \frac{0.15 \text{ m}^3}{0.00457 \text{ m}^3/\text{kg}} = 32.8 \text{ kg}$$

The specific volume of oxygen remains constant during this process, $v_2 = v_1$. Thus,

$$\left.\begin{aligned} T_{R2} &= \frac{T_2}{T_{cr}} = \frac{250}{154.8} = 1.615 \\ v_{R2} &= \frac{v_2}{RT_{cr}/P_{cr}} = \frac{0.00457\text{m}^3\text{/kg}}{(0.2598\text{kPa}\cdot\text{m}^3\text{/kg}\cdot\text{K})(154.8\text{K})/(5080\text{kPa})} = 0.577 \end{aligned}\right\} \begin{aligned} Z_2 &= 0.87 \\ Z_{h2} &= 1.0 \\ P_{R2} &= 2.4 \end{aligned}$$

$$P_2 = P_{R2}P_{cr} = (2.4)(5080) = \mathbf{12{,}190 kPa}$$

(*b*) The energy balance relation for this closed system can be expressed as

$$E_{\text{in}} - E_{\text{out}} = \Delta E_{\text{system}}$$

$$W_{\text{in}} = \Delta U = m(u_2 - u_1)$$

$$W_{\text{in}} = m[h_2 - h_1 - (P_2v_2 - P_1v_1)] = m[h_2 - h_1 - R(Z_2T_2 - Z_1T_1)]$$

where $h_2 - h_1 = RT_{cr}(Z_{h1} - Z_{h2}) + (h_2 - h_1)_{\text{ideal}}$
$= (0.2598)(154.8)(1.15 - 1) + (7275 - 6404)/32 = 33.25 \text{ kJ/kg}$

Substituting,

$$W_{\text{in}} = (32.8\text{kg})[33.25 - (0.2598\text{kJ/kg}\cdot\text{K})\{(0.87)(250) - (0.80)(220)\}\text{K}] = \mathbf{737 kJ}$$

Review Problems

11-68 For $\beta \geq 0$, it is to be shown that at every point of a single-phase region of an *h-s* diagram, the slope of a constant-pressure line is greater than the slope of a constant-temperature line, but less than the slope of a constant-volume line.

Analysis It is given that $\beta > 0$.

Using the Tds relation: $dh = T\,ds + v\,dP \longrightarrow \frac{dh}{ds} = T + \frac{dP}{ds}$

(1) P = constant: $\left(\frac{\partial h}{\partial s}\right)_P = T$

(2) T = constant: $\left(\frac{\partial h}{\partial s}\right)_T = T + v\left(\frac{\partial P}{\partial s}\right)_T$

But the 4th Maxwell relation: $\left(\frac{\partial P}{\partial s}\right)_T = -\left(\frac{\partial T}{\partial v}\right)_P$

Substituting: $\left(\frac{\partial h}{\partial s}\right)_T = T - v\left(\frac{\partial T}{\partial v}\right)_P = T - \frac{1}{\beta}$ (Eq.11-47)

Therefore, the slope of P = constant lines is **greater** than the slope of T = constant lines.

(3) v = constant: $\left(\frac{\partial h}{\partial s}\right)_v = T + v\left(\frac{\partial P}{\partial s}\right)_v$ (*a*)

From the ds relation: $ds = \frac{C_v}{T}dT + \left(\frac{\partial P}{\partial T}\right)_v dv$

Divide by dP holding v constant: $\left(\frac{\partial s}{\partial P}\right)_v = \frac{C_v}{T}\left(\frac{\partial T}{\partial P}\right)_v$ or $\left(\frac{\partial P}{\partial s}\right)_v = \frac{T}{C_v}\left(\frac{\partial P}{\partial T}\right)_v$ (*b*)

Using the properties P, s, v, the cyclic relation can be expressed as

$$\left(\frac{\partial P}{\partial T}\right)_v\left(\frac{\partial T}{\partial v}\right)_P\left(\frac{\partial v}{\partial P}\right)_T = -1 \longrightarrow \left(\frac{\partial P}{\partial T}\right)_v = -\left(\frac{\partial v}{\partial T}\right)_P\left(\frac{\partial P}{\partial v}\right)_T = (-\beta v)\left(\frac{1}{-\alpha v}\right) = \frac{\beta}{\alpha} \quad (c)$$

where we used the definitions of α and β. Substituting (*b*) and (*c*) into (*a*),

$$\left(\frac{\partial h}{\partial s}\right)_v = T + v\left(\frac{\partial P}{\partial s}\right)_v = T + \frac{T\beta v}{C_v\alpha} > T$$

Here α is positive for all phases of all substances. T is the absolute temperature that is also positive, so is C_v. Therefore, the second term on the right is always a positive quantity since β is given to be positive. Then we conclude that the slope of P = constant lines is **less** than the slope of v = constant lines.

11-69 Using the cyclic relation and the first Maxwell relation, the other three Maxwell relations are to be obtained.

Analysis (1) Using the properties *P*, *s*, *v*, the cyclic relation can be expressed as

$$\left(\frac{\partial P}{\partial s}\right)_v\left(\frac{\partial s}{\partial v}\right)_P\left(\frac{\partial v}{\partial P}\right)_T=-1$$

Substituting the first Maxwell relation, $\left(\frac{\partial T}{\partial v}\right)_s=-\left(\frac{\partial P}{\partial s}\right)_v$,

$$-\left(\frac{\partial T}{\partial v}\right)_s\left(\frac{\partial s}{\partial v}\right)_P\left(\frac{\partial v}{\partial P}\right)_s=-1\longrightarrow\left(\frac{\partial T}{\partial P}\right)_s\left(\frac{\partial s}{\partial v}\right)_P=1\longrightarrow\left(\frac{\partial T}{\partial P}\right)_s=\left(\frac{\partial v}{\partial s}\right)_P$$

(2) Using the properties *T*, *v*, *s*, the cyclic relation can be expressed as

$$\left(\frac{\partial T}{\partial v}\right)_s\left(\frac{\partial v}{\partial s}\right)_T\left(\frac{\partial s}{\partial T}\right)_v=-1$$

Substituting the first Maxwell relation, $\left(\frac{\partial T}{\partial v}\right)_s=-\left(\frac{\partial P}{\partial s}\right)_v$,

$$-\left(\frac{\partial P}{\partial s}\right)_v\left(\frac{\partial v}{\partial s}\right)_T\left(\frac{\partial s}{\partial T}\right)_v=-1\longrightarrow\left(\frac{\partial P}{\partial T}\right)_v\left(\frac{\partial v}{\partial s}\right)_T=1\longrightarrow\left(\frac{\partial s}{\partial v}\right)_T=\left(\frac{\partial P}{\partial T}\right)_v$$

(3) Using the properties *P*, *T*, *v*, the cyclic relation can be expressed as

$$\left(\frac{\partial P}{\partial T}\right)_v\left(\frac{\partial T}{\partial v}\right)_P\left(\frac{\partial v}{\partial P}\right)_T=-1$$

Substituting the third Maxwell relation, $\left(\frac{\partial s}{\partial v}\right)_T=\left(\frac{\partial P}{\partial T}\right)_v$,

$$\left(\frac{\partial s}{\partial v}\right)_T\left(\frac{\partial T}{\partial v}\right)_P\left(\frac{\partial v}{\partial P}\right)_T=-1\longrightarrow\left(\frac{\partial s}{\partial P}\right)_T\left(\frac{\partial T}{\partial v}\right)_P=-1\longrightarrow\left(\frac{\partial s}{\partial P}\right)_T=-\left(\frac{\partial v}{\partial T}\right)_P$$

11-70 It is to be shown that the slope of a constant-pressure line on an *h-s* diagram is constant in the saturation region and increases with temperature in the superheated region.

Analysis For P = constant, $dP = 0$ and the above relation reduces to $dh = Tds$, which can also be expressed as

$$\left(\frac{\partial h}{\partial s}\right)_P=T$$

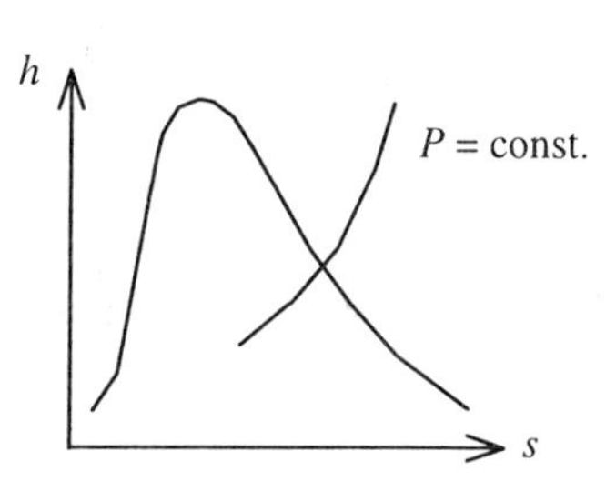

Thus the slope of the P = constant lines on an *h-s* diagram is equal to the temperature.

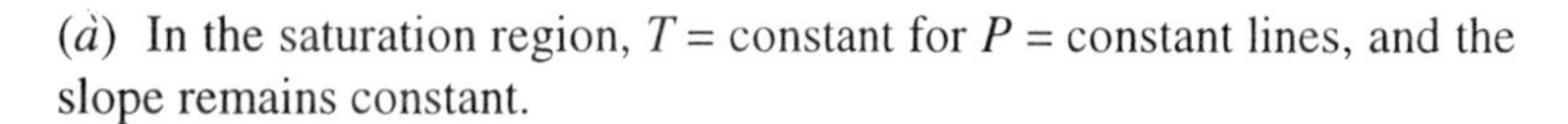

(*a*) In the saturation region, T = constant for P = constant lines, and the slope remains constant.

(*b*) In the superheat region, the slope increases with increasing temperature since the slope is equal temperature.

11-71 The relations for Δu, Δh, and Δs of a gas that obeys the equation of state $(P+a/v^2)v = RT$ for an isothermal process are to be derived.

Analysis (*a*) For an isothermal process $dT = 0$ and the general relation for Δu reduces to

$$\Delta u = u_2 - u_1 = \int_{T_1}^{T_2} C_v dT + \int_{v_1}^{v_2} \left(T\left(\frac{\partial P}{\partial T}\right)_v - P \right) dv = \int_{v_1}^{v_2} \left(T\left(\frac{\partial P}{\partial T}\right)_v - P \right) dv$$

For this gas the equation of state can be expressed as

$$P = \frac{RT}{v} - \frac{a}{v^2} \longrightarrow \left(\frac{\partial P}{\partial T}\right)_v = \frac{R}{v}$$

Thus,

$$T\left(\frac{\partial P}{\partial T}\right)_v - P = \frac{RT}{v} - \frac{RT}{v} + \frac{a}{v^2} = \frac{a}{v^2}$$

Substituting,

$$\Delta u = \int_{v_1}^{v_2} \frac{a}{v^2} dv = a\left(\frac{1}{v_1} - \frac{1}{v_2}\right)$$

(*b*) The enthalpy change Δh is related to Δu through the relation

$$\Delta h = \Delta u + P_2 v_2 - P_1 v_1$$

where

$$Pv = RT - \frac{a}{v}$$

Thus,

$$P_2 v_2 - P_1 v_1 = \left(RT - \frac{a}{v_2} \right) - \left(RT - \frac{a}{v_1} \right) = a\left(\frac{1}{v_1} - \frac{1}{v_2}\right)$$

Substituting,

$$\Delta h = 2a\left(\frac{1}{v_1} - \frac{1}{v_2}\right)$$

(*c*) For an isothermal process $dT = 0$ and the general relation for Δs reduces to

$$\Delta s = s_2 - s_1 = \int_{T_1}^{T_2} \frac{C_v}{T} dT + \int_{v_1}^{v_2} \left(\frac{\partial P}{\partial T}\right)_v dv = \int_{v_1}^{v_2} \left(\frac{\partial P}{\partial T}\right)_v dv$$

Substituting $(\partial P/\partial T)_v = R/v$,

$$\Delta s = \int_{v_1}^{v_2} \frac{R}{v} dv = R\ln\frac{v_2}{v_1}$$

11-72 It is to be shown that

$$C_v = -T\left(\frac{\partial v}{\partial T}\right)_s \left(\frac{\partial P}{\partial T}\right)_v \quad \text{and} \quad C_p = T\left(\frac{\partial P}{\partial T}\right)_s \left(\frac{\partial v}{\partial T}\right)_P$$

Analysis Using the definition of C_v ,

$$C_v = T\left(\frac{\partial s}{\partial T}\right)_v = T\left(\frac{\partial s}{\partial P}\right)_v \left(\frac{\partial P}{\partial T}\right)_v$$

Substituting the first Maxwell relation $\left(\frac{\partial s}{\partial P}\right)_v = -\left(\frac{\partial v}{\partial T}\right)_s$,

$$C_v = -T\left(\frac{\partial v}{\partial T}\right)_s \left(\frac{\partial P}{\partial T}\right)_v$$

Using the definition of C_p,

$$C_p = T\left(\frac{\partial s}{\partial T}\right)_P = T\left(\frac{\partial s}{\partial v}\right)_P \left(\frac{\partial v}{\partial T}\right)_P$$

Substituting the second Maxwell relation $\left(\frac{\partial s}{\partial v}\right)_P = \left(\frac{\partial P}{\partial T}\right)_s$,

$$C_p = T\left(\frac{\partial P}{\partial T}\right)_s \left(\frac{\partial v}{\partial T}\right)_P$$

11-73 The C_p of nitrogen at 300 kPa and 400 K is to be estimated using the relation given and its definition, and the results are to be compared to the value listed in Table A-2*b*.

Analysis (*a*) We treat nitrogen as an ideal gas with R = 0.297 kJ/kg· K and k = 1.397. Note that $PT^{-k/(k-1)}$ = C = constant for the isentropic processes of ideal gases. The C_p relation is given as

$$C_p = T\left(\frac{\partial P}{\partial T}\right)_s \left(\frac{\partial v}{\partial T}\right)_P$$

$$v = \frac{RT}{P} \longrightarrow \left(\frac{\partial v}{\partial T}\right)_P = \frac{R}{P}$$

$$P = CT^{k/(k-1)} \longrightarrow \left(\frac{\partial P}{\partial T}\right)_s = \frac{k}{k-1}CT^{k/(k-1)-1} = \frac{k}{k-1}\left(PT^{-k/(k-1)}\right)T^{k/(k-1)-1} = \frac{kP}{T(k-1)}$$

Substituting,

$$C_p = T\left(\frac{kP}{T(k-1)}\right)\left(\frac{R}{P}\right) = \frac{kR}{k-1} = \frac{1.397(0.297\text{kJ/kg}\cdot\text{K})}{1.397-1} = \mathbf{1.045kJ/kg\cdot K}$$

(*b*) The C_p is defined as $C_p = \left(\frac{\partial h}{\partial T}\right)_P$. Replacing the differentials by differences,

$$C_p \cong \left(\frac{\Delta h}{\Delta T}\right)_{P=300\text{kPa}}$$

$$= \frac{h(410K) - h(390\text{K})}{(410-390)\text{K}}$$

$$= \frac{(11{,}932 - 11{,}347)/28.0\text{kJ/kg}}{(410-390)\text{K}} = \mathbf{1.045kJ/kg\cdot K}$$

(Compare: Table A-2*b* at 400 K → C_p = 1.044 kJ/kg· K)

11-74 The temperature change of steam and the average Joule-Thompson coefficient during a throttling process are to be estimated.

Analysis The enthalpy of steam at 4.5 MPa and T = 400°C is h = 3204.7 kJ/kg. Now consider a throttling process from this state to 3.5 MPa. The temperature of the steam at the end of this throttling process will be

$$\left.\begin{array}{l} P = 3.5\text{kPa} \\ h = 3204.7\text{kJ/kg} \end{array}\right\} T_2 = 392.56°\text{C}$$

Thus the temperature drop during this throttling process is

$$\Delta T = T_2 - T_1 = 392.56 - 400 = \mathbf{-7.44°C}$$

The average Joule-Thomson coefficient for this process is determined from

$$\mu = \left(\frac{\partial T}{\partial P}\right)_h \cong \left(\frac{\Delta T}{\Delta P}\right)_{h=3204.7\text{kJ/kg}} = \frac{(392.56-400)°\text{C}}{(3.5-4.5)\text{MPa}} = \mathbf{7.44°C/MPa}$$

11-75 The initial state and the final temperature of argon contained in a rigid tank are given. The mass of the argon in the tank, the final pressure, and the heat transfer are to be determined using the generalized charts.

Analysis (*a*) The compressibility factor of argon at the initial state is determined from the generalized chart to be

$$\left.\begin{aligned} T_{R_1} &= \frac{T_1}{T_{cr}} = \frac{173}{151.0} = 1.146 \\ P_{R_1} &= \frac{P_1}{P_{cr}} = \frac{1}{4.86} = 0.206 \end{aligned}\right\} Z_1 = 0.95 \text{ and } Z_{h_1} = 0.18$$

Ar
-100 °C
1 MPa

Q

Then,

$$Pv = ZRT \longrightarrow v = \frac{ZRT}{P} = \frac{(0.95)(0.2081\,\text{kPa}\cdot\text{m}^3/\text{kg}\cdot\text{K})(173\,\text{K})}{1000\,\text{kPa}} = 0.0342\,\text{m}^3/\text{kg}$$

$$m = \frac{V}{v} = \frac{1.2\,\text{m}^3}{0.0342\,\text{m}^3/\text{kg}} = \mathbf{35.1\,kg}$$

(*b*) The specific volume of argon remains constant during this process, $v_2 = v_1$. Thus,

$$\left.\begin{aligned} T_{R_2} &= \frac{T_2}{T_{cr}} = \frac{273}{151.0} = 1.808 \\ v_{R_2} &= \frac{v_2}{RT_{cr}/P_{cr}} = \frac{0.0342\text{m}^3/\text{kg}}{(0.2081\text{kPa}\cdot\text{m}^3/\text{kg}\cdot\text{K})(151\text{K})(4860\text{kPa})} = 5.29 \end{aligned}\right\} \begin{aligned} P_{R_2} &= 0.315 \\ Z_2 &= 0.99 \\ Z_{h_2} &\cong 0 \end{aligned}$$

$$P_2 = P_{R_2} P_{cr} = (0.315)(4860) = \mathbf{1531 kPa}$$

(*c*) The energy balance relation for this closed system can be expressed as

$$E_{in} - E_{out} = \Delta E_{system}$$
$$Q_{in} = \Delta U = m(u_2 - u_1)$$
$$Q_{in} = m[h_2 - h_1 - (P_2 v_2 - P_1 v_1)] = m[h_2 - h_1 - R(Z_2 T_2 - Z_1 T_1)]$$

where

$$h_2 - h_1 = RT_{cr}(Z_{h_1} - Z_{h_2}) + (h_2 - h_1)_{\text{ideal}}$$
$$= (0.2081)(151)(0.18 - 0) + 0.5203(0 - (-100)) = 57.69\text{kJ/kg}$$

Thus,

$$Q_{in} = (35.1\text{kg})[57.69 - (0.2081\text{kJ/kg}\cdot\text{K})[(0.99)(273) - (0.95)(173)]\text{K}]$$
$$= \mathbf{1251 kJ}$$

11-76 Argon enters a turbine at a specified state and leaves at another specified state. Power output of the turbine and exergy destruction during this process are to be determined using the generalized charts.

Properties The gas constant and critical properties of Argon are R = 0.2081 kJ/kg.K, T_{cr} = 151 K, and P_{cr} = 4.86 MPa (Table A-1).

Analysis (*a*) The enthalpy and entropy departures of argon at the specified states are determined from the generalized charts to be

$$\left.\begin{aligned} T_{R_1} &= \frac{T_1}{T_{cr}} = \frac{600}{151} = 3.97 \\ P_{R_1} &= \frac{P_1}{P_{cr}} = \frac{7}{4.86} = 1.44 \end{aligned}\right\} Z_{h_1} \cong 0 \text{ and } Z_{s_1} \cong 0$$

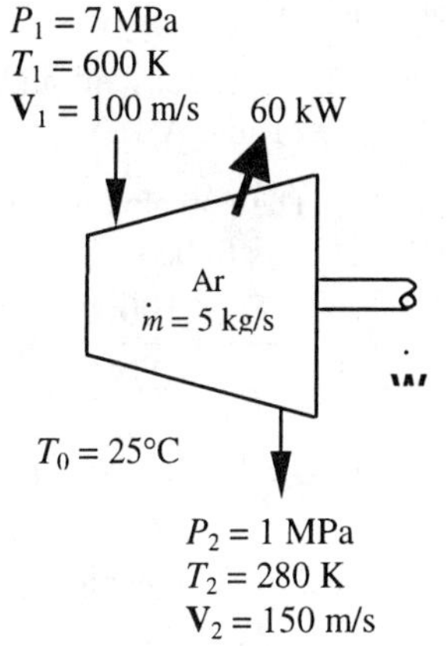

Thus argon behaves as an ideal gas at turbine inlet. Also,

$$\left.\begin{aligned} T_{R_2} &= \frac{T_2}{T_{cr}} = \frac{280}{151} = 1.85 \\ P_{R_2} &= \frac{P_2}{P_{cr}} = \frac{1}{4.86} = 0.206 \end{aligned}\right\} Z_{h_2} = 0.04 \text{ and } Z_{s_2} = 0.02$$

P_1 = 7 MPa
T_1 = 600 K
$\mathbf{V}_1$ = 100 m/s
60 kW
Ar
$\dot{m}$ = 5 kg/s
$\dot{W}$
T_0 = 25°C
P_2 = 1 MPa
T_2 = 280 K
$\mathbf{V}_2$ = 150 m/s

Thus,

$$\begin{aligned} h_2 - h_1 &= RT_{cr}\left(Z_{h_1} - Z_{h_2}\right) + \left(h_2 - h_1\right)_{\text{ideal}} \\ &= (0.2081)(151)(0 - 0.04) + 0.5203(280 - 600) = -167.8 \text{kJ/kg} \end{aligned}$$

The power output of the turbine is to be determined from the energy balance equation,

$$\dot{E}_{\text{in}} - \dot{E}_{\text{out}} = \Delta \dot{E}_{\text{system}} = 0 \text{ (steady)} \rightarrow \dot{E}_{\text{in}} = \dot{E}_{\text{out}}$$

$$\dot{m}(h_1 + \mathbf{V}_1^2/2) = \dot{m}(h_2 + \mathbf{V}_2^2/2) + \dot{Q}_{\text{out}} + \dot{W}_{\text{out}}$$

$$\dot{W}_{\text{out}} = -\dot{m}\left[(h_2 - h_1) + \frac{\mathbf{V}_2^2 - \mathbf{V}_1^2}{2}\right] - \dot{Q}_{\text{out}}$$

Substituting,

$$\dot{W}_{\text{out}} = -(5\text{kg/s})\left(-167.8 + \frac{(150\text{m/s})^2 - (100\text{m/s})^2}{2}\left(\frac{1\text{kJ/kg}}{1000\text{m}^2/\text{s}^2}\right)\right) - 60\text{kJ/s} = \mathbf{747.8kW}$$

(*b*) Under steady conditions, the rate form of the entropy balance for the turbine simplifies to

$$\dot{S}_{\text{in}} - \dot{S}_{\text{out}} + \dot{S}_{\text{gen}} = \Delta \dot{S}_{\text{system}}^{\nearrow 0} = 0$$

$$\dot{m}s_1 - \dot{m}s_2 - \frac{\dot{Q}_{\text{out}}}{T_{b,\text{out}}} + \dot{S}_{\text{gen}} = 0 \rightarrow \dot{S}_{\text{gen}} = \dot{m}(s_2 - s_2) + \frac{\dot{Q}_{\text{out}}}{T_0}$$

The exergy destroyed during a process can be determined from an exergy balance or directly from its definition $X_{\text{destroyed}} = T_0 S_{\text{gen}}$,

$$\dot{X}_{\text{destroyed}} = T_0 \dot{S}_{\text{gen}} = T_0\left(\dot{m}(s_2 - s_2) + \frac{\dot{Q}_{\text{out}}}{T_0}\right)$$

where $\quad s_2 - s_1 = R\left(Z_{s_1} - Z_{s_2}\right) + \left(s_2 - s_1\right)_{\text{ideal}}$

and $\quad \left(s_2 - s_1\right)_{\text{ideal}} = C_p \ln\frac{T_2}{T_1} - R\ln\frac{P_2}{P_1} = 0.5203\ln\frac{280}{600} - 0.2081\ln\frac{1}{7} = 0.0084 \text{kJ/kg} \cdot \text{K}$

Thus, $\quad s_2 - s_1 = R\left(Z_{s_1} - Z_{s_2}\right) + \left(s_2 - s_1\right)_{\text{ideal}} = (0.2081)[0 - (0.02)] + 0.0084 = 0.0042 \text{kJ/kg} \cdot \text{K}$

Substituting, $\quad \dot{X}_{\text{destroyed}} = (298\text{K})\left((5\text{kg/s})(0.0042\text{kJ/kg} \cdot \text{K}) + \frac{60\text{kW}}{298\text{K}}\right) = \mathbf{66.3kW}$

11-77 EES solution of this (and other comprehensive problems designated with the *computer icon*) is available to instructors at the *Instructor Manual* section of the *Online Learning Center* (OLC) at www.mhhe.com/cengel-boles. See the Preface for access information.

11-78E Argon gas enters a turbine at a specified state and leaves at another specified state. The power output of the turbine and the exergy destruction associated with the process are to be determined using the generalized charts.

Properties The gas constant and critical properties of argon are R = 0.04971 Btu/lbm.R, T_{cr} = 272 R, and P_{cr} = 705 psia (Table A-1E).

Analysis (*a*) The enthalpy and entropy departures of argon at the specified states are determined from the generalized charts to be

$$\left.\begin{array}{l} T_{R_1} = \dfrac{T_1}{T_{cr}} = \dfrac{1000}{272} = 3.68 \\ P_{R_1} = \dfrac{P_1}{P_{cr}} = \dfrac{1000}{705} = 1.418 \end{array}\right\} Z_{h_1} \cong 0 \text{ and } Z_{s_1} \cong 0$$

P_1 = 1000 psia
T_1 = 1000 R
$\mathbf{V}_1$ = 300 ft/s
80 Btu/s
Ar
$\dot{m}$ = 12 lbm/s
W
P_2 = 150 psia
T_2 = 500 R
$\mathbf{V}_2$ = 450 ft/s

Thus argon behaves as an ideal gas at turbine inlet. Also,

$$\left.\begin{array}{l} T_{R_2} = \dfrac{T_2}{T_{cr}} = \dfrac{500}{272} = 1.838 \\ P_{R_2} = \dfrac{P_2}{P_{cr}} = \dfrac{150}{705} = 0.213 \end{array}\right\} Z_{h_2} = 0.04 \text{ and } Z_{s_2} = 0.02$$

Thus ,

$$\begin{aligned} h_2 - h_1 &= RT_{cr}(Z_{h_1} - Z_{h_2}) + (h_2 - h_1)_{\text{ideal}} \\ &= (0.04971)(272)(0 - 0.04) + 0.1253(500 - 1000) = -63.2\text{Btu/lbm} \end{aligned}$$

The power output of the turbine is to be determined from the energy balance equation,

$$\dot{E}_{\text{in}} - \dot{E}_{\text{out}} = \Delta\dot{E}_{\text{system}} = 0\,(\text{steady}) \rightarrow \dot{E}_{\text{in}} = \dot{E}_{\text{out}}$$

$$\dot{m}(h_1 + \mathbf{V}_1^2/2) = \dot{m}(h_2 + \mathbf{V}_2^2/2) + \dot{Q}_{\text{out}} + \dot{W}_{\text{out}}$$

$$\dot{W}_{\text{out}} = -\dot{m}\left[(h_2 - h_1) + \frac{\mathbf{V}_2^2 - \mathbf{V}_1^2}{2}\right] - \dot{Q}_{\text{out}}$$

$$\dot{W}_{\text{out}} = -(12\text{lbm/s})\left(-63.2 + \frac{(450\text{ft/s})^2 - (300\text{ft/s})^2}{2}\left(\frac{1\text{Btu/lbm}}{25{,}037\text{ft}^2/\text{s}^2}\right)\right) - 80\text{Btu/s}$$

$$= \mathbf{651.4Btu/s = 922hp}$$

(*b*) Under steady conditions, the rate form of the entropy balance for the turbine simplifies to

$$\dot{S}_{\text{in}} - \dot{S}_{\text{out}} + \dot{S}_{\text{gen}} = \Delta\dot{S}_{\text{system}}^{\;\nearrow 0} = 0$$

$$\dot{m}s_1 - \dot{m}s_2 - \frac{\dot{Q}_{\text{out}}}{T_{b,\text{out}}} + \dot{S}_{\text{gen}} = 0 \quad \rightarrow \quad \dot{S}_{\text{gen}} = \dot{m}(s_2 - s_2) + \frac{\dot{Q}_{\text{out}}}{T_0}$$

The exergy destroyed during a process can be determined from an exergy balance or directly from its definition $X_{\text{destroyed}} = T_0 S_{\text{gen}}$,

$$\dot{X}_{\text{destroyed}} = T_0\dot{S}_{\text{gen}} = T_0\left(\dot{m}(s_2 - s_2) + \frac{\dot{Q}_{\text{out}}}{T_0}\right)$$

where $\qquad s_2 - s_1 = R(Z_{s_1} - Z_{s_2}) + (s_2 - s_1)_{\text{ideal}}$

and

$$(s_2 - s_1)_{\text{ideal}} = C_p \ln\frac{T_2}{T_1} - R\ln\frac{P_2}{P_1} = 0.1253\ln\frac{500}{1000} - 0.04971\ln\frac{150}{1000} = 0.00745\text{Btu/lbm}\cdot\text{R}$$

Thus,

$$s_2 - s_1 = R(Z_{s_1} - Z_{s_2}) + (s_2 - s_1)_{\text{ideal}} = (0.04971)[0-(0.02)] + 0.00745 = 0.00646\text{Btu/lbm}\cdot\text{R}$$

Substituting, $\dot{X}_{\text{destroyed}} = (535\text{R})\left((12\text{lbm/s})(0.00646\text{Btu/lbm}\cdot\text{R}) + \dfrac{80\text{Btu/s}}{535\text{R}}\right) = \mathbf{121.5Btu/s}$

11-79 An adiabatic storage tank that is initially evacuated is connected to a supply line that carries nitrogen. A valve is opened, and nitrogen flows into the tank. The final temperature in the tank is to be determined by treating nitrogen as an ideal gas and using the generalized charts, and the results are to be compared to the given actual value.

Assumptions **1** Uniform flow conditions exist. **2** Kinetic and potential energies are negligible.

Analysis We take the tank as the system, which is a control volume since mass crosses the boundary. Noting that the microscopic energies of flowing and nonflowing fluids are represented by enthalpy h and internal energy u, respectively, the mass and energy balances for this uniform-flow system can be expressed as

Mass balance: $m_{in} - m_{out} = \Delta m_{system} \rightarrow m_i = m_2 \quad (\text{since } m_{out} = m_{initial} = 0)$

Energy balance: $E_{in} - E_{out} = \Delta E_{system} \rightarrow 0 + m_i h_i = m_2 u_2$

Combining the two balances: $u_2 = h_i$

(*a*) From the ideal gas property table of nitrogen, at 225 K we read

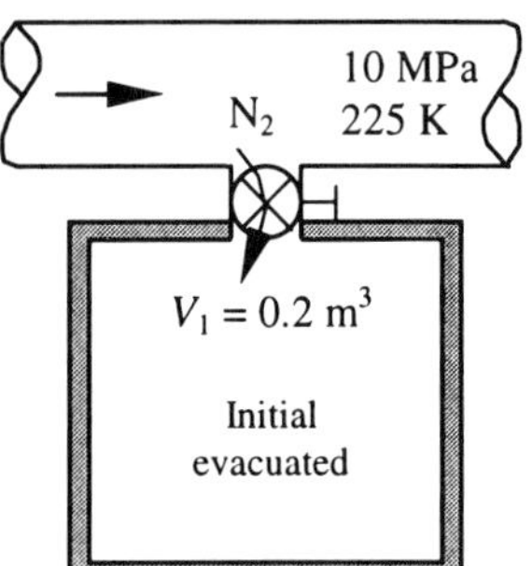

$$\bar{u}_2 = \bar{h}_i = \bar{h}_{@\,225\text{ K}} = 6{,}537\text{ kJ/kmol}$$

The temperature that corresponds to this $\bar{u}_2$ value is

$$T_2 = \mathbf{314.8\ K} \qquad (7.4\%\text{ error})$$

(*b*) Using the generalized enthalpy departure chart, h_i is determined to be

$$\left.\begin{aligned} T_{R,i} &= \frac{T_i}{T_{cr}} = \frac{225}{126.2} = 1.78 \\ P_{R,i} &= \frac{P_i}{P_{cr}} = \frac{10}{3.39} = 2.95 \end{aligned}\right\} Z_{h,i} = \frac{\bar{h}_{i,\text{ideal}} - \bar{h}_i}{R_u T_{cr}} = 0.9 \qquad \text{(Fig. A-31)}$$

Thus,

$$\bar{h}_i = \bar{h}_{i,\text{ideal}} - 0.9 R_u T_{cr} = 6{,}537 - (0.9)(8.314)(126.2) = 5{,}593\text{kJ/kmol}$$

and

$$\bar{u}_2 = \bar{h}_i = 5{,}593\text{kJ/kmol}$$

Try $T_2 = 280$ K. Then at $P_{R2} = 2.95$ and $T_{R2} = 2.22$ we read $Z_2 = 0.98$ and $(\bar{h}_{2,ideal} - \bar{h}_2)/R_u T_{cr} = 0.55$

Thus,

$$\bar{h}_2 = \bar{h}_{2,\text{ideal}} - 0.55 R_u T_{cr} = 8{,}141 - (0.55)(8.314)(126.2) = 7{,}564\text{kJ/kmol}$$
$$\bar{u}_2 = \bar{h}_2 - Z R_u T_2 = 7{,}564 - (0.98)(8.314)(280) = 5{,}283\text{kJ/kmol}$$

Try $T_2 = 300$ K. Then at $P_{R2} = 2.95$ and $T_{R2} = 2.38$ we read $Z_2 = 1.0$ and $(\bar{h}_{2,ideal} - \bar{h}_2)/R_u T_{cr} = 0.50$

Thus,

$$\bar{h}_2 = \bar{h}_{2,\text{ideal}} - 0.50 R_u T_{cr} = 8{,}723 - (0.50)(8.314)(126.2) = 8{,}198\text{kJ/kmol}$$
$$\bar{u}_2 = \bar{h}_2 - Z R_u T_2 = 8{,}198 - (1.0)(8.314)(300) = 5{,}704\text{kJ/kmol}$$

By linear interpolation,

$$T_2 = \mathbf{294.7\ K} \qquad (0.6\%\text{ error})$$

11-80 It is to be shown that $\frac{dv}{v} = \beta dT - \alpha dP$. Also, a relation is to be obtained for the ratio of specific volumes v_2/v_1 as a homogeneous system undergoes a process from state 1 to state 2.

Analysis We take $v = v(P, T)$. Its total differential is

$$dv = \left(\frac{\partial v}{\partial T}\right)_P dT + \left(\frac{\partial v}{\partial P}\right)_T dP$$

Dividing by v,

$$\frac{dv}{v} = \frac{1}{v}\left(\frac{\partial v}{\partial T}\right)_P dT + \frac{1}{v}\left(\frac{\partial v}{\partial P}\right)_T dP$$

Using the definitions of α and β,

$$\frac{dv}{v} = \beta dT - \alpha dP$$

Taking α and β to be constants, integration from 1 to 2 yields

$$\ln\frac{v_2}{v_1} = \beta(T_2 - T_1) - \alpha(P_2 - P_1)$$

which is the desired relation.

11-81 It is to be shown that $\frac{dv}{v} = \beta dT - \alpha dP$. Also, a relation is to be obtained for the ratio of specific volumes v_2/v_1 as a homogeneous system undergoes an isobaric process from state 1 to state 2.

Analysis We take $v = v(P, T)$. Its total differential is

$$dv = \left(\frac{\partial v}{\partial T}\right)_P dT + \left(\frac{\partial v}{\partial P}\right)_T dP$$

which, for a constant pressure process, reduces to

$$dv = \left(\frac{\partial v}{\partial T}\right)_P dT$$

Dividing by v,

$$\frac{dv}{v} = \frac{1}{v}\left(\frac{\partial v}{\partial T}\right)_P dT$$

Using the definition of β,

$$\frac{dv}{v} = \beta dT$$

Taking β to be a constant, integration from 1 to 2 yields

$$\ln\frac{v_2}{v_1} = \beta(T_2 - T_1) =$$

or

$$\frac{v_2}{v_1} = \exp[\beta(T_2 - T_1)]$$

which is the desired relation.

11-82 The volume expansivity of water is given. The change in volume of water when it is heated at constant pressure is to be determined.

Properties The volume expansivity of water is given to be $0.207\times10^{-6}\ \text{K}^{-1}$ at 20°C.

Analysis We take $v = v(P, T)$. Its total differential is

$$dv = \left(\frac{\partial v}{\partial T}\right)_P dT + \left(\frac{\partial v}{\partial P}\right)_T dP$$

which, for a constant pressure process, reduces to

$$dv = \left(\frac{\partial v}{\partial T}\right)_P dT$$

Dividing by v and using the definition of β,

$$\frac{dv}{v} = \frac{1}{v}\left(\frac{\partial v}{\partial T}\right)_P dT = \beta dT$$

Taking β to be a constant, integration from 1 to 2 yields

$$\ln\frac{v_2}{v_1} = \beta(T_2 - T_1)$$

or

$$\frac{v_2}{v_1} = \exp[\beta(T_2 - T_1)]$$

Substituting the given values and noting that for a fixed mass $V_2/V_1 = v_2/v_1$,

$$V_2 = V_1\exp[\beta(T_2 - T_1)] = (1\text{m}^3)\exp[(0.207\times10^{-6}\ \text{K}^{-1})(25-15)°\text{C}]$$
$$= 1.000002\text{m}^3$$

Therefore,

$$\Delta V = V_2 - V_1 = 1.000002 - 1 = 0.000002\ \text{m}^3 = \mathbf{2\ cm^3}$$

11-83 The volume expansivity of copper is given at two temperatures. The percent change in the volume of copper when it is heated at atmospheric pressure is to be determined.

Properties The volume expansivity of copper is given to be 49.2×10^{-6} K^{-1} at 300 K, and be 54.2×10^{-6} K^{-1} at 500 K

Analysis We take $v = v(P, T)$. Its total differential is

$$dv = \left(\frac{\partial v}{\partial T}\right)_P dT + \left(\frac{\partial v}{\partial P}\right)_T dP$$

which, for a constant pressure process, reduces to

$$dv = \left(\frac{\partial v}{\partial T}\right)_P dT$$

Dividing by v and using the definition of β,

$$\frac{dv}{v} = \frac{1}{v}\left(\frac{\partial v}{\partial T}\right)_P dT = \beta dT$$

Taking β to be a constant, integration from 1 to 2 yields

$$\ln\frac{v_2}{v_1} = \beta(T_2 - T_1)$$

or

$$\frac{v_2}{v_1} = \exp[\beta(T_2 - T_1)]$$

The average value of β is

$$\beta_{\text{ave}} = (\beta_1 + \beta_2)/2 = \left(49.2\times10^{-6} + 54.2\times10^{-6}\right)/2 = 51.7\times10^{-6}\,\text{K}^{-1}$$

Substituting the given values,

$$\frac{v_2}{v_1} = \exp[\beta(T_2 - T_1)] = \exp\left[\left(51.7\times10^{-6}\,\text{K}^{-1}\right)(500-300)\text{K}\right] = 1.0104$$

Therefore, the volume of copper block will increase by **1.04 percent**.

11-84 It is to be shown that the position of the Joule-Thompson coefficient inversion curve on the *T-P* plane is given by $(\partial Z/\partial T)_P = 0$.

Analysis The inversion curve is the locus of the points at which the Joule-Thompson coefficient μ is zero,

$$\mu = \frac{1}{C_p}\left(T\left(\frac{\partial v}{\partial T}\right)_P - v\right) = 0$$

which can also be written as

$$T\left(\frac{\partial v}{\partial T}\right)_P - \frac{ZRT}{P} = 0 \qquad (a)$$

since it is given that

$$v = \frac{ZRT}{P} \qquad (b)$$

Taking the derivative of (b) with respect to T holding P constant gives

$$\left(\frac{\partial v}{\partial T}\right)_P = \left(\frac{\partial (ZRT/P)}{\partial T}\right)_P = \frac{R}{P}\left(T\left(\frac{\partial Z}{\partial T}\right)_P + Z\right)$$

Substituting in (a),

$$\frac{TR}{P}\left(T\left(\frac{\partial Z}{\partial T}\right)_P + Z\right) - \frac{ZRT}{P} = 0$$

$$\frac{TR}{P}\left(T\left(\frac{\partial Z}{\partial T}\right)_P + Z - Z\right) = 0$$

$$\left(\frac{\partial Z}{\partial T}\right)_P = 0$$

which is the desired relation.

11-85 It is to be shown that for an isentropic expansion or compression process Pv^k = constant. It is also to be shown that the isentropic expansion exponent k reduces to the specific heat ratio C_p/C_v for an ideal gas.

Analysis We note that ds = 0 for an isentropic process. Taking $s = s(P, v)$, the total differential ds can be expressed as

$$ds = \left(\frac{\partial s}{\partial P}\right)_v dP + \left(\frac{\partial s}{\partial v}\right)_P dv = 0 \qquad (a)$$

We now substitute the Maxwell relations below into (a)

$$\left(\frac{\partial s}{\partial P}\right)_v = -\left(\frac{\partial v}{\partial T}\right)_s \text{ and } \left(\frac{\partial s}{\partial v}\right)_P = \left(\frac{\partial P}{\partial T}\right)_s$$

to get

$$-\left(\frac{\partial v}{\partial T}\right)_s dP + \left(\frac{\partial P}{\partial T}\right)_s dv = 0$$

Rearranging,

$$dP - \left(\frac{\partial T}{\partial v}\right)_s \left(\frac{\partial P}{\partial T}\right)_s dv = 0 \longrightarrow dP - \left(\frac{\partial P}{\partial v}\right)_s dv = 0$$

Dividing by P,

$$\frac{dP}{P} - \frac{1}{P}\left(\frac{\partial P}{\partial v}\right)_s dv = 0 \qquad (b)$$

We now define *isentropic expansion exponent* k as

$$k = -\frac{v}{P}\left(\frac{\partial P}{\partial v}\right)_s$$

Substituting in (b),

$$\frac{dP}{P} + k\frac{dv}{v} = 0$$

Taking k to be a constant and integrating,

$$\ln P + k \ln v = \text{constant} \qquad \longrightarrow \qquad \ln Pv^k = \text{constant}$$

Thus,

$$Pv^k = \text{constant}$$

To show that $k = C_p/C_v$ for an ideal gas, we write the cyclic relations for the following two groups of variables:

$$(s,T,v) \longrightarrow \left(\frac{\partial s}{\partial T}\right)_v \left(\frac{\partial v}{\partial s}\right)_T \left(\frac{\partial T}{\partial v}\right)_s = -1 \longrightarrow \frac{C_v}{T}\left(\frac{\partial v}{\partial s}\right)_T \left(\frac{\partial T}{\partial v}\right)_s = -1 \quad (c)$$

$$(s,T,P) \longrightarrow \left(\frac{\partial s}{\partial T}\right)_P \left(\frac{\partial P}{\partial s}\right)_T \left(\frac{\partial T}{\partial P}\right)_s = -1 \longrightarrow \frac{C_p}{T}\left(\frac{\partial P}{\partial s}\right)_T \left(\frac{\partial T}{\partial P}\right)_s = -1 \quad (d)$$

where we used the relations

$$C_v = T\left(\frac{\partial s}{\partial T}\right)_v \text{ and } C_p = T\left(\frac{\partial s}{\partial T}\right)_P$$

Setting Eqs. (*c*) and (*d*) equal to each other,

$$\frac{C_p}{T}\left(\frac{\partial P}{\partial s}\right)_T\left(\frac{\partial T}{\partial P}\right)_s = \frac{C_v}{T}\left(\frac{\partial v}{\partial s}\right)_T\left(\frac{\partial T}{\partial v}\right)_s$$

or,

$$\begin{aligned}\frac{C_p}{C_v} &= \left(\frac{\partial s}{\partial P}\right)_T\left(\frac{\partial P}{\partial T}\right)_s\left(\frac{\partial v}{\partial s}\right)_T\left(\frac{\partial T}{\partial v}\right)_s \\ &= \left(\frac{\partial s}{\partial P}\frac{\partial v}{\partial s}\right)_T\left(\frac{\partial P}{\partial T}\frac{\partial T}{\partial v}\right)_s \\ &= \left(\frac{\partial v}{\partial P}\right)_T\left(\frac{\partial P}{\partial v}\right)_s\end{aligned}$$

but

$$\left(\frac{\partial v}{\partial P}\right)_T = \left(\frac{\partial (RT/P)}{\partial P}\right)_T = -\frac{v}{P}$$

Substituting,

$$\frac{C_p}{C_v} = -\frac{v}{P}\left(\frac{\partial P}{\partial v}\right)_s = k$$

which is the desired relation.

Fundamentals of Engineering (FE) Exam Problems

11-86 A substance whose Joule-Thomson coefficient is negative is throttled to a lower pressure. During this process, (select the correct statement)

(a) the temperature of the substance will increase.
(b) the temperature of the substance will decrease.
(c) the entropy of the substance will remain constant.
(d) the entropy of the substance will decrease.
(e) the enthalpy of the substance will decrease.

Answer (a) the temperature of the substance will increase.

11-87 Consider the liquid-vapor saturation curve of a pure substance on the *P-T* diagram. The magnitude of the slope of the tangent line to this curve at a temperature *T* (in Kelvin) is

(a) proportional to the enthalpy of vaporization h_{fg} at that temperature,
(b) proportional to the temperature *T*,
(c) proportional to the square of the temperature *T*,
(d) proportional to the volume change v_{fg} at that temperature,
(e) inversely proportional to the entropy change s_{fg} at that temperature,

Answer (a) proportional to the enthalpy of vaporization h_{fg} at that temperature,

11-88 Based on the generalized charts, the error involved in the enthalpy of CO_2 at 350 K and 10 MPa if it is assumed to be an ideal gas is

(a) 0 (b) 20% (c) 35% (d) 50% (e) 65%

Answer (d) 50%

Solution Solved by EES Software. Solutions can be verified by copying-and-pasting the following lines on a blank EES screen. (Similar problems and their solutions can be obtained easily by modifying numerical values).

```
T=350 "K"
P=10000 "kPa"

Pcr=P_CRIT(CarbonDioxide)
Tcr=T_CRIT(CarbonDioxide)

Tr=T/Tcr
Pr=P/Pcr
Z=COMPRESS(Tr, Pr)
hR=ENTHDEP(Tr, Pr)
```

11-89 Based on data from the refrigerant-134a tables, the Joule-Thompson coefficient of refrigerant-134a at 0.6 MPa and 100°C is approximately

(a) 0 (b) -5°C/MPa (c) 11°C/MPa (d) 8°C/MPa (e) 26°C/MPa

Answer (c) 11°C/MPa

Solution Solved by EES Software. Solutions can be verified by copying-and-pasting the following lines on a blank EES screen. (Similar problems and their solutions can be obtained easily by modifying numerical values).

```
T1=100 "C"
P1=600 "kPa"
h1=ENTHALPY(R134a,T=T1,P=P1)

Tlow=TEMPERATURE(R134a,h=h1,P=P1+100)
Thigh=TEMPERATURE(R134a,h=h1,P=P1-100)

JT=(Tlow-Thigh)/200
```

11-90 For a gas whose equation of state is $P(v - b) = RT$, the specific heat difference $C_p - C_v$ is equal to

(a) R (b) $R - b$ (c) $R + b$ (d) 0 (e) $R(1 + v/b)$

Answer (a) R

Solution The general relation for the specific heat difference C_p - C_v is

$$C_P - C_v = -T\left(\frac{\partial v}{\partial T}\right)_P^2\left(\frac{\partial P}{\partial v}\right)_T$$

For the given gas, $P(v - b) = RT$. Then,

$$v = \frac{RT}{P} - b \longrightarrow \left(\frac{\partial v}{\partial T}\right)_P = \frac{R}{P}$$

$$P = \frac{RT}{v-b} \longrightarrow \left(\frac{\partial P}{\partial v}\right)_T = -\frac{RT}{(v-b)^2} = -\frac{P}{v-b}$$

Substituting,

$$C_P - C_v = -T\left(\frac{R}{P}\right)^2\left(-\frac{P}{v-b}\right) = \frac{TR^2}{P(v-b)} = \mathbf{R}$$

11-91 · · · 11-93 Design and Essay Problems

Chapter 12
GAS MIXTURES

Composition of Gas Mixtures

12-1C It is the average or the equivalent gas constant of the gas mixture. No.

12-2C No. We can do this only when each gas has the same mole fraction.

12-3C It is the average or the equivalent molar mass of the gas mixture. No.

12-4C The mass fractions will be identical, but the mole fractions will not.

12-5C Yes.

12-6C The ratio of the mass of a component to the mass of the mixture is called the mass fraction (*mf*), and the ratio of the mole number of a component to the mole number of the mixture is called the mole fraction (*y*).

12-7C From the definition of mass fraction,

$$mf_i = \frac{m_i}{m_m} = \frac{N_i M_i}{N_m M_m} = y_i \left(\frac{M_i}{M_m} \right)$$

12-8C Yes, because both CO_2 and N_2O has the same molar mass, $M = 44$ kg/kmol.

12-9 A mixture consists of two gases. Relations for mole fractions when mass fractions are known are to be obtained .

Analysis The mass fractions of *A* and *B* are expressed as

$$mf_A = \frac{m_A}{m_m} = \frac{N_A M_A}{N_m M_m} = y_A \frac{M_A}{M_m} \quad \text{and} \quad mf_B = y_B \frac{M_B}{M_m}$$

Where *m* is mass, *M* is the molar mass, *N* is the number of moles, and *y* is the mole fraction. The apparent molar mass of the mixture is

$$M_m = \frac{m_m}{N_m} = \frac{N_A M_A + N_B M_B}{N_m} = y_A M_A + y_B M_B$$

Combining the two equation above and noting that $y_A + y_B = 1$ gives the following convenient relations for converting mass fractions to mole fractions,

$$y_A = \frac{M_B}{M_A(1/mf_A - 1) + M_B} \quad \text{and} \quad y_B = 1 - y_A$$

which are the desired relations.

12-10 The molar fractions of the constituents of moist air are given. The mass fractions of the constituents are to be determined.

Assumptions The small amounts of gases in air are ignored, and dry air is assumed to consist of N_2 and O_2 only.

Properties The molar masses of N_2, O_2, and H_2O are 28.0, 32.0, and 18.0 kg/kmol, respectively (Table A-1).

Analysis The molar mass of moist air is

$$M = \sum y_i M_i = 0.78 \times 28.0 + 0.20 \times 32.0 + 0.02 \times 18 = 28.6 \text{ kg/kmol}$$

Moist air
78% N_2
20% O_2
2% H_2O
(Mole fractions)

Then the mass fractions of constituent gases are determined to be

N_2: $$w_{N_2} = y_{N_2}\frac{M_{N_2}}{M} = (0.78)\frac{28.0}{28.6} = \mathbf{0.764}$$

O_2: $$w_{O_2} = y_{O_2}\frac{M_{O_2}}{M} = (0.20)\frac{32.0}{28.6} = \mathbf{0.224}$$

H_2O: $$w_{H_2O} = y_{H_2O}\frac{M_{H_2O}}{M} = (0.02)\frac{18.0}{28.6} = \mathbf{0.012}$$

Therefore, the mass fractions of N_2, O_2, and H_2O in the air are 76.4%, 22.4%, and 1.2%, respectively.

12-11 The molar fractions of the constituents of a gas mixture are given. The gravimetric analysis of the mixture, its molar mass, and gas constant are to be determined.

Properties The molar masses of N_2, and CO_2 are 28.0 and 44.0 kg/kmol, respectively (Table A-1)

Analysis Consider 100 kmol of mixture. Then the mass of each component and the total mass are

$$N_{N_2} = 60\text{kmol} \longrightarrow m_{N_2} = N_{N_2} M_{N_2} = (60\text{kmol})(28\text{kg/kmol}) = 1{,}680\text{kg}$$

$$N_{CO_2} = 40\text{kmol} \longrightarrow m_{CO_2} = N_{CO_2} M_{CO_2} = (40\text{kmol})(44\text{kg/kmol}) = 1{,}760\text{kg}$$

$$m_m = m_{N_2} + m_{CO_2} = 1{,}680 \text{ kg} + 1{,}760 \text{ kg} = 3{,}440 \text{ kg}$$

Then the mass fraction of each component (gravimetric analysis) becomes

mole

60% N_2
40% CO_2

$$mf_{N_2} = \frac{m_{N_2}}{m_m} = \frac{1{,}680 \text{ kg}}{3{,}440 \text{ kg}} = 0.488 \quad \text{or} \quad \mathbf{48.8\%}$$

$$mf_{CO_2} = \frac{m_{CO_2}}{m_m} = \frac{1{,}760 \text{ kg}}{3{,}440 \text{ kg}} = 0.512 \quad \text{or} \quad \mathbf{51.2\%}$$

The molar mass and the gas constant of the mixture are determined from their definitions,

$$M_m = \frac{m_m}{N_m} = \frac{3{,}440 \text{ kg}}{100 \text{ kmol}} = \mathbf{34.40 \text{ kg/kmol}}$$

and

$$R_m = \frac{R_u}{M_m} = \frac{8.314 \text{ kJ/kmol}\cdot\text{K}}{34.4 \text{ kg/kmol}} = \mathbf{0.242 \text{ kJ/kg}\cdot\text{K}}$$

12-12 The molar fractions of the constituents of a gas mixture are given. The gravimetric analysis of the mixture, its molar mass, and gas constant are to be determined.

Properties The molar masses of O_2 and CO_2 are 32.0 and 44.0 kg/kmol, respectively (Table A-1)

Analysis Consider 100 kmol of mixture. Then the mass of each component and the total mass are

$$N_{O_2} = 60\text{kmol} \longrightarrow m_{O_2} = N_{O_2} M_{O_2} = (60\text{kmol})(32\text{kg/kmol}) = 1{,}920\text{kg}$$

$$N_{CO_2} = 40\text{kmol} \longrightarrow m_{CO_2} = N_{CO_2} M_{CO_2} = (40\text{kmol})(44\text{kg/kmol}) = 1{,}760\text{kg}$$

$$m_m = m_{O_2} + m_{CO_2} = 1{,}920\text{ kg} + 1{,}760\text{ kg} = 3{,}680\text{ kg}$$

Then the mass fraction of each component (gravimetric analysis) becomes

$$mf_{O_2} = \frac{m_{O_2}}{m_m} = \frac{1{,}920\text{ kg}}{3{,}680\text{ kg}} = 0.522 \quad \text{or} \quad \mathbf{52.2\%}$$

$$mf_{CO_2} = \frac{m_{CO_2}}{m_m} = \frac{1{,}760\text{ kg}}{3{,}680\text{ kg}} = 0.478 \quad \text{or} \quad \mathbf{47.8\%}$$

mole

60% O_2
40% CO_2

The molar mass and the gas constant of the mixture are determined from their definitions,

$$M_m = \frac{m_m}{N_m} = \frac{3{,}680\text{ kg}}{100\text{ kmol}} = \mathbf{36.80\ kg/kmol}$$

and

$$R_m = \frac{R_u}{M_m} = \frac{8.314\text{ kJ/kmol}\cdot\text{K}}{36.8\text{ kg/kmol}} = \mathbf{0.226\ kJ/kg\cdot K}$$

12-13 The masses of the constituents of a gas mixture are given. The mass fractions, the mole fractions, the average molar mass, and gas constant are to be determined.

Properties The molar masses of O_2, N_2, and CO_2 are 32.0, 28.0 and 44.0 kg/kmol, respectively (Table A-1)

Analysis (*a*) The total mass of the mixture is

$$m_m = m_{O_2} + m_{N_2} + m_{CO_2} = 5 \text{ kg} + 8 \text{ kg} + 10 \text{ kg} = 23 \text{ kg}$$

5 kg O_2
8 kg N_2
10 kg CO_2

Then the mass fraction of each component becomes

$$mf_{O_2} = \frac{m_{O_2}}{m_m} = \frac{5 \text{ kg}}{23 \text{ kg}} = \mathbf{0.217}$$

$$mf_{N_2} = \frac{m_{N_2}}{m_m} = \frac{8 \text{ kg}}{23 \text{ kg}} = \mathbf{0.348}$$

$$mf_{CO_2} = \frac{m_{CO_2}}{m_m} = \frac{10 \text{ kg}}{23 \text{ kg}} = \mathbf{0.435}$$

(*b*) To find the mole fractions, we need to determine the mole numbers of each component first,

$$N_{O_2} = \frac{m_{O_2}}{M_{O_2}} = \frac{5 \text{ kg}}{32 \text{ kg/kmol}} = 0.156 \text{ kmol}$$

$$N_{N_2} = \frac{m_{N_2}}{M_{N_2}} = \frac{8 \text{ kg}}{28 \text{ kg/kmol}} = 0.286 \text{ kmol}$$

$$N_{CO_2} = \frac{m_{CO_2}}{M_{CO_2}} = \frac{10 \text{ kg}}{44 \text{ kg/kmol}} = 0.227 \text{ kmol}$$

Thus,

$$N_m = N_{O_2} + N_{N_2} + N_{CO_2} = 0.156 \text{ kmol} + 0.286 \text{ kmol} + 0.227 \text{ kmol} = 0.669 \text{ kmol}$$

and

$$y_{O_2} = \frac{N_{O_2}}{N_m} = \frac{0.156 \text{ kmol}}{0.699 \text{ kmol}} = \mathbf{0.233}$$

$$y_{N_2} = \frac{N_{N_2}}{N_m} = \frac{0.286 \text{ kmol}}{0.669 \text{ kmol}} = \mathbf{0.428}$$

$$y_{CO_2} = \frac{N_{CO_2}}{N_m} = \frac{0.227 \text{ kmol}}{0.669 \text{ kmol}} = \mathbf{0.339}$$

(*c*) The average molar mass and gas constant of the mixture are determined from their definitions:

$$M_m = \frac{m_m}{N_m} = \frac{23 \text{ kg}}{0.669 \text{ kmol}} = \mathbf{34.4 \text{ kg/kmol}}$$

and

$$R_m = \frac{R_u}{M_m} = \frac{8.314 \text{ kJ/kmol} \cdot \text{K}}{34.4 \text{ kg/kmol}} = \mathbf{0.242 \text{ kJ/kg} \cdot \text{K}}$$

12-14 The mass fractions of the constituents of a gas mixture are given. The mole fractions of the gas and gas constant are to be determined.

Properties The molar masses of CH_4, and CO_2 are 16.0 and 44.0 kg/kmol, respectively (Table A-1)

Analysis For convenience, consider 100 kg of the mixture. Then the number of moles of each component and the total number of moles are

$$m_{CH_4} = 60\text{ kg} \longrightarrow N_{CH_4} = \frac{m_{CH_4}}{M_{CH_4}} = \frac{60\text{ kg}}{16\text{ kg/kmol}} = 3.750\text{ kmol}$$

$$m_{CO_2} = 40\text{ kg} \longrightarrow N_{CO_2} = \frac{m_{CO_2}}{M_{CO_2}} = \frac{40\text{ kg}}{44\text{ kg/kmol}} = 0.909\text{ kmol}$$

$$N_m = N_{CH_4} + N_{CO_2} = 3.750\text{ kmol} + 0.909\text{ kmol} = 4.659\text{ kmol}$$

mass

60% CH_4
40% CO_2

Then the mole fraction of each component becomes

$$y_{CH_4} = \frac{N_{CH_4}}{N_m} = \frac{3.750\text{ kmol}}{4.659\text{ kmol}} = 0.804 \quad \text{or} \quad \mathbf{80.4\%}$$

$$y_{CO_2} = \frac{N_{CO_2}}{N_m} = \frac{0.909\text{ kmol}}{4.659\text{ kmol}} = 0.195 \quad \text{or} \quad \mathbf{19.5\%}$$

The molar mass and the gas constant of the mixture are determined from their definitions,

$$M_m = \frac{m_m}{N_m} = \frac{100\text{ kg}}{4.659\text{ kmol}} = 21.46\text{ kg/kmol}$$

and

$$R_m = \frac{R_u}{M_m} = \frac{8.314\text{ kJ/kmol}\cdot\text{K}}{21.46\text{ kg/kmol}} = \mathbf{0.387\text{ kJ/kg}\cdot\text{K}}$$

12-15 The mole numbers of the constituents of a gas mixture are given. The mass of each gas and the apparent gas constant are to be determined.

Properties The molar masses of H_2, and N_2 are 2.0 and 28.0 kg/kmol, respectively (Table A-1)

Analysis The mass of each component is determined from

$$N_{H_2} = 8\text{kmol} \longrightarrow m_{H_2} = N_{H_2} M_{H_2} = (8\text{kmol})(2.0\text{kg/kmol}) = \mathbf{16kg}$$

$$N_{N_2} = 2\text{kmol} \longrightarrow m_{N_2} = N_{N_2} M_{N_2} = (2\text{kmol})(28\text{kg/kmol}) = \mathbf{56kg}$$

8 kmol H_2
2 kmol N_2

The total mass and the total number of moles are

$$m_m = m_{H_2} + m_{N_2} = 16\text{ kg} + 56\text{ kg} = 72\text{ kg}$$

$$N_m = N_{H_2} + N_{N_2} = 8\text{ kmol} + 2\text{ kmol} = 10\text{ kmol}$$

The molar mass and the gas constant of the mixture are determined from their definitions,

$$M_m = \frac{m_m}{N_m} = \frac{72\text{ kg}}{10\text{ kmol}} = 7.2\text{ kg/kmol}$$

and

$$R_m = \frac{R_u}{M_m} = \frac{8.314\text{ kJ/kmol}\cdot\text{K}}{7.2\text{ kg/kmol}} = \mathbf{1.155\text{ kJ/kg}\cdot\text{K}}$$

12-16E The mole numbers of the constituents of a gas mixture are given. The mass of each gas and the apparent gas constant are to be determined.

Properties The molar masses of H_2, and N_2 are 2.0 and 28.0 lbm/lbmol, respectively (Table A-1E).

Analysis The mass of each component is determined from

$$N_{H_2} = 5 \text{lbmol} \longrightarrow m_{H_2} = N_{H_2} M_{H_2} = (5 \text{ lbmol})(2.0 \text{ lbm/lbmol}) = \mathbf{10 \text{ lbm}}$$

$$N_{N_2} = 3 \text{lbmol} \longrightarrow m_{N_2} = N_{N_2} M_{N_2} = (3 \text{ lbmol})(28 \text{ lbm/lbmol}) = \mathbf{84 \text{ lbm}}$$

5 lbmol H_2
3 lbmol N_2

The total mass and the total number of moles are

$$m_m = m_{H_2} + m_{N_2} = 10 \text{ lbm} + 84 \text{ lbm} = 94 \text{ lbm}$$

$$N_m = N_{H_2} + N_{N_2} = 5 \text{ lbmol} + 3 \text{ lbmol} = 8 \text{ lbm}$$

The molar mass and the gas constant of the mixture are determined from their definitions,

$$M_m = \frac{m_m}{N_m} = \frac{94 \text{ lbm}}{8 \text{ lbmol}} = 11.75 \text{ lbm/lbmol}$$

and

$$R_m = \frac{R_u}{M_m} = \frac{1.986 \text{ Btu/lbmol} \cdot \text{R}}{11.75 \text{ lbm/lbmol}} = \mathbf{0.169 \text{ Btu/lbm} \cdot \text{R}}$$

P-v-T Behavior of Gas Mixtures

12-17C Normally yes. Air, for example, behaves as an ideal gas in the range of temperatures and pressures at which oxygen and nitrogen behave as ideal gases.

12-18C The pressure of a gas mixture is equal to the sum of the pressures each gas would exert if existed alone at the mixture temperature and volume. This law holds exactly for ideal gas mixtures, but only approximately for real gas mixtures.

12-19C The volume of a gas mixture is equal to the sum of the volumes each gas would occupy if existed alone at the mixture temperature and pressure. This law holds exactly for ideal gas mixtures, but only approximately for real gas mixtures.

12-20C The P-v-T behavior of a component in an ideal gas mixture is expressed by the ideal gas equation of state using the properties of the individual component instead of the mixture, $P_i v_i = R_i T_i$. The *P-v-T* behavior of a component in a real gas mixture is expressed by more complex equations of state, or by $P_i v_i = Z_i R_i T_i$, where Z_i is the compressibility factor.

12-21C Component pressure is the pressure a component would exert if existed alone at the mixture temperature and volume. Partial pressure is the quantity $y_i P_m$, where y_i is the mole fraction of component *i*. These two are identical for ideal gases.

12-22C Component volume is the volume a component would occupy if existed alone at the mixture temperature and pressure. Partial volume is the quantity $y_i V_m$, where y_i is the mole fraction of component *i*. These two are identical for ideal gases.

12-23C The one with the highest mole number.

12-24C The partial pressures will decrease but the pressure fractions will remain the same.

12-25C The partial pressures will increase but the pressure fractions will remain the same.

12-26C No. The correct expression is "the volume of a gas mixture is equal to the sum of the volumes each gas would occupy if existed alone at the mixture temperature and pressure."

12-27C No. The correct expression is "the temperature of a gas mixture is equal to the temperature of the individual gas components."

12-28C Yes, it is correct.

12-29C With Kay's rule, a real-gas mixture is treated as a pure substance whose critical pressure and temperature are defined in terms of the critical pressures and temperatures of the mixture components as

$$P'_{cr,m} = \sum y_i P_{cr,i} \qquad \text{and} \qquad T'_{cr,m} = \sum y_i T_{cr,i}$$

The compressibility factor of the mixture (Z_m) is then easily determined using these pseudo-critical point values.

12-30 A tank contains a mixture of two gases of known masses at a specified pressure and temperature. The volume of the tank is to be determined.

Assumptions Under specified conditions both O_2 and CO_2 can be treated as ideal gases, and the mixture as an ideal gas mixture.

Analysis The total number of moles is

$$N_m = N_{O_2} + N_{CO_2} = 8\,\text{kmol} + 10\,\text{kmol} = 18\,\text{kmol}$$

Then

$$V_m = \frac{N_m R_u T_m}{P_m} = \frac{(18\,\text{kmol})(8.314\,\text{kPa}\cdot\text{m}^3/\text{kmol}\cdot\text{K})(290\,\text{K})}{150\,\text{kPa}} = \mathbf{289.3\,m^3}$$

8 kmol O_2
10 kmol CO_2

290 K
150 kPa

12-31 A tank contains a mixture of two gases of known masses at a specified pressure and temperature. The volume of the tank is to be determined.

Assumptions Under specified conditions both O_2 and CO_2 can be treated as ideal gases, and the mixture as an ideal gas mixture.

Analysis The total number of moles is

$$N_m = N_{O_2} + N_{CO_2} = 8\,\text{kmol} + 10\,\text{kmol} = 18\,\text{kmol}$$

Then

$$V_m = \frac{N_m R_u T_m}{P_m} = \frac{(18\,\text{kmol})(8.314\,\text{kPa}\cdot\text{m}^3/\text{kmol}\cdot\text{K})(350\,\text{K})}{150\,\text{kPa}} = \mathbf{349.2\,m^3}$$

8 kmol O_2
10 kmol CO_2

350 K
150 kPa

12-32 A tank contains a mixture of two gases of known masses at a specified pressure and temperature. The mixture is now heated to a specified temperature. The volume of the tank and the final pressure of the mixture are to be determined.

Assumptions Under specified conditions both Ar and N_2 can be treated as ideal gases, and the mixture as an ideal gas mixture.

Analysis The total number of moles is

$$N_m = N_{Ar} + N_{N_2} = 0.5\text{ kmol} + 2\text{ kmol} = 2.5\text{ kmol}$$

And

$$V_m = \frac{N_m R_u T_m}{P_m} = \frac{(2.5\text{ kmol})(8.314\text{ kPa}\cdot\text{m}^3/\text{kmol}\cdot\text{K})(280\text{ K})}{250\text{ kPa}} = \mathbf{23.3\,m^3}$$

0.5 kmol Ar
2 kmol N_2

280 K
250 kPa

Q

Also,

$$\frac{P_2 V_2}{T_2} = \frac{P_1 V_1}{T_1} \longrightarrow P_2 = \frac{T_2}{T_1} P_1 = \frac{400\text{ K}}{280\text{ K}}(250\text{ kPa}) = \mathbf{357.1\,kPa}$$

12-33 The masses of the constituents of a gas mixture at a specified pressure and temperature are given. The partial pressure of each gas and the apparent molar mass of the gas mixture are to be determined.

Assumptions Under specified conditions both CO_2 and CH_4 can be treated as ideal gases, and the mixture as an ideal gas mixture.

Properties The molar masses of CO_2 and CH_4 are 44.0 and 16.0 kg/kmol, respectively (Table A-1)

Analysis The mole numbers of the constituents are

1 kg CO_2
3 kg CH_4

300 K
200 kPa

$$m_{CO_2} = 1 \text{ kg} \longrightarrow N_{CO_2} = \frac{m_{CO_2}}{M_{CO_2}} = \frac{1 \text{ kg}}{44 \text{ kg/kmol}} = 0.0227 \text{ kmol}$$

$$m_{CH_4} = 3 \text{ kg} \longrightarrow N_{CH_4} = \frac{m_{CH_4}}{M_{CH_4}} = \frac{3 \text{ kg}}{16 \text{ kg/kmol}} = 0.1875 \text{ kmol}$$

$$N_m = N_{CO_2} + N_{CH_4} = 0.0227 \text{ kmol} + 0.1875 \text{ kmol} = 0.2102 \text{ kmol}$$

$$y_{CO_2} = \frac{N_{CO_2}}{N_m} = \frac{0.0227 \text{ kmol}}{0.2102 \text{ kmol}} = 0.108$$

$$y_{CH_4} = \frac{N_{CH_4}}{N_m} = \frac{0.1875 \text{ kmol}}{0.2102 \text{ kmol}} = 0.892$$

Then the partial pressures become

$$P_{CO_2} = y_{CO_2} P_m = (0.108)(200 \text{ kPa}) = \mathbf{21.6 \text{ kPa}}$$
$$P_{CH_4} = y_{CH_4} P_m = (0.892)(200 \text{ kPa}) = \mathbf{178.4 \text{ kPa}}$$

The apparent molar mass of the mixture is

$$M_m = \frac{m_m}{N_m} = \frac{4 \text{ kg}}{0.2102 \text{ kmol}} = \mathbf{19.03 \text{ kg/kmol}}$$

12-34E The masses of the constituents of a gas mixture at a specified pressure and temperature are given. The partial pressure of each gas and the apparent molar mass of the gas mixture are to be determined.

Assumptions Under specified conditions both CO_2 and CH_4 can be treated as ideal gases, and the mixture as an ideal gas mixture.

Properties The molar masses of CO_2 and CH_4 are 44.0 and 16.0 lbm/lbmol, respectively (Table A-1E)

Analysis The mole numbers of gases are

1 lbm CO_2
3 lbm CH_4

600 R
20 psia

$$m_{CO_2} = 1 \text{ lbm} \longrightarrow N_{CO_2} = \frac{m_{CO_2}}{M_{CO_2}} = \frac{1 \text{ lbm}}{44 \text{ lbm/lbmol}} = 0.0227 \text{ lbmol}$$

$$m_{CH_4} = 3 \text{ lbm} \longrightarrow N_{CH_4} = \frac{m_{CH_4}}{M_{CH_4}} = \frac{3 \text{ lbm}}{16 \text{ lbm/lbmol}} = 0.1875 \text{ lbmol}$$

$$N_m = N_{CO_2} + N_{CH_4} = 0.0227 \text{ lbmol} + 0.1875 \text{ lbmol} = 0.2102 \text{ lbmol}$$

$$y_{CO_2} = \frac{N_{CO_2}}{N_m} = \frac{0.0227 \text{ lbmol}}{0.2102 \text{ lbmol}} = 0.108$$

$$y_{CH_4} = \frac{N_{CH_4}}{N_m} = \frac{0.1875 \text{ lbmol}}{0.2102 \text{ lbmol}} = 0.892$$

Then the partial pressures become

$$P_{CO_2} = y_{CO_2} P_m = (0.108)(20\text{psia}) = \mathbf{2.16 \text{ psia}}$$

$$P_{CH_4} = y_{CH_4} P_m = (0.892)(20\text{psia}) = \mathbf{17.84 \text{ psia}}$$

The apparent molar mass of the mixture is

$$M_m = \frac{m_m}{N_m} = \frac{4 \text{ lbm}}{0.2102 \text{ lbmol}} = \mathbf{19.03 \text{ lbm/lbmol}}$$

12-35 The masses of the constituents of a gas mixture at a specified temperature are given. The partial pressure of each gas and the total pressure of the mixture are to be determined.

Assumptions Under specified conditions both N_2 and O_2 can be treated as ideal gases, and the mixture as an ideal gas mixture.

Analysis The partial pressures of constituent gases are

0.3 m^3

0.6 kg N_2
0.4 kg O_2

300 K

$$P_{N_2} = \left(\frac{mRT}{V}\right)_{N_2} = \frac{(0.6\text{kg})(0.2968\text{kPa} \cdot \text{m}^3/\text{kg} \cdot \text{K})(300\text{K})}{0.3\text{m}^3} = \mathbf{178.1 \text{ kPa}}$$

$$P_{O_2} = \left(\frac{mRT}{V}\right)_{O_2} = \frac{(0.4\text{kg})(0.2598\text{kPa} \cdot \text{m}^3/\text{kg} \cdot \text{K})(300\text{K})}{0.3\text{m}^3} = \mathbf{103.9 \text{ kPa}}$$

and

$$P_m = P_{N_2} + P_{O_2} = 178.1 \text{ kPa} + 103.9 \text{ kPa} = \mathbf{282.0 \text{ kPa}}$$

12-36 The volumetric fractions of the constituents of a gas mixture at a specified pressure and temperature are given. The mass fraction and partial pressure of each gas are to be determined.

Assumptions Under specified conditions all N_2, O_2 and CO_2 can be treated as ideal gases, and the mixture as an ideal gas mixture.

Properties The molar masses of N_2, O_2 and CO_2 are 28.0, 32.0, and 44.0 kg/kmol, respectively (Table A-1)

Analysis For convenience, consider 100 kmol of mixture. Then the mass of each component and the total mass are

$$N_{N_2} = 65 \text{ kmol} \longrightarrow m_{N_2} = N_{N_2} M_{N_2} = (65 \text{ kmol})(28 \text{ kg/kmol}) = 1820 \text{ kg}$$
$$N_{O_2} = 20 \text{ kmol} \longrightarrow m_{O_2} = N_{O_2} M_{O_2} = (20 \text{ kmol})(32 \text{ kg/kmol}) = 640 \text{ kg}$$
$$N_{CO_2} = 15 \text{ kmol} \longrightarrow m_{CO_2} = N_{CO_2} M_{CO_2} = (15 \text{ kmol})(44 \text{ kg/kmol}) = 660 \text{ kg}$$
$$m_m = m_{N_2} + m_{O_2} + m_{CO_2} = 1{,}820 \text{ kg} + 640 \text{ kg} + 660 \text{ kg} = 3{,}120 \text{ kg}$$

65% N_2
20% O_2
15% CO_2

350 K
300 kPa

Then the mass fraction of each component (gravimetric analysis) becomes

$$mf_{N_2} = \frac{m_{N_2}}{m_m} = \frac{1{,}820 \text{ kg}}{3{,}120 \text{ kg}} = 0.583 \text{ or } \mathbf{58.3\%}$$
$$mf_{O_2} = \frac{m_{O_2}}{m_m} = \frac{640 \text{ kg}}{3{,}120 \text{ kg}} = 0.205 \text{ or } \mathbf{20.5\%}$$
$$mf_{CO_2} = \frac{m_{CO_2}}{m_m} = \frac{660 \text{ kg}}{3{,}120 \text{ kg}} = 0.212 \text{ or } \mathbf{21.2\%}$$

For ideal gases, the partial pressure is proportional to the mole fraction, and is determined from

$$P_{N_2} = y_{N_2} P_m = (0.65)(300 \text{ kPa}) = \mathbf{195 \text{ kPa}}$$
$$P_{O_2} = y_{O_2} P_m = (0.20)(300 \text{ kPa}) = \mathbf{60 \text{ kPa}}$$
$$P_{CO_2} = y_{CO_2} P_m = (0.15)(300 \text{ kPa}) = \mathbf{45 \text{ kPa}}$$

12-37 The masses, temperatures, and pressures of two gases contained in two tanks connected to each other are given. The valve connecting the tanks is opened and the final temperature is measured. The volume of each tank and the final pressure are to be determined.

Assumptions Under specified conditions both N_2 and O_2 can be treated as ideal gases, and the mixture as an ideal gas mixture

Properties The molar masses of N_2 and O_2 are 28.0 and 32.0 kg/kmol, respectively (Table A-1)

Analysis The volumes of the tanks are

$$V_{N_2} = \left(\frac{mRT}{P}\right)_{N_2} = \frac{(2\text{kg})(0.2968\text{kPa}\cdot\text{m}^3/\text{kg}\cdot\text{K})(298\text{K})}{200\text{kPa}} = \mathbf{0.884\ m^3}$$

$$V_{O_2} = \left(\frac{mRT}{P}\right)_{O_2} = \frac{(3\text{kg})(0.2598\text{kPa}\cdot\text{m}^3/\text{kg}\cdot\text{K})(298\text{K})}{500\text{kPa}} = \mathbf{0.465\ m^3}$$

2 kg N_2, 25°C, 200 kPa | 3 kg O_2, 25°C, 500 kPa

$$V_{\text{total}} = V_{N_2} + V_{O_2} = 0.884\ \text{m}^3 + 0.465\ \text{m}^3 = 1.349\ \text{m}^3$$

Also,

$$m_{N_2} = 2\ \text{kg} \longrightarrow N_{N_2} = \frac{m_{N_2}}{M_{N_2}} = \frac{2\ \text{kg}}{28\ \text{kg/kmol}} = 0.0714\ \text{kmol}$$

$$m_{O_2} = 3\ \text{kg} \longrightarrow N_{O_2} = \frac{m_{O_2}}{M_{O_2}} = \frac{3\ \text{kg}}{32\ \text{kg/kmol}} = 0.0938\ \text{kmol}$$

$$N_m = N_{N_2} + N_{O_2} = 0.0714\ \text{kmol} + 0.0938\ \text{kmol} = 0.1652\ \text{kmol}$$

Thus,

$$P_m = \left(\frac{NR_uT}{V}\right)_m = \frac{(0.1652\text{kmol})(8.314\text{kPa}\cdot\text{m}^3/\text{kmol}\cdot\text{K})(298\text{K})}{1.349\text{m}^3} = \mathbf{303.4\ kPa}$$

12-38 The volumes, temperatures, and pressures of two gases forming a mixture are given. The volume of the mixture is to be determined using three methods.

Analysis (*a*) Under specified conditions both O_2 and N_2 will considerably deviate from the ideal gas behavior. Treating the mixture as an ideal gas,

$$N_{O_2} = \left(\frac{PV}{R_uT}\right)_{O_2} = \frac{(8000\text{kPa})(0.3\text{m}^3)}{(8.314\text{kPa}\cdot\text{m}^3/\text{kmol}\cdot\text{K})(200\text{K})} = 1.443\text{kmol}$$

$$N_{N_2} = \left(\frac{PV}{R_uT}\right)_{N_2} = \frac{(8000\text{kPa})(0.5\text{m}^3)}{(8.314\text{kPa}\cdot\text{m}^3/\text{kmol}\cdot\text{K})(200\text{K})} = 2.406\text{kmol}$$

0.3 m³ O_2, 200 K, 8 MPa → $N_2 + O_2$, 200 K, 8 MPa ← 0.5 m³ N_2, 200 K, 8 MPa

$$N_m = N_{O_2} + N_{N_2} = 1.443\ \text{kmol} + 2.406\ \text{kmol} = 3.849\ \text{kmol}$$

$$V_m = \frac{N_m R_u T_m}{P_m} = \frac{(3.849\ \text{kmol})(8.314\ \text{kPa}\cdot\text{m}^3/\text{kmol}\cdot\text{K})(200\ \text{K})}{8000\ \text{kPa}} = \mathbf{0.8\ m^3}$$

(*b*) To use Kay's rule, we need to determine the pseudo-critical temperature and pseudo-critical pressure of the mixture using the critical point properties of O_2 and N_2 from Table A-1. But we first need to determine the Z and the mole numbers of each component at the mixture temperature and pressure,

O_2:
$$\left.\begin{aligned} T_{R,O_2} &= \frac{T_m}{T_{cr,O_2}} = \frac{200\text{K}}{154.8\text{K}} = 1.292 \\ P_{R,O_2} &= \frac{P_m}{P_{cr,O_2}} = \frac{8\text{MPa}}{5.08\text{MPa}} = 1.575 \end{aligned}\right\} Z_{O_2} = 0.77$$

N_2:
$$\left.\begin{aligned} T_{R,N_2} &= \frac{T_m}{T_{cr,N_2}} = \frac{200\text{K}}{126.2\text{K}} = 1.585 \\ P_{R,N_2} &= \frac{P_m}{P_{cr,N_2}} = \frac{8\text{MPa}}{3.39\text{MPa}} = 2.360 \end{aligned}\right\} Z_{N_2} = 0.863$$

$$N_{O_2} = \left(\frac{PV}{ZR_uT}\right)_{O_2} = \frac{(8000\text{kPa})(0.3\text{m}^3)}{(0.77)(8.314\text{kPa}\cdot\text{m}^3/\text{kmol}\cdot\text{K})(200\text{K})} = 1.874\text{kmol}$$

$$N_{N_2} = \left(\frac{PV}{ZR_uT}\right)_{N_2} = \frac{(8000\text{kPa})(0.5\text{m}^3)}{(0.863)(8.314\text{kPa}\cdot\text{m}^3/\text{kmol}\cdot\text{K})(200\text{K})} = 2.787\text{kmol}$$

$$N_m = N_{O_2} + N_{N_2} = 1.874\text{ kmol} + 2.787\text{ kmol} = 4.661\text{ kmol}$$

The mole fractions are

$$y_{O_2} = \frac{N_{O_2}}{N_m} = \frac{1.874\text{ kmol}}{4.661\text{ kmol}} = 0.402 \quad \text{and} \quad y_{N_2} = \frac{N_{N_2}}{N_m} = \frac{2.787\text{ kmol}}{4.661\text{ kmol}} = 0.598$$

$$\begin{aligned} T'_{cr,m} &= \sum y_i T_{cr,i} = y_{O_2}T_{cr,O_2} + y_{N_2}T_{cr,N_2} \\ &= (0.402)(154.8\text{ K}) + (0.598)(126.2\text{ K}) = 137.7\text{ K} \end{aligned}$$

$$\begin{aligned} P'_{cr,m} &= \sum y_i P_{cr,i} = y_{O_2}P_{cr,O_2} + y_{N_2}P_{cr,N_2} \\ &= (0.402)(5.08\text{ MPa}) + (0.598)(3.39\text{ MPa}) = 4.07\text{ MPa} \end{aligned}$$

Then,

$$\left.\begin{aligned} T_R &= \frac{T_m}{T'_{cr,O_2}} = \frac{200\text{K}}{137.7\text{K}} = 1.452 \\ P_R &= \frac{P_m}{P'_{cr,O_2}} = \frac{8\text{MPa}}{4.07\text{MPa}} = 1.966 \end{aligned}\right\} Z_m = 0.82 \qquad \text{(Fig. A-30b)}$$

Thus, $$V_m = \frac{Z_m N_m R_u T_m}{P_m} = \frac{(0.82)(4.661\text{ kmol})(8.314\text{ kPa}\cdot\text{m}^3/\text{kmol}\cdot\text{K})(200\text{ K})}{8000\text{ kPa}} = \mathbf{0.79\ m^3}$$

(*c*) To use the Dalton's law for this real gas mixture, we first need the Z of each component at the mixture temperature and pressure, which are determined in part (*b*). Then,

$$Z_m = \sum y_i Z_i = y_{O_2}Z_{O_2} + y_{N_2}Z_{N_2} = (0.402)(0.77) + (0.598)(0.863) = 0.83$$

Thus, $$V_m = \frac{Z_m N_m R_u T_m}{P_m} = \frac{(0.83)(4.661\text{ kmol})(8.314\text{ kPa}\cdot\text{m}^3/\text{kmol}\cdot\text{K})(200\text{ K})}{8000\text{ kPa}} = \mathbf{0.80\ m^3}$$

12-39 [*Also solved by EES on enclosed CD*] The mole numbers, temperatures, and pressures of two gases forming a mixture are given. The final temperature is also given. The pressure of the mixture is to be determined using two methods.

Analysis (*a*) Under specified conditions both Ar and N_2 will considerably deviate from the ideal gas behavior. Treating the mixture as an ideal gas,

$$\left.\begin{array}{l}\text{Initial state}: P_1V_1 = N_1R_uT_1 \\ \text{Final state}: P_2V_2 = N_2R_uT_2\end{array}\right\} P_2 = \frac{N_2T_2}{N_1T_1}P_1 = \frac{(4)(200\text{ K})}{(1)(220\text{ K})}(5\text{ MPa}) = \mathbf{18.2\ MPa}$$

(*b*) Initially,

$$\left.\begin{array}{l} T_R = \dfrac{T_1}{T_{cr,\text{Ar}}} = \dfrac{220\text{K}}{151.0\text{K}} = 1.457 \\ P_R = \dfrac{P_1}{P_{cr,\text{Ar}}} = \dfrac{5\text{MPa}}{4.86\text{MPa}} = 1.0278 \end{array}\right\} Z_{\text{Ar}} = 0.90$$

1 kmol Ar
220 K
5 MPa

3 kmol N_2
190 K
8 MPa

Then the volume of the tank is

$$V = \frac{ZNR_uT}{P} = \frac{(0.90)(1\text{ kmol})(8.314\text{ kPa}\cdot\text{m}^3/\text{kmol}\cdot\text{K})(220\text{ K})}{5000\text{ kPa}} = 0.33\text{ m}^3$$

After mixing,

Ar:

$$\left.\begin{array}{l} T_{R,\text{Ar}} = \dfrac{T_m}{T_{cr,\text{Ar}}} = \dfrac{200\text{K}}{151.0\text{K}} = 1.325 \\ v_{R,\text{Ar}} = \dfrac{v_{\text{Ar}}}{R_uT_{cr,\text{Ar}}/P_{cr,\text{Ar}}} = \dfrac{V_m/N_{\text{Ar}}}{R_uT_{cr,\text{Ar}}/P_{cr,\text{Ar}}} \\ \quad = \dfrac{(0.33\text{m}^3)/(1\text{kmol})}{(8.314\text{kPa}\cdot\text{m}^3/\text{kmol}\cdot\text{K})(151.0\text{K})/(4860\text{kPa})} = 1.278 \end{array}\right\} P_R = 0.90$$

N_2:

$$\left.\begin{array}{l} T_{R,\text{N}_2} = \dfrac{T_m}{T_{cr,\text{N}_2}} = \dfrac{200\text{K}}{126.2\text{K}} = 1.585 \\ v_{R,\text{N}_2} = \dfrac{v_{\text{N}_2}}{R_uT_{cr,\text{N}_2}/P_{cr,\text{N}_2}} = \dfrac{V_m/N_{\text{N}_2}}{R_uT_{cr,\text{N}_2}/P_{cr,\text{N}_2}} \\ \quad = \dfrac{(0.33\text{m}^3)/(3\text{kmol})}{(8.314\text{kPa}\cdot\text{m}^3/\text{kmol}\cdot\text{K})(126.2\text{K})/(3390\text{kPa})} = 0.355 \end{array}\right\} P_R = 3.75$$

Thus,

$$P_{\text{Ar}} = (P_RP_{cr})_{\text{Ar}} = (0.90)(4.86\text{ MPa}) = 4.37\text{ MPa}$$
$$P_{\text{N}_2} = (P_RP_{cr})_{\text{N}_2} = (3.75)(3.39\text{ MPa}) = 12.7\text{ MPa}$$

and $$P_m = P_{\text{Ar}} + P_{\text{N}_2} = 4.37\text{ MPa} + 12.7\text{ MPa} = \mathbf{17.1\ MPa}$$

12-40 EES solution of this (and other comprehensive problems designated with the *computer icon*) is available to instructors at the *Instructor Manual* section of the *Online Learning Center* (OLC) at www.mhhe.com/cengel-boles. See the Preface for access information.

12-41E The mole numbers, temperatures, and pressures of two gases forming a mixture are given. For a specified final temperature, the pressure of the mixture is to be determined using two methods.

Properties The critical properties of Ar are T_{cr} = 272 R and P_{cr} = 705 psia. The critical properties of N_2 are T_{cr} = 227.1 R and P_{cr} = 492 psia (Table A-1E).

Analysis (*a*) Under specified conditions both Ar and N_2 will considerably deviate from the ideal gas behavior. Treating the mixture as an ideal gas,

$$\left.\begin{array}{l}\text{Initial state}: P_1V_1 = N_1R_uT_1 \\ \text{Final state}: P_2V_2 = N_2R_uT_2\end{array}\right\} P_2 = \frac{N_2T_2}{N_1T_1}P_1 = \frac{(4)(360\text{ R})}{(1)(400\text{ R})}(750\text{ psia}) = \mathbf{2700\ psia}$$

(*b*) Initially,

$$\left.\begin{array}{l}T_R = \dfrac{T_1}{T_{cr,\text{Ar}}} = \dfrac{400\text{R}}{272\text{R}} = 1.47 \\ P_R = \dfrac{P_1}{P_{cr,\text{Ar}}} = \dfrac{750\text{psia}}{705\text{psia}} = 1.07\end{array}\right\} Z_{\text{Ar}} = 0.90$$

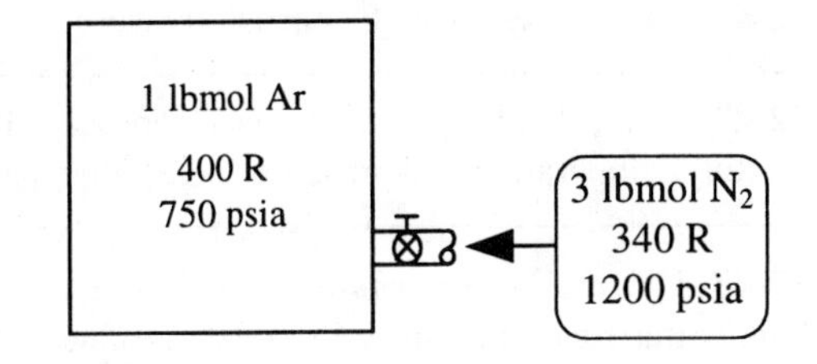

Then the volume of the tank is

$$V = \frac{ZNR_uT}{P} = \frac{(0.90)(1\text{ lbmol})(10.73\text{ psia}\cdot\text{ft}^3/\text{lbmol}\cdot\text{R})(400\text{ R})}{750\text{ psia}} = 5.15\text{ ft}^3$$

After mixing,

$$\text{Ar:}\quad\left.\begin{array}{l}T_{R,\text{Ar}} = \dfrac{T_m}{T_{cr,\text{Ar}}} = \dfrac{360\text{R}}{272\text{R}} = 1.324 \\ v_{R,\text{Ar}} = \dfrac{v_{\text{Ar}}}{R_uT_{cr,\text{Ar}}/P_{cr,Ar}} = \dfrac{V_m/N_{\text{Ar}}}{R_uT_{cr,\text{Ar}}/P_{cr,\text{Ar}}} \\ \quad = \dfrac{(5.15\text{ft}^3)/(1\text{lbmol})}{(10.73\text{psia}\cdot\text{ft}^3/\text{lbmol}\cdot\text{R})(272\text{R})/(705\text{psia})} = 1.244\end{array}\right\} P_R = 0.82$$

$$\text{N}_2\text{:}\quad\left.\begin{array}{l}T_{R,\text{N}_2} = \dfrac{T_m}{T_{cr,\text{N}_2}} = \dfrac{360\text{R}}{227.1\text{R}} = 1.585 \\ v_{R,\text{N}_2} = \dfrac{v_{\text{N}_2}}{R_uT_{cr,\text{N}_2}/P_{cr,\text{N}_2}} = \dfrac{V_m/N_{\text{N}_2}}{R_uT_{cr,\text{N}_2}/P_{cr,\text{N}_2}} \\ \quad = \dfrac{(5.15\text{ft}^3)/(3\text{lbmol})}{(10.73\text{psia}\cdot\text{ft}^3/\text{lbmol}\cdot\text{R})(227.1\text{R})/(492\text{psia})} = 0.347\end{array}\right\} P_R = 3.85$$

Thus,

$$P_{\text{Ar}} = (P_RP_{cr})_{\text{Ar}} = (0.82)(705\text{ psia}) = 578\text{ psia}$$
$$P_{\text{N}_2} = (P_RP_{cr})_{\text{N}_2} = (3.85)(492\text{ psia}) = 1{,}894\text{ psia}$$

and

$$P_m = P_{\text{Ar}} + P_{\text{N}_2} = 578\text{psia} + 1{,}894\text{psia} = \mathbf{2{,}472\ \ psia}$$

Properties of Gas Mixtures

12-42C Yes. Yes (extensive property).

12-43C No (intensive property).

12-44C The answers are the same for entropy.

12-45C Yes. Yes (conservation of energy).

12-46C We have to use the partial pressure.

12-47C No, this is an approximate approach. It assumes a component behaves as if it existed alone at the mixture temperature and pressure (i.e., it disregards the influence of dissimilar molecules on each other.)

12-48 The moles, temperatures, and pressures of two gases forming a mixture are given. The mixture temperature and pressure are to be determined.

Assumptions **1** Under specified conditions both CO_2 and H_2 can be treated as ideal gases, and the mixture as an ideal gas mixture. **2** The tank is insulated and thus there is no heat transfer. **3** There are no other forms of work involved.

Properties The molar masses and specific heats of CO_2 and H_2 are 44.0 kg/kmol, 2.0 kg/kmol, 0.657 kJ/kg.°C, and 10.183 kJ/kg.°C, respectively. (Tables A-1 and A-2b).

Analysis (*a*) We take both gases as our system. No heat, work, or mass crosses the system boundary, therefore this is a closed system with $Q = 0$ and $W = 0$. Then the energy balance for this closed system reduces to

$$E_{in} - E_{out} = \Delta E_{system}$$

$$0 = \Delta U = \Delta U_{CO_2} + \Delta U_{H_2}$$

$$0 = [mC_v(T_m - T_1)]_{CO_2} + [mC_v(T_m - T_1)]_{H_2}$$

CO_2	H_2
0.5 kmol	7.5 kmol
200 kPa	400 kPa
27°C	40°C

Using C_v values at room temperature and noting that $m = NM$, the final temperature of the mixture is determined to be

$$(0.5 \times 44\text{kg})(0.657\text{kJ/kg}\cdot°\text{C})(T_m - 27°\text{C}) + (7.5 \times 2\text{kg})(10.183\text{kJ/kg}\cdot°\text{C})(T_m - 40°\text{C}) = 0$$

$$T_m = \mathbf{38.9°C}\ (311.9\text{K})$$

(*b*) The volume of each tank is determined from

$$V_{CO_2} = \left(\frac{NR_uT_1}{P_1}\right)_{CO_2} = \frac{(0.5\text{kmol})(8.314\text{kPa}\cdot\text{m}^3/\text{kmol}\cdot\text{K})(300\text{K})}{200\text{kPa}} = 6.24\text{m}^3$$

$$V_{H_2} = \left(\frac{NR_uT_1}{P_1}\right)_{H_2} = \frac{(7.5\text{kmol})(8.314\text{kPa}\cdot\text{m}^3/\text{kmol}\cdot\text{K})(313\text{K})}{400\text{kPa}} = 48.79\text{m}^3$$

Thus,

$$V_m = V_{CO_2} + V_{H_2} = 6.24\text{ m}^3 + 48.79\text{ m}^3 = 55.03\text{ m}^3$$

$$N_m = N_{CO_2} + N_{H_2} = 0.5\text{ kmol} + 7.5\text{ kmol} = 8.0\text{ kmol}$$

and

$$P_m = \frac{N_mR_uT_m}{V_m} = \frac{(8.0\text{ kmol})(8.314\text{ kPa}\cdot\text{m}^3/\text{kmol}\cdot\text{K})(311.9\text{ K})}{55.03\text{ m}^3} = \mathbf{377\ kPa}$$

12-49 The temperatures and pressures of two gases forming a mixture are given. The final mixture temperature and pressure are to be determined.

Assumptions **1** Under specified conditions both Ne and Ar can be treated as ideal gases, and the mixture as an ideal gas mixture. **2** There are no other forms of work involved.

Properties The molar masses and specific heats of Ne and Ar are 20.18 kg/kmol, 39.95 kg/kmol, 0.6179 kJ/kg.°C, and 0.3122 kJ/kg.°C, respectively. (Tables A-1 and A-2b).

Analysis The mole number of each gas is

$$N_{\text{Ne}} = \left(\frac{P_1 V_1}{R_u T_1}\right)_{\text{Ne}} = \frac{(100\text{kPa})(0.45\text{m}^3)}{(8.314\text{kPa}\cdot\text{m}^3/\text{kmol}\cdot\text{K})(293\text{K})} = 0.0185\text{kmol}$$

$$N_{\text{Ar}} = \left(\frac{P_1 V_1}{R_u T_1}\right)_{\text{Ar}} = \frac{(200\text{kPa})(0.45\text{m}^3)}{(8.314\text{kPa}\cdot\text{m}^3/\text{kmol}\cdot\text{K})(323\text{K})} = 0.0335\text{kmol}$$

Ne	Ar
100 kPa	200 kPa
20°C	50°C

15 kJ

Thus,

$$N_m = N_{\text{Ne}} + N_{\text{Ar}} = 0.0185 \text{ kmol} + 0.0335 \text{ kmol} = 0.0520 \text{ kmol}$$

(*a*) We take both gases as the system. No work or mass crosses the system boundary, therefore this is a closed system with $W = 0$. Then the conservation of energy equation for this closed system reduces to

$$E_{\text{in}} - E_{\text{out}} = \Delta E_{\text{system}}$$

$$-Q_{\text{out}} = \Delta U = \Delta U_{\text{Ne}} + \Delta U_{\text{Ar}} \longrightarrow -Q_{\text{out}} = [mC_v(T_m - T_1)]_{\text{Ne}} + [mC_v(T_m - T_1)]_{\text{Ar}}$$

Using C_v values at room temperature and noting that m = NM, the final temperature of the mixture is determined to be

$$-15\text{kJ} = (0.0185 \times 20.18\text{kg})(0.6179\text{kJ/kg}\cdot°\text{C})(T_m - 20°\text{C})$$
$$+ (0.0335 \times 39.95\text{kg})(0.3122\text{kJ/kg}\cdot°\text{C})(T_m - 50°\text{C})$$
$$T_m = \mathbf{16.2°C} \ (289.2\text{K})$$

(*b*) The final pressure in the tank is determined from

$$P_m = \frac{N_m R_u T_m}{V_m} = \frac{(0.052 \text{ kmol})(8.314 \text{ kPa}\cdot\text{m}^3/\text{kmol}\cdot\text{K})(289.2 \text{ K})}{0.9 \text{ m}^3} = \mathbf{138.9 \ kPa}$$

12-50 The temperatures and pressures of two gases forming a mixture are given. The final mixture temperature and pressure are to be determined.

Assumptions **1** Under specified conditions both Ne and Ar can be treated as ideal gases, and the mixture as an ideal gas mixture. **2** There are no other forms of work involved.

Properties The molar masses and specific heats of Ne and Ar are 20.18 kg/kmol, 39.95 kg/kmol, 0.6179 kJ/kg.°C, and 0.3122 kJ/kg.°C, respectively. (Tables A-1 and A-2b).

Analysis The mole number of each gas is

$$N_{\text{Ne}} = \left(\frac{P_1 V_1}{R_u T_1}\right)_{\text{Ne}} = \frac{(100\text{kPa})(0.45\text{m}^3)}{(8.314\text{kPa}\cdot\text{m}^3/\text{kmol}\cdot\text{K})(293\text{K})} = 0.0185\text{kmol}$$

$$N_{\text{Ar}} = \left(\frac{P_1 V_1}{R_u T_1}\right)_{\text{Ar}} = \frac{(200\text{kPa})(0.45\text{m}^3)}{(8.314\text{kPa}\cdot\text{m}^3/\text{kmol}\cdot\text{K})(323\text{K})} = 0.0335\text{kmol}$$

Ne	Ar
100 kPa	200 kPa
20°C	50°C

8 kJ

Thus,

$$N_m = N_{\text{Ne}} + N_{\text{Ar}} = 0.0185 \text{ kmol} + 0.0335 \text{ kmol} = 0.0520 \text{ kmol}$$

(*a*) We take both gases as the system. No work or mass crosses the system boundary, therefore this is a closed system with $W = 0$. Then the conservation of energy equation for this closed system reduces to

$$E_{\text{in}} - E_{\text{out}} = \Delta E_{\text{system}}$$

$$-Q_{\text{out}} = \Delta U = \Delta U_{\text{Ne}} + \Delta U_{\text{Ar}} \longrightarrow -Q_{\text{out}} = \left[mC_v(T_m - T_1)\right]_{\text{Ne}} + \left[mC_v(T_m - T_1)\right]_{\text{Ar}}$$

Using C_v values at room temperature and noting that $m = NM$, the final temperature of the mixture is determined to be

$$-8\text{kJ} = (0.0185 \times 20.18\text{kg})(0.6179\text{kJ/kg}\cdot°\text{C})(T_m - 20°\text{C})$$
$$+ (0.0335 \times 39.95\text{kg})(0.3122\text{kJ/kg}\cdot°\text{C})(T_m - 50°\text{C})$$
$$T_m = \mathbf{27.0°C} \; (300.0\text{K})$$

(*b*) The final pressure in the tank is determined from

$$P_m = \frac{N_m R_u T_m}{V_m} = \frac{(0.052\text{ kmol})(8.314\text{ kPa}\cdot\text{m}^3/\text{kmol}\cdot\text{K})(300.0\text{ K})}{0.9\text{ m}^3} = \mathbf{144.1\text{ kPa}}$$

12-51 [*Also solved by EES on enclosed CD*] The temperatures and pressures of two gases forming a mixture in a mixing chamber are given. The mixture temperature and the rate of entropy generation are to be determined.

Assumptions **1** Under specified conditions both C_2H_6 and CH_4 can be treated as ideal gases, and the mixture as an ideal gas mixture. **2** The mixing chamber is insulated and thus there is no heat transfer. **3** There are no other forms of work involved. **3** This is a steady-flow process. **4** The kinetic and potential energy changes are negligible.

Properties The specific heats of C_2H_6 and CH_4 are 1.7662 kJ/kg.°C and 2.2537 kJ/kg.°C, respectively. (Table A-2b).

Analysis (*a*) The enthalpy of ideal gases is independent of pressure, and thus the two gases can be treated independently even after mixing. Noting that $\dot{W} = \dot{Q} = 0$, the steady-flow energy balance equation reduces to

$$\dot{E}_{in} - \dot{E}_{out} = \Delta \dot{E}_{system}^{\nearrow 0 \text{ (steady)}} = 0$$

$$\dot{E}_{in} = \dot{E}_{out}$$

$$\sum \dot{m}_i h_i = \sum \dot{m}_e h_e$$

$$0 = \sum \dot{m}_e h_e - \sum \dot{m}_i h_i = \dot{m}_{C_2H_6}(h_e - h_i)_{C_2H_6} + \dot{m}_{CH_4}(h_e - h_i)_{CH4.5}$$

$$0 = \left[\dot{m}C_p(T_e - T_i)\right]_{C_2H_6} + \left[\dot{m}C_p(T_e - T_i)\right]_{CH_4}$$

20°C
9 kg/s C_2H_6
200 kPa
45°C
CH

Using C_p values at room temperature and substituting, the exit temperature of the mixture becomes

$$0 = (9\text{kg/s})(1.7662\text{kJ/kg}\cdot°\text{C})(T_m - 20°\text{C}) + (4.5\text{kg/s})(2.2537\text{kJ/kg}\cdot°\text{C})(T_m - 45°\text{C})$$

$$T_m = \mathbf{29.7°C} \ (302.7\text{K})$$

(*b*) The rate of entropy change associated with this process is determined from an entropy balance on the mixing chamber,

$$\dot{S}_{in} - \dot{S}_{out} + \dot{S}_{gen} = \Delta \dot{S}_{system}^{\nearrow 0} = 0$$

$$[\dot{m}(s_1 - s_2)]_{C_2H_6} + [\dot{m}(s_1 - s_2)]_{CH_4} + \dot{S}_{gen} = 0 \rightarrow \dot{S}_{gen} = [\dot{m}(s_2 - s_1)]_{C_2H_6} + [\dot{m}(s_2 - s_1)]_{CH_4}$$

The molar flow rate of the two gases in the mixture is

$$\dot{N}_{C_2H_6} = \left(\frac{\dot{m}}{M}\right)_{C_2H_6} = \frac{9\text{kg/s}}{30\text{kg/kmol}} = 0.3\text{kmol/s}$$

$$\dot{N}_{CH_4} = \left(\frac{\dot{m}}{M}\right)_{CH_4} = \frac{4.5\text{kg/s}}{16\text{kg/kmol}} = 0.2813\text{kmol/s}$$

Then the mole fraction of each gas becomes

$$y_{C_2H_6} = \frac{0.3}{0.3 + 0.2813} = 0.516$$

$$y_{CH_4} = \frac{0.2813}{0.3 + 0.2813} = 0.484$$

Thus,

$$(s_2 - s_1)_{C_2H_6} = \left(C_p \ln \frac{T_2}{T_1} - R \ln \frac{yP_{m,2}}{P_1} \right)_{C_2H_6} = \left(C_p \ln \frac{T_2}{T_1} - R \ln y \right)_{C_2H_6}$$

$$= (1.7662 \text{kJ/kg} \cdot \text{K}) \ln \frac{302.7 \text{K}}{293 \text{K}} - (0.2765 \text{kJ/kg} \cdot \text{K}) \ln 0.516$$

$$= 0.240 \text{kJ/kg} \cdot \text{K}$$

$$(s_2 - s_1)_{CH_4} = \left(C_p \ln \frac{T_2}{T_1} - R \ln \frac{yP_{m,2}}{P_1} \right)_{CH_4} = \left(C_p \ln \frac{T_2}{T_1} - R \ln y \right)_{CH_4}$$

$$= (2.2537 \text{kJ/kg} \cdot \text{K}) \ln \frac{302.7 \text{K}}{318 \text{K}} - (0.5182 \text{kJ/kg} \cdot \text{K}) \ln 0.484$$

$$= 0.265 \text{kJ/kg} \cdot \text{K}$$

Noting that $P_{m,2} = P_{i,1} = 200$ kPa and substituting,

$$\dot{S}_{gen} = (9 \text{kg/s})(0.240 \text{kJ/kg} \cdot \text{K}) + (4.5 \text{kg/s})(0.265 \text{kJ/kg} \cdot \text{K}) = \mathbf{3.353 \ kW/K}$$

12-52 EES solution of this (and other comprehensive problems designated with the *computer icon*) is available to instructors at the *Instructor Manual* section of the *Online Learning Center* (OLC) at www.mhhe.com/cengel-boles. See the Preface for access information.

12-53 An equimolar mixture of helium and argon gases expands in a turbine. The isentropic work output of the turbine is to be determined.

Assumptions **1** Under specified conditions both He and Ar can be treated as ideal gases, and the mixture as an ideal gas mixture. **2** The turbine is insulated and thus there is no heat transfer. **3** This is a steady-flow process. **4** The kinetic and potential energy changes are negligible.

Properties The molar masses and specific heats of He and Ar are 4.0 kg/kmol, 40.0 kg/kmol, 5.1926 kJ/kg.°C, and 0.5203 kJ/kg.°C, respectively. (Table A-1 and Table A-2b).

Analysis The C_p and k values of this equimolar mixture are determined from

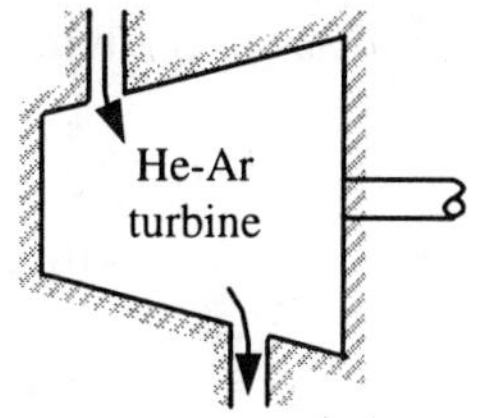

1.2 MPa
1300 K

He-Ar
turbine

200 kPa

$$M_m = \sum y_i M_i = y_{He} M_{He} + y_{Ar} M_{Ar} = 0.5 \times 4 + 0.5 \times 40 = 22 \text{ kg/kmol}$$

$$mf_i = \frac{m_i}{m_m} = \frac{N_i M_i}{N_m M_m} = \frac{y_i M_i}{M_m}$$

$$C_{p,m} = \sum mf_i C_{p,i} = \frac{y_{He} M_{He}}{M_m} C_{p,He} + \frac{y_{Ar} M_{Ar}}{M_m} C_{p,Ar}$$

$$= \frac{0.5 \times 4 \text{ kg/kmol}}{22 \text{ kg/kmol}} (5.1926 \text{ kJ/kg} \cdot \text{K}) + \frac{0.5 \times 40 \text{ kg/kmol}}{22 \text{ kg/kmol}} (0.5203 \text{ kJ/kg} \cdot \text{K})$$

$$= 0.945 \text{ kJ/kg} \cdot \text{K}$$

and

$k_m = 1.667$ since $k = 1.667$ for both gases.

Therefore, the He-Ar mixture can be treated as a single ideal gas with the properties above. For isentropic processes,

$$T_2 = T_1 \left(\frac{P_2}{P_1} \right)^{(k-1)/k} = (1200\text{K}) \left(\frac{200\text{kPa}}{1300\text{kPa}} \right)^{0.667/1.667} = 567.4\text{K}$$

From an energy balance on the turbine,

$$\dot{E}_{in} - \dot{E}_{out} = \Delta \dot{E}_{system}^{\nearrow 0 \text{ (steady)}} = 0$$

$$\dot{E}_{in} = \dot{E}_{out}$$

$$h_1 = h_2 + w_{out}$$

$$w_{out} = h_1 - h_2$$

$$w_{out} = C_p (T_1 - T_2) = (0.945\text{kJ/kg} \cdot \text{K})(1200 - 567.4)\text{K} = \mathbf{597.8 \ kJ/kg}$$

12-54E [*Also solved by EES on enclosed CD*] A gas mixture with known mass fractions is accelerated through a nozzle from a specified state to a specified pressure. For a specified isentropic efficiency, the exit temperature and the exit velocity of the mixture are to be determined.

Assumptions **1** Under specified conditions both N_2 and CO_2 can be treated as ideal gases, and the mixture as an ideal gas mixture. **2** The nozzle is adiabatic and thus heat transfer is negligible. **3** This is a steady-flow process. **4** Potential energy changes are negligible.

Properties The specific heats of N_2 and CO_2 are $C_{p,N2}$ = 0.248 Btu/lbm.R, $C_{v,N2}$ = 0.177 Btu/lbm.R, $C_{p,CO2}$ = 0.203 Btu/lbm.R, and $C_{v,CO2}$ = 0.158 Btu/lbm.R. (Table A-2Eb).

Analysis (*a*) Under specified conditions both N_2 and CO_2 can be treated as ideal gases, and the mixture as an ideal gas mixture. The C_p, C_v, and k values of this mixture are determined from

$$C_{p,m} = \sum mf_i C_{p,i} = mf_{N_2} C_{p,N_2} + mf_{CO_2} C_{p,CO_2}$$
$$= (0.8)(0.248) + (0.2)(0.203) = 0.239 \text{Btu/lbm} \cdot \text{R}$$
$$C_{v,m} = \sum mf_i C_{v,i} = mf_{N_2} C_{v,N_2} + mf_{CO_2} C_{v,CO_2}$$
$$= (0.8)(0.177) + (0.2)(0.158) = 0.173 \text{Btu/lbm} \cdot \text{R}$$
$$k_m = \frac{C_{p,m}}{C_{v,m}} = \frac{0.239 \text{Btu/lbm} \cdot \text{R}}{0.173 \text{Btu/lbm} \cdot \text{R}} = 1.382$$

90 psia
1800 R
80% N_2
20% CO_2
12 psia

Therefore, the N_2-CO_2 mixture can be treated as a single ideal gas with above properties. Then the isentropic exit temperature can be determined from

$$T_{2s} = T_1 \left(\frac{P_2}{P_1}\right)^{(k-1)/k} = (1{,}800 \text{R})\left(\frac{12 \text{psia}}{90 \text{psia}}\right)^{0.382/1.382} = 1031.3 \text{R}$$

From the definition of adiabatic efficiency,

$$\eta_N = \frac{h_1 - h_2}{h_1 - h_{2s}} = \frac{C_p(T_1 - T_2)}{C_p(T_1 - T_{2s})} \longrightarrow 0.92 = \frac{1{,}800 - T_2}{1{,}800 - 1031.3} \longrightarrow T_2 = \mathbf{1092.8\ R}$$

(*b*) Noting that, $q = w = 0$, from the steady-flow energy balance relation,

$$\dot{E}_{\text{in}} - \dot{E}_{\text{out}} = \Delta \dot{E}_{\text{system}}^{\not 0 \text{ (steady)}} = 0$$
$$\dot{E}_{\text{in}} = \dot{E}_{\text{out}}$$
$$h_1 + \mathbf{V}_1^2/2 = h_2 + \mathbf{V}_2^2/2$$
$$0 = C_p(T_2 - T_1) + \frac{\mathbf{V}_2^2 - \mathbf{V}_1^{2 \not 0}}{2}$$
$$\mathbf{V}_2 = \sqrt{2C_p(T_1 - T_2)} = \sqrt{2(0.239 \text{Btu/lbm} \cdot \text{R})(1{,}800 - 1{,}092.8)\text{R}\left(\frac{25{,}037 \text{ft}^2/\text{s}^2}{1 \text{Btu/lbm}}\right)} = \mathbf{2{,}909\ ft/s}$$

12-55E EES solution of this (and other comprehensive problems designated with the *computer icon*) is available to instructors at the *Instructor Manual* section of the *Online Learning Center* (OLC) at www.mhhe.com/cengel-boles. See the Preface for access information.

12-56 A piston-cylinder device contains a gas mixture at a given state. Heat is transferred to the mixture. The amount of heat transfer and the entropy change of the mixture are to be determined.

Assumptions **1** Under specified conditions both H_2 and N_2 can be treated as ideal gases, and the mixture as an ideal gas mixture. **2** Kinetic and potential energy changes are negligible.

Properties The constant pressure specific heats of H_2 and N_2 at 450 K are 14.501 kJ/kg.K and 1.039 kJ/kg.K, respectively. (Table A-2b).

Analysis (*a*) Noting that $P_2 = P_1$ and $V_2 = 2V_1$,

0.2 kg H_2
1.6 kg N_2
100 kPa
300 K
Q

$$\frac{P_2V_2}{T_2} = \frac{P_1V_1}{T_1} \longrightarrow T_2 = \frac{2V_1}{V_1}T_1 = 2T_1 = (2)(300\text{K}) = 600\text{K}$$

Also P = constant. Then from the closed system energy balance relation,

$$E_{\text{in}} - E_{\text{out}} = \Delta E_{\text{system}}$$

$$Q_{\text{in}} - W_{b,\text{out}} = \Delta U \quad \rightarrow \quad Q_{\text{in}} = \Delta H$$

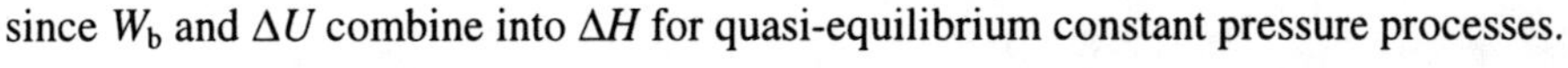

since W_b and ΔU combine into ΔH for quasi-equilibrium constant pressure processes.

$$\begin{aligned} Q_{\text{in}} &= \Delta H = \Delta H_{H_2} + \Delta H_{N_2} = \left[mC_{p,\text{ave}}(T_2 - T_1)\right]_{H_2} + \left[mC_{p,\text{ave}}(T_2 - T_1)\right]_{N_2} \\ &= (0.2\text{kg})(14.501\text{kJ/kg}\cdot\text{K})(600-300)\text{K} + (1.6\text{kg})(1.049\text{kJ/kg}\cdot\text{K})(600-300)\text{K} \\ &= \mathbf{1{,}374\ kJ} \end{aligned}$$

(*b*) Noting that the total mixture pressure, and thus the partial pressure of each gas, remains constant, the entropy change of the mixture during this process is

$$\begin{aligned} \Delta S_{H_2} &= \left[m(s_2 - s_1)\right]_{H_2} = m_{H_2}\left(C_p \ln\frac{T_2}{T_1} - R\ln\frac{P_2}{P_1}^{\nearrow 0}\right)_{H_2} = \left(C_p \ln\frac{T_2}{T_1}\right)_{H_2} \\ &= (0.2\text{kg})(14.501\text{kJ/kg}\cdot\text{K})\ln\frac{600\text{K}}{300\text{K}} \\ &= 2.01\text{kJ/K} \end{aligned}$$

$$\begin{aligned} \Delta S_{N_2} &= \left[m(s_2 - s_1)\right]_{N_2} = m_{N_2}\left(C_p \ln\frac{T_2}{T_1} - R\ln\frac{P_2}{P_1}^{\nearrow 0}\right)_{N_2} = \left(C_p \ln\frac{T_2}{T_1}\right)_{N_2} \\ &= (1.6\text{kg})(1.049\text{kJ/kg}\cdot\text{K})\ln\frac{600\text{K}}{300\text{K}} \\ &= 1.160\text{kJ/K} \end{aligned}$$

$$\Delta S_{\text{total}} = \Delta S_{H_2} + \Delta S_{N_2} = 2.01\text{kJ/K} + 1.16\text{kJ/K} = \mathbf{3.17\ kJ/K}$$

12-57 The states of two gases contained in two tanks are given. The gases are allowed to mix to form a homogeneous mixture. The final pressure, the heat transfer, and the entropy generated are to be determined.

Assumptions **1** Under specified conditions both O_2 and N_2 can be treated as ideal gases, and the mixture as an ideal gas mixture. **2** The tank containing oxygen is insulated. **3** There are no other forms of work involved.

Properties The constant volume specific heats of O_2 and N_2 are 0.658 kJ/kg.°C and 0.743 kJ/kg.°C, respectively. (Table A-2b).

Analysis (*a*) The volume of the O_2 tank and mass of the nitrogen are

$$V_{1,O_2} = \left(\frac{mRT_1}{P_1}\right)_{O_2} = \frac{(1\ \text{kg})(0.2598\ \text{kPa}\cdot\text{m}^3/\text{kg}\cdot\text{K})(288\ \text{K})}{300\ \text{kPa}} = 0.25\ \text{m}^3$$

$$m_{N_2} = \left(\frac{P_1V_1}{RT_1}\right)_{N_2} = \frac{(500\ \text{kPa})(2\ \text{m}^3)}{(0.2968\ \text{kPa}\cdot\text{m}^3/\text{kg}\cdot\text{K})(323\ \text{K})} = 10.43\ \text{kg}$$

$$V_{\text{total}} = V_{1,O_2} + V_{1,N_2} = 0.25\ \text{m}^3 + 2.0\ \text{m}^3 = 2.25\ \text{m}^3$$

Also,

$$m_{O_2} = 1\ \text{kg} \longrightarrow N_{O_2} = \frac{m_{O_2}}{M_{O_2}} = \frac{1\ \text{kg}}{32\ \text{kg/kmol}} = 0.03125\ \text{kmol}$$

$$m_{N_2} = 10.43\ \text{kg} \longrightarrow N_{N_2} = \frac{m_{N_2}}{M_{N_2}} = \frac{10.43\ \text{kg}}{28\ \text{kg/kmol}} = 0.3725\ \text{kmol}$$

$$N_m = N_{N_2} + N_{O_2} = 0.3725\ \text{kmol} + 0.03125\ \text{kmol} = 0.40375\ \text{kmol}$$

Thus,

$$P_m = \left(\frac{NR_uT}{V}\right)_m = \frac{(0.40375\ \text{kmol})(8.314\ \text{kPa}\cdot\text{m}^3/\text{kmol}\cdot\text{K})(298\ \text{K})}{2.25\ \text{m}^3} = \mathbf{444.6\ kPa}$$

(*b*) We take both gases as the system. No work or mass crosses the system boundary, and thus this is a closed system with $W = 0$. Taking the direction of heat transfer to be from the system (will be verified), the energy balance for this closed system reduces to

$$E_{\text{in}} - E_{\text{out}} = \Delta E_{\text{system}}$$

$$-Q_{\text{out}} = \Delta U = \Delta U_{O_2} + \Delta U_{N_2} \longrightarrow Q_{out} = [mC_v(T_1 - T_m)]_{O_2} + [mC_v(T_1 - T_m)]_{N_2}$$

Using C_v values at room temperature (Table A-2b), the heat transfer is determined to be

$$Q_{\text{out}} = (1\ \text{kg})(0.658\ \text{kJ/kg}\cdot°\text{C})(15-25)°\text{C} + (10.43\ \text{kg})(0.743\ \text{kJ/kg}\cdot°\text{C})(50-25)°\text{C}$$
$$= \mathbf{187.2\ kJ}\ \ (\text{from the system})$$

(*c*) For and *extended system* that involves the tanks and their immediate surroundings such that the boundary temperature is the surroundings temperature, the entropy balance can be expressed as

$$S_{\text{in}} - S_{\text{out}} + S_{\text{gen}} = \Delta S_{\text{system}}$$

$$-\frac{Q_{\text{out}}}{T_{b,\text{surr}}} + S_{\text{gen}} = m(s_2 - s_1) \rightarrow S_{\text{gen}} = m(s_2 - s_1) + \frac{Q_{\text{out}}}{T_{\text{surr}}}$$

The mole fraction of each gas is

$$y_{O_2} = \frac{N_{O_2}}{N_m} = \frac{0.0625}{0.8075} = 0.077$$

$$y_{N_2} = \frac{N_{N_2}}{N_m} = \frac{0.745}{0.8075} = 0.923$$

Thus,

$$(s_2 - s_1)_{O_2} = \left(C_p \ln \frac{T_2}{T_1} - R \ln \frac{yP_{m,2}}{P_1} \right)_{O_2}$$
$$= (0.918 \text{ kJ/kg}\cdot\text{K}) \ln \frac{298 \text{ K}}{288 \text{ K}} - (0.2598 \text{ kJ/kg}\cdot\text{K}) \ln \frac{(0.077)(444.6 \text{ kPa})}{300 \text{ kPa}}$$
$$= 0.5952 \text{ kJ/kg}\cdot\text{K}$$

$$(s_2 - s_1)_{N_2} = \left(C_p \ln \frac{T_2}{T_1} - R \ln \frac{yP_{m,2}}{P_1} \right)_{N_2}$$
$$= (1.039 \text{ kJ/kg}\cdot\text{K}) \ln \frac{298 \text{ K}}{323 \text{ K}} - (0.2968 \text{ kJ/kg}\cdot\text{K}) \ln \frac{(0.923)(444.6 \text{ kPa})}{500 \text{ kPa}}$$
$$= -0.0251 \text{ kJ/kg}\cdot\text{K}$$

Substituting,

$$S_{\text{gen}} = (1 \text{ kg})(0.5952 \text{ kJ/kg}\cdot\text{K}) + (10.43 \text{ kg})(-0.0251 \text{ kJ/kg}\cdot\text{K}) + \frac{187.2 \text{ kJ}}{298 \text{ K}} = \mathbf{0.962 \text{ kJ/K}}$$

12-58 EES solution of this (and other comprehensive problems designated with the *computer icon*) is available to instructors at the *Instructor Manual* section of the *Online Learning Center* (OLC) at www.mhhe.com/cengel-boles. See the Preface for access information.

12-59 Heat is transferred to a gas mixture contained in a piston cylinder device. The initial state and the final temperature are given. The heat transfer is to be determined for the ideal gas and non-ideal gas cases.

Properties The molar masses of H_2 and N_2 are 2.0, and 28.0 kg/kmol. (Table A-1).

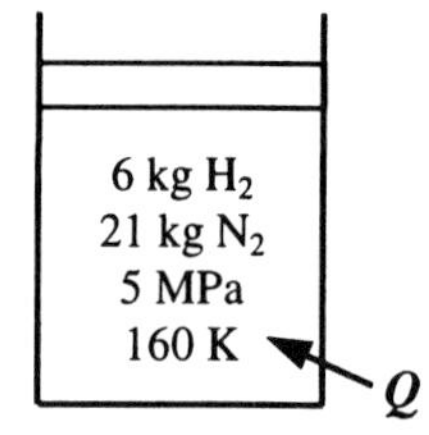

Analysis From the energy balance relation,

$$E_{in} - E_{out} = \Delta E$$

$$Q_{in} - W_{b,out} = \Delta U$$

$$Q_{in} = \Delta H = \Delta H_{H_2} + \Delta H_{N_2} = N_{H_2}\left(\bar{h}_2 - \bar{h}_1\right)_{H_2} + N_{N_2}\left(\bar{h}_2 - \bar{h}_1\right)_{N_2}$$

since W_b and ΔU combine into ΔH for quasi-equilibrium constant pressure processes

$$N_{H_2} = \frac{m_{H_2}}{M_{H_2}} = \frac{6\text{ kg}}{2\text{ kg/kmol}} = 3\text{ kmol}$$

$$N_{N_2} = \frac{m_{N_2}}{M_{N_2}} = \frac{21\text{ kg}}{28\text{ kg/kmol}} = 0.75\text{ kmol}$$

(*a*) Assuming ideal gas behavior, the inlet and exit enthalpies of H_2 and N_2 are determined from the ideal gas tables to be

$$H_2:\quad \bar{h}_1 = \bar{h}_{@\,160\text{ K}} = 4{,}535.4\text{ kJ/kmol},\qquad \bar{h}_2 = \bar{h}_{@\,200\text{ K}} = 5{,}669.2\text{ kJ/kmol}$$

$$N_2:\quad \bar{h}_1 = \bar{h}_{@\,160\text{ K}} = 4{,}648\text{ kJ/kmol},\qquad \bar{h}_2 = \bar{h}_{@\,200\text{ K}} = 5{,}810\text{ kJ/kmol}$$

Thus, $Q_{ideal} = 3 \times (5{,}669.2 - 4{,}535.4) + 0.75 \times (5{,}810 - 4{,}648) = \mathbf{4273\ kJ}$

(*b*) Using Amagat's law and the generalized enthalpy departure chart, the enthalpy change of each gas is determined to be

$$H_2:\quad \left.\begin{aligned} T_{R_1,H_2} &= \frac{T_{m,1}}{T_{cr,H_2}} = \frac{160}{33.3} = 4.805 \\ P_{R_1,H_2} &= P_{R_2,H_2} = \frac{P_m}{P_{cr,H_2}} = \frac{5}{1.30} = 3.846 \\ T_{R_2,H_2} &= \frac{T_{m,2}}{T_{cr,H_2}} = \frac{200}{33.3} = 6.006 \end{aligned}\right\} \begin{aligned} Z_{h_1} &\cong 0 \\ Z_{h_2} &\cong 0 \end{aligned}$$

Thus H_2 can be treated as an ideal gas during this process.

$$N_2:\quad \left.\begin{aligned} T_{R_1,N_2} &= \frac{T_{m,1}}{T_{cr,N_2}} = \frac{160}{126.2} = 1.27 \\ P_{R_1,N_2} &= P_{R_2,N_2} = \frac{P_m}{P_{cr,N_2}} = \frac{5}{3.39} = 1.47 \\ T_{R_2,N_2} &= \frac{T_{m,2}}{T_{cr,N_2}} = \frac{200}{126.2} = 1.58 \end{aligned}\right\} \begin{aligned} Z_{h_1} &= 1.3 \\ Z_{h_2} &= 0.7 \end{aligned}$$

Therefore,

$$\left(\bar{h}_2 - \bar{h}_1\right)_{H_2} = \left(\bar{h}_2 - \bar{h}_1\right)_{H_2,ideal} = 5{,}669.2 - 4{,}535.4 = 1{,}133.8\text{kJ/kmol}$$

$$\begin{aligned}\left(\bar{h}_2 - \bar{h}_1\right)_{N_2} &= R_u T_{cr}\left(Z_{h_1} - Z_{h_2}\right) + \left(\bar{h}_2 - \bar{h}_1\right)_{ideal} \\ &= (8.314\text{kPa}\cdot\text{m}^3\text{/kmol}\cdot\text{K})(126.2\text{K})(1.3 - 0.7) + (5{,}810 - 4{,}648)\text{kJ/kmol} \\ &= 1{,}791.5\text{kJ/kmol}\end{aligned}$$

Substituting, $Q_{in} = (3\text{kmol})(1{,}133.8\text{kJ/kmol}) + (0.75\text{kmol})(1{,}791.5\text{kJ/kmol}) = \mathbf{4745kJ}$

12-60 Heat is transferred to a gas mixture contained in a piston cylinder device discussed in previous problem. The total entropy change and the exergy destruction are to be determined for two cases.

Analysis The entropy generated during this process is determined by applying the entropy balance on an *extended system* that includes the piston-cylinder device and its immediate surroundings so that the boundary temperature of the extended system is the environment temperature at all times. It gives

$$S_{in} - S_{out} + S_{gen} = \Delta S_{system}$$

$$\frac{Q_{in}}{T_{boundary}} + S_{gen} = \Delta S_{water} \rightarrow S_{gen} = m(s_2 - s_1) - \frac{Q_{in}}{T_{surr}}$$

Then the exergy destroyed during a process can be determined from its definition $X_{destroyed} = T_0 S_{gen}$.

(*a*) Noting that the total mixture pressure, and thus the partial pressure of each gas, remains constant, the entropy change of a component in the mixture during this process is

$$\Delta S_i = m_i \left(C_p \ln\frac{T_2}{T_1} - R\ln\frac{P_2}{P_1}^{\phi 0} \right)_i = m_i C_{p,i} \ln\frac{T_2}{T_1}$$

Assuming ideal gas behavior and using C_p values at the average temperature, the ΔS of H_2 and N_2 are determined from

$$\Delta S_{H_2,ideal} = (6\text{kg})(13.60\text{kJ/kg}\cdot\text{K})\ln\frac{200\text{K}}{160\text{K}} = 18.21\text{kJ/K}$$

$$\Delta S_{N_2,ideal} = (21\text{kg})(1.039\text{kJ/kg}\cdot\text{K})\ln\frac{200\text{K}}{160\text{K}} = 4.87\text{kJ/K}$$

and

$$S_{gen} = 18.21\text{kJ/K} + 4.87\text{kJ/K} - \frac{4{,}273\text{kJ}}{298\text{K}} = \mathbf{8.74\ kJ/K}$$

$$X_{destroyed} = T_0 S_{gen} = (298\text{K})(8.74\text{kJ/K}) = \mathbf{2{,}605\ kJ}$$

(*b*) Using Amagat's law and the generalized entropy departure chart, the entropy change of each gas is determined to be

H_2:

$$\left.\begin{aligned} T_{R_1,H_2} &= \frac{T_{m,1}}{T_{cr,H_2}} = \frac{160}{33.3} = 4.805 \\ P_{R_1,H_2} = P_{R_2,H_2} &= \frac{P_m}{P_{cr,H_2}} = \frac{5}{1.30} = 3.846 \\ T_{R_2,H_2} &= \frac{T_{m,2}}{T_{cr,H_2}} = \frac{200}{33.3} = 6.006 \end{aligned}\right\} \begin{aligned} Z_{s_1} &\cong 1 \\ Z_{s_2} &\cong 1 \end{aligned}$$

Thus H_2 can be treated as an ideal gas during this process.

N_2:

$$\left.\begin{aligned} T_{R_1,N_2} &= \frac{T_{m,1}}{T_{cr,N_2}} = \frac{160}{126.2} = 1.268 \\ P_{R_1,N_2} = P_{R_2,N_2} &= \frac{P_m}{P_{cr,N_2}} = \frac{5}{3.39} = 1.475 \\ T_{R_2,N_2} &= \frac{T_{m,2}}{T_{cr,N_2}} = \frac{200}{126.2} = 1.585 \end{aligned}\right\} \begin{aligned} Z_{s_1} &= 0.8 \\ Z_{s_2} &= 0.4 \end{aligned}$$

Therefore,

$$\Delta S_{H_2} = \Delta S_{H_2,\text{ideal}} = 18.21\text{kJ/K}$$

$$\begin{aligned}\Delta S_{N_2} &= N_{N_2} R_u \left(Z_{s_1} - Z_{s_2}\right) + \Delta S_{N_2,\text{ideal}} \\ &= (0.75\text{kmol})(8.314\text{kPa}\cdot\text{m}^3/\text{kmol}\cdot\text{K})(0.8-0.4) + (4.87\text{kJ/K}) \\ &= 7.37\text{kJ/K}\end{aligned}$$

$$\Delta S_{\text{surr}} = \frac{Q_{\text{surr}}}{T_0} = \frac{-4{,}745\text{kJ}}{298\text{K}} = -15.92\text{kJ/K}$$

and

$$S_{\text{gen}} = 18.21\text{kJ/K} + 7.37\text{kJ/K} - \frac{4{,}745\text{kJ}}{298\text{K}} = \mathbf{9.66\ kJ/K}$$

$$X_{\text{destroyed}} = T_0 S_{\text{gen}} = (298\text{K})(9.66\text{kJ/K}) = \mathbf{2{,}879\ kJ}$$

12-61 Air is compressed isothermally in a steady-flow device. The power input to the compressor and the rate of heat rejection are to be determined for ideal and non-ideal gas cases.

Assumptions **1** This is a steady-flow process. **2** The kinetic and potential energy changes are negligible.

Properties The molar mass of air is 29.0 kg/kmol. (Table A-1).

Analysis The mass flow rate of air can be expressed in terms of the mole numbers as

$$\dot{N} = \frac{\dot{m}}{M} = \frac{2.90\ \text{kg/s}}{29.0\ \text{kg/kmol}} = 0.10\ \text{kmol/s}$$

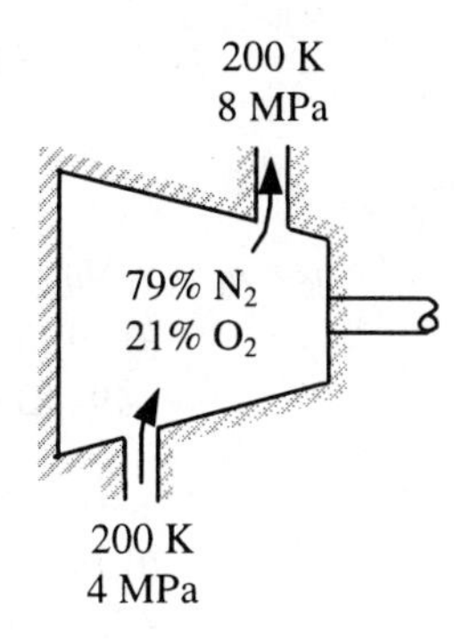

(*a*) Assuming ideal gas behavior, the Δh and Δs of air during this process is

$$\Delta\bar{h} = 0\ (\text{isothermal process})$$

$$\Delta\bar{s} = \bar{C}_p \ln\frac{T_2}{T_1}^{\nearrow 0} - R_u \ln\frac{P_2}{P_1} = -R_u \ln\frac{P_2}{P_1}$$

$$= -(8.314\text{kJ/kg}\cdot\text{K})\ln\frac{8\text{MPa}}{4\text{MPa}} = -5.763\text{kJ/kmol}\cdot\text{K}$$

Disregarding any changes in kinetic and potential energies, the steady-flow energy balance equation for the isothermal process of an ideal gas reduces to

$$\dot{E}_{\text{in}} - \dot{E}_{\text{out}} = \Delta\dot{E}_{\text{system}}^{\nearrow 0\ \text{(steady)}} = 0$$

$$\dot{E}_{\text{in}} = \dot{E}_{\text{out}}$$

$$\dot{W}_{\text{in}} + \dot{N}\bar{h}_1 = \dot{Q}_{\text{out}} + \dot{N}\bar{h}_2$$

$$\dot{W}_{\text{in}} - \dot{Q}_{\text{out}} = \dot{N}\Delta\bar{h}^{\nearrow 0} = 0 \longrightarrow \dot{W}_{\text{in}} = \dot{Q}_{\text{out}}$$

Also for an isothermal, internally reversible process the heat transfer is related to the entropy change by $Q = T\Delta S = NT\Delta\bar{s}$,

$$\dot{Q} = \dot{N}T\Delta\bar{s} = (0.10\text{kmol/s})(220\text{K})(-5.763\text{kJ/kmol}\cdot\text{K}) = -126.8\text{kW} \rightarrow \dot{Q}_{\text{out}} = 126.8\text{kW}$$

Therefore,

$$\dot{W}_{\text{in}} = \dot{Q}_{\text{out}} = \mathbf{126.8kW}$$

(*b*) Using Amagat's law and the generalized charts, the enthalpy and entropy changes of each gas are determined from

$$\bar{h}_2 - \bar{h}_1 = R_u T_{cr}(Z_{h_1} - Z_{h_2}) + (\bar{h}_2 - \bar{h}_1)_{\text{ideal}}^{\nearrow 0}$$

$$\bar{s}_2 - \bar{s}_1 = R_u(Z_{s_1} - Z_{s_2}) + (\bar{s}_2 - \bar{s}_1)_{\text{ideal}}$$

where

N_2:

$$\left.\begin{aligned} P_{R_1} &= \frac{P_{m,1}}{P_{cr,N_2}} = \frac{4}{3.39} = 1.18 \\ T_{R_1} = T_{R_2} &= \frac{T_m}{T_{cr,N_2}} = \frac{220}{126.2} = 1.74 \\ P_{R_2} &= \frac{P_{m,2}}{P_{cr,N_2}} = \frac{8}{3.39} = 2.36 \end{aligned}\right\} \begin{aligned} &Z_{h_1} = 0.4,\ Z_{s_1} = 0.2 \\ &Z_{h_2} = 0.8,\ Z_{s_2} = 0.35 \end{aligned}$$

O_2:

$$P_{R_1} = \frac{P_{m,1}}{P_{cr,O_2}} = \frac{4}{5.08} = 0.787$$

$$T_{R_1} = T_{R_2} = \frac{T_m}{T_{cr,O_2}} = \frac{220}{154.8} = 1.421$$

$$P_{R_2} = \frac{P_{m,2}}{P_{cr,O_2}} = \frac{8}{5.08} = 1.575$$

$$Z_{h_1} = 0.4,\ Z_{s_1} = 0.25$$

$$Z_{h_2} = 1.0,\ Z_{s_2} = 0.5$$

Then,

$$\begin{aligned}\bar{h}_2 - \bar{h}_1 &= y_i \Delta\bar{h}_i = y_{N_2}(\bar{h}_2 - \bar{h}_1)_{N_2} + y_{O_2}(\bar{h}_2 - \bar{h}_1)_{O_2} \\ &= (0.79)(8.314)(126.2)(0.4 - 0.8) + (0.21)(8.314)(126.2)(0.4 - 1.0) + 0 \\ &= -494 \text{ kJ/kmol}\end{aligned}$$

$$\begin{aligned}\bar{s}_2 - \bar{s}_1 &= y_i \Delta\bar{s}_i = y_{N_2}(\bar{s}_2 - \bar{s}_1)_{N_2} + y_{O_2}(\bar{s}_2 - \bar{s}_1)_{O_2} \\ &= (0.79)(8.314)(0.2 - 0.35) + (0.21)(8.314)(0.25 - 0.5) + (-5.763) \\ &= -7.18 \text{ kJ/kmol}\cdot\text{K}\end{aligned}$$

Thus,

$$\dot{Q}_{out} = -\dot{N}T\Delta\bar{s} = -(0.10\text{kmol/s})(220\text{K})(-7.18\text{kJ/kmol}\cdot\text{K}) = \mathbf{158.0\ kW}$$

$$\dot{E}_{in} - \dot{E}_{out} = \Delta\dot{E}_{system}^{\nearrow 0\ (steady)} = 0$$

$$\dot{E}_{in} = \dot{E}_{out}$$

$$\dot{W}_{in} + \dot{N}\bar{h}_1 = \dot{Q}_{out} + \dot{N}\bar{h}_2$$

$$\dot{W}_{in} = \dot{Q}_{out} + \dot{N}(\bar{h}_2 - \bar{h}_1) \longrightarrow \dot{W}_{in} = 158\text{ kW} + (0.10\text{kmol/s})(-494\text{kJ/kmol}) = \mathbf{108.6\ kW}$$

12-62 EES solution of this (and other comprehensive problems designated with the *computer icon*) is available to instructors at the *Instructor Manual* section of the *Online Learning Center* (OLC) at www.mhhe.com/cengel-boles. See the Preface for access information.

Special Topic: Chemical Potential and the Separation Work of Mixtures

12-63C

12-64C

12-65C

12-66C

12-67 Brackish water is used to produce fresh water. The minimum power input and the minimum height the brackish water must be raised by a pump for reverse osmosis are to be determined.

Assumptions **1** The brackish water is an ideal solution since it is dilute. **2** The total dissolved solids in water can be treated as table salt (NaCl). **3** The environment temperature is also 12°C.

Properties The molar masses of water and salt are M_w = 18.0 kg/kmol and M_s = 58.44 kg/kmol. The gas constant of pure water is R_w = 0.4615 kJ/kg·K (Table A-1). The density of fresh water is 1000 kg/m^3.

Analysis First we determine the mole fraction of pure water in brackish water using Eqs. 12-4 and 12-5. Noting that $\text{mf}_s = 0.00078$ and $\text{mf}_w = 1 - \text{mf}_s = 0.99922$,

$$M_m = \frac{1}{\sum \frac{\text{mf}_i}{M_i}} = \frac{1}{\frac{\text{mf}_s}{M_s} + \frac{\text{mf}_w}{M_w}} = \frac{1}{\frac{0.00078}{58.44} + \frac{0.99922}{18.0}} = 18.01 \text{ kg/kmol}$$

$$y_i = mf_i \frac{M_m}{M_i} \rightarrow y_w = mf_w \frac{M_m}{M_w} = (0.99922)\frac{18.01 \text{ kg/kmol}}{18.0 \text{ kg/kmol}} = 0.99976$$

The minimum work input required to produce 1 kg of freshwater from brackish water is

$$w_{\text{min, in}} = R_w T_0 \ln(1/y_w) = (0.4615 \text{ kJ/kg}\cdot\text{K})(285.15 \text{ K})\ln(1/0.99976) = 0.03164 \text{ kJ/kg fresh water}$$

Therefore, 0.03164 kJ of work is needed to produce 1 kg of fresh water is mixed with seawater reversibly. Therefore, the required power input to produce fresh water at the specified rate is

$$\dot{W}_{\text{min, in}} = \rho \dot{V} w_{\text{min, in}} = (1000 \text{ kg/m}^3)(0.280 \text{ m}^3/\text{s})(0.03164 \text{ kJ/kg})\left(\frac{1 \text{ kW}}{1 \text{ kJ/s}}\right) = \mathbf{8.86\ kW}$$

The minimum height to which the brackish water must be pumped is

$$\Delta z_{\text{min}} = \frac{w_{\text{min,in}}}{g} = \left(\frac{8.86 \text{ kJ/s}}{280 \text{ kg/s}}\right)\left(\frac{0.03164 \text{ kJ/kg}}{9.81 \text{ m/s}^2}\right)\left(\frac{1 \text{ kg.m/s}^2}{1 \text{ N}}\right)\left(\frac{1000 \text{ N.m}}{1 \text{ kJ}}\right) = \mathbf{3.23\ m}$$

12-68 A river is discharging into the ocean at a specified rate. The amount of power that can be generated is to be determined.

Assumptions **1** The seawater is an ideal solution since it is dilute. **2** The total dissolved solids in water can be treated as table salt (NaCl). **3** The environment temperature is also 15°C.

Properties The molar masses of water and salt are M_w = 18.0 kg/kmol and M_s = 58.44 kg/kmol. The gas constant of pure water is R_w = 0.4615 kJ/kg·K (Table A-1). The density of river water is 1000 kg/m³.

Analysis First we determine the mole fraction of pure water in ocean water using Eqs. 12-4 and 12-5. Noting that $\text{mf}_s = 0.035$ and $\text{mf}_w = 1 - \text{mf}_s = 0.965$,

$$M_\text{m} = \frac{1}{\sum \dfrac{\text{mf}_i}{M_i}} = \frac{1}{\dfrac{\text{mf}_s}{M_s} + \dfrac{\text{mf}_w}{M_w}} = \frac{1}{\dfrac{0.035}{58.44} + \dfrac{0.965}{18.0}} = 18.45 \text{ kg/kmol}$$

$$y_i = mf_i \frac{M_m}{M_i} \quad \rightarrow \quad y_w = mf_w \frac{M_m}{M_w} = (0.965)\frac{18.45 \text{ kg/kmol}}{18.0 \text{ kg/kmol}} = 0.9891$$

The maximum work output associated with mixing 1 kg of seawater (or the minimum work input required to produce 1 kg of freshwater from seawater) is

$$w_\text{max, out} = R_w T_0 \ln(1/y_w) = (0.4615 \text{ kJ/kg}\cdot\text{K})(288.15 \text{ K})\ln(1/0.9891) = 1.46 \text{ kJ/kg fresh water}$$

Therefore, 1.46 kJ of work can be produced as 1 kg of fresh water is mixed with seawater reversibly. Therefore, the power that can be generated as a river with a flow rate of 400,000 m³/s mixes reversibly with seawater is

$$\dot{W}_\text{max out} = \rho \dot{V} w_\text{max out} = (1000 \text{ kg/m}^3)(4\times 10^5 \text{ m}^3\text{/s})(1.47 \text{ kJ/kg})\left(\frac{1 \text{ kW}}{1 \text{ kJ/s}}\right) = \mathbf{582 \times 10^6 \text{ kW}}$$

Discussion This is more power than produced by all nuclear power plants (112 of them) in the US., which shows the tremendous amount of power potential wasted as the rivers discharge into the seas.

12-69 EES solution of this (and other comprehensive problems designated with the *computer icon*) is available to instructors at the *Instructor Manual* section of the *Online Learning Center* (OLC) at www.mhhe.com/cengel-boles. See the Preface for access information.

12-70E Brackish water is used to produce fresh water. The mole fractions, the minimum work inputs required to separate 1 lbm of brackish water and to obtain 1 lbm of fresh water are to be determined.

Assumptions **1** The brackish water is an ideal solution since it is dilute. **2** The total dissolved solids in water can be treated as table salt (NaCl). **3** The environment temperature is equal to the water temperature.

Properties The molar masses of water and salt are M_w = 18.0 lbm/lbmol and M_s = 58.44 lbm/lbmol. The gas constant of pure water is R_w = 0.1102 Btu/lbm·R (Table A-1E).

Analysis (*a*) First we determine the mole fraction of pure water in brackish water using Eqs. 12-4 and 12-5. Noting that mf_s = 0.0012 and mf_w = 1- mf_s = 0.9988,

$$M_m = \frac{1}{\sum \frac{mf_i}{M_i}} = \frac{1}{\frac{mf_s}{M_s} + \frac{mf_w}{M_w}} = \frac{1}{\frac{0.0012}{58.44} + \frac{0.9988}{18.0}} = 18.015 \text{ lbm/lbmol}$$

$$y_i = mf_i \frac{M_m}{M_i} \rightarrow y_w = mf_w \frac{M_m}{M_w} = (0.9988)\frac{18.015 \text{ lbm/lbmol}}{18.0 \text{ lbm/lbmol}} = \mathbf{0.99963}$$

$$y_s = 1 - y_w = 1 - 0.99963 = \mathbf{0.00037}$$

(*b*) The minimum work input required to separate 1 lbmol of brackish water is

$$\begin{aligned} w_{\text{min,in}} &= -R_u T_0 (y_w \ln y_w + y_s \ln y_s) \\ &= -(0.1102 \text{ Btu/lbmol.R})(525 \text{ R})[0.99963 \ln(0.99963) + 0.00037 \ln(0.00037)] \\ &= -0.191 \text{ Btu/lbm brackish water} \end{aligned}$$

(*c*) The minimum work input required to produce 1 lbm of freshwater from brackish water is

$$w_{\text{min, in}} = R_w T_0 \ln(1/y_w) = (0.1102 \text{ Btu/lbm}\cdot\text{R})(525 \text{ R})\ln(1/0.99963) = 0.0214 \text{ Btu/lbm fresh water}$$

Discussion Note that it takes about 9 times work to separate 1 lbm of brackish water into pure water and salt compared to producing 1 lbm of fresh water from a large body of brackish water.

12-71 A desalination plant produces fresh water from seawater. The second law efficiency of the plant is to be determined.

Assumptions **1** The seawater is an ideal solution since it is dilute. **2** The total dissolved solids in water can be treated as table salt (NaCl). **3** The environment temperature is equal to the seawater temperature.

Properties The molar masses of water and salt are M_w = 18.0 kg/kmol and M_s = 58.44 kg/kmol. The gas constant of pure water is R_w = 0.4615 kJ/kg·K (Table A-1). The density of river water is 1000 kg/m^3.

Analysis First we determine the mole fraction of pure water in seawater using Eqs. 12-4 and 12-5. Noting that $\text{mf}_s = 0.032$ and $\text{mf}_w = 1 - \text{mf}_s = 0.968$,

$$M_m = \frac{1}{\sum \frac{\text{mf}_i}{M_i}} = \frac{1}{\frac{\text{mf}_s}{M_s} + \frac{\text{mf}_w}{M_w}} = \frac{1}{\frac{0.032}{58.44} + \frac{0.968}{18.0}} = 18.41 \text{ kg/kmol}$$

$$y_i = mf_i \frac{M_m}{M_i} \rightarrow y_w = mf_w \frac{M_m}{M_w} = (0.968)\frac{18.41 \text{ kg/kmol}}{18.0 \text{ kg/kmol}} = 0.9900$$

The maximum work output associated with mixing 1 kg of seawater (or the minimum work input required to produce 1 kg of freshwater from seawater) is

$$w_{\text{max, out}} = R_w T_0 \ln(1/y_w) = (0.4615 \text{ kJ/kg}\cdot\text{K})(283.15 \text{ K})\ln(1/0.990) = 1.307 \text{ kJ/kg fresh water}$$

The power that can be generated as 1.4 m^3/s fresh water mixes reversibly with seawater is

$$\dot{W}_{\text{max out}} = \rho \dot{V} w_{\text{max out}} = (1000 \text{ kg/m}^3)(1.4 \text{ m}^3\text{/s})(1.307 \text{ kJ/kg})\left(\frac{1 \text{ kW}}{1 \text{ kJ/s}}\right) = 1.83 \text{ kW}$$

Then the second law efficiency of the plant becomes

$$\eta_{II} = \frac{\dot{W}_{\text{min,in}}}{\dot{W}_{\text{in}}} = \frac{1.83 \text{ MW}}{8.5 \text{ MW}} = 0.215 = \mathbf{21.5\%}$$

12-72 The power consumption and the second law efficiency of a desalination plant are given. The power that can be produced if the fresh water produced is mixed with the seawater reversibly is to be determined.

Assumptions **1** This is a steady-flow process. **2** The kinetic and potential energy changes are negligible.

Analysis From the definition of the second law efficiency

$$\eta_{II} = \frac{\dot{W}_{\text{rev}}}{\dot{W}_{\text{actual}}} \rightarrow 0.18 = \frac{\dot{W}_{\text{rev}}}{5.3 \text{ MW}} \rightarrow \dot{W}_{\text{rev}} = \mathbf{0.954 \text{ MW}}$$

which is the maximum power that can be generated.

Review Problems

12-73 The molar fractions of constituents of air are given. The gravimetric analysis of air and its molar mass are to be determined.

Assumptions All the constituent gases and their mixture are ideal gases.

Properties The molar masses of O_2, N_2, and Ar are 32.0, 28.0, and 40.0 kg/kmol. (Table A-1).

Analysis For convenience, consider 100 kmol of air. Then the mass of each component and the total mass are

$$N_{O_2} = 21\text{kmol} \longrightarrow m_{O_2} = N_{O_2}M_{O_2} = (21\text{kmol})(32\text{kg/kmol}) = 672\text{kg}$$
$$N_{N_2} = 78\text{kmol} \longrightarrow m_{N_2} = N_{N_2}M_{N_2} = (78\text{kmol})(28\text{kg/kmol}) = 2{,}184\text{kg}$$
$$N_{Ar} = 1\text{kmol} \longrightarrow m_{Ar} = N_{Ar}M_{Ar} = (1\text{kmol})(40\text{kg/kmol}) = 40\text{kg}$$

$$m_m = m_{O_2} + m_{N_2} + m_{Ar} = 672\text{ kg} + 2{,}184\text{ kg} + 40\text{ kg} = 2{,}896\text{ kg}$$

AIR

21% O_2
78% N_2
1% Ar

Then the mass fraction of each component (gravimetric analysis) becomes

$$mf_{O_2} = \frac{m_{O_2}}{m_m} = \frac{672\text{ kg}}{2{,}896\text{ kg}} = 0.232 \quad \text{or} \quad \mathbf{23.2\%}$$

$$mf_{N_2} = \frac{m_{N_2}}{m_m} = \frac{2{,}184\text{ kg}}{2{,}896\text{ kg}} = 0.754 \quad \text{or} \quad \mathbf{75.4\%}$$

$$mf_{Ar} = \frac{m_{Ar}}{m_m} = \frac{40\text{ kg}}{2{,}896\text{ kg}} = 0.014 \quad \text{or} \quad \mathbf{1.4\%}$$

The molar mass of the mixture is determined from its definitions,

$$M_m = \frac{m_m}{N_m} = \frac{2{,}896\text{ kg}}{100\text{ kmol}} = \mathbf{28.96\ kg/kmol}$$

12-74 Using Amagat's law, it is to be shown that $Z_m = \sum_{i=1}^{k} y_i Z_i$ for a real-gas mixture.

Analysis Using the compressibility factor, the volume of a component of a real-gas mixture and of the volume of the gas mixture can be expressed as

$$V_i = \frac{Z_i N_i R_u T_m}{P_m} \quad \text{and} \quad V_m = \frac{Z_m N_m R_u T_m}{P_m}$$

Amagat's law can be expressed as $V_m = \sum V_i(T_m, P_m)$. Substituting,

$$\frac{Z_m N_m R_u T_m}{P_m} = \sum \frac{Z_i N_i R_u T_m}{P_m}$$

Simplifying,

$$Z_m N_m = \sum Z_i N_i$$

Dividing by N_m,

$$Z_m = \sum y_i Z_i$$

where Z_i is determined at the mixture temperature and pressure.

12-75 Using Dalton's law, it is to be shown that $Z_m = \sum_{i=1}^{k} y_i Z_i$ for a real-gas mixture.

Analysis Using the compressibility factor, the pressure of a component of a real-gas mixture and of the pressure of the gas mixture can be expressed as

$$P_i = \frac{Z_i N_i R_u T_m}{V_m} \quad \text{and} \quad P_m = \frac{Z_m N_m R_u T_m}{V_m}$$

Dalton's law can be expressed as $P_m = \sum P_i(T_m, V_m)$. Substituting,

$$\frac{Z_m N_m R_u T_m}{V_m} = \sum \frac{Z_i N_i R_u T_m}{V_m}$$

Simplifying,

$$Z_m N_m = \sum Z_i N_i$$

Dividing by N_m,

$$Z_m = \sum y_i Z_i$$

where Z_i is determined at the mixture temperature and volume.

12-76 The mole numbers, pressure, and temperature of the constituents of a gas mixture are given. The volume of the tank containing this gas mixture is to be determined using three methods.

Analysis (*a*) Under specified conditions both N_2 and CH_4 will considerably deviate from the ideal gas behavior. Treating the mixture as an ideal gas gives

2 kmol N_2
6 kmol CH_4

200 K
12 MPa

$$N_m = N_{N_2} + N_{CH_4} = 2\,\text{kmol} + 6\,\text{kmol} = 8\,\text{kmol}$$

and

$$V_m = \frac{N_m R_u T_m}{P_m} = \frac{(8\,\text{kmol})(8.314\,\text{kPa}\cdot\text{m}^3/\text{kmol}\cdot\text{K})(200\,\text{K})}{12{,}000\,\text{kPa}} = \mathbf{1.11\ m^3}$$

(*b*) To use Kay's rule, we first need to determine the pseudo-critical temperature and pseudo-critical pressure of the mixture using the critical point properties of N_2 and CH_4 from Table A-1,

$$y_{N_2} = \frac{N_{N_2}}{N_m} = \frac{2\text{ kmol}}{8\text{ kmol}} = 0.25 \quad \text{and} \quad y_{CH_4} = \frac{N_{CH_4}}{N_m} = \frac{6\text{ kmol}}{8\text{ kmol}} = 0.75$$

$$T'_{cr,m} = \sum y_i T_{cr,i} = y_{N_2} T_{cr,N_2} + y_{CH_4} T_{cr,CH_4}$$
$$= (0.25)(126.2\text{ K}) + (0.75)(191.1\text{ K}) = 174.9\text{ K}$$

$$P'_{cr,m} = \sum y_i P_{cr,i} = y_{N_2} P_{cr,N_2} + y_{CH_4} P_{cr,CH_4}$$
$$= (0.25)(3.39\text{ MPa}) + (0.75)(4.64\text{ MPa}) = 4.33\text{ MPa}$$

Then,

$$\left.\begin{aligned} T_R &= \frac{T_m}{T'_{cr,m}} = \frac{200}{174.9} = 1.144 \\ P_R &= \frac{P_m}{P'_{cr,m}} = \frac{12}{4.33} = 2.77 \end{aligned}\right\} Z_m = 0.47 \qquad \text{(Fig. A-30b)}$$

Thus,

$$V_m = \frac{Z_m N_m R_u T_m}{P_m} = Z_m V_{ideal} = (0.47)(1.11\,\text{m}^3) = \mathbf{0.52\,m^3}$$

(*c*) To use the Amagat's law for this real gas mixture, we first need to determine the Z of each component at the mixture temperature and pressure,

N_2:

$$\left.\begin{aligned} T_{R,N_2} &= \frac{T_m}{T_{cr,N_2}} = \frac{200}{126.2} = 1.585 \\ P_{R,N_2} &= \frac{P_m}{P_{cr,N_2}} = \frac{12}{3.39} = 3.54 \end{aligned}\right\} Z_{N_2} = 0.85 \qquad \text{(Fig. A-30b)}$$

CH_4:

$$\left.\begin{aligned} T_{R,CH_4} &= \frac{T_m}{T_{cr,CH_4}} = \frac{200}{191.1} = 1.047 \\ P_{R,CH_4} &= \frac{P_m}{P_{cr,CH_4}} = \frac{12}{4.64} = 2.586 \end{aligned}\right\} Z_{CH_4} = 0.37 \qquad \text{(Fig. A-30b)}$$

Mixture: $$Z_m = \sum y_i Z_i = y_{N_2} Z_{N_2} + y_{CH_4} Z_{CH_4} = (0.25)(0.85) + (0.75)(0.37) = 0.49$$

Thus, $$V_m = \frac{Z_m N_m R_u T_m}{P_m} = Z_m V_{ideal} = (0.49)(1.11\,\text{m}^3) = \mathbf{0.544\ m^3}$$

12-77 A stream of gas mixture at a given pressure and temperature is to be separated into its constituents steadily. The minimum work required is to be determined.

Assumptions **1** Both the N_2 and CO_2 gases and their mixture are ideal gases. **2** This is a steady-flow process. **3** The kinetic and potential energy changes are negligible.

Properties The molar masses of N_2 and CO_2 are 28.0 and 44.0 kg/kmol. (Table A-1).

Analysis The minimum work required to separate a gas mixture into its components is equal to the reversible work associated with the mixing process, which is equal to the exergy destruction (or irreversibility) associated with the mixing process since

$$X_{\text{destroyed}} = W_{\text{rev,out}} - W_{act,u}^{\nearrow 0} = W_{\text{rev,out}} = T_0 S_{\text{gen}}$$

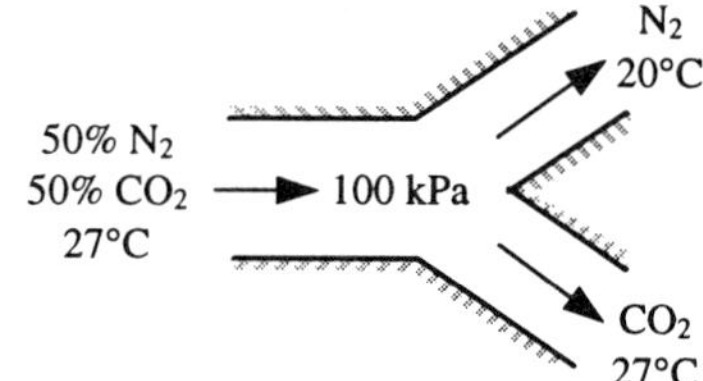

where S_{gen} is the entropy generation associated with the steady-flow mixing process. The entropy change associated with a constant pressure and temperature adiabatic mixing process is determined from

$$\bar{s}_{\text{gen}} = \sum \Delta\bar{s}_i = -R_u \sum y_i \ln y_i = -(8.314\text{kJ/kmol}\cdot\text{K})[0.5\ln 0.5 + 0.5\ln 0.5]$$
$$= 5.763\text{kJ/kmol}\cdot\text{K}$$

$$M_m = \sum y_i M_i = (0.5)(28\text{kg/kmol}) + (0.5)(44\text{kg/kmol}) = 36\text{kg/kmol}$$

$$s_{gen} = \frac{\bar{s}_{\text{gen}}}{M_m} = \frac{5.763\text{kJ/kmol}\cdot\text{K}}{36\text{kg/kmol}} = 0.160\text{kJ/kg}\cdot\text{K}$$

$$x_{\text{destroyed}} = T_0 s_{\text{gen}} = (300\text{K})(0.160\text{kJ/kg}\cdot\text{K}) = \mathbf{48.0\ kJ/kg}$$

12-78 A gas mixture is heated during a steady-flow process. The heat transfer is to be determined using two approaches.

Assumptions **1** Steady flow conditions exist. **2** Kinetic and potential energy changes are negligible.

Analysis Noting that there is no work involved, the energy balance for this gas mixture can be written, on a unit mole basis, as

$$\dot{E}_{in} - \dot{E}_{out} = \Delta\dot{E}_{system}^{\nearrow 0\ (steady)} = 0$$

$$\dot{E}_{in} = \dot{E}_{out}$$

$$\bar{q}_{in} + \bar{h}_1 = \bar{h}_2$$

$$\bar{q}_{in} = \Delta\bar{h}$$

180 K → $1O_2$+$3N_2$ 8 MPa 210 K, Q

Also, $y_{O_2} = 0.25$ *and* $y_{N_2} = 0.75$.

(*a*) Assuming ideal gas behavior, the inlet and exit enthalpies of O_2 and N_2 are determined from the ideal gas tables to be

$$O_2: \quad h_1 = h_{@180\text{ K}} = 5{,}239.6\text{ kJ/kmol}, \qquad h_2 = h_{@210\text{ K}} = 6{,}112.9\text{ kJ/kmol}$$

$$N_2: \quad h_1 = h_{@180\text{ K}} = 5{,}229\text{ kJ/kmol}, \qquad h_2 = h_{@210\text{ K}} = 6{,}100.5\text{ kJ/kmol}$$

Thus,

$$\begin{aligned}\bar{q}_{in,\,ideal} &= \sum y_i \Delta\bar{h}_i = y_{O_2}(\bar{h}_2 - \bar{h}_1)_{O_2} + y_{N_2}(\bar{h}_2 - \bar{h}_1)_{N_2} \\ &= (0.25)(6{,}112.9 - 5{,}239.6) + (0.75)(6{,}100.5 - 5{,}229) \\ &= \mathbf{872.0\ kJ/kmol}\end{aligned}$$

(*b*) Using the Kay's rule, the gas mixture can be treated as a pseudo-pure substance whose critical temperature and pressure are

$$\begin{aligned}T'_{cr,m} &= \sum y_i T_{cr,i} = y_{O_2} T_{cr,O_2} + y_{N_2} T_{cr,N_2} \\ &= (0.25)(154.8\text{ K}) + (0.75)(126.2\text{ K}) = 133.4\text{ K}\end{aligned}$$

$$\begin{aligned}P'_{cr,m} &= \sum y_i P_{cr,i} = y_{O_2} P_{cr,O_2} + y_{N_2} P_{cr,N_2} \\ &= (0.25)(5.08\text{ MPa}) + (0.75)(3.39\text{ MPa}) = 3.81\text{ MPa}\end{aligned}$$

Then,

$$\left.\begin{aligned}T_{R,1} &= \frac{T_{m,1}}{T_{cr,m}} = \frac{180}{133.4} = 1.349 \\ P_{R,1} = P_{R,2} &= \frac{P_m}{P_{cr,m}} = \frac{8}{3.81} = 2.100 \\ T_{R,2} &= \frac{T_{m,2}}{T_{cr,m}} = \frac{210}{133.4} = 1.574\end{aligned}\right\} \begin{aligned}Z_{h_1} &= 1.4 \\ Z_{h_2} &= 1.1\end{aligned}$$

The heat transfer in this case is determined from

$$\begin{aligned}\bar{q}_{in} &= \bar{h}_2 - \bar{h}_1 = R_u T_{cr}(Z_{h_1} - Z_{h_2}) + (\bar{h}_2 - \bar{h}_1)_{ideal} \\ &= R_u T_{cr}(Z_{h_1} - Z_{h_2}) + \bar{q}_{ideal} \\ &= (8.314\text{kJ/kmol}\cdot\text{K})(133.4\text{K})(1.4 - 1.1) + (872\text{kJ/kmol}) \\ &= \mathbf{1{,}205\ kJ/kmol}\end{aligned}$$

12-79 EES solution of this (and other comprehensive problems designated with the *computer icon*) is available to instructors at the *Instructor Manual* section of the *Online Learning Center* (OLC) at www.mhhe.com/cengel-boles. See the Preface for access information.

12-80 A gas mixture is heated during a steady-flow process, as discussed in the previous problem. The total entropy change and the exergy destruction are to be determined using two methods.

Analysis The entropy generated during this process is determined by applying the entropy balance on an *extended system* that includes the piston-cylinder device and its immediate surroundings so that the boundary temperature of the extended system is the environment temperature at all times. It gives

180 K → 8 MPa, $1O_2$+$3N_2$ → 210 K; Q

$$S_{in} - S_{out} + S_{gen} = \Delta S_{system}$$

$$\frac{Q_{in}}{T_{boundary}} + S_{gen} = \Delta S_{system} \rightarrow S_{gen} = m(s_2 - s_1) - \frac{Q_{in}}{T_{surr}}$$

Then the exergy destroyed during a process can be determined from its definition $X_{destroyed} = T_0 S_{gen}$.

(*a*) Noting that the total mixture pressure, and thus the partial pressure of each gas, remains constant, the entropy change of a component in the mixture during this process is

$$\Delta \bar{s}_i = \left(\bar{C}_p \ln\frac{T_2}{T_1} - R_u \ln\frac{P_2}{P_1}^{\nearrow 0} \right)_i = MC_p \ln\frac{T_2}{T_1}$$

Assuming ideal gas behavior and C_p values at the average temperature, the $\Delta\bar{s}$ of O_2 and N_2 are determined from

$$\Delta \bar{s}_{O_2,\text{ideal}} = (32\text{kg/kmol})(0.918\text{kJ/kg}\cdot\text{K})\ln\frac{210\text{K}}{180\text{K}} = 4.52\text{kJ/kmol}\cdot\text{K}$$

$$\Delta \bar{s}_{N_2,\text{ideal}} = (28\text{kg/kmol})(1.039\text{kJ/kg}\cdot\text{K})\ln\frac{210\text{K}}{180\text{K}} = 4.48\text{kJ/kmol}\cdot\text{K}$$

$$\begin{aligned}\Delta \bar{s}_{\text{sys,ideal}} &= \sum y_i \Delta\bar{s}_i = y_{O_2}\Delta\bar{s}_{O_2} + y_{N_2}\Delta\bar{s}_{N_2} \\ &= (0.25)(4.52\text{kJ/kmol}\cdot\text{K}) + (0.75)(4.48\text{kJ/kmol}\cdot\text{K}) \\ &= 4.49\text{kJ/kmol}\cdot\text{K}\end{aligned}$$

and

$$\bar{s}_{\text{gen}} = 4.49\text{kJ/kmol}\cdot\text{K} - \frac{872\text{kJ/kmol}}{298\text{K}} = \mathbf{1.56\ kJ/kmol\cdot K}$$

$$\bar{x}_{\text{destroyed}} = T_0\bar{s}_{gen} = (303\text{K})(1.56\text{kJ/kmol}\cdot\text{K}) = \mathbf{473\ kJ/kmol}$$

(*b*) Using the Kay's rule, the gas mixture can be treated as a pseudo-pure substance whose critical temperature and pressure are

$$T'_{cr,m} = \sum y_i T_{cr,i} = y_{O_2} T_{cr,O_2} + y_{N_2} T_{cr,N_2}$$
$$= (0.25)(154.8 \text{ K}) + (0.75)(126.2 \text{ K}) = 133.4 \text{ K}$$

$$P'_{cr,m} = \sum y_i P_{cr,i} = y_{O_2} P_{cr,O_2} + y_{N_2} P_{cr,N_2}$$
$$= (0.25)(5.08 \text{ MPa}) + (0.75)(3.39 \text{ MPa}) = 3.81 \text{ MPa}$$

Then,

$$\left.\begin{aligned} T_{R,1} &= \frac{T_{m,1}}{T_{cr,m}} = \frac{180}{133.4} = 1.349 \\ P_{R,1} = P_{R,2} &= \frac{P_m}{P_{cr,m}} = \frac{8}{3.81} = 2.100 \\ T_{R,2} &= \frac{T_{m,2}}{T_{cr,m}} = \frac{210}{133.4} = 1.574 \end{aligned}\right\} \begin{aligned} Z_{s_1} &= 0.8 \\ Z_{s_2} &= 0.45 \end{aligned}$$

Thus,

$$\Delta \bar{s}_{sys} = R_u(Z_{s_1} - Z_{s_2}) + \Delta \bar{s}_{sys,\,ideal}$$
$$= (8.314 \text{ kJ/kmol} \cdot \text{K})(0.8 - 0.45) + (4.49 \text{ kJ/kmol} \cdot \text{K})$$
$$= 7.39 \text{ kJ/kmol} \cdot \text{K}$$

and

$$\bar{s}_{gen} = 7.39 \text{kJ/kmol} \cdot \text{K} - \frac{1{,}204.7 \text{kJ/kmol}}{298 \text{K}} = \mathbf{3.35 \ kJ/kmol \cdot K}$$

$$\bar{x}_{destroyed} = T_0 \bar{s}_{gen} = (303 \text{K})(3.35 \text{kJ/kmol} \cdot \text{K}) = \mathbf{1015 \ kJ/kmol}$$

12-81 The masses, pressures, and temperatures of the constituents of a gas mixture in a tank are given. Heat is transferred to the tank. The final pressure of the mixture and the heat transfer are to be determined.

Assumptions He is an ideal gas and O_2 is a nonideal gas.

Properties The molar masses of He and O_2 are 4.0 and 32.0 kg/kmol. (Table A-1)

Analysis (*a*) The number of moles of each gas is

$$N_{\text{He}} = \frac{m_{\text{He}}}{M_{\text{He}}} = \frac{4 \text{ kg}}{4.0 \text{ kg/kmol}} = 1 \text{ kmol} \quad \text{and} \quad N_{O_2} = \frac{m_{O_2}}{M_{O_2}} = \frac{8 \text{ kg}}{32 \text{ kg/kmol}} = 0.25 \text{ kmol}$$

$$N_m = N_{\text{He}} + N_{O_2} = 1 \text{ kmol} + 0.25 \text{ kmol} = 1.25 \text{ kmol}$$

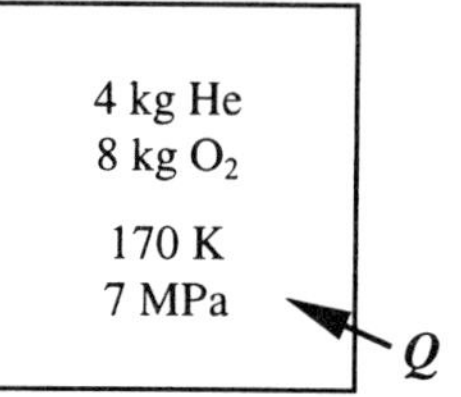

Then the partial volume of each gas and the volume of the tank are

He: $$V_{\text{He}} = \frac{N_{\text{He}} R_u T_1}{P_{m,1}} = \frac{(1 \text{ kmol})(8.314 \text{ kPa}\cdot\text{m}^3/\text{kmol}\cdot\text{K})(170 \text{ K})}{7{,}000 \text{ kPa}} = 0.202 \text{ m}^3$$

O_2: $$\left.\begin{aligned} P_{R_1} &= \frac{P_{m,1}}{P_{cr,O_2}} = \frac{7}{5.08} = 1.38 \\ T_{R_1} &= \frac{T_1}{T_{cr,O_2}} = \frac{170}{154.8} = 1.10 \end{aligned}\right\} Z_1 = 0.53$$

$$V_{O_2} = \frac{Z N_{O_2} R_u T_1}{P_{m,1}} = \frac{(0.53)(0.25 \text{ kmol})(8.314 \text{ kPa}\cdot\text{m}^3/\text{kg}\cdot\text{K})(170 \text{ K})}{7{,}000 \text{ kPa}} = 0.027 \text{ m}^3$$

$$V_{\text{tank}} = V_{\text{He}} + V_{O_2} = 0.202 \text{ m}^3 + 0.027 \text{ m}^3 = 0.229 \text{ m}^3$$

The partial pressure of each gas and the total final pressure is

He: $$P_{\text{He},2} = \frac{N_{\text{He}} R_u T_2}{V_{\text{tank}}} = \frac{(1 \text{ kmol})(8.314 \text{ kPa}\cdot\text{m}^3/\text{kmol}\cdot\text{K})(220 \text{ K})}{0.229 \text{ m}^3} = 7{,}987 \text{ kPa}$$

O_2: $$\left.\begin{aligned} T_{R_2} &= \frac{T_2}{T_{cr,O_2}} = \frac{220}{154.8} = 1.42 \\ v_{R,O_2} &= \frac{\bar{v}_{O_2}}{R_u T_{cr,O_2}/P_{cr,O_2}} = \frac{V_m/N_{O_2}}{R_u T_{cr,O_2}/P_{cr,O_2}} \\ &= \frac{(0.229\text{m}^3)/(0.25\text{kmol})}{(8.314\text{kPa}\cdot\text{m}^3/\text{kmol}\cdot\text{K})(154.8\text{K})/(5{,}080\text{kPa})} = 3.616 \end{aligned}\right\} P_R = 0.39$$

$$P_{O_2} = (P_R P_{cr})_{O_2} = (0.39)(5{,}080\text{kPa}) = 1981\text{kPa} = 1.981\text{MPa}$$

$$P_{m,2} = P_{\text{He}} + P_{O_2} = 7.987\text{MPa} + 1.981\text{MPa} = \mathbf{9.97\ MPa}$$

(*b*) We take both gases as the system. No work or mass crosses the system boundary, therefore this is a closed system with no work interactions. Then the energy balance for this closed system reduces to

$$E_{\text{in}} - E_{\text{out}} = \Delta E_{\text{system}}$$

$$Q_{\text{in}} = \Delta U = \Delta U_{\text{He}} + \Delta U_{O_2}$$

He: $$\Delta U_{\text{He}} = mC_v(T_m - T_1) = (4\text{kg})(3.1156\text{kJ/kg}\cdot\text{K})(220 - 170)\text{K} = 623.1\text{kJ}$$

O_2:
$$\left.\begin{array}{l} T_{R_1} = 1.10 \\ P_{R_1} = 1.38 \end{array}\right\} Z_{h_1} = 2.2 \quad \text{and} \quad \left.\begin{array}{l} T_{R_2} = 1.42 \\ P_{R_2} = \dfrac{9.97}{5.08} = 1.963 \end{array}\right\} Z_{h_2} = 0.6$$

$$\bar{h}_2 - \bar{h}_1 = R_u T_{cr}(Z_{h_1} - Z_{h_2}) + (\bar{h}_2 - \bar{h}_1)_{\text{ideal}}$$
$$= (8.314 \text{ kJ/kmol} \cdot \text{K})(154.8 \text{ K})(2.2 - 0.6) + (6{,}404 - 4{,}949) \text{ kJ/kmol} = 3{,}514 \text{ kJ/kmol}$$

Also,

$$P_{\text{He},1} = \frac{N_{\text{He}} R_u T_1}{V_{\text{tank}}} = \frac{(1 \text{ kmol})(8.314 \text{ kPa} \cdot \text{m}^3/\text{kg} \cdot \text{K})(170 \text{ K})}{0.229 \text{ m}^3} = 6{,}172 \text{ kPa}$$
$$P_{O_2,1} = P_{m,1} - P_{\text{He},1} = 7{,}000 \text{ kPa} - 6{,}172 \text{ kPa} = 828 \text{ kPa}$$

Thus,

$$\Delta U_{O_2} = N_{O_2}(\bar{h}_2 - \bar{h}_1) - (P_2 V_2 - P_1 V_1) = N_{O_2}(\bar{h}_2 - \bar{h}_1) - (P_{O_2,2} - P_{O_2,1}) V_{\text{tank}}$$
$$= (0.25 \text{ kmol})(3{,}514 \text{ kJ/kmol}) - (1{,}981 - 828)(0.229) \text{ kPa} \cdot \text{m}^3 = 614.5 \text{ kJ}$$

Substituting,

$$Q_{\text{in}} = 623.1 \text{ kJ} + 614.5 \text{ kJ} = \mathbf{1{,}238 \text{ kJ}}$$

12-82 · · · 12-84 EES solutions of these comprehensive problems are available to instructors at the *Instructor Manual* section of the *Online Learning Center* (OLC) at www.mhhe.com/cengel-boles. See the Preface for access information.

Fundamentals of Engineering (FE) Exam Problems

12-85 An ideal gas mixture whose apparent molar mass is 42 kg/kmol consists of nitrogen N_2 and three other gases. If the mole fraction of nitrogen is 0.30, its mass fraction is

(a) 0.10 (b) 0.20 (c) 0.30 (d) 0.50 (e) 0.70

Answer (b) 0.20

Solution Solved by EES Software. Solutions can be verified by copying-and-pasting the following lines on a blank EES screen. (Similar problems and their solutions can be obtained easily by modifying numerical values).

```
M_mix=42 "kg/kmol"
M_N2=28 "kg/kmol"

y_N2=0.3
mf_N2=(M_N2/M_mix)*y_N2

"Some Wrong Solutions with Common Mistakes:"
W1_mf = y_N2 "Taking mass fraction to be equal to mole fraction"
W2_mf= y_N2*(M_mix/M_N2) "Using the molar mass ratio backwords"
W3_mf= 1-mf_N2 "Taking the complement of the mass fraction"
```

12-86 An ideal gas mixture consists of 2 kmol of N_2 and 6 kmol of CO_2. The mass fraction of CO_2 in the mixture is

(a) 0.175 (b) 0.250 (c) 0.500 (d) 0.750 (e) 0.825

Answer (e) 0.825

Solution Solved by EES Software. Solutions can be verified by copying-and-pasting the following lines on a blank EES screen. (Similar problems and their solutions can be obtained easily by modifying numerical values).

```
N1=2 "kmol"
N2=6 "kmol"
N_mix=N1+N2

MM1=28 "kg/kmol"
MM2=44 "kg/kmol"

m_mix=N1*MM1+N2*MM2
mf2=N2*MM2/m_mix

"Some Wrong Solutions with Common Mistakes:"
W1_mf = N2/N_mix "Using mole fraction"
W2_mf = 1-mf2 "The wrong mass fraction"
```

12-87 An ideal gas mixture consists of 2 kmol of N_2 and 6 kmol of CO_2. The apparent gas constant of the mixture is

(a) 0.208 kJ/kg·K (b) 0.231 kJ/kg·K (c) 0.531 kJ/kg·K (d) 0.875 kJ/kg·K (e) 1.24 kJ/kg·K

Answer (a) 0.208 kJ/kg·K

Solution Solved by EES Software. Solutions can be verified by copying-and-pasting the following lines on a blank EES screen. (Similar problems and their solutions can be obtained easily by modifying numerical values).

```
Ru=8.314 "kJ/kmol.K"
N1=2 "kmol"
N2=6 "kmol"

MM1=28 "kg/kmol"
MM2=44 "kg/kmol"
R1=Ru/MM1
R2=Ru/MM2

N_mix=N1+N2
y1=N1/N_mix
y2=N2/N_mix

MM_mix=y1*MM1+y2*MM2
R_mix=Ru/MM_mix

"Some Wrong Solutions with Common Mistakes:"
W1_Rmix =(R1+R2)/2 "Taking the arithmetic average of gas constants"
W2_Rmix= y1*R1+y2*R2 "Using wrong relation for Rmixture"
```

12-88 A rigid tank is divided into two compartments by a partition. One compartment contains 3 kmol of N_2 at 600 kPa pressure and the other compartment contains 7 kmol of CO_2 at 200 kPa. Now the partition is removed, and the two gases form a homogeneous mixture at 300 kPa. The partial pressure of N_2 in the mixture is

(a) 75 kPa (b) 90 kPa (c) 150 kPa (d) 175 kPa (e) 225 kPa

Answer (b) 90 kPa

Solution Solved by EES Software. Solutions can be verified by copying-and-pasting the following lines on a blank EES screen. (Similar problems and their solutions can be obtained easily by modifying numerical values).

```
P1 = 600 "kPa"
P2 = 200 "kPa"
P_mix=300 "kPa"
N1=3 "kmol"
N2=7 "kmol"

MM1=28 "kg/kmol"
MM2=44 "kg/kmol"

N_mix=N1+N2
y1=N1/N_mix
y2=N2/N_mix

P_N2=y1*P_mix

"Some Wrong Solutions with Common Mistakes:"
W1_P1= P_mix/2 "Assuming equal partial pressures"
W2_P1= mf1*P_mix; mf1=N1*MM1/(N1*MM1+N2*MM2) "Using mass fractions"
W3_P1 = P_mix*N1*P1/(N1*P1+N2*P2) "Using some kind of weighed averaging"
```

12-89 A 100-L rigid tank contains an ideal gas mixture of 5 g of N_2 and 5 g of CO_2 at a specified pressure and temperature. If N_2 were separated from the mixture and stored at mixture temperature and pressure, its volume would be

(a) 35 L (b) 39 L (c) 50 L (d) 61 L (e) 65 L

Answer (d) 61 L

Solution Solved by EES Software. Solutions can be verified by copying-and-pasting the following lines on a blank EES screen. (Similar problems and their solutions can be obtained easily by modifying numerical values).

```
V_mix=100 "L"
m1=5 "g"
m2=5 "g"

MM1=28 "kg/kmol"
MM2=44 "kg/kmol"

N1=m1/MM1
N2=m2/MM2

N_mix=N1+N2
y1=N1/N_mix

V1=y1*V_mix  "L"

"Some Wrong Solutions with Common Mistakes:"
W1_V1=V_mix*m1/(m1+m2)  "Using mass fractions"
W2_V1= V_mix "Assuming the volume to be the mixture volume"
```

12-90 An ideal gas mixture consists of 3 kg of Ar and 6 kg of CO_2 gases. The mixture is now heated at constant volume from 250 K to 350 K. The amount of heat transfer is

(a) 374 kJ (b) 436 kJ (c) 488 kJ (d) 525 kJ (e) 664 kJ

Answer (c) 488 kJ

Solution Solved by EES Software. Solutions can be verified by copying-and-pasting the following lines on a blank EES screen. (Similar problems and their solutions can be obtained easily by modifying numerical values).

```
T1=250 "K"
T2=350 "K"

Cv1=0.3122; Cp1=0.5203 "kJ/kg.K"
Cv2=0.657; Cp2=0.846 "kJ/kg.K"

m1=3 "kg"
m2=6 "kg"

MM1=39.95 "kg/kmol"
MM2=44 "kg/kmol"

"Applying Energy balance gives Q=DeltaU=DeltaU_Ar+DeltaU_CO2"

Q=(m1*Cv1+m2*Cv2)*(T2-T1)

"Some Wrong Solutions with Common Mistakes:"
W1_Q = (m1+m2)*(Cv1+Cv2)/2*(T2-T1)  "Using arithmetic average of properties"
W2_Q = (m1*Cp1+m2*Cp2)*(T2-T1)"Using Cp instead of Cv"
W3_Q = (m1*Cv1+m2*Cv2)*T2 "Using T2 instead of T2-T1"
```

12-91 An ideal gas mixture consists of 30% helium and 70% argon gases by mass. The mixture is now expanded isentropically in a turbine from 400°C and 1.2 MPa to a pressure of 100 kPa. The mixture temperature at turbine exit is

(a) -217°C (b) -25°C (c) 72°C (d) 148°C (e) 256°C

Answer (b) -25°C

Solution Solved by EES Software. Solutions can be verified by copying-and-pasting the following lines on a blank EES screen. (Similar problems and their solutions can be obtained easily by modifying numerical values).

```
T1=400+273"K"
P1=1200 "kPa"
P2=100."kPa"

mf_He=0.3
mf_Ar=0.7

k1=1.667
k2=1.667

"The specific heat ratio k of the mixture is also 1.667 since k=1.667 for all componet
gases"
k_mix=1.667

T2=T1*(P2/P1)^((k_mix-1)/k_mix)-273

"Some Wrong Solutions with Common Mistakes:"
W1_T2 = (T1-273)*(P2/P1)^((k_mix-1)/k_mix) "Using C for T1 instead of K"
W2_T2 = T1*(P2/P1)^((k_air-1)/k_air)-273; k_air=1.4 "Using k value for air"
W3_T2 = T1*P2/P1 "Assuming T to be proportional to P"
```

12-92 One compartment of an insulated rigid tank contains 2 kmol of CO_2 at 20°C and 150 kPa while the other compartment contains 5 kmol of H_2 gas at 35°C and 300 kPa. Now the partition between the two gases is removed, and the two gases form a homogeneous ideal gas mixture. The temperature of the mixture is

(a) 25°C (b) 29°C (c) 22°C (d) 32°C (e) 34°C

Answer (b) 29°C

Solution Solved by EES Software. Solutions can be verified by copying-and-pasting the following lines on a blank EES screen. (Similar problems and their solutions can be obtained easily by modifying numerical values).

```
N_H2=5 "kmol"
T1_H2=35 "C"
P1_H2=300 "kPa"

N_CO2=2 "kmol"
T1_CO2=20 "C"
P1_CO2=150 "kPa"

Cv_H2=10.183; Cp_H2=14.307 "kJ/kg.K"
Cv_CO2=0.657; Cp_CO2=0.846 "kJ/kg.K"

MM_H2=2 "kg/kmol"
MM_CO2=44 "kg/kmol"

m_H2=N_H2*MM_H2
m_CO2=N_CO2*MM_CO2

"Applying Energy balance gives 0=DeltaU=DeltaU_H2+DeltaU_CO2"

0=m_H2*Cv_H2*(T2-T1_H2)+m_CO2*Cv_CO2*(T2-T1_CO2)

"Some Wrong Solutions with Common Mistakes:"
0=m_H2*Cp_H2*(W1_T2-T1_H2)+m_CO2*Cp_CO2*(W1_T2-T1_CO2) "Using Cp instead of Cv"
0=N_H2*Cv_H2*(W2_T2-T1_H2)+N_CO2*Cv_CO2*(W2_T2-T1_CO2) "Using N instead of mass"
W3_T2 = (T1_H2+T1_CO2)/2 "Assuming averate temperature"
```

12-93 A piston-cylinder device contains an ideal gas mixture of 3 kmol of He gas and 7 kmol of Ar gas at 30°C and 300 kPa. Now the gas expands at constant pressure until its volume doubles. The amount of heat transfer to the gas mixture is

(a) 5.3 MJ (b) 5.8 MJ (c) 32 MJ (d) 41 MJ (e) 63 MJ

Answer (e) 63 MJ

Solution Solved by EES Software. Solutions can be verified by copying-and-pasting the following lines on a blank EES screen. (Similar problems and their solutions can be obtained easily by modifying numerical values).

```
N_He=3 "kmol"
N_Ar=7 "kmol"

T1=30+273 "C"
P1=300 "kPa"
P2=P1

"T2=2T1 since PV/T=const for ideal gases and it is given that P=constant"
T2=2*T1 "K"

MM_He=4 "kg/kmol"
MM_Ar=39.95 "kg/kmol"

m_He=N_He*MM_He
m_Ar=N_Ar*MM_Ar

Cp_Ar=0.5203; Cv_Ar = 3122 "kJ/kg.C"
Cp_He=5.1926; Cv_He = 3.1156 "kJ/kg.K"

"For a P=const process, Q=DeltaH since DeltaU+Wb is DeltaH"
Q=m_Ar*Cp_Ar*(T2-T1)+m_He*Cp_He*(T2-T1)

"Some Wrong Solutions with Common Mistakes:"
W1_Q =m_Ar*Cv_Ar*(T2-T1)+m_He*Cv_He*(T2-T1) "Using Cv instead of Cp"
W2_Q=N_Ar*Cp_Ar*(T2-T1)+N_He*Cp_He*(T2-T1) "Using N instead of mass"
W3_Q=m_Ar*Cp_Ar*(T22-T1)+m_He*Cp_He*(T22-T1); T22=2*(T1-273)+273 "Using C for
T1"
W4_Q=(m_Ar+m_He)*0.5*(Cp_Ar+Cp_He)*(T2-T1) "Using arithmetic averate of Cp"
```

12-94 An ideal gas mixture of helium and argon gases with identical mass fractions enters a turbine at 1200 K and 1 MPa at a rate of 0.3 kg/s, and expands isentropically to 200 kPa. The power output of the turbine is

(a) 359 kW (b) 488 kW (c) 619 kW (d) 823 kW (e) 1630 kW

Answer (b) 488 kW

Solution Solved by EES Software. Solutions can be verified by copying-and-pasting the following lines on a blank EES screen. (Similar problems and their solutions can be obtained easily by modifying numerical values).

```
m=0.3 "kg/s"
T1=1200 "K"
P1=1000 "kPa"
P2=200 "kPa"

mf_He=0.5
mf_Ar=0.5

k_He=1.667
k_Ar=1.667

Cp_Ar=0.5203
Cp_He=5.1926
Cp_mix=mf_He*Cp_He+mf_Ar*Cp_Ar

"Specific heat ratio of mixture is also 1.667 since k=1.667 for all component gases"
k_mix=1.667

T2=T1*(P2/P1)^((k_mix-1)/k_mix)
-W_out=m*Cp_mix*(T2-T1)

"Some Wrong Solutions with Common Mistakes:"
W1_Wout= - m*Cp_mix*(T22-T1); T22 = (T1-273)*(P2/P1)^((k_mix-1)/k_mix)+273 "Using C
for T1 instead of K"
W2_Wout= - m*Cp_mix*(T222-T1); T222 = T1*(P2/P1)^((k_air-1)/k_air)-273; k_air=1.4
"Using k value for air"
W3_Wout= - m*Cp_mix*(T2222-T1); T2222 = T1*P2/P1 "Assuming T to be proportional to
P"
W4_Wout= - m*0.5*(Cp_Ar+Cp_He)*(T2-T1) "Using arithmetic average for Cp"
```

12-95 Design and Essay Problems

Chapter 13
GAS-VAPOR MIXTURES AND AIR CONDITIONING

Dry and Atmospheric Air, Specific and Relative Humidity

13-1C Yes; by cooling the air at constant pressure.

13-2C Yes.

13-3C Specific humidity will decrease but relative humidity will increase.

13-4C Dry air does not contain any water vapor, but atmospheric air does.

13-5C Yes, the water vapor in the air can be treated as an ideal gas because of its very low partial pressure.

13-6C The partial pressure of the water vapor in atmospheric air is called vapor pressure.

13-7C The same. This is because water vapor behaves as an ideal gas at low pressures, and the enthalpy of an ideal gas depends on temperature only.

13-8C Specific humidity is the amount of water vapor present in a unit mass of dry air. Relative humidity is the ratio of the actual amount of vapor in the air at a given temperature to the maximum amount of vapor air can hold at that temperature.

13-9C The specific humidity will remain constant, but the relative humidity will decrease as the temperature rises in a well-sealed room.

13-10C The specific humidity will remain constant, but the relative humidity will decrease as the temperature drops in a well-sealed room.

13-11C A tank that contains moist air at 3 atm is located in moist air that is at 1 atm. The driving force for moisture transfer is the vapor pressure difference, and thus it is possible for the water vapor to flow into the tank from surroundings if the vapor pressure in the surroundings is greater than the vapor pressure in the tank.

13-12C Insulations on *chilled water lines* are always wrapped with *vapor barrier jackets* to eliminate the possibility of vapor entering the insulation. This is because moisture that migrates through the insulation to the cold surface will condense and remain there indefinitely with no possibility of vaporizing and moving back to the outside.

13-13C When the temperature, total pressure, and the relative humidity are given, the vapor pressure can be determined from the psychrometric chart or the relation $P_v = \phi P_{sat}$ where P_{sat} is the saturation (or boiling) pressure of water at the specified temperature and ϕ is the relative humidity.

13-14 A tank contains dry air and water vapor at specified conditions. The specific humidity, the relative humidity, and the volume of the tank are to be determined.

Assumptions The air and the water vapor are ideal gases.

Analysis (*a*) The specific humidity can be determined form its definition,

21 kg dry air
0.3 kg H_2O vapor
30°C
100 kPa

$$\omega = \frac{m_v}{m_a} = \frac{0.3 \text{ kg}}{21 \text{ kg}} = \mathbf{0.0143 \text{ kg } H_2O / \text{kg dry air}}$$

(*b*) The saturation pressure of water at 30°C is

$$P_g = P_{\text{sat @ } 30°\text{C}} = 4.246 \text{ kPa}$$

Then the relative humidity can be determined from

$$\phi = \frac{\omega P}{(0.622 + \omega) P_g} = \frac{(0.0143)(100 \text{ kPa})}{(0.622 + 0.0143)4.246 \text{ kPa}} = \mathbf{52.9\%}$$

(*c*) The volume of the tank can be determined from the ideal gas relation for the dry air,

$$P_v = \phi P_g = (0.529)(4.246 \text{ kPa}) = 2.246 \text{ kPa}$$

$$P_a = P - P_v = 100 - 2.246 = 97.754 \text{ kPa}$$

$$V = \frac{m_a R_a T}{P_a} = \frac{(21 \text{ kg})(0.287 \text{ kJ / kg}\cdot\text{K})(303 \text{ K})}{97.754 \text{ kPa}} = \mathbf{18.7 \text{ m}^3}$$

13-15 A tank contains dry air and water vapor at specified conditions. The specific humidity, the relative humidity, and the volume of the tank are to be determined.

Assumptions The air and the water vapor are ideal gases.

Analysis (*a*) The specific humidity can be determined form its definition,

21 kg dry air
0.3 kg H_2O vapor
35°C
100 kPa

$$\omega = \frac{m_v}{m_a} = \frac{0.3 \text{ kg}}{21 \text{ kg}} = \mathbf{0.0143 \text{ kg } H_2O / \text{kg dry air}}$$

(*b*) The saturation pressure of water at 30°C is

$$P_g = P_{\text{sat @ } 35°\text{C}} = 5.628 \text{ kPa}$$

Then the relative humidity can be determined from

$$\phi = \frac{\omega P}{(0.622 + \omega) P_g} = \frac{(0.0143)(100 \text{ kPa})}{(0.622 + 0.0143)5.628 \text{ kPa}} = \mathbf{39.9\%}$$

(*c*) The volume of the tank can be determined from the ideal gas relation for the dry air,

$$P_v = \phi P_g = (0.399)(5.628 \text{ kPa}) = 2.246 \text{ kPa}$$

$$P_a = P - P_v = 100 - 2.246 = 97.754 \text{ kPa}$$

$$V = \frac{m_a R_a T}{P_a} = \frac{(21 \text{ kg})(0.287 \text{ kJ/kg}\cdot\text{K})(308 \text{ K})}{97.754 \text{ kPa}} = \mathbf{19.0 \text{ m}^3}$$

13-16 A room contains air at specified conditions and relative humidity. The partial pressure of air, the specific humidity, and the enthalpy per unit mass of dry air are to be determined.

Assumptions The air and the water vapor are ideal gases.

Analysis (*a*) The partial pressure of dry air can be determined from

$$P_v = \phi P_g = \phi P_{sat\ @\ 20°C} = (0.85)(2.339\text{ kPa}) = 1.99\text{ kPa}$$

$$P_a = P - P_v = 98 - 1.99 = \mathbf{96.01\ kPa}$$

(*b*) The specific humidity of air is determined from

$$\omega = \frac{0.622\,P_v}{P - P_v} = \frac{(0.622)(1.99\text{ kPa})}{(98 - 1.99)\text{ kPa}} = \mathbf{0.0129\ kg\ H_2O\,/\,kg\ dry\ air}$$

(*c*) The enthalpy of air per unit mass of dry air is determined from

$$h = h_a + \omega h_v \cong C_p T + \omega h_g$$

$$= (1.005\text{ kJ/kg}\cdot°\text{C})(20°\text{C}) + (0.0129)(2538.1\text{ kJ/kg}) = \mathbf{52.84\ kJ\,/\,kg\ dry\ air}$$

13-17 A room contains air at specified conditions and relative humidity. The partial pressure of air, the specific humidity, and the enthalpy per unit mass of dry air are to be determined.

Assumptions The air and the water vapor are ideal gases.

Analysis (*a*) The partial pressure of dry air can be determined from

AIR
20°C
85 kPa
85% RH

$$P_v = \phi P_g = \phi P_{sat\ @\ 20°C} = (0.85)(2.339\text{ kPa}) = 1.99\text{ kPa}$$

$$P_a = P - P_v = 85 - 1.99 = \mathbf{83.01\ kPa}$$

(*b*) The specific humidity of air is determined from

$$\omega = \frac{0.622\,P_v}{P - P_v} = \frac{(0.622)(1.99\text{ kPa})}{(85 - 1.99)\text{ kPa}} = \mathbf{0.0149\ kg\ H_2O\,/\,kg\ dry\ air}$$

(*c*) The enthalpy of air per unit mass of dry air is determined from

$$h = h_a + \omega h_v \cong C_p T + \omega h_g$$

$$= (1.005\text{ kJ/kg}\cdot°\text{C})(20°\text{C}) + (0.0149)(2538.1\text{ kJ/kg}) = \mathbf{57.95\ kJ\,/\,kg\ dry\ air}$$

13-18E A room contains air at specified conditions and relative humidity. The partial pressure of air, the specific humidity, and the enthalpy per unit mass of dry air are to be determined.

Assumptions The air and the water vapor are ideal gases.

Analysis (*a*) The partial pressure of dry air can be determined from

AIR
70°F
14.6 psia
85% RH

$$P_v = \phi P_g = \phi P_{sat\ @\ 70°F} = (0.85)(0.3632\text{ psia}) = 0.309\text{ psia}$$

$$P_a = P - P_v = 14.6 - 0.309 = \mathbf{14.291\ psia}$$

(*b*) The specific humidity of air is determined from

$$\omega = \frac{0.622\,P_v}{P - P_v} = \frac{(0.622)(0.309\text{ psia})}{(14.6 - 0.309)\text{ psia}} = \mathbf{0.0134\ lbm\ H_2O\,/\,lbm\ dry\ air}$$

(*c*) The enthalpy of air per unit mass of dry air is determined from

$$h = h_a + \omega h_v \cong C_p T + \omega h_g$$
$$= (0.24 \text{ Btu/lbm}\cdot{}^\circ\text{F})(70^\circ \text{F}) + (0.0134)(1092 \text{ Btu/lbm}) = \mathbf{31.43 \text{ Btu/lbm dry air}}$$

13-19 The masses of dry air and the water vapor contained in a room at specified conditions and relative humidity are to be determined.

Assumptions The air and the water vapor are ideal gases.

Analysis The partial pressure of water vapor and dry air are determined to be

$$P_v = \phi P_g = \phi P_{\text{sat @ 23°C}} = (0.50)(2.837 \text{ kPa}) = 1.42 \text{ kPa}$$
$$P_a = P - P_v = 98 - 1.42 = 96.58 \text{ kPa}$$

ROOM
240 m^3
23°C
98 kPa
50% RH

The masses are determined to be

$$m_a = \frac{P_a V}{R_a T} = \frac{(96.58 \text{ kPa})(240 \text{ m}^3)}{(0.287 \text{ kPa}\cdot\text{m}^3/\text{kg}\cdot\text{K})(296 \text{ K})} = \mathbf{272.9 \text{ kg}}$$

$$m_v = \frac{P_v V}{R_v T} = \frac{(1.42 \text{ kPa})(240 \text{ m}^3)}{(0.4615 \text{ kPa}\cdot\text{m}^3/\text{kg}\cdot\text{K})(296 \text{ K})} = \mathbf{2.49 \text{ kg}}$$

Dew-point, Adiabatic Saturation, and Wet-bulb Temperatures

13-20C Dew-point temperature is the temperature at which condensation begins when air is cooled at constant pressure.

13-21C Andy's. The temperature of his glasses may be below the dew-point temperature of the room, causing condensation on the surface of the glasses.

13-22C The outer surface temperature of the glass may drop below the dew-point temperature of the surrounding air, causing the moisture in the vicinity of the glass to condense. After a while, the condensate may start dripping down because of gravity.

13-23C When the temperature falls below the dew-point temperature, dew forms on the outer surfaces of the car. If the temperature is below 0°C, the dew will freeze. At very low temperatures, the moisture in the air will freeze directly on the car windows.

13-24C When the air is saturated (100% relative humidity).

13-25C These two are approximately equal at atmospheric temperatures and pressure.

13-26 A house contains air at a specified temperature and relative humidity. It is to be determined whether any moisture will condense on the inner surfaces of the windows when the temperature of the window drops to a specified value.

Assumptions The air and the water vapor are ideal gases.

Analysis The vapor pressure P_v is uniform throughout the house, and its value can be determined from

25°C

$\phi = 65\%$

10°C

$$P_v = \phi P_{g\,@\,25^\circ C} = (0.65)(3.169 \text{ kPa}) = 2.06 \text{ kPa}$$

The dew-point temperature of the air in the house is

$$T_{dp} = T_{sat\,@\,P_v} = T_{sat\,@\,2.06\,kPa} = \mathbf{17.9^\circ C}$$

That is, the moisture in the house air will start condensing when the temperature drops below 17.9°C. Since the windows are at a lower temperature than the dew-point temperature, some moisture **will condense** on the window surfaces.

13-27 A person wearing glasses enters a warm room at a specified temperature and relative humidity from the cold outdoors. It is to be determined whether the glasses will get fogged.

Assumptions The air and the water vapor are ideal gases.

Analysis The vapor pressure P_v of the air in the house is uniform throughout, and its value can be determined from

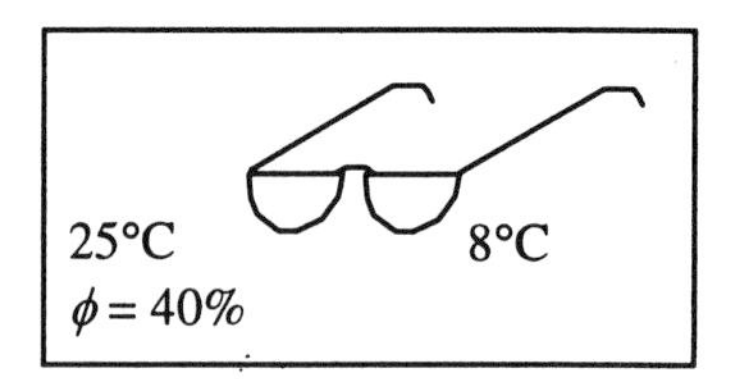

$$P_v = \phi P_{g\,@\,25°C} = (0.40)(3.169\text{ kPa}) = 1.268\text{ kPa}$$

The dew-point temperature of the air in the house is

$$T_{dp} = T_{sat\,@\,P_v} = T_{sat\,@\,1.268\text{ kPa}} = \mathbf{10.2°C}$$

That is, the moisture in the house air will start condensing when the air temperature drops below 10.2°C. Since the glasses are at a lower temperature than the dew-point temperature, some moisture will condense on the glasses, and thus they **will get fogged.**

13-28 A person wearing glasses enters a warm room at a specified temperature and relative humidity from the cold outdoors. It is to be determined whether the glasses will get fogged.

Assumptions The air and the water vapor are ideal gases.

Analysis The vapor pressure P_v of the air in the house is uniform throughout, and its value can be determined from

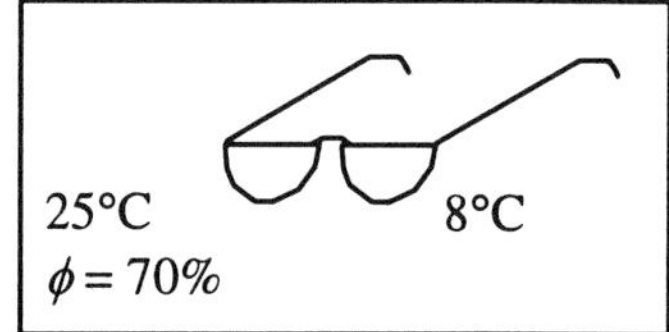

$$P_v = \phi P_{g\,@\,25°C} = (0.70)(3.169\text{ kPa}) = 2.22\text{ kPa}$$

The dew-point temperature of the air in the house is

$$T_{dp} = T_{sat\,@\,P_v} = T_{sat\,@\,2.22\text{ kPa}} = \mathbf{19.1°C}$$

That is, the moisture in the house air will start condensing when the air temperature drops below 19.1°C. Since the glasses are at a lower temperature than the dew-point temperature, some moisture will condense on the glasses, and thus they **will get fogged.**

13-29E A woman drinks a cool canned soda in a room at a specified temperature and relative humidity. It is to be determined whether the can will sweat.

Assumptions The air and the water vapor are ideal gases.

Analysis The vapor pressure P_v of the air in the house is uniform throughout, and its value can be determined from

80°F
50% RH
Cola
40°F

$$P_v = \phi P_{g@\,80°F} = (0.50)(0.5073\text{ psia}) = 0.254\text{ psia}$$

The dew-point temperature of the air in the house is

$$T_{dp} = T_{sat\,@\,P_v} = T_{sat\,@\,0.254\text{ psia}} = \mathbf{59.7°F}$$

That is, the moisture in the house air will start condensing when the air temperature drops below 59.7°C. Since the canned drink is at a lower temperature than the dew-point temperature, some moisture will condense on the can, and thus it **will sweat.**

13-30 The dry- and wet-bulb temperatures of atmospheric air at a specified pressure are given. The specific humidity, the relative humidity, and the enthalpy of air are to be determined.

Assumptions The air and the water vapor are ideal gases.

Analysis (*a*) The specific humidity ω_1 is determined from

$$\omega_1 = \frac{C_p(T_2 - T_1) + \omega_2 h_{fg2}}{h_{g1} - h_{f2}}$$

95 kPa
25°C
$T_{wb} = 20°C$

where T_2 is the wet-bulb temperature, and ω_2 is determined from

$$\omega_2 = \frac{0.622\, P_{g2}}{P_2 - P_{g2}} = \frac{(0.622)(2.339 \text{ kPa})}{(95 - 2.339) \text{ kPa}} = 0.0157 \text{ kg H}_2\text{O / kg dry air}$$

Thus,

$$\omega_1 = \frac{(1.005 \text{ kJ / kg·°C})(20 - 25)\text{°C} + (0.0157)(2454.1 \text{ kJ / kg})}{(2547.2 - 83.96) \text{ kJ / kg}} = \mathbf{0.0136 \text{ kg H}_2\text{O / kg dry air}}$$

(*b*) The relative humidity ϕ_1 is determined from

$$\phi_1 = \frac{\omega_1 P_1}{(0.622 + \omega_1) P_{g1}} = \frac{(0.0136)(95 \text{ kPa})}{(0.622 + 0.0136)(3.169 \text{ kPa})} = 0.641 \text{ or } \mathbf{64.1\%}$$

(*c*) The enthalpy of air per unit mass of dry air is determined from

$$h_1 = h_{a1} + \omega_1 h_{v1} \cong C_p T_1 + \omega_1 h_{g1} = (1.005 \text{ kJ / kg·°C})(25\text{°C}) + (0.0136)(2547.2 \text{ kJ / kg})$$
$$= \mathbf{59.8 \text{ kJ / kg dry air}}$$

13-31 The dry- and wet-bulb temperatures of air in room at a specified pressure are given. The specific humidity, the relative humidity, and the dew-point temperature are to be determined.

Assumptions The air and the water vapor are ideal gases.

Analysis (*a*) The specific humidity ω_1 is determined from

100 kPa
22°C
T_{wb} = 16°C

$$\omega_1 = \frac{C_p(T_2 - T_1) + \omega_2 h_{fg2}}{h_{g1} - h_{f2}}$$

where T_2 is the wet-bulb temperature, and ω_2 is determined from

$$\omega_2 = \frac{0.622\,P_{g2}}{P_2 - P_{g2}} = \frac{(0.622)(1.83\text{ kPa})}{(100 - 1.83)\text{ kPa}} = 0.0116\text{ kg H}_2\text{O / kg dry air}$$

Thus,

$$\omega_1 = \frac{(1.005\text{ kJ / kg}\cdot{}^\circ\text{C})(16 - 22)^\circ\text{C} + (0.0116)(2463.5\text{ kJ / kg})}{(2541.7 - 67.18)\text{ kJ / kg}}$$
$$= \mathbf{0.0091\text{ kg H}_2\text{O / kg dry air}}$$

(*b*) The relative humidity ϕ_1 is determined from

$$\phi_1 = \frac{\omega_1 P_1}{(0.622 + \omega_1)P_{g1}} = \frac{(0.0091)(100\text{ kPa})}{(0.622 + 0.0091)(2.67\text{ kPa})} = 0.540 \text{ or } \mathbf{54.0\%}$$

(*c*) The vapor pressure at the inlet conditions is

$$P_{v1} = \phi_1 P_{g1} = \phi_1 P_{\text{sat @ 22°C}} = (0.540)(2.67\text{ kPa}) = 1.442\text{ kPa}$$

Thus the dew-point temperature of the air is

$$T_{dp} = T_{\text{sat @ }P_v} = T_{\text{sat @ 1.442 kPa}} = \mathbf{12.3^\circ C}$$

13-32 EES solution of this (and other comprehensive problems designated with the *computer icon*) is available to instructors at the *Instructor Manual* section of the *Online Learning Center* (OLC) at www.mhhe.com/cengel-boles. See the Preface for access information.

13-33E The dry- and wet-bulb temperatures of air in room at a specified pressure are given. The specific humidity, the relative humidity, and the dew-point temperature are to be determined.

Assumptions The air and the water vapor are ideal gases.

Analysis (*a*) The specific humidity ω_1 is determined from

14.7 psia
70°F
T_{wb} = 60°F

$$\omega_1 = \frac{C_p(T_2 - T_1) + \omega_2 h_{fg2}}{h_{g1} - h_{f2}}$$

where T_2 is the wet-bulb temperature, and ω_2 is determined from

$$\omega_2 = \frac{0.622\,P_{g2}}{P_2 - P_{g2}} = \frac{(0.622)(0.2563\text{ psia})}{(14.7 - 0.2563)\text{ psia}} = 0.0110\text{ lbm H}_2\text{O / lbm dry air}$$

Thus,

$$\omega_1 = \frac{(0.24\text{ Btu / lbm}\cdot{}^\circ\text{F})(60 - 70)^\circ\text{F} + (0.0110)(1059.6\text{ Btu / lbm})}{(1092.0 - 28.08)\text{ Btu / lbm}}$$
$$= \mathbf{0.0087\ lbm\ H_2O\ /\ lbm\ dry\ air}$$

(*b*) The relative humidity ϕ_1 is determined from

$$\phi_1 = \frac{\omega_1 P_1}{(0.622 + \omega_1)P_{g1}} = \frac{(0.0087)(14.7\text{ psia})}{(0.622 + 0.0087)(0.3632\text{ psia})} = 0.558 \text{ or } \mathbf{55.8\%}$$

(*c*) The vapor pressure at the inlet conditions is

$$P_{v1} = \phi_1 P_{g1} = \phi_1 P_{\text{sat @ 70°F}} = (0.558)(0.3632\text{ psia}) = 0.2027\text{ psia}$$

Thus the dew-point temperature of the air is

$$T_{dp} = T_{\text{sat @ }P_v} = T_{\text{sat @ 0.2027 psia}} = \mathbf{53.2^\circ F}$$

Psychometric Chart

13-34C They are very nearly parallel to each other.

13-35C The saturation states (located on the saturation curve).

13-36C By drawing a horizontal line until it intersects with the saturation curve. The corresponding temperature is the dew-point temperature.

13-37C No, they cannot. The enthalpy of moist air depends on ω, which depends on the total pressure.

13-38 [*Also solved by EES on enclosed CD*] The pressure, temperature, and relative humidity of air in a room are specified. Using the psychrometric chart, the specific humidity, the enthalpy, the wet-bulb temperature, the dew-point temperature, and the specific volume of the air are to be determined.

Analysis From the psychometric chart we read

$$(a)\ \omega = 0.0181 \text{ kg H}_2 \text{ / kg dry air}$$
$$(b)\ h = 78.4 \text{ kJ / kg dry air}$$
$$(c)\ T_{wb} = 25.5°\text{C}$$
$$(d)\ T_{dp} = 23.3°\text{C}$$
$$(e)\ v = 0.890 \text{ m}^3 \text{ / kg dry air}$$

13-39 EES solution of this (and other comprehensive problems designated with the *computer icon*) is available to instructors at the *Instructor Manual* section of the *Online Learning Center* (OLC) at www.mhhe.com/cengel-boles. See the Preface for access information.

13-40 The pressure, temperature, and relative humidity of air in a room are specified. Using the psychrometric chart, the specific humidity, the enthalpy, the wet-bulb temperature, the dew-point temperature, and the specific volume of the air are to be determined.

Analysis From the psychometric chart we read

$$(a)\ \omega = 0.0148 \text{ kg H}_2 \text{ / kg dry air}$$
$$(b)\ h = 63.9 \text{ kJ / kg dry air}$$
$$(c)\ T_{wb} = 21.9°\text{C}$$
$$(d)\ T_{dp} = 20.1°\text{C}$$
$$(e)\ v = 0.868 \text{ m}^3 \text{ / kg dry air}$$

13-41 EES solution of this (and other comprehensive problems designated with the *computer icon*) is available to instructors at the *Instructor Manual* section of the *Online Learning Center* (OLC) at www.mhhe.com/cengel-boles. See the Preface for access information.

13-42E The pressure, temperature, and relative humidity of air in a room are specified. Using the psychrometric chart, the specific humidity, the enthalpy, the wet-bulb temperature, the dew-point temperature, and the specific volume of the air are to be determined.

Analysis From the psychometric chart we read

(*a*) $\omega = 0.0165 \text{ lbm H}_2\text{O / lbm dry air}$

(*b*) $h = 37.8 \text{ Btu / lbm dry air}$

(*c*) $T_{wb} = 74.3°\text{F}$

(*d*) $T_{dp} = 71.3°\text{F}$

(*e*) $v = 14.0 \text{ ft}^3 \text{ / lbm dry air}$

13-43E EES solution of this (and other comprehensive problems designated with the *computer icon*) is available to instructors at the *Instructor Manual* section of the *Online Learning Center* (OLC) at www.mhhe.com/cengel-boles. See the Preface for access information.

13-44 The pressure and the dry- and wet-bulb temperatures of air in a room are specified. Using the psychrometric chart, the specific humidity, the enthalpy, the relative humidity, the dew-point temperature, and the specific volume of the air are to be determined.

Analysis From the psychometric chart we read

(*a*) $\omega = 0.0092 \text{ kg H}_2 \text{ / kg dry air}$

(*b*) $h = 47.6 \text{ kJ / kg dry air}$

(*c*) $\phi = 49.6\%$

(*d*) $T_{dp} = 12.8°\text{C}$

(*e*) $v = 0.855 \text{ m}^3 \text{ / kg dry air}$

13-45 EES solution of this (and other comprehensive problems designated with the *computer icon*) is available to instructors at the *Instructor Manual* section of the *Online Learning Center* (OLC) at www.mhhe.com/cengel-boles. See the Preface for access information.

Human Comfort and Air-Conditioning

13-46C It humidifies, dehumidifies, cleans and even deodorizes the air.

13-47C (*a*) Perspires more, (*b*) cuts the blood circulation near the skin, and (*c*) sweats excessively.

13-48C It is the direct heat exchange between the body and the surrounding surfaces. It can make a person feel chilly in winter, and hot in summer.

13-49C It affects by removing the warm, moist air that builds up around the body and replacing it with fresh air.

13-50C The spectators. Because they have a lower level of activity, and thus a lower level of heat generation within their bodies.

13-51C Because they have a large skin area to volume ratio. That is, they have a smaller volume to generate heat but a larger area to lose it from.

13-52C It affects a body's ability to perspire, and thus the amount of heat a body can dissipate through evaporation.

13-53C Humidification is to add moisture into an environment, dehumidification is to remove it.

13-54C The metabolism refers to the burning of foods such as carbohydrates, fat, and protein in order to perform the necessary bodily functions. The metabolic rate for an average man ranges from 108 W while reading, writing, typing, or listening to a lecture in a classroom in a seated position to 1250 W at age 20 (730 at age 70) during strenuous exercise. The corresponding rates for women are about 30 percent lower. Maximum metabolic rates of trained athletes can exceed 2000 W. We are interested in metabolic rate of the occupants of a building when we deal with heating and air conditioning because the metabolic rate represents the rate at which a body generates heat and dissipates it to the room. This body heat contributes to the heating in winter, but it adds to the cooling load of the building in summer.

13-55C The metabolic rate is proportional to the size of the body, and the metabolic rate of women, in general, is lower than that of men because of their smaller size. Clothing serves as insulation, and the thicker the clothing, the lower the environmental temperature that feels comfortable.

13-56C Sensible heat is the energy associated with a temperature change. The sensible heat loss from a human body increases as (*a*) the skin temperature increases, (*b*) the environment temperature decreases, and (*c*) the air motion (and thus the convection heat transfer coefficient) increases.

13-57C Latent heat is the energy released as water vapor condenses on cold surfaces, or the energy absorbed from a warm surface as liquid water evaporates. The latent heat loss from a human body increases as (*a*) the skin wetness increases and (*b*) the relative humidity of the environment decreases. The rate of evaporation from the body is related to the rate of latent heat loss by $\dot{Q}_{\text{latent}} = \dot{m}_{\text{vapor}} h_{fg}$ where h_{fg} is the latent heat of vaporization of water at the skin temperature.

13-58 An average person produces 0.25 kg of moisture while taking a shower. The contribution of showers of a family of four to the latent heat load of the air-conditioner per day is to be determined.

Assumptions All the water vapor from the shower is condensed by the air-conditioning system.

Properties The latent heat of vaporization of water is given to be 2450 kJ/kg.

Analysis The amount of moisture produced per day is

$$\begin{aligned}\dot{m}_{\text{vapor}} &= \text{(Moisture produced per person)(No. of persons)}\\ &= (0.25\ \text{kg / person})(4\ \text{persons / day}) = 1\ \text{kg / day}\end{aligned}$$

Then the latent heat load due to showers becomes

$$\dot{Q}_{\text{latent}} = \dot{m}_{\text{vapor}} h_{fg} = (1\ \text{kg / day})(2450\ \text{kJ / kg}) = \mathbf{2450\ kJ / day}$$

13-59 There are 100 chickens in a breeding room. The rate of total heat generation and the rate of moisture production in the room are to be determined.

Assumptions All the moisture from the chickens is condensed by the air-conditioning system.

Properties The latent heat of vaporization of water is given to be 2430 kJ/kg. The average metabolic rate of chicken during normal activity is 10.2 W (3.78 W sensible and 6.42 W latent).

Analysis The total rate of heat generation of the chickens in the breeding room is

$$\dot{Q}_{\text{gen, total}} = \dot{q}_{\text{gen, total}}(\text{No. of chickens}) = (10.2\ \text{W / chicken})(100\ \text{chickens}) = \mathbf{1020\ W}$$

The latent heat generated by the chicken and the rate of moisture production are

$$\begin{aligned}\dot{Q}_{\text{gen, latent}} &= \dot{q}_{\text{gen, latent}}(\text{No. of chickens})\\ &= (6.42\ \text{W/chicken})(100\ \text{chickens}) = 642\ \text{W}\\ &= 0.642\ \text{kW}\end{aligned}$$

$$\dot{m}_{\text{moisture}} = \frac{\dot{Q}_{\text{gen, latent}}}{h_{fg}} = \frac{0.642\ \text{kJ / s}}{2430\ \text{kJ / kg}} = 0.000264\ \text{kg / s} = \mathbf{0.264\ g / s}$$

13-60 A department store expects to have a specified number of people at peak times in summer. The contribution of people to the sensible, latent, and total cooling load of the store is to be determined.

Assumptions There is a mix of men, women, and children in the classroom.

Properties The average rate of heat generation from people doing light work is 115 W, and 70% of is in sensible form (see Sec. 13-6).

Analysis The contribution of people to the sensible, latent, and total cooling load of the store are

$$\dot{Q}_{\text{people, total}} = (\text{No. of people}) \times \dot{Q}_{\text{person, total}} = 135 \times (115\text{ W}) = \mathbf{15{,}525\ W}$$

$$\dot{Q}_{\text{people, sensible}} = (\text{No. of people}) \times \dot{Q}_{\text{person, sensible}} = 135 \times (0.7 \times 115\text{ W}) = \mathbf{10{,}868\ W}$$

$$\dot{Q}_{\text{people, latent}} = (\text{No. of people}) \times \dot{Q}_{\text{person, latent}} = 135 \times (0.3 \times 115\text{ W}) = \mathbf{4658\ W}$$

13-61E There are a specified number of people in a movie theater in winter. It is to be determined if the theater needs to be heated or cooled.

Assumptions There is a mix of men, women, and children in the classroom.

Properties The average rate of heat generation from people in a movie theater is 105 W, and 70 W of it is in sensible form and 35 W in latent form (Table 12-8).

Analysis Noting that only the sensible heat from a person contributes to the heating load of a building, the contribution of people to the heating of the building is

$$\dot{Q}_{\text{people, sensible}} = (\text{No. of people}) \times \dot{Q}_{\text{person, sensible}} = 500 \times (70\text{ W}) = 35{,}000\text{ W} = \mathbf{119{,}420\ Btu/h}$$

since 1 W = 3.412 Btu/h. The building needs to be heated since the heat gain from people is less than the rate of heat loss of 120,000 Btu/h from the building.

13-62 The infiltration rate of a building is estimated to be 1.2 ACH. The sensible, latent, and total infiltration heat loads of the building at sea level are to be determined.

Assumptions **1** Steady operating conditions exist. **2** The air infiltrates at the outdoor conditions, and exfiltrates at the indoor conditions. **3** Excess moisture condenses at 5°C. **4** The effect of water vapor on air density is negligible.

Properties The gas constant and the specific heat of air are R = 0.287 kPa.m^3/kg.K and C_p = 1.0 kJ/kg·°C (Tables A-1 and A-3). The heat of vaporization of water at 5°C is $h_{fg} = h_{fg\,@\,5°C} = 2490\text{ kJ/kg}$ (Table A-4). The properties of the ambient and room air are determined from the psychrometric chart (Fig. A-33) to be

$$\left.\begin{aligned} T_{\text{ambient}} &= 32°\text{C} \\ \phi_{\text{ambient}} &= 50\% \end{aligned}\right\} w_{\text{ambient}} = 0.0150\text{ kg/kg dryair}$$

$$\left.\begin{aligned} T_{\text{room}} &= 24°\text{C} \\ \phi_{\text{room}} &= 50\% \end{aligned}\right\} w_{\text{room}} = 0.0093\text{ kg/kg dryair}$$

Analysis Noting that the infiltration of ambient air will cause the air in the cold storage room to be changed 0.8 times every hour, the air will enter the room at a mass flow rate of

$$\rho_{\text{ambient}} = \frac{P_0}{RT_0} = \frac{101.325\text{ kPa}}{(0.287\text{ kPa.m}^3/\text{kg.K})(32+273\text{ K})} = 1.16\text{ kg/m}^3$$

$$\dot{m}_{\text{air}} = \rho_{\text{ambient}} V_{\text{room}}\text{ACH} = (1.16\text{ kg/m}^3)(20\times 13\times 3\text{ m}^3)(1.2\text{ h}^{-1}) = 1086\text{ kg/h} = 0.3016\text{ kg/s}$$

Then the sensible, latent, and total infiltration heat loads of the room are determined to be

$$\dot{Q}_{\text{infiltration, sensible}} = \dot{m}_{\text{air}} C_p (T_{\text{ambient}} - T_{\text{room}}) = (0.3016\text{ kg/s})(1.0\text{ kJ/kg.°C})(32-24)°\text{C} = \mathbf{2.41\text{ kW}}$$

$$\dot{Q}_{\text{infiltration, latent}} = \dot{m}_{\text{air}} (w_{\text{ambient}} - w_{\text{room}}) h_{fg} = (0.3016\text{ kg/s})(0.0150-0.0093)(2490\text{ kJ/kg}) = \mathbf{4.28\text{ kW}}$$

$$\dot{Q}_{\text{infiltration, total}} = \dot{Q}_{\text{infiltration, sensible}} + \dot{Q}_{\text{infiltration, latent}} = 2.41+4.28 = \mathbf{6.69\text{ kW}}$$

Discussion The specific volume of the dry air at the ambient conditions could also be determined from the psychrometric chart at ambient conditions.

13-63 The infiltration rate of a building is estimated to be 1.8 ACH. The sensible, latent, and total infiltration heat loads of the building at sea level are to be determined.

Assumptions **1** Steady operating conditions exist. **2** The air infiltrates at the outdoor conditions, and exfiltrates at the indoor conditions. **3** Excess moisture condenses at 5°C. **4** The effect of water vapor on air density is negligible.

Properties The gas constant and the specific heat of air are R = 0.287 kPa.m^3/kg.K and C_p = 1.0 kJ/kg·°C (Tables A-1 and A-3). The heat of vaporization of water at 5°C is $h_{fg} = h_{fg\,@\,5°C} = 2490\text{ kJ/kg}$ (Table A-4). The properties of the ambient and room air are determined from the psychrometric chart (Fig. A-33) to be

$$\left.\begin{array}{l} T_{\text{ambient}} = 32°\text{C} \\ \phi_{\text{ambient}} = 50\% \end{array}\right\} w_{\text{ambient}} = 0.0150\text{ kg/kg dryair}$$

$$\left.\begin{array}{l} T_{\text{room}} = 24°\text{C} \\ \phi_{\text{room}} = 50\% \end{array}\right\} w_{\text{room}} = 0.0093\text{ kg/kg dryair}$$

Analysis Noting that the infiltration of ambient air will cause the air in the cold storage room to be changed 1.8 times every hour, the air will enter the room at a mass flow rate of

$$\rho_{\text{ambient}} = \frac{P_0}{RT_0} = \frac{101.325\text{ kPa}}{(0.287\text{ kPa.m}^3/\text{kg.K})(32+273\text{ K})} = 1.16\text{ kg/m}^3$$

$$\dot{m}_{\text{air}} = \rho_{\text{ambient}} V_{\text{room}} \text{ACH} = (1.16\text{ kg/m}^3)(20\times 13\times 3\text{ m}^3)(1.8\text{ h}^{-1}) = 1629\text{ kg/h} = 0.4524\text{ kg/s}$$

Then the sensible, latent, and total infiltration heat loads of the room are determined to be

$$\dot{Q}_{\text{infiltration, sensible}} = \dot{m}_{\text{air}} C_p (T_{\text{ambient}} - T_{\text{room}}) = (0.4524\text{ kg/s})(1.0\text{ kJ/kg.°C})(32-24)°\text{C} = \mathbf{3.62\text{ kW}}$$

$$\dot{Q}_{\text{infiltration, latent}} = \dot{m}_{\text{air}} (w_{\text{ambient}} - w_{\text{room}}) h_{fg} = (0.4524\text{ kg/s})(0.0150-0.0093)(2490\text{ kJ/kg}) = \mathbf{6.42\text{ kW}}$$

$$\dot{Q}_{\text{infiltration, total}} = \dot{Q}_{\text{infiltration, sensible}} + \dot{Q}_{\text{infiltration, latent}} = 3.62 + 6.42 = \mathbf{10.04\text{ kW}}$$

Discussion The specific volume of the dry air at the ambient conditions could also be determined from the psychrometric chart at ambient conditions.

Simple Heating and cooling

13-64C Relative humidity decreases during a simple heating process and increases during a simple cooling process. Specific humidity, on the other hand, remains constant in both cases.

13-65C Because a horizontal line on the psychometric chart represents a ω = constant process, and the moisture content ω of air remains constant during these processes.

13-66 Air enters a heating section at a specified state and relative humidity. The rate of heat transfer in the heating section and the relative humidity of the air at the exit are to be determined.

Assumptions **1** This is a steady-flow process and thus the mass flow rate of dry air remains constant during the entire process. **2** Dry air and water vapor are ideal gases. **3** The kinetic and potential energy changes are negligible.

Analysis (*a*) The amount of moisture in the air remains constant ($\omega_1 = \omega_2$) as it flows through the heating section since the process involves no humidification or dehumidification. The inlet state of the air is completely specified, and the total pressure is 95 kPa. The properties of the air are determined to be

$$P_{v1} = \phi_1 P_{g1} = \phi_1 P_{sat\,@\,15°C} = (0.3)(1.7051\text{ kPa}) = 0.51\text{ kPa}$$

$$P_{a1} = P_1 - P_{v1} = 95 - 0.51 = 94.49\text{ kPa}$$

$$v_1 = \frac{R_a T_1}{P_{a1}} = \frac{(0.287\text{ kPa}\cdot\text{m}^3/\text{kg}\cdot\text{K})(288\text{ K})}{94.49\text{ kPa}} = 0.875\text{ m}^3/\text{kg dry air}$$

Heating coils

95 kPa
15°C
30% RH
AIR
Heat
25°C

$$\omega_1 = \frac{0.622\,P_{v1}}{P_1 - P_{v1}} = \frac{0.622(0.51\text{ kPa})}{(95-0.51)\text{ kPa}} = 0.00336\text{ kg H}_2\text{O/kg dry air} (= \omega_2)$$

$$h_1 = C_p T_1 + \omega_1 h_{g1} = (1.005\text{ kJ/kg}\cdot°\text{C})(15°\text{C}) + (0.00336)(2528.9\text{ kJ/kg}) = 23.57\text{ kJ/kg dry air}$$

and

$$h_2 = C_p T_2 + \omega_2 h_{g2} = (1.005\text{ kJ/kg}\cdot°\text{C})(25°\text{C}) + (0.00336)(2547.2\text{ kJ/kg}) = 33.68\text{ kJ/kg dry air}$$

Also,

$$\dot{m}_{a1} = \frac{\dot{V}_1}{v_1} = \frac{4\text{ m}^3/\text{min}}{0.875\text{ m}^3/\text{kg dry air}} = 4.571\text{ kg/min}$$

Then the rate of heat transfer to the air in the heating section is determined from an energy balance on air in the heating section to be

$$\dot{Q}_{in} = \dot{m}_a(h_2 - h_1) = (4.571\text{ kg/min})(33.68 - 23.57)\text{ kJ/kg} = \mathbf{46.2\text{ kJ/min}}$$

(*b*) Noting that the vapor pressure of air remains constant ($P_{v2} = P_{v1}$) during a simple heating process, the relative humidity of the air at leaving the heating section becomes

$$\phi_2 = \frac{P_{v2}}{P_{g2}} = \frac{P_{v2}}{P_{sat\,@\,25°C}} = \frac{0.51\text{ kPa}}{3.169\text{ kPa}} = 0.161 \quad \text{or} \quad \mathbf{16.1\%}$$

13-67E Air enters a heating section at a specified pressure, temperature, velocity, and relative humidity. The exit temperature of air, the exit relative humidity, and the exit velocity are to be determined.

Assumptions **1** This is a steady-flow process and thus the mass flow rate of dry air remains constant during the entire process $(\dot{m}_{a1} = \dot{m}_{a2} = \dot{m}_a)$. **2** Dry air and water vapor are ideal gases. **3** The kinetic and potential energy changes are negligible.

Analysis (*a*) The amount of moisture in the air remains constant ($\omega_1 = \omega_2$) as it flows through the heating section since the process involves no humidification or dehumidification. The inlet state of the air is completely specified, and the total pressure is 1 atm. The properties of the air at the inlet state are determined from the psychometric chart (Figure A-33) to be

$$h_1 = 15.3 \text{ Btu / lbm dry air}$$
$$\omega_1 = 0.0030 \text{ lbm H}_2\text{O / lbm dry air } (= \omega_2)$$
$$v_1 = 12.9 \text{ ft}^3 \text{ / lbm dry air}$$

50°F
1 atm
AIR
40% RH
25 ft/s
D = 15 in
4 kW

The mass flow rate of dry air through the heating section is

$$\dot{m}_a = \frac{1}{v_1}\mathbf{V}_1 A_1 = \frac{1}{(12.9 \text{ ft}^3 / \text{lbm})}(25 \text{ ft / s})(\pi \times (15/12)^2 / 4 \text{ ft}^2) = 2.38 \text{ lbm / s}$$

From the energy balance on air in the heating section,

$$\dot{Q}_{\text{in}} = \dot{m}_a (h_2 - h_1)$$
$$4 \text{ kW}\left(\frac{0.9478 \text{ Btu/s}}{1 \text{ kW}}\right) = (2.38 \text{ lbm/s})(h_2 - 15.3)\text{Btu/lbm}$$
$$h_2 = 16.9 \text{ Btu/lbm dry air}$$

The exit state of the air is fixed now since we know both h_2 and ω_2. From the psychometric chart at this state we read

$$T_2 = \mathbf{56.8°F}$$

(*b*) $$\phi_2 = \mathbf{30.8\%}$$
$$v_2 = 131 \text{ ft}^3 \text{/lbm dry air}$$

(*c*) The exit velocity is determined from the conservation of mass of dry air,

$$\dot{m}_{a1} = \dot{m}_{a2} \longrightarrow \frac{\dot{V}_1}{v_1} = \frac{\dot{V}_2}{v_2} \longrightarrow \frac{\mathbf{V}_1 A}{v_1} = \frac{\mathbf{V}_2 A}{v_2}$$

Thus,

$$\mathbf{V}_2 = \frac{v_2}{v_1}\mathbf{V}_1 = \frac{13.1}{12.9}(25 \text{ ft / s}) = \mathbf{25.4 \text{ ft / s}}$$

13-68 Air enters a cooling section at a specified pressure, temperature, velocity, and relative humidity. The exit temperature, the exit relative humidity of the air, and the exit velocity are to be determined.

Assumptions **1** This is a steady-flow process and thus the mass flow rate of dry air remains constant during the entire process ($\dot{m}_{a1} = \dot{m}_{a2} = \dot{m}_a$). **2** Dry air and water vapor are ideal gases. **3** The kinetic and potential energy changes are negligible.

Analysis (*a*) The amount of moisture in the air remains constant ($\omega_1 = \omega_2$) as it flows through the cooling section since the process involves no humidification or dehumidification. The inlet state of the air is completely specified, and the total pressure is 1 atm. The properties of the air at the inlet state are determined from the psychometric chart (Figure A-33) to be

$$h_1 = 55.0 \text{ kJ / kg dry air}$$
$$\omega_1 = 0.0089 \text{ kg H}_2\text{O / kg dry air } (= \omega_2)$$
$$v_1 = 0.877 \text{ m}^3 \text{ / kg dry air}$$

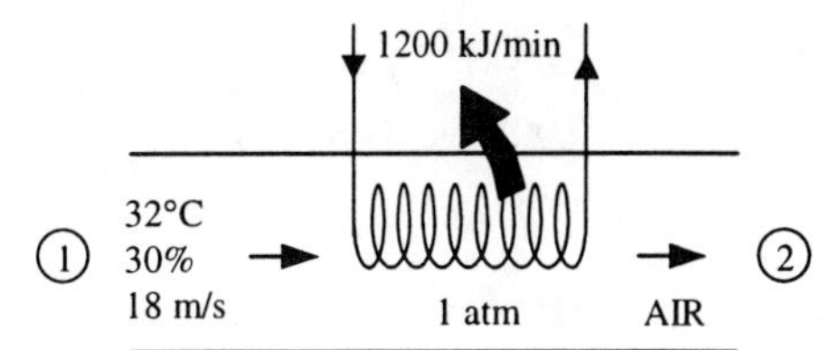

The mass flow rate of dry air through the cooling section is

$$\dot{m}_a = \frac{1}{v_1}\mathbf{V}_1 A_1 = \frac{1}{(0.877 \text{ m}^3/\text{kg})}(18 \text{ m/s})(\pi \times 0.4^2/4 \text{ m}^2) = 2.58 \text{ kg/s}$$

From the energy balance on air in the cooling section,

$$-\dot{Q}_{out} = \dot{m}_a(h_2 - h_1)$$
$$-1200/60 \text{ kJ/s} = (2.58 \text{ kg/s})(h_2 - 55.0) \text{ kJ/kg}$$
$$h_2 = 47.2 \text{ kJ/kg dry air}$$

The exit state of the air is fixed now since we know both h_2 and ω_2. From the psychometric chart at this state we read

$$T_2 = \mathbf{24.4°C}$$

(*b*)
$$\phi_2 = \mathbf{46.6\%}$$
$$v_2 = 0.856 \text{ m}^3 \text{ / kg dry air}$$

(*c*) The exit velocity is determined from the conservation of mass of dry air,

$$\dot{m}_{a1} = \dot{m}_{a2} \longrightarrow \frac{\dot{V}_1}{v_1} = \frac{\dot{V}_2}{v_2} \longrightarrow \frac{\mathbf{V}_1 A}{v_1} = \frac{\mathbf{V}_2 A}{v_2}$$

$$\mathbf{V}_2 = \frac{v_2}{v_1}\mathbf{V}_1 = \frac{0.856}{0.877}(18 \text{ m/s}) = \mathbf{17.6 \text{ m/s}}$$

13-69 Air enters a cooling section at a specified pressure, temperature, velocity, and relative humidity. The exit temperature, the exit relative humidity of the air, and the exit velocity are to be determined.

Assumptions **1** This is a steady-flow process and thus the mass flow rate of dry air remains constant during the entire process ($\dot{m}_{a1} = \dot{m}_{a2} = \dot{m}_a$). **2** Dry air and water vapor are ideal gases. **3** The kinetic and potential energy changes are negligible.

Analysis (*a*) The amount of moisture in the air remains constant ($\omega_1 = \omega_2$) as it flows through the cooling section since the process involves no humidification or dehumidification. The inlet state of the air is completely specified, and the total pressure is 1 atm. The properties of the air at the inlet state are determined from the psychometric chart (Figure A-33) to be

$$h_1 = 55.0 \text{ kJ / kg dry air}$$
$$\omega_1 = 0.0089 \text{ kg H}_2\text{O / kg dry air } (= \omega_2)$$
$$v_1 = 0.877 \text{ m}^3 \text{ / kg dry air}$$

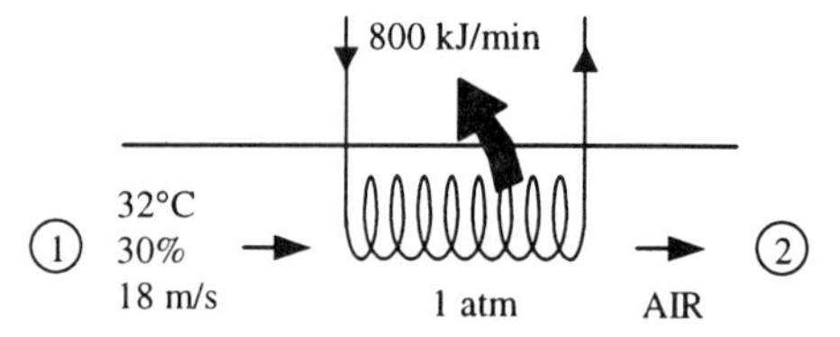

The mass flow rate of dry air through the cooling section is

$$\dot{m}_a = \frac{1}{v_1}\mathbf{V}_1 A_1 = \frac{1}{(0.877 \text{ m}^3\text{ / kg})}(18 \text{ m/s})(\pi \times 0.4^2 / 4 \text{ m}^2) = 2.58 \text{ kg/s}$$

From the energy balance on air in the cooling section,

$$-\dot{Q}_{out} = \dot{m}_a(h_2 - h_1)$$
$$-800/60 \text{ kJ/s} = (2.58 \text{ kg/s})(h_2 - 55.0) \text{ kJ/kg}$$
$$h_2 = 49.8 \text{ kJ/kg dry air}$$

The exit state of the air is fixed now since we know both h_2 and ω_2. From the psychometric chart at this state we read

$$T_2 = \mathbf{26.9°C}$$

(*b*) $$\phi_2 = \mathbf{40.0\%}$$

$$v_2 = 0.862 \text{ m}^3\text{ /kg dry air}$$

(*c*) The exit velocity is determined from the conservation of mass of dry air,

$$\dot{m}_{a1} = \dot{m}_{a2} \longrightarrow \frac{\dot{V}_1}{v_1} = \frac{\dot{V}_2}{v_2} \longrightarrow \frac{\mathbf{V}_1 A}{v_1} = \frac{\mathbf{V}_2 A}{v_2}$$

$$\mathbf{V}_2 = \frac{v_2}{v_1}\mathbf{V}_1 = \frac{0.862}{0.877}(18 \text{ m/s}) = \mathbf{17.7 \text{ m/s}}$$

Heating with Humidification

13-70C To achieve a higher level of comfort. Very dry air can cause dry skin, respiratory difficulties, and increased static electricity.

13-71 Air is first heated and then humidified by water vapor. The amount of steam added to the air and the amount of heat transfer to the air are to be determined.

Assumptions **1** This is a steady-flow process and thus the mass flow rate of dry air remains constant during the entire process $(\dot{m}_{a1} = \dot{m}_{a2} = \dot{m}_a)$. **2** Dry air and water vapor are ideal gases. **3** The kinetic and potential energy changes are negligible.

Properties The inlet and the exit states of the air are completely specified, and the total pressure is 1 atm. The properties of the air at various states are determined from the psychometric chart (Figure A-33) to be

$$h_1 = 31.1 \text{ kJ / kg dry air}$$
$$\omega_1 = 0.0064 \text{ kg H}_2\text{O / kg dry air} \ (= \omega_2)$$
$$h_2 = 36.2 \text{ kJ / kg dry air}$$
$$h_3 = 58.1 \text{ kJ / kg dry air}$$
$$\omega_3 = 0.0129 \text{ kg H}_2\text{O / kg dry air}$$

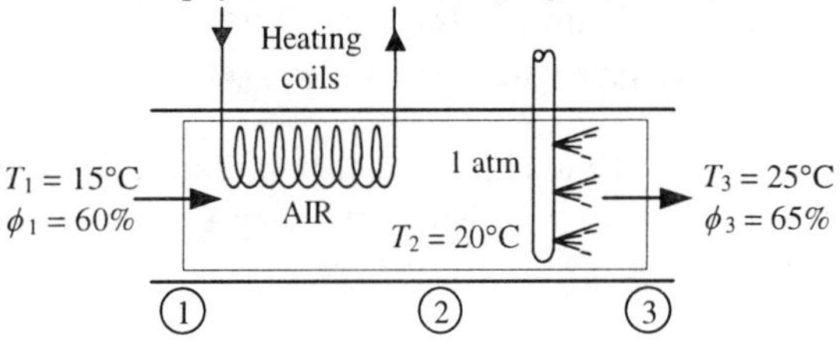

Heating coils
1 atm
AIR
$T_1 = 15°C$
$\phi_1 = 60\%$
$T_2 = 20°C$
$T_3 = 25°C$
$\phi_3 = 65\%$
(1) (2) (3)

Analysis (*a*) The amount of moisture in the air remains constant it flows through the heating section ($\omega_1 = \omega_2$), but increases in the humidifying section ($\omega_3 > \omega_2$). The amount of steam added to the air in the heating section is

$$\Delta\omega = \omega_3 - \omega_2 = 0.0129 - 0.0064 = \mathbf{0.0065 \text{ kg H}_2\text{O / kg dry air}}$$

(*b*) The heat transfer to the air in the heating section per unit mass of air is

$$q_{in} = h_2 - h_1 = 36.2 - 31.1 = \mathbf{5.1 \text{ kJ / kg dry air}}$$

13-72E Air is first heated and then humidified by water vapor. The amount of steam added to the air and the amount of heat transfer to the air are to be determined.

Assumptions **1** This is a steady-flow process and thus the mass flow rate of dry air remains constant during the entire process ($\dot{m}_{a1} = \dot{m}_{a2} = \dot{m}_a$). **2** Dry air and water vapor are ideal gases. **3** The kinetic and potential energy changes are negligible.

Properties The inlet and the exit states of the air are completely specified, and the total pressure is 1 atm. The properties of the air at various states are determined from the psychometric chart (Figure A-33E) to be

$$h_1 = 19.2 \text{ Btu / lbm dry air}$$
$$\omega_1 = 0.0055 \text{ lbm H}_2\text{O / lbm dry air}$$
$$h_2 = 23.3 \text{ Btu / lbm dry air}$$
$$\omega_2 = \omega_1 = 0.0055 \text{ lbm H}_2\text{O / lbm dry air}$$
$$h_3 = 31.2 \text{ Btu / lbm dry air}$$
$$\omega_3 = 0.0121 \text{ lbm H}_2\text{O / lbm dry air}$$

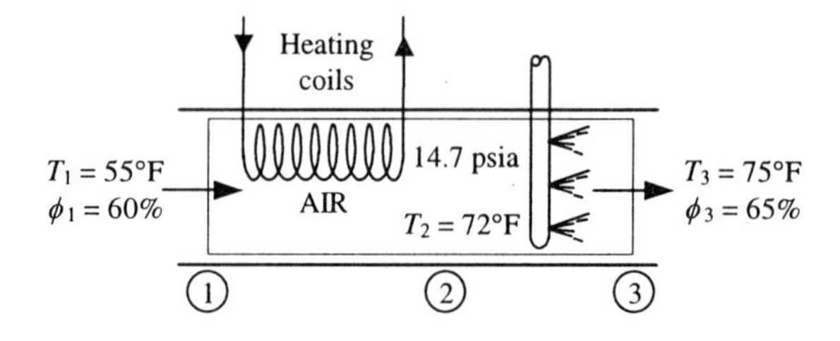

Analysis (*a*) The amount of moisture in the air remains constant it flows through the heating section ($\omega_1 = \omega_2$), but increases in the humidifying section ($\omega_3 > \omega_2$). The amount of steam added to the air in the heating section is

$$\Delta\omega = \omega_3 - \omega_2 = 0.0121 - 0.0055 = \mathbf{0.0066 \text{ lbm H}_2\text{O / lbm dry air}}$$

(*b*) The heat transfer to the air in the heating section per unit mass of air is

$$q_{in} = h_2 - h_1 = 23.3 - 19.2 = \mathbf{4.1 \text{ Btu / lbm dry air}}$$

13-73 Air is first heated and then humidified by wet steam. The temperature and relative humidity of air at the exit of heating section, the rate of heat transfer, and the rate at which water is added to the air are to be determined.

Assumptions **1** This is a steady-flow process and thus the mass flow rate of dry air remains constant during the entire process $(\dot{m}_{a1} = \dot{m}_{a2} = \dot{m}_a)$. **2** Dry air and water vapor are ideal gases. **3** The kinetic and potential energy changes are negligible.

Properties The inlet and the exit states of the air are completely specified, and the total pressure is 1 atm. The properties of the air at various states are determined from the psychometric chart (Figure A-33) to be

$$h_1 = 23.5 \text{ kJ/kg dry air}$$
$$\omega_1 = 0.0053 \text{ kg H}_2\text{O/kg dry air} (= \omega_2)$$
$$v_1 = 0.809 \text{ m}^3\text{/kg dry air}$$

$$h_3 = 42.3 \text{ kJ/kg dry air}$$
$$\omega_3 = 0.0088 \text{ kg H}_2\text{O/kg dry air}$$

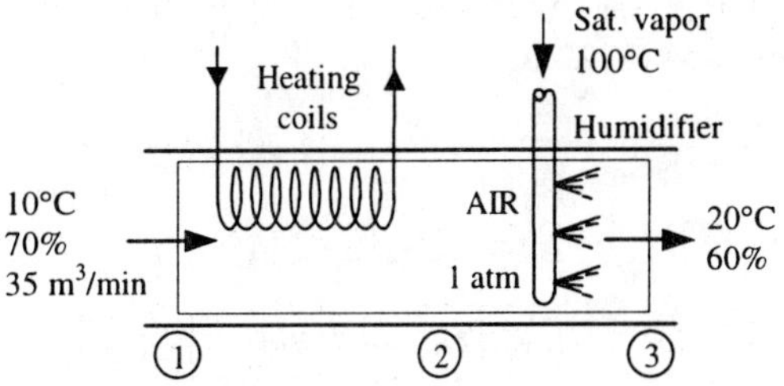

Sat. vapor
100°C
Heating coils
Humidifier
10°C
70%
35 m³/min
AIR
1 atm
20°C
60%
1
2
3

Analysis (*a*) The amount of moisture in the air remains constant it flows through the heating section ($\omega_1 = \omega_2$), but increases in the humidifying section ($\omega_3 > \omega_2$). The mass flow rate of dry air is

$$\dot{m}_a = \frac{\dot{\mathbf{V}}_1}{v_1} = \frac{35 \text{ m}^3\text{/min}}{0.809 \text{ m}^3\text{/kg}} = 43.3 \text{ kg/min}$$

Noting that $Q = W = 0$, the energy balance on the humidifying section can be expressed as

$$\dot{E}_{\text{in}} - \dot{E}_{\text{out}} = \Delta \dot{E}_{\text{system}}^{\nearrow 0 \text{ (steady)}} = 0$$
$$\dot{E}_{\text{in}} = \dot{E}_{\text{out}}$$
$$\sum \dot{m}_i h_i = \sum \dot{m}_e h_e \longrightarrow \dot{m}_w h_w + \dot{m}_{a2} h_2 = \dot{m}_a h_3$$
$$(\omega_3 - \omega_2) h_w + h_2 = h_3$$

Solving for h_2,

$$h_2 = h_3 - (\omega_3 - \omega_2) h_{g@100°C} = 42.3 - (0.0088 - 0.0053)(2676.1) = 32.9 \text{ kJ/kg dry air}$$

Thus at the exit of the heating section we have $\omega = 0.0053$ kg H_2O dry air and $h_2 = 32.9$ kJ/kg dry air, which completely fixes the state. Then from the psychometric chart we read

$$T_2 = \mathbf{19.4°C}$$
$$\phi_2 = \mathbf{37.8\%}$$

(*b*) The rate of heat transfer to the air in the heating section is

$$\dot{Q}_{\text{in}} = \dot{m}_a (h_2 - h_1) = (43.3 \text{ kg/min})(32.9 - 23.5) \text{ kJ/kg} = \mathbf{407 \text{ kJ/min}}$$

(*c*) The amount of water added to the air in the humidifying section is determined from the conservation of mass equation of water in the humidifying section,

$$\dot{m}_w = \dot{m}_a (\omega_3 - \omega_2) = (43.3 \text{ kg/min})(0.0088 - 0.0053) = \mathbf{0.15 \text{ kg/min}}$$

13-74 Air is first heated and then humidified by wet steam. The temperature and relative humidity of air at the exit of heating section, the rate of heat transfer, and the rate at which water is added to the air are to be determined.

Assumptions **1** This is a steady-flow process and thus the mass flow rate of dry air remains constant during the entire process ($\dot{m}_{a1} = \dot{m}_{a2} = \dot{m}_a$). **2** Dry air and water vapor are ideal gases. **3** The kinetic and potential energy changes are negligible.

Analysis (*a*) The amount of moisture in the air also remains constant it flows through the heating section ($\omega_1 = \omega_2$), but increases in the humidifying section ($\omega_3 > \omega_2$). The inlet and the exit states of the air are completely specified, and the total pressure is 95 kPa. The properties of the air at various states are determined to be

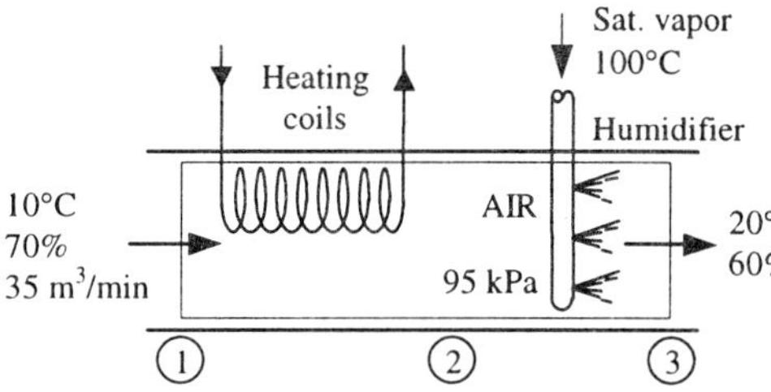

$$P_{v1} = \phi_1 P_{g1} = \phi_1 P_{\text{sat @ }10^\circ\text{C}} = (0.70)(1.2276\ \text{kPa}) = 0.859\ \text{kPa}\ (= P_{v2})$$

$$P_{a1} = P_1 - P_{v1} = 95 - 0.859 = 94.141\ \text{kPa}$$

$$v_1 = \frac{R_a T_1}{P_{a1}} = \frac{(0.287\ \text{kPa}\cdot\text{m}^3/\text{kg}\cdot\text{K})(283\ \text{K})}{94.141\ \text{kPa}} = 0.863\ \text{m}^3/\text{kg dry air}$$

$$\omega_1 = \frac{0.622 P_{v1}}{P_1 - P_{v1}} = \frac{0.622(0.859\ \text{kPa})}{(95 - 0.859)\ \text{kPa}} = 0.00568\ \text{kg H}_2\text{O/kg dry air}\ (= \omega_2)$$

$$h_1 = C_p T_1 + \omega_1 h_{g1} = (1.005\ \text{kJ/kg}\cdot^\circ\text{C})(10^\circ\text{C}) + (0.00568)(2519.8\ \text{kJ/kg}) = 24.36\ \text{kJ/kg dry air}$$

$$P_{v3} = \phi_3 P_{g3} = \phi_3 P_{\text{sat @ }20^\circ\text{C}} = (0.60)(2.339\ \text{kPa}) = 1.40\ \text{kPa}$$

$$\omega_3 = \frac{0.622 P_{v3}}{P_3 - P_{v3}} = \frac{0.622(1.40\ \text{kPa})}{(95 - 1.40)\ \text{kPa}} = 0.00930\ \text{kg H}_2\text{O/kg dry air}$$

$$h_3 = C_p T_3 + \omega_3 h_{g3} = (1.005\ \text{kJ/kg}\cdot^\circ\text{C})(20^\circ\text{C}) + (0.0093)(2538.1\ \text{kJ/kg}) = 43.70\ \text{kJ/kg dry air}$$

Also,

$$\dot{m}_a = \frac{\dot{V}_1}{v_1} = \frac{35\ \text{m}^3/\text{min}}{0.863\ \text{m}^3/\text{kg}} = 40.6\ \text{kg/min}$$

Noting that $Q = W = 0$, the energy balance on the humidifying section can be expressed as

$$\dot{E}_{in} - \dot{E}_{out} = \Delta\dot{E}_{system}^{\nearrow 0\ \text{(steady)}} = 0$$

$$\dot{E}_{in} = \dot{E}_{out}$$

$$\sum \dot{m}_e h_e = \sum \dot{m}_i h_i \longrightarrow \dot{m}_w h_w + \dot{m}_{a2} h_2 = \dot{m}_a h_3$$

$$(\omega_3 - \omega_2) h_w + h_2 = h_3$$

Solving for h_2,

$$h_2 = h_3 - (\omega_3 - \omega_2) h_{g\text{ @ }100^\circ\text{C}} = 43.7 - (0.0093 - 0.00568) \times 2676.1 = 34.0\ \text{kJ/kg dry air}$$

Thus at the exit of the heating section we have ω = 0.00568 kg H_2O dry air and h_2 = 34.0 kJ/kg dry air, which completely fixes the state. The temperature of air at the exit of the heating section is determined from the definition of enthalpy,

$$h_2 = C_p T_2 + \omega_2 h_{g2} \cong C_p T_2 + \omega_2 (2501.3 + 1.82 T_2)$$

$$34.0 = (1.005) T_2 + (0.00568)(2501.3 + 1.82 T_2)$$

Solving for h_2, yields $\qquad T_2 = \mathbf{19.5°C}$

The relative humidity at this state is

$$\phi_2 = \frac{P_{v2}}{P_{g2}} = \frac{P_{v2}}{P_{sat\,@\,19.5°C}} = \frac{0.859\text{ kPa}}{2.276\text{ kPa}} = 0.377 \text{ or } \mathbf{37.7\%}$$

(*b*) The rate of heat transfer to the air in the heating section becomes

$$\dot{Q}_{in} = \dot{m}_a (h_2 - h_1) = (40.6\text{ kg/min})(34.0 - 24.36)\text{ kJ/kg} = \mathbf{391\ kJ/min}$$

(*c*) The amount of water added to the air in the humidifying section is determined from the conservation of mass equation of water in the humidifying section,

$$\dot{m}_w = \dot{m}_a (\omega_3 - \omega_2) = (40.6\text{ kg/min})(0.0093 - 0.00568) = \mathbf{0.147\ kg/min}$$

Cooling with Dehumidification

13-75C To drop its relative humidity to more desirable levels.

13-76 Air is cooled and dehumidified by a window air conditioner. The rates of heat and moisture removal from the air are to be determined.

Assumptions **1** This is a steady-flow process and thus the mass flow rate of dry air remains constant during the entire process ($\dot{m}_{a1} = \dot{m}_{a2} = \dot{m}_a$). **2** Dry air and water vapor are ideal gases. **3** The kinetic and potential energy changes are negligible.

Properties The inlet and the exit states of the air are completely specified, and the total pressure is 1 atm. The properties of the air at various states are determined from the psychometric chart (Figure A-33) to be

$$h_1 = 86.4 \text{ kJ / kg dry air}$$
$$\omega_1 = 0.0212 \text{ kg H}_2\text{O / kg dry air}$$
$$v_1 = 0.894 \text{ m}^3 \text{ / kg dry air}$$

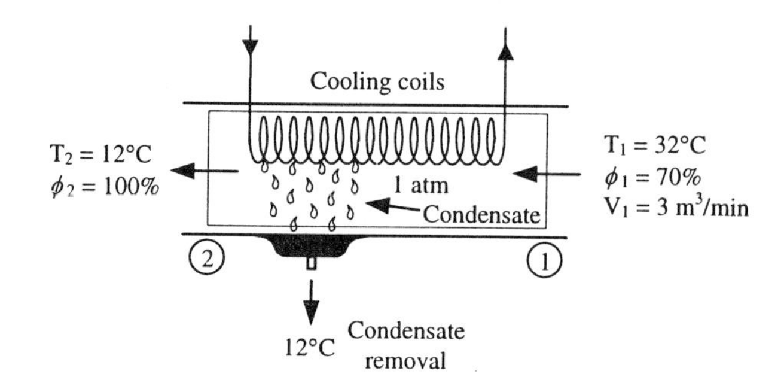

and

$$h_2 = 34.1 \text{ kJ / kg dry air}$$
$$\omega_2 = 0.0087 \text{ kg H}_2\text{O / kg dry air}$$

Also,

$$h_w \cong h_{f\,@12^\circ\text{C}} = 50.4 \text{ kJ / kg}$$

Analysis (a) The amount of moisture in the air decreases due to dehumidification ($\omega_2 < \omega_1$). The mass flow rate of air is

$$\dot{m}_{a1} = \frac{\dot{V}_1}{v_1} = \frac{3 \text{ m}^3 \text{ / min}}{0.894 \text{ m}^3 \text{ / kg dry air}} = 3.356 \text{ kg/min}$$

Applying the water mass balance and energy balance equations to the combined cooling and dehumidification section,

Water Mass Balance: $\sum \dot{m}_{w,i} = \sum \dot{m}_{w,e} \longrightarrow \dot{m}_{a1}\omega_1 = \dot{m}_{a2}\omega_2 + \dot{m}_w$

$$\dot{m}_w = \dot{m}_a(\omega_1 - \omega_2) = (8.95 \text{ kg/min})(0.0212 - 0.0087) = \mathbf{0.112\ kg/min}$$

Energy Balance:

$$\dot{E}_{\text{in}} - \dot{E}_{\text{out}} = \Delta \dot{E}_{\text{system}}^{\nearrow 0 \text{ (steady)}} = 0$$
$$\dot{E}_{\text{in}} = \dot{E}_{\text{out}}$$
$$\sum \dot{m}_i h_i = \dot{Q}_{out} + \sum \dot{m}_e h_e$$
$$\dot{Q}_{out} = \dot{m}_{a1}h_1 - (\dot{m}_{a2}h_2 + \dot{m}_w h_w) = \dot{m}_a(h_1 - h_2) - \dot{m}_w h_w$$

$$\dot{Q}_{out} = (3.356 \text{ kg/min})(86.4 - 34.1)\text{kJ/kg} - (0.112 \text{ kg/min})(50.4 \text{ kJ/kg})$$
$$= \mathbf{170\ kJ/min}$$

13-77 Air is first cooled, then dehumidified, and finally heated. The temperature of air before it enters the heating section, the amount of heat removed in the cooling section, and the amount of heat supplied in the heating section are to be determined.

Assumptions **1** This is a steady-flow process and thus the mass flow rate of dry air remains constant during the entire process ($\dot{m}_{a1} = \dot{m}_{a2} = \dot{m}_a$). **2** Dry air and water vapor are ideal gases. **3** The kinetic and potential energy changes are negligible.

Analysis (*a*) The amount of moisture in the air decreases due to dehumidification ($\omega_3 < \omega_1$), and remains constant during heating ($\omega_3 = \omega_2$). The inlet and the exit states of the air are completely specified, and the total pressure is 1 atm. The intermediate state (state 2) is also known since ϕ_2 = 100% and $\omega_2 = \omega_3$. Therefore, we can determined the properties of the air at all three states from the psychometric chart (Table A-33) to be

$$h_1 = 95.2 \text{ kJ / kg dry air}$$
$$\omega_1 = 0.0238 \text{ kg H}_2\text{O / kg dry air}$$

and

$$h_3 = 43.1 \text{ kJ / kg dry air}$$
$$\omega_3 = 0.0082 \text{ kg H}_2\text{O / kg dry air } (= \omega_2)$$

Also,

$$h_w \cong h_{f\,@10°\text{C}} = 42.0 \text{ kJ / kg}$$
$$h_2 = 31.8 \text{ kJ / kg dry air}$$
$$T_2 = \mathbf{11.1°C}$$

(*b*) The amount of heat removed in the cooling section is determined from the energy balance equation applied to the cooling section,

$$\dot{E}_{\text{in}} - \dot{E}_{\text{out}} = \Delta\dot{E}_{\text{system}} \nearrow^{0 \text{ (steady)}} = 0$$
$$\dot{E}_{\text{in}} = \dot{E}_{\text{out}}$$
$$\Sigma \dot{m}_i h_i = \Sigma \dot{m}_e h_e + \dot{Q}_{\text{out,cooling}}$$
$$\dot{Q}_{\text{out,cooling}} = \dot{m}_{a1}h_1 - (\dot{m}_{a2}h_2 + \dot{m}_w h_w) = \dot{m}_a(h_1 - h_2) - \dot{m}_w h_w$$

or, per unit mass of dry air,

$$q_{\text{out,cooling}} = (h_1 - h_2) - (\omega_1 - \omega_2)h_w$$
$$= (95.2 - 31.8) - (0.0238 - 0.0082)42.0$$
$$= \mathbf{62.7 \text{ kJ / kg dry air}}$$

(*c*) The amount of heat supplied in the heating section per unit mass of dry air is

$$q_{\text{in,heating}} = h_3 - h_2 = 43.1 - 31.8 = \mathbf{11.3 \text{ kJ / kg dry air}}$$

13-78 [*Also solved by EES on enclosed CD*] Air is cooled by passing it over a cooling coil through which chilled water flows. The rate of heat transfer, the mass flow rate of water, and the exit velocity of airstream are to be determined.

Assumptions **1** This is a steady-flow process and thus the mass flow rate of dry air remains constant during the entire process. **2** Dry air and water vapor are ideal gases. **3** The kinetic and potential energy changes are negligible.

Analysis (*a*) The dew point temperature of the incoming air stream at 35°C is

$$T_{dp} = T_{sat\,@\,P_v} = T_{sat\,@\,0.6\times 5.628\text{ kPa}} = 25.9°\text{C}$$

since air is cooled to 20°C, which is below its dew point temperature, some of the moisture in the air will condense.

The amount of moisture in the air decreases due to dehumidification $(\omega_2 < \omega_1)$. The inlet and the exit states of the air are completely specified, and the total pressure is 1 atm. Then the properties of the air at both states are determined from the psychometric chart (Table A-33) to be

$$h_1 = 90.3\text{ kJ / kg dry air}$$
$$\omega_1 = 0.0215\text{ kg H}_2\text{O / kg dry air}$$
$$v_1 = 0.904\text{ m}^3\text{ / kg dry air}$$

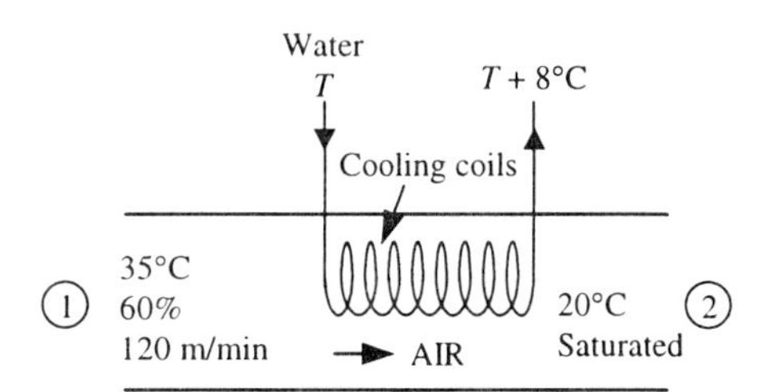

and

$$h_2 = 57.5\text{ kJ / kg dry air}$$
$$\omega_2 = 0.0147\text{ kg H}_2\text{O / kg dry air}$$
$$v_2 = 0.851\text{ m}^3\text{ / kg dry air}$$

Also,

$$h_w \cong h_{f\,@\,20°\text{C}} = 83.96\text{ kJ / kg} \qquad \text{(Table A-4)}$$

Then,

$$\dot{V}_1 = \mathbf{V}_1 A_1 = \mathbf{V}_1 \frac{\pi D^2}{4} = (120\text{ m/min})\left(\frac{\pi(0.3\text{ m})^2}{4}\right) = 8.48\text{ m}^3\text{ / min}$$

$$\dot{m}_{a1} = \frac{\dot{V}_1}{v_1} = \frac{8.48\text{ m}^3\text{ / min}}{0.904\text{ m}^3\text{ / kg dry air}} = 9.38\text{ kg/min}$$

Applying the water mass balance and the energy balance equations to the combined cooling and dehumidification section (excluding the water),

Water Mass Balance: $\quad \sum \dot{m}_{w,i} = \sum \dot{m}_{w,e} \longrightarrow \dot{m}_{a1}\omega_1 = \dot{m}_{a2}\omega_2 + \dot{m}_w$

$$\dot{m}_w = \dot{m}_a(\omega_1 - \omega_2) = (9.38\text{ kg/min})(0.0215 - 0.0147) = 0.064\text{ kg/min}$$

Energy Balance:

$$\dot{E}_{in} - \dot{E}_{out} = \Delta\dot{E}_{system}^{\nearrow 0\text{ (steady)}} = 0$$
$$\dot{E}_{in} = \dot{E}_{out}$$
$$\sum \dot{m}_i h_i = \sum \dot{m}_e h_e + \dot{Q}_{out} \rightarrow \dot{Q}_{out} = \dot{m}_{a1}h_1 - (\dot{m}_{a2}h_2 + \dot{m}_w h_w) = \dot{m}_a(h_1 - h_2) - \dot{m}_w h_w$$

$$\dot{Q}_{out} = (9.38\text{ kg/min})(90.3 - 57.5)\text{ kJ/kg} - (0.064\text{ kg/min})(83.96\text{ kJ/kg})$$
$$= \mathbf{302.3\text{ kJ / min}}$$

(*b*) Noting that the heat lost by the air is gained by the cooling water, the mass flow rate of the cooling water is determined from

$$\dot{Q}_{\text{cooling water}} = \dot{m}_{\text{cooling water}} \Delta h = \dot{m}_{\text{cooling water}} C_p \Delta T$$

$$\dot{m}_{\text{cooling water}} = \frac{\dot{Q}_w}{C_p \Delta T} = \frac{302.3 \text{ kJ/min}}{(4.18 \text{ kJ/kg}\cdot°\text{C})(8°\text{C})} = \mathbf{9.04 \text{ kg/min}}$$

(*c*) The exit velocity is determined from the conservation of mass of dry air,

$$\dot{m}_{a1} = \dot{m}_{a2} \longrightarrow \frac{\dot{V}_1}{v_1} = \frac{\dot{V}_2}{v_2} \longrightarrow \frac{\mathbf{V}_1 A}{v_1} = \frac{\mathbf{V}_2 A}{v_2}$$

$$\mathbf{V}_2 = \frac{v_2}{v_1}\mathbf{V}_1 = \frac{0.851}{0.904}(120 \text{ m/min}) = \mathbf{113 \text{ m/min}}$$

13-79 EES solution of this (and other comprehensive problems designated with the *computer icon*) is available to instructors at the *Instructor Manual* section of the *Online Learning Center* (OLC) at www.mhhe.com/cengel-boles. See the Preface for access information.

13-80 Air is cooled by passing it over a cooling coil. The rate of heat transfer, the mass flow rate of water, and the exit velocity of airstream are to be determined.

Assumptions **1** This is a steady-flow process and thus the mass flow rate of dry air remains constant during the entire process. **2** Dry air and water vapor are ideal gases. **3** The kinetic and potential energy changes are negligible.

Analysis (*a*) The dew point temperature of the incoming air stream at 35°C is

$$P_{v1} = \phi_1 P_{g1} = \phi_1 P_{\text{sat @ 35°C}} = (0.6)(5.628 \text{ kPa}) = 3.38 \text{ kPa}$$

$$T_{dp} = T_{\text{sat @ } P_v} = T_{\text{sat @ 3.38 kPa}} = 25.9°\text{C}$$

Since air is cooled to 20°C, which is below its dew point temperature, some of the moisture in the air will condense.

The amount of moisture in the air decreases due to dehumidification $(\omega_2 < \omega_1)$. The inlet and the exit states of the air are completely specified, and the total pressure is 95 kPa. Then the properties of the air at both states are determined to be

$$P_{a1} = P_1 - P_{v1} = 95 - 3.38 = 91.62 \text{ kPa}$$

$$v_1 = \frac{R_a T_1}{P_{a1}} = \frac{(0.287 \text{ kPa}\cdot\text{m}^3/\text{kg}\cdot\text{K})(308 \text{ K})}{91.62 \text{ kPa}} = 0.965 \text{ m}^3/\text{kg dry air}$$

$$\omega_1 = \frac{0.622\, P_{v1}}{P_1 - P_{v1}} = \frac{0.622(3.38 \text{ kPa})}{(95 - 3.38) \text{ kPa}} = 0.0229 \text{ kg H}_2\text{O/kg dry air}$$

$$h_1 = C_p T_1 + \omega_1 h_{g1} = (1.005 \text{ kJ/kg}\cdot°\text{C})(35°\text{C}) + (0.0229)(2565.3 \text{ kJ/kg})$$
$$= 93.92 \text{ kJ/kg dry air}$$

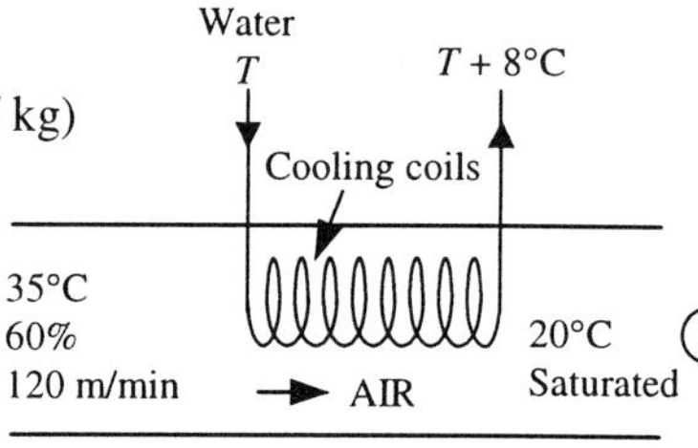

Water
T
$T + 8°\text{C}$
Cooling coils
35°C
60%
120 m/min
①
AIR
20°C
Saturated
②

and

$$P_{v2} = \phi_2 P_{g2} = (1.00) P_{\text{sat @ 20°C}} = 2.339 \text{ kPa}$$

$$v_2 = \frac{R_a T_2}{P_{a2}} = \frac{(0.287 \text{ kPa}\cdot\text{m}^3/\text{kg}\cdot\text{K})(293 \text{ K})}{(95 - 2.339) \text{ kPa}} = 0.908 \text{ m}^3/\text{kg dry air}$$

$$\omega_2 = \frac{0.622\, P_{v2}}{P_2 - P_{v2}} = \frac{0.622(2.339 \text{ kPa})}{(95 - 2.339) \text{ kPa}} = 0.0157 \text{ kg H}_2\text{O/kg dry air}$$

$$h_2 = C_p T_2 + \omega_2 h_{g2} = (1.005 \text{ kJ/kg}\cdot°\text{C})(20°\text{C}) + (0.0157)(2538.1 \text{ kJ/kg})$$
$$= 59.95 \text{ kJ/kg dry air}$$

Also,

$$h_w \cong h_{f\text{ @ 20°C}} = 83.96 \text{ kJ/kg} \qquad \text{(Table A-4)}$$

Then,

$$\dot{V}_1 = \mathbf{V}_1 A_1 = \mathbf{V}_1 \frac{\pi D^2}{4} = (120 \text{ m/min})\left(\frac{\pi (0.3 \text{ m})^2}{4}\right) = 8.48 \text{ m}^3/\text{min}$$

$$\dot{m}_{a1} = \frac{\dot{V}_1}{v_1} = \frac{8.48 \text{ m}^3/\text{min}}{0.965 \text{ m}^3/\text{kg dry air}} = 8.79 \text{ kg/min}$$

Applying the water mass balance and energy balance equations to the combined cooling and dehumidification section (excluding the water),

Water Mass Balance: $\sum \dot{m}_{w,i} = \sum \dot{m}_{w,e} \longrightarrow \dot{m}_{a1}\omega_1 = \dot{m}_{a2}\omega_2 + \dot{m}_w$

$$\dot{m}_w = \dot{m}_a(\omega_1 - \omega_2) = (8.79\ \text{kg/min})(0.0229 - 0.0157) = 0.0633\ \text{kg/min}$$

Energy Balance:

$$\dot{E}_{in} - \dot{E}_{out} = \Delta\dot{E}_{system}{}^{\nearrow 0\ (\text{steady})} = 0$$

$$\dot{E}_{in} = \dot{E}_{out}$$

$$\sum\dot{m}_i h_i = \sum\dot{m}_e h_e + \dot{Q}_{out} \rightarrow \dot{Q}_{out} = \dot{m}_{a1}h_1 - (\dot{m}_{a2}h_2 + \dot{m}_w h_w) = \dot{m}_a(h_1 - h_2) - \dot{m}_w h_w$$

$$\dot{Q}_{out} = (8.79\ \text{kg/min})(93.92 - 59.95)\text{kJ/kg} - (0.0633\ \text{kg/min})(83.96\ \text{kJ/kg}) = \mathbf{293.3\ kJ/min}$$

(*b*) Noting that the heat lost by the air is gained by the cooling water, the mass flow rate of the cooling water is determined from

$$\dot{Q}_{\text{cooling water}} = \dot{m}_{\text{cooling water}}\Delta h = \dot{m}_{\text{cooling water}} C_p \Delta T$$

$$\dot{m}_{\text{cooling water}} = \frac{\dot{Q}_w}{C_p\Delta T} = \frac{293.3\ \text{kJ/min}}{(4.18\ \text{kJ/kg}\cdot^\circ\text{C})(8^\circ\text{C})} = \mathbf{8.77\ kg/min}$$

(*c*) The exit velocity is determined from the conservation of mass of dry air,

$$\dot{m}_{a1} = \dot{m}_{a2} \longrightarrow \frac{\dot{V}_1}{v_1} = \frac{\dot{V}_2}{v_2} \longrightarrow \frac{\mathbf{V}_1 A}{v_1} = \frac{\mathbf{V}_2 A}{v_2}$$

$$\mathbf{V}_2 = \frac{v_2}{v_1}\mathbf{V}_1 = \frac{0.908}{0.965}(120\ \text{m/min}) = \mathbf{113\ m/min}$$

13-81E Air is cooled by passing it over a cooling coil through which chilled water flows. The rate of heat transfer, the mass flow rate of water, and the exit velocity of airstream are to be determined.

Assumptions **1** This is a steady-flow process and thus the mass flow rate of dry air remains constant during the entire process. **2** Dry air and water vapor are ideal gases. **3** The kinetic and potential energy changes are negligible.

Analysis (*a*) The dew point temperature of the incoming air stream at 90°F is

$$T_{dp} = T_{sat\,@\,P_v} = T_{sat\,@\,0.6\times 0.6988\text{ psia}} = 74°\text{ F}$$

Since air is cooled to 70°F, which is below its dew point temperature, some of the moisture in the air will condense.

The amount of moisture in the air decreases due to dehumidification $(\omega_2 < \omega_1)$.The inlet and the exit states of the air are completely specified, and the total pressure is 14.7 psia. Then the properties of the air at both states are determined from the psychometric chart (Table A-33E) to be

$$h_1 = 41.8\text{ Btu/lbm dry air}$$
$$\omega_1 = 0.0183\text{ lbm H}_2\text{O/lbm dry air}$$
$$v_1 = 14.26\text{ ft}^3\text{/lbm dry air}$$

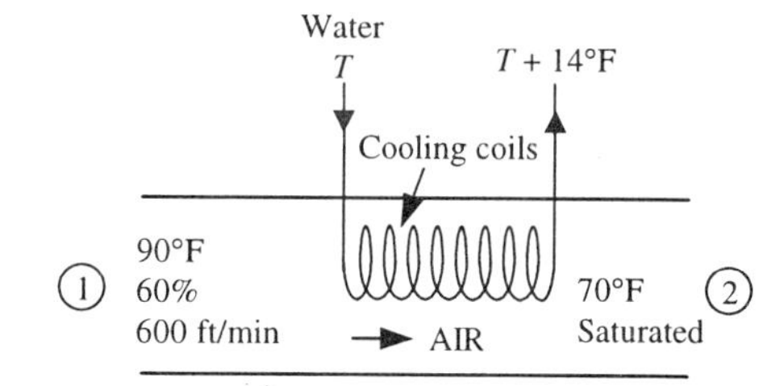

and

$$h_2 = 34.1\text{ Btu/lbm dry air}$$
$$\omega_2 = 0.0158\text{ lbm H}_2\text{O/lbm dry air}$$
$$v_2 = 13.68\text{ ft}^3\text{/lbm dry air}$$

Also,

$$h_w \cong h_{f\,@\,70°\text{F}} = 38.09\text{ Btu/lbm} \qquad \text{(Table A-4E)}$$

Then,

$$\dot{V}_1 = \mathbf{V}_1 A_1 = \mathbf{V}_1 \frac{\pi D^2}{4} = (600\text{ ft/min})\left(\frac{\pi(1\text{ ft})^2}{4}\right) = 471\text{ ft}^3\text{/min}$$

$$\dot{m}_{a1} = \frac{\dot{V}_1}{v_1} = \frac{471\text{ ft}^3\text{/min}}{14.26\text{ ft}^3\text{/lbm dry air}} = 33.0\text{ lbm/min}$$

Applying the water mass balance and the energy balance equations to the combined cooling and dehumidification section (excluding the water),

Water Mass Balance: $\sum \dot{m}_{w,i} = \sum \dot{m}_{w,e} \longrightarrow \dot{m}_{a1}\omega_1 = \dot{m}_{a2}\omega_2 + \dot{m}_w$

$$\dot{m}_w = \dot{m}_a(\omega_1 - \omega_2) = (33.0\text{ lbm/min})(0.0183 - 0.0158) = 0.083\text{ lbm/min}$$

Energy Balance:

$$\dot{E}_{in} - \dot{E}_{out} = \Delta\dot{E}_{system}^{\nearrow 0\text{ (steady)}} = 0$$

$$\dot{E}_{in} = \dot{E}_{out}$$

$$\sum \dot{m}_i h_i = \sum \dot{m}_e h_e + \dot{Q}_{out} \rightarrow \dot{Q}_{out} = \dot{m}_{a1}h_1 - (\dot{m}_{a2}h_2 + \dot{m}_w h_w) = \dot{m}_a(h_1 - h_2) - \dot{m}_w h_w$$

$$\dot{Q}_{out} = (33.0\text{ lbm/min})(41.8 - 34.1)\text{ Btu/lbm} - (0.083\text{ lbm/min})(38.09\text{ Btu/lbm})$$
$$= \mathbf{250.9\ Btu/min}$$

(*b*) Noting that the heat lost by the air is gained by the cooling water, the mass flow rate of the cooling water is determined from

$$\dot{Q}_{\text{cooling water}} = \dot{m}_{\text{cooling water}} \Delta h = \dot{m}_{\text{cooling water}} C_p \Delta T$$

$$\dot{m}_{\text{cooling water}} = \frac{\dot{Q}_w}{C_p \Delta T} = \frac{250.9 \text{ Btu/min}}{(1.0 \text{ Btu/lbm}\cdot^\circ\text{F})(14^\circ\text{F})} = \mathbf{17.9 \text{ lbm/min}}$$

(*c*) The exit velocity is determined from the conservation of mass of dry air,

$$\dot{m}_{a1} = \dot{m}_{a2} \longrightarrow \frac{\dot{V}_1}{v_1} = \frac{\dot{V}_2}{v_2} \longrightarrow \frac{\mathbf{V}_1 A}{v_1} = \frac{\mathbf{V}_2 A}{v_2}$$

$$\mathbf{V}_2 = \frac{v_2}{v_1}\mathbf{V}_1 = \frac{13.68}{14.26}(600 \text{ ft/min}) = \mathbf{576 \text{ ft/min}}$$

13-82E EES solution of this (and other comprehensive problems designated with the *computer icon*) is available to instructors at the *Instructor Manual* section of the *Online Learning Center* (OLC) at www.mhhe.com/cengel-boles. See the Preface for access information.

13-83E Air is cooled by passing it over a cooling coil through which chilled water flows. The rate of heat transfer, the mass flow rate of water, and the exit velocity of airstream are to be determined.

Assumptions **1** This is a steady-flow process and thus the mass flow rate of dry air remains constant during the entire process. **2** Dry air and water vapor are ideal gases. **3** The kinetic and potential energy changes are negligible.

Analysis (*a*) The dew point temperature of the incoming air stream at 90°F is

$$P_{v1} = \phi_1 P_{g1} = \phi_1 P_{sat\,@\,90°F} = (0.6)(0.6988\text{ psia}) = 0.42\text{ psia}$$

$$T_{dp} = T_{sat\,@\,P_v} = T_{sat\,@\,0.42\text{ psia}} = 74°\text{ F}$$

Since air is cooled to 70°F, which is below its dew point temperature, some of the moisture in the air will condense.

The mass flow rate of dry air remains constant during the entire process $(\dot{m}_{a1} = \dot{m}_{a2} = \dot{m}_a)$, but the amount of moisture in the air decreases due to dehumidification $(\omega_2 < \omega_1)$. The inlet and the exit states of the air are completely specified, and the total pressure is 14.4 psia. Then the properties of the air at both states are determined to be

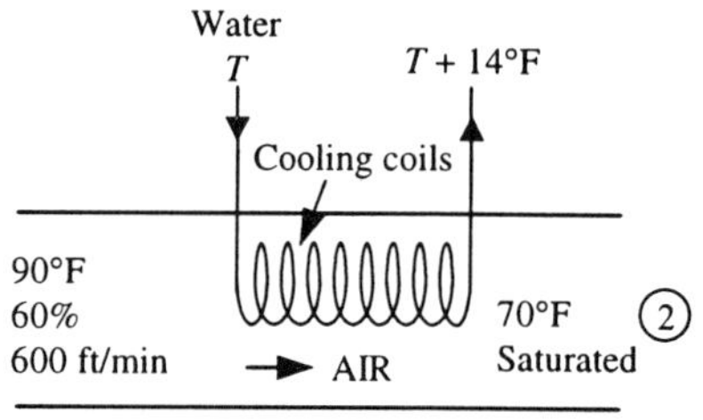

Water
T
T + 14°F
Cooling coils
90°F
60%
600 ft/min
①
AIR
70°F
Saturated
②

$$P_{a1} = P_1 - P_{v1} = 14.4 - 0.42 = 13.98\text{ psia}$$

$$\text{v}_1 = \frac{R_a T_1}{P_{a1}} = \frac{(0.3704\text{ psia}\cdot\text{ft}^3/\text{lbm}\cdot\text{R})(550\text{ R})}{13.98\text{ psia}} = 14.57\text{ ft}^3/\text{lbm dry air}$$

$$\omega_1 = \frac{0.622\,P_{v1}}{P_1 - P_{v1}} = \frac{0.622(0.42\text{ psia})}{(14.4 - 0.42)\text{ psia}} = 0.0187\text{ lbm H}_2\text{O}/\text{lbm dry air}$$

$$h_1 = C_p T_1 + \omega_1 h_{g1} = (0.24\text{ Btu/lbm}\cdot°\text{F})(90°\text{ F}) + (0.0187)(1100.7\text{ Btu/lbm})$$
$$= 42.18\text{ Btu/lbm dry air}$$

and

$$P_{v2} = \phi_2 P_{g2} = (1.00)\,P_{sat\,@\,70°F} = 0.36\text{ psia}$$

$$P_{a2} = P_2 - P_{v2} = 14.4 - 0.36 = 14.04\text{ psia}$$

$$\upsilon_2 = \frac{R_a T_2}{P_{a2}} = \frac{(0.3704\text{ psia}\cdot\text{ft}^3/\text{lbm}\cdot\text{R})(530\text{ R})}{14.04\text{ psia}} = 14.0\text{ ft}^3/\text{lbm dry air}$$

$$\omega_2 = \frac{0.622\,P_{v2}}{P_2 - P_{v2}} = \frac{0.622(0.36\text{ psia})}{(14.4 - 0.36)\text{ psia}} = 0.0159\text{ lbm H}_2\text{O}/\text{lbm dry air}$$

$$h_2 = C_p T_2 + \omega_2 h_{g2} = (0.24\text{ Btu/lbm}\cdot°\text{F})(70°\text{ F}) + (0.0159)(1092.0\text{ Btu/lbm})$$
$$= 34.16\text{ Btu/lbm dry air}$$

Also,

$$h_w \cong h_{f\,@\,70°F} = 38.09\text{ Btu/lbm} \qquad \text{(Table A-4E)}$$

Then,

$$\dot{V}_1 = \mathbf{V}_1 A_1 = \mathbf{V}_1 \frac{\pi D^2}{4} = (600\text{ ft/min})\left(\frac{\pi(1\text{ ft})^2}{4}\right) = 471\text{ ft}^3/\text{min}$$

$$\dot{m}_{a1} = \frac{\dot{V}_1}{\text{v}_1} = \frac{471\text{ ft}^3/\text{min}}{14.57\text{ ft}^3/\text{lbm dry air}} = 32.3\text{ lbm/min}$$

Applying the water mass balance and energy balance equations to the combined cooling and dehumidification section (excluding the water),

Water Mass Balance: $\sum \dot{m}_{w,i} = \sum \dot{m}_{w,e} \longrightarrow \dot{m}_{a1}\omega_1 = \dot{m}_{a2}\omega_2 + \dot{m}_w$

$$\dot{m}_w = \dot{m}_a(\omega_1 - \omega_2) = (32.3\text{ lbm/min})(0.0187 - 0.0159) = 0.0904\text{ lbm/min}$$

Energy balance:

$$\dot{E}_{in} - \dot{E}_{out} = \Delta \dot{E}_{system}^{\nearrow 0\text{ (steady)}} = 0$$

$$\dot{E}_{in} = \dot{E}_{out}$$

$$\sum \dot{m}_i h_i = \sum \dot{m}_e h_e + \dot{Q}_{out} \rightarrow \dot{Q}_{out} = \dot{m}_{a1}h_1 - (\dot{m}_{a2}h_2 + \dot{m}_w h_w) = \dot{m}_a(h_1 - h_2) - \dot{m}_w h_w$$

$$\dot{Q}_{out} = (32.3\text{ lbm/min})(42.18 - 34.16)\text{ Btu/lbm} - (0.0904\text{ lbm/min})(38.09\text{ Btu/lbm})$$
$$= \mathbf{255.6\ Btu/min}$$

(*b*) Noting that the heat lost by the air is gained by the cooling water, the mass flow rate of the cooling water is determined from

$$\dot{Q}_{\text{cooling water}} = \dot{m}_{\text{cooling water}}\Delta h = \dot{m}_{\text{cooling water}} C_p \Delta T$$

$$\dot{m}_{\text{cooling water}} = \frac{\dot{Q}_w}{C_p \Delta T} = \frac{255.6\text{ Btu/min}}{(1.0\text{ Btu/lbm}\cdot{}^\circ\text{F})(14^\circ\text{F})} = \mathbf{18.3\ lbm/min}$$

(*c*) The exit velocity is determined from the conservation of mass of dry air,

$$\dot{m}_{a1} = \dot{m}_{a2} \longrightarrow \frac{\dot{V}_1}{v_1} = \frac{\dot{V}_2}{v_2} \longrightarrow \frac{\mathbf{V}_1 A}{v_1} = \frac{\mathbf{V}_2 A}{v_2}$$

$$\mathbf{V}_2 = \frac{v_2}{v_1}\mathbf{V}_1 = \frac{14.0}{14.57}(600\text{ ft/s}) = \mathbf{577\ ft/min}$$

Evaporative Cooling

13-84C In steady operation, the mass transfer process does not have to involve heat transfer. However, a mass transfer process that involves phase change (evaporation, sublimation, condensation, melting etc.) must involve heat transfer. For example, the evaporation of water from a lake into air (mass transfer) requires the transfer of latent heat of water at a specified temperature to the liquid water at the surface (heat transfer).

13-85C During evaporation from a water body to air, the latent heat of vaporization will be equal to *convection* heat transfer from the air when *conduction* from the lower parts of the water body to the surface is negligible, and temperature of the surrounding surfaces is at about the temperature of the water surface so that the *radiation* heat transfer is negligible.

13-86C Evaporative cooling is the cooling achieved when water evaporates in dry air. It will not work on humid climates.

13-87 Air is cooled by an evaporative cooler. The exit temperature of the air and the required rate of water supply are to be determined.

Analysis (*a*) From the psychometric chart (Table A-33) at 36°C and 20% relative humidity we read

$$T_{wb1} = 19.5°\text{C}$$
$$\omega_1 = 0.0074 \text{ kg H}_2\text{O /kg dry air}$$
$$v_1 = 0.887 \text{ m}^3 \text{/kg dry air}$$

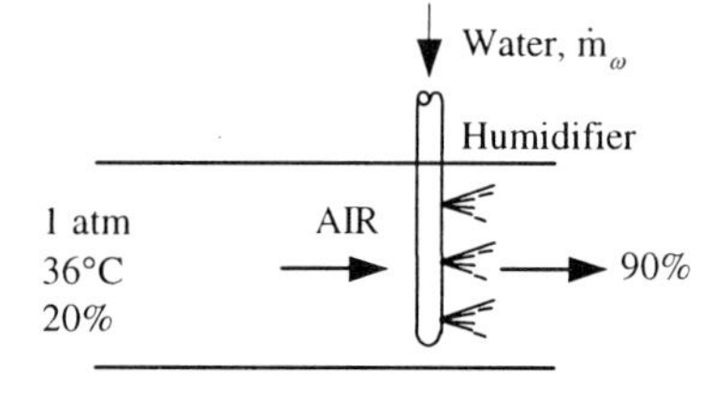

Assuming the liquid water is supplied at a temperature not much different than the exit temperature of the air stream, the evaporative cooling process follows a line of constant wet-bulb temperature. That is,

$$T_{wb2} \cong T_{wb1} = 19.5°\text{C}$$

At this wet-bulb temperature and 90% relative humidity we read

$$T_2 = \mathbf{20.5°\,C}$$
$$\omega_2 = 0.0137 \text{ kg H}_2\text{O / kg dry air}$$

Thus air will be cooled to 20.5°C in this evaporative cooler.

(*b*) The mass flow rate of dry air is

$$\dot{m}_a = \frac{\dot{V}_1}{v_1} = \frac{4 \text{ m}^3\text{/min}}{0.887 \text{ m}^3 \text{/kg dry air}} = 4.51 \text{ kg /min}$$

Then the required rate of water supply to the evaporative cooler is determined from

$$\begin{aligned}\dot{m}_{supply} &= \dot{m}_{w2} - \dot{m}_{w1} = \dot{m}_a(\omega_2 - \omega_1) \\ &= (4.51 \text{ kg /min})(0.0137 - 0.0074) \\ &= \mathbf{0.028 \text{ kg / min}}\end{aligned}$$

13-88E Air is cooled by an evaporative cooler. The exit temperature of the air and the required rate of water supply are to be determined.

Analysis (*a*) From the psychometric chart (Table A-33E) at 90°F and 20% relative humidity we read

$$T_{wb1} = 62.8°\text{F}$$
$$\omega_1 = 0.0060 \text{ lbm H}_2\text{O/lbm dry air}$$
$$v_1 = 14.0 \text{ ft}^3\text{/lbm dry air}$$

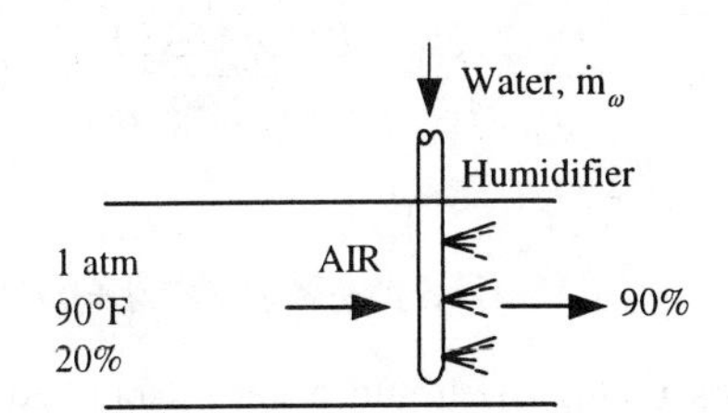

Assuming the liquid water is supplied at a temperature not much different than the exit temperature of the air stream, the evaporative cooling process follows a line of constant wet-bulb temperature. That is,

$$T_{wb2} \cong T_{wb1} = 62.8°\text{F}$$

At this wet-bulb temperature and 90% relative humidity we read

$$T_2 = \mathbf{64°F}$$
$$\omega_2 = 0.0116 \text{ lbm H}_2\text{O / lbm dry air}$$

Thus air will be cooled to 64°F in this evaporative cooler.

(*b*) The mass flow rate of dry air is

$$\dot{m}_a = \frac{\dot{V}_1}{v_1} = \frac{150 \text{ ft}^3\text{/min}}{14.0 \text{ ft}^3\text{/lbm dry air}} = 10.7 \text{ lbm/min}$$

Then the required rate of water supply to the evaporative cooler is determined from

$$\dot{m}_{supply} = \dot{m}_{w2} - \dot{m}_{w1} = \dot{m}_a(\omega_2 - \omega_1)$$
$$= (10.7 \text{ lbm/min})(0.0116 - 0.0060)$$
$$= \mathbf{0.06 \text{ lbm/min}}$$

13-89 Air is cooled by an evaporative cooler. The exit temperature of the air is to be determined.

Analysis The enthalpy of air at the inlet is determined from

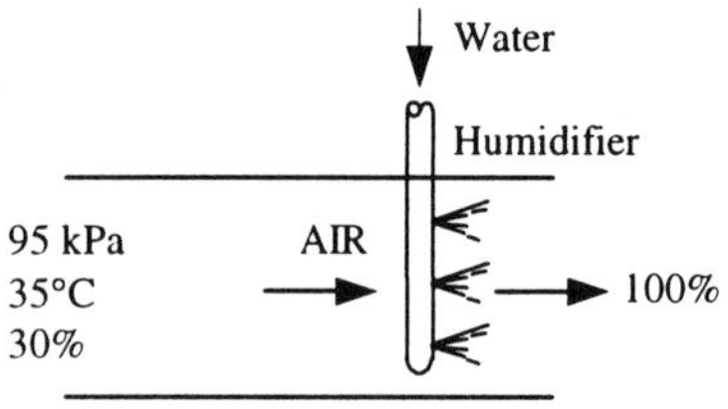

Water

Humidifier

95 kPa

35°C

30%

AIR

100%

$$P_{v1} = \phi_1 P_{g1} = \phi_1 P_{sat\,@\,35°C} = (0.30)(5.628\text{ kPa}) = 1.69\text{ kPa}$$

$$\omega_1 = \frac{0.622\,P_{v1}}{P_1 - P_{v1}} = \frac{0.622(1.69\text{ kPa})}{(95-1.69)\text{ kPa}} = 0.0113\text{ kg H}_2\text{O/kg dry air}$$

$$h_1 = C_p T_1 + \omega_1 h_{g1}$$

$$= (1.005\text{ kJ/kg}\cdot°\text{C})(35°\text{C}) + (0.0113)(2565.3\text{ kJ/kg}) = 64.2\text{ kJ/kg dry air}$$

Assuming the liquid water is supplied at a temperature not much different than the exit temperature of the air stream, the evaporative cooling process follows a line of constant wet-bulb temperature, which is almost parallel to the constant enthalpy lines. That is,

$$h_2 \cong h_1 = 64.2\text{ kJ / kg dry air}$$

Also,

$$\omega_2 = \frac{0.622\,P_{v2}}{P_2 - P_{v2}} = \frac{0.622\,P_{g2}}{95 - P_{g2}}$$

since air leaves the evaporative cooler saturated. Substituting this into the definition of enthalpy, we obtain

$$h_2 = C_p T_2 + \omega_2 h_{g2} \cong C_p T_2 + \omega_2 (2501.3 + 1.82 T_2)$$

$$64.2\text{ kJ / kg} = (1.005\text{ kJ / kg}\cdot°\text{C})\,T_2 + \frac{0.622\,P_{g2}}{95 - P_{g2}}(2501.3 + 1.82\,T_2)\text{kJ / kg}$$

By trial and error, the exit temperature is determined to be $T_2 = \mathbf{21.1°C}$.

13-90E Air is cooled by an evaporative cooler. The exit temperature of the air is to be determined.

Analysis The enthalpy of air at the inlet is determined from

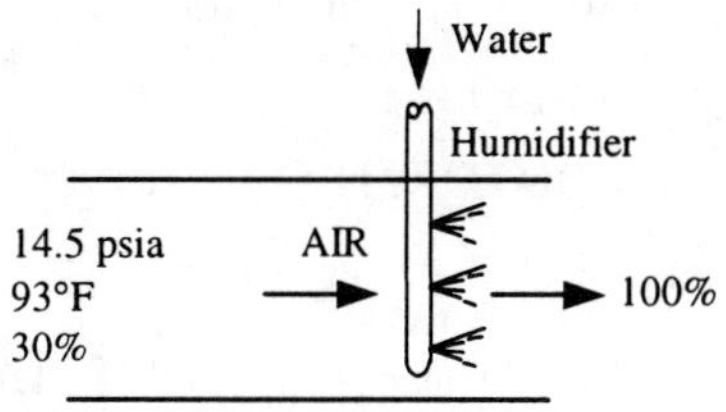

$$P_{v1} = \phi_1 P_{g1} = \phi_1 P_{sat\,@\,93^\circ F} = (0.30)(0.774\text{ psia}) = 0.232\text{ psia}$$

$$\omega_1 = \frac{0.622\,P_{v1}}{P_1 - P_{v1}} = \frac{0.622(0.232\text{ psia})}{(14.5-0.232)\text{ psia}}$$

$$= 0.0101\text{ lbm H}_2\text{O/lbm dry air}$$

$$h_1 = C_p T_1 + \omega_1 h_{g1}$$

$$= (0.24\text{ Btu/lbm}\cdot^\circ\text{F})(93^\circ\text{F}) + (0.0101)(1102.0\text{ Btu/lbm})$$

$$= 33.45\text{ Btu/lbm dry air}$$

Water

Humidifier

14.5 psia

93°F

30%

AIR

100%

Assuming the liquid water is supplied at a temperature not much different than the exit temperature of the air stream, the evaporative cooling process follows a line of constant wet-bulb temperature, which is almost parallel to the constant enthalpy lines. That is,

$$h_2 \cong h_1 = 33.45\text{ Btu / lbm dry air}$$

Also,

$$\omega_2 = \frac{0.622\,P_{v2}}{P_2 - P_{v2}} = \frac{0.622\,P_{g2}}{14.5 - P_{g2}}$$

since air leaves the evaporative cooler saturated. Substituting this into the definition of enthalpy, we obtain

$$h_2 = C_p T_2 + \omega_2 h_{g2} \cong C_p T_2 + \omega_2 (1061.5 + 0.435 T_2)$$

$$33.45\text{ Btu / lbm} = (0.24\text{ Btu / lbm}\cdot^\circ\text{F})\,T_2 + \frac{0.622\,P_{g2}}{14.5 - P_{g2}}(1061.5 + 0.435\,T_2)$$

By trial and error, the exit temperature is determined to be $T_2 = \mathbf{69.0^\circ F}$.

13-91 Air is cooled by an evaporative cooler. The final relative humidity and the amount of water added are to be determined.

Analysis (*a*) From the psychometric chart (Table A-33) at 32°C and 30% relative humidity we read

$$T_{wb1} = 19.4°\text{C}$$
$$\omega_1 = 0.0089 \text{ kg H}_2\text{O /kg dry air}$$
$$v_1 = 0.877 \text{ m}^3\text{ /kg dry air}$$

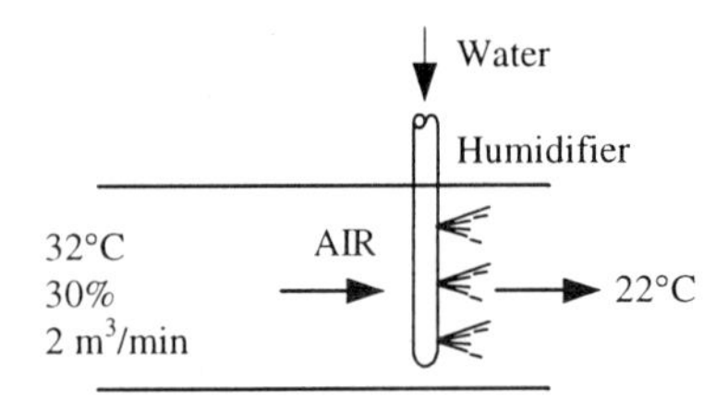

Assuming the liquid water is supplied at a temperature not much different than the exit temperature of the air stream, the evaporative cooling process follows a line of constant wet-bulb temperature. That is,

$$T_{wb2} \cong T_{wb1} = 19.4°\text{C}$$

At this wet-bulb temperature and 22°C temperature we read

$$\phi_2 = 77\%$$
$$\omega_2 = 0.0130 \text{ kg H}_2\text{O /kg dry air}$$

(*b*) The mass flow rate of dry air is

$$\dot{m}_a = \frac{\dot{V}_1}{v_1} = \frac{2 \text{ m}^3\text{ /min}}{0.877 \text{ m}^3\text{ /kg dry air}} = 2.28 \text{ kg /min}$$

Then the required rate of water supply to the evaporative cooler is determined from

$$\dot{m}_{supply} = \dot{m}_{w2} - \dot{m}_{w1} = \dot{m}_a(\omega_2 - \omega_1)$$
$$= (2.28 \text{ kg /min})(0.0130 - 0.0089)$$
$$= \mathbf{0.0094 \text{ kg / min}}$$

13-92 Air enters an evaporative cooler at a specified state and relative humidity. The lowest temperature that air can attain is to be determined.

Analysis From the psychometric chart (Table A-33) at 29°C and 40% relative humidity we read

$$T_{wb1} = 19.3°\text{C}$$

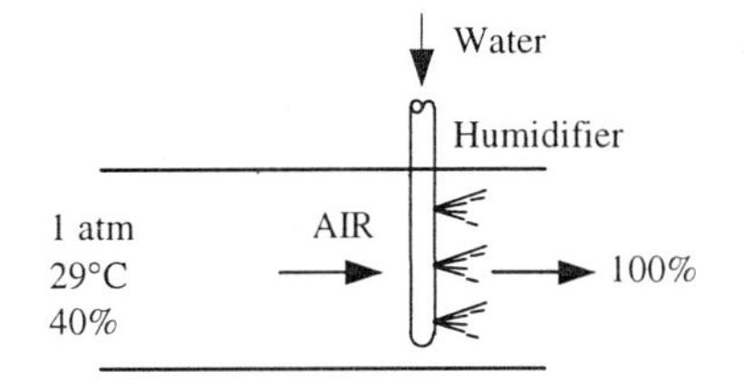

Assuming the liquid water is supplied at a temperature not much different than the exit temperature of the air stream, the evaporative cooling process follows a line of constant wet-bulb temperature, which is the lowest temperature that can be obtained in an evaporative cooler. That is,

$$T_{min} = T_{wb1} = \mathbf{19.3°C}$$

13-93 Air is first heated in a heating section and then passed through an evaporative cooler. The exit relative humidity and the amount of water added are to be determined.

Analysis (*a*) From the psychometric chart (Table A-33) at 15°C and 60% relative humidity we read

$$\omega_1 = 0.00635 \text{ kg H}_2\text{O / kg dry air}$$

The specific humidity ω remains constant during the heating process. Therefore, $\omega_2 = \omega_1 = 0.00635$ kg H_2O / kg dry air. At this ω value and 30°C we read $T_{wb2} = 16.6°C$.

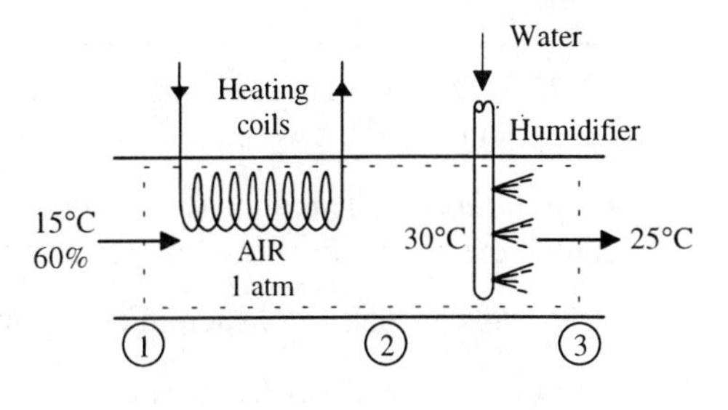

Assuming the liquid water is supplied at a temperature not much different than the exit temperature of the air stream, the evaporative cooling process follows a line of constant wet-bulb temperature. That is, $T_{wb3} \cong T_{wb2} = 16.6°C$. At this T_{wb} value and 25°C we read

$$\phi_3 = \mathbf{42.3\%}$$
$$\omega_3 = 0.00836 \text{ kg H}_2\text{O / kg dry air}$$

(*b*) The amount of water added to the air per unit mass of air is

$$\Delta\omega_{23} = \omega_3 - \omega_2 = 0.00836 - 0.00635 = \mathbf{0.00201 \text{ kg H}_2\text{O / kg dry air}}$$

Adiabatic Mixing of Airstreams

13-94C This will occur when the straight line connecting the states of the two streams on the psychometric chart crosses the saturation line.

13-95C Yes.

13-96 Two airstreams are mixed steadily. The specific humidity, the relative humidity, the dry-bulb temperature, and the volume flow rate of the mixture are to be determined.

Assumptions **1** Steady operating conditions exist **2** Dry air and water vapor are ideal gases. **3** The kinetic and potential energy changes are negligible. **4** The mixing section is adiabatic.

Properties Properties of each inlet stream are determined from the psychometric chart (Table A-33) to be

$$h_1 = 62.7 \text{ kJ/kg dry air}$$
$$\omega_1 = 0.0119 \text{ kg H}_2\text{O/kg dry air}$$
$$v_1 = 0.882 \text{ m}^3\text{/kg dry air}$$

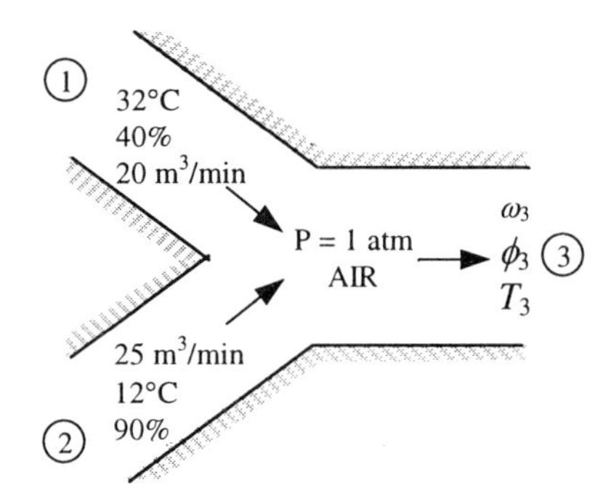

and

$$h_2 = 31.9 \text{ kJ/kg dry air}$$
$$\omega_2 = 0.0079 \text{ kg H}_2\text{O/kg dry air}$$
$$v_2 = 0.819 \text{ m}^3\text{/kg dry air}$$

Analysis The mass flow rate of dry air in each stream is

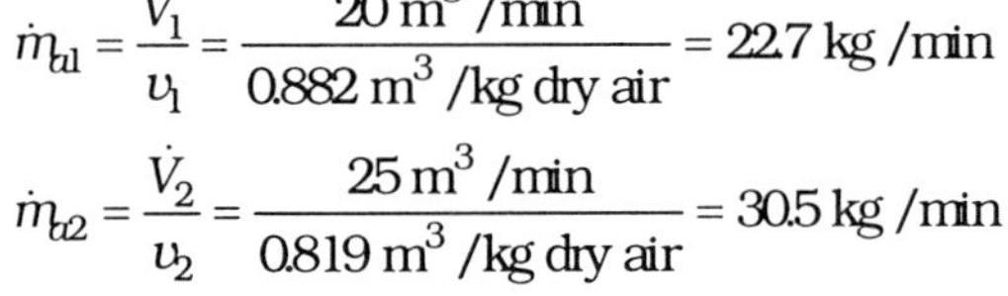

$$\dot{m}_{a1} = \frac{\dot{V}_1}{v_1} = \frac{20 \text{ m}^3\text{/min}}{0.882 \text{ m}^3\text{/kg dry air}} = 22.7 \text{ kg/min}$$

$$\dot{m}_{a2} = \frac{\dot{V}_2}{v_2} = \frac{25 \text{ m}^3\text{/min}}{0.819 \text{ m}^3\text{/kg dry air}} = 30.5 \text{ kg/min}$$

From the conservation of mass,

$$\dot{m}_{a3} = \dot{m}_{a1} + \dot{m}_{a2} = (22.7 + 30.5) \text{ kg/min} = 53.2 \text{ kg/min}$$

The specific humidity and the enthalpy of the mixture can be determined from Eqs. 13-24, which are obtained by combining the conservation of mass and energy equations for the adiabatic mixing of two streams:

$$\frac{\dot{m}_{a1}}{\dot{m}_{a2}} = \frac{\omega_2 - \omega_3}{\omega_3 - \omega_1} = \frac{h_2 - h_3}{h_3 - h_1}$$

$$\frac{22.7}{30.5} = \frac{0.0079 - \omega_3}{\omega_3 - 0.0119} = \frac{31.9 - h_3}{h_3 - 62.7}$$

which yields,

$$\omega_3 = \mathbf{0.0096 \text{ kg H}_2\text{O/kg dry air}}$$
$$h_3 = 45.0 \text{ kJ/kg dry air}$$

These two properties fix the state of the mixture. Other properties of the mixture are determined from the psychometric chart:

$$T_3 = \mathbf{20.6°C}$$
$$\phi_3 = \mathbf{63.4\%}$$
$$v_3 = 0.845 \text{ m}^3\text{/kg dry air}$$

Finally, the volume flow rate of the mixture is determined from

$$\dot{V}_3 = \dot{m}_{a3} v_3 = (53.2 \text{ kg/min})(0.845 \text{ m}^3\text{/kg}) = \mathbf{45.0 \text{ m}^3\text{/min}}$$

13-97 Two airstreams are mixed steadily. The specific humidity, the relative humidity, the dry-bulb temperature, and the volume flow rate of the mixture are to be determined.

Assumptions **1** Steady operating conditions exist **2** Dry air and water vapor are ideal gases. **3** The kinetic and potential energy changes are negligible. **4** The mixing section is adiabatic.

Analysis The properties of each inlet stream are determined to be

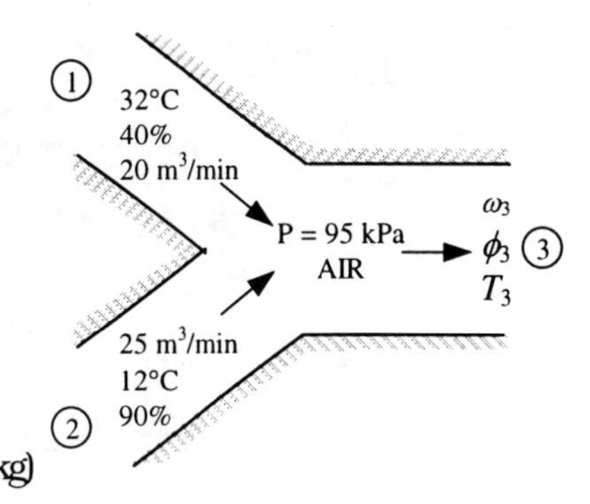

$$P_{v1} = \phi_1 P_{g1} = \phi_1 P_{sat\,@\,32°C} = (0.40)(4.800\ \text{kPa}) = 1.92\ \text{kPa}$$

$$P_{a1} = P_1 - P_{v1} = 95 - 1.92 = 93.08\ \text{kPa}$$

$$v_1 = \frac{R_a T_1}{P_{a1}} = \frac{(0.287\ \text{kPa}\cdot\text{m}^3/\text{kg}\cdot\text{K})(305\ \text{K})}{93.08\ \text{kPa}}$$

$$= 0.940\ \text{m}^3/\text{kg dry air}$$

$$\omega_1 = \frac{0.622\,P_{v1}}{P_1 - P_{v1}} = \frac{0.622(1.92\ \text{kPa})}{(95-1.92)\ \text{kPa}} = 0.0128\ \text{kg H}_2\text{O/kg dry air}$$

$$h_1 = C_p T_1 + \omega_1 h_{g1} = (1.005\ \text{kJ/kg}\cdot°\text{C})(32°\text{C}) + (0.0128)(2559.9\ \text{kJ/kg})$$

$$= 64.9\ \text{kJ/kg dry air}$$

and

$$P_{v2} = \phi_2 P_{g2} = \phi_2 P_{sat\,@\,12°C} = (0.90)(1.4186\ \text{kPa}) = 1.277\ \text{kPa}$$

$$P_{a2} = P_2 - P_{v2} = 95 - 1.277 = 93.723\ \text{kPa}$$

$$v_2 = \frac{R_a T_2}{P_{a2}} = \frac{(0.287\ \text{kPa}\cdot\text{m}^3/\text{kg}\cdot\text{K})(285\ \text{K})}{93.723\ \text{kPa}} = 0.873\ \text{m}^3/\text{kg dry air}$$

$$\omega_2 = \frac{0.622\,P_{v2}}{P_2 - P_{v2}} = \frac{0.622(1.277\ \text{kPa})}{(95-1.277)\ \text{kPa}} = 0.0085\ \text{kg H}_2\text{O/kg dry air}$$

$$h_2 = C_p T_2 + \omega_2 h_{g2} = (1.005\ \text{kJ/kg}\cdot°\text{C})(12°\text{C}) + (0.0085)(2523.4\ \text{kJ/kg}) = 33.5\ \text{kJ/kg dry air}$$

Then the mass flow rate of dry air in each stream is

$$\dot{m}_{a1} = \frac{\dot{V}_1}{v_1} = \frac{20\ \text{m}^3/\text{min}}{0.940\ \text{m}^3/\text{kg dry air}} = 21.3\ \text{kg/min}$$

$$\dot{m}_{a2} = \frac{\dot{V}_2}{v_2} = \frac{25\ \text{m}^3/\text{min}}{0.873\ \text{m}^3/\text{kg dry air}} = 28.6\ \text{kg/min}$$

From the conservation of mass,

$$\dot{m}_{a3} = \dot{m}_{a1} + \dot{m}_{a2} = (21.3 + 28.6)\ \text{kg/min} = 49.9\ \text{kg/min}$$

The specific humidity and the enthalpy of the mixture can be determined from Eqs. 13-24, which are obtained by combining the conservation of mass and energy equations for the adiabatic mixing of two streams:

$$\frac{\dot{m}_{a1}}{\dot{m}_{a2}} = \frac{\omega_2 - \omega_3}{\omega_3 - \omega_1} = \frac{h_2 - h_3}{h_3 - h_1}$$

$$\frac{21.3}{28.6} = \frac{0.0085 - \omega_3}{\omega_3 - 0.0128} = \frac{33.5 - h_3}{h_3 - 64.9}$$

which yields,

$$\boldsymbol{\omega_3 = 0.0103\ \text{kg H}_2\text{O/kg dry air}}$$

$$h_3 = 46.9\ \text{kJ/kg dry air}$$

These two properties fix the state of the mixture. Other properties are determined from

$$h_3 = C_p T_3 + \omega_3 h_{g3} \cong C_p T_3 + \omega_3 (2501.3 + 1.82 T_3)$$

$$46.9 \text{ kJ/kg} = (1.005 \text{ kJ/kg}\cdot°\text{C})\, T_3 + (0.0103)(2501.3 + 1.82\, T_3) \text{ kJ/kg}$$

$$T_3 = \mathbf{20.6°C}$$

$$\omega_3 = \frac{0.622\, P_{v3}}{P_3 - P_{v3}} \longrightarrow 0.0103 = \frac{0.622\, P_{v3}}{95 - P_{v3}} \longrightarrow P_{v3} = 1.55 \text{ kPa}$$

$$\phi_3 = \frac{P_{v3}}{P_{g3}} = \frac{P_{v3}}{P_{sat\,@\,T_3}} = \frac{1.55 \text{ kPa}}{2.44 \text{ kPa}} = 0.635 \text{ or } \mathbf{63.5\%}$$

Finally,

$$P_{a3} = P_3 - P_{v3} = 95 - 1.55 = 93.45 \text{ kPa}$$

$$v_3 = \frac{R_a T_3}{P_{a3}} = \frac{(0.287 \text{ kPa}\cdot\text{m}^3/\text{kg}\cdot\text{K})(293.6 \text{ K})}{93.45 \text{ kPa}} = 0.902 \text{ m}^3/\text{kg dry air}$$

$$\dot{V}_3 = \dot{m}_{a3} v_3 = (49.9 \text{ kg/min})(0.902 \text{ m}^3/\text{kg}) = \mathbf{45.0 \text{ m}^3/\text{min}}$$

13-98E Two airstreams are mixed steadily. The temperature, the specific humidity, and the relative humidity of the mixture are to be determined.

Assumptions **1** Steady operating conditions exist **2** Dry air and water vapor are ideal gases. **3** The kinetic and potential energy changes are negligible. **4** The mixing section is adiabatic.

Properties The properties of each inlet stream are determined from the psychometric chart (Table A-33E) to be

$$h_1 = 19.9 \text{ Btu/lbm dry air}$$
$$\omega_1 = 0.0039 \text{ lbm H}_2\text{O/lbm dry air}$$
$$v_1 = 13.30 \text{ ft}^3\text{/lbm dry air}$$

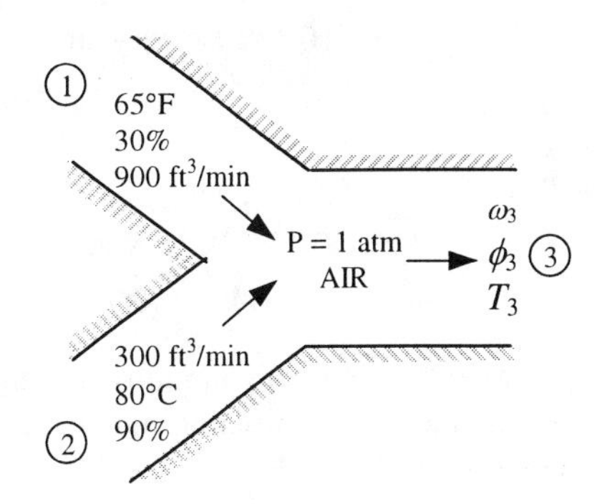

and

$$h_2 = 41.1 \text{ Btu/lbm dry air}$$
$$\omega_2 = 0.0200 \text{ lbm H}_2\text{O/lbm dry air}$$
$$v_2 = 14.03 \text{ ft}^3\text{/lbm dry air}$$

Analysis The mass flow rate of dry air in each stream is

$$\dot{m}_{a1} = \frac{\dot{V}_1}{v_1} = \frac{900 \text{ ft}^3\text{/min}}{13.30 \text{ ft}^3\text{/lbm dry air}} = 67.7 \text{ lbm/min}$$

$$\dot{m}_{a2} = \frac{\dot{V}_2}{v_2} = \frac{300 \text{ ft}^3\text{/min}}{14.03 \text{ ft}^3\text{/lbm dry air}} = 21.4 \text{ lbm/min}$$

The specific humidity and the enthalpy of the mixture can be determined from Eqs. 13-24, which are obtained by combining the conservation of mass and energy equations for the adiabatic mixing of two streams:

$$\frac{\dot{m}_{a1}}{\dot{m}_{a2}} = \frac{\omega_2 - \omega_3}{\omega_3 - \omega_1} = \frac{h_2 - h_3}{h_3 - h_1}$$

$$\frac{67.7}{21.4} = \frac{0.0200 - \omega_3}{\omega_3 - 0.0039} = \frac{41.1 - h_3}{h_3 - 19.9}$$

which yields,

(*a*) $\omega_3 = \mathbf{0.0078 \text{ lbm H}_2\text{O/lbm dry air}}$

$$h_3 = 25.0 \text{ Btu/lbm dry air}$$

These two properties fix the state of the mixture. Other properties of the mixture are determined from the psychometric chart:

(*b*) $T_3 = \mathbf{69.5°F}$

(*c*) $\phi_3 = \mathbf{49.0\%}$

13-99E EES solution of this (and other comprehensive problems designated with the *computer icon*) is available to instructors at the *Instructor Manual* section of the *Online Learning Center* (OLC) at www.mhhe.com/cengel-boles. See the Preface for access information.

13-100 A stream of warm air is mixed with a stream of saturated cool air. The temperature, the specific humidity, and the relative humidity of the mixture are to be determined.

Assumptions **1** Steady operating conditions exist **2** Dry air and water vapor are ideal gases. **3** The kinetic and potential energy changes are negligible. **4** The mixing section is adiabatic.

Properties The properties of each inlet stream are determined from the psychometric chart (Table A-33) to be

$$h_1 = 110.3 \text{ kJ / kg dry air}$$
$$\omega_1 = 0.0272 \text{ kg H}_2\text{O / kg dry air}$$

and

$$h_2 = 50.9 \text{ kJ / kg dry air}$$
$$\omega_2 = 0.0130 \text{ kg H}_2\text{O / kg dry air}$$

Analysis The specific humidity and the enthalpy of the mixture can be determined from Eqs. 13-24, which are obtained by combining the conservation of mass and energy equations for the adiabatic mixing of two streams:

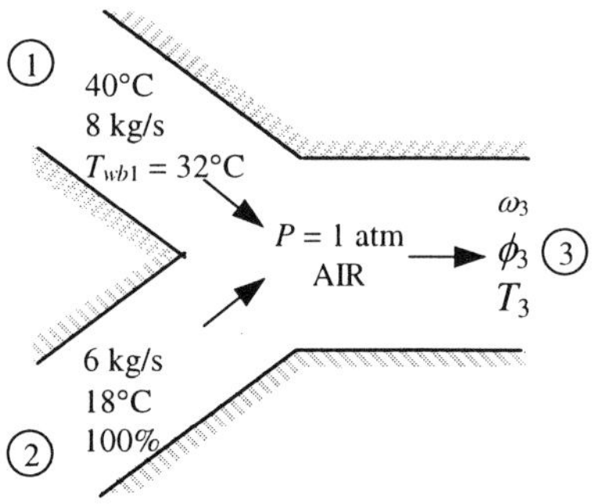

$$\frac{\dot{m}_{a1}}{\dot{m}_{a2}} = \frac{\omega_2 - \omega_3}{\omega_3 - \omega_1} = \frac{h_2 - h_3}{h_3 - h_1}$$

$$\frac{8.0}{6.0} = \frac{0.0130 - \omega_3}{\omega_3 - 0.0272} = \frac{50.9 - h_3}{h_3 - 110.3}$$

which yields,

(*b*) $\omega_3 = \mathbf{0.0211 \text{ kg H}_2\text{O / kg dry air}}$

$$h_3 = 84.8 \text{ kJ /kg dry air}$$

These two properties fix the state of the mixture. Other properties of the mixture are determined from the psychometric chart:

(*a*) $T_3 = \mathbf{30.7°C}$

(*c*) $\phi_3 = \mathbf{75.1\%}$

13-101 EES solution of this (and other comprehensive problems designated with the *computer icon*) is available to instructors at the *Instructor Manual* section of the *Online Learning Center* (OLC) at www.mhhe.com/cengel-boles. See the Preface for access information.

Wet Cooling Towers

13-102C The working principle of a natural draft cooling tower is based on buoyancy. The air in the tower has a high moisture content, and thus is lighter than the outside air. This light moist air rises under the influence of buoyancy, inducing flow through the tower.

13-103C A spray pond cools the warm water by spraying it into the open atmosphere. They require 25 to 50 times the area of a wet cooling tower for the same cooling load.

13-104 Water is cooled by air in a cooling tower. The volume flow rate of air and the mass flow rate of the required makeup water are to be determined.

Assumptions **1** Steady operating conditions exist and thus mass flow rate of dry air remains constant during the entire process. **2** Dry air and water vapor are ideal gases. **3** The kinetic and potential energy changes are negligible. **4** The cooling tower is adiabatic.

Analysis (*a*) The mass flow rate of dry air through the tower remains constant $(\dot{m}_{a1} = \dot{m}_{a2} = \dot{m}_a)$, but the mass flow rate of liquid water decreases by an amount equal to the amount of water that vaporizes in the tower during the cooling process. The water lost through evaporation must be made up later in the cycle to maintain steady operation. Applying the mass and energy balances yields

Dry Air Mass Balance:

$$\sum \dot{m}_{a,i} = \sum \dot{m}_{a,e} \longrightarrow \dot{m}_{a1} = \dot{m}_{a2} = \dot{m}_a$$

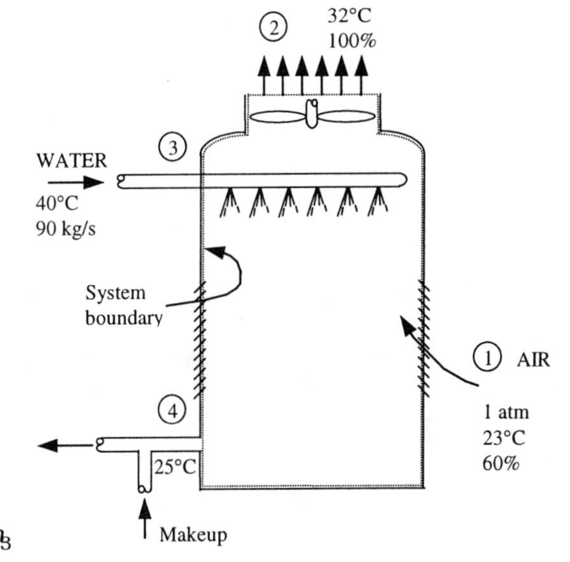

Water Mass Balance:

$$\sum \dot{m}_{w,i} = \sum \dot{m}_{w,e} \longrightarrow \dot{m}_3 + \dot{m}_{a1}\omega_1 = \dot{m}_4 + \dot{m}_{a2}\omega_2$$

$$\dot{m}_3 - \dot{m}_4 = \dot{m}_a(\omega_2 - \omega_1) = \dot{m}_{\text{makeup}}$$

Energy Balance:

$$\dot{E}_{in} - \dot{E}_{out} = \Delta\dot{E}_{system}^{\nearrow 0\ (\text{steady})} = 0$$

$$\dot{E}_{in} = \dot{E}_{out}$$

$$\sum \dot{m}_i h_i = \sum \dot{m}_e h_e \quad \text{since } \dot{Q} = \dot{W} = 0$$

$$0 = \sum \dot{m}_e h_e - \sum \dot{m}_i h_i$$

$$0 = \dot{m}_{a2} h_2 + \dot{m}_4 h_4 - \dot{m}_{a1} h_1 - \dot{m}_3 h_3$$

$$0 = \dot{m}_a(h_2 - h_1) + (\dot{m}_3 - \dot{m}_{\text{makeup}})h_4 - \dot{m}_3 h_3$$

Solving for $\dot{m}_a$,

$$\dot{m}_a = \frac{\dot{m}_3(h_3 - h_4)}{(h_2 - h_1) - (\omega_2 - \omega_1)h_4}$$

From the psychometric chart (Table A-33),

$$h_1 = 50.3 \text{ kJ/kg dry air}$$
$$\omega_1 = 0.0106 \text{ kg H}_2\text{O/kg dry air}$$
$$v_1 = 0.854 \text{ m}^3\text{/kg dry air}$$

and

$$h_2 = 110.8 \text{ kJ / kg dry air}$$
$$\omega_2 = 0.0307 \text{ kg H}_2\text{O / kg dry air}$$

From Table A-4,

$$h_3 \cong h_{f\,@\,40°C} = 167.57 \text{ kJ / kg H}_2\text{O}$$

$$h_4 \cong h_{f\,@\,25°C} = 104.89 \text{ kJ / kg H}_2\text{O}$$

Substituting,

$$\dot{m}_a = \frac{(90 \text{ kg/s})(167.57 - 104.89) \text{kJ/kg}}{(110.8 - 50.3) \text{ kJ/kg} - (0.0307 - 0.0106)(104.89) \text{ kJ/kg}} = 96.6 \text{ kg/s}$$

Then the volume flow rate of air into the cooling tower becomes

$$\dot{V}_1 = \dot{m}_a \text{v}_1 = (96.6 \text{ kg/s})(0.854 \text{ m}^3\text{/kg}) = \mathbf{82.5 \text{ m}^3/s}$$

(*b*) The mass flow rate of the required makeup water is determined from

$$\dot{m}_{makeup} = \dot{m}_a(\omega_2 - \omega_1) = (96.6 \text{ kg/s})(0.0307 - 0.0106) = \mathbf{1.942 \text{ kg/s}}$$

13-105E Water is cooled by air in a cooling tower. The volume flow rate of air and the mass flow rate of the required makeup water are to be determined.

Assumptions **1** Steady operating conditions exist and thus mass flow rate of dry air remains constant during the entire process. **2** Dry air and water vapor are ideal gases. **3** The kinetic and potential energy changes are negligible. **4** The cooling tower is adiabatic.

Analysis (*a*) The mass flow rate of dry air through the tower remains constant ($\dot{m}_{a1} = \dot{m}_{a2} = \dot{m}_a$), but the mass flow rate of liquid water decreases by an amount equal to the amount of water that vaporizes in the tower during the cooling process. The water lost through evaporation must be made up later in the cycle to maintain steady operation. Applying the mass balance and the energy balance equations yields

Dry Air Mass Balance:

$$\sum \dot{m}_{a,i} = \sum \dot{m}_{a,e} \longrightarrow \dot{m}_{a1} = \dot{m}_{a2} = \dot{m}_a$$

Water Mass Balance:

$$\sum \dot{m}_{w,i} = \sum \dot{m}_{w,e} \longrightarrow \dot{m}_3 + \dot{m}_{a1}\omega_1 = \dot{m}_4 + \dot{m}_{a2}\omega_2$$

$$\dot{m}_3 - \dot{m}_4 = \dot{m}_a(\omega_2 - \omega_1) = \dot{m}_{\text{makeup}}$$

Energy Balance:

$$\dot{E}_{in} - \dot{E}_{out} = \Delta \dot{E}_{system}^{\nearrow 0 \text{ (steady)}} = 0$$

$$\dot{E}_{in} = \dot{E}_{out}$$

$$\sum \dot{m}_i h_i = \sum \dot{m}_e h_e \quad (\text{since } \dot{Q} = \dot{W} = 0)$$

$$0 = \sum \dot{m}_e h_e - \sum \dot{m}_i h_i$$

$$0 = \dot{m}_{a2} h_2 + \dot{m}_4 h_4 - \dot{m}_{a1} h_1 - \dot{m}_3 h_3$$

$$0 = \dot{m}_a (h_2 - h_1) + (\dot{m}_3 - \dot{m}_{\text{makeup}}) h_4 - \dot{m}_3 h_3$$

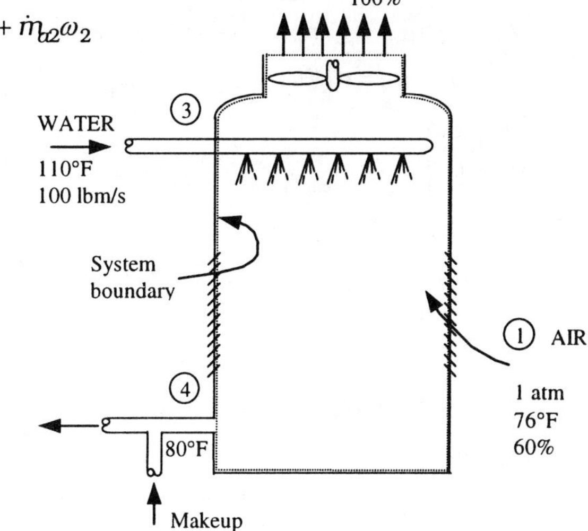

Solving for $\dot{m}_a$,

$$\dot{m}_a = \frac{\dot{m}_3 (h_3 - h_4)}{(h_2 - h_1) - (\omega_2 - \omega_1) h_4}$$

From the psychometric chart (Table A-33),

$$h_1 = 30.9 \text{ Btu/lbm dry air}$$
$$\omega_1 = 0.0115 \text{ lbm H}_2\text{O/lbm dry air}$$
$$v_1 = 13.76 \text{ ft}^3\text{/lbm dry air}$$

and

$$h_2 = 63.2 \text{ Btu/lbm dry air}$$
$$\omega_2 = 0.0366 \text{ lbm H}_2\text{O/lbm dry air}$$

From Table A-4E,

$$h_3 \cong h_{f@110°F} = 78.02 \text{ Btu/lbm H}_2\text{O}$$
$$h_4 \cong h_{f@80°F} = 48.09 \text{ Btu/lbm H}_2\text{O}$$

Substituting,

$$\dot{m}_a = \frac{(100 \text{ lbm/s})(78.02 - 48.09) \text{Btu/lbm}}{(63.2 - 30.9) \text{ Btu/lbm} - (0.0366 - 0.0115)(48.09) \text{ Btu/lbm}} = 96.3 \text{ lbm/s}$$

Then the volume flow rate of air into the cooling tower becomes

$$\dot{V}_1 = \dot{m}_a v_1 = (96.3 \text{ lbm/s})(13.76 \text{ ft}^3\text{/lbm}) = \mathbf{1325 \text{ ft}^3\text{/s}}$$

(*b*) The mass flow rate of the required makeup water is determined from

$$\dot{m}_{\text{makeup}} = \dot{m}_a(\omega_2 - \omega_1) = (96.3 \text{ lbm/s})(0.0366 - 0.0115) = \mathbf{2.42 \text{ lbm/s}}$$

13-106 Water is cooled by air in a cooling tower. The volume flow rate of air and the mass flow rate of the required makeup water are to be determined.

Assumptions **1** Steady operating conditions exist and thus mass flow rate of dry air remains constant during the entire process. **2** Dry air and water vapor are ideal gases. **3** The kinetic and potential energy changes are negligible. **4** The cooling tower is adiabatic.

Analysis (*a*) The mass flow rate of dry air through the tower remains constant ($\dot{m}_{a1} = \dot{m}_{a2} = \dot{m}_a$), but the mass flow rate of liquid water decreases by an amount equal to the amount of water that vaporizes in the tower during the cooling process. The water lost through evaporation must be made up later in the cycle to maintain steady operation. Applying the mass and energy balances yields

Dry Air Mass Balance:

$$\sum \dot{m}_{a,i} = \sum \dot{m}_{a,e} \longrightarrow \dot{m}_{a1} = \dot{m}_{a2} = \dot{m}_a$$

Water Mass Balance:

$$\sum \dot{m}_{w,i} = \sum \dot{m}_{w,e} \longrightarrow \dot{m}_3 + \dot{m}_{a1}\omega_1 = \dot{m}_4 + \dot{m}_{a2}\omega_2$$

$$\dot{m}_3 - \dot{m}_4 = \dot{m}_a(\omega_2 - \omega_1) = \dot{m}_{\text{makeup}}$$

Energy Balance:

$$\dot{E}_{in} - \dot{E}_{out} = \Delta \dot{E}_{system}^{\nearrow 0\ (\text{steady})} = 0$$

$$\dot{E}_{in} = \dot{E}_{out}$$

$$\sum \dot{m}_i h_i = \sum \dot{m}_e h_e \quad (\text{since } \dot{Q} = \dot{W} = 0)$$

$$0 = \sum \dot{m}_e h_e - \sum \dot{m}_i h_i$$

$$0 = \dot{m}_{a2} h_2 + \dot{m}_4 h_4 - \dot{m}_{a1} h_1 - \dot{m}_3 h_3$$

$$0 = \dot{m}_a (h_2 - h_1) + (\dot{m}_3 - \dot{m}_{\text{makeup}}) h_4 - \dot{m}_3 h_3$$

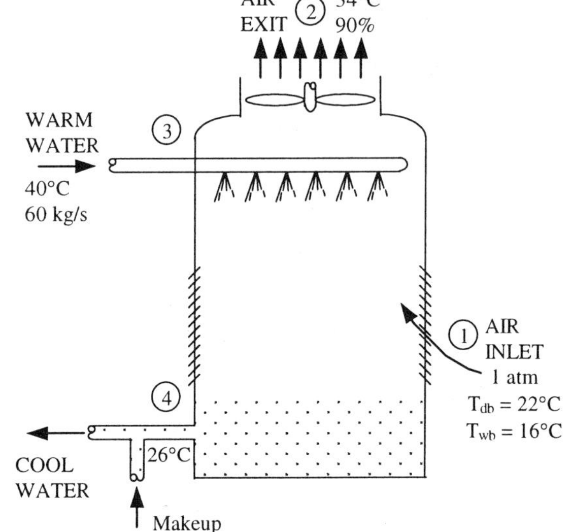

Solving for $\dot{m}_a$,

$$\dot{m}_a = \frac{\dot{m}_3 (h_3 - h_4)}{(h_2 - h_1) - (\omega_2 - \omega_1) h_4}$$

From the psychometric chart (Table A-33),

$$h_1 = 44.7 \text{ kJ/kg dry air}$$

$$\omega_1 = 0.0089 \text{ kg H}_2\text{O/kg dry air}$$

$$v_1 = 0.849 \text{ m}^3\text{/kg dry air}$$

and

$$h_2 = 113.5 \text{ kJ/kg dry air}$$

$$\omega_2 = 0.0309 \text{ kg H}_2\text{O/kg dry air}$$

From Table A-4,

$$h_3 \cong h_{f\,@\,40°C} = 167.57 \text{ kJ/kg H}_2\text{O}$$

$$h_4 \cong h_{f\,@\,26°C} = 109.07 \text{ kJ/kg H}_2\text{O}$$

Substituting,

$$\dot{m}_a = \frac{(60 \text{ kg/s})(167.57 - 109.07)\text{kJ/kg}}{(113.5 - 44.7) \text{ kJ/kg} - (0.0309 - 0.0089)(109.07) \text{ kJ/kg}} = 52.9 \text{ kg/s}$$

Then the volume flow rate of air into the cooling tower becomes

$$\dot{V}_1 = \dot{m}_a v_1 = (52.9 \text{ kg/s})(0.849 \text{ m}^3\text{/kg}) = \mathbf{44.9\ m^3/s}$$

(*b*) The mass flow rate of the required makeup water is determined from

$$\dot{m}_{\text{makeup}} = \dot{m}_a(\omega_2 - \omega_1) = (52.9 \text{ kg/s})(0.0309 - 0.0089) = \mathbf{1.16\ kg/s}$$

13-107 Water is cooled by air in a cooling tower. The volume flow rate of air and the mass flow rate of the required makeup water are to be determined.

Assumptions **1** Steady operating conditions exist and thus mass flow rate of dry air remains constant during the entire process. **2** Dry air and water vapor are ideal gases. **3** The kinetic and potential energy changes are negligible. **4** The cooling tower is adiabatic.

Analysis (*a*) The mass flow rate of dry air through the tower remains constant $(\dot{m}_{a1} = \dot{m}_{a2} = \dot{m}_a)$, but the mass flow rate of liquid water decreases by an amount equal to the amount of water that vaporizes in the tower during the cooling process. The water lost through evaporation must be made up later in the cycle to maintain steady operation. Applying the mass and energy balances yields

Dry Air Mass Balance:

$$\sum \dot{m}_{a,i} = \sum \dot{m}_{a,e} \longrightarrow \dot{m}_{a1} = \dot{m}_{a2} = \dot{m}_a$$

Water Mass Balance:

$$\sum \dot{m}_{w,i} = \sum \dot{m}_{w,e} \longrightarrow \dot{m}_3 + \dot{m}_{a1}\omega_1 = \dot{m}_4 + \dot{m}_{a2}\omega_2$$

$$\dot{m}_3 - \dot{m}_4 = \dot{m}_a(\omega_2 - \omega_1) = \dot{m}_{\text{makeup}}$$

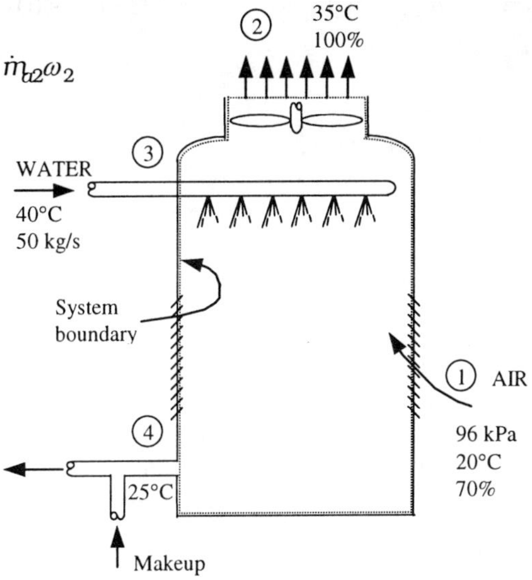

Energy Balance:

$$\dot{E}_{in} - \dot{E}_{out} = \Delta \dot{E}_{system}^{\nearrow 0 \text{ (steady)}} = 0$$

$$\dot{E}_{in} = \dot{E}_{out}$$

$$\sum \dot{m}_i h_i = \sum \dot{m}_e h_e \quad (\text{since } \dot{Q} = \dot{W} = 0)$$

$$0 = \sum \dot{m}_e h_e - \sum \dot{m}_i h_i$$

$$0 = \dot{m}_{a2} h_2 + \dot{m}_4 h_4 - \dot{m}_{a1} h_1 - \dot{m}_3 h_3$$

$$0 = \dot{m}_a (h_2 - h_1) + (\dot{m}_3 - \dot{m}_{\text{makeup}}) h_4 - \dot{m}_3 h_3$$

WATER 40°C 50 kg/s; ① AIR 96 kPa 20°C 70%; ② 35°C 100%; ③; ④ 25°C; System boundary; Makeup

Solving for $\dot{m}_a$,

$$\dot{m}_a = \frac{\dot{m}_3(h_3 - h_4)}{(h_2 - h_1) - (\omega_2 - \omega_1)h_4}$$

The properties of air at the inlet and the exit of the tower are calculated to be

$$P_{v1} = \phi_1 P_{g1} = \phi_1 P_{\text{sat @ }20°\text{C}} = (0.70)(2.339 \text{ kPa}) = 1.637 \text{ kPa}$$

$$P_{a1} = P_1 - P_{v1} = 96 - 1.637 = 94.363 \text{ kPa}$$

$$v_1 = \frac{R_a T_1}{P_{a1}} = \frac{(0.287 \text{ kPa}\cdot\text{m}^3/\text{kg}\cdot\text{K})(293 \text{ K})}{94.363 \text{ kPa}} = 0.891 \text{ m}^3/\text{kg dry air}$$

$$\omega_1 = \frac{0.622 P_{v1}}{P_1 - P_{v1}} = \frac{0.622(1.637 \text{ kPa})}{(96 - 1.637) \text{ kPa}} = 0.0108 \text{ kg H}_2\text{O/kg dry air}$$

$$h_1 = C_p T_1 + \omega_1 h_{g1} = (1.005 \text{ kJ/kg}\cdot°\text{C})(20°\text{C}) + (0.0108)(2538.1 \text{ kJ/kg})$$
$$= 47.5 \text{ kJ/kg dry air}$$

and

$$P_{v2} = \phi_2 P_{g2} = \phi_2 P_{\text{sat @ }35°\text{C}} = (1.00)(5.628 \text{ kPa}) = 5.628 \text{ kPa}$$

$$\omega_2 = \frac{0.622 P_{v2}}{P_2 - P_{v2}} = \frac{0.622(5.628 \text{ kPa})}{(96 - 5.628) \text{ kPa}} = 0.0387 \text{ kg H}_2\text{O/kg dry air}$$

$$h_2 = C_p T_2 + \omega_2 h_{g2} = (1.005 \text{ kJ/kg}\cdot°\text{C})(35°\text{C}) + (0.0387)(2565.3 \text{ kJ/kg})$$
$$= 134.5 \text{ kJ/kg dry air}$$

From Table A-4,

$$h_3 \cong h_{f\,@40°C} = 167.57 \text{ kJ/kg } H_2O$$

$$h_4 \cong h_{f\,@25°C} = 104.89 \text{ kJ/kg } H_2O$$

Substituting,

$$\dot{m}_a = \frac{(50 \text{ kg/s})(167.57 - 104.89) \text{kJ/kg}}{(134.5 - 47.5) \text{ kJ/kg} - (0.0387 - 0.0108)(104.89) \text{ kJ/kg}} = 37.3 \text{ kg/s}$$

Then the volume flow rate of air into the cooling tower becomes

$$\dot{V}_1 = \dot{m}_a v_1 = (37.3 \text{ kg/s})(0.891 \text{ m}^3\text{/kg}) = \mathbf{33.2\ m^3/s}$$

(*b*) The mass flow rate of the required makeup water is determined from

$$\dot{m}_{makeup} = \dot{m}_a(\omega_2 - \omega_1) = (33.2 \text{ kg/s})(0.0387 - 0.0108) = \mathbf{1.04\ kg/s}$$

Review Problems

13-108 Air is compressed by a compressor and then cooled to the ambient temperature at high pressure. It is to be determined if there will be any condensation in the compressed air lines.

Assumptions The air and the water vapor are ideal gases.

Properties The saturation pressure of water at 25°C is 3.169 kPa (Table A-4)..

Analysis The vapor pressure of air before compression is

$$P_{v1} = \phi_1 P_g = \phi_1 P_{sat\,@\,25°C} = (0.40)(3.169\text{ kPa}) = 1.27\text{ kPa}$$

The pressure ratio during the compression process is (800 kPa)/(92 kPa) = 8.70. That is, the pressure of air and any of its components increases by 8.70 times. Then the vapor pressure of air after compression becomes

$$P_{v2} = P_{v1} \times (\text{Pressure ratio}) = (1.27\text{ kPa})(8.70) = 11.0\text{ kPa}$$

The dew-point temperature of the air at this vapor pressure is

$$T_{dp} = T_{sat\,@\,P_{v2}} = T_{sat\,@\,11.0\text{ kPa}} = 47.4°\text{C}$$

which is greater than 25°C. Therefore, part of the moisture in the compressed air will **condense** when air is cooled to 25°C.

13-109E The error involved in assuming the density of air to remain constant during a humidification process is to be determined.

Properties The density of moist air before and after the humidification process is determined from the psychrometric chart (Table A-33E) to be

$$\left.\begin{array}{l} T_1 = 80°\text{F} \\ \phi_1 = 30\% \end{array}\right\} \rho_{air,1} = 0.0727\text{ lbm/ft}^3 \quad \text{and} \quad \left.\begin{array}{l} T_1 = 80°\text{F} \\ \phi_1 = 90\% \end{array}\right\} \rho_{air,2} = 0.07117\text{ lbm/ft}^3$$

Analysis The error involved as a result of assuming constant air density is then determined to be

$$\%\text{Error} = \frac{\Delta\rho_{air}}{\rho_{air,1}} \times 100 = \frac{0.0727 - 0.0712\text{ lbm/ft}^3}{0.0727\text{ lbm/ft}^3} \times 100 = \mathbf{2.1\%}$$

which is acceptable for most engineering purposes.

13-110 Dry air flows over a water body at constant pressure and temperature until it is saturated. The molar analysis of the saturated air and the density of air before and after the process are to be determined.

Assumptions The air and the water vapor are ideal gases.

Properties The molar masses of N_2, O_2, Ar, and H_2O are 28.0, 32.0, 39.9 and 18 kg / kmol, respectively (Table A-1). The molar analysis of dry air is given to be 78.1 percent N_2, 20.9 percent O_2, and 1 percent Ar. The saturation pressure of water at 25°C is 3.169 kPa (Table A-4). Also, 1 atm = 101.325 kPa.

Analysis (*a*) Noting that the total pressure remains constant at 101.32 kPa during this process, the partial pressure of air becomes

$$P = P_{\text{air}} + P_{\text{vapor}} \rightarrow P_{\text{air}} = P - P_{\text{vapor}} = 101.325 - 3.169 = 98.156 \text{ kPa}$$

Then the molar analysis of the saturated air becomes

$$y_{H_2O} = \frac{P_{H_2O}}{P} = \frac{3.169}{101.325} = \mathbf{0.0313}$$

$$y_{N_2} = \frac{P_{N_2}}{P} = \frac{y_{N_2,\text{dry}} P_{\text{dry air}}}{P} = \frac{0.781(98.156 \text{ kPa})}{101.325} = \mathbf{0.7566}$$

$$y_{O_2} = \frac{P_{O_2}}{P} = \frac{y_{O_2,\text{dry}} P_{\text{dry air}}}{P} = \frac{0.209(98.156 \text{ kPa})}{101.325} = \mathbf{0.2025}$$

$$y_{Ar} = \frac{P_{Ar}}{P} = \frac{y_{Ar,\text{dry}} P_{\text{dry air}}}{P} = \frac{0.01(98.156 \text{ kPa})}{101.325} = \mathbf{0.0097}$$

Air
1 atm
25°C
Lake

(*b*) The molar masses of dry and saturated air are

$$M_{\text{dry air}} = \sum y_i M_i = 0.781 \times 28.0 + 0.209 \times 32.0 + 0.01 \times 39.9 = 29.0 \text{ kg / kmol}$$

$$M_{\text{sat. air}} = \sum y_i M_i = 0.7566 \times 28.0 + 0.2025 \times 32.0 + 0.0097 \times 39.9 + 0.0313 \times 18 = 28.62 \text{ kg /kmol}$$

Then the densities of dry and saturated air are determined from the ideal gas relation to be

$$\rho_{\text{dry air}} = \frac{P}{(R_u / M_{\text{dry air}})T} = \frac{101.325\text{kPa}}{[(8.314\text{kPa} \cdot \text{m}^3/\text{kmol} \cdot \text{K})/29.0\text{kg/kmol}](25+273)\text{K}} = \mathbf{1.186 \text{ kg/m}^3}$$

$$\rho_{\text{sat. air}} = \frac{P}{(R_u / M_{\text{sat air}})T} = \frac{101.325\text{kPa}}{[(8.314\text{kPa} \cdot \text{m}^3/\text{kmol} \cdot \text{K})/28.62\text{kg/kmol}](25+273)\text{K}} = \mathbf{1.170 \text{ kg/m}^3}$$

Discussion We conclude that the density of saturated air is less than that of the dry air, as expected. This is due to the molar mass of water being less than that of dry air.

13-111E The mole fraction of the water vapor at the surface of a lake and the mole fraction of water in the lake are to be determined and compared.

Assumptions **1** Both the air and water vapor are ideal gases. **2** Air is weakly soluble in water and thus Henry's law is applicable.

Properties The saturation pressure of water at 60°F is 0.2563 psia (Table A-4E). Henry's constant for air dissolved in water at 60ºF (289 K) is given in Table 11-6 to be H = 62,000 bar.

Analysis The air at the water surface will be saturated. Therefore, the partial pressure of water vapor in the air at the lake surface will simply be the saturation pressure of water at 15°C,

$$P_{vapor} = P_{sat\,@\,60°F} = 0.2563 \text{ psia}$$

Assuming both the air and vapor to be ideal gases, the mole fraction of water vapor in the air at the surface of the lake is determined to be

$$y_{vapor} = \frac{P_{vapor}}{P} = \frac{0.2563 \text{ psia}}{13.8 \text{ psia}} = \mathbf{0.0186 \text{ (or 1.86 percent)}}$$

Air
13.8 psi
60°F
Lake

The partial pressure of dry air just above the lake surface is

$$P_{dry\ air} = P - P_{vapor} = 13.8 - 0.2563 = 13.54 \text{ psia}$$

Then the mole fraction of air in the water becomes

$$y_{dry\ air,liquid\ side} = \frac{P_{dry\ air,gas side}}{H} = \frac{13.54 \text{ psia}(1 \text{ atm}/14.696 \text{ psia})}{62{,}000 \text{ bar}(1 \text{ atm}/1.01325 \text{ bar})} = 1.51\times10^{-5}$$

which is very small, as expected. Therefore, the mole fraction of water in the lake near the surface is

$$y_{water,liquid\ side} = 1 - y_{dry\ air,\ liquid\ side} = 1 - 1.51\times10^{-5} = \mathbf{0.9999}$$

Discussion The concentration of air in water just below the air-water interface is 1.51 moles per 100,000 moles. The amount of air dissolved in water will decrease with increasing depth.

13-112 The mole fraction of the water vapor at the surface of a lake at a specified temperature is to be determined.

Assumptions **1** Both the air and water vapor are ideal gases. **2** Air at the lake surface is saturated.

Properties The saturation pressure of water at 12°C is 1.4186 kPa (Table A-4).

Analysis The air at the water surface will be saturated. Therefore, the partial pressure of water vapor in the air at the lake surface will simply be the saturation pressure of water at 15°C,

$$P_{vapor} = P_{sat\,@\,15°C} = 1.4186 \text{ kPa}$$

Air
100 kPa
12°C
Lake

Assuming both the air and vapor to be ideal gases, the partial pressure and mole fraction of dry air in the air at the surface of the lake are determined to be

$$P_{dry\ air} = P - P_{vapor} = 100 - 1.4186 = 98.5814 \text{ kPa}$$

$$y_{dry\ air} = \frac{P_{dry\ air}}{P} = \frac{98.5814 \text{ kPa}}{100 \text{ kPa}} = \mathbf{0.986} \text{ (or 98.6\%)}$$

Therefore, the mole fraction of dry air is 98.6 percent just above the air-water interface.

13-113E A room is cooled adequately by a 7500 Btu/h air-conditioning unit. If the room is to be cooled by an evaporative cooler, the amount of water that needs to be supplied to the cooler is to be determined.

Assumptions **1** The evaporative cooler removes heat at the same rate as the air conditioning unit. **2** Water evaporates at an average temperature of 70°F.

Properties The enthalpy of vaporization of water at 70°F is 1054 Btu/lbm (Table A-4E).

Analysis Noting that 1 lbm of water removes 1054 Btu of heat as it evaporates, the amount of water that needs to evaporate to remove heat at a rate of 5000 Btu/h is determined from $\dot{Q} = \dot{m}_{\text{water}} h_{fg}$ to be

$$\dot{m}_{\text{water}} = \frac{\dot{Q}}{h_{fg}} = \frac{7500 \text{ Btu/h}}{1054 \text{ Btu/lbm}} = \mathbf{7.12 \ lbm/h}$$

13-114E The required size of an evaporative cooler in cfm (ft^3/min) for an 8-ft high house is determined by multiplying the floor area of the house by 4. An equivalent rule is to be obtained in SI units.

Analysis Noting that 1 ft = 0.3048 m and thus 1 ft^2 = 0.0929 m^2 and 1 ft^3 = 0.0283 m^3, and noting that a flow rate of 4 ft^3/min is required per ft^2 of floor area, the required flow rate in SI units per m^2 of floor area is determined to

$$1 \text{ ft}^2 \leftrightarrow 4 \text{ ft}^3/\text{min}$$
$$0.0929 \text{ m}^2 \leftrightarrow 4 \times 0.0283 \text{ m}^3/\text{min}$$
$$1 \text{ m}^2 \leftrightarrow 1.22 \text{ m}^3/\text{min}$$

Therefore, a flow rate of **1.22 m^3/min** is required per m^2 of floor area.

13-115 A cooling tower with a cooling capacity of 440 kW is claimed to evaporate 15,800 kg of water per day. It is to be determined if this is a reasonable claim.

Assumptions **1** Water evaporates at an average temperature of 30°C. **2** The coefficient of performance of the air-conditioning unit is COP = 3.

Properties The enthalpy of vaporization of water at 30°C is 2430.5 kJ/kg (Table A-4).

Analysis Using the definition of COP, the electric power consumed by the air conditioning unit when running is

$$\dot{W}_{\text{in}} = \frac{\dot{Q}_{\text{cooling}}}{\text{COP}} = \frac{440 \text{ kW}}{3} = 146.7 \text{ kW}$$

Then the rate of heat rejected at the cooling tower becomes

$$\dot{Q}_{\text{rejected}} = \dot{Q}_{\text{cooling}} + \dot{W}_{in} = 440 + 146.7 = 586.7 \text{ kW}$$

Noting that 1 kg of water removes 2430.5 kJ of heat as it evaporates, the amount of water that needs to evaporate to remove heat at a rate of 586.7 kW is determined from $\dot{Q}_{\text{rejected}} = \dot{m}_{\text{water}} h_{fg}$ to be

$$\dot{m}_{\text{water}} = \frac{\dot{Q}_{\text{rejected}}}{h_{fg}} = \frac{586.7 \text{ kJ/s}}{2430.5 \text{ kJ/kg}} = 0.241 \text{ kg/s} = 869 \text{ kg/h} = 20{,}856 \text{ kg/day}$$

In practice, the air-conditioner will run intermittently rather than continuously at the rated power, and thus the water use will be less. Therefore, the claim amount of 15,800 kg per day is **reasonable**.

13-116 It is estimated that 190,000 barrels of oil would be saved per day if the thermostat setting in residences in summer were raised by 6°F (3.3°C). The amount of money that would be saved per year is to be determined.

Assumptions The average cooling season is given to be 120 days, and the cost of oil to be $20/barrel.

Analysis The amount of money that would be saved per year is determined directly from

$$(190{,}000\ \text{barrel}/\text{day})(120\ \text{days}/\text{year})(\$20/\text{barrel}) = \mathbf{\$456{,}000{,}000}$$

Therefore, the proposed measure will save about half-a-billion dollars a year.

13-117E Wearing heavy long-sleeved sweaters and reducing the thermostat setting 1°F reduces the heating cost of a house by 4 percent at a particular location. The amount of money saved per year by lowering the thermostat setting by 4°F is to be determined.

Assumptions The household is willing to wear heavy long-sleeved sweaters in the house, and the annual heating cost is given to be $600 a year.

Analysis The amount of money that would be saved per year is determined directly from

$$(\$600/\text{year})(0.04/^\circ\text{F})(4^\circ\text{F}) = \mathbf{\$96/year}$$

Therefore, the proposed measure will save the homeowner about $100 during a heating season..

13-118 Shading the condenser can reduce the air-conditioning costs by up to 10 percent. The amount of money shading can save a homeowner per year during its lifetime is to be determined.

Assumptions It is given that the annual air-conditioning cost is $500 a year, and the life of the air-conditioning system is 20 years.

Analysis The amount of money that would be saved per year is determined directly from

$$(\$500/\text{year})(20\ \text{years})(0.10) = \mathbf{\$1000}$$

Therefore, the proposed measure will save about $1000 during the lifetime of the system.

13-119 A tank contains saturated air at a specified state. The mass of the dry air, the specific humidity, and the enthalpy of the air are to be determined.

Assumptions The air and the water vapor are ideal gases.

Analysis (*a*) The air is saturated, thus the partial pressure of water vapor is equal to the saturation pressure at the given temperature,

$$P_v = P_g = P_{\text{sat @ 25°C}} = 3.169 \text{ kPa}$$

$$P_a = P - P_v = 97 - 3.169 = 93.831 \text{ kPa}$$

5 m³
25°C
97 kPa

Treating air as an ideal gas,

$$m_a = \frac{P_a V}{R_a T} = \frac{(93.831 \text{ kPa})(5 \text{ m}^3)}{(0.287 \text{ kPa}\cdot\text{m}^3/\text{kg}\cdot\text{K})(298 \text{ K})} = \mathbf{5.49 \text{ kg}}$$

(*b*) The specific humidity of air is determined from

$$\omega = \frac{0.622 P_v}{P - P_v} = \frac{(0.622)(3.169 \text{ kPa})}{(97 - 3.169) \text{ kPa}} = \mathbf{0.0210 \text{ kg } H_2O/\text{kg dry air}}$$

(*c*) The enthalpy of air per unit mass of dry air is determined from

$$h = h_a + \omega h_v \cong C_p T + \omega h_g = (1.005 \text{ kJ/kg}\cdot°\text{C})(25°\text{C}) + (0.0210)(2547.2 \text{ kJ/kg})$$
$$= \mathbf{78.62 \text{ kJ/kg dry air}}$$

13-120 EES solution of this (and other comprehensive problems designated with the *computer icon*) is availablé to instructors at the *Instructor Manual* section of the *Online Learning Center* (OLC) at www.mhhe.com/cengel-boles. See the Preface for access information.

13-121E Air at a specified state and relative humidity flows through a circular duct. The dew-point temperature, the volume flow rate of air, and the mass flow rate of dry air are to be determined.

Assumptions The air and the wate. vapor are ideal gases.

Analysis (*a*) The vapor pressure of air is

$$P_v = \phi P_g = \phi P_{\text{sat @ 60°F}} = (0.50)(0.2563 \text{ psia}) = 0.128 \text{ psia}$$

AIR
15 psia
50 f/s
60°F, 50%

Thus the dew-point temperature of the air is

$$T_{dp} = T_{\text{sat @ }P_v} = T_{\text{sat @ 0.128 psia}} = \mathbf{41.2°F}$$

(*b*) The volume flow rate is determined from

$$\dot{V} = \mathbf{V}A = \mathbf{V}\frac{\pi D^2}{4} = (50 \text{ ft/s})\left(\frac{\pi \times (8/12 \text{ ft})^2}{4}\right) = \mathbf{17.45 \text{ ft}^3/\text{s}}$$

(*c*) To determine the mass flow rate of dry air, we first need to calculate its specific volume,

$$P_a = P - P_v = 15 - 0.128 = 14.872 \text{ psia}$$

$$v_1 = \frac{R_a T_1}{P_{a1}} = \frac{(0.3704 \text{ psia}\cdot\text{ft}^3/\text{lbm}\cdot\text{R})(520 \text{ R})}{14.872 \text{ psia}} = 12.95 \text{ ft}^3/\text{lbm dry air}$$

Thus, $$\dot{m}_{a1} = \frac{\dot{V}_1}{v_1} = \frac{17.45 \text{ ft}^3/\text{s}}{12.95 \text{ ft}^3/\text{lbm dry air}} = \mathbf{1.35 \text{ lbm/s}}$$

13-122 Air enters a cooling section at a specified pressure, temperature, and relative humidity. The temperature of the air at the exit and the rate of heat transfer are to be determined.

Assumptions **1** This is a steady-flow process and thus the mass flow rate of dry air remains constant during the entire process $(\dot{m}_{a1} = \dot{m}_{a2} = \dot{m}_a)$. **2** Dry air and water vapor are ideal gases. **3** The kinetic and potential energy changes are negligible.

Analysis (*a*) The amount of moisture in the air also remains constant $(\omega_1 = \omega_2)$ as it flows through the cooling section since the process involves no humidification or dehumidification. The total pressure is 97 kPa. The properties of the air at the inlet state are

$$P_{v1} = \phi_1 P_{g1} = \phi_1 P_{\text{sat @ 35°C}} = (0.2)(5.628 \text{ kPa}) = 1.13 \text{ kPa}$$

$$P_{a1} = P_1 - P_{v1} = 97 - 1.13 = 95.87 \text{ kPa}$$

$$v_1 = \frac{R_a T_1}{P_{a1}} = \frac{(0.287 \text{ kPa·m}^3/\text{kg·K})(308 \text{ K})}{95.87 \text{ kPa}} = 0.922 \text{ m}^3/\text{kg dry air}$$

Cooling coils
35°C
20%
9 m^3/min
97 kPa
AIR
(1)
(2)

$$\omega_1 = \frac{0.622 P_{v1}}{P_1 - P_{v1}} = \frac{0.622(1.13 \text{ kPa})}{(97 - 1.13) \text{ kPa}} = 0.0073 \text{ kg H}_2\text{O/kg dry air} \; (= \omega_2)$$

$$h_1 = C_p T_1 + \omega_1 h_{g1} = (1.005 \text{ kJ/kg°C})(35\text{°C}) + (0.0073)(2565.3 \text{ kJ/kg}) = 53.90 \text{ kJ/kg dry air}$$

The air at the final state is saturated and the vapor pressure during this process remains constant. Therefore, the exit temperature of the air must be the dew-point temperature,

$$T_{dp} = T_{\text{sat @ } P_v} = T_{\text{sat @ 1.13 kPa}} = \mathbf{8.6°C}$$

(*b*) The enthalpy of the air at the exit is

$$h_2 = C_p T_2 + \omega_2 h_{g2} = (1.005 \text{ kJ/kg·°C})(8.6\text{°C}) + (0.0073)(2517.1 \text{ kJ/kg}) = 27.02 \text{ kJ/kg dry air}$$

Also,

$$\dot{m}_a = \frac{\dot{V}_1}{v_1} = \frac{9 \text{ m}^3/\text{s}}{0.922 \text{ m}^3/\text{kg dry air}} = 9.76 \text{ kg/min}$$

Then the rate of heat transfer from the air in the cooling section becomes

$$\dot{Q}_{out} = \dot{m}_a(h_1 - h_2) = (9.76 \text{ kg/min})(53.90 - 27.02) \text{ kJ/kg} = \mathbf{262.4 \text{ kJ/min}}$$

13-123 The outdoor air is first heated and then humidified by hot steam in an air-conditioning system. The rate of heat supply in the heating section and the mass flow rate of the steam required in the humidifying section are to be determined.

Assumptions **1** This is a steady-flow process and thus the mass flow rate of dry air remains constant during the entire process $(\dot{m}_{a1} = \dot{m}_{a2} = \dot{m}_a)$. **2** Dry air and water vapor are ideal gases. **3** The kinetic and potential energy changes are negligible.

Properties The amount of moisture in the air also remains constants it flows through the heating section $(\omega_1 = \omega_2)$, but increases in the humidifying section $(\omega_3 > \omega_2)$. The inlet and the exit states of the air are completely specified, and the total pressure is 1 atm. The properties of the air at various states are determined from the psychometric chart to be

$$h_1 = 17.7 \text{ kJ/kg dry air}$$
$$\omega_1 = 0.0030 \text{ kg H}_2\text{O/kg dry air} (= \omega_2)$$
$$v_1 = 0.807 \text{ m}^3\text{/kg dry air}$$

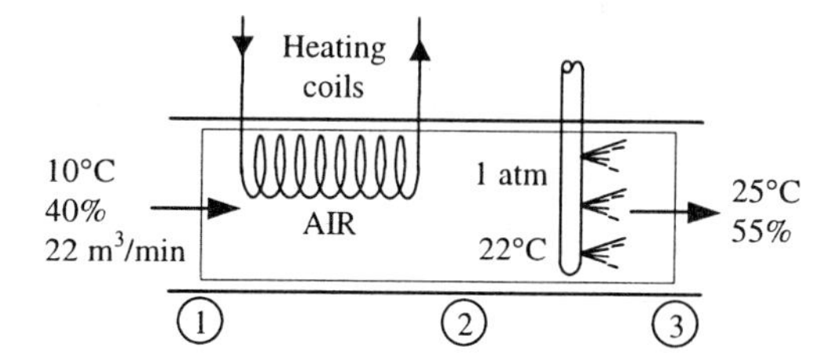

$$h_2 = 29.8 \text{ kJ / kg dry air}$$
$$\omega_2 = \omega_1 = 0.0030 \text{ kg H}_2\text{O / kg dry air}$$

$$h_3 = 52.9 \text{ kJ / kg dry air}$$
$$\omega_3 = 0.0109 \text{ kg H}_2\text{O / kg dry air}$$

Analysis (*a*) The mass flow rate of dry air is

$$\dot{m}_a = \frac{\dot{V}_1}{v_1} = \frac{22 \text{ m}^3\text{/min}}{0.807 \text{ m}^3\text{/kg}} = 27.3 \text{ kg/min}$$

Then the rate of heat transfer to the air in the heating section becomes

$$\dot{Q}_{in} = \dot{m}_a(h_2 - h_1) = (27.3 \text{ kg/min})(29.8 - 17.7) \text{kJ/kg} = \mathbf{330.3 \text{ kJ / min}}$$

(*b*) The conservation of mass equation for water in the humidifying section can be expressed as

$$\dot{m}_{a2}\omega_2 + \dot{m}_w = \dot{m}_{a3}\omega_3 \quad \text{or} \quad \dot{m}_w = \dot{m}_a(\omega_3 - \omega_2)$$

Thus,

$$\dot{m}_w = (27.3 \text{ kg/min})(0.0109 - 0.0030) = \mathbf{0.216 \text{ kg / min}}$$

13-124 Air is cooled and dehumidified in an air-conditioning system with refrigerant-134a as the working fluid. The rate of dehumidification, the rate of heat transfer, and the mass flow rate of the refrigerant are to be determined.

Assumptions **1** This is a steady-flow process and thus the mass flow rate of dry air remains constant during the entire process ($\dot{m}_{a1} = \dot{m}_{a2} = \dot{m}_a$). **2** Dry air and water vapor are ideal gases. **3** The kinetic and potential energy changes are negligible.

Analysis (*a*) The dew point temperature of the incoming air stream at 30°C is

$$T_{dp} = T_{sat\,@\,P_v} = T_{sat\,@\,0.7\times4.246\text{ kPa}} = 24^\circ\text{C}$$

Since air is cooled to 20°C, which is below its dew point temperature, some of the moisture in the air will condense.

The mass flow rate of dry air remains constant during the entire process, but the amount of moisture in the air decreases due to dehumidification ($\omega_2 < \omega_1$). The inlet and the exit states of the air are completely specified, and the total pressure is 1 atm. Then the properties of the air at both states are determined from the psychometric chart (Table A-33) to be

$$h_1 = 78.3 \text{ kJ/kg dry air}$$
$$\omega_1 = 0.0188 \text{ kg H}_2\text{O/kg dry air}$$
$$v_1 = 0.885 \text{ m}^3\text{/kg dry air}$$

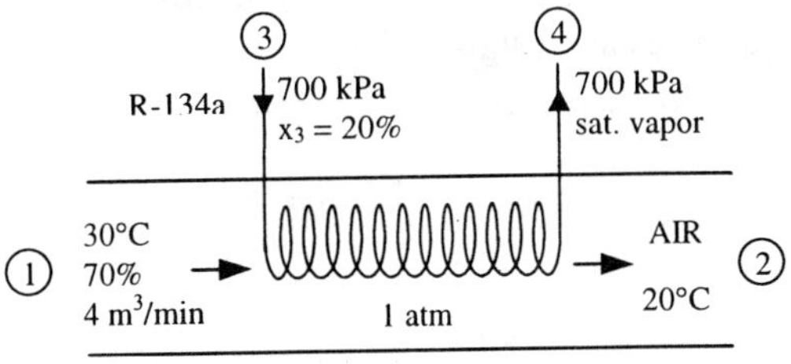

R-134a
3: 700 kPa, x_3 = 20%
4: 700 kPa, sat. vapor
1: 30°C, 70%, 4 m³/min
AIR
2: 20°C
1 atm

and

$$h_2 = 57.5 \text{ kJ/kg dry air}$$
$$\omega_2 = 0.0147 \text{ kg H}_2\text{O/kg dry air}$$

Also,

$$h_w \cong h_{f@20^\circ\text{C}} = 83.96 \text{ kJ/kg}$$

Then,

$$\dot{m}_{a1} = \frac{\dot{V}_1}{v_1} = \frac{4 \text{ m}^3\text{/min}}{0.885 \text{ m}^3\text{/kg dry air}} = 4.52 \text{ kg/min}$$

Applying the water mass balance and the energy balance equations to the combined cooling and dehumidification section (excluding the refrigerant),

Water Mass Balance: $\sum \dot{m}_{w,i} = \sum \dot{m}_{w,e} \longrightarrow \dot{m}_{a1}\omega_1 = \dot{m}_{a2}\omega_2 + \dot{m}_w$

$$\dot{m}_w = \dot{m}_a(\omega_1 - \omega_2) = (4.52 \text{ kg/min})(0.0188 - 0.0147) = \mathbf{0.0185 \text{ kg/min}}$$

(*b*) *Energy Balance*:

$$\dot{E}_{in} - \dot{E}_{out} = \Delta\dot{E}_{system}^{\nearrow 0\text{ (steady)}} = 0$$

$$\dot{E}_{in} = \dot{E}_{out}$$

$$\sum \dot{m}_i h_i = \dot{Q}_{out} + \sum \dot{m}_e h_e \longrightarrow \dot{Q}_{out} = \dot{m}_{a1}h_1 - (\dot{m}_{a2}h_2 + \dot{m}_w h_w) = \dot{m}_a(h_1 - h_2) - \dot{m}_w h_w$$

$$\dot{Q}_{out} = (4.52 \text{ kg/min})(78.3 - 57.5) \text{ kJ/kg} - (0.0185 \text{ kg/min})(83.96 \text{ kJ/kg}) = \mathbf{92.5 \text{ kJ/min}}$$

(*c*) The inlet and exit enthalpies of the refrigerant are

$$h_3 = h_g + x_3 h_{fg} = 86.78 + 0.2 \times 175.07 = 121.79 \text{ kJ/kg}$$

$$h_4 = h_{g\,@\,700\text{ kPa}} = 261.85 \text{ kJ/kg}$$

Noting that the heat lost by the air is gained by the refrigerant, the mass flow rate of the refrigerant becomes

$$\dot{Q}_R = \dot{m}_R(h_4 - h_3) \rightarrow \dot{m}_R = \frac{\dot{Q}_R}{h_4 - h_3} = \frac{92.5 \text{ kJ/min}}{(261.85 - 121.79) \text{ kJ/kg}} = \mathbf{0.66 \text{ kg/min}}$$

13-125 Air is cooled and dehumidified in an air-conditioning system with refrigerant-134a as the working fluid. The rate of dehumidification, the rate of heat transfer, and the mass flow rate of the refrigerant are to be determined.

Assumptions **1** This is a steady-flow process and thus the mass flow rate of dry air remains constant during the entire process. **2** Dry air and water vapor are ideal gases. **3** The kinetic and potential energy changes are negligible.

Analysis (*a*) The dew point temperature of the incoming air stream at 30°C is

$$P_{v1} = \phi_1 P_{g1} = \phi_1 P_{sat\,@\,30°C} = (0.7)(4.246\text{ kPa}) = 2.97\text{ kPa}$$

$$T_{dp} = T_{sat\,@\,P_v} = T_{sat\,@\,2.97\text{ kPa}} = 24°\text{C}$$

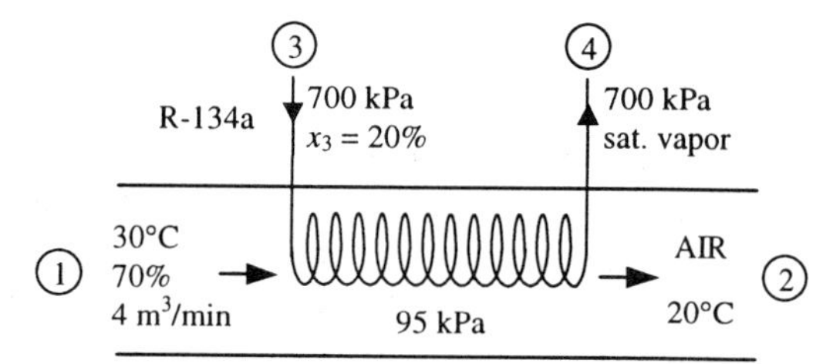

Since air is cooled to 20°C, which is below its dew point temperature, some of the moisture in the air will condense.

The amount of moisture in the air decreases due to dehumidification $(\omega_2 < \omega_1)$. The inlet and the exit states of the air are completely specified, and the total pressure is 95 kPa. The properties of the air at both states are determined to be

$$P_{a1} = P_1 - P_{v1} = 95 - 2.97 = 92.03\text{ kPa}$$

$$v_1 = \frac{R_a T_1}{P_{a1}} = \frac{(0.287\text{ kPa}\cdot\text{m}^3/\text{kg}\cdot\text{K})(303\text{ K})}{92.03\text{ kPa}} = 0.945\text{ m}^3/\text{kg dry air}$$

$$\omega_1 = \frac{0.622\,P_{v1}}{P_1 - P_{v1}} = \frac{0.622(2.97\text{ kPa})}{(95-2.97)\text{ kPa}} = 0.0201\text{ kg H}_2\text{O/kg dry air}$$

$$h_1 = C_p T_1 + \omega_1 h_{g1} = (1.005\text{ kJ/kg}\cdot°\text{C})(30°\text{C}) + (0.0201)(2556.3\text{ kJ/kg})$$
$$= 81.53\text{ kJ/kg dry air}$$

and

$$P_{v2} = \phi_2 P_{g2} = (1.00)P_{sat\,@\,20°C} = 2.339\text{ kPa}$$

$$\omega_2 = \frac{0.622\,P_{v2}}{P_2 - P_{v2}} = \frac{0.622(2.339\text{ kPa})}{(95-2.339)\text{ kPa}} = 0.0157\text{ kg H}_2\text{O /kg dry air}$$

$$h_2 = C_p T_2 + \omega_2 h_{g2} = (1.005\text{ kJ/kg}\cdot°\text{C})(20°\text{C}) + (0.0157)(2538.1\text{ kJ/kg})$$
$$= 59.95\text{ kJ/kg dry air}$$

Also,

$$h_w \cong h_{f@20°C} = 83.96\text{ kJ/kg} \qquad \text{(Table A-4)}$$

Then,

$$\dot{m}_{a1} = \frac{\dot{V}_1}{v_1} = \frac{4\text{ m}^3/\text{min}}{0.945\text{ m}^3/\text{kg dry air}} = 4.23\text{ kg/min}$$

Applying the water mass balance and the energy balance equations to the combined cooling and dehumidification section (excluding the refrigerant),

Water Mass Balance: $\quad \sum \dot{m}_{w,i} = \sum \dot{m}_{w,e} \longrightarrow \dot{m}_{a1}\omega_1 = \dot{m}_{a2}\omega_2 + \dot{m}_w$

$$\dot{m}_w = \dot{m}_a(\omega_1 - \omega_2) = (4.23\text{ kg/min})(0.0201 - 0.0157) = \mathbf{0.0186\text{ kg/min}}$$

(*b*) *Energy Balance*:

$$\dot{E}_{in} - \dot{E}_{out} = \Delta\dot{E}_{system}^{\nearrow 0\,(steady)} = 0$$

$$\dot{E}_{in} = \dot{E}_{out}$$

$$\sum \dot{m}_i h_i = \dot{Q}_{out} + \sum \dot{m}_e h_e \longrightarrow \dot{Q}_{out} = \dot{m}_{a1} h_1 - (\dot{m}_{a2} h_2 + \dot{m}_w h_w) = \dot{m}_a (h_1 - h_2) - \dot{m}_w h_w$$

$$\dot{Q}_{out} = (4.23\text{ kg/min})(81.53 - 59.95)\text{ kJ/kg} - (0.0186\text{ kg/min})(83.96\text{ kJ/kg}) = \mathbf{89.7\text{ kJ/min}}$$

(*c*) The inlet and exit enthalpies of the refrigerant are

$$h_3 = h_g + x_3 h_{fg} = 86.78 + 0.2 \times 175.07 = 121.79\text{ kJ/kg}$$

$$h_4 = h_{g@\,700\text{ kPa}} = 261.85\text{ kJ/kg}$$

Noting that the heat lost by the air is gained by the refrigerant, the mass flow rate of the refrigerant is determined from

$$\dot{Q}_R = \dot{m}_R (h_4 - h_3)$$

$$\dot{m}_R = \frac{\dot{Q}_R}{h_4 - h_3} = \frac{89.7\text{ kJ/min}}{(261.85 - 121.79)\text{ kJ/kg}} = \mathbf{0.640\text{ kg/min}}$$

13-126 Air is heated and dehumidified in an air-conditioning system consisting of a heating section and an evaporative cooler. The temperature and relative humidity of the air when it leaves the heating section, the rate of heat transfer in the heating section, and the rate of water added to the air in the evaporative cooler are to be determined.

Assumptions **1** This is a steady-flow process and thus the mass flow rate of dry air remains constant during the entire process ($\dot{m}_{a1} = \dot{m}_{a2} = \dot{m}_a$). **2** Dry air and water vapor are ideal gases. **3** The kinetic and potential energy changes are negligible.

Analysis (*a*) Assuming the wet-bulb temperature of the air remains constant during the evaporative cooling process, the properties of air at various states are determined from the psychometric chart (Table A-33) to be

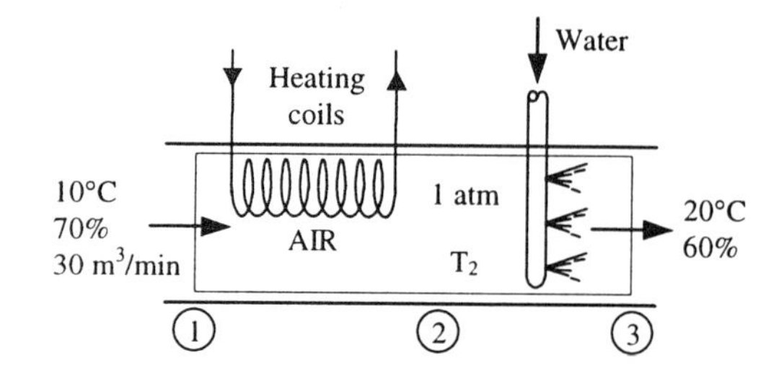

$$\left.\begin{array}{l} T_1 = 10°\text{C} \\ \phi_1 = 70\% \end{array}\right\} \begin{array}{l} h_1 = 23.5 \text{ kJ/kg dry air} \\ \omega_1 = 0.00532 \text{ kg/ H}_2\text{O/kg dry air} \\ v_1 = 0.810 \text{ m}^3/\text{kg} \end{array}$$

$$\left.\begin{array}{l} \omega_2 = \omega_1 \\ T_{wb2} = T_{wb3} \end{array}\right\} \begin{array}{l} T_2 = \mathbf{28.3°C} \\ \phi_2 = \mathbf{23.0\%} \\ h_2 \cong h_3 = 42.3 \text{ kJ/kg dry air} \end{array}$$

$$\left.\begin{array}{l} T_3 = 20°\text{C} \\ \phi_3 = 60\% \end{array}\right\} \begin{array}{l} h_3 = 42.3 \text{ kJ/kg dry air} \\ \omega_3 = 0.00875 \text{ kg/ H}_2\text{O/kg dry air} \\ T_{wb3} = 15.1°\text{C} \end{array}$$

(*b*) The mass flow rate of dry air is

$$\dot{m}_a = \frac{\dot{V}_1}{v_1} = \frac{30 \text{ m}^3/\text{min}}{0.810 \text{ m}^3/\text{kg dry air}} = 37.0 \text{ kg/min}$$

Then the rate of heat transfer to air in the heating section becomes

$$\dot{Q}_{in} = \dot{m}_a(h_2 - h_1) = (37.0 \text{ kg/min})(42.3 - 23.5) \text{ kJ/kg} = \mathbf{696 \text{ kJ/min}}$$

(*c*) The rate of water added to the air in evaporative cooler is

$$\begin{aligned} \dot{m}_{w,\text{added}} &= \dot{m}_{w3} - \dot{m}_{w2} = \dot{m}_a(\omega_3 - \omega_2) \\ &= (37.0 \text{ kg/min})(0.00875 - 0.00532) \\ &= \mathbf{0.127 \text{ kg/min}} \end{aligned}$$

13-127 EES solution of this (and other comprehensive problems designated with the *computer icon*) is available to instructors at the *Instructor Manual* section of the *Online Learning Center* (OLC) at www.mhhe.com/cengel-boles. See the Preface for access information.

13-128 Air is heated and dehumidified in an air-conditioning system consisting of a heating section and an evaporative cooler. The temperature and relative humidity of the air when it leaves the heating section, the rate of heat transfer in the heating section, and the rate at which water is added to the air in the evaporative cooler are to be determined.

Assumptions **1** This is a steady-flow process and thus the mass flow rate of dry air remains constant during the entire process ($\dot{m}_{a1} = \dot{m}_{a2} = \dot{m}_a$). **2** Dry air and water vapor are ideal gases. **3** The kinetic and potential energy changes are negligible.

Analysis (*a*) Assuming the wet-bulb temperature of the air remains constant during the evaporative cooling process, the properties of air at various states are determined to be

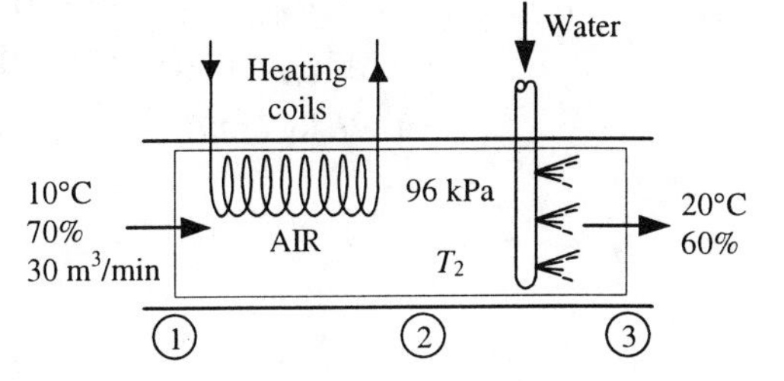

$$P_{v1} = \phi_1 P_{g1} = \phi_1 P_{sat\,@\,10°C} = (0.70)(1.2276\text{ kPa}) = 0.86\text{ kPa}$$

$$P_{a1} = P_1 - P_{v1} = 96 - 0.86 = 95.14\text{ kPa}$$

$$v_1 = \frac{R_a T_1}{P_{a1}} = \frac{(0.287\text{ kPa}\cdot\text{m}^3/\text{kg}\cdot\text{K})(283\text{ K})}{95.14\text{ kPa}}$$
$$= 0.854\text{ m}^3/\text{kg dry air}$$

$$\omega_1 = \frac{0.622\,P_{v1}}{P_1 - P_{v1}} = \frac{0.622(0.86\text{ kPa})}{(96-0.86)\text{ kPa}} = 0.00562\text{ kg H}_2\text{O/kg dry air}$$

$$h_1 = C_p T_1 + \omega_1 h_{g1} = (1.005\text{ kJ/kg}\cdot°\text{C})(10°\text{C}) + (0.00562)(2519.8\text{ kJ/kg})$$
$$= 24.21\text{ kJ/kg dry air}$$

and

$$P_{v3} = \phi_3 P_{g3} = \phi_3 P_{sat\,@\,20°C} = (0.60)(2.339\text{ kPa}) = 1.40\text{ kPa}$$

$$P_{a3} = P_3 - P_{v3} = 96 - 1.40 = 94.60\text{ kPa}$$

$$\omega_3 = \frac{0.622\,P_{v3}}{P_3 - P_{v3}} = \frac{0.622(1.40\text{ kPa})}{(96-1.40)\text{ kPa}} = 0.00921\text{ kg H}_2\text{O/kg dry air}$$

$$h_3 = C_p T_3 + \omega_3 h_{g3} = (1.005\text{ kJ/kg}\cdot°\text{C})(20°\text{C}) + (0.00921)(2538.1\text{ kJ/kg})$$
$$= 43.48\text{ kJ/kg dry air}$$

Also,

$$h_2 \cong h_3 = 43.48\text{ kJ/kg}$$
$$\omega_2 = \omega_1 = 0.00562\text{ kg H}_2\text{O/kg dry air}$$

Thus,

$$h_2 = C_p T_2 + \omega_2 h_{g2} \cong C_p T_2 + \omega_2(2501.3 + 1.82T_2) 43.48\text{ kJ/kg}$$
$$= (1.005\text{ kJ/kg}\cdot°\text{C})T_2 + (0.00562)(2501.3 + 1.82T_2)\text{ kJ/kg}$$

Solving for T_2,

$$T_2 = \mathbf{29.0°C} \longrightarrow P_{g2} = P_{sat\,@\,29°C} = 4.031\text{ kPa}$$

Thus,

$$\phi_2 = \frac{\omega_2 P_2}{(0.622 + \omega_2)P_{g2}} = \frac{(0.00562)(96)}{(0.622 + 0.00562)(4.031)} = 0.213 \text{ or } \mathbf{21.3\%}$$

(*b*) The mass flow rate of dry air is

$$\dot{m}_a = \frac{\dot{V}_1}{v_1} = \frac{30\ \text{m}^3/\text{min}}{0.854\ \text{m}^3/\text{kg dry air}} = 35.1\ \text{kg/min}$$

Then the rate of heat transfer to air in the heating section becomes

$$\dot{Q}_{in} = \dot{m}_a(h_2 - h_1) = (35.1\ \text{kg/min})(43.48 - 24.21)\ \text{kJ/kg} = \mathbf{676\ kJ/min}$$

(*c*) The rate of water addition to the air in evaporative cooler is

$$\begin{aligned}\dot{m}_{w,\text{added}} &= \dot{m}_{w3} - \dot{m}_{w2} = \dot{m}_a(\omega_3 - \omega_2)\\ &= (35.1\ \text{kg/min})(0.00921 - 0.00562)\\ &= \mathbf{0.126\ kg/min}\end{aligned}$$

13-129 Conditioned air is to be mixed with outside air. The ratio of the dry air mass flow rates of the conditioned- to-outside air, and the temperature of the mixture are to be determined.

Assumptions **1** Steady operating conditions exist. **2** Dry air and water vapor are ideal gases. **3** The kinetic and potential energy changes are negligible. **4** The mixing chamber is adiabatic.

Properties The properties of each inlet stream are determined from the psychometric chart (Table A-33) to be

$$h_1 = 34.3 \text{ kJ / kg dry air}$$
$$\omega_1 = 0.0084 \text{ kg H}_2\text{O / kg dry air}$$

and

$$h_2 = 68.5 \text{ kJ / kg dry air}$$
$$\omega_2 = 0.0134 \text{ kg H}_2\text{O / kg dry air}$$

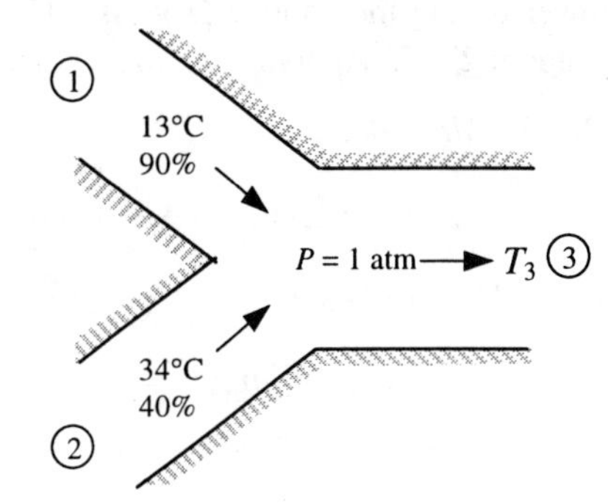

Analysis The ratio of the dry air mass flow rates of the Conditioned air to the outside air can be determined from

$$\frac{\dot{m}_{a1}}{\dot{m}_{a2}} = \frac{\omega_2 - \omega_3}{\omega_3 - \omega_1} = \frac{h_2 - h_3}{\omega_3 - h_1}$$

But state 3 is not completely specified. However, we know that state 3 is on the straight line connecting states 1 and 2 on the psychometric chart. At the intersection point of this line and $\phi = 60\%$ line we read

(*b*) $\quad T_3 = \mathbf{23.5°C}$

$$\omega_3 = 0.0109 \text{ kg H}_2\text{O /kg dry air}$$
$$h_3 = 51.3 \text{ kJ /kg dry air}$$

Therefore, the mixture will leave at 23.5°C. The $\dot{m}_{a1} / \dot{m}_{a2}$ ratio is determined by substituting the specific humidity (or enthalpy) values into the above relation,

(*a*) $$\frac{\dot{m}_{a1}}{\dot{m}_{a2}} = \frac{0.0134 - 0.0109}{0.0109 - 0.0084} = \mathbf{1.00}$$

Therefore, the mass flow rate of each stream must be the same.

13-130 EES solution of this (and other comprehensive problems designated with the *computer icon*) is available to instructors at the *Instructor Manual* section of the *Online Learning Center* (OLC) at www.mhhe.com/cengel-boles. See the Preface for access information.

13-131 [*Also solved by EES on enclosed CD*] Waste heat from the cooling water is rejected to air in a natural-draft cooling tower. The mass flow rate of the cooling water, the volume flow rate of air, and the mass flow rate of the required makeup water are to be determined.

Assumptions **1** Steady operating conditions exist. **2** Dry air and water vapor are ideal gases. **3** The kinetic and potential energy changes are negligible. **4** The cooling tower is adiabatic.

Analysis (*a*) The mass flow rate of dry air through the tower remains constant $(\dot{m}_{a1} = \dot{m}_{a2} = \dot{m}_a)$, but the mass flow rate of liquid water decreases by an amount equal to the amount of water that vaporizes in the tower during the cooling process. The water lost through evaporation is made up later in the cycle using water at 27°C. Applying the mass balance and the energy balance equations yields

Dry Air Mass Balance:

$$\sum \dot{m}_{a,i} = \sum \dot{m}_{a,e} \longrightarrow \dot{m}_{a1} = \dot{m}_{a2} = \dot{m}_a$$

Water Mass Balance:

$$\sum \dot{m}_{w,i} = \sum \dot{m}_{w,e} \longrightarrow \dot{m}_3 + \dot{m}_{a1}\omega_1 = \dot{m}_4 + \dot{m}_{a2}\omega_2$$

$$\dot{m}_3 - \dot{m}_4 = \dot{m}_a(\omega_2 - \omega_1) = \dot{m}_{\text{makeup}}$$

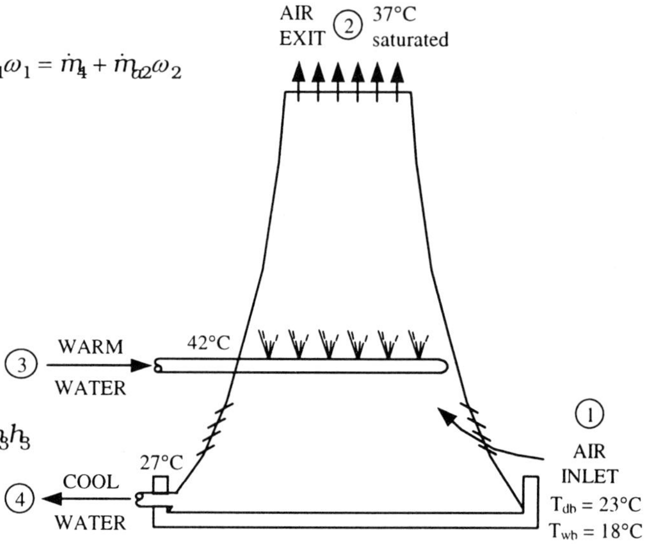

Energy Balance:

$$\dot{E}_{in} - \dot{E}_{out} = \Delta\dot{E}_{system}^{\nearrow 0\text{ (steady)}} = 0$$

$$\dot{E}_{in} = \dot{E}_{out}$$

$$\sum \dot{m}_i h_i = \sum \dot{m}_e h_e \quad (\text{since } \dot{Q} = \dot{W} = 0)$$

$$0 = \sum \dot{m}_e h_e - \sum \dot{m}_i h_i$$

$$0 = \dot{m}_{a2}h_2 + \dot{m}_4 h_4 - \dot{m}_{a1}h_1 - \dot{m}_3 h_3$$

$$0 = \dot{m}_a(h_2 - h_1) + (\dot{m}_3 - \dot{m}_{\text{makeup}})h_4 - \dot{m}_3 h_3$$

Solving for $\dot{m}_a$,

$$\dot{m}_a = \frac{\dot{m}_3(h_3 - h_4)}{(h_2 - h_1) - (\omega_2 - \omega_1)h_4}$$

From the psychometric chart (Table A-33),

$$h_1 = 50.8 \text{ kJ/kg dry air}$$
$$\omega_1 = 0.0109 \text{ kg H}_2\text{O/kg dry air}$$
$$v_1 = 0.854 \text{ m}^3\text{/kg dry air}$$

and

$$h_2 = 143.0 \text{ kJ/kg dry air}$$
$$\omega_2 = 0.0412 \text{ kg H}_2\text{O/kg dry air}$$

From Table A-4,

$$h_3 \cong h_{f\,@42°\text{C}} = 175.92 \text{ kJ/kg H}_2\text{O}$$
$$h_4 \cong h_{f\,@27°\text{C}} = 113.25 \text{ kJ/kg H}_2\text{O}$$

Substituting,

$$\dot{m}_a = \frac{\dot{m}_3(175.92 - 113.25)\text{kJ/kg}}{(143.0 - 50.8)\text{ kJ/kg} - (0.0412 - 0.0109)(113.25)\text{ kJ/kg}} = 0.706\,\dot{m}_3$$

The mass flow rate of the cooling water is determined by applying the steady flow energy balance equation on the cooling water,

$$\dot{Q}_{\text{waste}} = \dot{m}_3 h_3 - (\dot{m}_3 - \dot{m}_{\text{makeup}})h_4 = \dot{m}_3 h_3 - [\dot{m}_3 - \dot{m}_a(\omega_2 - \omega_1)]h_4$$
$$= \dot{m}_3 h_3 - \dot{m}_3[1 - 0.706(0.0412 - 0.0109)]h_4 = \dot{m}_3(h_3 - 0.9786h_4)$$

$$50{,}000 \text{ kJ/s} = \dot{m}_3(175.92 - 0.9786 \times 113.25) \text{ kJ/kg}$$
$$\dot{m}_3 = \mathbf{768.1\ kg/s}$$

and

$$\dot{m}_a = 0.706\dot{m}_3 = (0.706)(768.1 \text{ kg/s}) = 542.3 \text{ kg/s}$$

(*b*) Then the volume flow rate of air into the cooling tower becomes

$$\dot{V}_1 = \dot{m}_a v_1 = (542.3 \text{ kg/s})(0.854 \text{ m}^3/\text{kg}) = \mathbf{463.1\ m^3/s}$$

(*c*) The mass flow rate of the required makeup water is determined from

$$\dot{m}_{\text{makeup}} = \dot{m}_a(\omega_2 - \omega_1) = (542.3 \text{ kg/s})(0.0412 - 0.0109) = \mathbf{16.4\ kg/s}$$

13-132 EES solution of this (and other comprehensive problems designated with the *computer icon*) is available to instructors at the *Instructor Manual* section of the *Online Learning Center* (OLC) at www.mhhe.com/cengel-boles. See the Preface for access information.

Fundamentals of Engineering (FE) Exam Problems

13-133 A room is filled with saturated moist air at 30°C and a total pressure of 100 kPa. If the mass of dry air in the room is 100 kg, the mass of water vapor is

(a) 0.52 kg (b) 1.84 kg (c) 2.64 kg (d) 2.76 kg (e) 4.25 kg

Answer (d) 2.76 kg

Solution Solved by EES Software. Solutions can be verified by copying-and-pasting the following lines on a blank EES screen. (Similar problems and their solutions can be obtained easily by modifying numerical values).

```
T1=30 "C"
P=100 "kPa"
m_air=100 "kg"
RH=1

P_g=PRESSURE(Steam_NBS,T=T1,x=0)
RH=P_v/P_g
P_air=P-P_v
w=0.622*P_v/(P-P_v)
w=m_v/m_air

"Some Wrong Solutions with Common Mistakes:"
W1_vmass=m_air*w1; w1=0.622*P_v/P "Using P instead of P-Pv in w relation"
W2_vmass=m_air "Taking m_vapor = m_air"
W3_vmass=P_v/P*m_air "Using wrong relation"
```

13-134 A room contains 50 kg of dry air and 0.6 kg of water vapor at 25°C and 95 kPa total pressure. The relative humidity of air in the room is

(a) 1.2% (b) 18.4% (c) 56.7% (d) 65.2% (e) 78.0%

Answer (c) 56.7%

Solution Solved by EES Software. Solutions can be verified by copying-and-pasting the following lines on a blank EES screen. (Similar problems and their solutions can be obtained easily by modifying numerical values).

```
T1=25 "C"
P=95 "kPa"
m_air=50 "kg"
m_v=0.6 "kg"
w=0.622*P_v/(P-P_v)
w=m_v/m_air

P_g=PRESSURE(Steam_NBS,T=T1,x=0)
RH=P_v/P_g

"Some Wrong Solutions with Common Mistakes:"
W1_RH=m_v/(m_air+m_v) "Using wrong relation"
W2_RH=P_g/P "Using wrong relation"
```

13-135 A 40-m^3 room contains air at 30°C and a total pressure of 85 kPa with a relative humidity of 85%. The mass of dry air in the room is

(a) 25.7 kg (b) 37.4 kg (c) 39.1 kg (d) 41.8 kg (e) 45.3 kg

Answer (b) 37.4 kg

Solution Solved by EES Software. Solutions can be verified by copying-and-pasting the following lines on a blank EES screen. (Similar problems and their solutions can be obtained easily by modifying numerical values).

```
V=40 "m^3"
T1=30 "C"
P=85 "kPa"
RH=0.85

P_g=PRESSURE(Steam_NBS,T=T1,x=0)
RH=P_v/P_g
P_air=P-P_v

R_air=0.287 "kJ/kg.K"
m_air=P_air*V/(R_air*(T1+273))

"Some Wrong Solutions with Common Mistakes:"
W1_mass=P_air*V/(R_air*T1) "Using C instead of K"
W2_mass=P*V/(R_air*(T1+273)) "Using P instead of P_air"
W3_mass=m_air*RH "Using wrong relation"
```

13-136 A room contains air at 30°C and a total pressure of 92.0 kPa with a relative humidity of 85%. The partial pressure of dry air is

(a) 82.0 kPa (b) 85.8 kPa (c) 88.4 kPa (d) 90.6 kPa (e) 91.2 kPa

Answer (c) 88.4 kPa

Solution Solved by EES Software. Solutions can be verified by copying-and-pasting the following lines on a blank EES screen. (Similar problems and their solutions can be obtained easily by modifying numerical values).

```
T1=30 "C"
P=92 "kPa"
RH=0.85

P_g=PRESSURE(Steam_NBS,T=T1,x=0)
RH=P_v/P_g
P_air=P-P_v

"Some Wrong Solutions with Common Mistakes:"
W1_Pair=P_v "Using Pv as P_air"
W2_Pair=P-P_g "Using wrong relation"
W3_Pair=RH*P "Using wrong relation"
```

13-137 The air in a house is at 25°C and 65 percent relative humidity. Now the air is cooled at constant pressure. The temperature at which the moisture in the air will start condensing is

(a) 8.7°C (b) 11.3°C (c) 14.1°C (d) 17.9°C (e) 21.5°C

Answer (d) 17.9°C

Solution Solved by EES Software. Solutions can be verified by copying-and-pasting the following lines on a blank EES screen. (Similar problems and their solutions can be obtained easily by modifying numerical values).

```
T1=25 "C"
RH1=0.65
P_g=PRESSURE(Steam_NBS,T=T1,x=0)
RH1=P_v/P_g
T_dp=TEMPERATURE(Steam_NBS,x=0,P=P_v)

"Some Wrong Solutions with Common Mistakes:"
W1_Tdp=T1*RH1 "Using wrong relation"
W2_Tdp=(T1+273)*RH1-273 "Using wrong relation"
W3_Tdp=WETBULB(AirH2O,T=T1,P=P1,R=RH1); P1=100 "Using wet-bulb temperature"
```

13-138 On the psychrometric chart, a cooling and dehumidification process appears as a line that is
(a) horizontal to the left
(b) vertical downward
(c) diagonal upwards to the right (NE direction)
(d) diagonal upwards to the left (NW direction)
(e) diagonal downwards to the left (SW direction)

Answer (e) diagonal downwards to the left (SW direction)

13-139 On the psychrometric chart, a heating and humidification process appears as a line that is
(a) horizontal to the right
(b) vertical upward
(c) diagonal upwards to the right (NE direction)
(d) diagonal upwards to the left (NW direction)
(e) diagonal downwards to the right (SE direction)

Answer (c) diagonal upwards to the right (NE direction)

13-140 An air stream at a specified temperature and relative humidity undergoes evaporative cooling by spraying water into it at about the same temperature. The lowest temperature the air stream can be cooled to is
(a) the dry bulb temperature at the given state
(b) the wet bulb temperature at the given state
(c) the dew point temperature at the given state
(d) the saturation temperature corresponding to the humidity ratio at the given state
(e) the triple point temperature of water

Answer (b) the wet bulb temperature at the given state

13-141 Air is cooled and dehumidified as it flows over the coils of a refrigeration system at 85 kPa from 30°C and a humidity ratio of 0.023 kg/kg dry air to 15°C and a humidity ratio of 0.015 kg/kg dry air. If the mass flow rate of dry air is 0.7 kg/s, the rate of heat removal from the air is

(a) 5 kJ/s (b) 10 kJ/s (c) 15 kJ/s (d) 20 kJ/s (e) 25 kJ/s

Answer (e) 25 kJ/s

Solution Solved by EES Software. Solutions can be verified by copying-and-pasting the following lines on a blank EES screen. (Similar problems and their solutions can be obtained easily by modifying numerical values).

```
P=85 "kPa"
T1=30 "C"
w1=0.023

T2=15 "C"
w2=0.015
m_air=0.7 "kg/s"

m_water=m_air*(w1-w2)
h1=ENTHALPY(AirH2O,T=T1,P=P,w=w1)
h2=ENTHALPY(AirH2O,T=T2,P=P,w=w2)
h_w=ENTHALPY(Steam_NBS,T=T2,x=0)

Q=m_air*(h1-h2)-m_water*h_w

"Some Wrong Solutions with Common Mistakes:"
W1_Q=m_air*(h1-h2) "Ignoring condensed water"
W2_Q=m_air*Cp_air*(T1-T2)-m_water*h_w; Cp_air = 1.005 "Using dry air enthalpies"
W3_Q=m_air*(h1-h2)+m_water*h_w "Using wrong sign"
```

13-142 Air at a total pressure of 90 kPa, 15°C, and 60 percent relative humidity is heated and humidified to 25°C and 60 percent relative humidity by introducing water vapor. If the mass flow rate of dry air is 2.2 kg/s, the rate at which steam is added to the air is

(a) 0.014 kg/s (b) 0.032 kg/s (c) 0.11 kg/s (d) 0.18 kg/s (e) 0.28 kg/s

Answer (a) 0.014 kg/s

Solution Solved by EES Software. Solutions can be verified by copying-and-pasting the following lines on a blank EES screen. (Similar problems and their solutions can be obtained easily by modifying numerical values).

```
P=90 "kPa"
T1=15 "C"
RH1=0.60

T2=25 "C"
RH2=0.60
m_air=2.2 "kg/s"

w1=HUMRAT(AirH2O,T=T1,P=P,R=RH1)
w2=HUMRAT(AirH2O,T=T2,P=P,R=RH2)
m_water=m_air*(w2-w1)

"Some Wrong Solutions with Common Mistakes:"
W1_mv=0 "sine RH = constant"
W2_mv=w2-w1 "Ignoring mass flow rate of air"
W3_mv=RH1*m_air "Using wrong relation"
```

13-143 · · · 13-147 Design and Essay Problems

Chapter 14
CHEMICAL REACTIONS

Fuels and Combustion

14-1C Gasoline is C_8H_{18}, diesel fuel is $C_{12}H_{26}$, and natural gas is CH_4.

14-2C Nitrogen, in general, does not react with other chemical species during a combustion process but its presence affects the outcome of the process because nitrogen absorbs a large proportion of the heat released during the chemical process.

14-3C Moisture, in general, does not react chemically with any of the species present in the combustion chamber, but it absorbs some of the energy released during combustion, and it raises the dew point temperature of the combustion gases.

14-4C The dew-point temperature of the product gases is the temperature at which the water vapor in the product gases starts to condense as the gases are cooled at constant pressure. It is the saturation temperature corresponding to the vapor pressure of the product gases.

14-5C The number of atoms are preserved during a chemical reaction, but the total mole numbers are not.

14-6C Air-fuel ratio is the ratio of the mass of air to the mass of fuel during a combustion process. Fuel-air ratio is the inverse of the air-fuel ratio.

14-7C No. Because the molar mass of the fuel and the molar mass of the air, in general, are different.

Theoretical and Actual Combustion Processes

14-8C The causes of incomplete combustion are insufficient time, insufficient oxygen, insufficient mixing, and dissociation.

14-9C CO. Because oxygen is more strongly attracted to hydrogen than it is to carbon, and hydrogen is usually burned to completion even when there is a deficiency of oxygen.

14-10C It represent the amount of air that contains the exact amount of oxygen needed for complete combustion.

14-11C No. The theoretical combustion is also complete, but the products of theoretical combustion does not contain any uncombined oxygen.

14-12C Case (b).

14-13 Methane is burned with the stoichiometric amount of air during a combustion process. The AF and FA ratios are to be determined.

Assumptions **1** Combustion is complete. **2** The combustion products contain CO_2, H_2O, and N_2 only.

Properties The molar masses of C, H_2, and air are 12 kg/kmol, 2 kg/kmol, and 29 kg/kmol, respectively (Table A-1).

Analysis This is a theoretical combustion process since methane is burned completely with stoichiometric amount of air. The stoichiometric combustion equation of CH_4 is

$$CH_4 + a_{th}[O_2 + 3.76N_2] \longrightarrow CO_2 + 2H_2O + 3.76a_{th}N_2$$

CH₄

Air

Products

O_2 balance: $\quad a_{th} = 1+1 \longrightarrow a_{th} = 2$

Substituting, $\quad CH_4 + 2[O_2 + 3.76N_2] \longrightarrow CO_2 + 2H_2O + 7.52N_2$

The air-fuel ratio is determined by taking the ratio of the mass of the air to the mass of the fuel,

$$\mathrm{AF} = \frac{m_{\mathrm{air}}}{m_{\mathrm{fuel}}} = \frac{(2 \times 4.76\,\mathrm{kmol})(29\,\mathrm{kg/kmol})}{(1\,\mathrm{kmol})(12\,\mathrm{kg/kmol}) + (2\,\mathrm{kmol})(2\,\mathrm{kg/kmol})} = \mathbf{17.3\ kg\,air/kg\,fuel}$$

The fuel-air ratio is the inverse of the air-fuel ratio,

$$\mathrm{FA} = \frac{1}{\mathrm{AF}} = \frac{1}{17.3\,\mathrm{kg\,air/kg\,fuel}} = \mathbf{0.0578\ kg\,fuel/kg\,air}$$

14-14 Propane is burned with 75 percent excess air during a combustion process. The AF ratio is to be determined.

Assumptions **1** Combustion is complete. **2** The combustion products contain CO_2, H_2O, O_2, and N_2 only.

Properties The molar masses of C, H_2, and air are 12 kg/kmol, 2 kg/kmol, and 29 kg/kmol, respectively (Table A-1).

Analysis The combustion equation in this case can be written as

$$C_3H_8 + 1.75a_{th}[O_2 + 3.76N_2] \longrightarrow 3CO_2 + 4H_2O + 0.75a_{th}O_2 + (1.75\times 3.76)a_{th}N_2$$

where a_{th} is the stoichiometric coefficient for air. We have automatically accounted for the 75% excess air by using the factor $1.75a_{th}$ instead of a_{th} for air. The stoichiometric amount of oxygen ($a_{th}O_2$) will be used to oxidize the fuel, and the remaining excess amount ($0.75a_{th}O_2$) will appear in the products as free oxygen. The coefficient a_{th} is determined from the O_2 balance,

O_2 balance: $\quad 1.75a_{th} = 3 + 2 + 0.75a_{th} \longrightarrow a_{th} = 5$

Substituting, $\quad C_3H_8 + 8.75[O_2 + 3.76N_2] \longrightarrow 3CO_2 + 4H_2O + 3.75O_2 + 32.9N_2$

The air-fuel ratio is determined by taking the ratio of the mass of the air to the mass of the fuel,

$$AF = \frac{m_{air}}{m_{fuel}} = \frac{(8.75\times 4.76\text{kmol})(29\text{kg/kmol})}{(3\text{kmol})(12\text{kg/kmol}) + (4\text{kmol})(2\text{kg/kmol})} = \mathbf{27.5\ kg\,air/kg\,fuel}$$

14-15 Acetylene is burned with the stoichiometric amount of air during a combustion process. The AF ratio is to be determined on a mass and on a mole basis.

Assumptions **1** Combustion is complete. **2** The combustion products contain CO_2, H_2O, and N_2 only.

Properties The molar masses of C, H_2, and air are 12 kg/kmol, 2 kg/kmol, and 29 kg/kmol, respectively (Table A-1).

Analysis This is a theoretical combustion process since C_2H_2 is burned completely with stoichiometric amount of air. The stoichiometric combustion equation of C_2H_2 is

$$C_2H_2 + a_{th}[O_2 + 3.76N_2] \longrightarrow 2CO_2 + H_2O + 3.76a_{th}N_2$$

C_2H_2

Products

100%
theoretical air

O_2 balance: $\quad a_{th} = 2 + 0.5 \longrightarrow a_{th} = 2.5$

Substituting,

$$C_2H_2 + 2.5[O_2 + 3.76N_2] \longrightarrow 2CO_2 + H_2O + 9.4N_2$$

The air-fuel ratio is determined by taking the ratio of the mass of the air to the mass of the fuel,

$$AF = \frac{m_{air}}{m_{fuel}} = \frac{(2.5\times 4.76\text{kmol})(29\text{kg/kmol})}{(2\text{kmol})(12\text{kg/kmol}) + (1\text{kmol})(2\text{kg/kmol})} = \mathbf{13.3\ kg\,air/kg\,fuel}$$

On a mole basis, the air-fuel ratio is expressed as the ratio of the mole numbers of the air to the mole numbers of the fuel,

$$AF = \frac{m_{air}}{m_{fuel}} = \frac{2.5\times 4.76\text{ kmol}}{1\text{ kmol fuel}} = \mathbf{11.9\ kmol\ air\ /\ kmol\ fuel}$$

14-16 Ethane is burned with an unknown amount of air during a combustion process. The AF ratio and the percentage of theoretical air used are to be determined.

Assumptions **1** Combustion is complete. **2** The combustion products contain CO_2, H_2O, O_2, and N_2 only.

Properties The molar masses of C, H_2, and air are 12 kg/kmol, 2 kg/kmol, and 29 kg/kmol, respectively (Table A-1).

Analysis (*a*) The combustion equation in this case can be written as

$$C_2H_6 + a[O_2 + 3.76N_2] \longrightarrow 2CO_2 + 3H_2O + 2O_2 + 3.76aN_2$$

O_2 balance: $$a = 2 + 1.5 + 2 \longrightarrow a = 5.5$$

Substituting, $$C_2H_6 + 5.5[O_2 + 3.76N_2] \longrightarrow 2CO_2 + 3H_2O + 2O_2 + 20.68N_2$$

The air-fuel ratio is determined by taking the ratio of the mass of the air to the mass of the fuel,

$$\text{AF} = \frac{m_{\text{air}}}{m_{\text{fuel}}} = \frac{(5.5 \times 4.76\text{kmol})(29\text{kg/kmol})}{(2\text{kmol})(12\text{kg/kmol}) + (3\text{kmol})(2\text{kg/kmol})} = \mathbf{25.3\ \text{kgair/kgfuel}}$$

(*b*) To find the percent theoretical air used, we need to know the theoretical amount of air, which is determined from the theoretical combustion equation of C_2H_6,

$$C_2H_6 + a_{th}[O_2 + 3.76N_2] \longrightarrow 2CO_2 + 3H_2O + 3.76a_{th}N_2$$

O_2 balance: $$a_{th} = 2 + 1.5 \longrightarrow a_{th} = 3.5$$

Then,

$$\text{Percent theoretical air} = \frac{m_{\text{air, act}}}{m_{\text{air, th}}} = \frac{N_{\text{air, act}}}{N_{\text{air, th}}} = \frac{a}{a_{th}} = \frac{5.5}{3.5} = \mathbf{157\%}$$

14-17E Ethylene is burned with 200 percent theoretical air during a combustion process. The AF ratio and the dew-point temperature of the products are to be determined.

Assumptions **1** Combustion is complete. **2** The combustion products contain CO_2, H_2O, O_2, and N_2 only. **3** Combustion gases are ideal gases.

Properties The molar masses of C, H_2, and air are 12 lbm/lbmol, 2 lbm/lbmol, and 29 lbm/lbmol, respectively (Table A-1E).

Analysis (*a*) The combustion equation in this case can be written as

$$C_2H_4 + 2a_{th}[O_2 + 3.76N_2] \longrightarrow 2CO_2 + 2H_2O + a_{th}O_2 + (2\times 3.76)a_{th}N_2$$

where a_{th} is the stoichiometric coefficient for air. It is determined from

O_2 balance: $\quad 2a_{th} = 2 + 1 + a_{th} \longrightarrow a_{th} = 3$

Substituting, $\quad C_2H_4 + 6[O_2 + 3.76N_2] \longrightarrow 2CO_2 + 2H_2O + 3O_2 + 22.56N_2$

The air-fuel ratio is determined by taking the ratio of the mass of the air to the mass of the fuel,

$$\mathrm{AF} = \frac{m_{air}}{m_{fuel}} = \frac{(6\times 4.76\,\mathrm{lbmol})(29\,\mathrm{lbm/lbmol})}{(2\,\mathrm{lbmol})(12\,\mathrm{lbm/lbmol}) + (2\,\mathrm{lbmol})(2\,\mathrm{lbm/lbmol})} = \mathbf{29.6\ lbmair/lbmfuel}$$

(*b*) The dew-point temperature of a gas-vapor mixture is the saturation temperature of the water vapor in the product gases corresponding to its partial pressure. That is,

$$P_v = \left(\frac{N_v}{N_{prod}}\right)P_{prod} = \left(\frac{2\,\mathrm{lbmol}}{29.56\,\mathrm{lbmol}}\right)(14.5\,\mathrm{psia}) = 0.981\,\mathrm{psia}$$

Thus,

$$T_{dp} = T_{sat\,@\,0.981\,psia} = \mathbf{100.9°F}$$

14-18 Propylene is burned with 50 percent excess air during a combustion process. The AF ratio and the temperature at which the water vapor in the products will start condensing are to be determined.

Assumptions **1** Combustion is complete. **2** The combustion products contain CO_2, H_2O, O_2, and N_2 only. **3** Combustion gases are ideal gases.

Properties The molar masses of C, H_2, and air are 12 kg/kmol, 2 kg/kmol, and 29 kg/kmol, respectively (Table A-1).

Analysis (*a*) The combustion equation in this case can be written as

$$C_3H_6 + 1.5a_{th}[O_2 + 3.76N_2] \longrightarrow 3CO_2 + 3H_2O + 0.5a_{th}O_2 + (1.5 \times 3.76)a_{th}N_2$$

where a_{th} is the stoichiometric coefficient for air. It is determined from

O_2 balance: $\quad 1.5a_{th} = 3 + 1.5 + 0.5a_{th} \longrightarrow a_{th} = 4.5$

Substituting, $\quad C_3H_6 + 6.75[O_2 + 3.76N_2] \longrightarrow 3CO_2 + 3H_2O + 2.25O_2 + 25.38N_2$

The air-fuel ratio is determined by taking the ratio of the mass of the air to the mass of the fuel,

$$AF = \frac{m_{air}}{m_{fuel}} = \frac{(6.75 \times 4.76 \text{kmol})(29 \text{kg/kmol})}{(3 \text{kmol})(12 \text{kg/kmol}) + (3 \text{kmol})(2 \text{kg/kmol})} = \mathbf{22.2 \text{ kgair/kgfuel}}$$

(*b*) The dew-point temperature of a gas-vapor mixture is the saturation temperature of the water vapor in the product gases corresponding to its partial pressure. That is,

$$P_v = \left(\frac{N_v}{N_{prod}}\right)P_{prod} = \left(\frac{3 \text{kmol}}{33.63 \text{kmol}}\right)(110 \text{kPa}) = 9.813 \text{kPa}$$

Thus,

$$T_{dp} = T_{sat@9.813 \text{kPa}} = \mathbf{45.4°C}$$

14-19 Octane is burned with 250 percent theoretical air during a combustion process. The AF ratio and the dew-pint temperature of the products are to be determined.

Assumptions **1** Combustion is complete. **2** The combustion products contain CO_2, H_2O, O_2, and N_2 only. **3** Combustion gases are ideal gases.

Properties The molar masses of C, H_2, and air are 12 kg/kmol, 2 kg/kmol, and 29 kg/kmol, respectively (Table A-1).

Analysis (*a*) The combustion equation in this case can be written as

$$C_8H_{18} + 2.5a_{th}[O_2 + 3.76N_2] \longrightarrow 8CO_2 + 9H_2O + 1.5a_{th}O_2 + (2.5\times 3.76)a_{th}N_2$$

where a_{th} is the stoichiometric coefficient for air. It is determined from

C8H18 → Combustion chamber P = 1 atm → Products; Air 25°C →

O_2 balance: $\quad 2.5a_{th} = 8 + 4.5 + 1.5a_{th} \longrightarrow a_{th} = 12.5$

Substituting,

$$C_8H_{18} + 31.25[O_2 + 3.76N_2] \longrightarrow 8CO_2 + 9H_2O + 18.75O_2 + 117.5N_2$$

Thus,

$$AF = \frac{m_{air}}{m_{fuel}} = \frac{(31.25\times 4.76\text{kmol})(29\text{kg/kmol})}{(8\text{kmol})(12\text{kg/kmol}) + (9\text{kmol})(2\text{kg/kmol})} = \mathbf{37.8\ kgair/kgfuel}$$

(*b*) The dew-point temperature of a gas-vapor mixture is the saturation temperature of the water vapor in the product gases corresponding to its partial pressure. That is,

$$P_v = \left(\frac{N_v}{N_{prod}}\right)P_{prod} = \left(\frac{9\text{ kmol}}{153.25\text{ kmol}}\right)(101.325\text{ kPa}) = 5.951\text{ kPa}$$

Thus,

$$T_{dp} = T_{sat@5.951\text{ kPa}} = \mathbf{35.7°C}$$

14-20 Gasoline is burned steadily with air in a jet engine. The AF ratio is given. The percentage of excess air used is to be determined.

Assumptions **1** Combustion is complete. **2** The combustion products contain CO_2, H_2O, and N_2 only.

Properties The molar masses of C, H_2, and air are 12 kg/kmol, 2 kg/kmol, and 29 kg/kmol, respectively (Table A-1).

Analysis The theoretical combustion equation in this case can be written as

$$C_8H_{18} + a_{th}[O_2 + 3.76N_2] \longrightarrow 8CO_2 + 9H_2O + 3.76a_{th}N_2$$

where a_{th} is the stoichiometric coefficient for air. It is determined from

O_2 balance: $\quad a_{th} = 8 + 4.5 \longrightarrow a_{th} = 12.5$

The air-fuel ratio for the theoretical reaction is determined by taking the ratio of the mass of the air to the mass of the fuel for,

$$\mathrm{AF_{th}} = \frac{m_{\mathrm{air,th}}}{m_{\mathrm{fuel}}} = \frac{(12.5 \times 4.76\,\mathrm{kmol})(29\,\mathrm{kg/kmol})}{(8\,\mathrm{kmol})(12\,\mathrm{kg/kmol}) + (9\,\mathrm{kmol})(2\,\mathrm{kg/kmol})} = 15.14\,\mathrm{kg\,air/kg\,fuel}$$

Then the percent theoretical air used can be determined from

$$\text{Percent theoretical air} = \frac{\mathrm{AF_{act}}}{\mathrm{AF_{th}}} = \frac{21\,\mathrm{kg\,air/kg\,fuel}}{15.14\,\mathrm{kg\,air/kg\,fuel}} = \mathbf{139\%}$$

14-21 Ethane is burned with air steadily. The mass flow rates of ethane and air are given. The percentage of excess air used is to be determined.

Assumptions **1** Combustion is complete. **2** The combustion products contain CO_2, H_2O, and N_2 only.

Properties The molar masses of C, H_2, and air are 12 kg/kmol, 2 kg/kmol, and 29 kg/kmol, respectively (Table A-1).

Analysis The theoretical combustion equation in this case can be written as

$$C_2H_6 + a_{th}[O_2 + 3.76N_2] \longrightarrow 2CO_2 + 3H_2O + 3.76a_{th}N_2$$

where a_{th} is the stoichiometric coefficient for air. It is determined from

O_2 balance: $\quad a_{th} = 2 + 1.5 \longrightarrow a_{th} = 3.5$

The air-fuel ratio for the theoretical reaction is determined by taking the ratio of the mass of the air to the mass of the fuel for,

$$\mathrm{AF_{th}} = \frac{m_{\mathrm{air,th}}}{m_{\mathrm{fuel}}} = \frac{(3.5 \times 4.76\,\mathrm{kmol})(29\,\mathrm{kg/kmol})}{(2\,\mathrm{kmol})(12\,\mathrm{kg/kmol}) + (3\,\mathrm{kmol})(2\,\mathrm{kg/kmol})} = 16.1\,\mathrm{kg\,air/kg\,fuel}$$

The actual air-fuel ratio used is

$$\mathrm{AF_{act}} = \frac{\dot{m}_{\mathrm{air}}}{\dot{m}_{\mathrm{fuel}}} = \frac{132\ \mathrm{kg/h}}{6\ \mathrm{kg/h}} = 22\ \mathrm{kg\ air/kg\ fuel}$$

Then the percent theoretical air used can be determined from

$$\text{Percent theoretical air} = \frac{\mathrm{AF_{act}}}{\mathrm{AF_{th}}} = \frac{22\,\mathrm{kg\,air/kg\,fuel}}{16.1\,\mathrm{kg\,air/kg\,fuel}} = 137\%$$

Thus the excess air used during this process is **37%**.

14-22 Butane is burned with air. The masses of butane and air are given. The percentage of theoretical air used and the dew-point temperature of the products are to be determined.

Assumptions **1** Combustion is complete. **2** The combustion products contain CO_2, H_2O, and N_2 only. **3** Combustion gases are ideal gases.

Properties The molar masses of C, H_2, and air are 12 kg/kmol, 2 kg/kmol, and 29 kg/kmol, respectively (Table A-1).

Analysis (*a*) The theoretical combustion equation in this case can be written as

$$C_4H_{10} + a_{th}[O_2 + 3.76N_2] \longrightarrow 4CO_2 + 5H_2O + 3.76a_{th}N_2$$

where a_{th} is the stoichiometric coefficient for air. It is determined from

O_2 balance: $\quad a_{th} = 4 + 2.5 \longrightarrow a_{th} = 6.5$

The air-fuel ratio for the theoretical reaction is determined by taking the ratio of the mass of the air to the mass of the fuel for,

$$\text{AF}_{th} = \frac{m_{air,th}}{m_{fuel}} = \frac{(6.5 \times 4.76\text{kmol})(29\text{kg/kmol})}{(4\text{kmol})(12\text{kg/kmol}) + (5\text{kmol})(2\text{kg/kmol})} = 15.5\text{kgair/kgfuel}$$

The actual air-fuel ratio used is

$$\text{AF}_{act} = \frac{m_{air}}{m_{fuel}} = \frac{25\text{ kg}}{1\text{ kg}} = 25\text{ kg air / kg fuel}$$

Then the percent theoretical air used can be determined from

$$\text{Percent theoretical air} = \frac{\text{AF}_{act}}{\text{AF}_{th}} = \frac{25\text{ kg air / kg fuel}}{15.5\text{ kg air / kg fuel}} = \mathbf{161\%}$$

(*b*) The combustion is complete, and thus products will contain only CO_2, H_2O, O_2 and N_2. The air-fuel ratio for this combustion process on a mole basis is

$$\overline{\text{AF}} = \frac{N_{air}}{N_{fuel}} = \frac{m_{air}/M_{air}}{m_{fuel}/M_{fuel}} = \frac{(25\text{ kg})/(29\text{ kg/kmol})}{(1\text{ kg})/(58\text{ kg/kmol})} = 50\text{ kmolair/kmolfuel}$$

Thus the combustion equation in this case can be written as

$$C_4H_{10} + (50/4.76)[O_2 + 3.76N_2] \longrightarrow 4CO_2 + 5H_2O + 4.0O_2 + 39.5N_2$$

The dew-point temperature of a gas-vapor mixture is the saturation temperature of the water vapor in the product gases corresponding to its partial pressure. That is,

$$P_v = \left(\frac{N_v}{N_{prod}}\right)P_{prod} = \left(\frac{5\text{ kmol}}{52.5\text{kmol}}\right)(90\text{ kPa}) = 8.571\text{ kPa}$$

Thus,

$$T_{dp} = T_{sat\,@8.571\text{ kPa}} = \mathbf{42.7°C}$$

14-23E Butane is burned with air. The masses of butane and air are given. The percentage of theoretical air used and the dew-point temperature of the products are to be determined.

Assumptions **1** Combustion is complete. **2** The combustion products contain CO_2, H_2O, and N_2 only. **3** Combustion gases are ideal gases.

Properties The molar masses of C, H_2, and air are 12 lbm/lbmol, 2 lbm/lbmol, and 29 lbm/lbmol, respectively (Table A-1).

Analysis (*a*) The theoretical combustion equation in this case can be written as

$$C_4H_{10} + a_{th}[O_2 + 3.76N_2] \longrightarrow 4CO_2 + 5H_2O + 3.76a_{th}N_2$$

where a_{th} is the stoichiometric coefficient for air. It is determined from

O_2 balance: $\quad a_{th} = 4 + 2.5 \longrightarrow a_{th} = 6.5$

The air-fuel ratio for the theoretical reaction is determined by taking the ratio of the mass of the air to the mass of the fuel for,

$$AF_{th} = \frac{m_{air,th}}{m_{fuel}} = \frac{(6.5 \times 4.76 \text{ lbmol})(29 \text{ lbm/lbmol})}{(4 \text{ lbmol})(12 \text{ lbm/lbmol}) + (5 \text{ lbmol})(2 \text{ lbm/lbmol})} = 15.5 \text{ lbmair/lbmfuel}$$

The actual air-fuel ratio used is

$$AF_{act} = \frac{m_{air}}{m_{fuel}} = \frac{25 \text{ lbm}}{1 \text{ lbm}} = 25 \text{ lbm air / lbm fuel}$$

Then the percent theoretical air used can be determined from

$$\text{Percent theoretical air} = \frac{AF_{act}}{AF_{th}} = \frac{25 \text{ lbm air / lbm fuel}}{15.5 \text{ lbm air / lbm fuel}} = \mathbf{161\%}$$

(*b*) The combustion is complete, and thus products will contain only CO_2, H_2O, O_2 and N_2. The air-fuel ratio for this combustion process on a mole basis is

$$\overline{AF} = \frac{N_{air}}{N_{fuel}} = \frac{m_{air} / M_{air}}{m_{fuel} / M_{fuel}} = \frac{(25 \text{ lbm})/(29 \text{ lbm/lbmol})}{(1 \text{ lbm})/(58 \text{ lbm/lbmol})} = 50 \text{ lbmolair/lbmolfuel}$$

Thus the combustion equation in this case can be written as

$$C_4H_{10} + (50/4.76)[O_2 + 3.76N_2] \longrightarrow 4CO_2 + 5H_2O + 4O_2 + 39.5N_2$$

The dew-point temperature of a gas-vapor mixture is the saturation temperature of the water vapor in the product gases corresponding to its partial pressure. That is,

$$P_v = \left(\frac{N_v}{N_{prod}}\right)P_{prod} = \left(\frac{5 \text{ lbmol}}{52.5 \text{ lbmol}}\right)(14.7 \text{ psia}) = 1.4 \text{ psia}$$

Thus,

$$T_{dp} = T_{sat@1.4 \text{ psia}} = \mathbf{113.0°F}$$

14-24 The volumetric fractions of the constituents of a certain natural gas are given. The AF ratio is to be determined if this gas is burned with the stoichiometric amount of dry air.

Assumptions **1** Combustion is complete. **2** The combustion products contain CO_2, H_2O, and N_2 only.

Properties The molar masses of C, H_2, N_2, O_2, and air are 12 kg/kmol, 2 kg/kmol, 28 kg/kmol, 32 kg/kmol, and 29 kg/kmol, respectively (Table A-1).

Analysis Considering 1 kmol of fuel, the combustion equation can be written as

$$(0.65CH_4 + 0.08H_2 + 0.18N_2 + 0.03O_2 + 0.06CO_2) + a_{th}(O_2 + 3.76N_2) \longrightarrow xCO_2 + yH_2O + zN_2$$

The unknown coefficients in the above equation are determined from mass balances,

$$\begin{aligned}
&C: && 0.65 + 0.06 = x && \longrightarrow x = 0.71 \\
&H: && 0.65 \times 4 + 0.08 \times 2 = 2y && \longrightarrow y = 1.38 \\
&O_2: && 0.03 + 0.06 + a_{th} = x + y/2 && \longrightarrow a_{th} = 1.31 \\
&N_2: && 0.18 + 3.76a_{th} = z && \longrightarrow z = 5.106
\end{aligned}$$

Thus,

$$\begin{aligned}
&(0.65CH_4 + 0.08H_2 + 0.18N_2 + 0.03O_2 + 0.06CO_2) + 1.31(O_2 + 3.76N_2) \\
&\qquad \longrightarrow 0.71CO_2 + 1.38H_2O + 5.106N_2
\end{aligned}$$

The air-fuel ratio for the this reaction is determined by taking the ratio of the mass of the air to the mass of the fuel,

$$m_{air} = (1.31 \times 4.76 \text{ kmol})(29 \text{ kg/kmol}) = 180.8 \text{ kg}$$
$$m_{fuel} = (0.65 \times 16 + 0.08 \times 2 + 0.18 \times 28 + 0.03 \times 32 + 0.06 \times 44)\text{kg} = 19.2 \text{ kg}$$

and

$$AF_{th} = \frac{m_{air,\,th}}{m_{fuel}} = \frac{180.8 \text{ kg}}{19.2 \text{ kg}} = \mathbf{9.42 \text{ kg air / kg fuel}}$$

14-25 The composition of a certain natural gas is given. The gas is burned with stoichiometric amount of moist air. The AF ratio is to be determined.

Assumptions **1** Combustion is complete. **2** The combustion products contain CO_2, H_2O, and N_2 only.

Properties The molar masses of C, H_2, N_2, O_2, and air are 12 kg/kmol, 2 kg/kmol, 28 kg/kmol, 32 kg/kmol, and 29 kg/kmol, respectively (Table A-1).

Analysis The fuel is burned completely with the stoichiometric amount of air, and thus the products will contain only H_2O, CO_2 and N_2, but no free O_2. The moisture in the air does not react with anything; it simply shows up as additional H_2O in the products. Therefore, we can simply balance the combustion equation using dry air, and then add the moisture to both sides of the equation. Considering 1 kmol of fuel, the combustion equation can be written as

$$(0.65CH_4 + 0.08H_2 + 0.18N_2 + 0.03O_2 + 0.06CO_2) + a_{th}(O_2 + 3.76N_2) \longrightarrow xCO_2 + yH_2O + zN_2$$

The unknown coefficients in the above equation are determined from mass balances,

$$\begin{aligned}
&C: && 0.65 + 0.06 = x && \longrightarrow x = 0.71 \\
&H: && 0.65\times 4 + 0.08\times 2 = 2y && \longrightarrow y = 1.38 \\
&O_2: && 0.03 + 0.06 + a_{th} = x + y/2 && \longrightarrow a_{th} = 1.31 \\
&N_2: && 0.18 + 3.76a_{th} = z && \longrightarrow z = 5.106
\end{aligned}$$

Thus,

$$(0.65CH_4 + 0.08H_2 + 0.18N_2 + 0.03O_2 + 0.06CO_2) + 1.31(O_2 + 3.76N_2) \longrightarrow 0.71CO_2 + 1.38H_2O + 5.106N_2$$

Next we determine the amount of moisture that accompanies $4.76a_{th} = (4.76)(1.31) = 6.24$ kmol of dry air. The partial pressure of the moisture in the air is

$$P_{v,\text{in}} = \phi_{\text{air}} P_{\text{sat @ 25°C}} = (0.85)(3.169 \text{ kPa}) = 2.694 \text{ kPa}$$

Assuming ideal gas behavior, the number of moles of the moisture in the air ($N_{v,\text{in}}$) is determined to be

$$N_{v,\text{in}} = \left(\frac{P_{v,\text{in}}}{P_{\text{total}}}\right) N_{\text{total}} = \left(\frac{2.694 \text{ kPa}}{101.325 \text{ kPa}}\right)\left(6.24 + N_{v,\text{in}}\right) \longrightarrow N_{v,\text{air}} = 0.17 \text{ kmol}$$

The balanced combustion equation is obtained by substituting the coefficients determined earlier and adding 0.17 kmol of H_2O to both sides of the equation,

$$(0.65CH_4 + 0.08H_2 + 0.18N_2 + 0.03O_2 + 0.06CO_2) + 1.31(O_2 + 3.76N_2) + 0.17H_2O \longrightarrow 0.71CO_2 + 1.55H_2O + 5.106N_2$$

The air-fuel ratio for the this reaction is determined by taking the ratio of the mass of the air to the mass of the fuel,

$$m_{\text{air}} = (1.31 \times 4.76 \text{ kmol})(29 \text{ kg/kmol}) + (0.17 \text{ kmol} \times 18 \text{ kg/kmol}) = 183.9 \text{ kg}$$

$$m_{\text{fuel}} = (0.65 \times 16 + 0.08 \times 2 + 0.18 \times 28 + 0.03 \times 32 + 0.06 \times 44)\text{kg} = 19.2 \text{ kg}$$

and

$$\text{AF}_{\text{th}} = \frac{m_{\text{air, th}}}{m_{\text{fuel}}} = \frac{183.9 \text{ kg}}{19.2 \text{ kg}} = \mathbf{9.58 \text{ kg air / kg fuel}}$$

14-26 The composition of a gaseous fuel is given. It is burned with 130 percent theoretical air. The AF ratio and the fraction of water vapor that would condense if the product gases were cooled are to be determined.

Assumptions **1** Combustion is complete. **2** The combustion products contain CO_2, H_2O, O_2, and N_2 only.

Properties The molar masses of C, H_2, N_2, and air are 12 kg/kmol, 2 kg/kmol, 28 kg/kmol, and 29 kg/kmol, respectively (Table A-1).

Analysis (*a*) The fuel is burned completely with excess air, and thus the products will contain H_2O, CO_2, N_2, and some free O_2. Considering 1 kmol of fuel, the combustion equation can be written as

$$(0.60CH_4 + 0.30H_2 + 0.10N_2) + 1.3a_{th}(O_2 + 3.76N_2) \longrightarrow xCO_2 + yH_2O + 0.3a_{th}O_2 + zN_2$$

The unknown coefficients in the above equation are determined from mass balances,

$$\text{C:} \quad 0.60 = x \longrightarrow x = 0.60$$

$$\text{H:} \quad 0.60 \times 4 + 0.30 \times 2 = 2y \longrightarrow y = 1.50$$

$$O_2\text{:} \quad 1.3a_{th} = x + y/2 + 0.3a_{th} \longrightarrow a_{th} = 1.35$$

$$N_2\text{:} \quad 0.10 + 3.76 \times 1.3a_{th} = z \longrightarrow z = 6.70$$

Thus,

$$(0.60CH_4 + 0.30H_2 + 0.10N_2) + 1.755(O_2 + 3.76N_2) \longrightarrow 0.6CO_2 + 1.5H_2O + 0.405O_2 + 6.7N_2$$

The air-fuel ratio for the this reaction is determined by taking the ratio of the mass of the air to the mass of the fuel,

$$m_{air} = (1.755 \times 4.76 \text{ kmol})(29 \text{ kg/kmol}) = 242.3 \text{ kg}$$

$$m_{fuel} = (0.6 \times 16 + 0.3 \times 2 + 0.1 \times 28)\text{kg} = 13.0 \text{ kg}$$

and

$$AF_{th} = \frac{m_{air,\,th}}{m_{fuel}} = \frac{242.3 \text{ kg}}{13.0 \text{ kg}} = \mathbf{18.6\ kg\ air / kg\ fuel}$$

(*b*) For each kmol of fuel burned, 0.6 + 1.5 + 0.405 + 6.7 = 9.205 kmol of products are formed, including 1.5 kmol of H_2O. Assuming that the dew-point temperature of the products is above 20°C, some of the water vapor will condense as the products are cooled to 20°C. If N_w kmol of H_2O condenses, there will be 1.5 - N_w kmol of water vapor left in the products. The mole number of the products in the gas phase will also decrease to 9.205 - N_w as a result. Treating the product gases (including the remaining water vapor) as ideal gases, N_w is determined by equating the mole fraction of the water vapor to its pressure fraction,

$$\frac{N_v}{N_{prod.\,gas}} = \frac{P_v}{P_{prod}} \longrightarrow \frac{1.5 - N_w}{9.205 - N_w} = \frac{2.339 \text{ kPa}}{101.325 \text{ kPa}} \longrightarrow N_w = 1.32 \text{ kmol}$$

since $P_v = P_{sat\,@\,20°C}$ = 2.339 kPa. Thus the fraction of water vapor that condenses is 1.32/1.5 = 0.88 or **88%**.

14-27 EES solution of this (and other comprehensive problems designated with the *computer icon*) is available to instructors at the *Instructor Manual* section of the *Online Learning Center* (OLC) at www.mhhe.com/cengel-boles. See the Preface for access information.

14-28 The composition of a certain coal is given. The coal is burned with 50 percent excess air. The AF ratio is to be determined.

Assumptions **1** Combustion is complete. **2** The combustion products contain CO_2, H_2O, O_2, N_2, and ash only.

Properties The molar masses of C, H_2, O_2, and air are 12 kg/kmol, 2 kg/kmol, 32 kg/kmol, and 29 kg/kmol, respectively (Table A-1).

Analysis The composition of the coal is given on a mass basis, but we need to know the composition on a mole basis to balance the combustion equation. Considering 1 kg of coal, the numbers of mole of the each component are determined to be

$$N_C = (m/M)_C = 0.82/12 = 0.0683\text{kmol}$$
$$N_{H_2O} = (m/M)_{H_2O} = 0.05/18 = 0.0028\text{kmol}$$
$$N_{H_2} = (m/M)_{H_2} = 0.02/2 = 0.01\text{kmol}$$
$$N_{O_2} = (m/M)_{O_2} = 0.01/32 = 0.00031\text{kmol}$$

Considering 1 kg of coal, the combustion equation can be written as

$$(0.0683C + 0.0028H_2O + 0.01H_2 + 0.00031O_2 + \text{ash}) + 1.5a_{th}(O_2 + 3.76N_2)$$
$$\longrightarrow xCO_2 + yH_2O + 0.5a_{th}O_2 + 1.5\times 3.76a_{th}N_2 + \text{ash}$$

The unknown coefficients in the above equation are determined from mass balances,

$$\text{C:} \quad 0.0683 = x \longrightarrow x = 0.0683$$
$$\text{H:} \quad 0.0028\times 2 + 0.01\times 2 = 2y \longrightarrow y = 0.0128$$
$$O_2\text{:} \quad 0.00031 + 1.5a_{th} = x + y/2 + 0.5a_{th} \longrightarrow a_{th} = 0.0744$$

Thus,

$$(0.0683C + 0.0028H_2O + 0.01H_2 + 0.00031O_2 + \text{ash}) + 0.1116(O_2 + 3.76N_2)$$
$$\longrightarrow 0.0683CO_2 + 0.0128H_2O + 0.0372O_2 + 0.4196N_2 + \text{ash}$$

The air-fuel ratio for the this reaction is determined by taking the ratio of the mass of the air to the mass of the coal, which is taken to be 1 kg,

$$m_{air} = (0.1116\times 4.76\text{ kmol})(29\text{ kg/kmol}) = 15.4\text{ kg}$$
$$m_{fuel} = 1\text{ kg}$$

and

$$\text{AF}_{th} = \frac{m_{air,th}}{m_{fuel}} = \frac{15.4\text{ kg}}{1\text{ kg}} = \mathbf{15.4\text{ kg air/kg fuel}}$$

14-29 Octane is burned with dry air. The volumetric fractions of the products are given. The AF ratio and the percentage of theoretical air used are to be determined.

Assumptions **1** Combustion is complete. **2** The combustion products contain CO_2, CO, H_2O, O_2, and N_2 only.

Properties The molar masses of C, H_2, and air are 12 kg/kmol, 2 kg/kmol, and 29 kg/kmol, respectively (Table A-1).

Analysis Considering 100 kmol of dry products, the combustion equation can be written as

$$xC_8H_{18} + a[O_2 + 3.76N_2] \longrightarrow 9.21CO_2 + 0.61CO + 7.06O_2 + 83.12N_2 + bH_2O$$

The unknown coefficients *x*, *a*, and *b* are determined from mass balances,

$$N_2: \quad 3.76a = 83.12 \longrightarrow a = 22.11$$

$$C: \quad 8x = 9.21 + 0.61 \longrightarrow x = 1.23$$

$$H: \quad 18x = 2b \longrightarrow b = 11.05$$

$$(\text{Check } O_2: \quad a = 9.21 + 0.305 + 7.06 + b/2 \longrightarrow 22.11 \cong 22.10)$$

Thus,

$$1.23C_8H_{18} + 22.11[O_2 + 3.76N_2] \longrightarrow 9.21CO_2 + 0.61CO + 7.06O_2 + 83.12N_2 + 11.05H_2O$$

The combustion equation for 1 kmol of fuel is obtained by dividing the above equation by 1.23,

$$C_8H_{18} + 18.0[O_2 + 3.76N_2] \longrightarrow 7.50CO_2 + 0.50CO + 5.75O_2 + 67.68N_2 + 9H_2O$$

(*a*) The air-fuel ratio is determined by taking the ratio of the mass of the air to the mass of the fuel,

$$AF_{th} = \frac{m_{air,th}}{m_{fuel}} = \frac{(18.0 \times 4.76\ \text{kmol})(29\ \text{kg/kmol})}{(8\ \text{kmol})(12\ \text{kg/kmol}) + (9\ \text{kmol})(2\ \text{kg/kmol})} = \mathbf{21.8\ kg\ air/kg\ fuel}$$

(*b*) To find the percent theoretical air used, we need to know the theoretical amount of air, which is determined from the theoretical combustion equation of the fuel,

$$C_8H_{18} + a_{th}[O_2 + 3.76N_2] \longrightarrow 8CO_2 + 9H_2O + 3.76a_{th}N_2$$

$$O_2: \quad a_{th} = 8 + 4.5 \longrightarrow a_{th} = 12.5$$

Then,

$$\text{Percent theoretical air} = \frac{m_{air,act}}{m_{air,th}} = \frac{N_{air,act}}{N_{air,th}} = \frac{(18.0)(4.76)\text{kmol}}{(12.5)(4.76)\text{kmol}} = \mathbf{144\%}$$

14-30 Carbon is burned with dry air. The volumetric analysis of the products is given. The AF ratio and the percentage of theoretical air used are to be determined.

Assumptions **1** Combustion is complete. **2** The combustion products contain CO_2, CO, O_2, and N_2 only.

Properties The molar masses of C, H_2, and air are 12 kg/kmol, 2 kg/kmol, and 29 kg/kmol, respectively (Table A-1).

Analysis Considering 100 kmol of dry products, the combustion equation can be written as

$$xC + a[O_2 + 3.76N_2] \longrightarrow 10.06CO_2 + 0.42CO + 10.69O_2 + 78.83N_2$$

The unknown coefficients x and a are determined from mass balances,

$$N_2: \quad 3.76a = 78.83 \longrightarrow a = 20.965$$

$$C: \quad x = 10.06 + 0.42 \longrightarrow x = 10.48$$

$$(\text{Check } O_2: \quad a = 10.06 + 0.21 + 10.69 \longrightarrow 20.96 = 20.96)$$

Thus,

$$10.48C + 20.96[O_2 + 3.76N_2] \longrightarrow 10.06CO_2 + 0.42CO + 10.69O_2 + 78.83N_2$$

The combustion equation for 1 kmol of fuel is obtained by dividing the above equation by 10.48,

$$C + 2.0[O_2 + 3.76N_2] \longrightarrow 0.96CO_2 + 0.04CO + 1.02O_2 + 7.52N_2$$

(a) The air-fuel ratio is determined by taking the ratio of the mass of the air to the mass of the fuel,

$$AF_{th} = \frac{m_{air,th}}{m_{fuel}} = \frac{(2.0 \times 4.76\text{kmol})(29\text{kg/kmol})}{(1\text{kmol})(12\text{kg/kmol})} = \mathbf{23.0 \ kgair/kgfuel}$$

(b) To find the percent theoretical air used, we need to know the theoretical amount of air, which is determined from the theoretical combustion equation of the fuel,

$$C + 1[O_2 + 3.76N_2] \longrightarrow CO_2 + 3.76N_2$$

Then,

$$\text{Percent theoretical air} = \frac{m_{air,act}}{m_{air,th}} = \frac{N_{air,act}}{N_{air,th}} = \frac{(2.0)(4.76)\text{kmol}}{(1.0)(4.76)\text{kmol}} = \mathbf{200\%}$$

14-31 Methane is burned with dry air. The volumetric analysis of the products is given. The AF ratio and the percentage of theoretical air used are to be determined.

Assumptions **1** Combustion is complete. **2** The combustion products contain CO_2, CO, H_2O, O_2, and N_2 only.

Properties The molar masses of C, H_2, and air are 12 kg/kmol, 2 kg/kmol, and 29 kg/kmol, respectively (Table A-1).

Analysis Considering 100 kmol of dry products, the combustion equation can be written as

$$x\text{CH}_4 + a[\text{O}_2 + 3.76\text{N}_2] \longrightarrow 5.20\text{CO}_2 + 0.33\text{CO} + 11.24\text{O}_2 + 83.23\text{N}_2 + b\text{H}_2\text{O}$$

The unknown coefficients x, a, and b are determined from mass balances,

$$\text{N}_2: \quad 3.76a = 83.23 \longrightarrow a = 22.14$$

$$\text{C}: \quad x = 5.20 + 0.33 \longrightarrow x = 5.53$$

$$\text{H}: \quad 4x = 2b \longrightarrow b = 11.06$$

$$(\text{Check O}_2: \quad a = 5.20 + 0.165 + 11.24 + b/2 \longrightarrow 22.14 = 22.14)$$

Thus,

$$5.53\text{CH}_4 + 22.14[\text{O}_2 + 3.76\text{N}_2] \longrightarrow 5.20\text{CO}_2 + 0.33\text{CO} + 11.24\text{O}_2 + 83.23\text{N}_2 + 11.06\text{H}_2\text{O}$$

The combustion equation for 1 kmol of fuel is obtained by dividing the above equation by 5.53,

$$\text{CH}_4 + 4.0[\text{O}_2 + 3.76\text{N}_2] \longrightarrow 0.94\text{CO}_2 + 0.06\text{CO} + 2.03\text{O}_2 + 15.05\text{N}_2 + 2\text{H}_2\text{O}$$

(*a*) The air-fuel ratio is determined from its definition,

$$\text{AF}_{\text{th}} = \frac{m_{\text{air,th}}}{m_{\text{fuel}}} = \frac{(4.0 \times 4.76\text{ kmol})(29\text{ kg/kmol})}{(1\text{ kmol})(12\text{ kg/kmol}) + (2\text{ kmol})(2\text{ kg/kmol})} = \mathbf{34.5\ kgair/kgfuel}$$

(*b*) To find the percent theoretical air used, we need to know the theoretical amount of air, which is determined from the theoretical combustion equation of the fuel,

$$\text{CH}_4 + a_{\text{th}}[\text{O}_2 + 3.76\text{N}_2] \longrightarrow \text{CO}_2 + 2\text{H}_2\text{O} + 3.76a_{\text{th}}\text{N}_2$$

$$\text{O}_2: \quad a_{\text{th}} = 1 + 1 \longrightarrow a_{\text{th}} = 2.0$$

Then,

$$\text{Percent theoretical air} = \frac{m_{\text{air,act}}}{m_{\text{air,th}}} = \frac{N_{\text{air,act}}}{N_{\text{air,th}}} = \frac{(4.0)(4.76)\text{kmol}}{(2.0)(4.76)\text{kmol}} = \mathbf{200\%}$$

Enthalpy of Formation and Enthalpy of Combustion

14-32C For combustion processes the enthalpy of reaction is referred to as the enthalpy of combustion, which represents the amount of heat released during a steady-flow combustion process.

14-33C Enthalpy of formation is the enthalpy of a substance due to its chemical composition. The enthalpy of formation is related to elements or compounds whereas the enthalpy of combustion is related to a particular fuel.

14-34C The heating value is called the higher heating value when the H_2O in the products is in the liquid form, and it is called the lower heating value when the H_2O in the products is in the vapor form. The heating value of a fuel is equal to the absolute value of the enthalpy of combustion of that fuel.

14-35C If the combustion of a fuel results in a single compound, the enthalpy of formation of that compound is identical to the enthalpy of combustion of that fuel.

14-36C Yes.

14-37C No. The enthalpy of formation of N_2 is simply assigned a value of zero at the standard reference state for convenience.

14-38C 1 kmol of H_2. This is evident from the observation that when chemical bonds of H_2 are destroyed to form H_2O a large amount of energy is released.

14-39 The enthalpy of combustion of methane at a 25°C and 1 atm is to be determined using the data from Table A-26 and to be compared to the value listed in Table A-27.

Assumptions The water in the products is in the liquid phase.

Analysis The stoichiometric equation for this reaction is

$$CH_4 + 2[O_2 + 3.76N_2] \longrightarrow CO_2 + 2H_2O(\ell) + 7.52N_2$$

Both the reactants and the products are at the standard reference state of 25°C and 1 atm. Also, N_2 and O_2 are stable elements, and thus their enthalpy of formation is zero. Then the enthalpy of combustion of CH_4 becomes

$$h_C = H_P - H_R = \sum N_P \bar{h}^\circ_{f,P} - \sum N_R \bar{h}^\circ_{f,R} = \left(N\bar{h}^\circ_f\right)_{CO_2} + \left(N\bar{h}^\circ_f\right)_{H_2O} - \left(N\bar{h}^\circ_f\right)_{C_8H_{18}}$$

Using $\bar{h}^\circ_f$ values from Table A-26,

$$\begin{aligned} h_C &= (1\text{kmol})(-393{,}520\text{kJ/kmol}) + (2\text{kmol})(-285{,}830\text{kJ/kmol}) \\ &\quad - (1\text{kmol})(-74{,}850\text{kJ/kmol}) \\ &= \mathbf{-890{,}330kJ}\ (\text{perkmolCH}_4) \end{aligned}$$

The listed value in Table A-27 is -890,360 kJ/kmol, which is almost identical to the calculated value. Since the water in the products is assumed to be in the liquid phase, this h_c value corresponds to the higher heating value of CH_4.

14-40 EES solution of this (and other comprehensive problems designated with the *computer icon*) is available to instructors at the *Instructor Manual* section of the *Online Learning Center* (OLC) at www.mhhe.com/cengel-boles. See the Preface for access information.

14-41 The enthalpy of combustion of gaseous ethane at a 25°C and 1 atm is to be determined using the data from Table A-26 and to be compared to the value listed in Table A-27.

Assumptions The water in the products is in the liquid phase.

Analysis The stoichiometric equation for this reaction is

$$C_2H_6 + 3.5[O_2 + 3.76N_2] \longrightarrow 2CO_2 + 3H_2O(\ell) + 13.16N_2$$

Both the reactants and the products are at the standard reference state of 25°C and 1 atm. Also, N_2 and O_2 are stable elements, and thus their enthalpy of formation is zero. Then the enthalpy of combustion of C_2H_6 becomes

$$h_C = H_P - H_R = \sum N_P \bar{h}^\circ_{f,P} - \sum N_R \bar{h}^\circ_{f,R} = \left(N\bar{h}^\circ_f\right)_{CO_2} + \left(N\bar{h}^\circ_f\right)_{H_2O} - \left(N\bar{h}^\circ_f\right)_{C_8H_{18}}$$

Using $\bar{h}^\circ_f$ values from Table A-26,

$$\begin{aligned} h_C &= (2\text{ kmol})(-393{,}520\text{ kJ/kmol}) + (3\text{ kmol})(-285{,}830\text{ kJ/kmol}) \\ &\quad - (1\text{ kmol})(-84{,}680\text{ kJ/kmol}) \\ &= \mathbf{-1{,}559{,}850\ \ kJ}\ (\text{per kmolC}_2\text{H}_6) \end{aligned}$$

The listed value in Table A-27 is -1,559,900 kJ/kmol, which is almost identical to the calculated value. Since the water in the products is assumed to be in the liquid phase, this h_c value corresponds to the higher heating value of C_2H_6.

14-42 The enthalpy of combustion of liquid octane at a 25°C and 1 atm is to be determined using the data from Table A-26 and to be compared to the value listed in Table A-27.

Assumptions The water in the products is in the liquid phase.

Analysis The stoichiometric equation for this reaction is

$$C_8H_{18} + 12.5[O_2 + 3.76N_2] \longrightarrow 8CO_2 + 9H_2O(\ell) + 47N_2$$

Both the reactants and the products are at the standard reference state of 25°C and 1 atm. Also, N_2 and O_2 are stable elements, and thus their enthalpy of formation is zero. Then the enthalpy of combustion of C_8H_{18} becomes

$$h_C = H_P - H_R = \sum N_P \bar{h}^\circ_{f,P} - \sum N_R \bar{h}^\circ_{f,R} = \left(N\bar{h}^\circ_f\right)_{CO_2} + \left(N\bar{h}^\circ_f\right)_{H_2O} - \left(N\bar{h}^\circ_f\right)_{C_8H_{18}}$$

Using $\bar{h}^\circ_f$ values from Table A-26,

$$\begin{aligned} h_C &= (8\text{kmol})(-393{,}520\text{kJ/kmol}) + (9\text{kmol})(-285{,}830\text{kJ/kmol}) \\ &\quad - (1\text{kmol})(-249{,}950\text{kJ/kmol}) \\ &= \mathbf{-5{,}470{,}680\ kJ} \end{aligned}$$

The listed value in Table A-27 is -5,512,200 kJ/kmol for gaseous octane. The h_c value for liquid octane is obtained by adding $\bar{h}_{fg}$ = 41,460 kJ/kmol to it, which yields -5,470,740 kJ. Thus the two values are practically identical. Since the water in the products is assumed to be in the liquid phase, this h_c value corresponds to the higher heating value of C_8H_{18}.

First Law Analysis of Reacting Systems

14-43C In this case $\Delta U + W_b = \Delta H$, and the conservation of energy relation reduces to the form of the steady-flow energy relation.

14-44C The heat transfer will be the same for all cases. The excess oxygen and nitrogen enters and leaves the combustion chamber at the same state, and thus has no effect on the energy balance.

14-45C For case (*b*), which contains the maximum amount of nonreacting gases. This is because part of the chemical energy released in the combustion chamber is absorbed and transported out by the nonreacting gases.

14-46 Methane is burned completely during a steady-flow combustion process. The heat transfer from the combustion chamber is to be determined for two cases.

Assumptions **1** Steady operating conditions exist. **2** Air and combustion gases are ideal gases. **3** Kinetic and potential energies are negligible. **4** Combustion is complete.

Analysis The fuel is burned completely with the stoichiometric amount of air, and thus the products will contain only H_2O, CO_2 and N_2, but no free O_2. Considering 1 kmol of fuel, the theoretical combustion equation can be written as

$$CH_4 + a_{th}(O_2 + 3.76N_2) \longrightarrow CO_2 + 2H_2O + 3.76a_{th}N_2$$

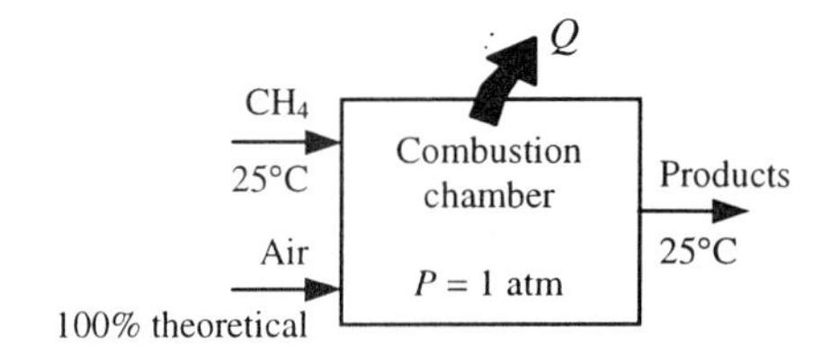

where a_{th} is determined from the O_2 balance,

$$a_{th} = 1 + 1 = 2$$

Substituting,

$$CH_4 + 2(O_2 + 3.76N_2) \longrightarrow CO_2 + 2H_2O + 5.64N_2$$

The heat transfer for this combustion process is determined from the energy balance $E_{in} - E_{out} = \Delta E_{system}$ applied on the combustion chamber with $W = 0$. It reduces to

$$-Q_{out} = \sum N_P\left(\bar{h}_f^\circ + \bar{h} - \bar{h}^\circ\right)_P - \sum N_R\left(\bar{h}_f^\circ + \bar{h} - \bar{h}^\circ\right)_R \longrightarrow -Q_{out} = \sum N_P \bar{h}_{f,P}^\circ - \sum N_R \bar{h}_{f,R}^\circ$$

since both the reactants and the products are at 25°C and both the air and the combustion gases can be treated as ideal gases. From the tables,

Substance	$\bar{h}_f^\circ$ **kJ/kmol**
CH_4	-74,850
O_2	0
N_2	0
$H_2O\ (\ell)$	-285,830
CO_2	-393,520

Thus,

$$-Q_{out} = (1)(-393{,}520) + (2)(-285{,}830) + 0 - (1)(-74{,}850) - 0 - 0 = -890{,}330 \text{ kJ / kmol CH}_4$$

or

$$Q_{out} = \mathbf{890{,}330\ kJ / kmol\ CH_4}$$

If combustion is achieved with 100% excess air, the answer would still be the same since it would enter and leave at 25°C, and absorb no energy.

14-47 Hydrogen is burned completely during a steady-flow combustion process. The heat transfer from the combustion chamber is to be determined for two cases.

Assumptions **1** Steady operating conditions exist. **2** Air and combustion gases are ideal gases. **3** Kinetic and potential energies are negligible. **4** Combustion is complete.

Analysis The H_2 is burned completely with the stoichiometric amount of air, and thus the products will contain only H_2O and N_2, but no free O_2. Considering 1 kmol of H_2, the theoretical combustion equation can be written as

$$H_2 + a_{th}(O_2 + 3.76N_2) \longrightarrow H_2O + 3.76a_{th}N_2$$

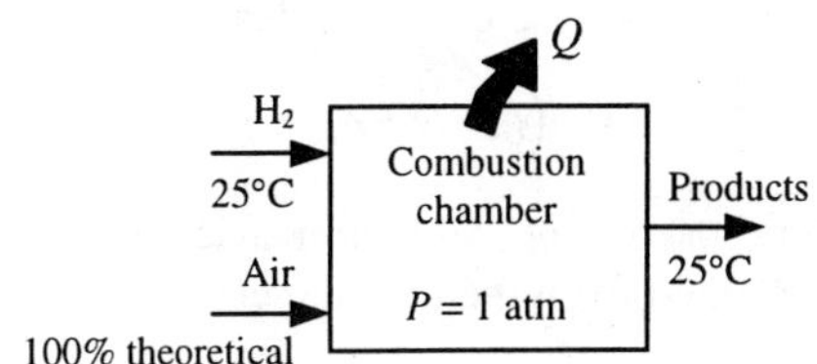

where a_{th} is determined from the O_2 balance to be $a_{th} = 0.5$. Substituting,

$$H_2 + 0.5(O_2 + 3.76N_2) \longrightarrow H_2O + 1.88N_2$$

The heat transfer for this combustion process is determined from the energy balance $E_{in} - E_{out} = \Delta E_{system}$ applied on the combustion chamber with $W = 0$. It reduces to

$$-Q_{out} = \sum N_P\left(\bar{h}_f^\circ + \bar{h} - \bar{h}^\circ\right)_P - \sum N_R\left(\bar{h}_f^\circ + \bar{h} - \bar{h}^\circ\right)_R \longrightarrow -Q_{out} = \sum N_P \bar{h}_{f,P}^\circ - \sum N_R \bar{h}_{f,R}^\circ$$

since both the reactants and the products are at 25°C and both the air and the combustion gases can be treated as ideal gases. From the tables,

Substance	$\bar{h}_f^\circ$ **kJ/kmol**
H_2	0
O_2	0
N_2	0
H_2O (ℓ)	-285,830

Substituting,

$$-Q_{out} = (1)(-285{,}830) + 0 - 0 - 0 - 0 = -285{,}830 \text{ kJ/kmol } H_2$$

or

$$Q_{out} = \mathbf{285{,}830 \text{ kJ/kmol } H_2}$$

If combustion is achieved with 80% excess air, the answer would still be the same since it would enter and leave at 25°C, and absorb no energy.

14-48 Liquid propane is burned with 150 percent excess air during a steady-flow combustion process. The mass flow rate of air and the rate of heat transfer from the combustion chamber are to be determined.

Assumptions **1** Steady operating conditions exist. **2** Air and combustion gases are ideal gases. **3** Kinetic and potential energies are negligible. **4** Combustion is complete.

Properties The molar masses of propane and air are 44 kg/kmol and 29 kg/kmol, respectively (Table A-1).

Analysis The fuel is burned completely with excess air, and thus the products will contain only CO_2, H_2O, N_2, and some free O_2. Considering 1 kmol of C_3H_8, the combustion equation can be written as

$$C_3H_8(\ell) + 2.5a_{th}(O_2 + 3.76N_2) \longrightarrow 3CO_2 + 4H_2O + 1.5a_{th}O_2 + (2.5)(3.76a_{th})N_2$$

where a_{th} is the stoichiometric coefficient and is determined from the O_2 balance,

$$2.5a_{th} = 3 + 2 + 1.5a_{th} \longrightarrow a_{th} = 5$$

C_3H_8 25°C | Air 12°C | Combustion chamber | P = 1 atm | $\dot{Q}$ | Products 1200 K

Thus,

$$C_3H_8(\ell) + 12.5(O_2 + 3.76N_2) \longrightarrow 3CO_2 + 4H_2O + 7.5O_2 + 47N_2$$

(a) The air-fuel ratio for this combustion process is

$$\mathrm{AF} = \frac{m_{air}}{m_{fuel}} = \frac{(12.5 \times 4.76\,\mathrm{kmol})(29\,\mathrm{kg/kmol})}{(3\,\mathrm{kmol})(12\,\mathrm{kg/kmol}) + (4\,\mathrm{kmol})(2\,\mathrm{kg/kmol})} = 39.2\,\mathrm{kg\,air/kg\,fuel}$$

Thus, $\dot{m}_{air} = (\mathrm{AF})(\dot{m}_{fuel}) = (39.2\,\mathrm{kg\,air/kg\,fuel})(1.2\,\mathrm{kg\,fuel/min}) = \mathbf{47.1\,kg\,air/min}$

(b) The heat transfer for this combustion process is determined from the energy balance $E_{in} - E_{out} = \Delta E_{system}$ applied on the combustion chamber with $W = 0$. It reduces to

$$-Q_{out} = \sum N_P\left(\bar{h}_f^\circ + \bar{h} - \bar{h}^\circ\right)_P - \sum N_R\left(\bar{h}_f^\circ + \bar{h} - \bar{h}^\circ\right)_R$$

Assuming the air and the combustion products to be ideal gases, we have h = h(T). From the tables,

Substance	$\bar{h}_f^\circ$ kJ/kmol	$\bar{h}_{285\,K}$ kJ/kmol	$\bar{h}_{298\,K}$ kJ/kmol	$\bar{h}_{1200\,K}$ kJ/kmol
C_3H_8 (ℓ)	-118,910	---	---	---
O_2	0	8296.5	8682	38,447
N_2	0	8286.5	8669	36,777
H_2O (g)	-241,820	---	9904	44,380
CO_2	-393,520	---	9364	53,848

The $\bar{h}_f^\circ$ of liquid propane is obtained by adding $\bar{h}_{fg}$ of propane at 25°C to $\bar{h}_f^\circ$ of gas propane. Substituting,

$$\begin{aligned}-Q_{out} &= (3)(-393{,}520 + 53{,}848 - 9364) + (4)(-241{,}820 + 44{,}380 - 9904) + (7.5)(0 + 38{,}447 - 8682) \\ &\quad + (47)(0 + 36{,}777 - 8669) - (1)(-118{,}910 + h_{298} - h_{298}) - (12.5)(0 + 8296.5 - 8682) \\ &\quad - (47)(0 + 8286.5 - 8669) \\ &= -190{,}464\,\mathrm{kJ/kmol\,C_3H_8}\end{aligned}$$

or

$$Q_{out} = 190{,}464\,\mathrm{kJ/kmol\,C_3H_8}$$

Then the rate of heat transfer for a mass flow rate of 0.1 kg/min for the propane becomes

$$\dot{Q}_{out} = \dot{N}Q_{out} = \left(\frac{\dot{m}}{N}\right)Q_{out} = \left(\frac{1.2\,\mathrm{kg/min}}{44\,\mathrm{kg/kmol}}\right)(190{,}464\,\mathrm{kJ/kmol}) = \mathbf{5194\ kJ/min}$$

14-49E Liquid propane is burned with 150 percent excess air during a steady-flow combustion process. The mass flow rate of air and the rate of heat transfer from the combustion chamber are to be determined.

Assumptions **1** Steady operating conditions exist. **2** Air and combustion gases are ideal gases. **3** Kinetic and potential energies are negligible. **4** Combustion is complete.

Properties The molar masses of propane and air are 44 lbm/lbmol and 29 lbm/lbmol, respectively (Table - 1E).

Analysis The fuel is burned completely with the excess air, and thus the products will contain only CO_2, H_2O, N_2, and some free O_2. Considering 1 kmol of C_3H_8, the combustion equation can be written as

$$C_3H_8(\ell)+2.5a_{th}(O_2+3.76N_2)\longrightarrow 3CO_2+4H_2O+1.5a_{th}O_2+(2.5)(3.76a_{th})N_2$$

where a_{th} is the stoichiometric coefficient and is determined from the O_2 balance,

$$2.5a_{th}=3+2+1.5a_{th}\longrightarrow a_{th}=5$$

C_3H_8 77°F → Combustion chamber P = 1 atm → Products 1800 R; Air 40°F; $\dot{Q}$

Thus,

$$C_3H_8(\ell)+12.5(O_2+3.76N_2)\longrightarrow 3CO_2+4H_2O+7.5O_2+47N_2$$

(*a*) The air-fuel ratio for this combustion process is

$$\mathrm{AF}=\frac{m_{air}}{m_{fuel}}=\frac{(12.5\times 4.76\ \text{lbmol})(29\ \text{lbm/lbmol})}{(3\ \text{lbmol})(12\ \text{lbm/lbmol})+(4\ \text{lbmol})(2\ \text{lbm/lbmol})}=39.2\ \text{lbmair/lbmfuel}$$

Thus, $\dot{m}_{air}=(\mathrm{AF})(\dot{m}_{fuel})=(39.2\ \text{lbm air/lbm fuel})(0.3\ \text{lbm fuel/min})=\mathbf{11.8\ lbm\ air\ /\ min}$

(*b*) The heat transfer for this combustion process is determined from the energy balance $E_{in}-E_{out}=\Delta E_{system}$ applied on the combustion chamber with W = 0. It reduces to

$$-Q_{out}=\sum N_P\left(\bar{h}_f^\circ+\bar{h}-\bar{h}^\circ\right)_P-\sum N_R\left(\bar{h}_f^\circ+\bar{h}-\bar{h}^\circ\right)_R$$

Assuming the air and the combustion products to be ideal gases, we have $h = h(T)$. From the tables,

Substance	$\bar{h}_f^\circ$ Btu/lbmol	$\bar{h}_{500R}$ Btu/lbmol	$\bar{h}_{537\ R}$ Btu/lbmol	$\bar{h}_{1800\ R}$ Btu/lbmol
C_3H_8 (ℓ)	-51,160	---	---	---
O_2	0	3466.2	3725.1	13,485.8
N_2	0	3472.2	3729.5	12,956.3
CO_2	-169,300	---	4027.5	18,391.5
H_2O (*g*)	-104,040	---	4258.0	15,433.0

The $\bar{h}_f^\circ$ of liquid propane is obtained by adding the $\bar{h}_{fg}$ of propane at 77°F to the $\bar{h}_f^\circ$ of gas propane. Substituting,

$$\begin{aligned}-Q_{out}&=(3)(-169{,}300+18{,}391.5-4027.5)+(4)(-104{,}040+15{,}433-4258)+(7.5)(0+13{,}485.8-3725.1)\\&\quad+(47)(0+12{,}959.3-3729.5)-(1)(-51{,}160+h_{537}-h_{537})-(12.5)(0+3466.2-3725.1)\\&\quad-(47)(0+3472.2-3729.5)\\&=-262{,}773\ \text{Btu / lbmol}\ C_3H_8\end{aligned}$$

or

$$Q_{out}=262{,}773\ \text{Btu / lbmol}\ C_3H_8$$

Then the rate of heat transfer for a mass flow rate of 0.1 kg/min for the propane becomes

$$\dot{Q}_{out}=\dot{N}Q_{out}=\left(\frac{\dot{m}}{N}\right)Q_{out}=\left(\frac{0.3\ \text{lbm/min}}{44\ \text{lbm/lbmol}}\right)(262{,}773\ \text{Btu/lbmol})=\mathbf{1792\ Btu/min}$$

14-50 Acetylene gas is burned with 20 percent excess air during a steady-flow combustion process. The AF ratio and the heat transfer are to be determined.

Assumptions **1** Steady operating conditions exist. **2** Air and combustion gases are ideal gases. **3** Kinetic and potential energies are negligible. **4** Combustion is complete.

Properties The molar masses of C_2H_2 and air are 26 kg/kmol and 29 kg/kmol, respectively (Table A-1).

Analysis The fuel is burned completely with the excess air, and thus the products will contain only CO_2, H_2O, N_2, and some free O_2. Considering 1 kmol of C_2H_2, the combustion equation can be written as

$$C_2H_2 + 1.2a_{th}(O_2 + 3.76N_2) \longrightarrow 2CO_2 + H_2O + 0.2a_{th}O_2 + (1.2)(3.76a_{th})N_2$$

where a_{th} is the stoichiometric coefficient and is determined from the O_2 balance,

$$1.2a_{th} = 2 + 0.5 + 0.2a_{th} \longrightarrow a_{th} = 2.5$$

Q

C_2H_2

25°C

Air

20% excess air

25°C

Combustion chamber

P = 1 atm

Products

1500 K

Thus,

$$C_2H_2 + 3(O_2 + 3.76N_2) \longrightarrow 2CO_2 + H_2O + 0.5O_2 + 11.28N_2$$

(*a*)
$$\text{AF} = \frac{m_{\text{air}}}{m_{\text{fuel}}} = \frac{(3 \times 4.76\text{kmol})(29\text{kg/kmol})}{(2\text{kmol})(12\text{kg/kmol}) + (1\text{kmol})(2\text{kg/kmol})} = \mathbf{15.9\ \ kgair/kgfuel}$$

(*b*) The heat transfer for this combustion process is determined from the energy balance $E_{in} - E_{out} = \Delta E_{system}$ applied on the combustion chamber with $W = 0$. It reduces to

$$-Q_{out} = \sum N_P\left(\bar{h}_f^\circ + \bar{h} - \bar{h}^\circ\right)_P - \sum N_R\left(\bar{h}_f^\circ + \bar{h} - \bar{h}^\circ\right)_R \longrightarrow -Q_{out} = \sum N_P\left(\bar{h}_f^\circ + \bar{h} - \bar{h}^\circ\right)_P - \sum N_R \bar{h}_{f,R}^\circ$$

since all of the reactants are at 25°C. Assuming the air and the combustion products to be ideal gases, we have $h = h(T)$. From the tables,

Substance	$\bar{h}_f^\circ$ kJ/kmol	$\bar{h}_{298\text{ K}}$ kJ/kmol	$\bar{h}_{1500\text{ K}}$ kJ/kmol
C_2H_2	226,730	---	---
O_2	0	8682	49,292
N_2	0	8669	47,073
H_2O (*g*)	-241,820	9904	57,999
CO_2	-393,520	9364	71,078

Thus,

$$\begin{aligned} -Q_{out} &= (2)(-393{,}520 + 71{,}078 - 9364) + (1)(-241{,}820 + 57{,}999 - 9904) + (0.5)(0 + 49{,}292 - 8682) \\ &\quad + (11.28)(0 + 47{,}073 - 8669) - (1)(226{,}730) - 0 - 0 \\ &= -630{,}565\text{kJ/kmol}C_2H_2 \end{aligned}$$

or

$$Q_{out} = \mathbf{630{,}565\ kJ/kmol\ C_2H_2}$$

14-51E Liquid octane is burned with 180 percent theoretical air during a steady-flow combustion process. The AF ratio and the heat transfer from the combustion chamber are to be determined.

Assumptions **1** Steady operating conditions exist. **2** Air and combustion gases are ideal gases. **3** Kinetic and potential energies are negligible. **4** Combustion is complete.

Properties The molar masses of C_3H_{18} and air are 54 kg/kmol and 29 kg/kmol, respectively (Table A-1).

Analysis The fuel is burned completely with the excess air, and thus the products will contain only CO_2, H_2O, N_2, and some free O_2. Considering 1 kmol of C_2H_2, the combustion equation can be written as

$$C_8H_{18}(\ell)+1.8a_{th}(O_2+3.76N_2)\longrightarrow 8CO_2+9H_2O+0.8a_{th}O_2+(1.8)(3.76a_{th})N_2$$

where a_{th} is the stoichiometric coefficient and is determined from the O_2 balance,

$$1.8a_{th}=8+4.5+0.8a_{th} \longrightarrow a_{th}=12.5$$

C_8H_{18} 77°F → Combustion chamber, P = 1 atm → Products 2500 R; Air, 80% excess air, 77°F; **Q**

Thus,

$$C_8H_{18}(\ell)+22.5(O_2+3.76N_2)\longrightarrow 8CO_2+9H_2O+10O_2+84.6N_2$$

(*a*)
$$\text{AF}=\frac{m_{\text{air}}}{m_{\text{fuel}}}=\frac{(22.5\times 4.76\text{ lbmol})(29\text{ lbm/lbmol})}{(8\text{ lbmol})(12\text{ lbm/lbmol})+(9\text{ lbmol})(2\text{ lbm/lbmol})}=\mathbf{27.2\ lbmair/lbmfuel}$$

(*b*) The heat transfer for this combustion process is determined from the energy balance $E_{in}-E_{out}=\Delta E_{system}$ applied on the combustion chamber with W = 0. It reduces to

$$-Q_{\text{out}}=\sum N_P\left(\bar{h}_f^\circ+\bar{h}-\bar{h}^\circ\right)_P-\sum N_R\left(\bar{h}_f^\circ+\bar{h}-\bar{h}^\circ\right)_R\longrightarrow -Q_{\text{out}}=\sum N_P\left(\bar{h}_f^\circ+\bar{h}-\bar{h}^\circ\right)_P-\sum N_R\bar{h}_{f,R}^\circ$$

since all of the reactants are at 77°F. Assuming the air and the combustion products to be ideal gases, we have $h = h(T)$. From the tables,

Substance	$\bar{h}_f^\circ$ Btu/lbmol	$\bar{h}_{537\text{ R}}$ Btu/lbmol	$\bar{h}_{2500\text{ R}}$ Btu/lbmol
C_8H_{18} (ℓ)	-107,530	---	---
O_2	0	3725.1	19,443
N_2	0	3729.5	18,590
CO_2	-169,300	4027.5	27,801
H_2O (*g*)	-104,040	4258.0	22,735

Thus,

$$\begin{aligned}-Q_{\text{out}}&=(8)(-169{,}300+27{,}801-4027.5)+(9)(-104{,}040+22{,}735-4258)+(10)(0+19{,}443-3725.1)\\&\quad+(84.6)(0+18{,}590-3729.5)-(1)(-107{,}530)-0-0\\&=-412{,}372\text{ Btu/lbmol}C_8H_{18}\end{aligned}$$

or

$$Q_{\text{out}}=\mathbf{412{,}372\ Btu/lbmol\ C_8H_{18}}$$

14-52 Benzene gas is burned with 95 percent theoretical air during a steady-flow combustion process. The mole fraction of the CO in the products and the heat transfer from the combustion chamber are to be determined.

Assumptions **1** Steady operating conditions exist. **2** Air and combustion gases are ideal gases. **3** Kinetic and potential energies are negligible.

Analysis (*a*) The fuel is burned with insufficient amount of air, and thus the products will contain some CO as well as CO_2, H_2O, and N_2. The theoretical combustion equation of C_6H_6 is

$$C_6H_6 + a_{th}(O_2 + 3.76N_2) \longrightarrow 6CO_2 + 3H_2O + 3.76a_{th}N_2$$

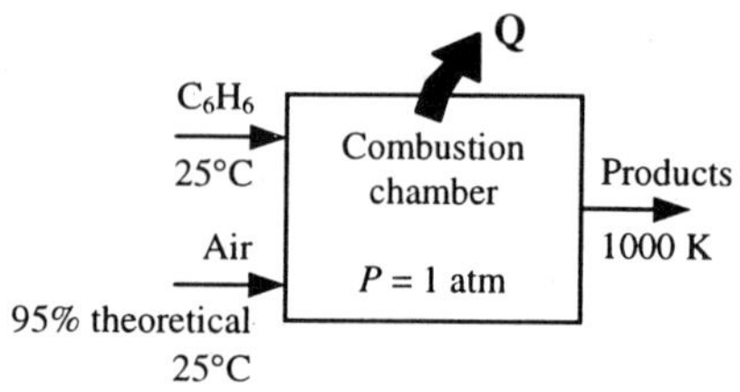

where a_{th} is the stoichiometric coefficient and is determined from the O_2 balance,

$$a_{th} = 6 + 1.5 = 7.5$$

Then the actual combustion equation can be written as

$$C_6H_6 + 0.95 \times 7.5(O_2 + 3.76N_2) \longrightarrow xCO_2 + (6-x)CO + 3H_2O + 26.79N_2$$

O_2 balance: $\quad 0.95 \times 7.5 = x + (6-x)/2 + 1.5 \longrightarrow x = 5.25$

Thus, $\quad C_6H_6 + 7.125(O_2 + 3.76N_2) \longrightarrow 5.25CO_2 + 0.75CO + 3H_2O + 26.79N_2$

The mole fraction of CO in the products is

$$y_{CO} = \frac{N_{CO}}{N_{total}} = \frac{0.75}{5.25 + 0.75 + 3 + 26.79} = 0.021 \text{ or } \mathbf{2.1\%}$$

(*b*) The heat transfer for this combustion process is determined from the energy balance $E_{in} - E_{out} = \Delta E_{system}$ applied on the combustion chamber with $W = 0$. It reduces to

$$-Q_{out} = \sum N_P\left(\bar{h}_f^\circ + \bar{h} - \bar{h}^\circ\right)_P - \sum N_R\left(\bar{h}_f^\circ + \bar{h} - \bar{h}^\circ\right)_R \longrightarrow -Q_{out} = \sum N_P\left(\bar{h}_f^\circ + \bar{h} - \bar{h}^\circ\right)_P - \sum N_R \bar{h}_{f,R}^\circ$$

since all of the reactants are at 25°C. Assuming the air and the combustion products to be ideal gases, we have $h = h(T)$. From the tables,

Substance	$\bar{h}_f^\circ$ kJ/kmol	$\bar{h}_{298\ K}$ kJ/kmol	$\bar{h}_{1000\ K}$ kJ/kmol
C_6H_6 (*g*)	82,930	---	---
O_2	0	8682	31,389
N_2	0	8669	30,129
H_2O (*g*)	-241,820	9904	35,882
CO	-110,530	8669	30,355
CO_2	-393,520	9364	42,769

Thus,

$$\begin{aligned}-Q_{out} &= (5.25)(-393{,}520 + 42{,}769 - 9364) + (0.75)(-110{,}530 + 30{,}355 - 8669) \\ &\quad + (3)(-241{,}820 + 35{,}882 - 9904) + (26.79)(0 + 30{,}129 - 8669) - (1)(82{,}930) - 0 - 0 \\ &= -2{,}112{,}779 \text{ kJ / kmol } C_6H_6\end{aligned}$$

or $\quad \dot{Q}_{out} = \mathbf{2{,}112{,}800\ kJ/kmol\ C_6H_6}$

14-53 Diesel fuel is burned with 20 percent excess air during a steady-flow combustion process. The required mass flow rate of the diesel fuel to supply heat at a specified rate is to be determined.

Assumptions **1** Steady operating conditions exist. **2** Air and combustion gases are ideal gases. **3** Kinetic and potential energies are negligible. **4** Combustion is complete.

Analysis The fuel is burned completely with the excess air, and thus the products will contain only CO_2, H_2O, N_2, and some free O_2. Considering 1 kmol of $C_{12}H_{26}$, the combustion equation can be written as

$$C_{12}H_{26} + 1.2a_{th}(O_2 + 3.76N_2) \longrightarrow 12CO_2 + 13H_2O + 0.2a_{th}O_2 + (1.2)(3.76a_{th})N_2$$

where a_{th} is the stoichiometric coefficient and is determined from the O_2 balance,

$\dot{Q}$ = 3000 kJ/s

$C_{12}H_{26}$ 25°C → Combustion chamber P = 1 atm → Products 500 K

Air 20% excess air 25°C

$$1.2a_{th} = 12 + 6.5 + 0.2a_{th} \longrightarrow a_{th} = 18.5$$

Substituting,

$$C_{12}H_{26} + 22.2(O_2 + 3.76N_2) \longrightarrow 12CO_2 + 13H_2O + 3.7O_2 + 83.47N_2$$

The heat transfer for this combustion process is determined from the energy balance $E_{in} - E_{out} = \Delta E_{system}$ applied on the combustion chamber with $W = 0$. It reduces to

$$-Q_{out} = \sum N_P\left(\bar{h}_f^\circ + \bar{h} - \bar{h}^\circ\right)_P - \sum N_R\left(\bar{h}_f^\circ + \bar{h} - \bar{h}^\circ\right)_R \longrightarrow -Q_{out} = \sum N_P\left(\bar{h}_f^\circ + \bar{h} - \bar{h}^\circ\right)_P - \sum N_R\bar{h}_{f,R}^\circ$$

since all of the reactants are at 25°C. Assuming the air and the combustion products to be ideal gases, we have $h = h(T)$. From the tables,

Substance	$\bar{h}_f^\circ$ kJ/kmol	$\bar{h}_{298\ K}$ kJ/kmol	$\bar{h}_{500\ K}$ kJ/kmol
$C_{12}H_{26}$	-291,010	---	---
O_2	0	8682	14,770
N_2	0	8669	14,581
H_2O (*g*)	-241,820	9904	16,828
CO_2	-393,520	9364	17,678

Thus,

$$\begin{aligned}-Q_{out} &= (12)(-393{,}520 + 17{,}678 - 9364) + (13)(-241{,}820 + 16{,}828 - 9904) \\ &\quad + (3.7)(0 + 14{,}770 - 8682) + (83.47)(0 + 14{,}581 - 8669) - (1)(-291{,}010) - 0 - 0 \\ &= -6{,}869{,}110\ \text{kJ/kmol}\ C_{12}H_{26}\end{aligned}$$

or $\quad Q_{out} = 6{,}869{,}110\ \text{kJ/kmol}\ C_{12}H_{26}$

Then the required mass flow rate of fuel for a heat transfer rate of 3000 kJ/s becomes

$$\dot{m} = \dot{N}M = \left(\frac{\dot{Q}_{out}}{Q_{out}}\right)M = \left(\frac{3000\ \text{kJ/s}}{6{,}869{,}110\ \text{kJ/kmol}}\right)(170\ \text{kg/kmol}) = 0.0742\ \text{kg/s} = \mathbf{74.2\ g/s}$$

14-54E Diesel fuel is burned with 20 percent excess air during a steady-flow combustion process. The required mass flow rate of the diesel fuel for a specified heat transfer rate is to be determined.

Assumptions **1** Steady operating conditions exist. **2** Air and combustion gases are ideal gases. **3** Kinetic and potential energies are negligible. **4** Combustion is complete.

Analysis The fuel is burned completely with the excess air, and thus the products will contain only CO_2, H_2O, N_2, and some free O_2. Considering 1 kmol of $C_{12}H_{26}$, the combustion equation can be written as

$$C_{12}H_{26} + 1.2a_{th}(O_2 + 3.76N_2) \longrightarrow 12CO_2 + 13H_2O + 0.2a_{th}O_2 + (1.2)(3.76a_{th})N_2$$

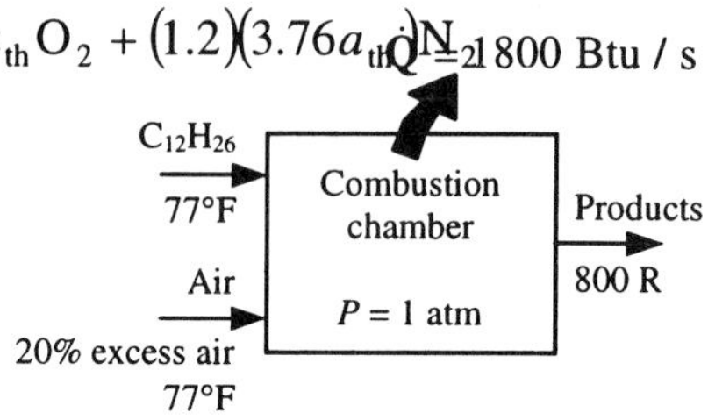

where a_{th} is the stoichiometric coefficient and is determined from the O_2 balance,

$$1.2a_{th} = 12 + 6.5 + 0.2a_{th} \longrightarrow a_{th} = 18.5$$

$\dot{Q} = 1800$ Btu/s

$C_{12}H_{26}$

77°F

Air

20% excess air

77°F

Combustion chamber

$P = 1$ atm

Products

800 R

Substituting,

$$C_{12}H_{26} + 22.2(O_2 + 3.76N_2) \longrightarrow 12CO_2 + 13H_2O + 3.7O_2 + 83.47N_2$$

The heat transfer for this combustion process is determined from the energy balance $E_{in} - E_{out} = \Delta E_{system}$ applied on the combustion chamber with $W = 0$. It reduces to

$$-Q_{out} = \sum N_P\left(\bar{h}_f^\circ + \bar{h} - \bar{h}^\circ\right)_P - \sum N_R\left(\bar{h}_f^\circ + \bar{h} - \bar{h}^\circ\right)_R \longrightarrow -Q_{out} = \sum N_P\left(\bar{h}_f^\circ + \bar{h} - \bar{h}^\circ\right)_P - \sum N_R \bar{h}_{f,R}^\circ$$

since all of the reactants are at 77°F. Assuming the air and the combustion products to be ideal gases, we have $h = h(T)$. From the tables,

Substance	$\bar{h}_f^\circ$ Btu/lbmol	$\bar{h}_{537\ R}$ Btu/lbmol	$\bar{h}_{800\ R}$ Btu/lbmol
$C_{12}H_{26}$	-125,190	---	---
O_2	0	3725.1	5602.0
N_2	0	3729.5	5564.4
H_2O (*g*)	-104,040	4258.0	6396.9
CO_2	-169,300	4027.5	6552.9

Thus,

$$\begin{aligned} -Q_{out} &= (12)(-169{,}300 + 6552.9 - 4027.5) + (13)(-104{,}040 + 6396.9 - 4258) \\ &\quad + (3.7)(0 + 5602.0 - 3725.1) + (83.47)(0 + 5564.4 - 3729.5) - (1)(-125{,}190) - 0 - 0 \\ &= -3{,}040{,}716 \text{ Btu/lbmol}C_{12}H_{26} \end{aligned}$$

or $\qquad Q_{out} = 3{,}040{,}716 \text{ Btu/lbmol } C_{12}H_{26}$

Then the required mass flow rate of fuel for a heat transfer rate of 1800 Btu/s becomes

$$\dot{m} = \dot{N}M = \left(\frac{\dot{Q}}{Q}\right)M = \left(\frac{1800 \text{ Btu/s}}{3{,}040{,}716 \text{ Btu/lbmol}}\right)(170 \text{ lbm/lbmol}) = \mathbf{0.1006\ \ lbm/s}$$

14-55 [*Also solved by EES on enclosed CD*] Octane gas is burned with 30 percent excess air during a steady-flow combustion process. The heat transfer per unit mass of octane is to be determined.

Assumptions **1** Steady operating conditions exist. **2** Air and combustion gases are ideal gases. **3** Kinetic and potential energies are negligible. **4** Combustion is complete.

Properties The molar mass of C_8H_{18} is 114 kg/kmol (Table A-1).

Analysis The fuel is burned completely with the excess air, and thus the products will contain only CO_2, H_2O, N_2, and some free O_2. The moisture in the air does not react with anything; it simply shows up as additional H_2O in the products. Therefore, for simplicity, we will balance the combustion equation using dry air, and then add the moisture to both sides of the equation. Considering 1 kmol of C_8H_{18}, the combustion equation can be written as

$$C_8H_{18}(g)+1.3a_{th}(O_2+3.76N_2)\longrightarrow 8CO_2+9H_2O+0.3a_{th}O_2+(1.3)(3.76a_{th})N_2$$

where a_{th} is the stoichiometric coefficient for air. It is determined from

$$O_2 \text{ balance: } 1.3a_{th}=8+4.5+0.3a_{th} \longrightarrow a_{th}=12.5$$

C_8H_{18}
25°C
Air
30% excess air
Combustion chamber
P = 1 atm
q
Products
600 K

Thus,

$$C_8H_{18}(g)+16.25(O_2+3.76N_2)\longrightarrow 8CO_2+9H_2O+3.75O_2+61.1N_2$$

Therefore, 16.25 × 4.76 = 77.35 kmol of dry air will be used per kmol of the fuel. The partial pressure of the water vapor present in the incoming air is

$$P_{v,in}=\phi_{air}P_{sat@25°C}=(0.60)(3.169\text{ kPa})=1.901\text{ kPa}$$

Assuming ideal gas behavior, the number of moles of the moisture that accompanies 77.35 kmol of incoming dry air is determined to be

$$N_{v,in}=\left(\frac{P_{v,in}}{P_{total}}\right)N_{total}=\left(\frac{1.901\text{ kPa}}{101.325\text{ kPa}}\right)(77.35+N_{v,in})\longrightarrow N_{v,in}=1.48\text{ kmol}$$

The balanced combustion equation is obtained by adding 1.48 kmol of H_2O to both sides of the equation,

$$C_8H_{18}(g)+16.25(O_2+3.76N_2)+1.48H_2O\longrightarrow 8CO_2+10.48H_2O+3.75O_2+61.1N_2$$

The heat transfer for this combustion process is determined from the energy balance $E_{in}-E_{out}=\Delta E_{system}$ applied on the combustion chamber with W = 0. It reduces to

$$-Q_{out}=\sum N_P\left(\bar{h}_f^\circ+\bar{h}-\bar{h}^\circ\right)_P-\sum N_R\left(\bar{h}_f^\circ+\bar{h}-\bar{h}^\circ\right)_R\longrightarrow -Q_{out}=\sum N_P\left(\bar{h}_f^\circ+\bar{h}-\bar{h}^\circ\right)_P-\sum N_R\bar{h}_{f,R}^\circ$$

since all of the reactants are at 25°C. Assuming the air and the combustion products to be ideal gases, we have $h = h(T)$. From the tables,

Substance	$\bar{h}_f^\circ$ kJ/kmol	$\bar{h}_{298\text{ K}}$ kJ/kmol	$\bar{h}_{600\text{ K}}$ kJ/kmol
C_8H_{18} (*g*)	-208,450	---	---
O_2	0	8682	17,929
N_2	0	8669	17,563
H_2O (*g*)	-241,820	9904	20,402
CO_2	-393,520	9364	22,280

Substituting,

$$\begin{aligned}-Q_{\text{out}} &= (8)(-393{,}520+22{,}280-9364)+(10.48)(-241{,}820+20{,}402-9904)\\&\quad+(3.75)(0+17{,}929-8682)+(61.1)(0+17{,}563-8669)\\&\quad-(1)(-208{,}450)-(1.48)(-241{,}820)-0-0\\&= -4{,}324{,}643\ \text{kJ/kmol}\,C_8H_{18}\end{aligned}$$

Thus 4,324,643 kJ of heat is transferred from the combustion chamber for each kmol (114 kg) of C_8H_{18}. Then the heat transfer per kg of C_8H_{18} becomes

$$q = \frac{Q_{out}}{M} = \frac{4{,}324{,}643\ \text{kJ}}{114\ \text{kg}} = \mathbf{37{,}935\ kJ\,/\,kg\ C_8H_{18}}$$

14-56 EES solution of this (and other comprehensive problems designated with the *computer icon*) is available to instructors at the *Instructor Manual* section of the *Online Learning Center* (OLC) at www.mhhe.com/cengel-boles. See the Preface for access information.

14-57 Ethane gas is burned with stoichiometric amount of air during a steady-flow combustion process. The rate of heat transfer from the combustion chamber is to be determined.

Assumptions **1** Steady operating conditions exist. **2** Air and combustion gases are ideal gases. **3** Kinetic and potential energies are negligible. **4** Combustion is complete.

Properties The molar mass of C_2H_6 is 30 kg/kmol (Table A-1).

Analysis The theoretical combustion equation of C_2H_6 is

$$C_2H_6 + a_{th}(O_2 + 3.76N_2) \longrightarrow 2CO_2 + 3H_2O + 3.76a_{th}N_2$$

where a_{th} is the stoichiometric coefficient and is determined from the O_2 balance,

$$a_{th} = 2 + 1.5 = 3.5$$

Then the actual combustion equation can be written as

$$C_2H_6 + 3.5(O_2 + 3.76N_2) \longrightarrow 1.9CO_2 + 0.1CO + 3H_2O + 0.05O_2 + 13.16N_2$$

The heat transfer for this combustion process is determined from the energy balance $E_{in} - E_{out} = \Delta E_{system}$ applied on the combustion chamber with $W = 0$. It reduces to

$$-Q_{out} = \sum N_P\left(\bar{h}_f^\circ + \bar{h} - \bar{h}^\circ\right)_P - \sum N_R\left(\bar{h}_f^\circ + \bar{h} - \bar{h}^\circ\right)_R$$

Assuming the air and the combustion products to be ideal gases, we have $h = h(T)$. From the tables,

Substance	$\bar{h}_f^\circ$ kJ/kmol	$\bar{h}_{500\text{ K}}$ kJ/kmol	$\bar{h}_{298\text{ K}}$ kJ/kmol	$\bar{h}_{800\text{ K}}$ kJ/kmol
C_2H_6 (*g*)	-84,680	---	---	---
O_2	0	14,770	8682	24,523
N_2	0	14,581	8669	23,714
H_2O (*g*)	-241,820	---	9904	27,896
CO	-110,530	---	8669	23,844
CO_2	-393,520	---	9364	32,179

Thus,

$$\begin{aligned}-Q_{out} &= (1.9)(-393{,}520 + 32{,}179 - 9364) + (0.1)(-110{,}530 + 23{,}844 - 8669)\\ &\quad + (3)(-241{,}820 + 27{,}896 - 9904) + (0.05)(0 + 24{,}523 - 8682) + (13.16)(0 + 23{,}714 - 8669)\\ &\quad - (1)(-84{,}680 + h_{298} - h_{298}) - (3.5)(0 + 14{,}770 - 8682) - (13.16)(0 + 14{,}581 - 8669)\\ &= -1{,}201{,}005\text{ kJ / kmol } C_2H_6\end{aligned}$$

or $$Q_{out} = 1{,}201{,}005 \text{ kJ / kmol } C_2H_6$$

Then the rate of heat transfer for a mass flow rate of 3 kg/h for the ethane becomes

$$\dot{Q}_{out} = \dot{N}Q_{out} = \left(\frac{\dot{m}}{N}\right)Q_{out} = \left(\frac{5\text{ kg/h}}{30\text{ kg/kmol}}\right)(1{,}201{,}005\text{ kJ/kmol}) = \mathbf{200{,}170\ kJ/h}$$

14-58 [*Also solved by EES on enclosed CD*] A mixture of methane and oxygen contained in a tank is burned at constant volume. The final pressure in the tank and the heat transfer during this process are to be determined.

Assumptions **1** Air and combustion gases are ideal gases. **2** Combustion is complete.

Properties The molar masses of CH_4 and O_2 are 16 kg/kmol and 32 kg/kmol, respectively (Table A-1).

Analysis (*a*) The combustion is assumed to be complete, and thus all the carbon in the methane burns to CO_2 and all of the hydrogen to H_2O. The number of moles of CH_4 and O_2 in the tank are

$$N_{CH_4} = \frac{m_{CH_4}}{M_{CH_4}} = \frac{0.12 \text{ kg}}{16 \text{ kg/kmol}} = 7.5 \times 10^{-3} \text{ kmol} = 7.5 \text{ mol}$$

$$N_{O_2} = \frac{m_{O_2}}{M_{O_2}} = \frac{0.6 \text{ kg}}{32 \text{ kg/kmol}} = 18.75 \times 10^{-3} \text{ kmol} = 18.75 \text{ mol}$$

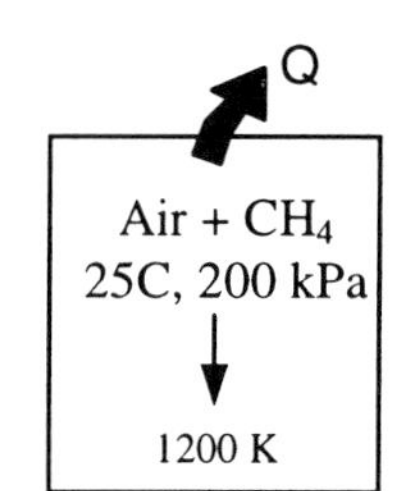

Then the combustion equation can be written as

$$7.5CH_4 + 18.75O_2 \longrightarrow 7.5CO_2 + 15H_2O + 3.75O_2$$

At 1200 K, water exists in the gas phase. Assuming both the reactants and the products to be ideal gases, the final pressure in the tank is determined to be

$$\left. \begin{array}{l} P_R V = N_R R_u T_R \\ P_P V = N_P R_u T_P \end{array} \right\} P_P = P_R \left(\frac{N_P}{N_R} \right) \left(\frac{T_P}{T_R} \right)$$

Substituting,

$$P_P = (200 \text{ kPa}) \left(\frac{26.25 \text{ mol}}{26.25 \text{ mol}} \right) \left(\frac{1200 \text{ K}}{298 \text{ K}} \right) = \mathbf{805 \text{ kPa}}$$

which is relatively low. Therefore, the ideal gas assumption utilized earlier is appropriate.

(*b*) The heat transfer for this constant volume combustion process is determined from the energy balance $E_{in} - E_{out} = \Delta E_{system}$ applied on the combustion chamber with $W = 0$. It reduces to

$$-Q_{out} = \sum N_P \left(\bar{h}_f^\circ + \bar{h} - \bar{h}^\circ - P\bar{v} \right)_P - \sum N_R \left(\bar{h}_f^\circ + \bar{h} - \bar{h}^\circ - P\bar{v} \right)_R$$

Since both the reactants and products are assumed to be ideal gases, all the internal energy and enthalpies depend on temperature only, and the $P\bar{v}$ terms in this equation can be replaced by $R_u T$. It yields

$$-Q_{out} = \sum N_P \left(\bar{h}_f^\circ + \bar{h}_{1200K} - \bar{h}_{298K} - R_u T \right)_P - \sum N_R \left(\bar{h}_f^\circ - R_u T \right)_R$$

since the reactants are at the standard reference temperature of 25°C. From the tables,

Substance	$\bar{h}_f^\circ$ kJ/kmol	$\bar{h}_{298\text{ K}}$ kJ/kmol	$\bar{h}_{1200\text{ K}}$ kJ/kmol
CH_4	-74,850	---	---
O_2	0	8682	38,447
H_2O (*g*)	-241,820	9904	44,380
CO_2	-393,520	9364	53,848

Thus,

$$\begin{aligned} -Q_{out} &= (7.5)(-393{,}520 + 53{,}848 - 9364 - 8.314 \times 1200) \\ &+ (15)(-241{,}820 + 44{,}380 - 9904 - 8.314 \times 1200) \\ &+ (3.75)(0 + 38{,}447 - 8682 - 8.314 \times 1200) \\ &- (7.5)(-74{,}850 - 8.314 \times 298) - (18.75)(-8.314 \times 298) \\ &= -5{,}251{,}791 \text{ J} = -5{,}252 \text{ kJ} \end{aligned}$$

Thus Q_{out} = **5252 kJ** of heat is transferred from the combustion chamber as 120 g of CH_4 burned in this combustion chamber.

14-59 EES solution of this (and other comprehensive problems designated with the *computer icon*) is available to instructors at the *Instructor Manual* section of the *Online Learning Center* (OLC) at www.mhhe.com/cengel-boles. See the Preface for access information.

14-60 A stoichiometric mixture of octane gas and air contained in a closed combustion chamber is ignited. The heat transfer from the combustion chamber is to be determined.

Assumptions **1** Both the reactants and products are ideal gases. **2** Combustion is complete.

Analysis The theoretical combustion equation of C_8H_{18} with stoichiometric amount of air is

$$C_8H_{18}(g)+a_{th}(O_2+3.76N_2)\longrightarrow 8CO_2+9H_2O+3.76a_{th}N_2$$

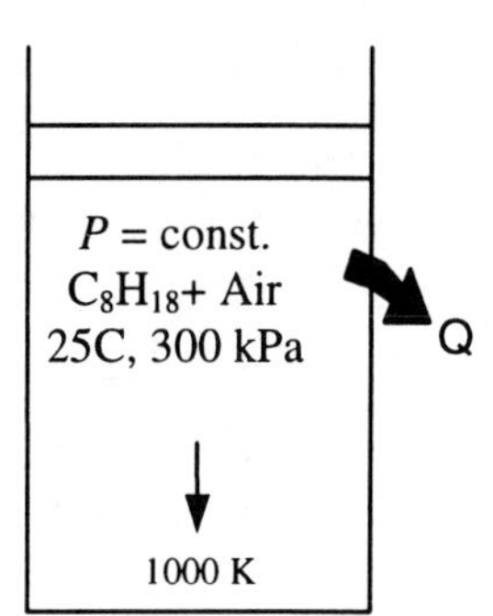

where a_{th} is the stoichiometric coefficient and is determined from the O_2 balance,

$$a_{th}=8+4.5=12.5$$

Thus,

$$C_8H_{18}(g)+12.5(O_2+3.76N_2)\longrightarrow 8CO_2+9H_2O+47N_2$$

The heat transfer for this constant volume combustion process is determined from the energy balance $E_{in}-E_{out}=\Delta E_{system}$ applied on the combustion chamber with $W_{other}=0$,

$$-Q_{out}=\sum N_P\left(\bar{h}_f^\circ+\bar{h}-\bar{h}^\circ-P\bar{v}\right)_P-\sum N_R\left(\bar{h}_f^\circ+\bar{h}-\bar{h}^\circ-P\bar{v}\right)_R$$

For a constant pressure quasi-equilibrium process $\Delta U + W_b = \Delta H$. Then the first law relation in this case is

$$-Q_{out}=\sum N_P\left(\bar{h}_f^\circ+\bar{h}_{1000\text{K}}-\bar{h}_{298\text{K}}\right)_P-\sum N_R\bar{h}_{f,R}^\circ$$

since the reactants are at the standard reference temperature of 25°C. Since both the reactants and the products behave as ideal gases, we have $h = h(T)$. From the tables,

Substance	$\bar{h}_f^\circ$ kJ/kmol	$\bar{h}_{298\text{ K}}$ kJ/kmol	$\bar{h}_{1000\text{ K}}$ kJ/kmol
C_8H_{18} (*g*)	-208,450	---	---
O_2	0	8682	31,389
N_2	0	8669	30,129
H_2O (*g*)	-241,820	9904	35,882
CO_2	-393,520	9364	42,769

Thus,

$$\begin{aligned}-Q_{out}&=(8)(-393{,}520+42{,}769-9364)+(9)(-241{,}820+35{,}882-9904-8.314\times1000)\\&\quad+(47)(0+30{,}129-8669)-(1)(-208{,}450)-0-0\\&=-3{,}681{,}254\text{kJ}(\text{per kmol of }C_8H_{18})\end{aligned}$$

or $\quad Q_{out}=3{,}681{,}254\text{kJ}(\text{per kmol of }C_8H_{18})$.

Total mole numbers initially present in the combustion chamber is determined from the ideal gas relation,

$$N_1=\frac{P_1V_1}{R_uT_1}=\frac{(300\text{kPa})(0.8\text{m}^3)}{(8.314\text{kPa}\cdot\text{m}^3/\text{kmol}\cdot\text{K})(298\text{K})}=0.0969\text{kmol}$$

Of these, 0.0969 / (1 + 12.5×4.76) = 1.601×10^{-3} kmol of them is C_8H_{18}. Thus the amount of heat transferred from the combustion chamber as 1.601×10^{-3} kmol of C_8H_{18} is burned is

$$Q_{out}=(1.601\times10^{-3}\text{kmol}C_8H_{18})(3{,}681{,}254\text{ kJ/kmol}C_8H_{18})=\mathbf{5894\ kJ}$$

14-61 A mixture of benzene gas and 30 percent excess air contained in a constant-volume tank is ignited. The heat transfer from the combustion chamber is to be determined.

Assumptions **1** Both the reactants and products are ideal gases. **2** Combustion is complete.

Analysis The theoretical combustion equation of C_6H_6 with stoichiometric amount of air is

$$C_6H_6(g) + a_{th}(O_2 + 3.76N_2) \longrightarrow 6CO_2 + 3H_2O + 3.76a_{th}N_2$$

Q

C_6H_6+ Air
25C, 1 atm

1000 K

where a_{th} is the stoichiometric coefficient and is determined from the O_2 balance,

$$a_{th} = 6 + 1.5 = 7.5$$

Then the actual combustion equation with 30% excess air becomes

$$C_6H_6(g) + 9.75(O_2 + 3.76N_2) \longrightarrow 5.52CO_2 + 0.48CO + 3H_2O + 2.49O_2 + 36.66N_2$$

The heat transfer for this constant volume combustion process is determined from the energy balance $E_{in} - E_{out} = \Delta E_{system}$ applied on the combustion chamber with $W = 0$. It reduces to

$$-Q_{out} = \sum N_P\left(\bar{h}_f^\circ + \bar{h} - \bar{h}^\circ - P\bar{v}\right)_P - \sum N_R\left(\bar{h}_f^\circ + \bar{h} - \bar{h}^\circ - P\bar{v}\right)_R$$

Since both the reactants and the products behave as ideal gases, all the internal energy and enthalpies depend on temperature only, and the $P\bar{v}$ terms in this equation can be replaced by R_uT.

It yields

$$-Q_{out} = \sum N_P\left(\bar{h}_f^\circ + \bar{h}_{1000K} - \bar{h}_{298K} - R_uT\right)_P - \sum N_R\left(\bar{h}_f^\circ - R_uT\right)_R$$

since the reactants are at the standard reference temperature of 25°C. From the tables,

Substance	$\bar{h}_f^\circ$ kJ/kmol	$\bar{h}_{298\ K}$ kJ/kmol	$\bar{h}_{1000\ K}$ kJ/kmol
C_6H_6 (*g*)	82,930	---	---
O_2	0	8682	31,389
N_2	0	8669	30,129
H_2O (*g*)	-241,820	9904	35,882
CO	-110,530	8669	30,355
CO_2	-393,520	9364	42,769

Thus,

$$\begin{aligned}
-Q_{out} &= (5.52)(-393{,}520 + 42{,}769 - 9364 - 8.314 \times 1000) \\
&\quad + (0.48)(-110{,}530 + 30{,}355 - 8669 - 8.314 \times 1000) \\
&\quad + (3)(-241{,}820 + 35{,}882 - 9904 - 8.314 \times 1000) \\
&\quad + (2.49)(0 + 31{,}389 - 8682 - 8.314 \times 1000) \\
&\quad + (36.66)(0 + 30{,}129 - 8669 - 8.314 \times 1000) \\
&\quad - (1)(82{,}930 - 8.314 \times 298) - (9.75)(4.76)(-8.314 \times 298) \\
&= -2{,}200{,}529 \text{ kJ}
\end{aligned}$$

or $\quad Q_{out} = \mathbf{2{,}200{,}529\ kJ}$

14-62E A mixture of benzene gas and 30 percent excess air contained in a constant-volume tank is ignited. The heat transfer from the combustion chamber is to be determined.

Assumptions **1** Both the reactants and products are ideal gases. **2** Combustion is complete.

Analysis The theoretical combustion equation of C_6H_6 with stoichiometric amount of air is

$$C_6H_6(g) + a_{th}(O_2 + 3.76N_2) \longrightarrow 6CO_2 + 3H_2O + 3.76a_{th}N_2$$

Q

C_6H_6+ Air
77F, 1 atm

1800 R

where a_{th} is the stoichiometric coefficient and is determined from the O_2 balance,

$$a_{th} = 6 + 1.5 = 7.5$$

Then the actual combustion equation with 30% excess air becomes

$$C_6H_6(g) + 9.75(O_2 + 3.76N_2) \longrightarrow 5.52CO_2 + 0.48CO + 3H_2O + 2.49O_2 + 36.66N_2$$

The heat transfer for this constant volume combustion process is determined from the energy balance $E_{in} - E_{out} = \Delta E_{system}$ applied on the combustion chamber with W = 0. It reduces to

$$-Q_{out} = \sum N_P\left(\bar{h}_f^\circ + \bar{h} - \bar{h}^\circ - P\bar{v}\right)_P - \sum N_R\left(\bar{h}_f^\circ + \bar{h} - \bar{h}^\circ - P\bar{v}\right)_R$$

Since both the reactants and the products behave as ideal gases, all the internal energy and enthalpies depend on temperature only, and the $P\bar{v}$ terms in this equation can be replaced by R_uT.

It yields

$$-Q_{out} = \sum N_P\left(\bar{h}_f^\circ + \bar{h}_{1800R} - \bar{h}_{537R} - R_uT\right)_P - \sum N_R\left(\bar{h}_f^\circ - R_uT\right)_R$$

since the reactants are at the standard reference temperature of 77°F. From the tables,

Substance	$\bar{h}_f^\circ$ Btu/lbmol	$\bar{h}_{537\text{ R}}$ Btu/lbmol	$\bar{h}_{1800\text{ R}}$ Btu/lbmol
C_6H_6 (*g*)	35,6860	---	---
O_2	0	3725.1	13,485.8
N_2	0	3729.5	12,956.3
H_2O (*g*)	-104,040	4258.0	15,433.0
CO	-47,540	3725.1	13,053.2
CO_2	-169,300	4027.5	18,391.5

Thus,

$$\begin{aligned}
-Q_{out} &= (5.52)(-169,300 + 18,391.5 - 4027.5 - 1.986\times 1800) \\
&+ (0.48)(-47,540 + 13,053.2 - 3725.1 - 1.986\times 1800) \\
&+ (3)(-104,040 + 15,433.0 - 4258.0 - 1.986\times 1800) \\
&+ (2.49)(0 + 13,485.8 - 3725.1 - 1.986\times 1800) \\
&+ (36.66)(0 + 12,956.3 - 3729.5 - 1.986\times 1800) \\
&- (1)(35,680 - 1.986\times 537) - (9.75)(4.76)(-1.986\times 537) \\
&= -946,870 \text{ Btu}
\end{aligned}$$

or $\quad Q_{out} = \mathbf{946,870\ Btu}$

Adiabatic Flame Temperature

14-63C For the case of stoichiometric amount of pure oxygen since we have the same amount of chemical energy released but a smaller amount of mass to absorb it.

14-64C Under the conditions of complete combustion with stoichiometric amount of air.

14-65 [*Also solved by EES on enclosed CD*] Hydrogen is burned with 20 percent excess air during a steady-flow combustion process. The exit temperature of product gases is to be determined.

Assumptions **1** Steady operating conditions exist. **2** Air and combustion gases are ideal gases. **3** Kinetic and potential energies are negligible. **4** There are no work interactions. **5** The combustion chamber is adiabatic.

Analysis Adiabatic flame temperature is the temperature at which the products leave the combustion chamber under adiabatic conditions (Q = 0) with no work interactions (W = 0). Under steady-flow conditions the energy balance $E_{in} - E_{out} = \Delta E_{system}$ applied on the combustion chamber reduces to

$$\sum N_P\left(\bar{h}_f^\circ + \bar{h} - \bar{h}^\circ\right)_P = \sum N_R\left(\bar{h}_f^\circ + \bar{h} - \bar{h}^\circ\right)_R$$

H$_2$ 7°C → Combustion chamber → Products T_P; Air 20% excess air 7°C

The combustion equation of H_2 with 20% excess air is

$$H_2 + 0.6(O_2 + 3.76N_2) \longrightarrow H_2O + 0.1O_2 + 2.256N_2$$

From the tables,

Substance	$\bar{h}_f^\circ$ kJ/kmol	$\bar{h}_{280\text{ K}}$ kJ/kmol	$\bar{h}_{298\text{ K}}$ kJ/kmol
H_2	0	7945	8468
O_2	0	8150	8682
N_2	0	8141	8669
H_2O (*g*)	-241,820	9296	9904

Thus,

$$(1)(-241,820 + \bar{h}_{H_2O} - 9904) + (0.1)(0 + \bar{h}_{O_2} - 8682) + (2.256)(0 + \bar{h}_{N_2} - 8669)$$
$$= (1)(0 + 7945 - 8468) + (0.6)(0 + 8150 - 8682) + (2.256)(0 + 8141 - 8669)$$

It yields $\bar{h}_{H_2O} + 0.1\bar{h}_{O_2} + 2.256\bar{h}_{N_2} = 270{,}116\text{ kJ}$

The adiabatic flame temperature is obtained from a trial and error solution. A first guess is obtained by dividing the right-hand side of the equation by the total number of moles, which yields 270,116/(1 + 0.1 + 2.256) = 80,488 kJ/kmol. This enthalpy value corresponds to about 2400 K for N_2. Noting that the majority of the moles are N_2, T_P will be close to 2400 K, but somewhat under it because of the higher specific heat of H_2O.

At 2300 K:
$$\bar{h}_{H_2O} + 0.1\bar{h}_{O_2} + 2.256\bar{h}_{N_2} = (1)(98{,}199) + (0.1)(79{,}316) + (2.256)(75{,}676)$$
$$= 276{,}856\text{kJ (Higher than 270,116 kJ)}$$

At 2250 K:
$$\bar{h}_{H_2O} + 0.1\bar{h}_{O_2} + 2.256\bar{h}_{N_2} = (1)(95{,}562) + (0.1)(77{,}397) + (2.256)(73{,}856)$$
$$= 269{,}921\text{kJ (Lower than 270,116 kJ)}$$

By interpolation, T_P = **2251.4 K**

14-66 EES solution of this (and other comprehensive problems designated with the *computer icon*) is available to instructors at the *Instructor Manual* section of the *Online Learning Center* (OLC) at www.mhhe.com/cengel-boles. See the Preface for access information.

14-67E Hydrogen is burned with 20 percent excess air during a steady-flow combustion process. The exit temperature of product gases is to be determined.

Assumptions **1** Steady operating conditions exist. **2** Air and combustion gases are ideal gases. **3** Kinetic and potential energies are negligible. **4** There are no work interactions. **5** The combustion chamber is adiabatic.

Analysis Adiabatic flame temperature is the temperature at which the products leave the combustion chamber under adiabatic conditions (Q = 0) with no work interactions (W = 0). Under steady-flow conditions the energy balance $E_{in} - E_{out} = \Delta E_{system}$ applied on the combustion chamber reduces to

$$\sum N_P\left(\bar{h}_f^\circ + \bar{h} - \bar{h}^\circ\right)_P = \sum N_R\left(\bar{h}_f^\circ + \bar{h} - \bar{h}^\circ\right)_R$$

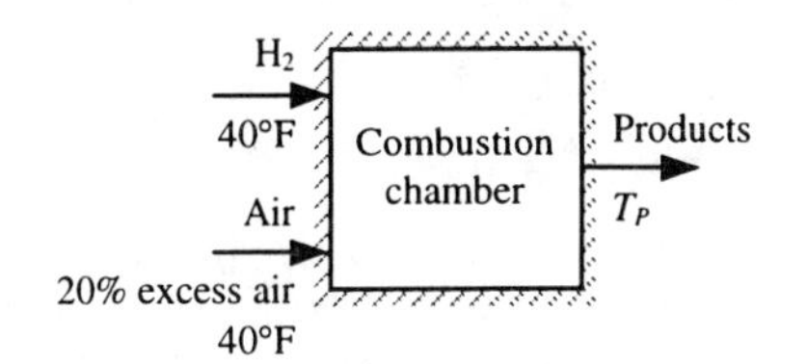

The combustion equation of H_2 with 20% excess air is

$$H_2 + 0.6(O_2 + 3.76N_2) \longrightarrow H_2O + 0.1O_2 + 2.256N_2$$

From the tables,

Substance	$\bar{h}_f^\circ$ Btu/lbmol	$\bar{h}_{500\ R}$ Btu/lbmol	$\bar{h}_{537\ R}$ Btu/lbmol
H_2	0	3386.1	3640.3
O_2	0	3466.2	3725.1
N_2	0	3472.2	3729.5
H_2O (*g*)	-104,040	3962.0	4258.0

Thus,

$$\begin{aligned}(1)(-140{,}040 + \bar{h}_{H_2O} - 4258) + (0.1)(0 + \bar{h}_{O_2} - 3725.1) + (2.256)(0 + \bar{h}_{N_2} - 3729.5)\\ = (1)(0 + 3386.1 - 3640.3) + (0.6)(0 + 3466.2 - 3725.1) + (2.256)(0 + 3472.2 - 3729.5)\end{aligned}$$

It yields $\bar{h}_{H_2O} + 0.1\bar{h}_{O_2} + 2.256\bar{h}_{N_2} = 152{,}094\ \text{Btu}$

The adiabatic flame temperature is obtained from a trial and error solution. A first guess is obtained by dividing the right-hand side of the equation by the total number of moles, which yields 152,094/(1 + 0.1+ 2.256) = 45,320 Btu/lbmol. This enthalpy value corresponds to about 5600 R for N_2. Noting that the majority of the moles are N_2, T_P will be close to 5600 R, but somewhat under it because of the higher specific heat of H_2O.

At 5100 R: $\bar{h}_{H_2O} + 0.1\bar{h}_{O_2} + 2.256\bar{h}_{N_2} = (1)(54{,}640) + (0.1)(43{,}021) + (2.256)(40{,}962)$
$= 151{,}352\ \text{Btu}\ (\text{Lower than } 152{,}094\ \text{Btu})$

At 5200 R: $\bar{h}_{H_2O} + 0.1\bar{h}_{O_2} + 2.256\bar{h}_{N_2} = (1)(55{,}957) + (0.1)(43{,}974) + (2.256)(41{,}844)$
$= 154{,}754\ \text{Btu}\ (\text{Higher than } 152{,}094\ \text{Btu})$

By interpolation, T_P = **5122 R**

14-68 Acetylene gas is burned with 30 percent excess air during a steady-flow combustion process. The exit temperature of product gases is to be determined.

Assumptions **1** Steady operating conditions exist. **2** Air and combustion gases are ideal gases. **3** Kinetic and potential energies are negligible. **4** There are no work interactions.

Analysis The fuel is burned completely with the excess air, and thus the products will contain only CO_2, H_2O, N_2, and some free O_2. Considering 1 kmol of C_2H_2, the combustion equation can be written as

$$C_2H_2 + 1.3a_{th}(O_2 + 3.76N_2) \longrightarrow 2CO_2 + H_2O + 0.3a_{th}O_2 + (1.3)(3.76)a_{th}N_2$$

where a_{th} is the stoichiometric coefficient and is determined from the O_2 balance,

$$1.3a_{th} = 2 + 0.5 + 0.3a_{th} \longrightarrow a_{th} = 2.5$$

75,000 kJ/kmol

C_2H_2

25°C

Air

30% excess air

27°C

Combustion chamber

Products

T_P

Thus,

$$C_2H_2 + 3.25(O_2 + 3.76N_2) \longrightarrow 2CO_2 + H_2O + 0.75O_2 + 12.22N_2$$

Under steady-flow conditions the energy balance $E_{in} - E_{out} = \Delta E_{system}$ applied on the combustion chamber with $W = 0$ reduces to

$$-Q_{out} = \sum N_P\left(\bar{h}_f^\circ + \bar{h} - \bar{h}^\circ\right)_P - \sum N_R\left(\bar{h}_f^\circ + \bar{h} - \bar{h}^\circ\right)_R$$

Assuming the air and the combustion products to be ideal gases, we have $h = h(T)$. From the tables,

Substance	$\bar{h}_f^\circ$ kJ/kmol	$\bar{h}_{298\text{ K}}$ kJ/kmol	$\bar{h}_{300\text{ K}}$ kJ/kmol
C_2H_2	226,730	---	---
O_2	0	8682	8736
N_2	0	8669	8723
H_2O (*g*)	-241,820	9904	---
CO_2	-393,520	9364	---

Thus,

$$\begin{aligned} -75{,}000 = &(2)(-393{,}520 + \bar{h}_{CO_2} - 9364) + (1)(-241{,}820 + \bar{h}_{H_2O} - 9904) \\ &+ (0.75)(0 + \bar{h}_{O_2} - 8682) + (12.22)(0 + \bar{h}_{N_2} - 8669) - (1)(226{,}730) \\ &- (3.25)(0 + 8736 - 8682) + (12.22)(0 + 8723 - 8669) \end{aligned}$$

It yields $\quad 2\bar{h}_{CO_2} + \bar{h}_{H_2O} + 0.75\bar{h}_{O_2} + 12.22\bar{h}_{N_2} = 1{,}321{,}184 \text{ kJ}$

The temperature of the product gases is obtained from a trial and error solution. A first guess is obtained by dividing the right-hand side of the equation by the total number of moles, which yields 1,321,184/(2 + 1 + 0.75 + 12.22) = 82,729 kJ/kmol. This enthalpy value corresponds to about 2500 K for N_2. Noting that the majority of the moles are N_2, T_P will be close to 2500 K, but somewhat under it because of the higher specific heats of CO_2 and H_2O.

At 2350 K:

$$\begin{aligned} 2\bar{h}_{CO_2} + \bar{h}_{H_2O} + 0.75\bar{h}_{O_2} + 12.22\bar{h}_{N_2} &= (2)(122{,}091) + (1)(100{,}846) + (0.75)(81{,}243) + (12.22)(77{,}496) \\ &= 1{,}352{,}961 \text{kJ (Higher than } 1{,}321{,}184 \text{ kJ)} \end{aligned}$$

At 2300 K:

$$\begin{aligned} 2\bar{h}_{CO_2} + \bar{h}_{H_2O} + 0.75\bar{h}_{O_2} + 12.22\bar{h}_{N_2} &= (2)(119{,}035) + (1)(98{,}199) + (0.75)(79{,}316) + (12.22)(75{,}676) \\ &= 1{,}320{,}517 \text{kJ (Lowerthan} 1{,}321{,}184 \text{kJ)} \end{aligned}$$

By interpolation, $\quad T_P$ = **2301 K**

14-69 A mixture of hydrogen and the stoichiometric amount of air contained in a constant-volume tank is ignited. The final temperature in the tank is to be determined.

Assumptions **1** The tank is adiabatic. **2** Both the reactants and products are ideal gases. **3** There are no work interactions. **4** Combustion is complete.

Analysis The combustion equation of H_2 with stoichiometric amount of air is

$$H_2 + 0.5(O_2 + 3.76N_2) \longrightarrow H_2O + 1.88N_2$$

H₂, AIR
25°C, 1 atm
T_P

The final temperature in the tank is determined from the energy balance relation $E_{in} - E_{out} = \Delta E_{system}$ for reacting closed systems under adiabatic conditions ($Q = 0$) with no work interactions ($W = 0$),

$$\sum N_P\left(\bar{h}_f^\circ + \bar{h} - \bar{h}^\circ - P\bar{v}\right)_P = \sum N_R\left(\bar{h}_f^\circ + \bar{h} - \bar{h}^\circ - P\bar{v}\right)_R$$

Since both the reactants and the products behave as ideal gases, all the internal energy and enthalpies depend on temperature only, and the $P\bar{v}$ terms in this equation can be replaced by R_uT.

It yields

$$\sum N_P\left(\bar{h}_f^\circ + \bar{h}_{T_P} - \bar{h}_{298K} - R_uT\right)_P = \sum N_R\left(\bar{h}_f^\circ R_uT\right)_R$$

since the reactants are at the standard reference temperature of 25°C. From the tables,

Substance	$\bar{h}_f^\circ$ kJ/kmol	$\bar{h}_{298\text{ K}}$ kJ/kmol
H_2	0	8468
O_2	0	8682
N_2	0	8669
H_2O (*g*)	-241,820	9904

Thus,

$$(1)(-241{,}820 + \bar{h}_{H_2O} - 9904 - 8.314 \times T_P) + (1.88)(0 + \bar{h}_{N_2} - 8669 - 8.314 \times T_P)$$
$$= (1)(0 - 8.314 \times 298) + (0.5)(0 - 8.314 \times 298) + (1.88)(0 - 8.314 \times 298)$$

It yields $\quad \bar{h}_{H_2O} + 1.88\bar{h}_{N_2} - 23.94 \times T_P = 259{,}648 \text{ kJ}$

The temperature of the product gases is obtained from a trial and error solution,

At 3050 K: $\quad \bar{h}_{H_2O} + 1.88\bar{h}_{N_2} - 23.94 \times T_P = (1)(139{,}051) + (1.88)(103{,}260) - (23.94)(3050)$
$$= 260{,}162 \text{kJ (Higher than 259,648 kJ)}$$

At 3000 K: $\quad \bar{h}_{H_2O} + 1.88\bar{h}_{N_2} - 23.94 \times T_P = (1)(136{,}264) + (1.88)(101{,}407) - (23.94)(3000)$
$$= 255{,}089 \text{kJ (Lower than 259,648 kJ)}$$

By interpolation, $\quad T_P = \mathbf{3045\ K}$

14-70 Octane gas is burned with 30 percent excess air during a steady-flow combustion process. The exit temperature of product gases is to be determined.

Assumptions **1** Steady operating conditions exist. **2** Air and combustion gases are ideal gases. **3** Kinetic and potential energies are negligible. **4** There are no work interactions. **5** The combustion chamber is adiabatic.

Analysis Under steady-flow conditions the energy balance $E_{in} - E_{out} = \Delta E_{system}$ applied on the combustion chamber with $Q = W = 0$ reduces to

$$\sum N_P\left(\bar{h}_f^\circ + \bar{h} - \bar{h}^\circ\right)_P = \sum N_R\left(\bar{h}_f^\circ + \bar{h} - \bar{h}^\circ\right)_R \longrightarrow \sum N_P\left(\bar{h}_f^\circ + \bar{h} - \bar{h}^\circ\right)_P = \sum N_R \bar{h}_{f,R}^\circ$$

since all the reactants are at the standard reference temperature of 25°C. Then,

$$C_8H_{18}(g) + 1.3a_{th}(O_2 + 3.76N_2) \longrightarrow 8CO_2 + 9H_2O + 0.3a_{th}O_2 + (1.3)(3.76)a_{th}N_2$$

where a_{th} is the stoichiometric coefficient and is determined from the O_2 balance,

$$1.3a_{th} = 8 + 4.5 + 0.3a_{th} \longrightarrow a_{th} = 12.5$$

Thus,

$$C_8H_{18}(g) + 16.25(O_2 + 3.76N_2) \longrightarrow 8CO_2 + 9H_2O + 3.75O_2 + 61.1N_2$$

Therefore, 16.25×4.76 = 77.35 kmol of dry air will be used per kmol of the fuel. The partial pressure of the water vapor present in the incoming air is

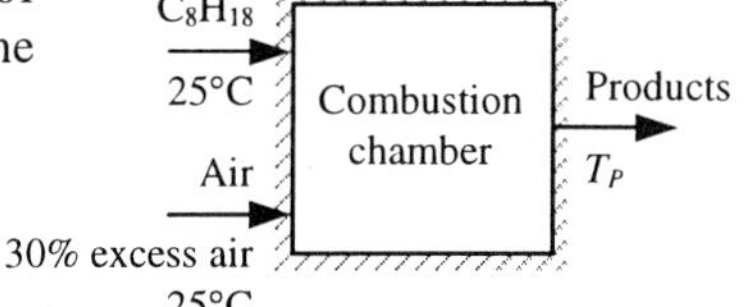

$$P_{v,in} = \phi_{air} P_{sat@25^\circ C} = (0.60)(3.169\text{kPa}) = 1.901\text{kPa}$$

Assuming ideal gas behavior, the number of moles of the moisture that accompanies 77.35 kmol of incoming dry air is determined to be

$$N_{v,in} = \left(\frac{P_{v,in}}{P_{total}}\right)N_{total} = \left(\frac{1.901\text{kPa}}{101.325\text{kPa}}\right)(77.35 + N_{v,in}) \longrightarrow N_{v,in} = 1.48\text{kmol}$$

The balanced combustion equation is obtained by adding 1.48 kmol of H_2O to both sides of the equation,

$$C_8H_{18}(g) + 16.25(O_2 + 3.76N_2) + 1.48H_2O \longrightarrow 8CO_2 + 10.48H_2O + 3.75O_2 + 61.1N_2$$

From the tables,

Substance	$\bar{h}_f^\circ$ kJ/kmol	$\bar{h}_{298\text{ K}}$ kJ/kmol
C_8H_{18} (*g*)	-208,450	---
O_2	0	8682
N_2	0	8669
H_2O (*g*)	-241,820	9904
CO_2	-393,520	9364

Thus,

$$(8)(-393{,}520 + \bar{h}_{CO_2} - 9364) + (10.48)(-241{,}820 + \bar{h}_{H_2O} - 9904) + (3.75)(0 + \bar{h}_{O_2} - 8682)$$
$$+ (61.1)(0 + \bar{h}_{N_2} - 8669) = (1)(-208{,}450) + (1.48)(-241{,}820) + 0 + 0$$

It yields $\quad 8\bar{h}_{CO_2} + 10.48\bar{h}_{H_2O} + 3.75\bar{h}_{O_2} + 61.1\bar{h}_{N_2} = 5{,}857{,}029 \text{ kJ}$

The adiabatic flame temperature is obtained from a trial and error solution. A first guess is obtained by dividing the right-hand side of the equation by the total number of moles, which yields 5,857,029/(8 + 10.48 + 3.75 + 61.1) = 70,287 kJ/kmol. This enthalpy value corresponds to about 2150 K for N_2. Noting that the majority of the moles are N_2, T_P will be close to 2150 K, but somewhat under it because of the higher specific heat of H_2O.

At 2000 K:

$$8\bar{h}_{CO_2} + 10.48\bar{h}_{H_2O} + 3.75\bar{h}_{O_2} + 61.1\bar{h}_{N_2} = (8)(100,804) + (10.48)(82,593) + (3.75)(67,881) + (61.1)(64,810)$$
$$= 5,886,451 \text{kJ (Higher than 5,857,029kJ)}$$

At 1980 K:

$$8\bar{h}_{CO_2} + 10.48\bar{h}_{H_2O} + 3.75\bar{h}_{O_2} + 61.1\bar{h}_{N_2} = (8)(99,606) + (10.48)(81,573) + (3.75)(67,127) + (61.1)(64,090)$$
$$= 5,819,358 \text{kJ (Lower than 5,857,029kJ)}$$

By interpolation, T_P = **1991 K**

14-71 EES solution of this (and other comprehensive problems designated with the *computer icon*) is available to instructors at the *Instructor Manual* section of the *Online Learning Center* (OLC) at www.mhhe.com/cengel-boles. See the Preface for access information.

Entropy Change and Second Law Analysis of Reacting Systems

14-72C Assuming the system exchanges heat with the surroundings at T_0, the increase-in-entropy principle can be expressed as

$$S_{gen} = \sum N_P \bar{s}_P - \sum N_R \bar{s}_R + \frac{Q_{out}}{T_0}$$

14-73C By subtracting $R\ln(P/P_0)$ from the tabulated value at 1 atm. Here P is the actual pressure of the substance and P_0 is the atmospheric pressure.

14-74C It represents the reversible work associated with the formation of that compound.

14-75 Hydrogen is burned steadily with oxygen. The reversible work and exergy destruction (or irreversibility) are to be determined.

Assumptions **1** Combustion is complete. **2** Steady operating conditions exist. **3** Air and the combustion gases are ideal gases. **4** Changes in kinetic and potential energies are negligible.

Analysis The combustion equation is $\quad H_2 + 0.5O_2 \longrightarrow H_2O.$

The H_2, the O_2, and the H_2O are at 25°C and 1 atm, which is the standard reference state and also the state of the surroundings. Therefore, the reversible work in this case is simply the difference between the Gibbs function of formation of the reactants and that of the products,

$$W_{rev} = \sum N_R \bar{g}^\circ_{f,R} - \sum N_P \bar{g}^\circ_{f,P} = N_{H_2}\bar{g}^\circ_{f,H_2}{}^{\not 0} + N_{O_2}\bar{g}^\circ_{f,O_2}{}^{\not 0} - N_{H_2O}\bar{g}^\circ_{f,H_2O} = -N_{H_2O}\bar{g}^\circ_{f,H_2O}$$
$$= -(1\text{kmol})(-237{,}180\text{kJ/kmol}) = \mathbf{237{,}180kJ} \quad (\text{per kmol of } H_2)$$

since the $\bar{g}^\circ_f$ of stable elements at 25°C and 1 atm is zero. Therefore, 237,180 kJ of work could be done as 1 kmol of H_2 is burned with 0.5 kmol of O_2 at 25°C and 1 atm in an environment at the same state. The reversible work in this case represents the exergy of the reactants since the product (the H_2O) is at the state of the surroundings.

This process involves no actual work. Therefore, the reversible work and exergy destruction are identical,

$X_{destruction}$ = **237,180 kJ** (per kmol of H_2)

We could also determine the reversible work without involving the Gibbs function,

$$W_{rev} = \sum N_R\left(\bar{h}^\circ_f + \bar{h} - \bar{h}^\circ - T_0\bar{s}\right)_R - \sum N_P\left(\bar{h}^\circ_f + \bar{h} - \bar{h}^\circ - T_0\bar{s}\right)_P$$
$$= \sum N_R\left(\bar{h}^\circ_f - T_0\bar{s}\right)_R - \sum N_P\left(\bar{h}^\circ_f - T_0\bar{s}\right)_P$$
$$= N_{H_2}\left(\bar{h}^\circ_f - T_0\bar{s}^\circ\right)_{H_2} + N_{O_2}\left(\bar{h}^\circ_f - T_0\bar{s}^\circ\right)_{O_2} - N_{H_2O}\left(\bar{h}^\circ_f - T_0\bar{s}^\circ\right)_{H_2O}$$

Substituting,

$$W_{rev} = (1)(0 - 298\times130.68) + (0.5)(0 - 298\times205.04) - (1)(-285{,}830 - 298\times69.92)$$
$$= 237{,}173\text{kJ}$$

which is almost identical to the result obtained before.

14-76 Ethylene gas is burned steadily with 20 percent excess air. The temperature of products, the entropy generation, and the exergy destruction (or irreversibility) are to be determined.

Assumptions **1** Combustion is complete. **2** Steady operating conditions exist. **3** Air and the combustion gases are ideal gases. **4** Changes in kinetic and potential energies are negligible.

Analysis (*a*) The fuel is burned completely with the excess air, and thus the products will contain only CO_2, H_2O, N_2, and some free O_2. Considering 1 kmol of C_2H_4, the combustion equation can be written as

$$C_2H_4(g) + 1.2a_{th}(O_2 + 3.76N_2) \longrightarrow 2CO_2 + 2H_2O + 0.2a_{th}O_2 + (1.2)(3.76)a_{th}N_2$$

where a_{th} is the stoichiometric coefficient and is determined from the O_2 balance,

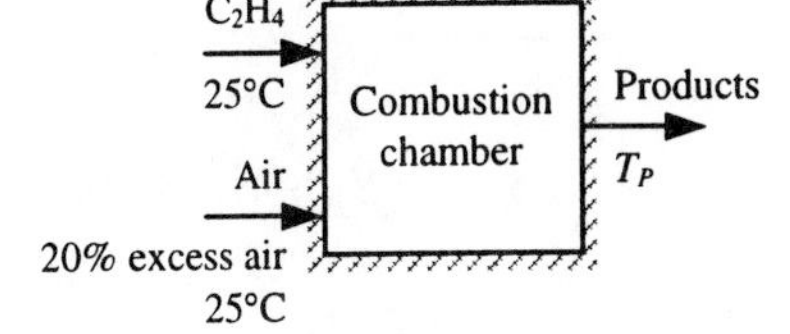

$$1.2a_{th} = 2 + 1 + 0.2a_{th} \longrightarrow a_{th} = 3$$

Thus,

$$C_2H_4(g) + 3.6(O_2 + 3.76N_2) \longrightarrow 2CO_2 + 2H_2O + 0.6O_2 + 13.54N_2$$

Under steady-flow conditions, the exit temperature of the product gases can be determined from the steady-flow energy equation, which reduces to

$$\sum N_P(\overline{h}_f^\circ + \overline{h} - \overline{h}^\circ)_P = \sum N_R \overline{h}_{f,R}^\circ = (N\overline{h}_f^\circ)_{C_2H_4}$$

since all the reactants are at the standard reference state, and for O_2 and N_2. From the tables,

Substance	$\overline{h}_f^\circ$ kJ/kmol	$\overline{h}_{298\text{ K}}$ kJ/kmol
C_2H_4 (*g*)	52,280	---
O_2	0	8682
N_2	0	8669
H_2O (*g*)	-241,820	9904
CO_2	-393,520	9364

Substituting,

$$(2)(-393{,}520 + \overline{h}_{CO_2} - 9364) + (2)(-241{,}820 + \overline{h}_{H_2O} - 9904)$$
$$+ (0.6)(0 + \overline{h}_{O_2} - 8682) + (13.54)(0 + \overline{h}_{N_2} - 8669) = (1)(52{,}280)$$

or,

$$2\overline{h}_{CO_2} + 2\overline{h}_{H_2O} + 0.6\overline{h}_{O_2} + 13.54\overline{h}_{N_2} = 1{,}484{,}083 \text{ kJ}$$

By trial and error, T_P = **2269.6 K**

(*b*) The entropy generation during this adiabatic process is determined from

$$S_{gen} = S_P - S_R = \sum N_P \overline{s}_P - \sum N_R \overline{s}_R$$

The C_2H_4 is at 25°C and 1 atm, and thus its absolute entropy is 219.83 kJ/kmol K (Table A-26). The entropy values listed in the ideal gas tables are for 1 atm pressure. Both the air and the product gases are at a total pressure of 1 atm, but the entropies are to be calculated at the partial pressure of the components which is equal to $P_i = y_i P_{total}$, where y_i is the mole fraction of component *i*. Also,

$$S_i = N_i \overline{s}_i(T, P_i) = N_i\left(\overline{s}_i^\circ(T, P_0) - R_u \ln(y_i P_m)\right)$$

The entropy calculations can be presented in tabular form as

	N_i	y_i	$\bar{s}_i^\circ(T,1atm)$	$R_u \ln(y_i P_m)$	$N_i \bar{s}_i$
C_2H_4	1	1.00	219.83	---	219.83
O_2	3.6	0.21	205.14	-12.98	784.87
N_2	13.54	0.79	191.61	-1.96	2620.94
					S_R = 3625.64 kJ/K
CO_2	2	0.1103	316.881	-18.329	670.42
H_2O	2	0.1103	271.134	-18.329	578.93
O_2	0.6	0.0331	273.467	-28.336	181.08
N_2	13.54	0.7464	256.541	-2.432	3506.49
					S_P = 4936.92 kJ/K

Thus,

$$S_{\text{gen}} = S_P - S_R = 4936.92 - 3625.64 = \mathbf{1311.28\ kJ/kmol \cdot K}$$

and

(*c*) $$X_{\text{destroyed}} = T_0 S_{\text{gen}} = (298\text{K})(1311.28\text{kJ/kmol} \cdot \text{K} C_2H_4) = \mathbf{390{,}760\ kJ/K}\ (\text{per kmol } C_2H_4)$$

14-77 Liquid octane is burned steadily with 50 percent excess air. The heat transfer rate from the combustion chamber, the entropy generation rate, and the reversible work and exergy destruction rate are to be determined.

Assumptions **1** Combustion is complete. **2** Steady operating conditions exist. **3** Air and the combustion gases are ideal gases. **4** Changes in kinetic and potential energies are negligible.

Analysis (*a*) The fuel is burned completely with the excess air, and thus the products will contain only CO_2, H_2O, N_2, and some free O_2. Considering 1 kmol C_8H_{18}, the combustion equation can be written as

$$C_8H_{18}(\ell) + 1.5a_{th}(O_2 + 3.76N_2) \longrightarrow 8CO_2 + 9H_2O + 0.5a_{th}O_2 + (1.5)(3.76)a_{th}N_2$$

where a_{th} is the stoichiometric coefficient and is determined from the O_2 balance,

$$1.5a_{th} = 8 + 4.5 + 0.5a_{th} \longrightarrow a_{th} = 12.5$$

T_0 = 298 K

$\dot{Q}$

C_8H_{18} (ℓ)

25°C

Air

50% excess air

25°C

Combustion chamber

Products

25°C

Thus,

$$C_8H_{18}(\ell) + 18.75(O_2 + 3.76N_2) \longrightarrow 8CO_2 + 9H_2O + 6.25O_2 + 70.5N_2$$

Under steady-flow conditions the energy balance $E_{in} - E_{out} = \Delta E_{system}$ applied on the combustion chamber with $W = 0$ reduces to

$$-Q_{out} = \sum N_P\left(\bar{h}_f^\circ + \bar{h} - \bar{h}^\circ\right)_P - \sum N_R\left(\bar{h}_f^\circ + \bar{h} - \bar{h}^\circ\right)_R \longrightarrow -Q_{out} = \sum N_P \bar{h}_{f,P}^\circ - \sum N_R \bar{h}_{f,R}^\circ$$

since all of the reactants are at 25°C. Assuming the air and the combustion products to be ideal gases, we have $h = h(T)$. From the tables,

Substance	$\bar{h}_f^\circ$ kJ/kmol
C_8H_{18} (ℓ)	-249,950
O_2	0
N_2	0
H_2O (l)	-285,830
CO_2	-393,520

Substituting,

$$-Q_{out} = (8)(-393{,}520) + (9)(-285{,}830) + 0 + 0 - (1)(-249{,}950) - 0 - 0 = -5{,}470{,}680 \text{ kJ / kmol of } C_8H_{18}$$

or $$Q_{out} = 5{,}470{,}680 \text{ kJ/kmol of } C_8H_{18}$$

The C_8H_{18} is burned at a rate of 0.4 kg/min or

$$\dot{N} = \frac{\dot{m}}{M} = \frac{0.4 \text{ kg/min}}{[(8)(12) + (18)(1)]\text{kg/kmol}} = 3.51 \times 10^{-3} \text{ kmol/min}$$

Thus,

$$\dot{Q}_{out} = \dot{N}Q_{out} = (3.51 \times 10^{-3} \text{ kmol/min})(5{,}470{,}680 \text{ kJ/kmol}) = \mathbf{19{,}195 \text{ kJ/min}}$$

The heat transfer for this process is also equivalent to the enthalpy of combustion of liquid C_8H_{18}, which could easily be de determined from Table A-27 to be $\bar{h}_C$ = 5,470,740 kJ/kmol C_8H_{18}.

(*b*) The entropy generation during this process is determined from

$$S_{\text{gen}} = S_P - S_R + \frac{Q_{\text{out}}}{T_{\text{surr}}} \longrightarrow S_{\text{gen}} = \sum N_P \bar{s}_P - \sum N_R \bar{s}_R + \frac{Q_{\text{out}}}{T_{\text{surr}}}$$

The C_8H_{18} is at 25°C and 1 atm, and thus its absolute entropy is $\bar{s}_{C_8H_{18}}$ = 360.79 kJ/kmol.K (Table A-26). The entropy values listed in the ideal gas tables are for 1 atm pressure. Both the air and the product gases are at a total pressure of 1 atm, but the entropies are to be calculated at the partial pressure of the components which is equal to $P_i = y_i P_{\text{total}}$, where y_i is the mole fraction of component i. Also,

$$S_i = N_i \bar{s}_i(T, P_i) = N_i\left(\bar{s}_i^\circ(T, P_0) - R_u \ln(y_i P_m)\right)$$

The entropy calculations can be presented in tabular form as

	N_i	y_i	$\bar{s}_i^\circ(T,1atm)$	$R_u \ln(y_i P_m)$	$N_i \bar{s}_i$
C_8H_{18}	1	1.00	360.79	---	360.79
O_2	18.75	0.21	205.14	-12.98	4089.75
N_2	70.50	0.79	191.61	-1.96	13646.69
					S_R = 18,097.23 kJ/K
CO_2	8	0.0944	213.80	-19.62	1867.3
H_2O (ℓ)	9	---	69.92	---	629.3
O_2	18.75	0.0737	205.04	-21.78	1417.6
N_2	70.50	0.8319	191.61	-1.64	13,624.1
					S_P = 17,538 kJ/K

Thus,

$$S_{\text{gen}} = S_P - S_R + \frac{Q_{\text{surr}}}{T_{\text{surr}}} = 17{,}538 - 18{,}097 + \frac{5{,}470{,}680 \text{ kJ}}{298 \text{ K}} = 17{,}799 \text{ kJ/kmol}\cdot\text{K}$$

and

$$\dot{S}_{\text{gen}} = \dot{N} S_{\text{gen}} = \left(3.51\times10^{-3} \text{ kmol/min}\right)\left(17{,}799 \text{ kJ/kmol}\cdot\text{K}\right) = \mathbf{62.47 \text{ kJ/min}\cdot\text{K}}$$

(*c*) The exergy destruction rate associated with this process is determined from

$$\dot{X}_{\text{destroyed}} = T_0 \dot{S}_{\text{gen}} = (298 \text{ K})(62.47 \text{ kJ/min}\cdot\text{K}) = 18{,}617 \text{ kJ/min} = \mathbf{310.3 \text{ kW}}$$

14-78 Acetylene gas is burned steadily with 20 percent excess air. The temperature of the products, the total entropy change, and the exergy destruction are to be determined.

Assumptions **1** Combustion is complete. **2** Steady operating conditions exist. **3** Air and the combustion gases are ideal gases. **4** Changes in kinetic and potential energies are negligible.

Analysis (*a*) The fuel is burned completely with the excess air, and thus the products will contain only CO_2, H_2O, N_2, and some free O_2. Considering 1 kmol C_2H_2, the combustion equation can be written as

$$C_2H_2(g) + 1.2a_{th}(O_2 + 3.76N_2) \longrightarrow 2CO_2 + H_2O + 0.2a_{th}O_2 + (1.2)(3.76)a_{th}N_2$$

where a_{th} is the stoichiometric coefficient and is determined from the O_2 balance,

$$1.2a_{th} = 2 + 0.5 + 0.2a_{th} \longrightarrow a_{th} = 2.5$$

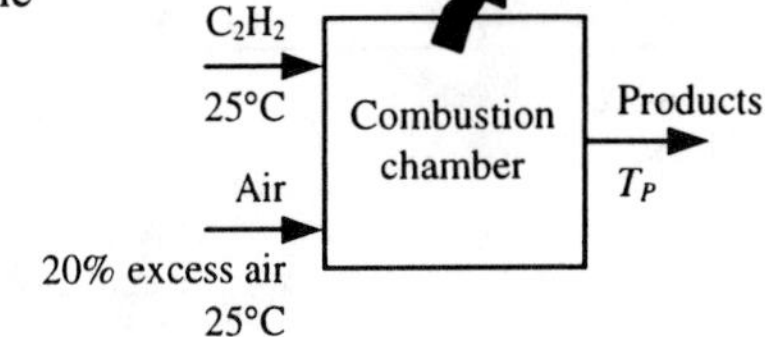

Substituting,

$$C_2H_2(g) + 3(O_2 + 3.76N_2) \longrightarrow 2CO_2 + H_2O + 0.5O_2 + 11.28N_2$$

Under steady-flow conditions the exit temperature of the product gases can be determined from the energy balance $E_{in} - E_{out} = \Delta E_{system}$ applied on the combustion chamber, which reduces to

$$-Q_{out} = \sum N_P\left(\bar{h}_f^\circ + \bar{h} - \bar{h}^\circ\right)_P - \sum N_R \bar{h}_{f,R}^\circ = \sum N_P\left(\bar{h}_f^\circ + \bar{h} - \bar{h}^\circ\right)_P - \left(N\bar{h}_f^\circ\right)_{C_2H_2}$$

since all the reactants are at the standard reference state, and $\bar{h}_f^\circ = 0$ for O_2 and N_2. From the tables,

Substance	$\bar{h}_f^\circ$ kJ/kmol	$\bar{h}_{298\ K}$ kJ/kmol
C_2H_2 (*g*)	226,730	---
O_2	0	8682
N_2	0	8669
H_2O (*g*)	-241,820	9904
CO_2	-393,520	9364

Substituting,

$$-300{,}000 = (2)\left(-393{,}520 + \bar{h}_{CO_2} - 9364\right) + (1)\left(-241{,}820 + \bar{h}_{H_2O} - 9904\right)$$
$$+ (0.5)\left(0 + \bar{h}_{O_2} - 8682\right) + (11.28)\left(0 + \bar{h}_{N_2} - 8669\right) - (1)(226{,}730)$$

or,
$$2\bar{h}_{CO_2} + \bar{h}_{H_2O} + 0.5\bar{h}_{O_2} + 11.28\bar{h}_{N_2} = 1{,}086{,}349 \text{ kJ}$$

By trial and error, T_P = **2062.1 K**

(*b*) The entropy generation during this process is determined from

$$S_{gen} = S_P - S_R + \frac{Q_{out}}{T_{surr}} = \sum N_P \bar{s}_P - \sum N_R \bar{s}_R + \frac{Q_{out}}{T_{surr}}$$

The C_2H_2 is at 25°C and 1 atm, and thus its absolute entropy is $\bar{s}_{C_2H_2} = 200.85 \text{ kJ/kmol}\cdot\text{K}$ (Table A-26). The entropy values listed in the ideal gas tables are for 1 atm pressure. Both the air and the product gases are at a total pressure of 1 atm, but the entropies are to be calculated at the partial pressure of the components which is equal to $P_i = y_i P_{total}$, where y_i is the mole fraction of component *i*. Also,

$$S_i = N_i \bar{s}_i(T, P_i) = N_i\left(\bar{s}_i^\circ(T, P_0) - R_u \ln(y_i P_m)\right)$$

The entropy calculations can be presented in tabular form as

	N_i	y_i	$\bar{s}_i^\circ(T,1atm)$	$R_u\ln(y_iP_m)$	$N_i\bar{s}_i$
C_2H_2	1	1.00	200.85	---	200.85
O_2	0.3	0.21	205.04	-12.98	654.06
N_2	11.28	0.79	191.61	-1.96	2183.47
				S_R =	3038.38 kJ/K
CO_2	2	0.1353	311.054	-16.630	655.37
H_2O	1	0.0677	266.139	-22.387	288.53
O_2	0.5	0.0338	269.810	-28.162	148.99
N_2	11.28	0.7632	253.068	-2.247	2879.95
				S_P =	3972.84 kJ/K

Thus,

$$S_{gen} = S_P - S_R + \frac{Q_{surr}}{T_{surr}} = 3972.84 - 3038.38 + \frac{+300{,}000\text{kJ}}{298\text{K}} = \mathbf{1941.2\ kJ/kmol \cdot K}$$

(*c*) The exergy destruction rate associated with this process is determined from

$$X_{destruction} = T_0 S_{gen} = (298\text{K})(1941.2\text{kJ/kmol}\cdot\text{K}) = \mathbf{578{,}469\ kJ}\ (\text{per kmol } C_2H_2)$$

14-79 CO gas is burned steadily with air. The heat transfer rate from the combustion chamber and the rate of exergy destruction are to be determined.

Assumptions **1** Combustion is complete. **2** Steady operating conditions exist. **3** Air and the combustion gases are ideal gases. **4** Changes in kinetic and potential energies are negligible.

Properties The molar masses of CO and air are 28 kg/kmol and 29 kg/kmol, respectively (Table A-1).

Analysis (*a*) We first need to calculate the amount of air used per kmol of CO before we can write the combustion equation,

$$v_{CO} = \frac{RT}{P} = \frac{(0.2968\text{kPa}\cdot\text{m}^3/\text{kg}\cdot\text{K})(310\text{K})}{(110\text{kPa})} = 0.836\text{m}^3/\text{kg}$$

$$\dot{m}_{CO} = \frac{\dot{V}_{CO}}{v_{CO}} = \frac{0.4\text{m}^3/\text{min}}{0.836\text{m}^3/\text{kg}} = 0.478\text{kg/min}$$

800 K

$\dot{Q}$

CO

37°C

Air

25°C

110 kPa

Products

900 K

Then the molar air-fuel ratio becomes

$$\overline{\text{AF}} = \frac{N_{\text{air}}}{N_{\text{fuel}}} = \frac{\dot{m}_{\text{air}}/M_{\text{air}}}{\dot{m}_{\text{fuel}}/M_{\text{fuel}}} = \frac{(1.5\text{kg/min})/(29\text{kg/kmol})}{(0.478\text{kg/min})/(28\text{kg/kmol})} = 3.03\text{kmolair/kmolfuel}$$

Thus the number of moles of O_2 used per mole of CO is 3.03/4.76 = 0.637. Then the combustion equation in this case can be written as

$$\text{CO} + 0.637(\text{O}_2 + 3.76\text{N}_2) \longrightarrow \text{CO}_2 + 0.137\text{O}_2 + 2.40\text{N}_2$$

Under steady-flow conditions the energy balance $E_{in} - E_{out} = \Delta E_{system}$ applied on the combustion chamber with $W = 0$ reduces to

$$-Q_{out} = \sum N_P(\bar{h}_f^\circ + \bar{h} - \bar{h}^\circ)_P - \sum N_R(\bar{h}_f^\circ + \bar{h} - \bar{h}^\circ)_R$$

Assuming the air and the combustion products to be ideal gases, we have $h = h(T)$. From the tables,

Substance	$\bar{h}_f^\circ$ kJ/kmol	$\bar{h}_{298\ K}$ kJ/kmol	$\bar{h}_{310\ K}$ kJ/kmol	$\bar{h}_{900\ K}$ kJ/kmol
CO	-110,530	8669	9014	27,066
O_2	0	8682	---	27,928
N_2	0	8669	---	26,890
CO_2	-393,520	9364	---	37,405

Substituting,

$$\begin{aligned} -Q_{out} &= (1)(-393{,}520 + 37{,}405 - 9364) + (0.137)(0 + 27{,}928 - 8682) \\ &\quad + (2.4)(0 + 26{,}890 - 8669) - (1)(-110{,}530 + 9014 - 8669) - 0 - 0 \\ &= -208{,}929\text{kJ/kmolofCO} \end{aligned}$$

Thus 208,929 kJ of heat is transferred from the combustion chamber for each kmol (28 kg) of CO. This corresponds to 208,929/28 = 7462 kJ of heat transfer per kg of CO. Then the rate of heat transfer for a mass flow rate of 0.478 kg/min for CO becomes

$$\dot{Q}_{out} = \dot{m}q_{out} = (0.478\text{ kg/min})(7462\text{kJ/kg}) = \mathbf{3567\ \ kJ/min}$$

(b) This process involves heat transfer with a reservoir other than the surroundings. An exergy balance on the combustion chamber in this case reduces to the following relation for reversible work,

$$W_{\text{rev}} = \sum N_R\left(\bar{h}_f^\circ + \bar{h} - \bar{h}^\circ - T_0\bar{s}\right)_R - \sum N_P\left(\bar{h}_f^\circ + \bar{h} - \bar{h}^\circ - T_0\bar{s}\right)_P - Q_{out}\left(1 - T_0/T_R\right)$$

The entropy values listed in the ideal gas tables are for 1 atm = 101.325 kPa pressure. The entropy of each reactant and the product is to be calculated at the partial pressure of the components which is equal to $P_i = y_i P_{\text{total}}$, where y_i is the mole fraction of component i, and $P_m = 110/101.325 = 1.0856$ atm. Also,

$$S_i = N_i\bar{s}_i(T,P_i) = N_i\left(\bar{s}_i^\circ(T,P_0) - R_u \ln(y_i P_m)\right)$$

The entropy calculations can be presented in tabular form as

	N_i	y_i	$\bar{s}_i^\circ(T,1atm)$	$R_u \ln(y_i P_m)$	$N_i\bar{s}_i$
CO	1	1.00	198.678	0.68	198.00
O_2	0.637	0.21	205.04	-12.29	138.44
N_2	2.400	0.79	191.61	-1.28	462.94
					S_R = 799.38 kJ/K
CO_2	1	0.2827	263.559	-9.821	273.38
O_2	0.137	0.0387	239.823	-26.353	36.47
N_2	2.400	0.6785	224.467	-2.543	544.82
					S_P = 854.67 kJ/K

The rate of exergy destruction can be determined from

$$\dot{X}_{\text{destroyed}} = T_0\dot{S}_{\text{gen}} = T_0\dot{m}\left(S_{\text{gen}}/M\right)$$

where

$$S_{\text{gen}} = S_P - S_R + \frac{Q_{\text{out}}}{T_{\text{res}}} = 854.67 - 799.38 + \frac{208{,}929\text{kJ}}{800\text{K}} = 316.5\text{kJ/kmol}\cdot\text{K}$$

Thus,

$$\dot{X}_{\text{destroyed}} = (298\text{K})(0.478\text{kg/min})(316.5/28\text{kJ/kmol}\cdot\text{K}) = \mathbf{1610kJ/min}$$

14-80E Benzene gas is burned steadily with 95 percent theoretical air. The heat transfer rate from the combustion chamber and the exergy destruction are to be determined.

Assumptions **1** Steady operating conditions exist. **2** Air and the combustion gases are ideal gases. **3** Changes in kinetic and potential energies are negligible.

Analysis (*a*) The fuel is burned with insufficient amount of air, and thus the products will contain some CO as well as CO_2, H_2O, and N_2. The theoretical combustion equation of C_6H_6 is

$$C_6H_6 + a_{th}(O_2 + 3.76N_2) \longrightarrow 6CO_2 + 3H_2O + 3.76a_{th}N_2$$

Q
C_6H_6
77°F
Air
95% theoretical
77°F
Combustion chamber
Products
1500 R

where a_{th} is the stoichiometric coefficient and is determined from the O_2 balance,

$$a_{th} = 6 + 1.5 = 7.5$$

Then the actual combustion equation can be written as

$$C_6H_6 + (0.95)(7.5)(O_2 + 3.76N_2) \longrightarrow xCO_2 + (6-x)CO + 3H_2O + 26.79N_2$$

The value of x is determined from an O_2 balance,

$$(0.95)(7.5) = x + (6-x)/2 + 1.5 \longrightarrow x = 5.25$$

Thus,

$$C_6H_6 + 7.125(O_2 + 3.76N_2) \longrightarrow 5.25CO_2 + 0.75CO + 3H_2O + 26.79N_2$$

Under steady-flow conditions the energy balance $E_{in} - E_{out} = \Delta E_{system}$ applied on the combustion chamber with $W = 0$ reduces to

$$-Q_{out} = \sum N_P\left(\bar{h}_f^\circ + \bar{h} - \bar{h}^\circ\right)_P - \sum N_R\left(\bar{h}_f^\circ + \bar{h} - \bar{h}^\circ\right)_R \longrightarrow -Q_{out} = \sum N_P\left(\bar{h}_f^\circ + \bar{h} - \bar{h}^\circ\right)_P - \sum N_R \bar{h}_{f,R}^\circ$$

since all of the reactants are at 77°F. Assuming the air and the combustion products to be ideal gases, we have $h = h(T)$.

From the tables,

Substance	$\bar{h}_f^\circ$ Btu/lbmol	$\bar{h}_{537\ R}$ Btu/lbmol	$\bar{h}_{1500\ R}$ Btu/lbmol
C_6H_6 (*g*)	35,680	---	---
O_2	0	3725.1	11,017.1
N_2	0	3729.5	10,648.0
H_2O (*g*)	-104,040	4258.0	12,551.4
CO	-47,540	3725.1	10,711.1
CO_2	-169,300	4027.5	14,576.0

Thus,

$$\begin{aligned}-Q_{out} &= (5.25)(-169{,}300 + 14{,}576 - 4027.5) + (0.75)(-47{,}540 + 10{,}711.1 - 3725.1) \\ &\quad + (3)(-104{,}040 + 12{,}551.4 - 4258) + (26.79)(0 + 10{,}648 - 3729.5) - (1)(35{,}680) - 0 - 0 \\ &= \mathbf{-1{,}001{,}434\ Btu/lbmol\ of\ C_6H_6}\end{aligned}$$

(*b*) The entropy generation during this process is determined from

$$S_{gen} = S_P - S_R + \frac{Q_{out}}{T_{surr}} = \sum N_P \bar{s}_P - \sum N_R \bar{s}_R + \frac{Q_{out}}{T_{surr}}$$

The C_6H_6 is at 77°F and 1 atm, and thus its absolute entropy is $\bar{s}_{C_6H_6}$ =269.20 Btu/lbmol R (Table A-26E). The entropy values listed in the ideal gas tables are for 1 atm pressure. Both the air and the product gases are at a total pressure of 1 atm, but the entropies are to be calculated at the partial pressure of the components which is equal to $P_i = y_i P_{total}$, where y_i is the mole fraction of component *i*. Also,

$$S_i = N_i \bar{s}_i(T, P_i) = N_i\left(\bar{s}_i^\circ(T, P_0) - R_u \ln(y_i P_m)\right)$$

The entropy calculations can be presented in tabular form as

	N_i	y_i	$\bar{s}_i^\circ(T,1atm)$	$R_u \ln(y_i P_m)$	$N_i \bar{s}_i$
C_6H_6	1	1.00	64.34	---	64.34
O_2	7.125	0.21	49.00	-3.10	371.21
N_2	26.79	0.79	45.77	-0.47	1238.77
					S_R = 1674.32 Btu/R
CO_2	5.25	0.1467	61.974	-3.812	345.38
CO	0.75	0.0210	54.665	-7.672	46.75
H_2O (*g*)	3	0.0838	53.808	-4.948	176.27
N_2	26.79	0.7485	53.071	-0.575	1437.18
					S_P = 2005.58 Btu/R

Thus,

$$S_{gen} = S_P - S_R + \frac{Q_{out}}{T_{surr}} = 1674.32 - 2005.58 + \frac{+1{,}001{,}434}{537} = 1533.6\ \text{Btu/R}$$

Then the exergy destroyed is determined from

$$X_{destroyed} = T_0 S_{gen} = (537\text{R})(1533.6\text{Btu/lbmol}\cdot\text{R}) = \mathbf{823{,}547\ Btu/R}\ (\text{per lbmol}C_6H_6)$$

14-81 [*Also solved by EES on enclosed CD*] Liquid propane is burned steadily with 150 percent excess air. The mass flow rate of air, the heat transfer rate from the combustion chamber, and the rate of entropy generation are to be determined.

Assumptions **1** Combustion is complete. **2** Steady operating conditions exist. **3** Air and the combustion gases are ideal gases. **4** Changes in kinetic and potential energies are negligible.

Properties The molar masses of C_3H_8 and air are 44 kg/kmol and 29 kg/kmol, respectively (Table A-1).

Analysis (*a*) The fuel is burned completely with the excess air, and thus the products will contain only CO_2, H_2O, N_2, and some free O_2. Considering 1 kmol of C_3H_8, the combustion equation can be written as

$$C_3H_8(\ell) + 2.5a_{th}(O_2 + 3.76N_2) \longrightarrow 3CO_2 + 4H_2O + 1.5a_{th}O_2 + (2.5)(3.76)a_{th}N_2$$

where a_{th} is the stoichiometric coefficient and is determined from the O_2 balance,

$$2.5a_{th} = 3 + 2 + 1.5a_{th} \quad \longrightarrow \quad a_{th} = 5$$

Substituting,

$$C_3H_8(\ell) + 12.5(O_2 + 3.76N_2) \longrightarrow 3CO_2 + 4H_2O + 7.5O_2 + 47N_2$$

The air-fuel ratio for this combustion process is

$$AF = \frac{m_{air}}{m_{fuel}} = \frac{(12.5 \times 4.76\text{kmol})(29\text{kg/kmol})}{(3\text{kmol})(12\text{kg/kmol}) + (4\text{kmol})(2\text{kg/kmol})} = 39.2\text{kg air/kg fuel}$$

Thus, $\dot{m}_{air} = (AF)(\dot{m}_{fuel}) = (39.2\text{kg air/kg fuel})(0.4\text{kg fuel/min}) = \mathbf{15.7kg\ air/min}$

(*b*) Under steady-flow conditions the energy balance $E_{in} - E_{out} = \Delta E_{system}$ applied on the combustion chamber with $W = 0$ reduces to

$$-Q_{out} = \sum N_P(\bar{h}_f^\circ + \bar{h} - \bar{h}^\circ)_P - \sum N_R(\bar{h}_f^\circ + \bar{h} - \bar{h}^\circ)_R$$

Assuming the air and the combustion products to be ideal gases, we have $h = h(T)$. From the tables, (The $\bar{h}_f^\circ$ of liquid propane is obtained by adding the h_{fg} at 25°C to $\bar{h}_f^\circ$ of gaseous propane).

Substance	$\bar{h}_f^\circ$ kJ/kmol	$\bar{h}_{285\text{ K}}$ kJ/kmol	$\bar{h}_{298\text{ K}}$ kJ/kmol	$\bar{h}_{1200\text{ K}}$ kJ/kmol
C_3H_8 (ℓ)	-118,910	---	---	---
O_2	0	8296.5	8682	38,447
N_2	0	8286.5	8669	36,777
H_2O (*g*)	-241,820	---	9904	44,380
CO_2	-393,520	---	9364	53,848

Thus,

$$\begin{aligned}-Q_{out} &= (3)(-393{,}520 + 53{,}848 - 9364) + (4)(-241{,}820 + 44{,}380 - 9904) \\ &\quad + (7.5)(0 + 38{,}447 - 8682) + (47)(0 + 36{,}777 - 8669) - (1)(-118{,}910 + h_{298} - h_{298}) \\ &\quad - (12.5)(0 + 8296.5 - 8682) - (47)(0 + 8286.5 - 8669) \\ &= -190{,}464\text{kJ/kmol of }C_3H_8\end{aligned}$$

Thus 190,464 kJ of heat is transferred from the combustion chamber for each kmol (44 kg) of propane. This corresponds to 190,464/44 = 4328.7 kJ of heat transfer per kg of propane. Then the rate of heat transfer for a mass flow rate of 0.4 kg/min for the propane becomes

$$\dot{Q}_{out} = \dot{m}q_{out} = (0.4\text{kg/min})(4328.7\text{kJ/kg}) = \mathbf{1732kJ/min}$$

(*c*) The entropy generation during this process is determined from

$$S_{gen} = S_P - S_R + \frac{Q_{out}}{T_{surr}} = \sum N_P \bar{s}_P - \sum N_R \bar{s}_R + \frac{Q_{out}}{T_{surr}}$$

The C_3H_8 is at 25°C and 1 atm, and thus its absolute entropy for the gas phase is $\bar{s}_{C_3H_8} = 269.91 \text{ kJ/kmol K}$ (Table A-26). Then the entropy of $C_3H_8(\ell)$ is obtained from

$$s_{C_3H_8}(\ell) \cong s_{C_3H_8}(g) - s_{fg} = s_{C_3H_8}(g) - \frac{\bar{h}_{fg}}{T} = 269.91 - \frac{15{,}060}{298.15} = 219.4 \text{ kJ/kmol} \cdot \text{K}$$

The entropy values listed in the ideal gas tables are for 1 atm pressure. Both the air and the product gases are at a total pressure of 1 atm, but the entropies are to be calculated at the partial pressure of the components which is equal to $P_i = y_i P_{total}$, where y_i is the mole fraction of component *i*. Then,

$$S_i = N_i \bar{s}_i (T, P_i) = N_i \left(\bar{s}_i^\circ (T, P_0) - R_u \ln(y_i P_m) \right)$$

The entropy calculations can be presented in tabular form as

	N_i	y_i	$\bar{s}_i^\circ(T, 1\text{atm})$	$R_u \ln(y_i P_m)$	$N_i \bar{s}_i$
C_3H_8	1	---	219.40	---	219.40
O_2	12.5	0.21	203.70	-12.98	2708.50
N_2	47	0.79	190.18	-1.96	9030.58
				S_R =	11,958.48 kJ/K
CO_2	3	0.0488	279.307	-25.112	913.26
H_2O (*g*)	4	0.0650	240.333	-22.720	1052.21
O_2	7.5	0.1220	249.906	-17.494	2005.50
N_2	47	0.7642	234.115	-2.236	11108.50
				S_P =	15,079.47 kJ/K

Thus,

$$S_{gen} = S_P - S_R + \frac{Q_{out}}{T_{surr}} = 15{,}079.47 - 11{,}958.48 + \frac{190{,}464}{298} = 3760.1 \text{ kJ/K (per kmol } C_3H_8)$$

Then the rate of entropy generation becomes

$$\dot{S}_{gen} = (\dot{N})(S_{gen}) = \left(\frac{0.4}{44} \text{ kmol/min}\right)(3760.1 \text{ kJ/kmol} \cdot \text{K}) = \mathbf{34.2 \text{ kJ/min} \cdot K}$$

14-82 EES solution of this (and other comprehensive problems designated with the *computer icon*) is available to instructors at the *Instructor Manual* section of the *Online Learning Center* (OLC) at www.mhhe.com/cengel-boles. See the Preface for access information.

Review Problems

14-83 A sample of a certain fluid is burned in a bomb calorimeter. The heating value of the fuel is to be determined.

Properties The specific heat of water is 4.18 kJ/kg.°C (Table A-3).

Analysis We take the water as the system, which is a closed system, for which the energy balance on the system $E_{in} - E_{out} = \Delta E_{system}$ with $W = 0$ can be written as

$$Q_{in} = \Delta U$$

or

$$\begin{aligned} Q_{in} &= mC\Delta T \\ &= (3\text{kg})(4.18\text{kJ/kg}\cdot{}^\circ\text{C})(1.5^\circ\text{C}) \\ &= 18.81\text{kJ}\ (\text{per gram of fuel}) \end{aligned}$$

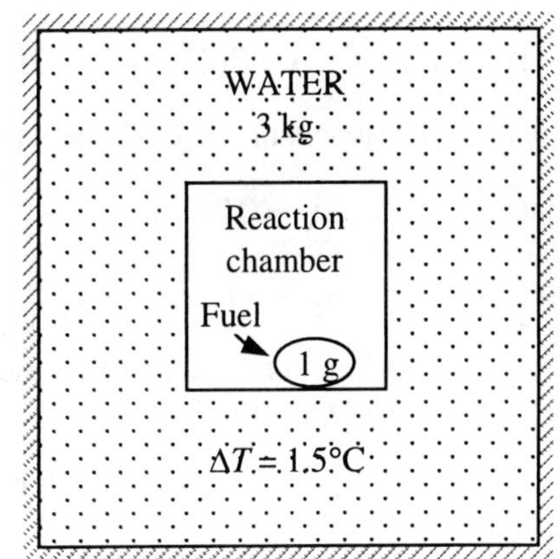

Therefore, heat transfer per kg of the fuel would be **18,810 kJ/kg fuel**. Disregarding the slight energy stored in the gases of the combustion chamber, this value corresponds to the heating value of the fuel.

14-84E Hydrogen is burned with 100 percent excess air. The AF ratio and the volume flow rate of air are to be determined.

Assumptions **1** Combustion is complete. **2** Air and the combustion gases are ideal gases.

Properties The molar masses of H_2 and air are 2 kg/kmol and 29 kg/kmol, respectively (Table A-1).

Analysis (*a*) The combustion is complete, and thus products will contain only H_2O, O_2 and N_2. The moisture in the air does not react with anything; it simply shows up as additional H_2O in the products. Therefore, for simplicity, we will balance the combustion equation using dry air, and then add the moisture to both sides of the equation. The combustion equation in this case can be written as

$$\mathrm{H_2} + 2a_{\text{th}}(\mathrm{O_2} + 3.76\mathrm{N_2}) \longrightarrow \mathrm{H_2O} + a_{\text{th}}\mathrm{O_2} + (2)(3.76)a_{\text{th}}\mathrm{N_2}$$

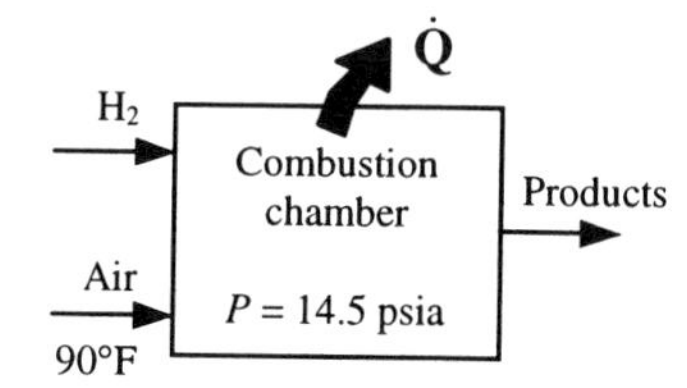

where a_{th} is the stoichiometric coefficient for air. It is determined from

O_2 balance: $\quad 2a_{\text{th}} = 0.5 + a_{\text{th}} \longrightarrow a_{\text{th}} = 0.5$

Substituting, $\quad \mathrm{H_2} + (\mathrm{O_2} + 3.76\mathrm{N_2}) \longrightarrow \mathrm{H_2O} + 0.5\mathrm{O_2} + 3.76\mathrm{N_2}$

Therefore, 4.76 lbmol of dry air will be used per kmol of the fuel. The partial pressure of the water vapor present in the incoming air is

$$P_{v,\text{in}} = \phi_{\text{air}} P_{\text{sat@90°F}} = (0.60)(0.6988\text{psia}) = 0.419\text{psia}$$

The number of moles of the moisture that accompanies 4.76 lbmol of incoming dry air ($N_{v,\text{ in}}$) is determined to be

$$N_{v,\text{in}} = \left(\frac{P_{v,\text{in}}}{P_{\text{total}}}\right) N_{\text{total}} = \left(\frac{0.419\text{psia}}{14.5\text{psia}}\right)(4.76 + N_{v,\text{in}}) \longrightarrow N_{v,\text{in}} = 0.142\text{lbmol}$$

The balanced combustion equation is obtained by substituting the coefficients determined earlier and adding 0.142 lbmol of H_2O to both sides of the equation,

$$H_2 + (O_2 + 3.76N_2) + 0.142H_2O \longrightarrow 1.142H_2O + 0.5O_2 + 3.76N_2$$

The air-fuel ratio is determined by taking the ratio of the mass of the air to the mass of the fuel,

$$\text{AF} = \frac{m_{\text{air}}}{m_{\text{fuel}}} = \frac{(4.76\text{lbmol})(29\text{lbm/lbmol}) + (0.142\text{lbmol})(18\text{lbm/lbmol})}{(1\text{lbmol})(2\text{lbm/lbmol})} = \mathbf{70.3\ \ lbm\ air/lbm\ fuel}$$

(*b*) The mass flow rate of H_2 is given to be 10 lbm/h. Since we need 70.3 lbm air per lbm of H_2, the required mass flow rate of air is

$$\dot{m}_{\text{air}} = (\text{AF})(\dot{m}_{\text{fuel}}) = (70.3)(25\text{ lbm/h}) = 1757\text{ lbm/h}$$

The mole fractions of water vapor and the dry air in the incoming air are

$$y_{\mathrm{H_2O}} = \frac{N_{\mathrm{H_2O}}}{N_{\text{total}}} = \frac{0.142}{4.76 + 0.142} = 0.029 \quad \text{and} \quad y_{\text{dry air}} = 1 - 0.029 = 0.971$$

Thus,

$$M = (yM)_{\mathrm{H_2O}} + (yM)_{\text{dryair}} = (0.029)(18) + (0.971)(29) = 28.7\text{lbm/lbmol}$$

$$v = \frac{RT}{P} = \frac{(10.73/28.7\text{psia}\cdot\text{ft}^3/\text{lbm}\cdot\text{R})(550\text{R})}{14.5\text{psia}} = 14.18\text{ft}^3/\text{lbm}$$

$$\dot{V} = \dot{m}v = (1757\text{lbm/h})(14.18\text{ft}^3/\text{lbm}) = \mathbf{24{,}916\ \ ft^3/h}$$

14-85 The composition of a gaseous fuel is given. The fuel is burned with 120 percent theoretical air. The AF ratio and the volume flow rate of air intake are to be determined.

Assumptions **1** Combustion is complete. **2** Air and the combustion gases are ideal gases.

Properties The molar masses of C, H_2, N_2, O_2, and air are 12, 2, 28, 32, and 29 kg/kmol (Table A-1).

Analysis (*a*) The fuel is burned completely with excess air, and thus the products will contain H_2O, CO_2, N_2, and some free O_2. The moisture in the air does not react with anything; it simply shows up as additional H_2O in the products. Therefore, we can simply balance the combustion equation using dry air, and then add the moisture to both sides of the equation. Considering 1 kmol of fuel, the combustion equation can be written as

$$(0.80\text{CH}_4 + 0.15\text{N}_2 + 0.05\text{O}_2) + 1.2a_{\text{th}}(\text{O}_2 + 3.76\text{N}_2) \longrightarrow x\text{CO}_2 + y\text{H}_2\text{O} + 0.2a_{\text{th}}\text{O}_2 + z\text{N}_2$$

The unknown coefficients in the above equation are determined from mass balances,

$$\text{C}:\ 0.80 = x \longrightarrow x = 0.80$$

$$\text{H}:\ (0.80)(4) = 2y \longrightarrow y = 1.6$$

$$\text{O}_2:\ 0.05 + 1.2a_{\text{th}} = x + y/2 + 0.2a_{\text{th}} \longrightarrow a_{\text{th}} = 1.55$$

$$\text{N}_2:\ 0.15 + (1.2)(3.76)a_{\text{th}} = z \longrightarrow z = 7.14$$

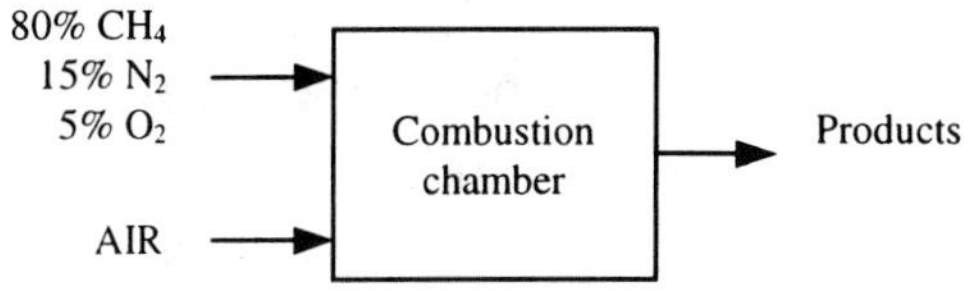

80% CH_4
15% N_2
5% O_2
AIR
Combustion chamber
Products

Next we determine the amount of moisture that accompanies $4.76\times1.2a_{\text{th}}$ = $4.76\times1.2\times1.55$ = 8.85 kmol of dry air. The partial pressure of the moisture in the air is

$$P_{v,\text{in}} = \phi_{\text{air}} P_{\text{sat @ 30°C}} = (0.60)(4.246\text{kPa}) = 2.548\text{kPa}$$

The number of moles of the moisture in the air ($N_{v,\text{ in}}$) is determined to be

$$N_{v,\text{in}} = \left(\frac{P_{v,\text{in}}}{P_{\text{total}}}\right)N_{\text{total}} = \left(\frac{2.548\text{kPa}}{100\text{kPa}}\right)(8.85 + N_{v,\text{in}}) \longrightarrow N_{v,\text{in}} = 0.23\text{kmol}$$

The balanced combustion equation is obtained by substituting the coefficients determined earlier and adding 0.23 kmol of H_2O to both sides of the equation,

$$(0.80\text{CH}_4 + 0.15\text{N}_2 + 0.05\text{O}_2) + 1.86(\text{O}_2 + 3.76\text{N}_2) + 0.23\text{H}_2\text{O} \longrightarrow 0.8\text{CO}_2 + 1.83\text{H}_2\text{O} + 0.31\text{O}_2 + 7.14\text{N}_2$$

The air-fuel ratio for the this reaction is determined by taking the ratio of the mass of the air to the mass of the fuel,

$$m_{\text{air}} = [(1.86)(4.76)\text{kmol}](29\text{kg/kmol}) + (0.23\text{kg})(18\text{kg/kmol}) = 260.9\text{kg}$$

$$m_{\text{fuel}} = [(0.8)(16) + (0.15)(28) + (0.05)(32)]\text{kg} = 18.6\text{kg}$$

and $$\text{AF} = \frac{m_{\text{air}}}{m_{\text{fuel}}} = \frac{260.9\text{kg}}{18.6\text{kg}} = \mathbf{14.0\ \ kgair/kgfuel}$$

(*b*) The mass flow rate of the gaseous fuel is given to be 1 kg/min. Since we need 14.0 kg air per kg of fuel, the required mass flow rate of air is

$$\dot{m}_{\text{air}} = (\text{AF})(\dot{m}_{\text{fuel}}) = (14.0)(1\text{kg/min}) = 14.0\text{kg/ min}$$

The mole fractions of water vapor and the dry air in the incoming air are

$$y_{\text{H}_2\text{O}} = \frac{N_{\text{H}_2\text{O}}}{N_{\text{total}}} = \frac{0.23}{9.85 + 0.23} = 0.023 \quad \text{and} \quad y_{\text{dry air}} = 1 - 0.023 = 0.977$$

Thus,

$$M = (yM)_{\text{H}_2\text{O}} + (yM)_{\text{dryair}} = (0.023)(18) + (0.977)(29) = 28.7\text{kg/kmol}$$

$$v = \frac{RT}{P} = \frac{(8.314/28.7\text{kPa}\cdot\text{m}^3/\text{kg}\cdot\text{K})(303\text{K})}{100\text{kPa}} = 0.878\text{m}^3/\text{kg}$$

$$\dot{V} = \dot{m}v = (14.0\text{kg/min})(0.878\text{m}^3/\text{kg}) = \mathbf{12.3\ \ m^3/min}$$

14-86 A gaseous fuel with a known composition is burned with dry air, and the volumetric analysis of products gases is determined. The AF ratio, the percent theoretical air used, and the volume flow rate of air are to be determined.

Assumptions **1** Combustion is complete. **2** Air and the combustion gases are ideal gases.

Properties The molar masses of C, H_2, N_2, O_2, and air are 12, 2, 28, 32, and 29 kg/kmol, respectively (Table A-1).

Analysis Considering 100 kmol of dry products, the combustion equation can be written as

$$x(0.80CH_4 + 0.15N_2 + 0.05O_2) + a(O_2 + 3.76N_2)$$
$$\longrightarrow 3.36CO_2 + 0.09CO + 14.91O_2 + 81.64N_2 + bH_2O$$

The unknown coefficients x, a, and b are determined from mass balances,

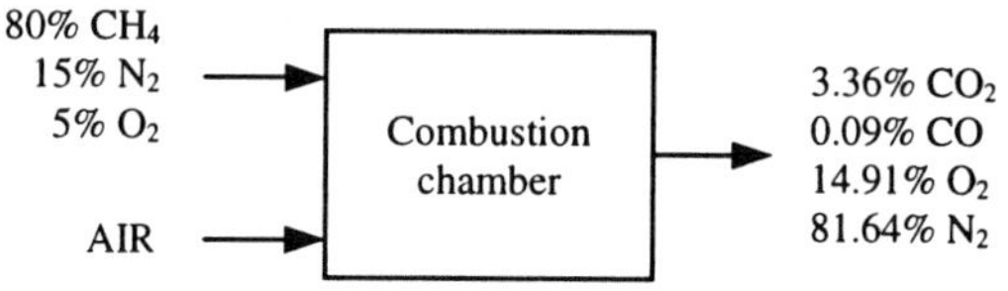

80% CH_4
15% N_2
5% O_2
AIR
Combustion chamber
3.36% CO_2
0.09% CO
14.91% O_2
81.64% N_2

$$C: \quad 0.80x = 3.36 + 0.09 \longrightarrow x = 4.31$$
$$H: \quad 3.2x = 2b \longrightarrow b = 6.90$$
$$N_2: \quad 0.15x + 3.76a = 81.64 \longrightarrow a = 21.54$$

$$\left[\text{Check } O_2: \quad 0.05x + a = 3.36 + 0.045 + 14.91 + b/2 \longrightarrow a = 21.78\right]$$

Thus,

$$4.31(0.80CH_4 + 0.15N_2 + 0.05O_2) + 21.54(O_2 + 3.76N_2)$$
$$\longrightarrow 3.36CO_2 + 0.09CO + 14.91O_2 + 81.64N_2 + 6.9H_2O$$

The combustion equation for 1 kmol of fuel is obtained by dividing the above equation by 4.31,

$$(0.80CH_4 + 0.15N_2 + 0.05O_2) + 5.0(O_2 + 3.76N_2)$$
$$\longrightarrow 0.78CO_2 + 0.02CO + 3.46O_2 + 18.95N_2 + 1.6H_2O$$

(*a*) The air-fuel ratio is determined from its definition,

$$AF = \frac{m_{air}}{m_{fuel}} = \frac{(5.0 \times 4.76\text{kmol})(29\text{kg/kmol})}{(18.6\text{kg/kmol})} = \mathbf{37.1\ kg\ air/kgfuel}$$

(*b*) To find the percent theoretical air used, we need to know the theoretical amount of air, which is determined from the theoretical combustion equation of the fuel,

$$(0.80CH_4 + 0.15N_2 + 0.05O_2) + a_{th}(O_2 + 3.76N_2) \longrightarrow 0.8CO_2 + 1.6H_2O + (0.15 + 3.76a_{th})N_2$$

$$O_2: \ 0.05 + a_{th} = 0.8 + 0.8 \longrightarrow a_{th} = 1.55$$

Then,

$$\text{Percenttheoreticalair} = \frac{m_{air,act}}{m_{air,th}} = \frac{N_{air,act}}{N_{air,th}} = \frac{(5.0)(4.76)\text{kmol}}{(1.55)(4.76)\text{kmol}} = \mathbf{323\%}$$

(*c*) The specific volume, mass flow rate, and the volume flow rate of air at the inlet conditions are

$$v = \frac{RT}{P} = \frac{(0.287\text{ kPa}\cdot\text{m}^3/\text{kg}\cdot\text{K})(298\text{ K})}{100\text{ kPa}} = 0.855\text{ m}^3/\text{kg}$$

$$\dot{m}_{air} = (AF)\dot{m}_{fuel} = (37.1\text{ kg air/kg fuel})(2.2\text{ kg fuel/min}) = 81.62\text{ m}^3/\text{min}$$

$$\dot{V}_{air} = (\dot{m}v)_{air} = (81.62\text{ kg/min})(0.855\text{ m}^3/\text{kg}) = \mathbf{69.8\ m^3/min}$$

14-87 CO gas is burned with air during a steady-flow combustion process. The rate of heat transfer from the combustion chamber is to be determined.

Assumptions **1** Steady operating conditions exist. **2** Air and combustion gases are ideal gases. **3** Kinetic and potential energies are negligible. **4** There are no work interactions. **5** Combustion is complete.

Properties The molar masses of CO and air are 28 kg/kmol and 29 kg/kmol, respectively (Table A-1).

Analysis We first need to calculate the amount of air used per kmol of CO before we can write the combustion equation,

$$v_{CO} = \frac{RT}{P} = \frac{(0.2968\text{kPa}\cdot\text{m}^3/\text{kg}\cdot\text{K})(310\text{K})}{(110\text{kPa})} = 0.836\text{m}^3/\text{kg}$$

$$\dot{m}_{CO} = \frac{\dot{V}_{CO}}{v_{CO}} = \frac{0.4\text{m}^3/\text{min}}{0.836\text{m}^3/\text{kg}} = 0.478\text{kg/min}$$

CO
37°C
Air
25°C
Combustion chamber
Q
Products
900 K

Then the molar air-fuel ratio becomes

$$\overline{\text{AF}} = \frac{N_{\text{air}}}{N_{\text{fuel}}} = \frac{\dot{m}_{\text{air}}/M_{\text{air}}}{\dot{m}_{\text{fuel}}/M_{\text{fuel}}} = \frac{(1.5\text{kg/min})/(29\text{kg/kmol})}{(0.478\text{kg/min})/(28\text{kg/kmol})} = 3.03\text{kmol air/kmol fuel}$$

Thus the number of moles of O_2 used per mole of CO is 3.03/4.76 = 0.637. Then the combustion equation in this case can be written as

$$CO + 0.637(O_2 + 3.76N_2) \longrightarrow CO_2 + 0.137O_2 + 2.40N_2$$

Under steady-flow conditions the energy balance $E_{\text{in}} - E_{\text{out}} = \Delta E_{\text{system}}$ applied on the combustion chamber with $W = 0$ reduces to

$$-Q_{\text{out}} = \sum N_P\left(\bar{h}_f^\circ + \bar{h} - \bar{h}^\circ\right)_P - \sum N_R\left(\bar{h}_f^\circ + \bar{h} - \bar{h}^\circ\right)_R$$

Assuming the air and the combustion products to be ideal gases, we have $h = h(T)$. From the tables,

Substance	$\bar{h}_f^\circ$ kJ/kmol	$\bar{h}_{298\text{ K}}$ kJ/kmol	$\bar{h}_{310\text{ K}}$ kJ/kmol	$\bar{h}_{900\text{ K}}$ kJ/kmol
CO	-110,530	8669	9014	27,066
O_2	0	8682	---	27,928
N_2	0	8669	---	26,890
CO_2	-393,520	9364	---	37,405

Thus,

$$\begin{aligned} -Q_{\text{out}} &= (1)(-393,520 + 37,405 - 9364) + (0.137)(0 + 27,928 - 8682) \\ &\quad + (2.4)(0 + 26,890 - 8669) - (1)(-110,530 + 9014 - 8669) - 0 - 0 \\ &= -208,929\text{kJ/kmol of CO} \end{aligned}$$

Then the rate of heat transfer for a mass flow rate of 0.956 kg/min for CO becomes

$$\dot{Q}_{\text{out}} = \dot{N}Q_{\text{out}} = \left(\frac{\dot{m}}{N}\right)Q_{\text{out}} = \left(\frac{0.478\text{ kg/min}}{28\text{ kg/kmol}}\right)(208,929\text{ kJ/kmol}) = \mathbf{3567\ \ kJ/min}$$

14-88 Methane gas is burned steadily with dry air. The volumetric analysis of the products is given. The percentage of theoretical air used and the heat transfer from the combustion chamber are to be determined.

Assumptions **1** Steady operating conditions exist. **2** Air and combustion gases are ideal gases. **3** Kinetic and potential energies are negligible. **4** There are no work interactions.

Analysis (*a*) Considering 100 kmol of dry products, the combustion equation can be written as

$$x\mathrm{CH_4} + a[\mathrm{O_2} + 3.76\mathrm{N_2}] \longrightarrow 5.20\mathrm{CO_2} + 0.33\mathrm{CO} + 11.24\mathrm{O_2} + 83.23\mathrm{N_2} + b\mathrm{H_2O}$$

The unknown coëfficients x, a, and b are determined from mass balances,

$$\mathrm{N_2}: \quad 3.76a = 83.23 \longrightarrow a = 22.14$$

$$\mathrm{C}: \quad x = 5.20 + 0.33 \longrightarrow x = 5.53$$

$$\mathrm{H}: \quad 4x = 2b \longrightarrow b = 11.06$$

$$(\text{Check } \mathrm{O_2}: \quad a = 5.20 + 0.165 + 11.24 + b/2 \longrightarrow 22.14 = 22.14)$$

Thus, $$5.53\mathrm{CH_4} + 22.14[\mathrm{O_2} + 3.76\mathrm{N_2}] \longrightarrow 5.20\mathrm{CO_2} + 0.33\mathrm{CO} + 11.24\mathrm{O_2} + 83.23\mathrm{N_2} + 11.06\mathrm{H_2O}$$

The combustion equation for 1 kmol of fuel is obtained by dividing the above equation by 5.53

$$\mathrm{CH_4} + 4[\mathrm{O_2} + 3.76\mathrm{N_2}] \longrightarrow 0.94\mathrm{CO_2} + 0.06\mathrm{CO} + 2.03\mathrm{O_2} + 15.04\mathrm{N_2} + 2\mathrm{H_2O}$$

To find the percent theoretical air used, we need to know the theoretical amount of air, which is determined from the theoretical combustion equation of the fuel,

$$\mathrm{CH_4} + a_{th}[\mathrm{O_2} + 3.76\mathrm{N_2}] \longrightarrow \mathrm{CO_2} + 2\mathrm{H_2O} + 3.76a_{th}\mathrm{N_2}$$

$$\mathrm{O_2}: \quad a_{th} = 1 + 1 \longrightarrow a_{th} = 2.0$$

Then, $$\text{Percent theoretical air} = \frac{m_{\text{air,act}}}{m_{\text{air,th}}} = \frac{N_{\text{air,act}}}{N_{\text{air,th}}} = \frac{(4.0)(4.76)\text{ kmol}}{(2.0)(4.76)\text{ kmol}} = \mathbf{200\%}$$

(*b*) Under steady-flow conditions, energy balance applied on the combustion chamber reduces to

$$-Q_{\text{out}} = \sum N_P\left(\bar{h}_f^\circ + \bar{h} - \bar{h}^\circ\right)_P - \sum N_R\left(\bar{h}_f^\circ + \bar{h} - \bar{h}^\circ\right)_R$$

Assuming the air and the combustion products to be ideal gases, we have $h = h(T)$. From the tables,

Substance	$\bar{h}_f^\circ$ kJ/kmol	$\bar{h}_{290\text{ K}}$ kJ/kmol	$\bar{h}_{298\text{ K}}$ kJ/kmol	$\bar{h}_{700\text{ K}}$ kJ/kmol
$\mathrm{CH_4}$ (*g*)	-74,850	---	---	---
$\mathrm{O_2}$	0	8443	8682	21,184
$\mathrm{N_2}$	0	8432	8669	20,604
$\mathrm{H_2O}$ (*g*)	-241,820	---	9904	24,088
CO	-110,530	---	8669	20,690
$\mathrm{CO_2}$	-393,520	---	9364	27,125

Thus,

$$\begin{aligned}-Q_{\text{out}} &= (0.94)(-393,520 + 27,125 - 9364) + (0.06)(-110,530 + 20,690 - 8669) \\ &\quad + (2)(-241,820 + 24,088 - 9904) + (2.03)(0 + 21,184 - 8682) + (15.04)(0 + 20,604 - 8669) \\ &\quad - (1)(-74,850 + h_{298} - h_{298}) - (4)(0 + 8443 - 8682) - (15.04)(0 + 8432 - 8669) \\ &= -530,144 \text{ kJ/kmol CH}_4\end{aligned}$$

or $$Q_{\text{out}} = \mathbf{530,144\ kJ/kmol\ CH_4}$$

14-89 A mixture of hydrogen and the stoichiometric amount of air contained in a rigid tank is ignited. The fraction of H_2O that condenses and the heat transfer from the combustion chamber are to be determined.

Assumptions **1** Steady operating conditions exist. **2** Air and combustion gases are ideal gases. **3** Kinetic and potential energies are negligible. **4** There are no work interactions. **5** Combustion is complete.

Analysis The theoretical combustion equation of H_2 with stoichiometric amount of air is

Q

H₂, Air
25°C

25°C

$$H_2 + a_{th}(O_2 + 3.76N_2) \longrightarrow H_2O + 3.76a_{th}N_2$$

where a_{th} is the stoichiometric coefficient and is determined from the O_2 balance,

$$a_{th} = 0.5$$

Thus,

$$H_2 + 0.5(O_2 + 3.76N_2) \longrightarrow H_2O + 1.88N_2$$

(*a*) At 25°C part of the water (say, N_w moles) will condense, and the number of moles of products that remains in the gas phase will be 2.88 - N_w. Neglecting the volume occupied by the liquid water and treating all the product gases as ideal gases, the final pressure in the tank can be expressed as

$$P_f = \frac{N_{f,\text{gas}} R_u T_f}{V} = \frac{(2.88 - N_w\,\text{kmol})(8.314\text{ kPa}\cdot\text{m}^3/\text{kg}\cdot\text{K})(298\text{ K})}{6\text{ m}^3}$$
$$= 412.9(2.88 - N_w)\text{kPa}$$

Then,

$$\frac{N_v}{N_{\text{gas}}} = \frac{P_v}{P_{\text{total}}} \longrightarrow \frac{1 - N_w}{2.88 - N_w} = \frac{3.169\text{ kPa}}{412.9(2.88 - N_w)\text{kPa}} \longrightarrow N_w = 0.992\text{ kmol}$$

Thus **99.2%** of the H_2O will condense when the products are cooled to 25°C.

(*b*) The energy balance $E_{in} - E_{out} = \Delta E_{\text{system}}$ applied for this constant volume combustion process with $W = 0$ reduces to

$$-Q_{\text{out}} = \sum N_P\left(\overline{h}_f^\circ + \overline{h} - \overline{h}^\circ - P\overline{v}\right)_P - \sum N_R\left(\overline{h}_f^\circ + \overline{h} - \overline{h}^\circ - P\overline{v}\right)_R$$

With the exception of liquid water for which the $P\overline{v}$ term is negligible, both the reactants and the products are assumed to be ideal gases, all the internal energy and enthalpies depend on temperature only, and the $P\overline{v}$ terms in this equation can be replaced by R_uT. It yields

$$-Q_{\text{out}} = \sum N_P\left(\overline{h}_f^\circ - R_uT\right)_P - \sum N_R\left(\overline{h}_f^\circ - R_uT\right)_R$$
$$= \sum N_P\overline{h}_{f,P}^\circ - \sum N_R\overline{h}_{f,R}^\circ + R_uT\left(\sum N_{P,gas} - \sum N_R\right)_R$$

since the reactants are at the standard reference temperature of 25°C. From the tables,

Substance	$\overline{h}_f^\circ$ **kJ/kmol**
H_2	0
O_2	0
N_2	0
H_2O (*g*)	-241,820
H_2O (ℓ)	-285,830

Thus,

$$-Q_{\text{out}} = (0.008)(-241{,}820) + (0.992)(-285{,}830) + 0 - 0 - 0 - 0 + 8.314 \times 298(1.89 - 3.38)$$
$$= -289{,}170\text{kJ} \quad (\text{perkmolH}_2)$$

or $\quad Q_{out} =$ **289,170 kJ** (per kmol H_2)

14-90 Propane gas is burned with air during a steady-flow combustion process. The adiabatic flame temperature is to be determined for different cases.

Assumptions **1** Steady operating conditions exist. **2** Air and combustion gases are ideal gases. **3** Kinetic and potential energies are negligible. **4** There are no work interactions. **5** The combustion chamber is adiabatic.

Analysis Adiabatic flame temperature is the temperature at which the products leave the combustion chamber under adiabatic conditions (Q = 0) with no work interactions (W = 0). Under steady-flow conditions the energy balance $E_{in} - E_{out} = \Delta E_{system}$ applied on the combustion chamber reduces to

$$\sum N_P\left(\bar{h}_f^\circ + \bar{h} - \bar{h}^\circ\right)_P = \sum N_R\left(\bar{h}_f^\circ + \bar{h} - \bar{h}^\circ\right)_R \longrightarrow \sum N_P\left(\bar{h}_f^\circ + \bar{h}_T - \bar{h}^\circ\right)_P = \left(N\bar{h}_f^\circ\right)_{C_3H_8}$$

since all the reactants are at the standard reference temperature of 25°C, and $\bar{h}_f^\circ = 0$ for O_2 and N_2.

(a) The theoretical combustion equation of C_3H_8 with stoichiometric amount of air is

$$C_3H_8(g) + 5(O_2 + 3.76N_2) \longrightarrow 3CO_2 + 4H_2O + 18.8N_2$$

From the tables,

Substance	$\bar{h}_f^\circ$ kJ/kmol	$\bar{h}_{298\ K}$ kJ/kmol
C_3H_8 (g)	-103,850	---
O_2	0	8682
N_2	0	8669
H_2O (g)	-241,820	9904
CO	-110,530	8669
CO_2	-393,520	9364

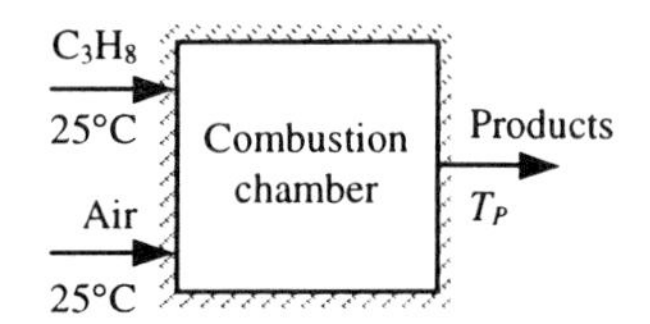

Thus,

$$(3)(-393{,}520 + \bar{h}_{CO_2} - 9364) + (4)(-241{,}820 + \bar{h}_{H_2O} - 9904) + (18.8)(0 + \bar{h}_{N_2} - 8669) = (1)(-103{,}850)$$

It yields

$$3\bar{h}_{CO_2} + 4\bar{h}_{H_2O} + 18.8\bar{h}_{N_2} = 2;274{,}675 \text{ kJ}$$

The adiabatic flame temperature is obtained from a trial and error solution. A first guess is obtained by dividing the right-hand side of the equation by the total number of moles, which yields 2,274,675 / (3 + 4 + 18.8) = 88,165 kJ/kmol. This enthalpy value corresponds to about 2650 K for N_2. Noting that the majority of the moles are N_2, T_P will be close to 2650 K, but somewhat under it because of the higher specific heats of CO_2 and H_2O.

At 2400 K: $$3\bar{h}_{CO_2} + 4\bar{h}_{H_2O} + 18.8\bar{h}_{N_2} = (3)(125{,}152) + (4)(103{,}508) + (18.8)(79{,}320) = 2{,}280{,}704 \text{kJ (Higher than 2,274,675kJ)}$$

At 2350 K: $$3\bar{h}_{CO_2} + 4\bar{h}_{H_2O} + 18.8\bar{h}_{N_2} = (3)(122{,}091) + (4)(100{,}846) + (18.8)(77{,}496) = 2{,}226{,}582 \text{kJ (Lower than 2,274,675kJ)}$$

By interpolation, T_P = **2394 K**

(b) The balanced combustion equation for complete combustion with 300% theoretical air is

$$C_3H_8(g) + 15(O_2 + 3.76N_2) \longrightarrow 3CO_2 + 4H_2O + 10O_2 + 56.4N_2$$

Substituting known numerical values,

$$(3)(-393{,}520+\bar{h}_{CO_2}-9364)+(4)(-241{,}820+\bar{h}_{H_2O}-9904)$$
$$+(10)(0+\bar{h}_{O_2}-8682)+(56.4)(0+\bar{h}_{N_2}-8669)=(1)(-103{,}850)$$

which yields

$$3\bar{h}_{CO_2}+4\bar{h}_{H_2O}+10\bar{h}_{O_2}+56.4\bar{h}_{N_2}=2{,}687{,}450\text{ kJ}$$

The adiabatic flame temperature is obtained from a trial and error solution. A first guess is obtained by dividing the right-hand side of the equation by the total number of moles, which yields 2,687,449 / (3 + 4 + 10 + 56.4) = 36,614 kJ/kmol. This enthalpy value corresponds to about 1200 K for N_2. Noting that the majority of the moles are N_2, T_P will be close to 1200 K, but somewhat under it because of the higher specific heats of CO_2 and H_2O.

At 1160 K:

$$3\bar{h}_{CO_2}+4\bar{h}_{H_2O}+10\bar{h}_{O_2}+56.4\bar{h}_{N_2}=(3)(51{,}602)+(4)(42{,}642)+(10)(37{,}023)+(56.4)(35{,}430)$$
$$=2{,}693{,}856\text{kJ (Higherthan2,687,450kJ)}$$

At 1140 K:

$$3\bar{h}_{CO_2}+4\bar{h}_{H_2O}+10\bar{h}_{O_2}+56.4\bar{h}_{N_2}=(3)(50{,}484)+(4)(41{,}780)+(10)(36{,}314)+(56.4)(34{,}760)$$
$$=2{,}642{,}176\text{kJ (Lowerthan2,687,450kJ)}$$

By interpolation, T_P = **1158 K**

(*c*) The balanced combustion equation for incomplete combustion with 95% theoretical air is

$$C_3H_8(g)+4.75(O_2+3.76N_2)\longrightarrow 2.5CO_2+0.5CO+4H_2O+17.86N_2$$

Substituting known numerical values,

$$(2.5)(-393{,}520+\bar{h}_{CO_2}-9364)+(0.5)(-110{,}530+\bar{h}_{CO}-8669)$$
$$+(4)(-241{,}820+\bar{h}_{H_2O}-9904)+(17.86)(0+\bar{h}_{N_2}-8669)=(1)(-103{,}850)$$

which yields

$$2.5\bar{h}_{CO_2}+0.5\bar{h}_{CO}+4\bar{h}_{H_2O}+17.86\bar{h}_{N_2}=2{,}124{,}684\text{ kJ}$$

The adiabatic flame temperature is obtained from a trial and error solution. A first guess is obtained by dividing the right-hand side of the equation by the total number of moles, which yields 2,124,684 / (2.5 + 4 + 0.5 + 17.86) = 85,466 kJ/kmol. This enthalpy value corresponds to about 2550 K for N_2. Noting that the majority of the moles are N_2, T_P will be close to 2550 K, but somewhat under it because of the higher specific heats of CO_2 and H_2O.

At 2350 K:

$$2.5\bar{h}_{CO_2}+0.5\bar{h}_{CO}+4\bar{h}_{H_2O}+17.86\bar{h}_{N_2}=(2.5)(122{,}091)+(0.5)(78{,}178)+(4)(100{,}846)+(17.86)(77{,}496)$$
$$=2{,}131{,}779\text{kJ (Higherthan2,124,684kJ)}$$

At 2300 K:

$$2.5\bar{h}_{CO_2}+0.5\bar{h}_{CO}+4\bar{h}_{H_2O}+17.86\bar{h}_{N_2}=(2.5)(119{,}035)+(0.5)(76{,}345)+(4)(98{,}199)+(17.86)(75{,}676)$$
$$=2{,}080{,}129\text{kJ (Lowerthan2,124,684kJ)}$$

By interpolation, T_P = **2343 K**

14-91 The highest possible temperatures that can be obtained when liquid gasoline is burned steadily with air and with pure oxygen are to be determined.

Assumptions **1** Steady operating conditions exist. **2** Air and combustion gases are ideal gases. **3** Kinetic and potential energies are negligible. **4** There are no work interactions. **5** The combustion chamber is adiabatic.

Analysis The highest possible temperature that can be achieved during a combustion process is the temperature which occurs when a fuel is burned completely with stoichiometric amount of air in an adiabatic combustion chamber. It is determined from

$$\sum N_P\left(\bar{h}_f^\circ + \bar{h} - \bar{h}^\circ\right)_P = \sum N_R\left(\bar{h}_f^\circ + \bar{h} - \bar{h}^\circ\right)_R \longrightarrow \sum N_P\left(\bar{h}_f^\circ + \bar{h}_T - \bar{h}^\circ\right)_P = \left(N\bar{h}_f^\circ\right)_{C_8H_{18}}$$

since all the reactants are at the standard reference temperature of 25°C, and for O_2 and N_2. The theoretical combustion equation of C_8H_{18} air is

$$C_8H_{18} + 12.5(O_2 + 3.76N_2) \longrightarrow 8CO_2 + 9H_2O + 47N_2$$

From the tables,

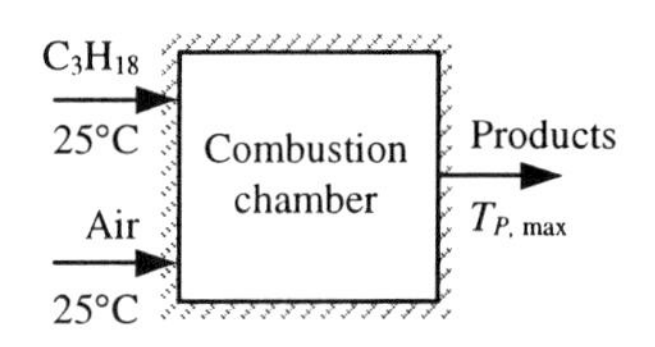

Substance	$\bar{h}_f^\circ$ kJ/kmol	$\bar{h}_{298\text{ K}}$ kJ/kmol
C_8H_{18} (ℓ)	-249,950	---
O_2	0	8682
N_2	0	8669
H_2O (*g*)	-241,820	9904
CO_2	-393,520	9364

Thus,

$$(8)\left(-393{,}520 + \bar{h}_{CO_2} - 9364\right) + (9)\left(-241{,}820 + \bar{h}_{H_2O} - 9904\right) + (47)\left(0 + \bar{h}_{N_2} - 8669\right) = (1)(-249{,}950)$$

It yields

$$8\bar{h}_{CO_2} + 9\bar{h}_{H_2O} + 47\bar{h}_{N_2} = 5{,}646{,}081 \text{ kJ}$$

The adiabatic flame temperature is obtained from a trial and error solution. A first guess is obtained by dividing the right-hand side of the equation by the total number of moles, which yields 5,646,081/(8 + 9 + 47) = 88,220 kJ/kmol. This enthalpy value corresponds to about 2650 K for N_2. Noting that the majority of the moles are N_2, T_P will be close to 2650 K, but somewhat under it because of the higher specific heat of H_2O.

At 2400 K:
$$8\bar{h}_{CO_2} + 9\bar{h}_{H_2O} + 47\bar{h}_{N_2} = (8)(125{,}152) + (9)(103{,}508) + (47)(79{,}320) = 5{,}660{,}828 \text{kJ (Higher than } 5{,}646{,}081 \text{kJ)}$$

At 2350 K:
$$8\bar{h}_{CO_2} + 9\bar{h}_{H_2O} + 47\bar{h}_{N_2} = (8)(122{,}091) + (9)(100{,}846) + (47)(77{,}496) = 5{,}526{,}654 \text{kJ (Lower than } 5{,}646{,}081 \text{kJ)}$$

By interpolation, T_P = **2395 K**

If the fuel is burned with stoichiometric amount of pure O_2, the combustion equation would be

$$C_8H_{18} + 12.5O_2 \longrightarrow 8CO_2 + 9H_2O$$

Thus,

$$(8)\left(-393{,}520 + \bar{h}_{CO_2} - 9364\right) + (9)\left(-241{,}820 + \bar{h}_{H_2O} - 9904\right) = (1)(-249{,}950)$$

It yields

$$8\bar{h}_{CO_2} + 9\bar{h}_{H_2O} = 5{,}238{,}638 \text{ kJ}$$

The adiabatic flame temperature is obtained from a trial and error solution. A first guess is obtained by dividing the right-hand side of the equation by the total number of moles, which yields 5,238,638/(8 + 9) = 308,155 kJ/kmol. This enthalpy value is higher than the highest enthalpy value listed for H_2O and CO_2. Thus an estimate of the adiabatic flame temperature can be obtained by extrapolation.

At 3200 K: $8\bar{h}_{CO_2} + 9\bar{h}_{H_2O} = (8)(174{,}695) + (9)(147{,}457) = 2{,}724{,}673 \text{kJ}$

At 3250 K: $8\bar{h}_{CO_2} + 9\bar{h}_{H_2O} = (8)(177{,}822) + (9)(150{,}272) = 2{,}775{,}272 \text{kJ}$

By extrapolation, T_P = **3597 K**

14-92E The work potential of diesel fuel at a given state is to be determined.

Assumptions **1** Steady operating conditions exist. **2** Air and combustion gases are ideal gases. **3** Kinetic and potential energies are negligible.

Analysis The work potential or availability of a fuel at a specified state is the reversible work that would be obtained if that fuel were burned completely with stoichiometric amount of air and the products are returned to the state of the surroundings. It is determined from

$$W_{rev} = \sum N_R\left(\bar{h}_f^\circ + \bar{h} + \bar{h}^\circ - T_0\bar{s}\right)_R - \sum N_P\left(\bar{h}_f^\circ + \bar{h} + \bar{h}^\circ - T_0\bar{s}\right)_P$$

or,

$$W_{rev} = \sum N_R\left(\bar{h}_f^\circ - T_0\bar{s}\right)_R - \sum N_P\left(\bar{h}_f^\circ - T_0\bar{s}\right)_P$$

since both the reactants and the products are at the state of the surroundings. Considering 1 kmol of $C_{12}H_{26}$, the theoretical combustion equation can be written as

$$C_{12}H_{26} + a_{th}(O_2 + 3.76N_2) \longrightarrow 12CO_2 + 13H_2O + 3.76a_{th}N_2$$

where a_{th} is the stoichiometric coefficient and is determined from the O_2 balance,

$$a_{th} = 12 + 6.5 \longrightarrow a_{th} = 18.5$$

Substituting,

$$C_{12}H_{26} + 18.5(O_2 + 3.76N_2) \longrightarrow 12CO_2 + 13H_2O + 69.56N_2$$

For each lbmol of fuel burned, 12 + 13 + 69.56 = 94.56 lbmol of products are formed, including 13 lbmol of H_2O. Assuming that the dew-point temperature of the products is above 77°F, some of the water will exist in the liquid form in the products. If N_w lbmol of H_2O condenses, there will be 13 - N_w lbmol of water vapor left in the products. The mole number of the products in the gas phase will also decrease to 94.56 - N_w as a result. Treating the product gases (including the remaining water vapor) as ideal gases, N_w is determined by equating the mole fraction of the water vapor to pressure fraction,

$$\frac{N_v}{N_{prod,\,gas}} = \frac{P_v}{P_{prod}} \longrightarrow \frac{13 - N_w}{94.56 - N_w} = \frac{0.4641\text{ psia}}{14.7\text{ psia}} \longrightarrow N_w = 10.34\text{ lbmol}$$

since $P_v = P_{sat\,@\,77°F} = 0.4641$ psia. Then the combustion equation can be written as

$$C_{12}H_{26} + 18.5(O_2 + 3.76N_2) \longrightarrow 12CO_2 + 10.34H_2O(\ell) + 2.66H_2O(g) + 69.56N_2$$

The entropy values listed in the ideal gas tables are for 1 atm pressure. Both the air and the product gases are at a total pressure of 1 atm, but the entropies are to be calculated at the partial pressure of the components which is equal to $P_i = y_iP_{total}$, where y_i is the mole fraction of component i. Also,

$$S_i = N_i\bar{s}_i(T, P_i) = N_i\left(\bar{s}_i^\circ(T, P_0) - R_u\ln(y_iP_m)\right)$$

The entropy calculations can be presented in tabular form as

	N_i	y_i	$\bar{s}_i^\circ(77°F, 1atm)$	$R_u\ln(y_iP_m)$	$\bar{s}_i$	$\bar{h}_f^\circ$, Btu / lbmol
$C_{12}H_{26}$	1	---	148.86	---	148.86	-125,190
O_2	18.5	0.21	49.00	-3.10	52.10	0
N_2	69.56	0.79	45.77	-0.47	46.24	0
CO_2	12	0.1425	51.07	-3.870	54.94	-169,300
H_2O (g)	2.66	0.0316	45.11	-6.861	51.97	-104,040
H_2O (ℓ)	10.34	---	16.71	---	16.71	-122,970
N_2	69.56	0.8259	45.77	-0.380	46.15	0

Substituting,

$$\begin{aligned} W_{\text{rev}} &= (1)(-125{,}190 - 537 \times 148.86) + (18.5)(0 - 537 \times 52.10) + (69.56)(0 - 537 \times 46.24) \\ &\quad - (12)(-169{,}300 - 537 \times 54.94) - (2.66)(-104{,}040 - 537 \times 51.97) \\ &\quad - (10.34)(-122{,}970 - 537 \times 16.71) - (69.56)(0 - 537 \times 46.15) \\ &= \mathbf{3{,}375{,}000\ Btu}\ (\text{per lbmol } C_{12}H_{26}) \end{aligned}$$

14-93 Liquid octane is burned with 200 percent excess air during a steady-flow combustion process. The heat transfer rate from the combustion chamber, the power output of the turbine, and the reversible work and exergy destruction are to be determined.

Assumptions **1** Combustion is complete. **2** Steady operating conditions exist. **3** Air and the combustion gases are ideal gases. **4** Changes in kinetic and potential energies are negligible.

Properties The molar mass of C_8H_{18} is 114 kg/kmol (Table A-1).

Analysis (*a*) The fuel is burned completely with the excess air, and thus the products will contain only CO_2, H_2O, N_2, and some free O_2. Considering 1 kmol of C_8H_{18}, the combustion equation can be written as

$$C_8H_{18} + 3a_{th}(O_2 + 3.76N_2) \longrightarrow 8CO_2 + 9H_2O + 2a_{th}O_2 + (3)(3.76a_{th})N_2$$

where a_{th} is the stoichiometric coefficient and is determined from the O_2 balance,

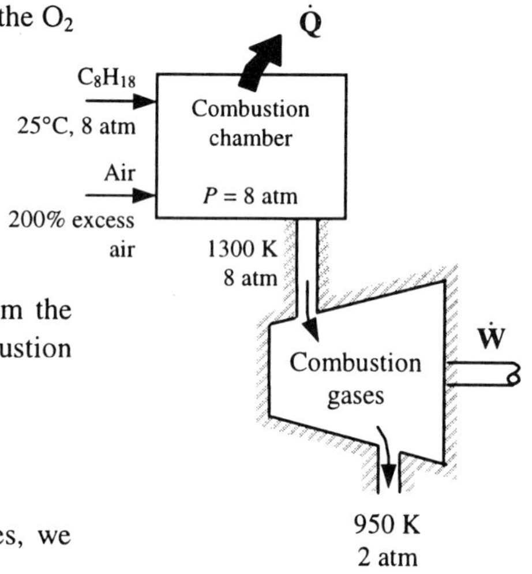

$$3a_{th} = 8 + 4.5 + 2a_{th} \longrightarrow a_{th} = 12.5$$

Substituting,

$$C_8H_{18} + 37.5(O_2 + 3.76N_2) \longrightarrow 8CO_2 + 9H_2O + 25O_2 + 141N_2$$

Q̇
C_8H_{18}
25°C, 8 atm
Air
200% excess air
Combustion chamber
P = 8 atm
1300 K
8 atm
Combustion gases
Ẇ
950 K
2 atm

The heat transfer for this combustion process is determined from the energy balance $E_{in} - E_{out} = \Delta E_{system}$ applied on the combustion chamber with $W = 0$,

$$-Q_{out} = \sum N_P\left(\bar{h}_f^\circ + \bar{h} - \bar{h}^\circ\right)_P - \sum N_R\left(\bar{h}_f^\circ + \bar{h} - \bar{h}^\circ\right)_R$$

Assuming the air and the combustion products to be ideal gases, we have $h = h(T)$. From the tables,

Substance	$\bar{h}_f^\circ$ kJ/kmol	$\bar{h}_{500\ K}$ kJ/kmol	$\bar{h}_{298\ K}$ kJ/kmol	$\bar{h}_{1300\ K}$ kJ/kmol	$\bar{h}_{950\ K}$ kJ/kmol
C_8H_{18} (ℓ)	-249,950	---	---	---	---
O_2	0	14,770	8682	42,033	26,652
N_2	0	14,581	8669	40,170	28,501
H_2O (*g*)	-241,820	---	9904	48,807	33,841
CO_2	-393,520	---	9364	59,552	40,070

Substituting,

$$\begin{aligned}-Q_{out} &= (8)(-393{,}520 + 59{,}522 - 9364) + (9)(-241{,}820 + 48{,}807 - 9904) \\ &\quad + (25)(0 + 42{,}033 - 8682) + (141)(0 + 40{,}170 - 8669) \\ &\quad - (1)(-249{,}950 + h_{298} - h_{298}) - (37.5)(0 + 14{,}770 - 8682) - (141)(0 + 14{,}581 - 8669) \\ &= -109{,}675 \text{kJ/kmol} C_8H_{18}\end{aligned}$$

The C_8H_{18} is burned at a rate of 0.8 kg/min or

$$\dot{N} = \frac{\dot{m}}{M} = \frac{0.8 \text{ kg/min}}{((8)(12) + (18)(1)) \text{ kg/kmol}} = 7.018 \times 10^{-3} \text{ kmol/min}$$

Thus,

$$\dot{Q}_{out} = \dot{N}Q_{out} = (7.018 \times 10^{-3} \text{ kmol/min})(109{,}675 \text{ kJ/kmol}) = \mathbf{770\ \ kJ/min}$$

(*b*) Noting that no chemical reactions occur in the turbine, the turbine is adiabatic, and the product gases can be treated as ideal gases, the power output of the turbine can be determined from the steady-flow energy balance equation for nonreacting gas mixtures,

$$-W_{out} = \sum N_P(\bar{h}_e - \bar{h}_i) \longrightarrow W_{out} = \sum N_P(\bar{h}_{1300K} - \bar{h}_{950K})$$

Substituting,

$$\begin{aligned} W_{out} &= (8)(59{,}522 - 40{,}070) + (9)(48{,}807 - 33{,}841) + (25)(42{,}033 - 29{,}652) + (141)(40{,}170 - 28{,}501) \\ &= 2{,}320{,}164 \text{ kJ/kmol } C_8H_{18} \end{aligned}$$

Thus the power output of the turbine is

$$\dot{W}_{out} = \dot{N}W_{out} = (7.018 \times 10^{-3} \text{ kmol/min})(2{,}320{,}164 \text{ kJ/kmol}) = 16{,}283 \text{ kJ/min} = \mathbf{271.4 \text{ kW}}$$

(*c*) The entropy generation during this process is determined from

$$S_{gen} = S_P - S_R + \frac{Q_{out}}{T_{surr}} = \sum N_P \bar{s}_P - \sum N_R \bar{s}_R + \frac{Q_{out}}{T_{surr}}$$

where the entropy of the products are to be evaluated at the turbine exit state. The C_8H_{18} is at 25°C and 1 atm, and thus its absolute entropy is $\bar{s}_{C_8H_{18}}$ =360.79 kJ/kmol· K (Table A-26). The entropy values listed in the ideal gas tables are for 1 atm pressure. The entropies are to be calculated at the partial pressure of the components which is equal to $P_i = y_i P_{total}$, where y_i is the mole fraction of component *i*. Also,

$$S_i = N_i \bar{s}_i(T, P_i) = N_i(\bar{s}_i^\circ(T, P_0) - R_u \ln(y_i P_m))$$

The entropy calculations can be presented in tabular form as

	N_i	y_i	$\bar{s}_i^\circ(T,1\text{ atm})$	$R_u \ln(y_i P_m)$	$N_i \bar{s}_i$
C_8H_{18}	1	1.00	360.79	17.288	343.50
O_2	37.5	0.21	220.589	4.313	8,110.34
N_2	141	0.79	206.630	15.329	26,973.44
				S_R =	35,427.28 kJ/K
CO_2	8	0.0437	266.444	-20.260	2,293.63
H_2O	9	0.0490	230.499	-19.281	2,248.02
O_2	25	0.1366	241.689	-10.787	6,311.90
N_2	141	0.7705	226.389	3.595	31,413.93
				S_P =	42,267.48 kJ/K

Thus,

$$S_{gen} = 42{,}267.48 - 35{,}427.28 + \frac{109{,}675 \text{ kJ}}{298 \text{ K}} = 7208.2 \text{ kJ/K (per kmol)}$$

Then the rate of entropy generation becomes

$$\dot{S}_{gen} = \dot{N}S_{gen} = (7.018 \times 10^{-3} \text{kmol/min})(7208.2 \text{ kJ/kmol} \cdot \text{K}) = 50.59 \text{ kJ/min} \cdot \text{K}$$

and

$$\dot{X}_{destruction} = T_0 \dot{S}_{gen} = (298 \text{ K})(50.59 \text{ kJ/min} \cdot \text{K}) = 15{,}075 \text{ kJ/min} = \mathbf{251.2 \text{ kW}}$$

$$\dot{W}_{rev} = \dot{W} + \dot{X}_{destruction} = 271.4 + 251.2 = \mathbf{522.6 \text{ kW}}$$

14-94 Methyl alcohol vapor is burned with the stoichiometric amount of air in a combustion chamber. The maximum pressure that can occur in the combustion chamber if the combustion takes place at constant volume and the maximum volume of the combustion chamber if the combustion occurs at constant pressure are to be determined.

Assumptions **1** Combustion is complete. **2** Air and the combustion gases are ideal gases. **4** Changes in kinetic and potential energies are negligible.

Analysis (*a*) The combustion equation of $CH_3OH(g)$ with stoichiometric amount of air is

$$CH_3OH + a_{th}(O_2 + 3.76N_2) \longrightarrow CO_2 + 2H_2O + 3.76a_{th}N_2$$

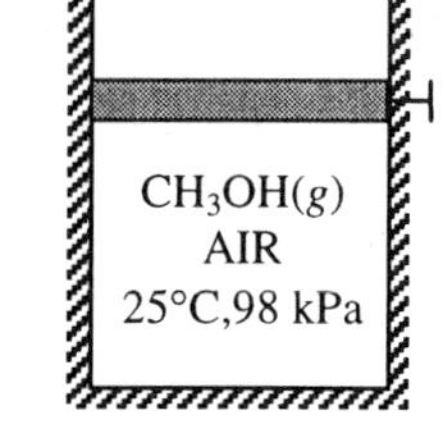

where a_{th} is the stoichiometric coefficient and is determined from the O_2 balance,

$$1 + 2a_{th} = 2 + 2 \quad \longrightarrow \quad a_{th} = 1.5$$

Thus,

$$CH_3OH + 1.5(O_2 + 3.76N_2) \longrightarrow CO_2 + 2H_2O + 5.64N_2$$

The final temperature in the tank is determined from the energy balance relation $E_{in} - E_{out} = \Delta E_{system}$ for reacting closed systems under adiabatic conditions ($Q = 0$) with no work interactions ($W = 0$),

$$0 = \sum N_P\left(\bar{h}_f^\circ + \bar{h} - \bar{h}^\circ - P\bar{v}\right)_P - \sum N_R\left(\bar{h}_f^\circ + \bar{h} - \bar{h}^\circ - P\bar{v}\right)_R$$

Assuming both the reactants and the products to behave as ideal gases, all the internal energy and enthalpies depend on temperature only, and the $P\bar{v}$ terms in this equation can be replaced by R_uT. It yields

$$\sum N_P\left(\bar{h}_f^\circ + \bar{h}_{T_P} - \bar{h}_{298K} - R_uT\right)_P = \sum N_R\left(\bar{h}_f^\circ - R_uT\right)_R$$

since the reactants are at the standard reference temperature of 25°C. From the tables,

Substance	$\bar{h}_f^\circ$ kJ/kmol	$\bar{h}_{298\ K}$ kJ/kmol
CH_3OH	-200,670	---
O_2	0	8682
N_2	0	8669
H_2O (*g*)	-241,820	9904
CO_2	-393,520	9364

Thus,

$$\begin{aligned}&(1)(-393{,}520 + \bar{h}_{CO_2} - 9364 - 8.314 \times T_P) + (2)(-241{,}820 + \bar{h}_{H_2O} - 9904 - 8.314 \times T_P)\\ &+ (5.64)(0 + \bar{h}_{N_2} - 8669 - 8.314 \times T_P) = (1)(-200{,}670 - 8.314 \times 298) + (1.5)(0 - 8.314 \times 298)\\ &+ (5.64)(0 - 8.314 \times 298)\end{aligned}$$

It yields

$$\bar{h}_{CO_2} + 2\bar{h}_{H_2O} + 5.64\bar{h}_{N_2} - 71.833 \times T_P = 734{,}388 \text{ kJ}$$

The temperature of the product gases is obtained from a trial and error solution,

At 2850 K:

$$\bar{h}_{CO_2} + 2\bar{h}_{H_2O} + 5.64\bar{h}_{N_2} - 71.833 \times T_P = (1)(152{,}908) + (2)(127{,}952) + (5.64)(95{,}859) - (71.833)(2850)$$
$$= 744{,}733 \text{ kJ (Higher than } 734{,}388 \text{ kJ)}$$

At 2800 K:

$$\bar{h}_{CO_2} + 2\bar{h}_{H_2O} + 5.64\bar{h}_{N_2} - 71.833 \times T_P = (1)(149{,}808) + (2)(125{,}198) + (5.64)(94{,}014) - (71.833)(2800)$$
$$= 729{,}311 \text{ kJ (Lower than } 734{,}388 \text{ kJ)}$$

By interpolation $\quad T_P = 2816 \text{ K}$

Since both the reactants and the products behave as ideal gases, the final (maximum) pressure that can occur in the combustion chamber is determined to be

$$\frac{P_1 V}{P_2 V} = \frac{N_1 R_u T_1}{N_2 R_u T_2} \longrightarrow P_2 = \frac{N_2 T_2}{N_1 T_1} P_1 = \frac{(8.64 \text{ kmol})(2816 \text{ K})}{(8.14 \text{ kmol})(298 \text{ K})}(98 \text{ kPa}) = \mathbf{983 \ kPa}$$

(*b*) The combustion equation of $CH_3OH(g)$ remains the same in the case of constant pressure. Further, the boundary work in this case can be combined with the *u* terms so that the first law relation can be expressed in terms of enthalpies just like the steady-flow process,

$$Q = \sum N_P \left(\bar{h}_f^\circ + \bar{h} - \bar{h}^\circ\right)_P - \sum N_R \left(\bar{h}_f^\circ + \bar{h} - \bar{h}^\circ\right)_R$$

Since both the reactants and the products behave as ideal gases, we have $h = h(T)$. Also noting that $Q = 0$ for an adiabatic combustion process, the 1st law relation reduces to

$$\sum N_P \left(\bar{h}_f^\circ + \bar{h}_{T_P} - \bar{h}_{298K}\right)_P = \sum N_R \left(\bar{h}_f^\circ\right)_R$$

since the reactants are at the standard reference temperature of 25°C. Then using data from the mini table above, we get

$$(1)(-393{,}520 + \bar{h}_{CO_2} - 9364) + (2)(-241{,}820 + \bar{h}_{H_2O} - 9904) + (5.64)(0 + \bar{h}_{N_2} - 8669) =$$
$$(1)(-200{,}670) + (1.5)(0) + (5.64)(0)$$

It yields

$$\bar{h}_{CO_2} + 2\bar{h}_{H_2O} + 5.64\bar{h}_{N_2} = 754{,}555 \text{ kJ}$$

The temperature of the product gases is obtained from a trial and error solution,

At 2350 K: $\quad \bar{h}_{CO_2} + 2\bar{h}_{H_2O} + 5.64\bar{h}_{N_2} = (1)(122{,}091) + (2)(100{,}846) + (5.64)(77{,}496)$
$$= 760{,}860 \text{ kJ (Higher than } 754{,}555 \text{ kJ)}$$

At 2300 K: $\quad \bar{h}_{CO_2} + 2\bar{h}_{H_2O} + 5.64\bar{h}_{N_2} = (1)(119{,}035) + (2)(98{,}199) + (5.64)(75{,}676)$
$$= 742{,}246 \text{ kJ (Lower than } 754{,}555 \text{ kJ)}$$

By interpolation, $\quad T_P = 2333 \text{ K}$

Treating both the reactants and the products as ideal gases, the final (maximum) volume that the combustion chamber can have is determined to be

$$\frac{PV_1}{PV_2} = \frac{N_1 R_u T_1}{N_2 R_u T_2} \longrightarrow V_2 = \frac{N_2 T_2}{N_1 T_1} V_1 = \frac{(8.64 \text{ kmol})(2333 \text{ K})}{(8.14 \text{ kmol})(298 \text{ K})}(0.8 \text{ L}) = \mathbf{6.65 \ L}$$

14-95 EES solution of this (and other comprehensive problems designated with the *computer icon*) is available to instructors at the *Instructor Manual* section of the *Online Learning Center* (OLC) at www.mhhe.com/cengel-boles. See the Preface for access information.

14-96 Methane is burned with the stoichiometric amount of air in a combustion chamber. The maximum pressure that can occur in the combustion chamber if the combustion takes place at constant volume and the maximum volume of the combustion chamber if the combustion occurs at constant pressure are to be determined.

Assumptions **1** Combustion is complete. **2** Air and the combustion gases are ideal gases. **4** Changes in kinetic and potential energies are negligible.

Analysis (*a*) The combustion equation of $CH_4(g)$ with stoichiometric amount of air is

$$CH_4 + a_{th}(O_2 + 3.76N_2) \longrightarrow CO_2 + 2H_2O + 3.76a_{th}N_2$$

where a_{th} is the stoichiometric coefficient and is determined from the O_2 balance,

$$a_{th} = 1 + 1 \longrightarrow a_{th} = 2$$

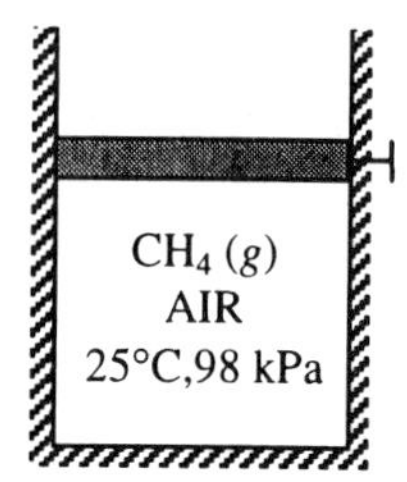

Thus,

$$CH_4 + 2(O_2 + 3.76N_2) \longrightarrow CO_2 + 2H_2O + 7.52N_2$$

The final temperature in the tank is determined from the energy balance relation $E_{in} - E_{out} = \Delta E_{system}$ for reacting closed systems under adiabatic conditions ($Q = 0$) with no work interactions ($W = 0$),

$$0 = \sum N_P(\bar{h}_f^\circ + \bar{h} - \bar{h}^\circ - P\bar{v})_P - \sum N_R(\bar{h}_f^\circ + \bar{h} - \bar{h}^\circ - P\bar{v})_R$$

Since both the reactants and the products behave as ideal gases, all the internal energy and enthalpies depend on temperature only, and the $P\bar{v}$ terms in this equation can be replaced by R_uT. It yields

$$\sum N_P(\bar{h}_f^\circ + \bar{h}_{T_P} - \bar{h}_{298K} - R_uT)_P = \sum N_R(\bar{h}_f^\circ - R_uT)_R$$

since the reactants are at the standard reference temperature of 25°C. From the tables,

Substance	$\bar{h}_f^\circ$ kJ/kmol	$\bar{h}_{298\,K}$ kJ/kmol
CH_4	-74,850	---
O_2	0	8682
N_2	0	8669
H_2O (*g*)	-241,820	9904
CO_2	-393,520	9364

Thus,

$$(1)(-393{,}520 + \bar{h}_{CO_2} - 9364 - 8.314 \times T_P) + (2)(-241{,}820 + \bar{h}_{H_2O} - 9904 - 8.314 \times T_P)$$
$$+ (7.52)(0 + \bar{h}_{N_2} - 8669 - 8.314 \times T_P) = (1)(-74{,}850 - 8.314 \times 298) + (2)(0 - 8.314 \times 298)$$
$$+ (7.52)(0 - 8.314 \times 298)$$

It yields

$$\bar{h}_{CO_2} + 2\bar{h}_{H_2O} + 7.52\bar{h}_{N_2} - 87.463 \times T_P = 870{,}609 \text{ kJ}$$

The temperature of the product gases is obtained from a trial and error solution,

At 2850 K:

$$\bar{h}_{CO_2} + 2\bar{h}_{H_2O} + 7.52\bar{h}_{N_2} - 87.463 \times T_P = (1)(152{,}908) + (2)(127{,}952) + (7.52)(95{,}859) - (87.463)(2850)$$
$$= 880{,}402 \text{kJ} \; (\text{Higher than } 870{,}609 \text{kJ})$$

At 2800 K:

$$\bar{h}_{CO_2} + 2\bar{h}_{H_2O} + 7.52\bar{h}_{N_2} - 87.463 \times T_P = (1)(149{,}808) + (2)(125{,}198) + (7.52)(94{,}014) - (87.463)(2800)$$
$$= 862{,}293 \text{ kJ (Lower than 870,609 kJ)}$$

By interpolation, T_P = 2823 K

Treating both the reactants and the products as ideal gases, the final (maximum) pressure that can occur in the combustion chamber is determined to be

$$\frac{P_1 V}{P_2 V} = \frac{N_1 R_u T_1}{N_2 R_u T_2} \longrightarrow P_2 = \frac{N_2 T_2}{N_1 T_1} P_1 = \frac{(10.52 \text{ kmol})(2823 \text{ K})}{(10.52 \text{ kmol})(298 \text{ K})}(98 \text{ kPa}) = \mathbf{928 \text{ kPa}}$$

(*b*) The combustion equation of $CH_4(g)$ remains the same in the case of constant pressure. Further, the boundary work in this case can be combined with the *u* terms so that the first law relation can be expressed in terms of enthalpies just like the steady-flow process,

$$Q = \sum N_P \left(\bar{h}_f^\circ + \bar{h} - \bar{h}^\circ\right)_P - \sum N_R \left(\bar{h}_f^\circ + \bar{h} - \bar{h}^\circ\right)_R$$

Again since both the reactants and the products behave as ideal gases, we have $h = h(T)$. Also noting that $Q = 0$ for an adiabatic combustion process, the energy balance relation reduces to

$$\sum N_P \left(\bar{h}_f^\circ + \bar{h}_{T_P} - \bar{h}_{298K}\right)_P = \sum N_R \left(\bar{h}_f^\circ\right)_R$$

since the reactants are at the standard reference temperature of 25°C. Then using data from the mini table above, we get

$$(1)(-393{,}520 + \bar{h}_{CO_2} - 9364) + (2)(-241{,}820 + \bar{h}_{H_2O} - 9904) + (7.52)(0 + \bar{h}_{N_2} - 8669) =$$
$$(1)(-74{,}850) + (2)(0) + (7.52)(0)$$

It yields

$$\bar{h}_{CO_2} + 2\bar{h}_{H_2O} + 7.52\bar{h}_{N_2} = 896{,}673 \text{ kJ}$$

The temperature of the product gases is obtained from a trial and error solution,

At 2350 K: $$\bar{h}_{CO_2} + 2\bar{h}_{H_2O} + 7.52\bar{h}_{N_2} = (1)(122{,}091) + (2)(100{,}846) + (7.52)(77{,}496)$$
$$= 906{,}553 \text{ kJ (Higher than 896,673 kJ)}$$

At 2300 K: $$\bar{h}_{CO_2} + 2\bar{h}_{H_2O} + 7.52\bar{h}_{N_2} = (1)(119{,}035) + (2)(98{,}199) + (7.52)(75{,}676)$$
$$= 884{,}517 \text{ kJ (Lower than 896,673 kJ)}$$

By interpolation, T_P = 2328 K

Treating both the reactants and the products as ideal gases, the final (maximum) volume that the combustion chamber can have is determined to be

$$\frac{PV_1}{PV_2} = \frac{N_1 R_u T_1}{N_2 R_u T_2} \longrightarrow V_2 = \frac{N_2 T_2}{N_1 T_1} V_1 = \frac{(10.52 \text{ kmol})(2328 \text{ K})}{(10.52 \text{ kmol})(298 \text{ K})}(0.8 \text{ L}) = \mathbf{6.25 \text{ L}}$$

14-97 It is to be shown that the work output of the Carnot engine will be maximum when $T_p = \sqrt{T_0 T_{af}}$. It is also to be shown that the maximum work output of the Carnot engine in this case becomes $w = CT_{af}\left(1 - \frac{\sqrt{T_0}}{\sqrt{T_{af}}}\right)^2$.

Analysis The combustion gases will leave the combustion chamber and enter the heat exchanger at the adiabatic flame temperature T_{af} since the chamber is adiabatic and the fuel is burned completely. The combustion gases experience no change in their chemical composition as they flow through the heat exchanger. Therefore, we can treat the combustion gases as a gas stream with a constant specific heat C_p. Noting that the heat exchanger involves no work interactions, the energy balance equation for this single-stream steady-flow device can be written as

$$\dot{Q} = \dot{m}(h_e - h_i) = \dot{m}C(T_p - T_{af})$$

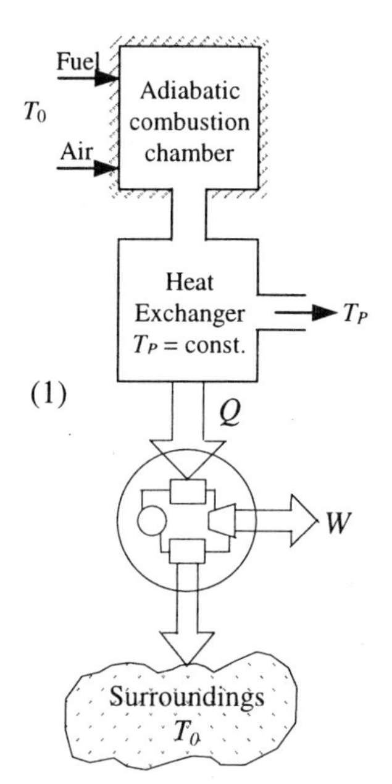

where $\dot{Q}$ is the negative of the heat supplied to the heat engine. That is,

$$\dot{Q}_H = -\dot{Q} = \dot{m}C(T_{af} - T_p)$$

Then the work output of the Carnot heat engine can be expressed as

$$\dot{W} = \dot{Q}_H\left(1 - \frac{T_0}{T_p}\right) = \dot{m}C(T_{af} - T_p)\left(1 - \frac{T_0}{T_p}\right) \quad (1)$$

Taking the partial derivative of $\dot{W}$ with respect to T_p while holding T_{af} and T_0 constant gives

$$\frac{\partial W}{\partial T_p} = 0 \longrightarrow -\dot{m}C\left(1 - \frac{T_0}{T_p}\right) + \dot{m}C(T_p - T_{af})\frac{T_0}{T_p^2}$$

Solving for T_p we obtain

$$T_p = \sqrt{T_0 T_{af}}$$

which the temperature at which the work output of the Carnot engine will be a maximum. The maximum work output is determined by substituting the relation above into Eq. (1),

$$\dot{W} = \dot{m}C(T_{af} - T_p)\left(1 - \frac{T_0}{T_p}\right) = \dot{m}C\left(T_{af} - \sqrt{T_0 T_{af}}\right)\left(1 - \frac{T_0}{\sqrt{T_0 T_{af}}}\right)$$

It simplifies to

$$\dot{W} = \dot{m}CT_{af}\left(1 - \frac{\sqrt{T_0}}{\sqrt{T_{af}}}\right)^2 \quad \text{or} \quad w = CT_{af}\left(1 - \frac{\sqrt{T_0}}{\sqrt{T_{af}}}\right)^2$$

which is the desired relation.

14-98 It is to be shown that the work output of the reversible heat engine operating at the specified conditions is $\dot{W}_{\text{rev}} = \dot{m}CT_0\left(\frac{T_{\text{af}}}{T_0} - 1 - \ln\frac{T_{\text{af}}}{T_0}\right)$. It is also to be shown that the effective flame temperature T_e of the furnace considered is $T_e = \frac{T_{\text{af}} - T_0}{\ln(T_{\text{af}}/T_0)}$.

Analysis The combustion gases will leave the combustion chamber and enter the heat exchanger at the adiabatic flame temperature T_{af} since the chamber is adiabatic and the fuel is burned completely. The combustion gases experience no change in their chemical composition as they flow through the heat exchanger. Therefore, we can treat the combustion gases as a gas stream with a constant specific heat C_p. Also, the work output of the reversible heat engine is equal to the reversible work W_{rev} of the heat exchanger as the combustion gases are cooled from T_{af} to T_0. That is,

$$\begin{aligned}\dot{W}_{\text{rev}} &= \dot{m}(h_i - h_e - T_0(s_i - s_e)) \\ &= \dot{m}C\left(T_{\text{af}} - T_0 - T_0\left(C\ln\frac{T_{\text{af}}}{T_0} - R\ln\frac{P_{\text{af}}}{P_0}\right)\right) \\ &= \dot{m}C\left(T_{\text{af}} - T_0 - T_0C\ln\frac{T_{\text{af}}}{T_0}\right)\end{aligned}$$

which can be rearranged as

$$\dot{W}_{\text{rev}} = \dot{m}CT_0\left(\frac{T_{\text{af}}}{T_0} - 1 - \ln\frac{T_{\text{af}}}{T_0}\right) \quad \text{or} \quad w_{\text{rev}} = CT_0\left(\frac{T_{\text{af}}}{T_0} - 1 - \ln\frac{T_{\text{af}}}{T_0}\right) \quad (1)$$

which is the desired result.

The effective flame temperature T_e can be determined from the requirement that a Carnot heat engine which receives the same amount of heat from a heat reservoir at constant temperature T_e produces the same amount of work. The amount of heat delivered to the heat engine above is

$$\dot{Q}_H = \dot{m}(h_i - h_e) = \dot{m}C(T_{\text{af}} - T_0)$$

A Carnot heat engine which receives this much heat at a constant temperature T_e will produce work in the amount of

$$\dot{W} = \dot{Q}_H \eta_{\text{th,Carnot}} = \dot{m}C(T_{\text{af}} - T_0)\left(1 - \frac{T_0}{T_e}\right) \quad (2)$$

Setting equations (1) and (2) equal to each other yields

$$\dot{m}CT_0\left(\frac{T_{\text{af}}}{T_0} - 1 - \ln\frac{T_{\text{af}}}{T_0}\right) = \dot{m}C(T_{\text{af}} - T_0)\left(1 - \frac{T_0}{T_e}\right)$$

$$T_{\text{af}} - T_0 - T_0\ln\frac{T_{\text{af}}}{T_0} = T_{\text{af}} - T_{\text{af}}\frac{T_0}{T_e} - T_0 + T_0\frac{T_0}{T_e}$$

Simplifying and solving for T_e, we obtain

$$T_e = \frac{T_{\text{af}} - T_0}{\ln(T_{\text{af}}/T_0)}$$

which is the desired relation.

14-99 · · · 14-108 EES solution of this (and other comprehensive problems designated with the *computer icon*) is available to instructors at the *Instructor Manual* section of the *Online Learning Center* (OLC) at www.mhhe.com/cengel-boles. See the Preface for access information.

Fundamentals of Engineering (FE) Exam Problems

14-109 A fuel is burned with 80 percent theoretical air. This is equivalent to

(a) 20% excess air (b) 80% excess air (c) 20% deficiency of air (d) 80% deficiency of air (e) stoichiometric amount of air

Answer (c) 20% deficiency of air

Solution Solved by EES Software. Solutions can be verified by copying-and-pasting the following lines on a blank EES screen. (Similar problems and their solutions can be obtained easily by modifying numerical values).

```
air_th=0.8
"air_th=air_access+1"
air_th=1-air_deficiency
```

14-110 Propane C_3H_8 is burned with 150 percent theoretical air. The air-fuel mass ratio for this combustion process is

(a) 5.3 (b) 10.5 (c) 15.7 (d) 23.4 (e) 39.3

Answer (d) 23.4

Solution Solved by EES Software. Solutions can be verified by copying-and-pasting the following lines on a blank EES screen. (Similar problems and their solutions can be obtained easily by modifying numerical values).

```
n_C=3
n_H=8

m_fuel=n_H*1+n_C*12
a_th=n_C+n_H/4
coeff=1.5 "coeff=1 for theoretical combustion, 1.5 for 50% excess air"

n_O2=coeff*a_th
n_N2=3.76*n_O2
m_air=n_O2*32+n_N2*28
AF=m_air/m_fuel
```

14-111 One kmol of ethane (C_2H_6) is burned with an unknown amount of air during a combustion process. If the combustion is complete and there are 2 kmol of free O_2 in the products, the air-fuel mass ratio is

(a) 25.2 (b) 21.2 (c) 17.6 (d) 14.9 (e) 12.1

Answer (a) 25.2

Solution Solved by EES Software. Solutions can be verified by copying-and-pasting the following lines on a blank EES screen. (Similar problems and their solutions can be obtained easily by modifying numerical values).

```
"The Fuel is C2H6"
n_C=2
n_H=6

m_fuel=n_H*1+n_C*12
a_th=n_C+n_H/4
(coeff-1)*a_th=2  "O2 balance: Coeff=1 for theoretical combustion, 1.5 for 50% excess
air"

n_O2=coeff*a_th
n_N2=3.76*n_O2
m_air=n_O2*32+n_N2*28
AF=m_air/m_fuel

"Some Wrong Solutions with Common Mistakes:"
W1_AF=1/AF "Taking the inverse of AF"
W2_AF=n_O2+n_N2 "Finding air-fuel mole ratio"
W3_AF=AF/coeff "Ignoring excess air"
```

14-112 A fuel is burned steadily in a combustion chamber. The combustion temperature will be the highest except when
(a) the fuel is preheated.
(b) the fuel is burned with a deficiency of air.
(c) the air is dry.
(d) the combustion chamber is well insulated.
(e) the combustion is complete.

Answer (b) the fuel is burned with a deficiency of air.

14-113 An equimolar mixture of carbon dioxide and water vapor at 1 atm and 50°C enter a dehumidifying section where the entire water vapor is condensed and removed from the mixture, and the carbon dioxide leaves at 1 atm and 50°C. The entropy change of carbon dioxide in the dehumidifying section is

(a) –5.8 kJ/kg·K (b) –0.13 kJ/kg·K (c) 0 (d) 0.13 kJ/kg·K (e) 5.8 kJ/kg·K

Answer (b) –0.13 kJ/kg·K

Solution Solved by EES Software. Solutions can be verified by copying-and-pasting the following lines on a blank EES screen. (Similar problems and their solutions can be obtained easily by modifying numerical values).

```
Cp_CO2=0.846
R_CO2=0.1889
T1=50+273 "K"
T2=T1
P1= 1 "atm"
P2=1 "atm"
y1_CO2=0.5; P1_CO2=y1_CO2*P1
y2_CO2=1; P2_CO2=y2_CO2*P2
Ds_CO2=Cp_CO2*ln(T2/T1)-R_CO2*ln(P2_CO2/P1_CO2)

"Some Wrong Solutions with Common Mistakes:"
W1_Ds=0 "Assuming no entropy change"
W2_Ds=Cp_CO2*ln(T2/T1)-R_CO2*ln(P1_CO2/P2_CO2) "Using pressure fractions backwards"
```

14-114 Methane (CH_4) is burned completely with 80% excess air during a steady-flow combustion process. If both the reactants and the products are maintained at 25°C and 1 atm and the water in the products exists in the liquid form, the heat transfer from the combustion chamber per unit mass of methane is

(a) 890 MJ/kg (b) 802 MJ/kg (c) 75 MJ/kg (d) 56 MJ/kg (e) 50 MJ/kg

Answer (d) 56 MJ/kg

Solution Solved by EES Software. Solutions can be verified by copying-and-pasting the following lines on a blank EES screen. (Similar problems and their solutions can be obtained easily by modifying numerical values).

```
T= 25 "C"
P=1 "atm"
EXCESS=0.8

"Heat transfer in this case is the HHV at room temperature,"
HHV_CH4 =55.53 "MJ/kg"
LHV_CH4 =50.05 "MJ/kg"

"Some Wrong Solutions with Common Mistakes:"
W1_Q=LHV_CH4 "Assuming lower heating value"
W2_Q=EXCESS*hHV_CH4 "Assuming Q to be proportional to excess air"
```

14-115 The higher heating value of a hydrocarbon fuel C_nH_m with $m = 6$ is given to be 1558 MJ/kmol of fuel. Then its lower heating value is

(a) 1426 MJ/kmol (b) 1466 MJ/kmol (c) 1502 MJ/kmol (d) 1514 MJ/kmol (e) 1551 MJ/kmol

Answer (a) 1426 MJ/kmol

Solution Solved by EES Software. Solutions can be verified by copying-and-pasting the following lines on a blank EES screen. (Similar problems and their solutions can be obtained easily by modifying numerical values).

```
HHV=1558 "MJ/kmol fuel"
h_fg=2.4423 "MJ/kg, Enthalpy of vaporization of water at 25C"
n_H=6
n_H2O=n_H/2
m_water=n_H2O*18
LHV=HHV-h_fg*m_water

"Some Wrong Solutions with Common Mistakes:"
W1_LHV=HHV - h_fg*n_water  "Using mole numbers instead of mass"
W2_LHV= HHV - h_fg*m_water*2 "Taking mole numbers of H2O to be m instead of m/2"
W3_LHV= HHV - h_fg*n_water*2 "Taking mole numbers of H2O to be m instead of m/2, and using mole numbers"
```

14-116 Acetylene gas (C_2H_2) is burned completely during a steady-flow combustion process. The fuel and the air enter the combustion chamber at 25°C, and the products leave at 1500 K. If the enthalpy of the products relative to the standard reference state is –404 MJ/kmol of fuel, the heat transfer from the combustion chamber is

(a) 177 MJ/kmol (b) 227 MJ/kmol (c) 404 MJ/kmol (d) 631 MJ/kmol (e) 751 MJ/kmol

Answer (d) 631 MJ/kmol

Solution Solved by EES Software. Solutions can be verified by copying-and-pasting the following lines on a blank EES screen. (Similar problems and their solutions can be obtained easily by modifying numerical values).

```
hf_fuel=226730/1000 "MJ/kmol fuel"
H_prod=-404 "MJ/kmol fuel"

H_react=hf_fuel
Q_out=H_react-H_prod

"Some Wrong Solutions with Common Mistakes:"
W1_Qout= -H_prod "Taking Qout to be H_prod"
W2_Qout= H_react+H_prod "Adding enthalpies instead of subtracting them"
```

14-117 Benzene gas (C_6H_6) is burned with 95 percent theoretical air during a steady-flow combustion process. The mole fraction of the CO in the products is

(a) 1.6% (b) 2.1% (c) 2.5% (d) 4.2% (e) 5%

Answer (b) 2.1%

Solution Solved by EES Software. Solutions can be verified by copying-and-pasting the following lines on a blank EES screen. (Similar problems and their solutions can be obtained easily by modifying numerical values).

```
n_C=6
n_H=6

a_th=n_C+n_H/4
coeff=0.95 "coeff=1 for theoretical combustion, 1.5 for 50% excess air"

"Assuming all the H burns to H2O, the combustion equation is
   C6H6+coeff*a_th(O2+3.76N2)----- (n_CO2) CO2+(n_CO)CO+(n_H2O) H2O+(n_N2)
N2"

n_O2=coeff*a_th
n_N2=3.76*n_O2
n_H2O=n_H/2
n_CO2+n_CO=n_C
2*n_CO2+n_CO+n_H2O=2*n_O2 "Oxygen balance"
n_prod=n_CO2+n_CO+n_H2O+n_N2 "Total mole numbers of product gases"
y_CO=n_CO/n_prod "mole fraction of CO in product gases"

"Some Wrong Solutions with Common Mistakes:"
W1_yCO=n_CO/n1_prod; n1_prod=n_CO2+n_CO+n_H2O "Not including N2 in
n_prod"
W2_yCO=(n_CO2+n_CO)/n_prod "Using both CO and CO2 in calculations"
```

14-118 A fuel is burned during a steady-flow combustion process. Heat is lost to the surroundings at 280 K at a rate of 1280 kW. The entropy of the reactants entering per unit time is 17 kW/K and that of the products is 15 kW/K. The total rate of exergy destruction during this combustion process is

(a) 720 kW (b) 936 kW (c) 1840 kW (d) 2140 kW (e) 2439 kW

Answer (a) 720 kW

Solution Solved by EES Software. Solutions can be verified by copying-and-pasting the following lines on a blank EES screen. (Similar problems and their solutions can be obtained easily by modifying numerical values).

```
To=280 "K"
Q_out=1280 "kW"
S_react=17 "kW'K"
S_prod= 15 "kW/K"

S_react-S_prod-Q_out/To+S_gen=0 "Entropy balance for steady state operation, Sin-Sout+Sgen=0"
X_dest=To*S_gen

"Some Wrong Solutions with Common Mistakes:"
W1_Xdest=S_gen "Taking Sgen as exergy destruction"
W2_Xdest=To*S_gen1; S_react-S_prod-S_gen1=0 "Ignoring Q_out/To"
```

14-119 . . . 14-124 Design and Essay Problems

14-120 Constant-volume vessels that store flammable gases are to be designed to withstand the rising pressures in case of an explosion. The safe design pressures for (*a*) acetylene, (*b*) propane, and (*c*) n-octane are to be determined for storage pressures slightly above the atmospheric pressure.

Analysis (*a*) The final temperature (and pressure) in the tank will be highest when the combustion is complete, adiabatic, and stoichiometric. In addition, we assume the atmospheric pressure to be 100 kPa and the initial temperature in the tank to be 25°C. Then the initial pressure of the air-fuel mixture in the tank becomes 125 kPa.

The combustion equation of $C_2H_2(g)$ with stoichiometric amount of air is

$$C_2H_2 + a_{th}(O_2 + 3.76N_2) \longrightarrow 2CO_2 + H_2O + 3.76a_{th}N_2$$

where a_{th} is the stoichiometric coefficient and is determined from the O_2 balance,

$$a_{th} = 2 + 0.5 \longrightarrow a_{th} = 2.5$$

Thus,

$$C_2H_2 + 2.5(O_2 + 3.76N_2) \longrightarrow 2CO_2 + H_2O + 9.40N_2$$

The final temperature in the tank is determined from the energy balance relation $E_{in} - E_{out} = \Delta E_{system}$ for reacting closed systems under adiabatic conditions ($Q = 0$) with no work interactions ($W = 0$),

$$0 = \sum N_P\left(\bar{h}_f^\circ + \bar{h} - \bar{h}^\circ - P\bar{v}\right)_P - \sum N_R\left(\bar{h}_f^\circ + \bar{h} - \bar{h}^\circ - P\bar{v}\right)_R$$

Assuming both the reactants and the products to behave as ideal gases, all the internal energy and enthalpies depend on temperature only, and the $P\bar{v}$ terms in this equation can be replaced by R_uT. It yields

$$\sum N_P\left(\bar{h}_f^\circ + \bar{h}_{T_P} - \bar{h}_{298K} - R_uT\right)_P = \sum N_R\left(\bar{h}_f^\circ - R_uT\right)_R$$

since the reactants are at the standard reference temperature of 25°C. From the tables,

Substance	$\bar{h}_f^\circ$ kJ/kmol	$\bar{h}_{298\,K}$ kJ/kmol
C_2H_2	226,730	---
O_2	0	8682
N_2	0	8669
H_2O (*g*)	-241,820	9904
CO_2	-393,520	9364

Thus,

$$(2)(-393{,}520 + \bar{h}_{CO_2} - 9364 - 8.314 \times T_P) + (1)(-241{,}820 + \bar{h}_{H_2O} - 9904 - 8.314 \times T_P)$$
$$+ (9.40)(0 + \bar{h}_{N_2} - 8669 - 8.314 \times T_P) = (1)(226{,}730 - 8.314 \times 298) + (2.5)(0 - 8.314 \times 298) + (9.40)(0 - 8.314 \times 298)$$

It yields

$$2\bar{h}_{CO_2} + \bar{h}_{H_2O} + 9.40\bar{h}_{N_2} - 103.094 \times T_P = 1{,}333{,}750 \text{ kJ}$$

The temperature of the product gases is obtained from a trial and error solution,

At 3200 K:

$$2\bar{h}_{CO_2} + \bar{h}_{H_2O} + 9.40\bar{h}_{N_2} - 103.094 \times T_P = (2)(174{,}695) + (1)(147{,}457) + (9.40)(108{,}830) - (103.094)(3200)$$
$$= 1{,}189{,}948 \text{kJ (Lower than 1,333,750kJ)}$$

At 3250 K:

$$2\bar{h}_{CO_2} + \bar{h}_{H_2O} + 9.40\bar{h}_{N_2} - 103.094 \times T_P = (2)(177{,}822) + (1)(150{,}272) + (9.40)(110{,}690) - (103.094)(3250)$$
$$= 1{,}211{,}347 \text{kJ} \ (\text{Lower than } 1{,}333{,}750 \text{kJ})$$

By extrapolation, $T_P = 3536$ K

Treating both the reactants and the products as ideal gases, the final (maximum) pressure that can occur in the combustion chamber is determined to be

$$\frac{P_1 V}{P_2 V} = \frac{N_1 R_u T_1}{N_2 R_u T_2} \longrightarrow P_2 = \frac{N_2 T_2}{N_1 T_1} P_1 = \frac{(12.40 \text{ kmol})(3536 \text{ K})}{(12.90 \text{ kmol})(298 \text{ K})}(125 \text{ kPa}) = 1426 \text{ kPa}$$

Then the pressure the tank must be designed for in order to meet the requirements of the code is

$$P = (4)(1426 \text{ kPa}) = \mathbf{5704 \ kPa}$$

14-120b The final temperature (and pressure) in the tank will be highest when the combustion is complete, adiabatic, and stoichiometric. In addition, we assume the atmospheric pressure to be 100 kPa and the initial temperature in the tank to be 25°C. Then the initial pressure of the air-fuel mixture in the tank becomes 125 kPa.

The combustion equation of $C_3H_8(g)$ with stoichiometric amount of air is

$$C_3H_8 + a_{th}(O_2 + 3.76N_2) \longrightarrow 3CO_2 + 4H_2O + 3.76a_{th}N_2$$

where a_{th} is the stoichiometric coefficient and is determined from the O_2 balance,

$$a_{th} = 3 + 2 \longrightarrow a_{th} = 5$$

Thus,

$$C_3H_8 + 5(O_2 + 3.76N_2) \longrightarrow 3CO_2 + 4H_2O + 18.80N_2$$

The final temperature in the tank is determined from the energy balance relation $E_{in} - E_{out} = \Delta E_{system}$ for reacting closed systems under adiabatic conditions ($Q = 0$) with no work interactions ($W = 0$),

$$0 = \sum N_P\left(\bar{h}_f^\circ + \bar{h} - \bar{h}^\circ - P\bar{v}\right)_P - \sum N_R\left(\bar{h}_f^\circ + \bar{h} - \bar{h}^\circ - P\bar{v}\right)_R$$

Assuming both the reactants and the products to behave as ideal gases, all the internal energy and enthalpies depend on temperature only, and the $P\bar{v}$ terms in this equation can be replaced by R_uT. It yields

$$\sum N_P\left(\bar{h}_f^\circ + \bar{h}_{T_P} - \bar{h}_{298\text{K}} - R_uT\right)_P = \sum N_R\left(\bar{h}_f^\circ - R_uT\right)_R$$

since the reactants are at the standard reference temperature of 25°C. From the tables,

Substance	$\bar{h}_f^\circ$ kJ/kmol	$\bar{h}_{298\text{ K}}$ kJ/kmol
C_3H_8	-103,850	---
O_2	0	8682
N_2	0	8669
H_2O (*g*)	-241,820	9904
CO_2	-393,520	9364

Thus,

$$(3)(-393{,}520+\bar{h}_{CO_2}-9364-8.314\times T_P)+(4)(-241{,}820+\bar{h}_{H_2O}-9904-8.314\times T_P)$$
$$+(18.80)(0+\bar{h}_{N_2}-8669-8.314\times T_P)=(1)(-103{,}850-8.314\times 298)+(5)(0-8.314\times 298)+(18.80)(0-8.314\times 298)$$

It yields

$$3\bar{h}_{CO_2}+4\bar{h}_{H_2O}+18.80\bar{h}_{N_2}-214.50\times T_P=2{,}213{,}231\text{ kJ}$$

The temperature of the product gases is obtained from a trial and error solution,

At 2950 K:

$$3\bar{h}_{CO_2}+4\bar{h}_{H_2O}+18.80\bar{h}_{N_2}-214.50\times T_P=(3)(159{,}117)+(4)(133{,}486)+(18.80)(99{,}556)-(214.50)(2950)$$
$$=2{,}250{,}173\text{kJ (Higherthan}2{,}213{,}231\text{kJ)}$$

At 2900 K:

$$3\bar{h}_{CO_2}+4\bar{h}_{H_2O}+18.80\bar{h}_{N_2}-214.50\times T_P=(3)(156{,}009)+(4)(130{,}717)+(18.80)(97{,}705)-(214.50)(2900)$$
$$=2{,}205{,}699\text{kJ (Lowerthan}2{,}213{,}231\text{kJ)}$$

By interpolation, T_P = 2908 K

Treating both the reactants and the products as ideal gases, the final (maximum) pressure that can occur in the combustion chamber is determined to be

$$\frac{P_1V}{P_2V}=\frac{N_1R_uT_1}{N_2R_uT_2}\longrightarrow P_2=\frac{N_2T_2}{N_1T_1}P_1=\frac{(25.80\text{ kmol})(2908\text{ K})}{(24.80\text{ kmol})(298\text{ K})}(125\text{kPa})=1269\text{ kPa}$$

Then the pressure the tank must be designed for in order to meet the requirements of the code is

$$P=(4)(1269\text{ kPa})=\mathbf{5076\ kPa}$$

14-120c The final temperature (and pressure) in the tank will be highest when the combustion is complete, adiabatic, and stoichiometric. In addition, we assume the atmospheric pressure to be 100 kPa and the initial temperature in the tank to be 25°C. Then the initial pressure of the air-fuel mixture in the tank becomes 125 kPa.

The combustion equation of $C_8H_{18}(g)$ with stoichiometric amount of air is

$$C_8H_{18}+a_{th}(O_2+3.76N_2)\longrightarrow 8CO_2+9H_2O+3.76a_{th}N_2$$

where a_{th} is the stoichiometric coefficient and is determined from the O_2 balance,

$$a_{th}=8+4.5\longrightarrow a_{th}=12.5$$

Thus,

$$C_8H_{18}+12.5(O_2+3.76N_2)\longrightarrow 8CO_2+9H_2O+47.0N_2$$

The final temperature in the tank is determined from the energy balance relation $E_{in}-E_{out}=\Delta E_{system}$ for reacting closed systems under adiabatic conditions (Q = 0) with no work interactions (W = 0),

$$0=\sum N_P\left(\bar{h}_f^\circ+\bar{h}-\bar{h}^\circ-P\bar{v}\right)_P-\sum N_R\left(\bar{h}_f^\circ+\bar{h}-\bar{h}^\circ-P\bar{v}\right)_R$$

Assuming both the reactants and the products to behave as ideal gases, all the internal energy and enthalpies depend on temperature only, and the $P\bar{v}$ terms in this equation can be replaced by R_uT. It yields

$$\sum N_P\left(\overline{h}_f^\circ + \overline{h}_{T_P} - \overline{h}_{298\text{K}} - R_u T\right)_P = \sum N_R\left(\overline{h}_f^\circ - R_u T\right)_R$$

since the reactants are at the standard reference temperature of 25°C. From the tables,

Substance	$\overline{h}_f^\circ$ kJ/kmol	$\overline{h}_{298\text{ K}}$ kJ/kmol
C_8H_{18}	-208,450	---
O_2	0	8682
N_2	0	8669
H_2O (*g*)	-241,820	9904
CO_2	-393,520	9364

Thus,

$$(8)\left(-393{,}520 + \overline{h}_{CO_2} - 9364 - 8.314\times T_P\right) + (9)\left(-241{,}820 + \overline{h}_{H_2O} - 9904 - 8.314\times T_P\right)$$
$$+ (47.0)\left(0 + \overline{h}_{N_2} - 8669 - 8.314\times T_P\right) = (1)(-208{,}450 - 8.314\times 298) + (12.5)(0 - 8.314\times 298) + (47.0)(0 - 8.314\times 298$$

It yields

$$8\overline{h}_{CO_2} + 9\overline{h}_{H_2O} + 47.0\overline{h}_{N_2} - 532.10\times T_P = 5{,}537{,}688\text{ kJ}$$

The temperature of the product gases is obtained from a trial and error solution,

At 2950 K:

$$8\overline{h}_{CO_2} + 9\overline{h}_{H_2O} + 47.0\overline{h}_{N_2} - 532.10\times T_P = (8)(159{,}117) + (9)(133{,}486) + (47.0)(99{,}556) - (532.10)(2950)$$
$$= 5{,}583{,}747\text{kJ (Higherthan5,534,220kJ)}$$

At 2900 K:

$$8\overline{h}_{CO_2} + 9\overline{h}_{H_2O} + 47.0\overline{h}_{N_2} - 532.10\times T_P = (8)(156{,}009) + (9)(130{,}717) + (47.0)(97{,}705) - (532.10)(2900)$$
$$= 5{,}473{,}570\text{kJ (Lowerthan5,534,220kJ)}$$

By interpolation, T_P = 2929 K

Treating both the reactants and the products as ideal gases, the final (maximum) pressure that can occur in the combustion chamber is determined to be

$$\frac{P_1 V}{P_2 V} = \frac{N_1 R_u T_1}{N_2 R_u T_2} \longrightarrow P_2 = \frac{N_2 T_2}{N_1 T_1} P_1 = \frac{(64.0\text{ kmol})(2929\text{ K})}{(60.5\text{ kmol})(298\text{ K})}(125\text{ kPa}) = 1300\text{ kPa}$$

Then the pressure the tank must be designed for in order to meet the requirements of the code is

$P = (4)(1300\text{ kPa}) = \mathbf{5200\ kPa}$

14-121 A certain industrial process generates a liquid solution of ethanol and water as the waste product. The solution is to be burned using methane. A combustion process is to be developed to accomplish this incineration process with minimum amount of methane.

Analysis The mass flow rate of the liquid ethanol-water solution is given to be 10 kg/s. Considering that the mass fraction of ethanol in the solution is 0.2,

$$\dot{m}_{\text{enthanol}} = (0.2)(10\text{kg/s}) = 2\text{kg/s}$$
$$\dot{m}_{\text{water}} = (0.8)(10\text{kg/s}) = 8\text{kg/s}$$

Noting that the molar masses M_{ethanol} = 46 and M_{water} = 18 kg/kmol and that mole numbers $N = m/M$, the mole flow rates become

$$\dot{N}_{\text{ethanol}} = \frac{\dot{m}_{\text{ethanol}}}{M_{\text{ethanol}}} = \frac{2 \text{ kg/s}}{46 \text{ kg/kmol}} = 0.04348 \text{ kmol/s}$$
$$\dot{N}_{\text{water}} = \frac{\dot{m}_{\text{water}}}{M_{\text{water}}} = \frac{8 \text{ kg/s}}{18 \text{ kg/kmol}} = 0.44444 \text{ kmol/s}$$

Note that

$$\frac{\dot{N}_{\text{water}}}{\dot{N}_{\text{ethanol}}} = \frac{0.44444}{0.04348} = 10.222 \text{ kmol H}_2\text{O/kmol C}_2\text{H}_5\text{OH}$$

That is, 10.222 moles of liquid water is present in the solution for each mole of ethanol.

Assuming complete combustion, the combustion equation of C_2H_5OH (ℓ) with stoichiometric amount of air is

$$C_2H_5OH(\ell) + a_{th}(O_2 + 3.76N_2) \longrightarrow 2CO_2 + 3H_2O + 3.76a_{th}N_2$$

where a_{th} is the stoichiometric coefficient and is determined from the O_2 balance,

$$1 + a_{th} = 4 + 3 \longrightarrow a_{th} = 3$$

Thus,

$$C_2H_5OH(\ell) + 3(O_2 + 3.76N_2) \longrightarrow 2CO_2 + 3H_2O + 11.28N_2$$

Noting that 10.222 kmol of liquid water accompanies each kmol of ethanol, the actual combustion equation can be written as

$$C_2H_5OH(\ell) + 3(O_2 + 3.76N_2) + 10.222H_2O(\ell) \longrightarrow 2CO_2 + 3H_2O(g) + 11.28N_2 + 10.222H_2O(\ell)$$

The heat transfer for this combustion process is determined from the steady-flow energy balance equation with $W = 0$,

$$Q = \sum N_P\left(\bar{h}_f^\circ + \bar{h} - \bar{h}^\circ\right)_P - \sum N_R\left(\bar{h}_f^\circ + \bar{h} - \bar{h}^\circ\right)_R$$

Assuming the air and the combustion products to be ideal gases, we have $h = h(T)$. We assume all the reactants to enter the combustion chamber at the standard reference temperature of 25°C. Furthermore, we assume the products to leave the combustion chamber at 1400 K which is a little over the required temperature of 1100°C. From the tables,

Substance	$\bar{h}_f^\circ$ kJ/kmol	$\bar{h}_{298\ K}$ kJ/kmol	$\bar{h}_{1400\ K}$ kJ/kmol
C_2H_5OH (ℓ)	-277,690	---	---
CH_4	-74,850	---	---
O_2	0	8682	45,648
N_2	0	8669	43,605
H_2O (*g*)	-241,820	9904	53,351
H_2O (ℓ)	-285,830	---	---
CO_2	-393,520	9364	65,271

Thus,

$$\begin{aligned} Q &= (2)(-393{,}520 + 65{,}271 - 9364) + (3)(-241{,}820 + 53{,}351 - 9904) \\ &\quad + (11.28)(0 + 43{,}605 - 8669) - (1)(-277{,}690) - 0 - 0 \\ &\quad + (10.222)(-241{,}820 + 53{,}351 - 9904) - (10.222)(-285{,}830) \\ &= 295{,}409 \text{ kJ/kmol of } C_2H_5OH \end{aligned}$$

The positive sign indicates that 295,409 kJ of heat must be supplied to the combustion chamber from another source (such as burning methane) to ensure that the combustion products will leave at the desired temperature of 1400 K. Then the rate of heat transfer required for a mole flow rate of 0.04348 kmol C_2H_5OH/s CO becomes

$$\dot{Q} = \dot{N}Q = (0.04348 \text{ kmol/s})(295{,}409 \text{ kJ/kmol}) = 12{,}844 \text{ kJ/s}$$

Assuming complete combustion, the combustion equation of $CH_4(g)$ with stoichiometric amount of air is

$$CH_4 + a_{th}(O_2 + 3.76N_2) \longrightarrow CO_2 + 2H_2O + 3.76a_{th}N_2$$

where a_{th} is the stoichiometric coefficient and is determined from the O_2 balance,

$$a_{th} = 1 + 1 \longrightarrow a_{th} = 2$$

Thus,

$$CH_4 + 2(O_2 + 3.76N_2) \longrightarrow CO_2 + 2H_2O + 7.52N_2$$

The heat transfer for this combustion process is determined from the steady-flow energy balance $E_{in} - E_{out} = \Delta E_{system}$ equation as shown above under the same assumptions and using the same mini table:

$$\begin{aligned} Q &= (1)(-393{,}520 + 65{,}271 - 9364) + (2)(-241{,}820 + 53{,}351 - 9904) \\ &\quad + (7.52)(0 + 43{,}605 - 8669) - (1)(-74{,}850) - 0 - 0 \\ &= -396{,}790 \text{ kJ/kmol of } CH_4 \end{aligned}$$

That is, 396,790 kJ of heat is supplied to the combustion chamber for each kmol of methane burned. To supply heat at the required rate of 12,844 kJ/s, we must burn methane at a rate of

$$\dot{N}_{CH_4} = \frac{\dot{Q}}{Q} = \frac{12{,}844 \text{ kJ/s}}{396{,}790 \text{ kJ/kmol}} = 0.03237 \text{ kmolCH}_4\text{/s}$$

or,

$$\dot{m}_{CH_4} = M_{CH_4}\dot{N}_{CH_4} = (16 \text{ kg/kmol})(0.03237 \text{ kmolCH}_4\text{/s}) = \mathbf{0.5179\ kg/s}$$

Therefore, we must supply methane to the combustion chamber at a minimum rate 0.5179 kg/s in order to maintain the temperature of the combustion chamber above 1400 K.

Chapter 15
CHEMICAL AND PHASE EQUILIBRIUM

The K_p and Equilibrium Composition of Ideal Gases

15-1C Because when a reacting system involves heat transfer, the increase-in-entropy principle relation requires a knowledge of heat transfer between the system and its surroundings, which is impractical. The equilibrium criteria can be expressed in terms of the properties alone when the Gibbs function is used.

15-2C No, the wooden table is NOT in chemical equilibrium with the air. With proper catalyst, it will reach with the oxygen in the air and burn.

15-3C They are

$$K_p = \frac{P_C^{\nu_C} P_D^{\nu_D}}{P_A^{\nu_A} P_B^{\nu_B}}, \quad K_p = e^{-\Delta G^*(T)/R_u T} \quad \text{and} \quad K_p = \frac{N_C^{\nu_C} N_D^{\nu_D}}{N_A^{\nu_A} N_B^{\nu_B}} \left(\frac{P}{N_{\text{total}}} \right)^{\Delta\nu}$$

where $\Delta\nu = \nu_C + \nu_D - \nu_A - \nu_B$. The first relation is useful in partial pressure calculations, the second in determining the K_p from gibbs functions, and the last one in equilibrium composition calculations.

15-4C (*a*) K_{p1}, (*b*) $1/K_{p1}$, (*c*) K_{p1}, (*d*) K_{p1}, (*e*) K_{p1}^2.

15-5C (*a*) K_{p1}, (*b*) $1/K_{p1}$, (*c*) K_{p1}^2, (*d*) K_{p1}, (*e*) $1/K_{p1}^3$.

15-6C (*a*) No, because K_p depends on temperature only.

(*b*) Yes, because the total mixture pressure affects the mixture composition. The equilibrium constant for the reaction $CO + \frac{1}{2}O_2 \Leftrightarrow CO_2$ can be expressed as

$$K_p = \frac{N_{CO_2}^{\nu_{CO_2}}}{N_{CO}^{\nu_{CO}} N_{O_2}^{\nu_{O_2}}} \left(\frac{P}{N_{\text{total}}} \right)^{(\nu_{CO_2} - \nu_{CO} - \nu_{O_2})}$$

The value of the exponent in this case is 1-1-0.5=-0.5, which is negative. Thus as the pressure increases, the term in the brackets will decrease. The value of K_p depends on temperature only, and therefore it will not change with pressure. Then to keep the equation balanced, the number of moles of the products (CO_2) must increase, and the number of moles of the reactants (CO, O_2) must decrease.

15-7C (*a*) No, because K_p depends on temperature only.

(*b*) In general, the total mixture pressure affects the mixture composition. The equilibrium constant for the reaction $N_2 + O_2 \Leftrightarrow 2NO$ can be expressed as

$$K_p = \frac{N_{NO}^{\nu_{NO}}}{N_{N_2}^{\nu_{N_2}} N_{O_2}^{\nu_{O_2}}} \left(\frac{P}{N_{total}} \right)^{(\nu_{NO} - \nu_{N_2} - \nu_{O_2})}$$

The value of the exponent in this case is 2-1-1 = 0. Therefore, changing the total mixture pressure will have no effect on the number of moles of N_2, O_2 and NO.

15-8C (*a*) The equilibrium constant for the reaction $CO + \frac{1}{2}O_2 \Leftrightarrow CO_2$ can be expressed as

$$K_p = \frac{N_{CO_2}^{\nu_{CO_2}}}{N_{CO}^{\nu_{CO}} N_{O_2}^{\nu_{O_2}}} \left(\frac{P}{N_{total}} \right)^{(\nu_{CO_2} - \nu_{CO} - \nu_{O_2})}$$

Judging from the values in Table A-28, the K_p value for this reaction decreases as temperature increases. That is, the indicated reaction will be less complete at higher temperatures. Therefore, the number of moles of CO_2 will decrease and the number moles of CO and O_2 will increase as the temperature increases.

(*b*) The value of the exponent in this case is 1-1-0.5=-0.5, which is negative. Thus as the pressure increases, the term in the brackets will decrease. The value of K_p depends on temperature only, and therefore it will not change with pressure. Then to keep the equation balanced, the number of moles of the products (CO_2) must increase, and the number of moles of the reactants (CO, O_2) must decrease.

15-9C (*a*) The equilibrium constant for the reaction $N_2 \Leftrightarrow 2N$ can be expressed as

$$K_p = \frac{N_{N}^{\nu_{N}}}{N_{N_2}^{\nu_{N_2}}} \left(\frac{P}{N_{total}} \right)^{(\nu_{N} - \nu_{N_2})}$$

Judging from the values in Table A-28, the K_p value for this reaction increases as the temperature increases. That is, the indicated reaction will be more complete at higher temperatures. Therefore, the number of moles of N will increase and the number moles of N_2 will decrease as the temperature increases.

(*b*) The value of the exponent in this case is 2-1 = 1, which is positive. Thus as the pressure increases, the term in the brackets also increases. The value of K_p depends on temperature only, and therefore it will not change with pressure. Then to keep the equation balanced, the number of moles of the products (N) must decrease, and the number of moles of the reactants (N_2) must increase.

15-10C The equilibrium constant for the reaction $CO + \frac{1}{2}O_2 \Leftrightarrow CO_2$ can be expressed as

$$K_p = \frac{N_{CO_2}^{\nu_{CO_2}}}{N_{CO}^{\nu_{CO}} N_{O_2}^{\nu_{O_2}}} \left(\frac{P}{N_{total}} \right)^{(\nu_{CO_2} - \nu_{CO} - \nu_{O_2})}$$

Adding more N_2 (an inert gas) at constant temperature and pressure will increase N_{total} but will have no direct effect on other terms. Then to keep the equation balanced, the number of moles of the products (CO_2) must increase, and the number of moles of the reactants (CO, O_2) must decrease.

15-11C The values of the equilibrium constants for each dissociation reaction at 3000 K are, from Table A-28,

$$N_2 \Leftrightarrow 2N \Leftrightarrow \ln K_p = -22.359$$

$$H_2 \Leftrightarrow 2H \Leftrightarrow \ln K_p = -3.685 \quad \text{(greater than -22.359)}$$

Thus H_2 is more likely to dissociate than N_2.

15-12 The equilibrium constant of the reaction H_2 + 1/2O_2 ↔ H_2O is listed in Table A-28 at different temperatures. The data are to be verified at two temperatures using Gibbs function data.

Analysis (*a*) The K_p value of a reaction at a specified temperature can be determined from the Gibbs function data using

$$K_p = e^{-\Delta G^*(T)/R_u T} \quad \text{or} \quad \ln K_p = -\Delta G^*(T)/R_u T$$

where

$$\Delta G^*(T) = \nu_{H_2O}\bar{g}^*_{H_2O}(T) - \nu_{H_2}\bar{g}^*_{H_2}(T) - \nu_{O_2}\bar{g}^*_{O_2}(T)$$

$H_2 + \frac{1}{2}O_2 \Leftrightarrow H_2O$

25°C

At 25°C,

$$\Delta G^*(T) = 1(-228{,}590) - 1(0) - 0.5(0) = -228{,}590 \text{ kJ/kmol}$$

Substituting,

$$\ln K_p = -(-228{,}590 \text{ kJ/kmol})/[(8.314 \text{ kJ/kmol}\cdot\text{K})(298 \text{ K})] = 92.26$$

or

$$K_p = \mathbf{1.12 \times 10^{40}} \quad \text{(Table A-28: } \ln K_p = 92.21)$$

(*b*) At 2000 K,

$$\begin{aligned}
\Delta G^*(T) &= \nu_{H_2O}\bar{g}^*_{H_2O}(T) - \nu_{H_2}\bar{g}^*_{H_2}(T) - \nu_{O_2}\bar{g}^*_{O_2}(T) \\
&= \nu_{H_2O}(\bar{h} - T\bar{s})_{H_2O} - \nu_{H_2}(\bar{h} - T\bar{s})_{H_2} - \nu_{O_2}(\bar{h} - T\bar{s})_{O_2} \\
&= \nu_{H_2O}[(\bar{h}_f + \bar{h}_{2000} - \bar{h}_{298}) - T\bar{s}]_{H_2O} \\
&\quad - \nu_{H_2}[(\bar{h}_f + \bar{h}_{2000} - \bar{h}_{298}) - T\bar{s}]_{H_2} \\
&\quad - \nu_{O_2}[(\bar{h}_f + \bar{h}_{2000} - \bar{h}_{298}) - T\bar{s}]_{O_2} \\
&= 1\times(-241{,}820 + 82{,}593 - 9904 - 2000\times 264.571) \\
&\quad - 1\times(0 + 61{,}400 - 8468 - 2000\times 188.297) \\
&\quad - 0.5\times(0 + 67{,}881 - 8682) - 2000\times 268.655) \\
&= -135{,}556 \text{ kJ/kmol}
\end{aligned}$$

Substituting,

$$\ln K_p = -(-135{,}556 \text{ kJ/kmol})/[(8.314 \text{ kJ/kmol}\cdot\text{K})(2000 \text{ K})] = 8.152$$

or $$K_p = \mathbf{3471} \quad \text{(Table A-28: } \ln K_p = 8.145)$$

15-13E The equilibrium constant of the reaction $H_2 + 1/2O_2 \leftrightarrow H_2O$ is listed in Table A-28 at different temperatures. The data are to be verified at two temperatures using Gibbs function data.

Analysis (*a*) The K_p value of a reaction at a specified temperature can be determined from the Gibbs function data using

$$K_p = e^{-\Delta G^*(T)/R_uT} \quad \text{or} \quad \ln K_p = -\Delta G^*(T)/R_uT$$

$H_2 + \frac{1}{2}O_2 \Leftrightarrow H_2O$
537 R

where

$$\Delta G^*(T) = \nu_{H_2O}\bar{g}^*_{H_2O}(T) - \nu_{H_2}\bar{g}^*_{H_2}(T) - \nu_{O_2}\bar{g}^*_{O_2}(T)$$

At 537 R,

$$\Delta G^*(T) = 1(-98{,}350) - 1(0) - 0.5(0) = -98{,}350 \text{ Btu/lbmol}$$

Substituting,

$$\ln K_p = -(-98{,}350 \text{ Btu/lbmol})/[(1.986 \text{ Btu/lbmol}\cdot\text{R})(537 \text{ R})] = 92.22$$

or

$$K_p = \mathbf{1.12\times 10^{40}} \quad (\text{Table A-28: } \ln K_p = 92.21)$$

(*b*) At 3240 R,

$$\begin{aligned}
\Delta G^*(T) &= \nu_{H_2O}\bar{g}^*_{H_2O}(T) - \nu_{H_2}\bar{g}^*_{H_2}(T) - \nu_{O_2}\bar{g}^*_{O_2}(T) \\
&= \nu_{H_2O}(\bar{h} - T\bar{s})_{H_2O} - \nu_{H_2}(\bar{h} - T\bar{s})_{H_2} - \nu_{O_2}(\bar{h} - T\bar{s})_{O_2} \\
&= \nu_{H_2O}[(\bar{h}_f + \bar{h}_{3240} - \bar{h}_{537}) - T\bar{s}]_{H_2O} \\
&\quad - \nu_{H_2}[(\bar{h}_f + \bar{h}_{3240} - \bar{h}_{298}) - T\bar{s}]_{H_2} \\
&\quad - \nu_{O_2}[(\bar{h}_f + \bar{h}_{3240} - \bar{h}_{298}) - T\bar{s}]_{O_2} \\
&= 1\times(-104{,}040 + 31{,}204.5 - 4258 - 3240\times 61.948) \\
&\quad - 1\times(0 + 23{,}484.7 - 3640.3 - 3240\times 44.125) \\
&\quad - 0.5\times(0 + 25{,}972 - 3725.1 - 3240\times 63.224) \\
&= -63{,}385 \text{ Btu/lbmol}
\end{aligned}$$

Substituting,

$$\ln K_p = -(-63{,}385 \text{ Btu/lbmol})/[(1.986 \text{ Btu/lbmol.R})(3240 \text{ R})] = 9.85$$

or

$$K_p = \mathbf{1.90\times 10^4} \quad (\text{Table A-28: } \ln K_p = 9.83)$$

15-14 The equilibrium constant of the reaction $CH_4 + 2O_2 \leftrightarrow CO_2 + 2H_2O$ at 25°C is to be determined.

Analysis The K_p value of a reaction at a specified temperature can be determined from the Gibbs function data using

$$K_p = e^{-\Delta G^*(T)/R_uT} \quad \text{or} \quad \ln K_p = -\Delta G^*(T)/R_uT$$

$CH_4 + 2O_2 \Leftrightarrow CO_2 + 2H_2O$
25°C

where

$$\Delta G^*(T) = \nu_{CO_2}\bar{g}^*_{CO_2}(T) + \nu_{H_2O}\bar{g}^*_{H_2O}(T) - \nu_{CH_4}\bar{g}^*_{CH_4}(T) - \nu_{O_2}\bar{g}^*_{O_2}(T)$$

At 25°C,

$$\Delta G^*(T) = 1(-394{,}360) + 2(-228{,}590) - 1(-50{,}790) - 2(0) = -800{,}750 \text{ kJ/kmol}$$

Substituting,

$$\ln K_p = -(-800{,}750 \text{ kJ/kmol})/[(8.314 \text{ kJ/kmol}\cdot\text{K})(298 \text{ K})] = 323.04$$

or

$$K_p = \mathbf{1.96 \times 10^{140}}$$

15-15 The equilibrium constant of the reaction $CO_2 \leftrightarrow CO + 1/2O_2$ is listed in Table A-28 at different temperatures. It is to be verified using Gibbs function data.

Analysis (*a*) The K_p value of a reaction at a specified temperature can be determined from the Gibbs function data using

$$K_p = e^{-\Delta G^*(T)/R_u T} \quad \text{or} \quad \ln K_p = -\Delta G^*(T)/R_u T$$

$CO_2 \Leftrightarrow CO + \frac{1}{2}O_2$
298 K

where

$$\Delta G^*(T) = \nu_{CO}\bar{g}^*_{CO}(T) + \nu_{O_2}\bar{g}^*_{O_2}(T) - \nu_{CO_2}\bar{g}^*_{CO_2}(T)$$

At 298 K,

$$\Delta G^*(T) = 1(-137{,}150) + 0.5(0) - 1(-394{,}360) = 257{,}210 \text{ kJ/kmol}$$

Substituting,

$$\ln K_p = -(257{,}210 \text{ kJ/kmol})/[(8.314 \text{ kJ/kmol}\cdot\text{K})(298 \text{ K})] = -103.81$$

or

$$K_p = \mathbf{8.20 \times 10^{-46}} \quad (\text{Table A-28: } \ln K_p = -103.76)$$

(*b*) At 1800 K,

$$\begin{aligned}
\Delta G^*(T) &= \nu_{CO}\bar{g}^*_{CO}(T) + \nu_{O_2}\bar{g}^*_{O_2}(T) - \nu_{CO_2}\bar{g}^*_{CO_2}(T) \\
&= \nu_{CO}(\bar{h} - T\bar{s})_{CO} + \nu_{O_2}(\bar{h} - T\bar{s})_{O_2} - \nu_{CO_2}(\bar{h} - T\bar{s})_{CO_2} \\
&= \nu_{CO}[(\bar{h}_f + \bar{h}_{1800} - \bar{h}_{298}) - T\bar{s}]_{CO} \\
&\quad + \nu_{O_2}[(\bar{h}_f + \bar{h}_{1800} - \bar{h}_{298}) - T\bar{s}]_{O_2} \\
&\quad - \nu_{CO_2}[(\bar{h}_f + \bar{h}_{1800} - \bar{h}_{298}) - T\bar{s}]_{CO_2} \\
&= 1 \times (-110{,}530 + 58{,}191 - 8669 - 1800 \times 254.797) \\
&\quad + 0.5 \times (0 + 60{,}371 - 8682 - 1800 \times 264.701) \\
&\quad - 1 \times (-393{,}520 + 88{,}806 - 9364 - 1800 \times 302.884) \\
&= 127{,}240.2 \text{ kJ/kmol}
\end{aligned}$$

Substituting,

$$\ln K_p = -(127{,}240.2 \text{ kJ/kmol})/[(8.314 \text{ kJ/kmol}\cdot\text{K})(1800 \text{ K})] = -8.502$$

or

$$K_p = \mathbf{2.03 \times 10^{-4}} \quad (\text{Table A-28: } \ln K_p = -8.497)$$

15-16 The equilibrium constant of the reaction $H_2O \leftrightarrow 1/2H_2 + OH$ is listed in Table A-28 at different temperatures. It is to be verified at a given temperature using Gibbs function data.

Analysis The K_p value of a reaction at a specified temperature can be determined from the Gibbs function data using

$$K_p = e^{-\Delta G^*(T)/R_uT} \quad \text{or} \quad \ln K_p = -\Delta G^*(T)/R_uT$$

$H_2O \Leftrightarrow \frac{1}{2}H_2 + OH$

25°C

where

$$\Delta G^*(T) = \nu_{H_2}\bar{g}^*_{H_2}(T) + \nu_{OH}\bar{g}^*_{OH}(T) - \nu_{H_2O}\bar{g}^*_{H_2O}(T)$$

At 298 K,

$$\Delta G^*(T) = 0.5(0) + 1(34{,}280) - 1(-228{,}590) = 262{,}870 \text{ kJ/kmol}$$

Substituting,

$$\ln K_p = -(262{,}870 \text{ kJ/kmol})/[(8.314 \text{ kJ/kmol}\cdot\text{K})(298 \text{ K})] = -106.10$$

or

$$K_p = \mathbf{8.34 \times 10^{-47}} \quad (\text{Table A-28: } \ln K_p = -106.21)$$

15-17 The temperature at which 5 percent of diatomic oxygen dissociates into monatomic oxygen at a specified pressure is to be determined.

Assumptions **1** The equilibrium composition consists of O_2 and O. **2** The constituents of the mixture are ideal gases.

Analysis The stoichiometric and actual reactions can be written as

Stoichiometric: $O_2 \Leftrightarrow 2O$ (thus $\nu_{O_2} = 1$ and $\nu_O = 2$)

Actual: $O_2 \Leftrightarrow \underbrace{0.95O_2}_{react.} + \underbrace{0.1O}_{prod.}$

$O_2 \Leftrightarrow 2O$

5 %

3 atm

The equilibrium constant K_p can be determined from

$$K_p = \frac{N_O^{\nu_O}}{N_{O_2}^{\nu_{O_2}}}\left(\frac{P}{N_{total}}\right)^{\nu_O - \nu_{O_2}} = \frac{0.1^2}{0.95}\left(\frac{3}{0.95+0.1}\right)^{2-1} = 0.0301$$

From Table A-28, the temperature corresponding to this K_p value is

$$T = \mathbf{3133\ K}$$

15-18 The temperature at which 5 percent of diatomic oxygen dissociates into monatomic oxygen at a specified pressure is to be determined.

Assumptions **1** The equilibrium composition consists of O_2 and O. **2** The constituents of the mixture are ideal gases.

Analysis The stoichiometric and actual reactions can be written as

Stoichiometric: $O_2 \Leftrightarrow 2O \quad (\text{thus } \nu_{O_2} = 1 \text{ and } \nu_O = 2)$

Actual: $O_2 \Leftrightarrow \underbrace{0.95O_2}_{react.} + \underbrace{0.1O}_{prod.}$

$O_2 \Leftrightarrow 2O$
5 %
10 atm

The equilibrium constant K_p can be determined from

$$K_p = \frac{N_O^{\nu_O}}{N_{O_2}^{\nu_{O_2}}}\left(\frac{P}{N_{Total}}\right)^{\nu_O-\nu_{O_2}} = \frac{0.1^2}{0.95}\left(\frac{10}{0.95+0.1}\right)^{2-1} = 0.1003$$

From Table A-28, the temperature corresponding to this K_p value is

$T =$ **3336 K**

15-19 [*Also solved by EES on enclosed CD*] Carbon monoxide is burned with 100 percent excess air. The temperature at which 97 percent of CO burn to CO_2 is to be determined.

Assumptions **1** The equilibrium composition consists of CO_2, CO, O_2, and N_2. **2** The constituents of the mixture are ideal gases.

Analysis Assuming N_2 to remain as an inert gas, the stoichiometric and actual reactions can be written as

Stoichiometric: $CO + \frac{1}{2}O_2 \Leftrightarrow CO_2$ (thus $\nu_{CO_2} = 1, \nu_{CO} = 1$, and $\nu_{O_2} = \frac{1}{2}$)

Actual: $$CO + 1(O_2 + 3.76N_2) \longrightarrow \underbrace{0.97CO_2}_{\text{product}} + \underbrace{0.03CO + 0.515O_2}_{\text{reactants}} + \underbrace{3.76N_2}_{\text{inert}}$$

The equilibrium constant K_p can be determined from

$$K_p = \frac{N_{CO_2}^{\nu_{CO_2}}}{N_{CO}^{\nu_{CO}} N_{O_2}^{\nu_{O_2}}}\left(\frac{P}{N_{total}}\right)^{(\nu_{CO_2}-\nu_{CO}-\nu_{O_2})}$$
$$= \frac{0.97}{0.03 \times 0.515^{0.5}}\left(\frac{1}{0.97+0.03+0.515+3.76}\right)^{1-1.5}$$
$$= 103.48$$

$CO + \frac{1}{2}O_2 \Leftrightarrow CO_2$
97 %
1 atm

From Table A-28, the temperature corresponding to this K_p value is T = **2276 K**

15-20 EES solution of this (and other comprehensive problems designated with the *computer icon*) is available to instructors at the *Instructor Manual* section of the *Online Learning Center* (OLC) at www.mhhe.com/cengel-boles. See the Preface for access information.

15-21 Carbon monoxide is burned with 100 percent excess air. The temperature at which 97 percent of CO burn to CO_2 is to be determined.

Assumptions **1** The equilibrium composition consists of CO_2, CO, O_2, and N_2. **2** The constituents of the mixture are ideal gases.

Analysis Assuming N_2 to remain as an inert gas, the stoichiometric and actual reactions can be written as

Stoichiometric: $CO + \frac{1}{2}O_2 \Leftrightarrow CO_2$ (thus $\nu_{CO_2} = 1, \nu_{CO} = 1$, and $\nu_{O_2} = \frac{1}{2}$)

Actual: $$CO + 1(O_2 + 3.76N_2) \longrightarrow \underbrace{0.97CO_2}_{\text{product}} + \underbrace{0.03CO + 0.515O_2}_{\text{reactants}} + \underbrace{3.76N_2}_{\text{inert}}$$

The equilibrium constant K_p can be determined from

$$K_p = \frac{N_{CO_2}^{\nu_{CO_2}}}{N_{CO}^{\nu_{CO}} N_{O_2}^{\nu_{O_2}}}\left(\frac{P}{N_{total}}\right)^{(\nu_{CO_2}-\nu_{CO}-\nu_{O_2})}$$
$$= \frac{0.97}{0.03 \times 0.515^{0.5}}\left(\frac{1}{0.97+0.03+0.515+3.76}\right)^{1-1.5}$$
$$= 103.48$$

$CO + \frac{1}{2}O_2 \Leftrightarrow CO_2$
97 %
1 atm

From Table A-28, the temperature corresponding to this K_p value is T = **2276 K** = **4097 R**

15-22 Hydrogen is burned with 150 percent theoretical air. The temperature at which 98 percent of H_2 will burn to H_2O is to be determined.

Assumptions **1** The equilibrium composition consists of H_2O, H_2, O_2, and N_2. **2** The constituents of the mixture are ideal gases.

Analysis Assuming N_2 to remain as an inert gas, the stoichiometric and actual reactions can be written as

Stoichiometric: $H_2 + \tfrac{1}{2}O_2 \Leftrightarrow H_2O$ (thus $\nu_{H_2O} = 1, \nu_{H_2} = 1$, and $\nu_{O_2} = \tfrac{1}{2}$)

Actual: $$H_2 + 0.75(O_2 + 3.76N_2) \longrightarrow \underbrace{0.98\,H_2O}_{\text{product}} + \underbrace{0.02\,H_2 + 0.26\,O_2}_{\text{reactants}} + \underbrace{2.82\,N_2}_{\text{inert}}$$

The equilibrium constant K_p can be determined from

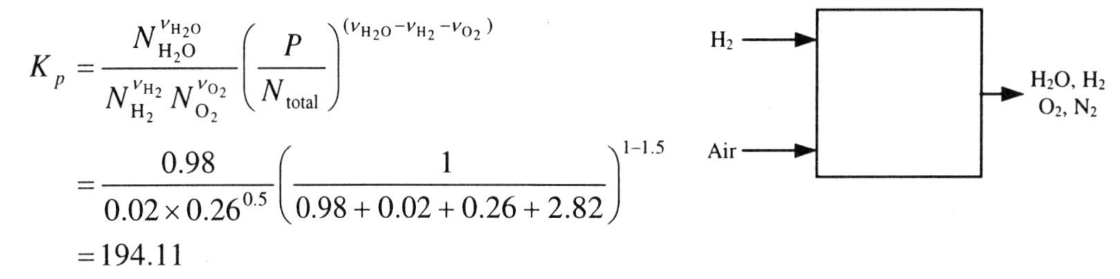

$$K_p = \frac{N_{H_2O}^{\nu_{H_2O}}}{N_{H_2}^{\nu_{H_2}} N_{O_2}^{\nu_{O_2}}} \left(\frac{P}{N_{\text{total}}} \right)^{(\nu_{H_2O} - \nu_{H_2} - \nu_{O_2})}$$

$$= \frac{0.98}{0.02 \times 0.26^{0.5}} \left(\frac{1}{0.98 + 0.02 + 0.26 + 2.82} \right)^{1-1.5}$$

$$= 194.11$$

From Table A-28, the temperature corresponding to this K_p value is T = **2472 K**.

15-23 Air is heated to a high temperature. The equilibrium composition at that temperature is to be determined.

Assumptions **1** The equilibrium composition consists of N_2, O_2, and NO. **2** The constituents of the mixture are ideal gases.

Analysis The stoichiometric and actual reactions in this case are

AIR
2000 K
2 atm

Stoichiometric: $\frac{1}{2}N_2 + \frac{1}{2}O_2 \Leftrightarrow NO$ (thus $\nu_{NO} = 1, \nu_{N_2} = \frac{1}{2}$, and $\nu_{O_2} = \frac{1}{2}$)

Actual: $3.76N_2 + O_2 \longrightarrow \underbrace{xNO}_{prod.} + \underbrace{yN_2 + zO_2}_{reactants}$

N balance: $7.52 = x + 2y$ or $y = 3.76 - 0.5x$

O balance: $2 = x + 2z$ or $z = 1 - 0.5x$

Total number of moles: $N_{total} = x + y + z = x + 4.76 - x = 4.76$

The equilibrium constant relation can be expressed as

$$K_p = \frac{N_{NO}^{\nu_{NO}}}{N_{N_2}^{\nu_{N_2}} N_{O_2}^{\nu_{O_2}}} \left(\frac{P}{N_{total}} \right)^{(\nu_{NO} - \nu_{N_2} - \nu_{O_2})}$$

From Table A-28, ln K_p = -3.931 at 2000 K. Thus K_p = 0.01962. Substituting,

$$0.01962 = \frac{x}{(3.76 - 0.5x)^{0.5}(1 - 0.5x)^{0.5}} \left(\frac{2}{4.76} \right)^{1-1}$$

Solving for x,

$$x = 0.0376$$

Then,

$$y = 3.76\text{-}0.5x = 3.7412$$

$$z = 1\text{-}0.5x = 0.9812$$

Therefore, the equilibrium composition of the mixture at 2000 K and 2 atm is

$$0.0376NO + 3.7412N_2 + 0.9812O_2$$

The equilibrium constant for the reactions $O_2 \Leftrightarrow 2O$ (ln K_p = -14.622) and $N_2 \Leftrightarrow 2N$ (ln K_p = -41.645) are much smaller than that of the specified reaction (ln K_p = -3.931). Therefore, it is realistic to assume that no monatomic oxygen or nitrogen will be present in the equilibrium mixture. Also the equilibrium composition is in this case is independent of pressure since $\Delta\nu = 1 - 0.5 - 0.5 = 0$.

15-24 Hydrogen is heated to a high temperature at a constant pressure. The percentage of H_2 that will dissociate into H is to be determined.

Assumptions **1** The equilibrium composition consists of H_2 and H. **2** The constituents of the mixture are ideal gases.

Analysis The stoichiometric and actual reactions can be written as

Stoichiometric: $H_2 \Leftrightarrow 2H$ (thus $\nu_{H_2} = 1$ and $\nu_H = 2$)

Actual: $H_2 \longrightarrow \underbrace{xH_2}_{react.} + \underbrace{yH}_{prod.}$

H_2
3200 K
8 atm

H balance: $2 = 2x + y$ or $y = 2 - 2x$

Total number of moles: $N_{total} = x + y = x + 2 - 2x = 2 - x$

The equilibrium constant relation can be expressed as

$$K_p = \frac{N_H^{\nu_H}}{N_{H_2}^{\nu_{H_2}}}\left(\frac{P}{N_{total}}\right)^{\nu_H - \nu_{H_2}}$$

From Table A-28, ln K_p = -2.534 at 3200 K. Thus K_p = 0.07934. Substituting,

$$0.07934 = \frac{(2-2x)^2}{x}\left(\frac{8}{2-x}\right)^{2-1}$$

Solving for x,

$$x = 0.95$$

Thus the percentage of H_2 which dissociates to H at 3200 K and 10 atm is

$$1 - 0.95 = 0.05 \text{ or } \mathbf{5.0\%}$$

15-25 Carbon dioxide is heated to a high temperature at a constant pressure. The percentage of CO_2 that will dissociate into CO and O_2 is to be determined.

Assumptions **1** The equilibrium composition consists of CO_2, CO, and O_2. **2** The constituents of the mixture are ideal gases.

Analysis The stoichiometric and actual reactions in this case are

Stoichiometric: $CO_2 \Leftrightarrow CO + \frac{1}{2}O_2$ (thus $\nu_{CO_2} = 1, \nu_{CO} = 1$, and $\nu_{O_2} = \frac{1}{2}$)

Actual: $CO_2 \longrightarrow \underbrace{xCO_2}_{\text{react.}} + \underbrace{yCO + zO_2}_{\text{products}}$

CO₂
2800 K
3 atm

C balance: $1 = x + y \longrightarrow y = 1 - x$

O balance: $2 = 2x + y + 2z \longrightarrow z = 0.5 - 1.5x$

Total number of moles: $N_{total} = x + y + z = 1.5 - 1.5x$

The equilibrium constant relation can be expressed as

$$K_p = \frac{N_{CO}^{\nu_{CO}} N_{O_2}^{\nu_{O_2}}}{N_{CO_2}^{\nu_{CO_2}}} \left(\frac{P}{N_{total}}\right)^{(\nu_{CO}+\nu_{O_2}-\nu_{CO_2})}$$

From Table A-28, $\ln K_p = -1.894$ at 2800 K. Thus $K_p = 0.1505$. Substituting,

$$0.1505 = \frac{(1-x)(0.5-1.5x)^{1/2}}{x} \left(\frac{3}{1.5-1.5x}\right)^{1.5-1}$$

Solving for x,

$$x = 0.332$$

Thus the percentage of CO_2 which dissociates into CO and O_2 is

1-0.332 = 0.668 or **66.8%**

15-26 A mixture of CO and O_2 is heated to a high temperature at a constant pressure. The equilibrium composition is to be determined.

Assumptions **1** The equilibrium composition consists of CO_2, CO, and O_2. **2** The constituents of the mixture are ideal gases.

Analysis The stoichiometric and actual reactions in this case are

Stoichiometric: $CO_2 \Leftrightarrow CO + \frac{1}{2}O_2$ (thus $\nu_{CO_2} = 1, \nu_{CO} = 1$, and $\nu_{O_2} = \frac{1}{2}$)

Actual: $CO + 3O_2 \longrightarrow \underbrace{xCO_2}_{\text{react.}} + \underbrace{yCO + zO_2}_{\text{products}}$

1 CO
3 O_2
2200 K
2 atm

C balance: $1 = x + y \longrightarrow y = 1 - x$

O balance: $7 = 2x + y + 2z \quad or \quad z = 3 - 0.5x$

Total number of moles: $N_{total} = x + y + z = 4 - 0.5x$

The equilibrium constant relation can be expressed as

$$K_p = \frac{N_{CO_2}^{\nu_{CO_2}}}{N_{CO}^{\nu_{CO}} N_{O_2}^{\nu_{O_2}}} \left(\frac{P}{N_{total}} \right)^{(\nu_{CO_2} - \nu_{CO} - \nu_{O_2})}$$

From Table A-28, $\ln K_p = 5.120$ at 2200 K. Thus $K_p = 167.34$. Substituting,

$$167.34 = \frac{x}{(1-x)(3-0.5x)^{0.5}} \left(\frac{2}{4-0.5x} \right)^{1-1.5}$$

Solving for x,

$x = 0.995$

Then,

$y = 1 - x = 0.005$

$z = 3 - 0.5x = 2.5025$

Therefore, the equilibrium composition of the mixture at 2200 K and 2 atm is

$\mathbf{0.995CO_2 + 0.005CO + 2.5025O_2}$

15-27E A mixture of CO, O_2, and N_2 is heated to a high temperature at a constant pressure. The equilibrium composition is to be determined.

Assumptions **1** The equilibrium composition consists of CO_2, CO, O_2, and N_2. **2** The constituents of the mixture are ideal gases.

Analysis The stoichiometric and actual reactions in this case are

2 CO
2 O_2
6 N_2
4320 R
3 atm

Stoichiometric: $CO + \frac{1}{2}O_2 \Leftrightarrow CO_2$ (thus $\nu_{CO_2} = 1, \nu_{CO} = 1$, and $\nu_{O_2} = \frac{1}{2}$)

Actual: $$2CO + 2O_2 + 6N_2 \longrightarrow \underbrace{xCO_2}_{\text{products}} + \underbrace{yCO + zO_2}_{\text{reactants}} + \underbrace{6N_2}_{\text{inert}}$$

C balance: $$2 = x + y \longrightarrow y = 2 - x$$

O balance: $$6 = 2x + y + 2z \longrightarrow z = 2 - 0.5x$$

Total number of moles: $$N_{total} = x + y + z + 6 = 10 - 0.5x$$

The equilibrium constant relation can be expressed as

$$K_p = \frac{N_{CO_2}^{\nu_{CO_2}}}{N_{CO}^{\nu_{CO}} N_{O_2}^{\nu_{O_2}}} \left(\frac{P}{N_{total}} \right)^{(\nu_{CO_2} - \nu_{CO} - \nu_{O_2})}$$

From Table A-28, $\ln K_p = 3.860$ at $T = 4320\ R = 2400$ K. Thus $K_p = 47.465$. Substituting,

$$47.465 = \frac{x}{(2-x)(2-0.5x)^{0.5}} \left(\frac{3}{10-0.5x} \right)^{1-1.5}$$

Solving for x,

$$x = 1.930$$

Then,

$$y = 2 - x = 0.070$$

$$z = 2 - 0.5x = 1.035$$

Therefore, the equilibrium composition of the mixture at 2400 K and 3 atm is

$$\mathbf{1.930CO_2 + 0.070CO + 1.035O_2 + 6N_2}$$

15-28 A mixture of N_2, O_2, and Ar is heated to a high temperature at a constant pressure. The equilibrium composition is to be determined.

Assumptions **1** The equilibrium composition consists of N_2, O_2, Ar, and NO. **2** The constituents of the mixture are ideal gases.

Analysis The stoichiometric and actual reactions in this case are

3 N_2
1 O_2
0.1 Ar
2400 K
10 atm

Stoichiometric: $\frac{1}{2}N_2 + \frac{1}{2}O_2 \Leftrightarrow NO$ (thus $\nu_{NO} = 1, \nu_{N_2} = \frac{1}{2}$, and $\nu_{O_2} = \frac{1}{2}$)

Actual: $$3N_2 + O_2 + 0.1Ar \longrightarrow \underbrace{xNO}_{\text{prod.}} + \underbrace{yN_2 + zO_2}_{\text{reactants}} + \underbrace{0.1Ar}_{\text{inert}}$$

N balance: $$6 = x + 2y \longrightarrow y = 3 - 0.5x$$

O balance: $$2 = x + 2z \longrightarrow z = 1 - 0.5x$$

Total number of moles: $N_{\text{total}} = x + y + z + 0.1 = 4.1$

The equilibrium constant relation becomes,

$$K_p = \frac{N_{NO}^{\nu_{NO}}}{N_{N_2}^{\nu_{N2}} N_{O_2}^{\nu_{O2}}} \left(\frac{P}{N_{\text{total}}} \right)^{(\nu_{NO} - \nu_{N2} - \nu_{O2})} = \frac{x}{y^{0.5} z^{0.5}} \left(\frac{P}{N_{\text{total}}} \right)^{1-0.5-0.5}$$

From Table A-28, $\ln K_p = -3.019$ at 2400 K. Thus $K_p = 0.04885$. Substituting,

$$0.04885 = \frac{x}{(3-0.5x)^{0.5}(1-0.5x)^{0.5}} \times 1$$

Solving for x,

$$x = 0.0823$$

Then,

$$y = 3 - 0.5x = 2.9589$$

$$z = 1 - 0.5x = 0.9589$$

Therefore, the equilibrium composition of the mixture at 2400 K and 10 atm is

$$\mathbf{0.0823NO + 2.9589N_2 + 0.9589O_2 + 0.1Ar}$$

15-29 The mole fraction of sodium that ionizes according to the reaction Na ⇔ Na^+ + e^- at 2000 K and 0.5 atm is to be determined.

Assumptions All components behave as ideal gases.

Analysis The stoichiometric and actual reactions can be written as

Stoichiometric: $Na \Leftrightarrow Na^+ + e^-$ (thus $\nu_{Na} = 1,\ \nu_{Na^+} = 1$ and $\nu_{e^-} = 1$)

Actual: $Na \longrightarrow \underbrace{xNa}_{react.} + \underbrace{yNa^+ + ye^-}_{products}$

Na ⇔ Na^+ + e^-
2000 K
0.5 atm

Na balance: $1 = x + y \ \text{or}\ y = 1 - x$

Total number of moles: $N_{total} = x + 2y = 2 - x$

The equilibrium constant relation becomes,

$$K_p = \frac{N_{Na^+}^{\nu_{Na^+}} N_{e^-}^{\nu_{e^-}}}{N_{Na}^{\nu_{Na}}}\left(\frac{P}{N_{total}}\right)^{(\nu_{Na^+}+\nu_{e^-}-\nu_{Na})} = \frac{y^2}{x}\left(\frac{P}{N_{total}}\right)^{1+1-1}$$

Substituting,

$$0.668 = \frac{(1-x)^2}{x}\left(\frac{0.5}{2-x}\right)$$

Solving for x,

$$x = 0.244$$

Thus the fraction of Na which dissociates into Na^+ and e^- is

1 - 0.244 = 0.756 or **75.6%**

15-30 Liquid propane enters a combustion chamber. The equilibrium composition of product gases and the rate of heat transfer from the combustion chamber are to be determined.

Assumptions **1** The equilibrium composition consists of CO_2, H_2O, CO, N_2, and O_2. **2** The constituents of the mixture are ideal gases.

Analysis (*a*) Considering 1 kmol of C_3H_8, the stoichiometric combustion equation can be written as

$$C_3H_8(\ell) + a_{th}(O_2 + 3.76N_2) \longrightarrow 3CO_2 + 4H_2O + 3.76a_{th}N_2$$

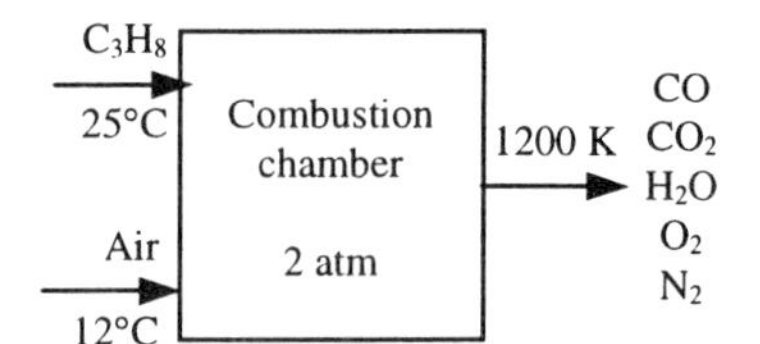

where a_{th} is the stoichiometric coefficient and is determined from the O_2 balance,

$$2.5a_{th} = 3 + 2 + 1.5a_{th} \longrightarrow a_{th} = 5$$

Then the actual combustion equation with 150% excess air and some CO in the products can be written as

$$C_3H_8(\ell) + 12.5(O_2 + 3.76N_2) \longrightarrow xCO_2 + (3-x)CO + (9-0.5x)O_2 + 4H_2O + 47N_2$$

After combustion, there will be no C_3H_8 present in the combustion chamber, and H_2O will act like an inert gas. The equilibrium equation among CO_2, CO, and O_2 can be expressed as

$$CO_2 \Leftrightarrow CO + \tfrac{1}{2}O_2 \quad (\text{thus } \nu_{CO_2} = 1, \nu_{CO} = 1, \text{ and } \nu_{O_2} = \tfrac{1}{2})$$

and

$$K_p = \frac{N_{CO}^{\nu_{CO}} N_{O_2}^{\nu_{O_2}}}{N_{CO_2}^{\nu_{CO_2}}} \left(\frac{P}{N_{total}} \right)^{(\nu_{CO} + \nu_{O_2} - \nu_{CO_2})}$$

where

$$N_{total} = x + (3-x) + (9-0.5x) + 4 + 47 = 63 - 0.5x$$

From Table A-28, $\ln K_p = -17.871$ at 1200 K. Thus $K_p = 1.73 \times 10^{-8}$. Substituting,

$$1.73 \times 10^{-8} = \frac{(3-x)(9-0.5x)^{0.5}}{x} \left(\frac{2}{63-0.5x} \right)^{1.5-1}$$

Solving for *x*,

$$x = 2.9999999 \cong 3.0$$

Therefore, the amount CO in the product gases is negligible, and it can be disregarded with no loss in accuracy. Then the combustion equation and the equilibrium composition can be expressed as

$$C_3H_8(\ell) + 12.5(O_2 + 3.76N_2) \longrightarrow 3CO_2 + 7.5O_2 + 4H_2O + 47N_2$$

and

$$\mathbf{3CO_2 + 7.5O_2 + 4H_2O + 47N_2}$$

(*b*) The heat transfer for this combustion process is determined from the steady-flow energy balance $E_{in} - E_{out} = \Delta E_{system}$ on the combustion chamber with $W = 0$,

$$-Q_{out} = \sum N_P \left(\bar{h}_f^\circ + \bar{h} - \bar{h}^\circ \right)_P - \sum N_R \left(\bar{h}_f^\circ + \bar{h} - \bar{h}^\circ \right)_R$$

Assuming the air and the combustion products to be ideal gases, we have $h = h(T)$. From the tables, (The $\bar{h}_f^\circ$ of liquid propane is obtained by adding the h_{fg} at 25°C to $\bar{h}_f^\circ$ of gaseous propane).

Substance	$\bar{h}_f^\circ$ kJ/kmol	$\bar{h}_{285\text{ K}}$ kJ/kmol	$\bar{h}_{298\text{ K}}$ kJ/kmol	$\bar{h}_{1200\text{ K}}$ kJ/kmol
C_3H_8 (ℓ)	-118,910	---	---	---
O_2	0	8696.5	8682	38,447
N_2	0	8286.5	8669	36,777
H_2O (g)	-241,820	---	9904	44,380
CO_2	-393,520	---	9364	53,848

Substituting,

$$\begin{aligned}-Q_{out} &= 3(-393{,}520+53{,}848-9364)+4(-241{,}820+44{,}380-9904)\\&\quad+7.5(0+38{,}447-8682)+47(0+36{,}777-8669)\\&\quad-1(-118{,}910+h_{298}-h_{298})-12.5(0+8296.5-8682)\\&\quad-47(0+8186.5-8669)\\&=-185{,}764\text{ kJ/kmol of }C_3H_8\end{aligned}$$

or $\qquad Q_{out} = 185{,}764\text{ kJ/kmol of }C_3H_8$

The mass flow rate of C_3H_8 can be expressed in terms of the mole numbers as

$$\dot{N} = \frac{\dot{m}}{M} = \frac{1.2\text{ kg/min}}{44\text{ kg/kmol}} = 0.02727\text{ kmol/min}$$

Thus the rate of heat transfer is

$$\dot{Q}_{out} = \dot{N}\times Q_{out} = (0.02727\text{ kmol/min})(185{,}746\text{ kJ/kmol}) = \mathbf{5066\text{ kJ/min}}$$

The equilibrium constant for the reaction $\frac{1}{2}N_2 + \frac{1}{2}O_2 \Leftrightarrow NO$ is $\ln K_p = -7.569$, which is very small. This indicates that the amount of NO formed during this process will be very small, and can be disregarded.

15-31 EES solution of this (and other comprehensive problems designated with the *computer icon*) is available to instructors at the *Instructor Manual* section of the *Online Learning Center* (OLC) at www.mhhe.com/cengel-boles. See the Preface for access information.

15-32E A steady-flow combustion chamber is supplied with CO and O_2. The equilibrium composition of product gases and the rate of heat transfer from the combustion chamber are to be determined.

Assumptions **1** The equilibrium composition consists of CO_2, CO, and O_2. **2** The constituents of the mixture are ideal gases.

Analysis (*a*) We first need to calculate the amount of oxygen used per lbmol of CO before we can write the combustion equation,

$$v_{CO} = \frac{RT}{P} = \frac{(0.3831 \text{ psia}\cdot\text{ft}^3/\text{lbm}\cdot\text{R})(560 \text{ R})}{16 \text{ psia}} = 13.41 \text{ ft}^3/\text{lbm}$$

$$\dot{m}_{CO} = \frac{\dot{V}_{CO}}{v_{CO}} = \frac{12.5 \text{ ft}^3/\text{min}}{13.41 \text{ ft}^3/\text{lbm}} = 0.932 \text{ lbm/min}$$

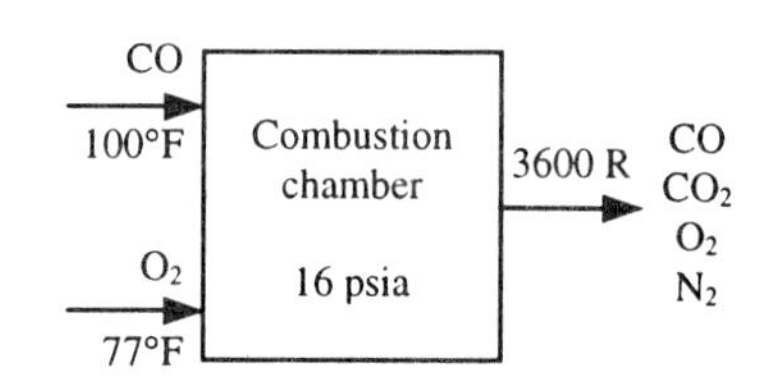

Then the molar air-fuel ratio becomes (it is actually O_2-fuel ratio)

$$\overline{AF} = \frac{N_{O_2}}{N_{fuel}} = \frac{\dot{m}_{O_2}/M_{O_2}}{\dot{m}_{fuel}/M_{fuel}} = \frac{(0.7 \text{ lbm/min})/(32 \text{ lbm/lbmol})}{(0.932 \text{ lbm/min})/(28 \text{ lbm/lbmol})} = 0.657 \text{ lbmol } O_2/\text{lbmol fuel}$$

Then the combustion equation can be written as

$$CO + 0.657O_2 \longrightarrow xCO_2 + (1-x)CO + (0.657-0.5x)O_2$$

The equilibrium equation among CO_2, CO, and O_2 can be expressed as

$$CO_2 \Leftrightarrow CO + \tfrac{1}{2}O_2 \quad (\text{thus } \nu_{CO_2} = 1, \nu_{CO} = 1, \text{ and } \nu_{O_2} = \tfrac{1}{2})$$

and

$$K_p = \frac{N_{CO}^{\nu_{CO}} N_{O_2}^{\nu_{O_2}}}{N_{CO_2}^{\nu_{CO_2}}} \left(\frac{P}{N_{\text{total}}}\right)^{(\nu_{CO}+\nu_{O_2}-\nu_{CO_2})}$$

where

$$N_{\text{total}} = x + (1-x) + (0.657-0.5x) = 1.657-0.5x$$
$$P = 16/14.7 = 1.088 \text{ atm}$$

From Table A-28, $\ln K_p = -6.635$ at $T = 3600 \text{ R} = 2000 \text{ K}$. Thus $K_p = 1.314\times10^{-3}$. Substituting,

$$1.314\times10^{-3} = \frac{(1-x)(0.657-0.5x)^{0.5}}{x}\left(\frac{1.088}{1.657-0.5x}\right)^{1.5-1}$$

Solving for *x*,

$$x = 0.9966$$

Then the combustion equation and the equilibrium composition can be expressed as

$$CO + 0.657O_2 \longrightarrow 0.9966CO_2 + 0.0034CO + 0.1587O_2$$

and

$$\mathbf{0.9966CO_2 + 0.0034CO + 0.1587O_2}$$

(*b*) The heat transfer for this combustion process is determined from the steady-flow energy balance $E_{in} - E_{out} = \Delta E_{system}$ on the combustion chamber with $W = 0$,

$$-Q_{out} = \sum N_P \left(\bar{h}_f^\circ + \bar{h} - \bar{h}^\circ\right)_P - \sum N_R \left(\bar{h}_f^\circ + \bar{h} - \bar{h}^\circ\right)_R$$

Assuming the air and the combustion products to be ideal gases, we have $h = h(T)$. From the tables,

Substance	$\bar{h}_f^\circ$ **Btu/lbmol**	$\bar{h}_{537\,R}$ **Btu/lbmol**	$\bar{h}_{560\,R}$ **Btu/lbmol**	$\bar{h}_{3600\,R}$ **Btu/lbmol**
CO	-47,540	3725.1	3889.5	28,127
O_2	0	3725.1	---	29,174
CO_2	-169,300	4027.5	---	43,411

Substituting,

$$\begin{aligned}-Q_{out} &= 0.9966(-169{,}300 + 43{,}411 - 4027.5)\\ &\quad + 0.0034(-47{,}540 + 28{,}127 - 3725.1)\\ &\quad + 0.1587(0 + 29{,}174 - 3725.1)\\ &\quad - 1(-47{,}540 + 3889.5 - 3725.1) - 0\\ &= -78{,}139 \text{ Btu/lbmol of CO}\end{aligned}$$

or $\quad Q_{out} = 78{,}139 \text{ Btu/lbm of CO}$

The mass flow rate of CO can be expressed in terms of the mole numbers as

$$\dot{N} = \frac{\dot{m}}{M} = \frac{0.932 \text{ lbm/min}}{28 \text{ lbm/lbmol}} = 0.0333 \text{ lbmol/min}$$

Thus the rate of heat transfer is

$$\dot{Q}_{out} = \dot{N} \times Q_{out} = (0.0333 \text{ lbmol/min})(78{,}139 \text{ Btu/lbmol}) = \mathbf{2601\ Btu/min}$$

15-33 Oxygen is heated during a steady-flow process. The rate of heat supply needed during this process is to be determined for two cases.

Assumptions **1** The equilibrium composition consists of O_2 and O. **2** All components behave as ideal gases.

Analysis (*a*) Assuming some O_2 dissociates into O, the dissociation equation can be written as

$$O_2 \longrightarrow xO_2 + 2(1-x)O$$

The equilibrium equation among O_2 and O can be expressed as

$$O_2 \Leftrightarrow 2O \quad (\text{thus } \nu_{O_2} = 1 \text{ and } \nu_O = 2)$$

Q̇

O₂ 298 K → O₂, O 3000 K

Assuming ideal gas behavior for all components, the equilibrium constant relation can be expressed as

$$K_p = \frac{N_O^{\nu_O}}{N_{O_2}^{\nu_{O_2}}}\left(\frac{P}{N_{total}}\right)^{\nu_O - \nu_{O_2}}$$

where $N_{total} = x + 2(1-x) = 2 - x$

From Table A-28, $\ln K_p = -4.357$ at 3000 K. Thus $K_p = 0.01282$. Substituting,

$$0.01282 = \frac{(2-2x)^2}{x}\left(\frac{1}{2-x}\right)^{2-1}$$

Solving for x gives $\qquad x = 0.943$

Then the dissociation equation becomes

$$O_2 \longrightarrow 0.943O_2 + 0.114O$$

The heat transfer for this combustion process is determined from the steady-flow energy balance $E_{in} - E_{out} = \Delta E_{system}$ on the combustion chamber with $W = 0$,

$$Q_{in} = \sum N_P\left(\bar{h}_f^\circ + \bar{h} - \bar{h}^\circ\right)_P - \sum N_R\left(\bar{h}_f^\circ + \bar{h} - \bar{h}^\circ\right)_R$$

Assuming the O_2 and O to be ideal gases, we have $h = h(T)$. From the tables,

Substance	$\bar{h}_f^\circ$ kJ/kmol	$\bar{h}_{298\,K}$ kJ/kmol	$\bar{h}_{3000\,K}$ kJ/kmol
O	249,190	6852	63,425
O_2	0	8682	106,780

Substituting,

$$Q_{in} = 0.943(0 + 106{,}780 - 8682) + 0.114(249{,}190 + 63{,}425 - 6852) - 0 = 127{,}363 \text{ kJ/kmol } O_2$$

The mass flow rate of O_2 can be expressed in terms of the mole numbers as

$$\dot{N} = \frac{\dot{m}}{M} = \frac{0.3 \text{ kg/min}}{32 \text{ kg/kmol}} = 0.009375 \text{ kmol/min}$$

Thus the rate of heat transfer is

$$\dot{Q}_{in} = \dot{N} \times Q_{in} = (0.009375 \text{ kmol/min})(127{,}363 \text{ kJ/kmol}) = \mathbf{1194 \text{ kJ/min}}$$

(*b*) If no O_2 dissociates into O, then the process involves no chemical reactions and the heat transfer can be determined from the steady-flow energy balance for nonreacting systems to be

$$\dot{Q}_{in} = \dot{m}(h_2 - h_1) = \dot{N}(\bar{h}_2 - \bar{h}_1) = (0.009375 \text{ kmol/min})(106{,}780 - 8682) \text{ kJ/kmol} = \mathbf{919.7 \text{ kJ/min}}$$

Simultaneous Reactions

15-34C It can be expresses as "$(dG)_{T,P} = 0$ for each reaction." Or as "the K_p relation for each reaction must be satisfied."

15-35C The number of K_p relations needed to determine the equilibrium composition of a reacting mixture is equal to the difference between the number of species present in the equilibrium mixture and the number of elements.

15-36 Two chemical reactions are occurring in a mixture. The equilibrium composition at a specified temperature is to be determined.

Assumptions **1** The equilibrium composition consists of H_2O, OH, O_2, and H_2. **2** The constituents of the mixture are ideal gases.

Analysis The reaction equation during this process can be expressed as

$$H_2O \longrightarrow xH_2O + yH_2 + zO_2 + wOH$$

$H_2O \Rightarrow$ H_2O, OH, O_2, H_2
3400 K
1 atm

Mass balances for hydrogen and oxygen yield

H balance: $2 = 2x + 2y + w$ (1)

O balance: $1 = x + 2z + w$ (2)

The mass balances provide us with only two equations with four unknowns, and thus we need to have two more equations (to be obtained from the K_p relations) to determine the equilibrium composition of the mixture. They are

$$H_2O \Leftrightarrow H_2 + \tfrac{1}{2}O_2 \qquad \text{(reaction 1)}$$

$$H_2O \Leftrightarrow \tfrac{1}{2}H_2 + OH \qquad \text{(reaction 2)}$$

The equilibrium constant for these two reactions at 3400 K are determined from Table A-28 to be

$$\ln K_{P1} = -1.891 \longrightarrow K_{P1} = 0.15092$$

$$\ln K_{P2} = -1.576 \longrightarrow K_{P2} = 0.20680$$

The K_p relations for these two simultaneous reactions are

$$K_{P1} = \frac{N_{H_2}^{\nu_{H2}} N_{O_2}^{\nu_{O2}}}{N_{H_2O}^{\nu_{H2O}}}\left(\frac{P}{N_{\text{total}}}\right)^{(\nu_{H2}+\nu_{O2}-\nu_{H2O})} \quad \text{and} \quad K_{P2} = \frac{N_{H_2}^{\nu_{H2}} N_{OH}^{\nu_{OH}}}{N_{H_2O}^{\nu_{H2O}}}\left(\frac{P}{N_{\text{total}}}\right)^{(\nu_{H2}+\nu_{OH}-\nu_{H2O})}$$

where $N_{\text{total}} = N_{H_2O} + N_{H_2} + N_{O_2} + N_{OH} = x + y + z + w$

Substituting,

$$0.15092 = \frac{(y)(z)^{1/2}}{x}\left(\frac{1}{x+y+z+w}\right)^{1/2} \qquad (3)$$

$$0.20680 = \frac{(w)(y)^{1/2}}{x}\left(\frac{1}{x+y+z+w}\right)^{1/2} \qquad (4)$$

Solving Eqs. (1), (2), (3), and (4) simultaneously for the four unknowns x, y, z, and w yields

$x = 0.574$ $\quad y = 0.308$ $\quad z = 0.095$ $\quad w = 0.236$

Therefore, the equilibrium composition becomes

$$\mathbf{0.574H_2O + 0.308H_2 + 0.095O_2 + 0.236OH}$$

15-37 Two chemical reactions are occurring in a mixture. The equilibrium composition at a specified temperature is to be determined.

Assumptions **1** The equilibrium composition consists of CO_2, CO, O_2, and O. **2** The constituents of the mixture are ideal gases.

Analysis The reaction equation during this process can be expressed as

$$2CO_2 + O_2 \longrightarrow xCO_2 + yCO + zO_2 + wO$$

CO_2, CO, O_2, O
3200 K
1 atm

Mass balances for carbon and oxygen yield

C balance: $2 = x + y$ (1)

O balance: $6 = 2x + y + 2z + w$ (2)

The mass balances provide us with only two equations with four unknowns, and thus we need to have two more equations (to be obtained from the K_P relations) to determine the equilibrium composition of the mixture. They are

$$CO_2 \Leftrightarrow CO + \tfrac{1}{2}O_2 \qquad \text{(reaction 1)}$$

$$O_2 \Leftrightarrow 2O \qquad \text{(reaction 2)}$$

The equilibrium constant for these two reactions at 3200 K are determined from Table A-28 to be

$$\ln K_{P1} = -0.429 \longrightarrow K_{P1} = 0.65116$$

$$\ln K_{P2} = -3.072 \longrightarrow K_{P2} = 0.04633$$

The K_P relations for these two simultaneous reactions are

$$K_{P1} = \frac{N_{CO}^{\nu_{CO}} N_{O_2}^{\nu_{O_2}}}{N_{CO_2}^{\nu_{CO_2}}} \left(\frac{P}{N_{total}} \right)^{(\nu_{CO}+\nu_{O_2}-\nu_{CO_2})}$$

$$K_{P2} = \frac{N_O^{\nu_O}}{N_{O_2}^{\nu_{O_2}}} \left(\frac{P}{N_{total}} \right)^{\nu_O-\nu_{O_2}}$$

where

$$N_{total} = N_{CO_2} + N_{O_2} + N_{CO} + N_O = x + y + z + w$$

Substituting,

$$0.65116 = \frac{(y)(z)^{1/2}}{x} \left(\frac{1}{x+y+z+w} \right)^{1/2} \qquad (3)$$

$$0.04633 = \frac{w^2}{z} \left(\frac{1}{x+y+z+w} \right)^{2-1} \qquad (4)$$

Solving Eqs. (1), (2), (3), and (4) simultaneously for the four unknowns x, y, z, and w yields

$x = 0.947$ $y = 1.053$ $z = 1.289$ $w = 0.475$

Thus the equilibrium composition is

$$\mathbf{0.947CO_2 + 1.053CO + 1.289O_2 + 0.475O}$$

15-38 Two chemical reactions are occurring at high-temperature air. The equilibrium composition at a specified temperature is to be determined.

Assumptions **1** The equilibrium composition consists of O_2, N_2, O, and NO. **2** The constituents of the mixture are ideal gases.

Analysis The reaction equation during this process can be expressed as

$$O_2 + 3.76N_2 \longrightarrow xN_2 + yNO + zO_2 + wO$$

Heat

AIR → Reaction chamber → O_2, N_2, O, NO 3000 K

Mass balances for nitrogen and oxygen yield

N balance: $7.52 = 2x + y$ (1)

O balance: $2 = y + 2z + w$ (2)

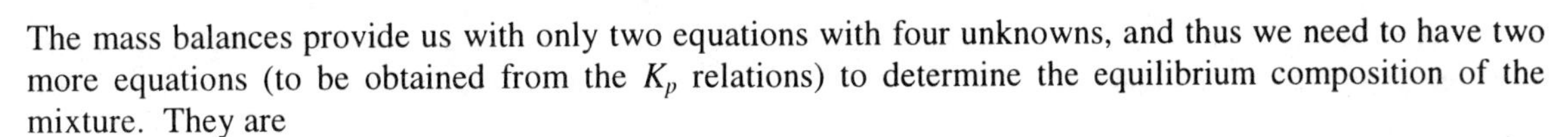

The mass balances provide us with only two equations with four unknowns, and thus we need to have two more equations (to be obtained from the K_p relations) to determine the equilibrium composition of the mixture. They are

$$\tfrac{1}{2}N_2 + \tfrac{1}{2}O_2 \Leftrightarrow NO \qquad \text{(reaction 1)}$$

$$O_2 \Leftrightarrow 2O \qquad \text{(reaction 2)}$$

The equilibrium constant for these two reactions at 3000 K are determined from Table A-28 to be

$$\ln K_{P1} = -2.114 \longrightarrow K_{P1} = 0.12075$$

$$\ln K_{P2} = -4.357 \longrightarrow K_{P2} = 0.01282$$

The K_P relations for these two simultaneous reactions are

$$K_{P1} = \frac{N_{NO}^{\nu_{NO}}}{N_{N_2}^{\nu_{N_2}} N_{O_2}^{\nu_{O_2}}} \left(\frac{P}{N_{total}}\right)^{(\nu_{NO} - \nu_{N_2} - \nu_{O_2})}$$

$$K_{P2} = \frac{N_O^{\nu_O}}{N_{O_2}^{\nu_{O_2}}} \left(\frac{P}{N_{total}}\right)^{\nu_O - \nu_{O_2}}$$

where $N_{total} = N_{N_2} + N_{NO} + N_{O_2} + N_O = x + y + z + w$

Substituting,

$$0.12075 = \frac{y}{x^{0.5} z^{0.5}} \left(\frac{1}{x + y + z + w}\right)^{1-0.5-0.5} \qquad (3)$$

$$0.01282 = \frac{w^2}{z} \left(\frac{1}{x + y + z + w}\right)^{2-1} \qquad (4)$$

Solving Eqs. (1), (2), (3), and (4) simultaneously for the four unknowns x, y, z, and w yields

$x = 3.731$ $\quad y = 0.058$ $\quad z = 0.855$ $\quad w = 0.232$

Thus the equilibrium composition is

$$\mathbf{3.731N_2 + 0.058NO + 0.855O_2 + 0.232O}$$

The equilibrium constant of the reaction $N_2 \Leftrightarrow 2N$ at 3000 K is $\ln K_P = -22.359$, which is much smaller than the K_P values of the reactions considered. Therefore, it is reasonable to assume that no N will be present in the equilibrium mixture.

15-39E [*Also solved by EES on enclosed CD*] Two chemical reactions are occurring in air. The equilibrium composition at a specified temperature is to be determined.

Assumptions **1** The equilibrium composition consists of O_2, N_2, O, and NO. **2** The constituents of the mixture are ideal gases.

Analysis The reaction equation during this process can be expressed as

$$O_2 + 3.76N_2 \longrightarrow xN_2 + yNO + zO_2 + wO$$

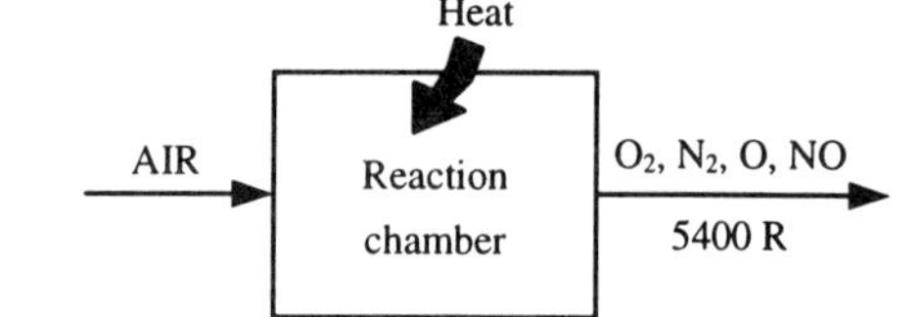

Mass balances for nitrogen and oxygen yield

N balance: $7.52 = 2x + y$ (1)

O balance: $2 = y + 2z + w$ (2)

The mass balances provide us with only two equations with four unknowns, and thus we need to have two more equations (to be obtained from the K_p relations) to determine the equilibrium composition of the mixture. They are

$$\tfrac{1}{2}N_2 + \tfrac{1}{2}O_2 \Leftrightarrow NO \qquad \text{(reaction 1)}$$

$$O_2 \Leftrightarrow 2O \qquad \text{(reaction 2)}$$

The equilibrium constant for these two reactions at T = 5400 R = 3000 K are determined from Table A-28 to be

$$\ln K_{P1} = -2.114 \longrightarrow K_{P1} = 0.12075$$

$$\ln K_{P2} = -4.357 \longrightarrow K_{P2} = 0.01282$$

The K_P relations for these two simultaneous reactions are

$$K_{P1} = \frac{N_{NO}^{\nu_{NO}}}{N_{N_2}^{\nu_{N_2}} N_{O_2}^{\nu_{O_2}}} \left(\frac{P}{N_{total}} \right)^{(\nu_{NO} - \nu_{N_2} - \nu_{O_2})}$$

$$K_{P2} = \frac{N_O^{\nu_O}}{N_{O_2}^{\nu_{O_2}}} \left(\frac{P}{N_{total}} \right)^{\nu_O - \nu_{O_2}}$$

where $N_{total} = N_{N_2} + N_{NO} + N_{O_2} + N_O = x + y + z + w$

Substituting,

$$0.12075 = \frac{y}{x^{0.5} z^{0.5}} \left(\frac{1}{x + y + z + w} \right)^{1-0.5-0.5} \qquad (3)$$

$$0.01282 = \frac{w^2}{z} \left(\frac{1}{x + y + z + w} \right)^{2-1} \qquad (4)$$

Solving Eqs. (1), (2), (3), and (4) simultaneously for the four unknowns x, y, z, and w yields

$x = 3.731$ $\quad y = 0.058$ $\quad z = 0.855$ $\quad w = 0.232$

Thus the equilibrium composition is

$$\mathbf{3.731N_2 + 0.058NO + 0.855O_2 + 0.232O}$$

The equilibrium constant of the reaction $N_2 \Leftrightarrow 2N$ at 5400 R is $\ln K_P$ = -22.359, which is much smaller than the K_P values of the reactions considered. Therefore, it is reasonable to assume that no N will be present in the equilibrium mixture.

14-40E EES solution of this (and other comprehensive problems designated with the *computer icon*) is available to instructors at the *Instructor Manual* section of the *Online Learning Center* (OLC) at www.mhhe.com/cengel-boles. See the Preface for access information.

15-41 Water vapor is heated during a steady-flow process. The rate of heat supply for a specified exit temperature is to be determined for two cases.

Assumptions **1** The equilibrium composition consists of H_2O, OH, O_2, and H_2. **2** The constituents of the mixture are ideal gases.

Analysis (*a*) Assuming some H_2O dissociates into H_2, O_2, and O, the dissociation equation can be written as

$$H_2O \longrightarrow xH_2O + yH_2 + zO_2 + wOH$$

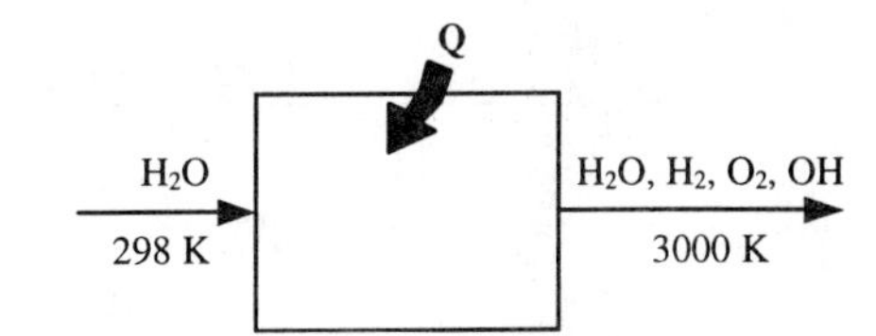

Mass balances for hydrogen and oxygen yield

H balance: $2 = 2x + 2y + w$ (1)

O balance: $1 = x + 2z + w$ (2)

The mass balances provide us with only two equations with four unknowns, and thus we need to have two more equations (to be obtained from the K_P relations) to determine the equilibrium composition of the mixture. They are

$$H_2O \Leftrightarrow H_2 + \tfrac{1}{2}O_2 \qquad \text{(reaction 1)}$$

$$H_2O \Leftrightarrow \tfrac{1}{2}H_2 + OH \qquad \text{(reaction 2)}$$

The equilibrium constant for these two reactions at 3000 K are determined from Table A-28 to be

$$\ln K_{P1} = -3.086 \longrightarrow K_{P1} = 0.04568$$

$$\ln K_{P2} = -2.937 \longrightarrow K_{P2} = 0.05302$$

The K_P relations for these three simultaneous reactions are

$$K_{P1} = \frac{N_{H_2}^{\nu_{H_2}} N_{O_2}^{\nu_{O_2}}}{N_{H_2O}^{\nu_{H_2O}}}\left(\frac{P}{N_{total}}\right)^{(\nu_{H_2}+\nu_{O_2}-\nu_{H_2O})}$$

$$K_{P2} = \frac{N_{H_2}^{\nu_{H_2}} N_{OH}^{\nu_{OH}}}{N_{H_2O}^{\nu_{H_2O}}}\left(\frac{P}{N_{total}}\right)^{(\nu_{H_2}+\nu_{O_2}-\nu_{H_2O})}$$

where

$$N_{total} = N_{H_2O} + N_{H_2} + N_{O_2} + N_{OH} = x + y + z + w$$

Substituting,

$$0.04568 = \frac{(y)(z)^{1/2}}{x}\left(\frac{1}{x+y+z+w}\right)^{1/2} \qquad (3)$$

$$0.05302 = \frac{(w)(y)^{1/2}}{x}\left(\frac{1}{x+y+z+w}\right)^{1/2} \qquad (4)$$

Solving Eqs. (1), (2), (3), and (4) simultaneously for the four unknowns *x*, *y*, *z*, and *w* yields

$x = 0.784$ $\quad y = 0.162$ $\quad z = 0.054$ $\quad w = 0.108$

Thus the balanced equation for the dissociation reaction is

$$H_2O \longrightarrow 0.784H_2O + 0.162H_2 + 0.054O_2 + 0.108OH$$

The heat transfer for this dissociation process is determined from the steady-flow energy balance $E_{in} - E_{out} = \Delta E_{system}$ with $W = 0$,

$$Q_{in} = \sum N_P \left(\bar{h}_f^\circ + \bar{h} - \bar{h}^\circ\right)_P - \sum N_R \left(\bar{h}_f^\circ + \bar{h} - \bar{h}^\circ\right)_R$$

Assuming the O_2 and O to be ideal gases, we have $h = h(T)$. From the tables,

Substance	$\bar{h}_f^\circ$ kJ/kmol	$\bar{h}_{298\text{ K}}$ kJ/kmol	$\bar{h}_{3000\text{ K}}$ kJ/kmol
H_2O	-241,820	9904	136,264
H_2	0	8468	97,211
O_2	0	8682	106,780
OH	39,460	9188	98,763

Substituting,

$$\begin{aligned} Q_{in} &= 0.784(-241{,}820 + 136{,}264 - 9904) \\ &\quad + 0.162(0 + 97{,}211 - 8468) \\ &\quad + 0.054(0 + 106{,}780 - 8682) \\ &\quad + 0.108(39{,}460 + 98{,}763 - 9188) - (-241{,}820) \\ &= 184{,}909 \text{ kJ/kmol } H_2O \end{aligned}$$

The mass flow rate of H_2O can be expressed in terms of the mole numbers as

$$\dot{N} = \frac{\dot{m}}{M} = \frac{0.2 \text{ kg/min}}{18 \text{ kg/kmol}} = 0.01111 \text{ kmol/min}$$

Thus,

$$\dot{Q}_{in} = \dot{N} \times Q_{in} = (0.01111 \text{ kmol/min})(184{,}909 \text{ kJ/kmol}) = \mathbf{2055 \text{ kJ/min}}$$

(*b*) If no dissociates takes place, then the process involves no chemical reactions and the heat transfer can be determined from the steady-flow energy balance for nonreacting systems to be

$$\begin{aligned} \dot{Q}_{in} &= \dot{m}(h_2 - h_1) = \dot{N}(\bar{h}_2 - \bar{h}_1) \\ &= (0.01111 \text{ kmol/min})(136{,}264 - 9904) \text{ kJ/kmol} \\ &= \mathbf{1404 \text{ kJ/min}} \end{aligned}$$

15-42 EES solution of this (and other comprehensive problems designated with the *computer icon*) is available to instructors at the *Instructor Manual* section of the *Online Learning Center* (OLC) at www.mhhe.com/cengel-boles. See the Preface for access information.

Variations of K_p with Temperature

15-43C It enables us to determine the enthalpy of reaction $\overline{h}_R$ from a knowledge of equilibrium constant K_P.

15-44C At 2000 K since combustion processes are exothermic, and exothermic reactions are more complete at lower temperatures.

15-45 The $\overline{h}_R$ at a specified temperature is to be determined using the enthalpy and K_P data.

Assumptions Both the reactants and products are ideal gases.

Analysis (*a*) The complete combustion equation of CO can be expressed as

$$CO + \tfrac{1}{2}O_2 \Leftrightarrow CO_2$$

The $\overline{h}_R$ of the combustion process of CO at 2200 K is the amount of energy released as one kmol of CO is burned in a steady-flow combustion chamber at a temperature of 2200 K, and can be determined from

$$\overline{h}_R = \sum N_P \left(\overline{h}_f^\circ + \overline{h} - \overline{h}^\circ\right)_P - \sum N_R \left(\overline{h}_f^\circ + \overline{h} - \overline{h}^\circ\right)_R$$

Assuming the CO, O_2 and CO_2 to be ideal gases, we have $h = h(T)$. From the tables,

Substance	$\overline{h}_f^\circ$ kJ/kmol	$\overline{h}_{298\text{ K}}$ kJ/kmol	$\overline{h}_{2200\text{ K}}$ kJ/kmol
CO_2	-393,520	9364	112,939
CO	-110,530	8669	72,688
O_2	0	8682	75,484

Substituting,

$$\begin{aligned}\overline{h}_R &= 1(-393{,}520 + 112{,}939 - 9364) \\ &\quad - 1(-110{,}530 + 72{,}688 - 8669) \\ &\quad - 0.5(0 + 75{,}484 - 8682) \\ &= \mathbf{-276{,}835\ kJ/kmol}\end{aligned}$$

(*b*) The $\overline{h}_R$ value at 2200 K can be estimated by using K_P values at 2000 K and 2400 K (the closest two temperatures to 2200 K for which K_P data are available) from Table A-28,

$$\ln\frac{K_{P2}}{K_{P1}} \cong \frac{\overline{h}_R}{R_u}\left(\frac{1}{T_1} - \frac{1}{T_2}\right) \quad \text{or} \quad \ln K_{P2} - \ln K_{P1} \cong \frac{\overline{h}_R}{R_u}\left(\frac{1}{T_1} - \frac{1}{T_2}\right)$$

$$3.860 - 6.635 \cong \frac{\overline{h}_R}{8.314\ \text{kJ/kmol}\cdot\text{K}}\left(\frac{1}{2000\ \text{K}} - \frac{1}{2400\ \text{K}}\right)$$

$$\overline{h}_R \cong \mathbf{-276{,}856\ kJ/kmol}$$

15-46E The $\bar{h}_R$ at a specified temperature is to be determined using the enthalpy and K_P data.

Assumptions Both the reactants and products are ideal gases.

Analysis (*a*) The complete combustion equation of CO can be expressed as

$$CO + \tfrac{1}{2}O_2 \Leftrightarrow CO_2$$

The $\bar{h}_R$ of the combustion process of CO at 3960 R is the amount of energy released as one kmol of H_2 is burned in a steady-flow combustion chamber at a temperature of 3960 R, and can be determined from

$$\bar{h}_R = \sum N_P\left(\bar{h}_f^\circ + \bar{h} - \bar{h}^\circ\right)_P - \sum N_R\left(\bar{h}_f^\circ + \bar{h} - \bar{h}^\circ\right)_R$$

Assuming the CO, O_2 and CO_2 to be ideal gases, we have $h = h(T)$. From the tables,

Substance	$\bar{h}_f^\circ$ Btu/lbmol	$\bar{h}_{537\,R}$ Btu/lbmol	$\bar{h}_{3960\,R}$ Btu/lbmol
CO_2	-169,300	4027.5	48,647
CO	-47,540	3725.1	31,256.5
O_2	0	3725.1	32,440.5

Substituting,

$$\begin{aligned}\bar{h}_R &= 1(-169{,}300 + 48{,}647 - 4027.5)\\ &\quad - 1(-47{,}540 + 31{,}256.5 - 3725.1)\\ &\quad - 0.5(0 + 32{,}440.5 - 3725.1)\\ &= \mathbf{-119{,}030\ Btu/lbmol}\end{aligned}$$

(*b*) The $\bar{h}_R$ value at 3960 R can be estimated by using K_P values at 3600 R and 4320 R (the closest two temperatures to 3960 R for which K_P data are available) from Table A-28,

$$\ln\frac{K_{P2}}{K_{P1}} \cong \frac{\bar{h}_R}{R_u}\left(\frac{1}{T_1} - \frac{1}{T_2}\right) \quad \text{or} \quad \ln K_{P2} - \ln K_{P1} \cong \frac{\bar{h}_R}{R_u}\left(\frac{1}{T_1} - \frac{1}{T_2}\right)$$

$$3.860 - 6.635 \cong \frac{\bar{h}_R}{1.986\ \text{Btu/lbmol}\cdot\text{R}}\left(\frac{1}{3600\ \text{R}} - \frac{1}{4320\ \text{R}}\right)$$

$$\bar{h}_R \cong \mathbf{-119{,}041\ Btu/lbmol}$$

15-47 The K_P value of the combustion process $H_2 + 1/2O_2 \Leftrightarrow H_2O$ is to be determined at a specified temperature using $\bar{h}_R$ data and K_P value .

Assumptions Both the reactants and products are ideal gases.

Analysis The $\bar{h}_R$ and K_P data are related to each other by

$$\ln\frac{K_{P2}}{K_{P1}} \cong \frac{\bar{h}_R}{R_u}\left(\frac{1}{T_1}-\frac{1}{T_2}\right) \quad \text{or} \quad \ln K_{P2} - \ln K_{P1} \cong \frac{\bar{h}_R}{R_u}\left(\frac{1}{T_1}-\frac{1}{T_2}\right)$$

The $\bar{h}_R$ of the specified reaction at 2400 K is the amount of energy released as one kmol of H_2 is burned in a steady-flow combustion chamber at a temperature of 2400 K, and can be determined from

$$\bar{h}_R = \sum N_P\left(\bar{h}_f^\circ + \bar{h} - \bar{h}^\circ\right)_P - \sum N_R\left(\bar{h}_f^\circ + \bar{h} - \bar{h}^\circ\right)_R$$

Assuming the H_2O, H_2 and O_2 to be ideal gases, we have $h = h\,(T)$. From the tables,

Substance	$\bar{h}_f^\circ$ kJ/kmol	$\bar{h}_{298\text{ K}}$ kJ/kmol	$\bar{h}_{2400\text{ K}}$ kJ/kmol
H_2O	-241,820	9904	103,508
H_2	0	8468	75,383
O_2	0	8682	83,174

Substituting,

$$\begin{aligned}\bar{h}_R &= 1(-241{,}820 + 103{,}508 - 9904)\\ &\quad - 1(0 + 75{,}383 - 8468)\\ &\quad - 0.5(0 + 83{,}174 - 8682)\\ &= -252{,}377 \text{ kJ / kmol}\end{aligned}$$

The K_P value at 2600 K can be estimated from the equation above by using this $\bar{h}_R$ value and the K_P value at 2200 K which is $\ln K_{P1} = 6.768$,

$$\ln K_{P2} - 6.768 \cong \frac{-252{,}377 \text{ kJ/kmol}}{8.314 \text{ kJ/kmol}\cdot\text{K}}\left(\frac{1}{2200\text{ K}} - \frac{1}{2600\text{ K}}\right)$$

$$\ln K_{P2} = 4.645 \quad (\text{Table A-28: } \ln K_{P2} = 4.648)$$

or

$$K_{P2} = \mathbf{104.1}$$

15-48 The $\bar{h}_R$ value for the dissociation process $CO_2 \Leftrightarrow CO + 1/2O_2$ at a specified temperature is to be determined using enthalpy and K_p data.

Assumptions Both the reactants and products are ideal gases.

Analysis (*a*) The dissociation equation of CO_2 can be expressed as

$$CO_2 \Leftrightarrow CO + \tfrac{1}{2}O_2$$

The $\bar{h}_R$ of the dissociation process of CO_2 at 2200 K is the amount of energy absorbed or released as one kmol of CO_2 dissociates in a steady-flow combustion chamber at a temperature of 2200 K, and can be determined from

$$\bar{h}_R = \sum N_P\left(\bar{h}_f^\circ + \bar{h} - \bar{h}^\circ\right)_P - \sum N_R\left(\bar{h}_f^\circ + \bar{h} - \bar{h}^\circ\right)_R$$

Assuming the CO, O_2 and CO_2 to be ideal gases, we have $h = h(T)$. From the tables,

Substance	$\bar{h}_f^\circ$ kJ/kmol	$\bar{h}_{298\ K}$ kJ/kmol	$\bar{h}_{2200K}$ kJ/kmol
CO_2	-393,520	9364	112,939
CO	-110,530	8669	72,688
O_2	0	8682	75,484

Substituting,

$$\begin{aligned}\bar{h}_R &= 1(-110{,}530 + 72{,}688 - 8669)\\ &\quad + 0.5(0 + 75{,}484 - 8682)\\ &\quad - 1(-393{,}520 + 112{,}939 - 9364)\\ &= \mathbf{276{,}835\ kJ/kmol}\end{aligned}$$

(*b*) The $\bar{h}_R$ value at 2200 K can be estimated by using K_P values at 2000 K and 2400 K (the closest two temperatures to 2200 K for which K_P data are available) from Table A-28,

$$\ln\frac{K_{P2}}{K_{P1}} \cong \frac{\bar{h}_R}{R_u}\left(\frac{1}{T_1} - \frac{1}{T_2}\right) \quad \text{or} \quad \ln K_{P2} - \ln K_{P1} \cong \frac{\bar{h}_R}{R_u}\left(\frac{1}{T_1} - \frac{1}{T_2}\right)$$

$$-3.860 - (-6.635) \cong \frac{\bar{h}_R}{8.314\ \text{kJ/kmol}\cdot\text{K}}\left(\frac{1}{2000\ \text{K}} - \frac{1}{2400\ \text{K}}\right)$$

$$\bar{h}_R \cong \mathbf{276{,}856\ kJ/kmol}$$

15-49 The $\overline{h}_R$ value for the dissociation process $O_2 \Leftrightarrow 2O$ at a specified temperature is to be determined using enthalpy and K_P data.

Assumptions Both the reactants and products are ideal gases.

Analysis (*a*) The dissociation equation of O_2 can be expressed as

$$O_2 \Leftrightarrow 2O$$

The $\overline{h}_R$ of the dissociation process of O_2 at 2900 K is the amount of energy absorbed or released as one kmol of O_2 dissociates in a steady-flow combustion chamber at a temperature of 2900 K, and can be determined from

$$\overline{h}_R = \sum N_P \left(\overline{h}_f^\circ + \overline{h} - \overline{h}^\circ\right)_P - \sum N_R \left(\overline{h}_f^\circ + \overline{h} - \overline{h}^\circ\right)_R$$

Assuming the O_2 and O to be ideal gases, we have $h = h(T)$. From the tables,

Substance	$\overline{h}_f^\circ$ kJ/kmol	$\overline{h}_{298\text{ K}}$ kJ/kmol	$\overline{h}_{2900\text{ K}}$ kJ/kmol
O	249,190	6852	61,332
O_2	0	8682	102,793

Substituting,

$$\begin{aligned}\overline{h}_R &= 2(249{,}190 + 61{,}332 - 6852) - 1(0 + 102{,}793 - 8682) \\ &= \mathbf{513{,}229\ kJ/kmol}\end{aligned}$$

(*b*) The $\overline{h}_R$ value at 2900 K can be estimated by using K_P values at 2800 K and 3000 K (the closest two temperatures to 2900 K for which K_P data are available) from Table A-28,

$$\ln\frac{K_{P2}}{K_{P1}} \cong \frac{\overline{h}_R}{R_u}\left(\frac{1}{T_1} - \frac{1}{T_2}\right) \quad \text{or} \quad \ln K_{P2} - \ln K_{P1} \cong \frac{\overline{h}_R}{R_u}\left(\frac{1}{T_1} - \frac{1}{T_2}\right)$$

$$-4.357 - (-5.826) \cong \frac{\overline{h}_R}{8.314\text{ kJ/kmol}\cdot\text{K}}\left(\frac{1}{2800\text{ K}} - \frac{1}{3000\text{ K}}\right)$$

$$\overline{h}_R \cong \mathbf{512{,}957\ kJ/kmol}$$

Phase Equilibrium

15-50C No. Because the specific gibbs function of each phase will not be affected by this process; i.e., we will still have $g_f = g_g$.

15-51C Yes. Because the number of independent variables for a two-phase (PH=2), two-component (C=2) mixture is, from the phase rule,

$$IV = C - PH + 2 = 2 - 2 + 2 = 2$$

Therefore, two properties can be changed independently for this mixture. In other words, we can hold the temperature constant and vary the pressure and still be in the two-phase region. Notice that if we had a single component (C=1) two phase system, we would have IV=1, which means that fixing one independent property automatically fixes all the other properties.

11-52C Using solubility data of a solid in a specified liquid, the mass fraction *w* of the solid *A* in the liquid at the interface at a specified temperature can be determined from

$$w_A = \frac{m_{\text{solid}}}{m_{\text{solid}} + m_{\text{liquid}}}$$

where m_{solid} is the maximum amount of solid dissolved in the liquid of mass m_{liquid} at the specified temperature.

11-53C The molar concentration C_i of the gas species *i* in the solid at the interface $C_{\text{i, solid side}}(0)$ is proportional to the *partial pressure* of the species *i* in the gas $P_{\text{i, gas side}}(0)$ on the gas side of the interface, and is determined from

$$C_{\text{i, solid side}}(0) = S \times P_{\text{i, gas side}}(0) \qquad (\text{kmol/m}^3)$$

where S is the *solubility* of the gas in that solid at the specified temperature.

11-54C Using Henry's constant data for a gas dissolved in a liquid, the mole fraction of the gas dissolved in the liquid at the interface at a specified temperature can be determined from Henry's law expressed as

$$y_{\text{i, liquid side}}(0) = \frac{P_{\text{i, gas side}}(0)}{H}$$

where *H* is *Henry's constant* and $P_{\text{i, gas side}}(0)$ is the partial pressure of the gas *i* at the gas side of the interface. This relation is applicable for dilute solutions (gases that are weakly soluble in liquids).

15-55 It is to be shown that a mixture of saturated liquid water and saturated water vapor at 100°C satisfies the criterion for phase equilibrium.

Analysis Using the definition of Gibbs function and enthalpy and entropy data from Table A-4,

$$g_f = h_f - Ts_f = (419.04 \text{ kJ/kg}) - (373.15 \text{ K})(1.3069 \text{ kJ/kg}\cdot\text{K}) = -68.6 \text{ kJ/kg}$$

$$g_g = h_g - Ts_g = (2676.1 \text{ kJ/kg}) - (373.15 \text{ K})(7.3549 \text{ kJ/kg}\cdot\text{K}) = -68.4 \text{ kJ/kg}$$

which are sufficiently close. Therefore, the criteria for phase equilibrium is satisfied.

15-56 It is to be shown that a mixture of saturated liquid water and saturated water vapor at 300 kPa satisfies the criterion for phase equilibrium.

Analysis The saturation temperature at 300 kPa is 406.7 K. Using the definition of Gibbs function and enthalpy and entropy data from Table A-5,

$$g_f = h_f - Ts_f = (561.47\text{ kJ/kg}) - (406.7\text{ K})(1.8607\text{ kJ/kg}\cdot\text{K}) = -185\text{ kJ/kg}$$
$$g_g = h_g - Ts_g = (2725.3\text{ kJ/kg}) - (406.7\text{ K})(6.9919\text{ kJ/kg}\cdot\text{K}) = -118.3\text{ kJ/kg}$$

which are sufficiently close. Therefore, the criteria for phase equilibrium is satisfied.

15-57 It is to be shown that a saturated liquid-vapor mixture of refrigerant-134a at 0°C satisfies the criterion for phase equilibrium.

Analysis Using the definition of Gibbs function and enthalpy and entropy data from Table A-12,

$$g_f = h_f - Ts_f = (50.02\text{ kJ/kg}) - (273.15\text{ K})(0.1970\text{ kJ/kg}\cdot\text{K}) = -3.791\text{ kJ/kg}$$
$$g_g = h_g - Ts_g = (247.23\text{ kJ/kg}) - (273.15\text{ K})(0.9190\text{ kJ/kg}\cdot\text{K}) = -3.795\text{ kJ/kg}$$

which are sufficiently close. Therefore, the criteria for phase equilibrium is satisfied.

15-58 The number of independent properties needed to fix the state of a mixture of oxygen and nitrogen in the gas phase is to be determined.

Analysis In this case the number of components is C = 2 and the number of phases is PH = 1. Then the number of independent variables is determined from the phase rule to be

$$\text{IV} = \text{C} - \text{PH} + 2 = 2 - 1 + 2 = 3$$

Therefore, three independent properties need to be specified to fix the state. They can be temperature, the pressure, and the mole fraction of one of the gases.

15-59 A liquid-vapor mixture of ammonia and water in equilibrium at a specified temperature is considered. The composition of the liquid phase is given. The composition of the vapor phase is to be determined.

Assumptions The mixture is ideal and thus Raoult's law is applicable.

Properties At 30°C, $P_{\text{sat,H}_2\text{O}} = 4.246\text{ kPa}$ and $P_{\text{sat,NH}_3} = 1116.5\text{ kPa}$.

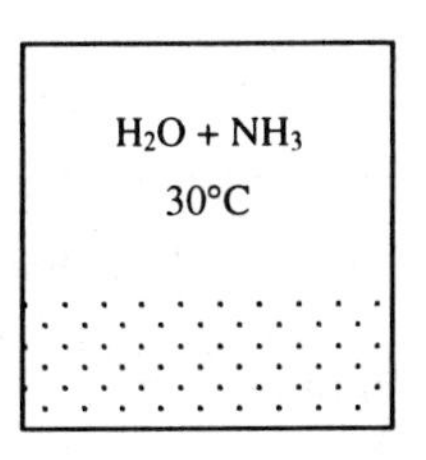

Analysis The vapor pressures are

$$P_{\text{H}_2\text{O}} = y_{f,\text{H}_2\text{O}} P_{\text{sat,H}_2\text{O}}(T) = 0.40(4.246\text{ kPa}) = 1.70\text{ kPa}$$
$$P_{\text{NH}_3} = y_{f,\text{NH}_3} P_{\text{sat,NH}_3}(T) = 0.60(1116.5\text{ kPa}) = 669.90\text{ kPa}$$

Thus the total pressure of the mixture is

$$P_{\text{total}} = P_{\text{H}_2\text{O}} + P_{\text{NH}_3} = (1.70 + 669.90)\text{ kPa} = 671.6\text{ kPa}$$

Then the mole fractions in the vapor phase become

$$y_{g,\text{H}_2\text{O}} = \frac{P_{\text{H}_2\text{O}}}{P_{\text{total}}} = \frac{1.7\text{ kPa}}{671.6\text{ kPa}} = \mathbf{0.0025} \quad \text{or} \quad 0.25\%$$

$$y_{g,\text{NH}_3} = \frac{P_{\text{NH}_3}}{P_{\text{total}}} = \frac{669.9\text{ kPa}}{671.6\text{ kPa}} = \mathbf{0.9975} \quad \text{or} \quad 99.75\%$$

15-60 A liquid-vapor mixture of ammonia and water in equilibrium at a specified temperature is considered. The composition of the liquid phase is given. The composition of the vapor phase is to be determined.

Assumptions The mixture is ideal and thus Raoult's law is applicable.

Properties At 20°C, $P_{sat,H_2O} = 2.339 \text{ kPa}$ and $P_{sat,NH_3} = 857.1 \text{ kPa}$.

Analysis The vapor pressures are

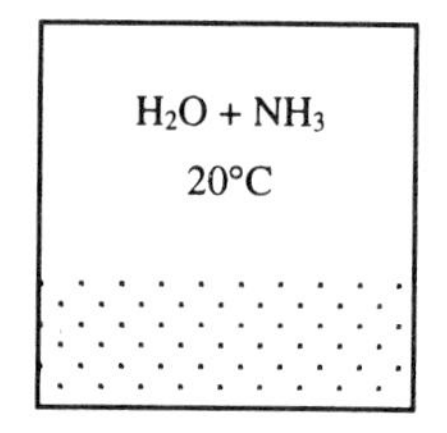

$$P_{H_2O} = y_{f,H_2O} P_{sat,H_2O}(T) = 0.50(2.339 \text{ kPa}) = 1.17 \text{ kPa}$$
$$P_{NH_3} = y_{f,NH_3} P_{sat,NH_3}(T) = 0.50(857.1 \text{ kPa}) = 428.55 \text{ kPa}$$

Thus the total pressure of the mixture is

$$P_{total} = P_{H_2O} + P_{NH_3} = (1.17 + 428.55) \text{ kPa} = 429.72 \text{ kPa}$$

Then the mole fractions in the vapor phase become

$$y_{g,H_2O} = \frac{P_{H_2O}}{P_{total}} = \frac{1.17 \text{ kPa}}{429.72 \text{ kPa}} = \mathbf{0.0027} \quad \text{or} \quad 0.27\%$$

$$y_{g,NH_3} = \frac{P_{NH_3}}{P_{total}} = \frac{428.55 \text{ kPa}}{429.72 \text{ kPa}} = \mathbf{0.9973} \quad \text{or} \quad 99.73\%$$

15-61 A liquid-vapor mixture of ammonia and water in equilibrium at a specified temperature is considered. The composition of the vapor phase is given. The composition of the liquid phase is to be determined.

Assumptions The mixture is ideal and thus Raoult's law is applicable.

Properties At 50°C, $P_{sat,H_2O} = 12.349 \text{ kPa}$ and $P_{sat,NH_3} = 2032.6 \text{ kPa}$.

Analysis We have $y_{f,H_2O} = 1\%$ and $y_{f,NH_3} = 99\%$. For an ideal two-phase mixture we have

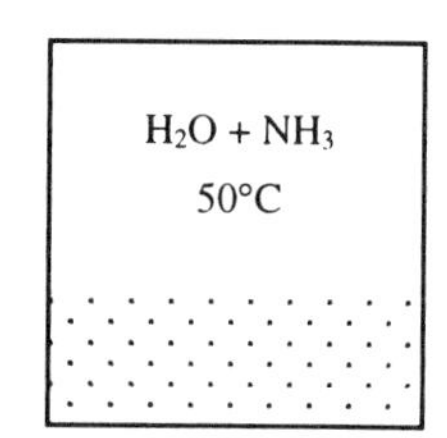

$$y_{g,H_2O} P_m = y_{f,H_2O} P_{sat,H_2O}(T)$$
$$y_{g,NH_3} P_m = y_{f,NH_3} P_{sat,NH_3}(T)$$
$$y_{f,H_2O} + y_{f,NH_3} = 1$$

Solving for y_{f,H_2O},

$$y_{f,H_2O} = \frac{y_{g,H_2O} P_{sat,NH_3}}{y_{g,NH_3} P_{sat,H_2O}}(1 - y_{f,H_2O}) = \frac{(0.01)(2032.6 \text{ kPa})}{(0.99)(12.369 \text{ kPa})}(1 - y_{f,H_2O})$$

It yields $y_{f,H_2O} = \mathbf{0.624}$ and $y_{f,NH_3} = \mathbf{0.376}$

15-62 Using the liquid-vapor equilibrium diagram of an oxygen-nitrogen mixture, the composition of each phase at a specified temperature and pressure is to be determined.

Analysis From the equilibrium diagram we read

Liquid: $37\% N_2$ and $63\% O_2$

Vapor: $10\% N_2$ and $90\% O_2$

15-63 Using the liquid-vapor equilibrium diagram of an oxygen-nitrogen mixture, the composition of each phase at a specified temperature and pressure is to be determined.

Analysis From the equilibrium diagram we read

$$\text{Liquid: } 68\%\text{N}_2 \text{ and } 32\%\text{ O}_2$$

$$\text{Vapor: } 32\%\text{N}_2 \text{ and } 68\%\text{ O}_2$$

15-64 Using the liquid-vapor equilibrium diagram of an oxygen-nitrogen mixture at a specified pressure, the temperature is to be determined for a specified composition of the vapor phase.

Analysis From the equilibrium diagram we read $T = \mathbf{88.5\ K}$.

15-65 Using the liquid-vapor equilibrium diagram of an oxygen-nitrogen mixture at a specified pressure, the temperature is to be determined for a specified composition of the liquid phase.

Analysis From the equilibrium diagram we read $T = \mathbf{79.5\ K}$.

15-66 A rubber plate is exposed to nitrogen. The molar and mass density of nitrogen in the iron at the interface is to be determined.

Assumptions Rubber and nitrogen are in thermodynamic equilibrium at the interface.

Properties The molar mass of nitrogen is M = 28.0 kg/kmol (Table A-1). The solubility of nitrogen in rubber at 298 K is 0.00156 kmol/m^3·bar (Table 15-3).

Analysis Noting that 250 kPa = 2.5 bar, the molar density of nitrogen in the rubber at the interface is determined to be

$$\begin{aligned} C_{\text{N}_2,\text{ solid side}}(0) &= S \times P_{\text{N}_2,\text{ gas side}} \\ &= (0.00156\text{ kmol/m}^3.\text{bar})(2.5\text{ bar}) \\ &= \mathbf{0.0039\ kmol/m^3} \end{aligned}$$

It corresponds to a mass density of

$$\begin{aligned} \rho_{\text{N}_2,\text{ solid side}}(0) &= C_{\text{N}_2,\text{ solid side}}(0)\, M_{\text{N}_2} \\ &= (0.0039\text{ kmol/m}^3)(28\text{ kg/kmol}) \\ &= \mathbf{0.1092\ kg/m^3} \end{aligned}$$

That is, there will be 0.0039 kmol (or 0.1092 kg) of N_2 gas in each m^3 volume of iron adjacent to the interface.

11-67 A rubber wall separates O_2 and N_2 gases. The molar concentrations of O_2 and N_2 in the wall are to be determined.

Assumptions The O_2 and N_2 gases are in phase equilibrium with the rubber wall.

Properties The molar mass of oxygen and nitrogen are 32.0 and 28.0 kg/kmol, respectively (Table A-1). The solubility of oxygen and nitrogen in rubber at 298 K are 0.00312 and 0.00156 kmol/m^3·bar, respectively (Table 15-3).

Analysis Noting that 500 kPa = 5 bar, the molar densities of oxygen and nitrogen in the rubber wall are determined to be

$$C_{O_2,\text{ solid side}}(0) = \mathcal{S} \times P_{O_2,\text{ gas side}}$$
$$= (0.00312 \text{ kmol/m}^3.\text{bar})(5 \text{ bar})$$
$$= \mathbf{0.0156 \text{ kmol/m}^3}$$

$$C_{N_2,\text{ solid side}}(0) = \mathcal{S} \times P_{N_2,\text{ gas side}}$$
$$= (0.00156 \text{ kmol/m}^3.\text{bar})(5 \text{ bar})$$
$$= \mathbf{0.0078 \text{ kmol/m}^3}$$

Rubber plate
O_2
25°C
500 kPa
C_{O2}
C_{N2}
N_2
25°C
500 kPa

That is, there will be 0.0156 kmol of O_2 and 0.0078 kmol of N_2 gas in each m^3 volume of the rubber wall.

15-68 A glass of water is left in a room. The mole fraction of the water vapor in the air and the mole fraction of air in the water are to be determined when the water and the air are in thermal and phase equilibrium.

Assumptions **1** Both the air and water vapor are ideal gases. **2** Air is saturated since the humidity is 100 percent. **3** Air is weakly soluble in water and thus Henry's law is applicable.

Properties The saturation pressure of water at 20°C is 2.339 kPa (Table A-9). Henry's constant for air dissolved in water at 20°C (293 K) is given in Table 15-2 to be H = 65,600 bar. Molar masses of dry air and water are 29 and 18 kg/kmol, respectively (Table A-1).

Analysis (*a*) Noting that air is saturated, the partial pressure of water vapor in the air will simply be the saturation pressure of water at 20°C,

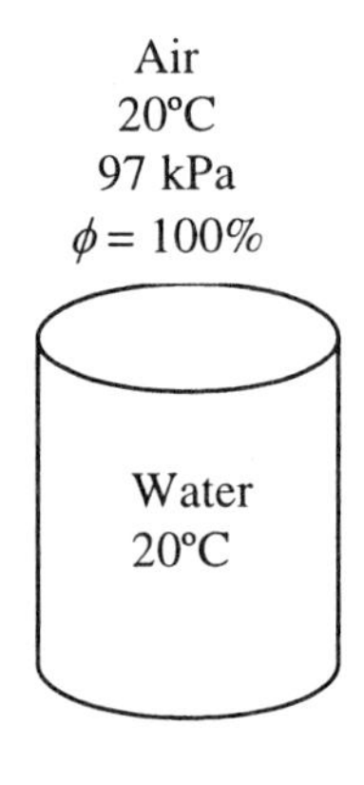

$$P_{\text{vapor}} = P_{\text{sat @ 20°C}} = 2.339 \text{ kPa}$$

Assuming both the air and vapor to be ideal gases, the mole fraction of water vapor in the air is determined to be

$$y_{\text{vapor}} = \frac{P_{\text{vapor}}}{P} = \frac{2.339 \text{ kPa}}{97 \text{ kPa}} = \mathbf{0.0241}$$

(*b*) Noting that the total pressure is 97 kPa, the partial pressure of dry air is

$$P_{\text{dry air}} = P - P_{\text{vapor}} = 97 - 2.339 = 94.7 \text{ kPa} = 0.947 \text{ bar}$$

From Henry's law, the mole fraction of air in the water is determined to be

$$y_{\text{dry air, liquid side}} = \frac{P_{\text{dry air, gas side}}}{H} = \frac{0.947 \text{ bar}}{65{,}600 \text{ bar}} = \mathbf{1.44 \times 10^{-5}}$$

Discussion The amount of air dissolved in water is very small, as expected.

15-69E Water is sprayed into air, and the falling water droplets are collected in a container. The mass and mole fractions of air dissolved in the water are to be determined.

Assumptions **1** Both the air and water vapor are ideal gases. **2** Air is saturated since water is constantly sprayed into it. **3** Air is weakly soluble in water and thus Henry's law is applicable.

Properties The saturation pressure of water at 80°F is 0.5073 psia (Table A-9E). Henry's constant for air dissolved in water at 80ºF (300 K) is given in Table 15-2 to be H = 74,000 bar. Molar masses of dry air and water are 29 and 18 lbm / lbmol, respectively (Table A-1).

Analysis Noting that air is saturated, the partial pressure of water vapor in the air will simply be the saturation pressure of water at 80°F,

$$P_{\text{vapor}} = P_{\text{sat @ 80°F}} = 0.5073 \text{ psia}$$

Then the partial pressure of dry air becomes

$$P_{\text{dry air}} = P - P_{\text{vapor}} = 14.3 - 0.5073 = 13.79 \text{ psia}$$

From Henry's law, the mole fraction of air in the water is determined to be

$$y_{\text{dry air, liquid side}} = \frac{P_{\text{dry air, gasside}}}{H} = \frac{13.79 \text{ psia}(1 \text{ atm}/14.696 \text{ psia})}{74{,}000 \text{ bar}(1 \text{ atm}/1.01325 \text{ bar})} = \mathbf{1.29 \times 10^{-5}}$$

which is very small, as expected.

The mass and mole fractions of a mixture are related to each other by

$$w_i = \frac{m_i}{m_m} = \frac{N_i M_i}{N_m M_m} = y_i \frac{M_i}{M_m}$$

where the apparent molar mass of the liquid water - air mixture is

$$M_m = \sum y_i M_i = y_{\text{liquid water}} M_{\text{water}} + y_{\text{dry air}} M_{\text{dry air}}$$
$$\cong 1 \times 29.0 + 0 \times 18.0 \cong 29.0 \text{ kg/kmol}$$

Then the mass fraction of dissolved air in liquid water becomes

$$w_{\text{dry air, liquid side}} = y_{\text{dry air, liquid side}}(0)\frac{M_{\text{dry air}}}{M_m} = 1.29 \times 10^{-5}\frac{29}{29} = \mathbf{1.29 \times 10^{-5}}$$

Discussion The mass and mole fractions of dissolved air in this case are identical because of the very small amount of air in water.

15-70 A carbonated drink in a bottle is considered. Assuming the gas space above the liquid consists of a saturated mixture of CO_2 and water vapor and treating the drink as a water, determine the mole fraction of the water vapor in the CO_2 gas and the mass of dissolved CO_2 in a 200 ml drink are to be determined when the water and the CO_2 gas are in thermal and phase equilibrium.

Assumptions **1** The liquid drink can be treated as water. **2** Both the CO_2 and the water vapor are ideal gases. **3** The CO_2 gas and water vapor in the bottle from a saturated mixture. **4** The CO_2 is weakly soluble in water and thus Henry's law is applicable.

Properties The saturation pressure of water at 27°C is 3.60 kPa (Table A-9). Henry's constant for CO_2 dissolved in water at 27°C (300 K) is given in Table 15-2 to be H = 1710 bar. Molar masses of CO_2 and water are 44 and 18 kg/kmol, respectively (Table A-1).

Analysis (*a*) Noting that the CO_2 gas in the bottle is saturated, the partial pressure of water vapor in the air will simply be the saturation pressure of water at 27°C,

$$P_{\text{vapor}} = P_{\text{sat @ 27°C}} = 3.60 \text{ kPa}$$

Assuming both CO_2 and vapor to be ideal gases, the mole fraction of water vapor in the CO_2 gas becomes

$$y_{\text{vapor}} = \frac{P_{\text{vapor}}}{P} = \frac{3.60 \text{ kPa}}{130 \text{ kPa}} = \mathbf{0.0277}$$

(*b*) Noting that the total pressure is 130 kPa, the partial pressure of CO_2 is

$$P_{\text{CO}_2\text{ gas}} = P - P_{\text{vapor}} = 130 - 3.60 = 126.4 \text{ kPa} = 1.264 \text{ bar}$$

From Henry's law, the mole fraction of CO_2 in the drink is determined to be

$$y_{\text{CO}_2\text{,liquid side}} = \frac{P_{\text{CO}_2\text{,gas side}}}{H} = \frac{1.264 \text{ bar}}{1710 \text{ bar}} = \mathbf{7.39 \times 10^{-4}}$$

Then the mole fraction of water in the drink becomes

$$y_{\text{water, liquid side}} = 1 - y_{\text{CO}_2\text{, liquid side}} = 1 - 7.39 \times 10^{-4} = 0.9993$$

The mass and mole fractions of a mixture are related to each other by

$$w_i = \frac{m_i}{m_m} = \frac{N_i M_i}{N_m M_m} = y_i \frac{M_i}{M_m}$$

where the apparent molar mass of the drink (liquid water - CO_2 mixture) is

$$M_m = \sum y_i M_i = y_{\text{liquid water}} M_{\text{water}} + y_{\text{CO}_2} M_{\text{CO}_2} = 0.9993 \times 18.0 + (7.39 \times 10^{-4}) \times 44 = 18.02 \text{ kg/kmol}$$

Then the mass fraction of dissolved CO_2 gas in liquid water becomes

$$w_{\text{CO}_2\text{, liquid side}} = y_{\text{CO}_2\text{, liquid side}}(0) \frac{M_{\text{CO}_2}}{M_m} = 7.39 \times 10^{-4} \frac{44}{18.02} = 0.00180$$

Therefore, the mass of dissolved CO_2 in a 200 ml ≈ 200 g drink is

$$m_{\text{CO}_2} = w_{\text{CO}_2} m_m = 0.00180(200 \text{ g}) = \mathbf{0.360 \text{ g}}$$

Review Problems

15-71 The equilibrium constant of the dissociation process $O_2 \leftrightarrow 2O$ is given in Table A-28 at different temperatures. The value at a given temperature is to be verified using Gibbs function data.

Analysis The K_P value of a reaction at a specified temperature can be determined from the Gibbs function data using

$$K_p = e^{-\Delta G^*(T)/R_u T} \quad \text{or} \quad \ln K_p = -\Delta G^*(T)/R_u T$$

$O_2 \Leftrightarrow 2O$

2000 K

where

$$\begin{aligned}\Delta G^*(T) &= \nu_O \bar{g}_O^*(T) - \nu_{O_2} \bar{g}_{O_2}^*(T) \\ &= \nu_O(\bar{h} - T\bar{s})_O - \nu_{O_2}(\bar{h} - T\bar{s})_{O_2} \\ &= \nu_O[(\bar{h}_f + \bar{h}_{2000} - \bar{h}_{298}) - T\bar{s}]_O - \nu_{O_2}[(\bar{h}_f + \bar{h}_{2000} - \bar{h}_{298}) - T\bar{s}]_{O_2} \\ &= 2 \times (249{,}190 + 42{,}564 - 6852 - 2000 \times 201.135) \\ &\quad -1 \times (0 + 67{,}881 - 8682 - 2000 \times 268.655) \\ &= 243{,}375 \text{ kJ/kmol}\end{aligned}$$

Substituting,

$$\ln K_p = -(243{,}375 \text{ kJ/kmol})/[(8.314 \text{ kJ/kmol}\cdot\text{K})(2000 \text{ K})] = -14.636$$

or

$$K_p = \mathbf{4.4 \times 10^{-7}} \qquad \text{(Table A-28: } \ln K_P = -14.622)$$

15-72 A mixture of H_2 and Ar is heated is heated until 15% of H_2 is dissociated. The final temperature of mixture is to be determined.

Assumptions **1** The constituents of the mixture are ideal gases. **2** Ar in the mixture remains an inert gas.

Analysis The stoichiometric and actual reactions can be written as

Stoichiometric: $H_2 \Leftrightarrow 2H$ (thus $\nu_{H_2} = 1$ and $\nu_H = 2$)

Actual: $H_2 + Ar \longrightarrow \underbrace{0.3H}_{\text{prod}} + \underbrace{0.85H_2}_{\text{react.}} + \underbrace{Ar}_{\text{inert}}$

$H_2 \Leftrightarrow 2H$

Ar

1 atm

The equilibrium constant K_P can be determined from

$$K_p = \frac{N_H^{\nu_H}}{N_{H_2}^{\nu_{H_2}}}\left(\frac{P}{N_{\text{total}}}\right)^{\nu_H - \nu_{H_2}} = \frac{0.3^2}{0.85}\left(\frac{1}{0.85 + 0.3 + 1}\right)^{2-1} = 0.0492$$

From Table A-28, the temperature corresponding to this K_P value is T = **3117 K**.

15-73 A mixture of H_2O, O_2, and N_2 is heated to a high temperature at a constant pressure. The equilibrium composition is to be determined.

Assumptions **1** The equilibrium composition consists of H_2O, O_2, N_2 and H_2. **2** The constituents of the mixture are ideal gases.

Analysis The stoichiometric and actual reactions in this case are

1 H_2O
2 O_2
5 N_2
2200 K
5 atm

Stoichiometric: $H_2O \Leftrightarrow H_2 + \frac{1}{2}O_2$ (thus $\nu_{H_2O} = 1, \nu_{H_2} = 1$, and $\nu_{O_2} = \frac{1}{2}$)

Actual: $H_2O + 2O_2 + 5N_2 \longrightarrow \underbrace{xH_2O}_{\text{react.}} + \underbrace{yH_2 + zO_2}_{\text{products}} + \underbrace{5N_2}_{\text{inert}}$

H balance: $2 = 2x + 2y \longrightarrow y = 1 - x$

O balance: $3 = x + 2z \longrightarrow z = 1.5 - 0.5x$

Total number of moles: $N_{\text{total}} = x + y + z + 5 = 7.5 - 0.5x$

The equilibrium constant relation can be expressed as

$$K_p = \frac{N_{H_2}^{\nu_{H_2}} N_{O_2}^{\nu_{O_2}}}{N_{H_2O}^{\nu_{H_2O}}} \left(\frac{P}{N_{\text{total}}} \right)^{(\nu_{H_2} - \nu_{O_2} - \nu_{H_2O})} = \frac{yz^{0.5}}{x} \left(\frac{P}{N_{\text{total}}} \right)^{1+0.5-1}$$

From Table A-28, $\ln K_P = -6.768$ at 2200 K. Thus $K_P = 0.00115$. Substituting,

$$0.00115 = \frac{(1-x)(1.5-0.5x)^{0.5}}{x} \left(\frac{5}{7.5-0.5x} \right)^{0.5}$$

Solving for x,

$x = 0.999$

Then,

$y = 1 - x = 0.001$

$z = 1.5 - 0.5x = 1.0005$

Therefore, the equilibrium composition of the mixture at 2200 K and 5 atm is

$$\mathbf{0.999H_2O + 0.001H_2 + 1.0005O_2 + 5N_2}$$

The equilibrium constant for the reaction $H_2O \Leftrightarrow OH + \frac{1}{2}H_2$ is $\ln K_P = -7.148$, which is very close to the K_P value of the reaction considered. Therefore, it is not realistic to assume that no OH will be present in equilibrium mixture.

15-74 The mole fraction of argon that ionizes at a specified temperature and pressure is to be determined.

Assumptions All components behave as ideal gases.

Analysis The stoichiometric and actual reactions can be written as

Stoichiometric: $Ar \Leftrightarrow Ar^+ + e^-$ (thus $\nu_{Ar} = 1, \nu_{Ar^+} = 1$ and $\nu_{e^-} = 1$)

Actual: $Ar \longrightarrow \underbrace{xAr}_{\text{react.}} + \underbrace{yAr^+ + ye^-}_{\text{products}}$

Ar balance: $1 = x + y$ or $y = 1 - x$

Total number of moles: $N_{total} = x + 2y = 2 - x$

The equilibrium constant relation becomes

$$K_p = \frac{N_{Ar^+}^{\nu_{Ar^+}} N_{e^-}^{\nu_{e^-}}}{N_{Ar}^{\nu_{Ar}}} \left(\frac{P}{N_{total}} \right)^{(\nu_{Ar^+} + \nu_{e^-} - \nu_{Ar})} = \frac{y^2}{x} \left(\frac{P}{N_{total}} \right)^{1+1-1}$$

Substituting,

$$0.00042 = \frac{(1-x)^2}{x} \left(\frac{0.2}{2-x} \right)$$

Solving for x,

$$x = 0.963$$

Thus the fraction of Ar which dissociates into Ar^+ and e^- is

$$1 - 0.963 = 0.037 \quad \text{or} \quad \mathbf{3.7\%}$$

$Ar \Leftrightarrow Ar^+ + e^-$
10,000 K
0.2 atm

15-75 [*Also solved by EES on enclosed CD*] Methane gas is burned with stoichiometric amount of air during a combustion process. The equilibrium composition and the exit temperature are to be determined.

Assumptions **1** The product gases consist of CO_2, H_2O, CO, N_2, and O_2. **2** The constituents of the mixture are ideal gases. **3** This is an adiabatic and steady-flow combustion process.

Analysis (*a*) The combustion equation of CH_4 with stoichiometric amount of O_2 can be written as

$$CH_4 + 2(O_2 + 3.76N_2) \longrightarrow xCO_2 + (1-x)CO + (0.5-0.5x)O_2 + 2H_2O + 7.52N_2$$

After combustion, there will be no CH_4 present in the combustion chamber, and H_2O will act like an inert gas. The equilibrium equation among CO_2, CO, and O_2 can be expressed as

$$CO_2 \Leftrightarrow CO + \tfrac{1}{2}O_2 \quad (\text{thus } \nu_{CO_2} = 1, \nu_{CO} = 1, \text{ and } \nu_{O_2} = \tfrac{1}{2})$$

and

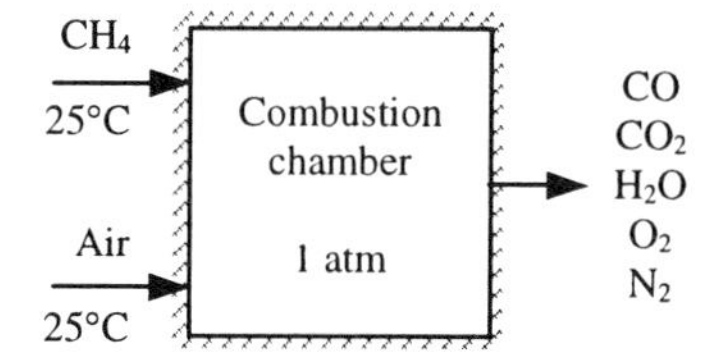

$$K_p = \frac{N_{CO}^{\nu_{CO}} N_{O_2}^{\nu_{O_2}}}{N_{CO_2}^{\nu_{CO_2}}} \left(\frac{P}{N_{total}} \right)^{(\nu_{CO}+\nu_{O_2}-\nu_{CO_2})}$$

where

$$N_{total} = x + (1-x) + (1.5-0.5x) + 2 + 7.52 = 12.02 - 0.5x$$

Substituting,

$$K_p = \frac{(1-x)(0.5-0.5x)^{0.5}}{x} \left(\frac{1}{12.02-0.5x} \right)^{1.5-1}$$

The value of K_P depends on temperature of the products, which is yet to be determined. A second relation to determine K_P and x is obtained from the steady-flow energy balance expressed as

$$0 = \sum N_P \left(\bar{h}_f^\circ + \bar{h} - \bar{h}^\circ \right)_P - \sum N_R \left(\bar{h}_f^\circ + \bar{h} - \bar{h}^\circ \right)_R \longrightarrow 0 = \sum N_P \left(\bar{h}_f^\circ + \bar{h} - \bar{h}^\circ \right)_P - \sum N_R \bar{h}_{f\,R}^\circ$$

since the combustion is adiabatic and the reactants enter the combustion chamber at 25°C. Assuming the air and the combustion products to be ideal gases, we have $h = h(T)$. From the tables,

Substance	$\bar{h}_f^\circ$ kJ/kmol	$\bar{h}_{298\,K}$ kJ/kmol
CH_4(g)	-74,850	--
N_2	0	8669
O_2	0	8682
H_2O(g)	-241,820	9904
CO	-110,530	8669
CO_2	-393,520	9364

Substituting,

$$\begin{aligned} 0 = {} & x(-393{,}520 + \bar{h}_{CO_2} - 9364) + (1-x)(-110{,}530 + \bar{h}_{CO} - 8669) \\ & + 2(-241{,}820 + \bar{h}_{H_2O} - 9904) + (0.5-0.5x)(0 + \bar{h}_{O_2} - 8682) \\ & + 7.52(0 + \bar{h}_{N_2} - 8669) - 1(-74{,}850 + h_{298} - h_{298}) - 0 - 0 \end{aligned}$$

which yields

$$x\bar{h}_{CO_2} + (1-x)\bar{h}_{CO} + 2\bar{h}_{H_2O} + (0.5-0.5x)\bar{h}_{O_2} + 7.52\bar{h}_{N_2} - 279{,}344x = 617{,}329$$

Now we have two equations with two unknowns, T_P and x. The solution is obtained by trial and error by assuming a temperature T_P, calculating the equilibrium composition from the first equation, and then checking to see if the second equation is satisfied. A first guess is obtained by assuming there is no CO in the products, i.e., x = 1. It yields T_P = 2328 K. The adiabatic combustion temperature with incomplete combustion will be less.

$$\text{Take } T_p = 2300 \text{ K} \longrightarrow \ln K_p = -4.49 \longrightarrow x = 0.870 \longrightarrow RHS = 641{,}093$$

$$\text{Take } T_p = 2250 \text{ K} \longrightarrow \ln K_p = -4.805 \longrightarrow x = 0.893 \longrightarrow RHS = 612{,}755$$

By interpolation, $T_p = \mathbf{2258\ K}$ and $x = 0.889$

Thus the composition of the equilibrium mixture is

$$\mathbf{0.889CO_2 + 0.111CO + 0.0555O_2 + 2H_2O + 7.52N_2}$$

15-76 EES solution of this (and other comprehensive problems designated with the *computer icon*) is available to instructors at the *Instructor Manual* section of the *Online Learning Center* (OLC) at www.mhhe.com/cengel-boles. See the Preface for access information.

15-77 A mixture of H_2 and O_2 in a tank is ignited. The equilibrium composition of the product gases and the amount of heat transfer from the combustion chamber are to be determined.

Assumptions **1** The equilibrium composition consists of H_2O, H_2, and O_2. **2** The constituents of the mixture are ideal gases.

Analysis (*a*) The combustion equation can be written as

H_2O, H_2, O_2
2800 K
5 atm

$$H_2 + 0.5O_2 \longrightarrow xH_2O + (1-x)H_2 + (0.5-0.5x)O_2$$

The equilibrium equation among H_2O, H_2, and O_2 can be expressed as

$$H_2O \Leftrightarrow H_2 + \tfrac{1}{2}O_2 \quad (\text{thus } \nu_{H_2O} = 1, \nu_{H_2} = 1, \text{ and } \nu_{O_2} = \tfrac{1}{2})$$

Total number of moles: $N_{total} = x + (1-x) + (0.5-0.5x) = 1.5 - 0.5x$

The equilibrium constant relation can be expressed as

$$K_p = \frac{N_{H_2}^{\nu_{H_2}} N_{O_2}^{\nu_{O_2}}}{N_{H_2O}^{\nu_{H_2O}}} \left(\frac{P}{N_{total}}\right)^{(\nu_{H_2}+\nu_{O_2}-\nu_{H_2O})}$$

From Table A-28, $\ln K_P$ = -3.812 at 2800 K. Thus K_P = 0.02210. Substituting,

$$0.0221 = \frac{(1-x)(0.5-0.5x)^{0.5}}{x}\left(\frac{5}{1.5-0.5x}\right)^{1+0.5-1}$$

Solving for x, $x = 0.944$

Then the combustion equation and the equilibrium composition can be expressed as

$$H_2 + 0.5O_2 \longrightarrow 0.944H_2O + 0.056H_2 + 0.028O_2$$

and $\mathbf{0.944H_2O + 0.056H_2 + 0.028O_2}$

(*b*) The heat transfer can be determined from

$$-Q_{out} = \sum N_P\left(\bar{h}_f^\circ + \bar{h} - \bar{h}^\circ - P\bar{v}\right)_P - \sum N_R\left(\bar{h}_f^\circ + \bar{h} - \bar{h}^\circ - P\bar{v}\right)_R$$

Since $W = 0$ and both the reactants and the products are assumed to be ideal gases, all the internal energy and enthalpies depend on temperature only, and the $P\bar{v}$ terms in this equation can be replaced by R_uT. It yields

$$-Q_{out} = \sum N_P\left(\bar{h}_f^\circ + \bar{h}_{2800K} - \bar{h}_{298K} - R_uT\right)_P - \sum N_R\left(\bar{h}_f^\circ - R_uT\right)_R$$

since reactants are at the standard reference temperature of 25°C. From the tables,

Substance	$\bar{h}_f^\circ$ kJ/kmol	$\bar{h}_{298\,K}$ kJ/kmol	$\bar{h}_{2800\,K}$ kJ/kmol
H_2	0	8468	89,838
O_2	0	8682	98,826
H_2O	-241,820	9904	125,198

Substituting,

$$\begin{aligned}-Q_{out} &= 0.944(-241{,}820 + 125{,}198 - 9904 - 8.314 \times 2800)\\ &\quad + 0.056(0 + 89{,}838 - 8468 - 8.314 \times 2800)\\ &\quad + 0.028(0 + 98{,}826 - 8682 - 8.314 \times 2800)\\ &\quad - 1(0 - 8.314 \times 298) - 0.5(0 - 8.314 \times 298)\\ &= -132{,}574 \text{ kJ/kmol } H_2\end{aligned}$$

or $\mathbf{Q_{out} = 132{,}574 \text{ J/mol } H_2}$

The equilibrium constant for the reaction $H_2O \Leftrightarrow OH + \frac{1}{2}H_2$ is $\ln K_P$ = -3.763, which is very close to the K_P value of the reaction considered. Therefore, it is not realistic to assume that no OH will be present in equilibrium mixture.

15-78 A mixture of H_2O and O_2 is heated to a high temperature. The equilibrium composition is to be determined.

Assumptions **1** The equilibrium composition consists of H_2O, OH, O_2, and H_2. **2** The constituents of the mixture are ideal gases.

Analysis The reaction equation during this process can be expressed as

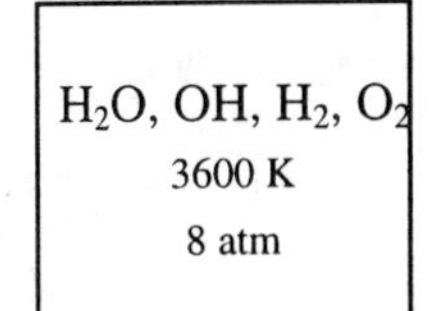

$$2H_2O + 3O_2 \longrightarrow xH_2O + yH_2 + zO_2 + wOH$$

Mass balances for hydrogen and oxygen yield

H balance: $4 = 2x + 2y + w$ (1)

O balance: $8 = x + 2z + w$ (2)

The mass balances provide us with only two equations with four unknowns, and thus we need to have two more equations (to be obtained from the K_P relations) to determine the equilibrium composition of the mixture. They are

$$H_2O \Leftrightarrow H_2 + \tfrac{1}{2}O_2 \quad \text{(reaction 1)}$$

$$H_2O \Leftrightarrow \tfrac{1}{2}H_2 + OH \quad \text{(reaction 2)}$$

The equilibrium constant for these two reactions at 3600 K are determined from Table A-28 to be

$$\ln K_{P1} = -1.392 \longrightarrow K_{P1} = 0.24858$$

$$\ln K_{P2} = -1.088 \longrightarrow K_{P2} = 0.33689$$

The K_P relations for these two simultaneous reactions are

$$K_{P1} = \frac{N_{H_2}^{\nu_{H2}} N_{O_2}^{\nu_{O2}}}{N_{H_2O}^{\nu_{H2O}}} \left(\frac{P}{N_{total}}\right)^{(\nu_{H_2} + \nu_{O_2} - \nu_{H_2O})}$$

$$K_{P2} = \frac{N_{H_2}^{\nu_{H2}} N_{OH}^{\nu_{OH}}}{N_{H_2O}^{\nu_{H2O}}} \left(\frac{P}{N_{total}}\right)^{(\nu_{H_2} + \nu_{OH} - \nu_{H_2O})}$$

where

$$N_{total} = N_{H_2O} + N_{H_2} + N_{O_2} + N_{OH} = x + y + z + w$$

Substituting,

$$0.24858 = \frac{(y)(z)^{1/2}}{x} \left(\frac{8}{x + y + z + w}\right)^{1/2} \quad (3)$$

$$0.33689 = \frac{(w)(y)^{1/2}}{x} \left(\frac{8}{x + y + z + w}\right)^{1/2} \quad (4)$$

Solving Eqs. (1), (2), (3), and (4) simultaneously for the four unknowns x, y, z, and w yields

$x = 1.371$ $y = 0.1646$ $z = 2.85$ $w = 0.928$

Therefore, the equilibrium composition becomes

$$\mathbf{1.371H_2O + 0.165H_2 + 2.85O_2 + 0.928OH}$$

15-79 A mixture of CO_2 and O_2 is heated to a high temperature. The equilibrium composition is to be determined.

Assumptions **1** The equilibrium composition consists of CO_2, CO, O_2, and O. **2** The constituents of the mixture are ideal gases.

Analysis The reaction equation during this process can be expressed as

$$3C_2O + 3O_2 \longrightarrow xCO_2 + yCO + zO_2 + wO$$

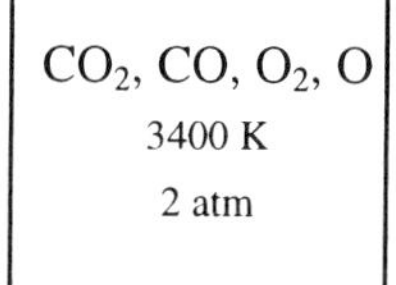

Mass balances for carbon and oxygen yield

C balance: $3 = x + y$ (1)

O balance: $12 = 2x + y + 2z + w$ (2)

The mass balances provide us with only two equations with four unknowns, and thus we need to have two more equations (to be obtained from the K_P relations) to determine the equilibrium composition of the mixture. They are

$$CO_2 \Leftrightarrow CO + \tfrac{1}{2}O_2 \quad \text{(reaction 1)}$$

$$O_2 \Leftrightarrow 2O \quad \text{(reaction 2)}$$

The equilibrium constant for these two reactions at 3400 K are determined from Table A-28 to be

$$\ln K_{P1} = 0.169 \longrightarrow K_{P1} = 1.1841$$

$$\ln K_{P2} = -1.935 \longrightarrow K_{P2} = 0.1444$$

The K_P relations for these two simultaneous reactions are

$$K_{P1} = \frac{N_{CO}^{\nu_{CO}} N_{O_2}^{\nu_{O_2}}}{N_{CO_2}^{\nu_{CO_2}}} \left(\frac{P}{N_{total}} \right)^{(\nu_{CO} + \nu_{O_2} - \nu_{CO_2})}$$

$$K_{P2} = \frac{N_O^{\nu_O}}{N_{O_2}^{\nu_{O_2}}} \left(\frac{P}{N_{total}} \right)^{\nu_O - \nu_{O_2}}$$

where

$$N_{total} = N_{CO_2} + N_{O_2} + N_{CO} + N_O = x + y + z + w$$

Substituting,

$$1.1841 = \frac{(y)(z)^{1/2}}{x} \left(\frac{2}{x + y + z + w} \right)^{1/2} \quad (3)$$

$$0.1444 = \frac{w^2}{z} \left(\frac{2}{x + y + z + w} \right)^{2-1} \quad (4)$$

Solving Eqs. (1), (2), (3), and (4) simultaneously for the four unknowns x, y, z, and w yields

$x = 1.313$ $y = 1.687$ $z = 3.187$ $w = 1.314$

Thus the equilibrium composition is

$$\mathbf{1.313CO_2 + 1.687CO + 3.187O_2 + 1.314O}$$

15-80 EES solution of this (and other comprehensive problems designated with the *computer icon*) is available to instructors at the *Instructor Manual* section of the *Online Learning Center* (OLC) at www.mhhe.com/cengel-boles. See the Preface for access information.

15-81 The $\overline{h}_R$ at a specified temperature is to be determined using enthalpy and K_p data.

Assumptions Both the reactants and products are ideal gases.

Analysis (*a*) The complete combustion equation of H_2 can be expressed as

$$H_2 + \tfrac{1}{2}O_2 \Leftrightarrow H_2O$$

The $\overline{h}_R$ of the combustion process of H_2 at 2400 K is the amount of energy released as one kmol of H_2 is burned in a steady-flow combustion chamber at a temperature of 2400 K, and can be determined from

$$\overline{h}_R = \sum N_P\left(\overline{h}_f^\circ + \overline{h} - \overline{h}^\circ\right)_P - \sum N_R\left(\overline{h}_f^\circ + \overline{h} - \overline{h}^\circ\right)_R$$

Assuming the H_2O, H_2, and O_2 to be ideal gases, we have $h = h(T)$. From the tables,

Substance	$\overline{h}_f^\circ$ kJ/kmol	$\overline{h}_{298\text{ K}}$ kJ/kmol	$\overline{h}_{2400\text{ K}}$ kJ/kmol
H_2O	-241,820	9904	103,508
H_2	0	8468	75,383
O_2	0	8682	83,174

Substituting,

$$\begin{aligned}\overline{h}_R &= 1(-241{,}820 + 103{,}508 - 9904) \\ &\quad - 1(0 + 75{,}383 - 8468) \\ &\quad - 0.5(0 + 83{,}174 - 8682) \\ &= \mathbf{-252{,}377\ kJ/kmol}\end{aligned}$$

(*b*) The $\overline{h}_R$ value at 2400 K can be estimated by using K_P values at 2200 K and 2600 K (the closest two temperatures to 2400 K for which K_P data are available) from Table A-28,

$$\ln\frac{K_{P2}}{K_{P1}} \cong \frac{\overline{h}_R}{R_u}\left(\frac{1}{T_1} - \frac{1}{T_2}\right) \quad \text{or} \quad \ln K_{P2} - \ln K_{P1} \cong \frac{\overline{h}_R}{R_u}\left(\frac{1}{T_1} - \frac{1}{T_2}\right)$$

$$4.648 - 6.768 \cong \frac{\overline{h}_R}{8.314\text{ kJ/kmol}\cdot\text{K}}\left(\frac{1}{2200\text{ K}} - \frac{1}{2600\text{ K}}\right)$$

$$\overline{h}_R \cong \mathbf{-252{,}047\ kJ/kmol}$$

15-82 EES solution of this (and other comprehensive problems designated with the *computer icon*) is available to instructors at the *Instructor Manual* section of the *Online Learning Center* (OLC) at www.mhhe.com/cengel-boles. See the Preface for access information.

15-83 The K_P value of the dissociation process $O_2 \Leftrightarrow 2O$ at a specified temperature is to be determined using the $\bar{h}_R$ data and K_P value at a specified temperature.

Assumptions Both the reactants and products are ideal gases.

Analysis The $\bar{h}_R$ and K_P data are related to each other by

$$\ln \frac{K_{P2}}{K_{P1}} \cong \frac{\bar{h}_R}{R_u}\left(\frac{1}{T_1} - \frac{1}{T_2}\right) \quad \text{or} \quad \ln K_{P2} - \ln K_{P1} \cong \frac{\bar{h}_R}{R_u}\left(\frac{1}{T_1} - \frac{1}{T_2}\right)$$

The $\bar{h}_R$ of the specified reaction at 2800 K is the amount of energy released as one kmol of O_2 dissociates in a steady-flow combustion chamber at a temperature of 2800 K, and can be determined from

$$\bar{h}_R = \sum N_P\left(\bar{h}_f^\circ + \bar{h} - \bar{h}^\circ\right)_P - \sum N_R\left(\bar{h}_f^\circ + \bar{h} - \bar{h}^\circ\right)_R$$

Assuming the O_2 and O to be ideal gases, we have $h = h(T)$. From the tables,

Substance	$\bar{h}_f^\circ$ kJ/kmol	$\bar{h}_{298\text{ K}}$ kJ/kmol	$\bar{h}_{2800\text{ K}}$ kJ/kmol
O	249,190	6852	59,241
O_2	0	8682	98,826

Substituting,

$$\begin{aligned}\bar{h}_R &= 2(249{,}190 + 59{,}241 - 6852) - 1(0 + 98{,}826 - 8682)\\ &= 513{,}014 \text{ kJ/kmol}\end{aligned}$$

The K_P value at 3000 K can be estimated from the equation above by using this $\bar{h}_R$ value and the K_P value at 2600 K which is $\ln K_{P1}$ =-7.521,

$$\ln K_{P2} - (-7.521) = \frac{513{,}014 \text{ kJ/kmol}}{8.314 \text{ kJ/kmol}\cdot\text{K}}\left(\frac{1}{2600 \text{ K}} - \frac{1}{3000 \text{ K}}\right)$$

$$\ln K_{P2} = -4.357 \quad (\text{Table A-28: } \ln K_{P2} = -4.357)$$

or

$$K_{P2} = \mathbf{0.0128}$$

15-84 It is to be shown that when the three phases of a pure substance are in equilibrium, the specific Gibbs function of each phase is the same.

Analysis The total Gibbs function of the three phase mixture of a pure substance can be expressed as

$$G = m_s g_s + m_\ell g_\ell + m_s g_s$$

where the subscripts s, ℓ, and g indicate solid, liquid and gaseous phases. Differentiating by holding the temperature and pressure (thus the Gibbs functions, g) constant yields

$$dG = g_s dm_s + g_\ell dm_\ell + g_s dm_s$$

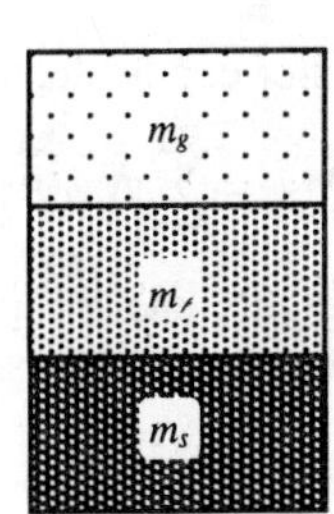

From conservation of mass,

$$dm_s + dm_\ell + dm_g = 0 \longrightarrow dm_s = -dm_\ell - dm_g$$

Substituting,

$$dG = -g_s(dm_\ell + dm_g) + g_\ell dm_\ell + g_g dm_g$$

Rearranging,

$$dG = (g_\ell - g_s)dm_\ell + (g_g - g_s)dm_g$$

For equilibrium, $dG = 0$. Also dm_ℓ and dm_g can be varied independently. Thus each term on the right hand side must be zero to satisfy the equilibrium criteria. It yields

$$g_\ell = g_s \text{ and } g_g = g_s$$

Combining these two conditions gives the desired result,

$$g_\ell = g_s = g_s$$

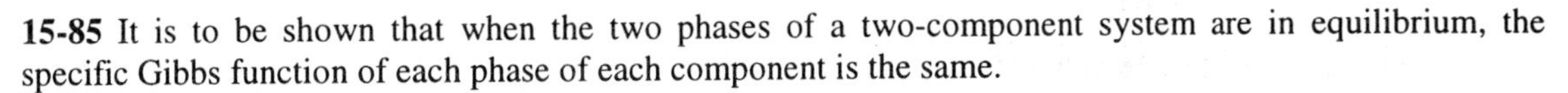

15-85 It is to be shown that when the two phases of a two-component system are in equilibrium, the specific Gibbs function of each phase of each component is the same.

Analysis The total Gibbs function of the two phase mixture can be expressed as

$$G = (m_{\ell 1} g_{\ell 1} + m_{g1} g_{g1}) + (m_{\ell 2} g_{\ell 2} + m_{g2} g_{g2})$$

where the subscripts ℓ and g indicate liquid and gaseous phases. Differentiating by holding the temperature and pressure (thus the Gibbs functions) constant yields

$$dG = g_{\ell 1} dm_{\ell 1} + g_{g1} dm_{g1} + g_{\ell 2} dm_{\ell 2} + g_{g2} dm_{g2}$$

From conservation of mass,

$$dm_{g1} = -dm_{\ell 1} \text{ and } dm_{g2} = -dm_{\ell 2}$$

Substituting,

$$dG = (g_{\ell 1} - g_{g1})dm_{\ell 1} + (g_{\ell 2} - g_{g2})dm_{\ell 2}$$

For equilibrium, $dG = 0$. Also $dm_{\ell 1}$ and $dm_{\ell 2}$ can be varied independently. Thus each term on the right hand side must be zero to satisfy the equilibrium criteria. Then we have

$$g_{\ell 1} = g_{g1} \text{ and } g_{\ell 2} = g_{g2}$$

which is the desired result.

15-86 A mixture of CO and O_2 contained in a tank is ignited. The final pressure in the tank and the amount of heat transfer are to be determined.

Assumptions **1** The equilibrium composition consists of CO_2 and O_2. **2** Both the reactants and the products are ideal gases.

Analysis The combustion equation can be written as

CO_2, CO, O_2
25°C
3 atm

$$CO + 3O_2 \longrightarrow CO_2 + 2.5O_2$$

The heat transfer can be determined from

$$-Q_{out} = \sum N_P \left(\bar{h}_f^\circ + \bar{h} - \bar{h}^\circ - P\bar{v}\right)_P - \sum N_R \left(\bar{h}_f^\circ + \bar{h} - \bar{h}^\circ - P\bar{v}\right)_R$$

Both the reactants and the products are assumed to be ideal gases, and thus all the internal energy and enthalpies depend on temperature only, and the $P\bar{v}$ terms in this equation can be replaced by $R_u T$. It yields

$$-Q_{out} = \sum N_P \left(\bar{h}_f^\circ + \bar{h}_{500K} - \bar{h}_{298K} - R_u T\right)_P - \sum N_R \left(\bar{h}_f^\circ - R_u T\right)_R$$

since reactants are at the standard reference temperature of 25°C. From the tables,

Substance	$\bar{h}_f^\circ$ kJ/kmol	$\bar{h}_{298\text{ K}}$ kJ/kmol	$\bar{h}_{500\text{ K}}$ kJ/kmol
CO	-110,530	8669	14,600
O_2	0	8682	14,770
CO_2	-393,520	9364	17,678

Substituting,

$$\begin{aligned}-Q_{out} &= 1(-393{,}520 + 17{,}678 - 9364 - 8.314 \times 500) \\ &\quad + 2.5(0 + 14{,}770 - 8682 - 8.314 \times 500) \\ &\quad - 3(0) - 1(-110{,}530 - 8.314 \times 298) \\ &= \mathbf{-271{,}528\ kJ/kmol\ CO}\end{aligned}$$

or $\quad Q_{out} = \mathbf{271{,}528\ kJ/kmol\ CO}$

The final pressure in the tank is determined from

$$\frac{P_1 V}{P_2 V} = \frac{N_1 R_u T_1}{N_2 R_u T_2} \longrightarrow P_2 = \frac{N_2 T_2}{N_1 T_1} P_1 = \frac{3.5}{4} \times \frac{500\text{ K}}{298\text{ K}}(3\text{ atm}) = \mathbf{4.40\ atm}$$

The equilibrium constant for the reaction $CO + \frac{1}{2}O_2 \Leftrightarrow CO_2$ is $\ln K_P = 57.62$, which is much greater than 7. Therefore, it is not realistic to assume that no CO will be present in equilibrium mixture.

15-87 Using Henry's law, it is to be shown that the dissolved gases in a liquid can be driven off by heating the liquid.

Analysis Henry's law is expressed as

$$y_{i,\text{ liquid side}}(0) = \frac{P_{i,\text{ gas side}}(0)}{H}$$

Henry's constant H increases with temperature, and thus the fraction of gas i in the liquid $y_{i,\text{ liquid side}}$ decreases. Therefore, heating a liquid will drive off the dissolved gases in a liquid.

15-88 A glass of water is left in a room. The mole fraction of the water vapor in the air at the water surface and far from the surface as well as the mole fraction of air in the water near the surface are to be determined when the water and the air are at the same temperature.

Assumptions **1** Both the air and water vapor are ideal gases. **2** Air is weakly soluble in water and thus Henry's law is applicable.

Properties The saturation pressure of water at 25°C is 3.169 kPa (Table A-9). Henry's constant for air dissolved in water at 25°C (298 K) is given in Table 15-2 to be H = 71,600 bar. Molar masses of dry air and water are 29 and 18 kg/kmol, respectively (Table A-1).

Analysis (*a*) Noting that the relative humidity of air is 70%, the partial pressure of water vapor in the air far from the water surface will be

$$P_{v,\text{room air}} = \phi P_{\text{sat @ 25°C}} = (0.7)(3.169\text{ kPa}) = 2.218\text{ kPa}$$

Air-water interface

Water 25°C

Assuming both the air and vapor to be ideal gases, the mole fraction of water vapor in the room air is

$$y_{\text{vapor}} = \frac{P_{\text{vapor}}}{P} = \frac{2.218\text{ kPa}}{100\text{ kPa}} = \mathbf{0.0222} \quad \textbf{(or 2.22\%)}$$

(*b*) Noting that air at the water surface is saturated, the partial pressure of water vapor in the air near the surface will simply be the saturation pressure of water at 25°C, $P_{v,\text{interface}} = P_{\text{sat @ 25°C}} = 3.169\text{ kPa}$. Then the mole fraction of water vapor in the air at the interface becomes

$$y_{v,\text{ surface}} = \frac{P_{v,\text{ surface}}}{P} = \frac{3.169\text{ kPa}}{100\text{ kPa}} = \mathbf{0.0317} \quad \textbf{(or 3.17\%)}$$

(*c*) Noting that the total pressure is 100 kPa, the partial pressure of dry air at the water surface is

$$P_{\text{air, surface}} = P - P_{v,\text{ surface}} = 100 - 3.169 = 96.831\text{ kPa}$$

From Henry's law, the mole fraction of air in the water is determined to be

$$y_{\text{dry air,liquid side}} = \frac{P_{\text{dry air,gas side}}}{H} = \frac{(96.831/100)\text{ bar}}{71{,}600\text{ bar}} = \mathbf{1.35 \times 10^{-5}}$$

Discussion The water cannot remain at the room temperature when the air is not saturated. Therefore, some water will evaporate and the water temperature will drop until a balance is reached between the rate of heat transfer to the water and the rate of evaporation.

15-89 A glass of water is left in a room. The mole fraction of the water vapor in the air at the water surface and far from the surface as well as the mole fraction of air in the water near the surface are to be determined when the water and the air are at the same temperature.

Assumptions **1** Both the air and water vapor are ideal gases. **2** Air is weakly soluble in water and thus Henry's law is applicable.

Properties The saturation pressure of water at 25°C is 3.169 kPa (Table A-9). Henry's constant for air dissolved in water at 25°C (298 K) is given in Table 15-2 to be H = 71,600 bar. Molar masses of dry air and water are 29 and 18 kg/kmol, respectively (Table A-1).

Analysis (*a*) Noting that the relative humidity of air is 40%, the partial pressure of water vapor in the air far from the water surface will be

$$P_{v,\text{room air}} = \phi P_{\text{sat @ 25°C}} = (0.4)(3.169\text{ kPa}) = 1.268\text{ kPa}$$

Assuming both the air and vapor to be ideal gases, the mole fraction of water vapor in the room air is

$$y_{\text{vapor}} = \frac{P_{\text{vapor}}}{P} = \frac{1.268\text{ kPa}}{100\text{ kPa}} = \mathbf{0.0127} \quad \textbf{(or 1.27\%)}$$

(*b*) Noting that air at the water surface is saturated, the partial pressure of water vapor in the air near the surface will simply be the saturation pressure of water at 25°C, $P_{v,\text{interface}} = P_{\text{sat @ 25°C}} = 3.169\text{ kPa}$. Then the mole fraction of water vapor in the air at the interface becomes

$$y_{v,\text{ surface}} = \frac{P_{v,\text{ surface}}}{P} = \frac{3.169\text{ kPa}}{100\text{ kPa}} = \mathbf{0.0317} \quad \textbf{(or 3.17\%)}$$

(*c*) Noting that the total pressure is 100 kPa, the partial pressure of dry air at the water surface is

$$P_{\text{air, surface}} = P - P_{v,\text{ surface}} = 100 - 3.169 = 96.831\text{ kPa}$$

From Henry's law, the mole fraction of air in the water is determined to be

$$y_{\text{dry air,liquid side}} = \frac{P_{\text{dry air,gas side}}}{H} = \frac{(96.831/100)\text{ bar}}{71{,}600\text{ bar}} = \mathbf{1.35 \times 10^{-5}}$$

Discussion The water cannot remain at the room temperature when the air is not saturated. Therefore, some water will evaporate and the water temperature will drop until a balance is reached between the rate of heat transfer to the water and the rate of evaporation.

15-90 A 2-L bottle is filled with carbonated drink that is fully charged (saturated) with CO_2 gas. The volume that the CO_2 gas would occupy if it is released and stored in a container at room conditions is to be determined.

Assumptions **1** The liquid drink can be treated as water. **2** Both the CO_2 gas and the water vapor are ideal gases. **3** The CO_2 gas is weakly soluble in water and thus Henry's law is applicable.

Properties The saturation pressure of water at 17°C is 1.96 kPa (Table A-4). Henry's constant for CO_2 dissolved in water at 17°C (290 K) is H = 1280 bar (Table 15-2). Molar masses of CO_2 and water are 44.01 and 18.015 kg/kmol, respectively (Table A-1). The gas constant of CO_2 is 0.1889 kPa.m^3/kg.K. Also, 1 bar = 100 kPa.

Analysis In the charging station, the CO_2 gas and water vapor mixture above the liquid will form a saturated mixture. Noting that the saturation pressure of water at 17°C is 1.96 kPa, the partial pressure of the CO_2 gas is

$$P_{CO_2,\text{ gas side}} = P - P_{\text{vapor}} = P - P_{\text{sat @ 17°C}} = 600 - 1.96 = 598.04 \text{ kPa} = 5.9804 \text{ bar}$$

From Henry's law, the mole fraction of CO_2 in the liquid drink is determined to be

$$y_{CO_2,\text{liquid side}} = \frac{P_{CO_2,\text{gas side}}}{H} = \frac{5.9804 \text{ bar}}{1280 \text{ bar}} = 0.00467$$

Then the mole fraction of water in the drink becomes

$$y_{\text{water, liquid side}} = 1 - y_{CO_2,\text{ liquid side}} = 1 - 0.00467 = 0.99533$$

The mass and mole fractions of a mixture are related to each other by

$$w_i = \frac{m_i}{m_m} = \frac{N_i M_i}{N_m M_m} = y_i \frac{M_i}{M_m}$$

where the apparent molar mass of the drink (liquid water - CO_2 mixture) is

$$M_m = \sum y_i M_i = y_{\text{liquid water}} M_{\text{water}} + y_{CO_2} M_{CO_2}$$
$$= 0.99533 \times 18.015 + 0.00467 \times 44.01 = 18.14 \text{ kg/kmol}$$

Then the mass fraction of dissolved CO_2 in liquid drink becomes

$$w_{CO_2,\text{ liquid side}} = y_{CO_2,\text{ liquid side}}(0) \frac{M_{CO_2}}{M_m} = 0.00467 \frac{44.01}{18.14} = 0.0113$$

Therefore, the mass of dissolved CO_2 in a 2 L ≈ 2 kg drink is

$$m_{CO_2} = w_{CO_2} m_m = 0.0113(2 \text{ kg}) = 0.0226 \text{ kg}$$

Then the volume occupied by this CO_2 at the room conditions of 25°C and 100 kPa becomes

$$V = \frac{mRT}{P} = \frac{(0.0226 \text{ kg})(0.1889 \text{ kPa}\cdot\text{m}^3/\text{kg}\cdot\text{K})(298 \text{ K})}{100 \text{ kPa}} = \mathbf{0.0127 \text{ m}^3 = 12.7 \text{ L}}$$

Discussion Note that the amount of dissolved CO_2 in a 2-L pressurized drink is large enough to fill 6 such bottles at room temperature and pressure. Also, we could simplify the calculations by assuming the molar mass of carbonated drink to be the same as that of water, and take it to be 18 kg/kmol because of the very low mole fraction of CO_2 in the drink.

Fundamentals of Engineering (FE) Exam Problems

15-91 If the equilibrium constant for the reaction H_2 + ½O_2 → H_2O is K, the equilibrium constant for the reaction $2H_2O$ → $2H_2$ + O_2 at the same temperature is

(a) $1/K$ (b) $1/(2K)$ (c) $2K$ (d) K^2 (e) $1/K^2$

Answer (e) $1/K^2$

15-92 If the equilibrium constant for the reaction CO + ½O_2 → CO_2 is K, the equilibrium constant for the reaction CO_2 + $3N_2$ → CO + ½O_2 + $3N_2$ at the same temperature is

(a) $1/K$ (b) $1/(K+3)$ (c) $4K$ (d) K (e) $1/K^2$

Answer (a) $1/K$

15-93 The equilibrium constant for the reaction H_2 + ½O_2 → H_2O at 1 atm and 1500°C is given to be K. Of the reactions given below, all at 1500°C, the reaction that has a different equilibrium constant is

(a) H_2 + ½O_2 → H_2O at 5 atm,

(b) $2H_2$ + O_2 → $2H_2O$ at 1 atm,

(c) H_2 + O_2 → H_2O+ ½O_2 at 2 atm,

(d) H_2 + ½O_2 + $3N_2$ → H_2O+ $3N_2$ at 5 atm,

(e) H_2 + ½O_2 + $3N_2$ → H_2O+ $3N_2$ at 1 atm,

Answer (b) $2H_2$ + O_2 → $2H_2O$ at 1 atm,

15-94 Of the reactions given below, the reaction whose equilibrium composition at a specified temperature is not affected by pressure is

(a) H_2 + ½O_2 → H_2O

(b) CO + ½O_2 → CO_2

(c) N_2 + O_2 → 2NO

(d) N_2 → 2N

(e) all of the above.

Answer (c) N_2 + O_2 → 2NO

15-95 Of the reactions given below, the reaction whose number of moles of products increases by the addition of inert gases into the reaction chamber at constant pressure and temperature is

(a) $H_2 + ½O_2 \rightarrow H_2O$

(b) $CO + ½O_2 \rightarrow CO_2$

(c) $N_2 + O_2 \rightarrow 2NO$

(d) $N_2 \rightarrow 2N$

(e) none of the above.

Answer (d) $N_2 \rightarrow 2N$

15-96 Moist air is heated to a very high temperature. If the equilibrium composition consists of H_2O, O_2, N_2, OH, H_2, and NO, the number of equilibrium constant relations needed to determine the equilibrium composition of the mixture is

(a) 1 (b) 2 (c) 3 (d) 4 (e) 5

Answer (c) 3

15-97 Propane C_3H_8 is burned with air, and the combustion products consist of CO_2, CO, H_2O, O_2, N_2, OH, H_2, and NO. The number of equilibrium constant relations needed to determine the equilibrium composition of the mixture is

(a) 1 (b) 2 (c) 3 (d) 4 (e) 5

Answer (d) 4

15-98 Consider a gas mixture that consists of two components. The number of independent variables that need to be specified to fix the state of the mixture is

(a) 1 (b) 2 (c) 3 (d) 4 (e) 5

Answer (c) 3

15-99 The value of Henry's constant for CO_2 gas dissolved in water at 290 K is 12.8 MPa. Consider water exposed to atmospheric air at 100 kPa that contains 2 percent CO_2 by volume. Under phase equilibrium conditions, the mole fraction of CO_2 gas dissolved in water at 290 K is

(a) 1.6×10^{-4} (b) 3.2×10^{-4} (c) 0.80×10^{-4} (d) 2.2×10^{-4} (e) 5.6×10^{-4}

Answer (a) 1.6×10^{-4}

Solution Solved by EES Software. Solutions can be verified by copying-and-pasting the following lines on a blank EES screen. (Similar problems and their solutions can be obtained easily by modifying numerical values).

```
H=12.8 "MPa"
P=0.1 "MPa"
y_CO2_air=0.02
P_CO2_air=y_CO2_air*P
y_CO2_liquid=P_CO2_air/H

"Some Wrong Solutions with Common Mistakes:"
W1_yCO2=P_CO2_air*H "Multiplying by H instead of dividing by it"
W2_yCO2=P_CO2_air "Taking partial pressure in air"
```

15-100 The solubility of nitrogen gas in rubber at 25°C is 0.00156 kmol/m^3·bar. When phase equilibrium is established, the density of nitrogen in a rubber piece placed in a nitrogen gas chamber at 500 kPa is

(a) 0.011 kg/m^3 (b) 0.22 kg/m^3 (c) 0.32 kg/m^3 (d) 0.56 kg/m^3 (e) 0.78 kg/m^3

Answer (b) 0.22 kg/m^3

Solution Solved by EES Software. Solutions can be verified by copying-and-pasting the following lines on a blank EES screen. (Similar problems and their solutions can be obtained easily by modifying numerical values).

```
T=25 "C"
S=0.00156 "kmol/bar.m^3"
MM_N2=28 "kg/kmol"
S_mass=S*MM_N2 "kg/bar.m^3"
P_N2=5 "bar"
rho_solid=S_mass*P_N2

"Some Wrong Solutions with Common Mistakes:"
W1_density=S*P_N2 "Using solubility per kmol"
```

15-101 and 15-102 Design and Essay Problems

Chapter 16
THERMODYNAMICS OF HIGH-SPEED GAS FLOW

Stagnation Properties

16-1C The temperature of the air will rise as it approaches the nozzle because of the stagnation process.

16-2C Stagnation enthalpy combines the ordinary enthalpy and the kinetic energy of a fluid, and offers convenience when analyzing high-speed flows. It differs from the ordinary enthalpy by the kinetic energy term.

16-3C Dynamic temperature is the temperature rise of a fluid during a stagnation process.

16-4C No. Because the velocities encountered in air-conditioning applications are very low, and thus the static and the stagnation temperatures are practically identical.

16-5 The state of air and its velocity are specified. The stagnation temperature and stagnation pressure of air are to be determined.

Assumptions **1** The stagnation process is isentropic. **2** Air is an ideal gas.

Properties The properties of air at room temperature are C_p = 1.005 kJ/kg.°C and k = 1.4 (Table A-2a).

Analysis The stagnation temperature of air is determined from

$$T_0 = T + \frac{\mathbf{V}^2}{2C_p} = 245.9\text{ K} + \frac{(470\text{ m/s})^2}{2\times 1.005\text{ kJ/kg}\cdot\text{K}}\left(\frac{1\text{ kJ/kg}}{1000\text{ m}^2/\text{s}^2}\right) = \mathbf{355.8\text{ K}}$$

Other stagnation properties at the specified state are determined by considering an isentropic process between the specified state and the stagnation state,

$$P_0 = P\left(\frac{T_0}{T}\right)^{k/(k-1)} = (44\text{ kPa})\left(\frac{355.8\text{ K}}{245.9\text{ K}}\right)^{1.4/(1.4-1)} = \mathbf{160.3\text{ kPa}}$$

16-6 Air at 300 K is flowing in a duct. The temperature that a stationary probe inserted into the duct will read is to be determined for different air velocities.

Assumptions The stagnation process is isentropic.

Properties The specific heat of air at room temperature is C_p = 1.005 kJ/kg.°C (Table A-2a).

Analysis The air which strikes the probe will be brought to a complete stop, and thus it will undergo a stagnation process. The thermometer will sense the temperature of this stagnated air, which is the stagnation temperature, T_0. It is determined from

$$T_0 = T + \frac{\mathbf{V}^2}{2C_p}$$

AIR
300 K
V

(*a*) $$T_0 = 300\text{ K} + \frac{(1\text{ m/s})^2}{2\times 1.005\text{ kJ/kg}\cdot\text{K}}\left(\frac{1\text{ kJ/kg}}{1000\text{ m}^2/\text{s}^2}\right) = \mathbf{300.00\ K}$$

(*b*) $$T_0 = 300\text{ K} + \frac{(10\text{ m/s})^2}{2\times 1.005\text{ kJ/kg}\cdot\text{K}}\left(\frac{1\text{ kJ/kg}}{1000\text{ m}^2/\text{s}^2}\right) = \mathbf{300.05\ K}$$

(*c*) $$T_0 = 300\text{ K} + \frac{(100\text{ m/s})^2}{2\times 1.005\text{ kJ/kg}\cdot\text{K}}\left(\frac{1\text{ kJ/kg}}{1000\text{ m}^2/\text{s}^2}\right) = \mathbf{304.98\ K}$$

(*d*) $$T_0 = 300\text{ K} + \frac{(1000\text{ m/s})^2}{2\times 1.005\text{ kJ/kg}\cdot\text{K}}\left(\frac{1\text{ kJ/kg}}{1000\text{ m}^2/\text{s}^2}\right) = \mathbf{797.51\ K}$$

16-7 The states of different substances and their velocities are specified. The stagnation temperature and stagnation pressures are to be determined.

Assumptions **1** The stagnation process is isentropic. **2** Helium and nitrogen are ideal gases.

Analysis (*a*) Helium can be treated as an ideal gas with C_p = 5.1926 kJ/kg °C and k = 1.667. Then the stagnation temperature and pressure of helium are determined from

$$T_0 = T + \frac{\mathbf{V}^2}{2C_p} = 50°\text{C} + \frac{(240\text{ m/s})^2}{2\times 5.1926\text{ kJ/kg}\cdot°\text{C}}\left(\frac{1\text{ kJ/kg}}{1000\text{ m}^2/\text{s}^2}\right) = \mathbf{55.5°C}$$

$$P_0 = P\left(\frac{T_0}{T}\right)^{k/(k-1)} = (0.25\text{ MPa})\left(\frac{328.7\text{ K}}{323.2\text{ K}}\right)^{1.667/(1.667-1)} = \mathbf{0.261\ MPa}$$

(*b*) Nitrogen can be treated as an ideal gas with C_p = 1.039 kJ/kg °C and k =1.400. Then the stagnation temperature and pressure of nitrogen are determined from

$$T_0 = T + \frac{\mathbf{V}^2}{2C_p} = 50°\text{C} + \frac{(300\text{ m/s})^2}{2\times 1.039\text{ kJ/kg}\cdot°\text{C}}\left(\frac{1\text{ kJ/kg}}{1000\text{ m}^2/\text{s}^2}\right) = \mathbf{93.3°C}$$

$$P_0 = P\left(\frac{T_0}{T}\right)^{k/(k-1)} = (0.15\text{ MPa})\left(\frac{366.5\text{ K}}{323.2\text{ K}}\right)^{1.4/(1.4-1)} = \mathbf{0.233\ MPa}$$

(*c*) The enthalpy and entropy of the steam at the specified state are h = 3137 kJ/kg and s = 6.9563 kJ/kg K. Then the stagnation enthalpy of the steam is

$$h_0 = h + \frac{\mathbf{V}^2}{2} = (3137.0\text{ kJ/kg}) + \frac{(480\text{ m/s})^2}{2}\left(\frac{1\text{ kJ/kg}}{1000\text{ m}^2/\text{s}^2}\right) = \mathbf{3252.2\ kJ/kg}$$

Assuming the stagnation process to be isentropic, the entropy of the steam at the stagnation state will be $s_0 = s$ = 6.9563. Thus the stagnation state is fixed, and the T and P at this state are, from Table A-6,

T_0 = **409.4°C** and P_0 = **2.975 MPa**

16-8 The inlet stagnation temperature and pressure and the exit stagnation pressure of air flowing through a compressor are specified. The power input to the compressor is to be determined.

Assumptions **1** The compressor is isentropic. **2** Air is an ideal gas.

Properties The properties of air at room temperature are C_p = 1.005 kJ/kg.°C and k = 1.4 (Table A-2a)

Analysis The exit stagnation temperature of air T_{02} is determined from

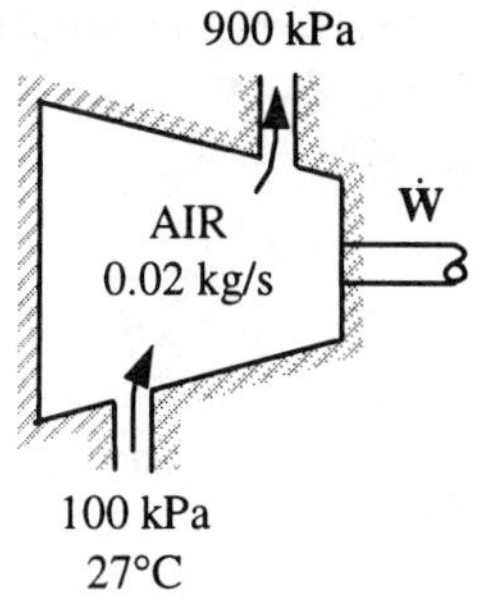

900 kPa

$$T_{02} = T_{01}\left(\frac{P_{02}}{P_{01}}\right)^{(k-1)/k} = (300.2\text{ K})\left(\frac{900}{100}\right)^{(1.4-1)/1.4} = 562.4\text{ K}$$

AIR
0.02 kg/s

$\dot{W}$

From the energy balance on the compressor,

$$\dot{W}_{\text{in}} = \dot{m}(h_{20} - h_{01})$$

100 kPa
27°C

or,

$$\dot{W}_{\text{in}} = \dot{m}C_p(T_{02} - T_{01}) = (0.02\text{ kg/s})(1.005\text{ kJ/kg}\cdot\text{K})(562.4 - 300.2)\text{K} = \mathbf{5.27\text{ kW}}$$

16-9E Steam flows through a device. The stagnation temperature and pressure of steam and its velocity are specified. The static pressure and temperature of the steam are to be determined.

Assumptions **1** The stagnation process is isentropic.

Properties The enthalpy and entropy of the steam at the specified stagnation state are h_0 = 1378.2 Btu/1bm and s_0 = 1.7825 Btu/1bm.R.

Analysis The static enthalpy of the steam becomes

$$h = h_0 - \frac{\mathbf{V}^2}{2} = (1378.2\text{ Btu/1bm}) - \frac{(900\text{ ft/s})^2}{2}\left(\frac{1\text{ Btu/1bm}}{25{,}037\text{ ft}^2/\text{s}^2}\right) = 1362.0\text{ Btu/1bm}$$

Assuming the stagnation process to be isentropic, the entropy of the steam at the static will be $s = s_0$ = 1.7825. Thus the static state fixed. Then the temperature and pressure at this state are determined from Table A-6E to be

T = **665.9°F** and P = **105.2 psia**

16-10 The inlet stagnation temperature and pressure and the exit stagnation pressure of products of combustion flowing through a gas turbine are specified. The power output of the turbine is to be determined.

Assumptions **1** The expansion process is isentropic. **2** Products of combustion are ideal gases.

Properties The properties of products of combustion are C_p = 1.157 kJ/kg.K, R = 0.287 kJ/kg.K, and k = 1.33.

Analysis The exit stagnation temperature T_{02} is determined to be

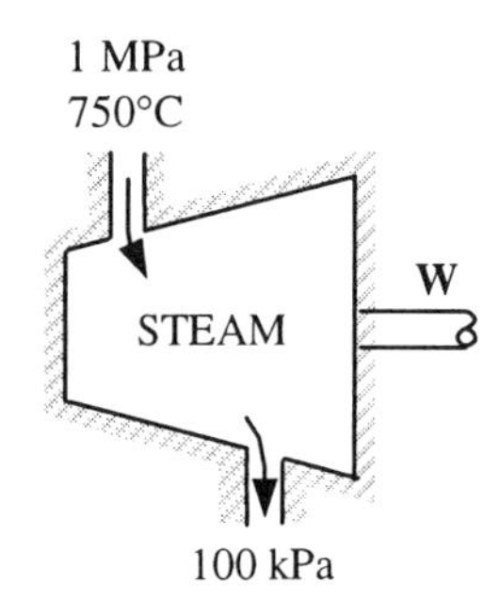

$$T_{02} = T_{01}\left(\frac{P_{02}}{P_{01}}\right)^{(k-1)/k} = (1023.2\text{ K})\left(\frac{0.1}{1}\right)^{(1.33-1)/1.33} = 577.9\text{ K}$$

Also,

$$C_p = kC_v = k(C_p - R) \longrightarrow C_p = \frac{kR}{k-1} = \frac{1.33(0.287\text{ kJ/kg}\cdot\text{K})}{1.33-1} = 1.157\text{ kJ/kg}\cdot\text{K}$$

From the energy balance on the turbine,

$$-w_{out} = (h_{20} - h_{01})$$

or,

$$w_{out} = C_p(T_{01} - T_{02}) = (1.157\text{ kJ/kg}\cdot\text{K})(1023.2 - 577.9)\text{ K} = \mathbf{515.2\text{ kJ/kg}}$$

16-11 Air flows through a device. The stagnation temperature and pressure of air and its velocity are specified. The static pressure and temperature of air are to be determined.

Assumptions **1** The stagnation process is isentropic. **2** Air is an ideal gas.

Properties The properties of air at an anticipated average temperature of 600 K are C_p = 1.051 kJ/kg.K and k = 1.376 (Table A-2b).

Analysis The static temperature and pressure of air are determined from

$$T = T_0 - \frac{\mathbf{V}^2}{2C_p} = 673.2 - \frac{(570\text{ m/s})^2}{2\times 1.051\text{ kJ/kg}\cdot\text{K}}\left(\frac{1\text{ kJ/kg}}{1000\text{ m}^2/\text{s}^2}\right) = \mathbf{518.6\text{ K}}$$

and

$$P_2 = P_{02}\left(\frac{T_2}{T_{02}}\right)^{k/(k-1)} = (0.6\text{ MPa})\left(\frac{518.6\text{ K}}{673.2\text{ K}}\right)^{1.376/(1.376-1)} = \mathbf{0.23\text{ MPa}}$$

Velocity of Sound and Mach Number

16-12C Sound is an infinitesimally small pressure wave. It is generated by a small disturbance in a medium. It travels by wave propagation. Sound waves cannot travel in a vacuum.

16-13C Yes, it is. Because the amplitude of ordinary sound wave is very small, and it does not cause any significant change in temperature and pressure.

16-14C The sonic velocity in a medium depends on the properties of the medium, and it changes as the properties of the medium changes.

16-15C In warm (higher temperature) air since $C = \sqrt{kRT}$

16-16C Helium, since $C = \sqrt{kRT}$ and helium has the highest kR value. It is 0.4 for air, 0.35 for argon and 3.46 for helium.

16-17C Air at specified conditions will behave like an ideal gas, and the speed of sound in an ideal gas depends on temperature only. Therefore, the speed of sound will be the same in both mediums.

16-18C In general, no. Because the Mach number also depends on the speed of sound in gas, which depends on the temperature of the gas. The Mach number will remain constant if the temperature is maintained constant.

16-19 The Mach number of an aircraft and the velocity of sound in air are to be determined at two specified temperatures.

Assumptions Air is an ideal gas with constant specific heats at room temperature.

Analysis (*a*) At 300 K air can be treated as an ideal gas with R = 0.287 kJ/kg K and k = 1.4. Thus

$$C = \sqrt{kRT} = \sqrt{(1.4)(0.287\ \text{kJ/kg}\cdot\text{K})(300\ \text{K})\left(\frac{1000\ \text{m}^2/\text{s}^2}{1\ \text{kJ/kg}}\right)} = \mathbf{347.2\ m/s}$$

and $$M = \frac{\mathbf{V}}{C} = \frac{240\ \text{m/s}}{347.2\ \text{m/s}} = \mathbf{0.69}$$

(*b*) At 1000 K air can be treated as an ideal gas with R = 0.287 kJ/kg K and k = 1.336. Thus,

$$C = \sqrt{kRT} = \sqrt{(1.336)(0.287\ \text{kJ/kg}\cdot\text{K})(1000\ \text{K})\left(\frac{1000\ \text{m}^2/\text{s}^2}{1\ \text{kJ/kg}}\right)} = \mathbf{619.2\ m/s}$$

and $$M = \frac{\mathbf{V}}{C} = \frac{240\ \text{m/s}}{619.2\ \text{m/s}} = \mathbf{0.39}$$

16-20 Carbon dioxide flows through a nozzle. The inlet temperature and velocity and the exit temperature of CO_2 are specified. The Mach number is to be determined at the inlet and exit of the nozzle.

Assumptions **1** CO_2 is an ideal gas. **2** This is a steady-flow process.

Properties The gas constant for carbon dioxide is R = 0.1889 kJ/kg K. Its specific heat ratio is listed in Table A-2b to be k_2 = 1.252, but there is no data listed above 1000 K. Therefore, we determine the k of CO_2 at 1200 K from Table A-20 using

$$k_1 = \frac{C_{p1}}{C_{v1}} = \frac{h_1}{u_1} = \frac{53{,}848 \text{ kJ/kmol}}{43{,}871 \text{ kJ/kmol}} = 1.227$$

1200 K
50 m/s
Carbon dioxide
400 K

Analysis (*a*) At the inlet

$$C_1 = \sqrt{k_1 RT_1} = \sqrt{(1.227)(0.1889 \text{ kJ/kg}\cdot\text{K})(1200 \text{ K})\left(\frac{1000 \text{ m}^2/\text{s}^2}{1 \text{ kJ/kg}}\right)} = 527.4 \text{ m/s}$$

Thus,

$$M_1 = \frac{\mathbf{V}_1}{C_1} = \frac{50 \text{ m/s}}{527.4 \text{ m/s}} = \mathbf{0.095}$$

(*b*) At the exit,

$$C_2 = \sqrt{k_2 RT_2} = \sqrt{(1.252)(0.1889 \text{ kJ/kg}\cdot\text{K})(400 \text{ K})\left(\frac{1000 \text{ m}^2/\text{s}^2}{1 \text{ kJ/kg}}\right)} = 307.6 \text{ m/s}$$

The nozzle exit velocity is determined from the steady-flow energy balance relation,

$$0 = h_2 - h_1 + \frac{\mathbf{V}_2^2 - \mathbf{V}_1^2}{2}$$

$$0 = (13{,}372 - 53{,}848)/44 \text{ kJ/kg} + \frac{\mathbf{V}_2^2 - (50 \text{ m/s})^2}{2}\left(\frac{1 \text{ kJ/kg}}{1000 \text{ m}^2/\text{s}^2}\right) \longrightarrow \mathbf{V}_2 = 1357.3 \text{ m/s}$$

Thus,

$$M_2 = \frac{\mathbf{V}_2}{C_2} = \frac{1357.3 \text{ m/s}}{307.6 \text{ m/s}} = \mathbf{4.41}$$

16-21 Nitrogen flows through a heat exchanger. The inlet temperature, pressure, and velocity and the exit pressure and velocity are specified. The Mach number is to be determined at the inlet and exit of the heat exchanger.

Assumptions **1** N_2 is an ideal gas. **2** This is a steady-flow process. **3** The potential energy change is negligible.

Properties The properties of N_2 are R = 0.2968 kJ/kg K and at T = 10°C = 283 K, k_1 = 1.400 (Table A-1 and A-2b).

Analysis
$$C_1 = \sqrt{k_1 R T_1} = \sqrt{(1.400)(0.2968\ \text{kJ/kg}\cdot\text{K})(283\ \text{K})\left(\frac{1000\ \text{m}^2/\text{s}^2}{1\ \text{kJ/kg}}\right)} = 342.9\ \text{m/s}$$

Thus,

$$M_1 = \frac{\mathbf{V}_1}{C_1} = \frac{100\ \text{m/s}}{342.9\ \text{m/s}} = \mathbf{0.292}$$

120kJ/k
150 kPa
10°C
100 m/s
Nitrogen
100 kPa
200 m/s

From the energy balance on the heat exchanger,

$$q_{in} = C_p(T_2 - T_1) + \frac{\mathbf{V}_2^2 - \mathbf{V}_1^2}{2}$$

$$120\ \text{kJ/kg} = (1.039\ \text{kJ/kg.°C})(T_2 - 10°\text{C}) + \frac{(200\ \text{m/s})^2 - (100\ \text{m/s})^2}{2}\left(\frac{1\ \text{kJ/kg}}{1000\ \text{m}^2/\text{s}^2}\right)$$

It yields

$$T_2 = 111.1°\text{C} = 384.3\ \text{K} \longrightarrow k_2 = 1.398$$

$$C_2 = \sqrt{k_2 R T_2} = \sqrt{(1.398)(0.2968\ \text{kJ/kg}\cdot\text{K})(384.3\ \text{K})\left(\frac{1000\ \text{m}^2/\text{s}^2}{1\ \text{kJ/kg}}\right)} = 399.3\ \text{m/s}$$

Thus,

$$M_2 = \frac{\mathbf{V}_2}{C_2} = \frac{200\ \text{m/s}}{399.3\ \text{m/s}} = \mathbf{0.501}$$

16-22 The velocity of sound in refrigerant-134a at 1 MPa and 60°C is to be determined.

Analysis Replacing the differentials by differences, the velocity of sound can be expressed approximately as

$$C^2 = \left(\frac{\partial P}{\partial \rho}\right)_s \cong \left(\frac{\Delta P}{\Delta(1/v)}\right)_s = \left(\frac{P_2 - P_1}{\dfrac{1}{v_2} - \dfrac{1}{v_2}}\right)_s$$

The entropy of refrigerant-134a at the given state (1 MPa, 60°C) is s = 0.9768 kJ/kg K. By interpolation, the specific volumes of the refrigerant-134a at this entropy and the listed pressures just below and just above the specified pressure (0.9 and 1.2 MPa) are determined to be 0.0256 m³/kg and 0.0191 m³/kg, respectively. Substituting,

$$C = \sqrt{\frac{(1200 - 900)\ \text{kPa}}{\left(\dfrac{1}{0.0191} - \dfrac{1}{0.0256}\right)\text{kg/m}^3}\left(\frac{1000\ \text{m}^2/\text{s}^2}{1\ \text{kPa}\cdot\text{m}^3/\text{kg}}\right)} = \mathbf{150.2\ m/s}$$

16-23 The maximum Mach number of a passenger plane for specified operating conditions is to be determined.

Assumptions Air is an ideal gas with variable specific heats.

Properties The gas constant for air is R = 0.287 kJ/kg K (Table A-2b).

Analysis Using the enthalpy and internal energy values from Table A-17 at –60°C (213 K), the specific heat ratio is

$$k = \frac{c_p}{c_v} = \frac{h_{@\,213\,\text{K}}}{u_{@\,213\,\text{K}}} = \frac{212.97\ \text{kJ/kg}}{151.83\ \text{kJ/kg}} = 1.403$$

and

$$C = \sqrt{kRT} = \sqrt{(1.403)(0.287\ \text{kJ/kg}\cdot\text{K})(213\ \text{K})\left(\frac{1000\ \text{m}^2/\text{s}^2}{1\ \text{kJ/kg}}\right)} = 292.9\ \text{m/s}$$

Thus,

$$M_{\max} = \frac{\mathbf{V}_{\max}}{C} = \frac{(945/3.6)\ \text{m/s}}{292.9\ \text{m/s}} = \mathbf{0.896}$$

16-24E Steam flows through a device. Its state and velocity are specified. The Mach number of steam is to be determined using two methods.

Analysis (*a*) Replacing the differentials by differences, the velocity of sound can be expressed approximately as

$$C^2 = \left(\frac{\partial P}{\partial r}\right)_s \cong \left(\frac{DP}{D(1/v)}\right)_s = \left(\frac{P_2 - P_1}{\dfrac{1}{v_2} - \dfrac{1}{v_2}}\right)_s$$

The entropy of steam at the given state (120 psia, 700°F) is s = 1.7825 Btu/1bm R. By interpolation, the specific volumes of the steam at this entropy and the listed pressures just below and just above the specified pressure (100 and 140 psia) are determined to be 6.549 ft^3/1bm and 5.046 ft^3/1bm, respectively. Substituting,

$$C = \sqrt{\frac{(140-100)\ \text{psia}}{\left(\dfrac{1}{5.046} - \dfrac{1}{6.549}\right)\text{1bm/ft}^3}\left(\frac{25{,}037\ \text{ft}^2/\text{s}^2}{1\ \text{Btu/1bm}}\right)\left(\frac{1\ \text{Btu}}{5.4039\ \text{ft}^3\cdot\text{psia}}\right)} = 2018.6\ \text{ft/s}$$

and $$M = \frac{\mathbf{V}}{C} = \frac{900\ \text{ft/s}}{2018.6\ \text{ft/s}} = \mathbf{0.446}$$

(*b*) Assuming ideal gas behavior

$$C = \sqrt{kRT} = \sqrt{(1.3)(0.1102\ \text{Btu/lbm}\cdot\text{R})(1160\ \text{R})\left(\frac{25{,}037\ \text{ft}^2/\text{s}^2}{1\ \text{Btu/lbm}}\right)} = 2039.8\ \text{ft/s}$$

Thus,

$$M = \frac{\mathbf{V}}{C} = \frac{900\ \text{ft/s}}{2039.8\ \text{ft/s}} = \mathbf{0.441}$$

Thus the ideal gas assumption is a reasonable one for steam at this state.

16-25E EES solution of this (and other comprehensive problems designated with the *computer icon*) is available to instructors at the *Instructor Manual* section of the *Online Learning Center* (OLC) at www.mhhe.com/cengel-boles. See the Preface for access information.

16-26 The expression for the velocity of sound for an ideal gas is to be obtained using the isentropic process equation and the definition of the velocity of sound.

Analysis The isentropic relation $Pv^k = A$ where A is a constant can also be expressed as

$$P = A\left(\frac{1}{v}\right)^k = A r^k$$

Substituting into the relation for the speed of sound,

$$C^2 = \left(\frac{\partial P}{\partial \rho}\right)_s = \left(\frac{\partial (A\rho)^k}{\partial \rho}\right)_s = kA\rho^{k-1} = k(A\rho^k)/\rho = k(P/\rho) = kRT$$

since for an ideal gas $P = \rho RT$ or $RT = P/\rho$. Therefore,

$$C = \sqrt{kRT}$$

16-27 The inlet state and the exit pressure of air are given for an isentropic expansion process. The ratio of the initial to the final velocity of sound is to be determined.

Assumptions Air is an ideal gas.

Properties The properties of air are R = 0.287 kJ/kg K and k = 1.4 (Table A-1 and A-2a). The specific heat ratio k varies with temperature, but in our case this change is very small and can be disregarded.

Analysis The final temperature of air is determined from the isentropic relation of ideal gases,

$$T_2 = T_{01}\left(\frac{P_2}{P_1}\right)^{(k-1)/k} = (333.2\text{ K})\left(\frac{0.4}{1.5}\right)^{(1.4-1)/1.4} = 228.4\text{ K}$$

Treating k as a constant, the ratio of the initial to the final velocity of sound can be expressed as

$$\text{Ratio} = \frac{C_2}{C_1} = \frac{\sqrt{k_1RT_1}}{\sqrt{k_2RT_2}} = \frac{\sqrt{T_1}}{\sqrt{T_2}} = \frac{\sqrt{333.2}}{\sqrt{228.4}} = \mathbf{1.21}$$

16-28 The inlet state and the exit pressure of helium are given for an isentropic expansion process. The ratio of the initial to the final velocity of sound is to be determined.

Assumptions Helium is an ideal gas.

Properties The properties of helium are R = 2.0769 kJ/kg K and k = 1.667 (Table A-1 and A-2a).

Analysis The final temperature of helium is determined from the isentropic relation of ideal gases,

$$T_2 = T_{01}\left(\frac{P_2}{P_1}\right)^{(k-1)/k} = (333.2\text{ K})\left(\frac{0.4}{1.5}\right)^{(1.667-1)/1.667} = 196.3\text{ K}$$

The ratio of the initial to the final velocity of sound can be expressed as

$$\text{Ratio} = \frac{C_2}{C_1} = \frac{\sqrt{k_1 RT_1}}{\sqrt{k_2 RT_2}} = \frac{\sqrt{T_1}}{\sqrt{T_2}} = \frac{\sqrt{333.2}}{\sqrt{196.3}} = \mathbf{1.30}$$

16-29E The inlet state and the exit pressure of air are given for an isentropic expansion process. The ratio of the initial to the final velocity of sound is to be determined.

Assumptions Air is an ideal gas.

Properties The properties of air are R = 0.06855 Btu/lbm R and k = 1.4 (Table A-1E and A-2Ea). The specific heat ratio k varies with temperature, but in our case this change is very small and can be disregarded.

Analysis The final temperature of air is determined from the isentropic relation of ideal gases,

$$T_2 = T_1\left(\frac{P_2}{P_1}\right)^{(k-1)/k} = (659.7\text{ R})\left(\frac{60}{170}\right)^{(1.4-1)/1.4} = 489.\mathbf{9}\text{ R}$$

Treating k as a constant, the ratio of the initial to the final velocity of sound can be expressed as

$$\text{Ratio} = \frac{C_2}{C_1} = \frac{\sqrt{k_1 RT_1}}{\sqrt{k_2 RT_2}} = \frac{\sqrt{T_1}}{\sqrt{T_2}} = \frac{\sqrt{659.7}}{\sqrt{489.9}} = \mathbf{1.16}$$

One Dimensional Isentropic Flow

16-30C (*a*) The exit velocity remain constant at sonic velocity, (*b*) the mass flow rate through the nozzle decreases because of the reduced flow area.

16-31C (*a*) The velocity will decrease, (*b*), (*c*), (*d*) the temperature, the pressure, and the density of the fluid will increase.

16-32C (*a*) The velocity will increase, (*b*), (*c*), (*d*) the temperature, the pressure, and the density of the fluid will decrease.

16-33C (*a*) The velocity will increase, (*b*), (*c*), (*d*) the temperature, the pressure, and the density of the fluid will decrease.

16-34C (*a*) The velocity will decrease, (*b*), (*c*), (*d*) the temperature, the pressure and the density of the fluid will increase.

16-35C They will be identical.

16-36C No, it is not possible.

16-37 Air enters a converging-diverging nozzle at specified conditions. The lowest pressure that can be obtained at the throat of the nozzle is to be determined.

Assumptions **1** Air is an ideal gas. **2** Flow through the nozzle is steady, one-dimensional, and isentropic.

Properties The specific heat ratio of air at room temperature is $k = 1.4$ (Table A-1).

Analysis The lowest pressure that can be obtained at the throat is the critical pressure P^*, which is determined from

$$P^* = P_0\left(\frac{2}{k+1}\right)^{k/(k-1)} = (1.2\text{ MPa})\left(\frac{2}{1.4+1}\right)^{1.4/(1.4-1)} = \mathbf{0.634\ MPa}$$

16-38 Helium enters a converging-diverging nozzle at specified conditions. The lowest temperature and pressure that can be obtained at the throat of the nozzle are to be determined.

Assumptions **1** Helium is an ideal gas. **2** Flow through the nozzle is steady, one-dimensional, and isentropic.

Properties The properties of helium are $k = 1.667$ and $C_p = 5.1926$ kJ/kg K (Table A-2a)

Analysis The lowest temperature and pressure that can be obtained at the throat are the critical temperature T^* and critical pressure P^*. First we determine the stagnation temperature T_0 and stagnation pressure P_0,

$$T_0 = T + \frac{\mathbf{V}^2}{2C_p} = 800\text{ K} + \frac{(100\text{ m/s})^2}{2 \times 5.1926\text{ kJ/kg}\cdot°\text{C}}\left(\frac{1\text{ kJ/kg}}{1000\text{ m}^2/\text{s}^2}\right) = 801\text{ K}$$

$$P_0 = P\left(\frac{T_0}{T}\right)^{k/(k-1)} = (0.7\text{ MPa})\left(\frac{801\text{ K}}{800\text{ K}}\right)^{1.667/(1.667-1)} = 0.702\text{ MPa}$$

Thus,

$$T^* = T_0\left(\frac{2}{k+1}\right) = (801\text{ K})\left(\frac{2}{1.667+1}\right) = \mathbf{600.7\ K}$$

and

$$P^* = P_0\left(\frac{2}{k+1}\right)^{k/(k-1)} = (0.702\text{ MPa})\left(\frac{2}{1.667+1}\right)^{1.667/(1.667-1)} = \mathbf{0.342\ MPa}$$

16-39 The critical temperature, pressure, and density of air and helium are to be determined at specified conditions.

Assumptions Air and Helium are ideal gases.

Properties The properties of air at room temperature are R = 0.287 kJ/kg K, k = 1.4, and C_p = 1.005 kJ/kg K. The properties of helium at room temperature are R = 2.0769 kJ/kg K, k = 1.667, and C_p = 5.1926 kJ/kg K. (Table A-1 and A-2a)

Analysis (*a*) Before we calculate the critical temperature T^*, pressure P^*, and density ρ^*, we need to determine the stagnation temperature T_0, pressure P_0, and density ρ_0.

$$T_0 = 100°\text{C} + \frac{\mathbf{V}^2}{2C_p} = 100 + \frac{(250\text{ m/s})^2}{2\times 1.005\text{ kJ/kg}\cdot°\text{C}}\left(\frac{1\text{ kJ/kg}}{1000\text{ m}^2/\text{s}^2}\right) = 131.1°\text{C}$$

$$P_0 = P\left(\frac{T_0}{T}\right)^{k/(k-1)} = (200\text{ kPa})\left(\frac{404.3\text{ K}}{373.2\text{ K}}\right)^{1.4/(1.4-1)} = 264.7\text{ kPa}$$

$$\rho_0 = \frac{P_0}{RT_0} = \frac{264.7\text{ kPa}}{(0.287\text{ kPa}\cdot\text{m}^3/\text{kg}\cdot\text{K})(404.3\text{ K})} = 2.281\text{ kg/m}^3$$

Thus,

$$T^* = T_0\left(\frac{2}{k+1}\right) = (404.3\text{ K})\left(\frac{2}{1.4+1}\right) = \mathbf{336.9\ K}$$

$$P^* = P_0\left(\frac{2}{k+1}\right)^{k/(k-1)} = (264.7\text{ kPa})\left(\frac{2}{1.4+1}\right)^{1.4/(1.4-1)} = \mathbf{139.8\ kPa}$$

$$\rho^* = \rho_0\left(\frac{2}{k+1}\right)^{1/(k-1)} = (2.281\text{ kg/m}^3)\left(\frac{2}{1.4+1}\right)^{1/(1.4-1)} = \mathbf{1.446\ kg/m^3}$$

(*b*) For helium, $$T_0 = T + \frac{\mathbf{V}^2}{2C_p} = 40 + \frac{(300\text{ m/s})^2}{2\times 5.1926\text{ kJ/kg}\cdot°\text{C}}\left(\frac{1\text{ kJ/kg}}{1000\text{ m}^2/\text{s}^2}\right) = 48.7°\text{C}$$

$$P_0 = P\left(\frac{T_0}{T}\right)^{k/(k-1)} = (200\text{ kPa})\left(\frac{321.9\text{ K}}{313.2\text{ K}}\right)^{1.667/(1.667-1)} = 214.2\text{ kPa}$$

$$\rho_0 = \frac{P_0}{RT_0} = \frac{214.2\text{ kPa}}{(2.0769\text{ kPa}\cdot\text{m}^3/\text{kg}\cdot\text{K})(321.9\text{ K})} = 0.320\text{ kg/m}^3$$

Thus,

$$T^* = T_0\left(\frac{2}{k+1}\right) = (321.9\text{ K})\left(\frac{2}{1.667+1}\right) = \mathbf{241.4\ K}$$

$$P^* = P_0\left(\frac{2}{k+1}\right)^{k/(k-1)} = (200\text{ kPa})\left(\frac{2}{1.667+1}\right)^{1.667/(1.667-1)} = \mathbf{97.4\ kPa}$$

$$\rho^* = \rho_0\left(\frac{2}{k+1}\right)^{1/(k-1)} = (0.320\text{ kg/m}^3)\left(\frac{2}{1.667+1}\right)^{1/(1.667-1)} = \mathbf{0.208\ kg/m^3}$$

16-40 Stationary carbon dioxide at a given state is accelerated isentropically to a specified Mach number. The temperature and pressure of the carbon dioxide after acceleration are to be determined.

Assumptions Carbon dioxide is an ideal gas with constant specific heats.

Properties The specific heat ratio of the carbon dioxide at 400 K is k = 1.252 (Table A-2b).

Analysis The inlet temperature and pressure in this case is equivalent to the stagnation temperature and pressure since the inlet velocity of the carbon dioxide said to be negligible. That is, $T_0 = T_i$ = 400 K and $P_0 = P_i$ = 800 kPa. Then,

$$T = T_0\left(\frac{2}{2+(k-1)\mathrm{M}^2}\right) = (400\text{ K})\left(\frac{2}{2+(1.252-1)(0.6)^2}\right) = \mathbf{382.6\ K}$$

and

$$P = P_0\left(\frac{T}{T_0}\right)^{k/(k-1)} = (800\text{ kPa})\left(\frac{382.6\text{ K}}{400\text{ K}}\right)^{1.252/(1.252-1)} = \mathbf{641\ kPa}$$

16-41 Air flows through a duct. The state of the air and its Mach number are specified. The velocity and the stagnation pressure, temperature, and density of the air are to be determined.

Assumptions Air is an ideal gas.

Properties The properties of air are R = 0.287 kPa.m³/kg.K and k = 1.4 (Table A-1 and Table A-2a).

Analysis The speed of sound in air at the specified conditions is

$$C = \sqrt{kRT} = \sqrt{(1.4)(0.287\text{ kJ/kg}\cdot\text{K})(373.2\text{ K})\left(\frac{1000\text{ m}^2/\text{s}^2}{1\text{ kJ/kg}}\right)} = 387.2\text{ m/s}$$

Thus,

AIR

$$\mathbf{V} = M \times C = (0.8)(387.2\text{ m/s}) = \mathbf{309.8\ m/s}$$

Also,

$$\rho = \frac{P}{RT} = \frac{200\text{ kPa}}{(0.287\text{ kPa}\cdot\text{m}^3/\text{kg}\cdot\text{K})(373.2\text{ K})} = 1.867\text{ kg/m}^3$$

Then the stagnation properties are determined from

$$T_0 = T\left(1+\frac{(k-1)\mathrm{M}^2}{2}\right) = (373.2\text{ K})\left(1+\frac{(1.4-1)(0.8)^2}{2}\right) = \mathbf{421.0\ K}$$

$$P_0 = P\left(\frac{T_0}{T}\right)^{k/(k-1)} = (200\text{ kPa})\left(\frac{421.0\text{ K}}{373.2\text{ K}}\right)^{1.4/(1.4-1)} = \mathbf{304.9\ kPa}$$

$$\rho_0 = \rho\left(\frac{T_0}{T}\right)^{1/(k-1)} = (1.867\text{ kg/m}^3)\left(\frac{421.0\text{ K}}{373.2\text{ K}}\right)^{1/(1.4-1)} = \mathbf{2.523\ kg/m^3}$$

16-42 EES solution of this (and other comprehensive problems designated with the *computer icon*) is available to instructors at the *Instructor Manual* section of the *Online Learning Center* (OLC) at www.mhhe.com/cengel-boles. See the Preface for access information.

16-43E Air flows through a duct. The state of the air and its Mach number are specified. The velocity and the stagnation pressure, temperature, and density of the air are to be determined.

Assumptions Air is an ideal gas.

Properties The properties of air are R = 0.06855 Btu/lbm.R and k = 1.4 (Table A-1E and Table A-2a).

Analysis The speed of sound in air at the specified conditions is

$$C = \sqrt{kRT} = \sqrt{(1.4)(0.06855\ \text{Btu/1bm}\cdot\text{R})(671.7\ \text{R})\left(\frac{25{,}037\ \text{ft}^2/\text{s}^2}{1\ \text{Btu/1bm}}\right)} = 1270.4\ \text{ft/s}$$

Thus,

$$\mathbf{V} = M \times C = (0.8)(1270.4\ \text{ft/s}) = \mathbf{1016.3\ ft/s}$$

Also,

$$\rho = \frac{P}{RT} = \frac{30\ \text{psia}}{(0.3704\ \text{psia}\cdot\text{ft}^3/\text{lbm}\cdot\text{R})(671.7\ \text{R})} = 0.1206\ \text{lbm/ft}^3$$

Then the stagnation properties are determined from

$$T_0 = T\left(1 + \frac{(k-1)\text{M}^2}{2}\right) = (671.7\ \text{R})\left(1 + \frac{(1.4-1)(0.8)^2}{2}\right) = \mathbf{757.7\ R}$$

$$P_0 = P\left(\frac{T_0}{T}\right)^{k/(k-1)} = (30\ \text{psia})\left(\frac{757.7\ \text{R}}{671.7\ \text{R}}\right)^{1.4/(1.4-1)} = \mathbf{45.7\ psia}$$

$$\rho_0 = \rho\left(\frac{T_0}{T}\right)^{1/(k-1)} = (0.1206\ \text{1bm/ft}^3)\left(\frac{757.7\ \text{R}}{671.7\ \text{R}}\right)^{1/(1.4-1)} = \mathbf{0.163\ 1bm/ft^3}$$

16-44 An aircraft is designed to cruise at a given Mach number, elevation, and the atmospheric temperature. The stagnation temperature on the leading edge of the wing is to be determined.

Assumptions Air is an ideal gas.

Properties The properties of air are R = 0.287 kPa.m^3/kg.K, C_p = 1.005 kJ/kg K, and k = 1.4 (Table A-1 and Table A-2a).

Analysis The speed of sound in air at the specified conditions is

$$C = \sqrt{kRT} = \sqrt{(1.4)(0.287\ \text{kJ/kg}\cdot\text{K})(236.15\ \text{K})\left(\frac{1000\ \text{m}^2/\text{s}^2}{1\ \text{kJ/kg}}\right)} = 308.0\ \text{m/s}$$

Thus,

$$\mathbf{V} = M \times C = (1.4)(308.0\ \text{m/s}) = 431.2\ \text{m/s}$$

Then,

$$T_0 = T + \frac{\mathbf{V}^2}{2C_p} = 236.15 + \frac{(431.2\ \text{m/s})^2}{2 \times 1.005\ \text{kJ/kg}\cdot\text{K}}\left(\frac{1\ \text{kJ/kg}}{1000\ \text{m}^2/\text{s}^2}\right) = \mathbf{328.7\ \ K}$$

Isentropic Flow Through Nozzles

16-45C (*a*) The exit velocity will reach the sonic velocity, (*b*) the exit pressure will equal the critical pressure, and (*c*) the mass flow rate will reach the maximum value.

16-46C (*a*) None, (*b*) None, and (*c*) None.

16-47C They will be the same.

16-48C Maximum flow rate through a nozzle is achieved when $M = 1$ at the exit of a subsonic nozzle. For all other M values the mass flow rate decreases. Therefore, the mass flow rate would decrease if hypersonic velocities were achieved at the throat of a converging nozzle.

16-49C M^* is the local velocity non-dimensionalized with respect to the sonic velocity at the throat, whereas M is the local velocity non-dimensionalized with respect to the local sonic velocity.

16-50C The fluid would accelerate even further instead of decelerating.

16-51C The fluid would decelerate instead of accelerating.

16-52C (*a*) The velocity will decrease, (*b*) the pressure will increase, and (*c*) the mass flow rate will remain the same.

16-53C No.

16-54 It is to be explained why the maximum flow rate per unit area for a given gas depends only on $P_0/\sqrt{T_0}$. Also for an ideal gas, a relation is to be obtained for the constant a in $\dot{m}_{max}/A^* = a\left(P_0/\sqrt{T_0}\right)$.

Properties The properties of the gas considered are $R = 0.287$ kPa.m³/kg.K and $k = 1.4$.

Analysis The maximum flow rate is given by

$$\dot{m}_{max} = A^* P_0 \sqrt{k/RT_0}\left(\frac{2}{k+1}\right)^{(k+1)/2(k-1)}$$

or

$$\dot{m}_{max}/A^* = \left(P_0/\sqrt{T_0}\right)\sqrt{k/R}\left(\frac{2}{k+1}\right)^{(k+1)/2(k-1)}$$

For a given gas, k and R are fixed, and thus the mass flow rate will depend on the parameter $P_0/\sqrt{T_0}$.

$\dot{m}_{max}/A^*$ can be expressed as a $\left(P_0/\sqrt{T_0}\right)$ where

$$a = \sqrt{k/R}\left(\frac{2}{k+1}\right)^{(k+1)/2(k-1)} = \sqrt{\frac{1.4}{(0.287\ \text{kJ/kg.K})\left(\frac{1000\ \text{m}^2/\text{s}^2}{1\ \text{kJ/kg}}\right)}}\left(\frac{2}{1.4+1}\right)^{2.4/0.8} = 0.0404\ (\text{m/s})\sqrt{\text{K}}$$

16-55 For an ideal gas, an expression is to be obtained for the ratio of the velocity of sound where $M = 1$ to the velocity of sound based on the stagnation temperature, C^*/C_0.

Analysis For an ideal gas the velocity of sound is expressed as $C = \sqrt{kRT}$. Thus,

$$\frac{C^*}{C_0} = \frac{\sqrt{kRT^*}}{\sqrt{kRT_0}} = \left(\frac{T^*}{T_0}\right)^{1/2} = \left(\frac{2}{k+1}\right)^{1/2}$$

16-56 For subsonic flow at the inlet, the variation of pressure, velocity, and Mach number along the length of the nozzle are to be sketched for an ideal gas under specified conditions.

Analysis

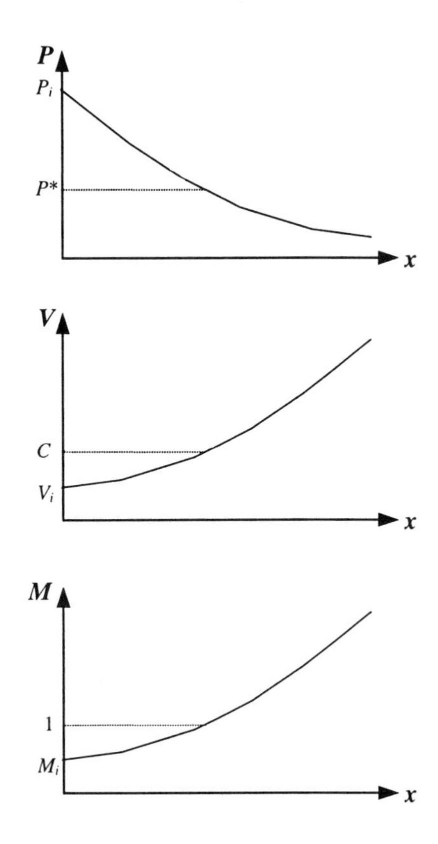

16-57 For supersonic flow at the inlet, the variation of pressure, velocity, and Mach number along the length of the nozzle are to be sketched for an ideal gas under specified conditions.

Analysis

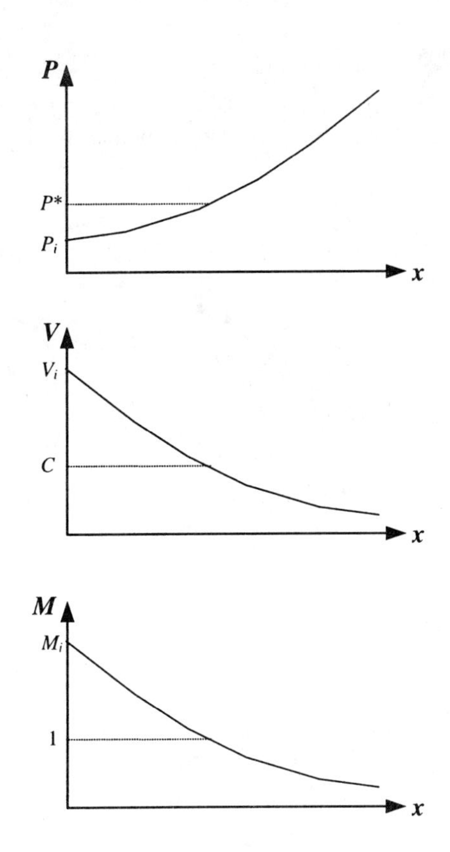

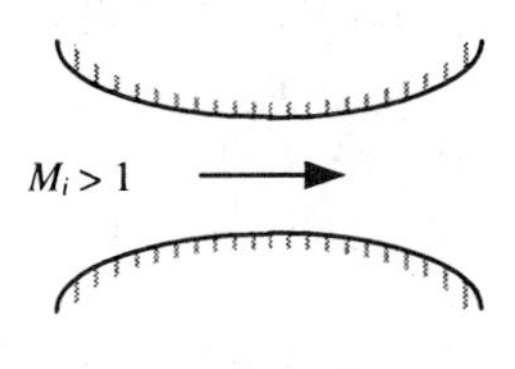

16-58 Air enters a nozzle at specified temperature, pressure, and velocity. The exit pressure, exit temperature, and exit-to-inlet area ratio are to be determined for a Mach number of $M = 1$ at the exit.

Assumptions **1** Air is an ideal gas with constant specific heats at room temperature. **2** Flow through the nozzle is steady, one-dimensional, and isentropic.

Properties The properties of air are $k = 1.4$ and $C_p = 1.005$ kJ/kg K (Table A-2a).

Analysis The properties of the fluid at the location where $M = 1$ are the critical properties, denoted by superscript *. We first determine the stagnation temperature and pressure, which remain constant throughout the nozzle since the flow is isentropic.

$$T_0 = T_i + \frac{\mathbf{V}_i^2}{2C_p} = 350\text{ K} + \frac{(150\text{ m/s})^2}{2\times 1.005\text{ kJ/kg}\cdot\text{K}}\left(\frac{1\text{ kJ/kg}}{1000\text{ m}^2/\text{s}^2}\right) = 361.2\text{ K}$$

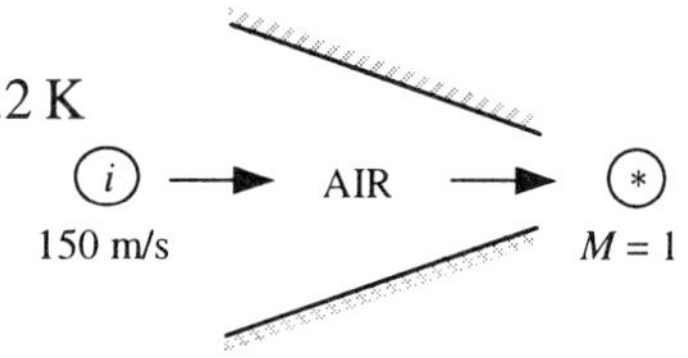

i → AIR → *

150 m/s

$M = 1$

and

$$P_0 = P_i\left(\frac{T_0}{T_i}\right)^{k/(k-1)} = (0.2\text{ MPa})\left(\frac{361.2\text{ K}}{350\text{ K}}\right)^{1.4/(1.4-1)} = 0.223\text{ MPa}$$

From Table A-15 (or from Eqs. 16-18 and 16-19) at $M = 1$, we read $T/T_0 = 0.83333$, $P/P_0 = 0.52828$. Thus,

$$T = 0.83333T_0 = 0.83333(361.2\text{ K}) = \mathbf{301.0\text{ K}}$$

and

$$P = 0.52828P_0 = 0.52828(0.223\text{ MPa}) = \mathbf{0.118\text{ MPa}}$$

Also,

$$C_i = \sqrt{kRT_i} = \sqrt{(1.4)(0.287\text{ kJ/kg}\cdot\text{K})(350\text{ K})\left(\frac{1000\text{ m}^2/\text{s}^2}{1\text{ kJ/kg}}\right)} = 375.0\text{ m/s}$$

and

$$M_i = \frac{\mathbf{V}_i}{C_i} = \frac{150\text{ m/s}}{375\text{ m/s}} = 0.40$$

From Table A-15 at this Mach number we read $A_i/A^* = 1.5901$. Thus the ratio of the throat area to the nozzle inlet area is

$$\frac{A^*}{A_i} = \frac{1}{1.5901} = \mathbf{0.629}$$

16-59 Air enters a nozzle at specified temperature and pressure with low velocity. The exit pressure, exit temperature, and exit-to-inlet area ratio are to be determined for a Mach number of $M = 1$ at the exit.

Assumptions **1** Air is an ideal gas. **2** Flow through the nozzle is steady, one-dimensional, and isentropic.

Properties The specific heat ratio of air is $k = 1.4$ (Table A-2a).

Analysis The properties of the fluid at the location where $M = 1$ are the critical properties, denoted by superscript *. The stagnation temperature and pressure in this case are identical to the inlet temperature and pressure since the inlet velocity is negligible. They remain constant throughout the nozzle since the flow is isentropic.

$$T_0 = T_i = 350 \text{ K}$$

$$P_0 = P_i = 0.2 \text{ MPa}$$

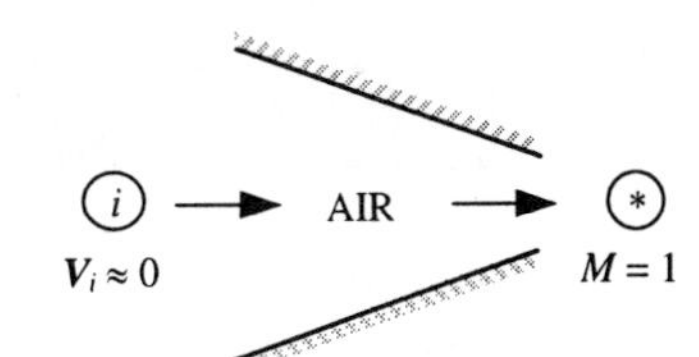

From Table A-15 (or from Eqs. 16-18 and 16-19) at $M = 1$, we read $T/T_0 = 0.83333$, $P/P_0 = 0.52828$. Thus,

$$T = 0.83333T_0 = 0.83333(350 \text{ K}) = \mathbf{291.7 \text{ K}}$$

and

$$P = 0.52828P_0 = 0.52828(0.2 \text{ MPa}) = \mathbf{0.106 \text{ MPa}}$$

The Mach number at the nozzle inlet is $M = 0$ since $V_i \cong 0$. From Table A-15 at this Mach number we read $A_i/A^* = \infty$. Thus the ratio of the throat area to the nozzle inlet area is

$$\frac{A^*}{A_i} = \frac{1}{\infty} = \mathbf{0}$$

16-60E Air enters a nozzle at specified temperature, pressure, and velocity. The exit pressure, exit temperature, and exit-to-inlet area ratio are to be determined for a Mach number of $M = 1$ at the exit.

Assumptions **1** Air is an ideal gas with constant specific heats at room temperature. **2** Flow through the nozzle is steady, one-dimensional, and isentropic.

Properties The properties of air are $k = 1.4$ and $C_p = 0.240$ Btu/lbm R (Table A-2Ea).

Analysis The properties of the fluid at the location where $M = 1$ are the critical properties, denoted by superscript *. We first determine the stagnation temperature and pressure, which remain constant throughout the nozzle since the flow is isentropic.

$$T_0 = T + \frac{\mathbf{V}_i^2}{2C_p} = 630\,\text{R} + \frac{(450\,\text{ft/s})^2}{2 \times 0.240\,\text{Btu/lbm}\cdot\text{R}}\left(\frac{1\,\text{Btu/1bm}}{25{,}037\,\text{ft}^2/\text{s}^2}\right) = 646.9\,\text{R}$$

i → AIR → *
450 ft/s
M = 1

and

$$P_0 = P_i\left(\frac{T_0}{T_i}\right)^{k/(k-1)} = (30\,\text{psia})\left(\frac{646.9\,\text{K}}{630\,\text{K}}\right)^{1.4/(1.4-1)} = 32.9\,\text{psia}$$

From Table A-15 (or from Eqs. 16-18 and 16-19) at $M = 1$, we read $T/T_0 = 0.83333$, $P/P_0 = 0.52828$. Thus,

$$T = 0.83333T_0 = 0.83333(646.9\text{ R}) = \mathbf{539.1\ R}$$

and

$$P = 0.52828P_0 = 0.52828(32.9\text{ psia}) = \mathbf{17.38\ psia}$$

Also,

$$C_i = \sqrt{kRT_i} = \sqrt{(1.4)(0.06855\,\text{Btu/1bm}\cdot\text{R})(630\,\text{R})\left(\frac{25{,}037\,\text{ft}^2/\text{s}^2}{1\,\text{Btu/1bm}}\right)} = 1230.4\text{ ft/s}$$

and

$$M_i = \frac{\mathbf{V}_i}{C_i} = \frac{450\text{ ft/s}}{1230.4\text{ ft/s}} = 0.3657$$

From Table A-15 at this Mach number we read $A_i/A^* = 1.7426$. Thus the ratio of the throat area to the nozzle inlet area is

$$\frac{A^*}{A_i} = \frac{1}{1.7426} = \mathbf{0.574}$$

16-61 Air enters a converging-diverging nozzle at a specified pressure. The back pressure that will result in a specified exit Mach number is to be determined.

Assumptions **1** Air is an ideal gas. **2** Flow through the nozzle is steady, one-dimensional, and isentropic.

Properties The specific heat ratio of air is $k = 1.4$ (Table A-2a).

Analysis The stagnation pressure in this case is identical to the inlet pressure since the inlet velocity is negligible. It remains constant throughout the nozzle since the flow is isentropic,

$$P_0 = P_i = 0.8\text{ MPa}$$

From Table A-15 at $M_e = 1.8$, we read $P_e/P_0 = 0.17404$.

Thus, $P = 0.17404P_0 = 0.17404(0.8\text{ MPa}) = \mathbf{0.1392\ MPa}$

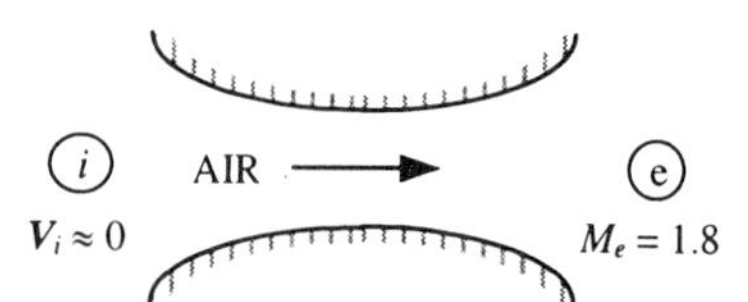

16-62 Nitrogen enters a converging-diverging nozzle at a given pressure. The critical velocity, pressure, temperature, and density in the nozzle are to be determined.

Assumptions **1** Nitrogen is an ideal gas. **2** Flow through the nozzle is steady, one-dimensional, and isentropic.

Properties The properties of nitrogen are $k = 1.4$ and $R = 0.2968$ kJ/kg K (Table A-1 and A-2a).

Analysis The stagnation pressure in this case are identical to the inlet properties since the inlet velocity is negligible. They remain constant throughout the nozzle,

$$P_0 = P_i = 700 \text{ kPa}$$

$$T_0 = T_i = 400 \text{ K}$$

$$\rho_0 = \frac{P_0}{RT_0} = \frac{700 \text{ kPa}}{(0.2968 \text{ kPa}\cdot\text{m}^3/\text{kg}\cdot\text{K})(400 \text{ K})} = 5.896 \text{ kg/m}^3$$

Critical properties are those at a location where the Mach number is $M = 1$. From Table A-15 at $M = 1$, we read $T/T_0 = 0.83333$, $P/P_0 = 0.52828$, and $\rho/\rho_0 = 0.63394$. Then the critical properties become

$$T^* = 0.83333T_0 = 0.83333(400 \text{ K}) = \mathbf{333.3\ K}$$

$$P^* = 0.52828P_0 = 0.52828(700 \text{ kPa}) = \mathbf{369.8\ MPa}$$

$$\rho^* = 0.63394\rho_0 = 0.63394(5.896 \text{ kg/m}^3) = \mathbf{3.738\ kg/m^3}$$

Also,

$$V^* = C^* = \sqrt{kRT^*} = \sqrt{(1.4)(0.2968 \text{ kJ/kg}\cdot\text{K})(333.3 \text{ K})\left(\frac{1000 \text{ m}^2/\text{s}^2}{1 \text{ kJ/kg}}\right)} = \mathbf{372.1\ m/s}$$

16-63 An ideal gas is flowing through a nozzle. The flow area at a location where $M = 2.5$ is specified. The flow area where $M = 1.2$ is to be determined.

Assumptions Flow through the nozzle is steady, one-dimensional, and isentropic.

Analysis The flow is assumed to be isentropic, and thus the stagnation and critical properties remain constant throughout the nozzle. The flow area at a location where $M_2 = 1.2$ is determined using A/A^* data from Table A-15 to be

$$M_1 = 2.5: \quad \frac{A_1}{A^*} = 2.6367 \quad \longrightarrow \quad A^* = \frac{A_1}{2.6367} = \frac{25 \text{ cm}^2}{2.6367} = 9.482 \text{ cm}^2$$

$$M_2 = 1.2: \quad \frac{A_2}{A^*} = 1.03044 \quad \longrightarrow \quad A_2 = (1.03044)A^* = (1.03044)(9.482 \text{ cm}^2) = \mathbf{9.77\ cm^2}$$

16-64 An ideal gas is flowing through a nozzle. The flow area at a location where $M = 2.5$ is specified. The flow area where $M = 1.2$ is to be determined.

Assumptions Flow through the nozzle is steady, one-dimensional, and isentropic.

Analysis The flow is assumed to be isentropic, and thus the stagnation and critical properties remain constant throughout the nozzle. The flow area at a location where $M_2 = 1.2$ is determined using the A/A^* relation,

$$\frac{A}{A^*} = \frac{1}{M}\left\{\left(\frac{2}{k+1}\right)\left(1+\frac{k-1}{2}M^2\right)\right\}^{(k+1)/2(k-1)}$$

For $k = 1.33$,

$$\frac{A}{A^*} = \frac{1}{M}\left\{\left(\frac{2}{1.33+1}\right)\left(1+\frac{1.33-1}{2}M^2\right)\right\}^{2.33/2\times 0.33} = \frac{1}{M}\left(0.85837\times(1+0.165M^2)\right)^{3.53}$$

Thus,

At $\quad M_1 = 2.5, \quad \frac{A_1}{A^*} = \frac{1}{2.5}\left(0.85837\times(1+0.165\times 2.5^2)\right)^{3.53} = 2.8467$

and, $\quad A^* = \frac{A_1}{2.8467} = \frac{25\text{ cm}^2}{2.8467} = 8.78\text{ cm}^2$

At $\quad M_2 = 1.2, \quad \frac{A_2}{A^*} = \frac{1}{1.2}\left(0.85837\times(1+0.165\times 1.2^2)\right)^{3.53} = 1.03157$

and $\quad A_2 = (1.03157)A^* = (1.03157)(8.78\text{ cm}^2) = \mathbf{9.06\ cm^2}$

16-65 [*Also solved by EES on enclosed CD*] Air enters a converging nozzle at a specified temperature and pressure with low velocity. The exit pressure, the exit velocity, and the mass flow rate versus the back pressure are to be calculated and plotted.

Assumptions **1** Air is an ideal gas with constant specific heats at room temperature. **2** Flow through the nozzle is steady, one-dimensional, and isentropic.

Properties The properties of air are $k = 1.4$, $R = 0.287$ kJ/kg K, and $C_p = 1.005$ kJ/kg K (Table A-2a).

Analysis The stagnation properties in this case are identical to the inlet properties since the inlet velocity is negligible. They remain constant throughout the nozzle since the flow is isentropic.,

$$P_0 = P_i = 900 \text{ kPa}$$

$$T_0 = T_i = 400 \text{ K}$$

The critical pressure is determined to be

$$P^* = P_0\left(\frac{2}{k+1}\right)^{k/(k-1)} = (900 \text{ kPa})\left(\frac{2}{1.4+1}\right)^{1.4/0.4} = 475.5 \text{ kPa}$$

i → AIR → e

$V_i \approx 0$

Then the pressure at the exit plane (throat) will be

$P_e = P_b$ for $P_b \geq 475.5$ kPa

$P_e = P^* = 475.5$ kPa for $P_b < 475.5$ kPa (choked flow)

Thus the back pressure will not affect the flow when $100 < P_b < 475.5$ kPa. For a specified exit pressure P_e, the temperature, the velocity and the mass flow rate can be determined from

Temperature $$T_e = T_0\left(\frac{P_e}{P_0}\right)^{(k-1)/k} = (400 \text{ K})\left(\frac{P_e}{900}\right)^{0.4/1.4}$$

Velocity $$V = \sqrt{2C_p(T_0 - T_e)} = \sqrt{2(1.005 \text{ kJ/kg}\cdot\text{K})(400 - T_e)\left(\frac{1000 \text{ m}^2/\text{s}^2}{1 \text{ kJ/kg}}\right)}$$

Density $$\rho_e = \frac{P_e}{RT_e} = \frac{P_e}{(0.287 \text{ kPa}\cdot\text{m}^3/kg\cdot K)T_e}$$

Mass flow rate $$\dot{m} = \rho_e V_e A_e = \rho_e V_e (0.001 \text{ m}^2)$$

The results of the calculations can be tabulated as

P_b, kPa	P_e, kPa	T_e, K	V_e, m/s	ρ_e, kg/m^3	$\dot{m}$ kg/s
900	900	400	0	7.840	0
800	800	386.8	162.9	7.206	1.174
700	700	372.3	236.0	6.551	1.546
600	600	356.2	296.7	5.869	1.741
500	500	338.2	352.4	5.151	1.815
475.5	475.5	333.3	366.2	4.971	1.820
400	475.5	333.3	366.2	4.971	1.820
300	475.5	333.3	366.2	4.971	1.820
200	475.5	333.3	366.2	4.971	1.820
100	475.5	333.3	366.2	4.971	1.820

P_e vs P_b

V_e, C vs P_b

$\dot{m}$, $\dot{m}_{max}$ vs P_b

100 475.5 900 kPa

16-66 EES solution of this (and other comprehensive problems designated with the *computer icon*) is available to instructors at the *Instructor Manual* section of the *Online Learning Center* (OLC) at www.mhhe.com/cengel-boles. See the Preface for access information.

16-67E Air enters a converging-diverging nozzle at a specified temperature and pressure with low velocity. The pressure, temperature, velocity, and mass flow rate are to be calculated in the specified test section.

Assumptions **1** Air is an ideal gas. **2** Flow through the nozzle is steady, one-dimensional, and isentropic.

Properties The properties of air are k = 1.4 and R = 0.06855 Btu/lbm R = 0.3704 psia ft^3/lbm R (Table A-2Ea).

Analysis The stagnation properties in this case are identical to the inlet properties since the inlet velocity is negligible. They remain constant throughout the nozzle since the flow is isentropic.

$$P_0 = P_i = 150 \text{ psia}$$

$$T_0 = T_i = 100°\text{F} = 560 \text{ R}$$

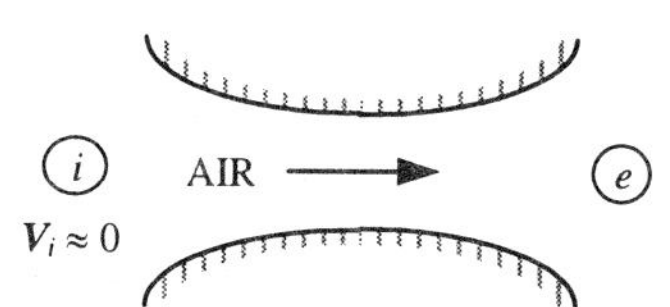

Then,

$$T_e = T_0\left(\frac{2}{2+(k-1)M^2}\right) = (560 \text{ R})\left(\frac{2}{2+(1.4-1)2^2}\right) = \mathbf{311\ R}$$

and

$$P_e = P_0\left(\frac{T}{T_0}\right)^{k/(k-1)} = (150 \text{ psia})\left(\frac{311}{560}\right)^{1.4/0.4} = \mathbf{19.1\ psia}$$

Also,

$$\rho_e = \frac{P_e}{RT_e} = \frac{19.1 \text{ psia}}{(0.3704 \text{ psia.ft}^3/\text{lbm.R})(311 \text{ R})} = 0.166 \text{ lbm/ft}^3$$

The nozzle exit velocity can be determined from $\mathbf{V}_e = M_e C_e$, where C_e is the velocity of sound at the exit conditions,

$$\mathbf{V}_e = M_e C_e = M_e\sqrt{kRT_e} = (2)\sqrt{(1.4)(0.06855 \text{ Btu/1bm.R})(311 \text{ R})\left(\frac{25{,}037 \text{ ft}^2/\text{s}^2}{1 \text{ Btu/1bm}}\right)} = \mathbf{1729\ ft/s}$$

Finally,

$$\dot{m} = \rho_e A_e \mathbf{V}_e = (0.166 \text{ 1bm/ft}^3)(5 \text{ ft}^2)(1729 \text{ ft/s}) = \mathbf{1435\ 1bm/s}$$

Air must be very dry in this application because the exit temperature of air is extremely low, and any moisture in the air will turn to ice particles.

Normal Shocks in Nozzle Flow

16-68C No.

16-69C The Fanno line represents the states which satisfy the conservation of mass and energy equations. The Rayleigh line represents the states which satisfy the conservation of mass and momentum equations. The intersections points of these lines represents the states which satisfy the conservation of mass, energy, and momentum equations.

16-70C No.

16-71C (*a*) decreases, (*b*) increases, (*c*) remains the same, (*d*) increases, and (*e*) decreases.

16-72 For an ideal gas flowing through a normal shock, a relation for $\mathbf{V}_y/\mathbf{V}_x$ in terms of k, M_x, and m_y is to be developed.

Analysis The conservation of mass relation across the shock is $\rho_x \mathbf{V}_x = \rho_y \mathbf{V}_y$ and it can be expressed as

$$\frac{\mathbf{V}_y}{\mathbf{V}_x} = \frac{\rho_x}{\rho_y} = \frac{P_x / RT_x}{P_y / RT_y} = \left(\frac{P_x}{P_y}\right)\left(\frac{T_y}{T_x}\right)$$

From Eqs. 16-35 and 16-38,

$$\frac{\mathbf{V}_y}{\mathbf{V}_x} = \left(\frac{1 + kM_y^2}{1 + kM_x^2}\right)\left(\frac{1 + M_x^2(k-1)/2}{1 + M_y^2(k-1)/2}\right)$$

16-73 Air flowing through a converging-diverging nozzle experiences a normal shock at the exit. The effect of the shock wave on various properties is to be determined.

Assumptions **1** Air is an ideal gas. **2** Flow through the nozzle is steady, one-dimensional, and isentropic before the shock occurs. **3** The shock wave occurs at the exit plane.

Properties The properties of air are $k = 1.4$ and $R = 0.287$ kJ/kg K (Table A-1 and Table A-2a).

Analysis The inlet stagnation properties in this case are identical to the inlet properties since the inlet velocity is negligible. Then,

$$P_{0x} = P_i = 1 \text{ MPa}$$

$$T_{0x} = T_i = 300 \text{ K}$$

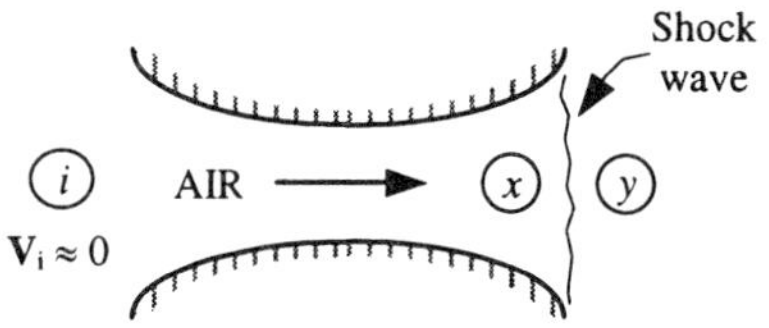

Then,

$$T_x = T_{0x}\left(\frac{2}{2+(k-1)M_x^2}\right) = (300 \text{ K})\left(\frac{2}{2+(1.4-1)2^2}\right) = 166.7 \text{ K}$$

and

$$P_x = P_{0x}\left(\frac{T_x}{T_0}\right)^{k/(k-1)} = (1 \text{ MPa})\left(\frac{166.7}{300}\right)^{1.4/0.4} = 0.1278 \text{ MPa}$$

The fluid properties after the shock (denoted by subscript y) are related to those before the shock through the functions listed in Table A-16. For $M_x = 2.0$ we read

$$M_y = 0.5774, \quad \frac{P_{0y}}{P_{0x}} = 0.7209, \quad \frac{P_y}{P_x} = 4.5000, \text{ and } \frac{T_y}{T_x} = 1.6875$$

Then the stagnation pressure P_{0y}, static pressure P_y, and static temperature T_y, are determined to be

$$P_{0y} = 0.7209 P_{0x} = (0.7209)(1.0 \text{ MPa}) = \mathbf{0.7209 \text{ MPa}}$$

$$P_y = 4.5000 P_x = (4.5000)(0.1278 \text{ MPa}) = \mathbf{0.5751 \text{ MPa}}$$

$$T_y = 1.6875 T_x = (1.6875)(166.7 \text{ K}) = \mathbf{281.3 \text{ K}}$$

The air velocity after the shock can be determined from $\mathbf{V}_y = M_y C_y$, where C_y is the velocity of sound at the exit conditions after the shock,

$$V_y = M_y C_y = M_y\sqrt{kRT_y} = (0.5774)\sqrt{(1.4)(0.287 \text{ kJ/kg}\cdot\text{K})(281.3 \text{ K})\left(\frac{1000 \text{ m}^2/\text{s}^2}{1 \text{ kJ/kg}}\right)} = \mathbf{194.1 \text{ m/s}}$$

16-74 Air enters a converging-diverging nozzle at a specified state. The required back pressure that produces a normal shock at the exit plane is to be determined for the specified nozzle geometry.

Assumptions **1** Air is an ideal gas. **2** Flow through the nozzle is steady, one-dimensional, and isentropic before the shock occurs. **3** The shock wave occurs at the exit plane.

Analysis The inlet stagnation pressure in this case is identical to the inlet pressure since the inlet velocity is negligible. Since the flow before the shock to be isentropic,

$$P_{0x} = P_i = 2 \text{ MPa}$$

It is specified that A/A^* =3.5. From Table A-15, Mach number and the pressure ratio which corresponds to this area ratio are the M_x =2.80 and P_x/P_{0x} = 0.03685. The pressure ratio across the shock for this M_x value is, from Table A-16, P_y/P_x = 8.98. Thus the back pressure, which is equal to the static pressure at the nozzle exit, must be

$$P_y = 8.98P_x = 8.98\times0.03685P_{0x} = 8.98\times0.03685\times(2 \text{ MPa}) = \mathbf{0.662\ MPa}$$

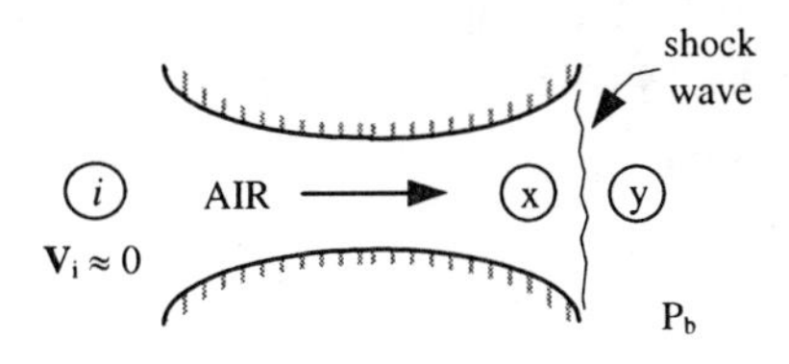

16-75 Air enters a converging-diverging nozzle at a specified state. The required back pressure that produces a normal shock at the exit plane is to be determined for the specified nozzle geometry.

Assumptions **1** Air is an ideal gas. **2** Flow through the nozzle is steady, one-dimensional, and isentropic before the shock occurs.

Analysis The inlet stagnation pressure in this case is identical to the inlet pressure since the inlet velocity is negligible. Since the flow before the shock to be isentropic,

$$P_{0x} = P_i = 2 \text{ MPa}$$

It is specified that A/A^* = 2. From Table A-15, the Mach number and the pressure ratio which corresponds to this area ratio are the M_x =2.20 and P_x/P_{0x} = 0.09352. The pressure ratio across the shock for this M_x value is, from Table A-16, P_y/P_x = 5.48. Thus the back pressure, which is equal to the static pressure at the nozzle exit, must be

$$P_y = 5.48P_x = 5.48\times0.09352P_{0x} = 5.48\times0.09352\times(2 \text{ MPa}) = \mathbf{1.025\ MPa}$$

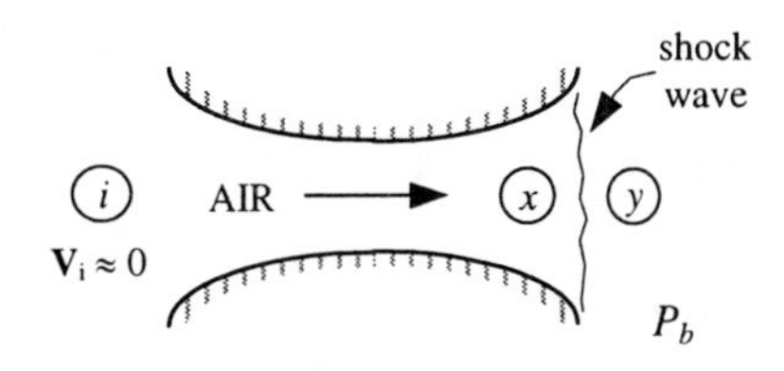

16-76 Air flowing through a nozzle experiences a normal shock. The effect of the shock wave on various properties is to be determined. Analysis is to be repeated for helium under the same conditions.

Assumptions **1** Air and helium are ideal gases. **2** Flow through the nozzle is steady, one-dimensional, and isentropic before the shock occurs.

Properties The properties of air are k = 1.4 and R = 0.287 kJ/kg K, and the properties of helium are k = 1.667 and R = 2.0769 kJ/kg K(Table A-1 and Table A-2a).

Analysis The air properties upstream the shock are

$M_x = 2.5$, $P_x = 61.64$ kPa, and $T_x = 262.15$ K

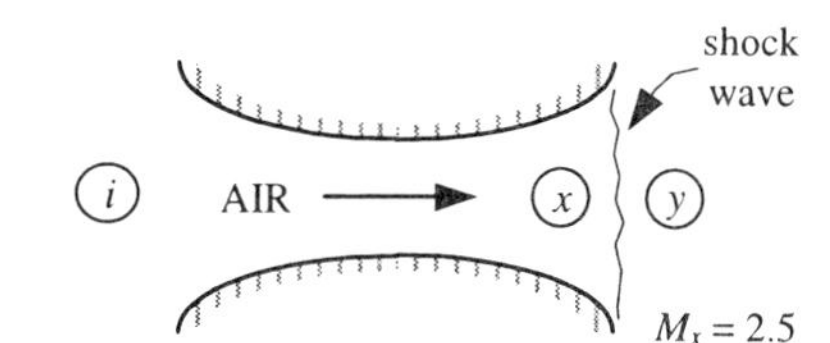

The fluid properties after the shock (denoted by subscript y) are related to those before the shock through the functions listed in Table A-16. For $M_x = 2.5$ we read

$$M_y = \mathbf{0.513}, \quad \frac{P_{0y}}{P_x} = 8.5262, \quad \frac{P_y}{P_x} = 7.125, \quad \text{and} \quad \frac{T_y}{T_x} = 2.1375$$

Then the stagnation pressure P_{0y}, static pressure P_y, and static temperature T_y, are determined to be

$P_{0y} = 8.5262P_x = (8.5262)(61.64 \text{ kPa}) = \mathbf{525.6\ kPa}$

$P_y = 7.125P_x = (7.125)(61.64 \text{ kPa}) = \mathbf{439.2\ kPa}$

$T_y = 2.1375T_x = (2.1375)(262.15 \text{ K}) = \mathbf{560.3\ K}$

The air velocity after the shock can be determined from $\mathbf{V}_y = M_yC_y$, where C_y is the velocity of sound at the exit conditions after the shock,

$$\mathbf{V}_y = M_yC_y = M_y\sqrt{kRT_y}$$

$$= (0.513)\sqrt{(1.4)(0.287 \text{ kJ/kg}\cdot\text{K})(560.3 \text{ K})\left(\frac{1000 \text{ m}^2/\text{s}^2}{1 \text{ kJ/kg}}\right)} = \mathbf{243.4\ m/s}$$

We now repeat the analysis for helium. This time we cannot use the tabulated values in Table A-16 since k is not 1.4. Therefore, we have to calculate the desired quantities using the analytical relations,

$$M_y = \left(\frac{M_x^2 + 2/(k-1)}{2M_x^2k/(k-1) - 1}\right)^{1/2} = \left(\frac{2.5^2 + 2/(1.667-1)}{2\times 2.5^2 \times 1.667/(1.667-1) - 1}\right)^{1/2} = \mathbf{0.553}$$

$$\frac{P_y}{P_x} = \frac{1+kM_x^2}{1+kM_y^2} = \frac{1+1.667\times 2.5^2}{1+1.667\times 0.553^2} = 7.5632$$

$$\frac{T_y}{T_x} = \frac{1+M_x^2(k-1)/2}{1+M_y^2(k-1)/2} = \frac{1+2.5^2(1.667-1)/2}{1+0.553^2(1.667-1)/2} = 2.7989$$

$$\frac{P_{0y}}{P_x} = \left(\frac{1+kM_x^2}{1+kM_y^2}\right)\left(1+(k-1)M_y^2/2\right)^{k/(k-1)}$$

$$= \left(\frac{1+1.667\times 2.5^2}{1+1.667\times 0.553^2}\right)\left(1+(1.667-1)\times 0.553^2/2\right)^{1.667/0.667} = 9.641$$

Thus, $P_{0y} = 11.546P_x = (9.641)(61.64 \text{ kPa}) = \mathbf{594.3\ kPa}$

$P_y = 7.5632P_x = (7.5632)(61.64 \text{ kPa}) = \mathbf{466.2\ kPa}$

$T_y = 2.7989T_x = (2.7989)(262.15 \text{ K}) = \mathbf{733.7\ K}$

$$\mathbf{V}_y = M_yC_y = M_y\sqrt{kRT_y} = (0.553)\sqrt{(1.667)(2.0769 \text{ kJ/kg}\cdot\text{K})(733.7 \text{ K})\left(\frac{1000 \text{ m}^2/\text{s}^2}{1 \text{ kJ/kg}}\right)} = \mathbf{881.4\ m/s}$$

16-77 Air flowing through a nozzle experiences a normal shock. The entropy change of air across the normal shock wave is to be determined.

Assumptions **1** Air and helium are ideal gases. **2** Flow through the nozzle is steady, one-dimensional, and isentropic before the shock occurs.

Properties The properties of air are $R = 0.287$ kJ/kg K and $C_p = 1.005$ kJ/kg K, and the properties of helium are $R = 2.0769$ kJ/kg K and $C_p = 5.1926$ kJ/kg K (Table A-1 and Table A-2a).

Analysis The entropy change across the shock is determined to be

$$s_y - s_x = C_p \ln\frac{T_y}{T_x} - R\ln\frac{P_y}{P_x}$$
$$= (1.005 \text{ kJ/kg}\cdot\text{K})\ln(2.1375) - (0.287 \text{ kJ/kg}\cdot\text{K})\ln(7.125) = \mathbf{0.200\ kJ/kg\cdot K}$$

For helium, the entropy change across the shock is determined to be

$$s_y - s_x = C_p \ln\frac{T_y}{T_x} - R\ln\frac{P_y}{P_x}$$
$$= (5.1926 \text{ kJ/kg}\cdot\text{K})\ln(2.7989) - (2.0769 \text{ kJ/kg}\cdot\text{K})\ln(7.5632) = \mathbf{1.142\ kJ/kg\cdot K}$$

16-78E [*Also solved by EES on enclosed CD*] Air flowing through a nozzle experiences a normal shock. The effect of the shock wave on various properties is to be determined. Analysis is to be repeated for helium under the same conditions.

Assumptions **1** Air and helium are ideal gases. **2** Flow through the nozzle is steady, one-dimensional, and isentropic before the shock occurs.

Properties The properties of air are k = 1.4 and R = 0.06855 Btu/lbm R, and the properties of helium are k = 1.667 and R = 0.4961 Btu/lbm R (Table A-1E and Table A-2Ea).

Analysis The air properties upstream the shock are

$M_x = 2.5$, $P_x = 10$ psia, and $T_x = 440.5$ R

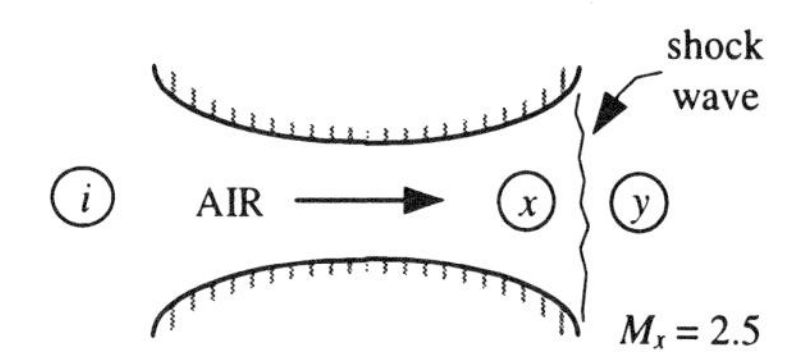

The fluid properties after the shock (denoted by subscript y) are related to those before the shock through the functions listed in Table A-16. For $M_x = 2.5$ we read

$$M_y = \mathbf{0.513}, \quad \frac{P_{0y}}{P_x} = 8.5262, \quad \frac{P_y}{P_x} = 7.125, \quad \text{and} \quad \frac{T_y}{T_x} = 2.1375$$

Then the stagnation pressure P_{0y}, static pressure P_y, and static temperature T_y, are determined to be

$$P_{0y} = 8.5262P_x = (8.5262)(10 \text{ psia}) = \mathbf{85.262\ psia}$$
$$P_y = 7.125P_x = (7.125)(10 \text{ psia}) = \mathbf{71.25\ psia}$$
$$T_y = 2.1375T_x = (2.1375)(440.5 \text{ R}) = \mathbf{941.6\ R}$$

The air velocity after the shock can be determined from $\mathbf{V}_y = M_y C_y$, where C_y is the velocity of sound at the exit conditions after the shock,

$$\mathbf{V}_y = M_y C_y = M_y \sqrt{kRT_y}$$
$$= (0.513)\sqrt{(1.4)(0.06855 \text{ Btu/1bm}\cdot\text{R})(941.6 \text{ R})\left(\frac{25{,}037 \text{ ft}^2/\text{s}^2}{1 \text{ Btu/1bm}}\right)} = \mathbf{771.6\ ft/s}$$

We now repeat the analysis for helium. This time we cannot use the tabulated values in Table A-16 since k is not 1.4. Therefore, we have to calculate the desired quantities using the analytical relations,

$$M_y = \left(\frac{M_x^2 + 2/(k-1)}{2M_x^2 k/(k-1) - 1}\right)^{1/2} = \left(\frac{2.5^2 + 2/(1.667-1)}{2\times 2.5^2 \times 1.667/(1.667-1) - 1}\right)^{1/2} = \mathbf{0.553}$$

$$\frac{p_y}{p_x} = \frac{1+kM_x^2}{1+kM_y^2} = \frac{1+1.667\times 2.5^2}{1+1.667\times 0.553^2} = 7.5632$$

$$\frac{T_y}{T_x} = \frac{1+M_x^2(k-1)/2}{1+M_y^2(k-1)/2} = \frac{1+2.5^2(1.667-1)/2}{1+0.553^2(1.667-1)/2} = 2.7989$$

$$\frac{p_{0y}}{p_x} = \left(\frac{1+kM_x^2}{1+kM_y^2}\right)\left(1+(k-1)M_y^2/2\right)^{k/(k-1)}$$

$$= \left(\frac{1+1.667\times 2.5^2}{1+1.667\times 0.553^2}\right)\left(1+(1.667-1)\times 0.553^2/2\right)^{1.667/0.667} = 9.641$$

Thus,
$$P_{0y} = 11.546P_x = (9.641)(10 \text{ psia}) = \mathbf{594.3\ psia}$$
$$P_y = 7.5632P_x = (7.5632)(10 \text{ psia}) = \mathbf{75.63\ psia}$$
$$T_y = 2.7989T_x = (2.7989)(440.5 \text{ R}) = \mathbf{1232.9\ R}$$

$$\mathbf{V}_y = M_y C_y = M_y\sqrt{kRT_y} = (0.553)\sqrt{(1.667)(0.4961 \text{ Btu/1bm.R})(1232.9 \text{ R})\left(\frac{25{,}037 \text{ ft}^2/\text{s}^2}{1 \text{ Btu/1bm}}\right)} = \mathbf{2794\ ft/s}$$

16-79E EES solution of this (and other comprehensive problems designated with the *computer icon*) is available to instructors at the *Instructor Manual* section of the *Online Learning Center* (OLC) at www.mhhe.com/cengel-boles. See the Preface for access information.

16-80 Air flowing through a nozzle experiences a normal shock. Various properties are to be calculated before and after the shock.

Assumptions **1** Air is an ideal gas. **2** Flow through the nozzle is steady, one-dimensional, and isentropic before the shock occurs.

Properties The properties of air are $k = 1.4$, $R = 0.287$ kJ/kg K, and $C_p = 1.005$ kJ/kg K (Table A-1 and Table A-2a).

Analysis The stagnation temperature and pressure before the shock are

$$T_{0x} = T_x + \frac{\mathbf{V}_x^2}{2C_p} = 217 + \frac{(680\text{ m/s})^2}{2(1.005\text{ kJ/kg}\cdot\text{K}}\left(\frac{1\text{ kJ/kg}}{1000\text{ m}^2/\text{s}^2}\right) = 447.0\text{ K}$$

$$P_{0x} = P_x\left(\frac{T_{0x}}{T_x}\right)^{k/(k-1)} = (22.6\text{ kPa})\left(\frac{447.0\text{ K}}{217\text{ K}}\right)^{1.4/(1.4-1)} = 283.6\text{ kPa}$$

shock wave

AIR → ⓧ ⓨ

The velocity and the Mach number before the shock are determined from

$$C_x = \sqrt{kRT}_x = \sqrt{(1.4)(0.287\text{ kJ/kg}\cdot\text{K})(217.0\text{ K})\left(\frac{1000\text{ m}^2/\text{s}^2}{1\text{ kJ/kg}}\right)} = \mathbf{295.3\text{ m/s}}$$

and

$$M_x = \frac{\mathbf{V}_x}{C_x} = \frac{680\text{ m/s}}{295.3\text{ m/s}} = \mathbf{2.30}$$

The fluid properties after the shock (denoted by subscript y) are related to those before the shock through the functions listed in Table A-16. For $M_x = 2.30$ we read

$$M_y = \mathbf{0.5344}, \quad \frac{P_{0y}}{P_x} = 7.2937, \quad \frac{P_y}{P_x} = 6.005, \quad \text{and} \quad \frac{T_y}{T_x} = 1.9468$$

Then the stagnation pressure P_{0y}, static pressure P_y, and static temperature T_y, are determined to be

$$P_{0y} = 7.2937P_x = (7.2937)(22.6\text{ kPa}) = \mathbf{164.8\text{ kPa}}$$

$$P_y = 6.005P_x = (6.005)(22.6\text{ kPa}) = \mathbf{135.7\text{ kPa}}$$

$$T_y = 1.9468T_x = (1.9468)(217\text{ K}) = \mathbf{422.5\text{ K}}$$

The air velocity after the shock can be determined from $\mathbf{V}_y = M_yC_y$, where C_y is the velocity of sound at the exit conditions after the shock,

$$\mathbf{V}_y = M_yC_y = M_y\sqrt{kRT_y}$$

$$= (0.5344)\sqrt{(1.4)(0.287\text{ kJ/kg.K})(422.5\text{ K})\left(\frac{1000\text{ m}^2/\text{s}^2}{1\text{ kJ/kg}}\right)} = \mathbf{220.2\text{ m/s}}$$

16-81 Air flowing through a nozzle experiences a normal shock. The entropy change of air across the normal shock wave is to be determined.

Assumptions **1** Air is an ideal gas. **2** Flow through the nozzle is steady, one-dimensional, and isentropic before the shock occurs.

Properties The properties of air are $R = 0.287$ kJ/kg K and $C_p = 1.005$ kJ/kg K (Table A-1 and Table A-2a).

Analysis The entropy change across the shock is determined to be

$$s_y - s_x = C_p \ln\frac{T_y}{T_x} - R\ln\frac{P_y}{P_x}$$
$$= (1.005\ \text{kJ/kg}\cdot\text{K})\ln(1.9468) - (0.287\ \text{kJ/kg}\cdot\text{K})\ln(6.005)$$
$$= \mathbf{0.155\ kJ/kg\cdot K}$$

16-82 The entropy change of air across the shock for upstream Mach numbers between 0.5 and 1.5 is to be determined and plotted.

Assumptions **1** Air is an ideal gas. **2** Flow through the nozzle is steady, one-dimensional, and isentropic before the shock occurs.

Properties The properties of air are $k = 1.4$, $R = 0.287$ kJ/kg K, and $C_p = 1.005$ kJ/kg K (Table A-1 and Table A-2a).

Analysis The entropy change across the shock is determined to be

$$s_y - s_x = C_p \ln\frac{T_y}{T_x} - R\ln\frac{P_y}{P_x}$$

where

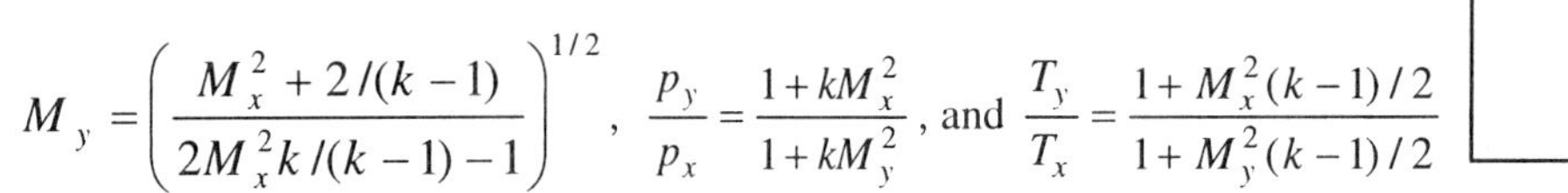
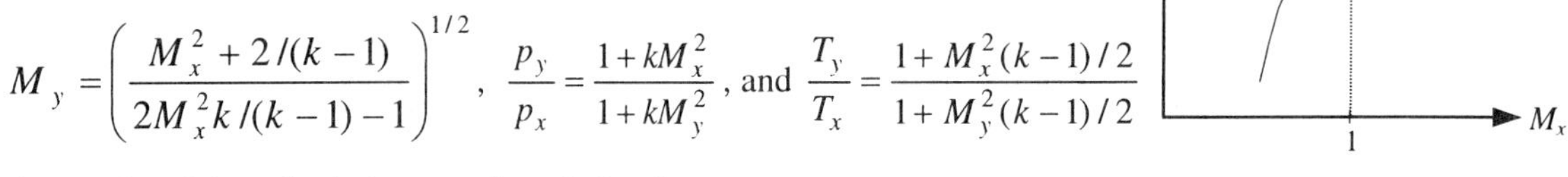

$$M_y = \left(\frac{M_x^2 + 2/(k-1)}{2M_x^2 k/(k-1) - 1}\right)^{1/2}, \quad \frac{P_y}{P_x} = \frac{1+kM_x^2}{1+kM_y^2}, \text{ and } \frac{T_y}{T_x} = \frac{1+M_x^2(k-1)/2}{1+M_y^2(k-1)/2}$$

The results of the calculations can be tabulated as

M_x	M_y	T_y/T_x	P_y/P_x	$s_y - s_x$
0.5	2.6458	0.1250	0.4375	-1.853
0.6	1.8778	0.2533	0.6287	-1.247
0.7	1.5031	0.4050	0.7563	-0.828
0.8	1.2731	0.5800	0.8519	-0.501
0.9	1.1154	0.7783	0.9305	-0.231
1.0	1.0000	1.0000	1.0000	0.0
1.1	0.9118	1.0649	1.2450	0.0003
1.2	0.8422	1.1280	1.5133	0.0021
1.3	0.7860	1.1909	1.8050	0.0061
1.4	0.7397	1.2547	2.1200	0.0124
1.5	0.7011	1.3202	2.4583	0.0210

The total entropy change is negative for upstream Mach numbers M_x less than unity. Therefore, normal shocks cannot occur when $M_x < 1$.

Nozzle and Diffuser Efficiencies

16-83C The velocity coefficient C_v and the efficiency η_N of a nozzle are related to each other by $C_v = \sqrt{\eta_N}$, and both quantities are always less than one. Therefore, C_v is always greater than η_N.

16-84C The pressure rise coefficient of a diffuser represents the increase in the static pressure of a fluid relative to the maximum possible pressure rise. The pressure recovery factor, on the other hand, represents the actual stagnation pressure of a fluid at the diffuser exit relative to the maximum possible stagnation pressure.

16-85 Atmospheric air enters a 95 percent efficient diffuser. The temperature, pressure, and flow area at the diffuser exit are to be determined.

Assumptions **1** Air is an ideal gas with constant specific heats at room temperature. **2** Flow through the diffuser is steady and one-dimensional. **3** The diffuser is adiabatic.

Properties The properties of air at room temperature are $k = 1.4$ and $R = 0.287$ kJ/kg·K (Tables A-1, A-2a).

Analysis The stagnation temperature and pressure at the diffuser inlet are determined from

$$T_{01} = T_1 + \frac{\mathbf{V}_1^2}{2C_p} = 260 + \frac{(250\text{ m/s})^2}{2(1.005\text{ kJ/kg}\cdot\text{K})}\left(\frac{1\text{ kJ/kg}}{1000\text{ m}^2/\text{s}^2}\right) = 291.1\text{ K}$$

$$P_{01} = P_1\left(\frac{T_{01}}{T_1}\right)^{k/(k-1)} = (90\text{ kPa})\left(\frac{291.1\text{ K}}{260\text{ K}}\right)^{1.4/(1.4-1)} = 133.7\text{ kPa}$$

Assuming the diffuser to be adiabatic, the energy equation reduces to $h_{01} = h_{02}$. Noting that $h = C_pT$ and the specific heats are assumed to be constant, we have

$$T_{01} = T_{02} = T_0 = 291.1\text{ K}$$

For an ideal gas with constant specific heats, the diffuser efficiency relation can be expressed as

$$\eta_D = \frac{h_{02s} - h_1}{h_{01} - h_1} = \frac{T_{02s} - T_1}{T_{01} - T_1}$$

Then the temperature T_{02s} becomes

$$0.95 = \frac{T_{02s} - 260}{291.1 - 260} \longrightarrow T_{02s} = 289.5\text{ K}$$

90 kPa
260 K
250 m/s
AIR
$\eta_D = 95\%$
80 m/s

Noting that $P_{02} = P_{02s}$,the isentropic relation between states 1 and 02s gives

$$P_{02} = P_{02s} = P_1\left(\frac{T_{02s}}{T_1}\right)^{k/(k-1)} = (90\text{ kPa})\left(\frac{289.5\text{ K}}{260\text{ K}}\right)^{1.4/(1.4-1)} = 131.1\text{ kPa}$$

Thus, $$T_2 = T_{02} - \frac{\mathbf{V}_2^2}{2C_p} = (291.1\text{ K}) - \frac{(80\text{ m/s})^2}{2(1.005\text{ kJ/kg}\cdot\text{K})}\left(\frac{1\text{ kJ/kg}}{1000\text{ m}^2/\text{s}^2}\right) = \mathbf{287.9\text{ K}}$$

Then the static exit pressure becomes

$$P_2 = P_{02}\left(\frac{T_2}{T_{02}}\right)^{k/(k-1)} = (131.1\text{ kPa})\left(\frac{287.9\text{ K}}{291.1\text{ K}}\right)^{1.4/(1.4-1)} = \mathbf{126.1\text{ kPa}}$$

Also, $$\rho_2 = \frac{P_2}{RT_2} = \frac{126.1\text{ kPa}}{(0.287\text{ kPa}\cdot\text{m}^3/\text{kg}\cdot\text{K})(287.9\text{ K})} = 1.526\text{ kg/m}^3$$

Thus the exit area is

$$A_2 = \frac{\dot{m}}{\rho_2\mathbf{V}_2} = \frac{15\text{ kg/s}}{(1.526\text{ kg/m}^3)(80\text{ m/s})} = \mathbf{0.123\text{ m}^2}$$

16-86 Atmospheric air enters a 95 percent efficient diffuser. The pressure rise coefficient and the pressure recovery factor are to be determined.

Assumptions **1** Air is an ideal gas with constant specific heats at room temperature. **2** Flow through the diffuser is steady and one-dimensional. **3** The diffuser is adiabatic.

Analysis The various pressures were determined in Problem 16-85 to be P_1 = 90, P_{01} = 133.7, P_2 = 126.1 and P_{02} = 131.1 kPa. Then the pressure rise coefficient and pressure recovery factor are determined from their definitions to be

Pressure rise coefficient: $$C_{PR} = \frac{P_2 - P_1}{P_{01} - P_1} = \frac{126.1 - 90.0}{133.7 - 90.0} = \mathbf{0.826}$$

Pressure recovery factor: $$F_P = \frac{P_{02}}{P_{01}} = \frac{131.1}{133.7} = \mathbf{0.981}$$

16-87E Atmospheric air enters a 90 percent efficient diffuser. The temperature, pressure, and flow area at the diffuser exit are to be determined.

Assumptions **1** Air is an ideal gas with constant specific heats at room temperature . **2** Flow through the diffuser is steady and one-dimensional. **3** The diffuser is adiabatic.

Properties The properties of air at room temperature are k = 1.4, R = 0.287 kJ/kg·K, and C_p = 0.240 Btu/lbm·R (Tables A-1E and A-2Ea).

Analysis The stagnation temperature and pressure at the diffuser inlet are determined from

$$T_{01} = T_1 + \frac{\mathbf{V}_1^2}{2C_p} = 450 + \frac{(800\text{ ft/s})^2}{2(0.240\text{ Btu/1bm}\cdot\text{R})}\left(\frac{1\text{ Btu/1bm}}{25{,}037\text{ ft}^2/\text{s}^2}\right) = 503.3\text{ R}$$

$$P_{01} = P_1\left(\frac{T_{01}}{T_1}\right)^{k/(k-1)} = (14.0\text{ psia})\left(\frac{503.3\text{ R}}{450\text{ R}}\right)^{1.4/(1.4-1)} = 20.7\text{ psia}$$

Assuming the diffuser to be adiabatic, the energy equation reduces to $h_{01} = h_{02}$. Noting that $h = C_pT$ and the specific heats are assumed to be constant, we have

$$T_{01} = T_{02} = T_0 = 503.3\text{ R}$$

For an ideal gas with constant specific heats, the diffuser efficiency relation can be expressed as

$$\eta_D = \frac{h_{02s} - h_1}{h_{01} - h_1} = \frac{T_{02s} - T_1}{T_{01} - T_1}$$

Then the temperature T_{02s} becomes

14 psia
450 R
800 m/s
AIR
η_D = 90%
250 ft/s

$$0.95 = \frac{T_{02s} - 450}{503.3 - 450} \longrightarrow T_{02s} = 500.6\text{ R}$$

Noting that $P_{02} = P_{02s}$, the isentropic relation between states 1 and 02s gives

$$P_{02} = P_{02s} = P_1\left(\frac{T_{02s}}{T_1}\right)^{k/(k-1)} = (14.0\text{ psia})\left(\frac{500.6\text{ R}}{450\text{ R}}\right)^{1.4/(1.4-1)} = 20.3\text{ psia}$$

Thus,

$$T_2 = T_{02} - \frac{\mathbf{V}_2^2}{2C_p} = (503.3\text{ R}) - \frac{(250\text{ ft/s})^2}{2(0.240\text{ Btu/1bm}\cdot\text{R})}\left(\frac{1\text{ Btu/1bm}}{25{,}037\text{ ft}^2/\text{s}^2}\right) = \mathbf{498.1\text{ R}}$$

Then the static exit pressure becomes

$$P_2 = P_{02}\left(\frac{T_2}{T_{02}}\right)^{k/(k-1)} = (20.3\text{ psia})\left(\frac{498.1\text{ R}}{503.3\text{ R}}\right)^{1.4/(1.4-1)} = \mathbf{19.6\text{ psia}}$$

Also,

$$\rho_2 = \frac{P_2}{RT_2} = \frac{19.6\text{ psia}}{(0.3704\text{ psia}\cdot\text{ft}^3/\text{1bm}\cdot\text{R})(498.1\text{ R})} = 0.106\text{ 1bm/ft}^3$$

Thus the exit area is

$$A_2 = \frac{\dot{m}}{\rho_2\mathbf{V}_2} = \frac{22\text{ lbm/s}}{(0.106\text{ lbm/ft}^3)(250\text{ ft/s})} = \mathbf{0.830\text{ ft}^2}$$

16-88 Air entering a 90 percent efficient diffuser experiences a normal shock at the diffuser entrance. The static pressure rise across the diffuser is to be determined.

Assumptions **1** Air is an ideal gas with constant specific heats at room temperature. **2** Flow through the diffuser is steady and one-dimensional. **3** The diffuser is adiabatic.

Properties Air properties at room temperature are $k = 1.4$, $R = 0.287$ kJ/kg·K, and $C_p = 1.005$ kJ/kg·K.

Analysis The speed of sound and the Mach number at the diffuser inlet before the shock are

$$C_x = \sqrt{kRT}_x = \sqrt{(1.4)(0.287\text{ kJ/kg}\cdot\text{K})(255.7\text{ K})\left(\frac{1000\text{ m}^2/\text{s}^2}{1\text{ kJ/kg}}\right)} = 320.5\text{ m/s}$$

$$M_x = \frac{\mathbf{V}_x}{C_x} = \frac{640\text{ m/s}}{320.5\text{ m/s}} = 2.0$$

AIR

x ① ②

$\eta_D = 90\%$

640 m/s

200 m/s

Using the values listed in Table A-16 for $M_x = 2.0$,

$$M_y = 0.5774, \quad \frac{\rho_y}{\rho_x} = \frac{\mathbf{V}_x}{\mathbf{V}_y} = 2.6666, \quad \frac{P_y}{P_x} = 4.5000, \quad \text{and} \quad \frac{T_y}{T_x} = 1.6875$$

the fluid properties after the shock at the diffuser inlet are determined to be

$$\mathbf{V}_1 = \mathbf{V}_y = \mathbf{V}_x / 2.6666 = (640\text{ m/s})/2.6666 = 240.0\text{ m/s}$$

$$P_1 = P_y = 4.5000 P_x = (4.5000)(54.0\text{ kPa}) = 243.0\text{ kPa}$$

$$T_1 = T_y = 1.6875 T_x = (1.6875)(255.7\text{ K}) = 431.5\text{ K}$$

Then the stagnation temperature and pressure at the diffuser inlet before the shock are determined from

$$T_{01} = T_1 + \frac{\mathbf{V}_1^2}{2C_p} = 431.5 + \frac{(240\text{ m/s})^2}{2(1.005\text{ kJ/kg}\cdot\text{K}}\left(\frac{1\text{ kJ/kg}}{1000\text{ m}^2/\text{s}^2}\right) = 460.2\text{ K}$$

$$P_{01} = P_1\left(\frac{T_{01}}{T_1}\right)^{k/(k-1)} = (243.0\text{ kPa})\left(\frac{460.2\text{ K}}{431.5\text{ K}}\right)^{1.4/(1.4-1)} = 304.4\text{ kPa}$$

Assuming the diffuser to be adiabatic, the energy equation reduces to $h_{01} = h_{02}$. Noting that $h = C_pT$ and the specific heats are assumed to be constant, we have

$$T_{01} = T_{02} = T_0 = 460.2\text{ K}$$

For an ideal gas with constant specific heats, the diffuser efficiency relation can be expressed as

$$\eta_D = \frac{h_{02s} - h_1}{h_{01} - h_1} = \frac{T_{02s} - T_1}{T_{01} - T_1}$$

Then the temperature T_{02s} becomes

$$0.90 = \frac{T_{02s} - 431.5}{460.2 - 431.5} \longrightarrow T_{02s} = 457.3\text{ K}$$

Noting that $P_{02} = P_{02s}$, the isentropic relation between states 1 and 02s gives

$$P_{02} = P_{02s} = P_1\left(\frac{T_{02s}}{T_1}\right)^{k/(k-1)} = (243.0\text{ kPa})\left(\frac{457.3\text{ K}}{431.5\text{ K}}\right)^{1.4/(1.4-1)} = 297.8\text{ kPa}$$

Then the static exit temperature and pressure become

$$T_2 = T_{02} - \frac{\mathbf{V}_2^2}{2C_p} = (460.2\text{ K}) - \frac{(200\text{ m/s})^2}{2(1.005\text{ kJ/kg}\cdot\text{K})}\left(\frac{1\text{ kJ/kg}}{1000\text{ m}^2/\text{s}^2}\right) = 440.3\text{ K}$$

Then the static exit pressure becomes

$$P_2 = P_{02}\left(\frac{T_2}{T_{02}}\right)^{k/(k-1)} = (297.8\text{ kPa})\left(\frac{440.3\text{ K}}{460.2\text{ K}}\right)^{1.4/(1.4-1)} = 255.1\text{ kPa}$$

Thus the static pressure rise across the diffuser is

$$\Delta P = P_2 - P_x = 255.1 - 54.0 = \mathbf{201.1\text{ kPa}}$$

16-89 Products of combustion enter the nozzle of a gas turbine engine. It is to be determined whether the nozzle is converging or converging-diverging. The exit velocity and the exit area are also to be determined.

Assumptions **1** The combustion gases behave as an ideal gas with specified properties. **2** Flow through the nozzle is steady, one-dimensional, and isentropic. **3** The nozzle is adiabatic.

Properties The properties of combustion gases are given to be $k = 1.34$ and $C_p = 1.16$ kJ/kg·K.

Analysis Before we can determine whether the nozzle is designed to be converging or converging-diverging, we need to calculate the critical pressure, P^*. The stagnation temperature and pressure at the nozzle inlet are determined from

$$T_{01} = T_1 + \frac{\mathbf{V}_1^2}{2C_p} = 1000 + \frac{(200\ \text{m/s})^2}{2(1.16\ \text{kJ/kg}\cdot\text{K})}\left(\frac{1\ \text{kJ/kg}}{1000\ \text{m}^2/\text{s}^2}\right) = 1017.2\ \text{K}$$

$$P_{01} = P_1\left(\frac{T_{01}}{T_1}\right)^{k/(k-1)} = (400\ \text{kPa})\left(\frac{1017.2\ \text{K}}{1000\ \text{K}}\right)^{1.34/0.34} = 427.8\ \text{kPa}$$

Combustion Gases
1 → 2
200 m/s
$\eta_N = 100\%$

These stagnation temperature and pressure values remain constant throughout the nozzle since the flow is assumed to be isentropic. That is,

$$T_0 = T_{01} = 1017.2\ \text{K}$$
$$P_0 = P_{01} = 427.8\ \text{kPa}$$

The critical pressure is determined from

$$P^* = P_0\left(\frac{2}{k+1}\right)^{k/(k-1)} = (427.8\ \text{kPa})\left(\frac{2}{1.34+1}\right)^{1.34/0.34} = 230.4\ \text{kPa}$$

The nozzle exit pressure is above the critical pressure, and thus the flow is not choked. Therefore, the flow cannot be supersonic at the nozzle exit. Thus the nozzle we have must be a **converging** one.

Since the flow is isentropic, the exit temperature can be determined from

$$\frac{T_{2s}}{T_{01}} = \left(\frac{P_2}{P_{01}}\right)^{(k-1)/k} \longrightarrow T_{2s} = (1017.2\ \text{K})\left(\frac{270\ \text{kPa}}{427.8\ \text{kPa}}\right)^{0.34/1.34} = 905.1\ \text{K}$$

Then the exit velocity is determined from the steady-flow energy balance $\dot{E}_{in} = \dot{E}_{out}$ with $q = w = 0$,

$$h_1 + \mathbf{V}_1^2/2 = h_2 + \mathbf{V}_2^2/2 \longrightarrow 0 = h_2 - h_1 + \frac{\mathbf{V}_2^2 - \mathbf{V}_1^{2\nearrow 0}}{2}$$

Solving for $\mathbf{V}_2$,

$$\mathbf{V}_2 = \sqrt{2C_p(T_1 - T_2)} = \sqrt{2(1.16\ \text{kJ/kg}\cdot\text{K})(1000-905.1)\text{K}\left(\frac{1000\ \text{m}^2/\text{s}^2}{1\ \text{kJ/kg}}\right)} = \mathbf{469.2\ m/s}$$

Also, $$R = C_p - C_v = C_p - \frac{C_p}{k} = C_p(1 - 1/k) = 1.16(1 - 1/1.34) = 0.294\ \text{kJ/kg}\cdot\text{K}$$

$$\rho_2 = \frac{P_2}{RT_2} = \frac{270\ \text{kPa}}{(0.294\ \text{kPa}\cdot\text{m}^3/\text{kg}\cdot\text{K})(905.1\ \text{K})} = 1.015\ \text{kg/m}^3$$

Then the exit area is determined from mass flow relation to be

$$A_2 = \frac{\dot{m}}{\rho_2 \mathbf{V}_2} = \frac{3\ \text{kg/s}}{(1.015\ \text{kg/m}^3)(469.2\ \text{m/s})} = 63.0\times10^{-4}\ \text{m}^2 = \mathbf{63.0\ cm^2}$$

16-90 EES solution of this (and other comprehensive problems designated with the *computer icon*) is available to instructors at the *Instructor Manual* section of the *Online Learning Center* (OLC) at www.mhhe.com/cengel-boles. See the Preface for access information.

16-91 Products of combustion enter the 95 percent efficient nozzle of a gas turbine. It is to be determined whether the nozzle is converging or converging-diverging. The exit velocity and the exit area are also to be determined.

Assumptions **1** The combustion gases behave as an ideal gas with specified properties. **2** Flow through the nozzle is steady and one-dimensional. **3** The nozzle is adiabatic.

Properties The properties of combustion gases are given to be $k = 1.34$ and $C_p = 1.16$ kJ/kg·K.

Analysis Before we can determine whether the nozzle is designed to be converging or converging-diverging, we need to calculate the critical pressure, P^*. The stagnation temperature and pressure at the nozzle inlet are determined from

$$T_{01} = T_1 + \frac{\mathbf{V}_1^2}{2C_p} = 1000 + \frac{(200\ \text{m/s})^2}{2(1.16\ \text{kJ/kg}\cdot\text{K})}\left(\frac{1\ \text{kJ/kg}}{1000\ \text{m}^2/\text{s}^2}\right) = 1017.2\ \text{K}$$

Combustion Gases

1 → 200 m/s

$\eta_N = 95\%$ → 2

$$P_{01} = P_1\left(\frac{T_{01}}{T_1}\right)^{k/(k-1)} = (400\ \text{kPa})\left(\frac{1017.2\ \text{K}}{1000\ \text{K}}\right)^{1.34/0.34} = 427.8\ \text{kPa}$$

Assuming the nozzle to be adiabatic, the energy equation reduces to $h_{01} = h_{02}$. Noting that $h = C_pT$ and the specific heats are assumed to be constant, we have

$$T_{01} = T_{02} = T_0 = 1017.2\ \text{K}$$

We can determine whether the nozzle is converging or converging-diverging by simply calculating the critical pressure P^* at the nozzle exit and comparing it to the specified exit pressure P_2. If $P_2 < P^*$ the nozzle is converging-diverging, otherwise it is converging.

The critical pressure P^* is the actual pressure at the actual throat when sonic velocity is reached at the throat ($M^* = 1$). The critical temperature T^* is related to the stagnation temperature T_0, and is determined from

$$\frac{T^*}{T_0} = \frac{2}{k+1} \longrightarrow T^* = T_0\left(\frac{2}{k+1}\right) = (1017.2\ \text{K})\left(\frac{2}{1.34+1}\right) = 869.4\ \text{K}$$

For an ideal gas with constant specific heats the nozzle efficiency between the inlet and the throat can be expressed as

$$\eta_N = \frac{h_{01} - h_t}{h_{01} - h_{ts}} = \frac{C_p(T_{01} - T_t)}{C_p(T_{01} - T_{ts})} = \frac{T_{01} - T^*}{T_{01} - T_{ts}}$$

Thus,

$$0.95 = \frac{1017.2 - 869.4}{1017.2 - T_{ts}} \longrightarrow T_{ts} = 861.6\ \text{K}$$

The critical (throat) pressure P^* for both the actual and the isentropic cases are the same, and is determined from

$$P^* = P_{01}\left(\frac{T_s^*}{T_{01}}\right)^{k/(k-1)} = (427.8\ \text{kPa})\left(\frac{861.6\ \text{K}}{1017.2\ \text{K}}\right)^{1.34/0.34} = 222.4\ \text{kPa}$$

The nozzle exit pressure is above the critical pressure, and thus the flow is not choked. Therefore, the flow cannot be supersonic at the nozzle exit. Thus the nozzle we have must be a **converging** one. Then,

$$P_2 = P_{2s} = 270\ \text{kPa}$$

The isentropic exit temperature can be determined from

$$\frac{T_{2s}}{T_{01}} = \left(\frac{P_2}{P_{01}}\right)^{(k-1)/k} \longrightarrow T_{2s} = (1017.2\text{ K})\left(\frac{270\text{ kPa}}{427.8\text{ kPa}}\right)^{0.34/1.34} = 905.1\text{ K}$$

The actual exit temperature is determined from

$$\eta_N = \frac{h_{01} - h_2}{h_{01} - h_{2s}} = \frac{T_{01} - T_2}{T_{01} - T_{2s}} \longrightarrow 0.95 = \frac{1017.2 - T_2}{1017.2 - 905.1} \longrightarrow T_2 = 910.7\text{ K}$$

Then the exit velocity is determined from the steady-flow energy balance $\dot{E}_{in} = \dot{E}_{out}$ with $q = w = 0$,

$$h_1 + \mathbf{V}_1^2/2 = h_2 + \mathbf{V}_2^2/2 \longrightarrow 0 = h_2 - h_1 + \frac{\mathbf{V}_2^2 - \mathbf{V}_1^2 \nearrow^0}{2}$$

Solving for $\mathbf{V}_2$

$$V_2 = \sqrt{2C_p(T_1 - T_2)} = \sqrt{2(1.16\text{ kJ/kg}\cdot\text{K})(1000 - 910.7)\text{K}\left(\frac{1000\text{ m}^2/\text{s}^2}{1\text{ kJ/kg}}\right)} = \mathbf{455.2\text{ m/s}}$$

Also,

$$R = C_p - C_v = C_p - \frac{C_p}{k} = C_p(1 - 1/k) = 1.16(1 - 1/1.34) = 0.294\text{ kJ/kg}\cdot\text{K}$$

$$\rho_2 = \frac{P_2}{RT_2} = \frac{270\text{ kPa}}{(0.294\text{ kPa}\cdot\text{m}^3/\text{kg}\cdot\text{K})(910.7\text{ K})} = 1.008\text{ kg/m}^3$$

Then the exit area is determined from mass flow relation to be

$$A_2 = \frac{\dot{m}}{\rho_2 \mathbf{V}_2} = \frac{3\text{ kg/s}}{(1.008\text{ kg/m}^3)(455.2\text{ m/s})} = 65.4 \times 10^{-4}\text{ m}^2 = \mathbf{65.4\text{ cm}^2}$$

Steam Nozzles

16-92C The delay in the condensation of the steam is called supersaturation. It occurs in high-speed flows where there isn't sufficient time for the necessary heat transfer and the formation of liquid droplets.

16-93 Steam enters a converging nozzle with a low velocity. The exit velocity, mass flow rate, and exit Mach number are to be determined for isentropic and 90 percent efficient nozzle cases.

Assumptions **1** Flow through the nozzle is steady and one-dimensional. **2** The nozzle is adiabatic.

Analysis (*a*) The inlet stagnation properties in this case are identical to the inlet properties since the inlet velocity is negligible. Thus $h_{01} = h_1$.

At the inlet,
$$\left.\begin{array}{l} P_1 = P_{01} = 3\text{ MPa} \\ T_1 = T_{01} = 600°\text{C} \end{array}\right\} \begin{array}{l} h_1 = h_{01} = 3682.3\text{ kJ/kg} \\ s_1 = s_{2s} = 7.5085\text{ kJ/kg}\cdot\text{K} \end{array}$$

At the exit,
$$\left.\begin{array}{l} P_2 = 1.8\text{ MPa} \\ s_2 = 7.5085\text{ kJ/kg}\cdot\text{K} \end{array}\right\} \begin{array}{l} h_2 = 3491.2\text{ kJ/kg} \\ v_2 = 0.1981\text{ m}^3\text{/kg} \end{array}$$

① → STEAM → ②
$V_i \approx 0$
a) $\eta_N = 100\%$
b) $\eta_N = 90\%$

Then the exit velocity is determined from the steady-flow energy balance $\dot{E}_{in} = \dot{E}_{out}$ with $q = w = 0$,

$$h_1 + V_1^2/2 = h_2 + V_2^2/2 \longrightarrow 0 = h_2 - h_1 + \frac{V_2^2 - V_1^{2 \nearrow 0}}{2}$$

Solving for V_2,

$$V_2 = \sqrt{2(h_1 - h_2)} = \sqrt{2(3682.3 - 3491.2)\text{ kJ/kg}\left(\frac{1000\text{ m}^2/\text{s}^2}{1\text{ kJ/kg}}\right)} = \mathbf{618.2\ m/s}$$

The mass flow rate is determined from

$$\dot{m} = \frac{1}{v_2}A_2V_2 = \frac{1}{0.1981\text{ m}^3/\text{kg}}(24\times10^{-4}\text{ m}^2)(618.2\text{ m/s}) = \mathbf{7.49\ kg/s}$$

The velocity of sound at the exit of the nozzle is determined from

$$C = \left(\frac{\partial P}{\partial \rho}\right)_s^{1/2} \cong \left(\frac{\Delta P}{\Delta(1/v)}\right)_s^{1/2}$$

The specific volume of steam at s_2 = 7.5085 kJ/kg·K and at pressures just below and just above the specified pressure (1.6 and 2.0 MPa) are determined to be 0.2172 and 0.1825 m^3/kg. Substituting,

$$C_2 = \sqrt{\frac{(2000 - 1600)\text{ kPa}}{\left(\frac{1}{0.1825} - \frac{1}{0.2172}\right)\text{kg/m}^3}\left(\frac{1000\text{ m}^2/\text{s}^2}{1\text{ kPa.m}^3}\right)} = 676.0\text{ m/s}$$

Then the exit Mach number becomes

$$M_2 = \frac{V_2}{C_2} = \frac{618.2\text{ m/s}}{676.0\text{ m/s}} = \mathbf{0.914}$$

(*b*) The inlet stagnation properties in this case are identical to the inlet properties since the inlet velocity is negligible. Thus $h_{01} = h_1$.

At the inlet,
$$\left.\begin{array}{l} P_1 = P_{01} = 3\text{ MPa} \\ T_1 = T_{01} = 600°\text{C} \end{array}\right\} \begin{array}{l} h_1 = h_{01} = 3682.3\text{ kJ/kg} \\ s_1 = s_{2s} = 7.5085\text{ kJ/kg}\cdot\text{K} \end{array}$$

At state 2s,
$$\left.\begin{array}{l} P_{2s} = 1.8\text{ MPa} \\ s_{2s} = 7.5085\text{ kJ/kg}\cdot\text{K} \end{array}\right\} h_{2s} = 3491.2\text{ kJ/kg}$$

The enthalpy of steam at the actual exit state is determined from

$$\eta_N = \frac{h_{01} - h_2}{h_{01} - h_{2s}} \longrightarrow 0.90 = \frac{3682.3 - h_2}{3682.3 - 3491.2} \longrightarrow h_2 = 3510.3\text{ kJ / kg}$$

Therefore,

$$\left.\begin{array}{l} P_2 = 1.8\text{ MPa} \\ h_2 = 3510.3\text{ kJ/kg} \end{array}\right\} \begin{array}{l} v_2 = 0.2003\text{ m}^3\text{/kg} \\ s_2 = 7.5317\text{ kJ/kg}\cdot\text{K} \end{array}$$

Then the exit velocity is determined from the steady-flow energy balance $\dot{E}_{in} = \dot{E}_{out}$ with $q = w = 0$,

$$h_1 + \mathbf{V}_1^2/2 = h_2 + \mathbf{V}_2^2/2 \longrightarrow 0 = h_2 - h_1 + \frac{\mathbf{V}_2^2 - \mathbf{V}_1^{2\nearrow 0}}{2}$$

Solving for $\mathbf{V}_2$,

$$\mathbf{V}_2 = \sqrt{2(h_1 - h_2)} = \sqrt{2(3682.3 - 3510.3)\text{ kJ/kg}\left(\frac{1000\text{ m}^2/\text{s}^2}{1\text{ kJ/kg}}\right)} = \mathbf{586.5\text{ m/s}}$$

The mass flow rate is determined from

$$\dot{m} = \frac{1}{v_2}A_2\mathbf{V}_2 = \frac{1}{0.2003\text{ m}^3\text{ / kg}}(24\times10^{-4}\text{ m}^2)(586.5\text{ m / s}) = \mathbf{7.03\text{ kg / s}}$$

The velocity of sound at the exit of the nozzle is determined from

$$C = \left(\frac{\partial P}{\partial \rho}\right)_s^{1/2} \cong \left(\frac{\Delta P}{\Delta(1/v)}\right)_s^{1/2}$$

The specific volume of steam at s_2 = 7.5317 kJ/kg·K and at pressures just below and just above the specified pressure (1.6 and 2.0 MPa) are determined to be 0.2196 and 0.1845 m^3/kg. Substituting,

$$C_2 = \sqrt{\frac{(2000 - 1600)\text{ kPa}}{\left(\frac{1}{0.1845} - \frac{1}{0.2196}\right)\text{kg/m}^3}\left(\frac{1000\text{ m}^2/\text{s}^2}{1\text{ kPa.m}^3}\right)} = 679.5\text{ m/s}$$

Then the exit Mach number becomes

$$M_2 = \frac{\mathbf{V}_2}{C_2} = \frac{586.5\text{ m / s}}{679.5\text{ m / s}} = \mathbf{0.863}$$

16-94E Steam enters a converging nozzle with a low velocity. The exit velocity, mass flow rate, and exit Mach number are to be determined for isentropic and 90 percent efficient nozzle cases.

Assumptions **1** Flow through the nozzle is steady and one-dimensional. **2** The nozzle is adiabatic.

Analysis (*a*) The inlet stagnation properties in this case are identical to the inlet properties since the inlet velocity is negligible. Thus $h_{01} = h_1$.

At the inlet,
$$\left.\begin{array}{l} P_1 = P_{01} = 450 \text{ psia} \\ T_1 = T_{01} = 900°\text{F} \end{array}\right\} \begin{array}{l} h_1 = h_{01} = 1468.3 \text{ Btu/lbm} \\ s_1 = s_{2s} = 1.7113 \text{ Btu/lbm}\cdot\text{R} \end{array}$$

At the exit,
$$\left.\begin{array}{l} P_2 = 275 \text{ psia} \\ s_{2s} = 1.7113 \text{ Btu/lbm}\cdot\text{R} \end{array}\right\} \begin{array}{l} h_2 = 1400.8 \text{ Btu/lbm} \\ v_2 = 2.5749 \text{ ft}^3\text{/lbm} \end{array}$$

1 → STEAM → 2

$V_i \approx 0$

a) $\eta_N = 100\%$
b) $\eta_N = 90\%$

Then the exit velocity is determined from the steady-flow energy balance $\dot{E}_{in} = \dot{E}_{out}$ with $q = w = 0$,

$$h_1 + V_1^2/2 = h_2 + V_2^2/2 \longrightarrow 0 = h_2 - h_1 + \frac{V_2^2 - V_1^{2} {}^{\nearrow 0}}{2}$$

Solving for V_2,

$$V_2 = \sqrt{2(h_1 - h_2)} = \sqrt{2(1468.3 - 1400.8) \text{ Btu/lbm}\left(\frac{25,037 \text{ ft}^2/\text{s}^2}{1 \text{ Btu/lbm}}\right)} = \mathbf{1838.5 \text{ ft/s}}$$

Then,

$$\dot{m} = \frac{1}{v_2} A_2 V_2 = \frac{1}{2.5749 \text{ ft}^3/\text{lbm}}(3.75/144 \text{ ft}^2)(1838.5 \text{ ft/s}) = \mathbf{18.6 \text{ lbm/s}}$$

The velocity of sound at the exit of the nozzle is determined from

$$C = \left(\frac{\partial P}{\partial \rho}\right)_s^{1/2} \cong \left(\frac{\Delta P}{\Delta(1/v)}\right)_s^{1/2}$$

The specific volume of steam at s_2 = 1.7113 Btu/lbm·R and at pressures just below and just above the specified pressure (250 and 300 psia) are determined to be 2.7726 and 2.4055 ft^3/lbm. Substituting,

$$C_2 = \sqrt{\frac{(300 - 250) \text{ psia}}{\left(\frac{1}{2.4055} - \frac{1}{2.7726}\right) \text{lbm/ft}^3}\left(\frac{25,037 \text{ ft}^2/\text{s}^2}{1 \text{ Btu/lbm}}\right)\left(\frac{1 \text{ Btu}}{5.4039 \text{ ft}^3\cdot\text{psia}}\right)} = 2051.5 \text{ ft/s}$$

Then the exit Mach number becomes

$$M_2 = \frac{V_2}{C_2} = \frac{1838.5 \text{ ft/s}}{2051.5 \text{ ft/s}} = \mathbf{0.896}$$

(*b*) The inlet stagnation properties in this case are identical to the inlet properties since the inlet velocity is negligible. Thus $h_{01} = h_1$.

At the inlet,
$$\left.\begin{array}{l} P_1 = P_{01} = 450 \text{ psia} \\ T_1 = T_{01} = 900°\text{F} \end{array}\right\} \begin{array}{l} h_1 = h_{01} = 1468.3 \text{ Btu/1bm} \\ s_1 = s_{2s} = 1.7113 \text{ Btu/1bm} \cdot \text{R} \end{array}$$

At state 2s,
$$\left.\begin{array}{l} P_{2s} = 275 \text{ psia} \\ s_{2s} = 1.7113 \text{ Btu/lbm} \cdot \text{R} \end{array}\right\} h_{2s} = 1400.8 \text{ Btu/lbm}$$

The enthalpy of steam at the actual exit state is determined from

$$\eta_N = \frac{h_{01} - h_2}{h_{01} - h_{2s}} \longrightarrow 0.90 = \frac{1468.3 - h_2}{1468.3 - 1400.8} \longrightarrow h_2 = 1407.6 \text{ Btu / 1bm}$$

Therefore,

$$\left.\begin{array}{l} P_2 = 275 \text{ psia} \\ h_2 = 1407.6 \text{ Btu/lbm} \end{array}\right\} \begin{array}{l} v_2 = 2.6052 \text{ ft}^3/\text{1bm} \\ s_2 = 1.7169 \text{ Btu/1bm} \cdot \text{R} \end{array}$$

Then the exit velocity is determined from the steady-flow energy balance $\dot{E}_{in} = \dot{E}_{out}$ with $q = w = 0$,

$$h_1 + \mathbf{V}_1^2 / 2 = h_2 + \mathbf{V}_2^2 / 2 \longrightarrow 0 = h_2 - h_1 + \frac{\mathbf{V}_2^2 - \mathbf{V}_1^{2} \nearrow 0}{2}$$

Solving for $\mathbf{V}_2$,

$$\mathbf{V}_2 = \sqrt{2(h_1 - h_2)} = \sqrt{2(1468.3 - 1407.6) \text{ Btu/1bm}\left(\frac{25{,}037 \text{ ft}^2/\text{s}^2}{1 \text{ Btu/1bm}}\right)} = \mathbf{1743.4 \text{ ft/s}}$$

Then,

$$\dot{m} = \frac{1}{v_2} A_2 \mathbf{V}_2 = \frac{1}{2.6052 \text{ ft}^3 / \text{1bm}} (3.75 / 144 \text{ ft}^2)(1743.4 \text{ ft / s}) = \mathbf{17.4 \text{ 1bm / s}}$$

The velocity of sound at the exit of the nozzle is determined from

$$C = \left(\frac{\partial P}{\partial \rho}\right)_s^{1/2} \cong \left(\frac{\Delta P}{\Delta(1/v)}\right)_s^{1/2}$$

The specific volume of steam at s_2 = 1.7169 Btu/lbm·R and at pressures just below and just above the specified pressure (250 and 300 psia) are determined to be 2.8057 and 2.4331 ft^3/lbm. Substituting,

$$C_2 = \sqrt{\frac{(300 - 250) \text{ psia}}{\left(\frac{1}{2.4331} - \frac{1}{2.8057}\right) \text{lbm/ft}^3} \left(\frac{25{,}037 \text{ ft}^2 / \text{s}^2}{1 \text{ Btu/lbm}}\right)\left(\frac{1 \text{ Btu}}{5.4039 \text{ ft}^3 \cdot \text{psia}}\right)} = 2060.2 \text{ ft/s}$$

Then the exit Mach number becomes

$$M_2 = \frac{\mathbf{V}_2}{C_2} = \frac{1743.4 \text{ ft / s}}{2060.2 \text{ ft / s}} = \mathbf{0.846}$$

16-95 Steam enters a converging-diverging nozzle with a low velocity. The exit area and the exit Mach number are to be determined.

Assumptions Flow through the nozzle is steady, one-dimensional, and isentropic.

Analysis The inlet stagnation properties in this case are identical to the inlet properties since the inlet velocity is negligible. Thus $h_{01} = h_1$.

At the inlet,

$$\left.\begin{array}{l} P_1 = P_{01} = 1\text{ MPa} \\ T_1 = T_{01} = 500°\text{C} \end{array}\right\} \begin{array}{l} h_1 = h_{01} = 3478.5\text{ kJ/kg} \\ s_1 = s_{2s} = 7.7622\text{ kJ/kg}\cdot\text{K} \end{array}$$

At the exit,

$$\left.\begin{array}{l} P_2 = 0.2\text{ MPa} \\ s_2 = 7.7622\text{ kJ/kg}\cdot\text{K} \end{array}\right\} \begin{array}{l} h_2 = 3000.4\text{ kJ/kg} \\ v_2 = 1.2330\text{ m}^3\text{/kg} \end{array}$$

Steam → ① ② $V_i \approx 0$ $\eta_N = 100\%$

Then the exit velocity is determined from the steady-flow energy balance $\dot{E}_{in} = \dot{E}_{out}$ with $q = w = 0$,

$$h_1 + \mathbf{V}_1^2/2 = h_2 + \mathbf{V}_2^2/2 \longrightarrow 0 = h_2 - h_1 + \frac{\mathbf{V}_2^2 - \mathbf{V}_1^2 {}^{\nearrow 0}}{2}$$

Solving for $\mathbf{V}_2$,

$$\mathbf{V}_2 = \sqrt{2(h_1 - h_2)} = \sqrt{2(3478.5 - 3000.4)\text{ kJ/kg}\left(\frac{1000\text{ m}^2/\text{s}^2}{1\text{ kJ/kg}}\right)} = 977.9\text{ m/s}$$

The exit area is determined from

$$A_2 = \frac{\dot{m}v_2}{\mathbf{V}_2} = \frac{(2.5\text{ kg/s})(1.2330\text{ m}^3/\text{kg})}{(977.9\text{ m/s})} = 31.5\times10^{-4}\text{ m}^2 = \mathbf{31.5\text{ cm}^2}$$

The velocity of sound at the exit of the nozzle is determined from

$$C = \left(\frac{\partial P}{\partial \rho}\right)_s^{1/2} \cong \left(\frac{\Delta P}{\Delta(1/v)}\right)_s^{1/2}$$

The specific volume of steam at s_2 = 7.7622 kJ/kg·K and at pressures just below and just above the specified pressure (0.1 and 0.3 MPa) are determined to be 2.0951 and 0.9036 m³/kg. Substituting,

$$C_2 = \sqrt{\frac{(300 - 100)\text{ kPa}}{\left(\frac{1}{0.9036} - \frac{1}{2.0951}\right)\text{kg/m}^3}\left(\frac{1000\text{ m}^2/\text{s}^2}{1\text{ kPa}\cdot\text{m}^3}\right)} = 563.7\text{ m/s}$$

Then the exit Mach number becomes

$$M_2 = \frac{\mathbf{V}_2}{C_2} = \frac{977.9\text{ m/s}}{563.7\text{ m/s}} = \mathbf{1.735}$$

16-96 Steam enters a converging-diverging nozzle with a low velocity. The exit area and the exit Mach number are to be determined.

Assumptions Flow through the nozzle is steady and one-dimensional.

Analysis The inlet stagnation properties in this case are identical to the inlet properties since the inlet velocity is negligible. Thus $h_{01} = h_1$.

At the inlet,

$$\left.\begin{array}{l} P_1 = P_{01} = 1\text{ MPa} \\ T_1 = T_{01} = 500°\text{C} \end{array}\right\} \begin{array}{l} h_1 = h_{01} = 3478.5\text{ kJ/kg} \\ s_1 = s_{2s} = 7.7622\text{ kJ/kg}\cdot\text{K} \end{array}$$

At state 2s,

$$\left.\begin{array}{l} P_2 = 0.2\text{ MPa} \\ s_2 = 7.7622\text{ kJ/kg}\cdot\text{K} \end{array}\right\} h_2 = 3000.4\text{ kJ/kg}$$

The enthalpy of steam at the actual exit state is determined from

$$\eta_N = \frac{h_{01} - h_2}{h_{01} - h_{2s}} \longrightarrow 0.95 = \frac{3478.5 - h_2}{3478.5 - 3000.4} \longrightarrow h_2 = 3024.3\text{ kJ / kg}$$

Therefore,

$$\left.\begin{array}{l} P_2 = 0.2\text{ MPa} \\ h_2 = 3024.3\text{ kJ/kg} \end{array}\right\} \begin{array}{l} v_2 = 1.2609\text{ m}^3\text{/kg} \\ s_2 = 7.8059\text{ kJ/kg}\cdot\text{K} \end{array}$$

Then the exit velocity is determined from the steady-flow energy balance $\dot{E}_{in} = \dot{E}_{out}$ with $q = w = 0$,

$$h_1 + \mathbf{V}_1^2/2 = h_2 + \mathbf{V}_2^2/2 \longrightarrow 0 = h_2 - h_1 + \frac{\mathbf{V}_2^2 - \mathbf{V}_1^{2\nearrow 0}}{2}$$

Solving for $\mathbf{V}_2$, $$\mathbf{V}_2 = \sqrt{2(h_1 - h_2)} = \sqrt{2(3478.5 - 3024.3)\text{ kJ/kg}\left(\frac{1000\text{ m}^2\text{/s}^2}{1\text{ kJ/kg}}\right)} = 953.1\text{ m/s}$$

The exit area is determined from

$$A_2 = \frac{\dot{m}v_2}{\mathbf{V}_2} = \frac{(2.5\text{ kg / s})(1.2609\text{ m}^3\text{ / kg})}{(953.1\text{ m / s})} = 33.1\times 10^{-4}\text{ m}^2 = \mathbf{33.1\text{ cm}^2}$$

The velocity of sound at the exit of the nozzle is determined from

$$C = \left(\frac{\partial P}{\partial \rho}\right)_s^{1/2} \cong \left(\frac{\Delta P}{\Delta(1/v)}\right)_s^{1/2}$$

The specific volume of steam at s_2 = 7.8059 kJ/kg·K and at pressures just below and just above the specified pressure (0.1 and 0.3 MPa) are determined to be 2.1417 and 0.9243 m^3/kg. Substituting,

$$C_2 = \sqrt{\frac{(300 - 100)\text{ kPa}}{\left(\frac{1}{0.9243} - \frac{1}{2.1417}\right)\text{kg/m}^3}\left(\frac{1000\text{ m}^2\text{ / s}^2}{1\text{ kPa}\cdot\text{m}^3}\right)} = 570.3\text{ m/s}$$

Then the exit Mach number becomes

$$M_2 = \frac{\mathbf{V}_2}{C_2} = \frac{953.1\text{ m / s}}{570.3\text{ m / s}} = \mathbf{1.671}$$

Review Problems

16-97 A 3-mm diameter leak develops in an automobile tire as a result of an accident. The initial mass flow rate of air through the leak is to be determined.

Assumptions **1** Air is an ideal gas with constant specific properties. **2** Flow of air through the hole is isentropic.

Properties The gas constant of air is R = 0.287 kPa.m^3/kg.K (Table A-1). The specific heat ratio of air at room temperature is k = 1.4 (Table A-2).

Analysis The absolute pressure in the tire is

$$P = P_{\text{gage}} + P_{\text{atm}} = 220 + 94 = 314 \text{ kPa}$$

The critical pressure is, from Table 16-2,

$$P^* = 0.5283 P_0 = (0.5283)(314 \text{ kPa}) = 166 \text{ kPa} > 94 \text{ kPa}$$

Therefore, the flow is choked, and the velocity at the exit of the hole is the sonic velocity. Then the flow properties at the exit becomes

$$\rho_0 = \frac{P_0}{RT_0} = \frac{314 \text{ kPa}}{(0.287 \text{ kPa}\cdot\text{m}^3/\text{kg}\cdot\text{K})(298 \text{ K})} = 3.671 \text{ kg/m}^3$$

$$\rho^* = \rho_0 \left(\frac{2}{k+1}\right)^{1/(k-1)} = (3.671 \text{ kg/m}^3)\left(\frac{2}{1.4+1}\right)^{1/(1.4-1)} = 2.327 \text{ kg/m}^3$$

$$T^* = \frac{2}{k+1} T_0 = \frac{2}{1.4+1}(298 \text{ K}) = 248.3 \text{ K}$$

$$\mathbf{V} = C = \sqrt{kRT^*} = \sqrt{(1.4)(0.287 \text{ kJ/kg}\cdot\text{K})\left(\frac{1000 \text{ m}^2/\text{s}^2}{1 \text{ kJ/kg}}\right)(248.3 \text{ K})} = 315.9 \text{ m/s}$$

Then the initial mass flow rate through the hole becomes

$$\dot{m} = \rho A \mathbf{V} = (2.327 \text{ kg/m}^3)[\pi(0.004 \text{ m})^2/4](315.9 \text{ m/s}) = 0.00924 \text{ kg/s} = \mathbf{0.554 \text{ kg/min}}$$

Discussion The mass flow rate will decrease with time as the pressure inside the tire drops.

16-98 The thrust developed by the engine of a Boeing 777 is about 380 kN. The mass flow rate of air through the nozzle is to be determined.

Assumptions **1** Air is an ideal gas with constant specific properties. **2** Flow of combustion gases through the nozzle is isentropic. **3** Choked flow conditions exist at the nozzle exit. **4** The velocity of gases at the nozzle inlet is negligible.

Properties The gas constant of air is R = 0.287 kPa.m^3/kg.K (Table A-1), and it can also be used for combustion gases. The specific heat ratio of combustion gases is k = 1.33 (Table 16-2).

Analysis The velocity at the nozzle exit is the sonic velocity, which is determined to be

$$\mathbf{V} = C = \sqrt{kRT} = \sqrt{(1.33)(0.287\ \text{kJ/kg}\cdot\text{K})\left(\frac{1000\ \text{m}^2/\text{s}^2}{1\ \text{kJ/kg}}\right)(295\ \text{K})} = 335.6\ \text{m/s}$$

Noting that thrust F is related to velocity by $F = \dot{m}\mathbf{V}$, the mass flow rate of combustion gases is determined to be

$$\dot{m} = \frac{F}{\mathbf{V}} = \frac{380{,}000\ \text{N}}{335.6\ \text{m/s}}\left(\frac{1\ \text{kg.m/s}^2}{1\ \text{N}}\right) = \mathbf{1132.4\ kg/s}$$

16-99 A stationary temperature probe is inserted into an air duct reads 85°C. The actual temperature of air is to be determined.

Assumptions **1** Air is an ideal gas with constant specific properties at room temperature. **2** The stagnation process is isentropic.

Properties The specific heat of air at room temperature is C_p = 1.005 kJ/kg.K (Table A-2).

Analysis The air that strikes the probe will be brought to a complete stop, and thus it will undergo a stagnation process. The thermometer will sense the temperature of this stagnated air, which is the stagnation temperature. The actual air temperature is determined from

$$T = T_0 - \frac{\mathbf{V}^2}{2C_p} = 85°\text{C} - \frac{(250\ \text{m/s})^2}{2\times 1.005\ \text{kJ/kg}\cdot\text{K}}\left(\frac{1\ \text{kJ/kg}}{1000\ \text{m}^2/\text{s}^2}\right) = \mathbf{53.9°C}$$

T

250 m/s

16-100 Nitrogen flows through a heat exchanger. The stagnation pressure and temperature of the nitrogen at the inlet and the exit states are to be determined.

Assumptions **1** Nitrogen is an ideal gas with constant specific properties. **2** Flow of nitrogen through the heat exchanger is isentropic.

Properties The properties of air are C_p = 1.039 kJ/kg.K and k = 1.4 (Table A-2).

Analysis The stagnation temperature and pressure of nitrogen at the inlet and the exit states are determined from

$$T_{01} = T_1 + \frac{\mathbf{V}_1^2}{2C_p} = 10°\text{C} + \frac{(100\text{ m/s})^2}{2\times 1.039\text{ kJ/kg}\cdot°\text{C}}\left(\frac{1\text{ kJ/kg}}{1000\text{ m}^2/\text{s}^2}\right) = \mathbf{14.8°C}$$

$$P_{01} = P_1\left(\frac{T_{01}}{T_1}\right)^{k/(k-1)} = (150\text{ kPa})\left(\frac{288.0\text{ K}}{283.2\text{ K}}\right)^{1.4/(1.4-1)} = \mathbf{159.1\,kPa}$$

Q_{in}

150 kPa
10°C
100 m/s

Nitrogen

100 kPa
200 m/s

From the energy balance relation $E_{in} - E_{out} = \Delta E_{system}$ with $w = 0$

$$q_{in} = C_p(T_2 - T_1) + \frac{\mathbf{V}_2^2 - \mathbf{V}_1^2}{2} + \Delta pe^{\nearrow 0}$$

$$150\text{ kJ/kg} = (1.039\text{ kJ/kg}\cdot°\text{C})(T_2 - 10°\text{C}) + \frac{(200\text{ m/s})^2 - (100\text{ m/s})^2}{2}\left(\frac{1\text{ kJ/kg}}{1000\text{ m}^2/\text{s}^2}\right)$$

$$T_2 = 139.9°\text{C}$$

and,

$$T_{02} = T_2 + \frac{\mathbf{V}_2^2}{2C_p} = 139.9°\text{C} + \frac{(200\text{ m/s})^2}{2\times 1.039\text{ kJ/kg}\cdot°\text{C}}\left(\frac{1\text{ kJ/kg}}{1000\text{ m}^2/\text{s}^2}\right) = \mathbf{159.1°C}$$

$$P_{02} = P_2\left(\frac{T_{02}}{T_2}\right)^{k/(k-1)} = (100\text{ kPa})\left(\frac{432.3\text{ K}}{413.1\text{ K}}\right)^{1.4/(1.4-1)} = \mathbf{117.2\,kPa}$$

16-101 An expression for the velocity of sound based on van der Waals equation of state is to be derived. Using this relation, the velocity of sound in carbon dioxide is to be determined and compared to that obtained by ideal gas behavior.

Properties The properties of CO_2 are R = 0.1889 kJ/kg· K and k = 1.279 at T = 50°C = 323.2 K (Tables A-1 and A-2).

Analysis Van der Waals equation of state can be expressed as $P = \dfrac{RT}{v-b} - \dfrac{a}{v^2}$.

Differentiating, $\left(\dfrac{\partial P}{\partial v}\right)_T = \dfrac{RT}{(v-b)^2} + \dfrac{2a}{v^3}$

Noting that $\rho = 1/v \longrightarrow d\rho = -dv/v^2$, the speed of sound relation becomes

$$C^2 = k\left(\frac{\partial P}{\partial r}\right)_T = v^2 k\left(\frac{\partial P}{\partial v}\right)_T$$

Substituting,

$$C^2 = \frac{v^2 kRT}{(v-b)^2} - \frac{2ak}{v}$$

Using the molar mass of CO_2 (M = 44 kg/kmol), the constant a and b can be expressed per unit mass as

$$a = 0.1882 \text{ kPa}\cdot\text{m}^6/\text{kg}^2 \quad \text{and} \quad b = 9.70\times10^{-4} \text{ m}^3/\text{kg}$$

The specific volume of CO_2 is determined to be

$$200 \text{ kPa} = \frac{(0.1889 \text{ kPa}\cdot\text{m}^3/\text{kg}\cdot\text{K})(323.2 \text{ K})}{\text{v} - 0.000970 \text{ m}^3/\text{kg}} - \frac{2\times 0.1882 \text{ kPa}\cdot\text{m}^6/\text{kg}^2}{v^2} \rightarrow v = 0.300 \text{ m}^3/\text{kg}$$

Substituting,

$$\text{C} = \left(\begin{array}{c}\dfrac{(0.300\,\text{m}^3/\text{kg})^2(1.279)(0.1889\,\text{kJ/kg}\cdot\text{K})(323.2\,\text{K})}{(0.300-0.000970\,\text{m}^3/\text{kg})^2}\left(\dfrac{1000\,\text{m}^2/\text{s}^2}{1\,\text{kJ/kg}}\right) \\ -\dfrac{2(0.1882\,\text{kPa.m}^6/\text{kg}^3)(1.279)}{(0.300\,\text{m}^3/\text{kg})^2}\left(\dfrac{1000\,\text{m}^2/\text{s}^2}{1\,\text{kPa}\cdot\text{m}^3/\text{kg}}\right)\end{array}\right)^{1/2} = 270.6\,\text{m/s}$$

If we treat CO_2 as an ideal gas, the velocity of sound becomes

$$C = \sqrt{kRT} = \sqrt{(1.279)(0.1889\,\text{kJ/kg}\cdot\text{K})(323.2\,\text{K})\left(\frac{1000\,\text{m}^2/\text{s}^2}{1\,\text{kJ/kg}}\right)} = \mathbf{279.4\ m/s}$$

16-102 The equivalent relation for the speed of sound is to be verified.

Analysis The two relations are $C^2 = \left(\frac{\partial P}{\partial \rho}\right)_s$ and $C^2 = k\left(\frac{\partial P}{\partial \rho}\right)_T$

From $r = 1/v \longrightarrow dr = -dv/v^2$. Thus,

$$C^2 = \left(\frac{\partial P}{\partial r}\right)_s = -v^2\left(\frac{\partial P}{\partial v}\right)_s = -v^2\left(\frac{\partial P}{\partial T}\frac{\partial T}{\partial v}\right)_s = -v^2\left(\frac{\partial P}{\partial T}\right)_s\left(\frac{\partial T}{\partial v}\right)_s$$

From the cyclic rule,

$$(P,T,s):\left(\frac{\partial P}{\partial T}\right)_s\left(\frac{\partial T}{\partial s}\right)_P\left(\frac{\partial s}{\partial P}\right)_T = -1 \longrightarrow \left(\frac{\partial P}{\partial T}\right)_s = -\left(\frac{\partial s}{\partial T}\right)_P\left(\frac{\partial P}{\partial s}\right)_T$$

$$(T,v,s):\left(\frac{\partial T}{\partial v}\right)_s\left(\frac{\partial v}{\partial s}\right)_T\left(\frac{\partial s}{\partial T}\right)_v = -1 \longrightarrow \left(\frac{\partial T}{\partial v}\right)_s = -\left(\frac{\partial s}{\partial v}\right)_T\left(\frac{\partial T}{\partial s}\right)_v$$

Substituting,

$$C^2 = -v^2\left(\frac{\partial s}{\partial T}\right)_P\left(\frac{\partial P}{\partial s}\right)_T\left(\frac{\partial s}{\partial v}\right)_T\left(\frac{\partial T}{\partial s}\right)_v = -v^2\left(\frac{\partial s}{\partial T}\right)_P\left(\frac{\partial T}{\partial s}\right)_v\left(\frac{\partial P}{\partial s1}\right)_T$$

Recall that

$$\frac{C_p}{T} = \left(\frac{\partial s}{\partial T}\right)_P \quad \text{and} \quad \frac{C_v}{T} = \left(\frac{\partial s}{\partial T}\right)_v$$

Substituting,

$$C^2 = -v^2\left(\frac{C_p}{T}\right)\left(\frac{T}{C_v}\right)\left(\frac{\partial P}{\partial v}\right)_T = -v^2k\left(\frac{\partial P}{\partial v}\right)_T$$

Replacing $-dv/v^2$ by $d\rho$,

$$C^2 = k\left(\frac{\partial P}{\partial \rho}\right)_T$$

16-103 For ideal gases undergoing isentropic flows, expressions for P/P^*, T/T^*, and ρ/ρ^* as functions of k and M are to be obtained.

Analysis Equations 16-18 and 16-21 are given to be $\frac{T_0}{T} = \frac{2+(k-1)M^2}{2}$ and $\frac{T^*}{T_0} = \frac{2}{k+1}$

Multiplying the two, $$\left(\frac{T_0}{T}\frac{T^*}{T_0}\right) = \left(\frac{2+(k-1)M^2}{2}\right)\left(\frac{2}{k+1}\right)$$

Simplifying and inverting, $$\frac{T}{T^*} = \frac{k+1}{2+(k-1)M^2} \qquad (1)$$

From $$\frac{P}{P^*} = \left(\frac{T}{T^*}\right)^{k/(k-1)} \longrightarrow \frac{P}{P^*} = \left(\frac{k+1}{2+(k-1)M^2}\right)^{k/(k-1)} \qquad (2)$$

From $$\frac{\rho}{\rho^*} = \left(\frac{\rho}{\rho^*}\right)^{k/(k-1)} \longrightarrow \frac{\rho}{\rho^*} = \left(\frac{k+1}{2+(k-1)M^2}\right)^{k/(k-1)} \qquad (3)$$

16-104 It is to be verified that for the steady flow of ideal gases $dT_0/T = dA/A + (1\text{-}M^2)\, dV/V$. The effect of heating and area changes on the velocity of an ideal gas in steady flow for subsonic flow and supersonic flow are to be explained.

Analysis We start with the relation $\frac{\mathbf{V}^2}{2} = C_p(T_0 - T)$, (1)

Differentiating, $\mathbf{V}\, d\mathbf{V} = C_p(dT_0 - dT)$ (2)

We also have $\frac{d\rho}{\rho} + \frac{dA}{A} + \frac{d\mathbf{V}}{\mathbf{V}} = 0$ (3)

and $\frac{dP}{\rho} + \mathbf{V}\, d\mathbf{V} = 0$ (4)

Differentiating the ideal gas relation $P = \rho RT$, $\frac{dP}{P} = \frac{d\rho}{\rho} + \frac{dT}{T} = 0$ (5)

From the speed of sound relation, $C^2 = kRT = (k-1)C_p T = kP/\rho$ (6)

Combining Eqs. (3) and (5), $\frac{dP}{P} - \frac{dT}{T} + \frac{dA}{A} + \frac{d\mathbf{V}}{\mathbf{V}} = 0$ (7)

Combining Eqs. (4) and (6), $\frac{dP}{\rho} = \frac{dP}{kP/C^2} = -\mathbf{V}\, d\mathbf{V}$

or, $\frac{dP}{P} = -\frac{k}{C^2}\mathbf{V}\, d\mathbf{V} = -k\frac{\mathbf{V}^2}{C^2}\frac{d\mathbf{V}}{\mathbf{V}} = -kM^2\frac{d\mathbf{V}}{\mathbf{V}}$ (8)

Combining Eqs. (2) and (6), $dT = dT_0 - \mathbf{V}\frac{d\mathbf{V}}{C_p}$

or, $\frac{dT}{T} = \frac{dT_0}{T} - \frac{\mathbf{V}^2}{C_p T}\frac{d\mathbf{V}}{\mathbf{V}} = \frac{dT}{T} = \frac{dT_0}{T} - \frac{\mathbf{V}^2}{C^2/(k-1)}\frac{d\mathbf{V}}{\mathbf{V}} = \frac{dT_0}{T} - (k-1)M^2\frac{d\mathbf{V}}{\mathbf{V}}$ (9)

Combining Eqs. (7), (8), and (9), $-(k-1)M^2\frac{d\mathbf{V}}{\mathbf{V}} - \frac{dT_0}{T} + (k-1)M^2\frac{d\mathbf{V}}{\mathbf{V}} + \frac{dA}{A} + \frac{d\mathbf{V}}{\mathbf{V}} = 0$

or, $\frac{dT_0}{T} = \frac{dA}{A} + \left[-kM^2 + (k-1)M^2 + 1\right]\frac{d\mathbf{V}}{\mathbf{V}}$

Thus, $\frac{dT_0}{T} = \frac{dA}{A} + (1 - M^2)\frac{d\mathbf{V}}{\mathbf{V}}$ (10)

Differentiating the steady-flow energy equation $q = h_{02} - h_{01} = C_p(T_{02} - T_{01})$

$$\delta q = C_p dT_0 \qquad (11)$$

Eq. (11) relates the stagnation temperature change dT_0 to the net heat transferred to the fluid. Eq. (10) relates the velocity changes to area changes dA, and the stagnation temperature change dT_0 or the heat transferred.

(*a*) When $M < 1$ (subsonic flow), the fluid will accelerate if the duck converges ($dA < 0$) or the fluid is heated ($dT_0 > 0$ or $\delta q > 0$). The fluid will decelerate if the duck converges ($dA < 0$) or the fluid is cooled ($dT_0 < 0$ or $\delta q < 0$).

(*b*) When $M > 1$ (supersonic flow), the fluid will accelerate if the duck diverges ($dA > 0$) or the fluid is cooled ($dT_0 < 0$ or $\delta q < 0$). The fluid will decelerate if the duck converges ($dA < 0$) or the fluid is heated ($dT_0 > 0$ or $\delta q > 0$).

16-105 A pitot tube measures the difference between the static and stagnation pressures for a subsonic airplane. The speed of the airplane and the flight Mach number are to be determined.

Assumptions **1** Air is an ideal gas with constant specific heat ratio. **2** The stagnation process is isentropic.

Properties The properties of air are R = 0.287 kJ/kg.K and k = 1.4 (Table A-1 Table A-2).

Analysis The stagnation pressure of air at the specified conditions is

$$P_0 = P + \Delta P = 70.109 + 22 = 92.109 \text{ kPa}$$

Then,

$$\frac{P_0}{P} = \left(1 + \frac{(k-1)M^2}{2}\right)^{k/k-1} \longrightarrow \frac{92.109}{70.109} = \left(1 + \frac{(1.4-1)M^2}{2}\right)^{1.4/0.4}$$

It yields $\qquad$ **M = 0.637**

The speed of sound in air at the specified conditions is

$$C = \sqrt{kRT} = \sqrt{(1.4)(0.287 \text{ kJ/kg} \cdot \text{K})(268.65 \text{ K})\left(\frac{1000 \text{ m}^2/\text{s}^2}{1 \text{ kJ/kg}}\right)} = 328.5 \text{ m/s}$$

Thus,

$$\text{V} = M \times C = (0.637)(328.5 \text{ m/s}) = \mathbf{209.3 \text{ m/s}}$$

16-106 The mass flow parameter $\dot{m}\sqrt{RT_0}/(AP_0)$ versus the Mach number for k = 1.2, 1.4, and 1.6 in the range of $0 \le M \le 1$ is to be plotted.

Analysis The mass flow rate parameter $(\dot{m}\sqrt{RT_0})/P_0A$ can be expressed as

$$\frac{\dot{m}\sqrt{RT_0}}{P_0A} = M\sqrt{k}\left(\frac{2}{2+(k-1)M^2}\right)^{(k+1)/2(k-1)}$$

Thus,

M	k = 1.2	k = 1.4	k = 1.6
0.0	0	0	0
0.1	0.1089	0.1176	0.1257
0.2	0.2143	0.2311	0.2465
0.3	0.3128	0.3365	0.3582
0.4	0.4015	0.4306	0.4571
0.5	0.4782	0.5111	0.5407
0.6	0.5411	0.5763	0.6077
0.7	0.5894	0.6257	0.6578
0.8	0.6230	0.6595	0.6916
0.9	0.6424	0.6787	0.7106
1.0	0.6485	0.6847	0.7164

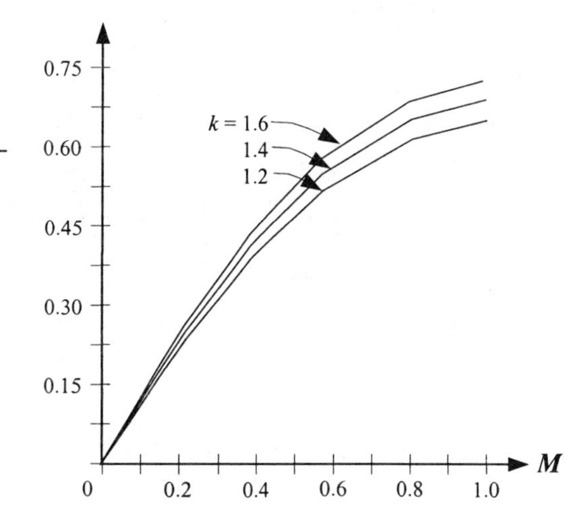

16-107 Helium gas is accelerated in a nozzle. The pressure and temperature of helium at the location where $M = 1$ and the ratio of the flow area at this location to the inlet flow area are to be determined.

Assumptions **1** Helium is an ideal gas with constant specific properties. **2** Flow through the nozzle is steady, one-dimensional, and isentropic.

Properties The properties of helium are $R = 2.0769$ kJ/kg.K, $C_p = 5.1926$ kJ/kg.K, and $k = 1.667$ (Tables A-1 and A-2).

Analysis The properties of the fluid at the location where $M = 1$ are the critical properties, denoted by superscript *. We first determine the stagnation temperature and pressure, which remain constant throughout the nozzle since the flow is isentropic.

$$T_0 = T_i + \frac{\mathbf{V}_i^2}{2C_p} = 500\text{ K} + \frac{(120\text{ m/s})^2}{2\times 5.1926\text{ kJ/kg}\cdot\text{K}}\left(\frac{1\text{ kJ/kg}}{1000\text{ m}^2/\text{s}^2}\right) = 501.4\text{ K}$$

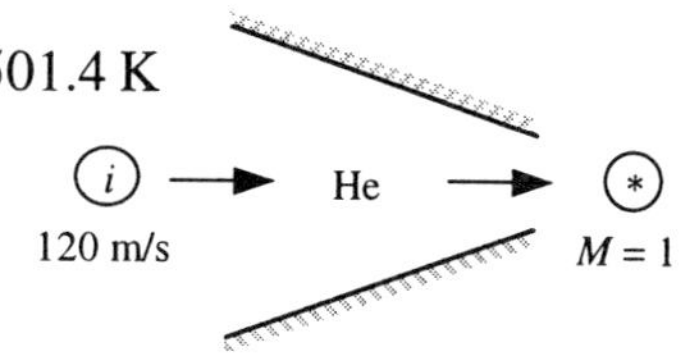

and

$$P_0 = P_i\left(\frac{T_0}{T_i}\right)^{k/(k-1)} = (0.8\text{ MPa})\left(\frac{501.4\text{ K}}{500\text{ K}}\right)^{1.667/(1.667-1)} = 0.806\text{ MPa}$$

The Mach number at the nozzle exit is given to be $M = 1$. Therefore, the properties at the nozzle exit are the *critical properties* determined from

$$T^* = T_0\left(\frac{2}{k+1}\right) = (501.4\text{ K})\left(\frac{2}{1.667+1}\right) = \mathbf{376.0\ K}$$

$$P^* = P_0\left(\frac{2}{k+1}\right)^{k/(k-1)} = (0.806\text{ MPa})\left(\frac{2}{1.667+1}\right)^{1.667/(1.667-1)} = \mathbf{0.393\ MPa}$$

The velocity of sound and the Mach number at the nozzle inlet are

$$C_i = \sqrt{kRT_i} = \sqrt{(1.667)(2.0769\text{ kJ/kg}\cdot\text{K})(500\text{ K})\left(\frac{1000\text{ m}^2/\text{s}^2}{1\text{ kJ/kg}}\right)} = 1316\text{ m/s}$$

$$M_i = \frac{\mathbf{V}_i}{C_i} = \frac{120\text{ m/s}}{1316\text{ m/s}} = 0.0912$$

The ratio of the entrance-to-throat area is

$$\frac{A_i}{A^*} = \frac{1}{M_i}\left[\left(\frac{2}{k+1}\right)\left(1+\frac{k-1}{2}M_i^2\right)\right]^{(k+1)/[2(k-1)]}$$

$$= \frac{1}{0.0912}\left[\left(\frac{2}{1.667+1}\right)\left(1+\frac{1.667-1}{2}(0.0912)^2\right)\right]^{2.667/(2\times 0.667)}$$

$$= 6.20$$

Then the ratio of the throat area to the entrance area becomes

$$\frac{A^*}{A_i} = \frac{1}{6.20} = \mathbf{0.161}$$

16-108 Helium gas enters a nozzle with negligible velocity, and is accelerated in a nozzle. The pressure and temperature of helium at the location where $M = 1$ and the ratio of the flow area at this location to the inlet flow area are to be determined.

Assumptions **1** Helium is an ideal gas with constant specific properties. **2** Flow through the nozzle is steady, one-dimensional, and isentropic. **3** The entrance velocity is negligible.

Properties The properties of helium are $R = 2.0769$ kJ/kg.K, $C_p = 5.1926$ kJ/kg.K, and $k = 1.667$ (Tables A-1 and A-2).

Analysis We treat helium as an ideal gas with $k = 1.667$. The properties of the fluid at the location where $M = 1$ are the critical properties, denoted by superscript *.

The stagnation temperature and pressure in this case are identical to the inlet temperature and pressure since the inlet velocity is negligible. They remain constant throughout the nozzle since the flow is isentropic.

$$T_0 = T_i = 500 \text{ K}$$

$$P_0 = P_i = 0.8 \text{ MPa}$$

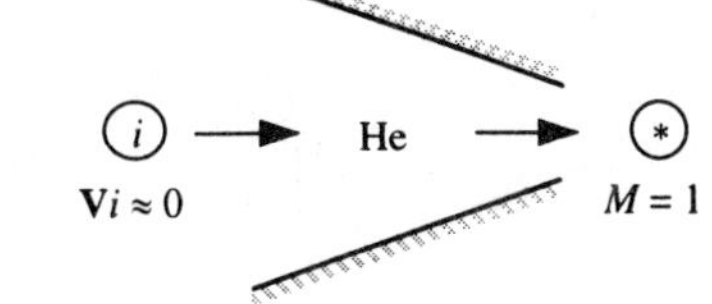

The Mach number at the nozzle exit is given to be $M = 1$. Therefore, the properties at the nozzle exit are the *critical properties* determined from

$$T^* = T_0\left(\frac{2}{k+1}\right) = (500\text{ K})\left(\frac{2}{1.667+1}\right) = \mathbf{375.0\ K}$$

$$P^* = P_0\left(\frac{2}{k+1}\right)^{k/(k-1)} = (0.8\text{ MPa})\left(\frac{2}{1.667+1}\right)^{1.667/(1.667-1)} = \mathbf{0.390\ MPa}$$

The ratio of the nozzle inlet area to the throat area is determined from

$$\frac{A_i}{A^*} = \frac{1}{M_i}\left[\left(\frac{2}{k+1}\right)\left(1+\frac{k-1}{2}M_i^2\right)\right]^{(k+1)/[2(k-1)]}$$

But the Mach number at the nozzle inlet is $M = 0$ since $\mathbf{V}_i \cong 0$. Thus the ratio of the throat area to the nozzle inlet area is

$$\frac{A^*}{A_i} = \frac{1}{\infty} = \mathbf{0}$$

16-109 Air enters a converging nozzle. The mass flow rate, the exit velocity, the exit Mach number, and the exit pressure-stagnation pressure ratio versus the back pressure-stagnation pressure ratio for a specified back pressure range are to be calculated and plotted.

Assumptions **1** Air is an ideal gas with constant specific heats at room temperature. **2** Flow through the nozzle is steady, one-dimensional, and isentropic.

Properties The properties of air at room temperature are R = 0.287 kJ/kg.K, C_p = 1.005 kJ/kg.K, and k = 1.4 (Tables A-1 and A-2).

Analysis The stagnation properties remain constant throughout the nozzle since the flow is isentropic. They are determined from

$$T_0 = T_i + \frac{\mathbf{V}_i^2}{2C_p} = 400\text{ K} + \frac{(180\text{ m/s})^2}{2\times 1.005\text{ kJ/kg}\cdot\text{K}}\left(\frac{1\text{ kJ/kg}}{1000\text{ m}^2/\text{s}^2}\right) = 416.1\text{ K}$$

and

$$P_0 = P_i\left(\frac{T_0}{T_i}\right)^{k/(k-1)} = (900\text{ kPa})\left(\frac{416.1\text{ K}}{400\text{ K}}\right)^{1.4/(1.4-1)} = 1033.3\text{ kPa}$$

The critical pressure is determined to be

$$P^* = P_0\left(\frac{2}{k+1}\right)^{k/(k-1)} = (1033.3\text{ kPa})\left(\frac{2}{1.4+1}\right)^{1.4/0.4} = 545.9\text{ kPa}$$

Then the pressure at the exit plane (throat) will be

$P_e = P_b$	for	$P_b \geq 545.9$ kPa
$P_e = P^* = 545.9$ kPa	for	$P_b < 545.9$ kPa (choked flow)

Thus the back pressure will not affect the flow when $100 < P_b < 545.9$ kPa. For a specified exit pressure P_e, the temperature, the velocity and the mass flow rate can be determined from

Temperature $$T_e = T_0\left(\frac{P_e}{P_0}\right)^{(k-1)/k} = (416.1\text{ K})\left(\frac{P_e}{1033.3}\right)^{0.4/1.4}$$

Velocity $$\mathbf{V} = \sqrt{2C_p(T_0 - T_e)} = \sqrt{2(1.005\text{ kJ/kg}\cdot\text{K})(416.1 - T_e)\left(\frac{1000\text{ m}^2/\text{s}^2}{1\text{ kJ/kg}}\right)}$$

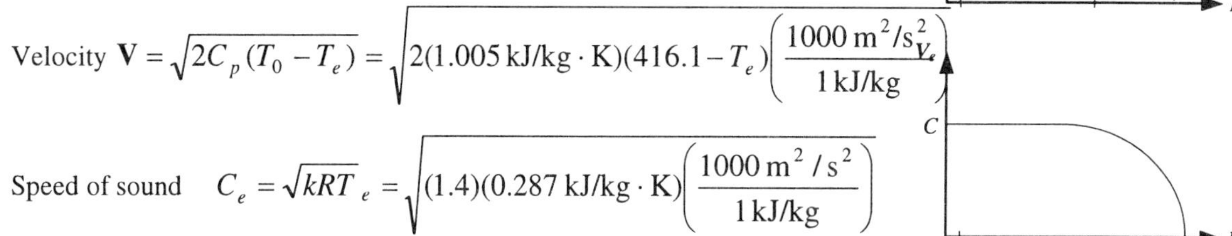

Speed of sound $$C_e = \sqrt{kRT_e} = \sqrt{(1.4)(0.287\text{ kJ/kg}\cdot\text{K})\left(\frac{1000\text{ m}^2/\text{s}^2}{1\text{ kJ/kg}}\right)}$$

Mach number $$M_e = \mathbf{V}_e / C_e$$

Density $$\rho_e = \frac{P_e}{RT_e} = \frac{P_e}{(0.287\text{ kPa}\cdot\text{m}^3/\text{kg}\cdot\text{K})T_e}$$

Mass flow rate $$\dot{m} = \rho_e\mathbf{V}_e A_e = \rho_e\mathbf{V}_e(0.001\text{ m}^2)$$

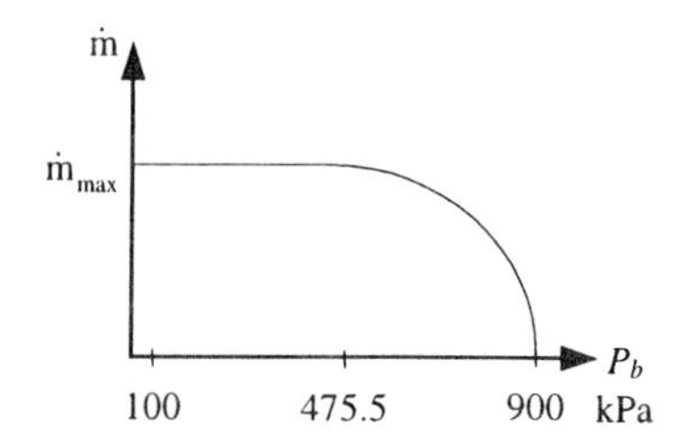

The results of the calculations can be tabulated as

P_b, kPa	P_b, P_0	P_e, kPa	P_b, P_0	T_e, K	V_e, m/s	M	ρ_e, kg/m^3	$\dot{m}$, kg / s
900	0.871	900	0.871	400.0	180.0	0.45	7.840	0
800	0.774	800	0.774	386.8	162.9	0.41	7.206	1.174
700	0.677	700	0.677	372.3	236.0	0.61	6.551	1.546
600	0.581	600	0.581	356.2	296.7	0.78	5.869	1.741
545.9	0.528	545.9	0.528	333.3	366.2	1.00	4.971	1.820
500	0.484	545.9	0.528	333.2	366.2	1.00	4.971	1.820
400	0.387	545.9	0.528	333.3	366.2	1.00	4.971	1.820
300	0.290	545.9	0.528	333.3	366.2	1.00	4.971	1.820
200	0.194	545.9	0.528	333.3	366.2	1.00	4.971	1.820
100	0.097	545.9	0.528	333.3	366.2	1.00	4.971	1.820

16-110 Steam enters a converging nozzle. The exit pressure, the exit velocity, and the mass flow rate versus the back pressure for a specified back pressure range are to be plotted.

Assumptions **1** Steam is to be treated as an ideal gas with constant specific heats. **2** Flow through the nozzle is steady, one-dimensional, and isentropic. **3** The nozzle is adiabatic.

Properties The ideal gas properties of steam are R = 0.462 kJ/kg.K, C_p = 1.872 kJ/kg.K, and k = 1.3.

Analysis The stagnation properties in this case are identical to the inlet properties since the inlet velocity is negligible. Since the flow is isentropic, they remain constant throughout the nozzle,

$$P_0 = P_i = 6 \text{ MPa}$$

$$T_0 = T_i = 700 \text{ K}$$

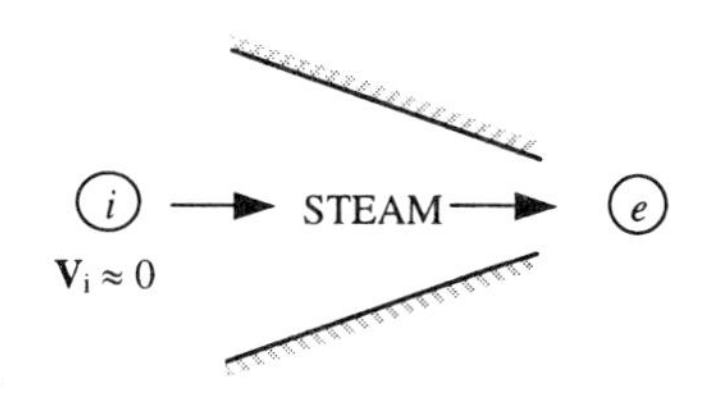

The critical pressure is determined from to be

$$P^* = P_0\left(\frac{2}{k+1}\right)^{k/(k-1)} = (6 \text{ MPa})\left(\frac{2}{1.3+1}\right)^{1.3/0.3} = 3.274 \text{ MPa}$$

Then the pressure at the exit plane (throat) will be

$P_e = P_b$ for $P_b \geq 3.274$ MPa

$P_e = P^* = 3.274$ MPa for $P_b < 3.274$ MPa (choked flow)

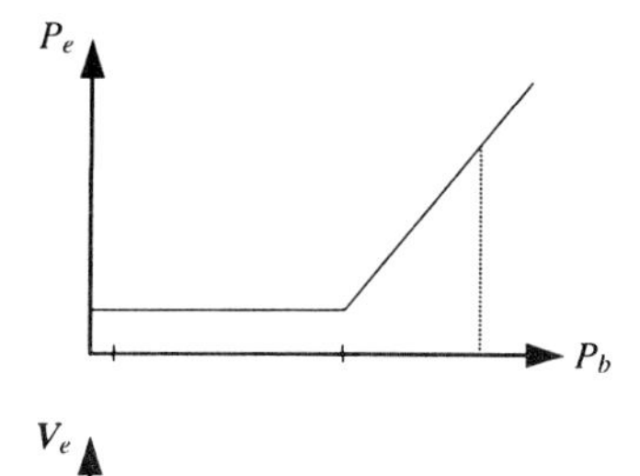

Thus the back pressure will not affect the flow when 3 < P_b < 3.274 MPa. For a specified exit pressure P_e, the temperature, the velocity and the mass flow rate can be determined from

Temperature $$T_e = T_0\left(\frac{P_e}{P_0}\right)^{(k-1)/k} = (700 \text{ K})\left(\frac{P_e}{6}\right)^{0.3/1.3}$$

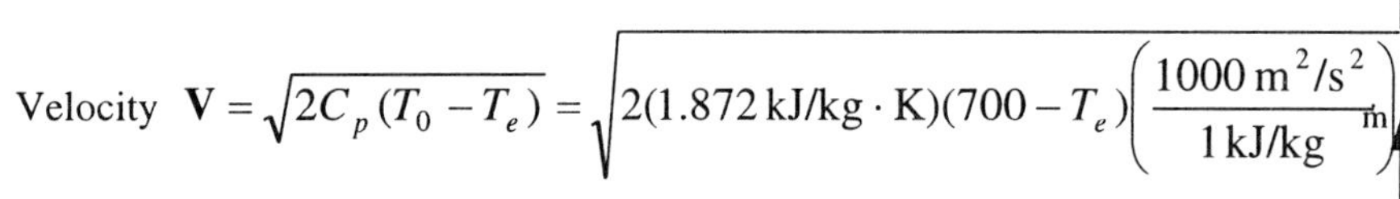

Velocity $$\mathbf{V} = \sqrt{2C_p(T_0 - T_e)} = \sqrt{2(1.872 \text{ kJ/kg}\cdot\text{K})(700 - T_e)\left(\frac{1000 \text{ m}^2/\text{s}^2}{1 \text{ kJ/kg}}\right)}$$

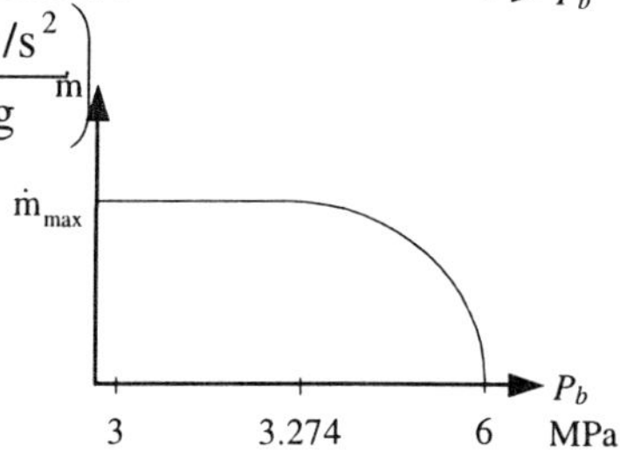

Density $$\rho_e = \frac{P_e}{RT_e} = \frac{P_e}{(0.462 \text{ kPa}\cdot\text{m}^3/\text{kg}\cdot\text{K})T_e}$$

Mass flow rate $$\dot{m} = \rho_e \mathbf{V}_e A_e = \rho_e \mathbf{V}_e (0.0008 \text{ m}^2)$$

The results of the calculations can be tabulated as follows:

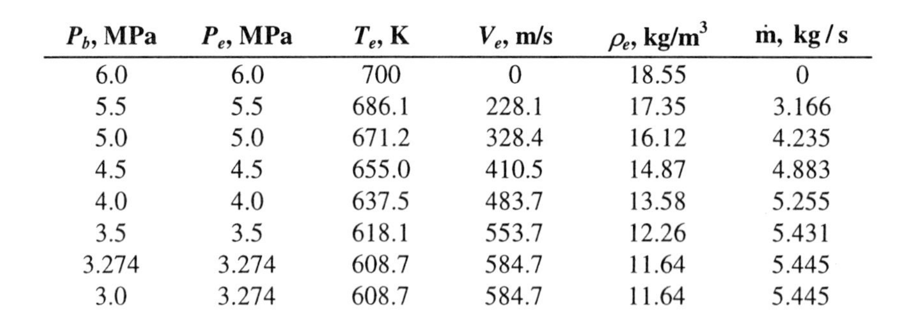

P_b, MPa	P_e, MPa	T_e, K	V_e, m/s	ρ_e, kg/m^3	$\dot{m}$, kg/s
6.0	6.0	700	0	18.55	0
5.5	5.5	686.1	228.1	17.35	3.166
5.0	5.0	671.2	328.4	16.12	4.235
4.5	4.5	655.0	410.5	14.87	4.883
4.0	4.0	637.5	483.7	13.58	5.255
3.5	3.5	618.1	553.7	12.26	5.431
3.274	3.274	608.7	584.7	11.64	5.445
3.0	3.274	608.7	584.7	11.64	5.445

16-111 An expression for the ratio of the stagnation pressure after a shock wave to the static pressure before the shock wave as a function of k and the Mach number upstream of the shock wave is to be found.

Analysis The relation between P_x and P_y is

$$\frac{p_y}{p_x} = \frac{1 + kM_y^2}{1 + kM_x^2} \longrightarrow p_y = p_x\left(\frac{1 + kM_x^2}{1 + kM_y^2}\right)$$

Substituting this into the isentropic relation

$$\frac{P_{0y}}{P_y} = \left(1 + (k-1)M_y^2/2\right)^{k/(k-1)}$$

Then,

$$\frac{P_{0y}}{P_x} = \left(\frac{1 + kM_x^2}{1 + kM_y^2}\right)\left(1 + (k-1)M_y^2/2\right)^{k/(k-1)}$$

where

$$M_y^2 = \frac{M_x^2 + 2/(k-1)}{2kM_y^2/(k-1) - 1}$$

Substituting,

$$\frac{P_{0y}}{P_x} = \left(\frac{(1 + kM_x^2)(2kM_x^2 - k + 1)}{kM_x^2(k+1) - k + 3}\right)\left(1 + \frac{(k-1)M_x^2/2 + 1}{2kM_x^2/(k-1) - 1}\right)^{k/(k-1)}$$

16-112 Nitrogen entering a converging-diverging nozzle experiences a normal shock. The pressure, temperature, velocity, Mach number, and stagnation pressure downstream of the shock are to be determined. The results are to be compared to those of air under the same conditions.

Assumptions **1** Nitrogen is an ideal gas. **2** Flow through the nozzle is steady, one-dimensional, and isentropic. **3** The nozzle is adiabatic.

Properties The properties of nitrogen are R = 0.297 kJ/kg.K and k = 1.4 (Table A-1 and Table A-2).

Analysis The inlet stagnation properties in this case are identical to the inlet properties since the inlet velocity is negligible. Assuming the flow before the shock to be isentropic,

$$P_{0x} = P_i = 700 \text{ kPa}$$

$$T_{0x} = T_i = 300 \text{ K}$$

Then,

$$T_x = T_{0x}\left(\frac{2}{2+(k-1)M_x^2}\right) = (300\text{ K})\left(\frac{2}{2+(1.4-1)3^2}\right) = 107.1\text{ K}$$

and

$$P_x = P_{0x}\left(\frac{T_x}{T_{0x}}\right)^{k/(k-1)} = (700\text{ kPa})\left(\frac{107.1}{300}\right)^{1.4/0.4} = 19.06\text{ kPa}$$

The fluid properties after the shock (denoted by subscript y) are related to those before the shock through the functions listed in Table A-35. For M_x = 3.0 we read

$$M_y = \mathbf{0.4752}, \quad \frac{P_{0y}}{P_{0x}} = 0.32834, \quad \frac{P_y}{P_x} = 10.333, \quad \text{and} \quad \frac{T_y}{T_x} = 2.679$$

Then the stagnation pressure P_{0y}, static pressure P_y, and static temperature T_y, are determined to be

$$P_{0y} = 0.32834 P_{0x} = (0.32834)(700 \text{ kPa}) = \mathbf{229.8\ kPa}$$

$$P_y = 10.333 P_x = (10.333)(19.06 \text{ kPa}) = \mathbf{196.9\ kPa}$$

$$T_y = 2.679 T_x = (2.679)(107.1 \text{ K}) = \mathbf{286.9\ K}$$

The velocity after the shock can be determined from $\mathbf{V}_y = M_y C_y$, where C_y is the velocity of sound at the exit conditions after the shock,

$$\mathbf{V}_y = M_y C_y = M_y\sqrt{kRT_y} = (0.4752)\sqrt{(1.4)(0.297\text{ kJ/kg}\cdot\text{K})(286.9\text{ K})\left(\frac{1000\text{ m}^2/\text{s}^2}{1\text{ kJ/kg}}\right)} = \mathbf{164.1\ m/s}$$

For air at specified conditions k = 1.4 (same as nitrogen) and R = 0.287 kJ/kg· K. Thus the only quantity which will be different in the case of air is the velocity after the normal shock, which happens to be 161.3 m/s.

16-113 A 92 percent efficient diffuser of an aircraft is considered. The static pressure rise across the diffuser and the exit area are to be determined.

Assumptions **1** Air is an ideal gas with constant specific heats at room temperature. **2** Flow through the diffuser is steady and one-dimensional. **3** The diffuser is adiabatic.

Properties Air properties at room temperature are $R = 0.287$ kJ/kg.K, $C_p = 1.005$ kJ/kg· K, and $k = 1.4$.

Analysis The inlet velocity is determined from

$$\mathbf{V}_1 = M_1 C_1 = M_1\sqrt{kRT_1} = (0.8)\sqrt{(1.4)(0.287\text{ kJ/kg}\cdot\text{K})(242.7\text{ K})\left(\frac{1000\text{ m}^2/\text{s}^2}{1\text{ kJ/kg}}\right)} = 249.8\text{ m/s}$$

Then the stagnation temperature and pressure at the diffuser inlet become

$$T_{01} = T_1 + \frac{\mathbf{V}_1^2}{2C_p} = 242.7 + \frac{(249.8\text{ m/s})^2}{2(1.005\text{ kJ/kg}\cdot\text{K})}\left(\frac{1\text{ kJ/kg}}{1000\text{ m}^2/\text{s}^2}\right) = 273.7\text{ K}$$

AIR

① $M_1 = 0.8$ $\quad\eta_D = 92\%\quad$ ② $M_2 = 0.3$

$$P_{01} = P_1\left(\frac{T_{01}}{T_1}\right)^{k/(k-1)} = (41.1\text{ kPa})\left(\frac{273.7\text{ K}}{242.7\text{ K}}\right)^{1.4/(1.4-1)} = 62.6\text{ kPa}$$

Assuming the diffuser to be adiabatic, the energy equation reduces to $h_{01} = h_{02}$. Noting that $h = C_pT$ and the specific heats are assumed to be constant, we have

$$T_{01} = T_{02} = T_0 = 273.7\text{ K}$$

For an ideal gas with constant specific heats, the diffuser efficiency relation can be expressed as

$$\eta_D = \frac{h_{02s} - h_1}{h_{01} - h_1} = \frac{T_{02s} - T_1}{T_{01} - T_1}$$

Then the temperature T_{02s} becomes

$$0.92 = \frac{T_{02s} - 242.7}{273.7 - 242.7} \longrightarrow T_{02s} = 271.2\text{ K}$$

Noting that $P_{02} = P_{02s}$, the isentropic relation between states 1 and 02s gives

$$P_{02} = P_{02s} = P_1\left(\frac{T_{02s}}{T_1}\right)^{k/(k-1)} = (41.1\text{ kPa})\left(\frac{271.2\text{ K}}{242.7\text{ K}}\right)^{1.4/(1.4-1)} = 60.6\text{ kPa}$$

The exit velocity can be expressed as

$$\mathbf{V}_2 = M_2 C_2 = M_2\sqrt{kRT_2} = (0.3)\sqrt{(1.4)(0.287\text{ kJ/kg}\cdot\text{K})\,T_2\left(\frac{1000\text{ m}^2/\text{s}^2}{1\text{ kJ/kg}}\right)} = 6.01\sqrt{T_2}$$

Thus $\quad T_2 = T_{02} - \dfrac{\mathbf{V}_2^2}{2C_p} = (273.7) - \dfrac{6.01^2 T_2\text{ m}^2/\text{s}^2}{2(1.005\text{ kJ/kg}\cdot\text{K})}\left(\dfrac{1\text{ kJ/kg}}{1000\text{ m}^2/\text{s}^2}\right) = 268.9\text{ K}$

Then the static exit pressure becomes

$$P_2 = P_{02}\left(\frac{T_2}{T_{02}}\right)^{k/(k-1)} = (60.6\text{ kPa})\left(\frac{268.9\text{ K}}{273.7\text{ K}}\right)^{1.4/(1.4-1)} = 57.0\text{ kPa}$$

Thus the static pressure rise across the diffuser is $\quad \Delta P = P_2 - P_1 = 57.0 - 41.1 = \mathbf{15.9\text{ kPa}}$

Also, $\quad \rho_2 = \dfrac{P_2}{RT_2} = \dfrac{57.0\text{ kPa}}{(0.287\text{ kPa}\cdot\text{m}^3/\text{kg}\cdot\text{K})(268.9\text{ K})} = 0.739\text{ kg/m}^3$

$$\mathbf{V}_2 = 6.01\sqrt{T_2} = 6.01\sqrt{268.9} = 98.6\text{ m/s}$$

Thus $\quad A_2 = \dfrac{\dot{m}}{\rho_2\mathbf{V}_2} = \dfrac{65\text{ kg/s}}{(0.739\text{ kg/m}^3)(98.6\text{ m/s})} = \mathbf{0.892\text{ m}^2}$

16-114 Helium gas is accelerated in a nozzle. For a specified mass flow rate, the throat and exit areas of the nozzle are to be determined for the cases of isentropic and 97% efficient nozzles.

Assumptions **1** Helium is an ideal gas with constant specific properties. **2** Flow through the nozzle is steady and one-dimensional. **3** The nozzle is adiabatic.

Properties The properties of helium are R = 2.0769 kJ/kg.K, C_p = 5.1926 kJ/kg.K, and k = 1.667 (Tables A-1 and A-2).

Analysis (*a*) The inlet stagnation properties in this case are identical to the inlet properties since the inlet velocity is negligible,

$$T_{01} = T_1 = 500 \text{ K}$$

$$P_{01} = P_1 = 1.0 \text{ MPa}$$

The flow is assumed to be isentropic, thus the stagnation temperature and pressure remain constant throughout the nozzle,

$$T_{02} = T_{01} = 500 \text{ K}$$

$$P_{02} = P_{01} = 1.0 \text{ MPa}$$

1 He * → 2
$V_i \approx 0$
a) η_N = 100%
b) η_N = 97%

The critical pressure and temperature are determined from

$$T^* = T_0\left(\frac{2}{k+1}\right) = (500 \text{ K})\left(\frac{2}{1.667+1}\right) = 375.0 \text{ K}$$

$$P^* = P_0\left(\frac{2}{k+1}\right)^{k/(k-1)} = (1.0 \text{ MPa})\left(\frac{2}{1.667+1}\right)^{1.667/(1.667-1)} = 0.487 \text{ MPa}$$

$$\rho^* = \frac{P^*}{RT^*} = \frac{487 \text{ kPa}}{(2.0769 \text{ kPa}\cdot\text{m}^3/\text{kg}\cdot\text{K})(375 \text{ K})} = 0.625 \text{ kg/m}^3$$

$$\mathbf{V}^* = C^* = \sqrt{kRT^*} = \sqrt{(1.667)(2.0769 \text{ kJ/kg}\cdot\text{K})(375 \text{ K})\left(\frac{1000 \text{ m}^2/\text{s}^2}{1 \text{ kJ/kg}}\right)} = 1139.4 \text{ m/s}$$

Thus the throat area is

$$A^* = \frac{\dot{m}}{\rho^*\mathbf{V}^*} = \frac{0.25 \text{ kg/s}}{(0.625 \text{ kg/m}^3)(1139.4 \text{ m/s})} = 3.51\times10^{-4}\text{m}^2 = \mathbf{3.51\ cm^2}$$

At the nozzle exit the pressure is P_2 = 0.1 MPa. Then the other properties at the nozzle exit are determined to be

$$\frac{P_0}{P_2} = \left(1+\frac{k-1}{2}M_2^2\right)^{k/(k-1)} \longrightarrow \frac{1.0 \text{ MPa}}{0.1 \text{ MPa}} = \left(1+\frac{1.667-1}{2}M_2^2\right)^{1.667/0.667}$$

It yields M_2 = 2.130, which is greater than 1. Therefore, the nozzle must be converging-diverging.

$$T_2 = T_0\left(\frac{2}{2+(k-1)M_2^2}\right) = (500 \text{ K})\left(\frac{2}{2+(1.667-1)\times 2.13^2}\right) = 199.0 \text{ K}$$

$$\rho_2 = \frac{P_2}{RT_2} = \frac{100 \text{ kPa}}{(2.0769 \text{ kPa}\cdot\text{m}^3/\text{kg}\cdot\text{K})(199 \text{ K})} = 0.242 \text{ kg/m}^3$$

$$\mathbf{V}_2 = M_2C_2 = M_2\sqrt{kRT_2} = (2.13)\sqrt{(1.667)(2.0769 \text{ kJ/kg}\cdot\text{K})(199 \text{ K})\left(\frac{1000 \text{ m}^2/\text{s}^2}{1 \text{ kJ/kg}}\right)} = 1768.0 \text{ m/s}$$

Thus the exit area is

$$A_2 = \frac{\dot{m}}{\rho_2\mathbf{V}_2} = \frac{0.25 \text{ kg/s}}{(0.242 \text{ kg/m}^3)(1768 \text{ m/s})} = 5.84\times10^{-4}\text{m}^2 = \mathbf{5.84\ cm^2}$$

(*b*) We treat helium as an ideal gas with $k = 1.667$, $R = 2.0769$ kJ/kg·K and $C_p = 5.1926$ kJ/kg·K. The inlet stagnation properties in this case are identical to the inlet properties since the inlet velocity is negligible,

$$T_{01} = T_1 = 500 \text{ K}$$
$$P_{01} = P_1 = 1.0 \text{ MPa}$$

Assuming the nozzle to be adiabatic, the energy equation reduces to $h_{01} = h_{02}$. Noting that $h = C_pT$ and the specific heats are assumed to be constant, we have

$$T_{01} = T_{02} = T_0 = 500 \text{ K}$$

The critical pressure P* is the actual pressure at the nozzle throat when sonic velocity is reached at the throat ($M^* = 1$). The critical temperature T^* is related to stagnation temperature T_0, and is determined from

$$\frac{T^*}{T_0} = \frac{2}{k+1} \longrightarrow T^* = T_0\left(\frac{2}{k+1}\right) = (500 \text{ K})\left(\frac{2}{1.667+1}\right) = 375.0 \text{ K}$$

For an ideal gas with constant specific heats, the nozzle efficiency between the inlet and the throat is

$$\eta_N = \frac{h_{01} - h_t}{h_{01} - h_{ts}} = \frac{C_p(T_{01} - T_t)}{C_p(T_{01} - T_{ts})} = \frac{T_{01} - T^*}{T_{01} - T_s^*}$$

Thus $$0.97 = \frac{500 - 375.0}{500 - T_s^*} \longrightarrow T_s^* = 371.1 \text{ K}$$

The critical (throat) pressure P^* for both the actual and isentropic cases are same, and is determined from

$$P^* = P_{01}\left(\frac{T_2^*}{T_{01}}\right)^{k/(k-1)} = (1.0 \text{ MPa})\left(\frac{371.1 \text{ K}}{500 \text{ K}}\right)^{1.667/0.667} = 0.475 \text{ MPa}$$

$$\rho^* = \frac{P^*}{RT^*} = \frac{475 \text{ kPa}}{(2.0769 \text{ kPa}\cdot\text{m}^3/\text{kg}\cdot\text{K})(375 \text{ K})} = 0.610 \text{ kg/m}^3$$

$$\mathbf{V}^* = C^* = \sqrt{kRT^*} = \sqrt{(1.667)(2.0769 \text{ kJ/kg}\cdot\text{K})(375 \text{ K})\left(\frac{1000 \text{ m}^2/\text{s}^2}{1 \text{ kJ/kg}}\right)} = 1139.4 \text{ m/s}$$

Thus the throat area is

$$A^* = \frac{\dot{m}}{\rho^*\mathbf{V}^*} = \frac{0.25 \text{ kg/s}}{(0.610 \text{ kg/m}^3)(1139.4 \text{ m/s})} = 3.60 \times 10^{-4} \text{ m}^2 = \mathbf{3.60 \text{ cm}^2}$$

At the nozzle exit the pressure is $P_2 = 0.1$ MPa. The isentropic exit temperature can be determined from

$$\frac{T_{2s}}{T_{01}} = \left(\frac{P_2}{P_{01}}\right)^{(k-1)/k} \longrightarrow T_{2s} = (500 \text{ K})\left(\frac{0.1 \text{ MPa}}{1.0 \text{ MPa}}\right)^{(1.667-1)/1.667} = 199.0 \text{ K}$$

The actual exit temperature is determined from

$$\eta_N = \frac{h_{01} - h_2}{h_{01} - h_{2s}} = \frac{T_{01} - T_2}{T_{01} - T_{2s}} \longrightarrow 0.97 = \frac{500 - T_2}{500 - 199} \longrightarrow T_2 = 208.0 \text{ K}$$

Then the exit velocity is determined from the steady-flow energy balance to be

$$h_1 + \frac{\mathbf{V}_1^2}{2} = h_2 + \frac{\mathbf{V}_2^2}{2} \rightarrow 0 = h_2 - h_1 + \frac{\mathbf{V}_2^2 - \mathbf{V}_1^2}{2} \rightarrow 0 = C_p(T_2 - T_1) + \frac{\mathbf{V}_2^2 - \mathbf{V}_1^2}{2}$$

Solving for $\mathbf{V}_2$,

$$\mathbf{V}_2 = \sqrt{2C_p(T_1 - T_2)} = \sqrt{2(5.1926 \text{ kJ/kg}\cdot\text{K})(500 - 208) \text{ K}\left(\frac{1000 \text{ m}^2/\text{s}^2}{1 \text{ kJ/kg}}\right)} = 1741.2 \text{ m/s}$$

Also $$\rho_2 = \frac{P_2}{RT_2} = \frac{100 \text{ kPa}}{(2.0769 \text{ kPa}\cdot\text{m}^3/\text{kg}\cdot\text{K})(208 \text{ K})} = 0.231 \text{ kg/m}^3$$

and $$A_2 = \frac{\dot{m}}{\rho_2\mathbf{V}_2} = \frac{0.25 \text{ kg/s}}{(0.231 \text{ kg/m}^3)(1741.2 \text{ m/s})} = 6.20 \times 10^{-4} \text{ m}^2 = \mathbf{6.20 \text{ cm}^2}$$

16-115E Helium gas is accelerated in a nozzle. For a specified mass flow rate, the throat and exit areas of the nozzle are to be determined for the cases of isentropic and 97% efficient nozzles.

Assumptions **1** Helium is an ideal gas with constant specific properties. **2** Flow through the nozzle is steady and one-dimensional. **3** The nozzle is adiabatic.

Properties The properties of helium are R = 0.4961 Btu/lbm·R = 2.6809 psia·ft³/lbm·R, C_p = 1.25 Btu/lbm·R, and k = 1.667 (Tables A-1E and A-2E).

Analysis (*a*) The inlet stagnation properties in this case are identical to the inlet properties since the inlet velocity is negligible,

$$T_{01} = T_1 = 900 \text{ R}$$

$$P_{01} = P_1 = 150 \text{ psia}$$

The flow is assumed to be isentropic, thus the stagnation temperature and pressure remain constant throughout the nozzle,

$$T_{02} = T_{01} = 900 \text{ R}$$

$$P_{02} = P_{01} = 150 \text{ psia}$$

(1) He (*) → (2)
$V_i \approx 0$
a) $\eta_N = 100\%$
b) $\eta_N = 97\%$

The critical pressure and temperature are determined from

$$T^* = T_0\left(\frac{2}{k+1}\right) = (900 \text{ R})\left(\frac{2}{1.667+1}\right) = 674.9 \text{ R}$$

$$P^* = P_0\left(\frac{2}{k+1}\right)^{k/(k-1)} = (150 \text{ psia})\left(\frac{2}{1.667+1}\right)^{1.667/(1.667-1)} = 73.1 \text{ psia}$$

$$\rho^* = \frac{P^*}{RT^*} = \frac{73.1 \text{ psia}}{(2.6809 \text{ psia}\cdot\text{ft}^3/\text{lbm}\cdot\text{R})(674.9 \text{ R})} = 0.0404 \text{ 1bm/ft}^3$$

$$\mathbf{V}^* = C^* = \sqrt{kRT^*} = \sqrt{(1.667)(0.4961 \text{ Btu/lbm}\cdot\text{R})(674.9 \text{ R})\left(\frac{25{,}037 \text{ ft}^2/\text{s}^2}{1 \text{ Btu/1bm}}\right)} = 3738 \text{ ft/s}$$

and
$$A^* = \frac{\dot{m}}{\rho^*\mathbf{V}^*} = \frac{0.2 \text{ 1bm/s}}{(0.0404 \text{ 1bm/ft}^3)(3738 \text{ ft/s})} = \mathbf{0.00132 \text{ ft}^2}$$

At the nozzle exit the pressure is P_2 = 15 psia. Then the other properties at the nozzle exit are determined to be

$$\frac{p_0}{p_2} = \left(1 + \frac{k-1}{2}M_2^2\right)^{k/(k-1)} \longrightarrow \frac{150 \text{ psia}}{15 \text{ psia}} = \left(1 + \frac{1.667-1}{2}M_2^2\right)^{1.667/0.667}$$

It yields M_2 = 2.130, which is greater than 1. Therefore, the nozzle must be converging-diverging.

$$T_2 = T_0\left(\frac{2}{2+(k-1)M_2^2}\right) = (900 \text{ R})\left(\frac{2}{2+(1.667-1)\times 2.13^2}\right) = 358.1 \text{ R}$$

$$\rho_2 = \frac{P_2}{RT_2} = \frac{15 \text{ psia}}{(2.6809 \text{ psia}\cdot\text{ft}^3/\text{lbm}\cdot\text{R})(358.1 \text{ R})} = 0.0156 \text{ 1bm/ft}^3$$

$$\mathbf{V}_2 = M_2C_2 = M_2\sqrt{kRT_2} = (2.13)\sqrt{(1.667)(0.4961 \text{ Btu/lbm}\cdot\text{R})(358.1 \text{ R})\left(\frac{25{,}037 \text{ ft}^2/\text{s}^2}{1 \text{ Btu/1bm}}\right)} = 5800 \text{ ft/s}$$

Thus the exit area is

$$A_2 = \frac{\dot{m}}{\rho_2\mathbf{V}_2} = \frac{0.2 \text{ lbm/s}}{(0.0156 \text{ lbm/ft}^3)(5800 \text{ ft/s})} = \mathbf{0.00221 \text{ ft}^2}$$

(*b*) We treat helium as an ideal gas with k = 1.667, R = 0.4961 Btu/lbm·R = 2.6809 psia·ft^3/lbm·R, and C_p = 1.25 Btu/lbm·R. The inlet stagnation properties in this case are identical to the inlet properties since the inlet velocity is negligible,

$$T_{01} = T_1 = 900 \text{ R}$$

$$P_{01} = P_1 = 150 \text{ psia}$$

Assuming the nozzle to be adiabatic, the energy equation reduces to $h_{01} = h_{02}$. Noting that $h = C_pT$ and the specific heats are assumed to be constant, we have

$$T_{01} = T_{02} = T_0 = 900 \text{ R}$$

The critical pressure P* is the actual pressure at the nozzle throat when sonic velocity is reached at the throat ($M^* = 1$). The critical temperature T^* is related to the stagnation temperature T_0, and is determined from

$$\frac{T^*}{T_0} = \frac{2}{k+1} \longrightarrow T^* = T_0\left(\frac{2}{k+1}\right) = (900 \text{ R})\left(\frac{2}{1.667+1}\right) = 674.9 \text{ R}$$

For an ideal gas with constant specific heats the nozzle efficiency between the inlet and the throat can be expressed as

$$\eta_N = \frac{h_{01} - h_t}{h_{01} - h_{ts}} = \frac{C_p(T_{01} - T_t)}{C_p(T_{01} - T_{ts})} = \frac{T_{01} - T^*}{T_{01} - T_s^*}$$

Thus $\quad 0.97 = \dfrac{900 - 674.9}{900 - T_s^*} \longrightarrow T_s^* = 667.9 \text{ R}$

The critical (throat) pressure P^* for both the actual and isentropic cases are same, and is determined from

$$P^* = P_{01}\left(\frac{T_2^*}{T_{01}}\right)^{k/(k-1)} = (150 \text{ psia})\left(\frac{667.9 \text{ R}}{900 \text{ R}}\right)^{1.667/0.667} = 71.2 \text{ psia}$$

$$\rho^* = \frac{P^*}{RT^*} = \frac{71.2 \text{ psia}}{(2.6809 \text{ psia}\cdot\text{ft}^3/\text{lbm}\cdot\text{R})(674.9 \text{ R})} = 0.0393 \text{ lbm/ft}^3$$

$$\mathbf{V}^* = C^* = \sqrt{kRT^*} = \sqrt{(1.667)(0.4961 \text{ Btu/lbm}\cdot\text{R})(674.9 \text{ R})\left(\frac{25{,}037 \text{ ft}^2/\text{s}^2}{1 \text{ Btu/lbm}}\right)} = 3738 \text{ ft/s}$$

Thus the throat area is

$$A^* = \frac{\dot{m}}{\rho^*\mathbf{V}^*} = \frac{0.2 \text{ lbm/s}}{(0.0393 \text{ lbm/ft}^3)(3738 \text{ ft/s})} = \mathbf{0.00136 \ ft^2}$$

At the nozzle exit the pressure is P_2 = 15 psia. The isentropic exit temperature can be determined from

$$\frac{T_{2s}}{T_{01}} = \left(\frac{P_2}{P_{01}}\right)^{(k-1)/k} \longrightarrow T_{2s} = (900 \text{ R})\left(\frac{15 \text{ psia}}{150 \text{ psia}}\right)^{(1.667-1)/1.667} = 358.2 \text{ R}$$

The actual exit temperature is determined from

$$\eta_N = \frac{h_{01} - h_2}{h_{01} - h_{2s}} = \frac{T_{01} - T_2}{T_{01} - T_{2s}} \longrightarrow 0.97 = \frac{900 - T_2}{900 - 358.2} \longrightarrow T_2 = 374.5 \text{ R}$$

Then the exit velocity is determined from the steady-flow energy balance to be

$$h_1 + \frac{\mathbf{V}_1^2}{2} = h_2 + \frac{\mathbf{V}_2^2}{2} \rightarrow 0 = h_2 - h_1 + \frac{\mathbf{V}_2^2 - \mathbf{V}_1^2}{2} \rightarrow 0 = C_p(T_2 - T_1) + \frac{\mathbf{V}_2^2 - \mathbf{V}_1^2}{2}$$

Solving for $\mathbf{V}_2$,

$$\mathbf{V}_2 = \sqrt{2C_p(T_1 - T_2)} = \sqrt{2(1.25 \text{ Btu/lbm}\cdot\text{R})(900 - 374.5)\text{R}\left(\frac{25{,}037 \text{ ft}^2/\text{s}^2}{1 \text{ Btu/lbm}}\right)} = 5735 \text{ ft/s}$$

Also, $$\rho_2 = \frac{P_2}{RT_2} = \frac{15 \text{ psia}}{(2.6811 \text{ psia}\cdot\text{ft}^3/\text{lbm}\cdot\text{R})(374.5 \text{ R})} = 0.0149 \text{ lbm/ft}^3$$

and $$A_2 = \frac{\dot{m}}{\rho_2 \mathbf{V}_2} = \frac{0.2 \text{ lbm/s}}{(0.0149 \text{ lbm/ft}^3)(5735 \text{ ft/s})} = \mathbf{0.00234 \text{ ft}^2}$$

16-116 Saturated steam enters a converging-diverging nozzle with a low velocity. The throat area, exit velocity, mass flow rate, and exit Mach number are to be determined for isentropic and 90 percent efficient nozzle cases.

Assumptions **1** Flow through the nozzle is steady and one-dimensional. **2** The nozzle is adiabatic.

Analysis (*a*) The inlet stagnation properties in this case are identical to the inlet properties since the inlet velocity is negligible. Thus $h_{10} = h_1$. At the inlet,

$$h_1 = (h_f + x_1 h_{fg})_{@3\text{ MPa}} = 1008.42 + 0.95 \times 1795.7 = 2714.3 \text{ kJ/kg}$$

$$s_1 = (s_f + x_1 s_{fg})_{@3\text{ MPa}} = 2.6457 + 0.95 \times 3.5412 = 6.0098 \text{ kJ/kg} \cdot \text{K}$$

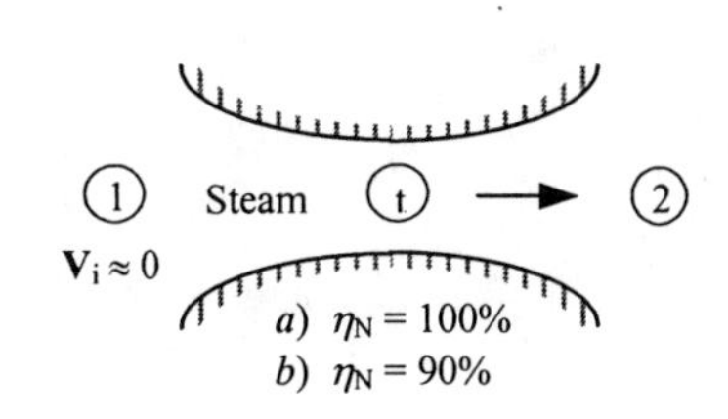

At the exit, P_2 = 1.2 MPa and $s_2 = s_{2s} = s_1$ = 6.0098 kJ/kg·K. Thus,

$$s_2 = s_f + x_2 s_{fg} \rightarrow 6.0098 = 2.2166 + x_2(4.3067) \rightarrow x_2 = 0.881$$

$$h_2 = h_f + x_2 h_{fg} = 798.65 + 0.881 \times 1986.2 = 2548.5 \text{ kJ/kg}$$

$$v_2 = v_f + x_2 v_{fg} = 0.001139 + 0.881 \times 0.1622 = 0.1440 \text{ m}^3\text{/kg}$$

Then the exit velocity is determined from the steady-flow energy balance to be

$$h_1 + \frac{\mathbf{V}_1^2}{2} = h_2 + \frac{\mathbf{V}_2^2}{2} \rightarrow 0 = h_2 - h_1 + \frac{\mathbf{V}_2^2 - \mathbf{V}_1^2}{2}$$

Solving for $\mathbf{V}_2$,

$$\mathbf{V}_2 = \sqrt{2(h_1 - h_2)} = \sqrt{2(2714.3 - 2548.5)\text{kJ/kg}\left(\frac{1000 \text{ m}^2/\text{s}^2}{1 \text{ kJ/kg}}\right)} = \mathbf{575.8 \text{ m/s}}$$

The mass flow rate is determined from

$$\dot{m} = \frac{1}{v_2} A_2 \mathbf{V}_2 = \frac{1}{0.1440 \text{ m}^3\text{/kg}}(24 \times 10^{-4} \text{m}^2)(575.8 \text{ m/s}) = \mathbf{9.60 \text{ kg/s}}$$

The velocity of sound at the exit of the nozzle is determined from

$$C = \left(\frac{\partial P}{\partial r}\right)_s^{1/2} \cong \left(\frac{DP}{D(1/v)}\right)_s^{1/2}$$

The specific volume of steam at s_2 = 6.0098 kJ/kg·K and at pressures just below and just above the specified pressure (1.1 and 1.3 MPa) are determined to be 0.1556 and 0.1341 m^3/kg. Substituting,

$$C_2 = \sqrt{\frac{(1300 - 1100)\text{kPa}}{\left(\frac{1}{0.1341} - \frac{1}{0.1556}\right)\text{kg/m}^3}\left(\frac{1000 \text{ m}^2/\text{s}^2}{1 \text{ kPa} \cdot \text{m}^3}\right)} = 440.6 \text{ m/s}$$

Then the exit Mach number becomes

$$M_2 = \frac{\mathbf{V}_2}{C_2} = \frac{575.8 \text{ m/s}}{440.6 \text{ m/s}} = \mathbf{1.307}$$

The steam is saturated, and thus the critical pressure which occurs at the throat is taken to be

$$P_t = P^* = 0.576 \times P_{01} = 0.576 \times 3 = 1.728 \text{ MPa}$$

Then at the throat,

$$P_t = 1.728 \text{ MPa} \quad and \quad s_t = s_1 = 6.0098 \text{ kJ/kg} \cdot \text{K}$$

Thus,

$$s_t = s_f + x_t s_{fg} \longrightarrow 6.0098 = 2.3789 + x_2(4.0154) \longrightarrow x_2 = 0.904$$

$$h_t = h_f + x_t h_{fg} = 875.54 + 0.904 \times 1920.5 = 2611.7 \text{ kJ/kg}$$

$$v_t = v_f + x_t v_{fg} = 0.001165 + 0.904 \times 0.1139 = 0.1042 \text{ m}^3\text{/kg}$$

Then the throat velocity is determined from the steady-flow energy balance,

$$h_1 + \frac{\mathbf{V}_1^2}{2}^{\nearrow 0} = h_t + \frac{\mathbf{V}_t^2}{2} \rightarrow 0 = h_t - h_1 + \frac{\mathbf{V}_t^2}{2}$$

Solving for $\mathbf{V}_t$,

$$\mathbf{V}_t = \sqrt{2(h_1 - h_t)} = \sqrt{2(2714.3 - 2611.7)\text{kJ/kg}\left(\frac{1000\ \text{m}^2/\text{s}^2}{1\ \text{kJ/kg}}\right)} = 453.0\ \text{m/s}$$

Thus the throat area is

$$A_t = \frac{\dot{m}v_2}{\mathbf{V}_t} = \frac{(9.6\ \text{kg/s})(0.1042\ \text{m}^3/\text{kg})}{(453.0\ \text{m/s})} = 22.1 \times 10^{-4}\text{m}^2 = \mathbf{22.1\ cm^2}$$

(*b*) The inlet stagnation properties in this case are identical to the inlet properties since the inlet velocity is negligible. Thus $h_{10} = h_1$. At the inlet,

$$h_1 = (h_f + x_1 h_{fg})_{@3\ \text{MPa}} = 1008.42 + 0.95 \times 1795.7 = 2714.3\ \text{kJ/kg}$$
$$s_1 = (s_f + x_1 s_{fg})_{@3\ \text{MPa}} = 2.6457 + 0.95 \times 3.5412 = 6.0098\ \text{kJ/kg} \cdot \text{K}$$

At state 2s, $P_2 = 1.2$ MPa and $s_2 = s_{2s} = s_1 = 6.0098$ kJ/kg·K. Thus,

$$s_{2s} = s_f + x_{2s} s_{fg} \longrightarrow 6.0098 = 2.2166 + x_{2s}(4.3067) \longrightarrow x_{2s} = 0.881$$
$$h_{2s} = h_f + x_{2s} h_{fg} = 798.65 + 0.881 \times 1986.2 = 2548.5\ \text{kJ/kg}$$

The enthalpy of steam at the actual exit state is determined from

$$\eta_N = \frac{h_{01} - h_2}{h_{01} - h_{2s}} \longrightarrow 0.90 = \frac{2714.3 - h_2}{2714.3 - 2548.5} \longrightarrow h_2 = 2565.1\ \text{kJ/kg}$$

Therefore at the exit, $P_2 = 1.2$ MPa and $h_2 = 2565.1$ kJ/kg·K. Thus,

$$h_2 = h_f + x_2 h_{fg} \longrightarrow 2565.1 = 798.65 + x_2(1986.2) \longrightarrow x_2 = 0.889$$
$$s_2 = s_f + x_2 s_{fg} = 2.2166 + 0.889 \times 4.3067 = 6.0453$$
$$v_2 = v_f + x_2 v_{fg} = 0.001139 + 0.889 \times 0.1622 = 0.1453\ \text{kJ/kg}$$

Then the exit velocity is determined from the steady-flow energy balance to be

$$h_1 + \frac{\mathbf{V}_1^2}{2} = h_2 + \frac{\mathbf{V}_2^2}{2} \rightarrow 0 = h_2 - h_1 + \frac{\mathbf{V}_2^2 - \mathbf{V}_1^2}{2}$$

Solving for $\mathbf{V}_2$,

$$\mathbf{V}_2 = \sqrt{2(h_1 - h_2)} = \sqrt{2(2714.3 - 2565.1)\text{kJ/kg}\left(\frac{1000\ \text{m}^2/\text{s}^2}{1\ \text{kJ/kg}}\right)} = \mathbf{546.3\ m/s}$$

The mass flow rate is determined from

$$\dot{m} = \frac{1}{v_2} A_2 \mathbf{V}_2 = \frac{1}{0.1453\ \text{m}^3/\text{kg}} (24 \times 10^{-4}\text{m}^2)(546.3\ \text{m/s}) = \mathbf{9.02\ kg/s}$$

The velocity of sound at the exit of the nozzle is determined from

$$C = \left(\frac{\partial P}{\partial \rho}\right)_s^{1/2} \cong \left(\frac{\Delta P}{\Delta(1/v)}\right)_s^{1/2}$$

The specific volume of steam at s_2 = 6.0453 kJ/kg·K and at pressures just below and just above the specified pressure (1.1 and 1.3 MPa) are determined to be 0.1570 and 0.1353 m³/kg. Substituting,

$$C_2 = \sqrt{\frac{(1300-1100)\text{kPa}}{\left(\frac{1}{0.1353}-\frac{1}{0.1570}\right)\text{kg/m}^3}\left(\frac{1000\text{ m}^2/\text{s}^2}{1\text{ kPa}\cdot\text{m}^3}\right)} = 442.5\text{ m/s}$$

Then the exit Mach number becomes

$$M_2 = \frac{\mathbf{V}_2}{C_2} = \frac{546.3\text{ m/s}}{442.5\text{ m/s}} = \mathbf{1.235}$$

The steam is saturated, and thus the critical pressure which occurs at the throat is taken to be

$$P_t = P^* = 0.576\times P_{01} = 0.576\times 3 = 1.728\text{ MPa}$$

At state 2ts, P_{ts} = 1.728 MPa and $s_{ts} = s_1$ = 6.0098 kJ/kg·K. Thus,

$$s_{ts} = s_f + x_{ts}s_{fg} \longrightarrow 6.0098 = 2.3789 + x_{2s}(4.0154) \longrightarrow x_{2s} = 0.904$$

$$h_{ts} = h_f + x_{ts}h_{fg} = 875.54 + 0.904\times 1920.5 = 2611.7\text{ kJ/kg}$$

The actual enthalpy of steam at the throat is

$$\eta_N = \frac{h_{01}-h_t}{h_{01}-h_{ts}} \longrightarrow 0.90 = \frac{2714.3-h_t}{2714.3-2611.7} \longrightarrow h_t = 2622.0\text{ kJ/kg}$$

Therefore at the throat, $P_2 = 1.728$ MPa and $h_2 = 2622.0$ kJ/kg. Thus,

$$h_t = h_f + x_t h_{fg} \longrightarrow 2622.0 = 875.54 + x_t(1920.5) \longrightarrow x_t = 0.909$$

$$\text{v}_t = \text{v}_f + x_t\text{v}_{fg} = 0.001165 + 0.909\times 0.1139 = 0.1047\text{ m}^3/\text{kg}$$

Then the throat velocity is determined from the steady-flow energy balance,

$$h_1 + \frac{\mathbf{V}_1^{2\ \nearrow 0}}{2} = h_t + \frac{\mathbf{V}_t^2}{2} \rightarrow 0 = h_t - h_1 + \frac{\mathbf{V}_t^2}{2}$$

Solving for $\mathbf{V}_t$,

$$\mathbf{V}_t = \sqrt{2(h_1-h_t)} = \sqrt{2(2714.3-2622.0)\text{kJ/kg}\left(\frac{1000\text{ m}^2/\text{s}^2}{1\text{ kJ/kg}}\right)} = 429.7\text{ m/s}$$

Thus the throat area is

$$A_t = \frac{\dot{m}\text{v}_2}{\mathbf{V}_t} = \frac{(9.02\text{ kg/s})(0.1047\text{ m}^3/\text{kg})}{(429.7\text{ m/s})} = 22.0\times 10^{-4}\text{m}^2 = \mathbf{22.0\text{ cm}^2}$$

16-117 and 6-118 There problems are solved by EES on enclosed CD.

16-119 The critical temperature, pressure, and density of an equimolar mixture of oxygen and nitrogen for specified stagnation properties are to be determined.

Assumptions Both oxygen and nitrogen are ideal gases with constant specific heats at room temperature.

Properties The specific heat ratio and molar mass are k = 1.395 and M = 32 kg/kmol for oxygen, and k = 1.4 and M = 28 kg/kmol for nitrogen (Tables A-1 and A-2).

Analysis The gas constant of the mixture is

$$M_m = y_{O_2} M_{O_2} + y_{N_2} M_{N_2} = 0.5 \times 32 + 0.5 \times 28 = 30 \text{ kg/kmol}$$

$$R_m = \frac{R_u}{M_m} = \frac{8.314 \text{ kJ/kmol} \cdot \text{K}}{30 \text{ kg/kmol}} = 0.2771 \text{ kJ/kg.K}$$

The specific heat ratio is 1.4 for nitrogen, and nearly 1.4 for oxygen. Therefore, the specific heat ratio of the mixture is also 1.4. Then the critical temperature, pressure, and density of the mixture become

$$T^* = T_0 \left(\frac{2}{k+1} \right) = (800 \text{ K}) \left(\frac{2}{1.4+1} \right) = \mathbf{667 \text{ K}}$$

$$P^* = P_0 \left(\frac{2}{k+1} \right)^{k/(k-1)} = (500 \text{ kPa}) \left(\frac{2}{1.4+1} \right)^{1.4/(1.4-1)} = \mathbf{264 \text{ kPa}}$$

$$\rho^* = \frac{P^*}{RT^*} = \frac{264 \text{ kPa}}{(0.2771 \text{ kPa} \cdot \text{m}^3/\text{kg} \cdot \text{K})(667 \text{ K})} = \mathbf{1.43 \text{ kg/m}^3}$$

Discussion If the specific heat ratios k of the two gases were different, then we would need to determine the k of the mixture from $k = C_{p,m}/C_{v,m}$ where the specific heats of the mixture are determined from

$$C_{p,m} = \text{mf}_{O_2} C_{p,O_2} + \text{mf}_{N_2} C_{p,N_2} = (y_{O_2} M_{O_2} / M_m) C_{p,O_2} + (y_{N_2} M_{N_2} / M_m) C_{p,N_2}$$

$$C_{v,m} = \text{mf}_{O_2} C_{v,O_2} + \text{mf}_{N_2} C_{v,N_2} = (y_{O_2} M_{O_2} / M_m) C_{v,O_2} + (y_{N_2} M_{N_2} / M_m) C_{v,N_2}$$

In this case it would give

$$C_{p,m} = (0.5 \times 32/30) \times 0.918 + (0.5 \times 28/30) \times 1.039 = 0.974 \text{ kJ/kg.K}$$

$$C_{p,m} = (0.5 \times 32/30) \times 0.658 + (0.5 \times 28/30) \times 0.743 = 0.698 \text{ kJ/kg.K}$$

and $k = 0.974/0.698 = 1.40$

16-120 ··· 16-124 EES solutions of these comprehensive problems are available to instructors at the *Instructor Manual* section of the *Online Learning Center* (OLC) at www.mhhe.com/cengel-boles. See the Preface for access information.

Fundamentals of Engineering (FE) Exam Problems

16-125 An aircraft is cruising in still air at 5°C at a velocity of 300 m/s. The air temperature at the nose of the aircraft where stagnation occurs is

(a) 5°C (b) 10°C (c) 20°C (d) 35°C (e) 50°C

Answer (e) 50°C

Solution Solved by EES Software. Solutions can be verified by copying-and-pasting the following lines on a blank EES screen. (Similar problems and their solutions can be obtained easily by modifying numerical values).

```
k=1.4
Cp=1.005 "kJ/kg.K"
T1=5 "C"
Vel1= 300 "m/s"
T1_stag=T1+Vel1^2/(2*Cp*1000)

"Some Wrong Solutions with Common Mistakes:"
W1_Tstag=T1 "Assuming temperature rise"
W2_Tstag=Vel1^2/(2*Cp*1000) "Using just the dynamic temperature"
W3_Tstag=T1+Vel1^2/(Cp*1000) "Not using the factor 2"
```

16-126 Air is flowing in a wind tunnel at 15°C, 80 kPa, and 200 m/s. The stagnation pressure at a probe
inserted into the flow stream is

(a) 82 kPa (b) 91 kPa (c) 96 kPa (d) 101 kPa (e) 114 kPa

Answer (d) 101 kPa

Solution Solved by EES Software. Solutions can be verified by copying-and-pasting the following lines on a blank EES screen. (Similar problems and their solutions can be obtained easily by modifying numerical values).

```
k=1.4
Cp=1.005 "kJ/kg.K"

T1=15 "K"
P1=80 "kPa"
Vel1= 200 "m/s"
T1_stag=(T1+273)+Vel1^2/(2*Cp*1000) "C"
T1_stag/(T1+273)=(P1_stag/P1)^((k-1)/k)

"Some Wrong Solutions with Common Mistakes:"
T11_stag/T1=(W1_P1stag/P1)^((k-1)/k); T11_stag=T1+Vel1^2/(2*Cp*1000) "Using deg. C
for temperatures"
T12_stag/(T1+273)=(W2_P1stag/P1)^((k-1)/k); T12_stag=(T1+273)+Vel1^2/(Cp*1000) "Not
using the factor 2"
T13_stag/(T1+273)=(W3_P1stag/P1)^(k-1); T13_stag=(T1+273)+Vel1^2/(2*Cp*1000)
"Using wrong isentropic relation"
```

16-127 An aircraft is reported to be cruising in still air at 10°C and 72 kPa at a Mach number of 0.86. The velocity of the aircraft is

(a) 9.2 m/s (b) 55 m/s (c) 186 m/s (d) 290 m/s (e) 321 m/s

Answer (d) 290 m/s

Solution Solved by EES Software. Solutions can be verified by copying-and-pasting the following lines on a blank EES screen. (Similar problems and their solutions can be obtained easily by modifying numerical values).

```
k=1.4
Cp=1.005 "kJ/kg.K"
R=0.287 "kJ/kg.K"

T1=10+273 "K"
P1=72 "kPa"
Mach=0.86

VS1=SQRT(k*R*T1*1000)
Mach=Vel1/VS1

"Some Wrong Solutions with Common Mistakes:"
W1_vel=Mach*VS2; VS2=SQRT(k*R*(T1-273)*1000) "Using C for temperature"
W2_vel=VS1/Mach "Using Mach number relation backwards"
W3_vel=Mach*VS3; VS3=k*R*T1 "Using wrong relation"
```

16-128 Air is flowing in a wind tunnel at 12°C and 66 kPa at a velocity of 230 m/s. The Mach number of the flow is

(a) 0.54 (b) 0.87 (c) 3.3 (d) 0.36 (e) 0.68

Answer (e) 0.68

Solution Solved by EES Software. Solutions can be verified by copying-and-pasting the following lines on a blank EES screen. (Similar problems and their solutions can be obtained easily by modifying numerical values).

```
k=1.4
Cp=1.005 "kJ/kg.K"
R=0.287 "kJ/kg.K"

T1=12+273 "K"
P1=66 "kPa"
Vel1=230 "m/s"

VS1=SQRT(k*R*T1*1000)
Mach=Vel1/VS1

"Some Wrong Solutions with Common Mistakes:"
W1_Mach=Vel1/VS2; VS2=SQRT(k*R*(T1-273)*1000) "Using C for temperature"
W2_Mach=VS1/Vel1 "Using Mach number relation backwards"
W3_Mach=Vel1/VS3; VS3=k*R*T1 "Using wrong relation"
```

16-129 Consider a converging nozzle with a low velocity at the inlet and sonic velocity at the exit plane. Now the nozzle exit diameter is reduced by half while the nozzle inlet temperature and pressure are maintained the same. The nozzle exit velocity will

(a) remain the same. (b) double. (c) quadruple. (d) go down by half. (e) go down to one-fourth.

Answer (a) remain the same.

16-130 Air is approaching a converging-diverging nozzle with a low velocity at 15°C and 95 kPa, and it leaves the nozzle at a supersonic velocity. The velocity of air at the throat of the nozzle is

(a) 10 m/s (b) 71 m/s (c) 311 m/s (d) 364 m/s (e) 412 m/s

Answer (c) 311 m/s

Solution Solved by EES Software. Solutions can be verified by copying-and-pasting the following lines on a blank EES screen. (Similar problems and their solutions can be obtained easily by modifying numerical values).

```
k=1.4
Cp=1.005 "kJ/kg.K"
R=0.287 "kJ/kg.K"

"Properties at the inlet"
T1=15+273 "K"
P1=95 "kPa"
Vel1=0 "m/s"

To=T1 "since velocity is zero"
Po=P1

"Throat properties"
T_throat=2*To/(k+1)
P_throat=Po*(2/(k+1))^(k/(k-1))

"The velocity at the throat is the velocity of sound,"
V_throat=SQRT(k*R*T_throat*1000)

"Some Wrong Solutions with Common Mistakes:"
W1_Vthroat=SQRT(k*R*T1*1000) "Using T1 for temperature"
W2_Vthroat=SQRT(k*R*T2_throat*1000); T2_throat=2*(To-273)/(k+1) "Using C for
temperature"
W3_Vthroat=k*R*T_throat "Using wrong relation"
```

16-131 Argon gas is approaching a converging-diverging nozzle with a low velocity at 20°C and 120 kPa, and it leaves the nozzle at a supersonic velocity. If the cross-sectional area of the throat is 0.015 m^2, the mass flow rate of argon through the nozzle is

(a) 0.41 kg/s (b) 3.4 kg/s (c) 5.3 kg/s (d) 17 kg/s (e) 22 kg/s

Answer (c) 5.3 kg/s

Solution Solved by EES Software. Solutions can be verified by copying-and-pasting the following lines on a blank EES screen. (Similar problems and their solutions can be obtained easily by modifying numerical values).

```
k=1.667
Cp=0.5203 "kJ/kg.K"
R=0.2081 "kJ/kg.K"
A=0.015 "m^2"

"Properties at the inlet"
T1=20+273 "K"
P1=120 "kPa"
Vel1=0 "m/s"

To=T1 "since velocity is zero"
Po=P1

"Throat properties"
T_throat=2*To/(k+1)
P_throat=Po*(2/(k+1))^(k/(k-1))
rho_throat=P_throat/(R*T_throat)

"The velocity at the throat is the velocity of sound,"
V_throat=SQRT(k*R*T_throat*1000)

m=rho_throat*A*V_throat

"Some Wrong Solutions with Common Mistakes:"
W1_mass=rho_throat*A*V1_throat; V1_throat=SQRT(k*R*T1_throat*1000);
T1_throat=2*(To-273)/(k+1) "Using C for temp"
W2_mass=rho2_throat*A*V_throat; rho2_throat=P1/(R*T1) "Using density at inlet"
```

16-132 Carbon dioxide enters a converging-diverging nozzle at 80 m/s, 350°C, and 300 kPa, and it leaves the nozzle at a supersonic velocity. The velocity of carbon dioxide at the throat of the nozzle is

(a) 12 m/s (b) 274 m/s (c) 312 m/s (d) 365 m/s (e) 386 m/s

Answer (d) 365 m/s

Solution Solved by EES Software. Solutions can be verified by copying-and-pasting the following lines on a blank EES screen. (Similar problems and their solutions can be obtained easily by modifying numerical values).

```
k=1.289
Cp=0.846 "kJ/kg.K"
R=0.1889 "kJ/kg.K"

"Properties at the inlet"
T1=350+273 "K"
P1=300 "kPa"
Vel1=80 "m/s"

To=T1+Vel1^2/(2*Cp*1000)
To/T1=(Po/P1)^((k-1)/k)

"Throat properties"
T_throat=2*To/(k+1)
P_throat=Po*(2/(k+1))^(k/(k-1))

"The velocity at the throat is the velocity of sound,"
V_throat=SQRT(k*R*T_throat*1000)

"Some Wrong Solutions with Common Mistakes:"
W1_Vthroat=SQRT(k*R*T1*1000) "Using T1 for temperature"
W2_Vthroat=SQRT(k*R*T2_throat*1000); T2_throat=2*(T_throat-273)/(k+1) "Using C for
temperature"
W3_Vthroat=k*R*T_throat "Using wrong relation"
```

16-133 Consider gas flow through a converging-diverging nozzle. Of the five statements below, select the one that is incorrect:

(a) The fluid velocity at the throat can never exceed the speed of sound.

(b) If the fluid velocity at the throat is below the speed of sound, the diverging section will act like a diffuser.

(c) If the fluid enters the diverging section with a Mach number greater than one, the flow at the nozzle exit will be supersonic.

(d) There will be no flow through the nozzle if the back pressure equals the stagnation pressure.

(e) The fluid velocity decreases, the entropy increases, and stagnation enthalpy remains constant during flow through a normal shock.

Answer (c) If the fluid enters the diverging section with a Mach number greater than one, the flow at the nozzle exit will be supersonic.

16-134 Combustion gases with k = 1.33 enter a converging nozzle at stagnation temperature and pressure of 400°C and 900 kPa, and are discharged into the atmospheric air at 20°C and 100 kPa. The lowest pressure that will occur within the nozzle is

(a) 32 kPa (b) 100 kPa (c) 321 kPa (d) 486 kPa (e) 672 kPa

Answer (d) 486 kPa

Solution Solved by EES Software. Solutions can be verified by copying-and-pasting the following lines on a blank EES screen. (Similar problems and their solutions can be obtained easily by modifying numerical values).

```
k=1.33
Po=900 "kPa"

"The critical pressure is"
P_throat=Po*(2/(k+1))^(k/(k-1))

"The lowest pressure that will occur in the nozzle is the higher of the critical or atmospheric pressure."

"Some Wrong Solutions with Common Mistakes:"
W1_Pthroat=900 "Assuming constant pressure"
W2_Pthroat=Po*(1/(k+1))^(k/(k-1)) "Using wrong relation"
W3_Pthroat=100 "Assuming atmospheric pressure"
```

16-135 ··· 16-137 Design and Essay Problems